D1683915

**SEAS AT THE MILLENNIUM:
AN ENVIRONMENTAL EVALUATION**

SEAS AT THE MILLENNIUM:
AN ENVIRONMENTAL EVALUATION

Edited by

Charles R.C. Sheppard
*Department of Biological Sciences,
University of Warwick,
Coventry, U.K.*

**Volume I
Regional Chapters:
Europe, The Americas and West Africa**

2000
PERGAMON
An imprint of Elsevier Science

AMSTERDAM – LAUSANNE – NEW YORK – OXFORD – SHANNON – SINGAPORE – TOKYO

ELSEVIER SCIENCE Ltd
The Boulevard, Langford Lane
Kidlington, Oxford OX5 1GB, UK

© 2000 Elsevier Science Ltd. All rights reserved.

This work is protected under copyright by Elsevier Science, and the following terms and conditions apply to its use:

Photocopying
Single photocopies of single chapters may be made for personal use as allowed by national copyright laws. Permission of the Publisher and payment of a fee is required for all other photocopying, including multiple or systematic copying, copying for advertising or promotional purposes, resale, and all forms of document delivery. Special rates are available for educational institutions that wish to make photocopies for non-profit educational classroom use.

Permissions may be sought directly from Elsevier Science Global Rights Department, PO Box 800, Oxford OX5 1DX, UK; phone: (+44) 1865 843830, fax: (+44) 1865 853333, e-mail: permissions@elsevier.co.uk. You may also contact Global Rights directly through Elsevier's home page (http://www.elsevier.nl), by selecting 'Obtaining Permissions'.

In the USA, users may clear permissions and make payments through the Copyright Clearance Center, Inc., 222 Rosewood Drive, Danvers, MA 01923, USA; phone: (978) 7508400, fax: (978) 7504744, and in the UK through the Copyright Licensing Agency Rapid Clearance Service (CLARCS), 90 Tottenham Court Road, London W1P 0LP, UK; phone: (+44) 20 7631 5555; fax: (+44) 20 7631 5500. Other countries may have a local reprographic rights agency for payments.

Derivative Works
Tables of contents may be reproduced for internal circulation, but permission of Elsevier Science is required for external resale or distribution of such material.
Permission of the Publisher is required for all other derivative works, including compilations and translations.

Electronic Storage or Usage
Permission of the Publisher is required to store or use electronically any material contained in this work, including any chapter or part of a chapter.

Except as outlined above, no part of this work may be reproduced, stored in a retrieval system or transmitted in any form or by any means, electronic, mechanical, photocopying, recording or otherwise, without prior written permission of the Publisher.
Address permissions requests to: Elsevier Science Global Rights Department, at the mail, fax and e-mail addresses noted above.

Notice
No responsibility is assumed by the Publisher for any injury and/or damage to persons or property as a matter of products liability, negligence or otherwise, or from any use or operation of any methods, products, instructions or ideas contained in the material herein. Because of rapid advances in the medical sciences, in particular, independent verification of diagnoses and drug dosages should be made.

First edition 2000

Library of Congress Cataloging in Publication Data
Seas at the millennium: an environmental evaluation / edited by Charles Sheppard. -- 1st ed.
   p. cm.
   Includes bibliographical references.
   ISBN 0-08-043207-7 (alk. paper)
   1. Marine ecology. I. Sheppard, Charles (Charles R.C.)
QH541.5.S3 S35 2000
577.7--dc21

00-034738

British Library Cataloguing in Publication Data
Seas at the millennium: an environmental evaluation
1. Marine ecology 2. Coastal ecology 3. Marine resources conservation
I. Sheppard, Charles R.C.
577.7

ISBN: 0-08-043207-7

∞ The paper used in this publication meets the requirements of ANSI/NISO Z39.48-1992 (Permanence of Paper).

Printed in The Netherlands.

CONTENTS

List of Authors xi

Acknowledgements xxi

Volume I. Regional Chapters: Europe, The Americas and West Africa

INTRODUCTION. 1
    Charles Sheppard

Chapter 1. THE SEAS AROUND GREENLAND 5
    Frank Riget, Poul Johansen, Henning Dahlgaard, Anders Mosbech, Rune Dietz and Gert Asmund

Chapter 2. THE NORWEGIAN COAST 17
    Jens Skei, Torgeir Bakke and Jarle Molvaer

Chapter 3. THE FAROE ISLANDS 31
    Maria Dam, Grethe Bruntse, Andrias Reinert and Jacob Pauli Joensen

Chapter 4. THE NORTH SEA 43
    Jean-Paul Ducrotoy, Mike Elliott and Victor N. de Jonge

Chapter 5. THE ENGLISH CHANNEL 65
    Alan D. Tappin and P. Chris Reid

Chapter 6. THE IRISH SEA 83
    Richard G. Hartnoll

Chapter 7. THE BALTIC SEA, ESPECIALLY SOUTHERN AND EASTERN REGIONS 99
    Jerzy Falandysz, Anna Trzosinska, Piotr Szefer, Jan Warzocha and Bohdan Draganik

Chapter 8. THE BALTIC SEA, INCLUDING BOTHNIAN SEA AND BOTHNIAN BAY. 121
    Lena Kautsky and Nils Kautsky

Chapter 9. THE NORTH COAST OF SPAIN 135
    Isabel Díez, Antonio Secilla, Alberto Santolaria and José María Gorostiaga

Chapter 10. SOUTHERN PORTUGAL: THE TAGUS AND SADO ESTUARIES 151
    Graça Cabeçadas, Maria José Brogueira and Leonor Cabeçadas

Chapter 11. THE ATLANTIC COAST OF SOUTHERN SPAIN 167
    Carlos J. Luque, Jesús M. Castillo and M. Enrique Figueroa

Chapter 12. THE CANARY ISLANDS 185
    Francisco García Montelongo, Carlos Díaz Romero and Ricardo Corbella Tena

Chapter 13. THE AZORES 201
    Brian Morton and Joseph C. Britton

Chapter 14. THE SARGASSO SEA AND BERMUDA 221
    Anthony H. Knap, Douglas P. Connelly and James N. Butler

Chapter 15. THE AEGEAN SEA 233
    Manos Dassenakis, Kostas Kapiris and Alexandra Pavlidou

Chapter 16. THE COAST OF ISRAEL, SOUTHEAST MEDITERRANEAN 253
    Barak Herut and Bella Galil

Chapter 17. THE ADRIATIC SEA AND THE TYRRHENIAN SEA 267
    Giuseppe Cognetti, Claudio Lardicci, Marco Abbiati and Alberto Castelli

Chapter 18. THE BLACK SEA 285
    Gülfem Bakan and Hanife Büyükgüngör

Chapter 19. THE GULF OF MAINE AND GEORGES BANK. 307
    Jack B. Pearce

Chapter 20. THE NEW YORK BIGHT 321
    Jack B. Pearce

Chapter 21. CHESAPEAKE BAY: THE UNITED STATES' LARGEST ESTUARINE SYSTEM 335
    Kent Mountford

Chapter 22. NORTH AND SOUTH CAROLINA COASTS 351
    Michael A. Mallin, JoAnn M. Burkholder, Lawrence B. Cahoon and Martin H. Posey

Chapter 23. THE GULF OF ALASKA. 373
    Bruce A. Wright, Jeffrey W. Short, Tom J. Weingartner and Paul J. Anderson

Chapter 24. SOUTHERN CALIFORNIA 385
    Kenneth C. Schiff, M. James Allen, Eddy Y. Zeng and Steven M. Bay

Chapter 25. FLORIDA KEYS. 405
    Phillip Dustan

Chapter 26. THE BAHAMAS. 415
    Kenneth C. Buchan

Chapter 27. THE NORTHERN GULF OF MEXICO 435
    Mark E. Pattillo and David M. Nelson

Chapter 28. COASTAL MANAGEMENT IN LATIN AMERICA. 457
    Alejandro Yáñez-Arancibia

Chapter 29. SOUTHERN GULF OF MEXICO 467
    Felipe Vázquez, Ricardo Rangel, Arturo Mendoza Quintero-Marmol, Jorge Fernández, Eduardo Aguayo, E.A. Palacio and Virender K. Sharma

Chapter 30. THE PACIFIC COAST OF MEXICO. 483
    Alfonso V. Botello, Alejandro O. Toledo, Guadalupe de la Lanza-Espino and Susana Villanueva-Fragoso

Chapter 31. BELIZE 501
    Alastair R. Harborne, Melanie D. McField and E. Kate Delaney

Chapter 32. NICARAGUA: CARIBBEAN COAST. 517
    Stephen C. Jameson, Lamarr B. Trott, Michael J. Marshall and Michael J. Childress

Chapter 33. NICARAGUA: PACIFIC COAST. 531
    Stephen C. Jameson, Vincent F. Gallucci and José A. Robleto

Chapter 34. EL SALVADOR 545
    Linos Cotsapas, Scott A. Zengel and Enrique J. Barraza

Chapter 35. JAMAICA 559
    Marjo Vierros

Chapter 36. PUERTO RICO 575
    Jack Morelock, Jorge Capella, Jorge Garcia and Maritza Barreto

Chapter 37. THE TURKS AND CAICOS ISLANDS 587
    Gudrun Gaudian and Paul Medley

Chapter 38. THE DUTCH ANTILLES 595
    Adolphe O. Debrot and Jeffrey Sybesma

Chapter 39. UK OVERSEAS TERRITORIES IN THE NORTHEAST CARIBBEAN: ANGUILLA, BRITISH VIRGIN ISLANDS, MONTSERRAT 615
    Fiona Gell and Maggie Watson

Chapter 40. THE LESSER ANTILLES, TRINIDAD AND TOBAGO. 627
    John B.R. Agard and Judith F. Gobin

Chapter 41. VENEZUELA 643
    Pablo E. Penchaszadeh, César A. Leon, Haymara Alvarez, David Bone, P. Castellano, María M. Castillo, Yusbelly Diaz, María P. Garcia, Mairin Lemus, Freddy Losada, Alberto Martin, Patricia Miloslavich, Claudio Paredes, Daisy Perez, Miradys Sebastiani, Dennise Stecconi, Victoriano Roa and Alicia Villamizar

Chapter 42. THE CARIBBEAN COAST OF COLOMBIA 663
    Leonor Botero and Ricardo Alvarez-León

Chapter 43. THE PACIFIC COAST OF COLOMBIA 677
    Alonso J. Marrugo-González, Roberto Fernández-Maestre and Anders A. Alm

Chapter 44. PERU 687
    Guadalupe Sanchez

Chapter 45. THE CHILEAN COAST 699
    Ramón B. Ahumada, Luis A. Pinto and Patricio A. Camus

Chapter 46. TROPICAL COAST OF BRAZIL 719
    Zelinda M.A.N. Leão and José M.L. Dominguez
Chapter 47. SOUTHERN BRAZIL 731
    Eliete Zanardi Lamardo, Márcia Caruso Bícego, Belmiro Mendes de Castro Filho, Luiz Bruner
    de Miranda and Valéria Aparecida Prósperi
Chapter 48. THE ARGENTINE SEA: THE SOUTHEAST SOUTH AMERICAN SHELF MARINE ECOSYSTEM . 749
    José L. Esteves, Nestor F. Ciocco, Juan C. Colombo, Hugo Freije, Guillermo Harris, Oscar Iribarne,
    Ignacio Isla, Paulina Nabel, Marcela S. Pascual, Pablo E. Penchaszadeh, Andrés L. Rivas and Norma
    Santinelli
Chapter 49. THE GULF OF GUINEA LARGE MARINE ECOSYSTEM 773
    Nicholas J. Hardman-Mountford, Kwame A. Koranteng and Andrew R.G. Price
Chapter 50. GUINEA 797
    Ibrahima Cisse, Idrissa Lamine Bamy, Amadou Bah, Sékou Balta Camara and Mamba Kourouma
Chapter 51. CÔTE D'IVOIRE 805
    Ama Antoinette Adingra, Robert Arfi and Aka Marcel Kouassi
Chapter 52. SOUTHWESTERN AFRICA: NORTHERN BENGUELA CURRENT REGION 821
    David Boyer, James Cole and Christopher Bartholomae

Index 841

Volume II. Regional Chapters: The Indian Ocean to The Pacific

Chapter 53. THE ARABIAN GULF 1
    D.V. Subba Rao and Faiza Al-Yamani
Chapter 54. NORTHWEST ARABIAN SEA AND GULF OF OMAN 17
    Simon C. Wilson
Chapter 55. THE RED SEA 35
    Charles R.C. Sheppard
Chapter 56. THE GULF OF ADEN 47
    Simon C. Wilson and Rebecca Klaus
Chapter 57. THE INDIAN OCEAN COAST OF SOMALIA 63
    Federico Carbone and Giovanni Accordi
Chapter 58. TANZANIA 83
    Martin Guard, Aviti J. Mmochi and Chris Horrill
Chapter 59. MOZAMBIQUE 99
    Michael Myers and Mark Whittington
Chapter 60. MADAGASCAR 113
    Andrew Cooke, Onésime Ratomahenina, Eulalie Ranaivoson and Haja Razafindrainibe
Chapter 61. SOUTH AFRICA 133
    Michael H. Schleyer, Lynnath E. Beckley, Sean T. Fennessy, Peter J. Fielding, Anesh Govender, Bruce
    Q. Mann, Wendy D. Robertson, Bruce J. Tomalin and Rudolph P. van der Elst
Chapter 62. THE NORTHWEST COAST OF THE BAY OF BENGAL AND DELTAIC SUNDARBANS 145
    Abhijit Mitra
Chapter 63. SOUTHEAST INDIA 161
    Sundararajan Ramachandran
Chapter 64. SRI LANKA 175
    Arjan Rajasuriya and Anil Premaratne
Chapter 65. THE ANDAMAN, NICOBAR AND LAKSHADWEEP ISLANDS 189
    Sundararajan Ramachandran
Chapter 66. THE MALDIVES 199
    Andrew R.G. Price and Susan Clark
Chapter 67. THE CHAGOS ARCHIPELAGO, CENTRAL INDIAN OCEAN 221
    Charles R.C. Sheppard
Chapter 68. THE SEYCHELLES 233
    Miles Gabriel, Suzanne Marshall and Simon Jennings

Chapter 69. THE COMOROS ARCHIPELAGO 243
    Jean-Pascal Quod, Odile Naim and Fouad Abdourazi
Chapter 70. THE MASCARENE REGION 253
    John Turner, Colin Jago, Deolall Daby and Rebecca Klaus
Chapter 71. THE BAY OF BENGAL 269
    D.V. Subba Rao
Chapter 72. BANGLADESH 285
    Dihider Shahriar Kabir, Syed Mazharul Islam, Md. Giasuddin Khan, Md. Ekram Ullah and Dulal C. Halder
Chapter 73. THE GULF OF THAILAND 297
    Manuwadi Hungspreugs, Wilaiwan Utoomprurkporn and Charoen Nitithamyong
Chapter 74. THE MALACCA STRAITS 309
    Chua Thia-Eng, Ingrid R.L. Gorre, S. Adrian Ross, Stella Regina Bernad, Bresilda Gervacio and Ma. Corazon Ebarvia
Chapter 75. MALACCA STRAIT INCLUDING SINGAPORE AND JOHORE STRAITS 331
    Poh Poh Wong
Chapter 76. EAST COAST OF PENINSULAR MALAYSIA 345
    Zelina Z. Ibrahim, Aziz Arshad, Lee Say Chong, Japar Sidik Bujang, Law Ah Theem, Nik Mustapha Raja Abdullah and Maged Mahmoud Marghany
Chapter 77. BORNEO 361
    Steve Oakley, Nicolas Pilcher and Elizabeth Wood
Chapter 78. CONTINENTAL SEAS OF WESTERN INDONESIA 381
    Evan Edinger and David R. Browne
Chapter 79. THE PHILIPPINES 405
    Gil S. Jacinto, Porfirio M. Aliño, Cesar L. Villanoy, Liana Talaue-McManus and
    Edgardo D. Gomez
Chapter 80. THE CORAL, SOLOMON AND BISMARCK SEAS REGION 425
    Michael E. Huber and Graham B.K. Baines
Chapter 81. ASIAN DEVELOPING REGIONS: PERSISTENT ORGANIC POLLUTANTS IN THE SEAS 447
    Shinsuke Tanabe
Chapter 82. SEA OF OKHOTSK 463
    Victor V. Lapko and Vladimir I. Radchenko
Chapter 83. SEA OF JAPAN 473
    Anatoly N. Kachur and Alexander V. Tkalin
Chapter 84. THE YELLOW SEA 487
    Suam Kim and Sung-Hyun Kahng
Chapter 85. TAIWAN STRAIT 499
    Woei-Lih Jeng, Chang-Feng Dai and Kuang-Lung Fan
Chapter 86. XIAMEN REGION, CHINA 513
    Chua Thia-Eng and Ingrid Rosalie L. Gorre
Chapter 87. HONG KONG 535
    Bruce J. Richardson, Paul K.S. Lam and Rudolf S.S. Wu
Chapter 88. SOUTHERN CHINA, VIETNAM TO HONG KONG 549
    Zhang Gan, Zou Shicun and Yan Wen
Chapter 89. VIETNAM AND ADJACENT BIEN DONG (SOUTH CHINA SEA) 561
    Dang Duc Nhan, Nguyen Xuan Duc, Do Hoai Duong, Nguyen The Tiep and Bui Cong Que
Chapter 90. CAMBODIAN SEA 569
    Touch Seang Tana
Chapter 91. THE AUSTRALIAN REGION: AN OVERVIEW 579
    Leon P. Zann
Chapter 92. TORRES STRAIT AND THE GULF OF PAPUA 593
    Michael E. Huber
Chapter 93. NORTHEASTERN AUSTRALIA: THE GREAT BARRIER REEF REGION 611
    Leon P. Zann
Chapter 94. THE EASTERN AUSTRALIAN REGION: A DYNAMIC TROPICAL/TEMPERATE BIOTONE 629
    Leon P. Zann

Chapter 95. THE TASMANIAN REGION ... 647
    Christine M. Crawford, Graham J. Edgar and George Cresswell
Chapter 96. VICTORIA PROVINCE, AUSTRALIA ... 661
    Tim D. O'Hara
Chapter 97. THE GREAT AUSTRALIAN BIGHT ... 673
    Karen Edyvane
Chapter 98. THE WESTERN AUSTRALIAN REGION ... 691
    Diana I. Walker
Chapter 99. THE SOUTH WESTERN PACIFIC ISLANDS REGION ... 705
    Leon P. Zann and Veikila Vuki
Chapter 100. NEW CALEDONIA ... 723
    Pierre Labrosse, Renaud Fichez, Richard Farman and Tim Adams
Chapter 101. VANUATU ... 737
    Veikila C. Vuki, Subashni Appana, Milika R. Naqasima and Maika Vuki
Chapter 102. THE FIJI ISLANDS ... 751
    Veikila C. Vuki, Leon P. Zann, Milika Naqasima and Maika Vuki
Chapter 103. THE CENTRAL SOUTH PACIFIC OCEAN (AMERICAN SAMOA) ... 765
    Peter Craig, Suesan Saucerman and Sheila Wiegman
Chapter 104. THE MARSHALL ISLANDS ... 773
    Andrew R.G. Price and James E. Maragos
Chapter 105. HAWAIIAN ISLANDS (U.S.A.) ... 791
    James E. Maragos
Chapter 106. FRENCH POLYNESIA ... 813
    Pat Hutchings and Bernard Salvat

Index ... 827

Volume III. Global Issues and Processes

Chapter 107. GLOBAL STATUS OF SEAGRASSES ... 1
    Ronald C. Phillips and Michael J. Durako
Chapter 108. MANGROVES ... 17
    Colin D. Field
Chapter 109. CORAL REEFS: ENDANGERED, BIODIVERSE, GENETIC RESOURCES ... 33
    Walter H. Adey, Ted A. McConnaughey, Allegra M. Small and Don M. Spoon
Chapter 110. WORLD-WIDE CORAL REEF BLEACHING AND MORTALITY DURING 1998:
    A GLOBAL CLIMATE CHANGE WARNING FOR THE NEW MILLENNIUM? ... 43
    Clive R. Wilkinson
Chapter 111. SEA TURTLES ... 59
    Jeanne A. Mortimer, Marydele Donnelly and Pamela T. Plotkin
Chapter 112. WHALES AND WHALING ... 73
    Sidney Holt
Chapter 113. SMALL CETACEANS: SMALL WHALES, DOLPHINS AND PORPOISES ... 89
    Kieran Mulvaney and Bruce McKay
Chapter 114. SEABIRDS ... 105
    W.R.P. Bourne and C.J. Camphuysen
Chapter 115. FISHERIES EFFECTS ON ECOSYSTEMS ... 117
    Raquel Goñi
Chapter 116. BY-CATCH: PROBLEMS AND SOLUTIONS ... 135
    Martin A. Hall, Dayton L. Alverson and Kaija I. Metuzals
Chapter 117. FISHERIES MANAGEMENT AS A SOCIAL PROBLEM ... 153
    Douglas C. Wilson
Chapter 118. FARMING OF AQUATIC ORGANISMS, PARTICULARLY THE CHINESE AND THAI
    EXPERIENCE ... 165
    Krishen J. Rana and Anton J. Immink

Chapter 119. CLIMATIC CHANGES: GULF OF ALASKA 179
    Howard Freeland and Frank Whitney

Chapter 120. EFFECTS OF CLIMATE CHANGE AND SEA LEVEL ON COASTAL SYSTEMS 187
    Shiao-Kung Liu

Chapter 121. PARTICLE DRY DEPOSITION TO WATER SURFACES: PROCESSES AND CONSEQUENCES . . 197
    Sara C. Pryor and Rebecca J. Barthelmie

Chapter 122. MARINE ECOSYSTEM HEALTH AS AN EXPRESSION OF MORBIDITY, MORTALITY
    AND DISEASE EVENTS 211
    Benjamin H. Sherman

Chapter 123. EFFECT OF MINE TAILINGS ON THE BIODIVERSITY OF THE SEABED:
    EXAMPLE OF THE ISLAND COPPER MINE, CANADA 235
    Derek V. Ellis

Chapter 124. MARINE ANTIFOULANTS 247
    Stewart M. Evans

Chapter 125. EUTROPHICATION OF MARINE WATERS: EFFECTS ON BENTHIC MICROBIAL
    COMMUNITIES 257
    Lutz-Arend Meyer-Reil and Marion Köster

Chapter 126. PERSISTENCE OF SPILLED OIL ON SHORES AND ITS EFFECTS ON BIOTA 267
    Gail V. Irvine

Chapter 127. REMOTE SENSING OF TROPICAL COASTAL RESOURCES: PROGRESS AND
    FRESH CHALLENGES FOR THE NEW MILLENNIUM 283
    Peter J. Mumby

Chapter 128. SATELLITE REMOTE SENSING OF THE COASTAL OCEAN: WATER QUALITY
    AND ALGAE BLOOMS 293
    Bertil Håkansson

Chapter 129. ENERGY FROM THE OCEANS: WIND, WAVE AND TIDAL 303
    Rebecca J. Barthelmie, Ian Bryden, Jan P. Coelingh and Sara C. Pryor

Chapter 130. MULTINATIONAL TRAINING PROGRAMMES IN MARINE ENVIRONMENTAL SCIENCE . . . 323
    G. Robin South

Chapter 131. GLOBAL LEGAL INSTRUMENTS ON THE MARINE ENVIRONMENT AT THE YEAR 2000 . . . 331
    Milen F. Dyoulgerov

Chapter 132. COASTAL MANAGEMENT IN THE FUTURE 349
    Derek J. McGlashan

Chapter 133. SUSTAINABILITY OF HUMAN ACTIVITIES ON MARINE ECOSYSTEMS. 359
    Paul Johnston, David Santillo, Julie Ashton and Ruth Stringer

Chapter 134. MARINE RESERVES AND RESOURCE MANAGEMENT 375
    Michael J. Fogarty, James A. Bohnsack and Paul K. Dayton

Chapter 135. THE ECOLOGICAL, ECONOMIC, AND SOCIAL IMPORTANCE OF THE OCEANS 393
    Robert Costanza

Index 405

CONTRIBUTING AUTHORS

Marco Abbiati
Dipartimento di Biologia Evoluzionistica, Università di Bologna, Via Tombesi dell'Ova 55, I-48100 Ravenna, Italy

Fouad Abdourazi
AIDE, Minizi, Mavouna, Moroni, Comoros

Nik Mustapha Raja Abdullah
Universiti Putra Malaysia, 43400 UPM Serdang, Malaysia

Giovanni Accordi
Centro di Studio per il Quaternario e l'Evoluzione Ambientale, C.N.R., Dipartimento di Scienze della Terra, Università degli Studi "La Sapienza", P. Aldo Moro, 5, 00185 Roma, Italy

Tim Adams
Secretariat of the Pacific Community (SPC), B.P. D5, 98848 Nouméa Cedex, New Caledonia

Walter H. Adey
Marine Systems Laboratory, Smithsonian Institution, NHB E-117, MRC 164, Washington, DC 20560, U.S.A.

Ama Antoinette Adingra
Centre de Recherches Océanologiques, BP V18, Abidjan, Côte d'Ivoire

John B.R. Agard
Department of Life Sciences, The University of the West Indies, St. Augustine, Trinidad and Tobago

Eduardo Aguayo
Instituto de Ciencias del Mar y Limnología, UNAM, Cd. Universitaria, A.P. 70-305, Mexico City, C.P. 04510 Mexico

Ramón B. Ahumada
Facultad de Ciencias, Universidad Católica de la Santísima Concepción, Campus San Andrés, Paicaví 3000, Casilla 297, Concepción, Chile

Faiza Al-Yamani
Kuwait Institute of Scientific Research, P.O. Box 1638, 22017 Salmiya, Kuwait

Porfirio M. Aliño
Marine Science Institute, University of the Philippines, 1101 Diliman, Quezon City, Philippines

M. James Allen
Southern California Coastal Water Research Project, 7171 Fenwick Lane, Westminster, CA 92683, U.S.A.

Anders A. Alm
Universidad de Cartagena, Facultad de Ciencias Químicas y Farmacéuticas, Zaragocilla, AA 1661, Cartagena, Colombia

Ricardo Alvarez-León
Universidad de la Sabana, Depto. Ciencias de la Vida, Campus Universitario Puente del Común, Edif. E-2, of. 232, Chía, Cundi., Colombia

Haymara Alvarez
Universidad Simon Bolivar, Apdo 89000, Caracas 1080-A, Venezuela

Dayton L. Alverson
Natural Resources Consultants, 1900 West Nickerson St., Suite 207, Seattle, WA 98119, U.S.A.

Paul J. Anderson
Kodiak Laboratory, National Marine Fisheries Service, National Oceanic and Atmospheric Administration, P.O. Box 1638, Kodiak, AK 99615, U.S.A.

Subashni Appana
Marine Studies Programme, University of the South Pacific, P.O. Box 1168, Suva, Fiji

Robert Arfi
Centre de Recherches Océanologiques, BP V18, Abidjan, Côte d'Ivoire

Aziz Arshad
Universiti Putra Malaysia, 43400 UPM Serdang, Malaysia

Julie Ashton
Greenpeace Research Laboratories, University of Exeter, Exeter EX4 4PS, U.K.

Gert Asmund
National Environmental Research Institute, Department of Arctic Environment, P.O. Box 358, DK-4000 Roskilde, Denmark

Amadou Bah
The National Center of Halieutic Sciences of Boussoura, B.P. 3060, Conakry, Republic of Guinea

Graham B.K. Baines
Environment Pacific, 3 Pindari St., The Gap, Brisbane, QLD 4061, Australia

Gülfem Bakan
Ondokuz Mayis University, Faculty of Engineering, Department of Environmental Engineering, 55139 Kurupelit, Samsun, Turkey

Torgeir Bakke
Norwegian Institute for Water Research (NIVA), P.O. Box 173 Kjelsaas, 0411 Oslo, Norway

Idrissa Lamine Bamy
The National Center of Halieutic Sciences of Boussoura, B.P. 3060, Conakry, Republic of Guinea

Enrique J. Barraza
Ministry of the Environment, San Salvador, El Salvador

Maritza Barreto
University of Puerto Rico Rio Piedras, Geography Department, Rio Piedras, Puerto Rico

Rebecca J. Barthelmie
Department of Wind Energy and Atmospheric Physics, Risø National Laboratory, DK-4000 Roskilde, Denmark

Christopher Bartholomae
National Marine Information and Research Centre, P.O. Box 912, Swakopmund, Namibia

Steven M. Bay
Southern California Coastal Water Research Project, 7171 Fenwick Lane, Westminster, CA 92683, U.S.A.

Lynnath E. Beckley
SAAMBR, P.O. Box 10712, Marine Parade, KwaZulu-Natal 4056, South Africa

Stella Regina Bernad
GEF/UNDP/IMO Regional Programme on Partnerships in Environmental Management for the Seas of East Asia (PEMSEA), P.O. Box 2502, Quezon City, 1165 Metro Manila, Philippines

Márcia Caruso Bícego
Universidade de São Paulo, Dept. Oceanografia Física do Instituto Oceanográfico, Pca do Oceanográfico, 191, Cidade Universitária, SP, 05508-900, Brazil

James A. Bohnsack
National Marine Fisheries Service, Southeast Fisheries Science Center, 75 Virginia Beach Drive, Miami, FL 33149, U.S.A.

David Bone
Universidad Simon Bolivar, Apartado 89000, Caracas 1080-A, Venezuela

Alfonso V. Botello
Institute for Marine and Limnology Sciences, National Autonomous University of Mexico, Marine Pollution Laboratory, P.O. Box 70305, México City 04510 D.F., Mexico

Leonor Botero
COLCIENCIAS, Trans. 9A # 133-28, Santafé de Bogotá, Colombia

W.R.P. Bourne
Department of Zoology, Aberdeen University, Tillydrone Avenue, Aberdeen AB24 2TZ, Scotland

David Boyer
National Marine Information and Research Centre, P.O. Box 912, Swakopmund, Namibia

Joseph C. Britton
Department of Biology, Texas Christian University, Fort Worth, Texas 76129, U.S.A.

Maria José Brogueira
Instituto de Investigação das Pescas e do Mar (IPIMAR), DAA, Av. Brasilia, 1400 Lisboa, Portugal.

David R. Browne
Department of Biology, McGill University, 1205 Docteur Penfield Avenue, Montreal, PQ, H3A 1B1 Canada

Grethe Bruntse
Kaldbak Marine Biological Laboratory, FO-180 Kaldbak, Faroe Islands

Ian Bryden
The Robert Gordon University, School of Mechanical and Offshore Engineering, Aberdeen, AB10 1FR, Scotland

Kenneth C. Buchan
Bahamian Field Station, San Salvador, Bahamas

Japar Sidik Bujang
Universiti Putra Malaysia, 43400 UPM Serdang, Malaysia

JoAnn M. Burkholder
Department of Botany, North Carolina State University, Raleigh, NC 27695-7612, U.S.A.

James N. Butler
Harvard University, Cambridge, Massachussetts, U.S.A.

Hanife Büyükgüngör
Ondokuz Mayis University, Faculty of Engineering, Department of Environmental Engineering, 55139 Kurupelit, Samsun, Turkey

Graça Cabeçadas
Instituto de Investigação das Pescas e do Mar (IPIMAR), DAA, Av. Brasilia, 1400 Lisboa, Portugal.

Leonor Cabeçadas
Direcção Geral do Ambiente (D.G.A.), R. da Murgueira, Zambujal, 2720 Amadora, Portugal

Lawrence B. Cahoon
Department of Biological Sciences, University of North Carolina-Wilmington, Wilmington, NC 28403, U.S.A.

Sékou Balta Camara
The National Center of Halieutic Sciences of Boussoura, B.P. 3060, Conakry, Republic of Guinea

C.J. Camphuysen
CSR Consultancy, Ankerstraat 20, 1794 BJ Oosterend, Texel, The Netherlands

Patricio A. Camus
Facultad de Ciencias, Universidad Católica de la Santísima Concepción, Campus San Andrés, Paicaví 3000, Casilla 297, Concepción, Chile

Jorge Capella
University of Puerto Rico R.U.M., Department of Marine Sciences, Mayagüez, Puerto Rico

Federico Carbone
Centro di Studio per il Quaternario e l'Evoluzione Ambientale, C.N.R., Dipartimento di Scienze della Terra, Università degli Studi "La Sapienza", P. Aldo Moro, 5, 00185 Roma, Italy

P. Castellano
Centro de Procesamiento de Imagenes Digitales (CPDI), Sartenejas, Caracas, Venezuela

Alberto Castelli
Dipartimento di Zoologia ed Antropologia, Corso Margherita di Savoia 15, I-17100 Sassari, Italy

Jesús M. Castillo
Departamento de Biología Vegetal y Ecología, Facultad de Biología, Universidad de Sevilla, Apdo 1095, 41080 Sevilla, Spain

María M. Castillo
Universidad Simon Bolivar, Apdo 89000, Caracas 1080-A, Venezuela

Michael J. Childress
Department of Biological Sciences, Idaho State University, Pocatello, ID 83209-8007, U.S.A.

Lee Say Chong
National Hydraulic Research Institute of Malaysia, Km 7 Jalan Ampang, 68000 Ampang, Malaysia

Nestor F. Ciocco
CENPAT-CONICET, Bv. Brown 3000, (9120) Puerto Madryn, Chubut, Argentina

Ibrahima Cisse
The National Center of Halieutic Sciences of Boussoura, B.P. 3060, Conakry, Republic of Guinea

Susan Clark
Department of Marine Sciences and Coastal Management, University of Newcastle upon Tyne, Newcastle upon Tyne, U.K.

Jan P. Coelingh
ECOFYS Energy and Environment, P.O. Box 8408, NL-3503 RK Utrecht, The Netherlands.

Giuseppe Cognetti
Dipartimento di Scienze dell'Uomo e dell'Ambiente, Università di Pisa, Via Volta 6, I-56124 Pisa, Italy

James Cole
2 Dolphin Cottage, 31 Penny St., Portsmouth PO1 2NH, U.K.

Juan C. Colombo
Química Ambiental y Bioquímica, Facultad de Ciencias Naturales y Museo, Universidad Nacional de La Plata, Paseo del Bosque s/n, (1900) La Plata, Argentina

Douglas P. Connelly
Bermuda Biological Station for Research, 17 Biological Station Lane, St. Georges, Bermuda, GE 01

Andrew Cooke
Cellule Environnement Marin et Côtier, Office National pour l'Environnement, B.P. 822, Antananarivo, Madagascar

Robert Costanza
Center for Environmental Science and Biology Department, and Institute for Ecological Economics, University of Maryland, Box 38, Solomons, MD 20688-0038, U.S.A.

Linos Cotsapas
Research Planning, Inc., 1121 Park St., Columbia, SC 29201, U.S.A.

Peter Craig
National Park of American Samoa, Pago Pago, American Samoa 96799, U.S.A.

Christine M. Crawford
Tasmanian Aquaculture and Fisheries Institute, University of Tasmania, Nubeena Crescent, Taroona, Tasmania 7053, Australia

George Cresswell
CSIRO Marine Research, Castray Esplanade, Hobart, Tasmania 7000, Australia

John Croxall
British Antarctic Survey, Natural Environment Research Council, High Cross, Madingley Road, Cambridge CB3 0ET, U.K.

Deolall Daby
Faculty of Science, University of Mauritius, Reduit, Mauritius

Henning Dahlgaard
Risø National Laboratory, DK-4000 Roskilde, Denmark

Chang-Feng Dai
Institute of Oceanography, National Taiwan University, Taipei, Taiwan, Republic of China

Maria Dam
Food and Environmental Agency, Debesartrøð, FO-100 Tórshavn, Faroe Islands

Manos Dassenakis
University of Athens, Department of Chemistry, Division III, Inorganic and Environmental Chemistry, Panepistimiopolis, Kouponia, Athens 15771, Greece

Paul K. Dayton
Scripps Institution of Oceanography, 9500 Gilman Dr., La Jolla, CA 92093, U.S.A.

Adolpe O. Debrot
Carmabi Foundation, Piscaderabaai, P.O. Box 2090, Curaçao, Netherlands Antilles

E. Kate Delaney
Department of Geography, University of Southampton, Southampton, SO17 1BJ, U.K.

Tom Dahmer
Hyder Consulting Ltd, Hong Kong; Ecosystems Ltd, Hong Kong

Yusbelly Diaz
Universidad Simon Bolivar, Apdo. 89000, Caracas 1080-A, Venezuela

Rune Dietz
National Environmental Research Institute, Department of Arctic Environment, P.O. Box 358, DK-4000 Roskilde, Denmark

Isabel Díez
Departamento de Biología Vegetal y Ecología, Facultad de Ciencias, Universidad del País Vasco, Apdo. 644, Bilbao 48080, Spain

José M.L. Dominguez
Laboratório de Estudos Costeiros, Centro de Pesquisa em Geofísica e Geologia, Universidade Federal da Bahia, Rua Caetano Moura 123, Federação, Salvador, 40210-340, Bahia, Brazil

Marydele Donnelly
IUCN/SSC Marine Turtle Specialist Group, 1725 DeSales St. NW #600, Washington, DC 20036, U.S.A.

Bohdan Draganik
Sea Fisheries Institute, 1 Kollataja Str., PL 81-332 Gdynia, Poland

Nguyen Xuan Duc
Institute of Ecology and Biological Resources, Nghia Do, Hanoi, Vietnam

Jean-Paul Ducrotoy
University College Scarborough, CERCI, Scarborough YO11 3AZ, U.K.

Do Hoai Duong
Institute of Hydrometeorology, Lang Thuong, Dong Da, Hanoi, Vietnam

Michael J. Durako
Center for Marine Science, The University of North Carolina at Wilmington, Wilmington, NC 28403, U.S.A.

Phillip Dustan
Department of Biology, University of Charleston, Charleston, SC 29424, U.S.A.

Milen F. Dyoulgerov
International Program Office, National Ocean Service, National Oceanic and Atmospheric Administration, 1305 East West Highway, Silver Spring, MD 20910, U.S.A.

Ma. Corazon Ebarvia
GEF/UNDP/IMO Regional Programme on Partnerships in Environmental Management for the Seas of East Asia (PEMSEA), P.O. Box 2502, Quezon City, 1165 Metro Manila, Philippines

Graham J. Edgar
Zoology Department, Tasmanian Aquaculture and Fisheries Institute, University of Tasmania, GPO Box 252-05, Hobart, Tasmania 7011, Australia

Evan Edinger
Department of Geology, St. Francis Xavier University, P.O. Box 5000, Antigonish, Nova Scotia B2G 2W5, Canada. Present address: Department of Earth Sciences, Laurentian University, Ramsey Lake Road, Sudbury, Ontario, P3E 2C6, Canada

Karen Edyvane
SA Research and Development Institute, P.O. Box 120, Henley Beach, South Australia 5022, Australia

Mike Elliott
IECS, University of Hull, Hull HU6 7RX, U.K.

Derek V. Ellis
Biology Department, University of Victoria, P.O. Box 3020, Victoria, B.C., V8W 3N5, Canada

Rudolph P. van der Elst
SAAMBR, P.O. Box 10712, Marine Parade, KwaZulu-Natal 4056, South Africa

Paul R. Epstein
Center for Health and the Global Environment, Harvard Medical School, Boston MA 02115, U.S.A

Mark V. Erdmann
Dept. of Integrative Biology, University of California, Berkeley, Berkeley, CA 94720, U.S.A.

José L. Esteves
CENPAT-CONICET, Bv. Brown 3000, (9120) Puerto Madryn, Chubut, Argentina

S.M. Evans
Dove Marine Laboratory (Department of Marine Sciences and Coastal Management, Newcastle University), Cullercoats, Tyne & Wear, NE30 4PZ, UK

Jerzy Falandysz
University of Gdansk, 18 Sobieskiego Str., PL 80-952 Gdansk, Poland

Kuang-Lung Fan
Institute of Oceanography, National Taiwan University, Taipei, Taiwan, Republic of China

Richard Farman
Southern Province, Department of Natural Resources, B.P. 3718, 98846 Nouméa Cedex, New Caledonia

Sean T. Fennessy
SAAMBR, P.O. Box 10712, Marine Parade, KwaZulu-Natal 4056, South Africa

Roberto Fernández-Maestre
Universidad de Cartagena, Facultad de Ciencias Químicas y Farmacéuticas, Zaragocilla, AA 1661, Cartagena, Colombia

Jorge Fernández
PEMEX-Exploración-Producción-Región Marina Suroeste, Calle 33 S/N, Edif. Cantarell, Cd. del Carmen, Campeche, C.P. 24170 Mexico

Renaud Fichez
IRD (Institute of Research for Development), B.P. A5, 98848 Nouméa Cedex, New Caledonia

Colin D. Field
Faculty of Science (Gore Hill), University of Technology, Sydney, P.O. Box 123, Broadway NSW 2007, Australia

Peter J. Fielding
SAAMBR, P.O. Box 10712, Marine Parade, KwaZulu-Natal 4056, South Africa

M. Enrique Figueroa
Departamento de Biología Vegetal y Ecología, Facultad de Biología, Universidad de Sevilla, Apdo. 1095, 41080 Sevilla, Spain

Belmiro Mendes de Castro Filho
Universidade de São Paulo, Dept. Oceanografia Física do Instituto Oceanográfico, Pca do Oceanográfico, 191, Cidade Universitária, SP, 05508-900, Brazil

William S. Fisher
U.S. Environmental Protection Agency, National Health and Environmental Effects Laboratory, Gulf Ecology Division, One Sabine Island Drive, Gulf Breeze, FL 32561, U.S.A

Michael J. Fogarty
University of Maryland Center for Environmental Science, Chesapeake Biological Lab., Solomons, MD, U.S.A. Present address: National Marine Fisheries Service, Northeast Fisheries Science Center, 166 Water St., Woods Hole, MA 02543, U.S.A.

Mark S. Fonseca
NOAA/National Ocean Service, Center for Coastal Fisheries and Habitat Research, 101 Pivers Island Road, Beaufort, NC 28516-9722, U.S.A.

Howard Freeland
Institute of Ocean Sciences, P.O. Box 6000, Sidney, B.C., V8L 4B2, Canada

Hugo Freije
Universidad Nacional del Sur, Química Ambiental, Av. Alem 1253, (8000) Bahía Blanca, Argentina

Miles Gabriel
Inter-consult Nambia (Pty) Ltd., P.O. Box 20690, Windhoek, Namibia

Bella Galil
Israel Oceanographic and Limnological Research, National Institute of Oceanography, P.O.Box 8030, Haifa 31080, Israel

Vincent F. Gallucci
University of Washington, School of Fisheries, Seattle, WA 98195, U.S.A.

Zhang Gan
Guangzhou Institute of Geochemistry, Chinese Academy of Sciences, Guangzhou 510640, People's Republic of China

María P. Garcia
Universidad Simon Bolivar, Apdo. 89000, Caracas 1080-A, Venezuela

Jorge Garcia
University of Puerto Rico R.U.M., Department of Marine Sciences, Mayagüez, Puerto Rico

Gudrun Gaudian
Sunny View, Main Street, Alne, N. Yorks, YO61 1RT, U.K.

Fiona Gell
ICLARM Caribbean and Eastern Pacific Office, PMB 158, Inland Messenger Service, Road Town, Tortola, British Virgin Islands

Bresilda Gervacio
GEF/UNDP/IMO Regional Programme on Partnerships in Environmental Management for the Seas of East Asia (PEMSEA), P.O. Box 2502, Quezon City, 1165 Metro Manila, Philippines

Ed Gmitrowicz
Hyder Consulting Ltd, Hong Kong; Ecosystems Ltd, Hong Kong

Judith F. Gobin
Institute of Marine Affairs, Hilltop Lane, Chaguaramas, Port of Spain, Trinidad and Tobago

Edgardo D. Gomez
Marine Science Institute, University of the Philippines, 1101 Diliman, Quezon City, Philippines

Raquel Goñi
Centro Oceanografico de Baleares Muelle de Poniente s/n, Apdo. 291, 07080 Palma de Mallorca, Spain

José María Gorostiaga
Departamento de Biología Vegetal y Ecología, Facultad de Ciencias, Universidad del País Vasco, Apdo. 644, Bilbao 48080, Spain

Ingrid Rosalie L. Gorre
GEF/UNDP/IMO Regional Programme on Partnerships in Environmental Management for Seas of East Asia (PEMSEA), P.O. Box 2502, Quezon City, 1165 Metro Manila, Philippines

Anesh Govender
SAAMBR, P.O. Box 10712, Marine Parade, KwaZulu-Natal 4056, South Africa

Stephen L. Granger
University of Rhode Island, Graduate School of Oceanography, Narragansett, Rhode Island 02882, USA

Martin Guard
Zoology Department, University of Aberdeen, Tillydrone Road, Aberdeen AB24 2TZ, U.K. and Department of Zoology and Marine Biology, University of Dar es Salaam, P.O. Box 35091, Dar es Salaam, Tanzania

Bertil Håkansson
Swedish Meteorological and Hydrological Institute, S-60176 Norrköping, Sweden

Dulal C. Halder
Independent University, Bangladesh (IUB), Plot 3 & 8, Road 10, Baridhara, Dhaka-1212, Bangladesh

Martin A. Hall
Inter-American Tropical Tuna Commission, 8604 La Jolla Shores Dr., La Jolla, CA 92037, U.S.A.

Alastair R. Harborne
Coral Cay Conservation, 154 Clapham Park Road, London, SW4 7DE, U.K.

Nicholas J. Hardman-Mountford
Centre for Coastal and Marine Sciences, Plymouth Marine Laboratory, Plymouth, U.K.

Guillermo Harris
Fundación Patagonia Natural, Marcos A. Zar 760, (9120) Puerto Madryn, Chubut, Argentina

Richard G. Hartnoll
Port Erin Marine Laboratory, University of Liverpool, Port Erin, Isle of Man IM9 6JA, British Isles

Frank Hawkins
Projet ZICOMA, BirdLife International, B.P. 1074, Antananarivo, Madagascar

Barak Herut
Israel Oceanographic and Limnological Research, National Institute of Oceanography, P.O.Box 8030, Haifa 31080, Israel

Sidney Holt
Hornbeam House, 4 Upper House Farm, Crickhowell, Powys, NP8 1BP, U.K.

Chris Horrill
Tanga Coastal Zone Conservation and Development Project, P.O. Box 5036, Tanga, Tanzania

Vicki Howe
Department of Maritime Studies and International Transport, University of Cardiff, Cardiff, Wales, U.K.

Michael E. Huber
Global Coastal Strategies, P.O. Box 606, Wynnum, QLD 4178, Australia

Manuwadi Hungspreugs
Department of Marine Science, Chulalongkorn University, Bangkok 10330, Thailand

George L. Hunt Jr.
Department of Ecology and Evolutionary Biology, University of California, Irvine, Irvine, CA 92697, U.S.A.

Pat Hutchings
The Australian Museum, Sydney, NSW 2010, Australia

Zelina Z. Ibrahim
National Hydraulic Research Institute of Malaysia, Km 7 Jalan Ampang, 68000 Ampang, Malaysia

Anton J. Immink
Fisheries Department, FAO, Rome, Italy

Oscar Iribarne
Universidad Nacional de Mar del Plata, Biologia, CC 573, Correo Central, (7600) Mar del Plata, Argentina

Gail V. Irvine
U.S. Geological Survey, Alaska Biological Science Center, 1011 E. Tudor Rd., Anchorage, AK 99503, U.S.A.

Ignacio Isla
Universidad Nacional de Mar del Plata, Centro de Geología de Costas

Syed Mazharul Islam
School of Liberal Arts and Science, Independent University, Bangladesh (IUB), Plot # 3 & 8, Road 10, Baridhara, Dhaka-1212, Bangladesh

Gil S. Jacinto
Marine Science Institute, University of the Philippines, 1101 Diliman, Quezon City, Philippines

Colin Jago
School of Ocean Sciences, University of Wales Bangor, LL59 5EY, U.K.

Stephen C. Jameson
Coral Seas Inc. – Integrated Coastal Zone Management, 4254 Hungry Run Road, The Plains, VA 20198-1715, U.S.A.

Woei-Lih Jeng
Institute of Oceanography, National Taiwan University, Taipei, Taiwan, Republic of China

Simon Jennings
CEFAS, Fisheries Laboratory, Lowestoft, NR33 OHT, U.K.

Jacob Pauli Joensen
Food and Environmental Agency, Debesartrøð, FO-100 Tórshavn, Faroe Islands

Poul Johansen
National Environmental Research Institute, Department of Arctic Environment, P.O. Box 358, DK-4000 Roskilde, Denmark

Paul Johnston
Greenpeace Research Laboratories, University of Exeter, Exeter EX4 4PS, U.K.

Victor N. de Jonge
National Institute for Coastal and Marine Management, Rijkswaterstaat, Ministry of Transport, Public Works and Water Management, P.O. Box 207, 9750 AE Haren, The Netherlands

Dihider Shahriar Kabir
School of Environmental Science and Management, Independent University, Bangladesh (IUB), Plot # 3 & 8, Road 10, Baridhara, Dhaka-1212, Bangladesh

Anatoly N. Kachur
Pacific Geographical Institute, Far East Branch, Russian Academy of Sciences, Vladivostok 690022, Russia

Sung-Hyun Kahng
Korea Ocean Research and Development Institute, Ansan P.O. Box 29, Seoul, 425-600, Korea

Kostas Kapiris
University of Athens, Department of Biology, Division of Zoology and Marine Biology, Panepistimiopolis, Kouponia, Athens 15784, Greece

Lena Kautsky
Department of Botany, Stockholm University, S-106 91 Stockholm, Sweden

Nils Kautsky
Department of Systems Ecology, Stockholm University, S-106 91 Stockholm, Sweden

W. Judson Kenworthy
NOAA/National Ocean Service, Center for Coastal Fisheries and Habitat Research, 101 Pivers Island Road, Beaufort, NC 28516-9722, U.S.A.

Oleg Khalimonov
International Maritime Organization, London, UK

Md. Giasuddin Khan
Centre for Environment and Geographical Information Systems Support, House 49, Road 27, Banani, Dhaka-1212, Bangladesh

Ruy K.P. Kikuchi
Departamento de Ciências Exatas, Universidade Estadual de Feira de Santana, BR-116, Campus Universitário, Feira de Santana, 44031-160, Bahia, Brazil

Suam Kim
Korea Ocean Research and Development Institute, Seoul, Korea. Present address: Dept. of Marine Biology, Pukyong National University, 599-1 Daeyeon 3-Dong, Nam-Gu, Pusan, 608-737, Korea

Rebecca Klaus
Department of Biological Sciences, University of Warwick, Coventry CV4 7RU, U.K.

Anthony H. Knap
Bermuda Biological Station for Research, 17 Biological Station Lane, St. Georges, Bermuda

Kwame A. Koranteng
Marine Fisheries Research Division, Ministry of Food and Agriculture, Ghana

Marion Köster
Institut für Ökologie der Ernst-Moritz-Arndt-Universität Greifswald, Schwedenhagen 6, D-18565 Kloster/Hiddensee, Germany

Aka Marcel Kouassi
Centre de Recherches Océanologiques, BP V18, Abidjan, Côte d'Ivoire

Mamba Kourouma
The National Center of Halieutic Sciences of Boussoura, B.P. 3060, Conakry, Republic of Guinea

Andreas Kunzmann
ZMT Bremen, Germany

Pierre Labrosse
Secretariat of the Pacific Community (SPC), B.P. D5, 98848 Nouméa Cedex, New Caledonia

Paul K.S. Lam
Department of Biology and Chemistry, City University of Hong Kong, 83 Tat Chee Avenue, Kowloon, Hong Kong

Eliete Zanardi Lamardo
University of Miami – RSMAS, Dept. Marine and Atmospheric Chemistry, 4600 Rickenbacker Causeway, Miami, FL 33149, U.S.A.

Guadalupe de la Lanza-Espino
Institute for Biological Sciences, National Autonomous University of Mexico, Marine Ecology Laboratory, P.O. Box 70233, México City 04515 D.F., México

Victor V. Lapko
Pacific Fisheries Research Centre, TINRO Centre, Vladivostok, Russia

Claudio Lardicci
Dipartimento di Scienze dell'Uomo e dell'Ambiente, Università di Pisa, Via Volta 6, I-56124 Pisa, Italy

Zelinda M.A.N. Leão
Laboratório de Estudos Costeiros, Centro de Pesquisa em Geofísica e Geologia, Universidade Federal da Bahia, Rua Caetano Moura 123, Federação, Salvador, 40210-340, Bahia, Brazil

Mairin Lemus
Instituto Oceanografico de Venezuela, Universidad de Oriente, Cumana, Venezuela

César A. Leon
Universidad Simon Bolivar, Apdo. 89000, Caracas 1080-A, Venezuela

Shiao-Kung Liu
Systems Research Institute, 3706 Ocean Hill Way, Malibu, CA, 90265, U.S.A.

Freddy Losada
Universidad Simon Bolivar, Apdo. 89000, Caracas 1080-A, Venezuela

Carlos J. Luque
Departamento de Biología Vegetal y Ecología, Facultad de Biología, Universidad de Sevilla, Apdo. 1095, 41080 Sevilla, Spain

Anmarie J. Mah
Vancouver Aquarium, P.O. Box 3232, Vancouver, British Columbia, Canada V6B 3X8

Michael A. Mallin
Center for Marine Science Research, University of North Carolina-Wilmington, Wilmington, NC 28403, U.S.A.

Bruce Q. Mann
SAAMBR, P.O. Box 10712, Marine Parade, KwaZulu-Natal 4056, South Africa

James E. Maragos
U.S. Fish and Wildlife Service, Pacific Islands Ecoregion, 300 Ala Moana Blvd., Box 50167, Honolulu, HI 96850, U.S.A.

Maged Mahmoud Marghany
Universiti Putra Malaysia, 43400 UPM Serdang, Malaysia

Alonso J. Marrugo-González
Universidad de Cartagena, Facultad de Ciencias Químicas y Farmacéuticas, Zaragocilla, AA 1661, Cartagena, Colombia

Michael J. Marshall
Coastal Seas Consortium, 5503 40th Avenue East, Bradenton, FL 34208, U.S.A.

Suzanne Marshall
Environment Agency, Kings Meadow Road, Reading, RG1 8DQ, U.K.

Alberto Martin
Universidad Simon Bolivar, Apdo. 89000, Caracas 1080-A, Venezuela

Ted A. McConnaughey
Marine Systems Laboratory, Smithsonian Institution, NHB E-117, MRC 164, Washington, DC 20560, U.S.A.

Melanie D. McField
Department of Marine Science, University of South Florida, 140 Seventh Ave South, St. Petersburg, FL 33701, U.S.A. and P.O. Box 512, Belize City, Belize

Derek J. McGlashan
Graduate School of Environmental Studies, Wolfson Centre, 106 Rottenrow East, University of Strathclyde, Glasgow, G4 0NW, U.K.

Bruce McKay
4058 Rue Dorion, Montreal, PQ H2K 4B9, Canada

Paul Medley
Sunny View, Main Street, Alne, N. Yorks, YO61 1RT, U.K.

Kaija I. Metuzals
Biological Sciences, University of Warwick, Coventry, U.K.

Lutz-Arend Meyer-Reil
Institut für Ökologie der Ernst-Moritz-Arndt-Universität Greifswald, Schwedenhagen 6, D-18565 Kloster/Hiddensee, Germany

Patricia Miloslavich
Universidad Simon Bolivar, Apartado 89000, Caracas 1080-A, Venezuela

Luiz Bruner de Miranda
Universidade de São Paulo, Dept. Oceanografia Física do Instituto Oceanográfico, Pca do Oceanográfico, 191, Cidade Universitária, SP, 05508-900, Brazil

Abhijit Mitra
Department of Marine Science, University of Calcutta 35, B.C Road, Calcutta 700 019, West Bengal, India.

Aviti J. Mmochi
Marine Environmental Chemistry, Institute of Marine Sciences, P.O. Box 668, Zanzibar, Tanzania

Jarle Molvaer
Norwegian Institute for Water Research (NIVA), P.O. Box 173 Kjelsaas, 0411 Oslo, Norway

Francisco García Montelongo
Department of Analytical Chemistry, Nutrition and Food Sciences, University of La Laguna, 38071 La Laguna, Spain

Jack Morelock
University of Puerto Rico R.U.M., Department of Marine Sciences, P.O. Box 3200, Lajas, Puerto Rico 00667

Jeanne A. Mortimer
Department of Zoology, University of Florida, Gainesville, FL 32611-8525, U.S.A. and Marine Conservation Society of Seychelles, P.O. Box 445, Victoria, Mahe, Seychelles

Brian Morton
The Swire Institute of Marine Science and Department of Ecology and Biodiversity, The University of Hong Kong, Hong Kong

Anders Mosbech
National Environmental Research Institute, Department of Arctic Environment, P.O. Box 358, DK-4000 Roskilde, Denmark

Kent Mountford
US Environmental Protection Agency, Chesapeake Bay Program, 410 Severn Ave., Suite 109, Annapolis, MD 21403, U.S.A.

Kieran Mulvaney
1219 W. 6th Avenue, Anchorage, AK 99501, U.S.A.

Peter J. Mumby
Centre for Tropical Coastal Management Studies, Department of Marine Sciences & Coastal Management, Ridley Building, The University, Newcastle upon Tyne, NE1 7RU, U.K.

Michael Myers
TCMC, University of Newcastle upon Tyne, Newcastle upon Tyne NE1 7RU, U.K.

Paulina Nabel
Museo Argentino de Ciencias Naturales-CONICET, Av. A. Gallardo 470, (1405) Buenos Aires, Argentina

Odile Naim
Laboratoire d'Ecologie Marine, Université de la Réunion, BP 9151,Saint-Denis messag 9, Réunion,France

Milika Naqasima
Marine Studies Programme, University of the South Pacific, P.O. Box 1168, Suva, Fiji

Milika Naqasima
Marine Studies Programme, University of the South Pacific, P.O. Box 1168, Suva, Fiji

David M. Nelson
National Ocean Service, 1305 East-West Highway, Silver Spring, MD 20910, U.S.A

Dang Duc Nhan
Institute of Nuclear Sciences and Techniques, P.O. Box 5T-160, Hoang Quoc Viet, Hanoi, Vietnam

Charoen Nitithamyong
Department of Marine Science, Chulalongkorn University, Bangkok 10330, Thailand

Scott W. Nixon
University of Rhode Island, Graduate School of Oceanography, Narragansett, Rhode Island 02882, USA

Steve Oakley
Institute of Biodiversity and Environmental Conservation, University of Malaysia, Kota Samarahan 93400, Sarawak, Malaysia

Tim D. O'Hara
Zoology Department, University of Melbourne, Parkville, Vic., Australia

E.A. Palacio
Instituto de Ingeniería, UNAM, Cd. Universitaria, Mexico City, C.P. 04510 México

Claudio Paredes
Universidad Simon Bolivar, Apartado 89000, Caracas 1080-A, Venezuela

Marcela S. Pascual
Instituto de Biología Marina y Pesquera "A. Storni", (8520) San Antonio Oeste, Río Negro, Argentina

Mark E. Pattillo
U.S. Army Corps of Engineers, Galveston District, P.O. Box 1229, Galveston, TX 77551-1229, U.S.A.

Alexandra Pavlidou
University of Athens, Department of Chemistry, Division III, Inorganic and Environmental Chemistry, Panepistimiopolis, Kouponia, Athens 15771, Greece

Jack B. Pearce
NMFS/NOAA, NE Fisheries Center, Woods Hole, MA 02543, U.S.A.

Pablo E. Penchaszadeh
Universidad Simon Bolivar, Apartado 89000, Caracas 1080-A, Venezuela

Daisy Perez
Universidad Simon Bolivar, Apartado 89000, Caracas 1080-A, Venezuela

Ronald C. Phillips
Commission of Environmental Research, Emirates Heritage Club, Abu Dhabi, United Arab Emirates. Correspondence: 1597 Meadow View Drive, Hermiston, OR 97838, U.S.A.

Niphon Phongsuwan
Phuket Marine Biological Center, P.O. Box 60, Phuket, 83000, Thailand

Nicolas Pilcher
Institute of Biodiversity and Environmental Conservation, University of Malaysia, Kota Samarahan 93400, Sarawak, Malaysia

Luis A. Pinto
Facultad de Ciencias, Universidad Católica de la Santísima Concepción, Campus San Andrés, Paicaví 3000, Casilla 297, Concepción, Chile

Pamela T. Plotkin
Center for Marine Conservation, 1725 DeSales St. NW #600, Washington, DC 20036, U.S.A.

Martin H. Posey
Center for Marine Science Research, University of North Carolina-Wilmington Wilmington, NC 28403, U.S.A.

Anil Premaratne
Coast Conservation Department, Sri Lanka

Andrew R.G. Price
Ecology and Epidemiology Group, Department of Biological Sciences, University of Warwick, Coventry, U.K.

Valéria Aparecida Prósperi
CETESB – Companhia de Tecnologia de Saneamento Ambiental, Setor de Ictiologia e Bioensaios com organismos aquáticos, Av. Prof. Frederico Hermann Jr., 345, Alto de Pinheiros, SP 05489-900, Brazil

Sara C. Pryor
Atmospheric Science Program, Department of Geography, Indiana University, Bloomington, IN 47405, U.S.A.

Bui Cong Que
Institute of Oceanology, Hoang Quoc Viet, Nghia Do, Hanoi, Vietnam

Arturo Mendoza Quintero-Marmol
PEMEX-Exploracíon-Produccion-Región Marina Noreste, Calle 31, Esq. Periferica, Cd. del Carmen, Campeche, C.P. 24170 Mexico

Jean-Pascal Quod
ARVAM, 14, Rue du stade de l'Est, 97490 Réunion, France

Vladimir I. Radchenko
Pacific Fisheries Research Centre, TINRO Centre, Vladivostok, Russia

Arjan Rajasuriya
National Aquatic Resources Research and Development Agency, Colombo, Sri Lanka

Sundararajan Ramachandran
Institute for Ocean Management, Anna University, Chennai 600 025, India

Krishen J. Rana
Fisheries Department, FAO, Rome, Italy

Eulalie Ranaivoson
Institut Halieutique et des Sciences Marines, B.P. 141, Toliara, Madagascar

Bemahafaly J. de D. Randriamanantsoa
Cellule des Océanographes de l'Université de Toliara (COUT), IHSM, B.P. 141, Toliara, Madagascar

Ricardo Rangel
Instituto de Ciencias del Mar y Limnología, UNAM, Cd. Universitaria, A.P. 70-305, Mexico City, C.P. 04510 Mexico

Onésime Ratomahenina
Ministère de la Recherche Scientifique, Direction Générale de la Recherche, B.P. 4258, Antananarivo, Madagascar

Haja Razafindrainibe
Cellule Environnement Marin et Côtier, Office National pour l'Environnement, B.P. 822, Antananarivo, Madagascar

P. Chris Reid
Sir Alister Hardy Foundation for Ocean Science, 1 Walker Terrace, The Hoe, Plymouth, PL1 3BN, U.K.

Andrias Reinert
Aquaculture Research Station of the Faroes, við Áir, FO-430 Hvalvík, Faroe Islands

Louise Richards
Hyder Consulting Ltd, Hong Kong; Ecosystems Ltd, Hong Kong

Bruce J. Richardson
Department of Biology and Chemistry, City University of Hong Kong, 83 Tat Chee Avenue, Kowloon, Hong Kong

Frank Riget
National Environmental Research Institute, Department of Arctic Environment, P.O. Box 358, DK-4000 Roskilde, Denmark

Andrés L. Rivas
CENPAT-CONICET, Bv. Brown 3000, (9120) Puerto Madryn, Chubut, Argentina

Victoriano Roa
Universidad Simon Bolivar, Apartado 89000 Caracas 1080-A, Venezuela

Wendy D. Robertson
SAAMBR, P.O. Box 10712, Marine Parade, KwaZulu-Natal 4056, South Africa

José A. Robleto
University of Mobile, Latin American Campus, San Marcos, Carazo, Nicaragua

Carlos Díaz Romero
Department of Analytical Chemistry, Nutrition and Food Sciences, University of La Laguna, 38071 La Laguna, Spain

S. Adrian Ross
GEF/UNDP/IMO Regional Programme on Partnerships in Environmental Management for the Seas of East Asia (PEMSEA), P.O. Box 2502, Quezon City, 1165 Metro Manila, Philippines

Bernard Salvat
Ecole Pratique des Hautes Etudes, URA CNRS 1453, Université de Perpignan, France, and Centre de Recherches Insulaires et Observatoire de l'Environnement, BP 1013, Moorea, Polynésia Française

Guadalupe Sanchez
Instituto del Mar del Peru, Callao, Peru

David Santillo
Greenpeace Research Laboratories, University of Exeter, Exeter EX4 4PS, U.K.

Norma Santinelli
Universidad Nacional de la Patagonia, Belgrano 504, (9100) Trelew, Chubut, Argentina

Alberto Santolaria
Departamento de Biología Vegetal y Ecología, Facultad de Ciencias, Universidad del País Vasco, Apdo. 644, Bilbao 48080, Spain

Suesan Saucerman
Environmental Protection Agency, EPA Region IX – WTR-5, 75 Hawthorne St., San Francisco, CA 94105-3901, U.S.A.

Kenneth C. Schiff
Southern California Coastal Water Research Project, 7171 Fenwick Lane, Westminster, CA 92683, U.S.A.

Michael H. Schleyer
SAAMBR, P.O. Box 10712, Marine Parade, KwaZulu-Natal 4056, South Africa

E.A. Schreiber
National Museum of Natural History, Smithsonian Institution, NHB MRC 116, Washington D.C. 20560, U.S.A.

Miradys Sebastiani
 Universidad Simon Bolivar, Apdo. 89000, Caracas 1080-A, Venezuela

Antonio Secilla
 Departamento de Biología Vegetal y Ecología, Facultad de Ciencias, Universidad del País Vasco, Apdo. 644, Bilbao 48080, Spain

Virender K. Sharma
 Chemistry Department, Florida Tech., 150 West University Blvd., Melbourne, FL 32901-6975, U.S.A.

Charles R.C. Sheppard
 Department of Biological Sciences, University of Warwick, Coventry CV4 7AL, U.K.

Benjamin H. Sherman
 Climate Change Research Center, Institute for the Study of Earth Oceans and Space, OSP HEED MMED program, 206 Nesmith Hall, University of New Hampshire, Durham, NH 03824, U.S.A.

Zou Shicun
 School of Chemistry and Chemical Engineering, Zhangshan University, Guangzhou 510301, People's Republic of China

Frederick T. Short
 Jackson Estuarine Laboratory, University of New Hampshire, 85 Adams Point Road, Durham, NH 03824, U.S.A.

Jeffrey W. Short
 Auke Bay Laboratory, National Marine Fisheries Service, National Oceanic and Atmospheric Administration, 11305 Glacier Highway, Juneau, AK 99801, U.S.A.

Jens Skei
 Norwegian Institute for Water Research (NIVA), P.O. Box 173 Kjelsaas, 0411 Oslo, Norway

Allegra M. Small
 Marine Systems Laboratory, Smithsonian Institution, NHB E-117, MRC 164, Washington, DC 20560, U.S.A.

G. Robin South
 International Ocean Institute – Pacific Islands, The University of the South Pacific, P.O. Box 1168, Suva, Republic of the Fiji Islands

Don M. Spoon
 Marine Systems Laboratory, Smithsonian Institution, NHB E-117, MRC 164, Washington, DC 20560, U.S.A.

Dennise Stecconi
 Centro de Procesamiento de Imagenes Digitales (CPDI), Sartenejas, Caracas. Venezuela

Ruth Stringer
 Greenpeace Research Laboratories, University of Exeter, Exeter EX4 4PS, U.K.

D.V. Subba Rao
 Mariculture and Fisheries Department, Kuwait Institute For Scientific Research, P.O. Box 1638, Salmiya 22017, Kuwait

Jeffrey Sybesma
 University of the Netherlands Antilles, Jan Noorduynweg 111, P.O. Box 3059, Curaçao, Netherlands Antilles

Piotr Szefer
 Medical University of Gdansk, 107 Gen. Hallera Ave., 80-416 Gdansk, Poland

Liana Talaue-McManus
 Marine Science Institute, University of the Philippines, 1101 Diliman, Quezon City, Philippines

Touch Seang Tana
 Department of Fisheries, 186 Norodom Blvd., P.O. Box 582, Phnom Penh, Cambodia

Shinsuke Tanabe
 Center for Marine Environmental Studies, Ehime University, Tarumi 3-5-7, Matsuyama 790-8566, Japan

Alan D. Tappin
 Centre for Coastal and Marine Science, Plymouth Marine Laboratory, Prospect Place, Plymouth PL1 3DH, U.K.

Ricardo Corbella Tena
 Department of Analytical Chemistry, Nutrition and Food Sciences, University of La Laguna, 38071 La Laguna, Spain

Gordon W. Thayer
 NOAA/National Ocean Service, Center for Coastal Fisheries and Habitat Research, 101 Pivers Island Road, Beaufort, NC 28516-9722, U.S.A.

Law Ah Theem
 Universiti Kolej Terengganu, Universiti Putra Malaysia, 21030 Kuala Terengganu, Malaysia

Chua Thia-Eng
 GEF/UNDP/IMO Regional Programme on Partnerships in Environmental Management for the Seas of East Asia (PEMSEA), P.O. Box 2502, Quezon City, 1165 Metro Manila, Philippines

Nguyen The Tiep
 Institute of Oceanology, Hoang Quoc Viet, Nghia Do, Hanoi, Vietnam

Alexander V. Tkalin
 Far Eastern Regional Hydrometeorological Research Institute (FERHRI), Russian Academy of Sciences, 24 Fontannaya Street, Vladivostok 690600, Russia

Alejandro O. Toledo
 Institute for Marine and Limnology Sciences, National Autonomous University of Mexico, Marine Pollution Laboratory, P.O. Box 70305, México City 04510 D.F., México

Bruce J. Tomalin
 SAAMBR, P.O. Box 10712, Marine Parade, KwaZulu-Natal 4056, South Africa

Tomas Tomascik
 Parks Canada – WCSC, 300-300 West Georgia St., Vancouver, British Columbia, Canada V6B 6B4

Michael S. Traber
 University of Rhode Island, Graduate School of Oceanography, Narragansett, RI 02882, USA

Lamarr B. Trott
 National Oceanic and Atmospheric Administration, National Marine Fisheries Service, 1315 East West Highway, Silver Spring, MD 20910, U.S.A.

Anna Trzosinska
 Institute of Meteorology and Water Management, 42 Waszyngtona Str., PL 81-342 Gdynia, Poland

Caroline Turnbull
 Department of Biological Sciences, University of Warwick, Coventry, UK

John Turner
 School of Ocean Sciences, University of Wales Bangor, Marine Science Laboratories, Anglesey, Gwynedd LL59 5EY, U.K.

Md. Ekram Ullah
 Environment Section, Water Resources and Planning Organisation (WARPO), House 4 A, Road 22, Gulshan-1, Dhaka-1212, Bangladesh

Wilaiwan Utoomprurkporn
Department of Marine Science, Chulalongkorn University, Bangkok 10330, Thailand

Marieke M. van Katwijk
University of Nijmegen, The Netherlands

Felipe Vázquez
Instituto de Ciencias del Mar y Limnología, UNAM, Cd. Universitaria, A.P. 70-305, Mexico City, C.P. 04510 Mexico

Marjo Vierros
UNEP-CAR/RCU, 14–20 Port Royal St., Kingston, Jamaica, Correspondence: Rosentiel School of Marine and Atmospheric Science, University of Miami, Dept. of Marine Geology & Geophysics, 4600 Rickenbacker Causeway, Miami, FL 33149-1098, U.S.A.

Alicia Villamizar
Universidad Simon Bolivar, Apartado 89000 Caracas 1080-A, Venezuela

Cesar L. Villanoy
Marine Science Institute, University of the Philippines, 1101 Diliman, Quezon City, Philippines

Susana Villanueva-Fragoso
Institute for Marine and Limnology Sciences, National Autonomous University of Mexico, Marine Pollution Laboratory, P.O. Box 70305, México City 04510 D.F., México

Maika Vuki
Chemistry Department, University of the South Pacific, P.O. Box 1168, Suva, Fiji

Veikila C. Vuki
Marine Studies Programme, University of the South Pacific, P.O. Box 1168, Suva, Fiji

Greg Wagner
Department of Zoology and Marine Biology, University of Dar es Salaam, Dar es Salaam, Tanzania

Diana I. Walker
Department of Botany, The University of Western Australia, Perth, WA 6907, Australia

Jan Warzocha
Sea Fisheries Institute, 1 Kollataja Str., PL 81-332 Gdynia, Poland

Maggie Watson
ICLARM Caribbean and Eastern Pacific Office, PMB 158, Inland Messenger Service, Road Town, Tortola, British Virgin Islands

Tom S. Weingartner
University of Alaska Fairbanks, Institute of Marine Science, School of Fisheries and Ocean Sciences, Fairbanks, AK 99775-7220, U.S.A.

Yan Wen
South China Sea Institute of Oceanology, Chinese Academy of Sciences, Guangzhou 510301, People's Republic of China

Frank Whitney
Institute of Ocean Sciences, P.O. Box 6000, Sidney, B.C., V8L 4B2, Canada

Mark Whittington
Frontier International, Leonard St., London EC2A 4QS, U.K.

Sheila Wiegman
American Samoa Environmental Protection Agency, Pago Pago, American Samoa 96799, U.S.A.

Clive R. Wilkinson
Australian Institute of Marine Science, PMB No. 3, Townsville MC 4810, Australia

Simon C. Wilson
Department of Biological Sciences, Warwick University, Coventry, CV4 7RU, U.K. and P.O. Box 2531, CPO 111, Seeb, Oman

Douglas C. Wilson
Institute for Fisheries Management and Coastal Community Development, P.O. Box 104, DK-9850 Hirtshals, Denmark

Poh-Poh Wong
Department of Geography, National University of Singapore, Singapore 119260

Elizabeth Wood
Marine Conservation Society, 9 Gloucester Road, Ross on Wye HR9 5BU, U.K.

Bruce A. Wright
Alaska Region, National Marine Fisheries Service, National Oceanic and Atmospheric Administration, 11305 Glacier Highway, Juneau, AK 99801, U.S.A.

Rudolf S.S. Wu
Department of Biology and Chemistry, City University of Hong Kong, 83 Tat Chee Avenue, Kowloon, Hong Kong

Sandy Wyllie-Echeverria
School of Marine Affairs, University of Washington, Seattle, WA 98105, U.S.A.

Alejandro Yáñez-Arancibia
Department of Coastal Resources, Institute of Ecology A.C., Km 2.5 Antigua Carretera Coatepec, P.O. Box 63, Xalapa 91000, Veracruz, México

Leon P. Zann
School of Resource Science and Management, Southern Cross University, P.O. Box 57, Lismore, NSW 2480, Australia

Eddy Y. Zeng
Southern California Coastal Water Research Project, 7171 Fenwick Lane, Westminster, CA 92683, U.S.A.

Scott A. Zengel
Research Planning, Inc., 1121 Park St., Columbia, SC 29201, U.S.A.

ACKNOWLEDGEMENTS

Several people greatly facilitated the logistical and editorial work of this series of 136 chapters. I am very grateful to Professor Leon Zann, in Australia, and Dr Jack Pearce in the USA who greatly assisted in the process of identifying sensible and manageable regions in their own respective parts of the world, and who helped to identify excellent people or groups of people to write about them. Both of them also contributed more than one excellent chapter themselves. Jack Pearce was also the co-editor of a more-or-less random collection of 16 of these chapters for a special issue of *Marine Pollution Bulletin* published simultaneously in 2000.

Bathymetry in the figures for these volumes was taken from 'GEBCO-97: The 1997 Edition of the GEBCO Digital Atlas'. This excellent product is published on behalf of the Intergovernmental Oceanographic Commission (of UNESCO) and the International Hydrographic Organisation as part of the General Bathymetric Chart of the Oceans (GEBCO) by the British Oceanographic Data Centre, Birkenhead. Further details are available at www.bodc.ac.uk. I wish to thank those who produced this invaluable digital data set for granting permission to use it. Coastlines, political boundaries and other cartographic details including some of the place names were taken from Europa Technologies 'Map elements: International & global map data components'. These GIS products were used in most 'Figure 1' sketch maps, and in many others. Early guidelines to authors bravely said that we would prepare 'standard' style maps from rough materials supplied by authors. The phrase 'rough materials' was taken rather literally by many, and for preparing the maps I am especially grateful to Anne Sheppard, whose interpretations of alleged draft maps often required prolonged diligence and clairvoyance. I thank Rebecca Klaus, who set up the GIS system, and I am also grateful to Olivia Langmead and Sheila O'Sullivan who provided editorial help with several draft chapters by converging numerous and variable dialects of the English language towards a common format. In the whole production process, Justinia Seaman in Elsevier provided masterful co-ordination of the huge project, and finally, when I thought that the editing process was complete, I had cause to be most grateful to Pam Birtles, whose production and copy editing skills added far more than just a final polish and a checking of references.

The final product is, of course, a credit to more than 350 authors. My emphasis throughout this series, particularly in the first two volumes, was on the lesser developed countries, precisely those areas with the least available information and commonly with the most pressing environmental difficulties. The breadth of material asked for in each chapter is considerable, and many large areas have very few resident marine scientists, who are in any case very over-stretched. Some had the data but insufficient time or resources to easily compile a review for a 'foreign language book'. Several had difficulty persuading their governments to allow them to do so, or to present data which might be embarrassing to their own employers. Several chapters have a long and interesting tale behind their gestation. That so many people did write is gratifying: an 'ordinary' chapter from some areas is anything but an ordinary achievement, when a range of obstacles conspired to prevent it. From personal experience, I know that in several countries these obstacles can include considerable censure and risk when describing, for example, environmental problems; which is understandable when it is realised that continuing aid may depend on the government pretending to comply with imposed 'sustainable use' measures. We should all be grateful to these authors, and I am also grateful for their subsequent acceptance of my sometimes drastic editorial changes done in the interests of brevity (mostly), format (usually) and language (sometimes). I hope that one of the benefits of this series will lie in its provision of information, so that one place may learn from the problems of another before repeating the same mistakes, and thus avoid impoverishing yet another bit of the world's coast and its people and, hopefully, reversing many of the problems.

Charles Sheppard

INTRODUCTION

Charles Sheppard

This series of articles reviews the condition of our coastal and marine environment at the turn of the millennium. The recent past has been a period when unprecedented demands have been made on global resources, generated both by higher expectations and soaring populations. It has also been a time when we started to become aware that biological resources are far from being unlimited, and that they can even be destroyed locally. As a result of that realisation, the last few decades of the 20th century also became the time when we started, sometimes with faltering and feeble steps, to try and manage the ways in which we use and interact with the ecosystems on which we ultimately depend.

This series aims to take stock of the condition of our marine and coastal habitats, and to summarise the extent to which they remain intact or otherwise, the ways in which they are being impacted or reduced, and the actions which we are taking to ensure that they may, by and large, remain in a functioning, sustainable and stable condition. Most chapters offer reviews of different regions, travelling around the world roughly from north to south or west to east. Following the area reviews, a second smaller section then examines some of the major global issues, processes and species groups.

DIVIDING UP THE OCEANS

Historically, the World's oceans have been divided up in many different ways. Political divisions came first, of course. Trade and influence lay behind the divisions and, from a 'modern' Eurocentric point of view anyway, may have started with the Treaty of Tordesillas in 1494 which drew a longitudinal line in the Atlantic 370 leagues (about 1800 km) west of the Cape Verde Islands (Whitfield, 1998). Land discovered to its west would be allocated to Spain, and land to the east to Portugal. Because this arrangement was even supported by a 'Papal Bull', it must have come as something of a shock for these early Europeans to discover, when they sailed to southeast Asia to claim that area, for example, that thriving maritime trades already existed and that many countries had their own areas of influence there already. The legacy of that system is clear in South America which was later found to span this line, but its ruling later disintegrated on the 'other side' of the world. In any case, other emerging European powers, specifically the Dutch and English, were not impressed by their own exclusion from that particular partition.

More recently there are the EEZs. A modern equivalent in some ways, Exclusive Economic Zones are blocks of sea over which countries claim jurisdiction. These come in several versions and with various sizes or extensions from the shore. Running quite contrary to van der Groot's (Grotius, 1609) old concept of the freedom of the oceans, these areas are viewed like national private property to varying degrees. They may cause friction where interests clash, in the same way that other types of property ownership may cause dispute, but they have one major advantage as can be seen in these volumes, which is that ownership may also lead to a desire to conserve, in ways that a 'free commons' never did.

Biological divisions are more recent, and a little more unusual. They are, in many cases, a much more sensible way of division. One of the first biological divisions is that by FAO into its marine fishery statistical areas. No ownership issue is implied here; these are used as part of the global and regional data-gathering programmes. Areas are large and are divided partly along major longitudinal and political lines, not necessarily reflecting the most sensible ecological regions. They are administrative conveniences. In the Indian Ocean, for example, the east and west sides are split, yet we know that the fishing fleets traverse the whole ocean, some crossing the boundary more often than remaining within one side or another. The UNEP Regional Seas Programme was also primarily arranged around geo-political units, though for reasons based on conservation needs, recognising the inter-connectedness of adjacent areas and the need for effective conservation to be 'regional'. Some regions, such as the South Pacific Region, are enormous while others such as the Kuwait Action Plan Region are relatively very small. Size alone does not matter, of course, and has not made too much difference to effectiveness. But, being primarily geo-political, the fact that some regions encompass discrete ecological areas too is sometimes largely due more to fortune than design. This is not a criticism; the Programme was pioneering and achieved much, especially in areas such as the Mediterranean. And some regions such as the Mediterranean or the Red Sea, Black Sea, or Arabian Gulf do indeed include much more closely defined ecosystems.

The biome scheme (Longhurst, 1998) attempts to divide the oceans into a larger number of more ecologically sensible divisions. It covers pelagic as well as coastal zones. Another scheme of great importance to coastal areas is that of the Large Marine Ecosystems, or LMEs (Sherman, 1995). This introduced about 50 regions determined according to climatic, oceanographic, and ecosystem criteria. Some very large areas of the Regional Seas Programme have been divided into six, seven or even more separate LMEs. This scheme is appropriate here because it deals with the coastal, or at least inshore, part of the oceans, and in this series of three volumes, emphasis is also on the inshore region. Yet several, if not all, LMEs are very all-embracing, and many include wide ranges of habitats and a range of quite different ecosystems within them.

THE AREA CHAPTERS

What method of division can a series of three volumes use which is scientifically sensible and physically possible? What limitations, other than space, apply? The chapters here divide the oceans into over one hundred discrete sections. Ecological criteria for the divisions were paramount, and as far as has been possible these have been followed. But two major factors conspire to let national boundaries intrude into this ideal scheme.

Firstly, in many parts of the world, marine scientists work in strictly nationalist ways. By this I mean that the marine scientists may be, for example, fisheries scientists working for their particular government; so work within their own national borders is all that they do. This limitation is more evident in poorer countries. Even when, for instance, an aid or development programme funds visiting scientists, the work is likely to simply aid marine resource exploitation (i.e. aid the fisheries scientist mentioned above) or may assist in mitigation of coastal damage in that particular country, rather than contributing to a wider programme of research. And when that country is relatively tiny or undeveloped so that there may be only one marine scientist employed who has to cover numerous and wide-ranging subjects, not to mention immediate crises, then it is clearly difficult to retain a broad perspective anyway.

The second impediment to a purely ecological or scientific division comes from the important fact that, even at this turn of millennium, so much of the world's coastline remains relatively unknown, in a scientific sense. Several parts of Africa for example have an enormous length of coast, but apart from fisheries work perhaps, little has been done in many of them. Certainly, only little coastal ecosystem work may have been done, and perhaps even less in terms of assessments of condition and quality in the coastal zone.

It would have been easy to be very unbalanced in approach. The number of scientific publications written about the North Sea, for example, must be one thousand times greater than the number published on the same length of coast in Somalia, yet I did not allow the 'North Sea' to receive very many more pages than 'Somalia'. I followed a slight tendency to create smaller divisions where there was a lot of information, not excessively so, but enough to ensure manageable units suitable for one chapter. The balance has not been easy, but that there is a fair balance for most places is a tribute to the willing scientists, about five hundred in all, who have contributed to these volumes.

Some areas are missing. Sometimes this was because it was simply not possible to find a scientist willing to write about a particular coastline, usually one of the lesser known areas (scientifically speaking) of the world. Sometimes, the reason was because political control greatly restricts the freedom of some scientists to act, or write, objectively. One chapter was pulled, due to political pressure, after being written and edited. In several areas communications are extremely difficult, and even in today's era of phones and e-mail, a simple lack of electricity or shortage of fax paper (yes, really!) were sufficient to exhaust attempts to encourage a chapter from a relatively unknown but doubtless interesting location. Added to this sad indictment of imbalance we may also remember that the areas of coast involved are almost certainly being affected, impacted, fished and built upon to a considerable extent, causing further impoverishment. These gaps need to be filled. Another reason for gaps is that, having agreed with some scientists that they would cover an area, the chapter did not get written, leaving it too late to find a substitute; this was particularly unfortunate, but only a few places were so affected.

GLOBAL ISSUES

Following the regional chapters are nearly 30 on global issues. Here I invited authors to describe and discuss matters which are not constrained by geography. One section is on important and globally widespread habitats such as seagrasses, mangroves or coral reefs, or on major species groups such as sea-birds, turtles and cetaceans. Various processes are also examined.

Perhaps unusually for ecology and 'environmental' volumes, there is a small section on fisheries and its problems. A very few years ago fisheries was viewed as being one branch of marine science compatible with environmental issues and ecology. Today, with collapsing fish stocks in every ocean, it is viewed increasingly as a science of efficient extraction—protein extraction—driven by economics as much as it is driven by the need for protein. Ever since it was shown that the economics of whaling were such that unsustainable use is financially preferable to sustainable use (Clark, 1973) the dice has been loaded against fish stocks and other marine resources too. This might be summed up by a fishery accountant's phrase: 'someone else will get them first if we don't, and if the fish disappear we can always put our money into microchips' (the accountant is hypothetical though the sentiment is not). The quantities of by-catch today are astonishingly wasteful, as is physical damage caused by unsustainable methods of extraction which destroy habitat along with substantial parts of our integrated food webs. As predicted by Clark (1999), consequences of over-fishing will be a major subject for all marine science journals in the very near future. Artisanal fishing is not so different either. The old idea that artisanal communities, especially in the good old days, lived in harmony with nature are becoming a little jaded now, and is going the way of older anthropological work which taught us about the 'noble savage'. Dynamite, DDT bombs and many other destructive methods, multiplied by several million people, cause no less havoc than do western industrial trawlers arriving to claim what they have bought. We cannot blame hungry people, which means only that the problem is bigger.

The pressures caused by these issues are raised and discussed in most of the individual area chapters as well as in this general section. Education, legal issues, training, valuation of what we have, energy, pollutant pathways, remote sensing, antifoulant effects, dumping and so on, are also all issues which are specifically addressed in this final section.

LEARNING FROM OTHERS

I asked all authors to include notes of optimism, positive ways forward, and to highlight successes and constructive activities, and indicate the benefits obtained. As we all know, learning comes from experience, from the carrot as much as the stick. You can see in these chapters where the successes lie and how they came about. Some authors in some areas found it quite difficult to be optimistic, however. We may hope that we can learn from experience, but in editing this series, it has been clear that it is also commonly the case that collectively we do not learn very quickly, sometimes for the silliest of reasons, one being that information obtained the hard way in one country may not get learned in the next, so that it then embarks on exactly the same sequences of activities. In this way, actions are repeated again and again which have been shown to have had debilitating consequences in other countries. A major role of these reviews is to try and finally reverse the waste that this causes.

But even within one country, some coastal activities, often grant-aided, may become necessary to reduce the impact of, or repair, other activities which also were grant-aided. Again, for many areas, the excellent chapters may help avoid this too. Management is supposed to take care of such matters, but 'management' is seen in too many cases to be a euphemism for merely tinkering blindly with complex systems whose interactions between themselves and with us are poorly understood. However, even though many groups of authors found it difficult to be optimistic, the sum total as recorded here offers an unusually good opportunity for learning more about what can go right as well as what can go wrong.

It is also an opportunity to see in a regional sense what can be most successful. Possibly the information contained can provide material for later 'meta-analysis' too. This should further help us put a perspective on things which a single country approach, or even a single region approach, cannot do. To give one simple example of what I mean, many poor countries seem to feel that tourism, sometimes wishfully called eco-tourism, will supply the cash needed for their sustainable development. Within one particular country, this may seem reasonable. But add together all the places in poor countries that are pinning their hopes on such tourism for their salvation, and we probably would find that for this to work, all the people in the rich countries would need to be permanently on vacation! In other words, such solutions may not 'scale up' from one country to the world generally. Like ecology generally, scale is everything.

The emphasis throughout has been on the coastal zone: the area where over a quarter of mankind lives, down to the edge of the continental shelf where man's activities and effects are substantial. It does not ignore the deep water areas of the oceans, but these deep areas are not focused upon here. A fourth volume and a different editor would be needed for that. Most chapters do, of course, explain very clearly the importance and effects of offshore areas and events, of ocean currents, climate and of upwelling, of other oceanographic features, of warming and of El Niño, and of all their interactions on the ecology of their area, and on the way that mankind uses and interacts with the sea.

Hundreds of authors have been involved in these volumes. Many are leading scientists, whose names will be familiar. Many others, perhaps from some lesser known areas, have previously had little experience of publishing in an international forum, so while my trepidation was great when I started this idea, the result has been most gratifying. The knowledge of all has come through authoritatively and clearly, and all have ensured that these chapters are, at the very least, interesting. But they are more than that. They are valuable, as they focus our attention onto many of the issues which simply must be resolved if we are to successfully and sustainably use our seas and coastal areas in the decades to come.

REFERENCES

Clark, C.W. (1973) Profit maximization and the extinction of animal species. *Journal of Political Economy* **81**, 950–961.

Clark, R.B. (1999) Looking back. *Marine Pollution Bulletin* **38**, 1057–1058.

Grotius (1609) (van der Groot). *Mare Liberum (The Freedom of the Sea)*. English translation: Clarendon Press, Oxford.

Longhurst, A. (1998) *Ecological Geography of the Sea*. Academic Press, San Diego. 398 pp.

Sherman, K. (1995) Achieving regional cooperation in the management of marine ecosystems: The use of the large marine ecosystem approach. *Ocean & Coastal Management* **29**, 165–185

Whitfield, P. (1998) *New Found Lands. Maps in the History of Exploration*. The British Library, London. 200 pp.

THE EDITOR

Charles Sheppard
Department of Biological Sciences
University of Warwick
Coventry CV4 7AL, U.K.

Chapter 1

THE SEAS AROUND GREENLAND

Frank Riget, Poul Johansen, Henning Dahlgaard, Anders Mosbech,
Rune Dietz and Gert Asmund

The seas around Greenland cover an area from the High Arctic in the north to the Subarctic in the south. The area is characterised by a cold climate with extreme variation in light and temperature, short summers and extensive ice cover in winter. Some areas are ice-covered all year round. Water masses from the Polar region enter the areas through the Fram Strait in the east and the Nares Strait and Canadian Archipelago in the west. At southeast Greenland, a branch of the relatively warm North Atlantic Current joins the East Greenland Current from the north. The currents flow around Cape Farewell and northward along the West Greenland coast, the most productive sea in Greenland.

Only 56,000 people live in Greenland. The seas are mainly used for fishing and hunting. Today, fishing is mainly carried out by a modern trawler fleet. The fish species in the area are heavily exploited, as are some of the marine mammals and seabird populations. Both onshore and offshore mineral exploration activities are currently undertaken; however, at present, no minerals are exploited in Greenland.

Contaminant levels in the Greenland marine ecosystem are relatively low compared with more southern latitudes. Exceptions are the high Cd and Hg concentrations found in the higher trophic levels, as a result of biomagnification. Although levels of persistent organic pollutants are relatively low compared with industrialised areas, these compounds are of concern because of their ability to biomagnify, and because in some areas marine mammals and seabirds constitute a significant part of the human diet.

The main sources of contamination are believed to be the industrialised areas in Europe, Russia and USA. Pollutants are transported to the areas around Greenland by the atmosphere and by the marine currents; however, transportation by ice may also play a role. The prevailing wind patterns, especially in winter, transport air masses from industrialised areas to the Arctic. The cold Arctic climate seems to create a sink for certain compounds, resulting in an accumulation. Local sources of pollutants in the marine environment around Greenland play a minor role. Three major mines have been in production in Greenland, and elevated heavy metal levels have been observed in fjord areas within approximately 40 km from the mine sites.

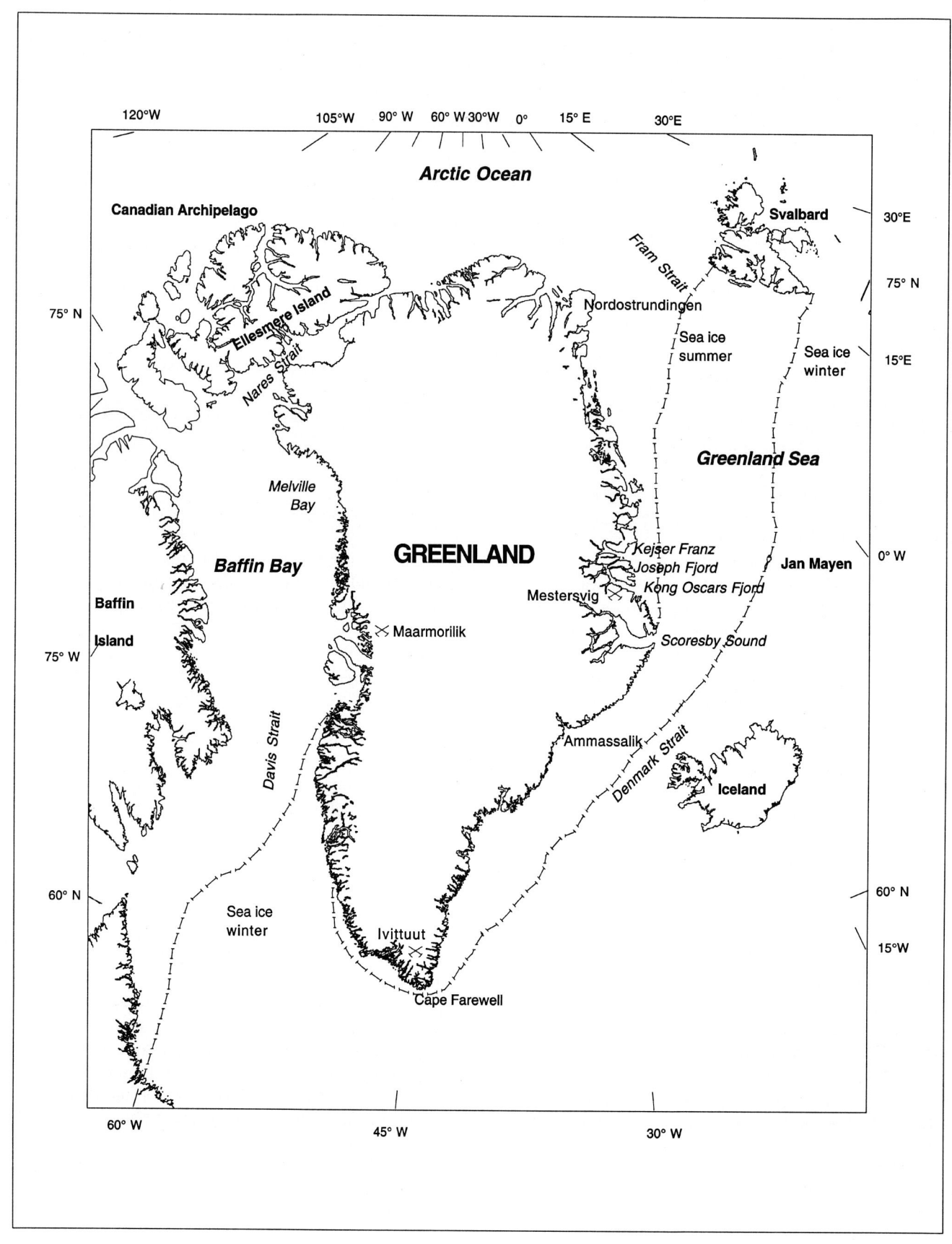

Fig. 1. Map showing Greenland and surrounding seas.

OCEANOGRAPHY

The physical oceanography of the Greenland waters has been described in detail by Buch (1990), and Valeur et al. (1996).

The Greenland Sea

Waters from the Arctic Basin are transported southward through the Fram Strait along the east coast of Greenland to the Greenland Sea (Figs. 1 and 2). The East Greenland Current flows over the Greenland shelf. During spring and early summer, it carries large amounts of pack ice along with it. The surface layer in the eastern part of the Greenland Sea is dominated by the northward-flowing Norwegian Atlantic Current—an extension of the North Atlantic Current.

Cooling of the surface waters during the winter period brings about a vertical exchange of water masses. The top layer of water sinks from the surface to the deeper water and is replaced by the warmer bottom water. This process of deep water formation may be one of the most important driving forces of current systems in the northern hemisphere. Furthermore, deep water formation has attracted special attention in climate modelling and prediction. The Greenland Sea is one of the places where the increasing carbon dioxide pollution in the atmosphere can be transferred to the deep sea.

In the Greenland Sea ice extends seaward successively during the winter. In some places isolated areas of open water, known as "polynyas", are found throughout the winter. The Northeast Water at Nordostrundingen is such a polynya. Winds, sea currents, tidewater currents and upwelling are all important factors affecting the complex dynamics of polynyas. These areas are important wintering areas for some marine mammals, and may also be important areas for the interchange of semi-volatile contaminants between water and air.

The Denmark Strait

The Denmark Strait is the passage between Greenland and Iceland. There is a submarine ridge between Greenland and Iceland. The East Greenland Current flows southward along the coast of East Greenland and rounds Cape Farewell. A branch of the North Atlantic Current, known as the Irminger Current, turns westward along the west coast of Iceland. Part of the current turns further towards Greenland, where it flows southward parallel to the East Greenland Current down to Cape Farewell, where it joins the East Greenland Current (Fig. 2).

The Davis Strait

The water masses flowing northward along the West Greenland coast originate partly from the cold East Greenland Current, and partly from the warmer Irminger Current. These two water masses mix intensely. The hydrographic conditions along West Greenland depend greatly on the relative strengths of these two currents. The West Greenland Current, which flows over the Greenland shelf, continues northward until it reaches a latitude of about 65–66°N in the Davis Strait. At this point, a part of it turns westward and unites with the south-flowing Baffin Current along the Canadian east coast, and a part continues northward in Baffin Bay. There is a submarine ridge approximately 700 m deep between Greenland and Baffin Island at about 67°N. South of the ridge, temperatures are between 3° and 4°C, while north of the ridge they are about zero or lower at depths of up to 1000 m and more.

A number of banks are found along the continental shelf of the west coast. The hydrographic conditions in these coastal areas are characterised by upwelling of nutrients from the bottom waters and stable stratification of water masses during summer making these waters among the most productive in Greenland. Productivity is also high in comparison with temperate seas such as the Danish waters and the northern North Sea (Smidt, 1979).

The marine climate off West Greenland has undergone drastic changes during the recent century. A strong and rapid increase in temperatures (1.5–2°C) started around 1920, and temperatures remained high until the late 1960s, when the first of three cold periods was observed. The following two cold periods were experienced in the early 1980s and the early 1990s. There is evidence that the distribution of some fish species has moved to more southerly areas during the latest colder period, especially on the Canadian side of the Davis Strait (NAFO, 1994).

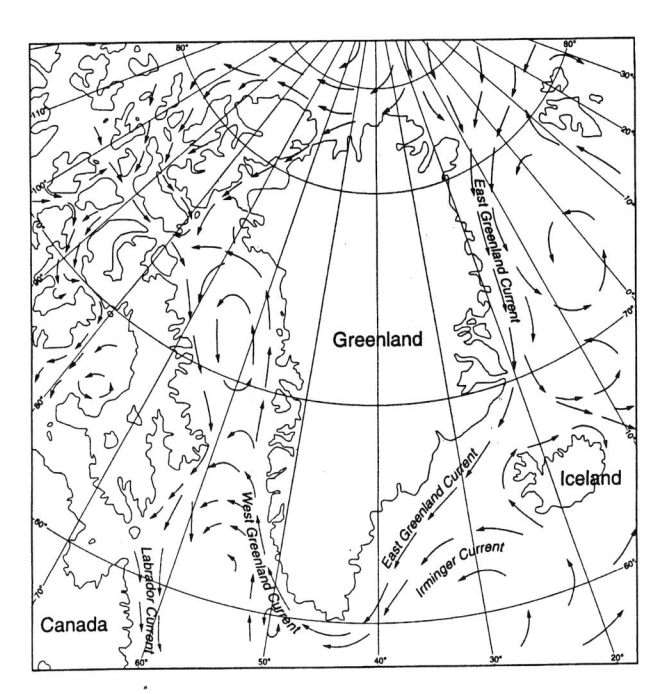

Fig. 2. General ocean surface circulation around Greenland. Modified from Oceanographic Atlas of the Polar Seas (US Navy Hydrographic Office, 1958).

Baffin Bay

Baffin Bay receives polar water from the Arctic Ocean through the Nares Strait and the Canadian Archipelago. This polar water flows southward along the eastern Canadian coast. Baffin Bay is covered by ice during winter, and in very cold winters, the ice can cover the whole Davis Strait. In summer the ice breaks up and drifts south along Canada's east coast (see Fig. 1). An important polynya called the North Water is found during winter in Smith Sound and the northernmost Baffin Bay.

Coastal Areas around Greenland

The coast around Greenland is dominated by bedrock shorelines with many skerries and several archipelagos. Very large differences in depths can be found within a short distance in the coastal zone. Some of the world's largest fjord complexes are found in Greenland, e.g. Kejser Franz Josephs Fjord and Scoresby Sund, leading north out of the Denmark Strait. In several places the icecap reaches the coast as glaciers at the heads of fjords—so-called icefjords. Deep fjords often continue as deep channels outside the coastal line, dividing the offshore banks.

POPULATION AND MARINE RESOURCES

Greenland has a population of 56,000, 13,000 of whom live in the capital, Nuuk. Eighty percent of the population lives in coastal towns and settlements in Southwest Greenland and the Disko Bay, where most of the commercial fishing also takes place and the fish processing plants are located. Outside this area subsistence hunting and fishing are predominant.

Fishery

The seas in Southeast and West Greenland, especially in the Davis Strait, have been important fishing grounds in the 20th century. Today, commercial fishing and the fishing industry constitute the most important sources of capital for the Greenland economy. Before the fishing zone around Greenland was extended in 1977, from 12 to 200 nautical miles, fishing was mainly carried out by foreign nations. Today, however, almost all fisheries are conducted by Greenlanders. The important fish species have been: halibut at the beginning of the century, cod in the middle of the century, and shrimp and Greenland halibut in modern times.

The most important fjord and coastal fishing is for Greenland halibut. The icefjords in Northwest Greenland are especially important in this regard. This fishery is mainly longline fishing which takes place from small open boats in summer and through the ice from dog sledges in winter. In recent years the annual inshore catch of Greenland halibut has been approximately 10,000 tons. The Disko Bay, situated in the northern part of the Davis Strait, is an area of relatively uniform depth which is an important area for shrimp fishing. Along the coastal area of the Davis Strait, there is traditional small-scale fishing targeting such species as cod, Greenland cod, wolf-fish and capelin.

A number of banks are found along the continental shelf of the Davis Strait and the Irminger Sea. During the 1950s and the 1960s, 300,000–400,000 tons of cod were fished annually, mainly from these banks. Today, however, cod has almost disappeared, probably as a result of a combination of too great a fishing pressure and declining sea temperatures. During this century there seems to have been a correlation between cod abundance in the Greenland part of the Davis Strait and the temperature (Buch et al., 1993). The slopes of the banks in the Greenland part of the Davis Strait and in the Irminger Sea (at depths between 300 and 600 m) are the main shrimp fishing grounds. Shrimp fishing in the Davis Strait is one of the world's largest cold-water shrimp fisheries, with an annual catch of about 70,000 tons in recent years, corresponding to an area of 16,000 km^2 trawled each year.

A bottom trawler fishery for Greenland halibut at greater depths, down to about 1000 m off the continental slopes has recently been established in Davis Strait.

In the Irminger Sea, redfish is exploited both by pelagic and bottom trawlers both on the shelf and in the area between Greenland and Iceland. Further north in the area between East Greenland and Iceland and Jan Mayen there is an industrial fishery based on maturing capelin. An annual catch of about 1 million tons has been reported in recent decades.

The fish resources in the Davis Strait and Denmark Strait may be considered to be heavily exploited. Fisheries are regulated on the basis of advice provided by the Northwest Atlantic Fisheries Organization (NAFO) and the International Council for the Exploration of the Sea (ICES). The catch figures presented in this section were taken from the catch statistics released by these two organizations.

Hunting

Hunting has always been important in Greenland where the rich wildlife, especially marine mammals, constitute the traditional food. Commercial fishing has become increasingly important to the Greenland economy throughout the 20th century, but traditional subsistence hunting and fishing is still the main occupation in the northern municipalities.

Ringed seals (*Phoca hispida*) and harp seals (*Pagophilus groenlandicus*) are generally the most important animals hunted, with sustainable annual catch-levels of approximately 75,000 and 20,000–50,000 animals, respectively (Kapel and Rosing-Asvid, 1996; Anon., 1995). There are, however, a number of examples of marine mammal and seabird stocks in West Greenland having suffered a severe decline during the last century as a result of increased hunting, egging and disturbance. There are documented examples of changes in distribution, where the most disturbed areas have been abandoned, and examples of a reduction in total numbers, although this is more difficult to document. Well documented examples of changes in distribution and numbers include the total abandonment of terrestrial walrus (*Odobenus rosmarus*) haul-out sites in West Greenland (Born et al., 1994), the disappearance of the beluga (*Delphinapterus leucas*) population wintering in Nuuk fjord (Heide-Jørgensen, 1994), the reduced number of active harbour seal (*Phoca vitulina*) haul-outs (Teilmann & Dietz, 1994), the reduced number of breeding thick-billed murre (*Uria lomvia*) including the abandonment of colonies up to 100,000 birds (Kamp et al., 1994), severely reduced numbers of common eiders (*Somateria mollissima*) in former large colonies (Boertman et al., 1996), and reduced numbers of king eiders (*Somateria spectabilis*) at moulting sites (Frimer, 1993; Mosbech and Boertmann, 1999).

Other factors such as climate changes (Vibe, 1967) have had ecological implications which have contributed to the observed changes in the numbers and distributions of seabirds and marine mammals. However, it is beyond doubt that the increased range and speed of motorboats and better hunting equipment has had a profound effect on the marine mammal and seabird fauna during the last century.

A number of conservation measures have been initiated to reverse this development, based mainly on hunting seasons, quotas and protected areas, as well as public education. A hunting permit has been required to hunt in Greenland since 1993. Today there exist detailed hunting regulations on all wildlife in Greenland.

PRESENT CONTAMINANTS LEVELS

Several studies of baseline levels of heavy metals in the marine food chains and trace metals in marine sediments in the Greenland seas have been made, whereas only few studies have been made on persistent organic pollutants (POPs) and radionuclides. A recent comprehensive assessment of contaminants in the marine ecosystem of Greenland has been made in connection with the Arctic Monitoring and Assessment Programme (AMAP) (Aarkrog et al., 1997; Johansen et al., 2000).

Metals

As there is no doubt that POPs and radionuclides in the marine ecosystem are derived from sources far away from Greenland, it is difficult to distinguish anthropogenic inputs of metals from the inputs from natural processes. Most studies deal with geographical, species-specific and tissue differences of contaminant levels. Very few studies have focused on possible biological effects in the marine ecosystem.

Dietz et al. (1996) compiled the available knowledge of lead (Pb), cadmium (Cd), mercury (Hg) and selenium (Se) in Greenland marine animals. Cd levels in seabirds and mammals are highest in the Smith Sound area, whereas for Hg the highest levels are found in the Denmark Strait. Neither Pb nor Se levels appear to increase in higher trophic levels, whereas this is clearly the case for Cd and Hg (Figs. 3 and 4). Because of this biomagnification, Cd and Hg levels in the top of the marine food chain are high. In many cases levels of Cd in seabirds and seals are higher in the Greenland marine ecosystem than in temperate areas, whereas this is not the case for Hg. Levels of Cd in some seabirds and marine mammals from Northwest Greenland might be high enough to cause kidney damage, however no such effects have been documented (Nilsson, 1997). The effects of Hg on the animals are difficult to assess. In all marine animal populations, even the most exposed ones, Se is abundant enough to detoxify Hg.

Information regarding metal concentrations in the marine environment over time is very sparse and often too short for making firm conclusions. Analysing sediment cores can give information about time trends in the seas. Samples throughout the Arctic, including Baffin Bay and the Davis Strait, have shown an increase in Hg concentrations in the upper sediment layers (Asmund and Nielsen, 2000). It is, however, uncertain as to whether diagenic processes can cause the movement of certain metals within the sediment column thereby affecting concentrations, and overall the chemical composition reflects the textural and mineralogical local differences (Loring and Asmund, 1996).

Indirect evidence of contamination of the Greenland marine environment can be obtained by analysing ancient material such as feathers of seabirds, human mummies and ice. Black guillemot (*Cepphus grylle*) is a stationary fish-eating seabird living in coastal zones all over Greenland. The Hg concentrations found in feathers of museum specimens indicate a slight increase in contamination over a period of about 100 years (Appelquist et al., 1985). Most of the food consumed by local people living along the coast of the Davis Strait is derived from the marine environment (seal, whale and birds). An increase has been found in the levels of Pb and Hg in hair and Pb in human bone tissues when compared with mummies from about the year 1500 (Grandjean, 1989; Hansen et al., 1989). There is also a documented increase for elements such as Pb, Cd, Zn and Cu in snow and ice cores taken on the Greenland ice cap (Boutron et al., 1995). Thus, there are indications that industrialisation has affected levels of contaminants found in Greenland far away from the sources.

Persistent Organic Pollutants

The concentration of POPs such as polychlorinated biphenyls (PCBs) and the chlorinated pesticide DDT also increase in higher trophic levels. A comparison of PCB and DDT levels in animals from the seas around Greenland indicates a decreasing trend from east Greenland to southwest Greenland and further north of the west coast (Fig. 5). In general,

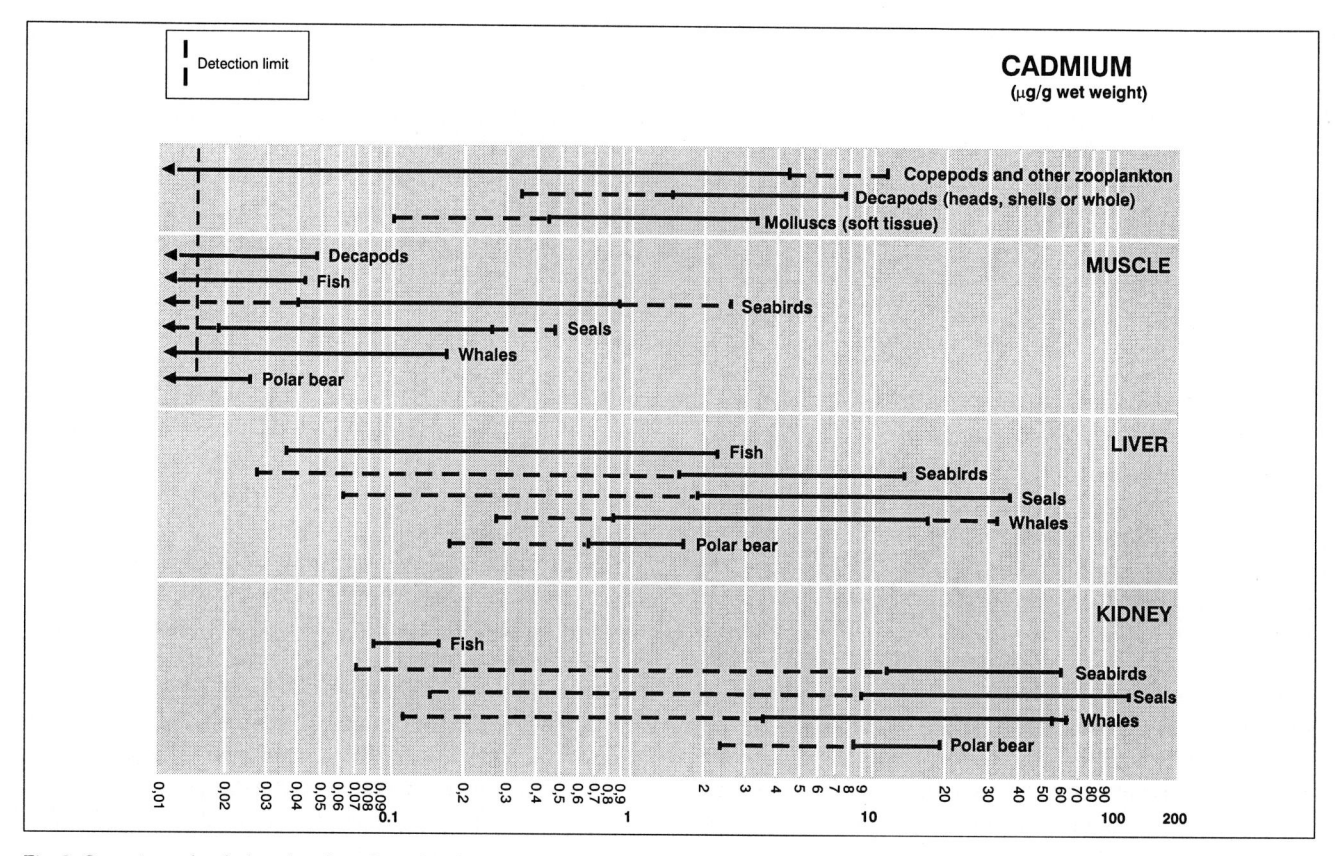

Fig. 3. Overview of cadmium levels in Greenland marine organisms (modified from Dietz et al., 1998). Solid bars show range of geometric means found among species analysed from Greenland and dotted bars ranges among species in the whole Arctic.

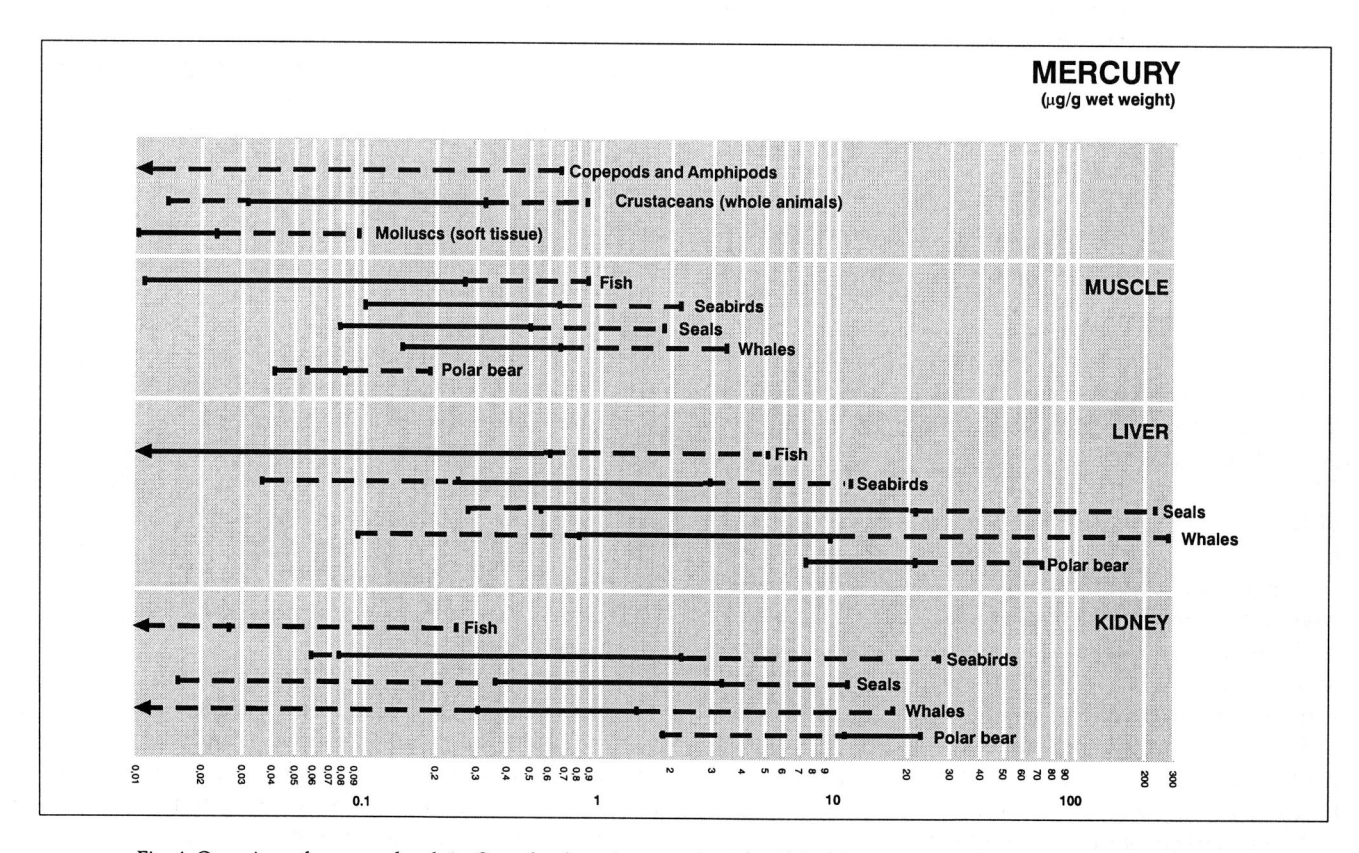

Fig. 4. Overview of mercury levels in Greenland marine organisms (modified from Dietz et al., 1998). Legend as in Fig. 3.

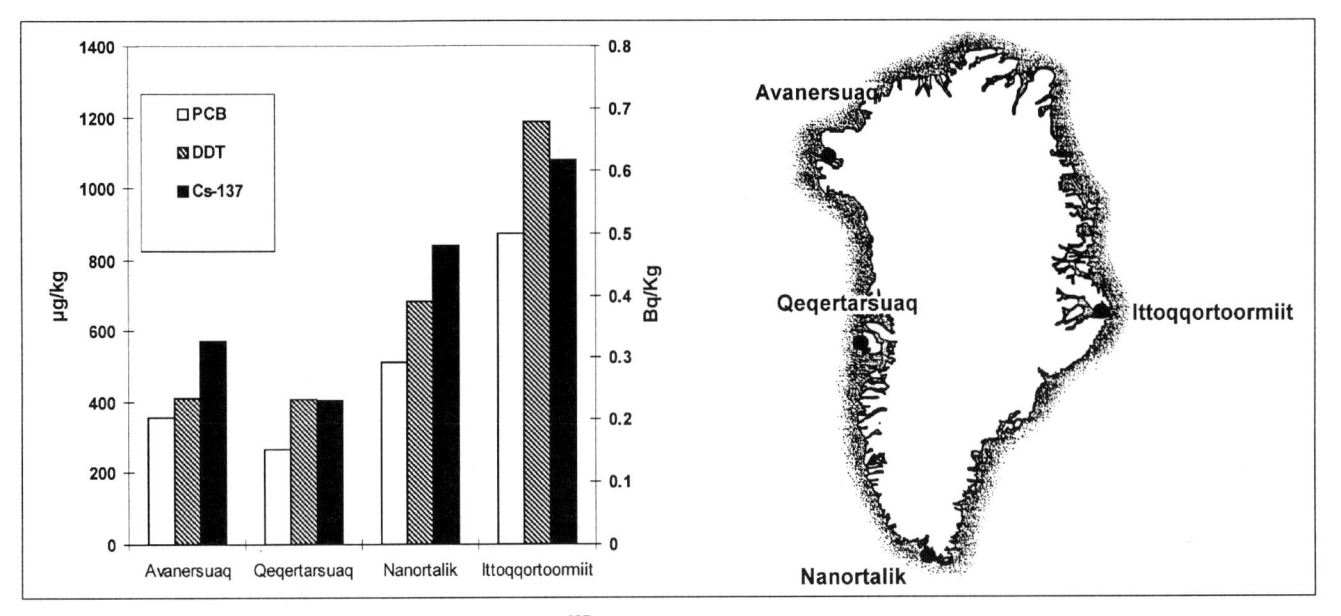

Fig. 5. PCB and DDT in blubber and ^{137}Cs in meat of ringed seals (data from Aarkrog et al. 1997).

the POP levels found in the seas of Greenland are lower than the levels found in northern Europe. However, since marine mammals and seabirds constitute major food items for the people of Greenland, there is concern that POPs may affect human health.

There are no studies of POPs in the Greenland marine ecosystem over time. However, in the Canadian Arctic area a few long-term studies have shown a decline of PCB and DDT levels in the Arctic environment over the last twenty years, probably because the use of these compounds was discontinued in the western world during the early 1970s (Muir et al., 1997). Less is known about time trends of other organic compounds.

Radionuclides

The radionuclides caesium-137(^{137}Cs), strontium-90(^{90}Sr), technetium (^{99}Tc) and plutonium (^{239,240}Pu) are, in practice, anthropogenic, as their existence in nature before the detonation of nuclear weapons was limited to extremely low levels. The main source was initially global fallout from atmospheric nuclear experiments 1955–1962, especially for ^{99}Tc and ^{137}Cs. This was later supplemented by long-distance marine transport of European coastal discharges from Sellafield and ^{137}Cs from the Chernobyl accident.

In Bylot Sound in the Thule district, the marine benthic environment was locally contaminated with approximately 0.5 kg plutonium (1 TBq) in 1968 after a nuclear weapon accident (Aarkrog et al., 1987). The main pollution is observed in the sediments and in the benthic fauna at 180–230 m.

Human exposure to radiation from radioactive contaminants in the Greenland marine environment is estimated to be insignificant (Aarkrog et al., 1997).

Levels of fallout of radionuclide ^{90}Sr have decreased in the Greenland waters since the mid 1960s, having a half-life of 10–12 years. This was also the case for ^{137}Cs, until the European discharges and Chernobyl perturbated the decline. As for ^{99}Tc, a significant increase in concentrations is expected from around 2001 due to increased discharges from Sellafield from 1994 and onwards.

SOURCES

Pollutants enter the marine environment from local sources and from sources outside Greenland via the atmosphere and ocean currents. The importance of sources and routes is not fully understood, but in general the anthropogenic contribution to the contaminant levels is dominated by sources outside Greenland. Local sources are, however, important in areas where mining has taken place (see below). Unregulated waste dumps in towns and settlements have caused local marine pollution from untreated percolate run-off and atmospheric fallout from uncontrolled waste burning (Anon., 1996). Potential effects have not been studied; however, these are expected to have been very localised because the cities and settlements have no large industries, and because the population is small and is scattered over a vast area. An action plan for improved waste handling was initiated in 1996 (Anon., 1996).

Mines

By far the most important source of pollution in Greenland to date is pollution by heavy metals from mining activities. Three major mines have been in production in Greenland: a cryolite mine in south Greenland, a lead–zinc mine in east Greenland and a lead–zinc mine in west Greenland. All are closed today. Table 1 gives some statistics of the mining operations.

Table 1
Major production statistics from Greenland mines.

Area	Ivittuut	Mestersvig	Maarmorilik
Product	cryolite	lead and zinc	lead and zinc
Operating years	1854–1987	1956–1962	1973–1990
Tonnes of ore produced	3,300,000	545,000	11,313,000
Percent zinc in ore	0.7	9.9	12.4
Percent lead in ore	0.3	9.7	4.2

Environmental monitoring of the pollution from these mines has been conducted (Johansen et al., 1985; Asmund et al., 1991; Johansen et al., 1991). A recent review of the operations, the studies conducted, an evaluation of pollutant sources and the mitigative measures taken is presented by Johansen and Asmund (1999).

Pollution of the sea with lead and zinc has been the main environmental problem at the three mines. In all three cases the pollution has been "local", affecting fjord areas within about 40 km of the mine sites. Although the three mines in Greenland are no longer operating, lead and zinc pollution still affects the fjords in the vicinity of the sites. However, heavy metal levels at the mine sites are declining.

Different sources of the lead and zinc pollution from the mines have been identified. The main source at the cryolite mine in Ivittuut is land-fill areas at the coast. The land-fill is composed of waste rock, containing lead and zinc, which are released into the sea when tidal water enters the land-fill areas. At the lead–zinc mine at Mestersvig two main sources have been identified: the main source of lead pollution seems to be spillage of concentrate on the coast, while for zinc the main source appears to be tailings which have been washed into a river and thereafter to the river delta and a nearby fjord. Several sources were identified at the lead–zinc mine at Maarmorilik: tailings, waste rock, ore crushing and transport of concentrate were the most important. In terms of the amounts of lead and zinc released into the marine environment, tailings were the most significant source of pollution. The tailings were discharged into a fjord; however, some lead and zinc dissolved in sea water and dispersed to other nearby fjords. While the mine was in production, an estimated 10–30 tonnes of lead and 30–55 tonnes of zinc dissolved in the fjord each year.

The release of heavy metals has caused significant elevations of lead and zinc in sea water, sediments and marine biota at all three sites, particularly in the intertidal zone. Brown seaweed has been used to monitor lead and zinc pollution at all sites. This has made it possible to make a direct comparison of the areas affected, as shown in Fig. 6. From the most recent studies, it appears that lead and zinc pollution is affecting the largest coastlines at Maarmorilik and Mestersvig, whereas only few kilometres of coast at Ivittuut have levels above background. A significant decline of levels is obvious at Maarmorilik. Although not shown, this is also the case for the other two sites.

In the intertidal zone blue mussels have also elevated concentrations of lead. Very high levels have been observed—up to 4000 µg/g (dry weight basis) in the most affected areas. Elevated lead concentrations have also been found in livers of stationary fish species (sculpins and catfish) and in shrimp.

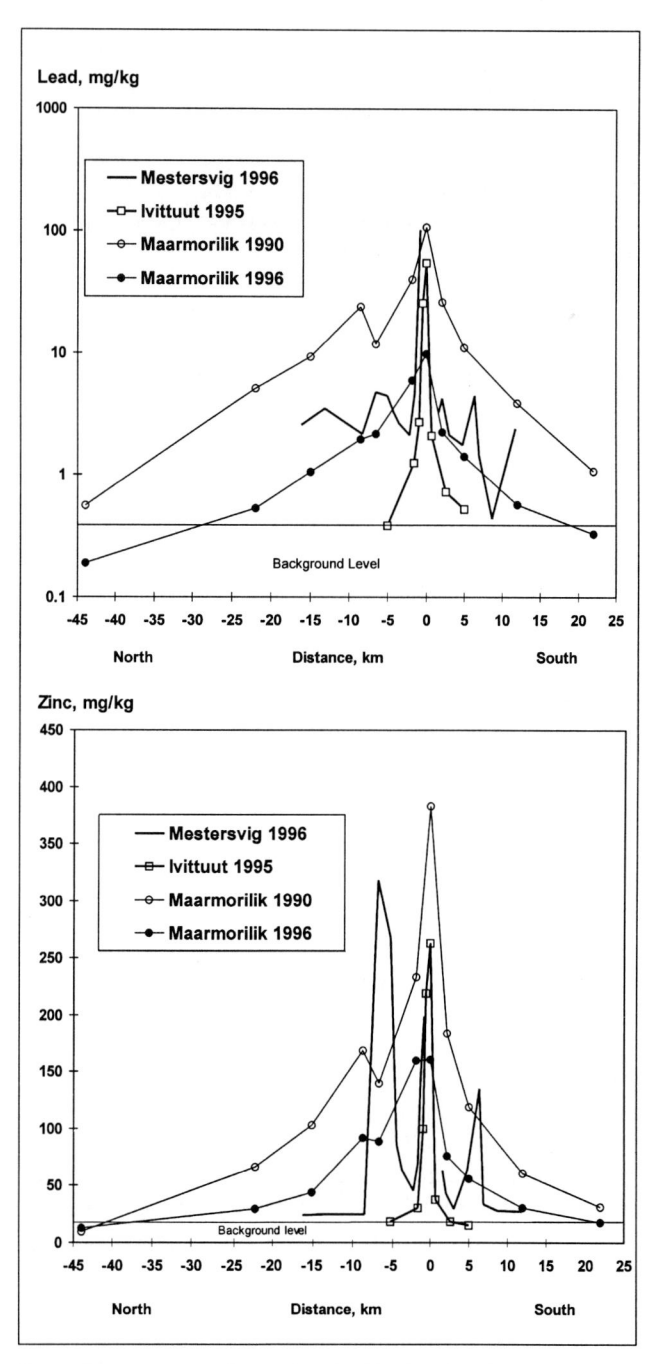

Fig. 6. Concentrations (logarithmic) of lead and zinc in new growth tips of seaweed (*Fucus spp.*) around the three major Greenlandic mines. Also shown are the intervals for background concentrations found elsewhere in Greenland.

Long-Distance Transport of Pollutants

Atmospheric Transport

The levels of especially persistent organic pollutants in the marine food chains around Greenland cannot be explained by local sources, and contaminants are likely to be derived from far away sources. One of the pathways for these compounds is the atmosphere, being the fastest and most direct route from the source.

Measurements of contaminants in the air have been carried out at research stations near the Greenland Sea, at northeast Greenland and Svalbard, near the Denmark Strait, at Iceland and near the Nares Strait, at Alert, Ellesmere Island. The contamination of the air in the Polar area is known as Arctic haze. Arctic haze is found all over the Polar area. It occurs mainly in the dark winter periods as a result of strong south to north air transport and weak pollutant transformation (Barrie and Barrie, 1990). Furthermore, during winter, the Arctic front is pushed far to the south by the prevailing high-pressure systems over the continents, and, as a result, the most polluted areas in Europe and Russia are included within the Arctic air mass (Fig. 7).

The atmospheric transport pathway is divided into two types; "one-hop" pathways and "multi-hop" pathways (Barrie et al., 1997). One-hop contaminants start their journey riding on the winds, and once landed, they usually settle. Heavy metals, with the exception of mercury and involatile organic compounds belong to the one-hop contaminants. Multi-hop contaminants are contaminants which are transported as gases. After landing, it is possible for them to become volatile again with higher temperatures in summer, and be transported further. Most organic compounds, many polycyclic aromatic hydrocarbons (PAHs) and mercury belong to this group. The chemical nature of POPs allows them to be systematically transferred via the atmosphere from warmer to colder areas, accumulating in the Arctic. This is a process called the cold-condensation effect (Shearer and Murray, 1997).

Contaminants in the air reach the sea by either wet deposition (rain, snow) or dry deposition (adhesion to particles). Contaminants from the terrestrial ecosystem may also end in the sea due to run-off from rivers. For some contaminants, concentrations are higher than in the air, suggesting that the sea may therefore act as a reservoir.

Marine Current Transport

Long-distance transport of radionuclides with the marine currents from northwest European coastal waters to the Arctic Ocean and further to Greenland has been demonstrated in several papers (Aarkrog et al., 1983; Dahlgaard, 1995; Kershaw and Baxter, 1995). This has been possible because discharges of some radionuclides, ^{137}Cs, ^{134}Cs, and ^{99}Tc, in particular, from the nuclear fuel reprocessing plant Sellafield in the UK have been high compared with other sources. The pathway for the transport of contaminants from Europe is shown in Fig. 8. The European contaminants are first transferred to the Arctic Ocean and from there they flow with the East Greenland Current. Accordingly, higher radionuclide concentrations are observed in East Greenland compared with West Greenland. It has been shown quantitatively, that a discharge of one ton per year in the Irish Sea leads to a concentration of 1–2 ng m^{-3} in the East Greenland Current 7–10 years later (Dahlgaard, 1995).

Caesium and technetium behave conservatively, i.e., their affinity to particles is sufficiently low to assure that the major fraction of the tracers stay in the water column during the approximately 10-year-long transport pathway. There is no reason to believe that other contaminants added to the European coastal waters which behave similarly would not undergo the same transportation. This is also seen with clearly non-conservative contaminants such as plutonium. Part of the Sellafield discharge for example has been transported to the Arctic (Kershaw and Baxter, 1995). East Greenland levels of POPs, like ^{137}Cs, ^{90}Sr and ^{99}Tc tend to be higher than comparable levels in West Greenland (Fig 5). If the main source of contamination was found to be in European coastal waters, as is the case for ^{99}Tc, levels in West Norway could be expected to be 5–50 times

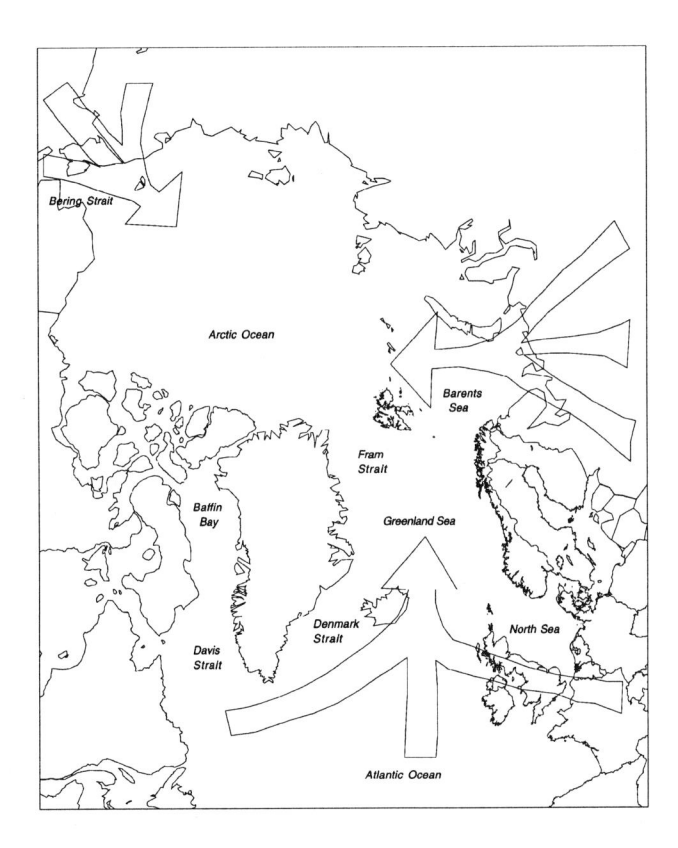

Fig. 7. Main winter transport routes (arrows) for air masses over the Arctic. Modified from Nilsson (1997).

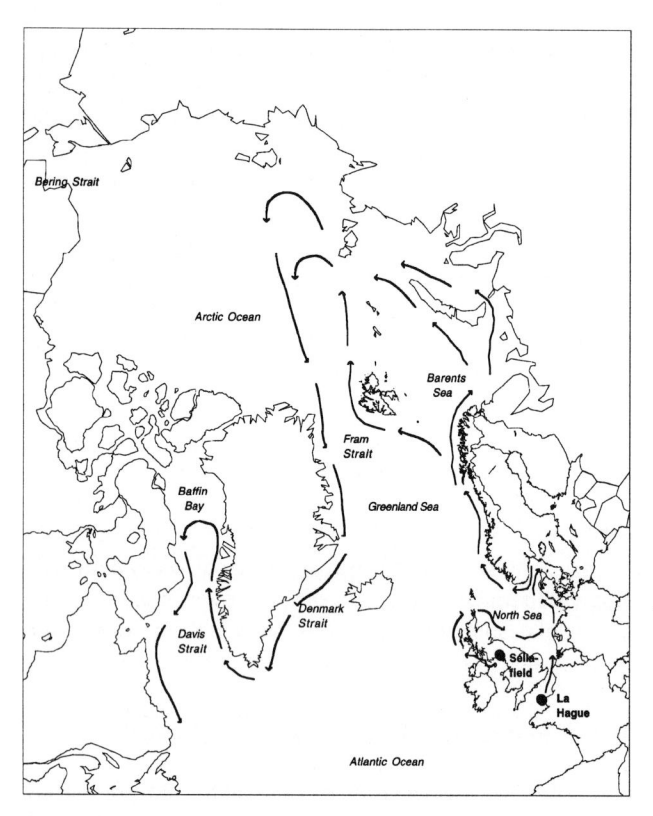

Fig. 8. Transport routes for European anthropogenic radionuclides from Sellafield and La Hague to the Arctic Ocean and the seas around Greenland.

higher than in East Greenland (Dahlgaard, 1995). PCB levels in harbour seals from the southern coast of Norway are approximately half the levels seen in seals from the north western coast of Norway (Skaare et al., 1990). A further decline has been seen in ringed seals in East Greenland, where the levels are seven times lower than at the Norwegian northwestern coast (Nilsson, 1997). This is an indication that marine transport of PCB from European waters to the Arctic is an important long-distance pathway. Siberian rivers, shelf seas and Arctic Ocean sea ice combined with atmospheric deposition may also play a significant role.

Sea Ice transport

The East Greenland Current, which originates mainly in the Arctic Ocean, flows from north to south along the entire East Greenland coast and contributes significantly to the water masses on the West Greenland coast. It has been suggested, that the vast amounts of ice in this transport system could be responsible for contaminant transport (Pfirman et al., 1995). This is consistent with the observation of higher levels in East Greenland. It is well documented that the Sellafield-derived radionuclides are transported with marine currents, but for other contaminants ice could be responsible for at least part of the long-distance transport, e.g. after deposition from Arctic haze on the Arctic Ocean ice cover (Pfirman et al., 1995).

FUTURE THREATS AND EXPECTATIONS FOR THE MARINE ENVIRONMENT

Oil Exploration

In recent years there has been an increase in offshore oil exploration activity in Greenland waters. Regional seismic programmes have been carried out over the Greenland parts of Baffin Bay, Davis Strait and Greenland Sea. More detailed seismic programmes have been conducted in the central Davis Strait (64°N), where exploratory drilling is expected to start in 2000. As early as 1976–77, five exploratory wells were drilled in the Davis Strait; however, these were found to be dry and exploration activities were ceased at that time.

The environmental impact of the regional seismic programmes is considered to have been minimal. Of greatest concern is the risk of accidental oil spills associated with oil exploration in the Greenland marine environment. Oil spills are difficult to control in icy waters, and may represent a threat, especially to seabird populations and coastal resources such as intertidal and subtidal spawning areas for capelins and lumpsucker (Mosbech et al., 1996).

Contaminants in the Marine Environment

Human activities significantly affect remote areas in the Arctic, largely as a result of distant emissions of pollutants into the atmosphere and the sea in industrialised parts of the world. The future trends of contaminants in the Greenland marine ecosystem depends on the future use of the compounds and control of their emission both on a local and a global scale.

Hg and Cd are the metals of greatest concern in the Greenland marine ecosystem. Although it may be difficult to assess the importance of the anthropogenic sources in relation to natural sources, there is evidence of increasing atmospheric Hg concentrations due to human activities. The increasing trend of Hg in biological samples and sediments, as discussed earlier, is also of concern and additional temporal studies of trends are needed. The flux of Cd through the atmosphere from human activities is small and is unlikely to have much impact on the natural cycles (Barrie et al., 1997). However, the high levels found in some marine animals are of concern.

The time trend of Pb in the Greenland snow indicates that control of emissions can be effective for the reduction of environmental contamination. Up until the mid-1960s there was a pronounced increase in Pb concentrations. This was followed by a marked decrease to present times because of the strong decline in the use of Pb alkyl additives in gasoline during the past two decades (Boutron et al., 1995).

Huge amounts of radioactive waste stored in interim depots in the north of Russia is cause for great concern for contamination of the marine environment (Sneve, 1997). The seas around Greenland are among the most threatened as the marine currents lead directly from the Russian shelf across the Arctic Ocean into the East Greenland Current.

REFERENCES

Aarkrog, A., Dahlgaard, H., Hallstadius, L., Hansen, H. and Holm, E. (1983). Radiocaesium from Sellafield effluents in Greenland waters. *Nature* **304**, 49–51.

Aarkrog, A., Boelskifte, S., Dahlgaard, H., Duniec, S., Holm, E. and Smith, J.N. (1987) Studies of Transuranics in an Arctic Marine Environment. *Journal of Radioanalytical and Nuclear Chemistry—Articles* **115**, 39–50.

Aakrog, A., Aastrup, P., Asmund, G., Bjerregaard, P., Boertmann, D., Carlsen, L., Christensen, J., Cleemann, M., Dietz, R., Fromberg, A., Storr-Hansen, E., Heidam, N.Z., Johansen, P., Larsen, H., Paulsen, G.B., Petersen, H., Pilegaard, K., Poulsen, M.E., Pritzl, G., Riget, F., Skov, H., Spliid, H., Weihe, P. and Wåhlin, P. (1997) AMAP Greenland 1994–1996. Arctic Monitoring and Assessment Programme (AMAP). Danish Environmental Protection Agency. Environmental Project No. 356, 788 pp.

Anon. (1995) Jagtinformation og fangstregistrering. *Greenland Home Rule*. 33 pp.

Anon. (1996) Handlingsplan for den fremtidige affaldshåndtering i byerne i Grønland. Report from Carl Bro as. 68 pp.

Appelquist, H., Drabæk, I. and Asbirk, S. (1985) Variation in mercury content of Guillemot feathers over 150 years. *Marine Pollution Bulletin* **16**, 244–248.

Asmund, G., Johansen, P. and Fallis, B.W. (1991) Disposal of mine wastes containing Pb and Zn near the ocean: an assessment of associated environmental implications in the Arctic. *Chemistry and Ecology* **5**, 1–15.

Asmund, G. and Nielsen, S.P. (2000) Mercury in dated Greenland marine sediments. *The Science of the Total Environment* **245**, 61–72.

Barrie, L.A. and Barrie, M.L. (1990) Chemical components of lower troposperic aerosols in the high Arctic: Six years of observations. *Journal of Atmospheric Chemistry* **11**, 211–226.

Barrie, L., Macdonald, R., Bidleman, T., Diamond, M., Gregor, D., Semkin, R., Strachan, W., Alaee, M., Backus, S., Bewers, M., Gobeil, C., Halsall, C., Hoff, J., Li, A., Lockhart, L., Mackay, D., Muir, D., Pudykiewicz, J., Reimar, K., Smith, J., Stern, G., Schroeder, W., Wagemann, R., Wania, F., and Yunker, M. (1997) Chapter 2. Sources, occurrence and Pathways. In: Jensen, J., Adare, K. and Shearer, R. (ed.), Canadian Arctic Contaminants Assessment Report, Indian and Northern Development, Canada, Ottawa, 1997, 41–63.

Born, E.W., Heide-Jørgensen, M.P. and Davis, R.A. (1994) The Atlantic walrus (*Odobenus rosmarus rosmarus*) in West Greenland. *Meddr. Grønland, Biosci.*, **40**, 33 pp. Copenhagen

Boertmann, D., Mosbech, A., Falk, K. and Kampp, K. (1996) Seabird colonies in western Greenland. National Environmental Research Institute, Denmark. NERI Technical Report No. 170, 98 pp.

Boutron, C.F., Candelone, J.-P. and Hong, S. (1995) Greenland snow and ice cores: unique archives of large-scale pollution of the troposphere of the Northern Hemisphere by lead and other heavy metals. *Science of the Total Environment* **160**, 233–241.

Buch, E. (1990) *A Monograph on the Physical Oceanography of the Greenland Waters*. Greenland Fisheries Resource Institute. 405 pp.

Buch, E., Horsted, Sv.Aa. and Hovgård, H. (1993) Fluctuations in the occurrence of cod at Greenland and their possible causes. *ICES Marine Science Symposium* **198**, 158–174.

Dahlgaard, H. (1995) Transfer of European coastal pollution to the Arctic: Radioactive tracers. *Marine Pollution Bulletin* **31**, 3–7.

Dietz, R., Riget, F. and Johansen, P. (1996) Lead, cadmium, mercury and selenium in Greenland marine animals. *Science of the Total Environment* **186**, 67–93.

Dietz, R., Pacyna, J, Thomas, D.J, Asmund, G., Gordeev, V., Johansen, P., Kimstach, V., Lockhart, L., Pfirman, S.L., Riget, F.F., Shaw, G., Wagemann, R. and White, M. (1998). Heavy Metals. Arctic Monitoring and Assessment Programme. AMAP Assessment Report: Arctic Pollution Issues. Oslo, Norway, pp. 373–524..

Frimer, O. (1993) Occurrence and distribution of king eiders *Somateria spectabilis* and common eiders *S. mollissima* at Disko, West Greenland. *Polar Research* **12(2)**, 111–116.

Grandjean, P. (1989) Bone analysis: silent testimony of lead exposures in the past. *Meddr. Grønland, Man Soc.* **12**, 156–160.

Hansen, J.C., Toribara, T.Y. and Muhs, A.G. (1989) Trace metals in human and animal hair from the 15th century graves at Qilakitsoq compared with recent samples. *Meddr. Grønland, Man Soc.* **12**, 161–167.

Heide-Jørgensen, M.P. (1994) Distribution, exploitation and population status of white whales (*Delphinapterus leucas*) and narwhals (*Monodon monoceros*) in West Greenland. *Meddr. Grønland, Biosci.*, **39**, 135–149.

Johansen, P., Hansen, M.M. and Asmund, G. (1985) Heavy metal pollution from mining in Greenland. In: Progr. Mine Water Congress, Granada, Spain, pp. 685–693.

Johansen, P., Hansen, M.M., Asmund, G. and Nielsen, P.B. (1991) Marine organisms as indicators of heavy metal pollution—experience from 16 years of monitoring at a lead zinc mine in Greenland. *Chemistry and Ecology* **5**, 35–55.

Johansen, P. and Asmund, G. (1999) Pollution from mining in Greenland—monitoring and mitigation of environmental impacts. In: Azcue, J.M. (ed.) *Environmental Impacts of Mining Activities—Emphasis on Mitigation and Remedial Measures*. Springer-Verlag, pp. 245–262.

Johansen, P., Muir, D.C.G. and Law, R.J. (eds.) (2000) Contaminants in the Greenland Environment. *The Science of the Total Environment*, Special Issue **245**, 277 pp.

Kamp, K., Nettleship, D.N. and Evans, P.G.H. (1994) Thickbilled murres of Greenland: status and prospects. In: Nettleship, D.N., Burger, J. and M. Gochfield (eds.) *Seabirds on Islands, Threats, Case Studies and Action Plans*. BirdLife International, Conservation Series no. 1., Pp. 133–154

Kapel, F.O. and Rosing-Asvid, A. (1996) Seal hunting statistics for 1993 and 1994, According to a new system of collecting information, compared to the previous list-of-game. *NAFO Scientific Council Studies* **26**, 71–86.

Kershaw, P. and Baxter, A. (1995) The transfer of reprocessing wastes from north-west Europe to the Arctic. *Deep-Sea Research Part II—Topical Studies in Oceanography* **42**, 1413–1448.

Loring, D.H. and Asmund, G. (1996) Major and trace-metal geochemistry of Greenland coastal and fjord sediments. *Environmental Geology* **28**, 2–11.

Mosbech, A. and Boertmann, D. (1999) Distribution, abundance and reaction to aerial surveys of post-breeding King Eiders (*Somateria spectabilis*) in Western Greenland. *Arctic* **52**, 188–203.

Mosbech, A., Dietz, R., Boermann, D. and Johansen, P. (1996) Oil exploration in the Fylla Area, an initial assessment of potential environmental impacts. National Environmental Research Institute, Denmark. NERI Technical Report No. 156, 92 pp.

Muir, D.C.G., Braune, B. DeMarch, B., Norstrom, R.J., Wagemann, R., Gamberg, M., Poole, K., Addison, R., Bright, D., Dodd, M., Duschenko, W., Eamer, J., Evans, M., Elkin, B., Grundy, S., Hargrave, B., Johnston, R., Kidd, K., Koening, B., Lockhart, L., Payene, J., Peddle, J. and Reimer, K. (1997) Chapter 3. Ecosystem Uptake and Effects. In: Jensen, J., Adare, K. and Shearer, R. (eds.), Canadian Arctic Contaminants Assessment Report, Indian and Northern Development, Canada, Ottawa, 1997, 183–294.

NAFO (1994) Report of the symposium Impact of Anomalous Oceanographic Conditions at the Beginning of the 1990s in the Northwest Atlantic on the Distribution and Behavior of Marine Life. *NAFO Science Council Studies* **24**, 3–5.

Nilsson, A. (1997) Arctic Pollution Issues: A State of the Arctic Environment Report. AMAP. Arctic Monitoring and Assessment Programme. Oslo 1997. 188 pp.

Pfirman, S.L., Eicken, H., Bauch, D. and Weeks, W.F. (1995) The Potential Transport of Pollutants by Arctic Sea Ice. *Science of the Total Environment* **159**, 129–146.

Shearer, R. and Murray, J.L. (1997) Chapter 1. Introduction. In: Jensen, J., Adare, K. and Shearer, R. (ed.), Canadian Arctic Contaminants Assessment Report, Indian and Northern Development, Canada, Ottawa, 1997, pp. 15–24.

Skaare, J.U., Markussen, N.H., Norheim, G., Haugen, S. and Holt, G. (1990) Levels of polychlorinated biphenyls, organochlorine pesticides, mercury, cadmium, copper, selenium, arsenic, and zinc in the Harbour Seal, *Phoca vitulina*, in Norwegian waters. *Environmental Pollution* **66**, 309–324.

Smidt, E.L.B. (1979) Annual Cycles of Primary Production and of Zooplankton at Southwest Greenland. *Meddr. Grønland, Biosci.*, **1**, 55 p.

Sneve, M.K. (1997) The *Lepse* floating base project. Int. Symp. Radionuclides in the Oceans, RADOC 96–97, Part 2, Norwich, UK.

Teilmann, J. and Dietz, R. (1994) Status of the Harbour Seal, *Phoca vitulina*, in Greenland. *Canadian Field-Naturalist* **108**(2), 139–155.

Valeur, H.H., Hansen, C., Hansen, K.Q., Rasmussen, L. and Thingvad, N. (1996) Weather, sea and ice conditions in eastern Baffin Bay, offshore Northwest Greenland., A review. Danish Meteorological Institute. Technical Report No. 96–12. 36 pp.

Vibe, C. (1967) Arctic animals in relation to climatic fluctuations. *Meddr. Grønland*, **170**(5), 227 pp.

THE AUTHORS

Frank Riget
*National Environmental Research Institute,
Department of Arctic Environment,
P.O. Box 358, DK-4000 Roskilde, Denmark*

Poul Johansen
*National Environmental Research Institute,
Department of Arctic Environment,
P.O. Box 358, DK-4000 Roskilde, Denmark*

Henning Dahlgaard
*Risø National Laboratory,
DK-4000 Roskilde, Denmark*

Anders Mosbech
*National Environmental Research Institute,
Department of Arctic Environment,
P.O. Box 358, DK-4000 Roskilde, Denmark*

Rune Dietz
*National Environmental Research Institute,
Department of Arctic Environment,
P.O. Box 358, DK-4000 Roskilde, Denmark*

Gert Asmund
*National Environmental Research Institute,
Department of Arctic Environment,
P.O. Box 358, DK-4000 Roskilde, Denmark*

Chapter 2

THE NORWEGIAN COAST

Jens Skei, Torgeir Bakke and Jarle Molvaer

A dominant feature of the Norwegian coastline is the deep Norwegian Trench to the south, outside a transition zone of skerries and islands and shallow silled fjords. The latter frequently contain anoxic water. To the west and north the coast is characterised by a shallow shelf, islands and deep and long fjords dissecting the coastline. These fjords are usually well ventilated due to deep sills. The long coastline encompasses gradients in climate, depth regimes, physio-chemical factors and varied biological habitats. The majority of the human population is situated along the coast and the environmental impact from domestic and industrial waste is particularly noticeable on the southern coastline where the fjords are not very suitable as recipients of waste. As a consequence of the pollution, health warnings have been issued concerning consumption of seafood in more than 10 fjords due to high levels of contaminants (dioxins, PCBs, PAHs, cadmium). Remedial actions have been taken to reduce contaminant input during the last decade, and one of the main environmental challenges at the turn of the millennium will be to reduce the impacts from secondary sources (contaminated sediments and dump sites). Offshore the main challenge will be to assess the potential effects of large volumes of contaminated produced water, and management of the many drilling waste piles on the seabed when the fields are closed down. An overall challenge for the Norwegian coast is the potential impact of climatic changes altering the major current systems, the fresh water input and the transboundary impacts from downstream effects of atmospheric and water transport of contaminants.

Seas at The Millennium: An Environmental Evaluation (Edited by C. Sheppard)
© 2000 Elsevier Science Ltd. All rights reserved

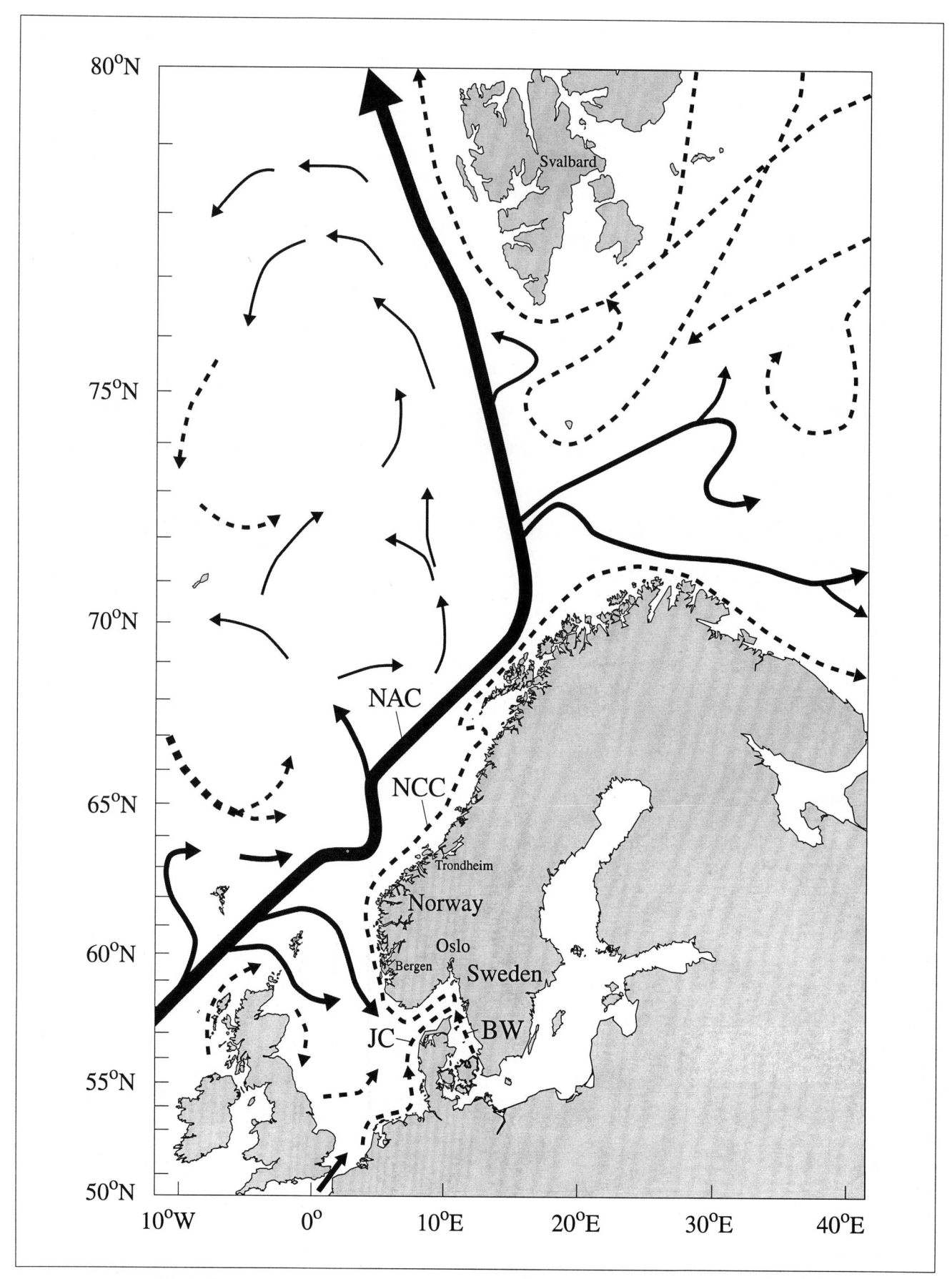

Fig. 1. The Norwegian coast. Arrows indicate major current systems (directions and magnitudes).

THE DEFINED REGION – ENVIRONMENTAL SETTING

The Norwegian coastline is the most fjord-dissected of any country in the world (Gregory, 1913), and appropriately the term fjord is of Norwegian origin. There are some 200 principal fjords along the mainland coast between 58°N and 71°N and another 35 along the Svalbard islands. The total coastline is about 21,000 km long (Fig. 1).

A fjord is a type of estuary: a semi-enclosed coastal body of water which has a free connection with the open sea and within which sea water is measurably diluted with fresh water derived from land drainage (Pritchard, 1967).

There are two major fjord regions of the world, a belt north of 43°N and a belt south of 42°S. These fjord coasts are commensurate with areas that were previously or are presently glaciated (Syvitski et al., 1987). The environmental setting of fjords is closely interrelated with respect to geomorphology, climatic conditions, water circulation, sediment sources and inputs and biotic factors. Most fjords have their main fluvial input at their head, with minor contributions at widely spaced point sources from their side-entry drainage basins. Norwegian mainland fjords locally receive meltwater from glaciers. Annual precipitation is variable along the coast (<1 m yr^{-1} to >2 m yr^{-1}), being greatest along the SW coast. The regulation of river flow for hydroelectric power greatly influences the regular seasonal effects of freshwater and sediment discharge into many Norwegian fjords.

Sills are commonly found at the mouth or within fjords, separating them into several basins. They may result from glacial deepening of the fjord basin compared with the adjacent shelf. Along the southern coast of Norway, fjords are short, relatively shallow, and with shallow sills, frequently containing stagnant bottom water, intermittently or permanently (Strøm, 1936). Minimal tides (<0.5 m amplitude) also contribute to bottom water stagnancy. Along the south-east Skagerrak coast, the embayments are small and irregular. The west coast fjords are long, narrow, steep sided, deep, and often have relatively deep sills (Holtedahl, 1967). Sognefjord (200 km long and 1300 m deep) and Hardangerfjord (180 km long and 900 m deep) are among the world's largest and deepest fjords. In the north, the mainland fjords are somewhat shorter, wider, and often without pronounced sills. Larger tidal amplitudes provide efficient flushing of fjord basins, and stagnant bottom water is rarely recorded. The Svalbard fjords are wider still, somewhat longer (five exceed 100 km) and most are shallower than 300 m.

Another geomorphological feature of the Norwegian coastline is the strandflat; the rim of ice-scoured uneven rocky ground which occurs along large parts of the Norwegian west and north-west coast to the seaward of the higher land. A great part of the strandflat is made up of low islands and skerries, and it continues below sea level in innumerable shallow rocks, until the sea-bottom drops, often very abruptly, towards deeper water (Holtedahl, 1960). Striking features are hat-like islands with a strandflat zone encircling a fairly high mountain.

Climate is of great significance to present-day and continuing modification of fjord morphology and sediment input. Climate has a major effect on the rates of weathering of rocks, and hence on sediment yields. Since climate governs the amount and type of precipitation, run-off, and water temperature, it also determines the style of estuarine circulation (Syvitski et al., 1987). Although Norway is within the same latitudinal range as Greenland or the Canadian Archipelago, the Gulf Stream or the North Atlantic Current conveys warm water from the Gulf of Mexico along the entire length of the Norwegian coast, creating a far warmer climate than on the Western side of the North Atlantic.

A pattern of two-layer flows with entrainment of marine water into the surface plume has become synonymous with fjord circulation: an outward-flowing surface layer and an inward-moving compensating current, compensating for the loss of salt entrained in the surface layer. Deep water circulation in fjords is to a large extent governed by density variations in coastal water. The frequency of these deep water renewals is variable, depending on meteorological conditions.

The typical feature of the Norwegian coastline is the tremendous north—south gradient due to climatic and geomorphological factors, including fjords at Svalbard at latitudes between 76°N and 80°N and fjords on the mainland between 58°N and 71°N (arctic to boreal climate). Additionally, there are strong gradients in salinity, nutrients, primary production, and in some cases oxygen, going from the continental shelf towards the heads of fjords.

The circulation in fjords also creates sedimentological, biogeochemical and biological gradients of environmental significance. Fjord circulation also influences the distribution and dispersion of contaminants. Environmental concerns are more important with respect to Norwegian fjords than for any other fjord country, as most of the population is situated at the coast and main cities at the head of fjords. As with coastal waters in most parts of the world, fjords are vulnerable to man's influences.

Due to geomorphological and hydrological factors (sills and restricted water circulation) fjords, being the buffer zone between land and the coast, act as traps for contaminants, where retention in sediments plays a major role (Skei, 1981). As a consequence local environmental problems occur, particularly in fjords acting as recipients for industrial and urban waste.

NATURAL ENVIRONMENTAL VARIABLES

Physical Forces

The annual runoff from the mainland of Norway for the standard normal period of 1931–1960 is estimated to be approximately 13000 m^3/s, of which 12%, 33%, 47% and 8%

enters the Skagerrak, North Sea, Norwegian Sea and the Barents Sea, respectively. Typical variations include nival regimes bordering the Skagerrak and the Barents Sea and more pluvio-nival regimes along the coastal and fjord districts bordering the North Sea and the Norwegian Sea.

The Norwegian coastal water is basically a mixture of two water masses: Atlantic water (salinity >35) and river water. Most of the Atlantic water enters the North Sea through the passages between the Faroe Islands and Scotland and between the Faroe Islands and Norway (Fig. 1). Most of the fresh water comes from three sources: local runoff to the coast, the Baltic Sea and from the large rivers draining to the southern part of the North Sea. These water masses combine to form the Norwegian Coastal Current (NCC) which is a dominant feature of the Norwegian coastline from the Swedish border to Stad at 62°N. It is characterised by high variability in time and space, and has a wedge-like appearance. Due to the Coriolis effect, the NCC follows the coastline, but the abrupt turn of the coastline at the southern tip of Norway combined with periods with northern and western winds, creates an area where the NCC moves away from the coast, with upwelling near the shore.

The volume transport of the NCC along the Norwegian Skagerrak coast is typically 200–250 000 m^3/s, showing a significant temporal variation (Anon., 1997).

North of Stad the NCC mixes with the warm saline water of the North Atlantic Current creating the "Norwegian Atlantic Current" which follows the Norwegian coastline to Nordkapp, where it divides into a northerly and an easterly branch (Fig. 1).

Tidal variations are semidiurnal, but due to an amphidromic point in the North Sea off the coast of southern Norway, the amplitude increases typically from 0.1–0.15 m along the Norwegian Skagerrak coast to 1–1.5 m in northern Norway.

Biotic Factors

The coastal primary production shows a strong seasonality. In November to January the light intensity is too low to support any production. Strong winds and low temperature mix the waters vertically. This prevents phytoplankton from staying in the surface layer, but also brings inorganic nutrients up from deeper waters (upwelling). In spring, when enhanced sunlight and density stabilisation of the surface waters occurs, this load of nutrients triggers a strong spring phytoplankton bloom.

The typical spring bloom starts in March at the south coast of Norway and in April along the northern coast. In Atlantic waters outside the coastal current the bloom starts as late as in May or June since stabilisation of the surface here is dependent solely on temperature. In the fjords the spring bloom may occur as early as February due to an earlier density stability caused by the reduced surface salinity. The fjords may also experience one or more early summer blooms triggered by nutrients in river runoff, due to the melting of snow in the mountains.

The spring bloom culminates as the nutrients are depleted, and during summer only a low level of production is sustained based on nutrients regenerated within the stable surface layer. In the autumn the stability of the surface water starts to break down due to wind and reduced temperatures, which may bring new nutrients to the surface when the light is still strong enough for production to occur. This supports another bloom in the southern coastal areas, but usually not as strong as the spring bloom.

In the last 20 years there seems to have been an increase in blooms of harmful algae, primarily along the south coast, but also in local areas further north. Local or regional incidences of toxic mussels have been frequent, and in later years some of these blooms have caused considerable mortality to salmon and trout in fish farming cages as well.

In spring 1988 a strong bloom of *Chrysochromulina polylepis* started in Kattegat and eastern Skagerrak and spread westwards with the coastal current before it culminated in south-western Norway. The bloom originated in water with elevated nutrient concentrations and altered nutrient ratios, suggesting a connection between this bloom and eutrophication (Skjoldal and Dundas, 1989). The bloom caused mass mortality in fish cages as well as among natural populations of marine organisms, and gained a tremendous public attention. In subsequent years there have been several blooms of this species without any discernible toxic effects as seen in 1988.

The strong seasonality in hydrography and primary production triggers a seasonality in other elements of the coastal ecosystem. Several grazers on the phytoplankton show rhythms in migration, growth and reproduction which optimise their chances of utilising this food source, and this again has repercussions for the higher levels of the food web. For instance, important fish stocks such as herring and cod spawn in spring which improves the chance of their larvae to match peaks in occurrence of their food: the zooplankton larvae. The production seasonality is also expressed in a seasonal input of organic matter to the bottom, which thus also elicits seasonality in the biology of sediment living organisms.

The sill separation of the deep basins of several fjords from the coast outside isolates, to some extent, the fjord community from the coastal community. The separation may have several consequences. In some deeper fjords with permanently cold bottom water and shallow sills the bottom fauna contains arctic species, suggesting an earlier zoogeographical connection, which has been cut off due to land elevation or climatic change. One example is Sørsalten in Nordland county, where the fauna at deep bottoms (100–150 m) contains certain species of bivalves which are only found elsewhere in the high Arctic (Haugen et al., 1988).

It has also been shown that the typical molluscan fauna in soft bottoms in some larger fjords on the west coast contains species which are not found at the same depth outside, but which are found along the coast of Ireland, France and Spain (Buhl-Mortensen and Høisæter, 1993). One hypothesis is that the fjord bottom may function as a refugia for species which were formerly widely distributed during periods of warm climate.

Deep Water Renewal

One of the most discussed aspects of silled fjords is the periodic flushing or renewal of their basin waters with adjacent coastal or shelf waters. Such interest may reflect: (1) the direct influence of renewal on fjord circulation and sedimentation; (2) the results of basin stagnation and anoxia and consequent destruction of sea life; and (3) the importance of renewal on the biogeochemical environment of basin waters and sediments (Syvitski et al., 1987). Renewal events occur during situations when the coastal water density at and above the fjord sill is higher than the density of the basin water. Renewals therefore depend on the combined effect of density variations in the coastal water where important factors are freshwater runoff variations, wind strength and direction, tidal activity and atmospheric pressure, and on the slow reduction of density from vertical diffusion in the basin water. A comprehensive discussion of the hydrodynamics of basin water exchange processes is given by Gade and Edwards (1980).

The fjord sill and the cold and dense bottom water in the deeper fjord basins prevent frequent exchange with outside water. Depending on the fjord topography and the depth and number of sills, the residence time of the basin water will typically vary from months to tens of years. The longest residence time is usually found in fjord basins which have several shallow sills separated by deep basins between themselves and the coastal water.

Over time, basin water will often become depleted in oxygen. The depletion is dependent on several factors such as time between water exchange events, volumes exchanged, characteristics of exchanged water, and input of organic (oxygen consuming) matter to the basins. Some fjords are by nature permanently anoxic (Skei, 1988), others face this situation intermittently. Basin water renewals may be partial or complete. Complete renewals in fjord basins with oxygen depletion and shallow sills are often accompanied by damage to the shallow biological communities, including fish kills (Strøm, 1936), as the deep basin water will be vertically displaced by the inflow of dense coastal water.

Sedimentation

Sediment transport and input to Norwegian fjords and coastal areas are seasonally dependent. Coastal areas of Norway have a thin soil cover, and the extent of erosion and sediment transport is moderate. There are two primary peaks of sediment input to the coastal water; during early spring (low land flood) and during early summer (high land flood due to snow melting in the mountains). During these periods the turbidity in fjords increases as the sediments are transported in the brackish surface layer. Most of the sedimentation takes place in the fjord basins, acting as efficient sediment traps. In some areas, rivers enter downstream lakes or dams before entering the fjords and most of the sediment load is retained in the lacustrine environment.

Besides fluvial input of sediments, episodic slumping may take place along steep sides of fjords and at fjord head deltas. Slumping events may also trigger turbidity currents, transporting large amount of sediments long distances in a short time. Fifty percent of the sediment column within the basins of Hardangerfjord is a result of slumps and turbidity currents (Holtedahl, 1965).

THE MAJOR SHALLOW WATER MARINE AND COASTAL HABITATS

Hard Bottom Habitats

The dominant shallow water habitat type along most of the coast is bedrock with dense growth of macroalgae (seaweeds and kelp) and associated sessile and mobile fauna. This habitat is characteristic both of the intertidal zone and to variable depths below this. In Norway, subtidal kelp communities extend to about 30–40 m depth in the coastal areas, but to less than 10 m in some fjords with turbid water. On the mid-west coast the kelp may be up to 3–4 m high and form dense forests hosting a highly diverse community of algae and animals (Fig. 2.) On one kelp plant 35–40 different species of epigrowth algae may be observed together with a similar number of sessile animal species. The fauna is especially rich in the holdfast of the kelp. In addition, one plant may host several thousand individuals of motile animals, primarily small crustaceans. At depths below the kelp zone, a less diverse, hard bottom community with sparse growth of encrusting algae, sponges, brachiopods and other sessile and motile animals may extend to considerable depths, especially along the fjord sides.

In certain regions, vast areas of former kelp forests have been denuded by the grazing effect of sea urchins. These areas seem to be very stable, and it is assumed that the large stocks of urchins continue to prevent kelp recolonization.

Coral Reefs

A particular type of hard bottom habitat is the deep water coral communities found most frequently at the offshore shelf edge outside mid Norway at 200 to 400 m depth, but also locally in some fjords from about 50 m depth and downwards. The basis for the reef formation is the coral *Lophelia pertusa*, but a few other species of corals are also

Fig. 2. A rich forest of the kelp *Laminaria hyperborea* at the Norwegian west coast.

present. These reefs, which may be several km long and more than 20 m high, host a large variety of associated fauna (Mortensen et al., 1995). Remote operated vehicle investigations in later years have shown considerable damage to some of the offshore reefs due to trawling activities.

Spatial Habitat Gradients

The hard bottom habitats vary in species composition and biodiversity due to a range of factors. The three main gradients of change are with depth, from the fjord head to the coast, and from south to north along the coast.

The rocky shore communities in the tidal zone show a typical vertical gradient in composition of sessile organisms as a function of tidal range, duration of air exposure and biological interaction. In some regions, especially in the Oslofjord area and in northern Norway, ice scouring may from time to time partially wipe out these communities. Below the tidal zone the vertical community gradient is influenced by light penetration, and in the fjords also by salinity. In the fjord basins, oxygen depletion may also influence the depth distribution.

The change in shallow water communities from the fjord head to the coast is stronger than the change from north to south along the coast. Although several factors influence the horizontal fjord gradient, the most important is the change in salinity from brackish water at the head to marine conditions at the coast. Shallow water species, which do not tolerate low salinity, may still extend far into the fjords, but are normally found below the brackish layer. Community structure in the tidal and upper subtidal zone is also influenced by wave intensity, which may change from the open, exposed coast to the sheltered fjord heads, and even locally, e.g., from the open to the sheltered side of islands and headlands.

In general, biodiversity increases from the fjord head to the coast. In the second longest fjord on the west coast, Hardangerfjorden, the diversity of benthic algae was shown to increase from 59 species in the innermost part to 166 species at the mouth of the fjord (Jorde and Klavestad, 1963). Also the diversity of soft bottom deep water fauna is lower in the fjord basins than on the shelf outside (e.g. Buhl-Mortensen and Høisæter, 1993), mainly as a result of finer sediments and higher levels of organic matter, possibly also recruitment isolation (Fig. 3). Fine grained sediments, which are most typical in fjord basins, tend to have lower fauna diversity than coarser sediments (e.g. Pedersen et al., 1995). However, grain size also appears to be a function of depth, and other factors linked to depth (e.g., flow velocities, organic matter supply (Flach and Thomsen, 1998)) may have a confounding influence on biodiversity.

The north–south gradient in marine communities along the coast is less than one might expect from the vast latitudinal distance covered (2000 km). The marine flora and fauna is in general dominated by cold–temperate species, and there is a considerable overlap in species composition all along the coast. Much of this is due to the Norwegian Coastal Current and the Norwegian Atlantic Current flowing along the coast, which tends to even out

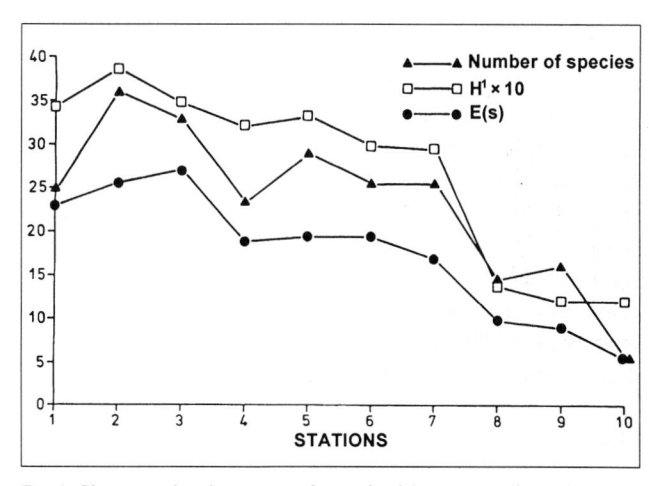

Fig. 3. Change in biodiversity indices of soft bottom molluscs from the shelf (stations 1–3) via the coast (stations 4–5) to fjord habitats (stations 6–10) at the Norwegian west coast. H': Shannon Wiener diversity; E(s): expected number of species per 200 individuals. From Buhl-Mortensen and Høisæter (1993).

the climatic difference, hence also the differences in shallow water flora and fauna. Still, there are several species which have their limit of distribution somewhere along the coast.

In south Norway there is also an east–west gradient in hard and soft bottom communities expressed as a decrease in species numbers and biodiversity towards the east (Moy et al., 1996). Also the individual size, abundance, and age of kelp decrease towards the east. This gradient is primarily due to changes in natural conditions along the coast, but also an effect of eutrophication in the east (Anon., 1997).

OFFSHORE SYSTEMS

Environmental Setting

The continental shelf is relatively narrow (approx. 20 km wide) along the south coast of Norway, where it is less than 200 m deep and bordered on the outside by the deeper Norwegian Trench. The shallow eastern part of the North Sea Plateau is also considered part of this shelf. The surface sediments are primarily fine to coarse sand.

Outside mid Norway, the shelf widens and is more irregular both in extent (60–200 km) and depth (45–500 m). In this area we find some of the main fishing banks along the coast. The bottom substrate in this region varies from silty depressions, wide bands of sorted sand and gravel, to rocks and boulders. Plough-marks from ancient icebergs are conspicuous. The narrowest part of the shelf (2 km wide) is outside Andøya in northern Norway. Outside Finnmark there is no clear distinction between the shelf and the shallow Barents Sea bottom.

The transport of water masses from south to north is a dominant feature. South of 62°N the Norwegian Coastal Current (NCC) flows northwards containing a mixture of Baltic water, Jutland current water, Norwegian fresh water and water of Atlantic origin. North of 62°N the NCC joins the saline and relatively warm North Atlantic Current forming a massive northward flowing coastal current (see Fig. 1). The NCC also carries a large nutrient load, and off the Norwegian Skagerrak coast the annual transport has been calculated to 1.7 million tonnes of nitrogen and 140,000–150,000 tonnes of phosphorus (Fig. 4). The corresponding nutrient transport along the Norwegian west coast and further north is not known, but is certainly significantly greater than off the Norwegian Skagerrak coast. In comparison the total discharges from the mainland were 90,000 tonnes of nitrogen and 3400 tonnes of phosphorus (Holtan et al., 1997).

The Bottom Fauna

The bottom fauna communities of the shelf are primarily known from the monitoring around Norwegian offshore petroleum sites in the eastern North Sea, and more recently also the Haltenbanken region. The shelf along the

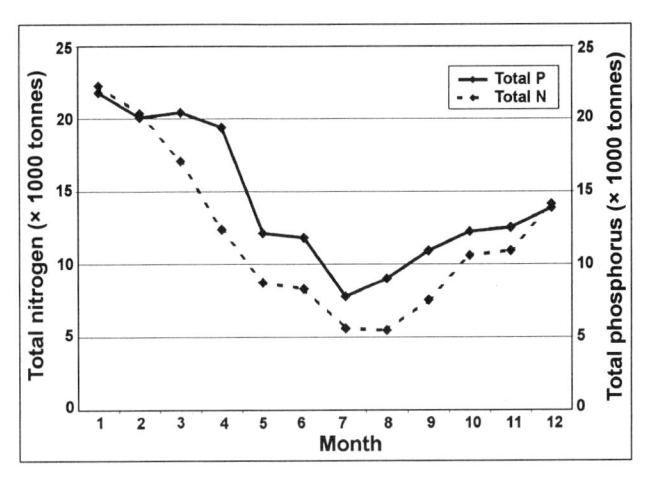

Fig. 4. Average monthly transport of nutrients (phosphorus and nitrogen) in the North Atlantic Current (NCC) off the Norwegian Skagerrak coast, based on data from 1990–1994 (from Anon., 1997).

Skagerrak and West Norway coast is also covered by the coastal monitoring programme, and by several local and regional surveys performed over the years. The faunal diversity is high, both on the coastal side of the Norwegian Trench and at the North Sea Plateau, and individual densities are normally in the range 1000–5000 individuals m^{-2} (Moy et al., 1996; Mannvik et al., 1998). Surface feeding bristle worms are in general the dominant animal group, followed by bivalves, crustaceans and brittle stars. The dominant species seem to fluctuate widely from one year to another in response to natural environmental factors (Pearson and Mannvik, 1998).

Seasonality in reproduction and recruitment may be strong and during May–June the sediment fauna may be temporarily dominated by very young individuals of a few species. In the Haltenbanken region the total faunal densities are in general less than in the North Sea (<1000 ind m^{-2}, Jensen et al., 1998), but the numbers of species are high, and the diversity is possibly higher than at the shelf further south. Also in this region bristle worms, crustaceans and bivalves dominate the bottom fauna.

Oil and Gas

Offshore oil and gas exploration and production have affected the bottom habitats locally, primarily due to the drilling activities. Until 1993, when the Norwegian authorities in practice prohibited the discharge of oil based drill cuttings to the sea, vast amounts of oil-contaminated drilling waste had been deposited around the platforms. Contaminated seabed sediments were detected as far as 15 km from the dumping sites, and effects on the fauna were detected out to a radius of 3 km around older fields (Olsgard and Gray, 1995). In general, the effects were reduced species diversity through loss of sensitive organisms, and increased dominance by a few opportunistic species. The new legislation elicited new technology in drilling waste management (re-injection, transport to shore, new drilling fluids) resulting in cleaner and less material deposited on the seabed. With few exceptions the contaminated areas around the fields have been significantly reduced in later years (Gray et al., 1999). A corresponding recovery of the bottom fauna also seems to occur, but apparently with a time lag.

Today, the main environmental challenge around the Norwegian offshore petroleum fields is to assess the potential effects of the large volumes of produced water which are separated from the oil and gas and discharged to the sea. These volumes are expected to increase considerably in the years to come as the fields get older, and are already the major contributor to hydrocarbon input to the sea from the offshore activities. In 1996 oil in produced water represented 80% of the total oil discharge to the sea at the Norwegian shelf (SFT, 1998a). A basic problem in assessing the threat of this pollution source to the marine resources is that satisfactory methods of monitoring biological effects in the pelagic habitat are lacking.

Another challenge linked to the aging of oil exploration sites is how to deal with the large deposits of older oil-contaminated drill cuttings on the seabed when the fields are closed down. There is likely to be an excess of 100 cuttings piles of various sizes distributed in the Norwegian sector of the North Sea (Cordah, 1998). Some of these may be more than 100 m in diameter and 5 m in height. The oil industry is at present preparing guidelines on how to obtain necessary information on the piles (extent, volume, pollutant content, stability, etc.) to handle them in a safe way at decommissioning. The options for dealing with the piles are *in situ* bioremediation, mechanical dispersal, recovery for dumping or onshore treatment, capping, reinjection, or leaving in place (Gerrard et al., 1998).

ANTHROPOGENIC INFLUENCES

Norway has a population of 4.2 million (1990) and an average population density of 13.8 per km^2. A total of 72% of the population lives in densely populated areas and these areas are to a large extent close to the coastline. Ultimately, the anthropogenic pressure on the coast is large and actions have been taken during the last decade to reduce discharges of harmful substances to the sea. These actions are partly a result of international agreements, declarations and conventions (i.e., the Oslo and Paris Commissions). The objectives of the North Sea declarations of 50% reduction in inputs of phosphorous and nitrogen were not reached by Norway in 1995 (45 and 30% respectively). It is particularly difficult to control the input of nitrogen from agriculture. The objectives of 50–70% reductions in input of hazardous substances were mostly reached by 1995. Complex and unpredictive sources of chemical contaminants are secondary sources (leakage from dumpsites, surface runoff from industrial sites and contaminated marine sediments).

Significant human alteration of the coastline started, to a large extent, at the turn of the century, with construction of hydroelectric power plants at the head of fjords. Following the development of this energy supply, heavy industry requiring hydroelectric power was established. Due to lack of flat land for waste disposal, fjords were the only real alternative for disposal. Old industry established at the beginning of the century has contaminated the Norwegian coast for almost 100 years (Skei and Molvær, 1988). Environmental awareness arose in the 1970s and establishment of monitoring programmes documented serious contamination problems locally in the vicinity of industrial activities.

The principal categories of anthropogenic impact on fjords and coastal water in Norway are:
- effluents from the pulp and paper industry (drastically reduced during the last decade);
- urban sewage and non-point source discharges (runoff from landfills, secondary input from contaminated sediments etc.);
- effluents from smelters and metal refining industry, chemical industry and mines (tailings).

Beside heavy industrial development, other human activities such as fishery and fish processing industry, aquaculture, kelp harvesting, shipping, harbour development, dredging and dumping, dumping of mine waste, road and tunnel building, and sand and gravel excavation are examples of activities which locally have created environmental concerns.

Eutrophication

The environmental focus has been on eutrophication in fjords with restricted water circulation and discharges of sewage. Oxygen depletion occurs in the bottom water of many fjord basins, and anoxic conditions are observed in fjords with shallow sills in the southern part of Norway, permanently or intermittently. By the introduction of sewage treatment in the 1980s and 1990s the situation with respect to high nutrient concentrations, algal growth and oxygen depletion has improved (Fig. 5). The state of eutrophication in the coastal water of southern Norway depends on the transboundary load carried by the NCC into Norwegian waters from "upstream countries", and moderate effects from regional eutrophication are clearly seen along the Norwegian Skagerrak coast (Anon., 1997).

Chemical Pollution

A serious concern about the environmental conditions in fjords and near coastal water is related to chemical contaminants in industrial effluents. Old smelters located at the head of fjords have caused contamination of polycyclic aromatic hydrocarbons (PAHs) and heavy metals in bottom sediments and organisms (Naes et al., 1995). The trapping of contaminants in the sediments has caused strong gradients and extremely high levels close to the point sources

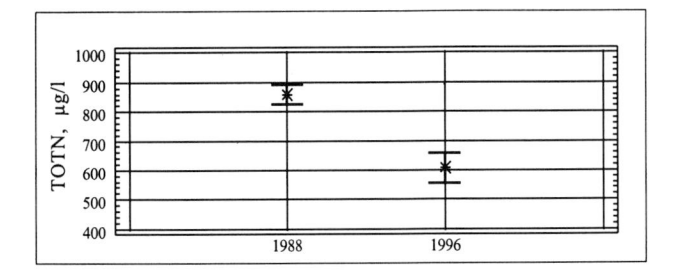

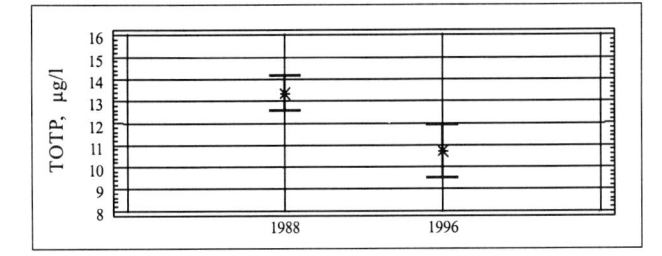

Fig. 5. Average concentrations of total nitrogen (TOTN) and total phosphorus (TOTP) in the brackish layer of Frierfjorden, southern Norway, in 1988–89 (1988 in figure) and 1996–97 (1996 in figure), following significant reductions of the nutrient load from municipal and industrial waste water. The concentrations during 1996–97 were statistically significant ($p < 0.05$) lower than during 1988–89.

(Skei, 1981). Additionally, the level of contaminants in fish and shellfish used for human consumption exceeds safety levels established by WHO in 14 fjord locations in Norway. These restrictions are partly due to elevated levels of PCB in fish liver, PAHs in shellfish (blue mussels) and a single case where the levels of cadmium in mussels and fish liver exceeds the safety limits.

The discharges from point sources are now being reduced due to strict governmental regulations. Consequently, secondary sources like waste disposal sites on land close to the shore, and hot spot sediments are becoming predominant sources. These sources are unpredictable and variable with time depending on rainfall, shore erosion and physical disturbance of bottom sediments. This causes a problem with respect to interpretation of data from monitoring of water quality, in particular. The environmental implications of hot spot sediments has been much in focus, particularly related to remobilization and bioavailability of contaminants (Skei, 1992). Fjord sediments near point sources are often more contaminated than sediments in other types of estuaries, due to low input of natural sediments. Rivers entering fjords often pass through lakes where much of the sediment load is emptied. As a consequence the contaminated fjord sediments are to a little extent diluted.

A particular problem of concern has been dioxin on the SW coast of Norway due to discharge from a magnesium plant near Frierfjord for more than 40 years. The annual discharge of dioxin during the time period between 1965 and 1975 was 5–10 kg. Today, this discharge has been reduced to 1–2 g per year. Despite this reduction, the levels of dioxin in crabs are elevated within an area up to 300–400 km from the point source (Knutzen et al., 1998). One hypothesis is that this is partly due to a time lag as a consequence of contamination in the food chain, and partly due to remobilization of dioxin in contaminated sediments and uptake in sediment living organisms.

In certain areas of the Norwegian west coast DDT was used as a pesticide in fruit-growing districts prior to 1970. In some of these areas elevated levels of DDT and its metabolites (DDD and DDE) are still observed in bottom sediments, in blue mussels, and in fish liver (Green et al., 1999). The reason why this is still an environmental problem is not known in detail. Several alternative explanations may be given:
- DDT stored in the soil in the fruit districts is washed back into the sea;
- DDT was buried in the ground when the banning of DDT in 1970 was effective and is now being washed into the sea;
- DDT is still being used illegally

In recent years determination of imposex (the induction of male sex characters in females) in populations of dogwhelks (*Nucella lapillus*) along the Norwegian coast has been associated with unintentional effects of the antifouling agent tributyltin (TBT) in ship paint (Følsvik et al., 1999). The occurrence of imposex in dogwhelk was investigated in 41 populations sampled in 1993–1995 along the Norwegian coast (Følsvik et al., 1999). Except in four populations in northern Norway, some degree of imposex occurred at all locations, indicating that the impact of TBT is a regional problem associated with the ship traffic along the coast.

Fishery and Fish Farming

In the coastal region, especially in western and northern Norway, fisheries and fish farming are the backbones of local and regional employment, and also important for the national economy. In 1997 the export value from fish and fish products were the second largest contributor to the total national export after oil and gas (Statistics Norway). About 10% of the total fish volume landed in 1995 came from fish farming (Statistics Norway).

Apart from the effects of fisheries on the fish stock sizes, which are regulated through national and international agreements, the traditional fishing activities have their own environmental effects. In particular, modern offshore trawling by large vessels is expected to cause physical disturbance to considerable areas of the shelf sea floor. It is known to have an impact on benthic communities, selectively removing larger animals on the sediment surface and smothering or damaging fauna in the sediments (Bergman and Hup, 1992). The same disturbing effects must be expected locally from intensive shrimp trawling along many fjord bottoms.

Fish farming is an expanding industry in Norway. The total number of active coastal fish farms have increased gradually from 256 in 1980 to 660 in 1995, with an even more intensive increase in produced fish from 8000 tonnes to 277,000 tonnes (Statistics Norway). The majority of fish farms lie on the west coast. The farming is almost entirely based on fish cages deployed in the sea.

During the late 80s, severe outbursts of epidemic diseases in the densely populated cages became a serious problem, and intensive use of antibiotics in these open systems, to counteract disease, became a matter of public concern (Braaten, 1991). Through improved farm maintenance routines, and strict regulation of maximum biomass, distance between farms, and transport of fish, most of the disease problems have been overcome. Today, the main environmental concern regarding the rapidly growing fish farming industry is the release of nutrients and organic matter, causing eutrophication in the local environment (Braaten, 1991). Another serious concern is the accidental release of live fish which may interbreed with wild fish and thereby alter the genetic composition of natural salmon and trout in Norwegian fjords and rivers (ICES, 1997).

Kelp and Seaweed Harvesting

Along the coast of mid Norway there is also a considerable harvesting of seaweed and kelp for industrial purposes. The annual harvest in recent years has been 165,000 tonnes, which is far more than in any other country in the NE Atlantic region (Ireland: 9500 tonnes, Scotland: 3000 tonnes, France: 490 tonnes) (Fosså, 1995). In particular the kelp harvesting has caused public concern. The highly diverse kelp forests function as nursery areas for several commercial fish species feeding on the associated fauna of the kelp. Although the harvesting is regulated to ensure regrowth of the kelp, the frequency of harvesting is too high (every fifth year in the same area) to allow the epigrowth community to develop completely (Christie et al., 1998). Another concern, still not sufficiently substantiated, is the function of the kelp forests as buffers against wave erosion of the shore. In some shallow areas it may be expected that the kelp dampens wave action and thus prevents loss of sandy shore areas lying behind the forests.

Sediments

In 1998, The Norwegian Pollution Control Authority presented an overview of sites along the Norwegian coast where sediments are severely contaminated by chemicals (SFT, 1998b). This is the basis for an action plan to clean up the most contaminated sites. High priority is given to 18 sites, seven of them being harbours. The main concern is contaminated sediments in shallow areas, exposed to resuspension due to waves, currents and propeller erosion and the bioavailability of contaminants residing in sediments. Experimental work on contaminated sediments has shown that trace metals are being released when a sediment is physically disturbed, either by physical forces (waves and currents) or by bioturbating organisms (Skei, 1992).

Oil Spills

In spite of heavy coastal ship traffic, Norway has experienced few serious oil spill incidents. The largest in recent times are 'Deifovos' at the Helgeland coast in 1981 (1000 tonnes of fuel oil), 'Mercantil Marica' in 1989 at Sognesjøen (420 tonnes fuel and diesel oil), 'Arisan' in 1992 close to the bird cliffs at Runde on the west coast (150 tonnes fuel oil), and 'Leros Strength' outside Stavanger in 1997 (150 tonnes fuel and diesel oil). In all incidents local damage to shoreline organisms and seabirds was recorded.

Introduction of New Marine Species

The introduction of foreign marine species to the coastal ecosystem with bilge and ballast water from ships, or attached to ship hulls, is a matter of public concern. A typical example is the Asian microalga *Biddulphia sinensis* which was introduced to the North Sea just after the turn of the century (Ostenfeld, 1908), and is now a typical member of the plankton community. Although there is no solid evidence that such introductions have occurred in Norway, it has been suggested that certain incidences of toxic algal blooms have been due to ballast water. Examples are the frequent occurrence of *Prymnesium parvum* on the west coast, and a strong bloom of *Chrysochromulina leadbeteri* in Vestfjorden, Nordland in 1991. The UN International Maritime Organization (IMO) is at present preparing improved regulations for ballast water management to minimise the risk of foreign species transfer by ships (IMO, 1998).

MONITORING PROGRAMMES, ENVIRONMENTAL QUALITY CRITERIA AND PROTECTIVE MEASURES

To a large extent the framework for the protection of the marine environment has been created through the four International conferences on the Protection of the North Sea, the OSPAR Convention covering the north-east Atlantic, and the Arctic Monitoring and Assessment Programme (AMAP) covering Arctic areas north of 62°N.

Regarding nutrients the Norwegian obligation is a reduction of 50% of nutrient load to the North Sea compared with 1985, of which the target for nitrogen will not be achieved until after the year 2000. Parallel to their efforts to fulfil these international obligations the Norwegian authorities apply a strategy based on local environmental quality goals and knowledge of the local recipients. This strategy includes various combinations of nutrient load reduction by means of sewage treatment plants and dilution through deep water outfalls in areas with high water exchange.

Monitoring of the quality of water, sediments and biota in fjords and coastal water started on a regular basis in about 1980 in Norway, initiated by the Norwegian Pollution Control Authority. Later on, monitoring programmes related to the Oslo Paris Commissions (JMP, later JAMP) were initiated to satisfy international requirements. A large database has been established containing data on contaminants in organisms (fish and shellfish) and in bottom sediments from localities along the Norwegian coast, where point sources have been identified (Green et al., 1999). Data from the Norwegian Arctic coast (Svalbard) have been collected within the framework of AMAP (Arctic Monitoring Assessment Programme).

Monitoring of the bottom conditions around Norwegian offshore petroleum installations have been conducted on a regular basis by the oil companies since the mid 1980s. The reduced discharges of contaminated drilling waste and the improvement in bottom conditions experienced in recent years are results of legislations imposed on the basis of this monitoring (Gray et al., 1999). Today, the oil companies, the authorities and the scientific community cooperate very well to ensure coordinated and high quality surveys are performed on a regional basis over the whole shelf area.

Environmental quality criteria for use in fjords and coastal water were established in 1993 and later revised in 1997 (Molvaer et al., 1997). These contain a numerical classification system for impact from nutrients, organic matter, micro pollutants and faecal bacteria, as well as a system for assessment of water quality suitability for various uses. The environmental quality criteria are a tool or set of guidelines used in monitoring of fjords and coastal water. Based on these criteria, decisions are being made regarding waste water treatment and environmental remediation.

In Norway there are three major cities situated on the coast—Oslo (0.5 million), Bergen (0.25 million) and Trondheim (0.25 million). Tertiary waste water treatment (including removal of nitrogen) for domestic waste is implemented in Oslo. In Bergen and Trondheim primary and secondary sewage treatment have just been implemented.

The city of Oslo is situated at the head of Oslofjord, which has a shallow sill at the entrance, restricting water circulation. As a consequence of incomplete deep water renewal and input of organic matter from sewage treatment plants, and runoff from land, several deep basins in Oslofjord turn anoxic for long periods. Large bottom areas are devoid of life due to the oxygen conditions (Fig. 6). Following a deep water renewal, the oxygen is quickly consumed and anoxic conditions are re-established, normally after a few months (Fig. 7).

The sediments of the harbour of Oslo are heavily contaminated by PCB, PAHs and heavy metals. As a result, the environmental authorities have not allowed dredging since 1992. At present there is great demand for dredging to maintain sufficient water depths for navigation. A major dredging operation is now planned, hydraulically removing 500,000–750,000 m³ of contaminated material. The

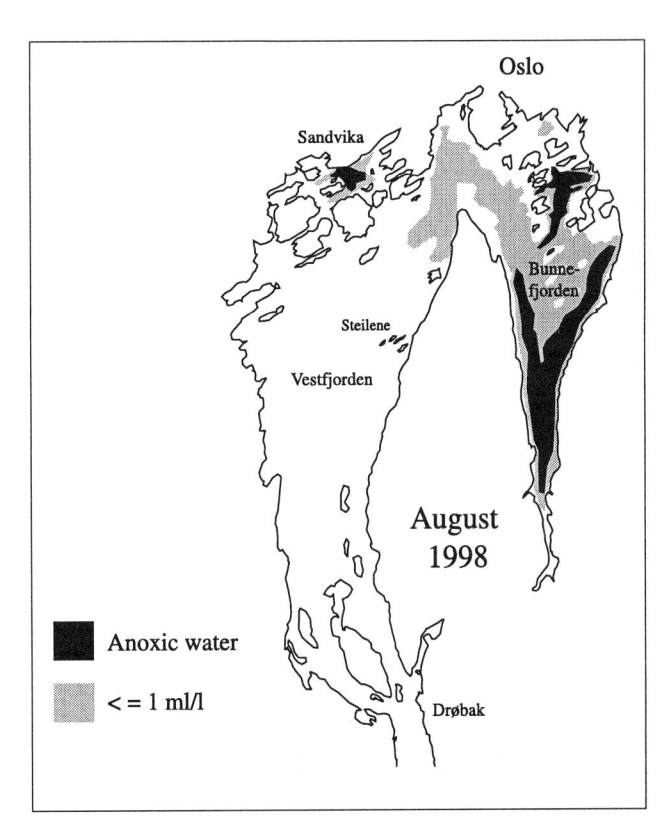

Fig. 6. Oxygen and hydrogen sulphide at 60 m depth in Oslofjord during a period of 10 years (1987–1997) (Jan Magnusson, pers. comm.).

dredged spoil may be pumped to a small anoxic basin and capped, about 3 km away from the harbour. The use of confined, natural anoxic basins as dumpsites for contaminated dredged sediments may be an environmentally acceptable low-cost alternative to treatment (Førstner and Calmano, 1998). By capping the material with clean sediments, an extra assurance is introduced, taking into account the possibility that the basin may be oxic and the sediments recolonized by organisms in the future.

The harbour of Bergen is also contaminated by organic waste and micro pollutants. There are no shallow sills and due to a tidal amplitude about twice that of Oslo the water circulation is better and oxic conditions prevail in the bottom water. The main contamination problem in the Bergen area is elevated levels of PCB in fish and shellfish, and restrictions on human consumption were introduced in 1992. A naval base situated close to Bergen has been singled out as a dominant PCB source. The levels of PCB measured in sediments near the naval base were extremely high, exceeding 20 mg/kg PCB (Konieczny, 1994) and it was decided to remove the contaminated sediments and dispose of the material in a confined dike facility. The objective of the clean-up was to reduce the exposure of PCB-contaminated sediments to benthic organisms and further food chain transfer to fish and shellfish.

Similar to Oslo and Bergen, the harbour of Trondheim is also contaminated, although the environmental conditions

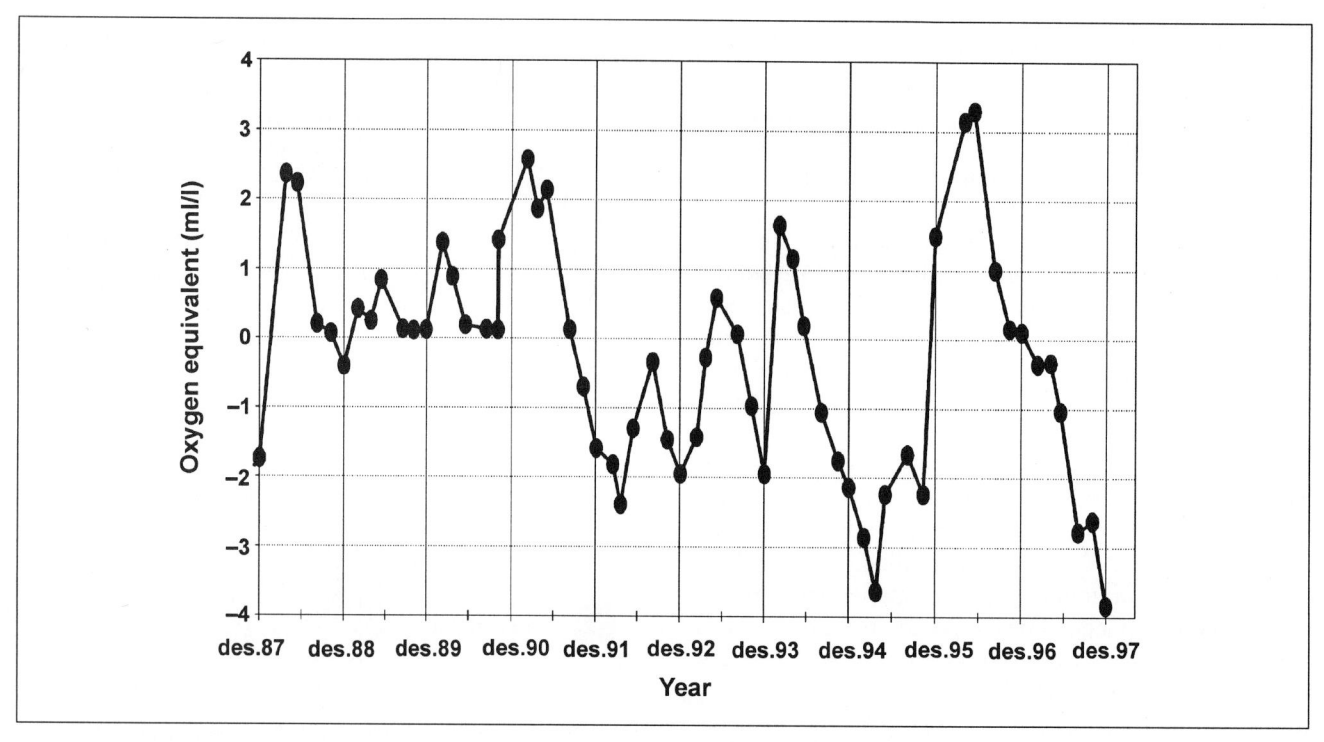

Fig. 7. Areas of oxygen depletion at the bottom in Oslofjord, August 1998 (Jan Magnusson, pers. comm.).

have been investigated less. Trondheimsfjorden has a deep sill (320 m) and the maximum water depth is 600 m. There are no oxygen problems in the main part of Trondheimsfjorden, but at the head of the fjord there are several isolated basins with shallow sills which are anoxic. In the vicinity of the city of Trondheim industrial activities (mining and smelters) have caused high levels of trace metals in sediments.

CHALLENGES AT THE TURN OF THE MILLENNIUM

The Norwegian coastline is sparsely populated compared with other coastlines in the world and the environmental situation should be of minor concern. However, the majority of the Norwegian population and industrial activity are closely associated with the coast, imposing environmental pressures. During the last couple of decades, considerable effort has been made to combat environmental problems by introducing treatment plants in industry and municipal works. Despite this effort, the environmental situation is not satisfactory in several fjords and harbours. The greatest challenge appears to be the impact from contaminated sediments and input of contaminants from on-shore dumpsites situated within reach of tidal activity. Underwater disposal sites containing drilling waste near off-shore oil installations and decommissioning of old oil fields represent other challenges. These sources of contamination are historical in origin, but will impose an environmental threat in the years to come. Remedial action will be very resource-demanding, and only the sites given highest priority are likely to be dealt with.

A lot of uncertainty is related to potential climatic changes and environmental consequences. A shift in the major current systems, including the North Atlantic Current, which has a great significance for the climate in Norway, may have dramatic consequences. A change in the fresh water supply will likewise have an impact on water exchange in fjords and water quality. A challenge will be to predict the consequences and to determine how to reduce them.

Fish farming in Norway is an expanding industry and a challenge is to utilise the maximum potentials of harvesting the sea in harmony with the environment and to maintain a healthy ecosystem at the turn of the millennium. This can only be achieved by an integrated coastal management and a management based on an ecosystem approach.

REFERENCES

Anon. (1997) The Norwegian North Sea coastal water. Eutrophication. Status and Trends. English summary report from the Norwegian Pollution Control Authority Expert Group on Eutrophication, December 1997. 90 pp.

Bergman, M.J.N. and Hup, M. (1992) Direct effects of beam trawling on macrofauna in sandy sediment in the southern North Sea. *ICES Journal of Marine Science* **49**, 5–11.

Braaten, B. (1991) Impact of pollution from aquaculture in six Nordic countries. Release of nutrients, effects, and waste water treatment. In *Aquaculture and the Environment*, eds. N. De Pauw and J.

Joyce, pp. 79–101. European Aquaculture Society Special Publication No. 16, Gent, Belgium.

Buhl-Mortensen, L. and Høisæter, T. (1993) Mollusc fauna along an offshore–fjord gradient. *Marine Ecology Progress Series* **97**, 209–224.

Christie, H., Fredriksen, S. and Rinde, E. (1998) Regrowth of kelp and colonization of epiphyte and fauna community after kelp trawling at the coast of Norway. *Hydrobiologia* **375/376**, 49–58.

Cordah (1998) The present status and effects of drill cuttings piles in the North Sea. Report no. Cordah/ODCP.004/1998.

Flach, E. and Thomsen, L. (1998) Do physical and chemical factors structure the macrobenthic community at a continental slope in the NE Atlantic? *Hydrobiologia* **375/376**, 265–285.

Følsvik, N., Berge, J.A., Brevik, E.M. and Walday, M. (1999) Quantification of organotin compounds and determination of imposex in population of dogwhelks (*Nucella lapillus*) from Norway. *Chemosphere* **38**, 681–691.

Förstner, V. and Calmano, W. (1998) Characterization of dredged materials. *Water Science and Technology* **38**, 149–157.

Fosså, J.H. (1995) Forvaltning av stortare. Prioriterte forskningsoppgaver (Management of kelp. Prioritized research activities). Institute of Marine Research Report, Bergen, Norway. 102 pp (in Norwegian).

Fosså, J.H. and Sjøtun, K. (1993) Tareskogsøkologi, fisk og taretråling (Kelp forest ecology, fish and kelp harvesting). *Fiskets Gang* **2**, 16–26 (in Norwegian).

Gade, H.G. and Edwards, A. (1980). Deep water renewals in fjords. In *Fjord Oceanography*, eds. H.J. Freeland, D.M. Farmer and C.D. Levings, pp. 453–489. Plenum Press, New York.

Gerrard, S., Grant, A., London, C. and Marsh, R. (1998) Understanding the relative impacts of cutting piles for different management options. University of East Anglia Research Report no. 31, UK. 189 pp.

Gray, J.S., Bakke, T., Beck, H.J. and Nilssen, I. (1999) Managing the environmental effects of the Norwegian oil and gas industry: from conflict to consensus. *Marine Pollution Bulletin* **38**, 525–530.

Green, N.W., Berge, J.A., Helland, A., Hylland, K., Knutzen, J. and Walday, M. (1999) Joint Assessment and Monitoring Programme (JAMP). National Comments regarding the Norwegian data for 1997. Norwegian Institute for Water Research (NIVA) report no. 3980-99, 144 pp.

Gregory, J.W. (1913) *The Nature and Origin of Fjords*. John Murray, London, 542 pp.

Haugen, I.N., Efraimsen, U., Golmen, L., Rygg, B. and Wikander, P.B. (1988) Undersøkelser i Sørsalten og Kolvereidvågen i Nærøy kommune, Nord-Trøndelag—Resipientundersøkelse (Recipient investigations in Sørsalten and Kolvereidvågen in Nærøy kommune, Nord-Trøndelag). Norwegian Institute for Water Research (NIVA) Report no. 2144, 41 pp.

Holtan, G., Berge, D., Holtan, H. and Hopen, T. (1997). Paris Convention. Annual report on direct and riverine inputs to Norwegian coastal waters during the year 1996. A. Principles, results and discussion. B. Data report. Norwegian Pollution Control Authority, Report 715/97. 138 pp.

Holtedahl, H. (1965) Recent turbidites in Hardangerfjord, Norway. *Colston Research Soceity Proceedings Bristol* **17**, 107–141.

Holtedahl, H. (1967) Notes on the formation of fjord and fjord valleys. *Geogr. Ann. Ser. A* **49**, 188–203.

Holtedahl, O. (1960) Geology of Norway. Norges Geologiske Undersøkelser Nr 208, 540 pp.

ICES (1997) Interactions between Salmon Culture and Wild Stocks of Atlantic Salmon: The Scientific and Management Issues. *ICES Journal of Marine Science* **54**, 963–1225.

IMO (1998) Marine Environment Protection Committee Protocol 42nd Session: 2–6 November 1998.

Jensen, T., Gjøs, N., Bakke, S.M. and Nøland, S.-A.(1998) Miljøundersøkelsen Haltenbanken. Region IV (Environmental investigation Haltenbanken. Region IV). Det Norske Veritas Report no. 98-3106, 159 pp (in Norwegian with English summary).

Jorde, I. and Klavestad, N.(1963) The natural history of the Hardangerfjord. 4. The benthonic algal vegetation. *Sarsia* **9**, 1–99.

Knutzen, J., Becher, G., Schlabach, M. and Utne Skaare, J. (1998) PCDF/PCDDDs, dioxin like PCBs, PCNs and Toxaphene in the edible crab (*Cancer pagurus*) from reference localities in Norway 1996. *Organohalogen Compounds* **39**, 292–298.

Konieczny, R. (1994) Priority hazardous substances in marine sediments in Norway. A status report. State Pollution Control Authority (SFT). Report no. TA 1119/1994, 96 pp. (in Norwegian).

Mannvik, H.P., Pearson, T., Carrol, M., Pettersen, A., Lie Gabrielsen, K. and Palerud, R (1998) Environmental monitoring survey. Region II 1997. Akvaplan-niva report no 411.98.1224-1, 250 pp.

Molvaer, J., Knutzen, J., Magnusson, J., Rygg, B., Skei, J. and Sörensen, J. (1997) Classification of environmental quality in fjords and coastal waters. A guide. State Pollution Control Authority (SFT). Report no. TA 1467/1997, 36 pp. (in Norwegian).

Mortensen, P.B., Hovland, M., Brattegard, T. and Farestveit, R. (1995) Deep water bioherms of the scleractinian coral *Lophelia pertusa* (L.) at 64°N on the Norwegian shelf: structure and associated megafauna. *Sarsia* **80**, 145–158.

Moy, F.E., Fredriksen, S., Gjøsæter, J., Hjolman, S., Jacobsen, T., Johannessen, T., Lein, T.E., Oug, E. and Tvedten, Ø.F. (1996) Utredning om benthossamfunn på kyststrekningen Fulehuk – Stad (Assessment of the benthic communities along the coast Fulehuk – Stad). Norwegian Institute for Water Research (NIVA) report no. 3551-96, 84 pp (in Norwegian).

Naes, K., Knutzen, J. and L.Berglind (1995) Occurrence of PAH in marine organisms and sediments from smelter discharge in Norway. *Science of the Total Environment* **163**, 93–106.

Olsgard, F. and Gray, J.S. (1995) A comprehensive analysis of the effects of offshore oil and gas exploration and production on the benthic communities of the Norwegian continental shelf. *Marine Ecology Progress Series* **122**, 277–306.

Ostenfeld, C.H. (1908). Meddelelser Fra Kommissionen For Danmarks Fiskeri- og Havundersøgelser. *Serie Plankton* **1**, No. 6. 44 pp.

Pearson, T. and Mannvik, H.P. (1998) Long-term changes in the diversity and faunal structure of benthic communities in the northern North Sea: natural variability or induced instability? *Hydrobiologia* **375/376**, 317–329.

Pedersen, A., Aure, J., Dahl, E., Green, N.W., Johnsen, T., Magnusson, J., Moy, F., Rygg, B. and Walday, M. (1995) Langtidsovervåking av miljøkvaliteten i kystområdene av Norge. Fem års undersøkelser: 1990–1994. Hovedrapport (long term monitoring of the environmental quality in the coastal area of Norway. Five years surveys: 1990–1995. Main report). Norwegian Institute for Water Research (NIVA) report no. 3332, 115 pp (in Norwegian).

Pritchard, D.W. (1967) Observations of circulation in coastal plain estuaries. In *Estuaries*, ed. G.H. Lauff, pp. 37–44. American Association for the Advancement of Science, Publ. 83, Washington D.C.

SFT (1998a) Discharges of oil and chemicals on the Norwegian continental shelf 1996. State Pollution Control Authority (SFT) Report no TA 1470/1997, 32 pp (in Norwegian).

SFT (1998b) Contaminated marine sediments. Status overviews and priorities. State Pollution Control Authority (SFT). Report no. TA 1547/1998, 74 pp. (in Norwegian).

Skei, J.M. (1981) The entrapment of pollutants in Norwegian fjord sediments—A beneficial situation for the North Sea. *Special Publs. Int. Assoc. Sediment* **5**, 461–468.

Skei, J.M. (1988) Framvaren—environmental setting. *Marine Chemistry* **3–4**, 209–218.

Skei, J.M. (1992) A review of assessment and remediation strategies for hot spot sediments. *Hydrobiologia* **235/236**, 629–638.

Skei, J.M. and Molvær, J. (1988) Fjords. In *Pollution of the North Sea. An*

Assessment, eds. W. Salomons, B.L. Bayne, E.K. Duursma and U. Førstner, pp. 100–110. Springer-Verlag, Berlin, Heidelberg, New York, London, Paris, Tokyo.

Skjoldal, H.R. and Dundas, I. (ed.) (1989) The *Chrysochromulina polylepis* bloom in the Skagerrak and the Kattegat in May–June 1988: Environmental conditions, possible causes, and effects. *ICES C.M.*, 1989/L. **18**, 1–60.

Strøm, K.M (1936) Land-locked water. Hydrography and bottom deposits in badly ventilated Norwegian fjords with remarks upon sedimentation under anaeorobic conditions. *Skrift. Norsk Vidensk. Akad.*, Oslo, No. 7, 85 pp.

Syvitski, J.P.M., Burrell, D.C. and Skei, J.M. (1987) *Fjords. Processes and Products*. Springer-Verlag, New York, Berlin, Heidelberg, London, Paris, Tokyo. 379 pp.

THE AUTHORS

Jens Skei
Norwegian Institute for Water Research (NIVA),
P.O. Box 173 Kjelsaas,
0411 Oslo, Norway

Torgeir Bakke
Norwegian Institute for Water Research (NIVA),
P.O. Box 173 Kjelsaas,
0411 Oslo, Norway

Jarle Molvaer
Norwegian Institute for Water Research (NIVA),
P.O. Box 173 Kjelsaas,
0411 Oslo, Norway

Chapter 3

THE FAROE ISLANDS

Maria Dam, Grethe Bruntse, Andrias Reinert and Jacob Pauli Joensen

---

The Faroe Islands are situated on the Greenland–Scotland Ridge from which they extend above sea level. This submarine ridge has sill levels between 400 and 600 m depth along most of its path. Flow across it therefore occurs in the upper layers (down to approx. 500 m) while the ridge blocks exchange of the deepest layers. The ridge thus acts as a partial barrier between the 'Arctic–Mediterranean' ocean north of the Faroes and the rest of the World ocean. In the areas north of the ridge, cooling and brine rejection in freezing water in the upper layers increase water density sufficiently to allow it to sink to intermediate (500–1000 m) or deep levels. This creates a pressure gradient which in the intermediate and deep layers drives a flow southwestwards over the ridge, while in the upper layers a compensating northeastward flow is induced. On the Faroe Plateau, sea temperatures are mainly influenced by the North Atlantic Water, reaching an average of 10°C in summer and decreasing to an average of 6°C in winter.

The 1100 km coastline mainly consists of bedrock and bedrock boulders. Only small areas, usually at the bottom of the fjords, are covered by sand or mud. Most parts of the hard substrate coast is steep cliffs. These and the relatively small tidal amplitude results in a narrow littoral zone. Many of the species resemble those found in the littoral zone of the adjacent areas in the North Atlantic. There are certain differences though, like the scarcity of *Fucus serratus*, and the lack of *Littorina littorea*.

The Faroese economy is strongly dependent on fishery and aquaculture, with fish and fish products accounting for over 99% of the gross value of the exported products in the period 1992–1997.

Chemical pollution is very limited, a consequence of its rather isolated position in the North Atlantic away from the dense populations and large industries of central Europe, and of favourable ocean currents. This does not, however, make the Faroese people a 'reference population' with respect to environmental pollution loads, as one would expect, because the intake of pollutants with marine mammals which are consumed is marked. The best-studied species is probably the long-finned pilot whale, *Globicephala melas*. Warnings to restrict the consumption of pilot whale meat were first issued in 1977, and this was apparently also the first time mercury was ever measured in Faroese pilot whale meat. Since then, heavy metals and persistent synthetic pollutants in these mammals have been measured. Mercury is of most interest in the islands, mainly due to the high concentrations carried in whale meat, but also since mercury tends to be associated with fish.

Seas at The Millennium: An Environmental Evaluation (Edited by C. Sheppard)
© 2000 Elsevier Science Ltd. All rights reserved

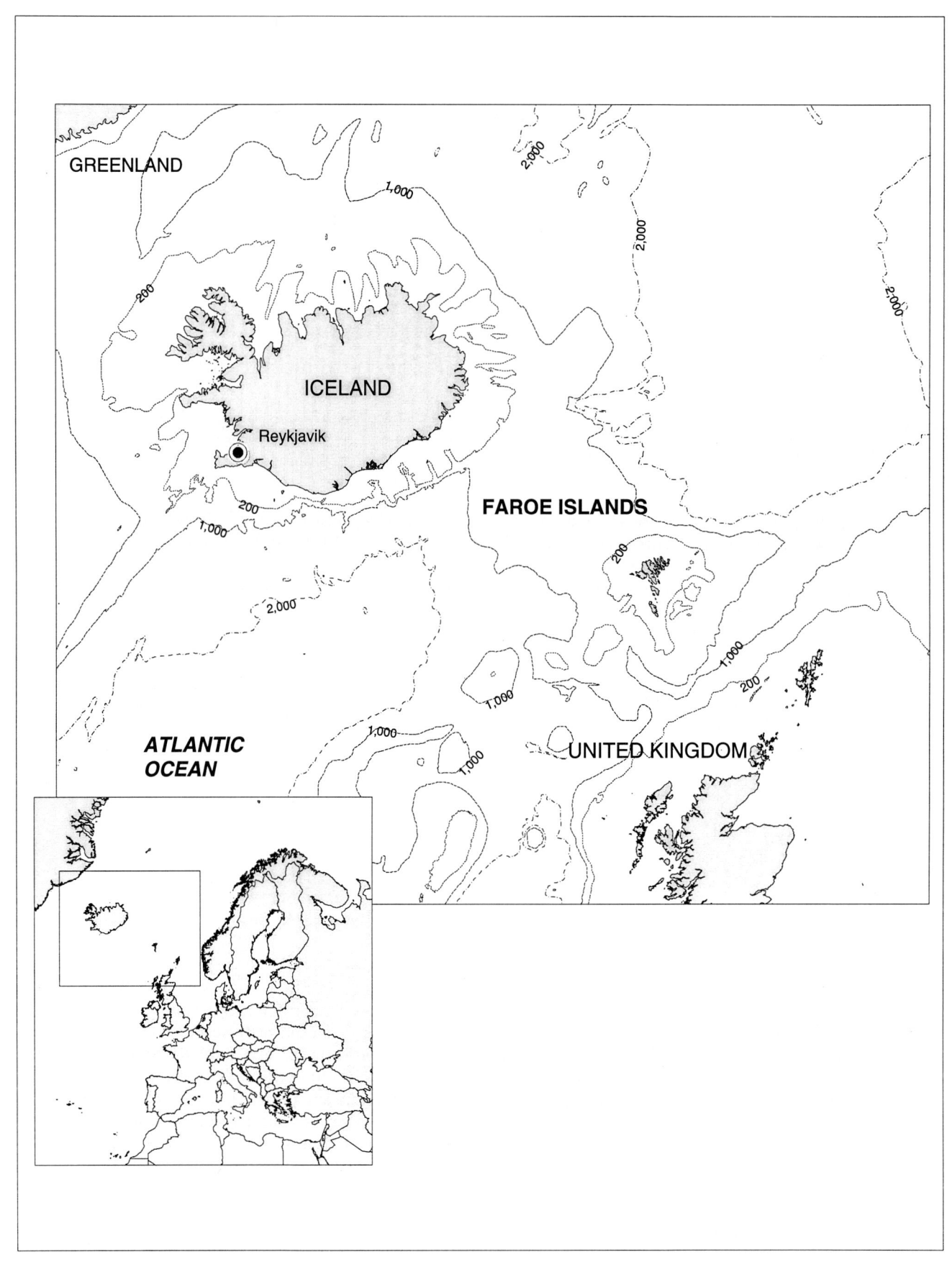

Fig. 1. The Faroe Islands with bottom topography. Adapted from Hansen, 1999.

THE REGION

The Faroe Islands (the Faroes) are situated on the Greenland–Scotland Ridge from which they, like Iceland, extend above sea level (Fig. 1). This submarine ridge has sills between 400 and 600 m deep along most of its path. The upper water layers (down to approx. 500 m) can flow across it, while exchange of water in the deepest layers is blocked by the ridge, which thus acts as a partial barrier between the Arctic Mediterranean north of the Faroes and the rest of the World's ocean, and this influences the region decisively. (The Arctic Mediterranean is the name commonly used to denote all the area north of the Greenland–Scotland Ridge. It includes the Arctic Ocean, but also the "Nordic Seas": the Norwegian Sea, Iceland Sea, and Greenland Sea, in addition to the Barents Sea.)

In the areas north of the ridge, cooling and brine rejection due to freezing induces water in the upper layers to increase its density sufficiently to allow it to sink to intermediate (500–1000 m) or deep levels. This creates a pressure gradient which in the intermediate and deep layers drives a flow southwestwards over the ridge while in the upper layers, a compensating northeastward flow is induced. This gives the ocean areas surrounding the Faroes a special character (Hansen and Østerhus, 1998).

In the upper layers, the water flowing northeastwards (Fig. 2a) derives from more southerly and warmer areas in the North Atlantic Current. In their upper layers, Faroese waters are therefore considerably warmer than the waters typically found at these latitudes. In deeper layers, on the other hand, the cold water that has sunk at higher latitudes passes the islands on its way towards the Atlantic. The deepest passage across the Greenland–Scotland Ridge is a channel which intrudes between the Faroes and Shetland (the Faroe–Shetland Channel) and continues in the narrow Faroe Bank Channel due west of the Faroes. Through this channel system, a continuous flow of sub-zero degree water passes and almost circles the islands (Fig. 2b). At levels deeper than about 500 m, the Faroes are therefore surrounded by cold water in almost all directions. The strong currents found at almost all levels also transport nutrients and organisms, with profound implications for the marine ecosystem in Faroese waters.

The formation of intermediate and deep waters in the north is an integral part of the global thermohaline circulation. With the climatic changes expected from the anthropogenic releases of greenhouse gases, changes in the thermohaline circulation have been predicted (IPCC, 1995). If these predictions are realized, they could lead to a reduced deep water formation in the Arctic Mediterranean area and hence also a reduced flow of warm Atlantic water past the Faroes.

THE COAST AND SHALLOW WATERS

The 1100 km coastline mainly consists of bedrock and bedrock boulders. Only small areas, usually at the bottom of the fjords, are occupied by sand or mud. Most of the hard substrate coast is steep, as the cliffs are alternating layers of volcanic basalt and tuff. The tuff is soft and is easily eroded and washed away, leaving weakened basalt to break off and fall into the sea (Rasmussen 1982).

Tidal amplitude varies within the islands from virtually none in the central region in the vicinity of the capital Tórshavn, to approximately 2–2.5 m in the outer parts of the group of islands to the west. The steep cliffs and the relatively small tidal amplitude results in a narrow littoral zone, especially in the sheltered and moderately exposed parts of the coast. In the exposed areas, waves reach much higher up the shore, moistening the cliffs, thus extending the area where marine life, algae and invertebrates can live. Sea temperatures on the Faroe Plateau are mainly influenced by North Atlantic water, reaching an average of 10°C in summer and decreasing to an average of 6°C in winter (Hansen, 1997).

For centuries scientists have been working in the area, and many older documents can be found on the species living here, their distribution and species diversity. In the past decade the Biofar programmes and the Farcos programme have investigated species and communities from the littoral zone to a depth of 1500–2000 m. Many samples have been taken and scientists from all over the world have participated in the identification of species. Several species new to science have been found and described during this extensive work and many species not previously known to live in this area have been registered.

The littoral communities and their response to wave exposure have been studied (Figs. 3 and 4) and a biological exposure scale (Dalby et al., 1978; Aarrestad and Lein, 1993) has been developed for the area (Bruntse et al., 1999b). Many of the species and their response to wave action resemble what can be observed in adjacent areas in the North Atlantic, with different Fucoids such as *Ascophyllum nodosum*, *Fucus evanescens*, *F. vesiculosus*, and *F. spiralis* being the most dominant species of algae on the sheltered and moderately exposed coasts, and *Aglaothamnion sepositum*, *Alaria esculenta*, *Corallina officinalis*, *Himanthalia elongata* and *Porphyra umbilicalis* being characteristic of the exposed coasts. On the Faroes, however, some species grow further into sheltered areas than seen on other North Atlantic coasts. One of the reasons for this may be the strictly oceanic climate of the Faroe Islands with little difference in summer and winter temperatures and high humidity caused by rain and fog (Bruntse et al., 1999b; Børgesen, 1902; Tittley et al., 1982).

Another difference, when comparing the Faroes with the adjacent coasts of, for example, Norway and Shetland Islands, is the scarcity of *Fucus serratus*, which has only been found at a few localities in one fjord in the Faroe Islands, and the lack of *Littorina littorea*, a common predator on Fucoids on other North Atlantic coasts (Bruntse et al., 1999b).

Below sea level, forests of the large kelp *Laminaria hyperborea* are found. They provide substrate and shelter for

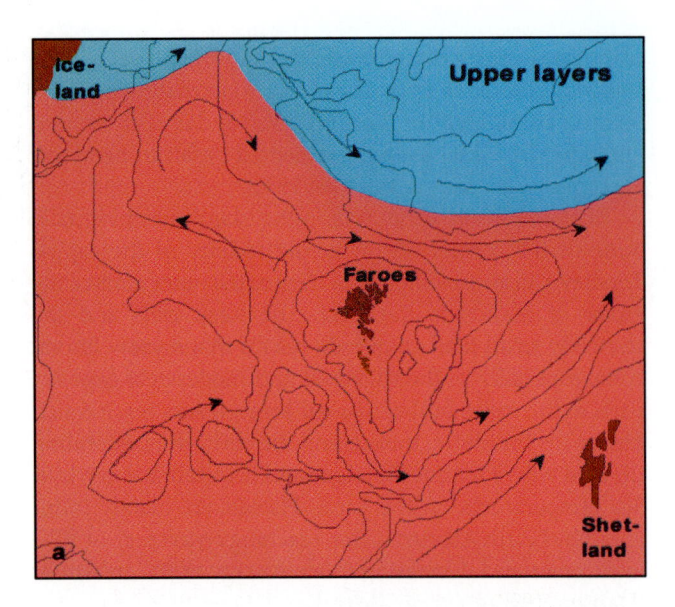

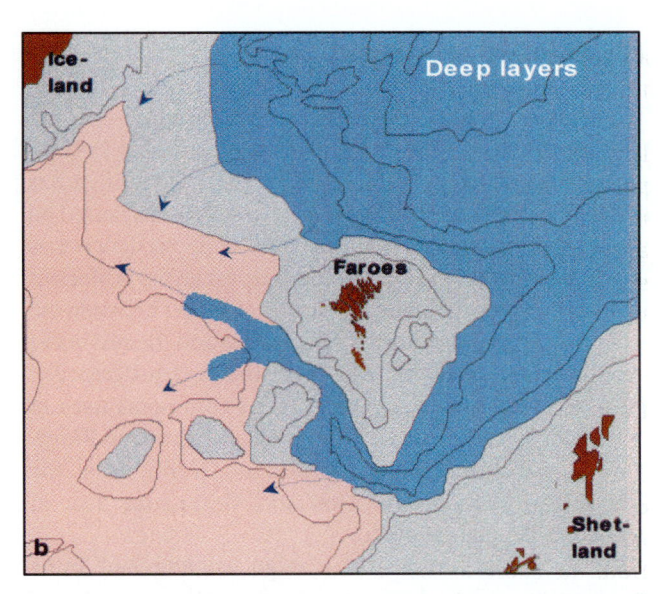

Fig. 2. Typical water flow in the (a) upper and (b) deeper layers indicated by arrows. The colours indicate water temperature with warm waters shown in red and cold waters in blue (adapted from Hansen, 1999).

many species. The large fronds of the kelp hold substrate for many colonies of bryozoans and hydroids, as well as mobile species such as the snail *Lacuna divariecata* and others, some eating the frond and others using it only as an elevated substrate (Bruntse et al., 1999a).

Below the fronds the stiff, erect and rough stipes are usually completely covered by different foliose and bushy algae such as *Phycodrys rubens*, *Plocamium cartilagineum*, and by different species of *Ceramium*. These epiphytes give shelter to thousands of small snails, polychaetes, amphipods and other crustaceans (Price and Farnham, 1982; Bruntse et al., 1999a).

The holdfasts securing the large kelp to the seabed are labyrinths of hapterons which harbour another very different community. Here tube-building polychaetes such as *Pomatoceros triqueter* and *Spirorbis spirorbis* dominate, together with different crustaceans and molluscs, both snails, notably *Gibbula cineraria*, and mussels, *Hiatella arctica*

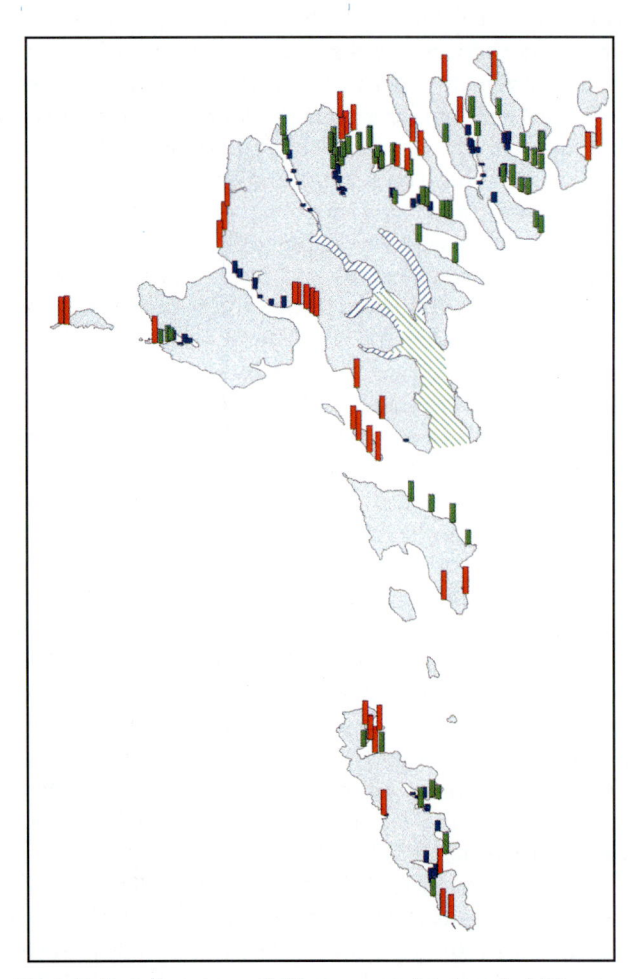

Fig. 3. Biological exposure. Red bars: exposed stations (0–2.75), green bars: moderately exposed stations (3–5.75), blue bars: sheltered stations (6–9). Heights of the columns indicate biological exposure grade, with the tallest being the most exposed. Blue and green hatching respectively indicate sheltered and moderately exposed areas with low tidal amplitude.

Fig. 4. The littoral of an exposed coast. Lower cliffs are covered by *Alaria esculenta* (photo: T.E. Lien).

and *Modiolus modiolus*. Many of the animals found here are juveniles using the extra substrate provided.

The kelp forest and its many niches also harbour many larger species of crustaceans such as *Hyas* sp. and *Cancer pagurus*, and echinoderms such as the sea star *Asterias rubens* and the two common sea-urchins *Echinus esculentus* and *Strongylocentrotus drobachiensis*, as well as numerous species of fish which either live here or find their way through while foraging (Bruntse et al., 1999a).

POPULATIONS AFFECTING THE AREA

The Faroese population of 45,000 inhabitants are highly dependent on fish and on the export of fish products. There is very little other industrial activity; the major discharge from the shore to the sea is therefore domestic effluent. Because the population is small and the ocean surrounding it is vast and has strong currents which promote rapid dilution, the entire area except for the most closed fjord systems may be seen as being only slightly contaminated. Certainly, the waters are much less affected than those near the more densely populated and industrialised parts of the northern hemisphere. In other words, the Faroese offshore waters may be regarded as being an area with mainly diffuse or background pollution levels.

AGRICULTURE

Agriculture is restricted in the Faroe Islands due to the climatic and topographic conditions. The short, cold summer does not allow production of barley or wheat for instance, so every grain or cereal product is imported. Potatoes, rhubarb and a few other temperature-tolerant root vegetables are grown in the islands, but not in quantities sufficient to meet the home market demand.

MARICULTURE

Fish farming is expanding rapidly at present, not least because of a new, more favourable trading agreement with the European Union. The fish farming is done in the fjords, and there is an urgent need to determine the carrying capacity for the most dense production areas. The production especially of Atlantic salmon, *Salmo salar*, and to a minor extent the Rainbow trout, *Oncorhynchus mykiss*, has increased to an anticipated annual production of 30,000 to 35,000 metric tons of round fish in 1999 since the initiation of mariculture activities in the early eighties. The produced salmonids are nearly all exported. At present there are 15 companies which own 26 marine installations in all (Reinert, 1999a), and in addition there are 18 companies directly involved in the production of smolt, at 19 localities (Reinert, 1999b). The demand for freshwater smolt production is secured by placing the tanks in artificial streams or, less desirable, directly into natural sections of rivers. The pharmaceuticals used in the industry are well regulated and subject to controls, using the regulations in the other Nordic countries as a model. The smolt is vaccinated and its skin disinfected before being transferred into cages, to reduce the risk of disease. In the mid 1990s, epidemics of the BKD (kidney disease) was the most severe medical problem. The frequency of outbreaks was less in 1998, but in 1999 are on the rise again (Reinert, 1999b). The other main indication treated with pharmaceuticals is the salmon lice, *Lepeoptherius salmonis* and *Caligus elongatus*. The origin of the breed of atlantic salmon now in production in the Faroes was roe imported from Norway in the period 1978–1984.

OFFSHORE RESOURCES

Fisheries

The Faroese economy is strongly dictated by its fishery, with fish and fish products accounting for over 99% of the gross value of the exported products in the period 1992–1997 (Hagstova Føroya, 1998). The importance of the traditional fishery for wild fish in the Faroese economy became obvious after the fishery collapse in the early 1990s, in particular in 1993. The fishery has, however, increased up to the "old" levels again, making 1998 a year of exceptionally high income for the Faroese fishery. The Faroese fishery zone was extended to 200 miles in 1977 and since then most of the ground-fish fishery (about 98% of the tonnage) in the area has been done by Faroese fishermen. The most valued fish species is the cod, *Gadus morhua*, which is taken in amounts of approximately 25,000 tons per year at the Landgrunnurin, whereas the annual catch at the Foroya Banki has been approximately 2000 tons per year in the period 1965–1989. Since then a marked decrease has occurred (Kristiansen, 1996). Favoured fish species also include saithe, *Gadus virens* (var. *upsi*), and haddock, *Gadus aeglefinus* (var. *hýsa*), each being taken in annual average catches of approximately 40,000 tons and 15,000 tons in the 1980s (Nicolajsen, 1996; Reinert, 1996). The catch of haddock plunged to a mere 4000 tons in 1993, the lowest in this century if the years of World War II are disregarded (Reinert, 1996). The average annual catch since 1978 of Redfish, *Sebastes marinus*, (var. *stóri kongafiskur*) at the Landgrunnurin is approximately 4000 tons (Reinert, 1996).

In 1997 the total catch of wild fish exported as either round fish or as fillets were just above 110,000 tons. In addition to this is approximately the same tonnage of fish used for industrial processing to flour etc. The export of prawns and shellfish combined amounts to about 10,000 tons/yr. The economical importance of the various groups of exported fish products is not mirrored in the tonnage: the export of salmon and trout which in 1997 amounted to approximately 15% that of the white wild fish in tonnage, amounted in value to more than 25% of that of the wild (Hagstova Føroya, 1998).

The fishery in the Faroese waters is regulated by the Føroya Landsstýri which sets the quotas with reference to advice issued by the International Council for the Exploration of the Seas (ICES). Denmark is, on behalf of the Faroe Islands and Greenland, a member of the Northeast Atlantic Fisheries Commission, NEAFO, and the Northwest Atlantic Fisheries Commissions, NAFO. Also Denmark has ratified the agreement on protecting the North Atlantic salmon under NASCO on behalf of both the Faroe Islands and Greenland.

Whaling

In contrast to the fishery, which has enormous economic importance, whaling is a non-commercial hunt only intended for domestic use. Whaling does, however, represent a valuable food source, which contributes on average approximately 30% of domestic meat production (Anon., 1999). The annual average catch of the long-finned pilot whale is variable, but the average number taken annually during the last 300 years has been around 850 animals (Bloch, 1999). The catch is organised according to laws and regulations adjusted from time to time, and the conservation of marine mammals in the Faroe Islands is secured through membership in the North Atlantic Marine Mammal Conservation Organisation (NAMMCO).

Chemical Pollution

The situation with chemical pollution in the Faroe Island seas are best described using some results from studies in the last decade. The best-studied species or matrix is probably the marine mammals, in particular the long-finned pilot whale, *Globicephala melas*. The substance most focused upon in the last few years has been mercury, where recent research shows that in the worst case the pilot whale meat may contain 6 mg/kg mercury (Dam, unpublished). Of this a major, but variable part is methyl mercury, a potent nerve toxin. Recent research involving 1000 new-born Faroese children (Weihe et al., 1996; Grandjean et al., 1997) suggests that intake of this mercury, especially when transferred to the unborn from the mothers, may give rise to adverse but subtle effects on the nervous development in the most heavily exposed individuals. The first time a warning to restrict the consumption of pilot whale meat was issued was in 1977, and this was apparently also the first time mercury was ever measured in Faroese pilot whale meat. Since then, several scientists, especially from other European countries, have screened heavy metals and synthetic persistent pollutants in these mammals (Aguilar et al., 1993; Borrell, 1993; Borrell and Aguilar, 1993; Caurant and Navarro, 1994; Simmonds et al., 1994; Borrell et al., 1995; Caurant and Amiard-Triquet, 1995; Caurant et al., 1993,

1994, 1996). However, attempts to link whale meat consumption to the mercury content of hair samples from volunteers have not been successful (Weihe 1998, pers. comm.). There are several reasons for this apparent discrepancy. Perhaps the most important is the large variation in mercury concentration among the individual whales; studies have shown that for whales the old truth still holds: mercury tends to concentrate with age and hence old individuals contain higher concentrations than young.

The synthetic pollutants most often reported on are the polychlorinated biphenyls (PCB), but other pesticides or metabolites of these are found in high concentrations (Borrell, 1993; Borrell and Aguilar, 1993; Simmonds et al., 1994). Recently, analyses for polybrominated diphenylethers (PBDE) were done on samples of whale blubber, and revealed high concentrations of these compounds especially in the tissue of young animals and males (Lindström et al., 1999) The PBDE, which are often referred to as flame retarders, that were found in the highest concentrations were the tetra and penta-brominated compounds. The maximum concentrations were recorded for 2,2',4,4'-TeBDE (TeBDE #47) and 2,2',4,4'-5-PeBDE (PeBDE #99) with 1782 ng/g lipid and 604 ng/g lipid respectively in a pooled blubber sample representing 13 juvenile males (Lindström et al., 1999). Later studies have confirmed the initial results, though levels as high as were found in the Vestmanna June 1996 *grind* were not detected (van Bavel et al., 1999).

In the 1980s a large study on the biology of long-finned pilot whales was carried out. The study involved specialists from many countries and data from more than 3400 individual whales. The basic results appeared in Donovan et al. (1993) and an overview of the work was compiled by Desportes et al. (1992).

The pilot whales are, however, transitory in that they only visit the Faroese waters for part of the year, so the most reliable results on local contamination would be those for stationary fish species, like cod (Table 1) or sediments. Seabirds also bear evidence of the local state of pollution (Table 2) and studies involving large numbers of species like black guillemots, common eiders and fulmars have been done lately. The diet is the major route of exposure of many organisms to environmental pollutants, therefore the above-mentioned studies included food choice studies as well as the chemical analyses of pollutants like the trace metals, PCBs, pesticides and polycyclic aromatic hydrocarbons (Dam, 1998a,b; Larsen and Dam, 1999).

Trace Metals

In the fjords and in the vicinity of marinas, production facilities or shipyards, a widespread occurrence of imposex in the dogwhelk *Nucella lapillus* was detected in a survey in 1996 (Følsvik et al., 1998). In the survey, sterile females were detected in one out of six sampled populations. This phenomenon has been shown to stem from the leakage of

Table 1

Concentrations of PCB and DDT in cod, *Gadus morhua*, and dab, *Limanda limanda*, from the Faroe Islands. Units are ng/g w.w. when not otherwise stated.

Species	Cod	Cod	Cod	Cod	Cod*	Dab
Tissue	Fillet	Fillet	Liver	Liver	Liver	Liver
Location	Føroya-banki	Munka-grunnin	Føroya-bankin	Munka-grunnin	Norðhavet	Brandansvík
Year of sampling	1995	1995	1995	1995	1994	1996/97
Length (cm)	75–86	49–57	75–86	49–57	46–59	20–30
Age (years)	3–4	3–4	3–4	3–4	2–3	2–6
No. ind.	6	3	7	6	25	153
Fat %	0.2	0.4	49.8	41.4	61	5.6–21.5
CB 28	0.2	<0.1	4.8	5	1.1–1.3	
CB 52	<0.1	<0.1	5.3	10.2	3.7–4.7	
CB 101	<0.1	0.2	9.3	21.5	7.4–11.1	
CB 118	<0.1	<0.1	12.1	22.9	6.4–8.5	2–7
CB 138	0.1	0.3	16.2	27.6	8.8–11.5	3–12
CB 153	<0.1	0.3	23.4	45.1	14.6–20.9	5–19
CB 180	<0.1	<0.1	7.8	13	2.8–4.4	
Sum PCB 7 (mg/kg lipids)			0.16	0.35	0.09–0.35	0.16–0.32
p,p'-DDE (ng/g w.w.)					26.8–31.9	7–32
Sum p,p'-DDT (mg/kg lipids)					0.07–0.08	0.11–0.28

Sources: Larsen and Dam, 1999b (cod), Stange et al., 1996 (cod*), Gunnarson et al., in prep. (dab).

Table 2

Concentrations of PCB and DDT in Faroese seabirds (Σ PCB 7 = CBs 28, 52, 101, 118, 138, 153 and 180).

Species	Year of sampling	Adults/mixed* or chicks	No. ind.	Σ PCB 7 (mg/kg lipids)	Tissue	Dry weight (%)	Lipids (%)	Σ PCB 7 (ng/g w.w.)	CB 153 (ng/g w.w.)	Sum** p,p-DDT (ng/g w.w.)	p,p-DDE (ng/g w.w.)
Fulmar	1998	mixed	25	10.8	liver	30.75	4.8	529	230	392	357
	1997	chicks	40	2.1	liver	29.15	6.4	131	55	86	62
	1998	mixed	24	19.3	subcut. fat	87.50	83.3	15969	6995	8081	6788
	1997	chicks	37	1.2	subcut. fat	87.20	87.2	1009	401	1350	1024
Black guillemots	1995/96/97	adults	56	1.1	liver		3.98	45	20		14
Com. eider	1996	adults	35	0.9	liver		2.44	21	10	14***	10

*Mixed = adults and juveniles. **Sum p,p'-DDT = p,p'-DDT + p,p'-DDE + p,p'-DDD.
***Analysed in a sub-sample of 4 males and 9 females (FEA, 1996).
Sources: Larsen and Dam, 1999b (fulmars) and Dam, 1998b (black guillemots and common eiders).

tributyltin compounds from the anti-fouling paint used on ship hulls. Also, elevated concentrations of some of the heavy metals have been found in such areas, either stemming from the use of lead-rich ship paint, from the anti-fouling formulations used on nets in mariculture, or other unknown sources (Býarverkfrødingurin, 1988; Christensen, 1992; FEA, 1996). With the exception of the lead pollution in an highly localised area, this pollution may be described as moderate, with levels up to 0.14 mg/kg d.w. sediment for Hg and 20 mg/kg ww for Cu in blue mussels (Food and Env. Agency, unpublished). The results of metal analyses on sediment samples from two fjords and one offshore station are given in Table 3, normally the results refer to analysis of the uppermost 2 cm of the sediments, this also with the organic pollutants.

Organic Pollutants

The background concentrations of the offshore area may be found in earlier surveys of the North Atlantic Ocean, when sediment sampled at Skeivibanki, near 61°40'N, 7°50'W, at 350 m depth was analysed (Stange et al., 1996; Magnusson et al., 1996) (Tables 4 and 5). In the study of Stange et al. (1996), the single PCB congeners analysed were not detected at <0.02–0.1 ng/g d.w. The other studies (Magnusson et al., 1996; Granmo, 1996) encountered results in the same

Table 3

Lead, copper and mercury concentrations in sediments in Faroese fjords and in one sample from an offshore station, Skeivibanki.

Location	Year	Pb (mg/kg d.w.)	Cu (mg/kg d.w.)	Hg (mg/kg d.w.)	dw (%)	Loss on ignition (%)	Reference
Skálafjørður	1992	15.0–38.6	61.2–131.3	0.08–0.17	32.9–50.0	6.4–12.6	Christensen (1992)
Skálafjørður	1996	20.4–27.7	70.1–115.9	0.09–0.14		8.4–11.8	FEA (1996)
Tangafjørður	1994	16.5	21.2	<0.04			Granmo (1996)
Skeivibanki	1994	7.3	59.3	<0.06	63	0.58*	Stange et al. (1996)

*Total Organic Content.

Table 4

Pesticides in sediments, in unit ng/g organic carbon. Σ PCB 7 = CBs 28, 52, 118, 101, 153, 138 and 180.

	Tangafjørður 1994 (Granmo 1996)	Skeivibanki 1991 (Magnusson et al. 1996)
Org. C (%)	1.93	0.61
Σ PCB 7	62	67*
HCB	1.04	–
α-HCH	0.52	1.31
β-HCH	1.55	1.15
α-HCH (lindane)	1.55	0.82
p,p´-DDT	0.52	8.90
p,p´-DDE	5.70	8.40
p,p´-DDD	5.18	0.00
α-Chlordane	0.52	0.98
γ-Chlordane	0.52	0.82
trans nonachlor	0.52	1.31

*Converted from Total PCB by using an empirically derived factor from a parallel study where Total PCB = 4.2 * Σ PCB 7.

range but were able to quantify at this level with the dominating contribution to the Σ PCB 7 (= sum of the IUPAC CBs no. 28, 52, 118, 101, 153, 138 and 180) being from CB 153 and 138 each at approx. 0.1 ng/g d.w., amounting to just above 50% of the Σ PCB 7 (Granmo, 1996). A curious result for the pesticide DDT is seen in the offshore location compared to that from the fjord, where the concentration of the original pesticide p,p´-DDT is actually higher in the offshore site than in the fjord. In addition to the exact value, which may be very sensitive to analytical conditions and the content of organic carbon, the relative concentrations between the p,p´-DDT and the metabolite p,p´-DDE is interesting. In the fjord sample the concentration of DDT is one tenth that of the metabolised compound, whereas in the Skeivibanki sample the concentration of the original product is equal to the metabolised one. This concentration of DDT is higher than in many locations in the Kattegat (Magnusson et al., 1996) but this may be more a question of slow degradation than high input because the concentration of the metabolised product in the Kattegat stations were all several times higher; 7–10 fold on a dry weight basis. In general this is the trend seen with the other pesticides analysed: the concentrations in the Faroese sediments were one half to one tenth those detected in the Kattegat and Skagerrak area. With the oil-derived substances found in the fjords or on the Skeivibanki, the difference with the North Sea area is still greater.

PROTECTIVE MEASURES AND THE FUTURE

The Faroe Islands are a self-governing part of the Kingdom of Denmark. In 1948 the Home Rule Act transferred the mandate of legislation to the Faroese Parliament in all areas defined to be issues of self-government. Hence, regulations issued in Denmark or the international conventions ratified by her are not automatically in force also in the Faroe Islands.

The Marine Environmental Act was first issued in 1985 and renewed by Act no. 646 of 12th August 1999. This Act is based on the London convention, adopted on 26th November 1976, the MARPOL convention, which was adopted by Parliament on 18th September 1984 and the new OSPAR convention which was ratified by the Faroese Parliament in October 1998. The Marine Environmental Act applies to the territorial waters and to the continental shelf.

The general Environmental Act came into force in 1989 and applies primarily to land and near shore areas such as fjords and sounds. There are also a number of other Acts that relate to environmental issues, such as protection of natural resources and heritage.

The Environmental Department at the Food and Environmental Agency is the central body of administration of the Environmental Act and the new Marine Environmental Act. The coastguard, The Faroes Inspection and Rescue Service, oversees compliance with the Marine Environmental Act.

Environmental issues were highlighted in the 1980s with the increase in domestic refuse. Today the Faroe Islands have a waste disposal system, covering all households and enterprises down to the tiniest village and island settlement. In the late 1980s modern waste incinerators and controlled dumping mechanisms and sites on the main islands were established. In 1993 a regulation was issued concerning sludge from domestic wastewater treatment. Earlier practice involved disposal of sludge at sea in places

Table 5

Polycyclic aromatic hydrocarbons, PAH, in sediments, in ng/g dw and in fish liver samples, in ng/g ww. The fish species analysed were cod, Gadus morhua, dab Limanda limanda, and long rough dab, Hippoglossoides platessoides.

Matrix	Sedim.	Sedim.	Sedim.	Cod	Cod	Dab	Dab	L. r. dab	L. r. dab
Location	Skeivi-banki	Skeivi-banki	Tanga-fjørður	Mýlings-grunnur	Mýlings-grunnur	Brand-ansvík	Brand-ansvík	61°40'N-7°45'W	62°53'N-9°06'W
Year of sampling	1994	1991	1994	Oct. 1997	Oct. 1997	Mar 96	Aug/Sep 96	Mar 96	Sep 96
Depth of sampling (m)	347	350						118–360	406–426
Dry weight or lipids (%)	63					4.38	20.5	7.52	19.7
Total organic carbon (%) / fish age (yrs)	13.55	0.61	1.93	4	4	3	4	510	
< 63 μm / no. of fish in sample	65%			24	25	21	39	21	34
Fluorene	0.2			3	2	1.8	1	2.9	0.5
Phenanthrene	1.5	2.0	7.9	4	3	3.3	1.7	4.2	1.2
Anthracene	0.8	1.0	3.1	<0.2	<0.2	<0.2	<0.2	<0.2	<0.2
Fluoranthene	4.3	4.0	21.0	<0.5	<0.5	0.7	0.6	0.6	<0.2
Pyrene	2.8	2.0	39.0	0.3	<0.2	0.6	<0.2	<0,2	<0.2
Benz(a)anthracene	2.1	3.0	11.0	<0.1	<0.1	<0.2	0.2	<0.2	<0.2
Chrysene	3.3	3.0	12.9	<0.2	0.4	0.2	0.5	0.2	<0.2
Benz(b+j+k) fluoranthenes	11.6	28.0*	55.0*	<0.9	<0.9	0.3	0.9	<0.2	<0.2
Benz(e)pyrene	3.7			<0.8	<0.8	0.2	0.4	<0.2	<0.2
Benz(a)pyrene	1.9	6.0	22.0	<0.1	<0.1	0.2	0.3	<0.2	<0.2
Perylene	nd					<0.2	<0.2	<0.2	<0.2
Indeno(1,2,3-cd) pyrene	11.3	33.0	55.0	<1	<1	<0.2	<0.2	<0.2	<0.2
Benzo(ghi) perylene	8.5	13.0	80.1	<0.5	<0.5	<0.2	0.4	<0.2	<0.2
Dibenz(a,c+a,h) anthracenes	nd			<1	<1	<0.2	<0.2	<0.2	<0.2

*Benz(b+k)fluoranthenes.
References: Sediments: Granmo 1996 (Tangafjørður), Magnusson et al. 1996 (Skeivabanki 1991) and Stange et al. 1996 (Skeivabanki 1994). Fish: Larsen and Dam, 1999a (cod), Dam, 1998a (dab and l.r. dab).

with strong outgoing currents; now, sludge is collected and treated with calcium and placed in sludgebeds.

The administrative system of environmental protection in the Faroe Islands is divided between the local authorities (municipalities) and the central authority (the Food and Environmental Agency and the Minister of Environment). In general, the local authority is responsible for every environmental issue which is not clearly defined by law to be regulated by the central authority.

The responsibility of the Food and Environmental Agency is to define the terms for operation of the potentially highly polluting companies, and if these terms are met satisfactorily, to grant permissions for operation. It is also the responsibility of the central administration to ensure that the local authorities have a high environmental perspective in their planning.

The history of the state of the environment in the context of chemical pollution is difficult to track without long time series of pollutant analyses. One study has been done on bird feathers from museum specimens. This study, on common and black guillemots (Somer, 1981), showed a slightly increased mercury concentration in the black guillemot feathers from the Faroe Islands from the turn of the century until the mid 1970s. The trend, however, was not quantified, probably due to individual variations. The present liver mercury concentration of these birds is 0.77 mg/kg (Dam, 1998b), which corresponds well with the reported approximately 2 mg/kg mercury in feathers in 1973 (Somer, 1981; Appelquist et al., 1985; Westermark et al., 1975; Pfeiffer-Madsen et al., 1978). Another way of doing time trend studies is to analyse material held in storage by nature itself, in the depths of the peat bogs or in the sediments. Such studies have not as yet been carried out in the Faroes, but are planned to take place in year 2000, at least with regard to mercury in sediment cores.

Mercury is the pollutant of most interest in these islands, mainly due to the high concentrations carried in the much-favoured meat from pilot whales, but also since mercury tends to be associated with fish, which are a major food source. The studies on pilot whale meat are not useful for extracting trends in the mercury concentrations, because individual variations among the whales are too high, and because the time span from when the first analyses were done until today is too short. With the persistent organic pollutants there is growing evidence that the restrictions on their use that were put into action some decades ago have had their effect. There have been marked reductions in the last decade, particularly with the pesticide DDT (Dam et al., unpublished). Still, when the analysed biota is scrutinised with respect to chemical pollutants, the result is often that the concentrations of pollutants within them gives reason for concern when compared to limits defined by food health authorities. In themselves, these wildlife species may represent a minor contribution to the overall intake of pollutants with the diet; but a world in which the only acceptable foodstuff is that produced in modern facilities it is both distressing and unacceptable, especially for people in "remote" areas who still utilise and value caught species for food. It is therefore a high priority to keep fighting the spreading of pollutants into the environment.

In recent years the prospect of finding oil reserves off the Faroe Islands has spurred the interest of the oil-producing companies, and a licensing round allotting exploration permits is expected to be opened in late 1999. At present there are several projects in the planning stage that will elucidate the environmental condition with particular reference to pollutants derived from oil production, both the chemical pollutants and those that may be described as causing physical disturbances. The fossil fuel search will, if realised, be a millennium project, and a very major one indeed for the Faroese population.

REFERENCES

Aarrestad, K. and T.E. Lein (1993) A computer program (Expon) for calculation of a biologically based exposure scale. Institutt for Fiskeri-og Marinbiologi. Rapport 5. University of Bergen.

Aguilar, A., Jover, L. and Borrell, A. (1993) Heterogeneties in organochlorine profiles of Faroese long-finned pilot whales: indication of segregation between pods? *Report of the International Whaling Commission* (Special Issue **14**), 359–367.

Anon. (1999) Whaling in the Faroe Islands—An Introduction. Office of the Prime Minister of the Faroe Islands, Department of Foreign Affairs.

Appelquist, H., Drabæk, I. and Asbirk, S. (1985) Variation in mercury content of guillemot feathers over 150 years. *Marine Pollution Bulletin* **16**, 244–248.

Bloch, D. (1999) 10 Tey bestu tølini fyri grindadráp. *Frøði 1*, vol. 5. (In Faroese).

Børgesen, F. (1902) Marine algae. In Botany of the Faroes based upon Danish Investigations, ed. E. Warming, Part II, pp. 339–532. Copenhagen.

Borrell, A. (1993) PCB and DDTs in blubber of cetaceans from the Northeastern North Atlantic. *Marine Pollution Bulletin* **26**, 146–151.

Borrell, A. and Aguilar, A. (1993) DDT and PCB pollution in blubber and muscle of long-finned pilot whales from the Faroe Islands. *Report international Whaling Commission* (Special Issue **14**), 351–358.

Borrell, A., Bloch, D. and Desportes, G. (1995) Age trends and reproductive transfer of organochlorine compounds in long-finned pilot whales from the Faroe Islands. *Environmental Pollution* **88**, 283–292.

Bruntse, G., Kongsrud, J.A. and Worsaae, K. (1999a) Species associated with stipes and holdfast of *Laminaria hyperborea*. In Marine Benthic Algae and Invertebrate Communities from Shallow Waters of the Faroe Islands, eds. G. Bruntse, T.E. Lein and R. Nielsen. Kaldbak Marine Biological Laboratory, ISBN 99918-3-065-0, pp. 50–60.

Bruntse, G., Lein, T.E., Nielsen, R. and Gunnarsson, K. (1999b) Response to wave exposure by littoral species in the Faroe Islands. *Fróðskaparrit* **47**, 181–198.

Býarverkfrøðingurin (1988) Kanning av dálkingarstøðuni á vágni og teimum størru áunum í Tórshavn, Býarverkfrøðingurin í Tórshavnar Kommunu. Faroe Islands (in Faroese).

Caurant, F. (1994) Bioaccumulation de quelques éléments traces (As, Cd, Cu, Hg, Se, Zn) chez le Globicéphale noir (*Globicephala melas*, Delphinidé) pêché au large des Iles Féroé. PhD thesis, University of Nantes.

Caurant, F. and Amiard-Triquet, C. (1995) Cadmium contamination in pilot whales Globicephala melas: source and potential hazard to the species. *Marine Pollution Bulletin* **30**, 207–210.

Caurant, F., Amiard-Triquet, C. and Amiard, J.-C. (1993) Factors influencing the accumulation of metals in pilot whales (*Globebicephala melas*) off the Faroe Islands. *Report International Whaling Commission* (Special Issue **14**), 369–390.

Caurant, F., Amiard, J.C., Amiard-Triquet, C. and Sauriau, P.G. (1994) Ecological and biological factors controlling the concentrations of trace elements (As, Cd, Cu, Hg, Se, Zn) in delphinids (*Globicephala melas*) from the North Atlantic Ocean. *Marine Ecology Progress Series* **103**, 207–219.

Caurant, F. and Navarro, M. (1994) Cadmium and mercury transfer during pregnancy in the marine mammal, long-finned pilot whale (*Globicephala melas*) of the Faroe Islands. European Research on Cetaceans. Proceedings from 8th Annual Conference of European Cetacean Society, Montpellier, France, 2–5 March, 1994, 226.

Caurant, F., Navarro, M. and Amiard, J.-C. (1996) Mercury in pilot whales: possible limits to the detoxification process. *The Science of the Total Environment* **186**, 95–104.

Christensen, L.Ole, 1992. "Skálafjordundersøgelsen" (Tungmetallbelastningen i fjorden 1992). Food and Environmental Agency, Faroe Islands, 18 pp. (in Danish).

Dalby, D.H., Cowell, E.B., Syrat, W.J. and Crothers, J.H. (1978) An exposure scale for marine shores of western Norway. *Journal of the Marine Biology Association, U.K.* **58**, 975–996.

Dam, M., 1998a. Målinger af miljøgifte i et udvalg af indikatorarter i det færøske marine miljø. Food and Environmental Agency, Faroe Islands, 1998:1, 92 pp. (in Danish).

Dam, M., 1998b. Hvad spiser tejst, edderfugl og topskarv på Færøerne, og hvad er indholdet af miljøgifte i disse fugle? Food and Environmental Agency, Faroe Islands, 1998:2, 87 pp. (in Danish).

Donovan, G.P., Lockyer, C.H. and Martin, A.R., eds. (1993). Biology of Northern Hemisphere Pilot Whales. *Report International Whaling Commission* (Special Issue **14**), 1–479.

Desportes, G., Bloch, D., Andersen, L.W. and Mouritsen, R. (1992) The international research programme on the ecology and status of the long-finned pilot whale off the Faroe Islands: Presentation, results and reference. *Fróðskaparrit* **40**, 9–29.

FEA, 1996. Food and Environmental Agency, Faroe Islands, not published.

Følsvik, N., Brevik, E.M., Berge, J.A. and Dam, M. (1998) Levels and effects of TBT in the Faroese coastal zone. *Fróðskaparrit*, Faroe Islands, 20 pp.

Grandjean, P., Weihe, P., White, R.F., Debes, F., Araki, S., Yokoyama, K., Murata, K., Sørensen, N., Dahl, R. and Jørgensen, P.J. (1997) Cognitive deficit in 7-year-old children with prenatal exposure to methylmercury. *Neurotoxicology and Teratology* **6**, 417–428.

Granmo, Å. (1996) Kristineberg Marine Research Station, Sweden (personal communication).

Gunnarsson, K., Skarpheðindóttir, H. and Dam, M. (in prep.) Integrated ecological monitoring of the coastal zone; environmental pollutants. A report of a Nordic project with participation from Iceland, The Faroe Islands and Norway.

Hagstova Føroyar (1998) Árbók fyri Føroyar, Hagstova Føroyar, Føroyar Skúlabókagrunnur, Tórshavn (in Faroese).

Hansen, B., 1997. Havið kring Föroyar. In *Fiskastovnar og umhvörvi 1997*, ed. H. í Jakupsstovu. Föroya Skúlabókagrunnur, Tórshavn.

Hansen, B. (1999) Havið. Føroya Skúlabókagrunnur.

Hansen, B. and Østerhus, S. (eds), 1998. *ICES Cooperative Research Report*, **225**, 246 pp.

Heinesen, S.P. and Johannesen, M. (1996) Vatnstðumátingar, 1989–1995. Landsverkfrøðingurin, Tórshavn.

IPCC 1995. *Climate Change 1995. The Science of Climate Change*. Cambridge University Press 1996, 572 pp.

Irvine, D.E.G. (1982) Seaweeds of the Faroes. 1: The flora. *Bulletin of the British Museum of Natural History (Bot.)* **10**, 109–131.

Kristianse, A. (1996) Chapters 4.2 and 4.3. In *Fiskastovnar og umhvørvi, 1996*, eds. Hjalti í Jákupsstova. Fisheries laboratory, Faroe Islands. 88 pp. (in Faroese).

Larsen, R.B. (1999) Food and Environmental Agency, Faroe Islands. Unpublished.

Lindström, G., Wingfors, H., Dam, M. and van Bavel, B. (1999) Identification of 19 brominated diphenylethers (PBDEs) in long-finned pilot whale (*Globicephala melas*) from the Atlantic. *Archives of Environmental Contamination and Toxicology* **36**, 355–363.

Magnusson, K., Ekelund, R., Dave, G., Granmo, Å., Förlin, L., Wennberg, l., Samuelsson, M.-O. Berggren M. and Brorström-Lundén, E. (1996) Contamination and correlation with toxicity of sediment samples from the Kattegat and Skagerrak. *Journal of Sea Research* **35** (1–3), 223–234.

Nicolajsen, Á. (1996) Chapter 4.5. In *Fiskastovnar og umhvørvi, 1996*, eds. Hjalti í Jákupsstova. Fisheries laboratory, Faroe Islands. 88 pp. (in Faroese).

Pfeiffer-Madsen, P., Sevel, T., Dahl-Madsen, K.I. and Møller, B. (1978) Measurement, monitoring and control of pollution caused by families and groups of dangerous substances of List I into the aquatic environment of the community. Report to the Environment and Consumer Protection Service, Commission of European Communities.

Price, J.H. and Farnham, W.F. (1982) Seaweeds of the Faroes. 3: Open Shores. *Bulletin of the British Museum of Natural History (Botany)* **10**, 153–225.

Rasmussen, J., (1982) Færöernes geologi. In *Danmarks Natur*, eds. A. Nörrevang and J. Lundö. Politikkens Forlag, Köbenhavn.

Reinert, A. (1999a) Alitíðindi, no. 2, Aquaculture Research Station of the Faroes, Hvalvík, Faroe Islands.

Reinert, A. (1999b) "Country Report" and "Production Report", Meeting papers of the International Salmon Farmers Association, at the General Meeting, Bergen, Norway, August 9th, 1999.

Reinert, J. (1996) Chapters 4.4 and 4.6. In *Fiskastovnar og umhvørvi, 1996*, eds: Hjalti í Jákupsstova. Fisheries laboratory, Faroe Islands. 88 pp. (in Faroese).

Simmonds, M.P., Johnston, P.A., Frensh, M.C., Reeves, R. and Hutchinson, J.D. (1994) Organochlorines and mercery in pilot whale blubber consumed by Faroe Islanders. *The Science of the Total Environment* **149**, 97–111.

Somer, E. (1981) Mercury: Interaction of scientific experiences and administrative regulations in Denmark, First conference on the scientific bases for environmental regulatory actions "Prevention of pollution by substances derived from wastes", Rome, May 11–13.

Stange, K., Maage, A. and Klungsøyr, J. (1996) Contaminants in fish and sediments in the North Atlantic ocean. *TemaNord* 1996, 552.

Tittley, I., Farnham, W.F. and Gray, P.W.G. (1982) Seaweeds of the Faroes. 2: Sheltered Fjoyrds and Sounds. *Bulletin of the British Museum of Natural History (Botany)*, **10**, 133–151.

Van Bavel, B., Sundelin, E., Lillbäck, J., Dam, M. and Lindström, G. (1999) Supercritical fluid extraction of polybrominated diphenyl ethers, PBDEs, from long-finned pilot whale (*Globicephala melas*) from the Atlantic. Presented at "Dioxin 1999" in Venice.

Weihe, P., Grandjean, P., Debes, F. and White, R. (1996) Health implications for Faroe Islanders of heavy metals and PCBs from pilot whales. *The Science of the Total Environment* **186**, 141–148.

Westermark, T., Odsjö, T. and Johnels, A. (1975) Mercury content of bird feathers before and after Swedish ban on alkyl mercury in agriculture. *Ambio* **4**, 87–92.

THE AUTHORS

Maria Dam
*Food and Environmental Agency,
Debesartrøð,
FO-100 Tórshavn, Faroe Islands*

Grethe Bruntse
*Kaldbak Marine Biological Laboratory,
FO-180 Kaldbak, Faroe Islands*

Andrias Reinert
*Aquaculture Research Station of the Faroes,
við Áir, FO-430 Hvalvík, Faroe Islands*

Jacob Pauli Joensen
*Food and Environmental Agency,
Debesartrøð,
FO-100 Tórshavn, Faroe Islands*

Chapter 4

THE NORTH SEA

Jean-Paul Ducrotoy, Mike Elliott and Victor N. de Jonge

The North Sea is a semi-enclosed, epi-continental large marine ecosystem of northwestern Europe. It is relatively shallow (average depth 90 m), and extends north to the Norwegian Trench (700 m). The climate is temperate. Because of highly developed industry and agriculture in its watershed, pollution from contaminants and nutrients has been a major environmental issue for several decades. Atmospheric pathways of contaminant inputs are also important. Fisheries and the protection of species and habitats have become major concerns, and legislation is still developing to address threats to biological diversity, especially of coastal areas which are under pressure from numerous activities.

The regulatory framework for the management of the North Sea is constantly changing. The 1992 'Paris Convention' came into force under the auspices of the Oslo and Paris Commission (OSPARCOM) in 1998. With the increasing influence of the European Union, there is considerable overlap between the EU and OSPAR, leading to duplication between these and other organisations such as the International Conferences for the Protection of the North Sea and the International Council for the Exploration of the Sea. Although a great deal of scientific research has been carried out in this region, the need for good scientific data is still a crucial precursor to management.

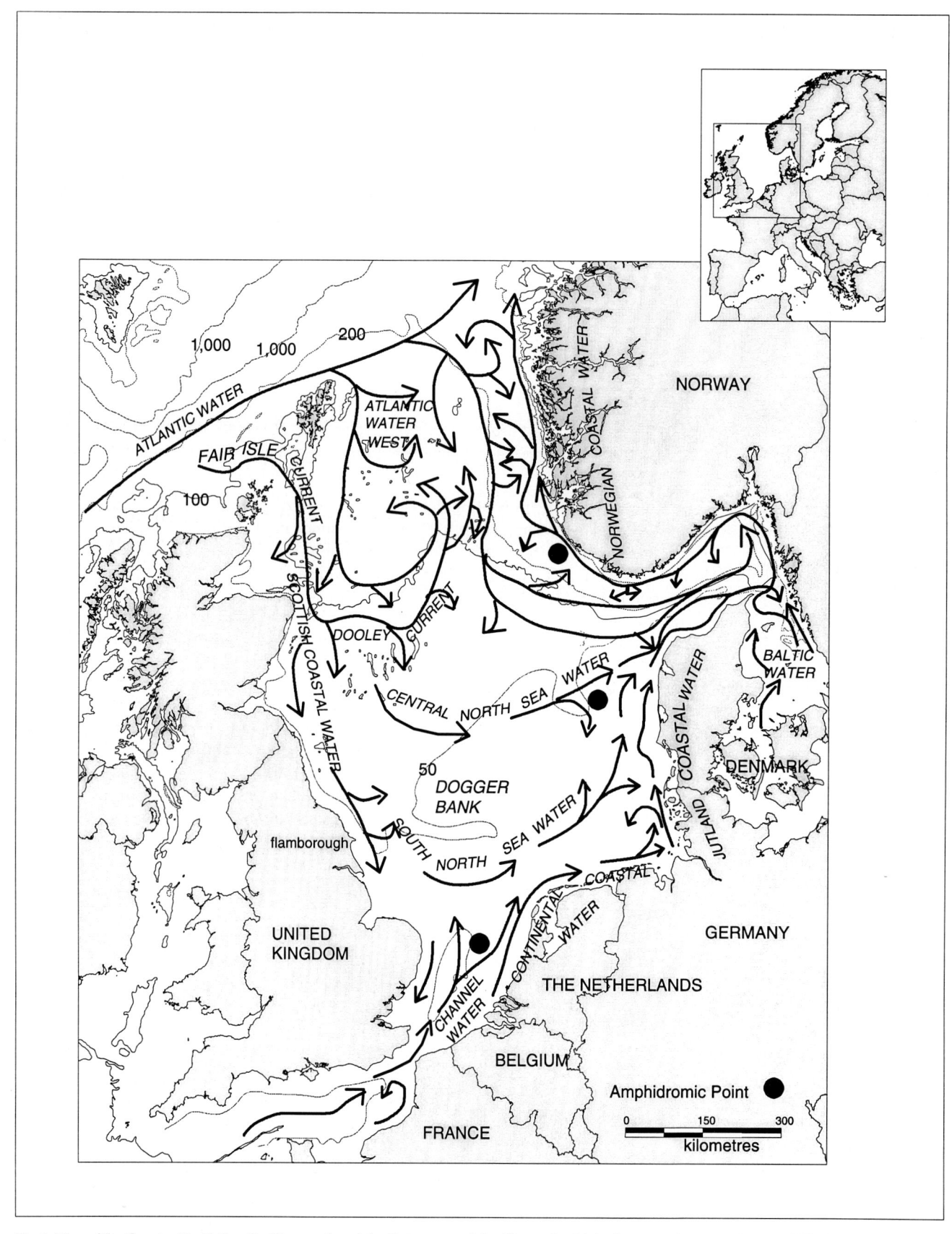

Fig. 1. Map of the Greater North Sea, the Skagerrak and the Kattegat, and the Channel, with bathymetry in metres (source: NSTF 1994) and general circulation (after Turrell et al., 1992). Amphidromic points are also shown.

THE NORTH SEA REGION

The North Sea as defined by the North Sea Task Force includes the area south of 62°N (NSTF, 1994), the Scandinavian Straits (the Skagerrak and the Kattegat) and the English Channel. This chapter concentrates on the region north of the narrowest constriction near Dover (Fig. 1).

The large, semi-enclosed, epi-continental North Sea (750,000 km^2, 94,000 km^3), has a mean depth of 90 m. It is a typical large marine ecosystem, but opens largely to the Atlantic, and it receives low-salinity water from the Baltic through the Kattegat and the Skagerrak, and from numerous rivers. An important part is the Wadden Sea whose area covers just 1% of the total, about half of which are tidal flats. Such fringes are important production and nursery areas.

The North Sea is one of only few major marine ecosystems to have been formed by flooding through the Holocene. The basin is shallow, deepening to the north, and is now heavily sedimented. Its western and northern coasts contain a variety of rocky and mountainous shores, sandy beaches, estuaries, and fjords, while the British east coast is bordered by cliffs, at places protected by pebbles, with a diversity of geological landscapes. On the eastern side, sand beaches and dunes prevail. The Wadden Sea offers large intertidal areas contained behind a girdle of barrier islands (Nienhuis, 1996).

In terms of species, the sea is still being colonised via the Dover Strait and around Scotland. The shallow southern part of the eastern coastline has been strongly impacted by development.

SEASONALITY AND NATURAL ENVIRONMENTAL VARIABLES

Climate

Climate (see Backhaus, 1989, 1996; Damm et al., 1994) is strongly influenced by an inflow of oceanic water from the Atlantic and by prevailing westerly winds, producing both short-period variability from strong continental influences and longer-term and generally weaker oceanic influences. Predominantly southwest winds alternate with easterly winds. The shallow nature of the basin, especially in the south, produces vertical thermal stratification.

During the past two decades, the North Sea has experienced exceptional climatic conditions; the period before 1988 had a series of severe winters, followed by particularly mild winters, and accompanying these changes, high salinities and increased storminess were noted (Becker et al., 1997). These changes were driven by the North Atlantic Oscillation, a periodic change in pressure between Iceland and Portugal, which determines the strength of the prevailing westerlies and ocean surface currents.

Warmer water enters through the Channel (Fig. 2), which means that temperature amplitudes of 8°C are found in the south, compared with 2°C in the north (Becker and Pauly, 1996). Salinity displays few variations in the open North Sea (35 ppt), though in the Scandinavian Straits it falls to 25 ppt and in the Wadden Sea salinities are usually lower than 30 ppt. Multi-year cycles of river discharges have been identified, with low salinities in the 1970s–1980s, and higher values in 1989–1995; reduced run-off leads to increased inflow of Atlantic water (Prandle et al., 1997). In future, increasing irrigation and abstraction of fresh water in the watershed may combine with effects of climate change to magnify changes already observed.

Winds, mainly westerly, increase with distance from the coast. They are seasonal, with high winds in spring. Winds were weaker in the 1960s than more recently (Coeling et al., 1996), a change which has affected water circulation, which in turn has increased cloudiness.

Tides are semi-diurnal (Pugh, 1987; Hardisty, 1990). The dominant lunar tide (M_2) and solar tide (S_2) are both sinusoidal. Tides mainly enter around the northern British coast but are then constrained by the narrowing and shallowing of the sea. This results in three well-developed amphidromic points (see Fig. 1). Spring tidal currents increase southwards, reaching 2–4 m s^{-1} near the southern coasts. Interactions between amphidromic systems cause the tidal range along the Dutch coast to decrease northwards (from 3 to 1.25 m) and then to increase again (to 2.5 m) along the Wadden Sea. Some mega-tidal estuaries and bays are found (up to 12 m in the Baie du Mont Saint Michel), which has important effects on the stability of the substrate.

The topography of the North Sea produces circulation influenced by the Coriolis force. Water entering from the north moves to the western coast while that entering from the Channel moves to the Dutch/Belgian coast (Turrell et al., 1992) creating a cyclonic (anti-clockwise) circulation. Currents are guided by topography and by depths in the

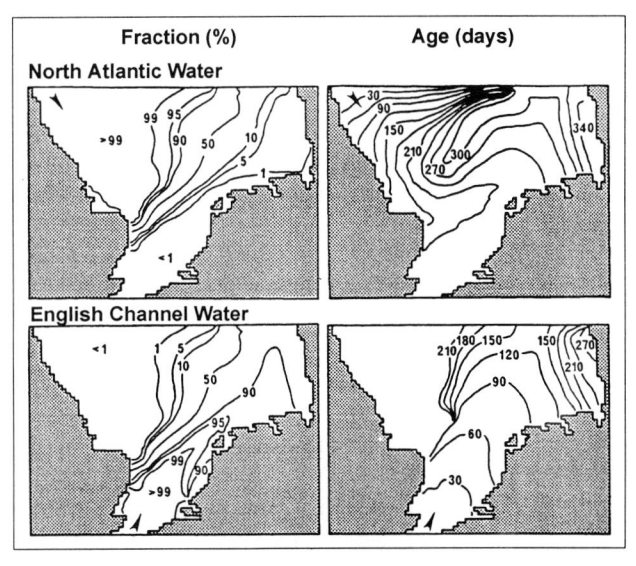

Fig. 2. Proportion and age of Atlantic water flowing into the North Sea (Reid et al., 1989). Arrows indicate the sources.

stratified northern North Sea. Depressed oxygen levels are found under 70 m where there is no wind influence. Flushing time for the entire North Sea has been estimated as one year (Otto et al., 1990) to 500 days. Eddies are a common feature and transient phenomena are generated along frontal boundaries, notably along the Flamborough front off northeast England, and the Frisian front off The Netherlands. Gyres and fronts are an important cause of patchiness amongst the biota.

Freshwater-induced circulation, especially seen as surface plumes, is evident off major estuaries such as the Rhine, Humber and Thames. However, given the well mixed nature of the region, some plumes may not be well developed.

Although there is no indication of increased storminess in the northeast Atlantic, there has been a steady increase of significant wave heights in last 30 years by about 2 to 3 cm per year (Storch, 1996). Extreme waves can reach 30 m and strongly contribute to resuspension of sediments from intertidal flats, and consequently may regulate the turbidity of the entire coastal or estuarine system (De Jonge, 1995; in press).

Surge is an important feature in this semi-enclosed marine ecosystem (Bijl, 1997). Relatively small increases of the intensity of depressions in the southern North Sea will result in a relatively large impact on the storm surges. The characteristic shape of the basin can produce strong storm surges, which increase in effect southwards. The most notable surge, in 1953, produced local sea levels up to 3 m higher than normal and caused a large loss of life in the Netherlands and, to a lesser extent, eastern England (Nienhuis and Smaal, 1994). This has led to the development of coastal protection works and surge barriers across several estuaries of the region.

Rivers and Stratification

The total catchment area of the North Sea is about 840,000 km^2, with a total annual input of about 300–350 km^3 of fresh water, a third of which comes from melt water from Scandinavia (ca. 15,000 m^3 s^{-1}) Wulff et al., 1990; Ehlin, 1981). In the North Sea proper, fresh water inputs, in particular the Rhine, are important (Table 1). The high outflow of Rhine water into the shallow southern region leads to vertical density distributions and sharp frontal structures around the river mouth (De Kok, 1996).

The region overall is well mixed in winter, except in deeper areas. In the north, horizontal thermal stratification occurs in summer, with a thermocline appearing in late spring, typically at a depth of 50 m (Warrach, 1998). Internal waves of tidal period are generated by the interaction of the tide with the Dogger Bank (Vested et al., 1996).

Vertical fronts result in upwelling along the coasts in stratified areas, i.e. the Kattegat, the Skagerrak, and off Norway (Becker, 1990). Proctor and James (1996) demonstrated a link between the horizontal and vertical distribution of water masses in the southern North Sea, showing that the onset of the spring thermocline was linked to the formation of the tidal front. This demonstrates the partitioning of the Sea into the stratified northern part and the well-mixed southern part along the Flamborough front. The deeper northern part remains stratified during the summer and has colder conditions. In contrast, the shallow southern areas remain well-mixed and warmer. These characteristics, across the Flamborough–Heligoland front, have been used to distinguish benthic community patterns (Glémarec, 1973).

Sediments

Intensive sediment movements take place in shallow parts. The anticlockwise gyre is a major mechanism for sediment distribution, but it is now accepted that only a finite stock of sediments is in transit (Guillen and Hoekstra, 1997). Net sediment transport paths indicate sediment deposition in the shallower, especially southern areas (Open University, 1989). The primary sediment sinks are concentrated along the North Sea margins—in the southeastern English estuaries, the Wash and Humber estuary systems (Doody et al., 1993), the Wadden Sea and the Dutch Delta areas. In the Wadden Sea, the concentration of suspended material is mainly governed by wind-induced waves and tidal currents. At increasing wind speed, waves resuspend fine sediment from the intertidal flats while the tidal currents redistribute the material. During periods of lower wind speeds, deposition occurs (De Jonge, 1995). Over a tidal cycle, changing currents may alternately erode or cause deposition of the channel bed (Postma, 1967). The complex sediment and bed-form distribution within the North Sea area, especially the southern basin, can be explained by the interaction of differing tidal components (Hardisty, 1990). The large-scale gyre operating especially in the southern area, for example, has produced the shallow Dogger Bank.

Table 1

Freshwater inputs to the North Sea (modified from OSPAR, in press)

Coastal areas	Run-off (km^3 yr^{-1})	Catchment area (km^2)
Scotland (including Forth)	16	41,000
Norway	58–70	45,500
Skagerrak and Kattegat	58–70	102,200
East coast of England (including Tyne, Tees, Humber and Thames)	32	74,500
Denmark and Germany (including Wadden Sea)	32	219,900
The Netherlands and Belgium (including Wadden Sea, Rhine, Meuse and Scheldt)	91–97	221,400
English Channel (including Seine)	9–37	137,000
Total North Sea	296–354	841,500

Seasonality

Both temperature and salinity show remarkable variability at annual, inter-annual and decadal scales (Becker et al., 1997). Several periods can be recognised. The period 1942 and 1977–1979 displayed the coldest winters in the second half of the 20th century, whereas the period 1989 and 1990 has had the mildest winters of the last 50 years. Changes in air–sea exchange processes and advection of heat and salt from the North Atlantic appear to dominate those climatic fluctuations. Even, if, at present, there are no indications of rapid warming except in the deepest parts of the North Sea, higher transport rates seem to have taken place in the period 1989–1994, with a higher transit through the English Channel since 1977.

Trends in salinity have not been found over 120 years of observation (Laane et al., 1996). Prandle et al. (1997) have summarised the large data set available from the 15 consecutive monthly surveys of the U.K. North Sea Project. Correlations between nitrate, nitrite, and ammonium correspond to the inter-conversion of these compounds. Oceanic and riverine inflows of phosphate, nitrate and silicate are shown to be insufficient to support their seasonal variability, suggesting that the seasonal cycle is maintained by internal recycling.

Climate Change

A general sea-level rise of 50 cm in 2100 has been predicted for the North Sea (IPCC, 1995; Watson et al., 1997). The spatial pattern over the region is affected by eustatic movements causing a tilt of the European landmass. Weaker or even negative trends in sea-level rise have been observed in the north (Scotland, Sweden). The southern part of the British Isles, especially the eastern coast, and parts of the Netherlands and Germany are undergoing isostatic rebound (e.g. Hardisty, 1990), such that large elevations have occurred in the German Bight, southern England and mid-Channel ports (NSTF, 1994).

On the German coastline, an increase in frequency, height and duration of storm episodes has been modelled (Toppe and Führböter 1994), which is linked to a stronger atmospheric circulation which is a result of a higher temperature gradient between the equator and the Arctic.

THE MAJOR SHALLOW WATER MARINE AND COASTAL HABITATS

Most of the North Sea is shallow and coastal, with extensive sedimentary areas, subtidal sandbanks and muddy areas (Rees et al., 1999; Elliott et al., 1998), many of which are rich feeding areas for higher trophic levels.

Open Sea Banks

Most open sea banks are found in the south. Deposition areas in particular seem to have a special ecology; the Dogger Bank is a good example (Kröncke and Knust, 1995). There, the biological situation differs from other areas in that phytoplankton production occurs throughout the year, connected with near-stable abundances of macrofauna and fish.

The hydrocarbon-rich substrata have created surface features associated with the deposits and their extraction. Of note are unique gas-seep pockmarked areas (Dando et al., 1991) which support high biodiversity and a reliance on chemosynthetic production, which is uncommon elsewhere in this area.

Intertidal Rocky and Sedimentary Areas

Rocky areas are the most biodiverse. For example, the macrofauna of the Steingrund, a stony, boulder and pebble area off Helgoland, comprises at least 289 taxa, dominated by several polychaete species (Kuhne and Rachor, 1996).

Intertidal sedimentary areas, such as the Wadden Sea and river estuaries, are highly influenced by sedimentary dynamics (De Jonge, in press). The abundance and vertical distribution of macrobenthic fauna is highly affected by erosion and deposition of fine sediments (Zuhlke and Reise, 1994; Sylvand, 1995). Sediment drift tends to increase from high tide to low tide level, while abundance of most species decreases, possibly due to changes in sediment stability. In mega-tidal estuaries, sediment mobility has crucial effects on the zonation on the tidal flats (Ducrotoy, 1998).

Biodiversity/Alien Species

Because of effects from pollution, brown and red seaweed diversity has decreased since the early 1930s in several English estuaries despite clean-up operations (Hardy et al., 1993). Some disturbances may have an opposite effect; at one subtidal site in the Thames estuary, over 200 invertebrate species were recorded in a dredged shipping channel (Attrill et al., 1996), creating a species richness far higher than elsewhere. However, this remains exceptional.

In the past, various alien seaweeds have found their way to the North Sea: *Sargassum muticum* and *Undaria pinnatifida* (1973), *Grateloupia doryphora* and *G. filicina* (1990), and *Heterosigma akashiwo* (1996). The brown alga *Sargassum muticum*, has now begun to colonise mussel beds in Germany (Schories et al., 1997), resulting in the displacement of a native species.

Hybridisation of salt-marsh plants has also taken place in the 20th century. *Spartina alterniflora* from America has cross-fertilised *S. maritima* to produce a polypoid hybrid *S. anglica*.

Various alien animal species have also been introduced to the North Sea and the English Channel. *Marenzelleria viridis* is a recent invader of European brackish waters from America (Rohner et al., 1996). *M. viridis* survives extreme environmental conditions, such as low oxygen, and competes well against other species. The introduction of the slipper limpet *Crepidula fornicata* has altered benthic habitats and competes with sought after species such as the scallop *Pecten maximus*.

Even if, in some instances, non-native species may increase the total biological diversity, native organisms may be eliminated by them. As stressed by ICES (1995) and Hiscock (1997), it is therefore sensible to prevent the introduction of non-natives, even if some effects can be positive, as in the case of *Ficopomatus enigmatus*, known to improve water quality from the filtering action of its settlements.

Plankton

The Continuous Plankton Recorder (CPR) survey provides a unique data set on the abundance of plankton in the North Sea and North Atlantic, and has yielded maps of the tows made and details of the species identified since 1931 (Warner and Hays, 1994). Correlations were found between the biota, salinity and temperature. In general, phytoplankton production is not as high in the open North Sea as in the coastal zone (Joint and Pomroy, 1993; Joint, 1997), though the Dogger Bank, for example, may exhibit higher primary production and more efficient energy transfer than most of the open North Sea (Heip et al., 1992). The tidal front off the northeast coast of the UK and the front between the low salinity water in the Kattegat and the higher salinity water in the Skagerrak are clear boundaries for plankton assemblages.

The Benthos

Depending on climate, a variable part of the microphytobenthos is suspended in the water column (De Jonge and Van Beusekom, 1995). This makes their communities difficult to understand and species assemblages of intertidal flats have been little investigated since Van Den Hoek et al. (1979). In general, spatial and temporal distributions of benthic diatom species are governed by physical processes, which abundances are more determined by substrate distribution and composition (De Jonge, 1992).

Nematodes are the dominant meiofaunal taxon overall. They are least abundant in the sandy sediments of the Southern Bight, increase to a maximum around 53°30'N and slowly decrease again towards the north. Copepod density and diversity are highest in the Southern Bight, due to the presence of many interstitial species; about 1500 species are thought to occur. Copepods show very distinct assemblages according to water depth and sediment type (Huys et al., 1992; Heip and Craeymeersch, 1995).

Macrobenthic assemblages have been recognised from the early 20th century, and in 1986, a synoptic survey of the North Sea was undertaken by a group of ten laboratories from seven countries (Heip et al., 1992; Kunitzer et al., 1992). The analysis of the macrofauna is based on over 700 taxa. In general, the North Sea macrofauna consists of northern species extending south to the northern margins of the Dogger Bank, and southern species extending north to the 100 m depth contour. The central North Sea is an area of

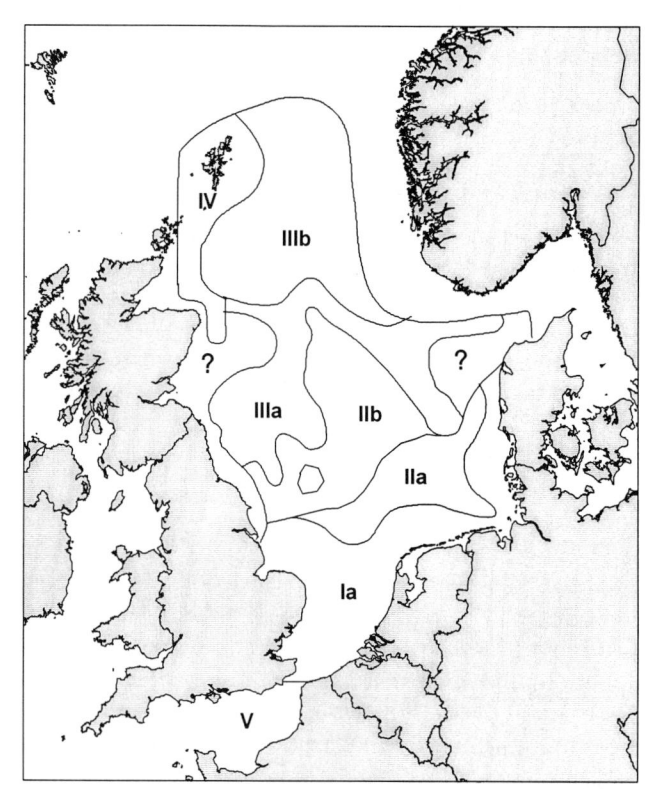

Fig. 3. Distribution of benthic assemblages in the North Sea, based on species abundance data (sources: Künitzer et al., 1992; Holme, 1966). The classification shows environmental factors and the indicator species. (I) mainly < 30 m, coarse sediments; (Ia) *Nephtys cirrosa, Echinocardium cordatum, Urothoe poseidonis*. (Ib) *Aonides paucibranchiata, Phoxocephalus holbolli, Pisione remota*. (IIb) 30–50 m, muddy fine sands, *Nucula nitosida, Callianassa subterranea, Eudorella trunculata*. (IIb) 50–70 m, fine sand, *Ophelia borealis, Nephtys longosetosa*. (IIIa) 70–100 m, fine sediments, no indicators. (IIIb) > 100 m, fine sediments, *Minuspio cirrifera, Thyasira* sp., *Aricidea catherinae, Exogone verugera*. (IV) < 100 m, coarse sediments, *Ophelia borealis, Exogone verugera, Spiophanes bombyx, Polycirrus* sp. (V) 30–100 m, gravels, *Nucula hanleyi* and *Venerupis rhomboides*, often associated with *Ophiotrix fragilis* on hard substrata or *Echinocardium cordatum* in muds.

overlap, especially around the 70 m depth contour (Fig. 3). Since then, the benthos has been related to both spatial and temporal patterns (e.g. Jennings et al., 1999; Basford et al., 1996), which illustrate both the large-scale sedimentary nature of the basin and the influences of Atlantic water inflow. Macrofaunal abundance and diversity both increase linearly towards the north. Macrofaunal biomass for the whole area averages 7 g dwt m^{-2} and decreases from south to north. This contrasting pattern can be only partially explained. Latitude and sediment characteristics, and water depth, determine part of the variance in species composition, density and biomass, but the patterns differ for different benthic groups, being related variously to current patterns, annual temperature variations, and availability of food. Large parts of the variance in many parameters remain unexplained (Heip and Craeymeersch, 1995), and use of these parameters in simple monitoring requires care.

Several types of plants have apparently disappeared from some coasts in the last few decades (see Ducrotoy, 1999b; NSTF, 1994). Red macroalgae disappeared from tidal creeks of the German Wadden Sea progressively up to the 1980s but knowledge is still incomplete. At Konigshafen Bay, an extensively studied part of the Wadden Sea, a total of 46 green algal species, 36 brown algal species and 26 red algal species have now been recorded within the last 120 years on soft and hard substrata. Significant long-term changes in species abundances have occurred in all three groups of algae.

Seagrass was reduced drastically in the 1930s by the so-called 'wasting disease' (Den Hartog, 1987) and has not re-colonised. In the past, *Zostera marina* covered a sub-littoral area of 65–150 km^2 in the western Dutch Wadden Sea (De Jonge et al., 1996). In the 1960s, it also declined on the Danish and French coasts. Reasons may be a change in currents, effects of cockle and mussel fisheries and competition by green algae (Den Hartog, 1994). Associated animals (snails, shrimps, and filefish) disappeared too (Reise et al., 1989).

Salt marshes are found in the upper reaches of sheltered intertidal areas. In deposition areas, colonisation has increased, but most marshes have declined (Rauss, 1998). These two situations can occur close to each other.

Fish

About 224 species of fish are found in the North Sea ranging in size from 5 cm gobies (*Pomatoschistus* sp.) to the 10 m basking shark (Cethorhinidae), but the great majority of information relates to commercial species. Most common species are typical of a semi-enclosed shelf-sea, although deep-water species are found along the northern shelf edge and in the deep waters of the Norwegian Trench and the Skagerrak. In general, less than 20% of species make up 95% of biomass. Different assemblages are found in relation to depth and substrate types (Potts and Swaby, 1991). In the northern North Sea, on the slope edge, the community is dominated by saithe *Pollachius virens* and haddock *Melanogrammus aeglefinus* which both represent more than half of the fish biomass there. Norway pout *Trisopteris esmarkii*, whiting *Merlangius merlangus*, blue whiting *Micromesistius poutassou* and cod *Gadus morhua* occur also.

In the central North Sea, at depths of 50 to 200 m, the association is comparable but haddock dominates, together with whiting and cod. In the shallower North Sea and the Channel, common dab *Limanda limanda* and whiting together account for almost half of the biomass. Grey gurnard *Eutrigla gurnardus* and plaice *Pleuronectes platessa* constitute about a fifth of the biomass. Horse mackerel *Trachurus trachurus* and sand-eels *Ammodytes marinus* and *Hyperoplus lanceolatus* also make a significant contribution to eastern and southern populations. Herring *Clupea harengus* use localised spawning and feeding grounds where they take advantage of well oxygenated gravel beds (Daan et al., 1996). Herring and mackerel *Scomber scombrus* show an interesting migratory pattern for spawning between the North Sea and adjacent seas such as the Baltic Sea or North East Atlantic, where they use the continental slope (ICES, 1990).

Elasmobranchs seem particularly sensitive to recent ecological changes. Two species of dogfish *Scyliorhinus canicula* and *Mustelus mustelus* have decreased over the last few decades, as have rays (Rajidae), conger eel *Conger conger*, sturgeon *Acipenser sturio* and allis shad *Alosa alosa*. Extinction of certain species has taken place at some coastal sites, i.e. the greater weaver *Trachinus draco* (Bergman et al. 1991).

Estuaries surrounding the North Sea are characteristic feeding, overwintering and nursery areas (e.g. Elliott et al., 1990; Elliott and Hemingway, in press). Hence the protection of the estuarine and wetland areas is of great importance for the sustainability of marine fisheries (Marshall and Elliott, 1998).

Birds

The bird population of the North Sea area is of global importance and great numbers of seabirds use the area for feeding and breeding. Twenty-eight species constituting 4.25 million seabirds breed, and a further 6 species feed here. There is an increasing concentration on rocky northern coasts, and there are important populations along the Wadden coasts and in Danish waters. Intertidal areas are of high conservation value; 12 million birds are found, for instance in the Wadden Sea. Waders and wildfowl rely on shallow, inshore and estuarine areas for feeding during their migrations (Tasker and Pienkowski, 1987), and the North Sea is an important part of the eastern Atlantic flyway (Davidson et al., 1991).

In the open North Sea, 10 million sea birds spend most of the year offshore, notably fulmars *Fulmarus glacialis*, guillemots *Uria aalge*, kittiwakes *Rissa tridactyla*, and lesser black-backed gulls *Larus fuscus*. Inshore species include several gulls and terns. The most important factors in their distribution are distance to land/nearest colony; fishing trawler abundance; and a factor linked to water stratification and surface temperature (Garthe and Damm, 1997). Densities in the vicinity of trawlers are particularly high in the case of gulls, showing the importance of fishery discards.

Pollution has an increasing impact on sea birds. For instance, an increase of organochlorines and mercury levels in the common guillemot during winter (Joiris et al., 1997) is explained by prey contamination. There has been an increase in mercury concentrations in seabirds of the German North Sea coast during the last 100 years, especially high levels during the 1940s, and reduced contamination more recently (Furness and Camphuysen, 1997), predominantly due to changes in local pollution inputs, but also from global atmospheric inputs.

Mammals

The harbour seal *Phoca vitulina* and the grey seal *Halichoerus grypus* breed along the coasts of the North Sea. An epidemic of phocine distemper virus in 1988 significantly reduced their numbers, but by the late 1990s, numbers recovered. The harbour porpoise *Phocoena phocoena* is the most common cetacean, and is known to be incidentally taken by fishing vessels (Kock and Benke, 1996), possibly with marked consequences for populations (Walton, 1997).

Contaminants (organochlorines, PCBs, and petroleum oil) find their way to marine mammals along food chains (Kleivane et al., 1995; Jensen, 1996). As a result of chronic low-level pollution from ships and discharges from offshore petroleum activity, about 50% of seal pups at the largest breeding colony in Norway are polluted each year by oil, though this causes little visible disturbance to the seal's behaviour and there has been little mortality. More serious are effects following spills, where animals may be affected by inhalation and suffer physiological damage. Exposure to these persistent compounds is also suspected to be the cause of decline of Baltic grey seals. There are indications that thyroid hormone and vitamin A status of grey seal pups are affected by the low exposure concentrations experienced at the Norwegian coast.

Dolphins (white-beaked dolphin *Lagenorhynchus albirostris*, the less common bottlenose dolphin *Tursiops truncatus*) and the minke whale *Balenoptera acutorostrata* are sighted regularly, but an increase in cetacean strandings has been noted in the 1990s. Winter-stranded sperm whales show an absence of food in the gut, weight loss and blubber thickness reductions compatible with an extended presence in the North Sea where food is inadequate (Jauniaux et al., 1998). Also, the coastal configuration of the southern North Sea makes it a trap for sperm whales (Ross, 1996).

Modelling Communities

Given the large amount of data accumulated for the North Sea, many models have arisen in order to produce a predictive capability. The UK NERC models for the southern part are producing good agreement for hydrographic processes and the exchange and dispersal of materials, including contaminants (Prandle, 1996). Other models for nutrients, eutrophication, contaminant fluxes and pelagic ecosystem functioning have also been created (see Andersen et al., 1996). Ecosystem modelling of smaller geographical areas has also been performed, for example the BOEDE model for the Ems–Dollard estuary (Baretta and Ruardij, 1988). Such models are providing, firstly a mathematical description of processes and, secondly, some capability for predicting changes to these highly variable systems (Carpentier et al., 1997).

POPULATIONS AFFECTING THE NORTH SEA

Approximately, 185 million people live in the highly industrialised countries which either have coasts along the North Sea or which have part or all of their territory in its catch-

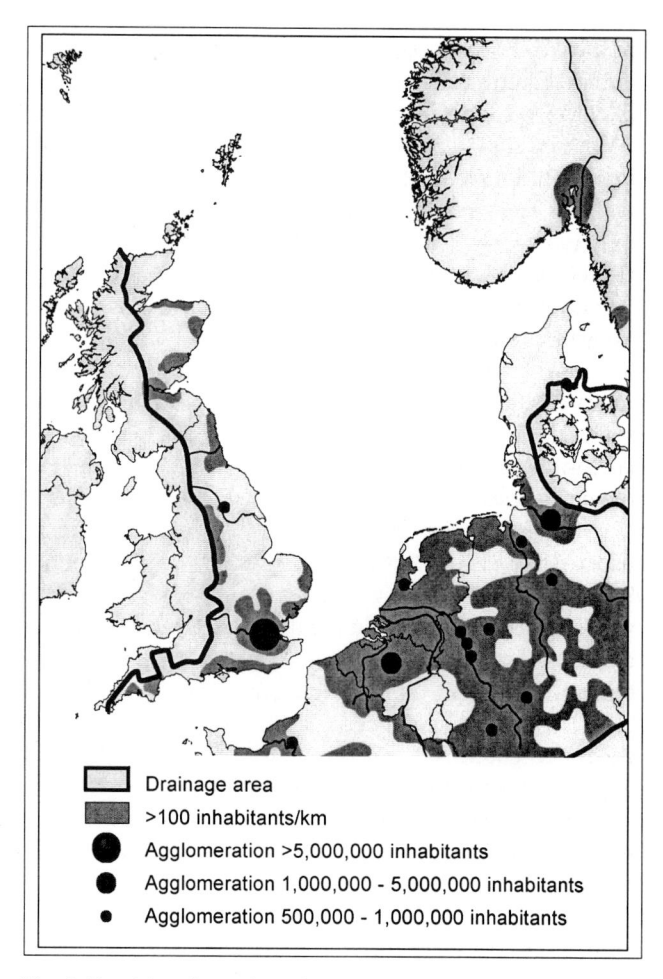

Fig. 4. Densities of people within catchment of the North Sea and main agglomerations (source: NSTF, 1994).

ment area (Fig. 4). An additional 85,000,000 inhabitants occupy the Baltic catchment which flows into the North Sea. Densities of populations differ greatly, with over 1000 inhabitants per km^2 on the coast of Belgium and the Netherlands to less than 50 inhabitants per km^2 along the coasts of Norway and Scotland. Despite a declining birth rate in these countries, the population was still increasing in the late 1990s due to immigration. Tourism induces large-scale seasonal migrations to the coast, amounting to millions of people on a single day. The end of the west–east divide has meant that trade and transport have increased too.

Historical Factors

During the 19th century considerable scientific and technological progress accompanied industrialisation. Scientists, both professionals and amateurs, gained a better knowledge of the North Sea environment, which led to the establishment of several marine stations which developed into reputable institutions. In response to requirements for better fisheries management, ICES was established in 1906, but it was not until 1986 that the Council gained an interest in the wider environment. The first reliable baseline surveys were obtained early on (e.g. Petersen 1914, 1918), but in the late 1960s onwards a true need for scientific surveys and monitoring arose following environmental disasters like the *Stella Maris* or the *Torrey Canyon* grounding (Weichart, 1973). Studies in marine ecology were largely incidental or accidental to studies of coastal habitats and birds up to that time. Internationally co-ordinated studies only took off in the 1980s with projects such as the European COST 647 programme (Lewis, 1984; 1997).

The adoption of EEZs in the North Sea is recent (1992). They have enabled States to adopt certain laws and regulations giving effect to international rules and standards for the protection of the marine environment. Nevertheless, some boundaries between States have been agreed only lately and more experience is needed in the application of the new zones.

Authority for the protection of species and habitats and fisheries is confused. It would seem that the EU is competent for the protection of species and habitats in territorial waters and for fisheries, whereas, the Oslo and Paris Commission is responsible for the protection of species and habitats in the EEZs.

Oil and gas exploration, primarily since the 1970s, has generated about 500 platforms. Sooner or later most will require decommissioning, disposal or re-use. The removal of the Brent Spar (1997) and its suggested disposal into deep water in the Atlantic caused a particularly well known controversy (Elliott, 1996).

Tourism and Recreation

Tourism is largely seasonal in this region. Bathing water quality has long been an issue of concern on the coasts of the North Sea but has recently improved due to action taken at European level. Impacts include those expected from high human concentrations as well as a pressure on habitats with natural features.

Fishing

The rich North Sea supports one of the world's most active fisheries (Table 2). The highest landings per unit area in the North East Atlantic are attained here, as the intensive fishery removes 30 to 40% of the biomass of exploited fish species each year (Gislason, 1992). Landings were highest in 1996 with 3.5 million tonnes (MT) taken. A decrease followed in 1997–1998 to 3 MT. The extensive overfishing of

Table 2

Landings of fish and shellfish by North Sea states in 1995 (source: IMM, 1997)

Denmark	45%	Germany	5%
Norway	22%	Sweden	4%
United Kingdom	12%	France	4%
Netherlands	7%	Belgium	1%

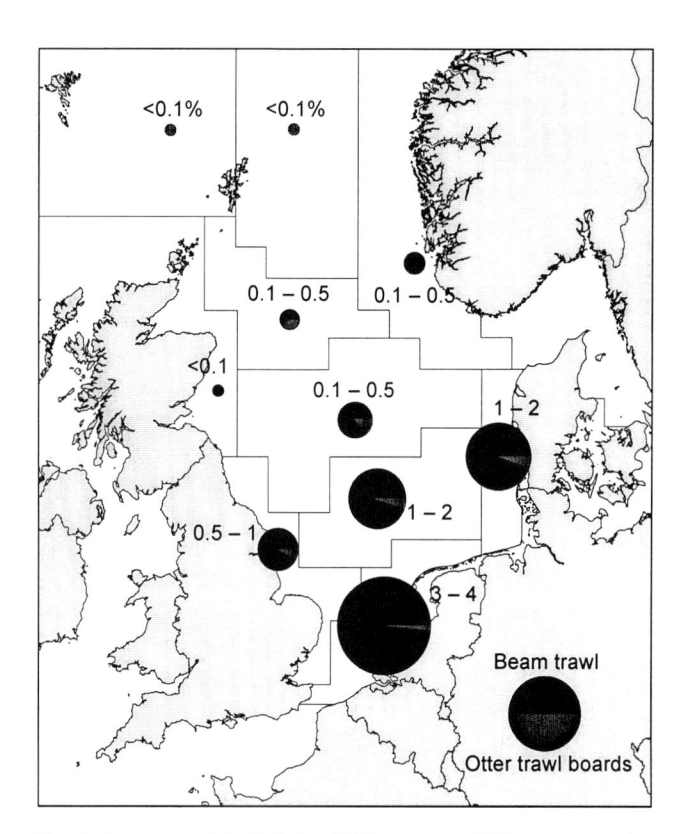

Fig. 5. Area swept in 1989 by fishing gears which penetrate the sediment of the North Sea, relative to the size of the corresponding NSTF areas. (source: ICES, 1994).

the North Sea, especially from beam trawling and other demersal methods, has had an effect on the structure of the fish and benthic communities and on trophic interactions (Jennings and Kaiser, 1998; Bergman et al., 1996). In particular, there are high levels of seabirds supported by discards and offal, and cause for concern about the loss of juvenile fishes. The extensive fishing activity, especially in the south, results in many areas of the bed being trawled several times each year (Fig. 5).

The high fishing pressure has given cause for concern regarding both the loss of mature and breeding fish populations and ecosystem changes (e.g. Jennings and Kaiser, 1998; Blaber et al., in press; IMM, 1997; Pope, 1989). These changes are coupled with indications of global warming which appear to be reducing fish distributions and their spawning patterns, for example of cod (Pope, 1989). These changes are amplified by pollution effects on the fish (Elliott et al., 1988).

Shellfish landings amount to approximately 250,000 t/year. Shellfish are gathered by dredges or trawls, and effects of the shell fishery depend on the total stock available. In the Wadden Sea approximately 11,000 ha are devoted to mussel culture; total production by the mussel fishery varies, but is on average about 50,000 tons wet weight.

The cockle fishery varies greatly, because it relies only on natural resources and is not cultured. For The Netherlands and Germany together, yields range from 100s to over 8500 tonnes fresh meat weight per annum. The cockle fishery is strongly regulated at present because birds and fishermen are competing for them; the birds always lose this contest if unregulated, due to the effectiveness of the suction dredging techniques used for cockles.

With between 30 and 40% of the biomass of commercially exploited fish in the North Sea being caught each year, stocks of herring, cod, mackerel and plaice have been fished beyond safe limits, and some stocks have been permanently depleted since the 1960s. Fishing also causes mortality of non-target species of benthos, fish, seabirds and mammals. The total mass of discards in the German Bight in 1991 was estimated at 36,000 t fish, 58,000 t starfish and 800 t swimming crabs (Garthe and Damm, 1997). Heavy towed gears disturb the uppermost layer of the seabed, while gillnets accidentally entangle seabirds and marine mammals (Kaiser and Spencer, 1996). Unwanted catch is usually returned to the sea where it is eaten by scavenging species, including seabirds (Gislason, 1994). It has proved difficult to isolate the consequences of these impacts, and until more is known about the environmental impact of fisheries management, action (or no action) will have to be agreed upon in the light of considerable scientific uncertainty.

Aquaculture/Mariculture

This is a minor activity throughout most of the North Sea, although there are areas of shellfish cultivation. In Norway and to a smaller extent Scotland, the coastline is suitable for cultivation of salmon *Salmon salar* and trout *Onchorynchus mykiss*. France and the Netherlands also produce quantities of oysters, scallops, and blue mussels. Problems have arisen, or are expected, with the introduction of alien species and with the use of pharmaceuticals. Seaweeds, in particular large browns such as *Laminaria* species, are exploited for alginates, fertilisation, and pharmaceuticals. High production is attained in the UK, France, and Norway.

Coastal Industry

The basins of the Seine, Rhine Meuse complex, Scheldt, Weser, Elbe, Humber and Thames are among the most heavily industrialised regions in the world. Industries of all types are located in estuaries and bays rather than along the open coast, especially in the Southern Bight. They most often coincide with large cities, or at locations with offshore hydrocarbon production. New gas-processing facilities have been opened in the late 1990s in Norway (Firierfjord), The Netherlands (Ems estuary), Germany (Ems estuary) and in England (Atwick, Humberside) with daily production of millions of m^3 of gas.

Military Activities

Military activities take place in various dedicated areas in the North Sea and include, for example, a bombing range on sandflats at Donna Nook at the mouth of the Humber estuary, eastern England, and underwater weapons and mine-sweeping test areas off Scotland. Dumping of munitions has been carried out for many years. Practices include detonating obsolete ammunition in estuaries along the French coast, including old chemical warfare agents.

Nature Conservation

The EC Directive on the Conservation of Wild Birds (the 'Birds Directive') (Council of the European Communities, 1979) complements the EC Directive on the Conservation of Natural Habitats and of Wild Fauna and Flora (the 'Habitats Directive') (Council of the European Communities, 1992). Both contribute to the conservation of biodiversity by requiring Member States to take measures to maintain or restore natural habitats and wild species. Together, they give effect both to site protection and species protection objectives. A major European initiative is the creation of protected sites in the *Natura 2000* network. These designations are to be added to other designations, for example RAMSAR sites. The EU Directives involve creation of Special Protected Areas (SPA) and Special Areas of Conservation (SAC) (Elliott et al., 1999). The SACs proposed by member states have not yet been confirmed, but include, for example, estuarine and coastal habitats at England's Flamborough Head and north Norfolk Coasts, and on the Dutch Wadden Sea (see Elliott et al., 1998).

Degree of Development and Trends

Due to international co-operation, and despite a new wave of economic growth, some results in pollution abatement have been achieved. Unfortunately, few positive trends in the quality of the environment have been recorded by scientists, but the cessation of dumping of sewage sludge in 1998 and of industrial wastes in 1993 has resulted in some decreases of concentrations of cadmium and mercury, PAHs and organochlorines. Due to new regulations, occasional decreases of PCBs have been detected. A reduction of a number of chemicals used in mariculture has not yet shown clear benefits, and no change has yet been demonstrated with TBT, for example. More stringent measures have to be enforced, and one of the main disappointments is an increase in nutrient concentrations despite regulatory action to halve discharges in the 1990s.

Economic activity is steadily increasing around and in the North Sea. Oil exploitation is increasing with new pipelines and platforms, with a move to exploit deeper resources as exploration moves further north and west to the North Atlantic Frontier. The gas-rich southern sector and the oil-rich northern sector still have reserves, although several fields have been exhausted. In 1993 (NSTF, 1994), 300 platforms produced 183 MT of oil when, in 1996, 573 platforms produced 325 MT. Much longer pipelines are being constructed: for example an 810-km pipeline links the

Norwegian field to Dunkirk (France). Yet, fields in environmentally sensitive coastal areas, such as the Wadden Sea may not be developed because of environmental restrictions (see Elliott et al., 1999). Environmental effects of drilling muds and seismic exploration are now more fully understood and require mitigation in developing new fields.

EUTROPHICATION

A general change in nutrient discharges has taken place recently, derived mainly from rivers, run-off and atmosphere. River inputs of phosphorus and nitrogen to the North Sea south of 56°N are estimated to be 127×10^3 and 1100×10^3 tonnes per year respectively (Gerlach, 1990; Portman, 1989). Whereas in some turbid areas such as estuaries, the high turbidity maintains a hyper-nutrified but non-eutrophic system; in clearer parts, symptoms of eutrophication are manifest (Kramer and Duinker 1988 in Salomons et al., 1989).

Changes in nutrient fluxes have caused various changes in nutrients along the European continental coasts (Radach and Lenhart, 1995; Beddig et al., 1997). For the rivers Rhine and Elbe, a strong increase and then recent decrease in dissolved nutrients have been demonstrated (Table 3).

Nutrient distribution in the coastal plume of the Humber Estuary was used to quantify the transfer of land-derived material from rivers and estuaries to coastal seas. Depending on the season, the estuary operated as a source or sink of nutrients (Morris et al., 1995) and a comparable situation was described for the continental Ems estuary (Van Beusekom and De Jonge, 1998).

Table 3
Nitrogen emissions from the different pathways into German river basins in the periods 1983–1987 and 1993–1997 (source: Behrendt, 1999)

	Nitrogen emissions (%)			
	1983–1987		1993–1997	
	Rhine	Elbe	Rhine	Elbe
Direct industrial discharge	19.0	18.2	4.8	7.7
Waste water treatment plants	30.6	21.0	32.2	19.5
Urban areas	3.0	6.3	2.9	7.1
Surface runoff	2.2	0.4	2.9	0.6
Erosion	1.3	2.0	1.8	2.8
Atmospheric deposition	1.0	1.7	1.2	1.6
Tile drainage	7.1	20.4	7.8	17.5
Groundwater	35.8	30.0	46.4	43.2
Percentage	100	100	100	100
Total in tonnes of nitrogen per annum	569,000	328,500	400,000	233,800

The extensive inputs of nutrients and the restricted nature of the North Sea circulation have led to increasing eutrophication events (e.g. Brockmann et al., 1988; Richardson, 1989; Cadée, 1992; Richardson and Heilmann, 1995). Many of the larger rivers produce detectable coastal nutrient plumes. Exceptional algal blooms have been produced and, while some have been a nuisance (e.g. *Phaeocystis* sp.), others have been toxic (e.g. *Gonyaulax* sp.) or have produced exceptional organic enrichment (e.g. *Chrysochromulina* sp.) (Richardson, 1989; Dippner, 1998). Some intertidal and shallow areas are experiencing macroalgal mats as a consequence of high nutrients and poor water exchange (den Hartog, 1994), e.g. *Enteromorpha* mats in the Tees estuary (northeast England) (Scott et al., 1999) and *Ulva* mats along the French, Dutch and German coasts (Piriou et al., 1991; Reise et al., 1994). A general shift from long-lived macrophytes to short-lived nuisance algal species has occurred at various sites (NSTF, 1994), requiring that about 50,000 tons are removed each year from North Sea beaches. Each year, up to 15% of the Wadden Sea is covered by algae, though with an apparent positive effect on zoobenthos productivity (Beukema and Cadee, 1986). While these algal mats generally are unsightly, they can also result in anaerobic conditions, the increase of opportunistic infaunal populations and the prevention of feeding by wading birds and fishes (Scott et al., 1999). Oxygen depletion may also adversely affect deposition areas like the German Bight (Brockmann et al., 1988), with consequent benthos and fish kills (Desprez et al., 1992). A more sustainable management of the river systems, balancing the interests of their legitimate users, is required (Edwards et al., 1997). Offshore deposition areas such as the Dogger Bank also seem to be affected by eutrophication (Kroncke and Knust, 1995).

The European Directive on the control of nitrates (Elliott et al., 1999) has required riparian states to ensure reductions in nitrate discharges and to consider the deleterious effects of such inputs. This has involved the creation of nitrate-vulnerable zones and concomitant changes in agricultural practices. These measures have led to notable decreases in inputs in some areas (Zevenboom and de Vries, 1996).

COASTAL EROSION AND LANDFILL

Extensive sand and gravel beds in the southern area and along the English eastern coast are used as sources of building aggregates. Building and development causing impact to coastal integrity is seen mostly in the shallow southern sector where dykes, artificially maintained dunes, and underwater barriers have been made. In addition, an increase in 'soft-engineering' techniques such as beach nourishment has led to shifts in sand. For example, the Race Bank area off eastern England provides material for beach replenishment along the English Lincolnshire coast, and intense longshore drift in the Netherlands requires constant beach nourishment. The marine aggregate extraction industry in the North Sea is well established, growing from 34 to 40 million m³ between 1992 and 1996. Impacts are site-specific and depend on numerous factors, including extraction method, sediment type and mobility, bottom topography, and bottom current strength (De Groot, 1996). The desired materials are unevenly distributed and should be considered as finite, except for additions from coastal erosion and, locally, river discharge (Arthurton et al., 1997). Long-term forecasts for supply are required and international co-operation is needed for the sustainable use of this irreplaceable resource.

Habitat infilling, in particular of salt marshes, has taken place for centuries almost everywhere in estuaries, intertidal bays and inlets throughout the region. Construction of artificial islands is now being considered in the southern North Sea, for instance in the Netherlands for the installation of a future airport. This is a highly political issue. Changes to the shoreline have been extensive in recent decades, and threats from rising sea levels and sinking landmasses have required new strategies. For example, water storage schemes in the Western Scheldt and managed retreat schemes along the English Essex coastline have been proposed and enacted as soft-engineering works, considered to be environmentally friendly and sustainable methods of dealing with long-term problems.

EFFECTS OF URBAN AND INDUSTRIAL ACTIVITIES

Artisanal and Industrial Use of the Coast

Coastal, conventional power stations are common around the North Sea margin. Cooling water intakes affect fishes, while antifoulants produce halogenated by-products. The eastern British coast alone supports 17 coastal power stations (Turnpenny and Coughlan, 1992). Whilst these are fuelled by oil, gas, coal and nuclear fuel, they all are direct-cooled by seawater. This represents possibly 500 cumecs of heated effluent being discharged to the coastal area.

When the 1974 economic crisis halted economic growth briefly in the region, nuclear energy spread in the 1980s to replace oil. At the end of the 20th century, France produces 75% of its electricity from nuclear power stations, Belgium 60%, Sweden 45%, Germany 35%, and the UK 25%. Use of tidal power is very limited, as the technology for wave energy has not been developed sufficiently for industrial exploitation (Delouche, 1992), and the tidal range around the North Sea in most cases is insufficient. Moreover, while tidal power generation has been proposed for several estuaries, further development is unlikely because of potential environmental effects, such as increased siltation and tidal range decrease behind the tidal energy barrage.

Wind use is slowly developing. Huge developments are planned offshore (about 10 km from the coast). While shoreline wind turbines are common, e.g. at Rotterdam, Zeebrugge and Blythe (northeast England), there are few offshore sites. The most developed is Tunø Knob in Denmark, although there are developments offshore of Blythe.

Metals

Metals have entered the North Sea ecosystem since historical times, naturally or due to human activities. Pollution from metal mining in the Pennine ore fields has induced contaminant fluxes to the North Sea for centuries. Industrial pollution and run-off from urban areas are important contributors of metal inputs and atmospheric sources (Macklin et al., 1997), but all processes need to be taken into consideration when modelling riverine metal fluxes to the North Sea, to avoid overestimation of metal contaminants from industrial and urban sources, and underestimation of contaminants from mining-related sources. Millward and Glegg (1997) suggested that the Humber estuary is a trap for fine, metal-contaminated sediments. For the German Bight a cadmium budget, set up with data from several experiments during the 1990s (Radach and Heyer, 1997), showed that the largest pools were contained in the sediments in deposition areas and were ingested by the benthos (Hall et al., 1996). It seems that less cadmium leaves the area than enters, thus increasing the load. Phytoplankton may absorb cadmium, and as a result, significant concentrations of cadmium (and mercury) are found in kidneys and liver of top predators.

Organics

It is unclear how organic synthetic chemicals affect the benthos but PCBs, for instance, are frequently found in fish liver, seal blubber, bird eggs and human fat in the North Sea, particularly the southern part. Octachlorostyrenes (OCSs) were found in benthic organisms from the international North Sea incineration area (Dethlefsen et al., 1996); OCS concentrations could be taken as an indication of incomplete combustion. The incineration was terminated in February 1991.

Concentrations of HCHs, PCBs, and triazines have been determined in the German Bight within the water column and rain water, and HCHs and PCBs in sediment samples (Huhnerfuss et al., 1997). Concentrations of insecticides and PCBs in sediment from the Thames estuary have been associated with sewage sludge dumping. The mass movement of water into the southern North Sea via the English Channel might also enhance organochlorines in sediments of the Essex salt marshes (Scrimshaw et al., 1996).

Radionuclides

Human-made radionuclides are clearly detectable, in particular from reprocessing plants at La Hague and Sellafield. The La Hague contribution to radioactivity, from antimony and caesium, in Norwegian waters (1992–1995) and the Barents Sea (1992–1997) was appreciable (Guéguéniat et al., 1997). Nevertheless, natural radioactivity should also be taken into account; Polonium$_{210}$ from phosphate ore adds its effects from nuclear power stations. Effects on biota are not known.

The Chernobyl explosion in 1985 contributed significantly to the direct contamination of the North Sea. Nevertheless, the Baltic Sea was affected to a much higher level and its outflow into the North Sea still supplies more Caesium$_{137}$ than the above-mentioned plants.

Cities

Disposal of dredged material into the North Sea amounted to approximately 70 million tonnes per year in the 1990s. The release of inert material such as minestone is decreasing and the discharge of industrial wastes has stopped. Vessel and aircraft scrap dumping will be forbidden by 2005. Good progress has therefore been achieved but a full strategy is still needed for the decommissioning of oil structures.

Officially, the discharge of sewage sludge ceased in 1998 but organic pollution is still high from the fisheries industry through fish processing. Litter and garbage disposal from ships overboard and from tourism is estimated at 600,000 m^3 per year. The same amount could be presently resting on the seabed.

The EU Directive designed to increase urban wastewater treatment, especially that of large population centres, will further decrease direct inputs of organic matter and nutrients. However, given the derogation clauses, it is likely that organic inputs will continue, especially in those waters deemed to have sufficient carrying capacity to degrade, disperse and assimilate the materials (Elliott et al., 1999; Ducrotoy and Elliott, 1997). It remains to be seen whether this has a net effect of reducing inputs to the North Sea.

Shipping and Offshore Accidents and Impacts

MARPOL is the main legal instrument. Shipping in the North Sea is the most intense in the world and the area is a major navigation route for some of the world's most developed and highly-populated economies. There is the continued development of major hub ports such as the Humber complex and the Harwich–Felixstowe complex in eastern England, together with Le Havre, Rotterdam and Hamburg. From these ports there will be the further development of short-sea transport in order to reduce the need for further, deep water ports (see Elliott and Ducrotoy, 1994).

Organotins

The effects of TBT, the active constituent of anti-foulant paints, on marine fauna have been extensively demonstrated with work done in this region and adjacent coasts. A correlation of the imposex effect with shipping intensities was established by Ten Hallers-Tjabbes et al. (1994). Policies to reduce TBT use were developed in a step-wise process: restrictions for small vessels came first, following alarm about coastal dogwhelks, while policy initiatives for offshore shipping started after findings of imposex in relation to general shipping in the North Sea. Global policies were further inspired by a consequent inventory in tropical offshore waters. At present, global policies, within the framework of IMO, move towards a total ban of TBT within the next decade (in 2008) together with a strong stimulation of development and application of environmentally safe alternatives.

Hydrocarbon Extraction

Environmental impacts occur at all stages of oil and gas production, They are the result of prospecting activities, of the physical impact due to the installation of rigs, of operational discharges when production starts and of accidental spills. Nihoul and Ducrotoy (1994) have estimated the input of oil to the North Sea from the offshore industry at 29% of the total. Operational discharges consist of production water and drilling cuttings. An overview over 10 years shows that, although the amount of oil discharged via production water is increasing as platforms are getting older, cuttings still account for 75% of the oil entering the sea as a result of normal operations (Fig. 6). Overall, the North Sea as a whole remains largely affected by discharges of oil from offshore activities and illegal discharges from ships, but even if spills represent a relatively small source of oil, they directly affect birds and mammals (Comphuysen and Franeker, 1992).

The effects on the marine environment of discharges of production water and of discarded oily cuttings have been extensively studied. Despite uncertainty about possible long-term effects, problems with oiled cuttings (drilling

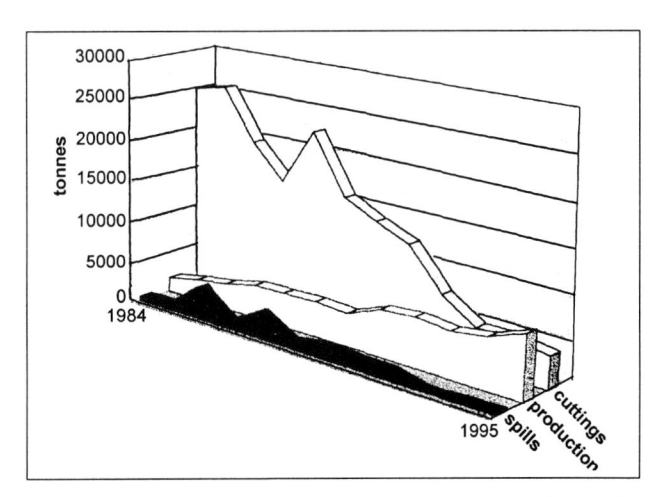

Fig. 6. Contribution of different sources of inputs of oil from the offshore installations (source: OSPAR, 1997).

muds) are acknowledged. PAHs originate from various sources such as flaring or engines, and their major sources are from platforms and ships (Laane et al., 1999).

Atmospheric Pathways

There are large uncertainties in results of different studies on atmospheric input of trace elements into the North Sea (Injuk and Vangrieken, 1995). Annual wet deposition flux of 1.08 ng Hg cm^{-2} yr^{-1} was estimated for the North Sea by Leermakers et al. (1995). Due to variation in dry deposition, the atmospheric input of lead can differ by a factor of 3. Total deposition values and thus atmospheric input data can differ by a factor of 10 (Schlunzen et al., 1997).

Persistent organic pollutants transport and deposition in Europe have been investigated using high-resolution 1990 emission estimates (Vanjaarsveld et al., 1997). Lindane (gamma-HCH) and benzo(a)pyrene were chosen to represent components found predominantly in the gas and particle phases, respectively. Atmospheric lifetimes and dispersion of gas-phase POPs are primarily determined by their solubility in water and degradation rate in soils and vegetation; they have much longer lifetimes over land than over water surfaces. Calculations indicate that gamma-HCH has the potential for dispersion throughout the hemisphere, with most of it ending up in large water bodies such as the North Sea.

Temporal and Spatial Trends

The input and transport of contaminants is widely discussed in Sundermann (1989), Andersen et al. (1996) and Salomons et al. (1989). Similarly, the biological responses to those contaminants are discussed, for example by the authors in Andersen et al. (1996) and contributors to the 1993 and 2000 QSRs. Pollution control measures and the implementation of international agreements and European legislation are designed to further reduce the inputs of conservative pollutants, and there is some evidence that inputs are decreasing (OSPAR, 1998). However, the nature of estuarine and coastal sediments dictates that they will act as sinks for contaminants. There have been wide geographical surveys of biological responses, such as EROD induction in marine flatfish, imposex in dogwhelks or 'Scope-for-Growth' in marine mussels. As an indication of biological responses of environmental quality changes, Widdows et al. (1995) showed that there were changes in Scope-for-Growth in marine mussels along the North Sea British coast and linked reductions in their fitness for survival to concentrations of contaminants in the water column. In particular, the reduction in quality with a progression southwards was linked to a greater bioavailability of pollutants.

Through international and European agreements, certain polluting inputs have now been stopped, notably incineration at sea and vessel disposal of industrial waste as well as sea dumping of sewage sludge. Despite this, aerial inputs of pollutants from the surrounding landmasses together with land-based pipeline discharges will still continue to input contaminants to the North Sea.

Table 4

Examples of international arrangements which affect the protection of the North Sea in the 1990s (from Ducrotoy and Elliott, 1997)

London Convention (1972)	Ban on sea dumping of radioactive waste: Amendment of the Annex (1993, for 25 years)
MARPOL (1973/78)	Generalisation of limit on oil discharges from ships: Amendment of Annex 1 (1993)
Convention on Biological Diversity (1993) 'Green Lungs' of Europe (1993)	Impacts of human activities other than pollution

PROTECTIVE MEASURES

Global/International

UNCLOS (United Nations, 1982) provides a two-fold framework for the protection of species and habitats through tackling pollution and through conservation and management measures. Its approach is reflected in the regional agreements concerning Europe, the North East Atlantic and the North Sea at regional and sub-regional levels (Table 4).

Recently, a series of global programmes targeting better understanding of marine ecology, including ocean circulation, have found developments in Europe. Future plans include the establishment of a Global Ocean Observation System (GOOS), which would offer the prospect of using data from monitoring programmes in conjunction with basin-scale models to improve operational oceanography and to provide forecasts of ocean climate. EMECS (Environmental Management of Enclosed Coastal Seas) have devoted several sessions to the North Sea. Research projects co-funded by the European Union's MAST programme and others co-ordinated by ICES, or conducted by individual nations, continue to add significantly to the understanding of the North Sea.

The European Union

Since the advent of the European Community, the question of competence in the exploitation and protection of marine natural resources has never been made clear. There are areas of 'shared competency', where, in certain matters, the individual Member States have sovereignty and, in others, it is the EU which makes decisions.

Ducrotoy and Elliott (1997) have discussed the degree to which a substantial number of various directives are relevant to the issue of biodiversity, especially for conservation of certain species or habitats in the relevant areas. The 'Birds Directive' and 'Habitats Directive' are probably going to remain as essential pieces of EU legislation in the foreseeable future. Nature conservation falls into the shared competency category. Whether they apply beyond 12-mile limits is debated (Ducrotoy, 1999a) but a 1999 ruling in the British High Court that the UK Government will be required to implement the European Nature Conservation Directives will dictate that the designations can be extended throughout the territorial waters (to 200 miles or the North Sea mid-line) of the northern European states. Member States have given competence to the EU in a number of other areas, including management of the troubled North Sea fisheries stocks through the Common Fisheries Policy.

Regional: The North Sea in the Framework of the North East Atlantic

The legislative framework has developed extensively during the past decade (Ducrotoy and Elliott, 1997). Taking the UK as a case study, Ducrotoy and Pullen (1998) examined the commitments and developments in integrated coastal zone management from an international and European perspective. They concluded that despite an array of legislative tools, there was still confusion regarding several aspects of conservation.

Tables 5 to 9 give the various articles and conventions of environmental importance. Of considerable importance is a new Convention for the Protection of the Marine Environment of the North East Atlantic, still known as the OSPAR Convention, based on the merged Oslo and Paris Conventions.

International Conferences for the protection of the North Sea, known as the International North Sea Conferences (INSC) were installed in the early 1980s (Ijlstra, 1990). They contributed to the establishment of the North Sea Task Force in 1987, which demonstrated that a holistic approach, when applied to the management of the marine environment, enhances policy-making at global and international levels, especially when linked to a marine ecosystem approach (Ducrotoy, 1997; 1998). The NSTF approach has been incorporated in the new ASMO (Assessment and Monitoring) Working Group of OSPAR.

The OSPAR Convention is central to co-operation (Nihoul, 1992). It acts as an interface between periodic ministerial meetings of the North Sea Conferences and several regional and sub-regional agreements; but, the success of the North Sea Conferences depends on the extent to which the EU and other bodies act on its recommendations (Pallemaerts, 1992, in Broadus et al., 1993). Even though OSPAR has been described as a framework model for the protection of other regional seas (Tromp and Wieriks, 1994), this issue remains a source of debate by scientists, some of whom feel that scientific findings are often not correctly translated into appropriate management (Ducrotoy and Elliott, 1997). There is also considerable overlap between

Table 5

Examples of arrangements for protecting European Seas in the 1990s (from Ducrotoy and Elliott 1997)

EU Directives	
Regulation of hazardous chemicals	Plant Protection Products Directive (1991)
	Existing Substances Regulation (1993)
	Pesticides Authorisation Directive (1993)
	Classification, Packaging and Labelling of Dangerous Substances Directive (amended 1993)
Measures against eutrophication	Urban Waste Water Treatment Directive (1991)
	Nitrates Directive (1991)
Microbiological water quality	Quality of Shellfish Harvesting (1991)
New basis for Common Fisheries Policy	Common Fisheries Policy Regulation (1992)
Protection of wildlife within territorial waters	Habitats Directive (1992)
	Natura 2000 (co-ordinated ecological network) (1995)
	Birds Directive (1992)
	Ecological Quality of Surface Water (proposed)

European Environment Agency (1993)

Table 6

Arrangements for protecting the North East Atlantic in the 1990s (from Ducrotoy and Elliott 1997). ASMO: Assessment and Monitoring; INSC: International North Sea Conference

OSPAR Convention for the Protection of the Marine Environment of the North-East Atlantic (ratified in 1998)	Precautionary principle, sustainable development, best environmental practice. ASMO, Annex V on the Protection of Species and Habitats in non-territorial waters
International Conferences on the Protection of the North Sea	INSC3 (1990)
	INSC4 (1995)

organisations. The protection framework in the North East Atlantic includes complex mechanisms incorporated within national, regional, European and global initiatives. The multiple legal tools most often duplicate themselves and create confusion. The control and assessment of introduced harmful substances and environmental protection in the North Sea have benefited from an increasing degree of legislation on species and habitats but here, also, with some confusion arising.

Subregional

National governments have greater legislative authority than local governments in marine environmental protec-

Table 7

Duplication of responsibility for the North Sea. UNCLOS: United Nations Conference on the Law of the Sea; OSPAR: Oslo and Paris Conventions; EU: European Union: UNCED: United Nations Conference on Environmental Development; INSC4: Fourth International Conferences on the Protection of the North Sea; IMO: International Maritime Organisation; QSR: Quality Status Reports; UWWTD: Urban Waste Water Treatment Directive, EIA: Environmental Impact Assessment (from Ducrotoy and Elliott 1997)

Global / International	Regional	Supranational
UNCLOS	OSPAR	EU Directives, etc.
UNCED	INSC4	
IMO/MARPOL		
Framework and agreement; contracting parties	framework and agreement; collation of data, preparation of QSR; recommendations for research/monitoring	*precise statutes* (e.g. TiO_2, Shellfish Growing and Harvest); *frameworks* (e.g. Dangerous Substances); *process control* (e.g. UWWTD, EIA); *wide-ranging* (e.g. Ecological Quality – draft)
	Grey, Black and Red lists; Nutrients, eutrophication	List I and II; Nitrates Directive and UWWTD

Table 8

Protection of marine species and habitats (instruments and science required): duplication or complementary?. EU: European Union; EEA: European Environmental Agency; OSPAR: Oslo and Paris Conventions; ICES: International Council for the Exploration of the Sea; ASMO: Assessment and Monitoring; INSCs: International Conferences on the Protection of the North Sea; UNCED: United Nations Conference on Environmental Development; NSTF: North Sea Task Force (from Ducrotoy and Elliott 1997)

Sub-Regional Agreements	EU/EEA	International Agreements	OSPAR (ICES)	INSCs
Conservation of Small Cetaceans; Trilateral Wadden Sea Co-operation	Common Fisheries Policy; Wild Birds Dir.; Habitats Dir.; Environmental Sensitive Areas; Natura 2000	Bern Convention; Bonn Agreement; Ramsar Convention; UNCED	ASMO -Regional Task Team (RTTs)	Recommendation to EU Recommendation to/from OSPAR
Management and action plans, ecological targets, Red List (species and biotopes); scientific knowledge (justification for management proposals)	basis for fishing yields (justification for policy), designation of areas, identification of habitats,	identification and protection of species; protection of isolated habitats	information on impacts other than pollution	classification system of habitats; co-ordination within and outside territorial waters; need for GIS, habitat evaluation, ecological objectives

Table 9

Examples of arrangements for protecting the North Sea at sub-regional level in the 1990s (from Ducrotoy and Elliott 1997)

Management plan, ecological targets, red list:	Convention on the International Commission for the Protection of the Elbe (1990)
	The Rhine Action Programme (1990)
Regional councils:	KIMO (Kommunenes Internasjonale Miljoorganisasjon)
	North Sea Commission (Environment Group) (1992)
Action plan	Agreement on the Conservation of Small Cetaceans of the Baltic and the North Seas (1993)
	Wadden Sea Trilateral Co-operation(1990)

tion, but local governments may invest more money in environmental protection and often execute more environmental policies (Rinne, 1994; de Jong, 1994). When national governments have established regional programmes, final remediation planning responsibility has often been delegated to local authorities (Table 9). Agenda 21 on Protection of the Oceans gives little attention to the primary role presently being played by the local authorities in controlling marine pollution. The role of local authorities in achieving these objectives must be better defined. There is a plethora of organisations concerned with environmental consequences of human activities, and the use, users, and problems within the semi-enclosed North Sea dictate the need for co-operation between local municipalities. Because of this, several sub-regional organisations such as KIMO (Kommunenes Internasjonale Miljoorganisasjon, Local Authorities Environmental Organisation) have been created. Aspects such as pollution control and best environmental practices in the North Sea area are considered by KIMO within local, national and international frameworks.

The North Sea has problems created by international rivers, for example the Rhine and Elbe, both of which contribute major amounts to the loading into the North Sea (Salomons et al., 1989). Initiatives during the past decade have led to reductions in those loads. The river Rhine has suffered because of numerous drastic environmental changes—for example, the regulation of the river bed and the construction of weirs and dams. Furthermore, discharges of agricultural, industrial and municipal wastewater have caused a deterioration in the water quality. After the Sandoz accident in Basle in November 1986, the states bordering the River Rhine agreed the Rhine Action Programme for its ecological rehabilitation (Vandijk et al., 1995). This programme, which should be realised by the year 2000, aims to (1) create conditions which will enable the return of higher species (such as salmon); safeguard Rhine water as a source for drinking water; (3) abate the contamination of sediments due to toxic compounds; and (4) fulfil the requirements of the North Sea Action Plan, as the river Rhine flows into the North Sea.

PERSPECTIVE

Major changes are taking place in approaching the control of marine pollution in the North Sea compared to those of a decade earlier. The international context is evolving constantly and is now oriented to "protect" the marine environment. Simplification of the regulatory framework is suggested. Despite the fact that a great deal of scientific research has been carried out in the North Sea, the need for obtaining good scientific data is still crucial as a precursor to management. Wider international collaboration is recommended, notably on biological monitoring and on the production of periodic assessments of the quality of the environment.

International reflection is therefore required to collect more relevant data, to produce scientific information and to develop criteria adapted to this enclosed coastal sea. In particular, further studies are needed to identify the role of global climatic change versus local disturbances, and this will have to operate through an international network. Such was the ultimate aim of the original COST 647 project (Lewis, 1996; 1999). With global warming now so much to the fore, it is surely time to establish baselines that will facilitate reliable distinction between human-made and natural changes in 20–50 years time.

In the 2000s, some marine environmental problems, such as pollutant inputs, may be brought under control with existing mechanisms. However, greater threats come from habitat change and loss, for example through dredging and aggregate removal and land reclamation. Whereas contamination can be reduced, the re-creation or restoration of habitat is much more difficult. Nevertheless, the North Sea management still represents a sectoral approach to problems despite the fact that it is becoming more widely recognised that an integrated and ecosystemic form of management is required. This is increasingly possible with some of the instruments now enacted (Ducrotoy and Elliott, 1997).

A future major challenge will be marine environmental planning in order to reconcile and achieve sustainability amongst the many users and uses of the North Sea and its coastal regions. The use of Shoreline Management Plans for coastal defences against a background of relative sea-level rise in many parts of the North Sea may, for instance, conflict with the adoption of habitat and wildlife management plans designed to protect the natural integrity of areas. At present most countries bordering the North Sea have a plethora of management plans designed to protect their coasts, but they will have to increase these to cover the whole sea area. With time, the overlapping jurisdictions of the European Union, OSPAR Commission, ICES and the North Sea Ministerial Meetings, together with the fact that they are all taking a wider and holistic view of marine management, dictates that marine management of the North Sea will become integrated.

REFERENCES

Andersen, J., Karup, H. and Nielsen, U.B. (eds.) (1996) *Scientific Symposium on the North Sea, Quality Status Report, 18–21 April 1994, Ebeltoft, Denmark.* Danish Environmental Protection Agency, Copenhagen.

Arthurton R.S., Harrison, D.J., Laban, C., Leth, J. and Lillywhite, R.P. (1997) Marine sand and gravel in North west Europe—a regional view of demand and resources. *ECMP '97 European Conference on Mineral Planning*, Zwolle, the Netherlands.

Attrill, M.J., Ramsay, P.M., Thomas, R.M. and Trett M.W. (1996) An estuarine biodiversity hot spot. *Journal of the Marine Biologists Association, UK* **76** (1), 161–175.

Backhaus, J.O. (1996) Climate sensitivity of European marginal seas, derived from the interpretation of modelling studies. *Journal of Marine Systems* **7** (2–4), 361–382.

Backhaus, J.O. (1989) The North Sea and the climate. *Dana, a Journal of Fisheries and Marine Research* **8**, 69–82.

Baretta, J. and Ruardij, P. (eds.) (1988) *Tidal Flat Estuaries, Simulation and Analysis of the Ems Estuary.* Springer-Verlag, Berlin.

Basford, D.J., Moore, D.C. and Eleftheriou, A.S. (1996) Variations in benthos in the North-Western North Sea in relation to the inflow of Atlantic Water, 1980–1984. *ICES Journal of Marine Science* **53**, 957–963.

Becker, G.A. and Pauly, M. (1996) Sea surface temperature changes in the North Sea and their causes. *ICES Journal of Marine Science* **53** (6), 887–898.

Becker, G.A. (1990) Die Nord Zee als physikalisches System. In *Warnsignale aus des Mordseewissenschaftliche Fakten*, eds. J.L. Lozán, W. Lenz. B. Rachor, B. Watermann and H. Westernhagen H (eds), Paul Parey, Berlin and Hamburg, 428 pp.

Becker, G.A., Frohse, A. and Damm, P. (1997) The North Western European shelf temperature and salinity variability. *Deutsche Hydrographische Zeitschrift* **49** (2–3), 145–161.

Beddig, S., Brockmann, U., Dannecker, W., Korner, D., Pohlmann, T., Puls, W., Radach, G., Rebers, A., Rick, H.J., Schatzmann, M., Schlunzen, H. and Schulz, M. (1997) Nitrogen fluxes in the German Bight. *Marine Pollution Bulletin* **34** (6), 382–394.

Behrendt, H. (1999) Estimation of the nutrient inputs into medium and large river basins—a case study for German rivers. *LOICZ Newsletter* **12**, 1–3.

Bergman, M.J.N., Fonds, M., De Groot, S.J. and Van Santbrink, J.W. (1996) Direct effects of beam trawl fishery on bottom fauna in the Southern North Sea. In *Scientific Symposium on the North Sea, Qual-*

ity Status Report, 18–21 April 1994, Ebeltoft, Denmark, eds. J. Andersen, H. Karup and U.B. Nielsen (eds.) Danish Environmental Protection Agency, Copenhagen., pp. 204–209

Beukema, J.J. and Cadee, G.C. (1986) Zoobenthos responses to eutrophication of the Dutch Wadden Sea. *Ophelia* 26, 55–64.

Bijl, W. (1997) Impact of wind climate change on the surge in the southern North Sea. *Climate Research* 8 (1), 45–59.

Blaber, S.J.M., Albaret, J.-J., Chong Ving Ching, Cyrus, D.P., Day, J.W., Elliott, M., Fonseca, D., Hoss, J., Orensanz, J., Potter, I.C. and Silvert, W. (1999) Effects of fishing on the structure and functioning of estuarine and nearshore ecosystems. *ICES Journal of Marine Science*, in press.

Broadus. J.M., Demish, S., Gjerge, K., Haas, P., Kaoru, Y., Peet, G., Repetto, S. and Roginko, A. (1993) Comparative assessment of regional international programs to control land-based marine pollution, the Baltic, North Sea, and Mediterranean. Woods Hole Oceanographic Institution, USA, 109 pp.

Brockmann, U., Billen, G. and Gieskes, W.W.C. (1988) North Sea Nutrients and Eutrophication. In: *Pollution of the North Sea, An Assessment*, eds. W. Salomons, B.L. Bayne, E.K. Duursma and U. Förstner. Springer-Verlag, Berlin, pp. 348–389.

Cadee, G.C. (1992) Trends in Marsdiep phytoplankton. In: *Present and Future Conservation of the Wadden Sea, Proceedings of the 7th International Wadden Sea Symposium, Ameland 1990*, eds. N. Danker, C.J. Smit and M. Scholl. Netherlands Institute of Sea Research Publication, Series 20.

Carpentier, P., Dewarumez, J.-M. and Leprêtre, A. (1997) Long term variability of the *Abra alba* community in the southern bight of the North Sea. *Oceanologica Acta* 20 (1), 283–290.

Coeling, J.P., Vanwijk A.J.M. and Holtslag A.A.M. (1996) Analysis of winds speed observations over the North Sea. *Journal of Wind Engineering* 61 (1), 51–69.

Comphuysen, C.J. and Franeker, J.A. (1992) The value of beached bird surveys in monitoring marine oil pollution. Technisch Rapport Vogelbescherming, 10 pp.

CPR Survey Team (1992). Continuous plankton records, the North Sea in the 1980s. *ICES Marine Science Symposia* 195, 243–248.

Daan, R., Booij, K., Mulder, M. and Vanweerlee, E.M. (1996) Environmental effects of a discharge of drill cuttings contaminated with ester-based drilling muds in the North Sea. *Environmental Toxicology and Chemistry* 15, 10.

Damm, P., Hinzpeter, H., Luthardt, H. and Terzenbach, U. (1994) Seasonal and interannual variability in the atmosphere and in the sea. In *Circulation and Contaminant Fluxes in the North Sea*, ed. Sundermann. Springer-Verlag, Berlin, pp. 11–55.

Dando, P.R., Austen, M.C., Burke, M.A. Jr, Kendall, M.A., Kennicut, M.C II, Judd, A.G., Moore, D.C., O'Hara, S.C.M., Schmalljohan, R. and Southward, A.J. (1991) Ecology of a North Sea pockmark with an active methane seep. *Marine Ecology Progress Series* 70, 49–63.

Davidson, N.C., Laffoley, D.D'A., Doody, J.P., Way, L.S., Key, R. Drake, C.M., Pienkowski, M.W., Mitchell, R. and Duff, K.L. (1991) Nature Conservation and Estuaries in Great Britain, Nature Conservancy Council, Peterborough, UK.

De Groot, S.J. (1996) The physical impact of marine aggregate extraction in the North Sea. *ICES Journal of Marine Science* 53 (6), 1051–1053.

De Jong, F. (1994) International environmental protection of the Wadden Sea and the North Sea and the integration of pollution and conservation policies. *Ophelia* 56, 37–45.

De Jonge, V.N. (1992) Physical processes and dynamics of microphytobenthos in the Ems estuary, the Netherlands. Doctorate Thesis, Rijks Universitat, Groningen, the Netherlands, 176 pp.

De Jonge, V.N. (1995) Wind driven tidal and annual gross transports of mud and microphytobenthos in the Ems estuary, and its importance for the ecosystem. In: *Changes in Fluxes in Estuaries*, eds. K.R. Dyer and C.F. D'Elia, pp. 29–40. Olsen & Olsen, Fredensborg, Denmark.

De Jonge, V.N. (1997) High remaining productivity in the Dutch western Wadden Sea despite decreasing nutrient inputs from riverine sources. *Marine Pollution Bulletin* 34, 427–436.

De Jonge, V.N. Relevant scales in bridging fundamental scientific information and management needs in coastal areas. *Continental Shelf Research*, in press.

De Jonge, V.N., De Jong, D.J. and Van Den Berg, S J. (1996) Reintroduction of eelgrass (*Zostera marina*) in the Dutch Wadden Sea: review of research and suggestions for management measures. *Journal of Coastal Conservation* 2 (2), 149–158.

De Jonge, V.N. and Van Beusekom, J.E.E. (1995) Wind and tide induced resuspension of sediment and microphytobenthos from tidal flats in the Ems estuary. *Limnology and Oceanography* 40, 766–778.

De Kok, J.M. (1996) A 2-layer model of the Rhine plume. *Journal of Marine Systems* 8 (3–4), 269–284.

Delouche, F. (1992) *Illustrated History of Europe, A Unique Portrait of Europe's Common History*. Hachette, Paris, 384 pp.

Den Hartog, C. (1987) 'Wasting disease' and other dynamic phenomena in *Zostera* beds. *Aquatic Botany* 27, 3–14.

Den Hartog, C. (1994) Suffocation of a littoral *Zostera* bed by *Enteromorpha radiata*. *Aquatic Botany* 47, 21–28.

Desprez, M., Rybarczyk, H., Wilson, J.G., Ducrotoy, J.-P. and Olivesi, R. (1992) Biological impact of eutrophication in the Bay of Somme and the induction and impact of anoxia. *Netherlands Journal of Sea Research* 30, 149–159.

Dethlefsen, V., Soffker, K., Buther, H. and Damm, U. (1996) Organochlorine compounds in marine organisms from the international North Sea incineration area. *Archive of Fishery and Marine Research* 44 (3) 215–242.

Dippner, J.W. (1998). Competition between different groups of phytoplankton for nutrients in the southern North Sea. *Journal of Marine Systems* 14 (1–2) 181–198.

Doody, J.P., Johnston, C. and Smith, B. (eds.) (1993) *Directory of the North Sea Coastal Margin*. Joint Nature Conservation Committee, Peterborough, UK.

Ducrotoy, J.-P. (1997) Scientific Management in Europe, The Case of the North Sea. In: *Saving the Seas, Values, Scientists and International Governance*, eds. L.A. Brooks and S.D. Van Deveer. Maryland Sea Grant, Washington, pp. 175–192.

Ducrotoy, J.-P. (1998) *Qualité écologique des milieux estuariens et littoraux, application à la Manche et à la mer du Nord*. Mémoire d'Habilitation à Diriger des Recherches, Université de Caen, 2 volumes

Ducrotoy, J.-P. (1999a) Protection, conservation and biological diversity in the North East Atlantic. *Aquatic Conservation, Marine and Freshwater Ecosystems* 9, 313–325.

Ducrotoy, J.-P. (1999b) Indication of changes in the marine flora of the North Sea in the 1990s. *Marine Pollution Bulletin* 38 (8) 646–654.

Ducrotoy, J.-P. and Elliott, M. (1997) Inter-relations between science and policy-making in the North Sea. *Marine Pollution Bulletin* 34 (9) 686–701.

Ducrotoy, J.-P. and Pullen, S. (1998) Integrated coastal zone management, Commitments and developments from an international, European and United Kingdom perspective. *Ocean & Coastal Management* 42, 1–18.

Edwards, A.M.C., Freestone, R.J. and Crockett, C.P. (1997) River management in the Humber catchment. *Science of the Total Environment* 194, 235–246.

Ehlin, U. (1981) Hydrology of the Baltic Sea. In *The Baltic Sea*, ed. A. Voipio. Elsevier, Amsterdam, pp. 123–134.

Elliott, M. and Ducrotoy, J.-P. (1994) Environmental perspectives for the Northern Seas—Summary. *Marine Pollution Bulletin* 29 (6–12), 647–660.

Elliott, M. and Hemingway, K.L. (Eds) *Fishes in Estuaries* (Provisional title) Fishing News Books, in press.

Elliott, M. (1996) *The 'Brent Spar Debacle'—The Role and Responsibility of*

Marine Scientists Sherkin Comment (Environmental Quarterly of Sherkin Island Marine Station, Ireland) September 1996 Issue No. 21 (pp. 21, 24)

Elliott, M. and Ducrotoy, J.-P. (1994) Environmental perspectives for the Northern Seas—summary. *Marine Pollution Bulletin* **29** (6–12) 647–660.

Elliott, M., Griffiths, A.H. and Taylor, C.T.L. (1988) The role of fish studies in estuarine pollution assessment. *Journal of Fish Biology* **33A**, 51–62.

Elliott, M., O'Reilly, M.G. and Taylor, C.J.L. (1990) The Forth Estuary, A nursery and overwintering area for North Sea fish species. *Hydrobiologia* **195**, 89–103.

Elliott, M., Nedwell, S., Jones, N.V., Read, S., Cutts, N.D. and Hemingway, K.L. (1998) Intertidal sand and mudflats and subtidal mobile sandbanks. Report to Scottish Association for Marine Science, Oban, for the UK Marine SAC project, 134 pp.

Elliott, M., Fernandes, T.F. and De Jonge, V.N. (1999) The impact of recent European Directives on estuarine and coastal science and management. *Aquatic Ecology* (in press).

Furness R.W. and Camphuysen, C.J. (1997) Seabirds as monitors of the marine environment. *ICES Journal of Marine Science* **54** (4), 726–737.

Garthe, S. and Damm, U. (1997) Discards from beam trawl fisheries in the German Bight (North Sea). *Archive of Fishery and Marine Research* **45** (3), 223–242.

Gerlach, S.A. (1990) Nitrogen, phosphorus, plankton and oxygen deficiency in the German Bight and in Kiel Bay. *Kieler Meeresforschung. Sonderh.* **7**, 1–341.

Gislason, H. (1992) Ecosystem impacts of fisheries in the North Sea. *Marine Pollution Bulletin* **29**, 520–127.

Glémarec, M. (1973) The benthic communities of the European North Atlantic continental shelf. *Oceanog. Mar. Biol. Ann. Rev.* **11** 263–289.

Guéguéniat, P., Kershaw, P., Hermann, J. and Dubois, P.B. (1997) New estimation of La Hague contribution to the artificial radioactivity of Norwegian waters (1992–1995) and Barents Sea (1992–1997). *Science of the Total Environment* **202**, 1–3, 249–266

Guillen, J. and Hoekstra, P. (1997) Sediment distribution in the nearshore zone, Grainsize evolution in response to shore-surface nourishment. *Estuarine Costal and Shelf Science* **45** (5), 639–652.

Hall, J.A., Frid, C.L.J. and Proudfoot, R.K. (1996) Effects of metal contamination on macrobenthos of two North Sea estuaries. *ICES Journal of Marine Science* **53** (6), 1014–1023

Hardisty, J (1990) *The British Seas*. Routledge, London.

Hardy, F.G., Evans, S.M. and Tremayne, M.A. (1993) Long-term changes in the marine macroalgae of three polluted estuaries in north-east England. *Journal of Experimental Marine Biology and Ecology* **172**, 81–92.

Heip, C. and Craeymeersch, J.A. (1995) Benthic community structures in the North Sea. *Helgolander Meeresuntersuchungen* **49** (1–4), 313–328.

Heip, C., Basford, D., Craeymeresch, J.A., Dewarumez, J.M., Dorjes, J., De Wilde, P., Duineveld, G., Eleftheriou, A., Hermann, P.M., Nierman, U., Kingston, P., Kunitzer, P., Rachor, A., Ruhmor, H., Soetaert, K. and Soltwedel, T. (1992) Trends in biomass, density and diversity of North sea macrofauna. *ICES Journal of Marine Science* **49**, 13–22.

Hiscock, K. (1997) Conserving biodiversity in North-East Atlantic marine ecosystems. In *Marine Biodiversity, Patterns and Processes*, eds. R.F.G. Ormond, J.D. Gage and M.V. Angel. Cambridge University Press, 18, pp. 415–427.

Holme, N.A. (1966) The bottom fauna of the Channel, part II. *Journal of the Marine Biological Association of the UK* **46**, 401–493.

Huhnerfuss, H., Bester, K., Landgraff, O., Pohlmann, T. and Selke, K. (1997) Annual balances of hexachlorocyclohexanes, polychlorinated biphenyls and triazines in the German Bight. *Marine Pollution Bulletin* **34** (6), 419–426

Huys, R., Herman, P.M.J, Heip, C.H.R. and Soetaert, K. (1992) The meiobenthos of the North Sea, density,biomass trends and distribution of copepod communities. *ICES Journal of Marine Science* **49** (1), 23–44.

ICES (1990) Report of the Mackerel Working Group, ICES CM 1990/Assess, 19.

ICES (1994) Report of the Study Group on Ecosystem Effects of Fishing Activities. ICES Cooperative Research Report no. 202, 15 pp.

ICES (1995) ICES code of practice on the introduction and transfers of marine organisms. ICES Cooperative Research Report no. 204, 5 pp.

Ijlstra, T. (1990) The third International Conferences for the Protection of the North Sea. *Marine Pollution Bulletin* **21** (5), 223–226.

IMM (Inter Ministerial Meeting) (1997) Assessment report on fisheries and related species and habitat issues. Danish Environmental Protection Agency, Copenhagen.

Injuk, J. and Vangrieken, R. (1995) Atmospheric concentrations and deposition of heavy metals over the North Sea. *Journal of Atmospheric Chemistry* **20** (2), 179–212.

IPCC (International Panel on Climate Change) (1995) Second assessment report: Climate change 1995, IPCC, Geneva, Switzerland, 64 pp.

Jauniaux, T., Brosens, L., Jacquinet, E., Lambrigtd, D., Addink, Smeenk, C. and Coignoul, F. (1998) Postmortem investigations on winter stranded sperm whales from the coasts of Belgium and the Netherlands. *Journal of Wildlife Diseases* **34** (1), 99–109.

Jennings, S. and Kaiser, M.J. (1998) The effects of fishing on marine ecosystems. *Advances in Marine Biology* **34**, 201–352

Jennings, S., Lancaster, J., Woolmer, A. and Cotter, J. (1999) Distribution, diversity and abundance of epibenthic fauna in the North Sea. *J. Mar. Biol. Ass. UK* **79**, 385–399.

Jensen, B.M. (1996) An overview of exposure to, and effects of, petroleum oil and organochlorine pollution in grey seals *Halochoerus grypus*. *Science of the Total Environment* **186** (1–2), 109–118.

Joint, I. (1997) Phytoplankton production in the North Sea. *Land Ocean Interaction in the Coastal Zone Report*, **29**, 27.

Joint, I. and Pomroy, A. (1993). Phytoplankton biomass and production in the North Sea. *Marine Ecology Progress Series* **99**, 169–182.

Joiris, C.R., Tapia, G. and Holsbeek, L. (1997) Increase of organochlorines and mercury levels in common guillemots *Uria aalge* during winter in the southern North Sea. *Marine Pollution Bulletin* **34** (12), 1049–1057.

Kaiser, M.J. and Spencer, B.F. (1996) The effects of beam trawl disturbance on infaunal communities in different habitats. *Journal of Animal Ecology* **65**, 348–358

Kleivane, L., Skaare, J.U., Bjorge, A., Deruiter, E. and Reijnders, P.J.H. (1995) Organochlorine pesticide residue and PCBs in Harbor porpoise *Phocoena phocoena* incidentally caught in Scandinavian waters. *Environmental Pollution* **89** (2), 137–146.

Kock, K.H. and Benke, H. (1996) On the by-catch of the harbour porpoise *Phocoena phocoena* in German fisheries in the Baltic and the North Sea. *Archive of Fishery and Marine Research* **44** (1–2), 95–114.

Kroncke, I. and Knust, C. (1995) The Dogger Bank, A special ecological region in the central North Sea. *Helgolander Meeresuntersuchungen* **49** (1–4), 335–353.

Kuhne, S. and Rachor, E. (1996) The macrofauna of a stony area in the German Bight (North Sea). *Helgolander Meeresuntersuchungen* **50** (4), 433–452.

Kunitzer, A., Basford, D., Craeymersch, J.A., Dewarumez, J.-M., Dorjes, J., Duineveld, G.C.A., Eleftheriou, A., Heip, C., Herman, P., Kingston, P., Niermann, U., Rachor, E., Rumohr, H. and De Wilde, P.A. (1992) The benthic fauna of the North Sea, species distribution and assemblages. *ICES Journal of Marine Science* **49** (2), 127–144.

Laane, R.W.P.M., Southward, A.J., Slinn, D.J., Allen, J., Groeneveld, G., De Vries, A. (1996) Changes and causes of variability in salinity and dissolved inorganic phosphate in the Irish Sea, English Channel and Dutch coastal zone. *ICES Journal of Marine Science* **53** (6), 933–944.

Laane, R.W.P.M., Sonneveldt, H., Van Der Weyden, A., Loch, J. and Groeneveld, G. (1999) Trends in the spatial and temporal distribution of metals and organic compounds in the Dutch coastal zone sediments from 1981–1996. *J. Sea Research* **41**, 1–17.

Leermakers, M., Baeyens, W., Ebinghaus, R., Kuballa, J. and Kock, H.H. (1995) Determination of atmospheric mercury during the North Sea experiment. *Water, Air and Soil Pollution* **97** (3–4), 257–263.

Lewis, J.R. (1984) Temporal and spatial changes in benthic coastal communities, COST 47 approach. *Marine Pollution Bulletin* **15**, 397–402.

Lewis, J. (1996) Coastal benthos and global warming, Strategies and problems. *Marine Pollution Bulletin* **32** (10), 698–700.

Lewis, J.R. (1997) Temporal and spatial data from the COST 647 rocky littoral programme. In Ecosystems Research Report No 16. In *Change in Marine Benthos, The Case for Long-term Studies*. E C. Brussels. pp. 1–13

Lewis, J.R. (1999) Coastal zone conservation and management, A biological indicator of climatic influences. *Aqua. Conserv.* **9**, 401–405.

Macklin, M.G., Hudson Edwards, K.A. and Dawson, E.J. (1997) The significance of pollution from historic metal mining in the Pennine ore fields on river sediment contaminant fluxes to the North Sea. *Science of the Total Environment* **194**, 391–397

Marshall, S. and Elliott, M. (1998) Environmental influences on the fish assemblage of the Humber estuary, UK. *Estuarine Coastal and Shelf Science* **46** 175–184.

Millward, G.E. and Glegg, G.A. (1997) Fluxes and retention of trace metals in the Humber Estuary. *Estuarine Coastal and Shelf Science* **44**, 97–105.

Morris, A.W., Allen, J.I., Howland, R.J.M. and Wood, R.G. (1995) The estuary plume zone - source or sink for land-derived nutrient discharges. *Estuarine Coastal and Shelf Science* **40** (4), 387–402.

Nienhuis, P.H. (1996) The North Sea coasts of Denmark, Germany and The Netherlands. In Marine Benthic Vegetation, eds. Schramm and Nienhuis. *Ecological Studies* **128**, 187–221.

Nienhuis, P.H. and Smaal, A.C. (1994) *The Oosterschelde—A Case Study of a Changing Estuary*. Kluwer, Dordrecht.

Nihoul, C. and Ducrotoy, J.-P. (1994) Impact of oil on the marine environment, Policy of the Paris Commission on operational discharges from the offshore industry. *Marine Pollution Bulletin* **29** (6–12), 323–329.

Nihoul, C. (1992) From the Oslo and Paris Conventions to the Convention for the Protection of the Marine Environment of the North-East Atlantic. *North Sea Task Force News*, May 1992, 3.

NSTF (North Sea Task Force) (1994) *The 1993 Quality Status Report of the North Sea*. Oslo and Paris Commissions/International Council for the Exploration of the Sea, eds. J.-P. Ducrotoy, J. Pawlak and S. Wilson. Olsen & Olsen, 132 pp.

Open University (1989) *Waves, Tides and Shallow-water Processes*. The Open University Course Team, Pergamon Press, Oxford.

Oslo and Paris Commissions (1997) Report on discharges, waste handling and air emissions from offshore installations for 1984–1995, Part 2. OSPAR, London, pp. 45–66.

Oslo and Paris Commissions (1998) Integrated assessment of inputs to the OSPAR Convention area 1990–1996. INPUT Special Assessment Workshop, the Hague, 26–27 March 1998. ASMO, London, 122 pp.

Otto, L., Zimmerman, J.T.F., Furnes, G.K., Mork, M., Soetre, R. and Becker, G. (1990) Review of the physical oceanography of the North Sea. *Netherlands Journal of Sea Research* **26** (2–4), 161–238.

Petersen, C.G.J. (1914) Valuation of the sea. II. The animal communities of the sea bottom and their importance for marine zoogeography. *Rep. Dan. Biol. Stn.* **21**, 1–44.

Petersen, C.G.J. (1918) The sea bottom and its production of fish food. *Rep. Dan. Biol. Stn.* **25**, 1–62.

Piriou, J.-Y., Menesguen, A. and Salomon, J.-C. (1991). Les marées vertes à ulves, conditions nécessaires, évolution et comparaison de sites. In *Estuaries and Coasts, Spatial and Temporal Intercomparisons*, eds. M. Elliott and J.-P. Ducrotoy. International Symposium Series, Olsen & Olsen, Fredensborg, pp. 117–122.

Pope, J. (1989) Fisheries research and management for the North Sea, the next hundred year. *Dana, A Journal of Fisheries and Marine Research* **8**, 33–44.

Portman, J.E. (1989) The chemical pollution status of the North Sea. *Dana, A Journal of Fisheries and Marine Research* **8**, 95–108.

Postma, H. (1967) Sediment transport and transportation in the estuarine environment. In: *Estuaries*, ed. G.H. Lauff. American Association for the Advancement of Science, Publication 83, pp. 158–184.

Potts, G.W. and Swaby, S.E. (1991) Evaluation of rare British marine fishes. Nature Conservation Council, CSD Report, 1220.

Prandle, D., Hydes, D.J., Jarvis, J. and MacManus, J. (1997) The seasonal cycles of temperature, salinity, nutrients and suspended sediments in the southern North Sea in 1988 and 1989. *Journal of Coastal and Shelf Science* **45** (5), 669–680.

Prandle, D. (1996) Modelling and Measuring, what we can and can't do and future prospects. In *Scientific Symposium on the North Sea Quality Status Report, 18–21 April 1994, Ebeltoft, Denmark*, eds. J. Andersen, H. Karup and U.B. Nielsen. Danish Environmental Protection Agency, Copenhagen, pp. 33–40.

Proctor, R. and James, I.D. (1996) A fine resolution 3D model of the Southern North Sea. *Journal of Marine Systems* **8** (3–4), 285–295.

Pugh, D. (1987) *Tides, Surges and Mean Sea Level*. Wiley & Son, Chichester.

Radach, G. and Heyer, K. (1997) cadmium budget for the German Bight in the North Sea. *Marine Pollution Bulletin* **34** (6) 375–381.

Radach, G. and Lenhart, H.J. (1995) Nutrient dynamics in the North Sea—Fluxes and budgets in the water column derived from ERSEM. *Netherlands Journal of Sea Research* **33** (3–4), 301–335.

Rauss, I. (1998) Marais salés de la baie des Veys. Rapport, GEMEL, 13 pp.

Rees, H.L., Pendle, M.A., Waldock, R., Limpenny, D.S. and Boyd, S.E. (1999) A comparison of benthic biodiversity in the North Sea, the English Channel and the Celtic Seas. *ICES Journal of Marine Science* **56**, 228–246.

Reid, P.C., Taylor, A.H. and Stevens, J. (1989) The hydrography and hydrographic balances of the North Sea. In *Pollution of the North Sea and Assessment*, eds. E. Salomons, B.L. Bayne, E.K. Duursma and U. Forstner. Springer-Verlag, Berlin.

Reise, K., Herre, E. and Sturm, M. (1989). Historical changes in the benthos of the Wadden Sea around the island of Sylt in the North Sea. *Helgoländer Meeresunters*, **43** 417–433.

Reise K., Kolbe, K. and De Jonge, V.N. (1994) Makroalgen und Seegrasbestände im Wattenmeer. In: *Warnsignale aus dem Wattenmeer—Wissenschaftliche Fakten*, eds. J.L. Lozán, E. Rachor, K. Reise, P. von Westernhagen and W. Lenz (eds.). Blackwell, Oxford, pp. 90–100.

Richardson, K. and Heilmann, J.P. (1995). Primary production in the Kattegat—past and present. *Ophelia*, **41**, 317–328.

Richardson, K. (1989) Algal blooms in the North Sea, the good, the bad and the ugly. *Dana, A Journal of Fisheries and Marine Research* **8** 83–94.

Rinne, P. (1994) Environment responsibility and co-operation between Baltic and North sea municipalities. 1. The Baltic eco-cities programme and its pilot function for municipalities bordering the North Sea. *Marine Pollution Bulletin* **29** (6–12), 627–630.

Rohner, M., Bastrop, R. and Jurss, K. (1996) Genetic differentiation in *Hediste diversicolor* for the North Sea and the Baltic Sea. *Marine Biology* **130** (2), 171–180

Ross, A (1996) Conservation of vulnerable species in a degraded environment—the case of the harbour porpoise (*Phocoena phocoena*) in the North Sea. In *Scientific Symposium on the North Sea Quality Status Report, 18–21 April 1994, Ebeltoft, Denmark*, eds. J. Andersen, H. Karup and U.B. Nielsen. Danish Environmental Protection Agency, Copenhagen, pp. 279–285.

Salomons, E., Bayne, B.L., Duursma, E.K. and Forstner, U. (eds.) (1989) *Pollution of the North Sea and Assessment*. Springer Verlag, Berlin.

Schlunzen, K.H., Stahlschmidt, T., Rebers, A., Niemeier, U., Kriews, M. and Dennecker, W. (1997) Atmospheric input of lead into the German Bight. *Marine Ecology Progress Series* **156**, 299–309.

Schories, D., Albrecht, A. and Lotze, H. (1997). Historical changes and inventory of macroalgae from Konigshafen Bay in the northern Wadden Sea. *Helgolander Meeresuntersuchungen* **51** (3), 321–341.

Scott, C.R., Hemingway, K.L., Elliott, M., De Jonge, V.N., Pethick, J.S., Malcolm, S. and Wilkinson, M. (1999) Impact of nutrients in estuaries. Report to the Environment Agency and English Nature, Cambridge Coastal Research Unit, University of Cambridge, and CEFAS, Lowestoft 250 pp.

Scrimshaw, M.D., Bubb, J.M. and Lester, J.N. (1996) Organochlorine contamination of UK Essex coast salt-marsh sediments. *Journal of Coastal Research* **12** (1) 246–255.

Storch, von H. (1996) The WASA project, Changing storm and wave climate in the North East Atlantic and adjacent seas. GKSS 96/E/61, 16 pp.

Sündermann, J. (Ed.) (1989) *Circulation and Contaminant Fluxes in the North Sea*. Springer-Verlag, Berlin.

Sylvand, B. (1995) La Baie des Veys Littoral occidental de la baie de Seine, Manche) 1972–1993, structure et évolution à long-terme d'un écosytème benthique intertidal de substrat meuble sous influence estuarienne. Thèse de Doctorat d'Etat de l'Université de Caen, 397 pp.

Tasker, M.L. and Pienkowski, M.W. (1987) *Vulnerable Concentrations of Birds in the North Sea*. Nature Conservancy Council, Peterborough.

Ten Hallers-Tjabbe, S.C.C., Kemp, J.F. and Boon, J.P. (1994) Imposex in whelks (*Buccinum undatum*) from the open North Sea—relation to shipping traffic intensities. *Marine Pollution Bulletin* **28** (5) 311–313.

Toppe, A. and Fuhrboter, A. (1994) Recent anomalies in mean and high tidal water levels at the German North Sea coastline. *Journal of Coastal Research* **10** (1), 206–209.

Tromp, D. and Wiericks, K. (1994) The OSPAR Convention, 25 years of North Sea protection. *Marine Pollution Bulletin* **29** (6–12), 622–626.

Turnpenny, A.W.H. and Coughlan, J. (1992) Power generation on the British coast, thirty years of marine biological research. *Hydroécologie Appliquée* Tome 4, **1**, 1–11

Turrell, W.R., Hendersen, E.W., Slesser, G., Payne, R. and Adams, R.D. (1992) Seasonal changes in the circulation of the northern North Sea. *Continental Shelf Research* **12** (2–3), 257–286.

United Nations (1982) *Law of the Sea Convention*. United Nations, New York.

Van Beusekom, J.E.E. and De Jonge, V.N. (1998) Retention of phosphorus and nitrogen in the Ems estuary. *Estuaries* **21**, 527–539.

Van Den Hoek, C., Admiraal, W., Colijn, F. and De Jonge, V.N. (1979) The role of algae and seagrasses in the ecosystem of the Wadden Sea, A review. In *Flora and Vegetation of the Wadden Sea*, ed. W.J. Wolff (ed.), pp. 9–118. Report of the Wadden Sea Working Group, Leiden, 3, 9–118.

Vandijk, G.M., Marteijn, E.C.L. and Schultewulwerleiding, A. (1995) Ecological rehabilitation of the Rhine—Plans, progress and perspectives. *Regulated rivers research and Management* **11** (4–4), 377–388.

Vanjaarsveld, J.A., Vanpul, C.W.A.J. and Deleeuw, F.A.A.M. (1997) Modelling transport and deposition of persistent organic pollutants in the European region. *Atmospheric Environment* **31** (7), 1011–1024.

Vested, H.J., Bareta, J.W., Ekebjaerg, L.C. and Labrosse, A. (1996) Coupling of hydrodynamical transport and ecological models for 2D horizontal flow. *Journal of Marine Systems* **8** (3–4), 255–267.

Walton, M.J. (1997) Population structure of harbour porpoises *Phocoena phocoena* in the seas around the UK and adjacent waters. *Proceedings of the Royal Society of London B* **264** (1378), 89–94.

Warner, A.J. and Hays, G.C. (1994) Sampling by the continuous plankton recorder survey. *Progress in Oceanography* **34** (2–3), 237–256.

Warrach, K. (1998) Modelling the thermal stratification of the North Sea. *Journal of Marine Systems* **14** (1–2), 151–165.

Watson, R.T., Zinyowera, M.C. and Moss, R.H. (eds.) (1997) *The Regional Impact of Climate Change: An Assessment of Vulnerability*. Cambridge University Press, Cambridge.

Weichart, G. (1973) Pollution of the North Sea. *Ambio* **2** (40), 99–106.

Widdows, J., Donkin, P., Brinsley, M., Evans, S.V., Salkeld, P.N., Franklin, A., Law, R. and Waldock, M.J. (1995) Scope for Growth and contaminant levels in North Sea mussels (*Mytilus edulis*). *Marine Ecology Progress Series* **127**, 131–148.

Wulff, F., Stigebrandt, A. and Rahm, L. (1990) Nutrient dynamics of the Baltic Sea. *Ambio* **19**, 126–133.

Zevenboom, W. and De Vries, I. (1996) Assessment of Eutrophication and perspectives. In *1996 Scientific Symposium on the North Sea Quality Status Report, 18–21 April 1994, Ebeltoft, Denmark*, eds. J. Andersen, H. Karup and U.B. Nielsen. Danish Environmental Protection Agency, Copenhagen, pp. 169–177.

Zuhlke, R. and Reise, K. (1994) Response of macrofauna to drifting tidal sediments. *Helgolander Meeresuntersuchungen* **48** (2–3), 277–289.

THE AUTHORS

Jean-Paul Ducrotoy
University College Scarborough, CERCI, Scarborough YO11 3AZ, U.K.

Mike Elliott
IECS, University of Hull, Hull HU6 7RX, U.K.

Victor N. de Jonge
National Institute for Coastal and Marine Management, Rijkswaterstaat, Ministry of Transport, Public Works and Water Management, P.O. Box 207, 9750 AE Haren, The Netherlands

Chapter 5

THE ENGLISH CHANNEL

Alan D. Tappin and P. Chris Reid

The English Channel is a marginal coastal sea located between the south coast of England and the northern coast of France. It is joined to the Celtic Sea (and hence the North Atlantic Ocean) in the west, and to the North Sea at its northeastern corner. Maximum depths range from 40 m in the east to 100 m in the west, extending to 160 m in the Hurd Deep.

The most important factor influencing the occurrence, abundance and distribution of intertidal and marine plant and animal species is the temperature of the water column and its variation. The sensitivity of species to temperature may be enhanced as the Channel lies at the boundary of two biogeographical provinces, the warm Lusitanean and the colder Boreal. Other important factors for benthic species include bed characteristics and physical disturbance.

The coastlines of France and England sustain a wide variety of natural habitats, including cliffs, rocky and sandy shores, shingle, dunes, salt marshes and saline lagoons. There are also a large number of estuaries, with freshwater inflow dominated by the River Seine. Extensive stretches of the coast are rural and support low-intensity arable farming. Interspersed along the coast are urban and industrial centres; the total population of the Channel coast is approximately 8 million, although this can increase significantly during the summer tourist season.

The Channel is one of the world's busiest areas for shipping. The region also supports a commercially important fishery (both fish and shellfish) and is a nationally important region for sand and gravel extraction.

Human activities on the coastal and marine environment have manifested themselves in a number of ways, including land reclamation, chemical pollution and fishing. These pressures remain palpable, although today a large number of national and transnational protective measures are in place. Indeed, natural factors may exert greater stresses on the coastline in the near future as the combined effects of sea-level rise and increasing storminess become more apparent.

Within the last decade there has been a realisation that an important way to protect the coastline in a sustainable manner is for all parties with a stake in the coastal and marine environment to coordinate their activities within an ongoing economic, management and legislative framework. This Integrated Coastal Zone Management is being actively adopted within this region, and is likely to be the key to the long-term protection of the English Channel.

Seas at The Millennium: An Environmental Evaluation (Edited by C. Sheppard)
© 2000 Elsevier Science Ltd. All rights reserved

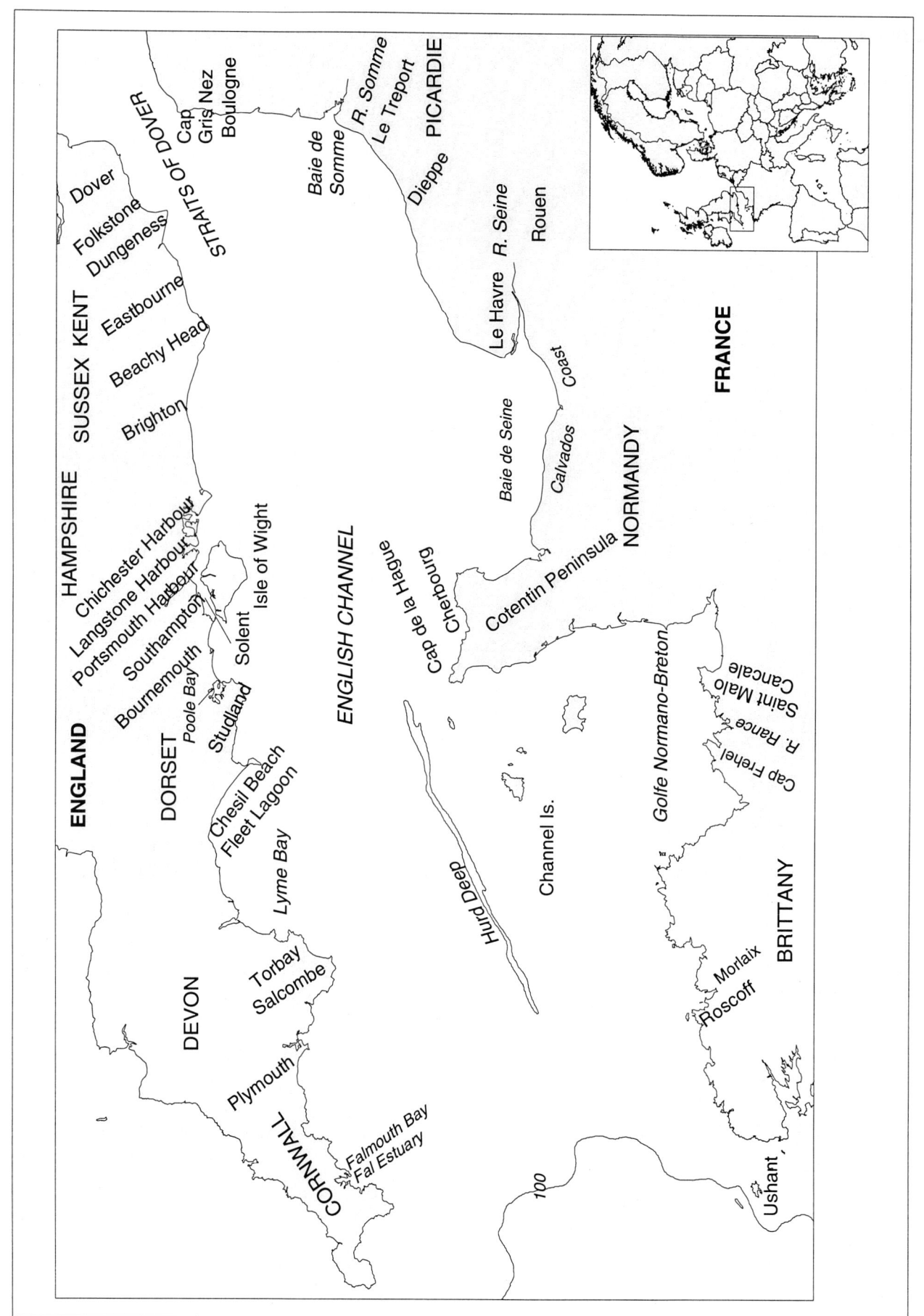

Fig. 1. Map of the English Channel.

THE DEFINED REGION

The Channel was formed following the rise in sea level at the end of the last Ice Age, 8000 years ago. Its northern flank along the south coast of England is approximately 650 km, while its southern, French coastline is about 1100 km in length. At the western end, where the Channel meets the Celtic Sea, the French and English coasts are 150 km apart. Its connection to the North Sea is by the Straits of Dover, only 37 km wide (Fig. 1). Eighty nine estuaries enter the Channel, draining both rural, urban and industrialised hinterlands. The eastern sector of the English coast is subsiding, leading to sea-level rise (Environment Agency, 1999) estimates ranging from +5.4 mm per year at Folkestone, to +5.1 mm at Southampton, +4.1 mm at Eastbourne and –0.3 mm at Salcombe (Carter 1988, cited in Environment Agency, 1999), though these estimates will also include effects from global warming (Bray et al., 1994).

Water depths in the Channel are 40 m at the Straits of Dover, increasing to 100 m at the western entrance. The seabed is generally flat, except at coastal boundaries, and there is a trough, the Hurd Deep, at the western margin which exceeds 160 m depth. The seabed is a mixture of bare rock, gravels, sands, and locally, silt and mud deposits (Fig. 2), the nature and extent of the covering generally reflecting the degree of tide- and wind-induced scour.

GENERAL PHYSICAL AND BIOLOGICAL ENVIRONMENT

Wind stress and barotropic tidal flows dominate water circulation, and since the former has a seasonal component, this is reflected in a change from generally anticyclonic circulation in winter to cyclonic in summer. Overall, wind stress dictates the residual flow of water, and as winds prevail from the south/southwest, net water movement is eastwards from the Atlantic into the Channel, and then into the North Sea (Pingree, 1980). Residual flows are higher in winter than summer since southwesterly gales are stronger and far more frequent (Cooper, 1967; Prandle, 1978; Pingree, 1980). However, significant amounts of water can occasionally flow into the Channel from the North Sea (Prandle, 1978). Radionuclide tracer studies have shown mean west–east currents with speeds of 0.02–0.03 m s^{-1} in the central eastern Channel. The mean annual export of water to the North Sea has been estimated to be ca. 114,000 m^3 s^{-1} (Salomon et al., 1993), with a range of 97,000–195,000 m^3 s^{-1} (Bailly du Bois et al., 1995), of which ca. 50% is wind driven (Prandle, 1993). Water transit times from Cap de la Hague (Cotentin Peninsula) to the Straits of Dover are of the order 110–152 days (Bailly du Bois et al., 1995), although along the coast transit times are longer (Dahlgaard, 1995). In contrast to the dispersion of water (equivalent to ~1000 km a^{-1}), the maximum residual transport of resuspendable sedimentary particles (size range 0.17-0.50 mm) has been estimated to be only a few tens of km a^{-1} (Boust et al., 1997).

Waters of the Channel can be divided into eight zones (Fig. 3) on the basis of both hydrographic properties and phytoplankton distribution and abundance (Reid et al., 1993). Each of these bodies of water is generally vertically mixed, although during summer the western Channel, where surface current speeds are relatively low (<1 m s^{-1}; Pingree, 1980), becomes thermally stratified. Coastal zones are fresher than central waters because of river run-off, and are more turbid in the east. Haline stratification may occur in the Baie de Seine because of the large flows from the River Seine. The Nord-Pas de Calais coast is rather distinct, being influenced by the fresher water flowing from the Baie de Seine, and to a lesser extent from the River Somme, which is trapped against the coast as it flows northwards to the Dover Strait.

Seasonal variations in the vertical stability of the western Channel waters have a marked influence on the growth of phytoplankton. A diatom bloom occurs during spring at the onset of stratification, lasts for approximately three weeks, and is terminated by a lack of nutrients, principally dissolved silicon. The spring bloom is followed by a flagellate-dominated community during the summer. Over this period, the phytoplankton are located at the pycnocline separating nutrient-poor and illuminated surface waters from nutrient-rich deep waters. During the autumn, increased wind stress and surface cooling promotes vertical mixing, bringing nutrients to the surface. This triggers a second diatom bloom, although it is shorter and less intense because of deeper mixing and less favourable light conditions. Records of primary production (estimated by fixation of ^{14}C) in the western Channel over the period 1964–74 showed little inter-annual variability, although the relative importance of contributory species did vary markedly. It has been suggested that primary production has little or no role to play in the long-term state of the Channel pelagic or benthic communities (e.g. Southward, 1980, 1983; Holme, 1983; Dauvin et al., 1993; Fromentin et al., 1997; Davoult et al., 1998).

A number of studies have concluded that the overriding factor influencing the occurrence, abundance and distribution of species is the mean temperature of the water column and the seasonal variability at any particular site. The distributions of benthic flora and fauna are also dependent on bed sediment characteristics, which in turn are influenced by water depth, current velocities and wave action. Southward and his colleagues (e.g. Southward, 1980, 1983, 1995; Southward et al., 1995; Holme, 1966, 1983) were among the first to recognise the role of decadal variations in climate and temperature on species abundance and distribution within pelagic, benthic and intertidal environments of the western Channel. Some of this work gave rise to the concept of the 'Russell cycle'. Results from the eastern Channel have confirmed this view (Fromentin and Ibanez, 1994; Alheit and Hagen, 1997; Fromentin et al., 1997). These investigations have clearly shown the importance of long time-series data in studies of environmental variability. It

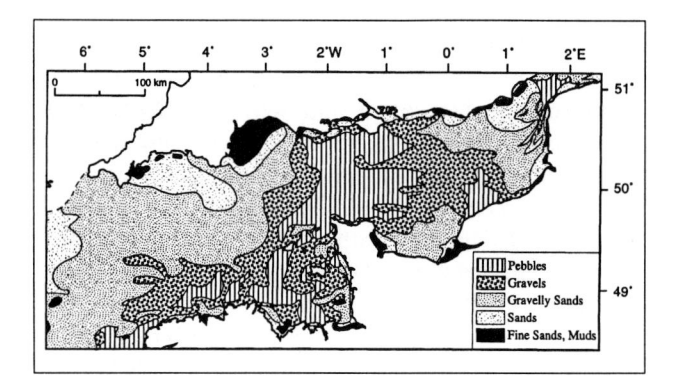

Fig. 2. Distribution of surface sediments in the Channel (Larsonneur et al., 1982).

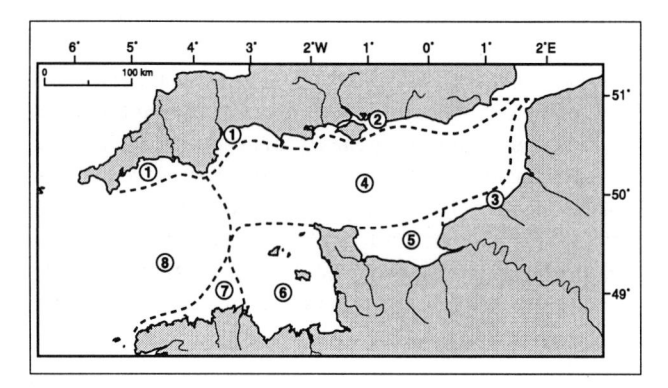

Fig. 3. Predominant water masses within the Channel. (1) Western English Channel Water; (2) Eastern English Channel Water; (3) Nord-Pas de Calais Coastal Water; (4) Central English Channel Water; (5) Baie de Seine Water; (6) Golfe Normano-Breton Water; (7) North Brittany Coastal Water; (8) Central Western Channel Water. (Reid et al., 1993).

may be that the Channel is particularly sensitive to climate fluctuations since it lies at the boundary of the Boreal (cold temperate) and Lusitanean (warm temperate) biogeographical provinces (Doody et al., 1993). Other factors may also influence species distributions, including pollution, biological competition/predation and physical disturbance, for example (Holme, 1966, 1983; Hawkins et al., 1994; Sanvicente-Añorve et al., 1996; Vallet and Dauvin, 1998; Dauvin, 1998; Kaiser et al., 1998).

COASTAL AND MARINE HABITATS

The English Coast

The habitats of this region have been comprehensively described (Barne et al., 1996a,b,c; 1998a,b; Pye and French, 1993a; English Nature, 1996). Cliffs, of which many exceed 20 m in height, are found along large stretches of the Channel coast (Fig. 4). Soft rocks (poorly consolidated marine, estuarine and fluvial sediments, periglacial head deposits and mudrocks) predominate in the central regions, whilst in the west and east, granitic and chalk rocks, respectively, are more common. Those in the west are highly exposed to wind and wave action. Many of the softer cliffs are subject to erosion, including those in Torbay, west Dorset (Christchurch Bay), the Isle of Wight and Kent (Folkestone Warren), although a number of chalk cliffs undermined by wave action are also vulnerable. Erosion is likely to accelerate not only because of sea-level rise, but also because of apparent increases in mean wave height in recent decades (Jelliman et al. 1991, cited in Environment Agency, 1999). There are approximately 34 major salt marsh localities along the coast, covering ca. 5200 ha, with the most extensive areas in Hampshire, Dorset and west Sussex. Extensive change to saltmarsh vegetation has been caused by the spread of the common cord grass *Spartina anglica*, which was first recorded in Southampton Water in the late 19th century (Tubbs, 1994). It now accounts for more than 75% of the saltmarsh vegetation along some south coast English estuaries, including Southampton Water, although in some areas the *Spartina* is dying because of the anaerobic soils with a high sulphide content that are produced by this plant. Marshes at a number of localities are either stable or accreting, although in Hampshire (i.e. the Solent, comprising Southampton Water, and Langstone, Chichester and Portsmouth Harbours), west Sussex and Kent, they are eroding (Fig. 5).

Many other estuarine habitats are also found in the Solent, including significant areas (ca. 4800 ha) of intertidal mud flats. Important intertidal sand flats are found in Dorset (ca. 2400 ha), particularly in Poole Bay, as well as in Sussex, Devon and Cornwall. Changes in the distribution and loss of these sediments over the next few decades are difficult to estimate, but may be extensive along the Channel coast because of the effects of projected rises in sea level (Pye and French, 1993a). One of the largest areas of shingle in Europe, at over 2700 ha, is situated at Dungeness. There are a number of other smaller shingle areas, including nationally important sites on the Isles of Scilly, and the Rye Harbour complex (720 ha) and Chesil Beach (250 ha). The latter feature separates the Fleet, the largest saline lagoon on the UK coast at 450 ha, from Channel waters. The number of lagoons in southern England doubled over the

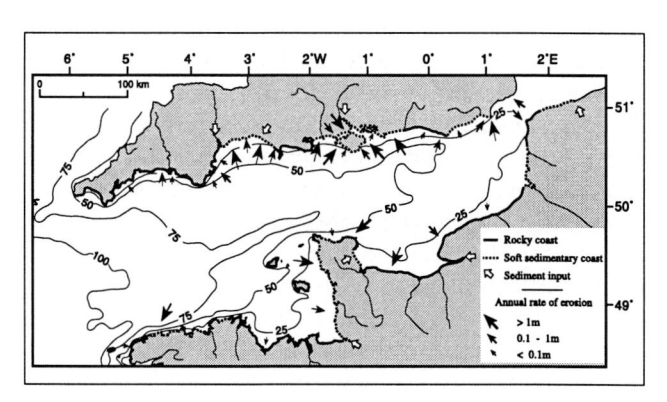

Fig. 4. Distributions of hard and soft coastlines along the French and English coasts of the Channel and their relative rates of erosion (Reid et al., 1993).

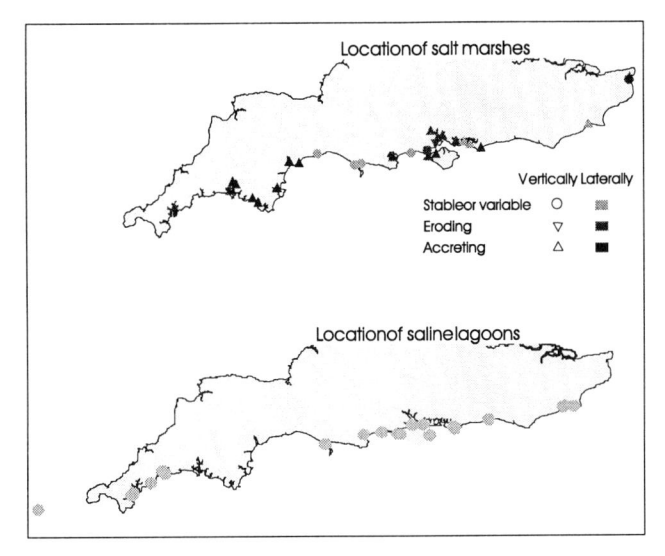

Fig. 5. Saltmarsh stability and saline lagoons along the Channel coast of England (Pye and French, 1993b; Barnes, 1989).

period 1800–1974, largely as a result of gravel extraction (Gubbay, 1988). There are a large number of sand dunes along the Devon and Cornwall coasts, and there is a dune system of note at Studland Bay in Dorset.

The French Coast

Sea cliffs are concentrated between Cap Gris Nez and Le Havre, although they also occur along the northern part of the Cotentin Peninsula, on the coast between Cancale and Saint Malo and in the Cap Frehel region. They are important habitats for seabirds (Reid et al., 1993). Rocky shores are mainly encountered in Brittany (Fig. 4) and on offshore islands. Pebble dominated habitats exist but are now endangered by recent increases in mean high tide level (Reid et al., 1993).

Extensive salt marshes occur on the Picardie and Cotentin coasts and in Mont Saint Michel Bay; a number of these areas are vulnerable to erosion (Fig. 4). In the Baie de Somme, where marshes cover 1250 ha, there are many artificial ponds, dug out for shooting water fowl. The upper areas of the marsh are given over to grass meadow dominated by *Puccinellia* spp. In the Marquenture, 3000 ha have been poldered and are separated from the sea by dunes. More than 3900 ha of marsh covered by samphire (*Salicornia* spp.), Sea Poa (*Puccinellia maritima*) and Red Fescue (*Festuca rubra*) are found at the head of Mont Saint Michel Bay. Sandy beaches are found mainly to the west of Le Havre and also to the north of the Somme, and are associated with both dune systems and rocky shores. The dunes are endangered by buildings, recreation and sand exhaustion, although protection has been provided at many sites through planting of *Pya* spp. The beaches are important nurseries for fish (e.g. turbot) and shrimp and as nesting sites for birds (Skov et al., 1995), and are also used extensively for mariculture (oysters and mussels).

Offshore Habitats

On the central and eastern side of the Channel coast of England, the seabed is generally shallow sloping, and there are extensive intertidal chalk platforms, sands and gravels. To the west, muddy sands are more common (e.g. in Lyme Bay), together with sands and some gravel areas. Off the coasts of Devon and Cornwall, coarse muddy sands and sands are dominant, except where currents are higher, when gravels predominate. On the French coast, silts and muds are also concentrated in bays and estuaries, particularly at the mouth of the Seine (Fig. 2) (Larsonneur et al., 1982). Offshore, the seafloor can be broadly divided into three zones: the western Channel where calcareous biogenic debris dominates, a central zone between the Cotentin Peninsula and the Isle of Wight characterised by extensive pebble deposits, and the eastern Channel where gravelly sands predominate (Fig. 2). Much of the area is covered in lag gravels, presently immobile. Locally, the gravel is overlain by finer, mobile deposits of sand, such as off the Devon coast and south of Beachy Head, with active tidal sand ridges south west of the Dover Strait. Rocky outcrops occur on the floor of the Straits of Dover, north of the Pays de Caux and off the Brittany coast between Ushant and the Channel Islands (Hamblin et al., 1992).

Biodiversity

Flora

The distribution and abundance of phytoplankton in the water column is to a large extent controlled by the hydrographic regime (Reid et al., 1993). Exceptional blooms of algae can occur if conditions are favourable, including *Gyrodinium aureolum* (which has given rise to fish and shellfish mortalities), *Phaeocystis* sp. (foam formation, shellfish mortalities) and coccolithophores (milky water). Algae giving rise to paralytic shellfish poisoning (PSP; *Alexandrium tamarensis*) and diarrhetic shellfish poisoning (DSP; *Dinophysis* spp.) have also occurred in the Channel. PSP was first recorded in 1988 on the north Brittany coast, and DSP in 1984 at Antifer, north of the Seine, and along the Calvados coast. The first DSP human intoxications were also recorded in 1984. Toxic algae are rare along the English coast, but between 1993–95, records of toxic algal blooms increased from one to fifteen, and from June to July 1997 there was a voluntary closure of the Fal Estuary shellfishery because PSP poisons were detected (Environment Agency, 1999). The extent to which these increased sightings are a consequence of a higher awareness of blooms or a real increase is not clear.

Deposits of a benthic calcareous algae, known as maërl, form a rare habitat with a rich associated fauna in the Golfe Normano-Breton and, less extensively, elsewhere (Baie de Seine, Falmouth Bay). There are two main types of maërl, one associated with a bed of fine sands and muds and the other with coarse sand, gravels and cockles. The maërl grow slowly, 1 mm yr^{-1}, as a surface crust. There is a commercial exploitation of these algae for carbonate, but this may be unsustainable and poses a risk to the long-term future of this habitat. There is little information on other benthic microalgae of the Channel (Reid et al., 1993). There has been some work in estuaries (e.g. Joint, 1978), where muds may support chlorophyll *a* levels up to 300 μg g^{-1}. Within deeper waters there are no systematic observations, although it is known that light can penetrate to a depth of 70 m in low-turbidity waters.

The Channel has ca. 700 species of benthic macroalgae, and is thus one of the most diversified habitats in the northern hemisphere (Cabioch et al., 1992). On moderately sloping shores with gullies and pools, typical of the Channel on both sides, fucoid algae are common. In the intertidal, *Pelvetia canaliculata* may be found, with *Fucus* spp. often lower on the slope, itself associated with a diverse underflora. In sheltered areas, *Fucus* spp. growth may be particularly developed. Sometimes the brown alga *Ascophyllum nodosum* may be dominant. On all shores, *Laminaria* species dominate the shallow (<15 m) subtidal zone. Growth of *Laminaria* species may be so profuse that commercial harvesting for alginate takes place. On exposed and sloping southwestern shores, the wave climate reduces the diversity of algal species. Elsewhere, extensive growth of green algae can occur, usually where beaches are sandy, gently sloping, of large surface area and in receipt of sufficient inputs of nitrogen. Examples include *Ulva* spp. and *Enteromorpha* spp. There are a number of introduced species within the Channel, including *Sargassum muticum* and *Undaria pinnatifada*. The former was introduced on oysters from Japan or Canada about 30 years ago and is now a widespread pest species, displacing native species, and fouling boat propellers, intakes, fishing nets and oyster beds (Environment Agency, 1999). The English coast is the southern limit for the alga *Ptilota plumosa*, and the northern limit for *L. ochroleuca*. Differences in temperatures between the eastern and western Channel, and the annual minima, are thought to explain the confinement of the brown alga *Cystoseria tamariscifolia* and the red alga *Gigartina acicularis* to the western Channel. Beds of the seagrasses *Zostera marina* and *Z. noltii* are found along the English (e.g. Portsmouth, Langstone and Chichester Harbours, Southampton Water, the Solent and the Isle of Wight) and French (Roscoff) coasts in shallow waters. *Z. marina* was nearly wiped out in the 1930s by an unspecified agent, and its regrowth in recent years may be due to natural recovery or associated with increased inputs of sewage. Detailed reviews of the flora within the intertidal and shallow subtidal waters along the English coast are given in Covey (1998) and Davies (1998).

Fauna

The general patterns of zooplankton follow those for the phytoplankton (Reid et al., 1993). In the western Channel copepod species are important, with highest numbers in the central areas from June to October. Some species are relatively scarce in English coastal waters (e.g. *Calanus helgolandicus*), becoming more abundant within frontal waters around Ushant and towards the French coast. Larvae of benthic organisms (decapods, echinoderms and bivalves) show peak abundance in July and August, a pattern that is similar to chlorophyll. The highest concentration of herbivorous zooplankton is found in the stratified waters close to the Ushant Front. In the Golfe Normano-Breton there is a gradient in both zooplankton type and biomass; the coastal zone is dominated by small holoplankton species (appendicularians and mysids, larvae of polychaetes, molluscs and crustaceans) whereas zooplankton in the frontal zone close to the Channel Islands are dominated by large copepods. Little is known about zooplankton populations in the English eastern Channel waters, although in the central eastern Channel, the summer maximum in biomass is dominated by copepods. Waters within the Baie de Seine are dominated by coastal species whilst the front in the waters off the Pas de Calais generally separates offshore species (e.g. *Calanus, Corycaeus, Centropages*) from inshore zooplankton such as *Cyclopina littoralis* and *Saphirella* sp. In these waters, seasonal changes in species composition have been linked to annual variations in both river flow from the Seine and air temperature.

The large variety of fish species can be divided into a number of groups (Reid et al., 1993). These include occasional visitors (e.g. pilchard which are at their northern or western limit), seasonal migrants (such as mackerel and cod), species such as the dab which live in the Channel throughout their life-cycle, resident species largely restricted to the coastal zone (e.g. dragonet, lesser weaver), estuarine species, including gobies and eelpout, and anadromous and catadromous species that move through the area to and from their spawning grounds (e.g. eels, shads, lampreys, salmonids). In addition, there are several species of shark including the Basking shark (*Cetorhinus maximus*), porbeagle (*Lamna nasus*) and thresher shark (*Alopias vulpinus*) (Potts and Swaby, 1996). Fish inhabiting English estuaries are described by Potts and Swaby (1993), and Pawson (1995) provides a synthesis of the known status of the stocks of the twenty most economically important fish in the Channel.

The most regularly observed cetaceans include the common dolphin, the bottle-nosed dolphin, and, in the western Channel only, the long-finned pilot whale. The striped dolphin is occasionally seen, and there are rare sightings of other cetaceans (minke whale, Atlantic white-sided dolphin, white-beaked dolphin, killer whale, false killer whale, fin whale, Risso's dolphin, humpbacked whale). The harbour porpoise is occasionally seen, and is now listed as a priority species by the UK Biodiversity Group (Environment Agency, 1999). Both grey seals and common seals occur, and the former are known to breed in the sea-caves of Cornwall and the Isles of Scilly. Common seals breed in the Baie de Somme, Baie de l'Orne and Baie de Veys. The

Greenland seal and mottled seal are more rare, although sightings have occurred along the French coast at Pas de Calais. In 1988, 48 dead seals (90% grey) were reported along the English south coast, having succumbed to the Phocine Distemper Virus that killed 1450 seals in the UK as a whole.

A recent description of the benthic macrofaunal community over the Channel bed (Hiscock, 1998) draws largely on the work of Holme (1961, 1966) and Cabioch et al. (1976). Macrofaunal associations within shallow coastal and estuarine waters of the English coast have been summarised by Doody et al. (1993), English Nature (1996), Covey (1998) and Davies (1998). Species distributions correlate with sediment type in both shallow and deep waters, although there is much variability. There are five 'boreal offshore associations'. A general pattern of sediment distribution and its associated fauna can also be described on the basis of depth and thermal characteristics, and there are twice as many species off western Brittany as there are in the Straits of Dover. The western species include widely distributed Atlantic stenothermal species, warm temperate species at their northern limit and some boreal stenothermal species. The eastern group, in contrast, is dominated by boreal eurythermal species. Marked temporal changes in benthic communities over the last century have been explained by changes in temperature with concomitant influxes of species from the Celtic and North Seas, with local variations due to dinoflagellate blooms, oil spills (both the oil and the detergents used for clean-up) and trawling (Holme, 1983). The macrofaunal populations within the Channel support an important commercial fishery.

The macrofauna of the rocky intertidal zone of the Channel is typical of much of the northeast Atlantic (Southward and Southward, 1978). On the most exposed shores the upper two-thirds are dominated by barnacles and limpets. As the wave energy reduces, limpets become increasingly important and the dogwhelk (*Nucella lapillus*) can be found on open rock. On more sheltered shores there are also found large numbers of mobile gastropods such as *Littorina littorea*. The distribution of the fauna reflects west to east gradients in physical parameters (Crisp and Southward, 1958; Lewis, 1964), and several species important in the western Channel are absent from the east. These tend to be animals approaching their northern limit of distribution. Examples include a number of barnacle species (*Balanus perforatus, Chthamalus montagui* and *C. stellatus*), the limpet *Patella depressa* and the gastropods *Monodonta lineata* and *Gibbula umbilicalis*. The western Channel is also close to the southern limit of the distribution of the barnacle *Semibalanus balanoides*, the dominant species of other UK rocky shores.

Seabirds can be divided into three groups depending on the primary foraging area of each species (Doody et al., 1993). Offshore species feed regularly more than 30 km from land, inshore species feed usually within sight of land and coastal species usually feed intertidally or very close to the shore. The Channel is used by seabirds throughout the year, but numbers of each group vary with the seasons. In general, numbers increase markedly during autumn and winter. For example, migrating guillemots, razorbills and kittiwakes feed off sprat between October and March. The terns leave in winter, with some large flocks using the area during spring and autumn migration periods. Relatively large numbers of divers and grebes move into the area in winter, particularly if conditions in Denmark and Germany are unfavourable. Surveys of breeding species over the last two decades have found that in general, numbers of most breeding species have been stable or have increased slightly (e.g. Tasker, 1996). The Scilly Isles is the foremost breeding area along the English coast. However, the UK Channel coast does not support major breeding numbers of offshore species at present, although it once did. On the Channel Islands, manx shearwaters and storm petrels breed in low numbers. Significant colonies of inshore species are found on the larger intertidal areas of the English coast, such as Chesil Beach, Poole Harbour, the salt marshes of the western Solent, Chichester and Langstone Harbours, and at Dungeness. Eight estuarine or other coastal areas are of international importance to migrating wildfowl in that they provide forage at times for more than 1% of the northwest European population of at least one species.

On the French coast, Mont Saint Michel Bay is an important location for shelducks, oystercatchers, grey plover, dunlin and sharp-tailed sandpiper, and migratory birds including Brent geese, widgeon and ringed plover (Reid et al., 1993).

URBAN AND RURAL POPULATIONS

Demography

Approximately 3.88 million people live in the English coastal region, at densities ranging from 0.8 to 46.0 persons per hectare. A further 4.97 million people live in the corresponding French area, primarily in 3700 towns of which only four have more than 100,000 inhabitants (Reid et al., 1993). There has been little change in land use along the French coast over a recent 5–10 year period. Exceptions included some urban increase, an urban decrease in the vicinity of Rouen (Seine), together with an increase in forest in the same area (European Commission, 1999a). Along the English coast there has been a net loss of 0.1–0.2% of rural land between 1990–93, with higher losses on the Isle of Wight (0.6%) (Environment Agency, 1999). Significant stretches of both coasts have a high development pressure. The number of retired people migrating to coastal areas, particularly in England, is creating demands for new residential areas and amenities. For example, 36% of the coastal population in Devon are over 60 years of age (Barne et al., 1996a). Resources are further stretched during the tourist season when, locally, populations can increase by a factor of 6–7 (Environment Agency, 1999).

Industry

Industry along the English coast is concentrated around the main urban centres and ports. Southampton is by far the largest port, and there is a large oil refinery at nearby Fawley. Dover is a significant port for passengers and containers.

On the French coast, diverse industrial development is located on the Seine Estuary. The most important industries include titanium dioxide processing plants in Le Havre, with discharges into the Baie de Seine of ~ 14000 m^{-3} d^{-1} sulphuric acid waste in 1993. Localised effects of this on the benthos come from both the high acidity and smothering by iron hydroxide which forms on contact with the seawater (Reid et al., 1993). The annual metal inputs have been substantial, although in recent years these have decreased. A major phosphoric acid manufacturing plant is situated on the Seine at Rouen. Until the mid-1980s this plant discharged several thousands of tonnes per day of phosphogypsum waste into the Baie de Seine and adjacent coastal seas, although the waste is now stored on land.

Agriculture

Agriculture on both sides of the Channel is predominantly a mixture of arable and pasture for animal grazing. Much of the farming is low intensity, although along the eastern coast of the UK and within the Seine and Calais regions of France, arable farming is relatively intensive (European Environment Agency, 1995). High input animal production (cattle, poultry, pigs) is also found in Normandy and Brittany.

Tourism

Tourism has been an important activity along the French and English coasts since the 19th century. Major English resorts include the Brighton, Bournemouth and Torbay conurbations and the Kent, Dorset and southeast Cornish coasts. In France, the Brittany and Normandy coasts host similar activities and there are pressures on port and resort areas like Calais, Boulogne and St. Malo. Pressures on coastal development and infrastructure have also arisen as traditional hotel and beach-based recreation has changed to encompass wider recreational needs, including countryside access, watersports, and changing trends in ferry activity.

There has been an extensive development of marinas and mooring facilities for small boats along much of these coasts in recent years. These facilities have a localised effect on water movements, act as traps for fine sediments, are a source of contaminants from antifouling paints, waste water and sewage, and have caused habitat loss (Reid et al., 1993; Environment Agency, 1999). In 1992, the two coasts offered moorings for 20,000 boats, with the highest concentrations in the Solent and Côtes d'Armor areas. The use of antifouling paints containing tributyltin has been, and continues to be, a major pollution problem (see later).

Energy

There are two operational nuclear power plants in southeast England, at Dungeness, and four along the French coast, at Paluel, Penly, Flamanville and Gravelines. At certain times of the year chlorine is used as a biocide, but its environmental impact is assessed to be minimal. A nuclear fuel reprocessing plant, built in 1959, is located at Cap de la Hague on the Cotentin Peninsula. Radionuclides contained in the liquid effluents released into the Channel from this plant have been used as tracers of water masses. Wytch Farm in Dorset, owned by BP, produces 90% of the onshore oil in the UK and is the largest onshore production site in Europe. The oil field extends under Poole Harbour and Poole Bay, and is estimated to hold 53 Mt, of which a half have been used to date (Department of Trade and Industry, 1997). There are no production platforms for oil or gas in the English or French sectors of the Channel at present. The world's first operational tidal barrage power station is located on the Rance Estuary. However, there are few detailed plans to add further alternative energy production capacity, including tidal barrages, tidal currents, wave devices and wind turbines, although wind and wave energies are very high and tidal ranges large (Environment Agency, 1999). In a recent study, local authorities from Kent and Nord-Pas de Calais have begun evaluating the potential for offshore wind generation in the Transmanche region (European Commission, 1999b).

Shipping

The Channel is one of the most frequently used shipping areas in the world, and carries large container ships, oil tankers and ferries. Because the region also supports an important commercial fishery there are in addition numerous fishing boats. Shipping is particularly heavy through the Dover Strait with approximately 150 ships per day sailing in each direction in addition to some 300 ferry movements daily; there is therefore more than an average risk of collisions (Lloyds Register of Shipping, 1998). Information on UK oil pollution events, primarily from ships, are given annually, and many reports arise from the observations of slicks caused by the illegal discharge of ballast water, which leads to chronic contamination of the coastlines. Over the last three decades there have been a number of oil spills leading to observable effects on the local environment. The most important included those from the Torrey Canyon (1976), the Olympic Bravery and the Boehlen (1976), the Amoco Cadiz (1978), the Tanio (1980), and the Rose Bay (1990). Concerns have also been raised over spills of toxic chemicals (Johnston et al., 1997), such as the loss of 5.8 t of lindane from the MV Perintis in March 1989 and 7.2 t of a carbamate and furathiocarb containing seed treatment

Table 1

Total landings of 20 species of fish by French, English and Channel Island vessels from the English Channel in 1989 (from Pawson, 1998).

Species		Landings (tonnes)
Common name	Scientific name	
Spurdog*	*Squalus acanthias*	621
Thornback ray	*Raja clavata*	{1975
Cuckoo ray	*Raja naevus*	
Herring*	*Clupea harengus*	10732
Sprat	*Sprattus sprattus*	1584
Pilchard	*Sardina pilchardus*	1899
Cod*	*Gadus morhua*	5711
Whiting	*Merlangius merlangus*	6008
Pollack	*Pollachius pollachius*	2982
Hake	*Merluccius merluccius*	932
Ling	*Molva molva*	2310
Bass*	*Dicentrachus labrax*	1311
Black bream	*Spondyliosoma cantharus*	640
Mackerel*	*Scomber scombrus*	16203
Scad	*Trachurus trachurus*	3354
Red gunard	*Aspitrigla cuculus*	4193
Plaice*	*Pleuronectes platessa*	8687
Lemon sole	*Microstomus kitt*	1143
Sole*	*Solea solea*	3022
Angler	*Lophius piscatorius*	3810

*Species for which sufficient information exists for stock management purposes

from the Sherbro in December 1993 (Law and Campbell, 1998). In both cases there were no observable ecological effects, although in the first example the central Channel fishery was disrupted for 3–4 weeks, whilst in the latter sachets of pesticide were washed up hundreds of km downstream (Law and Campbell, 1998).

Fisheries

The Channel supports a large fishery of round, flat and pelagic fish (Fig. 6). Of the 4000 boats that fish in the Channel, ~90% operate within 10 km of the coast (Pascoe, 1997). Fishing methods include beam trawling, otter trawling, pelagic/midwater trawling, dredging, line fishing, potting and netting (Pascoe and Robinson, 1997). In tonnage, herring, cod, whiting, mackerel and plaice number among the main exploited species from a total catch of ca. 77,000 t (Table 1). In 1994 the English catch was estimated to be worth ~£100 million, and directly employed 4800 people (Pascoe, 1997).

There is also extensive fishing for crustaceans and molluscs, and the activity is both socially and economically significant; indeed this fishery is the main income for coastal fishing boats on the French Channel coast. The annual reported catch of crustaceans was 14,663 t in 1989 (Pawson, 1995). The edible crab (*Cancer pagarus*), spider crab (*Maja squinado*) and lobster (*Homarus gammarus*), are in ton-

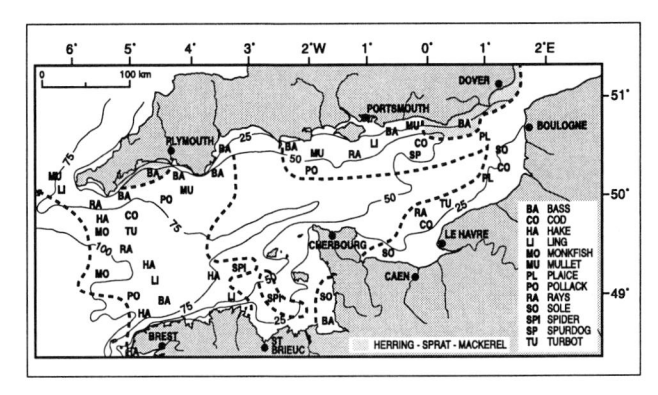

Fig. 6. The distribution of the main fishing areas and fish species caught in the Channel (Reid et al., 1993).

nage, as well as value, the principal exploited species, with larger populations and catches in the western Channel (Fig. 7a). The lobsters and spider crabs breed in shallow water and so are susceptible to coastal environmental impacts. The principle species of molluscs exploited include scallop (*Pecten* and *Chlamys*) and cuttlefish (*Sepia*), the catch of which was 22,735 t in 1989 (Pawson, 1995), although significant quantities of dog cockle (*Glycimeris*), cockle (*Cerastoderma*) and hard clam (*Venus*) are also landed (Fig. 7b). The farming of oysters (*Crassotrea*, with some *Ostrea edulis*) and mussels occurs extensively along the Normandy and northern Brittany coasts, having expanded markedly over the last 20–30 years, and to a lesser extent along the English coast. In 1990, French production of oysters was 41,000 t and mussels 26,500 t, dwarfing English tonnage. In 1991, mussel production was extended to offshore floating structures, adding a further 25 000 t for that year.

A number of small, but nevertheless important, fisheries for salmon and sea trout exist, especially in rivers and estuaries. A recent reduction in catches of these fish is believed primarily to have been due to overfishing and lon-

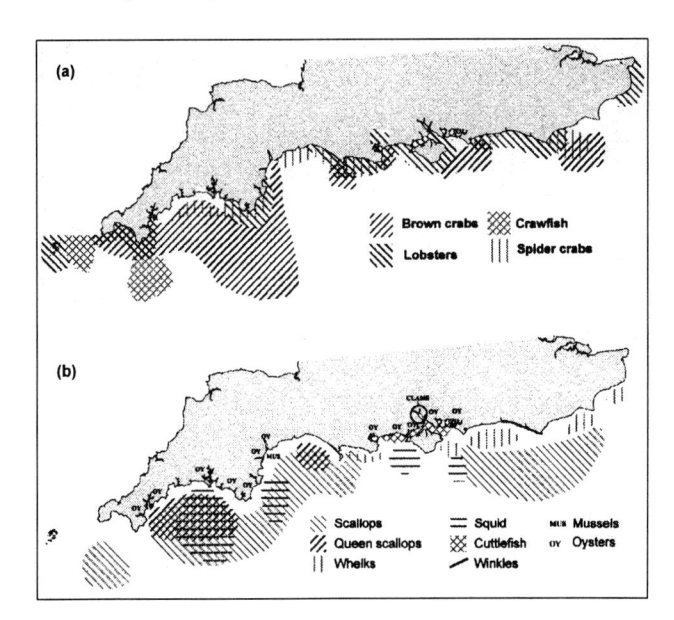

Fig. 7. Major crustacean and molluscan fisheries along the English coast of the Channel (Gray, 1995).

ger-term changes in the hydrometeorology of the North Atlantic Ocean, as well as deterioration of spawning areas in the upper reaches of some rivers (Reid and Planque, 1999).

Sand and Gravel Extraction

The Channel has significant reserves of sand and gravel. It is the second most important area within Britain, in terms of tonnage, with an annual production in excess of 5 M t. Dredging takes place predominantly in the eastern Channel to the south and west of the Isle of Wight from a number of licensed areas. French resources of sand and gravel are estimated at 5.5 M t (siliceous) and 3.7 M t (calcareous). In the eastern Channel these are not exploited significantly. In the central Channel off Dieppe and Le Havre, 0.82 M t is extracted per annum. In Brittany, sands are exploited by many outlets on a small scale to give an annual estimated production of 0.36 M t and 0.61 M t for siliceous sands, and calcareous sands and maërl, respectively. The area of greatest anticipated demand for marine aggregates is the eastern English Channel, where reserves are large (Environment Agency, 1999).

ANTHROPOGENIC IMPACTS ON THE CHANNEL

Loss of Coastal Habitats

Coastal habitats, and especially estuaries, are an important wildlife resource. Whilst changes to them by natural factors remains important (e.g. erosion), degradation by human activity, which has gone on since at least Roman times, remain a palpable threat (Davidson et al., 1995). These include land reclamation, aggregate extraction, dredging, spoil dumping, waste tipping, pollution and effects from coastal defences (Pye and French, 1993a; European Commission, 1998a). Along the English coast, developments including Portsmouth Harbour, numerous dock areas and construction of motorways have resulted in significant losses of salt marsh. Within Southampton Water in the 1950s to 1970s, extensive erosion occurred following oil-induced death of vegetation (Carpenter and Pye, 1996). It is suggested that a further 2% nationally of salt marsh will be lost to development in the next 20 years, with most loss near to conurbations such as Portsmouth and Southampton (Pye and French, 1993a; Davidson et al., 1995).

Reductions in the areas of saltmarsh and intertidal muds have a number of consequences that have rarely been quantified. These include (a) alterations to estuarine shape, and hence tidal currents and sediment transport (b) a reduction in the productivity and biomass of plants and animals, which may affect the quantity of fish and birds the estuary can support (c) reductions in the number of wintering wildfowl, particularly if alternative sites have also been restricted, and (d) reduced opportunities for birds to feed (Davidson et al., 1995).

Chemical Contamination and Impacts

Nutrients and Eutrophication

During winter the distributions of nitrate and phosphate throughout Channel waters are inversely correlated with salinity (Tappin et al., 1993), reflecting the mixing of river waters with oceanic waters. During the spring and summer these relationships are weakened as source signals change, and *in situ* nutrient cycling becomes important (Hoch and Ménesguen, 1997; Gentilhomme and Lizon, 1998; Jordan and Joint, 1998). The main anthropogenic sources of nutrients to the Channel include those from sewage treatment plants and industry, agricultural and urban run-off, including storm water overflows, and inputs from intensive aquaculture. For example, there are approximately 200 consented sewage discharges (>1000 m^3 d^{-1} effluent) along the UK Channel coast. The control on inputs of nutrients to coastal waters is regulated, although the relevant Directives may be poorly implemented (European Commission, 1998b). The direct input of total nitrogen from England for 1992 was estimated to be 9600 t (Oslo and Paris Commissions, 1995), which was approximately half the estimated river input of 20,000 t. For phosphate, sewage and riverine inputs were similar, at 1700 and 1300 t a^{-1}, respectively. The larger part of the direct nutrient input from the UK is into the eastern Channel, reflecting the greater population density in southeastern England (Reid et al., 1993). There is little evidence that the nutrient inputs give rise to widespread problems of eutrophication, except in three areas of restricted water exchange, Chichester and Langstone Harbours, and the Fal Estuary, which are now designated as 'eutrophic sensitive areas' under the Urban Waste Water Treatment Directive (Environment Agency, 1999). The most recent data for France show that in 1990, total nitrogen inputs to the Channel and North Sea (although most will be to the Channel) equalled 112,000 t, of which 99% was riverine (Oslo and Paris Commissions, 1992). There are also three areas on the coast of France where increased inputs of nutrients from agriculture may have contributed to eutrophication, the northern Brittany coast ('green' tides on beaches, estuarine blooms), Artois Picardie (*Phaeocystis* blooms linked to fish and shellfish mortality) and the mouth of the Seine (discoloured waters from algal blooms). In the latter case, concentrations of riverine nitrate have increased sharply since the mid-1960s, and are correlated with rates of application of nitrogen fertilisers; indeed it has been estimated that ca. 80% of the riverine nitrate originates from agricultural sources (Reid et al., 1993).

Metals

Metal concentrations of waters and suspended sediments are generally low, and reflect extensive dilution with relatively clean Atlantic water (e.g. Cossa and Fileman, 1991;

Tappin et al., 1993; Achterberg et al., 1999); concentrations generally increase in the vicinity of estuaries and the nearshore. Throughout the Channel, levels of metals in sediments appear to be largely controlled by the dilution of fine clay-type minerals high in metals with coarser material (silts, sands and gravels), except for cadmium, for which carbonate may be important (Reid et al., 1993). At a small number of estuarine sites levels within surficial sediments are particularly high. These may be due to a natural phenomenon such as the influence of metalliferous catchments (Fal Estuary/Restronguet Creek; copper, zinc, arsenic), Looe Estuary (silver, lead), past activities such as mining, or they may be the result of more recent contamination, such as dumping (Baie de Seine; cadmium, mercury, lead, zinc), oil refinery discharges (Southampton Water; copper) and urban runoff (Portsmouth Harbour, Southampton Water; lead). Data on the concentrations of metals in the nearshore biota generally reflect the loadings in the sediments (Burt et al., 1992). Tolerance to zinc has been developed by some animals in heavily contaminated sites such as Restronguet Creek, and larval settlement of bivalves in the Fal Estuary was observed to be inhibited because of the high concentrations of sedimentary zinc and copper; however, other toxicological responses have not been observed. In addition, studies on fish and shellfish throughout the Channel have shown no evidence of serious contamination (Reid et al., 1993). The one exception is found with tributyl tin (TBT), an antifoulant used on the hulls of small boats.

Tributyl tin is one of a growing class of substances that disrupt hormonal activity within organisms to the extent that acute physiological damage can occur (see Hawkins et al., 1994). Concentrations of TBT as low as $5\ ng\ l^{-1}$ can lead to sterility (described as imposex) and eventual extinction of local populations of the dog whelk (Gibbs et al., 1991), and levels in the mid-1980s reached $1000\ ng\ l^{-1}$ in some localities. The link between TBT and imposex led to a banning of the use of TBT antifouling paints on boats under 25 m length throughout the European Union, including France and the UK, in 1987. In the Solent, water column concentrations of TBT, and also those in sediments and clams, in the Hamble and Itchen over the period 1986–92 had declined, whereas in the Test Estuary, still used by larger vessels exempt from the ban, concentrations were stable or had increased (Langston et al., 1994). Imposex has recently been observed in female dog whelks examined as far apart as Cornwall and Margate in Kent (Langston et al., 1997). It was concluded that a number of dog whelk populations remained vulnerable to further losses or extinction ten years after the TBT ban was implemented. In addition, a number of sites along the western Channel coast of England in 1997 failed to meet their environmental quality standard for TBT (Environment Agency, 1999). The IMOs marine environmental protection committee has recommended a complete ban on the use of TBT from 2003 and its removal from all ships by 2008.

Organic Materials

Measurements for micro-organic compounds in the water column include those for hexachlorocyclohexanes (HCHs) and the s-triazines, simazine, atrazine and Irgarol 1051 (the latter is a booster biocide) (Law and Allchin, 1994; Law et al., 1994; Gough et al., 1994). Some measurements of polychlorinated biphenyls have also been undertaken [references cited in Reid et al. (1993)]. A number of these measurements were carried out as part of an internationally co-ordinated Monitoring Master Plan initiated by the North Sea Task Force (Law et al., 1994). In the early 1990s, concentrations of HCH in the Channel were significantly below the UK EQS value ($100\ ng\ l^{-1}$), whilst atrazine and simazine concentrations were sufficiently low that further significant monitoring effort beyond estuarine waters was not considered worthwhile (Law et al., 1994). In the Seine Estuary, concentrations of lindane have been recorded as high as $50\ ng\ l^{-1}$ (Abarnou, 1988, cited in Reid et al., 1993). Regular monitoring between 1989 and 1993 for lindane following the sinking of MV Perintis in 1987 showed concentrations generally $<1\ ng\ l^{-1}$, with a maximum of $1.8\ ng\ l^{-1}$ (Law and Allchin, 1994). Contamination of sediments by micro-organics, including PCBs, lindane, DDT and the drins are generally low or not detectable, with slightly elevated values found near some estuaries and ports (Reid et al., 1993). As with the metals, MAFF studies have shown no serious contamination of biota by micro-organic compounds in the Channel. Nevertheless, a number of studies have shown significantly elevated levels of PCBs and organochlorines in the meio- and macrofauna, including shellfish and fish, of the Baie de Seine (Goerke and Weber, 1998), and monitoring of ecotoxicological effects is ongoing. For PCBs, concentrations in fish and shellfish in the Baie de Seine have in the recent past approached the maximum allowed for human consumption, although they are now declining.

Hydrocarbons, Including Oil

Concentrations of total hydrocarbons in offshore waters collected between 1990–92 were low, and less than $1\ \mu g\ l^{-1}$ Ekofisk crude oil equivalents (Law et al., 1994), although because of the complexity of these mixtures the toxicological effects are difficult to assess. Enhanced values of total hydrocarbons have been observed close to Southampton and Portsmouth (adjacent to the Fawley oil refinery) and in the Seine Estuary (Marchand and Caprais, 1985, cited in Reid et al., 1993). Levels of total hydrocarbons in surficial sediments of the Channel are generally $<10\ \mu g\ g^{-1}$ and are therefore considered uncontaminated. As in the water column, increased concentrations were found in sediments sampled near the Fawley refinery (Fileman and Law, 1988), and also near Kimmeridge Bay (Dorset), where there are natural oil seepages.

The major impacts of hydrocarbons have followed oil spills from ships. The largest spill, from the oil tanker

Amoco Cadiz, in March 1978, contaminated 350 km of the Brittany coast when approximately 64,000 t of crude oil came ashore during the four weeks following the accident. At the time of the spill it was estimated that 30% of the benthic fauna and associated organisms and 5% of the macroalgae were killed in affected areas (Cabioch et al., 1978). Concentrations of hydrocarbons in the intertidal sediments decreased several-fold over the following three to four years, but were still measurable a decade later (Dauvin, 1998; Mille et al., 1998). Six to seven years after the spill, in the sector most exposed to the sea, the affected ecosystem had returned to normal although in more enclosed bays and in cohesive sediment effects were still visible 10–15 years on (Dauvin, 1998). Two years prior to the grounding of the Amoco Cadiz, the Torrey Canyon spilled 10,000 t of oil onto the Cornish coast. However, the major damage to the intertidal ecology came not from the oil but from the 14,000 t of dispersant used to remove the oil from the shores. The resultant widespread extinction of the 'keystone' species, the grazer *Patella*, initiated the extensive ecological disruption that followed. Only after 10–15 years did the ecology return to normal (Hawkins et al., 1994), thus again emphasising the importance of long-term observations following major pollution incidents.

Radioactivity

The major source of anthropogenic radionuclides to the Channel is from the nuclear waste reprocessing plant at Cap de la Hague (European Commission, 1999c). The principal nuclides include ^{3}H, ^{106}Ru, ^{60}Co, ^{90}Sr, ^{99}Tc and ^{125}Sb. There has been no measurable radiological impact on human or marine organisms from either the liquid or aerosol releases (Germain and Guegueniat, 1993).

Microbiological Contamination and Impacts

Shellfish

The Channel is home to a substantial shellfish industry, particularly on the French coast. Marketing of shellfish has historically been affected by pollution as shellfish retain and concentrate microbial pathogens derived principally from sewage-contaminated waters. For example, a recent study linked higher winter levels of pathogens in farmed shellfish to increased runoff from agricultural catchments in Brittany and Normandy where animal rearing is intensive (Dupont et al., 1992). Gastroenteritis and hepatitis A are associated with the consumption of raw or undercooked bivalves. The Shellfish Waters Directive provides for controls with respect to water quality, whilst a Shellfish Hygiene Directive relates to the quality of the animals themselves. In the latter case, the Directive demands that the shellfish be sufficiently clear of faecal coliforms, including *Escherischia coli*, before they can be sold (Category A) and also free of *Salmonella*. Shellfish of Category B or C can be transferred to Category A following an approved heat treatment, or if this is not possible, then depuration in tanks or relaying to cleaner waters. Category D animals are deemed unfit for human consumption under any circumstances. Most fisheries are Category B or C, but in both France and England there have been a number of closures of shellfisheries because of gross contamination by pathogens.

Bathing Water Quality

Concerns over the quality of coastal bathing waters with respect to aesthetics and microbiological contamination, and thus contraction of disease, led to the introduction of one of the earliest (1975) European Community Directives on the environment, the Bathing Waters Directive (76/160/EEC). Water quality, with respect to nineteen microbiological and physicochemical parameters, must be regularly monitored during the holiday season, with coliform counts generally used to assess compliance. In 1998, for example, >95% of French and UK Channel waters complied with the Directive limits. Information on the bathing water quality of the Channel is continually updated.

Impacts Arising from Fishing

Important changes have taken place in the species caught and the levels of exploitation of fish stocks over the last century as boats, gear and fish location equipment have developed. The scale of the exploitation of the stock and the deployment of bottom-towed gear in particular is likely to have had important and lasting effects on the Channel environment and ecosystems. A correct evaluation of the effects of the fishery is difficult, however, because it is likely to be one of several anthropogenic and natural pressures (Rogers et al., 1998), and in any case there are few historical measurements against which any effects can be assessed. In principle, the fishery can have direct effects, including (a) the mortality of the target fish and biota caught incidentally, (b) increasing the availability of food to other species as discards, waste and dead/damaged animals, (c) the disturbance of the seabed and (d) the production of litter. Indirect effects can include modifications both to predator–prey relationships and energy flows. Available data on fish growth, fecundity and disease (Ministry of Agriculture, Fisheries and Food, 1995) are all indicative of healthy fish populations. However, knowledge of the status of many stocks is poor, with sufficient information for management purposes only available for seven of the 20 species listed in Table 1. There is strong evidence for declining stocks of sole, cod, whiting, angler fish and herring (International Council for the Exploration of the Sea, 1998), with high fishing mortalities an important factor (Mellon, 1998). A number of areas within the Channel and its estuaries have fishing restrictions on them to protect stocks (Rogers, 1997). Bioeconomic modelling of the fishery for management purposes, linking biological and economic understanding, is still at an early stage (Pascoe, 1997; Pascoe and Robinson, 1997).

It was suggested in the 1960s (Holme, 1966) that heavy bottom gear towed by boats was having a detrimental effect on the benthos of the Channel. Holme (1983) reiterated this argument, and concluded that fishing disturbance was a major determinant of the ecology of the benthic fauna in the western Channel, which hitherto had been shaped by natural factors. Studies along the French and English coasts have shown that trawling may have an important effect on the juvenile stages of scallops in a number of areas (e.g. Kaiser et al., 1998). There have been no scientific assessments for management purposes of crab, spider crab and lobster populations. These stocks are all likely to be exploited at unsustainable levels. Indiscriminate fishing methods contribute to stock reduction through physical disturbance or loss of undersized animals e.g. spider crabs, clams (Reid et al., 1993).

Impacts arising from Sand and Gravel Extraction, and Dumping of Waste

There is little evidence of detrimental effects due to aggregate extraction, probably because the areas utilised are geographically restricted (Reid et al., 1993). Furthermore, any effects are likely to be masked by changes arising from benthic trawling.

Approximately 5 Mt a year of dredge spoil from six French harbours is dumped at licensed sites off Caen-Ouistreham, Grandcamp, Maisy, Rouen, Le Havre, Dieppe, le Tréport and Boulogne. More than half comes from the navigation channel of the Seine Estuary. Dredge spoil from harbours along the English south coast totals 2.5–3.5 M t per year and is dumped at 25 licensed sites. These sites are all located in high dispersion areas, and for this reason there appears to be no long-term effects on the local environment (Reid et al., 1993). Sewage sludge was dumped at four monitored sites off the English coast (~40000 t a^{-1}), but this ceased at the end of 1998 in response to the OSPAR Convention of 1992 and European legislative requirements. The dumping of solid and liquid industrial wastes in the Channel has been prohibited by the Oslo and Paris Convention since 1992.

PROTECTIVE AND REMEDIATION MEASURES

A number of provisions exist at the national and international levels to monitor and protect coastal and offshore habitats of the English Channel from the effects of both natural and human-induced changes to the environment. The provisions (Table 2), encompass those relating to habitat and wildlife conservation and pollution prevention. Issues relating to coastal defence, in contrast, are dealt with at the national level only by the relevant government and maritime agencies.

Within the UK, coastal habitats are afforded protection at the local level through their designation as National Nature Reserves, Areas of Outstanding Natural Beauty and Sites of Special Scientific Interest. In France, important sites have been given the status of Réserves Naturelles (e.g. Sept Ilse, Baie du Mont Saint Michel, Baie du Somme) or Zones de Protection Spéciale. Other locations have also been designated by the Conservatoire de l'Espace Littoral et des Rivages Lacustres to afford enhanced protection. A number of estuaries and coastal regions qualify as Special Protection Areas (SPAs) under the EC Directive on the Conservation of Wild Birds, including the Exe Estuary, and Chesil Beach and the Fleet, among others (Doody et al., 1993). Several marine areas will also become Special Areas of Conservation (SAC) under the EC Directive on the Conservation of Natural Habitats and of Wild Flora and Fauna. By 2004 the designation of SACs and SPAs across Europe will be complete and will form the Natura 2000 network. Coastal habitats will also be given priority status under national Biodiversity Action Plans derived from the EC biodiversity strategy and the UN Convention on Biodiversity (European Commission, 1998b). The linked coast of Dorset and East Devon (UK) is being considered as a World Heritage Site for designation during 2000. Figure 8 provides an overview of the habitats of major importance for nature conservation and their location.

The discharge of contaminants and waste materials to the Channel, and contaminant monitoring, are regulated by global (e.g. London, Bonn and Lisbon Conventions), regional (e.g. Oslo and Paris Convention, North Sea Task Force) and European initiatives (the latter including the Bathing Waters, Urban Waste Water Treatment and Nitrate Directives, as described in earlier sections). Monitoring of contaminants in sediments, biota and waters is carried out under the Oslo and Paris Convention of 1992, and also by the North Sea Task Force (NSTF), of which the Oslo and Paris Commissions (OSPARCOM) is a member. OSPARCOM established a Joint Monitoring Programme (JMP) in 1978, and incorporated the Monitoring Master Plan of the NSTF in 1989. Recently, the JMP has been super-

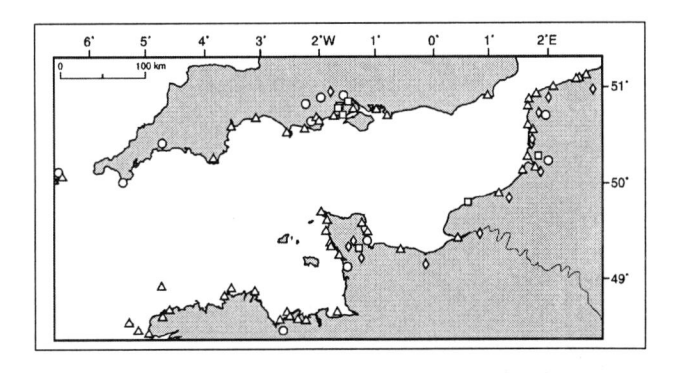

Fig. 8. Sites of major importance for nature conservation, by habitat type. Triangles: coastal and halophytic communities. Squares: bogs and marshes. Circles: forests. Diamonds: non-marine waters (European Commission, 1999a).

Table 2

International Conventions and Agreements and European Commission Directives relevant to protection of the water quality, wildlife and habitats of the English Channel.

International Conventions and Agreements
Convention on the Prevention of Marine Pollution by the Dumping of Wastes and Other Matter (1972, London)
Convention on the Prevention of Marine Pollution from Ships (1973, 1978)
Convention for the Prevention of Marine Pollution from Land-Based Sources (1974, Paris)
Protocol amending the Protocol for the Prevention of Marine Pollution from Land-Based Sources (1986, Paris)
Convention on the Conservation of Migratory Species of Wild Animals (1979, Bonn) (UNEP)
Convention on the Conservation of European Wildlife and Natural Habitats (1979, Bern) (Council of Europe)
Agreement for Cooperation in Dealing With Pollution of the North Sea by Oil and Other Harmful Substances (1983, Bonn)
Cooperation Agreement for the Protection of the Coasts and Waters of the North-East Atlantic against Accidental Pollution (1990, Lisbon)
Convention for the Protection of the North East Atlantic (Revised Oslo/Paris Convention) (1992, Paris)
UN Convention on the Law of the Sea (1982, Montego Bay, Jamaica)
UN Agreement on the Conservation of Small Cetaceans of the Baltic and North Seas (1992, New York)
UN Convention on Biodiversity (1992, Rio de Janeiro)
Ministerial Declarations from North Sea Conferences

EC Directive		Directive number
Dangerous Substances	Pollution caused by the discharge of certain dangerous substances into the aquatic environment	76/464/EEC
	Limit values and quality objectives for mercury discharges by the chlor-alkali electrolysis industry	82/176/EEC
	Limit values and quality objectives for mercury discharges by sectors other than the chlor-alkali electrolysis industry	84/156/EEC
	Limit values and quality objectives for cadmium discharges	83/513/EEC
	Limit values and objectives for discharges of hexachlorocyclohexane	84/491/EEC
	Limit values and objectives for discharges of certain dangerous substances included in List I of the Annex to Directive 76/464/EEC	86/280/EEC
	Amending Annex II to Directive 86/280/EEC on limit values and quality objectives for discharges of certain dangerous substances included in List I of the Annex to Directive 76/464/EEC	88/347/EEC
	Amending Annex II to Directive 86/280/EEC on limit values and quality objectives for discharges of certain dangerous substances included in List I of the Annex to Directive 76/464/EEC	90/415/EEC
Bathing Waters		76/160/EEC
Quality Required for Shellfish Waters		79/923/EEC
Shellfish Hygiene		91/492/EEC
Urban Waste Water Treatment		91/271/EEC (Amended by 98/15/EC)
Protection of Waters against Pollution caused by Nitrates from Agricultural Sources		91/676/EEC
The Conservation of Natural Habitats and of Wild Fauna and Flora		92/43/EEC
Conservation of Wild Birds		79/409/EEC
Integrated Pollution Prevention and Control		96/61/EEC

seded by the Joint Assessment and Monitoring Programme (JAMP). The issues identified by JAMP as important in the context of human impacts fall under the following broad headings: contaminants, eutrophication, litter, fisheries, mariculture and habitats and ecosystem health (Izzo et al., 1997). The national agencies that contribute significantly to the JAMP monitoring exercises include the Ministry of Agriculture, Fisheries and Food and the Environment Agency in the UK, and the Institut Français de Recherche pour l'Exploitation de la Mer (IFREMER) in France. IFREMER manages three French national monitoring networks, namely those for water quality (Reseau national d'observation de la qualité de milieu marin, RNO), microbiological contamination (Reseau de surveillance microbiologique du littoral français, REMI) and phytoplankton populations (Surveillance du phytoplankton du littoral français, REPHY) (European Environment Agency, 1999).

Coastal defence works have been undertaken for many years along the Channel coast in order to preserve human life and property, agricultural land and industrial zones. Because of increasing sea levels and storminess in the region, coastal defence work will continue and probably increase in importance. Structures will range from simple wooden groynes installed on beaches, soft engineering options such as beach recharge, and major engineering works, including breakwaters and seawalls. In many cases, these defences will also contribute to the protection of important natural habitats. However, future defence works will require a sound economic justification and so replacement of deteriorating defences may not be assured. The alternative in these situations is to allow a managed retreat of the coastline to a point where hard defences are economic. It is not clear to what extent this will occur along the French and UK Channel coasts. An advantage of this approach, nevertheless, is that new natural habitats can be created (Pye and French, 1993).

Because of the diverse range of human activities (economic, recreational) located on the coast, and the myriad of regulatory bodies, it became clear a decade ago that there was a need for integrated management and planning if the development and use of the coastal zone were to be sustainable. This approach led to the concept of Integrated Coastal Zone Management or ICZM (e.g. European Commission, 1999a,b), the object of which is to ensure that all land and sea use issues are co-ordinated, including development, conservation, waste disposal, fisheries, transport and coastal protection. Within the UK for example, ICZM was reflected in the formulation of local management plans for several regions, including Portsmouth and Chichester Harbours, and a number of estuaries (Environment Agency, 1999). In 1995 the European Commission initiated a demonstration programme to show what practical conditions should be met if sustainable development is to be achieved in coastal zones. The programme was predicated on three main ideas; (i) that improved co-operation between all interested parties is the basis for sustainable development (ii) that such co-operation can develop full, comprehensible information on the state of the environment, the origin of the changes affecting it, the implications of policies and measures at the various levels, and the options, and (iii) that co-operation must be organised and maintained (European Commission, 1999d). Thirty-five demonstration sites were chosen throughout Europe, including several along the Channel coast (the Cote d'Opale near Dunkirk in France, and the Isle of Wight and the Kent, Dorset, and Devon and Cornwall coasts in England). The programme finished in 1999, and the final conclusions and recommendations for a European Strategy for Integrated Coastal Zone Management are being compiled for consideration by the European Parliament and Council in the year 2000.

ACKNOWLEDGEMENTS

We would like to thank Kate Davis for redrawing the figures. We also thank Blackwells for permission to reproduce Figure 2 (© 1982), MAFF for the upper part of Figure 5 (© Crown), Elsevier Science for the lower part of the same figure (© 1989) and CEFAS for Figure 7 (© British Crown 1995).

REFERENCES

Achterberg, E.P., Colombo, C. and van den Berg, C.M.G. (1999) The distribution of dissolved Cu, Zn, Ni, Co and Cr in English coastal surface waters. *Continental Shelf Research* **19**, 537–558.

Alheit, J. and Hagen, E. (1997) Long-term climate forcing of European herring and sardine populations. *Fisheries Oceanography* **6**, 130–139.

Bailly du Bois, P., Salomon, J.C., Gandon, R. and Guegueniat, P. (1995) A quantitative estimate of English Channel water fluxes from 1987 to 1992 based on radiotracer distribution. *Journal of Marine Systems* **6**, 457–481.

Barne, J.H., Robson, C.F., Kaznowska, S.S., Doody, J.P., Davidson, N.C. and Buck, A.L. (eds) (1996a) Coasts and Seas of the United Kingdom. Region 10, South-west England: Seaton to the Roseland Peninsula. Coastal Directory Series. Joint Nature Conservation Committee, Peterborough.

Barne, J.H., Robson, C.F., Kaznowska, S.S., Doody J.P. and Davidson, N.C. (eds) (1996b) Coasts and Seas of the United Kingdom. Region 9, Southern England: Hayling Island to Lyme Regis. Coastal Directory Series. Joint Nature Conservation Committee, Peterborough.

Barne, J.H., Robson, C.F., Kaznowska, S.S., Doody, J.P., Davidson N.C. and Buck, A.L. (eds) (1996c) Coasts and Seas of the United Kingdom. Region 11, The Western Approaches: Falmouth Bay to Kenfig. Coastal Directory Series. Joint Nature Conservation Committee, Peterborough.

Barne, J.H., Robson, C.F., Kaznowska, S.S., Doody, J.P., Davidson N.C. and Buck, A.L. (eds) (1998a) Coasts and Seas of the United Kingdom. Region 7, South-east England: Lowestoft to Dungeness. Coastal Directory Series. Joint Nature Conservation Committee, Peterborough.

Barne, J.H., Robson, C.F., Kaznowska, S.S., Doody, J.P., Davidson, N.C. and Buck, A.L. (eds) (1998b) Coasts and Seas of the United Kingdom. Region 8, Sussex: Rye Bay to Chichester Harbour. Coastal Directory Series. Joint Nature Conservation Committee, Peterborough.

Barnes, R.S.K. (1989) The coastal lagoons of Britain: an overview and conservation appraisal. *Biological Conservation* **49**, 295–313.

Boust, D., Colin, C., Leclerc, G. and Baron, Y. (1997) Distribution and transit times of plutonium-bearing particles throughout the Channel. *Radioprotection* **32**, 123–128.

Bray, J.M., Hooke, J.M. and Carter, D.J. (1994) Tidal Information: Improving the Understanding of Relative Sea-Level Rise on the South Coast of England. Report to the Standing Conference on Problems associated with the Coastline. The River and Coastal Environments Research Group, University of Portsmouth, February 1994.

Burt, G.R., Bryan, G.W., Langston, W.J. and Hummerstone, L.G. (1992) Mapping the Distribution of Metal Contamination in United Kingdom Estuaries. Report of the Plymouth Marine Laboratory.

Cabioch, L., Dauvin, J.-C. and Gentil, F. (1978) Preliminary observations on pollution of the sea bed and disturbance of sub-littoral communities in northern Brittany by oil from the 'Amoco Cadiz'. *Marine Pollution Bulletin* **9**, 303–307.

Cabioch, L., Gentil, F., Glaçon, R. and Rotiere, C. (1976) Le Macrobenthos des Fonds Meubles de la Manche: Distribution Generale et Ecologie. In: *Biology of Benthic Organisms*, eds. B.F. Keegan, P. O'Ceidigh and P.J.S. Boaden. Pergamon Press. pp. 115–128.

Cabioch, J., Floch, J.-Y., Le Toquin, A., Boudouresque, C., Meinesz, C.F. and Verlaque, M., (1992) *Guides des Algues des Mers d'Europe*. Delachaux et Niestlé.

Carpenter, K.E. and Pye, K. (1996) Saltmarsh Change in England and Wales—Its History and Causes. R&D Technical Report W12, HR Wallingford Ltd., Wallingford, UK.

Cooper, L.H.N. (1967) The physical oceanography of the Celtic Sea. *Oceanography and Marine Biology Review* **5**, 99–110.

Cossa, D. and Fileman, C. (1991) Mercury concentrations in surface waters of the English Channel: a co-operative study. *Marine Pollution Bulletin* **22**, 197–200.

Covey, R. (1998) Eastern Channel (Folkestone to Durleston Head)

(MNCR Sector 7). In: *Marine Nature Conservation Review. Benthic Marine Ecosystems of Great Britain and the North-East Atlantic*, ed. K. Hiscock. Joint Nature Conservation Committee, Peterborough. pp. 199–218.

Crisp, D.J. and Southward, A.J. (1958) The distribution of intertidal organisms along the coasts of the English Channel. *Journal of the Marine Biological Association of the UK* 37, 157–208.

Dahlgaard, H. (1995) Radioactive tracers as a tool in coastal oceanography: an overview of the MAST-52 project. *Journal of Marine Ecosystems* 6, 381–389.

Dauvin, J.-C. (1998) The fine sand *Abra alba* community of the Bay of Morlaix twenty years after the Amoco Cadiz oil spill. *Marine Pollution Bulletin* 36, 669–676.

Dauvin, J.-C., Dewarumez, J.-M., Elkaim, B., Bernardo, D., Fromentin, J.-M. and Ibanez, F. (1993) Cinétique de *Abra alba* (mollusque bivalve) de 1977 à 1991 en Manche-Mer du Nord, relation avec les facteurs climatiques. *Oceanologica Acta* 16, 413–422.

Davidson, N.C., Laffoley, D. d'A. and Doody, J.P. (1995) Land-claim on British Estuaries: Changing Patterns and Conservation Implications. In: *Coastal Zone Topics: Process, Ecology & Management. 1. The Changing Coastline*, ed. N.V. Jones. Joint Nature Conservancy Council, Peterborough. pp. 68–80.

Davies, J. (1998) Western Channel (Durleston Head to Cape Cornwall, including the Isles of Scilly) (MNCR Sector 8). In: *Marine Nature Conservation Review. Benthic Marine Ecosystems of Great Britain and the North-East Atlantic*, ed. K. Hiscock. Joint Nature Conservation Committee, Peterborough. pp. 219–253.

Davoult, D., Dewarumez, J.-M. and Migné, A. (1998) Long-term changes (1979–1994) in two coastal benthic communities (English Channel): analysis of structural developments. *Oceanologica Acta* 21, 609–617.

Department of Trade and Industry (1997) *The Energy Report: Oil and Gas Resources of the United Kingdom. Volume Two*. The Stationary Office, London.

Doody, J.P., Johnston, C. and Smith, B. (eds.) (1993) *Directory of the North Sea Coastal Margin*. Joint Nature Conservation Committee, Peterborough.

Dupont, J., Jehl-Pietri, C., Hervé, C. and Ménard, D. (1992) Comparative study of bacterial and viral faecal contamination in shellfish: demonstration of seasonal variations. *Biomedical Letters* 47, 329–335.

English Nature (1996) *Important Areas for Marine Wildlife Around England*. English Nature, Peterborough.

Environment Agency (1999) *The State of the Environment of England and Wales: Coasts*. The Stationary Office, London.

European Commission (1998a) First Report on the Implementation of the Convention on Biological Diversity by the European Community. Office of Official Publications of the European Communities, Luxembourg.

European Commission (1998b) The Implementation of Council Directive 91/676/EEC concerning the Protection of Waters against Pollution caused by Nitrates from Agricultural Sources. Report of the Commission to the Council and European Parliament. Office of Official Publications of the European Communities, Luxembourg.

European Commission (1999a) Towards a European Integrated Coastal Zone Management (ICZM) Strategy. Office for Official Publications of the European Communities, Luxembourg.

European Commission (1999b) Lessons From the European Commission's Demonstration Programme on Integrated Coastal Zone Management (ICZM). Office for Official Publications of the European Communities, Luxembourg.

European Commission (1999c) Radiation Protection 104. Radioactive Effluents from Nuclear Power Stations and Nuclear Fuel Reprocessing Plants in the European Community, 1991–95. Office for Official Publications of the European Communities, Luxembourg.

European Commission (1999d) Internet site http:\\europa.eu.int/comm/dg11/iczm/.

European Environment Agency (1995) *Europe's Environment. The Dobriš Assessment*, eds. D. Stanners and P. Boudreau. Office of Official Publications of the European Communities, Luxembourg.

European Environment Agency (1999) Internet site http://themes.eea.eu.int/.

Fileman, T.W. and Law, R.J. (1988) Hydrocarbon concentrations in sediments and water from the English Channel. *Marine Pollution Bulletin* 19, 390–393.

Fromentin, J.M. and Ibanez, F. (1994) Year-to-year changes in meteorological features of the French coast area during the last half-century. Examples of two biological responses. *Oceanologica Acta* 17, 285–296.

Fromentin, J.M., Dauvin, J.C., Ibanez, F., Dewarumez, J.M. and Elkaim, B. (1997) Long-term variations of four macrobenthic community structures. *Oceanologica Acta* 20, 43–53.

Gentilhomme, V. and Lizon, F. (1998) Seasonal cycle of nitrogen and phytoplankton biomass in a well-mixed coastal system (Eastern English Channel). *Hydrobiologica* 361, 191–199.

Germain, P. and Guegueniat, P. (1993) Impact of industrial nuclear releases into the English Channel. *Radioprotection* 28, 271–275.

Gibbs, P.E., Bryan, G.W. and Spencer, S. (1991) The impact of tributyl tin (TBT) pollution on *Nucella lapillus* (Gastropoda) populations around the coast of south-east England. *Oceanologica Acta Special Publication* 11, 157–162.

Goerke, H. and Weber, K. (1998) The bioaccumulation pattern of organochlorine residues in *Lanice conchilega* (Polychaeta) and its geographical variation between the English Channel and the German Bight. *Chemosphere* 37, 1283–1298.

Gough, M.A., Fothergill, J. and Hendrie, J.D. (1994) A survey of southern England coastal waters for the s-triazine antifouling compound Irgarol 1051. *Marine Pollution Bulletin* 28, 613–620.

Gray, M.J. (1995) The coastal fisheries of England and Wales. Part III: A review of their status 1992–1994. Fisheries Technical Report No. 100. MAFF Directorate of Fisheries Research, Lowestoft.

Gubbay, S. (1988) *A Coastal Directory for Marine Nature Conservation*. Marine Nature Conservation Society, Ross-on-Wye.

Hamblin, R.J.O., Crosby, A., Balson, P.S., Jones, S.M., Chadwick, R.A., Penn, I.E. and Arthur, M.J. (1992) *United Kingdom Offshore Regional Report Series; the Geology of the English Channel*. HMSO, London.

Hawkins, S.J., Proud, S.V., Spence, S.K. and Southward, A.J. (1994) From the Individual to the Community and Beyond: Water Quality, Stress Indicators and Key Species in Coastal Systems. In *Water Quality and Stress Indicators in Marine and Freshwater Ecosystems: Linking Levels of Organisation (Individuals, Populations, Communities)*, ed. D.W. Sutcliffe. Freshwater Biological Association, Ambleside, Cumbria. pp. 35–62.

Hiscock, K. (1998) Introduction and Atlantic-European Perspective. In *Benthic Marine Ecosystems of Great Britain and the North-East Atlantic. Marine Nature Conservation Review*, ed. K. Hiscock. Joint Nature Conservation Committee, Peterborough, pp. 3–70.

Hoch, T. and Ménesguen, A. (1997) Modelling the biogeochemical cycles of elements limiting primary production in the English Channel. II. Sensitivity analyses. *Marine Ecology Progress Series* 146, 189–205.

Holme, N.A. (1961) The bottom fauna of the English Channel. *Journal of the Marine Biological Association of the UK* 41, 397–461.

Holme, N.A. (1966) The bottom fauna of the English Channel. II. *Journal of the Marine Biological Association of the UK* 46, 401–493.

Holme, N.A. (1983) Fluctuations in the benthos of the western English Channel. Proceedings 17th European Biology Symposium, Brest, France, 27 September–1 October, 1982. *Oceanologica Acta Special Publication*, 121–124.

International Council for the Exploration of the Sea (1998) Report of the ICES Advisory Committee on Fishery Management 1997 Parts 1 and 2. ICES Cooperative Research Report No 223.

Izzo, G., Chaussepied, M., Bokn, T. and Manzella, G.M.R. (1997) Data Collected Within the Framework of the Regional European Sea Conventions. European Environment Agency Technical Report No. 3. European Environment Agency, Copenhagen.

Johnston, P., Marquardt, S., Keys, J. and Jewell, T. (1997) Shipping and handling of pesticide cargoes; the need for change. *Journal of the Chartered Institute for Water and Environmental Managers* **11**, 157–163

Joint, I. (1978) Microbial production of an estuarine mudflat. *Estuarine, Coastal and Marine Science* **7**, 185–195.

Jordan, M.B. and Joint, I. (1998) Seasonal variation in nitrate:phosphate ratios in the English Channel 1923–1987. *Estuarine, Coastal and Shelf Science* **46**, 157–164.

Kaiser, M.J., Armstrong, P.J., Dare, P.J. and Flatt, R.P. (1998) Benthic communities associated with a heavily fished scallop ground in the English Channel. *Journal of the Marine Biological Association of the UK* **78**, 1045–1059.

Langston, W.J., Bryan, G.W., Burt, G.R. and Pope, N.D. (1994) Effects of Benthic Metals on Estuarine Benthic Organisms. National Rivers Authority Report 105, R&D Note 203. National Rivers Authority and NERC Plymouth Marine Laboratory.

Langston, W., Gibbs, P., Livingstone, D., Burt, G., O'Hara, S., Pope, N., Bebianno, M., Coelho, M., Porte, C., Bayona, J., McNulty, M., Lynch, G. and Keegan, B. (1997) Risk Assessment of Organotin Antifoulings on Key Benthic Organisms of European Coastal Habitats (BOATS). NERC Plymouth Marine Laboratory Final Report on Contract No. MAS2-CT94-0099.

Larsonneur, C., Bouysse, P. and Auffret, J.-P. (1982) The superficial sediment of the English Channel and the Western Approaches. *Sedimentology* **29**, 851–864.

Law, R.J. and Allchin, C.R. (1994) Hexachlorohexanes in seawater in the English Channel 1989–1993, following the loss of MV Perintis. *Marine Pollution Bulletin* **28**, 704–706.

Law, R.J. and Campbell, J.A. (1998) The effects of oil and chemical spillages at sea. *Journal of the Chartered Institute of Water and Environmental Management* **12**, 245–249.

Law, R.J., Waldock, M.J., Allchin, C.R., Laslett, R.E. and Bailey, K.J. (1994) Contaminants in seawater around England and Wales: results from monitoring surveys, 1990–1992. *Marine Pollution Bulletin* **28**, 668–675.

Lewis, J.R. (1964) *The Ecology of Rocky Shores*. English Universities Press, London.

Lloyds Register of Shipping (1998) UK Coastguard Agency -Risk Analysis of Spills of Bunker Fuel Oils, Refined Products and Vegetable Oils in UK waters. Lloyd's Register, System Integrity & Risk Management.

Mellon, C. (1998) Les stocks de poissons démersaux sous quota en Manche est et Mer du Nord. *Cybium* **22**, 357–369.

Mille, G., Munoz, D., Jacquot, F., Rivet, L. and Bertrand, J.-C. (1998) The Amoco Cadiz oil spill: evolution of petroleum hydrocarbons in the Ile Grande salt marshes (Brittany) after a 13-year period. *Estuarine, Coastal and Shelf Science* **47**, 547–559.

Ministry of Agriculture, Fisheries and Food (1995) Monitoring for Diseases in Marine and Freshwater Fish, 1992. Aquatic Environment Monitoring Report Number 46. Directorate of Fisheries Research, Lowestoft.

Oslo and Paris Commissions (1992) *Monitoring and Assessment*. Chameleon Press, London.

Oslo and Paris Commissions (1995) Data Report on Riverine and Direct Inputs of Contaminants to the Maritime Area of the Paris Convention in 1992.

Pascoe, S. (1997) A Preliminary Bioeconomic Model of the UK Component of the Fisheries of the English Channel. Research Paper 112. Centre for the Economics and Management of Aquatic Resources, University of Portsmouth.

Pascoe, S. and Robinson, C. (1997) Bioeconomic Modelling of the Fisheries of the English Channel: Overview Report. Report 45. Centre for the Economics and Management of Aquatic Resources, University of Portsmouth.

Pawson, M.G. (1995) Biogeographical Identification of English Channel Fish and Shellfish Stocks. Fisheries Research Technical Report No. 99. Directorate of Fisheries Research, Lowestoft.

Pingree, R.D. (1980) Physical Oceanography of the Celtic Sea and English Channel. In: *The North West European Shelf Seas: the Sea Bed and the Sea in Motion, Volume II*, eds. F.T. Banner, M.B. Collins and K.S. Massie. Elsevier, Amsterdam. pp. 415–465.

Potts, G.W. and Swaby, S.E. (1993) Review of the Status of Estuarine Fishes. English Nature Research Report No. 34. English Nature, Peterborough.

Potts, G.W. and Swaby, S.E. (1996) Fish: Other species. In *Coasts and Seas of the United Kingdom. Region 9, Southern England: Hayling Island to Lyme Regis*, eds. J.H. Barne, C.F. Robson, S.S. Kaznowska, J.P. Doody and N.C. Davidson. Joint Nature Conservation Committee, Peterborough. pp. 119–121.

Prandle D. (1978) Monthly mean residual flows through the Dover Strait 1949–1972. *Journal of the Marine Biological Association of the UK* **58**, 965–973.

Prandle, D. (1993) Year-long measurements of flow through the Dover Strait by H.F. Radar and acoustic Doppler current profilers (ADCP). *Oceanologica Acta* **16**, 457–468.

Pye, K. and French, P.W. (1993a) *Targets for Habitat Re-creation*. English Nature Science Series. English Nature, Peterborough.

Pye, K. and French, P.W. (1993b) Erosion and accretion processes on British saltmarshes, Vol. 3. Final Report to MAFF, CSA Contract 1976.

Reid, P.C., Auger, C., Chaussepied, M. and Burn, M. (1993) Quality Status Report of the North Sea 1993. Report on subregion 9, The Channel. UK Department of the Environment, London.

Reid, P.C. and Planque, B. (1999) Long-term planktonic variations and the climate of the North Atlantic. In *Ocean Life of Atlantic Salmon*, ed. D. Mills. Fishing News Books, Oxford. pp. 153–169.

Rogers, S.I. (1997) A Review of Closed Areas in the United Kingdom Exclusive Economic Zone. Science Series Technical Report Number 106. Centre for Environment, Fisheries and Aquaculture Science, Lowestoft.

Rogers, S.I., Rijnsdorp, A.D., Damm, U. and Vanhee, W. (1998) Demersal fish populations in the coastal waters of the UK and continental NW Europe from beam trawl survey data collected from 1990 to 1995. *Journal of Sea Research* **39**, 9–102.

Salomon, J.-C., Breton, M. and Guegueniat, P. (1993) Computed residual tidal flow through the Dover Strait. *Oceanologica Acta* **16**, 449–455.

Sanvicente-Añorve, L., Leprêtre, A. and Davoult, D. (1996) Large-scale spatial pattern of the macrobenthic diversity in the eastern English Channel. *Journal of the Marine Biological Association of the UK* **76**, 153–160.

Skov, H., Durinck, J., Leopold, M.F. and Tasker, M.L. (1995) *Important Bird Areas for Seabirds in the North Sea*. Birdlife International, Cambridge.

Southward, A.J. (1980) The western English Channel—an inconstant ecosystem? *Nature* **285**, 361–366.

Southward, A.J. (1983) Fluctuations in the ecosystem of the western Channel: a summary of studies in progress. Proceedings 17th European Biology Symposium, Brest, France, 27 September–1 October, 1982. *Oceanologica Acta Special Publication*, 187–189.

Southward, A.J. (1995) The importance of long-time series in understanding the variability of natural systems. *Helgoländer Meeresuntersuchungen* **49**, 329–333.

Southward, A.J., Hawkins, S.J. and Burrows, M.T. (1995) Seventy years' observations of changes in distribution and abundance of zooplankton and intertidal organisms in the western English Channel in relation to rising sea temperature. *Journal of Thermal Biology* **20**, 127–155.

Southward, A.J. and Southward, E.C. (1978) Recolonization of rocky shores in Cornwall after use of toxic dispersants to clean up the Torrey Canyon spill. *Journal of the Fisheries Research Board of Canada* **35**, 682–706.

Tappin, A.D., Hydes, D.J., Burton, J.D. and Statham, P.J. (1993) Concentrations, distributions and seasonal variability of dissolved Cd, Co, Cu, Mn, Ni, Pb and Zn in the English Channel. *Continental Shelf Research* **13**, 941–969.

Tasker, M.L. (1996) Seabirds. In *Coasts and Seas of the United Kingdom. Region 9, Southern England: Hayling Island to Lyme Regis*, eds. J.H. Barne, C.F. Robson, S.S. Kaznowska, J.P. Doody and N.C. Davidson. Joint Nature Conservation Committee, Peterborough. pp. 122–124.

Tubbs, C.R. (1994) *Spartina* on the South Coast: an Introduction. In *Spartina anglica in Great Britain*. Focus on Nature Conservation No. 5, ed. J.P. Doody. Nature Conservancy Council, Peterborough.

Vallet, C. and Dauvin, J.-C. (1998) Composition and diversity of the benthic boundary layer macrofauna from the English Channel. *Journal of the Marine Biological Association of the UK* **78**, 387–409.

THE AUTHORS

Alan D. Tappin
Centre for Coastal and Marine Science,
Plymouth Marine Laboratory,
Prospect Place,
Plymouth PL1 3DH, U.K.

P. Chris Reid
Sir Alister Hardy Foundation for Ocean Science,
1 Walker Terrace,
The Hoe,
Plymouth, PL1 3BN, U.K.

Chapter 6

THE IRISH SEA

Richard G. Hartnoll

In global terms, the Irish Sea is relatively small, enclosed and sheltered. It is characterised by strong tidal currents and a well mixed water body, but exchange with the Celtic Sea is limited and its water has a considerable residence time exceeding one year. Temperature, nutrient levels and primary production all show clear seasonality. Levels of primary production are low compared with comparable neighbouring areas. Anthropogenic input is greatest in the north east sector, reducing water quality, but never so severely that the region ceases to be productive and support fisheries.

There are several problems currently affecting the Irish Sea: (a) Nutrients have roughly doubled over the past forty years, reaching levels which pose a risk of eutrophication. The incidence of algal blooms is increasing, and there is limited scope for remedial action. This must pose the major threat. (b) All of the major fisheries are being exploited at levels substantially above the optimum. Current regulations are of limited success. (c) There are several anthropogenic inputs which are of concern and require continued monitoring—sewage, heavy metals, organic compounds and radionuclides. None currently have widespread severe impact, and most inputs are being reduced.

The overall prognosis for the Irish Sea is one of cautious optimism.

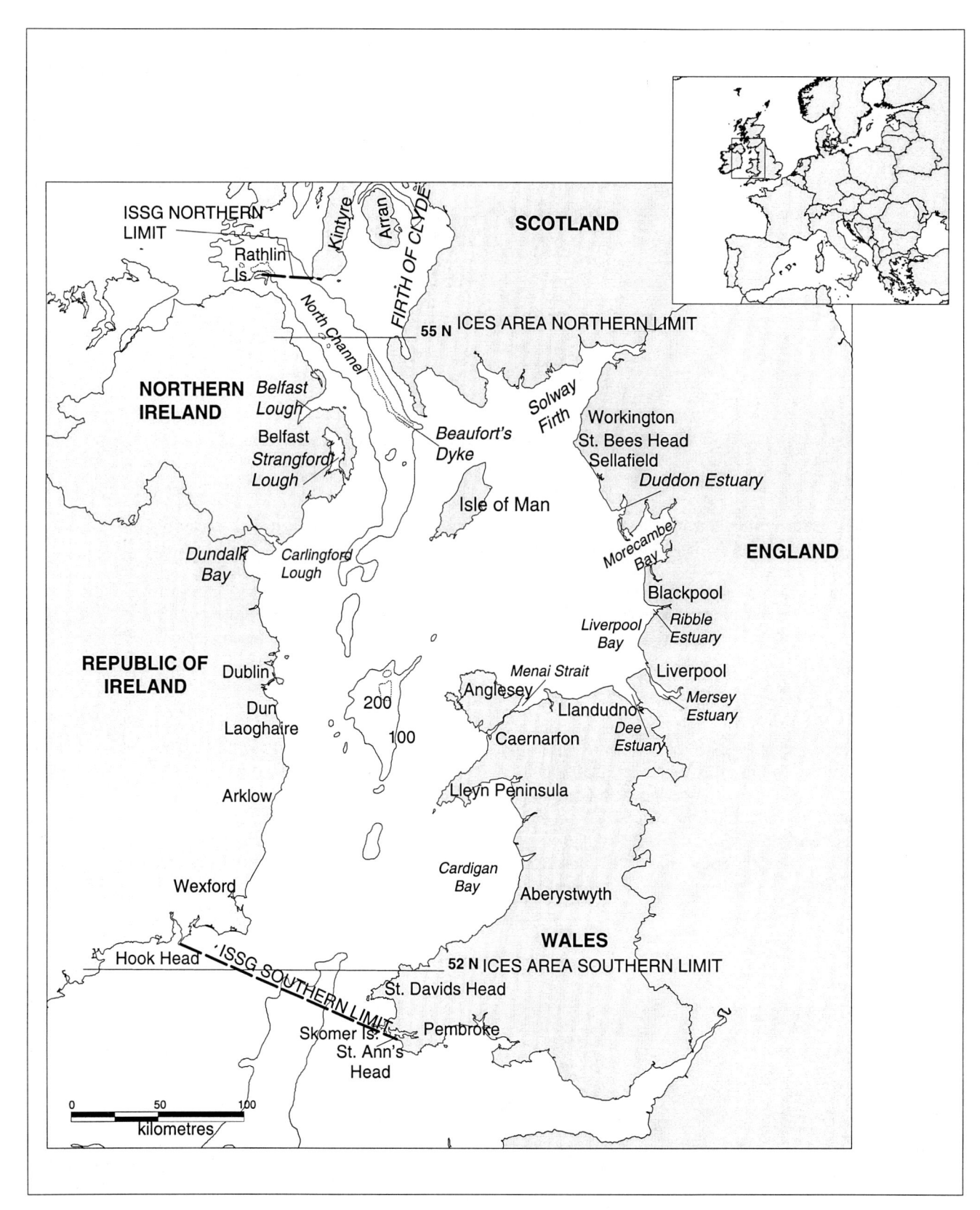

Fig. 1. A general map of the Irish Sea showing accepted boundaries (ISSG definition, and ICES Area boundaries), bathymetry (contours at 50 m, 100 m and 200 m), major urban areas, and locations and geographic features mentioned in the text.

THE DEFINED REGION

The Irish Sea lies within the OSPAR Convention Region III, the 'Celtic Seas'. The Irish Sea is quite well defined, with an area of approximately 45,000 km^2 and a volume of 2,430 km^3, and bounded by the coastlines of England, Wales, Scotland and Ireland. It comprises territorial waters of the United Kingdom, the Republic of Ireland, and the Isle of Man. It is connected to the Celtic Sea portion of the Atlantic Ocean by St Georges Channel to the south, and by the North Channel to the north. There are no absolute boundaries to the Irish Sea. The Irish Sea Study Group (ISSG) defined the limits as a line between Rathlin Island and the Mull of Kintyre to the north, and one from Hook Head to St Anne's Head in the South (ISSG, 1990a). The ICES Fisheries Statistical Area VIIa covers a similar area, being bounded on the north by the 55°N parallel and on the south by the 52°N parallel. The northern boundary of the ICES area is to the south of the ISSG limit, and is probably more appropriate as it unequivocally excludes the Clyde. The two southern boundaries are very similar (Fig. 1). The margins of the Irish Sea comprise on the east side the Marine Nature Conservation Review (MNCR) British coastal sectors 10 (Cardigan Bay and North Wales), 11 (Liverpool Bay and the Solway) and 12 (Clyde Sea). On the western side are the BioMar Irish coastal sectors IR2 (North Channel) and IR3 (East).

Despite having two substantial connections to the Celtic Sea, the Irish Sea has a strong individuality which sets it apart from the adjacent areas and justifies its consideration as a distinct entity. Prime reasons for this are:
– its enclosed and comparatively sheltered nature;
– a relatively long residence time for much of its water mass (Dickson and Boelens, 1988);
– shallow depths over much of its extent;
– substantial land-based inputs.

NATURAL ENVIRONMENTAL VARIABLES

Topography and Sediments of the Irish Sea

The Irish Sea is composed of two constituent parts, a deeper north/south trough extending up the western side, and a series of shallow embayments to the east (Fig. 1). Along the axis of the trough there is a minimum depth of 80 m, with maximum depths exceeding 275 m in parts of the North Channel along Beaufort's Dyke. The eastern embayments are less than 50 m depth, and there are two main areas. In the north is the region to the east of the Isle of Man which includes the Solway, Morecambe Bay and Liverpool Bay. To the south is Cardigan Bay. This depth pattern has a major influence on the nature of sediments, current flow, and stratification and nutrient flux: consequently it is a major determinant of the distribution of biotic communities.

The distribution of bottom sediments is related to depth and to the strength of tidal currents (see below). Coarser gravelly sediments are widespread in the southern Irish Sea in St George's Channel and Cardigan Bay, and extend up through the central area past the Isle of Man. Most of the other areas to the east and west are sandy. There are three substantial muddy regions—in the areas of weak tidal currents to the southwest of the Isle of Man and off St Bees Head, and in a deep area off Holyhead (Fig. 2a). Fine muddy sediments also occur in most estuaries.

Tides and Currents

The Irish Sea experiences semi-diurnal tides which differ greatly in range. Along the Cumbrian and Lancashire coasts are found the largest tides with mean spring ranges exceeding 8 m. The range generally exceeds 4 m through most of the Irish Sea, but declines in the North Channel and along the Irish coast south of Arklow. Thus Ballycastle on the coast of Northern Ireland has a spring range of about 1 m, and at Wexford on the Irish coast it is less than 2 m.

These large tides generate strong tidal currents. The strongest currents are in the North Channel and St Georges Channel and off Anglesey, where they exceed 1 m/s during average spring tides. Close to headlands and in confined channels they can exceed 4 m/s. In contrast in areas extending southwest of the Isle of Man to the Irish coast, and eastwards to St Bees Head, currents do not exceed 0.25 m/s. The influence of these tidal currents on sediment distribution has already been noted.

The residual currents in the Irish Sea are complex and variable. Near surface and near bottom currents differ, and the effects of wind can have a predominant effect in the short term, reversing the long term trends for periods of months. The basic long-term flow, averaged over a year or more, is from south to north (Fig. 3). The flow rate averages between 2 and 8 km/day on a yearly basis, so that water will take at least a full year to travel the entire length. The eastern Irish Sea and Liverpool Bay form a series of eddy systems within which there is a considerably longer residence time. There is a southward flowing countercurrent on the western side of the North Channel which can penetrate south as far as Dublin. It has recently been shown (Hill et al., 1996) that there is a cyclonic gyre in the western basin of the Irish sea each spring and summer, which probably acts as a retention mechanism for plankton. This has potential importance for the recruitment of commercial species (Dickey-Collas et al., 1997).

Temperature and Salinity

The average monthly seawater temperature in the central Irish Sea ranges from a maximum of about 13–14°C in July/August to a minimum of 6–7°C in February/March (Shammon et al., 1998). In the shallow eastern embayments ranges are more extreme, with higher summer temperatures and lower winter ones. The water column is generally well mixed, but the area to the west of the Isle of Man becomes thermally stratified in summer, with the surface

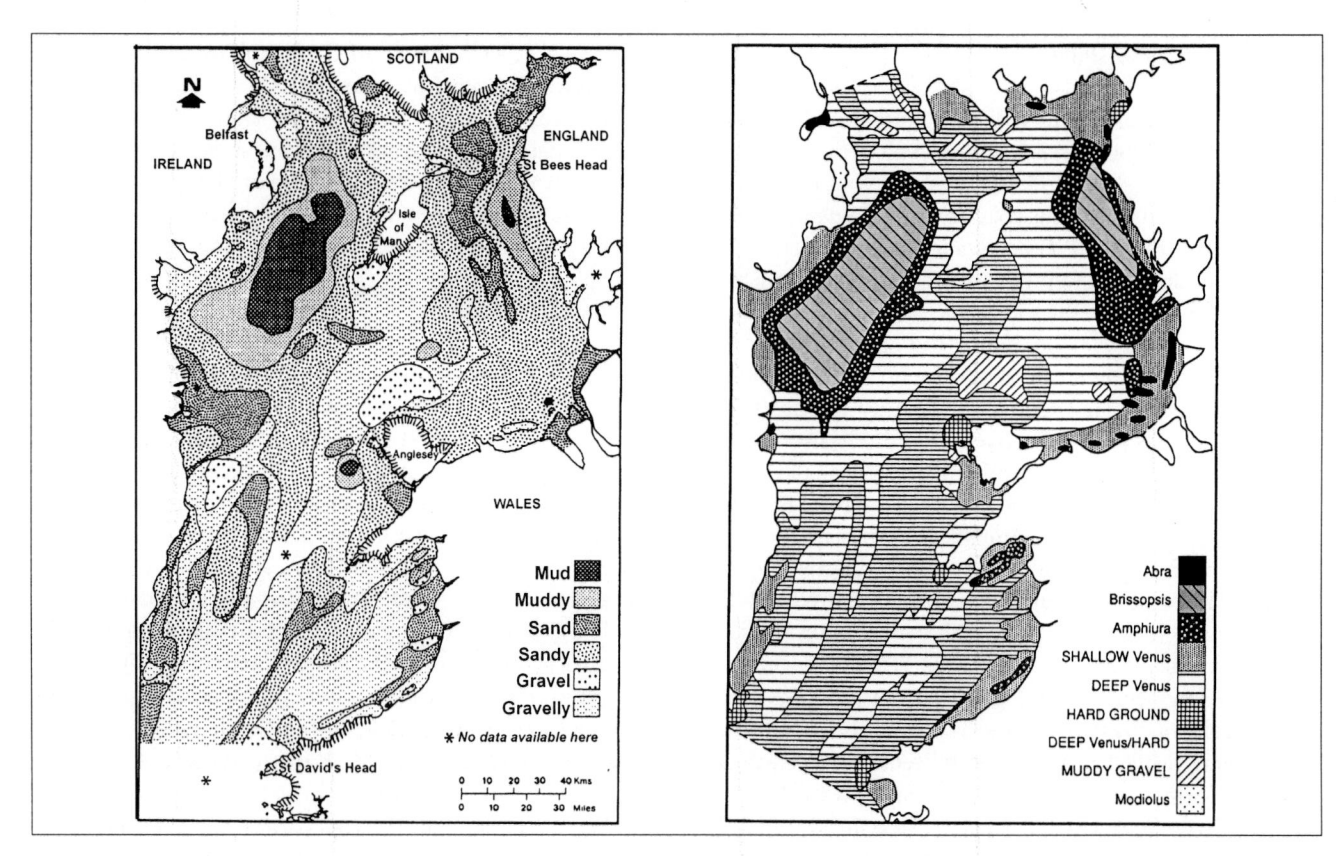

Fig. 2. (a) Superficial sea bed sediments of the Irish Sea, and major areas of rocky coastline—hatched shoreline (partly after ISSG, 1990a). (b) General distribution of major macrobenthic communities in the Irish sea (after ISSG, 1990a).

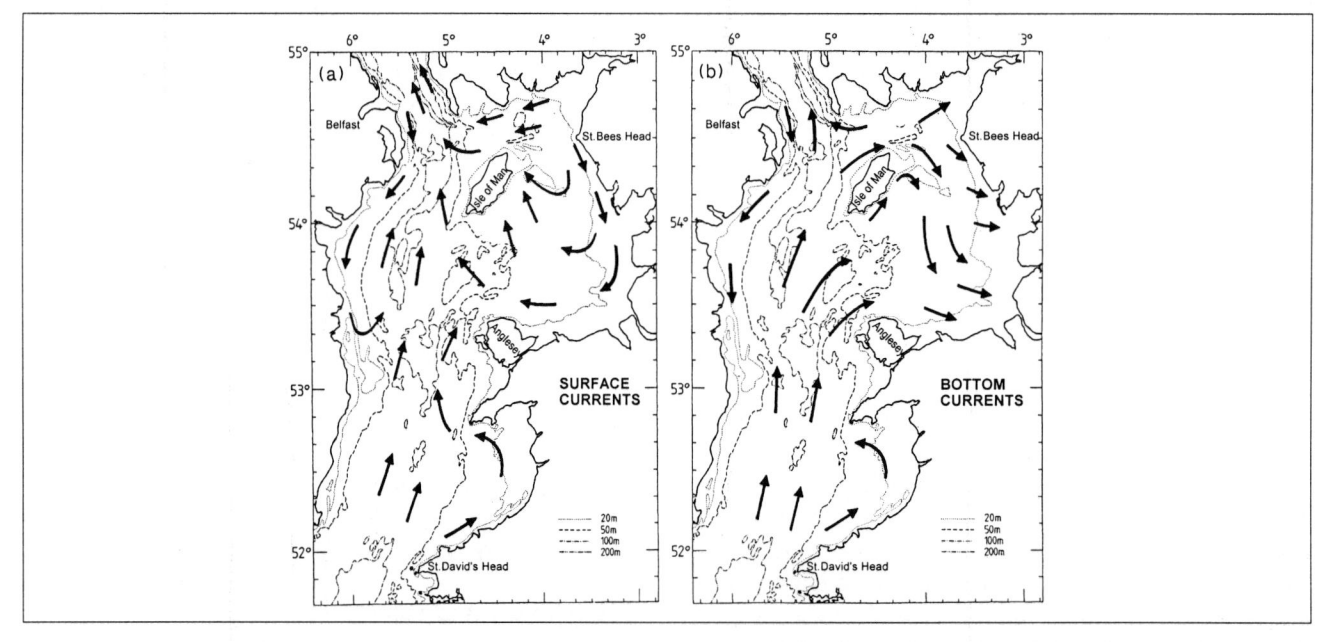

Fig. 3. General pattern of (a) surface and (b) bottom residual currents of the Irish Sea (after ISSG, 1990b).

layers up to 5°C warmer than the deeper parts. There are other areas of less distinctly stratified 'transitional' waters (see Fig. 7). The transitional waters off the Lancashire and Cumbrian coasts can show haline stratification in winter and spring. Distinct fronts separate the stratified and well mixed regions, and are characterised by high biological activity. There is a ninety year surface temperature data set for Port Erin in the Isle of Man (Fig. 4). There is evidence of a small increase in mean annual temperature over this period (0.5–1.0°C), but it is not possible at present to decide whether this is part of a natural long-term cycle, or evidence for the anthropogenic effect of global warming.

Salinity in central areas varies between 34 and 35, with minor fluctuations over the year—this low value reflects

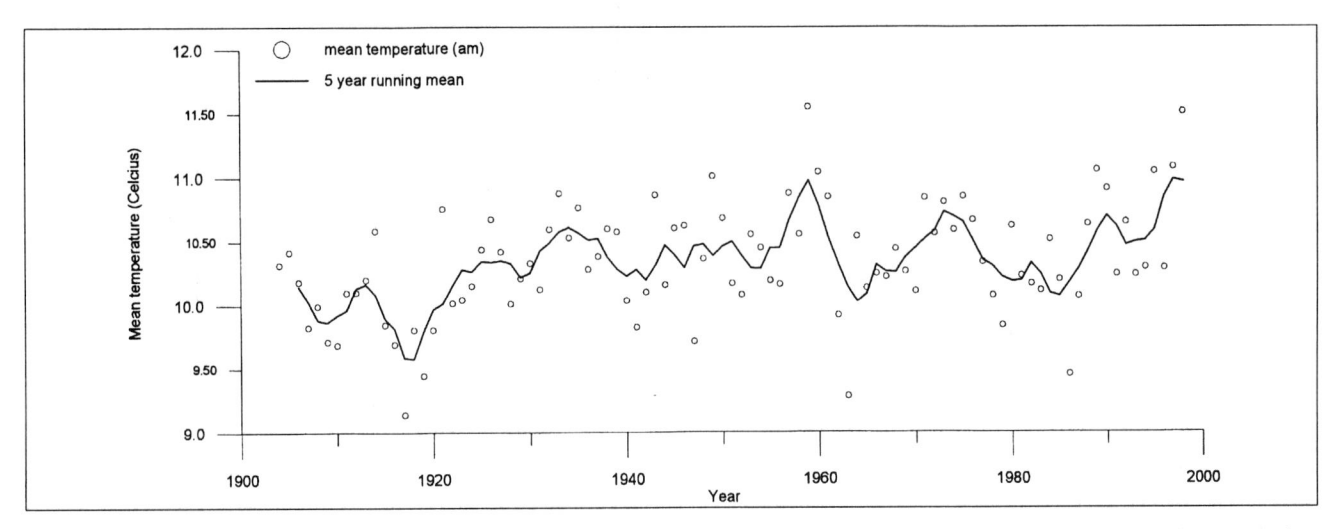

Fig. 4. Annual mean surface seawater temperature at Port Erin, Isle of Man from 1904 to 1998 inclusive. The open circles are annual means, the line is the five year running mean (after Shammon et al., 1998).

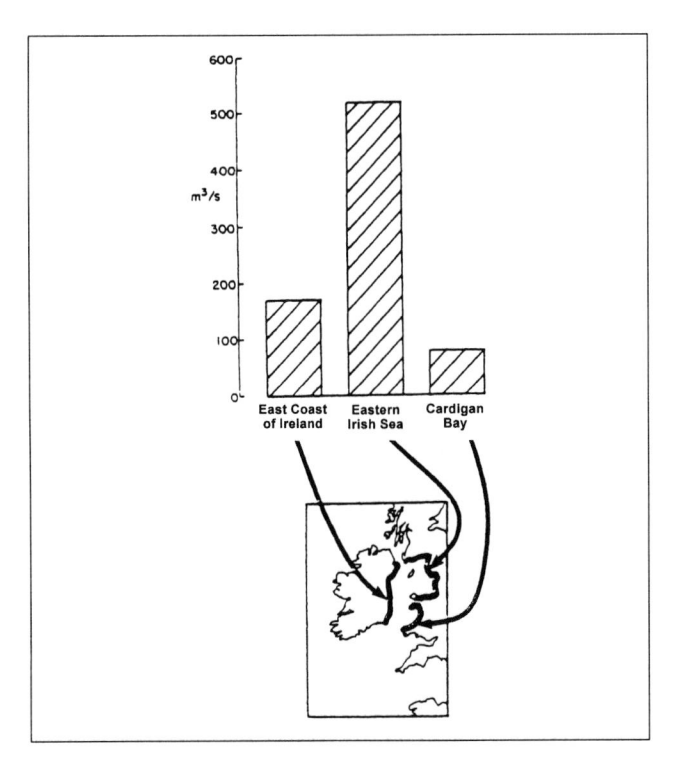

Fig. 5. Average river discharge levels into the Irish Sea from major regions of the coastline (after ISSG, 1990b).

the semi-enclosed nature of the sea. In the shallower eastern region it is lower, and may fall below 33 in the winter. The reason for this salinity reduction is the pattern of river discharges (Fig. 5). Some two-thirds of the discharge is into the north-eastern Irish Sea, with the Solway being the single largest source. This input pattern also has implications for the input of pollutants.

Nutrients

The important nutrients are nitrate, phosphate and silicate - they are essential for the primary plant growth upon which the productivity of the ecosystem depends. During the winter the water column is generally well mixed and nutrients are evenly distributed throughout the offshore areas. Typical winter levels are: total oxidised nitrogen 8 μg-atN/L; soluble reactive phosphate 0.5 μg-at P/L; silicate 5.0 μg-atSi/L. These levels are higher near to land as a result of freshwater runoff and anthropogenic inputs, especially on the English side where the major riverine input occurs (Fig. 5). During summer nutrient levels are depleted as the nutrients are taken up by growing phytoplankton: depletion occurs throughout the water column in the mixed areas, and down to the thermocline at about 20 m in the stratified areas. Typical summer surface levels are: total oxidised nitrogen 0.5 μg-at N/L; soluble reactive phosphate 0.1 μg-at P/L; silicate 1.5 μg-at Si/L.

There is clear evidence that nutrient levels have increased in the central Irish Sea over the last forty years (Allen et al., 1998). The levels of nitrate and phosphate have more or less doubled (Fig. 6), whilst the level of silicate has increased slightly. The Si:N ratio at the winter maximum has declined from around 1.25 to 0.75. It is assumed that this is due to anthropogenic inputs. These are all signs of possible eutrophication, and areas near the English coast show very clear indications of this—winter TON concentrations up to 29 μg-at N/L, and N:Si ratios up to 2.6. The Comprehensive Studies Task Team of the Marine Pollution Monitoring Group (CSTT, 1997) have suggested values exceeding 12 and 2 respectively as indicating waters liable to eutrophication.

THE MAJOR BENTHIC MARINE AND COASTAL HABITATS

Intertidal Habitats

The Irish Sea has a wide range of eroding and depositing shores, providing a diverse and rich assemblage of species and communities. The main areas of rocky coastline (Fig.

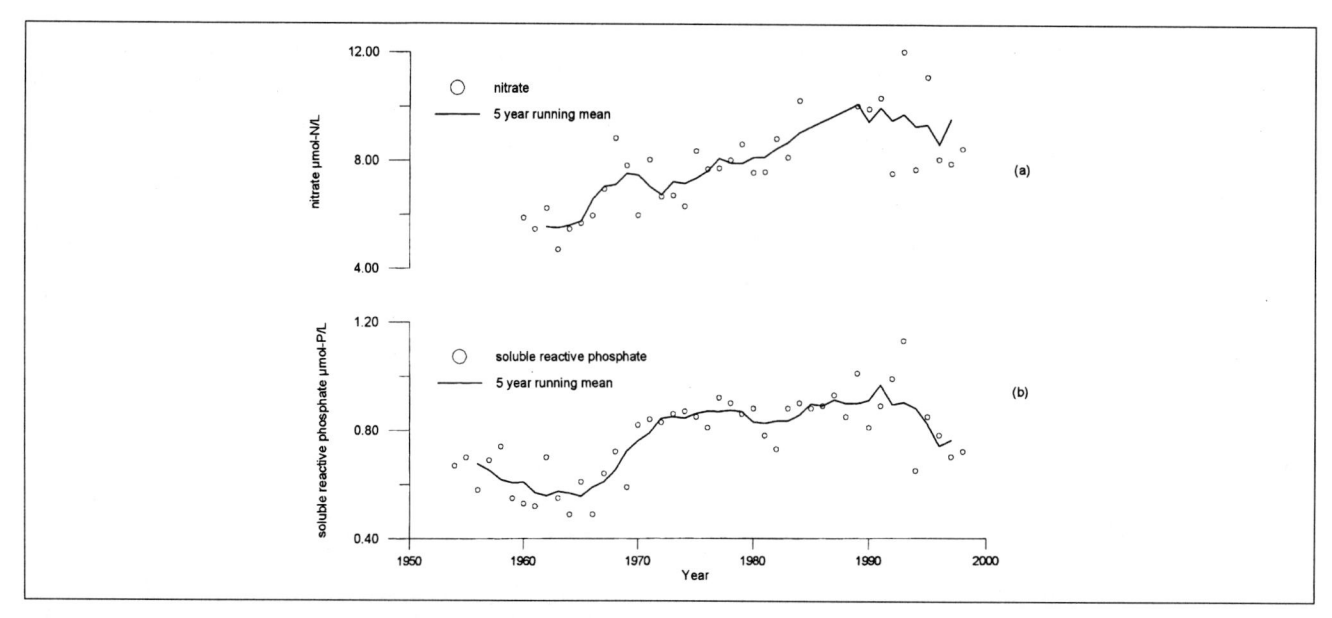

Fig. 6. Annual winter (December to March inclusive) maxima of (a) nitrate and (b) soluble reactive phosphate in the surface waters off the Isle of Man. For 1954 to 1998 inclusive, with five-year running means (after Shammon et al., 1998).

2a) are as follows, proceeding anti-clockwise from St Annes head:
- the north Pembrokeshire coast,
- the Lleyn Peninsula,
- Anglesey,
- Dumfries & Galloway,
- Kintyre,
- Co Antrim and North Down,
- Co. Dublin,
- Hook Head,
- the southern coast of the Isle of Man.

There are, of course, many smaller areas of rocky shore, often with very well developed communities. Most of the above also include small areas of soft shoreline. Rocky shore communities vary with the degree of wave action. Under strong wave exposure communities are animal-dominated, with barnacles, mussels and limpets predominating. As shelter from wave action increases the amount of macro-algae becomes greater, and very sheltered shores have dense algal canopies.

Most of the remaining areas are dominated by eroding shores, varying from shingle to mud, and usually with scattered areas of hard substrate. Sediments become finer in sheltered areas, inlets and estuaries. Thus open coastlines, such as Cardigan Bay, are predominantly shingle and coarse sand shores, whilst more sheltered areas such as Morecambe Bay consist of sand and mud. The organisms inhabiting these shores are almost all burrowing, and show signs of activity only when covered by the tide. The composition of the community varies with the grade of the sediment, which in turn is determined mainly by wave action and currents: more water movement means coarser sediments. The coarser sediments have a predominance of more mobile organisms including crustaceans, bivalves and polychaetes. Fine sediments have an increasing preponderance of species inhabiting permanent tubes, largely worms.

The distribution of intertidal habitats around the Irish Sea, and of the communities found in them, have been described in very considerable detail (ISSG, 1990; Huckbody et al., 1992; Taylor and Parker, 1993; Connor et al., 1997a; Hiscock, 1998). No attempt will be made here to summarise this very substantial body of information, but some aspects of particular interest will be highlighted.

The geographical distribution of intertidal species has been studied in detail, and has shown that a number of southern species reach the limits of their geographical distribution in or near the Irish Sea. Some of these examples are reviewed in Lewis (1964), and include the barnacle *Balanus perforatus*, the limpet *Patella depressa*, the top shell *Monodonta lineata* and the alga *Bifurcaria bifurcaria*. Such species are of special interest on two counts. Firstly since they are surviving at the limit of their range they will be sensitive to any increased stress, such as from pollution. Secondly, as they are presumably limited by cold tolerance, any global warming (for which there is perhaps evidence in the Irish Sea—see earlier section on "Temperature and Salinity") could result in extension of their range. The topshell *Monodonta* had its northern limit on Anglesey trimmed back by a severe winter (Crisp, 1964). It has been shown that changes in the dominance of the southern barnacle *Chthamalus* and the northern *Semibalanus* in the Irish Sea have been correlated with temperature fluctuations (Southward and Crisp, 1956). A comprehensive review of such changes in the area immediately to the south of the Irish Sea has been produced by Southward et al., 1995), and other examples (e.g. limpets, topshells, algae) of changes in limits correlated with temperature are reviewed by Hill et al. (1998). The future geographic distribution and abundance of southern and northern intertidal species will need monitoring if global warming meets expectations.

Another aspect of geographical distribution, also linked to global warming, is the spread and impact of introduced species. The Australian barnacle *Elminius modestus* arrived in the U.K. in the forties, and was established in the Irish Sea in 1948 (see review in Lewis, 1964). It competes with *Semibalanus*, and whilst the impact has been limited to date, this may be increased if temperatures rise. The Japanese brown alga *Sargassum muticum* has become established in Strangford Lough. Another Japanese brown alga, *Undaria*, has yet to reach the Irish Sea, but warming will make this more likely. Environmental impacts could be substantial.

Subtidal Habitats

Since the Irish Sea is all relatively shallow in global terms, there is no point in treating shallow and deep habitats separately. The basic division in the Irish Sea is between hard and soft subtidal substrates.

Hard substrates are essentially rock, with occasional artificial structures. They are predominantly confined to the coastal regions where rocky shorelines extend beneath low water, and their distribution is broadly similar to that of the rocky shorelines mentioned above—overall they comprise a small proportion of the sea bed of the Irish Sea. The biota are typical of this environment around the British Isles, and their composition is largely determined by a combination of depth and water movement. The shallower infralittoral region, down to about 15 m (but varying with water quality from 5 to 20 m) is dominated by algae, particularly the large kelp *Laminaria hyperborea*. The deeper circalittoral has mainly animal-dominated communities, where a variety of sponges and coelenterates are often prominent. Particularly diverse communities flourish in areas of strong tidal currents which keep the substrate and organisms free of fine sediment, and provide an abundant supply of suspended food for filter feeders. Detailed descriptions of hard substrate communities are available (Hiscock and Mitchell, 1980; Connor et al., 1997b; Hartnoll, 1998; Birkett et al., 1998).

The major part of the sea bed of the Irish Sea is covered by sediments and, as described earlier, the grade of sediment is determined by a combination of depth and current speed. The composition of the benthic communities is very closely linked to sediment type as Jones (1951) has demonstrated for the area around the Isle of Man. Because of this, and because the sediment types are generally distributed over large areas (Fig. 2a), the main macrobenthic communities are similarly distributed. Benthic communities are generally characterised on the basis of prominent species (see Jones, 1950; Thorson, 1957; Glemarec, 1973), and the general distribution of such communities is shown in Fig. 2b. The general agreement with the distribution of sediment types is clear.

A general description of these offshore communities in the Irish Sea is provided in Mackie (1990). He points out that these communities appear very similar to those occurring in similar environments in other areas around Britain such as the North Sea and Celtic Sea, but at the same time emphasises our lack of detailed information. Mackie also lists the considerable number of studies on which this synthesis of bottom community composition and distribution is based. The major recent study is the BIOMÔR 1 study of the sediments and benthic communities in the eastern past of the southern Irish Sea (Mackie et al., 1995). This examines rigorously the association of species groups with sediment types, and questions the validity of the traditional 'fixed' scheme used in Fig. 2b.

A number of threats to the benthos have been identified, which will be elaborated in later sections. Some are localised, and these are indicated in Fig. 8. They include the dumping of sewage sludge or of dredging spoil, the extraction of aggregate, the extraction of oil and gas, and the release of radioactive effluent. Other threats are more widespread. These include the release of sewage effluent, the effects of fishing (on both target and non-target species), and the effects of shipping operations.

OFFSHORE SYSTEMS

Plankton and Nekton

Here we will consider the planktonic systems, and the pelagic organisms which depend upon them. The general consensus from numerous recent studies is that primary production in the Irish Sea is lower than in most other U.K. coastal waters (see discussion and references in Huckbody et al., 1992 and Taylor and Parker, 1993). The annual phytoplankton bloom commences later and declines earlier than in other U.K. areas, and a clear autumn bloom is rarely observed (Savidge and Kain, 1990). The length of the phytoplankton production season ranges from six months in coastal regions north of Dublin to as little as two months in the North Channel (Gowen et al., 1995). Possible factors responsible for this late blooming are strong tidal mixing and low solar radiation in spring (Gowen et al., 1995). There are major year-to-year variations in both the timing and the magnitude of the spring bloom (Slinn and Eastham, 1984). The spring bloom is dominated by diatoms. However, in terms of total annual primary production the larger phytoplankton (cells >20 μm) contribute less than the nanophytoplankton (cells 5–20 μm) (Graziano, 1988). Higher levels of primary production occur in some of the inshore areas adjacent to regions of high nutrient input, such as in Liverpool Bay (Foster et al., 1982).

An important factor in relation to primary production is the stratification of the water masses and the distribution of fronts. In the summer there are well developed stratified areas in the western Irish Sea, separated from the mixed

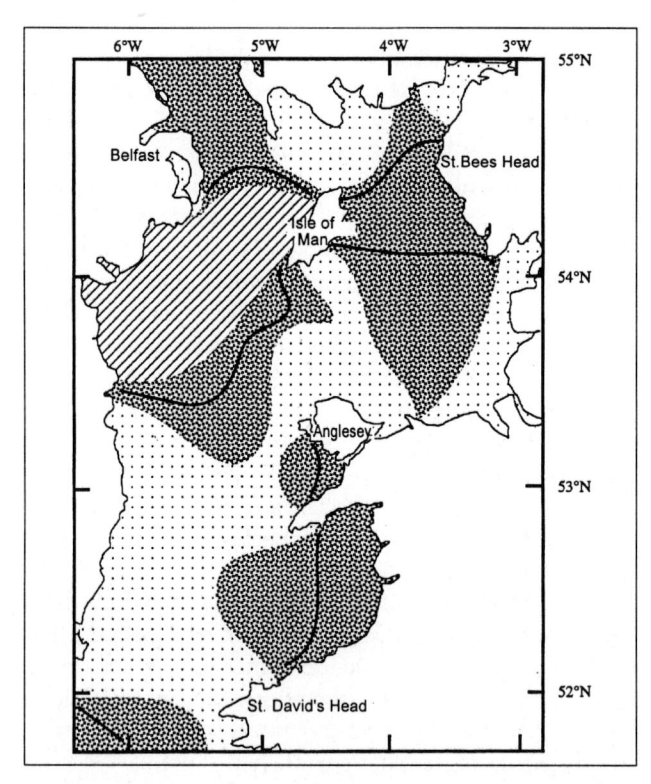

Fig. 7. Summer hydrographic conditions in the Irish Sea to show areas of mixed (fine dots), transitional (heavy dots) and stratified (diagonal lines) waters (based on Pingree and Griffiths, 1978). Approximate position of fronts indicated by bold lines (after Brand and Wilson, 1996).

areas by well defined fronts (Fig. 7). These frontal areas are regions of increased phytoplankton concentration (Savidge, 1976), though not to the extent observed in some other European seas. Nevertheless they are important in the distribution of some species (see below). However, there do not seem to be consistent differences in primary productivity between the stratified and mixed areas (Savidge, 1976; Richardson et al., 1985), though regional differences in production do occur. Production ranges from 155 g C m^{-2} yr in coastal areas north of Dublin to 66 g C m^{-2} yr in areas to the south east of the Isle of Man (Gowen et al., 1995).

Apart from the effect upon primary productivity, a major concern in relation to phytoplankton and the status of the Irish Sea is the occurrence of algal blooms. As discussed above, nutrient levels have roughly doubled over the past forty years, providing a clear index of eutrophication (Allen et al., 1998). There are now areas of the Irish Sea which are formally classified as hypernutrified (CSTT, 1997). One of the most obvious, and potentially damaging, effects of eutrophication is an increased incidence of phytoplankton blooms, which may be classified as 'nuisance' or 'toxic'. Nuisance algae include *Phaeocystis pouchettii*, *Noctiluca scintillans* and *Gyrodinium aureolum*, all of which are responsible for producing 'red tides' in the Irish Sea. These algae are not toxic, but discolour the water, produce odours, and can be aesthetically unpleasant in bathing areas. If present in high density, especially in enclosed areas, they can cause deoxygenation and fish kills. Red tides have been known in inshore areas of the Irish Sea for many years (Jones and Haq, 1963), but their incidence is increasing, especially in the offshore regions.

Toxic algae present a greater potential problem, since they can pass through the food chain and produce gastrointestinal (DSP) or neurological (PSP) illness. Red tides of toxic species have not been recorded in the Irish Sea, but a variety of toxic species are present giving the potential for such to occur. These species include *Dinophysis acuta*, *D. acuminata*, *D. rotundata*, *Pseudonitzchia seriata* and *Alexandrium tamarense*. Generally concentrations have been low, but *Dinophysis* levels have exceeded the MAFF and EA guideline action levels, though shellfish samples have so far proved on analysis clear of its toxin (Shammon et al., 1998). In 1999 azaspira acid, a suspected phytoplankton toxin, was detected in shellfish from the northern Irish Sea. Also in that year a large area of the Scottish scallop fishery, extending close to the northern boundary of the Irish Sea, was closed on account of Amnesic Shellfish Toxin contamination. Clearly the situation in the Irish Sea will need careful monitoring, and must constitute a cause for concern.

Zooplankton populations are generally typical of those of the western seaboard of Britain (Williamson, 1952; Scrope-Howe and Jones, 1985), and like the phytoplankton show substantial variations in abundance from year to year. However, the levels of zooplankton in the Irish Sea are even lower than those of phytoplankton when compared to areas such as the North Sea (Savidge and Kain, 1990). The 'excess' phytoplankton perhaps passes directly to the benthic community—on the basis of shellfish landings benthic productivity appears to be high (see Fisheries section). The relative scarcity of zooplankton is a suggested cause of the low fish production in the Irish Sea, which on an area basis is only about 40% that in the North Sea (Brander and Dickson, 1984): it is postulated that fish recruitment is inhibited by the short and late plankton production cycle in the Irish Sea, which reduces the food supply for larval fish.

The only large pelagic species of commercial importance are herring (*Clupea harengus*) and squid (*Loligo forbesi*), considered below. However, a variety of other large pelagic species are found. The occurrence of cetaceans in the Irish Sea is reviewed by Northridge (1990), who lists fifteen species. The majority of these are occasional vagrants, but a few merit more consideration. The harbour porpoise, *Phocoena phocoena*, is probably the commonest cetacean in the area, and is often associated with fronts between the stratified and mixed areas (Jones, 1986). The larger common dolphin, *Delphinus delphis*, is probably the most often sighted due to its habit of congregating in large schools. One of the only two British resident populations of the bottlenose dolphin, *Tursiops truncatus*, occurs in Cardigan Bay. For this reason Cardigan Bay has been designated a candidate Special Area of Conservation for marine interest (see Protective Measures section) (Fig. 10). Two species of pinniped occur regularly and breed in the Irish Sea (Northridge, 1990). The common seal, *Phoca vitulina*, is centred along the coast of north eastern Ireland where its breeding sites occur. The more abundant grey seal, *Halichoerus grypus*, occurs generally around the Irish Sea, with breeding sites on the Welsh and Irish coasts, and on the Isle of Man. A threat to marine mammals in the Irish Sea as top predators is that they tend to accumulate toxins in their blubber—high levels of heavy metals and PCBs have been recorded (Baines, 1998).

A particularly interesting pelagic species in the Irish Sea is the basking shark, *Cetorhinus maximus*. This is the second largest fish species, growing to over 10 m length, and considerable numbers appear in the Irish Sea each summer. They are particularly found in the area to the west of the Isle of Man associated with the frontal systems between the stratified and mixed waters (Fig. 7)—it is assumed that they favour these areas because of the concentrations there of plankton on which they feed. Little is known of their biology, and they disappear each winter to unknown destinations. Populations of the species appear to be generally in decline worldwide, in part due to exploitation, and the local population may also be declining. Hence the Irish Sea population merits interest and conservation. The species received full protection in Isle of Man waters in 1990 (IOM Wildlife Act), and in British waters in 1998 (Wildlife and Countryside Act).

Fisheries

It is generally accepted that Irish Sea fish stocks are exploited at a fishing pressure above the optimum, and that only reductions in permitted quotas will effect a cure (Lockwood, 1990). The balance of the fishery has undergone various historical changes. The more striking changes in recent times include the decline of the herring fishery from a high in the seventies (though there had been other lean periods in the past), and the development of fisheries for scallops, queen scallops and *Nephrops* over the last forty years.

The main species fished in the Irish Sea, with details of landings by quantity and value, are listed in Table 1. The information is for the ICES Fisheries Statistical Area VIIa, taken from Wood (1998), and there are a number of interesting features. Of the fish, the herring is still the largest landing by quantity, but only eighth by value: it certainly no longer dominates the fishery as it did at times in the past. Within the demersal fishery the landings are spread over a wide range of species, with whiting having the largest landings, but cod contributing the greatest value to the fishery. Despite its relatively small quantity, the high price of soles lifts them to third place in value. However, the most striking feature of the table is the dominance of shellfish, which easily exceed the combined demersal and pelagic fish landings in both quantity and value. Of the shellfish,

Table 1

Fish landings in the Irish Sea (ICES Fisheries Statistical Area VIIa) in 1997, using data from Wood (1998). Only the more important species in each category are listed separately. Scientific names given where category is comprised of a single species.

Species		Quantity (tonnes)	Value (£ 000)
Demersal			
Whiting	*Merlangius merlangus*	3,183	1,376
Cod	*Gadus morhua*	2,451	2,766
Dogfish		1,369	965
Skates and rays		898	832
Plaice	*Pleuronectes platessa*	822	833
Haddock	*Melanogrammus aeglefinus*	768	605
Saithe	*Pollachius virens*	662	270
Monks or anglers	*Lophius piscatorius*	377	526
Conger eels	*Conger conger*	287	223
Soles	*Solea solea*	197	1,199
Pollack (Lythe)	*Pollachius pollachius*	171	137
Witches	*Glyptocephalus cynoglossus*	143	107
Total demersal		12,581	11,944
Pelagic			
Herring	*Clupea harengus*	5,341*	559*
Horse mackerel	*Trachurus trachurus*	588	70
Total pelagic		5,934	631
Shellfish			
Nephrops	*Nephrops norvegicus*	7,181	11,575
Mussels	*Mytilus edulis*	5,651	2,102
Queens	*Aequipecten opercularis*	5,414	1,788
Scallops	*Pecten maximus*	1,454	2,149
Whelks	*Buccinum undatum*	790	211
Cockles	*Cerastoderma edule*	738	434
Crabs	*Cancer pagurus*	401	326
Squid	*Loligo forbesi*	127	246
Lobsters	*Homarus gammarus*	75	629
Total shellfish		22,160	20,069

*Data for area VII (N) only in this case.

the scampi or Dublin Bay prawn (*Nephrops norvegicus*) is the most important, comprising over half of the shellfish value. The fact that some species are relatively low on the Irish Sea list (e.g. whelks, cockles, crabs and lobsters) tends to conceal the fact that they are often very important in a local context. These fisheries still operate at a largely artisanal level and contribute substantially to employment and the economy in certain coastal communities. Thus on the west coast of Wales the most valuable fishery is for lobsters, and in the Dee Estuary and Morecambe Bay for shrimps (*Crangon crangon*) and cockles (*Cerastoderma edule*). Details of the distribution of catches within the Irish Sea of the more important species are given in Hillis and Grainger (1990).

The relatively low fish production, and high benthic production, in the Irish Sea were mentioned earlier. The figures in Table 1 support these observations, and an

Table 2

The landings of demersal fish, pelagic fish and shellfish expressed as a percentage of total landings by quantity and by value. Values are given for the Irish Sea (ICES area VIIa), for the North Sea (ICES area IV), and for the United Kingdom as a whole.

	Demersal		Pelagic		Shellfish	
	Quantity	Value	Quantity	Value	Quantity	Value
Irish Sea	31%	37%	15%	2%	54%	61%
North Sea	68%	74%	16%	6%	16%	20%
U.K.	55%	61%	22%	6%	23%	33%

interesting comparison can be made with the North Sea (Table 2). Whilst the pelagic fish comprise similar proportions of the landing by quantity in both areas, in the North Sea demersal fish are far more important than shellfish, whereas the opposite is true in the Irish Sea. The North Sea is more representative of the partitioning of the landings in the United Kingdom overall.

All fisheries in the Irish Sea are operating intensively, and in most cases there is clear evidence that fishing pressure is above the optimum. For a number of species for which data have been collected there has been a decline in spawning stock biomass (SSB) over time (Wood, 1998). For cod the SSB declined from 16,000 t in 1968 to 8,000 t in 1996. Plaice SSB fell from 8,000 t in 1964 to 4,000 t in 1995, and sole SSB from 6,000 t in 1970 to 3,500 t in 1995. The most abundant species in the landings, whiting, fell from 17,000 t in 1980 to 11,000 t in 1996. For all of these species there were fluctuations in SSB over the periods covered, but the general declines were obvious. Figures for recruitment at age 0 or 1 are much more variable, and do not correlate well with SSB, but they also display a general decline. There is no reason to attribute these declines to environmental factors such as pollution or climatic change—fishing pressure is clearly the primary cause. The stock of sole in the Irish Sea is considered by the ICES/ACFM to be 'close to or outside safe biological limits'. However, regulation of fishing levels in the Irish Sea is a complex problem for several reasons. Firstly, the area includes territorial waters of the United Kingdom, Ireland and the Isle of Man. Secondly, there is the impact of EU legislation on the regulation of the fisheries. Thirdly, there is the mixed nature of most fisheries, so that regulatory measures controlling one-species-fishery will have impacts upon other species. This will certainly be so whilst the main regulatory instrument is Total Allowable Catch (TAC) under the EU Common Fishery Policy, which invariably results in extensive discard of some species once their TAC is exceeded. There will be no simple solutions.

This intensive fishing within the Irish Sea is bound to have environmental impacts: one fishery will have impacts upon other fisheries (Lockwood, 1990), and fisheries will have impacts upon the non-target species and the environment.

Impacts on fisheries include situations where one fishery may catch juveniles of other commercial species as a by-catch, causing high mortalities. Thus inshore areas which are important nursery sites for plaice and sole (e.g. Solway Firth, Morecambe Bay, North Wales Coast) are also sites of significant shrimp (*Crangon crangon*) fisheries. The small mesh shrimp nets catch many small flatfish, many of which die even if released. There is a similar effect of the *Nephrops* fishery on whiting and cod. The small meshed *Nephrops* trawls catch large numbers of juvenile whiting and cod which will all die as a result—for every tonne of *Nephrops* landed about half a tonne of whiting is discarded. A more complex relationship occurs between cod and *Nephrops*, where there are not only by-catch effects, but the fact that cod is a major predator of *Nephrops* (Brander and Bennett, 1986). The conclusion of multi-species modelling is that, to maximise the fishery profit, pressure should be reduced on *Nephrops* but increased on cod—clearly problems are complex.

Fishing also affects the environment (Hartnoll, 1990), which is not surprising since the area trawled or dredged annually in the Irish Sea is 2.5 times the sea bed area. It has long been assumed that this activity must have major effects on the benthos, but until recently the evidence has been largely anecdotal. However there is now increasing hard evidence of impacts in the Irish Sea, based upon both time series studies and experimental programmes (see Hill et al., 1999 for references and discussion). Observed changes include a decrease in fragile and epifaunal species, an increase in epifaunal scavenger/predator species, and an increase in the polychaete: mollusc ratio. In particular, the species-rich *Modiolus* community must be considered at risk. It is also suspected that intensive dredging will coarsen the grade of the sediment. In the intertidal zone cockle (*Cerastoderma edule*) harvesting—especially by mechanical means (tractors, hydraulic dredges)—disrupts the environment. The cockle is preyed on heavily by the oystercatcher (*Haematopus ostralegus*), which has resulted in culling programmes of the birds in Morecambe Bay.

POPULATIONS AFFECTING THE AREA

The two largest coastal cities on the Irish Sea are Dublin and Liverpool, both with populations exceeding 500,000. The only other town with a population exceeding 100,000 is Belfast (Fig. 1). Centred on the two major cities are the regions of greatest population density: Merseyside and the greater Dublin area. Within these areas population density exceeds 2,000 per km^2, but otherwise outside towns levels are generally below 300 per km^2. For the U.K. census statistics are not helpful for calculating populations within a given distance of the coast. However for Ireland it is possible to calculate that 52% of the national population live within 10 km of the coast.

Notable recent population trends include an increase of over 15% in the last ten years in the area to the north of

Dublin City, and a smaller increase in the coastal regions of Northern Ireland. In contrast, a decrease has occurred in the Merseyside area.

In addition to the resident coastal population, there are numerous resort towns around the Irish Sea which attract large numbers of visitors, particularly during the summer: populations may increase more than threefold during such periods. Major Irish Sea resorts include Aberystwyth, Llandudno and Blackpool. There is increasing coastal tourism away from established resorts, and such areas include Cardigan Bay, Colwyn Bay, the Isle of Man, the Cumbrian coast, Strangford and Carlingford Loughs, and the south-east coast of Ireland. Coastal tourism has major economic importance for the areas concerned: in Blackpool tourist expenditure is around £550 million annually. However, it can overload the infrastructure, accelerate coastal development, and increase pollution problems.

RURAL FACTORS

Much of the coastal area of the Irish Sea is fertile and subject to relatively intensive agriculture. The major impact of this activity on the marine environment is by way of material entering the river systems and thus the sea, and the main substances of concern are plant nutrients. These arise from agricultural waste, and from the run-off of fertilisers from the fields. On average, about 3% of the nitrogen and phosphorus applied as fertiliser runs off to the water system. Table 3 itemises the sources of nutrient input to the Irish Sea.

Riverine input is clearly the major source of nitrogen, and atmospheric input (which is often ignored) is the other major source. The contribution from domestic sewage is relatively small, so changes in treatment regime, including the introduction of tertiary treatment with nutrient scrubbing, would have limited impact on the Irish Sea nitrogen budget. Rivers are also a more important source of phosphorus than domestic sewage, but both are dwarfed by the industrial loading. However, this is predominantly from a single source where the discharge regime has recently changed (see earlier discussion).

It is unlikely that agricultural practices can be substantially changed in the short term, so the nutrient inputs

Table 3

Sources of nutrient input to the Irish Sea in tonnes per year (after ISSG, 1990b).

Source	Nitrogen	Phosphorus
Atmosphere	43,000	2,000
Rivers	76,400	6,120
Domestic	10,700	1,920
Industry	2,640	16,260*
Sludge	3,840	720

*Relates largely to a single industrial source. Recent changes are discussed in the section on Industrial Uses of the Coast.

from riverine sources to the Irish Sea are unlikely to change. There is also little that can be dome at present to moderate atmospheric input. Consequently the problems of potential eutrophication highlighted earlier can be expected to persist.

Rural activities otherwise have little direct impact on the coastal ecosystems. Artisanal usage of the shore is restricted, as discussed above.

COASTAL EROSION AND LANDFILL

In some areas of the Irish Sea coastal erosion is a problem. In the northwestern coast of the Isle of Man erosion rates of 1.2 m/yr have been observed, up to 1.5 m/yr between the Dee and Mersey estuaries, and up to 2 m/yr in sand dune areas north of Liverpool. To limit erosion coastal protection work has been undertaken in a number of areas, e.g., the southwestern Wirral Peninsula, and on the Irish coast between Larne and Belfast Lough. In industrialised areas or centres of population protective sea walls have been established in estuaries or along the coast. Thus much of the coastline of the Ribble and Mersey estuaries, Belfast Lough and around Dublin is now artificial. The main environmental effect is to reduce wetland areas which provide important feeding areas for birds.

There have been no major reclamation programmes in the area. A proposal to erect a barrage across the mouth of Morecambe Bay was mooted, but not pursued—there were major environmental objections. However there has been a piecemeal reclamation of land, particularly in estuarine areas, resulting in a loss of coastal habitat. Major sites of such reclamation are Morecambe Bay and the Dee Estuary, where these processes have been continuing over the past centuries. Details of the history of reclamation in the Dee Estuary are given in Taylor and Parker (1993).

EFFECTS FROM URBAN AND INDUSTRIAL ACTIVITIES

Artisanal and Non-industrial Use of the Coast

Artisanal uses of the coastline are limited in the Irish Sea. Some of the inshore fisheries are basically artisanal, being locally based and operating from small boats or from the shore. These include the gathering of algae and gastropods intertidally for food—both very limited. More extensive are the fisheries for crabs and lobsters, shrimp, and for cockles. Only the last of these has substantial environmental impacts, as discussed earlier. In some areas fixed nets are operated from the shore to catch migrating salmon (*Salmo salar*). Mariculture is very limited in the Irish Sea.

The main non-industrial use of the coastline has been for residential purposes, which has affected large areas in the past. However, many coastal areas are now under some form of protection (see Protective Measures section), and planning restrictions on near-shore development have everywhere become stringent.

Industrial Uses of the Coast

Industrial inputs to the Irish Sea from urban sources are discussed below. This section covers a small number of major industrial activities which are not in urban areas. Along the Cumbrian coast between Whitehaven and Maryport large quantities of blast furnace spoil have been dumped in the past covering the natural shoreline with a layer of solidified 'slagcrete', which now supports intertidal communities. There are however two important current activities.

Radioactive discharges are an emotive issue in the Irish Sea, which has several times been denigrated as 'the most radioactively contaminated sea in the world', a statement which must, however, have further qualification. In 1988 the total content of man-made radionuclides in the sediments and water of the Irish Sea was 20,000 TBq, whilst the natural content of the water alone was 30,000 Tbq (ISSG, 1990b). These radioactive discharges arise from various sources (Fig. 8). There are several smaller ones—nuclear power stations at Heysham, Trawsfyydd and Wylfa, and British Nuclear Fuels (BNFL) establishments at Capenhurst, Chapelcross and Springfields. However, the major source is the BNFL fuel reprocessing plant at Sellafield on the Cumbrian coast. For all of these sources discharge limits are set by Government Authorising Departments, and have been complied with. Different radionuclides behave and disperse differently after release. Some, such as caesium^{137}, are highly soluble and are widely dispersed in the Irish Sea—concentrations in water (Fig. 9) and organisms fall with distance from Sellafield. Others, such as plutonium and americium, tend to accumulate in muddy sediments of the Sellafield coastal zone, and remain more localised. Studies have been done of the dosage received by consumers of fish in the Sellafield area, and even the most at risk consumed less than 20% of the ICRP principal dose limit of 1 millisievert per year. Since a high in the mid-seventies, discharges from Sellafield have fallen consistently. However, whilst radioactive output declines, nitrogen output has increased: it was 1500 tonnes nitrogen in 1996, and is expected to rise, making Sellafield a significant contributor to the nutrient budget.

The other major industrial input of concern arises from phosphate rock processing by the Albright and Wilson Marchon works on the Cumbrian coast just to the north of St Bees Head (Fig. 1). This plant has been responsible for a major input of phosphorus to the Irish Sea, though changes in raw materials and in manufacturing methods have improved effluent quality in recent years. Thus input from the plant has fallen from 5,600 tonnes phosphorus per annum in 1991 to around 1,800 tonnes in 1997, though the latter value still represented 45% of phosphorus inputs from land to the north east Irish Sea. Elevated phosphate

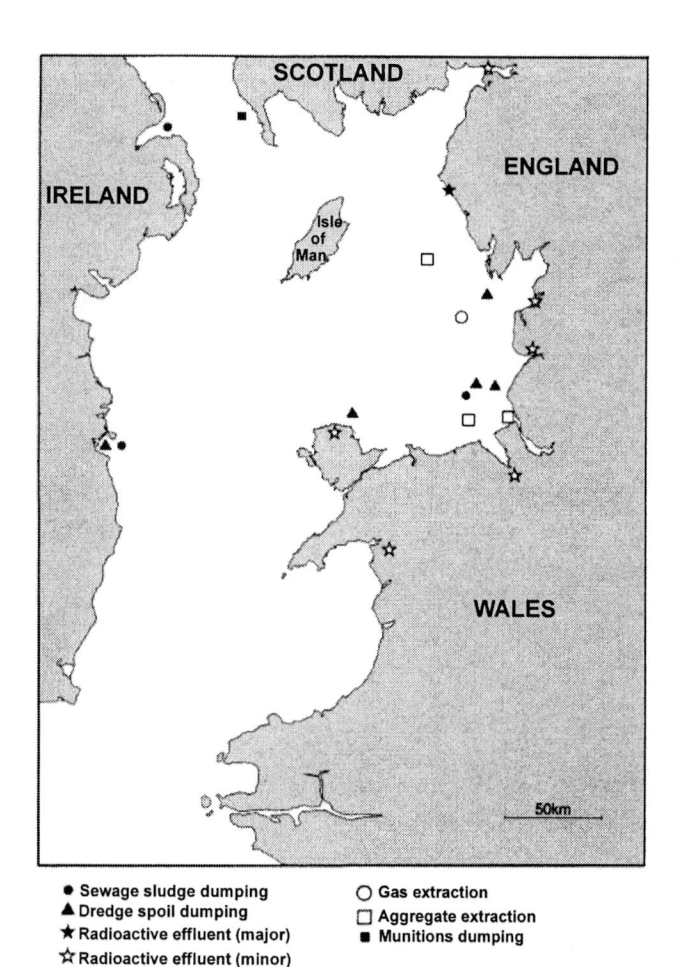

● Sewage sludge dumping ○ Gas extraction
▲ Dredge spoil dumping □ Aggregate extraction
★ Radioactive effluent (major) ■ Munitions dumping
☆ Radioactive effluent (minor)

Fig. 8. The major sites for industrial effluent release, waste disposal and resource exploitation in the Irish Sea.

levels occur in sea water around St Bees Head (Kennington et al., 1998). This plant will have contributed to the rise in nutrient levels in the Irish Sea over the last forty years. Although it is too early to be certain, there are indications of a fall of dissolved phosphorus levels in recent years (Fig. 6), paralleling the fall in output from the plant. Implications for the potential eutrophication of the area will need to be evaluated. However, as the system appears to be nitrate-limited at present (Kennington et al., 1999), beneficial effects may be minimal. The plant also releases cadmium and other heavy metals, so that whilst most heavy metals show their highest concentrations in Liverpool Bay, the highest levels of cadmium are found off the Cumbrian coast (ISSG, 1990b). Elevated levels have been recorded in shore biota, but have fallen recently as effluent quality improves.

Cities

Apart from the building up of the coastal areas, the major environmental impact of cities is the generation within a concentrated area of sewage and other wastes which are discharged to the sea. A primary concern is the biodegradable component of these discharges, and this is summarised in Table 4. The bulk of these discharges occur into the

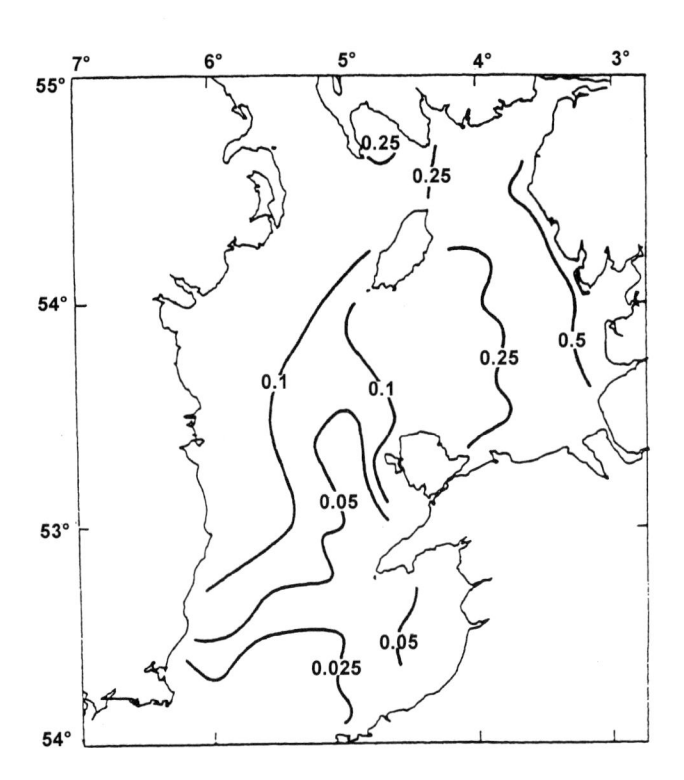

Fig. 9. The concentration of caesium^{137} (Bq/kg) in filtered seawater from the Irish Sea in 1987 (after ISSG, 1990b).

eastern side of the sea, especially into the Liverpool Bay area. The table also includes dumped material, but these are deposited further offshore (see next section) and do not have the same local environmental impact. Domestic sewage is the single largest source of BOD, and reflects the fact that there is still much sewage discharged untreated, or with only primary treatment, into the Irish Sea. Industrial and riverine inputs are also major contributors. Despite this level of input there is no significant deoxygenation of bottom waters in the Irish Sea, partly due to the high level of tidal mixing. This is so even in high input areas like Liverpool and Dublin Bays.

An indication of water quality as a result of sewage effluent discharge is the level of compliance with the EU Directive (76/160/EEC) on quality of bathing water based on number of total and faecal coliforms. Generally, compliance has been good around the Irish Sea—the main areas with failing beaches have been around Dublin, the Isle of Man, and the coasts of Lancashire and Cumbria: untreated discharges and short sea outfalls are the prime reasons. The more advanced sewage treatments being implemented in all these areas should improve the situation.

Other components of the discharges, particularly the non-biodegradable ones, are a cause for greater concern. Of these heavy metals and synthetic organic compounds present the major problems. Heavy metals are discharged mainly into the Liverpool Bay area from the heavily industrialised Merseyside region (see previous section for discussion of release of cadmium on the Cumbrian coast). This result in higher levels in both water column and sediments than in the rest of the Irish Sea, but these do not exceed the UK Environmental Quality Standards (ISSG, 1990b). Mercury has been seen as a particular risk, but measures to limit discharges have resulted in a reduction of 50% in sediment concentrations over recent years. Average levels in fish do not reach the EU EQS of 0.3 mg per kg fish flesh, but this can be exceeded in some large specimens. Alkyl lead has been responsible for bird deaths in the Mersey area, and organo-tin (TBT) has reduced populations of the whelk *Nucella lapillus* in many Irish Sea harbours (Gibbs et al., 1987), though this problem is now declining following the restrictions on the use of TBT in ships' antifouling.

Synthetic organic compounds enter the Irish Sea in river inflows and in dumped sludge. Organochlorines, probably from pesticide residues, are highest in Liverpool Bay but the contamination seems not to be of recent origin, and levels appear to be declining. PCBs are also causing concern, but there is little information on their distribution. High levels of PCBs have been found in marine mammal species in the Irish Sea (Morris et al., 1989), sufficient to place them 'at risk'. Levels of PCBs in some fish from Liverpool Bay are unacceptably high.

Shipping and Offshore Accidents and Impacts

There are seven major ports along the Irish Sea coast of England and Wales, which handle over 77 million tonnes of traffic and about 19,000 ship movements annually. Dublin and Dun Laoghaire handle some 9 million tonnes annually. The shipping consists partly of cargo vessels, but also of the various ferry routes connecting Great Britain, the Isle of Man and Ireland. There have been no major spills involving shipping in the Irish Sea. However, on many beaches a considerable proportion of the debris is related to shipping.

Aggregate extraction from offshore sites can only be carried out under license, and is currently restricted to four sites (Fig. 8) in the north-east Irish sea (Bellamy, 1997). Two are at the mouth of the Mersey, one in Liverpool Bay

Table 4

Discharges to the Irish Sea in tonnes BOD per year (after ISSG, 1990b).

	Republic of Ireland	Great Britain and Northern Ireland
Discharge		
Domestic sewage	22,500	98,900
Industrial discharge	11,200	35,000
Riverine inputs	25,200	36,200
Dumping		
Sewage sludge	8,800	12,500
Dredged material	8,400	93,800

('Hilbre Swash'), and one between the Isle of Man and Cumbria. These areas are licensed for 1.5 million tonnes per year, but this limit is not currently reached. The sand and gravel extracted are used in the construction industry, and also for beach nourishment; 200,000 m³ of sand from the Hilbre Swash was used for beach nourishment at Prestatyn on the North Wales coast. Studies have been done on the biological impact of extraction (Kenny, 1997). The general conclusion is that most extraction areas already experience high natural disturbance, and that physical and biological recovery is relatively quick (2–3 years).

Sludge generated by coastal sewage treatment plants may be dumped offshore. In the Irish Sea sludge dumping occurs in Liverpool Bay, in outer Belfast Lough, and outer Dublin Bay (Fig. 8)—adjacent to the three largest centres of population. However, this activity is declining. Dumping in Dublin Bay is planned to cease, and dumping at the Liverpool Bay and Belfast Lough sites fell from 4 to <2.5 million tonnes wet weight between 1988 and 1996. Material dredged from ports and navigation channels is also dumped offshore—in the Irish Sea this originates mainly from the Mersey and from Dublin.

A cause for concern has been the dumping of munitions and related materials after both world wars in the Beaufort's Dyke area of the North Channel (Figs. 1 and 8). Between 1945 and 1963 approximately 1 million tonnes of munitions were dumped. These materials have been washed ashore following disturbance during the laying of gas pipelines. Investigations have failed to detect contamination in sediments and commercial species from the area (SOAEFD, 1996). However, it was shown that materials were present over a wider area than suspected.

There have been a number of exploratory drilling programmes for oil and gas in the Irish Sea licensed under the UK Petroleum (Production) Act, but the majority have been unproductive and have had no long-term environmental impact. All exploratory drilling must be preceded by an environmental appraisal. The only commercial operation which has been developed is the Morecambe Bay gas field (Fig. 8), 35 km west of Blackpool. This consists of a central processing and accommodation complex, linked by pipelines to drilling platforms. To date there have been no environmental problems.

PROTECTIVE MEASURES

Some of the protective measures in place have already been outlined—such as those for fisheries, effluents, dumping and extraction. The fisheries regulations cannot be regarded as very successful at present—exploitation of most stocks is still too high. Regulations on discharges and other sea bed uses are generally operating successfully. In this section protected areas will be considered.

There are several international legislations affecting coastal areas extending seawards to high water, or in some cases to low water mark. These apply to all countries bordering the Irish Sea except the Isle of Man (which is not an EC member). Of relevance to the Irish Sea are RAMSAR sites (Convention of Wetlands of International Importance 1971) to protect important wetlands, of which a number have designated or candidate status around the Irish Sea: e.g., Morecambe Bay and the Dee Estuary. There are also Special Protected Areas (SPAs) designated under the EU Directive on Conservation of Wild Birds (EEC/79/409) for the protection of birds, especially migratory species. SPAs are often also RAMSAR sites, and are normally already SSSIs (see below), and protected by national legislation. There are a considerable number of SPAs around the Irish Sea.

Protected coastal areas under national and local legislation are numerous and diverse—too much so to attempt to itemise here. A detailed inventory, which includes the Irish Sea, is provided by Gubbay (1988; 1993). In the U.K. the main designation is as a Site of Special Scientific Interest (SSSI) under the Wildlife and Countryside Act, 1981. These are far too numerous to list; there are, for example, seventeen coastal SSSIs around Anglesey alone. Other measures of national protection include National Trust properties, Heritage Coast designation, and areas within National Parks. Fisher and Bolt (1990) list 32 types of protective designation for coastal areas in the various Irish Sea bordering countries.

The development of marine (i.e. subtidal) protected areas has been a much slower process. In the Irish Republic the Wildlife Act 1976 has the power to create marine reserves, but to date none have been designated in the Irish Sea. However, the Lambay Islands, just north of Dublin, are a Proposed Area of Scientific Interest. In Great Britain the Wildlife and Countryside Act 1981 includes legislation to designate Marine Nature Reserves (MNRs). Similar legislation has been passed in Northern Ireland (1985 Northern Ireland Amenity Lands Order) and the Isle of Man (Wildlife Act 1990). However, progress has been very slow, in part due to the extensive consultation processes required under these legislations with only two MNRs established so far in the Irish Sea (Fig. 10). In Great Britain Skomer Island, off the coast of Pembrokeshire and just within the Irish Sea under the ISSG definition, was declared an MNR in 1990 (see Hiscock, 1998 for details). In Northern Ireland Strangford Lough was declared an MNR in 1995. In Wales the Menai Strait, between Anglesey and the mainland, is proposed as an MNR and consultations are in progress.

A catalyst to the designation of marine reserves has been the EU 'Habitats Directive' (92/43/EEC), which requires the creation of marine Special Areas of Conservation (marine SACs). A number of areas in the Irish Sea have been declared candidate SACs (Fig. 10) because they are particularly rich in the following habitats (Annex I) or species (Annex II) specified within the Habitats Directive: BD – bottlenose dolphin *Tursiops truncatus*; E – estuaries; GS – grey seal *Halichoerus grypus*; I – large shallow inlets and bays; M – mud and sand flats not covered by sea water at low tide; R – reefs; S – sandbanks which are slightly covered

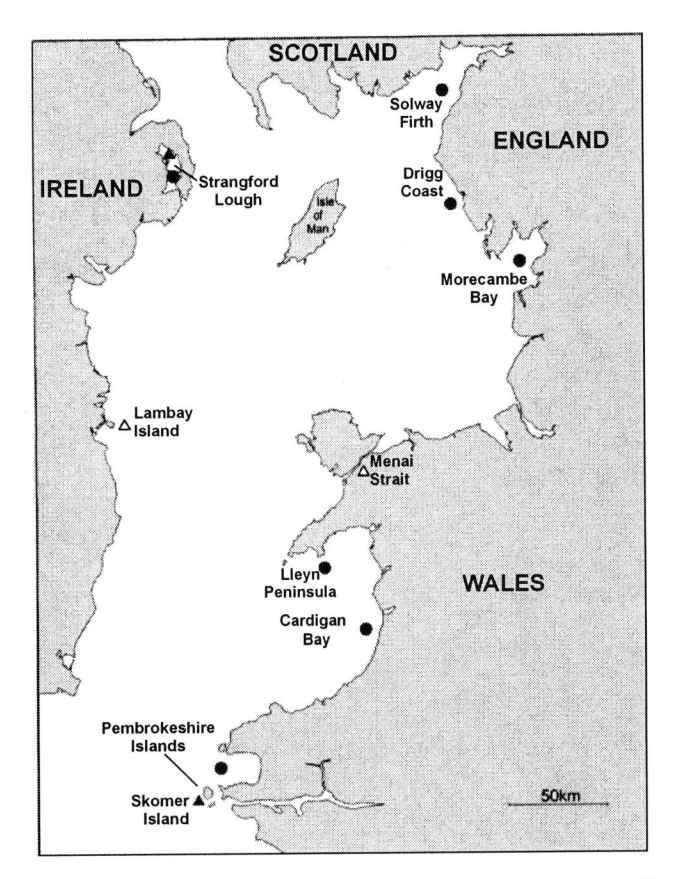

Fig. 10. The location of designated and proposed marine protected areas in the Irish Sea. Solid triangles: Designated Marine Nature Reserves; open triangles: proposed Marine Nature Reserves; solid circles: candidate marine Special Areas of Conservation.

by seawater at all times. The candidate sites and the features responsible for their designation are:
- Pembrokeshire Islands (E, I, R, GS)
- Cardigan Bay (BD)
- Lleyn Peninsula and the Sarnau (E, R)
- Morecambe Bay (M, I)
- Drigg Coast (E)
- Solway Firth (S, E, M)
- Strangford Lough (I)

Management plans for these candidate SACs are currently being developed.

REFERENCES

Allen, J.R., Slinn, D.J., Shammon, T.M., Hartnoll, R.G. and Hawkins, S.J. (1998) Evidence for eutrophication of the Irish Sea over four decades. *Limnology and Oceanography* **43**, 1970–1974.

Baines, M.E. (1998) Marine mammals of the Irish Sea: a review of species and conservation issues. *Irish Sea Forum Seminar Report* **15**, 49–52.

Bellamy, A.G. (1997) The marine aggregate dredging industry in the Irish Sea: the need for and economic importance of marine sand and gravel. *Irish Sea Forum Seminar Report* **17**, 7–13.

Birkett, D.A., Maggs, C.A., Dring, M.J., Boaden, P.J.S. and Seed, R. (1998) *Infralittoral Reef Biotopes with Kelp Species (Vol. VII). An overview of dynamics and sensitivity characteristics for conservation management of marine SACs*. Scottish Association for Marine Science (UK Marine SACs Project). 174 pp.

Brand, A.R. and Wilson, U.A.W. (1996) Seismic surveys and scallop fisheries. A report on the impact of a seismic survey on the 1994 Isle of Man queen scallop fishery. Port Erin Marine Laboratory, Isle of Man.

Brander, K.M. and Bennett, D.B. (1986) Interactions between Norway lobster (*Nephrops norvegicus*) and cod (*Gadus morhua*) and their fisheries in the Irish Sea. In North Pacific workshop on stock assessment and management of invertebrates, G.S. Jamieson and N. Bourne (eds). *Can. spec. Publ. Fish. Aquatic Science* **92**, 269–281.

Brander, K.M. and Dickson, R.R. (1984) An investigation of the low level of fish production in the Irish Sea. *Rapp. P.-v. Réun. Cons. Int. Explor. Mer* **183**, 234–242.

CSST (Comprehensive Studies Task Team) (1997) Comprehensive studies for the purposes of articles 6 and 8.5 of DIR91/271 EEC. The Urban Waste Water Treatment Directive. The Department of the Environment for Northern Ireland, The Environment Agency, the Scottish Environment Protection Agency and the Water Services Association.

Connor, D.W., Brazier, D.P., Hill, T.O. and Northen, K.O. (1997a) Marine Nature Conservation review: marine biotope classification for Britain and Ireland. Volume 1. Littoral biotopes. Version 97.06. *JNCC Report*, No. **229**, 1–362.

Connor, D.W., Dalkin, M.J., Hill, T.O., Holt, R.H.F. and Sanderson, W.G. (1997b) Marine Nature Conservation review: marine biotope classification for Britain and Ireland. Volume 2. Sublittoral biotopes. Version 97.06. *JNCC Report*, No. **230**, 1–448.

Crisp, D.J. (1964) The effects of the severe winter of 1962–63 on marine life in Great Britain. *Journal of Animal Ecology* **33**, 165–210.

Dickey-Collas, M., Brown, J., Fernand, L., Hill, A.E., Horsburgh, K.J. and Garvine, R.W. (1997) Does the western Irish Sea influence the distribution of pelagic juvenile fish? *Journal of Fish Biology* **51** (suppl. A), 206–229.

Dickson, R.R. and Boelens, R.G.V. (1988) The status of current knowledge on anthropogenic influences in the Irish Sea. *ICES, Copenhagen. Coop. Res. Rep.* **155**, 1–88.

Fisher, E. and Bolt, S. (1990) Marine legislation. In: Irish Sea Study Group. *The Irish Sea: An Environmental Review. Part 1: Nature Conservation*. Liverpool University Press. pp 335–352.

Foster, P., Voltalina, D. and Beardall, J. (1982) A seasonal study of the distribution of surface state variables in Liverpool Bay. IV. The spring bloom. *Journal of Experimental Marine Biology and Ecology* **62**, 93–115.

Gibbs, P.E., Bryan, G.W., Pascoe, P.L. and Burt, G.R. (1987) The use of the dogwhelk, Nucella lapillus, as an indicator of tributyltin (TBT) contamination. *Journal of the Marine Biology Association of the U.K.* **67**, 507–523.

Glemarec, M. (1973) The benthic communities of the European North Atlantic continental shelf. *Oceanography and Marine Biology* **11**, 263–289.

Gowen, R.J., Stewart, B.M., Mills, D.K. and Elliott, P. (1995) Regional differences in stratification and its effect on phytoplankton production and biomass in the northwestern Irish Sea. *Journal of Plankton Research* **17**, 753–769.

Graziano, C. (1988) Some observations on the plankton of the north Irish Sea. Ph.D. thesis, University of Liverpool. 121 pp.

Gubbay, S. (1988) *A Coastal Directory for Marine Nature Conservation*. Marine Conservation Society, Ross-on-Wye. 319 pp.

Gubbay, S. (1993) *Marine Protected Areas in European Waters. The British Isles*. Marine Conservation Society, Ross-on-Wye. 160 pp.

Hartnoll, R.G. (1990) Environmental effects of harvesting. In: Irish Sea Study Group. *The Irish Sea: An Environmental Review. Part 3: Exploitable Living Resources*. Liverpool University Press. pp 139–154.

Hartnoll, R.G. (1998) *Circalittoral Faunal Turf Biotopes (Vol. VIII). An overview of dynamics and sensitivity characteristics for conservation *management of marine SACs.* Scottish Association for Marine Science (UK Marine SACs Project). 109 pp.

Hill, A.E., Brown, J. and Fernand, L. (1996) The western Irish sea gyre: a retention system for Norway lobster. *Oceanologica Acta* **19**, 357–368.

Hill, A.S., Veale, L.O., Pennington, D., Whyte, S.G., Brand, A.R. and Hartnoll, R.G. (1999) Changes in Irish Sea benthos: possible effects of forty years of dredging. *Estuarine and Coastal Shelf Science* **48**, 739–750.

Hill, S., Burrows, M.T. and Hawkins, S.J. (1998) *Intertidal Reef Biotopes (Vol. VI). An overview of dynamics and sensitivity characteristics for conservation management of marine SACs.* Scottish Association for Marine Science (UK Marine SACs Project). 84 pp.

Hillis, J.P. and Grainger, R.J.R. (1990) The species exploited. In: Irish Sea Study Group. *The Irish Sea: An Environmental Review. Part 3: Exploitable living resources.* Liverpool University Press. pp 83–125.

Hiscock, K. (ed). (1998) *Marine Nature Conservation Review. Benthic Marine Ecosystems of Great Britain and the North-east Atlantic.* Peterborough, Joint Nature Conservation Committee. 404 pp.

Hiscock, K. and Mitchell, R. (1980) The description and classification of sublittoral epibenthic ecosystems. In *The Shore Environment: 2 Ecosystems*, J.H. Price, D.E.G. Irvine and W.F. Farnham (eds.). Academic Press, London, pp. 323–370.

Huckbody, A.J., Taylor, P.M., Hobbs, G. and Elliott, R. (eds). (1992) *Caernarfon and Cardigan Bays. An Environmental Appraisal.* Hamilton Oil Company, London. 71 pp.

ISSG (Irish Sea Study Group) (1990a) *The Irish Sea—An Environmental Review. Part 1. Nature Conservation.* Liverpool University Press. 403 pp.

ISSG (Irish Sea Study Group) (1990b) *The Irish Sea—An Environmental Review. Part 2. Waste Inputs and Pollution.* Liverpool University Press. 165 pp.

Jones, N.S. (1950) Marine bottom communities. *Biology Review* **25**, 283–313.

Jones, N.S. (1951) The bottom fauna off the south of the Isle of Man. *Journal of Animal Ecology* **20**, 132–144.

Jones, P.G.W. and Haq, S.M. (1963) The distribution of Phaeocystis in the eastern Irish Sea. *J. Conseil* **28**, 8–20.

Jones, P.H. (1986) Cetaceans seen in the Irish Sea and approaches, late summer 1983. *Nature in Wales (New Series)* **3**, 62–64.

Kennington, K., Allen, J.R., Shammon, T.M., Hartnoll, R.G., Wither, A. and P. Jones. (1998) The distribution of phytoplankton and nutrients in the North East Irish Sea during 1997. Environment Agency R & D Technical Report E55, pp 1–38, 111 figs.

Kennington, K., Allen, J.R., Wither, A., Shammon, T.M. and Hartnoll, R.G. (1999) Phytoplankton and nutrient dynamics in the northeast Irish Sea. In Biological, Physical and Geochemical Features of Enclosed and Semi-enclosed Marine Systems, E.M. Blomqvist, E. Bonsdorff and K. Essink (eds). *Hydrobiologia* **393**, 57–67.

Kenny, A.J. (1997) The biology of marine aggregates and the impacts of commercial dredging. *Irish Sea Forum Seminar Report* **17**, 31–40.

Lewis, J.R. (1964) *The Ecology of Rocky Shores.* English Universities Press, London. 323 pp.

Lockwood, S.J. (1990a) Factors influencing stocks. In: Irish Sea Study Group. *The Irish Sea: An Environmental Review. Part 3: Exploitable living resources.* Liverpool University Press. pp. 45–65.

Lockwood, S.J. (1990b) Fisheries interactions. In: Irish Sea Study Group. *The Irish Sea: An Environmental Review. Part 3: Exploitable living resources.* Liverpool University Press. pp. 126–137.

Mackie, A.S.Y. (1990) 3. Offshore benthic communities of the Irish Sea. In The Irish Sea—an Environmental Review. *Irish Sea Study Group Report Part 1. Nature Conservation.* Liverpool University Press. pp. 169–218.

Mackie, A.S.Y., Oliver, P.G. and Rees, I.S. (1995) Benthic biodiversity in the Southern Irish Sea. Studies in Marine Biodiversity and Systematics from the National Museum of Wales. *BIOMÔR Reports* **1**, 1–263.

Morris, R.J., Law, R.J., Allchin, C.R., Kelly, C.A. and Fileman, C.F. (1989) Metals and organochlorines in dolphins and porpoises of Cardigan Bay. *Marine Pollution Bulletin* **20**, 512–523.

Northridge, S. (1990) Mammals in the Irish Sea. In: Irish Sea Study Group. *The Irish Sea: An Environmental Review. Part 1: Nature Conservation.* Liverpool University press. pp. 325–333.

Pingree, R.D. and Griffiths, K.D. (1978) Tidal fronts on the shelf seas around the British Isles. *Journal of Geophysical Research* **83**, 4615–4622.

Richardson, K., Lavin-Peregrina, M.F., Mitchelson, E.G. and Simpson, J.H. (1985) Seasonal distribution of chlorophyll a in relation to physical structure in the western Irish Sea. *Oceanologica Acta* **8**, 77–86.

Savidge, G. (1976) A preliminary study of the distribution of chlorophyll a in the vicinity of fronts in the Celtic and western Irish Seas. *Estuarine and Coastal Marine Science* **4**, 617–625.

Savidge, G. and Kain, J.M. (1990) Productivity of the Irish Sea. In *The Irish Sea: an Environmental Review. Part 3: Exploitable Living Resources*, T.A. Norton and A.J. Geffen (eds.). Liverpool University Press. pp. 9–43.

Scrope-Howe, S. and Jones, D.A. (1985) Biological studies in the vicinity of a shallow-sea tidal mixing front. V. Composition, abundance and distribution of zooplankton in the western Irish Sea, April 1980 to November 1981. *Philosophical Transactions of the Royal Society of London B* **310**, 501–519.

Shammon, T.M., Slinn, D.J. and R.G. Hartnoll. (1998) Seventh annual report. Long term studies of the Irish Sea: environmental monitoring and contamination. Report to the Department of Local Government and the Environment, Isle of Man. 34 pp.

Slinn, D.J. and Eastham, J.F. (1984) Routine hydrographic observations in the Irish Sea off Port Erin, during 1972–1981 inclusive. *Annals of Biology, Copenhagen* **38**, 42–44.

SOAEFD (1996) Surveys of the Beaufort's Dyke explosives disposal site November 1995–July 1996. Final Report. *Scottish Office Agriculture, Environment and Fisheries Department, Fisher. Res. Serv. Rept* **15/96**, 1–24.

Southward, A.J. and Crisp, D.J. (1956) Fluctuations in the distribution and abundance of intertidal barnacles. *Journal of the Marine Biology Association of the U.K.* **35**, 211–229.

Southward, A.J., Hawkins, S.J. and Burrows, M.T. (1995). 70 Years observations of changes in distribution and abundance of zooplankton and intertidal organisms in the western English Channel in relation to rising sea temperature. *Journal of Thermal Biology* **20**, 127–155.

Taylor, P.M. and Parker, J.G. (eds). (1993). *The Coast of North Wales and North West England. An Environmental Appraisal.* Hamilton Oil Company, London. 80 pp.

Thorson, G. (1957) Bottom communities (sublittoral or shallow shelf). In J.W. Hedgepeth (ed.), Treatise on Marine Ecology and Paleoecology. Volume 1. Ecology. *Mem. Geological Society of America* **67**, 461–534.

Williamson, D.I. (1952) Distribution of plankton in the Irish Sea in 1949 and 1950. *Proc. Transactions of the Liverpool Biological Society* **58**, 1–46.

Wood, I. (1998) *United Kingdom Sea Fishery Statistics 1997.* London, The Stationery Office. 148 pp.

THE AUTHOR

Richard G. Hartnoll
*Port Erin Marine Laboratory,
University of Liverpool,
Port Erin, Isle of Man IM9 6JA, British Isles*

Chapter 7

THE BALTIC SEA, ESPECIALLY SOUTHERN AND EASTERN REGIONS

Jerzy Falandysz, Anna Trzosinska, Piotr Szefer, Jan Warzocha and Bohdan Draganik

This chapter describes the ecological situation in the Baltic Proper, focusing on the southern and eastern coasts, and describes the main problems which have arisen there in the second half of the 20th century.

Eutrophication of the Baltic Proper is an on-going process which has been well documented for more than 40 years and which increased very rapidly during the 1970s. At present the nutrient concentrations in the photic zone are stable, though at a level sufficiently high to support intensive primary production. Extremely large loads of organic matter, nitrogen and phosphorus from land-based sources as well as large anoxic areas of the seafloor have a great impact on the cycling of nutrients. The most important effects of eutrophication are increasing primary production, decrease in water transparency and increase in organic matter sedimentation, and accelerating depletion of oxygen in the deep layers. Most typical marine species do not occur in the Baltic Proper, or else occur here at the edge of their ecological range. Thus even small changes in environmental conditions influence their spatial distribution. The main factors limiting the biodiversity and the immigration of marine organisms are low salinity and low water temperature.

There are four groups of natural immigrants in the Baltic flora and fauna: the Northwest European euryhaline marine and brackish-water species, the freshwater species, and (third and fourth groups) glacial relicts, which reached the Baltic either through ice-dammed lakes from Siberia or by a westerly marine route.

Because of its semi-enclosed nature, the Baltic Sea is vulnerable to adverse impacts of large-scale anthropogenic inputs. Significant quantities of chemical substances have been discharged into the Baltic Sea from industrial wastewaters, municipal sewage, agricultural run-off, atmospheric fallout, marine paints, dumping of wastes, and chemical warfare agents as well as sediments dredged from the port canals during the 20th century. Pollution of the Baltic Sea with persistent, toxic and bioaccumulative compounds has become a real threat to Baltic wildlife, notably some birds of prey and marine mammals. Following a ban on use of some organochlorine pesticides and alkyl mercury in the 1960s–70s and the restrictions on the use of some organochlorinated industrial chemicals, their concentrations now appear to be decreasing in all types of samples and at all locations, and there is also an improvement in the population status of higher predators in recent years.

Seas at The Millennium: An Environmental Evaluation (Edited by C. Sheppard)
© 2000 Elsevier Science Ltd. All rights reserved

Fig. 1. Map of the Baltic Sea (from Bergström and Carlsson, 1993; modified) showing its large drainage basin.

REGIONAL SETTING

The Baltic Sea is a young postglacial inland sea, with a drainage basin over four times its sea area (Fig. 1). The surrounding land is densely inhabited, urbanized, and used mainly for agricultural and industrial purposes. It is connected to the North Sea via the Kattegat and narrow inlets of the Belt Sea and Sound—the transition zone. The main problems which have arisen during the past decades in the Baltic Proper are attributed to the rapidly increasing loads of nutrients, organic matter and other harmful substances discharged from the land-based sources via rivers and the atmosphere. The impacts of these on the environmental conditions and biota have become evident, first and foremost in the coastal area.

Although the Baltic Sea is divided into natural basins by bottom topography and economic sectors, it is largely an integrated system, highly sensitive to events in the adjacent North Sea, the land and the atmosphere. Polluted areas are related partly to distance from the North Sea, local hydrologic conditions, the catchment area of adjacent rivers and the extent of conservation measures in the surrounding countries.

At the end of the 1960s, there was deep concern about the marked deterioration of water and biota in the Baltic Sea. This resulted in the Convention on the Protection of the Marine Environment of the Baltic Sea Area—Helsinki Convention—by all riparian countries. Considering the political situation, the Helsinki Convention of 1974 was a unique international agreement, though not until 1992 were the coastal zones included. All Baltic countries assisted in its implementation by performing joint monitoring programmes and, starting in 1981, a series of the assessments has been published by the Helsinki Commission (HELCOM). These summarise scientific results from the beginning of the century and reflect the present status of knowledge. In this chapter, these collective works are often referred to under the name HELCOM.

The Baltic Proper is the largest subdivision of the Baltic Sea. Its area of 211,069 km^2 is 51% of the whole sea and its volume of 13,045 km^3 is 60% of the total (Melvasalo et al., 1981; HELCOM, 1990 and 1996). It extends from the Darss Sill (18 m depth) in the transition zone to the entrances to the Gulfs of Bothnia, Finland and Riga. Several regions are distinguished based on bottom topography: the Arkona Basin, the Bornholm Basin and the Gotland Basin (Fig. 2), the last being subdivided into eastern and western parts. The Gdansk Basin is a southward extension of the Eastern Gotland Basin, and is frequently treated as a separate natural region because the Gdansk Deep (max. depth 118 m) acts as a sink for suspended matter carried by the Vistula River, the largest river draining into the Baltic Proper.

Continuous inflow of more saline water from the North Sea is hampered by shallow sills. Only major inflows, approximately 100 km^3 in volume, reach the Bornholm Basin. To renew the deep or intermediate water layers in the Gotland and Gdansk Basins, even greater volumes of dense oceanic water of high salinity, low temperature and high oxygen concentration are required. These proceed in cascades eastward and northward through the Slupsk Furrow which has a sill depth of approximately 60 m. Major inflows occur at irregular intervals, mostly in winter (Fig. 3). Their impact depends not only on the volume of water but

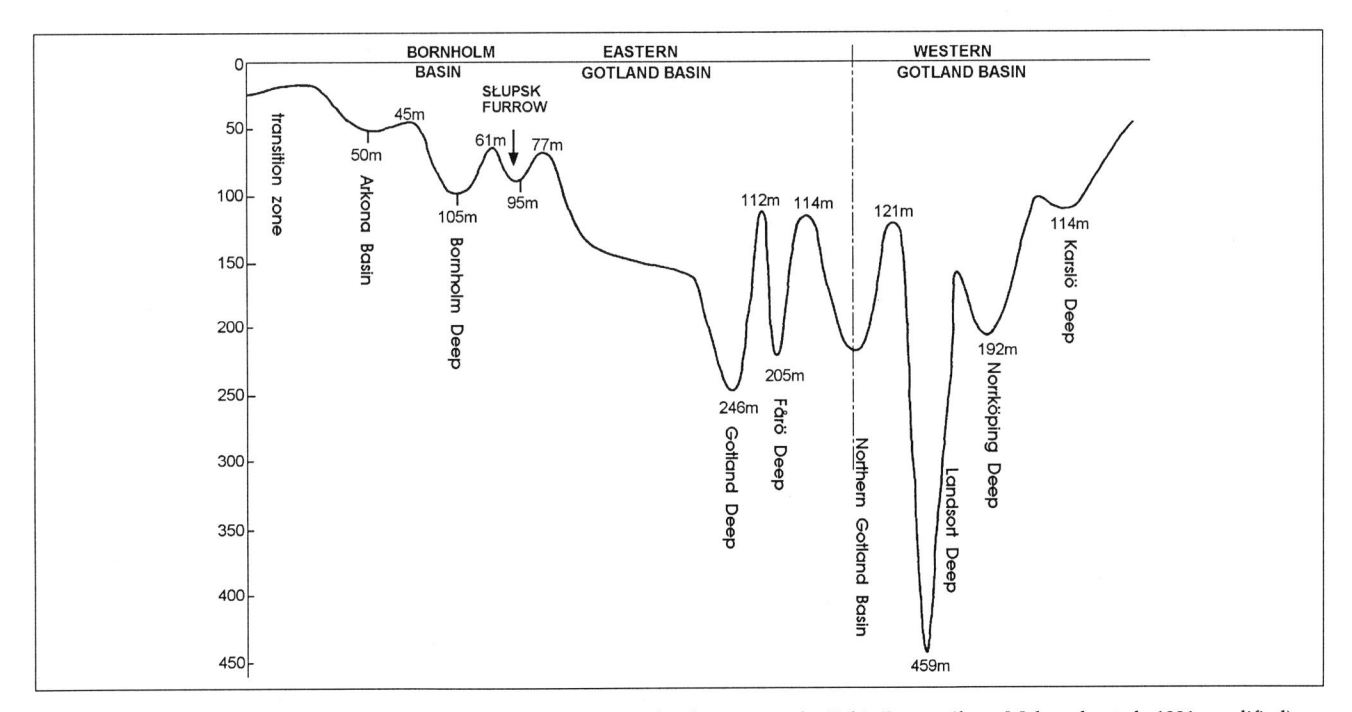

Fig. 2. Schematic display of the bottom profiles along the longitudinal section in the Baltic Proper (from Melvasalo et al., 1981; modified).

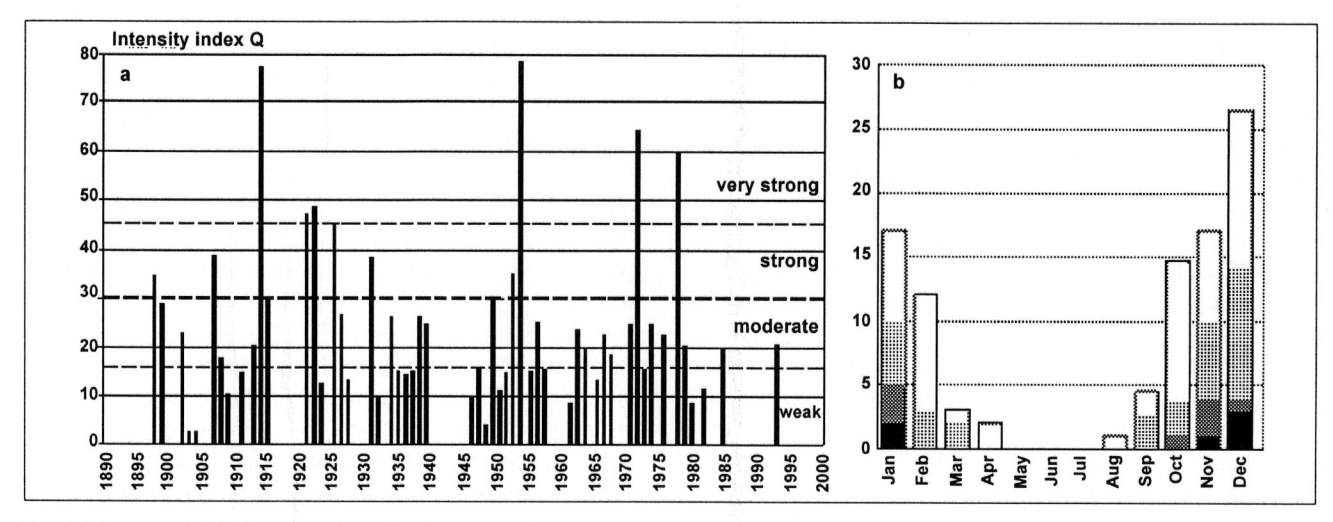

Fig. 3. Major oceanic inflows into the Baltic Sea during the present century (a) and their seasonal distribution (b). Intensity index Q takes into account the duration of the inflow and salinity of the inflowing water (from Matthäus and Francke, 1992 and HELCOM, 1996).

also on its salinity and the duration of the event. The causes of these inflowing waters are not well understood but meteorological and hydrological conditions are important.

There are pronounced horizontal salinity gradients in the surface layers due to extensive river run-off (Fig. 4). The salinity of surface water is spatially variable, and in the Baltic Proper, ranges from about 1 psu in estuarine areas up to 9 psu in the western region.

Cyberski (1995) reported statistically significant long-term trends in seasonal river outflows draining into the Baltic. Seasonal changes began in the 1920s and have accelerated since the 1970s, coinciding with the energy crisis and with attempts to improve water storage facilities. Seasonal variations in river outflow to the Baltic Sea as well as recent climatic changes may affect different elements of the water balance and of the distribution of species.

A horizontal salinity gradient exists in the deep waters of the Baltic Proper. Fonselius et al. (1984) studied 100-year series of salinity data, finding that salinity varied from over 14 psu to about 21 psu in the near-bottom layer of the Bornholm Deep, whereas in the southern and northern basins these variations were less, e.g. from over 11 to 14 psu in the Gotland Deep. Changes in surface water temperature are governed by the increased continental influence in the east as well as latitude (Melvasalo et al., 1981). In the Baltic Proper, the average winter sea surface temperature is around 2°C. The extent of ice cover is very variable (Majewski and Lauer, 1994). During August, mean sea surface temperature is 16–18°C in the southern part, about 16°C in the central part and 15–16°C in the northern part of the Baltic Proper. Between 1989–1993, the mild winters caused positive water temperature anomalies (HELCOM,

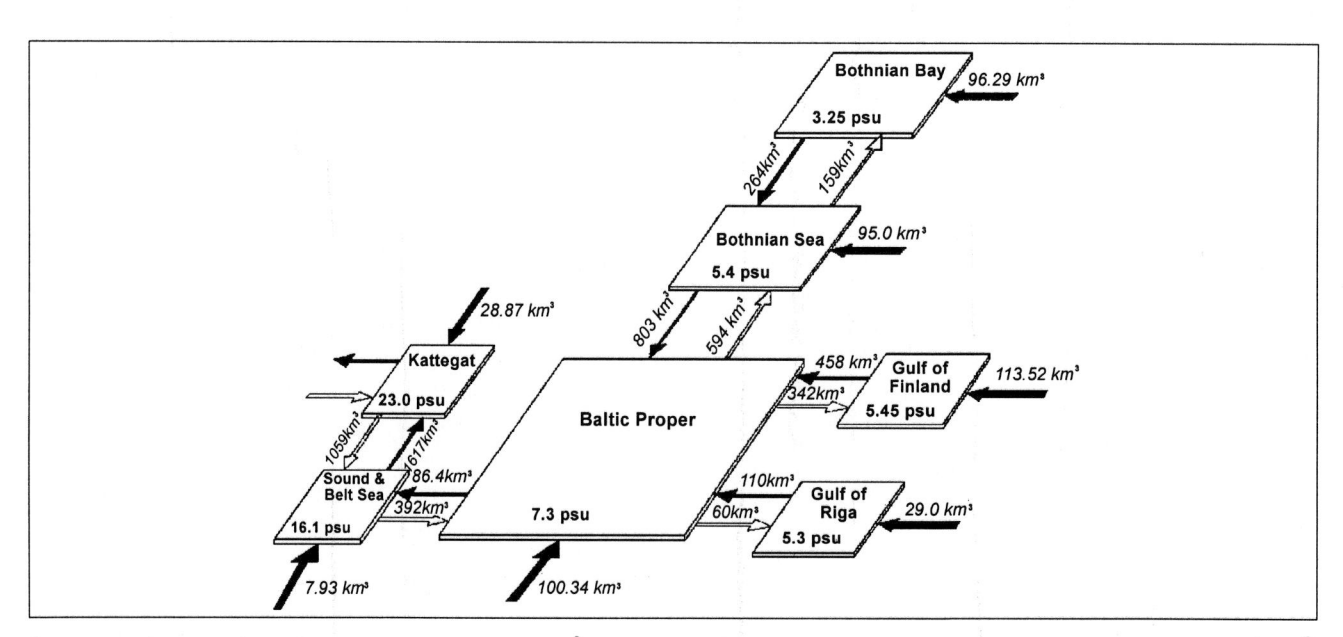

Fig. 4. Annual water exchange between the Baltic regions (km^3), mean long-term salinity of surface water (psu) and regional riverine inflow (km^3, thick arrows) (from HELCOM, 1986; Mikulski, 1991; and Cyberski, 1995; modified).

1996). Deep waters have more or less stable temperatures (5–8°C), which are influenced by the frequency and season of the major inflows.

The water budget and seasonal variations in water temperature result in marked vertical gradients in water density. In summer, warm surface water is separated from cold deeper water by a thermocline at approximately 20 m depth. The main barrier between the low salinity upper (isohaline) layers and higher salinity (heterohaline) deep layers occurs, on the average, at 40–70 m, depending on the region and the season. Major inflows of water from the North Sea significantly change the depth of the permanent halocline as well as the relative volumes of the isohaline and heterohaline layers.

The residence time of Baltic Sea water varies spatially but is estimated from the salinity distribution to be in the range 20–35 years. Important biogeochemical elements spend a much shorter time in the Baltic; Wulff et al. (1990) calculated that the average residence times for silicate, phosphorus and nitrogen compounds are 13, 11 and 5 years respectively.

CONDITION OF THE BALTIC PROPER

Eutrophication

Research and monitoring programmes have concentrated much effort on oxygen and nutrients. In the Baltic Proper, the occurrence of these components is closely linked, both in the photic zone and in the deep water layers.

Eutrophication of the Baltic Proper, defined as an increasing rate of input of organic matter into the ecosystem, is an on-going process that has been well documented for more than 40 years, increasing very rapidly during the 1970s. The concurrence of increasing loads of nitrogen, phosphorus and organic matter with high frequency of major oceanic inflows seems to be responsible for this change in trophic status.

Inputs from the Land and the Atmosphere

Most pollutants enter the Baltic Sea in river waters; the very few exceptions include organohalogenated compounds, butyltins, some radionuclides and, probably, lead. The Baltic Proper receives 21% of the total run-off into the Baltic Sea. Three main rivers: the Wisla (Vistula), the Nemunas (Neman) and the Odra (Oder) provide 72% of this, with mean long-term flow rates of 1081, 664 and 574 $m^3\ s^{-1}$, respectively. The fresh water from these three rivers and from several dozen smaller ones enters the sea along the southern coast and is incorporated into the surface layer characterized by a prevailing counter-clockwise current system. Some rivers such as the Neman and Oder (and a considerable amount of waste water) enter the sea through lagoons and coastal lakes which have retention times of several weeks to several months. These reservoirs serve as natural purification basins and are, as a consequence, seriously degraded. However, a substantial reduction of the pollution load to the Baltic takes place here, although this is usually disregarded in input compilations.

Comparison of the pollution loads entering the Baltic Sea in 1990 and 1995 shows an increase in nitrogen and organic matter by 15 and 10%, respectively, but a decrease in phosphorus by 18% (HELCOM, 1993a and 1998). However, because of considerable interannual variations in riverine run-off, these estimates are hardly comparable. More complete and representative data were collected during the 1995 trial.

Poland is the main contributor of organic matter, nitrogen and phosphorus into the Baltic Proper (Table 1). Almost the entire Polish territory (99.7%) is included in the drainage basin of the Baltic Proper, and about 90% of the Polish runoff is carried by the Vistula and Odra (Oder) rivers. The high phosphorus load in the Odra River (40% of the total) has decreased markedly since 1988 (Niemirycz et al., 1996).

Poland has a population of 38 million with a population density of 123 inhabitants/km^2 (i.e., approximately 45% of the population of the Baltic countries). About 62% of the

Table 1

Distribution of the drainage area and the riverine and direct point source loads of organic matter (given as BOD_7), total nitrogen and total phosphorus discharged into the Baltic Proper in 1995 (HELCOM 1998)

Region/ Country	Drainage area km^2	BOD_7 load		Nitrogen load		Phosphorus load	
		Total t/a	Area specific (kg/km^2)	Total t/a	Area specific (kg/km^2)	Total t/a	Area specific (kg/km^2)
Baltic Proper	537,328	492,860	917	316,223*	589*	17,799	33
Poland	331,196 (61.6%)	288,605 (58.5%)	871	214,747 (67.9%)	648	14,208 (79.8%)	43
Lithuania	98,890 (18.4%)	96,416 (19.6%)	975	36,824 (11.6%)	372	1,405 (7.9%)	14
Sweden	67,766 (12.6%)	34,863 (7.1%)	514	36,421 (11.5%)	537	1,011 (5.7%)	15
Other countries**	39,476 (7.4%)	72,975 (14.8%)	1849	28,234* (8.9%)	715*	1,174 (6.6%)	30

*Data not complete. **Denmark, Estonia, Germany, Latvia and Russia.

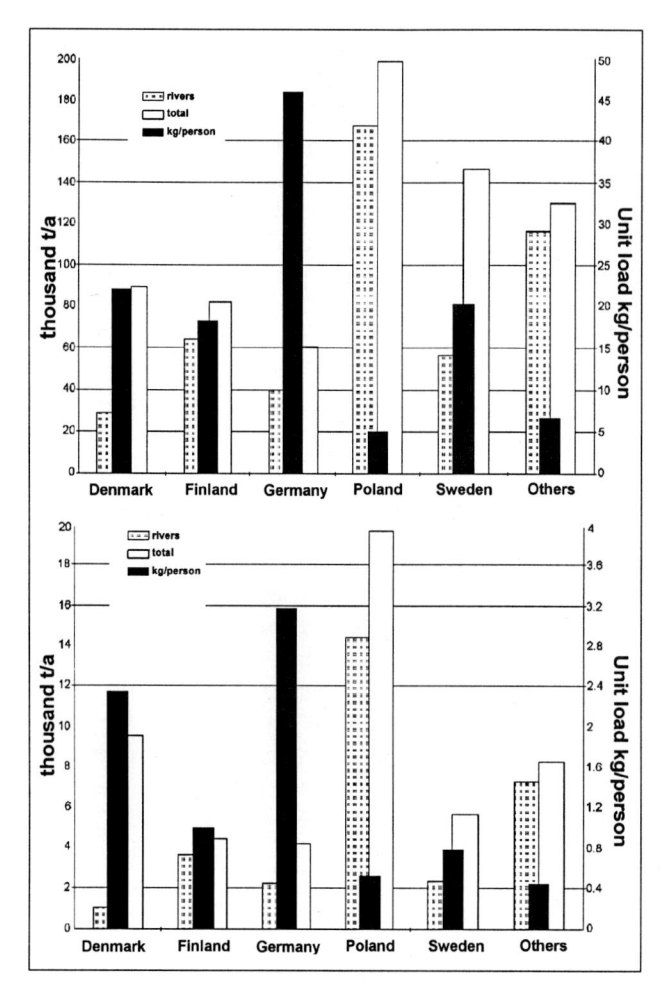

Fig. 5. Nitrogen (top) and the phosphorus (bottom) loads discharged into the Baltic Sea by countries in 1990 (from Niemirycz et al., 1996; modified).

Polish population is concentrated in urban areas. Over 60% of the land area is agricultural, and has a poor soil permeability, which has serious consequences for the pollution load (Fig. 5). In contrast with the total loads, the pollution loads per capita and per unit area arable land in Poland are the lowest among the Baltic countries.

Nearly 83% of the Lithuanian territory, the second greatest contributor of organic matter and nutrients, is in the Baltic Proper drainage area (Table 1). This territory is dominated by agriculture (54%) and forests (31%). The population is about 3.5 million, which corresponds to 57 inhabitants/km² (HELCOM, 1998).

Only 19% of Swedish territory forms part of the drainage area to the Baltic Proper. This territory is heavily forested (52%) and has a population of 4.1 million (48 inhabitants/km²). Agricultural land takes up 16% of the catchment area, wetlands and lakes 13%. The Norrström is the major river with mean long-term flow rate of 166 m³ s⁻¹.

In other countries in the watershed of the Baltic Proper, population density ranges from 9 inhabitants/km² in Estonia to 86 in Germany. The proportion of arable land is also very variable: 14% in Estonia to about 70% in Germany.

In the case of organic matter, rivers contribute 88% to the total load of the Baltic Proper (Table 1), while direct municipal and industrial sources contribute 8% and 4%, respectively. Similar ratios are observed for phosphorus, and total nitrogen.

According to HELCOM (1998), 80–95% of nitrogen in the rivers discharging into the Baltic Proper is of anthropogenic origin, mainly derived from diffuse sources. This reveals serious limitations in the implementation of the 1988 Ministerial Declaration which recommended a 50% overall reduction of pollution input. In fact, there were considerable decreases in organic matter and phosphorus loads in Poland in 1995 as compared with the 1988–1989 period (Rybinski et al., 1992) but the decrease of nitrogen was negligible. Nevertheless, environmental investment in Poland, which increased from 0.6% of GDP in 1989 to 1.3% in 1993, did result in a substantial reduction of nitrogen and phosphorus from point sources (Report, 1995). Input of airborne nitrogen has gradually increased during the 20th century, reaching its highest level in the 1980s. The mean deposition of total nitrogen in the entire Baltic was estimated as 324 000 tonnes annually in 1986–1990 (HELCOM, 1991). The value for the Baltic Proper was 100,000 t/a, almost a quarter of the total annual load of the riverine, direct point sources and airborne nitrogen. It follows from the combined emission/input models that only 65% of the airborne nitrogen originated from countries bordering the Baltic Sea.

Recent estimates show that there was a decrease in the nitrogen deposition into the Baltic Proper by about 20–30% and into its drainage area by about 10–25% from the mid-1980s to 1995 (HELCOM, 1997). The main reason was a decline of approximately 20% in the emission of nitrogen oxides and ammonia from the five largest contributors, viz. Denmark, Germany, Poland, Sweden and United Kingdom. Based on measurements carried out in the Polish coastal zone, the flux of the airborne assimilable nitrogen compounds decreased by 35% for the oxidized forms and by 50% for ammonia during 1987–1994 (IMGW, 1987–1998).

Oxygen

Seasonal fluctuations in temperature and primary productivity result in fluctuations in the oxygen concentration and the degree of oxygen saturation. Salinity variations exert a small effect on the solubility of oxygen in the isohaline layer of the Baltic Proper.

During the cold season, homogeneous isohaline waters contain some 8-9 cm³ dm⁻³ of dissolved oxygen, which corresponds to a saturation of around 100±4%. In spring, when water temperature and photosynthetic activity both rise rapidly, oxygen concentrations remain at almost the same level, but long-term means of water saturation values increase to 110–115% in the open sea and to about 120% in coastal zones (Majewski and Lauer, 1994). At the peak of the phytoplankton blooms in areas with abundant nutrient supply, supersaturation in the euphotic layer has recently

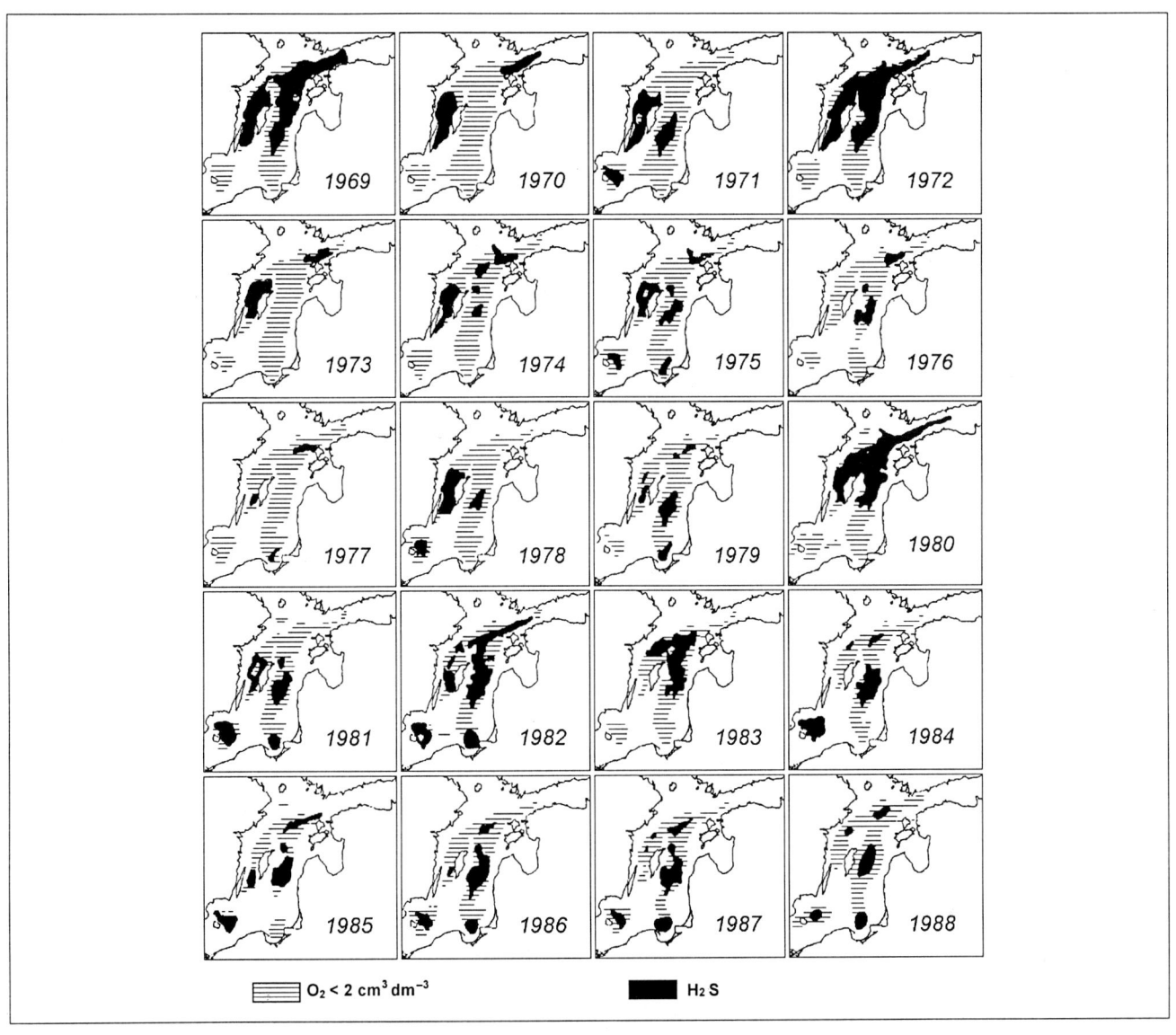

Fig. 6. Oxygen deficiency and hydrogen sulphide in the near-bottom water of the Baltic Proper (from Majewski and Lauer 1994; according to Andersin and Sandler, 1989; IMGW 1987-1989 and HELCOM 1990).

reached even 200% (IMGW, 1987–1998). In early summer, the isohaline waters become usually divided by the thermocline into the uppermost layer with an oxygen concentration of 6–7 cm^3 dm^{-3} and the remnant winter waters of permanently high oxygen concentrations.

High supersaturation of oxygen extended over the entire warm period can be used as a measure of on-going eutrophication in the Baltic Proper. For instance, in the surface water of the Gulf of Gdansk, near the Vistula River mouth, a significant long-term increase in oxygen saturation of 1.4% per year was observed from 1979 to 1993 (HELCOM, 1996).

Deep water below the pycnocline cannot equilibrate with the atmosphere. Oxygen conditions there are mainly dependent on the strength and frequency of oceanic inflows and on the intensity of bacterial degradation of organic matter. Other factors affecting the extent and persistence of anoxic zones in the Baltic Proper are temperature of the major inflows and magnitude of the density gradients in the water column (HELCOM, 1990, 1996). During stagnation periods, after the complete exhaustion of oxygen supplied by the inflows, hydrogen sulphide is formed (Fig. 6). Andersin and Sandler (1989) found that the bottom areas suitable for benthic fauna had not diminished significantly in the central Baltic during 1963–1987. However, a clear deterioration had taken place in the Arkona and Bornholm Basins because periods with favourable oxygen conditions have become shorter. Recent investigations have confirmed a serious deterioration of oxygen conditions in the southern Baltic with oxygen-deficient areas spreading towards the coasts, to the relatively shallow Hanö and Pomeranian Bays, and the Gdansk Basin as well as towards the transition zone (HELCOM, 1990, 1996; Lindahl, 1998; Report, 1998).

There was an exceptionally long period of stagnation from 1978 to 1993. In the first phase, anoxia spread widely over the bottom in the Baltic Proper. Then, during the second phase from 1985, as a consequence of reduced salinity in the deep waters and considerable sinking of the halocline, there was more efficient vertical mixing, and oxygen penetrated more deeply into intermediate layers at around 100 m depth, improving conditions at the sea floor. In 1993–1994, three successive oceanic inflows renewed the deep waters in the Baltic Proper. Hydrogen sulphide disappeared from all deeps, including the Gotland Deep where reducing conditions have been recorded since the 1950s (Fonselius, 1969). These salt water inflows have suppressed the mixing across the halocline and limited the downward transport of oxygen. In 1994–1996, the deep basins of the Baltic Proper became anoxic again, first of all in the southern regions. These frequently alternating oxygen conditions are attributed to the growing pool of organic matter, which has accumulated at the bottom as a result of the intensive phytoplankton blooms and the input of suspended matter from rivers.

Temporal and Spatial Variability in Nutrients

Seasonal and annual variations have been widely studied. Because of the differences in climate and bathymetry within the Baltic Sea, they are usually referred to particular regions and/or water bodies (Melvasalo et al., 1981; HELCOM, 1987, 1990, 1993b and 1996).

Seasonal fluctuations in nutrient concentrations of surface waters of the Bornholm and Gdansk Deeps and the southern part of the Gotland Basin, averaged over 20 years, show distinct temporal and spatial differences of accumulation in winter and uptake by autotrophic organisms in spring. There is a time lag of about 2–4 weeks in the accumulation and assimilation peaks, when moving from the Arkona Basin toward the northern Baltic. Another time lag of about 1–2 weeks occurs between the coastal zone and offshore areas.

In the 1990s, winter nutrient concentrations in the photic layer became much more consistent throughout the offshore area of the Baltic Proper. However, exceptions were found in the northern Baltic (the Landsort Deep had elevated phosphate and nitrate contents), as well as in the southern Baltic (the Gdansk Deep had elevated nitrate). Compared with the 1960s, an overall increase took place: 1.5–5 fold for nitrate and 2–3.5 fold for phosphate, depending on the region.

During vernal phytoplankton blooms, the pool of assimilable nitrogen and phosphorus was already consumed by June–July in all areas except estuaries. Nitrate depletion in warm water creates conditions promoting growth of blue-green algae that fix several hundred thousand tons of nitrogen in the Baltic Proper. From summer to December, nitrogen is a limiting nutrient in most parts, though in the Pomeranian Bay and the most inner part of the Gulf of Gdansk phosphorus has been a limiting nutrient at the beginning of summer since the 1980s (Trzosinska, 1992).

Silicate has never been the limiting factor in the Baltic Proper, though since the 1980s, almost complete silicate consumption has occasionally occurred following vast phytoplankton blooms.

The accumulation of nutrients starts in January–February. At the peak of nutrient concentration during winter, the mean molar ratio of nitrate to phosphate is approximately 7 in the Bornholm Deep and the Gotland Basin, but as high as 10 in the Gdansk Deep. When compared with the 1960s, this means an increase in the N/P ratio by a few percent for the offshore regions and by 50% for the Gdansk Basin. Before eutrophication accelerated in the 1970s, the N/P ratios in the trophic zone of the Baltic Proper were significantly lower than the Redfield ratio (16:1), which reflected the steady state relations between the environment and the biota in the ocean. In the Bornholm Basin during the vernal phytoplankton bloom, the uptake ratio was 15:1 (HELCOM, 1987) while for the spring/summer species in the southern Baltic it was somewhat lower at 14:1 (HELCOM, 1996). The mean uptake ratio of silicate versus phosphate was close to the Redfield ratio; it ranged from 13:1 in the Gotland Basin to 18:1 in the Bornholm Deep.

Oxygen saturation near the bottom reflects seasonality in respiration and remineralization, though these may be overwhelmed by occasional oceanic inflows, slow water advection, vertical density gradient weakening northwards and the long stagnation period. Phosphate fluctuations are connected with resuspension or remobilization, and silicate also accumulates in deep waters whenever oxygen declines. On the other hand, decreasing redox potential promotes the denitrification activity. Rönner (HELCOM, 1990) calculated that denitrification is responsible for the overall nitrogen loss of 470 000 tons annually.

The variety of the input and sink mechanisms, with temporal and spatial differences in their efficiency, does not permit any realistic mass balance calculations. Nevertheless, nutrient budgets calculated by Wulff et al. (1990) are very impressive and contain some management implications regarding the desired reduction in pollution loads (Fig. 8).

The first signs of increasing fertility were reported in the mid-1970s (Melvasalo et al., 1981; HELCOM, 1987). Long-term positive trends, calculated by means of approximately 20-year data series, were in most cases highly significant. In surface water of the Baltic Proper, mean annual accumulation rates of phosphate during the winter seasons ranged from 0.015 to 0.26 mmol/m^3 and of nitrate from 0.17 to 0.34 mmol/m^3, depending on the region. A rate 2–4 times higher was found for phosphate in deep water. In spite of anoxic conditions, nitrate accumulated in some water layers of the Baltic deep basins (Nehring, 1989).

In the 1980s, when external inputs were still high, the rate of eutrophication slowed down. The most characteristic feature of that period was the long-lasting stagnation in the Baltic deep waters, the longest ever observed during the 20th century. As a result of diminishing salinity and increas-

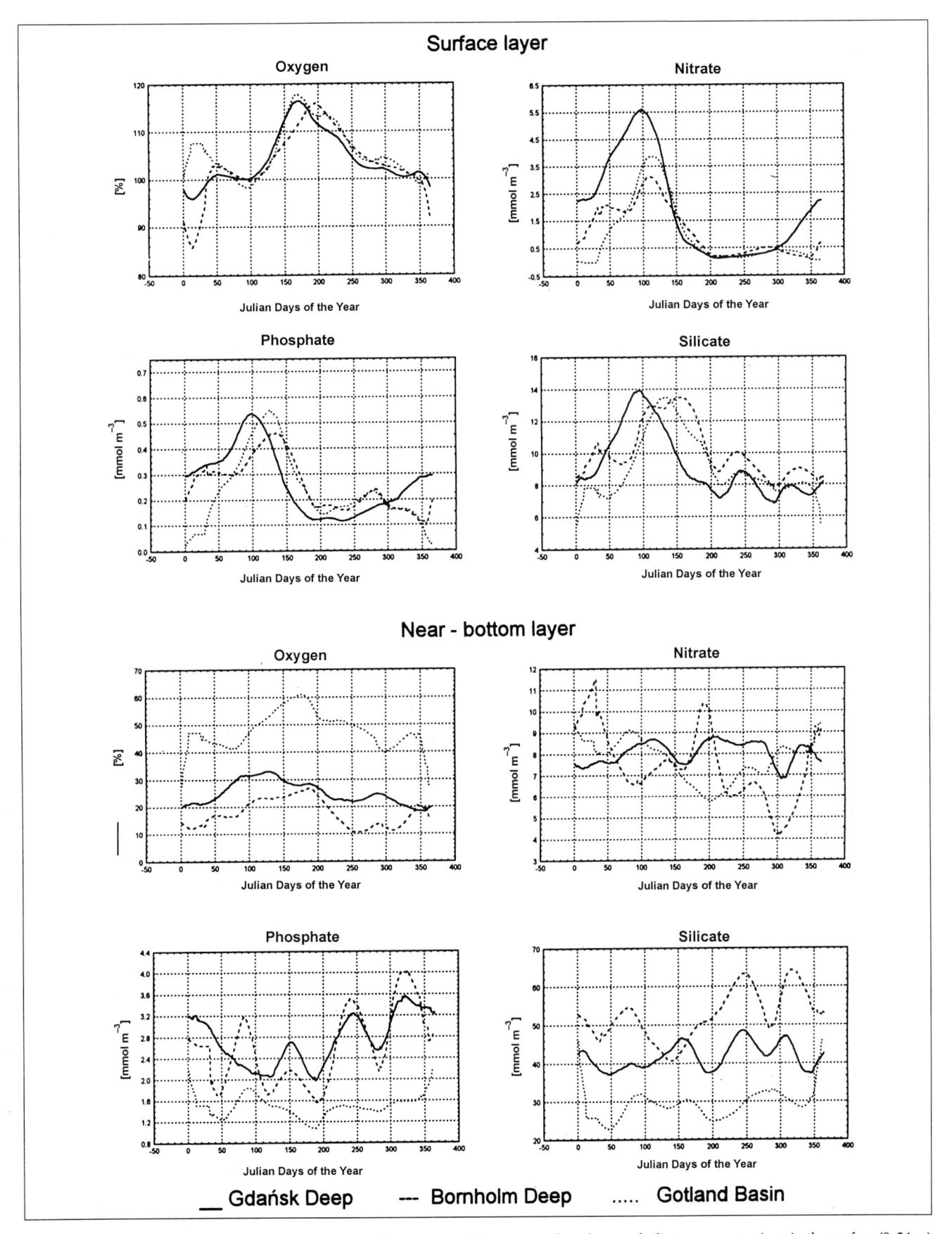

Fig. 7. Seasonal development of the water saturation with oxygen and the nitrate, phosphate and silicate concentrations in the surface (0–24 m) and near-bottom layers of the Bornholm and Gdansk Deeps and the southern part of the Gotland Basin, averaged over 1979–1998 (from Lysiak-Pastuszak, 1999).

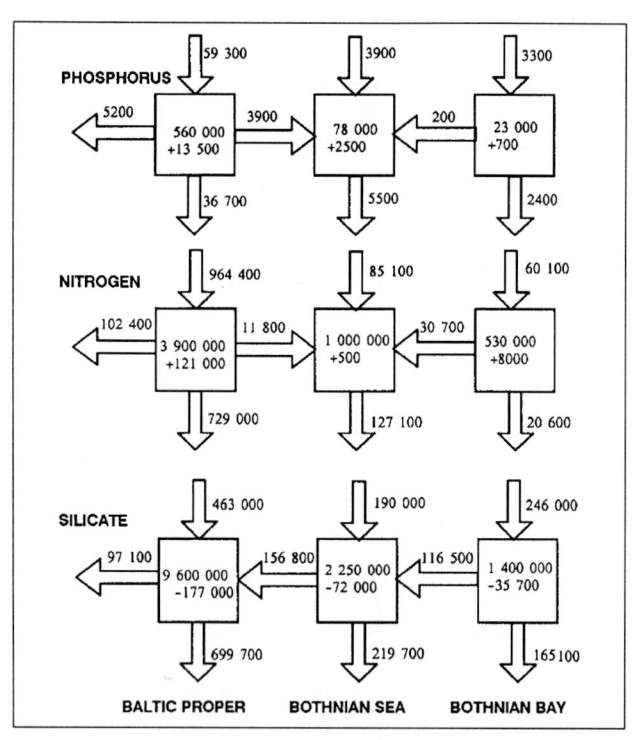

Fig. 8. Nutrient budgets calculated for the sub-basins of the Baltic Sea, 1971–1981. Storages (tonnes) and annual changes (tonnes/annually) are shown in each box. Net flows between basins and the atmosphere and internal sinks are shown as arrows between, above and below boxes (Wulff et al., 1990).

ing temperature of the deep waters, the weakening vertical density gradient supported downward transport of oxygen and upward transport of nutrients over a vast area of bottom at intermediate water depths (HELCOM, 1990). The long-term increase in the phosphate and nitrate concentrations continued, but was interrupted by periods with decreasing concentrations. Trzosinska (HELCOM, 1990) found almost cyclic behaviour in the phosphate and nitrate accumulation in the Gdansk Deep of three and six to seven years, probably caused by variations in the atmospheric circulation affecting both riverine run-off and oceanic inflows.

At present, concentrations of assimilable compounds of phosphorus, nitrogen and silicates in the photic zone of the Baltic Proper are at stable levels, though sufficiently high to support intensive primary production. During the last few decades, the phytoplankton primary production has almost doubled in some areas, with a resultant doubling of phytoplankton biomass and its subsequent sedimentation (Ambio, 1990).

Secondary effects of the eutrophication included drastic changes in water transparency in the Baltic Proper. Over 30 years, this decreased from about 8–10 m to about 5–7 m, most of the reduction being in the 1980s (HELCOM, 1993b). In estuarine areas, no direct change could be expected, but in the offshore areas of the Gdansk Basin, a decline in water transparency was demonstrated, and recent calculations over a 40-year period both confirmed this and revealed a steeper decline in the Bornholm Basin during the warm seasons.

This may adversely affect phytoplankton production and species succession. In the southern Baltic, average primary productivity, although increasing by 2–3% annually (Renk, 1990; IMGW, 1997–1998), is not as high as could be expected from the rich nutrient supply. Elmgren (1989) estimated that eutrophication increased pelagic primary production in the period 1900–1980 by 30–70%, and sedimentation by 70–190%.

Because diatoms dominate the phytoplankton, silicates are good indicators of eutrophication in the Baltic Proper. Silicates have decreased in the whole water column (HELCOM, 1990, 1996), although there was no evidence of a significant decline in the riverine outflow. Differences between the mean silicate concentrations in surface water in 1954–1969 and 1970–1991 ranged during the productive seasons from a few to several dozen mmol/m^3 in the southern Baltic and the northern Baltic, including the Bothnian Sea (Majewski and Lauer, 1994). Wulff et al. (1990) found that silicate storage diminished by 177 000 tonnes annually in the Baltic Proper during 1971–1981.

BIOTA IN THE BALTIC PROPER

The main limitation on species in the Baltic is low salinity, and there is a decreasing number of marine species along the diminishing salinity gradient. The least number of species in the Baltic occur in a salinity of 5 psu at the northern end. Low temperature is also an important factor, as is the relatively young age of the Baltic which has been a brackish sea for only 6000 years. There are, therefore, not many species which can be regarded as typical Baltic, brackish-water species, most having immigrated from adjacent seas and freshwater bodies during different periods of its evolution, beginning with the last glacial period about 12,000 years ago.

There are four groups of natural immigrants. The first group consists of Northwest European euryhaline marine and brackish-water species, e.g. the molluscs *Macoma balthica* and fish *Clupea harengus*, and the second are freshwater species, e.g. *Theodoxus fluviatilis* and *Perca fluviatilis*. The third and fourth groups include glacial relicts that reached the Baltic either through ice-dammed lakes from Siberia, e.g. the isopod *Saduria entomon* and the mysid *Mysis relicta*, or by a westerly route through the sea, e.g. the bivalve *Astrate borealis* and amphipod *Pontoporeia femorata*. This migration process still continues (Dahl, 1956; Segerstråle, 1957, 1972; Remane, 1958; Jansson, 1972; Elmgren, 1984).

The Main Coastal and Marine Biotopes

Sandy coasts, dunes and moraine landscapes dominate the shores of Germany, Poland, Lithuania, Russia, Latvia as well as southern Sweden. Dunes occur in various stages of succession, and dunes covered by forests—e.g. Leba in

Poland—are typical of such coasts. High active cliffs built of clays and sands are also present, as are cliff and rocky coasts in the western part (e.g. Rügen Island). In the southern part of the Baltic Proper, lagoons and coastal lakes are characteristic: Szczecin Lagoon (Oder Haff), Vistula Lagoon and Curonian Lagoon. There are several types of coastal salty meadows, coastal bogs and marshes, and large peat bog complexes are located along the southern coasts (e.g. along Lebsko Lake in Poland).

Offshore the biotopes can be divided into those above and below the halocline. Sandy sediments mixed with gravel deposits dominate the sea floor of the coastal zone. In the deep water zone, silty sediments prevail (Lozán et al., 1996; HELCOM, 1998a).

Pelagic and Benthic Organisms

The most abundant phytoplankton species are Chlorophyceae, and diatoms. Phytoplankton composition and dominance changes between the three blooms in spring, summer and autumn. Copepods dominate the zooplankton, while in the summer season *Cladocera* become more abundant. Rotifera form a major portion of the mesozooplankton in summer. The macrozooplankton consists of a few species permanently present, e.g. *Aurelia aurita*. Some species are only observed occasionally, when introduced with saline water inflows from the North Sea, e.g. *Pleurobranchia pileus*, *Cyanea capillata* and *Sagitta elegans*.

The Baltic Proper, especially the southern part with its predominantly sandy bottoms, does not favour development of much macroalgae. In the littoral zone some seagrass and reeds are found such as *Zostera marina*, *Chara*, *Potamogeton*, and *Phragmites communis*. Species of freshwater origin grow on bottoms where wave action is limited. The most typical brown algal species is *Fucus vesiculosus*, though in some areas (Gulf of Gdansk) filamentous brown algae *Ectocarpus* sp. and *Pilayella littoralis* are very abundant. *Ceramium* sp. is a very common red algae on underwater piles, stones and other plants.

Bivalves dominate the macrozoobenthos of the southern Baltic, in particular *Macoma balthica*, *Mya arenaria*, *Mytilus edulis*, and *Cardium glaucum*. Crustaceans and polychaetes are also typical species. The greatest diversity occurs in shallow littoral, sandy and sandy-muddy bottoms with greater habitat diversity and is much less on deep muddy bottoms, though in the latter biomass is relatively high, reaching 300–500 g wet weight m^{-2} due to the great abundance of dominant bivalves *M. balthica*. However, decreasing species diversity, abundance and biomass of macrofauna with depth is seen in the Bornholm, Gdansk and Gotland Basins because of the oxygen deficit.

Baltic marine fish diversity decreases from 57 species in Arkona Basin to 22 in the Gulf of Finland. Cod, herring, sprat, plaice and brill are typical marine species spawning in the Baltic Proper. Other marine species occur here sporadically, e.g. anchovy, whiting, horse mackerel and mackerel, but do not spawn in the Baltic. The high variability in marine fish species numbers is a consequence of different intensities of saline water inflows from the North Sea. This is especially true for species with pelagic embryonic stages, where salinity and oxygen conditions determine their embryonic survival. A characteristic feature of this region is the abundance of many freshwater species, e.g. perch *Perca fluviatilis*, in coastal waters (Segesträle, 1957; Jansson, 1972; Maagard and Rheinheimer, 1974; Augustowski, 1987; Lozán et al., 1996; HELCOM, 1996; Rheinheimer, 1998).

Birds and Mammals

A relatively small number of birds breed on southern coasts of the Baltic Proper, but larger breeding-grounds are located in mouths of rivers, within lakes and coastal lagoons, e.g. various gulls and ringed plover. Various water birds (especially Anseriformes) gather in large numbers on shallow coastal waters and lagoons during migrations. The most important Baltic wintering sites are located in the Baltic Proper along the German–Polish coast (Vorpommern and Szczecin lagoons), Pomeranian Bay between Denmark, Germany and Poland, and the Gulf of Riga. At each of the above sites around a million birds gather. The most abundant wintering species are the tufted duck *Aythya fuligula*, pochard *Aythya ferina*, scaup *Aythya marila*, goosander *Mergus merganser*, long-tailed duck *Clangula hyemalis*, velvet scoter *Melanitta fusca*, common scoter *Melanitta nigra*, black-throated diver *Gavia arctica*, red-throated diver *Gavia stellata* and slavonian grebe *Podiceps auritus*. Shores along river mouths are important stopover sites for migratory waders (*Charadrii*) too. Intensive migrations of passerines also take place along the southern coast of the Baltic.

Marine mammals are only occasionally observed in the Baltic Proper. The most commonly occurring species is the largest Baltic seal, the grey seal *Halichoreus grypeus*, which migrates mainly from Estonia, the nearest breeding area. The ringed seal *Pusa hispida* is rare, while the common seal *Phoca vitulina* occurs mainly in the Kattegat and Skagerrak. There is little information on the Baltic porpoise, *Phoceona phoceona*, which occurs mainly in the west, although it is sometimes found in fishing nets elsewhere (Augustowski, 1987; Andell et al., 1994; HELCOM, 1996; Lozán et al., 1996).

Biological Effects of Eutrophication

Eutrophication is the main anthropogenic factor here. There is a large natural annual phytoplankton variability, but intensity of phytoplankton blooms, including those of toxic algae, may be a general indicator of primary production increase. In the Baltic Proper, no major effects from harmful algae have been observed, although species toxic to mammals have been found. Distinctive, often drastic, changes were observed in benthic macroalgae and vascular plant composition and distribution, during the 1970s, possibly reflecting eutrophication, and a decrease in water transparency may explain their decrease in depth range. Such changes were observed along the coasts of Latvia, Lithuania, Russia, Poland, Germany and the southern coast of Sweden. *Fucus vesiculosus* communities underwent drastic changes, with this community disappearing in some regions. In the shallow littoral zone, many species of red and brown algae have become extinct, while others, including seagrass, became more restricted. In their place, opportunistic green algae and filamentous red algae became dominant.

The most drastic, adverse changes to benthic fauna occurred below the halocline. Long-term oxygen deficits caused changes including, in some cases, even the total disappearance of macroscopic life on the bottom. In the first half of the 20th century, Bornholm, Gdansk and Gotland Basins were inhabited by numerous benthic species. The total extinction of macrozoobenthos on the Bornholm Basin bottom was observed for the first time in the early 1950s. Today, the bottom of deeps below 70–80 m depth show no signs of macroscopic life, and sediments are covered by anaerobic bacteria. There is a lot to suggest that oxygen deficiency in the deep water has contributed to low effectiveness of cod spawning. Cod may hatch only in waters of 10–11 psu minimum salinity, which allows spawn to float. In less saline waters the eggs fall to the bottom and die. In the Bornholm Basin, where waters are sufficiently saline for effective spawning, oxygen deficits have become a limiting factor below about 60 m. In the shallow littoral zone (e.g. Pomeranian Bay, Gulf of Gdansk), increasing sedimentation of organic matter together with a lack of water mixing contribute to summer oxygen deficiencies with consequent effects on benthic species (Maagard and Rheinheimer, 1974; Jansson, 1972; Jarvekulg, 1979; Kautsky et al., 1986; Elmgren, 1989; Cederwall and Elmgren, 1990; Andell et al., 1994; HELCOM, 1996 and 1998a; Lozán et al., 1996).

Alien Species

Alien species have been introduced by ballast waters and the construction of artificial waterways which have destroyed natural biogeographical barriers. During the 20th century, more than 60 new species have been introduced, some becoming common in the southern Baltic: the polychaeta *Marenzelleria viridis* from North America, the bivalve *Dreissena polymorpha* from the Caspian Sea, crustaceans *Eriocheir sinensis* from China and *Rhithropanopeus harrisi tridentata*, and the barnacle *Balanus improvisus*. Some now dominate, e.g. *M. arenaria* and *M. viridis*, and a few may influence native species with the same ecological niche. A round goby *Neogobius melanostomus*, introduced to the Gulf of Gdansk from the Caspian Sea, is an example of a newly occurring fish species. It was observed for the first time in the Gulf of Gdansk in 1990. Since then, it has shown a constant increase and now occurs in increasing numbers in catches, and may compete for food with native fish species. The low species diversity of the Baltic and resulting weak inter-species competition favour alien species migration. Also, because of the relatively short time during which the Baltic Sea has existed and the lack of endemic fauna, the introduction of new species should be seen as a continuous migration process. Some species considered to be newly introduced may be found on endangered species lists, e.g. *Mya arenaria*, *Dreissena polymorpha* and *Rhitropanopeus harrisi* (Hagerhall, 1994; Skóra and Stolarski, 1993; HELCOM, 1996).

FISH STOCKS AND FISHERIES

For centuries, the renewable living resources of the Baltic Sea have provided food for the people inhabiting the coastal zone. Archaeological excavations show that molluscs 10,000 years ago were an important food source for people along the western coast of the Baltic. Taking 56°30'N as the northern boundary for the Baltic Proper, the fish populations from this area support a landing of 0.5 million tonnes of fish annually. Three species: herring *Clupea harengus*, sprat *Sprattus sprattus* and cod *Gadus morhua* contribute most of the catch. Others, like the eel *Anguilla anguilla* and salmon *Salmo salar*, though highly priced, are fished in small quantities. The Baltic cod stock has been the most important fishery resource during the last fifty years. However, since the reproductive success of this stock heavily depends on the Atlantic water inflow from the North Sea, a repetition of the situation in the 1980–1985 period, when cod catches reached their peak, can hardly be expected.

Fish abundance along the southern coast attracted settlers, and fishing villages were set up close to each other. Where fishing proved to be the only means of making a living, unique fishermen's associations were developed which assumed the responsibility of organising members' fishing activities and their compliance with fishing regulations. These associations called "maszoperie" were less important in those localities where alternative occupations existed.

In the second half of the 20th century, the administrations of countries being politically, military and economically dependent on the USSR, with the exception of Poland, closed down their private fishing sectors. The private fishery was restored in those countries after 1990, but tough working conditions, catch seasonality and low income has rendered this occupation unattractive. Consequently, the number of coastal villages with operating fishing boats has decreased in the last 50-year period.

Offshore fishing is carried out with cutters ranging in length from 15 to 30 m. Larger vessels are rare. Between 1950 and 1990 the bulk of the landed catch was taken with trawls, but increasing fuel prices and dispersed cod concentrations forced the Baltic fishers to switch to gill nets.

Herring has proved to be very adaptive and has the greatest biomass of all Baltic fish. The price of Baltic herring has been lower than that of North Sea herring since the fifteenth century (Ropelewski, 1963). Herring together with sprat constitute the main food item for Baltic Sea predatory fish like salmon, cod and trout. Salmon *Salmo salar*, together with eel, one of the most expensive fish in the Baltic countries, was in danger of being exterminated in the Baltic due to its requirement of unpolluted fresh running water for spawning. Efforts focused on rearing fertilised salmon eggs, fry, and smolts for stocking of the rivers draining into the Baltic Sea, have had a spectacular success in restoring the population of the Baltic salmon. The salmon fishery, mainly controlled by Danish fishermen in the southern Baltic until the late 80s, has lately seen a considerable increase in the share allocated to the Polish, Lithuanian and Latvian fishermen.

The economic success of the Baltic fishery depends on cod. Cod catches reached a peak in 1984 with 440,000 tonnes, only to decrease to 66,400 tonnes just 10 years later. This drastic drop without recovery resulted from the lack of a strong cohort able to replenish the stock as well as high fishing intensity. The stock size of the Baltic plaice, *Pleuronectes platessa*, has decreased since the end of the 1980s to a level that makes the fish insignificant in catches from the western part of the area. Until the mid 1970s, plaice composed 30% of the flatfish catches from the Bornholm basin (west of 18°E), while in the latter half of the 1990s it has declined to 4%. The northern boundary for plaice follows the line running from the Gulf of Riga to Gotland.

Turbot *Psetta maxima* inhabits European coastal waters, and is the largest (70 cm) Baltic flatfish. Turbot prey on other fish, and belong, with salmon and cod, to the group of top predators in the Baltic Sea ecosystem. The fish had very limited importance for the Baltic fishery till 1990, at least for Polish, Lithuanian and Latvian fisheries. The easing of trade rules after 1990 in the former Eastern bloc countries, resulted in higher prices being offered for turbot, which led to an increase in catch to 1,000 tonnes annually, compared to 160 tonnes previously. For a fish with a low reproduction efficiency and slow growth, the outcome of rapidly increasing fishing pressure is likely to lead to overfishing, as is demonstrated by a decrease in the annual catch and smaller average sizes.

An increase in salinity recorded in the 1920s caused the appearance of mackerel *Scomber scombrus* and garfish *Belone belone*. They spawn in the Baltic waters occasionally and are fished when available. Lesser sand-eel *Ammodytes tobianus*, sand-eel *Hyperoplus lanceolatus* and four-bearded rockling *Enchelyopus cimbrius*, although common, are only fished at some localities on the southern coast. Eelpout or viviparous blenny *Zoarces viviparus*, a unique viviparous Baltic fish species, is also common and abundant in some shallow coastal waters and is exploited mainly by the Latvian fishery. The lumpsucker, *Cyclopterus lumpus*, inhabits places with hard bottom along the Polish coast, and during the last decade has been recorded more frequently in catches.

The density distribution of fish within the southern Baltic is not uniform. It depends on environmental factors, mostly those determining primary production. One factor influencing the latter is an upwelling noticeable in the southern Baltic. The northern coast of the Hel Peninsula is one of the places where these effects are most marked, and a map produced by Hensen (1875) shows the distribution of fishing effort along the southern coast more than 100 years ago, which shows that in the Gulf of Gdansk, highest boat density distribution in the 1870s overlapped with the area of most intensive upwelling.

A large number of freshwater species spawn in the rivers and channels connecting lakes and lagoons with the sea. These are limited to inshore waters. Of the typical freshwater species, pike-perch *Stizostedion lucioperca*, perch *Perca fluviatilis*, and bream *Abramis brama* are widely distributed in the southern Baltic. The number of freshwater fish species found in the sea increases eastward. These species only have importance for the local fishery, and their total catch does not exceed 20,000 tonnes for the entire Baltic area. Zährte *Vimba vimba* is a freshwater fish, spawning in the middle and upper parts of the Baltic rivers but undertakes a long migration to feeding grounds in estuaries and bays. It is one of the most valuable species of the carp family. The zährte population of the Vistula River system supported annual catches of 160 tonnes during the period 1950–1965, but the dam constructed on the Vistula River in Wloclawek in 1968 has impeded fish migrations to the spawning grounds located in the upper course of the river, so today the fish is practically extinct in the Gulf of Gdansk. In order to rebuild the population to the levels recorded in the 1950s, the river needs to be stocked with 100 tonnes of fry of 2 g individual weight (Backiel and Bontemps, 1995). Catches of zährte in the Curonian lagoon did not exceed 3 tonnes annually during the last decade (Repecka, 1998).

The three largest lagoons—Szczecin, Vistula and Curonian lagoons—extend over 3100 km² and are very productive areas for fisheries. Their shallow brackish waters with a maximum salinity of 4 psu are favourable for freshwater fish and, in the areas adjacent to the outlets to the sea, for marine species. The number of fish species inhabiting these waters ranges from 44 to 52 (Dunin-Kwinta, 1995; Repecka, 1998), though some of these spend only a part of their life there. The last records of sturgeon *Acipenser sturio* in the lagoons date from the first decade of the 20th century (Bartel et al., 1996). In the Vistula Lagoon, 29 fish species were fished during 1890–1920, but during the last decade, 19 species only were fished. However, some species disappear for long periods only to reappear again at a later date. Sabre carp *Pelecus cultratus*, was considered endangered in Polish fresh water and was not recorded in the Vistula Lagoon after 1920, but it reappeared again in the 1980s in such quantities as to ensure a catch of over 120 tonnes (Keida, 1998); however current catches are not exceeding 60 tonnes.

An increase in the living standard in Poland has influenced the nature of fishing. Species such as pike-perch, eel, bream and perch are suffering the highest fishing pressure, whereas land ruffe *Gymnocephalus cernuus*, silver bream *Blicca bjoerkna* and roach *Rutilus rutilus*, are abundant in the lagoons, but are avoided by fishermen. The three-spined stickleback, *Gasterosteus aculeatus*, a small fish occasionally reaching a maximum length of 10 cm, is common in the coastal waters but has no consumer value. It is a frequent prey item of larger predator fish and used to be fished for reduction (oil) in quantities up to 6,000 tonnes annually. Since 1945, the stickleback population has not been a target for fishing activities in the Gulf of Gdansk and the Vistula Lagoon, and during the last decade of the century it has become a food competitor for other valuable fish species. To suppress the stickleback population, the Puck Bay waters were stocked with rainbow trout, which proved to be very efficient predators of stickleback abundance.

During the last 25 years, herring, common in Baltic lagoons, was fished in largest quantities in the Vistula Lagoon. The fishing season is limited to few weeks and depends on the number of spawners entering the Lagoon's water after the ice has melted. The largest herring catches in the Lagoon were recorded in 1950–1952 (12,000–17,0000 tonnes). In 1979–1989 they varied at 8000 tonnes. Since 1988, a sharply decreasing trend has been observed with only 2000 tons being caught in 1996 (Krasovskaya, 1998). The other fish in the herring family, twaite shad (*Alosa fallax*), has been absent from the Vistula Lagoon since 1920 (Bartel et al., 1996), and in the Curonian Lagoon since 1960. It reappeared in 1995 in the Curonian Lagoon, however, and has been regularly recorded in gill-net catches since (Maksimov and Toliušis, 1998).

Fish assemblages in the Szczecin and Vistula lagoons can hardly be considered stable. Some have disappeared in both lagoons within the last thirty years (1965–1994), possibly partly a consequence of pollution.

Aquaculture is not an attractive enterprise. Stocking the sea with valuable fish such as salmon, trout and whitefish *Coregonus lavaretus* is therefore a unique activity to the southern Baltic.

Management of Fishing

The self-regulatory fishery model is based, in principle, on records of past catch volume taken by fishermen, and is currently accepted by the International Baltic Sea Fishery Commission, but it can hardly be acknowledged as the best model. In forthcoming years, its suitability will be appraised, especially in the light of requirements such as maintenance of biodiversity, development of tourism, relating fishing effort to the stock productivity, and in determining the total expenditure on stocking spent from a country's central budgets.

The main management tool of exploited Baltic fish is the Total Allowable Catch (TAC) assigned for each fish stock. The assessments are performed under the auspices of ICES. TAC volumes, once accepted, become the annual catch quotas and are divided among the member countries, the administration for which is further assigned to ship owners. Alongside the catch quota system, several regulatory measures curb the amount of fishing effort expended. These include a minimum landable length size of fish, minimum allowable mesh size, and closed seasons.

Research effort is not uniform. The fish species being the object of most intensive studies are those which contribute to the bulk of catch taken annually from the Baltic Sea. In 1988–1995 64% of the published scientific papers concerning Baltic fish were focused on four species: herring, cod, sprat and salmon. With regards to 23 other fish species, the number of publications was five or less on each (Draganik, 1996).

ENVIRONMENTAL POLLUTANTS

Organic Materials

The semi-enclosed Baltic Sea is vulnerable to the significant quantities of chemical substances discharged into it in the 20th century from industrial wastewaters, municipal sewage, agricultural run-off, atmospheric fallout, marine paints, dumping of wastes and chemical warfare agents as well as from sediments dredged from port channels. Pollution with persistent, toxic and bioaccumulative compounds has become a real threat to wildlife, and a huge range of harmful organic pollutants is involved, including various organochlorine pesticides and organohalogenated industrial chemicals, as well as pulp-mill related polychlorinated organic compounds and other chlorinated and non-chlorinated aromatic compounds.

The concentrations of polychlorinated biphenyls (PCBs), DDT and its analogues (DDTs), isomers of hexachlorocyclohexane (γ, α and β; HCHs), hexachlorobenzene (HCBz), dieldrin and chlordanes (CHLs) were higher in fish caught from the southern and western Baltic than in the North Sea, Norwegian Sea or the Shelf of Iceland in the 1980s (Fig. 9). PCBs and polycyclic aromatic hydrocarbons (PAHs) from sediments in the 1990s showed geographic variations, along with many other persistent, toxic and bio-accumulative chemicals in surface sediment and biota. Highest concentrations were more frequently recorded in the southern Baltic Sea. Herring and perch (Figs. 10 and 11) caught in the southern part in 1992 indicated a somewhat higher concentration of DDTs than in specimens from various sites in the northern waters of the Bothnian Bay and Bothnian Sea. This seems to have been a consequence of the heavy use of DDT in Poland from the 1950s to the 1970s, and in the former East Germany until the 1980s (Strandberg et al., 1998a; Falandysz et al., 1999). The concentrations of PCBs in herring and perch from the Gulf of Gdansk in the 1990s are similar to those in fish from various sites in the northern Baltic Sea, while the concentrations of

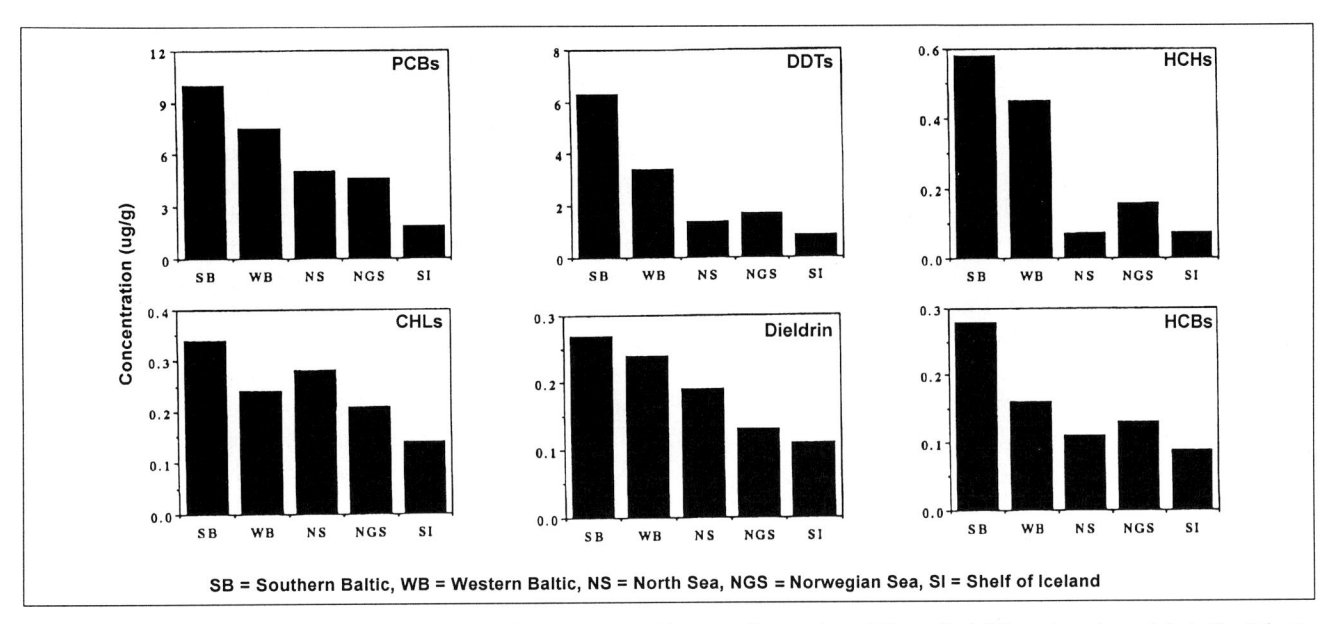

Fig. 9. Geographical distribution and concentrations of major organochlorine pollutants in cod-liver oil of different marine origin in the Atlantic Ocean (from Falandysz et al., 1994).

Table 2
Concentrations of persistent organochlorine contaminants in biota (ng/g lipid wt) from the Baltic south coast*

Substance	Plankton	Blue mussel	Herring[a]	Flounder[a]	Cod[a]	Black cormorant[a]	Harbour porpoise[a]	White-tailed sea eagle[a]	
								Coast	Inland
HCHs	35	39	75	51	130	430	140	42	57
HCBz	12	7.9	41	20	14	470	200	2200	530
PCBz	ND	ND	8.3	8.0	8.3	29	ND	270	88
DDTs	300	910	1300	1600	1200	9500	7800	780000	130000
TCPM-H/OH	19	ND	3.7	22	16	230	21	16000	460
PCBs	210	1600	1300	5000	1400	34000	10000	1100000	15000
PCNs	18	110	33	64	130	210	2.6	960	73
CHLs	8.7	12	49	18	19	100	800	6100	210
Endosulfan 1 and 2	ND	ND	ND	ND	ND	ND	ND	ND	ND
Endrin	ND	ND	ND	ND	ND	ND	ND	ND	ND
Isodrin	ND	ND	ND	ND	ND	ND	ND	ND	ND
Aldrin	ND	ND	ND	ND	ND	ND	ND	ND	ND
Dieldrin	32	15	70	27	51	190	740	3300	190
Mirex	ND	ND	ND	<0.15	ND	ND	15	680	65
PCDDs[b]	12	170	38	3.7	3.6	50	3.2	230	NA
PCDFs[b]	7.9	140	29	4.7	4.2	58	2.4	240	NA
Lipids (%)	1.8	1.5	9.0	4.8	6.0	4.5	87.0	8.8	11.2
No[c]	4	2 (700)	1 (3)	3 (15)	2 (8)	3	4	10	3

*Data from Falandysz, 1998; Strandberg et al., 1998a,b; Falandysz et al., 1998 and 1999; Falandysz, unpublished.
[a]Whole fish, blubber of harbour porpoise and breast muscles of cormorant and sea eagle were analysed.
[b]TCDD TEQs (2,3,7,8-tetrachloro dibenzo-p-dioxin toxic equivalents) in pg/g lipid weight.
[c]Number of samples and number of animals (in parentheses).
NA, not analysed; ND, not detected.

HCHs, HCBz, CHLs and dieldrin are much lower or quite similar to concentrations in the Gulf of Bothnia. There is also a vast amount of data available for many other similar compounds, many of which are still ubiquitous pollutants in the Baltic Sea environment in the 1990s. Some of those chemicals are found in relatively high concentrations in marine animals higher up in the food web along the Baltic south coast (Table 2).

Atmospheric concentrations of individual congeners of chlorobiphenyl in the Baltic Sea region appear to be influ-

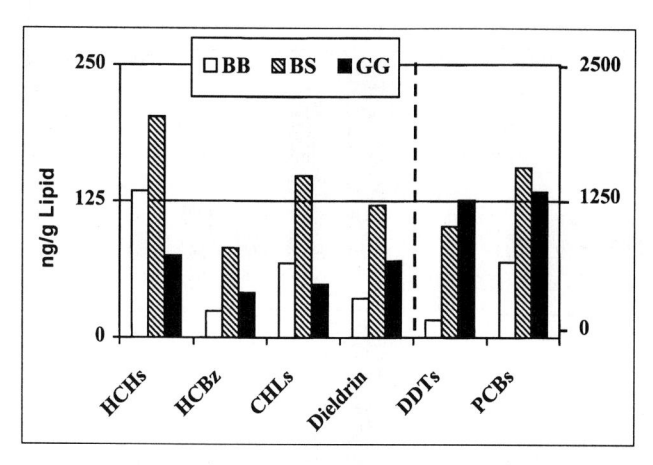

Fig. 10. Spatial variation of organochlorines in herring (*Clupea harengus*) from the Bothnian Bay (BB), Bothnian Sea (BS) and Gulf of Gdansk (GG) (from Strandberg et al., 1998a).

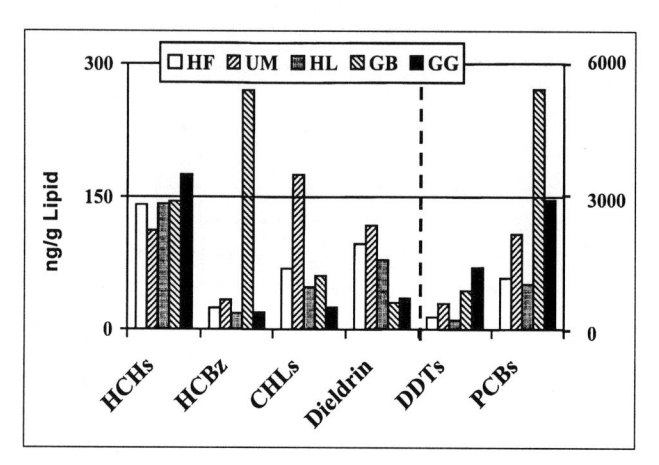

Fig. 11. Spatial variation of organochlorines in perch (*Perca fluviatilis*) from station HF in the Bothnian Bay (BB), UM, HL and GB in the Bothnian Sea (BS) and Gulf of Gdansk (GG) (from Strandberg et al., 1998a).

enced by their physical–chemical properties, ambient temperature, and geographical location. A latitudinal gradient with higher levels in the south was found for the total PCBs and for individual congeners. Consequently, the most volatile congeners are transported more easily and fixed in colder regions (Agrell et al., 1999). Concentrations of many of these organic substances have decreased continuously over the last two decades in all types of samples and at all locations in the Baltic Sea, although the rates differ (HELCOM, 1996; Bignert et al., 1998; Kjeller and Rappe, 1995) but there is no clear trend for organobrominated chemicals such as polybrominated diphenyl ethers (PBDEs) and hexabromocyclododecane (HBCD) (Bernes, 1998; Kierkegaard et al., 1999).

The structure of some of the persistent, toxic and bioaccumulative organohalogenated compounds that contaminate the Baltic biota still remain unknown. Recently a relatively new but poorly known group of chlorinated polycyclic aromatic hydrocarbons (Cl-PAHs) such as mono- to heptachlorosubstituted fluorenes (Cl-Flo), phenanthrenes/anthracenes (Cl-Phe/Ant) and pyrenes/fluoranthenes (Cl-Pyr/Flu) were identified and quantified in various types of non-biological material in the Baltic Sea region, including particulate matter settling in the Stockholm archipelago and sediment from the Baltic Sea (Ishaq et al., 1999).

The people and biota around the Baltic Sea have been exposed for many years to complex mixtures of environmental pollutants. Small doses of daily ingested methylmercury and organochlorines such as DDTs and PCBs are blamed for reproductive disturbances and failure in birds of prey such as white-tailed sea-eagle *Haliaeetus albicilla* and osprey *Pandion haliaeetus*. Following the ban on the use of DDT, cyclodiene pesticides and alkyl mercury in the 1960s–70s and restrictions on the use of PCNs, PCBs and PCTs, these bird populations are recovering. The seals of the Baltic Sea suffered from reproductive impairment and increased prevalence of jaw damage coincident with changes in concentrations of some persistent pollutants that peaked in the 1970s. Methylsulphone metabolites of DDE and PCBs ($MeSO_2$–DDE/PCB) are considered to be the most likely cause due to their accumulation in the adrenal cortex (Bernes, 1998).

Metallo-organic Environmental Pollutants

Tributyltin (TBT) in antifouling paints seems to constitute the highest risk among the organometallic environmental pollutants in the Baltic Sea. In 1988 the Helsinki Commission recommended that member countries restrict pollution caused by antifouling paints containing organotin compounds by 1991. However, these bans were primarily applied for boats less than 25 m in length, so at present TBT remains an important pollutant in areas with high shipping traffic. In addition, the restrictions of TBT use did not greatly reduce its consumption due to a resulting change in application from antifouling paint to wood preservation, PCV plastic stabilisation and catalyst in the synthesis of silicones (Fent, 1996).

Very few monitoring studies on butyltin or its effects in the Baltic Sea have been carried out (HELCOM, 1990; Kannan and Falandysz, 1997, 1999; Szpunar et al., 1997; Senthilkumar et al., 1999). They were found in relatively high concentrations in surface sediment and in muscle tissue, liver and eggs of all fish species examined from Polish coastal waters (Table 3). Especially high concentrations of TBT, DBT and MBT were found in sediments collected from shipyards, seaports and marinas. Available data suggest that BTs contamination of fish and sediments from the Baltic south coast continues to be high.

Metals

Poland is a rich mining area with abundant Zn-Pb and Cu deposits situated mainly in Silesia (Osika, 1986). Ag and Cd are associated with the Zn-Pb ores (Osika, 1986), and Poland has the highest abundance of Ag per unit area of

Table 3
Concentrations of butyltins in muscle tissue of fish (ng/g wet weight) and surface sediments (ng/g dry weight) from the Baltic south coast*

Matrix and site	Year	MBT	DBT	TBT	BTs
Fish					
Gulf of Gdansk					
Flounder	1990				316
Flounder	1997	12	20	51	83
Turbot	1990				39
Turbot	1997	11	17	73	100
Herring	1990				40
Eel	1990				190
Sea trout	1990				51
Cod	1990				19
Eelpout	1990				130
Pikeperch	1990				460
Puck Bay					
Roach	1997	43	530	2700	3300
Brown trout	1997	15	8.0	56	78
Herring	1997	11	10	57	78
Ruff	1997	8.0	19	17	44
Smelt	1997	8.0	14	150	170
Vistula River					
Roach	1997	14	14	74	100
Sediment					
Open Sea	1994–95				<1.0
Gdynia Seaport	1994				1200–7300
Puck Bay	1994				24
Shipyards, Gdansk	1998	1200–46000	2000–42000	2600–40000	5800–130000
Dead Vistula River	1993–94				120
Szczecin Lagoon	1994				15–27

*Data from Kannan and Falandysz, 1997; Szpunar et al., 1997 and Senthilkumar et al., 1999, respectively.

any country (Singer, 1995). In 1991, 5.3×10^6 tonnes of Zn-Pb ores and 31×10^6 tonnes of Cu ores were mined (Helios Rybicka, 1996). These mining operations have caused very significant environmental contamination. For example, Helios Rybicka (1996) reported that, on average, 3331 tonnes of Zn, 448 tonnes of Pb, 515 tonnes of Cu, 40 tonnes of Cd, 443 tonnes of Cr and 284 tonnes of Ni are discharged into the Baltic Sea each year by rivers from the Polish drainage basin.

The highest loads of metals entering the Baltic Sea come from tributaries in Poland and the Baltic countries (Matschullat, 1997). Other important point sources are single industries along the Finnish and Swedish coasts (Bruneau, 1980; Matschullat, 1997). The Vistula River is responsible for ca 10% of the input of Pb, Cd, Cr, Cu and Zn. This river and its tributaries traverse highly industrialized and mining areas. Agricultural activity also contributes nutrients and eroded soil to the river waters. In 1989, the Vistula was estimated to have carried 2930 tonnes/year Zn, 12.8 tonnes/year Cd, 196 tonnes/year Pb, 233 tonnes/year Cu and 15.5 tonnes/year Hg. Until recently, municipal systems in Poland discharged 900,000 m^3 of untreated sewage and 1,400,000 m^3 of partially treated sewage, much of which would have ended up in the Vistula. The Gdansk region has been designated an ecologically endangered area in Poland and one of the pollution "hot spots" in the Baltic.

The Szczecin Lagoon and Pomeranian Bay (southern Baltic) are supplied by the Oder River, the second largest river in Poland, which drains the heavily industrialized heartland of Silesia. It transports 90 tonnes/year Cu, 792 tonnes/year Zn, 15.5 tonnes/year Cd and 104 tonnes/year Pb into the Baltic (Neumann et al., 1996). Near the mouth of the Odra/Oder River, the <2 μm size fraction of the riverine sediments contains 426 mg/kg Cu, 3114 mg/kg Zn, 20 mg/kg Cd and 1132 mg/kg Pb (Helios Rybicka, 1996).

In addition, atmospheric transport of metals is significant (Smal and Salomons, 1995; Helios Rybicka, 1996; Szefer et al., 1996; Verner et al., 1996), and evidence has been presented for the long-range transport of heavy metals to the Baltic from the heavily industrialized regions of Central and Eastern Europe. According to Matschullat (1997) metals such as Pb, Cd, Cu and Zn show similar atmospheric and riverine inputs but Hg, As, Co, Cr and Ni are mainly transported to the Baltic Sea by rivers. In 1985, 1388 tonnes Pb were introduced into the Baltic from atmospheric routes; this load decreased in 1990 to 627 tonnes (Krüger, 1996). Using the best available technology in the non-ferrous metals industry and using only unleaded gasoline in Europe, it was estimated that the atmospheric Pb input could be reduced to 190 tonnes/year. For Cd, atmospheric input into the Baltic is about 19 tonnes/year. A reasonable estimate for the Pb deposition during the latter half of the 1980s, derived from both measurements and models, seems to be 1300 tonnes/year, with about 70% of this coming from the region (HELCOM, 1997).

Atmospheric and Riverine Fluxes

Most Cu, Zn, Ag, Cd and Pb in the sediments of the Gulf of Gdansk are of anthropogenic origin (Szefer et al., 1993, 1995, 1996). According to Renner et al. (1998) Cu, Zn and Ag principally enter via the Vistula River while Cd and Pb are introduced, in part, by atmospheric transport, though a significant proportion of the latter in the southern Baltic also enters from rivers (Table 4) (HELCOM, 1998). The dual source of Cd and Pb may explain the complex inter-element relationships displayed by these elements in the sediments of this region. Silver, on the other hand, enters with sewage sludge (Ravizza and Bothner, 1996).

Table 4

Riverine and direct point source metals load entering the Baltic Proper in 1995 (HELCOM, 1998)

Hg		Cd		Zn		Cu		Pb	
Total load (kg/a)	Area specific load (kg/km^2)	Total load (kg/a)	Area specific load (kg/km^2)	Total load (kg/a)	Area specific load (kg/km^2)	Total load (kg/a)	Area specific load (kg/km^2)	Total load (kg/a)	Area specific load (kg/km^2)
9909	0.0184	10881	0.020	1198882	2.23	229543	0.43	153416	0.286

Table 5

Total annual deposition of metals to the Baltic Proper in respect to their total deposition in the Baltic Sea in 1991–1995 (HELCOM, 1997)

	Pb (100 kg/yr)				Cd (10 kg/yr)			
	South Baltic Proper	Baltic Sea Deposition			South Baltic Proper	Baltic Sea Deposition		
		Wet	Dry	Total		Wet	Dry	Total
1991	1626			6280	651			2605
1992	1692			6024	698			2572
1993	1907			7026	749			2880
1994	2366			6774	968			2875
1995	1888			5769	779			2516
Average	1896	5858	517	6375	769	2354	336	2690
CV (%)	14			7	14			

Analysis of land mosses as biomonitors indicated that the highest atmospheric metallic loads are mainly in the southern part of the Baltic catchment (the Black Triangle, e.g. Poland, Czech Republic and Germany). It is estimated that in 1995 total depositions of Pb to the southern Baltic from Germany and Poland were 73.2 and 34.1 tonnes/year, respectively. For Cd annual loads amounted to 3.42 and 1.57 tonnes (HELCOM, 1997a, 1997b) (Table 5). For many metals, the Baltic is more polluted than the Black and Adriatic Seas (Seculiae and Vertacnik, 1997).

Biota

Metals such as Cd, Pb, Zn and Cu in biota may be classified as "biophile" or "anomalously enriched elements" (Li, 1981). There is widespread atmospheric contamination of Pb in the southern Baltic, transported mainly by air masses. However, the Vistula River is the major source of other metallic pollutants such as Cd (Szefer et al., 1996). Other elements, namely Co, Th, Ca, Mg, Ti, U and, to a lesser extent, Mn and Ni, have low biological affinity in biota in the Baltic Proper.

Fish are an important seafood in Baltic countries and have been extensively investigated for their concentrations of metals. In contrast to Pb, concentrations of metals such as Cd, Cu and Zn in fish increased during the 1980s. The decrease in Pb could be a result of the reduced air emissions from car traffic in Finland, Sweden, Germany and Denmark, where unleaded gasoline is mainly used (Jansson and Dahlberg, 1999).

Metals may concentrate along a food chain. However Szefer (1991) found that in general there is no biomagnification of several metals in the southern Baltic food chain. The lack of metal biomagnification is connected with generally higher levels in prey than in the predator since food is not the exclusive vector for element input (Fowler, 1982, 1985; Luoma, 1983). Moreover, factors such as incomplete absorption of metals across the gut, rapid excretion and dilution in muscle (Fowler, 1982), representing a large part of the total body weight, are responsible for lower pollution levels in the predator than its prey. The opposite trend is observed for Hg (Falandysz, pers. comm.) which reaches maximum concentration in higher trophic levels, e.g. in fish muscle and fish-eating birds.

Radionuclides

Benthic organisms such as bladder wrack (*Fucus vesiculosus*) can be useful as biomonitors of radionuclides (HELCOM, 1996). In seafood, concentrations of ^{137}Cs and ^{134}Cs in fish clearly reflected the distribution pattern of the Chernobyl fallout. Benthic sediments are also appropriate material for radiological studies because radioactive substances are adsorbed on suspended particles. The total inventory of

Table 6
Amounts of chemical munitions and warfare agents dumped in the Baltic Proper (Anon, 1993; HELCOM, 1996; Glasby, 1997a)

Area	Quantities of munitions (tonnes)	Estimated quantities of warfare agents (tonnes)	Types of warfare agents
Bornholm Basin	35,300 (certain) to 43,300 (uncertain)	5300–6 500	Mustard gas, viscous mustard gas, Clark I, Clark II, Adamsite, Chloroacetophenone and (less certain: phosgene, nitrogen mustard and tabun)
East of Bornholm	8000 (not verified)		Not known
Area SW of Bornholm	Up to 15,000 (uncertain)	2250	Not known
Gotland Basin	2000	1000	Mustard gas, Adamsite, chloroacetophenone

^{137}Cs in the Baltic sediments in 1991 was ca 1400 TBq and the corresponding value reported for the period before the Chernobyl accident, i.e. at the beginning of the 1980s, was 277 TBq (HELCOM, 1995).

Samples of Baltic seawater, biota and bottom sediments were analysed for contents of ^{137}Cs, $^{239+240}$Pu and ^{210}Po (Holm, 1995; Skwarzec, 1997). The activity concentration of ^{137}Cs in seawater of the Pomeranian Bay was high and was found to be similar to the rest of the Baltic Sea. It results from the radioactive contamination from Chernobyl in 1986. Prior to the accident, typical ^{137}Cs content in the Baltic ranged between 12 and 16 mBq/kg. As a result of the Chernobyl fallout, concentrations of this radionuclide in surface water increased to 100–1000 mBq/kg. This highly variable distribution pattern reached a certain degree of equilibrium by 1988, amounting to 105±15 mBq/kg. The 10% decrease in activity concentrations noted from 1988 to 1993 can be connected with very slow ^{137}Cs decay. According to Holm (1995), the major source of plutonium in the Baltic Sea is fallout from nuclear tests. The Chernobyl accident made a very small contribution to the total activity concentrations of ^{239,240}Pu, however contributions of ^{238}Pu and ^{241}Pu were more significant. It should be emphasized that discharges from the reprocessing plants at Sellafield and La Hague have also influenced concentration activity of radionuclides in the Baltic Sea by the inflow of contaminated waters through the Danish Straits. Fortunately, these discharges have recently been significantly reduced. The least significant source of artificial radionuclides are the discharges released from nuclear installations within the Baltic drainage area (HELCOM, 1995, 1996).

Radiation doses to man from consumption of seafood contaminated with ^{137}Cs and ^{90}Sr in the Baltic Sea have been evaluated (HELCOM, 1995, 1996, 1997). The total collective dose to man from contaminated Baltic seafood with the above mentioned radionuclides is estimated to be 2300 man sieverts (manSv). Approximately 60% of this dose corresponds to the Chernobyl fallout, 30% to nuclear weapons fallout, 10% to European reprocessing facilities and 0.03% to nuclear installations bordering the Baltic Sea (HELCOM, 1995, 1996).

Chemical Munitions in the Baltic Proper

Large quantities of chemical warfare agents were dumped in the Baltic Proper after World War II (Glasby, 1997a,b) (Table 6). Between 1947 and 1948 the Soviet Military Administration in Germany dumped 34,000 tonnes of chemical munitions containing approximately 12,000 tonnes of chemical warfare agents in the Bornholm and Eastern Gotland Basins. According to unconfirmed reports, the British Military Authority ordered the dumping of four ships with about 15,000 tonnes of chemical munitions southwest of Bornholm in 1946 (Anon, 1993; Theobald, 1994; Glasby, 1997a). Both these basins are characterised by stable stratification with anoxic conditions developing below the halocline, and usually have slight bottom currents. The Baltic Sea is a relatively shallow-water, semi-closed and dynamic basin and therefore is not an ideal place to dump toxic chemicals. During major storms, the anoxic basins like the Bornholm and Gotland Basins are flushed and in consequence adjacent bottom sediments are extensively reworked. Under the oxic conditions, the steel casings may have rusted. Owing to periodic major flushing, any toxic substances stored in the bottom of the basins would have been released and dispersed throughout the Baltic (Glasby, 1997a).

Although western scientists are relatively optimistic in their opinion concerning the status of this in the Baltic (Anon, 1993), Russian opinion appears much more negative (Glasby, 1997b). The metal casing of the disposed munitions are estimated to be 70–80% rusted through. It is possible that a massive gas infiltration will take place in Baltic waters within the next few years (Anon, 1993; Glasby, 1997b).

REFERENCES

Agrell, C., Orla, L., Larsson, P., Backe, C. and Wania, F. (1999) Evidence of latitudinal fractionation of polychlorinated biphenyl congeners along the Baltic Sea region. *Environmental Science and Technology* 33, 1149–1156.

Ambio (1990) Special report: Current status of the Baltic Sea. *Ambio* 7, 1–10.

Andell, P., Dürinck, J. and Skov, H. (1994) Baltic marine areas of outstanding importance for wintering seabirds. *WWF Bulletin* **5**, 1–8.

Andersin, A.B. and Sandler, H. (1989) Occurrence of hydrogen sulphide and low oxygen concentrations in the Baltic deep basins. *Proceedings of the 16th Conference of the Baltic Oceanographers*, Kiel, Vol. 1.

Anon. (1993) Chemical munitions in the southern and western Baltic Sea compilation, assessment and recommendations. *Bundesamt für Seeschiffahrt und Hydrographie*, Hamburg 60.

Augustowski, B. (1987) *Southern Baltic* (in Polish). Polish Academy of Sciences, Wroclaw. pp. 412.

Backiel, T. and Bontemps, S. (1995) Estimation by three methods of *Vimba vimba* population in the Vistula River system. *Archiwa Rybołówstwa Polskiego* **3**, 137–158.

Bartel, R., Wilkonska, H. and Borowski, W. (1996) Changes in the ichthyofauna in the Vistula Lagoon. In *Proceedings of the International Meeting on Baltic Network of Biodiversity and Productivity of Selected Species in Coastal Ecosystems*, ed. R. Volskis, 4–8 October 1995, Nida, Lithuania. UNESCO Regional Office for Science and Technology for Europe, Vilnius, pp. 12–24.

Bergström, S. and Carlsson, B. (1993) Hydrology of the Baltic Basin. *SMHI Reports Hydrology* **7**, 1–21.

Bernes, C. (1998) Persistent organic pollutants. A Swedish view of an international problem. *Monitor* **16**, Swedish Environmental Protection Agency. ISBN 91-620-1189-8.

Bignert, A., Olsson, M., Persson, W., Jensen, S., Zakrisson, S., Litzén, K., Eriksson, U., Häggberg, L. and Alsberg, T. (1998) Temporal trends of organochlorines in Northern Europe, 1967–1995. Relation to global fractionation, leakage from sediments and international measures. *Environmental Pollution* **99**, 177–198.

Bruneau, L. (1980) Pollution from industries in the drainage area of the Baltic. *Ambio* **9**, 145–152.

Cederwall, H. and Elmgren, R. (1990) Eutrophication of the Baltic Sea. *Ambio* **19**, 109–112.

Cyberski, J. (1995) Recently observed and prognostic changes in water balance and their impact on the salinity in the Baltic Sea (in Polish). Wydawnictwo Uniwersytetu Gdanskiego. *Rozprawy i monografie* **206**, 1–210.

Dahl, E. (1956) Ecological salinity boundaries in pikilohaline waters. *Oikos* **7**, 1–21.

Draganik B. (1996) Obiektywnosc i przydatnosc kryterium rzadkosci w decyzjach dotyczacych ochrony zagrozonych gatunków ryb. *Zoologica Poloniae* **41** supplement, 7–22.

Dunin-Kwinta, I. (1995) Fisheries of the Szczecin Lagoon. In *50 Years of the Polish Marine Management in the Western Pomerania* (in Polish). eds. I. Dunin-Kwinta and J. Stanielewicz. Szczecin, pp. 113–121.

Elmgren, R. (1984) Trophic dynamics in the enclosed, brackish Baltic Sea. *Rapports et Proces-Verbaux des Reunions, Conseil International pour l'Exploration de Mer* **183**, 152–169.

Elmgren, R. (1989) Man's impact on the ecosystem of the Baltic Sea: energy flow today and at the turn of the century. *Ambio* **18**, 326–332.

Falandysz, J. (1998) Polychlorinated naphthalenes: an environmental update. *Environmental Pollution* **101**, 77–90.

Falandysz, J., Kannan, K., Tanabe, S. and Tatsukawa, R. (1994) Organochlorine pesticides and polychlorinated biphenyls in cod-liver oils: North Atlantic, Norwegian Sea, North Sea and Baltic Sea. *Ambio* **23**, 288–293.

Falandysz, J., Strandberg, B., Strandberg, L., Bergqvist, P.-A. and Rappe, C. (1998) Concentrations and spatial distribution of chlordanes and some other cyclodiene pesticides in Baltic plankton. *The Science of the Total Environment* **215**, 253–258.

Falandysz, J., Strandberg, B., Strandberg, L. and Rappe, C. (1999) Tris(4-chlorophenyl)methane and tris(4-chlorophenyl)methanol in sediment and food webs from the Baltic south coast. *Environmental Science and Technology* **33**, 517–521.

Fent, K. (1996) Ecotoxicology of organotin compounds. *Critical Reviews in Toxicology* **26**, 1–117.

Fonselius S. (1969) Hydrography of the Baltic deep basins III. *Fishery Board of Sweden, Ser. Hydrogr.* No. **23**, 1–97.

Fonselius S., Szaron J., Öström B. (1984) Long-term salinity variations in the Baltic Sea deep water. *Rapports et Proces-Verbaux des Reunions, Conseil International pour l'Exploration de Mer* **185**, 140–149.

Fowler, S.W. (1982) Biological transfer and transport processes. In *Pollutant Transfer and Transport in the Sea. Vol. II*, ed. G. Kullenberg. CRC Press, Boca Raton, FL, pp. 1–65.

Fowler, S.W. (1985) Heavy metal and radionuclide transfer and transport by marine organisms. In *Heavy Metals in Water Organisms*, ed. J. Salánki. Symposia Biologia Hungarica, Vol. 29. Akadémiai Kiadó, Budapest, pp. 191–205.

Glasby, G.P. (1997a) Disposal of chemical weapons in the Baltic Sea. *The Science of the Total Environment* **206**, 267–273.

Glasby, G.P. (1997b) Sea-dumped chemical weapons: Aspects, Problems and Solutions. A.V. Kaffka (ed.), Kluwer Academic Publishers, Dordrecht, 1996; 170 pp. Book review; *The Science of the Total Environment* **208**, 145–147.

Hagerhall, A.B. (1994) Alien species in the Baltic marine environment. *WWF Bulletin* **4**, 10–14.

HELCOM (1986) Water balance of the Baltic Sea, ed. M. Falkenmark. *Baltic Sea Environment Proceedings* No. 16.

HELCOM (1987) First periodic assessment of the state of the marine environment of the Baltic Sea area, 1980–1985. *Baltic Sea Environment Proceedings* No. **17B**.

HELCOM (1990) Second periodic assessment of the state of the marine environment of the Baltic Sea, 1984–1988. *Baltic Sea Environment Proceedings* No. **35B**.

HELCOM (1991) Airborne pollution load to the Baltic Sea 1986–1990. *Baltic Sea Environment Proceedings* No. 39.

HELCOM (1993a) Second Baltic Sea pollution load compilation. *Baltic Sea Environment Proceedings* No. 45.

HELCOM (1993b) First assessment of the state of the coastal waters of the Baltic Sea. *Baltic Sea Environment Proceedings* No. 54.

HELCOM (1995) Radioactivity in the Baltic Sea 1984–1991. *Baltic Sea Environment Proceedings* No. 61.

HELCOM (1996) Third periodic assessment of the state of the marine environment of the Baltic Sea, 1989–1993; *Baltic Sea Environment Proceedings* No. **64B**.

HELCOM (1997) Airborne pollution load to the Baltic Sea 1991–1995. *Baltic Sea Environment Proceedings* No. 69.

HELCOM (1998) The third Baltic Sea pollution load compilation. *Baltic Sea Environment Proceedings* No. 70.

HELCOM (1998a). Red list of marine and coastal biotopes and biotope complexes of the Baltic Sea, Belt Sea and Kattegat. *Baltic Sea Environment Proceedings* No. 75.

Helios Rybicka, E. (1996) Impact of mining and metallurgical industries on the environment in Poland. *Applied Geochemistry* **11**, 3–9.

Hensen, V. (1875) Über die Befischung der deutschen Küsten. In *Jahresbericht der Commission zur wissenschaftlichen Untersuchung der deutschen Meere in Kiel für die Jahre 1872, 1873* (in German). Berlin, pp. 341–380.

Holm, E. (1995) Plutonium in the Baltic Sea. *Appl. Radiat. Isot.* **46**, 1225–1229.

IMGW (1987–1998) Environmental conditions in the Polish zone of the southern Baltic Sea (in Polish), eds. B. Cyberska, Z. Lauer and A. Trzosinska. Institute of Meteorology and Water Management, Maritime Branch Materials, Gdynia.

Ishaq, R., Näf, C., Zebhur, Y., Järnberg, U. and Broman, D. (1999) PCBs, PCNs, PCDD/Fs, PAHs and Cl-PAHs in air and water particulate samples—patterns and variations. *Environmental Science and Technology* (to be submitted).

Jansson, B.O. (1972) Ecosystem approach to the Baltic problem. *Bull. Ecol. Res. Comm. NFR* **16**, 82 pp.

Jansson, B.-O. and Dahlberg, K. (1999) The environmental status of the Baltic Sea in the 1940s, today, and in the future. *Ambio* **28**, 312–319.

Jarvekulg, A. (1979) Donnaja fauna vostocnoj casti Baltijskogo Morja. Sostav i ekologija respledelenija. Tallin.

Kannan, K. and Falandysz, J. (1997) Butyltin residues in sediment, fish, fish-eating birds, harbour porpoise and human tissues from the Polish coast of the Baltic Sea. *Marine Pollution Bulletin* **34**, 203–207.

Kannan, K. and Falandysz, J. (1999) Response to the comment on: Butyltin residues in sediment, fish, fish-eating birds, harbour porpoise and human tissues from the Polish coast of the Baltic Sea. *Marine Pollution Bulletin* **38**, 61–63.

Kautsky N., Kautsky, H., Kautsky, U. and Waern, M. (1986) Decreased depth penetration of *Fucus vesiculosus* (L.) since 1940. indicates eutrophication of the Baltic Sea. *Marine Ecology Progress Series* **28**,

Keida, M.E. (1998) Description of fishery and catch composition of roach (*Rutilus rutilus* (L.)), saberfish (*Pelecus cultratus* (L.)) and perch (*Perca fluviatilis* (L.)) in the Vistula Lagoon of the Baltic Sea. Proceedings of the Symposium „Freshwater fish and the herring populations in the Baltic coastal lagoons; environment and fisheries", MIR, Gdynia, pp. 109–118.

Kierkegaard, A., Sellström, U., Bignert, A., Olsson, M., Asplund, L., Jansson, B. and de Wit, C. (1999) Temporal trends of a polybrominated diphenyl ether (PBDE), a methoxylated PBDE, and hexabromocyclododecane (HBCD) in Swedish biota. *Organohalogen Compounds* **40**, 367–370.

Kjeller, L.-O. and Rappe, C. (1995) Time trends in levels, patterns, and profiles for polychlorinated dibenzo-p-dioxins, dibenzofurans, and biphenyls in a sediment core from the Baltic Proper. *Environmental Science and Technology* **29**, 346–352.

Krasovskaya, N.V. (1998) Reproduction and abundance dynamics of the Baltic herring (*Clupea hearengus membras*, L.) in the Vistula Lagoon of the Baltic Sea in 1980–1997. Proceedings of the Symposium „Freshwater fish and the herring populations in the Baltic coastal lagoons; environment and fisheries", MIR, Gdynia, pp. 128–142.

Krüger, O. (1996) Atmospheric deposition of heavy metals into North European marginal seas: Scenarios and trends for lead. *GeoJournal* **39**, 117–131.

Li, Y.-H. (1981) Geochemical cycles of elements in human perturbation. *Geochimica Cosmochimica Acta* **45**, 2073–2084.

Lindahl, A.H. (1998) Preliminary study of the conditions in some coastal parts of the Bothnian Sea. Ad hoc Expert Meeting on the Status of Eutrophication in the Baltic Marine Area, 2–4 February 1998, Berlin.

Lozán, J.L., Lampe, R. Matthäus, W., Rachor, E., Rumohr, H. and Westernhagen, H. (eds.) (1996) *Warnsignale aus der Ostsee: wissenschaftliche Fakten*. Parey, Berlin, pp. 1–385.

Luoma, S.N. (1983) Bioavailability of trace metals to aquatic organisms—a review. *The Science of the Total Environment* **28**, 1–22.

Lysiak-Pastuszak, E. (1999) Ph.D. thesis. An impact of the climatic and hydrological factors on the variability in the physical and chemical parameters in the marine environment of the southern Baltic. University of Gdansk (in Polish).

Majewski, A. and Lauer, Z. (eds.) (1994) *Atlas of the Baltic Sea* (in Polish). Institute of Meteorology and Water Management, Warszawa.

Magaard, L. and Rheinheimer, G. (1974) *Meereskunde der Ostsee*. Springer-Verlag, Berlin, pp. 171–260.

Maksimov, Yu. and Š. Toliušis, 1998 Water pollution effect on abundance of coastal herring (*Clupea harengus membras*) and twaite shad (*Alosa fallax* Lacepede) in the Curonian Lagoon. Proceedings of the Symposium Freshwater Fish and the Herring Populations in the Baltic Coastal Lagoons; Environment and Fisheries, MIR, Gdynia, pp.153–160.

Matschullat, J. (1997) Trace element fluxes to the Baltic Sea: problem of input budget. *Ambio* **26**, 363–368.

Matthäus, W. and Francke, H. (1992) Characteristics of major Baltic inflows—a statistical analysis. *Continental Shelf Research* **12**, 1375–1400.

Melvasalo, T., Pawlak, J., Grasshoff, K., Thorell, L. and Tsiban, A. (eds.) (1981) Assessment of the effects of pollution on the natural resources of the Baltic Sea, 1980. *Baltic Sea Environment Proceedings* No. **5B**, 1–426.

Mikulski, A. (1991) Comparative investigations on the water balance in the Baltic regions (in Polish). *Czasopismo Geograficzne* **63**, 177–185.

Nehring, D. (1989) Phosphate and nitrate trends and the ratio oxygen consumption to phosphate accumulation in central Baltic deep waters with alternating oxic and anoxic conditions. *Beitr. Meeresk.*, Berlin, **59**, 47–58

Neumann, T., Leipe, T., Brand, T. and Shimmield, G., (1996) Accumulation of heavy metals in the Oder Estuary and its off-shore basins. *Chem. Erde* **56**, 207–222.

Niemirycz, E., Bogacka, T. and Taylor, R. (1996) The Polish contribution to the total pollution input to the Baltic Sea (in Polish). *Wiadomosci IMGW* **39/40**, 63–84.

Osika, R. (1986) Poland. In *Mineral Deposits of Europe Volume 3: Central Europe*, eds. F.W. Dunning and A.M. Evans, pp. 55–97. The Institution of Mining and Metallurgy The Mineralogical Society, London.

Ravizza, G.E. and Bothner, M.H. (1996) Osmium isotopes and silver as tracers of anthropogenic metals in sediments from Massachusetts and Cape Cod bays. *Geochimica Cosmochimica Acta* **60**, 2753–2763.

Remane, A. (1958) Okologie der Brackwassers. In: Die Biologie des Brackwassers, eds. A. Remane and Schlipper. *Binnengewasser* **12**, 1–126.

Renk, H. (1990) Primary production of the southern Baltic in 1979–1983, *Oceanologia* **29**, 51–75

Renner, R.M., Glasby, G.P. and Szefer, P. (1998) Endmember analysis of heavy-metal pollution in surficial marine sediments from the Gulf of Gdansk and Southern Baltic Sea. *Applied Geochemistry* **13**, 313–318.

Repecka, R. (1998) Biology and resources of the main commercial fish species in the Lithuanian part of the Curonian Lagoon. Proceedings of the Symposium Freshwater Fish and the Herring Populations in the Baltic Coastal Lagoons; Environment and Fisheries, MIR, Gdynia, pp. 185–195.

Report (1995) Report on the status of the implementation of the Ministerial Declaration 1988 by the Republic of Poland. Draft, eds. E. Niemirycz and A. Trzosinska.

Report (1998) Short-term effects of 1997 flood on the marine environment of the Gulf of Gdansk and Pomeranian Bay (in Polish), A. Trzosinska, E. Andrulewicz (eds.), Morski Instytut Rybacki, Gdynia, pp. 1–76.

Rheinheimer, G. (1998) Pollution in the Baltic Sea. *Naturwissenschaften* **85**, 318–329.

Ropelewski, A. (1963) 1000 years of our fishery (in Polish). Wydawnictwo Morskie, Gdynia, pp 1–157.

Rybinski, J., Niemirycz, E. and Makowski, Z. (1992) Pollution load, In: An assessment of the effects of pollution in the Polish coastal area of the Baltic Sea 1984–1989. *Studia i Materialy Oceanologiczne* **61**, Marine Pollution (2), 21–52.

Segerstrále, S.G. (1957) Baltic Sea. *Mem. Geol. Soc. America* **67**, 751–800.

Segerstrále, S.G. (1972) The distribution of some malacostraca crustaceans in the Baltic Sea in relation to the temperature factor. *Merentutkimuslait. Julk.* **237**, 13–26.

Sekuliæ, B. and Vertaènik, A. (1997) Comparison of anthropogenical and "natural" input of substances through waters into Adriatic, Baltic and Black Sea. *Water Research* **31**, 3178–3182.

Senthilkumar, K., Duda, C.A., Villeneuve, D.L., Kannan, K., Falandysz, J., Giesy, J.P. (1999) Butyltin compounds in sediment and fish from the southern Baltic Sea, Poland, 1997–1998. *Environmental Science and Pollution Control* (in press).

Singer, D.A. (1995) World class base and precious metal deposits—a quantitative analysis. *Economic Geology* **90**, 88–104.

Skóra, K. and Stolarski, J. (1993) Round goby—a fishy invader. *WWF Bulletin* **1**.

Skwarzec, B. (1997) Polonium, uranium and plutonium in the southern Baltic Sea. *Ambio* **26**, 113–117.

Smal, H. and Salomons, W. (1995) Acidification and its long-term impact on metal mobility. In *Biogeodynamics of Pollutants in Soils and Sediments Risk Assessment of Delayed and Non-Linear Responses*, eds. W. Salomons and W.M. Stigliani, pp.193–212. Springer-Verlag, Berlin.

Strandberg, B., Strandberg, L., van Bavel, B., Bergqvist, P-A., Broman, D., Falandysz, J., Näf, C., Papakosta, O., Rolff, C. and Rappe, C. (1998a) Concentrations and spatial variations of cyclodienes and other organochlorines in herring and perch from the Baltic Sea. *The Science of the Total Environment* **215**, 69–83.

Strandberg, B., Strandberg, L., Berqvist, P.-A., Falandysz, J. and Rappe, C. (1998b) Concentrations and biomagnification of 17 chlordane compounds and other organochlorines in harbour porpoise *Phocoena phocoena* from southern Baltic Sea. *Chemosphere* **37**, 2513–2523.

Szefer P. (1991) Interphase and trophic relationships of metals in a southern Baltic ecosystem. *The Science of the Total Environment* **101**, 201–215.

Szefer, P., Pempkowiak, J. and Kaliszan, R. (1993) Distribution of elements in surficial sediments of the Southern Baltic determined by factor analysis. *Studia i Mateialy Oceanologiczne* **67**, *Marine Pollution* (3), 85–93.

Szefer, P., Glasby, G.P., Pempkowiak, J. and Kaliszan, R. (1995) Extraction studies of heavy-metal pollutants in surficial marine sediments of Gdansk Bay and the Southern Baltic Sea off Poland. *Chemical Geology* **120**, 111–126.

Szefer, P., Glasby, G.P., Szefer, K., Pempkowiak, J. and Kaliszan, R. (1996) Heavy-metal pollution in surficial marine sediments from the southern Baltic Sea off Poland. *Journal of Environmental Science and Health* **31A**, 2723–2754.

Szpunar, J., Falandysz, J., Schmitt, V.O. and Obrebska, E. (1997) Butylyins in marine and freshwater sediments of Poland. *Bulletin of Environmental Contamination and Toxicology* **58**, 859–864.

Theobald, N. (1994) HELCOM-Arbeitsgruppe 'Chemische Kampfstoffe'-Schlussfolgerungen und kunftige Aktivitäten. *Dtsch Hydrogr. Z. Suppl.* **2**, 133–138.

Trzosinska A. (1992) Nutrients, In An assessment of the effects of pollution in the Polish coastal area of the Baltic Sea, 1984–1989, *Studia i Materialy Oceanologiczne* **61**, *Marine Pollution* (2), 107–130.

Verner, J.F., Ramsey, M.H., Helios Rybicka, E. and Jedrzejczyk, B. (1996) Heavy metal contamination of soils around a Pb-Zn smelter in Bukowno, Poland. *Applied Geochemistry* **11**, 11–16.

Wulff, F., Stigebrandt, A. and Rahm, L. (1990) Nutrient dynamics of the Baltic Sea. *Ambio* **29**, 126–133.

THE AUTHORS

Jerzy Falandysz
University of Gdansk,
18 Sobieskiego Str.,
PL 80-952 Gdansk, Poland

Anna Trzosinska
Institute of Meteorology and Water Management,
42 Waszyngtona Str.,
PL 81-342 Gdynia, Poland

Piotr Szefer
Medical University of Gdansk,
107 Gen. Hallera Ave.,
80-416 Gdansk, Poland

Jan Warzocha
Sea Fisheries Institute,
1 Kollataja Str.,
PL 81-332 Gdynia, Poland

Bohdan Draganik
Sea Fisheries Institute,
1 Kollataja Str.,
PL 81-332 Gdynia, Poland

Chapter 8

THE BALTIC SEA, INCLUDING BOTHNIAN SEA AND BOTHNIAN BAY

Lena Kautsky and Nils Kautsky

The Baltic Sea region is one of the largest brackish water areas in the world with a salinity declining from about 10 PSU in the south in the Baltic Proper, declining through the Bothnian Sea to 2 PSU in the northern Bothnian Bay. It is surrounded by 14 densely populated and industrialised countries, where 90 million people live within the drainage area. The narrow and shallow Danish straits limit the water exchange and result in a residence time of the water of 25–30 years. The north–south salinity gradient restricts the northward penetration of marine organisms, and the few which do occur in the north often show dwarfism. A permanent halocline at about 60–80 m depth prevents vertical circulation, and oxygenation of the deep water is limited to a few events of salt-water inflow. About one third of the bottom area of the Baltic Proper is devoid of higher life due to low oxygen. The Baltic Sea has no tides.

The hard bottoms are dominated by communities of *Fucus vesiculosus* and *Mytilus edulis*, and some further 30 species. Soft-bottom communities are dominated by the bivalve *Macoma balthica* and three more species in the Baltic Proper, but both biomass and number of species decrease rapidly to the north. Production of pelagic fish like herring and sprat is high, but benthic species like cod and flounder are declining due to eutrophication. There is also a significant fishery of freshwater fish such as pike and perch in the coastal areas. The pollution load is high, affecting e.g. reproduction of top carnivores, although species such as seals and white-tailed sea eagle, which were threatened by DDT and PCB levels until recently, are now recovering. The main problem is eutrophication, favouring growth of nuisance algae, and a lowering of oxygen in the benthos. Since most organisms of this region are already living under severe physiological stress, they are sensitive to pollution, and the low number of species increases the risk that whole functional groups can be lost.

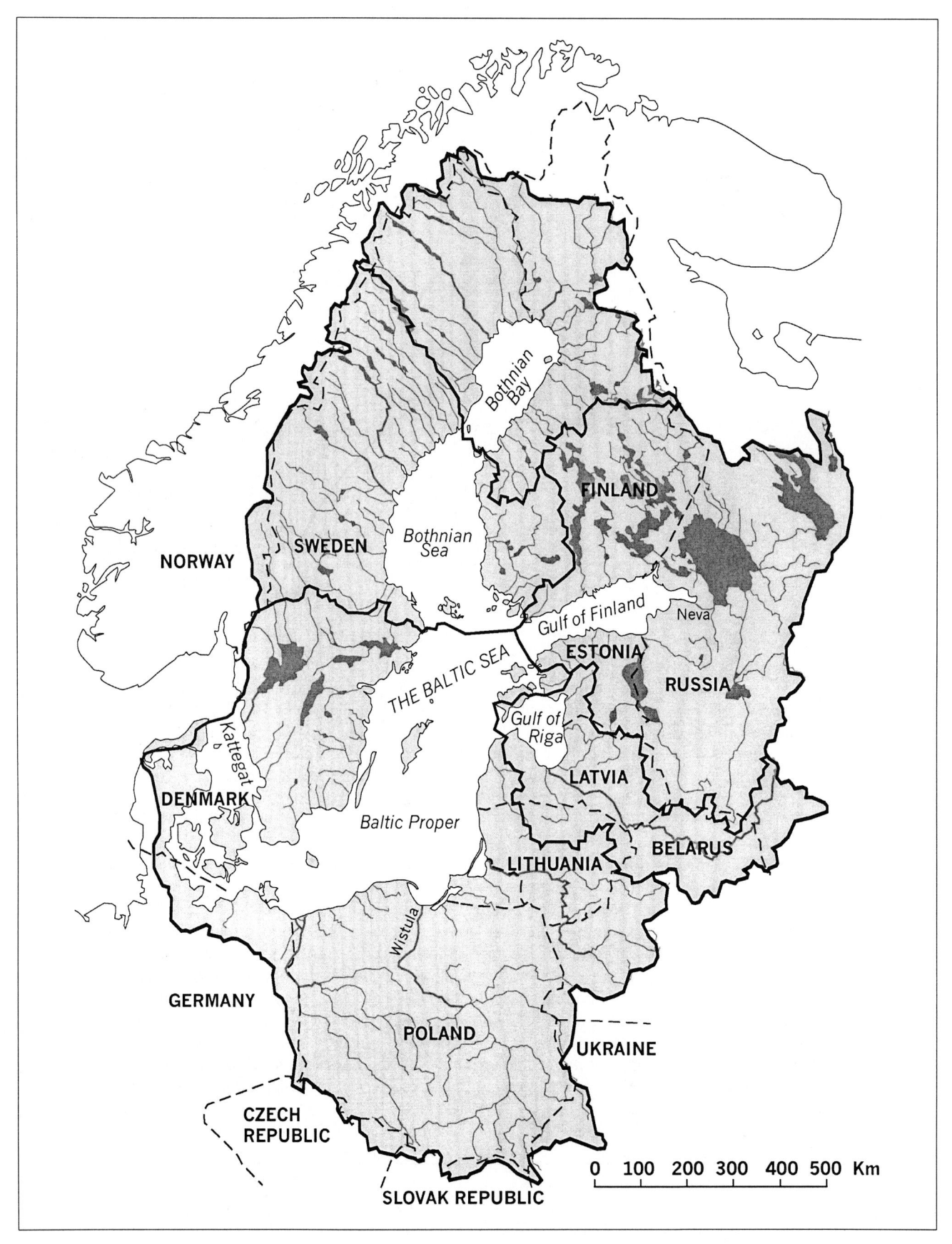

Fig. 1. The Baltic Sea and its drainage area divided into five larger sub-areas indicated by thick black lines. The major basins are the Bothnian Bay, the Bothnian Sea, the Baltic Proper and the Gulf of Finland and the Gulf of Riga. (Drawing by Robert Kautsky)

INTRODUCTION

The Baltic Sea region includes the Bothnian Sea and Bothnian Bay, together with the Gulf of Finland and Gulf of Riga (Fig. 1). The region has attracted much attention from marine scientists during the past century, and is one of the most well studied of the world. It links nine nations together along its coasts and 14 countries, with a total population of about 90 million people, lie within its drainage area. It is one of the largest brackish water areas in the world, characterised by a large input of freshwater from precipitation and many rivers, and may thus be looked upon as a gigantic estuary. Its five basins are separated by shallower areas or sills. The Baltic Proper is the largest and the deepest, with a maximum depth of 459 m. Sweden has the longest coastline of the countries surrounding the Baltic Sea, extending from the southern part of the Baltic Proper, the Öresund, to the northern Gulf of Bothnia.

The drainage area of the Baltic Sea is over 1.7 million km^2, which is more than 4 times larger than the water surface area (0.42 million km^2) (see Fig. 1). The largest rivers are the Neva, entering the Gulf of Finland, and Wistula, flowing into the Baltic Proper, but a large number of rivers enter the Bothnian Sea and Bothnian Bay. The climate is cold–temperate and rather humid with high land run-off and low evaporation. There are significant seasonal variations in river discharge, with a maximum run-off generally occurring in the spring following snow-melting. This gives the Baltic Sea a positive water balance with an annual freshwater surplus of 481 km^2 (Matthäus and Schinke, 1999). The most important obstacle for a free water exchange with the North Sea is the shallow depth across the Danish Straits.

The dominating water circulation in each of the three largest basins, i.e. the Bothnian Bay, Bothnian Sea and Baltic Proper, are anti-clockwise cells created by the rotation of the earth's Coriolis force. Thus, when a river enters the Baltic, the water turns to the right and follows the shore. This will result in a longer residence time for the heavily polluted water from the large rivers Wistula and Neva entering in the south and east, which is thus kept in the Baltic Proper for a longer period than otherwise would have been the case. The residence time of the water has been estimated to be 25–35 years.

ENVIRONMENTAL FACTORS AFFECTING THE BIOTA

The most important factor affecting life is salinity. Due to large volume, a positive water balance and a semi-enclosed topography, the Baltic shows very small variations in salinity at any given point (Ehlin, 1981). The salinity in the Northern Quark (a transition area between the Bothnian Bay and Bothnian Sea), varies between 3 and 4 PSU and increases in the Bothnian Sea to about 5 PSU (Fig. 2). The

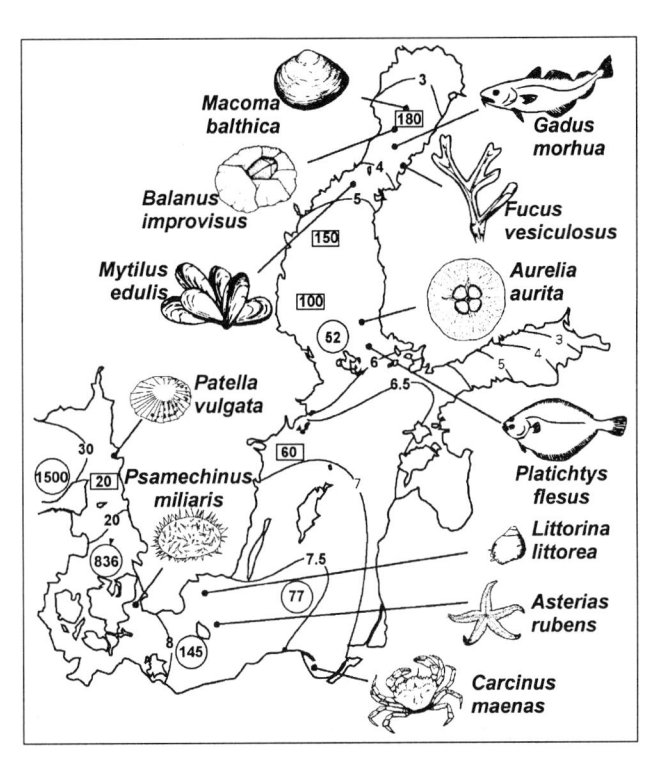

Fig. 2. Distribution limits for some common marine species in the Baltic Sea. Surface water isohalines. Salinity is presented as PSU. Numbers within squares: number of days with ice cover in the shallow parts of the area. Numbers within circles: number of macrofauna species. (After Jansson, 1972). Drawings by C. Bollner.

salinity increases rapidly from 6–8 PSU in the Baltic Proper through the Danish Sounds to about 15–20 PSU in the southern Kattegat. In the northern Kattegat and Skagerrak, salinities are more variable with a mean of about 25–30 PSU in the surface water. Reviews on the physical and biological characteristics of the Baltic Sea are given in Voipio (1981), Håkansson et al. (1996) and Kuparinen et al. (1996).

Annual solar radiation in the Bothnian Bay is about 30% and the number of degree-days in the depth interval 0–10 m is 65% of the values estimated for the southern part of the Baltic Proper. Thus, the salinity gradient is paralleled by a productive season of only 4–5 months in the far north of the Bothnian Bay, caused by up to six months of ice cover. In the southern part of the Baltic the productive season is much longer, i.e. 8–9 months, with open sea ice only in cold years, although coastal embayments have ice-cover more frequently (Fig. 2).

The central parts of the Baltic Proper are permanently stratified. The upper layer consists of brackish water with salinities of 6–8 PSU and a deeper water layer of 10–14 PSU. A combined halo- and thermocline at depths between 60 and 80 m prevents vertical circulation. During spring when the upper water mass is heated, a shallow thermocline develops at 15–30 m depth that further restricts vertical circulation during the summer months.

In the Baltic Proper, spring heating starts in March–April, while in the Gulf of Bothnia it is delayed by one to two months. Surface water temperatures in both north and south show about the same range from 0°C in the winter period to around 20°C during warm summers, but the length of the warm water period and growth season decreases northward. Surface water temperature reaches its maximum in August, nearly simultaneously over the entire Baltic, albeit with different values in the Bothnian Bay, the Bothnian Sea, and the Baltic Proper.

The tidal difference is about 20 cm on the Swedish North Sea coast, while within the Baltic Sea it is almost negligible: about 2 cm only. Changes in weather, however, cause large irregular differences in the water levels, up to 1 m, which have effects on shallow benthic communities and their zonation pattern (Johannesson, 1989). Ninety percent of the low-water periods occur in winter–early spring (i.e. October–March) and, during longer periods of low water, benthic animals and plants might be killed by freezing and/or desiccation.

Environmental conditions in the deep water of the Baltic Proper are strongly dependent on inflows through the Danish Sound of saline and oxygenated water from the North Sea. These inflows are restricted by the narrow channels and shallow sills, about 10–20 m deep. The main inflow of denser, more saline, and oxygenated water takes place in autumn and winter, and is mainly dependent on air pressure and storms. Since the mid 1970s only a few major inflows have occurred and in the ten-year period February 1983 to January 1993, there have been none, which has resulted in a reduction in salinity in the deeper parts of the Baltic Sea (Mattäus and Schinke, 1999). Higher runoff from the rivers is usually followed by lowered salinity, first in the surface waters of the coastal areas and then, with a delay of about 6 years, in the deeper waters of the Baltic.

COASTAL AND MARINE HABITATS

There are large differences in topographical, hydrodynamic and biological character between the coasts of the different regions. Archipelagos dominate the Swedish and the Finnish coastline, clint coasts can be found on the west coast of Gotland and in Estonia. The archipelagos in the northern Baltic Proper, the Åland Sea and the Archipelago Sea are all shallow areas with thousands of islands forming a huge maze with a variety of sounds and bays. The still-continuing land uplift, following the release of pressure from ice during the last glaciation period, is on average 1.5 mm per year and may lead to the isolation of new water bodies. The archipelago of Stockholm contributes 25% of the total coastline in Sweden. Similarly the archipelago of Åland and the Finnish Archipelago Sea are all very important areas for marine production, because they have a long coast line, and well developed marine communities.

In the middle and southern parts of the Baltic Proper there are several off-shore shallow (less than 20–25 m) bank areas consisting of rock or a mosaic of boulders, gravel and sand. These areas are valuable spawning grounds and nursery areas for young fish.

Other typical coastal habitats are the large coastal lagoons, which have many features in common, e.g a lower salinity than the open Baltic and a more limited water exchange which traps nutrients within them. These shallow lagoons have a high biomass production due to the good supply of nutrients, high temperature and light. The large supply of food attracts migratory birds, and the southern part of the Baltic coastline is part of the East-Atlantic Flyway for migrants. Many of the coastal lagoons have been made nature reserves and long-term management plans of these valuable south-eastern parts of the Baltic Sea are now being developed.

Biodiversity of the Baltic Sea

The Baltic Sea was formed only some 12,000 years ago when the ice-cap retreated after the last glaciation. Since then it has changed a number of times from a marine area to a large freshwater lake and then back again into a marine/brackish environment. The marine flora and fauna of the Baltic was probably recruited after the beginning of the Littorina Sea period (ca. 7500 BP), when the connection between Baltic and Atlantic was wider. After that there has been a progressive decline in salinity until ca. 3000 BP, after which it stabilised (Voipio, 1981).

Compared to fully marine environments, these brackish waters have a very species-poor flora and fauna. The total number of marine species decreases dramatically as one goes through the Danish Straits into the Baltic Proper and then continues to decrease through the Archipelago Sea into the Bothnian Sea, and finally the Bothnian Bay (Fig. 2). Of the 1500 macroscopic animals found off the Swedish North Sea coast, only some 70 species still occur in the central Baltic. Of approximately 320 benthic macroalgae along the Danish and Swedish Kattegat coasts, only about 90 remain in the northern part of the Baltic Proper, the Gulf of Finland and the Bothnian Sea (Fig. 3) (Wallentinus, 1991;

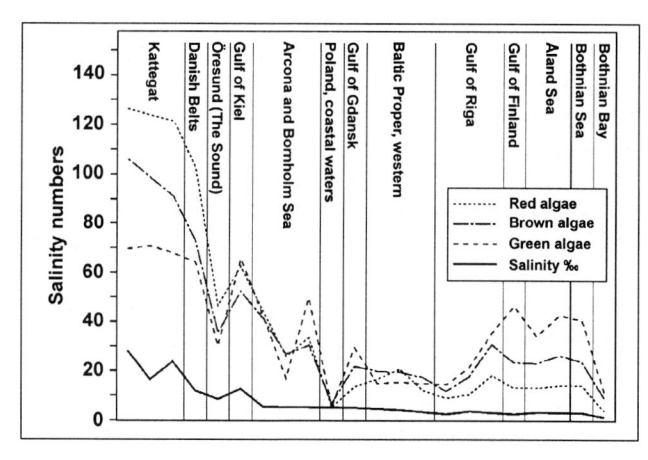

Fig. 3. Number of red, brown and green algae in different parts of the Baltic Sea, and (continuous line) salinity. (After Nielsen et al., 1995).

Nielsen et al., 1995). Two hundred species of bivalve molluscs are found in the North Sea while there are only four in the Gulf of Finland. In Bothnian Bay very few species are able to grow, due to the low salinity, and the proportion of freshwater species is high. However, for some marine organisms, low temperature rather than salinity seems to be the factor limiting further penetration into the region.

It seems unlikely that the Baltic Sea contains any unique or endemic species; the region has probably not existed long enough. In fact almost all of the organisms that once were suspected to be unique to the Baltic have later been shown to be adapted variants of species found elsewhere (see Nielsen et al., 1995). The candidates are all either just subspecies, or belong to little-studied or taxonomically difficult groups. The Baltic flora and fauna therefore sharply contrasts to, for example, the Mediterranean Sea, where 40% of the plant species are considered to be endemic.

The long-term adaptation to brackish-water conditions has, however, created considerable differences between e.g. macroalgal populations living in the Baltic and those found in normal seawater (Russell, 1988; Serrão et al., 1999). There is evidence that several marine macroalgae species in the Baltic may have undergone some ecotypic differentiation with regard to salinity tolerance. Also, most invertebrate species seem not to be physiologically fully adapted to the low saline environment. For example, the blue mussel *Mytilus edulis* is a marine intruder into the brackish environment and a euryhaline osmoconformer (Schlieper, 1971). In the Baltic Sea this species inhabits waters down to about 4‰S, although it has problems with water and energy balances at salinities below 18‰S (Schlieper, 1971), resulting in reduced size and growth rates compared to North Sea specimens (Remane and Schlieper, 1971; Kautsky, 1982). Due to very few predators or competitors for space in the Baltic Sea, this has become a dominant species despite these physiological problems, covering hard substrates from the water surface to more than 30 m depth (Jansson and Kautsky, 1977; Kautsky, 1981).

Immigrant species from fresh water, here represented by the gastropod snail *Theodoxus fluviatilis*, is one of the dominant macroscopic species in the Baltic littoral (Jansson and Kautsky, 1977). This species is thought to have some 400,000 years of evolutionary history under limnetic conditions (Russell-Hunter, 1964), and is a recent invader in the Baltic (ca. 7000 years ago; Hubendick, 1947). The size of this species decreases with increasing salinity in the Baltic Sea (Remane and Schlieper, 1971), and larger specimens are found in freshwater lakes than exist in the Baltic Proper (Kangas and Skoog, 1978). This is exactly the opposite to the change in size exhibited by marine mollusc species like *Mytilus edulis*.

Some organisms, in particular the glacial relict crustaceans, are well adapted to the low temperature conditions prevailing in winter. Thus, the large isopod *Saduria entomon*, the bottom-dwelling amphipod *Pontoporeia femorata* and the mysid shrimp *Mysis relicta* even breed under the winter ice. In this way they improve the possibilities for juveniles which are hatched during the productive peak of the pelagic ecosystem in early spring (Voipio, 1981).

Hard-bottom Communities

On hard substrate there are about 100 species of brown algae in the North Sea but only some 20 species in the Baltic Sea (Fig. 3). In the uppermost zone on rocky shores a typical succession of species occurs during the year. In spring when the ice cover is gone, the annual succession starts with a heavy bloom of sessile diatoms covering the bare rocks at the waterline. The diatoms are soon replaced by the filamentous brown algae *Pilayella littoralis*, followed by the green algae *Cladophora glomerata* during summer, and filamentous red algae *Ceramium tenuicorne* and brown algae during autumn and winter. This belt looks different in different parts of the Baltic. In the Bothnian Bay a perennial, i.e., long-lived, vegetation of *Cladophora aegagrophila* and the water moss *Fontinalis* starts at 2 m depth which is the normal lower limit of influence of the sea ice (Voipio, 1981). In the Bothnian Sea many fresh water species are still abundant. The vertical distribution of vegetation is dependent on light conditions. The benthic vegetation reaches down to about 10 m in the Bothnian Bay, to about 18–25 m in the Baltic Proper and to about 30 m in the Belt Sea area (Voipio, 1981).

Being the only large long-lived, belt-forming algae on hard bottoms *Fucus vesiculosus*, together with *Fucus serratus* in the southern and middle part of the Baltic Proper, are key species in these ecosystems. Their geographical distribution limit is, as for many other species, set by the salinity during the reproductive period (Serrão et al., 1999; Malm, 1999). *Fucus* populations within the Baltic are permanently submerged, due to the combined effects of lack of tides, ice scour in winter, and extended periods of low water levels in spring. The bladderwrack extends down to about 8 m depth in the northern regions of its distribution, while in the Danish straits in the south it is only found close to the surface. An exceptionally rich fauna, including mussels, snails, crustaceans, bryozoans and even insect larvae (Voipio, 1981) inhabits the *Fucus vesiculosus* stands. This is the most species-rich community in the Baltic Sea, containing some ten species of algae and 30 macroscopic animal species. Many fish species pass their larval stages in the *Fucus* belt feeding on the rich variety of small animals. During summer, *Fucus* can get heavily overgrown by several species of filamentous green and brown algae. Competition for light and nutrients by these filamentous algae are, in fact, a threat to its photosynthesis and growth. The isopod *Idothea* has been sometimes found to be able to graze heavily on *Fucus* and destroy the bladderwrack communities over large areas (Kangas et al., 1982; Malm, 1999).

At depths where scarcity of light limits algal growth, the blue mussel *Mytilus edulis* may dominate entirely. The mussel belt normally starts at a few metres of depth and often extends to 30 m or until the bottom gets too muddy to allow the mussel to attach. In the Baltic Proper the animal biomass is totally dominated by the blue mussel (*Mytilus edulis*), which accounts for more than 90% of the total animal biomass on hard bottoms. The total blue mussel population in the Baltic Proper including the Åland Sea down to 25 m depth is about 8 million tons of dry weight including shells (Kautsky, 1981). The main predators on the mussels are eider ducks and some fish, e.g. plaice (Kautsky, 1981, 1982).

The filtering capacity of blue mussels is very high; the total amount of mussels within 160 km^2 in the Askö area, northern Baltic Proper, are theoretically able to filter the whole overlying water mass (mean depth about 20 m) once in about two weeks in summer. For the total Baltic Sea, the mussels are able annually to filter the total volume of the Sea, which could be compared to natural residence time of about 25–30 years. Particulate matter is, in this way consumed, metabolically processed and inorganic salts excreted (Kautsky, 1980, 1981). The nitrogen and phosphorus excreted by the mussels fill the entire nutrient demands of the large macroalgae in the area where they grow. On top of that the *Mytilus* population also furnishes 5% of the nitrogen and 20% of the phosphorus needed to maintain the total phytoplankton biomass in the same area. In total, the blue mussels in the entire Baltic Sea excrete roughly 250,000 tons of inorganic nitrogen and 80,000 tons of inorganic phosphorus per year. These figures are easily as large as the total land-based inputs to the Baltic of these substances. However, it should be noted that the mussels contribution is a re-circulation of already existing nutrients, while the land inputs represent a new addition of nutrients to the ecosystem that may lead to eutrophication.

The low salinity, temperature and short growing season, all contribute to much lower standing stock, i.e. only about 1% of the biomass per m shore line in the Bothnian Bay (Fig. 4A) increasing to 10% in the Bothnian Sea (Fig. 4B) compared to the Baltic Proper (Fig. 4C). Total net primary production in the Baltic Sea has been estimated at 5.9×10^8 kg carbon per year of which 87% was found in the Baltic Proper, 11% in the Bothnian Sea and only 2% in the Bothnian Bay (Kautsky, 1995) (Fig. 4). In the Baltic Proper the filtering blue mussels consume about 35% of the pelagic production and a large proportion (33–42%) is channelled to the benthic consumers on deeper soft bottoms (Kautsky and Kautsky, 1991). In the Bothnian Bay and Bothnian Sea, filtering bivalves play a much smaller role, and only 5% of the pelagic primary production is consumed by filter feeders. In the Bothnian Bay, 33% of the total production of carbon is covered by allochthonous carbon from land run-off, which is four times higher than in other parts of the Baltic Sea (Elmgren, 1984).

Soft-bottom Communities

Most of the sea floor consists of muddy and sandy sediments. In the shallow protected bays on the coast the freshwater influence is obvious, resulting in extensive beds of reed, *Potamogeton* spp. and other rooted brackish and freshwater macrophytes (Münsterhjelm, 1987). Marine eelgrass (*Zostera marina*) may cover sandy bottoms rich in organic matter especially off the southern coasts of the Baltic Proper, but never reaches the same dominance and productivity as in fully marine waters.

The fauna of shallow bays has a corresponding freshwater character, where insect larvae of dragon flies and mayflies are found together with water bugs and freshwater fish like bream, rudd, tench, pike and perch. The production of this area is high, with a production outburst in spring when the shallow water is rapidly warmed up and the microflora blooms, stimulating secondary production. At this time many fish spawn here since potential larval food is abundant (Lappalainen et al., 1977). In fact much of the production goes to larval food for fish (Nellbring, 1988). Due to its high production, the shallow soft bottoms function as nutrient traps and much is stored as plant biomass, especially in the large archipelagos.

The main parts of the Baltic soft bottom have been classified as a *Macoma* community from the dominating marine mussel *Macoma balthica* (Voipio, 1981). The deep basins of the Baltic Proper have a greatly impoverished fauna, with almost no macroscopic animals (Voipio, 1981). In the Bothnian Bay and the central part of the Bothnian Sea a species-poor community of the isopod *Saduria entomon* and amphipod *Pontoporeia* spp. dominates. Also on the soft bottoms, primary production and animal biomass decreases from south to north, partly caused by the salinity decrease, but also by differences in growth season (Fig. 2) (Elmgren, 1984). The low values in the Bothnian Bay are mainly a result of the exclusion of the marine mussels. In the absence of filter-feeding mussels in the Bothnian Bay the decomposition of organic material is taken over by a variety of meiofauna living in the sediment. This results in a longer and therefore less efficient food chain before the macrofauna and fish levels are reached.

Pelagic Communities

Annual pelagic primary production is highest in the Baltic Proper and decreases to the north, being only some 10% of the values for the Baltic Proper in the Bothnian Bay (Fig. 4) (Elmgren, 1984). This annual production is a sum of several intensive so-called algal blooms during the year, varying from about nine in the hydrographically "noisy" belts in the south to three in the northern Baltic Proper, and only one in the Bothnian Bay. Here the spring and autumn blooms have merged together, which is a typical polar feature.

The spring bloom in the northern Baltic Proper starts in late February or early March. With increasing light and

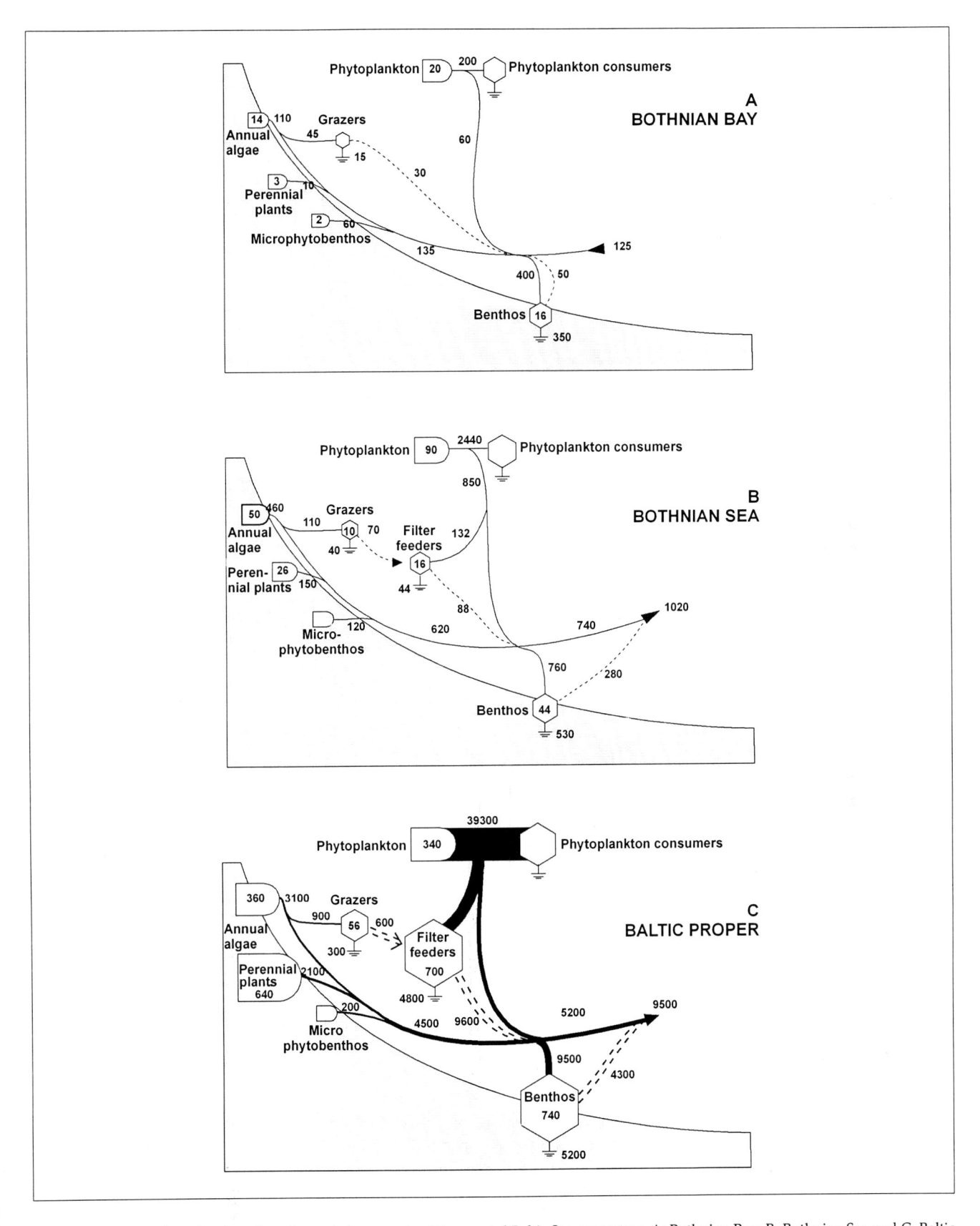

Fig. 4. Standing stock and carbon flow through three parts of the coastal Baltic Sea ecosystem, A: Bothnian Bay, B: Bothnian Sea and C: Baltic Proper. Bullet-shaped symbols are primary producers and hexagons are consumers. Phytoplankton consumers include both the microbial loop and the zooplankton. Downpointing arrows from the storages are respiration loss. Numbers inside the symbols are standing stock (10^5 kg carbon) and by the arrows are carbon flow (10^5 kg C y^{-1}). Black arrows show primary production and hatched arrows secondary production and faeces. (From Kautsky and Kautsky, 1991).

abundant nutrients, phytoplankton biomass rapidly increases, mainly composed of dinoflagellates, e.g. *Gonyaulax catenata* and diatoms *Thalassiosira baltica* and *Skeletonema costatum*. The fixation of solar energy is intense, primary production reaching values of 1.5 g carbon/m^2/day (Fig. 4). Bacterial populations rapidly increase and serve as food for pelagic ciliates, which promptly increase. These in turn are consumed by the rotifers *Synchaeta*, which also have a high potential for explosive growth and propagation.

The rapid sinking of diatoms, in combination with the absence of large herbivores, results in as much as 40% of the organic matter synthesised during the spring bloom sinking out of the pelagic zone to settle on the soft bottoms below (Fig. 4). In fact this constitutes as much as half of the annual requirement of food of the soft-bottom communities (see Kautsky, 1995).

In July–August a conspicuous bloom of blue-green algae, especially *Nodularia spumigena*, dominates the pelagic zone in the Baltic Proper. This nitrogen-fixing freshwater cyanobacterium produces gas vacuoles and sometimes these float in large quantities to the surface, where winds and currents arrange the thread-bundles in dense patches. In the usually nitrogen-starved Baltic waters, the process of nitrogen fixation is important and the estimated 100,000 tons of nitrogen fixed in the northern and central Baltic Proper are certainly important for the input of potential energy through stimulated fixation of solar energy. A problem of the *Nodularia* blooms is that they are potentially toxic to dogs, cattle and children that drink from the water.

FISH AND FISHERIES

As with other plants and animals, the fish fauna exhibits a mixture of species of freshwater and marine origin. Again, the salinity decrease causes the Baltic to be poor in species. In Kattegat there are 80 marine fish species, in the Sound only 55, in the Archipelago Sea about 20 and in the Gulf of Bothnia only eight, and a mix of freshwater and marine fish species such as perch, pike, cod and flounder can often be caught in the same net (Voipio, 1981).

The fishery in the Baltic Sea is rather high compared to other seas but varies in different parts. Freshwater fish are especially important in the catch from the northern and eastern parts. Leisure fishermen fish pike intensively.

The pelagic fish in the surface layers benefit from moderate eutrophication, and herring and sprat have greatly increased their populations in the Baltic for many years. The potential fish food on soft bottoms above the primary halocline has increased up to sevenfold since the 1920s, most probably due to eutrophication (Cederwall and Elmgren, 1980). The growth conditions for commercial fish living and feeding in deeper areas are less favourable in the Baltic than in the North Sea. Cod and flounder grow less, and the stagnant water in the deep basins has made the former foraging and reproduction areas in the southern

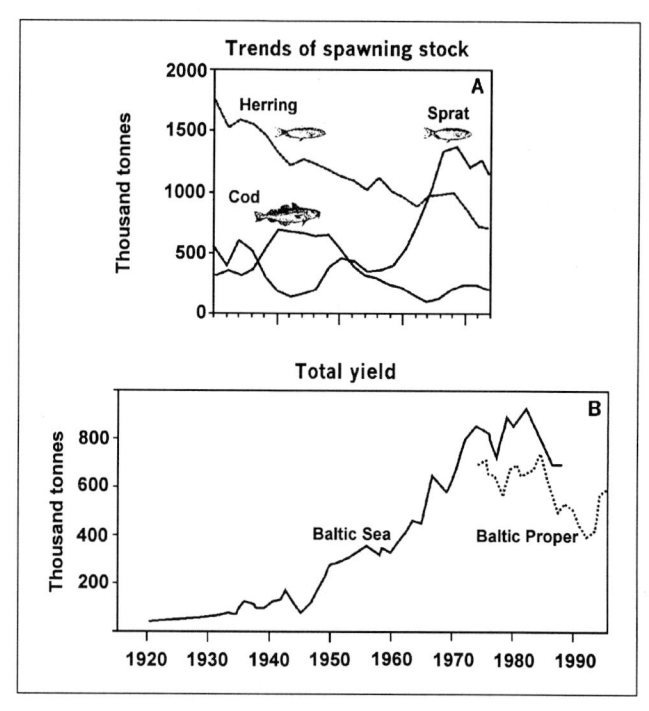

Fig. 5. A. Spawning stock of cod, herring and sprat in the Baltic Sea in 1000 tonnes (from ICES, 1998) B. Time series of total fish yield of cod, herring and sprat in the Baltic Sea (continuous line) and in the Baltic Proper (dotted line). The yield from the Baltic Proper dominates the fishery in the Baltic Sea (Thurow, 1989).

Baltic uninhabitable and unproductive. Cod eggs need a minimum of 10 PSU to develop successfully (Nissling and Westin, 1991), and the increased primary production since the 1950s has caused oxygen depletion in the deeper areas which has depressed reproductive success (Hansson and Rudstam, 1990). The critical salinity for cod reproduction is reached at 60–80 m depth in the Bornholm, Gotland and Arcona deep. However, single year-classes of cod may still be successful and give rise to intensive fishing as in 1979–80 (Thurow, 1980) (Fig. 5). As a result of the reduced cod population, the sprat and herring biomass has increased due to reduced predation.

The population of salmon in the Baltic Sea belongs to the Atlantic salmon species, *Salmo salar*. The Swedish stocks of Baltic salmon are mainly native to the big rivers in the northern part of the country. Originally some 50 Baltic Sea rivers accommodated spawning salmon populations (Voipio, 1981). Today very few still harbour reproducing stocks and much of the genetic variability has been lost. The building of dams for hydroelectric power plants since the 1940s has drastically changed and reduced the salmon populations. By establishing compensatory hatcheries at each river and mass rearing of smolts, which are released into the river, it has been possible to maintain stocks. The wild stocks now only contribute 10–15% of the total population, while the rest originates from artificially produced stocks with a reduced genetic variation (Ackefors et al., 1991).

Introduction of Non-native Species

Introductions of alien species may have great impact on the species living in the low diversity communities of the Baltic Sea (Leppäkoski, 1984). Through human activities many of the natural barriers have been weakened, e.g. by boat traffic. This has resulted in the unintentional importation of species via e.g. ballast water or as growth on ship hulls. In addition, intentional transplantation of commercial species of shellfish and fish, and associated organisms, has taken place with aquaculture. The number of introductions have increased drastically over the last 50–100 years. In addition, the slight increase in the salinity during recent decades has resulted in the spontaneous invasion of several euryhaline fauna species, a process known as "oceanization".

Today, more than 30 species (Fig. 6), mainly unintentionally introduced animals, have been reported from the central and northern Baltic coastal areas. A database containing introduced species is now kept by the Centre of Systems Analysis, University of Klaipéda.

The soft-shell clam *Mya arenaria* was introduced to European waters from North America in the 16th or 17th century with ballast water, and has been able to persist and thrive in the Baltic for 300–400 years (see Olenin and Leppäkoski, 1999). Some introduced species may be of economic interest, e.g. the Canada goose *Branta canadensis* which is hunted both in Finland and Sweden. Others are nuisance organisms: e.g. the Mitten crab, *Eriocheir sinensis* and the hydroid *Cordylophora caspia* that may damage fishing gear, the migrating mussel *Dreissena polymorpha* which may overgrow underwater constructions and clog cooling systems, and the muskrat which may damage constructions.

The introduction of non-native species has also increased the functional diversity of the benthic system in the Baltic Sea. Such changes in functions include *Dreissenia polymorpha* which filter-feeds in the low salinity areas where *Mytilus edulis* is absent, the barnacle *Balanus improvisus* which suspension-feeds in the uppermost hydrolittoral zone, the crab *Eriocheir sinensis*, a predator and scavenger, and the polychaete *Marenzelleria viridis* which is increasing bioturbation of sediment. *Mya arenaria* shells form a secondary hard substrate and the empty shells of *Balanus improvisus* serve as new microhabitat, increasing the habitat variability (Olenin and Leppäkoski, 1999). These examples show that non-native invertebrate species may modify their new habitats by increasing the functional diversity. These modifications include creating new microhabitats for associated fauna, broadening the food base and changing the energy flow of coastal ecosystems, and have been reported for coastal lagoons in the southern part of the Baltic Sea (Olenin and Leppäkoski, 1999). Young *Dreissena* are important food for certain fish and water fowl, and in the Gdansk Bay crabs feed as adults on *Dreissena* and as young on *Cordylophora*. The *Rithropanopeus* crab is, in its turn, a common food item for fish. In this way a whole new food chain has been constructed by introduced species.

POLLUTION SENSITIVITY OF THE BALTIC ECOSYSTEM

The drainage basin of the Baltic Sea is densely populated and heavily industrialised which adds a significant stress from pollution. There has long been a debate among marine biologists and environmental legislators concerning whether the Baltic or North Sea is the most sensitive to further pollution. Experimental studies have shown that a number of invertebrates from the Baltic are more sensitive in terms of physiological response and survival when exposed to toxic substances, compared to the same species from more saline environments, and there are no examples of the opposite (Tedengren et al., 1988).

The higher sensitivity to pollutants of Baltic populations can be explained by factors such as changes in the characteristics of many toxic substances (e.g. metals) with salinity. Also the relative ionic concentration of a certain amount of poisonous substance is higher in the low-saline Baltic Sea as compared to the North Sea. Furthermore, many toxicants are known to interact with membrane permeability and osmoregulatory mechanisms, that are already under strain at low salinities (Tedengren et al., 1988).

The generally lower genetic variation in Baltic Sea populations and the reduced ecosystem complexity will also increase vulnerability of the Baltic ecosystem. The risk that whole life functions may be excluded because of environmental disturbances is relatively high, since the low

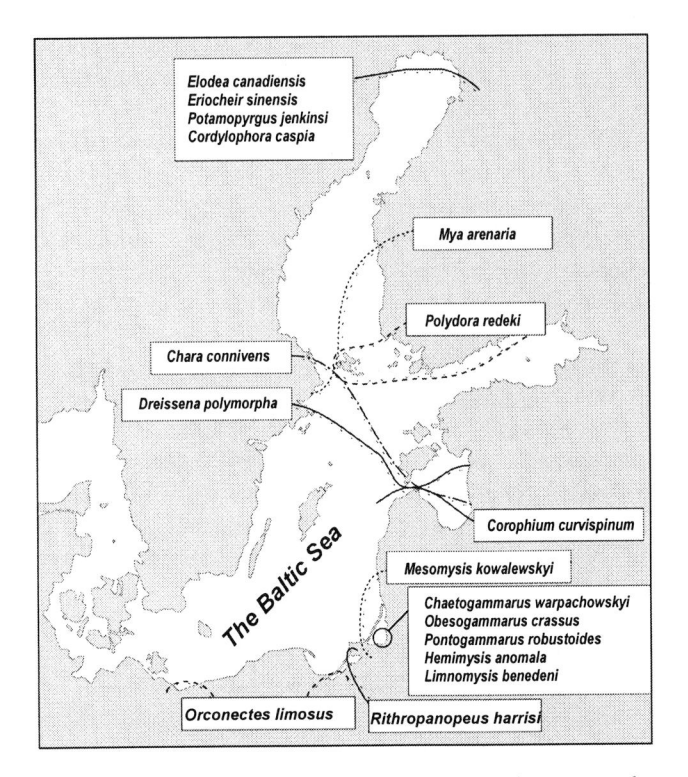

Fig. 6. Distribution of some of the newly introduced species in the Baltic Sea. (From Leppäkoski, 1984).

number of species in each functional group in the Baltic Sea increases the probability that whole functions are lost. The potential effects of this can be observed in the Bothnian Bay where filter-feeding mussels are absent due to low prevailing salinities. This has led to significant changes in energy flow compared to other parts of the Baltic and the North Sea where the energy pathways are in fact very similar (Elmgren, 1984).

EFFECTS OF POLLUTION

Eutrophication

One of the largest threats to the Baltic Sea is eutrophication. Inputs of nutrients have drastically increased due to changes in land use, excessive use of fertilisers in agriculture, loss of wetlands, sewage outlets and emissions of nitrogen from fossil fuel burning. The large inputs of nutrients, year after year, have resulted in increased concentrations of nitrogen and phosphorus in all basins (Fig. 7). The load of phosphorus has increased about five times and nitrogen load about three times since the 1940s. Nitrate has increased in all basins since the late 1960s. With the exception of the Bothnian Bay, this is also true for phosphate (Elmgren, 1989). The main part of the Baltic Sea

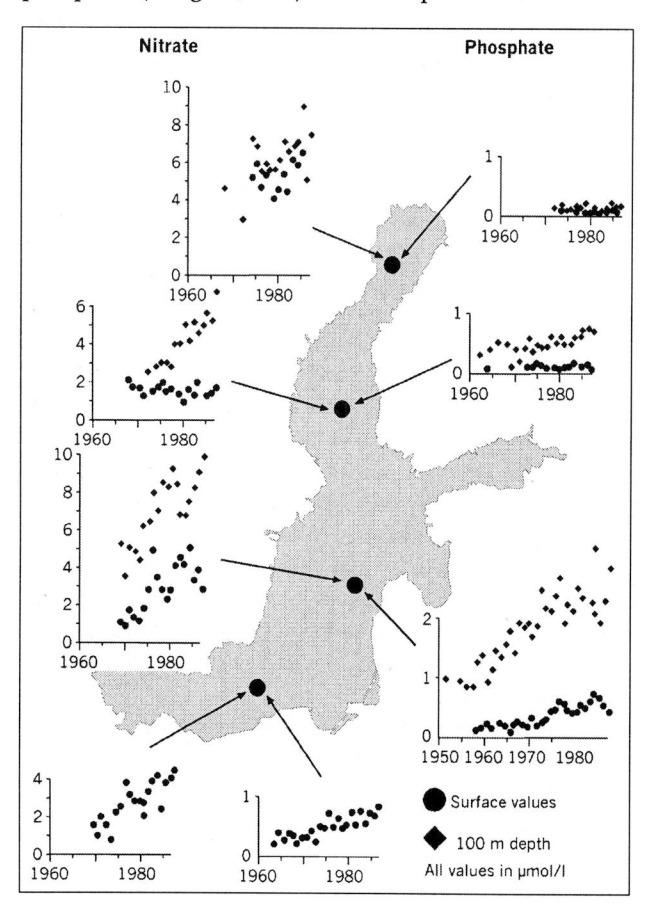

Fig. 7. Changes in nitrate and phosphate concentrations in surface water and at 100 m depth at four sites in the Baltic Sea. (From Elmgren, 1989).

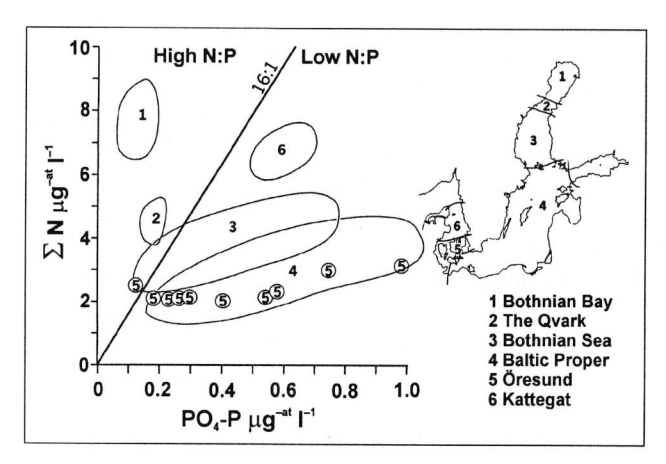

Fig. 8. Relationship between concentrations of phosphate and inorganic nitrogen in the different larger areas of the Baltic Sea during November–December 1975. (After Fonselius, 1978).

has inorganic N:P ratios below the Redfield ratios of 16:1. The Bothnian Sea and especially the Bothnian Bay have lower nutrient concentrations and higher N:P ratios (Fig. 8) (Fonselius, 1978). In the Baltic Proper the main limiting nutrient is nitrogen, except in the inner parts of the coastal areas influenced by land run-off, rivers and sewage where phosphorus may be limiting at least during part of the growth season (Larsson, 1984). The increasing nutrient load has resulted in a doubling of biological production in both the pelagic and the benthic communities since the turn of the century (Elmgren, 1989). Most of the changes have probably taken place since the 1940s.

The increased phytoplankton production makes the water less transparent, resulting in a reduction in mean Secchi depth in the Baltic Proper by about 3 m compared to the first half of this century (Sandén and Håkansson, 1996). The increased production in the pelagic system has further resulted in an increased sedimentation of organic carbon by 70–90% and a doubling of the macrobenthic production above the halocline (Cederwall and Elmgren, 1980). This increased production and sedimentation has in turn lead to anoxic conditions in the deep soft-bottom areas, and anoxic sediments have increased from about 20,000 km^2 in the 1940s to 70,000 km^2 today, resulting in about one third of the bottom of the Baltic Proper being devoid of higher life today (Jonsson et al., 1990).

Eutrophication has, in many coastal areas, resulted in changes in the macroalgal communities, i.e. the disappearance or decrease of long-lived *Fucus* spp. and red algae, and increases in the amounts of opportunistic filamentous algae such as *Pilayella littoralis* and *Ceramium tenuicorne* (Kautsky et al., 1992). The disappearance of *Fucus* may be a rather slow process, as is the case in increasingly eutrophicated areas where *Fucus* is shaded by epiphytic filamentous algae that have a more efficient nutrient uptake, or where re-colonisation is inhibited by increased sediment dust on rocky surfaces (Wallentinus, 1991). On average, the maximum depth distribution of *Fucus vesiculosus* in offshore areas has changed from about 11 m to about 8 m since the

1940s, due to reduced light penetration, which corresponds to a halving of its biomass (Kautsky et al., 1986). Some of the *Fucus* communities are now recovering, especially in areas that have now been cleaned. However, re-colonisation by *Fucus* is totally dependent on the success of sexual reproduction and free space being available for the attachment of the zygote to the surface of the rock (Serrão et al., 1999). The high abundance of filamentous algae found today (e.g. *Pilayella littoralis*, *Ceramium tenuicorne* and *Cladophora glomerata*), largely inhibits recruitment of fucoid zygotes by making attachment sites on hard substratum unavailable.

Today, drifting filamentous algal mats are often readily observed. These algal mats are a hindrance to fisheries and a nuisance to recreation activities in many areas, but also affect other parts of the coastal ecosystem such as the seagrass meadows and shallow sediment animal communities (Norkko and Bonsdorff, 1996). Eutrophication has favoured the blue mussel, as the increasing phytoplankton production provides food and limits the distribution of the bladderwrack. Thus the deeper rocky bottoms on which earlier the bladderwrack was growing, are now instead covered by the blue mussel (Kautsky et al., 1992).

The local disappearance of *Fucus* has been associated with declines in fish and fisheries (e.g. Kangas et al., 1982; Kautsky et al., 1992). The three-spined and nine-spined sticklebacks and minnow which were abundant and common in the Archipelago Sea (Seili area in the northern Baltic Proper) in the early 1970s and 1980 had almost disappeared in 1996, and shores that were earlier rich in fish fauna are now almost devoid of them (Rajasilta et al., 1999). Similar patterns have been found in many areas along the Swedish coast. The decline in bladderwrack in many coastal areas, both in Sweden and in Finland has negatively affected the available spawning and nursery grounds of several coastal fish species, e.g. perch and pike. In many enclosed bays, low oxygen due to eutrophication has also reduced the amount of animal species and biomass production (Norkko and Bonsdorff, 1996).

Pulp Mill Industry

The structure of industry is rather different in the different regions around the Baltic. In the northern parts, i.e in Sweden and Finland, the metal and pulp and paper industries are the main industries polluting the Bothnian Bay and Bothnian Sea. Although the general decline in *Fucus vesiculosus* has been related to eutrophication, in some areas the decline is rather due to point-source industrial discharges, mainly from pulp mills (Lehtinen et al., 1988). Field transplant experiments and toxicity tests pointed out chlorinated organic compounds from the bleaching process as the causal factor (Lehtinen et al., 1988). The highly toxic effluents produced during the bleaching process have also affected the fish communities, and strong pressure from the public has forced the pulp mill industry to change the bleaching process.

Organic Pollutants

A large number of halogenated organic pollutants have been identified in the Baltic Sea and levels in some fish, e.g. herring and cod liver, are so high that recommendations have been issued to restrict consumption. The use of PCB and DDT during the 1960s and 1970s and the ability of these substances to accumulate in the food chain has affected reproduction in top predators like the seals and the white-tailed sea eagle. Joint international measures to reduce inputs to the Baltic, have resulted in a decline in concentrations of PCB and DDT in herring muscle and guillemot eggs since the 1970s (Bernes, 1988; Jansson and Dahlberg, 1999). DDT is not used in the area any more but is still reaching the Baltic Sea via precipitation. The restrictions on DDT use in the Baltic Sea countries in the 1970s, in combination with supplying non-toxic food, saved from extinction the white-tailed eagle, of which only about 50 pairs remained in the 1970s. The reduction in PCB concentrations has also positively affected the seal populations (Bernes, 1988; Elmgren, 1989) in the Baltic Sea. Three species of seals live here: the ringed seal, mainly found in the Bothian Bay, Bothnian Sea, Gulf of Finland and Riga Bight, the grey seal and the harbour seal. The number of grey seals numbered more than 100,000 animals in 1900 and declined to numbers of less than 200 in the late 1970s, mainly as a result of PCB pollution (Helander, 1989). Now the stock is recovering, with about 8000 animals and an annual increase of about 8% per year.

Heavy Metals

The most important point sources of heavy metals and inorganic acids are found along the coast of the Bothnian Bay and Bothnian Sea. Although the loading factors are higher in the Baltic Proper, the highest levels in biota are found in the Bothnian Sea (Lithner et al., 1990). This may at least partly be explained by the lower salinity in the Bothnian Sea, since both uptake and toxicity of metals depend on the ionic form of metals which vary with salinity and the organic content in the water. The anthropogenic loads of cadmium, lead and mercury have increased 5–7 times compared to natural loads in the Baltic Proper, while copper and zinc loads have doubled (Lithner et al., 1990). Metal concentrations in the sediments increased from the 1950s until the 1970s, but now seem to be decreasing due to measures taken in the industry (Jonsson, 1992). Lead concentrations are today decreasing in fish, due to reduced air emissions from car traffic. Although industrial effluents are now being reduced, the deliberate spreading of copper and organic tin through use of antifouling paint, is becoming significant. This is serious, as copper and the herbicide Irgarol 1051 in antifouling paint have been shown to affect reproduction and the youngest stages of *Fucus vesiculosus* (Andersson, 1996).

PROTECTIVE MEASURES

The HELCOM Convention on the Baltic Marine Environment Protection strategy came into force in 1980 and, following the Brundtland Commission and the Rio Declaration, much effort has been made to improve the conditions of the Baltic Sea and protect the environment.

Fifteen marine protected areas have so far been established in Sweden, in most cases based on their ecological/biogeographical and social values (Nilsson, 1998). Since marine coastal areas are often strongly affected by activities on land, protection will only be efficient if the reserve is supported by active coastal and land use planning policy.

In 1988, the ministers of environment of Baltic countries declared that the nutrient load of nitrogen and phosphorus should be reduced by half within eight years, and in 1992, 132 "pollution hotspots" were identified and a joint comprehensive action program aiming at reducing pollution from these hotspots was adopted (HELCOM, 1988). However, the reduction target of nutrients by 50% has still not been reached due to the difficulties of reducing diffuse sources, such as leakage from agricultural soils and emissions by transportation. Still, the nutrient load to the Baltic Sea has levelled off, but part of the improvements achieved are due to the economic recession in the Baltic States, Poland and Russia. These states are the only ones that have reached a 50% reduction of nutrient leaching from agricultural soils due to the reduced use of fertilisers during the 1990s. Thus, there is a risk that the recovering economies in these densely populated countries may again lead to increasing nutrient loads, unless proper measures are taken.

After independence in 1990, the economy of the Baltic states changed from a state-controlled economy with huge collective farms to a market economy with privately owned family farms which drastically reduced the inputs of nitrogen and phosphorus (Löfgren et al., 1999). During the same period, a relatively high nutrient utilisation efficiency was achieved by high crop yields in Sweden. Still this does not seem to be enough to reduce the nutrient load to the Baltic to such an extent that the 50% reduction, agreed upon by the HELCOM states, can be reached. The most difficult problems are huge emissions from diffuse non-point sources and long-distance distribution via precipitation. An important step for the future development of the Baltic region, is the International Convention on Long-Range Transport of Air Pollutants, which was signed by about 30 countries, including Sweden, in December 1999. This will be especially important since between 20 and 40% of the total nitrogen load is entering the Baltic via precipitation, and about two-thirds comes from the countries surrounding the Baltic themselves. The long-range air pollution precipitated over the Baltic Sea does not only contain nitrogen but also heavy metals and persistent organic compounds such as DDT and PCB, which are also regulated in the Convention. The aim is to reduce emissions to half in 2010 compared to 1990.

Other complementary measures to reduce the nutrient load have also been taken, including increasing the plant cover, spring application of animal manure, increased use of catch crop and construction of wetlands and small dams (Löfgren et al., 1999). The effectiveness of these measures to reduce the nutrient load to the Baltic Sea still needs to be proven. A key to decreasing the amount of waste and sewage is collecting, sorting, and recycling. This will, however, not be enough and new technologies and infrastructures will need to be developed jointly with changes in attitudes and behaviour (Jansson and Dahlberg, 1999).

REFERENCES

Ackefors, H., Johansson, N. and Wahlberg, B. (1991) The Swedish compensatory programme for salmon in the Baltic; an action plan with biological and economical implications. *ICES Mar. Sci. Symp.* **192**, 109–119.

Andersson, S. (1996) The influence of salinity and antifouling agents on Baltic Sea *Fucus vesiculosus* with emphasis on reproduction. PhD thesis, Department of Systems ecology, Stockholm University, Sweden

Bernes, C. (1988) Monitor 1988. Sweden's Marine environment Ecosystems under pressure. National Swedish Environmental Protection Board, Stockholm pp. 1–207.

Cederwall, H. and Elmgren, R. (1980) Biomass increase of benthic macrofauna demonstrating eutrophication of the Baltic Sea. *Ophelia* Suppl. 1, 287–304.

Ehlin, U. (1981) Hydrology of the Baltic Sea. In The Baltic Sea, ed. A. Voipio. Elsevier, Amsterdam, pp. 123–134.

Elmgren, R. (1984) Trophic dynamics in the enclosed brackish Baltic Sea. *Rapports et Proces-Verbaux des Reunions, Conseil International pour l'Exploration de Mer* **183**, 152–169.

Elmgren, R. (1989) Man's impact on the ecosystem of the Baltic Sea: energy flows today and at the turn of the century. *Ambio* **18**, 326–332.

Fonselius, S.H. (1978) On nutrients and their role as production limiting factors in the Baltic. *Acta Hydrochimica et Hydrobiologica* **6**, 329–339.

Hansson, S. and Rudstam, L.G. (1990) Eutrophication and the Baltic fish communities. *Ambio* **19**, 123–125.

Helander, B. (1989) Survey of grey seal *Halichoerus grypus* and harbour seal *Phoca vitulina* along the Swedish Baltic coast 1975–1984. In *Proceedings of the Soviet–Swedish Symposium, Influence of Human Activities on the Baltic Ecosystem*, eds. A. Yablokov and M. Olsson, pp 10–21. Leningrad.

HELCOM (Helsinki Commission) (1988) 1988 Ministerial Declaration on the Protection of the Marine Environment of the Baltic Sea. HELCOM, Helsinki, Finland.

Hubendick, B. (1947) Die Verbreitungsverhältnisse der limnischen Gastropoden in Sudschweden. *Zoologiska Bidrag fran Uppsala* **24**, 419–559.

Håkansson, B., Alenius, P. and Brydsten, L. (1996) Physical environments in the Gulf of Bothnia. *Ambio* Spec. Rep. **8**, 5–12.

Jansson, B.O. (1972) Ecosystem approach to the Baltic problem. *Bulletin from the Ecological Research Committee, NFR* **16**, 1–82.

Jansson, A-M. and Kautsky, N. (1977) Quantitative survey of hard bottom communities in a Baltic archipelago. In *Biology of Benthic Organisms. 11th European Symposium of Marine Biology, Galway, Oct. 1976*, eds. B.F. Keegan, P. O'Ceidigh and P.J.S. Boaden, pp 359–366. Pergamon Press, Oxford.

Jansson, B.-O. and Dahlberg, K. (1999) The environmental status of the Baltic Sea in the 1940s, today and in the Future. *Ambio* **28**, 312–319.

Johannesson, K. (1989) The bare zone of Swedish rocky shores: why is it there? *Oikos* **54**, 77–86.

Jonsson, P. (1992) Large scale changes of contamination in the Baltic Sea sediments during the twentieth century. PhD thesis, Uppsala University, Sweden.

Jonsson, P., Carman, R. and Wulff, F. (1990) Laminated sediments in the Baltic a tool for evaluating nutrient mass balances. *Ambio* **19**, 152–158.

Kangas, P. and Skoog, G. (1978) Salinity tolerance of *Theodoxus fluviatilis* (Mollusca, Gastropoda) from freshwater and from different salinity regimes in the Baltic Sea. *Estuarine and Coastal Marine Science* 6: 409–416.

Kangas, P., Autio, H., Hällfors, G., Luther, H., Niemi, A. and Salemaa H. (1982) A general model of the decline of *Fucus vesiculosus* at Tvärminne, south coast of Finland in 1977–81. *Acta Botanica Fennica* **118**, 1–27.

Kautsky, N. (1981) On the trophic role of the blue mussel (*Mytilus edulis* L.) in a Baltic coastal ecosystem and the fate of the organic matter produced by the mussels. *Kieler Meeresforschungen, Sonderh.* **5**, 454–461.

Kautsky, N. (1982) Growth and size structure in a Baltic *Mytilus edulis* L. population. *Marine Biology* **68**, 117–133.

Kautsky, U. (1995) Ecosystem processes in coastal areas of the Baltic Sea. PhD thesis, Department of Systems Ecology, Stockholm Univ.

Kautsky, L. and Kautsky, H. (1989) Algal species diversity and dominance along gradients of stress and disturbance in marine environments. *Vegetatio* **83**, 259–267.

Kautsky, U. and Kautsky, H. (1991) Coastal production in the Baltic Sea. In: *The Biology and Ecology of Shallow Coastal Waters*, eds. A. Eletheriou et al. Olsen and Olsen Pub., Denmark, pp. 31–38.

Kautsky, N. and Wallentinus, I. (1980) Nutrient release from a Baltic *Mytilus*–red algal community and its role in benthic and pelagic productivity. *Ophelia* (Suppl. 1), 17–30.

Kautsky, H., Kautsky, L., Kautsky, N., Kautsky, U. and Lindblad, C. (1992) Studies on the *Fucus vesiculosus* community in the Baltic Sea. *Acta Phytogeographica Suecica* **78**, 33–48.

Kautsky, N., Kautsky, H., Kautsky, U. and Waern, M. (1986) Decreased depth penetration of *Fucus vesiculosus* since the 1940s indicates eutrophication of the Baltic Sea. *Marine Ecology Progress Series* **28**, 1–8.

Kuparinen, J., Leonardsson, K., Mattila, J. and Wikner, J. (1996) Food web structure and function in the Gulf of Bothnia, the Baltic Sea. *Ambio* Spec. Rep. **8**, 13–21.

Larsson, P.-O. (1984) Some characteristics of the Baltic Salmon (*Salmo salar* L.) population. Ph.D. Thesis, University of Stockholm, Stockholm, Sweden. 80 pp.

Lappalainen, A., Hällfors, G. and Kangas, P. (1977) Littoral benthos of the Northern Baltic Sea. IV. Pattern and dynamics of macrobenthos in a sandy-bottom *Zostera marina* community in Tvärminne. Internationale Revue der Gesamten Hydrobiologie **62** (4), 465–503.

Lehtinen, K.J., Notini, M., Mattson, J. and Landner, L. (1988) Disappearance of bladder wrack (*Fucus vesiculosus* L.) in the Baltic Sea: relation with pulp mill chlorate. *Ambio* **17**, 387–393.

Lithner, G., Borg, H., Grimås, U., Götberg, A., Neumann, G. and Wrådhe, H. (1990) Estimating the loads to the Baltic Sea. *Ambio* Spec. Rep. **7**, 7–9.

Leppäkoski, E. (1984) Introduced species in the Baltic Sea and its coastal ecosystems. *Ophelia* Suppl. 3, 123–135.

Löfgren, S., Gustafson, A., Steineck, S. and Ståhlnacke, P. (1999) Agricultural development and nutrient flows in the Baltic states and Sweden after 1988. *Ambio* **28**, 320–327.

Malm, T. (1999) Distribution patterns and ecology of *Fucus serratus* L. and *Fucus vesiculosus* L. in the Baltic Sea. PhD Thesis, Department of Botany, Stockholm University. 130 pp.

Mattäus, W. and Schinke, H. (1999) The influence of river run off on deep water conditions of the Baltic Sea. *Hydrobiologia* **393**, 1–10.

Münsterhjelm, R. (1987) Flads and gloes in the archipelago. *Geological Survey of Finland. Special paper* **2**, 55–61.

Nellbring, S. (1988) Quantitative and qualitative studies of fish in shallow water, northern Baltic Proper. Thesis. Department of Zoology, Stockholm University.

Nielsen, R., Kristiansen, A., Mathiesen L. and Mathiesen H. (1995) Distributional index of the benthic macroalgae of the Baltic Sea area. The Baltic Marine Biologists Publication No. 18. *Acta Botanica Fennica* **155**.

Nilsson, P. (1998) Criteria for the selection of marine protected areas —an analysis. Swedish EPA, Report 4834.

Nissling, A. and Westin, L. (1991) Egg buoyancy of Baltic cod (*Gadus morhua*) and its implications for cod stock fluctuations in the Baltic. *Marine Biology* **111**, 33–35.

Norkko, A. and Bonsdorff, E. (1996) Population responses of coastal zoobenthos to stress induced by drifting algal mats. *Marine Ecology Progress Series* **140**, 141–151.

Olenin, S. and Leppäkoski, E. (1999) Non-native animals in the Baltic Sea:alteration of benthic habitats in coastal inlets and lagoons. *Hydrobiologia* **393**, 233–243.

Rajasilta, M., Mankii, J., Ranta-Aho, K. and Vourinen, I. (1999) Littoral fish communities in the Archipelago Sea, SW Finland: a preliminary study of changes over 20 years. *Hydrobiologia* 393, 253–260.

Remane, A. and Schlieper, C. (1971) Biology of brackish water. Wiley Interscience, New York, 350 pp.

Russell, G. (1988) The seaweed flora of a young semi-enclosed sea: the Baltic. Salinity as a possible agent of flora divergence. *Helgoländer Meeresunters.* **42**, 243–250.

Russell-Hunter, W.D. (1964) Physiological aspects of ecology in nonmarine molluscs. In *Physiology of Mollusca*, Vol. 1, eds. K.M. Wilbur and C.M. Yonge, pp. 83–126. Academic Press, London.

Sandén, P. and Håkansson, B. (1996) Long-term trends in Secchi depth in the Baltic Sea. *Limnology and Oceanography* **41**, 346–351.

Schlieper, C. (1971) Physiology of brackish water. In Biology of Brackish Water, eds. A. Remane and C. Schlieper, pp. 211–350. Wiley Interscience, New York.

Serrão, E., Brawley, S.H., Hedman, J., Kautsky, L. and Samuelsson, G. (1999) Reproductive success of *Fucus vesiculosus* (Phaeophycea in the Baltic Sea. *Journal of Phycology* **35**, 254–269.

Tedengren, M., Arnér, M. and Kautsky, N. (1988) Ecophysiology and stress response of marine and brackish water *Gammarus* species (Crustacea, Amphipoda) to changes in salinity and exposure to cadmium and diesel-oil. *Marine Ecology Progress Series* **47**, 107–116.

Thurow, F. (1980) The state of fish stocks in the Baltic. *Ambio* **9**, 153–157.

Thurow, F. (1989) Fisheries resources of the Baltic region. In Comprehensive Security for the Baltic. An Environmental Approach, ed. A.H. Westing, pp. 54–61. PRIO, Sage Publ., London.

Voipio, A. (ed.) (1981) *The Baltic Sea*. Elsevier Oceanographic Series. Elsevier, Amsterdam. 418 pp.

Wallentinus, I. (1991) The Baltic Sea gradient. In *Ecosystems of the World 24: Intertidal and Littoral Ecosystems*, eds. A.C. Mathieson and P.H. Nienhuis, pp. 83–108. Elsevier, Amsterdam.

THE AUTHORS

Lena Kautsky
Department of Botany, Stockholm University,
S-106 91 Stockholm, Sweden

Nils Kautsky
Department of Systems Ecology, Stockholm University,
S-106 91 Stockholm, Sweden

Chapter 9

THE NORTH COAST OF SPAIN

Isabel Díez, Antonio Secilla, Alberto Santolaria and
José María Gorostiaga

The North coast of Spain is an open coast characterised by a narrow continental shelf where weak, variable and wind driven currents prevail. The annual phytoplankton and zooplankton cycle follows the pattern of a temperate sea with the highest planktonic biomass in spring followed by a secondary peak in autumn. During summer water stratification, this typical cycle can be modified by the development of shelf-edge fronts and coastal upwellings. Other important physical processes and hydrological structures of the area are the warming trend of the sea surface temperature (SST) (a mean increase of 1.4°C for the period 1972–1993), the poleward geostrophical warm current running northwards along the Portuguese–Spanish slope during winter, the development of slope water oceanic eddies, and annual warming of waters off the eastern coast as a result of the entrainment of water masses during the spring–summer period. Thus, SST in the Basque coast ranges from 11°C to 22°C, in winter and summer, respectively, whilst in Galicia it varies from 13.5°C to 19°C, respectively. This leads to a geographically varied flora and fauna, in which the Galician coast is enriched in septentrional species, whilst the Basque coast shows a remarkable resemblance to the Portugal and northern Atlantic Morocco communities. The great pressure of uses on the coastal fringe and the lack of enforcement of strict limits to human interventions have led to the environmental degradation of some parts. Estuaries have been specially impacted as a result of the concentration of industrial-urban centres in these areas since the end of the 19th century. Pollution and land reclamation process of estuarine intertidal areas and coastal wetlands have led these areas to be among the most threatened habitats in the north coast of Spain today. Despite protective measures, the landfill continues, mainly related to tourism. Urban centres have serious infrastructural deficits in waste-water treatment systems, such that only 20% of the total 12 million population equivalent conforms with the requirements of Directive 91/271/EC concerning urban waste-water treatment.

Nutrient enrichment from agriculture does not seem to have affected the general pattern of nutrient limitation during summer water stratification, but there is no information on pesticide impacts in coastal waters. On the other hand, environmental eutrophication resulting from aquaculture on the western coast (Rías Bajas) has been related to the appearance of toxic marine phytoplankton blooms. Mining, steel industries and metal treatments have been predominant industrial activities for decades, as is revealed by extremely high heavy metal concentrations in several coastal areas. The eastern part of the coast has been the most impacted region. Gross organic enrichment, water oxygen depletion, anaerobic sediment conditions, high heavy-metal concentration and trace organic pollutants (PAHs, PCBs, and pesticides) have led to extremely reduced fauna along the Nervión estuary (Bilbao).

With respect to navigation and routine port operations, a comprehensive impact assessment has not been carried out. However, the use of tributyltin (TBT) in antifouling paints for fishing

---

Seas at The Millennium: An Environmental Evaluation (Edited by C. Sheppard)
© 2000 Elsevier Science Ltd. All rights reserved

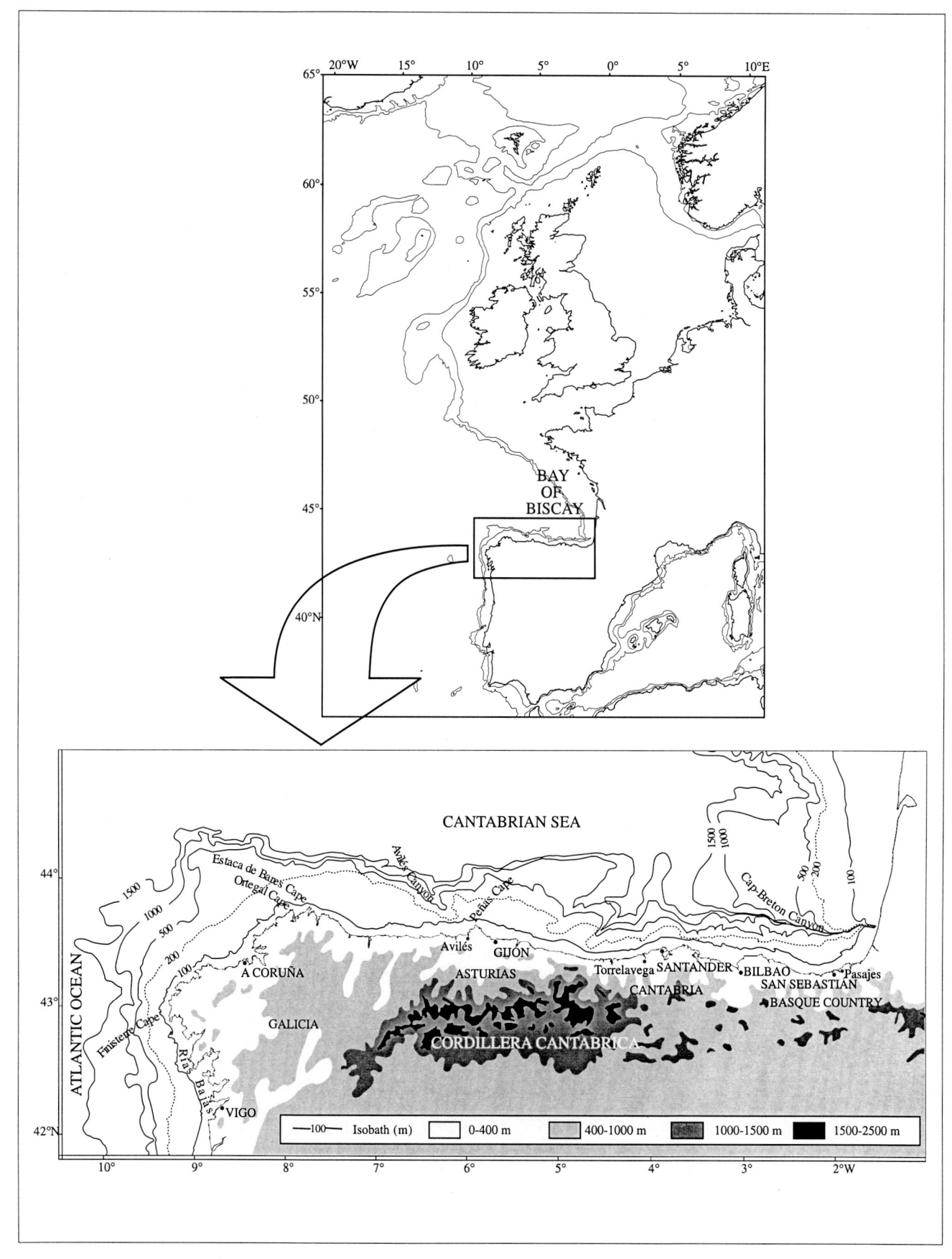

Fig. 1. Altitude and bathymetric map of the north coast of Spain showing the main industrial-urban centres along the coast.

and merchant fleets has been reported to have harmful effects on benthic communities. The northwestern coast of Spain is particularly exposed to oil spills. The most notable recent accidents have been the wrecks of 'Aegean Sea' (in 1992) and 'Monte Urquiola' (in 1976), with 79,096 and 100,000 tonnes of oil released into the sea, respectively.

Other environmental concerns arise from the over-exploitation of living marine resources. Hake, anglerfish, megrim and blue whiting, and the pelagic sardine, anchovy, mackerel, horse mackerel and albacore are the main species exploited in the area by coastal fisheries. Most of them are fully or over-exploited and discards represent a high percentage of the catches. In addition, fisheries have a major impact on shelf communities due to the degradation of sea bottom habitats by trawling. The increase of economic, human and legal efforts for the protection of natural resources carried out over the last few years is insufficient. In order to efficiently protect those still extensive well preserved shores and to recover the degraded ones, Spanish Authorities need to enforce strict limits. Likewise, they should formally establish an appropriate and global strategy to progressively reduce industrial and domestic waste-waters with the long-term goal of zero discharge. Finally, more environmental surveys should be carried out in order to manage properly the coastal habitats and to assess reliably their conservation state.

THE DEFINED REGION

The northern and northwestern coasts of Spain, approximately 2,000 km long, are geographically located between 41°53′ N to 43°40′ N and 01°40′ W to 09°20′ W at the southern Bay of Biscay (Fig. 1). From the classical biogeographical viewpoint it belongs to the Lusitanian province of the Mediterranean–Atlantic European region which extends from southern Ireland and Great Britain to the Rio de Oro on the African northwestern coast (van den Hoek, 1975). From the perspective of plankton biogeography it belongs to the Boreal Subtropical Transitional Zone (van der Spoel and Heyman, 1983). From a more ecological approach, taking into account features such as nutrient dynamics, continental shelf topography and algal blooms, Longhurst (1998) includes this area in two provinces: the Northeast Atlantic Shelf province (northern coast) and the Eastern Canary Coastal province (western coast of Galicia). Finally, with regard to the proposed management of the ecosystems, the northern coast of Spain is included in the Western Iberian large marine ecosystem (LME) (Sherman and Alexander, 1989), although the consideration of the Bay of Biscay as a LME is currently being studied.

Along the western coast of Galicia, the surface waters flow down from north to south under the effect of the North Atlantic Drift, forming part of the Canary Current System. The winter northwesterly winds cause this current to diverge at the Ortegal Cape generating a surface water movement from west to east along the north coast of Spain that moves up to the north at the Cap Breton canyon. During spring, this westerly current becomes weakened due to the change in direction of the prevailing winds. When its inertia is overcome, the surface water flow changes direction, remaining from east to west during summer.

The topography of the sea floor is characterised by a narrow continental shelf (from 5 to 50 miles) which becomes wider westwards, and by bathymetric singularities such as submarine canyons and marginal shelves. The slope is particularly steep (6° as a mean) with measured gradients of 20° (Montadert et al., 1974). In general, the sedimentary coverage is reduced, and the rocky bottom is almost continuous in shallow waters. The sea bed in the inner Cantabrian shelf is most commonly sandy with little organic matter, except for the most eastern region where silt can be found associated with the Cap Breton basin. In the Galician shelf, from Estaca de Bares to Finisterre, there are fine sediments associated with the upwelling events, whilst to the south of Finisterre (western Galician coast) sediments are mainly of a fine silty-sandy nature. The middle and outer western shelf shows a layer of silt rich in organic matter coming from the estuaries (Rías Bajas).

In winter, sea surface temperatures (SST) in eastern waters (11°C) are cooler than in the western region (13.5°C), whilst in August this SST gradient is inverted, ranging between 22°C to 19°C from east to west (with the exception of northwestern areas affected by upwelling events which can achieve 14°C in summer). Therefore, surface waters from the northwestern coast are cooler and show a narrower SST seasonal range. As a result, the Galician algal flora is similar to that found on the French Brittany coast; in contrast, benthic vegetation from the Cantabrian coast shows a remarkable resemblance to the Portugal and northern Atlantic Morocco floras. Likewise, the geographical location of the coast results in the presence of a classic temperate fauna with groups of boreal and subtropical affinity, resulting in a great biological richness.

The formation of a summer thermocline (50 m depth) limits nutrient transfer processes from deep to surface waters where the light and temperature conditions are more favourable for the development of primary producers. In addition to this limitation, nutrient supply from rivers is reduced (Prego and Vergara, 1998) and open waters do not show permanent symptoms of eutrophication. In this way, the excessive growth of macroalgae associated with eutrophic waters found on Danish, German or Dutch coasts, commonly termed 'green tides', have not been found along the northern Spanish Atlantic coast (Niell et al., 1996). However, eddies, small river plumes and shelf-edge fronts (Fernández et al., 1993), and upwelling events (Fraga, 1981; Botas et al., 1990; Borja et al., 1996) are important hydrological features that fertilise waters and keep the photosynthesis active along the annual cycle. These processes and mechanisms affect the distribution of marine organisms throughout the region so that the most productive and rich areas are associated with them. The maximum phytoplanktonic biomass is reached in spring followed by a second peak in autumn, with the exception of the upwelling areas from the northwestern and western coasts which show biomass peaks in summer (Varela, 1992). The annual cycle of zooplankton abundance, dominated by copepods (*Paracalanus parvus*, *Acartia clausi*), is typical of temperate areas showing spring and autumn peaks. Mesozooplankton biomass is highly variable, ranging from 1–70 mg m^{-3} dry weight (d.w.) (Valdés and Moral, 1998). These changes are related to seasonal and inter-annual periodic cycles of the environmental conditions and non-periodic events such as continental runoff, storms and upwellings, amongst others.

This planktonic richness is not only the food base for pelagic species such as anchovies, pilchards or mackerels, but also for a great number of benthic polychaetes, mollusca and crustacea groups, as well as other benthic infauna that form part of the diet of demersal fish (hake, blue whiting etc.) and of benthic fish such as sole and anglerfish (Fig. 2) (Olaso, 1990). Man has traditionally exploited numerous fisheries in these waters.

The northern coast of Spain belongs to Divisions VIIIc and IXa of the International Council for the Exploration of the Sea (ICES). In these Divisions, the main demersal fisheries are hake (*Merluccius merluccius*), anglerfish (*Lophius piscatorius* and *Lophius budegasa*), megrim (*Lepidorhombus whiffiagonis* and *Lepidorhombus boscii*) and blue whiting (*Micromesistius poutassou*), and the main pelagic fisheries are pilchard (*Sardina pilchardus*), anchovy (*Engraulis encrasicholus*), mackerel (*Scomber scombrus*) and horse mackerel (*Trachurus trachurus*).

SEASONALITY, CURRENTS AND NATURAL ENVIRONMENTAL VARIABLES

The coastline of northern Spain is mainly formed by cliffs which alternate with beaches and estuaries of small dimensions on the Cantabrian coast, and 'rías' (tectonic drowned valleys with positive estuarine circulation) on the west coast of Galicia. Numerous rivers with a very short trajectory flow into the coast. These have their origin on the Cordillera Cantábrica which is located very close to, and parallel with, the coast with summits above 2,500 m.

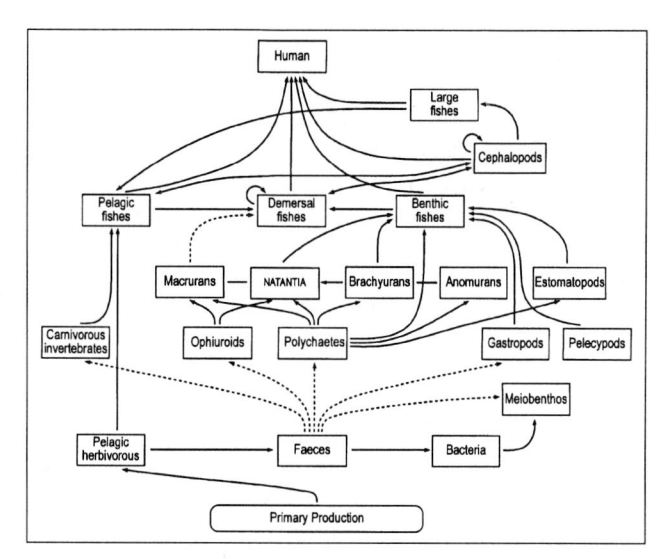

Fig. 2. Simplified trophic web of the Cantabrian shelf including the main invertebrates groups involved (modified from Olaso, 1990).

This is an open coast exposed to large fetches and with a narrow (5–50 km) continental shelf. The latter explains why its coastal dynamics depends mainly on the conditions of the oceanic waters. Sea swell reaches this coast from as far away as 1,500 km, from storm areas such as Gran Sol. These low pressure systems (resulting from the contact between the cold polar air mass and the temperate air mass from the Azores high pressure system) generate intense winds in the North Atlantic and strong waves, mainly from the northwest. According to Penin (1980), the maximum wave height is 2.4 m during a period of 30 days a year, between 3.5 m and 3.8 m during 5 days a year and 4.8 m once a year. Every 10 years, the height of the maximum wave of the year is 7.5 m. The most frequent wave periods registered in the area range between 8 s and 12 s.

This is a region of weak general currents. The surface water movement is variable and is determined by the direction and speed of the wind, although the general trend is the development of a west to east current along the northern coast during winter which reverses in summer. The central (oceanic) area of the Bay of Biscay is characterised by a weak anticyclonic circulation (Koutsikopoulos and Le Cann, 1996). As a result of continental margin current instabilities interacting with bottom topography, frequent cyclonic and anticyclonic eddies are developed (100 km diameter, with their nucleus located at 200 m). These structures, termed "SWODDIES" (Slope Water Oceanic Eddies) by Le Cann and Pingree (1994), are important in water mixing processes in this region.

In the littoral zone, the circulation variability is more noticeable and the west to east (and *vice versa*) oscillations are superimposed onto the tidal currents. Tides are semidiurnal. The tidal wave enters the Bay of Biscay from the west and propagates towards the east, increasing its amplitude as it enters the Bay. Thus, the mean tidal range in the Galician coast is approximately 1.2 m during neap tides and

3.5 m at spring tides, whilst in the Basque Country is 1.5 m and 4.5 m, respectively.

In the northwestern corner of the coast, water masses from the North Atlantic and the Bay of Biscay meet. According to Fraga (1989), south of the Finisterre Cape the sub-surface water is typically North Atlantic Central Water (NACW) whilst off the Ortegal Cape the upper part of NACW, just below the surface layer (100 m depth), disappears and the 100–400 m layer is occupied by Bay of Biscay Central Water (BBCW) (sigma-t range 27.17–27.23). Below both of them a more salty Mediterranean Water (MW) mass (sigma-t 27.75) can be found with its nucleus located at 1,200 m (Le Floch, 1968). Further down (1,800 m) water from the Labrador current (sigma-t 27.82) is present, whereas at 2,800 m the so-called Deep North Atlantic Water appears.

The surface waters experience a process of seasonal stratification between May and October in a layer of about 40–50 m depth, and a mixing process of the water column between November and April (Valencia, 1993; Lavín et al., 1998) with temperature differences between surface and 100 m depth waters of less than 1°C. As mentioned, the eastern part of the coast shows a wide SST range (from 11°C to 22°C in February and August, respectively) (Valencia, 1993). The warming process of waters off the Basque Country is the result of the entrainment of water masses in the southern corner of the Bay of Biscay due to the relaxation of wind and current regimes during summer (Borja et al., 1996). In contrast, the warmer waters of Galicia during winter (13.5°C) seem to be a manifestation of the poleward geostrophical warm current running northwards along the Portuguese and Spanish slope which is known as 'Corriente de Navidad' (Le Cann and Pingree, 1994).

Sea surface temperature in the southern Bay of Biscay shows a strong warming trend with a mean increase of 1.4°C for the period 1972–1993 (Koutsikopoulos et al., 1998; Lavín et al., 1998). The trend reduces with depth and disappears at 100 m depth. Several biological observations can be associated with this climate warming. Villate et al. (1997) detected changes in mesozooplankton communities (e.g., increase of *Temora stylifera* abundance) while Quéro et al. (1998) pointed out an increasing trend in the presence of tropical fishes which appeared recently and spread northwards.

Important hydrological structures of the area are the strong episodic upwelling events occurring in the northwestern corner during summer (April–October). Along the Galician coast, North Atlantic Central Water (NACW) upwells from 150 m (Fraga, 1981). This upwelling is controlled by the latitudinal displacement of the general upwelling produced by the North Atlantic anticyclonic gyre that extends from Galicia to Cape Vert (Wooster et al., 1976). In addition to the strong upwelling events of the Galician coast, less intense upwelling processes occur along the Cantabrian coast. The western component of the wind induces an inshore surface circulation with downwelling events, while easterly winds lead to an offshore transport of surface waters and subsequent upwelling conditions. These pulsing upwellings occur during summer and may be described as weak upwelling events since their effects are restricted to pushing the thermocline up close to the surface (Borja et al., 1996). In the case of the central Cantabrian upwelling (Botas et al., 1990), this is favoured by the topographic features of Peñas Cape, the Avilés submarine canyon and the internal waves action.

THE MAJOR SHALLOW WATER MARINE AND COASTAL HABITATS

Wetlands and Marshes

The flora and vegetation of Cantabrian–Atlantic coastal wetlands show a transitional character between those of North Atlantic and Mediterranean regions, which partially explains their high diversity (Bueno, 1997). Although these areas are not very extensive, they include a wide variety of habitats such as sandy shores, estuarine waters, intertidal mudflats and saltmarshes, permanent freshwater streams and brackish pools, sedge and reed beds, and a few coastal lagoons and dune systems. In permanently submerged areas, beds of *Zostera marina* grow, whilst in low intertidal areas, emerging during most low tides, are beds of *Zostera noltii* with monospecific beds of *Spartina maritima* in higher levels. Areas usually flooded during high tide support a saltmarsh community of *Halimione portulacoides, Inula crithmoides, Aster tripolium, Arthrocnemum perenne, Arthrocnemum fruticosum, Triglochin maritima, Limonium vulgare, Puccinellia maritima, Spergularia media, Sarcocornia perennis, Salicornia ramosissima, Suaeda maritima*, etc. Beds of *Scirpus maritimus, Juncus maritimus, Juncus gerardii* and *Carex extensa* are found in areas with a weak tidal influence and freshwater input. Close to the sea in the dune areas, *Ammophila arenaria, Euphorbia paralias, Erygium maritimum, Cakile maritima, Carex arenaria, Aetheorrhiza bulbosa* thrive.

Soft-bottom Communities

Sediment characteristics, tidal height and wave exposure are among the main environmental factors determining the distribution and abundance of soft-bottom faunal assemblages (Junoy and Viéitez, 1990). On beaches, polychaetes (*Scolaricia typica, Scolaricia mesnili, Nephthys cirrosa*) and crustaceans (*Bathyporeia pelagica, Urothoe brevicornis, Sphaeroma rugicauda-monodi*) are the most abundant groups. Molluscs are also common (*Tellina tenuis, Donax trunculus, Donax vittatus*). In sublittoral sheltered sands, the molluscs *Cerastoderma edule, Hydrobia ulvae* and *Angulus tenuis*, the polychaetes *S. typica* and *N. cirrosa*, and the crustaceans *U. brevicornis* and *Urothoe poseidonis* can achieve high densities. *Venus fasciata, Venus casina* or *Chamella gallina* are associated with coarse sediments. In midlittoral sheltered sands, crustaceans (*Eurydice pulchra, Bathyporeia sarsi-pilosa* and *Haustorius arenarius*) are dominant, accompanied by *C. edule* and *Ophelia bicornis*. Muddy sand flats are dominated by polychaetes (*Capitella capitata, Pygospio elegans, Spio martinensis-decoratus, Streblospio benedicti, Hediste diversicolor*) although other species such as *Scrobicularia plana* or *Cyathura carinata* can be abundant. High abundance of the mud-snail *Peringia ulvae* is found in seagrass meadows. Other species in these habitats are the polychaetes *C. capitata, Pygospio elegans* and *H. diversicolor*, the amphipod *Melita palmata* and the isopod *Idothea chelipes*.

Intertidal Rocky Communities

Intertidal communities show a gradient of vegetation from open marine coast to the inner estuarine waters. Estuarine coasts are characterised by the presence of *Fucus ceranoides, Monostroma oxyspermum, Enteromorpha intestinalis* and *Rizoclonium tortuosum* belts. Sheltered coasts show abundant *Fucus spiralis, Fucus vesiculosus, Catenella caespitosa, Gelidium pusillum, Ascophyllum nodosum,* and *Pelvetia canaliculata*. Coasts moderately exposed to wave action are highly diverse; among the most abundant species are *Mastocarpus stellatus, Caulacanthus ustulatus, Corallina elongata, Laurencia obtusa, Bifucaria bifurcata, Stypocaulon scoparium, Chondracanthus acicularis, Gelidium latifolium, Chondracanthus teedi, Chondrus crispus, Cystoseira baccata* and *Cystoseira tamariscifolia*. Finally, exposed coasts are characterised by *Corallina elongata, Litophyllum incrustans, Litophyllum lichenoides, Pterosiphonia complanata, Mesophyllum lichenoides,* and *Gelidium sesquipedale*. Likewise, along the shores of northern Spain, a change in the species composition of the phytobenthic communities occurs from the cooler northwestern coast to the warmer Basque coast (Anadón and Niell, 1981). So, species associated with cool water and which occur in Galicia, such as *Ascophyllum nodosum, Fucus serratus, Himanthalia elongata,* or *Chorda phyllum* gradually disappear eastwards.

Open coasts are characterised by the development of a barnacle belt (*Chthamalus montagui, Chthamalus stellatus*) at the mid-intertidal level where limpets (*Patella intermedia*) are widely spread and mussels (*Mytilus galloprovincialis*) are restricted to crevices together with the gastropod *Littorina neritoides*. At low intertidal levels, under the algae canopy, the barnacle *Balanus perforatus* can achieve significant abundances. Among caespitose vegetation, limpets (*Patella ulyssiponensis*), small gastropods (*Gibbula umbilicalis, Hinia incrassata, Bittium reticulatum*) and amphipods (*Hyale perieri*) exercise an intense herbivorous pressure. The bryozoan *Electra pilosa* is a frequent epiphyte on large perennial algae. Sea urchins (*Paracentrotus lividus*) are abundant in low-intertidal pools together with starfish (*Marthasterias glacialis*), anemones (*Anemonia viridis, Actinia equina*), hermit crabs (*Clibanarius* spp.), crabs (*Carcinus maenas, Pachygrapsus marmoratus, Eriphia verrucosa*), shrimps (*Palaemon serratus*) and small blennid and gobids fishes (*Coryphoblennius galerita, Blennius pholis, Gobius paganellus*) among others groups.

Subtidal Rocky Communities

Gelidium sesquipedale beds are the most significant subtidal vegetation along the Cantabrian coast, due to both the vast surface areas they occupy as well as their commercial exploitation. This vegetation uses the space through its complex vertical layering (Gorostiaga et al., 1998), and it is composed of a well-developed crustose layer of *Mesophyllum lichenoides* and *Zanardinia prototypus*, a poor understorey layer of *Pterosiphonia complanata, Corallina officinalis, Rhodymenia pseudopalmata* and *Cryptopleura ramosa*, a well-developed canopy of *Gelidium sesquipedale*, a poor epiphytic layer of *Plocamium cartilagineum* and *Dictyota dichotoma* (well developed in late spring). This vegetation thrives in habitats very exposed to wave action with little to moderate sand sedimentation levels. Other macrophytes achieving large populations are *Laminaria hyperborea, Laminaria ochroleuca, Saccorhiza polyschides, Cystoseira baccata, Cystoseira usneoides* and *Halidrys siliquosa*. Kelps are not abundant along the Cantabrian coast, whilst become common on the Galician coasts (Bárbara and Cremades, 1996). On sandy bottoms of northwestern coast, maerl vegetation composed of *Phymatolithon calcareum* and *Lithothamnion corallinoides* is found.

Among the most frequent invertebrates are the bryozoans (*Aetea anguina, Electra pilosa, Crisia eburnea, Scrupocellaria reptans, Turbicellepora magnicostata*) and hydrozoans (*Obelia geniculata, Sertularia distans, Aglaophenia pluma*) growing mainly on large macrophytes. Amphipods (*Caprella pennantis*) are abundant among algae, whereas below the macrophytes fronds a high specific richness composed of polychaetes (*Pomatoceros lamarcki, Salmacina disteri*), gastropods (*Haliotis tuberculata, Calliostoma zizyphinum, Bittium reticulatum, Aplysia* sp., *Hypselodoris cantabrica*), sea urchins (*Paracentrotus lividus, Sphaerochinus granularis, Arbacia lixula*), starfish (*Marthasterias glacialis, Echinaster sepositus*), brittle stars (*Ophioderma longicauda*), sea cucumbers (*Holothuria forskali*), crustaceans (*Balanus perforatus, Cancer pagurus, Maja squinado, Liocarcinus puber, Homarus gammarus, Palinurus elephas*) and cephalopods (*Octopus vulgaris*) is found. The perforated bivalve *Rocellaria dubia* is an abundant species in the infaunal community.

OFFSHORE SYSTEMS

The annual cycle of phytoplankton is typical of a temperate sea (Fernández et al., 1993). The water mixing process of winter is followed by a phytoplankton bloom (4–3 mg Chl m^{-3}, 0–20 m, March) during spring when surface waters receive increasing irradiance and heat. In summer, solar radiation gives rise to a thermal water column stratification. During this period, the thermocline prevents the supply of nutrients from deep layers to the surface, and phytoplankton biomass declines. Finally, in autumn the thermocline is destroyed as a result of surface water cooling, and the wind and wave intensity increases, causing a moderate phytoplankton bloom. This pattern presents some peculiarities with respect to other northern Bay of Biscay areas (Varela, 1996). The summer subsurface chlorophyll maximum (SCM) of the Cantabrian Sea is related to the 'nitracline', not to the thermocline, and the stratification process which started in spring is of a saline origin, not thermal. A particular aspect of phytoplankton ecology is the development of red tides in late summer in the western coast (Rías Bajas) (Fraga et al., 1988), which occasionally are composed of toxic marine phytoplankton.

The typical phytoplankton cycle can be modified by two oceanographic processes (Fernández et al., 1993; Varela, 1996). One of them is the development of a convergence front between the stratified coastal water and the off-shelf, well-mixed oceanic waters which originate in the Northern Azores and which flow northwards along the slope. Both sides of this saline front encourage different food webs. In the oceanic and more saline area characterised by microflagellates, the microbial loop is the main trophic web, whilst in the other area, where different populations of chain-forming diatoms dominate, the classical food webs prevail. The second oceanographic process is the coastal upwelling during summer stratification of strong intensity in the western and northwestern coast (Fraga, 1981; Prego and Varela, 1998) which is weaker along the north coast (Botas et al., 1990; Borja et al., 1996). High densities of large phytoplankton, mainly diatoms, are observed in upwelling conditions, in contrast to the condition in nutrient-poor waters where low densities are found with a predominance of dinoflagellates and microflagellates. The shelf-edge fronts and upwelling events have a strong influence on fisheries because the success of pelagic fishes recruitment depends on larvae finding appropriate food.

Phytoplankton blooms are followed by herbivorous zooplankton peaks, with maximum values in late spring extending into summer (30–50 mg d.w. m^{-3}) and a secondary peak in autumn. Copepods account for 68% of the total abundance of the neritic water zooplankton and up to 83% in the oceanic area (Valdés and Moral, 1998). *Paracalanus parvus*, *Clausocalanus* spp., *Acartia clausi*, *Calanus helgolandicus*, *Oithona plumifera*, *Temora longicornis*, *T. stylifera*, *Centropages typicus* and *C. chierchiae* account for more than 90% of the overall copepod abundance. Biomass is highly variable depending on seasonal and inter-annual periodic environmental cycles as well as non-periodic events such as continental runoff, storms or upwellings.

These plankton communities support important pelagic fisheries such as sardine (*Sardina pilchardus*) and anchovy (*Engraulis encrasicholus*). The fast-moving pelagic carnivores mackerel (*Scomber scombrus*) and horse mackerel (*Trachurus trachurus*) are targeted by the spring fishery in the Cantabrian Sea, whereas albacore (*Thunnus alalunga*) is mainly caught during summer when they migrate.

A high percentage of the primary production sinks and is exported to the benthic system, which is characterised by rich faunal communities composed of septentrional and meridional Atlantic species. Most megabenthic species are not abundant. The crustaceans *Polybius henslowi*, *Munida sarsi*, *Liocarcinus depurator*, *Munida intermedia* and *Pagurus prideauxi*, cephalopods *Illex coindetti*, *Todaropsis eblanae*, *Loligo forbesi*, and the anemone *Actinauge richardi* are the most abundant species (Olaso, 1990). According to Olaso (1990), the invertebrates of the megabenthos from soft grounds of the Cantabrian shelf are differentiated into five groups: (1) a coastal group (40–90 m) characterised by *Sepia officinalis*, *Loligo vulgaris*, *Alloteuthis media*, *Octopus vulgaris*, and *Pagurus bernhardus*; (2) a group of the upper edge of the shelf (90–140 m) characterised by *Sepia elegans*, *Loligo forbesi*, *Calliostoma granulatum*, *Marthasterias glacialis* and *Ophiura ophiura*; (3) a shelf group (140–300 m): *Actinauge richardi*, *Argobuccinun olearium*, *Eledone cirrhosa*, *Munida sarsi*, *Stichopus regalis*, etc.; (4) a group of the upper edge of the slope (300–500 m): *Pasiphaea sivado*, *Munida perarmata*, *Galeodea rugosa*, *Stichopus tremulus*, *Porania pulvillus*, *Bathynectes maravigna*, and *Nephrops norvegicus*; and (5) a deep slope group (500–700 m): *Phormosoma placenta*, *Nymphaster arenatus*, *Opistotheuthis agassizzii*, *Philocheras echinulatus*, *Bathypolipus sponsalis* and *Polycheles thyphlops*. At greater depth still, the density of megabenthos decreases. Suspension-feeders and detritivorous animals are the main trophic groups. Sibuet and Segonzac (1986) pointed out that at 2000 m depth, 64% of megabenthos density corresponds to detritivorous species, with 34% being suspension-feeders. Of this latter, 29% are echinoderms.

This rich system supports a great fish diversity. Crustaceans are the food base for alevins, juveniles and young demersal fishes, whereas molluscs and echinoderms are occasional prey or food for specialised species. Demersal fish communities from soft grounds change with depth. According to Sánchez (1993), the community associated with gravel and sand bottoms between 90 and 150 m depth is characterised by *Trisopterus luscus*, *Trisopterus minutus*, *Capros aper*, *Merluccius merluccius*, *Lophius piscatorius*, *Lepidorhombus whiffiagonis*, *Callionymus lyra* and *Scyliorhinus canicula*. On sandy and muddy bottoms from the outer shelf (150–250 m depth), the most representative species are *Lepidorhombus boscii*, *Lophius budegasa*, *Micromesistius variegatus*, *Conger conger* and *Gadiculus argenteus*. The edge shelf community (250–400 m) inhabiting muddy bottoms is characterised by *Malacocephalus laevis*, *Bathysolea profundicola*, *Chimaera monstrosa*, *Galeus melastomus*, *Phycis blennoides* and *Antonogadus macrophthalmus*. Finally, in the slope community (400–650 m), *Notacanthus bonapartei*, *Deania calceus*, *Etmopterus spinax*, *Trachyrhynchus trachyrhynchus* and *Lepidion eques* are found. Some of these species have been traditionally exploited, with hake (*M. merluccius*), anglerfish (*L. piscatorius* and *L. budegasa*, respectively), and megrim (*L. whiffiagonis* and *L. boscii*) as the main demersal fisheries.

POPULATIONS AFFECTING THE AREA

Until the end of the 19th century, human settlements on the coastal zone appeared to have exploited their natural resources in a relatively balanced way. This trend changed with the industrialisation process which demanded large volumes of water for cooling systems, a receiving medium for wastes, and ports. Industrial areas quickly became population centres which supplied the required labour. In recent years an added consideration which resulted in high population concentrations, has been the use of the coast for leisure activities as well as economic and social activities, especially around estuaries. From the administrative viewpoint, the northern coast of Spain is split into four autonomous communities (Galicia, Principado de Asturias, Cantabria and Basque Country), with approximately 6.5 million inhabitants. The main urban-industrial centres along the coast in a decreasing population order are: Bilbao, Vigo, A Coruña, Gijón-Avilés, Santander-Torrelavega and San Sebastián- Pasajes areas (Fig. 3a).

Until 1876, the Basque Country was mainly a rural population, but since then the strong expansion of mining and steel industries turned the region into one of the major mineral suppliers for Western Europe, thus becoming a heavily industrialised area. Its modern and competitive industry reached its production peak in the year 1906 with 6.5 million tons of iron ore per year. At the same time the activity of the ports of Bilbao (the most important communication and transport route of northern Spain) and Pasajes increased. Following successive stages of expansion and decline, a dense network of industrial activities was developed (chemical plants, metal treatments, paints, paper mills, cement kilns, petrochemicals etc.) and large shipyards were established. In 1973, the worldwide economic crisis was felt and the heavy industry of the Basque Country fell into a severe decline, leaving behind serious environmental damages (Azkona et al., 1984). However, since the early 1990s, the closure of mines and of some of the most polluting industries together with installation of waste-water treatment plants and corrective measures in industries, are resulting in a recovery of coastal water quality (Gorostiaga and Díez, 1996). Nevertheless, the pressure on the coastal zones continues to increase, and of the total Basque population (ca. 2,130,000 inhabitants), 56% is concentrated in the coastal strip with an average density of 1419 inhabitants/m^2. Its main urban conglomeration is the metropolitan area of Gran Bilbao (approx. one million inhabitants), followed by the San Sebastian-Pasajes area (200,000 inhabitants). Further, 33% of the industrial companies and 28% of the working population are concentrated on the coast.

The Galician population (currently 2,785,394 inhabitants), despite heavy emigration until the mid-70s, is still densely populated, as 49% of the population lives in the coastal strip with an average density of 332 inhabitants/km^2. The growth of the main urban centres began from 1950 onwards (A Coruña with 316,000 inhabitants and Vigo with 398,000 inhabitants) along with a decline in rural areas (Bosque-Maurel and Vilà-Valentí, 1990). Round the estuaries of the west coast there is an intense concentration of economic activities. The fishing sector stands out and makes Galicia one of the most important fishing regions of Europe, with a landed fishing average of 600,000 tons per year (54% of Spanish fishing). Along the coastal strip there are 54 fishing harbours, with the most important ones being Vigo, A Coruña, Ribeira and Marín. The importance of the fishing sector is highlighted since it, in turn, drives other associated activities such as canning and freezing, pre-cooked products, marine cultures, machinery, naval construction, financial activities, insurance, etc. The late industrialisation process of Galicia is also concentrated in two large urban centres, Vigo and A Coruña.

The population of the Principado de Asturias (around 1,100,000 inhabitants) is mainly concentrated in the central area of the territory, in the urban centres of Gijón, Oviedo and Avilés where 51% of the population is based resulting in an average density above 500 inhabitants/km^2. Asturias is characterised by its great contrasts, from semi-deserted rural areas, to coal mining areas and industrial areas, mainly steel works, which are currently in crisis, but which some decades ago were leaders in the Spanish steel sector. The main industrial centres, Gijón and Avilés, are both located by the coast at both sides of Peñas Cape; this is the part of the Asturian coast which has deteriorated most. The remaining coastal zone does not have a high population density and is in a good state of conservation when compared to other coastal regions. The Cantabria situation is similar (527,000 inhabitants): the population is concentrated in the capital, Santander, and on its industrial belt where 50% of the population is based. Its coast, well preserved over long stretches, is tourist orientated and subjected to a strong development pressure.

Industrial and urban development in these coastal areas has led to substantial environmental degradation of several stretches of the coast as a result of the discharge of waste waters, disposal of litter, draining, filling and land-filling of ecologically valuable systems, disposal of dredged material and sewage sludge on poorly known habitats and over-exploitation of non-living and living marine resources. Despite increasing environmental awareness in social and political sectors, the extensive shorelines which are still in a good condition remain subject to the risk of degradation.

RURAL FACTORS

Galicia, Principado de Asturias, Cantabria and the Basque Country have a total of 5,256,000 hectares of useful agricultural land (UAL). Despite intense forest clearance over centuries, woodlands still occupy large areas (2,955,041 ha) (Fig. 3b), but a high percentage of natural forestry resources have been replaced mainly by pine (*Pinus radiata*)

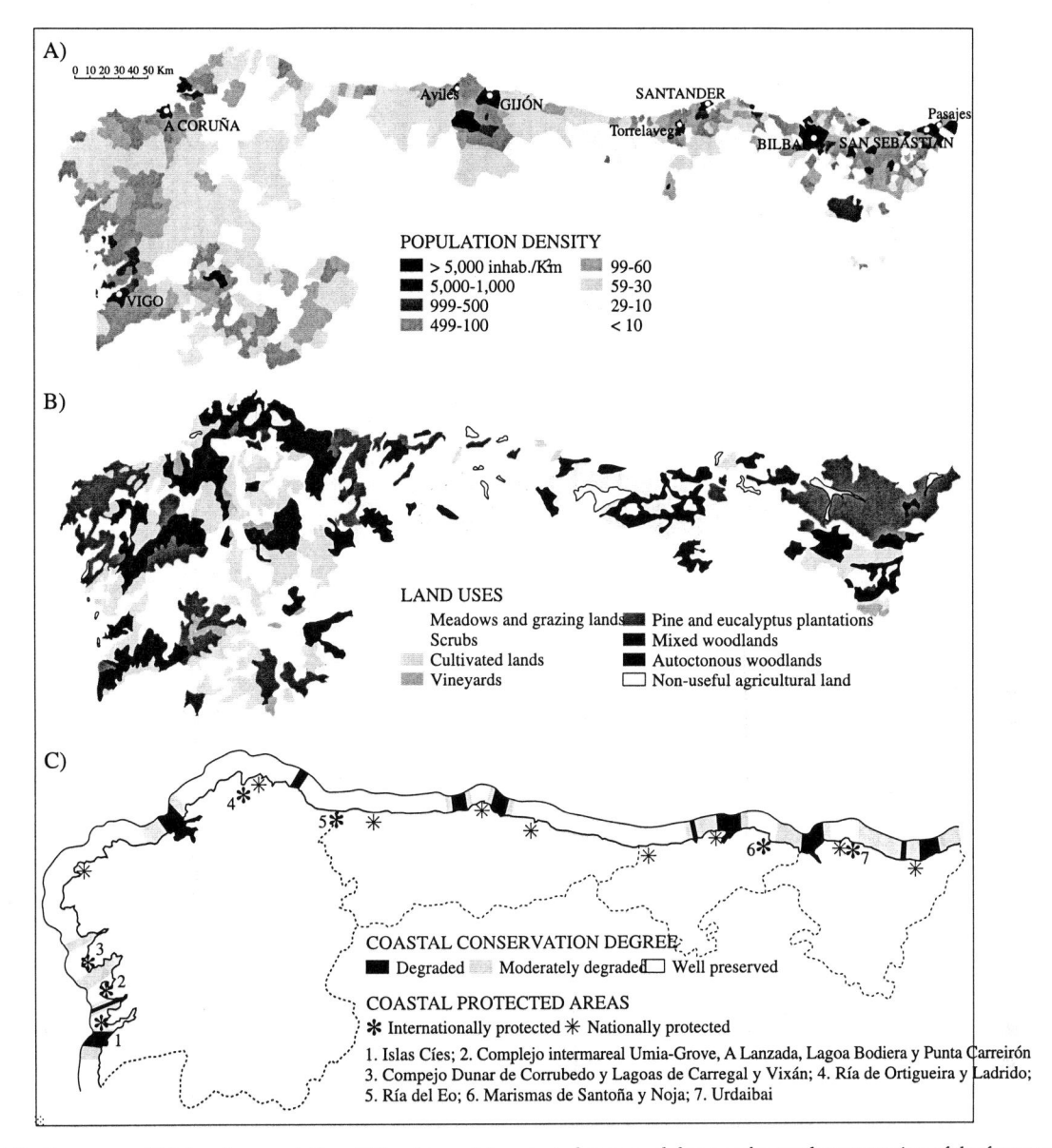

Fig. 3. Distribution maps of (a) density population, (b) land uses, (c) protected areas and degree of coastal preservation of the four north Spanish autonomous communities.

and eucalyptus (*Eucaliptus globulus*) plantations. These exotic trees, introduced in the 19th century, have spread widely over the past 50 years and they currently represent around 1,100,000 ha. Meadows and grazing lands occupy 896,551 ha (17% UAL), whereas cultivated lands only extend over 717,744 ha (14% UAL).

The land use involving the greatest soil loss—between 25 and 100 t ha^{-1} year^{-1}—is the forestry plantations of these trees. Their location on the steepest slopes (gently sloping lands are assigned to agricultural and industrial uses) and the plantation work are the main factors causing soil erosion. Run-off after high rainfall (annual mean of 1,000–1,608 mm in the north coast of Spain, increasing from west to east) transports the particulate matter to the coast through numerous short rivers; however, habitat alterations of rocky bottom communities by smothering have not been detected, which can be partially explained by the construction of dams in the upper course of the rivers and the still large extent of vegetation cover. Exceptions can be found at the mouth of the polluted rivers such as Nervion river (Bilbao), where sediment excess is one of the main destabilizing factors amongst hard bottom communities (Saiz-Salinas and Isasi-Urdangarin, 1994). However, together with soil loss, this particulate matter has its origin in several sources such as industrial and domestic waste-water effluents.

Very little data can be found on the impact on coastal waters of pesticides used in agricultural production. There are no indications that much effect has come from agricultural fertilizers transported to the coastal waters by rivers. According to Prego and Vergara (1998), the annual continental contributions of nutrients to the Bay of Biscay from the Cantabrian basin (524 m^3 s^{-1} of freshwater) may be estimated as 1.0×10^9 mol of N (dissolved inorganic nitrogen), 0.062×10^9 mol of phosphate and 1.2×10^9 mol of silicate with Nalón river (Asturias) followed by Nervión river (Basque Country) as the main sources. Cantabrian fluxes present very small values in comparison with those of the French basin, where only the Loire river annually transports 6.4×10^9 mol of N, 0.11×10^9 mol of P and 1.2×10^9 mol of Si. In addition, the nutrient fluxes are distributed among several small rivers all along the coast, and no large coastal plumes are formed. Therefore, coastal fertilisation due to continental waters could be considered negligible, only affecting estuarine zones at the mouth of the rivers in spring and summer.

Artisanal fishing and shellfish harvesting have been pursued by coastal communities all along the northern Spanish shoreline, especially in Galicia (a fishing power at European level) and in the Basque Country. The artisanal fleet is composed of small vessels, usually no more than 20 gross registered tons (range 5–100 GRT). Boats, belonging to family businesses, work close to the coast every day. Artisanal fishing uses a wide diversity of fishing gear such as surface, midwater and bottom long-line, handline, baskets and gillnets, and catches are distributed among a great variety of target species including hake, mackerel, horse mackerel, red mullets, anglerfish, sole, seabass, conger, red seabream, red scorpion fish and crustaceans (swimcrab, lobster) among others. This activity is characterised by its adaptability to variable situations, the varied fishing methods adaptable to catching different target species. Artisanal fishing and shellfish harvesting shows a clear decline as a consequence of the fall in target species abundance resulting from systematic overexploitation. Implementation of regulation measures has restrained fishermen from a indiscriminate catch, although furtive fishing is a widespread activity even in those polluted areas.

Aquaculture is not widespread along the Cantabrian coast. In some coastal areas the water quality is not acceptable (all industrialised estuaries), while in other cases the rough orography or the small extension of the estuaries are the reasons for its poor development. However, the 'Rías Bajas' in the western coast of Galicia shows very favourable conditions as its large sheltered areas support a great biological richness as a result of summer upwelling of deep nutrient-rich waters. Fish (turbot, salmon), clams, oysters, scallops, cockles and mainly mussels (*Mytilus galloprovincialis*) are cultivated in these rich areas. Mussel cultivation began in 1946 and nowadays its annual production is estimated to be 200,000 tons, being one the largest worldwide. The cultivation is carried out on floating platforms, locally termed "bateas", which consist of a rectangular wooden framework from 100 to 500 m² held up by steel floats covered with fibre-glass or polyester. The "batea" are fixed to the ground with steel chains and a ballast of 20 tons. From the wooden framework, 3 cm wide and 10–12 m long nylon ropes are hung, on which mussels are cultivated. There are currently more than 3,500 "bateas" in operation. Environmental eutrophication (ammonium increase) resulting from the aquaculture could be related to some deleterious biological observations. Since the cultivation of mussels became well developed in the mid-70s, toxic marine phytoplankton (*Gymnodinium catenatum*, *Alexandrium* sp., *Dinophysis acuta*, *Prorocentrum lima*) have been recorded. Wyatt and Reguera (1989) related this harmful algal blooms (HABs) to the hypertrophication of the environment as result of mussel production, agricultural fertilizers, sewage discharge from cities and increase in forest fires and soil transport to the sea, as well as a widespread climatic phenomenon.

COASTAL EROSION AND LANDFILL

Sediment transport and its redistribution along the coast has suffered some local alterations, mainly on the beaches around the urban-industrial and recreational centres. However, on the north coast of Spain, the coastal erosion problem is not as serious as it is on the Mediterranean Spanish coast. The Cantabrian coast is mainly formed by calcareous cliffs, where population and infrastructure density is low in comparison with the less extensive estuarine zones. Sea cliffs retreat as a result of wave-produced erosion; this is usually subtle and not easily perceived although sometimes it becomes evident when road sections close to the shoreline are destroyed.

Human pressure on estuarine and coastal wetland zones has started a landfill process which has led these areas to be amongst the most threatened habitats in the north coast of Spain (Bueno, 1997). Since the second half of the 19th century, in just the two eastern autonomous communities of Cantabria and Basque Country, 64 km² of intertidal and marsh areas have been lost by draining, filling and reclamation processes, representing 40% of the area which existed after the post-Flandrian retreat (Rivas and Cendrero, 1991). A similar trend can be expected for the two remaining regions, Asturias and Galicia. In some areas such as Bilbao and Pasajes, the intertidal zones have been totally occupied. The landfill has been related to the development of economic activities. At the beginning of the 19th century, commerce was the main activity and the predominant use of the reclaimed lands was urban and tidal mills. During the second half of that century, the reclamation of intertidal and marsh areas was made for agriculture and mining spoil disposal purposes. Since the beginning of 20th century, industrial installations and transport infrastructure became predominant, followed during the last three decades by construction of recreational and residential centres for tourism activity. Despite protective measures and increasing awareness of the ecological importance of these areas, reclamation processes continue, mainly related to tourism.

According to Rivas and Cendrero (1991), the distribution of land-use types in the newly artificially created lands from the Cantabrian and Basque Country is approximately 36%

agriculture, 22% industry, 14% residential, 13% others and 15% without use. In addition to a reduction of the biodiversity and impoverishment of landscapes, landfilling implies losses in terms of monetary value. From a biological productivity viewpoint, the substitution of intertidal or wetland areas by agricultural lands represent a loss of about $US 250,000 km^{-2} year^{-1}. In the same way, based on the 'willingness to pay' by population, the aesthetic qualities and recreational potential of these areas were estimated to be $US 200×10^6 per year. Reclamation and occupation of intertidal and marsh zones also represent a risk regarding an eventual rise of sea level. If the sea level rises in response to a global warming in the way predicted by several models, most of the filled areas, which are less than one meter above sea level (73 km^2), would be submerged. The economic impacts including the loss of buildings, roads, ports, industrial and agricultural installations, etc., were estimated at around $US 5,000×10^6 by Rivas and Cendrero (1991).

In addition to inundation, a sea level rise would cause erosion and the retreat of shorelines as a consequence of a sediment budget deficit, a salt intrusion into aquifers and an increasing risk of flooding and storm damage. Flooding processes are frequent in the northeastern Spanish coast due to the torrential character of the rivers. Every 25 years, rivers tend to flow over the whole flood plain and have catastrophic effects. In the Basque Country, the intense rainfall (350–400 mm) in 1983 resulted in a flood that caused 50 casualties and a capital loss of $US 900,000.

Great amounts of sediments are annually dredged to maintain navigation channels, for building or landfilling purposes. The most detrimental of these actions are those which occur on well preserved coastal zones and even on protected areas such as the 'Urdaibai Biosphere Reserve' where dredging is periodically carried out to serve a shipyard. The most direct impact is the destruction of the soft-bottom habitats and the removal of epifaunal and infaunal communities. The time for recolonization is variable depending on the environmental conditions of the area and the maturity of the community. López-Jamar and Mejuto (1988) reported for the harbour area of A Coruña Bay a six-month recolonization period after dredging operations. If there is rocky bottom in the dredging vicinity, the light attenuation caused by suspended sediments and the smothering effect when they are deposited, affect phytobenthic communities, especially the more sensitive species, leading to a proliferation of opportunistic ones such as blue-green algae (*Spirulina* sp.). Deleterious effects are intensified when dredging is carried out in harbour areas and polluted estuaries due to the potential introduction of contaminants. Although this type of action legally requires an environmental impact assessment, the latter are rarely comprehensive studies, most of the time being a mere formality. Together with biological and chemical impacts, dredging has an erosive effect on the outer beaches due to changes in the sediment budget. Other environmental concerns arise from the translocation of dredged spoils to dumpsites, wetlands and marshes or areas very close to the shoreline (<10 km away). Harmful effects resulting from sediment disposal are poorly understood and more effort must be made to manage these sediments properly, mainly when they come from harbours, embayments or polluted areas.

EFFECTS FROM URBAN AND INDUSTRIAL ACTIVITIES

Urban settlements of the north coast of Spain have a serious deficit of waste-water treatment systems. From the total of 11,979,138 population equivalent (p.e.) (1 p.e. means the organic biodegradable load having a five-day biochemical oxygen demand—BOD$_5$—of 60 g of oxygen per day) only 20% conforms to the requirements of the Directive 91/271/EC concerning urban waste-water treatment. In this respect, Cantabria and the Basque Country are the less favoured regions with only 5% and 17% of the population equivalent, respectively. From the main coastal urban centres, only Bilbao and Santander with primary sewage treatment, and Vigo (secondary treatment) have treatment plants. Unfortunately, at present, several projects of submarine outfalls (after a primary treatment) have been selected as corrective measures, following the erroneous principle of 'dilution'.

Industrial development this century has resulted in 479 km of shoreline with intense to moderate industrial use. The occupation of wetlands and marshes and the discharge of great amounts of waste waters are responsible for the loss of valuable habitats and the degradation of several coastal stretches (Fig. 3c). The Basque Country has been the most impacted region. From a total of 440,051,106 m^3/year land-to-sea sewage wastes reaching the Basque coast, 277,365,801 m^3/year are industrial effluents, 74,818,907 m^3/year are domestic waste-waters, 182,366 m^3/year are faecal waters and the rest are mixed industrial–urban–faecal waste-water effluents (Gobierno Vasco, 1995). Mining, steel industries and metal treatments have been predominant industrial activities along the north coast of Spain for decades. This is revealed by the extremely high heavy metal concentrations recorded in the surface waters and sediments of several coastal areas (Table 1). Guerrero-Pérez and co-workers (1988) found the highest values for zinc (230 µg l^{-1}) and lead (9.10 µg l^{-1}) close to Suances (Cantabria), which receives the industrial sewage wastes from Torrelavega and a mining area. At the Nervión river mouth (Bilbao), extremely high values of cadmium occur in surface waters (3,320 ng l^{-1}) and elevated levels were recorded for most heavy metals examined. Gross organic enrichment, water oxygen depletion and anaerobic sediment conditions are the dominant anthropogenic influences in the Nervión estuary according to Saiz-Salinas (1997), although trace organic pollutants (PAHs, PCBs, and pesticides) also show

Table 1

Dissolved heavy metals in surface coastal waters and concentrations found in the 63 μm fraction of surface sediment layer (dry weight) from several sites along the north coast of Spain (modified from Guerrero-Pérez et al., 1988).

Site	Coordinates		Depth (m)	Surface water			Sediment (heavy metals in ppm)						
	Lat. N	Long. W		Zn (μg/l)	Cd (ng/l)	Pb (μg/l)	Cu	Zn	Cd	Pb	Hg	Mn	Cr
Miño River mouth	41°52'6	8°56'6	54	8.2	30	0.55	21.2	94.7	0.087	21.8	0.16	256	21.7
Ría de Vigo	42°13'0	8°52'5	38	6.8	60	0.51	24.0	913.0	0.130	54.4	0.25	216	47.0
Ría de Pontevedra	42°22'0	8°54'0	47	10.7	200	1.64	27.9	107.5	0.071	37.9	0.31	195	30.6
Ría de Arousa	42°30'0	8°57'0	60	5.4	300	0.50	30.2	113.4	0.360	44.7	0.31	209	54.6
Ría de Muros	42°43'7	9°03'5	45	6.4	32	0.76	26.3	105.5	0.078	39.5	0.20	188	44.6
Pta. Rocundo	43°16'0	9°09'0	110	3.6	120	0.22	24.4	94.7	0.040	38.3	0.11	211	44.5
A Coruña	43°22'1	8°22'0	19	12.1	61	0.43	33.1	780.0	0.470	79.9	0.53	167	28.0
Avilés	42°36'7	5°57'7	36	2.7	61	0.18	-	-	-	-	-	-	-
Gijón	43°34'0	5°40'5	25	-	-	-	54.0	1936.0	1.550	169.9	3.74	801	46.4
Llanes	43°29'8	4°45'0	100	4.2	63	1.08	28.1	197.8	0.470	82.0	1.51	359	38.0
Suances-Torrelavega	43°27'0	4°02'8	28	230.0	770	9.10	-	-	-	-	-	-	-
Santander Bay	43°28'5	3°44'8	22	10.3	56	0.74	23.5	2422.0	2.840	136.0	0.91	308	14.7
Nervíon River mouth	42°20'0	3°01'6	15	98.2	3320	7.35	459.0	1721.0	6.620	840.0	4.05	742	144.1

high values. Saiz-Salinas (1997) reported a massive reduction of fauna in three-quarters of the estuary as a response to heavily polluted conditions, whereas Díez et al. (1999) pointed out that as far as 17 km away from the Nervión river mouth, the pollution still affects phytobenthic communities. At present, together with the ending of the sewerage plan, dredging the estuary is being considered to remove sequestrated pollutants in sediments, even though dredging can be a risky and potentially damaging remedy.

Shipping operations and accidents are other sources of pollution. The harmful effects of the use of organotin (OT) compounds (dibutyltin, DBT; tributyltin, TBT; etc.) in anti-fouling paints has been reported by Ruiz et al. (1998), who pointed out the frequent female sterilisation of the intertidal gastropod *Nucella lapillus* on the northwestern coast of Spain. The authors suggest fishing and merchant fleets are the dominant sources of OT pollution, with pleasure boats being of lesser importance. Likewise, the Galician coast is particularly exposed to oil spills as it is located in one of the most important sea lanes for petroleum transport. The last oil spill occurred on 3rd December 1992 when the Greek oil tanker 'Aegean Sea' ran aground at Torre de Hercules (near A Coruña harbour). The tanker, carrying 79,096 tonnes of Brent-type crude oil, broke up and exploded. Almost all the oil was released into the sea and over 200 km of shore were affected, with 4,000 m^3 of oil removed from the sea and beaches (MOPT, 1996). Sixteen years before, in 1976, the Spanish tanker 'Monte Urquiola', exploded not far from the Aegean Sea wreck. The 100,000-tonne spill damaged 210 km of coast (Palomo et al., 1978). Although navigation and routine ports operations are another clear source of marine pollution, a comprehensive impact assessment has not been carried out.

The fishing industry expands the list of negative human impacts in the north coast of Spain. Most of the traditional fishery resources of Divisions VIIIc and IXa of ICES are fully or over-exploited and several stocks are depleted (AZTI, 1999). Hake, anglerfish and megrim, the major demersal resource of this area, are out of the safe biological limits. The Advisory Committee for Fisheries Management (ACFM) advises that hake fishing mortality should be reduced by 20% for 1999, in order to allow recovery. In the same way, ICES advises that anglerfish and megrim fishing mortality should be reduced by 50% with respect to 1997, whereas black-bellied angler should be reduced by 30%. The total allowable catches (TACs) established for these species have been greater than the catches obtained in recent years.

In addition, Sánchez (1993) pointed out that fisheries have a major impact on the Cantabrian shelf ecosystem. Alteration and degradation of sea bottom habitats by trawling gears encourage the development of opportunistic species of wide distribution and high growth rates, whereas the most specialized slow-growing species are eliminated, resulting in a progressive simplification of the communities.

In relation to pelagic fisheries, sardine stock presents the most critical situation as spawning biomass shows an historical minimum (Fig. 4) and the population is out of safe biological limits. At the moment, TACs have not been applied to this fishery. Currently exploitation of mackerel is not compatible with the precautionary principle, and catches should be reduced by 30%. Anchovy from the Bay of Biscay and horse mackerel southern stock are considered to be within safe biological limits.

A large percentage of all the fish pulled from the sea is returned to it (Table 2). Discards consist of non-commercial species, fish that cannot be legally landed (too small speci-

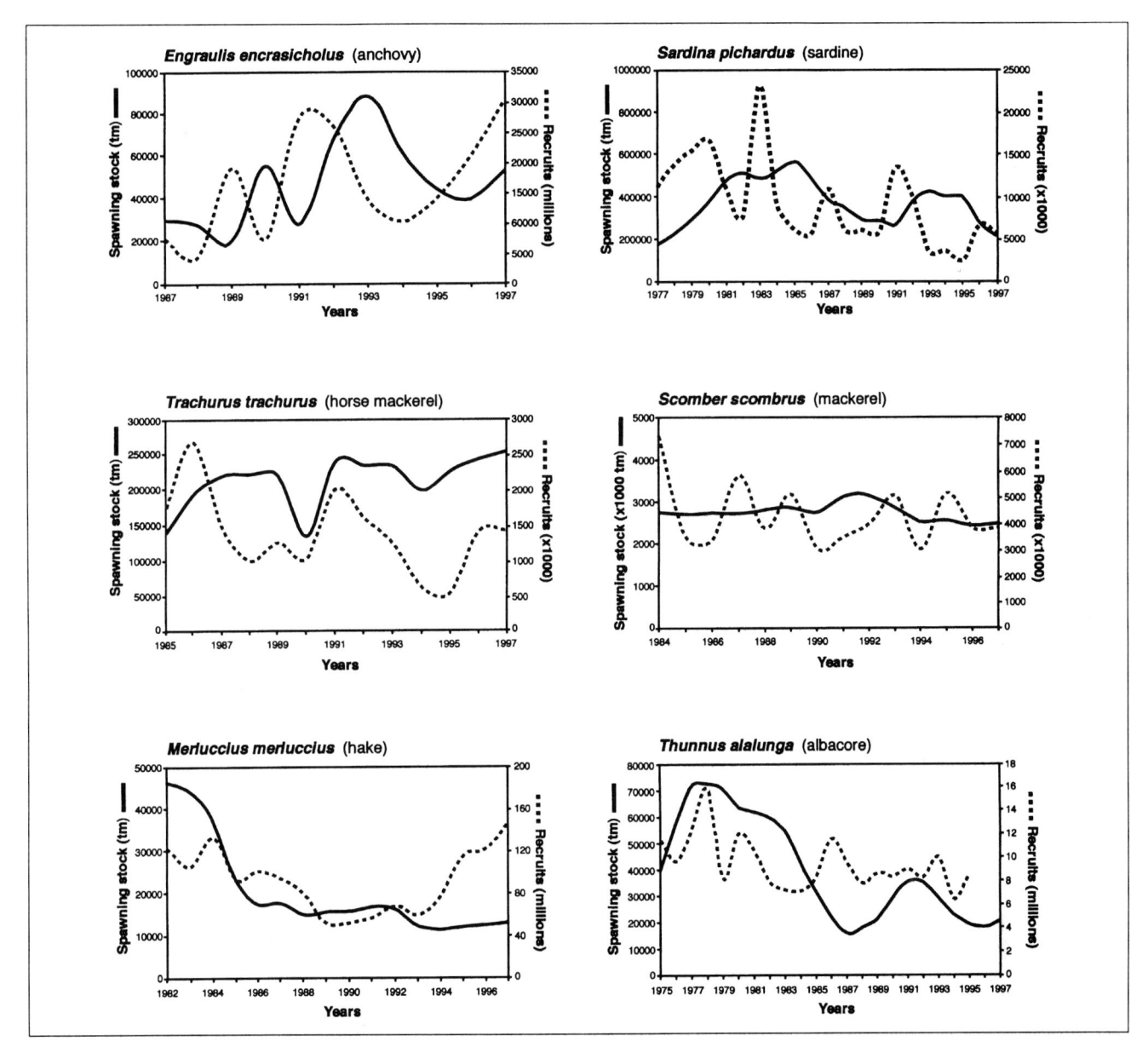

Fig. 4. Stock evolution in terms of spawning biomass and number of recruits of some of the main fishery resources exploited in the north coast of Spain (modified from AZTI, 1999).

mens, exceeded TAC) or specimens deteriorated during fishing operations (Pérez et al., 1996). Most of them do not survive, with the subsequent loss of food for the upper trophic web stages. Trawling is the fishing method causing highest discard rates.

PROTECTIVE MEASURES

Over the last few years the Spanish administration has increased its economic, human and legal efforts to stop the progressive deterioration of the coastline and to protect the natural resources and biodiversity of coastal ecosystems.

Spain has signed, amongst others, the following international agreements which have effects on the protection and conservation of the coastlines and seas: 'RAMSAR Convention' (Irán 1971) regarding Wetlands of International Importance, 'OSPAR Convention' (Paris 1992) for the Protection of the Marine Environment from the North-East Atlantic, 'MARPOL Convention' (London 73/79) for the Prevention of Pollution from Ships, 'United Nations Convention on the Law of the Sea (UNCLOS)' (Montego Bay 1982) and the 'United Nations Convention on Biological Diversity or Biodiversity' (Rio de Janeiro 1992).

At a national level, the administrative management of the coast is established by the Coastal Law 22/1988 of 28th of July, and its implementing Regulation of 1989. One of the most relevant and positive aspects of this law was the classification of the sea and its shore as a collective heritage. Likewise, this law sets a series of limitations to the ownership and use of land adjoining the sea shore and also controls land-to-sea waste disposal. Generally speaking,

Table 2

Spanish fleet discards in the Divisions VIIIc and IXa of ICES during 1994, classified according to the fishing method (modified from Pérez et al., 1996)

Fishing method	ICES Zone	Percentage (discard/catch)				Total percentage			Retained species	
		Quarter of the year				Discard/ catch	Discard/ retained catch	Discard/ 1st retained sp.		
		1st	2nd	3rd	4th					
Trawling	VIIIc	40.5	32.7	30.1	34.9	34.7	53.1	139.3	*Micromesistius poutassou*	38%
	IXa	28.7	67.3	73.8	54.9	59.2	145.1	807.9	*Trachurus trachurus*	35%
Long-lining	VIIIc	36.7	10.3	4.5	16.3	18.7	23.0	115.1	*Phycis biennoides*	20%
Gillnetting	VIIIc	11.0	9.3	47.9	39.3	25.3	33.9	101.2	*Merluccius merluccius*	33%
Purse seining	VIIIb,c East	1.2–4.5	15.1–33.0	30.0–35.0	–	23.0–30.4	32.8–43.6	70.0–94.0	*Trachurus mediterraneus*	46%
	VIIIc West	2.0	1.4	6.3	10.1	6.4	6.9	11.1	*Sardina pilchardus*	69%
	IXa North	0.0	0.2	0.8	2.1	0.9	0.9	1.4	*Trachurus trachurus*	63%

both the extent and severity of the enforcement of the coastal law have been limited. A clear example is the administrative process involved in the demarcation of land that falls under public ownership, which twelve years after the publishing of the Law, is yet to be concluded. In addition, on many occasions, the coastal strip which is directly affected by the Coastal Law has a very small extension compared with the coastal territories where the problems affecting the coast are felt.

Some autonomous communities have created a specific regulation to be enforced in the coastal zone in order to achieve an integrated management, as is the case with Asturias and the Basque Country, whilst Cantabria and Galicia have no specific legal instrument that establishes a regulation of the uses and activities of the coastal zone.

The declaration of reserves and natural protected areas falls under the Law 4/1989 of 27 of March for the Conservation of the Natural Areas and the Wild Fauna and Flora (in its Title no. III they refer to the coastal ecosystems), recently modified by the Law 41/1997 of 5th September. The legislative condition with maximum protection foreseen by this law is the 'National Park'. On the northern coast of Spain no marine national parks exist, although the Environmental Department of the Galicia Government (Consellería de Medio Ambiente de la Xunta de Galicia) has already initiated the procedures established by the Law 47/97 in order to create the first Coastal National Park, that of the 'Atlantic Islands'.

The central governmental law is supplemented at a regional level by other regulations such as the Law 16/94 of 30 June, for the Conservation of Nature in the Basque Country.

As a result of the enforcement of such laws a total of 16 protected coastal areas (41,216.4 ha) are acknowledged in northern Spain (by the end of 1999), which as a result of their high environmental value have been protected either at a national or at an international level. The legislative instruments under which these areas are protected are numerous such as: Natural Parks, Natural Reserves, Partial Natural Reserves, Protected Biospheres, Natural Monuments, Generally Protected Areas, etc. depending on the degree of protection, the size of the area and the local administration. Within these areas, traditional uses of the space compatible with the conservation of the natural resources are promoted, and access to visitors is facilitated.

Seven of these protected areas (Fig. 3c) have international protection status of one or more of the following: UNESCO "Man And Biosphere Reserve" (MaB), "Wetland of International Importance" of the Ramsar list, and "Special Protection Areas" (SPAs) from ECC Directive of April 1979 on the Conservation of Wild Birds: 1. Islas Cíes (SPA); 2. Complejo intermareal Umia-Grove, A Lanzada, Lagoa Bodiera y Punta Carreirón (SPA, Ramsar list); 3. Complejo Dunar de Corrubedo y Lagoas de Carregal y Vixán (Ramsar list); 4. Ría de Ortigueira y Ladrido (SPA, Ramsar list); 5. Ría del Eo (SPA, Ramsar list); 6. Marismas de Santoña y Noja (SPA, Ramsar list); and 7. Urdaibai (Man And Biosphere Reserve, SPA, Ramsar list).

The increase of economic, human and legal efforts for the protection of natural resources and biodiversity seems insufficient, since at present any protected area is still at risk of being seriously destroyed or deteriorated. Even those protected areas which are regarded as most notable for nature conservation such as Santoña, Victoria and Joyel complex, the most important wetland of the northern Iberian Peninsula for water birds with up to 86 species found and with the only breeding site in northern Spain for *Ardea pupurea*, *Ixobrychus minutus* and *Himantopus himantopus*, has been progressively degraded after the implementation of protective measures. The road building across the main wetland (Santoña), the development of several small settlements, scattered building, hotels, single-family dwellings, and a disco at the edge of the wetland result in a strong development pressure with highly adverse environmental implications for the wetlands.

Speculative activities, insufficient studies and the lack of a legal regulation to enforce strict limits to human interventions in protected areas are some of the reasons that explain the current failure of these protective measures. In addition, it is essential not to restrict conservation to specific areas subjected to protection schemes. It is equally important to consider the uses that are made of the non-protected areas.

Likewise, in order to preserve marine habitats, the administrative Spanish Authorities should formally establish an appropriate and global strategy to progressively reduce industrial and domestic waste-waters with the long-term goal of zero discharge. A good start would be to carry out a comprehensive inventory of pollution sources. At present, the lack of information does not allow a reliable report to be made of the conservation state of the coast, more environmental surveys being needed. Finally, environmental education is essential for conservation to be possible, and information should be public and easily available.

REFERENCES

Anadón, R. and Niell, F.X. (1981) Distribución longitudinal de macrófitos en la costa asturiana (N de España). *Inves. Pesq.* **45**, 143–156.

Azkona, A., Jenkins, S.H. and Roberts, H.M.G. (1984) Sources of pollution of the estuary of the river Nervion, Spain-a case study. *Water Science and Technology* **16**, 95–125.

AZTI (1999). *AZTI ARRANTZA 1999: Nuestros Recursos Pesqueros.* AZTI (ed.) Instituto Tecnológico Pesquero y Alimentario, Sukarrieta. pp. 55.

Bárbara, I. and Cremades, J. (1996) Seaweeds of the Ría de A Coruña (NW Iberian Peninsula, Spain). *Bot. Mar.* **39**, 371–388.

Borja, A., Uriarte, A., Valencia, V., Motos, L. and Uriarte, A. (1996) Relationships between anchovy (*Engraulis encrasicolus*) recruitment and the environment in the Bay of Biscay. *Sci. Mar.* **60** (suppl. 2): 179–192.

Bosque-Maurel, J. and Vilà-Valentí, J. (1990) *Geografía de España, 10 Vols.* Editorial Planeta, S.A., Barcelona, España.

Botas, J.A., Fernández, E., Bode, A. and Anadón, R. (1990) A persistent upwelling off the central Cantabrian coast (Bay of Biscay). *Estuarine and Coastal Shelf Science* **30**, 185–199.

Bueno, A. (1997) Flora y vegetación de los estuarios asturianos. *Cuadernos de Medio Ambiente, Naturaleza 3.* Servicio Central de Publicaciones del Principado de Asturias. pp. 352.

Díez, I., Secilla, A., Santolaria, A. and Gorostiaga, J.M. (1999) Phytobenthic intertidal community structure along an environmental pollution gradient. *Marine Pollution Bulletin* **38**, 463–472.

Fernández, E., Cabal, J., Acuña, J.L., Bode, A., Botas, J.A. and García-Soto, C. (1993) Plankton distribution across a slope current-induced front in the southern Bay of Biscay. *Journal of Plankton Research* **15**, 619–641.

Fraga, F. (1981) Upwelling off the Galician coast, northwest Spain. In: *Coastal Upwelling*, F.A. Richards (ed.), pp. 176–182. American Geophysical Union, Washington D.C.

Fraga, F. (1989) Circulación de las masas de agua en el Golfo de Vizcaya. In: *Oceanografía del Golfo de Vizcaya.* Univ. del País Vasco (ed.), San Sebastian, pp. 11–21.

Fraga, S., Anderson, D.M., Bravo, I., Reguera, B., Steidinger, K.A. and Yentsch, C.M. (1988) Influence of upwelling relaxation on dinoflagellates and shellfish toxicity in ría de Vigo, Spain. *Estuarine and Coastal Shelf Science* **27**, 349–361.

Gobierno Vasco (1995) *Inventario de vertidos tierra-mar del País Vasco.* Viceconsejería de Medio Ambiente. Vitoria-Gasteiz.

Gorostiaga, J.M. and Díez, I. (1996) Changes in the sublittoral benthic marine macroalgae in the polluted area of Abra de Bilbao and proximal coast (Northern Spain). *Marine Ecology Progress Series* **130**, 157–167.

Gorostiaga, J.M., Santolaria, A., Secilla, A. and Díez, I. (1998) Sublittoral benthic vegetation of eastern Basque coast (N Spain): structure and environmental factors. *Bot. Mar.* **41**, 455–465.

Guerrero-Pérez, J., Rodriguez-Puente, C. and Jornet-Sancho, A. (1988) Estudio de metales pesados en aguas y sedimentos superficiales en las costas cantábrica y gallega. *Inf. Técn. Inst. Esp. Oceanogr.* **64**, 16.

Junoy, J. and Viéitez, J.M. (1990) Macrozoobenthic community structure in the Ría de Foz, an intertidal estuary (Galicia, Northwest Spain). *Marine Biology* **107**, 329–339.

Koutsikopoulos, C. and Le Cann, B. (1996) Physical processes and hydrological structures related to the Bay of Biscay anchovy. *Sci. Mar.* **60** (suppl. 2), 9–19.

Koutsikopoulos, C., Beillois, P., Leroy, C. and Taillefer, F. (1998) Temporal trends and spatial structures of the sea surface temperature in the Bay of Biscay. *Oceanologica Acta* **21**, 335–344.

Lavín, A., Valdés, L., Gil, J. and Moral, M. (1998) Seasonal and interannual variability in properties of surface waters off Santander, Bay of Biscay, 1991–1995. *Oceanologica Acta* **21**, 179–190.

Le Cann, B. and Pingree, R. (1994) Circulation dans le golfe de Gascogne: une revue de travaux récents. In: *Actes IV Colloque International d'Océanographie du Golfe de Gascogne*, Cendrero O. and Olaso I. (eds.). Inst. Esp. Oceanogr., pp. 217–234.

Le Floch, J. (1968) Quelque propietés des eaux d'origine méditerranéene dans le golfe de Gascogne. *Travails Centre Recherche Etudes Océanographique* **7**, 25–36.

Longhurst, A. (1998) *Ecological Geography of the Sea.* Academic Press, London, pp. 398.

Lopez-Jamar, E. and Mejuto, J. (1988) Infaunal benthic recolonization after dredging operations in La Coruña Bay, NW Spain. *Cahiers de Biologie Marine* **29**, 37–49.

Montadert, L., Winnock, E., Delteil, J-R. and Grau, G. (1974) Continental margins of Galicia-Portugal and Bay of Biscay. In: *The Geology of Continental Margins*, C.A. Burk and C.L. Drake (eds.), pp. 323–342. Springer-Verlag. New York.

M.O.P.T. (1996) *Seguimiento de la contaminación producida por el accidente del buque Aegean Sea.* Servicio de Publicaciones del Ministerio de Obras Públicas y Transportes.

Niell, F.X., Fernández, C., Figueroa, F.L., Figueiras, F.G., Fuentes, J.M., Pérez-Llorens, Garcia-Sánchez, M.J., Hernández, I., Fernández, J.A., Espejo, M., Buela, J., Garcia-Jiménez, M.C., Clavero, V. and Jiménez, C. (1996) Spanish Atlantic Coasts. In: *Marine Benthic Vegetation: Recent Changes and the Effects of Eutrophication*, W. Schramm and P.H. Nienhuis (eds.), pp. 265–281. Springer-Verlag, Berlin.

Olaso, I. (1990) Distribución y abundancia del megabentos invertebrado en los fondos de la plataforma cantábrica. *Publ. Espec. Inst. Esp. Oceanogr.* **5**, pp. 128.

Palomo, C., Acosta, J., Andres, J.R., Herranz, P., Rey, J. and Sanz, J.L. (1978) Contribución al estudio del impacto causado en el litoral de La Coruña (España) por el derrame de crudos originado en el accidente del petrolero "Monte Urquiola" (mayo 1976). *Bol. Inst. Esp. Oceanogr.* **4**, 131–162

Penin, F. (1980) Le prisme littoral Aquitain histoire et evolution recente des environnements morphosedimentaires. Ph.D. Thesis, Université de Bordeaux 1.

Pérez, N., Pereda, P., Uriarte, A., Trujillo, V., Olaso, I. and Lens, S. (1996). Descartes de la flota española en el área del ICES. *Dat. Resúm. Inst. Esp. Oceanogr.*, no. 2, pp. 142.

Prego, R. and Varela, M. (1998) Hydrography of the Artabro Gulf in summer: western coastal limit of Cantabrian seawater and wind-induced upwelling at Prior Cape. *Oceanologica Acta* **21**, 145–155.

Prego, R. and Vergara, J. (1998) Nutrient fluxes to the Bay of Biscay from Cantabrian rivers (Spain). *Oceanologica Acta* **21**, 271–278.

Queró, J.C., Du Buit, M.H. and Vayne, J.J. (1998) Les observations de poissons tropicaux et le réchauffement des eaux dans l'Atlantique européen. *Oceanologica Acta* **21**, 345–352.

Rivas, V. and Cendrero, A. (1991) Use of natural and artificial accretion on the North Coast of Spain: historical trends and assessment of some environmental and economic consequences. *Journal of Coastal Research* **7**, 491–507.

Ruiz, J.M., Quintela, M. and Barreiro, R. (1998) Ubiquitous imposex and organotin bioaccumulation in gastropods *Nucella lapillus* from Galicia (NW Spain): a possible effect of nearshore shipping. *Marine Ecology Progress Series* **164**, 237–244.

Saiz-Salinas, J.I. and Isasi-Urdangarin, I. (1994) Response of sublittoral hard substrate invertebrates to estuarine sedimentation in the outer harbour of Bilbao (N. Spain). *P.S.Z.N.I.: Mar. Ecol.* **15**, 105–131.

Saiz-Salinas, J.I. (1997) Evaluation of adverse biological effects induced by pollution in the Bilbao estuary (Spain). *Environmental Pollution* **96**. 351–359.

Sánchez, F. (1993) Las comunidades de peces de la plataforma del Cantábrico. *Publ. Espec. Inst. Esp. Oceanogr.* **13**, pp. 137.

Sherman, K. and Alexander, L.M. (1989) Biomass Yields and Geography of Large Marine Ecosystems. *American Association for the Advancement of Science Selected Symp.* **111**, 1–493.

Sibuet, M. and Segonzac, M. (1986) Abondance et repartition de l'epifaune mégabenthique. In *Peuplements profonds du Golfe de Gascogne*. Campagnes Biogas, IFREMER, pp. 143–156.

Valdés, L. and Moral, M. (1998) Time-series analysis of copepod diversity and species richness in southern Bay of Biscay off Santander, Spain, in relation to environmental conditions. *ICES Journal of Marine Science* **55**, 783–792.

Valencia, V. (1993) Estudio de la variación temporal de la hidrografía y el plancton en la zona nerítica frente a San Sebastián entre 1988–1990. Inf. Técn. no. 52 Dep. Agric. y Pesca, Gob. Vasco. Vitoria.

Van den Hoek, C. (1975) Phytogeographic provinces along the coasts of the northern Atlantic Ocean. *Phycologia* **14**, 317–330.

Van der Spoel, S. and Heyman, R.P. (1983) *A Comparative Atlas of Zooplankton. Biological Patterns in the Oceans*. Springer-Verlag, Berlin. pp. 186.

Varela, M. (1992) Upwelling and Phytoplankton ecology in Galician (NW Spain) rías and shelf waters. *Bol. Inst. Esp. Oceanogr.* **8**, 57–74.

Varela, M. (1996) Phytoplankton ecology in the Bay of Biscay. *Sci. Mar.* **60** (suppl. 2): 45–53.

Villate F., Moral, M. and Valencia, V. (1997) Mesozooplankton community indicates climate changes in a shelf area of the inner Bay of Biscay throughout 1988 to 1990. *Journal of Plankton Research* **19**, 1617–1636.

Wooster, W.S., Bakun, A. and McLain, D.R. (1976) The seasonal upwelling cycle along the eastern boundary of the North Atlantic. *Journal of Marine Research* **34**, 131–141.

Wyatt, T. and Reguera, B. (1989) Historical trends in the red tide phenomenon in the Rías Bajas of northwest Spain. In: *Red Tides: Biology, Environmental Science, and Toxicology. Proc. First Int. Symp. on Red Tides* (November 10–14. Takamatsu, Japan). T. Okaichi et al. (eds.), pp. 33–36. Elsevier. New York.

THE AUTHORS

Isabel Díez
*Departamento de Biología Vegetal y Ecología,
Facultad de Ciencias,
Universidad del País Vasco, Apdo. 644,
Bilbao 48080, Spain*

Antonio Secilla
*Departamento de Biología Vegetal y Ecología,
Facultad de Ciencias,
Universidad del País Vasco, Apdo. 644,
Bilbao 48080, Spain*

Alberto Santolaria
*Departamento de Biología Vegetal y Ecología,
Facultad de Ciencias,
Universidad del País Vasco, Apdo. 644,
Bilbao 48080, Spain*

José María Gorostiaga
*Departamento de Biología Vegetal y Ecología,
Facultad de Ciencias,
Universidad del País Vasco, Apdo. 644,
Bilbao 48080, Spain*

Chapter 10

SOUTHERN PORTUGAL: THE TAGUS AND SADO ESTUARIES

Graça Cabeçadas, Maria José Brogueira and Leonor Cabeçadas

The coastal waters of southern Portugal, which include the Tagus and Sado estuaries, are influenced by submarine canyons, headlands and by intense discharges of freshwater (the annual average flow of the Tagus river is about 400 $m^3 s^{-1}$). The area is also affected by intense upwelling processes, and changes in the last 10–15 years in the upwelling pattern off Portugal are thought to be responsible for decreasing recruitment of sardine and horse-mackerel.

Spatial phytoplankton variability seems to be controlled, to some extent, by the establishment of a frontal boundary separating the river plume from tidally mixed nearshore waters; the boundary apparently functions as a barrier between two different phytoplankton populations, *Thalassiosira* spp. and *Detonula pumila*. Considerable work has been done in the Tagus estuary whose environment has been well characterised, particularly its dynamics related to the river regime and tidal conditions. The latter variables control the turbidity of the estuary and the materials deposited in the inner and outer shelf. Information on the biota of the zone—plankton and marine biological resources including fish, chephalopeds and benthic species—as well as on contamination by bacterial and pollutants, is fairly extensive. The system has a high biodiversity and, in general, is in reasonably good condition. However, in the inner parts of the Tagus estuary there are potential problems with high levels of Hg, Pb, TBTs and some bacterial contamination. The area is under considerable human pressure, being one of the most populated and industrialized regions of Portugal, and thus it receives considerable urban and industrial wastes. In addition to this there is intense maritime traffic and significant tourism in the area, and recent enterprises such as the EXPO Exhibition, the construction of the new bridge and several marinas have each led to specific but usually relatively minor environmental impacts. Dredging in the Tagus and Sado estuaries takes place regularly, and in the Tagus the amount of material removed increased by a additional 2.5 million tons during the construction of the new bridge, but monitoring studies undertaken during these operations apparently revealed no significant effects. In general, contaminants in fish, molluscs and bivalves are below those allowed in organisms for human consumption.

The relatively healthy conditions of these Portuguese coastal waters are a result partly of the hydrological and biological characteristics of the system, and also a consequence of recent installation of several waste treatment systems, removal of some industrial complexes from the area, creation of estuarine and marine protected areas and as well as being aided by implementation of environmental and nature conservation legislation.

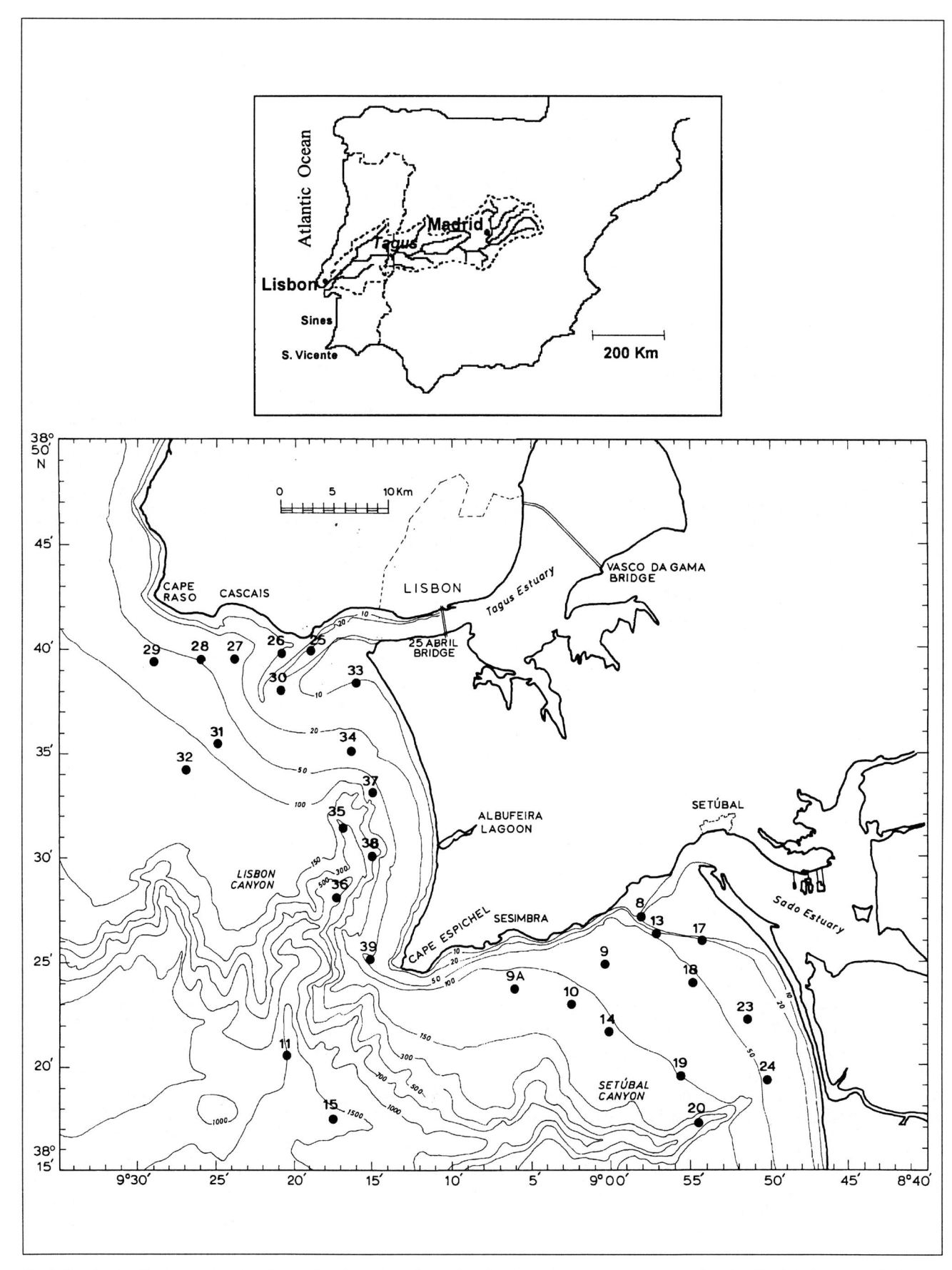

Fig. 1. The Iberian Peninsula showing Portugal and the Tagus hydrologic basin, the Tagus and Sado estuaries, the Albufeira lagoon and adjacent coastal zone.

THE DEFINED REGION

Portuguese Coastal Waters

The Portuguese west coast lies at the eastern boundary of the subtropical North Atlantic. This determines many of its atmospheric and oceanographic characteristics.

A surficial mixing layer is present along the Portuguese coast from the surface to 100–150 m depth showing an entrained salty water mass (salinity >36.3) extending from Cape S. Vicente to Lisbon (Fig. 1). As a result of river runoff, pronounced salinity gradients occur, indicating frontal structures. The characteristics of the central water of the eastern part of the North Atlantic, ENACW (Fiúza, 1984) determine the hydrologic characteristics of the coastal ocean of Portugal, which occupy all the shelf and slope from 100 m to 350–400 m depth. This central water consists of two water masses of different origins (Fiúza, 1984): a relatively warm and salty subtropical branch ($27.0 < \sigma_t < 27.1$) ($ENACW_{st}$) which moves to the north, and another branch, which is less saline and colder from subpolar origin ($ENACW_{sp}$) ($27.1 < \sigma_t < 27.3$) moving slowly southward.

The strong influence of Mediterranean Water (MW) is characterised by relatively high salinities and temperatures. Two main cores of MW can be distinguished in the Atlantic: an upper core centred at about 800 m, which has a relatively high temperature maximum and a lower core near 1200 m, which has a salinity maximum (Madelain, 1970). The influence of MW is evident on the slope especially in the canyons at depths greater than 400 m.

Seasonal wind-driven events occur along the Portuguese coast. During summer (April–September), more active and persistent coastal upwelling conditions prevail off the west coast of Portugal (Fiúza, 1984). However, upwelling activity has occurred in winter in recent years (Santos et al., 1998). The origin of the upwelling water is considered to be the ENACW, which varies significantly from north to south. North of 40°N, this water originates from $ENACW_{sp}$, while south of this latitude the upwelling source is predominantly $ENACW_{st}$ (Fiúza, 1984).

The Tagus and Sado Coastal Zone

The Tagus and Sado estuaries and adjacent waters (see Fig. 1) are located between 38°15′ and 38°50′N and between 8°40′ and 9°35′W. The coastal area is a complex system under the influence of the major submarine canyons of Lisbon and Setúbal. A well developed proto-fluvial delta forms a transition area between the Tagus estuary and the narrow continental shelf (3–30 km). The coastline of this region is interrupted by large Capes (Raso and Espichel) and by pronounced embayments (Lisbon and Setúbal). The morphology of the area is deeply marked by the intense discharge of the Tagus river. Figure 2 shows the values of Tagus river discharges and precipitation from 1980 to 1997.

The bathymetry and topography of the area, along with variations in bottom friction and wind forcing, largely

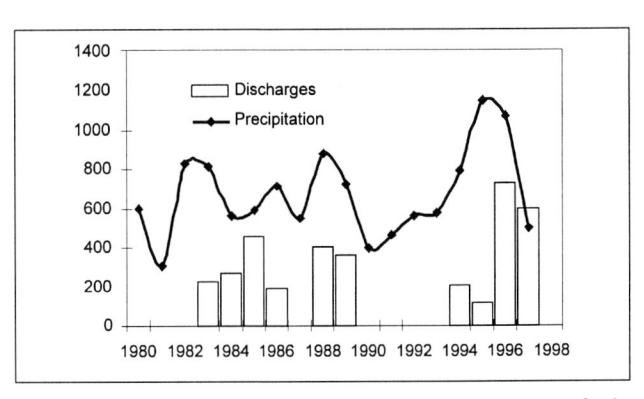

Fig. 2. Precipitation (mm) and mean Tagus river discharges ($m^3 s^{-1}$). Sources: Precipitation — Instituto de Meteorologia; mean Tagus river discharges — Instituto da Água.

control the three-dimensional structure of the upwelling over the area (Fiúza, 1984). Fiúza (1984) noted that the source of the upwelling occurred at less than 20 m depth at several points close to the coast, from Cape Espichel to Sado mouth and from Tróia to Sines. There were also indications of active upwelling near the head of the Setúbal canyon. Effects of the upwelling centres observed in the Espichel zone have been found southward of the cape, over the continental slope at the Sines area (Estrada, 1995).

The Tagus and Sado Estuaries

The Tagus estuary is one of the largest on the west coast of Europe, covering an area of 320 km² (Fig. 3). The main source of freshwater in the estuary is the Tagus river, the largest of the Iberian Peninsula draining about 81,000 km² (Fig. 1), producing an annual average flow of the order of 400 $m^3 s^{-1}$. As its discharge may vary monthly from 100 to 2200 $m^3 s^{-1}$, the residence time of freshwater in the estuary is highly variable and may range seasonally from 65 to 6 days (Martins et al., 1984 in Vale and Sundby, 1987). The estuary has two very distinct regions. The large and shallow upper estuary (over 10 km width in some places with maximum depths of 15 m), characterised by extensive tidal flats occupying approximately 40% of the total area, and salt marshes covering approximately 35% of the intertidal areas, is subjected to intense sediment deposition and erosion which is gradually changing the bathymetry. The lower estuary consists of a channel about 30 m deep, 2 km wide and 12 km long, which opens into a large bay by a canyon 15–20 m deep delimited by sand banks. The mouth of the estuary is characterised by two features, the island of Bugio and cape S. Julião.

The estuary is meso-tidal and its circulation is mainly tidally driven, the tides being semi-diurnal and varying between 0.75 m at neap tides in Cascais, and 4.3 m at spring tides in the upper estuary. The typical water velocity is of the order of 1 m s^{-1} but can exceed 2.5 m s^{-1} at the mouth of the estuary. Wind can be locally important but is of secondary importance in the global circulation. Stratification varies

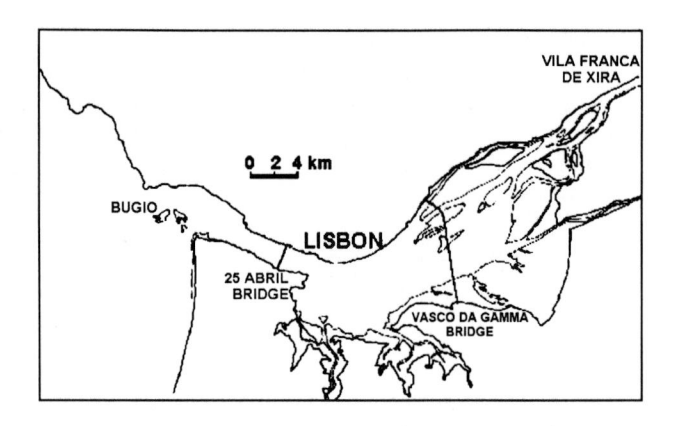

Fig. 3. Map of the Tagus Estuary.

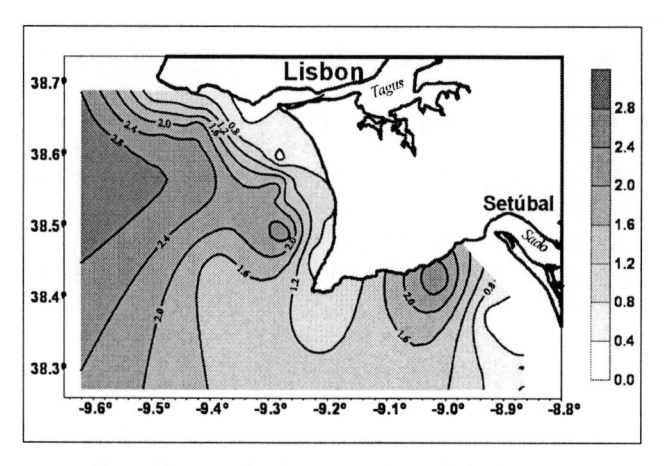

Fig. 4. Horizontal sedimentary C_{org} (%) distribution.

strongly with river flow and tidal conditions, the estuary being well mixed at spring tides but partially stratified in average conditions (Vale and Sundby, 1987). The amplitude of the tide appears, however, to be the controlling variable, and is responsible to a large extent for the turbidity of the Tagus estuary. Thus, a fortnightly erosion–deposition cycle of turbidity occurs in the estuary. During spring tides the turbidity zone may be expanded to the mouth of the estuary. Nevertheless, the fluctuation of this cycle may reach the same magnitude on a semidiurnal cycle. The concentration of suspended particles in the Tagus river may vary over a broad interval, ranging from as little as 4 mg l^{-1} at low river flow to 300 mg l^{-1} at river flood conditions (Vale and Sundby, 1987). Flow patterns at the mouth of the estuary are very different on ebb and flood. On ebb, a tidal jet forms along the channel, with two large eddies on each side, which get stronger and move further away from the coast on spring tides. On flood, as the currents over the sandbanks are larger, the two coastal eddies dissipate and a clockwise eddy forms near the southern margin of the channel (Fortunato et al., 1998). The strong mixing present in the mouth of the estuary is thus promoted by the "tidal pumping" and chaotic stirring. This induces marine waters to enter the estuary close to the coast, while enhancing flushing of the estuarine waters and reducing residence times (Fortunato et al., 1998).

Despite limited export of sediments to the coast—which is a result of the dams in the Tagus river—the main source of fine-grained sediments in the continental shelf is the river discharge, the deposition being 0.16 cm yr^{-1} in the inner shelf and 0.16 to 2.09 cm yr^{-1} in the outer shelf (Jouanneau et al., 1997). The materials deposited are mainly clay and enriched in organic material (Cabeçadas and Brogueira, 1997) (Fig. 4).

The Sado estuary is the second largest estuary in Portugal, but the Sado river has an annual mean flow of only approximately 5 $m^3 s^{-1}$, so a noticeable plume does not develop in adjacent waters. Furthermore, the narrowing of the shelf near the river mouth contributes to the rapid dilution of fresh water into the Atlantic.

Albufeira Lagoon

The Albufeira lagoon is located between the Tagus estuary and the Espichel cape (Fig. 1). It comprises two zones commonly known as the "lagoa grande" and the "lagoa pequena". The lagoon occupies an area of 1.32 km^2, and has a mean depth of 4.5 m and a maximum depth of 15 m. The system exhibits two hydrologic situations, communication and isolation from the sea. This latter condition leads occasionally to the environmental degradation of the system.

NATURAL ENVIRONMENTAL VARIABLES AND SEASONALITY

Early oceanographic work along the Portuguese coast and particularly in the area considered here was carried out in 1927 (Ramalho and Dentinho, 1928). From 1934 to 1936 Bôto (1945) studied seasonal variation of salinity, temperature, dissolved oxygen and nutrients in the area, including Cape Raso, Cape Espichel, Sesimbra and Lisbon-Cascais embayment. Since then several surveys of water quality have been carried out (Costa, 1957, Geada, 1972, Ataíde, 1975 in CNA, 1976). In 1982 and 1983, under the framework of the International Project "Estudo Ambiental do Estuário do Tejo" (UNDP/POR/77/016), surveys on the Tagus estuary conditions were conducted. Since then, regular studies are conducted, especially to characterise the water quality of the estuaries and coastal region. Additionally, a coupled physical-ecological model is being implemented (Ferreira, 1995), and a two-dimensional model of turbidity for the Tagus was developed by Portela and Neves (1997) which is used to reproduce the spatial and temporal variation of turbidity resulting from tidal flow.

Natural Environmental Variables

Biological productivity in Portuguese coastal waters is moderate. Values of chlorophyll *a* are usually below 5 mg m⁻³, although, occasionally, values as high as 35 mg m⁻³ have been recorded in the Tagus embayment (Cabeçadas et al., 1999). Levels of primary production of 1.5 g C m⁻² d⁻¹ and 0.2 g C m⁻² d⁻¹ were estimated for the Tagus adjacent waters in the summer of 1981 (Cabeçadas, 1983) and in the spring of 1994, respectively.

During upwelling in the area, diatoms are the major phytoplankton group, being mainly chain-forming species of *Pseudo-nitzschia*, *Chaetoceros*, *Thalassiosira* and *Thalassionema nitzschioides*, while during winter coccolithophorids dominate (Abrantes and Moita, 1999). With regard to nutrient conditions, concentrations of NO_3, NH_4, PO_4 and $Si(OH)_4$ lie between the values presented in Table 1.

Tagus and Sado Coastal Waters

Hydrographic and chemical structures of coastal areas adjacent to large rivers are often determined by the river discharges, even under upwelling conditions. This is illustrated by results obtained during a study conducted in the Tagus and Sado adjacent coastal area in March 1994 when steady persistent northerly winds (10–20 km h⁻¹) favourable to upwelling were present. During the sampling period the outbreak of an unusually intense spring phytoplankton bloom occurred in Tagus-dominated coastal waters (Cabeçadas et al., 1999). An examination of the physical and chemical patterns obtained for Tagus and Sado coastal waters revealed that in the Tagus embayment a frontal boundary is established, separating the river plume stratified water body from a mixed zone localised along the Lisbon–cape Espichel shore (Fig. 5).

The nutrient horizontal gradients in the embayments reflect, in general, the considerable loadings of N, P and Si discharged by the Tagus river, and concentrations above 10.0 μM NO_3, 0.6 μM PO_4 and 7.0 μM $Si(OH)_4$ were measured in the river plume, as shown in Fig. 6, relative to PO_4 pattern.

Vertically, the effect of the river discharge was noticed up to 10–15 m depth, paralleling PO_4 (Fig. 7). Lower nutrient concentrations are displayed, especially around the cape (PO_4 < 0.1 μM, NO_3 < 0.5 μM). Simultaneously, patterns of Chl *a* and phytoplankton densities show a marked gradient of increasing productivity around the Espichel shore (up to 15 mg m⁻³ Chl *a* and 0.7×10⁶ cells l⁻¹)

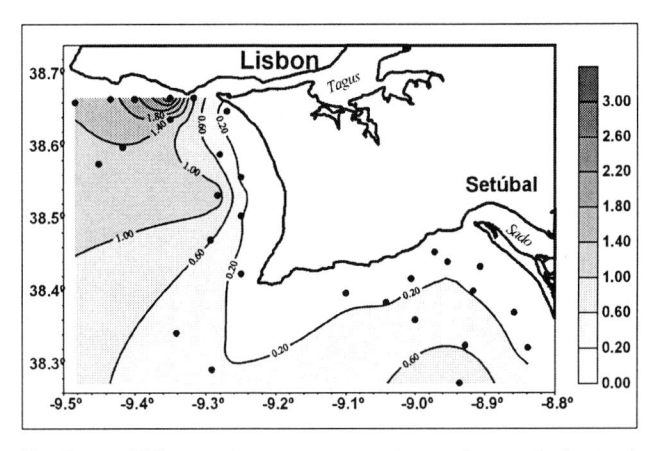

Fig. 5. $\Delta\sigma_t$ (difference between σ_t at surface and σ_t at the bottom) distribution in the Tagus and Sado embayments.

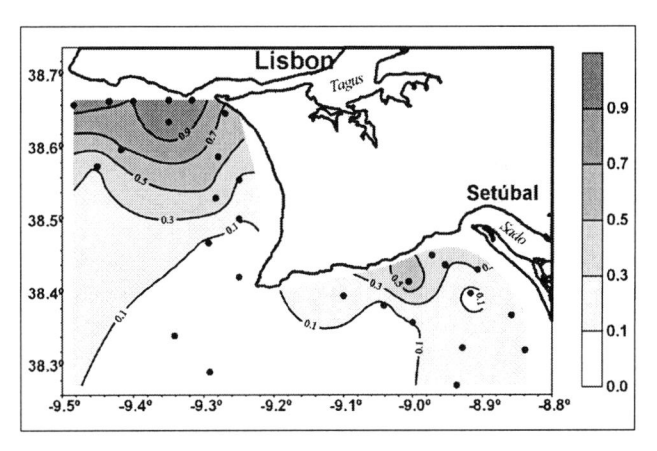

Fig. 6. Horizontal PO_4 (μM) distribution pattern in the Tagus and Sado embayments.

and maximum concentrations offshore within the Tagus stratified region (Chl *a* > 30 mg m⁻³ and 3×10⁶ cells l⁻¹). The small diatom *Thalassiosira* sp. (< 10 μm) was responsible for the bloom peaks while phytoplankton biomass close to the Espichel shore was dominated by the larger diatom *Detonula pumila* (> 0.1×10⁶ cells l⁻¹). A frontal boundary separated the river plume from the tidally mixed nearshore water (Fig. 5) and apparently functioned as a barrier between the two different phytoplankton populations—the nanoplanktonic population and the microplanktonic population. This seems to control, to some extent, the spatial phytoplankton variability and the size of the spring bloom in the area (Figs. 8a and b).

Besides the physical structure conditioning the phytoplankton variability in the area, chemical factors are also likely to limit phytoplankton growth, as the amounts of suspended particles ranging only between 1 and 12 mg l⁻¹, are not likely to induce light limitation to phytoplankton growth. Clear deviations of nutrient ratios from the Redfield values, as well as PO_4 levels below the half saturation constant for phytoplankton uptake, where phytoplankton peaks emerge, point to phosphorus limita-

Table 1

Nutrient levels in the Tagus and Sado embayments (0–100 m depth)

Nutrients (μM)	Winter	Summer
NO_3	0.3–15.0	0.3–10.0
NH_4	0.2–5.0	0.4–3.0
PO_4	0.05–9.0	0.05–0.8
$Si(OH)_4$	0.3–9.0	0.8–5.0

Sources: Bôto (1945); Cavaco (unpublished data); Silva et al. (1986); Cabeçadas (unpublished data).

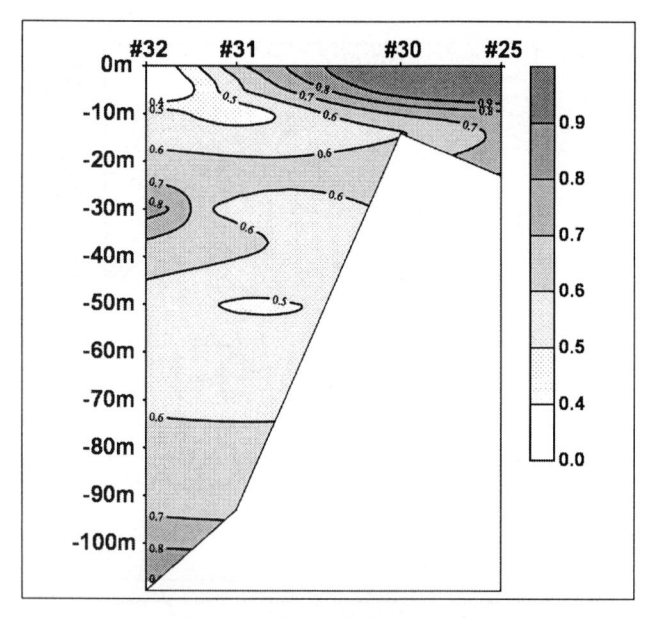

Fig. 7. Vertical PO$_4$ (μM) distribution along the transect #25 to #32.

tion by the spring. However, estimates of sedimentary mobile P and N present in the coastal area reveal that significant potential amounts of PO$_4$ can be released from the sediment to the water column in the Tagus and Sado adjacent area (Fig. 9). The potential pool of mobile P measured in the Tagus (159 mmol m^{-2}) was higher than in the Sado area (50 mmol m^{-2}) (Cabeçadas and Brogueira, 1997). Thus, it is likely that fluxes from the sediment play a significant role in the water column nutrient enrichment and in the biological productivity of the coastal waters, in particular when P limits the phytoplankton growth. In addition, the occurrence of seasonal upwelling in the area is expected to contribute to the transport of phosphorus released from the sediments to the productive zone. A shift to nitrogen limitation of phytoplankton productivity was thought to occur in the same area by late summer (Brogueira et al., 1994).

Compared with the Tagus embayment structure, waters adjacent to the Sado do not exhibit such marked stratified areas, as Sado river discharges are smaller. As a result, nutrient horizontal surficial gradients in Sado adjacent coastal waters are weaker (Fig. 6) and phytoplankton biomass (as Chl a), in general, does not attain values over 4 mg m^{-3}.

Albufeira lagoon

In Albufeira lagoon the phytoplankton community is predominantly composed of marine species, with diatoms dominating over the year. Ichthyofauna and benthic fauna present high species diversity. Mussel (*Mytilus edulis* L.) farming in suspended cultures on ropes began in 1976.

During a period of isolation from the sea, the lagoon exhibited a sharp stratification and high nutrient concentrations, as well as oxygen depletion near the bottom, favouring the occurrence of considerable fluxes of silicate

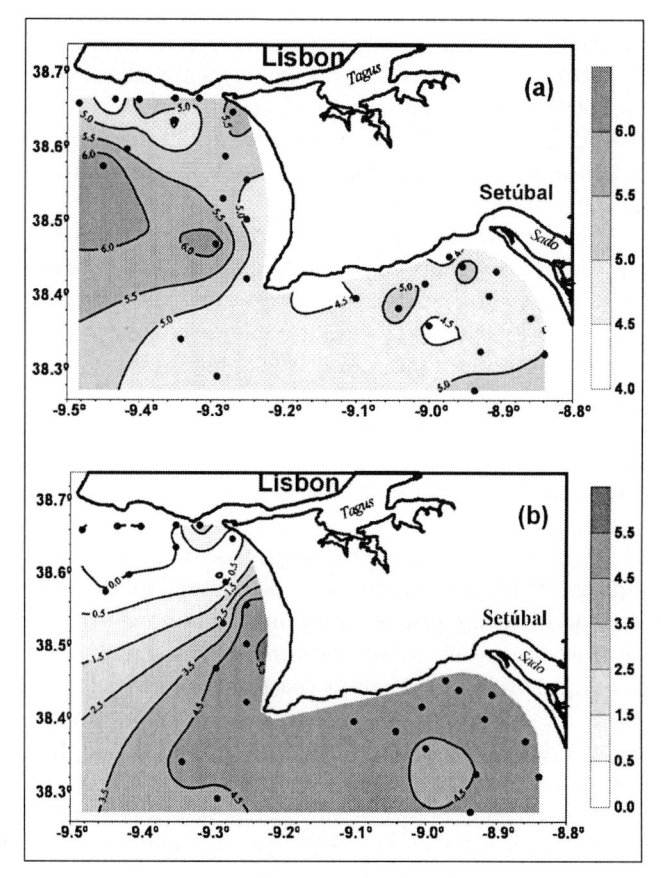

Fig. 8. Horizontal distribution of (a) *Thalassiosira* sp. (log$_{10}$ cells l^{-1}) and (b) *Detonula pumila* (log$_{10}$ cell l^{-1}).

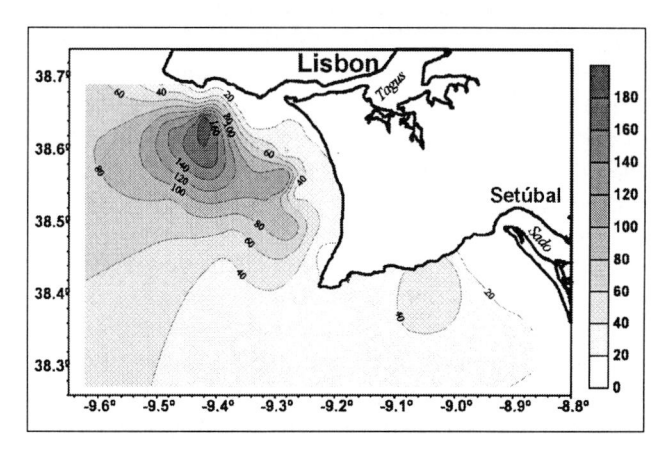

Fig. 9. Horizontal distribution of potential mobile sedimentary P (mmol m^{-2}).

(125 μmol m^{-2} d^{-1}) and ammonia (721 μmol m^{-2} d^{-1}) from the sediment (Nogueira et al., 1996). Consequently, planktonic and benthic species, as well as ichthyofauna diversity, were reduced and episodes of mussel mortality occurred. However, as soon as the connection with the ocean was established the system recovered immediately.

A study conducted by Coutinho (1998) indicates that phytoplankton mean diversity in the lagoon has not shown a significant change in recent years and species richness (mean values) even increased from 16 to 26 species.

SHALLOW MARINE HABITATS

Plankton

Diatoms were identified as being the most important plankton group present along the Tagus estuary and adjacent waters: *Skeletonema costatum*, *Cilindrotheca closterium*, *Pseudo-nitzschia seriata*, *Chaetoceros didymum*, *Amphiprora alata* and *Thalassiosira rotula* were the main marine species in the estuary (Sousa e Silva et al., 1969). Rodrigues and Moita (1979) studied the composition, abundance and distribution of phytoplankton in the estuary and reported that freshwater species were dominant in the upper estuary and had densities of up to 60×10^6 cells l^{-1} during summer. In the lower estuary and adjacent waters, densities were much lower (0.1–0.5×10^6 cells l^{-1}) and the species were predominantly marine although freshwater species were still present, showing the strong influence of the river in the adjacent coastal waters.

Studies of zooplankton communities in the Tagus estuary (Sobral, 1982) showed that the highest values of biomass were reached in winter and spring, Copepoda being the dominant group. The upper part of this system is mainly dominated by the freshwater species *Bosmina longirostris* (Cladocera) at low tide, and by *Calanipeda aquaedulcis* and *Acanthocyclops robustus* (Copepoda) at high tide. The highest biomass values were observed in the middle estuary which is an important concentration zone for *Mesopodopsis slabberi* (Mysidacea), *Acartia tonsa* and *Euterpina acutifrones* (Copepoda). The lower estuary has high biodiversity, but is less important in terms of biomass.

Tagus Salt Marshes

The southern and eastern parts of the Tagus estuary consist of extensive inter-tidal mudflat areas covered, in part, by halophytic vegetation inundated daily by the tide. These salt marshes are mainly colonised by *Spartina maritima*, *Halimione portulacoides* and *Arthrocnemum fruticosum*. Studies conducted by Madureira et al. (1997) revealed differences in sulphur speciation between non- vegetated and vegetated sediments; specifically Tagus sediments colonised by *S. maritima* have been shown to be dominated by S oxidised species. Because marsh plants in this estuary remain active almost all year round, the oxidation processes are only interrupted temporarily in summer when the temperature reaches 40°C, contrasting with some other European marsh plants which reduce their activity in winter, when oxidation processes stop. Therefore, seasonal variations of S speciation are hardly detected in this estuary. Caçador et al. (1996) also show evidence that plant roots in the Tagus area significantly modify sediment quality, eventually increasing metal sequestration in the rhizosphere, and conclude that marsh plants play an important role in sediment chemistry in this estuary and may function as an efficient natural means for keeping the ecosystem healthy. Studies by Sundby et al. (1998) show that the net result of the colonization by *S. maritima* is that the rhizosphere becomes oxidised and metals toxic to plants are immobilised within discrete microenvironments.

Primary Production in the Tagus

Measurements of primary production by microphytobenthos, phytoplankton and macrophyte algae, have been carried out in some areas of the Tagus estuary. Brotas (1995) estimated that there is 11,000 tonnes C yr^{-1} of the primary production of microphytobenthos in the intertidal area of Tagus, based on a mean production of 15 mg C m^{-2} h^{-1} for two sites. Phytoplankton production in the lower estuary was measured by Cabeçadas (1999) over the spring period; values obtained ranged from 3 to 15 mg C m^{-2} h^{-1}, which might be roughly converted to 5,720 tonnes C yr^{-1} for total Tagus estuary phytoplankton production. For the macrophyte algae, a production of 4,000 tonnes C yr^{-1} was calculated based on populations of *Ulva lactuca* (10–100 mg C m^{-2} h^{-1}) and *Fucus vesiculosus* (40–160 mg C m^{-2} h^{-1}) (Ferreira and Ramos, 1989).

Catarino et al. (1985) measured a rate of 324 mg C $m^{-2} h^{-1}$ for *Spartina maritima* and 35 mg C $m^{-2} h^{-1}$ for *Arthrocnemum fruticosum*. Based on the cover and the monthly biomass variation per dry weight of several species of the salt marsh vegetation, an annual production of 17,790 tonnes of biomass was estimated (Catarino, 1981). According to Simenstad and Costa (1997), the conditions of the Tagus estuary promote a detritus-based food web based mainly on epibenthic crustaceans. Phytoplankton-based pathways are not so significant, and sediment microalgae and marsh macrophytes may be more important to the Tagus food web.

Marine Biological Resources

The Tagus estuary is very rich in terms of diversity and densities of young fish, especially flat fish. The area is used by several species as a nursery ground, particularly soles (*Solea vulgaris* and *S. senegalensis*) and sea bass (*Dicentrarchus labrax*). This is mainly attributed to the high production of suitable food, and high temperatures (Costa and Bruxelas, 1989). Many other fish species are present, such as common two-banded seabream (*Diplodus vulgaris*), Lusitanian toadfish (*Halobatrachus didactylus*), and several species of gobies, particularly the sand goby (*Pomatochistus minutus*) (Costa et al., 1997).

Another species found is eel (*Anguilla anguilla*). Glass-eel represents an economically important resource of the Tagus estuary. Since 1987, legislation (D.R.43/87) concerned with restrictions of glass-eel fishing is being applied without much success.

Chephalopeds are also found, in particular common cuttlefish (*Sepia officinalis*) and common squid (*Alotheuthis subulata*). The economically more important crustaceans abundant in the Tagus estuary are the brown shrimp (*Crangon crangon*) and the green crab (*Carcinus maenas*) (Martins et al., 1994).

Three major benthic groups are distinguished: polychaetes, bivalves and crustaceans, in particular amphipods. The coastal areas adjacent to the Tagus and Sado estuaries are very important for banks of molluscs. Their commercial exploitation (such as of *Ensis siliqua*, *Callista chione*, *Chamelea striatula* and *Donax* spp.) in the intertidal coastal zone of the region started in 1984, and now has a considerable socio-economic importance (Dias et al., 1994). However, as a result of over-exploitation, a decline of the beds of *Ensis siliqua*, *Spisula solida* and *Venus striatula* has taken place. In 1996, the fisheries of these species in the Setúbal region were closed. A study carried out in 1997 by Gaspar et al. (1998) revealed that the respective beds showed no signs of recovery, despite the maintenance of the controls place on collection in the area. South of Cape Espichel an extensive bed of *Callista chione* has been detected (Gaspar et al., 1998).

Contamination

Bacterial contamination

Total coliforms in the Tagus estuary increased significantly from approximately 70 to 11,500 col ml^{-1} between 1953 and 1975 (CNA, 1976). Microbiological investigations in 1981 (Rheinheimer and Schmaljohann, 1983) showed values of 150 to 400 col l^{-1} close to the mouth of Tagus estuary, and values were much higher between Lisbon and Cascais (6,500–75,000 col l^{-1}).

Concerning the protection and quality of the coastal waters, the application of EU directive 76/160/CEE led to a monitoring programme which started in January 1993. This programme consisted of regular determinations of coliforms and some physical-chemical parameters during summer. Ten percent of the beaches in this area of Portugal were still contaminated in 1997 (Instituto da Cígua e Direcção-Geral da Saríde, 1998).

The contamination of marine molluscs from the Tagus estuary was surveyed from 1962 to 1970 (IBM, 1962–1970 in CNA, 1976). From 1991 to 1993, bivalves collected in the Tagus estuary, especially furrow shell (*Scrobicularia plana*) contained high values of coliform bacteria (790 to >110,000/ 100 g), with *Salmonella* being detected in 30% of analysed samples (Dias and Nunes, 1994). Contrasting with this situation, in the coastal area of Lisbon and Setúbal, samples of razor clam (*Ensis siliqua*) and smooth calista (*Callista chione*) had very low levels of coliform and no *Salmonella*. This indicated that bivalves in the coastal area were, in general, healthy (Dias and Nunes, 1994). Since 1995, control of the bacterial conditions of bivalves molluscs along the Portuguese coastal waters has led to the prohibition for human consumption of some species (common mussel, *Mytilus edulis*, furrow shell, *Scrobicularia plana* and common cockle, *Cerastoderma edule*) in the Tagus estuary, while the consumption of carpet shell (*Venerupis pullastra*) was allowed only after purification. Common mussels (*Mytilus edulis*) from Albufeira lagoon also have to be subjected to purification, while bivalves from the littoral area have no restrictions on their consumption (Pissarra et al., 1997). A monitoring programme is under way consisting of regular determinations of marine biotoxins, Amnesic Shellfish Poisoning (ASP), Diarrhetic Shellfish Poisoning (DSP) and Paralytic Shellfish Poisoning (PSP) as well as of mercury in the bivalves, and phytoplankton composition in seawater. This programme is carried out by IPIMAR (Instituto de Investigação das Pescas e do Mar) and derives from the implementation of EU directive (91/492/CEE) in Portugal.

Metals

Mendes and Vale (1984) measured the highest contents of Pb, Cd, Cr and Cu in mussels collected in the mouth of the Tagus. The sharp seaward decrease of metal contents (Fe, Mn, Pb, Cd, Cr and Cu) in mussels from the transition zone indicates the limit of the estuarine water influence on the adjacent coastal waters, although in winter a less clear influence of the estuary could be drawn from metal concentration in mussels along the shoreline. Contamination of organisms by Pb seems to be a problem in this area. Indeed, Mendes and Vale (1984) indicate that the levels of Pb in mussels (*Mytilus galloprovincialis*) were of the same order of magnitude as those from polluted areas. In specific areas of the Tagus estuary, two epibenthic species, brown shrimp *Crangon crangon* and rose shrimp *Palaemon serratus* were both contaminated by Pb, exceeding the recommended maximum values for human consumption of crustacea (Ribeiro, 1999).

TBT

Along the Portuguese coast, most of the populations of the biological indicator *Nucella lapillus* develop imposex, with particularly high incidences along the shore of Estoril, where imposex percentages of 100% are reached (Santos, 1996). Similarly high imposex percentages, both for *Nucella lapillus* and *Hinia reticulata* occur in Setúbal and Tróia (Guerra et al., 1999). These results must be related to the intense maritime traffic of Lisbon and Setúbal harbours, the shipyards of Lisnave and Setenave, and also to the nearby marinas. TBT levels in the Tagus and Sado estuaries and near harbours and shipyards were 0.02–0.4 μg Sn l^{-1} in water, 0.2–0.7 μg Sn g^{-1} in sediments (dry weight) and 0.5–5.0 μg Sn g^{-1} (dry weight) in various organisms (Coelho and Bebiano, 1998); these levels exceeded those used in several countries for the protection of marine species. Legislation imposing restrictions on the use of antifouling paints containing Sn was adopted in Portugal in 1993 (DL 54/93). As it applies only to small boats (< 25 m length) and to certain immersed equipment, these regulations seem inadequate, as much larger ships and oil tankers regularly use the estuaries in Portuguese offshore waters.

OFFSHORE SYSTEMS

Fisheries

There are rich fisheries in the plume zone of the Tagus estuary (Cascais/Lisbon embayment). European hake (*Merluccius merluccius*) and horse-mackerel (*Trachurus trachurus*)—both important commercial species—are present in the area, although snipefish (*Macrorhamphosus scolopax*)—a non commercial species—may be the most abundant. Many other fish species are found, namely silver scabbardfish (*Lepidopus caudatus*), golden mullet (*Liza aurata*), bastard sole (*Microchirus azevia*) and red mullets (*Mullus* spp.) (Martins et al., 1994). Some crustaceans and molluscs are abundant as well. In the coastal area of the Sado estuary species are especially abundant. The main catches are of horse-mackerel (*Trachurus trachurus*), followed by common mackerel (*Scomber japonicus*) and European pilchard (*Sardina pilchardus*). Sparidae is also an important group here, specifically white seabream (*Diplodus* spp.), axillary seabream (*Pagellus acarne*), bogue (*Boops boops*), salema (*Sarpa salpa*), black seabream (*Spondyliosoma cantharus*) and gilthead seabream (*Sparus aurata*) (Dias and Cabral, 1994).

In Portuguese continental waters the black scabbard fish (*Aphanopus carbo*) is mainly caught extensively, in a confined and deep area 1000–2000 m deep off Cape Espichel. This fishery, which started in 1983, operates all year round (Martins et al., 1989).

Marine Mammals

Systematic monitoring of marine mammals dependent on the Portuguese littoral has been conducted in recent years in order to assure their conservation. Cetaceans and pinnipeds were identified as the most abundant mammals and these are most frequently noticed between cape Roca and Sines (Sequeira et al., 1996). This seems to be linked to the characteristics of the area mentioned above, such as coastal upwelling events and discharges from the estuaries, both being conditions which favour high productivity. Common dolphin (*Delphinus delphis*) can be observed in high numbers along the coast and close to the Arrábida–Cape Espichel shore; a permanent population (about 30 individuals) of the bottlenose dolphin (*Tursiops truncatus*) is established as well. Mortalities of marine mammals along the Portuguese coast have been attributed to natural and accidental causes mainly due to the use of gillnets. Measures to protect numbers of these mammals since 1981 (D.L. 263/81) have been successful; no significant changes in distribution and occurrence patterns of these species have been noticed.

Macrofauna

Studies on the structure of the benthic fauna communities along the coast adjacent to the Tagus and Sado estuaries have shown that the macrofaunal populations reflect the intense hydrodynamic nature of the area as well as the granulometric structure of sediments. *Ophiura albida* is associated with fine sand in Tagus, and *Echinocordium cordatum* in Sado (Gaudêncio and Guerra, 1994). Bivalves are dominant in medium to coarse sand, while limpets are abundant on all intertidal rocky shores (Gaudêncio and Guerra, 1994). Analysis of distribution, abundance, habitats, size-structure and reproductive cycles of *Patella* spp. reveals that these populations have shorter life-spans and a smaller maximum size in Portugal than in NW Europe. Also, their reproductive behaviour seems to be influenced by the higher seawater and air temperatures of Portugal, resulting in more successful recruitment. On rocky shores with a dense algal cover and where wave energy is low, the species *Gibbula umbilicalis* is the most abundant. In Portugal, these species have longer spawning periods, smaller maximum sizes, shorter life spans and higher densities than in more northern countries. Another factor, which almost certainly limits both maximum size and life span of *Gibbula*, is very likely to be its collection for human food (Gaudêncio and Guerra, 1986).

Grain size of the sand-based habitat is the most important factor controlling the distribution and development of the dominant Amphipod *Ampelisca* spp. along the coast from Lisbon canyon to the north (Marques and Bellan-Santini, 1993). These Amphipods constitute an important food source for fishes, even being the primary food for demersal species. Consequently, *Ampelisca* species can be expected to play a significant role in the diet of many secondary consumers in the Portuguese coast.

BENTHIC VEGETATION

The coastal algal flora of Portugal is considered to be quite rich, with 460 species identified. Among the dominant members of more than forty species that reach southern Portugal are the brown algae *Laminaria saccharina*, *Laminaria hyperborea*, *Fucus serratus*, *Pelvetia canaliculata*, *Ascophyllum nodosum*, and *Himantalia elongata*, and the red algae *Chondrus crispus*, *Palmaria palmata* and *Ceramium shuttleworthianum*.

Most studies have concentrated mainly on the distribution, biomass and growth of *Gelidium sesquipedale*, the major species exploited for agar (Palminha et al., 1982; Oliveira, 1989 in Oliveira and Cabeçadas, 1996). One of its more important commercial beds is localised off Cape Espichel, where a potential for yield overharvesting has been noted (Santos, 1994).

There have been temporal changes in the density of communities of brown algae, Laminariales and Fucales, which have been attributed to natural causes such as variation in solar irradiance levels (Oliveira and Cabeçadas, 1996). These cyclic oscillations were also observed in *Porphyra* and *Gelidium* populations. Oliveira and Cabeçadas

(1996) remark that, in contrast, changes in the natural flora caused by pollution discharges have occurred only occasionally in lagoons and estuaries or near harbours or sewage discharges.

UPWELLING EFFECTS ON FISHERIES

A decrease in the Portuguese sardine fishery has been attributed to the increase in the April–September upwelling index observed from 1950 to 1980 (Dickson et al., 1988). This might have caused a delay in the onset of the spring bloom, and thus reduce the biomass of both phytoplankton and zooplankton during the growing season. Considering that April is a spawning time in this species, stronger upwelling may not permit the formation of a planktonic population suitable for the newly hatched sardines. Santos et al. (1998) observed changes in the upwelling pattern off Portugal, specifically an increase in the frequency of upwelling favourable northerly winds during winter (January–March). Modifications of the directions and intensity of the wind in winter was attributed to the strength, position and shape of the Azores High. Upwelling indices for the period 1987–1997 on the west coast of Portugal showed that decreasing trends in the recruitment of small pelagic populations of the region (sardine and horse-mackerel) in the last 10–15 years may be related to the increase of upwelling favourable conditions in winter. In fact, as the spawning for those species off Portugal takes place predominantly in winter, Figueiredo and Santos (1989) consider that this creates conditions for larval mortality, namely they are transported offshore into unfavourable feeding areas. Whatever the explanation, it seems that changing weather patterns have exerted a marked effect on the sardine stocks of Portugal.

Toxic Blooms and Bacteria

Nearshore outbreaks of *Dinophysis* were observed off the Portuguese coast, in summer and autumn. The Tagus/Sado embayments mark the southern limit of both *D. acuta* and *D. acuminata*. This distribution can be explained by differences in stratification patterns between the north-western and the south-western coast, associated with the northern wide and flat shelf, the lower salinities and the intense upwelling. South of Setúbal, the establishment of stratification conditions suitable for *Dinophysis* growth is, probably, not achieved, as the upwelling pattern is different due to the steep shelf and the weakly pronounced shelf break (Palma et al., 1998). In the area off Sines, the phytoplankton community present during the local upwelling consists mainly of diatoms, dominated by *Detonula pumila*, and includes typical red tide forming dinoflagellates (*Prorocentrum triestinum*, *Oblea rotunda*, *Prorocentrum rostratum* and *Gymnodinium catenatum*). Estrada (1995) suggest that the rapid appearance of *G. catenatum* in the area off Sines could be favoured by the advection of warm offshore water from a poleward surface current, induced by relaxation of coastal upwelling.

In Portuguese coastal waters, episodes of DSP and PSP have been associated with the presence of *Dinophysis acuminat* and *D. sacculus*, and *G. catenatum* (Sampayo, 1994). The dinoflagellate *G. catenatum* has been observed since 1981 along the coast of Portugal (Estrada, 1995) and the first confirmed occurrence of DSP associated with *Dinophysis* spp. was reported in 1987 (Anon., 1994 in Palma et al., 1998). In the area of the Tagus and Sado estuaries, Albufeira Lagoon and coastal zone, important bivalves resources have been affected by the biotoxins produced by these toxic species (Sampayo, 1994).

Diatom Sediment Records

The impact of the upwelling phenomena on the recent sediment on the Portuguese shelf has been investigated by Monteiro et al. (1983) and Abrantes (1988). Distribution patterns of diatoms abundance reflect the more intense, persistent and homogeneous upwelling north of Nazaré canyon and the "inner" upwelling front south of Lisbon. Resting spores, mainly of *Chaetoceros*, record the position of the zone where higher variability occurs. According to Abrantes (1988) this supports the hypothesis that diatom species distributions off Portugal are good indicators of the changing character of the upwelling system.

POPULATIONS AFFECTING THE AREA

Population Density

The Lisbon Area and Tagus Valley, which includes the Greater Lisbon Area (seven municipalities) and Setúbal Peninsula Area (nine municipalities), occupy an area of 11,934 km^2 and have a total of 3,299,050 inhabitants, corresponding to a population density of approx. 300 people km^{-2} (INE, 1993) (Table 2).

The predicted slight demographic increase in the region is mainly a result of an internal migratory process, with a flow towards the littoral and the Greater Lisbon area. In 1995/96, in preparation for 'Expo 98', redevelopment took place in the degraded industrial eastern part of Lisbon over 340 ha with 5 km along the river, which led to the relocation

Table 2

Demography trends

	Population present 1991*	Population estimated 1995**
Greater Lisbon	1,831,877	1,834,070
Setúbal Peninsula	640,493	658,320

Sources: *Instituto de Estatística (INE, 1993); **Marktest (1995).

of many industrial units. The new urban area for residential and commercial uses is predicted to lead to an increase of population in the near future.

The degree of development in the area has been intense during the last ten years. Ocean recreation has expanded, and maritime traffic has increased considerably, 4500 units/year in Lisbon harbour and 1237 in Setúbal harbour, in 1994 (modified from Anuário dos Portos de Portugal, 1994). Tourism is very important in the region, mainly in the Lisbon–Cascais axis, along the Tagus–Setúbal coast and in the Sado estuary. Aquaculture activities in Tagus do not involve a significant number of people, as this is still a poorly developed activity in this estuary.

RURAL FISHING

Artisanal fishing is an important activity in the Tagus estuary. According to Carneiro et al. (1997), there are 280 boats and 439 fishermen using trawlers, "botes" and "saveiros". The fishing gears mostly used are bottom and beam trawl, eel trap, longline, gillnet and trammel. Stow net is used for glass-eel and boat dredge for bivalves catches.

In the coastal waters adjacent to the Sado estuary, small 'surrounding' nets are very important between Cape da Roca and Sines. In 1993, catches using this fishing technique caught more than 20% of the total landings in Setúbal harbour (Dias and Cabral, 1994). This method has a negative impact on the juveniles of several species living in the bottom of coastal areas (Franca and Costa, 1984 in Dias and Cabral, 1994). Also, Costa et al. (1997) report that the reduction in the capture density of sole (*S. vulgaris* and *S. senegalensis*) in the Tagus estuary may be related to the fact that juveniles of these species, which have a high commercial value, are fished intensely and illegally in the upper areas of the estuary.

COASTAL EROSION AND LANDFILL

The main agent of littoral erosion on the Portuguese coast is wave action, but in the vicinity of the final reaches of the large estuaries and of tidal inlets, tidal currents are important, as the tidal range is approximately 3.5 m. The main sediment source is the Tagus river, though much is trapped in deposition zones in the final reach, and sediment discharge is also now reduced in dams and weirs (Abecassis, 1994). The areas between Cape Raso and the Tagus mouth, and between Cape Espichel and the Sado mouth have very little sediment input other than a few small creeks and direct wave action on the cliffs. The littoral transport in this areas is almost zero. The sections of coast between the Tagus mouth and Cape Espichel and the Sado mouth and Cape de Simes are quite similar from a physiographic point of view, being crenulate-shaped beaches, 28 km and 65 km long respectively (Abecassis, 1994).

DREDGING

Dredging in the Tagus estuary takes place in areas where sediments are enriched with organic matter and trace metals deriving from domestic and industrial sewage. Several areas of the Tagus estuary are dredged annually by the Port Authority, but the amounts of removed material increased up to 2.5 million tonnes with the dredging of a transversal channel during the construction of the New Bridge, Vasco da Gama, over the estuary (Vale et al., 1998). Dredged, resuspended material, which remains in the vicinity of the operations at neap tide, is dispersed over larger areas when currents are stronger. Although resuspended material has not increased metal contamination in suspended particles, laboratory experiments have demonstrated that they may accumulate in mussels (Fig. 10), and so in the Tagus, dredging operations may transfer contamination to living organisms (Vale et al., 1998). Nevertheless, levels were below those allowed in organisms for human consumption.

Impact on the Benthic Macrofauna and Fish

Surveys of the subtidal benthic macrofauna, also in the Tagus estuary (Vale et al., 1989) showed no major changes from the dredging, except in the central area where there was a decrease on species richness (ca. 60%), abundance and diversity, and the dominance of the amphipod *Corophium volutator*. A drop in the Shannon index of diversity from 2.3 to 3.3 in the surrounding area to 0.5–1.5 at the affected site was a result of an impoverishment in polychaete and bivalve species and a sharp increase of *C. volutator* abundance. Whether changes on the macrofauna structure can be attributed to pollutants solely or to other aspects related to disposal is not clear. In the areas of dredging, species abundance and richness were comparable to those from other estuarine zones.

Monitoring studies on fish fauna (Costa et al., 1997) showed that 1% of fish species were malformed, a slightly higher value that which may be considered acceptable.

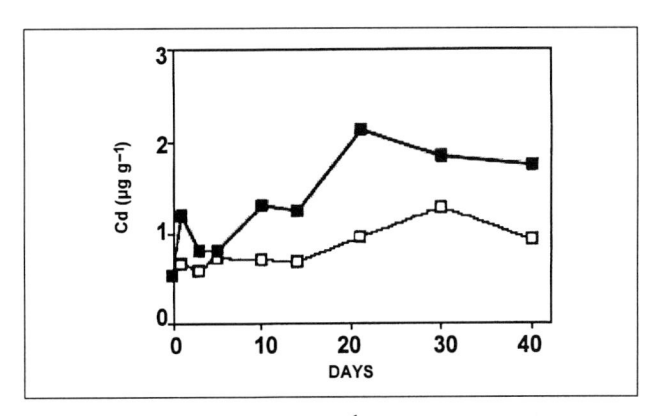

Fig. 10. Concentration of Cd ($\mu g\ g^{-1}$, dry weight) in *Mytillus galloprovincialis* of control (□) and contaminated (■) tanks during a 40-day experiment (from Vale et al., 1998).

Artisanal Uses

Since 1994 a programme for fisheries development (PROPESCA) has improved aquacultural production in Portugal. In Lisbon and the Tagus Valley production was 633 tonnes in 1997 (DGPA, Direcção Geral das Pescas e Aquicultura). Most development in the Tagus and Sado estuaries took place in old salines and tidal mill reservoirs. The production system is mainly extensive; only a few units are semi-intensive. Cultivated fishes are predominantly sea bream (*Dicentrarchus labrax*) and sea bass (*Sparus auratus*).

URBAN AND INDUSTRIAL EFFECTS

In this region, most pollutants are discharged directly into the central estuary, contrasting with most estuaries in Europe, where pollutants from industrial regions are discharged into rivers and brought to the estuaries via the rivers. Partially treated effluents from Greater Lisbon (which has approximately 2 million inhabitants) enter the Tagus estuary directly. Commercial and fishing activities have been adversely affected by the inputs of fertilizers and pesticides, specifically molinate, endolsulfan, chlorfenvinphos, and lindane as well. Many of the pollutants are incorporated into the sediments and tidally transported. There is an enrichment in Zn in the surface layer of the sediment (Vale, 1990; Paiva et al., 1997) attributed to human activities in the middle and lower parts of the estuary. Dissolved mercury comes particularly from a chloralkali plant, as a result of which the total dissolved Hg in the lower estuary (67 ng dm^{-3}) and in the outer estuary (56 ng dm^{-3}) in the 1980s, were much higher than upstream (12 ng dm^{-3}) (Figueres et al., 1985). Values found later in 1997 during dredging were of the same order of magnitude, varying from <1 to 60 ng dm^{-3}, although reaching 200 ng dm^{-3} at certain periods (Pereira et al., 1997). In biota, Lima (pers. comm.) recently measured Hg in several commercial fish species (*Trachurus trachurus*, *Merlucius merlucius*, *Pagellus erytrinus* and *Mullus barbatus*) and bivalves (*Mytilus edulis*, *Donax trunculus* and *Venerupis senegalensis*) finding, in general, concentrations below 500 $\mu g\ kg^{-1}$, the limit allowed for human consumption. Levels of Hg in brown shrimp (*Crangon crangon*) collected in specific areas of the estuary in 1997 (Simas, 1998) were also lower than the limits recommended.

The decline of oyster production in the Tagus estuary has also been associated with industrial development. The Portuguese oyster (*Crassostrea angulata*) was, for a long time, an important biological resource here (Vilela, 1954), but by the 1960s, it had sharply declined following the establishment of industry and the simultaneous occurrence of "gill disease". This massive mortality of oysters happened at the same time as in other European countries, and this resource has not recovered. Despite the appearance of another species *Ostrea edulis* in the estuary, Tagus oysters now are too small and have no commercial value. At the same time,

Table 3

Total discharges (direct and riverine) to the Tagus estuary

Year	SPM (tonne yr^{-1})	Total N (tonne yr^{-1})	Total P (tonne yr^{-1})	Tagus mean flow ($m^3 s^{-1}$)
1994*	170103	15702	14238	208
1996**	147806	17529 (NO_3)	2165	330
1997**	300464	26019 (NO_3)	4254	700

Sources: *INAG; **DRALVT (Direcção Regional do Ambiente de Lisboa e Vale do Tejo).

Costa et al. (1997) report a reduction in the capture density of two sole (*S. vulgaris* and *S. senegalensis*) in the Tagus estuary and suggest that this might be related to environmental factors, although the effects of intense and illegal fishing activity could not be excluded.

Wastes Disposal

Discharges of particulate material, phosphorus and nitrogen to the estuary, including the direct (industrial and domestic) and the riverine loads (Tagus, Sorraia and Trancão rivers) in 1994, 1996 and 1997 are presented in Table 3. A considerable reduction of phosphorus loads to the estuary is noticeable since 1996: this is likely to be a result of the installation of several waste treatment systems (Alcântara, Beirolas, Frielas).

Treated domestic effluents from four municipalities around Lisbon, serving an area of 22,000 ha with a population over 600,000, have been discharged since May 1994 through the largest outfall system in Portugal, located west of the bay of Cascais. A monitoring programme integrating chemical parameters, macrofaunal communities and sediment toxicity has been implemented by SANEST (Sistema de Saneamento da Costa do Estoril) since 1997.

PROTECTIVE MEASURES

This coastal area includes zones with different protection status within the Portuguese National Protected Areas system: the Tagus Estuary Nature Reserve, the Arrábida Natural Park, the Protected Landscape Area of Caparica Coast Fossil Cliffs, the Sado Estuary Nature Reserve and the Albufeira Lagoon Ramsar site (Fig. 11). Their protection status habitats are presented in Table 4. Portugal has signed and ratified several International Conventions such as the European OSPAR Convention for the protection of the sea, the Convention on Biological Diversity, which includes marine biological diversity and the Ramsar Convention on the conservation of internationally important wetlands, especially waterfoul habitats. Under the EC Bird Directive 79/409, Special Protected Areas (SPA) were designated and under the EC Habitats Directive 92/43, a National Sites List (NSL) was also designated, with the objective of joining the European Conservation Web NATURA 2000, in 2004.

Table 4

Protected areas in the Tagus and Sado adjacent coastal area

Classified area	Protection status/ European Directives/ International Conventions	Habitats	Birds
1. Tagus Estuary	Nature Reserve; SPA; NSL; RAMSAR site	Estuary; Intertidal mud-flats; Salt-marshes; Salines; Natural Oyster beds; Rice fields; Seagrass beds; Sandbanks	Breeding area for waterfowl species (1); Passing through and/or winter refuge for waterfowl species (2)
2. Fossil Cliff of Caparica Coast	Protected Landscape Area	White limestone cliffs with marine fossils; Dune fringe; Mediterranean scrubs; Pinewoods	Typical Mediterranean species (3)
3. Albufeira Coastal Lagoon	RAMSAR site	Coastal lagoon; Brackish shallow waters; Sandbanks; Reedbeds	Passing through area during migrations of waterfowl and trans-saharian passeriformes (4). Winter refuge (5)
4. Arrábida-Cape Espichel	Natural Park; SPA; NSL	Upwelling area; Fronts; Exposed Rocky Shores; Tidal-Swept; Communities; Caves; Sandy beaches; Embayments; Bays; Shallow marine waters; Deep marine waters; White and grey limestone mountain; Mediterranean maquis vegetation	Migration corridor for the important African-paleartic system (6). Sea and seashore species (7)
5. Sado Estuary	Nature Reserve; SPA; NSL; RAMSAR site	Estuary; Intertidal mud-flats; Salt Marshes; Natural Oyster beds; Salines; Shallow marine waters; Seagrass beds; Rice fields; Dune fringe; Sandbanks	Breeding area for waterfowl species (8); passing through and/or winter refuge for waterfowl species (9)

Sources: Instituto da Conservação da Natureza, ICN (1999).
SPA = Special Protected Areas, designated under the EC Habitats Directive 92/43/CEE.
NSL = National Sites List, designated under the Habitats Directive 92/43/CEE.
Ramsar Convention, Ramsar sites recognised as internationally important wetlands particularly for waterfowl habitats.
Species:
(1) *Ardea purpurea, Himantopus himantopus, Glareola pratincola, Circus aeruginosus, Charadrius alexandrinus.*
(2) *Anser anser, Anas crecca, Anas clypeata, Pluvialis squatarola, Calidris alpina, Limosa limosa, Recurvirostra avosetta* (reaching 75% of the European totals), *Phoenicopterus ruber, Tringa totanus.*
(3) *Sylvia undata, Lullula arborea, Cyanopica cyanus cooki, Merops piaster.*
(4) (5) Winter refuge for *Fulica atra; Ixobrychus minutus, Sterna hirundo, Sterna albifrons, Alcedo atthis, Falco peregrinus, Bubo bubo, Sylvia undata.*
(6) *Nycticorax nycticorax, Egretta garzetta, Ardea purpurea, Ciconia nigra, Pernis apivorus, Elanus, caeruleus, Gyps fulvus, Aegypius monachus, Circus pyga, Aquila adalberti, Aquila chrysaetos, Hieraaetus pennatus, Hieraaetus fasciatus, Falco naumanni, Falco peregrinus, Grus grus, Otis tarda, Burhinus oedicnemus, Pluvialis apricaria, Pterocles orientalis, Pterocles alchata, Bubo bubo, Coracias garrulus, Melanocorypha calandra, Galerida theklae, Lullula arborea, Oenanthe leucur, Sterna albifrons, Phalacrocorax crabo.*
(7) *Larus marinus, Sterna albifrons, Sterna albifrons, Phalacrocorax crabo.*
(8) *Ardea purpurea, Himantopus himantopus, Circus aeruginosus, Sterna albifrons.*
(9) *Phalacrocorax carbo, Anas clypeata, Calidris alpina, Recurvirostra avosetta, Phoenicopterus ruber, Pluvialis squatarola.*

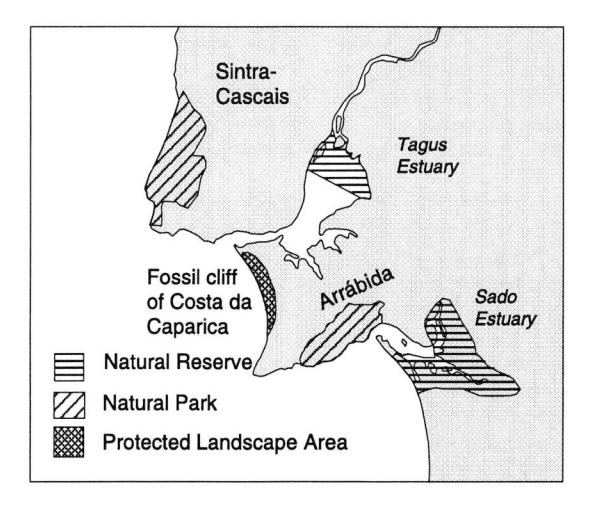

Fig. 11. Map of the zones with different protection status in the area under focus.

REFERENCES

Abecassis, F. (1994) Geomorphological characterization of the Portuguese coast. *Littoral* 4, Sept. 26–29, Lisbon, 25–29 pp.

Abrantes, F. and Moita, M.T. (1999) Water column and recent sediment data on diatom and coccolithophorids off Portugal confirm sediment record of upwelling events. *Oceanologica Acta* **22** (3), 319–336.

Abrantes, F. (1988) Diatom assemblages as upwelling indicators in surface sediments off Portugal. *Marine Geology* **85**(1), 15–39.

Bôto, R.G. (1945) Contribuição para os estudos de oceanografia ao longo da costa de Portugal. *Travaux de la Station de Biologie Maritime de Lisbonne* **49**, 102 pp.

Brogueira, M.J., Cabeçadas, G. and Rocha, C. (1994) Sado estuary coastal waters: thermohaline properties and chemical composition. *GAIA* 23–25.

Brotas, V. (1995) Distribuição espacial e temporal do microfitobentos no estuário do Tejo (Portugal): pigmentos fotossintéticos, povoamentos e produção primária. Ph.D., FCUL, 144p.

Cabeçadas, G. and Brogueira, M.J. (1997) Sediments in a Portuguese coastal area—pool sizes of mobile and immobile forms of nitrogen and phosphorus. *Marine and Freshwater Research* **48**, 559–563.

Cabeçadas, L. (1999). Phytoplankton production in the Tagus estuary (Portugal). *Oceanologica Acta* **22** (2), 205–214.

Cabeçadas, L., Brogueira, M.J. and Cabeçadas, G. (1999) Phytoplankton spring bloom in the Tagus coastal waters: hydrological and chemical conditions. *Aquatic Ecology* **33** (3), 243–250.

Caçador, I., Vale, C. and Catarino, F. (1996) The influence of plants on concentrations and fractionation of Zn, Pb and Cu in salt marsh sediments (Tagus Estuary, Portugal). *Journal of Aquatic Ecosystem Health* **5**, 193–198.

Carneiro, M., Martins, R. and Franca, L.P. (1997) A pesca artesanal local no Estuário do Tejo. *ECSA Local Meeting. The Tagus Estuary Adjoining Coastlines*. Lisbon, 25–27 June, 61–62 pp. (Abstract).

Catarino, F. (1981) O papel das zonas húmidas do tipo sapal na descontaminação das águas. *Ciência, IV série*, **1**(2), 9–16.

Catarino, F., Tenhhunen, J.D., Brotas, V. and Lange, O.L. (1985) Applications of CO_2-porometer methods to assessment of components of photosynthetic production in estuarine ecosystems. *Marine Biology* **89**, 37–43.

CNA (1976) Environmental study of the Tejo estuary. Summary of the preliminary report. Phase I. 44p.

Coelho, M.R. and Bebiano, M.J. (1998). TBT na costa portuguesa. *IX Simpósio Ibérico de Química Marinha*, Univ. de Aveiro, 16–19 de Abril 1998, no. 20.

Costa, M.J. and Bruxelas, A. (1989) The structure of fish communities in the Tagus Estuary, Portugal, and its role as a nursery for commercial fish species. Topics in Marine Biology. J.D. Ros (Ed.). *Scient. Mar.* **53**(2–3), 561–566.

Costa, M.J., Pereira, C.D., Teixeira, C.M., Jorge, F.M., Salgado, J.P., Costa, J.L., Gordo, L.S. and Correia, M.J. (1997) Monitoring of macrozoobenthic invertebrates and fish. In *Vasco da Gama Bridge—Lisbon, Environmental Monitoring Works During Construction*. CEMA, June. pp. 22–29.

Coutinho, M.T.P. (1998) Diversidade do fitoplâncton e comportamento do índice de Shannon numa lagoa costeira com cultivo de Mytilus. *Revista Biol. (Lisboa)* **16**, 17–30.

Dias, M.D.S. and Cabral, H.N. (1994) Actividade das pequenas cercadoras nas áreas adjacentes ao Estuário do Sado. *Publicações Avulsas do IPIMAR* **1**, 35–46.

Dias, M.D.S. and Nunes, M.C. (1994) Contaminação bacteriana de moluscos bivalves. *Publicações Avulsas do IPIMAR* **1**, 251–261.

Dias, M.D.S., Gaudêncio, M.J. and Guerra, M.T. (1994) Moluscos bivalves com importância economica nas areas costeiras adjacentes aos estuários dos rios Tejo e Sado. *Publicações Avulsas do IPIMAR* **1**, 47–58.

Dickson, R.R., Kelly, P.M., Colebrook, J.M., Wooster, W.S. and Cushing, D.H. (1988) North winds and production in the eastern North Atlantic. *Journal of Plankton Research* **10**, 151–169.

Estrada, M. (1995) Dinoflagellate assemblages in the Iberian upwelling area. In *Harmful Marine Algal Blooms*, eds. P. Lassus, G. Arzul, E. Erard, P. Gentien and C. Marcaillou. Techique et Documentation, Lavoisier, Intercept Ltd, pp. 157–162.

Ferreira, J.G. and Ramos, L. (1989) A model for the estimation of annual production rates of macrophyte algae. *Aquatic Botany* **33**, 53–70.

Ferreira, J.G. (1995) Ecowin — an object–oriented ecological model for aquatic ecosystems. *Ecological Modelling* **79**, 21–34.

Figueiredo, I.M. and Santos, A.M.P. (1989) Reproductive biology of *Sardina pilchardus* (Walb.): seasonal maturity evolution (1986 to 1988). ICES CM 1989/H: 40, 4 pp.

Figueres, G., Martin, J.M., Meybeck, M. and Seyler, P. (1985) A comparative study of mercury contamination in the Tagus estuary (Portugal) and major French estuaries (Gironde, Loire, Rhône). *Estuarine, Coastal and Shelf Science* **20**, 183–203.

Fiúza, A.F.G. (1984) Hidrologia e dinâmica das águas costeiras de Portugal (Hydrology and dynamics of the Portuguese coastal waters). Ph.D. dissertation, Universidade de Lisboa, 294 pp.

Fortunato, A.B., Baptista, A.M. and Luettich Jr, R.A. (1998) A three-dimensional model of tidal currents in the mouth of the Tagus estuary. *Continental Shelf Research* **17**(14), 1689–1714.

Gaspar, M.B, Dias, M.D. and Monteiro, C.C. (1998) A pescaria de bivalves do litoral oceânico da região de Setúbal. Situação actual dos principais bancos (Junho de 1997). *Relat. Cient. Téc. Inst. Invest. Pescas Mar* **33**, 12 p.

Gaudêncio, M.J. and Guerra, M. (1994) Povoamentos macrozoobentónicos das áreas costeiras adjacentes aos estuários dos rios Tejo e Sado (Portugal). *Publicações Avulsas do IPIMAR* **1**, 99–110.

Gaudêncio, M.J. and Guerra, M.T. (1986) Preliminary observations on *Gibbula umbilicalis* (da Costa, 1778) on the Portuguese coast. *Hydrobiologia* **142**, 23–30.

Guerra, M.T. and Gaudêncio, M.J. (1986) Aspects of the ecology of *Patella* spp. on the Portuguese coast. *Hydrobiologia* **142**, 57–69.

Guerra, M., Sacramento, C., Wergikoski, B. and Gaudêncio, M.J. (1999) Desenvolvimento de Metodologias para a avaliação dos efeitos de tri-butil-estanho (TBT) na costa portuguesa e corredores de navegação. *Relatório da 1ª Fase, Protocolo de Colaboração IPIMAR/DGA*, 17 pp.

INE (1993) Região de Lisboa e Vale do Tejo, Outubro, Vol. 3.

Instituto da Cígua e Direcção-Geral da Saríde (1998) Qualidade das Cíguas Balneares (dados da época balnear de 1997), 4 pp.

Jouanneau, J.M., Oliveira, A., Rodrigues, A., Weber, O., Garcia, C. and Dias J.A. (1997) Sedimentological evidence of present day trapping on the portuguese margin off Tagus estuary. *Third EU Conference on Exchange Process at the Continent/Ocean Margins in the North Atlantic, Vigo/Spain*. May 1997 (Abstract).

Madelain, F. (1970) Influence de la topographie du fond sur l'écoulement Méditerranéen entre le détroit de Gibraltar et le Cap Saint-Vincent. *Cahiers Océnographiques* **22** (1), 43–61.

Madureira, M.J., Vale, C. and Gonçalves, M.L. (1997) Effect of plants on sulphur geochemistry in the Tagus salt-marshes sediments. *Marine Chemistry* **58**, 27–37.

Marques, J.C. and Bellan-Santini, D. (1993) Biodiversity in the ecosystem of the Portuguese continental shelf: distributional ecology and the role of benthic amphipods. *Marine Biology* **115**, 555–564.

Martins, M.R., Cascalho, A., Franca, M.L. (1994) Prospecção e avaliação de recursos vivos (peixes, moluscos e crustáceos) de interesse económico no estuário do Tejo. *Publicações Avulsas do IPIMAR* **1**, 17–29.

Martins, M.R., Martins, M.M. and Cardador, F. (1989). Portuguese fishery of black scabbard fish (*Aphanopus carbo* Lowe, 1839) off Sesimbra Waters. ICES CM 1989/G: 38, 29 pp.

Mendes, R. and Vale, C. (1984) Geographic variation of heavy metal contents in mussels (*Mytilus galloprovincialis*) from the coastal area adjacent to the Tagus estuary, Portugal. M.M. 1984/E:38. Marine Environmental Quality Committee. 14 p.

Monteiro, J.H., Abrantes, F.G., Dias, J.M.A. and Gaspar, L.C. (1983) Upwelling records in recent sediments from southern Portugal: a reconnaissance survey. In *Coastal Upwelling, Its Sediment Record. Part B: Sedimentary Records of Ancient Coastal Upwelling*, ed. J. Thiede and E. Suess, pp. 145–162 pp. Plenum, New York.

Nogueira, M.C., Cabeçadas, G. and Brogueira, M.J. (1996) Resposta em termos ambientais da Lagoa de Albufeira a situações de isolamento e de comunicação com o mar. *Bol. Inst. Port. Invest. Marít., Lisboa* **2**, 49–54.

Oliveira, J.C. and Cabeçadas, G. (1996) The North Sea and the Atlantic Coasts of Western and Southern Europe. Portugal. In *Marine Benthic Vegetation*, eds. W. Schramm and P.H. Nienhuis. Ecological Studies Vol. 123. pp. 283–292. Springer-Verlag, Berlin and Heidelberg.

Paiva, P., Jouanneau, J.-M, Araújo, F., Weber, O., Rodrigues, A. and Dias J.M.A. (1997) Elemental distribution in a sedimentary deposit on the shelf off the Tagus estuary (Portugal). *Water, Air and Soil Pollution* **99**, 507–514.

Palma, A.S., Vilarinho, G. and Moita, M.T. (1998) Interannual trends in the longshore distribution of *Dinophysis* off the Portuguese coast. In *Harmful Algae*, eds. B. Reguera, J. Blanco, M.L. Fernández and T. Wyatt, pp. 124–127. Xunta de Galicia and International Oceanographic Commission of UNESCO.

Palminha, F. Melo, R. and Santos, R. (1982) A existencia de *Gelidium sesquipedale* (Clem) Bornet et Thuret, na costa sul do Algarve I: Distribuicão local. *Bol. Inst. Nac. Invest. Pescas* 8, 93–105.

Pereira, E., Vale, C., Ramalhosa, E. and Duarte, A. (1997) The relationship between mercury distribution coefficient and suspended particulate matter concentration in the Tagus estuary. *ECSA, The Tagus Estuary and Adjoining Coastlines*, pp. 38 (Abstract).

Pissarra, J., Sampayo, M.A., Nunes, M.C., Cachola, R., Dias, M.D., Vieira, A. and Lima, C. (1997) Classificação e controlo das zonas de produção de bivalves. *Divulgação IPIMAR* 1, 6 p.

Portela, L.I. and Neves, R. (1997) Distribution of phytoplankton and suspended sediment in the Tagus esyuary: implications for turbidity. *ECSA. The Tagus Estuary and Adjoining Coastlines.* pp. 66 (Abstract).

Ramalho, A. and Dentinho, L. (1928) Notas sobre as condições oceanográficas ao largo da costa de Portugal em 1927. *Travaux de la Station de Biologie Maritime de Lisbonne* 15, 20 p.

Rheinheimer, G. and Schmaljohann (1983) Investigations on the influence of coastal upwelling and polluted rivers on the microflora of the northeastern Atlantic off Portugal. I. Size and composition of the bacterial population. *Botanica Marina* XXVI, 137–152.

Ribeiro, A.P. (1999) Contaminação por metais no epibentos de Cala do Norte do Estuário do Tejo. ECOTEJO 98, Rel. A-8503-01-99-UNL, Ed. DCEA/FCT. 43 p.

Rodrigues, A. and Moita, M.T. (1979) Sistema dos sapróbios aplicado ao estudo do fitoplâncton, para uma possível classificação da água no estuário do Tejo. *Bol. Inst. Nac. Invest. Pescas* 2, 27–53.

Sampayo, M.A. (1994) Fitoplâncton tóxico e toxicidade em moluscos bivalves. In: Seminário sobre Recursos Haliêuticos, Ambiente, Aquacultura e Qualidade do Pescado da Península de Setúbal. *Publ. Avulsa IPIMAR* 1, 263–27

Santos, A.M.P., Borges, M.F., Pestana, G. and Groom, S. (1998) North Atlantic Oscillation (NAO) and small pelagic fish dynamics off Portugal. In: *Proceedings of GLOBEC First Open Science Meeting, Paris, 17–20 March 1998*, p. 72.

Santos, M. (1996) Contribuição para o estudo da biologia de *Nucella lapillus* L. *(Mollusca, Gastropoda)* na costa continental Portuguesa. Padrões de distribuição, variabilidade morfológica e efeitos de contaminação por TBT. Dissertação de mestrado, Fac. Ciênc., Univ. Porto, 45 pp.

Santos, R. (1994) Frond dynamics of the commercial seaweed *Gelidium sesquipedale*: effects of size of frond history. *Marine Ecology Progress Series* 107, 295–305.

Sequeira, M., Inácio, A., Silva, M.A. and Reiner, F. (1996) Arrojamentos de mamíferos marinhos na costa continental portuguesa entre 1989 e 1994. *Estudos de Biologia e Conservação da Natureza, ICN*, 19, 52 p.

Silva, M.C., Calvão, T. and Figueiredo, H. (1986) Controlo de qualidade de água. Resultados referentes os observaçoes realizadas en 1982 e 1983. Estudo Ambiental do Estuário do Tejo. SEARN Relatório Projecto Tejo No. 9, 167 pp.

Simas, M.T. (1998) Contaminação por mercúrio em alguns níveis da Cala do Norte do Estuário do Tejo. ECOTEJO 98, Rel. A-8503-01-98-UNL, Ed. DCEA/FCT. 42 p.

Simenstad, C.A. and Costa, M.J. (1997) Exogeneous and endogenous controls on food web structure supporting estuarine fishes: a comparison of the Columbia River (USA) and Tagus River (Portugal) Estuaries. *ECSA. The Tagus Estuary and Adjoining Coastlines.* pp. 43 (Abstract).

Sobral, P. (1982) Zooplâncton no estuário do Tejo. Distribuição e abundância. In *Estuarine Processes. An Application to the Tagus Estuary.* CNA. pp. 371–375.

Sousa e Silva, E., Assis, M.E. and Sampayo, M.A. (1969) Primary productivity in the Tagus and Sado estuaries from May 1967 to May 1968. *Notas e Estudos do Instituto de Biologia Maritima* 37, 30 pp.

Sundby, B., Vale, C., Caçador, I., Madureira, M.J. Catarino, F. and Caetano, M. (1998) Metal-rich concretions on the roots of salt marsh plants: mechanism and rate of formation. *Limnology and Oceanography* 43(2), 245–252.

Vale, C. and Sundby, B. (1987) Suspended sediment fluctuations in the Tagus estuary on semi diurnal and fortnightly times scales. *Estuarine and Coastal Shelf Science* 25, 495–508.

Vale, C. (1990) Temporal variations of particulate metals in the Tagus river estuary. *The Science of the Total Environment* 97/98, 137–154.

Vale, C., Ferreira, A.M., Micaelo, C., Caetano, M., Pereira, E., Madureira, M.J. and Ramalhosa, E. (1998) Mobility of contaminants in relation to dredging operations in a mesotidal estuary (Tagus estuary, Portugal). *Water Science and Technology* 37 (6–7), 25–31.

Vale, C., Gaudêncio, M.J. and Guerra, M.T. (1989) Evaluation of the ecological impact. In *Proceedings of the International Seminar on the Environmental Aspects of dredging Activities, Nantes, France, 22 Nov.–1 Dec.*, eds. C. Alzieu and B. Gallene, pp. 119–128.

Vilela, H. (1954) As ostras no consumo e na economia nacional. *Boletim das Pescas* 43, 24–45.

THE AUTHORS

Graça Cabeçadas
*Instituto de Investigação das Pescas e do Mar (IPIMAR), DAA, Av. Brasilia,
1400 Lisboa, Portugal.*

Maria José Brogueira
*Instituto de Investigação das Pescas e do Mar (IPIMAR), DAA, Av. Brasilia,
1400 Lisboa, Portugal.*

Leonor Cabeçadas,
*Direcção Geral do Ambiente (D.G.A.) R. da Murgueira, Zambujal,
2720 Amadora, Portugal*

Chapter 11

THE ATLANTIC COAST OF SOUTHERN SPAIN

Carlos J. Luque, Jesús M. Castillo and M. Enrique Figueroa

The Atlantic coast of Southern Spain (Gulf of Cadiz) is strongly influenced in terms of its oceanography and species composition by both the Atlantic Ocean and the Mediterranean Sea. Strong currents converge in this area due to the prevailing winds, and occasionally during autumn and winter, strong storms occur that can cause severe erosion of the coastline. A rich terrestrial and marine morphology exists in the Gulf of Cadiz, which allows a variety of coastal ecosystems. This, together with the mild climatic conditions, results in a relatively high biodiversity, and the oceanic currents also lead to relatively high productivity.

Up to now the coast has been fairly well conserved and is in generally good condition. Nevertheless it is subject to pressures from many socio-economic activities, both traditional (fishing, harbour, salt mines, etc.) and newly introduced (intensive fishing, industry, tourism, urban development, development of water resources, intensive agriculture and recreational harbours). The area contains several protected areas, including the famous Doñana National Park, though there are pressures on the borders of several and a certain amount of degradation in some. Infrastructure for seasonally elevated populations of visitors has generated some problems, and fisheries of a non-traditional nature are beginning to result in noticeable changes in catches on the productive and broad continental shelf. The area as a whole contains coastal habitats which are unusual in Europe, and the value of this is gradually being recognised by the authorities.

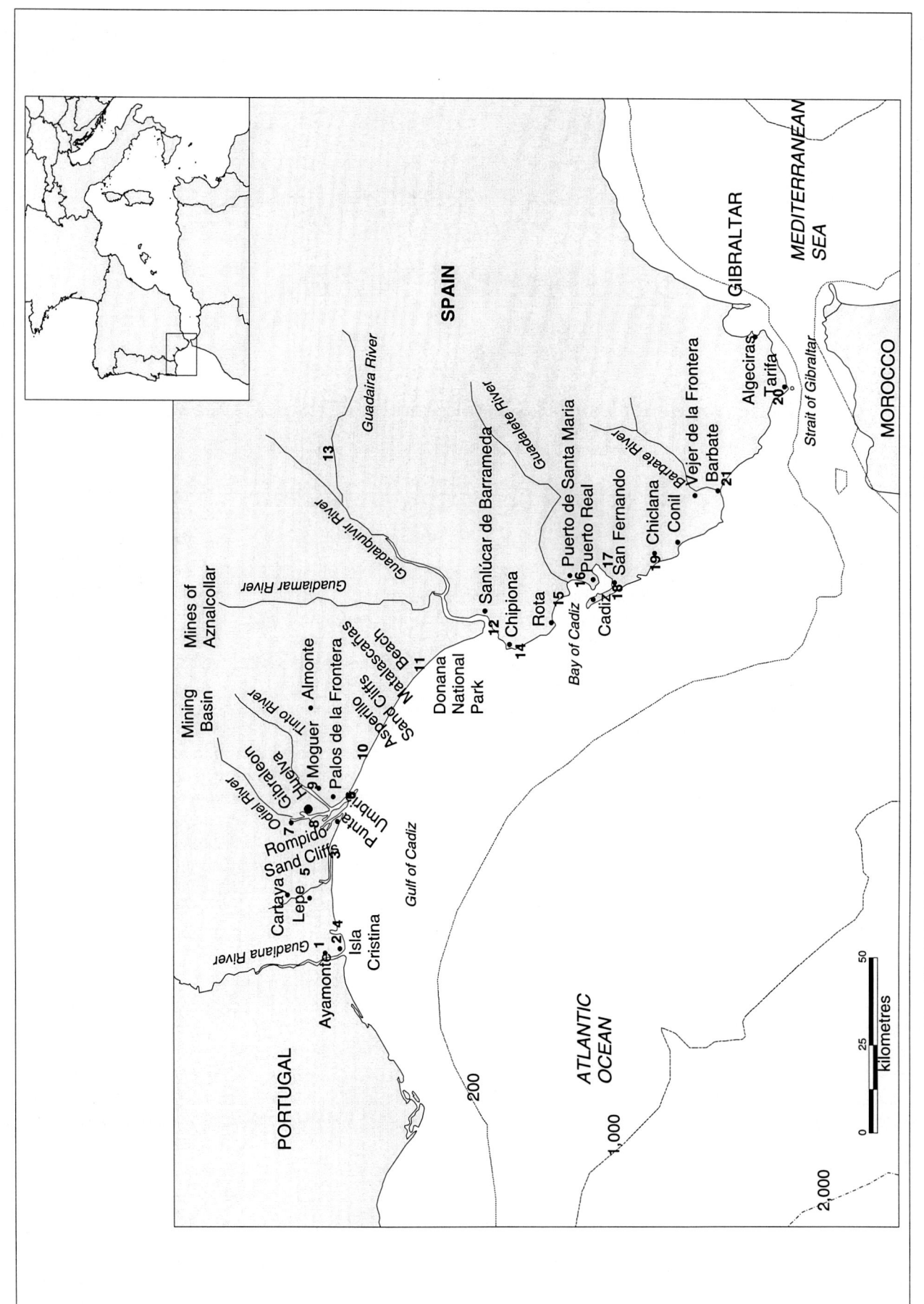

Fig. 1. The Gulf of Cadiz in the Atlantic coast of Southern Spain. Numbers related to sites named in Tables 12–14.

THE DEFINED REGION

The southern Atlantic coast of Spain, covered in this chapter, lies between Guadiana River to the West on the border with Portugal and Tarifa, which is the southernmost cape of the European mainland at 36°00'N. The water mass comprised within these limits is locally known as the Gulf of Cadiz (Fig. 1).

This coastline of Spain is 330 km long and includes the province of Huelva and the western part of the province of Cadiz. Its special geographical location, bordering the only natural entrance to the Mediterranean Sea, bestows a particular character onto its marine ecosystem, since it is influenced by the water masses of both the Atlantic Ocean and Mediterranean Sea. It is hence an important mixing area. The Strait of Gibraltar also acts as an effective bottleneck between Atlantic and Mediterranean waters.

SEASONALITY, CURRENTS AND NATURAL ENVIRONMENTAL VARIABLES

The mean annual temperature on the south-western coast is around 18°C. This area receives the highest radiation of the Iberian Peninsula, with 188 to 222 kJ cm^{-2} during autumn and winter and 423 to 444 kJ cm^{-2} during spring and summer. Annual values correspond to 2800 to 3200 hours of sun per year. Precipitation varies along the coast from below 500 mm in the western area to over 800 mm in the east, and long periods of rain alternating with long periods of drought may occur in the irregular Mediterranean climate. Highest rainfall occurs in November–December and March–April.

In summer, drought is extreme with less than 1 mm rain in July and August (Capel Molina, 1975, 1981; García Novo and Merino, 1993). Minor rivers may dry up during certain drought periods and especially during summer. This irregularity affects the biota of the region, which is well adapted to water shortages. Furthermore, the decrease in the quantity of water carried by the rivers modifies the hydrological dynamics of the estuaries.

Although its climate is Mediterranean, the southern Atlantic coast is predominantly influenced by the ocean, reducing extreme temperatures and favouring rainfall. Occasionally during autumn and winter, strong storms generate important erosive processes, especially in sand substratum areas, even damaging buildings and promenades on the coast. The oceanic influence increases towards the Strait of Gibraltar. Some of the physical phenomena, such as tidal amplitude and strong storms, are more related to the Atlantic. There is a drop in tidal amplitude from Huelva (spring tidal amplitude reaches a maximum of about 3.60 m) to Tarifa (1.30 m).

Two masses of water meet in the Gulf of Cadiz, near the Strait of Gibraltar: a surface and a deep one. The surface layer comes from the great North-Atlantic currents which split in two branches at the Cape of San Vicente (Portugal), one branch heading to the Canary Islands with the other entering the Gulf of Cadiz with an anticyclonic movement. Part of this mass of water enters the Mediterranean Sea, which has a water balance deficit (Fig. 2). This is a consequence of the high solar radiation, prevailing winds and insufficient water mass coming into this sea from rivers. The rest of the flow continues south towards the African coast (AMA, 1996). The estimated water volume that enters the Mediterranean Sea through the Strait of Gibraltar ranges between 63 and 146 km^3/day (Luengo Acosta et al., 1998). This surface water has a low relative salinity of about 35‰ down to 600 m depth (AMA, 1996). It has typical North-Atlantic characteristics, although it is warm due to its latitude, ranging between 10–15°C in winter and 18–25°C in summer.

However, deep, saline waters (36.5‰) flow out from the Mediterranean beneath the Atlantic water (AMA, 1996), spreading out in the Gulf of Cadiz. These currents result in the movement of animal and plant species through the Strait of Gibraltar in both directions, although this point is also the precise distribution limit for some species.

The minimum gap between Europe and Africa is about 14 km and its minimum depth is 300 m. This gap generates vigorous currents. Mediterranean waters are theoretically renewed every 97 years. Also, the orientation of the mountain ranges focuses the winds through this gap.

All these factors, together with the strong current parallel to the coast circulating from west to east, and south-easterly winds, considerably influence the coast.

Relatively large rivers (by Mediterranean standards) and the length of this coast, flow into this coastal arc. The main ones are the Guadiana (820 km long and with a basin of 67,500 km^2) and the river Guadalquivir (675 km long and a basin 57,000 km^2), the Piedras, the Tinto, the Odiel, the Guadalete and the Barbate. These rivers markedly influence the southern Atlantic coast of Spain, and drain a

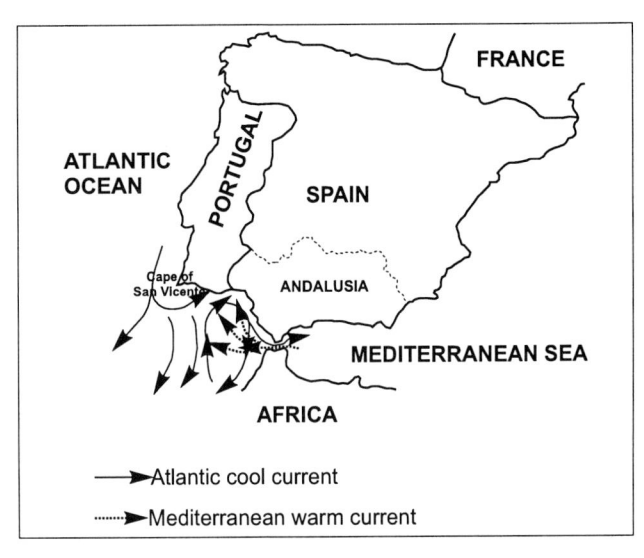

Fig. 2. Currents in the Gulf of Cadiz (from Ramos et al., 1996).

large amount of organic and inorganic material. Estuaries may be large; the influence of Guadalquivir estuary in particular, extends more than 100 km inland.

MAJOR MARINE AND COASTAL SHALLOW WATER HABITATS

Although human influence in this region has been intense at least since the 1 millennium B.C., the low fertility of sandy and gravel soils and the steep slopes of cliffs have preserved most of the coastal area from cultivation or intense exploitation (García Novo and Merino, 1993). The southern Atlantic coast of Spain still possesses varied ecosystems in a good state of conservation (Figueroa et al., 1980).

The town of Cadiz divides the coast into two sectors. In the Western sector continental detritic platforms on erosional coasts form sandstone cliffs and vast beaches. These alternate with shallow depressions of tectonic origin where rivers discharge, building up depositional coasts with marshes, sand banks, and spits with extensive dune systems (García Novo and Merino, 1993). In this area the relief is rather flat and the main geographical feature is a large coastal dune system interrupted by consolidated sand cliffs (El Rompido, El Asperillo) that are carved by an active wave pattern. At their foot is a soft sand platform. The raised coastal platform and its detritic cover now form a retreating coast, with cliffs ranging in height from 4 to 50 m. The rate of coastal erosion ranges from 0.5 m to 2 m yr^{-1} (Vanney and Menanteau, 1979).

Varied plant communities live in these coastal sandy locations and their distribution depends on several environmental factors. The following species are found on the beaches: *Cakile maritima* ssp. *aegyptiaca, Polygonum maritimum* and *Salsola kali* ssp. *kali*. Increased substrate stability at the rear part of the beach is indicated by *Crucianella maritima, Eryngium maritimum, Otanthus maritimum, Helichrysum picardii* among others. At the back of the beach, embryo dunes develop with *Ammophila arenaria* ssp. *arundinacea* as the building species. With increased surface stability, shrub species may become established, for example *Retama monosperma* forms dense stands at Punta Umbría, and this is associated with nitrophilic grassland species. Further up, if the surface of the sand plain lies high above the phreatic water table, a xerophytic shrub vegetation develops with *Cistus salvifolius, Halimium halimifolium, Helichrysum picardii, Lavandula stoechas, Rosmarinus officinalis*.

Extensive Quaternary sand deposits are widespread. Together with the beaches and sand plains they constitute over 60% of the coastal ecosystems in the western sector. *Pinus pinea* (umbrella pine) now forms extensive forests all over the southwestern coast of Spain. Planting of this species started in the 17th century and became widespread. Littoral dunes were extensively fixed during the first quarter of this century by planting *Ammophila arenaria* and *Pinus pinea* (Granados et al., 1984). *Juniperus oxicedrus* ssp. *macrocarpa* occurs in a few places either in pure patches or associated with other tree species. This juniper is now restricted to fixed dunes in a narrow band about 1 km wide near the sea. In some places *Quercus suber* woodland occurs (García Novo and Merino, 1993).

River estuaries range from continental marshes (Guadalquivir River) to tidal saltmarshes (Guadiana River, Piedras River, and Odiel-Tinto Estuary in Huelva). These cover large areas due to the tidal amplitude and to the flat terrain near the river mouths. The southern Atlantic saltmarshes of Spain are numerous in relation to the coast length. Strong coastal currents transport materials from retreating cliff sectors and seabeds to depressions, thus giving rise to sand banks and long spit formations. These large deposits at the mouths of the rivers create spectacular geo-morphological coastal formations, such as littoral sand spits (El Rompido), and large dune formations. Spits range from a few hundred metres in the Huelva estuary to huge sand accumulations over 30 km long and 4–6 km wide, such as in Doñana National Park (Dabrio González, 1980). The rate of land accretion on spits reaches 25 m yr^{-1} (García Novo and Merino, 1993).

The continental shelf of this sector is the largest of the Iberian Peninsula; the 100 m contour-line is 35 km from shore. The submerged river valleys are submarine canyons and the slope of the shelf is very smooth. The sandy sediment comes from eroded materials from the south coast of Portugal, as well as from the rivers. The muddy bottoms are less common and lie further from the coast (Fig. 3). They are mainly associated with the three large rivers, Guadiana, Odiel-Tinto and Guadalquivir. Few epibenthic organisms are found, due to the instability of the substrate caused by the continuous movement of the currents, whereas endobenthic organisms are more abundant, especially polychaetes, and bivalves, such as *Chamelea gallina* and *Donax trunculus*, which are commercially exploited. Other organisms are found, such as *Psammechinus miliaris, Astropecten* sp. and *Holothuria forskali*, as well as numerous fish species, many of which are commercially fished. Although rocky bottomed areas exist, they are scarce.

An area of great natural interest is the Odiel saltmarshes, one of the most extensive and best conserved tidal saltmarsh systems in Spain, located in the Tinto and Odiel rivers joint estuary. It contains a rich biological community, as well as particular geomorphological dynamics. The main halophyte species are *Zostera noltii, Spartina maritima* and *Sarcocornia perennis* in low saltmarsh areas, and *Sarcocornia fruticosa, Spartina densiflora,* and *Arthrocnemum macrostachyum* mainly in high saltmarshes. Many birds make use of the marshes since they are located on one of the main migratory routes between Europe and Africa. Among them, the most important are populations of the flamingo (*Phoenicopterus ruber*) and the nesting spoonbill colonies (*Platalea leucorodia*). The main environmental problem

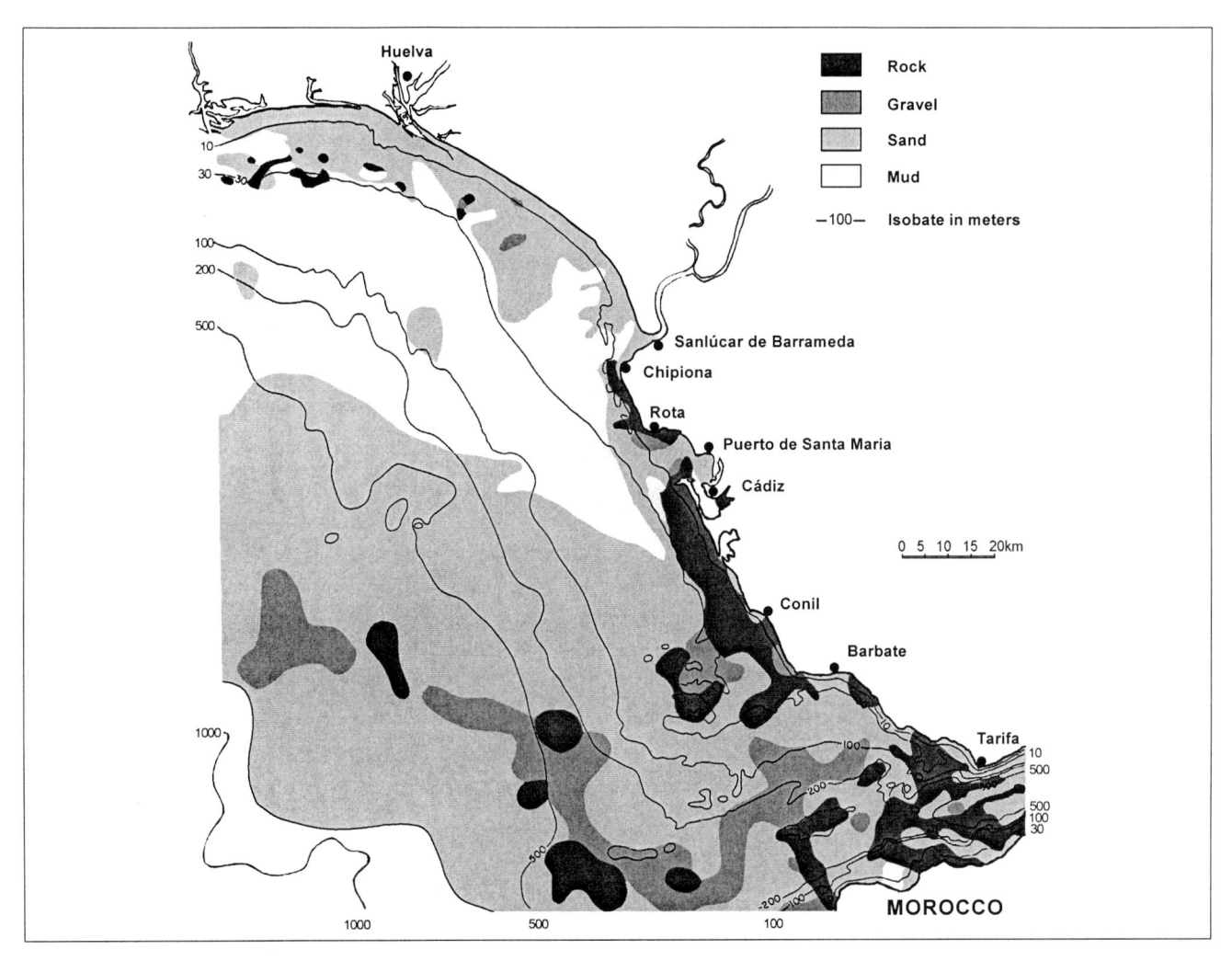

Fig. 3. Sediments of the Gulf of Cadiz (Ramos et al., 1996).

within this saltmarsh is the intense heavy metal contamination caused by the neighbouring industrial site of Huelva, and mining activities (Jiménez-Nieva and Luque, 1995). In spite of the heavy contamination load of this estuary, a great part of it is protected by the Government as a *Paraje Natural* (Natural Area) and as a Biosphere Reserve.

One of the most important natural areas in Europe is the Doñana National Park. It is located at the mouth of the Guadalquivir and comprises parts of the provinces of Huelva and Seville, covering 50,720 hectares. Doñana extends over two contrasting ecosystems: the sands (beach, active dunes and stabilized dunes) and the marshes. The dunes are 4–30 m high with advancing fronts under prevailing southwesterly winds, which may extend for over 2 km. Stable dunes have been generated under different climatic conditions and a much greater sand supply. They include large parabolic dunes 1–2 km long (García Novo and Merino, 1997). Sand vegetation varies greatly according to the distance from the water table. There are several notable Mediterranean scrub species; *Halimium halimifolium*, *H. commutatum*, *Cistus salvifolius*, *Stauracanthus genistoides*, *Calluna vulgaris*, *Erica scoparia*, *Ulex minor*, *Calluna vulgaris*, and others. Examples of the mature forest community that existed in the area before the sands were set in motion are; *Juniperus oxycedrus* subsp. *macrocarpa*, *Quercus suber* and *Olea europaea* var. *sylvestris*. On the active dunes of Doñana, *Pinus pinea* occurs mostly in the slacks with a few scattered trees. The marsh covers about 27,000 hectares. This ecosystem is flooded marsh in winter months and dried marsh during summer drought periods. Some of the plant species found in the area are: *Scirpus maritimus*, *Schoenoplectus lacustris*, *Sarcocornia fruticosa*, *Arthrocnemum macrostachyum*, *Salicornia ramosissima*. 'La Vera' separates marsh clay flooded soils from sandy soils. This is a long narrow stretch, with its width ranging to more than 1 km and serves as an ecotone between marshland and dunes.

The dune scrub landscape is very rich in mammals (*Cervus elaphus*, *Dama dama*, *Genetta gennetta*, *Sus scrofa*, *Vulpes vulpes*, *Oryctolagus cuniculus*, and others). Numerous bird species use Doñana as a breeding site and as a migratory stop, mainly on the marsh areas (*Phoenicopterus ruber*, *Platalea leucorodia*, *Anser anser*, *Ardea purpurea*, *Anas acuta*, *Anas penelope*, *Anas crecca*, *Limosa limosa*, *Recurvirostra avosetta*, and others). Some endangered species are found in the area, such as *Aquila adalberti* and *Lynx pardina*. Doñana, which has great ecological value, is subjected to immense pressures, mainly urban expansion and surrounding agricultural activities.

In the Eastern Sector is the Cadiz-Tarifa arc, dominated by rugged cliffs and rocky coastline (Peris-Mora, 1994). Parallel sierras of sandstone, calcareous rocks and schist reach the coast in large steps, forming capes with huge cliffs separated by bays with beaches and estuaries. Beaches are much smaller than in the western sector because no large source of detritic sediments exists, and a series of capes with high cliffs interrupts the coastline (García Novo and Merino, 1993).

Approaching the Strait of Gibraltar, more rugged bottoms with shallow rocky outcrops are found. In this area the continental platform narrows markedly, being only 2 or 3 km wide around Tarifa (Luengo Acosta et al., 1998). The benthos is predominantly rocky, due to rocky blocks coming off the cliffs, most of them lying on sand. The variety of bottom surfaces provides multiple habitats for the local communities of organisms. Epibenthic species are more abundant than endobenthic organisms. The upper illuminated levels are inhabited by seaweeds: *Phyllophora heredia, Halopitys pinastroides, Cystoseira fimbriata* and *C. tamariscifolia*. The lower shady levels are dominated by animal communities, such as the colonial anthozoans *Parazoanthus axinellae* and *Astroides calycularis*, which cover vast areas, and a great range of sponges, ascidians, tubificid polichaetes, and sea slugs. This is also home to a wide variety of fish, such as *Diplodus sargus, D. vulgaris, Sparus aurata, Argyrosomus regius*, etc. (Luengo Acosta et al., 1998).

The Strait of Gibraltar deserves separate consideration. It is located between two seas of great importance and the sea currents between them allow an important movement of nutrients. This, together with the great variety of habitats, and the variability of the physico-chemical conditions, such as temperature and water salinity, allows the area to maintain diverse animals and plant species.

Consequently, a rich coastal and marine morphology exists in the Gulf of Cadiz, which allows a variety of coastal ecosystems. The sea and coastal substratum is varied, ranging from muddy to rocky areas separated by sandy zones and from points exposed to the effects of the currents to more sheltered areas, and from saline to fresh water areas close to the mouths of the large rivers. Thus there are numerous different habitats. Furthermore, it is the point where two large well-defined systems meet, allowing the presence of Atlantic species in the Mediterranean and vice versa. All these circumstances, together with the optimal environmental conditions, result in a relatively high biodiversity.

OFFSHORE SYSTEMS

The hydrodynamic characteristics and the anticyclonic movement in the Gulf of Cadiz produce an upwelling movement in the westernmost part of this coast, which

Table 1

Fresh fish (metric tons) auctioned in the main markets of Huelva and Cadiz in 1996–1997 (IEA, 1998).

	1996			1997		
	Mollusc	Crustacean	Fish	Mollusc	Crustacean	Fish
Cadiz	3301	2263	34187	4477	2530	52237
Huelva	2199	331	2411	2271	819	8518

consequently increases primary productivity (AMA, 1996). In the Gulf of Cadiz, the highest rates of chlorophyll are found around the four large estuaries: Guadiana, Odiel-Tinto, Guadalquivir and Bay of Cadiz and the peninsular southern coast, with values around 30 mg chlorophyll-a m^{-3} (López Cotelo, 1993; Prieto et al., 1997). Primary production is high and maximum values are reached in summer. Atlantic waters contain a rich and varied plankton with a high number of taxa, with copepods the main components of the zooplankton (AMA, 1996).

Likewise, fisheries productivity is favoured by the mixing water masses, mild climate, the deposition of nutrients from large rivers and the vast continental platform. This is reflected in the important fishing tradition around the coasts of the Gulf of Cadiz, which is especially rich in historically important commercial species of molluscs, crustaceans and fish (Table 1) (López Cotelo, 1993).

Around 50 species of fish, molluscs and crustaceans are caught in this area, among them *Sardina pilchardus, Solea vulgaris, Dicologoglosa cuneata, Engraulis encrassicholus, Diplodus senegalensis, D. vulgaris* and *Merluccius merluccius*. The main cephalopods are *Octopus vulgaris, Loligo vulgaris* and *Sepia officinalis*; and among the bivalves *Chamelea gallina* and *Donax trunculus*, that gather in large shoals. The main crustaceans are *Penaeus kerathurus, Processa edulis, Palaeomon serratus, P. xiphias, Sycionia carinata, Carcinus maenas*, and *Macropipus arcuatus* (Muñoz Pérez and Sánchez de Lamadrid, 1994; López Cotelo, 1993). This last study pointed out that pelagic fish living on sandy and stony seabeds were mainly found near the coast, whereas cephalopods were the dominant species in deeper water throughout the year. It also indicates that the total biomass caught by fishing increases in relation to depth, cephalopods being the main species in this case. The pelagic fish catch represents only a small percentage of the total biomass.

POPULATIONS AFFECTING THE AREA

Huelva (10,128 km^2 and 454,735 inhabitants) and Cadiz (7,440 km^2 and 1,105,762 inhabitants) are located on the southwest coast of Spain. The entire coastline of Huelva faces the Atlantic Ocean, whereas only the western part of the province of Cadiz faces the Atlantic. There are numerous coastal municipalities (Table 2) which exploit the

Table 2

Registered population of Atlantic coast municipalities in 1996 (IEA, 1998)

Coastal Municipalities in Huelva	Population	Coastal Municipalities in Cadiz	Population
Almonte	16,264	Barbate	21,888
Ayamonte	17,566	Cadiz	145,595
Cartaya	11,435	Conil de la Frontera	16,687
Gibraleón	11,166	Chiclana de la Frontera	53,001
Huelva	140,675	Chipiona	15,518
Isla Cristina	17,310	Puerto de Santa María	72,460
Lepe	18,325	Puerto Real	33,069
Moguer	13,371	Rota	24,197
Palos de la Frontera	6,884	San Fernando	85,882
Punta Umbría	10,998	Sanlúcar de Barrameda	61,088
		Tarifa	14,993
		Vejer de la Frontera	12,823

Table 4

Seasonal nature of tourist use of the coast (population) (Consejería de Medio Ambiente, 1998a)

	Municipalities/Areas	Winter	Summer
Cadiz	Barbate	22,664	29,463
	Cadiz-San Fernando	243,026	315,934
	Conil de la Frontera	16,359	21,267
	Chiclana	50,697	65,906
	Chipiona	15,399	20,019
	Puerto de Santa María	69,656	90,553
	Puerto Real	31,086	40,412
	Rota	24,287	31,573
	Sanlúcar de Barrameda	59,780	77,714
	Tarifa	14,934	19,414
	Vejer de la Frontera	12,957	16,844
Huelva	Almonte/Matalascañas	17,107	70,000
	Ayamonte	15,796	50,000
	Cartaya/El Rompido	11,002	25,502
	Huelva	145,049	200,000
	Isla Cristina/La Antilla	17,729	134,700
	Lepe	17,185	17,185
	Moguer/Palos de la Frontera/Mazagón	19,931	49,000
	Punta Umbría	10,793	120,000
Total		815,437	1,395,486

resources. The southern Atlantic coast provides the population mainly with fishing and tourist activities.

Many rich fishing grounds exist here, allowing fishing to be one of the major economic activities in the Gulf of Cadiz. The production for 1991 was estimated at $260,000,000 involving around 15,000 workers (CAP, 1997). The percentage of the working population employed in the fishing sector in 1989 in the municipalities of Isla Cristina and Punta Umbría was 40%, and was 30% in Cartaya (Fraidias et al., 1997). A considerable fleet, including the traditional fishing fleet, operates near the coast and another one operates offshore along the African coast as far as the Indian Ocean (Table 3). Around 80% of the fishing boats work on the southern Atlantic coast.

Currently, fishing stocks are below the sustainable maximum rates due to excessive exploitation in recent years. Fishing effort has increased and catches decreased, greatly affecting fishermen's income.

Table 3

Fish catches in the provinces of Cadiz and Huelva (metric tons) (CAP, 1997).

Year	Cadiz	Huelva
1987	78,112	17,389
1988	61,411	15,236
1989	75,115	16,650
1990	63,033	15,183
1991	61,467	14,413
1992	51,053	14,940
1993	50,472	17,300
1994	57,036	16,467

Since the 1970s, tourism has become important. The large sandy beaches, together with a favourable climate, provided the coastal municipalities with an economically valuable resource. Almost all the southern Atlantic coastline offers excellent tourist resources, which coexist with magnificent natural beauty.

Growing economic activities related to tourism have caused a large population increase, swelled further in summer (Table 4). In Punta Umbría beach (Huelva), winter population is about 11,000 and summer population is 120,000. This has resulted in several environmental problems. One example is that the urban sewage farms are not able to cope with increased usage.

There is a tendency to build new housing estates in coastal areas, especially around sandy beaches. This results in the loss of natural habitats. There is a greater housing demand in towns within the provinces of Huelva and Cadiz and also in neighbouring provinces (Sevilla, Córdoba and Badajoz). The construction of the infrastructure for new urban sites, such as Matalascañas (Almonte), is also a problem. This is an important centre surrounded by unique natural areas, such as the Doñana National Park. Although it has a lower population in winter, in summer it reaches 50,000 inhabitants. Controversy has arisen over the construction and improvement of road works around this locality, which may damage the surrounding natural reserves, thus threatening species such as the lynx (*Linx pardina*).

Until the middle of this century industrial activity had not had any impact on this environment. The first important urban–industrial areas were developed in the 1960s, when two large industrial sites were created in Huelva and the Bay of Cadiz. The industrial activities at these sites were environmentally hazardous (oil, chemical, paper factories, etc.). Industrial waste was dumped directly into the estuaries and sea without any kind of control, with consequences still evident to this day (Consejería de Medio Ambiente, 1998a). The estuary of the Odiel has also received a great metal load transported by the rivers from mining zones north of the province. This situation was worsened by industrial contamination.

At the end of the 1970s and during the 1980s, environmental concerns began to develop. Research was carried out on this region's natural resources which led to further studies on local terrestrial, aquatic, coastal and fisheries resources. Studies were also carried out to assess the extent of pollution in areas of high urban and industrial concentration, such as the Bay of Cadiz and the Huelva estuary. The aim of these studies was to examine the consequences of environmental pollution on coastal resources and to determine the extent of their degradation.

RURAL FACTORS

Fertilisers and mechanical farming have created problems of erosion and soil contamination of water resources.

Intensive agriculture is an important economic activity along the Guadalquivir and other rivers flowing into the Gulf of Cadiz. This has caused pollution and eutrophication in rivers and subterranean waters.

Underground waters of the Gulf of Cadiz have contamination problems, not only from the intensive agriculture developed in these areas, but also due to tourist activities. The most affected area is between Ayamonte and Huelva. Aquifers such as the Aluvial of the Guadalquivir, in the province of Seville, and the Rota–Sanlúcar–Chipiona in Cadiz contain nitrate levels higher than the permitted standards (50 mg/l). This is usually caused by uncontrolled and excessive use of fertilisers.

Surface waters of coastal zones of the Gulf of Cadiz are affected by considerable agricultural contamination such as uncontrolled dumping of highly polluting olive oil waste in rivers and shores, and the excessive use of pesticides in rice fields (in Guadalquivir Marshes). An extreme example of this happened in 1998 in the Barbate River marshes (Cadiz), where large numbers of fish died due to massive amounts of agricultural chemical agents. This nitrate contamination is causing progressive eutrophication of rivers, reservoirs and coastal waters. Thus, the whole southern Atlantic Andalusian coast is regarded as an area vulnerable to nitrate contamination (Table 5).

Deforestation that took place in former times along the southern Atlantic rivers, together with excessive cattle-raising activities and fires (very frequent in Mediterranean ecosystems) have caused serious erosion (Table 6). For instance, large eucalyptus plantations were established on the northern mountains of Huelva for cellulose production. As a result, the original forest has gradually disappeared and has been replaced by pyrophyte scrub of poor ecological interest.

Sometimes fires are started deliberately in forest and scrub in order to increase land for crops or cattle grazing, especially on mountains. This causes erosion, particularly during periods of heavy rainfall.

A program to reforest the area with native species adapted to the soil and climate is currently taking place. These species are mainly holm oak (*Quercus rotundifolia*), cork oak (*Q. suber*) and several pine species. The aim is to regenerate the natural vegetation of the ecosystems and thereby prevent erosion and subsequent desertification.

Inappropriate techniques, such as terrace building for reforestation, have caused strong erosion on high slopes,

Table 5

Fertilizer consumption (metric tons) during 1996 in some provinces of Andalusia whose rivers flow into the Gulf of Cadiz (Consejería de Medio Ambiente, 1998a)

Province	With nitrogen	With phosphates	With potassium
Cadiz	11,039	1,784	1,461
Córdoba	66,381	20,738	10,053
Granada	34,727	11,573	10,495
Huelva	5,891	2,694	2,643
Jaén	29,738	8,455	6,029
Sevilla	87,191	34,196	19,301

Table 6

Number of fires and affected forest and scrub vegetation surface in 1998 (Consejería de Medio Ambiente, 1999)

	Cadiz	Huelva
Wooded Surface (hectares)	80	530
Scrub Surface (hectares)	531	427
Number of fires	119	295

Table 7

Instance of erosion rates in the province of Huelva in 1997 (Consejería de Medio Ambiente, 1998a)

Tm/ha/year	0–1	1–4	4–8	8–12	12–20	20–50	50–75	75–100	100–200	>200
%	36	8	6	5	7	14	6	4	7	7

and this, together with torrential rainfall, contributes to soil loss due to flowing water and large amounts of suspended solids (Table 7).

Suspended solid concentrations on the southern Atlantic coast are closely related to fluvial load. Maximum values are 23 mg/l around the mouth of the Guadalquivir River, and 18 mg/l around the Odiel River. Mean suspended solid concentrations in the Gulf of Cadiz in autumn are 10 mg/l, whereas summer values are 3.01 mg/l. The maximum rates (5.5 mg/l) are at the mouth of the Guadalquivir. Likewise maximum concentrations of nitrates, nitrites, ammonium and orthophosphates generally occur in autumn and are associated with the mouths of the large rivers (López Cotelo, 1993).

Most of the intensive crops of coastal zones of the Gulf of Cadiz are seasonal, and grow only during the summer. Hence strong erosion occurs during the period when the soil lacks vegetation, especially since this coincides with periods of heavy rainfall.

The southern Atlantic region of Spain has a serious water shortage. Therefore numerous reservoirs have been constructed. In the Guadiana basin there are more than 100 reservoirs, holding a water volume over 11,000 Hm^3 (Fraidias et al., 1997). In the Guadalquivir there are 54 reservoirs, with a water capacity of 6,427 Hm^3. These reservoirs extensively affect the coast by modifying the river dynamics and acting as sediment traps. More than 80% of the water is used in agriculture and only 11.7% is urban demand (Consejería de Medio Ambiente, 1999). A large area of the southern Atlantic river basins is devoted to irrigated crops, which increases the water shortage of the region. In the Guadalquivir basin, for instance, there is 443,000 ha of irrigated land. Irrigation water is inappropriately used. The most common technique is to completely flood the cultivated soil, and most of the water is ultimately wasted. These methods are currently being modernised. For example, with regard to tree crops, only the trees are watered and water is not wasted on the space between them.

For more than 3000 years, the local population has been linked with fishing in the Gulf of Cadiz. Archaeological remains of salting industries have been found in several locations. Until the last few decades, fishing had a traditional character and was sustainable. It was based on three factors. The fishing fleet was not oversized with respect to the fishing resources. Secondly, fishing activities were selective, without massive fishing of individuals or of species, and it avoided catching small-sized individuals. This allowed the resources to regenerate. Finally, fishing boats were small.

However, fishing today is no longer sustainable as a result of continuous over-exploitation. Over 80% of the fishing fleet operates within coastal waters (CAP, 1997) (Table 8). There is an increase in boat size, and many catch illegally, in closed seasons, in exclusion areas or with illegal net sizes.

Table 8

Fishing boats on southern Atlantic fishing grounds in 1994 (CAP, 1995)

Technique	No. of boats	Register Gross Tonnage
Traditional	664	2340
Trawling	240	6337
Cerco	76	3140
Paternoster line	18	368
Rastro	164	1168
Total	1162	13353

Traditional fishing in the Guadalquivir estuary has been very important. However, the amount of caught fish since the 1960s is diminishing. River pollution, the removal of young fish and over-exploitation are the main factors for this. For example, a large population of sturgeon (*Acipenser sturio*) lived in the area. Caviar from these fish achieved international fame and some companies were established to commercialise it. The last individual was caught in the 1970s.

Most widespread fishing techniques have a long tradition in the region. Many practices that exhaust resources have been banned and many traditional fishing techniques are still used in the Gulf of Cadiz. These include:

(1) Alcatruz, mainly used to catch octopus on the coast of Huelva. This takes advantage of the habit of the octopus of sheltering in caves and rocks. A rope is used to which clay pitchers are attached. The pitchers are submerged 10–60 cm deep until the octopus enters it. This is a passive and selective technique.

(2) Tablilla and Chivo, mainly used on the coast of Cadiz and employs fish hooks to catch octopus. The environmental impact is likewise minimal.

(3) Mollusc and other invertebrate fishing. Shellfishing is carried out in sandy beaches and intertidal saltmarsh plains. Species involved in shellfishing include: *Marphysa sanguinea, Diopatra neapolitana, Nereis diversicolor, Scrobicularia plana, Tapes decussatus, Tapes aureus, Cerastoderma edule, Murex brandaris, Monodonta turbinata, Upogebia deltaura, Palaemonetes varians, Uca tangeri* and *Carcinus maenas*. Fishing is mainly performed manually. The capture of *Donax trunculus* is worth highlighting. This fish is highly valued in the local cuisine and is caught with the help of a rake attached to a sack-like net dragged along the sandy bottom of the beach. The individuals are then selected according to the size established by the regulations. Thus selective fishing is achieved which causes little impact. This method is still in use, especially around the Doñana National Park. *Nereis diversicolor* is caught on intertidal plains, especially in saltmarshes, for use as bait for sport fishing. They are mainly hand-caught and individuals reach a high price. The environmental impact of this

technique is low on intertidal plains, but if performed on banks, considerable erosion takes place (Castillo et al., 2000).

(4) Jábega, a technique normally carried out on sandy beaches. It requires the use of a net thrown from a boat which is manually dragged. Its impact on the coastal habitats is not very high, although, if the net size is too small, many juveniles may be caught. However, this fishing method is not frequently used on the southwest coast of the Iberian Peninsula.

(5) Almadraba, a technique with a long historical usage, having been practised on the coasts of Huelva and Cadiz for about 2000 years to catch tuna (*Thunnus thynnus*) on its migratory route along the coast. It consists of static labyrinth-like nets (anchored to the bottom) that are placed near the coast. To avoid this obstacle, tuna move parallel to the net for about 2 km. When they find a gap in the net they swim through it and continue through several compartments of the net until they are eventually caught. This structure is installed in spring, when tuna heads from the Atlantic towards the Mediterranean to breed and it is dismantled in autumn when they return to the Atlantic (Suárez de Vivero, 1988). Nowadays it is only used in the localities of Barbate and Tarifa. Recently there has been a decrease in fishing quantities without any clear reason. It may be due to a change in the migratory routes or to uncontrolled fishing performed by the fleet of other countries in the Mediterranean (CAP, 1997).

(6) Corrales (stockyards), a technique practised in the provinces of Cadiz, Sanlúcar, Chipiona and Rota. Fishing using *corrales* appears to date back 2000 years. The *corrales* are stone walls built in intertidal areas that surround certain natural and artificial shelters. Fish and cephalopods trapped within these walls are caught during the two daily periods of low tide. Due to the continuous maintenance required by these structures, they are now used less frequently. A large range of species is caught in them, probably due to nearby seagrass prairies of *Cymodocea nodosa* (Bernal et al., 1997).

(7) Nasas (fish traps), small-sized static traps used to fish shores and channels of estuaries. The nets are conical, and held open at the mouth by a metallic ring. Inside, there are further nets that prevent the fish from escaping.

(8) Another technique worth mentioning is the use of a paternoster line, a main cord to which multiple hooks have been attached. Different species are caught according to the size of the hooks.

COASTAL EROSION AND LANDFILL

Building and development are mainly located in the harbour zones and estuaries. The construction of ports and docks, dikes, dredging and industrial waste have altered the sedimentary balance, resulting in erosion and siltation of numerous coastal locations, some of great ecological value.

The construction of harbour infrastructure has caused considerable direct impact. The area occupied by the two great commercial ports is 1540 hectares at Huelva and 326 hectares in the Bay of Cadiz (IEA, 1998). These have involved large-scale transformation of the coastal landscape. Furthermore, there are 10 fishing or recreational ports on this section of the coast, as well as the military port at Rota. To maintain access to these ports, continual dredging is required, since these coasts have highly dynamic sedimentary processes. The harbour of Huelva is a particular case. The problem there is that these dredged materials contain high levels of heavy metal concentrations (Luque et al., 1998) which need to be confined in isolated deposits.

The largest changes have been effected by the construction of structures such as dikes and break waters, generally related to commercial, fishing, recreational or military ports. As dikes interfere with the coastal drift, sedimentation is scarce at beaches located east of them. This explains why emerged beach surfaces are being lost in these areas. In addition, erosive effects of rough weather during winter are magnified in areas heavily influenced by man. The loss of sandy beaches, an important tourist resource, has prompted the authorities to artificially regenerate some areas by dredging the sand from coastal areas (Anfuso et al., 1997). This activity requires heavy investment and the effect of these large projects also affects the erosion of tidal saltmarshes of some estuaries (Castillo et al., 2000).

One of the most striking cases of this type occurred at the end of the 1970s, with the construction of a dike, around 15 km in length, on the Huelva estuary. It was constructed to stop the advance of sand towards the river mouth of the port, in order to prevent it closing. However, the dike has interrupted the transport of sediments by the coastal west-east drift, causing the deposition of large amounts of sand to the west leading to the formation of sandy beaches up to 5 km long, and the loss of sandy beaches to the east. Besides altering the coastline, the water and coastal sedimentary dynamics have also been considerably affected (Ojeda et al., 1995; Castillo et al., 2000). The implications are widespread. The beaches of the Gulf of Cadiz usually have a period of winter erosion, due to spells of rough weather, and a period of summer sedimentation (Reyes et al., 1996); a lack of sedimentary materials can change the balance.

Dredging has large environmental impacts such as silting of large areas of saltmarshes, the appearance of large isolated deposits or the re-suspension of heavy metals (Castillo et al., 2000). This is particularly apparent where the ports are located within estuaries, such as Huelva or Punta Umbría, which are highly contaminated (Luque, 1996; Luque et al., 1998). In the case of the former, the construction of the sand-holding dike did not prevent the navigation channel from silting up. Dredging reintroduces the previously isolated heavy metals from benthic sediments back into the estuarine trophic network, counteracting the role of the sedimentation process in removing them (Delaune and Gambrell, 1996). Until several decades ago, dredged materials containing contaminated sediments were dumped onto previously well-conserved saltmarshes, thereby causing their total destruction. These acid, salty sediments could only be colonised by a few species, such as *Spergularia rubra* subsp. *longipes* (Luque, 1996). Later on, dredged materials were dumped away from coastal areas. Sediment dumping is now prohibited, causing immense problems in finding suitable grounds to place these materials. Large deposits have been accumulating on the sides of navigable rivers, creating a serious environmental impact.

Sand loss, like contamination, is another environmental threat that affects beaches directly and, in this area, is of crucial social and economic importance. Sand is lost mostly during winter and autumn storms, and in extreme cases entails the physical disappearance of the beach.

The only solution is the artificial transfer of sediments to beaches. This is an extremely costly project, but essential to preserve the beach as a resource. This needs to be carried out periodically. As an example, it is estimated that for the section of coast between Rota and Chipiona between 20,000 and 40,000 m^3 of sand is required every year. In 1998 numerous beaches suffered from considerable sand loss. Several buildings have been flooded during winter storms at a new tourist resort in Isla Canela (Ayamonte), and 750,000 m^3 of sand have been transferred to the beach in Camposoto (San Fernando), on the coast of Cadiz. The construction of protection jetties has also been necessary (Consejería de Medio Ambiente, 1999).

Another activity worthy of mention is arid soil extraction, used to regenerate beaches that suffer erosion during winter and autumn storms. Although the location where this soil is obtained is studied prior to extraction, in order to avoid damage to the marine environment, it can still cause the disappearance of some seagrass prairies. This activity also contributes to the environmental damage of other coastal areas, as sand dredging destroys the habitat of certain species (CAP, 1997). It can also cause detrimental effects on marine habitats adjacent to the filled areas.

Industries located in estuarine areas have caused a considerable direct impact on the coastal environment for many years. They have also generated huge amounts of contaminated waste. Liquid waste has been spilled in the sea, and solid waste has been dumped in saltmarshes, regarded as worthless at the time. An example of this is the estuary of the Tinto and Odiel Rivers. Slightly radioactive industrial solid waste was dumped on 1500 hectares of saltmarsh (40% of the total surface of the Tinto saltmarshes in 1956).

The saltmarshes at the Gulf of Cadiz have undergone numerous transformations, mainly due to different ways of infilling and excavation, leading to their total devastation. Another factor contributing to their decline has been reclamation for agricultural purposes; these ecosystems contain high concentrations of nutrients, which led to dredging of these areas for fields.

The natural dynamics of most coastal dune systems have been irreversibly altered, mostly in areas used for tourism and second-residence construction. The effects of tourism have also been noticed, with additional disturbances from fires, hunting, exploitation of water resources and uncontrolled rubbish dumping (Menanteau and Martin Vicente, 1979). Only 60% of Huelva, and 75% of Cadiz coastal sand areas are maintained in a decent state.

EFFECTS FROM URBAN AND INDUSTRIAL ACTIVITIES

In recent decades, some traditional coastal practices have almost disappeared, for example, mills powered by tidal energy. Other traditional usage still persists, especially marine salt extraction. The favourable geological and climatic conditions have made the saltmarshes on the Gulf of Cadiz a centre of sea salt extraction for centuries. The muddy sediments allowed easy shaping, to channel and store seawater in pools, while the strong solar radiation and high wind frequency lead to fast evaporation. The traditional marine salt production areas of the Bay of Cadiz are at the mouth of the Guadiana River, at Huelva, and at the Guadalquivir River.

Throughout history the marine salt industry has experienced a succession of periods of expansion followed by periods of decline. The most spectacular development took place during the last century, when most existing marine salt-producing areas were ploughed and the marine salt industry became the most important in the area. This created the characteristic landscape still evident today. The main crisis started in the 1940s, caused by the difficulties in adapting new mechanisation techniques to salt extraction, and the loss of market following the development of cold food storage. Today, numerous marine salt production areas have been abandoned and disappeared. These areas have been drained and filled for industrial or construction purposes. Some abandoned pools in traditional marine salt production areas are used by birds, such as *Phoenicopterus ruber* and *Himantopus himantopus*.

Many traditional marine salt production areas keep their old structure. These are now often used for mariculture, especially given current over-exploitation of fishing resources. The coastal ecosystem also benefits from the reorganisation of activities, since it is not necessary to modify the new areas. These, and other new ones, have considerably enhanced fish farming. In 1997, 104 fish farming industries existed in the province of Cadiz and 13 in Huelva (IEA, 1998). The main species used are: *Sparus aurata, Dicentrarchus labrax, Solea senegalensis, Penaeus kerathurus, P. japonicus, Palaemon longirostris* and *Palaemonetes varians*. Results have not been very successful, often because of inappropriate techniques or use of non-native species.

Industrial exploitation of littoral resources has put pressure on the coastal fishing grounds, especially on the continental shelf. The richness of the area inspired uncontrolled exploitation, which led to a progressive impoverishment of shoals and fishing grounds (López Cotelo, 1993). Despite their capacity for recovery, they are still very affected by over-exploitation. Trawling is the main fishing technique used, which actively changes the nature of the seabed. An enormous amount of fish from many different species is caught. Furthermore, it may cause a large-scale ecological impact when carried out within protected areas. Vast rock-free areas enable trawling activities to take place all over the region, and the catch includes large amounts of cuttlefish and sea bass (López Cotelo, 1993). About 34 fishing grounds have been identified on the continental shelf and on the talus up to 800 m (Sobrino, 1996). Several factors may be contributing to the deterioration of the fishery:

- use of new technologies, which increase catch sizes;
- increase in number and quality of the fishing fleet;
- use of destructive techniques and practices;
- non-compliance with laws regulating trawling net use, and fishing during closed seasons and within breeding zones;
- urban and industrial pollution of coastal areas;
- catching juvenile fish, thereby preventing populations from regenerating (CAP, 1997).

In the north of Huelva and Sevilla there is considerable mining activity. The minerals are extracted from opencast mines, and include iron, copper, lead, zinc, gold, silver, arsenic, cobalt and mercury (IGME, 1982). Drainage from these areas carries sediments containing high heavy metals concentrations. The Odiel and Tinto drain a mining basin that has been exploited for more than 4000 years. pH levels down to pH 2 are caused by natural oxidation of sulphurous minerals, and these loads are transported, undiluted, to the mouth of the joint estuary, where they mix with seawater. These high levels of contamination render these two rivers among the most polluted rivers in the world (Luque, 1996; Luque et al., 1998).

The greatest ecological catastrophe of Spain's recent history took place in the mines of Aznalcóllar (Sevilla) in April 1998. The holding wall of the mineral waste pool broke, resulting in an enormous spillage that reached the Guadiamar River. This tributary of the Guadalquivir River is also a considerable water supply for the Doñana National Park. The wastewater contained solid sediments and had high concentrations of heavy metals in suspension, originating from the pyrite washing and treatment processes at the mine. The total amount of water and mud spilled into the Guadiamar River was estimated at 4.5 Hm^3 (Table 9). This caused the river to flood, reaching 200 m on both sides of its shores and covering an area of 4286 hectares. The flood reached up to 40 km from the mine pool. It was eventually controlled shortly before the flood could enter the National Park or the Guadalquivir River; 37.4 metric tons of dead fish were recorded during the first month after the event. Urgent action was taken to regenerate the whole river and other contaminated areas, such as the construction of a wall to hold the spilled waste, removal of contaminated mud and treatment of contaminated water by neutralising and precipitating out the heavy metals (Table 10). Water is also purified before entering the Guadalquivir River. A large program to evaluate the catastrophe has been instigated, but although great efforts have been made, it is not possible to return the site to its original state. It is feared that heavy metal contamination will persist in this river basin. Certain heavy metal loads may have reached the sea.

Table 9

Polluting element concentration in pyrite mud (Consejería de Medio Ambiente, 1998b)

Element	Concentration (mg/kg)
Zn	8000
Pb	8000
As	5000
Cu	2000
Co	90
Tl	55
Bi	70
Cd	28
Hg	15

Table 10

Maximum concentrations of heavy metals and other elements analysed on fluvial sediment samples (Guadiamar River) taken 15 days after the 1998 catastrophe (Consejería de Medio Ambiente, 1998b)

Element	Concentration (%)	Element	Concentration (mg/kg)
Al	8.08	Ba	441
Ca	7.18	Sr	235
P	0.09	Ag	61
Ti	0.34	Ni	68
Mg	3.38	Hg	16.6
K	2.57	Bi	87
Na	0.47	Cd	31
Mn	0.14	Ga	20
Fe	36.60	Ge	4
S	41.80	Mo	8.4
As	0.52	Sb	722
Cu	0.20	Se	12
Pb	1.22	Sn	23
Zn	0.89	Te	0.4
		Th	12
		Tl	55
		Co	98
		Cr	95
		V	121
		U	2.4

Table 11

Compounds and quantities dumped into the Huelva Estuary by the Huelva Industrial Centre over 1992–1993 (AMA, 1994)

Compound (Unit)	1992	1993	Compound (Unit)	1992	1993
Solids in suspension (Tm)	176,145	50,328	Oils and fats (Tm)	35.1	92.7
Phenols (kg)	33,637	31,874	Cadmium (kg)	1,138	2,526
Fluorides (Tm)	5,863	4,953	Copper (kg)	21,784	21,942
Ammonia (Tm)	1,062	858	Chromium (VI) (kg)	151.5	109
Total phosphorus (Tm)	29,575	20,706	Mercury (kg)	–	769.9
Sulphates (Tm)	–	37,266	Manganese (kg)	2,2743	97,652
Arsenic (Tm)	132.8	155	Nickel (kg)	98.3	1,329
Iron (Tm)	398	1,360	Titanium (kg)	3134	17,455
Acids (Tm)	3,117	1,704	Lead (kg)	3719	9,616
Cyanide (kg)	62.9	94.15	Zinc (kg)	80,617	96,930

The Guadalquivir estuary has an important fish farming industry, which takes place within the limits of the Doñana Natural Park. A vast saltmarsh region was transformed to establish large mariculture pools, destroying the saltmarsh in the process.

Both main industrial centres on the Gulf of Cadiz, the Bay of Cadiz and Huelva now suffer high rates of pollution, but were originally developed on coastal habitats. In the mid 1960s an industrial estate was built in Huelva, and mainly occupied by petrochemical factories. It is one of the major sites for this industry in Spain. Since its establishment, this site has caused many environmental problems, especially by dumping waste in the Odiel and Tinto estuary (Table 11).

Luque et al. (1999) reported heavy metal incorporation into the estuarine trophic web. Heavy metal accumulation has been recorded in halophyte vegetation. But despite the high concentrations of heavy metals, no lethal effects have been observed on local vegetation, nor constant mortalities in the animal populations. Nevertheless, isolated fish kills have been observed, probably related to occasional dumping of industrial waste, although it has not been proved. A Corrective Programme for Waste Disposal in the Huelva Estuary is the main initiative on the part of the Andalusian Public Administration in face of the constant episodes of industrial contamination produced in this region (Jiménez-Nieva and Luque, 1995). However, several factories on the Huelva estuary and the Bay of Cadiz still dump contaminated waste.

Contamination caused by urban wastewater flushing causes a major impact. There is a lack of infrastructure in many coastal towns for completely purifying wastewater, despite the EU regulation that will force every town with a population larger than 15,000 inhabitants to purify waste waters in 2001. This, combined with the strong increase in population during summer months due to tourism, causes sewage pollution of waters adjacent to most populated coastal areas. In 1998, major contamination incidents occurred at locations in the province of Cadiz, where non-purified urban wastewater (no waste disposal infrastructure existed there) was dumped directly into two streams that eventually flowed onto two environmentally valuable beaches. Other contamination was due to malfunctioning of the urban waste disposal network. A great effort is being made to build waste disposal infrastructure to purify the urban waste of the most populated municipalities (Consejería de Medio Ambiente, 1998a). Ten years from 1996, a major investment in the construction of sewage farms will be made following an agreement between the EU and the Andalusian Government. The Gulf of Cadiz is a priority area within this programme.

The Environmental Department of the Andalusian Regional Government has been monitoring the quality of coastal waters off the Gulf of Cadiz for the past 10 years through the so-called Water Police Plan program (Consejería de Medio Ambiente, 1998a). This is a coastal police plan that monitors the state of water quality through a periodic sample-gathering campaign. Fifty eight sampling sites on rivers have been established, with a further 33 on the coast (Consejería de Medio Ambiente, 1998a) (Tables 12, 13 and 14). According to these surveys, industrial contamination of the Gulf of Cadiz is closely related to large basic or heavy industrial activities that produce highly contaminating residues. These are mining industries in the north of the provinces of Huelva and Seville and other industries located on some of the main estuaries and bays of the region, such as the Tinto and Odiel estuary or the Bay of Cadiz. Urban waste in urban and tourist areas, such as the cities of Cadiz and Huelva together with numerous industrial estates, has created rivers among the most polluted in the region. However, coastal water has not deteriorated substantially, as the strong Atlantic currents swiftly disperse the contaminants.

One of the main problems yet to be faced in the Gulf of Cadiz comes from the intensive nature of the tourism along this coast. The huge demand for building land near the beaches has led to a considerable population increase along the coast. The urbanisation process has particularly taken

Table 12

Heavy metal concentration (micrograms/litre), pH and conductivity (mS/cm) in coastal and river waters on the Gulf of Cadiz in 1998. n = number of samplings carried out through the year

Location	Code*	n	Cu	Zn	Mn	Ni	Cr	Cd	Pb	As	Fe	pH	Conductivity
Guadiana river	1	4	2.0	11.2	12.0	<1	<1	<0.5	<5	2.0		7.9	29.0
Ayamonte	2	4	<1	13.7	13.7	<1	<1	<0.5	<5	1.6		8.0	
Lepe	3	4	<1	16.2	13.5	<1	<1	<0.5	<5	1.9		8.0	
Carreras river	4	1	1.0	8.0	59.0	<1	<1	<0.5	<5	1.9		7.8	53.3
Piedras river	5	1	4.0	14.0	38.0	<1	<1	<0.5	<5	3.2		7.9	53.0
Punta Umbría	6	4	4.5	20.7	10.5	1.0	<1	<0.5	<5	4.3		8.0	
Odiel estuary	7	8	69.6	647.9	283.5	4375.0	<1	5.2	<5	20.6	47.2	7.4	
Odiel river	8	4	5595.0	16875.0	9085.0	171.0	13.2	57.2	249.5	5.0	8285.0	3.3	1.6
Tinto river	9	4	16730.0	40125.0	7227.5	145.2	30.2	163.7	113.5	55.6	146900.0	2.6	2.8
Mazagón	10	4	4.7	23.5	7.0	1.0	<1	<0.5	<5	2.0		8.1	
Torre de la Higuera	11	4	12.7	16.5	4.5	<1	<1	<0.5	<5	2.0		8.1	
Sanlúcar de Barrameda	12	4	3.2	23.5	4.0	2.7	<1	<0.5	<5	2.1		7.9	
Guadaira river	13	4	4.5	35.2	171.5	11.2	<1	<0.5	<5	1.2		7.4	1.7
Chipiona	14	1	5.0	17.0	<1	2.0	<1	<0.5	<5	2.0		8.1	
Guadalete river	15	2	1.0	16.5	30.5	4.0	<1	<0.5	<5	1.1		7.9	30.6
Valdelagrana	16	4	<1	17.2	10.5	<1	<1	<0.5	<5	1.2		8.1	
Punta de San Luis	17	4	<1	13.2	10.7	<1	<1	<0.5	<5	1.2		8.1	
Cadiz	18	1	2.0	12.0	<1	<1	<1	<0.5	<5	1.8		8.1	
Conil	19	1	<1	13.0	3.0	<1	<1	<0.5	<5	1.6		8.2	
Ensenada de Bolonia	20	1	<1	13.0	4.0	1.0	<1	<0.5	<5	1.8		8.1	
Barbate	21	4	<1	10.2	3.0	<1	<1	<0.5	<5	1.1		8.1	

*Code: For location of sampling sites, see Fig. 1.

Table 13

Heavy metal concentrations (in mg/kg) in coastal and river sediments in the Gulf of Cadiz in 1998

Location	Code*	Cu	Zn	Mn	Ni	Cr	Cd	Pb	As	Hg
Guadiana river	1	34	179	712	30	44	<0.5	18	30	0.5
Ayamonte	2	20	117	368	19	26	<0.5	16	16	0.1
Lepe	3	31	162	308	11	30	<0.5	15	16	0.4
Carreras river	4	48	167	403	30	47	<0.5	23	19	0.8
Piedras river	5	42	163	309	30	40	<0.5	13	10	0.3
Punta Umbría	6	85	322	225	6	19	<0.5	19	23	0.2
Odiel estuary	7	1840	1860	565	37	66	2.5	624	577	5.8
Odiel river	8	616	364	314	26	49	<0.5	540	514	5.1
Tinto river	9	398	326	149	15	33	1.7	406	1150	1.1
Mazagón	10	105	468	209	6	22	<0.5	50	65	0.3
Torre de la Higuera	11	40	327	223	4	16	<0.5	26	24	0.3
Sanlúcar de Barrameda	12	35	109	449	23	41	<0.5	26	16	0.2
Guadaira river	13	35	128	504	30	56	<0.5	25	4.8	0.2
Chipiona	14	6	39	365	3	15	<0.5	6	6.3	0.1
Guadalete river	15	47	115	336	25	53	<0.5	28	5.5	0.4
Valdelagrana	16	4	37	343	3	19	<0.5	6	4.8	0.3
Punta de San Luis	17	48	169	431	29	46	<0.5	42	9.5	0.7
Cadiz	18	5	29	142	5	4	<0.5	2	3.4	<0.1
Conil	19	4	28	289	5	10	<0.5	3	4.7	0.3
Ensenada de Bolonia	20	3	20	341	8	18	<0.5	9	4.1	<0.1
Barbate	21	11	38	543	15	27	<0.5	9	7.8	0.2

*Code: For location of sampling sites, see Fig. 1.

Table 14

Suspended solids (SS), dissolved oxygen (O_2), chemical oxygen demand (COD), nitrites (NO_2^-), nitrates (NO_3^-), ammonium (NH_4^+) and phosphates (PO_4^{3-}) in coastal and river water in the Gulf of Cadiz in 1998 (mg/l).

Location	Code*	SS	O_2	COD	NO_2^-	NO_3^-	NH_4^+	PO_4^{3-}
Guadiana river	1	29.2	7.02	17.50	0.02	2.55	0.06	0.12
Ayamonte	2				<0.01	0.51	0.16	0.10
Lepe	3				<0.01	0.42	0.16	0.10
Carreras river	4	32.0	6.2	<15.00	<0.01	1.31	0.05	0.10
Piedras river	5	59.0	6.70	<15.00	0.02	0.90	0.07	0.10
Punta Umbría	6				0.02	0.42	0.16	0.45
Odiel estuary	7				0.15	1.27	0.38	3.15
Odiel river	8	6.0		<15.00			0.35	
Tinto river	9	14.5		16.25			0.24	
Mazagón	10				0.02	0.83	0.16	0.10
Torre de la Higuera	11				<0.01	0.75	0.16	0.10
Sanlúcar de Barrameda	12	625.5	7.32	17.50	0.13	11.14	0.10	0.25
Guadaira river	13	186.5	0.52	174.50	0.83	5.07	21.55	14.90
Chipiona	14				0.02	0.95	0.05	0.10
Guadalete river	15	54.0	6.15	17.00	0.38	4.25	1.19	0.20
Valdelagrana	16				0.02	0.74	0.08	0.10
Punta de San Luis	17				<0.01	0.59	0.05	0.10
Cadiz	18				<0.01	0.52	0.05	0.10
Conil	19				<0.01	0.40	0.05	0.10
Ensenada de Bolonia	20				<0.01	0.40	0.05	0.10
Barbate	21				<0.01	0.65	0.05	0.10

*Code: For location of sampling sites, see Fig. 1.

place on extensive beach areas from Ayamonte to Chiclana and the population density of these developments can threaten the natural coastal balance.

Vast extensions of Pinus pinea forests and scrub located on sandy coastal areas are being destroyed to build residential areas. These forests frequently contain protected or threatened species e.g. (Juniperus phoenicea ssp. turbinata), (Juniperus oxycedrus ssp. macrocarpa) which demonstrates a lack of social interest and awareness of the natural resources.

Today, despite numerous measures taken to preserve coastal environments, there is strong economic pressure to develop large building projects in areas of great natural beauty without the appropriate environmental protection. Coastal ecosystems face a bleak future, since they are still regarded as an appropriate choice for tourist settlements.

The Gulf of Cadiz, as the natural gate to the Mediterranean sea, plays an important role connecting Atlantic and Mediterranean trading ports as well as the Suez Canal. Many ships pass through the Strait of Gibraltar. This involves the risk of serious environmental damage, since strong winds, currents and storms are frequent in this area. There has been no major incident recently, but there have been isolated episodes of accidental spillage, such as one in the Bay of Cadiz in 1997. A ship, stranded because of a storm, spilled 300 tons of diesel oil into the sea (Consejería de Medio Ambiente, 1998a). Pollution from deliberate dumping of contaminants into the coastal waters by ships is scarce. Some ships perform illegal washing of their tanks here and some of this oil-based waste reaches the coast, causing tar deposits. It is almost impossible to identify which ship is responsible, due to heavy naval traffic.

Navigation around port areas may have other impacts. Shipping traffic within the estuary of the Tinto and Odiel has been regarded as one of the main causes of strong erosion of the channels and saltmarsh, which has led to the loss of vegetation over large areas of protected zones (Castillo et al., 2000).

An expensive extraction programme is being carried out to regenerate sandy beaches that suffer strong winter erosion. This seriously impacts animal and plant communities living in the habitat. In some of these beaches it is estimated that around 40,000 m³ of sand are needed every year to prevent them from disappearing (Consejería de Medio Ambiente, 1998a). Many of these regenerated beaches are not balanced with the coastal hydrodynamics, so if the beaches are regenerated with a slope higher than the original one they may disappear again (Anfuso et al., 1997; Benavente et al., 1997).

PROTECTION MEASURES

Different administrations are involved in the management of the coastal resources: State, regional and local. Each deals with a certain aspect and they generate different regulations. The southern Atlantic coastal area belongs to the Andalusian Regional Government. The Environmental Department of the Regional Government (Consejería de Medio Ambiente) has the main protection and conservation powers over the Atlantic coast of Andalusia. From its establishment, 20 years ago, it has demonstrated an understanding of coastal ecosystems and the importance of implementing conservation policies.

From the 1980s onwards there has been awareness of natural resources and of their conservation. Prior to this, little attention was paid to the protection of the marine environment, despite its economic and biological importance.

The current policy from government and official organisations is geared towards total protection of the coastal environment. The exploitation of coastal resources such as fishing, tourism or recreational activities is essential for the local economy. Hence the government attempts to maintain them sustainably.

Marine benthic protection is carried out in three legislative ways; fishing, territorial and environmental regulations. The first official initiative to protect the marine benthos did not come from the protection of natural areas but from the recuperation and restocking within the framework of fishing regulations.

The law on Andalusian Territory Regulation in 1985 includes the protection of these ecosystems, establishing measures and recommendations to guarantee the compatibility of their use and their exploitation with their protection and renewal (Molina Vázquez, 1998). Certain guidelines are determined:

1. Adoption of measures to protect communities from activities that may alter their biological performance.
2. Strong surveillance of areas of high biological interest, such as bays, coves, estuaries and reefs. These, together with seagrass prairies are essential to maintain the coast production potential.
3. Regulation on live resources in interior waters, establishing:
   (a) Banned areas for extracting activities.
   (b) Areas which can be used for resource extraction.
   (c) Protected areas for biological communities.
   (d) Areas of high biological value.
   (e) Types of fishing and fishing regulations.

There are 88 natural areas under protection in Andalusia (Consejería de Medio Ambiente, 1999) but only 14 of them include maritime–terrestrial areas that affect coastal resources of the southern Atlantic region (Fig. 4, Table 15).

Some of these are continental areas that border the sea, whereas others are estuarine areas whose waters are therefore protected. Only La Breña and Barbate Saltmarshes Natural Park and Doñana National Park include measures related to the sea in their planning and regulations. They have, therefore, power to regulate activities and practices in the sea, and to attempt to reconcile marine resources exploitation with their protection within the National Park. For example, within the Guiding Plan for the Use and Management of La Breña and Barbate saltmarshes Natural Park, permitted activities and practices are identified; so-called 'economic' fishing is restricted and the use of any kind of fishing techniques is banned within the influence area of bird nesting sites. Limited sport fishing is permitted. Submarine fishing and gathering of marine organisms is

Table 15

Protected natural areas with maritime–terrestrial areas in the Gulf of Cadiz

	Type of protected area	Name	Area (ha)
1	National Park	Doñana	50,720
2	Natural Park	La Breña and Barbate Saltmarsh	4,815
3	Natural Park	Bay of Cadiz	10,453
4	Natural Park	Doñana	55,327
5	Natural Area	Trocadero Island	525
6	Natural Area	Sancti Petri Saltmarsh	170
7	Natural Area	Los Lances Beach	226
8	Natural Area	Juniper sand dune woodland of Punta Umbría	162
9	Natural Area	Domingo Rubio Channel	480
10	Natural Area	Isla Cristina Saltmarsh	2,385
11	Natural Area	Odiel Saltmarsh	7,185
12	Natural Area	Rio Piedras Saltmarsh and Rompido Sand Spit	2,530
13	Natural Reserve	Enmedio Island	480
14	Natural Reserve	Burro Saltmarsh	597

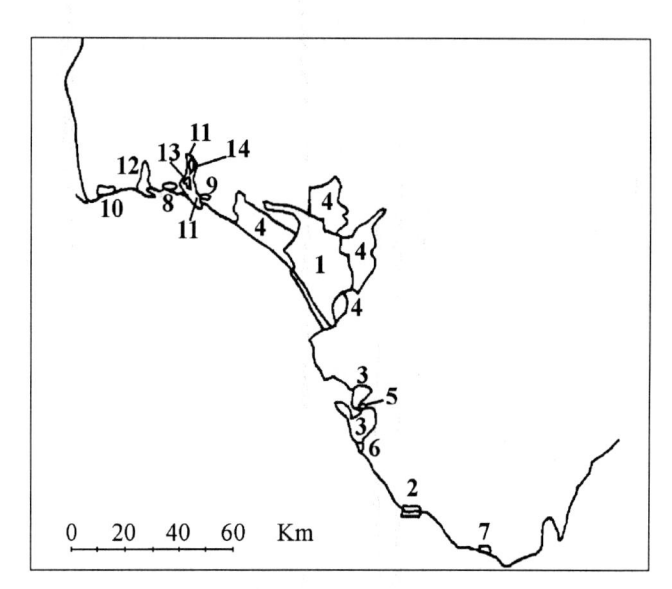

Fig. 4. Location of protected natural areas with maritime–terrestrial areas in the Gulf of Cadiz. 1. Doñana National Park. 2. La Breña and Barbate Saltmarsh Natural Park. 3. Bay of Cadiz Natural Park. 4. Doñana Natural Park. 5. Trocadero Island Natural Area. 6. Sancti Petri Saltmarsh Natural Area. 7. Los Lances Beach Natural Area. 8. Juniper sand dune woodland of Punta Umbria Natural Area. 9. Domingo Rubio Channel Natural Area. 10. Isla Cristina Saltmarsh Natural Area. 11. Odiel Saltmarsh Natural Area. 12. Rio Piedras Saltmarsh and Rompido Sand Spit Natural Area. 13. Enmedio Island Natural Reserve. 14. Burro Saltmarsh Natural Reserve.

prohibited (Molina Vázquez, 1998). To conclude, several practices and activities are restricted in the sea.

Protection measures include restricting operations of the large trawling fleet to within 6 nautical miles of the coast. Besides legal measures, artificial reefs have been installed since 1985 in several coastal areas, especially in areas of special ecological and fisheries interest, such as breeding grounds at the mouth of the main Atlantic rivers. Artificial reefs are concrete structures with varied shapes that would tear the trawling nets if used on them. At the same time these structures enhance fishery productivity and increase traditional fishing practices, which are more selective of fisheries resources (Sobrino, 1996; AMA, 1992) and which contribute to the protection of seagrass beds damaged through fishing activities. Several studies have proved the efficiency of artificial reefs in discouraging trawling.

Currently several Sector Plans for the Arrangement of Fishing Resources are being carried out. They include a strict regulation on activities, in order to prevent the depletion of fishing resources. These are still under strong pressure; at this moment measures exist to prohibit the catching of juvenile fish and the establishment of closed seasons for numerous commercial species. An important example is the Arrangement and Regulation plan for fishing activity in the estuary of the Guadalquivir River, one of the coastal areas of greatest interest in the Gulf of Cadiz, since its ecological value affects directly the southern Atlantic coast of Spain. This estuary is a habitat of great biological diversity, which serves as a breeding and feeding ground for juvenile fish. A variety of species have been exploited in a non-aggressive manner by traditional fishing. But lately fishing effort has intensified, especially of *Palaemon* sp. and baby eels (*Anguilla anguilla* juveniles) (species of great economic value), with the help of non-selective fishing gear which also catches juvenile fish of other commercially exploited species. This directly affects the availability of fishing resources on the coastal platform of Cadiz and Huelva.

Regarding the coastal environment, several measures have been developed to protect and plan the different littoral resources. Numerous programmes and actions affecting the whole Andalusian coast are being implemented. The following are the main ones:
- water treatment and sewerage programmes, to install sewage farms in coastal localities that presently dump their waste on protected natural areas;
- monitoring and control plans for industrial liquid waste to improve the quality of coastal waters, to be applied in the main contaminated areas of Huelva and Bay of Cadiz;
- coastal water regulations, with the purpose of controlling authorised dumping at sea, dumping quantities, surveillance and control of the coast (Consejería de Medio Ambiente, 1998a).

All current legal measures are not sufficient to conserve and protect natural areas of high interest. Surveillance and control measures are especially important to keep an adequate balance between conservation and exploitation in the next millennium, since many illegal activities are still being practised in all sectors affecting the environment.

REFERENCES

AMA (Agencia de Medio Ambiente: Environmental Agency) (1992) *Informe. Medio Ambiente en Andalucía.* Junta de Andalucía.

AMA (1994) *Informe, Medio Ambiente en Andalucía.* Junta de Andalucía.

AMA (1996) *Recursos Naturales de Andalucía.* Dirección General de Planificación. Junta de Andalucía. Sevilla.

Anfuso, G. Benavente, J., Peyes, J.L., Gracia, F.J., López-Aguayo, F. and Andrés, J. (1997) Towards a sustainable beach nourishment: goals and failures in the Bay of Cadiz (SW Spain). *VI International Europen Union for Coastal Conservation Conference (Abstract).* Italy.

Benavente, J., Reyes, J.L., Anfuso, G., Gracia, F.J. and López-Aguayo, F. (1997) Respuesta diferencial en playas regeneradas de la Bahía de Cadiz. *IV Jornadas españolas de ingeniería de puertos y costas. Volumen III.* Universidad Politécnica de Valencia. Valencia.

Bernal, M.A., Vallespin, C., Villar, N., Reglero, P., Esteban, C., Pumar, J.C., Bravo, R., Gómez-Cama, C., Soriguer, M.C. and Hernando, J.A. (1997) Macrofauna intermareal en los corrales de Rota (SW, de España). *2nd Symposium on the Atlantic Iberian Continental Margin.* Cadiz.

CAP (Consejería de Agricultura y Pesca: Agriculture and Fishing Department of the Regional Government) (1995) *La agricultura y pesca en Andalucía, Memoria 1994.* Junta de Andalucía. Sevilla.

CAP (1997) *Plan de modernización del sector pesquero andaluz.* Junta de Andalucía. Jaén.

Capel Molina, J.J. (1975) Tipos de tiempo de invierno en la Andalucía atlántica. *Bol. R. Soc. Geogr.,* **91** (1–2), 7–63.

Capel Molina, J.J. (1981) *Los Climas de España.* Oikos-Tau. Barcelona.

Castillo, J., Luque, C.J., Castellanos, E.M. and Figueroa, M.E. (2000) Causes and consequences of saltmarsh erosion in an Atlantic estuary, Huelva, Spain. *Journal of Coastal Conservation* (In press).

Consejería de Medio Ambiente: Environmental Department of the Regional Government (1998a) *Informe Ambiental 1997.* Medio Ambiente en Andalucía. Junta de Andalucía. Madrid.

Consejería de Medio Ambiente (1998b) *Informe al Parlamento de Andalucía sobre las consecuencias de la rotura de la balsa de estériles de las minas de Aznalcóllar.* Junta de Andalucía.

Consejería de Medio Ambiente (1999) *Informe Ambiental 1998.* Medio Ambiente en Andalucía. Junta de Andalucía. Madrid.

Dabrio González, C.J. (1980) Dinámica costera en el Golfo de Cadiz. Sus implicaciones en el desarrollo socioeconómico de la región. In: *Reunión Nacional de Geología Ambiental y Ordenación del Territorio.* Santander, pp. 337–356.

Delaune, R.D. and Gambrell, R.P. (1996) Role of sedimentation in isolating metal contaminants in wetland environments. *Journal of Environmental Science and Health* **31** (9), 2349–2362.

Figueroa, M.E., Cota, H. and García Novo, F. (1980) Conservación de ecosistemas del litoral suratlántico de España. *I Reunión Iberoam. Zool. Vert.* La Rábida, 1977.

Fraidias, J., Leal, A. and Serrano, J. (1997) *Modelo de gestión del estuario del Río Guadiana.* Consejería de Medio Ambiente. Junta de Andalucía. Sevilla.

García Novo, F. and Merino, J. (1993) Dry coastal ecosystems of southwestern Spain. In: *Dry Coastal Ecosystems. Polar Regions and Europe,* ed. E. van der Marel. Elsevier. London.

García Novo, F. and Merino, J. (1997) Pattern and process in the dune system of the Doñana National Park, southwestern Spain. In: *Ecosystems of the World 2C. Dry Coastal Ecosystems. General aspects,* ed. E. van der Marel. Elsevier. Amsterdam.

Granados, M., Martín, A., Fernández Alés, R. and García Novo, F. (1984) Etude diachronique dún écosystème à longue échelle. La pinède de Marismillas (Parc National de Doñana). *Mélanges de la Casa de Velazquez* **20**, 393–418.

IGME (1982) Mapa Geológico de España. Hoja de Nerva (no. 938). E. 1:50.000. Ministerio de Industria y Energía.

IEA (Instituto de Estadística de Andalucía: Statistical Institute of Andalusia) (1998) *Anuario Estadístico de Andalucía*. Junta de Andalucía. Sevilla.

Jiménez-Nieva, F.J. and Luque, C.J. (1995) Odiel Marshes. In: *Management of Mediterranean Wetlands*, eds. C. Morillo and J.L. González. Dirección General de Conservación de la Naturaleza, Mº de Medio Ambiente. Tomo 3. pp. 381–393.

López Cotelo, I. (1993) *Recursos marinos del Golfo de Cadiz. Litoral de Huelva*. Consejería de Agricultura y Pesca. Junta de Andalucía. Sevilla.

Luengo Acosta, F., Asensio Romero, B., Ávila Puyana, M.A. and Lozano Única, J. (1998) *Andalucía bajo el mar. Guía del buceador*. Junta de Andalucía. Málaga.

Luque, C.J. (1996) *Typification, cartography, and heavy metal content of plant communities in the Odiel Saltmarshes. Population dynamics of different species of the genus Spergularia*. Ph.D. Thesis, University of Seville, Spain. (In Spanish). (Unpublished).

Luque, C.J., Castellanos, E.M., Castillo, J.M., González, M., González-Vilches, M.C. and Figueroa, M.E. (1998) Distribución de metales pesados en sedimentos de las marismas del Odiel (Huelva, SO. España). *Cuaternario y Geomorfología* **12** (3–4), 77–85.

Luque, C.J., Castellanos, E,.M., Castillo, J.M., González, M., González Vilches, M.C. and Figueroa, M.E. (1999) Metals in halophytes of a contaminated estuary (Odiel Saltmarshes, SW Spain). *Marine Pollion Bulletin* **38** (1), 49–51.

Menanteau, L. and Martin Vicente, A. (1979) Environment et tourisme. Exemple de la Costa de la Luz (Andalousie atlantique). Tourisme et développement régional en Andalousie. *Casa de Velazquez, Série Rech. Sci. Soc.* **5**, 241–310.

Molina Vázquez, F. (1998) Espacios marítimos protegidos de Andalucía. In: *Regiones y ciudades enclaves, relaciones fronterizas, cooperación técnica y el desarrollo en Iberoamérica y Mar de Alborán*, ed. Rafael Cámara, pp. 111–128. Librería Andaluza. Sevilla.

Muñoz Pérez, J.L. and Sánchez de Lamadrid, A. (1994) *El medio físico y biológico en la bahía de Cadiz: Saco interior*. Ed. Junta de Andalucía-CAP. Sevilla.

Ojeda, J., Sánchez, E., Fernández-Palacios, A. and Moreira, J.M. 1995. Study of the dynamics of estuarine and coastal waters using remote sensing: the Tinto-Odiel estuary, SW Spain. *Journal of Coast. Conservation* **1**, 109–118.

Peris-Mora, E. (1994) *Contaminación y otras agresiones sobre el litoral de las diez comunidades costeras españolas*. Coastwatch-1993. Ed. Universidad Politécnica de Valencia.

Prieto, L., García, C.M., Corzo, A., Ruíz, J. and Echevarría, F. (1997) Estructura hidrológica, bacterioplancton, fitoplancton y actividad nitrato reductasa en el Estrecho de Gibraltar y áreas adyacentes en período de estratificación. 2nd Symposium on the Atlantic Iberian Continental Margin. Cadiz.

Reyes, J.L., Benavente, J., Gracia, F.J. and López-Aguayo, F. (1996) Efectos de los temporales sobre las playas de la Bahía de Cadiz. *IV reunión de Geomorfología*, eds. A. Grandal dAnglade and J. Pagés Valcarlos. Sociedad Española de Geomorfología. El Castro. La Coruña.

Ramos, F., Sobrino, I. and Jiménez, M.P. (1996) *Cartografía de especies y caladeros "Golfo de Cádiz"*. CAP. Junta de Andalucia.

Sobrino, I. (1996) *Estudio de los componentes de la flota de arrastre que opera en el Golfo de Cadiz. Cartografiado de los recursos explotados por dicha flota*. Informe final. Instituto Español de Oceanografía. Madrid.

Suárez de Vivero, J.L. (1988) Fondos marinos. In: *El Litoral*. CETU. Junta de Andalucía.

Vanney, J.R. and Menanteau, L. (1979) Types de reliefs littoraux et dunaires en Basse Andalousie. *Mélanges Casa Velázquez* **15**, 5–52.

THE AUTHORS

Carlos J. Luque
Departamento de Biología Vegetal y Ecología,
Facultad de Biología,
Universidad de Sevilla, Apdo 1095,
41080 Sevilla.
Spain

Jesús M. Castillo
Departamento de Biología Vegetal y Ecología,
Facultad de Biología,
Universidad de Sevilla, Apdo 1095,
41080 Sevilla, Spain

M. Enrique Figueroa
Departamento de Biología Vegetal y Ecología,
Facultad de Biología,
Universidad de Sevilla, Apdo 1095,
41080 Sevilla, Spain

Chapter 12

THE CANARY ISLANDS

Francisco García Montelongo, Carlos Díaz Romero and
Ricardo Corbella Tena

The Canary Islands of Spain are of volcanic origin, situated in the north Atlantic between 39°45′ and 14°49′N, and between 31°17′ and 13°20′W. They are markedly different from the nearest part of the continent of Africa in spite of their proximity to it, with rugged shorelines surrounded by an irregular oceanic seabed. Their sublittoral slopes are steep and great depths are found not far from the shoreline. Thus, the area for primary producers is small and production capacity is limited. Rocky seabeds are found everywhere, while sandy, clay or shell sediments are rare though they are more common in the eastern islands. The rocky bottoms are very uneven and craggy, with many shallows, cliffs and underwater caves.

Their position in a subtropical area near the west of Africa has resulted in a unique climate. They are alternatively under the influence of the subtropical hot high pressures causing a stable climate, and of less frequent polar low pressures, which bring unstable rainy weather. The Azores anticyclone is the primary factor resulting in the Trade Winds and stable weather, but when high pressure moves towards the west, outbreaks of colder air masses from the north and, less frequently, hotter outbreaks from the Sahara may reach the islands. This climate together with the influence of the Canary Island Stream gives the islands their mild temperatures and low daily variation which are so favoured by visitors.

The pelagic system is affected by the presence of different oceanic species from tropical and from temperate seas. This brings a relatively large marine biomass to the islands. The pelagic coastal system may also be classified into a group of species living nearer the coast, and another belonging to oceanic waters which go to coastal waters to reproduce and grow. Another unique feature of the Canarian coastal ecosystem is that many deep-water species live very close to the coast and rise to surface at night. These usually deep species have become an integral part of this shallow ecosystem. This phenomenon is almost unique in calm waters.

The most common communities in this area are those of rocky shores which are adapted to extremes of moisture, salinity, abrasion and light intensity. Production per unit area is high, though the total area is low. Fishing resources are made up of a great variety of species, most of them endemic, but low in number. Fishing for shellfish is a traditional activity even though production is relatively low. In some places this resource is over-fished, but most of these species reproduce in the winter and grow quickly, so that this over-exploitation may be replenished fairly quickly.

Development in the last decades has favoured intensive and indiscriminate use of the limited marine resources of these islands. Only in 1986 did the Local Government of the Canary Islands establish rules to protect and recover marine resources. However, after ten years, only a few protective actions are well established and results are scarce so far. Another important problem for the coastal ecosystem comes from displacement of the population due to the increasing

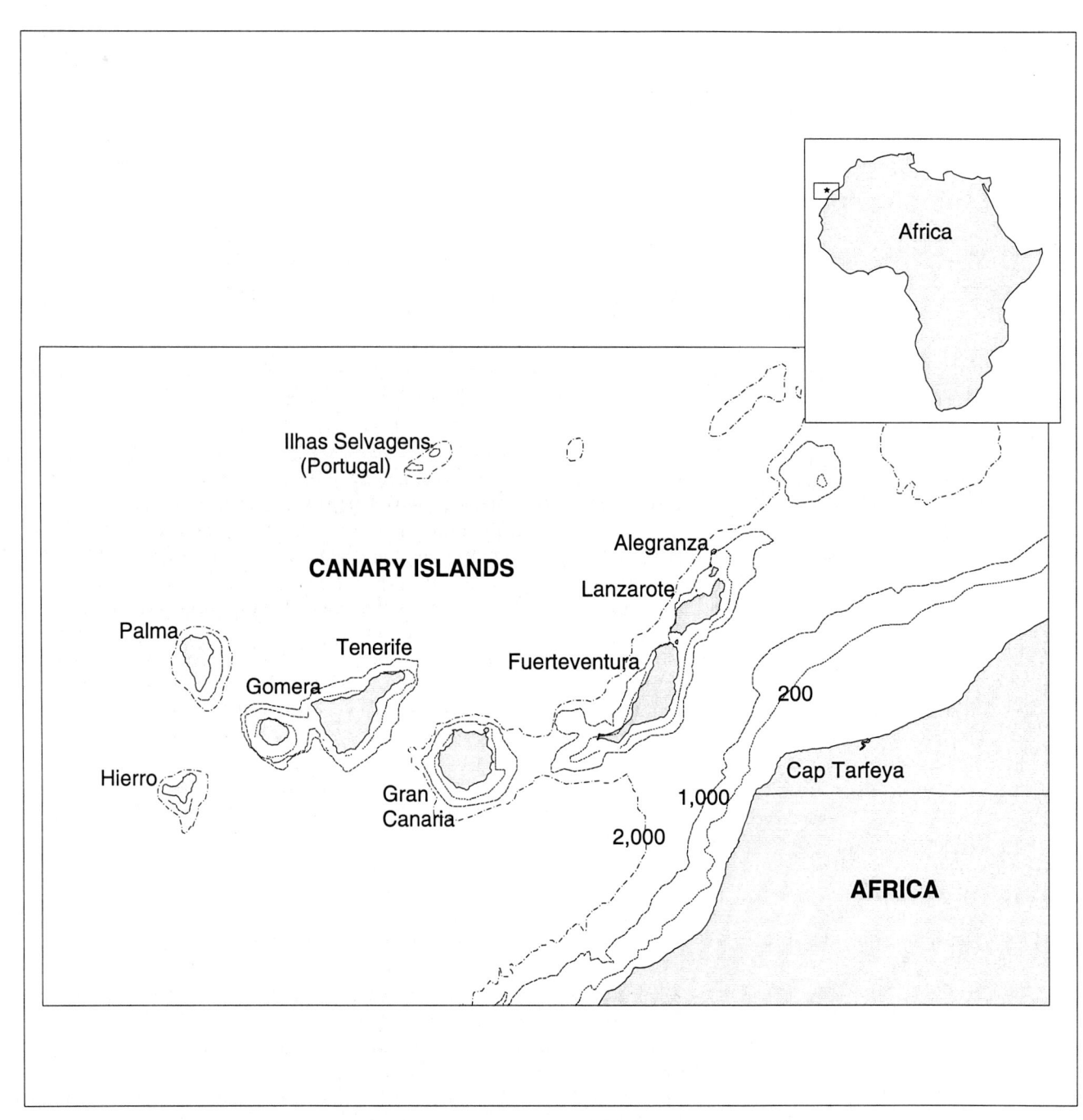

Fig. 1. Map of the Canary Islands, showing bathymetry.

demands and pressures of tourism in coastal tourist resorts. Sewage disposal occurs by outfalls into coastal waters, though in the last few years public opinion has demanded action to overcome often severe pollution of bathing beaches, so that the main cities such as Santa Cruz de Tenerife now have large coastal sewage treatment plants, whose treated waters are also used for agricultural purposes.

PHYSICAL AND GEOLOGICAL BACKGROUND TO THE REGION

The Canary Islands are a group of islands whose volcanic origin has created the rugged shoreline and the irregular oceanic seabed around them. They are situated 111 km off the northwestern coast of Africa (Fig. 1 and Table 1). These islands are markedly different from the nearest part of the continent of Africa in spite of their proximity to it. Their geological evolution is quite independent, not only in terms of time (they were formed a hundred million years after the African continent), but also in terms of their nature as the islands are totally volcanic in origin. The Canary Islands are oceanic islands and not continental ones (Bravo, 1954, 1970; Báez et al., 1982). The evolution of their flora and fauna is also independent of Africa's, and subject to special climatic conditions (Bacallado Aránega, 1984).

The Canary Islands' position between 27°37' and 29°25'N and 13°20' and 18°10'W means that it lies on the southern border of the temperate area of the Northern hemisphere on the Tropic of Cancer. There are seven major islands (Tenerife, Gran Canaria, Fuerteventura, Lanzarote, La Palma, La Gomera and El Hierro), several small islets (Alegranza, Graciosa, Montaña Clara and Lobos) and several rocks (Roque del Este, Roque del Oeste, Roques de Anaga, etc.), covering 7542 km² ranging in size from El Hierro 278 km² to Tenerife 2058 km². There is a total of 1490 km of coastline (Table 2). The highest peak in the islands, is Pico del Teide in Tenerife, which has an altitude of 3,717 m.

These islands have emerged from the ocean as a result of continental drift and the accumulation of lava from several volcanic eruptions (18 eruptions in the last 500 years), although there is still controversy about this (Afonso, 1989–1992). The islands have emerged independently from each other, and there are depths of 2,000 m in the channels between them, with the exception of Tenerife–La Gomera and Gran Canaria–Fuerteventura–Lanzarote, and between Lanzarote–Fuerteventura and the northwestern Africa coast where depths of up to 1,500 m can be found (Fig. 1).

The seabed between the Makaronesy Archipelagos is also deep. The deepest point between the Canaries and the Salvages is more than 3000 m, and between Canaries/Salvages and Madeira is more than 4500 m. In this area are other geological formations including banks and ridges (Dacia, Concepción) and underwater mountains (Great Meteor, Cruisier, Atlantis, Ampère, Seine, Josephine, Endeavour). Some of these are extremely steep with peaks not very far below the surface, i.e., Dacia (–86 m), Ampère (–60 m), Seine (–146 m), and Concepción (–161 m). The geomorphological characteristics of several of these structures suggest that they rose above sea-level in the distant past and disappeared under the surface as a consequence of wave erosion (Bravo, 1954, 1970; Alfonso, 1989–1992).

The small land area of these islands makes their slopes very steep and, therefore great depths can be found not far from the shoreline. Furthermore, their relatively short existence means that their surrounding, submerged platforms are small (Table 2), not more than 200 m wide around most of the islands, and these platforms have also been further reduced in size by later lava flows from successive volcanic eruptions. However, the older islands (Gran Canaria, Gomera, Fuerteventura and Lanzarote) have larger platforms towards their south-southeastern coasts, as much as 30 km wide around Fuerteventura. Thus, the area where primary producers live (algae and phanerogama) is small and production capacity is limited and different for every

Table 1

Physical data on the archipelagos of the Makaronesy archipelagoes (data for 1984 and *1996)

Name	Age ($\times 10^{-6}$ years)	Area (km²)	Maximum altitude (m)	Distance to nearest continent (km)	Inhabitants	Mean year temperature (°C)	Rain (mm/year)
Azores	3–7	2344	2351	1600	300,000	17.3	968
Salvajes	no data	4	154	350	0	no data	no data
Madeira	60–70	810	1816	700	280,000	18.8	602
Canaries	35–40	7542	3717	111	1,605,400*	21.1	290
Cape Verde	>100?	4033	2829	500	300,000	17.7	266

Table 2
Some geographical characteristics of the Canary Islands

Island	Area (km²)	Coast (km)	Coastal platform (km²)*	Age (×10⁻⁶ years)	Maximum height (m) /Name	Inhabitants (1996)
Tenerife	2,034	358	315	7.2	3,717/Pico del Teide	665,562
Fuerteventura	1,660	326	695	16.6	807/Pico de la Zarza	41,629
Gran Canaria	1,560	236	324	13.9	1,959/Pozo de las Nieves	714,139
Lanzarote	846	213	461	19.0	671/Peñas del Chache	77,233
La Palma	708	155	152	1.6	2,426/Roque de los Muchachos	81,521
La Gomera	360	97	216	12.0	1,487/Garajonay	16,978
El Hierro	268	166	93	0.7	1,501/Malpaso	8,338
La Graciosa	27	266	–	–	166/Las Agujas	500
Alegranza	12	289	–	–	269/La Caldera	–
Lobos	6	122	–	–	122/La Caldera	–
Montaña Clara	1	256	–	–	25/La Marina	–
Roque del Este	<1	–	–	–	–	–
Roque del Oeste	<1	–	–	–	–	–

*Down to –50 m.

island according to the size of its continental platform (Carrillo et al., 1985; Aguilera Klint, 1994).

Due to the steepness of the islands, rocky seabeds are found everywhere while sandy, clay or shell sediments are rare, though they are more common in the eastern islands. The rocky bottoms are very uneven and craggy, with many shallows, cliffs and underwater caves. Sandy and pebble beaches (18% of the coast) adjoin larger platforms, while most of the coast and seabed is rocky or cliff, generating a wide variety of ecosystems. In spite of the erosive action of the sea, the intertidal platforms are also very small, which is due not only to their youth but also to the effects of the most recent volcanic materials arriving onto the coast from the nearby volcanos (Bravo, 1970).

Several factors are responsible for the climate around of the Canary Islands: (a) the atmospheric conditions at these subtropical latitudes, (b) the influence of its own relief, (c) the presence of a cold oceanic stream and, (d) their proximity to a continent. Thus, their position in a subtropical area and near the west of Africa has resulted in a unique climate. Essentially, they are alternately under the influence of the subtropical hot high pressures (anticyclone), causing a stable climate, and of the less frequent polar low pressures, which bring unstable rainy weather. As a consequence of high pressures to the north of the islands, the Azores anticyclones, also well known as the Trade Winds, flow towards the Canaries. The Trade Winds present markedly seasonal variations. They predominate in summer (90–95%) and are less frequent in winter (50%), and result in stable weather with stratus-cumulus clouds to the north of the islands and clear skies toward the south. Another consequence of the Trade Winds is the stratification of the low troposphere into two layers, the lower layer being wet and fresh and the upper layer dry and hot. This thermal inversion layer is found between 950 m and 1500 m above sea-level and forms 'a sea of stratus-cumulus clouds' towards the north of the islands with the highest altitudes, where the altitude blocks the passage of wind (Fig. 2). The Azores anticyclone is the primary factor resulting in the Trade Winds and the stable weather, but when high pressure moves towards the west, outbreaks of colder air masses from the north and, less frequently, hotter outbreaks from the Sahara may reach the islands. This atmospheric phenomena, together with the influence of the Canary Island Stream, gives the islands their unique climate with mild temperatures and low daily variations (Table 3) (Afonso, 1989–1992).

Because of its peculiar climate, these islands are a very popular tourist destination, especially in autumn and winter. As an example, Tenerife, with 665,562 inhabitants, received as many as 4,125,468 tourists and the archipelago as a whole had as many as ten million tourists in 1996. This massive tourism has basically been concentrated in resorts

Fig. 2. The 'sea of clouds' which is the result of Trade Winds in the northern part of Tenerife.

Table 3

Maximum (M) and minimum (m) temperatures in °C at three locations in Tenerife

Location		J	F	M	A	M	J	J	A	S	O	N	D	Mean
S.C. Tenerife	M	20.4	20.6	21.7	22.7	23,8	26.0	28.3	28.8	27.7	26.0	23.6	21.4	24.2
(sea level)	m	14.4	14.3	14.7	15.7	16.9	18.5	20.1	20.7	20.5	19.4	17.4	15.5	17.3
Los Rodeos	M	15.1	15.5	16.8	17.1	18.1	19.3	21.9	23.3	22.9	21.4	18.5	15.6	18.7
(617 m)	m	8.4	8.2	8.7	9.5	11.2	12.6	14.3	15.2	14.9	13.7	11.5	9.4	11.5
Izaña	M	6.9	7.4	9.5	11.1	13.7	17.9	21.8	21.7	17.9	13.4	10.0	7.4	13.2
(2,367 m)	m	0.8	0.6	1.8	3.0	5.2	9.1	3.0	13.0	9.8	6.3	3.5	1.4	5.6

on the coastline which has therefore been over-exploited. During the same period as the development of the coastline, other socio-economic changes have taken place in the Archipelago. For economic reasons the population has shifted from traditional agricultural work to the new tourist-related activities. The effect of this has been to put great pressure on the coastal marine ecosystems as there are now new artificial beaches, sporting marinas, fishing harbours, and boat trips to see attractions such as the whale *Globicephala macrorinchas* and the dolphin communities found in the waters around the islands and cliff zones towards the northwestern of Tenerife. Only very recently have local authorities implemented measures to control this over-exploitation (Bacallado et al, 1989).

SEASONALITY, CURRENTS, AND NATURAL ENVIRONMENTAL VARIABLES

In spite of its geographical situation in the subtropics, the physico-chemical oceanographic parameters of the sea, such as salinity, temperature, etc., are quite different from other seas in the subtropics because the Islands are influenced by the Canary Island Stream as well as by the upwelling (Mascareño and Molina, 1970) which originates along the northwestern coast of Africa.

The anticyclonic current system in the North Atlantic Ocean—the Gulf Stream—carries warm waters towards Europe, but on turning south before the Bay of Biscay, it starts cooling relative to oceanic waters around it and becomes colder still as it reaches the Equator. Thus, when the Canary Island Stream reaches the islands it can be considered as a cold current (Fig. 3). This current flows mainly south-southwest, parallel to the African coast. However, around the islands, cyclonic eddy currents appear as the islands and banks obstruct the stream. Therefore, west of Tenerife, an eddy stream going north can be found, and generally there are major variations in the currents around the islands. On many occasions during the year, especially around Tenerife, Fuerteventura and Lanzarote currents move parallel to the shoreline (Mascareño, 1972; Molina and Laatzen, 1986; Zenk et al., 1986).

As a result of the effect of the islands, current speed is generally low (Fig. 4), especially in summer, when the speed reaches a maximum of 25 m s^{-1}. However, this speed can increase to 60 m s^{-1}, when currents coincide wth tide flows. As usual, current speed diminishes as depth increases. The flux of the Canary Island Stream close to the north of the islands can reach a maximum of 6×10^6 m^3 s^{-1} in summer (Zenk et al., 1986).

Observations carried out recently have revealed a subsurface current near the northwestern slope of Africa, between Cape Verde and Cape Bojador, flowing between depths of 200 and 300 m, with a maximum speed about 15 m s^{-1}, influencing the waters at 100–500 m and to a distance of about 40 km from the continental slope. This current, sometimes referred to as the Canary Islands Countercurrent because it can reach the islands, especially the most easterly ones. These kinds of current have also been observed near other upwelling zones (Mascareño and Molina, 1970; Mascareño, 1972; Molina and Laatzen, 1986).

The Canary Islands, in common with the adjacent coasts of Africa, experience two tides a day. Every 24 h 50 min, two high and two low tides occur with similar amplitudes between consecutive tides. Maximum amplitude reached during the summer and winter equinoxes (March and September) can be as large as 3 m in this area. During the solstices, the minimum amplitude is about 0.7 m. However, the influence of the seabed and coastal topography as well as meteorological conditions, may influence the tides, generating local exceptions.

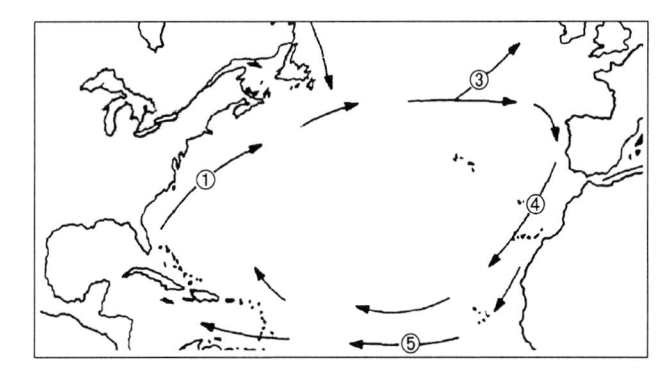

Fig. 3. Main surficial currents in the North Atlantic Ocean: (1) Gulf Stream (warm), (3) North Atlantic Stream, (4) Canary Islands Stream (cold), (5) North-equatorial Stream.

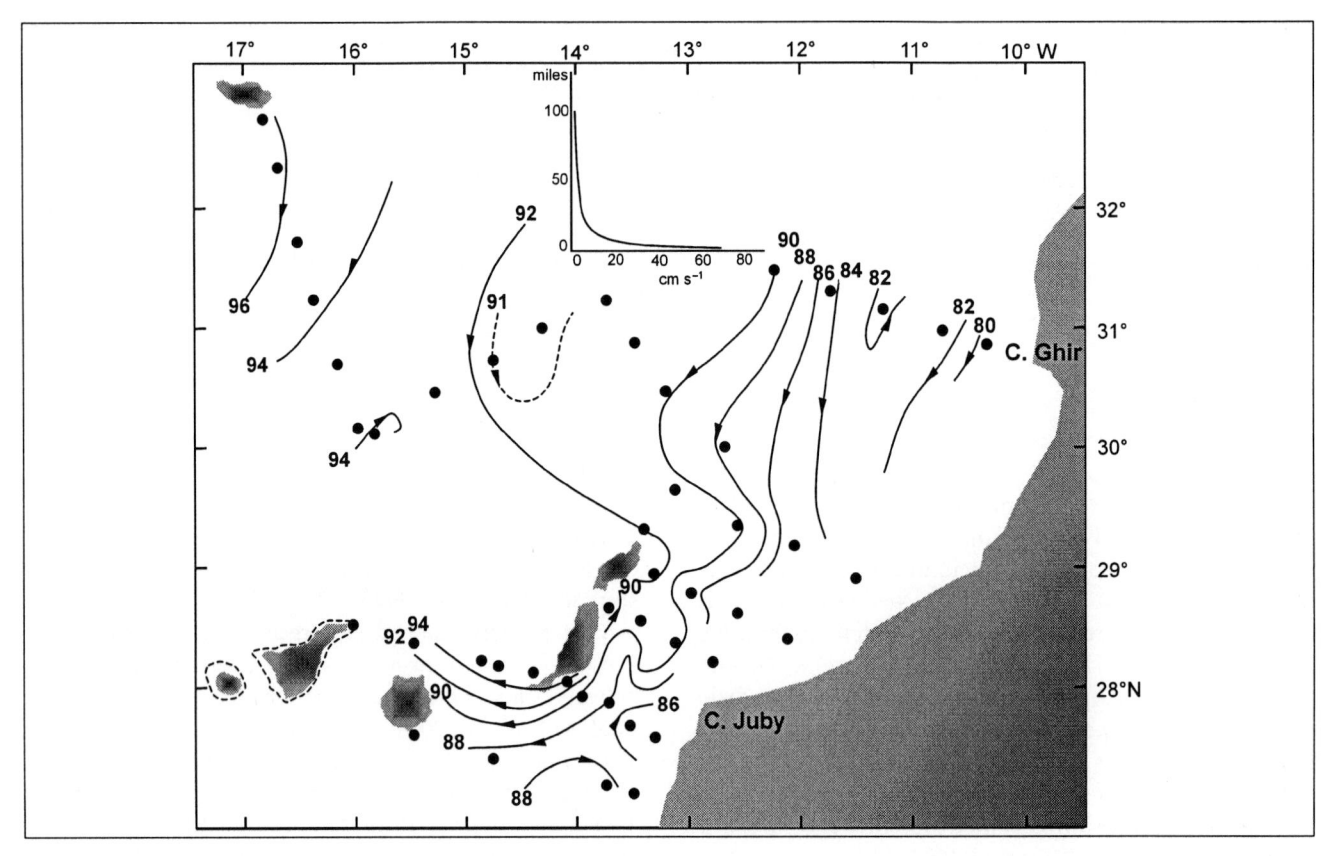

Fig. 4. Current speed around the Canaries.

In summer the windward coasts of the islands are continuously influenced by waves of not more than 3 m which arise from the action of the Trade Winds. In the autumn, as the Trade Winds weaken, these wind waves decrease and the water become very calm. However, North Atlantic lows travelling towards Europe cause waves up to 10 m high which may travel about 3500 km in 2.5–3.5 days from north-northwest to north-northeast, generating swell waves with an amplitude of about 3.5 m. When these lows travel close to the islands, they may cause infrequent southwesterly storms on their southwestern coasts (Afonso, 1989–1992).

On the northwestern coast of Africa and especially in summer, the Alisios Winds (Trade Winds) cause a net surface water movement perpendicular to the wind direction towards the open ocean which, in turn, causes cold water from as deep as 250 m to rise to the surface. This is known as the Northwestern African Upwelling, which is nutrient-rich and low in temperature and salinity. Unfortunately, the high nutrient content only influences biological activity near the African coast where there is an important fishing bank (the Saharan Fishing Bank). However, the low temperature and salinity influences the islands, with gradients parallel to the Equator (Afonso, 1989–1992).

Surface temperature, under the influence of this variable upwelling phenomenon, shows large spatial and temporal differences especially in the eastern islands nearest to the African coast. Generally, water down to a depth of 800 m shows a horizontal distribution of isothermal lines parallel to the African coast, with increasing temperatures towards the open sea. Thus temperature differences of up to 3°C can be found between the western and eastern waters, and differences exist of up to 5°C compared with adjacent African coastal waters. In open ocean surface water, the temperature reaches a maximum of about 25°C in September and October, while minimum temperatures of about 17°C occur in winter. Distribution of temperature with depth shows three intervals: (a) sea level down to 150 m with a large thermal variation over the year due to energy exchange with the atmosphere, (b) between 150 and 2000 m where temperatures go down from 15–19°C to 4°C, and (c) between 2000 and the bottom, with polar water inputs where temperatures are as low as 2.5°C (Figs. 5 and 7). At the end of spring and the beginning of summer, a seasonal thermocline at about 150–200 m can be observed, whose maximum is at the end of summer and which disappears by the end of autumn (Afonso, 1989–1992).

There is a slight variation in salinity in the first 100 m (36.10–36.90°). This is not as homogeneous as temperature, probably because of the input of subtropical surface waters coming from the Central North Atlantic Ocean, which reach the islands after some mixing and sinking. At depths of over 100 m there is a large decrease in salinity (35.20–35.50° at 800 m) which then increases to 35.30–35.70°

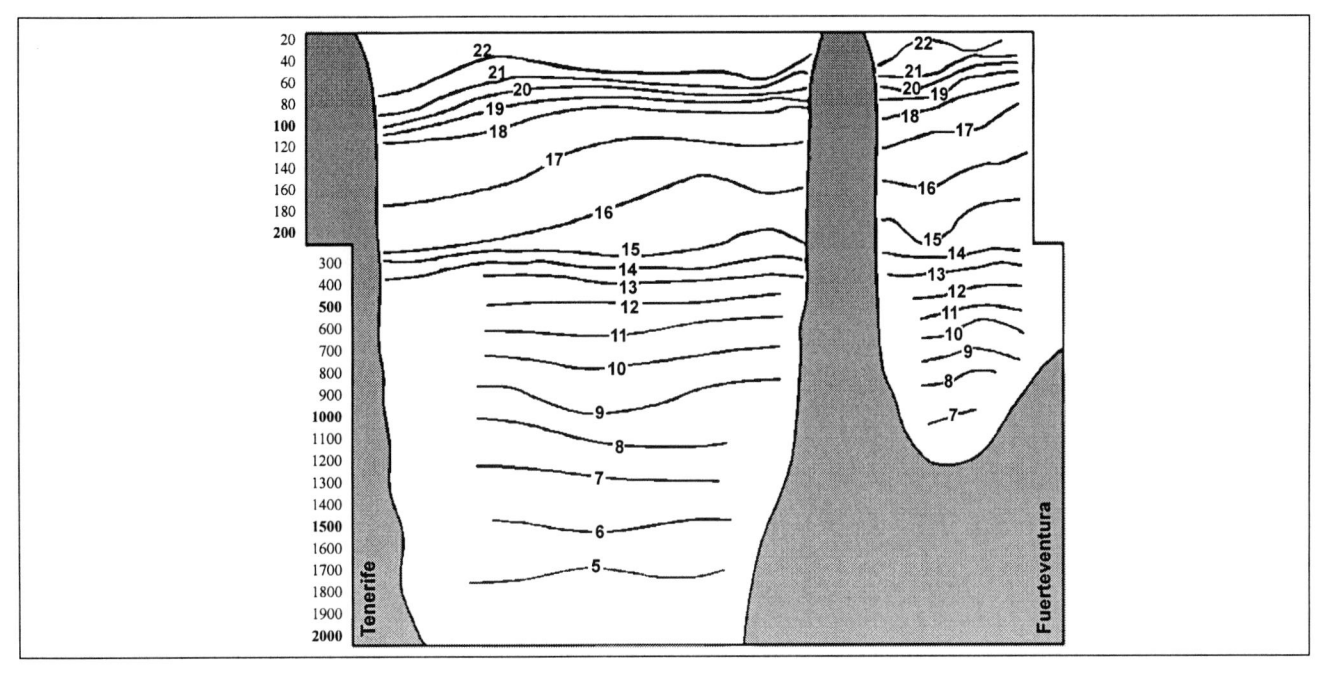

Fig. 5. Temperature variation with depth (September).

between 1,100 and 1,300 m caused by waters exiting the Mediterranean Sea. Salinity further decreases, especially below 2,000 m, with values as low as 34.90° near the ocean bed (Figs. 6 and 7) (Afonso, 1989–1992).

Dissolved oxygen, 5.00–5.50 cc O_2 l^{-1}, in these waters and down to 100 m deep, is supersaturated. Values decrease with depth and may fall to 3.5 cc O_2 l^{-1} at 700–1000 m, increasing again to levels similar to surface levels near the bottom (Fig. 8). This minimum level of dissolved oxygen is associated with the oxidation of suspended organic matter. However, a different situation is found between Lanzarote and Fuerteventura and the northwestern coast of Africa where the upwelling causes dissolved oxygen to be lower than corresponding levels near the western island waters.

The absence of a continental platform and the oceanic characteristics of these islands means that the surrounding waters are oceanic in their nature, and of course they are oligotrophic (Fig. 9). Nutrient concentrations found down to the seasonal thermocline range from 'not detected' to: phosphate 0.18 μatom-g P l^{-1}, silicate 2.0 μatom-g Si l^{-1}, and nitrate 2.5 μatom-g N l^{-1}. Normal nitrite levels are about 0.10 μatom-g N l^{-1} with a maximum level of about 0.30 μatom-g

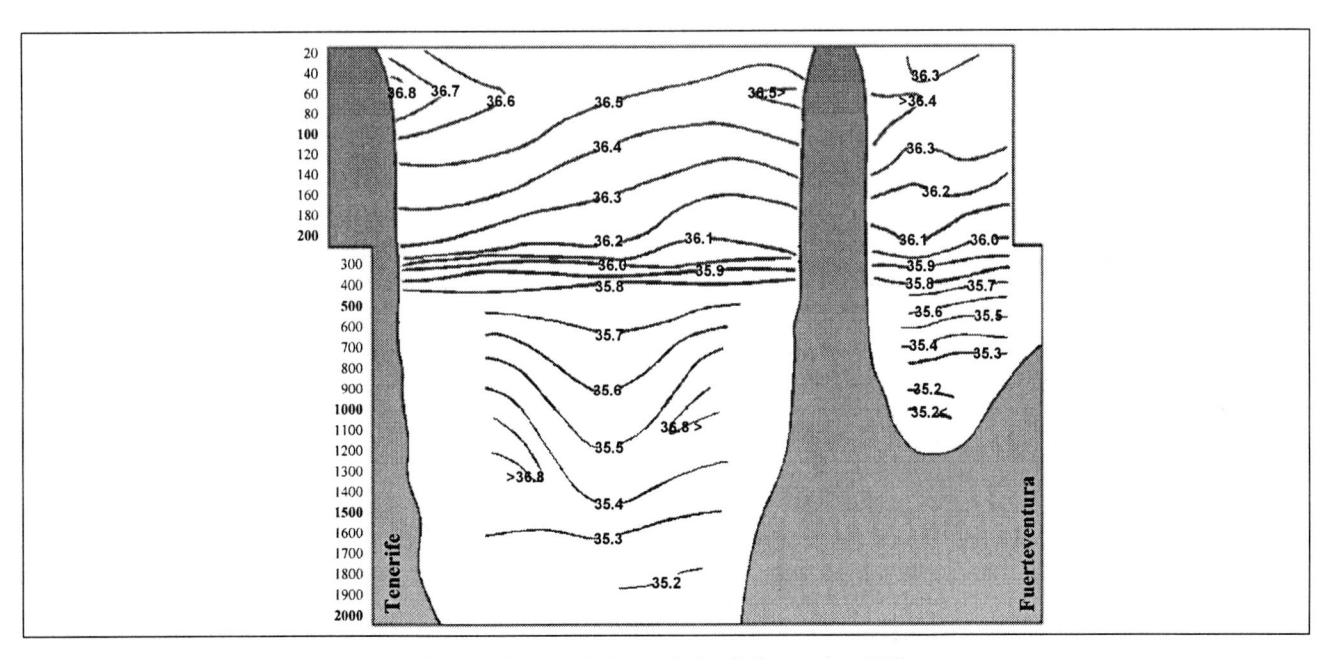

Fig. 6. Salinity variation with depth (September 1979).

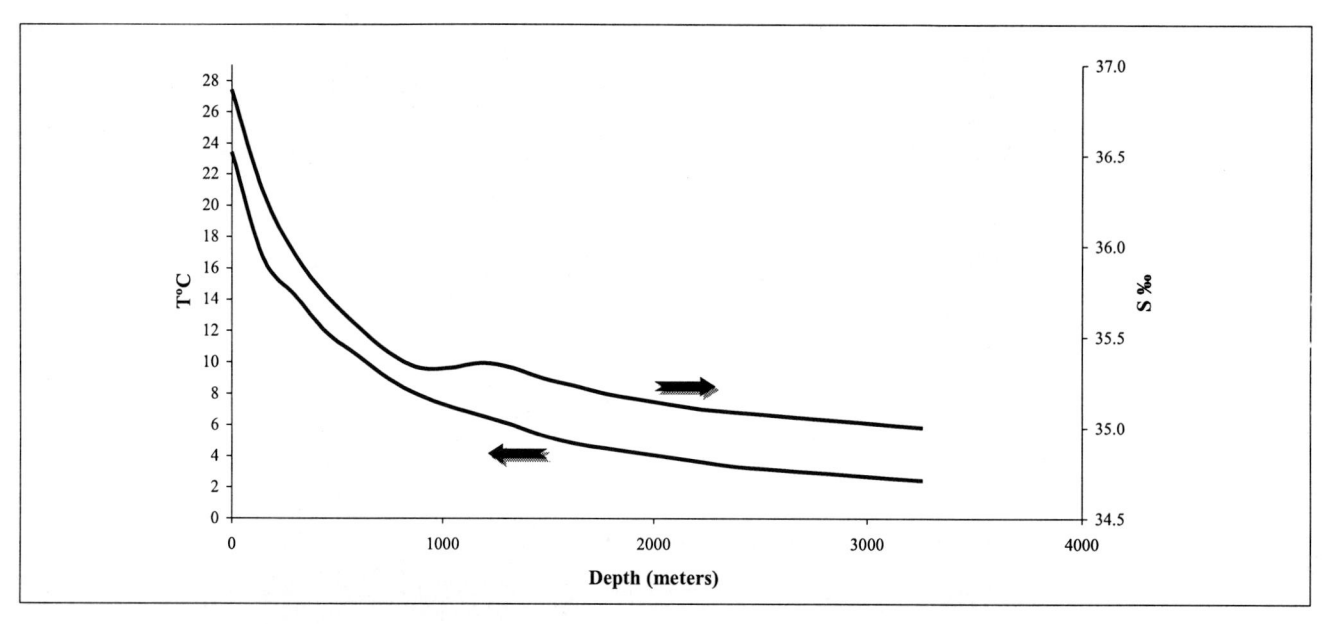

Fig. 7. Temperature and salinity variation with depth.

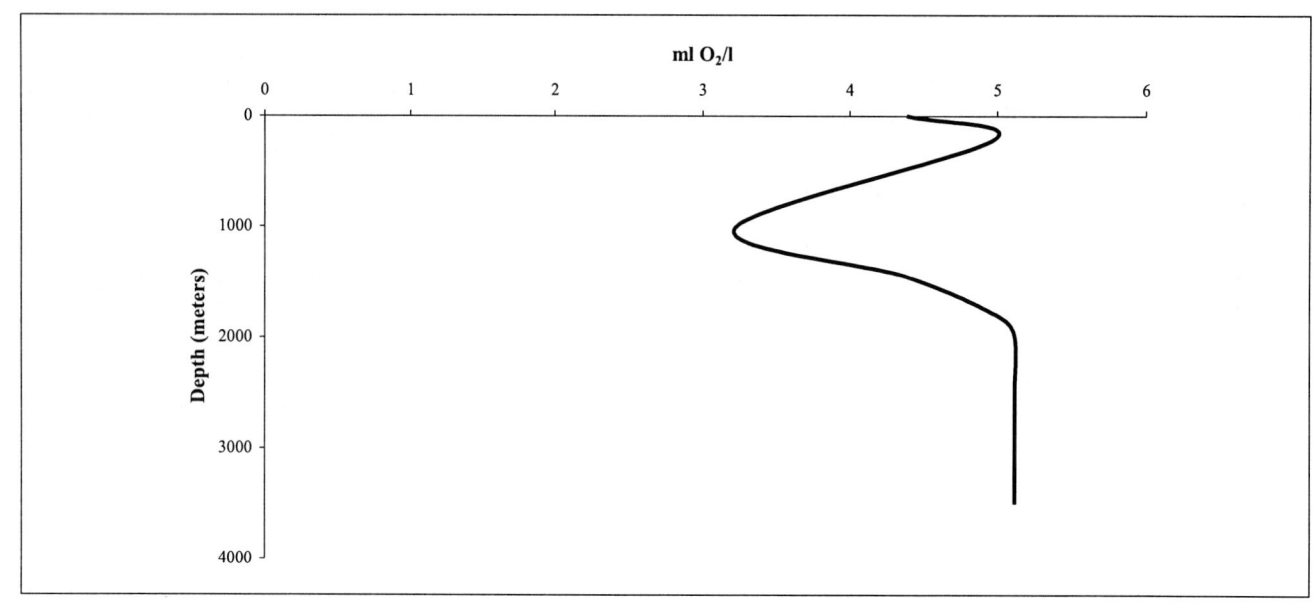

Fig. 8. Oxygen concentration variation with depth.

N l^{-1} associated with the seasonal thermocline (Fig. 9), levels which can be attributed to nitrification processes in the ocean. Ammonia is detected with a maximum level of 1.0 μatom-g N l^{-1} and its vertical distribution is very irregular. Under typical oceanic conditions, organic matter is found up to 115 mg-Cl^{-1}, of which 2.2% is phytoplankton, and 2.1% zooplankton. Taking 0.02% as fish content, 95.7% of the organic matter exists as suspended, microscopic particulate matter (Afonso, 1989–1992; Llinás, 1988; Llinás et al., 1996).

Phytoplankton around the Canary Islands is very variable on and near the surface as well as in the deep and depends on the seasons. Thus, in springtime when phytoplankton blooms, chlorophyll reaches a maximum, 1.0–1.5 mg ml^{-1}, which diminishes with depth (Fig. 10). As the bloom dies away, chlorophyll content in the surface waters goes down to about 0.1 mg ml^{-1}, and a maximum of 0.3–0.5 mg ml^{-1} is reached at the depth of the seasonal thermocline (Braun and Real, 1981). Most of the phytoplankton organisms, up to 80%, are smaller than 10 μ diameter. The number of zooplankton individuals are about 250 organisms m^{-3} in the first 200 m, and 170 organisms m^{-3} down to 500 m, with mean values of 1 mg-C m^{-3}. This is highly variable over the year, so primary production varies greatly over the year with a mean value of 300 mg-C m^{-2} d^{-1}, and a mean secondary production of zooplankton of 70 mg m^{-2} d^{-1}, giving a mean efficiency in the trophic web transfer of 23% (Braun, 1981; León and Braun, 1973).

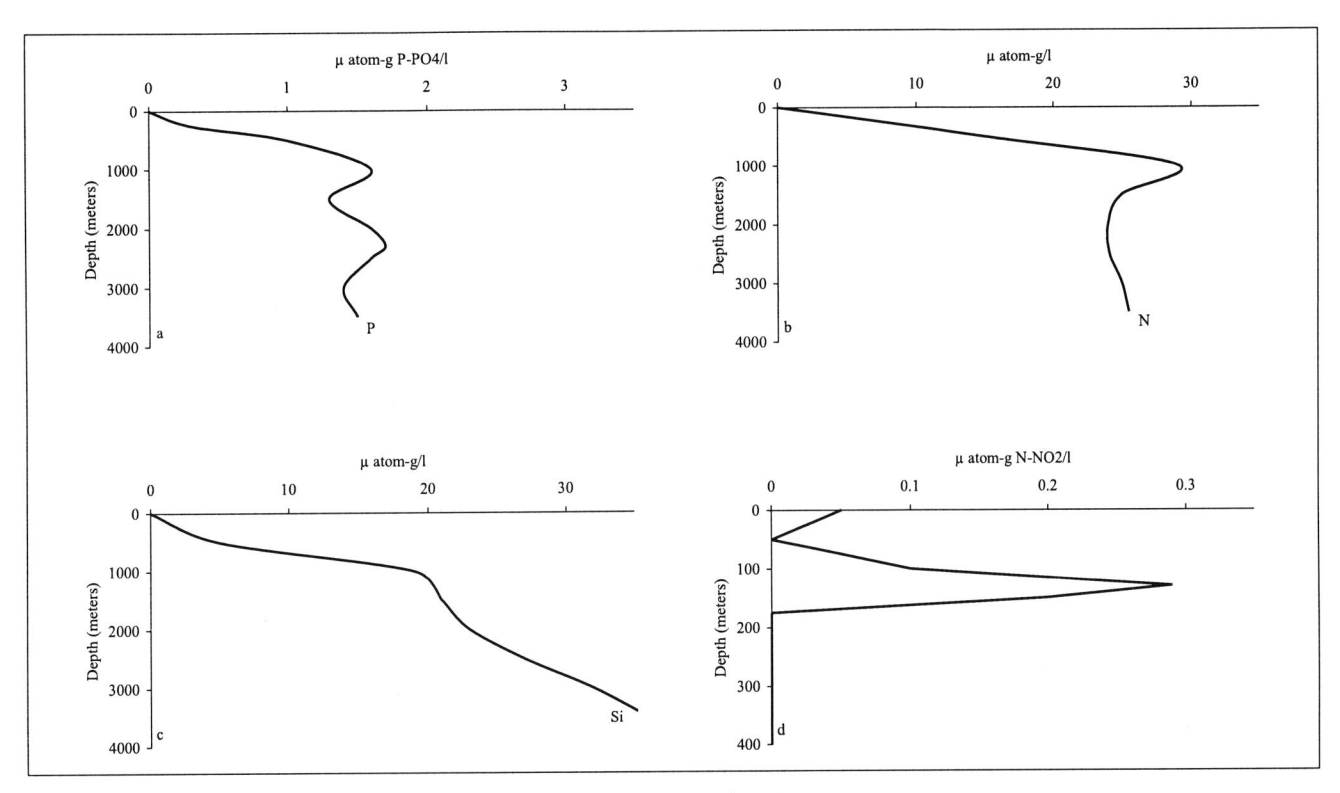

Fig. 9. Nutrients and nitrite.

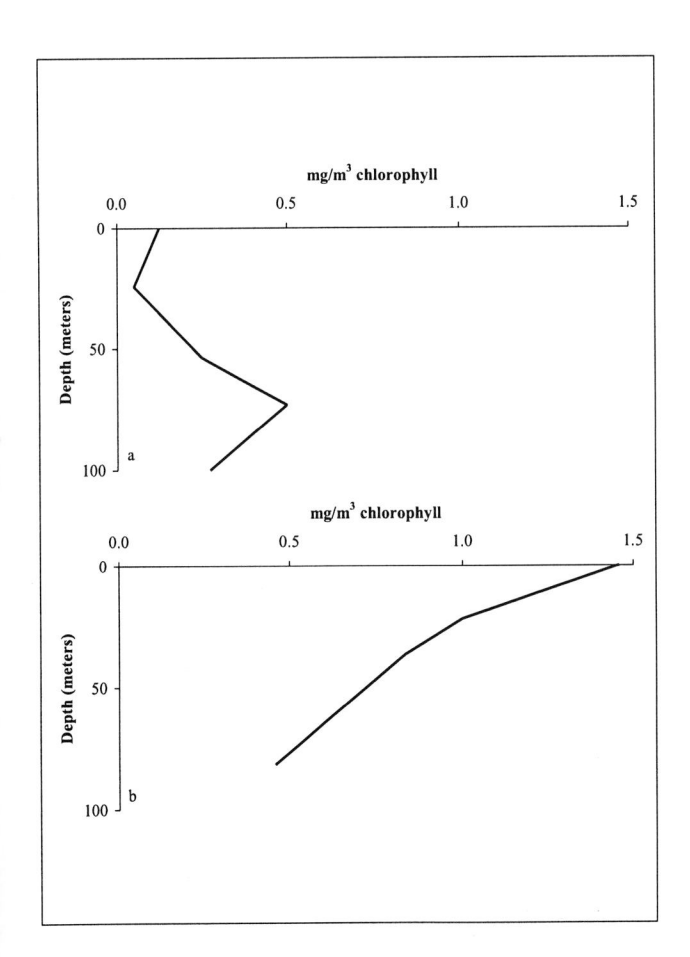

Fig. 10. Chlorophyll in summer (*top*) and spring (*bottom*).

MARINE ECOSYSTEMS

When studying the marine ecosystem in the Canary Island it is necessary to bear in mind two main factors: (a) the small dimension of the island platforms which limits the extent of the shallow benthic area, and (b) the fact that the sea around the Canaries is oligotrophic, low in biomass production. This makes the biomass for all the species small and so the productivity capacity of the whole ecosystem is very low. As a consequence, the Canaries are in a peculiar situation compared to other tropical and subtropical islands around the world, i.e., the islands' ecosystems are not only fairly diversified but also unique and fragile, easily vulnerable, because of the low numbers each species and the very complex inter-relationships between them (León and Braun, 1973).

Mean primary production of phytoplankton is calculated to be 300 mg-C m^{-2} d^{-1}, and mean secondary production of zooplankton is about 38.8 mg-C m^{-2} d^{-1}, with its maximum slightly behind the phytoplankton maximum. As the islands disturb wind and stream flows, there are higher leeward concentrations of zooplankton with corresponding increases in fish resources (Braun and Real, 1981; Braun, 1981).

The structure of the pelagic system is conditioned to a large degree by the presence of different oceanic pelagic species which reach these island when they migrate: fish, turtles, small cetaceans, sperm whales (*Physeter macrocephalus*), rorquals (*Balaenoptera physalus, B. acutorostrata, B.*

edeni), and tunnids. Some of the latter are tropical such as bigeye, skipjack and yellowfin, while others are from temperate seas (bluefin, albacore), which in turn bring a relatively large marine biomass to the islands. These species prey on coastal and littoral pelagic fish. The pelagic coastal system may be classified into two main groups: (1) those species living nearer the coast (i.e., *Atherina presbyter, Boops boops, Trachinotus ovatus*), and (2) those belonging to oceanic waters (i.e., *Scomber japonicus, Sardinella aurita, Sardina pilchardus, Trachurus* sp.) which go to the coastal waters to reproduce and grow. *Sphyraena viridensis, Seriola carpenteri* and *S. dumerili* are also found nearer the coast (Bravo de Laguna and Escánez, 1975; Brito, 1991).

Another unique feature of the Canarian coastal ecosystem is that many species which live in deep waters, which are very close to the coast, rise up to the surface at night. Therefore, these usually deep species have become an integral part of this shallow ecosystem. This does not happen in continental areas where such species are only found in deep waters several miles from the coast. Thus, organisms such as small fish, cephalopod, crustaceans, etc., are concentrated at about 500 m below the surface during the day, and rise to the surface at night, increasing the biomass on which larger fish, sea birds and cetaceans can prey. Hence, there are stable populations of, for example, the protected *Globicephala macrorinchus* to the south-southwest of Tenerife, where the seabed near the coast is very deep. This phenomenon is almost unique in calm waters and creates the circumstances for large populations of easily caught *Thodarodes sagittatus sagittatus*, which is the basic food for *Globicephala macrorinchus*. The co-existence of species from temperate and tropical waters also occurs in the ichthyofauna of the Canary Islands. An example of this is that two larger species of shark, pilgrim shark (*Cetorhinus maximus*) from cold temperate seas, co-exist with the whale shark (*Rhyncodon typus*) from tropical areas (Bravo de Laguna and Escánez, 1975).

In the supralittoral (Splash) area, the most common communities are found on the rocky shores, and are made up of a few species which have adapted to extremes of moisture, salinity, abrasion and light intensity. Greenbluish algae (*Calotrix* sp.) and lichens are the main vegetation, and *Littorina* gastropods are the main invertebrates. Other species found here include crustaceans such as *Ligia italica*, winkles (*Littorina striata, L. neritoides, L. punctata*), crabs (*Grapsus grapsus*) and detritivorous specimens of *Talitrus* sp., *Talorchestria* sp. and *Orchestia* sp. There is plenty of production by phytoplankton, phanerogamous (*Ruppia* sp.) and benthonic microalgae, where small invertebrates live, in the small coastal lagoons, such as El Golfo in Lanzarote (Bacallado Aránega, 1984; Afonso, 1989–1992).

The mesolittoral area is formed by fairly steep rocky cliffs. Here the biota falls into three main zones, defined by the resistance of the species to emersion. The highest level, which is most exposed to dryness, contains large colonies of *Chthamalus steliatus* (up to 10,000 specimens m^{-2}),

co-existing with the limpet *Patella piperata*. The middle level is characterised by less desiccation-resistant algae such as *Gelidium* sp., *Caulacanthus* sp., *Fucus* sp. and, depending on the plankton production, specimens of *Perna picta*, and other molluscs such as *Patella ulyssiponensis aspera* and *Patella candei, Gibbula* sp., and *Osilinus* sp. The lowest zone is mainly formed by strips of the algae *Jania* sp., *Corallina* sp., *Ulva* sp., *Cytoseria* sp., and has a richer associated fauna of amphipods, isopods, polychaetes and decapods. Rock fissures, with greater water retention, provide the habitat for many other species, the most common of which include *Paracentrotus lividus, Perna perna, Eriphia verrucosa* and *Actinia equina* (Bacallado Aránega, 1984; Afonso, 1989–1992).

When high tide recedes, the highest tidepools formed in depressions in the mesolittoral zone contain biota which has adapted to extreme environmental changes. These include *Cyanophytes, Cytoseria* and *Entheromorphae* algae with *Gobius madeirense, Blennius parcorvis*, and the crustacean *Palaemon elegans* as associated fauna. In the lower level tidepools is a richer fauna including anthozoae (*Anemonia sulcata, Aiptasia mutabilis, Polythoa canariensis*), molluscs (*Aplysia* sp., *Chiton canariensis, Clanculus bertheloti, Doris bertheloti*), crustaceans (*Xantho* sp., *Porcellana platycheles*), echinoderms (*Paracentrotus lividus*), sponges (*Clathrina coriacea, Plakortis simplex*), and fish (*Blennius trigloides, B. cristatus, Gobius paganellus, Lepadogaster lepadogaster purpurea, L. zebrina, L.candollei, Abudefduf luridus, Thalassoma pavo*) (Bacallado Aránega, 1984; Afonso, 1989–1992).

When the mesolittoral is a mobile one of sand, gravel and pebbles, biota are very scarce, due to both the mobility and to the low content of organic matter. Generally only amphipods such as *Talitrus* sp., *Talorchestria* sp. and *Orchestia* sp. can be found, though in some months specimens of the marine winkle *Osilinus attratus* can be seen in gravel and pebble strata (Bacallado Aránega, 1984; Afonso, 1989–1992).

The sublittoral area is usually covered with heavy communities of the alga *Cytoseria abies-marina* which, according to sea conditions, may be accompanied by other species such as *Gelidium* sp. and *Corallin*a sp., in which the crab *Plagusia depressa* and barnacles of the genus *Balanus* live. When the zone is affected by strong seas, heavy colonies of the barnacle *Pollicipes cornucopiae* are seen. Down to about 40 m are seagrasses (*Cymodocea nodosa*) and green algae (*Caulerpa prolifera*) whose growth appears to be strongly controlled by the presence of large colonies of the echinoderm *Diadema antillarum*, together with other grazing echinoderms such as *Paracentrotus lividus, Arbacia lixula*, and fish such as *Sarpa salpa* and *Kyphosus sectator*.

On rocky strata where algae are not very abundant, a wide range of fauna can be found: spongiidae (*Ircinia* spp., *Verongia aerophoba*), anemone (*Anemonia sulcata, Aiptasia mutabilis*), sea stars (*Martasterias glacialis, Coscinasterias tenuispina*), fish (*T. pavo, A. luridus, Centrolabrus trutta, Sparisoma cretensis, Spheroides spengleri, Chromis limbatus,*

Diplodus sp., *Canthigaster rostrata*, *Coris judis*, etc.). Caves and rock fractures in this area are the habitat or the refuge of shade-dwelling fauna such as corals (*Madracis asperula*, *Phyllangia mouchezii*, *Caryophyllia inornata*), spongiidae (*Spongionella pulchella*, *Axinella damicornis*), ascidiae (*Ascidia mentula*), anemones (*Telmatactis* sp.), crabs (*Stenorrynchus lanceolatus*), canarian lobster (*Scyllarides latus*), cephalopods (*Octopus vulgaris*), muraenidae (*Muraena helena*, *Lycodintis* sp., *Gymnothorax* sp.), and fish (*Scorpaena maderensis*, *Apogon emberbis*, *Priacanthus cruentatus*, *Epinephelus guaza*, *Mycteroperca rubra*, *Serranus cabrilla*, *S. atricauda*, *Argyrosomus regius*. Highly diversified and complex communities of invertebrates and fish are found on the rocky seabeds where pelagic and semipelagic fish interact with benthic species. Some authors have found as many as 132 different species of fish including grazers and other which feed on small crustaceans, such as *Abudefduf luridus*, *Thalassoma pavo*, *Sparisoma cretensis*, *Chromis limbatus*, *Canthigaster rostrata*, *Boops boops*. Predators such as *Sphyraena viridensis*, *Seriola* sp., *Pomatomus saltator*, *Mycteroperca rubra*, and *Epinephelus guaza* are common (Bacallado Aránega, 1984; Afonso, 1989–1992).

Sandy seabeds in the Canaries normally have little vegetation and productivity is based on detritus coming from algae and seagrasses as well as on microalgae developing as a thin layer on the sand. Higher production and complexity develops in the transition areas between rocky seabeds and seagrass meadows, where a large number of detritivorous crustaceans transfer organic matter between the two interacting systems. These transition seabeds are the habitat of *Mullus surmuletus* and *Pagellus erythrinus* and other species of fish. Seagrass meadows develop on sandy seabeds from several metres deep to about 35 m in sheltered areas, and are formed mainly of *Cymodocea nodosa*, sometimes mixed with the green algae *Caulerpa prolifera* which may go down to a depth of 50 m. Apart from stabilizing the sandy substrata, these beds support populations of fish such as *Sarpa salpa* and *Kyphosus sectator* together with many other young which feed on the microfauna inhabiting the plants. During the year these meadows also produce a large quantity of organic detritus which flows down-slope to deeper sites. These *Cymodocea-Caulerpa* communities play an important role on the southern and south-eastern coast of the islands where the development of algae on the rocky seabeds is otherwise very limited due to the grazing urchin *Diadema antillanum* (Bacallado Aránega, 1984; Afonso, 1989–1992).

A special ecosystem is created by the subterranean volcanic complex of lagoons and tunnels called Los Jameos del Agua in Lanzarote. As they are isolated, they have a very special fauna of endemic invertebrates such as *Munidopsis polymorpha* with a high ecological and biogeographical value. This is an oligotrophic ecosystem sustained on microalgae, mainly diatoms, developing in the outer lagoons which receive some light through breaks in the roof and on planktonic organisms which enter the system through the subterranean outlets to the sea (Bacallado Aránega, 1984; Afonso, 1989–1992).

FISHING RESOURCES

In contrast to the fishing on the Saharian Bank where there is upwelling and where there is a large biomass for commercial exploitation with few species, fishing resources in the Canary Islands are made up of a great variety of species, most of them endemic, but low in number (García Cabrera, 1970).

It is known that in the past turtles and cetaceans were fished. The cetacean fisheries have been tentatively studied, and the exploitation and disappearance of the monk seal (*Monachus monachus*) is well documented. These occurred during the conquest (15th–16th centuries). It is also known that when the Phoenicians arrived on the islands they were looking for *Thais haemastoma* from which they extracted a purple dye which was highly valued. *Spondylus senegalensis* shells were used as money by the Spanish and Portuguese in their commerce with traders from Guinea Gulf during the second half of the 15th century. A well known example of indirect exploitation in post-conquest years is ambergris arriving on the coast (Viera y Clavijo, 1982; Viera y Clavijo, 1942; Webb and Berthelot, 1836–1850). In contrast, algae have not been exploited except in a minor way as a bait for some herbivorous fish (i.e., *Sarpa salpa*) though there have been some failed attempts to use them as manure.

Fishing resources include: shellfish, pelagic oceanic fish (*Todarodes sagittatus sagittatus* and related species, tuna fish, sharks), coastal pelagic fish (sardines, mackerel, *Atherina presbyter*, *Trachurus* sp.), littoral demersal fish (*Sparisoma cretensis*, *Spondyliosoma cantharus*, *Mycteroperca rubra*, *Pagrus pagrus*, *Sphyraena viridensis*, *Sarpa salpa*, *Epinephelus* sp.), and deep-bottom demersal (*Merluccius merluccius*, *Polyprion americanus*) (Bravo de Laguna and Escánez, 1975; García Cabrera, 1970; Franquet and Brito, 1995; Franquet, 1985). Fishing for shellfish is a traditional activity even though production is relatively low (Fig. 11). In the past limpets were an important foodstuff, especially in pre-hispanic times and also in times of food shortages on the islands. Nowadays, fishing for shellfish mainly occurs in the summer and in some places this resource is over-fished. However, most of these species reproduce in the winter and grow quickly so that this over-exploitation may be replenished fairly quickly. The crabs (*Xanto poressa*, *Plagusia depressa*), urchin (*Paracentrotus lividus*) and polychaete (*Perinereis cultrifera*) are caught for bait (Franquet, 1985).

Fishing for pelagic fish is also a traditional activity which was introduced by the first European settlers using 'chinchorros' a kind of non-selective net drawn along the littoral and taken out of the sea on sandy beaches. Today this kind of net is banned and fishing is carried out using encirclement nets ('sardinales', 'traiñas') or with pelagic traps of different shapes according to the species to be

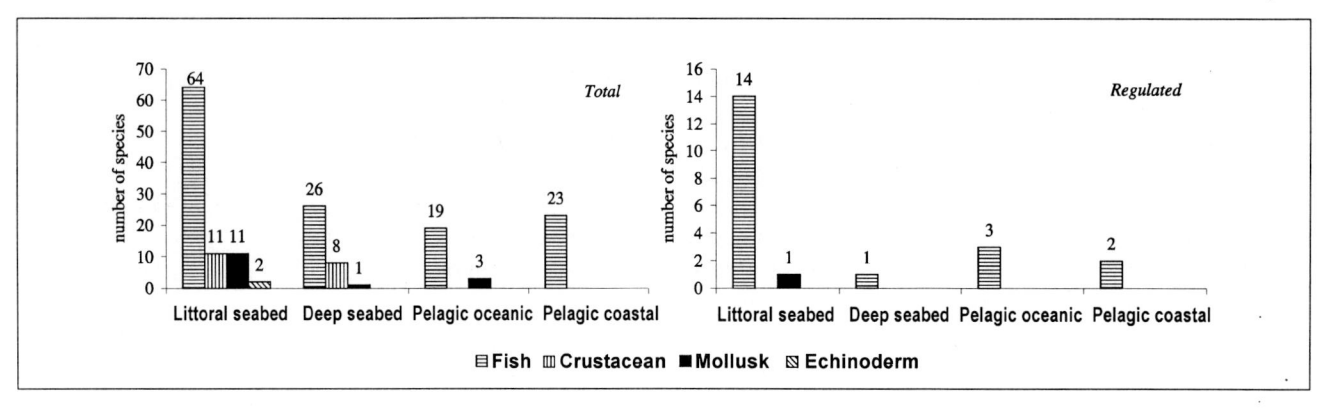

Fig. 11. Comparison between the number of species fished and those actually protected.

fished. This harvests an important fish resource, whose numbers tend to fluctuate (Franquet and Brito, 1995). Some of the targeted species are migratory fish reaching the coast.

Tuna are the most important fishing resource around these islands, even though the only hooks used are hand-made and were introduced in the 1960s to take advantage of the species' migratory characteristics. Skipjack, *Katsuwonus pelamis*, crossing the island seas from spring to autumn, accounts for more than half of the catches (Fig. 12). Tuna fish will be an important economic resource in the future (Afonso, 1989–1992). Another pelagic oceanic resource are *Todarodes sagittatus sagittatus* and related species which also seem to be under-exploited, along with pelagic sharks (*Isurus oxyrinchus, Carcharhinus branchyurus*). Marlin, swordfish and others are only caught for pleasure.

Fish living at depths of 100 m were in the past the main fish caught by the Canarian fishermen. They have produced very high yields. One part of the catch would be sold as fresh fish, and the rest salted, because at the time preservation techniques were fairly crude. However, the modern use of non-selective nets, improvements in navigation, the loss of the Saharan Bank and the increased demand for these fish caused by tourist development and demographic growth has led to the over-exploitation of an ecosystem whose inter-relations are very complex and species abundance relatively low. As a consequence, the yield started to decline and fishing moved to deeper waters in order to meet the demand. In 1970 came the first warnings that littoral fish down to 100 m deep were being over-exploited and that some kind of regulation would be necessary. There are no data on this over-exploitation but it is known that catches as well as fish sizes have markedly diminished. Symptoms of this over-exploitation are species substitution in the ecosystems and in fishing. As an example of this, *Chromis limbatus*, as well as other small-sized fish were favoured as their predators disappeared but, in turn, have started to become commercialised. Other symptoms of over-exploitation are the rapid development of species like the highly mobile and voracious urchin *Diadema antillanum* mentioned earlier, because its main predator *Chilo mycterus reticulatus* (today a protected species) has diminished (Franquet and Brito, 1995; Franquet, 1985).

Resources from deeper areas, as deep as 500 m, such as *Phycis phycis, Pagrus pagrus, Serranus* sp., and *Polyprion americanus*, have also been over-exploited. This can be seen in the cases of *Pagellus* sp. and *Mora moro* near Lanzarote where long hook-lines and modern techniques for fish localization have made an impact. On the other hand, other abundant species which were not very popular are no longer fished. Species like crabs, shrimps, prawns and several sharks species (*Centrophorus* sp., *Mustelus* sp., *Centroscymnus coelolepis*, etc.) are under-exploited (Franquet and Brito, 1995; Franquet, 1985).

DEVELOPMENT PRESSURES AND PROTECTIVE MEASURES

Before the 1950s, the interaction between man and the marine ecosystem in the Canaries was well balanced, but the most sensitive species, which were easy to fish and not very abundant, have been diminished. Today the limpet *Patella candei candei* is only found in Fuerteventura. Other species have even disappeared, like the monk-seal (*Monachus monachus*), which was reported to have existed in big colonies near the small islet of Lobos (Lobos: seals) at the beginning of this century, and the limpet-eater *Haematopus*

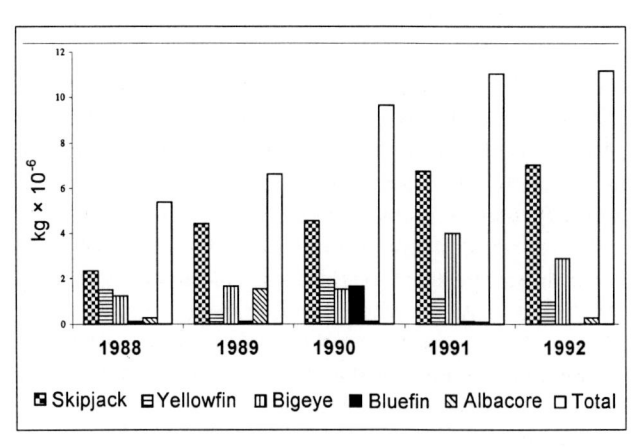

Fig. 12. Tuna fish catches between 1988 and 1992.

Table 4
Main characteristic of ecosystems and impacts on coastal communities

Community characteristics	Coastal ecosystems					
	Coastal lagoons	Intertidal rocky & pebble bottoms	Subtidal rocky bottoms	Subtidal sandy bottoms	Algae meadows	Pelagic zone
Primary producers	Phanerogams, unicellular and macroscopic algae	Benthic algae		Unicellular benthonic algae	Phanerogams, epiphytic and green algae	Phytoplankton
Secondary producers	Insects, crustaceans, molluscs, polichetae	Invertebrates and fish		Invertebrates	Fish, invertebrates	Zooplankton
Higher trophic levels	Birds, fish	Birds and fish	Fish, macro-invertebrates			Birds, fish, cetaceans
Main biologic and ecological values	High production important to birds	Several communities, some endemic, important production, growth zone	Several communities, some endemic, important production, reproduction and growth zone	Several simple communities	Important production for the ecosystem. Sand stabilisation, reproduction and growth zone	Basic production for the littoral ecosystem
Main human activities and impacts	Building, sewage outfalls, rubbish dumping	Building, new docks, breakwaters and artificial beaches, sewage and industrial outfalls, rubbish dumping, shellfishing	New docks, break-waters building, sewage and industrial outfalls, rubbish dumping, fishing and shellfishing	Sand extraction, new docks and break-waters building, sewage and industrial outfalls, rubbish dumping, fishing and shellfishing	Sand extraction, new docks and breakwater building, sewage and industrial outfalls, fishing	Sewage and industrial outfalls, turbidity originated by dredging and rubbish dumping, fishing

meadewaldoi an endemic bird (Viera y Clavijo, 1982; Webb and Berthelot, 1836–1850). Table 4 shows the main ecological features and the impacts on the coastal communities in the Canary Islands (Bacallado et al., 1989).

The model of development in the last decades has favoured the intensive and indiscriminate use of the limited marine resources of these islands, forcing a severe overexploitation in many marine zones. It was recognised that this could lead to an ecological disaster if protection measures were not taken.

Only in 1986 did the Local Government of the Canary Islands establish rules to protect and, if possible, to recover marine resources. Thus, minimum size for some twenty species, industrial use of traps and nets, sport fishing, beach sand extraction, and a Fishing Survey Service was started. However, after ten years, only a few protective actions are well established and results are scarce so far. Thus, among the hundred or so commercial species, only thirty are protected by local and national protective measures and it is necessary to improve the Fishing Survey Service (Bacallado et al., 1989).

A new problem is the increasing and uncontrolled development of sport fishing to meet the demands of increasing tourism, which competes even with commercial fishing. In addition, the use of modern, large commercial fishing boats competes with the small motor and rowing boats used by local fishermen for the big migratory tuna. However, marine protected areas where fishing is severely restricted have been recently established and, at most, only local fishermen have any access. Access by scuba fishers to many marine areas is strictly forbidden. Also, marine reserves have now been established (Fig. 13) to reduce fishing pressure and to serve as protected zones for reproduction and to allow marine resources to recover (Bacallado et al., 1989).

Another important problem for the coastal ecosystem comes from the displacement of the population due to the increasing demands and pressures of tourism in coastal tourist resorts. Sewage disposal occurs by outfalls into fairly shallow waters only due to the steep characteristics of the coast. However, in recent years, public opinion has increasingly demanded action to overcome often severe pollution of ocean bathing beaches and accumulation of contaminants in marine biota near these shoreline outfalls. Thus, small coastal sewage treatment plants are been completed to discharge only secondary or tertiary treated effluents to coastal waters (Fig. 14). The main cities like Santa Cruz de Tenerife now have large coastal sewage treatment plants where treated waters are now being used for agricultural purposes.

It is necessary to bear in mind that these islands have no rivers. Thus, the erosive ravines typical of these islands only contain water after rain. Potable and irrigation waters are mainly obtained through wells and 'galerías' (tunnels) drilled horizontally in the mountains, the excess being stored in open reservoirs and, more recently, in ponds built with special plastic liners which make use of old volcanic cones (Bravo, 1954, 1970).

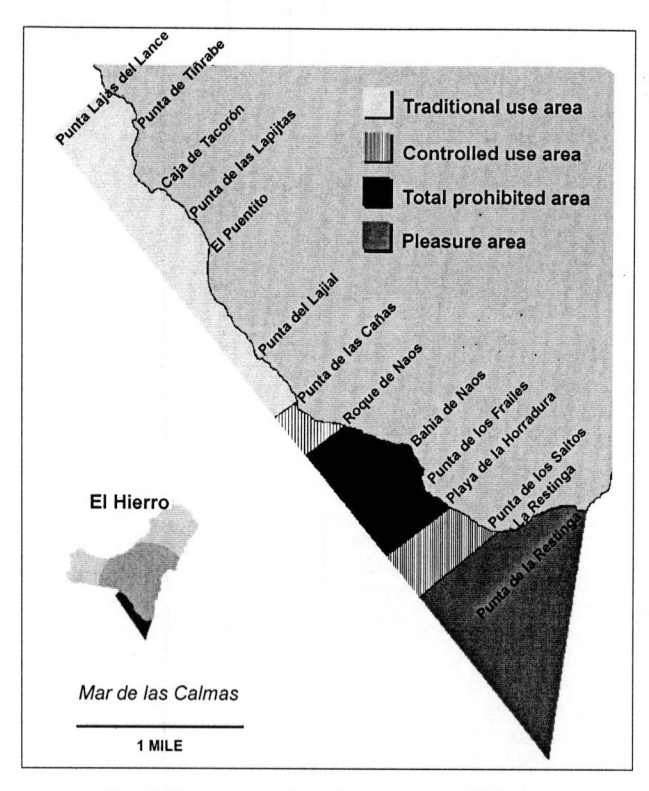

Fig. 13. The proposed marine reserve in El Hierro.

1996). However, studies on coastal biota contamination show that, in general, the pollution is moderate compared to most other European coastal areas (Corbella and García Montelongo, 1996; Díaz et al., 1989, 1990, 1994; García Montelongo et al., 1994; Hardisson et al., 1985a,b; Peña et al., 1996a,b).

Local and national governments are increasing financial support for seafood farming (*Sparus aurata*, lobster, *Dicentrarchus labrax*, *Sparisoma cretensis*, and others). The Marine Laboratory in Tenerife (Spanish Oceanographic Institute) has a wide research plan to study seafood farming of commercial endemic species in artificial pools, which will be set free as juveniles for restocking coastal areas.

REFERENCES

Afonso, L. (ed.) (1989–1992) *Geografía de Canarias*. Vols. 1–7, Editorial Interinsular Canaria, Tenerife.

Aguilera Klint, F. (ed.) (1994) *Canarias: Economía, Ecología y Medio Ambiente*. F. Lemus, La Laguna.

Bacallado Aránega, J.J. (ed.) (1984) *Fauna (marina y terrestre) del Archipiélago Canario*. Edirca, S.L., Las Palmas de Gran Canaria, 1984.

Bacallado, J.J., Cruz, T., Brito, A., Barquín, J. and Carrillo, M. (1989) *Reservas Marinas de Canarias*. Gobierno de Canarias. Santa Cruz de Tenerife.

Báez, M., Bravo, T. and Navarro, J.F. (1982) *Canarias: origen y poblamiento*. CESC, Madrid.

Braun, J.G. and Real, F. (1981). *The Vertical Distribution of Chlorophyll in Canary Islands Waters*. ICES, Biological Oceanography Commitee, 1981/L:7

Braun, J.G. (1981) Estudios de producción en aguas de las Islas Canarias.II. Producción de zooplancton. *Boletín Instituto Español de Oceanografia* 290, 90–96.

Bravo de Laguna, J. and Escánez, J. (1975) Informe sobre las posibilidades pesqueras de elasmobranquios en el Archipiélago Canaro. *Publicaciones Técnicas de la Dirección General de Pesca Maritima* 11, 169–192.

Bravo, T. (1954) *Geografía General de las Islas Canarias*. 2 vols. Goya Ediciones, Santa Cruz de Tenerife.

Bravo, T. (1970) *La situación de Canarias en la tectónica atlántica*. Universidad de La Laguna, La Laguna.

Brito, A. (1991) *Catálogo de los peces de las Islas Canarias*. F. Lemus Editor, La Laguna.

Carrillo, J., González, J.A., Castillo, J. and Gómez, J. (1985) Recursos demersales de Lanzarote y Fuerteventura. *Simp. Int. Afl. O. Afr., Instituto de Investigaciones Pesqueras* 2, 799–823.

Conde, J.E., Peña, E. and García Montelongo, F. (1996) Sources of tar balls and oil slicks on the coast of the Canary Islands. *International Journal of Environmental Analytical Chemistry* 62, 77–84.

Corbella, R. and García Montelongo, F. (1996) Aliphatic hydrocarbons in *Patella ullyssiponensis aspera* from the coast of Fuerteventura (Canary Islands). *Química Analitica* 5 (Suppl. 2), S275–S281.

Díaz, C. Galindo, L. and García Montelongo, F. (1994) Distribution of metals in some fishes from Santa Cruz de Tenerife, Canary Islands. *Bulletin of Environmental Contaminant Toxicology* 52, 374–381.

Díaz, C., Galindo, L. and García Montelongo, F. (1989) Mercury levels in limpets (*patella* sp.) from the coast of Santa Cruz de Tenerife, Canary Islands. *Archivio di Oceanografia e Limnologia* 21, 191–198.

Díaz, C., Galindo, L., García Montelongo, F., Larrechi, M.S. and Rius, F.X. (1990) Metals in coastal waters of Santa Cruz de Tenerife, Canary Islands. *Marine Pollution Bulletin* 21, 91–95.

Impacts from industrial pollution are not great due to the fact that there are no large industries around the islands. Only one refinery (Tenerife), which recently passed the ISO-9000 standards, power stations in every island, and small manufacturing industries exist. The main pollution sources are localised near commercial docks and the small industrial areas which have mandatory sewage treatment plants. Another pollution source is that which originates away from the islands but which is carried to them by the Canary Islands Stream and the Trade Winds (Conde et al.,

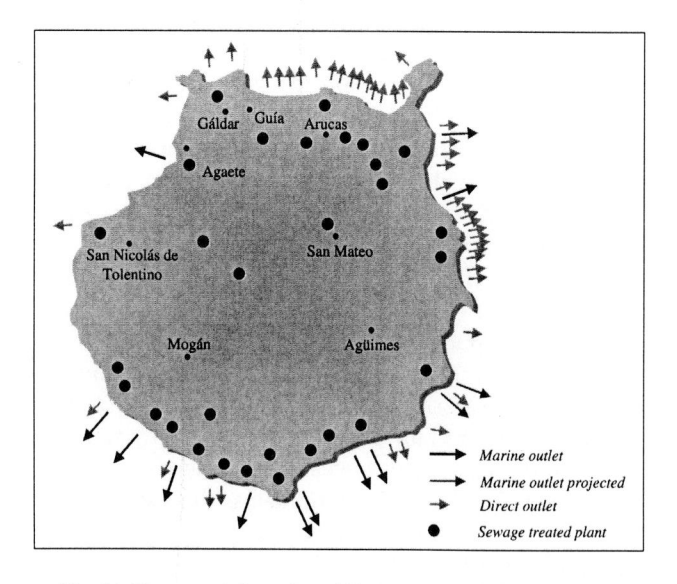

Fig. 14. Direct outfalls and outfall pipes around Gran Canaria.

Franquet, F. and Brito, A. (1995) *Especies de interés pesquero de Canarias*. Gobierno de Canarias, Santa Cruz de Tenerife.

Franquet, F. (1985) *Guía de peces, crustáceos y moluscos de interés comercial del Archipiélago Canario*, Gobierno de Canarias, Santa Cruz de Tenerife.

García Cabrera, C. (1970) *La pesca en Canarias y Banco Sahariano*. CESIC, Santa Cruz de Tenerife.

García Montelongo, F., Díaz, C., Galindo, L., Larrechi, M.S. and Rius, X. (1994) Heavy metals in three fish species from the coastal waters of Santa Cruz de Tenerife (Canary Islands). *Scientia Marina*. **58**, 179–183.

Hardisson, A., Díaz, C., Galán, J.M., Galindo, L., and García Montelongo, F. (1985a) Niveles de concentración de Pb, Cd, Cu, Zn y Fe en cefalópodos congelados y en conserva. *Revista de Toxicologia* **2**, 121–127.

Hardisson, A., Galindo, L. and García Montelongo, F. (1985b) Niveles de concentración de mercurio en pescados y cefalópodos frescos y enlatados. *Alimentaria* **161**, 79–82.

León, A.R. and Braun, J.G. (1973) Ciclo anual de la producción primaria y su relación con los nutrientes en aguas canarias. *Boletín del Instituto Español de Oceanografía* **167**.

Llinás, O., González, J.A. and Rueda, M.J. (eds.) (1996) *Oceanografía y recursos marinos del Atlántico centro-oriental*. Gobierno de Canarias, Las Palmas de Gran Canaria.

Llinás, O. (1988) *Análisis de la distribución de nutrientes en la masa de agua central noratlántica en las Islas Canarias*. PhD Thesis, Universidad de La Laguna.

Mascareño, D. and Molina, R. (1970) Contribution à l'étude de l'upwelling dans la zone canarienne africaine. *Rapports des Procès-Verbaux, Conseil International pour l'Exploration de la Mer* **159**, 61–73.

Mascareño, D. (1972) Algunas consideraciones oceanográficas de las aguas del archipiélago canario. *Boletín del Instituto Español de Oceanografía* **158**, 7–79.

Molina, R. and Laatzen, F.L. (1986) Corrientes en la región comprendida entre las Islas Canarias orientales, Marruecos y las Islas Madeira. Campaña 'Norcanarias I'. *Revista de Geofisica*. **42**, 41–52.

Peña Méndez, E.M., Astorga España, M.S. and García Montelongo, F. (1996a) Polychlorinated biphenyls in two mollusc species from the coast of Tenerife (Canary Islands, Spain). *Chemosphere* **32**, 2371–2380.

Peña, E.M., Conde, J.E. and García Montelongo, F. (1996b) Evaluation of *Osilinus attratus* as a bioindicator organism to monitor oil pollution in the Canary Islands. *Archives of Environmental Contamination and Toxicology* **31**, 444–452.

Viera y Clavijo, J. (1942) *Diccionario de Historia Natural de las Islas Canarias* (1866). 2 vols., Santa Cruz de Tenerife.

Viera y Clavijo, J. (1982) *Noticias de la Historia General de las Islas Canarias* (1866). 2 vols. Goya Ediciones, Santa Cruz de Tenerife.

Webb, P.B. and Berthelot, S. (1836–1850) *Histoire Naturelle des Iles Canaries*. 6 vols., Plon Ed. and Béthume Ed., Paris.

Zenk, W., Finke, M., Müller, T.J. and Llinás, O. (1986) The role of the Canary Current in the subtropical Atlantic gyre circulation. (S8.30) Joint Meeting of the E.G.S and E.S.C, Kiel, August, 1986.

THE AUTHORS

Francisco García Montelongo
Department of Analytical Chemistry,
Nutrition and Food Sciences,
University of La Laguna,
38071 La Laguna, Spain

Carlos Díaz Romero
Department of Analytical Chemistry,
Nutrition and Food Sciences,
University of La Laguna,
38071 La Laguna, Spain

Ricardo Corbella Tena
Department of Analytical Chemistry,
Nutrition and Food Sciences,
University of La Laguna,
38071 La Laguna, Spain

Chapter 13

THE AZORES

Brian Morton and Joseph C. Britton

The nine islands of the Azores, with a total land area of 2233 km², command a marine Exclusive Economic Zone of more than one million km². With a total population of only ~249,000, these remote islands, rising steeply from oceanic depths >4000 m on both sides of the mid-Atlantic Ridge, should be models of marine environmental health.

Climatic changes and the prevailing pattern of ocean circulation around the islands, however, bring trace levels of emission gasses and water-borne contaminants. With little industry and a sparse population, inshore pollution levels are low. Recreational beaches are classified as either good or satisfactory and most significant contaminant levels are restricted to major boat harbours. TBT-induced imposex in the predatory dogwhelk *Stramonita haemastoma* was, for example, only detectable at the popular trans-Atlantic yacht harbour at Horta, Faial. Similarly, heavy metal levels were generally higher in the harbour of the most populous city (65,000) of Ponta Delgada on São Miguel.

Most heavy-metal research has focused on seabirds, it being demonstrated, for example, that there has been a three-fold increase in mercury contamination over the last 100 years in the North Atlantic. High levels of mercury have also been recorded from cephalopods and mesopelagic fishes. Metal levels have also been examined in intertidal amphipods and barnacles and shown to be high near the harbour at Ponta Delgada and in some remote areas, as for example, near a coastal thermal spring. Cadmium levels of 167 μg g^{-1} in *Chthamalus stellatus* are the highest ever recorded, worldwide, for a barnacle but this represents natural bioavailability from the native lava rock. Elsewhere, such Cd levels have been shown to influence gastropod growth adversely. It seems clear that in the Azores natural leaching is more important than pollution in defining the health of local marine communities. However attractive the concept, therefore, it is unlikely that the Azores can serve as a 'clean', 'control' site for pollution studies elsewhere.

The greatest human threats to native marine communities are the introduction of exotic species and excessive exploitation of some fisheries resources. Regarding the former, the Azores have fared much better than other island groups, but with increasing interactions with world markets, the potential for detrimental exotic invasions remains high. Rats and feral cats have had profound adverse effects on seabird populations and Monteiro et al. (1996) have documented the decline in seabird numbers in the Azores. Over-exploitation of certain inshore fisheries, especially limpets and, perhaps, octopus, produces equally severe impacts on community ecology.

Seas at The Millennium: An Environmental Evaluation (Edited by C. Sheppard)
© 2000 Elsevier Science Ltd. All rights reserved

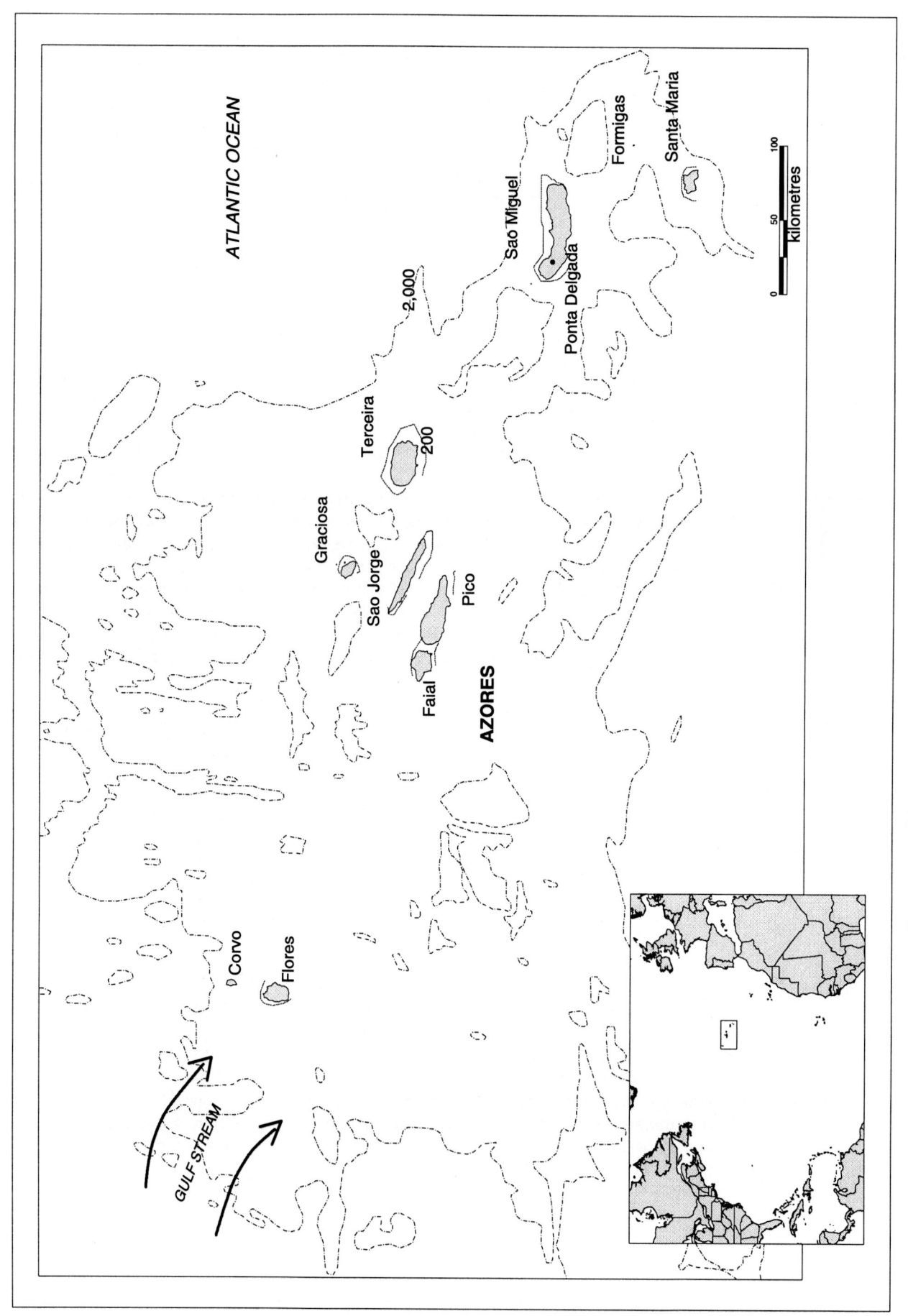

Fig. 1. A map of the Azorean Exclusive Economic Zone in the North Central Atlantic.

THE DEFINED REGION

The Exclusive Economic Zone of the Azores Archipelago (Fig. 1), including nine volcanic islands, numerous islets and several seamounts, straddles the Mid-Atlantic Ridge along a WNW–ESE axis in the north-central Atlantic Ocean between latitudes 37° and 40°N. Two of the islands, Flores and Corvo, at approximately 40°N 32°W, arise from the North American Plate on the western side of the ridge. The other islands on the eastern side are upon an unusual crustal feature, the Azores Microplate, positioned as a triangular wedge between the Eurasian and African Plates and extending ESE to approximately 37°N, 25°W. The Azores constitute the most remote islands in the North-central Atlantic Ocean, 1510 km from Lisbon, Portugal on the European continent and 3203 km from Boston on the North American continent, at approximately similar latitudes.

All of the islands are geologically young, consisting mostly of Pliocene and Quaternary igneous and sedimentary rocks (Azevedo et al., 1991), but some deposits on Santa Maria are from the Miocene, between eight to ten million years ago (Serralheiro and Madeira, 1993). The Exclusive Economic Zone of the archipelago, extending over an area of almost one million km^2, has been tectonically active and formative, with the islands emerging regularly since the Miocene. Episodic volcanic activity, erosion and sedimentary deposition have interacted to generate the modern landforms of the archipelago. Capelinhos Mountain, Faial, less than 50 years old and, hence, the most recent landform, represents a microcosm of the processes—both formative and destructive—which have come together to produce all of the Azores (Morton et al., 1998).

The youthfulness of the archipelago and its remote position in the central North Atlantic Ocean makes it a natural laboratory for studies of biogeography, colonisation processes and evolution. Prior to human intervention, the Azores islands were a natural meeting place for marine biota from different origins. Unlike the terrestrial flora, which claims several endemic plants, there are few endemic marine species, perhaps because of the youthfulness of the islands. Young (1998) considers the barnacle, *Tesseropora arnoldi* a Tethyan relict endemic, but Southward (1998) disputes this interpretation. Other possible endemics include certain algae (Prud'homme van Reine, 1988; Tittley and Neto, 1995), amphipods (Lopes et al., 1993) and gastropods (Gofas, 1990; Knudsen, 1995). Most of the present resident marine and maritime species of the islands were recruited from elsewhere and this process is continuing and, possibly, accelerating, with the human-assisted introduction of exotic species reviewed by Morton and Britton (1999). There are also other papers which focus on specific groups, notably the algae (South and Tittley, 1986; Prud'homme van Reine, 1988; Neto, 1994; Tittley and Neto, 1995), sponges (Boury-Esnault and Lopes, 1985), hydroids (Rees and White, 1966; Cornelius, 1992), amphipods (Lopes et al., 1993), barnacles (Young, 1998), rissoid gastropods (Gofas, 1989, 1990) and fishes (Santos et al., 1997), that illustrate the diverse origins of these components of the Azorean shallow water biota but, nevertheless, with strongest links to the shores of the European Eastern Atlantic at the same latitude, the Mediterranean and other Macaronesian (Madeira and Canaries) islands.

SEASONALITY, CURRENTS, NATURAL ENVIRONMENTAL VARIABLES

The Azores occupy the temperate zone of the northern hemisphere. The archipelago is unaffected by subtropical trade winds but polar low pressure fronts may sweep the islands at any time, especially in winter. Prolonged sunshine is uncommon. Considerable cloudiness and some rainfall occur in every season, with an average of 120 overcast days annually. February is the cloudiest, with overcast skies expected 70% of the time. Sunny skies are most frequent in August, occurring about half the time, but the islands experience only about ten clear days each year.

Heavy storms in the Azores are neither common nor prolonged, although occasional extra-tropical disturbances, originally born in the trade wind belt south of the Azores as tropical storms, or hurricanes, sometimes strike the islands with considerable fury. Only in July and August, when a high pressure cell normally sits directly upon the Azores, do the islands experience prolonged periods of fine weather. The ridge of high pressure migrates southward in winter, exposing the islands to occasional polar depressions which approach from the northwest. The Azores lie, however, near to the southern limits reached by most polar fronts. This, and the tempering influences of the eastern limbs of the Gulf Stream, provide the Azores with a pleasant, though somewhat rainy, climate throughout the year. Rainfall increases with latitude and altitude. Santa Maria is the southernmost and driest island. Mean annual rainfall at sea level varies from 86.4 cm at Ponta Delgada, São Miguel, to 144.8 cm at Santa Cruz, Flores. At Carvão, Terceira, at an elevation of 524 m, the mean annual rainfall is 226 cm. Mean annual air and water temperatures are strikingly similar, being 13°C and 16°C, respectively, in winter and 17°C and 19°C, respectively, in summer. February is the coldest month and August the warmest. At Ponta Delgada, the mean minimum winter and mean maximum summer air temperatures are 6°C and 28°C, respectively. Winter frosts are rare at sea level but occur at some of the higher elevations on each island. Snow, though uncommon, can blanket mountain tops briefly but usually melts within a few hours. Pico do Pico, at 2351 m, is the only mountain in the Azores where snow and ice persist on the ground for more than a few hours during winter.

Wind direction in the Azores is variable throughout the year whereas wind speed is decidedly seasonal. Prevailing westerly (anti-cyclonic) winds dominate in winter, with mean daily wind speeds ranging from 9 to 23 knots. In the summer, under the influence of seasonal, relatively stationary, high pressure cells, wind direction is more variable but chiefly from the northeast with mean daily speeds ranging from 4.5 to 13 knots. Harrison et al. (1996) measured ground-level concentrations of atmospheric particulate sulphate, nitrate, chloride, ammonium, iron, sulphur dioxide, gaseous nitric acid and total particle count on Santa Maria during two periods. Persistent anticyclonic conditions during the first period ensured that air sweeping across Santa Maria had not been in contact with continental regions for many days. Levels of the measured parameters were low and comparable to values recorded from other remote island locations. During the second period, however, the air over Santa Maria had crossed Great Britain only three days earlier. It contained substantially increased concentrations of sulphate, nitrate and nitric acid.

Gale-force winds, blowing mainly from the west and northwest, can be expected for three or four days each month during winter, but only one day in every two months during summer. Sea fog is relatively infrequent in the Azores and is usually confined to periods of early morning calm. Coastal fog is, otherwise, generally restricted to eight to ten days in June. In contrast, the higher slopes of every island are often shrouded with clouds and this can be expected in all seasons.

A few climatic studies have evaluated atmospheric pollutants in the vicinity of the Azores. In 1992, for example, mean concentrations of the anthropogenic radioactive oceanographic tracers ^{99}Tc, ^{90}Sr and ^{137}Cs were measured for waters east and northeast of the Azores (Dahlgaard et al., 1995). Half-lives of 20 years have been calculated for ^{90}Sr and ^{137}Cs in the northeast Atlantic surface water, corresponding to a mean residence time of between 80 and 100 years. The measured ^{137}Cs:^{90}Sr ratio was not, however, in agreement with published data on rainwater samples and not significantly different from that calculated for global fallout. With no significant additional sources of artificial radionuclides in the region, Azorean values are characteristic of 1960's levels of global fallout. Kasibhatla et al. (1996) studied large-scale patterns of tropospheric ozone over the North Atlantic in summer, with one focus over the Azores. By comparing present-day and pre-industrial simulations, they concluded that anthropogenic NO_x emissions have significantly perturbed ozone levels. Present day ozone levels above the North Atlantic were at least twice as high as pre-industrial levels.

As a result of their central position in the North Atlantic Ocean, the Azores are influenced by weak, but complex, surface oceanic currents (Fig. 2) and, perhaps, even more complex, deep water flows (Santos et al., 1995). The western limb of the clockwise-rotating North Atlantic gyre, the Gulf Stream, is a swift, strong, current which dominates the North Atlantic circulation. It arises from tropical latitudes and flows northeastward, delivering warm water well into the boreal zone. It sweeps northwest of Bermuda and continues eastward, losing strength slowly. It breaks into two limbs at about 40°N to form the North Atlantic Current and the more southerly Azores Current. The former continues towards Europe where it splits into two branches. The northern limb flows northeast and tempers the climate of the British Isles. The southern limb flows toward the southeast, but direction and intensity vary seasonally. During the summer, it tracks easterly at a considerable distance from the Azores (Fig. 2A) whereas in the winter it approaches the islands closely from the north (Fig. 2B). The Azores Current also splits into two branches. A northern limb flows eastward across Azorean southern shores while a southern one flows south at a considerable distance southwest of the archipelago. During the summer, the former continues eastward to Madeira where it joins the Madeira and Canary Currents, both of which flow south off the African coast. In winter, however, it divides into two branches, the southern one turning abruptly south shortly

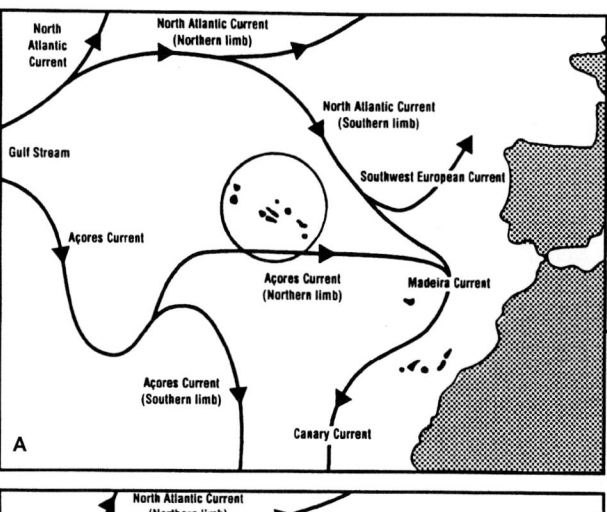

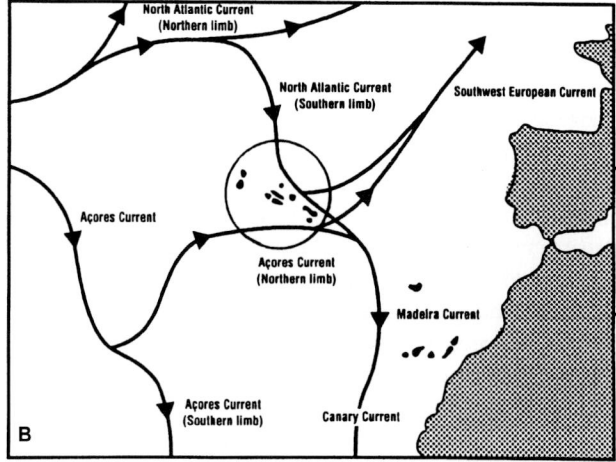

Fig. 2. The pattern of surface circulation in the North Atlantic in A, summer and B, winter (after Gofas, 1990).

after passing the Azores while the northern one joins the southerly element of the North Atlantic Current to form the Southwest European Current flowing to the northeast, towards the English Channel. All of these limbs continue to produce complex and unpredictable branches, meanders and eddies, even briefly reversing primary current directions, especially in autumn and winter (Krauss and Meincke, 1982; Krauss et al., 1990).

The foregoing, despite its complexity, still fails to sufficiently describe the North Atlantic current patterns, for it considers only surface flow. Other currents flow at depth, often in directions counter to those which move at the surface. For example, the gradual convergence of surface flow east of the Azores produces a downwelling of water which moves southward (Kaese and Siedler, 1982; Pollard and Pu, 1985). Similarly, there is a steady westward flow of Mediterranean water from the Straits of Gibraltar to the vicinity of the Azores at a depth of approximately 900 m, separated from surface water by a halocline (Gofas, 1990). When this water impacts the Mid-Atlantic Ridge, some of it is pushed to the surface.

A few studies have examined pollutants in the vicinity of the Azores which were, most likely, transported there from other regions by currents. Class and Ballschmiter (1988) measured levels of trace organics such as bromochlomethanes in the air and surface waters at various locations in the Atlantic Ocean and detected a correlation between high concentrations of such compounds and the occurrence of algae on the coastlines of the Azores, Bermuda and Tenerife (Canaries). Peroxyacetylnitrate (PAN) levels in the marine boundary layer at various locations in the Atlantic were measured by Mueller and Rudolph (1992). The mixing ratios of PAN varied from 2000 mg l^{-1} in the English Channel to <0.4 mg l^{-1} south of the Azores.

THE MAJOR SHALLOW WATER MARINE AND COASTAL HABITATS

The coastal ecology of the Azores has been described by Morton et al. (1998). All of the islands of this archipelago rise precipitously from the deep-sea floor and emerge generally without a significant subtidal shelf. Only in a few locations, including some swimming beaches, a few harbours, some islet lagoons and, especially, the strait between the islands of Faial and Pico, does the intertidal zone grade into a shallow, offshore, flat. The tidal pattern in the Azores is semidiurnal, with only a slight diurnal inequality (Fig. 3). The tidal range is less than one metre. Many times each year tidal patterns are masked by strong onshore winds and waves.

The prevailing intertidal habitat in the Azores is a hardshore usually dominated by black basaltic lava (Fig 4). The pattern of intertidal zonation on such a shore is illustrated in Fig. 5, along with representative species. Frequently, the land meets the sea as sheer, inaccessible, cliffs. Sometimes, however, either ancient lava flows or massive landslides (*fajãs*) produce more accessible coastal platforms which rise a few metres above the sea and extend toward it dozens, and sometimes hundreds, of metres (Fig. 6). The bases of these platforms, and, indeed, even the bases of the sheer vertical seacliffs are generally cloaked with a luxuriant cover of algae. Neto (1994) records 307 species of intertidal and shallow subtidal marine algae from the archipelago, including 193 species of Rhodophyta, 66 species of Phaeophyta and 44 species of Chlorophyta. Medium-sized macrophytes such as *Fucus spiralis, Gelidium latifolium, Pterocladia capillacea,* and *Cystoseira tamariscifolia* are among the conspicuous intertidal species, but larger, laminarian, brown algae are conspicuously absent. Most marine algae in the Azores are relatively small, filamentous, coralline, bushy or epiphytic and distinctly mat-forming. The sublittoral fringe and shallow subtidal shores throughout the Azores are carpeted by a thick, prickly, algal turf accommodating a distinctive suite of marine communities and protecting a rich, mostly cryptic, fauna which, except for the micromolluscs (Bullock et al., 1990; Bullock, 1995), is largely unstudied. The high percentage

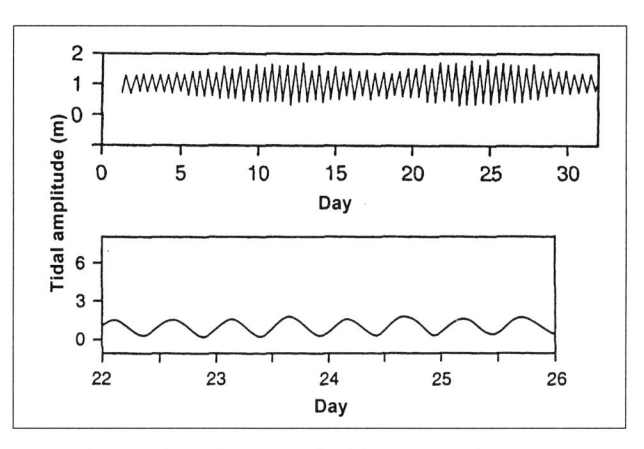

Fig. 3. The semidiurnal pattern of tidal change in the Azores over a period of one month (top) and four days (bottom).

Fig. 4. The wave-exposed shore at Barro Vermelho, Graciosa, Azores.

Fig. 5. The exposed basalt shore at Caloura, São Miguel, with significant zoning organisms identified and typical, representative, species illustrated. Scale is in metres. A, *Lacerta dugesi*; B, Buckshorn plantain, *Plantago coronopus*; C, Sea spleenwort, *Asplenium marinum*; D, Rock samphire, *Crithmum maritimum*; E, Common sea lavender, *Limonium vulgare*; F, *Melarhaphe neritoides*; G, *Littorina striata*; H, *Ligia italica*; I, *Chthamalus stellatus*; J, *Patella candei gomesii*; K, *Lyngbya confervoides*; L, *Enteromorpha intestinalis*; M, *Patella ulyssiponensis aspera*; N, *Fucus spiralis*; O, *Omalogyra atomus*; P, *Cladophora prolifera*; Q, *Corallina officinalis*; R, *Halopteris filicina*; S, *Jania rubens*; T, *Arundo donax*; U, *Gigartina acicularis*; V, *Cystoseira tamariscifolia*; W, *Ulva rigida*; X, *Jania adhaerens*; Y, *Lepidochitona simrothi*; Z, *Patella ulyssiponensis aspera*; A1, *Columbella adansoni*; B1, *Stramonita haemastoma*; C1, *Laurencia pinnatifida*; D1, *Callithamnion corymbosum*; E1, *Mitra nigra*; F1, *Chaetomorpha aerea*; G1, *Pterocladia capillacea*; H1, *Arbacia lixula*; I1, *Sphaerechinus granularis*; J1, *Gelidium pusillum* (after Morton et al., 1998).

algal cover on many Azorean shores may, however, reflect the local culinary predilection for limpets ('*lapas*'). Two such molluscan herbivores normally crop the algae. Morton et al. (1998) identified a shore on Santa Maria where limpet harvesting had not taken place. Here, limpets were abundant, the rocks were largely devoid of algae and a major limpet predator, *Stramonita haemastoma*, was also present in considerable numbers. Throughout much of the Azores, however, excessive harvesting of limpets, especially along easily accessible shores, has both decimated their populations and favoured the development of a thick algal turf, despite legislative protection and a 'closed' season for harvesting (Santos et al., 1995).

At another Santa Maria locality, São Lourenço, lies a shore sandwiched between outcrops of black lava, consisting of brownish orange, highly weathered, palagonitic tuff.

Fig. 6. Fajã dos Cubres from an overlook along the north side of São Jorge, Azores.

Here, as elsewhere in the islands where semilithified, friable, tuffs and volcanic ash deposits stand in the intertidal, the habitat is depauperate in comparison with basalt shores, there being fewer algae, limpets and barnacles, notably *Chthamalus stellatus*.

The second most common intertidal habitat in the Azores is characterized by large, rounded, mobile, beach stones, many of which can be called, appropriately, boulders. Azorean boulder or cobble shores are formed under a variety of conditions. Where recent, poorly lithified, ashfalls, reach the sea, as occurred following the 1957 volcanic eruptions at Capelinhos, Faial, the fine particles of ash were removed quickly from the intertidal zone by waves and currents, leaving behind a boulder-strewn cobble beach of unstable roll-stones. Alternatively, strong nearshore currents adjacent to a depositional environment can also produce such shores, as occurs at Fajã Grande, Flores. On a smaller scale, similar conditions can either deposit or expose boulders and cobbles between adjacent headlands, as, for example, at Barro Vermelho, Graciosa. Regardless of their origins, cobble shores, at first glance, seem the most depauperate, with the mobile rocks pulverizing any attached biota. Remarkably, however, deep within such beaches, between the mobile roll stones, is a modestly diverse community of fragile detritivores, including flatworms, amphipods, isopods and ellobiid gastropods (Martins, 1976; 1980). Most members of this fauna derive nutrition from allochthonous detritus which accumulates here, especially beached algal debris washed in from adjacent shores.

Sandy beaches are rare in the Azores and, where encountered, are of high recreational value. When and where present, they are influenced strongly by waves and nearshore currents. Seasonal wave patterns may modify their geomorphology overnight, with storm waves transforming sandy beaches into cobble shores. Accordingly, the fauna of Azorean sandy beaches appears spartan (Morton et al., 1998), but no meiofaunal studies have yet been undertaken. In contrast, offshore, subtidal, sands in harbours and bays are rich in life and may be dominated by one bivalve, *Ervilia castanea* (Morton, 1990a). They may become so abundant in such habitats that when red, empty, shells wash onto nearby sandy beaches, the shore may become markedly discoloured. Such an *E. castanea*-based community occupies the accessible lagoon crater of the drowned offshore volcanic islet of Ilhéu de Vila Franca, São Miguel, and where the only information on Azorean sand beach dynamics has been obtained (Morton, 1990b).

Several Azorean islands and associated, uninhabited, islets are seabird roosting and nesting sites. The most obvious birds occupying them are the Common tern *Sterna hirundo* and the possibly threatened Roseate tern *Sterna dougalli* (del Nevo et al., 1990). Also common, but rarely observed on shore even though it nests there, is Cory's shearwater *Calonectris diomedea*. It nests in a burrow on sea cliffs and within which a single egg is laid in summer. While fledglings are in the nest, adults spend the day fishing at sea, returning to feed their offspring at night. The eerie calls of adult birds coming ashore are the fabric of folklore and the basis of some claims that ghosts haunt the islands.

About 115 species of nearshore fishes inhabit the shallow waters surrounding the Azores (Santos et al., 1997). A significant number of them (80%) either also occur in the Mediterranean or have close affiliations with species there. A similar relationship exists with much of the inshore fauna of the archipelago (Morton and Britton, 2000). With no rivers in the Azores and only a few streams, many of which fall precipitously into the sea as coastal waterfalls, there is only one catadromous fish, the eel *Anguilla anguilla*, and no anadromous species.

Despite the paucity of flowing surface water on most islands, normally abundant precipitation ensures an ample supply of groundwater, especially on the more elevated portions of the islands. Often, this groundwater flows into the sea *via* subtidal fissures. At three places on the Azores, at Lajes, on Pico, and at Fajã de Santa Cristo and Fajã dos Cubres, São Jorge, are coastal platforms, each containing central lagoons, or submerged basins, which are protected by a seaward rampart of boulders. Seawater in all of the basins is diluted by incursing groundwater, producing miniature coastal wetlands of varying salinity, depending upon both the particular site and location within it (Morton et al., 1995; 1996; 1997). The basis for productivity also varies from place to place, being *Juncus* and drift algae at Lajes, the seagrass *Ruppia maritima* at Fajã dos Cubres and drift algae at Fajã de Santa Cristo (Morton et al., 1997). Such places are unique wetland habitats in the Azores and, we suspect, other Macaronesian islands. Calls have been made for their protection (Morton and Tristão da Cunha, 1993; Santos et al., 1995; Morton et al., 1995, 1996, 1997). The flora and fauna of the marsh behind the seaward boulder rampart at Lajes, Pico is illustrated in Fig. 7.

There was once a magnificent marsh at Paúl, Praia da Vitória, Terceira, but this has now been reclaimed at the expense of the endemic Moorhen *Gallinula chloropus*

Fig. 7. The *Juncus* marsh at Lajes do Pico with typical, representative species illustrated. A, *Enteromorpha linza*; B, *Asparagopsis armata*; C, *Gigartina acicularis*; D, *Orchestia gammarellus*; E, *Chaetomorpha linum*; F, *Ulva rigida*; G, *Tanais dulongii*; H, *Melita palmata*; I, Grey heron, *Ardea cinerea*; J, Wimbrel, *Numenius phaeopus*; K, Turnstone, *Arenaria interpres*; L, Little-ringed plover, *Charadrius dubius*; M, Grey wagtail, *Motacilla cinerea*; N, Sharp rush, *Juncus acutus*; O, *Phascolosoma granulatum*; P, *Lekanesphaera rugicauda*; Q, *Pseudomelampus exiguus*; R, *Myosotella myosotis*; S, *Ovatella vulcani*; T, *Auriculinella bidentata*; U, *Sphaeroma serratum*; V, *Cingula trifasciata*; W, *Spirorbis spirorbis*; X, *Manzonia unifasciata*; Y, *Amphipholis squamata*; Z, *Onchidella celtica*; A1, *Lasaea adansoni*; B1, *Nereis diversicolor*; C1, *Pachygrapsus marmoratus*; D1, *Littorina striata*; E1, *Pedipes pedipes*; F1, *Ligia italica*; G1, Marine rush, *Juncus maritimus*; H1, *Anemonia sulcata*; I1, *Euphrosyne foliosa*; J1, *Cardita calyculata*; K1, Nematoda sp.; L1, *Ligia oceanica*; M1, Açorean spurry, *Spergularia azorica*; N1, Rock samphire, *Crithmum maritimum*; O1, Portulaca, *Portulaca oleracea* (after Morton et al., 1998).

correiana which is likely extinct (Agostinho in Bannerman and Bannerman, 1966). At Cabo de Praia, adjacent to Praia da Vitória, a quarry has been excavated so deep and so close to the sea that tidal seawater now floods parts of its floor daily (Morton et al., 1997). Here, a human engineered mistake has created another coastal wetland, born within the last ten years and still in an early stage of evolution. Colonized by wind-blown spores and seeds and bird phoresy, productivity is based around the estuarine grass *Ruppia maritima* and green algae, with flies, *Psilopa nitidula*, being the primary consumer, interestingly, as they were at the original marsh at Paúl (Morton et al., 1997). Also at the quarry are the remnant representatives of the coastal avifauna which were once so abundant at Paúl (Agostinho in Bannerman and Bannerman, 1966). Morton et al. (1997) argue that the quarry wetland, regardless of the unusual nature of its origin, should now be considered a unique Azorean coastal wetland resource deserving of protection and conservation.

In addition to the few habitats and communities already enumerated, there are, in the Azores, many other foci of biotic potential, including intertidal and shallow subtidal pools, sea stacks, sea arches, subtidal overhangs, sea caves and thermal seeps associated with both hard shores and sands, gravel beds at the mouths of streams, semi-emergent islands such as the Formigas and submarine seamounts scattered throughout the archipelago. Of considerable importance, however, are Azorean harbours. Harbour walls are the most common sheltered, albeit artificial, rocky habitats in the archipelago, some old and of hewn rock, others new and of smooth concrete. Regardless of composition, the native biota has reacted to these artificial substrata in a variety of ways (Morton et al., 1998). The floors of the harbours comprise another habitat, accumulating not only muds and silts but also a time-sequence of macro- and micro-pollutants (the former including cans, bottles and shards; the latter including heavy metals, pesticides and other chemical debris) introduced there by man. Like all harbours throughout the world, Azorean ones are also the focus of a unique form of biopollution — they are foci of introduced, alien or exotic, marine species (Morton and Britton, 1999) that may constitute the greatest long-term threat to Azorean coastal ecosystems. Just as feral cats and rats have already had profound impacts upon seabird populations in the Azores (Le Grand et al., 1984), the arrival of several new exotic marine species with a present focus in harbours (Zibrowius and Bianchi, 1981; Knight-Jones et al., 1991; Wirtz and Martins, 1993; Knight-Jones and Perkins, 1998), may prove to be disadvantageous to the present Azorean marine shallow water biota, even though their present impacts seem inconsequential.

OFFSHORE SYSTEMS

Since the first colonization of the islands, the sea has provided Azoreans with a rich fisheries harvest. Of the 460 species of fishes recorded from the Azores (Santos et al., 1997), about 100 of these plus several invertebrates including the squid *Loligo forbesi*, octopus *Octopus vulgaris*, spiny lobster *Panlinurus elephas*, slipper lobster *Scyllarides latus*, limpets *Patella* spp. and a barnacle *Megabalanus azoricus* are exploited. Of the fish, however, Isidro and Pereira (1998) identify 43 commercially important species. These author's updated assessment of the Azorean fishing industry provides the basis for this section.

The Azorean fishing fleet can be divided into three categories: small, medium and large (tuna) vessels. There are about 1900 small boats (<12.5 m long) scattered throughout the islands at every coastal village. Crews of one or two men launch and return their vessels each day carrying a variety of man-made gear, including lines, pots and traps, designed to catch an array of species. Medium-sized vessels, between 12 and 25 m in length and 13–18 tonnes, have crews of 5–8 men and can remain at sea for up to ten days. They usually carry a variety of gear enabling capture of either demersal (bottom long lines) or epi-and mesopelagic species (tangle-nets or long lines). The catch is held at sea on crushed ice and returned to one of 49 registered island ports. The tuna fleet comprises of 27 larger vessels of between 24 and 32 m in length and 140–200 tonnes. Each has a crew of between 10 and 19 men, who fish for tuna using live bait and pole-and-line gear.

The tuna fishery serves canning industries on the islands of São Miguel, Terceira and Faial and most of the medium-sized vessels are also registered here, representing the major population foci.

Since the mid-1980s, the numbers of small, artisanal, vessels has decreased by some 29%. They remain important, however, away from the major population centres. The medium-sized and tuna vessels have been extensively modernized and their power and efficiency increased over the last two decades. Indeed, 25 of the 27 vessels of the tuna fleet were built since 1984.

The main fishery activities are focused around size of catch and habitat, i.e. large (mostly tuna) pelagic, small pelagic, offshore demersal and inshore fisheries (Fig. 8).

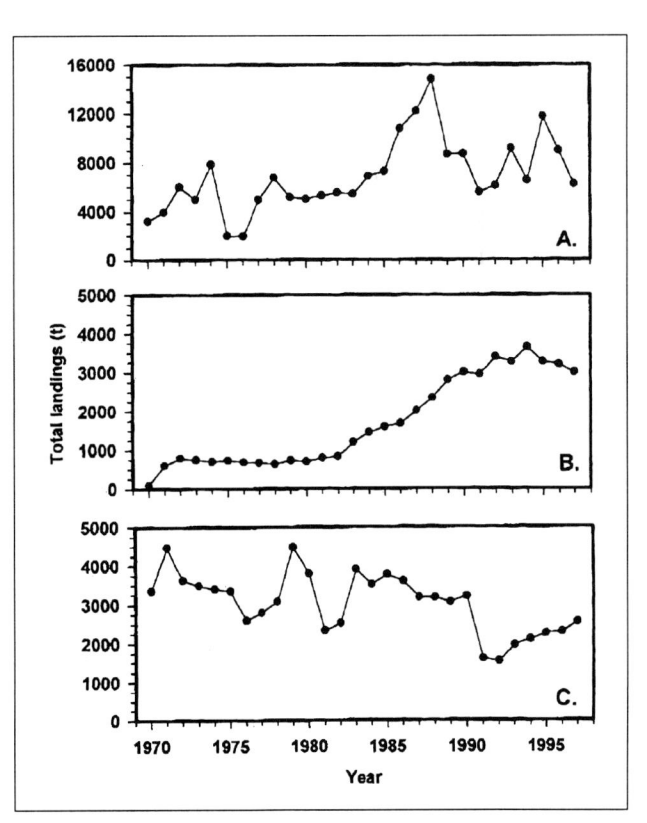

Fig. 8. Landings of the three main Azorean fisheries from 1970–1997. (A): The tuna fishery, including Bigeye, Albacore and Skipjack tuna. (B): The demersal fishery, including Red sea bream, Conger, Forkbeard, Blue-mouth rockfish, Rockfish, Alfonsim and Wreckfish. (C): The small pelagic fishery, including Blue-jack mackerel and Chub mackerel (after Isidro and Pereira, 1998).

Smaller-scale pelagic fisheries focus on seasonal target species such as Kitefish shark, Swordfish, Blue and Mako sharks and squid. The demersal fishery includes catches of Red sea bream, Conger, Forkbeard, Blue-mouth rockfish, Rockfish, Alfonsim and Wreckfish. Inshore fisheries target Barracudas, Mullets, Bogues and the several invertebrates listed previously. Changes in total catch between 1982 and 1996 reflect changes in the fishing fleet, with the small pelagic and inshore catch declining from 20% and 28% to 14% and 16%, respectively, of the total and the demersal and tuna components increasing from 9% and 43% to 20% and 50%, respectively. Porteiro (1994) has shown that the landings of the squid *Loligo forbesi* reached a maximum of 473 tonnes in 1989, but has declined in the 1990s. Silva and Krug (1992) used catch records and other data in a Virtual Population Analysis to estimate stock size of the Forkbeard, *Phycis phycis*. They concluded there had been a decrease in abundance from 8.59 million individuals in 1983 to about 4.63 million in 1990. During the same period, stock biomass decreased from 10,780 tonnes to 8,013 tonnes. Concurrent with these changes in catch, however, was an overall increase in fisheries income.

The tuna fishery mainly targets Albacore, Bigeye and Skipjack tuna, although a few Atlantic bluefin and Yellowfin are caught. The fishing season reflects the migratory behaviour of the target species, that is, between March and April and June and July for Bigeye; June, July and September for Skipjack and August, September, November and December for Albacore.

The demersal fishery is mostly a general one but the highly variable depth and bottom topography has promoted development of specialized fishing strategies and, sometimes, gear. As the medium-sized fisheries vessels have modernized, landings of demersal species have increased and some catch efforts have become focused for a particular catch, notably, Wreckfish and Silver scabbardfish.

The small pelagic fishery, targeting especially young Blue jack and Chub mackerel, has fluctuated over the last 30 years, without any significant increase in catch (Fig. 8B). The considerable variation in catch associated with this fishery is not explained easily, in part because the distribution and migratory behaviour of the primary target species, perhaps in association with climatic variation, are poorly known. The small-vessel, inshore fishery is, as noted earlier, in general decline. Due to its artisanal nature, catches have been difficult to quantify. This and the small pelagic fishery do, however, serve a useful purpose by providing bait fish for the tuna, squid and other fisheries (Martins, 1982; Pierce et al., 1994).

With no significant subtidal shelf and with seamounts constituting only 0.3% and 1.8% of the Azorean Exclusive Economic Zone down to 500 and 1000 m, respectively, natural productivity in this region is low. The probability that the present fishery can be maintained is low, in part because of its dependence on migratory species, which can be caught by other fleets elsewhere and, in part, because of its reliance on slow growing, long-lived, demersal species. Concern has been raised, not only about the present level of fisheries yield, but whether it is sustainable (Silva et al., 1994; Isidro and Pereira, 1998).

Some shore fisheries activities, for example, the limpet fishery, have had such profound ecological impacts upon rocky shore community ecology that a strictly enforced closed season for harvesting has been necessary (Santos et al., 1995). Similarly, the practice of overturning rocks for bait crabs, *Pachygrapsus* spp. and the artisanal *Octopus vulgaris* fishery keeps the resident community of this habitat depauperate. In this instance, humans are acting as a keystone predator to the shore community.

Historically, Cory's shearwater chicks, *Calonectris diomedea*, were minced alive for fish bait (LeGrand, 1984) and Azoreans were also whalers, mostly pursuing the Sperm whale, *Physeter macrocephalus*. Morton et al. (1998) provide a brief account of the Azorean whaling industry and see the "Protective Measures" Section below.

POPULATIONS AFFECTING THE AREA

A human population of about 250,000 occupies the Azores. It is distributed unevenly among the islands, with over half (132,000) living on São Miguel and half of those (65,000) inhabiting the city of Ponta Delgada. The population of Corvo, the smallest island, is about 450 (Santos et al., 1995).

Using population statistics, production data and theoretical calculations, Lobo (1987) rejected suggestions that the waters close to Angra do Heroismo, Terceira, were polluted by domestic sewage and industrial effluents. Twenty years later, in the only known study of its kind, Mendes et al. (1997) examined the sanitary quality of sands from Azorean beaches. They cultured a variety of bacterial and fungal indicators of sanitary health, including total faecal coliforms, faecal enterococci, *Clostridium*, *Pseudomonas aeruginosa*, and such fungi as yeasts and potential pathogenic and allergic saprophytes. They concluded from these assays, however, that all Azorean beaches should be classified as either environmentally good or satisfactory, according to proposed guidelines for the microbiological quality of sands, and should be awarded a local 'Blue Flag' of high environmental quality.

In the Azores, the neogastropod whelk *Stramonita haemastoma* shows varying degrees of imposex, that is, the superposition of male characters upon females, induced by tributyltin (TBT) contained in marine anti-fouling paints. The highest levels of imposex were recorded from harbours, as at Horta, Faial (Spence et al., 1990). Blaber (1970) first described the condition in the dogwhelk *Nucella lapillus* from Plymouth, England and it became known as 'imposex' following the study by Smith (1981) of the mud snail *Ilynassa obsoleta* from eastern North America. Unlike these species, *S. haemostoma* has a teleplanic larva so that recruitment can be sustained, even from distant sources, where local TBT pollution is high. Accordingly, Azorean rocky shore community ecology is relatively uninfluenced by variations in numbers of this 'keystone' predator, being more profoundly influenced by the non-sustainable harvesting by humans of its principal prey, the limpets *Patella* spp. (Hawkins et al., 1990).

RURAL FACTORS

The largest threat to Azorean seabirds does not appear to be from pollution *per se*, but impacts from humans and introduced feral species. In the 16th and 17th centuries, large populations of seabirds nested on every island. Today, surface-nesting seabirds are restricted mostly to offshore islets. Current threats to Azorean seabirds include predation by introduced cats and rats, human disturbance and exploitation, habitat loss to invasive alien plants, overgrazing of nesting sites by rabbits and domestic animals and competition with fisheries, although the latter is unquantified (Monteiro et al., 1996).

Artificial fertilizers are applied to fields throughout the Azores and are known to bring about eutrophication in ponds and lakes. There have been no studies of their effects upon the sea in the Azores.

COASTAL EROSION AND LANDFILL

On volcanic islands such as the Azores, coastal erosion commences immediately following the latest land-forming eruption. Significant coastal erosion has occurred since the 1957 eruption of Capelinhos Mountain, Faial, as evidenced by the broad boulder beach originally deposited as a thick blanket of volcanic ash and pyroclastic inclusions. Only a few studies of coastal landforms and erosion in the Azores have been made, among the earliest being the pioneer work of Berthois (1953) on the coastal morphology of São Miguel Island. Boothe et al. (1978) addressed recent volcanism on São Miguel, including some information on its effects on coastal landforms.

Tectonic activity, especially earthquakes, may also have profound consequences for the coastal zone. São Jorge, for example, is a high, elongate, island, most of which meets the sea as precipitous cliffs rising hundreds of metres above the surf. At several points on São Jorge, massive landslides have produced depositional coastal platforms (*fajãs*), especially along the northern coast. Fajã dos Cubres and Fajã de Santo Cristo are two of the largest, their geomorphology and ecology described in detail by Morton et al. (1995, 1996), respectively. Both were created from landslides produced by the earthquake of 9 July 1757, depositing a volume of debris estimated to be as much as six times greater than that presently in evidence (Forjaz and Fernandes, 1975) and, today, they constitute the most important coastal environments in the Azores.

Borges et al. (1997) used aerial photographs from 1955, 1974 and 1988 to assess coastal erosion of sea cliffs on the southern coast of São Miguel at Rocha dos Campos. Erosion rates varied from place to place, with a mean of $0.12\,m\,year^{-1}$. These authors also pointed out that many existing dwellings had been erected too close to eroding escarpments, as elsewhere, suggesting that the builders failed to anticipate the necessary distances needed for safety. Morton et al. (1998) addressed another high-risk coastal settlement at Relva, along the southwestern coast of São Miguel, where some dwellings occur within a few metres of a precipice. On the other hand, here and elsewhere in the Azores, the exotic, South African, Hottentot fig *Carpobrotus edulis*, has been utilized as a ground cover to stabilize sea cliffs.

Winter storms are especially erosive on some Azorean beaches. The storm that hit São Miguel during Christmas 1996, produced 12–14 m waves which broke over the outer wall of Ponta Delgada Harbour, ripping ships from their moorings and putting five ocean-going vessels onto rocks along the inner margin of the harbour. This same storm lashed Ribeira Grande on the northern coast of São Miguel causing a dramatic change in beach topography and appearance. Sand was removed from the beach by destructive waves, exposing the rock and cobble base (Morton et al., 1998). Such waves also render sea cliffs unstable, especially those composed of poorly lithified volcanic ash, by undercutting them.

Human-induced beach erosion can be seen between Rabo de Peixe and Ribeira Seca on the north shore of São Miguel. Here, a cottage industry has developed in which beach sand from the foot of already eroding sea cliffs is dug up and transported inland in panniers carried by horses (Morton et al., 1998), to be sold to the construction industry. Another cottage industry with the potential to enhance coastal erosion is the almost unregulated artisanal collection of the red alga *Pterocladia capillacea* for agar extraction and pesticide use (Depledge et al., 1992).

Perhaps the most dramatic example of both landfill and human-induced coastal erosion can be seen on the island of Terceira, and described earlier. The most extensive coastal wetland in the Azores once occurred at Paúl, near Praia da Vitória (Bannerman and Bannerman, 1966). During this century, the Paúl wetland has been gradually, but progressively, reclaimed (Morton et al., 1998). In the last stages of the Paúl wetland reduction, a coastal quarry at Cabo da Praia was excavated too closely to the sea and sufficiently deep to penetrate the pizometric level for tidal groundwater. As a result, a new tidal wetland is being created (Morton et al., 1997). As elsewhere, therefore, the coastal landform of the Azores is affected profoundly by natural and human-induced perturbations. At this mid-Atlantic site, however, where tectonic and climatic events interact most dramatically, the former overrides the latter. Notwithstanding, human effects have the power to be most significant on wetlands and Morton et al. (1999) have made an especial plea for their conservation in the Azores.

EFFECTS FROM URBAN AND INDUSTRIAL ACTIVITIES

Mercury levels in the feathers of seven species of Azorean seabirds were examined by Monteiro and Furness (1995). Median levels were much higher in petrels than in shearwaters and terns (Madeiran storm petrel = 12.5 $\mu g\ g^{-1}$; Bulwer's petrel = 22.1 $\mu g\ g^{-1}$; Little shearwater = 2.1 $\mu g\ g^{-1}$; Cory's shearwater = 6.0 $\mu g\ g^{-1}$; Roseate tern = 20 $\mu g\ g^{-1}$; Common tern = 2.3 $\mu g\ g^{-1}$). Mercury levels were significantly lower in shearwater and tern chicks and, among the latter, mercury levels decreased by 60–70% with increasing age. Mercury levels in Madeiran storm petrels also varied seasonally, being 50% lower in spring than in autumn breeders.

Levels of mercury in seabirds show a dose–response relationship, providing a means for monitoring ocean pollution by this metal. Monteiro and Furness (1995), measuring mercury levels in feathers from museum specimen seabirds representing broad geographic areas, suggested that contamination over the last 100 years has experienced a three-fold increase in the Northeastern Atlantic, but not, significantly, in the South Atlantic over the same period. Later, Monteiro and Furness (1997) examined mercury levels in bird features representing top predators of either epi- or mesopelagic fishes. In this study, the authors showed that over the last 100 years mercury levels had increased by 1.1–1.9% $year^{-1}$ in the epipelagic food chain and 3.5–4.8% $year^{-1}$ in the mesopelagic one, the latter rate being previously undetected.

Between 1989 and 1990, two scorpionfish species, *Helicolenus dactylus* and *Pontinus kuhlii*, were captured in Azorean waters and examined for mercury content (Monteiro et al., 1991). There were highly significant correlations between mercury levels in both species and sex, length, weight, age, growth and condition. Accumulation rates were significantly faster, for example, in females than males. It was concluded, therefore, that such benthic predators might constitute useful heavy metal biomonitors.

Cory's shearwater, *Calonectris diomedea*, from the Mediterranean and Salvage Islands were shown to have high concentrations of heavy metals in their tissues, attributable to accumulation from their prey, both fish and squid (Renzoni et al., 1986). Squids, for example, are known to accumulate high concentrations of cadmium in the digestive gland (Martin and Flegal, 1975). A recent study by Stewart et al. (1997) showed that fledgling (12-week-old) Cory's shearwater chicks from the Azores accumulated measurable amounts of cadmium, zinc and copper in both kidney and liver tissues. Levels of zinc and copper were similar to that reported for shearwaters and other seabirds from other islands (Renzoni et al., 1986), but virtually all of the cadmium present in the fledglings had accumulated after hatching, as eggs contained only trace amounts (Renzoni et al., 1986). Cadmium concentrations in Azorean fledgling shearwaters were lower than that reported from adults on other Atlantic and Mediterranean islands.

Several studies have examined levels of mercury in oceanic fishes taken from near the Azores. For example, total mercury in the muscle of the Swordfish *Xiphias gladius* ranged from 0.06 to 491 $\mu g\ g^{-1}$ dry weight (Monteiro and Lopes, 1990), with rates of accumulation significantly faster in males than females. Mean mercury levels, ranging from 57 to 377 ppb, in eight species of either epi- (<200 m) and mesopelagic (>300 m) fishes caught near the Azores were significantly and positively correlated with median diurnal depth (Monteiro et al., 1996). Mercury concentrations in fish tissues increased four-fold from the epipelagic to the upper mesopelagic, but with no additional increase within the middle mesopelagic down to a depth of 1200 m. Such depth-related bioaccumulation seems to be determined, first, by the availability of mercury in food and, eventually, by aspects of water chemistry which control its uptake and speciation as it is transferred from the base of the epipelagic food chain to the mesopelagic predators (Monteiro et al., 1996).

Closer to shore, total mercury and methylmercury levels were measured in 1993–1994 for a variety of commercial fish and shellfish from São Miguel (Andersen and Depledge, 1997). Mean concentrations of total mercury were highest in the gills and midgut gland (0.864 and 1.265 $\mu g\ g^{-1}$, respectively) of the decapod crab *Cancer pagurus*. Apart from the White seabream, *Diplodus sargus cadenati*, mean total concentrations in fish muscle tissues were low (0.043–0.371 $\mu g\ g^{-1}$) with more than 80% of the total mercury in the form of methylmercury. High levels in some White seabream individuals, up to 24.61 $\mu g\ g^{-1}$ wet weight, might have been derived from local hot spots with high background levels of mercury.

Mercury levels in mantle muscle tissues were determined for four species of Azorean cephalopods, *Loligo forbesi*, *Octopus vulgaris*, *Ommastrephes bartrami* and *Todarodes sagittatus* (Monteiro et al., 1992). Mercury concentrations increased exponentially with body size, but the squid *L. forbesi* had significantly higher levels than the other three species. Female *L. forbesi* generally had higher levels than males and mercury levels in both sexes, especially small and medium-sized individuals, varied seasonally. Of especial interest was that there were significantly higher mercury levels in the mantle muscles of the coastal octopus, *O. vulgaris*, collected near urban areas than those from more remote, less developed, shorelines.

When levels of organochlorinated pesticides and PCBs in fishes were compared from samples collected in mainland Portugal estuaries and from the Azores, only the former were considered to be very polluted by such compounds (Magalhaes and Barros, 1987). Concentrations of these pollutants in Azorean fishes were relatively low.

The whole body concentrations of copper and zinc in four coastal Azorean talitroid amphipods, i.e. *Talitroides topitotum* (euterrestrial), *Orchestia gammarelus* (maritime

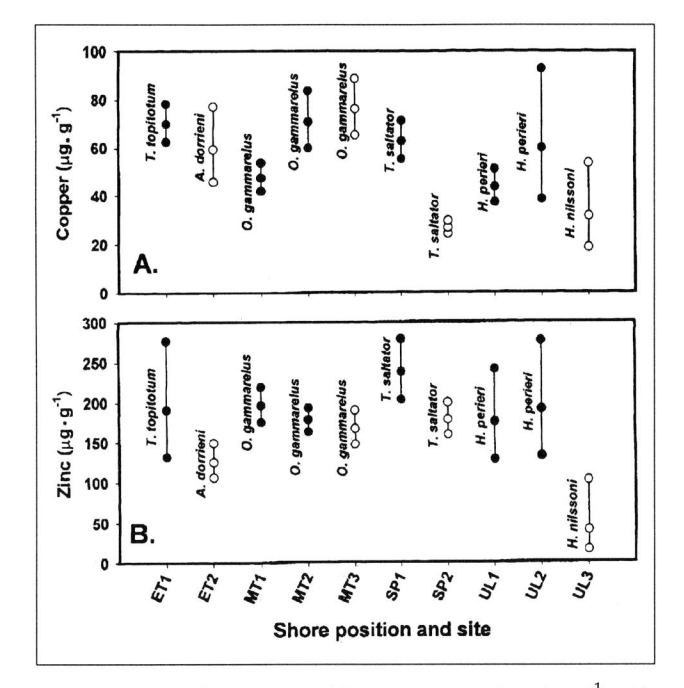

Fig. 9. Estimates of A, copper and B, zinc concentrations ($\mu g\ g^{-1}$) with 95% confidence limits for Azorean (solid circles) and Scottish (open circles) amphipods from different levels on the shore. Mean concentrations are for standard either 5 or 10 mg dry weight measurements derived from best fit double log regressions. All Azores amphipods were from the island of São Miguel. Shore positions, locations and full species citations are as follows: ET1, *Talitroides topitotum* from euterrestrial leaf litter at 290 m above sea level, São Miguel; ET2, *Arcitalitrus dorrieni* from a euterrestrial site on the island of Colonsay, inner Hebrides, Scotland; MT1, MT2 and MT3, *Orchestia gammarelus* from the upper maritime (30 m above sea level) on Ilhéu de Vila Franca, from the lower maritime/supralittoral fringe boundary at Calhetas, and from the upper supralittoral fringe at Millport, Scotland, respectively; SP1 and SP2, *Talitrus saltator* from the supralittoral fringe at Vila Franca do Campo and from a similar position at Millport, Scotland, respectively; UL1 and UL2, *Hyale perieri* from the upper eulittoral at Ferraria and Mosteiros, respectively and UL3, *Hyale nilssoni* from the upper eulittoral at Millport, Scotland. From Moore et al. (1995), who also provide other significant constraints and clarifications for these summarized data.

terrestrial), *Talitrus saltator* (supralittoral) and *Hyale pareiri* (high littoral) were examined by Moore et al. (1995) and we have summarized these data in Fig. 9. Trace metal concentrations from the marine amphipods seem somewhat elevated above those of the more maritime and euterrestrial species. For example, copper concentrations in *O. gammarelus* were significantly higher for more seaward populations than those occupying more euterrestrial sites (Fig. 9A). This might suggest a marine source of heavy metal contamination. There are, however, few industrial sources for such metals in the Azores. Even more enigmatic, concentrations of copper and zinc (Fig. 9B) in the Azorean amphipods were either higher or much higher than those recorded for either conspecifics or congeners occupying the industrialized Clyde Estuary (Millport), Scotland, where copper and zinc levels are low year round (Rainbow and Moore, 1990). Such results are difficult to interpret. It is likely that the Azorean sources are natural ones, reflecting enhanced bioavailabilities. Regardless of source, this casts doubt, as Moore et al. (1995) characterized it, on the 'simplistic' notion that an isolated, volcanic, mid-Atlantic shore community might provide a pristine 'control' environment and baseline measurements for studies of heavy metal contamination elsewhere.

Barnacles are good bioindicators of copper, zinc and cadmium concentrations (Rainbow and White, 1989). In a study of these metals in the barnacle *Chthamalus stellatus* from five shore locations around São Miguel, Weeks et al. (1995) showed that body copper concentrations varied about twofold between localities (30.2–12.4 $\mu g\ g^{-1}$) indicating local variation in bioavailability.

Zinc concentrations in *C. stellatus* varied considerably and were highest at a thermal spring site (2754 $\mu g\ g^{-1}$) and close to the city of Ponta Delgada (436 $\mu g\ g^{-1}$), other values being similar, i.e. ~200–300 $\mu g\ g^{-1}$. Cadmium concentrations were high at all sites, the highest reported value of 167 $\mu g\ g^{-1}$, again near Ponta Delgada, being the highest ever recorded for a barnacle. *Semibalanus balanoides* from the notoriously metal-rich 'hot spot' of Dulas Bay, England, had a cadmium tissue concentration of 60 $\mu g\ g^{-1}$ (Rainbow, 1987). Stenner and Nickless (1975), moreover, report cadmium concentrations of only 5.1 to 6.3 $\mu g\ g^{-1}$ in *C. stellatus* from the mainland coasts of Portugal and Spain. These data again, therefore, cast doubt on the usefulness of the Azores as an unpolluted, 'control', site for anthropogenic-induced metal bioavailabilities.

In a test of the possibility that elevated zinc levels in *Chthamalus stellatus* were the result of proximity to a thermal spring (Weeks et al., 1995), Vedel and Depledge (1995) measured copper and zinc concentrations in the limpet *Patella candei gomesi* from the same site and others on São Miguel, one close to Ponta Delgada. Zinc and copper concentrations were higher in limpets from the thermal spring site than the 'colder' others (Zinc, 41.5 vs. 35.5 $\mu g\ g^{-1}$; Copper, 5.59 vs. 4.29 $\mu g\ g^{-1}$), with all values within the ranges established by Eisler (1981) for these two metals as typical of 'clean' sites.

PROTECTIVE MEASURES

Santos et al. (1995) provided a comprehensive review of Azorean marine resources, a list of marine species currently receiving legislative protection and an inventory of established conservation sites in the Azores. Morton and Tristão da Cunha (1993) and Morton et al. (1995, 1996, 1997, 1998) added to and updated this review, especially by identifying additional sites and measures deserving attention. Here we provide only a brief summary of the information contained in these references.

Only a few marine animals are protected by legislation in the Azores. Whales, especially the Sperm whale, *Physeter macrocephalus*, were once among the most exploited Azorean marine vertebrates. Azorean whaling, always an artisanal, shore-based, activity, once consisted of up to 200

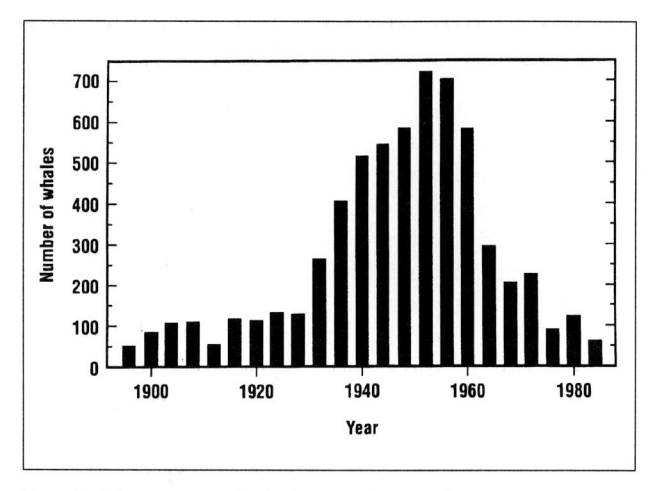

Fig. 10. The pattern of whaling catches in the Azores, from the beginning to the end of the industry (after Morton et al., 1998).

small boats designed to capture and return sperm whales to oil factories on several of the islands. Whaling began in the Azores in the latter years of the 19th century and reached its peak in the mid-1950s (Fig. 10). Thereafter, it began to decline and ceased in 1987. Today, all sea mammals are protected by the Azorean government and whaling has become 'whale-watching', with several commercially successful operations based mostly at Lajes, Pico.

Few additional pelagic animals are protected by legislation. The various fisheries conduct their harvesting almost without regulation, although catch records for most commercially important species are kept (Fig. 8). With the exception of one inshore species, the Dusky perch, *Epinephelus marginatus*, which cannot be taken by spearfishing, there are no special regulations concerned with minimum catch size and fishing season for Azorean fishes. There is, however, a special regulation limiting the size of nets used to catch live bait and a daily catch limit for spearfishing. In contrast, over 20 species of mainland Portugal marine fishes are protected by various regulations. Santos et al. (1995) called for legislative protection of additional Azorean fishes, to be determined on a species by species basis. Sea turtles are generally uncommon in the Azores, except the Loggerhead, *Caretta caretta*, which visits the islands during transatlantic migrations. No species is protected by specific legislation, nor are they particularly sought after, their paucity in the area itself rendering a certain measure of protection.

Azoreans consider limpets, *Patella* spp., a culinary delight and have denuded most accessible shore of them. Beginning in 1985, limpets received legislative protection, including a ban on their collection in some localities for several years, the prohibition of commercial collecting on all islands, the creation of protected areas and protected seasons, and, eventually, licence requirements for limpet collectors. The imposition of these regulations had a positive effect on limpet populations, with both Azorean species making noticeable recoveries. A second mollusc, the bivalve *Venerupis decussatus*, although almost certainly

introduced into the Azores from mainland Europe, has received considerable attention and protection, largely on the basis of its extremely limited distribution. In the Azores it is known only from the remote Fajã de Santa Cristo lagoon on São Jorge. Azoreans once would make the arduous journey to Santa Cristo by foot to collect and remove bucketfuls of clams. Today, both the site and its unique bivalve resident are protected by legislation, but there is an increasing demand by some to construct a road to Santa Cristo. Two lobsters, *Scyllarides latus* and *Palinurus elephas*, and a barnacle, *Megabalanus azoricus*, have also received various forms of protection from over-fishing, including limited closed seasons, minimum sizes and capture methods. The Common octopus, *Octopus vulgaris*, is heavily exploited throughout the islands, but appears to be holding its ground without protective regulations (Gonçalves, 1991).

Seabirds are among the most conspicuous elements of the marine and maritime fauna throughout the world and those in the Azores are no exception. Cory's shearwater, *Calonectris diomedea*, is the most abundant species and is joined by seven others, all of which are regular breeders on either some or all islands. The Azores is also the most important European breeding site for the Roseate tern, *Sterna dougallii*, comprising over 60% of this population (Santos et al., 1995). Fifteen major Azorean seabird sites have been identified at least as 'Special Protection Areas for Birds' under the European Union Wildbirds Directive. A few have received either additional local or regional legislative protection, being designated as either Nature Reserves, Marine Protected Areas or Special Ecological Areas, although these may have received such designations for reasons other than, or in addition to, seabird protection (Table 1). Despite these designations, the seabirds, themselves, have received remarkably little protective attention. So-called protected sites are abused widely from the standpoint of conservation and species protection. Ilhéu do Topo at the western end of São Jorge was designated a Nature Reserve in 1984, but Santos et al. (1995) described how cattle grazing on the islet affected nesting seabird colonies. They suggested that this problem was solved when the regional administration purchased Topo, yet in 1997 we observed that although the cows were gone, they had been replaced by sheep and goats. Ilhéu de Vila Franca, São Miguel, is a designated, locally regulated, Marine Protected Area, but this designation has not stopped local entrepreneurs constructing concrete concession stands and sunbathing sites for holiday visitors in blatant disregard of the damage this has done to the natural community. As the protection of seabirds, however ineffective, seems intimately linked to the recognition of specific coastal sites deserving protection, we will use this as a segue to a discussion of the latter.

The Azorean government has recognized nine Marine Protected Areas, one Protected Landscape and one Special Ecological Area on four islands and one shallow bank (Table 1). The level of protection afforded each by such designations has been minimal and, as evidenced at Ilhéu

Table 1

Existing marine protected areas in the Azores.

Island	Designation	Site	Remarks
Santa Maria	Marine Reserve	Baía da Maia	All established in 1987, mainly for recreation; all are also biologically and ecologically important
	Marine Reserve	Baía de São Lourenço	
	Marine Reserve	Baía dos Anjos	
	Marine Reserve	Baía da Praia	
Formigas	Marine Reserve	Formigas Islets and Dolabarat Bank	Barely emergent islets and shallow submarine bank between Santa Maria and São Miguel
São Miguel	Marine Reserve	Ilhéu de Vila Franca	Despite 'Reserve' status and an important shearwater nesting site, it experiences heavy recreational abuse
São Jorge	Natural Reserve	Ilhéu do Topo	Established in 1984, it continues to be abused by grazing livestock
	Special Ecological Area	Lagoa do Santo Cristo	Established in 1989 for protection of both the site and its unique bivalve resident, *Venerupis decussatus*
Faial	Protected Landscape	Monte da Guia	Terrestrial resources protected; limited protection for marine resources (see Table 2)

do Topo and Ilhéu de Vila Franca, either non-existent or an invitation for abuse. Part of the problem has been that, although these sites have been recognized for their uniqueness, no accompanying enforcement legislation has been enacted to ensure the conservation of their unique state. One might argue that until such enforcement legislation is enacted, the recognition of additional Marine Protected Areas would be an exercise in futility. Nevertheless, 31 other coastal sites have been proposed for protection (Table 2). In many instances, protection of seabird populations is the primary reason such a status has been suggested, but several sites have been recommended for protection for a variety of other reasons. The aesthetic quality of Caloura and Mosteiros, São Miguel, Monte Brasil, Terceira and Ponta dos Rosais, São Jorge, to name a few, is adequate justification for protection status. The tourism potential for such sites as Porto Formoso, São Miguel and Baixa do Canal between Faial and Pico is another. Protection of Mosteiros, São Miguel and Lajes do Pico could regulate already excessive human abuse of these fragile ecosystems.

Of especial interest are the extremely fragile coastal wetland environments of the Azores—environments rarely encountered on oceanic islands. A once-extensive marsh on Terceira at Paúl (Bannerman and Bannerman, 1966) has already been destroyed. There remains, however, four sites which can be properly called coastal wetlands: Lajes do Pico (Morton et al., 1996), Fajã do Santo Cristo (Morton and Tristão da Cunha, 1993) and Fajã dos Cubres (Morton et al., 1995) on São Jorge and a newly created tidal wetland at the bottom of a quarry at Cabo de Praia (Morton et al., 1997). Each of these sites share similar features but also remain unique from one another (Morton et al., 1999). All experience tidal fluctuations and all contain some water of noticeably reduced salinity. Except for a small pond at Lajes do Pico and some groundwater seeps along one side of the lagoon at Santo Cristo, São Jorge, these two sites are mostly true marine systems. At Fajã dos Cubres and the quarry wetland at Cabo de Praia, groundwater dilution of seawater is sufficiently great to produce a truly estuarine condition. All of these sites attract a coastal avifauna with a unique assemblage of between 10-20 species at each site, among which Turnstones *Arenaria interpres*, terns *Sterna* spp., Kentish plovers *Charadrius alexandrinus*, and Grey herons *Ardea cinerea* are especially common (Morton et al., 1999).

Long isolated from mainland European events and communication, the Azorean legacy is one of a brutish work ethic and the exploitation of all attainable reserves to sustain an island's people. As the second millennium closes, two events, Portugal's admission to the European Union and the increasing insular interaction with the mainland, have improved the Azorean quality of life greatly. A developing tourist industry, founded upon a more affluent clientele seeking to experience the Azorean cultural heritage, has stimulated a commercial environment whereby traditional foraging habits along the shore and the exploitation of all accessible marine resources for food are no longer either acceptable or necessary. It is also unlikely that, in the foreseeable future, the Azores will experience significant industrialization and, thus, pollution is likely to remain near background levels. Ironically, however, some geological influences elevate the background levels sufficiently that these islands cannot serve as a pristine 'control' site for pollution parameters monitored elsewhere.

The future thus lies with (i) the sustainable utilization of existing marine resources and their protection from foreign fleets and (ii) the protection and conservation of unique Azorean habitats which have been identified by Santos et al. (1995) and Morton et al. (1998) and in this paper. Both measures will serve local people well in the third millennium and ensure that important seascapes and their unique biotas are protected for aesthetic, recreational and educational reasons and scientific research. Identified sites also have great tourist potential. Most important, however,

Table 2

Marine sites in the Azores proposed for protection.

Island	Site	References	Remarks
Santa Maria	Lagoinhas	Santos et al., 1995	Islets; nesting Roseate terns
	Ilhéu da Vila	Santos et al., 1995	Islet; 7 species of resident seabirds
São Miguel	Caloura	Hamer et al., 1989; Hawkins et al., 1990; Britton, 1995; Santos et al., 1995; Morton et al., 1998	Spectacular seascapes; important ecological site
	Mosteiros	Santos et al., 1995	Popular inshore fishery site; broad subtidal platform; protected status would regulate the fishery
	Porto Formoso	Santos et al., 1995	Ecologically interesting sandy and rocky shores; tourism potential
	Nordeste	Santos et al., 1995	Rocky exposed shore; protected status would create a reserve for marine biota
	Ribeira Quente	Santos et al., 1995; Morton et al., 1998	Small estuarine system; hydrothermal vents on sandy beach and to depths of 15 m; regulations needed to prevent sand removal
Terceira	Cabo de Praia	Morton et al., 1997; Morton et al., 1998	New coastal wetland accidentally created in a quarry; shorebird and wetland protection
	Ilhéus das Cabras and Ilhéu dos Fradinhos	Santos et al., 1995	Seabird and marine fauna protection
	Ilhéu de Mós	Santos et al., 1995	Protection of breeding terns
	Monte Brasil	Santos et al., 1995	Extend existing terrestrial reserve to protect shore biota
	Serreta	Santos et al., 1995	Protection of littoral fish and crustaceans
Graciosa	Ilhéu da Praia	Grimmet and Jones, 1989; Santos et al., 1995	Diverse marine fauna and flora; important seabird site
	Ponta do Carapacho and Ilhéu de Baixo	Grimmet and Jones, 1989; Santos et al., 1995	Important subtidal habitats; breeding seabirds
	Ponta Branca and Ilhéu	Santos et al., 1995	Seabird site
	Baía da Vitória to Baía das Diagaves	Santos et al., 1995	Diverse marine flora and fauna
São Jorge	Fajã dos Cubres	Morton et al., 1995; Santos et al., 1995; Morton et al., 1998	Unique, low salinity, coastal wetland; important shorebird site
	Fajã do Santo Cristo	Morton and Tristão da Cunha, 1993; Santos et al., 1995; Morton et al., 1998	Already protected Special Ecological Area (Table 1); expand and extend protection to Fajã dos Cubres, including the coastline between the two.
	Morros das Velas	Santos et al., 1995	Important seabird site
	Ponta dos Rosais	Santos et al., 1995	Scenic; rich in fish
Pico	Ilhéus da Madalena and Baixa do Canal	Santos et al., 1995	Two islets with nesting seabirds at the western end and the shallow bank between Pico and Faial; underwater tourism
	Lajes do Pico	Santos et al., 1995; Morton et al., 1996; Morton et al., 1998	Littoral platform and coastal wetlands in urban setting; heavy human pressure
	Baía das Canas	Santos et al., 1995	Rich sublittoral fauna; nesting shearwaters
Faial	Capelinhos	Santos et al., 1995; Morton et al., 1998	Created by volcanic eruptions 1957-59; already a terrestrial Nature Preserve; extend protection to marine environment
	Morro de Castelo Branco	Santos et al., 1995	already a terrestrial Nature Preserve; extend protection to marine environment; nesting seabirds; dolphins; fishes
	Ponta dos Cedros to Ponta do Salão	Santos et al., 1995	Rich sublittoral fauna
	Feteira to Horta Harbour	Santos et al., 1995	Includes Monte da Guia, a protected Landscape (Table 1); proposal extends marine protection
Flores	Ponta Delgada to Santa Cruz	del Nevo et al., 1990; Santos et al., 1995	Rocky seacliffs, islets and stacks along NE coast shelter the largest breeding population of Roseate terns in Eastern Atlantic
	Ponta dos Bredos to Ponta Lopo Vaz	del Nevo et al., 1990; Santos et al., 1995	Another large concentration of Roseate terns along SW coast
	Baixa Rosa to Ponta Delgada	Santos et al., 1995; Morton et al., 1998	Rocky seacliffs, seaside waterfalls, islets and stacks; subtidal fauna; seabirds; includes Fazenda de Santa Cruz
Corvo	Entire island	Santos et al., 1995; Morton et al., 1998	Small in both size and human population (450), the entire island could be designated a coastal reserve

is the need to preserve the Azores as a great biological, biogeographical and ecological *in situ* experiment that can serve to enrich knowledge of remote island evolution. The scientific value of the Azores must be matched by a local need to protect the Azorean heritage. The path to be followed in the coming millennium, therefore, is one that sustains the islands, their biota and the people. One of the greatest threats to this course is the accidental and/or deliberate introduction of alien species which confounds attempts to understand natural biogeographical events. But, here too, there is the opportunity for research to achieve an understanding of the way the human species interacts with this special environment.

REFERENCES

Andersen, J.L. and Depledge, M.H. (1997) A survey of total mercury and methylmercury in edible fish and invertebrates from Azorean waters. *Marine Environmental Research*, **44**, 331–350.

Azevedo, J.M.M., Portugal Ferreira, M.R. and Martins, J.A. (1991) The emergent volcanism of Flores Island, Azores (Portugal). *Arquipél. Life Earth Sci.* **9**, 37–46.

Bannerman, D.A. and Bannerman, W.M. (1966) *Birds of the Atlantic Islands, Volume III, Azores*. Oliver & Boyd, London.

Berthois, L. (1953) Ile de São Miguel, géomorphologie littorale. *Communicações dos Serviços Geológicos de Portugal*, **34**, 41–71.

Blaber, S.J.M. (1970) The occurrence of a penis-like outgrowth behind the right tentacle in spent females of *Nucella lapillus* (L.). *Proc. Malacol. Soc. Lond.* **39**, 231–233.

Booth, B., Croasdale, R. and Walker, G.P.L. (1978) A quantitative study of five thousand years of volcanism on São Miguel, Azores. *Philosophical Transactions of the Royal Society London, Series A.* **288**, 271–319.

Borges, P.A., Cruz, J.V. and Andrade, C.F. (1997) O recuo da arriba littoral no local da Rocha dos Campos (Água d'Alto, São Miguel) um estudo do caso. *Açor.* **8**, 391–401.

Boury-Esnault, N. and Lopes, M.T. (1985) Les démosponges littorales de l'archipel des Açores. *Ann. Inst. Oceanogr., Paris, Nouv. Sér.* **61**, 149–225.

Britton, J.C. (1995) The relationship between position on shore and shell ornamentation in two size-dependent morphotypes of *Littorina striata* with an estimate of evaporative water loss in these morphotypes and *Melarhaphe neritoides*. In *Developments in Hydrobiology III. Advances in Littorinid Biology*, ed. J. Graham, P.J. Mill and D.G. Reid, pp. 129–142. Kluwer, Dordrecht, the Netherlands.

Bullock, R.C. (1995) The distribution of the molluscan fauna associated with the intertidal coralline algal turf of a partially submerged volcanic crater, the Ilhéu de Vila Franca, São Miguel, Azores. In *Proc. Second Internat. Workshop of Malacol. and Mar. Biol., Vila Franca do Campo, São Miguel, Azores, 1991*, ed. A.M.F. Martins, pp. 9–55. Açor., Suppl. 4.

Bullock, R.C., Turner, R.D. and Fralick, R.A. (1990) Species richness and diversity of algal-associated micromolluscan communities from Sao Miguel, Acores. In *Proc. First Internal. Workshop Malacol. Mar. Biol., Vila Franca do Campo, São Miguel, Azores, 1988*, ed. A.M.F. Martins, pp. 39–58. Açor., Supl., Outubro 1990.

Class, T. and Ballschmiter, K. (1988) Chemistry of organic traces in air. 8. Sources and distribution of bromo- and bromochloromethanes in marine air and surface water of the Atlantic Ocean. *Journal of Atmospheric Chemistry* **6**, 35–46.

Cornelius, P.F.S. (1992) The Azores hydroid fauna and its origin, with a discussion of rafting and medusa supression. *Arquipélago, Life and Earth Sciences* **10**, 75–99.

Dahlgaard, H., Chen, Q., Herrmann, J., Nies, H., Ibbett, R.D. and Kershaw, P.J. (1995) On the background level of ^{99}Tc, ^{90}Sr and ^{137}Cs in the North Atlantic. *Journal of Marine Systems* **6**, 571–578.

del Nevo, A.J., Dunn, E.E., Medeiros, F.M., LeGrand, G., Akers, P. Avery, M.I. and Monteiro, L.R. (1990) The status of Roseate terns *Sterna dougallii* and Common terns *Sterna hirundo* in the Azores. *Seabird* **15**, 30–37.

Depledge, M.H., Weeks, J.M., Frias-Martins, A., Tristao da Cunha, R. and Costa, A. (1992) The Azores. Exploitation and pollution of the coastal ecosystem. *Marine Pollution Bulletin* **24**, 433–435.

Eisler, R. (1981) *Trace Metal Concentrations in Marine Organisms*. Pergamon Press, Oxford.

Forjaz, V.H. and Fernandes, N.S.M. (1975) *Carta Geológica de Portugal na escalaq do 1/50,000. Noticia explicativa das folhas A e B da ilha de S. Jorge (Açores)*. Serviços Geológicos de Portugal, Lisbon.

Gofas, S. (1989) Two new species of *Alvania* (Rissoidae) from the Azores. *Publ. Ocas. Soc. Portuguesa Malacol.* **14**, 39–42.

Gofas, S. (1990) The littoral Rissoidae and Anabathridae of Sao Miguel, Azores. In *Proc. First Internal. Workshop Malacol. Mar. Biol., Vila Franca do Campo, São Miguel, Azores, 1988*, ed. A.M.F. Martins, pp. 97–134. Açor., Supl., Outubro 1990.

Gonçalves, J.M. (1991) The Octopoda (Mollusca:Cephalopoda) of the Azores. *Arquipélago, Life and Earth Sciences* **9**, 75–81.

Grimmet, R.F.A. and Jones, T.A. (1989) Important bird areas in Europe. *ICBP Technical Publication No. 9*. ICBP, Cambridge, 544–558.

Hamer, K.C., Baber, I., Furness, R.W., Klomp, N.I., Lewis, S.A., Stewart, F.M., Thompson, D.R. and Zonfrillo, B. (1989) *Ecology and Conservation of Azorean Seabirds. Report of the 1989 Glasgow University Expedition to the Azores Islands*, Glasgow, 63 pp.

Harrison, R.M., Peak, J.D. and Msibi, M.I. (1996) Measurements of airborne particulate and gaseous sulphur and nitrogen species in the area of the Azores, Atlantic Ocean. *Atmosphere and Environment* **30**, 133–143.

Hawkins, S. J., Burnay, L.P., Neto, A. I., Tristao da Cunha, R. and Martins, A.M.F. (1990) A description of the zonation patterns of molluscs and other biota on the south coast of São Miguel, Azores. In *Proc. First Internal. Workshop Malacol. Mar. Biol., Vila Franca do Campo, São Miguel, Azores, 1988*, ed. A.M.F. Martins, pp. 21–38. Açor., Supl., Outubro 1990.

Isidro, E.J. and Pereira, J.G. (1998) A pesca nos Açores, breve descrição; A brief description of the Azorean fisheries. In *Assoc. Internac. Urbanistas, Bol.—Especial "Açores, bem-vindo!"*, pp. 51–61.

Kaese, R.H. and Siedler, G. (1982) Meandering of the subtropical front south-east of the Azores. *Nature, London* **300**, 245–246.

Kasibhatla, L.P., Levy, H., Klonecki, A. and Chameides, W.L. (1996) Three-dimensional view of the large-scale tropospheric ozone distribution over the North Atlantic Ocean during summer. *Journal of Geophysical Research (D, Atmos.).* **101**, 29305–29316,

Knight-Jones, P. and Perkins, T.H. (1998) A revision of *Sabella, Bispira* and *Stylomma* (Polychaeta: Sabellidae). *Zoological Journal of the Linnean Society* **123**, 385–467.

Knight-Jones, P., Knight-Jones, E.W. and Buzhinskaya, G. (1991) Distribution and interrelationships of northern spirorbid genera. *Bulletin of Marine Science* **48**, 189–197.

Knudsen, J. (1995) Observations on reproductive strategy and zoogeography of some marine prosobranch gastropods (Mollusca) from the Azores. In *Proc. Second Internat. Workshop Malacol. Mar. Biol., Vila Franca do Campo, São Miguel, Azores, 1991*, ed. A.M.F. Martins, pp. 135–158. Açor., Supl. 4.

Krauss, W. and Meincke, J. (1982) Drifting Buoy Trajectories in the North Atlantic Current. *Nature, London.* **296**, 737–740.

Krauss, W., Kaese, R.H. and Hinrichsen, H.H. (1990) The branching of the Gulf Stream southeast of the Grand Banks. *Journal of Geophysical Research* **95**, 13089–13103.

LeGrand, G. (1984) *Ornithologie et Conservation aux Açores*, 246 pp. Universidade dos Açores, Ponta Delgada.

Le Grand, G., Emmerson, K. and Martin, A. (1984) The status and conservation of seabirds in the Macaronesian islands. In *Status and Conservation of the World's Seabirds*, eds. J.P. Croxall, P.G.H. Evans and R.W. Schreiber, pp. 377–391. ICBP Technical Publication No. 2.

Lobo, M.A.G. (1987) Theoretical polluted charges rejected by the town of Angra do Heroismo. *Recursos Hidricos* 8, 55–70.

Lopes, M.F.R., Marques, J.C. and Bellan-Santini, D. (1993) The benthic amphipod fauna of the Azores (Portugal): an up-to-date annotated list of species, and some biogeographic considerations. *Crustaceana* 65, 204–217.

Magalhães, M.J. and Barros, M.C. (1987) The contamination of fish with chlorinated hydrocarbons in Portugal: continental coast and Azores islands. *Environmental Monitoring and Assessment* 8, 37–57.

Martin, J.H. and Flegal, A.R. (1975) High copper concentrations in squid livers in association with elevated levels of silver, cadmium and zinc. *Marine Biology* 30, 51–55.

Martins, A.M.F. (1976) *Mollusques des Açores. Ilhéu de Vila Franca do Campo, 1. Basommatophora*, 31 pp., 5 plates. Ponta Delgada: Centenário da Fundação do Museu "Carlos Machado".

Martins, A.M.F. (1980) *Notes on the habitat of five halophile Ellobiidae in the Azores*. Ponta Delgada: Centenário da Fundação do Museu "Carlos Machado".

Martins, H.R. (1982) Biological studies of the exploited stock of *Loligo forbesi* (Mollusca: Cephalopoda) in the Azores. *Journal of the Marine Biological Association of the U.K.* 62, 799–808.

Mendes, B., Urbano, P., Alves, C., Lapa, N., Morais, J., Nascimento, J., Oliveira, J.F.S., Morris, R., Grabow, W.O.K. and Jofre, J. (eds.) (1997) Health-related water microbiology 1996. Sanitary quality of sands from beaches of Azores Islands. *Water Science and Technology* 35, 147–150.

Monteiro, L.R., Costa, V., Furness, R.W. and Santos, R.S. (1996) Mercury concentrations in prey fish indicate enhanced bioaccumulation in mesopelagic environments. *Marine Ecology Progress Series* 141, 21–25.

Monteiro, L.R. and Furness, R.W. (1995) Seabirds as monitors of mercury in the marine environment. *Water, Air and Soil Pollution* 80, 851–870.

Monteiro, L.R. and Furness, R.W. (1997) Accelerated increase in mercury contamination in North Atlantic mesopelagic food chains as indicated by time series of seabird feathers. *Environ. Toxicol. Chem.* 16, 2489–2493.

Monteiro, L.R. and Lopes, H.D. (1990) Mercury content of swordfish *Xiphias gladius* in relation to length, weight, age and sex. *Marine Pollution Bulletin* 21, 293–296.

Monteiro, L.R., Isidro, E.J. and Lopes, H.D. (1991) Mercury content in relation to sex, size, age and growth in two scorpionfish (*Helicolenus dactylopterus* and *Pontinus kuhlii*) from Azorean waters. *Water, Air and Soil Pollution* 56, 359–367.

Monteiro, L.R., Porteiro, F.M. and Gonçalves, J.M. (1992) Inter-and intra-specific variation of mercury levels in muscle of cephalopods from the Azores. *Arquipél., Life Earth Sci.* 10, 13–22.

Monteiro, L.R., Ramos, J.A. and Furness, R.W., (1996) Past and present status and conservation of the seabirds breeding in the Azores Archipelago. *Biological Conservation* 78, 319–328.

Moore, P.G., Rainbow, P.S., Weeks, J.M. and Smith, B. (1995) Observations on copper and zinc in an ecological series of talitroidean amphipods (Crustacea: Amphipoda) from the Azores. In *Proc. Second Internal. Workshop Malacol. Mar. Biol., Vila Franca do Campo, São Miguel, Azores, 1991*, ed. A.M.F. Martins, pp. 93–102. *Açor.*, Supl. 4.

Morton, B. (1990a) The biology and functional morphology of *Ervilia castanea* (Bivalvia: Tellinacea) from the Azores. In *Proc. First Internal. Workshop Malacol. Mar. Biol., Vila Franca do Campo, São Miguel, Azores, 1988*, ed. A.M.F. Martins, pp. 75–96. *Açor.*, Supl., Outubro 1990.

Morton, B. (1990b) The intertidal ecology of Ilheu de Vila Franca - a drowned volcanic crater in the Azores. In *Proc. First Internal. Workshop Malacol. Mar. Biol., Vila Franca do Campo, São Miguel, Azores, 1988*, ed. A.M.F. Martins, pp. 3–20. *Açor.*, Supl., Outubro 1990.

Morton, B. and Britton, J.C. (1999) Origins of the Açorean intertidal biota: the significance of introduced species, survivors of chance events. *Arquipél., Life Earth Sci. Suppl.* 2A, 29–51.

Morton, B. and Britton, J.C. (2000) The origins of the coastal and marine flora and fauna of the Azores. *Oceanography and Marine Biology: Annual Reviews.*, in press.

Morton, B. and Tristão da Cunha, R. (1993) The Fajã do Santo Cristo, São Jorge, revisited and a case for Azorean coastal conservation. *Açor.* 7, 539–553.

Morton, B., Britton, J.C. and Frias Martins, A.M. (1995) Fajã dos Cubres, São Jorge: a case for coastal conservation and the first record of *Ruppia maritima* Linnaeus (Monocotyledones; Ruppiaceae) from the Açores. *Açor.* 8, 11–30.

Morton, B., Britton, J.C. and Frias Martins, A.M. (1996) The Lajes do Pico marsh: a further case for coastal conservation in the Açores. *Açor.* 8, 183–200.

Morton, B., Britton, J.C. and Frias Martins, A.M. (1997) The former marsh at Paúl, Praia da Vitória, Terceira, Açores, and the case for the development of a new wetland by rehabilitation of the quarry at Cabo do Praia. *Açor.* 8, 285–307.

Morton, B., Britton, J.C. and Frias Martins, A.M. (1998.) *Coastal Ecology of the Açores*. Sociedade Afonso Chaves, Ponta Delgada.

Mueller, K.P. and Rudolph, J. (1992) Measurements of peroxyacetylnitrate in the marine boundary layer over the Atlantic. *Journal of Atmospheric Chemistry* 15, 361–367.

Neto, A.I. (1994) Checklist of the benthic marine algae of the Azores. *Arquipél., Life Mar. Sci.* 12A, 15–34.

Pierce, G.J., Boyle, P.R., Hastie, L.C. and Santos, M. (1994) Diets of squid *Loligo forbesi* and *Loligo vulgaris* in the Northeast Atlantic. *Fisheries Research* 21, 149–163.

Pollard, R.T. and Pu, S. (1985) Structure and circulation of the upper Atlantic Ocean northeast of the Azores. *Progress in Oceanography* 14, 443–462.

Porteiro, F.M. (1994) The present status of the squid fishery (*Loligo forbesi*) in the Azores Archipelago. *Fisheries Research* 21, 243–253.

Prud'homme van Reine, W.F. (1988) Phytogeography of seaweeds of the Azores. *Helgo. Meersunters.* 42, 165–185.

Rainbow, P.G. (1987) Heavy metals in barnacles. In *Barnacle Biology, Crustacean Issues*, ed. A.J. Southward, pp. 405–417. A.A. Balkema, Rotterdam.

Rainbow, P.S. and Moore, P.G. (1990) Seasonal variation in copper and zinc concentrations in three talitrid amphipods (Crustacea). *Hydrobiologia* 196: 65–72.

Rainbow, P.S. and White, S.L. (1989) Comparative strategies of heavy metal accumulation by crustaceans: zinc, copper and cadmium in a decapod, an amphipod and a barnacle. *Hydrobiologia* 174, 245–262.

Rees, W.J. and White, E. (1966) New records and fauna list of hydroids from the Azores. *Ann. Mag. Nat. Hist., Ser.* 13. 9, 271–284.

Renzoni, A., Focardi, S., Fossi, C., Leonzio, C. and Mayol, J. (1986) Comparison between concentrations of mercury and other contaminants in eggs and issues of Cory's shearwater *Calonectris diomedea*, collected on Atlantic and Mediterranean islands. *Environmental Pollution* 46: 263–295.

Santos, R.S., S.J. Hawkins, S.J., Monteiro, L.R., Alves, M. and Isidro, E.J. (1995) Marine research, reserves and conservation in the Azores. *Aquatic Conservation* 5, 311–354.

Santos, R.S., Porteiro, F.M. and Barreiros, J.P. (1997) Marine fishes of the Azores. Annotated checklist and bibliography. *Arquipél., Life Mar. Sci., Suppl.* 1, 244 pp.

Serralheiro, A. and Madeira, J. (1993) Stratigraphy and geochronology of Santa Maria Island (Azores). *Açor.* 7, 575–592.

Silva, H.M. and Krug, H.M. (1992) Virtual population analysis of the forkbeard, *Phycis phycis* (Linnaeus, 1766), in the Azores. *Arquipél., Life Earth Sci.* 10, 5–12.

Silva, H.M., Krug, H.M. and Meneses, G.M. (1994) Bases para a regulamentação da pescade demersais nos Açores. *Arqui. Depart. Oceanogr. Pes. Universid. Açores. Sér. estudos.* 7/94, 41 pp.

Smith, B.S. (1981) Tributyltin compounds induce male characteristics on female mud snails *Nassarius obsoletus* = *Ilyanassa obsoleta*. *Journal of Applied Toxicology* **1**, 141–144.

South, G.R. and Tittley, I. (1986) *A Checklist and Distributional Index of the Marine Benthic Algae of the North Atlantic Ocean*. Huntsman Marine Laboratory and British Museum (Natural History), St. Andrews and London. 76 pp.

Southward, A.J. (1998) Notes on Cirripedia of the Azores region. *Arquipél., Life Mar. Sci.* **16A**, 11–26.

Spence, S.K., Hawkins, S.J. and Santos, R.S. (1990) The mollusc *Thais haemastoma*—an exhibitor of 'imposex' and potential biological indicator of tributyltin pollution. *Marine Ecology* **11**, 147–156.

Stenner, R.D. and Nickless, G. (1975) Heavy metals in organisms of the Atlantic coast of S.W. Spain and Portugal. *Marine Pollution Bulletin* **6**, 89–92.

Stewart, F.M., Monteiro, L.R., and Furness, R.W. (1997) Heavy metal concentrations in Cory's shearwater, *Calonectris diomedea*, fledglings from the Azores, Portugal. *Bulletin of Environmental Contamination and Toxicology* **58**, 115–122.

Tittley, I. and Neto, A.I. (1995) The marine algal flora of the Azores and its biogeographical affinities. *Bol. Mus. Munic. Funchal*, Supl. 4, 747–766.

Vedel, G. and Depledge, M.H. (1995) Temperature tolerance and selected trace metal concentrations in some Azorean gastropod molluscs. In *Proc. Second Internal. Workshop Malacol. Mar. Biol., Vila Franca do Campo, São Miguel, Azores, 1991*, ed. A.M.F. Martins, pp. 113–124. *Açor.*, Suppl. 4.

Weeks, J.M., Rainbow, P.S. and Depledge, M.H. (1995) Barnacles (*Chthamalus stellatus*) as biomonitors of trace metal bioavailability in the waters of São Miguel (Azores). In *Proc. Second Internal. Workshop Malacol. Mar. Biol., Vila Franca do Campo, São Miguel, Azores, 1991*, ed. A.M.F. Martins, pp. 103–111. *Açor.*, Suppl. 4.

Wirtz, P. and Martins, H.R. (1993) Notes on some rare and little known marine invertebrates from the Azores, with a discussion of the zoogeography of the region. *Arquipél., Life Mar. Sci.* 11A, 55–63.

Young, P.S. (1998) Cirripedia (Crustacea) from the "Campagne Biaçores" in the Azores region, including a generic revision of Verrucidae. *Zoosystema* 20, 31–91.

Zibrowius, H. and Bianchi, C.N. (1981) *Spirorbis marioni* et *Pileolaria berkeleyana*, Spirorbidae exotiques dans les ports de la Mediterranée nord-ocidentale. *Rapp. Procès-verbaux Réunions. Commiss. Internat. Explor. sci. Mer Méditerr.* 27, 163–164.

THE AUTHORS

Brian Morton
*The Swire Institute of Marine Science and Department of Ecology and Biodiversity,
The University of Hong Kong, Hong Kong*

Joseph C. Britton
*Department of Biology,
Texas Christian University,
Fort Worth, Texas 76129, U.S.A.*

Chapter 14

THE SARGASSO SEA AND BERMUDA

Anthony H. Knap, Douglas P. Connelly and James N. Butler

The Sargasso Sea is a major oceanic gyre, with low nutrients as the main feature of the biogeochemistry of the area. In summer, the region is dominated by the Bermuda High which creates a barrier to further frontal passage and conditions for the formation of shallow, fresh, warm mixed layers, which typically shoal to less than 20 m. Temperatures peak at 28°C, and rain forms low salinity layers in the surface 10 m which could stimulate phytoplankton blooms. The area is notable for several features, including its rafts of pelagic *Sargassum*, the *Anguilla* eel and the Atlantic Humpback whale which is seen in relatively large numbers.

The only land mass in the Sargasso Sea is the small island of Bermuda. Surrounding it is a coral reef ecosystem which extends around the island up to 10 km from the islands. The limestone supports the most northerly coral reefs in the Atlantic, which have a reduced diversity of corals, but which are well adapted to the relatively high latitudes and clear seasonal changes. In some areas there are remarkable, nearly continuous linear sequences of emergent algal-vermetid reefs.

The greatest threat to the reefs of Bermuda is siltation, and locally derived contamination effects are minimal. Northwesterly winds from North America often contain contaminants such as anthropogenic sulphur and other trace substances, but apart from this, some tar contamination from shipping traffic has been of concern. Local issues revolve around excessive fishing, some locally generated air contamination and concerns about coral reefs due to dredging and ship grounding. Generally the area is relatively free of pollution.

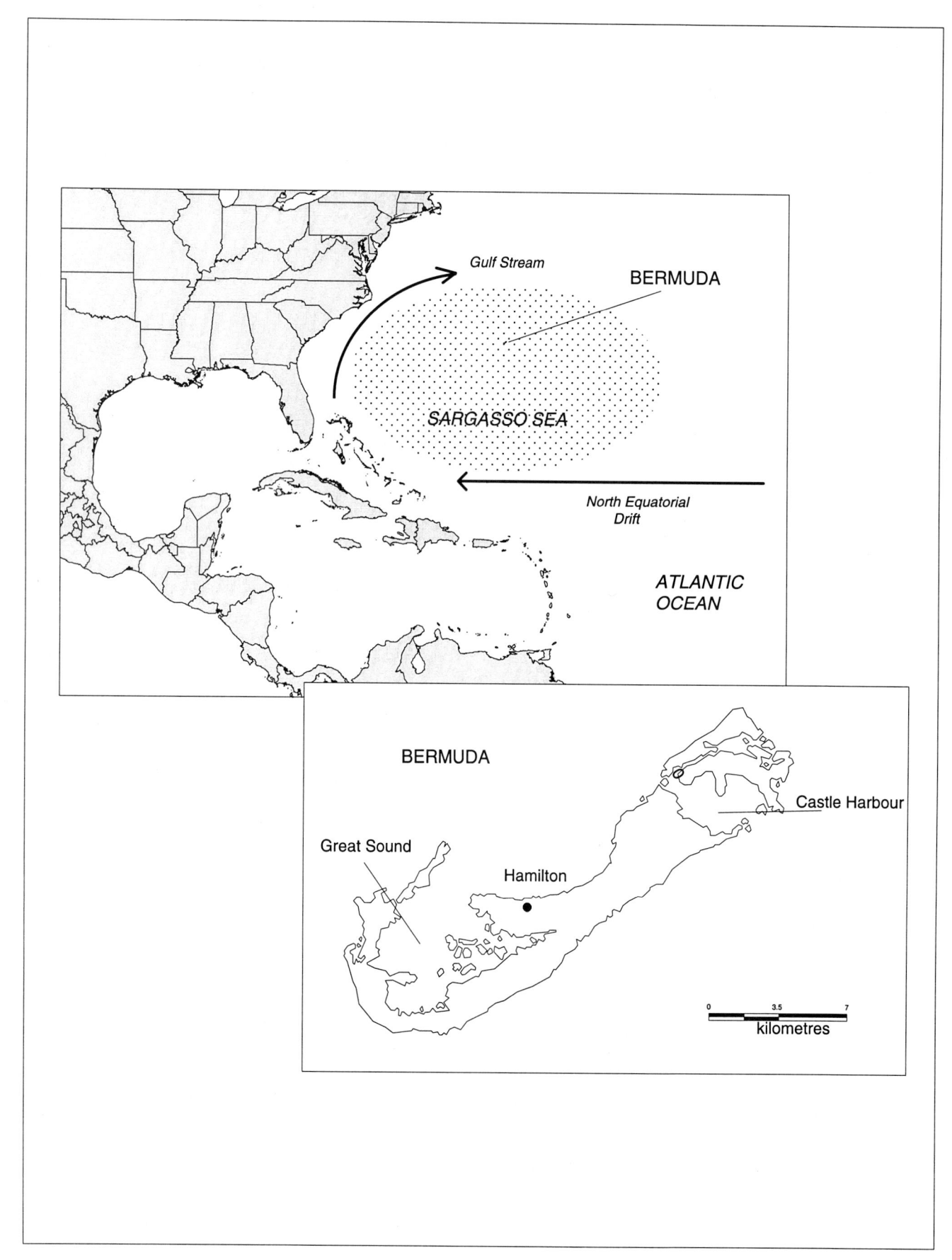

Fig. 1. The Sargasso Sea and Bermuda, including the main currents.

THE DEFINED REGION

The Sargasso Sea is an anticyclonic gyre in the central North Atlantic Ocean (Fig. 1). Traditionally the boundaries of the Sargasso Sea were defined by the presence of pelagic *Sargassum* species. However, now the Sargasso Sea is defined in terms of physical boundaries—the Gulf Stream to the west, the North Atlantic current to the north, the Canaries Current to the east, and the North Equatorial Current to the south (Dietrich, 1965). The only land mass in the Sargasso Sea is the island of Bermuda.

Sargassum may be found anywhere within these boundaries, and although it may be more consistently abundant in certain sectors, abundance at any single location is quite unpredictable. As Parr (1939) recognized, it is probably impossible to predict *Sargassum* quantities accurately on a local scale (Butler et al., 1983).

SEASONALITY, CURRENTS, NATURAL ENVIRONMENTAL VARIABLES

The region is characterized by weak geostrophic recirculation with net flow towards the southwest that drives net Ekman downwelling rates of ~4 cm day^{-1}, and by high eddy energetics. These mesoscale eddy phenomena include cold core rings and smaller cyclonic and anticyclonic eddies (McGillicuddy et al., 1998). Within the 25°N–32°N zonal band subtropical frontogenesis between westerlies and easterlies enhances the slope of the meridional seasonal thermocline and helps generate the baroclinic eddies.

Between 25°N and 32°N there is a transition region between: (1) relatively eutrophic (productive) waters to the north which is the formation site of subtropical mode water (STMW or 18°C mode water), subject to deep winter mixing and associated nutrient enrichment of the surface layer (Worthington, 1976); and (2) a relatively oligotrophic subtropical convergence zone to the south where nutrient-rich mode water underlies the permanently stratified euphotic zone for most of the year. Formation of subtropical mode water occurs each winter as the passage of cold fronts across the Sargasso Sea erodes the seasonal thermocline and causes convective mixing, resulting in mixed layers of 150–300 m depth. The subtropical mode water sinks and spreads southward separating the seasonal and permanent thermoclines, recognizable by a minimum in the profile of potential vorticity. In summer, the region is dominated by a high pressure system (known as the Bermuda High) which creates a barrier to further frontal passage. Conditions are then ripe for the formation of shallow, fresh, warm mixed layers, which typically shoal to less than 20 m, with temperatures peaking at 28°C. Wet deposition forms low salinity layers in the surface 10 m which could stimulate phytoplankton bloom formation (Michaels et al., 1993). Summer and early fall tropical storm activity brings increased wind, changing the thermal and physical structure and hence ecosystem dynamics. In intermediate waters of the Sargasso Sea, significant climatic events, such as shifts in the North Atlantic Oscillation which drives convection in the Labrador Sea, can be clearly observed with a six-year lag.

The formation of the Bermuda High is also responsible for the variable transport of atmospheric contaminants to the region. During the summer, primarily southwesterly winds bring air from the lower latitudes, generally transporting clean warmer air, while during the transitional period when the high weakens, cooler air is transported as Northwesterlies from North America. This air often contains contaminants such as anthropogenic sulphur and other trace contaminants (Jickells et al., 1982; Knap et al., 1988).

The other major factor in this area of the North Atlantic Ocean is the presence of Tropical Cyclones. The tropical cyclone season runs from June to November; however most of the activity is generally in August, September and October. For tropical cyclones to form and grow in these basins under the current climate state, the following environmental conditions must exist (Gray, 1968).

1. Ocean waters warmer than 26°C (80°F). Warm water is needed to fuel the heat engine of the tropical cyclone.
2. An unstable atmosphere that cools rapidly enough with height to allow for convection or thunderstorm activity.
3. Relatively moist layers near the mid-troposphere, about 5 km in altitude.
4. A minimum distance of at least 500 km from the equator in order for the Earth's rotation to impart a sizable spin.
5. A pre-existing disturbance near the Earth's surface with sufficient rotation and convergence. Tropical cyclones do not generate spontaneously.
6. Low values (less than about 30 k/h) of vertical wind shear between the surface and the upper troposphere (12 km).

The El Niño-Southern Oscillation (ENSO) affects tropical cyclone activity in the Atlantic as well as elsewhere. ENSO is a fluctuation on the scale of a few years in the ocean–atmosphere system involving large changes throughout the tropical Pacific. The state of ENSO can be characterized, among other features, by sea surface temperature (SST) anomalies in the eastern and central equatorial Pacific. Warmings in this region are referred to as El Niño events and coolings as La Niña events.

ENSO greatly alters global atmospheric circulation patterns. It affects tropical cyclone frequencies mainly by altering relative vorticity patterns in the lower atmosphere and by changing the vertical shear profile. High levels of relative vorticity impart spin to a tropical cyclone's inflowing winds, allowing an incipient storm to develop its characteristic pinwheel shape.

The Atlantic basin feels the effects of ENSO remotely through changes in vertical shear. Vertical shear increases during El Niño events, primarily due to increased westerly winds in the upper troposphere. During La Niña events, westerly winds at this level decrease, thus reducing vertical shear. Due to these effects, it is twice as likely for two or more landfalling hurricanes to strike the US during a neutral year than during an El Niño year.

Besides ENSO, another global parameter, the stratospheric Quasi-Biennial Oscillation (QBO), an east–west shift of stratospheric winds that encircle the globe near the equator also affects hurricane formation (Wallace, 1973). In the Atlantic, activity increases during the west phase. While the mechanism by which the stratospheric QBO influences tropical cyclones is uncertain, changes in vertical shear may play a role, as may changes in the stability of the upper troposphere.

In addition to the global effects of ENSO and QBO, local effects also appear to influence tropical cyclone frequency within individual basins. These include variations in local sea-level air pressures, sea surface temperatures (SSTs), and trade wind and monsoon circulation.

Over the Atlantic basin, monsoonal rainfall from June through September in Africa's Western Sahel shows a close association with the frequency of intense landfalling hurricanes. Wet years (e.g., 1988 and 1989) are marked by more intense hurricanes, while drought years (e.g., 1990–1993) are marked by fewer intense hurricanes. Variations in vertical shear and the intensity of African easterly waves may link the two phenomena.

Generally, an average of 9–10 tropical storms form annually in the Atlantic basin. Of these, 5–6 reach hurricane strength. Atlantic hurricanes constitute about 12% of the global total of hurricanes and typhoons. Analysis of the number of tropical storms in the Atlantic basin shows substantial yearly variability, but no significant trends. In contrast, the number of intense hurricanes decreased markedly from the 1940s through the 1990s, even when considering 1995 and 1996, years that nearly broke the record for number of intense hurricanes. The mean intensity of Atlantic tropical cyclones has also decreased, although there has been no significant change in the peak intensity reached by the strongest hurricane each year. Instead of a linear trend, we may more accurately depict the intense hurricane fluctuations as a very active regime in the 1940s through the late 1960s, followed by a repressed period in the 1970s through 1994, and concluded by a "spike" of activity in 1995 and 1996. The very active 1995 and 1996 seasons are more representative of seasons in the 1940s and 1950s.

These hurricanes have a remarkable effect on the Sargasso Sea and on the only land mass—Bermuda. Recently investigation of the effects of these hurricanes on ocean temperatures show that massive changes of temperature in the top layers of the ocean are affected after a hurricane passes, sometimes lowering the summer ocean surface temperatures by 4°C. This can persist in large areas for at least three weeks to a month (Nelson, 1998). Also, these tropical cyclones can be responsible for significant outgassing of carbon dioxide to the atmosphere (Bates et al., 1998).

On the local scale, hurricanes tend to hit the only land mass of Bermuda quite infrequently as it is a small land mass in this large ocean. These cyclones tend to cause less damage to the reef structure in Bermuda than in the Caribbean, as Bermuda coral species are morphologically adapted to dealing with large Atlantic storms.

THE MAJOR OPEN OCEAN HABITATS

The Sargasso Sea has typical abyssal depths of 4000 metres. The main organisms are pelagic and mid- and deep-water organisms of the Atlantic ocean. Many of these were described by Dr. William Beebe in his record bathyscaphe dives off Bermuda in 1932 which resulted in the first live radio broadcasts from the deep ocean. There are far too many species and systems to name in this review but there are some aspects of specific interest such as the general biogeochemistry of the area which defines the open ocean communities, the *Sargassum* Community, the yearly eel migration/spawning activity, as well as the migration of marine mammals. There are also many species of migrating fish which are part of a sport fishery in Bermuda.

Biogeochemistry of the Area around Bermuda

We probably know more about the area of the Sargasso Sea southeast of Bermuda than any other open ocean area in the world. This is the site of a number of long-term monitoring programs. The longest, located at Hydrostation S, is 22 km southeast of the island of Bermuda, was started by Henry Stommel and colleagues in 1954 (Schroeder and Stommel, 1969) and is the longest continuous running open ocean time-series in the world. Biweekly sampling of hydrography at this site continues to this day and has provided critical information on the physical structure of the ocean, as well as a decadal scale framework for the biogeochemically driven Bermuda Atlantic Time-series (BATS) program (Michaels and Knap, 1996). From 1957–1963, biological and chemical measurements were added to the Hydrostation S cruises. These included a time-series of nutrients, chlorophyll and primary productivity and these data resulted in classic papers on the seasonal cycle of biogeochemistry in the Sargasso Sea (Menzel and Ryther, 1960). In addition to Hydrosation S, a 23-year time series of deep-ocean sediment trap collections began in 1976 and continues today (Deuser and Ross, 1980; Conte et al., 2000). This study revealed the strong seasonality of deep ocean particle fluxes and has been a platform for numerous other investigations of particle composition and flux processes. A 19-year time series of atmospheric measurements (begun in 1980), including the Western Atlantic Ocean Experiment (WATOX) and the Air-Ocean Chemistry Experiment (AEROCE) studies (Galloway et al., 1992), provide a valuable context for interpreting air-sea exchanges which may be important for upper ocean biogeochemistry (Knap et al., 1986a).

The Hydrostation S program attracted other time-series investigations near Bermuda that began before the inception of BATS (see Michaels and Knap, 1996). These time series included studies of: radioactive tracers, coupled to measurements of oxygen and nutrients, which gave insights into the residence times of the water in the main thermocline and rates of biological activity; the upper ocean nitrogen cycle using nitrogen isotope ratios; benthic boundary layer fluxes and biogeochemistry at the sea floor, and dissolved inorganic carbon. Other time-series programs have been initiated since the inception of BATS in 1988. Ocean optics and remote sensing programs at BATS provide an important link between ocean biogeochemistry and satellite oceanography (Siegel et al., 2000). A testbed mooring program has been initiated near the BATS site in order to develop and test new instrumentation, to ground-truth satellite ocean colour data and to provide high temporal resolution physical, bio-optical and biogeochemical data in support of BATS.

The annual cycle of hydrography in the upper ocean at BATS is primarily driven by seasonal changes in surface heat flux and wind stress. Strong thermal stratification is present from April to October, largely due to higher heat fluxes and lower wind stresses. In the summer, mixed layer temperatures range from 27–29°C, and there is a fresher mixed layer, a subsurface salinity maximum, and strong density gradients in the upper 100 m. Warming and cooling of the surface layer (~1–2 m), associated with the diurnal thermal cycle, ranges from ~0.2°C to 2.5°C depending on day-to-day changes in net daytime surface heat fluxes (controlled by amount of cloudiness and wind stress) and mixed layer depth. In the winter, the mixed layer is more saline and temperatures range from ~19–21°C, while mixed layer depths vary from 150–300 m deep. Total carbon dioxide concentrations within the mixed layer are highest in winter due to vertical entrainment of TCO_2-rich subsurface water, while oxygen is typically oversaturated in the surface layer.

The annual cycle of biogeochemistry is intricately linked to this physical forcing, since vertical mixing brings nutrients into the euphotic zone during the winter. Typically, each year, the upward nutrient flux supports a short spring bloom period (January–March) of higher primary production rates, enhanced concentrations of chlorophyll biomass, and higher concentrations of suspended particulate organic carbon and nitrogen (Fig. 2). During thermal stratification in summer, measurable nutrients are absent within the euphotic zone, primary production rates are lower, a subsurface chlorophyll maximum is present, and TCO_2 concentrations reach their minima.

One of the interesting aspects of this BATS time-series shows that the cycling of C, N, and P at BATS is apparently not at the traditional Redfield ratios. The Redfield ratio (Redfield et al., 1963) is a fundamental assumption used to model ocean uptake of CO_2 on a global scale. The summer drawdown of dissolved inorganic carbon (DIC) by primary production in the surface waters, occurs in the absence of measurable nutrients (Bates et al., 1996) and is an extreme example of non-Redfield ratio DIC depletions in the North

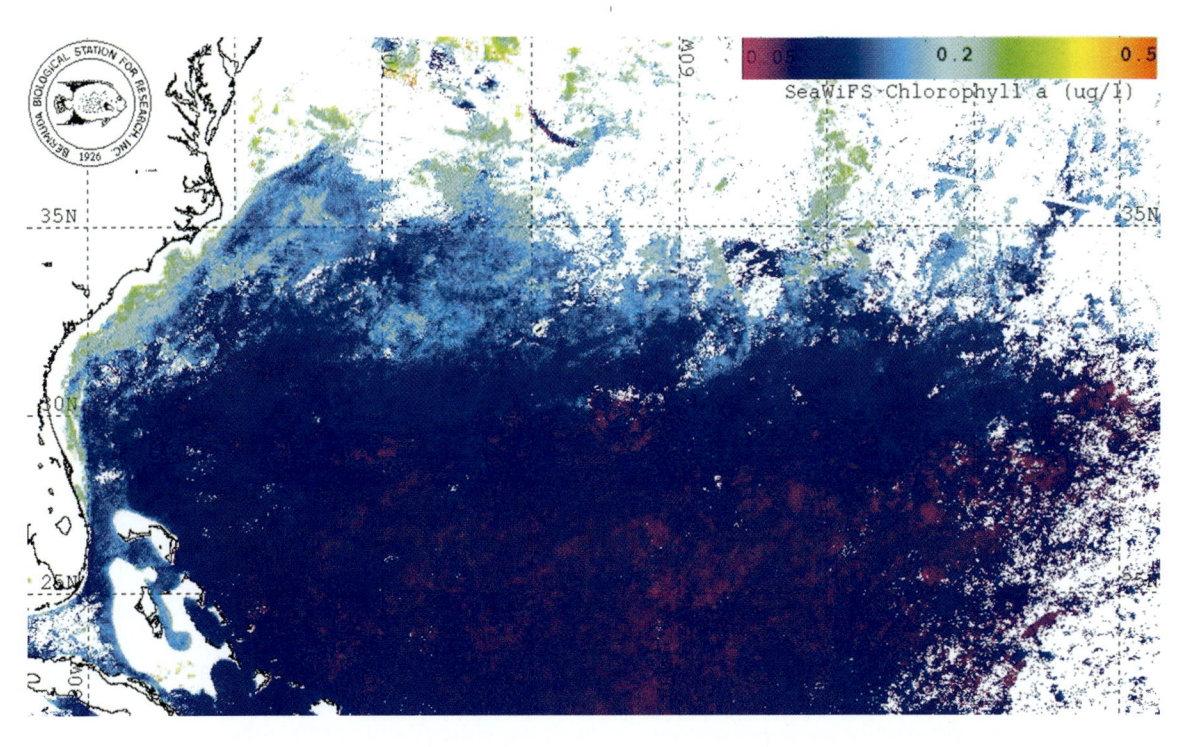

Fig. 2. SeaWifs image of Sargasso Sea, showing chlorophyll a concentrations (January 1999).

Atlantic. At the BATS station, nitrate:phosphate ratios in the upper thermocline consistently exceed the Redfield ratio of 16. This pattern differs from that found in much of the world's oceans where N:P ratios in surface waters are below Redfield, presumably because phosphate is labile and regenerated more quickly from organic matter than nitrate. This raises questions about what mechanisms cause the observed ratios. Michaels et al. (1994) speculate that the pattern may be due to one of many possible causes: an advected feature; heterotrophic uptake of phosphate below the euphotic zone; storage of phosphate by live algae such as diatoms as they sink through the thermocline; and vertical migration of nitrogen-fixing cyanobacterial colonies such as *Trichodesmium* sp., descending to depth to incorporate phosphate, and then ascending to fix carbon and nitrogen. Recent analyses of this anomaly indicate that nitrogen fixation may indeed be a likely source of new nitrogen to the Sargasso Sea. These unresolved paradoxes, common to most of the Sargasso Sea, are major conceptual uncertainties in our understanding of the oceanic carbon cycle and must be adequately explained to fully model the role of the oceans in the global carbon budget

Sargassum

Perhaps the most unique open-ocean habitat is the floating *Sargassum* community. It maintains its population entirely by vegetative growth. The two common species are *Sargassum natans* and *S. fluitans* which are thought to be derived from ancestral littoral stock. The alga is restricted to the surface where it maintains itself by gas-filled bladder floats. Growth and propagation result from seasonal fragmentation, usually during winter, of the larger and older plants. The animals inhabiting the *Sargassum* have been described as a displaced benthos; a majority of the species are derivative of littoral genera and species (Hedgepeth, 1957). Surprisingly the total organization of the Sargassum community is relatively unknown. A field guide of the community noted 101 species (Morris and Mogelberg, 1973) and as many as 54 species of fish spend their time associated with the *Sargassum* for some stage of their life cycle; however, only two species of fish spend their whole life cycle in *Sargassum*.

There has been concern lately that on Bermuda there is far less *Sargassum* present than in the past (Wingate, personal communication). *Sargassum* in early years was used as a fertilizer on Bermuda and there appears to be less around. Butler et al. (1983) carried out a study in 1981 and concluded that there was no significant change in *Sargassum* biomass from 1933 to 1981, except for an area northeast of the Antilles (10–25°N, 62–68°W) where measurements made in November 1977 and 1980 were 0.1% of values measured in February and March 1933. This difference may be the result of changes in seasonal distribution. Also, as *Sargassum* occupies the upper reaches of the ocean, it was impossible to collect samples of it without also collecting tar resulting from ballast water cleaning of tankers. It was this that led to a number of studies of tar in the open ocean at Bermuda since 1980. This community is an interesting one for future study as it occupies a fascinating niche in the open-ocean ecosystem.

Anguilla

One of the interesting mysteries of this part of the ocean is the spawning of the American eel (*Anguilla rostrata*) and the european eel (*Anguilla anguilla*). The capture of various sizes of American and European eel larvae were used to infer that adults of these species migrate to the Sargasso Sea and spawn and die, and the routes and mechanisms of the adult migration are still unknown. *Anguilla* occupy fresh water streams, rivers, brackish waters and the open ocean in various stages in their life cycle. They apparently spawn in the Sargasso Sea and ocean currents transport the developing larvae northward until the young metamorphose into juveniles and then move upstream along the continents. There is apparently no difference in the vertical distribution of the two species in the Sargasso Sea as larvae (leptocephali) of both species were found in equal abundance between 0 and 350 m. Developing eels apparently remain in fresh or brackish water for about 10 years before they return to the Sea to spawn. Although it is known that the larvae of both species have been positively identified by electrophoresis, no adult eels have been found and therefore where and what depth the adults occupy is still largely unknown, although Fricke and Kaese (1995) suggest that based on number of larvae present in their samples the European eel may spawn further south at 20°N latitude.

Humpback Whales

The Bermuda Islands are located about halfway between the major western North Atlantic humpback whale breeding areas in the Antilles and the northern feeding grounds extending from the New England coast to Iceland. Therefore they provide an excellent place for observation of this species (Katona et al., 1980). Adult whales reach up to 30 tons in weight and they range from 12–19 m in length. The flippers of these whales are unique as they measure up to one-third of the body weight; they are used as cooling planes through which a whale loses heat. This allows them to live closer to the warmer equatorial waters than other whales. They also use these large fins to defend themselves against killer whales, as well as using them to sweep krill into their mouths. Although there are many reports of the sighting of whales off Bermuda, going back to the early settlers, it was not until Verril in 1902 that there was a description, including a description of their seasonality in which they arrived off Bermuda late February or early March and left about the first of June, accompanied by suckling cubs (Verril, 1902). Although whaling began in 1911 in Bermuda, it finished in 1942. Various studies have

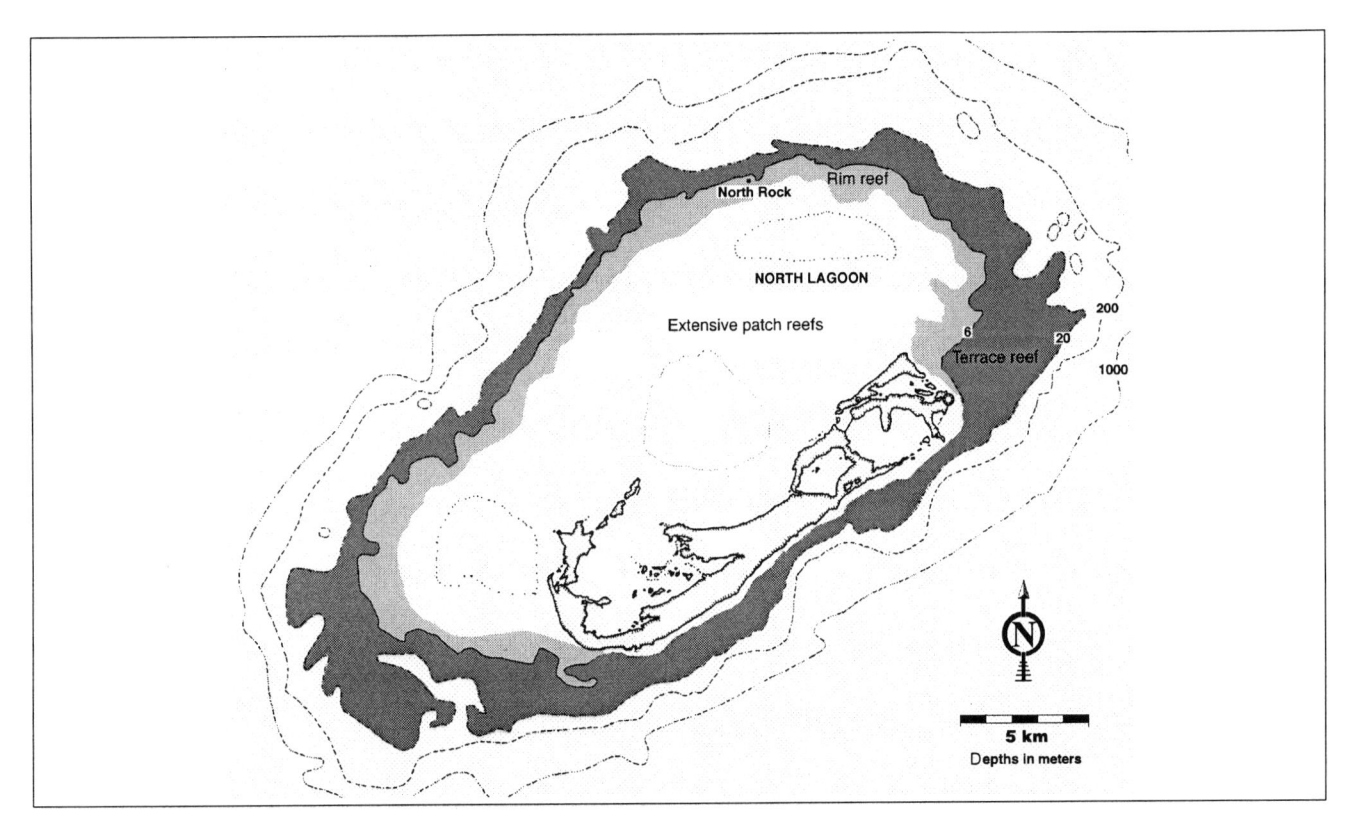

Fig. 3. Reef zones around Bermuda. Shading denotes shallow water and reefs.

been carried out using acoustic studies of humpbacks, and from 1967 to 1972 at least 10 whales were observed off Bermuda per day with a maximum of 25 in a single day. These whales also produce "songs" which last on average 15 minutes but can extend to 30 minutes (Payne and McVay, 1971). There is some suggestion that the humpback whale population off Bermuda has decreased since the 17th, 18th and 19th century (Stone et al., 1987) but this may be due to the better food availability in other areas. The US government acoustic arrays have also been used to determine the whale population in the Sargasso Sea and plans are underway to try and use these systems for this purpose (Knap, personal communication).

THE MAJOR SHALLOW WATER MARINE AND COASTAL HABITATS

Bermuda is a small, volcanic island system which rises dramatically from the deep abyssal plain of the Sargasso Sea and is located 32°20'N latitude and 64°45'W longitude. There is a coral reef ecosystem which extends around the island up to 10 km from the land mass. Water depths on the reef range from 0.5 m to about 30 m until the reef drops off very rapidly to abyssal depths; depths around the platform are over 2000 m. Further away the depths are over 4000 m. There are two extinct volcanoes which rise to within 30 m of the surface. These are 15 and 30 km southwest of Bermuda, called Challenger Bank and Plantagenet Bank respectively. Bermuda is made up of about 150 islands arranged in a narrow chain with a total land mass of 56 km². The total size of the Bermuda platform is about 775 km². Bermuda is a carbonate cap on an extinct volcano and the volcanic rock can be found about 80 m below the present surface. Bermuda is the most Northerly coral reef island in the Atlantic and, due to the presence of the Gulf Stream, nearby temperatures on the platform range from 16°C to 32°C, enabling corals to survive.

Extensive coral reef zones are developed at the margin and on the shallow flanks of the Bermuda Pedestal (Fig. 3). The rim reef is a shallow (3–10 m) zone, about 0.5–1 km wide that separates the ocean from the North Lagoon. This zone is reduced on the narrow southeastern edge of the pedestal adjacent to the islands, where a nearly continuous linear sequence of emergent algal-vermetid reefs ("boilers") separate the nearshore reef zone from the ocean. Seaward of the northern rim reef and southern boiler reefs is the extensive main terrace. Below the main terrace reefs is the deep fore-reef that slopes sharply off and terminates at about 60 m (Fricke and Meischner, 1985). Within the North Lagoon is an extensive array of patch reefs that vary in size and configuration (Garret et al., 1971), interspersed with deep (15–18 m) muddy basins.

The rim and terrace reefs have similar coral assemblages, made up of a few species, but coverage ranges from 25% in the former zone to 50% in the latter (Dodge et al., 1982; Smith and Musik, 1981). *Diploria strigosa, D. labyrinthiformis, Montastrea franksi, M. cavernosa,* and *Porites astreoides* are the dominant corals in these zones, along with the hydrozoan

Millepora alcicornis. Less common species include *Stephanocoenia michelini*, *Favia fragum*, *Agaricia fragilis*, *Madracis decactis*, *Siderastrea* spp., *Scolymia cubensis*, and *Isophyllia sinuosa*. The deep fore-reef community is composed of *Montastrea* spp., *A. fragilis*, *S. michelini*, and *Madracis* spp.

The lagoonal patch reefs have a similar *Diploria–Montastrea–Porites* community structure with lesser coverage (<20%). However, the patch reefs closer to the island and within Castle Harbour support a different community of primarily branched species (*Madracis decactis, M. mirabilis, Oculina diffusa*) that grow on the vertical sides of the reefs (Dryer and Logan, 1978). This reef community may have developed as the result of higher sedimentation rates close to shore and the degree of protection from wave energy. Coral diversity is greatest on these reefs, along with other sessile invertebrates and benthic algae.

Coral growth rates appear to be seasonal, with faster growth in the summer months, but reduced for some species compared to Caribbean congeners (Thomas and Logan, 1992). Higher growth rates for several coral species are found within the lagoonal reefs compared to the outer reef zones (Logan et al., 1994).

The threats to the reef environment in Bermuda are the effects of dredging, hurricanes, ship groundings, increased temperatures (Cook et al., 1993), various coral diseases, changes in the ecosystem due to overfishing (Butler et al., 1993) as well as some damage from pleasure boat fishing and diving. There is a no-collection policy for corals in Bermuda.

The greatest threat for the Bermuda reefs is siltation. This is usually caused by dredging or ship traffic. This decreases the clarity of the water, decreasing the productivity of symbiotic algae in the corals thus providing less nutrition to the corals. In Bermuda, dredging for the airport in the early 1940s destroyed a number of inshore reefs (Dodge and Vaisnys, 1977). However, it should be noted that for the most part, the reefs in Bermuda are quite healthy because of the large dilution factor of the Atlantic Ocean washing past the island continually.

The Bermuda inshore waters divide into two nutrient regimes. In the more open bodies of water such as the offshore reef areas, nutrient levels are very low and there is evidence that the benthic activity of the coral reef community reduces nutrient concentrations below that of the surrounding Sargasso Sea (Jickells and Knap, 1984). In the more enclosed inshore water such as the four basins of the Great Sound, Harrington Sound, Castle Harbor and St. Georges Harbor, restricted water exchange and increasing amounts of land inputs increase dissolved inorganic nitrogen. Also sewage discharge to the groundwater lens interacts with the carbonate sediments leading to a phosphate-limited ecosystem in the enclosed ecosystem. However, although this does lead to elevated primary productivity about the adjacent Sargasso Sea, only the most landward part of the Great Sound system, Hamilton Harbor, exhibits seasonal eutrophication.

These inshore areas are the home for seagrass communities, inshore water communities and in limited numbers, mangroves. There are only a few mangrove stands left on the island. There were some commercially harvestable clams and scallops in the inshore waters until the 1970s, but the numbers have dwindled over the years, primarily due to fishing pressure. These are now endangered. There is an active program of shellfish mariculture underway to bring these species back to sustainable numbers.

POPULATIONS AFFECTING THE AREA

Bermuda is a UK self-governing colony and, after the return of Hong Kong to China, it is now the largest British Colony. Bermuda has a population of about 65,000 people; approximately 5000 of these are expatriates on work permits. The two major businesses of the island are tourism and international business. Tourism numbers have been continuing to decline over the years primarily due to competition from other resorts, while international business, primarily re-insurance, and other capital market business is increasing.

There is no heavy industry on the island and therefore little industrial waste. There is a mass-burn incinerator which was built in the early 1990s to reduce the need to land-fill the Island's refuse. Electricity is generated by the use of fuel oil generators and there is one major and two minor exhaust stacks. One additional stack will be added for the new power plant in 1999. These stacks are monitored by the Bermuda Biological Station for Research under contract to the Ministry of the Environment of the Government of Bermuda.

RURAL FACTORS

There is little agricultural activity on the island and therefore little agricultural run-off. The only land run-off of any significance may be the use of pesticides, fungicides and nutrients from the golf courses of the island into the inshore waters of the island. Harrington Sound, the enclosed basin in the centre of the island, has a flushing time of 150 days (Morris et al., 1977), and this is the most sensitive to such land-based contaminants.

COASTAL EROSION AND LANDFILL

The Bermuda airport construction in the 1940s introduced a heavy suspended load of sediment in Castle Harbor, which greatly changed the ecology of the coral reefs (Dodge and Vaisnys, 1977). The dominant species changed from *Diploria strigosa* (brain coral) to *Porites* (finger coral). There is also a Government quarry that cuts carbonate stone for building material. This activity creates fine sediment which increases the turbidity of this inshore basin.

Turbidity is also created by the almost daily passage of ships into Bermuda. The cruise ships have drafts very close to the dredge limit of the island and therefore sediments are disturbed for short periods of time into the surface of the water, thus affecting turbidity (Sarkis, pers. comm.).

EFFECTS FROM URBAN AND INDUSTRIAL ACTIVITIES

There are two types of urban inputs to this area of the ocean: regional and local. Regional contaminants are oil and tar contamination as well as atmospheric inputs of organic and inorganic contaminants as a result of long-range atmospheric transport. There are two main reviews of this subject: on tar contamination (Butler et al., 1973) and long-range atmospheric transport (Knap, 1990).

Input of petroleum residues (pelagic tar) from tanker operations and accidents in the North Atlantic was highlighted by Thor Heyerdahl during his Ra expedition in the 1970s. In Bermuda, studies of zooplankton were hampered as pelagic tar was caught in the towed neuston nets to such a degree that the nets became useless (Morris, 1971). This started a long-term study of tar in the Sargasso Sea (Butler et al., 1973). Due to tanker routes crossing the Sargasso Sea and the past practices of cleaning tanks in the open ocean—primarily in the warmer waters of the Gulf Stream and Sargasso Sea—large quantities of petroleum residues were captured by the Sargasso Sea gyre. Generally tar in the open ocean lasts from months to a year before it breaks up into small particles. In some instances tar lumps the size of soccer balls were washed up on local beaches. Quantitative measurements of tar in the open ocean were made using neuston nets. An alternative method is to collect tar from oceanic beaches (Morris and Butler, 1973). Knap et al. (1980) repeated a Bermuda beach survey carried out in 1971–1972 and found a 15% increase in the mean over this period compared with a supposedly 27% decrease due to improvements in tanker operations. These surveys were continued, and Smith and Knap (1985) showed a significant decrease by a factor of 8 in the amount of tar stranding on Bermuda beaches. Tar studies have continued over the years and demonstrate a continual decline, indicating that newer operational methods for cleaning of tankers and international pollution agreements appear to be working.

Long-range atmospheric transport is responsible for the transfer of many types of contaminants to the Sargasso Sea. For organochlorine compounds such as polychlorinated biphenyls, pesticides, etc., atmospheric transport is the main route of these compounds to the ocean in this area. There have been many long-term studies of atmospheric transport to the Sargasso Sea using Bermuda as a base. In the 1970s various collections were carried out. Jickells et al. (1982) first reported "acid rain" in Bermuda and showed that rapid long-range transport from North America brought anthropogenic sulphate to the Sargasso Sea resulting in lowering pH in Bermuda rainwater from around 7 to 4.0. This prompted the start of the Western Atlantic Ocean Experiment (WATOX) which used a site at the west end of the island of Bermuda as well as ships of opportunity and aircraft flights. The aim of the program was to detail the mode and extent of transport of trace metals (Church et al., 1982), organics (Knap et al., 1988) and sulphur and nitrogen compounds (Galloway et al., 1992). This was followed by the AEROCE program which installed a 20 m walkup tower at the west end of Bermuda. This site is still operational.

There have been few measurements of organochlorines in the atmosphere of the North Atlantic ocean and certainly very few since major studies in the 1970s (Bidleman and Olney, 1974). Knap et al. (1988) measured trace organic in rain events on Bermuda and for some specific organic contaminants such as alpha and gamma hexachlorcyclohexane (HCH), concentrations were higher when airmasses originated in North America than when they originated in the Sargasso Sea. Concentrations of contaminants in precipitation were highly variable indicating that just a few events can be responsible for a large flux. Concentrations measured by aircraft over the Sargasso Sea showed no change with altitude and concentrations were quite low. Knap et al. (1986b) measured the flux of some organic compounds to the deep sea using a sediment trap 25 km south of the atmospheric site at 3200 m depth in the Sargasso Sea. They report a flux for chlordane and dieldrin, once two commonly used pesticides in the U.S., of 20 and 40 ng m^2 year^{-1} which is about 8 times lower than the wet flux. Unfortunately there are only a few measurements of organochlorines in this part of the Atlantic, and they are dated. When last reported, the concentrations of PCBs in open ocean water southeast of Bermuda was about 100 pg l^{-1}. There are also only a few data on the concentrations of these compounds in marine organisms in the Sargasso Sea. Knap and Jickells (1984) report concentrations of organochlorines such as DDT, DDE and total PCB concentrations determined in blubber, flesh, kidney and heart of the Goosebeaked whale. The highest concentrations were found in the blubber with total DDT concentrations from 12–20 mg g^{-1} wet weight and PCB concentrations from 8–12 mg g^{-1}.

There have been quite a few studies on trace metals in the Sargasso Sea (Church et al., 1982). It is fairly clear that the atmosphere above the Sargasso Sea is contaminated by emissions from North America and this has perturbed the oceanic and biochemical cycles of several trace metals in the North Atlantic. Boyle et al. (1996) have shown a significant decrease in lead in the surface Sargasso Sea over time, correlated with the decrease in lead used for gasoline additives in the United States. Jickells et al. (1984) have investigated the sedimentation rate of trace elements in the Sargasso sea using a sediment trap at 3200 m. The fluxes were variable, largely driven by the primary productivity cycle of the surface waters. These fluxes suggested an anthropogenic signal of copper, lead and zinc in the open ocean. Generally, other than these trace elements, there is little contamination of the waters of the Sargasso Sea. Knap and

Jickells (1984) report contamination of Cuvier beaked whales and report concentrations of 8 trace metals in blubber, flesh, liver, kidney and heart. The data seem to be in the range of other toothed whales except for cadmium, which in this study was higher than other contaminants in whales.

Local sources of contaminants in Bermuda are small. In Bermuda there is a mass burn incinerator and the emissions tend to be diluted by the large air mass around Bermuda and therefore go out to sea. In the 1980s there was significant lead contamination due to the use of leaded gasoline. Concerns about health issues led to the ban of leaded gasoline on the island, which resulted in a very rapid decline of lead in air around the island; there was a fall of >90% in a year. Generally, contamination levels in Bermuda are low (Jickells and Knap, 1984) with elevated concentrations near the main harbours and boat marinas. There is a waste dump site where concentrations of certain trace metals are higher than ambient. However, this is an area of good dispersal and so far there have not been significant impacts.

PROTECTIVE MEASURES

The Sargasso Sea is governed by international conventions regarding the discharge of waste from ships at sea. Bermuda does claim an Exclusive Economic Zone and does license vessels fishing within this zone. The Bermuda economic zone is governed by building codes, fisheries regulations and pollution regulations. The remainder of the Sargasso Sea is governed by international conventions (Hayward et al., 1981).

Bermuda has had a long history of conservation laws. As early as 1620 the killing of sea turtles was banned and is noted as the first conservation law in the "New World". The early settlers on Bermuda passed many laws to preserve cedar wood in 1632, and against exporting it in 1659. After the American Civil War there were a number of fire-arms in Bermuda so the Protection of Birds Act of 1870 was passed. Most of Bermuda planning is now protected under the Development and Planning Act of 1965 which means permission has to be sought for any change to the land or marine areas. The Oil Pollution Act of 1973 set maximum fines for oil pollution offences. There are now a number of Acts in Bermuda that protect the environment in many ways. Perhaps the most important marine Act lately was an amendment to the fisheries act which banned the use of fish pots for fishing. This has been reported by Butler et al. (1993) and has resulted in the reversal of the decline of the Bermuda fisheries.

ACKNOWLEDGEMENTS

The authors would like to thank Dr. David Malmquist, Dr. Debbie Steinberg, Dr. Robbie Smith. This work was funded by the Bermuda Government and the Zemurray Foundation. This is BBSR Contribution No. 1550.

REFERENCES

Bates, N.R., Michaels, A.F. and Knap, A.H. (1996) Seasonal and interannual variability of oceanic carbon dioxide species at the U.S. JGOFS Bermuda Atlantic Time-series Study (BATS) site. *Deep-Sea Research II* **43**, 347–383.

Bates, N.R., Michaels, A.F. and Knap, A.H. (1998) Contribution of hurricanes to local and global estimates of air-sea exchange of CO_2. *Nature* **395**, 58–61.

Bidleman, T.F. and Olney, C.E. (1974) Chlorinated hydrocarbons in the Sargasso Sea, atmosphere and surface waters. *Science* **183**, 516–518.

Boyle, E.A., Chapnick, S.D., Shen, G.T., Bacon, M.P. (1986) Temporal variability of lead in the western North Atlantic. *Journal of Geophysical Research* **91**, 8573–8593.

Butler, J.N., Morris, B.F., Cadwalleder, J. and Stoner, A.W. (1983) *Studies of Sargassum and the Sargassum Community*. Bermuda Biological Station Special Publication No. 22. 307 pp.

Butler, J.N., Burnett-Herkes, J., Barnes, J.A. and Ward, J. (1993) The Bermuda fisheries: A tragedy of the commons averted? *Environment* **35** (1), 6–15.

Butler, J.N., Morris, B.F. and Sass, J. (1973) *Pelagic Tar from Bermuda and the Sargasso Sea*. Special Publication 10, Bermuda Biological Station for Research, St. Georges West, Bermuda, 346 pp

Castonguay, M. and McCleave, J.D. (1987) Vertical distributions, diel and ontogenetic vertical migrations and net avoidance of Anguilla and other common species in the Sargasso Sea. *Journal of Plankton Research* **9** (1), 195–214.

Church, T.M., Tramontano, J.M., Scudlark, J.R., Jickells, T.D., Tokos, J.J., Knap, A.H. and Galloway, J. (1984) The wet deposition of trace metals to the western Atlantic Ocean at the mid-Atlantic coast and on Bermuda. *Atmospheric Environment* **18** (12), 2657–2664.

Conte, M.H., Deuser, W.G. and Ralph, N. (2000) Seasonal and interannual variability in deep ocean particle flux at the Ocean Flux Program (OFP)/Bermuda-Atlantic Time Series (BATS) site near Bermuda. *Deep-Sea Research II*, in press

Cook, C.B., Dodge, R.E. and Smith S.R. (1993) Fifty years of impacts on coral reefs in Bermuda. In *Global Aspects of Coral Reefs: Health, Hazards and History*, ed. R.N. Ginsburg, pp. F8–F14. RSMSAS, Miami, 420 pp.

Deuser, W.G. and Ross, E.H. (1980) Seasonal changes in the flux of organic carbon to the deep Sargasso Sea. *Nature* **283**, 364–365.

Dietrich, G. (1965) *General Oceanography*. Wiley-Interscience, N.Y.

Dodge, R.E. and Vaisnys, J.R. (1977) Coral populations and growth patterns: responses to sedimentation and turbidity associated with dredging. *Journal of Marine Research* **35** (4), 715–730.

Dodge, R.E., Logan, A. and Antonius, A. (1982) Quantitative reef assessment studies in Bermuda: A comparison of methods and preliminary results. *Bulletin of Marine Science* **32**, 745–760.

Dryer, S. and Logan, A. (1978) Holocene reefs and sediments of Castle Harbour, Bermuda. *Journal of Marine Research* **36**, 399–425.

Fricke, H. and Kaese, R. (1995) Tracking of artificially matured eels (*Anguilla anguilla*) in the Sargasso sea and the problem of the eels spawning site. *Naturwissenschaften* **82** (1), 32–36.

Fricke, H. and Meischner, D. (1985) Depth limits of Bermudian Scleractinian corals: a submersible survey. *Marine Biology* **88** (2), 175–187.

Galloway, J.N., Penner, J.E. and Atherton, C.S. (1992) Sulfur and nitrogen levels in the North Atlantic Ocean's atmosphere: A synthesis of field and modeling results. *Global Biogeochemical Cycles* **6**, 77–100.

Garret, P., Smith, D.L., Wilson, O.A. and Patriquin, D. (1971) Physiography, ecology and sediments of two Bermuda patch reefs. *Journal of Geology* **79**, 647–668.

Gray, W.M. (1968) A global view of the origin of tropical disturbances and storms. *Monthly Weather Review* **96**, 669–700.

Hedgepeth, J.W. (1957) Marine Biogeography. In *Treatise on marine Ecology and Paleoecology, Vol. 1. Marine Ecology*, ed. J.W. Hedgepeth. Memoir no. 67. Geological Society of America, 1296 pp.

Hayward, S., Gomez, V.H. and Sterrer, W. (1981) *Bermuda's Delicate Balance*. Bermuda Biological Station for Research, Inc. Special Publication No. 20, 402 pp.

Jickells, T.D., Knap, A.H. and Church, T.M. (1984) Trace metals in Bermuda rainwater. *Journal of Geophysics Research* **89** (D1), 1423–1428.

Jickells, T.D., Knap, A.H., Church, T.M., Galloway, J.G. and Miller, J.M. (1982) Acid rain in Bermuda. *Nature* **297**, 55–57.

Jickells, T.D. and Knap, A.H. (1984) Trace metals in Bermuda inshore waters. *Estuarine, Coastal and Shelf Science* **18** (3), 245–262.

Katona, S.K., Harcourt, P. Perkins, J.S. and Kraus, S.D. (1980). Humpback Whales. A catalogue of individuals identified by fluke photographs. College of the Atlantic, Bar Harbor, Maine.

Knap, A.H. (1990) In *The Long-Range Atmospheric Transport of Natural and Contaminant Substances*, ed. A.H. Knap. NATO ASI Series Vol. 297. Kluwer Academic Publishers, Netherlands. 321 pp.

Knap, A.H., Iliffe, T.M. and Butler, J.M. (1980) Has the amount of tar in the ocean changed in the past decade? *Marine Pollution Bulletin* **11**, 161–164.

Knap, A.H. and Jickells, T.D. (1984) Trace metals and organochlorines in the goosebeaked whale (*Ziphius cavirostris*). *Marine Pollution Bulletin* **14**, 271–274.

Knap, A.H., Jickells, T.D., Pszenny, A. and Galloway, J (1986a) The significance of atmospherically derived fixed nitrogen on the productivity of the Sargasso Sea. *Nature* **320**, 158–160.

Knap, A.H., Binkley, K.S. and Deuser (1986b) The flux of synthetic organic compounds to the Sargasso Sea. *Nature* **319** (6054), 572–574.

Knap, A.H., Binkley, K.S. and Artz, R.S. (1988) The occurrence and distribution of trace organic compounds in Bermuda precipitation. *Atmospheric Environment* **22** (7), 1411–1423.

Logan, A., Yang, L. and Tomascik, T. (1994) Linear extension rates in two species of *Diploria* from high latitude reefs in Bermuda. *Coral Reefs* **13**, 225–242.

McGillicuddy, D.J. Jr., Robinson, A.R., Seigel, D.A., Johnson, R., Dickey, T.D., Jannasch, H.W., McNeil, J., Michaels, A.F. and Knap, A.H. (1998) New evidence for the impact of mesoscale eddies on biogeochemical cycling in the Sargasso Sea. *Nature* **394**, 263–265.

Menzel, D.W. and Ryther, J.H. (1960) The annual cycle of primary production in the Sargasso Sea off Bermuda. *Deep-Sea Research* **6**, 351–367.

Michaels, A.F., Siegel, D.A., Johnson, R.J., Knap, A.H. and Galloway, J.N. (1993) Episodic inputs of atmospheric nitrogen to the Sargasso Sea: Contributions to new production and phytoplankton blooms. *Global Biogeochemical Cycles* **7**, 339–351.

Michaels, A.F., Bates, N.R., Buesseler, K.O., Carlson, C.A., Knap, A.H. (1994) Carbon-cycle imbalances in the Sargasso Sea. *Nature* **372**, 537–540.

Michaels, A.F. and Knap, A.H. (1996) Overview of the U.S.JGOFS BATS and Hydrostation S program. *Deep-Sea Research* **43** (2–3), 157–198.

Morris, B.F. (1971) Petroleum: tar quantities floating in the northwestern Atlantic take with a new quantitative neuston net. *Science, N.Y.* **173**, 430–432.

Morris, B.F. and Butler, J.N. (1973) Petroleum residues in the Sargasso Sea and on Bermuda beaches. *Proceedings of the Conference on Prevention and Control of Oil Spills*. American Petroleum Institute, Washington, DC, pp. 521–529.

Morris, B.F. and Mogelberg, D.D. (1973) *Identification Manual to the Pelagic Sargassum Fauna*. Bermuda Biological Station Special Publication, No. 11, 63 pp.

Morris, B., Barnes, J., Brown, F. and Markham, J. (1977) *The Bermuda Marine Environment*. Special Publication 15, Bermuda Biological Station for Research, 120 p.

Nelson, N.B. (1998) Spatial and temporal extent of sea surface temperature modifications by hurricanes in the Sargasso Sea during the 1995 season. *Monthly Weather Review* **126**, 1364–1368.

Parr, A.E. (1939) Quantitative observations on the pelagic Sargassum vegetation of the western North Atlantic. *Bulletin of the Bingham Oceanographic Collection, Peabody Museum of Natural History, Yale University* **6** (7), 1–94.

Payne R. and McVay, S. (1971) Songs of the humpback whale. *Science, N.Y.* **173**, 585–597.

Redfield, A.C., Ketchum, B.H. and Richards, F.A. (1963) The influence of organisms on the composition of sea water. In *The Sea*, ed. M.N. Hill. Wiley, New York, pp. 26–77.

Schroeder, E. and Stommel, H. (1969) How representative is the series of Panulirus stations of monthly mean conditions off Bermuda? *Progress in Oceanography* **5**, 31–40.

Siegel, D.A., Westberry, T.K., O'Brien, M.C., Nelson, N.B., Michaels, A.F., Morrison, J.R., Scott, A.J., Caporelli, E.A., Sorensen, J.C., Maritorena, S., Garver, S.A., Brody, E.A., Ubante, J. and Hammer, M.A. (2000) Bio-Optical modelling of primary production on regional scales: The Bermuda Bio-Optics Project. *Deep-Sea Research II*, in press.

Smith, S.R. and Musik, K.M. (1981) Coral Survey of Bermuda's reefs. Sixteenth Meeting of the Association of Island Marine Laboratories of the Caribbean. St Georges, Bermuda, GE 01. Vol. 16 p. 13.

Smith, S.R. and Knap, A.H. (1985) Significant decrease in the amount of tar stranding on Bermuda. *Marine Pollution Bulletin* **16**, 19–21.

Stone, G.S., Katona, S.K. and Tucker, E.B. (1987) History, Migration and present status of Humpback whales *Megaptera novaengliae* at Bermuda. *Biological Conservation* **42**, 122–145.

Thomas, M.L. and Logan, A. (1992) *A Guide to the Ecology of the Shoreline and Shallow-Water Marine Communities of Bermuda*. Special Publication 30, Bermuda Biological Station for Research, 345 pp.

Verril, A.E. (1902) The Bermuda Islands. *Transactions of the Conn. Academy of Arts and Sciences 1901–1903* **11**, 413–911.

Wallace, J.M. (1973) General circulation of the tropical lower stratosphere. *Review of Geophysics and Space Physics* **11**, 191–222.

Worthington, L.V. (1976) *On the North Atlantic Circulation*. The Johns Hopkins Oceanographic Studies 6, 110 pp.

THE AUTHORS

Anthony H. Knap
*Bermuda Biological Station for Research,
17 Biological Station Lane,
St. Georges, Bermuda*

Douglas P. Connelly
*Bermuda Biological Station for Research
17 Biological Station Lane,
St. Georges, Bermuda, GE 01*

James N. Butler
*Harvard University
Cambridge, Massachussetts, U.S.A.*

Chapter 15

THE AEGEAN SEA

Manos Dassenakis, Kostas Kapiris and Alexandra Pavlidou

The Aegean Sea is one of the Eastern Mediterranean sub-basins located between the Greek and the Turkish coasts and the islands of Crete and Rhodes. It has more than 2000 islands and a complex coastline with numerous gulfs. The main physical characteristics of this system are the general cyclonic movement of surface water, the high salinity, the dissolved oxygen saturation, the low biomass and the low availability of nutrients. It is affected also by the Black Sea water which comes into the Aegean through the Dardanelles.

Many human activities (urban, industrial, marine, rural, etc.) affect the water quality and threaten this fragile ecosystem. Most of the environmental problems are concentrated near the large cities where many scheduled works for pollution control are either delayed or are ineffectively designed. As a result, in many areas the level of heavy metals, organic pollutants and nutrient concentrations are significantly elevated. Some gulfs suffer seasonally from eutrophication and anoxic condition problems. Oil spills and litter have frequently been encountered on the beaches during the last few years. Furthermore, the excessive use of fertilisers and pesticides contributes significantly to the deterioration of many coastal ecosystems of the Aegean.

The Aegean coastline is one of the most valuable natural resources, as it generates value from tourism, recreation, transport, fisheries, aquaculture and salt extraction, but it is managed without any serious planning both in Greece and in Turkey. Social pressures for better environment quality are still low, although numerous economic activities are directly dependent on improving the environmental condition. It is still not clear to many that high environment quality is directly related to socio-economic development and prosperity.

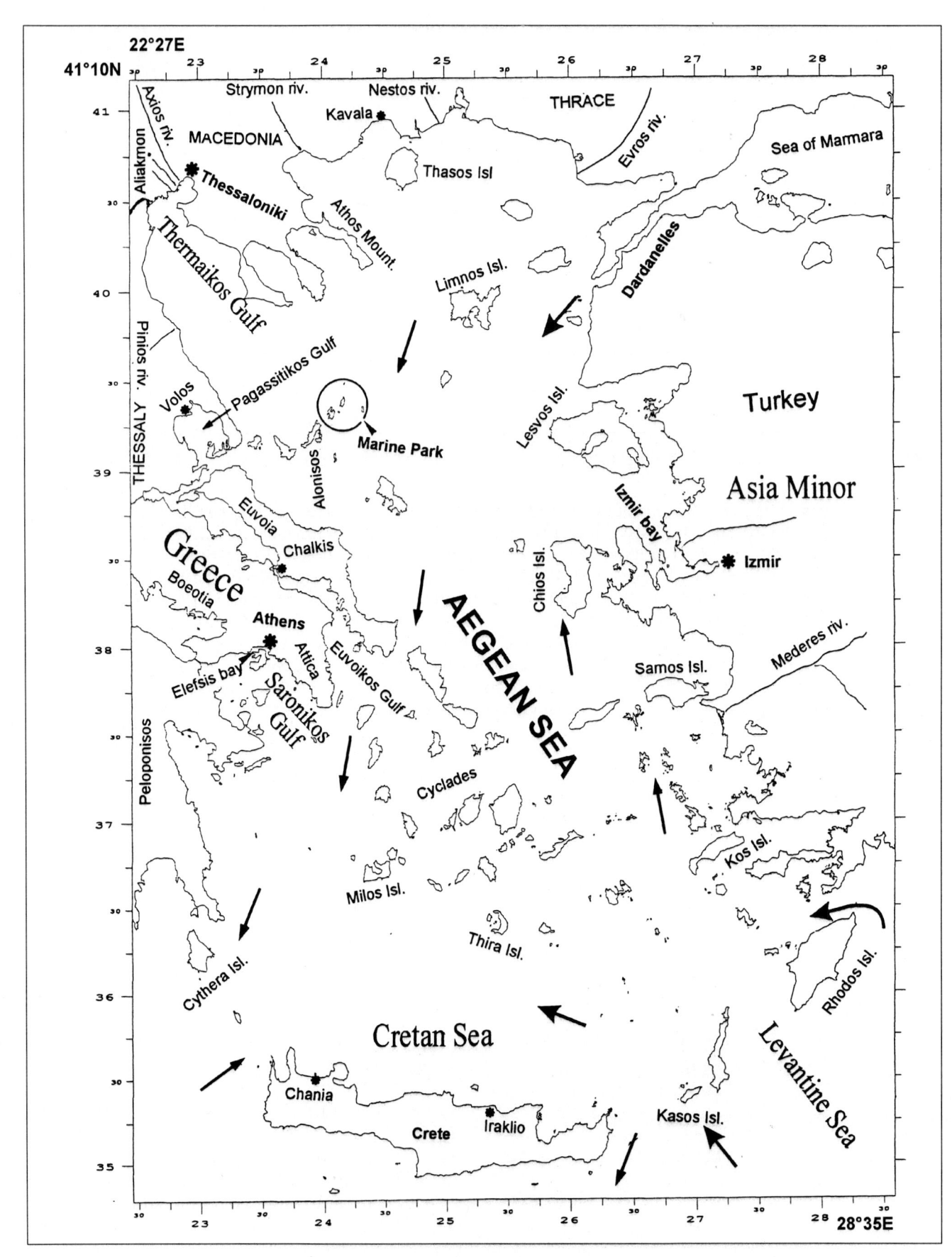

Fig. 1. The area studied. The main gulfs and rivers and the main direction of surface currents are presented.

THE DEFINED REGION

The Aegean Sea (Fig. 1) is one of the Eastern Mediterranean sub-basins located between the Greek and the Turkish coasts and the islands of Crete and Rhodes. It has more than 2000 islands forming small basins and narrow passages, and a very irregular coastline topography. Its area is 2×10^5 km^2, its volume 7.4×10^4 km^3 and its maximum depth 2500 m north of Crete. The Aegean extends about 640 km from Crete northward to the coast of Thrace, and its width ranges from 195 to 400 km. The Cythera–Crete and Crete–Kasos–Rodos straits in the southwest and southeast of the basin, respectively, are the boundaries of the Aegean Sea with the main body of the Eastern Mediterranean. Three major basins exist in the Aegean Sea:
1. in the Northern Aegean, the basin of the Mountain Athos (max depth 1500 m),
2. in the Central Aegean, the Chios basin (depths up the 1100 m), and
3. in the Southern Aegean Sea there is the deepest (max. depth 2500 m) and largest basin, north of the Cretan coast.

The Aegean Sea receives some water of relatively low salinity (about 16‰) from the Black Sea. The water flow from the Black Sea to the Aegean through the Dardanelles has been estimated to be 12,600 m^3/s (taking into account a reverse flow, from the Aegean to the Black Sea, to 6100 m^3/s). Black Sea water, after mixing over the sea of Marmara, generates a distinct surface water layer in the North Aegean (UNEP(OCA)/MED 1996a).

The coast of the Aegean is mountainous. There are extensive coastal plains only in Macedonia and Thrace, in the north. Some large rivers discharge into the sea. In the north, the Axios, Strimon, Nestos and Evros enter from the Balkan Peninsula, the Pinios enters from the Thessaly plain in the Greek coast and the Menderes enters from Anatolia. The most important ports that ring the Aegean littoral are Pireaeus and Thessaloniki in Greece and Izmir in Turkey.

The Aegean Basin consists of a submerged block of the Earth's crust across which long belts of folded rocks extend from the mountain of Greece to the mountains of Turkey. On the seafloor these belts form submarine ridges, separated by deep basins. The ridges, clearly traceable on the seafloor, are the foundation of most of the Aegean islands.

The Aegean sea is a region of active extensional tectonism within the overriding part of a convergent plate margin system (Africa with respect to Europe) in the Mediterranean. The Aegean has split a pre-existing tectonized continental crust and is also stretching in response to a general westward motion of the Turkish–Anatolian continental block imposed by the southern collision of the Arabian plate. Given the specific geodynamic setting, the Aegean has been considered a 'deforming microplate'. The so-called 'North Aegean trough' is interpreted as the westward extension of the Anatolian fault zone. In this plate tectonic framework it represents the northern transform type boundary of the Aegean microplate. The southern border of this microplate is formed by the Levantine segment (also of transform type) of the Greek subduction zone. The eastern and western microplate boundaries of the Aegean domain are less well defined (Dewey and Sengor, 1979).

The overall late Miocene–present evolution of the Aegean sea may be directly dependent on two geodynamic processes: to the southeast of the area, collision between Arabia and Turkey leads to the lateral (westward) expulsion of the Aegean continental crust, and to the west, Greek subduction is still active. Nearly all of the Greek area was land in the mid-Miocene. It was even connected by land to Asia Minor. As a result of sea regression at that time, the sea was situated south of Crete and of the Dodecanese islands. During the Upper Miocene the sea then occupied all the area of today's Aegean Sea. Only the Cyclades were still land and connected to Attica and Euvoia. The Mediterranean Sea at that time consisted of only a few salt-water basins with a small inflow of fresh water. In conjunction with the evaporation rates, this resulted in the development of large salt contents which caused the sedimentation of evaporates. At the end of the Miocene–Lower Pliocene, the Aegean Sea dried out. The sea regressed further to the southwest of Crete and the Ionian islands, while Paratethys penetrated the North Aegean area. The Island Arc and the internal sea of the Cretan Sea were formed during the Pliocene. At the end of the Pliocene, the sea regressed from the North and Central Aegean and gave its place to lakes with an Efxinocaspian fauna. The geodynamic evolution of the Greek area during recent time is mainly attributed to the collision of the African tectonic plate with the Euroasiatic tectonic plate. An important impact of this collision is the remarkable volcanic activity in the Greek Arc area. The South Aegean volcano arc includes many Plio-Quaternary volcanoes, i.e. Nisiros, Kos, Patmos, Thira, Milos, Antiparos, Krommionia, Methana, Lihades, Kammena Vourla, Psathoura, those on the northwest edge of the Almopian volcano arc, Koula and Aidini in West Turkey (Katsikatsos, 1992).

The main gulfs in the coasts of Aegean Sea which are under environmental pressure as a result of significant polluting human activities are as follows.

The Thermaikos Gulf, a semi-enclosed gulf in the northwest Aegean Sea, with the city of Thessaloniki (approximately one million inhabitants) on its northern coast. It is fairly long and shallow (depth not exceeding 50 m). The inner section of the gulf has a total surface of 300 km^2 and a maximum depth of 30 m.

The Saronikos Gulf, in the vicinity of Athens, is considered to be among the most polluted Greek gulfs. Its area is about 2600 km^2 and its maximum depth is 450 m. The Saronikos Gulf is subdivided into four sections on the basis of geomorphological criteria:
1. the South Gulf (external), with a maximum depth of 400 m,
2. the West Gulf, with depths between 70 and 90 m,

3. the East Gulf, with depths up to 100 m, and
4. the Elefsis Bay, between Salamis island and Attiki coast, with depths up to 30 m.

The Bay of Izmir is one of the largest bays of the Turkish Aegean coast, but is very shallow. It extends about 24 km in an east–west direction and its average width is approximately 5 km. It consists of three sections: the Inner Bay, the Middle Bay and the Outer Bay. The bottom slopes in the northern and northwestern coastlines of the Bay are mild, whereas those in the south are steep, reaching 20 m deep. The sea depth of Izmir Bay ranges from 0–20 m in the Inner Bay, 0–40 m in the Middle Bay and 40–60 m in the Outer Bay.

The Pagassitikos Gulf is a semi-enclosed inlet of sea situated on the east coast of central Greece, to the north of Euvoia island. It communicates with the open Aegean Sea through the Trickeri channel, 5.5 km wide and 80 m deep.

The Euvoikos Gulf is a shallow embayment of the Aegean Sea formed by the eastern coasts of Attica and Boeotia and the western coast of Euvoia island.

SEASONALITY, CURRENTS AND NATURAL ENVIRONMENTAL VARIABLES

The most sustained prevailing winds in the Aegean during all four seasons are those of the northern sector. The second most sustained are the winds of the southern sector. The eastern and western winds have much lower frequencies. In summer, dry winds of the northern sector—called Etesians—can reach gale force.

A simplified view of surface currents in the Aegean is presented in Fig. 1. The general circulation pattern is complex, having seasonal and local variability, consisting of basin, sub-basin and mesoscale features. The various flows in and out of the Aegean arise in response to the continuity requirements dictated by the circulation patterns on either side of the passages. The dynamical structure of the straits comes from the Rhodos gyre, the Asia Minor current, the anticyclones found to the southeast of Crete and the Eastern Cretan Sea (Theocharis et al., 1993).

The most permanent feature of the surface movements in the Aegean is probably the cyclonic movement. The surface layer circulation in the North Aegean Sea is dominated by the Black Sea water. Cyclonic surface circulation carries Black Sea water towards the west, and then follows a south-southwesterly route (Zodiatis, 1994). The complexity of the geometry of the North Aegean with respect to the main axis current direction forms small-scale cyclonic and anticyclonic flows along the general direction of movement (Lascaratos, 1992).

The salinity of the Aegean is rather high (mean value 38.9%, max values about 39.2% during early autumn). In the western part of the region the outflow of the rivers and the Black Sea water cause a slight decrease of the salinity. The deep water of the Aegean is characterised as one of the densest in the World Ocean, having temperatures of about 15°C in summer and 14°C in winter (Lascaratos, 1992).

Sea surface temperature range through the year in the Aegean is typically 8–26°C. The lower values are observed in the North Aegean and the higher in the southwest part. The temporal variations of temperatures are strongly influenced by the seasonal atmospheric conditions (Vlahakis and Pollatou, 1993)

The Aegean responds, even at great depths, to seasonal and/or interannual variations, through complex mechanisms. Within this region there is no climatic factor to dramatically influence the distribution of the organisms. Instead, they respond to the seasonal and local variations of the small-scale physical parameters.

A brief description of the behaviour of Aegean water masses, based on satellite images was given by LeVourch et al. (1992).

May marks the beginning of spring, with the warming of the Gulf of Salonica. The Black Sea water is still delineated by sharp fronts off the Dardanelles.

In June the shallow coastal waters of Greece and Turkey warm up, whereas the central Aegean water remains colder. A cold water area may be observed around Rhodes, contrasting with the warm water of the Asia Minor Current.

In July and August the summer thermal structures occur and the northerly winds blowing over the Aegean Sea produce upwelling zones along the western Turkish coasts as far as eastern Crete. Other zones, protected from the wind, create sharp thermal fronts. These northerlies also produce clearly delineated cold water areas on both sides of Crete. The Rhodes cold water is clearly apparent.

September still presents summer features although the gradients weaken. The cool water southwest and southeast of Crete as well those of Rhodes strongly contrasts with the surrounding water. Along the Turkish coast, given the weakening of the northerlies, the warm water can extend northwards up to Samos (occasionally Lesvos).

October and November represent a transition towards autumn dominated by a contrast between the cold central Aegean water and the warmer coastal areas. The cool water of Crete (southwest and east) remains distinct.

In December the whole western Aegean Sea is colder than the eastern part due to the influx of cold Black Sea water which contributes to the cooling of the northern part of the Aegean. The cold water near Rhodes remains well delineated, although the previously sharp contrast between this and the southwestern Crete cold water tends to weaken. Some cold river runoff may be observed in the Gulf of Salonica.

In January the west–east gradients remain strong, the southwest Crete cold spot disappears, while the contrasts around the eastern Crete cold water weaken.

In February the gradients attenuate and the cold water of the northern Aegean moves southwards through the Euvoia and Cyclades Straits. The water around Crete is usually homothermal.

In March and April the gradients are north–south, with a double frontal system: there is an important extension of the cold water of the northern Aegean Sea, from the Dardanelles to Euvoia, with an important contribution from the cold Black Sea water. The secondary system marks the limit, north of the Cyclades, of the relatively warm water of the South Aegean Sea.

The Aegean basin is partitioned by many islands which create local conditions which confuse the clear seasonal evolution. August may be considered as a typical summer month (cold in the east, warm in the west, cold patches at both ends of Crete). December can be understood as a typical autumn month (cold in the west, warm in the east, cold patch east of Crete). March may be interpreted as a typical winter month (cold in the north, warm in the south, no gradients, except between Euvoia and Dardanelles). June may be considered as a typical spring month (cold central water surrounded by warm coastal water).

The Saronikos Gulf has five distinct water masses: Elefsis Bay, the Inner, Central and Outer gulfs and the Western basin. The Inner and Central gulfs frequently exhibit similar water mass characteristics. In winter, the waters of the gulf are well mixed, down to a depth of about 40 m. The Elefsis Bay waters develop into a distinct water mass during spring. Winds develop two patterns of circulation in the Saronikos gulf, cyclonic and anticyclonic (Barbetseas, 1983).

Analysis of the water mass of Thermaikos Gulf characteristics reveals that there is a general anticlockwise pattern of water movement. Low salinity surface waters discharging from the rivers flow towards the south along the western coastline. Conversely, relatively high salinity waters, which are of southerly origin, from the open waters of the Aegean Sea, intrude along the eastern coastline (Balopoulos and Friligos, 1994).

Pagassitikos Gulf is generally characterised by weak currents (<10 cm/s). Residual currents in both the near-surface and near-bed layers of the western area are towards the southwest, whilst in the eastern basin, the upper-layer residual flow is predominantly towards the northeast. The wind condition is the most likely factor affecting the water movements whereas the tidally induced currents are rather small (Voutsinou-Taliadouri and Balopoulos, 1989).

The Euvoikos Gulf is naturally divided by the narrow straits of Euripos. The restricted area of the straits can be considered as a separate section of particular interest due to the significant tidal phenomena that are observed there. The morphology of the area causes the development of a strong (about 12 km/h at the narrowest position) tidal current that changes its direction every six hours. Such currents are very rare in the non-tidal Mediterranean. It causes the rapid transport and dispersion of pollutants, but also significantly affects the sedimentation processes, especially in the immediate vicinity of the Euripos straits where the fine-grained sediment fraction is eliminated. There are also some small bays where the efficiency of the current in removing any pollutants is rather limited since water velocity is usually less than 10 cm/s. Usual current velocities in the area are 6–18 cm/s and occasionally 20–30 cm/sec (Vlachakis and Tsiblis, 1993).

In Izmir Bay studies show that the principal cause of mixing in the upper layer is the wind, the surface currents are in the same direction as the wind, and the dominant current during the summer is toward the east, while during the winter the dominant current direction is toward the northwest.

The very low tidal variations in this bay result in little water currents while vertical mixing can be caused due to the density changes through warming and evaporation (UNEP, 1994).

THE MAJOR SHALLOW WATER ECOSYSTEMS AND BIOTIC COMMUNITIES

Plankton

The Aegean can be characterised as an oligotrophic area, having low biomass and low availability of nutrients and growth factors. With the exception of the areas around the mouth of rivers, where dense zooplankton communities are formed, the plankton biomass remains low at 500–1000 individuals per cubic metre of water. The Aegean Sea is 12–18 times lower than the Black Sea in terms of secondary productivity, four times lower than the Adriatic Sea, but similar in terms of productivity to the Tyrrhenian and Ligurian Seas. The northern part of the Aegean shows a tendency to be somewhat ecologically isolated while the South Aegean has the character of a transitional area (Ignatiades et al., 1994).

The oligotrophic character of the Aegean Sea is confirmed by the low abundance of phytoplankton, in combination with the relatively high number of species. Diatoms and dinoflagellates are the main species, but in the Central Aegean—especially during spring—there is abundance of coccolithophores that form, together with copilates, interesting prey–predator relations. There are significant differences in planktonic populations between spring and autumn (Pagou et al., 1990).

The majority of zooplankton species are small-sized Atlanto-Mediterranean forms known from the Western Mediterranean, while some species have migrated from the Black Sea and from the Red Sea. The zooplanktonic communities of the Aegean Sea are clearly dominated by copepods which usually constitute 75–95% of the total populations. Copilates are the second most abundant species group. The dominant species in the North Aegean are *Temora stylifera*, *Centropages typicus*, *Calanus minor*, and *Acartia clausi*, while dominant species in the South Aegean are *Temora stylifera*, *Clausocalanus arcuicornis*, *Oithona plumifera*, and *Acartia negligens*.

Many other differences have been observed between the copepod fauna of the two basins of the Aegean Sea. Four species of cladocerans have been found in the Aegean

Sea (*Evadne spinifera, E. tergestina, Penilia avirostris,* and *Podon intermedius*), while ten species of chaetognaths have been identified, the most abundant being *Sagitta minima* and *S. enflata*.

From the nature of the hydrological profiles in the South Aegean, Tselepides and Eleftheriou (1992) concluded that the oligotrophic water column gives rise to a dominance of small phytoplankters and a long 'microbial loop' type of food chain. Most of the biomass consists of bacteria and microzooplankton, and an inverted biomass pyramid better describes the system. An interesting implication of the latter is that fixed nitrogen is recycled many more times in oligotrophic waters, while the climax community is supported by new nutrients being injected from below the euphotic zone.

Pollution in the Saronikos Gulf influences the plankton abundance, composition and structure, but the intensity of the impact from contaminants and nutrients depends on the water circulation and topography of each area (UNEP, 1996a). Near Athens and in Elefsis Bay there is an excessive predominance of diatoms, dinoflagellates and coccolithophores. Peak abundance is observed in summer due to the impact of greater human activity in this season. Furthermore, pollution tends to reduce the number of species in periods when the abundance is greatly raised. *Coccollithus sp.* and *Thalassiosira aetivalis* predominate in this area, suggesting that pollution causes the bloom of species which depends on the local conditions and which are scarce in clean waters.

The Gulf of Elefsis is the most eutrophic area in the Saronikos Gulf, as nutrients are also released from sediments in anoxic periods. Pagou (1990) reported 10^7 cells/l of *Gymnodinium breve* in November 1977 and 29×10^6 cells/l of *Scripsiella trochoidea* in May of the same year. Concentrations higher than 2×10^7 cells/l of *Gymnodinium sp., Pyramimonas sp., Thalassiosira sp., Leptoxcylidrus danicus, Nitzschia delicatissima* and *Noctiluca scintilas* have also been observed along the Attiki coasts. Yannopoulos and Yannopoulos (1973) reported that >90% of zooplankton species consisted of the copepod *Acartia clausi*.

The increase in phytoplankton abundance in Thermaikos Gulf is correlated with increased phosphate and nitrate concentrations which come from industrial pollution and rivers. In the northern part of this Gulf, zooplankton is abundant and communities are characterised by dominance of opportunistic species (*Acartia clausi, Oithona nana,* and *Podon polyphemoides*). In contrast, the eastern region is strongly influenced by incoming water masses from the Aegean Sea and zooplankton communities reflect a much more diversified and stable situation (*Acartia clausi, Evadne spinifera,* and *Centropages typicus*) (Siokou-Frangou and Papathanassiou, 1991).

In Izmir Bay, interaction of inorganic phosphate and nitrogen with physical factors such as light and temperature has led to the development of dinoflagellate or diatom dominated blooms. When phosphate levels in the water column are adequate, *A. minutum, P. micans* and *S. trochoidea* dominate the dinoflagellate red tides which may be observed here. In contrast, diatoms such as *T. allenii, T. anguste-lineata,* the euglenoid *E. gymnastica* and the prasinophyte *P. propulsum* tend to dominate and bloom in nitrogen-rich waters, depending on light and temperature (Koray et al., 1996).

Red tide phenomena were recorded in Pagassitikos Gulf in July 1987 (and in summer 1997), caused by flagellates of the species *Gymodinium catenatum* (11,150,000 cells/l), *Cachonina niei* and *Gonyalax tamarensis* (Pagou, 1990).

Fish Fauna

Few studies exist on the fish fauna of the North Aegean Sea and most existing literature concerns the distribution of certain fish species. A total of 141 identified fish representing 101 genera are found there. This ichthyofauna has a mainly Atlanto-Mediterranean character with some boreal and truly tropical elements (84%). There are only four cosmopolitan species, nine worldwide species and nine species endemic in the Mediterranean (Papaconstantinou et al., 1994).

The cosmopolitan fishes are *Dalatias licha, Hoplostethus mediterraneus, Raja clavata* and *Squalus acanthias*. The more abundant worldwide species are *Scomber japonicus* and *Zeus faber,* while the more abundant Atlanto-Mediterranean species are *Coelorhynchus coelorynchus, Gadiculus argenteus, Micromesistius poutassou, Merluccius merluccius, Mullus barbatus* and *Serranus cabrilla*.

The more commercial fish species are *Merluccius merluccius, Mullus barbatus, Mullus surmuletus, Trigla lyra, Trisopterus minutus capellanus, Pagellus erythrinus* and *Micromesistius poutassou*.

The fish fauna of the South Aegean Sea is mainly made up from species of Atlanto-Mediterranean character (85%) while there are a 6% endemic species and a further 8% with cosmopolitan character. In this region there exist more thermophillic species than the North Aegean Sea, especially around the Dodecanessa islands.

In the Central Aegean the more abundant species are: *Serranus hepatus, Argentina sphyraena, Lepidotrigla cavillone, Capros aper, Dentex maroccanus* and *Mullus surmuletus*.

The Benthic Fauna

The benthic fauna of the Aegean Sea forms a separate subsystem of the Mediterranean fauna (Murdoch and Onuf, 1974). The native populations of the Aegean Sea are a source of enrichment for the Black Sea and are themselves influenced by the less saline waters of the Black Sea. Furthermore, they are enriched both with Indopacific immigrants and warm water species of Atlantic origin (Turkay et al., 1987). Despite easy access to the sea, regular research into this fauna began only recently, even though several studies had been conducted more than a century ago.

Using echinoderms as an example, a total of 107 species are reported from Greek seas. Thus 70.4% of the Mediterranean echinoderm fauna are present in the Aegean. Among them 73 species are common in both western and eastern Greece, two of them occur only in the west, one has been reported only in the sea of Kythira (South Aegean), 21 are known only in the east, eight are common for both the Aegean and Kythira seas and two have no locality specification (Pancucci-Papadopoulou, 1996). In the entire Mediterranean, five species of crinoids and 52 species of holothuroids have been found (Sinis, 1977), so that the holothuroid fauna of the Aegean Sea are 56% of that of the Mediterranean (Koukouras and Sinis, 1981).

There are 404 polychaetes (Annelida, Polychaeta) in Greek waters; of these, 56 were first reported in the Aegean. Their high diversity is favoured by substrate heterogeneity but is reduced by intense hydrodynamic conditions and especially by pollution. It has been suggested that polychaete communities alone can reflect, with high fidelity, the general diversity patterns of benthic ecosystems (Symboura, 1997).

From the 400 bivalvia species known in the Mediterranean, about 300 or more have been reported in Greek waters. This contrasts with the theory of a poor Eastern Mediterranean fauna. In fact, the number of bivalvia species constantly increases with new studies in either unexplored bathymetric zones or overlooked geographical areas. In the last ten years approximately 25 more bivalvia species have come to light (Zenetos, 1996).

The number of decapod species known in the Aegean Sea runs to 231 (74 Natantia, 23 Macrura Reptantia, 35 Anomoura and 99 Brachyura). These 231 species comprise the 70.64% of the Mediterranean decapod fauna and seven of these species are reported for the first time from the eastern Mediterranean and three from the Aegean Sea (Koukouras et al., 1992).

On the South Aegean continental slope, biodiversity and the total biomass of benthos declines with depth, especially between 300 and 400 m. Polychaetes are the most important group overall, together with crustaceans.

During the last fifty years in Elefsis Bay an obvious decrease in abundance of some invertebrates has been demonstrated following the start of the industrial development of the coast (i.e., Cephalopods, Decapods). Other species have increased in abundance and species not previously recorded in the Gulf have appeared (i.e., *Patella* sp., *Mytilus edulis*).

Marine Plants

There are about 1000 species of macroscopic plants in the Mediterranean of which 15–20% are endemic. From them, there are five Phanerogams marines, which are considered as endangered in the Aegean: *Posidonia oceanica*, *Cymodocea nodosa*, *Zostera marina*, *Zostera noltii* and *Halophila stipulacea*. In Greece, other species recommended for protection are: *Lithophyllum lichenoides*, *Schimmelmannia ornata* (Rhodophyta), *Cystoseira amentacea* (Fucophyceae), *Posidonia oceanica* and *Zostera marina* (Phanerogams).

The famous endemic seagrass species *Posidonia oceanica* is regressing under the impacts of pollution, coastal and river works, trawling, dredging and anchoring. The meadows in the Saronikos Gulf are well developed in the outer gulf but are absent in the eastern part. The reasons for this absence are pollution, topography and construction on the beach (Panayotidis, 1988).

In the North Aegean Sea at depths from 5 to 78 m, 21 Chlorophyceae, 29 Phaeophycae and 70 Rhodophycae were identified. In this region some *Cystoseira* communities were totally eliminated (Tsekos and Haritonidis, 1982). In general, *Cystoseira* forests are regressing under the negative impact of pollution.

In the Saronikos Gulf 197 species of algae have been identified. They are classified into four main groups: seven species in Chlorophyceae, 34 in Bryopsidophyceae, 37 in Phaeophycae and 119 in Rhodophycae. The ratio Rhodophycae/Phaeophycae is about 3.22, which suggests that the flora of the Saronikos Gulf is of the Atlantic subtropical type. The algae were classed into eight ecological groups, the main groups being sciophile algae and the photophile algae. The surface covered and the number of species are affected by polluted water. The lowest values are recorded in the vicinity of the sewage outfall of Athens (Diapoulis, 1988).

Marine Mammals

The mammals of Greece include a total of 116 species and can be divided into 31 families which belong to eight classes: Insectivora, Chiroptera, Lagomorpha, Rodentia, Carnivora, Perissodactyla, Artiodactyla and Cetacea. Nineteen of the 80 or more cetaceans of the world have been observed in the Mediterranean. Though several of these species are frequently seen and others have been washed ashore from time to time, there is really very little information available on species living in the eastern Mediterranean. The fact that Greek seas are oligotrophic, combined with overfishing, represents a serious threat to many species with small populations. The populations of the Mediterranean Monk Seal are continuously declining throughout its distribution area, partly as a result of getting caught and damaged in fishing nets. More than half of the surviving Monk Seals today live in Greece (Scoullos et al., 1994a). According to the Red Data Book (1992), seven cetacean species in Greece are included in the IUCN 'endangered' listings, two in the 'vulnerable' and five in the 'rare' categories.

Sea Birds

Drawing on the current available information, the Montpellier meeting of experts (1995) listed 15 species of birds as endangered or threatened—all having a direct relation with the marine environment. In Greece, there are 30 water birds (Red Data Book, 1992) of which the raptors (21 birds) are the most endangered group in Greece. The destruction by continuing human encroachment into the wetlands, combined with intense and often uncontrolled hunting has serious repercussions for these birds.

Marine Reptiles

The Greek marine herpetofauna is perhaps one of the richest among the European countries. It contains 16 species of amphibians and 28 species of reptiles. From these species, amphibians, two species of terrapin (*Emys arbicularis* and *Mauremys caspica*) and two species of water snake belonging to the genus *Natrix*, face significant problems from human activities, especially drainage, pollution and tourist development. The sea turtles are the most endangered. The species *Caretta caretta* is the only species which nests in Greece. It faces very serious problems arising from tourist development in areas where it lays its eggs. Greece is home to the majority of the loggerhead turtles in the Mediterranean, with the main nesting beaches in Zakynthos, Crete and Peloponnese. Of these, Zakynthos in the Ionian Sea is the single most important nesting area in the Mediterranean for the loggerheads. An average of 1300 nests are made each year and these are concentrated on six beaches totalling only 5 km in length.

OFFSHORE SYSTEMS

In the Aegean Sea, given the large number of islands, there are few extensive offshore areas. Among the principal basins of the Mediterranean, the intermediate and deep waters of the Aegean Sea have the lowest concentrations of nutrients and the highest of oxygen. Oceanographic research during 1994 has shown that dissolved oxygen is almost saturated in the surface layer (about 6 ml/l in winter and 4.8 ml/l in summer).

The surface layer is separated from the lower layers of intermediate and deep water by a transition layer of 100–200 m thickness, where the concentration of nutrients increases quickly. Dissolved oxygen shows a sharp decrease in this transitional layer. The most striking feature is the presence of the nutrient-rich (>5.8 µg/l for nitrates and >0.180 µg/l for phosphates) and oxygen-poor water mass of the Cretan Sea, which lies below 200 m depth (Souvermezoglou et al., 1997). Figure 2 shows a characteristic distribution of dissolved oxygen and nitrates along the 36° parallel in the Cretan Sea.

The Mediterranean is considered to differ from the large ocean basins in that phosphate is the main nutrient limiting the growth of phytoplankton. The eastern Mediterranean is strongly phosphorus limited. The N/P ratio increases eastwards (24–27 across the Cretan Sea). Massive transport of nutrients from deep layers to the impoverished euphotic

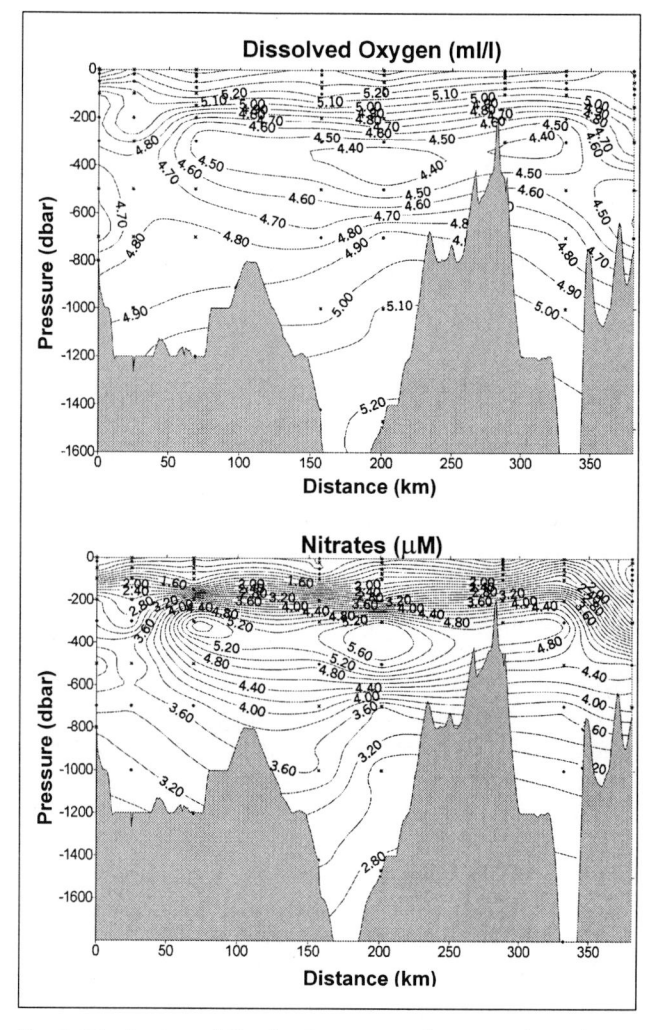

Fig. 2. Distribution of dissolved oxygen and nitrates along the 36° parallel.

zone occurs when vertical mixing processes are observed (Tselepides et al., 1997).

The chlorophyll in the southern Aegean Sea ranges between 0.1 and 0.4 mg/m^3 and reaches its maximum at about 100 m depth. The relevant concentrations in the northern Aegean ranges between 0.1 and 0.8 mg/m^3 with a maximum at about 40 m depth. The vertical distribution of chlorophyll in the northern, southern and central Aegean have a characteristic shape that is presented in Fig. 3 (Siokou-Frangou et al., 1997).

Deep-water Fauna

The deep Mediterranean fauna is relatively poor in animal species. The warm homothermy below 300 m ($T = 13°C$) is a limiting factor for many bathyo-abyssal animals and a number of important deep-sea groups are absent or poorly represented.

The vertical distribution of the Mediterranean copepods shows characteristic features of an isolated oceanic basin. In the deep layers of the Greek seas the density of the zooplankton is very low. Below 1000 m, only 0.1–0.2 ind./m^3

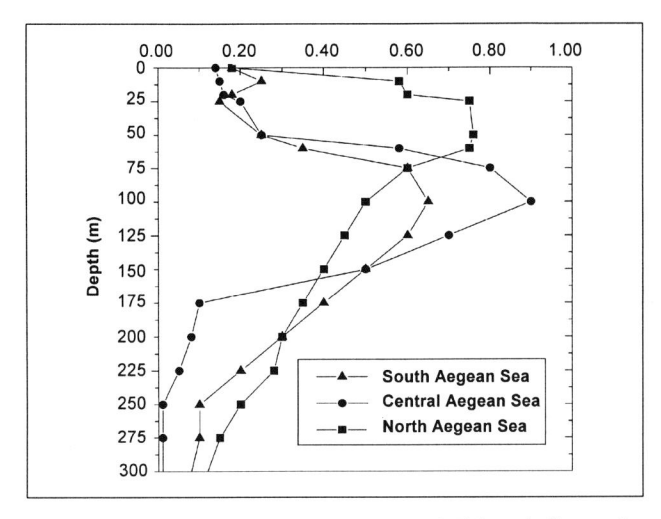

Fig. 3. Characteristic vertical distribution of Chlorophyll-a in the South, Central and North Aegean.

were found in the northwest Levantine Sea. Copepods were the most dominant constituent with percentages between 64 and 92% (Pancucci-Papadopoulou et al., 1990).

Some investigations into the benthic fauna of the Aegean Sea have demonstrated that below the 1000 m depth few organisms were encountered. Some unidentified amphipods, penaeids and macrourids fishes constituted the fauna. These organisms decreased with the depth. Polychaetes comprised 48% of the total fauna in the Cretan Sea followed by the Crustaceans, Molluscs and Sipunculids. The total biomass declined with depth especially between 300 and 400 m, while the diversity initially increased and after 300 m declined with depth (Tselepides and Eleftheriou, 1992)

Thirty-seven mesopelagic, bathypelagic and a small number of bathyal fish species have been identified in the Greek seas during the last few years. Zoogeographically, the highest proportion of these species belongs to the Atlanto-Mediterranean group (72.2%). Some of those are of a worldwide distribution (16.7%), while some others are endemic in the Mediterranean Sea (1.1%). It is suggested that the North Aegean could be recognised as a distinct faunistic "sub-province", with a good number of "endemic" fauna. In the North Aegean there is a decrease of the number of fish species in depths greater than 500 m in combination with an abrupt decrease of abundance beyond 750 m (Papaconstantinou et al., 1994).

HUMAN POPULATIONS AFFECTING THE AREA

(The data in this section is from UNEP (1989, 1996b), Environment Statistics (1996), Europe in Numbers (1995), National Statistic Service of Greece (1996), UNEP(OCA)/MED WG.111/Inf.5 (1996b).)

Greek mythology tells of a legendary king of Athens named Aegeus who threw himself into the sea when he was mistakenly led to the believe that his son Thesus had been killed by the Minotaur on the island of Crete. Known thereafter as the Aegean, that sea, its islands and its shores have been home to several civilisations for many centuries.

These Aegean cultures had their roots in the Neolithic period from the sixth through the fourth millennium BC, when the advent of agriculture led to increased settlement in the area. Significant civilisations were developed in the area, like the Cycladic in the islands of the Cyclades, the Trojan in western Anatolia, the Minoan on Crete and the Hellenic on mainland Greece. Later periods of Roman, Byzantine and Ottoman Empires were also significant for the Aegean. Many important cities were established near the Aegean coast, and many ships of different types sailed the Aegean for merchant and military activities. The Aegean is one of the most important areas for ancient world naval history.

Numerous ancient monuments are almost everywhere along the Aegean coasts but there are also significant underwater monuments that need specific protection from human activities. These include ruins of ancient civilisations which flourished there before the area was covered by sea-water (i.e. Kechreai) and items of archaeological value found on the seabed as a result of shipwrecks (i.e. Lavrio, Peristera, Dhokos).

Very little is known about how far environmental problems in the area may have been caused by poor management long ago. Deforestation and mining have long been main reasons for environmental degradation but the significance of the problems was lower as they were restricted to a small percentage of the coast.

Although the Aegean Sea is surrounded by Greece and Turkey, most of its islands belong to Greece. Therefore the total coastline of Greece is 7300 km of continental shore and 7700 km of island shore, whereas the coasts of Turkey are 4400 km continental and 800 km of islands. Thus about 75% of the Aegean coastline is Greek and 25% is Turkish.

The total population of Greece is approximately ten million while that of Turkey is approximately fifty-five million. In Greece 89.7% of the population lives in the coastal region which is 76% of the total area (population density is about 78 ind/km^2 or 0.6 ind/km of coast). In contrast, in Turkey the coastal area supports 15.7% of the total population and the coastal population is 20.3% of the total (population density, 63 ind/km^2 or 1.9 ind/km).

The major cities in the area are Athens (with approximately four million inhabitants), Thessaloniki (with approximately one million inhabitants) and Izmir (with approximately two million inhabitants), but near the coast there are numerous smaller cities and ports. The urban population in 1985 was 60% in Greece and 48% in Turkey, but it is estimated that in 2000 the percentages will be 68% for Greece and 54% for Turkey. The total population of the coastal cities in Greece is approximately 7,100,000 inhabitants which increases to approximately 10,600,000 in the summer months as a result of tourism. The corresponding population in Turkey is 3,500,000 which increases to 5,100,000.

Greece and Turkey are in a middle level with respect to economical development. The distribution of the labour force in 1980 was:
- agriculture: 37% in Greece, 54% in Turkey,
- industry: 28% in Greece, 13% in Turkey, and
- services: 35% in Greece, 33% in Turkey.

Marine activities like shipping, fishing and oil drilling are significant for the marine environment, especially from Greece.

Most industrial development in the area has occurred mainly since the Second World War but its degree remains low in comparison with the other countries of the western Mediterranean. The industries (and the pollution that they cause) are concentrated in gulfs (Saronikos, Thermaikos, Izmir) which are only a small percentage of the whole Aegean coastline. In contrast, the tourist activities, although having smaller environmental consequences in some respects, are dispersed throughout most coasts of the Aegean and may cause significant damage to the coastal ecosystems.

Up to 1986 there was only sporadic information concerning pollution of the Aegean Sea. McGill (1961) presented a summary of the seasonal patterns of oxygen and phosphorus distribution in the eastern and western Mediterranean Sea. During an expedition of the Oceanographic ship 'Academic Kovalevski' in October 1980, surface seawater samples were collected from eight stations in the Aegean Sea (Rozhanskaya, 1973). During two oceanographic expeditions covering the whole Aegean Sea, Romanov et al. (1977) studied the Cu, As, and Hg distributions. Huynh-Ngoc and Fukai (1978) reported trace element concentrations in surface seawater.

Oceanography in Greece was developed as a significant science mainly in the second half of the 1970s, when the initial research was restricted to nearshore areas as there were no oceanographic vessels for open-sea research. Oceanography in Turkey was less developed than in Greece and was also restricted to nearshore areas (Greek Oceanographers Association, 1994).

The first interdisciplinary oceanographic cruise in the open Aegean by Greek scientists took place during 1980 (UNEP, 1986b), with 'Nautilus', a ship of Greek Hydrographic Service. Physical, geological, chemical and biological data were collected, but this was not followed up by systematic studies.

The construction of the oceanographic ship 'Aegean' from the National Centre for Marine Research in Greece (and later of the smaller ship 'Filia' from the Institute of Marine Biology of Crete) gave the opportunity of open-sea environmental research. From 1986 until 1992, twelve seasonal cruises were performed in the region comprising the North and South Aegean Sea, Cretan Sea, passages of Cretan Arc, Northwest Levantine Sea as well as North and South Ionian Sea in the framework of three different projects. Since then, oceanographic research in Aegean has continued and now covers all the oceanographic sections (NCMR, 1996b).

RURAL FACTORS

Agriculture is a significant activity on the coasts of the Aegean Sea and significant quantities of fertilisers and pesticides are used as agricultural practices become increasingly intensive. The small-scale cultivations that occurred on so many of the islands have been abandoned as they were insufficiently economically profitable in comparison with the touristic activities that have commonly replaced them. The agricultural population in 1985 was 2,600,000 in Greece and 25,500,000 in Turkey.

Table 1 shows land use by main category in Greece (1994) and Turkey (1987). The irrigated land during the period 1988–90 was 30.3% for Greece (18.4% for 1968–70) and 8.4% for Turkey (6.2% for 1968–70).

The development of intensive agriculture was significant in both countries and the fertiliser consumption in t/ha/yr is shown in Table 2. Total consumption in Greece for 1994 was 535,000 t (nitrogen: 334,000 t N; phosphate: 144,000 t P_2O_5; potash: 57,000 t K_2O) whereas in Turkey for the 1988–90 period it was 1,766,000 t.

Significant nitrogen and phosphorus loads are also generated by the livestock population of the Mediterranean-bound basins of the two countries. These loads are estimated at 145,000 t N and 27,000 t P for Greece and 280,000 t N and 41,000 t P for Turkey. The quantities that are transferred into the Aegean are estimated at 5000–15,000 t P/yr and 30,000–130,000 t N/yr for Greece, and 25,000–60,000 t N/y and 6000–8000 t P/yr for Turkey.

The main rivers of the Balkan peninsula carry their load into the northern section of the Aegean Sea. The seven

Table 1

Land use by main category in Greece (1994) and Turkey (1987)

Greece	%	Turkey	%
Arable land	17.0	Arable land	31.8
Permanent crops	8.2	Permanent crops	3.8
Wooded area	22.3	Forests and woodlands	26.2
Permanent grassland	13.6	Land under permanent meadows and pastures	11.6
Other areas	38.9	Other land	26.6

Table 2

Fertiliser consumption

	Greece (t/ha/yr)	Turkey (t/ha/yr)
1964/66	66.0	6.0
1968/70	83.1	15.4
1974/76	119.0	31.0
1984/86	165.0	58.0
1988/90	171.3	63.4

Table 3

Nutrient concentrations in the Aegean Sea (all values in µgat/l)

		Nitrates	Nitrites	Ammonia	Phosphates	Silicates
East Aegean 1980 (1)	Mean	0.57	0.26	0.33	0.075	1.05
	Range	0.09–2.28	0.05–1.59	0.65–1.35	0.07–0.11	0.75–3.27
North Aegean 1994 (2)	Range	0.10–2.80	0.01–0.13	0.10–0.95	0.02–0.16	0.30–3.50
North Aegean 1992–94 (3)	Mean	0.84	0.04	0.34	0.07	1.65
South Aegean 1992–94 (3)	Mean	0.78	0.03	0.28	0.06	1.83
Evripos Straits 1997–98 (7)	Mean	5.38	0.09	3.90	0.35	7.83
	Range	0.01–68.2	0.01–0.59	0.88–27.2	0.02–1.32	0.92–123.7
Evripos Straits 1993 (6)	Mean	2.70	0.20	0.80	0.30	
South Evoikos 1980 (8)	Mean	0.49	0.08	0.23	0.18	1.72
North Evoikos (8)	Mean	4.27	0.08	0.60	0.34	16.1
Elefsis Gulf (10)	Mean	2.94	0.49	0.65	0.61	5.06
Saronikos Gulf (10)	Mean	1.90	0.21	1.19	0.29	2.65
Saronikos Gulf 1995 (5)	Range	0.05–9.12	0.02–1.46	0.33–15.27	0.02–3.40	0.36–477
Thessaloniki Gulf (10)	Mean	1.63	0.61	1.65	0.64	4.09
Thermaikos Gulf (10)	Mean	1.26	0.30	0.88	0.19	3.65
Izmir Bay 1987 (4)	Range	0.44–4.20	0.11–3.50		0.21–2.51	
Kavala 1975–76 (8)	Mean	0.59	0.08	0.36	0.22	1.82
Kavala 1982 (9)	Mean	0.33	0.04	0.28	0.18	1.74
Pagasitikos Gulf (10)	Mean	0.99	0.20	0.94	0.12	3.42
Mediterranean Background (8)		0.42	0.16	0.36	0.12	1.22

(1) UNEP (1986b). (2) Souvermezoglou et al. (1997b). (3) Kucuksezgin et al. (1995). (4) UNEP (1994). (5) Psyllidou et al. (1997). (6) Dassenakis and Kloukiniotou (1994). (7) Dassenakis et al. (1990). (8) Friligos (1987). (9) Friligos and Karydis (1988). (10) NCMR (1989).

main rivers that flow into the Aegean (six from Greece and one from Turkey) have an annual discharge of about 1000 m³/s and carry into the sea about 70,000 t N/yr, 3,000 t P/yr and 45–60 Mt of suspended sediments. In addition the Black Sea water which enters the Aegean through the Dardanelles also contributes nutrients to the Aegean. It is estimated that the export of total nitrogen, phosphorus and organic carbon from the Marmara Sea to the south is 4–5 times larger than the importation into the Black Sea from the Aegean, through the near-bottom current of opposite direction (Polat and Tugrul, 1996).

Some characteristic mean values and ranges of nutrient concentrations in the Aegean regions are shown in Table 3.

The consumption of pesticides during 1989 in Greece was 8151 t (fungicides: 1925 t; herbicides: 3031 t; insecticides: 2844 t; other: 11 t). There only a few data for the concentrations of pesticides in the marine environment and their effects in the marine ecosystems in this region.

Table 4

Values of selected pesticides from UNEP (1985) and Satsmadjis et al. (1988)

Mytilus galloprovincialis

ppDDT: 5 µg/kg (0–1014)   Hexachloro-cycloexane: 1.9 µg/kg (0.4–5)

ppDDD^{-1}: 7 µg/kg (0–45)   ppDDE: 10 µg/kg (1–255)

Mullus barbatus

ppDDT: 23 µg/kg (4–110)   Hexachloro-cycloexane: 5 µg/kg (0.8–50)

ppDDD^{-1}: 14 µg/kg (0–140)   ppDDE: 33 µg/kg (1–75)

p,p'-DDT: 2.0–4.0   p,p'-DDD: 1.1–3.2

Endrin: 0.1–0.3   a-BHC: 3.2–4.1

Dieldrin: 1.0–1.5   Heptachlor: 0.0–0.1 a-BHC: 0.2–0.3

Mean values and ranges of pesticide concentrations of two species considered to be pollution indicators in the Mediterranean which are monitored indicate that although the mean values are not significantly elevated, the very wide ranges show that in polluted areas bioaccumulation mechanisms may lead to dangerous levels (Table 4).

The Thermaikos and Pagassitikos Gulfs are strongly affected by agriculture runoff and rivers. The Gulfs of Izmir and Euvoikos are also influenced, whereas the Saronikos Gulf is affected only from domestic and industrial activities and has less surrounding agriculture.

Four major rivers (Aliakmon, Loudias, Axios and Gallikos) flow into the Thermaikos Gulf and contribute to an enrichment of organic carbon, nutrients and particulate matter. The total drainage area of the surrounding the Gulf region is 3600 km and the total annual discharge of water is 10.2×10^6 m^3. The total sediment supply is estimated at 3–4×10^6 t/yr (Voutsinou-Taliadori and Varnavas, 1995). Nutrient concentrations at the mouth of the rivers are high and the area of Themaikos Gulf show values of nutrients 2–6 times higher than a corresponding oligotrophic area (NO_3: 1.7 times; NO_2: 1.8; SiO_4: 2.6; NH_4: 6; PO_4: 5.5). As a consequence, red tides and phytoplankton blooms have been observed on occasion in the gulf and anoxic conditions develop during the summer period near the harbour of Thessaloniki. The blooms are generally caused by the diatoms *Nitzschia closterium*, *Cerataulina bergonii*, *Leptocylindrus minimus*, *Chetoceros socialis* and *Thalassiosira* sp. (Gotsis-Scretas and Friligos, 1988).

A study of the concentrations of organophosphorus pesticides, parathion and malathion, in Thermaikos has shown that these compounds are present at concentrations ranging from 2–46 ng/l in water samples and 3–35 ng/l in surface sediments. The higher values were measured near the Axios estuary due to extensive agricultural cultivations (Fytianos and Samanidou, 1991).

High trophic levels are found in the northern area of the bay of Volos due to nutrients from agricultural, domestic and industrial activities. The Pagassitikos Gulf is enriched in nutrients also through the channel which was used for the draining of Lake Karla, and this channel remains in use to protect the area against the floods. Concentrations of organochlorine pesticides in tissue of *Mullus barbatus* of this Gulf for the 1986–87 period are shown in Table 4.

High terrestrial inputs are also the main reasons for the increased nutrient concentrations in Izmir Bay. The total suspended solids concentration is so high that the water is turbid and the average depth of Secchi disk is 2–10 m. Eutrophication of the inner bay has already started with increasing levels of chlorophyll, nutrients and faecal coliforms. Poor visibility in the water is exacerbated by increasing occurrences of red tides and other noxious algal blooms. The primary productivity in the bay is 7–10 times higher than the offshore production in this region. The eutrophication is spreading progressively to the rest of the bay (Balci et al., 1995). All the pollution indicators monitored in the area are an order of magnitude higher than the coastal stations, while nutrient concentrations measured in the inner bay were several orders of magnitude higher than those found in clean waters nearby. The dissolved oxygen levels in the inner bay decrease to dangerously low levels (~3 mg/l), especially in the summer and dissolved oxygen disappears in some parts of the coastal section for extended periods of time (UNEP, 1994).

In the Euvoikos Gulf there are several nutrient sources which have similar significance: agriculture, domestic and industrial effluents, aquacultures, small rivers all contribute. Probably as a result of the tidal currents in the area, eutrophication phenomena are rare although the values near the sources are high. Phosphorus remains the limiting factor for phytoplankton growth (Dassenakis et al., 1999).

Fishing

From ancient times, fishing has played a very important role in the Aegean economy. However, the fishery sector makes up only 3% of the GNP, and only 20% of the revenues from the whole agricultural sector, which itself accounts for 13.7% of the GNP of Greece. The fish production in the Aegean for the 1970–1984 period is shown in Table 5. The total Greek fisheries production has levelled off and is now inadequate to meet local needs. As a result, net imports of fisheries products are currently over 40,000 t/yr (1993).

Aegean demersal/inshore and pelagic/semipelagic fisheries resources are overfished, as a result of the increased fishing effort, the intense modernization of the fleet, and the inadequacy of the management measures currently in use (Stergiou and Petrakis, 1993). The common gear used in fisheries in the Aegean are purse seines, trawlers, beach seines and nets. All fishing gear—in the broadest sense—are environmentally damaging to some degree. Much fishing gear is unselective with respect to both size and species in the catch. Compared with gill nets, for example, trammel nets are believed to be unselective in the way they catch a wide size range of individuals and a much larger number of different species. The use of other gear is prohibited but sporadic fishing with dynamite and some industrial chemicals in some areas is also obvious.

Recent evidence indicates that one of the most effective management techniques for multispecies fisheries is the

Table 5

Fish catches (UNEP, 1996b)

Year	Catch (t)
1970	48,200
1975	68,200
1980	85,200
1984	87,300
1997	177,400

creation of marine harvest refugia. Such protected areas, in which fishing is prohibited, provide both an abundance of recruits and a certain proportion of adult fishes in these areas.

The creation of refuges may be applicable for better management of Greek demersal and inshore fisheries because of the extreme multispecies and multigear nature of the latter.

COASTAL EROSION AND LANDFILL

The very long Aegean coastline is one of the most precious natural resources of the region, as it provides tourism, recreation, transport, fisheries, aquacultures and salt extraction, but as yet it has been managed without any serious planning. Over the last century—and since the 1960s in particular—a significant move of the population towards the coastal areas has taken place, as people observed a higher level of economic development, through tourism, industry, transport and agriculture. This expansion and intensification of human development on the coast has greatly increased the environmental problems and the threats to Greece's wildlife, leading to a considerable reduction in the number of many species. The most serious problem faced by fauna is the alteration and destruction of habitats. This is almost exclusively due to recent economic development and appears in various forms now, such as the draining of wetlands, extensive tree cutting, land clearing due to forest fires, development of coastal housing and tourist installations and building on mountains.

Data for 1987 concerning the significant problem of soil erosion (UNEP, 1996b) indicate that it is severe. It is excessive in 4,700,000 ha (35.6% of the total area) in Greece and 7,560,000 ha (36.9%) of the arable land in Turkey. Almost 54,000 ha/yr of arable land is lost through erosion.

It must be mentioned that the river outflow in the Aegean Sea has been significantly reduced during the last twenty years due to the construction of hydroelectric dams and the establishment of irrigation systems for the water needs of intensive cultivation. The irrigated land has almost doubled in the period 1970–1990. The annual fluctuations of the river flows have been disturbed so they no longer follow normal annual cycles. The reduction of sediment load coming from the rivers leads to increased beach erosion and loss of deltaic environments and loss of wetlands. Changes in offshore profiles and shelf transport processes also occur. At times, in the summer, the flow is low enough to permit the inflow of saline water into the river bed. The formation of very long and sharp salt wedges has been observed in many cases, which affects the estuarine ecosystems.

The area of the remaining wetlands of Greece has been estimated to range up to 190,000 ha, but this represents a loss of nearly 300,000 ha during this century (Psilovikos, 1992). Approximately 100,000 ha are included in 11 designated Ramsar sites, where 15 areas, apart from the Ramsar wetlands, have been classed as Special Protected Areas (SPA) in compliance with EEC Directive 79/409. Although large-scale drainage projects resulting from natural policy seem to have ceased, Greece's wetlands continue to be under considerable pressure and face new threats.

EFFECTS FROM URBAN AND INDUSTRIAL ACTIVITIES

(The data in this section is from UNEP (1989, 1990d, 1996b), and UNEP(OCA)/MED WG.111/Inf.5 (1996a).)

Urban and industrial activities are important pollution sources on Aegean coasts. They are closely associated, as all industrial areas are located close to the main coastal cities. The establishment of waste water treatment plants during the last decade has reduced the load of pollutants flowing to the sea but the percentage of population served by these plants is still small (11.4% in Greece during 1992). Data for the quantities of municipal waste waters and for their disposal are shown in Table 6.

Some characteristic mean values and ranges of trace metal, PAH, and PCB concentrations in seawater, sediments and organisms of the Aegean Sea are shown in Tables 7–10.

About 40% of Greek industry is located at the coast of Attiki, along the northern part of the Saronikos Gulf. A large amount of industrial effluents are discharged into the sea. Some of the main industries (oil refineries, shipyards, chemical plants, food, metal, cement industries etc.) are located within the Elefsis Bay. It is estimated that about 20,000 t of petroleum come into this gulf annually. During the period 1986–91, approximately 130 pollution events from ships and about 25 from refineries and industries were recorded in the gulf. For a long period of time the sewage of

Table 6

Municipal waste water 1996 and municipal waste disposal (data from UNEP 1989, 1990a, 1996b)

	Municipal waste water 1996 (10^6 m^3/yr)		
	Untreated	Primary treatment	Secondary treatment
Greece (520.26)	43.5%	42.2%	14.3%
Turkey (404.87)	25.5%	53.4%	21.1%

	Municipal waste disposal (10^6 m^3/yr)	
	Greece	Turkey
Into the sea or rivers through municipal sewer system	373.76	358.48
Onto land	0.29	0.12
Into sub-soil	136.33	–
In irrigation ponds	1.53	4.08
In recreation areas	–	42.20

Table 7

Concentrations of trace metals in Aegean seawater (unless otherwise indicated, all values are in µg/l)

		Cu		Pb		Zn		Ni		Cd		Mn		Hg (ng/l)
		D	P	D	P	D	P	D	P	D	P	D	P	D
Cretan Sea 1994 (1)	Mean	0.25		0.46				0.3		0.025		0.43		
	Range	0.04–0.36		0.07–0.87				0.10–0.50		0.005–0.05		0.01–0.72		
East Aegean Sea 1980 (2)	Mean	0.9	0.4	1.3	0.4	6.0	1.0	1.2	0.1			2.8	0.6	
	Range	0.5–2.3	0.2–0.8	0.9–2.6	0.2–0.9	4.0–9.5	0.3–3.0	0.8–2.2	0.1–0.5			0.8–5.5	0.2–1.4	
Open Aegean 1978 (3)	Mean	0.3				3.0				0.07				40
Saronikos Gulf 1986–93 (4)	Mean	1.25	0.27	0.94	0.15			3.07	0.70	0.113	0.009	1.01	2.85	
	Range	0.10–10.7	0.02–2.11	0.03–12.2	0.005–1.2			0.20–22.5	0.04–7.10	0.008–1.2	0.001–0.23	0.10–2.20	0.01–5.50	
Elefsis Gulf 1986–93 (4)	Mean	1.52	0.60	1.08	0.26			4.75	1.04	0.159	0.011	2.64	9.49	
	Range	0.30–9.50	0.04–8.52	0.05–11.0	0.01–3.30			0.40–41.9	0.03–18.1	0.014–2.3	0.001–0.14	0.12–13.0	0.01–230	
Euripos Straits 1993 (5)	Mean	4.50	1.86	1.60	8.85	19.4	16.1	1.79	1.35	0.13	0.016	1.56	5.12	
	Range	0.45–20.7	0.40–6.80	0.14–2.70	0.50–57.7	2.55–120	0.25–130	0.33–6.88	0.52–8.50	0.04–0.38	0.003–0.10	0.34–3.08	3.22–9.02	
Euripos Straits 1997 (7)	Mean	0.80	0.35	1.51	1.21	8.60	1.57							
	Range	0.17–2.0	0.09–2.93	0.02–12.1	0.08–0.52	1.15–38.1	0.13–8.48			0.04				
South Euvoikos 1980 (6)	Mean	1.8	1.1	2.2	1.5	9.2	3.2	1.4	0.3	<0.004–0.11		1.7	2.0	
	Range	0.7–5.4	0.4–3.8	1.4–3.5	0.4–5.2	5.2–20.6	0.7–15.1	1.0–1.7	0.1–0.5			0.5–4.9	0.7–3.6	
Izmir Gulf 1986 (8)	Range	1.1–14.5		2–14						0.43–1.13				
Open Mediterranean (9)	Range	0.10–0.40		<0.1		0.15–2.2		<0.5		<0.1				

D = dissolved, P = particulate.
(1) Souvermetzoglou et al. (1997b). (2) UNEP (1986b). (3) Huyn-Ngoc and Fukai (1978). (4) Scoullos et al. (1994b). (5) Dassenakis et al. (1996). (6) Scoullos and Dassenakis (1983b). (7) Dassenakis et al. (1999). (8) Environmental Problems of Turkey (1989). (9) UNEP (1990a).

Athens (about 600,000 m³/day) was untreated and was discharged to the shallow Keratsini Bay through the Central Sewage Outfall. The primary treatment of the wastes of Athens started in September 1994 in the Seawage Treatment Plant at Psytallia island. The total amount of the removed organic load is estimated at approximately 40%. The disposal of the effluents through two pipes placed at the bottom of the gulf has led to significant changes in the distributions of dissolved oxygen and nutrients in the gulf and intermittently anoxic conditions in Elefsis bay may soon extend to the other parts of the Saronikos Gulf if there is not a secondary treatment of the effluents.

Eight years (1986–1993) of monitoring trace metals in the Saronikos have shown that the Elefsis Gulf is the most polluted sub-region of the gulf with metals. Over the last few years some of the main industries have ceased, thus the concentrations of main pollutants have become reduced (Scoullos et al., 1994b).

Thermaikos Gulf is heavily polluted by sewage from Thessaloniki and the industrial influents of about 250 factories. The sewage waters are estimated at 150,000 m³/d of which only 30–40% are treated, while a new treatment plant is under construction. Planned outfall will discharge the waste water after treatment into the bay, at a depth of 20 m. The improvement of the city's waste water treatment plant may reduce the frequency of observation of phenomena such as anoxia or red tides (Latinopoulos et al., 1996).

The Euvoikos Gulf is significantly affected by large amounts of domestic and industrial waste. The waste water treatment plant of the town of Chalkis is located on a small island in the gulf and the treated effluents are disposed into the sea. Several industries such as cement, textile, paint, food, metal-forming and ceramic factories, shipyards etc., are located near the coastal zone (Dassenakis et al, 1999).

The population in the area of the Bay of Izmir exceeds 2,500,000 and extensive industrial development has been the driving force behind the vigorous economic growth of Izmir over the last 40 years. During this period, urban land expanded more than four times, with the rate of land conversion very often faster than the population growth.

Table 8
Trace metals in Aegean surface sediments. All values in μg/g.

Area		Copper T	Copper E	Lead T	Lead E	Zinc T	Zinc E	Nickel T	Nickel E	Manganese T	Manganese E	Mercury T
East Aegean Sea 1980 (1,2)	Mean	20	10.3		22.2	40	18	145	75	925	522	
	Range	5–30	3.8–17.2	10–25	14.4–31.0	22–55	6.8–45.4	60–305	43.2–130	270–2920	60–1950	
East Aegean Sea 1993 (3)	Range	3–77				19–162		11–344		113–2625		
Izmir Bay (4)	Range	14–870		20–280		53–860						
Saronikos Gulf 1988 (5)	Mean	91	47	112	87	175	78			725	520	
	Range	43–188	20–108	63–274	33–130	100–632	27–405			345–1223	220–1020	0.5–1 (1976) (9)
Gulf of Elefsis 1977–84 (6,7)	Mean	140	103	237	147	385	340		92	844	362	
1992 (5)	Range	25–150		160–500		125–1500		40–80		350–1000		
Thermaikos Gulf 1975–76 (8)	Range	7–69	4–37	16–268	13–228	39–560	23–299	75–440		340–2600		
Euripos Straits 1993 (10)	Mean	52.3	23.7	54.8	43.9	158.5	90.9		64.4	480		
	Range	28.4–80.4	1.3–41.8	27.5–110	15–100	77–377	24–200		52–100	351–676		
Euripos Straits 1997 (13)	Mean	57.7	18.5	82.3	51.8	124	63.2			502	297	
	Range	4.4–240	2.9–58.0	33.1–183	24.0–125	35.1–435	15.2–377			300–700	176–393	
North Euvoikos Gulf 1984 (8)	Range		0–28		0–27		9–58					
South Euvoikos Gulf 1980 (11)	Mean		7.4		28.3		14.1		50.2		315	
	Range		3.4–11		15.4–38.2		8.5–17.5		34.3–75.3		150–800	0.8–1.8 (1980) (9)
Pagassitikos Gulf 1975–76 (12)	Mean	30		30		130		165		1070		
	Range	10–55	0.5–16	15–50	5.3–28	30–945	1.1–30	55–420		255–4200		
Open Mediterranean (14)	Range	10–49		10–93		20–260		9–46		52–2560		

T = total, E = extractable.
(1) Scoullos and Dassenakis (1983a). (2) Voutsinou–Taliadouri (1982). (3) Ergin et al. (1993). (4) UNEP (1989). (5) Foufa (1993). (6) Scoullos (1979). (7) Voutsinou-Taliadouri et al. (1987). (8) Chester and Voutsinou (1981). (9) UNEP (1988). (10) Dassenakis et al. (1996). (11) Scoullos and Dassenakis (1983b). (12) Voutsinou-Taliadouri and Balopoulos (1989). (13) Dassenakis et al. (1999). (14) UNEP (1990a).

Approximately 44% of the total 26,000 ha of agricultural land has been converted to urban uses while approximately 14% has been neglected or degraded. Most of the industries in Izmir were established in the inner bay region. Their number (including small-scale units) is likely to have exceeded 6500. The main type of industries in Izmir are food, beverage manufacturing and bottling, tanneries, vegetable oil and soap production, chemical industries, paper and pulp factories, textile industries and metal processing. The wastes of Izmir, consisting of 100,000 m³/day of industrial discharges and 300,000 m³ of sewage, are dumped directly into the inner bay through 128 canals and 10 streams, without any pre-treatment. Domestic and industrial wastes represent 50% of the polluting load in the gulf (the remainder coming from rainfall 15%, ships 4%, rivers 10%, erosion 8%, agriculture 10%, other 3%) and contain 248,600 kg/day BOD, 213,000 TS, 149,000 N, and 3660 P. A waste water treatment plant is under construction which will be serving 2.4 million people initially and 4.85 million by 2020. The treated waste water, however, is expected to contain a relatively high concentration of suspended solids (120 mg/l), N (27 mg/l) and P (10 mg/l). This may still have the potential to trigger eutrophication processes (UNEP, 1994).

Samples of benthic fauna collected from the Saronikos Gulf (*Mytilus galloprovincialis, Parapenaeus longirostris*) showed that the organisms had not been affected significantly by the pollution of the gulf (Voutsinou-Taliadouri, 1982). Mussel samples from the Saronikos and Pagassitikos Gulfs showed that there is no evident difference between the two gulfs regarding the concentration of trace elements in their flesh (Grimanis et al., 1982) whereas the trace element level in *Parapenaeus longirostris* in the Saronikos Gulf is considerably lower than that found in other polluted areas of the Mediterranean Sea (Voutsinou-Taliadouri, 1980). The concentrations of PCBs and DDTs in *Mullus barbatus* of the Saronikos Gulf are lower compared with the corresponding areas elsewhere in the Mediterranean Sea, while the same species caught in the north near Alexandroupolis and the south in Crete did not show high concentrations (UNEP, 1996b). The Pagassitikos Gulf contained higher levels of chlorinated biphenyls than were found in the Saronikos

Table 9

Trace metals in organisms of the Aegean Sea (data from UNEP, 1985, 1989, 1990a, 1996b; Strogyloudi et al., 1997; NCMR, 1996a). All values in µg/g dry weight.

	Mytilus galloprovincialis (mollusc)				
	Aegean (1986)		Saronikos (1978)	Izmir (1978)	Saronikos 1996
	Range	Mean	Range	Range	Range
Cu	0.75–2.80	1.6	0.2–4.5	36–64	2.6–15.9
Zn	9.2–97.7	45	12–87	336–452	73–289
Pb	0.055–8.26	1.1		83–110	2.1–11.0
Cd	0.005–0.78	0.1	0.06–0.08	6.6–12	0.04–1.54
Hg		0.105	0.06–0.2	0.9–1.1	
Cr			0.11–7.8	26–55	0.8–27.6
Fe			17–32	308–356	78–789

	Mullus barbatus (fish)					
	Aegean (1986)		Alexand-roupoli (1994)	Pagassitikos (1994)	Chios (1994)	Saronikos (1995)
	Mean	Range	Mean	Mean	Mean	Range
Cu	0.6	0.22–1.47	2.15	1.33	1.92	0.92–1.75
Zn	3.5	2.57–6.89	47.9	15.1	41.0	11.8–17.8
Cd	0.047	0.015–0.162				0.001–0.011
Hg	0.175	0.015–1.4				
Cr			1.96	1.08	1.96	0.18–1.43

Gulf in *Mullus barbatus*, but these values seem to be lower than those reported in areas like the Adriatic Sea and France (Georgacopoulos-Gregoriades et al., 1991).

Shipping is also an important traditional activity for the Greek inhabitants of the coasts and islands of Aegean. This influences the marine environment mainly through the disposing into the sea of significant quantities of petroleum hydrocarbons.

The petroleum entering the sea from land-based sources and from marine activities has been reduced in the Mediterranean after the implementation of the MAR-POL convention, but some accidents have caused environmental damage. Since 1980, ten significant (more than 500 tonnes of petroleum in the sea) accidents have occurred in different regions of the Aegean. During a typical year (e.g. 1995) about 300 pollution events were recorded off the Greek coast (most of them small in size). Seventy-one of these came from ships, 70 from land-based sources, 9 from other sources and 145 were not identified. Two hundred and nine of them involved hydrocarbons, 61 urban effluents, 6 solid wastes, 5 other chemicals and 2 colour substances.

For 69 of these events, the Greek coastguard's anti-pollution equipment (ships, chemicals, skimmers, gathering systems, etc.) was needed to prevent damage to the marine environment. During the period 1991–93, the Greek courts fined those responsible for more than 1000 pollution events (650 from ships and 350 from land-based sources) more than a billion drachmas (about 400 million dollars).

The pollution of the Aegean with petroleum has led to the observation of tar balls on almost every Aegean coast. The wide range in the PAH concentrations measured in the Aegean (Table 10) indicates that significant pollution remains restricted to nearshore and enclosed areas. In Turkey it is estimated that 4.4×10^6 t/yr of oil discharges to the sea from refineries. The oil production in Greece from offshore drilling near the island of Thasos in northern Greece was 300,000 t in 1992. Pollution in the area is significantly elevated after the establishment of the drilling.

Regarding the quality of bathing water, it is well known that the Aegean is one of the main touristic areas of the Mediterranean and thus seawater quality has an important significance. There is systematic monitoring of the Greek coasts from April to October for total and faecal coliforms. For 1995 the sampling points which were not in compliance with the quality requirements of the E.U. were 0.7% of 1500 points in all the Greek coast.

The Aegean coasts have no nuclear power stations. The main source of marine pollution from radionuclides is the Black Sea water—mainly after the Chernobyl nuclear reactor accident which affected the Aegean Sea, after the radioactive cloud reached Greece on 3 May 1986. There was an immediate uptake by the marine organisms of the radio-

Table 10

Organic pollutants in the Aegean Sea (data from UNEP, 1986a, 1989, 1990b, 1994, 1996b; Balci, 1993; Botsou, 1997; Michopoulos, 1996; Sakellariadou et al., 1994; Dassenakis et al., 1999)

	Range	Mean
PAHs in Seawater (µg/l)		
East Aegean (1983–86)	0.09–5.9	
East Aegean (1986–89)	0.09–25.5	
East Aegean (1983)	0.14–1.39	0.86
East Aegean (1977–79)	0.8–11.5	
South Aegean (1977–79)	0.6–12.1	3.2
Aegean (1981)	2.9–13.7	
Saronikos (1981)	1.6–5.6	
Izmir bay (1983)	0.75–9.4	5.0
South Euvoikos (1997)	<0.04–0.49	0.17
North Crete (1997)	0.09–0.32	
Agios Nikolaos, Crete (1996)	0.02–3.47	1.6
Lesvos (1996)	0.09–9.6	2.05
Chios	1.3–2.5	
Rodos	2.0–3.7	
PCBs in Aegean		
Seawater (ng/l)	0.2–1.3	0.36
Sediments (µg/kg)	0.6–775	155
Mullus barbatus (µg/kg)	14–1613	432
Mytilus galloprovincialis (µg/kg)	40–80	62

active pollutants which were added to their environment. For ^{137}Cs, measured concentrations in marine samples appeared higher, up to one order of magnitude compared to those observed before the accident. Radionuclide concentrations in fresh water fish were 10–100 times higher than those observed in marine fish species (Florou et al., 1987). The annual outflow of ^{137}Cs from the Black Sea to the Mediterranean, through the straits of the Dardanelles during the period 1987–1992 was 1.6–1.8% of the total amount of this radionouclide present in the 0–50 m layer of the Black Sea. The respective value for ^{90}Sr was 0.8%. The transfer of the Black Sea water mass resulted in considerable increase of the radioactivity concentrations in the North Aegean. The activity concentration of ^{137}Cs in the surface layer in the mouth of the Dardanelles during 1993 was 118.1 ± 18.0 Bq/m^3 varying according to the surface current circulation pattern. This is quite high when compared with the average value for the Aegean Sea (20.7 ± 14.7 Bq/m^3) and the average value for the Ionian Sea (9.2 ± 2.5 Bq/m^3), and is six times higher than in the Adriatic Sea. Chernobyl debris is still detected in the impacted areas (Florou and Chaloulou, 1997).

PROTECTIVE MEASURES

Most of the efforts for the protection of the Aegean Sea are closely connected to the efforts for the protection of the Mediterranean Sea as a whole. Most derive from the establishment of MAP/UNEP and the Convention of Barcelona.

Environmental degradation in the area would be much more serious and, in many respects, irreversible if the countries around the Mediterranean had not, in 1975, committed themselves to work together to combat the different forms of pollution and to control development. This form of co-operation, launched and promoted under the aegis of the United Nations Environment Programme (UNEP), has kept the name it was originally given—the Mediterranean Action Plan (MAP). The MAP is an example for similar plans, for many reasons: its solid legal framework adopted in Barcelona in 1976 and since then complemented by new protocols, the diversity of its components and its dynamic Activity Centres, the protection measures taken by all parties involved and the prospective work to 2000 and 2025, and finally the first field activities to manage its coastal areas.

The MAP is one of the largest international programmes in the environmental field, and involves the interaction of eighteen countries and the E.U., practically all the major Organisations of the United Nations System, four Regional Activity Centres, a growing list of Non-Governmental Organisations and a vast number of scientists. Over 100 laboratories participating in the MEDPOL Programme have tracked down, identified and attempted to more fully understand both the sources of pollution and the processes which take place in the sea, the soil and the air. The contracting parties to the Barcelona Convention and Protocols, the Mediterranean States and the E.U. are, jointly, the final decision makers within the framework of the MAP. From the beginning of the Programme in 1975, the Mediterranean States had expressed the view that its headquarters should be located in a Mediterranean city. Athens was formally selected by the contracting parties in 1981 and the Co-ordinating Unit moved there on 1 July 1982 (UNEP, 1995)

The Barcelona Convention of 1976 for the Protection of the Mediterranean Sea against pollution in combination with the International MARPOL 73/78 convention for the prevention of pollution from ships have played a beneficial role in the reduction of the pollution of the Aegean Sea. A disadvantage in the efforts for the protection of marine environment in the area is the inadequate co-operation between Greece and Turkey in almost all aspects. The debate between the two countries has a long historical background but is also related to the continental shelf of the Aegean and the management of its wealth.

An important peculiarity of the Aegean is that it is affected by the pollution of the Black Sea which is not included in the MAP programme and thus a wider international co-operation is needed for its effective protection.

The establishment of monitoring systems for the measurement of the temporal and spatial variations and distributions of main pollutants is another important result of the MAP/MEDPOL programme. The Greek MEDPOL National Monitoring Programme includes 211 stations in the Aegean area, mainly at coastal sites, distributed in six regions: Saronikos Gulf, Thermaikos Gulf, Gulf of Kavala, Northern Crete, Island of Lesvos and Rhodos Island. These stations cover sources of pollution (20), general coastal, hot spot and estuarine areas (70), bathing areas (120) and airborne sources (1).

The main parameters selected are heavy metals (in sea water, effluent, biota, sediment, suspended matter and precipitation), halogenated hydrocarbons (in biota, effluent and sediment), petroleum hydrocarbons (in biota, sea water and sediment), polyaromatic hydrocarbons (in biota, sediment, effluent and sea water), total coliforms, faecal coliforms, faecal streptococci, nutrients (in sea water and effluent), chlorophyll in plankton, total organic carbon and calcium carbonate in sediment, dissolved ions in effluent, pesticides and polychlorinated biphenyls (in sea water, effluent, sediment and biota) as well as standard parameters in sea water. Meteorological parameters and ozone concentrations are also measured in air. The frequency of sampling varies depending on the pollution parameter.

As well as the UNEP/MAP efforts and the E.U. directives for the environmental protection, in this area there is also significant national legislation for the protection of the coasts, the fishery and for marine transports, but the services responsible for this subject are not well organised or sufficiently equipped to be fully effective (Greek Oceanographers Association, 1994).

In Greece there are 14 protected areas (63,000 ha). Most of those connected with the Aegean sea are the estuaries of the rivers Evros, Nestos, and Axios, which are wetlands of high ecological significance and which fall under the protection of the Ramsar Convention. In the Mediterranean part of Turkey there are six protected areas of 65,000 ha. In Greece there are 10 National parks of 350,000 ha, and in Turkey eight of 170,000 ha, but there is only one marine park. A marine park is a designated area, which usually includes a stretch of coast as well as sea and which constitutes a vigorous ecosystem with great biological diversity, interesting geological structure and/or important cultural elements. The park is subject to specific legislation which aims to protect and conserve rare habitats and threatened species encompassed in it.

The Marine Park of North Sporades is connected with the protection of the monk seal *Monachus monachus*. Efforts to protect this area began in the early 1970s when the local authorities passed the first regulations protecting the monk seal. This was followed in 1988 by legislation introduced jointly by the Ministers for the Environment, Agriculture and Merchant Marine. The legislation initially covered a two-year period, but was renewed in 1990. Finally in May 1992, the area was declared a National Park by Presidential Decree. The creation of the National Park has the following aims.

1. The protection conservation and management of the wildlife and landscape which constitute natural heritage and a valuable national natural resource, in extended terrestrial and sea areas of the North Sporades.
2. The protection of one of the most important habitats of the monk seal (*Monachus monachus*) which is high on the list of species threatened by extinction, both in Europe and worldwide.
3. The protection of other rare and threatened plant and animal species which find refuge in the islands.
4. The development of the region by the sustainable use of natural resources.

The basic philosophy of the organisation of the Park is its division into two main protection zones (A and B). Zone A (1587 km^2) is the strict protection zone and in some of its areas—chosen on the basis of the urgency for protection, uniqueness and wilderness of plant and animal life—special protection measures are in force. In Zone B (678 km^2) which includes inhabited areas, protection measures are less stringent. Alonnisos is the largest island in the Park which also encompasses six smaller islands and 22 uninhabited islands and rocky outcrops.

The increased interest of the local people in the environmental problems, the significant environmental education activities and the increasing societal pressure on the governments for effective environmental protection, in combination with sustainable forms of development, will all contribute to a better environmental future for the Aegean.

REFERENCES

Balci, A., Kucuksezgin, F., Kontas, A. and Altay, O. (1995) Eutrophication in Izmir Bay, eastern Aegean. *Toxicological and Environmental Chemistry* **48** (1–2), 31–48.

Balci, A. (1993) Petroleum hydrocarbons in East Aegean. *Marine Pollution Bulletin* **26** (4), 222–223.

Balopoulos, E.T. and Friligos, N.C. (1994) Water circulation and eutrophication in the NW Aegean Sea: Thermaikos Gulf. *Toxicological and Environmental Chemistry* **41** (3–4), 155–167.

Barbetseas, S. (1983) An analysis of the water masses in the gulf of Saronikos, Greece. Special RPT M-102, SACLANTSEN p. 57.

Botsou, F. (1997) Determination of PAHs in Lesvos island, B.Sc. Thesis, Univ. of Aegean.

Chester, R. and Voutsinou, F. (1981) The initial assessment of trace metal pollution in coastal sediments. *Marine Pollution Bulletin* **12**, 84.

Dassenakis, M. and Kloukiniotou, M. (1994) Trace metals and nutrients in the southern part of Evripos Straits. In *Proceedings of the International Conference on Restoration and Protection of the Environment II. Patra, August 1994*, pp. 280–287.

Dassenakis, M., Kloukiniotou, M. and Pavlidou, A. (1996) The effects of long existing pollution to trace metals levels in a small tidal Mediterranean bay. *Marine Pollution Bulletin* **32** (3), 275–282.

Dassenakis, M., Arsenikos, S., Botsou, F., Depiazi, G., Adrianos, H., Zaloumis, P. and Drosis, J. (1999) General trends in marine pollution of the central part of Euvoikos Gulf. *Proceedings of the 6th International Conference on Environmental Science and Technology, Samon, 30 Aug.–2 Sept. 1999*, Vol. 2, pp. 108–115.

Dewey, J.R. and Sengor A.M. (1979). Aegean and surrounding regions complex multiplate and continuum tectonics in a convergent zone. *Geological Society of American Bulletin* **90**, 84–92.

Diapoulis, A. (1988) The study of the marine algae in the Saronikos Gulf. N.C.M.R. Technical report, Athens, pp. 55–56

Environment Statistics (1996). Office des Publications Officieles des Communautes Europeennes, Luxemburg, 1997.

Environmental Problems Foundation of Turkey (1989) *Environmental Profile of Turkey*, Ankara, November 1989.

Ergin, M., Bodur, M., Ediger, V., Yemenicioglu, S., Okyar, M., Kubilay, N. (1993) Sources and dispersal of heavy metals in surface sediments along the eastern Aegean shelf. *Bolletino di Oceanologica Teorica ed Applicata* **XI** (1), 27–44.

Europe in Numbers, (1995). Office des Publications Officieles des Communautes Europeennes, Luxemburg, 4th edn.

Florou, H. and Chaloulou, Ch. (1997) Migration of the Black Sea radioactive pollution to the Aegean Sea via water mass circulation through the Straits of Dardanelles. In *Proceedings of the 5th National Symposium of Oceanography and Fisheries*, pp. 275–278.

Florou, H., Kritidis, P., Synetos, S. and Chaloulou, Ch. (1987) Some radioecological aspects on the radioactive impact of the Chernobyl reactor accident to the Greek marine environment. In *Proceedings of the Congres Franco-Italian, Roma*, pp. 124–126.

Foufa, E. (1993) Evolution of trace metal levels and magnetic properties in sediments of Saronikos gulf, Greece. M.Sc Thesis, University of Athens.

Friligos, N. (1987) Eutrophication assessment in Greek coastal waters. *Toxicological and Environmental Chemistry* **15**, 185–196.

Friligos, N. and Karydis, M. (1988) Nutrients and phytoplankton distributions during spring in the Aegean Sea. *Vie Milieu* **38** (2), 133–143.

Fytianos, N. and Samanidou, M. (1991). Study of biogeochemical cycle of organophosphorus pesticides in Thermaikos Gulf, Greece. In Final reports on Research Projects dealing with mercury, toxicity and analytical techniques. *MAP Technical Report Series No 51*, UNEP, Athens, pp. 133–140.

Georgakopoulos-Gregoriades, E., Vassilopoulou, V. and Stergiou, K. (1991) Multivariate analysis of organochlorines in red mullet

from Greek waters. *Marine Pollution Bulletin* **22** (5), 237–241.
Gotsis-Scretas, O. and Friligos, N. (1988) Eutrophication and phytoplankton ecology in the Thermaikos Gulf. *Rapp. Comm. Int. Mer Medit.* **31** (2), 297.
Greek Oceanographers Association (1994). The state of oceanography in Greece. *Mesopelaga* **1**, 6–7.
Grimanis A.P., Zafiropoulos, D., Papadopoulou, C., Economou, T., Vassilaki-Grimani, M. (1982) Trace elements in *Mytilus galloprovincialis* from three gulfs of Greece. *VIes Journees Etud. Pollutions, Cannes, CIESM*, 319–322.
Huynh-Ngoc, L., and Fukai, R. (1978) Levels of trace metals in open Mediterranean surface waters—A summary report. *IVes Journees Etud. Pollutions, CIESM, Antalya, Turkey*, pp. 171–175.
Ignatiades, L., Gotsis, O., Apostolopoulou-Moraitou, M., Pagou, K., Sassalou, V., Kapiris, K., Zervoudaki, S. and Psara, S. (1994) Plankton and productivity in S. Aegean Sea: MAST/MTP PELAGOS 1 CRUISE. *First workshop of the Mediterranean Targeted Project*, pp. 77–82.
Katsikatsos, G.X. (1992) *Geology of Greece*. University of Patra, pp. 99–110.
Koray, T., Buyukisik, B., Parlak, H. and Gokpinar, S. (1996) Eutrophication Processes and Algal Blooms in Izmir Bay. *MAP Technical Reports Series No. 104*.
Koukouras, A. and Sinis, A. (1981) Benthic fauna of the North Aegean Sea. II Crinoidea and Holothurioides (Echinodermata). *Vie Milieu* **31** (3–4), 271–281.
Koukouras, A., Dounas, C., Turkay, M. and Voutsiadou-Koukoura, E. (1992) Decapod crustacean of the Aegean Sea: New information, check list, affinities. *Senckenbergiana Maritima* **22** (3/6) 64–70.
Kucuksezgin, F., Balci, A., Kontas, A. and Altay, O. (1995) Distribution of Nutrients and Chlorophyll-a in the Aegean Sea. *Oceanologica Acta* **18** (3), 343–352.
Lascaratos, A. (1992) Hydrology of the Aegean Sea. Winds and currents of the Mediterranean basin. In *Proceedings of the Workshop held at Santa Teresa, Italy*, ed. H. Charnock, No 40, Vol. 1, pp. 313–334.
Latinopoulos, P.D., Krestenitis, Y.N. and Valioulis, I.A. (1996) A decision analysis approach to a coastal pollution problem: The sewage system of the City of Thessaloniki (Greece). In *Coastal Environment: Environmental Problems in Coastal Regions*, eds. A.J. Ferrante and C.A. Brebbia. Computational Mechanics Inc, Billerica, MA, USA, pp. 11–21.
Le Vourch, J., Millot, C., Castagne, N., Le Borgne, P. and Olry, J-P. (1992) Atlas of Thermal Fronts of the Mediterranean Sea Derived from Satellite Imagery. *Memoires de l'Institut Oceanographique, Monaco* **16**, 78–79.
McGill, D.A. (1961) A Preliminary study of the oxygen and phosphate distribution in the Mediterranean Sea. *Deep Sea Research* **9**, 259–269.
Michopoulos N. (1996) Study of the lake of Agios Nikolaos, Crete, M.Sc. Thesis in Oceanography, Univ. of Athens.
Murdoch W.W. and Onuf, C.P. 1974. The Mediterranean as a system. Part I. Large Ecosystems. *International Journal of Environmental Studies* **5**, 275–284.
NCMR (National Center for Marine Research) (1989) Chemical oceanographic study of Amvrakikos gulf. Study for the Ministry of Environment, Athens.
NCMR (National Center for Marine Research) (1996a) Report for MED-POL, Saronikos Project for 1995 period. Athens.
NCMR (National Center for Marine Research) (1996b). The 1994–99 Development plan, Athens.
National Statistic service of Greece (1996). Environmental Statistics, Year 1994 with comparative 1993 data, Athens, pp. 45–50.
Pagou, K. (1990) Eutrophication problems in Greece. *Water Pollution Research Reports* **16**, 97–114.
Pagou, K., Pancucci-Papadopoulou, A., Gialamas, B. and Siokou-Frangou, I. (1990) Distribution of plankton in Aegean Sea. In *Proceedings of the Third National Conference of Oceanography and Fisheries*, pp. 487–494.
Panayotidis P. (1988). Etude de l'impact de la pollution sur les herbiers de Posidonia oceanica, dans le golfe Saronikos. UNEP, *MAP Technical Reports Series No. 22*, pp. 85–104.
Pancucci-Papadopoulou, M.A. (1996) *The Echinodermata of Greece*. Hell. Zoological Society, 162 pp.
Pancucci-Papadopoulou, A., Siokou-Frangou, I. and Christou, E. (1990). On the vertical distribution and composition of deepwater Copepod populations in the Eastern Mediterranean Sea. *Rapp. Comm. Int. Mer Medit.* **32** (1), 199.
Papaconstantinou, C., Vassilopoulou, V., Petrakis, G., Caragitsou, E., Mytilineou, C., Fourtouni, A. and Politou, C.-Y. (1994) The demersal fishfauna of the North and West Aegean Sea. *Bios* **2**, 35–45.
Papaevagelou, S. (1993) The Chemical Oceanography of Pagassitikos Gulf. M.Sc Thesis, University of Athens.
Polat, C. and Tugrul, S. (1996) Chemical exchange between the Mediterranean and the Black Sea via the Turkish straits. Bulletin Institut Oceanographique, Monaco, No. special 17. *CIESM Science Series* **2**, 167–185.
Psilovikos, A. (1992) *Prospects for Wetlands and Waterfowl in Greece. Managing Mediterranean Wetlands and Their Birds*. IWRB Spec. Publ. No. 20, pp. 53–55.
Psyllidou-Giouranovits, R., Pavlidou, A. and Georgakopoulou-Gregoriadou, E. (1997) Recent measurements of nutrients and dissolved oxygen in the Saronikos and the Elefsis Gulfs (1995). In *Proceedings of the 5th National Conference of Oceanography and fisheries*, pp. 73–74.
Red Data Book (1992) *The Red Book Data of Threatened Vertebrates of Greece*. Greek Ornithological Society, 356 pp.
Romanov, A.S., Ryabinin, A.I., Lazareva, E.A. and Zhidkova, L.B. (1977) Copper, arsenic and mercury in Aegean sea waters (1974–75). *Oceanology* **17** (2), 160–162.
Rozhanskaya, L.I. (1973) Mg, Cu and Zn concentrations in Aegean and Ionian seawaters. Field studies in Mediterranean Sea during the 67th cruise of the R/V Akademik Kovalevsky in September–October 1970. Kiev, Nankova dumka.
Sakellariadou, F., Tselentis, V.S. and Tzannatos, E. (1994) Dissolved/dispersed petroleum hydrocarbons content in Greek coastal areas. *International symposium Pollution of the Mediterranean Sea WTSAC, Nicosia, Cyprus*, pp. 151–155.
Satsmadjis, J., Georgakopoulos-Gregoriades, E. and Voutsinou-Taliadouri, F. (1988) Monitoring of organochlorine pollution in the Pagassitikos gulf. N.C.M.R. Technical report, Athens, pp. 68–69.
Scoullos, M. (1979) Chemical studies of the Gulf of Elefsis, Greece, Ph.D. Thesis, Dept. of Oceanography. University of Liverpool.
Scoullos, M. and Dassenakis, M (1983a) Trace metals in suspended particles and sediments of the central and eastern Aegean Sea. *Second International Symposium on Environmental Pollution and its Impact on Life in the Mediterranean Region. Iraklion, Crete*.
Scoullos, M. and Dassenakis, M. (1983b) Trace metals in a Tidal Mediterranean Embayment. *Marine Pollution Bulletin* **14** (1), 24–29.
Scoullos, M., Mantzara, M. and Constantianos, V. (1994a) *The Book Directory for the Mediterranean Monk Seal* (Monachus monachus) *in Greece*. Contract with the C.E.U.,DG XI, 4-3010(92)7829. ed. M. Scoullos. Copyright Elliniki Etairia, Athens, Greece.
Scoullos, M., Dassenakis, M., Pavlidou, A., Matzara, B., and Bolkas, S. (1994b) A brief account on trace metal levels in the Saronikos Gulf, based on the 1986-1993 MED-POL monitoring programme. In *Proceedings of an International Conference on Restoration and Protection of the Environment II*, pp. 272–279.
Sinis, A. (1977) Keys for the identification of the Mediterranean Echinoderms 1. Crinoidea and Holothurioides (in Greek). *Sci. Annals, Fac. Phys. Mathem., Univ. Thessaloniki* **17** (2), 223–265.
Siokou-Frangou, I. and Papathanassiou, E. (1991) Differentiation of zooplankton populations in a polluted area. *Marine Ecology Progress Series* **76**, 41–51.

Siokou-Frangou, I., Christou, E.D., Gotsis-Scretas, O., Kontoyannis, H., Krasakopoulou, E., Pagou, K., Pavlidou, A., Souvermezoglou, E. and Theocharis, A. (1997) Impact of physical processes upon chemical and biological properties in the Rhodes gyre area. *International Conference on Progress in Oceanography of the Mediterranean Sea, Rome, November 17–19*, pp. 265–270.

Souvermezoglou, E., Krasakopoulou, E. and Pavlidou, A. (1997) Interannual changes of oxygen and nutrients in Cretan Sea. *5th National Conference of Oceanography and Fisheries*, pp. 185–188.

Stergiou, K. and Petrakis, G. (1993) Description, assessment of the state and management of the demersal and inshore fisheries resources in the Greek seas. *Fresh. Environ. Bulletin* **2**, 312–319.

Strogyloudi, E., Catsiki, V.A. and Galenou, E. (1997) Concentrations of Metals in Mullus Barbatus from North Aegean Sea. In *Proceedings of the 5th National Symposium of Oceanography and Fisheries*, pp. 279–281.

Symboura, N. (1997) Benthic Polychaetes (Polychaeta) as indicators of diversity in benthic ecosystems. *Fifth Hellenic Symposium on Oceanography and Fisheries, Kavala, Greece, April 15–18, 1997*, pp. 445–448.

Theocharis, A., Georgopoulos, D., Laskaratos, A. and Nittis, K. (1993) Water masses and circulation in the central region of the eastern Mediterranean: Eastern Ionian, South Aegean and northwest Levantine, 1987–87. *Deep Sea Research II Top. Stud. Oceanography* **40** (6) 1121–1142.

Tsekos, I. and Haritonidis, S. (1982) Contribution to the study of populations of benthic macroalgae of the Greek seas. *Thalassographica* **1** (5) 61–153.

Tselepides, A. and Eleftheriou, A. (1992) South Aegean Continental Slope. Benthos: Macroinfaunal–environmental Relationships. Deep-Sea Chains and the Global Carbon Cycle, ed. Rowe, Pariente. pp. 139–156.

Tselepides, A., Polychronaki, T., Dafnomili, F., Plaiti, W. Zivanovic, S. (1997). Distribution of nutrients, chloroplastic pigments, POC, PON and ATP in the Cretan Sea (NE Mediterranean) Seasonal and interannual variability. *5th National Conference of Oceanography and Fisheries*, pp. 189–192.

Turkay, M., Fisher, G. and Neumann, V. (1987) List of the marine Crustacea ecapoda of the Northern Sporades (Aegean Sea) with systematic and zoogeographic remarks. *Investigacion Pesquera* **51** (suppl. 1), 87–109.

UNEP(OCA)/MED WG.111/Inf.5 (1996a) Assessment of the state of Eutrophication in the Mediterranean Sea, UNEP, Athens, pp 80–90.

UNEP(OCA)/MED WG.111/Inf.9 (1996b) Survey of Pollutants from Land-based Sources in the Mediterranean. UNEP, Athens.

UNEP (1985) Report of the state of pollution of the Mediterranean Sea. UNEP/IG.56/INF. 4.

UNEP (1986a). Baseline studies and monitoring of oil and petroleum hydrocarbons in marine waters. *MAP Technical Reports Series No. 1.*

UNEP (1986b). Biogeochemical studies of selected pollutants in the open waters of the Mediterranean (MED POL VII) Addendum, Greek Oceanographic Cruise 1980. *MAP Technical Reports Series No. 8*, Addendum. Athens

UNEP (1988) M. Bernhard: Mercury in the Mediterranean. *UNEP Regional Seas Reports and Studies No. 98.*

UNEP (1989) State of the Mediterranean marine environment. *MAP Technical Reports Series No. 28*, UNEP, Athens.

UNEP (1990a) L. Jeftic: State of the marine environment in the Mediterranean region. *MAP Technical Reports No 28.*

UNEP (1990b). Assessment of the state of pollution of the Mediterranean Sea by organohalogen compounds. *MAP Technical Reports No. 39.* pp. 40–60.

UNEP (1994). Integrated management study for the area of Izmir. *MAP Technical Reports Series* **84**, Split.

UNEP (1995). The Mediterranean Action Plan: Saving our common heritage, UNEP, Athens.

UNEP (1996a). Pollution effects on plankton composition and spatial distribution, near the sewage outfall of Athens. *MAP Technical Reports Series* **96**, 121 p.

UNEP (1996b). State of the marine and coastal environment in the Mediterranean region. *MAP Technical Reports Series* **100**, UNEP, Athens.

Vlachakis, G.N and Tsimblis, M.N. (1993) Euripus problem: review and a proposal. In *Proceedings of the 4th National Symposium of Oceanography and Fisheries*. National Centre for Marine Research, pp. 156–159.

Vlahakis, G.N. and Pollatou, R.S. (1993) Temporal variability and spatial distribution of sea surface temperatures in the Aegean Sea. *Theoretical and Applied Climatology* **47**, 15–23.

Voutsinou-Taliadouri, F. (1980) Trace metals in marine organisms from the Saronikos gulf. *Ves Jour. Etud. Pollutions, Cagliari, CIESM*, pp. 275–280.

Voutsinou-Taliadouri, F. (1982) Metal concentration in polluted and unpolluted Greek sediments: a comparative study. *VIes Jour. Etud. Pollutions, Cannes, CIESM*, pp. 329–333.

Voutsinou-Taliadouri, F. and Satsmadjis, N. (1982) Trace metals in the Pagassitikos Gulf, Greece. *Estuarine Coastal and Shelf Science* **15**, 221–228.

Voutsinou-Taliadouri, F. and Balopoulos, E. (1989) Geochemical and water flow features in a semienclosed embayment of the western Aegean Sea (Pagassitikos Gulf) and physical oceanographic and geochemical conditions in Thermaikos Bay (Northwestern Aegean, Greece). *Water Science and Technology* **21** (12), 188.

Voutsinou-Taliadouri, F., Georgakopoulou, E. and Fragoudaki, S. (1987) Trace metal pollution in surface sediments of the Gulf of Elefsis, Greece. In *Proceedings of the 2nd Greek Symposium on Oceanography and Fisheries*. National Centre for Marine Research, pp. 99–108.

Voutsinou-Taliadouri, F. and Varnavas, S. (1995) Geochemical and sedimentological patterns in the Thermaikos Gulf, North-west Aegean Sea, formed from a multisource of elements. *Estuarine, Coastal and Shelf Science* **40**, 295–320.

Yannopoulos C. and Yannopoulos A. (1973) The Saronikos and the S. Evvoikos gulf, Aegean Sea. Zooplankton standing stock and environmental factors. *Pelagos* **IV** (2), 73–81.

Zenetos, A. (1996) *The Marine Bivalvia of Greece*. National Centre for Marine Research, Hell. Zool. Soc. (eds.), 318 pp.

Zodiatis, G. (1994) Advection of the Black Sea water in the North Aegean Sea. *Global Atmos. Ocean Syst.* **2** (1), 41–60.

THE AUTHORS

Manos Dassenakis
University of Athens, Department of Chemistry, Division III, Inorganic and Environmental Chemistry, Panepistimiopolis, Kouponia, Athens 15771, Greece

Kostas Kapiris
University of Athens, Department of Biology, Division of Zoology and Marine Biology, Panepistimiopolis, Kouponia, Athens 15784, Greece

Alexandra Pavlidou
University of Athens, Department of Chemistry, Division III, Inorganic and Environmental Chemistry, Panepistimiopolis, Kouponia, Athens 15771, Greece

Chapter 16

THE COAST OF ISRAEL, SOUTHEAST MEDITERRANEAN

Barak Herut and Bella Galil

The Israeli coast, at the southeastern corner of the Mediterranean, describes a slightly curved line, with Haifa Bay the sole embayment in the mostly sandy coast. The Nilotic sediments transported from the Nile delta northwards by the prevailing inner shelf and wave-induced longshore currents, produce a shallow shelf, narrowing considerably northwards. North of Haifa Bay, for lack of Nilotic sand, the coast is mostly rocky.

Although the Neolithic city of Jericho, nearly 9000 years old, marks the beginning of civilization, only in the late 19th century did anthropogenic changes begin affecting the environment. The opening of the Suez Canal in 1864—linking the Red Sea with the Mediterranean—allowed hundreds of Erythrean species to settle along the Levantine coasts. Some abundant invaders are exploited commercially, others constitute a nuisance or economic burden, and yet others outcompete native species.

The Levantine Basin is considered the most impoverished in the Mediterranean. The completion of the high dam at Aswan in the mid 1960s deprived the Levant of its influx of freshwater, nutrients and sediments, contributing to the diminishing sediment transport and negatively impacting fisheries.

The rapid increase in population density along the Israeli coastal plain in the past half-century, and its consequent urbanization, generated land reclamation schemes. Sand mining in the past and the existing marine structures along the coast have depleted sand reserves and increased coastal erosion.

Effluents, such as sewage, agricultural run-off or industrial wastes may increase nutrient loading locally, most notably in Haifa Bay and in some lower reaches of the coastal streams. In such nutrient-enhanced sites, appropriate ambient physical conditions may cause the development of toxic algal blooms. Yet, overall, levels of toxic contaminants are low, except for Haifa Bay with its concentration of heavy industries.

Protective measures and monitoring activities are implemented by several legislative / administrative systems, and by removal and abating land-base pollution sources. A modified National Plan for the Israeli Mediterranean coast—including fourteen marine reserves—is in the process of affirmation and should provide a tool for an integrated sustainable coastal zone management.

Seas at The Millennium: An Environmental Evaluation (Edited by C. Sheppard)
© 2000 Elsevier Science Ltd. All rights reserved

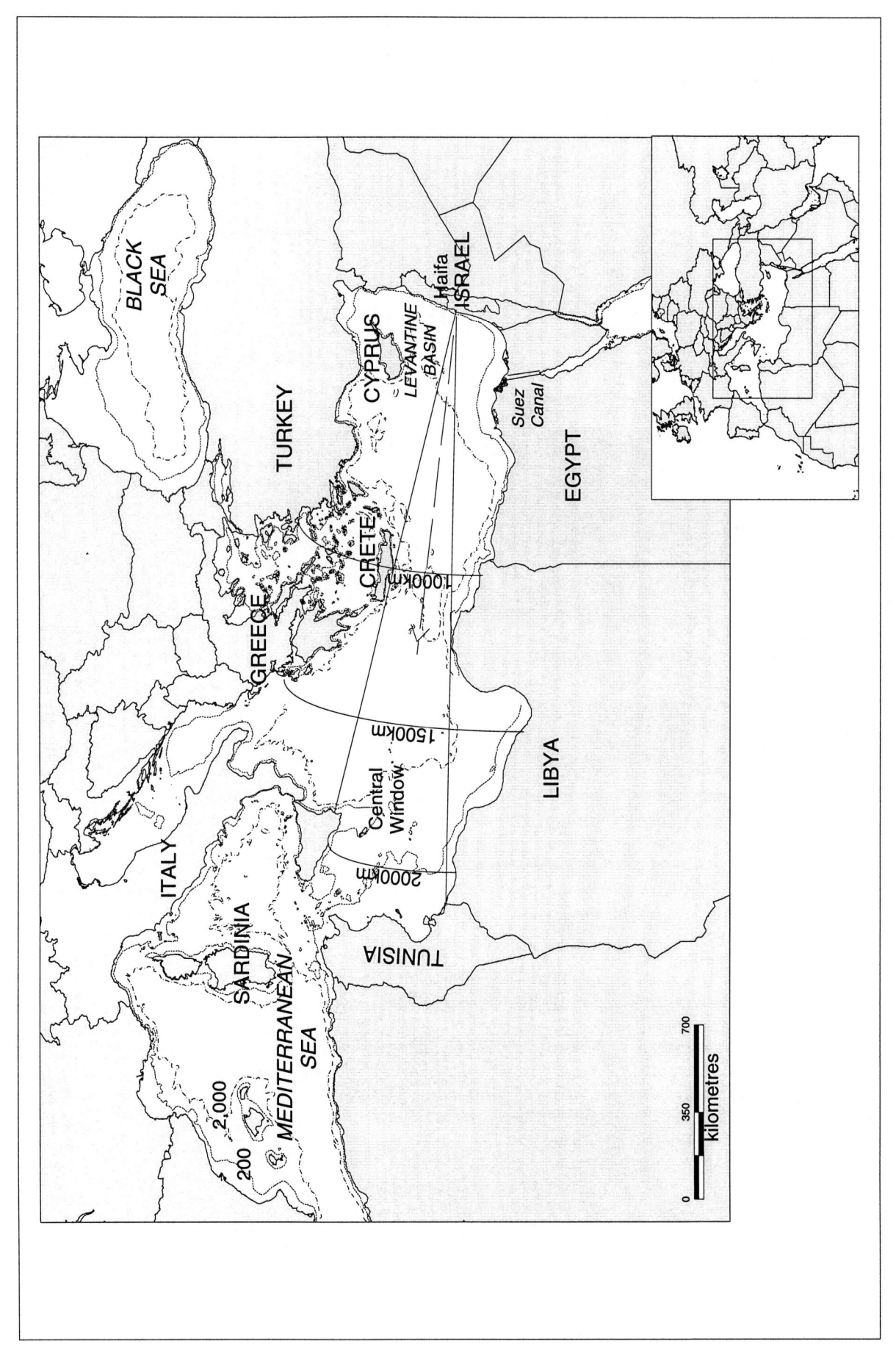

Fig. 1. Map of the East Mediterranean Basin. The fetch distances facing the Mediterranean coast of Israel are included (reprinted from Carmel et al, 1985).

LOCALE

The easternmost Mediterranean, from Rhodes to Cyreneica, is the Levant basin (Fig. 1). At the southeastern corner of the basin, the Israeli coast describes a slightly curved line, 180 km long, with Haifa Bay the sole embayment (Fig. 2). The shelf narrows gradually toward the north, extending to a distance of 10–20 km from the coast (Fig. 2). Haifa Bay, about 11 km long and 6 km wide at its southern end, is bordered in the west by three submerged sandstone ridges parallel to the coast. South of the Bay, the Israeli littoral is mostly sandy, with isolated sandstone outcrops. North of the bay, rocky shores predominate (Emery and Neev, 1960), but a continuous rocky substrate from the upper to the sublittoral zone is present in only a few places and the occurrence of vermetid reefs is rarer still (Safriel, 1966). Thirteen streams traverse the coastal plain, of which two, the Kishon and Naaman, flow into Haifa Bay (Fig. 3).

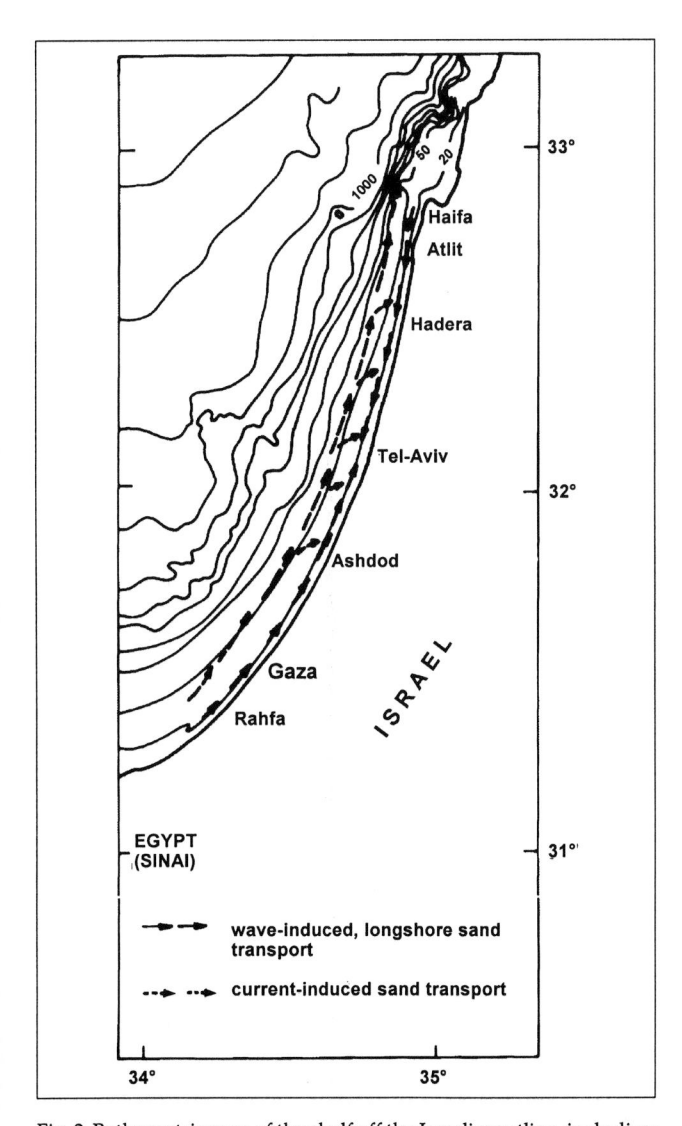

Fig. 2. Bathymetric map of the shelf off the Israeli coastline, including the sediment transport pattern according to Emery and Neev (1960).

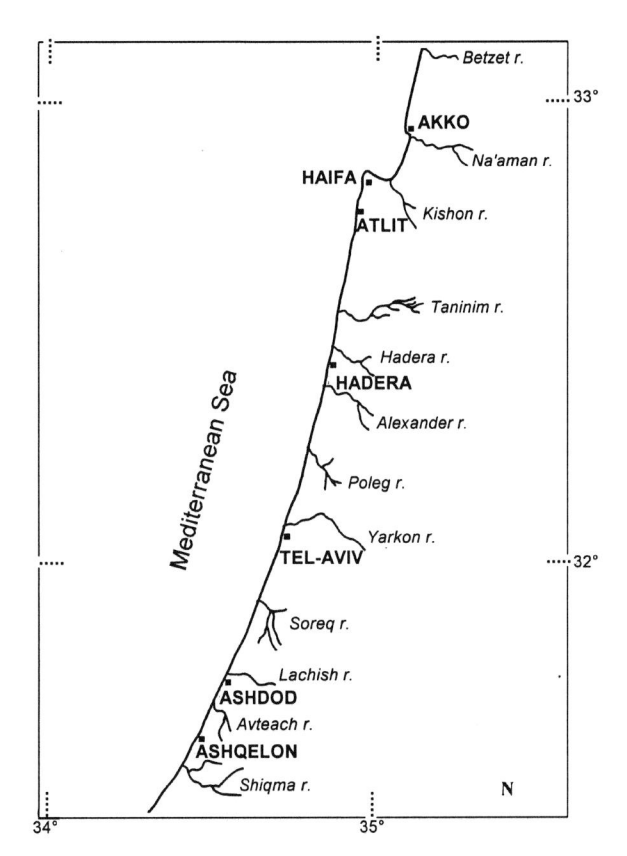

Fig. 3. Location map showing the coastal orientation, Haifa Bay and the main coastal streams which drain into the sea.

NATURAL CHARACTERISTICS

The vertical profiles of temperature and salinity in the southeastern Levantine waters reveal four water masses (Hecht et al., 1988): Levantine Surface Water (LSW) (S $\geq$ 38.95, 0–40 m), Atlantic Water (AW) (S $\leq$ 38.87, 65–95 m), Levantine Intermediate Water (LIW) (S $\geq$ 38.94, 200–310 m) and Deep Water (DW) (T <13.8°C and S <38.74, below 700 m). The formation of these water masses and their circulation in the eastern Mediterranean has been studied extensively (POEM, 1992). The two upper water masses are most affected by seasonality. Surface water temperatures show high seasonal variations and range from ~16°C during winter to ~28°C in summer. The upper ~100 m are well mixed in most winters and become stratified during the rest of the year with an upper mixed layer of about 25 m and a sharp halocline and thermocline below. The bottom of the seasonal thermocline is at about 100 m (Hecht et al., 1988).

The offshore waters are oligotrophic, with extremely low nutrient values most of the year (Berman et al., 1984; Krom et al., 1991; Yacobi et al., 1995). The concentration of phosphate in surface waters is close or below the detection limit (0.01 μM), increasing to about 0.28 μM in deep waters (>700 m depth). Nitrate concentrations show a similar trend with values between detection limit and approximately 6 μM, and silicic acid concentrations ranging from 1 to 12 μM. A typical vertical nutrient profile shows low concentrations in the upper 150 m (<0.05 and <1 μM phosphate and nitrate, respectively), a nutricline development between 150–200 m and increasing concentrations down to 600 m. Silicic acid, however, continuously increases with depth. Near surface chlorophyll-a (chl-a) concentrations vary from 0.4 μg/l near-shore and decrease westward to 0.05 at 20 km offshore (Berman et al., 1986; Gitelson et al., 1996; Yacobi et al., 1995). The seaward gradient is seasonal and higher in summer than in winter and early spring (Berman et al., 1986). The depth distribution of chl-a concentrations typically shows a deep chlorophyll maximum (DCM) between 75 and 130 m. The Levantine oligotrophy is attributed to the eastward surface flow of nutrient-depleted waters, further diluted by the nutrient-depleted LIW; the arid and semi-arid climate with low terrigeneous runoff of nutrient-rich waters; and the relatively narrow continental shelf off Israel and northward, allowing little benthic–pelagic coupling for nutrient recycling (Berman et al., 1984). In addition, the construction of the Aswan Dam arrested the flow of the major terrigeneous nutrients source (Azov, 1991).

The circulation on the shelf, dominated by the geostrophic current and shelf waves, is mostly northward (Rosentraub, 1995). Thus, the currents are mostly parallel to the coast, along depth contours, and effect particle transport along the shelf. The current fluctuations occur on both the daily (mainly sea breeze controlled) and synoptic (3–14 day) time scales. The strongest currents, predominately northward, occur in winter and summer, while in spring and autumn currents are weaker and alternate from north to south (Rosentraub, 1995). In the stratified summer the relatively strong currents and fluctuations are confined to the upper layer and increase towards the continental slope to mean speeds of 40 cm s^{-1}. In the mixed winter the currents distribution is uniform throughout the water column.

The coast is fanned by light winds (<10 knots) for 80% of the year, fresh winds (11–21 knots) for 18%, and by strong winds (>22 knots) for only 2% (Rosen, 1998). In winter and early spring the winds are mostly from a southwest direction attributed to eastward-moving depressions across the Levantine Basin. In summer and autumn the area is controlled by a northwest wind system on top of the local breeze. The wind transports sand particles from the beach inland, where gaps in the coastal cliff exist, forming sand dunes which extend to a distance of up to 10 km from shore. The highest, longest waves are westerly, due to the longer fetch (2400 km) (Fig. 1) (Carmel et al., 1985). Wave height distribution allows for low waves (<1 m) for almost half the time, moderate waves (1–2 m) for 25% of the time, tall waves (2–4 m) for 20%, and storm waves (>4 m) for only 5% of the time, mostly from the West (Rosen, 1998). The angle between the wave direction and the coastline orientation (beach normal) is the main factor that determines transport of sand by the wave-induced longshore currents (Emery and Neev, 1960). The tidal currents are weak, about 0.05 cm s^{-1}, and have only a minor effect on sediment flow. The extremely low natural flow in the coastal streams, except for occasional storm runoff, and their lower reaches bathymetry, permits the penetration of seawater inland.

The quartz sands making up the shelf, beaches and inland dunes of Israel originate in the Nile river and delta (Pomerancblum, 1966) and constitute a part of the Nile littoral cell which extends from the Nile delta to Akko (Inman and Jenkins, 1984). The Nile sediments are transported northward by currents on the upper continental shelf, and by the wave-induced longshore currents (Emery and Neev, 1960) (Fig. 2). The magnitude of the longshore sediment transport is estimated at 170,000–540,000 m^3 $year^{-1}$, decreasing northwards (Golik, 1997). However, the pattern of sand movement along the coast, and particularly along its northern part (from south of Hadera to Haifa) is not entirely clear, and may be either northward or southward (Golik, 1993, 1997). The contribution of local streams to the sand budget is negligible as their drainage basins are dominated by carbonate rocks.

The geomorphology of the beaches is diverse: from Gaza to Tel-Aviv, 30–50 m wide sandy beaches backed by large dune fields are prevalent; from Tel-Aviv to Hadera, the beaches are narrower (0–30 m), bordered by a calcareous cliff; from Hadera to Haifa the narrow beach lacks the cliff; Haifa Bay is bordered by a sandy beach; and from Akko to the Lebanese border the beach is mostly rocky. Ridges of eolianite sandstone (locally called "Kurkar") extend parallel to the coastline on the coastal plain and on the upper shelf.

THE SHALLOW MARINE HABITATS, THE RED SEA INVADERS

Community studies of the continental shelf of Israel have been conducted for nearly half a century. The earlier studies were mainly fishery research (Wirszubski, 1953; Gilat-Gottlieb, 1959), and it was only in the 1960s that an extensive program was undertaken by Gilat (1964) to describe the macrobenthal communities off the Israeli coast. Gilat, and later Galil and Lewinsohn (1981) and Tom and Galil (1990) recognized parallel dominant species with other parts of the Mediterranean. However, Galil and Lewinsohn (1981) noted that the sandy–mud associations at depths of 20–50 m have no known parallel outside the Levant.

Vermetid reefs occur in the narrow zone of the infralittoral fringe, on exposed rocky shores under high energy conditions. The rimmed reef platforms are formed by the gregarious sessile gastropod *Dendropoma petraeum* (Monterosato) and *Vermetus triqueter* (Bivona), endemic to the Mediterranean. The porous structure of the vermetid reef and the underlining aeolanithic sandstone allows for a rich community of endoliths. The resulting local organic enrichment coupled with structural complexity of the edifice, have led to the establishment of an extraordinarily dense ichthyofauna of great diversity—probably the most diverse ecosystem in the Mediterranean coast of Israel.

The total annual catch of the Mediterranean fisheries of Israel is estimated at 3000 tons. The fishery is divided into three categories according to the methods employed: inshore, employing trammel nets and hook-and-line, purse seine, and trawl fishery. Red Sea immigrants constitute 50% of the trawl catch.

The opening of the Suez Canal that joined the Red Sea and the Mediterranean initiated a remarkable faunal movement. Despite physical and hydrological impediments, hundreds of Erythrean species settled in the Mediterranean, forming thriving populations along the Levantine coasts. This extraordinary movement is considered to be "the most important biogeographic phenomenon witnessed in the contemporary oceans" (Por, 1989).

Many—but by no means all—immigrants have become ubiquitous. In fact, three modes of population gain can be discerned: persevering, exploding and ubiquitous. Many, if not most, of the immigrants have established small but stable populations off the Levantine coast.

Rather more attention has been given to those instances of explosive population gain followed by sharp decline. In the late 1940s the immigrant goldband goatfish, *Upeneus moluccensis*, made up 10–15% of the total mullid catches. Following the exceptionally warm winter of 1954–55, its percentage in the catch increased to 83% (Oren, 1957). Its share has since been reduced to 30% of the catch (Ben Tuvia, 1973). Following that same winter, the brushtooth lizardfish, *Saurida undosquamis*, became commercially important and its proportion in trawl fisheries catches rose to 20% in the late 1950s. The population then diminished and catches stabilized at about 5% of the total trawl catch (Ben Yami and Glazer, 1974). Similarly, the gastropod *Rhinoklavis kochi*, first reported in Haifa Bay in the mid-1960s, spread rapidly to become by the late 1970s one of the dominant species on sandy-mud bottoms between 20 and 60 m (Galil and Lewinsohn, 1981; Tom and Galil, 1990). Samples taken a decade later consisted mostly of empty shells (Galil, 1993). Then there are immigrants that are common and abundant. Immigrant fish now constitute nearly half of total fish biomass in commercial trawl catches along the Israeli coast (Golani and Ben Tuvia, 1995): *strombus decorus persicus*, littering the shallow sandy littoral: "one can speak of an invasion... hundreds of dead shells on the beaches and shoals of live *Strombus*, of all sizes, colors and patterns, feeding on the sea floor up to 20 meters depth" (Curini-Galleti, 1988). Another immigrant that proliferated in an astonishingly short time is the nomadic jellyfish, *Rhopilema nomadica* (Galil et al., 1990). It was first collected in the Mediterranean in 1977; by the mid-1980s huge swarms would appear each summer along the southeastern Levant coast, and by 1995 also off the southeastern coast of Turkey (Kideys and Gucu, 1995) and Cyprus. The massive swarms, the sizable biomass of these voracious planktotrophs must play havoc with the meagre resources of this oligotrophic sea, and when those shoals draw nearer shore, they impact fisheries, coastal installations and tourism.

Invasion is often followed by competition for resources or direct interference between native and invading species; the former outcompeted wholly, or in part, from their habitat space. Por (1978) maintained that "Other than the case of *Asterina gibbosa* there is no known case in which a lessepsian migrant (Red Sea invader) species has completely replaced a local one". Indeed, the decimation of the indigenous sea star, *A. gibbosa*, populations from the Israeli coast paralleled the rapid advent of its Red Sea congener *A. wega* (Achituv, 1973). But it is far from alone: there are other documented instances of an extreme change in abundance that can be attributed to the new competition. A native penaeid prawn, *Penaeus kerathurus*, was "very commonly caught by trawlers on Israel coastal shelf especially on sandy or sandy mud bottoms" according to Holthuis and Gottlieb (1958), and supported a commercial fishery throughout the 1950s. It has since nearly disappeared and its habitat overrun by the Red Sea penaeid prawns. The immigrant snapping shrimps *Alpheus inopinatus* and *A. edwardsi* are more common now in the rocky littoral than the native *A. dentipes* (Lewinsohn and Galil, 1982). The reduction in the numbers of the previously prevalent indigenous jellyfish, *Rhizostoma pulmo*, has coincided with the massive presence of *R. nomadica*, and may also be a case of competitive displacement. The local red mullet, *Mullus barbatus*, and the native hake, *Merluccius merluccius*, were both displaced into deeper, cooler waters by their respective Red Sea competitors, *Upeneus moluccensis* and *Saurida undosquamis* (Oren, 1957).

However, the immigrants' ascendancy resulted not only in displacement of some indigenous species but, increasingly, in displacement among the immigrant species themselves. *Trachypenaeus curvirostris* was first recorded in the Mediterranean in the late twenties (Steinitz, 1929); already so abundant, it was sold on the Haifa fish market and was the most common penaeid on sandy-mud bottoms (Galil, 1986). In 1987 another immigrant, *Metapenaeopsis aegyptia*, joined it on the sandy-mud bottoms (Galil and Golani, 1990). By 1993 in samples collected along the central Israeli coast at depth of 35 m, *M. aegyptia* outnumbered *T. curvirostris* by three to one, and by 1996 outnumbered it by 25 to one. That same year yet another migrant penaeid was recorded on the sandy-mud bottoms, *Metapenaeopsis moigensis consobrina* (Galil, 1997). *Charybdis longicollis* was first recorded in the Mediterranean in the mid 1950s (Holthuis, 1961), and has since dominated the macrobenthic fauna on silty-sand bottoms off the Israeli coast, forming up to 70% of the biomass at places (Galil, 1986). Of the thousands of specimens collected over three decades none was parasitized until, in 1992, a few parasitized crabs were collected; the parasite was identified as a sacculinid rhizocephalan,

Heterosaccus dollfusi—itself a Red Sea immigrant (Galil and Lutzen, 1995). Within three years it spread as far as the eastern Anatolian coast and the infection rate at Haifa Bay rose to 77%. In the past year, we find, for the first time since the invasion of *Charybdis*, large numbers of the indigenous portunid crab *Liocarcinus vernalis*.

The invasion of the Mediterranean by Red Sea fauna is a dynamic, ongoing, surprising process. The unique history of the easternmost Mediterranean—which left it warm, salty and impoverished—is the basis of a singular synergy between anthropogenic and natural environmental factors, past and present. Although the expected outcome of invasion is a reduction in diversity, we are witnessing an invasion that increases faunal diversity, and augments the local fisheries.

POPULATION AFFECTING THE AREA

From the Neolithic Age, nearly 9,000 years ago, until the late nineteenth century, the major population centres along the southeastern Levant were situated inland. However, the political realities that shaped the map of Israel forced a shift to the coastal plain, where the population expanded from 100,000 to four million within a century (out of a total of 5.8 million). Today, Israel is in a unique situation among the developed countries due to the combination of its high population density and high growth rate (Fig. 4). The population increase resulting from a high birth rate and massive immigration—three-quarters of a million in the early 1990s alone—subjects the coastal plain to increasing demands for housing, tourism, recreation, transportation, ports, energy and industrial facilities, sewer and effluent outlets, and marine farming. These pressures led to examination of land reclamation schemes, either adjacent to the coastline or as offshore artificial islands, for housing, airports and various infrastructure facilities (e.g. gas terminals, seawater desalinization plants). Thus, the coastal strip, which constitutes 5% of Israel's area, is and will be subjected to all those activities that may have a drastic short- or long-term impact on the marine systems.

COASTAL EROSION

Since the construction of the low Aswan dam in 1902 the Nile delta has retreated about 10 km—5 km since the completion of the high dam in 1964 alone. There is no concrete evidence yet that the negative sand budget in the delta has affected the Israeli coast.

The coastal sand reservoirs of Israel have been depleted by massive mining for construction, outlawed only in 1964, and by the construction since the late 1960s of coastal structures hindering its natural northward movement. The Israeli building industry utilizes at present approximately 8 million m³ annually; however, already planned marine structures (Table 1) would greatly increase the demand. To satisfy construction needs options such as offshore sand mining and sand importation are being examined (Golik, 1997). The effects of detached breakwaters and other marine structures on beach morphology were investigated by Nir (1982). He estimated that 600,000 m³ of sand were trapped by the detached breakwaters alone. In 35 years, Ashdod port breakwater alone trapped more than 4.5 million m³ of sand and has greatly affected the nearby shoreline and sea bottom (Golik et al., 1996). Golik (1997) estimated the total sand deficit due to human activity along the Israeli coast at 20 million m³—equal to at least 37 years of natural sand accretion.

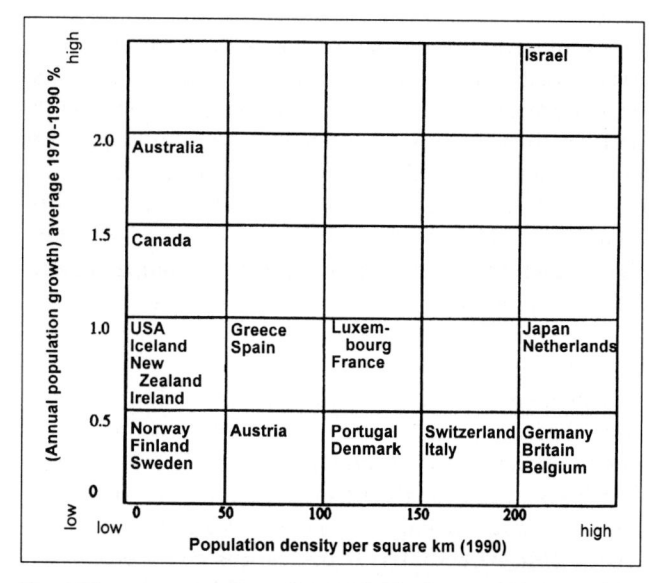

Fig. 4. The unique Israel situation with regard to population density and growth rate. Reprinted from Mazor (1993).

Table 1

Existing and future planned coastal structures along the Mediterranean coast of Israel

Marine structure	Present amount	Future development
Ports (Haifa and Ashdod)	2	large expansions of the existing ports and addition of a large port off Gaza
Marinas and small harbors	7	addition of at least one marina; expansion of harbors
Power plants + cooling basins	4	
Offshore unprotected coal unloading terminals	2	development of gas terminals
Breakwaters	at least 16	additions to protect beach and cliff erosion
Marine outfalls (excluding beach outfalls)	4	additional are planned
Mariculture farms	3	additional are planned
Artificial islands offshore	–	several are planned

Proof of increasing coastal erosion is also supplied by the recent submarine archeological finds: the exposure of a Neolithic (8,000–10,000 B.P.) human skeleton at a depth of 8 m off Atlit, whose fragile bones would not have survived the destructive power of the waves had they been long uncovered (Hershkovitz and Galili, 1990; Galili et al., 1993); or a 2000-year-old merchant ship, its wood in perfect condition, which was exposed at a depth of about 2 m south of Haifa (Linder, 1992).

EFFECTS OF LAND-BASED POLLUTION SOURCES

The main pollution sources are connected to the intense coastal metropolis and industrial activities. More than 100 industries, cities and small settlements discharge waste into the sea, either directly or via the coastal rivers. Since the early 1980s the overall pollution load into the coastal zone has declined due to the combination of the following factors: (a) the increase of public awareness and governmental actions, (b) the national commitment to implement international conventions on marine pollution prevention, and (c) the progress and expansion of the National Sewage Project and sewage treatment. At present only two cities (Akko and Naharya) discharge raw sewage (after primary filtration) into the sea.

The three major classes of pollutants affecting the area are nutrients, heavy metals and toxic organic compounds. Their introduction into the coastal zone is via both point and non-point sources. The main point source inputs include direct pipeline discharges and riverine input, and the main diffused pathways are atmospheric deposition and runoff. Insufficient data are available to allow a reliable estimation of the pollution inputs to the coastal zone. Table 2 presents a preliminary estimate of the pollution load entering the coastal zone, and its distribution among the main sources (Herut and Krom, 1996; UNEP, 1997; Herut et al., 1998, 1999a, in press). Nitrogen and phosphorous are introduced mainly through the Kishon River, Gush Dan outfall and all other coastal streams in approximately equal amounts. The atmospheric deposition is dominant in the open sea. The table probably represents minimal values because it does not include input of agricultural runoff into the whole coastal zone area, and because the riverine input does not include the particulate and dissolved organic nutrient loads (which are unknown). The main heavy metal and organic loads are industrial via the Kishon River and the Gush Dan outfall.

Since 1982 more than one million tons of coal fly ash (approximately 17% of the total amount generated) were dumped at a deep water site (1500 m, 200 km^2) located 70 km off the Israeli coast beyond the continental shelf (Kress et al., 1993). Industrial sludge was dumped as well at a nearby second deep-water site (Kress et al., 1998). The Israeli government has agreed to the cessation of waste disposal in those sites.

Haifa Bay

Haifa Bay is exposed to the highest pollution load along the Israeli coast. Two rivers flow into the bay, the Naaman at its northern part, and the Kishon in the south. The latter is regarded as the most polluted river in Israel (Cohen et al., 1993; Herut et al., 1993, 1994; Herut and Kress, 1997). At the northern part of Haifa Bay anthropogenic mercury is introduced from a chlor-alkali plant (Hornung et al., 1984; Herut et al., 1996). High loads of the heavy metals (Herut and Kress, 1997), nutrients (Kress and Herut, 1998; Herut et al., 1999b) and organics (Cohen et al., 1993) are introduced via the Kishon River. Most anthropogenic heavy metals, nitrates and phosphates are discharged by acidic industrial effluents (about 90% of the total amount), while organic matter and ammonium originate mainly from the Haifa District sewage treatment plant effluents (about 90 and 70% of the total amount, respectively). Petroleum-derived compounds are released mainly from the adjacent refineries.

Nutrients

In contrast to the oligotrophic open eastern Mediterranean waters, in Haifa Bay the dissolved nutrient and chl-a levels are due to anthropogenic nutrient input through the Kishon River. The nutrient and chl-a concentrations decrease by more than an order of magnitude from the Kishon estuary towards the bay and the open sea (Fig. 5). The o-phosphate concentrations vary by more than two orders of magnitude. The chl-a concentrations in the bay are higher by a factor of about four than the maximal concentrations reported for the Israeli continental shelf (Berman et al., 1984). The bay ecosystem is N-limited in contrast to the P-limitation on the outer shelf and in the deep

Table 2
Estimated loading of pollutants to the coastal waters

Parameter	Kishon river**	Coastal rivers	Gush Dan sewage sludge outfall	Ashdod outfall	Atmospheric input*
N (t/y)	3,450	1,800–3,670	2,900	600	1,110
P (t/y)	2,900	350–580	1,200	7	74
Cd (kg/y)	2,048	53	430	–	45
Cu (kg/y)	3,194		19,000	–	1,672
Pb (kg/y)	810	3,000	1,670	–	9,623
Hg (kg/y)	60 ?		60	–	–
Zn (t/y)	48		54	–	8,029
BOD (t/y)	5,371		13,286	2,630	–
COD (t/y)	6,225		40,665	12,150	–
Oils (t/y)	25		–	11	–

*Atmospheric input was calculated for area of 180 km length and 21.6 km width (12 miles) from shore. Nutrients load are wet deposition only.
**Only part of the load is transported into Haifa Bay.

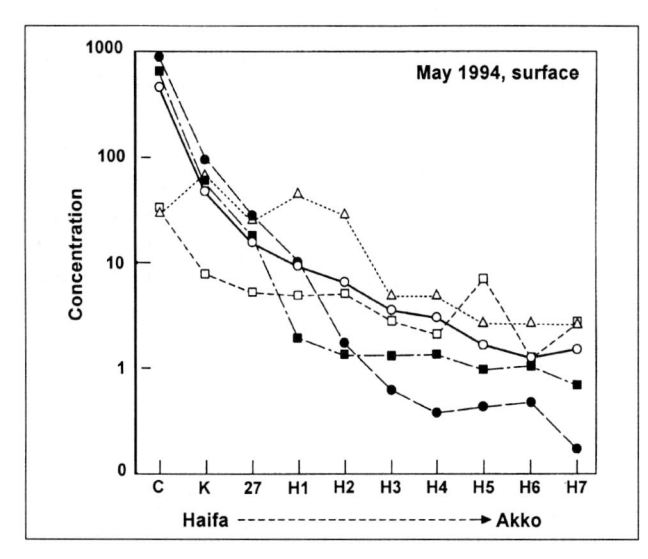

Fig. 5. The drastic decrease in the concentrations of some parameters in the surface waters along an S–N transect in Haifa Bay (from the Kishon estuary and Haifa port in the south to Akko at the north). Open circle: PO_4 (μM); open triangle: chlorophyll-a (mg l^{-1}); filled square: NH_4 (μM); open square: suspended particulate matter (mg l^{-1}). Reprinted from Kress and Herut (1998). Reproduced by permission of Academic Press.

sea (Krom et al., 1991). The physical conditions (low currents, warm weather, high radiation, seasonal stratification) and nutrient overloading in Haifa Bay may easily lead to the development of potentially harmful algal blooms. Indeed, sporadic events of toxic plankton blooms (Kress et al., 1995; Kimor et al., 1996; Kress and Herut, 1998), unusual fish mortality and complaints of eye rashes by swimmers, were attributed to unusually high amounts of anthropogenic nutrients introduced via the Kishon River.

Toxic Metal and Organic Pollutants

The sediments in Haifa Bay are contaminated by mercury from two sources, a chlor-alkali plant in the north and the Kishon River in the south, but the level of contamination is not high (Fig. 6). Most of the mercury in the sediments of the northern part of the bay accumulated prior to the introduction of waste treatment facilities by the chlor-alkali plant in 1976 (Fig. 7). Since the beginning of routine monitoring in the area in the early 1980s, the amount of mercury in the sediments has decreased continuously, probably due to wave-induced resuspension of contaminated particles and their subsequent seaward transport (Fig. 7) (Herut et al., 1996). The concentrations of mercury in benthic bivalves and fish sampled off the chlor-alkali plant and in the bay, respectively, also decreased during the same period (Fig. 7), and now approach background values. These data indicate that the bioavailable fraction of the mercury in the sediments of the northern bay is small, and therefore, that the remaining mercury reservoir in the sediments probably does not constitute a high ecological risk.

The sediments in the Kishon estuary are contaminated by several heavy metals and petroleum-derived compounds (Cohen et al., 1993). The degree of contamination in this area depends mainly on the hydrological conditions in the river and is maximal after heavy river floods (Herut and Kress, 1997). In recent years, mercury concentrations in the sediments of the Kishon estuary increased to higher levels than those found in the sediments of the northern part of the bay off the chlor-alkali plant. In 1992, cadmium and mercury concentrations in the sediments were much higher than those found in previous years (Fig. 8). This is a result of a major flood in the Kishon in 1992 which carried a large volume of contaminated river sediments towards

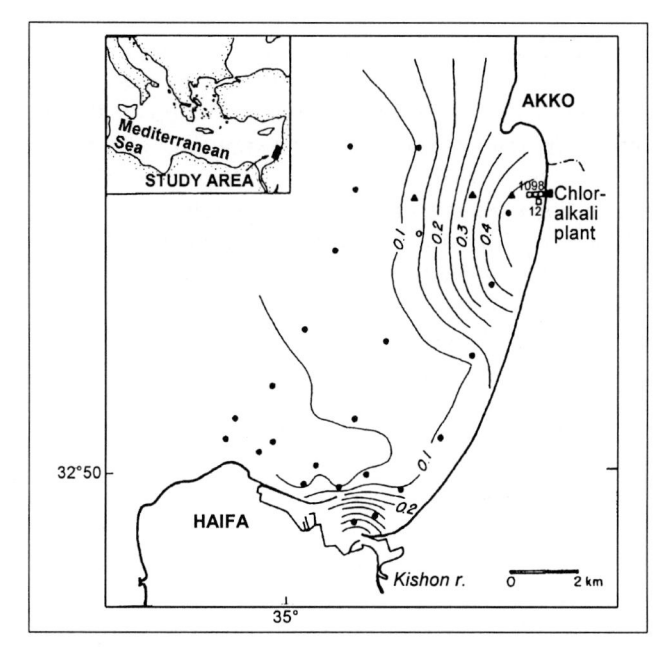

Fig. 6. Map of Haifa Bay showing the distribution of mercury concentrations (μg g^{-1} dry wt.) in the surficial sediments. From Herut et al. (1996).

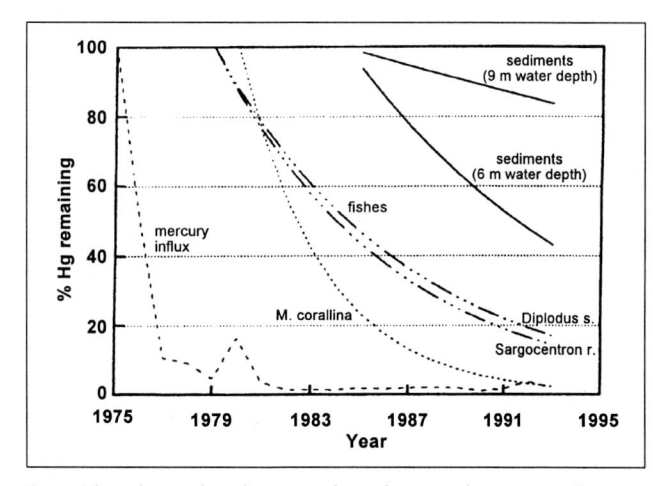

Fig. 7. The relationships between the reduction of mercury influx into northern Haifa Bay, the reduction in the amount of mercury in the top 50 cm of the sediments and the reduction of mercury concentrations in the biota (bivalves and fish) of Haifa Bay. Reprinted from Herut et al. (1996).

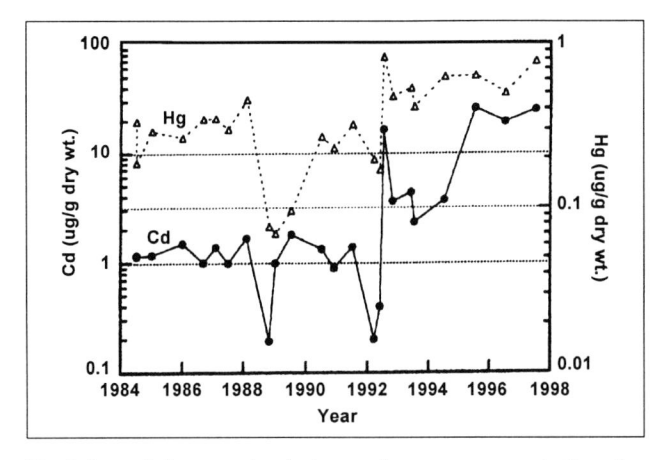

Fig. 8. Annual changes of cadmium and mercury concentrations ($\mu g\ g^{-1}$ dry wt.) in surficial sediments from the Kishon estuary.

Haifa Bay, while usually the lower river system acts as a trap for contaminants discharged by the adjacent industries (Hornung et al., 1989; Cohen et al., 1993).

The suspended particulate matter (SPM) in Haifa Bay contains relatively high concentrations of mercury, cadmium, copper and zinc (Herut and Kress, 1997). The main source of mercury in surface water SPM is the chlor-alkali plant in the northern part of the bay. In near-bottom SPM the main source of mercury is the Kishon estuary. The Kishon is also the source of SPM contaminated by the other toxic metals (Herut and Kress, 1997; Herut et al., 1999b).

The coastal zone outside Haifa Bay

In general, with regard to the environmental levels of potentially toxic heavy metals and organic contaminants, the status of the coastal zone is quite satisfactory. Heavy metals and organics do not significantly contaminate the offshore sediments outside Haifa Bay. However, relatively high concentrations of some metals (5–7 times higher than "background" levels) are present near a sewage sludge outfall (Gush Dan) 5 km from central Israel. Very slight enrichment is also apparent near an industrial wastewater outfall off southern Israel (Ashdod), off the estuary of the Yarkon stream (Tel-Aviv area) and near a coal terminal off central Israel.

The sediments and SPM in some of the lower reaches of the coastal streams are enriched by mercury, cadmium, copper, zinc and lead indicating anthropogenic contamination but the degree of contamination is not high (Herut et al., 1995).

The mercury concentrations in commercial-size specimens of fish caught off the Israeli coastline do not represent a risk to human health. In all fish examined, including those sampled in Haifa Bay, the mercury concentrations were well below the national safety limit for seafood. Other trace metal concentrations do not reveal any indication of contamination, and are below strict international sea-food standards.

In the offshore sediments, in benthic fauna and fish, there is no significant organic contamination. Recent screening of organic contaminants along the coastline revealed very low concentrations, and PCBs and dioxins could not be detected in the sediments (Herut et al., 1997).

The lower reaches of some coastal streams (Soreq, Poleg and Alexander) contain high nutrient concentrations (Herut et al., 1998). The high degree of contamination is attributed to the discharge of domestic and industrial wastes, and to agricultural runoff. The spatial and temporal variations of the nutrient and chl-a concentrations in the lower streams system and in the shallow water along the coast are poorly known. The phytoplankton species are also poorly known and a substantial effort to identify and quantify them is needed. The water quality is monitored for microbial indicators of fecal pathogens in authorized bathing beaches. Occasionally a few beaches are closed because of microbial contamination, but generally the status of the water quality is quite satisfactory. Other beaches, which are open to the public but are not declared as authorized bathing beaches and which extend over 100 km (out of 130 km of beaches), are not monitored at present.

PROTECTIVE MEASURES

At present, the marine environmental protective measures in Israel are encompassed within the framework of three main complementary legislative/administrative systems: the land-use planning system, the system of marine and pollution control and the system for nature protection. Recently, a national plan for an integrated sustainable coastal zone management was completed. It aims to guide the national coastal zone committee in a sustainable global planning policy, including nature protection.

The land-use planning system (under the Planning and Building law of 1965) incorporates regulations for environmental protection both in regional masterplans and in specific projects. An environmental impact statement (EIS) has been a statutory requirement since 1982 for development projects with anticipated significant environmental impacts. In 1983 a National Masterplan for the Mediterranean coast was affirmed. The plan includes regulations on the reservation and protection of the natural coastline and beaches, on the open view to the sea and on free public passage to the beach. An additional part of the program was submitted in 1991, and includes a more detailed land-use assessment based on a land and sea uses database prepared by a professional multidisciplinary team. This revised program has not yet been affirmed.

The Ministry of Environment and, for some aspects, the Ministry of Transportation and the Ministry of Interior, are responsible for legislation concerning marine pollution. This includes the prohibition of oil discharge into the sea from ships and land-based marine installations, the regulation of dumping at sea, the regulation of waste discharges into the sea (directly and indirectly) from land-based

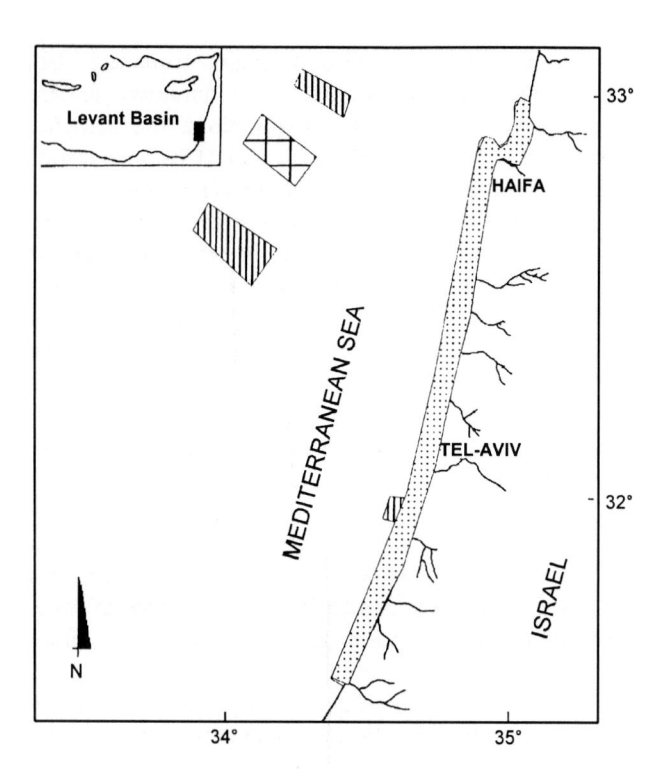

Fig. 9. Schematic map of sites monitored for heavy metal levels and biological structure at the Israeli Mediterranean continental margin. Dots: routine long-term monitoring activities; dark lines: waste disposal sites; squares: control site for deep sea monitoring.

sources and the maintenance of cleanliness in the public domain. Most of the related activities are financed by the "Sea Pollution Prevention Fund" which is based on taxes and penalization, and thus implements the "Polluter Pays" principle.

Israel is a party to MARPOL 73/78 on prevention of pollution from ships, and is also a party to the 1976 Barcelona Convention and its protocols, and has accepted the changes made in 1996. Incorporation of the new Barcelona 1996 principles in the law is now under way.

The Israeli Nature Reserves and National Parks Authority oversees 36 reserves and fourteen coastal parks along the Mediterranean coast of Israel. There are four types of nature reserve: marine reserves (proposed and declared), coastal reserves (proposed and declared), islet reserves, and protected natural assets belts. Declared reserves have full legal protection; proposed reserves have a limited level of protection until they are legally declared. All fourteen proposed marine reserves (area 2500 ha, shoreline 45921 m) come under the limited protection. Of the 20 coastal reserves (area 3500 ha, shoreline 45732 m), four are declared, and thus fully protected. The two islets reserves (33 ha), serving as important nesting sites, are also fully protected.

Several marine and coastal pollution-monitoring programs were undertaken in Israel since the 1970s (Fig. 9). Until the early 1970s, only the monitoring of fecal pathogens in authorized bathing beaches was carried out. Since then other monitoring programs, as well as compliance monitoring programs in outfalls and dumping sites, were held. The "status and trends" type of heavy metal monitoring has been carried out continuously since the early 1980s in Haifa Bay and from 1988 in the entire coastline, and constitutes a part of Israel's National Monitoring Program within the framework of MED POL (the international Mediterranean Pollution Research and Monitoring Program). In addition, several research projects aimed at understanding the pathways, fate and impact of anthropogenic metals in the marine environment were performed. The overall aim of the monitoring and related research activities is to provide a basis for decision-making on a variety of management issues such as marine waste disposal, pollution control and seafood safety. The specific objectives are to assess the status of the coastal zone with regard to pollutant contamination, to identify contamination sources and to detect early signs of potential health and ecological risks.

The results of the monitoring are used as a basis for environmental decisions, control and abatement activities, as well as for the assessment of their effectiveness (Cohen, 1995). Such a contribution is demonstrated by the following two examples related to the presence of tar balls on the coastline and the bioaccumulation of mercury in fish from Haifa Bay (Cohen, 1995).

Because of its heavy presence on the beaches, monitoring of tar quantities started in 1975. The origin of this tar was unknown and among several speculations government agencies tended to believe it originated mainly from oil discharged into the sea to the west of Israel (500–1500 km). Analyses of vanadium, nickel and sulphur indicated that most of the tar balls were formed from oil transported to

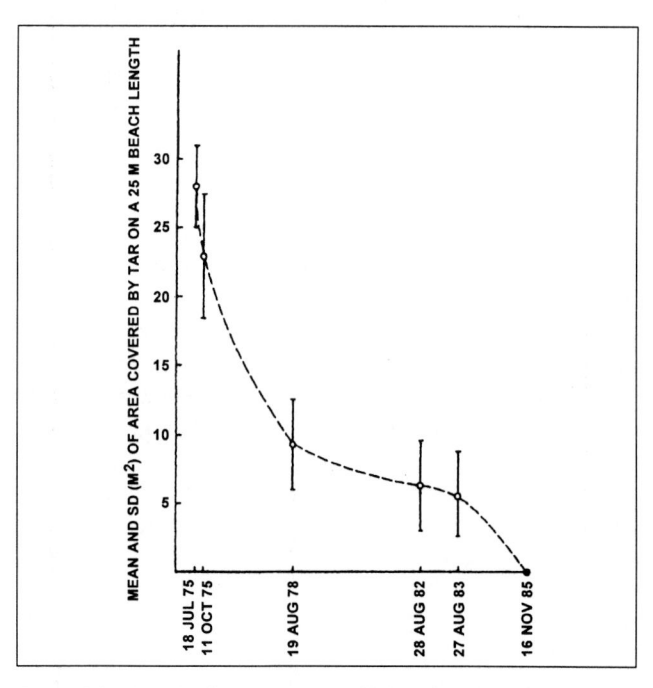

Fig. 10. Monitoring of tar quantity on the beach in the central coast of Israel (reprinted from Golik and Rosenberg, 1987).

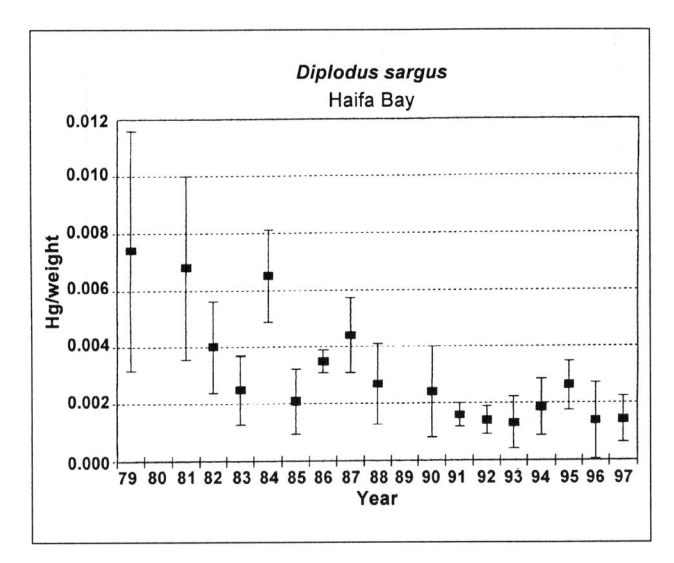

Fig. 11. Decrease of mercury concentration/body weight ratios (as a proxy to contamination level) in specimens of Diplodus sargus (annual average ± S.D.) from Haifa Bay.

and from Israel and discharged into the sea less than 100 km from the coast (Shekel and Ravid, 1976). As a result the Ministry of Environment implemented a national program and new legislation for the prevention of oil pollution. Further monitoring confirmed the drastic decrease of the tar balls on the beaches (Fig. 10) (Golik, 1982; Golik and Rosenberg, 1987).

Toward the late 1980s—in the framework of the toxic heavy metals monitoring programme in Haifa Bay—it was found that in large specimens of *Diplodus sargus* (a commercially important fish) the mercury concentrations exceeded the national safety limit (1 ppm) (Hornung et al., 1984; Krom et al., 1990). The main source of anthropogenic mercury was a chlor-alkali plant that had discharged mercury-containing effluents into the northern part of the bay since 1956. The drastic reduction (90%) of mercury discharge in 1976 due to the installation of treatment facilities in the plant was apparently not sufficient. As a result of the findings, the government prohibited the sale of *D. sargus* specimens of more than 100 g in size caught in Haifa Bay, and enforced the chlor-alkali plant management to further reduce its mercury discharge. A clear decreasing trend of the mercury levels (below safety limit for all market sizes) was then seen in further sampling (Fig. 11) (Herut et al., 1996; 1998). Thus, the long-term monitoring results contributed to the environmental management decisions and verified the effectiveness of the management.

SUMMARY

Most of the Israeli population, industry and agriculture are concentrated along the narrow coastal plain. The past decade has seen unprecedented growth, and the pressures to utilize coastal lands are increasing apace. Implementation of the present plans—many still mired in controversy —will dramatically change the coastline, in particular in and around the metropolitan areas. In addition there are proposals for marine constructions such as artificial islands for habitation, tourism or airfields, together with plans for a submarine gas pipeline, sand mining and marine farming. The planned enlargement of the Mediterranean ports and construction of additional marinas anticipate an increase in marine transport and leisure activities.

The National Plan has stated that marine and coastal resources should be utilized to the benefit of all citizens now and in the future, despite pressures to accommodate special interests. As rapid development and population growth continue along the coastal plain, increasingly heavy demands will be placed on the remaining natural habitats. The protection of the marine environment depends on careful adherence to the principle of sustainable development that combines environmental commitments within the marine and coastal resource development and management policy.

ACKNOWLEDGEMENTS

The authors would like to express their thanks to Dr. Yuval Cohen for his comments and suggestions.

REFERENCES

Achituv, Y. (1973) On the distribution and variability of the Indo-Pacific sea star Asterina wega (Echinodermata: Asteroidea) in the Mediterranean Sea. *Marine Biology* **18**, 333–336.

Azoz, Y. (1991) The eastern Mediterranean—a marine desert? *Marine Pollution Bulletin* **23**, 225–232.

Ben Tuvia, A. (1973) Man-made changes in the eastern mediterranean sea and their effect on the fishery resources. *Marine Biology* **19**, 197–203.

Ben Yami, M. and Glaser, T. (1974) The invasion of Saurida undosquamis (Richardson) into the Levant Basin—An example of biological effect of interoceanic canals. *Fishery Bulletin* **72**, 359–373.

Berman, T., Azov, Y., Schneller, A., Walline, P. and Townsend, D.W. (1986) Extent, transparency and phytoplankton distribution of the neritic waters overlying the Israeli coastal shelf. *Oceanologica Acta* **9**, 439–447.

Berman, T., Townsend, D.W., El Sayed, S.Z., Trees, C.C. and Azov, Y. (1984) Optical transparancy, chlorophyll and primary productivity in the Eastern Mediterranean near the Israeli coast. *Oceanologica Acta* **7**, 367–372.

Carmel, Z., Inman, D.L. and Golik, A. (1985) Directional wave measurements at Haifa, Israel, and sediment transport along the Nile littoral cell. *Coastal Engineering* **9**, 21–36.

Cohen, Y., Kress, N. and Hornung, H. (1993) Organic and trace metal pollution in the sediments of the Kishon river (Israel) and possible influence on the marine environment. *Water Science and Technology* **32**, 53–59.

Cohen, Y. (1995) The Israeli experience of using marine and coastal pollution monitoring as a basis for environmental management. *Water Science and Technology* **27**, 439–447.

Curini-Galletti, M. (1988) Notes and Tidings. *La Conchiglia* **14**, 232–233.

Emery, K.O. and Neev, D. (1960) Mediterranean beaches of Israel. *Israel Geological Survey Bulletin* **26**, 1–23.

Galil, B.S. and Lewinsohn, Ch. (1981) Macrobenthic communities of the Eastern Mediterranean continental shelf. PSZNI *Marine Ecology* 2, 343–352.

Galil, B.S. (1986) Red Sea decapods along the Mediterranean coast of Israel, ecology and distribution. In: Dubinsky Z. and Y. Steinberger (eds). *Environmental Quality and Ecosystem Stability*, Vol. III/A, Bar Ilan University Press, pp. 179–183.

Galil, B.S. and Golani, D. (1990) Two new migrant decapods from the eastern Mediterranean. *Crustaceana* 58(3), 229–236.

Galil, B.S., Spanier, E. and Ferguson, W.W. (1990) The Scyphomedusae of the Mediterranean coast of Israel, including two lesepsian migrants new to the Mediterranean. *Zoologische Mededelingen* 64, 95–105.

Galil, B.S. (1993) Lessepsian migration: New findings on the foremost anthropogenic change in the Levant Basin fauna. In: N.F.R. Della Croce (ed.), *Symposium Mediterranean Seas 2000*, pp. 307–318.

Galil, B.S. and Lutzen, J. (1995) Biological observations on *Heterosaccus dolfusi* Boschma (Cirripedia, Rhizocephala), a parasite of *Charybdis longicollis* Leene (Decapoda: Brachyura), a lessepsian migrant to the Mediterranean. *Journal of Crustacean Biology* 15(4), 659–670.

Galil, B.S. (1997) Two Lessepsian migrant decapods new to the coast of Israel. *Crustaceana* 70(1), 111–114.

Galili, E., Hershkovitz, I., Gopher, A., Weinstein-Evron, M., Lernou, H., Lerneou, O., Kislev, M. and Kolska-Hershkovitz, L. (1993) Atlit Yam, a submerged neolithic site off the Israeli coast. *Journal of Field Archeology* 20, 133–157.

Gilat, E. (1964) The macrobenthic animal communities of the Israeli continental shelf in the Mediterranean. *Rapports et Proces-Verbaux des Revmious de la CIESM* 17, 103–106.

Gilat-Gottlieb, E. (1959) Study of the benthos in Haifa Bay. Ecology and zoogeography of invertebrates. *Spec. Publ. Sea Fish. Res. Sta., Haifa*, 131 pp. (in Hebrew).

Gitelson, A., Karnieli, A., Goldman, N., Yacobi, Y.Z. and Mayo, M. (1996) Chlorophyll estimation in the Southeastern Mediterranean using CZCS images: adaptation of an algorithm and its validation. *Journal of Marine Systems* 9, 283–290.

Golani, D. and Ben Tuvia, A. (1995) Lessepsian migration and the Mediterranean Fisheries of Israel. In *Condition of the World's Aquatic Habitats. Proceedings of the World Fisheries Congress*, ed. N.B. Armantrout, pp. 279–289. Science Publishers, Lebanon, New Hampshire.

Golik, A. (1982) The distribution and behavior of tar balls along the Israeli coast. *Estuarine, Coastal and Shelf Science* 15, 267–276.

Golik, A. and Rosenberg, N. (1987) Quantitative evaluation of beach-stranded tar balls by means of air photographs. *Marine Pollution Bulletin* 18, 289–298.

Golik A. (1993) Indirect evidence for sediment transport on the continental shelf of Israel. *Geo-Marine Letters* 13: 159–164.

Golik, A., Rosen, D.S., Golan, A. and Shoshany, M. (1996) The effect of Ashdod port on the sorrounding seabed, shorline and sediments. IOLR Report H/02/96, 66 pp.

Golik, A. (1997) Dynamics and management of sand along the Israeli coastline. Bulletin de Institute Oceanographique, Monaco, special 18, *Ciesm Science Series* 3, 97–110.

Hecht, A.,Pinardi N. and Robinson A.R. (1988) Currents, water masses, eddies and jets in the Mediterranean Levantine Basin. *Journal of Physical Oceanography* 18, 1320–1353.

Hershkovitz, I. and Galili, E. (1990) 8000 year old human remains on the sea floor near Atlit, Israel. *Journal of Human Evolution* 5, 319–358.

Herut, B., Hornung H., Krom M.D., Kress N. and Cohen Y. (1993) Trace metals in shallow sediments from the Mediterranean coastal region of Israel. *Marine Pollution Bulletin* 26, 675–682.

Herut, B., Hornung H. and Kress N. (1994) Mercury, lead, copper, zinc and iron in shallow sediments of Haifa Bay, Israel. *Fresenius Environmental Bulletin* 3, 147–151.

Herut, B., Hornung, H., Kress, N., Krom, M.D. and Shirav, M. (1995) Trace metals in sediments at the lower reaches of Mediterranean coastal rivers, Israel. *Water Science and Technology* 32, 239–246.

Herut, B., Hornung, H., Kress, N. and Cohen Y. (1996) Environmental relaxation in response to reduced contaminant input: The case of mercury pollution in Haifa bay, Israel. *Marine Pollution Bulletin* 32, 366–373.

Herut, B., Hornung, H. and Kress, N. (1997) Long-term record mercury decline in Haifa Bay (Israel) shallow sediments. *Fresenius Environmental Bulletin* 6, 48–53.

Herut, B. and Kress, N. (1997) Particulate metals contamination in the Kishon River Estuary, Israel. *Marine Pollution Bulletin* 34, 706–711.

Herut, B. and Krom, M.D. (1996) Atmospheric input of nutrients and dust to the SE Mediterranean. In *Impact of Desert Dust from Northern Africa across the Mediterranean*, eds. S. Guerzoni and R. Chester. Kluwer, Dordrecht, pp. 349–358.

Herut, B., Hornung, H., and Kress, N. (1998) Monitoring of heavy metals along the Mediterranean coast of Israel in 1997. IOLR Rep. H18/98, 26 pp. (in Hebrew and executive summary in English).

Herut, B., Krom, M.D., Pan, G. and Mortimer, R. (1999a) Atmospheric input of nitrogen and phosphorus to the SE Mediterranean: sources, fluxes and possible impact. *Limnology and Oceanography* 44, 1683–1692.

Herut, B., Tibor, G., Yacobi, Y.Z. and Kress, N. (1999b) Synoptic measurements of chlorophyll-a and suspended particulate matter in a transitional zone from polluted to clean seawater utilizing airborne remote sensing and ground measurements, Haifa Bay (SE Mediterranean). *Marine Pollution Bulletin* 38, 762–772.

Herut, B., Kress, N. and Hornung, H. Nutrients pollution at the lower reaches of Mediterranean coastal rivers in Israel. *Water Science and Technology*, in press.

Holthuis, L.B. (1961) Report on a collection of Crustacea Decapoda and Stomatopoda from Turkey and the Balkans. *Zoologische Verhandelingen Leiden* 47, 1–67.

Holthuis, L.B. and Gottlieb, E. (1958) An annotated list of the decapod Crustacea of the Mediterranean coast of Israel, with an appendix listing the Decapoda of the Eastern Mediterranean. *Bulletin of the Research Council Israel* 7B, 1–126.

Hornung, H., Krumgalz, B. and Cohen, Y. (1984) Mercury pollution in sediments, benthic organisms and inshore fishes of Haifa Bay, Israel. *Marine Environmental Research* 12, 191–208.

Hornung, H., Krom, M.D. and Cohen, Y. (1989) Trace metal distribution in sediments and benthic fauna of Haifa Bay, Israel. *Estuarine, Coastal and Shelf Science* 29, 43–56.

Inman, D.L. and Jenkins, S.A. (1984) The Nile littoral cell and man's impact on the coastal zone of the southeastern Mediterranean. Scripps Institute of Oceanography, Ref. Series 84-31, 43 pp.

Kideys, A.E. and Gucu, A.C. (1995) Rhopilema nomadica: a lessepsian schyphomedusan new to the Mediterranean coast of Turkey. *Israel Journal of Zoology* 41, 615–617.

Kimor, B., Kress, N. and Herut, B. (1996) A short-lived toxic plankton bloom in Haifa Bay (Israel). ASLO Ocean Sciences Meeting, Feb. 12–16 1996, San Diego.

Kress, N. and Herut, B. (1998) Hypernutriphication in the oligotrophic Eastern Mediterranean. A study in Haifa Bay, Israel. *Estuarine, Coastal and Shelf Science* 46, 645–646.

Kress, N., Golik, A., Galil, B. and Krom, M.D. (1993) Monitoring the disposal of coal fly ash at a deep water site in the eastern Mediterranean Sea. *Marine Pollution Bulletin* 26(8), 447–456.

Kress, N., Herut, H. and Angel, D.L. (1995) Environmental conditions of the water column in Haifa Bay, Israel, during September–October 1993. *Water Science and Technology* 32, 57–64.

Kress, N., Hornung, H. and Herut, B. (1998) Concentrations of Hg, Cd, Cu, Zn, Fe and Mn in deep sea benthic fauna from the Southeastern Mediterranean Sea. A comparison study between fauna collected at a pristine area and at two waste disposal sites. *Marine Pollution Bulletin* 36, 911–921.

Krom, M.D., Hornung, H. and Cohen, Y. (1990) Determination of the environmental capacity of Haifa Bay with respect to the input of mercury. *Marine Pollution Bulletin* **21**, 349–354.

Krom, M.D., Kress, N., Brenner, S. and Gordon, L.I. (1991) Phosphorus limitation of primary productivity in the Eastern Mediterranean Sea. *Limnology and Oceanography* **36**, 424–432.

Lewinsohn, Ch. and Galil, B.S. (1982) Notes on species of Alpheus (Crustacea Decapoda) from the Mediterranean coast of Israel. *Quad. Lab. Tecno. Pesca, Ancona.* **3**(2–5), 207–210.

Linder, E. (1992) Ma'agan Michael ship wreck. Excavating an ancient merchanman. *Biblical Archeology Review* **18**, 24–35.

Mazor, A. (1993) *Israel 2020—Masterplan for Israel in the 21st Century.* Technion Research and Development Foundation. Haifa, Israel.

Muerdter, D.R., Kennett, J.P. and Thunell, R.C. (1984) Late Quaternary sapropel sediments in the Eastern Mediterranean Sea: Faunal variatins and chronology. *Quaternary Research* **21**, 1–69.

Nir, Y. (1982) Offshore artificial structures and their influence on the Israel and Sinai Mediterranean beaches. *Proceedings of the 18th International Conference on Coastal Engineering, ASCE*, pp. 1837–1856.

NOAA (1996) *NOAA's Estuarine Eutriphication Survey. Volume 1: South Atlantic Region.* Office of Ocean Resources Conservation Assessment, Silver Spring, MD. 50 pp.

Oren, O.H. (1957) Changes in the temperature of the Eastern Mediterranean Sea in relation to the catch of the Israel trawl fishery during the years 1954/55 and 1955/56. *Bulletin of the Institute of Oceanography of Monaco* **1102**, 1–12.

POEM Group (1992) General circulation of the eastern Mediterranean. *Earth Science Reviews* **32**, 285–309.

Pomerancblum, M. (1966) The distribution of heavy minerals and their hydraulic equivalents in sediments of the Mediterranean continental shelf of Israel. *Journal of Sedimentary Petrology* **36**, 161–174.

Por, F.D. (1978) *Lessepsian Migration—The influx of Red Sea biota into the Mediterranean by way of the Suez Canal.* Ecological Studies, 23. Springer-Verlag, 228 pp.

Por, F.D. (1989) *The Legacy of Tethys—An Aquatic Biogeography of the Levant.* Kluwer. Dordrecht.

Rosen S.D. (1998) Characterisation of mete-oceanographic climate in the study sector. IOLR Report H16/98.

Rosentraub Z. (1995) Circulation on the Mediterranean continental shelf and slope of Israel. IAEPSO 21 General Assembly, Honolulu, Hawaii, 5–12 August 1995.

Safriel, U.N. (1966) Recent vermetid formation on the Mediterranean shor of Israel. *Proceedings of the Malacological Society of London* **37**, 27–34.

Shekel, Y. and Ravid, R. (1976) Sources of tar ball pollution on Israel's beaches. Tech. Rep., Israel Institute of Petroleum and Energy, 45 pp.

Steinitz, W. (1929) Die Wanderung indopazifischer Arten ins Mittelmeer seit Beginn der Quartarperiode. *Inst. Rev. ges. Hydrobiol. Hydrog.* **22**, 1–90.

Tom, M. and Galil, B.S. (1990) The macrobenthic associations of Haifa Bay, Mediterranean coast of Israel. *PSZNI Marine Ecology* **12**(1), 75–86.

UNEP (1977) Identification of priority pollution hot spots and sensitive areas in the Mediterranean. Athens, Greece.

UNEP/UNESCO/FAO (1988) Eutriphication in the Mediterranean Sea. Receiving capacity and monitoring of long term effects. *MAP Technical Report Series* **21**, UNEP, Athens.

Wirszubski, A. (1953) On the biology and biotope of the Red Mullet, Mullus barbatus. *Bulletin Sea Fisheries Research, Haifa* **7**, 1–20.

Yacobi, Y., Zohari, T., Kress, N., Hecht, A., Robart, R.D., Wood, A.M. and Li, W.K.W. (1995) Chlorophyll distribution throughout the Southeastern Mediterranean in relation to the physical structure of the water mass. *Journal of Marine Systems* **6**, 179–190.

THE AUTHORS

Barak Herut
*Israel Oceanographic and Limnological Research,
National Institute of Oceanography,
P.O.Box 8030,
Haifa 31080, Israel*

Bella Galil
*Israel Oceanographic and Limnological Research,
National Institute of Oceanography,
P.O.Box 8030,
Haifa 31080, Israel*

Chapter 17

THE ADRIATIC SEA AND THE TYRRHENIAN SEA

Giuseppe Cognetti, Claudio Lardicci, Marco Abbiati and Alberto Castelli

The Adriatic Sea differs sharply from the Tyrrhenian Sea and the rest of the Mediterranean in its hydrographic and hydrobiological characteristics. Three different areas can be distinguished: the Northern, the Central and the Southern Adriatic. The first shows the most accentuated peculiarities: maximum depth does not exceed 40 meters, temperature may fluctuate between 5°C and 28°C during the year, salinity is highly variable, and tides may rise higher than 1 m. Organic matter input from rivers, and the resulting nutrient enrichment, leads to elevated primary productivity, particularly in the Northern and the Central Adriatic. Both the circulation and distribution of water masses in the Adriatic Sea are strongly influenced not only by the morphology of the three basins but also by the fresh water inflow of continental origin.

In the Tyrrhenian Sea the continental shelf extends only for a relatively short distance. The depth exceeds 2000 m throughout virtually the whole of the basin, with a maximum depth of 3840 m. Input from the continental waters is minimal so salinity remains constant at roughly 38 PSU. In winter the Tyrrhenian Sea maintains a constant temperature of about 13°C, while in summer the temperature is around 23–24°C. The mean value of primary productivity is much lower than in the Adriatic. The surface Atlantic current feeds a cyclonic gyre with a southern branch entering between Sardinia and Sicily and moving in a northeasterly direction, and the other towards the southwest. Levantine intermediate waters penetrate mainly through the strait of Sicily at a depth ranging between 300 and 400 m.

A clear distinction between coastal and off-shore habitats is found only in the Southern Adriatic where the bottom drops to considerable depths. Particularly significant is the beach-rocks habitat enclosed within the surrounding soft bottom. Bathial fauna are found only in the Southern Adriatic. The extensive coastal lagoons are of considerable ecological and biogeographic interest. Several endemisms are known, some of sarmatic origin. Only the Central and Southern Adriatic are characterised off-shore by an oceanic planktonic community. The Northern Adriatic is characterised by neritic plankton throughout its extension. Production of nanoplankton is dominant.

In the Tyrrhenian Sea the hard bottom communities are amongst the most important of the whole Mediterranean. Of particular interest are the 'trottoir' and the coralligenous formations typical of the Tyrrhenian Sea which can develop on rocky and sandy bottoms at depths ranging between 20 and 1230 m. Zooplankton shows some characteristics that distinguish it from other

Seas at The Millennium: An Environmental Evaluation (Edited by C. Sheppard)
© 2000 Elsevier Science Ltd. All rights reserved

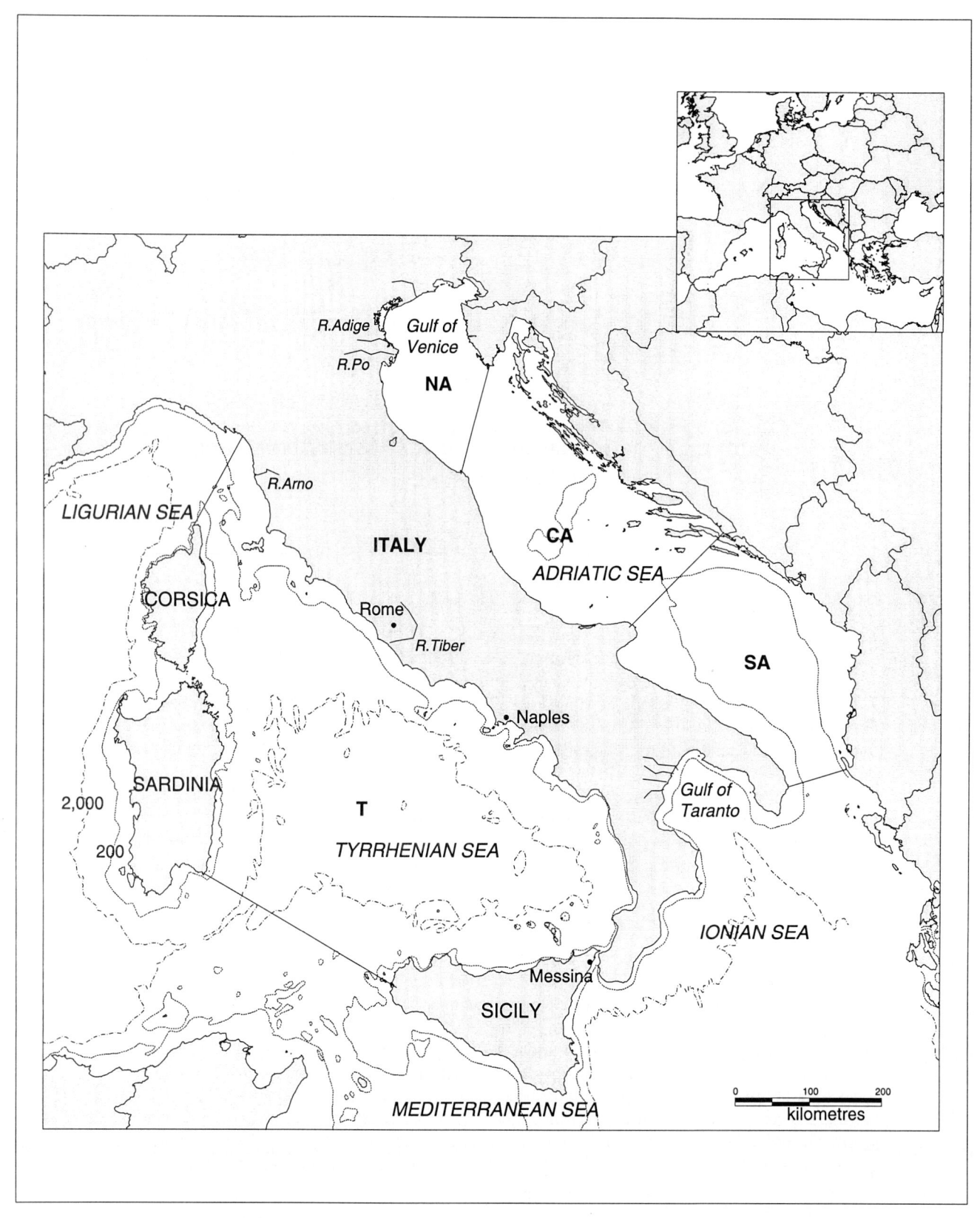

Fig. 1. Conventional boundaries of the Tyrrhenian Sea (T) and of the Adriatic Sea (NA = Northern Adriatic; CA = Central Adriatic; SA = Southern Adriatic).

areas of the Western Mediterranean. Abyssal fauna is present in the centre of the basin although it is considerably poorer than that of the Atlantic.

Along the coast of both seas there are important ports and numerous large urban centres whose population is substantially increased during the tourist season. The numerous industrial complexes co-exist with intense agriculture and animal-rearing activities. Maritime traffic is intense. The most important fishing ports are in the Central and Southern Adriatic and aquaculture is highly developed. In the Adriatic natural gas reserves and oil fields lie along the Italian coast: there are 70 off-shore platforms, and oil or gas flows to the mainland through pipelines. Water quality is impaired by a number of factors, particularly the excessive quantity of nitrogen and phosphorus. On the Tyrrhenian sea floor, several areas of interest for mineral extraction have been identified. The presence of cinnabar mines in Tuscany has increased the levels of mercury in the water and sediment in the Northern Tyrrhenian. Other metals are also present in considerable concentration in the waters of the Southern Tyrrhenian due to underwater volcanic activity.

Both the Adriatic and Tyrrhenian have been included within the framework of protective measures set up on the basis of international conventions. Considerable progress has been made in designing waste purification technology and measures designed to establish protected marine areas are already in place. Interdisciplinary research developed in the universities and in various research centres has long been a driving force and has made a crucial contribution to knowledge of oceanography and marine biology. Proposals based on data concerning the current ecological situation have also been put forward, aimed at water clean-up and improving fishery production.

THE DEFINED REGION

Adriatic Sea

The Adriatic Sea is a basin measuring roughly 138,000 km² extending between the Italian and the Balkan peninsulas. Approximately 800 km in length, it communicates with the Ionian Sea through the Otranto Channel (Fig. 1). At its broadest point, between Manfredonia and Drin, its width is 200 km, decreasing to just 71 km at the narrowest point in the Otranto Channel.

From a structural point of view, the Adriatic can be considered as a large geosyncline lying between the Apennines to the west, the Alps to the north, and the Dinaric system to the east. The gradual uplifting of these ridges has led to progressive sinking of the interposed geosyncline. Over the various geological periods the Adriatic has undergone variations in extension and depth. During the Pliocene it extended throughout virtually the whole of the Po plain, while during the Pleistocenic glacial expansion stages, following the lowering of the sea level, it occupied a considerably smaller area than it does at present (Brambati, 1992).

Three areas, with different biological, hydrologic and sedimentologic characteristics can be distinguished (Fig. 1).

In the Northern Adriatic the waters are relatively shallow. Between the estuaries of the Po and Istria, maximum depth does not exceed 40 m, and as far as Pesaro the 10 m isobath is located 4–6 km from the coast (the maximum depth of the Gulf of Trieste is 25 m). This is due to terrigenous sediment input, as shown by the advancement of the Po delta. In the Central Adriatic depth increases gradually, reaching over 300 m in parts of the central trough. The underwater sill of Pelagosa, which links the Gargano to the Dalmation coast and represents the area which the Tremiti Islands emerge from, forms the demarcation. Beyond this lies the Southern Adriatic, where the depth reaches 1590 m.

Along the western side the Northern Adriatic coastline is characterised by lagoons, coastal swamplands and ponds, and wide sandy beaches. In the Central Adriatic from Pesaro to the Gargano the Apennine ridges at times extend as far as the coast.

On the eastern side, from Istria to Dubrovnik, the coastline is steep and karstic, with peninsulas, inlets, deep coves and innumerable islands parallel to the coast. The Albanian coastline is flat, with few bays and extensive marshlands.

The Adriatic differs sharply from the rest of the Mediterranean by virtue of its hydrographic and hydrobiological characteristics. The temperature range in the most northerly areas is far greater than elsewhere in the Mediterranean. In winter, temperatures may fall below 5°C, especially in the presence of the bora, a strong wind blowing from the northeast. Salinity is highly variable throughout the northern basin. Tidal amplitudes show greater fluctuation than in any other region of the Mediterranean; for instance, in Venice and along the coast of Istria, tidal ranges exceed 1 m. In the Central and Southern Adriatic, on the other hand, the maximum range is similar to average levels found in the rest of the Mediterranean (roughly 30 cm) (Hopkins, 1985; Ott, 1992).

Organic matter input from river and the resulting nutrient enrichment leads to elevated primary productivity, making the Adriatic one of the richest fish stock areas of the Mediterranean (Margalef, 1985). Organic pollution is elevated along the western coast, due not only to river transport but also to intense human activity carried out along and near the coast. Pollution levels peak during the summer, especially in the northwestern sector (Marchetti, 1992). The benthic communities in certain coastal areas are typical of unstable environments (Crema et al., 1991).

Tyrrhenian Sea

The Tyrrhenian Sea lies between the Italian peninsula and the three large islands of Corsica, Sardinia and Sicily (Fig. 1), but its geographic boundaries are not well defined and are thus merely indicative. Its northern limit is generally considered to lie between the Cape Corso and the peninsula, following a line stretching roughly 90 km between Cape Corso, the Island of Elba and Piombino. This demarcation line, which separates it from the Ligurian Sea, is however used for administration purposes. The scientifically based distinction takes into account the substantially homogeneous character of the biota of the Tuscan Archipelago and places the line between Cape Corso and the estuary of the Magra River, thereby including the northern islands of the Archipelago. The southern demarcation is usually indicated by the line between Cape Carbonara (Southern Sardinia) and Cape Lilibeo (Western Sicily), which stretches for about 300 km. In addition, the Tyrrhenian Sea links with the Ionian Sea through the Strait of Messina, and with the Sea of Sardinia and the Sea of Corsica through the Bocche di Bonifacio.

The continental shelf extends only for a relatively short distance, and is particularly limited along the coasts of the large islands and Calabria. It extends further in the area of the Tuscan Archipelago and along some of the coastal areas of Latium and Campania. Depths are in excess of 2000 m throughout virtually the whole of the basin, with a maximum depth of 3840 m south of the Pontine Archipelago. The bottom is not uniform due to the presence of ridges and extinct volcanic structures. As well as sediment deposits there are also calcite and pyrite based clayey and sandy deposits, which testify to past volcanic activity (Maldonado, 1985).

The coastline is mainly high and jagged, with wide sandy beaches found only in Tuscany, Latium and Corsica. Vast coastal marshlands can also be found in Corsica, Tuscany, Latium, Campania and Sardinia.

Continental water input is minimal, so salinity remains constant at roughly 38 PSU. Surface water temperature in winter is generally around 14°C, rising to an average of 23°C in summer. This allows the water to exert a mitigating action on the coastal climate (Astraldi et al., 1995).

The waters of the Tyrrhenian Sea are famous for their crystal-clear quality. Even at depths of 100 m, photosynthetic algae can sometimes be found, since the absence of large rivers means that the organic matter load is low. Consequently, primary productivity is low. The biocenoses of the Tyrrhenian are those typical of the Western Mediterranean. However, it has been noted in recent years that warm-water species typical of the eastern and southern Mediterranean basins have tended to move towards the northern sector (Francour et al., 1994).

SEASONALITY, CURRENTS, NATURAL ENVIRONMENT VARIABLES

Adriatic Sea

The water temperature is subject to marked annual variations, particularly in the Northern Adriatic, where temperature fluctuates between 5 and 28°C at the surface and between 12 and 17°C at the bottom. During winter, temperature remains constant at around 5–8°C from the surface to the bottom owing to the shallow depth and the cold bora winds, so the northern coasts do not benefit from the mild temperatures characteristic of the Mediterranean climate. Both the circulation and distribution of water masses in the Adriatic Sea are strongly influenced not only by the morphology of the three Adriatic basins but also (particularly in the Northern Adriatic) by the freshwater inflow of continental origin (Fig. 2) (Franco and Michelato, 1992).

In the Central and Southern Adriatic the water column is divided into three levels: surface, composed of low density waters influenced by the effect of the rivers, especially along the western coasts; intermediate, with high salinity

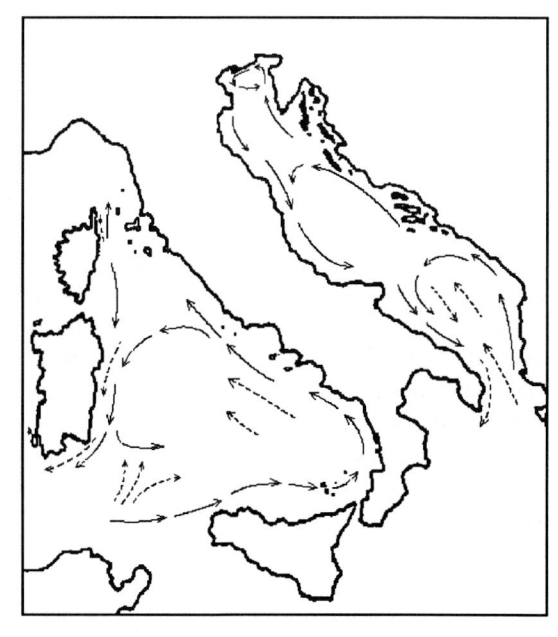

Fig. 2. The circulation in Tyrrhenian and Adriatic Sea. Solid arrows: surface currents; dotted arrows: Levantine intermediate waters.

water of Ionic origin; and deep, consisting of denser waters formed during the winter period. The intermediate level waters (at a depth of 200–600 m) undergo a marked influence from Levantine Intermediate Water (LIW), which has its origin during winter in the Western Mediterranean through the mingling of low temperature (15°C) and high salinity water (39.1 PSU) and subsequently spreads throughout the Adriatic (Buljan and Zore-Armanda, 1976).

In the Central and Southern Adriatic the surface circulation is always cyclonic, with northward and southward movement along the eastern and western coasts, respectively. This gyre is strongly stabilised vertically by the thermal flows at the surface and is sustained by the advection of eastward-moving diluted western waters, as well as by the jagged nature of the coastlines and by the southern waters of the intermediate current. In winter, on the other hand, the circulation generates two semi-stationary cyclonic gyres, one of which is situated north of the Gargano Promontory in the Central Adriatic, and one to the south of this promontory, in the Southern Adriatic. In both areas the pronounced evaporation activity caused by the northeasterly wind results in high density waters in the central area of the cyclonic gyres, with vertical convective processes (Franco and Michelato, 1992).

Salinity, which in the Otranto Channel is on average equal to that of the Mediterranean, namely about 38 PSU, decreases gradually proceeding northwards. In the Gulf of Venice salinity can decrease along the coast during the spring period to as little as 18–20 PSU. This is attributable to the inflow of river waters from the Alps and the Apennines and the phenomenon of underwater springs welling up from subterranean streams that flow down off the karstic plateau.

Nutrient concentration is low in surface waters, while the nitrate/nitrogen ratio is elevated, with a high concentration of silicates particularly in the Levantine intermediate current that flows into the Southern Adriatic and in the deep higher density waters. The main source of nutrients for the surface layers of the Southern Adriatic would appear to derive from convective transfer of Levantine Intermediate Waters originating from the Ionian Sea. Terrigenous sediment seems to play a less important role. Nutrients produced by mineralization processes in the deeper parts of the water column are transported into the Ionian Sea and the eastern basin of the Mediterranean by the deep circulation of denser waters (Buljan and Zore-Armanda, 1976).

In the Northern Adriatic, annual fluctuations in the formation and circulation of water masses are governed by two main factors: the considerable variation between winter and summer heat flows on the surface, and the marked input of freshwater masses into a basin with a rather shallow average depth (Hendershott and Rizzoli, 1976). During winter the cold coastal waters, diluted by river discharge and extremely rich in nutrients, form a band restricted to the coastal areas. This is separated from the off-shore waters by a frontal system, so coastal waters tend to move southwards. Salinity is high in the off-shore area of this basin, as the waters are advected from the southern basins and are actively mixed as a result of surface cooling by strong north winds. Given these conditions of marked instability of the vertical column, masses of high density water are generated. These move towards the Central Adriatic basin, particularly when driven by the bora wind. At the onset of spring a thermocline commences, which subsequently causes the freshwater masses of continental origin to enter the off-shore marine area. This dilution of surface waters intensifies their decrease in density. Water of this type then tends to spread throughout the surface area of the Northern Adriatic (Fonda Umani et al., 1992). The subsequent surface warming, together with the dilution processes and the inflow of water masses from the southern basins, lead to a strong vertical stabilization of the water column during summer, with three clearly distinct layers of water mass separated by pronounced density gradients. The increase in vertical stability of the water column, resulting from the pycnoclines, reduces vertical diffusion and thereby influences transport of dissolved and particulate organic matter and nutrients. Thus in spring and summer, diffusion transport of river and terrigenous input would appear to represent the main mechanism of nutrient transfer to the waters of the basin (Franco and Michelato, 1992).

Tyrrhenian Sea

In winter, the Tyrrhenian Sea, like virtually the whole of the Mediterranean, presents a homogeneous temperature of about 13°C extending from the surface to the bottom. Salinity, which has a mean value of 38 PSU, tends to increase slightly from the surface to the bottom. As the surface temperature increases to around 24–25°C, a surface thermocline is formed whose depth can vary according to the regime of winds and currents. When the warming process comes to an end and wave action intensifies as autumn approaches, the thermocline tends to become deeper and disappear. On account of the scarce inflow of continental waters, nutrient availability does not reach the levels observed in the Northern and Central Adriatic. According to Hopkins (1985), terrigenous phosphorus transfer constitutes roughly one-third of the new production (i.e., not linked to nutrients released through biological decomposition processes) in the Tyrrhenian Sea, while the remaining two-thirds are due to vertical phosphorus transport from deep water to surface layers, mainly as a result of mixing processes driven by convective motion and currents. The influence of freshwater masses of terrigenous origin leads to a marked nutrient increase in certain coastal areas such as the estuary of the Arno and the Tiber Rivers, or in rather enclosed bays or gulfs where there is strong pressure from man-related activities, such as the Gulfs of Naples and Palermo.

The surface Atlantic current feeds a cyclonic gyre in the Tyrrhenian Sea, with a southern branch entering between Sardinia and Sicily and moving northwards along the coastline of the peninsula (Fig. 2). The mass of Atlantic water entering into the Tyrrhenian circulation has been estimated to amount to one-third of the total mass originating from Gibraltar. This water from Gibraltar divides into two main branches when it reaches the Straight of Sicily, one moving in a northeasterly direction and the other towards the southwest (Hopkins, 1985). Upon reaching the Island of Elba the northeasterly current is deflected towards Corsica and travels northwards through the Corsican Channel (between the islands of Corsica and Capraia). The direction of flow in the latter channel is always towards the north, with intensity varying according to season and depth (Astraldi and Gasperini, 1992).

However, this picture only represents the general pattern, as the regime of winds is known to be capable of influencing currents on a small and medium scale. The motion of water masses may thus be complex and difficult to interpret. For instance, in the Corsican Channel, 85% of the total flow occurs during winter and spring, with a high of 1.3×10^6 m³/s as compared to a summer low of 0.2×10^6 m³/s. A close correspondence between wind direction and force in the Gulf of Lyon on the one hand, and flows in the Corsican Channel on the other, has also been demonstrated. As a result of this seasonal trend, the waters of the Tuscan Archipelago and the Tuscan coast come into contact with Tyrrhenian waters during winter, while during the summer, when the flow from the southerly direction is less intense, Ligurian water with a colder average temperature shifts towards these areas (Astraldi and Gasperini, 1992).

Levantine intermediate water penetrates into the Tyrrhenian Sea mainly through the Straight of Sicily, at a depth ranging between 300–400 m, with a northwards and northwestern movement as shown by the high salinity water isolines extending from Sicily to Sardinia. In addition, one branch of this current has been reported to exit from the Tyrrhenian basin off the coast of southeastern Sardinia.

Formation of deep waters comes about through the action of cold dry winds blowing from northern and eastern Europe, while the warm damp winds from North Africa play a major role in formation of high density waters. Such processes mainly involve the waters of the continental shelf, and lead to the formation of water masses which, following the continental slope, move towards the abyssal plain situated in the central–southern area. Evidence for this movement can be obtained from the distribution of water temperature and water oxygen values (Hopkins, 1985). On the assumption that higher temperatures and lower water oxygen values imply older deep water, this author deduced a not altogether surprising general movement towards the Alboran Sea from the Western Mediterranean. The Tyrrhenian deep waters have lower oxygen values, around 4.4 ppm, compared to the values of approximately 4.5 ppm found in the Balearic Basin and about 4.6 ppm in the Ligurio-Provençal basin. Tyrrhenian deep waters remain almost entirely unstratified vertically. A body of deep water flowing into the Tyrrhenian basin has been detected at a depth of roughly 2500 m off the coast of Sardinia (Hopkins, 1985).

THE MAJOR SHALLOW MARINE AND COASTAL HABITATS

Adriatic Sea

The strikingly different characteristics of the western and eastern Adriatic coasts, at least in the northern and central basins, result in a marked divergence between their benthic coastal communities (Vatova, 1949; Crema et al., 1991).

This means that in the Northern and Central Adriatic a fair degree of homogeneity is found. There is a clear-cut distinction between coastal and off-shore habitats found only in the Southern Adriatic and a few more northerly areas of the eastern coastline, where the bottom drops to considerable depths. Much of the Northern and Central Adriatic basin is characterised by sandy and muddy bottoms. Along the coastline of the Central and Northern Adriatic, the bottom presents an extensive area of sedimentation, influenced by continental input. Somewhat further off shore, an erosion zone is also present, in which 'relict sands' can be found (Stefanon, 1984). In the most northerly area there is greater biocenotic heterogeneity due to the presence of a hard bottom (beach-rocks), which hosts an interesting biocenosis, and of vast areas of coarse sand characterised by *Branchiostoma lanceolatum*.

The extensive coastal lagoons are of considerable ecological and biogeographic interest. Several endemisms are known, some of which are of sarmatic origin, such as the bryozoan *Tendra zostericola* and the crustacean *Heterotanais gurneyi*, both of which are typical of the Black Sea (Sacchi et al., 1983). In contrast, the area to the south of the Po mouth is characterised by a much more pronounced biocenotic continuity, in which the different habitats are distinguished primarily by the extent of coastal input. Thus there is a gradient extending outwards from the Po mouth in which the trophic conditions of the system gradually decrease in a southwards direction, modifying the characteristics of the coastal habitat (Fig. 3) (Marchetti, 1992).

The Northern Adriatic communities are characterised by the absence of typical warm-water Mediterranean species, presenting instead a number of other species that represent the relicts of temperate–cold affinity assemblages that invaded the Mediterranean during the last glacial age. Among these, mention should be made of the phaeophycean *Fucus virsoides*, the only species of the *Fucus* genus found in the Mediterranean, the gasteropod *Littorina saxatilis* and the proseriate *Monocelis tenuis*, which are common along the European Atlantic coasts but absent in other areas of the Mediterranean (Cognetti, 1994).

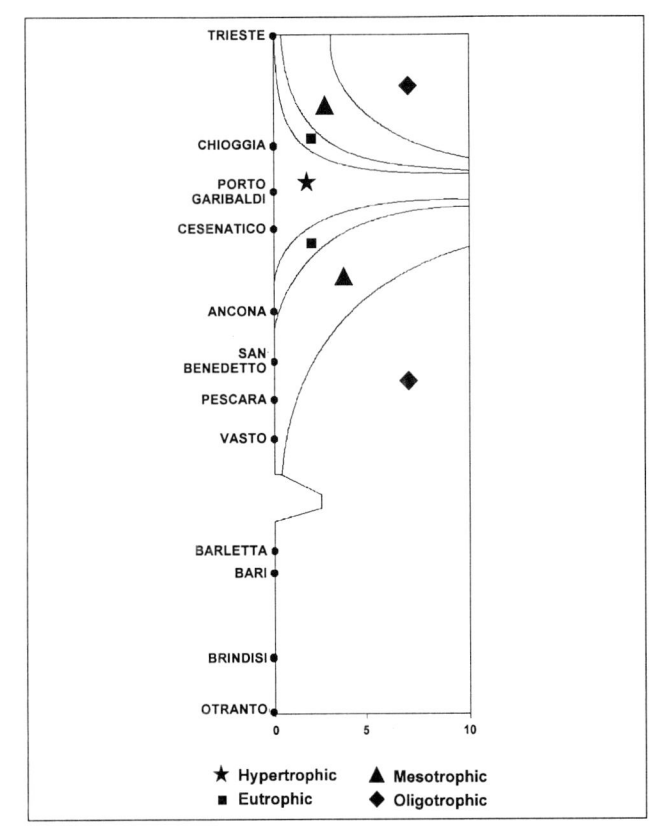

Fig. 3. Indicative trophic conditions of the seawater along the Adriatic coast from Trieste to Otranto.

★ Hypertrophic ▲ Mesotrophic
■ Eutrophic   ◆ Oligotrophic

themselves. The presence of islands of hard substrates with elevated levels of biodiversity on the vast expanse of muddy-sandy bottom creates abrupt discontinuities in the ecological pattern.

Tyrrhenian Sea

Although the Tyrrhenian Sea is not bounded by any clear-cut geographic demarcation, it is fairly isolated within the western basin of the Mediterranean on account of its limited exchanges, which take place mainly through the Corsican Channel and the Sardinian Channel (neither of which reach to great depths). A further separation is constituted by the mountain chains of the Alps and the Apennines, which protect it from the influence of meteorological events and keep the surface temperatures of this basin fairly elevated throughout the year. By virtue of these complex circumstances the Tyrrhenian Sea is the sector inhabited by the most typical Mediterranean biocenoses, characterised by the presence of warm-water species and a high percentage of subtropical affinity species (Astraldi et al., 1995). Species distribution is characterised by a marked latitudinal gradient; some species, such as the chlorophyta *Penicillus capitatus*, the cnidarian *Astroides calycularis*, the polychaete *Hermodice caruncolata* and the labride *Thalassoma pavo*, are common along the southern coasts, becoming rarer further towards the northern part of this basin. In recent years, however, some of these warm water species have been reported to be shifting gradually northwards, a phenomenon thought to be correlated with rising temperature in the waters of the western Mediterranean (Béthoux et al., 1990; Francour et al., 1994).

The Tyrrhenian coastline is characterised by an alternation between rocky and sandy habitats. In addition, there are certain areas with exclusive habitats that are quite unique: the sandbanks and vast shoals along the Tuscan coastline (Meloria and Vada), the numerous archipelagoes (Tuscan, Pontino, Eolian and La Maddalena), and the straits of Messina and Bonifacio. The hard bottom populations are undoubtedly among the most important in the whole of the Mediterranean, given the extension of rocky coastlines both along the peninsula itself and on the surrounding islands. Of particular interest is the 'trottoir', one of the most typical components of the wave exposed intertidal zone of the Tyrrhenian Sea. The 'trottoir' is formed predominantly by the calcareous matrix of the calcified alga *Lithophyllum tortuosum*, and is characterised by elevated biological diversity (Riggio, 1990).

Corralligenous formations, which represent another element typical of the Tyrrhenian Sea (Sarà, 1969), can develop on rocky or sandy bottoms, but equally on a pebbled bottom, at depths ranging between 20–130 m or even deeper, depending on the clarity of the water. In its diversity and wealth of different species, this habitat has been compared to the tropical coral reefs (Ros et al., 1985). A major role in these formations is played by bioconstructor calcareous algae such

In the southern basin, the overall hydrologic and climatic conditions together with the greater diversification among habitats result in a general picture more closely resembling that of the Tyrrhenian Sea. Rocky coastlines are interspersed with sandy shores, and the environmental features are therefore more varied, and also feature the increasingly influential *Posidonia oceanica* meadows and coralligenous biocenoses in the more southerly areas.

Particularly significant is the beach-rocks habitat (Stefanon, 1984). Stefanon reported the discovery of areas of rocky bottom enclosed within the surrounding soft bottom, labelling these rock formations as "beachrocks" due to their analogous nature with similar structures found along the Californian coast. The size and morphology of such beachrocks varies widely: some are extremely small, others are larger, at times extending for thousands of square meters. Stefanon suggested that beachrocks owe their origin to the cementation of sediments that contained embedded mollusc thanatocenoses, with an iso-orientation pointing to their origin as deposits of shoreline debris. Since the beachrocks appear to be arranged primarily in lines parallel to the coast, the hypothesis that has been put forward is that they represent the remnants of ancient coastlines. In addition to such structures, one also finds extensive areas with organogenic-detritic formations which, in contrast to beachrocks, are of biological origin and have arisen on the support provided by the beachrocks as *Mesophyllum lichenoides*, *Lithophyllum expansum* and *Peyssonnelia rosa-marina* and by sessile animals of various groups (Sarà, 1969; Sartoretto et al., 1996).

One of the most important elements of soft bottom habitats is represented by the meadows of *Posidonia oceanica*, endemic to the Mediterranean, and *Cymodocea nodosa* (Cinelli et al., 1995). These phanerogams live on soft bottoms from 1 m down to a depth of about 40 m. In many areas of the Tyrrhenian Sea, and in particular in the Tuscan Archipelago and the Gulf of Naples, *P. oceanica* grows in vast meadows which play an extremely important ecological role. Recent research has shown that many of the major meadows, especially in the coastal area, are affected by regression phenomena, partly attributable to increased water turbidity and to fine matter sedimentation (Marbà et al., 1996; Balestri et al., 1998). Moreover, typical sublittoral algal communities in many areas are being replaced by various species of algal felts (*Polysyphonia flocculosa*, *P. opaca*, *P. subulata*, *Boergeseniella fruticolosa* and *Acrothamnium reyssi*) that exhibit particularly elevated resistance to sedimentation (Verlaque, 1994; Airoldi et al., 1995).

The lagoon-like environments occurring along the Tyrrhenian coast constitute a unique scientific and economic heritage. The biocenoses of these basins display certain features attributable not only to the physical environment but also to the marine biocenoses that lie beyond the lagoon area. Thus certain species succeed in colonizing some of the lagoons, thereby giving rise to populations with their own distinct characteristics as compared to the original species. Many of these basins are highly polluted and are affected by severe environmental decay, as a result of inadequate control over use of the natural resources and irrational exploitation for production purposes (Cognetti 1994; Lardicci et al., 1997).

OFFSHORE SYSTEMS

Adriatic Sea

No offshore system can be defined for the Adriatic, given its geomorphological characteristics.

Study of the distribution of plankton in the Adriatic shows that only the central zone of the Central-Southern Adriatic is characterised by an 'oceanic' community. This community is subject to seasonal fluctuation, extending as far as the Gargano-Kotor line during summer, and up to the Ancona-Sibenik line during winter. Throughout the year the phytoplankton is dominated by diatoms, while in the microzooplanktonic community the relation between tintinnid and non-tintinnid ciliata is tilted in favour of the latter. In the mesozooplanktonic community, which is characterised by an elevated degree of diversity and stability, herbivorous copepoda (*Paracalanus parvus*, *Ctenocalanus vanus*, *Clausocalanus*) predominate throughout the year. The carnivorous species *Oithona helgolandica*, *Centrophages typicus*, *Oncaea* and *Corycaeus* are widely represented (Fonda Umani et al., 1992).

The Northern Adriatic is characterised by neritic plankton throughout its extension. Off-shore, the plankton presents a moderate biomass, while biomass is elevated near the coast but has low diversity. Only in special conditions do neritic species of the Southern Adriatic such as *Isaia clavipes* and *Codonellopsis schabii* populate the northern areas. *Tintinnopsis*, *Acartia* and *Penilia avirostris* are frequent. Throughout the area, production of nanoplankton is dominant and represents an important source of nutrition for filter feeders such as *Penilia avirostris* and Thaliacea (Fonda Umani, 1996).

Primary productivity is extremely high, especially in the north and centre, while values in the Southern Adriatic are similar to those found in the Tyrrhenian Sea. Productivity values for the whole basin reach a mean value of 240 gC/m^2 (Antoine et al., 1996).

Bathyal fauna is found only in the centre of the Southern Adriatic (maximum depth 1590 m), with characteristics similar to those of other Mediterranean basins. No abyssal fauna is present.

Tyrrhenian Sea

The zooplankton of the Tyrrhenian sea shows some characteristics that distinguish it from other areas of the Western Mediterranean. Some species are typical of the Southern Tyrrhenian area, such as *Euphausia krohnii*, while others are indicators of Eastern Mediterranean waters, such as *Stylocheiron suhmii*. Yet others, for instance *Pterosagitta draco* and *Acartia danae*, are of Atlantic origin. The influence of Atlantic waters can also be detected in the central basin of the Northern Tyrrhenian sea, where *Acartia danae* and *Centropages violaceus* are present. As regards copepoda, all the Mediterranean species abound.

Microzooplankton populations, over 50% of which are composed of tintinnids, present a fairly elevated biomass, at times greater than that recorded in the Central and Southern Adriatic (Fonda Umani and Monti, 1990).

Primary productivity values for the whole basin reach a mean value of 140 gC/m^2 (Morel and André, 1991). It was shown by Innamorati et al. (1998) that a generalised change in the Tyrrhenian waters has been taking place since 1991, with a shift from the nitrogen deficiency observed in the 1970s–1980s (N/P = 4–7) to phosphorus deficiency, due to an increase in nitrates and a slight decrease in phosphates (N/P = 20–30). This change probably derives from the inflow of Levantine Intermediate Waters (LIW, 200–800 m), in which the N/P ratio has risen to roughly 36 accompanied by an increase in temperature and salinity. Concomitantly with this new situation has come production of mucilaginous aggregates and a change in the regularity of seasonal phytoplankton blooming cycles.

In the Central-Southern Tyrrhenian Sea, depths are in excess of 3500 m. Abyssal fauna can therefore be found in the centre of the basin, although it is considerably poorer than that of the Atlantic (as is generally observed in the Mediterranean), in particular for the macrobenthos. Polychaetes are much better represented and consist of endemic Mediterranean species that live in the bathyal as well as the abyssal zone, i.e. *Aricidea abyssalis* and *Aedicira mediterranea*. Numerous bathyal species move into the abyssal depths because they are not held back by low temperatures, unlike in the Atlantic. The temperature is known to remain at a constant 13°C in the Mediterranean even at great depths. However, despite the presence of euribate species, the Tyrrhenian does have genuine abyssal fauna composed particularly of meiobenthic species (Pérès, 1985).

HUMAN POPULATIONS AFFECTING THE AREA

Along the Italian Adriatic coastline the most important urban areas, with populations of over 100,000, are Trieste, Venice, Rimini, Ancona, Pescara, Bari and Brindisi. Along the Balkan coast the main cities are Rijeka and Split. But it should be taken into account that the entire basin of the River Po, with over 20–30 million inhabitants, also gravitates towards the Adriatic.

The ancient Venetian Republic, which dominated the Adriatic for many centuries, has left remains of its settlements and constructions along the Dalmation coast. Similarly, the utilization of marine resources over the centuries by the coastal populations has shown the enduring influence of the ancient Venetian traditions, traces of which have come down to the present day both in fisheries and other activities linked to the marine environment. For instance, aquaculture boasts a long-standing tradition; currently there are important production centres in the Italian coastal areas of Friuli, Venetia and Emilia Romagna, as well as in the Limsky channel and the Molostan channel in Croatia. Albania has considerable potential for developing this sector, given the presence of numerous lagoons; contacts with Italy are currently being established in order to promote the modern development of aquaculture in these areas.

Commercial maritime traffic docks at the ports of Venice and Trieste, from whence goods are shipped to central Europe. Bari and Brindisi also provide passenger ferry traffic to the eastern Mediterranean. Rijeka, Dubrovnik and Split have predominantly local traffic. Ancona and San Benedetto del Tronto are the most important fishing ports.

Along the Tyrrhenian coast there are numerous large urban centres with a high density of population (Livorno, Rome, Naples and Palermo); the population also undergoes a substantial increase during the summer period due to the influx of tourists. However, the impact on the hydrographic basin is far lower than that of the Adriatic basin.

Commercial shipping activity in the area is intense, since the Tyrrhenian Sea lies at the centre of the main Mediterranean shipping routes, with major commercial ports at Livorno, Naples, and Palermo. Considerable passenger traffic between the peninsula and the various islands also occurs in other smaller ports. In addition, the numerous seaside resorts with well-equipped tourist harbours lead to very active pleasure-boat traffic.

RURAL FACTORS

The continual influx of nutrients into the marine environment through surface waters has long been a cause for concern, as levels have risen constantly, and frequently exceeded the limits set by the E.U. Nutrient enrichment is due above all to phosphate-based fertilizers, use of which has continuously increased throughout the E.U. Furthermore, effluent from animal-rearing establishments transports microelements such as copper and zinc as well as pesticide and veterinary drug residues. Research is currently being conducted in Italy to determine the effects of animal farming on water, with the aim of maintaining nutrient and micropollutant levels within the limits set by international conventions and safeguarding groundwater quality.

The problem of agricultural pollutants has come to attention ever more forcefully in recent decades, due to marked changes in cultivation and rearing techniques that have resulted in a pronounced improvement in production but have also had dramatic consequences for the natural environment. Attempts are therefore being made to devise a sustainable production system, so that a balance can be achieved between adequate productivity to satisfy the requirements of the population and the need to avoid impact on the aquatic ecosystem (Cianci, 1996).

Despite considerable progress in this direction, the agricultural system still exerts a strong polluting pressure on the Adriatic basin and, albeit to a lesser extent, on the Tyrrhenian basin. The greater pollution load in the Adriatic can be explained by the fact that the waters of the hydrographic Po basin, which drain into the Adriatic, transport pollutants from crops, intensive rearing establishments (dairy and meat cattle, hog farms, chicken farms) and from the related industrial activities also present in the Po basin. This leads to continuous discharge into the waters of substantial quantities of nitrogen of organic and inorganic origin, as well as soluble forms of phosphorus, potassium, sulphur, heavy metals (copper and zinc), pesticides and pharmaceuticals (drug residues and their metacatabolites). Furthermore, the great concentration of modern intensive animal-rearing establishments in Emilia Romagna has also had an adverse effect on the phreatic waters on account of nitrate leaching or percolation caused by rainfall.

The Tyrrhenian Sea, on the other hand, receives water from basins in which intensive agricultural and animal-rearing activities are less concentrated. In addition, its hydrologic condition contributes to limiting accumulation of the above type of pollutants.

In Italy efforts are now being directed towards developing methods to reduce the environmental impact by seeking to improve nutritional efficiency in animal production. Thus innovative procedures designed to modify protein fraction digestibility in animal feedstuffs, the availability of phosphorus of plant origin, together with the use of an appropriate amino-acid balance have been put in place. These have successfully reduced nitrogen and phosphorus discharge into waste waters by over 30% in some areas, with no untoward consequences on food production and quality (Cianci, 1996).

COASTAL EROSION AND LANDFILL

Adriatic Sea

The western coastline extends for roughly 1260 km, while the eastern coastline measures over 6300 km as a result of its extremely jagged geomorphological pattern.

Much of the Italian coastline has long been subjected to profound modifications linked to the very varied activities of the coastal populations. In some cases this has had drastic consequences on the landscape, and has wrought fundamental changes in the morphological characteristics of the coastline and the surrounding mainland. It should suffice to mention the demolition of the system of dunes that previously lined extended stretches of the coast and were removed in order to enlarge the beaches and make way for bathing establishments (Zunica, 1987).

Phenomena of coastal erosion have also been increasingly observed, and can be attributed to a number of causes, mainly brought about by human activity. Jetties and mooring facilities for boats and harbours of various sizes have been built, influencing the flow of currents along the coast, with the result that shorelines have receded or become silted up. Furthermore, numerous other activities such as embanking measures at river estuaries, extracting construction materials from river beds, regulating rivers further upstream, or building dams for purposes of irrigation or hydroelectric power have impaired sediment, sand and gravel transport to the sea. This has deprived the shore of many of the materials that are essential to the proper balance of a beach.

In Venetia and Emilia Romagna, vast expanses of swamps and lagoon areas in the alluvial land surrounding the coast have been subject to reclamation schemes. Particular interest centres on the evolution of the Po delta, which has been the focus of repeated and massive maintenance and water regulation engineering projects since the 17th century. As a result of these interventions, the delta has advanced by an average of 70 m a year. The material transported by the river, whose average flow stands at 1500 m^3/s, has been calculated to amount to 13 million tonnes per year. The entire area is characterised by swamps and lagoon fish hatcheries, and the continual construction of dikes has created levees so that the arms of the delta are now overhanging with respect to the sea. Soil subsidence has also occurred in the past few years, at times lowering the land by as much as a metre, with the consequence of water bursting from the soil.

Subsidence phenomena are linked to intensive exploitation of underground water resources and hydrocarbons (Brambati, 1983). In the province of Ravenna the land has sunk by roughly 11.6 mm a year since the 1950s.

Climate represents another important factor: over recent decades more severe storm patterns and the resulting greater imbalance in coastal regimes has aggravated erosion. The beaches of the Central Adriatic have been most severely affected. They exhibit generalised erosion caused mainly by the decreased transport of material from the numerous rivers, many of which are temporally intermittent streams, with alternating regimes of high and low streamflow. In many cases the river waters are utilised upstream; furthermore, the frequent practice of extracting sand and gravel from river beds close to the estuaries has deprived the shoreline of large quantities of sediment. In the Marche and Abruzzo erosion affects not only the sandy beaches of many seaside resorts but also the coastal highways and even the railway that follows the coastline (Zunica, 1987). The Albanian Adriatic coastline is low and characterised by a series of deltas, lagoons, and marshes that have so far been little utilised. No major evidence of erosion has been observed.

Tyrrhenian Sea

The Tyrrhenian coastline has a linear extension of roughly 3700 km, of which roughly 45% is composed of beaches. Thirty-seven percent of the beaches are undergoing erosion while only 6% are advancing. Erosion is often more pronounced in the vicinity of river estuaries, and is to be attributed to causes similar to those already pointed out for the Adriatic: extraction of material from riverbeds, a decrease in river flow due to interruptions located upstream (often involving reservoirs and dams, as is the case in Calabria and Campania), and destruction of the protective belt of dunes. For example, just south of the River Arno the erosion that began to be manifested towards the end of the last century has gradually extended down the coast as far south as Livorno, so that the shore is receding at a rate of roughly 10 m a year. The process of erosion may be aggravated by structures built out into the sea that alter the hydraulic conditions and thereby impede the natural shift of sediments along the coast (VVAA, 1984). Such a phenomenon is observed along the coastline to the north of Naples around the estuary of the River Volturno, or in Northern Tuscany where the port of Marina di Carrara impedes transport of detritus from the River Magra, thereby causing the shoreline south of the estuary to recede. In contrast, around the actual estuary of the Magra, a law forbidding extraction of inert material from the river has resulted in an advancement of the shoreline. Along the Sardinian and Corsican shorelines, with rare exceptions, erosion is virtually non-existent.

A general picture of erosion along the Italian coasts is given in Zunica (1987).

EFFECTS OF URBAN AND INDUSTRIAL ACTIVITIES

Adriatic Sea

Along the Italian Adriatic coast, particularly between Trieste and Rimini, numerous industrial complexes have been set up (shipyards, machine industry, chemical and food industries, petroleum refineries, power stations, etc.). These coexist with intense agricultural and animal-rearing activities, which have undergone particularly marked development in Emilia Romagna and Apulia. Tourism, mainly focusing on seaside resorts and bathing attractions, is widespread on both sides of the Adriatic coast, and on the Italian side it has led to substantial spin-offs in other economic sectors involving the manufacturing and craft industries (pleasure-boat shipyards, furniture-making, the garment trade, etc.), with a number of local specializations (Biagini, 1990).

The most important area for seaside and bathing resorts is situated in the northern provinces, from Monfalcone to Cattolica, where there is an uninterrupted stretch of long sandy beaches that have been developed for mass tourism. Improved communications with Central Europe favour a ceaseless flow of tourists, and routes have now been opened up to tourist flows from Eastern Europe as well. Along the central and southern coastline, the presence of Italian and international tourists has increased constantly in recent years. Overall, it has been calculated that the Adriatic coast accounts for one-third of total foreign presence in Italy (Lozato-Giotard, 1990).

The Dalmation coast also attracts intense tourist activity, partly drawing an elite form of tourism in search of local arts and crafts, with a notable component of Austrians and Germans. During the recent wartime period, Dalmation tourism, which was a strong rival to tourism on the Italian side of the Adriatic coast, went into a steep decline; more recent data suggest that it is now enjoying something of a revival.

Along the Italian coast, maritime traffic can count on a system of 35 ports. The shipping trade mainly involves crude oil and refinery products. Trade with Central Europe takes place above all through Trieste and Venice, and to a lesser extent through Rijeka and Koper. The main crude oil ports are Venice and Ravenna, where important refineries have also been set up.

One remarkable characteristic of the Northern and Central Adriatic is represented by the lagoon ports (Venice, Chioggia) and the canal harbours (Monfalcone, Ravenna, San Benedetto del Tronto, Pescara, etc.). The latter have

Table 1

Data pertaining to motorised fishing craft in 1994 (ISTAT, 1997)

Basin	No.	TSL
Northern Adriatic	2,175	19,806
Central Adriatic	2,232	48,634
Southern Adriatic	1,476	35,741
Adriatic Sea	5,883	104,181
Northern Tyrrhenian	769	11,780
Central Tyrrhenian	749	9,119
Southern Tyrrhenian	1,988	17,553
Tyrrhenian Sea	3,506	38,452
Italy	15,798	245,637

been created by digging canals to a depth needed to cater for the mooring of ships of varying tonnage. Such canal harbours require continual maintenance.

The vast expanse of the continental shelf, the nutrient input and the resulting elevated primary productivity make the Adriatic one of the richest fish stock areas of the Mediterranean. Table 1 shows the most recent data concerning the number of trawlers and their tonnage. The most highly developed sector, as regards both number and tonnage of motor boats (34.8% and 40% of the Adriatic fishing fleet, respectively), is the Central Adriatic, which offers fully equipped modern ports such as Ancona, San Benedetto del Tronto and Pescara. The Northern Adriatic has a greater number of craft (38.8%) but they are of lesser tonnage (8.7 t); the main fishing ports are Grado, Chioggia, Porto Garibaldi and Cesenatico. The Southern Adriatic shows the smallest number of craft (26.4%) but they are of greater tonnage, better equipped for preserving the catch and for operating in other seas as well; the major ports of this area are Manfredonia, Barletta, Trani, Molfetta and Bari.

Intensive and extensive aquaculture is highly developed, with coastal lagoons and small lakes stretching uninterrupted for hundreds of kilometres from Monfalcone to Ravenna and along the coast of Apulia. In the Gulf of Trieste vast marine areas are devoted to shellfish farming. Along the Dalmation coast there are innumerable small coves, and in the rias that stretch many kilometres into the interior, salinity varies on account of the inflow from waterways and the presence of freshwater springs on the bottom. These represent ideal environments for aquaculture. The most important establishments are located on the Limsky channel in Istria and in the Gulf of Malostrom near Dubrovnik. Overall, aquaculture is undertaken in Italy over an area of roughly 130,000 ha, two-thirds of which are within the Adriatic; in Croatia, an area of roughly 14,000 ha is devoted to aquaculture.

Total production of maritime and lagoon fishing in the Italian Adriatic in 1995 is shown in Table 2.

Table 2

Total production in quintals of maritime and lagoon fishing catch in the Adriatic and Tyrrhenian Sea in 1995 (ISTAT, 1997)

Basin	Fish	Molluscs	Crustaceans
Adriatic			
Northern	370,257	340,997	24,748
Central	306,044	196,252	37,441
Southern	505,706	87,923	31,036
Tyrrhenian			
Northern	82,528	97,335	27,845
Central	84,703	31,787	11,760
Southern	118,066	49,645	10,286

Beneath the Adriatic lie natural gas reserves and oilfields whose presence in the Northern Adriatic is due to the geologic unity of this basin. Oilfields are also found in the Central and Southern Adriatic.

Along the entire Italian Adriatic coast there are 70 offshore platforms, all of which are fully automated. Oil or gas flows to the mainland through pipelines fitted with automatic signalling devices, thus allowing continuous control. The boring of test wells and subsequent drilling operations make use of container ships to transport the production waters to specific appropriately equipped ports. So far, no incidents of particular impact on the environment have occurred. Two-thirds of the total quantity of methane extracted in Italy, accounting for 35% of total domestic consumption, derives from the Adriatic.

Economic development over the last 20 years has resulted in intense building activity, often carried out without any form of control along the entire length of the Italian coastline. This has led to severe environmental damage in many areas, blighting the landscape and causing an increase in organic pollution.

Water quality is impaired by a number of factors, above all the excessive quantity of nitrogen and phosphorus discharged into the sea by rivers (the Po alone has a hydrographic basin of 70,000 km^2), as well as the waste waters from the coast, especially during the summer months. These conditions, taken together with the hydrographic characteristics particularly in the Northern Adriatic, can lead to extensive dystrophic phenomena, causing very severe damage both to fishing and to tourism. The quantity of phytoplankton, which normally ranges between 1000 and 100,000 cells per litre in the Northern Adriatic, can reach a density of millions of cells per litre in periods of intense eutrophication, leading to the phenomenon of anoxia during the night. At times the massive presence of toxic species of Dinoflagellata can make it necessary to suspend the sale of bivalve molluscs in order to avoid the risk of food poisoning, which would have adverse economic consequences for the producers (Crisciani et al., 1991).

An additional phenomenon is the presence of mucilaginous aggregates (mucopolysaccharides secreted mainly during the reproduction of diatoms) which invade the Northern and Central Adriatic in certain years and render both fishing and bathing totally impossible. In normal conditions mucilage forms in small quantities on the seabed throughout the Mediterranean at the beginning of spring, and disappears at the beginning of summer. In some years, however, the metabolic activity of diatoms and the resulting production of mucilage continues intensely throughout the summer. This is probably a result of the abnormal increase in nutrients, but may also be related to the occurrence of dry winters and enduring conditions of absolute calms (Fonda Umani et al., 1992; Stachowitsch et al., 1990).

Overfishing can also be a cause of severe damage to benthic communities. An example is provided by the selective harvesting of molluscs, an important activity in this basin. This has caused profound changes in the sandy bottom where psammic bivalves such as *Chamelea gallina* are harvested from boats equipped with devices capable of aspirating the bottom. Equally severe damage has been wrought on the rocky bottom of the Southern Adriatic and the eastern coast, where the bottom has been virtually destroyed over extensive areas as a result of harvesting endolitic bivalves, in particular *Lithophaga lithophaga*. This has led to almost total desertification in some areas (Fanelli et al., 1994).

Tyrrhenian Sea

The main industrial centres are found in Tuscany and Northern Latium (shipyards, machine and chemical industries, petroleum refineries in Livorno, steel mills in Piombino, power stations at Civitavecchia). The Arno and Serchio basins are intensely industrialised (leather tanning, paper mills, furniture industry, etc.). Mining and quarrying industries are also well developed (marble from the Apuan Alps, mercury from Mt. Amiata, pyrite and various minerals from the Colline Metallifere). The southern and insular Tyrrhenian coasts, with the exception of the industrial installations located on the outskirts of Naples and Palermo, have no heavy industry.

Agricultural activity is highly developed above all in Tuscany, Latium and Campania. A number of speciality plants are grown, in addition to traditional crops.

Along the entire length of the coastline of the peninsula and the islands, tourism has reached intense levels of development, with positive spin-offs that have enhanced local arts and crafts as well as boatyards and ship-building for leisure sailing. Tuscany boasts vast expanses of sandy beaches (Viareggio, Forte dei Marmi) which, similar to the Northern Adriatic coast, have been developed for mass tourism and to some extent for package holidays. The smaller islands of the various archipelagoes are extremely popular holiday destinations (Elba, Capri, Ischia, Lipari, La Maddalena); similarly, the Sorrento Peninsula in Campania, the 'Smeralda' Coast in Sardinia, and the beaches of Corsica are renowned for their beauty. Famous throughout the world, these areas are frequented the whole year round and cater mainly for quality tourism, attracting an international élite (Lozato-Giotard, 1990).

Maritime traffic is intense throughout the Tyrrhenian Sea. Not only do numerous ferry companies ply between the various islands and the continent, but there is also flourishing commercial maritime traffic. Furthermore, the Tyrrhenian is traversed by the intensely travelled routes of oil tankers of varying tonnage. The main Tyrrhenian port is Livorno, which is the hub of many different types of trade (commercial, industrial, oil tankers and passenger ships). Additional ports are those of Piombino, Civitavecchia, Naples, Gioia Tauro, Messina, Palermo and Bastia. There are also numerous tourist harbours, some of which are of considerable importance, scattered along the coasts of the peninsula and the smaller and larger islands.

The most important fishing ports are in Sicily, but they are not situated on the coast that faces the Tyrrhenian Sea. Fishery production is low, with levels that fall notably below the production achieved in the Adriatic (Table 2). This is due to the fairly limited extension of the continental shelf and the scant availability of nutrients, which in turn can be attributed to the lack of any major inflow of river water.

Worth mentioning, however, is the sector of the coastal tuna and swordfish industry, which is well developed. In spring, tuna (*Thunnus thynnus*) move towards the coast in large shoals for reproduction. They can thus be caught in great webs of special nets known as 'tonnare volanti' utilised in the open sea. Fixed nets, 'tonnare', which once represented the basic fishing technique and were installed along the coast, have now been largely abandoned. Swordfish (*Xiphias gladius*) show a behavioural pattern similar to that of tuna, living at considerable depth and moving nearer to the coast for reproduction. Fishing activities centring around swordfish are found mainly in the Strait of Messina. Tuna fishing was widespread until quite recently, but over the last few years a steep decline in production has been observed. Swordfish are also becoming scarce, owing to the use of driftnets and illegal capture of undersized young fish.

Aquaculture, which is mainly intensive, is undertaken in Tuscany in the marshland near Castiglione della Pescaia, and in the Orbetello lagoon, as well as in the brackish coastal lakes and ponds of Latium, Campania and Sardinia. Sicily has a vigorous development both of intensive and extensive aquaculture in the semi-enclosed body of water in and around the inlet of Marsala on the west coast. In Corsica, aquaculture has been developed in the vast swampy lake of Biguglia.

Oil prospecting is carried out in several areas facing the coasts of Latium and Campania, and particularly also in the area bounded by the Strait of Sicily and the Tyrrhenian Sea on the continental shelf off northwestern Sicily.

Several areas of interest for mineral extraction have been identified on the Tyrrhenian bottom: ferromagnetic ore deposits near the Island of Elba, iron and manganese in the Eolian Archipelago, titanium-magnetite on the continental shelf of Latium, and metalliferous sands containing zirconium, rutile, and tin south of Sardinia and off the coast of Calabria.

Over the past few decades the Tyrrhenian coasts have experienced a sharp influx of population and a considerable increase in the size and number of settlements. Development has been intense, not only through the expansion of urban areas and resorts that have long exerted a powerful attraction as centres of tourism, but also through the creation of new settlements of considerable size in areas that were virtually uninhabited until very recently (tourist villages complexes, leisure-boat harbours, small towns). In certain cases, such as the 'Smeralda' Coast in Sardinia, these new settlements have achieved international renown. However, unauthorised construction and building development has often resulted in severe blighting of the landscape. In Corsica, on the other hand, exploitation of the coast does not appear to have given particular cause for environmental concern, partly because of the absence of major industrial installations and large cities, and partly by virtue of careful control over construction activity.

It is above all in summer that overcrowding in the coastal areas leads to an increase in the pollution load, which can reach unsustainable levels in some areas. Carefully targeted prevention measures are therefore urgently required. Nevertheless, the organic pollution generally remains localised, and thanks to the hydrographic characteristics of the Tyrrhenian Sea the generalised dystrophic situations observed in the Adriatic do not arise.

As the Tyrrhenian sea is traversed by many of the major oil tanker routes, the coasts are frequently blighted by petroleum pollution, caused mainly by illegal discharge of ballast water used to flush out the cisterns, and to a lesser extent by mistakes or accidents during loading and unloading operations at the ports.

Heavy metal and polychlorobiphenyl pollution naturally constitutes a risk in the heavily industrialised areas (Livorno, Piombino, Naples and Palermo) and around the estuaries of the various rivers (Magra, Serchio, Arno, Tiber, Garigliano and Volturno) that have been affected by the industrial, agricultural and animal-rearing activities present in the territory through which the waters flow. A degree of improvement has been noted in the last few years, since efficient depurators have entered into function.

The presence of cinnabar mines in Tuscany has increased the levels of mercury in the waters and sediment of the Northern Tyrrhenian. Concentration of this metal, as detected by methylmercury levels in deep benthic organisms, exceeds permitted levels. Other metals have also been found in considerable concentration in the waters around the Eolian Islands, where the phenomenon has been ascribed to underwater volcanic activity (Renzoni et al., 1990; Aubert, 1994).

PROTECTIVE MEASURES

Both the Adriatic and the Tyrrhenian have been included within the framework of protective measures set up on the basis of international conventions and implemented through joint actions by the countries bordering on the Mediterranean. Numerous different aspects of environmental protection are encompassed within such measures, which focus on defending the environment against oil spills or discharge of toxic pollutants into the sea, providing safeguards for endangered species, taking measures to counter the reduction of fish stocks, and so forth. Other projects set up mainly on a local basis are concerned with depurators, containment of coastal erosion and the establishment of protected marine areas.

Considerable progress has been made in designing waste-purification technology in the areas most heavily affected by human activity. Nitrogen and phosphorus abatement is achieved through biological purifying plants that make use of activated sludge. In recent years a substantial number of such plants have been set up, leading to a marked improvement in the general situation, particularly in the Northern Adriatic. In addition, the major industrial areas are now provided with multi-phase depurators in which effluent is subjected to several steps of chemical-physical treatment.

Of particular interest is the purification system that has been designed for the lower valley of the River Arno (Tuscany), the site of almost 1000 leather-tanning factories. The pollution load of leather-tanning wastes involves not only chromium and sulphides but also an elevated COD, partly characterised by bioresistant substances (polyphenols, sulphonates). The new purifying plants are among the most modern in the whole of Europe, and have an efficiency of over 98%. They are kept under continuous control, given the naturalistic importance of the Arno and its estuary (Taponeco et al., 1997).

The balance and dynamics of Italian coastal areas is regulated by Act of Law No. 183 of 18 May 1989, "Rules for the organizational and functional restructuring of land protection". This law concerns mining, gravel and sand extraction and provides for reconstruction of the dune belts.

In the Adriatic, interdisciplinary research has long been a driving force and has made a crucial contribution to knowledge of oceanography and marine biology. Proposals based on data concerning the current ecological situation have also been put forward, aimed at water clean-up and improving fishery production. The main academic institutions engaged at present in research on the Adriatic are the Universities of Trieste, Venice, Padua, Modena, Bologna, Bari, Zagreb and Vienna. There are also a number of important Research Stations forming part of other Italian bodies such as the National Council of Research in Trieste, Venice, Bologna, Ancona, Lesina; a similar function is fulfilled by the Institutes of Dubrovnik and of Roviny in Croatia, and that of Piran in Slovenia. For the Tyrrhenian basin, the research centres that have made the greatest contribution to scientific research in this field are the Zoological Station of Naples and the Thalassographic Institute of Messina. The University Institutes of Pisa and of the other universities in Tuscany, Liguria, Sicily, Latium, Sardinia and Corsica should also not be overlooked.

Both in the Adriatic and the Tyrrhenian areas, measures designed to establish protected marine areas have already been in place for some time. The aim pursued by such measures is to safeguard still intact biocenoses and to protect reproduction and recruitment areas of valuable species, as well as instituting biological rest areas to improve fishery resources (Cognetti, 1991).

Along the Italian Adriatic coast the only marine park that has so far entered fully into function is the one situated at Miramare in the Gulf of Trieste, which is managed by the WWF. Although well organised as an educational and marine environment control centre, it is of very small size (little more than 1 km along the promontory and reaching a depth of 18 m, which represents the limit of the external perimeter). Other marine parks are currently being set up on the Tremiti Islands and the coast of the Gargano Promontory.

On the eastern coast, Slovenia has an integral reserve at Strunjan, which includes the entire peninsula (5 km^2), and a marine area of 7 km^2 linked to the oceanographic station of Portorose. Along the Croatian coast there are numerous protected areas that have been set up in pursuit of various different goals: the special natural reserve of the Limsky canal, which includes the entire rias (10 km long and 600 m wide) and is designed to encourage production, the integral nature reserve of Lokrum to which the biological station of Dubrovnik is linked, the special nature reserve of Maloston, a long narrow rias devoted to bivalve rearing, and the Mljet park (Plavsich-Gojkovic, 1983; Cognetti, 1991).

A number of marine parks (the Park of the Tuscan Archipelago, the Eolian Islands Park, the La Maddalena Archipelago Park) have also been instituted along the Italian Tyrrhenian coast, but to date they have not been fully brought into function. The marine park of the island of Ustica, set up in 1986, constitutes a special case. It includes the marine area surrounding the island up to a distance of three miles from the coast, and is subdivided into an integral protection area, where all commercial activity is banned, and other areas with diversified regulations where a variety of ecocompatible commercial activities can be carried out. The island also houses a marine biology laboratory linked to the University of Palermo.

On the Tyrrhenian coast of Corsica lies the Lavezzi nature reserve, which embraces 5000 ha of sea and land. Set up in 1982, it resembles the Ustica park in having diversified regimes, and is also the site of a marine biology laboratory. In addition, the Lavezzi park presents one of the most typical French institutions, the 'cantonnement de pêche', marine areas protected specifically in order to allow repopulation of commercial species. The Tyrrhenian coast of Corsica has three such areas: the cantonnement of Bastia, extending over 791 ha, instituted in 1977; the cantonnement of Porto Vecchio, 1000 ha, instituted in 1978; the cantonnement of Bonifacio, which encompasses the entire bay and extends for a length of 8 km, instituted in 1983 (Augier, 1985).

A project is also under way for the creation of a transboundary park in the Bocche di Bonifacio, which would include the Lavezzi marine park and the recently instituted park of the La Maddalena Archipelago (Cognetti, 1993).

Despite the improved ecological conditions achieved over the last 10 years in a number of risk areas (thanks to the introduction of more efficient depurators and effective prevention measures), many dire problems remain. All too many issues crucial for safeguarding the marine environment and its resources frequently cannot be resolved due to the lack of rational management of the coasts. Certainly, there is growing awareness of the need for in-depth study of the ecological characteristics both of the Tyrrhenian and the Adriatic Sea, with investigation addressing virtually all branches of oceanography and basic and applied marine biology. High-quality research in these areas is being conducted by teams of scientists affiliated with the numerous academic and research institutions mentioned above. However, the management criteria and policies enacted for protection of the marine environment may be substantially biased by country-specific factors. For instance, in the case of Italy, the measures that have been taken for the conservation and management of protected areas are unsatisfactory. Thus dozens of new parks have been set up along the coastline, yet so far none of them are fully functioning, because no provision has been made to accommodate local needs. In other words, these conservation projects are not set within a broader context of environmental management that would make it possible to reconcile conservation with local requirements, in the overarching framework of sustainable development. A glance at the French approach and that of other countries bordering on the Mediterranean highlights the way in which proper management of the marine environment must be based on appropriate policy choices and on cooperation between research bodies and the public administration.

STRAIT OF MESSINA

The physical–chemical properties and certain biological aspects of the waters in the Strait of Messina are unique to this special part of the Mediterranean. The Strait presents a very limited coastal strip, with dramatic underwater morphological contrasts such as escarpments and faults. The Ionian and Tyrrhenian basins have opposite tidal cycles and the intense tidal currents alternately flow into both basins. These tidal currents can reach speeds of up to 5 knots on the bottom and 6–7 knots on the surface, creating an extremely complex hydrodynamic structure descending to a depth of over 1400 m. Moreover, powerful vertical currents of cold water from the bottom of the Ionian Sea (1000 m in depth) rise to 75–130 m below the surface along the walls of the sill of the Strait. On account of the strong currents, the bottom of the sill of the Strait, despite its great depth, shows many similarities with coastal deposits, including rocky pinnacles and coarse sediment (sands and gravels). Deep water movements result in classical upwelling phenomena, leading to nutrient enrichment. Seasonal proliferation of plankton is observed at the northern opening of the Strait, as well as the rise of abyssal species (e.g. *Myctophid* fishes), many of which are found beached. The vast areas of rocky bottom and the crystal-clear water allow algal assemblages to extend down to depths in excess of 100 m. Animal communities are dominated by epibenthic species, especially filter feeders that are arranged over several layers of epibiontic organisms (the coral *Errina aspera* and the cirriped *Pachylasma giganteum*). Many species are of great ecological and biogeographic interest. They reveal marked affinity with Atlantic populations and they could be considered as relict species. In addition, there is an interesting presence of Mediterranean endemisms and Atlanto-Mediterranean elements. The Atlantic affinity species are characterised by large-sized rheophile algal species such as *Desmarestia dresnayi*, *Laminaria ochroleuca* and *Saccorhiza polyschides*. There are also animal species with very limited distribution (e.g. *Errina aspera*, *Tubbrea micrometrica*). The algae probably penetrated

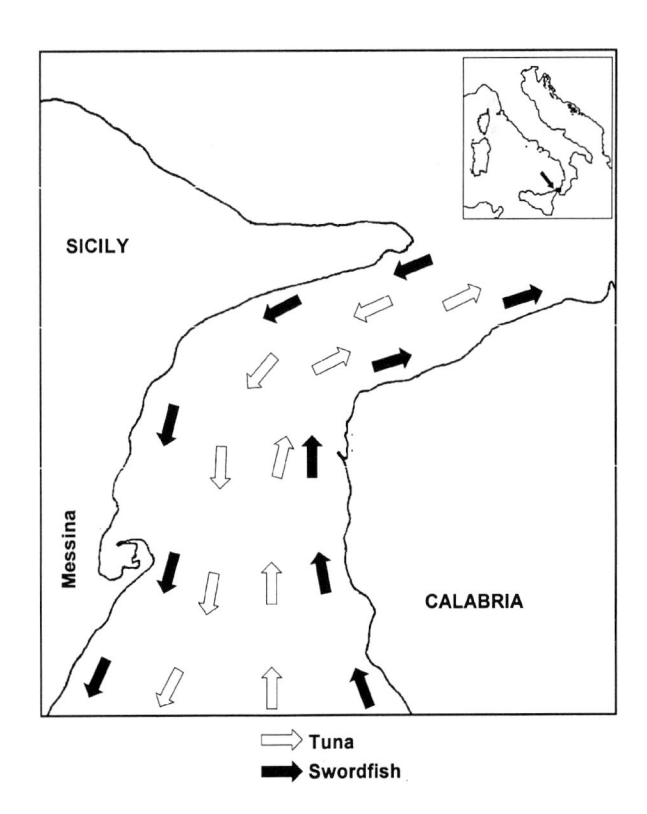

Fig. 4. Migration routes of tuna and swordfish in the Strait of Messina.

into the Mediterranean during the cold periods of the pleistocene, the peculiar hydrodynamic conditions subsequently allowing their persistence in the area of the Strait. Endemic species such as the molluscs *Jujubinus seguenzae* and *Alvania peloritana* also perform an important ecological function. Furthermore, on account of its rheologic features, the Strait plays a crucial role as a centre for dispersal of the reproductive elements of important species forming part of Mediterranean biocenoses (Cognetti et al., 1992; VVAA, 1995).

The Strait of Messina is a transit route for many cetaceans, and represents an obligatory route for the migrations of large pelagic fishes such as *Xiphias gladius*, *Thunnus thynnus* and *T. alalunga* (Fig. 4). In addition, it seems to be the main reproduction area for swordfish. Fishery activities centring around the latter species—a veritable symbol of the Mediterranean fishing tradition—date back to extremely ancient times and still represent an important source of fishing production in this area today (Cognetti et al., 1992).

EXOTIC SPECIES

Recent years have seen a mounting number of reports of species originating from other biogeographic regions in Adriatic and Tyrrhenian Sea (Zibrowius, 1991). One of the most striking examples is that of the bivalve *Scapharca inaequivalvis*, a species of Indo-Pacific origin, which first appeared in the 1980s along the coast of the Northern and Central Adriatic. It has since invaded the coasts and lagoons, developing into considerable biomasses. In numerous areas it has replaced many of the native species that were characteristic of soft bottom coastal biocenoses such as *Cerastoderma glaucum* and *Tapes decussatus*. This successful colonization by *S. inaequivalvis* is related to its high tolerance of the oxygen depletion that has occurred during the repeated dystrophic crises in the Northern Adriatic since the late 1980s. *S. inaequivalvis* is able to reduce its basal metabolism to minimum levels, and can thus survive even in conditions of drastically reduced oxygen availability, thanks to the fact that its respiratory pigment consists of myoglobin, a much more efficient oxygen transporter than the hemocyanins present in other Adriatic bivalve species (Cognetti and Cognetti, 1994).

Intrusion of allochthonous species has also affected Adriatic lagoon environments. For instance, the hydrozoan *Garveia franciscana* and the bryozoans *Tendra zoostericola* and *Tricellaria inopinata* have appeared in the Venice Lagoon, and some of these have succeeded in reproducing and expanding their colony to the point of becoming the dominant element of such communities (Sacchi et al., 1983). It has also been reported that species introduced into controlled lagoon areas and fishing grounds for purposes of aquaculture have at times become dispersed throughout these lagoons, and are now invading vast marine areas.

This is the case with the crustacean *Penaeus japonicus*, and also the bivalve *Crassostrea gigas*, the latter having by now replaced the native species *Ostrea edulis*. Equally harmful has been the introduction of the bivalve *Tapes philippinarum*. Initially used in preference to the native *T. decussatum* on account of the higher growth rate and greater size of *T. philippinarum* adults, in many areas this introduced species is now replacing the native species (Cognetti, 1994).

In the Tyrrhenian sea certain algal species of tropical origin are becoming widespread. Thus *Caulerpa taxifolia*, first reported along the coasts of Provence about fifteen years ago and probably introduced directly by man, is now spreading throughout the Tyrrhenian Sea, and more recently the Adriatic Sea, together with *Caulerpa racemosa*, a typically lessepsian species (Piazzi et al., 1994; Ceccherelli and Cinelli, 1997). *Polysiphonia setacea*, which is probably also of tropical origin, is present in the Tyrrhenian with populations capable of adapting to mean temperatures that are decidedly lower than in their native area. This species is now colonizing vast rocky areas at the expense of the typical native species (Airoldi et al., 1995). *Desdemona ornata*, an austral polychaete species, was first reported in the brackish microhabitats of the Island of Elba (Lardicci and Castelli, 1986) and is now spreading to other lagoon environments of Tyrrhenian and Adriatic.

REFERENCES

Airoldi, L., Rindi, F. and Cinelli, F. (1995) Structure, seasonal dynamics and reproductive phenology of a filamentous turf assemblage on a sediment influenced, rocky subtidal shore. *Botanica Marina* 38, 227–237.

Antoine, D., Morel, A. and Andrè, J.M. (1996) Algal pigment distribution and primary production in the eastern Mediterranean as derived from coastal zone color scanner observations. *Journal of Geophysical Research* 100, 16193–16209.

Astraldi, M. and Gasparini, G.P. (1992) The seasonal characteristics of the circulation in the North Mediterranean Basin and their relationship with the atmospheric–climatic conditions. *Journal of Geophysical Research* 97, 9531–9540.

Astraldi, M., Bianchi, C.N, Gasparini, G.P. and Morri, C. (1995) Climatic fluctuations, current variability and marine species distribution: a case study in the Ligurian Sea (north-west Mediterranean). *Oceanologica Acta* 18, 139–149.

Aubert, M. (1994) *La Méditerranée. La Mer et les Hommes*. Les Edition de l'Environment, Paris.

Augier, H. (1985) Protected Marine areas. The example of France: appraisal and prospects. *Council of Europe. Nature and Environmental Series* 31, 1–44.

Balestri, E., Piazzi, L. and Cinelli, F. (1998) Survival and growth of transplanted and natural seedlings of *Posidonia oceanica* (L.) Delile in a damaged coastal area. *Journal of Experimental Marine Biology and Ecology* 228, 209–225.

Béthoux, J.P., Gentili, P., Raunet, J. and Taillez, D. (1990) Warming trend in the western Mediterranean deep water. *Nature* 347, 660–662.

Biagini, E. (1990) *La riviera di Romagna: sviluppo di un sistema regionale turistico*. Patron ed. Bologna.

Brambati, A. (1983) I problemi del mare Adriatico. *Atti Convegno Internazionale Università Trieste*: 117–132.

Brambati, A., (1992) Origin and evolution of the Adriatic Sea. In: *Marine Eutrophication and Population Dynamics*, eds G. Colombo, I. Ferrari, V. Ceccherelli and R. Rossi, pp. 327–346. 25th European Marine Biology Symposium.

Buljan, M. and Zore-Armanda, M. (1976) Oceanographic properties of the Adriatic Seas. *Oceanography and Marine Biology Annual Revue* 14, 11–98.

Ceccherelli, G. and Cinelli, F. (1997) Short-term effects of nutrient enrichment of the sediment and interactions between the seagrass *Cymodocea nodosa* and the introduced green alga *Caulerpa taxifolia* in a Mediterranean bay. *Journal of Experimental Marine Biology and Ecology* 217, 165–177.

Cianci, D. (1996) Attività agricole e inquinamento delle acque. Recupero e gestione della fascia costiera. *Atti della VII Rassegna del mare, Mare Amico,* Roma, 135–140.

Cinelli, F., Fresi E., Lorenzi, C., and Mucedola, A. (1995) *La Posidonia oceanica*. Rivista Marittima. Roma.

Cognetti, G. (1991) Marine reserves and conservation of Mediterranean coastal habitats. *Council of Europe. Nature and Environmental Series* 50, 1–86.

Cognetti, G. (1993) Transboundary marine parks in the Mediterranean. *Marine Pollution Bulletin*, 26, 292–293.

Cognetti, G. (1994) Colonization of brackish waters. *Marine Pollution Bulletin* 10, 583–586.

Cognetti, G. and Cognetti, G. (1994) *Inquinamenti e protezione del mare.* Calderini ed., Bologna.

Cognetti, G., Cinelli, F., Castelli, A., Curini-Galletti, M., Lardicci, C., Abbiati, M., Maltagliati, F., Benedetti-Cecchi, L. and Airoldi, L. (1992) *Attraversamento in alveo dello Stretto di Messina. Studio di impatto ambientale. Vol. 4: Benthos*. Centro Interuniversitario di Biologia Marina di Livorno, ENI.

Crema, R., Castelli, A. and Prevedelli, D. (1991) Long term eutrophication effects on macrofaunal communities in Northern Adriatic Sea. *Marine Pollution Bulletin* 10, 503–508.

Crisciani, F., Ferraro, S. and Raicick, F. (1991) Climatological markers on the algal blooms episodes of 1729 and 1991 in the Northern Adriatic sea. *Bollettino di Oceanologia Teorica ed Applicata* 9, 367–370.

Fanelli, G., Piraino, S., Belmonte, G., Geraci, S. and Boero, F. (1994) Human predation along Apulian rocky coasts (SE Italy): desertification caused by *Lithophaga lithophaga* (Mollusca) fisheries. *Marine Ecology Progress Series* 110, 1–8.

Fonda Umani, S. (1996) Pelagic biomass and production in the Adriatic Sea. *Scientia Marina* 60, 65–77.

Fonda Umani, S. and Monti, M. (1990) Distribuzione dei popolamenti microzooplanctonici nell'arcipelago Toscano. In: *Progetto mare, ricerca sullo stato biologico chimico e fisico dell'Alto Tirreno Toscano,* ed Laboratorio di Ecologia, Dipartimento di Biologia Vegetale, pp. 157–260. Università di Firenze.

Fonda Umani, S., Franco, P., Ghirardelli, E. and Malej, A. (1992) Outline of oceanography and the plankton of the Adriatic Sea. In *Marine Eutrophication and Population Dynamics*, eds. G. Colombo, I. Ferrari, V. Ceccherelli and R. Rossi, pp. 347–366. 25th European Marine Biology Symposium.

Franco, P. and Michelato, A. (1992) Northern Adriatic Sea: Oceanography of the basin proper and of the western coastal zone. In: *Marine Coastal Eutrophication*, eds. R.A. Vollenweider, R. Marchetti and R. Viviani, pp. 35–62. Proceedings of an International Conference. Elsevier.

Francour, P., Boudouresque, C.F., Harmelin, J.G., Harmelin-Vivien, M.L. and Quignard, J.P. (1994) Are the Mediterranean waters becoming warmer? Information from biological indicators. *Marine Pollution Bulletin* 28, 523–526.

Hendershott, M.C. and Rizzoli, P. (1976) The winter circulation of the Adriatic Sea. *Deep Sea Research* 23, 353–370.

Hopkins, T.S. (1985) The physics of the sea. In: *Western Mediterranean,* ed. R. Margalef, pp. 100–125. Pergamon Press Ltd.

Innamorati, M., Melley, A. and Nuccio, C. (1998) Cambiamento generale delle caratteristiche trofiche del Mar Tirreno. *Biologia Marina Mediterranea* 5, 41–46.

ISTAT (Istituto Centrale di Statistica) (1997) Settore Agricoltura. Statistiche della caccia e della pesca. Annuario n. 10.

Lardicci, C. and Castelli, A. (1986) *Desdemona ornata* (Banse 1957) (Polychaeta, Sabellidae, Fabricinae) new record in the Mediterranean sea. *Oebalia* 13, 195–200.

Lardicci C., Rossi F. and Castelli A. (1997). Analysis of macrozoobenthic community structure after severe dystrophic crises in a Mediterranean coastal lagoon. *Marine Pollution Bulletin* 7, 536–547.

Lozato-Giotard, J.P. (1990) *Méditerranée et tourisme*. Masson, Paris.

Maldonado, A. (1985) Evolution of the Mediterranean basins and a detailed reconstruction of the cenozoic paleoceanography. In: *Western Mediterranean*, ed. R. Margalef, pp. 17–59, Pergamon Press Ltd.

Marbà, N., Duarte, C. M., Cebrian, J., Gallegos, M.E., Olesen, B. and Sand-Jensen, K. (1996) Growth and population dynamics of *Posidonia oceanica* on the Spanish Mediterranean coast: elucidating seagrass decline. *Marine Ecology Progress Series* 137, 203–213.

Marchetti, R. (1992) The problems of the Emilia-Romagna coastal waters: facts and interpretations. In *Marine Coastal Eutrophication*, eds. R.A. Vollenweider, R. Marchetti and R. Viviani, pp. 21–34, Proceedings of an International Conference. Elsevier.

Margalef, R. (1985) Introduction to the Mediterranean. In: *Western Mediterranean,* ed. R. Margalef, pp. 1–16, Pergamon Press Ltd.

Morel, A. and André, J.M. (1991) Pigment distribution and primary production in the Western Mediterranean as derived and modelled from Coastal Zone Color Scanner observations. *Journal of Geophysical Research* 96, 12685–12698.

Ott, J.A. (1992) The Adriatic benthos: problems and perspectives. In *Marine Eutrophication and Population Dynamics*, eds. G. Colombo, I. Ferrari, V. Ceccherelli and R. Rossi, pp. 367–378. 25th European Marine Biology Symposium.

Pérès, J.M. (1985) History of the Mediterranean biota and the colonization at depths. In: *Western Mediterranean,* ed. R. Margalef, pp. 198–232, Pergamon Press Ltd.

Piazzi, L., Balestri, E. and Cinelli, F. (1994) Presence of *Caulerpa racemosa* in the North-Western Mediterranean. *Cryptogamie, Algologie* 15, 183–189.

Plavsich-Gojkovic, N. (1983) Le parc national sur l'ile adriatique de Mljet. *Rapports et Procés–Verbaux des Commission pour l'exploration scientifique de la mer Méditerranée* 28, 95–96.

Renzoni, A., Chemello, G., Gaggi, C., Bargagli, R. and Bacci, E. (1990) *Methylmercury in deep sea organisms from the Mediterranean.* FAO/UNEP/IAEA Consultation Meeting. La Spezia.

Riggio, S. (1990) Criteri guida per la creazione di parchi marini ed istituzione di riserve costiere in Sicilia. In: *Parchi marini del Mediterraneo, aspetti naturalistici e gestionali*, ed. Chiarella, pp. 171–181. Atti del 1 Convegno Internazionale San Teodoro.

Ros, J.D., Romero, J., Ballesteros, E. and Gili, J.M. (1985) Diving in Blue Water. The Benthos. In *Western Mediterranean,* ed. R. Margalef, pp. 233–295, Pergamon Press Ltd.

Sacchi, C. F., Bianchi, C.N., Morri, C., Occhipinti, A. and Sconfietti, R. (1983) Noveaux éléments pour la zoogéographie lagunaire de la haute Adriatique. *Rapports et Procés–Verbaux des Commission pour l'exploration scientifique de la mer Méditerranée* 29, 163–166.

Sarà, M. (1969) Research on coralligenous formations: problems and perspectives. *Pubblicazioni della Stazione Zoologica di Napoli* 37, 124–134.

Sartoretto, S., Verlaque, M. and Laborel, J. (1996) Age of settlement and accumulation rate of submarine "coralligene" (–10 to –60 m) of the northwestern Mediterranean Sea; relation to Holocene rise in sea level. *Marine Geology* 130, 317–331.

Stachowitsch, M., Fanuko, N. and Richter, M. (1990) Mucus aggregates in the Adriatic Seas: an overview of stages and occurrences. *Pubblicazioni della Stazione Zoologica di Napoli: I. Marine Ecology* 11, 327–350.

Stefanon, A. (1984) Sedimentologia del Mare Adriatico: rapporti tra erosione e sedimentazione olocenica. *Bollettino di Oceanologia teorica Applicata* **2**, 281–324.

Taponeco, G., Giannini, G. and Giberti, G. (1997) *Normativa Ambientale*, ed. Felici, Camera di commercio, Pisa.

Vatova, A. (1949) La fauna bentonica dell'Alto e Medio Adriatico. *Nova Thalassia* **1**, 1–110.

Verlaque, M. (1994) Inventaire des plantes introduites en Méditerranée: origines et répercussions sur l'environment et les activités humaines. *Oceanologica Acta*, **17**, 1–23.

VVAA (1984) *Atlante delle spiaggie italiane*, ed. Selca. Consiglio Nazionale delle Ricerche, Dinamica dei litorali.

VVAA (1995) *The Straits of Messina Ecosystem*, eds. L. Guglielmo, A. Manganaro and E. de Domenico, University of Messina.

Zibrowius, H. (1991) Ongoing modification of the Mediterranean marine fauna and flora by the establishment of exotic species. *Mésogée* **51**, 83–107.

Zunica, M. (1987). *Lo spazio costiero italiano: dinamiche fisiche e umane*, ed. V. Levi, Rome.

THE AUTHORS

Giuseppe Cognetti
*Dipartimento di Scienze dell'Uomo e dell'Ambiente,
Università di Pisa,
Via Volta 6,
I-56124 Pisa, Italy*

Claudio Lardicci
*Dipartimento di Scienze dell'Uomo e dell'Ambiente,
Università di Pisa,
Via Volta 6,
I-56124 Pisa, Italy*

Marco Abbiati
*Dipartimento di Biologia Evoluzionistica,
Università di Bologna,
Via Tombesi dell'Ova 55.
I-48100 Ravenna, Italy*

Alberto Castelli
*Dipartimento di Zoologia ed Antropologia,
Corso Margherita di Savoia 15,
I-17100 Sassari, Italy*

Chapter 18

THE BLACK SEA

Gülfem Bakan and Hanife Büyükgüngör

The Black Sea is the world's largest landlocked inland sea. Almost one third of the entire land area of continental Europe drains into it and, during the last 30 years, the Black Sea environment has suffered a catastrophic degradation from the waterborne waste from 17 countries. Due to natural causes, the sub-halocline waters of the Black Sea are anoxic. In spite of this natural deficiency, the Black Sea has served mankind well in the past through its provision of food resources, as a natural setting for recreation and transportation and even as a disposal site for waste, including perhaps nuclear wastes. In return, it has been exploited and degraded in many ways. Unregulated and unplanned freshwater withdrawal for irrigation purposes, hydro- and thermal-power generation, the use of coastal areas for construction and the many untreated industrial and agricultural wastes discharged into the rivers that drain into the sea have all had detrimental effects on its health.

The large natural river supply of phosphorus and nitrogen, essential nutrients for marine plants and algae, has always made the Black Sea very fertile. The serious degradation it faces now can be explained by a variety of factors ranging from high pollution loads from the rivers that discharge into the sea to improper policies and inadequate management practices. Among the most serious problems is the high level of eutrophication by nutrients from land-based sources. Other factors in the degradation of the marine environment include the introduction of opportunistic species such as the comb jellyfish, *Mnemiopsis leydi*; changes in the hydrological balance caused by construction of dams on major rivers; chemical and microbiological pollution, synthetic organic contaminants, heavy metals, radionuclides, dumping, and oil pollution.

These have caused the environment of the Black Sea to deteriorate dramatically in terms of biodiversity, habitats, fisheries resources, aesthetic and recreational value and water quality. Land-based sources are identified as the primary factor causing the present crisis situation, and this is where research efforts have been targeted.

Black Sea riparian countries have committed themselves to prevent, reduce and control pollution from land-based sources in accordance with Article VII of the Bucharest Convention. Gathering information on the sources of pollution was one of the basic requirements of the Odessa Ministerial Declaration of the Black Sea countries. One of the objectives of the Program for Environmental Management and Protection in the Black Sea, known as the Black Sea Environmental Program, BSEP, is the preparation of a Black Sea Action Plan.

This chapter reviews the Black Sea in environmental terms, its current environmental status and major problems arising from human use of both the sea and its watershed. It also comments on major trends, problems and successes mainly in the light of the Black Sea Environmental Program (BSEP). Following a general review, the southern, Turkish coast is focused upon.

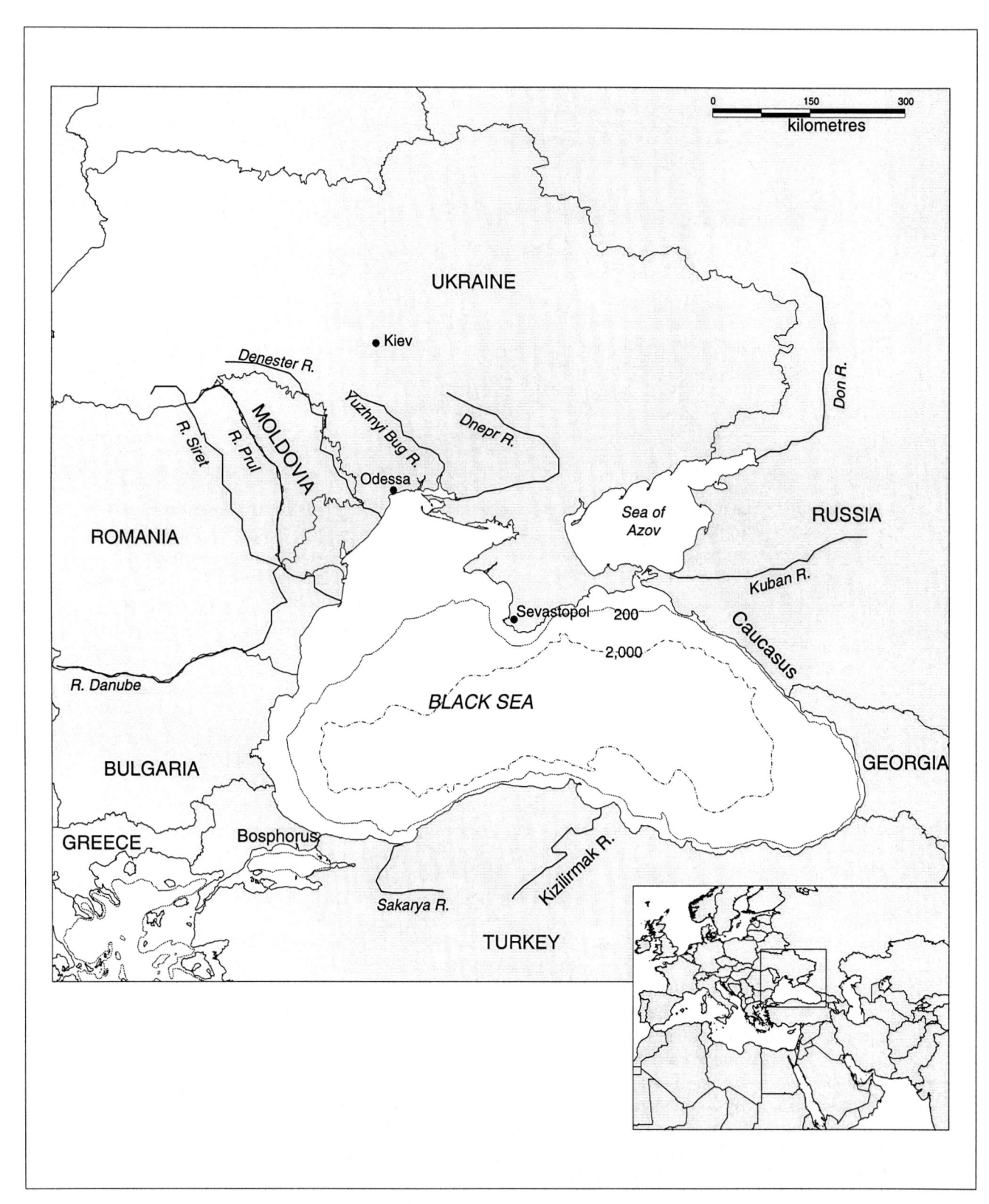

Fig. 1. The Black Sea and its drainage basin.

THE DEFINED REGION

The Black Sea (Fig. 1) is situated between 40°55' to 46°32'N and 27°27' to 41°32'E. To the south, it is connected to the Mediterranean through the Bosphorus, which is the world's narrowest strait, with an average width of 1.6 km, an average depth of 36 m and a total length of 31 km. To the north, the Black Sea is connected with the Sea of Azov through the shallow Kerch Strait, which has a depth of less than 20 m. The Black Sea is surrounded by six countries located in Europe and Asia: Bulgaria, Georgia, Romania, Russia, Turkey and Ukraine. In fact, the Black Sea is influenced by seventeen countries, thirteen capital cities and some 160 million people. Indeed, the second, third and the fourth major European rivers, the Danube, Dniper and Don, discharge into this sea and so almost one third of the entire land area of continental Europe affects it.

It is a semi-enclosed sea, whose only connection to the world's oceans is the narrow Bosphorus Channel. The Black Sea has an area of 4.2×10^5 km^2 with maximum and average depths of 2200 and 1240 m respectively. Further, the Black Sea contains the world's largest anoxic water mass. Below the layer of oxygenated surface water, hydrogen sulphide builds up in the deep water down to a maximum depth of 2200 m. The hydrographic regime is characterised by low-salinity surface water of river origin overlying high-salinity deep water of Mediterranean origin. A steep pycnocline centred at about 50 m is the primary physical barrier to mixing and is the origin of the stability of the anoxic interface (Murray et al., 1989). More precisely, 90% of its water mass is anoxic (Sorokin, 1983).

Murray et al. (1989) reported dramatic changes in the oceanographic characteristics of the anoxic interface of the Black Sea over decadal or shorter time scales. The anoxic, sulphide-containing interface has moved up in the water column since the US cruises in 1969 and 1975. In addition, a suboxic zone overlays the sulphide-containing deep water, and the expected overlap of oxygen and sulphide was not present. It was believed that this results from horizontal mixing or flushing events that inject denser, saltier water into the relevant part of the water column. It is possible that man-made reduction in freshwater inflow into the Black Sea could cause these changes, although natural variability cannot be discounted. Moreover, a long-term increase in nitrate concentration and a concomitant decrease in the silicate and ammonia concentrations in this upper layer are indicative of the considerable changes taking place in the biochemical regime of the Black Sea (Tugrul et al., 1992).

The main sources of pollution in the Black Sea are the rivers which flow into the region. The major rivers flowing into the Black Sea and their discharges are; Danube (203 km^3 yr^{-1}), Dniper (54 km^3 yr^{-1}), Dniesta (9.3 km^3 yr^{-1}), Don (28 km^3 yr^{-1}) and Kuban (13 km^3 yr^{-1}). In addition to these, a large number of smaller rivers along the Turkish and Bulgarian coasts contribute another 28 km^3 yr^{-1} to the water budget of the sea (Balkas et al., 1990).

SEASONALITY, CURRENTS, NATURAL ENVIRONMENTAL VARIABLES OF THE BLACK SEA

Temporal and spatial variability of atmospheric conditions is a distinguishing feature of the Black Sea climate. Meteorological conditions vary in both east–west and north–south directions, particularly during winter when the centre of maximum pressure over Siberia dominates the region. The average pressure is about 1020 mb in the north and 1016 mb in the south. The highest average wind speed, observed in January and February, is about 16 knots. There are significant seasonal changes in air temperatures: the highest average temperatures are 24°C in the central and northern parts and 22°C in the southern parts. The average daily minimum and maximum values in August are about 19°C and 28°C, respectively, along the southern coast of the Black Sea, where regional variations in air temperature can reach 8°C.

The Black Sea is located in a semi-arid climatic zone, and as a result, evaporation (332–392 km^3/yr) exceeds rainfall (225–300 km^3/yr). Runoff (350 km^3/yr) originating primarily from the humid zone to the north, leads to an excess of net freshwater inflow and subsequent dilution of surface sea waters. In the Mediterranean, in contrast, evaporative losses exceed net freshwater input. As a result, the relatively less saline and lighter waters of the Black Sea flow into the Mediterranean over the top of a more saline counterflow (Fig. 2) (see Ünlüata and Oguz 1988; Özsoy et al., 1986).

The predominant semi-permanent elements of the general circulation in the Black Sea consist of a cyclonic boundary current that essentially runs parallel to the basin's periphery, two cyclonic gyres that nearly split the basin area into two and a series of cyclonic and anticyclonic mesoscale eddies that appear to come from the larger-scale features (Fig. 3) (Neumann, 1942). Within the general atmospheric circulation, the Black Sea is under the effects of the Azores anticyclone and Persian Gulf cyclone in summer, and is affected by the arctic low pressure centre and the Asian high centre in winter (Tolmazin, 1985). Satellite imagery and high resolution hydrographic data along the Turkish coast indicate the presence of temporally and spatially variable mesoscale features, and specific organized forms of local nonstationary currents (such as jets, filaments and dipole eddies), embedded within the cyclonic general circulation (Oguz et al., 1992).

Circulation features of the Black Sea examined in September–October 1990 (Oguz et al., 1993) shows a circulation pattern for the upper 300–400 dbar of a cyclonically meandering Rim Current, a series of anticyclonic eddies between the coast and the Rim Current, and a basin-wide, multi-centred cyclonic cell in the interior. Deeper in the intermediate depth (defined here between 500 and 1000 dbar) circulations are weaker and reveal considerable structural variability, with counter-currents, moving eddy centres, coalescence of eddies, and associated recirculation cells. A descriptive synthesis of the upper layer circulation (Fig. 4),

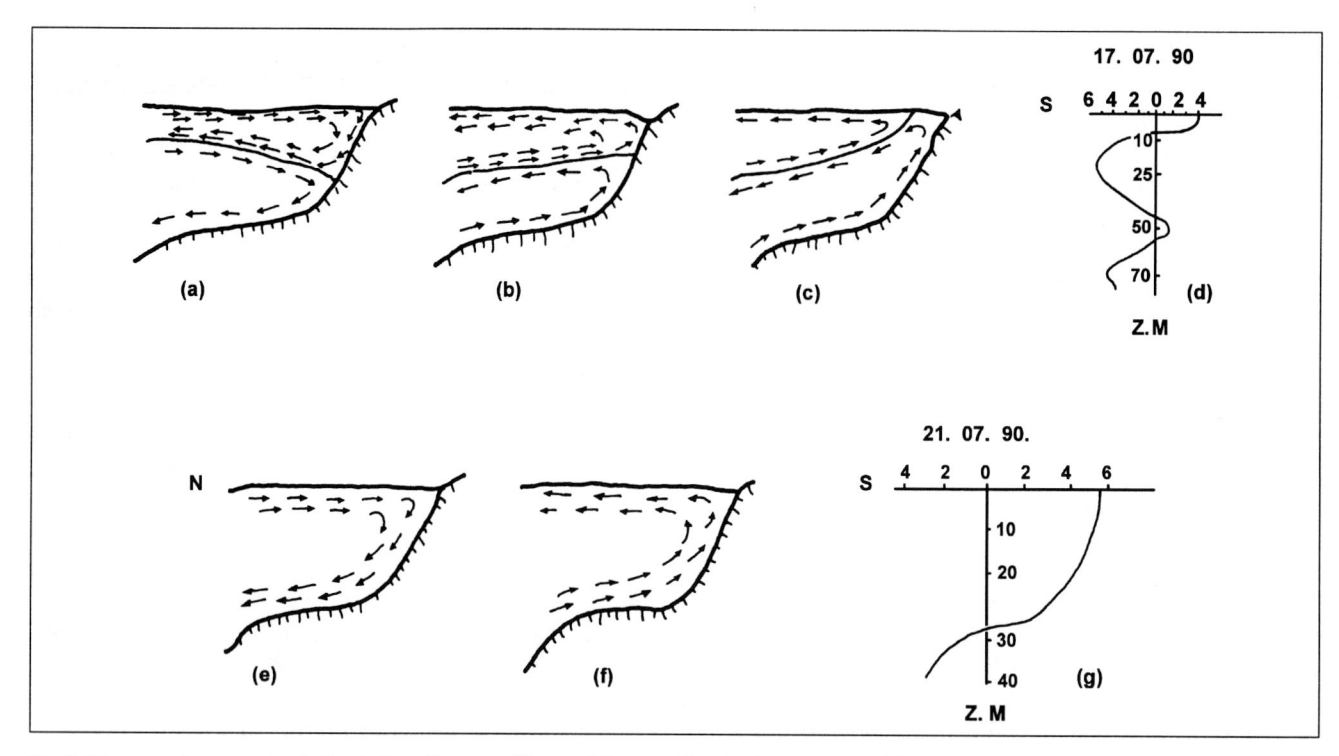

288 SEAS AT THE MILLENNIUM: AN ENVIRONMENTAL EVALUATION

Fig. 2. Diagrams for cross circulation induced by upwelling and downwelling in the presence of thermocline in summer (a, b, c, d) and in winter (e, f, g) (source IOC, 1993).

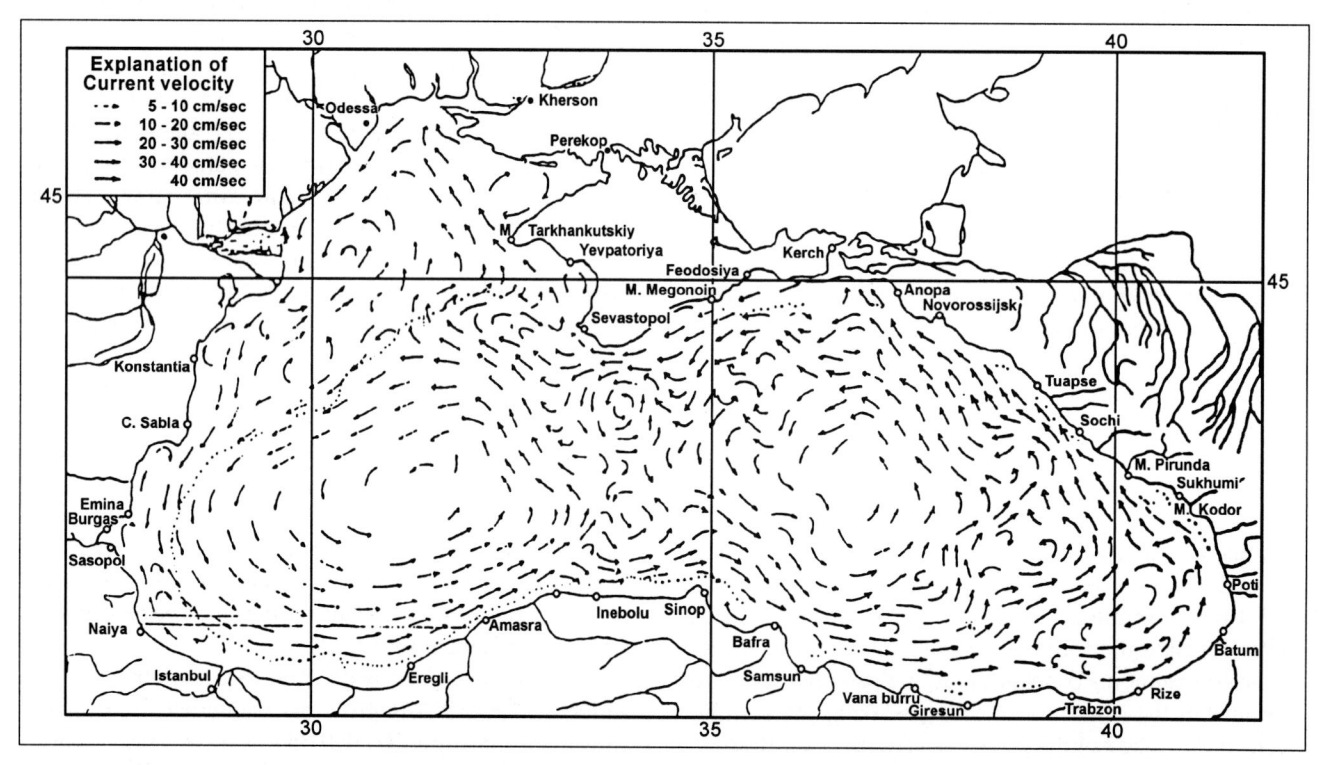

Fig. 3. General surface circulation (after Neumann, 1942; source Balkas et al., UNEP, 1990; reproduced by permission of Prof. Turgut Balkas).

combining the present results with earlier findings, identifies quasi-permanent and recurrent features even though the shape, position, strength of eddies and meander pattern of currents vary. The features discussed by Oguz et al. (1993) reflect the effect of multiple and variable forcings and modifications introduced by internal dynamic processes on the basin circulation.

A distinguishing characteristic of the Black Sea is the presence of a permanent halocline located between 100 and 200 m. The Black Sea is the world's largest body of water

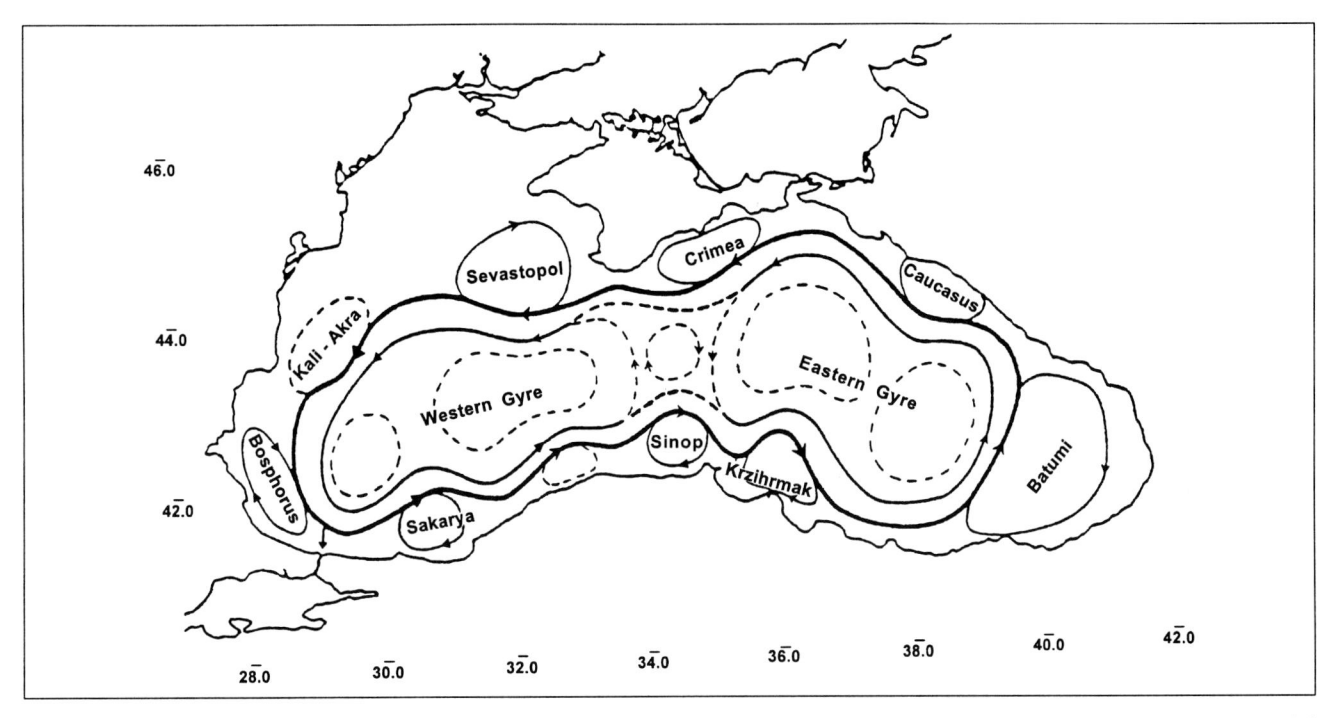

Fig. 4. Schematic of the main features of the upper layer general circulation emerging from synthesis of past studies and the SO90 survey. Solid (dashed) lines indicate quasi-permanent (recurrent) features of the general circulation (source Oguz et al., 1993).

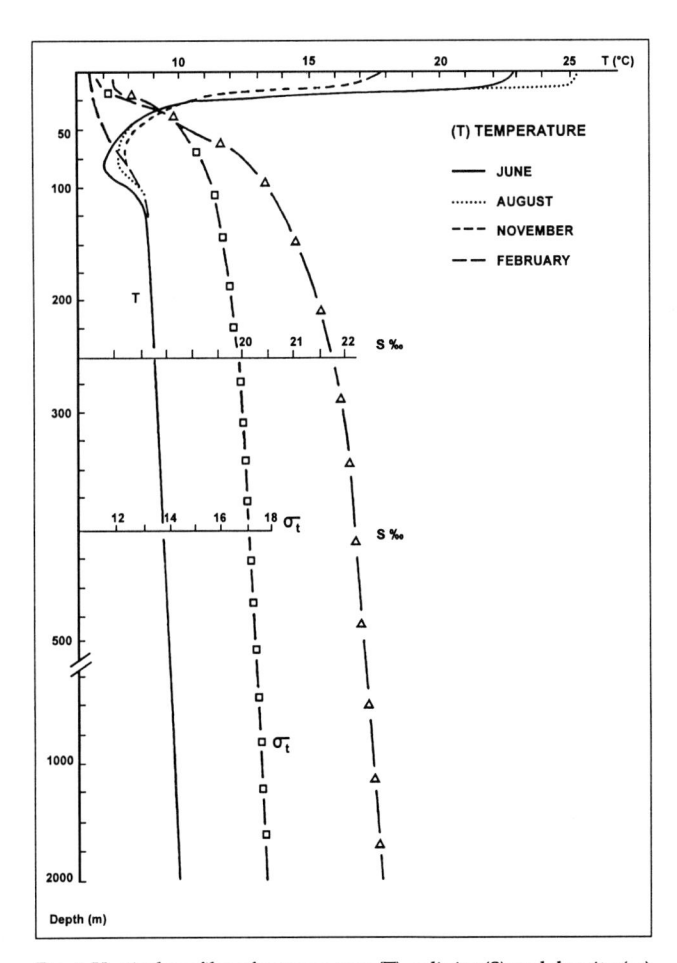

Fig. 5. Vertical profiles of temperature (T), salinity (S) and density (σ_t) in the Black Sea (source Balkas et al., UNEP, 1990; reproduced by permission of Prof. Turgut Balkas).

with this character. The stratification is generated by freshwater input and the Mediterranean inflow of water of a higher salinity. The vertical pattern of salinity is essentially the same everywhere in the basin, except in the coastal and nearby shelf regions located particularly in the northeastern part of the basin. The mean salinity of the deep waters in the Black Sea is 22.2–22.4‰ depending on the regional deep water convection and mixing characteristics. A representative salinity profile is shown in Fig. 5 including vertical profiles of temperature and density in the Black Sea (Balkas et al., 1990).

Salinity of surface waters varies depending on the amount of evaporation, precipitation and river run-off, but variations are almost absent below 200 m. Average surface salinities are about 18.0–18.5‰ during winter and are typically 1.0–1.5‰ higher than those observed in summer, particularly in the western and southeastern parts. Salinities in summer may attain much lower values even in the shelf and coastal areas of the north-northeastern Black Sea where values of about 14–16‰ are frequently observed due to run-off (Balkas et al., 1990).

Water temperature shows much more pronounced variations than salinity. Mean annual surface temperature varies from 16°C in the southern, to 13°C in the northeastern and 11°C in the north-western parts. While in the upper 50–70 m water layer the temperature has seasonal fluctuations and considerable vertical variations, the temperature of deeper waters remains constant through the year. Typically, the temperature at a depth of 1000 m is about 9°C and shows a slight increase of 0.1°C per 1000 m towards the bottom due to geothermal heat flux from the bottom (Fig. 5).

THE MAJOR SHALLOW WATER MARINE AND COASTAL HABITATS AND OFFSHORE SYSTEMS OF THE BLACK SEA

Anthropogenic eutrophication of shelves and coastal waters has been the most damaging of all the many harmful human influences on the oceans and sea, both in terms of its scale and its consequences. Phenomena such as the decrease in taxonomic diversity, population explosions of some species to the detriment of others, declining water transparency, the degradation of benthic macrophytocenoses, hypoxia in the bottom benthic layers, mass mortalities of the zoobenthos and nektobenthic fish are all directly or indirectly linked to the sharp increase in mineral and organic elements in the run-off, which has had a strong impact on rates of primary production.

Due to its geographic position and morphometric features, the Black Sea is a classical example of the above. Eutrophication, bottom trawling and accidental introduction of exotic species have led to profound changes in the biota. The most endangered are the largest assemblages of mussel, *Mytilus galloprovincialis* (containing more than 100 species of invertebrates, including endemic and relict species), brown alga *Cystoseira barbata* (60 species) and red algae *Phyllophora nervosa* (90 species). Introduced exotic marine organisms predated on indigenous bivalve molluscs (the Pacific snail *Rapana thomasiana*), and on zooplankton, pelagic fish eggs and larvae (Atlantic ctenophore *Mnemiopsis leidyi*) (Zaitsev and Manaev, 1997).

Results of a three-year Black Sea Environmental Programme, BSEP, are presented in Zaitsev and Mamaev (1997). Specific measures are proposed to protect Black Sea keystone species (*Mytilus, Cystoseria, Phyllophora*), forming nuclei of large bottom communities. The establishment of new protected areas, prohibition of bottom fishing gears in certain areas, construction of artificial reefs and ballast water control are among the recommendations.

According to most of the reports on the threats to the Black Sea, eutrophication is the Sea's primary problem. As a result of the nutrient flow, principally into the western and northern continental shelves, algal blooms have intensified over the past decades. This, in turn, has triggered some fundamental biological alterations to the upper water layers. In the first place, the algal blooms have reduced the amount of sunlight penetrating into the water column. This has resulted in the mass mortality of shallow water macrophytes, such as *Phyllophora* (a red algae) formerly an important component of Black Sea ecosystems and a major economic resource (Sorensen et al., 1997).

The water above the main pycnocline is where phytoplankton thrive in most marine basins. Particularly in winter and spring seasons, the main pycnocline in the Black Sea lies closer to the surface than in other seas. Winter blooms during winter convection and uniform surface temperatures above the pycnocline are characteristic of the open Black Sea. It may be assumed that in the open sea, annual phytoplankton production amounts to 200–250 mg C m^{-2}, characteristic of mesotrophic waters evolving to eutrophic conditions (Vinogradov and Tumantseva, 1993).

As a result of eutrophication, there have been changes in the composition of the phytoplankton community. In the early 1960s, the nutrient content was a limiting factor for phytoplankton development. Since 1970, the nutrients exceeded the phytoplankton requirements and stopped being a factor controlling primary production. A large amount of phytoplankton adds to the suspended matter, reducing light penetration. Phytoplankton blooms also deplete dissolved oxygen with the result that conditions become lethal to the majority of organisms (Balkaþ et al., 1990). In recent years such water blooms became characteristic of the northwestern part of the Black Sea, and the area covered by the blooms increased 10- to 30-fold in comparison with 1950–1960 (Zaitsev and Mamaev, 1997).

The dominant Black Sea mesoplankton are copepods, among them the interzonal *Calnus ponticus*. Until the summer of 1988, the biomass of *Sagitta setosa* was almost as high, but after rapid development of introduced macroplankton *Mnemiopsis*, the former decreased rapidly. Herbivorous mesoplankton biomass including consumers of nanoplankton, and *Noctiluca*, now constitutes 52–88% of the total mesoplankton mass. The Black Sea macroplankton includes three large species: the ctenophore *Mnemiopsis leidyi* and the two medusas *Aurelia aurita* and *Rhyzostoma pulmo*. The latter inhabit mainly contaminated coastal regions of the Black Sea and the Sea of Azov and almost never penetrate into the open sea. *Aurelia aurita* and *Mnemiopsis leidyi*, by contrast, inhabit cleaner coastal waters and open sea waters (Vinogradov and Tumantseva, 1993).

By the end of the 1980s, the total biomass of the ctenophore, *Mnemiopsis leidyi*, which was introduced by ships from the Atlantic coast of North America, was estimated at a billion tons. It is known that it is an active predator, feeding on zooplankton and crustaceans in particular, and also on pelagic eggs and fish larvae. The late 1980s were also marked by a sharp decline and then collapse of the catch of the main commercial fish, the anchovy. It may be that *Mnemiopsis* was not the major reason for the decline of the existing Black Sea fisheries, but the example of *Mnemiopsis* and other exotics demonstrates that biological diversity is very fragile (Zaitsev and Mamaev, 1997).

Long-term modifications on the Black Sea marine environment resulting from human activity has induced significant changes not only in phytoplankton but also in zooplankton and zoobenthos. The zooplankton community increased from 2.56 mg/m^3 in 1961 to 18.30 mg/m^3 in 1967 and 16.96 to 155.56 mg/m^3 during 1976–1977. In 1983 this was 8719 mg/m^3. *Noctiluca miliaris* had an explosive increase, reaching 15,712 individuals m^{-3}, *Acartia clausi* reached 5835 individuals m^{-3} and *Pleopsis polyphemoides* 1760 individuals m^{-3}. The diversity indices of zooplankton have low values due to a few opportunistic species such as *Pleopsis polyphemoides* and *Acartia clausii* that produce

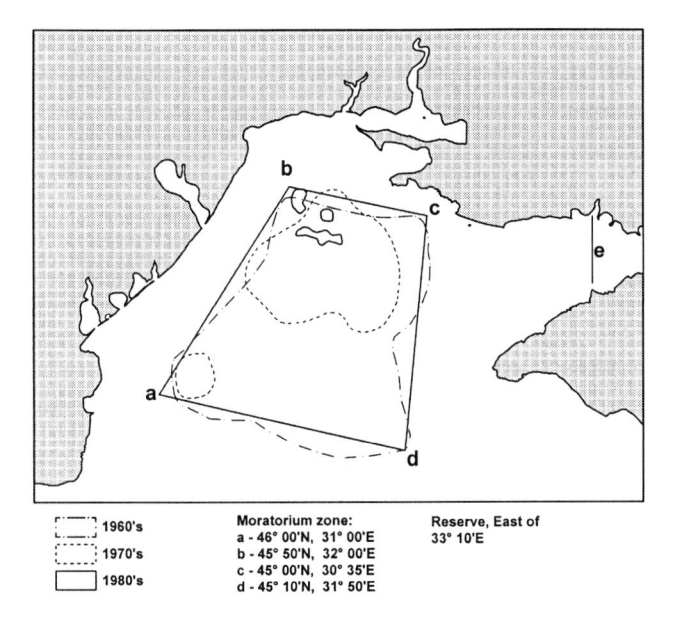

Fig. 6. Progressive reduction of Zernov's *Phyllophora* field on the NWS and proposed protection area (source Zaitsev and Mamaev, 1997).

monospecific populations (Balkas et al., 1990). When nanoplankton and algae die and fall to the bottom of the shelf floor they begin to decay, consuming huge quantities of oxygen in the process. Consequently, about 95% of the Ukrainian coastline and the entire Sea of Azov (shared by Ukraine and Russia) now suffer from hypoxia (Sorensen et al., 1997).

At the same time, transparency worsens as the result of greater quantities of phytoplankton, *Noctiluca*, jellyfish, detritus and other organisms and particles. By the 1980s transparency had fallen to 7–8 m and then 2–2.5 m and, during blooms, it declined to less than 1 m. In the 1960s the compensation point was about 45–55 m in the open water of the northwestern shelf and 18–20 m closer to the shore. In the 1980s the compensation point shrank to 20–25 m in the open sea and 6–8 m close to the shore. This means that by the 1980s, the bottom algae, which until the 1960s had grown close to the shore at depths over 6–8 m and in the open sea at depths of more than 20–25 m, were below the compensation point. The result has been a severe degradation of benthic algae. This process is graphically illustrated by the example of Zernov's *Phyllophora* field (Fig. 6) (Zaitsev and Mamaev, 1997).

Moreover, there have been spectacular changes in the biomass of Jellyfish (*Aurelia aurita*). In 1978, in the 0–10 m layer there were 47 million tons of *Aurelia* with a total biomass (wet weight) of 300–450 million tons for the whole of the Black Sea. This is much greater than the biomass of the anchovy. This phenomenon can be related to the development of phytoplankton and zooplankton as a result of eutrophication and to the considerable reduction in the number of some plankton-feeding trophic competitors of the jellyfish. Benthic fauna has also shown qualitative and quantitative changes. *Ostrea sublamellosa* and *Gibbula divaricata* are now extinct along the Romanian coast. As a consequence of the more frequent phytoplankton blooms, a great quantity of organic detritus is accumulating on the sea bed. This detritus is an ideal food for *Melinna*, which shows a great tolerance of large changes in dissolved oxygen concentrations and even temporary anaerobic conditions and the presence of hydrogen sulphide. Owing to these modified conditions, new organisms have penetrated into the Black Sea and formed large populations. These are *Polidora ciliata*, *Callinectes sapidus*, *Rapana thomasiana*, *Mya arenaria* and, more recently *Scapharca inequivalvis* (Balkas et al., 1990).

Thus, for the benthos, one of the most damaging effects of eutrophication has been the creation of zones in the bottom layers of water with reduced (hypoxia) or no (anoxia) dissolved oxygen. Hypoxia and mass mortalities became an annual occurrence from 1973 and the area of the northwestern shelf over which they extended increased to 30,000–40,000 km^2, depending on the particular features of each summer and autumn seasons (Zaitsev and Mamaev, 1997).

Recent widespread changes in the biological diversity in the Black Sea are largely due to effects of human activities. Loss or imminent loss of endangered species which have ecological and/or economic value for the Black Sea ecosystem, degradation of coastal wetlands, loss of habitats and communities and degradation of landscape are the most common responses of the Black Sea ecosystem to man's activities. Species listed in Table 1 are mainly keystone species at the centre of communities which are highly characteristic of the local environment, and include threatened endemic as well as relict species. Those communities have dramatically decreased due to the eutrophication caused by inflow of untreated sewage from point and non-point sources and from polluted rivers, from hypoxia caused by eutrophication, increased turbidity, including that induced by use of various types of bottom gear, toxic pollution, over-harvesting and destruction of breeding grounds (GEF, 1996).

There are currently three species of dolphin in the Black Sea: the common dolphin *Delphinus delphis*, the harbour porpoise *Phoceana phoceana* and the bottlenosed dolphin *Tursiops truncatus*. In 1950 the total dolphin population is thought to have been approximately 1 million. An aerial survey in 1983–84 estimated that there were 60,000–100,000 individuals in the waters of the former Soviet Union. A Turkish ship-based survey in 1987 put the total Black Sea population at 445,440. The true figure, however, is probably considerably less and there is an urgent need for an accurate stock assessment. The decline of the dolphin population in the Black Sea may be attributed to three main factors: accidental killings directly related to gill net fishing, destruction of the coastal ecosystem due to harmful or overfishing, and the direct effect of pollution (Öztürk, 1996).

The occurrence and distribution of the critically endangered Mediterranean monk seal in the Black Sea poses a

Table 1
Loss or imminent loss of endangered species in the Black Sea and its wetlands (modified from GEF Draft Report, 1996; reproduced by permission of BSEP)

Taxon	Main Species Groups	Population size and geographic range
Bottom plant community	Red algae (phyllophora) Brown algae	3% of reference level on Ukrainian shelf less than 1% of reference level on Romanian and Ukrainian shelf
Commercial and/or endemic animals	Molluscs * Mussel * Oyster * *Hypanis*	30% reference level on NWS less than 5% reference level 50% of reference level, brackish waters in Russia and Ukraine
Bottom crustaceans	Crabs (about 14 species)	30–50% of reference level,
	Shrimps (more than 20 species)	40% of reference level, basin wide
Fish	Gobiidae (20 species, 10 commercial, all endemic)	20% reference level, NWS
Wetlands communities	Consisting of plants, invertebrates, amphibians, reptiles, birds and mammals in all well over 2000 species	Variable, but declining
Marine mammals	3 dolphins Monk seal	5–10% of reference level, basin wide few specimens left

*As reference against which to evaluate present population sizes, the population sizes for the 1960s were used.

more intractable problem. The primary threat to the few remaining individuals appears to come from the numerous types of pollution, overfishing, food shortages and coastal ecosystem degradation.

Fisheries Issues Along the Black Sea Coast

Unlike the Mediterranean, the Black Sea was traditionally a rich fishing area, supporting some two million people. This economic resource has now been almost entirely destroyed. Biodiversity has also suffered and the future options for resource development of the Black Sea have been reduced. During the 1980s the Black Sea fishing industry collapsed. Total catches, estimated at 900,000 tons in 1986, fell to about 100,000 tons for all countries in 1992.

The loss of fisheries resources in the Black Sea is an issue which transcends the usual boundaries of stock management, which is commonly a process of managing the activities of the fishermen themselves. The declining fisheries are a clear consequence of the degradation of the ecosystem itself, which is in turn intimately related to land-based human activities.

Changes in the ichthyofaunal composition of the Black Sea have primarily involved alterations in the number of individuals in specific populations. For many species, fish populations have declined so sharply that they have lost their importance for commercial fishing, and remain within the Black Sea ichthyofauna only as zoological representatives of the species. In the period 1960–1970 there were 26 commercial fish species which were caught by the tens or even hundreds of thousands of tons. But by the 1980s only five species were left, although they were joined in the early 1990s by the introduced haarder *Mugil soiury* (Zaitsev and Mamaev, 1997; Pavlov et al., 1996).

The Turkish fishery industry, composed entirely of small-scale owner-operators, exploits the coastal waters of the country's extensive coastline. Within Turkish waters, the Black Sea is the most productive region, accounting for an average 85% of Turkey's total catch. Anchovy is the most abundant fish species. It is a winter seasonal fish, available in densest concentration from November to March. The fish descends during the day to a depth of 70 to 80 m, and rises during the night to the higher water layers thereby moving toward the coast into depths of 10–40 m. Purse-seining is the only fishing method used by Turkish fishermen to catch the anchovy, and since it is available in large quantities, it has a comparatively low market value, and thus has a high consumption rate especially in coastal residential areas. Anchovy is considered to be an important supplementary source of protein for the growing population (Dinçer et al., 1995).

It is clear that the main resources of the Black Sea are transboundary in nature and require cooperative action. The recent collapse of the fisheries is directly connected with degradation of the water quality and destruction of spawning grounds, unplanned development as well as uncontrolled fisheries practices. However, to date, there has been no international agreement on appropriate levels of fishing by each coastal state, and the current distribution of benefits does not well reflect the territorial distribution of resources. Although precise estimates of the socioeconomic impacts of the collapse of the fisheries are not available in terms of earnings and employment, it is clear that annual catch values for the fishery declined by at least $ 300 million from the mid 1980s to the early 1990s.

It must be stressed that cooperative management on transboundary fishery resources throughout the world is based on proper scientific information and analysis, and it is followed by negotiation on overall yield and fishing effort, and appropriate shares of the allowable catch between all parties. Further considerations have been developed from information collected, and evaluations performed during fisheries working groups of the General Fisheries Council for the Mediterranean, and through the GEF Fisheries Activity Centre in Romania. A brief summary table by ecological category of fish resource is given here in Table 2 (GEF, 1996).

Table 2

Summary table by ecological category of fish resource (modified from GEF Draft Report, 1996; reproduced by permission of BSEP)

Resource category	Species	State of stocks
Pelagic fish	Sprat, Whiting, Anchovy, Horse Mackerel	stock size of most small pelagics has partially recovered from Mnemiopsis-caused depletion which began in the early 90's.
Demersal fish	Turbot, Spiny dogfish	Decline in stocks in most areas
Anadromous fish	Sturgeons	Some species as giant sturgeon are endangered, others are depleted
	Shad	Recovering
Indicators of ecosystem health	Red algae, dolphins	
	Venus clam, blue mussels	Declining in many years
	Artificial reefs	maybe used to increase plant and animal diversity
	Native grey/golden mullets, red mullet, rays and skates, native salmon-trout	Depleted and rare
Exploitable species	*Mya* clam, blue crabs, sandlance & silversides	Unexploited, large biomass in some countries
	Rapana snail	Overexploited in Turkey and Bulgaria
	Haarder (exotic grey mullet)	Expanding stock area
Seasonal migrants	Mackerel, Bonito, Bluefish	Stock heavily fished throughout Mediterranean, Aegean, Marmara and drastic reduction of immigration into the Black Sea.

Changes in the Ecology of Coastal Wetlands of the Black Sea

The coastal zone wetland ecosystems occupy large areas and link the huge catchment area with the Black Sea itself. Wetlands are highly productive ecosystems whose formation, functioning and characteristics are determined by the water regime. They support a unique diversity of flora and fauna because of the constant inflow of water and alternating dry and wet periods. Limans, lagoons, estuaries and deltas are the most widespread types of wetlands in the Black Sea region.

Conservation areas in each country, existing or proposed, are listed in Table 3. The Kizilirmak delta near Samsun is one of the major coastal wetlands on the Black Sea coast of Turkey. The delta is approximately 10,000 hectares of natural area consisting of three main lakes surrounded by extensive reedbeds and marshes. Of the 420 bird species in Turkey, 309 have so far been recorded in the delta. Some of these are very rare and only occasionally seen in Turkey, such as the Olive-backed Pipit *Anthus hodgsoni* and the Little Bunting *Emberiza pusilla*. Others are threatened on a global scale, such as the Dalmatian Pelican *Pelecanus crispus*, the Pygmy Cormorant *Phalacrocorax pygmeus* and the Red-breasted Goose *Brenta ruficollis* (BSEP, 1994).

The main anthropogenic influences resulting in the loss and degradation of wetlands in the Black Sea region are draining of wetlands in the interests of agriculture, forestry and fishing or construction; dumping dredge spoils and solid waste; waste discharge (sewage and fertilizer-polluted water) causing eutrophication; dumping of toxic pollutants (pesticides, industrial and radioactive waste, heavy metals, organic compounds); overfishing, overhunting of birds and mammals, and overgrazing, which undermine biodiversity and reduce food resources and feeding possibilities for wild animals.

Table 3

Conservation areas (modified from GEF Draft Report, 1996; reproduced by permission of BSEP)

Country	Coverage by protected areas	Critical ecosystems, landscapes, habitats
Ukraine	Wetlands	
	Existing:	Danube Delta Nature Reserve
	Proposed:	Chermomorsky, Mys Martyan, Karadag, Cape Tarkhankut, Bolshoy Fontan Cape, Zmeiny Island
Bulgaria	Wetlands	
	Proposed:	Coketryse bank, Ahtopol Resova river, Primorsko Ropotamo river, Byala Shkorpilovtz sy, Cape Kaliakra Kamen Briag
Turkey	Wetlands	
	Existing:	Kizilirmak Delta, Yesilirmak
	Proposed:	Bosphorus
Romania	Wetland Coasts	
	Existing::	Danube Delta Biosphere reserve,
	Proposed: :	Siutghiol lake, Techirghiol lake, Cape Tuzla, Cape Midya, Mamaia Bay
Georgia	Wetlands	
	Existing :	Kolkheti Nature reserve,
	Wetlands Coast Marine Habitats	
	Proposed :	Kolkheti National Park
	Basin/catchment ecosystems and landscapes. **Proposed:**	Adjara National Park, Central Caucasus protected areas
Russia	Wetlands	
	Existing:	Lake Abrau
	Proposed:	Cape Utrish
	Basin /catchment ecosystems and landscapes. **Existing:**	Sochi National Park

Radiation and Oil in the Black Sea

The release of large quantities of radionuclides to the lower atmosphere from the Chernobyl nuclear power plant in 1986 has had implications on the health of the Black Sea. Clouds containing radioactive material reached the Black Sea area from 2 May 1986. In addition to direct fallout, the Black Sea is expected to have received an input from the Danube and the Dnepr, both of which drain watersheds heavily impacted by the Chernobyl fallout, and the Dnepr which drains Chernobyl itself. The ^{134}Cs, ^{137}Cs concentrations measured and ^{134}Cs/^{137}Cs ratios calculated from the data obtained during the cruise of the Turkish research vessel R/V PIRI REIS are presented in Table 4.

On 13 March 1994 the fully laden Cypriot oil tanker, the Nassia, collided with an empty freighter, the Ship Broker, shortly after entering the Bosphorus from the Black Sea. The Nassia disaster was just one of 200 collisions and groundings in the Turkish straits in the last 12 years, 138 of which have occurred since 1990. The incident highlighted the enormous demands facing the local emergency services. The six littoral states of the Black Sea officially inaugurated an Emergency Response Activity Centre (ERAC) at a meeting in Varna, Bulgaria, 16–18 May 1994. Total oil pollution input to the Black Sea is estimated at 110,840 t/yr. Of this, 48% is transported by the Danube river and most of the remainder is introduced from land-based sources through inadequate waste treatment and poor handling of oil and oil products (Table 5). Total oil discharges from the Black Sea coastal countries are 57,404 t/yr or about 52% of the total. Accidental oil spills average 136 t/yr, and the illegal discharges from shipping are excluded. The amount reaching the Black Sea from ballast water discharges by ships is unknown but thought to be considerable.

In pilot surveys, oil levels were measured in sediments and sea water. Sediment levels were found to be of concern near sea ports (Odessa and Socchi), but in open coast and the Bosphorus outflow areas, the levels were relatively low and correlated with lipid content. The levels of oil and petroleum hydrocarbons in general in sediments were comparable with those of the Mediterranean. In the EROS measurements of dissolved oil, rather high levels of fresh oil were observed, especially near the discharge of the River Danube. Concentrations of polyaromatic hydrocarbons (PAH), particularly toxic petroleum hydrocarbon compounds, are generally low and include contributions from petrogenic (oil) and pyrogenic (combustion products) sources.

THE SOUTHERN BLACK SEA, TURKEY

The oceanography of the Black Sea has been relatively well studied and documented. The same, however, cannot be said for documentation of the levels of marine pollution especially in coastal areas. Although a complete assessment of the pollution throughout the basin is not currently possible, the consequences of pollution, such as its effect on tourism in the region, loss of biodiversity, changes in the hydrological balance due to the construction of dams on rivers, reduction or even collapse of fisheries have been reported (Murray et al., 1989; Balkas et al., 1990; Knudsen, 1991; Mee, 1992).

The resources of the Black Sea and its problems are mainly shared by six coastal countries, Bulgaria, Georgia, Romania, Russia, Turkey and Ukraine. For these countries, there is a strong need for harmonizing legal and policy objectives and for developing common strategies for the control of pollution. For this purpose, representatives of the Black Sea countries drafted their own Convention for the Protection of the Black Sea against Pollution which was signed in Bucharest in April 1992. Later on, a ministerial Declaration on the Protection of the Black Sea Environment was signed by all six Ministers of the Environment in Odessa in April 1993 in order to set the goals, priorities and timetable needed to bring about environmental actions. In June 1993, a three-year Black Sea Environmental Programme (BSEP) was established.

Table 4
^{134}Cs and ^{137}Cs concentrations and ^{134}Cs/^{137}Cs ratios in surface waters (Livingstone et al., 1986)

Location	^{137}Cs (bg/m^3)	^{134}Cs (bg/m^3)	^{134}Cs/^{137}Cs
Black Sea	165±14	78±4	0.47
Black Sea	70±5	35±3	0.50
Black Sea	41±6	17±1	0.42
Bosphorus (mouth)	85±4	35±3	0.48
Bosphorus	74±4	35±3	0.46

Table 5
Oil pollution of the Black Sea (t/yr) (GEF Draft Report, 1996; reproduced by permission of BSEP)

Source of pollution	Bulgaria	Georgia	Romania	Russian Federation	Turkey	Ukraine	Total
Domestic	5,649.00	–	3,144.10	–	7.30	21,215.90	30,016.30
Industrial	2.72	78.00	4,052.50	52.78	752.86	10,441.00	15,379.86
Land–based	–	–	–	4,200.00	–	5,169.20	9,369.20
Rivers	1,000.00	–	–	165.70	–	1,473.00	2,638.70
Total	6,651.72	78.00	7,196.60	4,418.48	760.16	38,299.10	57,404.06

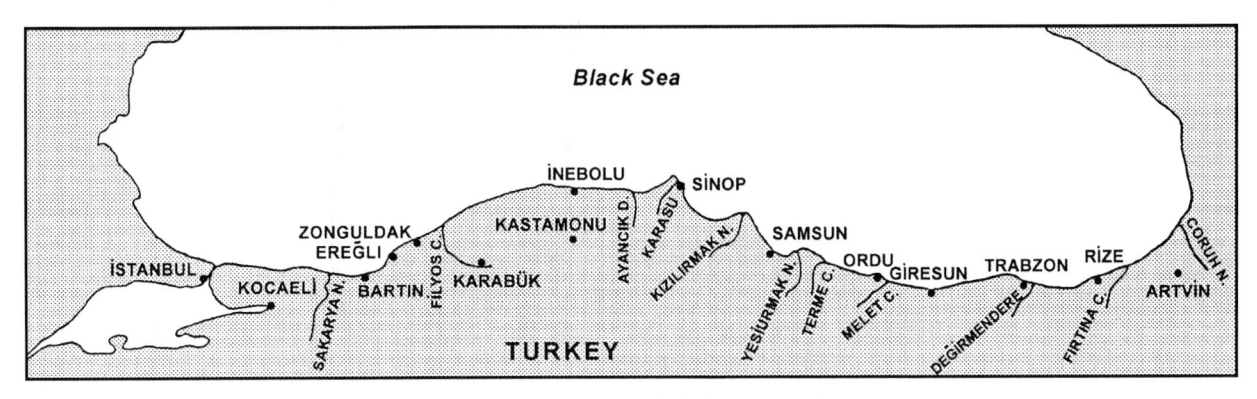

Fig. 7. Major Turkish rivers and Black Sea coastal towns.

At the end, a Strategic Action Plan was developed based on the recommendations in the Bucharest convention and Odessa declaration and was signed by the countries surrounding the Black Sea in Istanbul in October 1996. The signed convention amongst the Black Sea countries contains a legal framework for the establishment of a Black Sea Commission and provides protocols for protection against land-based sources of pollution, for regulating dumping and for emergency action in the case of spills of oil or other toxic substances. This Black Sea Action Plan, BSEP, formed the basis for the scientific and regulatory actions that are taken to protect the Black Sea.

The overall BSEP objectives stated in the GEF Black Sea Environmental Programme are: to improve the capacity of Black Sea countries to assess and manage the environment, to support the development and implementation of new environmental policies and laws and to facilitate the preparation of sound environmental investments. The first step in creating the Black Sea Action Plan was the completion of a systematic scientific analysis of the root causes of environmental degradation in the Black Sea.

The Black Sea Environmental Programme (BSEP) is financed under GEF which is managed by UNEP, UNDP and the World Bank. In the Black Sea Environmental Programme in order to improve the capacity and forge new linkages, a system of thematic Working Parties was established, based upon regional Activity Centres. Each Black Sea country agreed to host one of these Centres such as an Emergency Response to oil spills (Varna, Bulgaria); Fisheries (Constanta, Romania); Pollution Assessment (Odessa, Ukraine); Coastal Zone Management (Krasnodar, Russia); Biodiversity (Batumi, Georgia); and Pollution Control (Istanbul, Turkey) (Fig. 7). Then, corresponding National Focal Points were established for each Centre in each of the other countries such as the Ondokuz Mayýs University, Research Centre of Environment in Samsun, Turkey which served as the national focal point of the protocol for protection against land-based sources of pollution of the Black Sea coast of Turkey. The BSEP has also created a Black Sea Data System and a Black Sea Geographic Information System. For general programme coordination, a Programme Coordinating Unit (PCU) was also established in Istanbul.

General Characteristics and Populations of the Southern Black Sea Coast

The National Focal Point—Ondokuz Mayis University, gathered data for three years during 1995–1997, on land-based sources of pollution of the Black Sea coast of Turkey. In addition, it established a monitoring system for bathing water quality. A total of 33 land-based pollution source sampling stations and around 30 beaches for bathing water quality were selected (Final Project Report, 1996, 1997).

The Black Sea coast of Turkey is 1695 km long extending from the Bulgarian border in the west to the Georgia border in the east (Fig. 7). The area of the region is about 141,000 km^2 or about 18% of the total surface area of Turkey. The total population of the region is about 8,000,000 people (Table 6) divided into 14 city centres.

The topography of the region has a profound influence on the distribution of the population and thus pollution sources along the coast. The Ponthic mountains that extend along the coastline rise sharply within 30 km of the shore

Table 6

Cities at the Black Sea region of Turkey (UNCED, 1992)

City	Surface area (km^2)	Population (1990)
Amasya	5,520	285,729
Artvin	7,436	212,833
Bolu	11,051	536,869
Çorum	12,820	609,863
Giresun*	6,934	490,087
Gümüþhane	10,227	169,375
Kastamonu	13,108	423,611
Ordu*	6,001	830,105
Rize*	3,920	348,776
Samsun*	9,579	1,158,400
Sinop*	5,862	256,153
Tokat	9,958	719,251
Trabzon*	4,685	795,849
Zonguldak*	8,629	1,073,563

*Cities on the coastal zone of the region.

and do not allow for the development of large cities. These high and young mountains have played an important role in the civilization of the Black Sea region. Mountains are also an obstacle for transportation and so the fairly small population of the mountains live in small settlements rather than large cities. Although the climate is favourable, agriculture is limited, and there are few natural harbours.

The longest rivers are the Kizilirmak (1335 km), the Sakarya (824 km), Yesilirmak (519 km), Filyos (228 km) and Melet (165 km). Among these rivers, the Kizilirmak, Yesilirmak and Sakarya drain areas of 78, 65 and 58 thousand km^2, respectively. These flow year round, though after heavy rains and snow melting, the level of the water increases. However, the Black Sea region of Turkey is not rich in lakes; there are ice-lakes at the west side of the region but they are small. The only large lake is the Uzun (12 km^2).

The topography of the Black Sea region affects the direct discharges to the sea. Some cities and towns use sewerage systems, but small towns use septic tanks. Present sewerage systems are mainly old, but new systems are being constructed. For domestic and industrial pollutants, the streams and rivers are the main sources of pollutants into the Black Sea (Bakan et al., 1996).

Rural Factors Affecting the Southern Black Sea, Turkish Coast

The coastal zone supports extensive agriculture and includes the most productive lands of Turkey (Table 7). The most important cereals are wheat, rice and corn. The Black Sea region produces almost 9% of the country's total wheat production, about 10% of the country's total barley production, 56% of the country's total corn production and almost 34% of the country's total rice production. Rice is grown within the Yesilirmak, Gökirmak, Kizilirmak and Devrez deltas. Important industrial products produced in this region are tobacco and sugar beet and there is some flax seed production. This region also produces about 19% of the country's total tobacco, and 20% of the sugar beet. The most important pulse product in the Black Sea region is bean with 21% of the total and this region produces almost 32% of the country's total olives and 81% of the hazelnuts, 36% of the country's walnuts, and large quantities of fruit (UNCED, 1992).

Table 7

Breakdown of agricultural production in the Black Sea region of Turkey (UNCED, 1992)

Crop	Proportion in the total production (%)
Cereals	85.1
Industrial plants	5.8
Oily seeds	1.8
Pulses	4.3
Lump plants	3.0

Development of Coastal Areas in the Southern of Black Sea

In the past, the economy of the Turkish Black Sea coast, where nearly 20% of the population is located, was based on agriculture, and even though industry has been developing rapidly, agricultural activity is still important. Industrial development, supported by the government and private enterprise, is mostly concentrated in the Samsun area, but its distribution is influenced by the availability of agricultural products, by population density, and by the presence of transport facilities including ports and railways. Such industrial development puts heavy stress on valuable agricultural areas such as, for example, those where the best quality tobacco is grown. Some of these agricultural areas have already been replaced by industry and now suffer from industrial pollution. The Black Sea coast is poorly developed for tourism, mostly due to the less favourable climate compared with that of the Aegean and the Mediterranean coast. The government is making efforts to upgrade and promote tourism along the Black Sea (Balkas et al., 1990).

During recent developments, large amounts of material were dumped in coastal waters. This affected water quality, especially turbidity and transparency. Filter-feeding organisms were most affected; the macroflora in the rocky zones of shallow waters diminished greatly, both in quality and quantity, and the most important perennial brown alga *Cystoseria barbata*, once very abundant and sheltering a rich fauna, has suffered greatly. Human impact, combined with some harsh winters when the sea froze and high water turbidity, did not allow repopulation (Balkas et al., 1990).

The eastern Black Sea has been exposed to severe coastal erosion and shoreline recession for the last 30 years. One of the most important reasons for this is the response of the coast to man-made activities. As a result of sand mining by people and municipalities, the coastal dynamic was changed. Another important factor was the construction, by filled soil, of a highway near the shore. The wave energy increases by reflected waves from these slopes and causes a seaward sediment transport. Another important reason for erosion and shoreline recession is incorrect site selection, planning and design of coastal structures such as harbours and fishery harbours (Yüksek et al., 1995).

LAND-BASED POLLUTION

Monitoring

Inventories of land-based sources of pollution (LBS) and beaches in the Black Sea coast of Turkey were made in order to set up a valid monitoring programme, using rapid assessment methods. This permitted convenient assessment of the effectiveness of various pollution control options (WHO, 1993). In this major study, the physical

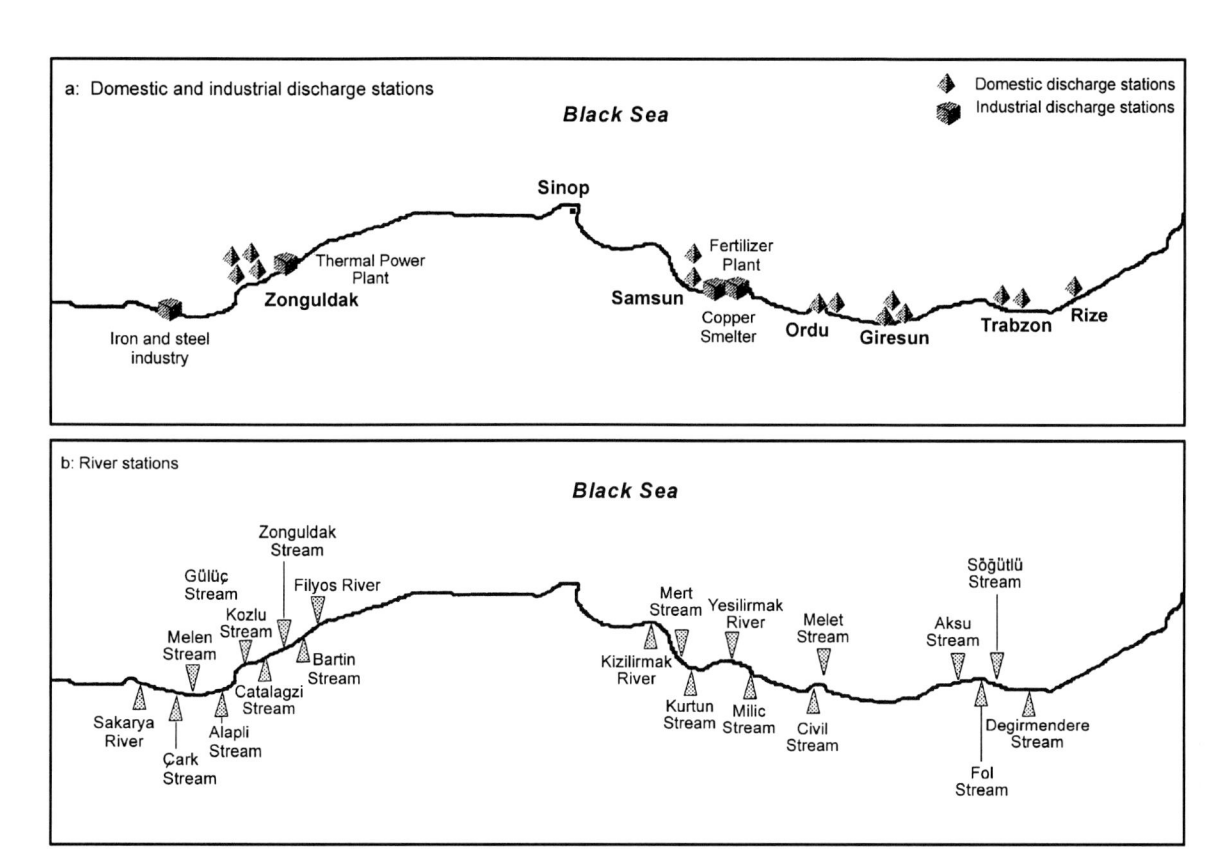

Fig. 8. Locations of the sampling stations.

Table 8
Inventory calculations for the domestic sources of each city along the Black Sea coast of Turkey

Location	Population*	BOD$_5$ (t/yr)	COD (t/yr)	TSS (t/yr)	Total N (t/yr)	Total P (t/yr)
Sinop	49,111	887	1,411	3,720	125	45
Samsun	530,508	9,250	14,700	36,223	1,324	803
Ordu	194,293	3,510	5,580	14,715	496	180
Giresun	420,941	7,604	11,995	31,880	1,076	391
Trabzon	457,152	8,258	14,011	36,364	1,160	424
Rize	50,365	1,954	3,108	8,194	276	152
Zonguldak	301,092	5,533	8,800	23,194	783	285

*Only the population on the coast.

boundary for the land-based sources inventory was determined as the borders of the municipalities on the Black Sea coast of Turkey (Fig. 8). Rivers were treated as 'point sources' but whole river basins were not considered. Diffuse sources and storm waters were not included in the rapid assessment part of the project, and only point sources that are subdivided into domestic and industrial sources were considered, and were estimated by applying the WHO Rapid Assessment technique (1993). The methods, results and the information gathered from municipalities were summarized in Bakan et al. (1996).

A summary of inventory calculations for the domestic sources of each city along the coast of the Black Sea of Turkey are given in Table 8. There were around 37,000 t/yr BOD$_5$, 60,000 t/yr COD, 154,000 t/yr TSS, 5240 t/yr total N and 2280 t/yr total P loads entering from the Black Sea coast of Turkey. Details on industrial discharges are shown in Table 9.

The region is not heavily industrialized. Although there are a number of small scale production facilities located around settlement areas, there are only four large industrial establishments along the coast: an iron and steel complex located approximately 40 km to the west of the city of Zonguldak which also has a central chemical treatment plant, a 600 MW thermal power plant located 20 km east of Zonguldak, and a fertiliser plant and a copper smelter 20 km east of Samsun. Pollution load calculations were performed for industries with liquid wastes, in using given waste load factors (WHO, 1993) for each raw material or process used. Liquid waste loads were classified as the conventional pollutants BOD$_5$, TSS, Total N and Total P, toxic and other important substances (Table 10). There are around 6000 t/yr BOD$_5$, 4200 t/yr TSS, 474 t/yr total N and 137 t/yr total P loads entering the Black Sea from the various industries located at each city.

Total domestic industrial and riverine pollution loads of the major pollutants such as BOD, TSS, TN and TP were calculated for each of the six Black Sea countries (Table 11). The data for annual loads of the other countries to the Black Sea are obtained from GEF (1996) which is the most recent comprehensive assessment of annual fluxes of pollutants.

Table 9

The major Turkish industries and their type of waste in the Black Sea region

Type of industry	Probable pollutants and characteristics of effluents	Location
Food manufacturing (Slaughtering, dairy products, canning of fruits / vegetables /fish, grain mill & bakery products, sugar factories, etc.)	BOD, COD, suspended material, chemical material, organic material, odour, specific pollutants from each type of manufacturing such as sugar, slaughtering.	Giresun, Ordu, Samsun, Sinop, Sakarya, Trabzon, Zonguldak
Manufacture of paper & paper products (pulp & paper)	pH change, high amount of suspended solids, colloidal & dissolved material, cellulose	Giresun, Zonguldak
Manufacture of other non-metallic mineral product (mainly cement factories)	heated cooling water, suspended solids, some inorganic salts	Ordu, Samsun, Trabzon
Manufacture of wood & cork products	Organic from staining and sealing wood products	Ordu, Sakarya
Non-ferrous metal basic industries	Acid, metals, toxic, low volume, mainly mineral matter	Samsun, Trabzon
Manufacture of industrial chemicals (manufacture of fertilizers and pesticides, resins & plastics)	Acids, mineral elements, suspended solids, caustic, phenols, formaldehyde	Samsun, Sakarya
Manufacture of textiles	highly alkaline, coloured, high BOD & temperature, high SS.	Samsun, Zonguldak
Non-metallic mineral products	red colour, alkaline nonsettleable SS	Samsun
Beverage industries (soft drinks)	increase in BOD, suspended material, precipitable solid material, fat & oil	Giresun, Ordu, Trabzon
Tea plant factories	wastes from treatment of tea leaves	Rize, Artvin, Trabzon, Giresun
Cigarettes	wastes from tobacco & its treatment	Samsun, Sinop, Trabzon
Coal mining	SiO_2, $CaCO_3$, Al_2O_3, cobalt, cadmium, lithium in coal ash	Zonguldak
Hazelnut	suspended material	Trabzon, Ordu, Giresun

Table 10

Waste load calculation totals for the important industries present in the Black Sea cities of Turkey

City	Discharge (t/yr)	BOD_5 (t/yr)	TSS (t/yr)	Total N (t/yr)	Total P (t/yr)
Sakarya	1,247	957	766	47.10	27.70
Zonguldak	126,292	262	150	0.46	0.03
Sinop	50	38	15	0.36	0.08
Samsun	5,950	2,706	2,514	419.41	108.40
Ordu	3,330	21	27	0.76	0.05
Giresun	17,416	1,975	616	0.05	0.04
Trabzon	3,232	159	143	6.37	1.12

The annual particulate load from Turkish domestic, industrial and national river sources into the Black Sea accounts for 7.94% of the total load into the basin including the international riverine input. The BOD contribution of Turkey is about 6.07% of the total in the Black Sea, with similar values for the other parameters (Table 12).

The pollutant exchange through the straits is not taken into account in the evaluation given above. Net influx from the Sea of Azov to the Black Sea via the Kerch Strait, and pollutant exchange via the Bosphorus Strait between the Marmara Sea and the Black Sea is presented in Table 13 (Delft Hydraulics Lab., 1996; Tugrul and Polat, 1995). There is an important pollutant input from the Sea of Azov to the Black Sea. These pollutants are carried through the Don, Kuban, Protoka and Kalmius Rivers. The nitrogen and

Table 11

Annual loads of pollutants from countries surrounding the Black Sea (GEF Draft Report, 1996; reproduced by permission of BSEP)

Countries	BOD_5 (t/yr)			TSS (t/yr)			Total N (t/yr)			Total P (t/yr)		
	(I)	(II)	Total	(I)	(II)	Total	(I)	(II)	Total	(I)	(II)	Total
Bulgaria	4,166	3,785	7,951	6,990	27,250	34,240	2,483	1,985	4,468	693	432	1,125
Georgia	6,434	2,180	8,614	8,830	650	9,480	1,584	1	1,585	435	–	435
Romania	39,775	–	39,775	67,310	–	67,310	89,671	–	89,671	515	–	515
Turkey	44,805	18,090	62,895	167,920	4,120,000	4,287,920	6,008	12,730	18,738	2,257	1,713	3,970
Russia	1,935	23,586	25,521	2,030	698,530	700,560	415	13,076	13,491	504	533	1,037
Ukrainian	56,261	8,030	64,291	104,300	652,630	756,930	39,866	1,895	41,761	4,271	1,163	5,434
Int. Rivers	–	–	844,573	–	–	4,8803,520	–	–	229,181	–	–	39,888

(I) = Domestic + industrial sources. (II) = National rivers.

Table 12

Distribution of pollutant load totals of all Black Sea countries calculated from the inventory studies in 1995 (GEF Draft Report, 1996; reproduced by permission of BSEP)

Component	Total BOD (t/yr)	Total SS (t/yr)	Total N (t/yr)	Total P (t/yr)
Domestic + industrial	153,378	356,932	147,968	8,675
International rivers	844,573	48,803,520	229,181	39,888
National rivers	37,684	4,816,511	14,715	2,713
Total	1,035,635	53,976,963	391,864	51,276
% of Turkey	6.07	7.94	4.78	7.74

Table 13

Pollutant exchange through straits (GEF Draft Report, 1996; reproduced by permission of BSEP)

	TN (t/yr)	TP (t/yr)	TOC (t/yr)	BOD (t/yr)	TSS (t/yr)
Kerch Strait					
Net influx from Sea of Azov	43,900	3,100	–	90,200	1,948,300
Bosphorus Strait					
Influx into the Marmara Sea	190,000	12,000	1,520,000	–	–
Influx into the Black Sea	60,000	10,000	350,000	–	–

phosphorus loads are 43,900 t/yr and 3100 t/yr, respectively (Sarikaya et al., 1999).

The Black Sea is thus a recipient of high quantities of pollutants from point and diffuse sources. This demonstrates the significance of control of pollution, not only from the six Black Sea riparian countries, but also from all of the countries within the Black Sea basin. In this respect, every country in the basin is responsible to some degree for the pollution of the Black Sea. Therefore, coordinated and joint efforts are essential. Fulfilment of the commitments and adherence to the recommendations of the Strategic Action Plans are essential to reach the goals of reduced pollution and improved water quality.

As a result of the evaluation of the domestic and industrial inventories, 33 routine monitoring sampling stations were selected in Turkey. Among these, six were industrial discharge channels, six were sewage outlets and 21 were rivers and streams (Fig. 8). Seasonal composite water samples were collected both from the rivers and industrial or sewage discharge points, and at each station, measurements were made of several parameters using the standard methods given by APHA, AWWA (1990). Annual discharges of pollutants by each river and stream are given in Table 14.

Although the concentrations of most of the inorganic pollutants are fairly low in rivers and streams compared to domestic and industrial discharges, total annual pollutant fluxes are dominated by rivers and streams. Industrial and domestic outfalls are important sources of local pollution, but they account for only a minute fraction of annual pollutant fluxes. This dominating influence of rivers and streams on annual loads is due to overwhelmingly high quantities of water being carried by the rivers and streams compared to domestic and industrial discharges. Without exception, annual loads of all pollutants flowing into the Black Sea showed strong correlation with discharge.

Among the rivers and streams, the Sakarya and Filyos rivers, which are located on the western part of the coast, and the Kizilirmak, Yesilirmak and Mert rivers, which are located on the eastern Black Sea coast, account for more than 70% of the total annual fluxes of most of the conventional pollutants (Figs. 9–12). This ranking is expected because these rivers are the ones with the highest annual water discharge into the Black Sea. The only exception to

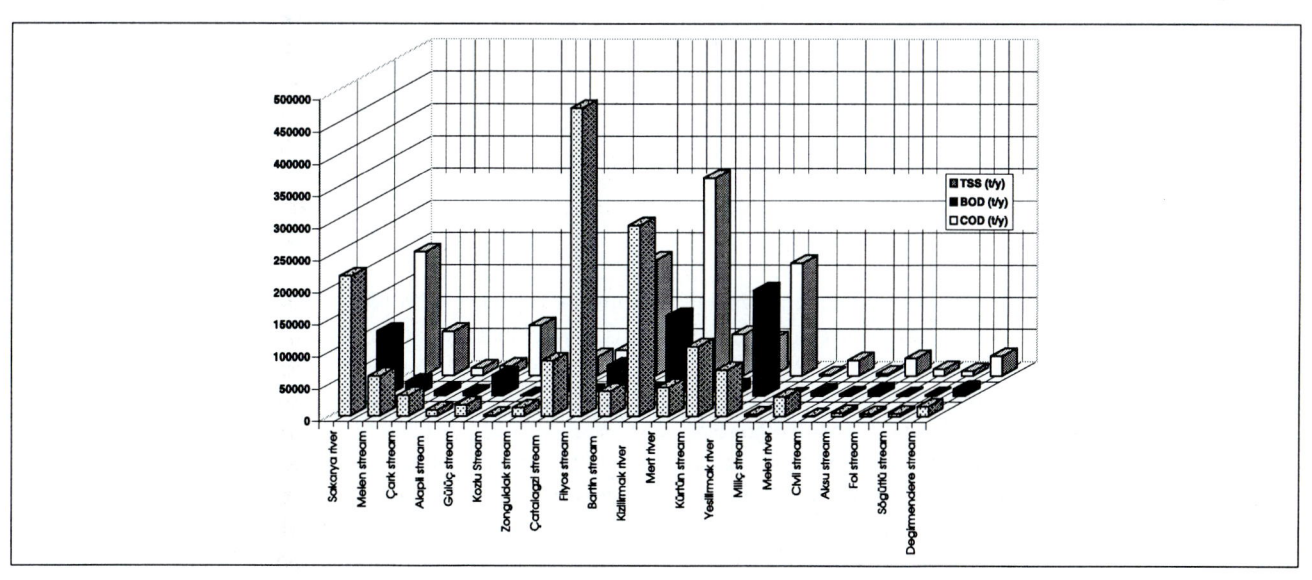

Fig. 9. Annual loads of TSS, BOD and COD from different sources along the Turkish Coast into the Black Sea.

Table 14

Annual load of pollutants from rivers and streams along the Turkish Black Sea coast

Name	Discharge (km³/yr)	TSS (t/yr)	BOD (t/yr)	COD (t/yr)	Ortho-P (t/yr)	Total P (t/yr)	NH₃-N (t/yr)	NO₃-N (t/yr)	NO₂-N (t/yr)	TKN (t/yr)	Detergent
Sakarya river	6.02	217,695	99,805	192,439	1,214.4	1,201.5	3,449	11,354	121	26,703	693.1
Melen stream	1.57	61,818	21,366	68,304	149.6	170.7	565	2,006	55	9,339	253.7
Çark stream	0.31	32,102	7,774	11,524	174.3	247.8	329	690	10	1,289	209.3
Alapli stream	0.27	9,328	4,460	14,539	44.4	60.7	67	550	4.8	647	88.6
Gülüç stream	1.19	17,413	32,214	77,277	43.6	77.5	1,459	5,530	24	3,206	180.4
Kozlu stream	0.02	1,438	291	864	10.9	12.4	96	71	1.7	76	4.4
Zonguldak stream	0.13	13,258	17,792	29,178	47.9	48.4	214	452	2.9	912	27.7
Çatalagzi stream	0.13	85,825	5,805	39,072	4.8	19.9	298	315	2.6	557	23.0
Filyos stream	3.22	478,764	46,779	180,102	566.9	574.6	554	2,152	93	4,777	614.8
Bartýn stream	0.36	38,636	7,367	19,812	28.7	36.5	102	81	8.9	394	57.1
Kizilirmak river	7.39	296,815	124,241	307,263	78.8	147.2	6,139	7,765	141	16,368	1,613.9
Mert river	1.06	44,848	20,996	64,010	371.7	473.7	1,178	1,694	384	441	970.5
Kürtün stream	0.16	108,245	14,772	56,106	157.8	45.8	55	231	10	654	524.8
Yesilirmak river	10.26	71,563	164,153	175,230	3,277.7	1,126.7	2,894	5,813	211	16,959	1,758.9
Miliç stream	0.43	2,666	378	1,601	153.9	65.6	6.3	57	4.3	500	524.6
Melet river	0.83	30,059	6,515	23,834	97.2	64.6	196	1,774	13	997	170.8
Civil stream	0.16	274	2,509	3,134	27.9	44.6	9.4	22	6.1	246	257.2
Aksu stream	0.97	5,233	9,073	27,115	84.3	41.2	98	1,282	12	640	220.1
Fol stream	0.20	3,469	1,471	10,091	67.8	67.4	100	483	8.1	158	138.3
Södütlü stream	0.12	4,270	1,478	7,137	28.7	9.4	98	480	2.8	158	73.4
Degirmendere stream	0.87	15,427	11,147	30,560	989.3	1,406.7	279	459	17.8	1,133	132.0

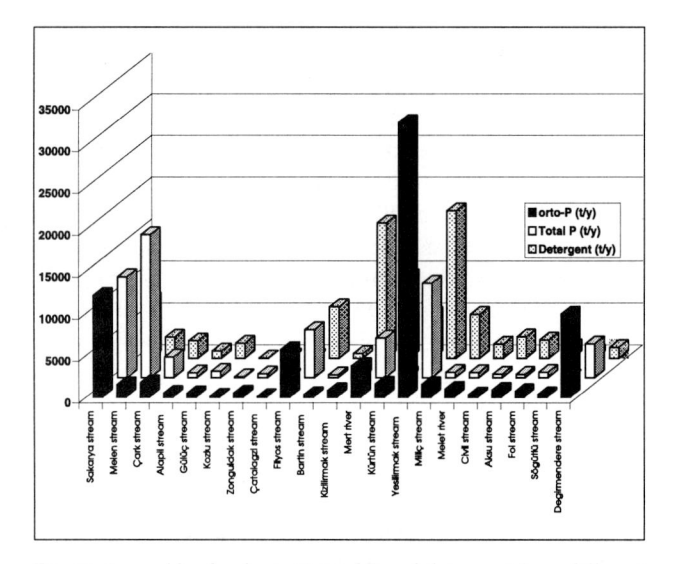

Fig. 10. Annual loads of orto-P, total P and detergent from different sources along the Turkish Coast into the Black Sea.

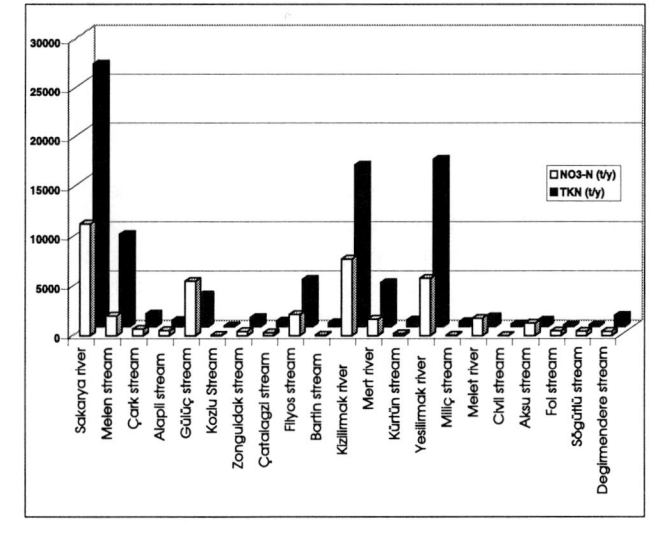

Fig. 11. Annual loads of NO₃-N and TKN from different sources along the Turkish Coast into the Black Sea.

the indicated trend was observed in the annual flux of total P (Fig. 10). In addition to the five main rivers indicated above, the Degirmendere stream accounts for 24% of the annual phosphorus discharge into the Black Sea.

Most of the rivers and streams flowing into the Black Sea are polluted with COD, BOD, NO_2, NH_3 and P. Organic pollution resulting from discharges of untreated domestic waste from the villages through which these rivers and streams flow appears to be the most important problem along the Black Sea coast. The sampled rivers and streams are not polluted with NO_3 (Fig. 11). Agricultural fertilizer use, which is the main source of NO_3 in river water, is not extensive along the Black Sea coast due to the low population density and lack of flat erodible land (Özdemir, 1997).

The annual load of pollutants directly from the sewerage system of cities located along the Turkish Black Sea coast are given in Table 15. Cities with high populations and discharge rates such as Samsun and Trabzon were the main domestic sources of pollution (Figs. 13–16). Inventory

Table 15
Annual load of pollutants from sewerage system of cities along the Turkish Black Sea Coast

Name	Discharge (km³/yr)	TSS (t/yr)	BOD (t/yr)	COD (t/yr)	orto-P (t/yr)	Total P (t/yr)	NH_3-N (t/yr)	NO_3-N (t/yr)	NO_2-N (t/yr)	TKN (t/yr)	Detergent (t/yr)
Sinop	0.004	596	827	1635	32.7	37.3	85.9	7.3	0.13	114.6	4.7
Samsun	0.008	1600	2054	3037	46.9	62.4	25.6	12.3	1.04	132.0	48.1
Ordu	0.010	886	1946	820	54.9	68.1	19.3	17.1	0.14	44.8	51.8
Giresun	0.004	473	2063	2249	27.9	50.8	16.5	9.5	0.13	128.5	36.2
Trabzon	0.010	1489	2099	2221	69.3	49.8	9.6	30.2	0.13	208.5	118.5
Rize	0.009	276	1477	1282	32.7	43.9	41.8	14.2	0.15	285.5	30.5

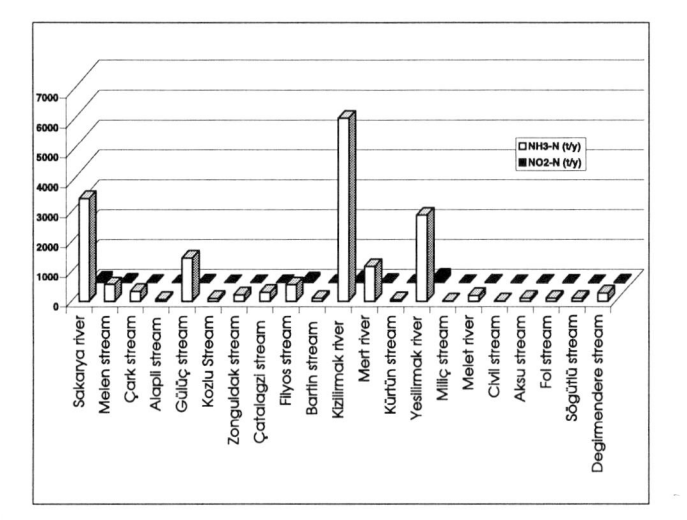

Fig. 12. Annual loads of NH_3-N and NO_2-N from different sources along the Turkish Coast into the Black Sea.

calculations (see Table 10) were higher than the monitoring results because they include all the cities, towns and related populations in their calculations whereas monitoring results were only the results of samples collected directly from the discharge point of sewerage systems of city centres.

In the study of Tuncer et al. (1998), concentrations of metals, 11 pesticides and PCBs measured in the Black Sea coast of Turkey were also given. Although industrial discharges do not make significant contributions to the fluxes of conventional pollutants, they account for a significant fraction of metal discharges into the Black Sea on an annual basis. This is particularly obvious for Cd, where industries account for approximately 90% of the annual Cd flux. Among the industries, the copper smelter located approximately 20 km east of the city of Samsun accounts for most of the metal discharges. The Kizilirmak, Yesilirmak, Sakarya, Filyos and Gülüç rivers are also important riverine sources for Cd, Cu, Pb and Zn.

In addition to inorganic pollutants, concentrations of 11 pesticides and PCBs, including lindane, heptachlor, heptachlor epoxy, aldrin, dieldrin, endrin, p,p'DDE, o,p'DDE, o,p'DDD, o,p'DDT and p,p'DDT were also measured (Tuncer et al., 1998). Except for lindane, aldrin and heptachlor epoxy, rivers dominate the fluxes of pesticides into the Black Sea. The highest flux was found for the o,p'DDE which is followed by the endrin and DDD. Each year 316 tons of o,p'DDE, 176 tonnes of endrin and 115 tonnes of DDD are being discharged from the Turkish coast into the Black Sea. Although the use of DDT in Turkey has been

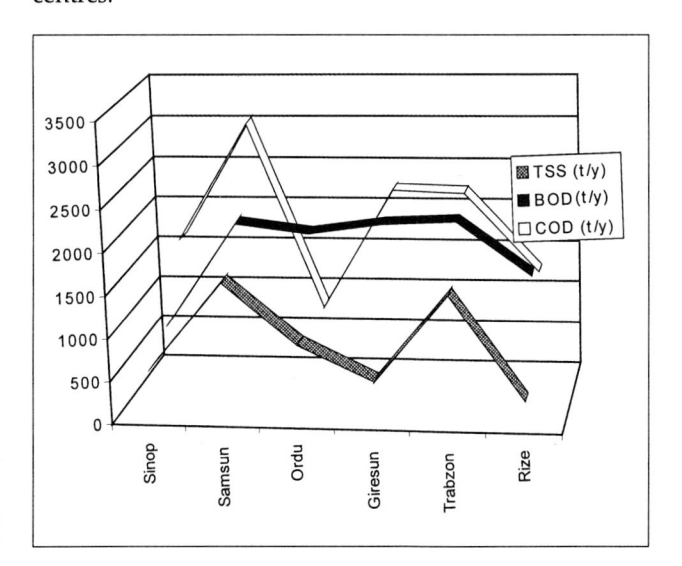

Fig. 13. Annual loads of TSS, BOD and COS from the cities along the Turkish coast into the Black Sea.

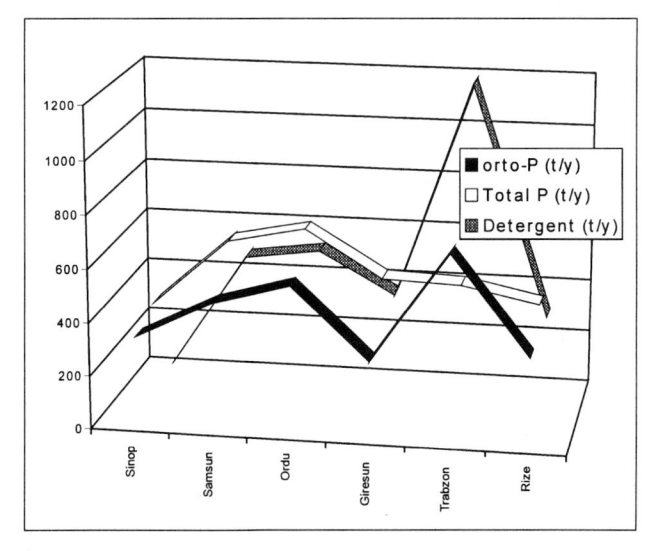

Fig. 14. Annual loads of orto-P, total P and detergent from the cities along the Turkish Coast into the Black Sea.

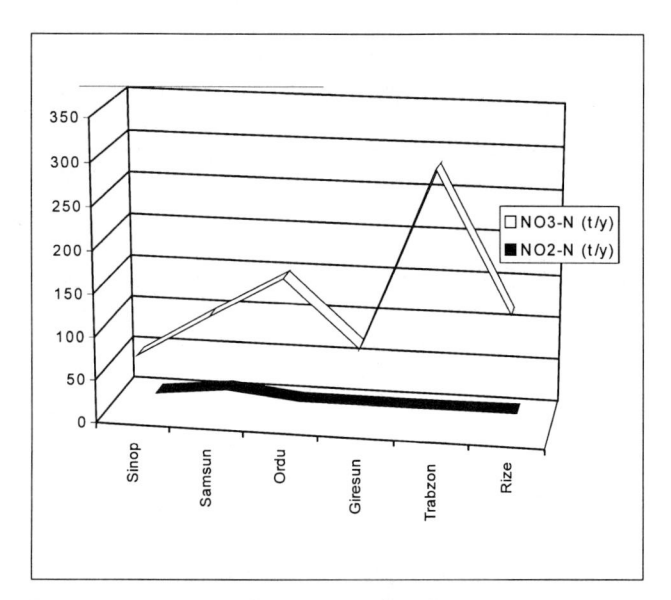

Fig. 15. Annual loads of NO_3-N and NO_2-N from the cities along the Turkish Coast into the Black Sea.

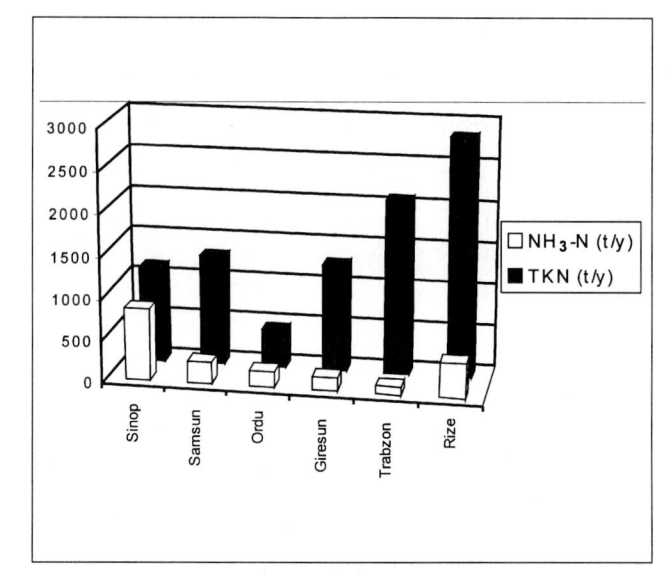

Fig. 16. Annual loads of NH_3-N and TKN from the cities along the Turkish Coast into the Black Sea.

banned for the last 30 years, approximately 100 tonnes of DDT are discharged via rivers suggesting illegal use of this attractive but dangerous pesticide. Special monitoring studies have also been started with regard to trace metals, PCBs, pesticides, PAH and related organic compounds, mainly in water, sediment and mussel tissues.

The local water pollution due to domestic waste discharges can be completely overcome with domestic waste treatment programmes, not only in large coastal cities, but also in cities along the banks of these polluted rivers and streams. Industries along the Black Sea coast have the potential to create local pollution problems for different pollutants. Among these industries, the Erregli iron and steel complex has an efficient, operational treatment plant and does not discharge significant quantities of any of the pollutants included here. However, wastes are discharged directly into the Black Sea without any treatment from the CATES thermal power plant in the Zonguldak area, the TUGSAS fertilizer plant and the KBI secondary copper smelter in the Samsun region. Among these, the CATES power plant is an important source of suspended particles. The TUGSAS and KBI plants are important sources of particularly toxic trace elements and to a certain extent nutrients as well. It should also be noted that the TUGSAS and KBI plants also generate substantial air pollution problems in the region with their high sulphuric acid emissions.

Beach Quality of the Southern Black Sea

Beach and bathing water quality inventory studies have also been completed for the Black Sea Coast of Turkey (Table 16) (Özkoç et al., 1997). Most beaches on the Black Sea coast are included in class 1b, which means that the beach and the bathing water quality are always of good quality. However, there is at present insufficient confidence in this good condition and they are not heavily used.

Table 16

Classification of beaches along the Black Sea coast of Turkey. The collection of water samples was done according to WHO (1995). Measurements were of microbiological and biochemical parameters, as well as physical (light transparency, water colour, surface flow velocity, surface flow direction) and visual (tar, foam, litter, algae) conditions as given in WHO (1995). (Final Project Report, 1997).

Beach City	Classification						
	1a	1b	2a	2b	3	4	5
Kocaeli	–	–	2	–	–	–	–
Sakarya	–	2	–	–	1	–	–
Bolu	–	1	1	–	–	–	–
Zonguldak	–	2	–	–	–	–	–
Bartin	–	–	–	–	1	–	–
Kastamonu	–	–	–	1	1	–	–
Sinop	–	3	–	–	–	–	–
Samsun	–	1	2	1	–	–	–
Ordu	–	1	–	–	1	–	–
Giresun	–	–	–	2	–	–	–
Trabzon	1	–	–	–	1	–	–
Rize	–	–	1	–	–	–	–
Artvin	–	–	2	–	1	–	–

Classification of beaches (WHO, 1995):
1a = Good quality, safe and fully used.
1b = Good quality, safe and for further development.
2a = Good quality, unsafe beach.
2b = Safe beach, bad quality of beach and water.
3 = Unsafe beach, bad quality of beach and water.
4 = Not suitable for tourists.
5 = Beach not used for bathing.

In the Black Sea coastal region, approximately 10,385,000 people have sewerage coverage and discharge an estimated 571,175,000 m^3/year. The state of recreational waters and beaches in the Black Sea coastal area is given in Table 17. Regular beach closures occur in many of the Black Sea

Table 17

State of recreational waters and beaches in the Black Sea coastal area
(GEF Draft Report, 1996; reproduced by permission of BSEP)

Country	Performed samples[1]	Number failed samples[2]	Percentage failed samples	Total length of beaches (km)
Bulgaria	859	45	5	150
Georgia	n.a.	n.a.	n.a.	160
Romania	1100	n.a.	n.a.	170
Russia	1800	200	11	600
Turkey	529	74	14	950
Ukraine	785	348	44	1400

1. Numbers of samples reported are partial and refer to some of the sample stations along the Black Sea coastal area.
2. Failure was judged by exceedence of 1/20 samples: faecal coliforms 2000/100 ml; faecal streptococci 400/100 ml.

countries and, although no cause–effect relationship has been clearly established, there are increasingly frequent outbreaks of serious waterborne diseases such as cholera and Hepatitis A. The need for better sewage treatment is evident, as is the need for greater transparency in the sharing of information on this subject.

CONCLUSIONS

The waters of the Black Sea have the unfortunate distinction of having the longest residence time among all enclosed coastal seas around the world. By comparison, even the enclosed Mediterranean Sea has the next longest residence time among enclosed coastal seas, approximately 90 to 100 years. The Mediterranean flushes and renews itself three and a half times during the time it takes the Black Sea to complete its very sluggish renewal cycle. Moreover, since the tides are small and currents and flushing are weak, inflows into the Black Sea remain relatively close to their source for long periods of time. Therefore, the pollutant loads from rivers and outfalls to the Sea move relatively slowly away from their origins, exacerbating the adverse impacts (Sezer, 1998).

Sustainable development of the Black Sea requires continued, even enhanced, international cooperation. The Black Sea Action Plan, BSEP, once adopted by the six coastal countries, together with the Bucharest Convention, formed a comprehensive framework for sustainable regional management. Special attention was given to domestic wastewater and toxicity, oil and grease and nutrient loads in determining the dominant point sources on the coast which affect human health, ecosystems, sustainability or economy. For the evaluation of priority areas, several effects have been considered. These are human health, effects on drinking water quality, natural aquatic life, effects on wetland and recreational areas, effects on other beneficial uses of the sea and effects on economy and welfare. For point sources which discharge oil and petrochemicals, risks of accidental pollution have also been considered.

Different types of pollutants in domestic and/or industrial discharges have different effects on human health and ecosystems at the point of discharge and in the surrounding environment. This surrounding environment may be very large and may extend beyond international borders. The risks increase proportionally with the quantity of the wastewater and concentration of the pollutant.

It was concluded in the study of BSEP that the Black Sea receives large quantities of domestic and industrial wastewater which is mostly untreated. The hydrological cycles of the northern rivers are being significantly manipulated and the wastewaters discharged into the basin adversely affect the Black Sea. It is expected that the Black Sea coast will continue to develop rapidly; new townships and industries, and increased wastes, will cause it further stress. As a result of the manipulation of the northern rivers, both the quantity and quality of the water reaching the Black Sea will continue to be considerably reduced. At present, about 50% of the water consumed goes back into the river systems without sufficient treatment.

The manipulation of rivers flowing into the northern Black Sea and the accompanying increase in the levels of pollution significantly affect fisheries, which are also faced with over-exploitation. Many economically important species of fish have disappeared regionally as well as from the entire Black Sea. Furthermore, increasing nutrient input through rivers and the alteration in the stratification of the coastal water masses have led to even more prolonged periods of eutrophication. Hypoxia has become a frequent phenomenon in certain areas in the northern region. Mass mortalities of the major species of the food chain accompany the hypoxia in water masses as large as 3500 km^3. The change in the ecology of the Black Sea is indicated by the explosion in numbers of jellyfish. On the other hand, significant amounts of oil emanating from some Black Sea countries are transported across the Black Sea. It is evident that this also poses a threat which requires serious monitoring efforts and regulatory measures.

In the 40-year period from 1950 to 1990 the phosphorus load transported by the Danube alone increased from 13,000 to 30,000 tons, while the nitrogen load soared from 140,000 to over 700,000 tons, most of it the result of intensified agricultural development and the widespread use of phosphate detergents. Over the same period, the amount of organic matter discharged from the Danube into the Black Sea increased five times, to 10 million metric tons. Since the 1950s, there has been a three-fold increase in nitrates and a seven-fold increase in phosphates in the Dnestr River as well. In 1991, the Danube contribution of inorganic nitrogen per year was more than double the load from the Rhine but less than half the total river input into the North Sea and four times that to the Baltic Sea (Mee, 1992).

According to GEF (1996) the total annual input of both phosphorus and nitrogen (dissolved and suspended) from the Danube, Dnestr, and the Dnieper is now on the decline. In 1995, measurements indicated an input from these three rivers of 40,000 metric tons of phosphorus and 264,000 metric tons of nitrogen.

In the 1950s, *Phyllophora* meadows covered an area the size of the Netherlands. The meadows were an important nursery for fish, providing nourishment and shelter from storms, and turned out to be vital to the maintenance of the sea's large stocks of anchovy, turbot, mackerel, flounder and sturgeon. They also helped maintain the chemical health of the sea's surface layers by emitting an estimated 2 million cubic meters of oxygen a year into the water. According to the "Saving the Black Sea Newsletter", the collapse of the productive fishery of the Black Sea since 1988 can largely be ascribed to the accidental introduction of the Ctenophore *Mnemiopsis* in the ballast water of ships. As regards the stocks of plankton-feeding fish, harvest rates as high as those in the late 1980s give pelagic fish little chance of competing with the invertebrate predator.

Riverine contributions of BOD, TSS and TP loads are about 85, 99 and 81% of the total. This demonstrates the significance of control of pollution, not only from the six Black Sea riparian countries, but also from all of the countries within the Black Sea basin. Although ranking changes with respect to the type of pollutant, every country in the basin is responsible. Therefore, coordinated and joint efforts are essential in order to effectively implement pollution control. Fulfilment of the commitments and adherence to the recommendations of the Strategic Action Plans are essential to reach the goals of reduced pollution and improved water quality in the Black Sea.

The solution of the Black Sea's environmental problems demands that uniform strict rules be adopted by each country. It means that the regulations should also cover those countries which influence the Black Sea environment through the rivers, mainly Danube, Dneper and Dnester. This in turn means joint solving of ecological problems over a very large territory covered by many states with varying standards of living. In this way the environment of the Black Sea is closely linked to the environmental problems of the large river catchments.

The inevitable involvement of economic and political interests could interfere with the creation of a general environmental policy for preservation of the Black Sea environment. Some good results have been achieved in this field so far (e.g. the Black Sea Environmental Programme), but the practical implementation and enforcement of the agreements through the hierarchical administrative channels turns out to be a very difficult and complicated task.

The fulfilment of the agreements requires that each country concerned creates an environmental policy compatible with international objectives. This means harmonisation of legislation and standards, preparation of effluent discharge inventories and mapping of major pollution sources, harmonisation of analytical and laboratory methods and establishment of water monitoring programmes. These components are stated in the activities of the Black Sea Environmental Programme, but the legislative frame for their realization still does not exist in all countries in the region.

ACKNOWLEDGEMENTS

Some part of this study has been supported through the GEF-BSEP Program and the Turkish Ministry of Environment.

REFERENCES

APHA, AWWA, WPCF (1990) *Standard Methods for the Examination of Water and Wastewater 18th Edition*. Copyright by American Public Health Association, Washington.

Arsov, R. et al. (eds.) (1995) Proceedings of the 1995 Black Sea Regional Conference on Environment Protection Technologies for Coastal Areas, Varna, Bulgaria. *Water Science and Technology* 32 (7).

Black Sea Transboundary Diagnostic Analysis, Report prepared by the GEF, BSEP Coordination Unit, Ýstanbul. (Level II), Marine Science Country Profiles, Unesco, Paris.

Bakan, G. et al. (1996) Evaluation of the Black Sea Land-Based Sources Inventory Results of the Coastal Region of Turkey, *Proc. of the Int. Workshop on MED & Black Sea ICZM*, Nov. 2–5, Sarýgerme, Turkey.

Balkas, T. et al. (1990) State of the Marine Environment in the Black Sea Region. *UNEP Regional Sea Reports and Studies No. 124*, UNEP, 41 pp.

BSEP (1994) Saving the Black Sea. *Official Newsletter of the GEF, Black Sea Environmental Programme*, Issue 1.

Delft Hydraulics Laboratory (1996) Integrated Water Resources Management Azov Sea.

Dinçer, A.C. Köse, E. and Durukanoglu, H.F. (1995) An economic evaluation of fishing vessels of the Black Sea. *MEDCOAST '95*, Spain, pp. 569–577.

Economopoulos, A.P. (1993) *Assessment of Sources of Air, Water and Land Pollution, Part One: Rapid Inventory Techniques in Environmental Pollution*. WHO, Geneva.

Final Project Report (1995) Black Sea Region Land-Based Routine Pollution Programme, Prepared by Research Center of Environment in Ondokuz Mayis University, Samsun, Turkey.

Final Project Report (1996) GEF National Black Sea Land-Based Pollution Routine Monitoring Project. Prepared by Research Center of Environment in Ondokuz Mayis University, Samsun, Turkey.

Final Project Report (1997) GEF National Protection and Management Routine Pollution Monitoring Project about the Monitoring of Land-Based Pollutants, Bathing Water and Drinking Water Quality. Prepared by Research Center of Environment in Ondokuz Mayis University, Samsun, Turkey.

GEF (The Global Environmental Facility, Black Sea Environmental Programme) (1996).

GEF (1996) Black Sea Transboundary Diagnostic Analysis, United Nations Development Programme, Istanbul.

GEF Black Sea Environmental Programme, CoMSBlack/Woods Hole Oceanographic Institution, The Intergovermental Oceanographic Commission of UNESCO (1996) *Black Sea Bibliography, 1974–1994*, ed. V.O. Mamaev, D.G. Aubrey and V.N. Eremeev).

GEF, Black Sea Environmental Programme, BSEP (1994) Joint First Meeting of the Working Parties for Routine and Special Pollution Monitoring, Summary Report.

Integrated Coastal Zone Management in the Mediterranean and Black Sea: Immediate needs for research, education/training, and implementation (MED and Black Sea ICZM 96), (Özhan, E., ed.), *Ocean and Coastal Management* **34** (3), 1997, 227–232.

Intergovernmental Oceanographic Commission (1993) Black Sea Research Country Profiles.

Knudsen, S. (1991) Change in the fishermen's adaptations along the eastern Black Sea coast of Turkey, Paper presented at the Symposium: Ecological Problems and Economical Prospects of the Black Sea, Istanbul, Turkey.

Livingstone, H.D. et al. (1986) Chernobyl fallout studies in the Black Sea and other ocean areas, New York, Department of Energy, Tech. Rep. EML-460.

Mee, L.D. (1992) The Black Sea in crisis: The need for concerted international action. *Ambio* **21** (4), 278–286.

Murray, J.W. et al. (1989) Unexpected changes in the oxic/anoxic interface in the Black Sea. *Nature* **338** (6214), 411–413.

Neumann, G. (1942) Die absolute Topographie des physikalishen Meerresniveaus und die Oberflachenstromungen des Schwarzen Meeres, *Ann. Hydrgr. Berl.* **70** (9), 265–282.

Oguz, T. et al. (1992) The Black Sea circulation: its variability as inferred from hydrographic and satellite observations *Journal of Geophysical Research* **97** (12), 569–584.

Oguz, T. et al. (1993) Circulation in the surface and intermediate layers of the Black Sea. *Deep Sea Research* **40** (8), 1597–1612.

Özdemir, A. et al. (1997) The nitrogen loads carried by rivers and streams to the Black Sea in Turkey. *MEDCOAST '97*, Malta, pp. 327–336.

Özhan, E. (ed.) (1996). Proceedings of the 2nd International Conference on the Mediterranean Coastal Environment, MEDCOAST 95, Oct. 24–27, 1995. Tarragona, Spain.

Özkoç, H.B. et al. (1997) Investigation of land-based pollution parameters in the surface waters of the Black Sea, *MEDCOAST '97*, pp. 315.

Özsoy, F. et al. (1986) On the physical oceanography of the Bosphorus, the Sea of Marmara and Dardanelles, Erdemli, Içel, Turkey, Middle East Technical University.

Öztürk, B. (1996) The Turkish national programme for the conservation of the Black Sea dolphins, *Int. Symp. on the Marine Mammals of the Black Sea*, 27–30 June, 1994, Istanbul, Turkey, pp. 108–110.

Öztürk, I. et al. (eds.) (1995) Selected Proceedings of 2 IAWQ Int. Symp. on Marine Disposal Systems held in Istanbul, Turkey, 9–11 Nov., 1994. *Water Science and Technology* **32** (2).

Öztürk, B. (ed.) (1994). Proceedings of the First International Symposium on the Marine Mammals of the Black Sea, 27–30 June, 1994, Istanbul, Turkey.

Öztürk, I., Eroglu, V., and Akkoyunlu, A., (1992) Marine outfall applications on the Turkish Coast of the Black Sea. *Water Science and Technology* **25** (9), 203–2220.

Pavlov, V. et al. (1996) Impact of fishing on Black Sea dolphins off the Crimea coasts, *Int. Symp. on the Marine Mammals of the Black Sea*, 27–30 June, 1994, Istanbul, Turkey, pp. 41–43.

Sarikaya, H.Z., Sevimli, M.F. and Çitil, E. (1999) Region-wide assessment of the land-based sources of pollution of the Black Sea. *Water Science and Technology* **39** (8), 193–200.

Sezer, S. (1998) Integrating Economics into Environmental Management Case Study: The Black Sea Environmental Programme. *Kriton Curi Int. Symp.*, Bogaziçi University, Istanbul.

Sorensen, J. et al. (1997) The Black Sea: Another Environmental Tragedy in Our Times? *MEDCOAST '97*, pp. 741.

Sorokin, Y.I. (1983) The Black Sea. In *Estuaries and Enclosed Seas. Ecosystems of the World*, Vol. 26, ed. B.H. Ketchum. Elsevier, Amsterdam, pp. 253–291.

Tolmazin, D. (1985) Changing coastal oceanography of the Black Sea. I. Northwestern shelf. *Progress in Oceanography* **15**, 217–276.

Tugrul, S. and Polat, C. (1995) Quantitative comparison of the influxes of nutrients and organic carbon into the Sea of Marmara both from anthropogenic sources and from the Black Sea. *Water Science and Technology* **32** (2), 115–121.

Tugrul, S. et al. (1992) Changes in the hydrochemistry of the Black Sea inferred from water density profiles. *Nature* **359** (6391), 137–139.

Tuncer, G. et al. (1998) Land-based sources of pollution along the Black Sea coast of Turkey: concentrations and annual loads to the Black Sea. *Marine Pollution Bulletin* **36** (6), 409–423.

UNCED (1992) Turkey: National Report to UNCED, United Nations Conference on Environment & Development, Ministry of Environment of Turkey. United Nations Publ., New York.

Ünlüata, U. and Oguz, T. (1988) A review of the dynamical aspects of the Bosphorus. In *On the Atmospheric and Oceanic Circulation in the Mediterranean*, ed. H. Charnock. Erdemli, Içel, Turkey, Middle East Technical University.

Vinogradov, M. and Tumantseva, N. (1993) Some results of investigations of the Black Sea Research. *IOC Marine Science Country Profile Series, Black Sea Research Country Profiles Level II*, 3, pp. 80–94.

WHO (1993) Assessment of sources of air, water and land pollution, Part 1: Rapid inventory techniques in environmental pollution. WHO, Geneva.

WHO (1995) *Manual for recreational water and beach quality monitoring and assessment*, Draft. WHO Regional Office for Europe, European Centre for Environmental and Health.

Yüksek, O. et al. (1995) Coastal erosion in eastern Black Sea region, Turkey. *Coastal Engineering* **26** (3–4), 225–239.

Zaitsev, Y. and Mamaev, V. (1997) Biological Diversity in the Black Sea: Main Changes and Ensuing Conservation Problems, *Proceedings of the Third Int. Conf. on the Med. Coastal Env., MEDCOAST '97*, pp. 171.

THE AUTHORS

Gülfem Bakan
Ondokuz Mayis University, Faculty of Engineering, Department of Environmental Engineering, 55139, Kurupelit, Samsun, Turkey

Hanife Büyükgüngör
Ondokuz Mayis University, Faculty of Engineering, Department of Environmental Engineering, 55139, Kurupelit, Samsun, Turkey

Chapter 19

THE GULF OF MAINE AND GEORGES BANK

Jack B. Pearce

With the exception of the Boston Harbor and one or two other port/industrial sites, the Gulf of Maine has remained relatively unimpacted by urban and industrial contamination compared to its earlier status and waters such as the New York Bight. This chapter reviews natural environmental variables, the status of shallow water embayments and coastal waters, and the condition of the contained living marine resources. Principal fisheries are in poor condition today, albeit more intense management of the "take" and habitat quality, are beginning to result in limited improvements to the status of the stocks and their habitat.

Seas at The Millennium: An Environmental Evaluation (Edited by C. Sheppard)
© 2000 Elsevier Science Ltd. All rights reserved

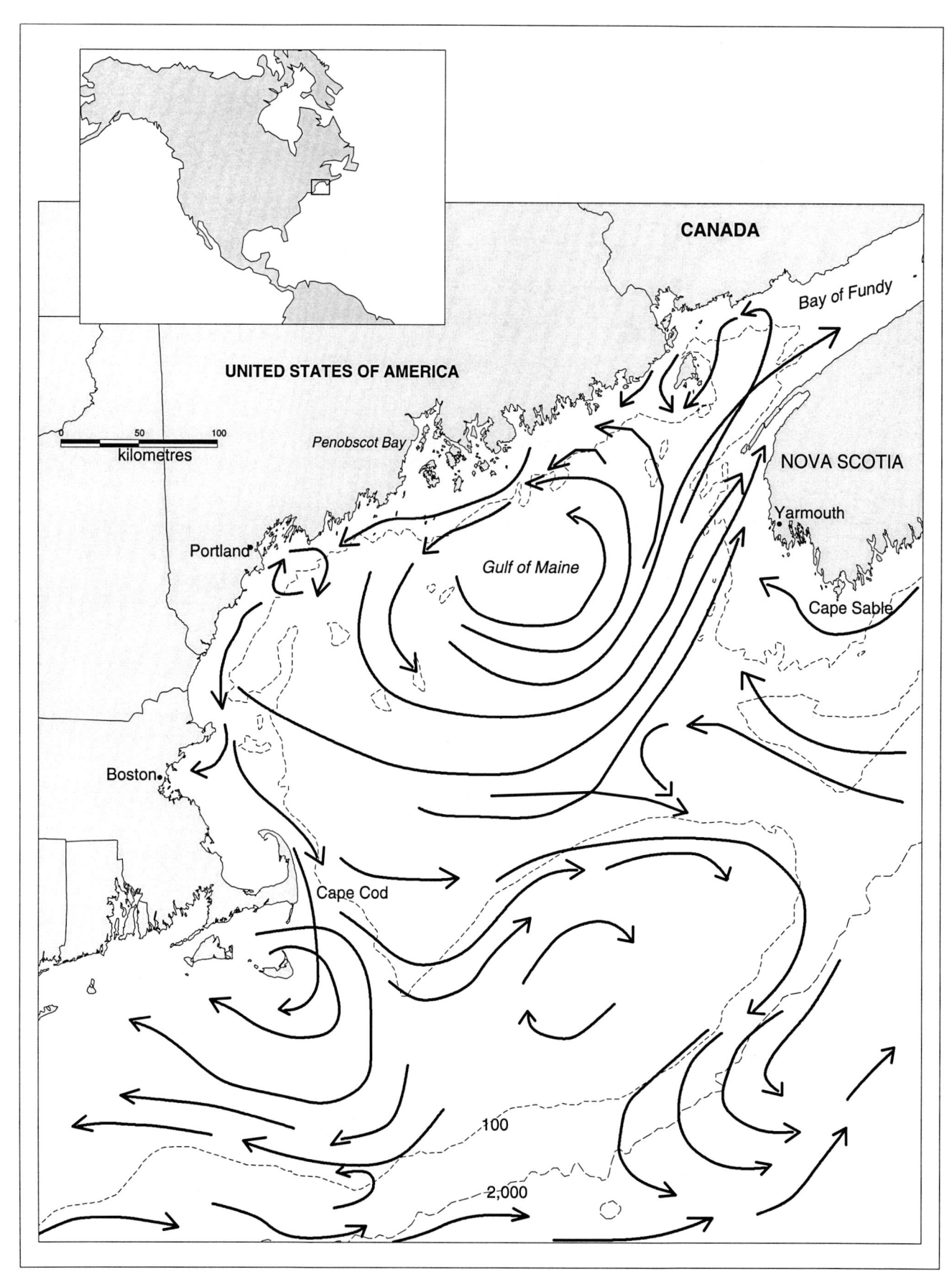

Fig. 1. Map of Gulf of Maine and Georges Bank. Bigelow's (1927) classical circulation schematic for the Gulf of Maine region in summer months, based on multiple experiments with surface drift bottles, hydrography, and plankton distributions.

INTRODUCTION

Unlike many of the coastal and shelf waters discussed in this volume, the Gulf of Maine has been well-studied during the past century (see Townsend, 1997). In part, this is because of its juxtaposition to a number of marine research centres (Huntsman Laboratory, Marine Biological Laboratory, National Marine Fisheries Service Laboratories, and Woods Hole Oceanographic Institution) and academic institutions, but also because of the curiosity of a wide range of scientists who understood early the dependency of the living marine resources, and the economics of coastal communities, on an extensive knowledge of the physical aspects of the embayment and the numerous species which inhabit it. Henry Bigelow was foremost among the early pioneers who dedicated much of their careers to investigations of the Gulf. Within a decade following the war to end all wars, Bigelow (1927) had published extensively on the circulation of the Gulf. His work was based on classical hydrographic measurements and drift-bottle observations. As a professor, Bigelow had considerable interests in the biology of marine fishes and their relationships to physical and chemical driving forces; his 1927 work was published in the *Fishery Bulletin* and still serves as a starting point for many oceanographers in eastern North America. Some four decades later, using basic drift-bottle data, Bumpus and Lauzier (1965) elaborated upon Bigelow's studies and suggested that seasonal changes in current strength were related to stratification. Since the sixties, scores, even hundreds, of papers and proceedings have resulted in several syntheses of the physical oceanography of the Gulf of Maine, most of these published between the late 1970s and mid 1980s. The first large compendium was the volume on Georges Bank, edited by Backus and Bourne (1987) and published by MIT Press. While this volume focused on the Bank, propinquity demanded that its authors draft an extensive coverage of the Gulf. The several sections and chapters on Physical Science remain as an authoritative narrative on the geology, climatology and meteorology, waves and tides, seasonal circulation, and erosion and scouring. The discipline of marine science is not likely to see, again, a comparable group of authors.

In addition to its coverage of the Physical Sciences, *Georges Bank* (Backus and Bourne, 1987) provided chapters on phytoplankton, primary production, and microbiology; zoology and secondary production; and the fisheries and some three score authors provided coverage of the realms of plankton, benthic communities, and fisheries, including the dynamics of food chains, man's impacts on the biota, and the futures of these systems. Entirely too large to constitute a summary, this volume of almost 600 pages is, in effect, a treatise on a significant portion of the boreal northwest Atlantic and a significant history.

In 1991 the Gulf of Maine Council on the Maine Environment (GOM Council) sponsored a Scientific Workshop on the Gulf. Once more, some 250 scientists met for three days to discuss, review, and document what is known about the biological, chemical, and physical components of the Gulf of Maine system. More important, the sessions were guided by a manifesto for "Provision of the Environmental Information Needed to Substantially Improve Our Ability to Understand, Monitor, and Manage the Marine Environment of the Gulf of Maine" (Wiggin and Mooers, 1992).

The deliberations of some twelve score scientists eventually led to the Proceedings of the Gulf of Maine Scientific Workshop (Wiggin and Mooers, 1992), again standing as a comprehensive (394 pp.) statement of what we thought we knew about the Gulf and what we yet needed to know in order to "effectively manage" this water body.

It was less than four years after the GOM Council Workshop that the Regional Association for Research on the Gulf of Maine sponsored a Scientific Symposium and Workshop which culminated in the Proceedings of the Gulf of Maine Ecosystem Dynamics (Wallace and Braasch, 1997). This Symposium was organized around (1) nine panel members who elaborated on a variety of management issues, (2) a score of presenters to the plenary sessions, and (3) four working group reports. Finally, the various aforementioned activities considered some ninety poster papers which were displayed during the Symposium.

At the same time as these workshops/symposia were being held, Conkling (1995) and others published the volume *From Cape Cod to the Bay of Fundy: An Environmental Atlas of the Gulf of Maine*. This uses several score satellite images, aerial photographs, and maps, along with photographs and tabular materials, to illustrate the physical environment and its contained biological stocks. The volume not only covers the shelf and coastal zone, it also illustrates "...the thin edge between land and sea", the estuaries, rivers, and coastal ponds, farmlands, and forests.

In addition to these aforementioned volumes, hundreds of individual papers have been written albeit most of the significant titles are referenced and well summarized in these volumes. The Gulf and Georges Bank generally occupy the area between 41° and 45°N, and 65° and 70°W. The Scotian shelf, Browns Bank, Georges Bank, Nantucket Shoals, and Cape Cod tend to separate the Gulf of Maine *per se* from the open ocean. Boreal waters from the north (Labrador and St. Lawrence seaways) flow into the Gulf over the Scotian shelf, through the Northeast Channel, and via advective processes inherent in the Georges Bank system. Finally, waters exit from the Great South Channel, to flow south into the Middle Atlantic Bight. This movement of water is updated and well summarized in Beardsley et al. (1997) and Townsend (1997).

There are three principal basins in the Gulf: Georges Basin to the east, Jordan Basin in the northeast, and Wilkinson Basin in the west. In addition, there are several ledges; Cashes and Jeffries are especially prominent. The basins are all deeper than 200 m and, with coastal estuaries and shoals and ledges, result in a highly diverse habitat which has, in past decades and centuries, constituted one of the continent's, if not the world's, more productive fishery grounds (see Boreman et al., 1997 and Steneck, 1997).

THE NATURAL ENVIRONMENTAL VARIABLES: CURRENTS, TIDES, WAVES AND NUTRIENTS

The Gulf has one of the planet's greatest measured tidal systems and, because of its relief, receives cold, somewhat reduced salinity boreal waters as well as nutrient-rich offshore waters. Because of several vertical exchange processes (see Smith, 1997), both horizontal and vertical compartments are affected by exchanges of different waters and nutrients at time scales of minutes to months, and, perhaps, decades, again accounting for the unusual productivity of the Gulf and its bordering shoals and banks.

In the most general sense, the Gulf has a counter-clockwise coastal circulation (Fig. 1) carrying inputs from the Scotian shoals and Northeast Channel, as well as riverine inputs to the Maine Coast and Bay of Fundy to the west, thence to the southwest to exit via the Great South Channel. As already suggested, the major currents can vary widely due to riverine input, winds, warm core ring phenomena, and the gross movement of offshore Labrador boreal waters (Beardsley et al., 1997).

Townsend (1997) says that Scotian shelf waters entering the Gulf at surface, and slope waters via the Northeast Channel, dominate the flux of nitrogen to surface waters, accounting for some 54 gCm^{-2} yr^{-1} of *new* primary production. There are other external sources of nutrients including atmospheric inputs, riverine exports, and certain anthropogenic activities.

The average rate of planktonic primary production in the Gulf is 290 gCm^{-2} yr^{-1}; however, there is considerable variation. For instance, the rate over Georges Bank is some 400 gCm^{-2} yr^{-1} (see O'Reilly et al., 1987). Even though primary production is greater over the Bank, secondary production is less, 18% of the phytoplankton production (Cohen and Grosslein, 1987). The same authors report secondary production in the Gulf to be 26% of phytoplankton production. This leads to an important thought earlier framed by Dugdale and Goering (1967) and cited by Townsend (1997): "...measurement of primary production alone is not enough to assess the capacity of a region to support production at higher tropic levels in the food-chain".

Townsend suggests that because the Gulf of Maine primary production consists of a larger proportion of "new" primary production, than on Georges Bank, therefore allowing for a relatively greater level of secondary production in the Gulf, there are considerable temporal and spatial differences in zooplankton between these two entities. Durbin (1997) notes four hydrologically different regions in the Gulf, each with a distinct, representative zooplankton community. These are the estuarine areas, affected by riverine transport, a well-mixed coastal region(s), the central Gulf, and shallow bordering banks (Georges and Brown) and shoals (Nantucket). The estuarine and coastal systems have seasonal changes in the relative abundance of certain key species. This phenomena is thought to be due to the extensive range of temperatures (well below 0°C to near 30°C) and salinities measured in coastal and estuarine waters relative to the central Gulf and the banks (Durbin, 1997). Other plankton and benthic species are precluded from coastal habitats by occasional, extremely low winter temperatures. For instance, the parasitic mussel crab, *Pinnotheres maculatus*, *never* occurs in intertidal or sublittoral habitats north of Cape Cod, even though its principal host, the blue mussel, *Mytilus edulis*, does; however, the crab has been regularly collected in mussels trawled from 200 m of water in the Central Gulf of Maine. Winter temperatures are relatively greater at depth.

In regard to long-term variation, there are not many extensive works documenting change in species other than fish. The latter, as discussed later, has shown ups and downs, but severe declines in standing stocks of fishes occurred because of overharvesting in recent decades (Boreman et al., 1997). Jossi and Goulet (1993) reported that the ecologically important copepod, *Calanus finmarchicus*, routinely collected from the Gulf, had an increasing trend between 1961 and 1989. During the past decade several exotic invertebrates have been introduced to the Gulf, or have spread to other areas once introduced to the Gulf of Maine. As an example, the grapsoid crab, *Hemigrapsus sanguineus*, invaded New Jersey from Japan about a decade ago (McDermott, 1998). By 1997 the crab had spread northward to the Woods Hole area and in 1998 it was reported several times in Cape Cod Bay, the southwestern extent of the Gulf of Maine. Just as the introduced (from Europe) green crab, *Carcinus maenus*, this exotic crab can cause extensive damage to shellfish beds and mariculture stocks.

MAJOR SHALLOW WATER MARINE AND COASTAL HABITATS

The Gulf of Maine is serviced by several moderate sized riverine systems, most of which have significant estuaries at their mouths (Fig. 2). Much of the coastline of the Gulf is characterized by rocky out-croppings, and headlands although there are some sandy beaches, especially at or near the mouths of rivers. Harvey et al. (1995) described in considerable detail the rocky intertidal zone and estuaries of the Gulf, with special attention to the salt marsh plants, seagrasses, and seaweeds and kelp. They note that river basins and runoff contribute 250 billion gallons of fresh waters to the system annually. There are a score of principal estuarine embayments which receive these freshwaters, mixing them with saline waters. Major rivers (and estuaries) include the St. John, St. Croix, Penobscot, Kennebec/Androscoggin complex, Piscataqua/Great Bay, Saco, and Merrimac.

A principal flowering plant, the marsh grass, *Spartina*, serves to filter and entrap sediments and organic debris being carried seaward, thus continually adding to the areal

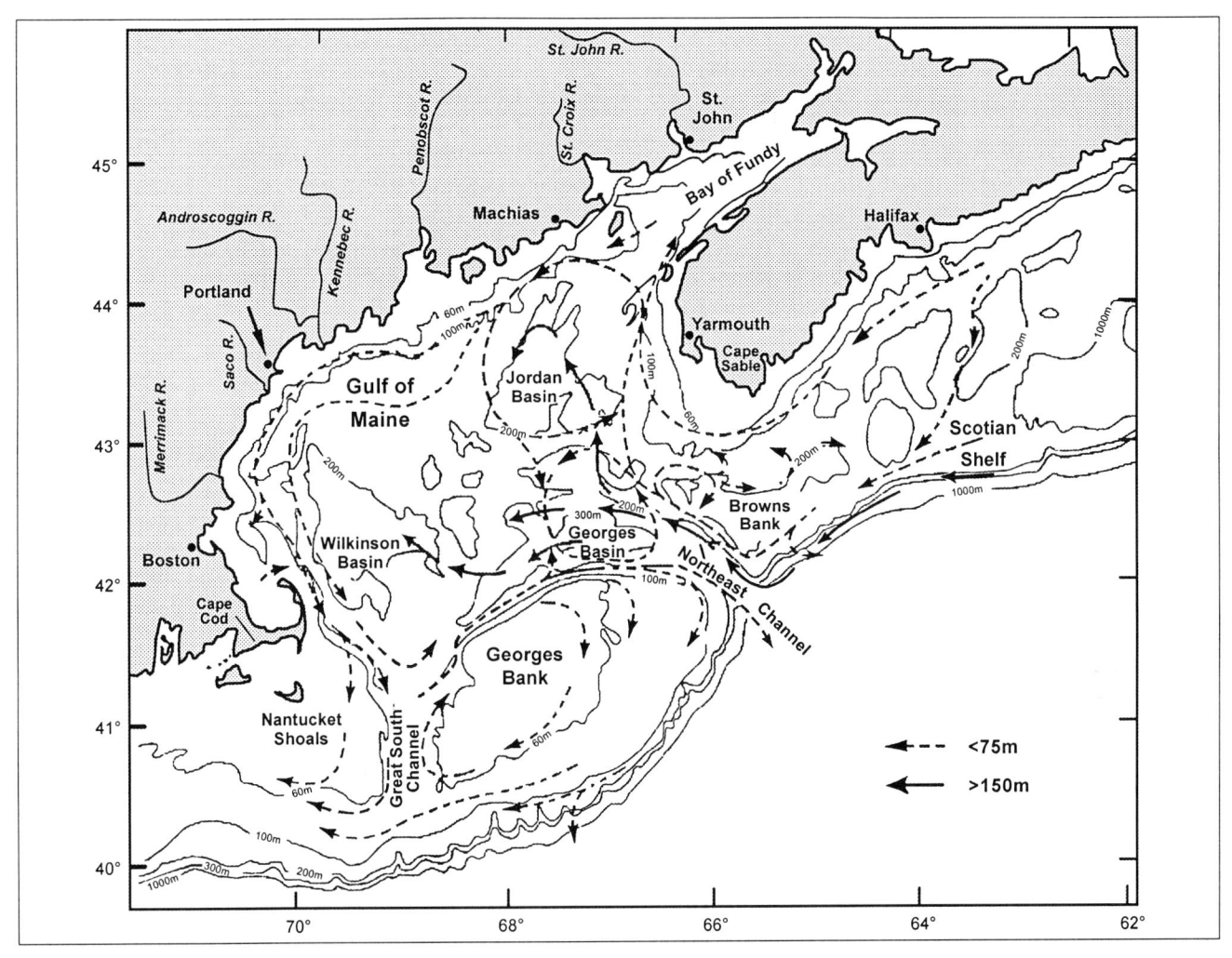

Fig. 2. Schematic of upper and lower layer circulation in the Gulf of Maine during the stratified season based on most recent moored, hydrographic, and drifter data. Major streams, channels, and cities and topographic features noted (from Beardsly et al., 1997).

extent of deltas and estuaries. Sizable salt marshes exist at the mouths of the aforementioned rivers as well as along the northern margins of Cape Cod and at the head of the Bay of Fundy. Further seaward, but "blending-in", another seagrass, *Zostera*, occurs in deeper coastal waters. Like the *Spartina*, it provides unique habitats for biota, protects the seafloor and mainlands from storm surges and erosion, and continuously entraps particulate materials, thus maintaining the basis for marshlands, seagrasses, and seaweed habitats.

In deeper waters, over rock outcroppings, and in areas where currents and waves scour away the soft sediments, a variety of brown, green, and red algaes provide the dominant floristic base for community structure (see Witman, 1996). To a great degree the algal zonation is comparable to that of Great Britain and northern Europe as described decades ago by Yonge (1949), Southward (1958), and Lewis (1964). *Fucus* and *Ascophyllum* are dominant above the low water line, the large kelps *Alaria* and *Laminaria* below it to depths of 25 meters. Both the rockweeds and kelps provide shelter, habitat, stability, oxygenation, and a significant basis (particulate debris) for food webs in the Gulf of Maine.

The sinuous coastlines of the Gulf, including the estuaries, river mouths, rocky coasts, offshore islands, and occasional gravel or sandy beaches, provide a diverse habitat for many boreal invertebrates, estuarine and marine fishes, marine mammals, sea birds, and turtles. The seagrasses, rockweeds and kelps, cobble and boulders, and rocky overhangs, as well as the muddy sea floors, all contribute shelter and habitat for those flora and fauna which can accommodate to a wide range of temperature and salinity, winter intertidal ice scouring, currents and waves, and solar radiation effects which occur during the exceedingly low tides in the Bay of Fundy and other parts of the Gulf. Larsen, Witman and coworkers have carried out extensive studies of the benthos and should be consulted (Leichter and Witman, 1997; Witman, 1996).

Among the lower invertebrates of ecological and commercial significance in the Gulf are a wide variety of polychaete worms (some, i.e. the nereids and bloodworms, having commercial value as baits); smaller crustaceans such as the amphipods and isopods; several taxa of shrimp; larger grapsoid and cancroid crabs; lobsters; bivalves, including hard and soft clams, mussels, oysters, and scallops;

several sea stars and brittle stars; a number of small and large limpets, snails, whelks, and moonsnails; and extensive beds of the green urchin. The latter are being increasingly harvested, principally for export to Japanese markets. Because they prey on the early stages of seaweeds and kelps, in areas where active harvesting has removed most urchins, kelp has grown luxuriantly relative to earlier years and decades. The invertebrate life forms or taxa include myriad numbers of species, many, if not most, similar to those reported upon by workers in the British Isles and northern Europe (again, see Yonge, 1949; Southward, 1958; Lewis, 1964). If, however, one compares the number of macrobenthic littoral and sublittoral species in the Gulf with lists garnered from studies in the English Channel, the sea lochs of the Scottish West Coast, or the North Sea there tends to be a far greater number in the northeast Atlantic. Certainly, if one reads *Between Pacific Tides* (Ricketts et al., 1985) there is an indication of a far larger number of both infaunal and epibenthic species in comparable habitats (Straits of Georgia, Puget Sound, Monterey Bay) along the North American west coast. Although not exhaustive, Smith (1964) and Pollock (1998) provide reasonable lists of marine animals from northeastern North America, i.e. the Northwest Atlantic.

Most important, however, is the fact that the species lists for many areas, including the Gulf of Maine, are growing. As previously noted, the Asian grapsoid crab, *Hemigrapus sanguineus*, was reported from the New Jersey coast over a decade ago and by 1997 it had spread to Cape Cod and the Gulf of Maine via the Cape Cod Canal. Besides the introductions of invertebrate species, several algae have been carried from Europe and Asia to North America, thence spreading to the Gulf of Maine. *Codium*, or dead-man's fingers is an example (Witman, 1996). The continual addition of new species to regional waters of the Gulf will increase diversity but will also have poorly understood biological and ecological consequences (Carlton and Geller, 1993). Recently, experiments have been used to demonstrate the possible longevity of megalopa and crab instars of *Cancer irroratus* (a species found in the GOM) in chemical tanker ballast water (Hamer et al., 1998). Such studies will help to demonstrate the possibility for fauna to be moved great distances.

During the past century there have been numerous taxonomic and biological studies of the megabenthic population of the Gulf but relatively few quantitative investigations or assessments of community structure and productivity. Langton and Uzman (1989) used photography to document the distributions and abundance of larger benthic organisms and Rowe et al. (1975, 1982) used traditional cores and grabs to assess the fauna of the deeper basins of the Gulf. While there is a paucity of information on the megafauna, a review of Gulf studies to date suggests "...a remarkable diversity of plant and animal species", relatively unpolluted conditions, and unusually high levels of biological productivity (Waterman, 1995). Again,

Witman (1996) and Leichter and Witman (1997) report on many aspects of the benthos but especially on the epibenthic forms on outcroppings, seawalls, and pinnacles.

THE OFFSHORE SYSTEMS

As noted in the introduction and section on natural environmental variables, the Gulf of Maine is effectively separated from the continental shelf and slope by land masses such as Nova Scotia and Cape Cod as well as by Georges Bank. The latter actually constitutes a part of the shelf and is characterized by the Canyons which are prevalent along the northeast coast of North America (see for example Dillon and Zimmerman (1970), Roberson (1964), and Uchupi and Austin (1987)). It is an environment that is only periodically affected and moderated by the Gulf Stream and the periodic intrusion of warm-core rings which spin off the Gulf Stream. These phenomena and related events are well treated in several chapters in Backus and Bourne (1987). While previous studies have shown toxic metal contaminant concentrations in the central Gulf, until recently the data have not been robust enough to draw conclusions from. The U.S. Geological Survey is presently mapping these data for use in interpreting possible effects (USGS, 1998).

POPULATIONS AFFECTING THE AREA

Fishing was, perhaps, the first industry practised in the Gulf; in fact, fishers sailed from and returned their catch to Europe well before the Pilgrims settled the Plymouth colony in the early 17th Century. Later a range of endeavours characterized the early industrial bases of the Gulf; by the late 18th Century, Boston was a major harbor, mercantile centre, and producer of hundreds of manufactured products ranging from hardware to ships. Very soon after, Portland (Maine), Portsmouth (New Hampshire), St. John (New Brunswick), and scores of smaller cities and towns followed suit.

Until recently the Gulf's special physical and chemical attributes resulted in fish stocks supporting some 20,000 fishers and support persons. They garnered their living from one half million metric tons of fin- and shellfish harvested annually, and worth $650 million (U.S.) in the early 1990s (Apollonio and Mann, 1995). As detailed in Boreman et al. (1997), a severe depletion of fish stocks due to overharvesting has resulted in "...severe economic and social dislocations in hundreds of communities..." of Newfoundland, other Atlantic provinces, and the New England states (see Apollonio and Mann, 1995; Boreman et al., 1997). Moreover, the intense trawling and dredging necessary to garnering the catches has "...decimated the food chain on the seafloor off New England" (Allen, 1998). The consequences of overfishing and trawling (Witman, 1998) are, today, probably the greatest ecological and economic issues in most New England states and the Canadian Maritime provinces. Nevertheless, annual revenues for fishers and fish farms climbed to $989 million in 1997 according to the National Marine Fisheries Service (Globe, 1998).

Principal species taken from the Gulf for human consumption have long included the members of the cod family, i.e. cod, haddock, red and white hake, and pollack; flounders, redfish, and shrimp and lobster have also been economically important. Pelagic forms, i.e. herring and mackerel, have also been an important source of revenue, as have the highly migratory pelagic great tunas and swordfishes. The Atlantic salmon was once an important commercial and recreational species but, because of damming of rivers, pollution, siltation, and other consequences of agriculture and forestry, it is all but extinct in its former home streams in New England and much of the Maritimes! Likewise, the redfish, which lives in the more or less thermally stable deeper waters, and grows and reproduces slowly, was long ago fished to very low levels. The prognosis for the overall recovery of fish stocks in Gulf of Maine waters is not at all clear; Canada has shut down most of its groundfisheries almost completely (Nickerson, 1998) and, relative to the 1970s–early 1990s, the U.S. has taken draconian management steps which have recently reduced fishing.

Using satellite imagery, Conkling (1995) depicts the development to date in the watersheds bordering the Gulf (for instance see his figure on p. XIX). What is obvious is that the early settlements grew and expanded at the expense of riversheds, estuaries, and coastal waterways. To a large degree, early logging practices allowed this early-on in the coastal zone, later in the more interior regions. Initially, trees were felled along streams and rivers to be floated to coastal embayments for sawing and other processes necessary to commercial lumber and woods. Once riparian habitats were "clear-cut" it was an easy matter to poke roadways further inland and to introduce farming where only a decade before virgin timber was dominant.

The forestlands that we see today are the consequences of almost half a millennium of use; the first ships to arrive in the Gulf undoubtedly had their masts replaced with timbers from the bounding watersheds. Dated carbon ashes show that early explorers left not only campfires but also burned over forests. Even today Maine remains the most densely forested state (90%), with New Hampshire a close second (86%) (Conkling et al., 1995). Yet, with few exceptions, most of these forests are secondary, even tertiary growth, regardless of the original species complex. Intense logging was initiated over two centuries ago and most of the prime stands and more valuable species removed to build ships, homes, and furniture or to fuel the growing industries of the 18th and 19th centuries. Salmon spawning beds were being eroded and silted even as the Revolution and War of 1812 were being fought. Later, paper mills sited on key rivers leading to the Gulf introduced chlorine and other bleaches for the first time, and small smelters and forges discharged a variety of toxic trace metals, oils and greases, and organic wastes to riparian habitats and estuaries. With the clearing of forests came farming and the beginning of the use of fertilizers which when carried to aquatic systems resulted in riverine, estuarine, and coastal enrichment, plankton blooms, and occasional hypoxia. As timberlands bordering the Gulf disappeared, new energy sources were sought to power the developing industrial empires of North America. My ancestors left Yorkshire in the 1840s to grind grains with dammed water power near Toronto; in spite of the economy of water power, a transition was made to steam, electricity, and thence back to water, again, as a way to address the energy issues of the late 20th century.

At this time, the urban areas sited on or near the Gulf of Maine are relatively stable. Boston, the largest city, consists of a metropolitan complex of some one million persons, while the Portland, Maine area is home for 400 thousand. Scores of cities, including Salem and Gloucester, Massachusetts and St. John, New Brunswick and Yarmouth, Nova Scotia have variously 50 to 150 thousand souls within their metropolitan jurisdiction. Maine remains famous for its potatoes and blueberries, and most river valleys have some farming, ranging from hay and straw to dairying. Generally, the stony soils of New England and the maritimes do not, however, lend themselves to large scale agriculture; heavy agrarian runoffs, such as found in Chesapeake Bay area and the North American mid-West, are not representative of the area.

While the area was once characterized by extensive biological productivity, i.e. forests and fishes, it *never* had uniquely valuable or extensive geological deposits, i.e. metals, coal, other minerals, or oil. Consequently, *truly* heavy industry has not been on the scene, ever. Some smelting and petrochemical plants grew up locally but turn of the century industry produced mostly fabrics, including cotton and woollen clothes, felt, and brass and silverware. Boston and Portland have no petrochemical complexes which stretch for miles along waterways, as exist in Linden and Camden, New Jersey, Philadelphia, Pennsylvania, and Wilmington, Delaware. Today Boston is famous for its computer software, military weaponry, razors, and academic endeavours, as well as hospitals and high technology brain trusts. Because of its products, much of its export is via Logan International Airport, not Boston Harbor. Much the same can be said for Portland, St. John, New Brunswick, and other smaller cities on embayments; while they may ship fish and forestry products, today, with some exception, many products travel by truck or plane-occasionally by surface vessels or train.

As highways, malls, and airports were built, construction *once* resulted in considerable erosion and siltation. Today, however, federal and state (provincial) regulations require blockage of eroded runoff, settling basins, and recycling of eroded sediments. Even the building of individual homes or the placement of curbing requires screens and devices to contain freed sediments. Sand and gravel trucks and train hoppers must have covers to prevent the blowing of dusts, and mining sites cannot, any longer, use open washing systems to separate various grain sizes or categories of mineral wastes.

EFFECTS FROM URBAN AND INDUSTRIAL ACTIVITIES

As noted above, the riparian lands, forests, and river valleys adjunct to the Gulf have seen the consequences of fishing, forestry, agriculture, water diversion, and "urbanization" for over two centuries (Van Dusen and Johnson-Hayden, 1989; Cerulo and Hancock, 1986). Early in this development, "urbanization" was characteristic of the major ports such as Boston, Portsmouth, New Hampshire; Portland, Maine; and St. John, New Brunswick. Hundreds of villages and towns sprung up along riparian waterways as forestry, agriculture, and industry developed to take advantage of nature's abundance and vast amounts of water power. However, by the turn of the 19th century the forest reserves had declined, far better fields had been found in the midwest prairies and western plains, and water power had been supplanted by steam, electricity, and diesel. A literal mass exodus occurred and, overnight, many small Gulf watershed communities disappeared into second and third growth woodlots and forests. As the hinterlands lost their denizens, however, coastal towns and cities gained them; the new mills of Fall River and New Bedford, Massachusetts, Lowell and Boston, Massachusetts, Manchester and Portsmouth, New Hampshire, and Portland, Maine, resulted in burgeoning populations and altogether new forms of stress, i.e. industrial pollution (PCBs from early electronic factories), road and railway development, harbor dredging, and contamination from the ever increasing maritime fleets. This later "history" of the Gulf of Maine is well treated and summarized in Van Dusen and Johnson-Hayden (1989). The coastal industries grew variously during World Wars I and II; large numbers of military shipyards and other factories opened in all the states and provinces bordering the Gulf. It is only after some 50 years of relative peace that these providers of plowshares and swords have either shut down or adopted adequate pollution controls.

In many instances, old mills have evolved into housing, facilities for light industry (computers, software, and electronics), shopping malls, or clothing outlets. While some counties continue to show a net population loss, many coastal counties had population increases of 15–25% during the period 1980–1986, and even greater increases in the present decade; Cape Cod (Barnstable County, Massachusetts) is the fastest growing county in the state, perhaps in all of New England. The total 1986 population (approximate) of the coastal counties of states bordering the Gulf was 4,593,000; 95,000 in Nova Scotia, 134,000 in New Brunswick, 320,000 in Maine, 167,000 in New Hampshire, and 3,877,000 in Massachusetts (Van Dusen and Johnson-Hayden, 1989. Much of the development in recent years has been on the landmass bordering the southwestern sectors of the Gulf, albeit tourism has often more than tripled human populations and development in the small cities and towns of coastal Maine and New Brunswick, such as Boothbay and Bar Harbor, Maine or St. Andrews, New Brunswick. Often such development has been at the expense of agricultural lands or forests.

As mentioned previously, research on the Gulf started with the Huntsman staff (St. Andrews) and Henry Bigelow (1927) just after the turn of the century. While there is a fairly long history of marine biology and oceanography, there has been precious little work done on pollution and contaminants, their sources, fates, and effects, in the Gulf. In recent reviews, Farrington (1997) discusses the sources, transport, and fate of chemicals (trace metals and organics) of environmental concern in the Gulf of Maine, and McDowell (1997) provides an historical overview for biological effects of toxic chemical contaminants. Beginning with noting the importance of information on the distribution(s) of contaminants in space and time, and understanding how they are sequestered or partitioned in water, sediments, and biota, McDowell (1997) suggested that, based on a few studies, i.e. Larsen (1992), we can judge that: "Trace metals, chlorinated pesticides, polychlorinated biphenyls (PCBs) and polycyclic aromatic hydrocarbons (PAHs) are found in sediments and biota throughout the Gulf of Maine ecosystem". Given this, we should look for the sources. Farrington, considering the more recent literature, states that many of the point-sources of metals and organics have been dealt with and that the non-point sources in the Gulf "should be targeted". The latter includes atmospheric inputs, urban and agrarian run off, and inputs from historically contaminated sediments to overlying waters. He suggested that we should attempt to ascertain the relative contributions from each of these sources. As for temporal and spatial variability, Farrington says we must look at human interventions, i.e. locations of outfalls, dump sites and dredging, seasonal biological activity(s), and the physical dynamics of *nearshore* and *offshore* waters, giving special attention to the physical couplings between these two. Since there has been a paucity of studies of the farther offshore areas, more sampling is needed as well as a greater effort to elucidate the transport routes from estuaries, rivers, and other inshore areas. McDowell (1997), based on NOAA National Status and Trends (NS&T) Program data, says that several Gulf sites were among the twenty most contaminated in U.S. coastal waters; this was true for both trace metals and organics. Boston and Salem Harbors and Penobscot Bay seem to be most heavily contaminated. Again, these long-term monitoring studies generally confirm reports by Pearce and Johnson-Hayden (1986), Larsen (1992), and Buchholtz ten Brink et al. (1996), to the

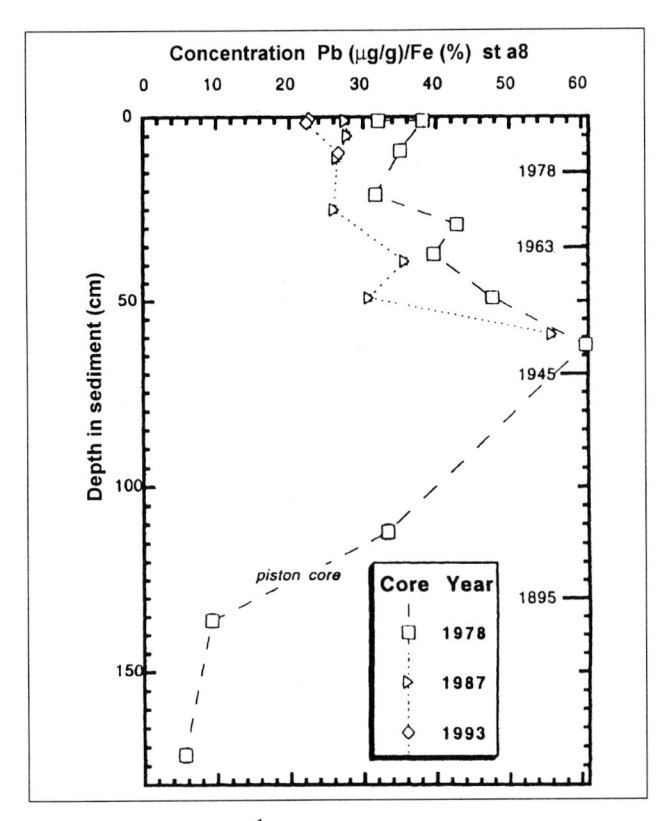

Fig. 3. Profiles of Pb (μg g^{-1})/Fe(%) concentration ratios with sediment depth in cores collected at station 8, between 1978 and 1993. Dates shown on the right axis apply only to the core collected in 1978 and are based on model calculations using zero age for the bottom of the mixed layer, 15 cm thick. Analytical error is typically within the size of the symbol. Sample deeper than 60 cm in 1978 were collected with a piston corer (from Bothner et al., 1998).

effect that the larger harbors (urban areas), and some smaller ones, are significantly polluted. The latter say: "The Gulf of Maine region has both some of the highest and lowest contaminant concentrations in the nation (U.S.)", with higher values tending to be reported in harbors or the more heavily used embayments such as Casco Bay. Even then they state: "Metals are all *below* [my emphasis] concentrations likely to cause effects in test organisms while PCB, DDT, and chorodane are mostly below levels suspected of evoking toxic biological responses". Citing the NS&T data (Gottholm and Turgeon, 1992), Buchholtz ten Brink et al. (1996) suggest that sediments from coastal areas of the northern Gulf are "relatively pristine", with increasing concentrations towards the more urbanized southwestern coastal areas. Very recent research results show that concentrations of metals in surface sediments of the Gulf have decreased during the period 1977–1993 (Bothner et al., 1998). These authors also looked at relative concentrations of selected metals at different core depths, or historical times. When these data are graphed or plotted a profile is produced (Fig. 3). They found that a history of lead in Boston Harbor sediments goes back to at least 1895! From that period to about 1945 there was a rapid increase in lead loading in sediments, to a maximum of 250 μg g^{-1}. The authors suggest this might have been due to ship building during World War II. Further accumulations of lead (via atmospheric inputs) occurred until 1973, when the nation switched to unleaded gasolines (Nriagu, 1989). The average lead in fine sediments in 1977–8 was 143 μg g^{-1}; in 1993 the average lead concentration was 62% the former level. The cause of such decreasing concentrations (for several metals) is thought to be a decrease in "...industrial point sources, sewage, street runoff, combined sewer overflows, and the atmosphere", probably reflecting better treatment of wastes, a reduction of lead in automobile exhausts, and the general application of technology and new regulations and legislation. Nevertheless, a 1993 study of relations between sediment concentrations of metals and biological effects suggested that some sediments in Boston Harbor might remain inimical to certain stages or instars of benthic organisms (Long et al., 1996). The concentrations of toxic metals and organics in sediments collected at smaller or less intensively developed embayments were usually far less (Fig. 4) than found in Boston Harbor (for example, see Pearce et al., 1985, and Reid et al., 1987).

When considering the sources, transport or fate, and effects of inorganic or organic contaminants, it is always of interest to know how much of a particular contaminant has become sequestered in the body tissues (flesh or muscle and various organs) of fishes and other marine organisms. Also, we want to know what the effects of this are. In the Gulf of Maine some information is known about body burdens in certain species, from specific localities. Most such areas are, however, generally inshore and in harbor or port situations where contaminants occur in greater concentrations. The National Status and Trends program (NS&T) has considerable data for such areas (Gottholm and Turgeon, 1992) (Fig 5). Again, biota from heavily contaminated habitats in the Gulf generally have greater body burdens. Data for fishes taken further offshore are quite rare, but the Northeast Monitoring Program (NEMP) collected eleven species of shelf and offshore fish in late 1979. These were analyzed for PCBs and petroleum hydrocarbons and the data reported as maps (Boehm and Hirtzer, 1982). Seven collecting stations could be considered as within the general bounds of the Gulf. Other stations were distributed over the shelf and south to Chesapeake Bay. In one species, the silver hake, 86% of the samples, contained relatively high levels of petroleum hydrocarbons (6–90 μg/g) and PCBs (0.1–0.5 ppm). Generally, higher values were found in fish collected from the New York Bight but some silver hake from the southern Gulf of Maine had significant levels of petroleum hydrocarbons and PCBs, especially when compared with red hake, haddock, and three species of flatfish. Since this preliminary but unique work was done, however, PCBs have been better managed, as have petroleum moieties, generally. One would expect that body burdens have dropped during the past two decades and it would be worth resurveying the area to ascertain what the levels currently are, and what the effects might be, experimentally.

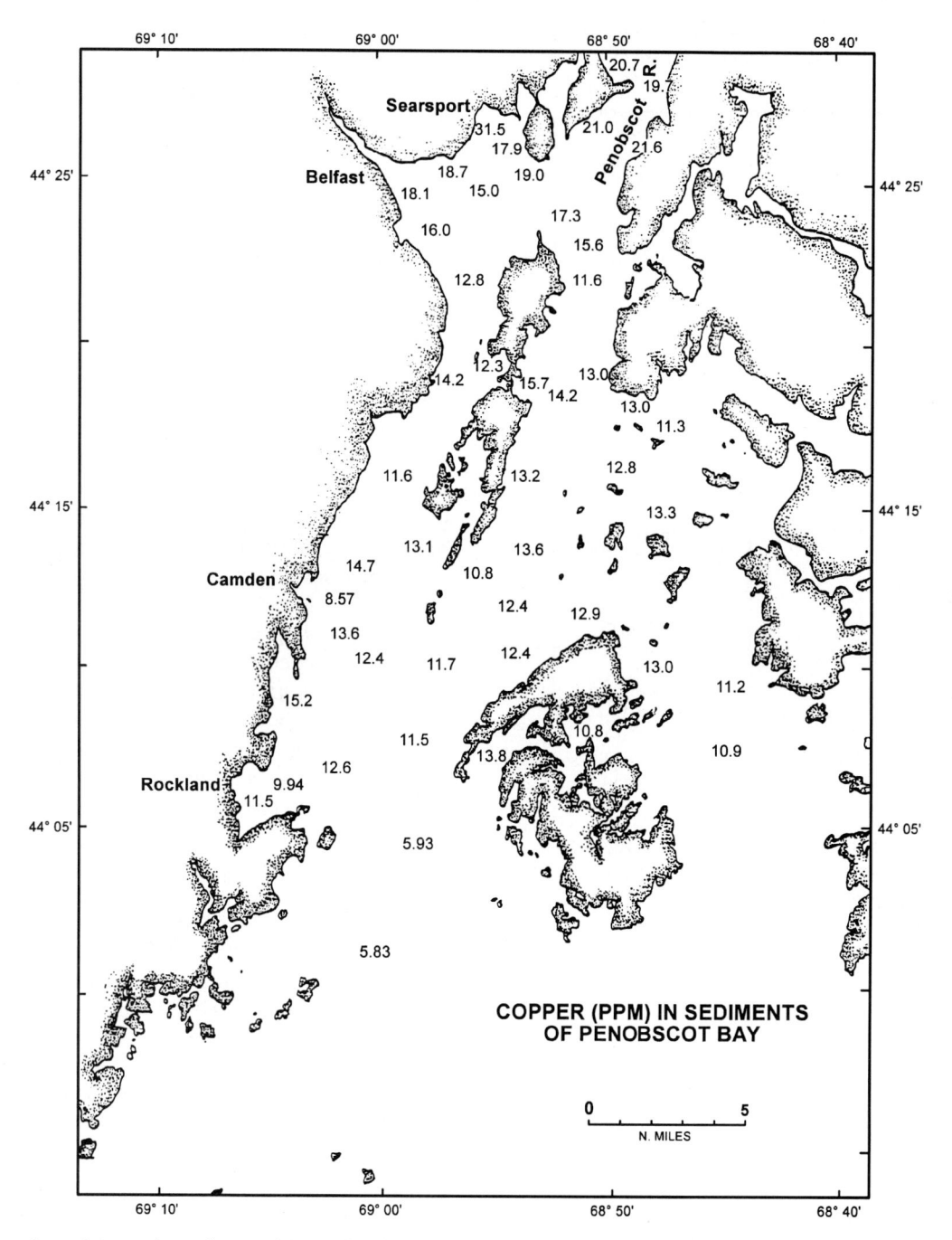

Fig. 4. Concentrations of copper (ppm, dry weight) in surficial sediments of Penobscot Bay, Maine, 1982 (from Pearce et al., 1985). Note increasing values for sediments from the inner bay.

Reflecting the concerns of Farrington (1997) and others, federal and state agencies have invested heavily in the development of coastal pollution discharge inventories. One of the first was developed for the Gulf of Maine; entitled "Gulf of Maine Point Source Inventory: A Summary by Watershed for 1991" (NOAA, 1994). It has already gone through significant change as clean water legislation has mandated local abatement or treatment at some 275 major and 1,655 minor "direct discharging point source facilities" in U.S. and Canadian watersheds draining to the Gulf.

At the same time as government agencies have identified major sources as well as the fates of certain specific contaminants, other nongovernment organizations have prepared literature compilations for the Gulf, as well as the several smaller embayments which form part of the overall system. Larsen and Webb (1997) prepared an environmental bibliography of information relevant to the marine ecosystem of Cobscook Bay, listing some six hundred annotated citations in 150 pp, including a list of keywords and an index. The large majority (430) of the titles were from 1970 to 1996. Other compendia were concerned

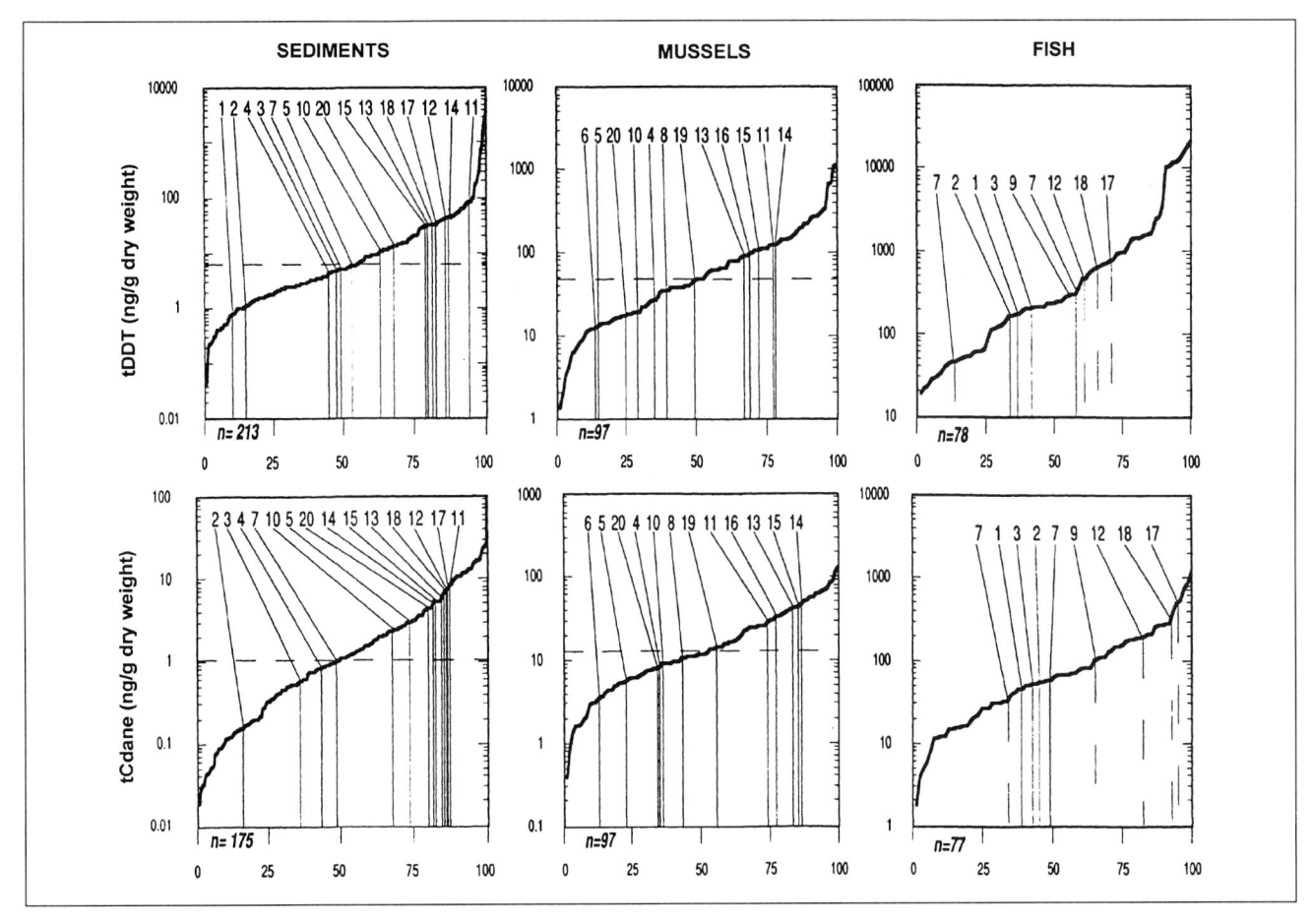

Fig. 5. Distributions of organics at Gulf of Maine sites in relation to nationwide (all NS&T sites) concentrations for sediments, mussel tissues, and fish livers (Vertical scale chemical concentrations are logarithmic; the horizontal scale is cumulative percent of national sites). Mean concentrations in sediments of total DDT (tDDT), an aggregated value of DDT and the metabolites DDD and DDE, are higher in the Salem and Boston Harbor area than at other Gulf sites. While tDDT concentrations in mussels from Salem and Boston Harbor area sites cluster around the 75th percentile, the remaining Gulf sites are below the national mean. Mean concentrations of tDDT in mussel tissue from the Folger Point and the Boston Harbor area sites are higher than other Gulf sites. The winter flounder livers collected from the Boston Harbor area sites have higher mean concentrations of tDDT than other Gulf area sites. Among NS&T winter flounder sites, Boston Harbor had the highest mean tDDT concentration (780 ppb) (from Gottholm and Turgeon, 1992).

with the Bay of Fundy (Plant, 1985) and the coast of Maine, especially near Eastport (Shenton and Horton, 1973). Many of the citations in these bibliographies are *apropos* the entire Gulf of Maine and the perspective student of the Gulf is urged to obtain copies, especially of Larsen and Webb (1997).

SUMMARY

The Gulf of Maine has been well studied in regard to most physical processes, and many biological and chemical systems. Sufficient research has been done to suggest that principal harbors and embayments historically have been contaminated, some heavily so. For instance, Boston Harbor has, on occasion, been found to be the most polluted in the northeast (Reid et al., 1987) and Casco Bay, especially the urban area around Portland, Maine, has sediments bearing significant levels of organic and inorganic contaminants (Reid et al., 1987; Buchholtz ten Brink et al., 1996). Larsen et al. (1985) found at least traces of PCBs at every station occupied in the Gulf. The sources included point discharges, urban and rural runoffs, dumping of dredged materials from contaminated harbor sediments, and atmospheric inputs (again, see Buchholtz ten Brink et al., 1996 for several figures and discussion). The question now is, has pollution abatement in the past two decades helped to improve demonstrably contaminated habitats? With recent regulation and legislation there has, in effect, been improvement; only "clean", relatively non-contaminated dredged materials can be dumped today. Moreover, as point sources of discharge are improved, harbors have shown reductions in sediment contaminant loads; in Boston Harbor, lead and other metals have decreased to half the levels found in 1977 (Bothner et al., 1998). This suggests that in future decades harbor management, i.e., channel deepening and widening, can be carried out without the same degree of concern about spreading contamination. The same is probably true for other harbors and coastal waters of the Gulf of Maine. As we improve the quality of air, waters, and sediments in and over river basins, harbors, and estuaries, the sources of many contaminants will be reduced. Moreover, subsequent biological effects will be lessened, especially when sediments might be eroded, resuspended, or dredged and spoiled.

While regional, national, and international progress is being made in (1) pollution abatement, (2) habitat restoration, and (3) physically disruptive processes, recent conferences, meetings, and symposia have emphasized that the understanding of habitat function is an important matter for future resource management (Pederson, 1997). And, in fact, the Working Group on Human Induced Biological Change, convened as a subgroup of the Symposium on Gulf of Maine Ecosystem Dynamics (see Wallace and Braasch, 1997), suggested that "When considering breadth of impacts, habitat alteration issues related to fishing are arguably the major environmental issues facing the Gulf of Maine" (Pederson, 1997). Given the foregoing, what are the ways ahead to remediate coastal marine habitats such as the Gulf of Maine, now that we have reasonably good data on the degree or severity of the problems? One working group session chair said that we must move ahead by generating proposals to address the specific issues given in the table of priority issues (Table 1) (Pederson, 1997).

Other conferences and symposia have had working group reports which have come to generally similar conclusions and recommendations, albeit with differing specific goals and tasks. For instance, the sediment and water quality working group for the Gulf of Maine Habitat Workshop, sponsored by the Regional Association for Research on the Gulf of Maine (RARGOM), arrived at the goals and associated tasks in Table 2.

One thing we can be certain of is that modern technology, including biotechnology, will be a major part of future monitoring, research, and management. Already geographic information systems (GIS) are being used to predict such things as the preferred habitats for North American right whales (Moses and Finn, 1997). While satellite imagery has been used for over two decades to quantify the spatial dimensions and rates of marine primary production, increasing uses may even include studies of the littoral and sublittoral benthos from the skies (Larsen et al., 1985). And while underwater photography and TV were used decades ago to census fish stocks and invertebrates (Langton and Uzman, 1989), increasing uses include assessments of damage done to the sea floor by mobile fishing gear (see Dorsey and Pederson, 1998).

Finally, because of past research and monitoring, many of the past major issues are now being managed and regulated, largely on the basis of peer-reviewed publication and subsequent reporting in the press and popularized scientific journals. As human populations increase in the Gulf, and as its resources are more intensively used and new schemes involving offshore mining or mariculture come into place, new issues will arise which can only be dealt with through science and technology. Fishing and

Table 1

Research Priorities for Habitat Degradation and Resource Exploitation: Specific issues are listed in order of priority within each major issue (from Pederson, 1997)

Habitat description issues:
– Define, map and identify economically and ecologically important habitats in the Gulf of Maine;
– Evaluate the nature of the risks to which these habitat are subjected, biological, chemical, and physical;
– Determine interactions among and between habitats; and
– Prioritize habitats for restoration and determine criteria to evaluate success of restoration efforts.

Habitat alteration issues:
– Develop tools to quantify the effects of mobile gear on the quantity and quality of important habitats in the Gulf of Maine;
– Determine thresholds beyond which normal functioning of habitats is impaired;
– Determine the effect of sediment transport on habitat form and function;
– Determine the effect of sea level rise on habitat quantity and quality; and
– Determine the role of exotic/invasive species on habitat form and function.

Science/Policy interface issues:
– Develop a rationale for long-term monitoring of habitat quantity and quality in the Gulf of Maine;
– Insure that the results of scientific research are framed in a format valuable to decision makers (i.e. legislatures, managers, and citizenry);
– Include public education as a tool for improved management of important habitats in the Gulf of Maine; and
– Revisit and draw on historical data and information to aid current habitat protection efforts.

Table 2

Primary research goals/tasks identified by the sediment and water quality working group, Gulf of Maine Habitat Workshop (from Buchholtz ten Brink et al., 1994)

In setting research priorities, identify the endpoint and keep goals continuously in mind.
Ultimate endpoint is zero toxic effects.
Goals:
– Most, if not all, waters to be fishable and swimmable,
– maintain ecological diversity and multiple human use in the Gulf of Maine,
– maintain healthy ecosystems, and
– manage the Gulf of Maine in a way that we progress towards pristine ecosystems. And:
– The links between potentially toxic contaminant concentrations and biotic effects must be better established.
– Transport paths must be studied to determine how contaminants move and become mobilized in the environment, and subsequently accessible to organisms.
– The effectiveness and net costs of remediation practices in meeting goals needs to be more clearly established (and more effective approaches developed if needed)

other resource-based activities in the Gulf of Maine will become an entirely different business than it was half a century ago, or is today. Mariculture is certain to grow in areal extent and the diversity of species used. The so-called ITQs, sanctuaries (McArdle, 1997), closed areas, habitat restoration, and underused species will all become parts of complex fisheries management plans. Some managers even feel that trawling gears will, to one degree or another, be supplanted by various fixed gears proven to be less damaging to habitats. And, future development in coastal zones and river basins will be managed so as to avoid contamination and physically degrading activities. As suggested earlier, much of this is already happening in an *ad hoc* fashion and more such efforts will evolve. In her recent paper, Hanna (1998) states that as the demands for marine resources grow (due to population increase and ever higher standards of living), an inevitable consequence will be the undermining of those institutions established to coordinate and constrain human actions. This suggests that in the Gulf of Maine, governments, agencies, *and* the citizenry must be driven towards establishing a "...compatibility between economic incentives and ecosystem objectives..." long before the various fabrics are rent.

REFERENCES

Allen, S. (1998) Bottoming out. *The Globe*, Boston, MA, 17 August 1998, pp. C1 and C3.

Apollonio and Mann (1995) A peculiar piece of water. In *From Cape Cod to the Bay of Fundy*, ed. P. Conkling. The MIT Press, Cambridge, MA, pp. 77–95.

Backus, R. and Bourne, D. (1987) *Georges Bank*. The MIT Press, Cambridge, MA, 593 pp.

Beardsley, R., Butman, B., Geyer, W. and Smith, P. (1997) Physical oceanography of the Gulf of Maine: An update. In *Proceedings of the Gulf of Maine Ecosystem Dynamics*, eds. G. Wallace and E. Braasch. MIT Press, Cambridge, MA, pp. 39–52.

Bigelow, H. (1927) Physical oceanography of the Gulf of Maine. *Bulletin of the United States Bureau of Fisheries* **40** (2), 511–1027.

Boehm, P. and Hirtzer, P. (1982) Gulf and Atlantic Survey for Selected Organic Pollutants in Finfish. NOAA Technical Memorandum NMFS-F/NEC-13, 68 pp.

Boreman, J., Nakashima, B., Wilson, J. and Kendell, R. (1997) *Northwest Atlantic Groundfish: Perspectives on a Fishery Collapse*. Am. Fish. Soc., Bethesda, MD, 242 pp.

Bothner, M., Buchholtz ten Brink, M. and Manheim, F. (1998) Metal concentrations in surface sediments of Boston Harbor. *Marine Environmental Research* **45** (2), 127–155.

Buchholtz ten Brink, M., Manheim, F. and Bothner, M. (1994) The contaminated sediment database: A tool for research and management in Massachusetts waters. Geol. Soc. America, Ann. Meeting Abstracts, 26(7)ISSN 0016-7592, AP.A-203.

Buchholtz ten Brink, M., Manheim, F. and Bothner, M. (1996) Contaminants in the Gulf of Maine: What's here and should we worry? In *The Health of the Gulf of Maine Ecosystem: Cumulative Impacts of Multiple Stressors*, eds. D. Dow and E. Braasch. RARGOM Report 96-1. Regional Association for Research on the Gulf of Maine, Portsmouth, NH, pp. 91–115.

Bumpus, D. and Lauzier, L. (1965) Surface circulation on the Continental shelf of eastern North America between Newfoundland and Florida. *American Geographical Society Serial Atlas of the Marine Environment, Folio 7*, Am. Geogr. Soc., New York, 8pp.

Carlton, J. and Geller, J. (1993) Ecological roulette: The global transport of nonindigenous marine organisms. *Science* **261**, 78–82.

Cerulo, M. and Hancock, W. (1986) The mixing of the waters. *Journal of the Maine Audubon Society* **3** (7), 16–21.

Cohen, E. and Grosslein, M. (1987) Production on Georges Bank compared with other shelf ecosystems. In *Georges Bank*, ed. R. Backus. MIT Press, Cambridge, MA, pp. 383–391.

Conkling, P. (ed.) (1995) *From Cape Cod to the Bay of Fundy. An Environmental Atlas of the Gulf of Maine*. The MIT Press, Cambridge, MA, 258 pp.

Conkling, P., Irland, L. and Harvey, J. (1995) Views of the forest: Timber, history, and wild lands of the Gulf of Maine watershed. In *From Cape Cod to the Bay of Fundy. An Environmental Atlas of the Gulf of Maine*, ed. P. Conkling, pp. 167–187. The MIT Press, Cambridge, MA.

Dillon, W. and Zimmerman, H. (1970) Erosion by biological activity in two New England submarine canyons. *Journal of Sedimentary Petrology* **40**, 542–547.

Dorsey, E. and J. Pederson (eds.) (1998). *Effects of Fishing Gear on the Sea Floor of New England*. Conservation Law Foundation, Boston, MA, 160 pp.

Dugdale, R. and Goering, J. (1967) Uptake of new and regenerated forms of nitrogen in primary productivity. *Limnology and Oceanography* **12**, 196–206.

Durbin, E. (1997) Zooplankton dynamics of the Gulf of Maine and Georges Bank Region. In: *Proceedings of the Gulf of Maine Ecosystems Dynamics*, eds. G. Wallace and E. Braasch. RARGOM Report 97-1, Regional Association for Research on the Gulf of Maine, Dartmouth, NH, pp. 53–67.

Farrington, J. (1997) Sources, transport, and fate of chemicals of environmental concern in the Gulf of Maine: Trace metals and organic compounds. In *Proceedings of the Gulf of Maine Ecosystem Dynamics*, eds. G. Wallace and E. Braasch. RARGOM Report 97-1. Regional Association for Research on the Gulf of Maine, Dartmouth, NH, pp. 135–139.

Globe (1998) Maine tops region in fishing revenue. *The Boston Globe* **26** (Sep. 1998), B10.

Gottholm, B. and Turgeon, D. (1992) Toxic contaminants in the Gulf of Maine. NOAA National Status and Trends Program, Silver Spring, MD, 14 pp.

Hamer, J., McCollin, T. and Lucas, I. (1998) Viability of decapod larvae in ships' ballast water. *Marine Pollution Bulletin* **36** (7), 646–647.

Hanna, S. (1998) Institutions for marine ecosystems: Economic incentives and fishery management. *Ecological Applications* **8** (1), 5170–5174.

Harvey, J., Mann, K., Podolsky, R. and Meyer, S. (1995) The thin edge between land and sea. In: *From Cape Cod to the Bay of Fundy. An Environmental Atlas of the Gulf of Maine*, ed. P. Conkling. The MIT Press, Cambridge, MA, pp. 121–143.

Jossi, J. and Goulet, J. (1993) Zooplankton trends: US north-west shelf ecosystem and adjacent regions differ from north-east Atlantic and North Sea. Internat'l Council Explor. Seas, *Journal of Marine Science* **50**, 303–313.

Langton, R. and Uzman, J. (1989) A photographic survey of the megafauna of the central and eastern Gulf of Maine. *Fisheries Bulletin* **87**, 945–954.

Larsen, P. (1992) Marine environmental quality in the Gulf of Maine: A Review. *Reviews in Aquatic Science* **6**, 67–87.

Larsen, P. (1992a) An overview of the environmental quality of the Gulf of Maine. In: *The Gulf of Maine*, eds. D. Townsend and P. Larsen, pp. 71–95. NOAA Coastal Ocean Program Regional Synthesis, No. 1. U.S. Department of Commerce, Washington, D.C.

Larsen, P., Gadbois, D. and Johnson, A. (1985) Distribution of PCBs in the surficial sediments of the deeper waters of the Gulf of Maine. *Marine Pollution Bulletin* **16**, 439–445.

Larsen, P. and Webb, R. (1997) Cobscook Bay: An Environmental Bibliography. Bigelow Laboratory Technical Report #100. Maine Chapter of the Nature Conservancy, Brunswick, ME, 150 pp.

Lewis, J. (1964) *The Ecology of Rocky Shores*. The English Universities Press, London, 323 pp.

Leichter, J. and Witman, J. (1997) Waterflow over subtidal rockwalls. *Journal of Experimental Marine Biology and Ecology* 209, 293–307.

Long, E., Sloane, G., Carr, R., Scott, K., Thursby, G. and Wade, T. (1996) Sediment toxicity in Boston Harbor: magnitude, extent, and relationships with chemical toxicants. NOAA Technical Memorandum NOS ORCA 96. Silver Spring, MD.

McArdle, D. (ed.) (1997) *California Marine Protected Areas*. California Sea Grant System (University of California at LaJolla). Publ. No. T-039, 268 pp.

McDermott, J. (1998) The Western Pacific brachyuran (*Hemigraphus sanguineus*: Grapsidae) in its new habitat along the Atlantic coast of the United States: geographic distribution and ecology. *ICES Journal of Marine Science* 55, 289–298.

McDowell, J. 1997. Biological effects of toxic chemical contaminants in the Gulf of Maine. In *Proceedings of the Gulf of Maine Ecosystem Dynamics*, eds. G. Wallace and E. Braasch. RARGOM Report 97-1. Regional association for Research on the Gulf of Maine, Dartmouth, NH, pp. 183–192.

Moses, E. and Finn, J. (1997). Using geographic information systems to predict North Atlantic right whale (*Eubalaena gracialis*) habitat. *Journal of Northwest Atlantic Fisheries Science* 22, 37–46.

Nickerson, C. (1998) Newfoundland, farewell. *The Boston Sunday Globe Magazine*, 20 Sep. 98, pp. 14–32.

NOAA (1994) Gulf of Maine Point Source Inventory: A Summary by Watershed for 1991. The National Coastal Pollutant Discharge Inventory. National Oceanic and Atmospheric Administration. Silver Spring, MD. 11 pp., 14 appendices.

Nriagu, J. (1989) The history of leaded gasoline. In *Proceedings of the International Conference, Heavy Metals in the Environment*, ed. J. Vernet. CEB Consultants, Edinburgh, pp. 361–366.

O'Reilly, J., Evans-Zetlin, C. and Busch, D. (1987) Primary Production. In *Georges Bank*, ed. R. Backus. The MIT Press, Cambridge, MA, pp. 220–233.

Pearce, J., Berman, C. and Rosen, M. (1985) Annual NEMP Report on the Health of the Northeast Coastal Waters. NOAA Technical Memorandum NMFS-F/NEC-35. Woods Hole, MA, 68 pp.

Pearce, J. and Johnson-Hayden, A. (1986) A not so distant warning. *Journal of the Maine Audubon Society* 3 (7), 30–33.

Pederson, J. (1997) Human induced biological change. In *Proceedings of the Gulf of Maine Ecosystems Dynamics*. MIT Press, Cambridge, MA, pp. 263–269.

Plant, S. (1985) *Bay of Fundy Environmental and Tidal Power Bibliography* (second edition). Can. Tech. Rept. Fish. Aquatic. Sci. (No. 1339), Bedford Institute of Oceanography, Dartmouth, N.S.

Pollock, L. (1998) *A Practical Guide to the Marine Animals of Northeastern North America*. Rutgers University Press, New Brunswick, NJ, 367 pp.

Reid, R., Ingham, M. and Pearce, J. (1987) NOAA's Northeast Monitoring Program (NEMP): A Report of Progress of the First Five Years (1979–84). NOAA Technical Memorandum NMFS-F/NEC44. Woods Hole, MA, 138 pp.

Ricketts, E., Calvin, J. and Hedgpeth, J. (1985) *Between Pacific Tides*. Stanford University Press, CA, 652 pp.

Roberson, M. (1964) Continuous seismic profiler survey of Oceanographer, Gilbert, and Lydonia submarine canyons, Georges Bank. *Journal of Geophysical Research* 69, 4779–4789.

Rowe, G., Polloni, P. and Haedrich, R. (1975) Quantitative biological assessment of the benthic fauna in deep basins of the Gulf of Maine. *Journal of the Fisheries Research Board of Canada* 32, 1805–1812.

Rowe, G., Polloni, P. and Haedrich, R. (1982) The deep-sea macrobenthos on the continental margin of the northwest Atlantic Ocean. *Deep-sea Research* 29, 257–278.

Shenton, E. and Horton, D. (1973) *Literature Review of the marine Environmental Data for Eastport, Maine*. TRIGOM Publication No. 2A, The Research Institute of the Gulf of Maine (contact RARGOM, Dartmouth, NH).

Smith, P. (1997) Vertical exchange processes in the Gulf of Maine. In *Proceedings of the Gulf of Maine Ecosystem Dynamics*, eds. G. Wallace and E. Braasch. MIT Press, Cambridge, MA, pp. 69–104.

Smith, R. (ed.) (1964) *Keys to Marine Invertebrates of the Woods Hole Region*. The Marine Biological Laboratory, Woods Hole, MA.

Southward, A. (1958) The zonation of plants and animals on rocky shores. *Biological Review* 33, 137–177.

Steneck, R. (1997) Fisheries induced biological changes to the structure and function of the Gulf of Maine ecosystem. In *Proceedings of the Gulf of Maine Ecosystem Dynamics*, eds., G. Wallace and E. Braasch. MIT Press, Cambridge, MA, pp. 153–167.

Townsend, D. (1997) Cycling of carbon and nitrogen in the Gulf of Maine. In *Proceedings of the Gulf of Maine Ecosystem Dynamics*, eds. G. Wallace and E. Braasch. RARGOM Report 97-1, Regional Association for Research on the Gulf of Maine, Dartmouth, NH, pp. 117–134.

Uchupi, E. and Austin, J. (1987) Morphology. In *Georges Bank*, eds. R. Backus and D. Bourne. The MIT Press, Cambridge, MA, pp. 24–30.

USGS (1998) Database of contaminated sediments for the Gulf of Maine. U.S. Geological Survey Information Sheet, May 1998, 2 pp.

Van Dusen, K. and Johnson-Hayden, A. (1989) *The Gulf of Maine: Sustaining Our Common Heritage*. Maine State Planning Office, Augusta, ME, 63 pp.

Wallace, G. and Braasch, E. (1997) *Proceedings of the Gulf of Maine Ecosystem Dynamics Scientific Symposium and Workshop*. RARGOM Report 97-1. Regional Association for Research on the Gulf of Maine, Dartmouth, NH, 352 pp.

Waterman, M. (1995) Marine protected areas in the Gulf of Maine. *Nat. Areas J.* 15 (1), 43–49.

Wiggin, J. and Mooers, C. (1992) *Proceedings of the Gulf of Maine Scientific Workshop, Woods Hole, MA, 8–10 January 1991*. Urban Harbors Institute, University of Massachusetts at Boston, 394 pp.

Witman, J. (1996) Dynamics of Gulf of Maine Benthic communities. In: D. Dow and E. Braasch, RARGOM Report 96-1, Dartmouth, NH, pp. 51–69.

Witman, J. (1998) Natural disturbance and colonization on subtidal hard substrates in the Gulf of Maine. In *Effects of Fishing Gear on the Sea Floor of New England*, eds. E. Dorsey and J. Pederson. MIT Press, Cambridge, MA, pp. 30–37.

Yonge, C. (1949) *The Sea Shore*. Collins, London, 311 pp.

THE AUTHOR

Jack B. Pearce
NMFS/NOAA, NE Fisheries Center
Woods Hole, MA 02543, U.S.A.

Chapter 20

THE NEW YORK BIGHT

Jack B. Pearce

This chapter provides an overview of the major physical, hydrographic, and chemical conditions in the New York Bight, as well as details of the principal shallow waters, embayments, and coastal habitats adjacent to the Bight, based on historical and recent data. Given these data, living resources at risk and the factors affecting them are reviewed. The principal conclusions are that, although the Bight and ancillary waters were once heavily contaminated, the ending of ocean disposal, upgrading of sewage treatment, and general overall improvement in the handling of wastes have resulted in significant improvement in habitat quality.

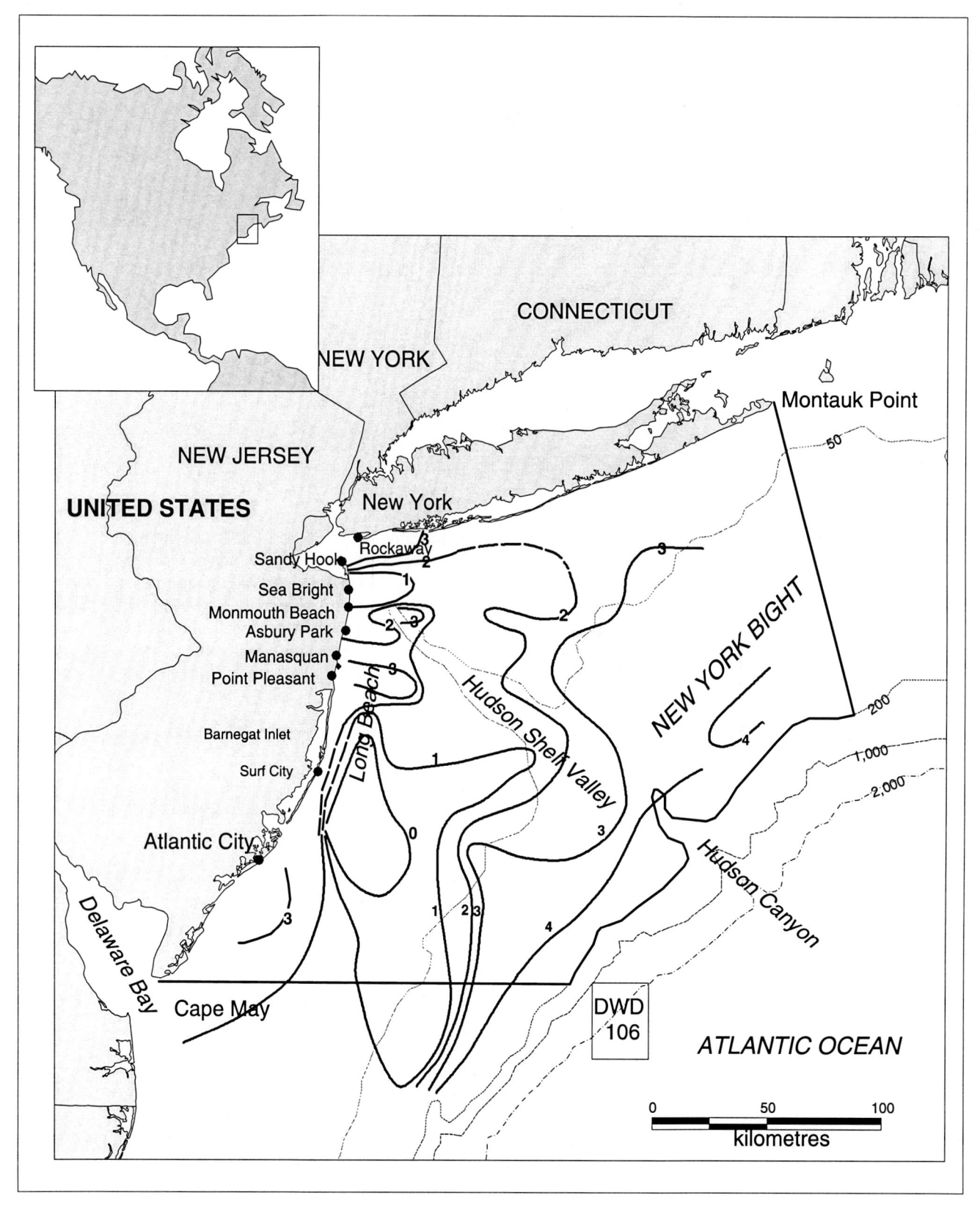

Fig. 1. The New York Bight outlined with vertical and horizontal lines from Montauk Pt. and Cape May respectively; isolines indicate the hypoxia situation in 1976, with values for dissolved oxygen (p.p.m.). The deep-water dump site 106 (DWD 106) shown.

THE REGION

The New York Bight is a portion of the Middle Atlantic Bight (sometimes defined as the Virginian Sea). The general boundaries of the latter are Cape Hatteras, North Carolina to the south, Cape Cod in the north. The New York Bight is always defined as bordered by the New Jersey coast to Cape May, New Jersey, to the south, and Long Island east to Montauk Point, New York. To seaward the boundary is the shelf slope break. Thus the New York Bight lies within 39° and 41°N, 72° and 75°W (see Fig. 1).

The New York Bight provides the major sea routes leading to and from Raritan Bay and the Port of New York/New Jersey, one of the busiest in North America (Hammon, 1976). An urban area extending from metropolitan Philadelphia to New York City includes a human population of 24 million and the land areas bounding the Bight are among the nation's most travelled (Brail and Hughes, 1977) and industrialized (Koebel and Kruekeberg, 1975). Much of the domestic and industrial waste waters ultimately flow into the Bight via the Delaware River/Bay, the Raritan and Hudson Rivers (Lower Bay Complex, see Duedall et al., 1979), and scores of smaller streams flowing to the New Jersey shore and off of Long Island. In addition, there are over two hundred outfalls carrying human and industrial wastes of various kinds, and with varying degrees of sewage treatment (Mueller and Anderson, 1978). Finally, the New York Bight has long had designated sites for receiving sewage sludges (since 1924), chemical wastes (the infamous acid waste grounds and 106-mile offshore site [Pearce et al., 1983]), heavily contaminated dredged materials, and, earlier, certain radioactive wastes.

At the same time, the recreational fisheries in the area are among the largest in North America; the commercial fisheries are also moderately large, including various shellfish, i.e. lobsters and surf clams and hard clams, and oysters and mussels inshore. Several species of groundfish and pelagic apex predators are taken and there are large fisheries for baitfish and industrial species.

Because of the intensive use made of the Bight and its contained living marine resources (LMR), these have been well described and discussed in the scientific literature over the past four decades (see Grosslein and Azarovitz, 1982; McHugh and Ginter, 1978). There have been well over two dozen major conferences and workshops which have resulted in significant proceedings and reports and commercially published volumes. Of these, the following have, in part, been important to the preparation of the following narrative: Pearce (1972), NOAA (1975, 1975a), O'Connor and Stanford (1979), McIntyre and Pearce (1980), GESAMP (1982, 1990, 1990a), Mayer (1982), Pearce et al., (1983), NOAA (1988), Pacheco (1988), Studholme et al. (1995), Sherman et al. (1996a), and Adams et al. (1998).

In addition, the extensive *MESA New York Bight Atlas Monograph Series* is acknowledged; this series originally was planned around 32 titles, most of which were published between the mid 1970s and early 1980s. Published through the New York Sea Grant Institute, the titles include subject matter important as to (1) how metropolitan New York City affected the Bight; (2) detailed studies of the fisheries impacted by man; and (3) their supporting food chains. Most had highly professional charts and maps showing the location of key activities and species and all the monographs were at the cutting edge of science. Most recent conference proceedings are based on data in them. They remain to this day a major product of the NOAA Marine Ecosystems Analysis (MESA) program. Those readers interested in obtaining copies should contact the New York Sea Grant Office, Albany, or this author.

Finally, during the earlier years of study, at least three extensive bibliographies were developed (MESA, 1974; Renwick, 1984; Tiedeman, 1984) which provided many early references to the New York Bight, some topical and important even today, and referenced later in this paper. The recent compilations by Mayer (1982), Studholme et al. (1995), and Sherman et al. (1996a) have extensive bibliographies which will bring the reader more or less up to date.

MAJOR PHYSICAL, HYDROGRAPHIC, AND CHEMICAL FACTORS

This topic has been well described recently by Mountain and Arlen (1995) and previously by Bowman and Wunderlich (1977). Brooks (1996) also has a brief overview which touches on the New York Bight. Generally, temperatures are moderate, rarely reaching freezing or exceeding 24°C. The Northeast Fisheries Science Center (NEFSC) has taken vertical temperature records biannually over some three decades and there are scores of other investigations doing the same. The water column is uniform top to bottom in winter because of surface cooling and mixing due to winds, upwellings and currents. In spring, however, surface warming results in a stratification which grows stronger in the later summer months. Strong northwest winds may drive surface waters offshore, resulting in cooler upwelling phenomenon along beaches on the inner shelf. Salinities tend to be about 32% but changes can occur quickly due to the direction of outflows from the Raritan Bay, Hudson River systems. During early spring, plumes of low-salinity water may occur, often extending southward. Likewise, lower salinity water may depart Delaware Bay but to the north, affecting the southern New Jersey coast. Again, long-term data for temperature, and salinity data, can be had by contacting the NEFSC, Woods Hole, as can information on unusual temperature and salinity conditions. Bottom currents and their relation to sediment resuspension and transport are well covered by Manning (1995). Because of up- and down-welling via the Hudson Shelf Valley (HSV), materials can be transported seaward in this system. Given winter storms, summer upwelling and several major current systems, very few events contribute

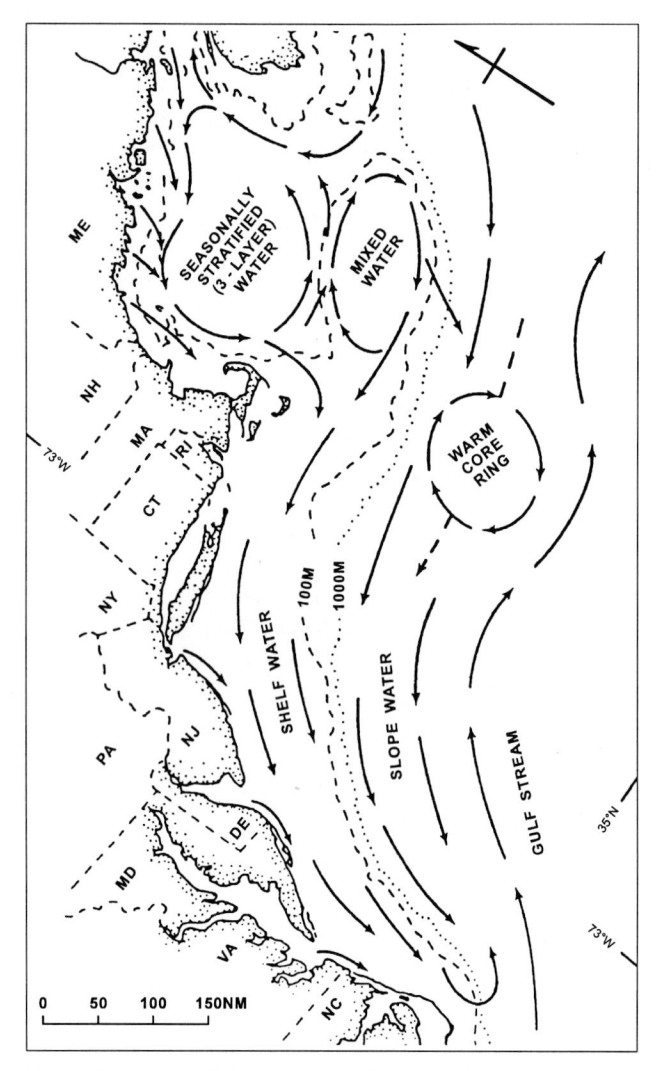

Fig. 2. Mean nontidal surface circulation in the Gulf of Maine, Georges Bank, southern New England, and the Middle Atlantic Bight. The northern sector of the shelf ecosystem is characterized by a cyclonic gyre and a seasonally stratified three-layered water-mass system over the deep basins of the Gulf of Maine and mixed water with an anticyclonic gyre over the shoal bottom of Georges Bank. Further south, the waters move southwesterly along the broad shelf of southern New England to the narrower, gently sloping shelf plain of the Middle Atlantic Bight. From Ingham et al. (1982).

to the annual transport at any one site. Movement down and up the HSV was discussed at length by Davis et al. (1995), and Bopp et al. (1995) cover what is known, generally, about the distribution of sediments and associated contaminants in the HSV and on nearby continental shelf habitats. Likewise, Stumpf and Biggs (1988) detail the surficial sediments and morphology of the continental shelf of the Middle Atlantic Bight, of which the New York Bight is a part.

In his summarization of the physical oceanography of the Middle Atlantic Bight, Cook (1988) uses the earlier works of Ingham et al. (1982), Saila et al. (1973), and others to detail what was known of the larger Middle Atlantic Bight, especially the transient effects of Gulf Stream meanders and warm-core rings (Fig. 2), features which can have far reaching effects on transport and survival of living marine resources (LMR), especially eggs and larvae. He also discussed the "cold pool", a more or less continuous body of subsurface water with connection to the Georges Bank, and south to Cape Hatteras. First described by Bigelow (1933), it may extend on occasion to Cape Hatteras. As the warm core rings, it may affect fish eggs and larvae and the general movement of estuarine and shelf waters.

Finally, Cook (1988) treats upwelling, another important feature of coastal/shelf waters in general, but especially the Jersey shore where cold, clean upwelled coastal waters can affect various amenities. Such phenomena may play transitory roles in anoxia events and other aspects of water quality, which will be discussed later.

There are moderate tidal currents throughout the New York Bight with the Shrewsbury and Navesink Rivers draining into the Sandy Hook Bay, the Raritan River and Arthur Kill draining into Raritan Bay, and these bays and the Hudson River flowing across the Sandy Hook-Rockaway Transect, thence generally south along the New Jersey coast. As already noted, wind from the northwest can drive stratified surface waters seaward, allowing the upwelling of colder waters along both the Long Island and New Jersey coastlines. Such sudden temperature changes are important to the LMR and the fisheries, per se. The entire area can be subject to strong "nor'Easters" and tropical depressions of hurricane force. Accompanying winds have seriously damaged thousands of resorts and domiciles in the past (see Yasso and Hartman, 1976).

Because much of the waters leaving urban and semi-rural areas via the Raritan and Hudson are substantially nutrient enriched, the inner New York Bight is often measured as having exceptional levels of nutrients and very high rates of primary production (Sherman et al., 1996b; O'Reilly and Busch, 1984). These elevated levels of nutrients (Fig. 3), along with wastes from outfall pipes and past ocean dumping, have been responsible for considerable loadings to the sea floor of the New York Bight, especially in the "Apex" and the Hudson Shelf Valley. Deshpande and Powell (1995) detail the differences at stations near the sludge and dredged material dumping sites and at "reference" stations. Likewise, Packer et al. (1995) studied the changes in total organic carbon (TOC) associated with the sediments as ocean dumping ceased. The rates of seabed oxygen consumption (SOC) decreased significantly following cessation of sludge dumping and concomitant decreases in TOC in 1986 (see Phoel et al., 1995).

Major questions in regard to nutrient enrichment and availability have been addressed by several teams: Mueller et al. (1976) reported that some 58% of the anthropogenic organic carbon in the Bight was from the Raritan–Hudson estuarine plume, 5% from dumped sewage sludge. Using satellite imagery to define the plume based on turbidity, Fedosh and Munday (1982) said that the estuarine plume intersects the basin where dumped sludge settles some 12%

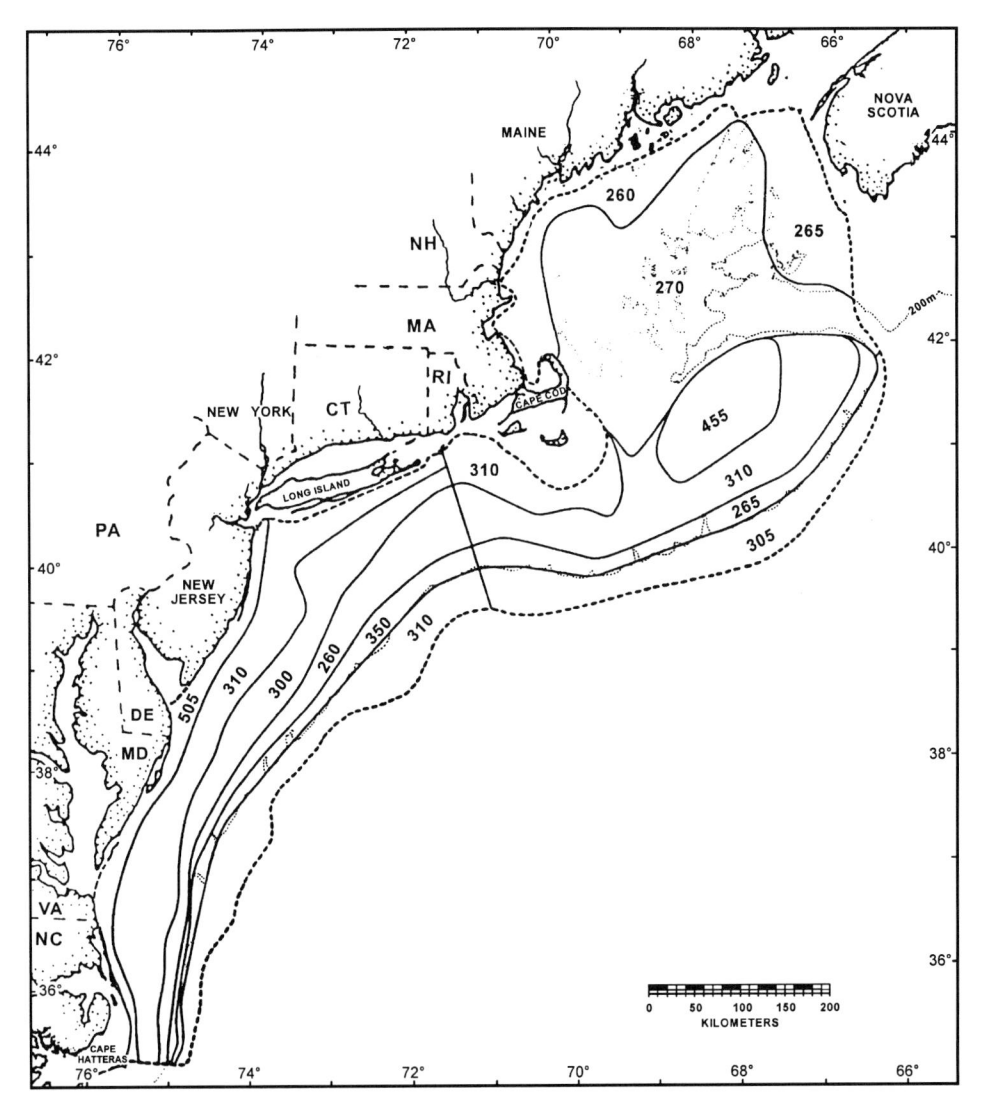

Fig. 3. Estimated annual total primary production (particulate and dissolved organic carbon), Cape Hatteras to Nova Scotia by subarea [g C m^{-2} yr^{-1} (1 g C = 10 kcal = 41.67 kJ)]. Some of the greatest production levels found off the New Jersey coast, probably a result of coastal loading by nutrients. From O'Reilly and Bush (1984).

of the time, and, thus, the plume (with many sources) may contribute to the highly contaminated Christiaensen Basin, along with dumped sludge materials. Because of the exceptional rates of nutrient input to the Bight, there have been numerous reviews and assessments of sewage-derived nutrients in the Bight. Malone (1982) discusses the factors influencing the fate of nutrients in the Bight and McLaughlin et al. (1982) review the importance of nutrients to phytoplankton production in the New York Harbor. In the *MESA New York Bight Atlas Monograph 12*, Yentsch (1977) discusses the basis for plankton production in the Bight, giving special attention to the role of vertical mixing. Finally, a paper by Mearns and Word (1982) and the volume by Swanson and Sindermann (1979) review the effects of nutrients and hydrological conditions on the communities and ecosystem of the New York Bight, especially the coastwide hypoxia experienced in 1976. While the last word is not in, nutrient loading from several sources has had measurable effects on habitat quality in the Bight and changes following cessation of sludge dumping suggest that reduced loading has been good for the LMR of the Bight (see Studholme et al., 1995).

MAJOR SHALLOW WATER MARINE AND COASTAL HABITATS

Almost without exception, the New York Bight is bordered with sandy coastal beaches, often with dunes standing between the beaches, and the lagoons or bays into which major rivers, or smaller streams, drain (Yasso and Hartman, 1976). The continental shelf is generally quite wide (over 200 km in many places) and often traversed by canyons and associated shelf valleys. The origins of submarine canyons, their role in the ecology of fishes, and the nature of surficial sediments are well reviewed by Cooper et al. (1987). These authors also provide exceptional photographs and line drawings of the fauna typical of these canyons, which are found in the deeper semi-boreal waters of Georges Bank

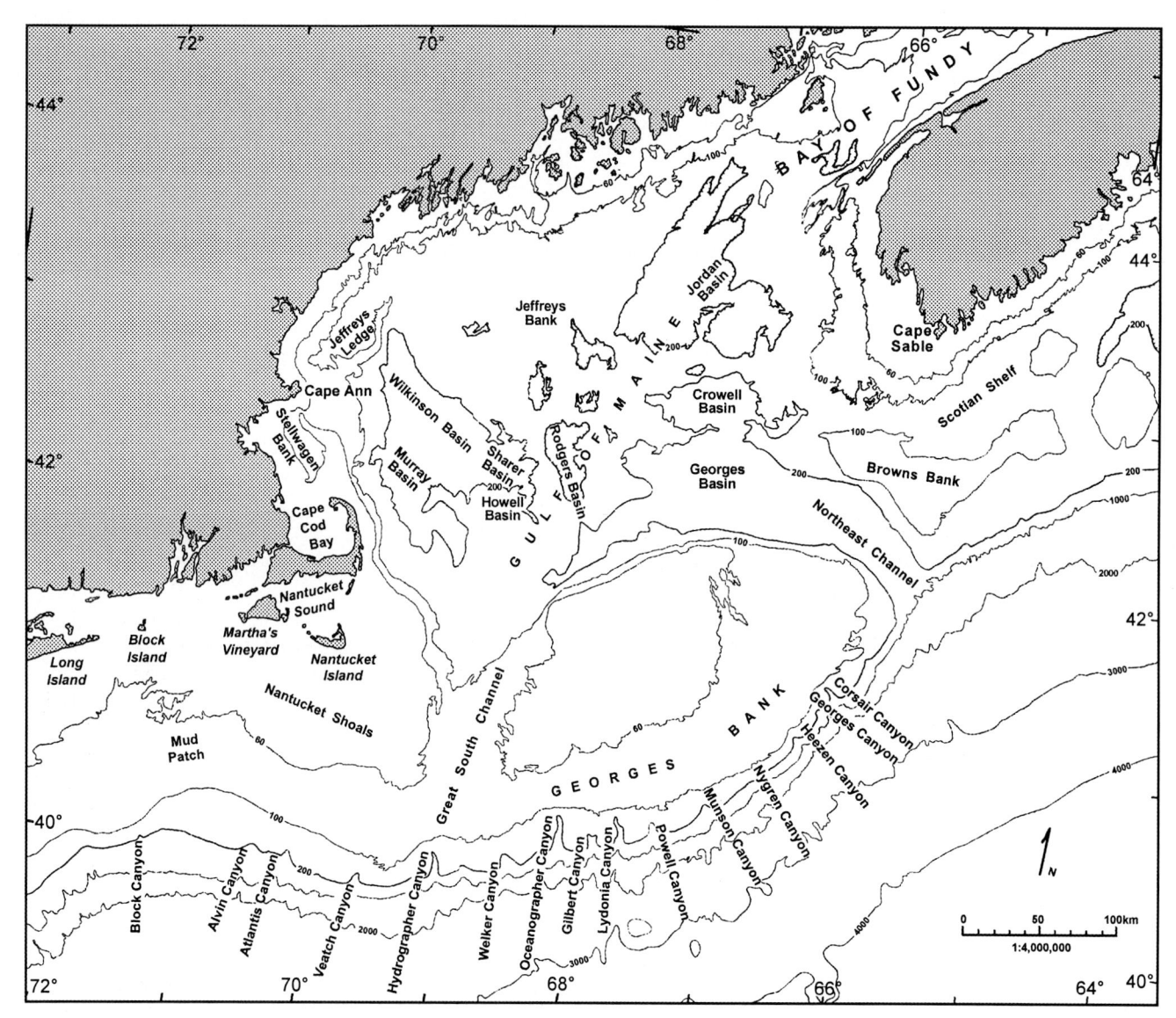

Fig. 4. Location of canyons at northeast margins of the New York Bight. See Fig. 1 for location of Hudson Canyon and Deepwater Dumpsite 106. From Cooper et al. (1987).

southward to Cape Hatteras. The Hudson Shelf Valley (HSV) begins in the apex of the New York Bight and continues across the Continental Shelf (Fig. 4); it is an exceptionally important geological feature, providing habitat for LMR and a conduit for on- and offshore movements of water and entrained substances.

Because the coasts of Long Island and New York consist of monotonous sands in the littoral and intertidal, there is a relatively poor intertidal fauna (see Pearce et al., 1981), showing a reduced species diversity relative to embayments or deeper waters. The few rocky outcroppings, Shrewsbury Rocks off Long Branch, New Jersey, are augmented by seawalls, groins, and piers and docks. Interestingly, even though such habitats are hundreds of kilometers away from other hard bottoms, i.e., the rocky Maine coasts, most of the epibenthic, or attached invertebrate fauna found on rocky outcroppings, artificial reefs, or submerged rocky coastal protection devices (groins, etc.) are representative of epibenthic faunas found both north and south of the New York Bight. This is testimony to the various massive water movements, i.e. the Labrador and Gulf Streams, capable of transporting larvae thousands of kilometers.

Many of the embayments behind the coastal dunes, i.e. Barnegat Bay, Great Bay, and Shark River Inlet along the New Jersey coast (see Fig. 1), are extremely important features in the life history of several species of fish and invertebrates. These areas are often referred to as "nursery grounds", semi-protected places where fishes spawn, eggs hatch, and larval and juvenile forms feed and grow. There are scores of papers and books on these subjects but the NOAA-published volume, *Our Changing Fisheries* (Shapiro, 1971), has a good early account of estuarine productivity and nursery grounds.

The estuaries of the New York Bight have been vastly altered from their original condition as viewed by Henry Hudson from the deck of the late 16th century *Half Moon*. At one time these benthic habitats were characterized by extensive wetlands covered with native *Spartina* grasses and other vegetation. Today most of the marshes have been altered by dredging and filling; installation of docks, piers, and seawalls; and the introduction of exotics such as *Phragmites*. Regular overharvesting of shellfish has often driven to absence, if not extinction, of many of the key invertebrate fauna which are important to the reworking and aeration of sediments, or as food for fishes and mankind. In his monograph on port facilities and commerce, Hammon (1976) details how the Ports of New York and New Jersey grew and what the ultimate consequences have been. His photographs on pages 33 and 35 state what no narrative ever could about the sad state of affairs in the New York Harbor of 1976. Moreover, the New York and New Jersey Hudson River harbors can serve as case studies of how man has affected the coastal habitats and resources of the entire New York Bight; yes, even the entire coastline of the US East Coast. There are two major histories (Burrows and Wallace, 1998; Albion, 1939) and one minor one (Squires, 1981) which treat the human colonization of New York harbor and provide considerable degree of understanding to the forces behind urbanization and estuarine degradation, as it exists today. Duedall et al. (1979) describe the Lower Bay Complex, or estuary, which in turn shares many of the features of the Boston, Philadelphia, and Baltimore Harbors. In recent years such harbors have been characterized by derelict piers, falling down factories and warehouses, and abandoned vessels of all sizes (see Duedall et al., 1979). During the later 1980s and the 1990s, efforts by port authorities, municipalities, and the US Corps of Engineers have resulted in considerable clean up efforts. These often came about *after* "coastal sweeps", where federal agencies and environmental organizations organized volunteer beach clean-ups. A report from the US Gulf of Maine Association (Hoagland and Kite-Powell, 1997) suggests that such activities, along with media publicity, can result in improvements in habitat quality for LMR as well as the public esthetics.

OFFSHORE SYSTEMS

Early reports about deep-water, alternative species for the fisheries resulted in considerable interest in tilefish, red crabs, and deeper water American lobster and their habitats (see Cooper et al., 1987). Also, in the 1970s various state and federal agencies were interested in locating alternative dumping sites for sewage sludges and dredged materials ("spoils"). Consequently, the National Marine Fisheries Service (NMFS) was asked to locate sites on the Continental Shelf, both north and south of the Hudson Shelf Valley (canyon), as well as well offshore at a site approximated 106 miles to sea from the entrance to New York Harbor. The NMFS, NOAA's MESA Program, the US Army Corps of Engineers, and others were asked to investigate, initially, the Continental Shelf Alternative (CSA) sites. While these were early on reported to have a paucity of LMR of importance to the commercial interests, later they were categorized as objectionable by fishers and the fisheries industries. Thus attention was turned to the deep-water dump site 106 (DWD-106), which was viewed more favorably as a site for future disposal of sewage sludges, industrial (acid) wastes, and highly contaminated dredged materials. The majority of the deep-water site was located deeper than 1,800 m, between latitudes 38°40'N and 39°00'N and longitudes 72°00'W and 72°30'W (see Pearce et al., 1979). In May 1974 NOAA conducted an environmental baseline investigation at the DWD-106 and surrounding areas. These studies included attention to: past radioactive waste disposal; hydrographic and physical oceanographic conditions (historical and recent); analyses of micronutrients; status of benthic community structure; assessment of deep-water fishes; a systematic analysis of mid-water fishes; and analyses of plankton, especially fish eggs and larvae and juveniles, as well as the invertebrate zooplankton. Preliminary analyses of the data forthcoming from these baseline cruises and investigations are given in separate papers collectively published by NOAA (1975a), with certain data on benthic fauna and heavy metal burdens given in Pearce et al. (1979). Generally it was reported that dumping of industrial wastes to 1974 had not impacted benthic community structure nor increased heavy metal burdens in the benthic macrofauna or nekton. These data did, however, result in "benchmarks" or "baselines" to be used in future monitoring activities by government agencies and several academic institutions. Moreover, a wide range of biological and hydrographic measurements, echo-soundings, and photography added greatly to what was *then* known about the deeper waters of the New York Bight.

By the early 1980s, the Congress and federal agencies were considering designating the DWD-106 an alternative site for permanent or temporary disposal of sewage sludge. The then NMFS, Northeast Fisheries Science Center, was requested to prepare an update for the status of the DWD-106, again considering the biology, physical factors, and degree of contamination. The report (Pearce et al., 1983) noted that given the dynamics of the system some 116,000 km^2 might be influenced by such future dumping. Once again, this report provided substantial background materials on the DWD-106 and the Continental Slope and abyssal waters off the New York Bight.

A review of the foregoing literature suggests that the deeper offshore systems of the Bight are consistent with those similar habitats off of Georges Bank, and the coastline to Cape Hatteras. There would not appear to be significant contaminant burdens in these deeper waters based on these earlier reports.

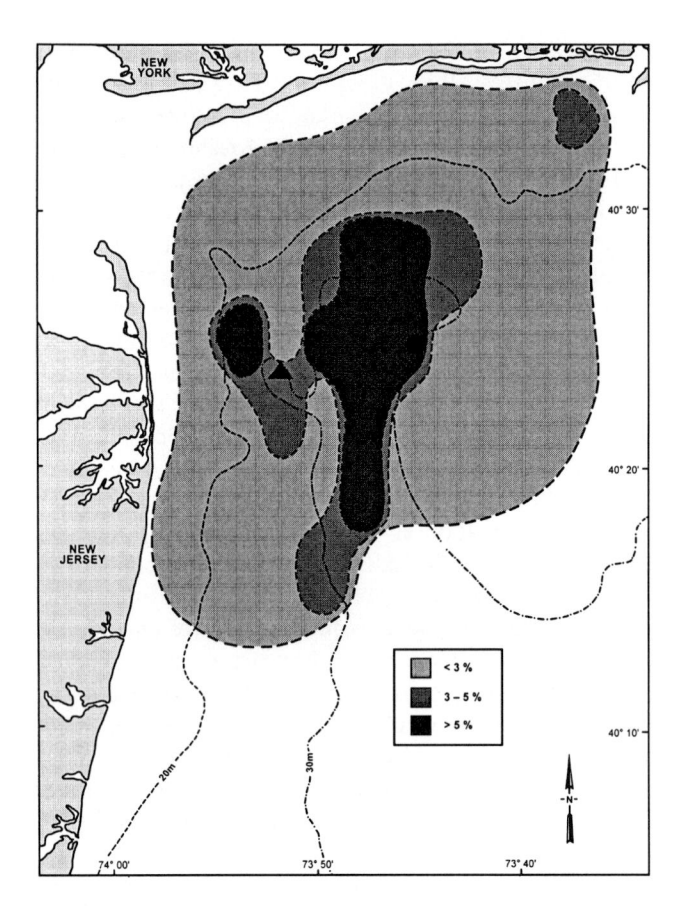

Fig. 5. Distribution of digestible organic carbon (average %) in apex sediments. Locations of dredge spoil (triangles) and sewage sludge (circles) dumpsites are indicated. From Mayer (1982), p. 217.

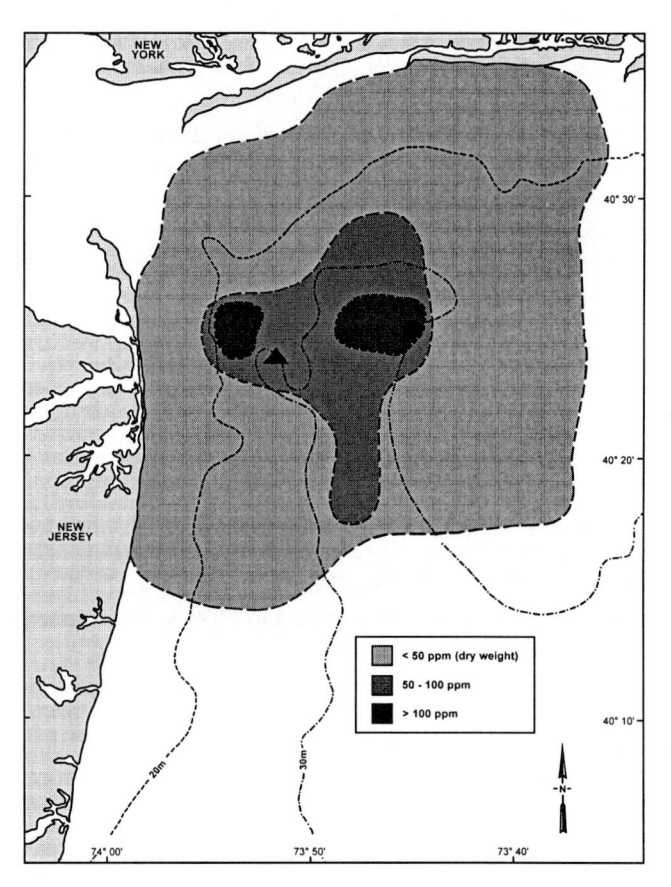

Fig. 6. Average concentrations of lead in the apex sediments. Location of dredge spoil (triangles) and sewage sludge (circles) dumpsites are indicated. From Mayer (1982), p. 218.

POPULATIONS AND CONDITIONS AFFECTING THE BIGHT

Even as the New York metropolitan area was being colonized by 17th century settlers, the Hudson and East Rivers and harbors were used for waste disposal. Rubbish and garbage were simply "tipped" off the riverbank into an ebbing tide. Later, as wharves and piers were built, these were used as locations from which to jettison wastes which were eventually carried to sea. As the area grew to its present proportions, eventually some 24 million souls got rid of much of their waste by direct dumping, but no longer just in the harbor. Garbage was first barged to the Bight and dumped, but, because it often floated back to land, even when dumped ten miles to sea, society started to use land fills on Staten Island and in the Jersey meadowlands. This early history is noted in the three recommended New York City histories (see Albion, 1939, and Hammon, 1976), and Mueller and Anderson (1978) review how New York City, as "the largest center of manufacturing in the United States", handled its industrial wastes in the late 1970s, while Gross (1976) detailed how the city rid itself of monumental amounts of solid waste as the City and the "tall ships" celebrated the Nation's two hundredth birthday. These solid wastes included dredge spoils from highly contaminated industrial harbors, rubble from demolition, and sewage and industrial sludges. Since the City is served by combined storm sewers (CSS), much of the sewage was often swept into the harbors when heavy storm run-off washed the streets into the sewers and carried residual sludges seaward via riverine transport, and ebbing tides. The eventual fate of the sludge and its effects were initially covered by Pearce (1972) in an early paper based on a report to the US Army Corps of Engineers (Pearce, 1969). Not only did such waste disposal practices impinge upon the harvesting of fishes and shellfish (O'Connor, 1996), it also affected recreation and a multibillion-dollar tourism industry in New York and New Jersey. Carls (1978) summarized components of the highly diversified recreational industry as it existed in the late seventies, and served millions, as well as how the waste disposal activities might affect tourism, while O'Connor et al. (1977) reviewed the water quality of the New York Bight. They stated: "It is clear that man's impact on the region has been significant as reflected in the analysis of the data presented here." The data included dissolved oxygen (a decrease in saturation levels from 67% in 1949, to 30% in 1974), light, temperature, salinity, pH, and various forms of nitrogen and phosphorus. Also, these authors considered trace metals, coliform bacteria, and phytoplankton chlorophyll. Many of these

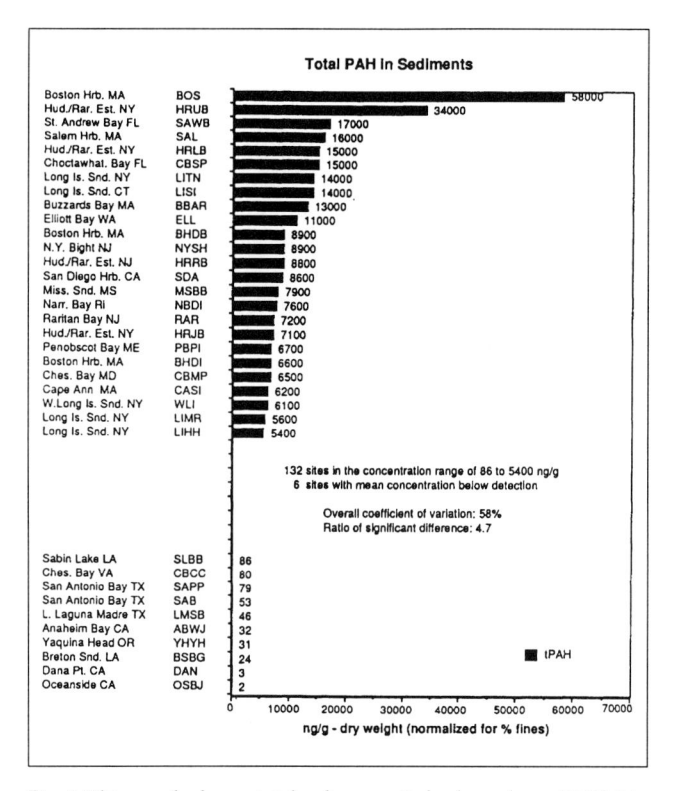

Fig. 7. This graph shows total polyaromatic hydrocarbons (PAHs) in sediments at sampling sites in New England and Middle-Atlantic states (northwest Atlantic). Where other organic contaminants as well as trace metals have been analyzed in the same sediments, similar high values were found at most of these stations. Likewise, microbial indicators of mammalian wastes also frequently are elevated at these sites. Finally, increased incidence of fish diseases have been observed at stations showing greatest contamination. From NOAA (1988a).

data were reconsidered and the values updated in Studholme et al. (1995) (see Figs. 5, 6, and 7).

While there are nearly 24 million persons living on or near the Bight, Squires (1981) reckons that some 18 million of them are dependent in one way or another on it. Wilk et al. (1998) said perhaps 40 million persons use the Bight and Raritan Bay, but many probably only occasionally or every few years as a fishing ground. However, New Jersey, a relatively small state (7.8 million persons), is one of the most densely settled, near 900 persons per square mile (US Bureau of the Census, 1980). But they are not uniformly distributed; the New Jersey pine barrens is almost devoid of people, whereas the area near the lower Hudson is extremely densely populated. This same area is, or was in 1981, the largest producer of chemicals in North America (Jones et al., 1975). Those populations in Pennsylvania and Delaware which border the Delaware Bay are, likewise, very densely settled, and associated with large petrochemical and pharmaceutical industries. As noted earlier, substantial riverine flows to and tides from Delaware Bay also carry possibly deleterious materials northward into the Bight.

These densely settled areas are all served principally by the auto as a prime person mover. Over 70% of all weekday trips in 1970 were via the automobile; only 13% were made by rapid transit (the subway), 12% by bus (Brail and Hughes, 1977). Since then the auto has increased its share. This is largely because of continued movement to suburban areas not well serviced by public transportation. The shift from public conveyances to the car greatly increases congestion and pollution.

Much of this pollution, for instance lead from petroleum combustion, has left a permanent mark within the vertical varves of the sediments in the Lower Bay Complex, the New York Bight, and the Hudson River Shelf Valley (see Bothner et al., 1998); by analyzing and aging deep cores it is possible to demonstrate when in history a particular contaminant first entered the sediments of the Boston Harbor or the New York Bight. The Port of New York, as many New England Harbors, has trace elements going back to pre-Civil War and Revolutionary times.

RESOURCES AT RISK

The New York Bight is a *temperate* water body, sometimes considered as the northern half of the "Virginian Sea". As such, it has resident finfish and invertebrate species as well as migratory taxa which move from the Gulf of Maine and other boreal habitats south to winter in the Bight, and from the Carolinian and Virginian waters northward in the summer. The latter taxa, such as striped bass, use the Bight and riverine systems as spawning grounds and nursery areas for juveniles.

Resident offshore shellfish species such as sea scallops and surf clams constitute a large commercial fishery. There are substantial bottomfish fisheries, especially for cod which move into the Bight in winter months. Large industrial fisheries are based on the menhaden, a herring-like fish. Finally, there are several very large recreational fisheries which target cod and other gadids, flatfish such as the summer and winter flounders, reef and hardbottom dwelling forms such as the serranid, the black sea bass, and the temperate water wrasses, the cunner and the tautog.

These several fisheries and the general biology (habitats) and distributions for over two score species have been well described by several authors including Breder (1948), Bigelow and Schroeder (1953), Freeman and Walford (1974), Grosslein and Azarovitz (1982), and McHugh and Ginter (1978). The latter three publications contain information specific to the New York Bight, as well as other LMR, and uses made of the Bight by fishes.

The Bight was one of the original fishing grounds of North America and has been an important area since the mid-seventeenth century. Reasonable records of commercial landings go back to the 1880's, and several fishing clubs and societies maintain records of recreational catches made before the turn of the century. The various states and the federal government have maintained quite reasonable statistical records for saltwater recreational catches from 1960 to the present. Many commercial records are based on landings or sales *per se* but the National Marine Fisheries Service has conducted annual seasonal assessment cruises since the 1960s to the present time. Along with the fisher's logbooks and port landings' data, they provide the basis for fish management in the northwest Atlantic.

At one time (1960s) "foreign fishing" was very intense in and off the New York Bight; this resulted in reduced domestic catches and was the stimulus for passage of the PL 94-265, the Magnuson Act, which ended unlimited foreign fishing inside the 200 nautical mile line (1 March 1977). Since then the size (tonnage) and number of domestic trawling vessels have increased to a point where the various stocks have been fished to historic low points (see Mayo et al., 1998). A recent (4 December 1998) newspaper article declared: "Another fishing ground collapses: Cod stocks drop in Gulf of Maine" (Allen, 1998). Both the Gulf of Maine and Georges Bank cod and haddock stocks are at low points (Mayo et al., 1998), as are most of the other groundfish species, i.e. various species of flounders. The large pelagic forms sought in the Bight, tuna, swordfish, and certain sharks, are also considered overfished and in danger. Likewise, sea and bay scallops are being intensely managed to prevent further declines in stocks and total collapse of the various shellfisheries.

Periodically, the NMFS develops background documents on the status of regional fisheries and the distribution and abundance of key species within a particular area. These documents are important to establishing habitat requirements (see Packer et al., 1998) for various fish and shellfish species and to assess or model the environmental consequences of dredging and spoiling operations in the New York Bight (see Long et al., 1995, and US Army Corps of Engineers, 1996, 1997). Wilk et al. (1998), for instance, provide data and charts showing the seasonal distributions and abundance of 26 species of finfish and shellfish which constitute the dominant (90% of total number of individuals and weight) taxa in the Hudson–Raritan Estuary, a principal embayment to the New York Bight. Such information becomes extremely important when an agency or organization proposes an operation likely to have deleterious effects on key living marine resources (LMR).

In addition to the LMR, society today demands continual supplies of various minerals, including sands and gravel. The New York Bight, Raritan Bay, and the lower Hudson estuary have extensive deposits of these which have been reviewed and discussed by Schlee and Sanko (1975), who note that 26,446 million short tons of sand and gravel occur in an area of 15,112 km^2 on the shelf off New Jersey. Moreover, they report that an average 5.5 millions yds^3/yr were dredged from the *lower bay complex* of New York Harbor during 1966–1974 and Kastens et al. (1978) reported 2.2 million mtons/yr, each year since 1946. These latter authors review what is known about the mineral resources and the associated fauna which might be "impacted by dredging operations". A more extensive review was provided by Brinkhuis (1980), who includes an assessment of the biological effects (on benthos) of sand mining.

A principal question then comes up. How have the various stresses affected the LMR, as well as mineral extraction and tourism and aesthetics? Early research in the 1960s found shellfish and habitats contaminated with bacteria. Subsequent research yielded data which showed that coastal pollution and ocean dumping had affected the distribution and abundance of common ocean species. The findings have been well summarized in Mayer (1982), Studholme et al. (1995), and Sherman et al. (1996). Generally it has been the bottom creatures, the benthos and demersal fishes, which have most frequently been reported affected. However, studies of genetics in fish stocks, reported upon in the aforementioned compilations, suggest that waste disposal and associated contaminants have impacted the gene systems of a variety of fishes, pelagic as well as demersal.

Also, various reports have dealt with the effects of wastes washed up on beaches, especially on bathing, diving (scuba), and recreational fishing (Pearce, 1988). By addressing these various issues and problems through well conceived research, many of the causes (ocean dumping, ocean outfalls, poorly managed landfills, etc.) have been illuminated and management steps taken to end the offending causes. For instance, in the New York Bight the Interstate Sanitation Commission has dealt with a wide range of outfalls, initiating pre-treatment, closing down derelict sources, and generally reducing inappropriate discharges. Through legislation, the U.S. Congress effectively ended the ocean disposal of sewage sludge in the New York Bight on 31 December 1987. Caspe (1995) provides the details as to how enabling legislation was finally enacted, and Studholme et al. (1995) provide the narrative from a score of authors indicating the changes which occurred after the cessation of dumping.

From all evidence there has been a progressive improvement over the past decade in the New York Bight as a habitat for man and fishes. Fishes apparently have less disease, benthic species seem to be returning to habitats from which they had been precluded, and beaches are more swimmable. Major Bight restoration plans (New York–New Jersey Harbor Estuary Program, 1996) have led to the initiation of new projects and programs, directly and through example. These deal with the *management information* needed for habitat and living resources; toxic contamination; dredged materials; pathogenic contamination; floatable debris; nutrients and organic enrichment; rainfall induced events; and appropriate public involvement and education. As this interstate, regional effort moves forward, the federal agencies continue to plan for further eventual management of dredged material disposal in the Bight, but with appropriate alternatives for highly contaminated spoils (see US EPA, 1997). The NOAA (1996, 1996a) continued research to better understand how various contaminants exist in the Bight, how these may have been absorbed by LMR, and what the health risks might be for people consuming fishes and shellfish.

While we make progress in dealing with these issues, society points out new issues or further concerns which then must be dealt with (Meador and Layher, 1998). These authors suggest that with a growing economy there is ever more construction, and pressures increase to mine streams and coastal waters. This is especially true in the New York metropolitan area where there are few land-based sources of aggregates. Likewise, there are few ways to dispose of dredged materials (spoils), industrial wastes, and sewage wastes. The New York Bight will constantly be turned to as a body of water to receive wastes, but also as a source of LMR, clean aggregates, and aesthetic pleasures, and as a site for harbor activities (shipping), manufacturing, and waters (cooling, dilution, drinking).

Many agencies, for instance, the Waste Reduction and Management Institute, Marine Science Research Center, State University of New York (Stony Brook), are considering "new ways forward" in terms of multiple uses of the seas and management of sewage materials and other wastes (see Swanson et al., 1998). Moreover, federal and state agencies, such as the National Oceanic and Atmospheric Administration (NOAA), are carrying out long-term monitoring programs (the National Status and Trends Program) designed to measure levels of contaminants at established sites (some 233 nationally), and to, periodically, monitor these to collect animal tissues and sediment, and then re-analyze these tissues as needed to demonstrate longer-term changes at stations which might be stressed by certain of man's activities (see O'Connor, 1995, for details). All of these programs grew out of the earlier research programs in the Bight, and recommendations provided by numerous conference proceedings (see Mayer, 1982, and Studholme et al., 1995).

Finally, the urban planners, zoning officials, industrialists, and city and sanitary engineering professions have turned their attention to developing better ways to design and build coastal waste disposal systems. Gunnerson and French (1996) have written a book which focuses on how best to prevent ecosystem damage and protect human health and habitat quality for Living Marine Resources (LMR). Entitled "Wastewater Management for Coastal Cities," the volume depends on case studies from the Yangtze and Thames River estuaries, Boston Harbor, and the Southern California Bight for examples and summarization and progress. As with the New York Bight there is evidence of progress in upgrading severely degraded systems in several parts of the world. Marine scientists and coastal zone managers would, however, do well to turn again to certain United Nations proceedings (GESAMP, 1982, 1990, 1990a) to see what was suggested as important some two decades ago, and how far we have since come. Most important, if we are in any way to live from the seas, we *must* understand that humankind in the future will have to appreciate the hundreds, even thousands, of ways that we collectively affect the sustainability of the World's oceans (Pearce, 1991).

REFERENCES

Adams, D., O'Connor, J. and Weisburg, S. (1998) Sediment Quality of the NY/NJ Harbor System. The U.S. EPA Environmental Monitoring and Assessment Program. Report EPA/902-R-98-001, March 1998. U.S. EPA Regional Office, Region 2, New York City.

Albion, R. (1939) *The Rise of New York Port (1815–1860)*. Charles Scribner's Sons, New York. 481 pp.

Allen, S. 1998. Another fishing ground collapses. *The Boston Globe* **4** (Dec.), pp. 1 and 26.

Bigelow, H. (1933) Studies of the waters on the continental shelf, Cape Cod to Chesapeake Bay. I. The cycle of temperature. *Papers in Physical Oceanography and Meteorology* **2** (4), 1–135.

Bigelow, H. and Schroeder, W. (1953) *Fishes of the Gulf of Maine*. Fishery Bulletin 74, US Fish & Wildlife Service, US Government Printing Office, Washington DC. 577 pp.

Bopp, R., Robinson, D., Simpson, H., Biscaye, P. and Anderson, R. (1995) Recent sediment and contaminant distributions in the Hudson Shelf Valley. In *Effects of the Cessation of Sewage Sludge Dumping at the 12-Mile Site*, eds. Studholme, O'Reilly and Ingham. U.S. Department of Commerce, SPO, Seattle WA. pp. 61–83.

Bothner, M., Buchholtz ten Brink, M. and Manheim, F. (1998) Metal concentration in surface sediments of Boston Harbor. *Marine Environmental Research* **45** (2), 127–155.

Bowman, M. and Wunderlich, L. (1977) *Hydrographic Properties*. MESA New York Bight Atlas Monograph No. 1. New York Sea Grant Institute, Albany. 78 pp.

Brail, R. and Hughes, J. (1977) *Transportation*. MESA New York Bight Atlas Monograph No. 24. New York Sea Grant Institute, Albany. 37 pp.

Breder, C. (1948) *Marine Fishes of the Atlantic Coast*. G.P. Putnam and Sons. New York. 332 pp.

Brinkhuis, B. (1980) Biological Effects of Sand and Gravel Mining in the Lower Bay of New York Harbor. An Assessment from the Literature. Spec. Rpt. 34, Reference No. 80-1. State University of New York, Mar. Sci. Res. Ctr., Stony Brook, NY, 193 pp.

Brooks, D. (1996) Physical oceanography of the shelf and slope seas from Cape Hatteras to Georges Bank: a brief overview. In *The Northeast Shelf Ecosystem*, eds. K. Sherman, N. Jaworksi, and T. Smayda. Blackwell Science, Cambridge, MA, pp. 47–74.

Burrows, E. and Wallace, M. (1998) *Gotham. A History of New York City to 1898*. Oxford University Press, New York City. 1416 pp.

Carls, E. (1978) *Recreation*. MESA New York Bight Atlas Monograph No. 19. New York Sea Grant Institute, Albany. 32 pp.

Caspe, R. (1995) Historical background of the 12-mile dumpsite. In *Effects of the Cessation of Sewage Sludge Dumping at the 12-mile Site*, eds. Studholme, O'Reilly, and Ingham. U.S. Dept. of Commerce, Seattle Printing Office, Washington State, pp. 9–11.

Cook, S. (1988) Physical oceanography of the Middle Atlantic Bight. In *Characterization of the Middle Atlantic Water Management Unit*, Pacheco (ed.). NOAA Technical Memorandum NMFS-F/NEC-46, Woods Hole MA, pp. 1–49.

Cooper, R., Valentine, P., Uzmann, J. and Slater, R. (1987) Submarine canyons. In *Georges Bank*, eds. R. Backus and D. Bourne. The MIT Press, Cambridge MA, pp. 52–63.

Davis, W., McKinney, R. and Watkins, W. (1995) Response of the Hudson Shelf Valley sewage sludge-sediment reservoir to cessation of disposal at the 12-mile site. In *Effects of the Cessation of Sewage Sludge Dumping at the 12-mile Site*, eds. Studholme, O'Reilly, and Ingham. U.S. Dept. of Commerce, Seattle Printing Office, Washington State, pp. 49–60.

Deshpande, A. and Powell, P. (1995) Organic contaminants in sediments of the New York Bight Apex associated with sewage sludge dumping. In *Effects of the Cessation of Sewage Sludge Dumping at the 12-mile Site*, eds. Studholme, O'Reilly, and Ingham. U.S. Dept. of Commerce, Seattle Printing Office, Washington State, pp. 101–112.

Duedall, I., O'Connor, H., Wilson, R. and Parker, J. (1979) *The Lower Bay Complex.* MESA New York Bight Atlas Monograph 29. New York Sea Grant Institute, Albany. 47 pp.

Fedosh, M.S. and Munday, J. (1982) Satellite analysis of estuarine plume behavior. In Proceedings, Oceans '88. IEEE, Washington, D.C., 20–22 Sept. 1982.

Freeman, B. and Walford, A. (1974) *The Angler's Guide to the United States Atlantic Coast, Section II. Block Island to Cape May, NJ.* National Marine Fisheries Service, Seattle Printing Office, WA.

GESAMP (1982) GESAMP: The Health of the Oceans. UNEP Regional Seas Reports and Studies, No. 16. United Nations Environment Programme, Regional Seas Programme Activity Centre, Geneva, Switzerland. 111 pp.

GESAMP (1990) GESAMP: Technical Annexes to the Report on the State of the Marine Environment. UNEP Regional Seas Reports and Studies No. 114/2. United Nations Environment Programme, Regional Seas Programme Activity Centre, Nairobi, Kenya, pp. 321–676.

GESAMP (1990a) GESAMP: The State of the Marine Environment. United Nations Environment Programme, Regional Seas Programme Activity Centre, Nairboi, Kenya, pp. 1–111.

Gross, M. (1976) *Waste Disposal.* MESA New York Bight Atlas Monograph 26. New York Sea Grant Institute, Albany. 32 pp.

Grosslein, M. and Azarovitz, T. (1982) *Fish Distribution.* MESA New York Bight Atlas Monograph 15. New York Sea Grant Institute, Albany. 182 pp.

Gunnerson, C. and French, J. (1996) *Wastewater Management for Coastal Cities. The Ocean Disposal Option.* Springer-Verlag, Berlin. 345 pp.

Hammon, A. (1976) *Port Facilities and Commerce.* MESA New York Bight Atlas Monograph 20. New York Sea Grant Institute, Albany. 41 pp.

Hoagland, P. and Kite-Powell, H. (1997) Characterization and Mitigation of Marine Debris in the Gulf of Maine. Prepared under Contract GM 97-13. Woods Hole Research Consortium, 168 Alden Street, Duxbury MA 02332-3836. 31 pp.

Ingham, M., Armstrong, R., Chamberlin, J., Cook, S., Mountain, D., Schlitz, R., Thomas, J., Bisagni, J., Paul, J. and Warsh, C. (1982) Summary of the physical oceanographic processes and features pertinent to pollution distribution in the coastal and offshore waters of the northeastern United States, Virginia to Maine. NOAA Technical Memorandum NMFS-F/NEC-17, Woods Hole MA. 166 pp.

Jones, H., Bronheim, H. and Palmedo, P. (1975) *Electricity Generation and Oil Refining.* MESA New York Bight Atlas Monograph 25. Albany. 58 pp.

Kastens, K., Fray, C. and Schubel, J. (1978) *Environmental Effects of Sand Miming in the Lower Bay of New York Harbor, Phase 1.* Special Rpt. 15, Reference 78-3. State University of NY, Mar. Sci. Res. Ctr., Stony Brook NY.

Koebel, C. and Krueckeberg, D. (1975) *Demographic Patterns.* MESA New York Bight Atlas Monograph 23. New York Sea Grant Institute, Albany. 54 pp.

Long, E., Wolf, D., Scott, K., Thursby, G., Stern, E., Peven, C. and Swartz, T. (1995) Magnitude and Extent of Sediment Toxicity in the Hudson-Raritan Estuary. U.S. Dept. of Commerce, NOAA/NOS/ORCA, National Status and Trends Program, NOAA Tech. Memo. 88. 230 pp.

Malone, T. (1982) Factors influencing the fate of sewage-derived nutrients in the lower Hudson estuary and New York Bight. In *Ecological Stress and the New York Bight*, ed. G. Mayer. Estuarine Research Federation, Columbia, SC, pp. 389–400.

Manning, J. (1995) Observations of bottom currents and estimates of resuspended sediment transport in the vicinity of the 12-mile dumpsite. In *Effects of the Cessation of Sewage Sludge Dumping at the 12-mile Site*, eds. Studholme, O'Reilly, and Ingham. U.S. Dept. of Commerce, Seattle Printing Office, Washington State, pp. 33–47.

Mayer, G. (ed.) (1982) *Ecological Stress and the New York Bight: Science and Management.* Estuarine Research Federation, Columbia, SC. 715 pp.

Mayo, R., O'Brien, L. and Wigley, S. (1998) Assessment of the Gulf of Maine Cod Stock. Northeast Fisheries Science Center Reference Document 98-13. November 1998, Woods Hole, MA. 16 pp.

McHugh, J. and Ginter, J. (1978) *Fisheries.* MESA New York Bight Atlas Monograph 16. New York Sea Grant Institute, Albany. 129 pp.

McIntyre, A. and Pearce, J. (1980). *Biological Effects of Marine Pollution and the Problems of Monitoring.* Rep. Proces-Verbaux Reunions, Vol. 179, International Council Explor. Seas, Copenhagen. 346 pp.

McLaughlin, J., Kleppel, G., Brown, M., Ingham, R. and Samuels, W. (1982) The importance of nutrients to phytoplankton production in New York Harbor. In *Ecological Stress and the New York Bight*, ed. G. Mayer. Estuarine Research Federation, Columbia, SC, pp. 469–479.

Meador, M. and Layher, A. (1998) Fisheries habitats—Instream sand and gravel mining. *Fisheries* **23** (11), 6–12.

Mearns, A. and Word, J. (1982) Forecasting effects of sewage solids on marine benthic communities. In *Ecological Stress and the New York Bight*, ed. G. Mayer. Estuarine Research Federation, Columbia, SC, pp. 495–512.

MESA (1974) *Bibliography of the New York Bight*, Parts 1 (citations) and 2 (indexes). Marine Ecosystems Analysis Program (MESA), NOAA, Rockville, MD. 184 pp and 493 pp, respectively.

Mountain, D. and Arlen, L. (1995) Oceanographic conditions in the inner New York Bight during the 12-mile dumpsite study. In *Effects of the Cessation of Sewage Sludge Dumping at the 12-mile Site*, eds. Studholme, O'Reilly, and Ingham. U.S. Dept. of Commerce, Seattle Printing Office, Washington, pp. 21–31.

Mueller, J., Anderson, A. and Jervis, J. (1976) Contaminants in the New York Bight: Sources, mass loads, significance. *Journal of Water Pollution Control Fed.* **48** (2309–2326).

Mueller, J. and Anderson, A. (1978) *Industrial Wastes.* MESA New York Bight Atlas Monograph 30. New York Sea Grant Institute, Albany. 37 pp.

NOAA (1975) Ocean Dumping in the New York Bight. NOAA Technical Report ERL 321-MESA2. NOAA Environmental Laboratories, Boulder CO. 78 pp.

NOAA (1975a) May 1975 Baseline Investigation of Deepwater Dumpsite 106. NOAA Dumpsite Evaluation Report 75-1. Rockville MD. 388 pp.

NOAA (1988) Hudson/Raritan Estuary: Issues, Resources, Status, and Management. NOAA Estuary-of-the-Month Seminar Series No. 9. NOAA Estuarine Program Office, Washington DC. 170 pp.

NOAA (1988a) A Summary of Selected Data on Chemical Contaminants in Sediments Collected During 1984, 1985, 1986, and 1987. NOAA Tech. Memo. NOS OMA 44. NOAA Office of Oceanography and Marine Assessment. National Status and Trends Program. Rockville MD. 88 pp.

NOAA (1996) Levels of Seventeen 2, 3, 7, 8–Chlorinated Dioxin and Furan Congeners in Muscle of Four Species of Recreational Fish from the New York Bight Apex. NOAA, Northeast Fisheries Science Center, Highlands NJ. 16 pp.

NOAA (1996a) Contaminant Levels in Muscle and Hepatic Tissue of Lobster from the New York Bight Apex. NOAA, Northeast Fisheries Science Center. Highlands NJ. 72 pp.

O'Connor, D., Thomann, R. and Salas, H. (1977) *Water Quality.* MESA New York Bight Atlas Monograph 27. New York Sea Grant Institute, Albany. 104 pp.

O'Connor, J. and Standford, H. (1979) Chemical Pollutants of the

New York Bight. NOAA, Marine Ecosystems Analysis Project (MESA), Stony Brook, Long Island NY. 217 pp.

O'Connor, T. (1996) Coastal sediment contamination in the northeast shelf large marine ecosystem. In *The Northeast Shelf Ecosystem*, eds. K. Sherman, N. Jaworksi and T. Smayda. Blackwell Science, Cambridge MA, pp. 239–257.

O'Reilly J. and Busch, D. (1984) Phytoplankton primary production on the Northwest Atlantic Shelf. *Rapp. P.-V. Reun. Cons. Int. Explor. Mer* **183**, 255–268.

Pacheco, A. (1988) Characterization of the Middle Atlantic Water Management Unit. NOAA Tech. Memo. NMFS-F/NEC-56, Woods Hole MA. 321 pp.

Packer, D., Finneran, T., Arlen, L., Koch, R., Fromm, S., Finn, J., Fromm, S. and Drexler, A. (1995) Fundamental and mass properties of surficial sediments in the innter New York Bight and responses to the abatement of sewage sludge dumping. In *Effects of the Cessation of Sewage Sludge Dumping at the 12-mile Site*, eds. Studholme, O'Reilly, and Ingham. U.S. Dept. of Commerce, Seattle Printing Office, Washington, pp. 155–170.

Packer, D., Cargnelli, L., Griesbach, S. and Shumway, S. (1998) Essential Fish Habitat Source Document: Sea Scallops, *Placopecten magellanicus* (Gmelin), Life History and Habitat Characteristics, August 1998. NOAA Tech. Memo. NMFS-NE (in press). 20 pp.

Pearce, J. (1969) The Effects of Waste Disposal in the New York Bight — Interim Report for January 1, 1970. To the U.S. Army Corps of Engineers. Sandy Hook Marine Laboratory, U.S. Bureau of Sport Fish and Wildlife. 61 pp.

Pearce, J. (1972) The effects of solid waste disposal on benthic communities in the New York Bight. In *Marine Pollution and Sea Life*. Food and Agriculture Organization of the United Nations. Fishery News (Books) Ltd., Surrey, England, pp. 404–411.

Pearce J. (chr.) (1988) The State of the Ocean: The Report by the (New Jersey Governor's) Blue Ribbon Panel on Ocean Incidents — 1987. New Jersey Department of Environmental Protection, Trenton NJ. 81 pp.

Pearce, J. (1991) Collective effects of development on the marine environment. In H. Chamley (ed.). Environment of the Epicontinental Seas. *Oceanologica Acta*, Spec. Issue No. **11**, 287–298.

Pearce, J., Caracciolo, J., Greig, R., Wenzloff, D. and Steimle, F. (1979) Benthic fauna and heavy metal burdens in marine organisms and sediments of a continental slope dumpsite off the northeast coast of the United States (Deepwater Dumpsite 106). *Ambio*, Special Report, No. 6, 101–104.

Pearce, J., Miller, D. and Berman, C. (eds.) (1983) 106-mile Site Characterization Update. NOAA Tech. Memo. NMFS-F/NEC-26. Woods Hole MA. 475 pp.

Pearce, J., Radosh, D., Caraciollo, J. and Steimle, F. (1981) *Benthic Fauna*. MESA New York Bight Atlas Monograph 14. New York Sea Grant Institute. Albany. 79 pp.

Phoel, W., Fromm, S., Sharack, K. and Zetlin, C. (1995) Changes in sediment oxygen consumption in relation to the phaseout and cessation of dumping at the New York Bight sewage sludge dumpsite. In *Effects of the Cessation of Sewage Sludge Dumping at the 12-mile Site*, eds. Studholme, O'Reilly, and Ingham. U.S. Dept. of Commerce, Seattle Printing Office, Washington State, pp. 145–154.

Renwick, H. (1984) Bibliography of Scientific Studies at Gateway National Recreation Area (draft). Rutgers University, Center for Coastal and Environmental Studies. New Brunswick, NJ. 20 pp.

Saila, S., Bumpus, D., Lynde, R., Shaw, D., Kester, D., Courant, R., Smayda, T., Jeffries, H., Johnson, W., Pratt, S., Pilson, M., Goldstein, E., Heppner, F. and Gould, L. (1973) *Coastal and Offshore Environmental Inventory—Cape Hatteras to Nantucket Shoals*. Marine Publication No. 2, University of Rhode Island, Kingston, RI. 702 pp.

Schlee, J. and Sanko, P. (1975) *Sand and Gravel*. MESA New York Bight Atlas Monograph. New York Sea Grant Institute, Albany. 26 pp.

Shapiro, S. (ed.) (1971) *Our Changing Fisheries*. U.S. Government Printing Office, Washington, DC. 534 pp.

Sherman, K., Jaworski, N. and Smayda, T. (eds.) (1996) *The Northeast Shelf Ecosystem: Assessment, Sustainability, and Management*. Blackwell Science, Cambridge, MA. 564 pp.

Sherman, K., Grosslein, M., Mountain, D., Busch, D., O'Reilly, J. and Theroux, R. (1996) The Northeast Shelf Ecosystem: An Initial Perspective. In *The Northeast Shelf Ecosystem*, eds. Sherman, Jaworski, and Smayda. Blackwell Science, Cambridge, MA, pp. 103–126

Squires, D. (1981) *The Bight of the Big Apple*. The New York Sea Grant Institute, Albany, NY. 84 pp.

Studholme, A., O'Reilly, J. and Ingham, M. (eds.) (1995) Effects of the Cessation of Sewage Sludge Dumping at the 12-mile site. U.S. Dept. of Commerce, Seattle Printing Office, Washington State. 255 pp.

Stumpf, R. and Biggs, R. (1988) Surficial morphology and sediments of the Continental Shelf. In *Characterization of the Middle Atlantic Water Management Unit*, ed. Pacheco. NOAA Tech. Memo. NMFS-F/NEC-56, Woods Hole, MA. 321 pp.

Swanson, L. and Sindermann, C. (eds.) (1979) *Oxygen Depletion and Associated Benthic Mortalities in New York Bight, 1976*. NOAA Professional Paper 11. NOAA, Rockville MD. 345 pp.

Swanson, L., Bortman, M., O'Connor, T. and Stanford, H. (1998) Management of Sewage Materials. Draft Manuscript. Waste Reduction and Management Institute, Marine Sciences Research Center, State University of New York, Stony Brook, NY. 39 pp.

Tiedemann, J. (1984) *The Marine Environments of New Jersey and New York. An Annotated Bibliography*. New Jersey Sea Grant Publication No. NJSG-84-131. New Jersey Marine Science Consortium, Ft. Hancock NJ. 108 pp.

U.S. Army Corps of Engineers (1996) Interim Report: Dredged Material Management Plan for the Port of New York and New Jersey. U.S. Army Corps of Engineers, New York District, New York, NY. 251 pp.

U.S. Army Corps of Engineers (1997) Progress Report: Dredge Material Management Plan for the Port of New York and New Jersey. U.S. Army Corps of Engineers, New York District, New York, NY. 44 pp.

U.S. Bureau of the Census (1980) Estimates of the Population of States, by Age: July 1, 1971 to 1979. US Government Printing Office, Washington DC. (selected pp.).

U.S. Environmental Protection Agency (1997) Supplement to the Environmental Impact Statement on the New York Dredged Material Disposal Site Designation of the Historic Area Remediation Site (HARS) in the New York Bight Apex. U.S. EPA, Region 2, New York.

Wilk, S., Pikanowski, P., McMillan, D. and MacHaffie, E. (1998) Seasonal Distribution and Abundance of 26 Species of Fish and Megainvertebrates Collected in the Hudson–Raritan Estuary, January 1992–December 1997. Northeast Fisheries Science Center Reference Document 98-10, Woods Hole MA. 145 pp.

Yasso, W. and Hartman (1976) *Beach Forms and Coastal Processes*. MESA New York Bight Atlas Monograph 11. New York Sea Grant Institute, Albany NY. 50 pp.

Yentsch, C. (1977) *Plankton Production*. MESA New York Bight Atlas Monograph 12. New York Sea Grant Institute, Albany NY. 25 pp.

THE AUTHOR

Jack B. Pearce
NMFS/NOAA, NE Fisheries Center
Woods Hole, MA 02543, U.S.A.

Chapter 21

CHESAPEAKE BAY: THE UNITED STATES' LARGEST ESTUARINE SYSTEM

Kent Mountford

The Chesapeake Bay is a large, partially mixed, estuarine system on the Eastern seaboard of the United States. It was formed over the last 10,000 years as sea level rose following the last North American ('Wisconsin') glaciation. The surrounding watershed drains 165,759 km^2 via nine major river systems. Native Americans occupied the landscape starting at least 12,000 BP, before sea level rise began inundating the ancestral river courses; most of their prehistory therefore, until the Late Woodland period (1250 BP) has been swallowed beneath a growing Chesapeake. The earliest recorded European explorations date from the mid-16th century. Permanent European settlement began in 1607, followed by major environmental disruptions from the 18th through the 20th centuries.

Deforestation, widespread agriculture, heavy harvesting of the Bay's natural resources and industrial development have played successive roles in stressing the ecosystem. The presence of two ports of world prominence and the nation's capital, Washington DC, engendered tremendous population growth following the two World Wars. The exponential increase in use of agricultural fertilizers and sewage discharges accelerated declines in water quality, especially underwater seagrass meadows and harvestable fish and shellfish resources. These losses stimulated a political process which, over the years 1983–1999, made significant investments in nutrient reductions and habitat improvement, achieving measurable progress in restoring the bay.

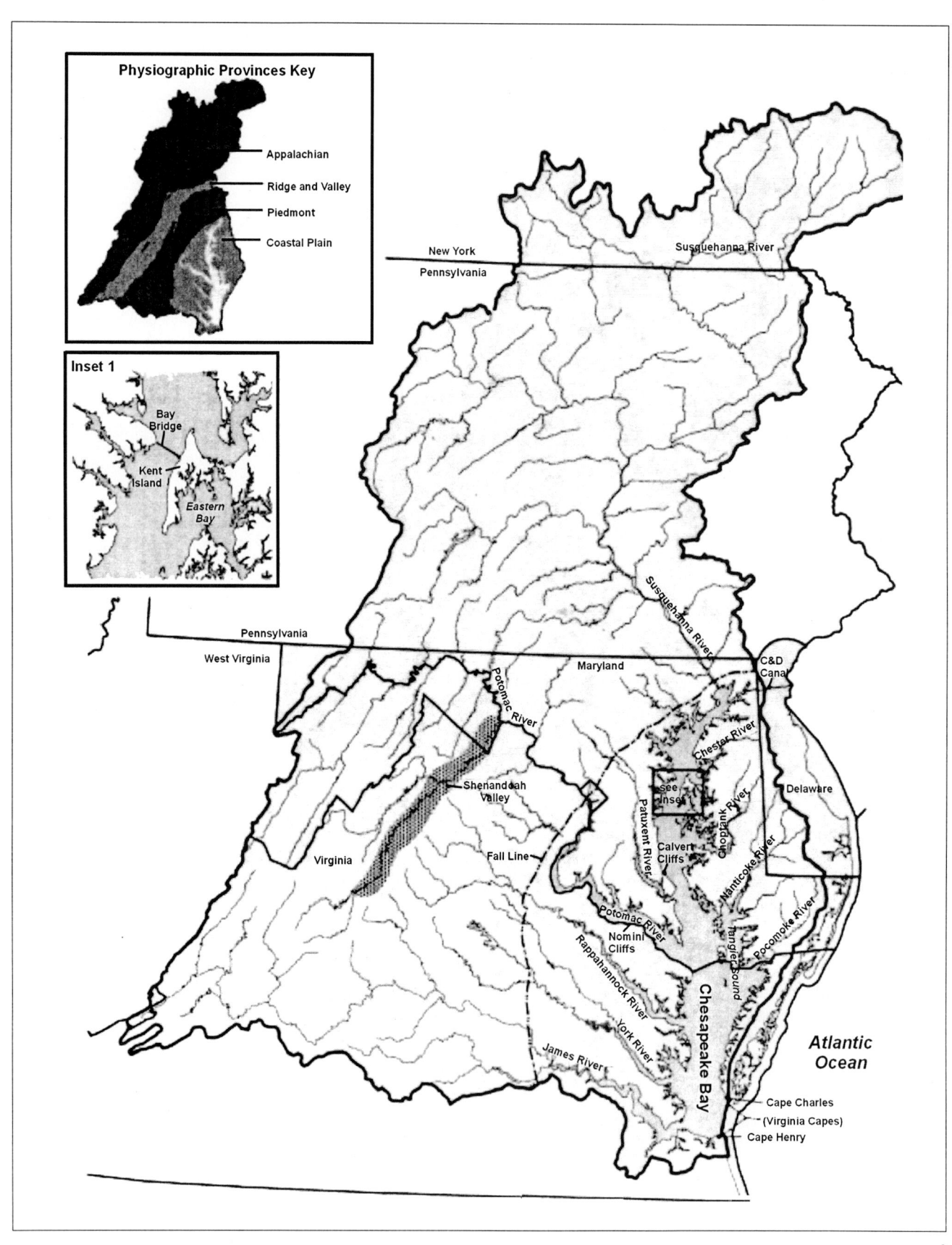

Fig. 1. Map of Chesapeake Bay watershed, the largest estuarine embayment in temperate North America. Surface area of the mainstem is 6495 km² (2507 miles²), and the tributaries below head-of-tide add 4933 km² (1904 miles²). The map is overlain with political boundaries for the states of New York, Pennsylvania, Delaware, Maryland, Virginia, West Virginia and the District of Columbia, of which all or parts drain to the Bay. Place and feature references correspond to the text.

THE DEFINED REGION

Chesapeake Bay opens to the North Atlantic Ocean at 37° North on the Eastern Seaboard of the United States. Eastern North America is richly indented with estuarine systems which cover a variety of latitudinal and tidal gradients. The Chesapeake is the nation's largest estuarine system on either coast and one of the largest in the world, draining a watershed of approximately 165,759 km^2 (64,000 square miles) (Fig. 1). The Bay is about 305 km long from north to south and its margins below the fall-line, richly indented with tributaries, have a shoreline extending some 11,845 km. The bay is relatively shallow, with an average depth of 6.7 m. When the vast area of its drainage basin is compared with its volume, the ratio for Chesapeake Bay is 2743:1. This ratio is by far the highest among 30 of the world's major gulfs and seas, the next nearest being the Gulf of Finland with a ratio of 382:1 (unpublished data of J. Bartholemew in Horton and Eichbaum, 1991). This indicates slow flushing capacity when compared to the other systems.

Geologically, the basin drains south and east across a Holocene landscape from the Appalachians, which are severely eroded mountains of Permian age (Cleves, 1989). The geometry of the coastal plain region seems to have been set up about 35 million years ago by the impact of a bolide (comet or meteor) which created a crater some 6400 km^2 in area, and threw out some 4400 km^3 of ejecta, much of which scattered north and east. This predisposed drainage southward through subsequently deposited layers of overlying sediments, and may have established the present orientation of Chesapeake Bay. The bolide also truncated all the existing deep aquifers with its brine and breccia-filled crater, placing significant limitations on well water resources in the modern, heavily developed, Norfolk–Hampton Roads area (Poag et al., 1997).

Chesapeake Bay is considered a "drowned river system" (Fig. 2). It was created by rising sea level over about ten millennia which have elapsed following the last great North American (or Wisconsin) continental glaciation. Rising water inundated largely Cretaceous sediments laid down on the floor of a much older, post-bolide coastal sea. The Bay became recognizable by its modern shape about 5000 BP, but many small tributary systems which were likewise drowned, remain entombed under later erosional deposits as "fossil" systems. These drainages can be imaged by sub-bottom profiling and mapped because of the seismic opacity of methane rising through the bottom sediments (Hill et al., 1992). Nonetheless, the "original" narrow late Pleistocene river channels drowned by sea level rise remain as deep axial channels in the bay preserving "deep troughs" from 21–53 m.

Erosional losses of fast land along the Bay have been a feature of Chesapeake Bay since its formation but apparently increasing sea-level rise rates during the past century have hastened the disappearance of several bay islands. Salt marshes are also decreasing, especially where the sources of accreting sediment and accumulating plant detritus are deficient. These losses have become both real estate and wildlife habitat issues. Together with the crowding of human financial interests and structures close to waterfront, pressure has increased for artificial shoreline stabilization, much of which destroys outright or further degrades habitat. Published data indicate that between 1978 and 1994, 447 km (278 miles) of Chesapeake shoreline was armored in Maryland alone (Titus, 1998). In 1994 the state ironically suspended gathering these statistics.

The bay, as defined by Pritchard (1955), is a partially mixed estuary, with fairly stable vertical salinity stratification (accentuated by a summer thermocline) and a long, if weakly defined, wedge of higher salinity water which intrudes three-quarters of the way from the Virginia Capes to the Susquehanna Flats. In most summers the pycnocline averages around 6.7 m depth, and places a cap on some 20×10^{12} m^3—40% of the bay's total volume. The upper and lower layers are thus separated and normally fail to completely mix for several months.

The "salt wedge" also intrudes into most of the bay's major tributaries and both here and in the mainstem, summer hypoxia or anoxia is common. Periodic reduced dissolved oxygen was probably characteristic of the deepest channels during the pre-European period, but geochemical evidence from deep cores suggests that only in the past century has anoxia become an annual feature. This corresponds to the period when post-World War II basin agriculture, development and industry began delivering massive nutrient loads to the bay (Brush et al., 1998).

Resultant increases in plankton blooms, a decrease in the euphotic zone and the subsequent rain of organic debris into subpycnocline waters now drives an annual decline of oxygen. Oxygen-consuming processes in the lower layer proceed without replenishment. This completely extinguishes the deep water benthos each summer and remobilizes stored nitrogen and phosphorus from the sediments to stimulate further algal growth. The subpycnocline hypoxic zone is usually stable from May into October. During this period strong lateral winds, which persist for more than 24 hours, drive warmer less salty water away from the windward shore and cause "upwelling", or more accurately tilting of the pycnocline, which advects deeper more saline and hypoxic or anoxic water up onto the bay's shallow flanks. Sanford et al. (1990) report that these wind (and tide) forced oscillations can occur at intervals of 2–6 days in summer and the pycnocline can move as much as 10 m vertically and 4 km laterally. This variable but repeated movement of water with low or zero oxygen over sessile benthos or territorial fish populations is stressful or lethal. Breitburg (1992) and others have documented substantial kills, or so-called "Jubilees" where lethargic crabs and fish, seeking dissolved oxygen, can be easily caught at the water's edge.

Ten principal rivers (Table 1) enter the Bay, flow being dominated at the North end by the Susquehanna, which

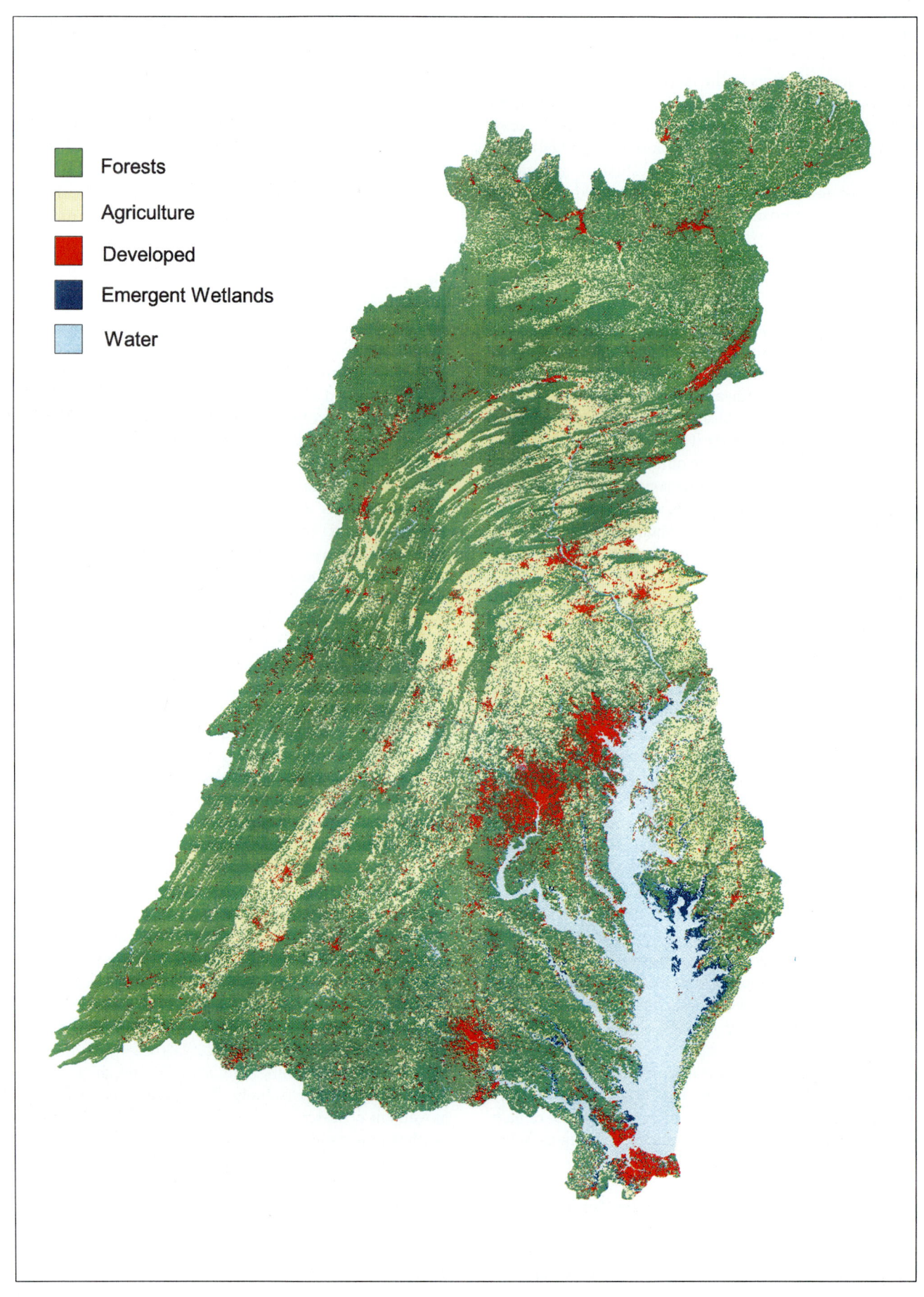

Fig. 2. A land use map of the Chesapeake Bay Basin, based on 1991–1993 MRLC, Landsat Imagery. Shown are basin physiographic provinces (upper left), forested, agricultural and urbanized areas.

Table 1

The ten principal rivers entering Chesapeake Bay, together with their drainage basin areas in km^2, percent of the total watershed, average flow in m^3 s^{-1} and percent of total flow. Totals are less than 100% because of other minor rivers entering the Bay. Flows are created using the Chesapeake Bay watershed model to sum measured discharge at the fall lines with estimated discharge from tributaries below fall line. Basin subareas are from the Chesapeake Bay Program GIS system.

Tributary river basin	Watershed area (km^2)	% of Total CB watershed	Av. flow (m^3 s^{-1})	% of Total CB flow
Western Shore Rivers N to S				
Susquehanna	71,218.43	42.80	1033	52.6
Patuxent	2,277.91	1.37	22	1.1
Potomac	36,715.33	22.07	312	15.9
Rappahannock	6,665.03	4.01	74	3.7
York	6,671.25	4.01	68	3.5
James	26,123.12	15.70	289	14.7
Eastern Shore Rivers N to S				
Chester	1,032.57	0.62	10	0.5
Choptank	1,558.27	0.94	17	0.9
Nanticoke	2,012.78	1.21	22	1.1
Pocomoke	1,867.75	1.12	15	0.8
Total (principal rivers only)	166,384.03	93.60	1847	94.9

Flows averaged for the period 1984–1995. Percentages do not total 100 because of numerous small, unlisted tributaries.

under average flows results in fresh water at the Bay's upper extremity. About midway to the Atlantic, the Potomac, second largest tributary, is confluent with the bay. The James River, settled by Europeans in the 17th century, joins the Bay near its mouth. These three rivers cut deeply into the continent and penetrate into all four biogeographic provinces of coastal plain, piedmont, "ridge and valley" and the Appalachian mountains. River systems north and south drain to other estuarine systems and those to the west drain to the great Mississippi Basin and southward to the Gulf of Mexico. Three smaller rivers, the Patuxent, Rappahannock and York drain only the piedmont and coastal plain biogeographic provinces.

Four major rivers enter from the Chesapeake's low relief and largely agricultural Eastern Shore, a peninsula which separates the Bay from the Atlantic. This "Delmarva" peninsula is partitioned among the three US States of Delaware, Maryland and Virginia. This portion of the basin at the time of settlement had large wetland and non-tidal wet woodland areas, many of which have since been drained for agriculture. The tributary streams of these rivers are closely anastomosed with farming and livestock activity and thus subject to nutrient and agrochemical impacts.

SEASONALITY, CURRENTS, NATURAL ENVIRONMENTAL VARIABLES

Seated squarely in temperate latitudes, the main stem of Chesapeake Bay stretches from 36°54′N to about 39°37′N, which is about the latitude of Spain and Portugal. The watershed reaches substantially farther into colder climates at latitude 42°45′N, this being parallel with the most southerly of the Great Lakes, North America's large freshwater "inland seas".

Salinity within Chesapeake Bay varies strongly with runoff from the basin (Figs. 3 and 4), but ranges from that of neritic ocean waters (ca. 30‰) at the mouth, to fresh where the Susquehanna discharges. During the last 15 years of the 20th century, salinity regimes were observed widely oscillating higher and lower in the bay and its tributaries as a result of freshwater flow. Oligohaline and tidal fresh habitat, while important in the rivers, occupies only a small area in the uppermost mainstem bay during wet years.

In the early 20th century, a series of navigation locks on a branch of the Elk River were excavated to sea level and the "C and D Canal" opened to the Delaware Estuary (see Fig. 1). This has been widened and deepened for shipping with resulting tidal exchange—of salinity, marine life and occasionally contaminants—between these two estuaries.

Coriolis effect drives incoming higher salinity waters against the Eastern Shore so isohalines tilt northward on that side, and the occurrence of higher salinity species thus differs from east to west, as well as from north to south during runs of wet and dry years. These up and down bay differences are especially true for benthic infauna for which there is an extensive long-term database (Holland et al., 1989).

The northern hemisphere winter-to-spring climate surrounding Chesapeake Bay is dominated by cold west and northwest continental weather "outbreaks" but modified by the Bay's water mass and proximity of the Atlantic Gulf

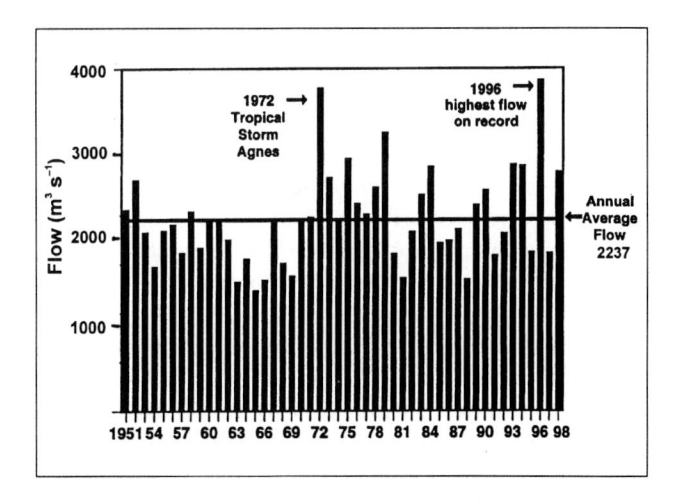

Fig. 3. Year to year variation in average river discharge rates to Chesapeake Bay, 1951 to 1998. The annual mean river flow includes the flow from all of the Bay's tributaries. Data from Bue (1968) and US Geological Survey, Towson, MD (Sylvester, 1999).

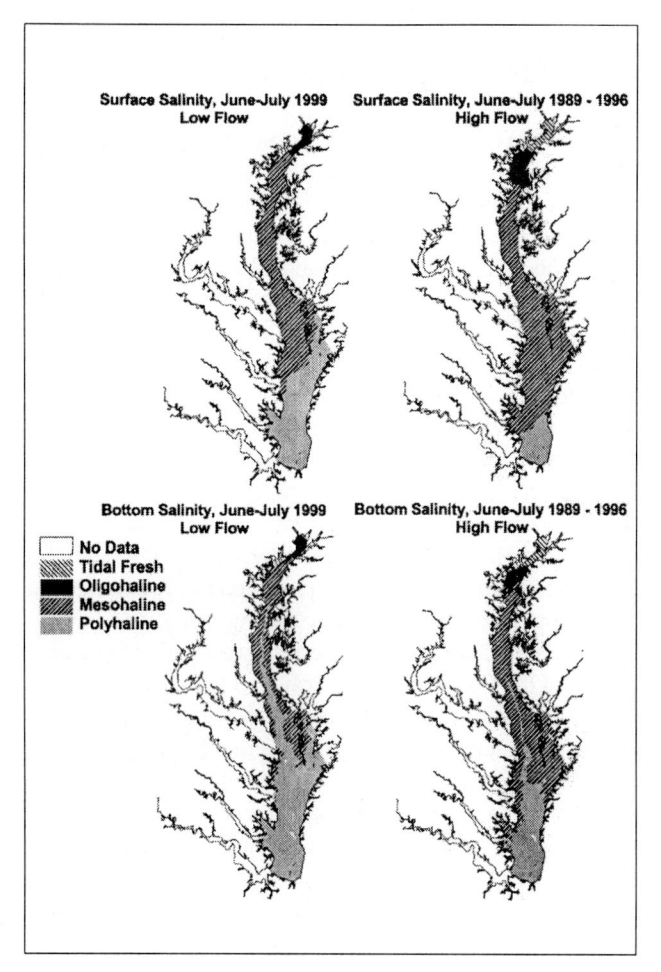

Fig. 4. Salinity, monitored by the Chesapeake Bay Program 1985–2000, shows that between wet and dry periods the surface and bottom areas covered by mesohaline and polyhaline habitat vary dramatically (David Jasinski, University of Maryland). (A) Surface Salinity June–July, 1999, a dry year. (B) Surface Salinity June–July, 1989 and 1996 wet years averaged. (C) Bottom Salinity June–July, 1999, a dry year. (D) Bottom Salinity June–July, 1989 and 1996, wet years averaged.

Stream. Winter storms bring substantial snowfalls during individual years and create significant snow-pack across much of the watershed, but mostly in the northern parts. Snow near the bay quickly melts in many winters. Delay of runoff by prolonged cold, or acceleration of runoff occasioned by snow melt, strongly alters the spring flow or "freshet" of Chesapeake Rivers. The freshet is a major figure of the annual hydrograph and it establishes conditions during the critical spring migration, spawning and growth periods for many living resources.

The palaeoecological record shows that major storm and freshet events have occurred for thousands of years, but without the sediment (and nutrient) deliveries observed from the late Colonial to modern periods. While overall rainfall water volumes might not have changed, freshet intensities over the Bay's four centuries of European settlement appear to have increased in magnitude as agriculture, deforestation, urbanization and impervious surface have proliferated. It is estimated that removal of 25–30% of forest cover produces permanent change in watersheds (Brush, pers. comm.; Brush et al., 1998).

The modern freshet period of Chesapeake Rivers, taken here as the average flow January through May, is very variable but two recent years have come close to the extremes for the period of record since 1951. Average annual flow for the entire year across the period of record 1951–98 was about 2215 $m^3\ s^{-1}$ (78,200 cfs) but the range for the monthly average flow during the *winter–spring freshet* period was from 2280 $m^3\ s^{-1}$ (80,500 cfs) in 1999, within a 16 month drought, and 4723 $m^3\ s^{-1}$ (166,780 cfs) in 1996, an extremely high flow period (US Geological Survey, 1999).

In colder winters, tributary creeks and the major rivers are ice-covered for periods of weeks to months. In the last decades of the 20th century, an overall warming of winter air (and Bay water) temperatures has been observed so that in some winters no significant ice-cover has formed even on tributary sub-estuaries of small size. In more rare winters (3–5 times in the most recent century) Chesapeake Bay freezes over, so that (usually unconsolidated and shifting) ice cover occludes free water surface. This can be an inhibitory factor for small shipping and the significant winter shellfisheries, notably those for oysters and crabs. Ice cover prevents warmer Bay waters from modifying the air, and the albedo of snow atop ice, prevents early spring sun from warming the bay. These factors combine to prolong cold conditions.

Summer through autumn is usually substantially drier, sometimes with serious multi-year droughts such as those experienced in the 1930s, 1960s and late 1990s. Strong summer warming of the bay above 25°C in the mainstem and above 30°C in quiescent small tributaries, decreases the land–water temperature differential which in spring drives convective onshore winds or "sea breezes". Normal summer winds are therefore light but summer climate is dominated by south and southwesterly flows and there are episodic severe thunderstorms. These are usually afternoon or evening events, with spectacular thunder, cloud to ground lightning and torrential downpours often exceeding rainfall rates of 25–50 mm/h, and can be associated with waterspouts or damaging tornados. The intensity of these events stunned early English explorers (Smith, 1624). Whitney (1996) estimates that in Northeastern North America the return frequency for forest canopy disruption from episodic storms is on the order of 500 years so, dramatic as it may be, this force has historically been a small component of deforestation.

Hurricanes from the Caribbean often have trajectories which impact the Chesapeake. Most by this latitude are degraded to "tropical storms" (winds under ca. 34 $m\ s^{-1}$) but these can still deposit rainfalls of 300 mm or more in a day or two, with catastrophic flooding results, and immense pulses of temporarily stored nutrients can be carried off the landscape and out of overloaded wastewater treatment facilities. Storm events also pulse sediments into the bay, and turbulence associated with high flows can resuspend nutrient-rich material from behind dams where storage has occurred during quiescent periods. Major dams like those on the Susquehanna appear to be nearing steady state for sediment retention after decades of net accumulation.

As larger areas of rainfall-impervious surfaces cover the watershed, flood frequency and intensity may be increasing. Some smaller watersheds have 40% or more cover by impervious (paved) surface. As shoreline development proliferates and hundreds of thousands of recreational boats are used annually on the bay, the frequency of human exposure to storm hazard escalates. With global warming fairly well documented, the potential for increased storm frequency for both tropical and extratropical cyclones is a major coastal management factor in the new millennium.

High humidity and hot, hazy conditions in summer contribute both to human discomfort and the persistence of air pollution in wide areas surrounding the Chesapeake's urban centres. While the bulk of airborne nitrogen deposition comes from an "airshed" covering all or parts of 11 states, most outside the drainage basin, the significance of vehicle miles travelled within the drainage area is great. The contribution of atmospheric pollution to total basin wide nutrient loads is on the order of 21–27%, but only about 6% is directly deposited to the bay's surface.

The usual presence of the stinging, and often large (20 cm × 1.5 m) estuarine jellyfish *Chrysaora quinquecirrha* from June or July through August or September, widely limits swimming and other water contact recreation during these uncomfortable periods. These "sea nettles" were not mentioned by early colonists, and only enter the historical record sporadically (1750, 1864, 1899) until the 20th century. Baird and Ulanowicz (1988) hypothesize that changes brought about in the food web by eutrophication have encouraged sea nettle growth relative to earlier centuries. In hotter and drier summers the occurrences are prolonged and the number of these organisms can increase locally to over 100 m^{-3}. The comb jelly *Mnemiopsis leidyi* is often abundant in spring and fall and, while benign to recreational swimmers, it is a voracious consumer of smaller zooplankton. Chrysaora is a predator of *Mnemiopsis*, which is usually scarce when many nettles are present. Chrysaora, by decreasing predation, might thus positively impact the abundance of other plankton species.

MAJOR SHALLOW WATER MARINE AND COASTAL HABITATS

Prominent among Chesapeake Bay's habitats at the time of European contact in the 17th century were its vast underwater seagrass meadows, occupying large areas from 1–3 m deep along tributaries, the flanks on either side of the channels, and around many marshy islands. These submerged aquatic vegetation habitats are variously estimated to have covered 1214–2428 km^2 (300–600,000 acres) and were extraordinarily valuable for the spawning and shelter of fish and shellfish, notably billions of the commercially valuable blue crab *Callinectes sapidus*.

Regular aerial surveys were not begun until bay grasses underwent a collapse after Tropical Storm "Agnes" in June, 1972 (Fig. 5), but from the 1938 and 1952 surveys, a number of aerial photographs taken by the U.S. Department of Agriculture Soil Conservation Service will enable modern specialists to reconstruct grass coverage from a half century ago for about three quarters of at least the Maryland portion of Chesapeake Bay (Naylor, 1999, pers. comm.). Though reduced in size to the order of 28,300 hectares (70,000 acres), these grass beds continue to be of great importance to the bay ecosystem. They vary in area from year to year, contracting in wet, turbid periods and expanding in dry years with clearer water, but over a decade and a half the trend has been for increasing coverage.

The earliest known pilot chart of lower Chesapeake Bay, drawn about 1607 by Robarte Tindall in 1608 (Haile, 1998) shows about 24 reefs drawn to help navigators avoid them. These were emergent intertidal oyster banks and represented a small part of an immense benthic stock of the shellfish *Crassostrea virginica* (Hargis and Haven, 1995). Newell (1988) estimated these animals could in several days filter through their gills a volume of water about equivalent to the entire bay. With substantially more oligotrophic conditions in the past, this is believed to have meant the bay was once much clearer than at present. Commercial oyster dredging under sail and steam, mostly during the 19th century, removed—essentially mined—a large part of this stock. Maryland alone harvested 15,000,000 bushels (528,000 m^3) in the year 1885. Newell estimates that the 1990s stock of oysters, perhaps 1% of its historic level, could take over 400 days to filter a comparable volume of water. The remaining oyster populations are still harvested annually by hand and mechanical pincer-like tongs and dredged

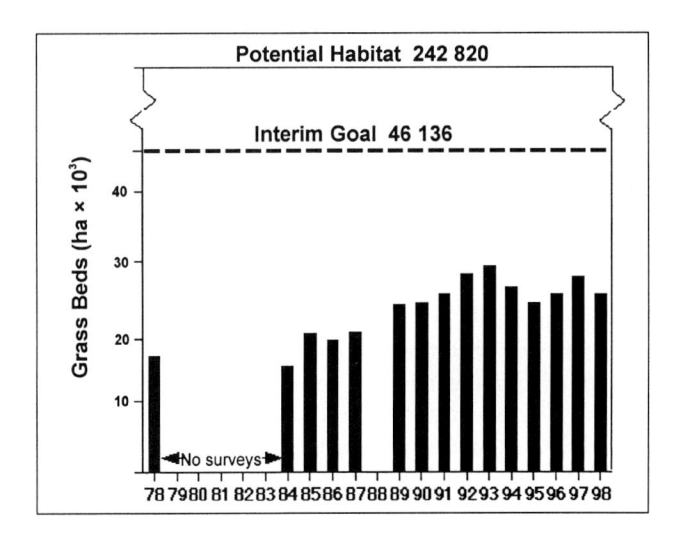

Fig. 5. Submerged Aquatic Vegetation area coverage from 1978 to 1998. Vertical axis shows the estimated original extent of these grass beds and an interim goal for year 2005, towards which restoration efforts are proceeding (Sylvester, 1999).

Fig. 6. A sailing "skipjack", one of a small remnant fleet still licensed to dredge oysters in Chesapeake Bay. Photo by Kent Mountford.

Table 2

Summary of the 4007 known aquatic species in Chesapeake Bay by major taxonomic groupings and habitat. Not included are macro-algae of which there are 52 spp. in the Potomac estuary alone (Lippson et al., 1979), or the 28 species of submerged aquatic plants in 11 families (Batiuk et al., 1992)

	Taxonomic Group					
	Benthos	Fish	Micro-zoo-plankton	Meso-zoo-plankton	Phyto-plankton	Total
Fresh	171	109	64	146	397	887
Olgiohaline	158	136	44	127	328	793
Mesohaline	324	175	46	183	386	793
Polyhaline	364	169	22	177	481	1213
Total	1017*	589	176	633	1592	4007

*Of these 106 are bivalves and 270 are crustaceans.
Source: Johnson, Jacqueline Interstate Commission on Potomac River Basin.

picturesquely by a small surviving fleet of sailing "skipjacks", remnants of a thousand such vessels (Fig. 6).

Holland et al. (1989), using estimates for the filtering ability of current infaunal benthos, also calculated that a large portion of the bay's water could still be filtered but this does not appear to have resulted in clearer water. Taxa of the benthos are roughly divided between suspension feeders in shallower waters and deposit feeders in deeper waters. Where dissolved oxygen falls to low levels, the benthos is extirpated in summer and the recolonizers each autumn are opportunistic and rarely achieve significant biomass before being killed by hypoxia the following spring. Holland et al.'s data, and that of others, nonetheless indicates that the modern benthos converts a large fraction of phytoplankton production to infaunal biomass and forms a major energy transfer path to higher trophic levels.

The Chesapeake has about 4000 aquatic species, including plankton but not macroalgae. These are summarized in Table 2. Many species habitat requirements have been described by Funderburk et al. (1991) and Lippson and Lippson (1997).

There are about 267 fish species recorded as occurring in estuarine Chesapeake Bay, allowing some uncertainty about including rare seasonal visitors (Murdy et al., 1997). While the bay anchovy *Anchoa mitchelli* is the most abundant fish in the bay, migratory species far exceed in biomass the 32 fully resident taxa.

There are only seven truly anadromous species: two sturgeon, the shortnosed *Acipenser brevirostris* and the Atlantic *A. oxyrhynchus*, were once a staple in 17th century colonial Virginia and in the 19th and early 20th centuries provided a significant commercial harvest for flesh and caviar. By the late 1920s they were, and are still, quite uncommon. Four migratory spawning alosines, shad and herring species, were once far more abundant than today. They have declined due to some 2500 blockages of the watershed's rivers and streams by dams or other structures, and as a consequence of overfishing early in the 19th and early 20th century. The peak harvest after 1890 was about 6.9 million kg (Fig. 7). Harvest of American shad within the Bay has been illegal in Maryland since 1980 and in Virginia since 1994. The offshore intercept fishery which captures gravid female American shad for their roe, a regional delicacy, will be closed by 2004.

The striped bass or "rockfish" *Morone saxatilis* is an anadromous fish which in adulthood can exceed 55 kg. It has been known from waters of the Virginian province since explorers described it in the 1580s. Very low harvests of this once-abundant commercial and recreational species caused concern starting in the 1970s (Fig. 8). A hard-fought temporary moratorium on fishing, some pollution control in spawning habitat and carefully adjusted harvest limits have largely restored spawning stock for this species to healthy levels. This is considered a significant management success story.

Juvenile habitat in the Bay's tributaries for striped bass overlaps somewhat in time and space with that of white

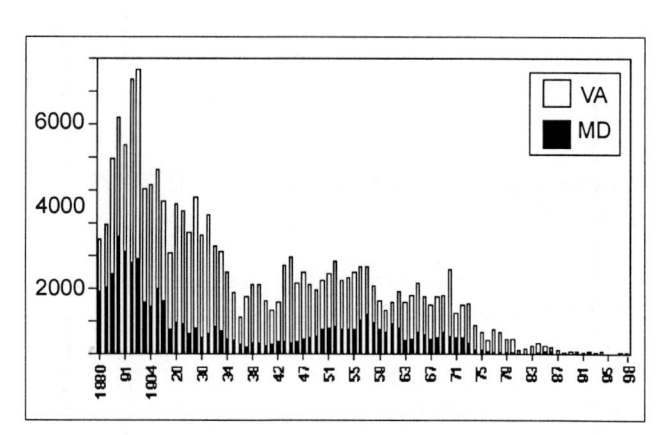

Fig. 7. Harvest statistics for American Shad in Virginia and Maryland portions of Chesapeake Bay 1880–1998 in millions of kilograms (Sylvester, 1999).

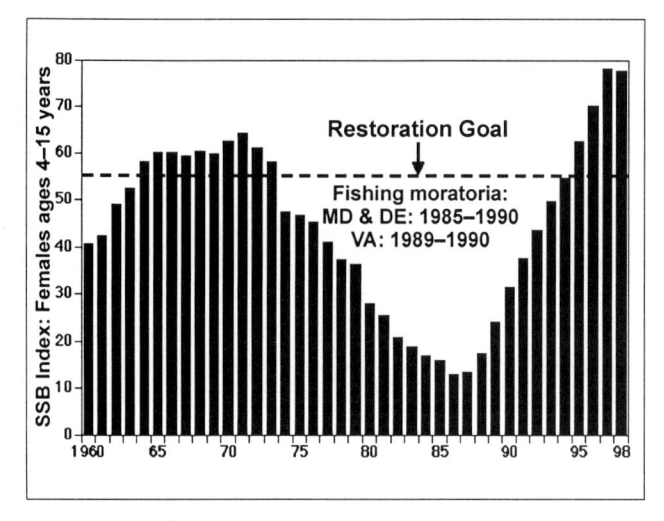

Fig. 8. Spawning stock estimates for the striped bass or "rockfish" *Morone saxatilis*, 1960–1998 (Sylvester, 1999).

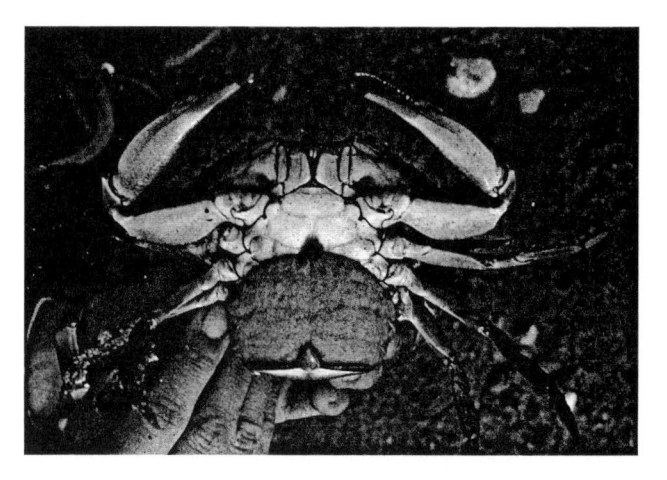

Fig. 9. A gravid female, or "sponge", blue crab *Callinectes sapidus*. Mated females migrate to deeper water farther down-bay and carry eggs over the winter. They begin spawning in spring and the zoea larvae are largely carried out to sea. See text (Photo by Kent Mountford).

perch *Morone americana* which is also an abundant and popular species. Resource competition between these species has been suggested. In 1999 there were concerns about the decline of some forage fish species like the bay anchovy *Anchoa mitchelli* and seasonally occurring menhaden *Brevoortia tyrannus* particularly as the population biomass of striped bass increased. Menhaden are also under substantial harvest pressure and, Atlantic coast-wide, catches fell during the last years of the 20th century. Menhaden provide the largest single commercial landing from Chesapeake Bay (180,707,000 kg in 1998), with a dockside value of US$ 13.9 million.

Several coastal ocean spawners seasonally occupy the Bay and some (for example, three species of *Cynoscion* the "sea trouts", croaker *Micropogon undulatus* and spot *Leiostomus xanthurus*) support significant recreational or commercial fisheries. Others like the bluefish *Pomatrix saltatrix* and spanish mackerel *Scomberomorus maculatus* are more episodic, but significant catches occur in times of abundance.

Declines in the Bay's unusually high historic fishery yields have transferred substantial pressure to the fishery for blue crabs *Callinectes sapidus* (Fig. 9). Annual harvest has fluctuated widely between 18,000 and 50,000 m tons (40–110 million pounds) since the 1950s, but in the late 1990s decreasing yields have caused concern among managers about finding risk-averse fishery thresholds.

Marine mammals are not viewed as common Chesapeake Bay fauna. Nonetheless, manatee *Trichechus manatus* have rarely but repeatedly been reported in Chesapeake tributaries from 1676 to 1995. Inshore dolphin *Tursiops* sp. utilize the lower Bay and occasionally the rarer harbor porpoise *Phocena phocena* move up the Eastern Shore as far as Eastern Bay (below Kent Island) by about August and down the Western shore into autumn. As many as 100, but more usually a dozen *Tursiops* have been seen in aerial surveys. Loggerhead turtles, *Caretta caretta*, in the aggregate

Fig. 10. Calvert Cliffs on the western shore of Chesapeake Bay erode continually to expose rich deposits of Miocene fossils, but private residences built imprudently atop them are often threatened by shoreline recession (photo by Kent Mountford).

number from about 3000 individuals up to a potential 15,000 individuals utilize the Bay mouth. Net captured, stranded, or ship-propeller killed, specimens are found as far north as Calvert Cliffs (Fig. 10). Kemp's ridley *Lepidochelys kempi* and the leatherback *Dermochelys coriacea* are sometimes encountered. Occasional humpback (1994) and minke (1999) whales are sighted in the Bay, in some cases as far north as the Chesapeake Bay Bridge (39°N lat.). The harp seal *Phoca groenlandica*, once decimated by hunting, has expanded its worldwide population to about 7 million and in 1999 one was recently found in Tangier Sound (Scofield, 1999, pers. comm.).

OFFSHORE SYSTEMS

The central and lower mainstem of Chesapeake Bay ranges in width from 21 to 32 km and the Bay mouth is about 24 km wide, so its connection with the coastal ocean is substantial. The Chesapeake Bay ebb tidal plume tends to fold southward along the North Carolina Coast and following major storm events like Tropical Storm "Agnes" in 1972, the salinity signature of Chesapeake water can be traced as far south as Oregon Inlet over a period of at least 41 days (NASA, 1981). Incoming tides also receive flow coming down the Atlantic Coast from farther North, part an inshore countercurrent circulation to the offshore, northward-flowing Gulf stream.

Large amounts of airborne nitrogen and other contaminants are deposited at sea over the continental shelf. These, added to waterborne contributions from other mid-Atlantic estuaries provide a reservoir which shares portions of these loads with the Chesapeake. Model studies designed to simulate processes within the Bay have required repeated extensions of boundary conditions into neritic waters to account for such processes. Such models, among the most complicated yet constructed, suggest that the coastal ocean is also a significant source of phosphorus to the Chesapeake (Linker, 1996).

In summer, subtropical species entrained in this major ocean current system, like the occasional triggerfish *Balistes carolinensis* and pompano *Trachinotus* spp. are recorded reaching the bay, though they are normally from outside the Virginian province.

The blue crab *Callinectes sapidus* spawns in multiple pulses from spring into summer around the lower Chesapeake and the bulk of zoea are carried to sea where they metamorphose through several stages into megalopa. The megalopa rely upon regional wind patterns and the return circulation of coastal ocean waters to carry them back into the estuary for growth to adulthood.

POPULATIONS AFFECTING THE AREA

Native American peoples have occupied the watershed for at least 12,000 years, and certainly since the end of the Wisconsin Glaciation or "Ice Age" when rising sea level slowly filled ancient river valleys to create today's Chesapeake. Native American populations at the time of European contact may have been lower than in earlier periods but are estimated at around 100,000 persons in the watershed, and about 30,000 persons in "tidewater", the portion of the basin below the major river fall lines. Native American peoples had significant impacts on their environment through selective game hunting, burning and, after about 3000 BP, agriculture. Today native Americans are a minute fraction of total population.

European adventurers visited the Bay in the mid to late 16th century and English colonists planted the first precarious but permanent colony at Jamestown in 1607. This commenced a long sequence during which the watershed population and its societal impacts overwhelmed those of Native Americans. Heavy emphasis on agriculture, especially tobacco grown on the coastal plain, benefitted in the short term from slave labour and hundreds of thousands of Africans were brought in to feed this demand until this practice ended following the American Civil War. African American citizens currently make an important contribution to total population and human resources.

Exploitation of the Bay's shell- and fin-fisheries was contemporary with heavy industrialization around the major ports at Norfolk (shipbuilding and naval stores) and Baltimore (steel and chemical manufacture). This brought many immigrant peoples other than English. Following the American Civil War, the World Wars, Korean and Vietnam Wars, immense bursts of population growth and the sprawl of development occurred across formerly forested or agricultural landscape. Latino and Asian-American people, many arriving to work in agricultural or service occupations, were forming an increasing proportion of population at the end of the millennium.

Growth and development, much of it concentrated along the interstate highway network linking Harrisburg, Baltimore, Washington, Richmond and Norfolk, has had profound, and probably irreversible effects on the watershed and consequently, Chesapeake Bay. The value of the Bay's fisheries has been estimated at $1 billion US annually. Overall, bay-related economic activity with its socioeconomic multipliers was estimated in 1989, at $678 billion annually (MD Dept. Econ. & Employment Development).

In 1985 total nutrient loads of 162.8 million kg (359 million pounds) of nitrogen and 11.8 million kg (26 million pounds) of phosphorus were estimated from the watershed. While the watershed's forests were estimated to release about 14% of the nitrogen and 4% of phosphorus, the bulk of nutrients came from point and non-point sources as indicated in Fig. 11. These largely result from human activity. Some 6.6 million m³ (1.5 billion gallons per day) of variously treated municipal sewage effluents enter the Bay and emerge through its tributaries daily.

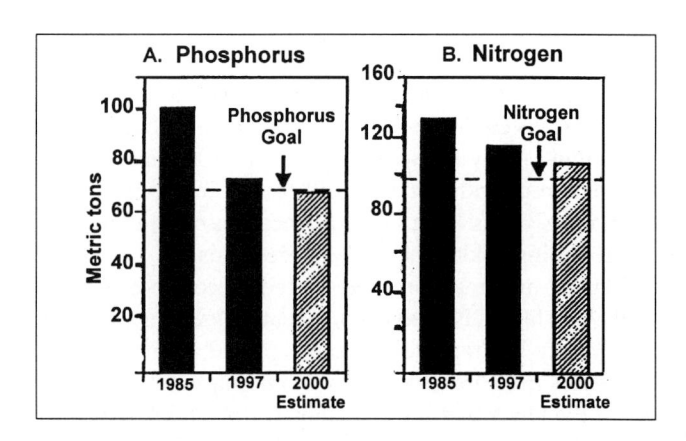

Fig. 11. Nutrient loads for phosphorus and nitrogen delivered to Chesapeake Bay in millions of kg yr^{-1}. (A) Phosphorus and (B) nitrogen, for a reference year 1985, 12 years' progress in 1997, and an estimate for year 2000. Point source phosphorus, by 1997, was reduced by 49% and nitrogen reduced by 18% (see text) (Sylvester, 1999).

Non-point, or diffuse, sources still dominate and of those agriculture contributes the largest share. Point sources, those from industrial and municipal generators, were very large in the past but have been more amenable to control using improved technology.

Atmospheric sources contribute about 21% of non-point nitrogen pollution. Inputs include those from power plant and industrial combustion sources, many of which arrive as atmospheric plumes from outside the bay watershed. The contributing "airshed" is considered to cover all or part of eleven states. Signal among atmospheric sources are the emissions from motor vehicles for which the aggregate "miles travelled annually" increased from 65 billion in 1970 to about 150 billion in 1999. Between 1997 and 2010, vehicle miles travelled are projected to increase 32% to almost 190 billion, while for the same period population is only expected to grow by 10%.

Population at the end of the 20th century was estimated at 15.7 million persons with a mean daily increase (in 1999) of some 311.2 persons (113,588 annually). Population is projected at 17.8 million in the year 2020 (Chesapeake Bay Program, 1999). More significant has been the rate of population sprawl, a phenomenon where urban centres decline and suburban and rural areas increase in population at radii farther and farther from the urban epicentre. Such sprawl development was also focusing along formerly bucolic rural water bodies in Maryland and Virginia. Overall, the number of persons per household is decreasing, and the lot size occupied by each dwelling is increasing. In the watershed 24.7% of housing units are outside the reach of municipal sewer systems. The fraction of non-point source nutrients contributed by their septic systems, which discharge nitrogen to groundwater, is also increasing.

RURAL FACTORS

At the time of European settlement in 1607 historians believe the Chesapeake watershed was about 95% forested. There appear to have also been significant grassland or "prairie" areas in the Shenandoah and Susquehannah River Valleys, and regions in Maryland of scrub forest on serpentine barrens with thin or absent soil. These may have totalled about 8300 km^2, colonists record that much of the coastal plain was in old-growth forest with some trees near 3 m in circumference, and many species had the potential to produce very tall and straight timber. Near the bay some areas had open park-like forest stands, often hardwoods but also with areas of pine (Middleton, 1994).

Only a few tiny remnants of forest purported to be uncut remain, all the rest being logged over once or more for firewood, fencing, charcoal or timber, reduced to agriculture, paved or put in urban land use. After the American Civil War, and during subsequent westward population expansion in the United States, substantial areas reverted to second (or third) growth woodland. Despite abandonment

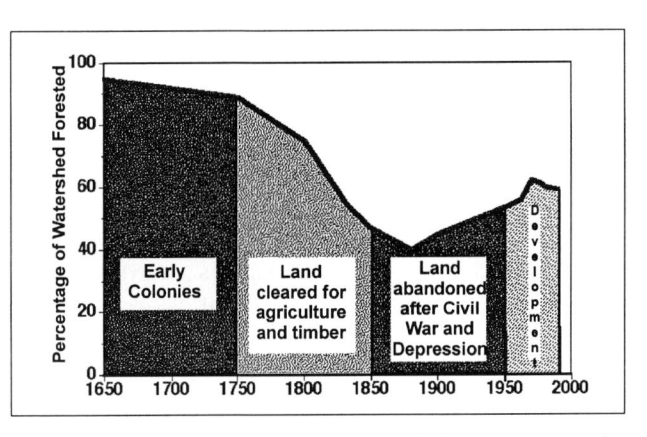

Fig. 12. Estimated time-course of overall forest cover in the Chesapeake Basin from 1650 onward. Rate of forest loss in 1999 was estimated at 52 ha d^{-1} (130 acres) (Sylvester, 1999).

of agricultural land, and temporary regrowth of woodlands, deforestation might have approached 80% near the end of the 19th century.

About 30% forest cover appears to have regrown since, but in the latter half of the 20th century forest loss resumed at a rate of at least 40.5 hectares (100 acres) d^{-1}. This loss continues, and may have accelerated to over 52 hectares d^{-1} (130 acres) by 1999, accompanied by the removal of streamside wooded buffers and forest fragmentation, both of which seriously degrade terrestrial and in-stream habitat quality. In 1990–91, forest in some successional stage covered about 59% of the watershed (Fig. 12).

Agricultural practice was initially conducted by girdling trees, slashing and burning understorey, and planting among the stumps, which mimicked Native American practices. This was suitable for tobacco culture, a major colonial cash crop for the Maryland and Virginia portions of the basin for nearly 200 years. Plow agriculture, characterized by deep furrow tillage for cereal and row-crops became widespread from the early 18th century onwards. As this technology was carried up onto piedmont and into Appalachian mountain topography, soil erosion became a major factor following rainfall events, and resulted in the filling of many small bay-side and tributary river shipping ports (Middleton, 1994).

Modern agricultural croplands cover 36,360 km^2 (9 million acres) in the basin. In aggregate, farming is the largest contributor of nitrogen and phosphorus in runoff to the bay. The resultant sediment, turbidity and algal standing crop increases are linked with broad declines in the areal extent of submerged aquatic vegetation. Much management attention is being focused on concentrated animal feeding operations for poultry and livestock, a segment of the industry which grew rapidly in the late 20th century. While direct causation has not been established, non-point nutrient sources along the Nanticoke and Pocomoke Rivers were linked with a six-year localized decline of submerged aquatic vegetation in Tangier Sound and with outbreaks of a toxic form of the dinoflagellate *Pfiesteria piscicida* during the summer of 1997.

COASTAL EROSION AND LANDFILL

As a result of sea-level rise, most of the fast shoreline, the margins of tributaries below head of tide, and virtually all island footprints in Chesapeake Bay are eroding, in some localized areas at over 3 m annually. There are limited areas where, as a result of alongshore migration of sediment eroded from cliffs and banks, accretion is occurring.

The Cliffs of Calvert in Maryland—at places over 30 m high—and those at Nomini in Virginia are many kilometres in length. Breakdowns from the upper portions of these cliffs, and subsequent toe erosion, exposes vast and scientifically valuable fossil deposits of Miocene and Eocene age. These breakdowns often carry away large trees, occasionally imprudently sited homes and rarely have killed fossil hunters below. Calvert Cliffs' fossil deposits are known worldwide for the thousands of shark teeth found. Single teeth of *Carcharodon megalodon* from the Miocene, are occasionally found as trophies exceeding 15 cm in size.

The loss of real estate, here and elsewhere, has stimulated erosion control programs employing a variety of shoreline hardening techniques ranging from the planting of marsh vegetation (chiefly *Spartina alterniflora* or American beachgrass *Ammophila breviligulata*), various fabricated matting and sediment containment products, offshore stone breakwaters parallel to the beach, "riprap" (a continuous stone revetment) and bulkheading of wood, steel or aluminum. The interruption of littoral sediment flow often accelerates erosion on adjacent parcels and stimulates further shoreline structures. Reflected wave energy from hardened shorelines frequently exacerbates erosion locally.

All of these structures cover or interrupt the function of natural shoreline habitat and adversely affect many species, aquatic and terrestrial. Examples include nesting territories of tiger beetles (*Cicindella* spp., one of which was on the endangered list in 1999), the diamondback terrapin (*Malaclemys terrapin*) which returns to natal sand beaches for nesting. These beaches are also often breeding habitats for the horseshoe crab (*Limulus polyphemus*).

PROTECTIVE MEASURES

Chesapeake Bay has historically been an immensely productive ecosystem. It maintained this production in the face of escalating human population from 100,000 to over six million persons and suffered depredations supported by increasingly sophisticated technology for about three centuries. Declines in major fish and shellfish resources began to generate serious public outcry in the last third of the of the 19th century, but the economic growth and development pressure surrounding and following major wars through the first half of the 20th century largely overwhelmed these concerns.

In June of 1972, Tropical Storm "Agnes", which had been a Caribbean hurricane, dumped rainfalls of 30 cm over large portions of the Chesapeake (Davis, 1974). This storm struck during fish migration and spawning and at a critical growth stage for submerged aquatic vegetation. Flooding, especially along the Susquehanna, was catastrophic. Tremendous loads of sediment accompanied the freshwater flows, and bay resources staggered. The effect on Chesapeake Bay was quickly perceived by the public and elected officials.

In the same timeframe, passage of the federal Clean Water Act in 1972 provided major resources nationwide for the cleanup of pollution and, following a fact-finding tour around the bay, a Republican Senator from Maryland, Charles McC. Mathias, shepherded through Congress a $25 million appropriation to study the Bay's problems over the subsequent five years (USEPA, 1983). Out of this work, made public in 1983 at a conference hosted by basin politicians, came a Federal–State Cooperative Agreement, and creation of the USEPA Chesapeake Bay Program, which at the Millennium was in its seventeenth year.

The Bay Program was, at its creation, unique in North America both in its commitment to base management decisions on science and on its cooperative, non-regulatory nature. Governors of the partner States and the District of Columbia were more likely to agree to and strive for more ambitious goals to which they agreed voluntarily, than to those imposed by regulation. The "Bay Program" model became a template for estuarine restoration programs nationwide, and to some extent worldwide.

A monitoring program was established in 1984 to track the bay's condition, and was sustained, largely intact, to the end of the century. It provides one of history's most comprehensive estuarine databases, at least for the suite of parameters integrating hydrography, nutrients, plankton and benthos. This monitoring served as a framework upon which an astonishing number of accessory studies covering physics, biology and estuarine processes were carried. A long-term citizen-based volunteer monitoring program in a number of tributaries was also established through the Alliance for Chesapeake Bay, a regional membership organization, and was also sustained from 1985–2000.

Subsequent to the original 1983 Chesapeake Bay Agreement, a second Agreement in 1987 set a number of goals to be reached by the millennium. Chief among these was a commitment to reduce incoming nutrients which are controllable by technical means to a level 40% lower than those reaching the Bay in 1985. Setting this goal, which involved acceptance and action by each of the basin states and District of Columbia governments, was a landmark process. To facilitate it, data from the comprehensive monitoring program was utilized in construction of an eventually complex and comprehensive three-dimensional hydrographic and water quality model (Linker, 1996). When projection scenarios reducing then-current inputs by 40% indicated measurable bay habitat improvement, regional political leaders agreed to underwrite the reduction goal.

As part of implementing bay recovery, a matching grant program was set up to encourage state non-point source pollution reductions. These grants initially focused on improving farm practices but have expanded to include urban programs and educational objectives. Deep tillage was substantially reduced, fertilizer use was brought more in line with crop needs, and winter cover crops to keep nitrogen from leaching to groundwater were greatly expanded. For farms above threshold livestock levels, nutrient management plans for manures were required by 1999 in Pennsylvania, Maryland, and Virginia. Integrated pest management was also expanded. With the intent of reaching the 40% reduction goal, each jurisdiction signing the agreement began putting in place detailed nutrient reduction plans called "Tributary Strategies". At the millennium these were in place for Maryland and Pennsylvania but still being developed in Virginia. In 2000, the phosphorus reduction has been met but the nitrogen goal was not attained.

Wastewater treatment nationwide was improving throughout the 1970s because of Clean Water Act resources, and this benefitted the Bay, but in the 1980s, EPA's single nutrient strategy was modified for estuaries like the Chesapeake to include nitrogen removal (Davidson et al., 1997). The introduction of biological nutrient removal ("BNR") technology made costly but practical wastewater nitrogen reductions possible. Effluent concentrations at some plants were reduced from 18 mg l^{-1} to as low as 3 mg and an expected average of 8 mg.

In year 2000, about 65 wastewater plants reduce nutrients using BNR technology, in which bacterial processes are manipulated to reduce both phosphorus and nitrogen in effluent (Randall et al., 1992), and within a few years that number should be 102 plants. Tributary strategies are to be fully implemented by 2020, at which time 64% of all wastewater entering the Bay should be treated using BNR technology. Actual municipal wastewater nutrient discharges from 1950 to 1997 and those projected for later years are shown in Fig. 13.

Agricultural best management practices (including reduced tillage of the soil, fertilizer and pesticide application based on soil testing and crop needs, the use of winter cover crops) have spread widely in the basin with nutrient management plans covering 1.7 million acres in 1997, rapidly increasing year to year towards a goal of 3.3 million acres in 2000.

Overall, since the commencement of the Chesapeake Bay Program, areal coverage of submerged aquatic vegetation has increased, and that has been associated with more areas meeting established water quality criteria (Batiuk et al., 1992). This is considered progress towards the interim goal for restoring submerged aquatic vegetation (see Fig. 5). Several extraordinary freshet years apparently introduced sufficient nutrients and turbidity to cause interannual variability in bay-wide vegetation coverage, and in the Tangier Sound area, where nitrogen loads from tributary rivers were still increasing, the areal extent of submerged vegetation declined for several years running in the late 1990s.

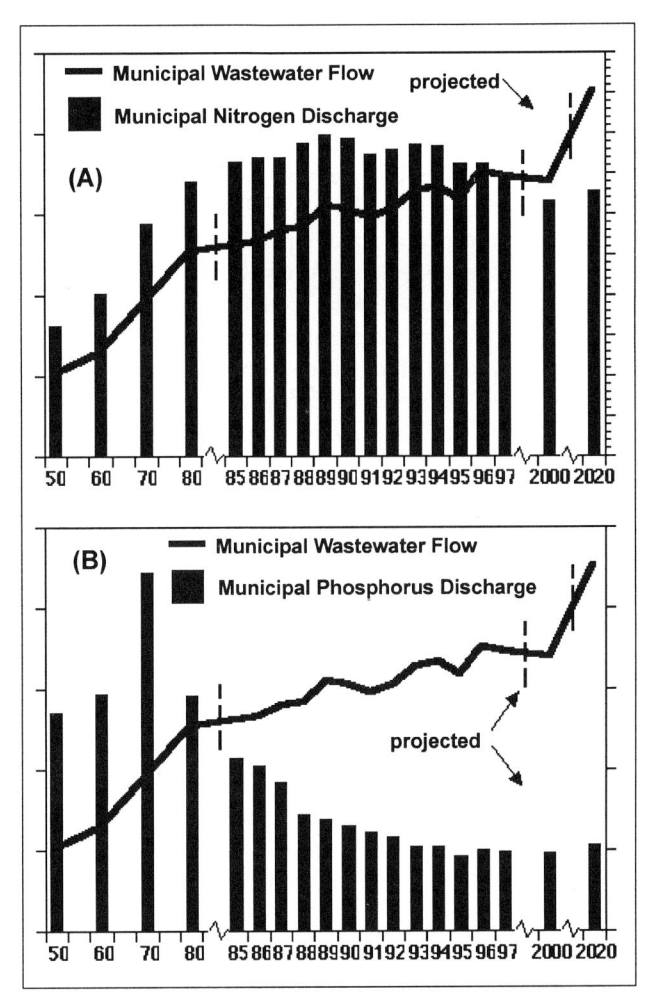

Fig. 13. Municipal wastewater flows and loads discharged to Chesapeake Bay for nitrogen (A) and phosphorus (B) for 1950 through 1997, with projections for 2000 and 2020, as population growth begins to reverse trend (Sylvester, 1999).

Habitat restoration is a high Chesapeake Bay Program priority and removal of blockages to resident fish movement and anadromous fish migration opened 1738 km of stream course. The target by year 2003 is 2985 km (1357 miles). The number of shad arriving at the Susquehanna to spawn increased from several hundred in the early 1980s to over 81,000 fish in 1999, though this is trivial against historic population levels. A goal for restoring forested stream buffers along Chesapeake watercourses where they are currently absent has been set at 4444 km, or "2020 miles by the year 2020".

Fossil oyster shell thousands of years old is "mined" by dredging from upper Chesapeake Bay and, together with other inert bulk materials, has been used to construct several artificial reef structures, emergent at low tide and reminiscent of those described by early colonists. These appear attractive for oyster spat settlement and growth, but oysters are still subject to annual mortalities from the combined effect of the protozoan diseases MSX and Dermo. Tidal wetland losses have slowed to a trickle but it must be remembered that overall 50% of wetlands present in North America at the time of European settlement have already been lost. The long-term ecological function has yet to be demonstrated for artificial wetlands, created as "mitigation" for damage done in other projects.

Significant management lessons were learned through the successful striped bass restoration exercise, and multi-species management is now considered a productive route to follow with other targeted species.

With a long history of industrial and general business activity focused near metropolitan areas like the seaports of Baltimore and Norfolk, with military activity and heavy development surrounding Washington, DC, substantial accumulations of toxic materials cause concern. Many contaminants are widely detectable in water and sediments across the bay and tributaries, and show significant profile changes with depth in sediment, and therefore over time. Analyses in the mid to late 1990s suggest that significant ecosystem effects concentrate around "regions of concern" in the Patapsco, Anacostia and Elizabeth Rivers.

Many toxic waste sites have been stabilized at significant public and industry expense. Following the principle of "pollution prevention" espoused by EPA, and coupled with regulatory and enforcement actions, large decreases in the amounts of known toxic materials discharged to the bay have been achieved. The Chesapeake Bay Program in 1999 estimated that from 1988 to 1997 overall toxic releases were reduced 67%.

Maintaining viability for the bay's two major seaports is high on the political agenda. Economic interests perceive the need for maintaining, altering and deepening major harbour approach channels and the Chesapeake and Delaware Canal, which connects from near the Susquehanna's mouth to the adjacent Delaware River. Each year about 10,000 vessels reportedly use Chesapeake ports, including Baltimore and Norfolk. Large quantities of dredged material are removed each year on a regular basis and additional dredging is proposed to accommodate vessels of still deeper draft. At the millennium, placement in containments (diked areas and artificial island structures) is favoured in comparison to overboard disposal, but both options result in the release of large quantities of nutrients into the water column.

Further commitments to continuing restoration of the estuary are put in place for the time-frames 2000–2030 and these must deal with continued population growth, pressures of the economy and significant changes in land use. There is clear recognition that fisheries cannot be sustained without managing them as associations of interdependent species. In 1999 there is much emphasis on sustainable rather than unbridled growth, with an eye to preserving natural landscape and its ecosystem services, managing impervious surfaces, encouraging sound agriculture and clustering development to control sprawl.

The success of these strategies in "capping" nutrient loads to the bay at the 40% reduction goal level, and the prospects for coping with population and economic growth pressures are serious challenges. There must be strong political willpower to make still greater reductions in nutrient loading if we are to continue improvements in the Chesapeake ecosystem. Whether such willpower can transcend individual state and federal administrations will determine success or stagnation in the early decades of this new millennium.

REFERENCES

Baird, D. and Ulanowicz, R. (1989) Seasonal dynamics of the Chesapeake Bay ecosystem. *Ecological Monographs* 59, 329–364.

Batiuk, R., Orth, R., Moore, K., Dennison, W., Stevenson, J., Staver, L., Carter, V., Rybicki, N., Hickman, R., Kollar, S., Bieber, S. and Heasly, P. (1992) Submerged aquatic vegetation habitat requirements and restoration targets: A technical synthesis. US Environmental Protection Agency, CBP/TRS 83/92, Annapolis, Maryland. 185 pp.

Breitburg, D.L. (1992) Episodic hypoxia in the Chesapeake Bay: Interacting effects of recruitment, behavior and physical disturbance. *Ecological Monographs* 62, 525–546.

Bue, C.D. (1968) Monthly Surface-water Inflow to Chesapeake Bay. US Geological Survey Open File Report, Arlington, Virginia. 45 pp. (This report is modified at USGS Website by hypertext at md.usgs.gov/publications/).

Brush, G., Hill, J. and Unger, M. (1998) Pollution History of Chesapeake Bay. NOAA Technical Memo NOS-ORCA 212. 75 pp. and 4 appendices.

Chesapeake Bay Program (1999) The State of the Chesapeake Bay, a report to citizens of the Bay region. EPA 903-R99-013, CBP/TRS 222/108. Chesapeake Bay Program, Annapolis, MD, 61 pp.

Cleves, E.T. (1989) Appalachian Piedmont Landscapes from the Permian to the Holocene. *Geomorphology* 2, 159–179.

Davidson, S., Merwin, J., Capper, J., Power, G. and Shivers, F. (1997) *Chesapeake Waters*. Tidewater Publ., Centreville, Maryland, 272 pp.

Davis, J. (ed.) (1974) Report on the Effects of Tropical Storm Agnes on the Chesapeake Bay Estuarine System. J. Davis, Chesapeake Research Consortium Contribution 34 for US Army Corps of Engineers. 59 pp. plus 6 Appendices.

Funderburk, S., Mihurski, J., Jordan, S. and Riley, D. (1991) Habitat Requirements for Chesapeake Bay Living Resources. Chesapeake Research Consortium, Solomons, Maryland, 300 pp.

Haile, E.W. (1998) *Jamestown Narratives, Eyewitness Accounts of the Virginia Colony*. Round House, Champlian, Virginia. 946 pp.

Hargis, W.J. and Haven, D. (1995) The Precarious State of the Chesapeake Bay Public Oyster Resource. Contrib. 1965. Chesapeake Research Consortium Inc., Edgewater, Maryland. 51 pp.

Hill, J.M., Halka, J., Conkwright, R., Koczot, K., Park, J. and Colman, S. (1992) Distribution and Effects of Shallow Gas on Bulk Estuarine Sediment Properties. *Continental Shelf Research* 12, 1219–1229.

Holland, A., Shaughnessy, A., Scott, L., Dickens, V., Gerritsen, J. and Ranasinghe, J. (1989) Long-term Benthic Monitoring and Assessment Program for the Maryland Portion of Chesapeake Bay: Interpretive Report. Maryland Power Plant Siting Program, CBRM-LTB/EST-89-2. Annapolis, Maryland. 112 pp. + appendices.

Horton, T. and Eichbaum, W. (1991) *Turning the Tide*. Island Press, Washington DC, 324 pp.

Johnson, J. (1999) Interstate Commission on the Potomac Basin, Rockville, Maryland. Personal communication.

Linker, L. (1996) Models of the Chesapeake Bay. *Sea Technology* 37, 49–55.

Lippson, A.J., Haire, M., Holland, A., Jacobs, F., Jensen, J., Moran-Johnson, R., Polgar, T. and Richkus, W. (1979) Environmental Atlas of the Potomac Estuary. Martin Marietta Corp. for Maryland Department of Natural Resources, 280 pp.

Lippson, R. and Lippson, A.J. (1997) *Life in the Chesapeake Bay*. Johns Hopkins Press, Baltimore, Maryland, 294 pp.

Maryland Department of Economic & Employment Development, Office of Research (1989) Economic Importance of the Chesapeake Bay. Baltimore, Maryland, 12 pp.

Middleton, A.P. (1994) *Tobacco Coast*. Johns Hopkins Univ. Press, Baltimore, Maryland, 508 pp.

Murdy, E., Birdsong, R. and Musick, J. (1997) Fishes of Chesapeake Bay, Smithsonian Institution, Washington, DC. 324 pp.

NASA (1981) Chesapeake Bay Plume Study, Superflux, 1980. NASA Conference Publication 2188. National Aeronautics and Space Administration, Washington, DC, 522 pp.

Naylor, M. (1999) Maryland Department of Natural Resources. Personal communication.

Newell, R. (1988) Ecological changes in the Chesapeake bay: are they the result of overharvesting the American oyster, *Crassostrea virginica*. In Understanding the Estuary: Advances in Chesapeake Bay Research, Conference Proceedings, 29–31 March 1988, Baltimore, MD. Chesapeake Research Consortium Pub. 129, Solomons, MD, pp. 536-546.

Poag, C.W. (1997) The Chesapeake Bay bolide impact: a convulsive event in Atlantic coastal plain evolution. *Sedimentary Geology* 108, 45–90.

Pritchard, D.W. (1955) Estuarine circulation patterns. *Proceedings American Society of Civil Engineers* 81, 1–11.

Randall, C., Barnard, J. and Stensel, H. (1992) *Design and Retrofit of Wastewater Treatment Plants for Biological Nutrient Removal*. Technomic Publications. Lancaster, Pennsylvania, 417 pp.

Sanford, L., Sellner, K. and Breitburg, D. (1990) Covariability of dissolved oxygen with physical processes in the summertime Chesapeake Bay. *Journal of Marine Research*, 48, 567–590.

Scofield, D. (1999) National Aquarium at Baltimore, Maryland. Personal communication.

Smith, J. (1624) (Facsimile) *A Generall Historie of Virginia, New England and the Summer Isles, Books I–IV*. Michael Sparkes, London, 248 pp.

Sylvester, N. (1999) Environmental Indicators: Measuring our Progress. US Environmental Protection Agency, Annapolis, MD, maintained as a dynamic web document at www/chesapeakbay.net on the "quick links" menu at Environmental Indicators.

Titus, J. (1998) Rising seas, coastal erosion and the Takings Clause: How to save wetlands and beaches without hurting property owners. *Maryland Law Review* 57, 1279–1399.

USEPA Chesapeake Bay Program (1983) Chesapeake Bay: A Profile of Environmental Change. US Environmental Protection Agency, Philadelphia, PA, 200 pp.

Whitney, G.G. (1996) *From Coastal Wilderness to Fruited Plain*. Cambridge Press, 451 pp.

THE AUTHOR

Kent Mountford

*US Environmental Protection Agency,
Chesapeake Bay Program,
410 Severn Ave., Suite 109,
Annapolis, MD 21403, U.S.A.*

Chapter 22

NORTH AND SOUTH CAROLINA COASTS

Michael A. Mallin, JoAnn M. Burkholder, Lawrence B. Cahoon and Martin H. Posey

This coastal region of North and South Carolina is a gently sloping plain, containing large riverine estuaries, sounds, lagoons, and salt marshes. The most striking feature is the large, enclosed sound known as the Albemarle–Pamlico Estuarine System, covering approximately 7530 km^2. The coast also has numerous tidal creek estuaries ranging from 1 to 10 km in length. This coast has a rapidly growing population and greatly increasing point and nonpoint sources of pollution. Agriculture is important to the region, swine rearing most notably increasing fourfold during the 1990s.

Estuarine phytoplankton communities in North Carolina are well studied; the most important taxonomic groups are diatoms, dinoflagellates, cryptomonads, and cyanobacteria. Several major poorly flushed estuaries are eutrophic due to nutrient inputs, and toxic dinoflagellates (*Pfiesteria* spp.) may reach high densities in nutrient-enriched areas. Fully marine waters are relatively oligotrophic. Southern species enter in subsurface intrusions, eddies, and occasional Gulf Stream rings, while cool water species enter with the flow of the Labrador Current to the Cape Hatteras region. The Carolinas have a low number of endemic macroalgae, but species diversity can be high in this transitional area, which represents the southernmost extension for some cold-adapted species and the northernmost extension of warm-adapted species. In North Carolina the dominant seagrass, *Zostera marina*, lies at its southernmost extension, while a second species, *Halodule wrightii* is at its northernmost extent. Widgeon-grass *Ruppia maritima* is common, growing in brackish water or low-salinity pools in salt marshes. Seagrasses are now much reduced, probably due to elevated nitrogen and increased sedimentation.

In sounds, numerically dominant benthic taxa include bivalves, polychaetes and amphipods, many showing gradients in community type from mesohaline areas of the eastern shore to near marine salinities in western parts. The semi-enclosed sounds have extensive shellfisheries, especially of blue crab, northern quahogs, eastern oysters, and shrimp. Problems include contamination of some sediments with toxic substances, especially of metals and PCBs at sufficiently high levels to depress growth of some benthic macroinvertebrates. Numerous fish kills have been caused by toxic dinoflagellate outbreaks, and fish kills and habitat loss have been caused by episodic hypoxia and anoxia in rivers and estuaries. Oyster beds currently are in decline because of overharvesting, high siltation and suspended particulate loads, disease, and coastal development. Fisheries monitoring which began in the late 1970s shows greatest recorded landings in 1978–1982; since then, harvests have declined by about a half.

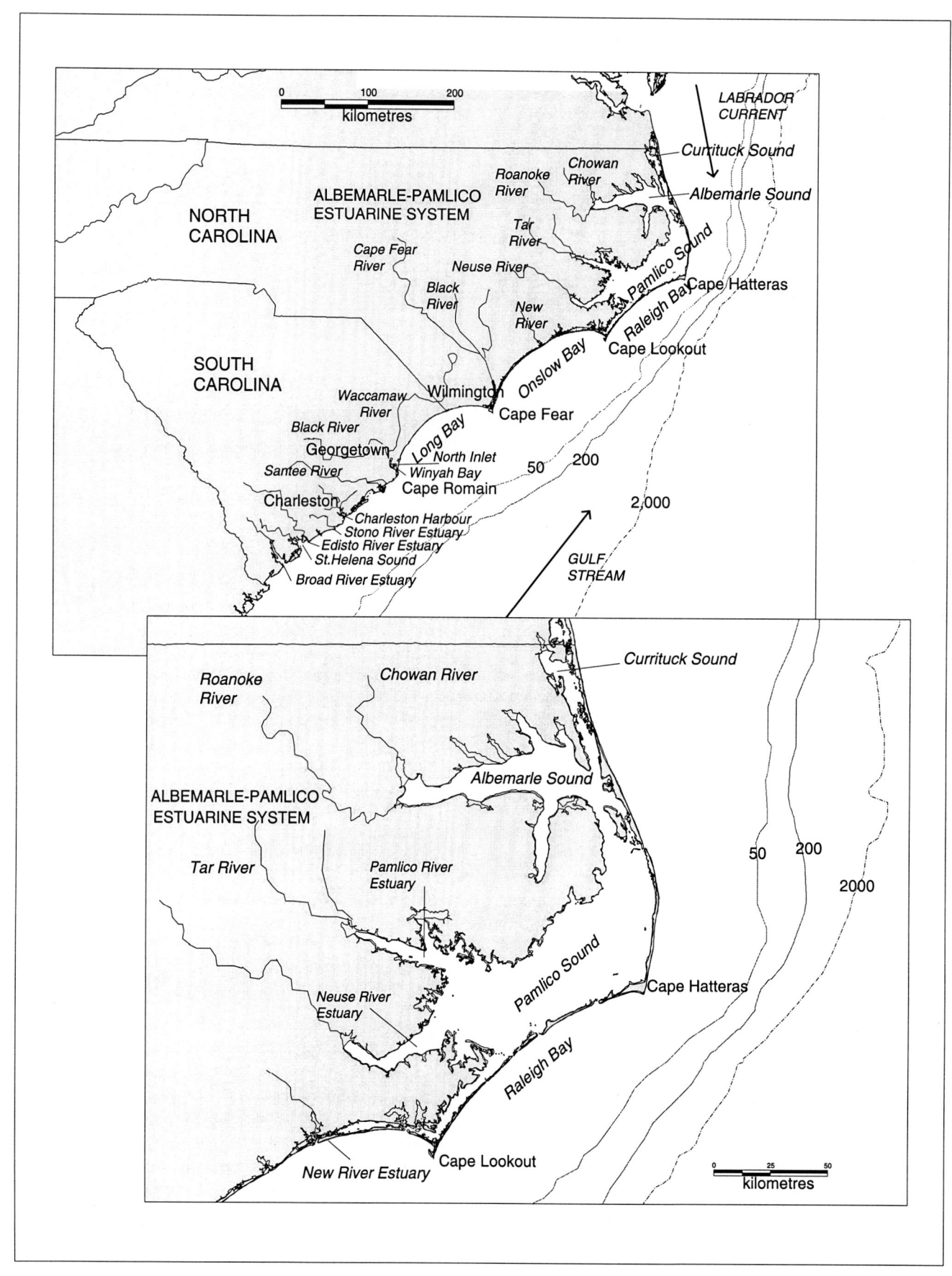

Fig. 1. The North and South Carolina coastal areas, southeastern United States.

Some management plans have been developed toward improving water quality and fisheries sustainability. Major challenges include; high coliform levels leading to closures of shellfish beds, a problem that has increased with urban development and increasing cover of watershed by impervious surfaces; high by-catch and heavy trawling activity; overfishing which has led to serious declines in many wild fish stocks; and eutrophication. Comprehensive plans limiting nutrient inputs are needed for all coastal rivers and estuaries, not only those that already exhibit problems. There is a critical need to improve management of nonpoint nutrient runoff, through increased use of streamside vegetated buffers, preservation of remaining natural wetlands, and construction of artificial wetlands. Improved treatment processes, based on strong incentive programs, should also be mandated for present and future industrial-scale animal operations.

PHYSICAL SETTING

This region includes the diverse collection of estuaries along the North and South Carolina coasts, and the continental shelf waters eastward to the Gulf Stream (Fig. 1), located approximately 76°W, 36°30′N in the north, to 80°42′W, 32°17′N in the south. This coastal region is a gently sloping plain, containing riverine estuaries, sounds, lagoons, and salt marshes. Most of it is bordered by barrier islands formed within the past 15,000 years during sea level rise, and these islands play a major role in hydrological and biological estuarine processes. From the barrier islands the continental shelf gradually deepens to approximately 50–60 m at the shelf break. The shelf is largely a soft-bottom system consisting of shallow (<1 m) relict sediments (primarily fine, medium and coarse sands with varying amounts of calcareous sands) overlying a series of sedimentary/calcareous lithofacies, which outcrop to form rock-reef structures with up to 5 m relief. These hard-bottom outcrops, sometimes termed "live bottoms", support benthic algae and invertebrate communities with a variety of associated fishes (Cahoon et al., 1990a).

From the shelf break seaward there is a rapid deepening, from approximately 50 km seaward of the capes, to about 100 km seaward from bays (Fig. 1). The position of the Gulf Stream roughly marks the shelf break, but this current is highly dynamic. Frictional forcing by the Gulf Stream drives the predominantly counter-clockwise circulation in bays, especially Raleigh, Onslow, and Long Bays. Filaments of offshore water sometimes move inshore along the south side of the shoals seaward of Cape Lookout and Cape Fear. Topographic features such as the "Charleston Bump", a rise in the continental slope off Charleston, and wind forcing, drive Gulf Stream meanders up to 30–40 km seaward of the shelf break or onto the shelf itself (Pietrafesa et al., 1985). Intrusions typically reach the mid-shelf, but may reach 10 km from the shore (Atkinson et al., 1980). Meanders are also associated with shelf-edge upwelling, which can advect nutrient-enriched slope water on to the shelf. Upwelled nutrients are a significant portion of total new nutrients entering the shelf ecosystem (Atkinson, 1985). Wind forcing, particularly winds associated with storms, can drive significant flows shoreward (Pietrafesa and Janowitz, 1988).

The most striking feature along this coast is the large, enclosed sound known as the Albemarle–Pamlico Estuarine System (APES) (Fig. 1). Covering an area of approximately 7530 km² this is the second largest estuary by surface area in the United States. This system is bounded by the Outer Banks of North Carolina, a series of heavily utilized barrier islands. The Chowan, Roanoke and other rivers feed the northernmost part, the oligohaline Albemarle Sound. The APES is constricted at Roanoke Island and widens to the south into the polyhaline Pamlico Sound which is fed by the Pamlico and the Neuse. South of the APES lies a series of narrow euhaline sounds between the mainland and barrier islands. The most significant estuaries in southern North Carolina are the lagoons that form the New River Estuary, and 65 km to the south of that system lies the Cape Fear Estuary, which drains the largest watershed in North Carolina. This is not constrained by barrier islands. Throughout the North Carolina coastline are numerous tidal estuaries ranging from 1 to 10 km in length.

Rivers originating in the piedmont feed most of the large estuaries in North Carolina. Exceptions are the New River, and the Black and Northeast Cape Fear Rivers, which are blackwater coastal plain rivers. Piedmont soils are largely clays that are reactive and bind with potential pollutants; thus, erosion and sedimentation in the piedmont can affect coastal water quality. Coastal Plain soils are generally less reactive and sandier, but the waters draining swamps here are darkly stained by dissolved organic matter.

In northern South Carolina, the largest estuary is Winyah Bay, fed by the blackwater Waccamaw and Black Rivers and the piedmont-derived PeeDee River. Adjoining it is North Inlet, a salt marsh estuary consisting of a maze of high salinity tidal creeks. To the south lie the North and South Santee River Estuaries, which have drainages arising in the piedmont. The principal urbanized estuarine system in South Carolina is the Charleston Harbor area, fed by the coastal plain-derived Ashley and Wando Rivers, and the piedmont-derived Cooper River. Just to the south lie the lowland Stono River Estuary, and the Edisto River Estuary which receives some drainage from the piedmont. Further south still lie St. Helena Sound and the Broad River Estuary, which contain numerous tidal creeks, islands, and salt marshes. In contrast to North Carolina, most of the large estuaries in South Carolina are open to the ocean.

Sources of Pollutants

These coasts have rapidly growing populations. The eight North Carolina counties bordering the Atlantic Ocean (Currituck, Dare, Hyde, Carteret, Onslow, Pender, New Hanover, and Brunswick) experienced a population increase of 32% in the period 1977–1997, from 345,200 to 504,700 (U.S. Census Bureau records). Certain coastal counties in South Carolina, especially Horry, Beaufort, and Berkeley Counties, are also growing rapidly (Bailey, 1996). Human and domestic animal wastes are sources of enteric microbes affecting shellfish beds and swimming beaches, while residential land development, commercial landscaping and golf courses are all sources of fertilizers, pesticides, herbicides, sedimentation and turbidity. Thus, urbanization and population growth lead to greatly increased nonpoint source pollution of coastal waters. Tourism also stresses natural resources and can add to pollutant loads (Bailey, 1996). North Carolina's Outer Banks, barrier island communities along the central and southern coasts of the state, and the Myrtle Beach Grand Strand and the Hilton Head area in South Carolina all experience extensive tourism, especially during warmer seasons (April through October).

In well-populated and/or industrialized areas, point source discharges may be substantial. For example, in the Cape Fear basin alone there are 641 licensed point source discharges (NC DWQ, 1996). Major industrial point sources include the world's largest phosphate mine (Pamlico Estuary), pulp and paper mills (Roanoke, Chowan, Neuse, Cape Fear, and Winyah Bay), a steel mill (Winyah Bay), a metal plating industry (Neuse), and textile manufacturers (Roanoke, Chowan, Pamlico, Neuse, Cape Fear).

Agriculture is important to the region. The 28 counties surrounding the APES include 45% of North Carolina's cropland (Copeland and Grey, 1989). About 24% of the Cape Fear watershed is devoted to cropland or pastureland (NC DWQ, 1996). Agriculture is also a major source of fertilizers, pesticides, sedimentation and faecal bacteria to these coastal waters.

During the 1980s and early 1990s, tobacco farming began to decline, along with an explosive growth in the swine population, from approximately 2.7 million in 1990 to over 10 million head in 1997 (Burkholder et al., 1997; Cahoon et al., 1999a). These hogs are intensively reared. Their wastes are hosed into 'lagoons', to be sprayed onto the surrounding fields, but this system has demonstrated a number of flaws. Accidents have released concentrated wastes into coastal streams and estuaries (Burkholder et al., 1997; Mallin et al., 1997a), while leakage, spray field runoff, and deposition of atmospheric ammonium have all released appreciable quantities of nutrients into coastal waters (Burkholder et al., 1997). In 1998, for example, the North Carolina Division of Air Quality (Raleigh, NC) reported that during the 1990s there had been a 30% increase in ammonia in the airshed of the eastern third of that state, nearly half of which had been contributed by sprayed swine wastes. Vast amounts of nutrients are shipped into eastern North Carolina in feeds (Cahoon et al., 1999a). In 1995 alone, about 90,700 tonnes of nitrogen and 29,930 tonnes of phosphorus were required in the Cape Fear basin (Glasgow and Burkholder 2000). Over 80% of these nutrients remain in the basin, and the watersheds are increasingly acting as nutrient 'sinks', some of which enter coastal waters. Further, an average pig produces the equivalent quantity of waste as three to four people, and pig waste is much richer in BOD materials (Dewi et al., 1994). The availability of excessive animal manure has not driven major changes in use of commercial fertilizers, so that the rise in livestock production has created a concomitant rise in total nutrient loadings within these basins (Cahoon et al., 1999a).

FLORA AND FAUNA OF NORTH AND SOUTH CAROLINA COASTAL WATERS

Estuarine Phytoplankton

Much is known about estuarine phytoplankton communities in North Carolina (Mallin, 1994), although little has been published for South Carolina. All of the systems studied are characterized by phytoplankton productivity increases that coincide with increasing water temperatures in late spring and summer, and subsequent decreases in productivity as the water cools. Certain North Carolina riverine estuaries, especially the Pamlico and Neuse, often experience additional large productivity pulses and dense algal blooms in late winter–early spring. These blooms consist primarily of the dinoflagellates *Heterocapsa triquetra*, *Prorocentrum minimum*, *Amphidinium* and *Gymnodinium* spp. They occur when elevated winter flows bring large concentrations of nutrients to the lower portions of estuaries. In the Neuse Estuary, for example, phytoplankton productivity has been significantly correlated with river flow which after a lag period, is correlated with rain events in the piedmont (Mallin et al., 1993). In general, riverine estuaries whose flushing is constrained by barrier islands, support the highest phytoplankton production. Well-flushed sounds or unconstrained rivers (such as the Cape Fear system) maintain much lower phytoplankton biomass and productivity; and complex tidal creek/salt marsh systems (i.e., North Inlet) exhibit moderate phytoplankton productivity (Tables 1 and 2) (Mallin, 1994; Bricker et al., 1999).

The most important taxonomic groups as reflected by cell counts are diatoms, dinoflagellates, cryptomonads, and

Table 1

Phytoplankton productivity in North and South Carolina estuaries (modified from Mallin, 1994)

Estuarine system	Volumetric ($g\ C\ m^{-3}$)	Areal ($g\ C\ m^{-2}$)
Beaufort Channel, NC (Williams and Murdoch, 1966)	17	68
Beaufort estuaries, NC (Thayer, 1971)	56	67
Calico Creek, NC (Sanders and Kuenzler, 1979)	315	145
Neuse Estuary, NC (Mallin et al., 1991)	75	280
Neuse Estuary, NC (Paerl et al., 1995)	108	370
Newport River, NC (Williams and Murdoch, 1966)	74	74
Pamlico River Estuary, NC (Kuenzler et al., 1979)	150	500
North Inlet, SC (Sellner et al., 1976)	n/a	259

n/a = Data not available.

Table 2

Phytoplankton abundance and chlorophyll *a* concentrations for North and South Carolina estuaries (updated and expanded from Mallin, 1994)

Estuarine system	Cells (no. ml^{-1})		Chlorophyll *a* ($\mu g\ l^{-1}$)	
	Mean	Range	Mean	Range
Beaufort Channel, NC				
Williams and Murdoch (1966)	2,000	130–5,400	4	2–9
Beaufort estuaries, NC				
Thayer (1971)	1,700	360–8,200	4	2–9
Calico Creek, NC				
Sanders and Kuenzler (1979)	n/a	1,000–1,000,000	n/a	6–140
Pamlico Estuary, NC				
Hobbie (1971)	n/a	1,000–340,000	11	1–48
Copeland et al. (1984)	n/a	n/a	n/a	5–100
Stanley (1992)	4,200	630–20,600	17	1–184
Neuse Estuary, NC				
Mallin et al. (1991)	1,600	210–4,200	12	2–23
Mallin and Paerl (1994)	1,700	560–4,400	14	2–65
Fensin (1997)	36,500*	27,100–46,200*	19	6–54
Pinckney and Paerl (1997)	n/a	n/a	14	n/a–90
Glasgow and Burkholder (2000)	n/a	n/a	23	2–220
New River Estuary, NC				
Mallin et al. (1997b) polyhaline	2,900	800–10,000	14	3–35
Mallin (this chapter) systemwide	n/a	n/a	20	1–379
Cape Fear Estuary, NC				
Carpenter (1971)	1,700	250–7,300	n/a	n/a
Mallin et al. (1999a)	n/a	n/a	7	1–33
Tidal Creeks, NC				
Mallin et al. (1999b) summer	7,900	1,440–26,440	14	1–114
North Inlet, SC				
Sellner et al. (1976)	1,600	540–3,400	n/a	n/a

n/a = Data not available.
*These high counts reflect the inclusion of picoplanktonic blue-green algae (cyanobacteria) which were not quantified in previous studies. See text for further explanation.

cyanophytes (Mallin, 1994; Fensin, 1997). In euhaline sounds and higher salinity regions, diatoms seasonally dominate, especially the centric diatoms *Thalassiosira* spp., *Skeletonema costatum*, *Cyclotella* spp., and the pennate diatoms *Nitzschia closterium* and *Navicula* spp. Winter-blooming dinoflagellates such as *Heterocapsa triquetra*, and other species such as *Ceratium* spp. and *Katodinium rotundatum* are often abundant, and dinoflagellates were reported to have been dominant, overall, in the Pamlico Estuary (Copeland et al., 1984). The toxic dinoflagellate, *Pfiesteria piscicida* reaches highest densities in nutrient-enriched areas during some fish kill/epizootic events—up to 13,000 zoospores ml^{-1}, with gametes and zoospores at densities up to 250,000 cells ml^{-1} in surface water and foam during fish kills (Burkholder et al., 1995; Burkholder and Glasgow, 1997; Glasgow et al., 1995). Cryptomonads are well represented in North Carolina estuaries, especially *Cryptomonas testaceae*, *Chroomonas minuta*, *Chroomonas amphioxiae* and *Hemiselmis virescens*. Other taxa that are sometimes abundant in these systems include the euglenoid *Eutreptia* sp., the chrysophyte *Calicomonas ovalis*, the chlorophytes *Chlamydomonas* spp., and the prasinophytes *Pyramimonas* spp.

Cyanophytes historically were considered rare in these estuaries, except for occasional summer appearances of filamentous *Phormidium* spp. and colonial *Microcystis marina*. However, picoplanktonic, cryptic forms (mostly as *Gloeothece* spp., *Lyngbya limnetica*, *Merismopedia punctata* and *Aphanothece microscopica*) in the mesohaline, eutrophic Neuse Estuary are now known to seasonally dominate the estuarine phytoplankton (Fensin, 1997). Cyanophytes were dominant among phytoplankton in cell number throughout the period from June 1994–October 1995, representing 50–65% of the total phytoplankton cells. On the basis of biovolume, during that period picoplanktonic cyanophytes were dominant except during spring; they comprised ca. 85% of the total phytoplankton biovolume in summer season, and their average contribution to total biovolume seasonally was ca. 61%. During winter their biovolume positively correlated with water-column ammonium concentrations. Their maximal cell numbers ($2.3–3.0 \times 10^5$ cells ml^{-1}, depending on the station sampled) were attained during spring, and were positively correlated with soluble reactive phosphate concentrations. Other studies also have now documented cyanophytes as being abundant (Pinckney et al., 1997), and it is likely that this group was previously overlooked because of their size and cryptic appearance (Burkholder and Cuker, 1991).

Phytoplankton typically reach maximum values during summer although late winter or spring blooms can occur. Several North Carolina estuaries can be considered moderately to highly eutrophic (Tables 1 and 2); these include the Neuse, Pamlico, and New River Estuaries (Mallin, 1994; NOAA, 1996; Bricker et al., 1999; Glasgow and Burkholder, 2000). Upper reaches of the smaller urban tidal creek estuaries can host dense algal blooms as well, especially during summer (Table 2). The eutrophic portions of these systems can be characterized by elevated nutrient loading and poor flushing. Well-flushed systems like euhaline sounds, and open rivers like the Cape Fear rarely host algal blooms. In the near-pristine tidal creeks of North Inlet, SC, anthropogenic nitrate loading is very limited, and summer phytoplankton abundance was limited by microzooplankton grazing, rather than nitrogen loading, while diatom growth in winter was stimulated by ammonium inputs (Lewitus et al., 1998).

Estuaries contain various microalgal taxa that have caused fin- or shellfish kills or impaired fish health in other regions, such as the dinoflagellates *Gymnodinium sanguineum*, *Gyrodinium galetheanum*, and *Prorocentrum minimum*; the chrysophyte *Phaeocystis*; the diatom *Pseudo-nitzschia australis*; and the raphidophytes *Heterosigma akashiwo* and *Chattonella antiqua* (Burkholder unpubl. data; Hallegraeff, 1993; Burkholder, 1998). The two known species of ichthyotoxic *Pfiesteria* also occur in Carolina estuaries (Burkholder, unpubl. data). Aside from the latter two *Pfiesteria* species, noxious algal blooms have not been reported in Carolina waters with exception of a bloom of the chrysophyte, *Phaeocystis globosa*, following a major swine effluent lagoon rupture (Burkholder et al., 1997). Elsewhere, *Phaeocystis* blooms also have been related to major inputs of raw sewage wastes into lower rivers and estuaries (Hallegraeff, 1993).

Only a little information is available on phytoplankton of North Carolina's expansive enclosed sounds (e.g. Copeland et al., 1983). Dinoflagellates were reported as dominant in the low-salinity Albemarle Sound (chlorophyll *a* range of 3–40 μg l^{-1}; Copeland et al., 1983). Salinity in Pamlico Sound is typically at 25–30 ppt salinity or higher (Epperly and Ross, 1986). Seasonal sampling over the past two years (Burkholder et al., unpublished) indicates that the phytoplankton of Pamlico Sound are dominated by estuarine and marine taxa (Marshall, 1976), predominantly diatoms.

Marine Phytoplankton

The marine waters are relatively oligotrophic, with low phytoplankton abundance except in a narrow inshore zone. Chlorophyll *a* concentrations are typically ca. 1 μg l^{-1} inshore, declining to 0.1–0.01 μg l^{-1} in the Gulf Stream (Cahoon and Cooke, 1992). The nearshore assemblage is dominated by small centric diatoms and flagellates, along with occasional blooms of larger centric diatoms (Marshall, 1969, 1971, 1976, 1978). These surveys have not yet been repeated to identify trends over the past 20–30 years. Available information indicates that warmer-water neritic associations often contain abundant diatom genera such as *Rhizosolenia*, *Hemiaulus*, and *Coscinodiscus*, together with *Skeletonema costatum* and *Thalassiosira* spp. (e.g., Marshall, 1978). Moving seaward, dinoflagellates and coccolithophorids tend to dominate. Winter assemblages also typically include various colder-water forms including diatoms such as *Amphiprora hyperborea*, *Biddulphia aurita*, *Chaetoceros* spp.; dinoflagellates such as *Ceratium* spp., *Dinophysis* spp., and certain gonyaulacoids, mixed with coccolithophorids such as *Cyclococcolithus leptoporus* and *Emeliana huxleyi*.

These trends are believed to arise, in part, because nearshore currents carry 'seed' populations of various taxa into Carolina coastal areas. Gulf Stream phytoplankton enter in Gulf Stream rings or eddies, and through this mechanism, tropical species may occur throughout an annual cycle in this area. Moreover, cool water species tend to move southward in a temporal pattern that follows the flow of the cold, offshore Labrador Current to the Cape Hatteras region. A major portion of the shelf water that flows southwest of New England originates in the Gulf of Maine and Georges Bank area, thus favouring transport of phytoplankton into neritic areas southward. In late spring as winter winds lessen and warmer water temperatures develop, successional patterns and seasonal assemblages re-establish along a gradient from Cape Hatteras northward. Cape Hatteras is thus a natural coastal feature associated with a geographic division of many phytoplankton (Marshall, 1978). Intrusions of nutrient-enriched slope water also occur, which can stimulate blooms and sometimes create a near-bottom chlorophyll *a* maximum.

High-clarity and low-nutrient concentrations of coastal waters along the Carolinas prevail under normal conditions, and incidences of harmful marine algal blooms have been rare. An exception was an incursion of the toxic dinoflagellate, *Gymnodinium breve* (formerly called *Ptychodiscus brevis*), in the fall of 1987 which persisted through winter 1988, following transport of bloom-containing eddies from the Gulf Stream in to shore (Tester et al., 1991). This event is believed to have occurred because extremely unusual and persistent weather conditions (warm, dry, calm with very little wind) allowed the eddies to maintain integrity, and the dinoflagellate populations to grow. The *G. breve* outbreak caused major losses to the commercial shellfish industry (NC DMF records, Morehead City, NC), and was strongly correlated with a catastrophic decline in bay scallop (*Argopecten irradians concentricus*) recruitment in succeeding years (Summerson and Peterson, 1990).

Macroalgae

In biogeographical terms, North Carolina's macroalgal communities lie along the border of the cold temperate

North Atlantic and warm temperate Carolina regions of species distributions, delineated at Cape Hatteras; South Carolina's macroalgae occur entirely within the latter (Lüning, 1990). At the general confluence of these two regions, the North Carolina coast is a transition zone for macroalgal communities (Schneider and Searles, 1991). The Carolinas have a low number of endemic macroalgae relative to the well-developed cool temperate flora to the north, and the rich tropical flora to the south. Species diversity can nonetheless be high in this transitional area, which represents the southernmost extension for certain cold-adapted species and the northernmost extension of some warm-adapted species. The latter influence appears to be more important in contributing to species diversity. At Cape Hatteras the Gulf Stream turns eastward, allowing many deep-water, warm-adapted macroalgal species to thrive (e.g., *Caulerpa* spp., *Dictyota dichotoma*, *Botryocladia occidentalis*, and many others of the ca. 800 macroalgal species of the western Atlantic tropical region (Lüning, 1990). Thus, macroalgal diversity increases from ca. 100 species along coastal Maryland and Virginia to ca. 300 on the North Carolina coast, because of this northern extension of the sublittoral Caribbean algal flora. The benthic macroalgal flora of the continental shelf of the Carolinas may represent the only truly subtropical biogeographic region in Atlantic North America (Kapraun, 1980).

The littoral macroalgal flora has pronounced seasonality, especially in North Carolina. In warmer seasons, tropical species of Caribbean affinity predominate, while in colder seasons species of cool temperate (New England) affinity are dominant. A third, less conspicuous assemblage is comprised of warm temperate species along with certain cosmopolitan species that have become adapted to warm temperate conditions. These generally attain highest abundance in spring and fall (Kapraun, 1980).

Macroalgal colonization in the Carolinas is limited by a paucity of solid substrata; indeed, most intertidal collections are made along artificial rock jetties. Water temperatures in the salt marshes can exceed 35°C, and a limited number of cyanophyte mat formers and rapidly growing forms such as *Enteromorpha* spp. and *Ectocarpus* spp. tend to predominate; these, along with *Cladophora* spp., can show undesirable growth in rapid response to nutrient enrichment (Burkholder et al., 1992a). Brackish *Chara* species can sometimes become abundant in salt marshes as well; and the vascular salt marsh macrophytes can be heavily colonized by small macroalgal taxa such as *Porphyra, Polysiphonia, Bryopsis,* and *Ulva* species.

The rich subtidal macroalgal communities occur mainly on rocky outcrops, stones, and organic concretions such as those 10–20 km off Cape Hatteras. These have relatively clear water at 15–60 m depths—temperature range ca. 10–24°C (Lüning, 1990), with most species restricted to depths of less than 40 m (Schneider, 1976). Macroalgae are important members of these 'live bottom' communities (Cahoon et al., 1990a; Thomas and Cahoon, 1993; Mallin et al., 1992).

Benthic Microalgae

Benthic microalgae are important primary producers in estuarine ecosystems, where these flora are exposed to high nutrient availability and light fluxes in shallow waters (Mallin et al., 1992; Cahoon et al., 1999b). This is certainly true in Carolina estuaries (e.g., Zingmark, 1986; Freeman, 1989; Pinckney and Zingmark, 1993; Coleman and Burkholder, 1995; Nearhoof, 1994; Cahoon et al., 1999b). Benthic microalgal biomass frequently exceeds phytoplankton biomass in estuarine water columns by factors of 10–100 (Cahoon and Cooke, 1992), and is often inversely related to integrated phytoplankton biomass (Fig. 2).

Seagrass epiphytes can contribute a significant proportion of total community production. Epiphytes on the historically dominant seagrass, *Zostera marina*, exhibit a bimodal curve in productivity with maxima during spring and fall (Penhale, 1977). Generally, however, little is known of the species composition and species-specific productivity of seagrass epiphyte communities. Their composition has been characterized in mesocosm (Burkholder et al., 1992a; Coleman and Burkholder, 1994) and field studies (Coleman and Burkholder, 1995); depending on the nutrient regime, epiphyte communities in spring were found to consist primarily of diatoms and cyanophytes, shifting to abundant cyanophytes, dinoflagellates, and cryptomonads in warmer months. During autumn the epiphytes of field communities were dominated by the crustose adnate red alga *Sahlingia subintegra*, with dinoflagellates (especially mixotrophic or heterotrophic *Amphidinium* sp., *Polykrikos* sp., and *Cochlodinium* sp.) co-dominant in biovolume contribution. Nitrate enrichment stimulated production of the adnate diatom *Cocconeis placentula*, several other diatoms, and cyanophytes.

The clear waters of the continental shelf also support significant benthic microalgal production (Cahoon and

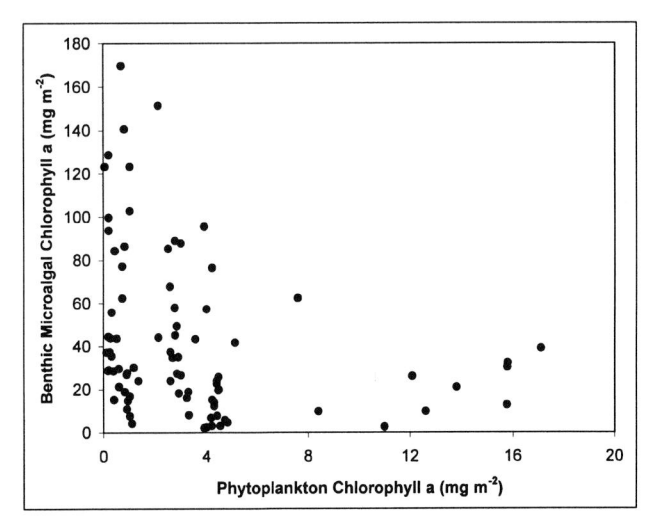

Fig. 2. Relationship between benthic microalgal chlorophyll *a* and chlorophyll *a* content of the overlying water column in North Carolina coastal waters.

Cooke, 1992) dominated by diatoms (Cahoon and Laws, 1993). As in estuarine waters, benthic microalgal biomass can exceed integrated phytoplankton biomass (Cahoon et al., 1990b); but benthic microalgal production in deeper shelf waters is more light-limited, and is approximately equal to phytoplankton production (Cahoon and Cooke, 1992). Benthic microalgae support production of fishes that forage over soft-bottom habitats (Thomas and Cahoon, 1993).

Seagrasses and Other Rooted Submersed Aquatic Vegetation (SAV)

Along the North Carolina coast the dominant seagrass, eelgrass (*Zostera marina*), lies at the southernmost extension of its geographic range on the U.S. Atlantic seaboard, where its growth is stressed in summer from the high water temperatures (Thayer et al., 1984a), Maximum shoot length of eelgrass in this region averages only about 40 cm. Temperatures above 30°C are detrimental to eelgrass from northern regions (Zimmerman et al., 1989). However, shallow water temperatures along the North Carolina coast typically reach 31–33°C (Burkholder et al., 1992), and the populations inhabiting these waters are believed to be a separate ecotype from more northern populations (Touchette, 1999). Primary productivity of eelgrass in this region is regarded as comparatively low: above-ground, 0.59–1.23 g C m^{-2} d^{-1}; below-ground, 0.15–0.28 g C m^{-2} d^{-1} (Penhale, 1977; Thayer et al., 1984b). Eelgrass in North Carolina has one growing season in winter-spring, and another in autumn when production of new shoots is higher (Burkholder et al., 1992a; 1994). The warm summer season is a time of reduced growth or dehiscence, in which *Z. marina* uses carbon stored in its rhizomes for sustenance. Metabolism is reduced and most leaves are sloughed (Burke et al., 1996; Touchette, 1999). Shoot production is especially pronounced during the fall, whereas leaf production and carbon reserves are more important in spring, apparently in preparation for the warm summer season (Burkholder et al., 1992a, 1994; Touchette, 1999).

North Carolina is the northernmost extension for a second seagrass species, shoalgrass (*Halodule wrightii*) (Thayer et al., 1984a). The two species co-occur to some extent, but tend to be temporally separated with *Z. marina* attaining maximal production in winter to early summer, and *H. wrightii* in late summer–early fall (Kenworthy, 1981; Burkholder et al., 1994). The latter species extends into the upper intertidal zone. In North Carolina it generally is found in euhaline regimes, although it is strongly influenced by the high turbidity and freshwater flow (Steel, 1991; Glasgow and Burkholder, 2000). *Zostera marina* tends to be dominant or co-dominant in euhaline environments (Thayer et al., 1984a).

A third estuarine/marine SAV species that is sometimes called a seagrass is the euryhaline angiosperm, widgeongrass (*Ruppia maritima*). This species responds most favourably to nutrient pollution (especially inorganic N; Burkholder et al., 1994). It is widespread, and grows in brackish water or low-salinity pools in salt marshes (Thayer et al., 1984a). Shoalgrass and widgeon-grass are sometimes exposed to air during low tide, whereas eelgrass is only rarely exposed (Thayer et al., 1984a, Touchette, 1999).

At present, the remaining seagrass meadows in North Carolina occur mostly in sandy or muddy sediments on the landward side of the Outer Banks, with sparse cover along most of the mainland (Ferguson et al., 1988). Yet, elderly fishermen and fishermen's journal accounts from the late 1800s describe extensive beds of such vegetation in many embayments along the mainland where it is now absent (pers. comm. to Burkholder). Several large eelgrass beds have disappeared within the past decade in the Intracoastal Waterway (Morehead City area, following extensive use of herbicides to control macroalgal growth) and near Harkers Island along the mainland (following construction activity; Burkholder, unpubl. data). Early data to resolve historical trends in seagrass distributions are sparse; the only complete maps in existence for this coast were compiled during the late 1980s (Ferguson et al., 1988).

Light availability has been considered a primary factor limiting seagrass distribution (Stevenson, 1988), effected partly through turbidity. Nutrient enrichment can stimulate epiphytic algal growth, but more recently nitrate enrichment has been related to declines in eelgrass as a direct physiological response (Burkholder et al., 1992a). The ability of *Z. marina* to take up nitrate through its leaves during periods of light or darkness (Touchette, 1999), and the apparent inability of plants grown in sandy nearshore sediments to 'shut off' water-column nitrate uptake leads to severe internal imbalances in carbon metabolism (Touchette, 1999). Although *Z. marina* shows this response in both its spring and fall growing seasons, water-column nitrate inhibition appears more pronounced during warm spring seasons, when the plants would otherwise allocate more of their carbon to storage in below-ground tissues (Burkholder et al., 1992). Elevated N can also render the plants more susceptible to pathogens such as the marine slime mould, *Labrynthula zosterae* (Muehlstein, 1992; Short and Wyllie-Echeverria, 1996).

Accelerating coastal development in North Carolina has been associated with increased nutrient (nitrate) loading from, among other sources, septic effluent leachate (Stanley, 1992). It is likely that eelgrass meadows along the mainland have disappeared because of reduced light availability in combination with increased water-column nitrate enrichment. The presence of the barrier islands further from mainland influences provides sheltered habitat for seagrass growth (Steel, 1991). Eelgrass appears to be a more 'oligotrophic' indicator, since it is the most sensitive of the three seagrass species to light reduction as well as water-column nitrate enrichment, while shoalgrass is sensitive to higher nitrate loading, and widgeon-grass is stimulated by high nitrate (Burkholder et al., 1994). The state's remaining seagrass beds are classified by the North Carolina Marine Fisheries Commission as critical habitat for many finfish and shellfish species, with *Z. marina* considered the most valuable (Thayer et al., 1984a).

Abundant meadows of freshwater/brackish SAV such as tapegrass or freshwater eelgrass (*Vallisneria americana*) hisorically were a prominent feature of the Pamlico Estuary (Copeland et al., 1984) and, to a lesser extent, the Neuse, with pondweeds (*Potamogeton* spp.) and widgeongrass as subdominants. This submersed vegetation significantly declined from the mid-1980s under excessive sediment loading, then experienced a resurgence as modest improvements in erosion control promoted an increase in water clarity (Steel, 1991; NC DEHNR, 1994). This information is derived mostly from many recent complaints about abundant SAV around dock areas (Neuse) and from visual observations and fishermen's anecdotal accounts (Pamlico).

Estuarine Zooplankton

The most common and abundant estuarine zooplankter in the region is the calanoid copepod *Acartia tonsa*, followed by the calanoid *Paracalanus crassirostris* and the cyclopoid copepod *Oithona colcarva* (Thayer et al., 1974; Lonsdale and Coull, 1977; Fulton, 1984; Mallin, 1991; Mallin and Paerl, 1994; Houser and Allen, 1996). Harpacticoid copepods can be abundant at times (i.e. *Microsetella norvegica, Euterpina acutifrons*, and others), and cladocerans (i.e. *Evadne nordmandii, Podon polyphemoides*, and *Penilla* spp.) are either rare or appear in periodic blooms. Species richness of copepods is low in mesohaline systems such as the Pamlico and Neuse estuaries. However, euhaline systems such as the Beaufort, NC area and North Inlet, SC also yield calanoids such as *Centropages* spp., *Labidocera aestiva* and *Psuedodiaptomus coronatus*, and cyclopoids such as *Corycaeus* spp., *Oncaea venusta* and *Saphirella* sp. (Thayer et al., 1974; Lonsdale and Coull, 1977; Fulton, 1984).

In North Carolina, copepods display highest peak densities in late spring–early summer in euhaline areas and in mid-to-late summer in mesohaline estuaries, while abundances are generally lowest in late winter. There is a positive correlation between zooplankton abundance and water temperature in North Carolina estuaries (Fulton, 1984; Mallin, 1991; Mallin and Paerl, 1994). Seasonal differences were less pronounced in North Inlet compared with the more northerly estuaries. Mesh size of nets is important (Table 3). Diel periodicity is evident; postnaupliar stages of several copepods, particularly *Acartia tonsa*, are more abundant in the water column at night than during the day (Fulton, 1984, Mallin and Paerl, 1994). However, copepod nauplii display no such periodicity. Houser and Allen (1996) also have found a tidal signal in North Inlet in which copepods were most abundant at high tide and least abundant during daytime low tides.

Table 3

Comparison of mean zooplankton abundance and biomass for estuaries in North and South Carolina (updated from Mallin, 1991)

Estuary	Net mesh (μm)	Density (number m^{-3})	Biomass (mg dry wt m^{-3})
Beaufort area, NC			
Thayer et al. (1974)	156	4,000 (1970)	14.0
		8,400 (1971)	21.0
Fulton (1984)*	76	21,900 (30 mo.)	47.8
Neuse, NC			
Mallin (1991)	76	34,530 (20 mo.)	15.3
		31,224 (1989)	17.2
Mallin and Paerl (1994)	60	137,150 (22 mo.)	38.7
North Inlet, SC			
Lonsdale and Coull (1977)	156	9,257 (20 mo.)	16.1
Houser and Allen (1996)	153	21,555 (6 mo. summer)	n/a

*Post-naupliar copepod data only.
n/a = Data not available.

Other organisms can occasionally be abundant. Using small mesh nets, Mallin and Paerl (1994) demonstrated periodic high densities of tintinnid protozoans and the large mixotrophic dinoflagellate *Polykrikos hartmanni* (Table 3). Meroplanktonic organisms, especially barnacle nauplii and polychaete larvae, may be abundant at times (Lonsdale and Coull, 1977; Mallin, 1991; Mallin and Paerl, 1994; Houser and Allen, 1996).

Zooplankton play a key role in transfer of energy through the food chain in these estuaries. Mallin and Paerl (1994) demonstrated that on an annual basis estuarine zooplankton grazed 38–45% of the daily phytoplankton production in the Neuse Estuary, ranging from 2% in winter to >100% in summer. Zooplankton grazing rates were positively correlated with phytoplankton productivity rates and abundance of centric diatoms and dinoflagellates. Zooplankton, especially copepods, are preyed upon extensively by larval and juvenile fish in these waters and their abundance patterns likely are controlled, in part, by such predation (Thayer et al., 1974; Kjelson et al., 1975; Fulton, 1985). In the mesohaline Neuse Estuary, zooplankton abundance and planktonic trophic transfer are greatest in late spring through summer, when anadromous fish larvae migrating from the open ocean reach these estuarine nursery areas (Mallin and Paerl, 1994). Most anadromous fish arrive when planktonic trophic coupling is strongest and depart in fall, when planktonic trophic transfer, zooplankton abundance, and phytoplankton productivity all decrease. Thus, in spring and summer there appears to be high trophic efficiency and tight planktonic food chain coupling in these waters.

Marine Zooplankton

The composition of the zooplankton community in coastal ocean waters is heavily influenced by the mixing of inshore waters with Gulf Stream waters. The inshore zooplankton community is dominated by small copepods, chaetognaths, ctenophores, and larval fishes with estuarine affinities. Offshore zooplankton assemblages include many species of both large and small copepods, gelatinous forms of several taxa, and larval fishes advected by the Gulf Stream and with more tropical affinities (Bowman, 1971; Fahay, 1975; Paffenhofer, 1980). Some of the more abundant taxa found include the cyclopoid copepods *Oncaea* spp., *Oithona* spp., and *Corycaeus* spp.; the calanoid copepods *Paracalanus* spp. and *Eucalanus pileatus*; and the ostracod *Conchoecia* spp.

Demersal forms in North Carolina shelf waters (Cahoon and Tronzo, 1990, 1992) are relatively distinct from the holozooplankton, and are dominated by harpacticoid and cyclopoid copepods, nematodes, amphipods, cumaceans, and others. These animals are closely associated with benthic substrata, and conventional sampling methods generally have not sampled them adequately, forcing use of other sampling techniques such as re-entry and emergence trapping (Cahoon and Tronzo, 1992). The abundance of demersal zooplankton (as numbers of animals m^{-2}) sometimes approximates the numbers of holozooplankton in the overlying water column (Cahoon and Tronzo, 1992), suggesting that a significant fraction of total zooplankton biomass in these coastal waters is under-sampled and poorly considered.

Coastal Benthic Invertebrate Communities

Many estuarine communities along the North Carolina and South Carolina coasts are separated from adjacent coastal oceans by extensive barrier islands with narrow inlets. These estuaries can be divided into three broad categories: (1) the extensive Pamlico–Albemarle–Currituck Sound system, (2) the smaller, barrier island systems to the south, and (3) the estuarine reaches of the several rivers that empty directly into the coastal ocean.

The key physical characteristics of the Albemarle–Pamlico Estuarine System (APES) that structure the benthic communities are inputs from several river systems, a broad, shallow sound, and relatively few, narrow inlet connections to the adjacent coastal ocean that effect a long residency time for waters in the sound. The eastern portion of Pamlico Sound and much of Albemarle and Currituck Sounds are oligohaline to mesohaline. The northernmost Currituck sound was linked by an inlet to the ocean until 1830, when it closed and the sound began to transform to a freshwater system. The APES region, in general, is characterized by low benthic diversity and high seasonality in abundances, especially in the lower reaches of the rivers emptying into the sound (Tenore, 1972; Chester et al., 1983). Numerically dominant taxa include the bivalves *Macoma balthica* and *Macoma mitchelli*, the polychaetes *Capitella* spp., *Mediomastus californiensis*, *Mediomastus ambiseta*, *Heteromastus filiformis*, *Streblospio benedicti*, *Maranzellaria*, and *Nereis* spp., and the amphipods *Ampelisca* and *Corophium*. Such opportunistic taxa are typical of many estuarine systems along the mid-Atlantic coast of North America. Abundances can be high during late winter and early spring, reaching densities of over 3500 per 100 cm^2 in some locations (Posey and Alphin, unpubl. data). However, many individuals are relatively small in size, emphasizing the importance of recent recruitment events and annual adult mortality. Abundances decline during mid to late summer, especially in deeper areas. This decline has been related to a variety of factors, including influx of predatory fish and decapods and summer bottom-water hypoxia (Tenore, 1972; Epperly and Ross, 1986; Pietrafesa et al., 1986; Lenihan and Peterson, 1998). The Pamlico and Neuse Estuaries support a small trawling fishery for blue crabs as well as an extensive and growing blue crab pot fishery. Crab pot fishing and trawling for shrimp occurs throughout much of the eastern Pamlico Sound.

The central Pamlico Sound system provides a gradient in community types from mesohaline areas of the eastern shore to near marine salinities in some parts of the western shore. Abundances are seasonal throughout, but seasonal variations in freshwater input have a greater effect along the east. Water in the Pamlico has a long residency time, which makes it behave as a semi-enclosed system, with larval retention and limited adult loss to the adjacent ocean. There is high internal productivity, but a greater residency of nutrients and pollutants. Albemarle and Currituck Sounds both drain into Pamlico Sound, and are dominated by oligohaline to mesohaline benthic communities with abundance patterns characteristic of low or pulsed planktonic recruitment.

The APES supports extensive shellfisheries. The blue crab fishery is the largest in terms of overall landings, and these sounds contribute almost 80% of North Carolina's catch for this species (Table 4) (McKenna et al., 1998). Shrimp, scallop, clam and oyster fisheries are also important, but most of these fisheries are either in decline or showing signs of stress (McKenna et al., 1998). Reasons include overfishing, eutrophication, sediment loading and other pollution, and disturbance to bottom habitats. Major environmental problems facing the bottom communities of Pamlico and Albemarle Sounds are related to the long flushing time (Epperley and Ross, 1986). Increased inputs of sediment and nutrients from agriculture and development have led to decreased water quality, greater turbidity, and increased sedimentation. This pollution has been linked to declines of SAV, and to changes in the benthic composition. About 22% of the Pamlico Sound can currently be classified as significantly impacted (Hackney et al., 1998) but, depending on the estuary, from 38–78% of the sediment areas in North Carolina were contaminated with toxic substances (especially nickel, arsenic, mercury, chromium, and

Table 4

Stock status of marine fish taxa or habitat groups in North Carolina estuarine and coastal waters with commercial catches worth more than $500,000 annually (NC DMF, 1999)

Taxa group	Status	Value of 1998 landings (US $)
Blue crab *Callinectes sapidus*	concern	44,952,300 (72¢/lb)
Shrimp *Peneaus aztecus, P. duorarum, P. seiferus*	viable	10,826,100 ($2.35/lb)
Southern flounder *Paralichthys lethostigma*	concern/overfished	7,117,300 ($1.80/lb)
Summer flounder *Paralichthys dentatus*	stressed/recovering	5,427,100 ($1.82/lb)
Atlantic menhaden *Brevoortia tyrannus*	viable	4,071,000 (7¢/lb)
Atlantic croaker *Micropogonias undulatus*	concern	3,424,800 (32¢/lb)
Reef fish (71 species)	concern/overfished	3,332,000 ($1.58/lb)
King mackerel *Scomberomorus cavalla*	viable	1,749,400 ($1.53/lb)
Weakfish/grey trout *Cynoscion regalis*	stressed/recovering	1,694,200 (51¢/lb)
Black sea bass *Centropristis striata*	concern/overfished	1,063,200 ($1.46/lb)
Striped mullet *Mugil cephalus*	concern	1,061,400 (48¢/lb)
Spot *Leiostomus xanthuris*	viable	1,001,700 (41¢/lb)
Eastern oyster *Crassostrea virginica*	overfished	974,400 ($4.13/lb)
Bluefish *Pomatomus saltatrix*	stressed/recovering	763,400 (26¢/lb)
Black (gag,grey) grouper *Mycteroperca bonaci*	overfished	742,100 ($2.50/lb)
Kingfishes (southern, northern, gulf or sea mullet)	unknown	742,100 ($1.04/lb)
Spiny dogfish (dogfish shark) *Squalus acanthius*	overfished	649,100 (13¢/lb)
Striped bass *Morone saxatilis*	viable	519,400 ($1.23/lb)

PCBs), at sufficiently high levels to depress growth of various benthic macroinvertebrates used as food by commercially valuable finfish. Benthic macroinvertebrates were present in areas with hypoxia; only areas where appreciable toxic substances occurred, with or without hypoxia, had depleted macroinvertebrates (Hackney et al., 1998).

Most of the other estuarine areas along the Carolina coasts are characterized by narrow barrier island sound systems, interspersed with rivers that empty almost directly into the coastal ocean. These relatively small estuarine areas are often less than 2–3 km in width but extending 10–20 km between inlets. They are exemplified by Masonboro Sound in North Carolina and North Inlet in South Carolina. They have extensive intertidal regions (often greater than 50–70% of the total area), large expanses of *Spartina alterniflora* salt marshes, and relatively high salinities (often reaching 35–37 ppt in the Masonboro Sound area), and are well flushed.

Communities in these areas are relatively diverse, with opportunistic taxa such as capitellid polychaetes, the polychaete *Streblospio*, and *Corophium* amphipods as well as more stable assemblages such as thalassinid shrimp, dense beds of hemichordates, maldanid polychaetes and arenicolids (Fox and Ruppert, 1985; Posey et al., 1995). Low dissolved oxygen is seldom a significant problem here. Seasonal maxima in abundance of many amphipods and polychaetes occur in winter and early spring, with summer declines related to predation and possibly temperature, and there is relative stability in species composition among years. However, increasing development of terrestrial areas adjacent to the sounds is likely to expose these benthic communities to increased sediment runoff as well as nutrient and organic inputs (Mallin et al., 1999b; Posey et al., 1999a).

Coastal rivers, as exemplified by the Cape Fear in North Carolina, are affected strongly by storms and freshwater runoff. The Cape Fear experiences periodic freshwater flow events related to major fronts and hurricanes, which can lead to significant salinity reductions, increases of agricultural runoff and hypoxic conditions. As a result, opportunistic taxa dominate, such as *Mediomastus*, *Maranzellaria* and *Corophium* (Mallin et al., 1999c). The strong flow and high silt load in these rivers generally reduces effects of chronic nutrient loading on planktonic productivity (Mallin et al., 1999a), but does inhibit many sedentary suspension feeders as well as submerged macrophytes.

Coastal rivers are particularly susceptible to upstream land-use practices. Poor sediment or nutrient retention practices, increasing development, and livestock farms have increased the potential for major effects on the benthic communities when major storms and associated flooding occur. The Cape Fear River may experience 90% declines in species abundances after the passage of a major storm (Mallin et al., 1999c).

Commercially important benthic shellfisheries in the barrier island and coastal river regions include blue crabs, northern quahogs (hard clams), eastern oysters, and shrimp (Table 4). These are affected by loss of seagrass beds south of the central North Carolina coast. Seagrass beds are recognized as a critical juvenile habitat for the blue crab, and there has been increasing interest in identifying alternative juvenile nursery areas for this species; ongoing work

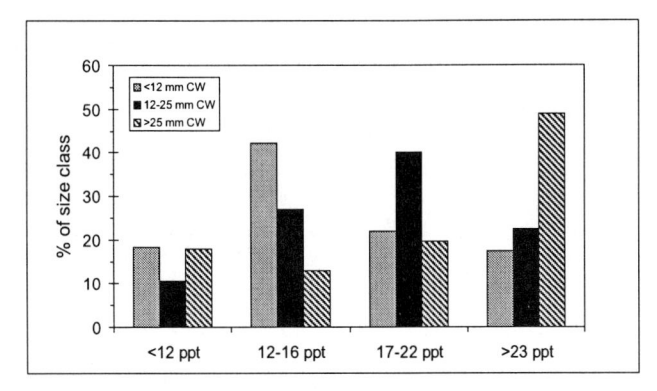

Fig. 3. Distribution of juvenile blue crabs by size class across the salinity gradient in the Cape Fear River estuary (Posey and Alphin, Univ. of North Carolina at Wilmington, unpublished data). Size classes for blue crabs reflect sizes that are thought to rely on seagrass habitat in other estuaries (<12 mm carapace width (CW); Pile et al. 1996) and the sizes at which they move into lower salinity marsh habitats (>25 mm CW).

(Posey and Alphin, unpublished; Posey et al., 1999b) has identified oyster beds and salt marshes, as well as lower salinity reaches of coastal rivers. For example, in the Cape Fear Estuary, the highest concentration of small juvenile blue crabs occurs in salinities <15 ppt, with larger size classes occurring in more saline portions of the estuary (Fig. 3). This distribution indicates habitat segregation along the salinity gradient in the coastal rivers of this region.

Oyster beds are present both in the sounds and near the mouths of rivers. They currently are in decline because of overharvesting, high siltation and suspended particulate loads, disease, and coastal development (Breitburg et al., in press). Considerable effort has been focused on maintaining the fishery through mariculture and transplant operations. However, oyster beds provide critical habitat for many fish species (Posey et al., 1999b) and may have significant impacts on water quality. Therefore, they may need to be managed as an essential habitat, much as seagrass beds are managed. Fishery managers recently have supported efforts to re-establish oyster beds in areas where they have been lost or degraded, as a method of habitat mitigation.

Offshore Benthic Communities

The benthic shelf habitats include a diverse mix of hard bottoms, hard-bottom veneer (shallow, mobile sands overlying rock), and deeper sand areas. Hard bottoms are productive fisheries habitats. Their benthic communities vary with depth and according to vertical relief (Posey and Ambrose, 1994). Low-relief hard bottoms are subject to periodic sand scour and possibly burial, so often harbor lower diversity, dominated by a few emergent octocorals and sponges. On high-relief hard bottoms, horizontal surfaces are usually dominated by macroalgae such as *Sargassum* and *Dictyota*, while vertical surfaces and overhangs are dominated by a diverse array of sponges, bryozoans, hydrozoans, and tunicates. Diversity is lowest inshore, where the water is colder in winter and where there are turbidity and salinity effects from adjacent estuaries. Offshore, at depths greater than 35–40 m, the Gulf Stream influence becomes stronger and the community may be dominated by a variety of tropical and subtropical Caribbean taxa. Large areas of offshore bottom habitat are veneer habitats that are often covered by a thin, mobile layer of sand. These do not have such a diverse array of encrusting taxa.

Soft-sediment communities are diverse. A recent monitoring study along the southern North Carolina coast found over 500 taxa, with over 20 taxa comprising at least 1% of the individuals. These communities exhibit seasonality, but there is considerable consistency between years. The habitats are dominated by polychaetes as well as a variety of small bivalves. Possible future environmental problems are related to mining or drilling activities, while deeper soft-substrate deposits have been targeted in some areas for beach 'renourishment', a practice which could cause habitat disruption (Posey and Alphin, 1999).

Estuarine and Coastal Finfish Communities

The Carolinas support diverse populations of fishes, ranging from estuarine assemblages to pelagic fishes. Estuarine communities consist of a mix of species with three types of life history strategy, including anadromous species that spend most of their lives in saltwater but return to freshwater streams to spawn (e.g., river herrings, shad), resident 'estuarine-dependent' species that spend their entire life in the estuary (e.g., white perch, catfish), and estuarine migratory species that spawn in the open ocean, around inlets, or near shore (e.g., spot, Atlantic menhaden, Atlantic croaker, weakfish, and flounders) (Table 4). These species dominate the estuarine finfisheries, and also emigrate to join nearshore stocks that migrate seasonally along the Atlantic Coast (NC DMF, 1993). Coastal populations also include migratory marine fishes (mackerels, bluefish); rock/reef-associated fishes (the snapper–grouper complex, sea basses, haemulids), and larger pelagic fishes such as wahoo, swordfish, marlins and sharks.

Commercial and recreational fisheries are important (Table 4). Considerable effort has been devoted to their management and enhancement, including construction of artificial reefs in estuarine and coastal ocean waters. North Carolina historically has had, and still maintains, one of the richest fishery resource bases in the nation, attributed in part to the mix of northern and southern species that tend to overlap in the Cape Hatteras area, and to the more extensive, dissected coastline and expansive habitats afforded by the Albemarle–Pamlico Estuarine System. South Carolina has a similar mix of warm-water species (Sedberry and Van Dolah, 1984) but is lacking species of northern origin. North Carolina's blue crab fishery is among the three largest in the nation, along with the blue crab fisheries in Chesapeake Bay and Louisiana (SC DNR, 1999). North Carolina also has the northernmost major shrimp fishery in the country in

Pamlico Sound (SC DNR, 1999). Shrimp (brown and white: *Panaeus aztecus* and *P. setiferus,* respectively) are the most important commercial fishery in South Carolina, with blue crabs secondary. Unlike the major inshore shrimp trawling activity allowed in North Carolina, shrimp trawlers in South Carolina are restricted to ocean waters except for limited periods during fall, when trawlers work the lower areas of Winyah and North Santee Bays (SC DNR, 1999).

Most fisheries monitoring programs began in the late 1970s–1980s, although some commercial landings data for North Carolina extend back to the 1880s. The five-year period with the greatest recorded landings was 1978–1982, with average landings of 1.62×10^8 kg (357 million pounds), and a maximum of 196×10^8 kg (NC DMF, 1993). Atlantic menhaden, flounders, weakfish, Atlantic croaker, and white perch, as well as shellfish such as northern quahogs, blue crabs, and shrimp established all-time records or reached levels not seen for many years. Since that period, total finfish and shellfish harvests have declined by about a half, down to 9×10^7 kg by the early 1990s (NC DMF, 1993).

Twenty-eight individual species and five other species groupings are listed as commercially valuable (NC DMF, 1999). The species groupings consist of reef fish (71 species); sharks (two dominant species, with an unknown number of other common to rare species); shrimp (three species), kingfish (three species), catfish (four species), and river herrings (two species), for a total of 107 species plus sharks. The most valuable fishery in North Carolina is the blue crab fishery, followed by shrimp, southern flounder, summer flounder and Atlantic menhaden (Table 4). In 1998 these yielded a dockside value of more than $72 million. Populations of the shrimp and Atlantic menhaden are evaluated by NC DMF as viable or in good condition; however, summer flounder has been seriously overfished and is listed as 'stressed but recovering.' Two species, blue crab and southern flounder, have been evaluated as stressed populations of 'concern' with clear signs of overfishing.

Fisheries that, as of 1998, remained viable with healthy stocks included shrimp, Atlantic menhaden, king mackerel, striped bass, spotted seatrout, bay scallops, Spanish mackerel, dolphin, spot, and five reef fish, for a total of 15 viable species. Three species: weakfish, bluefish, and summer flounder are 'stressed but recovering,' following additional regulations to reduce recruitment overfishing. A total of 56 species were listed as stressed populations of 'concern,' including blue crab and southern flounder as mentioned, 1 major stock of black sea bass, Atlantic croaker, 51 reef fish, and striped mullet. Ten species were listed as depressed, 'overfished' populations, including 1 stock of black sea bass, eastern oyster, black grouper, red drum, river herring (two species, one stock), spiny dogfish, monkfish, Atlantic sturgeon, scup, tautog, 15 reef fish, and 'sharks.' Ten species were listed as 'unknown' in status. A total of 26 species, including most of the commercially important fishes with the exception of the reef fish category, are strongly dependent on estuarine habitats during critical periods such as spawning or recruitment. Fish management plans have been developed at the federal and/or state levels within the past ca. 15 years toward improving their protection and sustainability (NC DMF, 1999).

ENVIRONMENTAL CONCERNS IN ESTUARINE AND COASTAL SYSTEMS

The majority of estuaries in this region show low to moderate eutrophic conditions (Bricker et al., 1999). However, three estuaries in North Carolina (the Pamlico River, Neuse River, and New River) ranked as highly eutrophic, are showing symptoms including phytoplankton blooms, bottom-water hypoxia and anoxia, fish kills, and loss of SAV (e.g., Ferguson et al., 1988; NOAA, 1996; Bricker et al., 1999).

Inorganic nutrient concentrations in the most productive areas of eutrophic estuaries in North Carolina are little different compared to those of a nearby non-eutrophic system (the Cape Fear River Estuary) (Table 5). However, chlorophyll *a* concentrations in the eutrophic systems far exceed those of the Cape Fear (Table 5). The major difference between the systems is morphometric, with the open Cape Fear Estuary well flushed by tidal action. This flushing suppresses phytoplankton bloom development. Light is more strongly attenuated in the Cape Fear, relative to the Neuse, Pamlico, and New River Estuaries (Table 5). Tidal dampening in the eutrophic systems allows turbidity to settle out, so the slower flowing waters provide an ideal environment for bloom formation. Estuaries in South Carolina all show either low or moderate eutrophication symptoms (NOAA, 1996; Bricker et al., 1999), and most are more geomorphologically similar to the Cape Fear than the Neuse or Pamlico systems.

The eutrophic environment has led to numerous estuarine fish kills in recent years (Table 6). The Neuse and Pamlico Estuaries in particular have suffered kills of up to one billion fish (Burkholder et al., 1995). The ichthyotoxic dinoflagellate, *Pfiesteria piscicida,* was discovered in these waters in the early 1990s (Burkholder et al., 1992b), and its

Table 5

Average inorganic nutrient (dissolved inorganic nitrogen, DIN, and dissolved inorganic phosphorus, DIP, μg l^{-1}) and chlorophyll *a* concentrations (μg l^{-1}) in the most productive areas of four North Carolina estuaries (the first three, eutrophic), 1993–1998. Light attenuation coefficient (*k*) values are given in the last column. (Data records from Burkholder, North Carolina State University; the North Carolina Division of Water Quality; and Mallin and Cahoon, University of North Carolina at Wilmington)

Estuary	DIN	DIP	Chl *a*	Chl *a* range	*k*
Pamlico	125	28	32	4–360	–
Neuse	86	46	21	1–210	1.80
New River	370	64	70	1–380	3.10
Cape Fear	360	36	9	1–33	4.04

Table 6

Total fish kills and *Pfiesteria*-induced fish kills in four North Carolina estuaries, 1991–1998. (Data records from Burkholder, North Carolina State University; the North Carolina Division of Water Quality; and Mallin, University of North Carolina at Wilmington)

Estuary	Fish kills	Toxic *Pfiesteria*-related fish kills
Pamlico	26	13
Neuse	124	65
New River	6	2
Cape Fear	3	0

lethality to over 30 species of finfish and shellfish has been documented (Burkholder et al., 1995; Burkholder and Glasgow, 1997). Blooms of *Pfiesteria piscicida* and a second toxic *Pfiesteria* species have accounted for up to 50% of the estuarine fish kills in the Pamlico, Neuse, and New River Estuaries on an annual basis in the past decade (Table 6). Many fish kills are primarily caused by low dissolved oxygen as well, and occasionally by temperature shock and toxicant (e.g., pesticide) exposure (NC DWQ fish kill database, Raleigh, NC). The extent to which these and other factors act in concert to impair fish health is beginning to be examined, especially for *Pfiesteria* and low dissolved oxygen (Burkholder unpubl. data). In contrast, the well-flushed Cape Fear system suffers few fish kills, with none attributed to *Pfiesteria* (Table 6). Many of the South Carolina estuaries are similar hydrologically to the Cape Fear.

In the eutrophic systems discussed above, bacterial decay of dead algal biomass creates a high BOD, leading to hypoxia/anoxia (Lenihan and Peterson, 1998; Burkholder et al., 1999). Hypoxia and anoxia can also be caused by BOD loads from run-off from poorly treated swine waste (Burkholder et al., 1997; Mallin et al., 1997a; Mallin et al., 1999c). Such incidents have caused massive fish and benthic invertebrate kills.

The Carolinas have experienced frequent hurricanes in recent years, including Hugo which primarily struck South Carolina (1988); Bertha and Fran (1996), Bonnie (1998) and Dennis and Floyd (1999) that struck North Carolina. These events caused power outages in wastewater treatment facilities that led to rerouting of untreated or partially treated sewage into rivers and estuaries. Also, the storms caused breaching of the walls and flooding of swine-waste lagoons. The BOD loads after Fran and Bonnie combined to cause anoxia and large-scale fish kills in the Cape Fear and Neuse River estuaries (Mallin et al., 1999c; Burkholder et al., 1999). Dennis and Floyd combined to produce the wettest September on record at several observing stations; the high volume of fresh water apparently pushed many fish populations down-estuary into the sounds and coastal ocean, so that major fish kills were not reported. Additionally, prolonged standing water on the floodplain likely allowed smaller fish access to food-rich refuge areas in riparian wetlands. The large volume of floodwaters apparently diluted inputs, so that nutrients, BOD, and faecal coliform bacterial concentrations were much lower following Floyd compared to levels recorded after Fran and Bonnie.

Following Hurricane Floyd, there were erroneous press reports of a massive 'dead zone' in Pamlico Sound on the basis of a relatively small area extending out from the mouth of the Neuse, where hypoxic conditions were found in the lower 1–2 m of a 6–7 m water column. There was no reported fish kill, and only a few dead blue crabs were found (NC DWQ records, Raleigh, NC). Within seven days, this bottom-water oxygen sag had recovered to >5 mg DO l^{-1}, and a 60 km transect across the sound from north to south indicated that DO was >5 mg l^{-1} throughout (Burkholder, unpubl. data). However, the massive flooding did inundate agricultural and industrial sources of numerous pollutants including nutrients, petrochemicals, metals, animal decomposition products, BOD, and microbial pathogens. Some of this material was likely transported into the sounds and adjacent marine coastal waters, as evidenced by obvious darkly coloured plumes of turbid material. Some fish populations in the Pamlico Sound developed obvious signs of disease for 6–8 weeks after the hurricane (NC DMF records; L.B. Crowder, Duke U., pers. comm.). The high volume of fresh water that was delivered into the sounds would have caused salinity stress which is known to weaken fish and render them more susceptible to attack by opportunistic pathogens (Couch and Fournie, 1993). The recent storm events have given notice that the sounds, while more distant from terrestrial pollution sources, are not immune to degradation and need to be monitored more closely.

Research in the Cape Fear River and Estuary indicates that much of its chronic BOD load comes from nonpoint runoff. These sources can include riparian forests, swine lagoon sprayfields and poultry litterfields, and manure from livestock grazing areas. We performed correlation analyses among physical and biological parameters at a sampling station representing inputs to the lower river/upper estuary, using 41 consecutive months of data (Table 7) and showed that there was a highly significant correlation between five-day BOD and faecal coliform counts, indicating that much of the material contributing to these parameters was derived from the same sources. However, there was a very strong correlation between turbidity and both faecal coliforms and river flow on the day of sample collection, and a significant correlation between turbidity and BOD (Table 7). Turbidity is often a strong signal of nonpoint source runoff. Since river flow is correlated with both turbidity and faecal coliforms, nonpoint source runoff appears to be important to both BOD and faecal coliform pollution of the estuary.

Closures of shellfish beds due to high bacterial counts are a widespread problem in North Carolina waters. The most productive clam and oyster harvest waters in North Carolina are located from lower Pamlico Sound south to the South Carolina border, an area undergoing rapid

Table 7
Correlation analyses among various physical and biological parameters at a station on the mainstem lower Cape Fear River ($n = 41$ data points; correlation coefficient (r)/probability (p); ns = not significant at $p < 0.05$)

	Faecal coliforms	Turbidity	Daily flow
BOD_5	0.518	0.444	ns
	0.001	0.005	
Turbidity	0.858	1.000	0.736
	0.001	0.0	0.001
Faecal coliforms	1.000	0.858	0.469
	0.0	0.001	0.005

urbanization (Preyer, 1994). Approximately 18% of these waters are permanently closed due to high coliform counts, and much of the remaining area becomes temporarily closed following rain events (P.K. Fowler, pers. comm.). Most urban areas of this coast were brought into centralized sewer systems in the 1970s and early 1980s; thus, most closures result from nonpoint pollution. In South Carolina, shellfish grounds in urbanized estuaries such as Charleston Harbor and Winyah Bay are closed to harvest, whereas much of the non-urbanized North Inlet is approved for shellfishing (DeVoe et al., 1992).

The rapid urbanization of these coastal areas is an important factor leading to shellfish bed closures. Increasing cover of impervious surfaces concentrates pollutants during dry periods, and then serves as a rapid conduit of these pollutants into water bodies during and following rain. In many coastal locations, stormwater runoff directly enters sensitive shellfishing beds without any pre-treatment. In a four-year study of five urbanized tidal creeks in southeastern North Carolina, Mallin et al. (2000) found a strong correlation ($r = 0.945$, $p = 0.015$) between average estuarine faecal coliform counts and percent of developed land in the watershed, and an even stronger correlation ($r = 0.975$, $p = 0.005$) between average counts and percent of the watershed covered by impervious surfaces. In this study, as with the Cape Fear River study, faecal coliform counts for the creeks showed a highly significant correlation with turbidity levels.

Turbidity not only serves as a carrier of pollutants, but also can be directly harmful to estuarine functioning. Turbidity can interfere with fish and shellfish feeding, can block solar irradiance, and lead to boating problems through sedimentation. Upstream of Charleston, SC, two reservoirs were created in the 1940s for power generation purposes. Water was diverted from the Santee River through these reservoirs into the Cooper River, then to Charleston Harbor. This increased the average monthly flow of the Cooper River from 12 $m^3 s^{-1}$ to 455 $m^3 s^{-1}$ (DeVoe et al., 1992). This flow dramatically increased the sedimentation (and shoaling) rate in Charleston Harbor, and dredging costs to maintain open ship channels became prohibitive. In the 1970s, rediversion of the water back to the Santee River was proposed and eventually completed, reducing the Cooper River's flow to 122 $m^3 s^{-1}$. With the rediversion, the lower Santee River is again experiencing fresh and oligohaline conditions, and the lower Cooper River is undergoing resalinization to more normal conditions (Hackney and Adams, 1992).

An issue is the environmental impact on fish health, and on the viability of fish populations. However, the destructive nature of certain fishing practices is also important. By-catch continues to be an extremely serious problem, with some of the highest wastes associated with shrimp trawling and the flynet fishery (NC DMF, 1993). The shallow estuaries of North Carolina, especially the Pamlico and the New River, sustain heavy trawling activity which disrupts and damages benthic habitats (Watling and Norse, 1998) while also resuspending sediments. Although certain fishing practices are not allowed in primary fish nursery areas, they are used immediately adjacent to them as well as in secondary fish nursery areas. The extent to which trawling and other destructive fishing practices contribute to the degradation and loss of critical habitats in these important fish nursery grounds needs to be critically evaluated, so that effective management strategies can be imposed.

Another component of the fisheries issue that needs to be strengthened is the economic assessment and valuation of both the fish and the natural habitat resources that sustain them. Both fish and water are greatly undervalued in North Carolina, as elsewhere in the nation. The present-day commercial fishing industry is, in fact, made possible largely through federal subsidies (Safina, 1995). Fishermen obtain remarkably little compensation for many of the fish species they sell dockside (Table 4), relative to the lucrative gains made by many other components of the seafood industry who use those fish. Public understanding of the underlying value of fish and supporting aquatic habitat, and the critical need to move toward increased sustainability of these resources, will be greatly improved when more realistic and effective means of economic valuation are practised and conveyed (Safina, 1995).

Offshore Environmental Concerns

Environmental issues in the coastal ocean off North and South Carolina may arise from utilization of economic resources in the U.S. EEZ, including hydrocarbons, sand and gravel, and phosphorite deposits (Cahoon et al., 1996). Proposals by Mobil Oil (in consortium with other companies) and Chevron to explore for oil and/or natural gas off the North Carolina coast have been withdrawn as of this writing, but several active lease blocks remain off this coast. It is likely that interest in this area will remain high, particularly as pressure increases to develop new oil and gas sources in the U.S. A more novel hydrocarbon resource, methane gas hydrates, is believed to be present in high and potentially extractable quantities in the surface sediments of the Blake Plateau. Demand for beach renourishment sand is likely to continue to rise, since beach erosion from hurricanes has become severe. Consequently, commercial extraction of sands and gravel from offshore sources is likely to begin soon, with environmental assessments and lease sales impending. A high-grade phosphorite ore deposit is also present in the surface sedimentary layers of Onslow Bay in quantities on the order of 10^9 tons. Extraction of any of these resources poses challenges to the integrity of benthic communities, and may also affect downstream plankton and fish populations. Possible impacts on other organisms, particularly protected seabird, sea turtle, and mammal species, remain to be assessed.

PROGNOSES FOR THE FUTURE

Several major critical habitat and resource issues remain to be resolved in the Carolinas. For example, there are no personnel in the Carolinas' environmental agencies who are responsible for tracking and periodically remapping critical seagrass habitat. Moreover, unlike management practices in the Chesapeake Bay region (Dennison et al., 1993), the Carolinas lack water quality regulations for sediment and nutrient loadings to protect seagrass habitat. In addition to pollution stressors on seagrass meadows in North Carolina, destructive fishing practices such as clam-tonging and clam-raking (which destroy the perennating root/rhizome complex) are legally permitted in seagrass beds (Peterson et al., 1983, 1987; NC MFC, 1998). For many years sediment was allowed to be dredged ca. 6 m in depth and several hectares in extent at each taking, from areas within the seagrass meadows on the landward side of the barrier islands. The affected area, called the "Canadian Hole", yielded sediment that has been used to 'nourish' or build the land supporting roads along the barrier islands as it is continuously and seriously eroded by storms. Most of these finely particulate sediments are removed from the 'nourished' areas by winds within a year or less, so that the process must be repeated. Such practices destroy seagrass habitat and should be prevented, for the protection of this critical habitat and the fisheries that depend upon them.

Like many regions worldwide, overfishing has led to serious declines in many wild fish stocks of the Carolinas (NC DMF, 1993, 1999). For some stocks, certain practices leading to recruitment overfishing are still allowed (Table 4), and need to be reduced or restricted. Certain North Carolina fish populations, severely overfished, are now showing signs of recovery following strengthened regulations to reduce the fishing pressure (Table 4) (NC DMF, 1999). The same cannot be said for many fish habitats that continue to be degraded by water pollution. For some commercially valuable wild fish stocks that will require a decade or more to recover, there is increasing interest in aquaculture (Defur and Rader, 1995), though this also requires good water quality and, in some cases (e.g., striped bass aquaculture in the Pamlico Estuary watershed) requires substantial use of freshwater aquifers. The Castle Hayne aquifer is already experiencing a cone of depression of more than 2 m per year from this and other increased uses, accompanied by saltwater intrusion that is damaging certain fish habitats.

Constructive measures have been taken to reduce eutrophication in North Carolina estuaries over the past decade. A 1988 statewide ban on phosphate-containing detergents led to decreased P loading from point sources, particularly in the Neuse (NC DEM, 1991). Blue-green algal blooms have shown a concurrent decrease in the Neuse River as well (Mallin, 1994). After 25 years, in 1993 the world's largest phosphate mine installed a waste treatment system that reduced effluent phosphate concentrations by 90% to the Pamlico Estuary (NC DEHNR, 1994). However, algal blooms and fish kills still commonly occur (Tables 5 and 6). As an example, the authors of this paper witnessed a slow but continuous (weeks long) kill on the Pamlico Estuary in July 1997, during which numerous dead fish bearing lesions floated to the surface while dissolved oxygen conditions and other factors were adequate to protect fish health (6.9 mg l^{-1}). *Pfiesteria* concentrations in the water were at levels known to cause death of fish under laboratory conditions; and the presence of an actively toxic population of *P. piscicida* was verified using fish bioassays (Burkholder and Glasgow, 1997; Burkholder et al., 1999).

In the New River estuary, major improvements to wastewater treatment facilities in 1997–1999 appear to be having positive effects, with generally lower nutrient and chlorophyll levels recorded in 1998–1999 compared with the previous four years (Mallin et al., unpublished). An intensive five-year study in the Neuse Estuary (Glasgow and Burkholder, 2000) found that total P significantly decreased (by 14% from 1993–1998), but inorganic N increased by 38% while total N also increased significantly (by 16%). Management strategies for the Neuse Estuary are now targeting a 30% reduction in nitrogen loading (NC DEHNR, 1997), but without consideration for additional management of P which can be critically important (Fisher et al., 1992; Paerl et al., 1995; Mallin et al., 1999a). While this action will probably yield some improvement, more stringent N and P reductions will be needed to reverse the eutrophication process, and to hold the problem in check as human population growth increases (Glasgow and Burkholder, 2000).

The recent rise in the intensive swine industry represents a massive potential nutrient source that will remain a significant threat. Recent hurricane-induced pollution resulted in a 30-month moratorium on licensing of new industrial hog farms in North Carolina, and a ban on future hog houses and waste lagoons on the 100-year floodplain. However, political factors and the sheer size of this industry

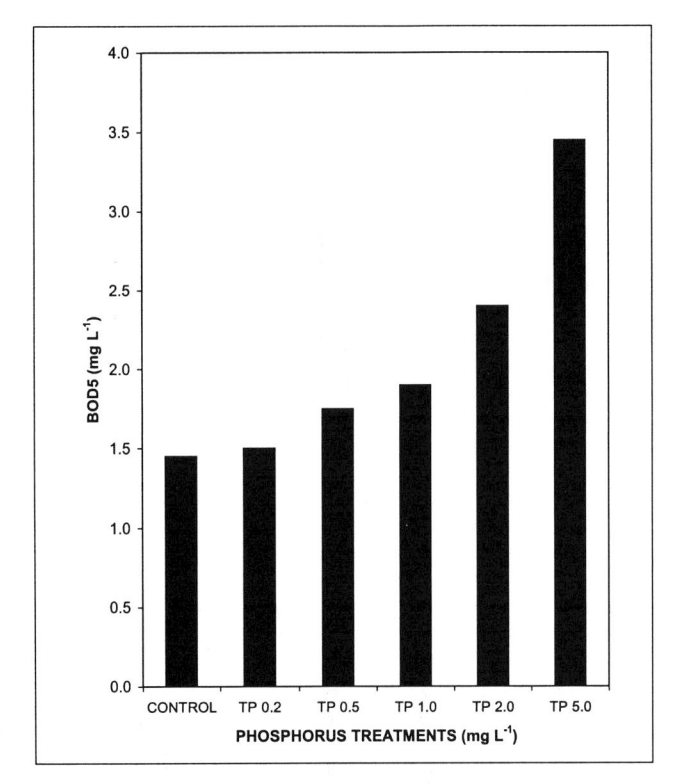

Fig. 4. Response of water samples from the Northeast Cape Fear River, a coastal blackwater stream, to additions of total phosphorus (50% orthophosphate + 50% glycerophosphate) over a gradient from zero to 5.0 mg l^{-1}, as measured by five-day biochemical oxygen demand.

suggest that nutrient imports will continue at very high levels. Atmospheric transport of nitrogen and runoff enriched in both nutrients are thus also likely to continue. For example, experiments by Mallin and Cahoon using water from coastal plain blackwater streams show that nutrient loading can cause direct increases in BOD without stimulating algal bloom formation first (Fig. 4). Thus, even if traditional eutrophication symptoms do not presently affect such systems, hypoxia/anoxia may be a future problem. Comprehensive plans limiting nutrient inputs are needed for all of the Carolinas' coastal rivers and estuaries; as a proactive measure, such planning should be extended to estuaries that currently do not sustain noxious algal blooms, rather than considering only estuaries that already have developed these characteristic signals of nutrient over-enrichment (e.g., Preyer, 1994). These plans should include discharge limits on total N and P for all municipal and private wastewater treatment systems, and phase-in of biological nutrient removal for all major wastewater treatment plants. Besides these point sources, there is a critical need to improve management of nonpoint nutrient runoff through increased use of streamside vegetated buffers, water level control structures, preservation of remaining natural wetlands, and construction of artificial wetlands. Improved treatment processes, based on strong incentive programs, should also be mandated for present and future industrial-scale animal operations.

Since estuaries reflect the impact of pollutant loading from distant sources as well as nearby sources, control measures that account for all contributing sources should be initiated on a watershed basis. General watershed management strategies are inadequate to protect these environmentally sensitive coastal resources unless accompanied by additional, localized coastal area-specific plans (Preyer, 1994). For example, tidal creeks in urbanized southeastern North Carolina sustain dense algal blooms in spring and summer, summer hypoxia, and the presence of *Pfiesteria* in nutrient-enriched areas (Mallin et al., 1999b). These systems generally are not impacted by point source inputs, but receive nutrient loading from urban and suburban lawns, gardens, and golf courses. Thus, nonpoint source runoff especially needs to be managed better in these smaller, urbanized watersheds. With steadily increasing urbanization along the South Atlantic seaboard, we predict that eutrophication symptoms in both large and small systems will significantly worsen unless aggressive steps are taken to alter current land development and nutrient management practices.

ACKNOWLEDGMENTS

For funding support we thank the Center for Marine Science Research at the University of North Carolina at Wilmington; the City of Wilmington; the Cooperative Institute of Fisheries Oceanography; the Department of Botany at North Carolina State University; the Lower Cape Fear River Program; the National Undersea Research Center; New Hanover County; the North Carolina General Assembly; the North Carolina Department of Environment & Natural Resources; the North Carolina Agricultural Research Foundation; the Northeast New Hanover Conservancy; the UNC Water Resources Research Institute; North Carolina Sea Grant and the U.S. Environmental Protection Agency. For their many helpful contributions in fieldwork, laboratory work, facilitation, and discussions about the writing and/or preparation of supporting materials, we thank T.D. Alphin, S.H. Ensign, P. Foster, H.B. Glasgow Jr., R.P. Lowe, D.B. Mayes, M.R. McIver, J.F. Merritt, J.E. Nearhoof, D.C. Parsons, R. Reed, and G.C. Shank.

REFERENCES

Atkinson, L.P. (1985) Hydrography and nutrients of the southeastern U.S. continental shelf. In *Oceanography of the Southeastern U.S. Continental Shelf*, eds. L.P. Atkinson, D.W. Menzel and K.A. Bush, pp. 77–92. American Geophysical Union, Washington, DC.

Atkinson, L.P., Singer, J.J. and Pietrafesa, L.J. (1980) Volume of subsurface intrusions into Onslow Bay, North Carolina. *Deep-Sea Research* 27, 421–434.

Bailey, W.P. (1996) Population trends in the coastal area, concentrating on South Carolina. In *Sustainable Development in the Southeast-*

ern Coastal Zone, eds. F.J. Vernberg, W.B. Vernberg and T. Siewicki, pp. 55–73. University of South Carolina Press, Columbia, SC.

Breitburg, D., Coen, L., Luckenbach, M., Mann, R., Posey, M. and Wesson, J. (in press). Oyster reef restoration: convergence of harvest and conservation strategies. *Journal of Shellfish Research.*

Bricker, S.B., Clement, C.G., Pirhalla, D.E., Orlando, S.P. and Farrow, D.R.G. (1999) National Estuarine Eutrophication Assessment: Effects of Nutrient Enrichment in the Nation's Estuaries. NOAA, National Ocean Service, Special Projects Office and the National Centers for Coastal Ocean Science. Silver Spring, MD, 71 pp.

Bowman, T.E. (1971) The distribution of calanoid copepods off the southeastern United States between Cape Hatteras and southern Florida. *Smithsonian Contributions in Zoology* #96.

Burke, M.K., Dennison, W.C. and Moore, K.A. (1996) Non-structural carbohydrate reserves of eelgrass *Zostera marina*. *Marine Ecology Progress Series* **137**, 195–201.

Burkholder, J.M. (1998) Implications of harmful microalgae and heterotrophic dinoflagellates in management of sustainable marine fisheries. *Ecological Applications* **8**, S37–S62.

Burkholder, J.M. and Cuker, B.E. (1991) Response of periphyton communities to clay and phosphate loading in a shallow reservoir. *Journal of Phycology* **27**, 373–384.

Burkholder, J.M. and Glasgow, Jr., H.B. (1997) *Pfiesteria piscicida* and other *Pfiesteria*-like dinoflagellates: Behavior, impacts, and environmental controls. *Limnology and Oceanography* **42**, 1052–1075.

Burkholder, J.M., Glasgow, Jr., H.B. and Cooke, J.E. (1994) Comparative effects of water-column nitrate enrichment on eelgrass *Zostera marina*, shoalgrass *Halodule wrightii*, and widgeongrass *Ruppia maritima*. *Marine Ecology Progress Series* **105**, 121–138.

Burkholder, J.M., Glasgow Jr., H.B. and Fensin, E.E. (1996) Physical, Chemical, and Biological Characteristics of Water Samples Collected from the Neuse Estuary in the Vicinity of Cherry Point, North Carolina, 1993–1995. Final report to the U.S. Marine Air Station, Cherry Point, NC.

Burkholder, J.M., Mason, K.M. and Glasgow Jr., H.B. (1992a) Water-column nitrate enrichment promotes decline of eelgrass *Zostera marina* L.: Evidence from seasonal mesocosm experiments. *Marine Ecology Progress Series* **81**, 163–178.

Burkholder, J.M., Noga, E.J., Hobbs, C.W., Glasgow Jr., H.B. and Smith, S.A. (1992b) New "phantom" dinoflagellate is the causative agent of estuarine fish kills. *Nature* **358**, 407–410; *Nature* **360**, 768.

Burkholder, J.M., Glasgow Jr., H.B. and Hobbs, C.W. (1995) Distribution and environmental conditions for fish kills linked to a toxic ambush predator dinoflagellate. *Marine Ecology Progress Series* **124**, 43–61.

Burkholder, J.M., Mallin, M.A., Glasgow Jr., H.B., Larsen, L.M., McIver, M.R., Shank, G.C., Deamer-Melia, N., Briley, D.S., Springer, J., Touchette, B.W. and Hannon, E.K. (1997) Impacts to a coastal river and estuary from rupture of a swine waste holding lagoon. *Journal of Environmental Quality* **26**, 1451–1466.

Burkholder, J.M., Mallin, M.A. and Glasgow, Jr., H.B. (1999) Fish kills, bottom-water hypoxia, and the toxic *Pfiesteria* complex in the Neuse River and Estuary. *Marine Ecology Progress Series* **179**, 301–310.

Cahoon, L.B. and Cooke, J.E. (1992) Benthic microalgal production in Onslow Bay, North Carolina. *Marine Ecology Progress Series* **84**, 185–196.

Cahoon, L.B. and Tronzo, C.R. (1990) New records of amphipods and cumaceans from Onslow Bay, North Carolina, in demersal zooplankton collections. *Journal of the Elisha Mitchell Scientific Society* **106**, 78–84.

Cahoon, L.B. and Tronzo, C.R. (1992) Quantitative estimates of demersal zooplankton abundance in Onslow Bay, North Carolina. *Marine Ecology Progress Series* **87**, 197–200.

Cahoon, L.B. and Laws, R.A. (1993) Benthic diatoms from the North Carolina continental shelf: Inner and mid shelf. *Journal of Phycology* **29**, 257–263.

Cahoon, L.B., Lindquist, D.G. and Clavijo, I.E. (1990a) "Live bottoms" in the continental shelf ecosystem: A misconception? In *Diving for Science*, ed. W. Jaap, pp. 39–47. American Academy of Underwater Sciences. Costa Mesa, CA.

Cahoon, L.B., Redman, R.L. and Tronzo, C.R. (1990b) Benthic microalgal biomass in sediments of Onslow Bay, North Carolina. *Estuarine and Coastal Shelf Science* **31**, 805–816.

Cahoon, L.B., Clark, W.F. and Crawford, K.P. (1996) North Carolina's Ocean Resources Plan. In *Proceedings of the Coastal Society's 15th International Conference*, eds. T.E. Bigford and R.H. Boyles, Jr., pp. 359–365. The Coastal Society, Alexandria, VA.

Cahoon, L.B., Mickucki, J.A. and Mallin, M.A. (1999a) Nutrient imports to the Cape Fear and Neuse River basins to support animal production. *Environmental Science and Technology* **33**, 410–415.

Cahoon, L.B., Nearhoof, J.E. and Tilton, C.L. (1999b) Sediment grain size effect on benthic microalgal biomass in shallow aquatic ecosystems. *Estuaries* **22**, 735–741.

Carpenter, E.J. (1971) Annual phytoplankton cycle of the Cape Fear River Estuary, North Carolina. *Chesapeake Science* **12**, 95–104.

Chester, A.J., Fergesun, R.L. and Thayer, G.W. (1983) Environmental gradients and benthic macroinvertebrate distributions in a shallow North Carolina estuary. *Bulletin of Marine Science* **33**, 282–295.

Coleman, V.L. and Burkholder, J.M. (1994) Community structure and productivity of epiphytic microalgae on eelgrass (*Zostera marina* L.) under water-column nitrate enrichment. *Journal of Experimental Marine Biology and Ecology* **179**, 29–48.

Coleman, V.L. and Burkholder, J.M. (1995) Response of microalgal epiphytes to nitrate enrichment in an eelgrass (*Zostera marina* L.) meadow. *Journal of Phycology* **31**, 36–43.

Copeland, B.J. and Grey, J. (1989) Albemarle–Pamlico Estuarine System: Preliminary Analysis of the Status and Trends. Albemarle–Pamlico Study Report 89-13A. North Carolina Department of Environment and Natural Resources, Raleigh, NC.

Copeland, B.J., Hodson, R.G. and Riggs, S.R. (1984) The Ecology of the Pamlico River, North Carolina: An Estuarine Profile. U.S. Fish and Wildlife Service, Office of Biological Services (Technical Report) FWS/OBS FWS/OBS/82-06, 1–83.

Copeland, B.J., Hodson, R.G., Riggs, S.R. and Easley Jr., J.E. (1983) *The Ecology of Albemarle Sound, North Carolina: An Estuarine Profile.* U.S. Fish and Wildlife Service, Office of Biological Services (Technical Report) FWS/OBS FWS/OBS/83-01, 1–68.

Couch, J.A. and Fournie, J.W. (eds.) (1993) *Pathobiology of Marine and Estuarine Organisms.* CRC Press, Boca Raton, FL.

Dame, R.F. and Allen, D.M. (1996) Between estuaries and the sea. *Journal of Experimental Marine Biology and Ecology* **200**, 169–185.

DeFur, P.L. and Rader, D.N. (1995) Aquaculture in estuaries: feast or famine? *Estuaries* **18**, 2–9.

Den Hartog, C. (1970) *The Sea-Grasses of the World.* North-Holland Publishing Co., Amsterdam, The Netherlands.

Dennison, W.C., Orth, R.J., Moore, K.A., Stevenson, J.C., Carter, V., Kollar, S., Bergstrom, P.W. and Batiuk, R.A. (1993) Assessing water quality with submersed aquatic vegetation—habitat requirements as barometers of Chesapeake Bay health. *BioScience* **43**, 86–94.

DeVoe, M.R., Davis, K.B. and Van Dolah, R.F. (1992) *Characterization of the Physical, Chemical and Biological Conditions and Trends in Three South Carolina Estuaries, 1970–1985.* South Carolina Sea Grant Consortium, Charleston, SC.

Dewi, L.A., Axford, R.F.E., Fayez, I., Marai, M. and Omed, H. (eds.) (1994) *Pollution in Livestock Production Systems.* CAB International, Wallingford, UK.

Epperley, S.P. and Ross, S.W. (1986) Characterization of the North Carolina Pamlico–Albemarle estuarine complex. NOAA Technical Memorandum NMFS-SEFC-175. NOAA, Washington, D.C., USA.

Fahay, M.P. (1975) An Annotated List of Larval and Juvenile Fishes Captured with a Surface-Towed Meter Net in the South Atlantic Bight During Four DOLPHIN Cruises Between May 1967 and February 1968. NOAA Tech. Rept. NMFS SSRF-685.

Fensin, E.E. (1997) Population Dynamics of *Pfiesteria*-like Dinoflagellates, and Environmental Controls in the Mesohaline Neuse Estuary, North Carolina. Master of Science thesis, Department of Botany, North Carolina State University, Raleigh, NC.

Ferguson, R.L., Rivera, J.A. and Wood, L.L. (1988) Submerged Aquatic Vegetation in the Albemarle–Pamlico Estuarine System. Report No. 88-10 to the Albemarle–Pamlico Estuarine Study. NC DEHNR and the U.S. Environmental Protection Agency – National Estuarine Program, Raleigh, NC.

Fisher, T.R., Peele, E.R., Ammerman, J.W. and Harding Jr., L.W. (1992) Nutrient limitation of phytoplankton in Chesapeake Bay. *Marine Ecology Progress Series* 82, 51–63.

Fox, R.S. and Ruppert, E.E. (1985). *Shallow-Water Marine Benthic Macroinvertebrates of South Carolina*. University of South Carolina Press, Columbia, SC, 329 pp.

Freeman, D.B. (1989) The Distribution and Trophic Significance of Benthic Microalgae in Masonboro Sound, North Carolina. Master of Science thesis, Department of Biology, UNC-Wilmington, Wilmington, NC.

Fulton, R.S. III (1984) Distribution and community structure of estuarine copepods. *Estuaries* 7, 38–50.

Fulton, R.S. III. (1985) Predator–prey relationships in an estuarine littoral zooplankton community. *Ecology* 66, 21–29.

Glasgow, H.B., Jr. and Burkholder, J.M. (2000) Water quality trends and management implications from a five-year study of a poorly-flushed, eutrophic estuary. *Ecological Applications*. (In press).

Glasgow, H.B. Jr., Burkholder, J.M., Morton, S.L. and Springer, J. (2000). A new species of toxic *Pfiesteria*. Ninth annual Conference on Harmful Algal Blooms, Hobart, Tasmania (abstract).

Glasgow, H.B. Jr., Burkholder, J.M., Schmechel, D.E., Tester, P.A. and Rublee, P.A. (1995) Insidious effects of a toxic dinoflagellate on fish survival and human health. *Journal of Toxicology and Environmental Health* 46, 101–122.

Hackney, C.T., Grimley, J., Posey, M., Alphin, T. and Hyland, J. (1998) Sediment contamination in North Carolina's estuaries. Center for Marine Sciences Research Publication #198. University of North Carolina-Wilmington, Wilmington, NC.

Hackney, C.T. and Adams, S.M. (1992) Aquatic communities of the Southeastern United States: past, present, and future. In *Biodiversity of Southeastern United States Aquatic Communities*, eds. C.T. Hackney, S.M. Marshall and W.M. Martin, pp. 747–760. John Wiley and Sons, Inc., New York, NY.

Hallegraeff, G.M. (1993) A review of harmful algal blooms and their apparent global increase. *Phycologia* 32, 79–99.

Hobbie, J.E. (1971) Phytoplankton species and populations in the Pamlico River Estuary of North Carolina. Report No. 56, Water Resources Research Institute, University of North Carolina, Raleigh, N.C.

Houser, D.S. and Allen, D.M. (1996) Zooplankton dynamics in an intertidal salt-marsh basin. *Estuaries* 19, 659–673.

Kapraun, D.F. (1980) *An Illustrated Guide to the Benthic Marine Algae of Coastal North Carolina. I. Rhodophyta*. The University of North Carolina Press, Chapel Hill, NC.

Kenworthy, W.J. (1981) The Interrelationship Between Seagrasses *Zostera marina* and *Halodule wrightii*, and the Physical and Chemical Properties of Sediments in a mid-Atlantic Coastal Plain Estuary Near Beaufort, North Carolina (U.S.A.). Master of Science thesis, University of Virginia, Charlottesville, VA.

Kjelson, M.A., Peters, D.S., Thayer, G.W. and Johnson, G.M. (1975) The general feeding ecology of postlarval fishes in the Newport River Estuary. *Fishery Bulletin* 73, 137–144.

Kuenzler, E.J., Stanley, D.W. and Koenigs, J.P. (1979). Nutrient Kinetics in the Pamlico River, North Carolina. Report No. 139. UNC Water Resources Research Institute, Raleigh, N.C.

Lenihan, H.S. and Peterson, C.H. (1998) How habitat degradation through fishery disturbance enhances impacts of hypoxia on oyster reefs. *Ecological Applications* 8, 128–140.

Lewitus, A.J., Koepfler, E.T. and Morris, J.T. (1998) Seasonal variation in the regulation of phytoplankton by nitrogen and grazing in a salt-marsh estuary. *Limnology and Oceanography* 43, 636–646.

Lonsdale, D.J. and Coull, B.C. (1977) Composition and seasonality of zooplanktion of North Inlet, South Carolina. *Chesapeake Science* 18, 272–283.

Lüning, K. (1990) *Seaweeds—Their Environment, Biogeography, and Ecophysiology*. John Wiley & Sons, Inc., New York, NY.

Mallin, M.A. (1991) Zooplankton abundance and community structure in a mesohaline North Carolina estuary. *Estuaries* 14, 481–488.

Mallin, M.A. (1994) Phytoplankton ecology of North Carolina estuaries. *Estuaries* 17, 561–574.

Mallin, M.A. and Paerl, H.W. (1994) Planktonic trophic transfer in an estuary: Seasonal, diel, and community structure effects. *Ecology* 75, 2168–2185.

Mallin, M.A., Paerl, H.W. and Rudek, J. (1991) Seasonal phytoplankton composition, productivity, and biomass in the Neuse River Estuary, North Carolina. *Estuarine, Coastal and Shelf Science* 32, 609–623.

Mallin, M.A., Burkholder, J.M. and Sullivan, M.J. (1992) Benthic microalgal contributions to coastal fishery yield. *Transactions of the American Fisheries Society* 121, 691–695.

Mallin, M.A., Paerl, H.W., Rudek, J. and Bates, P.W. (1993) Regulation of estuarine primary production by rainfall and river flow. *Marine Ecology Progress Series* 93, 199–203.

Mallin, M.A., Burkholder, J.M., McIver, M.R., Shank, G.C., Glasgow, Jr. H.B., Touchette, B.W. and Springer, J. (1997a) Comparative effects of poultry and swine waste lagoon spills on the quality of receiving streamwaters. *Journal of Environmental Quality* 26, 1622–1631.

Mallin, M.A., Cahoon, L.B., McIver, M.R., Parsons, D.C. and Shank, G.C. (1997b) Nutrient Limitation and Eutrophication Potential in the Cape Fear and New River Estuaries. Report No. 313. Water Resources Research Institute of the University of North Carolina, Raleigh, NC.

Mallin, M.A., Cahoon, L.B., McIver, M.R., Parsons, D.C. and Shank, G.C. (1999a) Alternation of factors limiting phytoplankton production in the Cape Fear Estuary. *Estuaries* 22, 985–996.

Mallin, M.A., Esham, E.C., Williams, K.E. and Nearhoof, J.E. (1999b) Tidal stage variability of fecal coliform and chlorophyll *a* concentrations in coastal creeks. *Marine Pollution Bulletin* 38, 414–422.

Mallin, M.A., Posey, M.H., Shank, G.C., McIver, M.R., Ensign, S.H. and Alphin, T.D. (1999c) Hurricane effects on water quality and benthos in the Cape Fear watershed: natural and anthropogenic impacts. *Ecological Applications* 9, 350–362.

Mallin, M.A., Williams, K.E., Esham, E.C. and Lowe, R.P. (2000) Effect of human development on bacteriological water quality in coastal watersheds. *Ecological Applications* (in press).

Marshall, H.G. (1969) Phytoplankton distribution off the North Carolina coast. *American Midland Naturalist* 81, 241–257.

Marshall, H.G. (1971) Composition of phytoplankton off the southeastern coast of the United States. *Bulletin of Marine Science* 21, 806–825.

Marshall, H.G. (1976) Phytoplankton distribution along the eastern coast of the U.S. I. Phytoplankton distribution. *Marine Biology* 38, 81–89.

Marshall, H.G. (1978) Phytoplankton distribution along the eastern coast of the USA. Part II. Seasonal assemblages north of Cape Hatteras, North Carolina. *Marine Biology* 45, 203–208.

McKenna, S.K., Henry, L.T. and Diably, S. (1998) *North Carolina Fishery Management Plan—Blue Crab*. North Carolina Department of

Environmental Protection and Natural Resources, Division of Marine Fisheries, Morehead City, NC.

Muehlstein, L.K. (1992) The host-pathogen interaction in the wasting disease of eelgrass *Zostera marina*. *Canadian Journal of Botany* **70**, 2081–2088.

Nearhoof, J.E. (1994) Effects of Water Depth and Clarity on the Distribution and Relative Abundance of Phytoplankton and Benthic Microalgae in North Carolina Estuaries. Master of Science thesis, Department of Biology, UNC-Wilmington, Wilmington, NC.

NC DEHNR (North Carolina Department of Environment, Health & Natural Resources now NC DENR) (1994) Water Quality Progress in North Carolina: 1992 – 1993 305(b) Report. NC DEHNR, Raleigh, NC.

NC DEM (North Carolina Division of Environmental Management) (1991) An evaluation of the effects of the North Carolina phosphate detergent ban. Report No. 91-04. NC DEM, Raleigh, NC.

NC DENR (North Carolina Department of Environment & Natural Resources) (1997) Neuse River Nutrient-Sensitive Waters (NSW) Management Strategy. NC DENR, Raleigh, NC.

NC DMF (North Carolina Division of Marine Fisheries) (1993) Description of North Carolina's Coastal Fishery Resources, 1972-1991. NC DEHNR–DMF, Morehead City, NC, 213 pp. and appendices.

NC DMF (North Carolina Division of Marine Fisheries) (1999) Stock Status of Important Coastal Fisheries in North Carolina, 1999. Available at website www.ncdmf.net/stocks. NC DMF, Morehead City, NC.

NC DWQ (North Carolina Division of Water Quality) (1996) Cape Fear River Basinwide Water Quality Management Plan. North Carolina Department of Environment, Health & Natural Resources, Division of Water Quality, Raleigh, NC.

NC MFC (North Carolina Marine Fisheries Commission) (1998) North Carolina Fisheries Rules for Coastal Waters 1998–1999. NCMFC and NC DENR-DMF, Morehead City, NC.

NOAA (National Oceanic & Atmospheric Administration) (1992) Status of Fishery Resources off the Southeastern United States for 1991. NOAA Technical Memorandum NMFS-SEFSC-306, Silver Spring, MD.

NOAA (National Oceanic & Atmospheric Administration) (1996) NOAA's Estuarine Eutrophication Survey, Volume 1: South Atlantic Region. NOAA, Office of Ocean Resources Conservation Assessment, Silver Spring, MD, 50 pp.

Paerl, H.W., Mallin, M.A., Donahue, C.A., Go, M. and Peierls, B.J. (1995) Nitrogen Loading Sources and Eutrophication of the Neuse River Estuary, North Carolina: Direct and Indirect Roles of Atmospheric Deposition. Report 291. UNC Water Resources Research Institute, Raleigh, NC.

Paffenhofer, G.-A. (1980) Zooplankton distribution as related to summer hydrographic conditions in Onslow Bay, North Carolina. *Bulletin of Marine Science* **30**, 819–832.

Penhale, P. (1977) Macrophyte-epiphyte biomass and productivity in an eelgrass (*Zostera marina* L.) community. *Journal of Experimental Marine Biology and Ecology* **26**, 211–224.

Peterson, C.H., Summerson, H.C. and Fegley, S.R. (1983) Relative efficiency of two clam rakes and their contrasting impacts on seagrass biomass. *Fishery Bulletin* **81**, 429–434.

Peterson, C.H., Summerson, H.C. and Fegley, S.R. (1987) Ecological consequences of mechanical harvesting of clams. *Fishery Bulletin* **85**, 281–298.

Peterson, C.H., Summerson, H.C. and Luettich, Jr. R.A. (1996) Response of bay scallops to spawner transplants: a test of recruitment limitation. *Marine Ecology Progress Series* **132**, 93–107.

Pietrafesa, L.J. and Janowitz, G.S. (1988) Physical oceanographic processes affecting larval transport around and through North Carolina inlets. *American Fisheries Society Symposium* **3**, 34–50.

Pietrafesa, L.J., Janowitz, G.S. and Wittman, P.A. (1985) Physical oceanographic processes in the Carolina Capes. In *Oceanography of the Southeastern U.S. Continental Shelf*, eds. L.P. Atkinson, D.W. Menzel and K.A. Bush, pp. 23–32. American Geophysical Union, Washington, DC.

Pietrafiesa, L.J., Janowitz, G.S., Miller, J.M., Noble, E.B., Ross, S.W. and Epperly, S.P. (1986) Abiotic factors influencing the spatial and temporal variability of juvenile fish in Pamlico Sound, North Carolina. In *Estuarine Variability*, ed. D.A. Wolfe, pp. 341–353. Academic Press, Inc.

Pile, A.J., Lipcius, R.N., van Montfrans, J. and Orth, R.J. (1996) Density-dependent settler–recruit–juvenile relationships in blue crabs. *Ecological Monographs* **66**, 277–300.

Pinckney, J.L. and Zingmark, R.G. (1993) Modeling the annual production of intertidal benthic microalgae in estuarine ecosystems. *Journal of Phycology* **29**, 396–407.

Pinckney, J.L., Millie, D.F., Vinyard, B.T. and Paerl, H.W. (1997) Environmental controls of phytoplankton bloom dynamics in the Neuse River Estuary, North Carolina, U.S.A. *Canadian Journal of Fisheries and Aquatic Sciences* **54**, 2491–2501.

Posey, M.H. and Ambrose, Jr. W.G. (1994) Effects of proximity to an offshore hard-bottom reef on infaunal abundances. *Marine Biology* **118**, 745–753.

Posey, M.H. and Alphin, T.D. (1999) Monitoring of benthic faunal responses to sediment removal associated with the Carolina Beach and vicinity—area south project. Final Report submitted to NOAA Sanctuaries and Reserves Division. 24 pp.

Posey, M.H., Powell, C., Cahoon, L. and Lindquist, D. (1995) Top down vs. bottom up control of benthic community composition on an intertidal tideflat. *Journal of Experimental Marine Biology and Ecology* **185**, 19–31.

Posey, M.H., Alphin, T.D., Cahoon, L., Lindquist, D. and Becker, M.E. (1999a) Interactive effects of nutrient additions and predation on infaunal communities. *Estuaries* **22**, 785–792.

Posey, M.H., Alpin, T.D., Powell, C.M. and Townsend, E. (1999b) Use of oyster reefs as habitat for epibenthic fish and decapods. In *Oyster Reef Habitat Restoration: A Synopsis and Synthesis of Approaches*, eds. M.W. Luckenbach and J.A. Wesson, pp. 229–238. Virginia Institute of Marine Science Press.

Preyer, L.R. (chair) (1994) *Charting a Future for Our Coast—A Report to the Governor of North Carolina*. North Carolina Coastal Futures Committee. North Carolina Department of Environment, Health & Natural Resources, Raleigh, NC, 106 pp.

Safina, C. (1995) The world's imperiled fish. *Scientific American* **273**, 46–53.

Sanders, J.G. and Kuenzler, E.J. (1979) Phytoplankton population dynamics and productivity in a sewage-enriched tidal creek in North Carolina. *Estuaries* **2**, 87–96.

SC DNR (South Carolina Dept. of Natural Resources) (1999) Marine Fisheries publications website: http://www.dnr.state.sc.us/marine/index.html

Schneider, C.W. (1976) Spatial and temporal distributions of benthic marine algae on the continental shelf of the Carolinas. *Bulletin of Marine Science* **26**, 133–151.

Schneider, C.W. and Searles, R.B. (1991) *Seaweeds of the Southeastern United States—Cape Hatteras to Cape Canaveral*. Duke University Press, Durham, NC, 553 pp.

Sedberry, G.R. and Van Dolah, R.F. (1984) Demersal fish assemblages associated with hard bottom habitats in the South Atlantic Bight of the U.S.A. *Environmental Biology of Fishes* **11**, 241–258.

Sellner, K.G., Zingmark, R.G. and Miller, T.G. (1976) Interpretations of the ^{14}C method of measuring the total annual production of phytoplankton in a South Carolina estuary. *Botanica Marina* **19**, 119–125.

Short, F.T. and Wylle-Echeverria, S. (1996) Natural and human-induced disturbance of seagrasses. *Environmental Conservation* **23**, 17–27.

Stanley, D.W. (1992). Historical Trends: Water Quality and Fisheries, Albemarle–Pamlico Sounds, with Emphasis on the Pamlico River

Estuary. North Carolina Sea Grant College Program Publication UNC-SG-92-04. Institute for Coastal and Marine Resources, East Carolina University, Greenville, NC, 215 pp.

Steel, J. (ed.) (1991) Status and Trends Report of the Albemarle–Pamlico Estuarine Study. Albemarle–Pamlico Estuarine Study, U.S. Environmental Protection Agency National Estuarine Program and the North Carolina Department of Environment, Health & Natural Resources, Raleigh, NC.

Stevenson, J.C. (1988) Comparative ecology of submersed grass beds in freshwater, estuarine, and marine environments. *Limnology and Oceanography* **33**, 867–893.

Summerson, H.C. and Peterson, C.H. (1990) Recruitment failure of the bay scallop, *Argopectin irradians concentricus*, during the first red tide, *Ptychodiscus brevis*, outbreak recorded in North Carolina. *Estuaries* **13**, 322–331.

Tenore, K.R. (1972) Macrobenthos of the Pamlico River Estuary, North Carolina. *Ecological Monographs* **42**, 51–69.

Tester, P.A., Stumpf, R.P., Vukovich, F.M., Fowler, P.K. and Turner, J.T. (1991) An expatriate red tide bloom: Transport, distribution, and persistence. *Limnology and Oceanography* **36**, 1053–1061.

Thomas, C.J. and Cahoon, L.B. (1993) Stable isotope analyses differentiate between different trophic pathways supporting rocky-reef fishes. *Marine Ecology Progress Series* **95**, 19–24.

Touchette, B.W. (1999) Physiological and Developmental Responses of Eelgrass (*Zostera marina* L.) to Increases in Water-Column Nitrate and Temperature. Ph.D. Thesis, Dept. of Botany, North Carolina State University, Raleigh, NC.

Thayer, G.W. (1971) Phytoplankton production and distribution of nutrients in a shallow unstratified estuarine system near Beaufort, N.C. *Chesapeake Science* **12**, 240–253.

Thayer, G.W., Hoss, D.E., Kjelson, M.A., Hettler, Jr. W.F. and Lacroix, M.W. (1974) Biomass of zooplankton in the Newport River Estuary and influence of postlarval fishes. *Chesapeake Science* **15**, 9–16.

Thayer, G.W., Kenworthy, W.J. and Fonseca, M.S. (1984a) The Ecology of Eelgrass Meadows of the Atlantic Coast: A Community Profile. U.S. Fish and Wildlife Service, Office of Biological Services [Temporary Report] FWS/OBS FWS/OBS/84-02, 1–147.

Thayer, G.W., Bjorndal, K.A., Ogden, J.C., Williams, S.L. and Zieman, J.C. (1984b) Role of larger herbivores in seagrass communities. *Estuaries* **7**, 351–376.

Watling, L. and Norse, E.A. (1998) Disturbance of the seabed by mobile fishing gear: a comparison to forest clearcutting. *Conservation Biology* **12**, 1180–1197.

Williams, R.B. and Murdoch, M.B. (1966) Phytoplankton production and chlorophyll concentration in the Beaufort Channel, North Carolina. *Limnology and Oceanography* **11**, 73–82.

Zimmerman, R.C., Smith, R.D. and Alberte, R.S. (1989) Thermal acclimation and whole-plant carbon balance in *Zostera marina* L. (eelgrass). *Journal of Experimental Marine Biology and Ecology* **130**, 93–109.

Zingmark, R.G. (1986) Production of microbenthic algae. In *South Carolina Coastal Wetlands and Impoundments: Ecological Characteristics, Management, Status, and Use. Vol. II: Technical Synthesis*, eds. R. Devoe and D. Baughman, pp. 179–194. SC Sea Grant Publication #SC-SG-TR-86-2, Charleston, SC.

THE AUTHORS

Michael A. Mallin
*Center for Marine Science Research,
University of North Carolina-Wilmington,
Wilmington, NC 28403, U.S.A.*

JoAnn M. Burkholder
*Department of Botany,
North Carolina State University,
Raleigh, NC 27695-7612, U.S.A.*

Lawrence B. Cahoon
*Department of Biological Sciences,
University of North Carolina-Wilmington,
Wilmington, NC 28403, U.S.A.*

Martin H. Posey
*Center for Marine Science Research,
University of North Carolina-Wilmington
Wilmington, NC 28403, U.S.A.*

Chapter 23

THE GULF OF ALASKA

Bruce A. Wright, Jeffrey W. Short, Tom J. Weingartner and
Paul J. Anderson

The Gulf of Alaska (GOA) is a large (336,000 km^2) and productive ecosystem, but is showing increasing signs of being influenced by global-scale events, including global warming and contamination. The natural beauty of the rugged coastline, the vast wilderness, and the pristine waters still represent an ecosystem that is clean and pure. Government agencies, both State of Alaska and federal, have been mostly successful in managing the GOA's resources. These organizations are attempting to apply an ecosystem management approach to protecting and utilizing the resources found in this northern ecosystem. However, ecosystem management requires a baseline of understanding of the natural history of the species that ply these waters, and little information exists. Research associated with the *Exxon Valdez* oil spill, and with the University of Alaska and resource management agencies has advanced the understanding of this region tremendously, but much more needs to be done.

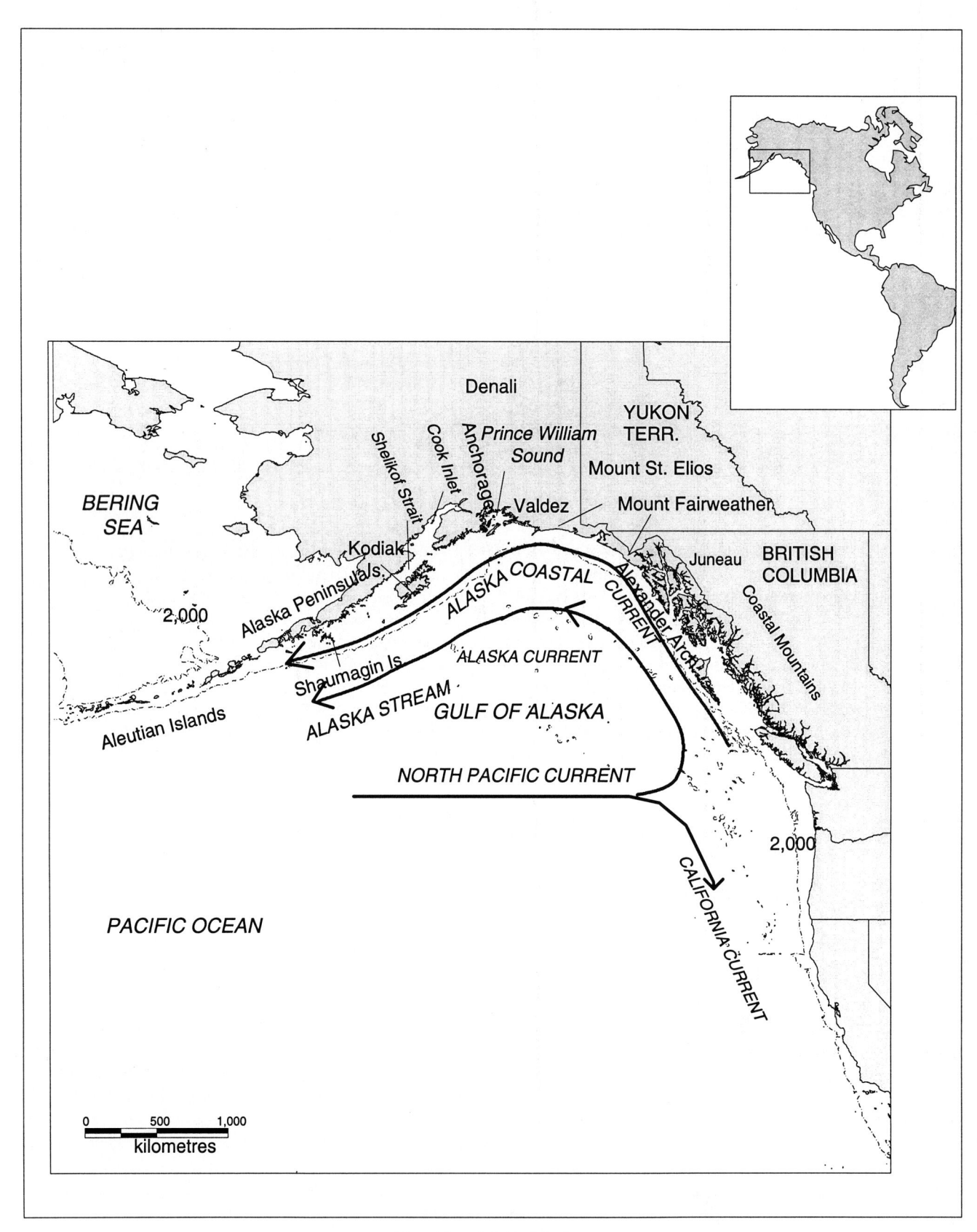

Fig 1. Map of Gulf of Alaska.

THE DEFINED REGION

The Gulf of Alaska (GOA) is defined as that area north of 52°N, between about 127°30'W on the east where it meets the British Columbia coast, and 176°W on the west including most of the Aleutian Islands. The GOA includes very rugged and extreme coasts with Cook Inlet in the north, Prince William Sound (PWS) just east of that, and Alexander Archipelago further southeast (Fig. 1). The western GOA includes the Alaska Peninsula coastline and the Shumagin and Kodiak Island groups. The arc formed between the extremes extends about 3600 kilometres. The GOA's continental shelf is estimated to be 336,000 km².

GEOGRAPHIC SETTING

The eastern and northern terrestrial margins of the GOA sustain very high natural erosion rates. The region is still undergoing de-glaciation from the last ice age, and contains the largest ice fields and glaciers in North America. The region also experiences rapid uplift from isostatic rebound following loss of ice field mass, and from continental uplift resulting from subduction of the northward-moving north Pacific plate. These processes have produced high mountain ranges immediately adjacent to the coast, with numerous short but high-volume and often high-velocity coastal streams and rivers that transport large sediment burdens to the GOA, especially where coastal glaciers are involved.

The mountains, many exceeding 2000 m, peak with Denali. As storms track across the GOA, the moisture-laden clouds are forced into this ring of mountains, which can produce localized annual rainfall exceeding 8 m per year or many times that amount in snow at the upper elevations. The GOA storms have no place to go as they pound against the mountains in what has been referred to as the graveyard of the North Pacific storms.

THE MAJOR SHALLOW WATER MARINE AND COASTAL HABITATS

The GOA shelf supports a diverse ecosystem that includes several commercially important fisheries such as crab, shrimp, pollock, salmon, and halibut. In aggregate these stocks imply that the GOA is amongst the world's largest fisheries with annual catches exceeding 300 g/1000 m³ (Brodeur and Ware, 1992).

The relative dominance of the commercially important fish species changed in the mid-1970s; crab and shrimp declined while salmon and groundfish populations increased (see box). These population shifts coincided with the beginning of a decadal North Pacific change in the atmosphere and ocean. From the human perspective these alterations required the commercial fishing industry to invest substantially in infrastructure adjustments so as to remain economically viable. Subsequent changes in this ecosystem followed in the 1980s with substantial declines in populations of sea lions and puffins. Dramatic though this "regime shift" was, Parker et al. (1995) show evidence that the abundance of halibut and other commercially important species varies on decadal time scales in conjunction with northern North Pacific Ocean temperatures. These correlations and the regime shift suggest that the GOA ecosystem is sensitive to climate variations on time scales ranging from the interannual to the interdecadal; however, the specific mechanisms linking climate to ecosystem alterations are unknown. Elucidation of these mechanisms requires an understanding of the seasonal cycle of the principal physical, chemical and biological variables.

PHYSICAL OCEANOGRAPHY

The alongshore flow on the shelf and slope of the GOA is generally cyclonic (Reed and Schumacher, 1986). Flow over the continental slope consists of the Alaska Current, a relatively broad, diffuse flow in the north and northeast GOA, and the Alaskan Stream, a swift, narrow, western boundary current in the west and northwest GOA (Fig. 1). Together these currents comprise the poleward limb of the North Pacific Ocean's subarctic gyre and they provide the oceanic connection between the GOA shelf and the Pacific Ocean. Reed and Schumacher (1986) suggest that flow in the Alaskan Stream is relatively constant year round. However, Musgrave et al. (1992) and Okkonen (1992) show that sometimes the Alaskan Stream captures large eddies or forms prominent meanders and Royer (1982) suggests that the seasonal signal in baroclinic transport is less than 10% of the mean flow. In the northeast GOA, the "Sitka Eddy" occasionally forms and slowly propagates westward across the GOA. To the extent that these low-frequency features impinge on the shelfbreak they could contribute to the shelf circulation and exchange of water masses.

The most striking feature of the shelf circulation is the Alaska Coastal Current (Fig. 1), a swift (0.2–1.8 m s^{-1}), coastally constrained flow, typically found within 35 km of the coast. This current persists throughout the year and circumscribes the GOA shelf for at least some 2500 km from where it originates on the northern British Columbia shelf (or possibly even the Columbia River depending on the season) to where it enters the Bering Sea in the western GOA. In contrast to the coastal current, the shelf flow between the offshore edge of the coastal current and the shelfbreak is weaker and more variable. The source of this variability is uncertain, but potential mechanisms include separation of the coastal current as it flows around coastal promontories; baroclinic instability of the coastal jet or meandering of the Alaska Current along the shelfbreak.

The dynamics of the basin and the shelf are closely coupled to the Aleutian Low pressure system. Storm systems propagate eastward into the GOA and are then

Trophic Shift in Marine Communities in the Gulf of Alaska

Recently there has been information presented that the Gulf of Alaska (GOA) ecosystem has undergone some abrupt and significant environmentally induced changes starting in the mid to late 1970s (Piatt and Anderson, 1996; Anderson et al., 1997; Anderson and Piatt, 1999). Most of the biological changes in the NE Pacific and the GOA follow from the lowest trophic levels through the benthic and pelagic fish stocks to sea birds and marine mammals. The extent and degree of these changes are now well documented and provide an important perspective in determining future strategies for management of the marine ecosystem in the GOA. Environmental change has been indicated as the greatest influence on the system, and it is likely that changes in the environment will continue to occur in the future as they have in the past. Probably one of the best long-term data series that is useful in describing environmentally induced changes in the GOA is the data collected from small-mesh trawl surveys conducted nearly continuously from 1953 to the present. These data are used here to demonstrate the great degree of control that the environment has on the ecological structure of the GOA.

GOA OCEAN/CLIMATE VARIABILITY

Perhaps the broadest measure of contemporary climate variability in the GOA is the North Pacific Pressure Index (NPPI) which represents a wide-spread and relatively low frequency climate signal that has significant impact on the GOA ocean circulation and temperature (Fig. 1a). The NPPI models the changing location and intensity of the Aleutian low. Persistent strong blocking high pressure ridges during winter over eastern Asia (Siberian high) displace the Aleutian low to the south and intensify it (Wilson and Overland, 1986). This leads to cold northwest winds in the western and central GOA which favour upwelling in nearshore areas along with rapid cooling of the water column. When the storm track follows a more northerly tack across the Pacific, the nearshore region of the GOA is more subject to vertical mixing and downwelling. These two semi-stable states generally typify the two climate regimes of the GOA, the first representing the cold regime and the latter the warm regime. Analysing dendroclimatic data, back to about 1500, it appears these two climate regimes have oscillated at an average frequency of 15 years with the longest warm period lasting 34 years and the longest cold period lasting 26 years (Ingraham et al., 1998). The current warm period should have reverted to the cold dominated regime in 1995 according to analysis by Ingraham et al. (1998). Wiles et al. (1998), found evidence of the ENSO effect in GOA coastal tree-ring chronologies as well as evidence for a bidecadal oscillation (19 years) in the climate tendency of the GOA region. It thus appears that the climate varies in the GOA in a somewhat predictable fashion with either the cold or warm regime lasting from one to three decades before reversal. Changes in the climatology of the NE Pacific and the GOA lead to changes in ocean conditions that have a direct effect on the marine ecosystem. Lower trophic level animals demonstrate a particularly rapid response to environmentally induced changes.

POPULATION CHANGES OF SHRIMP AND FORAGE FISH

Many shrimps and forage fish (mostly in the family Osmeridae) can be considered trophospecies; they share similar prey and predators. Shrimp and forage fish react very quickly to environmentally induced changes, owing to their low relative trophic level. These forage species groups declined from relatively high levels in abundance in the period 1970–1984 to uniformly low abundance across a broad region of the GOA after 1985 (Anderson et al., 1997). Capelin (*Mallotus villosus*) composed 84% of the osmerid biomass prior to 1981; recently the dominant species has shifted to Eulachon (*Thaleichthys pacificus*). A change in the composition and timing of zooplankton abundance that was observed in the GOA after the late 1970s (Mackas et al., 1998; Brodeur and Ware, 1992) could have adversely affected planktivores like shrimp and forage fish. These changes highlight some of the major changes in abundance and population structure that have recently been observed in the GOA after the climate change that occurred in 1977 (Fig. 1b).

MAJOR INCREASES IN COD AND GROUNDFISH

In stark contrast to the decline of crustacean and forage fish populations have been the increase of Pacific cod, walleye pollock, and several species of pleuronectid groundfish. Yearly CPUE of arrowtooth flounder (*Atheresthes stomias*) exhibited a pattern opposite to that of the pink shrimp, increasing from a low of 0.02 kg/km in 1971 to a high of 28.8 kg/km in 1992. Arrowtooth flounder were caught in inshore

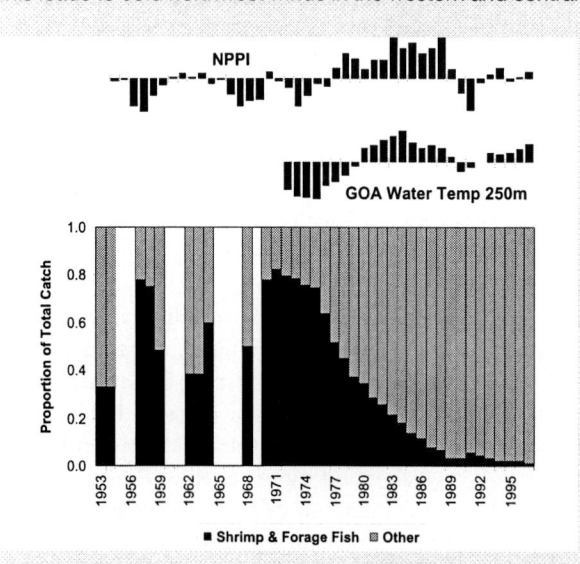

Fig. 1a. Normalised anomalies of North Pacific pressure Index (NPPI) and Gulf of Alaska water temperature at 250 m compared with long term composition of small-mesh trawl survey catch composition (3-year running average) in the Gulf of Alaska from 1953 to 1997 (from Anderson and Piatt, 1999).

bays near Kodiak Island during 1985–1995 where they were not present before. Mean CPUE of arrowtooth flounder increased in virtually every bay sampled. Areas of greatest increase in mean CPUE (up to 410.7 kg/km per grid cell) were east and north of Kodiak Island, and east of Afognak Island at Portlock Bank. Lesser increases were seen in bays west and south of Kodiak Island, and bays east of the Alaska Peninsula. Yearly CPUE of walleye pollock increased from a low of 17.5 kg/km in 1975 to a high of 284.8 kg/km in 1991. The greatest increase in mean CPUE (up to 410.7 kg/km) occurred in bays southeast of Kodiak Island and Portlock Bank. Lesser increases occurred in bays east of the Alaska Peninsula, south of Kodiak Island and southeast of Afognak Island.

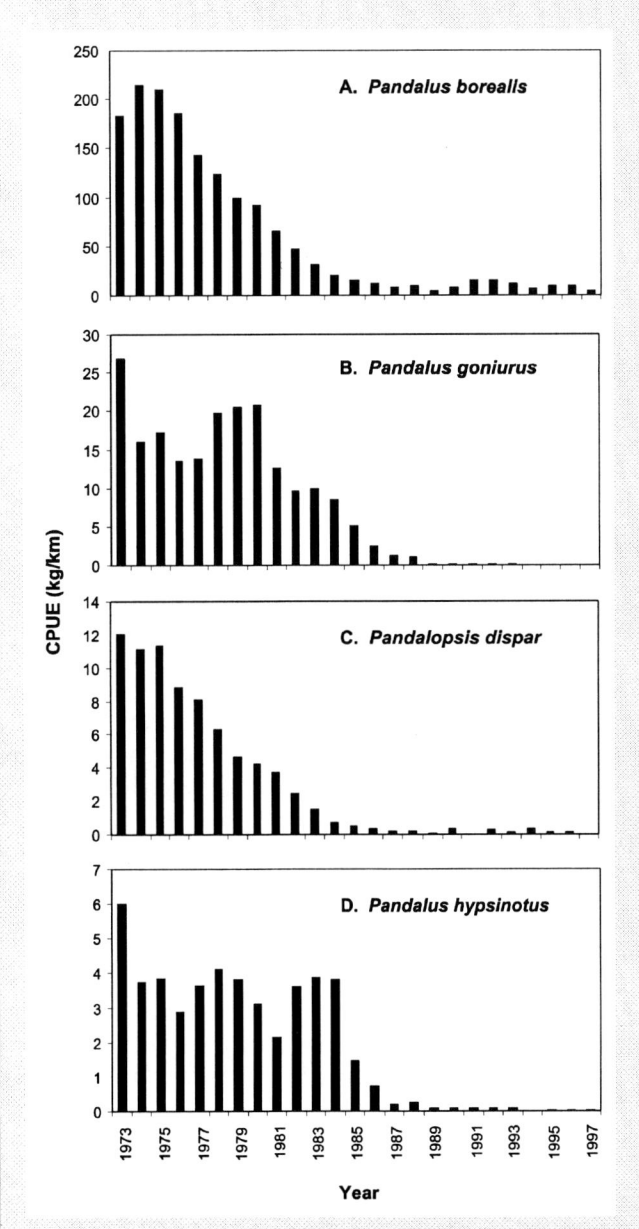

Fig. 1b. Relative abundance of (a) *Pandalus borealis*, (b) *P. goniurus*, (c) *Pandalopsis dispar* and (d) *Pandalus hypsinotus* in the Gulf of Alaska 1973–1997 (data smoothed 3-year running average).

CONCLUSION

As indicated by abrupt changes in NPPI, and sea temperature at depth (250 m) anomalies (Fig. 1a), average climate in the GOA shifted from cool to warm around 1977 (Mantua et al., 1997). Standardized trawl data collected from 1953 to 1997 reveal the ecological impact of this climate regime shift. Most of the increasing taxa are benthic or demersal in the adult stage (cod, pollock, flatfish, starfish), some are pelagic (jellyfish, cephalopods), and while most decreasing taxa are generally pelagic as adults, some are benthic (crab, sculpin). Although there were historically large fisheries for pandalid shrimp in many areas of the GOA (Orensanz et al., 1998), they also declined in areas that were seldom, if ever, fished (Anderson and Gaffney, 1977) and populations continued to decline even after fisheries were eliminated. Capelin have never been targeted by commercial fisheries in the GOA. Simultaneous declines were also noted in ecologically disparate taxa such as crabs, sculpins, herring, pricklebacks, eelpouts, snailfish, greenling, sablefish and Atka mackerel. The geographic and temporal coherence of the collapse of so many taxa argues for a large-scale common cause such as climate change, a conclusion also reached by Orensanz et al. (1998). The strength of association between shrimp catches and water temperature supports the hypothesis that the GOA ecosystem is regulated to a large degree by "bottom-up" processes (Francis et al., 1998, McGowan et al., 1998).

Decadal fluctuations in biomass and composition of fish communities apparently had a direct impact on sea birds and marine mammals that subsist on forage species and juvenile age groups of groundfish (Francis et al., 1998). During the cold regime prior to 1977, seabirds and marine mammals relied on fatty forage species such as capelin. As forage biomass declined in the early 1980s, forage species disappeared from bird and mammal diets and were replaced largely by juvenile pollock (Piatt and Anderson, 1996; Merrick et al., 1997). Declines in production and abundance of several sea bird and marine mammal populations in the GOA followed (Piatt and Anderson, 1996). Because juvenile pollock have low energy densities compared to fatty forage species such as capelin (Payne et al., 1999), some predator declines may be attributable to changes in diet composition. Perhaps more importantly, the total biomass of all forage taxa, including juvenile pollock, may now be limiting owing to the enormous food demands of adult groundfish. (Livingston, 1993; Yang, 1993; Hollowed et al., 1999). Marine mammal populations are, in general, thought to be food-limited (Estes, 1979). Population changes of marine mammals in the GOA most likely reflect the changed composition and abundance of preferred food sources. Climate-induced changes in the entire ecosystem from the lowest to the highest levels occur at different temporal scales, abrupt and rapid population change occurs at the low trophic levels while higher trophic levels are constrained by a mixture of "top-down" and "bottom-up" processes which take more time to manifest themselves.

REFERENCES

Anderson, P.J. and Gaffney, F. (1977) Shrimp of the Gulf of Alaska. *Alaska Seas and Coasts* 5(3), 1–3.

Anderson, P.J. and Piatt, J.F. (1999) Community reorganization in the Gulf of Alaska following ocean climate regime shift. *Marine Ecology Progress Series*, in press.

Anderson, P.J., Blackburn, J.E. and Johnson, B.A. (1997) Declines of forage species in the Gulf of Alaska, 1972–1995, as an indicator of regime shift. In *Forage Fishes in Marine Ecosystems*. Proceedings of the International Symposium on the Role of Forage Fishes in Marine Ecosystems. Alaska Sea Grant College Program Report No. 97-01 pp. 531–544.

Brodeur, R.D. and Ware, D.M. (1992) Long-term variability in zooplankton biomass in the subarctic Pacific Ocean. *Fisheries Oceanogr.* 1, 32–38.

Francis, R.C. and Hare, S.R. (1994) Decadal-scale regime shifts in the large marine ecosystems of the North-east Pacific: a case for historical science. *Fisheries Oceanogr.* 3 (4), 279–291.

Hollowed, A.B., Ianelli, J.N. and Livingston, P. (1998) Including predation mortality in stock assessments: A case study for Gulf of Alaska walleye pollock. *ICES Mar. Sci. Symp.*, In press.

Ingraham, W.J., Ebbesmeyer, C.C. and Hinrichsen, R.A. (1998) Imminent climate and circulation shift in the northeast Pacific Ocean could have a major impact on marine resources. *EOS, Transactions Am. Geo. Union* 79 (16), 199–201.

Livingston, P.A. (1993) The importance of predation by groundfish, marine mammals and birds on walleye pollock *Theragra chalcogramma* and Pacific herring *Clupea pallasi* in the eastern Bering Sea. *Marine Ecology Progress Series* 102, 205–215.

Mackas, D.L., Goldblatt, R. and Lewis, A.G. (1998) Interdecadal variation in developmental timing of *Neocalanus plumchrus* populations at Ocean Station P in the subarctic North Pacific. *Can. J. Fish. Aquat. Sci.* 55, 1878–1893.

Mantua, N.J., Hare, S.R., Zhang, Y., Wallace, J.M. and Francis, R.C. (1997) A Pacific interdecadal climate oscillation with impacts on salmon production. *Bulletin of the American Meteorological Society* 78, 1069–1079.

McGowan, J.A., Cayan, D.R. and Dorman, L.M. (1998) Climate-ocean variability and ecosystem response in the Northeast Pacific. *Science* 281, 210–217.

Merrick, R.L., Loughlin, T.R. and Calkins, D.G. (1987) Decline in the abundance of the northern sea lion, *Eumetopia jubatus*, in Alaska, 1956–1986. *U.S. Fish. Bulletin* 85, 351–365.

Orensanz, J.M., Armstrong, J., Armstrong, D. and Hilborn, R. (1998) Crustacean resources are vulnerable to serial depletion—the multifaceted decline of crab and shrimp fisheries in the greater Gulf of Alaska. *Rev. Fish Biol. Fisheries* 8, 117–176.

Payne, S.A., Johnson, B.A. and Otto, R.S. (1999). Proximate composition of some north-eastern Pacific forage fish species. *Fish. Oceanogr.* 8 (3), 159–177.

Piatt, J.F. and Anderson, P. (1996) Response to Common Murres to the *Exxon Valdez* oil spill and long-term changes in the Gulf of Alaska ecosystem. In *Exxon Valdez Oil Spill Symposium Proceedings*, eds. S.D. Rice, R.B. Spies, D.A. Wolfe and B.A. Wright. American Fisheries Symposium No. 18, pp. 720–737.

Wiles, G.C., D'Arrigo, R.D. and Jacoby, G.C. (1998) Gulf of Alaska atmosphere–ocean variability over recent centuries inferred from coastal tree-ring records. *Climate Change* 38, 289–306.

Wilson, J.G. and Overland, J.E. (1986) Meteorology. In *The Gulf of Alaska, Physical Environment and Biological Resources*, eds. D.W. Hood and S.T. Zimmerman. MMS/NOAA, Alaska Office, Anchorage, OCS Study MMS 86-0095, pp. 31–54.

Yang, M. (1993) Food habits of the commercially important groundfishes in the Gulf of Alaska in 1990. U.S. Dep. of Commerce, NOAA Tech. Memoran. MNMFS-AFSC-22.

---

blocked by the mountain ranges of Alaska and British Columbia. Thus the regional winds are strong and cyclonic and the precipitation rates are very high. The positive wind-stress curl forces cyclonic circulation in the deep GOA while on the shelf these winds impel an onshore surface Ekman drift and establish a cross-shore pressure gradient that forces the Alaska Coastal Current. The high rates of precipitation cause an enormous freshwater flux (~20% larger than the average Mississippi River discharge) that feeds the shelf as a "coastal line source" extending from Southeast Alaska to Kodiak Island (Royer, 1982). The seasonal variability in winds and freshwater discharge is large. The mean monthly "upwelling index" at locations on the GOA shelf is negative in most months indicating the prevalence of coastal convergence (e.g., this index is a measure of the strength of cyclonic wind stress in the GOA). Cyclonic winds are strongest from November through March and feeble or even weakly anticyclonic in summer when the Aleutian Low is displaced by the North Pacific High (Wilson and Overland, 1986). The seasonal runoff cycle exhibits slightly different phasing from the winds; it is maximum in early fall, decreases rapidly through winter when precipitation is stored as snow, and attains a secondary maximum in spring due to snowmelt.

The shelf hydrography and circulation vary seasonally and are linked to the annual cycles of wind and freshwater discharge. The cross-shore salinity structure mimics density on the GOA shelf in April and September. In April, the stratification and the offshore front (defined here to be the surface intersection of the 32.0 isohaline) are relatively weak. By contrast in September, a 25 km wide wedge of strongly stratified water lies adjacent to the coast and is bounded on the offshore side by a prominent front. Royer et al. (1979) showed that surface drifters released on the shelf but shoreward of the front drifted onshore in accordance with Ekman dynamics. Upon encountering the front the drifters moved in the alongfront (e.g. westward) direction consistent with the geostrophic tendency implied by the cross-shore density distributions. Royer et al. (1979) hypothesized that ageostrophic offshore spreading of the dilute surface layer occurred on the inshore side of the front. In their analysis of currents measured inshore of the front, Johnson et al. (1988) found that this is indeed the case and that surface offshore flow was positively, and significantly, correlated with discharge. These studies imply that near-surface waters converge from either side of the front. This pattern of cross-shelf circulation would tend to accumulate plankton which might then attract foraging fish. Moreover, the front and the region inshore of it might be an area of enhanced productivity because entrainment and/or frontal instability could resupply the surface layer with nutrients from depth. The alongshore transport appears to be important in advecting zooplankton to important juvenile fish foraging areas.

Near-bottom salinities are higher in fall than spring. Xiong and Royer (1984) showed that on average maximum bottom salinities occur in fall and are nearly coincident with minimum surface salinities and maximum inshore stratification. Although the surface waters are diluted by coastal discharge (which peaks in fall), the source of the high salinity water is the onshore intrusion of slope water in response to the seasonal relaxation (or reversal) in downwelling.

Royer's (1996) analysis of monthly anomalies from the GOA shelf show very low-frequency (interdecadal) variations in bottom water salinity that imply interannual variability in the onshore flux of slope water and/or differences in slope water properties. These differences likely result in differences in the onshore flux of nutrients to the GOA shelf.

PRIMARY PRODUCTIVITY AND NUTRIENT CYCLES

There are few primary production measurements from this region and those that were made were done so at widely varying locations and times. However, satellite images of phytoplankton indicate the GOA to generate a relatively high level of primary production along the continental shelf (Fig. 2). While both Sambrotto and Lorenzen (1986) and Parsons (1986) conclude that the largest production rates occur on the shelf, nothing can be said about interannual variability. An additional limitation in understanding production here is the lack of nutrient data, particularly from the shelf. We do know that the shelf's nutrient source must be from the deep ocean as the coastal runoff nutrient concentrations are very low. These low concentrations are not unexpected given the steep, mountainous coastline and the extensive snow fields. Conceivably the shelf euphotic zone, especially in inshore waters, becomes nutrient-depleted but we emphasize that this is not known.

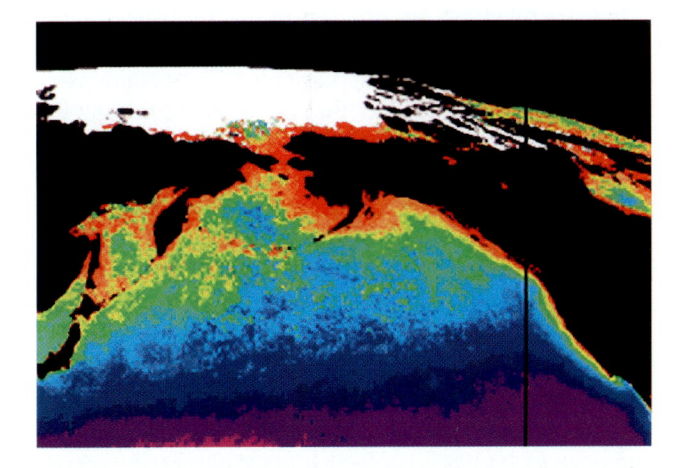

Fig. 2. Free-floating photosynthetic organisms (phytoplankton). This figure shows the Gulf of Alaska waters to be highly productive. Upon closer inspection the waters in Lower Cook Inlet are exceptionally productive. Image from NASA.

Although little is known about surface nutrient concentrations, there are suggestions of large year-to-year differences in subsurface nutrient concentrations. Incze and Ainair (1996) show large interannual differences in nutrient concentrations at depths >150 m along one section in Shelikof Strait (in the western GOA) that they occupied each spring between 1985 and 1989. Because of the unique bathymetry of this area it is unclear whether these differences apply to other GOA shelf regions. We speculate that the interannual salinity variations shown by Royer (1996) imply variability in deep water nutrient concentrations. These nutrient data are the only synoptic deep ocean and shelf nutrient data available for the northern GOA. The salinity–NO_3 relationship is correlated using data from between 125 and 450 m depth at stations within the Alaskan Stream and on the western shelf. This depth interval covers the range of bottom water salinities observed by Royer (1996) and Xiong and Royer (1984). The correlation appears to be good and we note that a change in salinity from 32.0 to 33.0 involves nearly a doubling in the NO_3 concentrations.

Zooplankton

Zooplankton are a critical link in the transfer of energy from primary producers to apex predators. Therefore, any process influencing the abundance and distribution of zooplankton can ultimately have an impact on fisheries. Zooplankton are therefore a critical component of attempts to understand the relationship of long-term climate variations to fish production.

The zooplankton community on the shelf of the GOA is dominated by a combination of oceanic and neritic herbivorous and omnivorous copepod stocks (Cooney, 1986a,b; Incze et al., 1996). The major oceanic species include *Neocalanus plumchrus, N. flemingeri, N. cristatus, Eucalanus bungii* and *Metridia pacifica*. Neritic taxa are dominated by *Pseudocalanus* spp. and *Calanus marshallae*, with lesser amounts of *Acartia* spp., *Centropages abdominalis* and *Calanus pacificus*. In addition to copepods, a number of micronektonic species contribute substantially to the overall density of forage for fish on the GOA shelf. The euphausiid species include primarily *Thysanoessa inermis, T. spinifera* and *Euphausia pacifica*, with lower densities of *Thysanoessa raschii, T. longipes, T. inspinata, Tessarabrachion oculatum* and *Euphausia pacifica*. Amphipods include *Cyphocaris challengeri, Parathemisto pacifica*, and *Primno macropa*. Oceanographic conditions affecting the transport and production of these taxa influence their absolute and relative densities and distribution over the shelf, and thus their availability to fish predators on the shelf.

During spring and summer, 25–78% of the copepod biomass over the shelf is dominated by the oceanic species complex. The distribution of oceanic relative to neritic copepods is determined to a large extent by cross-shelf transport (Cooney, 1986a) and water mass type (Incze et al., 1996; Napp et al., 1996). Although most of the copepod biomass in lower Shelikof Strait occurred consistently in the Alaska Coastal Current from 1986–1989; there was a fourfold (3–12 g C m^{-2}) interannual variation in maximum biomass (Incze et al., 1996; Napp et al., 1996). Zooplankton biomass on the shelf outside of Prince William Sound in May 1996 varied by up to an order of magnitude, with maximum values occurring in the shelf water offshore of the Alaska Coastal Current.

In addition to late copepodid stages of the major copepod taxa, the early nauplii stages are the primary forage for the first-feeding larval stages of a variety of fish. Based on water temperature, copepod development rates and flow rates of the Alaska Coastal Current, copepods producing the major cohort of nauplii stage larvae available to first-feeding pollock larvae in Shelikof Strait, originated during February–March on the shelf off Prince William Sound and east of GAK1 (GAK1 is a long-term oceanographic monitoring station just outside Resurrection Bay, Alaska). The nauplii consumed by first-feeding fish larvae are produced primarily by the neritic zooplankton community. Therefore, pre-bloom conditions on the north central GOA shelf might crucially influence survival of larval fish further downstream (west and south) near Kodiak Island.

No data are available on interannual differences in zooplankton biomass for the north central GOA shelf. However, a multi-year data set of zooplankton settled volumes measured during April and May near Ester Island, in the southern end of Prince William Sound, is available. In this part of Prince William Sound the zooplankton community is influenced primarily by advection from the GOA shelf. Cooney (pers. comm.) found a significant positive correlation between the logarithm of the average settled (zooplankton) volume for April and May and the average of the upwelling index on the northern GOA. The mechanistic link between these two variables is not obvious, or necessarily direct, as there are a number of possible explanations. In April and May oceanic species of the genus *Neocalanus* dominate zooplankton biomass suggesting a direct link that anomalously weak springtime downwelling enhances subsurface onshore transport of oceanic copepods from the shelfbreak. Alternatively, these same conditions could have elevated primary production (e.g., through onshore nutrient advection) during the spring months, thereby providing a more continuous and abundant zooplankton food supply. An anomalously positive April–May upwelling index implies reduced wind stress, precipitation rates, cloud cover and (perhaps) higher air temperatures. All these variables influence upper ocean stratification through wind mixing, surface heat flux and coastal discharge (note that cloud cover and air temperatures affect the springtime melt of snow accumulated through winter in the coastal mountain ranges). Stratification influences the vertical distribution of plant cells and, along with cloud cover (light availability), influences primary production rates. These physical variables through their influence on phytoplankton food quality and/or abundance would affect zooplankton.

If cross-shelf advection is a major source of zooplankton biomass on the shelf, then conditions that enhance zooplankton biomass at the shelfbreak should also enhance shelf zooplankton densities when favourable onshore transport conditions occur. Comparisons of zooplankton densities in the GOA between 1956–1962 and 1980–1989 show a doubling in average biomass around the GOA's perimeter since the early 1960s. The reason for this increase is uncertain, however, suggested hypotheses include increased primary productivity due to an increased winter wind stress and elevated summer winds producing a more northward displacement of the subarctic current into the GOA during the 1980s. The positive correlation between zooplankton densities and surface salinities suggests stronger vertical mixing (Brodeur and Ware, 1992) that led to enhanced new production and better feeding conditions for herbivorous zooplankton. Primary production rates were apparently three to four times higher in the GOA in 1987–1988 than earlier measurements indicated (Welschmeyer et al., 1993). Although the latter attribute the differences to methodology, the zooplankton and wind data cited above suggest that there might have been real decadal variation in annual production rates. Salmon probably benefited from these elevated zooplankton densities because salmon production nearly doubled between the 1950s and 1980s (Rogers, 1986). The major environmental shift suggested by the collapse of the crustacean fishery and its replacement by a ground fish fishery in the late 1970s and early 1980s could also be a consequence of enhanced zooplankton biomass because the early life history stages of demersal fishes feed on zooplankton.

Offshore Systems

The pelagic GOA is relatively productive biologically. The counterclockwise circulation at the margins produces continuous upwelling in the central GOA from Ekman pumping (Broecker, 1991), resulting in the continuous introduction of deep-water nutrients. This sustains year-round primary productivity that is modulated but still significant during winter, at an annual production rate as high as 170 g C m^{-2} y^{-1} (Welschmeyer et al., 1993). This comparatively high production concurrent with high nutrients (especially nitrate) and characteristically absent phytoplankton blooms (Parsons and Lalli, 1988) has been explained in part by micronutrient limitation by iron (Martin et al., 1991). In this scenario, low iron availability favours small phytoplankton cell sizes grazed continuously by both micro- and macrozooplankton (Miller, 1993). The doubling of zooplankton biomass, approximately doubled from the late 1950s to the 1980s, has led to coastal invasion by pelagic species, quite possibly in response to the climatically driven shift in the production regime during the mid-1970s (Brodeur and Ware, 1992).

HUMAN POPULATIONS AFFECTING THE AREA

About 500,000 people live in the terrestrial catchment basin draining into the GOA. More than half live in or near Anchorage in upper Cook Inlet, and nearly all of the remainder live in one of a dozen much smaller towns located along some 3000 km of coastline from Prince Rupert in the east to Kodiak in the west. Recent population growth has been slow, with a mean annual growth rate of less than 1% over the last 15 years. Previous population growth was more rapid, increasing from about 50,000 to 320,000 during the interval 1940 to 1980. The great majority of the significant human impacts in the region have occurred since 1940.

By far the most important primary industries in the region are related to natural resource extraction, followed by tourism. Manufacturing is negligible. Extraction-related industries include commercial fishing, logging, oil production and trans-shipment, and mining. Commercial fishing has been important since the late 19th century. Large-scale commercial logging of the coastal rainforest began in the 1950s and reached peak production in the 1980s, but has declined somewhat following closure of the two pulp mills in the region. Small-scale oil production occurred at Katalla during the first two decades of this century. Significant oil production began in Cook Inlet in the 1960s, and oil trans-shipments increased dramatically with the opening of the trans-Alaska pipeline terminal in 1977 at Valdez. Similarly, only two large-scale mining complexes have operated in the region: at Juneau (gold), and at McCarthy (copper), with ore from the latter exported as a concentrate. Another ore shipping terminal is located at Skagway for lead–zinc mines (currently closed) in the Yukon. Apart from these, other coastal mines have been scattered, small-scale operations, few of which currently operate.

BASELINE STUDIES—POLLUTANTS

The GOA is a candidate region for monitoring planet-wide dispersion of pollutants. Upwelling of deep oceanic water in the central GOA by Ekman pumping tends to divert surface waters away from the region, thereby exporting pollutants entrained in surface waters. The relatively small surface area of the terrestrial catchment basin provides few compartments for accumulation of pollutants before they are flushed into the Gulf by the heavy rainfall characteristic of the region, so terrestrial residence times are usually brief. The high erosion rates result in rapid burial of pollutants exported to coastal benthic sediments. Finally, the relatively sparse population concentrated in a few centres dispersed across a very long coastline, and the general absence of manufacturing, make local inputs negligible beyond the immediate vicinity of urbanization. Hence, pollution indicators that appear consistently at sampling sites remote from urbanized areas are *prima facie* evidence of long-distance pollutant transport and dispersion over a very large area.

Four baseline studies have been done to document pollutant levels or monitor trends. The earliest began in 1977 in Prince William Sound to document petroleum hydrocarbons in intertidal sediments and mussels prior to anticipated oil pollution along the marine oil-transport corridor through the PWS. This multi-seasonal study stopped in 1980 (Karinen et al., 1993), but resumed in 1989 just prior to beaching of oil spilled by the T/V *Exxon Valdez* (see box). The U.S. National Ocean Service, National Status and Trends (NS&T) program began monitoring intertidal sediments and mussels, and some benthic fishes in the region beginning in 1984 (in O'Connor and Pearce, 1999). Monitored analytes include polycyclic aromatic hydrocarbons (PAH), polychlorinated biphenyls (PCBs), DDT isomers, other chlorinated pesticides (listed in Brown et al., 1999, and Wade et al., 1999), butyltins and metals. Tributyltin concentrations were documented in mussels in 1987 (Short and Sharp, 1989). Finally, Regional Citizen's Advisory Councils (RCACs), which are federally-mandated but industry-funded semi-public oversight organizations, have commenced PAH monitoring in 1993 of marine sediments and mussels in Prince William Sound (Payne et al., 1998), and in Cook Inlet (Lees et al., 1999).

Results from these baseline studies indicate that the GOA is among the least polluted marine ecosystems on Earth. At sampling stations more than 10 to 20 km away from the outskirts of towns, pollutants are either near analytical detection limits, or else are geographically widespread within a low and relatively narrow concentration range. Pollutants in the first category include butyltins (Short and Sharp, 1989) and some of the less persistent, more recently introduced chlorinated pesticides (NS&T datasets). Pollutants in the second category include metals, PAHs and the persistent organic pollutants (POPs) such as PCBs, DDT-related compounds, and a few other chlorinated pesticides (Brown et al., 1999; Wade et al., 1999; NS&T datasets). PAHs in mussels are often not detected, and when detected are usually less than 20 ng/g dry weight at remote sites absent a clearly identifiable areawide pollution incident (such as the *Exxon Valdez* oil spill). A regional background of PAHs has been consistently detected in benthic sediments west of Yakutat Bay to Shelikof Strait, and has been attributed to oil seeps (Bence et al., 1996) but is more likely coal particles eroded from the extensive coastal deposits in the region (Short et al., 1999).

The consistent appearance of relatively narrow concentration ranges of POPs in sediments and mussels throughout the GOA coastline strongly suggests widespread dispersion in the upper water column. Concentrations of individual POPs in these matrixes are usually less than 1–2 ng/g. Benthic sediment core samples from the western GOA indicate accumulations of PCBs and DDTs beginning in the 1930s, and the other chlorinated organics one to two decades later (Iwata et al., 1994). These pollutants concentrate in higher trophic levels of the GOA food chain including the largely zooplanktivorous sockeye salmon (Ewald et

The *Exxon Valdez* Oil Spill

The *Exxon Valdez* oil spill (EVOS) is easily the single most devastating pollution event in the GOA to date. On March 24, 1989 the T/V *Exxon Valdez* struck Bligh Reef in Prince William Sound (PWS). At least 40,800 m^3 of crude oil, and possibly as much as three times that amount* spilled into PWS on the northern margin of the GOA. The spilled oil spread as a surface slick during the following three days, and was then mixed thoroughly with the seasurface layer by a high-energy storm over the next four days. Winds and currents carried the oil in a southwesterly direction over the next eight weeks, oiling shorelines intermittently along some 1750 km of coast extending up to 750 km from Bligh Reef. By the end of April 1989, about 40% of the spilled oil accumulated on beaches, 33% had dispersed into the water column, 20% evaporated, and the remainder was recovered, according to estimates based on the minimum spill volume (Wolfe et al., 1994). Ultimately, about 14% of the oil spilled was recovered.

The spill occurred at the worst time possible within the annual biological production cycle. The spring phytoplankton bloom typically begins in late March, with primary production near 300 g C/m^2/y over an extensive area within and outside PWS. This productivity supports a diverse food web, with a suite of apex predators attracted to the region to forage in support of further migrations or to reproduce. Avian and marine mammal predators that congregate in the area and associate with the sea surface were thus most affected. Even ten years after the EVOS there is continued evidence of cytochrome P450 elevation in intertidally dependent apex predators within the spill zone. This is strong evidence to suggest continued long-term exposure to *Exxon Valdez* oil via the intertidal food web to higher trophic levels, including sea otters and some sea ducks.

The effects of the EVOS on fish were less severe but at least as protracted as on marine mammals and birds. The most immediate effect on fish was through ingestion of oiled prey. The consequent poor growth especially of juveniles led to losses estimated at nearly 20% of the pink salmon year class (Geiger et al., 1996), which fortunately was largely mitigated by the hatcheries that had attained full-scale salmon production the same year. Nearly half the intertidal spawning habitat of Pacific herring was exposed to oil in PWS (Brown et al., 1996), which led to lower survival rates of larvae (Kocan et al., 1996). Oil that persisted in the intertidal over the next few years also caused longer term impacts to fish utilizing this habitat for reproduction. Both field and laboratory studies motivated by the EVOS found that fish embryos exposed to one part per billion concentrations of polycyclic aromatic hydrocarbons (PAH) exhibit a manifold of delayed effects that appear randomly among exposed populations throughout the life span duration (Carls et al., 1999; Heintz et al., 1999; Bue et al., 1998). Although these effects were not anticipated, population level impacts were small to negligible in the field, because of the limited persistence of the oil, the small proportion of available habitat affected and replenishment of affected populations by immigration.

Despite unprecedented efforts, attempts to clean the oiled beaches had very limited success. High-energy winter storms appeared to be at least as effective at re-introducing residual beach oil back into the water column, where it dispersed offshore into the Gulf of Alaska. By the end of 1992, beach cleaning efforts were terminated, and removal of the oil remaining was left to natural processes. Relatively unweathered oil was still present in 1998 in the interstices of cobble to boulder patches or beneath mussel beds on beaches that were heavily oiled initially (Irvine et al., 1999; Brodersen et al., 1999; Hayes and Michel, 1999; Babcock et al., 1996; Boehm et al., 1996). Oil in these remaining patches is substantially protected from further weathering or physical dispersion, and so is likely to persist on times scales of decades until dispersed by very high-energy storm events.

REFERENCES

Babcock, M.M., Irvine, G.V., Harris, P.M., Cusick, J.A. and Rice, S.D. (1996) Persistence of oiling in mussel beds three and four years after the *Exxon Valdez* oil spill. In *Proceedings of the Exxon Valdez Oil Spill Symposium*, eds. S.D. Rice, R B. Spies, D.A. Wolfe and B.A. Wright. Am. Fish. Soc. Symp. 18, pp. 286–29.

Boehm, P.D., Mankiewicz, P.J., Hartung, R., Neff, J.M., Page, D.S., Gilfillan, E.S., O'Reilly, J.E. and Parker, K.R. (1996) Characterization of mussel beds with residual oil and the risk to foraging wildlife 4 years after the *Exxon Valdez* oil spill. *Environmental Toxicology and Chemistry* **15**, 1289–1303.

Brodersen, C., Short, J., Holland, L., Carls, M., Pella, J., Larsen, M. and Rice, S. (1999) Evaluation of oil removal from beaches 8 years after the *Exxon Valdez* oil spill. Proc. 22nd Arctic and Marine Oilspill Program Technical Seminar, Calgary, Alberta, June 1999, pp. 325–336.

Brown, E.D., Baker, T.T., Hose, J.E., Kocan, R.M., Marty, G.D., McGurk, M.D., Norcross, B.L. and Short, J.W. (1996) Injury to the early life history stages of Pacific herring in Prince William Sound after the *Exxon Valdez* oil spill. In *Proceedings of the Exxon Valdez oil spill symposium*, eds. S.D. Rice, R B. Spies, D.A. Wolfe and B.A. Wright. Am. Fish. Soc. Symp. 18, pp. 449–462.

Bue, B.G., Sharr, S. and Seeb, J.E. (1998) Evidence of damage to pink salmon populations inhabiting Prince William Sound, Alaska, two generations after the *Exxon Valdez* oil spill. *Trans. Am. Fish. Soc.* **127**, 35–43.

Carls, M.G., Rice, S.D. and Hose, J.E. (1999) Sensitivity of fish embryos to weathered crude oil: Part I. Low-level exposure during incubation causes malformations, genetic damage, and mortality in larval Pacific herring (*Clupea pallasi*). *Environmental Toxicology and Chemistry* **18** (3), 481–493.

Geiger, H.J., Bue, B.G., Sharr, S., Wertheimer, A.C. and Willette, T.M. (1996) A life history approach to estimating damage to Prince William Sound pink salmon caused by the *Exxon Valdez* oil spill. In *Proceedings of the Exxon Valdez oil spill symposium*, eds. S.D. Rice, R.B. Spies, D.A. Wolfe and B.A. Wright. Am. Fish. Soc. Symp. 18, pp. 487–498.

Hayes, M.O. and Michel, J. (1999) Factors determining the long-term persistence of *Exxon Valdez* oil in gravel beaches. *Marine Pollution Bulletin* **38**, 92–101.

Heintz, R.A., Short, J.W. and Rice, S.D. (1999) Sensitivity of fish embryos to weathered crude oil: Part II. Increased mortality of pink salmon (*Oncorhynchus gorbuscha*) embryos incubation downstream from weathered *Exxon Valdez* crude oil. *Environmental Toxicology and Chemistry* **18** (3), 494–503.

Irvine, G.W., Mann, D.H. and Short, J.W. (1999) Multi-year persistence of oil mousse on high energy beaches distant from the *Exxon Valdez* spill origin. *Marine Pollution Bulletin* **38**, 572–584.

Kocan, R.M., Hose, J.E., Brown, E.D. and Baker, T.T. (1996) Pacific herring (*Clupea pallasi*) embryo sensitivity to Prudhoe Bay petroleum hydrocarbons: Laboratory evaluation and in situ exposure at oiled and unoiled sites in Prince William Sound. *Can. J. Fish. Aquat. Sci.* **53**, 2366–2375.

Wolfe, D.A., Hameedi, M.J., Galt, J.A., Watabayashi, G., Short, J., O'Clair, C., Rice, S., Michel, J., Payne, J.R., Braddock, J., Hanna, S. and Sale, D. (1994) The fate of the oil spilled from the *Exxon Valdez*. *Environmental Science and Technology* **28**, 561A–568A.

---

*The minimum release amount was provided by the spiller but was never independently verified, and the maximum corresponds with the capacity of the damaged cargo compartments. The actual amount spilled is likely somewhere between these extremes, because tidal and wave-driven oscillations of the impaled vessel probably replaced some of the oil with seawater in the hold.

al., 1998), sea otters and Bald Eagles (Estes et al., 1997) and killer whales (Matkin et al., 1999). Adult sockeye salmon migrating to freshwater spawning grounds have recently been identified as the major vector transporting these pollutants to anadromous streams and lakes (Ewald et al., 1998). Likely proximate sources of these pollutants include wet and dry atmospheric deposition to surface waters of the GOA, and east Asian runoff carried to North America by the North Pacific current (Iwata et al., 1994; Ewald et al., 1998).

Concentrations of many of the monitored pollutants increase as urbanized areas are approached, but the NS&T data indicate that marine pollution at even the most polluted coastal stations monitored in the GOA is usually moderate compared with stations along the coast of the conterminous U.S. (NS&T datasets).

RURAL FACTORS

Agriculture in the GOA region is very limited and declining. However, agricultural practices in other parts of the world may contribute to contaminant loads found in the GOA food web. Industrial logging has involved widespread clear cutting on both public and private lands. This kind of logging can lead to severe habitat degradation for fish and wildlife, including reduction of salmon spawning and rearing habitat. The U.S. Department of Agriculture has, in recent years, reduced industrial logging on the national forests in Alaska. Logging on Native selected lands continues. The State of Alaska has adopted Forest Practices Act regulations that protect salmon streams. Salmon stream buffers and maintenance of large woody debris in streams and rivers are techniques useful for protecting and promoting healthy streams and salmon runs.

Fishing practices have the potential to impact the health of the GOA ecosystem. High seas gillnetting affected the population of target and non-target species in the GOA waters outside of State of Alaska jurisdiction. Fortunately, this type of fishing was prohibited. The dramatic increase in salmon sharks in the GOA waters may be in response to elimination of this fishing technique. Shark abundances in the GOA near shore (salmon sharks, and Pacific sleeper sharks) are much more abundant than 15 years ago. This, in itself, can result in dramatic restructuring of the GOA ecosystem.

Long-line fishing and bottom trawling can impact important fish habitat. Researchers are beginning to study essential fish habitats, as required under federal legislation, in hopes of defining and protecting this vital habitat. Long-line fishing techniques have been adjusted to protect endangered bird species. Some seabirds will take the bait as long-line gear is set, drowning the bird. Several avoidance techniques have been successful in reducing this non-intended take. Trawling exclusion zones have been employed to protect the endangered GOA Steller sea lion population.

EFFECTS FROM URBAN AND INDUSTRIAL ACTIVITIES

Industrial uses of the coast associated with timber harvest and the fishing industry include localized smothering of marine habitat due to log rafting and fish processing. When logs are rafted in the marine environment, the bark and debris can accumulate on the bottom altering the habitat, making it unsuitable to fish and crabs. Outfalls of fish processing facilities can result in hundreds of square metres being smothered in putrefying wastes. In both of these cases, State of Alaska resource managers are researching and addressing this issue using best management practices.

Tourism has become an important industry in Alaska. Large-scale cruise liners bring more than a million visitors annually. This increase in activities in the marine environment has led to increasing water and air pollution, and impacts to marine resources. As the tourism industry continues to grow, the likelihood of greater impacts will increase.

The GOA remains relatively unperturbed by anthropogenic pollutants, and its geographical situation together with the low human population density suggest that in future most of the pollution input will arise from distant sources, particularly east Asia. Prevailing patterns of wind flows and ocean currents tend to convey east Asia atmospheric pollution and terrestrial run-off to the GOA. Once there, replacement of surface water by pristine deep oceanic water upwelled within the central GOA tends to depurate these pollutants. Hence, increasing trends of pollution burdens in compartments of the GOA may signal parallel trends within large proportions of the entire North Pacific ocean. For these reasons, the GOA is an especially appropriate region for monitoring pollution trends of the North Pacific.

REFERENCES

Bence, A.E., Kvenvolden, K.A. and Kennicutt, M.C. (1996) Organic geochemistry applied to environmental assessments of Prince William Sound, Alaska, after the *Exxon Valdez* oil spill—a review. *Organic Geochemistry* **24**, 7–42.

Brodeur, R.D. and Ware, D.M. (1992) Long-term variability in zooplankton biomass in the subarctic Pacific Ocean. *Fisheries Oceanogr.* **1**, 32–38.

Broecker, W.S. (1991) The great ocean conveyor. *Oceanago* **4**, 79–89.

Brown, D.W., McCain, B.B., Horness, B.H., Sloan, C.A., Tilbury, K.L., Pierce, S.M., Burrows, D.G., Chan, S.L., Landahl, J.T. and Krahn, M.M. (1999) Status, correlations and temporal trends of chemical contaminants in fish and sediment from selected sites on the Pacific coast of the USA. *Marine Pollution Bulletin* **37**, 67–85.

Cooney, R.T. (1986a) The seasonal occurrence of *Neocalanus cristatus*, *Neocalanus plumchrus* and *Eucalanus bungii* over the northern Gulf of Alaska. *Continental Shelf Research* **5**, 541–553.

Cooney, R.T. (1986b) Zooplankton. In *The Gulf of Alaska, Physical Environment and Biological Resources*, eds. D.W. Hood and S.T. Zimmerman. MMS/NOAA, Alaska Office, Anchorage, OCS Study MMS 86-0095, pp. 285–303.

Estes, J.A. (1997) Exploitation of marine mammals: r-selection of K-strategists? *Journal of the Fisheries Research Board of Canada* **36**, 1009–1017.

Ewald, G., Larson, P., Linge, H., Okla, L. and Szarzi, N. (1998) Biotransport of organic pollutants to an inland Alaska lake by migrating sockeye salmon (*Oncorhynchus nerka*). *Arctic* **51** (1), 40–47.

Incze, L.S. and Ainaire. T. (1996) Distribution and abundance of copepod nauplii and other small (40–300 μm) zooplankton during spring in Shelikof Strait, Alaska. *Fisheries Bulletin* **92**, 67–78.

Incze, L.S., Siefert, D.W. and Napp, J.M. (1996) Mesozooplankton of Shelikof Strait, Alaska: abundance and community composition. *Continental Shelf Research* **17**, 287–305.

Iwata, H., Tanabe, S., Aramoto, M., Sakai, N. and Tatsukawa, R. (1994) Persistent organochlorine residues in sediments from the Chukchi Sea, Bering Sea and Gulf of Alaska. *Marine Pollution Bulletin* **28**, 746–753.

Johnson, W.R., Royer, T.C. and Luick, J.L. (1988) On the seasonal variability of the Alaska Coastal Current, *Journal of Geophysical Research* **93**, 12423–12437.

Karinen, J.F., Babcock, M.M., Brown, D.W., MacLeod Jr. W.D., Ramos, L.S. and Short, J.W. (1993) (revised December 1994). Hydrocarbons in intertidal sediments and mussels from Prince William Sound, Alaska, 1977–1980: Characterization and probable sources. U.S. Dept. Commer., NOAA Tech. Memo. NMFS-AFSC-9, 70 p.

Lees, D.C., Payne, J.R. and Driskell, W.B. (1999) Technical evaluation of environmental monitoring program for Cook Inlet Regional Citizens' Advisory Council. Final Report prepared for the Cook Inlet Regional Citizens' Advisory Council, Kenai, Alaska 99611. CIRCAC Project No. 98-023E, January, 1999, 168 pp.

Martin, J.H., Gordon, R.M. and Fitzwater, S.E. (1991) The case for iron. *Limnology and Oceanography* **36**, 1793–1802.

Matkin, C.O., Scheel, D., Ellis, G., Barrett-Lennard, L., Jurk, H. and Saulitis, E. (1999) Comprehensive killer whale investigation, *Exxon Valdez* Oil Spill Trustee Council Annual Report (Restoration Project 98012), North Gulf Oceanic Society, Homer, AK.

Miller, C.B. (1993) Pelagic production processes in the Subarctic Pacific. *Progress in Oceanography* **32**, 1–15.

Musgrave, D., Weingartner, T. and Royer, T.C. (1992) Circulation and hydrography in the northwestern Gulf of Alaska. *Deep-Sea Research* **39**, 1499–1519.

Napp, J.M., Incze, L.S., Ortner, P.B., Siefert, D.L.W. and Britt, L. (1996) The plankton of Shelikof Strait, Alaska: standing stock, production, mesoscale variability and their relevance to larval fish survival. *Fish. Oceanography* **5** (suppl. 1), 19–38.

National Status and Trends Datasets. (Current). Available at ftp:// seaserver.nos.noaa.gov/datasets/nsandt/

O'Connor, T.P. and Pearce, J. eds. (1999) U.S. Coastal Monitoring: NOAA's National Status and Trends Results. *Marine Pollution Bulletin* **37**, 1–113.

Okkonen, S.R. (1992) The shedding of an anticyclonic eddy from the Alaskan Stream as observed by the GEOSAT altimeter. *Geophysics Research Letters* **19**, 2397–2400.

Parker, K.S., Royer, T.C. and Deriso, R.B. (1995) High-latitude climate forcing and tidal mixing by 18.6-year lunar nodal cycle and low-frequency recruitment trends in Pacific halibut (*Hippoglossus stenolepis*). In *Climate Change and Northern Fish Populations*, ed. R.J. Beamish. Can. Spec. Publ., Fish. Aquat. Sci., no. 121, pp. 449–459.

Parsons, T.R. (1986) Ecological relations. In *The Gulf of Alaska, Physical Environment and Biological Resources*, eds. D.W. Hood and S.T. Zimmerman. MMS/NOAA, Alaska Office, Anchorage, OCS Study MMS 86-0095, pp. 561–570.

Parsons, T.R. and Lalli, C.M. (1988) Comparative oceanic ecology of the plankton communities of the subarctic Atlantic and Pacific Oceans. *Oceanogr. Mar. Annu. Rev.* **26**, 317–359.

Payne, J.R., Driskell, W.B. and Lees, D.C. (1998) Long term environmental monitoring program data analysis of hydrocarbons in intertidal mussels and marine sediments, 1993–1996. Final report prepared for the Prince William Sound Regional Citizens Advisory Council, Anchorage, Alaska 99501. PWS RCAC Contract No. 611.98.1. March, 1998, 16 p.

Reed, R.K. and Schumacher, J.D. (1986) Physical Oceanography. In *The Gulf of Alaska, Physical Environment and Biological Resources*, eds. D.W. Hood and S.T. Zimmerman. MMS/NOAA, Alaska Office, Anchorage, OCS Study MMS 86-0095, pp. 57–76.

Rogers, D.E. (1986) Pacific Salmon. In *The Gulf of Alaska: Physical Environment, and Biological Resources*, eds. D.W. Hood and S.T. Zimmerman. MMS/NOAA, Alaska Office, Anchorage, OCS Study MMS 86-0095, pp. 561–476.

Royer, T.C. (1982) Coastal freshwater discharge in the Northeast Pacific, *Journal of Geophysical Research* **87**, 2017–2021.

Royer, T.C. (1996) Interdecadal hydrographic variability in the Gulf of Alaska, 1970–1995. *EOS, Transaction, AGU*, **77**, F368.

Royer, T.C., Hansen, D.V. and Pashinski, D.J. (1979) Coastal flow in the northern Gulf of Alaska as observed by dynamic topography and satellite-tracked drogued drift buoys. *Journal of Physical Oceanography* **9**, 785–801.

Sambrotto, R. and Lorenzen, C.J. (1986) Phytoplankton and Primary Production. In *The Gulf of Alaska, Physical Environment and Biological Resources*, eds. D.W. Hood and S.T. Zimmerman. MMS/NOAA, Alaska Office, Anchorage, OCS Study MMS 86-0095, pp. 249–282.

Short, J.W. and Sharp, J.L. (1989) Tributyltin in bay mussels (*Mytilus edulis*) of the Pacific coast of the United States. *Environmental Science and Technology* **23**, 740–743.

Short, J.W., Kvenvolden, K.A., Carlson, P.R., Hostettler, F.D., Rosenbauer, R.J. and Wright, B.A. (1999) Natural hydrocarbon background in benthic sediments of Prince William Sound, Alaska: Oil vs coal. *Environmental Science and Technology* **33**, 34–42.

Wade, T.L., Sericano, J.L., Gardinali, P.R., Wolff, G. and Chambers, L. (1999) NOAA's 'Mussel Watch' project: current use organic compounds in bivalves. *Marine Pollution Bulletin* **37**, 20–26.

Welschmeyer, N.A., Strom, S., Goericke, R., DiTullio, G., Belvin, M. and Petersen, W. (1993) Primary production in the subarctic Pacific Ocean: project SUPER. *Progress in Oceanography* **32**, 101–135.

Wilson, J.G. and Overland, J.E. (1986) Meteorology. In *The Gulf of Alaska, Physical Environment and Biological Resources*, eds. D.W. Hood and S.T. Zimmerman. MMS/NOAA, Alaska Office, Anchorage, OCS Study MMS 86-0095, pp. 31–54.

Xiong, Q. and Royer, T.C. (1984) Coastal temperature and salinity observations in the northern Gulf of Alaska, 1970–1982. *Journal of Geophysical Research* **89**, 8061–8068.

THE AUTHORS

Bruce A. Wright
Alaska Region, National Marine Fisheries Service,
National Oceanic and Atmospheric Administration,
11305 Glacier Highway, Juneau, AK 99801

Jeffrey W. Short
Auke Bay Laboratory, National Marine Fisheries Service,
National Oceanic and Atmospheric Administration,
11305 Glacier Highway, Juneau, AK 99801, U.S.A.

Tom S. Weingartner
University of Alaska Fairbanks,
Institute of Marine Science,
School of Fisheries and Ocean Sciences,
Fairbanks, AK 99775-7220, U.S.A.

Paul J. Anderson
Kodiak Laboratory, National Marine Fisheries Service,
National Oceanic and Atmospheric Administration,
P.O. Box 1638, Kodiak, AK 99615, U.S.A.

Chapter 24

SOUTHERN CALIFORNIA

Kenneth C. Schiff, M. James Allen, Eddy Y. Zeng and Steven M. Bay

---

The Southern California Bight (SCB) has undergone tremendous changes over the last 100 years resulting from natural and anthropogenic alteration of the coastal zone. A large influx of population during the 1900s has propelled the coastal community surrounding the SCB from a small pueblo (<2000 in 1900) to the largest metropolitan centre in the U.S. (>17 million in 1998). This rapid urbanization has placed extreme pressures on marine resources including loss of habitat, discharge of pollutants, and overfishing.

As the population has grown in the four counties bordering the shoreline of the SCB, so have discharges of pollutants to the ocean. The major source of pollutants in the early 1970s was publicly owned treatment works (POTWs). Regulation of these discharges has led to improved treatment, source control, and pretreatment programs. As a result, cumulative pollutant loads for POTWs have been reduced several fold, even orders of magnitude for some constituents. Similar to trends observed in many areas of the nation, non-point sources have become larger contributors of potential pollutants as POTWs have reduced their inputs. In the SCB, urban runoff contributes more trace metals (chromium, copper, lead, and zinc) and nutrients (nitrate and phosphorus) than all other sources combined.

As the inputs of pollutants have declined and the dominant sources have shifted, the fate and distribution of pollutants in the SCB has changed over the last 30 years. Studies have observed decreasing concentrations in water, sediments, and biota. For example, decreasing concentrations in near-surface sediments are recorded in sediment core profiles. Also, fish tissue concentrations have decreased compared to similar measurements made in the 1970s and 1980s. However, legacy inputs continue to place both the marine ecosystem and public health at risk. Among the most important constituents of concern in the SCB is total DDT (o,p' and p,p' isomers of DDT, DDE, and DDD). Total DDT is widely dispersed; it is measured in 89% of the SCB sediments and has contaminated nearly 100% of Pacific and Longfin sanddab populations. In regions where sediment concentrations of total DDT are highest (e.g. Palos Verdes Shelf), commercial fishing is prohibited and recreational anglers are warned about consuming tainted bottom-feeding fish.

Along with reductions in pollutant inputs over the last 30 years, scientists have observed the recovery of some marine ecosystems. Five phylogenetic groups are evaluated in this article including kelp (algae), benthic invertebrates, fish, seabirds, and marine mammals. Among the most-studied groups are benthic infauna and fish communities. In 1994 approximately 91% of the SCB mainland shelf contained benthic communities classified as "reference." Although fish diseases (e.g., fin rot and epidermal tumours) were common in the 1970s, their occurrence is currently at background levels. In almost all cases, interactions have occurred between natural and anthropogenic factors. For example, kelp beds near large POTW discharges that were a

Seas at The Millennium: An Environmental Evaluation (Edited by C. Sheppard)
© 2000 Elsevier Science Ltd. All rights reserved

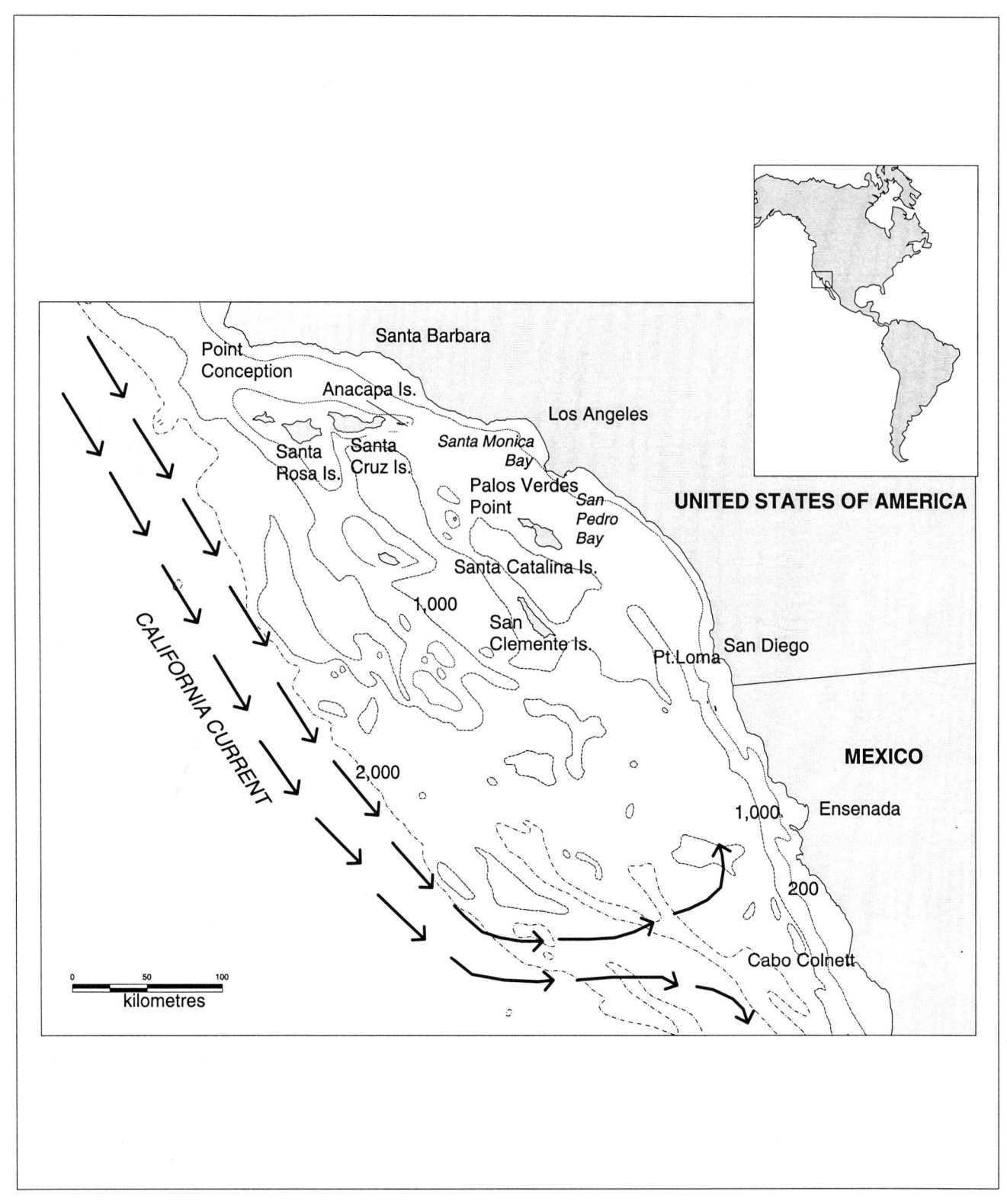

Fig. 1. Map of Southern California, including general circulation patterns in the Southern California Bight (after Hickey, 1993).

fraction of their historical extent in 1970 have shown exceptional recruitment and are currently flourishing. However, during the same time period, natural events such as the 1987–88 El Niño negatively impacted the kelp and reduced bed extent to levels not observed since 1970.

Ecosystem management of the SCB is improving as we enter the new millennium. The improvement began when resource managers recognized that traditional monitoring programs were not providing the information they needed to make responsible stewardship decisions. Regulatory and permitted discharge agencies have since created an open dialogue to identify the most important monitoring objectives. In addition, they have cooperatively designed and implemented a coordinated, integrated regional monitoring program. Regional monitoring has evaluated the full range of natural variability and cumulative impacts from multiple discharges, enabling assessment of the overall condition of the SCB.

PHYSICAL AND BIOLOGICAL SETTING

Geography and Oceanography

The Southern California Bight (SCB) is the oceanographic region off southern California in the U.S (Fig. 1). This area is formed where the coast makes a sharp bend to the east, causing the southward-flowing California Current to flow far offshore before intersecting the mainland again in northern Baja California. The SCB extends from Point Conception, California (lat. 34°30'N; long. 120°30'W), in the northwest to Cabo Colnett, Baja California (lat. 31°00'N; long. 116°20'W), in the southeast, and is bounded to the west by the California Current (SCCWRP, 1973; Dailey et al., 1993). It includes an area of approximately 78,000 km^2 with a shoreline distance of over 300 km (Dailey et al., 1993).

Surface waters of the SCB flow in a large, counter-clockwise eddy, where the warmer surface waters from the northerly flowing Davidson Countercurrent mix with the colder, southerly flowing California Current. Water temperatures range from 12 to 16°C north of Point Conception and >18°C in Baja California, whereas temperatures range from 14 to 20°C in southern California (Eber, 1977).

The oceanic environment off southern California varies decadally (Smith, 1995) and aperiodically, such as during El Niño (anomalously warm) and La Niña (anomalously cold) events (Murphree and Reynolds, 1995; Lynn et al., 1995). El Niño events occur when the location of atmospheric high and low pressure areas in the southern hemisphere shift (hence, El Niño-Southern Oscillation or ENSO). During an El Niño event, the California Current flow weakens, water temperatures increase, and the thermocline deepens as warm, saline, oligotrophic water moves north into the SCB (Lynn et al., 1995; Murphree and Reynolds, 1995). The reverse occurs during La Niña events. During normal periods, the California Current is strong, as is upwelling, and waters are cooler and more productive. Strong El Niño events affected the SCB in 1929–30, 1957–59, 1982–83 (Smith, 1995), and 1997–98. Strong La Niña events occurred in 1933, 1975–76, and (with less cold water) 1988–1989 (Smith, 1995). Water temperatures were cooler than normal from 1942–1976 (except for the 1957–59 El Niño event) and warmer than normal prior to 1942 and since 1976 (particularly since the 1982–83 El Niño event (Smith, 1995).

The waters of the SCB overlie the continental borderland of southern California (Emery, 1960; Dailey et al., 1993). The outer edge of the borderland is the Patton Escarpment, which lies some 250–300 km offshore, and is defined by a sharp change in slope at 1000 m. The continental borderland consists of a number of offshore islands, submerged banks, submarine canyons, and deep basins. The result is an unusually narrow mainland shelf, which averages 3 km in width (ranging from 1–20 km) and ends in waters of 200 m depth. Elsewhere in the U.S., the mainland shelf may be 10–200 times wider. The narrowness of the mainland shelf in the SCB makes it particularly susceptible to human activities.

The dominant bottom environment in the SCB consists of sandy and muddy sediments (Emery, 1960; Dailey et al., 1993). Sediments with high percentages of sand generally dominate the shelf, whereas sediments with high percentages of silt generally dominate the slope and basins. Along the mainland shelf, rocky bottoms are most commonly found inshore near rocky headlands, along edges of submarine canyons, and at the shelf break. Only rarely do outcrops occur in deep water (i.e., Santa Monica Bay and San Pedro Bay). Rocky bottom is more common along the shelf of the offshore islands and banks where the supply of sand and silt is minimal.

BIOGEOGRAPHIC PROVINCES AND HABITATS

The SCB is a rich ecosystem; over 5000 species of invertebrates, 480 species of fish, and 195 species of marine birds are found in this region (Dailey et al., 1993). The diversity found in the SCB is owed, in part, to its transitional zonation between two biogeographic provinces. The San Diegan Province to the south introduces sub-tropical species while the Oregonian Province to the north introduces temperate species into the SCB (Briggs, 1974). Each of these provinces has distinctive biota. For example, more than 70% of all algal species found in California occur in the SCB; half of these species have their northern or southern range endpoints located within the SCB (Murray and Bray, 1993).

A number of habitats are found in the SCB, based upon the physical structure or other properties of the environment. These include water-column, hard-bottom (and kelp beds), and soft-bottom (sand and mud) habitats. The intertidal zone is a unique habitat due to tidal influences. Estuaries are also unique as they respond to variability in salinities. Estuaries and lagoons in the SCB are typically small (<2 km^2) and have little natural runoff, except during winter storms. Salinity decreases dramatically to near zero during winter storms, whereas estuaries become hypersaline during hot, dry periods.

Hard-bottom habitat provides substrate for attachment of algae and sessile organisms, and crevices for refuge for mobile organisms. Hard-bottom habitats frequently have abundant algal cover in shallow water. Subtidal hard-bottom habitat occurs primarily near headlands and near the shelf break and outcroppings on the shelf. It is more abundant on the islands than on the mainland shelf.

Kelp beds (consisting largely of giant kelp, *Macrocystis pyrifera*) are usually attached to hard-bottom substrate and provide vertical structure of the habitat to the sea surface. This increases the complexity of the habitat, with tangled holdfasts at the bottom, columns of kelp stipes in midwater, and a dense canopy of kelp blades at the surface. For many invertebrates, this substrate also provides a source of food. Kelp beds are typically found at depths shallower than 30 m, being limited by light penetration (Quast, 1968). Kelp beds are patchily distributed along the coast, with large beds in the Santa Barbara area, on the Palos Verdes Shelf, near Point Loma, and on the Channel Islands (CSWQCB, 1964). Unlike hard-bottom habitats, the distribution and extent of kelp beds changes more readily, as beds are dislodged by storm swells and die off in periods of low nutrients (Stull, 1995).

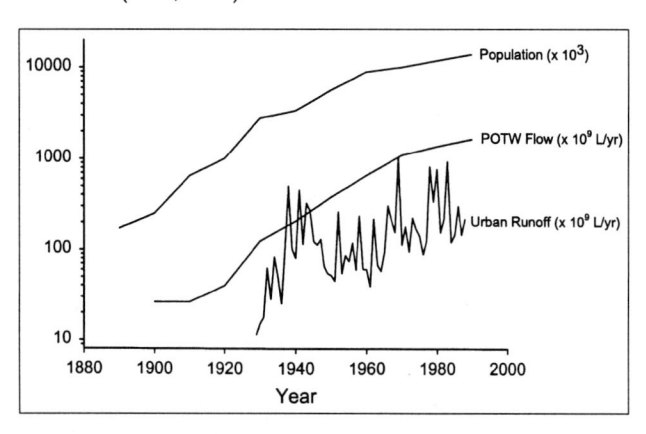

Fig. 2. Population growth and associated increases in wastewater flow and surface runoff in the Southern California Bight over the last 100 years. Wastewater flows are from the four largest facilities that represent approximately 95% of all wastewater flows in the region. Surface runoff flows are from the Los Angeles River that represents approximately 33% of the gauged runoff in the region.

Soft-bottom habitat is the most extensive benthic habitat, particularly on the mainland shelf, slopes, and basins. Soft-bottom habitat consists of sandy and muddy sediments with much less relief than hard-bottom habitat. Patches of sediment differing in grain size (e.g., gravel, sand, silt, clay) provide distinct habitats to different species of infaunal organisms; however, polychaetous annelids dominate in most unimpacted regions. Grain size naturally covaries with depth on the mainland shelf of the SCB. As a result, four soft-bottom communities are often defined by depth (Bergen et al., 1999; Jones, 1969). These consist of a shallow water assemblage (<30 m), a mid-depth assemblage (30–120 m), and two deep assemblages (120 to >200 m) segregated by grain size (stratified at approximately 40% fines).

HUMAN CONTRIBUTIONS AND ANTHROPOGENIC INPUTS

Southern California is home to over 17 million people and is among the most densely populated areas in the nation. Nearly 25% of the nation's coastal population lives in the four coastal counties that surround the SCB (Culliton et al., 1990). In fact, the population of the SCB has doubled in the last 30 years and is expected to increase another 2 million by 2010 (Fig. 2). Rapid urbanization represents risk to the coastal ecology by encroachment of habitat and by contributing anthropogenic contaminants to the coastal environment.

The loss of coastal habitat has reduced some ecosystems to critically small areas. In Santa Monica Bay alone, an estimated 95% of wetland area has been lost to dredge-and-fill operations for marinas and ports, diked for agriculture, or developed for coastal urbanization (SMBRP, 1988). In the SCB, coastal wetland ecosystems provide essential habitat for fish nursery areas, as well as for two marine birds on the endangered species list (California least tern and Belding's savannah sparrow). Moreover, wetlands in the SCB are an important rest stop on the Pacific Flyway, the major bird migratory route between northerly and southerly latitudes. A second habitat lost to urbanization is beaches. Coastal erosion has been attributed to reduced sediment delivery resulting from development within coastal watersheds. Sediment yields of 7.7 million tons were one-third lower than 10 years earlier (Rodolpho, 1970). However, coastal erosion programs in the SCB have not conclusively identified reduced sediment yields as the primary cause for the loss of this habitat.

Multiple sources discharge pollutants into the SCB (Fig. 3). Sources that contribute pollutants include publicly owned wastewater treatment plants (POTWs), surface runoff from urban and agricultural watersheds, disposal of contaminated dredged materials, industrial facilities, power generating stations, oil and gas production, vessel activities from recreational marinas and commercial ports,

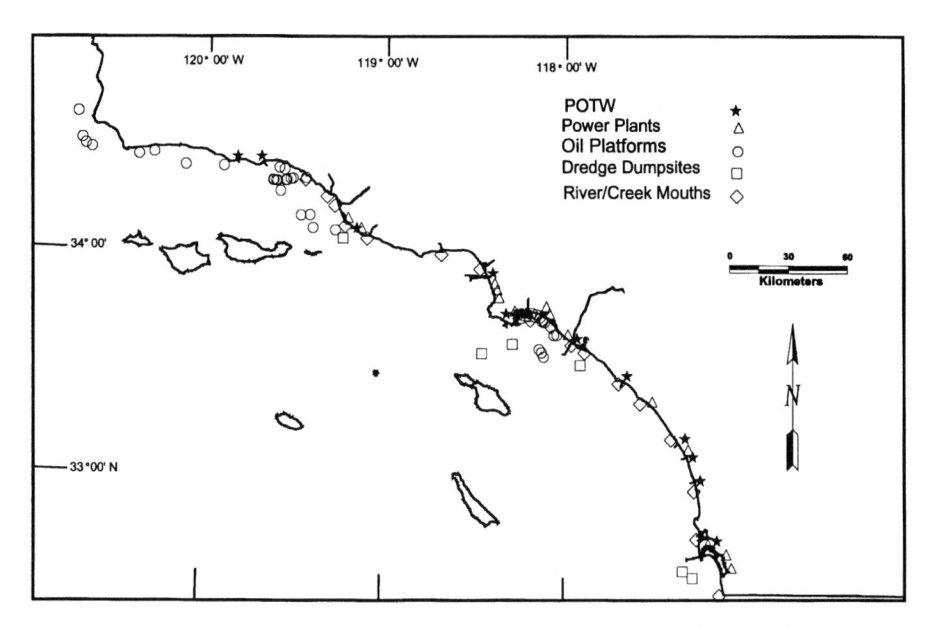

Fig. 3. Map of potential pollutant sources to the Southern California Bight.

aerial deposition, and hazardous material spills, among others. The types of constituents that have been identified as potential pollutants in the SCB include heavy metals; chlorinated hydrocarbons (e.g., pesticides, fungicides, herbicides); petroleum hydrocarbons (e.g., polycyclic aromatic hydrocarbons [PAHs]); nutrients (nitrogenous and phosphate compounds); bacteria; and oxygen-depleting substances (e.g., biosolids).

Although urbanization and population growth have been increasing over the last 30 years, the mass emission of potential pollutants has been decreasing. Overall, there has been a 70% reduction in contaminant inputs to SCB coastal waters from all sources. Between 1971 and 1996, general constituents (e.g., suspended solids and biological oxygen demand) have decreased by 50%; combined heavy metals have decreased by 90%; and chlorinated hydrocarbons have decreased by more than 99%. These reductions have been so significant that some constituents are no longer detected in any source (e.g., polychlorinated biphenyls [PCBs]) (Raco-Rands, 1999).

The reductions observed over the last 25 years have largely been the result of reductions from POTWs (Fig. 4). In 1971 the four largest POTWs, which serve the City and County of Los Angeles, County of Orange, and City of San Diego, cumulatively discharged 1035 mgd of wastewater effluent and contributed the vast majority of these potential pollutants to the SCB (SCCWRP, 1973). While POTW flows have been steadily increasing over time, dramatic reductions in mass emissions for all constituents have occurred. In some instances, the reductions were the result of wholesale bans (e.g., DDT and PCB); but for most constituents, the reductions have been the result of increased source control, pretreatment, reclamation, and treatment plant upgrades. In 1971, for example, 105 mgd (10%) of the

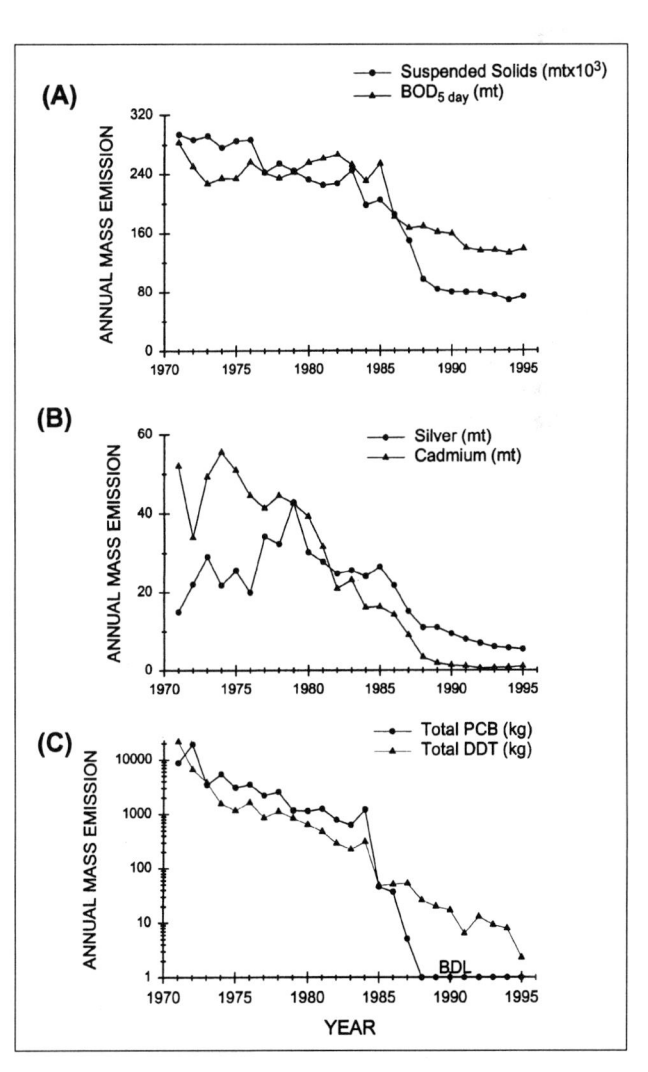

Fig. 4. Cumulative mass emissions of selected constituents from publicly owned treatment works in the Southern California Bight between 1971 and 1995.

> ### DDT Contamination in Southern California
>
> Patterns of contamination and effects in the SCB are dominated by the legacy of the historical production and use of the pesticide 1,1,1-trichloro-2,2-bis(*p*-chlorophenyl) ethane (DDT). Total DDT, which includes both the *ortho*- and *para*-substituted isomers of DDT and its metabolites DDE and DDD, are man-made chemicals that produce neurotoxic, estrogenic, and carcinogenic effects in animals. Originally used throughout the U.S. from the 1930s to 1971, DDT was hailed as a superior pesticide and helped to drastically reduce the incidence of malaria and other diseases passed by insects such as mosquitoes. Due to its resistance to chemical, physical, and biological degradation, however, DDT has been widely distributed in the global ecosystem (including Antarctica). Moreover, its lipophilic nature means that it is fat-soluble and has dramatically affected several non-target organisms including birds and marine mammals.
>
> The DDT contamination is particularly significant in the coastal regions of southern California because the world's largest DDT manufacturer, the Montrose Chemical Corporation, discharged large quantities of DDT-enriched wastes via the Los Angeles County Sanitation Districts' Joint Water Pollution Control Plant (JWPCP) outfall on the mainland shelf off Palos Verdes. An estimated 20 metric tons of total DDT were being discharged through the JWPCP outfall in 1971. Montrose was also discharging DDT wastes by ocean dumping at two locations in the Santa Monica and San Pedro Basins prior to 1970. In 1971, the use of DDT was banned in the U.S. by federal regulation and disposal of DDT residues into the sewer system followed not long afterwards. Currently, inputs of total DDT are extremely low; an estimated 1.4 kg of DDT was discharged from all POTWs in the SCB during 1997.
>
> Historic DDT inputs and their redistribution represent a potential source of contamination that is still impacting sensitive species. It remains unclear how much still exists in the SCB, but an estimated 156 metric tons of total DDT still existed in the sediments on the Palos Verdes Shelf in 1992. The Palos Verdes Shelf is now the site of a Superfund investigation and is the epicentre of the nation's largest environmental damage assessment lawsuit. Although DDT was banned in the U.S., it is still being used in other countries including Mexico.

combined wastewater flow discharged by POTWs underwent secondary treatment. In 1996, 506 mgd (49%) received secondary treatment. Capital improvements to POTWs throughout the SCB are estimated to have cumulatively cost over $5 billion.

While emissions of pollutants from POTWs have decreased over time due to improved controls and treatment, other sources have remained steady or increased. For example, surface runoff from urban and agricultural watersheds has increased due to larger flows and lack of significant concentration-reduction strategies. Surface runoff, which is much more variable than POTW flows, has increased over time (see Fig. 2). Larger flows have occurred as a consequence of the impervious materials used in urban areas (i.e., concrete) and inland discharges from municipal and industrial facilities. More than 95% of the flow in the Los Angeles River during dry weather is the result of effluent discharges upstream.

Surface runoff is currently one of the largest sources of pollutants to the SCB (Table 1). In 1995 runoff discharged more nutrients (nitrate and phosphate) and heavy metals (chromium, copper, lead, and zinc) than all other sources combined. Since stormwater sewers and sanitary sewers are not combined in the metropolitan areas of the SCB, surface runoff receives no treatment prior to discharge into coastal oceans, bays, and wetlands. Moreover, POTW discharges are located at great depths (60–100 m) and miles from shore, while storm drains discharge across the beach where potential human contact is great.

Regulation and control of stormwater discharges has lagged behind POTWs partly due to the infrequent rainfall the SCB receives each year. An average of 12–14 storm events occur per year, which typically last from 6–12 h. However, these short but intense rain events generate substantial increases in runoff. For example, flow in large concrete-lined flood control channels measuring more than 90 m in width can increase from <5 cfs to >20,000 cfs in less than 3 h. These flood events contribute more than 95% of the total runoff volume and pollutant load annually.

Measurements of runoff are tremendously variable and have only recently been addressed in a regulatory context. However, other potential sources remain unmonitored, such as aerial deposition of pollutants either directly to coastal water bodies or indirectly (whereby pollutants are deposited on terrestrial surfaces and then washed into the sea following rain events). Southern California has among the worst air quality in the nation (U.S. EPA, 1997), yet no formal monitoring program exists to examine this potential source of pollutants to aquatic environments; all air monitoring is used to support the U.S. EPA Clean Air Act Amendments that address human health impacts from respiration of contaminants.

DISTRIBUTION AND FATE OF ANTHROPOGENIC INPUTS

It is evident that the importance of contamination sources affecting the ecology of the SCB has been shifting over the last two decades. This shift has gradually affected the distribution patterns of contaminants in the coastal environment off southern California, which are considered in the examination of contaminant fate and distribution.

Contaminants entering into the SCB undergo a number of physical, chemical, and biological processes in the water

Table 1
Total mass emissions of selected constituents from various sources to the coastal oceans of the southern California Bight.

Constituent	Total load	Percent of total load							
		Urban runoff	Large POTWs	Small POTWs	Industrial facilities	Power plants	Oil platform	Ocean dumping	Hazardous material spills
Year of load estimate		1994–95	1997	1995	1995	1995	1990	1995	1990
Flow (1×10^9)	13,668	21.36	11.19	1.44	0.17	65.81	0.04	–	<0.01
Suspended solids (mt)	674,200	88.76	10.90	0.29	0.05	0.01	<0.01	–	–
BOD (mt)	140,541	–	98.19	1.68	0.13	<0.01	<0.01	–	–
Oil and grease (mt)	19,922	–	96.37	2.32	0.45	0.14	0.72	–	–
Nitrate-N (mt)	9,224	95.41	2.87	1.65	–	0.07	–	–	–
Nitrite-N (mt)	151	–	84.11	15.89	–	–	–	–	–
Ammonia-N (mt)	45,898	1.96	90.06	7.84	0.12	0.01	–	–	–
Organic N (mt)	5,880	–	99.00	1.00	–	–	–	–	–
Phosphate (mt)	4,702	61.68	38.32	–	–	–	–	–	–
Total phosphorus (mt)	1,841	–	100.0	–	–	–	–	–	–
Cyanide (kg)	8,026	–	80.99	18.71	<0.01	<0.01	0.30	–	–
Arsenic (kg)	5,723	–	87.37	6.67	4.11	1.00	0.86		
Cadmium (kg)	2,085	–	47.01	21.68	0.21	30.94	0.16	<0.01	–
Chromium (kg)	38,396	76.05	18.23	3.65	0.25	1.05	0.78	<0.01	–
Copper (kg)	149,464	58.61	35.46	4.53	0.03	1.31	0.06	<0.01	–
Lead (kg)	51,349	76.53	4.67	4.64	0.03	2.29	11.83	<0.01	–
Mercury (kg)	262	–	8.39	4.19	0.03	85.38	2.02	–	–
Nickel (kg)	91,572	63.67	32.53	2.96	0.15	0.01	0.69	<0.01	–
Selenum (kg)	9,212	–	84.67	8.48	6.85	<0.01	<0.01	–	–
Silver (kg)	6,031	<0.01	89.54	10.38	0.01	<0.01	0.07	–	–
Zinc (kg)	443,437	71.35	19.39	3.57	0.24	4.17	1.27	<0.01	–
Phenols (kg)	166,643	–	97.57	0.02	0.84	<0.01	1.57	–	–
Chlorinated	2,900	–	96.55	3.45	<0.01	<0.01	–	–	–
Nonchlorinated	94,966	–	99.83	0.17	<0.01	<0.01	–	–	–
Total DDT (kg)	3	–	91.18	8.82	<0.01	<0.01	–	–	–
Total PCB (kg)	<0.1	–	–	–	–	–	–	–	–

column, and settle into the sea floor and/or are accumulated by biota. A large amount of contaminants may be decomposed chemically or biologically or remain buried in sediments. However, some contaminants are likely to re-enter the water column and become available for redistribution or bioaccumulation. Such a cycling of contaminants plays an important role in modifying the spatial and temporal trends of contaminant distribution. Our discussion about fates of environmental contaminants will involve three compartments including water column, sediment, and biota. In each compartment we will examine: (1) the spatial and temporal distributions of contaminants; (2) contaminant correlations among compartments; and (3) mechanisms that influence transport among compartments.

Water Column

The waters of the SCB can be classified, based upon the transport dynamics, into three categories: near-surface waters (0–200 m), intermediate waters (200 m to basin sill depth), and deep basin waters (Eganhouse and Venkatesan, 1993). The near-surface waters are density stratified, resulting in much stronger vertical concentration gradients than horizontal gradients. On the other hand, the spatial distribution in the intermediate waters is dominated by advection with insignificant vertical concentration gradient. In the deep basin waters, both density stratification and horizontal advection are weak. Instead, eddy diffusion between the deep and upper waters is the major mechanism for water exchange.

Research in the area of near-surface waters has been focused on the influence of anthropogenic activities. Sea-surface microlayer contamination and bottom fluxes have been the main concerns. Since surface water microfilms have a greater affinity with particles than subsurface water, the sea-surface microlayer is enriched with organic and inorganic materials and is reflective of influences from anthropogenic activities. A survey conducted by Cross et al. (1987) in six locations of the SCB found that concentrations of DDTs, PCBs, PAHs, and a group of trace metals (silver, chromium, copper, iron, manganese, nickel, lead, and zinc) were two or three orders of magnitude higher in the harbour locations (Los Angeles and Long Beach Harbors) than in the nearshore locations (San Pedro Channel, Huntington Beach, Palos Verdes Shelf, and Redondo Harbor). In addition, low molecular weight PAHs were found relatively enriched in the harbour samples while high molecular weight PAHs were relatively dominant in the nearshore samples. This suggested that petroleum-related residues were the main source of contamination inside the harbour. Another study conducted in the coastal area off San Diego also obtained higher concentrations of PAHs and aliphatic hydrocarbons in microlayer samples collected from inside San Diego Bay than those collected in the nearshore stations (Zeng and Vista, 1997).

Bottom fluxes of particles and organic and inorganic materials were clearly correlated with sewage inputs in the SCB (Hendricks and Eganhouse, 1992). Sewage-derived contaminants appeared to be transported to basin waters (Crisp et al., 1979), probably via upper water advection. This transfer may have caused a widespread distribution of historically discharged contaminants, such as DDTs, throughout the entire SCB. Indeed, a number of sediments collected from Santa Monica and San Pedro Basins had percent of DDEs (%DDEs) in total DDTs values similar to those contained in nearshore sediments, leading to the conclusion that DDTs in basin sediments were originally derived from nearshore sediments (Zeng and Venkatesan, 1999).

Recent investigations of contaminant distribution in the water column of the SCB have made considerable progress with the application of an *in-situ* sampling technique (Green et al., 1986, Tran and Zeng, 1997). Unlike sediment traps typically used in the past to capture sinking particles in the water column, the *in-situ* sampling approach collects particles and dissolved materials separately. The capability of this approach to process a large quantity of water permits the detection of ultra-low levels of contaminants. A recent sampling on the Palos Verdes Shelf (heavily contaminated), off Newport Beach (moderately contaminated), and off Dana Point (lightly contaminated) (Fig. 1) using this technique acquired valuable information about the magnitudes of water column DDT, PCB, and PAH contamination (Tran and Zeng, 1997). Concentrations of DDTs at all the sampling stations (~1 m from the sea floor) were higher than the discharge limit established by the State of California (California State Water Resources Control Board, 1997). Another sampling on the Palos Verdes Shelf showed that the spatial distribution of DDTs (mostly p,p'-DDE) in the water column was similar to that of p,p'-DDE in the sediments. In addition, the vertical concentration of water column DDTs decreased with increasing distance from the sea floor. This evidence, when taken cumulatively, suggests that contaminated sediments remain a main source of DDT contamination to the water column of the Palos Verdes Shelf (Zeng et al., 1999).

Sediment

Sediments of the SCB have been studied extensively since the early 1970s (McDermott et al., 1974; Young et al., 1975; Young et al., 1976). As a depositing reservoir, sediments often provide critical links to the temporal trends of contaminant inputs as well as the current status of contamination in the marine environment. Earlier investigations mostly converged on "hot spots" (i.e., areas severely impacted by known sources). Sediments of the Palos Verdes Shelf and of Santa Monica Bay near sewage outfalls have long been recognized as such "hot spots." San Diego Bay is another location that has been deemed a "hot spot" and is on the State of California's list of impaired water bodies. Sediments contain high concentrations of PCBs and PAHs (Mearns et al., 1991), presumably due to discharges from U.S. Navy operations and commercial shipping activities. Newport Bay, another State-listed impaired water body, contains sediments that have high levels of DDTs, non-DDT chlorinated pesticides, PCBs, PAHs, and some trace metals (Phillips et al., 1998).

Scientists in the SCB have been attempting to use molecular markers to identify the source(s) of contaminants that comprise a "hot spot" and other locations with significant sediment concentrations (Eganhouse et al., 1988; Venkatesan and Kaplan, 1990; Sanudo-Wilhelmy and Flegal, 1992; Venkatesan, 1995; Zeng et al., 1997). Linear alkylbenzenes (LABs) and coprostanol are two commonly used tracers of sewage inputs. Iron is also used as a reference element for determining trace metal enrichment due to anthropogenic activities (Schiff and Weisberg, 1999). Phillips et al. (1997), using a combination of molecular markers and principal component analysis, identified three relatively distinct areas of the San Pedro Shelf impacted by wastewater discharge (near the outfall of the Orange County Sanitation District), riverine inputs (close to the mouths of the Santa Ana River and Newport Bay), and natural seepage and historical pesticide and hydrocarbon inputs (at a deep slope and canyon region). Sewage-derived contaminants are likely to be transported to basin sediments via current advection (mainly northwesterly in the nearshore region of the SCB), as corroborated by the presence of faecal sterols and trialkylamines (Venkatesan and Kaplan, 1990; Venkatesan, 1995) and LABs (Chalaux et al., 1992) in sediments of the Santa Monica and San Pedro Basins. It should be recognized that distinguishing contaminants from sources other than sewage inputs remains a difficult task, and much work is needed to develop appropriate tools for such purposes as non-sewage sources have become increasingly important in the SCB.

The vast majority of sediment chemistry evaluations have been performed in a very small portion of the SCB. Only rarely have Bight-wide surveys been conducted to examine the overall condition of the region (Word and Mearns, 1979; Thompson et al., 1987, 1993; SCBPP Steering Committee, 1998). The last survey, conducted in 1994,

Table 2

Percent of mainland shelf area with sediment contamination that was detectable, anthropogenically enriched, or above the sediment quality guidelines effects range-low, and effects range median. Sediment chemistry data from Schiff and Gossett (1998). Anthropogenic enrichment based upon Schiff and Weisberg (1999). Sediment quality guidelines from Long et al. (1995).

	Detectable	Enriched	Effects range low	Effects range median
Arsenic	100.0	6.8	1.5	0.0
Cadmium	99.1	31.2	2.1	0.0
Chromium	100.0	21.4	7.3	0.0
Copper	100.0	16.4	6.8	0.0
Lead	100.0	16.5	0.5	0.0
Mercury	95.7	–	6.3	0.0
Nickel	99.9	3.2	3.2	1.8
Silver	98.5	20.2	7.3	1.0
Zinc	100.0	16.5	2.7	0.0
LMW PAH	0.0	0.0	0.0	0.0
HMW PAH	0.0	0.0	0.0	0.0
Total DDT	81.8	81.8	63.7	10.4
Total PCB	45.6	45.6	15.3	0.7
Any trace metal	100.0	50.1	13.7	2.8
Any organic	82.1	82.1	63.7	10.4
Any contaminant	100.0	89.0	66.8	12.3

collected and analyzed about 250 sediment samples from Point Conception to the U.S.–Mexico International Border with water depths ranging from 30 to 200 m. This survey identified that 89% of the SCB sediments were anthropogenically contaminated (Table 2) (Schiff, 1999). When effects sediment quality guidelines were used to evaluate potential biological effects (Long et al., 1995), 12% of the SCB sediments contained at least one contaminant at a level where biological effects were likely, and about 66% of the SCB sediments contained contaminants at levels where biological effects may occasionally occur.

Total DDT is the most widespread contaminant of concern in the SCB (Fig. 5, Table 2) (Schiff and Gossett, 1998). In addition to the historical inputs of DDT-enriched wastes via sewage outfalls, acid wastes containing DDTs were also dumped at two locations in the Santa Monica and San Pedro Basins, respectively, until the late 1960s (Chartrand et al., 1985). The composition of originally discharged DDT residues has been well preserved under the anoxic conditions in the deep basin sediments (Venkatesan et al., 1996), which is sharply different from those found in the nearshore sediments (Zeng and Venkatesan, 1999). In two sediment cores taken in areas adjacent to the Santa Monica Basin and San Pedro Basin dump sites, the %DDEs varied with the time of deposition. At the sediment sections related to the periods during which dumping was active, %DDEs values were similar to those found in the dumped acid wastes. On the other hand, surface layer sediments had %DDEs values similar to those found in the nearshore sediments (i.e., Palos Verdes Shelf and Santa Monica Bay). These observations further verify that sediment DDTs are widely distributed due to continuing dispersal of older deposits to distant areas such as the Santa Monica and San Pedro Basins (Zeng and Venkatesan, 1999).

Sediments near sewage outfalls reflect the trend of mass emissions over the last 50 years, found by examining sediment core samples (Fig. 6) (Eganhouse et al., 1988; SCCWRP, 1995; Stull et al., 1996; Anderson et al., 1999). Concentrations of trace metals, PAHs, total organic carbon, and total extractable hydrocarbons experienced a steady

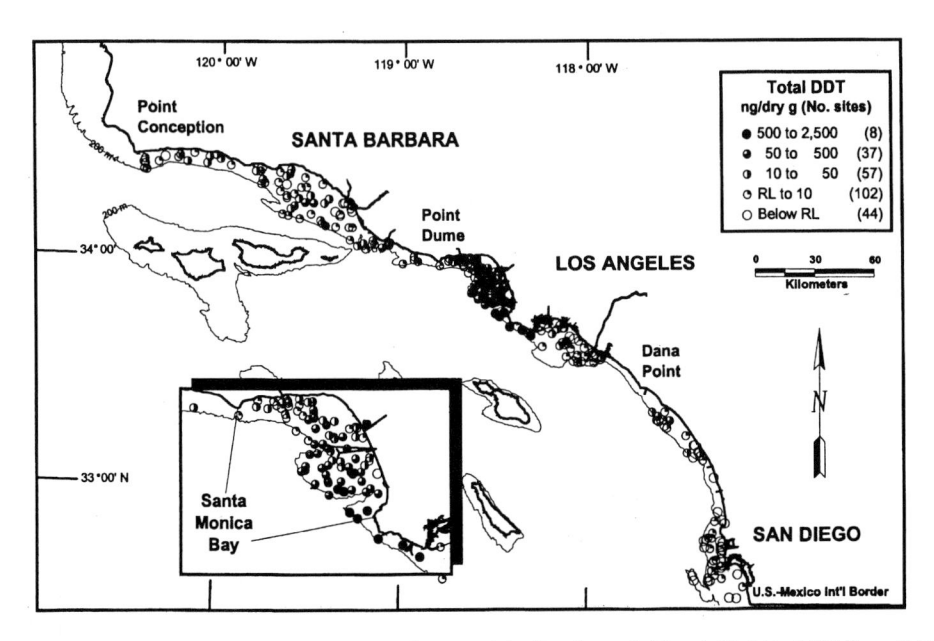

Fig. 5. Distribution of total DDT concentrations in sediment of the Southern California Bight in 1994 (from Schiff, 1999).

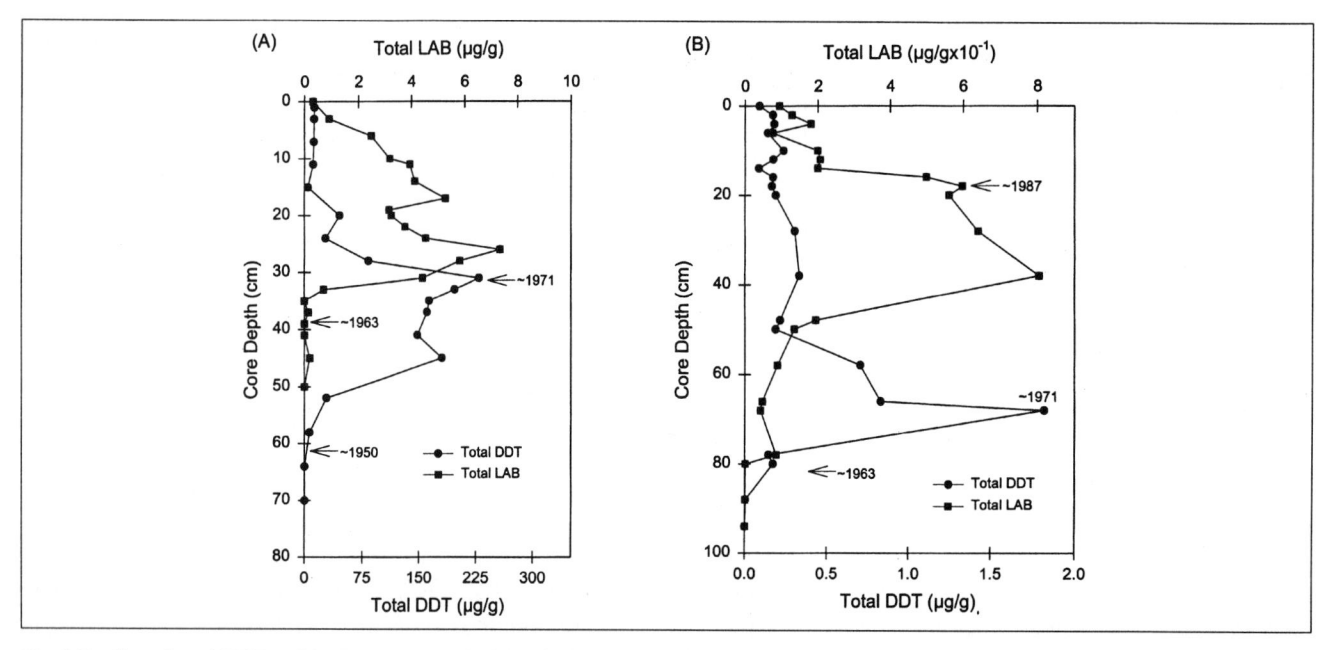

Fig. 6. Profiles of total DDT and LAB concentrations in sediment cores collected near sewage treatment plant outfalls from (a) the Palos Verdes Shelf and (b) Santa Monica Bay (SCCWRP, 1995).

increase prior to 1970 when mass emissions from POTWs were also increasing. After 1970, sediment concentrations decreased as the mass emissions of organic constituents and trace metals declined due to the constantly improved treatment methods and better source control. Concentrations of chlorinated hydrocarbons such as DDTs and PCBs peaked at core depths corresponding to 1970 when the mass emissions of these compounds peaked in POTW effluents in the SCB.

Biota

Biota capable of accumulating contaminants provide another avenue for redistribution of contaminants. Patterns of bioaccumulation in the SCB indicate that tissue residues of trace metals, organotins, DDTs, PCBs, PAHs, chlordane, and dieldrin are related to the sources of these constituents (Mearns et al., 1991). For example, increasing concentrations of chlorinated hydrocarbons have been observed in plankton, invertebrates (mussels and sand crabs), and a variety of fish species as sample locations approach the Palos Verdes Shelf. A Bight-wide survey in 1994 examined bioaccumulation of chlorinated hydrocarbons in three flatfishes (Pacific sanddab, Longfin sanddab, and Dover sole) and found that nearly 100% of each sanddab species and the majority of the sole were contaminated with total DDT (Schiff and Allen, 1999). Bioaccumulation of DDTs and PCBs has been measured in higher order predators such as fish-eating birds (e.g., brown pelican, bald eagle, and double-crested cormorant) and mammals (e.g., sea lion, dolphin, and sea otter). The spatial relationships to sources for these higher order predators are less clear-cut, however, due to their extended hunting range and potential depuration methods such as reproduction and nursing.

Bioaccumulation patterns in the SCB can be characterized as follows. First, the magnitude of bioaccumulation is generally consistent with the magnitude of sediment contamination. Bioaccumulation of chlorinated hydrocarbons shows the highest correlations with sediment contamination among all of the contaminant classes. Trace metals appear to be the only exception; increased trace metal tissue concentrations have been observed in fish caught from relatively clean areas such as Dana Point (Mearns et al., 1991). Second, concentrations of all contaminants in marine species have experienced a dramatic decline during the last three decades. In the 1994 Bight-wide survey, DDTs and PCBs were detected in 100% of the Pacific sanddab and Longfin sanddab individuals, but all 12 of the non-DDT pesticides were consistently not detected. Moreover, DDT and PCB concentrations in Pacific sanddab and Longfin sanddab collected from similar locations had decreased as much as two orders of magnitude from 1985 to 1994 (Schiff and Allen, 1999). Third, PAHs were rarely detected at low levels (Mearns et al., 1991). This is largely attributed to the ability of marine organisms to metabolize PAHs.

EFFECTS OF ANTHROPOGENIC ACTIVITIES ON MARINE BIOTA

Effects of human activities in the SCB are evaluated at five phylogenetic levels in this article, including kelp (algae), benthic invertebrates, fish, birds, and marine mammals. Effects on algal species in the SCB have focused largely on giant kelp because of its important role in SCB ecology as well as its value to humans as a recreational and commercial resource. Benthic invertebrates, predominantly soft-bottom infauna, have been studied in great detail owing to their "sentinel" status. Infauna live within the sediments where potential exposure to deposited contaminants is greatest, they are relatively sensitive to pollutant effects, and they are immobile so they cannot escape a large-scale human impact. Effects on fish communities in the SCB have been studied from two perspectives: fisheries-related and pollutant-related. Fisheries-related assessments focus on impacts to populations of specific species, mostly due to the commercial or recreational value to humans. Pollutant-related assessments focus more on impacts to marine ecosystems. This ecosystem evaluation has two components including effects on individual fish, such as impaired reproduction, or on fish assemblages, such as imbalanced communities.

Kelp

Kelp beds (*Macrocystis pyrifera*) are important habitats in the SCB because they provide habitat and food for rocky subtidal environments. Moreover, drift kelp provides food and habitat for a variety of pelagic and benthic organisms. Although kelp canopy extent in the SCB is estimated to be 88 km^2, less than 0.1% of the SCB area, kelp beds provide nearly 6% of the total energy input into the SCB (Hood 1993). Kelp growth rates have been measured up to 2 m per day. Therefore, impacts to kelp beds can exert large effects in the SCB.

Kelp beds are sensitive to both natural and anthropogenic impacts. Natural impacts occur largely as the result of fluctuations in water temperature, nutrient availability, and wave energy. The most severe natural disturbances occur during El Niño periods when warm, nutrient-poor waters negatively affect kelp growth and recruitment. In addition, unusually strong storms occur during El Niño periods that can tear kelp holdfasts from their rocky substratum. Anthropogenic impacts can also damage kelp beds. Such impacts have been attributed to increased turbidity that reduces light penetration (Dean et al., 1987) or to potentially toxic pollutants. In other cases, anthropogenic impacts such as inputs of nutrients from POTWs have been shown to stimulate kelp growth (Tegner et al., 1995). Kelp is currently used in many effluent monitoring programs for toxicity testing to protect ecosystem health (U.S. EPA, 1995).

The kelp beds near POTWs have been slowly returning to their historical areal extent since the 1950s, when extensive kelp beds, such as those near Palos Verdes, were dramatically reduced (Fig. 7). The decline of kelp beds has been attributed to a combination of natural and anthropogenic effects. In the late 1950s, the waters near Palos Verdes were experiencing an increase of pollutant inputs from a nearby sewage outfall while, at the same time, a strong El Niño had occurred in the SCB. In addition, herbivorous sea urchins were abundant as a result of the overfishing of their main competitors (abalone) and predators (lobsters and California sheephead) (Murray

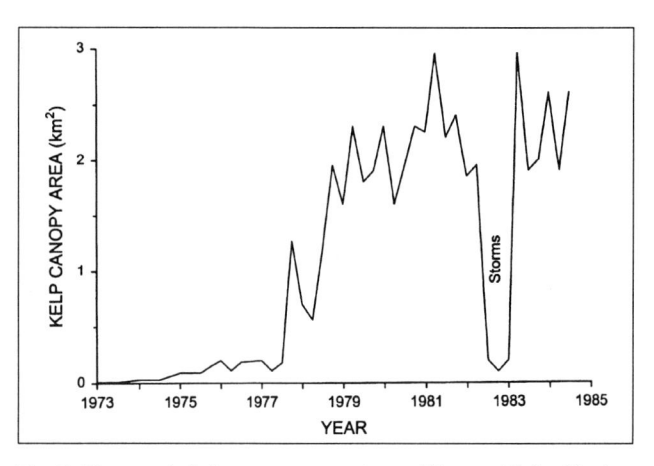

Fig. 7. Changes in kelp canopy over time offshore of Palos Verdes, California (from MBC, 1988).

and Bray, 1993). As pollutant inputs decreased, the kelp bed canopy returned and eventually flourished.

Benthos

Studies of benthic organisms have been used for over 30 years to assess environmental conditions in the SCB (Jones, 1969; Word and Mearns, 1979). Most of the benthic monitoring in southern California has examined the mainland shelf (to 100 m depth) with an emphasis on describing the localized effects of discharges from individual sources, including POTW outfalls (Stull et al., 1986; Zmarzly et al., 1994; Diener et al., 1995; Dorsey et al., 1995), industrial discharges (Southern California Edison Company, 1997), dredged material disposal (U.S. EPA 1987), and urban stormwater runoff (Bay et al., 1998).

The most consistent and severe impacts to the benthos have been associated with POTW discharges. Most of the sewage produced by the coastal population of southern California is treated and disposed of through four large ocean outfall systems, each discharging between 250 and 500 billion litres/day. Monitoring studies in the 1970s near these outfalls showed altered benthic communities characterized by low species diversity and reduced abundance in areas nearest the outfalls. These communities were dominated by deposit-feeding annelids and had markedly reduced abundances of key crustacean and echinoderm species, such as the ophiuroid *Amphiodia urtica*. Impacts co-occurred with elevated sediment concentrations of a number of discharge constituents, including organic carbon, chlorinated hydrocarbons, and trace metals.

Substantial improvements to the condition of the benthos near POTW outfalls have occurred in the last two decades, the result of improvements in effluent treatment and industrial source controls. Changes in the benthos off the Palos Verdes Peninsula illustrate this situation (Stull, 1995; Bergen et al., 1999). In the 1970s, degraded conditions (defaunation or the loss of major taxonomic groups) extended more than 15 km away from a large POTW outfall system operated by the County of Los Angeles. By 1990 conditions had improved to the extent that the most severe effects were absent, strong impacts were restricted to an area within 5 km of the outfall, and communities typical of undisturbed areas were present within the monitoring zone.

Areas of altered benthos due to environmental stress presently occupy a very small area of the mainland shelf of the SCB. A regional survey was conducted in 1994 that examined benthic communities at 251 sites between Point Conception and the U.S.–Mexico International Border (Bergen et al., 1999). Benthic communities were classified as typical of reference areas over approximately 91% of the shelf (Fig. 8), with an additional 8% of the area showing slight differences in community composition. Reduced benthic diversity was observed over approximately 2% of the shelf, with many of these areas located near river discharges (Fig. 8).

Aside from the areas near POTW outfalls, the relationship between benthic community alterations on the mainland shelf and anthropogenic activities is unclear. Over 89% of the SCB has evidence of chemical contamination (Schiff, 1999), yet most communities appear to be unaffected. Altered benthos near river discharges may reflect the combined effects of natural disturbances from flooding and transient contaminant inputs from agricultural and urban runoff (Bergen et al., 1999).

Sediment toxicity studies have also been used to assess sediment quality in the SCB. These studies, first conducted in the 1980s, show temporal improvements in sediment quality near POTW outfalls that correspond to reduced chemical mass emissions and improved benthic communities (Swartz et al., 1986; SCCWRP, 1992). Regional studies during 1992–94 have examined offshore areas as well as bays and estuaries. Sediment quality was highest on the mainland shelf (Bay, 1996), where no significant toxicity to benthic invertebrates was detected (Fig. 9). Toxicity was prevalent in enclosed bays and estuaries, however, with 14–66% of the area toxic to amphipod crustaceans (Long et al., 1996). The incidence of sediment toxicity was correlated with a number of contaminants, most notably copper, zinc, PAHs, PCBs, and Chlordane (Fairey et al., 1998). The spatial extent of sediment toxicity in southern California bays and estuaries is relatively high, compared to a national average of 11% (Long et al., 1996).

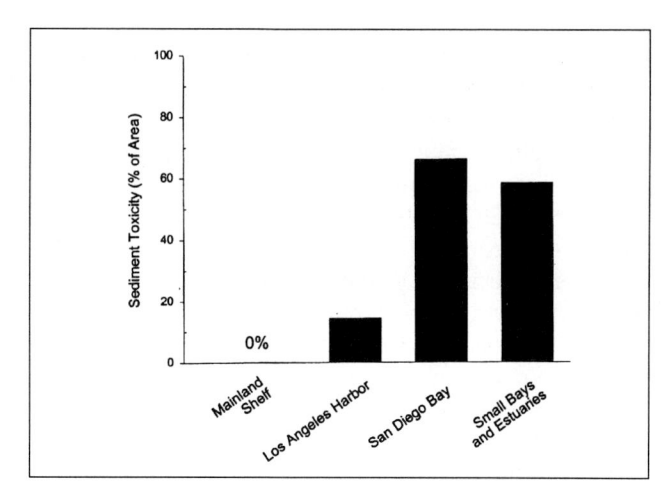

Fig. 9. Percent of area in different habitats of the Southern California Bight that exhibited sediment toxicity to amphipod crustaceans during 1994.

Commercial Fisheries

Commercial fishing is or has been conducted by nets (purse seines, lampara nets, otter trawls, gill-nets, seines), hook-and-line (longlines, rod-and-reel), and traps. Throughout much of the century, the pelagic wetfish fishery dominated the fisheries of the SCB (Frey, 1971). This fishery primarily targets four species including Pacific sardine (*Sardinops sagax*), northern anchovy (*Engraulis mordax*), chub (= Pacific) mackerel (*Scomber japonicus*), and jack mackerel (*Trachurus symmetricus*); all are primarily caught by purse seine. The Pacific sardine fishery dominated from the early 1920s to the early 1950s, but crashed completely from the 1950s through the 1970s (Wolf and Smith, 1992). Chub mackerel followed a similar pattern to Pacific sardine (Konno and Wolf, 1992). Northern anchovy and jack mackerel were the primary focus of the fishery from the 1950s through 1982. Chub mackerel and California market squid (*Loligo opalescens*) were dominant in the 1980s and California market squid has been the dominant fishery in the 1990s (CDFG, 1998).

The Pacific sardine fishery was reopened in 1986 and, since that time, landings have increased. Although intensive fishing (and perhaps increasing pollution) may have contributed to the crash in sardine populations, a historical record of relative abundance of scales of pelagic

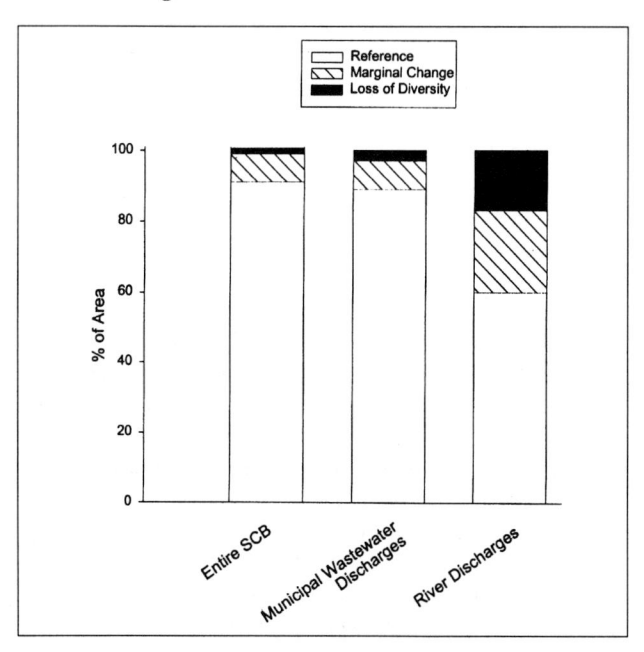

Fig. 8. Percent of area on the mainland shelf of the Southern California Bight with impacted benthic infaunal communities during 1994.

fishes from sediment cores in the Santa Barbara Basin suggests that sardines have undergone natural population explosions and crashes (20–80 years in duration) during the past 2,000 years (Soutar and Isaacs, 1969; Baumgartner et al., 1992).

Commercial trawl fishing is currently limited to the Santa Barbara Channel (off Santa Barbara and off the Channel Islands), and is excluded within 1.9 km (1 nautical mile) from shore (Barsky, 1990). Gillnet fishing was very important during the 1970s and 1980s, but was banned within 4.8 km (3 miles) of the coast in 1990. Set gillnets were banned due to potential impacts on populations of target species (California halibut, *Paralichthys californicus*, and white seabass, *Atractoscion nobilis*); drift gillnets were banned to protect incidental by-catch of non-target species including marine mammals and seabirds (Weber, 1997). Trap fisheries have come and gone; currently a trap fishery provides live fish for restaurants (CDFG, 1998). Red sea urchin (*Strongylocentrotus franciscanus*) was one of the most important fisheries by value in the late 1980s, but landings have decreased since 1988 (CDFG, 1998). Abalone (*Haliotis* spp.) populations were heavily fished in the 1970s and 1980s, and populations have decreased so dramatically that all fishing in the SCB has been banned since May of 1997 (CDFG, 1998).

Recreational Fisheries

Fishing is an important recreational activity in the SCB. Recreational landings averaged 14 million fish annually between 1981 and 1984 (Helvey et al., 1987). Recreational catches vary in species composition between fishing modes and over time. Barred sand bass (*Paralabrax nebulifer*), rockfishes (*Sebastes* spp.), and kelp bass (*Paralabrax clathratus*) have generally been among the most important species on commercial passenger fishing vessels (CDFG, 1998). Rockfishes (*Sebastes* spp.) are important during the fall to winter season, whereas basses (*Paralabrax* spp.) and warm-water migrants (Pacific barracuda, *Sphyraena argentea*; yellowtail, *Seriola lalandis*) are important during the summer season. Chub mackerel is one of the most important species taken by anglers on piers or boats (Allen et al., 1996). Rockfish abundance has decreased during the past two decades, probably due to decreased recruitment during the recent period of ocean warming and to overfishing (Love et al., 1998).

Pollution Effects on Fish

High levels of total DDT and total PCB have been implicated in reproductive impairment in white croaker (*Genyonemus lineatus*) sampled from San Pedro Bay (Cross and Hose, 1988; Hose et al., 1989; Hose and Cross, 1994). The bioaccumulation of these chlorinated hydrocarbons was correlated with decreased proportions of spawning females, decreased fecundity, and decreased fertilization success. Increased levels of these compounds have also been correlated to subcellular damage in white croaker and kelp bass. An increased frequency of micronuclei, a by-product of DNA damage, was observed in blood cells of these two fish species with increased concentrations of total DDT and total PCB. Other subcellular markers of contaminant exposure have been observed in additional fish species including Dover sole, hornyhead turbot (*Pleuronectes verticalis*), and English sole(*Pleuronectes vetulus*).

Fish diseases, such as fin erosion and epidermal tumours, are another mechanism by which impacts from human activities are manifested in individual fish. Fin erosion levels were very high in the early 1970s on the Palos Verdes Shelf near the POTW outfall, affecting 33 of 151 species examined, when mass emissions from POTWs were highest (Mearns and Sherwood, 1977). Dover sole was the species with the highest prevalence (30%) of fin erosion in the 1970s, almost all of which occurred on the Palos Verdes Shelf. By the middle 1980s, fin erosion had disappeared in Dover sole from that area (Stull, 1995) and was virtually absent in Dover sole throughout the SCB in 1994 (Allen et al., 1998). The cause of this disease was not determined, but was assumed to be related to sediment contamination, based upon its highest prevalence in benthic fishes found near the POTW outfall and its disappearance as wastewater treatment improved.

Epidermal tumours were found consistently in Dover sole sampled from the SCB in the early 1970s (Mearns and Sherwood, 1977). Prevalence was highest in juveniles (60–120 mm) and was similar among areas in the central mainland coast of the SCB. However, the disease had been found outside of the SCB as far back as 1946 (Mearns and Sherwood, 1977). Incidences of this disease have decreased since the 1970s and are presently at background levels (Allen et al., 1998). These x-cell pseudotumours are probably the result of amoeboid parasitism (Cross, 1988). Oral papillomas in white croaker and microscopic liver abnormalities in several species have been found near the Los Angeles area, but the prevalence of these abnormalities is low (Mearns and Sherwood, 1977; Malins et al., 1986).

Fish Community Impacts

Fish assemblages in the SCB vary by habitat and depth with distinct bay, rocky bottom, and soft-bottom assemblages (Allen, 1985). Demersal (soft-bottom) fish assemblages are the focus of studies of pollution impact because outfalls are on soft-bottom habitat and the species are relatively sedentary and respond to environmental stress. Demersal fish assemblages of the shelf vary in species composition by depth, with assemblages roughly corresponding to inner shelf (10–30 m), middle shelf (30–100 m), and outer shelf (100–200 m) zones, and with some distinct assemblages forming where these zones overlap (Allen, 1982; Allen and Moore, 1997; Allen et al., 1998; Allen et al., 1999). Fish assemblages also change with depth along the slope below

200 m and in the basins (Allen and Mearns, 1977; Cross, 1987).

Demersal fish communities have shown shifts in abundance, biomass, diversity, and species composition in response to wastewater discharges in the SCB. Some species typical of reference assemblages were absent from and other species were attracted to outfalls (Allen, 1977; Cross et al., 1985; Stull and Tang, 1996). Missing reference assemblage species included hornyhead turbot and California tonguefish (*Symphurus atricauda*), while species attracted to the outfall included white croaker, shiner perch (*Cymatogaster aggregata*), and curlfin sole (*Pleuronichthys decurrens*). The changes in species composition were attributable to food habits; as the benthic infaunal community changed from crustaceans to polychaetes, the fish species that preferred crustaceans also diminished. As wastewater treatment improved and benthic communities recovered, outfall fish assemblages became more similar to reference communities at the same depths (Stull and Tang, 1996; Allen et al., 1998).

Natural factors also play a role in changes to fish communities. For example, demersal fish communities changed somewhat in species composition between the 1970s and the 1990s, following a warming trend in SCB waters (Allen, 1982; Allen and Moore, 1997). Most changes occurred on the inner shelf (10–30 m) and middle shelf (30–100 m) regions, and were largely the result of increased abundances of warm-water species and decreased abundances of cool-water species.

Birds

Bird populations in the SCB have been impacted by three factors: contamination, loss of habitat, and changes in food availability. Three common species (the brown pelican, California least tern, and Belding's savannah sparrow) are currently listed as endangered species and receive federal protection under the U.S. EPA Endangered Species Act of 1973.

Dramatic declines in the populations of seabirds (brown pelican and double crested cormorant) and birds of prey (bald eagle and peregrine falcon) were observed during the late 1950s to 1970s (Anderson and Hickey, 1970). Discharges and biomagnification of DDTs (and possibly PCBs) during this time caused severe eggshell thinning and nest abandonment in these species and led to reproductive failure in breeding colonies. Diminished reproduction in seagulls due to the estrogenic effects of DDT was also observed (Fry et al., 1987). Reduction of DDT inputs resulted in a reduction of thinning and the recovery of seabird populations to historic levels (Gress, 1994). Temporal variations in seabird populations still occur in the SCB, but these changes appear to be associated with reductions in food supply caused by natural factors such as El Niño events (Table 3).

Table 3

Nesting, reproduction and productivity estimates for brown pelicans from the Channel Islands in the southern California Bight. Shaded rows indicate intervals when El Niño conditions were present, which were often associated with decreased reproductive success. Data from Gress (1994).

Year	Nest attempts	Young produced[a]	Productivity[b]
1969	750	4	0.01
1970	552	1	0.00
1971	540	7	0.01
1972	261	57	0.22
1973	247	34	0.14
1974	416	305	0.73
1975	292	256	0.88
1976	417	279	0.67
1977	76	39	0.51
1978	210	37	0.18
1979	1258	980	0.78
1980	2244	1515	0.68
1981	2946	1805	0.61
1982	1862	1175	0.63
1983	1877	1159	0.62
1984	628	530	0.84
1985	6194	7902	1.28
1986	7349	4601	0.63
1987	7167	4898	0.68
1988	2878	2500	0.87
1989	5959	3500	0.59
1990	2425	654	0.30
1991	6383	1792	0.28
1992	1752	394	0.22
1993	4157	2811	0.68

[a]Total number of young fledged from all nests.
[b]Number of young produced per nest attempt.

Bioaccumulation of DDTs and PCBs remains prevalent in many bird species and is still producing adverse biological effects in sensitive species. Reintroduction programs have reestablished breeding populations of bald eagles and peregrine falcons on the Channel Islands, but reproductive success is still reduced by eggshell thinning (Wiemeyer et al., 1993). The reproductive output of brown pelicans is still below levels considered necessary to maintain a stable population and eggshell thinning is still observed (Gress, 1994).

Development of coastal bays and estuaries has severely restricted the habitat available for breeding and foraging for several species, leading to population declines in species such as the California least tern and Belding's savannah sparrow. Between 75 and 90% of coastal wetland habitat in southern California has been dredged, filled, or otherwise modified. Restrictions on coastal development and remediation efforts in recent years have helped to stabilize the populations of these endangered species.

Marine Mammals

At least six species of pinnipeds occur in the SCB (Bonnell and Dailey, 1993) and some of the largest rookeries on the west coast of the U.S. are found along the Channel Islands (primarily San Miguel Island). The most common species in the SCB is the California sea lion (*Zalophus californianus*). In the early part of this century, the northern elephant seal (*Miroranga angustirostris*) was nearly extinct as a species. As a result of protection from harvesting, this species has become one of the more abundant pinnipeds off the California coast (Bonnell and Dailey, 1993).

Marine mammals that are resident in the SCB (e.g., sea lions and coastal dolphins) bioaccumulate high concentrations of DDTs and PCBs in their tissues. Tissue concentrations peaked in the 1970s and were associated with an increased incidence of premature births in sea lions (DeLong et al., 1973). Present-day tissue concentrations remain high in southern California marine mammals, but a causal link with adverse biological effects has not been established. The population of the California sea lion is increasing at an annual rate of 5–10%. Hundreds of young seals and sea lions strand themselves along the coast each year, but the principal factors in these strandings are lack of food due to unfavourable oceanographic conditions and parasitic infestations (Bonnell and Dailey, 1993).

Cetaceans are also diverse in the SCB. Among toothed whales, dolphins are the most prominent, with several species common to the area. The most commonly observed baleen whale in the area is the grey whale (*Eschrichtius robustus*), which makes seasonal migrations through the area (winter and spring) between the Bering Sea and the lagoons off southern Baja California (Bonnell and Dailey, 1993). Although populations were depleted in the early part of the century due to whaling, populations have rebounded sufficiently that this species is no longer listed as endangered.

HUMAN HEALTH CONCERNS

Human health concerns in the SCB focus on two distinct areas. The first is the risk of contracting illness through body-contact recreation (i.e., swimming, surfing, SCUBA-diving, etc.) along the shoreline. The second human health concern is the risk of long-term illness, including cancer, from consuming contaminated fish.

Shoreline Impacts

Beaches are among the most valuable recreational resources in the SCB. The U.S. Lifesaving Association estimated that 146 million beach-goers visited SCB beaches in 1997, more swimmers than visited beaches in the states of Florida, Hawaii, and New Jersey combined (Schiff et al., 1999). This influx justifies the tremendous quantity of recreational shoreline monitoring conducted in the SCB; an estimated $3 million is spent annually assessing the water quality along high-use beaches. This monitoring consists of sampling and measuring indicator bacteria such as total coliform, faecal coliform, and enterococcus.

In the summer of 1998, an integrated assessment of water quality along the SCB shoreline was conducted (Noble et al., 1999). Approximately 95% of the shoreline mile-days met water quality standards; only 5% of the shoreline during any summer day at any random location would have exceeded water quality thresholds for bacteria contamination. However, this level of water quality exceedence was not consistent among various shoreline types. The highest use sandy beaches (>50,000 swimmers per year) exceeded water quality thresholds less than 2% of their shoreline mile-days, while those adjacent to storm drain outfalls that discharged surface runoff exceeded thresholds as much as 60% of their shoreline mile-days. The density of bacteria that are measured in surface runoff can be quite high. Measurements of faecal coliforms in wet weather discharges routinely reach 10^5–10^6 cfu/100 mL and densities of indicator bacteria are highly correlated to rainfall. Moreover, spatial correlations have been observed as densities along the shoreline decrease with distance from surface runoff outfalls.

Bacteria such as coliforms and enterococcus are only indicators of the potential pathogens that can induce illness. Measurements of human enteric virus have shown that potential pathogens are frequently found in surface runoff, even in dry weather when natural flows do not exist. An epidemiological study conducted in Santa Monica Bay in 1994 indicated that swimmers are more likely to become ill after swimming near storm drain discharges (Haile et al., 1999). The incidence of illness increased from 88 to 305 out of 10,000 swimmers within 50 m of a storm drain. The types of illnesses most frequently observed included significant respiratory disease, fever, coughing with phlegm, chills, highly credible gastroenteritis, and ear discharge.

Seafood Consumption

Health risks from consumption of seafood organisms in the SCB come primarily from two carcinogens, DDTs and PCBs. Current sources of both carcinogens are sediments in the Los Angeles area. Some recreationally caught seafood organisms (e.g., white croaker) have sufficiently high levels that health advisories restricting consumption are posted in pertinent fishing areas (SCCWRP, 1994; Pollock et al., 1991). Advisories are currently posted when advisory tissue concentrations of DDTs and PCBs exceed 100 ppb wet weight. Advisories have been posted for white croaker, California corbina (*Menticirrhus undulatus*), queenfish (*Seriphus politus*), surfperches (Embiotocidae spp.), and California scorpionfish (*Scorpaena guttata*), with white croaker being the most restricted. Advisories have been posted in fishing areas along the Los Angeles County and Orange County coasts; most advisories were in the Palos Verdes and Los Angeles Harbor–Long Beach Harbor areas, where consumption of white croaker was not recommended. White croaker were nevertheless consumed in these areas, mostly (57%) by Hispanic anglers (Allen et al., 1996). Although commercial fishing for white croaker is banned in that area, commercially caught white croaker with high levels of DDT sometimes appear in local Asian markets (Gold et al., 1998).

MONITORING AND MANAGEMENT OF THE SCB

A tremendous amount of marine monitoring is conducted in the SCB, yet few of the results are used to assist environmental managers tasked with stewarding natural resources. More than 40 local, state or federal public agencies, private industry, and academic institutions spent an estimated $17 million in 1990 on marine monitoring (NRC, 1990). The majority of these funds were used to assess impacts from single sources in the immediate vicinity of a discharge in compliance with National Pollutant Discharge Elimination System (NPDES) permits.

Most marine monitoring in the SCB cannot be used to assist environmental managers in managing the coastal ecosystem (NRC, 1990) due to several fundamental reasons. First, management objectives are not clearly defined and managers are often unsure of what to do with the monitoring data once it has been collected. For example, many monitoring programs were designed in the early 1970s when little was known about the ocean environment and programs were based largely upon characterization of near-field and far-field conditions. After 30 years of monitoring, our understanding has improved and management questions required to maintain or improve beneficial uses of the coastal zone have changed; yet the monitoring programs have remained inflexible. Second, monitoring is inefficient due to poor sampling designs, a lack of monitoring objectives, and the failure of most programs to incorporate natural variability in space or time. For example, each of the NPDES monitoring programs has been designed by different individuals in separate jurisdictional agencies. The result is a series of disparate programs that cannot be integrated to make large-scale assessments of the overall condition of the SCB. Despite all of the effort that is expended, only 5% of the SCB mainland shelf is actually monitored.

Resource managers are attempting to address these limitations as we move into the new millennium by refining important management questions and redesigning monitoring programs. Both regulators and dischargers have agreed upon a set of four questions that embody the needs of management objectives:

– Is it safe to swim in the ocean?
– Is it safe to eat the local seafood?
– Is the health of the ecosystem being safeguarded?
– Are fisheries and other living resources being adequately protected?

Regional monitoring is one example of how SCB resource managers are answering some of these questions and addressing the issues that have plagued current programs (Cross and Weisberg, 1996). Regional monitoring in the SCB, unlike other areas of the nation, consists of local agencies participating in an integrated and coordinated effort; a minor proportion of effort comes from agencies outside of the SCB (e.g., the U.S. Environmental Protection Agency). Instead, local regulators and permittees work cooperatively to define monitoring objectives. Regulators provide a cost-neutral exchange of effort to participating agencies by relieving permittees of a subset of routine monitoring activities. Regional monitoring designs move away from traditional "end-of-pipe" stations and sample the entire SCB. This strategy enables assessments of the full range of natural variability, evaluation of the effect of cumulative impacts from multiple discharges, and determination of the overall condition of the SCB. The involvement of multiple agencies in regional monitoring provides benefits that extend beyond the single project, by setting minimum quality assurance goals for all programs that ensure comparability, establishing data management systems to share information, and establishing communication among regulatory jurisdictions and discharge agencies.

REFERENCES

Allen, L.G. (1985) A habitat analysis of the nearshore marine fishes from southern California. *Bulletin of the Southern Californian Academy of Science* 84(3), 133–155.

Allen, M.J. (1977) Pollution-related alterations of southern California demersal fish communities. *American Fisheries Society, Cal-Neva Wildlife Transactions* 1977, 103–107.

Allen, M.J. (1982) Functional structure of soft-bottom fish communities of the southern California shelf. Ph.D. dissertation. University of California, San Diego, La Jolla, California. 577 pp.

Allen, M.J., Diener, D., Mubarak, J., Weisberg, S.B. and Moore, S.L. (1999) Demersal fish assemblages of the mainland shelf of southern California in 1994. In Southern California Coastal Water Research Annual Report 1997–1998 (S.B. Weisberg and D. Hallock, eds.), pp. 101–112. S. Calif. Coastal Water Res. Proj., Westminster, California.

Allen, M.J. and Mearns, A.J. (1977) Bottomfish populations below 200 meters. In annual report for the year ended 30 June 1977, pp. 117–120. S. Calif. Coastal Water Res. Proj., El Segundo, California.

Allen, M.J. and Moore, S.L. (1997) Recurrent groups of demersal fishes on the mainland shelf of southern California in 1994.. In Southern California Coastal Water Research Project, Annual Report 1996 (S.B. Weisberg, C. Francisco and D. Hallock, eds.), pp. 122–128. So. Calif. Coastal Water Res. Proj., Westminster, California.

Allen, M.J., Moore, S.L., Schiff, K.C., Weisberg, S.B., Diener, D., Stull, J.K., Groce, A., Mubarak, J., Tang, C.L. and Gartman, R. (1998) Southern California Bight Pilot Project: V. Demersal fishes and megabenthic invertebrates. So. Calif. Coastal Water Res. Proj., Westminster, California. 324 pp.

Allen, M.J., Velez, P.V., Diehl, D.W., McFadden, S.E. and Kelsh, M. (1996) Demographic variability in seafood consumption rates among recreational anglers of Santa Monica Bay, California in 1991–1992. *Fishery Bulletin* 94, 597–610.

Anderson, D.W. and Hickey, J.J. (1970) Oological data on egg and breeding characteristics of brown pelicans. *Wilson Bulletin* 82, 14–28.

Anderson, J.W., Zeng, E.Y. and Jones, J.M. (1999) Correlation between response of human cell line and distribution of sediment polycyclic aromatic hydrocarbons and polychlorinated biphenyls on Palos Verdes Shelf, California, USA. *Environmental Toxicology and Chemistry* 18, 1506–1610.

Anderson, J.W., Reish, D.J., Spies, R.B., Brady, M.E. and Segelhorst, E.W. (1993) Human impacts. In *Ecology of the Southern California Bight: A Synthesis and Interpretation*, eds. M.D. Dailey, D.J. Reish and J.W. Anderson, pp. 682–766. University of California Press, Berkeley, California.

Barsky, K.C. (1990) History of the commercial California halibut fishery. In *The California Halibut, Paralichthys californicus, Resource and Fisheries*, ed. C.W. Haugen., pp. 217–227. California Department of Fish and Game, Fish Bulletin 174.

Baumgartner, T.R., Soutar, A. and Ferreira-Bartrina, V. (1992) Reconstruction of the history of Pacific sardine and northern anchovy populations over the past two millennia from sediments of the Santa Barbara Basin. California Cooperative Oceanic Fisheries Investigations Report 33, pp. 24–40.

Bay, S.M. (1996) Sediment toxicity on the mainland shelf of the southern California Bight in 1994. In Southern California Coastal Water Research Project Annual Report 1994–95 (M.J. Allen, C. Francisco and D. Hallock, eds.), pp. 128–136. Westminster, California.

Bay, S., Schiff, K., Greenstein, D. and Tiefenthaler, L. (1998) Stormwater runoff effects on Santa Monica Bay, toxicity, sediment quality, and benthic community impacts. In *California and the World Ocean '97*, Vol. 2, pp. 900–921, eds. O.T. Magoon, H. Converse, B. Baird and M. Miller-Henson. American Society of Civil Engineers, Reston, Virginia.

Beers, J.R. (1986) Organisms and the food web. In *Plankton Dynamics of the Southern California Bight*, ed. R.W. Eppley, pp. 84–175. Springer-Verlag, New York, New York.

Bergen, M., Cadien, D., Dalkey, A., Montagne, D., Smith, R.W., Stull, J.K., Velarde, R.G. and Weisberg, S.B. (1999) Assessment of benthic infauna on the mainland shelf of Southern California. In Southern California Coastal Water Research Project Annual Report 1997–98 (S.B. Weisberg and D. Hallock, eds.), pp. 143–154. Westminster, California.

Bonnell, M.L. and Dailey, M.D. (1993) Marine mammals. *Ecology of the Southern California Bight: A Synthesis and Interpretation*, eds. M.D. Dailey, D.J. Reish and J.W. Anderson, pp. 605–681. University of California Press, Berkeley, California.

Briggs, J.C. (1974) *Marine Zoogeography*. McGraw-Hill Book Co., New York, New York. 475 pp.

California State Water Resources Control Board (1997) *California Ocean Plan*, p. 25.

CDFG (California Department of Fish and Game) (1998) Review of some California fisheries for 1997. California Cooperative Oceanic Fisheries Investigations Report 39, pp. 9–24.

Chalaux, N, Bayona, J.M., Venkatesan, M.I. and Albaigés, J. (1992) Distribution of surfactant markers in sediments from Santa Monica basin, Southern California. *Marine Pollution Bulletin* 24, 403–407.

Chartrand, A.B., Moy, S., Safford, A.N., Yoshimura, T. and Schinazim L.A. (1985) Ocean dumping under Los Angeles Regional Water Quality Control Board Permit: A review of past practices, potential adverse impacts, and recommendations for future action. Report from California Regional Water Quality Control Board, 47 pp.

Crisp, P.T., Brenner, S., Ventkatesan, M.I., Ruth, E. and Kaplan, I.R. (1979) Organic chemical characterization of sediment-trap particulates from San Nicolas, Santa Barbara, Santa Monica and San Pedro basins, California. *Geochimica Cosmochimica Acta* 43, 1791–1801.

Cross, J.N. and Weisberg, S.B. (1996) The southern California Bight pilto project: an overview. In Southern California Coastal Water Research Project Annual Report 1994–95 (M.J. Allen, C. Francisco and D. Hallock, eds.), pp. 104–108. Westminster, California.

Cross, J.N. (1988) Fin erosion and epidermal tumors in demersal fish from southern California. In *Ocean Processes in Marine Pollution, Vol. 5, Urban Wastes in Coastal Marine Environments*, eds. D.A. Wolfe and T.P. O'Connor, pp. 57–64. Krieger Publishing Co., Malabar, Florida.

Cross, J.N. and Hose, J.E. (1988) Evidence of impaired reproduction in white croaker (*Genyonemus lineatus*) from contaminated areas off southern California. *Marine Environmental Research* 24, 185–188.

Cross, J.N. (1987) Demersal fishes of the upper continental slope off southern California. *California Cooperative Oceanic Fisheries Investigations Reports* 28, 155–167.

Cross, J.N., Hardy, J.T., Hose, J.E., Hershelman, G.P., Antrim, L.D., Gossett, R.W. and Crecelius, E.A. (1987) Contaminant concentrations and toxicity of sea-surface microlayer near Los Angeles, California. *Marine Environment Research* 23, 307–323.

Cross, J.N., Roney, J. and Kleppel, G.S. (1985) Fish food habits along a pollution gradient. *California Department of Fish and Game* 71(1), 28–39.

CSWQCB (California State Water Quality Control Board) (1964) An investigation of the effects of discharged waste on kelp. California State Water Quality Control Board, Sacramento, California. Publ. No. 26. 124 pp.

Culliton, T., Warren, M., Goodspeed, T., Remer, D., Blackwell, C. and McDonough III, J. (1990) Fifty years of population changes along the nation's coasts. Coastal Trends Series, Report No. 2, National Oceanic and Atmospheric Administration, Strategic Assessment Branch. Rockville, MD.

Dailey, M.D., Anderson, J.W., Reish, D.J. and Gorsline, D.S. (1993) The California Bight: Background and setting. In *Ecology of the Southern California Bight: A Synthesis and Interpretation*, eds. M.D. Dailey, D.J. Reish, and J.W. Anderson, pp. 1–18. University of California Press, Berkeley, California.

Dean, T.A., Schroeter, S. and Dixon, J. (1987) The effect of the San Onofre Nuclear Generating Station on the giant kelp, *Macrocystis pyrifera*. Report to the Marine Review Committee, Inc. Santa Barbara, CA.

DeLong, R.L., Gilmartin, W.G. and Simpson, J.G. (1973) Premature births in California Sea Lions: association with organochlorine pollutant residue levels. *Science* 181, 1168–1170.

Diener, D.R., Fuller, S.C., Lissner, A., Haydock, C.I., Maurer, D., Robertson, G. and Gerlinger, T. (1995) Spatial and temporal patterns of the infaunal community near a major ocean outfall in southern California. *Marine Pollution Bulletin* 30, 861–878.

Dorsey, J.H., Phillips, C.A., Dalkey, A., Roney, J.D. and Deets, G.B. (1995) Changes in assemblages of infaunal organisms around wastewater outfalls in Santa Monica Bay, California. *Bulletin of the Southern California Academy of Science* 94, 46–64.

Eber, L.E. (1977) Contoured depth-time charts (0–200 m, 1950–1966) of temperature, salinity, oxygen and sigma-t at 23 CalCOFI stations in the California Current. California Cooperative Oceanic Fisheries Investigations, Atlas 25. 231 pp.

Eganhouse, R.P., Olaguer, D.P., Gould, B.R. and Phinney, C.S. (1988) Use of molecular markers for the detection of municipal sewage sludge at sea. *Marine Environmental Research* 25, 1–22.

Eganhouse, R.P. and Venkatesan, M.I. (1993) Chemical oceanography and geochemistry. *Ecology of the Southern California Bight: A Synthesis and Interpretation*, eds. M.D. Dailey, D.J. Reish, and J.W.

Anderson, pp 71–174. University of California Press, Los Angeles, California.

Emery, K.O. (1960) *The Sea off Southern California, a Modern Habitat of Petroleum.* John Wiley and Sons, New York, New York. 366 pp.

Fairey, R., Roberts, C., Jacobi, M., Lamerdin, S., Clark, R., Downing, J., Long, J., Hunt, J., Anderson, B., Newman, J., Tjeerdema, R., Stephenson, M. and Wilson, C. (1998) Assessment of sediment toxicity and chemical concentrations in the San Diego Bay region, California, USA. *Environmental Toxicology and Chemistry* 17, 1570–1581.

Frey, H.T. (1971) California's living marine resources and their utilization. California Department of Fish and Game, Sacramento, California. 148 pp.

Fry, D.M., Toone, C.K., Speich, S.M., and Peard, R.J. (1987) Sex ratio skew and breeding patterns of gulls: demographic and toxicological considerations. *Studies in Avian Biology* 10, 13–22.

Gold, M.D., Alamillo, J., Fleischli, S., Forrest, J., Gorke, R., Heibshi, L. and Gossett, R. (1998) Let the buyer beware: A determination of DDT and PCB concentrations in commercially sold white croaker. Heal the Bay, Santa Monica, California. 18 pp.

Green, D.R., Stull, J.K. and Heesen, T.C. (1986) Determination of chlorinated hydrocarbons in coastal waters using a moored *in situ* sampler and transplanted live mussels. *Marine Pollution Bulletin* 17, 324–329.

Gress, F. (1994) Reproductive performance, eggshell thinning, and organochlorines in brown pelicans and double-crested cormorants breeding in the Southern California Bight. Report 18 in Southern California Bight Natural Resource Damage Assessment Expert Reports. U.S. Department of Justice, Environment and Natural Resources Division. Washington, D.C.

Haile, R.W., Witte, J.S., Gold, M., Cressey, R., McGee, C., Millikan, R.C., Glasser, A., Harawa, N., Ervin, C., Harmon, P., Harper, J., Dermand, J., Alamillo, J., Barrett, K., Nides, M. and G.-Y. Wang. (1999) The health effects of swimming in ocean water contaminated by storm drain runoff. *Epidemiology* 10(4), 355–363.

Helvey, M.S., Crooke, J. and Milone, P.A. (1987) Marine recreational fishing and associated state–Federal research in California, Hawaii and the Pacific Island Territories. *Marine Fisheries Review* 49 (2), 8–14.

Hendricks, T.J. and Eganhouse, R.P. (1992) Modification and verification of sediment deposition model. Technical Report 265, Southern California Coastal Water Research Project, Westminster, California.

Hickey, B.M. (1993) Physical oceanography. In *Ecology of the Southern California Bight: A Synthesis and Interpretation*, eds. M.D. Dailey, D.J. Reish and J.W. Anderson, pp. 19–70. University of California Press, Berkeley, California.

Hood. D.W. (1993) Ecosystem relationships. In *Ecology of the Southern California Bight: A Synthesis and Interpretation*, eds. M.D. Dailey, D.J. Reish and J.W. Anderson, pp. 782–835. University of California Press, Berkeley, California.

Hose, J.E. and J.N. Cross. (1994) Evaluation of evidence for reproductive impairment in white croaker induced by DDT and/or PCBs. Prepared for National Oceanic and Atmospheric Administration. Prepared by Occidental Colleg, VANTUNA Research Group, Los Angeles, California. 54 pp.

Hose, J.E., Cross, J.N., Smith, S.G. and Diehl, D. (1989) Reproductive impairment in a fish inhabiting a contaminated coastal environment off southern California. *Environmental Pollution* 57, 139–148.

Jones, G.F. (1969) The benthic macrofauna of the mainland shelf of southern California. *Allan Hancock Monographs in Marine Biology* 4, 1–219.

Konno, E.S. and Wolf, P. (1992) Pacific mackerel. In *California's Living Marine Resources and Their Utilization*, eds. W.S. Leet, C.M. Dewees and C.W. Haugen, pp. 91–93. University of California, Davis, Department of Wildlife and Fisheries Biology, Sea Grant Extension Program, Davis, California. Sea Grant Extension Publication, UCSGEP-92-12.

Long, E.R., MacDonald, D.D., Smith, S.L. and Calder, F.D. (1995) Incidence of adverse biological effects within ranges of chemical concentrations in marine and estuary sediments. *Environmental Management* 19, 81–97.

Long, E.R., Robertson, R., Wolfe, D.A., Hameedi, J. and Sloane, G.M. (1996) Estimates of the spatial extent of sediment toxicity in major U.S. estuaries. *Environmental Science Technology* 30, 3585–3592.

Love, M.S., Caselle, J.E. and Herbinson, K. (1998) Declines in nearshore rockfish recruitment and populations in the Southern California Bight as measured by impingement rates in coastal electrical power generating stations. *Fishery Bulletin* 96, 492–501.

Lynn, R.J., Schwing, F.B. and Hayward, T.L. (1995) The effect of the 1991–1993 ENSO on the California Current System. *Cooperative Oceanic Fisheries Investigations Report*, 36, 57–71.

Malins, D.C., McCain, B.B., Brown, D.W., Myers, M.S. and Chan, S.L. (1986) Marine pollution study: Los Angeles vicinity. Final report to CSWRQB, Sacramento, California. Prepared by National Oceanic and Atmospheric Administration, National Marine Fisheries Service, Seattle, WA. 64 pp.

MBC Applied Environmental Sciences (1988) The state of Santa Monica Bay. Part One: Assessment of conditions and pollution impacts. Prepared for Southern California Association of Governments, Los Angeles, CA. Prepared by MBC Applied Environmental Sciences, Costa Mesa, CA. 420 pp.

McDermott, D.J., Heesen T.C. and Young, D.R. (1974) DDT in bottom sediments around five southern California outfall systems. TM 217, Southern California Coastal Water Research Project, El Segundo, California, 54 pp.

Mearns, A.J., Matta, M., Shigenaka, G., MacDonald, D., Buchman, M., Harris, H., Golas, H. and Laurenstein, G. (1991) Contaminant trends in the Southern California Bight: Inventory and assessment. NOAA Technical Memorandum NOS ORCA 62. 413 pp.

Mearns, A.J. and Sherwood, M.J. (1977) Distribution of neoplasms and other diseases in marine fishes relative to the discharge of wastewater. *Annals of the New York Academy of Science* 298, 210–224.

Murphree, T. and Reynolds, C. (1995) El Niño and La Niña effects on the Northeast Pacific: The 1991–1993 and 1988–1989 events. *California Cooperative Oceanic Fisheries Investigations Report* 36, 45–56.

Murray, S. and Bray, R. (1993) Benthic macrophytes. In *Ecology of the Southern California Bight: A Synthesis and Interpretation*, eds. M.D. Dailey, D.J. Reish and J.W. Anderson, pp. 304–368. University of California Press, Berkeley, California.

Noble, R.T., Dorsey, J.H., Leecaster, M., Orozco-Borbon, V., Reid, D., Schiff, K. and Weisberg, S.B. (1999) A regional survey of the microbiological water quality along the shoreline of the southern California Bight. *Environmental Monitoring and Assessment*, in press.

NRC (1990) Monitoring southern California's coastal waters. National Research Council. National Academy Press, Washington, D.C. 154 pp.

Phillips, B., Anderson, B., Hunt, J., Newman, J., Tjeerdema, R., Wilson, C.J., Long, E.R., Stephenson, M., Puckett, M., Fairey, R., Oakden, J., Dawson, S. and Smythe, H. (1998) Sediment chemistry toxicity and benthic community conditions in selected water bodies of the Santa Ana Region. Report to the State Water Resources Control Board, Sacramento, CA, from University of California, Santa Cruz, CA.

Phillips, C.P., Venkatesan, M.I. and R. Bowen. (1997) Interpretations of contaminant sources to San Pedro Shelf using molecular markers and principal component analysis. In *Molecular Markers in Environmental Geochemistry*, ed. E.P. Eganhouse, pp. 242–260. American Chemical Society, Washington, D.C.

Pollack, G.A., Uhaa, I.J., Fan, A.M., Wisniewski, J.A. and Witherell, I. (1991) A study of chemical contamination of marine fish from southern California II: Comprehensive study. California Environmental Protection Agency, Office of Environmental Health Hazard Assessment. Sacramento, CA, 161 pp.

Quast, J.C. (1968) Some physical aspects of the inshore environment, particularly as it affects kelp-bed fishes. In *Utilization of Kelp-bed Resources in Southern California*, eds. W.J. North and C.L. Hubbs, pp. 25–34. California Department of Fish and Game, Fish Bulletin 139.

Raco-Rands, V. (1999) Characteristics of effluents from large municipal wastewater treatment facilities. In Southern California Coastal Water Research Project Annual Report 1997–1998 (S.B. Weisberg and D. Hallock, eds.),. Westminster, California.

Rodolpho, K.S. (1970) Annual suspended sediment supplied to the continental borderland by the southern California watershed. *Journal of Sedimentary Petrology* **40**, 666–671.

Sanudo-Wilhelmy, A.S. and Flegal, A.R. (1992) Anthropogenic silver in the Southern California Bight: A new tracer of sewage in coastal waters. *Environmental Science and Technology* **26**, 2147–2152.

SCBPP Steering Committee (1998) Southern California Bight 1994 Pilot Project: I Executive summary. Southern California Coastal Water Research Project, Westminster, California.

SCCWRP (Southern California Coastal Water Research Project) (1973) The ecology of the Southern California Bight: Implications for water quality management. Southern California Coastal Water Research Project, El Segundo, California. Technical Report 104. 531 pp.

SCCWRP (Southern California Coastal Water Research Project) (1992) Temporal and spatial changes in sediment toxicity in Santa Monica Bay. In Southern California Coastal Water Research Project Annual Report 1990–91 and 1991–92 (J.N. Cross and C. Francisco, eds.), pp. 81–87. Long Beach, California.

SCCWRP (Southern California Coastal Water Research Project) (1994) Contamination of recreational seafood organisms off Southern California. Southern California Coastal Water Research Annual Report 1992–1993 (J.N. Cross, C. Francisco and D. Hallock, eds.), pp. 100–110. S. Calif. Coastal Water Res. Proj. Auth., Westminster, California.

SCCWRP (Southern California Coastal Water Research Project) (1995) Post-depositional distribution of organic contaminants near the Hyperion 7-mile outfall. Southern California Coastal Water Research Annual Report 1993–1994 (J.N. Cross, C. Francisco and D. Hallock, eds.), pp. 43–54. S. Calif. Coastal Water Res. Proj. Auth., Westminster, California.

Schiff, K.C. (1999) Sediment chemistry on the mainland shelf of the southern California Bight. In Southern California Coastal Water Research Project Annual Report 1997–98 (S.B. Weisberg and D. Hallock, eds.), pp. 76–88. Westminster, California.

Schiff, K.C., Weisberg, S.B. and Dorsey, J.H. (1999) Microbiological monitoring of marine waters in southern California. In Southern California Coastal Water Research Project Annual Report 1997–98 (S.B. Weisberg and D. Hallock, eds.), pp. 179–186. Westminster, California.

Schiff, K.C. and Allen, M.J. (1999) Bioaccumulation of chlorinated hydrocarbons in livers of flatfishes from the Southern California Bight. *Environmental Toxicology and Chemistry* in press.

Schiff, K.C. and R.W. Gossett. (1998) Southern California Bight 1994 Pilot Project: III. Sediment chemistry. Southern California Coastal Water Research Project, Westminster, California.

Schiff, K.C. and Weisberg, S.B. (1999) Iron as a reference element for determining trace metal enrichment in Southern California coastal shelf sediments. *Marine Environmental Research* **48**, 161–176.

SMBRP (1988) State of the bay: scientific assessment. Santa Monica Bay Restoration Project, Monterey Park, CA.

Smith, P.E. (1995) A warm decade in the Southern California Bight. *California Cooperative Oceanic Fisheries Investigations Report*, **36**, 120–126.

Somner, H. and Clarke, F.N. (1946) Effect of red water on marine life in Santa Monica Bay, California. *California Fish and Game* **32**, 100–101.

Soutar, A. and Isaacs, J.D. (1969) History of fish populations inferred from fish scales in anaerobic sediments off California. *California Cooperative Oceanic Fisheries Investigations Report* **13**, 63–70.

Southern California Edison Company (1997) Annual marine environmental analysis and interpretation: Report on 1996 data, San Onofre nuclear generating station. Prepared by Southern California Edison, ECOSystems Management, and Ogden Environmental and Energy Services.

Stull, J.K. (1995) Two decades of marine biological monitoring, Palos Verdes, California, 1972 to 1992. *Bulletin of the Southern California Academy of Sciences* **94**(1), 21–45.

Stull, J., Haydock, C.I., Smith, R.W. and Montagne, D.E. (1986) Long-term changes in the benthic community on the coastal shelf of Palos Verdes, southern California. *Marine Biology* **91**, 539–551.

Stull, J.K., Swift, D.J.P. and Niedoroda, A.W. (1996) Contaminant dispersal on the Palos Verdes continental margin: I. Sediments and biota near a major California wastewater discharge. *Science Total Environment* **179**, 73–90.

Stull, J.K. and Tang, C.L. (1996) Demersal fish trawls off Palos Verdes, Southern California, 1973–1993. *California Cooperative Oceanic Fisheries Investigations Report* **37**, 211–240.

Swartz, R.C., Cole, F.A. Schultz, D.W. and DeBen, W.A. (1986) Ecological changes in the Southern California Bight near a large sewage outfall: benthic conditions in 1980 and 1983. *Marine Ecology Progress Series* **31**, 1–13.

Tegner, M.J., Dayton, P.K., Edwards, P.B., Riser, K.L., Chadwick, D.B., Dean, T.A. and Deysher, L. (1995) Effects of a large sewager spill on a kelp forest community: catastrophe or disturbance? *Marine Environmental Research* **40**, 181–224.

Thompson, B., Laughlin, J.D. and Tsukada, D. (1987) 1985 Reference site survey. SCCWRP Technical Report #221, Southern California Coastal Water Research Project. Long Beach, CA.

Thompson, B., Tsukada, D. and O'Donahue, D.. (1993) 1990 Reference site survey. SCCWRP Technical Report #269, Southern California Coastal Water Research Project. Long Beach, CA.

Tran, K. and Zeng, E. (1997) Laboratory and field testing on an INFILTREX 100 pump. In Southern California Coastal Water Research Project Annual Report 1996 (S.B. Weisberg, C. Francisco and D. Hallock, eds.), pp. 137–146. Westminster, California.

U.S. EPA. (1987) Environmental impact statement for San Diego (LA-5) ocean dredged material disposal site designation. U.S. Environmental Protection Agency, Region IX. San Francisco, California.

U.S. EPA. (1995) Short-term methods for estimating the chronic toxicity of effluents and receiving waters to west coast marine and estuarine organisms. U.S. Environmental Protection Agency, Office of Research and Development, Washington, D.C. EPA/600/R-95/136.

U.S. EPA. (1997) National air pollutant emission trends, 1900-1996. U.S. Environmental Protection Agency, Office of Air Quality, Research Triangle Park, NC EPA-454/R-97-011.

Venkatesan, M.I. (1995) Coprostanol as a Chemical Probe to Assess Sewage and Toxic Pollutant Inputs in Southern California Basins. Sea Grant Biennial Report, Project No. R/CZ-126PD, pp. 159–169.

Venkatesan, M.I., Greene, G.E., Ruth, E. and Chartrand, A.B. (1996) DDTs and dumpsite in the Santa Monica Basin, California. *Science of the Total Environment* **179**, 61–71.

Venkatesan, M.I. and Kaplan, I.R. (1990) Sedimentary coprostanol as an index of sewage addition in Santa Monica Basin, southern California. *Environment Science Technology* **24**, 208–214.

Weber, M.L. (1997) The marine fisheries of southern California: A briefing book. Cabrillo Marine Aquarium, San Pedro, California. 144 pp.

Wiemeyer, S.N., Bunck, C.M. and Stafford, C.J. (1993) Environmental contaminants in bald eagle eggs 1980–84 and further interpretations of relationships to productivity and shell thickness. *Arch. Environ. Contam. Toxicol.* **24**, 213–227.

Wolf, P. and Smith, P.E. (1992) Pacific sardine. In *California's Living Marine Resources and Their Utilization*, eds. W.S. Leet, C.M. Dewees and C.W. Haugen, pp. 83–86. University of California, Davis, Department of Wildlife and Fisheries Biology, Sea Grant Extension Program, Davis, California. Sea Grant Extension Publication, UCSGEP-92-12.

Word J.Q. and Mearns, A.J. (1979) 60-m control survey off southern California. Technical Memorandum 229. Southern California Coastal Water Research Project. Long Beach, California.

Young, D.R., McDermott, D.J. and Heesen, T.C. (1976) DDT in sediments and organisms around southern California outfalls. *Journal of the Water Pollution Control Federation* **48**, 1919–1928.

Young, D.R., McDermott, D.J., Heesen, T.C. and Hotchkiss, D.A. (1975) DDT residues in bottom sediments, crabs, and flatfish off southern California submarine outfalls. *Californian Water Pollution Control Association Bulletin* **12**, 62–66.

Zeng, E.Y., Khan, A.R. and Tran, K. (1997) Organic pollutants in the coastal environment off San Diego, California. 3. Using linear alkylbenzenes to trace sewage-derived organic materials. *Environmental Toxicology and Chemistry* **16**, 196–201.

Zeng, E.Y. and Venkatesan, M.I. (1999) Dispersion of sediment DDTs in the coastal ocean off southern California. *Science of the Total Environment* **229**, 195–208.

Zeng, E.Y. and Vista, C. (1997) Organic pollutants in the coastal environment off San Diego, California. 1. Source identification and assessment by compositional indices of polycyclic aromatic hydrocarbons. *Environmental Toxicology and Chemistry* **16**, 179–188.

Zeng, E.Y., Yu, C.C. and Tran, K. (1999) In situ measurements of chlorinated hydrocarbons in the water column off the Palos Verdes Peninsula, California. *Environmental Science and Technology* **33**, 392–398.

Zmarzly, D.L., Stebbins, R.D. Pasko, D., Duggan, R.M. and Barwick, K.L. (1994) Spatial patterns and temporal succession in soft-bottom macroinvertebrate assemblages surrounding an ocean outfall on the southern San Diego shelf: Relation to anthropogenic and natural events. *Marine Biology* **118**, 293–307.

THE AUTHORS

Kenneth C. Schiff
*Southern California Coastal Water Research Project,
7171 Fenwick Lane,
Westminster, CA 92683, U.S.A.*

M. James Allen
*Southern California Coastal Water Research Project,
7171 Fenwick Lane,
Westminster, CA 92683, U.S.A.*

Eddy Y. Zeng
*Southern California Coastal Water Research Project,
7171 Fenwick Lane,
Westminster, CA 92683, U.S.A.*

Steven M. Bay
*Southern California Coastal Water Research Project,
7171 Fenwick Lane,
Westminster, CA 92683, U.S.A.*

Chapter 25

FLORIDA KEYS

Phillip Dustan

The Florida Keys form an elongated chain of 822 low-lying islands, extending for over 220 miles south of Florida, and are the only living coral reefs in the continental United States. A wide shelf area populated with seagrass beds, patch reefs, and banks of carbonate sand separates outer reefs from the islands. The original terrestrial vegetation of the islands consisted of mixed tropical hardwood forest and extensive mangroves. Many of the trees were of Caribbean origin and not found elsewhere in the continental United States, but most of the larger trees have been logged and the vegetational composition highly altered.

Urbanization on the islands has been intense, mostly around Key Largo, in the Northern Keys, Marathon in the Middle Keys, and Key West in the South. The islands are connected by an overseas highway containing 19.3 miles of bridge spans; a water pipeline paralleling the roadway provides the only source of fresh water. The permanent resident population of the Florida Keys grew from 5,657 in 1870 to 53,058 in 1990, with about 30% of the residents living in Key West. In 1995–1996, there were 2.54 million visitors.

A significant portion of reef degradation here is probably related to watershed alterations from agriculture, development, and tourism in South Florida; efforts there to channel water flow have resulted in additional flow into the coastal waters of Florida Bay and the Keys, bringing increased sedimentation and nutrients. Ship groundings and anchor damage have been widespread and severe. As elsewhere, coral reef populations in the Keys are subject to stresses of global warming and events such as the Caribbean-wide mass mortality of *Diadema antellarum*. Monitoring has been most complete at Carysfort Reef, which has continued to decline, and is now entering a state of ecological collapse: similar ecological degradation has occurred on many reefs throughout the Florida Keys.

In 1990, the Florida Keys National Marine Sanctuary Act came into force and directed the U.S. Environmental Protection Agency to institute a water quality assurance and protection plan for the Florida Keys. Presently, management utilizes the concept of multi-use zoning to generate varying levels of resource protection.

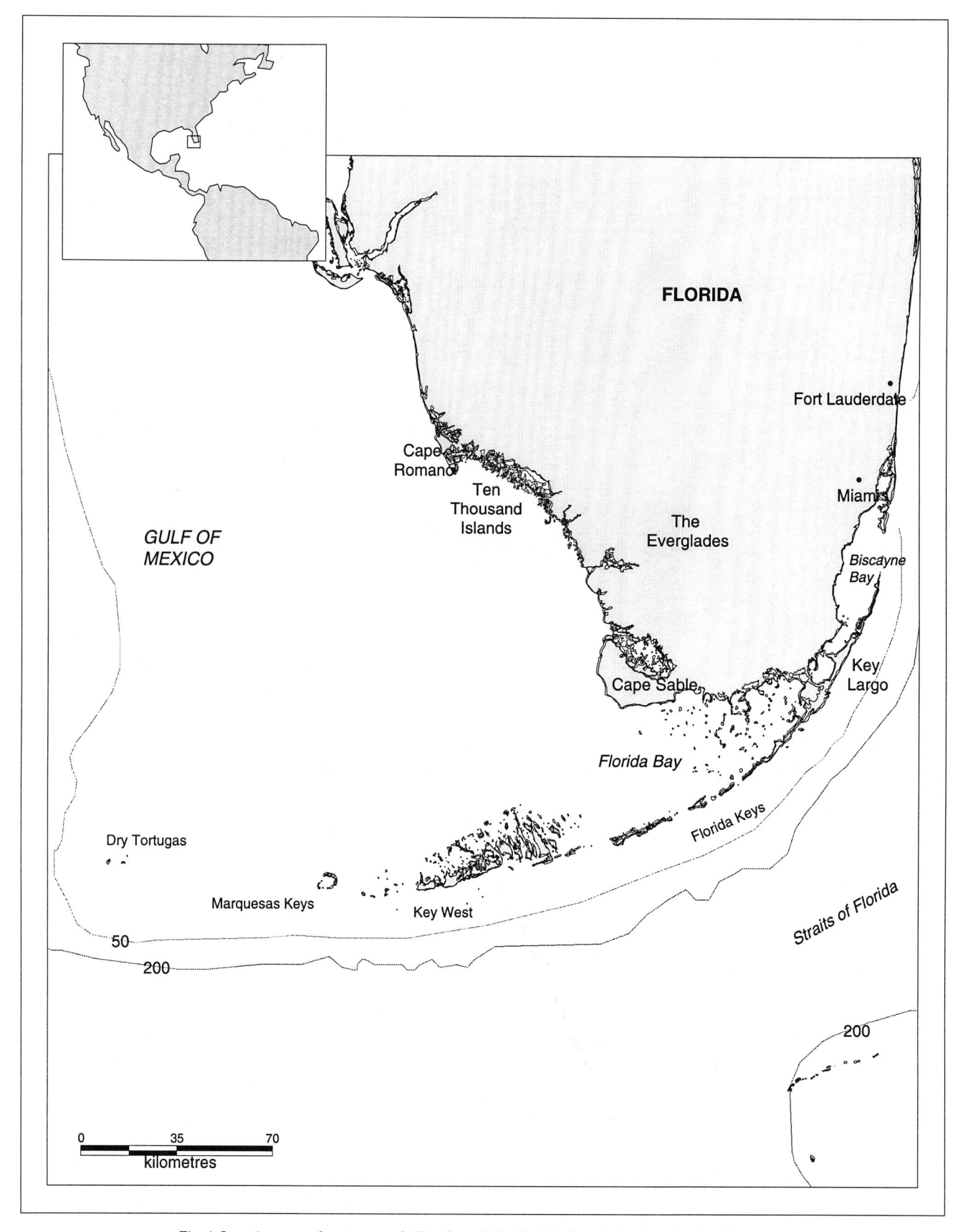

Fig. 1. Location map showing south Florida and the Florida Keys including the Dry Tortugas.

DEFINED REGION

The Florida Keys were discovered by Ponce De Leon on 12 May 1513 during his voyage of exploration in search of the Fountain of Youth. The islands form an elongated chain of 822 low-lying islands, extending for over 220 miles at the southern tip of the Florida peninsula. The islands extend from the southeastern tip of the Florida peninsula to the Dry Tortugas and lie between the Gulf of Mexico and the Atlantic Ocean (Figs. 1 and 2). The shallow offshore waters of the Keys support the only living coral reefs in the continental United States. The Florida Keys, excluding the Dry Tortugas, are separated from the mainland by shallow bays: Biscayne Bay, Barnes Sound, Blackwater Sound and Florida Bay. The Florida Current flows along the eastern edge of the Keys on its way to becoming the Gulf Stream at higher latitudes.

The climate of the Florida Keys is sub-tropical. Rain falls principally in spring and summer (approximately 1 m/year), and also with the passage of winter cold fronts which penetrate from northern temperate latitudes. Hurricanes, which occur from June to November, may cross the Keys causing intense damage to both terrestrial and marine communities.

MAJOR SHALLOW WATER, MARINE AND COASTAL HABITATS

Coral reefs are the best known marine habitat of this region. They exist at the northern extent of the tropical Western Atlantic. Mayer (1914, 1916) first suggested a "temperature divide" between Fowey Rocks and Carysfort Reef. Using data collected by lighthouse keepers, Mayer hypothesized that winter polar cold fronts lowered the water temperature in this region and north of it to values below the threshold necessary for reef development.

Coral reefs are most abundant in the upper and lower areas, which are separated by large tidal passes that connect Florida Bay with the Atlantic Ocean. Ginsburg and Shinn (1964) advanced the hypothesis that cold water from Florida Bay also had a controlling influence on the distribution of reefs within the Keys as there is a "sparsity of thriving reefs opposite the large tidal passes of the Middle Keys". Recently, this has become known as the "Florida Bay Hypothesis" which has been expanded to suggest that both the physical and chemical characteristics of Florida Bay strongly influence the health and vitality of reef-building corals in the Florida Keys (cf. LaPointe, 1999; Porter et al., 1999). The present ecology of the reef tract is probably

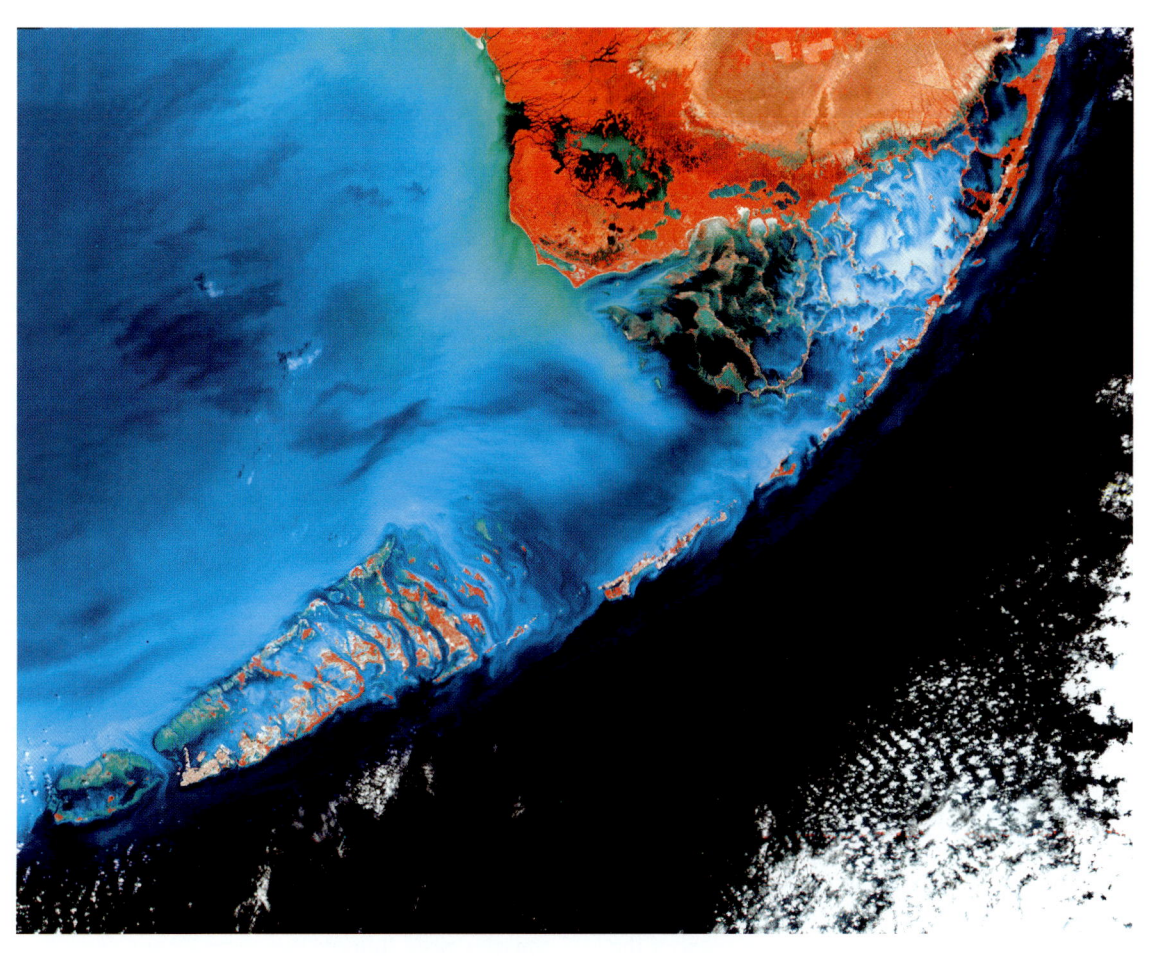

Fig. 2. Thematic Mapper MSS image of Florida Keys, 18 August 1972.

similar to that of the earlier Pleistocene reefs, as both are dominated by the same general species of corals and algae.

A wide shelf area separates the outer reefs from the islands of the Florida Keys. This area is populated with seagrass beds, patch reefs, and banks of carbonate sand. For more complete descriptions of the geological setting of the Florida Keys see Shinn et al. (1989), Jaap (1984), and Jaap and Hallock (1990).

The islands of the Florida Keys are composed of calcium carbonate rock that dates from an earlier high sea-level stand of the Pleistocene (80–120 kbp) and sits on top of the now submerged foothills of the very old Appalachian Mountains. The highest point in the Keys, only about 6 m above sea level, is found on Windley Key. The Keys can be divided into three major areas: Upper, Middle, and Lower Keys (Shinn et al., 1989). The Upper and Middle Keys are coralline limestone known as the Key Largo Formation. The Upper Keys are oriented nearly north–south and face into the prevailing seas. The Middle Keys face the east–southeast winds. The islands of the Lower Keys are constructed of oolitic limestone and lie parallel to the prevailing east-southeast winds (Shinn et al., 1989).

The terrestrial vegetation of the islands during the time of Ponce De Leon's discovery consisted of mixed tropical hardwood forest and extensive mangrove communities. Many of the trees native to the Keys are of Caribbean origin and are not found elsewhere in the continental United States. Most of the larger trees were logged by the Spanish and by later settlers. Since early times, numerous species have been introduced to the area and, today, there are very few, if any, stands of pre-European vegetation left.

POPULATIONS AFFECTING THE AREA

Urbanization along the island chain has been intense over the last 50 years. Most of the development has centred around Key Largo, in the Northern Keys, Marathon in the Middle Keys, and Key West in the South. Situated within Monroe County, this chain of islands is connected by an overseas highway containing 19.3 miles of bridge spans. The original roadway utilized 42 defunct railroad bridges between Key Largo and Key West. The bridges were replaced in the early 1980s and in many cases the old bridges still run parallel to the new with some utilized as fishing bridges. A pipeline carrying water parallels the roadway and provides the only source of fresh water for the entire Florida Keys. The original installation of these bridges by Henry Flagler in the early 1900s began the human modification of the Florida Keys by altering the flow of water through the tidal passes between the islands (Swart et al., 1996).

The permanent resident population of the Florida Keys grew from 5,657 in 1870 to 53,058 in 1990, with about 30% of the residents living in Key West. A very large part of the economy of the Florida Keys revolves around tourism, which provides the major source of employment for local residents. Other industries include retail services, commercial fishing and government institutions such as the Florida Keys National Marine Sanctuary. The seasonal influx of visitors virtually doubles the population of the Keys. For example, in 1990, the combined resident and visitor population of the region was 113,053, with visitors comprising 53% of the functional population (Monroe County Planning Department). From June 1995 to May 1996, 2.54 million visitors travelled to the Florida Keys, with a majority (1.4 million) visiting Key West. The Upper Keys received over 911,000; the Middle Keys, 697,000; and over 304,000 visited the Lower Keys excluding Key West. Two thirds of the visitors participated in at least one water related activity (28% snorkelling, 8% SCUBA diving, 21% fishing), providing $792 million of the estimated $1.2 billion generated by the tourist economy. Presently there are 24,800 boats registered in Monroe County, and approximately 14% of visitors (roughly 365,000 people) brought boats with them on their visits to the Keys.

STRESS TO THE ENVIRONMENT AND REEFS

In the Florida Keys, the question is frequently asked, which is the single factor mainly responsible for recent deterioration, sediments or nutrients? It may well be that the factor is the accumulation of a series of nested stresses which may be: as local as fishing and tourism; as regional as cities, agriculture, and industry; and as global as deforestation of rainforests, the hole in the ozone, and the greenhouse effect. Each factor compounds the others, and the vitality of the reef declines. On a large geographical scale, the addition of nutrients, organic carbon, and sediments from poor land-use practices are responsible for coastal hypoxia near river mouths and deltas. Many of these areas become 'dead zones' when the water becomes depleted of oxygen to the extent that it no longer supports aerobic metabolism (Cooper and Brush, 1991; Malakoff, 1998; Costanza et al., 1998). Coastal ocean current patterns circulate coastal pollutants and their effects throughout the seas (Fig. 2).

Some of the increased stresses that Florida Keys coral populations are now exposed to are simply amplifications of naturally occurring stress, while others are new in terms of presently living corals. The most important include increased sedimentation which smothers corals, increased nutrients which, in the absence of adequate herbivory, will result in algal overgrowth, elevated temperatures which promote bleaching, and diseases which seem to be more prevalent in areas close to centres of human habitation.

Reefs in all tropical seas are threatened by degraded ecological conditions that originate locally, regionally, and from the shores of distant continents (Bryant et al., 1998; Hatziolos et al., 1998). Many of these stresses are nested within each other which probably amplifies the negative consequences. Remote oceanic reefs are affected by global change such as elevated ocean temperatures and increased ultraviolet irradiance. Reefs in most coastal waters are affected by these factors and by additional stressors such as increased sediments, carbon, nutrients, and harvesting, especially when located near population centres. The geography of the Florida Keys places them at a location where many of these factors appear to have converged to produce a suite of stressors that currently are pushing the reefs into a state of ecological collapse.

Effects of different or individual stresses are often extremely difficult to tease apart. The impacts of each stress may be cumulative and probably synergistic. For example, corals with weakened immune systems are more susceptible to disease than are healthy corals, and in the Florida Keys, disease is a significant source of colony mortality. But coral recruitment and regeneration rates are low and appear to be decreasing, which will amplify the total decline.

AGRICULTURAL AND URBAN FACTORS

A significant portion of reef degradation in the Florida Keys is probably related to watershed lands that have been altered from their natural state. The increase in human manipulation of South Florida from Lake Okeechobee southward includes agriculture, development, and tourism. Generally, natural terrestrial ecosystems may export little in the way of nutrients, carbon, and sediments, but agriculture, urbanization and deforestation have reduced the capacity of the South Florida terrestrial ecosystems to trap and retain these materials. Efforts to channel water flow to prevent flooding have resulted in further flow of excess fresh water, nutrients, and sediments into the coastal waters, including Florida Bay (cf. Davis and Ogden, 1994). Coral reef ecosystems have, in general, evolved to be very efficient in trapping and retaining nutrients even in concentrations that are below levels of detection. This creates a situation in which materials from a diffuse array of sources contribute to pervasive levels of chronic stress to these reefs. Both point and non-point sources contribute to the hydraulic flow that pushes sediments, nutrients, and contaminants into the sea.

Locating the sources of increased nutrient and sediment levels and other stresses has proven to be elusive. Upstream of the Keys, the effluent of cities, towns, and farms slowly bleeds into the sea through numerous canals, rivers, and coastal bays. Point sources, such as sewage outfalls, deep injection well package plants or agricultural irrigation canals, are usually known and can be controlled through permitting processes, but their effluents are not easily traceable once they enter the tropical shallow water ecosystems.

Recent studies have demonstrated that the effluent from Class 5 shallow-well injection package plants migrates through the Key Largo limestone at rates of several metres per day. Since this is the preferred method for waste disposal by developments and hotels, this presents a significant pollution threat for the near-shore coastal environment of the area. Tidal pumping generates a hydraulic flow that pushes effluent eastward into the coastal Atlantic Ocean (Reich et al., 1999). The migration rates vary from 3 to 30 m/day, depending on tides and weather conditions. This effluent contributes to the nutrient enrichment of the very shallow coastal waters of the Florida reef tract but so far, has not been positively identified among the offshore reefs (Shinn, pers. comm.).

INDUSTRIAL STRESSES

Ship groundings and anchor damage have been widespread and severe throughout the Florida Keys for some years (Dustan, 1977a; Davis, 1977). Small boat groundings are common and some can be catastrophic for both reef and boat. In 1984, the freighter *Wellwood* grounded on Molasses Reef, destroying approximately 1500 m^2 of reef substrate (Hudson and Diaz, 1988). This large-scale grounding was followed by at least four major ship and numerous small boat groundings since 1985. Currently, the Florida Keys National Marine Sanctuary (FKNMS) assesses fines that are approximately $2500/$m^2$ of reef damaged. The frequency of boat and ship groundings in the Florida Keys continued to increase; in the fiscal year 1999, there were 540 small boat groundings documented in the FKNMS.

GLOBAL STRESSES

On a larger geographic scale, the Florida Keys are downstream from almost every source of sediment or nutrient in the Caribbean basin and Gulf of Mexico. Materials wash into the sea from the west and east coasts of the Florida peninsula and through the Everglades. The area of sediment influx extends to the watershed of the Mississippi River and continues throughout the Caribbean Sea. Sediments from as distant as the Orinoco and Amazon Rivers have been identified on Carysfort Reef (Dustan, unpublished). In 1993, floodwaters from the U.S. Midwest combined with the Loop Current, which flows into the Florida Current, and reduced salinity and oxygen concentrations along the reefs of the Keys. On 14 September, the SeaKeys Station, off Key Largo, recorded a decrease in salinity from 36 to below 32 ppt, and a decrease in dissolved oxygen from 7.1 to 5.2 on Molasses Reef (Fig. 3) (Porter, per comm.). This signal was unusually large due to the magnitude of the floods, and it clearly demonstrated the connectivity between interior watersheds of North America and the Florida Keys.

Coral and reef populations of the Florida Keys, like reefs everywhere, are also subject to global-scale stresses such as global warming and increased ultraviolet due to ozone thinning. Coral bleaching has been correlated to increased water temperatures in the late summer and early fall. Several particularly serious mass bleachings have occurred in the last 20 years. Most of these events were recorded in many areas of the wider Caribbean and thus have been

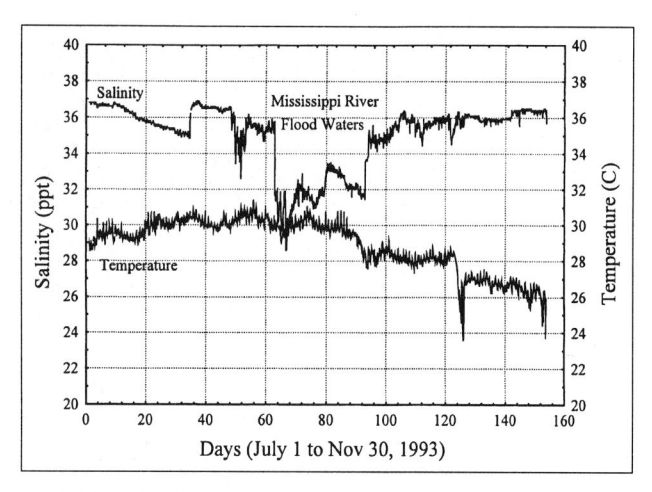

Fig. 3. Mississippi floods cause changes in temperature and salinity at Molasses Reef, Key Largo during summer 1993. Data from the SEAKEYS/C-MAN Station at Molasses Reef.

linked to warming seas. However, recent work (Shinn, 1998; 1999) points to another, previously unsuspected ecological stressor which parallels the impact of global warming: transatlantic African dustfall. Their most recent work has uncovered a soil fungus, *Aspergillus sydowii*, that causes a Caribbean-wide disease that occurs in sea fans. Isolated African dust collected in the Caribbean contains species of *Aspergillus* and there is an apparent correlation between increased amounts of dust and various disease outbreaks. The atmospheric distribution of African dust is a potential cause of other synchronous Caribbean-wide coral diseases, including those that have killed the staghorn coral *Acropora cervicornis* and the sea urchin *Diadema*. The near extinction of these organisms in 1983 correlates with the period of highest annual dust transport to the Caribbean since measurements began in 1965. In addition to *Aspergillus* spores, African dust is composed of the major crustal elements including iron, phosphorus, sulphate, aluminum, and silica. These elements, especially iron, may enhance the growth of tropical marine algae.

LOCALIZED ECOLOGICAL REEF STRESS

Starting in 1982, there was a mass mortality of *Diademia antillarun* throughout the Caribbean Sea (Lessios et al., 1984). Populations have still not recovered to anywhere near their pre-mortality densities (Lessios, 1988; pers. comm.). Urchins are still rare on Keys reefs, 16 years later. Following the mass mortality, macro-algal populations throughout the Caribbean soared as they were released from herbivory pressure. Thus, while nutrient levels were apparently rising due to increased urbanization and human population growth, levels of herbivory on the reef plummeted. At present, these two scenarios cannot be uncoupled and debate continues on which stressor is ecologically the most important (see Lapointe, 1999; Hughes et al., 1999).

CORAL VITALITY: LONG-TERM STUDY OF CARYSFORT REEF

Coral populations in the Florida Keys have declined precipitously since 1974 when the first quantitative monitoring of reefs began in the Florida Keys (Dustan, 1977b, 1985, 1999; Dustan and Halas, 1987). Coral recruitment was much lower than that found in other parts of the Caribbean such as Discovery Bay, Jamaica, and two coral diseases, Black Band Disease (Fig. 4) (Garrett and Ducklow, 1975; Antonius, 1977) and White Plague (Fig. 5) (Dustan, 1977b; Richardson, 1998) were just becoming significant. Algal-sediment encroachment comprised the major agents of mortality (Dustan, 1977b) in the 1970s. Between 1974 and 1982, cover and diversity on Carysfort Reef increased in shallow areas while the deeper, fore-reef terrace showed significant losses (Dustan and Halas, 1987). Change in shallow water seemed to be driven by the destruction of the dominant stands of *Acropora palmata* (elkhorn coral). Overall coral cover had appeared to increase because the previously rich, three-dimensional habitat had been reduced to planar rubble which covered more of the bottom while smaller colonizing species settled on open substrate. Deeper on the reef, colonies were dying from disease and sediment damage, and were no longer being replaced by recruitment.

In July 1984, observations on the phenotypic condition of over 9,800 corals on 19 different reefs in the Key Largo region revealed that 60% of colonies showed signs of physical or biological stress, 5–10% were infected with disease and about one third appeared healthy. Surprisingly, virtually all the reefs contained approximately the same proportion of unhealthy corals. This argues in favour of widespread stress factors such as water quality rather than effects that are local to specific reefs (Dustan, 1993). High rates of mortality continued to be documented elsewhere in the Florida Keys between 1984 and 1991 (Porter and Meier, 1992).

A second study site, in the Dry Tortugas, was initially used as a control site for the Carysfort Reef, off Key Largo. Coral development there was very rich, with little or no disease (Dustan, 1985; Jaap, et al., 1989). There was extensive anchor damage caused by shrimp boats seeking refuge from storms, but otherwise the corals were vibrant. The reefs of the Dry Tortugas, however, were ecologically very different from the northern Keys, so direct comparisons concerning reef health were imprecise. During the winter of 1976–77 an extreme cold front reduced water temperatures to 14°C on January 21, 1977 when 96% of the living coral cover at depths <2 m died (Porter et al., 1982).

Carysfort Reef has continued to decline in deep and shallow waters. By June 1998, coral cover had decreased to approximately 5% cover. During one dive on Carysfort Reef in July 1998, not one colony of star coral, *Montastrea annularis* species complex was seen that was not infected with White Plague disease. Large colonies (over 1 m

Fig. 4. Black Band Disease overgrowing large colony of the star coral, *Montastrea annularis* species complex. Insert shows close-up of the action of disease, a cyanophyte algal mat that can kill coral tissue at rates on the order of millimetres/day. A large coral colony, which may be 250 to 500 years old, can be killed in less than a year. Photo P. Dustan, Carysfort Reef, 2 m depth.

Fig. 5. Colony of *Mycetophyllia ferox* infected with the White Plague. Inset shows a close-up of the active disease, which is killing the live coral tissue at a rate of 1 to 3 mm/day. Photo P. Dustan, Carysfort Reef, 16 m.

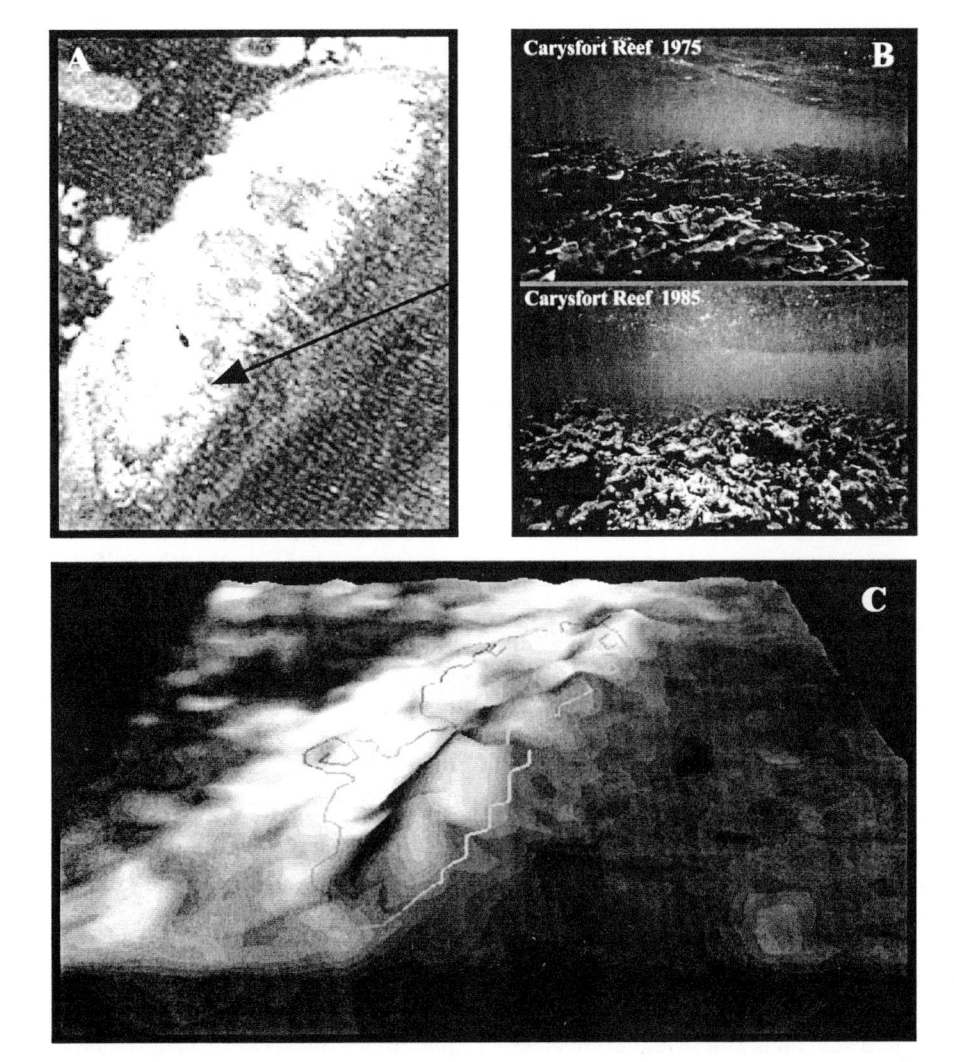

Fig. 6. Change on Carysfort Reef as detected in temporal texture analysis using Thematic Mapper imagery (Dustan, Dobson, Nelson, 1999). (A) Aerial photograph. (B) Two photographs, taken from the same vantage point 10 years apart in 1975 and 1985, illustrate the rapid degradation of the shallow *Acropora palmata* zone that has occurred on Carysfort Reef, Key Largo, FL, USA. Photos by Phillip Dustan. (C) 3-Dimensional temporal texture in which the vertical axis represents habitat variability.

diameter) were rapidly being overtaken by White Plague. Since the skeletal growth rate of *M. annularis* has been measured at 5–10 mm/year, it is estimated that these colonies are at least 100 years old (Dustan, 1975). Some colonies are at least twice this age, and White Plague can kill them in less than a single year. With such rapid mortality of large colonies, coral cover will fall below 5% cover, and corals will cease to provide any significant contribution to reef framework construction.

It is not an overstatement to suggest that this reef is entering a state of ecological collapse. Similar ecological degradation has occurred on many reefs throughout the Florida Keys, including Molasses Reef, Looe Key, and Sand Key. Carysfort, however, is the only reef where this long-term change has been documented with quantitative line transect studies. In fact, the change is so extensive that it can be detected in Landsat Thematic Mapper satellite imagery (Dustan, 1999 (Fig. 6).

Observations at the long-term site in the Dry Tortugas in June 1999 suggest that the reefs there are showing slower decline than in the Key West area. It would appear that corals in the Dry Tortugas are not stressed to the same degree as they are in the "mainland" Florida Keys. Coral cover on Bird Key has decreased to an estimated 20–25% as opposed to the 5–10% for Carysfort Reef. These reefs are buffered from Key West by 65 miles of ocean, which may help to explain why they remain somewhat healthier than the Keys reefs. Although the reefs in the Dry Tortugas are in marginally better condition, they too are experiencing considerable decline and there is cause for serious concern.

USEPA CORAL REEF MONITORING PROJECT

In 1990, the Florida Keys National Marine Sanctuary Act established the FKNMS and directed the U.S. Environmental Protection Agency (USEPA) to institute a water quality assurance and protection plan for the Florida Keys. This plan was to include monitoring of the status and trends of the seagrasses, coral reefs and hard-bottom communities and of the water quality. The result was the creation of a program called the Coral Reef Monitoring Project (CRMP) (Dustan et al., 1996). The CRMP team instituted repetitive underwater observations and video transects to provide estimates of biodiversity, distribution, and coverage of reef corals and associated benthic organisms. Starting in 1996, the CRMP annually sampled 160 stations at 40 sites on 32 reefs that are distributed throughout the Florida Keys. In June 1999, 10 more stations were added at three sites in the Dry Tortugas.

In the first three years of sampling, the CRMP has witnessed a significant increase in the geographical distribution of diseases which kill corals and in the number of species with diseases (Table 1) (see Peters, 1996 and Richardson, 1998 for reviews of coral diseases). Like disease, coral bleaching has become relatively common throughout the studied sites. Initial findings gave rise to a second project called the Florida Keys Coral Reef Disease Study, which is also funded by USEPA (D. Santavy, E. Mueller, and J. Porter, principal investigators). Similar trends in diseases are appearing in other marine organisms, most notably in the Florida Keys where gorgonians have been infected with a pathogen identified as *Aspergillus sydowii*, an opportunistic terrestrial soil fungus (Smith et al., 1996). Concomitant with increases in the distribution and type of diseases, have been apparent losses of species and coral cover on some reefs. For example, coral cover on the fore reef terrace of Carysfort Reef declined from 13.3% to 5.3% between 1996 and 1998 (Fig. 7).

Table 1

Station and species disease data from the Coral Reef Monitoring Project (Harvell et al., 1999)

	Number of Stations with Disease				
	WH	BB	OD	Diseased Stations	Percent Diseased
1996	7	7	16	26	16%
1997	61	11	66	95	59%
1998	97	28	92	131	82%
Increase n (96–98)	90	21	76	105	
Increase % (96–98)	1285%	300%	475%	404%	

	Number of Coral Species with Disease				
	WH	BB	OD	Diseased Species	Percent Diseased
1996	3	2	8	11	27%
1997	22	4	22	28	68%
1998	28	7	28	35	85%
Increase n (96–98)	25	5	20	24	
Increase % (96–98)	833%	250%	188%	218%	

WH = white plague BB = black band disease OD = other diseases.

PROTECTIVE MEASURES

The Florida Keys have been placed under state and federal protection programmes beginning with the creation of the Dry Tortugas National Monument in 1935 by President Franklin Roosevelt (Table 2). John Pennekamp Coral Reef State Park was established in 1960, and was the United States' first underwater coral reef park. Biscayne National Park was established in 1968. The Marine Protection, Research and Sanctuaries Act set forth the guidelines for establishing the Key Largo National Marine Sanctuary in 1975. The protected area was expanded in 1981 with the formation of Looe Key National Marine Sanctuary. The Florida Keys National Marine Sanctuary and Protection Act 1990 brought the entire Florida Keys under federal protection. In 1992 the Dry Tortugas National Monument was

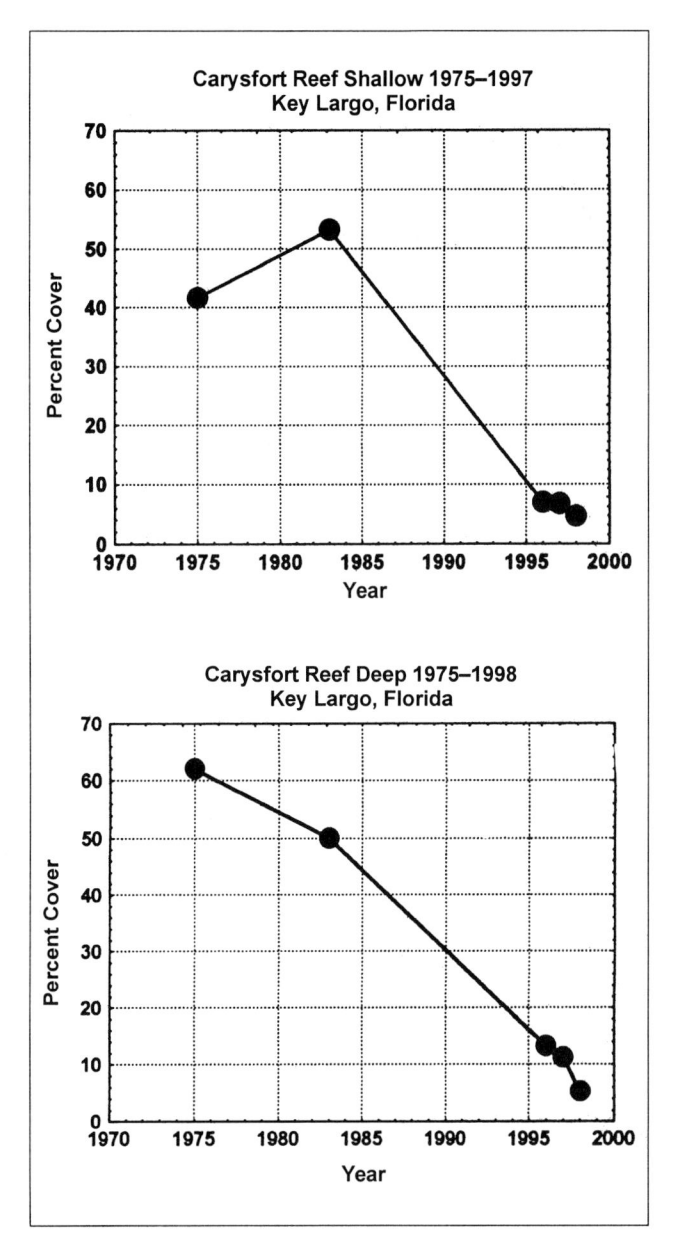

Fig. 7. Changes in coral cover on Carysfort Reef, 1974–1998.

Table 2
Florida Keys Coral Reef Ecosystem: Timeline

Year	Event
1935	Dry Tortugas National Monument
1960	John Pennekamp Coral Reef State Park established 10 December 1960
1968	Biscayne National Park established
1969	Skin Diver Magazine sounds alarm on reef degradation
1972	Marine Protection, Research and Sanctuaries Act
1973	Coral Diseases discovered in Key Largo (Antonius, 1974)
1974	Beginning of long-term reef monitoring at Carysfort Reef
1975	Key Largo National Marine Sanctuary established
1981	Looe Key National Marine Sanctuary established
1983	Significant reduction of corals at Carysfort Reef since 1974
1984	Key Largo Coral Vitality Study, 60% corals stressed
1987	Severe Coral Bleaching throughout Caribbean
1991	Florida Keys National Marine Sanctuary and Protection Act 1990
1992	Reef coral degradation estimated at 5% loss per year (Porter and Meier)
1995	USEPA Water Quality Protection Plan Coral Reef Monitoring project
1997	Outbreak of White Band Disease 2. pathogen identified (Richardson)
1998	First US Coral Reef Task Force Meeting at Biscayne National Park
1999	Carysfort Reef coral coverage below 5%
2000	Proposed Tortugas 2000 National Marine Sanctuary

redesignated as Dry Tortugas National Park to protect both the historical and natural features. Sanctuary designation for a larger portion of the Dry Tortugas banks and surrounding area is underway with the Tortugas 2000 project. This would protect a larger area of deep reef that has luxuriant coral development which is prime fisheries habitat that may serve as a recruitment area for the Florida Keys.

At present, the FKNMS utilizes the concept of multi-use zoning to generate varying levels of resource protection. Mooring buoys have proliferated throughout the Keys to protect reefs from anchor damage, and sanctuary officers patrol the waters. In an effort to reduce large vessel groundings, a series of radar beacons have been placed on outer reef lighthouses to help warn commercial shipping traffic of the reefs.

CONCLUSIONS

Data collected since 1974 presents a grim picture for the future of the reefs of the Florida Keys. Recovery from the present severe ecological degradation is probably not possible within a human lifetime. A coral reef is a structure of ancient ecological design. Its physical morphology, its orientation to the forces of the sea, and its community structure have been tested and moulded by time and natural selection. However, its design for ultimate conservation and use of nutrients through symbioses and detailed trophic interactions has made it vulnerable to changing environmental conditions, particularly changes in temperature, increases in sedimentation, increased nutrient concentrations, and over-harvesting. Ironically, the many values of coral reefs—as a fisheries resource, for coastal protection and building materials, and as tourist attractions—now are contributing to their steady and rapid decline in the Florida Keys. As clear as these facts are, we still lack specific knowledge concerning the identification of the factors and the magnitude of the relative contributions of each, though it is clear that in combination their effects are severe and unsustainable. Future research needs to be targeted at process studies. We are learning that coral reefs are indicators of overall oceanic health and global climate change, and a better understanding of their condition and responses may assist not only the reefs themselves and their marine environment, but also help our understanding of the complex relationships in our global ecosystems.

REFERENCES

Antonius, A. (1977) Mortality in reef corals. *Proc. 3rd I.C.R.S.* **2**, 617–624.

Bryant, D., et al. (1998) *Reefs at Risk*. World Resources Institute, Washington, DC.

Cooper, S. and Brush, G. (1991) Long-term history of Chesapeake Bay anoxia. *Science* **254**, 992–996.

Costanza, R., et al. (1998) Principles for Sustainable Governance of the oceans. *Science* **281**, 198–199.

Davis, G.E. (1977) Anchor damage to a coral reef on the coast of Florida. *Biological Conservation* **11**, 29–34.

Davis, S.M. and Ogden, J.C. (1994) *Everglades, The Ecosystem and its Restoration*. St. Lucie Press, Delray Beach, FL, 826 pp.

Dustan, P. (1975) Growth and form in the reef-building coral *Montastrea annularis*. *Marine Biology* **33**, 101–107.

Dustan, P. (1977a) Besieged reefs of the Florida Keys. *Natural History* **86** (4), 72–76.

Dustan, P. (1977b) Vitality of reef coral population off Key Largo, Florida: recruitment and mortality. *Environmental Geology* **2**, 51–58.

Dustan, P. (1985) Community structure of reef-building corals in the Florida Keys: Carysfort Reef, Key Largo and Long Key Reef, Dry Tortugas. Atoll Research Bulletin, No. 288.

Dustan, P. (1993) Developing Methods for Assessing Coral Reef Vitality: A Tale of Two Scales: Global Aspects of Coral Reefs, June 10–11. University of Miami. pp. M8-M14

Dustan, P. (1999) Coral reefs under stress: Sources of mortality in the Florida Keys. *United Nations Forum* **23**, 147–155

Dustan, P. and Halas, J. (1987) Changes in the reef-coral population of Carysfort Reef, Key Largo, Florida, 1975–1982. *Coral Reefs* **6**, 91–106.

Dustan, P., Jaap, W., Porter, J.W. and Wheaton, J. (1996) Coral Reef Monitoring In the Florida Keys (Abstract). Eight International Coral Reef Symposium, Panama, June 1996

Garrett, P. and Ducklow, H. (1975) Coral diseases in Bermuda. *Nature* **253**, 349–350.

Ginsburg and Shinn, E.A. (1964) Distribution of the reef-building community in Florida Bay and the Bahamas (abst.). *American Association of Petroleum Geologists Bulletin* **48**, 527.

Hatziolos, M., Hooten, A.J. and Fodor, M. (1998) *Coral Reefs: Challenges and Opportunities for Sustainable Development*. The World Bank, Washington, DC. 225 pp.

Hudson, J.H. and Diaz, R. (1988) Damage survey and restoration of M/V Wellwood grounding site, Molasses Reef, Key Largo national marine Sanctuary, Florida. *Proc. 6th I.C.R.S.*, Australia **2**, 231–236.

Hughes, T.P. and Connell, J.H. (1999) Multiple stressors on coral reefs: A long term perspective. *Limnology and Oceanography* **44** (3, part 2), 932–940.

Jaap, W.C. and Hallock, P. (1990) In *Coral Reefs. Ecosystems of Florida*, eds. R.L. Myers and J.S. Ewel, pp. 574–618. University of Central Florida Press, Orlando, FL.

Jaap, W.C. (1984) In The Ecology of the South Florida Coral Reefs: A community profile. U.S. Dept. of the Interior, Fish and Wildlife Service, Minerals Management Service, Washington, DC, 138 pp.

Jaap, W.C., Lyons, W.G., Dustan, P. and Halas, J.C. (1989) Stony Coral (*Scleractinia* and *Milleporina*) community structure at Bird Key Reef, Ft. Jefferson National Monument, Dry Tortugas, Florida. Fla. Mar. Res. Publ. No. 46, 31 pp.

Lapointe, B.E. (1999) Simultaneous top-down and bottom-up forces controlling macroalgal blooms on coral reefs (reply to comment by Hughes et al.) *Limnology and Oceanography* **44** (6), 1586–1592.

Lessios, H.A. (1988) Mass mortality of *Diadema antillarum* in the Caribbean: What have we learned. *Annual Reviews of Ecology and Systematics* **19**, 371–393.

Lessios, H.A., Robertson, D.R. and Cubit, J.D. (1984) Spread Of *Diadema* Mass Mortality through the Caribbean. *Science* **226**, 335–337.

Malakoff, D. (1998) Death by suffocation in the Gulf of Mexico. *Science* **281**, 190–192.

Mayer, A.G. (1914) The Effects of Temperature on Tropical Marine Animals. Carnegie Institute Wash. Pub. 183, V6, pp. 1–24.

Mayer, A.G. (1916) The Temperature of The Florida Coral-Reef Tract. National Academy of Science V2, pp. 97–98.

Peters, E.C. (1996) Diseases of coral reef organisms. In *Life and Death of Coral Reefs*, ed. C. Birkland, pp. 114–139. Chapman-Hall, New York.

Porter, J.W., Battey, J.F. and Smith, J. (1982). Perturbation and change in coral reef communities. *P.N.A.S.* **79**, 1678–1681.

Porter, J.W. and Meier, O. (1992) Quantification and loss and change in Floridian reef coral populations. *American Zoologist* **32**, 625–640.

Porter, J.W., Lewis, S.L. and Porter, K.G. (1999) The effect of multiple stressors on the Florida Keys coral reef ecosystem: a landscape hypothesis and a physiological test. *Limnology and Oceanography* **44** (3), 941–949.

Reich, C.D., Shinn, E.A., Hickey, T.D. and Tihansky, A.B. (1999) Hydrogeology of a dynamic marine system in a carbonate environment, Key Largo Limestone Formation, Florida Keys. USGS Open-File Report 99-181, pp. 88–89.

Richardson, L.L. (1998) Coral Diseases: What is Really Known? *TREE* **13** (11) 438–443.

Shinn, E.A. (1998) Did Dust Do It? SEPM Annual Meeting, Salt Lake City, Abstract.

Shinn, E.A. (1999) Water Quality in The Florida Keys: Schizophrenia in Paradise, Abstract. Geological Society of America Southeastern Section Meeting Athens, Georgia, Vol 31 no. 3 March, p. A-67.

Shinn, E.A., et al. (1989) Reefs of Florida and the Dry Tortugas. In Field Trip Guidebook T176, 28th International Geological Congress, Amer. Geophys. Union, Washington, DC, 53 pp.

Smith, G.W., Ives, L.D., Nagelkerken, I. and Ritchie, K.B. (1996) Caribbean sea-fan mortalities. *Nature* **383**, 487

Swart, P.K., Healy, G.F., Dodge, R.E., Kramer, P., Hudson, J.H., Halley, R.B. and Robblee, M.B. (1996) The stable oxygen and carbon isotopic record from a coral growing in Florida Bay: a 60 year record of climatic and anthropogenic influence. *Palaeogeography, Palaeoclimatology, Palaeoecology* **123** (1–4) 219–237.

THE AUTHOR

Phillip Dustan
*Department of Biology
University of Charleston,
Charleston, SC 29424, U.S.A.*

Chapter 26

THE BAHAMAS

Kenneth C. Buchan

The archipelago of the Bahamas contains the largest tropical shallow water area in the Western Atlantic. Located on the northern and eastern margins of two large submerged banks and a number of smaller more isolated banks, the Bahama Islands, of which there are over seven hundred, are low-lying and composed of limestone. A sub-tropical climate and a geographic position between two major warm ocean currents affect the region with seasonal variability which influences the biological communities inhabiting the ocean and coastal areas.

The Bahama Banks are separated from the North American continent by the Florida Straits and from each other by deep channels, some in excess of 2000 m deep. Two deep water channels cut into the larger Great Bahama Bank. Most of the marine area is shallow (<20 m), resulting in an extremely important marine resource with both ecological and economic value.

The Bahama Islands are dependent on their seas to maintain a GDP of US$ 2.7 billion through tourism and harvest of marine resources. To date, the fishing industry has benefited from the relatively high ecological productivity of the shallow banks and their related habitats. Commercially important fisheries resources include the Spiny Lobster, Conch and the Nassau Grouper which together make up the bulk of fisheries income.

Clear warm waters and white sand beaches, along with its close proximity to the USA, make the Bahamas a prime tourist destination. Tourism is the mainstay of the Bahamian economy accounting for 60% of the countries Gross Domestic Product.

Agricultural and forestry operations are limited and impacts in the coastal zone from these activities are negligible. However, land reclamation and construction for tourism development, along with sand mining, dredging, over-fishing, poor fishing practices and their respective impacts of habitat loss, beach erosion and over-exploitation of target and non-target marine resources are becoming increasingly apparent as development pressures grow.

Environmental regulations are in place through a number of parliamentary Acts. Management of established marine and coastal protected areas has been undertaken by the Bahamas National Trust who, along with other organisations, carry out environmental education programs to increase awareness and reduce impact on the marine and coastal areas of the archipelago.

Fig. 1. Map of the Commonwealth of the Bahamas.

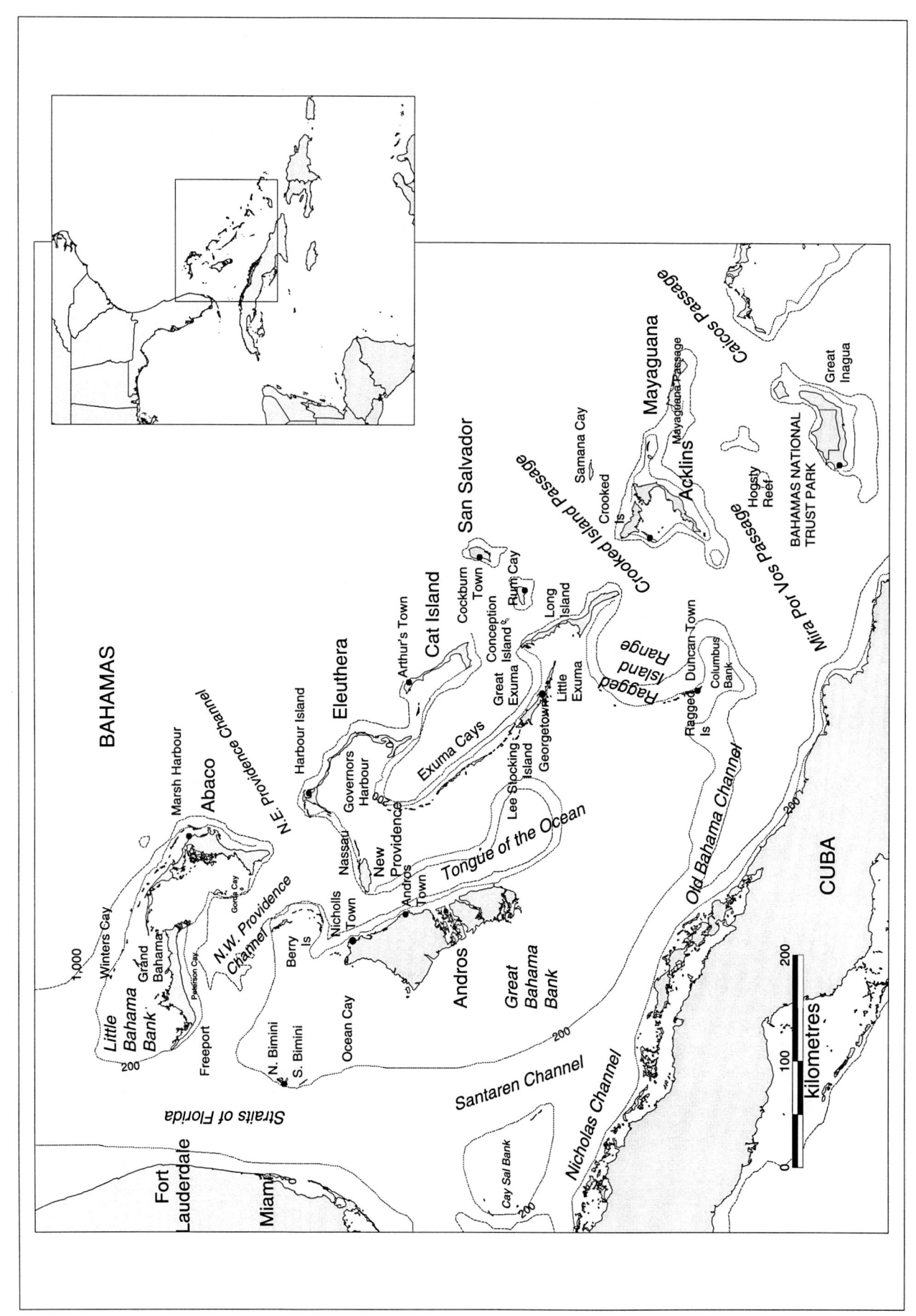

INTRODUCTION

The Bahama Islands form an archipelago in the tropical West Atlantic north of the Greater Antilles and southeast of Florida in the United States of America. Covering an area of 13,860 square km with a total land and sea area of approximately 300,000 km^2, the Bahamas consist of over 3000 low-lying carbonate islands, cays and rocks. The island territory extends from Grand Bahama on the Little Bahama Bank at 27.5°N, 1126 km southeast to Great Inagua (20°N), just north of Haiti (Fig. 1).

The northern and central islands are located on two vast carbonate platforms averaging 10 m in depth. The Little Bahamas Bank is located in the northern Bahamas while the Great Bahamas Bank begins approximately 100 km south, extending to the south and southeast. These Banks are separated by the Northwest and Northeast Providence Channels, and the Great Bahama Bank is split by two deep water channels. The first of these channels is the Tongue of the Ocean (approximately 1500–1800 m deep) which separates Andros Island from New Providence and the Exuma Cays. The second is the Exuma Sound which is similar in depth to the Tongue of the Ocean and forms a deep area to the east of the Exumas.

Beyond the central Bahamas to the southeast, the islands are located on a series of carbonate platforms beyond which the island territory of the Turks and Caicos begins.

The Cay Sal Bank is located approximately 50 km west of the Great Bahama Bank and covers an area of nearly 4000 km^2. Most of this bank is submerged beneath waters averaging 12 m in depth. A number of small islands, cays and rocks are scattered along the north, east and west margins of the Bank.

The Bahama Banks are isolated from neighbouring land on all sides, from the United States by the Florida Straits, to the North by the Atlantic Ocean, from Cuba in the south by the Old Bahama Channel and to the east by deep water between Mayaguana Island and Great Inagua beyond where the Caicos Bank begins.

THE ORIGIN OF THE BAHAMAS

The geological origin of the Bahamas Archipelago has been debated since 1853 when it was proposed that the Bahamas platform was a huge delta formed by the Gulf Stream as it met the Atlantic Ocean. The Bahamas began to develop around 200 million years ago, during the formation of the Atlantic Ocean. Later hypotheses on the formation of the Bahamas suggest that tectonic activity formed the deep channels and banks as grabens and horsts (Mullins and Lynts, 1977). Dietz et al. (1970) suggested that the channels and banks are the result of long-term depositional processes where carbonate production kept pace with subsidence and turbidity currents eroded the carbonate mass forming deep channels. Other schools of thought suggest the existence of a "megabank" which included Florida, Northern Cuba, the Turks and Caicos Islands and the Blake Plateau to the north. Around eighty million years ago this "megabank" was altered substantially by an event, perhaps the creation of the Gulf of Mexico (Sealey, 1994). This event led to flooding of the Blake Plateau (now at 900 m depth), the separation of the Bahamas from Cuba and Florida, the creation of small banks separated by deep water in the southeast and the creation of the troughs and channels within and between the Little and Great Bahama Banks.

It is generally accepted that the Bahama Banks developed in an area conducive to hermatypic activity and that skeletal remains were deposited as sediments. In addition, oolitic sediment precipitated from the oceanic waters of the Atlantic as it moved on to the warm shallow banks. Drill holes have indicated that carbonate deposits making up the platform reach depths of over 5.4 km (Meyerhoff and Hattin, 1974).

The islands of the Bahamas formed when the platform became exposed during sea-level lowstands caused by four major glacial events during the Pleistocene (Fig. 2) (Sealey, 1994). Sea level dropped as much as 120 m, exposing the Bank sediments. Fine, light rounded oolitic limestone was then blown by the trade winds to form dunes, which became lithified after flooding. During the interglacial periods sea level rose approximately 5–6 m above present-day sea level, and fossil reefs throughout the Bahamas have been used to date sea-level change (White et al., 1997). Erosional features are now common throughout the islands and the limestone rock has been weathered into karst formations such as caves, sink holes and solution pits (Gerace et al., 1998). The limestone is very porous and is permeated by subterranean conduits formed by rainwater. Consequently, rainwater is diverted underground, so there are no rivers and very little freshwater run-off from the

Fig. 2. The islands of the Bahamas are low-lying limestone structures, typically located on the northern and eastern margins of large shallow water banks. From the air the influence of currents on the bank sands can be seen in features such as channels and large rippled areas. (Photograph Matthew Robinson).

islands except during the heaviest of rains. For a more detailed account on the geology of the Bahamas (see Carew and Mylroie, 1997).

Much of the biota existing in the Bahamas today is of Caribbean origin and was introduced during the last glacial advance when sea level dropped almost 100 m. This facilitated the movement of plants and animals across the much reduced oceanic gap between the southern Bahamas and Cuba and Hispaniola.

SEASONALITY, CURRENTS, NATURAL ENVIRONMENTAL VARIABLES

Climate

The Bahamas Archipelago spans six degrees of latitude and nine degrees of longitude across the Tropic of Cancer, so there are regional variations in weather patterns and a mix of climatic conditions throughout the island chain. The climate of the Bahamas is sub-tropical and has distinct winter and summer regimes (Halkitis et al., 1982). During the winter, southward-moving cold polar air masses stream over the islands from the United States, and although moderated by the Gulf Stream, these fronts can reduce air temperature in the northwestern Bahamas significantly. In the summer, warm moist air moves northwards from the Caribbean (Fig. 3). The islands in the north on average exhibit cooler temperatures and higher precipitation than those in the south.

Precipitation figures (Fig. 4) indicate that average rainfall in Grand Bahama, the most northerly island, is twice that of Great Inagua in the South (1400 mm and 700 mm respectively). The temperature in Great Inagua is generally greater with a maximum mean monthly temperature of 29°C (84°F) compared to 27°C (81°F) in Grand Bahama and a

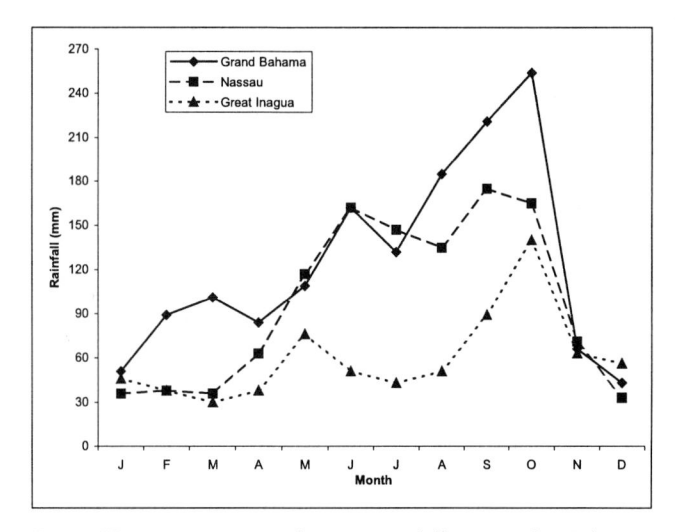

Fig. 4. Variation in annual mean rainfall across the Bahamas Archipelago using data from islands located in the northern, central and southern Bahamas. (Data from Shaklee, 1996).

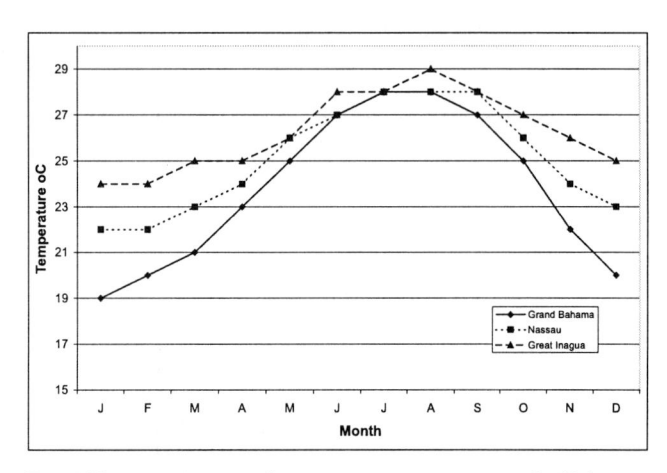

Fig. 5. Variation in annual mean temperatures across the Bahamas Archipelago using data from islands located in the northern, central and southern Bahamas. (Data from Shaklee, 1996).

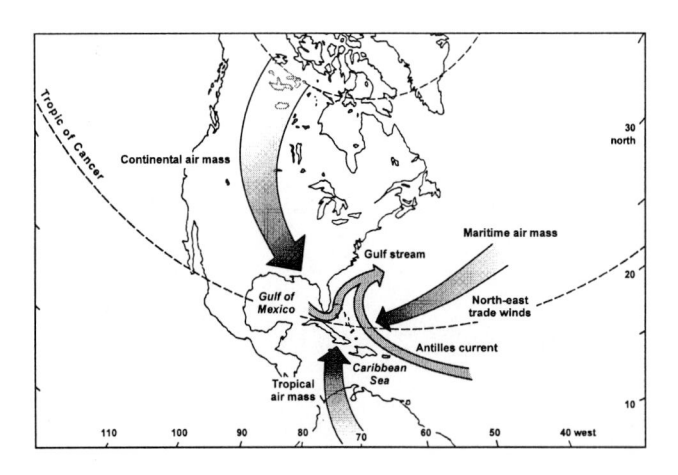

Fig. 3. Winds and ocean currents affecting The Bahamas. Cold winds from North America may be more northerly or westerly. The northeast trade winds curve around and blow from the east or south east in the summer. (From Sealey, 1994).

minimum mean monthly temperature of 25°C (77°F) in Great Inagua compared to 20°C (68°F) in Grand Bahama (Shaklee, 1996) (Fig. 5).

The Bahamas are exposed to significant hurricane and tropical storm activity during the months of August through October. The islands are located along the path taken by many North Atlantic hurricanes and are particularly susceptible to damage because they are low-lying. Hurricanes and storms impacting the Bahamas may also originate in the Gulf of Mexico and Caribbean Sea. On average three hurricanes can be expected to cross some portion of the Bahamas archipelago every four years (Shaklee, 1989), which accounts for approximately one in every seven which develops in the North Atlantic. Hurricanes affecting the Bahamas have been somewhat variable from year to year (Table 1). One of the worst was the "Great Bahama Hurricane" (Albury, 1975) which hit in September 1866; unlike most which affect some islands but not others, this

Table 1

Hurricane occurrence, by decades, from 1900–1999 in the north Atlantic and the Bahamas. (Modified from C.J. Neumann et al., 1987)

	1900–09	1910–19	1920–29	1930–39	1940–49	1950–59	1960–69	1970–79	1980–89	1990–99
North Atlantic	35	34	38	47	50	69	62	49	40	56
Bahamas	7	2	9	11	13	11	6	1	2	5
Percentage	20%	6%	24%	23%	26%	16%	19%	2%	5%	9%

hurricane worked its way up the chain of islands. Houses and farm crops were destroyed and all but one vessel in Nassau Harbour at the time were sunk or broken up.

Major hurricanes affecting the Bahamas in recent years include: Hurricane David in 1979 which caused massive beach erosion with the average beach profile loss of 2–2.5 m on Cabbage Beach in New Providence; Andrew in 1992, which caused a 7 m storm surge in Eleuthera carrying seawater approximately 1 mile inland; and Hurricane Lili in 1996 which caused major structural damage to residences on the island of San Salvador. Most recently, the category 4 hurricane Floyd caused major coastal flooding and damage to seawalls, roads and coastal residences on San Salvador, Cat Island, Eleuthera and Abaco in September 1999.

Large northerly swells are also known to occur on occasion as a result of particularly violent storms originating in the Northern Atlantic, causing shore erosion on the north coasts of many Bahama islands, as in October 1991.

Oceanography

Ocean currents and sea surface water temperatures influence temperature throughout the archipelago. Currents affecting the Bahamas originate from two places (see Fig. 3). The Gulf Stream moves between Florida and the Bahamas from the Caribbean Sea and the Gulf of Mexico, while the Antilles Current flows onto the Archipelago after it has moved westward across the Atlantic Ocean from the Coast of Africa, originally as part of the North Equatorial Current. The path of the Gulf Stream remains fairly constant in its position. However, the Antilles current shifts to the north in the summer creating warmer temperatures in the Northern Bahamas, and to the south during the winter months providing warmer temperatures to the southern islands (Shaklee, 1996).

Sea surface water temperatures are variable across the islands, which affects the biological components of the shallow water marine habitats. For five months of the year waters in the northern Bahamas are below optimum for coral growth. This factor combined with fluxes in salinity and turbidity from trade-wind-induced currents affects the abundance of coral (Newell and Imbrie, 1955). *Acropora cervicornis* is absent and growth of *A. palmata* is restricted in the northwestern Bahamas. Additionally, cold fronts from the United States during the winter months can last several days creating cold ocean currents which contribute to this reduction in growth and species diversity. Inshore shallow water temperature may be reduced very quickly during winter storms and has resulted in mortality of some fish species in bays and estuaries (Newell et al., 1959). In contrast, recent summer ocean temperatures in the Central Bahamas have exceeded 30°C for extended periods of time causing extensive coral bleaching (McGrath and Smith, 1999).

The tidal range throughout the Bahamas is approximately 1.5 m and is a semi-diurnal mixed type (Sullivan, 1991) with four tidal extremes. Salinity is fairly consistent along the platform margins of the Banks at around 35 ppt, but may be higher across the shallow banks due to evaporation.

THE MAJOR SHALLOW WATER MARINE AND COASTAL HABITATS

The islands of the Bahamas are located for the most part on the northern and eastern margins of the extensive platforms of the Banks. In general, coastal waters are warm and clear, lacking any impact from rivers or other terrestrial run-off. In contrast, the interiors of the platforms are generally turbid due to tidal circulation, and are more variable in temperature and salinity (Newell et al., 1959). Consequently, reef development on the interior of the platforms is inhibited.

The Bahama Bank Platforms

The marine sediments of the platforms, excluding any organic matter, are pure calcium carbonate in the form of aragonite with a lesser amount of calcite from coralline algae and foramniferans (Newell et al., 1959). Grain sizes range from medium-grained sand to silt and clay.

Recently, extensive work has been done on marine community classification. Sullivan (Sullivan, 1991; Sullivan-Sealey, 1999) describes soft sediment and hard substrate habitats of the Exuma Cays and presented two classifications schemes, one which describes marine communities throughout the Bahamas Archipelago and the other a more specific scheme to aid in the interpretation of aerial photographs of Montagu Bay, New Providence and the Exuma Cays. Table 2 summarizes the habitat classifications present in this region.

Table 2

Summary of Benthic community classifications for the Bahamas Archipelago (see Sullivan-Sealey, 1999)

Soft Sediment Communities	
Sand-Mud/Bare Bottom (Calcareous Muds)	(a) Mud bank and mud bottom often with polychaete or crustacean burrow mounds (b) Intertidal mud flats (c) Island moats (d) Anchialine ponds/saline land locked ponds (e) Mangrove channels/lagoons
Sand-Mud/Seagrass	(a) Sparse seagrass <30% cover (b) Moderate to dense seagrass community (c) Seagrass patches on matrix of soft sediment
Sand/Bare Bottom	(a) Sand beaches (b) Sandy shoals and sand bars
Sand/Seagrasses/Algal Canopy	(a) Sparse seagrass (b) Sandy algal canopy (dominant calcareous green algae) (c) Mixed algal canopy (sparse seagrasses, red algae and green algae) (d) Sand bioturbation zone, polychaete or crustacean burrows dominate. Small patches of algae (typically calcareous green) may be present
Rubble/Loosely Consolidated Hard Bottom	(a) Cobble rubble beaches (intertidal) (b) Reef rubble communities (c) Mollusc reefs (d) Serpulid worm (polychaete) reefs
Hard Substrate Communities	
Sparse Hard-bottom Communities	(a) Mixed coral/sponge/algae sparse hard bottom with less than 30% cover of the above (b) Algal dominated sparse hard bottom with <50% cover of algae and reduced sponge and coral coverage (c) Coral/octocoral-dominated sparse hard bottom reduced sponge and algal coverage
Dense Hard-bottom Communities	(a) Mixed coral/sponge/algae dense hard bottom >50% cover of the above. Low relief carbonate platform (b) Algal-dominated dense hard bottom reduced occurrence of sponges and corals (c) Octocoral-dominated dense hard-bottom: characterized by a visual dominance of octocorals on a low-relief limestone pavement with little or no sediment accumulation
Tidal Channel Communities	(a) Sparse hard-bottom: characterized by shallow depth <5 m sparse sand accumulation (b) Dense hard-bottom: similar to sparse tidal channel communities
Hard-bottom Seagrasses	(a) Dense seagrass patches on matrix of hard-bottom (seagrass > 50% of total area) (b) Hard-bottom matrix with dense seagrass patches (seagrass < 50% of the total area)
Hard-bottom/Coral Reef Communities	(a) Patch Reefs – Linear/bank patch reef – Dome-shaped/lagoonal/Channel patch reef (b) Fringing Reef Systems – Back reef rubble – High relief spur and groove – Transitional reefs – Low-relief spur and groove – Fore reef terrace – Deep reef resources or escarpments (c) Bank-barrier Reef Systems – Back reef rubble – High relief spur and groove – Transitional reefs – Low-relief spur and groove – Fore reef terrace – Deep reef resources or escarpments

Hard-bottom Nearshore Platform/ Rocky Intertidal	Zone A: Intertidal Spray Zone
	Zone B: Upper Intertidal Zone
	Zone C: Lower Intertidal Zone
	(a) Windward Rocky Community
	  Windward Rocky Intertidal
	  Windward Rocky Platform
	
	(b) Leeward Rocky Community
	  Leeward Rocky Intertidal
	  Leeward Rocky Platform

Important Shallow Marine Habitats

Coral Reefs

Coral reefs cover an area of just over 1800 km² of the Great Bahama Bank and approximately 324 km² of the Little Bahama Bank (Wells, 1988). These reefs are most prominent on the windward north and eastern sides of the islands and cays, developing best a short distance from shore. The occurrence of coral in the central area of the Bank platforms is limited, due to turbidity and variable physical conditions. The best development of coral reef is in association with islands along the margins of the platforms. Reefs are not restricted to the north and east and development of healthy systems is evident on the west coasts of many of the islands including the Exuma Cays and San Salvador. Possibly the third largest barrier reef in the world is located off the east coast of Andros, the largest of the Bahamian Islands. However, some researchers question the barrier reef classification because of the discontinuous nature of the reef.

There are about 30 species of hermatypic corals, of which only a few can be described as contributing significantly to the reef-building process. These are: *Montastrea annularis*, *Montastrea cavernosa*, *Acropora palmata*, *Acropora cervicornis*, *Siderastrea siderea*, *Diploria labyrinthiformis* and *Porites porites* (Squires, 1958; Newell et al., 1959). More recently extensive studies have focused on ecological factors such as community structure (Chiappone and Sullivan, 1991; Sullivan and Chiappone, 1992; Sullivan et al., 1994; Chiappone et al., 1996) and reef health (Curran et al., 1993; Lang et al., 1988; McGrath and Smith, 1999).

In the Exuma Cays, Sluka et al. (1996) described 53 species of algae, 49 sponges, 36 scleractinian corals, 29 octocorals, 3 black corals, 4 anemones, 2 zooanthids and 2 corallimorpharians. Several coral diseases have been described, including black and type II white band diseases (Ritchie and Smith, 1998), and Aspergillosis in *Gorgonia* spp. (Nagelkerken et al., 1997a,b), and a number of bleaching events have also been described (McGrath and Smith, 1999). Mass bleaching was observed around New Providence Island in August 1998 where approximately 60% of coral heads were bleached. Extensive bleaching was also reported at Walkers Cay in the northern Bahamas (Wilkinson, 1999).

The sea urchin *Diadema antillarum* died off throughout the Bahamas in 1983 as it did elsewhere in the Caribbean. In recent years the *Diadema* have been seen more frequently on the reefs, but in numbers far below the pre-die-off populations.

Marine fishes have been described in detail by Bohlke and Chaplin (1968) who describe species abundance and distribution throughout the Bahamas. Sluka et al. (1996) concentrated their efforts on describing the status of Groupers in the Exuma Cays Land and Sea Park and concluded that the park was indeed protecting the abundance, size and reproductive output of a number of Grouper species normally targeted by fishermen.

Seagrass

Seagrasses can be found on the Great and Little Bahama Banks, the Cay Sal Bank and in tidal estuaries, lagoons and sheltered bays of Islands and Cays across the archipelago. Three seagrass species are most commonly encountered: *Thalassia testudium*, *Syringodium filiforme* and *Halodule wrightii*. On San Salvador Island, Smith et al. (1990) concluded that low energy sites had a higher frequency of *Thalassia* with a lesser amount of *Syringodium* and *Halodule*, whereas high energy sites appeared to favour *Syringodium*. Seasonal variations in biomass also occurred with increases in *Syringodium* and *Halodule* and a decrease in *Thallasia* during the months of July to December. Leaf biomass and production of seagrasses in the Bahamas is shown in Table 3.

Seagrass meadows form important nursery areas and habitat for reef fish and invertebrates; they provide organic material to down stream habitats, stabilize sediments and act as a food source to herbivorous fishes and sea turtles.

Mangroves

Large areas of mangrove can be found along the margins of many sheltered bays, lagoons and tidal estuaries throughout the Bahamas. These trees, in particular the red mangrove with its many prop roots, provide stability to the coastal fringe and trap sediments forming overwash islands, and make a land-forming progression outward from the shore. Many species of juvenile fish utilize the area beneath the mangrove roots for foraging. The high organic content of the sediments due to leaf litter, and the algae-coated roots provide food for polychaete worms, crustaceans and other detritivores which young fish feed on. The mangrove root system also provides shelter from predation by larger carnivorous fishes. Species common to the mangrove areas in the Bahamas include Snappers (*Lutjanus* spp.), Grunts (*Haemulon* spp.), Parrotfishes (*Scarus* spp. and *Sparisoma* spp.), and Mojarra (*Gerres* spp. and *Eucinostomus* spp.), also in some areas the Nassau Grouper (*Epinephelus striatus*).

Mangroves throughout the Bahamas mostly have a typical zonation, with red mangrove (*Rhizophora mangle*) dominating the water's edge, then becoming interspersed

Table 3

Leaf biomass and production of seagrass species in the Bahamas (modified from Short, 1986)

Species	Biomass (g dry m^{-2})	Production (g C m^{-2} d^{-1})
Thalassia testudinum	5.3–200	0.14–2.10
Syringodium filiforme	7.6–159	0.01–2.00
Halodule wrightii	5.9	0.004

Fig. 6. A typical tidal creek fringed by red mangrove (*Rhizophora mangle*) with *Thallasia testudium* dominated seagrass beds adjacent. In the Bahamas the red mangrove tends to be dwarfed due to a lack of fresh water, low substrate fertility and occasional low temperatures. (Pigeon Creek, San Salvador Island, photo Matthew Robinson).

with black mangrove (*Avicennia germinans*) landward, eventually progressing toward an area dominated by white mangrove (*Laguncularia racemosa*) and Buttonwood (*Conocarpus erectus*) (Fig. 6). However, this classic zonation pattern is not exhibited in some of the inland lakes of San Salvador island (Godfrey et al., 1993), where red mangrove, which is typically more sensitive to environmental extremes of high salinity and periods of lower temperatures (Lugo, 1994) is replaced by black mangrove in the fringe closest to the water. This is also the case in some other sheltered locations.

The mangroves on San Salvador Island have been described as an ecosystem under stress (Lugo, 1994) and may be typical of other mangrove areas throughout the Bahamas. This stress is reflected most in the red mangrove whose trees are typically stunted, the leaves are small and inflexible, the tree canopy is thinned and the leaves are orientated upwards. There is also a high occurrence of albinism in *Rhizophora* seedlings (Godfrey and Klekowski, 1989). The sources of stress for mangrove can be attributed to a number of factors (Lugo, 1994). Being located just above the Tropic of Cancer, the northern Bahamas are sometimes exposed to lower temperatures, in combination with a low annual rainfall which can cause drought and an increase in salinity, subsequently restricting the development of mangrove stands. They grow optimally with some freshwater influence, but there are no rivers, few freshwater springs and little freshwater run-off to mangroves in the Bahamas, and the substrate has low fertility (Kass and Stephens, 1990; Kass et al., 1994).

In Bimini, as elsewhere in the Bahamas, much of the mangrove is stunted in its growth form, but mangrove roots are dense, forming a habitat for juvenile fishes. The Bimini lagoon which has been earmarked for development is an important habitat for Bone Fish and as a Lemon Shark nursery area (Correia et al., 1995).

The western half of Andros island is predominantly wetland with extensive areas of mangrove. This area forms an important habitat for birds such as cattle egrets and ospreys.

Mangrove habitat is under stress from coastal development and has been removed extensively in New Providence and Grand Bahama for coastal development purposes.

OFFSHORE SYSTEMS

The most striking deep-water features of the Bahamas are the channels which penetrate and surround the Banks. Although there is some debate about their origin, certain features of these canyons are that channels consist of a large U-shaped trough with a V-cut canyon on the trough floor. The steep sides of the troughs are constructional as opposed to erosional, being built up by depositional processes, and the V-shaped canyons are erosional features cut by extreme but occasional turbidity currents (Sealey, 1994).

The Tongue of the Ocean has been well studied by submersible explorations in the 1960s and 70s. The trough features include eroded cliffs down to about 400 m followed by gullied slopes extending to the trough floor which is flat except where the V-shaped canyons split the floor. These canyons rarely begin at the start of the trough but eventually spread open becoming part of the main trough floor. Turbidity currents are responsible for the submarine canyon formation. A build-up of sediment on the bank platform at the head of the trough will now and again be disturbed by storms or perhaps tectonic activity. The sediment cascades into the trough and is carried by submarine currents with velocities up to 45 mph which erodes the trough floor. The geological record shows that turbidity currents occur anywhere from every 500 to every 10,000 years (Sealey, 1994). Porter (1973) made some observations on the biota of the Tongue of the Ocean. As the submersible descended into the trough, genera and species declined, along with the total biomass. Near the surface, sponges replaced stony corals as the dominant species and living cover decreased from around 50–90% at the surface to less than 1% at 300 m.

The Bahamian Exclusive Economic Zone (EEZ) extends 200 miles into the Atlantic Ocean. Within these waters there is tremendous potential for commercial fishing as they encompass an area along the migration routes of some high-value commercial pelagic species such as yellow and blackfin tuna, and swordfish (BREEF and MacAlister, Elliott and Partners Ltd 1998). Bimini has developed a deep-sea sport fishing industry targeting fish which utilize the productive waters of the Gulf Stream (Campbell, 1978). The Bahamas is known worldwide for its deep-sea sportfishing. Sports fishermen utilize the deep-water areas, targeting pelagics such as blue marlin, white marlin, wahoo, dolphinfish and tunas.

In the 1970s and 80s, the Bahamas Department of Fisheries considered the potential of a commercial deep-water 50–300 m snapper and grouper fishery on the bank drop-offs. Although not exploited to any great extent so far, interest is increasing in this type of fishery utilizing fish traps.

POPULATIONS AFFECTING THE AREA

Before the arrival of Columbus on San Salvador Island in 1492, the Bahamas were inhabited by the Lucayan Indians. Historical records and archaeological studies have suggested that the population of the Bahamas was around 40,000 at this time (Sealey, 1990). The Lucayan people farmed the land and fished to sustain themselves, but they eventually died out early in the 1500s as a result of disease and slaving raids by the Spanish. By the 1780s the population of the Bahamas was around 4000. Populations fluctuated over the following years with a major increase at the end of the 1700s as those loyal to the British Crown looked for places to settle following the American War of Independence. These loyalists established plantations using slave labour. In 1807 the British government abolished trade in slaves so that the plantations could no longer operate. Many of the Loyalists left at this time, and by 1840 there were few plantations left (Sealey, 1990). The freed slaves remained and established themselves in New Providence, Andros, Grand Bahama, San Salvador, Rum Cay, Long Island and the Exuma Cays, and by 1843 the population had grown to around 25,000. This number slowly increased until the early 1900s when islanders left to seek employment elsewhere, and this lull continued until the 1920s when it started to grow once more. Since the 1950s there has been an increase in the population to the present day (Table 4).

The most recent official census of population in the Bahamas was made in 1990, though estimates for 1998 were 293,700 with a projection for the year 2000 of 302,800. The main population centres are New Providence (2152.5 individuals per square mile) and Grand Bahama (77.2 individuals per square mile).

Historically, the inhabitants from the Lucayan Indians to present day fishermen have always utilized the resources of its shallow marine environment. During this time significant fisheries have been established. Coastal areas on many of the islands, in particular New Providence and Grand Bahama, have been developed for tourism and residential areas.

THE TOURISM INDUSTRY AND ITS EFFECT ON THE POPULATION

In 1997, the islands of the Bahamas accommodated 1,617,595 stopover visitors, of whom 81% came from the United States of America, 6% from Canada, 8% from Europe and 5% from other countries. Approximately 74% of these were on vacation, and according to surveys on visitors to Nassau by the Bahamas Ministry of Tourism (1998) the primary reason for vacationing in the Bahamas was its beaches (27%) followed by the climate (17%). Twenty percent of those visiting Grand Bahama gave value for money as their main reason. Visitors to the out islands also indicated beaches (29%) as their primary reason for choosing the Bahamas, followed by sporting attractions (including SCUBA diving and snorkelling) (26%). With the coastal zone and marine environment being the main tourist attractions in the Bahamas, much coastal development has taken place.

Development in the tourism sector has led to numerous problems on some of the islands, such as waste management and the excessive use of freshwater resources.

AGRICULTURE IN THE BAHAMAS

In 1997, agriculture generated $56.44 million. Abaco, Andros, and Grand Bahama are the main centres for agriculture, on land mostly leased from the government to Bahamian farmers. This land makes up about 90% of agricultural land throughout the archipelago.

Presently, crop production for export is concentrated on Abaco, Andros, Grand Bahama and Eleuthera, and consists mainly of citrus fruits (grapefruit, lemons, limes and oranges), but also includes cucumbers, okra, avocados, papaya, squash, tomatoes and zucchini. About 95% of export is to the USA.

On many of the smaller Bahamian Islands small-scale subsistence agriculture is carried out. Limited soils which

Table 4
Population of the Bahamas (Official Census, 1980 and 1990)

Island	1980	1990
Abaco	7,271	10,061
Acklins	618	428
Andros	8,307	8,155
Berry Islands	509	634
Bimini	1,411	1,638
Cat Island	2,215	1,678
Crooked Island and Long Island	553	423
Eleuthera, Harbour Island and Spanish Wells	10,631	10,524
Exumas	3,670	3,539
Grand Bahama	33,102	41,035
Inagua	924	985
Long Island	3,404	3,107
Mayaguana	464	308
New Providence	135,437	171,542
Ragged Island	164	89
Rum Cay and San Salvador	825	539
Total	209,505	254,685

Table 5

Marine resources utilised throughout The Bahamas (modified from Sealey, 1990)

Edible	scalefish	grouper, snapper, hogfish, jacks, grunts
	shellfish	conch, whelks, chiton, spiny lobster, stone crab, queen helmet
	sportfish	marlin, tuna, wahoo, mackerel, kingfish, dolphin, swordfish, sailfish, bonefish
	other	green and loggerhead turtle
Non-edible	biological	turtle shell, conch shell, coral, black coral, shells, sponge
	chemical	salt, water from desalination
	mineral	aragonite, building sand, petroleum

Table 6

Summary of total recorded landings of marine products in The Bahamas during 1998 (Bahamas Department of Fisheries, 1999)

	Weight (lbs)	Value (US $)
Crawfish tails	5,478,508	53,364,247
Crawfish whole	215,144	776,233
Conch	1,477,374	3,651,628
Stone crab	85,126	609,001
Turtle (green)	5,072	6,571
Turtle (loggerhead)	2,052	3,693
Nassau grouper	1,125,817	2,674,401
Other grouper	228,235	460,581
Grouper fillet	108,803	327,422
Snappers	1,721,359	2,363,558
Jacks	202,411	216,381
Grunts	198,232	155,601
Sharks	4,312	10,248
Others	343,214	415,479
Total	11,195,659	65,035,044

are usually deepest in carbonate sink holes and other dissolution features are utilised to grow sweet potatoes, water melon and other produce. Fertilizers are rarely used, and in conjunction with a low rainfall, a very porous substrate and little topographic relief, agricultural run-off is negligible.

Golf course construction has taken place on a number of the larger islands. However, no information on fertilizer-related nutrient loading of coastal areas is available, and whether this is a particular problem for coastal waters is not clear.

ARTISANAL AND COMMERCIAL FISHERIES IN THE BAHAMAS

A diversity of resources are harvested from Bahamian waters. In economic terms the most important of these are shellfish and scalefish fisheries.

Approximately 100,000 sq. miles of the Bahamas territory is marine and around half of this area is shallow (less than 20 m) and very productive in fisheries terms (Table 5). Three main fisheries exist and contribute the most to an income of almost US $62 million per year or 2.25% of the country's Gross Domestic Product in 1997 (BREEF and MacAlister Elliott and Partners Ltd, 1998). These are the Spiny Lobster *Panulirus argus*, the Queen Conch *Strombus gigas* and the Nassau Grouper *Epinephalus striatus*.

Other species fished include Snappers (*Lutjanus* spp.), Grunts (*Haemulon* spp.), Jacks (*Caranx* spp.), other grouper, Green and Loggerhead Turtles, Shark, Stone Crabs, the Queen Helmet Shell (*Cassis madagascariensis*) and sponges (*Hippospongia lachne* and *Spongia* spp.). Most of the catch is landed on New Providence, Abaco and Eleuthera and to a lesser extent on Grand Bahama, Long Island and Andros (Fig. 7 and Table 6).

Spiny Lobster *Panulirus argus*

Of the three main species harvested in the Bahamas, the Spiny Lobster (*Panulirus argus*) or "Crawfish" as it is referred to in the Bahamas is the most important contributor to the Bahamian economy. Being the fourth largest spiny lobster fishery in the world (after Australia, Brazil and Cuba) almost 2600 metric tonnes of lobster tails (around 7000 tonnes live weight) were fished in 1997 which had a value of US $58.7 million (Department of Fisheries, 1998). Export of the Spiny Lobster is to the U.S. (60%), France (35%) and Canada (5%) with a small developing fraction going to Japan.

They are caught with three different fishing techniques. Traditionally, Spiny Lobster are speared or trapped. Spearfishing involves free diving or diving with the assistance of

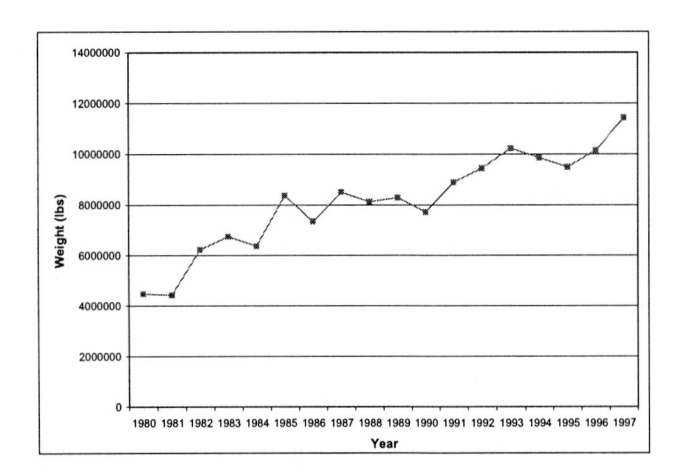

Fig. 7. Total recorded fisheries landings by weight from 1980 to 1997.

surface supply air, and hooking them out of their hiding places and spearing them. Destructive methods have also been used to force lobster into the open, such as use of bleach (Campbell, 1978). Although this destructive practice has reduced in recent times, (it is now illegal to carry bleach on fishing vessels) other substances have reportedly been utilised such as gasoline and detergent. In addition, traps are constructed from slats of wood and string which are baited with cowhide. Normally larger fishing vessels will employ this fishing technique and deploy up to about two thousand traps.

In the late 1980s, artificial habitats known as condos (or casitas elsewhere in the Caribbean) were introduced and are now being used widely throughout the Bahamas. These structures are constructed from large rectangular pieces of aluminium and have three wooden sides. The condos attract the lobster.

Commercial fishing vessels will remove most of the lobsters' tails and freeze them for storage. The tails are often soaked in preservative before freezing.

The pelagic larval phase of the Spiny Lobster is up to a year. The fishery in the Bahamas may therefore be dependent on upstream supply, although localised ocean circulation may replenish stocks from within the Bahamas. Considering the importance of upstream supply, concerns are directed at the status of regional stocks, in particular those from Turks and Caicos, the Eastern Caribbean and Brazil. Throughout the wider Caribbean it is generally accepted that most Spiny Lobster stocks are either fully or over exploited.

The Bahamas Department of Fisheries collect lobster landing data indirectly through licensed fish buyers and processors, and because there are only a few main buyers, the data collected is considered fairly comprehensive. There is less direct information, for example on catch per unit effort (CPUE). Analysis of the available data however, suggests recent declines in CPUE (BREEF and MacAllister, Elliot and Partners Ltd., 1998).

Queen Conch *Strombus gigas*

Conch landings in 1997 were approximately 1.43 million lbs (Bahamas Department of Fisheries, 1998). Less than a quarter of this was exported to the U.S. and Canada, generating around US $1 million in export sales. Most conch (60%) are fished during the closed lobster season, when full-time fishermen turn to this valuable alternative. Conch are usually fished while free-diving although it is permitted to use compressed air during the lobster season. Conch may sometimes be picked up from very shallow water, but these individuals tend to be juveniles.

Stock assessment is very difficult because much of the conch catch is not sold to licensed buyers, but is sold directly to businesses and consumers. Generally, inshore populations have become depleted throughout the Caribbean and fishermen are searching in deeper waters. Populations are in decline. Some countries have taken conservation measures by enforcing a complete ban on conch fishing, but this is not the case in the Bahamas although the Department of Fisheries are currently assessing stocks. BREEF/MacAllister, Elliot and Partners (1998) surmise that it is likely that the Bahamas could support a relatively large Conch fishery and that depleted stocks are mostly evident around population centres in the Northern and Central Bahamas.

Nassau Grouper *Epinephalus striatus*

Nassau Grouper landings for 1997 were approximately 1.13 million lbs, making this probably the largest Nassau Grouper landing in the world. Other grouper species caught totalled around 15% of the weight with similar amounts of grouper fillet being landed. In total 1.45 million lbs of grouper and filet were landed with a total value of around US $3.29 million. As with the Queen Conch, grouper landings are hard to assess because of direct sales.

Shallow water scalefish such as the Nassau Grouper are caught using spears, either while free-diving or using compressed air. Traps, hook and line and nets are also utilised. In December and January each year the Nassau Grouper form spawning aggregations which are also exploited by fishermen when they are easily caught. Exploitation of these spawning aggregations has led to major declines throughout the Caribbean. Indeed exploitation of the species is severe in most parts of the Caribbean, except those which control fishing during spawning periods. Catch per unit effort data from the Bahamas is inconclusive, but evidence from other parts of the Caribbean suggests that the Nassau Grouper is extremely vulnerable to even low fishing effort on spawning aggregations.

Exploitation of Other Scalefish

Although the major scalefish fishery is for Nassau Grouper, Snapper, Grunts, other Grouper, Hogfish and Jacks are also targeted. Fish-attracting devices are often used to encourage the aggregation of Snappers and Grunts. Scalefish exports (not including grouper) totalled approximately 200,000 lbs in 1997. A majority of this can be attributed to the Red Snapper and Mutton Snapper. Shallow water fishes are speared, trapped or netted.

Deep-water snapper and grouper are fished using traps; long lines with ten or more hooks are not permitted. Exploitation of deep-water fish and the techniques used to catch them have been of some concern for three main reasons: They have been overexploited in other countries because of a slow growth rate and sexual maturation, mechanical hoists used to recover traps can cause extensive damage to fish habitats, and traps may break free and continue to catch fish for extended periods in cases where the use of biodegradable trap materials is not enforced.

OTHER FISHERIES RESOURCES

From 1840 until 1940 the harvest of sponge from shallow marine areas was a booming industry. At one time there were 265 schooners, 322 sloops and 2808 open boats committed to sponging (Campbell, 1978). Sponges were exported to France, and nearly one third of the Bahamian work force made their living from the sponge fishery (Campbell, 1978). However the fishery came to a sudden halt in 1939 when a fungal disease killed around 90% of the harvestable sponge in two years. In modern times synthetic sponges have prevented the sponge fishery ever becoming a major economic force in the Bahamas again, although some is still harvested from the Acklins Bight and Central Andros (Sealey, 1994) and sent to Nassau for export. The value of sponge export during the 1980s was very low, ranging between US $29,000–375,000 per year. The higher value came about as a result of disease affecting the Mediterranean sponge production and harvest. Recent fisheries statistics from 1995 to 1997 show a marked increase in harvest, with export values averaging 126,854 lbs and US $932,301 respectively (Bahamas Department of Fisheries (DoF), 1998).

Three species of sea turtle are seen throughout the Bahamas, the Green Turtle *Chelonia Mydas*, the Hawksbill *Eretmochelys imbricata*, the Loggerhead *Caretta caretta* and on rare occasions the Leatherback *Dermochelys coriacea*. The Loggerhead is exploited less than the other species as its meat is not as palatable and its shell is of little commercial value. As a result it is the most common of the sea turtles in the Bahamas, and is still fished with restrictions on size. The Hawksbill has become seriously depleted in many parts of the world, the Bahamas being no exception. Its shell is of great economic value to markets in Europe and Asia and is used to make jewellery and other products. It is now prohibited to capture or be in the possession of a Hawksbill turtle in the Bahamas. The Leatherback is rarely seen in the Bahamas and is not specifically mentioned in legislation apart from generalised statements on closed seasons, and while the Green Turtle is still fished in the Bahamas, fishery regulations outline a minimal harvestable size.

Historically, since the 1860s when export of turtle shell from the Bahamas totalled approximately 20,000 lbs in weight (Campbell, 1978), exploitation of turtles has varied considerably. In 1984 almost 100,000 lbs of turtle were landed, since when landings have decreased. The Hawksbill turtle is no longer exploited and the combined weights of Loggerhead and Green Turtle landed in 1997 and 1998 were just over 7000 lbs (Fig. 8).

Little comparative data exists on turtle nesting throughout the Bahamas, but some studies on the beaches of the islands and cays on the Cay Sal bank indicate significant turtle nesting (Addison and Morford, 1996).

Fisheries Labour

The labour force employed in the fisheries sector has been estimated at around 9,300 persons, following a fisheries census in 1995. Approximately 95% of this number (8835) are fishermen with the remainder being employed in processing or buying stations (Bahamas Department of Fisheries, 1998). This accounts for 7.2% of the total labour force in the Bahamas. The contribution of the fisheries to the Gross Domestic Product of the Bahamas in relation to other sectors is shown in Table 7.

Table 7

Contribution to the GDP in 1995 of various sectors of the Bahamian Economy (from Department of Statistics figures presented in BREEF and MacAlister, Elliot and Partners Ltd., 1998)

Sector	Contribution to GDP (US $ million)	Contribution to GDP (%)
Wholesale and retail trade	409.70	14.90
Hotels	269.85	9.82
Real Estate	227.05	8.27
Communications	106.28	3.87
Manufacturing	85.11	3.10
Electricity	75.56	2.75
Business activity	74.27	2.70
Construction	73.31	2.67
Fisheries	61.67	2.25
Insurance	60.22	2.19
Restaurants	48.38	1.76
Banking	48.36	1.76
Air transport and allied services	45.87	1.67
Transport (excluding shipping and air)	42.37	1.54
Shipping and allied services	38.12	1.39
Mining and quarrying	25.76	0.94
Agriculture	25.75	0.94
Total*	2746.13	

*Also includes other elements such as the public sector.

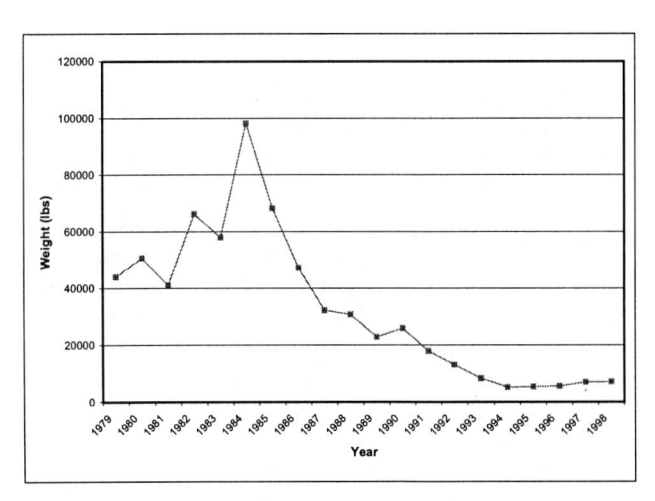

Fig. 8. Turtle landings in The Bahamas from 1979 to 1998.

There are approximately 4080 commercial fishing vessels, 646 of which are between 20 and 100 ft in length, with the remaining vessels ranging between 10 and 19.5 ft. Many smaller vessels listed within the total number of fishing vessels work in conjunction with larger vessels as smaller fishing tenders or platforms.

Fishery catch statistics are collected by the Bahamas Department of Fisheries. Only recently (1990) has catch per unit effort (CPUE) data been recorded on a computer database, and previous records are incomplete. Consequently, it is very difficult for fisheries officers to make an accurate assessment of the main Bahamas fisheries. Major fishing grounds in the Bahamas include the Great and Little Bahama Banks, the Cay Sal Bank, and the Crooked Island–Acklins Island Bank.

AQUACULTURE

Aquaculture operations in the Bahamas have been economically unsuccessful to date, in part due to transportation costs of equipment, supplies and the chosen culture organisms. Initial capital investment for these operations has often been too low for complete establishment.

Table 8 presents a summary of the aquaculture operations which have been established in the Bahamas outlining the species cultured, location and operational status. In Long Island the salt ponds from an abandoned salt production facility were utilised for shrimp production. Using approximately 100 acres of salt pond they produced about 300,000 lbs of shrimp per year. In addition, this company cultured Redfish quite successfully, but there was an inability to control rising salinity because there was no source of fresh water. Furthermore, logistical problems with transportation costs and the long distance from the United States subsequently resulted in closure of the site. Similarly, *Tilapia* were cultured in Freeport (Grand Bahama), Nassau and Lee Stocking island in the Exuma Cays. These are all operations which have since closed down, although some interest is being shown for the re-development of aquaculture activities on Lee Stocking Island.

On Walker's Cay in Abaco, 'Clown Fish' and other small reef fish were cultured for the US aquarium trade until 1997. The only remaining aquaculture operation is in Grand Bahama where white prawn is being cultured. No reclamation or excavation of mangrove areas has taken place to establish any aquaculture operations in the Bahamas.

Government incentives are high for the establishment of aquaculture operations, including attractive lease and purchase agreements, inter-agency cooperation to expedite applications, duty free concessions on equipment and supplies and no taxes on business profits. It has been shown that aquaculture production is possible in the Bahamas and that conditions are excellent for many cultures. Historically, the reasons for the failure of most of the aquaculture ventures has been a lack of full understanding of the

Table 8

The type and status of aquaculture operations in The Bahamas (information from Bahamas Department of Fisheries)

Location	Type	Species Cultured
Rudder Cut, Exuma	experimental shellfish culture	American oyster
Freeport, Grand Bahama	experimental shellfish culture	oysters and clams
Berry Islands	experimental shellfish culture	*Strombus gigas*
Walker's Cay, Abaco	commercial tropical aquarium fish farm	*Amphipnon* sp.; *Gobiosoma* sp.; Anemones; *Centropyge* sp.
Marsh Harbour, Abaco	spiny lobster ranching facility	*Panulirus argus*
Nassau, New Providence	pilot/commercial shrimp farm	*Penaeus vannamei*; *Penaeus monodon*; *Perna perna*; *Crassostrea*
Barbary Bay, Grand Bahama	pilot/commercial fish farm	*Tilapia* sp.; Tiger cichlids
Lee Stocking Island	experimental/ commercial	*Epinephulus striatus*; *Panulirus argus*; *Tilapia*
Freeport, Grand Bahama	pilot/commercial shrimp farm	*Penaeus vannamei*
Clarence Town, Long Island	commercial fish and shrimp farm	*Penaeus vannamei*; *Sciaenops ocellata*
Lynards Cay, Abaco	lobster/grouper ranching facility (experimental)	*Panulirus argus*; *Epinephulus striatus*
Nassau, New Providence	commercial fish	*Tilapia* sp.
Freeport, Grand Bahama*	commercial shrimp farm	*Penaeus vannamei*

*Active; all the others are now inactive.

culture process, poor on-site management and undercapitalization, an inability to control disease, and poor returns from the local market because of the availability of fresh fish. The Department of Fisheries are taking steps towards organising the proper training for Bahamians who wish to venture into the shrimp farming industry (Deleveaux, 1997).

EFFECTS FROM URBAN AND INDUSTRIAL ACTIVITIES

Impacts of Coastal Development

Deforestation

Historically, Mahogany, Cedar, Braziletto, *Lignum vitae* and Mastic along with some other species including pine, were used for construction and boat building. In modern times however the demand for this wood is very small and much of the broad leaf coppice from which the hard woods came has been completely removed. There is potential for commercial use of the remaining pine forests at sustainable levels, but forestry activities at present are insignificant.

The ground is poor, lacking sufficient amounts of soil, so limited deforestation is unlikely to increase the amount of soil run-off. Furthermore, rainwater is mostly diverted below ground. Even during the heaviest rainfall, storm run-off is relatively clear with little suspended matter, and poses a minimal threat to near-shore biological habitats. However, it is well documented that run-off was considerable following land clearances during the Loyalist period.

Destruction of Coastal Wetlands

Mangrove areas have been destroyed at a number of areas in the Bahamas, most notably in Nassau, (New Providence), Freeport (Grand Bahama), Marsh Harbour (Abaco) and George Town (Great Exuma). These wetlands are cleared for mosquito control and water front access. Plans for future tourism development in Bimini also include dredging, extraction and infilling of mangrove areas.

Sand Mining and Dredging

During the 1950s, four areas of the Bahamas Banks were identified as suitable for the extraction of sand (Sealey, 1994). These sites are associated with the Bank margins where precipitation of calcium carbonate occurs as cool oceanic waters move onto the shallow Banks. Sites were located at the ends of the Tongue of the Ocean and the Exuma Sound, one at Joulters Cay at the extreme North end of Andros Island and one in the Southern Bimini islands. Combined, these sites were estimated to have a resource of around 50–100 billion tonnes of oolitic sand. This is particularly pure (97% $CaCO_3$) and could be used as chemical-grade sand as well as in the construction industry. Having the advantage of being of uniform grain size, sieving and crushing operations were not necessary. Additionally, the sand was located in shallow water and was therefore easily extracted.

At present only the Bimini site is being worked. The development of this site was advantageous because of its close proximity to its main market, the USA. With the sand forming close to the margin of the Great Bahama Bank, only a short dredged channel was necessary to reach the deep shipping lanes of the Florida Straits. It was necessary for this mining operation to be large scale as the market value of the oolitic sand per tonne was very low (US $3). In the late 1970s a man-made island known as Ocean Cay was constructed from dredged sands and, with an area of 95 acres this island was equipped with a dock, airstrip, accommodation and power. Production went from almost 4 million tonnes in 1980 to 1.2 million tonnes in 1993. Extraction reduced due to competition from a similar operation off the Yucatan (Mexican) coast, and because periodic poor weather damaged the Ocean Cay facility. Hurricane Andrew in 1992 caused extensive damage.

A second area where dredging of sand occurs in the Bahamas is off Rose Island (Grand Bahama). There is great potential for other dredging operations to begin in the Bahamas. Beach erosion has become a major problem in South Florida and in 1995 a U.S. company approached the Bahamian Government seeking permission to dredge sand to reclaim two beaches.

On a smaller scale, particularly in the Family Islands, beaches are mined for building sand for local construction projects. This activity has altered the beach profile in some locations and resulted in reduced coastal protection and some beach erosion.

Coastal Erosion

Perhaps the most important tourism attribute of the Bahama Islands is its hundreds of white beaches. Most are prone to some degree of erosion during tropical weather disturbances, but generally normal geomorphological processes replenish such disturbed areas. Problems arise when external factors affect replenishment such as the dredging of sand from nearshore areas, the construction of hotels or breakwaters and any other intrusion on to the beach area.

There are numerous examples of beach erosion problems in the Bahamas, particularly on the more developed islands of New Providence and Grand Bahama. On the foreshore of Montagu Bay in Nassau, New Providence, the sand moved from the beach following dredging for the Paradise Island golf course. A similar situation occurred when sand was removed from Goodman's Bay for the Cable Beach golf course (Sealey, 1998, 1999). Other examples of beach erosion have been due to the intrusion of structures onto the beach. Part of a large hotel was constructed on Cable Beach resulting in a down current loss of sand. Efforts to reverse this process by constructing a dock and a concrete groyne caused further erosion.

Examples also exist of channels which have been cut into the coast to gain access to protected rock cut marinas. In South Bimini, such a channel was cut on a windward coastline breaching the northerly longshore drift. Consequently, the channel began to fill with sand and the existing beach on the down-current side of the marina began to erode, and continues to do so.

Sealey (1982) noted with concern severe erosion of parts of the north coast of New Providence, caused by large northerly waves which are remnants of huge storms which impact north-facing shores. This erosion is mainly due to the loss of sand dunes and back beach areas caused by cars and pedestrian traffic damaging the vegetation responsible for the consolidation of the beach sand (Fig. 9).

Destruction of the sand dunes at Delaporte on New Providence resulted in reduced coastal protection and

Fig. 9. A typical beach with low-level vegetation consolidating sand behind the beach, reducing wind and water erosion. In the background is an erosional remnant of an eolian dune complex. (Rice Bay and North Point, San Salvador Island, photo Matthew Robinson).

increased erosion. The front edges of the dunes receded following the removal of beach vegetation which had been viewed as a problem because it trapped unsightly litter left behind by beach users. At Orange Hill and Saunders Beach, roads have been constructed too close to the ocean, not allowing room for a full beach profile with the result that here, too, beach destruction has followed (Sealey, 1998).

Sources of Pollution and Their Effects on Water Quality and Near-shore Habitats

Sources of pollution include on-site disposal areas, domestic sewage soakaways (septic tanks), municipal injection wells, live-aboard boats and yachts, storm water run-off and dredging activities.

Sewage and Water Use

The geological characteristics of the Bahamas cause great problems for freshwater use and liquid waste disposal. This is compounded by tourism development and indeed has been a factor affecting the development of many of the outer islands.

Freshwater supply for residential areas is generally from subterranean freshwater lenses because desalination technology is expensive. Desalination operations are found in many of the hotels and other tourist facilities, but to cut costs some hotels process slightly saline ground water as opposed to seawater. The distribution and amount of freshwater is linked to specific factors such as island size, shape, climate and geology. On smaller islands with high populations water resources are limited. New Providence is a good example in that the freshwater lens is approximately 17,500 acres for a population of 171,542. In contrast, Andros Island's lens is 338,585 acres and the population is around 8155 (Cant, 1996). Forty percent of New Providence's fresh water is supplied by barge from Andros.

The porous limestone substrate presents a serious problem and makes groundwater lenses particularly vulnerable to contamination by liquid wastes and contaminated runoff. Domestic wastes are generally treated in septic tanks which are normally combined with a disposal well or drainage field. Many septic tanks do not conform to building regulations and do not work properly (Cant, 1996). In less developed areas shallow latrines may be used, and in some cases direct discharge into the sea still occurs.

In populated areas wastes are normally treated to primary or secondary levels of sewage treatment and then effluent is discharged into deep injection disposal wells. Wells designed for large amounts of waste are in excess of 600 ft in depth. In Paradise Island and some other resort areas, waste water is recycled for irrigation use on golf courses, but waste disposal techniques used in the Bahamas may be described as inadequate and contamination of groundwater is evident in many urban areas.

Pollution is also evident in seawater, particularly enclosed bays and harbours. The Bahamas are renowned for their excellent cruising environment and beautiful anchorages, but holding tanks are not required on sail boats and live-aboard vessels and pollution is evident in some anchorages. An average tidal range of around 50 cm creates very little tidal flow and minimal flushing of bay areas.

Few quantitative studies on the effects of pollution have been done in the Bahamas. Sullivan-Sealey (1999) explored the relationship between urban coastal development and the health of near-shore coral reefs. Using sites in the undeveloped Exuma Cays and a highly populated area of New Providence known as Montagu Bay, comparisons of physical and ecological factors were made. Water quality parameters including salinity, temperature, dissolved oxygen, turbidity, chlorophyll a, total nitrogen and total phosphorus were measured over two summer and winter sampling periods. The likely assumption was made that septic tanks less than 500 m from the seashore will leach nutrients into surface water and will eventually contaminate near-shore coral reef areas. Also, measurements of species richness, benthic coverage, coral density, coral size, coral recruitment rate and herbivore density and size were compared between study areas.

Little difference was seen in water and sediments between developed and undeveloped sites. Ecological parameters showed more distinct variation between sites, with more species of fleshy macroalgae (typically found in nutrient-rich areas) in Montagu Bay and although chemical analysis showed that nutrient levels were similar in each site, the authors suggest that the macroalgal growth may be a response to very subtle changes in nutrient concentrations and the amount of particulate matter in the water column. Furthermore, species composition in Montagu Bay includes a high coverage of zooanthids and anemones which are adapted to areas with high particulate matter. On the patch reefs of the Exuma Cays, some coral heads were particularly large in comparison to those in Montagu

Bay, although coral abundance was similar. There was low coral recruitment to both sites, and herbivorous fish assemblages were very similar, although Parrotfish and Surgeonfish were generally larger in the Exuma sites. Predatory fish were larger and, along with the Spiny Lobster, were more abundant at the undeveloped site.

Significant perturbations have occurred over the last fifty years in New Providence, in particular around the City of Nassau and its harbour which has been modified and expanded on a number of occasions. An analysis and interpretation of a series of aerial photographs dating back to 1943, allowed Sullivan-Sealey (1999) to detect major changes in coastal habitats as a result of construction of man-made cays, breakwaters and the dredging activities relating to these developments. Findings included the identification of two major sedimentation events from dredging and in-filling activities in 1967 and 1989, the physical removal of reefs that presented navigation hazards at each end of the harbour, and the alteration of water flow to reefs following breakwater construction and in-filling of the shoreline. In all, approximately 29 ha of seafloor were altered from 1943 to 1995.

In conclusion, much of the degradation and alteration of the Bay habitats was during more acute construction events and is likely to have had a far greater effect on the near-shore reefs than nutrient loading from local residences.

Cruiseship Discharge

There have been no notable instances of oil spill in Bahamian waters although tar balls are often found on beaches. Much debris is washed up on windward-facing shores around the islands, some of which is identifiable as coming from cruise ships. The larger islands are regularly visited by cruise ships and major harbour expansion has occurred in both Nassau and Freeport over the last 20 years. There have been some instances of garbage disposal and sewage holding tank flushing from these ships, though recently, cruise ship companies have established facilities on small islands and cays such as Gorda Cay, Little San Salvador and Little Stirrup Cay in the Berry Islands.

PROTECTIVE MEASURES

The Bahamian Government have ratified and initiated a number of preventative and protective legislative acts with regard to coastal areas and their territorial waters. Some of the more significant relate to the use of coastal and marine resources. Government policy restricts commercial fishing to the native population and, as a consequence, all vessels fishing within The Bahamas Exclusive Fishery Zone must be owned fully by a Bahamian citizen residing in the Bahamas (Bahamas National Trust, 1992).

The use of bleach or other noxious or poisonous substances for fishing, or possession of such substances on board a fishing vessel without the written approval of the Minister, is prohibited. Spear fishing within one mile of the coast of New Providence and Freeport and two hundred yards off the coast of all other Family Islands is prohibited, as is the use of firearms or explosives. For nets, a minimum mesh size of two inches is necessary, except when fishing goggle-eye or pilchard. Scale-fish traps are required to have self-destruct panels and minimum mesh sizes of 1×2 inches for rectangular wire mesh traps and 1.5 inches (greatest length of mesh) for hexagonal wire mesh traps. Those wishing to sell fish catches in New Providence must possess a permit. A permit is required to use air compressors for fishing purposes and the use of compressors is restricted to the period 1 August–31 March and to 10–20 m depth.

Harvesting of coral is prohibited, as is the construction of artificial reefs without permission from the Minister. There is specific legislation relating to the most commercially viable species. For example, the Spiny Lobster fishery has a closed season from 1 April–31 July. Individuals under $3\frac{3}{8}$ inches carapace length or six inches tail length may not be harvested and permits are necessary for any vessels trapping lobster. Specifications for lobster traps are outlined in the legislation; gravid females may not be taken and the stripping of eggs from the female is prohibited.

Conch is also protected and harvesting or possession of a shell without a well-formed lip is prohibited. Export of unprocessed conch meat from the Bahamas is not permitted.

The capture or possession of any hawksbill turtle is prohibited, and there is an annual closed season from 1 April–31 July for all other species. Size limits are imposed. All turtles captured must be landed whole, and the taking or possession of turtle eggs is prohibited.

Scalefish regulations are species-specific. The capture of Bonefish by nets is prohibited. The purchase or selling of Bonefish is illegal and the capture of Grouper and Rock fish weighing less than 3 lbs is prohibited. Bone Fish are commercially important as a sports fish throughout the Bahamas and are found in shallow lagoon areas and tidal creeks, notably in Andros, Crooked Island, Long Island and Bimini.

Stone Crab have an annual closed season from 1 June–15 October, and a minimum harvestable claw length of 4 inches. The harvest of female stone crabs is prohibited. The capture or molesting of marine mammals is prohibited. Persons may capture marine mammals for scientific, educational or exhibition purposes only with the written permission of the Minister. Finally, sponges are harvested with a minimum size limit of 5.5 inches for wool and grass sponge and 1 inch for hard head and reef sponge, and a permit is required to engage in aquaculture activities.

Many sports fishermen are drawn to the Bahamas each year to take part in tournaments and these events are also restricted by legislation. Sport fishing tournament directors must have the written approval of the Minister to organize

Table 9
Marine and Coastal National Parks of the Bahamas

Name	Area	Description
Inagua National Park	287 sq. miles	Located on Great Inagua this site hosts the world's largest breeding colony of West Indian Flamingos
Union Creek Reserve	7 sq. miles	Also on Great Inagua this park is a small enclosed tidal creek which is significant because of the Green Turtle population.
Exuma Cays Land and Sea Park	176 sq. miles	This Park when established in 1958 was the first marine fishery reserve in the Caribbean region. The Park encompasses the northern part of the Exuma Cays chain of islands.
Pelican Cays Land and Sea Park	850 hectares	Located in Great Abaco this was developed as a sister park to the Exuma Cays. The main features of the marine ecosystem are the submarine caves and coral reefs.
Peterson Cay National Park	13,000 hectares	This cay is a low-lying eroded fossil dune ridge and is the only Cay off the leeward coast of Grand Bahama. The Park originally only included the Cay and surrounding waters out to a quarter mile. Boundaries now include Barbary Beach on Grand Bahama and beyond the cay to the drop-off, and from Sharp Rocks point in the north-east to the Grand Lucayan Waterway in the south-west giving a total area of around 13,000 hectares. Notable features include a rich and diverse marine fauna and flora.
Conception Island National Park	850 hectares	Established in 1971. This uninhabited island is important for migratory and nesting birds and Green Turtles.
Black Sound Cay		Located off Green Turtle Cay in Abaco, this park is comprised of a thick mangrove stand of significance to wintering birds.

or hold tournaments. A permit is required for foreign vessels to engage in sport fishing, and under this permit certain restrictions are imposed. Fishing gear, unless otherwise authorised, is restricted to hook and line, and the number of lines in the water at any time unless otherwise authorised, is restricted to six. There is a maximum combined total of six fish per person on any vessel for King Fish, Dolphin Fish and Wahoo. All other migratory fish caught, unless it is to be used, should not be injured unnecessarily and should be returned to the sea alive.

Vessel bag limits for other fishery resources are 20 lbs of scale fish, 10 conch and 6 Spiny Lobster per person at any time. The possession of turtle is prohibited. The above amounts may also be exported by the vessel upon leaving The Bahamas.

Penalties totalling B $1000 (1 Bahamian $ is equal to 1 US $) or imprisonment for six months or both are imposed for resisting or obstructing a fisheries inspector. For contravention of regulations unless otherwise stipulated in the Act, a B $3000 fine or imprisonment for 1 year, or both, is imposed. All gear utilised in the contravention of fisheries regulations is confiscated and forfeited to the Crown.

The Bahamas have signed up to CITES, a convention covering 38 species of flora and fauna in the Bahamas. Of these listed species seven are marine and include Loggerhead, Hawksbill, Green and Leatherback Turtles, Whales and Dolphins, the West Indian Manatee and the Queen Conch.

In 1998, the Bahamian Government signed the RAMSAR Convention and designated Inagua National Park as a RAMSAR site. Table 9 summarizes the marine and coastal national parks of the Bahamas, whose management has been designated to the Bahamas National Trust.

A Land and Sea Park has been proposed for Andros island in conjunction with the Bahamas Reef Environment Education Foundation (BREEF), the Department of Fisheries and the National Trust.

Over and above fisheries legislation, additional regulations are imposed in relation to National Parks and are made under the Bahamas National Trust act, operating in conjunction with all other laws in the Bahamas. These are fairly widespread and cover hunting, capture, removal or damage of many species, whether alive or found dead. There are also strict regulations covering dumping and discharging wastes.

Permanent moorings may not be placed within any National Park and it is illegal to drop anchor on the coral reef, unless under emergency circumstances. Additionally, personal watercraft and air boats are not permitted within anchorage and creek areas. No vessel may remain for more than two weeks in any one anchorage on any visit to or voyage through the Exuma Cays Land and Sea Park.

Regulations are also imposed on persons who own land within the boundaries of any National Park. These relate to dock and breakwater construction, the construction of a sewerage outlet or overflow below the high water mark of the sea and the dredging of the sea bed adjacent to their land. Written permission is necessary from The Bahamas National Trust.

Perhaps the best evidence that protective measures are having a positive effect on the marine environment and its resources is found in the Exuma Cays Land and Sea Park, which was the first of its kind, being established in 1959. The biomass of Nassau Grouper was shown to be statistically greater inside, and within 5 km of the Park boundaries. Furthermore, reproductive output (egg production) was measured at six times more than that outside the Park (Bahamas National Trust, 1999). Studies have also shown that the concentration of conch inside the

ECLSP is approximately 47 times that found anywhere else in the world. Estimates have also been made on the export potential of conch eggs from the Park and it is suggested that as many as ten million exported eggs will reach adulthood elsewhere in the Bahamas.

The Bahamas National Trust (BNT) was created by an act of parliament in 1959. The Trust, a non-governmental organisation, was given the responsibility of managing the National Parks of the Bahamas. In more recent times the Bahamas National Trust has been one of the main organisations influencing environmental policy. On the recommendation of the BNT the Bahamas Environment Science and Technology (BEST) Commission was established in 1992 within the office of the Prime Minister to create environmental policy and programs. The National Trust works closely with the BEST commission in reviewing Environmental Impact Assessments. In 1999 a precedent was set by requiring independent professional review of an impact assessment for a proposed development on Clifton Cay, New Providence (Bahamas National Trust, 1999).

Marine environment awareness is being promoted by the National Trust through educational programs and activities. In recent years the Bahamas Reef Environment Education Foundation (BREEF) has been established. This organisation has concentrated on training Bahamian teachers to teach environmental education to students. BREEF have also compiled a fisheries management action plan for the Bahamas and are providing the expertise to create a proposal for the Andros Land and Sea Park.

ACKNOWLEDGEMENTS

I would like to gratefully acknowledge the following individuals and organisations who shared information about the many aspects of the Bahamas covered in this chapter: The Bahamian Field Station, the Bahamas Department of Fisheries, the Bahamas Department of Tourism, Sir Nicholas Nuttal of the Bahamas Reef Environment Education Foundation (BREEF), Dr. Donald Gerace, and Dr. Kathleen Sullivan-Sealey. In particular I would like to thank Mr. Neil Sealey for his valuable input and very useful comments on this chapter.

REFERENCES

Addison, D.S. and Morford, B. (1996) Sea Turtle nesting activity on the Cay Sal Bank. *Bahamas Journal of Science* 3 (3), 31–36.

Albury, P. (1975) *The Story of the Bahamas*. MacMillan Education Ltd, London and Basingstoke. pp. 294.

Bahamas Department of Fisheries (1998) Summary Report: An economic overview of the Bahamian commercial fishing industry. Prepared by the Bahamas Department of Fisheries.

Bahamas Department of Fisheries (1999) Summary of total recorded landings of marine products in the Bahamas during 1998. Prepared by Bahamas Department of Fisheries.

Bahamas Ministry of Tourism (1998) Tourism in the Islands of the Bahamas 1997 in Review.

Bahamas National Trust (1992) Summaries of Legislation affecting wildlife and national parks in the Bahamas. The Bahama Parrot Conservation Committee, Bahamas National Trust, 17 pp.

Bahamas National Trust (1999) Currents, Newsletter of the Bahamas National Trust Vol. 15 number 1, April 1999.

BREEF (Bahamas Reef Environment Educational Foundation) and MacAlister Elliot and Partners Ltd. (1998) Fisheries Management Action Plan for the Bahamas: Report to the Bahamas Department of Fisheries.

Bohlke J.E. and Chaplin C.C.G. (1968) *Fishes of the Bahamas*. Published for The Academy of Natural Sciences of Philadelphia by Livingstone Publishing Company, Wynnewood, PA. 771 pp.

Campbell, D.G. (1978) *The Ephemeral Islands: A Natural History of the Bahamas*. MacMillan Education Ltd. London and Basingstoke. 151 pages.

Cant, R.V. (1996) Water supply and sewerage in a small island environment: The Bahamian experience. In *Coastal and Estuarine Studies, Small Islands Marine Science and Sustainable Development*, ed. G.A. Maul, pp. 329–340. American Geophysical Union.

Carew, J.L. and Mylroie, J.E. (1997) Geology of the Bahamas. In *Geology and Hydrogeology of Carbonate Islands*, eds. H.L. Vacher and T. Quinn. Developments in Sedimentology, Vol. 54. Elsevier Science, Amsterdam, pp. 91–139.

Chiappone, M. and Sullivan K.M. (1991) A comparison of line transect and linear percentage sampling for evaluating stony coral (Scleratinia and Milleporina) community similarity and area coverage on reefs of the central Bahamas. *Coral Reefs* 10, 139–154.

Chiappone, M., and Sullivan, K.M. and Lott, C. (1996) Hermatypic scleractinian corals of the south eastern Bahamas: A comparison to western Atlantic reef systems. *Caribbean Journal of Science* 32 (1), 1–13.

Correia, J., de Marignac, J.R.C. and Gruber, S.H. (1995) Young lemon shark behavior in Bimini Lagoon. *Bahamas Journal of Science* 3 (1), 2–8.

Curran, H.A., Smith, D.P., Meigs, A.E., Pufall and Greer M.L. (1993) The health and short-term change of two coral patch reefs, Fernandez Bay, San Salvador Island, Bahamas. In *Colloquium on Aspects of Coral Reefs—Health, Hazards, and History*, University of Miami, Florida, pp. F1–F7.

Deleveaux, V.K.W. (1997) The history, status and potential of aquaculture in the Bahamas. Department of Fisheries, Ministry of Agriculture and Fisheries, Nassau, Bahamas.

Dietz, R.S., Holden, J.C. and Sproll, W.P. (1970) Geotectonic evolution and subsidence of the Bahama platform. *Geological Society of America Bulletin* 81, 1915–1928.

Gerace, D.T., Ostrander, G.K. and Smith, G.W. (1998). San Salvador, Bahamas. In *Caribbean Coastal Marine Productivity (CARICOMP): Coral Reef, Seagrass, and Mangrove Site Characteristics*, ed. B. Kjerfve. UNESCO, Paris, pp. 229–245.

Godfrey, P.J. and Klekowski (1989) Mutations for chlorophyll–deficiency ("albinism") in the red mangroves of San Salvador Island: Mendel's law in Bahamian swamps, pp. 25–39. In *Proceedings of the 3rd Symposium on the Botany of the Bahamas*, Bahamian Field Station, San Salvador, Bahamas.

Godfrey, P.J., Edwards, D.C., Davis, R.L. and Smith, R.R. (1993) Natural history of northeastern San Salvador island, a "new world" in the new world. Field guide, Bahamian Field Station, San Salvador Bahamas, 28 pp.

Halkitis, M., Smith, S. and Rigg, K. (1982) *The Climate of the Bahamas*. The Bahamas Geographical Association. Nassau, Bahamas.

Kass, L.B. and Stephens, L.J. (1990) The trees of the Mangrove swamp community of San Salvador Island, Bahamas and their "succession" patterns, pp. 53–65. In *Proceedings of the 3rd Symposium on the Botany of the Bahamas*, ed. R. Smith. Bahamian Field Station, San Salvador, Bahamas.

Kass, L.B., Stephens, L.J., Kozacko, M. and Carter, J. (1994) Continued studies of mangrove ecosystems on San Salvador Island, Bahamas, pp. 50–57. In *Proceedings of the 5th Symposium on the Natural History of the Bahamas*, ed. L.B. Kass.

Lang, J.C., Wicklund, R.I. and Dill, R.F. (1988) Depth and habitat related bleaching of zooxanthellate reef organisms near lee Stocking Island, Exuma Cays, Bahamas. *Proceedings of the Sixth International Coral Reef Symposium* 3, 269–274.

Lugo, A.E. (1994) San Salvador Mangroves: An ecosystem under chronic stress, pp. 60–63. In *The Proceedings of the Fifth Symposium on the Natural History of the Bahamas*, ed. L.B. Kass.

McGrath, T.A. and Smith, G.W. (1999). Monitoring the Coral Patch Reefs of San Salvador Island, Bahamas. *Proceedings of the 8th Symposium on the Natural History of the Bahamas*. Bahamian Field Station, San Salvador, Bahamas.

Meyerhoff, A.A. and Hatten, C.W. (1974) Bahamas salient of North America: Tectonic framework, stratigraphy, and petroleum potential. *American Association of Petroleum Geologists Bulletin* 58, 1201–1239.

Mullins, H.T. and Lynts, G.W. (1977) Origin of the northwestern Bahama platform: Review and reinterpretation. *Geological Society of America Bulletin* 88, 1447–1461.

Nagelkerken, I., Buchan, K.C., Smith, G.W., Bonair, K., Bush, P., Garzon-Ferreira, J., Botero, L., Gayle, P., Heberer, C., Petrovic, C., Pors, L. and Yoshioka, P. (1997a) Widespread disease in Caribbean sea fans: I. Spreading and general characteristics. *Proceedings of the 8th International Coral Reef Symposium* 1, 679–682.

Nagelkerken, I., Buchan, K. C., Smith, G.W., Bonair, K. Bush, P., Garzon-Ferreira J., Botero L., Gayle, P., Harvell, C.D., Heberer, C., Kim, K., Petrovic, C., Pors, L. and Yoshioka, P. (1997b) Widespread disease in Caribbean sea fans: II. Patterns of infection and tissue loss. *Marine Ecology Progress Series* 160, 255–263.

Nelson, R.J. (1853) On the geology of the Bahamas and on coral formation generally. *Quarterly Journal of the Geological Society London* 9, 200–215.

Neumann, C.J., Jarvinen, B.R., Pike, A.C. and Nelms, J.D. (1987) *Tropical Cyclones of the North Atlantic Ocean, 1871–1986*. National Climate Data Center, Asheville, NC.

Newell, N.D. and Imbrie, J. (1955) Biogeological reconnaissance in the Bimini area, Great Bahama Bank. *Transactions of the New York Academy of Science, Series 2* 18 (1), 3–14.

Newell, N.D., Imbrie, J., Purdy, E.G. and Thurber, D.L. (1959) Organism communities and bottom facies, Great Bahama Bank. *Bulletin of the American Museum of Natural History, New York* 117 (Article 4), pp. 181–240.

Porter, J.W. (1973) Ecology and composition of deep reef communities off the Tongue of the Ocean, Bahama Islands. *Discovery* 9 (1), 3–12.

Ritchie, K.B. and Smith, G.W. (1998) Type II White-Band Disease. *Revista de Biologia Tropical* 46 suppl. (5), 201–203.

Sealey, N.E. (1982) Conservation and the coast: potential hazards in the Bahamas. *Naturalist* 6 (2).

Sealey, N.E. (1990) *The Bahamas Today: An Introduction to the Human and Economic Geography of the Bahamas*. MacMillan Education Ltd., London and Basingstoke. 120 pp.

Sealey, N.E. (1994) *Bahamian Landscapes. An Introduction to the Geography of the Bahamas*. Second Edition. Media Enterprises Ltd., Nassau, Bahamas. 128 pp.

Sealey, N.E. (1998) The sand on our beaches. *The Nassau Tribune*, April 25, 1998.

Sealey, N.E. (1999) The threat to the environment from recent major developments in the Bahamas. Abstract for the 8th Symposium on the Natural History of the Bahamas, Bahamian Field Station, San Salvador, Bahamas.

Shaklee, R.V. (1989) *Hurricanes in the Bahamas*. The Bahamian Field Station Ltd., San Salvador, Bahamas. 82 pp.

Shaklee, R.V. (1996) *Weather and Climate, San Salvador Island, Bahamas*. The Bahamian Field Station Ltd., San Salvador, Bahamas. 67 pp.

Short, F.T. (1986) Nutrient ecology of Bahamian Seagrasses. In *Proceedings of the First Symposium on the Botany of the Bahamas*, ed. R.R. Smith.

Sluka, R., Chiappone, M., Sullivan, K.M. and Wright, R. (1996) Habitat and Life in the Exuma Cays, The Bahamas: The status of Groupers and coral reefs in the northern cays. The Nature Conservancy, University of Miami, FL, 83 pp.

Smith, G.W., Short, F.T. and Kaplan, D.I. (1990) Distribution and Biomass of seagrasses in San Salvador, Bahamas. In *Proceedings of the Third Symposium on the Botany of The Bahamas*, ed. R. Smith, pp. 67–77. Bahamian Field Station, San Salvador, Bahamas, 83 pp.

Squires, D.F. (1958) Stony corals from the vicinity of Bimini, Bahamas, British West Indies. *Bulletin of the American Museum of Natural History* 115, Article 4.

Sullivan, K.M. (ed.) (1991) Guide to the shallow-water marine habitats and benthic invertebrates of the Exuma Cays Land and Sea Park, Bahamas. Sea and Sky Foundation, Coral Gables, FL.

Sullivan, K.M. and Chiappone, M. (1992) A comparison of belt quadrat and species presence and absence sampling for evaluating stony coral (Scleractinia and Milleporina) and sponge species patterning on patch reefs of the central Bahamas. *Bulletin of Marine Science* 50 (3), 464–488.

Sullivan, K.M., Chiappone, M. and Lott, C. (1994) Abundance patterns of stony corals on platform margin reefs of the Caicos Bank. *Bahamas Journal of Science* 1 (3), 2–11.

Sullivan-Sealey, K.M. (ed.) (1999) Water quality and coral reefs: Temporal and spatial comparisons of changes with coastal development. The Nature Conservancy, Marine Conservation Science Center, University of Miami, Florida. 184 pp.

Wells, S.M. (1988) Bahamas. In *Coral Reefs of the World. Volume 1: Atlantic and Eastern Pacific*. Prepared by IUCN Conservation Monitoring Centre, Cambridge, U.K., pp. 13–28.

White, B., Curran, H.A. and Wilson, M.A. (1997) Bahamian Sangamonian coral reefs and sea level change. In *Proceeding of the 8th Symposium on the Geology of the Bahamas and Other Carbonate Regions*, ed. J. Carew. Bahamian Field Station, San Salvador, Bahamas.

Wilkinson, C. (1999) Coral bleaching reports from the Caribbean. *Caribbean Marine Science* Number 1, April 1999. Newsletter of the Association of Marine Laboratories of the Caribbean.

THE AUTHOR

Kenneth C. Buchan
Bahamian Field Station,
San Salvador, Bahamas

Chapter 27

THE NORTHERN GULF OF MEXICO

Mark E. Pattillo and David M. Nelson

The northern Gulf of Mexico is characterized by features such as bathymetry, hydrography, productivity and trophically dependent populations that make this a distinct region. The natural resources of this area are important to the economy of the coastal border states of the United States. Of particular importance in this area are the fish and fisheries resources, recreational facilities, offshore oil production, and the wide diversity of fish, bird, and mammal species. However, this area is increasingly being subjected to stress from a number of areas. Growing population pressures impose an increased use of and demand on Gulf resources. The large Gulf watershed introduces many environmental problems associated with runoff of pesticides, fertilizers, toxic substances, and trash. There is an increasing incidence and extent of harmful algal blooms, oxygen depletion events, pollution, loss of wetlands, and losses of fishery productivity and yield through overexploitation and by-catch discards. Efforts are presently underway to restore the ecological damage in the northern Gulf through the various local, state, federal, national, and international jurisdictions responsible for the management of the northern Gulf's rich biodiversity habitats, and other assets. In addition, these groups are working to develop more effective means for the governance of this area to assure that its users will be able to realize the long-term sustainable benefits to be derived from the Gulf's productive waters.

Seas at The Millennium: An Environmental Evaluation (Edited by C. Sheppard)
© 2000 Elsevier Science Ltd. All rights reserved

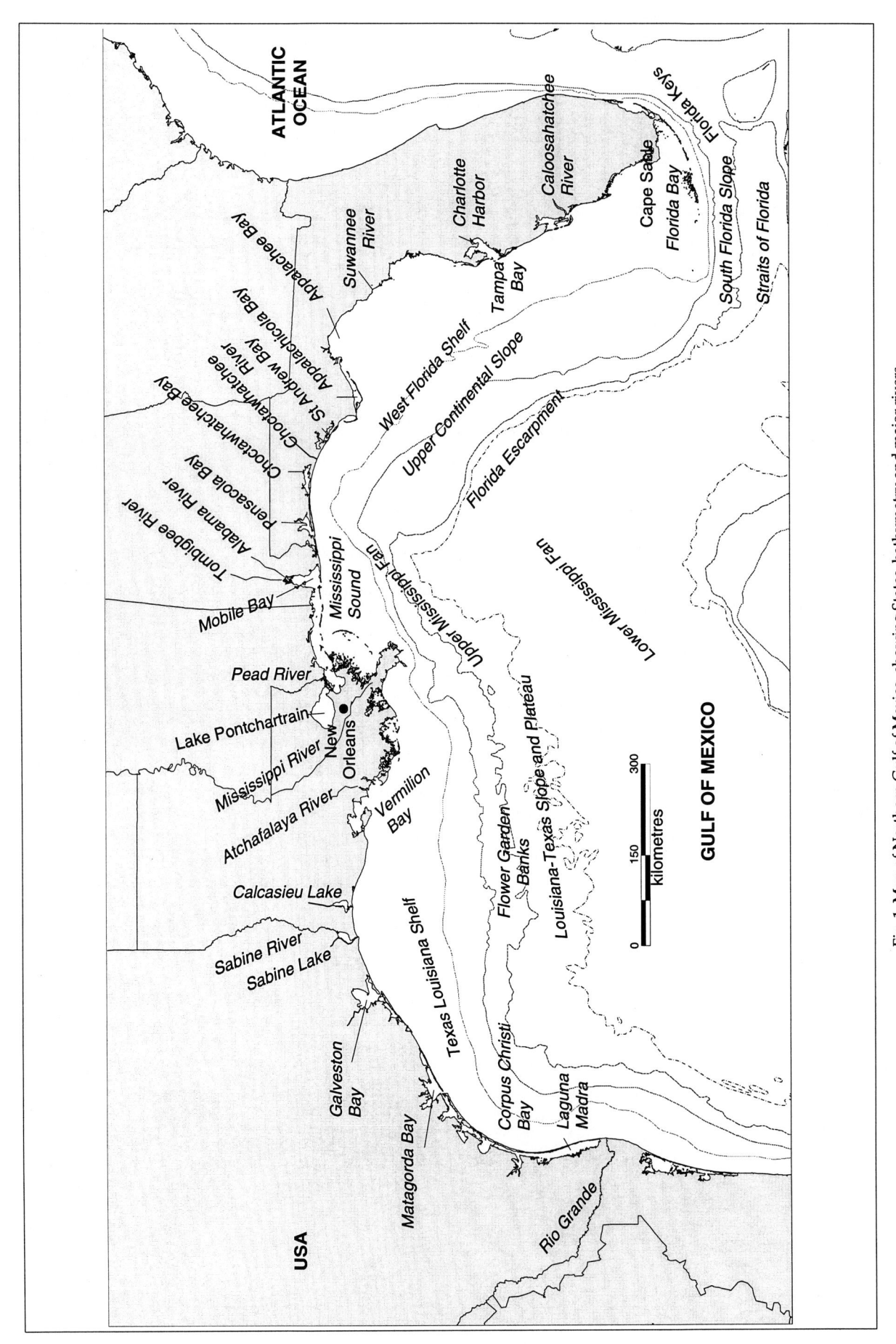

Fig. 1. Map of Northern Gulf of Mexico, showing States, bathymetry and major rivers.

THE DEFINED REGION

The Gulf of Mexico is the ninth largest body of water in the world, spanning nearly 5 degrees of latitude and 15 degrees of longitude (Kenworthy, 1992). It is a semi-enclosed, partially land-locked, marginal sea bounded by the southeastern periphery of the North American continent and the island of Cuba (Britton and Morton, 1989; Gore, 1992) (Fig 1). Its southeastern boundary extends from the northeastern tip of the Yucatan Peninsula at Cabo Catoche, Mexico to Key West in Florida. There is considerable interaction with the waters and biota of the Caribbean basin through the Yucatan Channel, and with the North Atlantic Ocean through the Straits of Florida. The Gulf basin is roughly oval in shape, approximately 1600 km east–west and 800 km north–south, covering 1.5 million km². The North American shore is approximately 25,000 km long, and a broad plain that extends from Florida to eastern Mexico dominates this coastal area. Shallow coastal waters contain an assortment of physicochemical environments including extensive barrier island lagoons, 33 major river systems and 207 estuaries (Kenworthy, 1992) in the largest concentration of such systems in the contiguous U.S. (NOAA, 1985).

The Gulf has a partially isolated environment which, when coupled with its distinct bathymetry, hydrogeography, productivity, and trophically dependent species, characterizes the Gulf of Mexico as a large marine ecosystem. The northern and southern portions of this system can be divided into two separate subsystems (Briggs, 1974; Hoese and Moore, 1998; Britton and Morton, 1989; Gore, 1992). Although there is a relatively high degree of habitat homogeneity and a moderate to high species similarity at corresponding latitudes, these drastically change from north to south. Differences occur largely due to climatic changes, but also due to the faunal distribution barrier imposed by the Straits of Florida, and changes in sediment sources between the two regions.

The Gulf of Mexico lies within a transition region between temperate and subtropical (Briggs, 1974; Ruffner, 1980; Hoese and Moore, 1998; Britton and Morton, 1989; Gore, 1992). The transition is more pronounced for shore biota than for offshore or continental shelf species. Northern shores lie within the warm-temperate Carolinian Province, and are isolated by tropical Florida from similar shores along the eastern seaboard of the USA. Warm tropical currents enter the Gulf of Mexico and are widely diffused into the northern Gulf during the summer. However, in the winter, these currents tend to circulate further south, probably as a result of frequent northerly winds during this season. Such winds also lower the surface temperature of onshore water until it is below the tolerance of most tropical organisms. Plant and animal communities of the northern Gulf undergo a distinct pattern of change in response to these seasonal temperature changes, in contrast with the time-stable, increasingly diverse communities of tropical southern Mexico and the West Indies.

The Florida Loop Current, passing through the Florida Straits, restricts tropical species from the West Indies from entering the northern Gulf (Briggs, 1974; Tomczak and Godfrey, 1994; Hoese and Moore, 1998). This current is not only very fast, ≥1 m/s at the surface, but its water mass properties preclude much mixing between it and surrounding tropical water. This current forms an effective barrier to dispersal of tropical species from the Caribbean.

Differences in sediments also separate the northern and southern Gulf of Mexico. Terrigenous sediments are found in the north, while biogenic carbonate sediments dominate southern shores (Briggs, 1974; Britton and Morton, 1989; Gore, 1992).

The northern Gulf can be divided into eastern, central, and western sectors separated by transitional zones (Rezak et al., 1990; Darnell, 1991; Hoese and Moore, 1998). These sectors include the western Gulf (south Texas from the Rio Grande to Matagorda Bay), central Gulf (Texas–Louisiana border to Perdido Bay, just east of Mobile Bay), and eastern Gulf (Apalachicola Bay to the Florida Keys). Transitional areas occur in upper Texas and the area around the head of DeSoto Canyon.

The western and central sectors are closely related. Estuaries, bays, and lagoons line almost the entire coastline (Darnell, 1991; Hoese and Moore, 1998). The Mississippi River carries large amounts of silt and clay westward, resulting in high nearshore turbidity with a persistent nepheloid layer over predominantly terrigenous bottom sediments. As a result, attached vegetation is rare, and seagrasses are typically found only in small stands in bays. Rocky outcrops are not as prominent here as in the eastern sector. The western sector tends to have higher temperatures and salinities in nearshore waters than the central sector due to lower rainfall, decreased influence from the Mississippi, and less estuarine circulation. In the more arid portion of the western Gulf (i.e. Texas), river flow is governed by surface runoff generated by storms and, therefore, is highly variable (Orlando et al., 1993). Along the upper Texas coast, the greater frequency and intensity of precipitation, coupled with the detention created by reservoirs leads to considerable overlap in individual storm impulses. As a result, freshwater inflow is a seasonal runoff surge lasting several weeks to a few months. In south Texas, freshwater impulses are more infrequent and typically occur as a series of nearly discrete flood pulses. These lower levels of freshwater input have led to lower biological diversity and productivity than that found in the central sector (Darnell, 1991; Hoese and Moore, 1998).

Bays and estuaries of the central Gulf as well as the nearshore area are strongly influenced by the variability of the Mississippi and other rivers (Rezak et al., 1990; Darnell, 1991; Hoese and Moore, 1998). Typically, turbidity is higher and salinity is lower than the other sectors. High nutrient loads brought by the Mississippi River raise biological productivity, and because of its more northerly location, winter temperatures tend to be colder here.

The eastern sector is distinct from the central and western areas (Rezak et al., 1990; Darnell, 1991; Hoese and Moore, 1998). Bays and estuaries are isolated from one another, winter temperatures remain higher, and outflow from the Mississippi River has very little influence. Except for Mobile Bay, Apalachicola Bay, and a few small estuaries, bay waters east of the Mississippi River are clear and salty, lacking the cloudy nepheloid layer, so seagrasses cover much of the bay bottoms. On the shelf, water clarity is very high and less of the coarser sand in offshore areas has been covered by mud. Carbonate sediments are dominant, and rocky outcrops, ledges, and hard bottoms are widespread. Biological diversity is quite high with many unique species, e.g. species of sea bass, snapper, angelfish, butterfly fish, etc. Surface currents facilitate the recruitment of tropical species and enhance the maintenance of populations of these species once established (Smith, 1976). Like the western Gulf, this is an arid region with precipitation levels declining as one proceeds southward from the panhandle region (Orlando et al., 1993). This has led to freshwater inflow conditions similar to the western sector, with a seasonal runoff surge occurring in northern Florida and fewer, more discrete flood pulses to the south.

SEASONALITY, CURRENTS, NATURAL ENVIRONMENTAL VARIABLES

The northern Gulf of Mexico has a wide diversity of watersheds and coastal geomorphology (Darnell, 1990; Hoese and Moore, 1998), influenced by seasonal and daily climatic changes, tidal regime, amount and distribution of rainfall, temperature ranges, and the type and distribution of sediments (EPA, 1994a). Catastrophic events, such as floods, cold front storms, droughts and hurricanes, superimpose more dramatic changes.

Several natural gradients contribute to the diversity of habitats found in the northern Gulf (Darnell, 1990; Mulholland et al., 1997; Hoese and Moore, 1998). Rainfall and freshwater inflow decrease and temperature slightly increases southwest from the Mississippi River to the Rio Grande River and southern Florida. These variations result in low salinities (<10 ppt) in the marshes and bayous of western Louisiana, the Mississippi River, and East Texas; moderate salinities along much of the central Texas coast, the east-central Louisiana coast, and the Florida panhandle, and often hypersaline conditions (>40 ppt) in bays and lagoons of south Texas and south Florida. Seasonal extremes also vary, with the southern parts of Texas and Florida remaining subtropical for most of the year while coastal marine waters from East Texas to the Florida panhandle are usually colder in winter.

Shallow, turbid embayments and extensive marsh and estuarine systems characterize the Central Gulf shores (NOAA, 1997b). The coastline is exposed to ocean waters without the protection of barrier islands. Semi-permanent currents, prevailing southeasterly winds and wave-driven currents control circulation patterns in the immediate nearshore areas of these estuaries. The large volume of freshwater from the Mississippi, Atchafalaya and other rivers brings about density gradients near the estuary mouths. The massive quantity of nutrients carried to the central Gulf by these rivers has resulted in plankton-rich waters that fuel extensive food webs, making this one of the most productive areas on Earth (Gunter, 1963; Darnell, 1990; Hoese and Moore, 1998). Longshore currents transport this river water westward and southwestward along the northern Gulf from October through March (Rabalais, 1990). Between June and September, this component weakens and reverses for short time periods to produce perpendicular movements of water across the shelf.

The northern Gulf shelf is further enriched by annual pulses of shrimps, crabs, and fish which migrate from these estuarine nursery areas back to the continental shelf. Large tracts of salt marshes have developed along this part of the Gulf coast in response to these conditions. As freshwater inflows taper off toward south Texas, coastal waters become less turbid; this and the increasing water salinity favours seagrasses. The large bays, lagoons and barrier islands of Texas are often hypersaline, especially in summer (NOAA, 1997b). Along the Florida shelf, east of Mobile Bay, estuarine nursery areas are more limited in number and size due to lower freshwater and sediment inflows (Darnell, 1990; Hoese and Moore, 1998) and faunal densities are more patchy. As in South Texas, the inner shelf here supports dense seagrass meadows. From Cape Romano northward, the shoreline consists of sandy beaches, some rocky areas, swamplands and tidal marshes (NOAA, 1997b). The coastline is partially enclosed and protected by barrier islands. South of Cape Romano, where freezes are absent, thick stands of mangroves, mangrove islands, tidal channels and extensive wetlands dominate. This area, which includes the Everglades, is highly affected by tidal action, weather-related events and canal structures.

A gradient in bottom composition comes from different levels of riverine input (Hoese and Moore, 1998). Coarser-grained sandy sediments are more common off the arid South Texas coast. Sediments become more muddy as one progresses away from the barrier islands, although considerable sand exists on the Louisiana shelf. Rocky reefs occur off of Texas, most of the continental shelf off Louisiana, especially near the continental slope, and along the west coast of Florida (Darnell, 1990; Hoese and Moore, 1998). These reefs provide habitat for tropical reef fish that are not commonly found in the inshore shallow zones.

East of the Mississippi Delta to near Pensacola, Florida, the shelf is largely composed of coarse sand, with many areas of hard bottoms and accumulations of shells, very different from that of most of the western Gulf. The Florida west coast consists largely of limestone and derived detrital sediments, which have favoured the spread of many coral reef fishes northward. Areas with fine sediment and mud occur in the vicinity of the larger rivers, contributing to habitat and species diversity.

There is a temperate to tropical latitudinal gradient (Ruffner, 1980; Britton and Morton, 1989; Rezak et al., 1990). The northern Gulf is generally humid and warm-temperate, but three distinctive climatic regions occur. North of a line between Cape Romano, near Naples, Florida, westward to Cape Rojo, near Tampico, Mexico, is the warm-temperate to marginally subtropical Carolinian Province. However, the influence of winter cold fronts is pronounced. Tropical fauna pushes up to near the northern shore of the Gulf, but is kept offshore by cold winters and possibly by competition with temperate fishes (Hoese and Moore, 1998). Ample rain (109–140 cm) produces a humid climate and diverse maritime flora which is primarily deciduous (Britton and Morton, 1989). Southward is a transitionally subtropical region, with an annual rainfall range of <52–75 cm. Hard winter freezes, although uncommon, are sufficiently frequent to exclude all but the hardiest tropical plants. The tip of Florida southward from Cape Romano, extends into tropical waters of the Caribbean Province (Briggs, 1974). Tropical fauna comes closer to shore, eventually displacing the temperate fauna (Hoese and Moore, 1998).

Southerly to southeasterly prevailing winds (Ward, 1980) are governed by the intensity of the Azores–Bermuda High and thus blow onshore for much of the Gulf coastline from approximately Corpus Christi Bay, Texas to Apalachee Bay, Florida. This flow is most pronounced in the summer. In winter, the High weakens and the Gulf falls under the increasing influence of continental weather patterns generated by extratropical cyclones (Leipper, 1954; Gore, 1992). Most cyclones take place during the winter, particularly January, and are responsible for the seasonal "northers", fierce storms that can produce large-scale erosion of beaches and destruction of coastal properties. Rainfall and, more importantly, snowfall may vary substantially from the more southerly areas, with more precipitation falling in winter than in summer, in contrast with southern Gulf areas. The western and northwestern Gulf also receive large amounts of precipitation from tropical disturbances in late summer and early fall (Ward, 1980).

Hurricanes can cause widespread and long-lasting ecological effects to Gulf habitats (Gore, 1992; Orlando et al., 1993; EPA, 1994a; Hoese and Moore, 1998). They occur primarily from June to October, but are most common in late summer and early fall. Heavy rains can cause inundation and erosion, and sea-surface waters may be abnormally cooled as much as 4°C. Pollutants may be injected in large pulses into estuaries, and sediments may remain suspended for weeks, limiting light penetration (Franceschini and El-Sayed, 1968; EPA, 1994a). Waves with heights of 6 m or greater may flood most coastal fresh and salt water wetland habitats, destroying huge areas of seagrass beds and other bottom communities. High winds devastate forested areas and mangroves. Depending on direction, storm surges can either inject large water volumes into coastal areas or flush water from estuaries through existing inlets, breaches, or overwashes through barrier islands (Orlando et al., 1993), massively altering salinity (EPA, 1994a) and causing severe impacts to marine fauna such as the important penaeid shrimp (Pattillo et al., 1997).

In the Gulf of Mexico, water circulation is dominated by a portion of the Gulf Stream System called the Florida Loop Current, which brings Caribbean waters into the Gulf through the Yucatan Channel (Briggs, 1974; Darnell and Defenbaugh, 1990; Hoese and Moore, 1998). Water circulation is one of the most important processes that determines local distributions of epibenthic fauna here (Continental Shelf Associates, 1998). After entering through the Yucatan Straits, the Loop Current flows northward into the central Gulf and then deflects east, travels south parallel to the Florida west coast and exits the Gulf through the Straits of Florida (Briggs, 1974; NOAA, 1985; Rabalais, 1990). This current has been known to intrude upon the northern Gulf east of the Mississippi River Delta (Darnell and Defenbaugh, 1990) while the southward-flowing arm may impinge upon the outer Florida shelf. It will periodically spin off large, clockwise-rotating rings, which proceed toward Mexico (Darnell and Defenbaugh, 1990; Rezak et al., 1990; Wiseman and Sturges, 1999), maintaining their structure for many months and carrying with them massive amounts of heat, salt, and water. They can also transport nutrient-laden water of the continental shelf hundreds of kilometres seaward (Wiseman and Sturges, 1999).

Most surface currents of the continental shelves are strongly coupled with the wind (Darnell and Defenbaugh, 1990; Rezak et al., 1990). Bottom currents are primarily tidal, but are complicated by inertial oscillations and shelf waves (Rezak et al., 1990). As the currents pass over irregularities on the bottom, turbulent flow resuspends fine sediment to form a nepheloid layer.

Natural extremes of temperature, salinity, or dissolved oxygen can have profound effects on the northern Gulf (Hoese and Moore, 1998; Pattillo et al., 1997). "Red tides" and more recently "brown tides" kill large numbers of fishes, increase water turbidity and stress seagrass beds. Resultant low oxygen levels cause narcotic effect or mortalities in fishes, while extremely high oxygen levels can cause mortalities from gas bubble disease.

Occasional prolonged freezes can also cause significant mortalities, especially in areas with water restriction (McEachron et al., 1994). Hard freezes in south Texas have reduced populations of many species, including important game fish (e.g. snook, spotted seatrout, red drum, black drum), to a point where recovery takes several years. Shifts in species distributions can occur during prolonged periods of drought or high freshwater inflow (Hoese and Moore, 1998; Pattillo et al., 1997). Salinities increase during droughts, displacing estuarine species further upstream. When high salinity waters are moved further offshore because of sustained high levels of freshwater inflow the reverse occurs. Large declines in harvestable oysters in Texas are invariably linked to such salinity perturbations (Stanley and Sellers, 1986; Britton and Morton, 1989).

THE MAJOR SHALLOW WATER MARINE AND COASTAL HABITATS

The United States Gulf of Mexico is a diverse and productive ecosystem which provides many benefits to society (Duke and Kruczynski, 1992; EPA, 1994a). The 25,000 km of the U.S. Gulf Coast is dominated by tidal marshes, mangroves, and submerged seagrass beds in a total of 207 estuaries. Major associated systems also include upland forests, forested wetlands, oyster reefs, and coral reefs. The estuaries provide habitats for 48% of the commercially important species of the Gulf and 45% of its important recreational species, as well as several threatened and endangered species. The fish and invertebrate harvest of the Gulf constitutes about 33% by weight and 29% by dollar value of the total U.S. harvest.

The Gulf coastal plain, extending from central Florida to east Texas, was once covered with upland forests consisting of loblolly, slash, and longleaf pines (Gore, 1992; EPA, 1994a). The slowly percolating, silty sandy loams are particularly suitable for herbaceous plants and shrubs which provide food and shelter for a variety of open land and woodland wildlife, including game birds and animals that are popular with recreational hunters.

Forested wetlands generally are hardwoods in low-lying areas, maintained by the regime of alternating wet and dry periods and by soils that are saturated or inundated for some of the growing season (Keeland et al., 1995). They include bottomland forested wetlands, cypress swamps, and mangrove forests (NOAA, 1985; EPA, 1994a; Cowardin et al., 1979). The largest areas extend from Florida Bay north to Cape Romano (NOAA, 1985), where red and black mangrove stands penetrate up to six miles inland (14 miles along river corridors) grading into brackish and freshwater wetlands, including the bald cypress and swamps of the Everglades. Virgin swamp forests, predominantly bald cypress and water tupelo, lie in the Mississippi Delta and along major rivers of the north central Gulf.

Bottomland forest wetlands occur adjacent to streams and drainage ways, as well as depressional areas, which may or may not be stream-fed (EPA, 1994a). The soils of this habitat are frequently wet, highly acidic, less permeable, usually infertile, and poorly suited for agricultural or commercial use. Common trees include water oak, overcup oak, sweet gum, sweet bay, black gum, tupelo gum, and loblolly and slash pines. Vegetation species in the understorey can include various sedges, joint grasses, ferns, shrubs (such as blueberries and hollies), and climbing vines (such as blackberries and catbriars). Cypress swamps occur in freshwater areas such as sluggish streams, shallow lake basins, and flat upland areas with waterlogged, semi-fluid clays that are frequently flooded with 0.3 to 2 m of water for most of the year (EPA, 1994a; Shaw and Fredine, 1956). These swamps support muskrat, opossum, raccoons, deer, and many small mammals.

Approximately 202,350 hectares of mangroves occur along the Gulf coast, almost exclusively in Florida (Beccasio et al., 1982; EPA, 1994a). Mangrove forests typically occur in brackish to saline tidal water on low relief, muddy shores along tropical and subtropical coastlines, particularly those with estuaries. They develop best in sheltered environments or in the lee of protective structures receiving fluvial and allochthonous sediment inputs, where rainfall is >200 cm per year and where there is no pronounced dry season, and most are found in tidal ranges from 0.5 to 3 m or more (Shaw and Fredine, 1956; Mitsch and Gosselink, 1993). Because they tolerate high salinity, mangroves form effective and lasting buffers against storm surges, and because of their high productivity, mangroves play a major role in the dynamics of estuaries. Many invertebrates and fishes derive energy from detritus and dissolved organic carbon contributed by mangroves, which also provide important nursery areas to larval and juvenile animals, many important to fisheries. Their productivity couples them to other systems through the export of high quality energy, and they are closely linked with seagrass beds and coral reefs too, where these systems coexist (Cintron-Molero, 1992).

The mangroves grade into coastal salt marshes which are found between the marine and terrestrial habitats of the Gulf region (Mitsch and Gosselink, 1993; EPA, 1994a). Salt marshes are predominately intertidal, being found in areas that are tidally flooded at irregular or frequent intervals. A shoreline slope is necessary to allow for tidal flooding and stability of the vegetation, and adequate protection from wave and storm energy is also a physical requirement. Their sediments originate from upland runoff, marine reworking of coastal shelf sediments, or from organic production with the marsh itself. On the Gulf coast, salt marshes develop mainly on low energy coastlines; the Mississippi River deltaic marshes are the major example of this type of environment. They are one of the most valuable biological resources, as is reflected in the fact that nearly all the commercial finfish and invertebrate fisheries in the northern Gulf are dependent on them. Productivity rivals that of seagrasses and mangroves in more southerly regions, and saltmarshes also provide many physical benefits including sediment entrapment, purification of overland runoff, storm surge buffer zones, and dry season reservoirs.

Seagrass meadows (Fig. 2) are recognized today as one of the most important communities in shallow coastal systems (Zieman and Zieman, 1989; Duke and Kruczynski, 1992; Fonseca, 1992; Gore, 1992; EPA, 1994b). With the exception of coral reefs, seagrass communities have the highest diversity in the marine environment, and their productivity may exceed that of the benthic algae and plankton combined. Seagrasses are an important food resource and refuge area and serve as nursery areas for the

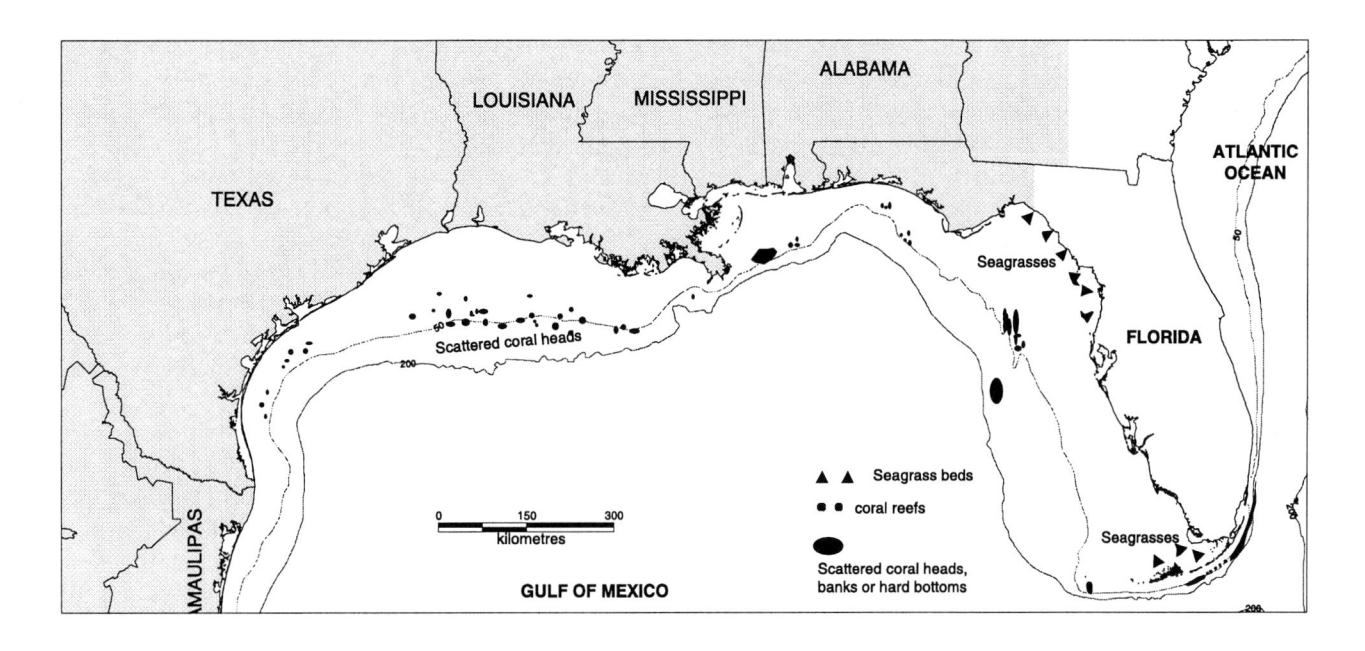

Fig. 2. Map of seagrasses, coral reefs and coral hard grounds in the northern Gulf of Mexico.

juveniles of a variety of finfish and shellfish of commercial and recreational importance. Although seagrasses usually do not provide food directly, indirect trophic linkages to other coastal systems, through a detrital food web and associated epiphytic and benthic algae, are enormous. Their rhizomes stabilize soils, thereby reducing erosion and preserving sediment microflora. They trap sediment and dampen wave energy, thereby contributing to the maintenance of good water quality. Their loss decreases bottom and shoreline stability, decreases faunal abundance, increases sediment suspension, and turbidity and decreases primary production (Kenworthy and Haunert, 1991).

Seagrasses occupy over 323,760 hectares within the estuaries and shallow near-coastal waters of the Gulf (EPA, 1994a). Approximately 95% of this is in Florida and Texas, where seagrasses occupy about 20% of the bay bottoms. Most species of seagrasses are predominantly found in shallow, clear, tropical or subtropical marine waters of moderate current strength (Cowardin et al., 1979; NOAA, 1985). In the Gulf, the most abundant species is turtlegrass, *Thalassia testudinum*. Other species are manatee grass (*Syringodium filiforme*), shoalgrass (*Halodule wrightii*), widgeon grass (*Ruppia maritima*), and two species of stargrass (*Halophila*). Despite the low diversity of species, seagrasses occupy a wide variety of habitats including but not restricted to, sand shoals, shallow muddy and sheltered lagoons, high-energy tidal channels, and relatively deep open-water continental shelves (Kenworthy, 1992). Their ability to grow in these very different environments results from their phenotypic plasticity and the wide diversity of morphology and life history strategies.

Fluctuations in salinity, water temperature and especially turbidity are detrimental to seagrass beds (Kenworthy and Haunert, 1991; Fonseca, 1992; NOAA, 1985; EPA, 1994a). Reduction in water clarity, as from dredging, nutrient loading, stormwater runoff, agricultural drainage and boating activities, upset the stability and function of established and developing seagrass areas (Kenworthy and Haunert, 1991; Fonseca, 1992). The nearly continuous distribution of seagrasses around the Gulf's periphery is interrupted from Alabama to the Laguna Madre, Texas due to low salinity and turbidity from the Mississippi River (NOAA, 1985; EPA, 1994a). There, only scattered patches of seagrass communities, mostly in bays, are found.

Many Gulf estuaries support extensive subtidal and intertidal oyster reefs dominated by the American oyster, *Crassostrea virginica* (EPA, 1994a). These generally occur in shallow, well-mixed estuaries, lagoons, and oceanic bays where they tolerate widely fluctuating water conditions (Butler, 1954; NOAA, 1985; Stanley and Sellers, 1986; Pattillo et al., 1997). Because of their tolerance, there are few places along the Gulf where at least a scattered growth of oysters is not found, and they may grow in waters ranging in depth from 0.3 m above to 12 m below mean low tide (Butler, 1954). Sands or very soft muds rarely support oyster reefs, but almost any other bottom type is suitable (Butler, 1954; Galtsoff, 1964; Britton and Morton, 1989). Oyster reefs occur where currents are sufficient to provide a continuous but nonturbulent flow of water that will deliver adequate food but not excessive turbidity. Accordingly, large crowded oyster reefs are usually located in or near rapid tidal streams.

Oyster reefs support a great diversity of highly productive benthic and epibenthic species, and so have an important ecological, environmental, and economic role.

They provide a stable substrate, reduce water turbidity and metabolise carbon at a high rate. The reef community is important in remineralizing organic matter and releasing nutrients, and it also provides concentrated food sources to estuarine fish, shellfish, and birds. Oyster harvests are significant to the fishing economies in Texas, Louisiana, and Florida, while in the vicinity of active delta formation, buried reefs can raise the subaqueous platform on which subaerial land can develop. They influence water flow within an estuary and stabilize intertidal sediment within estuarine environments (Dame and Patten, 1981).

Coral reefs and hard bottoms are found throughout the Gulf along the continental shelf (EPA, 1994a,b). Although they occur in relatively low nutrient waters, they are biologically productive and taxonomically very diverse, supporting thousands of plant and animal species. Coral reefs are closely interdependent with other marine and terrestrial communities in the Gulf. Energy, chemical constituents, and mobile species move between reef habitats and mangrove, seagrass, benthic, and hard-bottom habitats. Coral reefs are important economically in Florida for commercial and recreational fishing, boating, scuba diving, snorkelling, as well as educational activities.

Coral reefs are wave resistant carbonate structures (Britton and Morton, 1989; EPA, 1994b; NOAA, 1985) inhabiting only those regions where the water temperature remains above 18°C year-round. In the northern Gulf, reef building corals are not common (Fig. 2). In the eastern and northeastern sections of the Gulf, coral reefs have formed on limestone veneers and escarpments. In the northwest section, reefs have formed on isolated banks that protrude from a mud bottom on the seaward half of the continental shelf. Salt tectonics, diapirism, are responsible for producing these banks and creating a hard substrate on which the reef-building organisms can grow above the surrounding turbid water (Rezak et al., 1990). These areas are the northernmost for corals in this region, yet exhibit some of the most extensive biological zonation of corals in the Gulf. More than 130 such banks occur in the northwest Gulf, including the East and West Flower Garden Banks off the Texas–Louisiana coast, which have been designated as natural marine sanctuaries. Because corals are also sensitive to water turbidity, they are absent around river mouths or other areas with high loads of suspended materials. Below a depth of about 50 m most species in the Gulf disappear.

Hard bottoms are thin veneers of coral communities lacking the density, reef development, and species diversity of patch and outer bank reefs (NOAA, 1985; EPA, 1994a). They often occur in high energy environments on naturally occurring rock outcrops or shell middens, as well as on artificial substrates.

Natural crude oil seeps and brine pools provide some unique and interesting habitats of the Gulf of Mexico (EPA, 1994a; MacDonald, 1998). These are located near the bottom of the continental shelf from DeSoto Canyon westward to salt domes off Texas in water depths that range between 350 and 2200 m. Natural oil seeps are the result of salt tectonism that can trap pools of hydrocarbons and/or open large faults that extend from deeply buried reservoirs all the way to the surface, providing conduits through which petroleum can travel upward. The hydrocarbons released at the site of a seep provide a source of chemical energy for a variety of tubeworms, mussels, and clams through their symbiosis with chemosynthetic bacteria. These animals, in turn, support an abundant epifauna of fishes, shrimps, galatheid crabs, and other invertebrates that are commonly found in smaller numbers at shallower depths. The habitat area provided by a seep is generally small (less than 100 m^2) because the volume of seeping material is small and quickly dispersed (approximately 100 litres/day). Seawater percolating downward through limestone layers, dissolves underlying salt domes to form a brine mixture. This mixture is heavier and denser than the surrounding seawater and flows along the bottom, collecting in depressions. Brine pools form in cavities created by the expansion, eruption and subsequent collapse of a gaseous dome. Besides being extremely salty, they contain high levels of hydrogen sulphide and almost no dissolved oxygen. Bacteria that oxidize the hydrogen sulphide live on the interface between brine pools and "normal" seawater. Recently, these pools were found to contain beds of mussels tolerant to hypersaline conditions and an almost complete absence of dissolved oxygen (MacDonald, 1998). These molluscs also apparently rely on symbiotic bacteria.

OFFSHORE SYSTEMS

The pelagic zone of the Gulf is divided into the neritic province and the oceanic province. The neritic province generally includes all water less than 200 m in depth, or the continental shelf. In the northern Gulf, the continental shelf is fairly uniform with gentle slope (Galtsoff, 1964). Water density, circulation, and turbidity are the primary factors controlling vertically distributed water column habitats in the pelagic zone (EPA, 1994a). In shallow waters along the Gulf coast, water temperatures average 18°C in winter and 29°C in summer. Deeper water remains relatively constant at 17–18°C (Rabalais, 1990). The water is turbid to a depth of about 20 m, and in deeper water, turbidity evolves into a 5–25 m thick nepheloid layer that extends over the shelf break (EPA, 1994a).

In the oceanic province, the deepest portion is an irregularly shaped pit called the Mexican Basin (Lynch, 1954; Darnell and Defenbaugh, 1990; Gore, 1992) which includes the maximum recorded depth of 3850 m in the Sigsbee Deep. Abyssal bottom waters are 4.35°C and have a salinity of 34.97 ppt (Pequenat et al., 1990). The bottom of this area is very flat, but areas of noticeable relief do occur. In the western Gulf, the abyssal plain floor and the shelf surface are disrupted by salt diapirs that form gentle hills and ridges and provide habitats for bottom dwelling fauna.

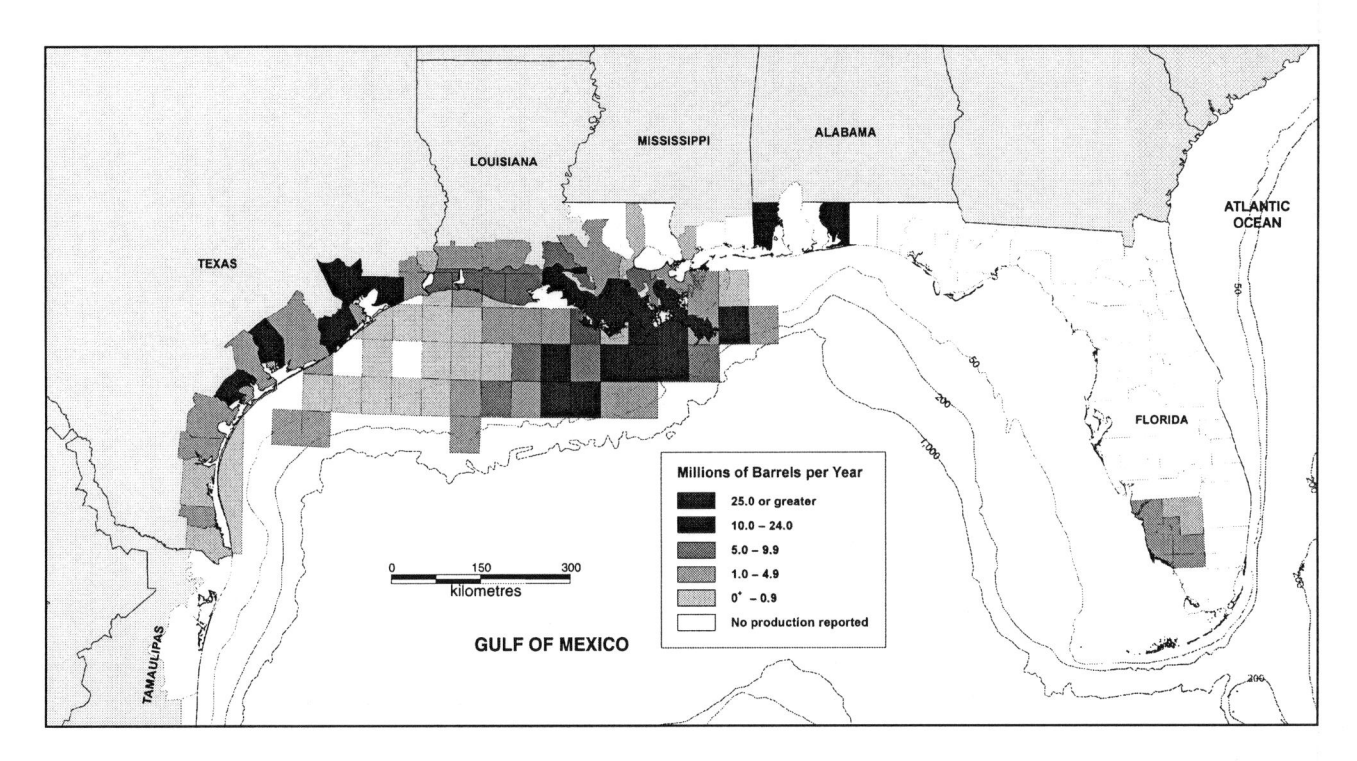

Fig. 3. Map of oil production in northern Gulf of Mexico.

A 9500 km² offshore area between the Mississippi and Sabine Rivers is subject to recurring incidents of oxygen depletion (Darnell, 1990; EPA, 1994b). The nation's richest and most extensive fishing grounds surround this area. Nutrient enrichment from the Mississippi River is a major cause of this, although naturally occurring water column density stratification may be a contributor. Recently, a small oxygen-depleted area was discovered off Florida.

Human activities in the oceanic realm are a major concern (Darnell, 1990) (Fig. 3). Modern offshore oil drilling platforms are self-contained, but the potential for accidental discharges is significant. The risk of a major tanker spill will increase as the number of tankers increases in response to oil import demands. Drilling support boats and commercial ships do not always follow prescribed regulations when discharging garbage and wastes. Pipeline burial can create significant short-term turbidity and disturb benthic communities.

POPULATIONS AFFECTING THE AREA

The northern Gulf of Mexico ranks fourth in total population among the five U.S. coastal regions, accounting for 13 percent of the nation's total coastal population (Culliton et al., 1990). This area is highly valued for residential and seasonal housing, especially in Florida. The coastal population has the second fastest rate of growth of all the major U.S. coastal regions, with the population projected to increase 144% between 1960 and 2010. The major population centres in the Gulf region include Houston, New Orleans, Tampa/St. Petersburg. Regionally, western Florida continues to be the most rapidly growing area followed by Texas. The population growth affects water and land use directly. In addition, population growth in the rest of the U.S. and the world results in increased demands for agricultural products, timber and minerals from the region surrounding the Gulf, placing increasing stresses on land and water resources (Mulholland et al., 1997). Increasing population pressures in this region mean an increasing use of and demands on the Gulf of Mexico.

RURAL FACTORS

The northern Gulf of Mexico receives the Mississippi and Atchafalaya Rivers, which drain 40 percent of the U.S. and parts of Canada (EPA, 1994b). This inflow dwarfs the input from any other Gulf Coast system, providing 79 percent of Gulf of Mexico freshwater inflow. Nutrient loadings of the Mississippi River and for the Gulf have risen dramatically over the last three decades due to increasing agricultural, commercial, and residential development, and are causing growing eutrophication problems (Day et al., 1995) (Fig. 4).

Over 70 percent of the Gulf coast is devoted to agriculture and forestry (McKinney, 1991). This has had a profound effect on many aspects of the coastal environment. Irrigation is increasingly important, and in many places is the largest use of water withdrawals (Mulholland et al., 1997). As ground water extraction grows, saltwater intrusion into coastal aquifers becomes a significant problem. Increasing fertilizer use is increasing nitrate

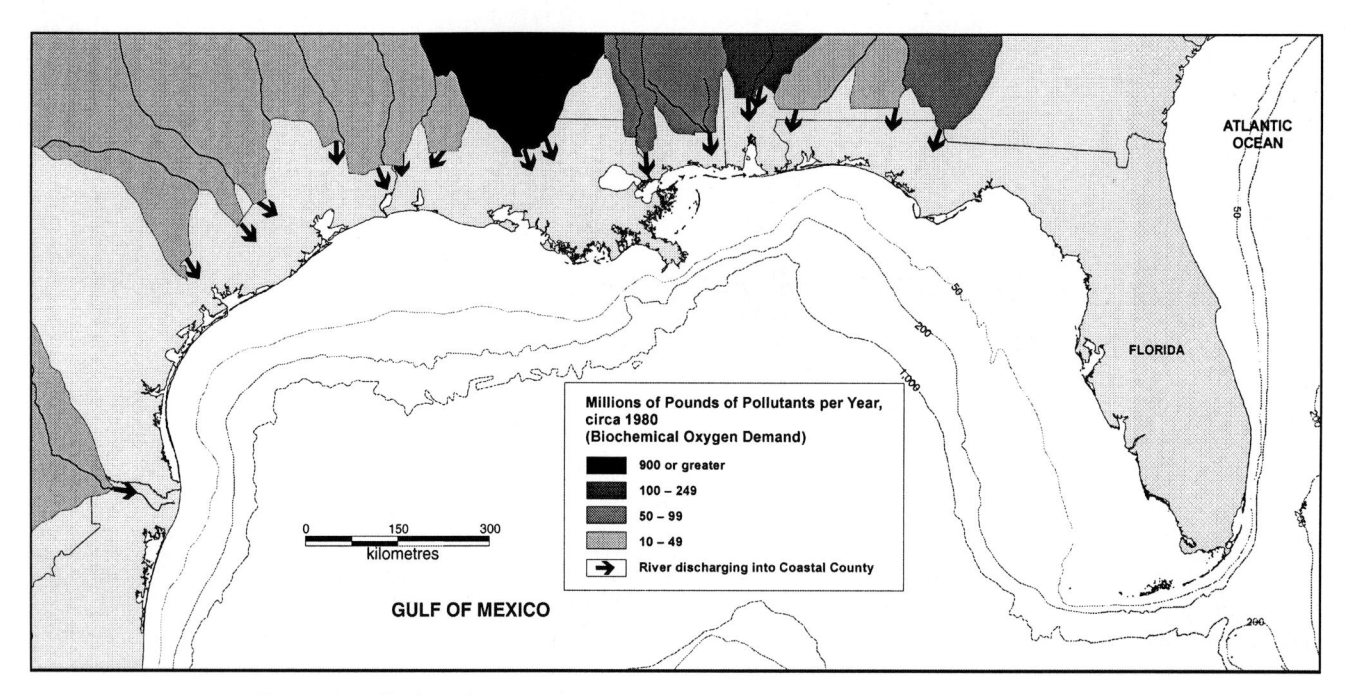

Fig. 4. Map of biological oxygen demand entering northern Gulf of Mexico via major rivers.

concentrations, and the dissolution of potentially toxic elements from soils and their accumulation in ecosystems are accelerated by irrigation (Schmitt and Bunck, 1995). Mercury accumulation in the Everglades, leached from soils and vegetation, has increased due to irrigation and other activities associated with agriculture. In 1978, Gulf coastal counties used an estimated 22 million pounds of pesticide (McKinney, 1991) which became concentrated in the bodies of both fish and shellfish (Hansen and Wilson, 1970). Oysters exposed to concentrations of DDT as low as 0.1 ppb in the surrounding water may concentrate up to 7 ppm in their bodies within a month; this and other trace pollutants may pose a threat to the reproduction, survival, or marketability of fish or shellfish. Impacts of this were most dramatically observed in the extirpation of brown pelicans from the northern Gulf, due to eating tainted fish (McKinney, 1991).

Many acres of pine forests that formerly grew in the Gulf coastal plain have been clearcut and converted to farmland or replanted for pulp wood (EPA, 1994a). Further losses have resulted from construction of flood-control structures and reservoirs, mining and petroleum extraction, and urban development (Keeland et al., 1995). Large-scale federal navigation, flood-control, and drainage projects have played a large role in these conversions by making previously flood-prone lands dry enough for planting crops. Alterations in hydrology and poor timber management have degraded the condition of many of the remaining forests. The effects of these factors on the ecosystems of streams that form the tributary system of the Gulf of Mexico are complex (EPA, 1994a). Pine forests take approximately 70 to 80 years to reach the maturity needed to provide the type of habitat many birds and animals need, but pine trees are typically harvested after only 50 to 60 years of growth. Timber harvesting may also change the distribution of precipitation and soil storage capacity, and lead to exposure of underlying soil, opening the land to erosion and nutrient leaching (Smith, 1980; Meehan, 1991). As trees are removed from an area, runoff increases, carrying increased amounts of sediments and nutrients away in streams and down to the Gulf.

Similar problems are caused from overgrazing by livestock, by elimination of riparian areas by channel widening, channel aggradation, and by lowering of the water table (Brinson, 1990; Armour et al., 1991; Platts, 1991). As the livestock industry has grown, the number of animals has increased far beyond the carrying capacity of available pasture. Livestock are attracted to the riparian areas causing alterations in these habitats and affecting tree reproduction. Without the recruitment of young trees, riverine forests develop an unstable age structure and suffer soil compaction and increases in surface runoff and erosion. As wind and water carry exposed topsoil away, terrestrial and aquatic productivity is lost and a larger sediment load is transported into the Gulf.

Instream dredging for construction materials can result in a number of effects (Meador and Layher, 1998). Channel degradation and erosion has led to increased water velocity and decreased sediment load. In areas of extensive mining, an upstream progression of degradation and erosion known as headcutting can occur. Headcuts can cause dramatic changes in a stream bank and channel that may affect water chemistry and temperature, bank stability, available cover, and siltation. The runoff and discharge from mining operations increases sedimentation and turbidity and may be responsible for decreases in dissolved oxygen and increases in temperature observed downstream from these sites.

Nutrient and sediment loading cause increases in water turbidity that can have serious impacts on habitats (NOAA, 1985; Kenworthy and Haunert, 1991; EPA, 1994a; NOAA, 1997a). Excess nutrients from numerous sources likewise contribute to seagrass declines, and excessive discharges also cause irreparable damage to productive oyster reefs (Galtsoff, 1964). Closings of bay areas to commercial harvest due to human and animal generated contaminants are common throughout the Gulf of Mexico, causing serious economic hardships for fishermen (EPA, 1994a; NOAA, 1997a).

Fisheries

As more and more people come to the Gulf region to enjoy or exploit this area, they have an increasing impact on coastal resources. Increasing demand for seafood has increased fishing pressure on Gulf of Mexico habitats (Bohnsack, 1992), and the advent of improved navigational aids, electronic fish-finding equipment, fishing gear and vessel technology has greatly expanded the ability of fishermen to find and catch fish. This has led to overfishing. For example, in the last 30 years, overfishing has driven down the average weight of swordfish caught at sea from 573 kg to 198 kg, a size at which swordfish are two years short of reproducing. Fishing typically targets top predators and keystone species important for maintaining community structure. Furthermore, by-catch discards are increasing, and process waste may also artificially support large numbers of opportunistic, mobile scavengers and predators, including other fish, crabs and seabirds. Overfishing is thought to be a major reason for the decline of reef and hard bottom communities, as well as vandalism and mistreatment by divers (EPA, 1994a; NOAA, 1998). Reef fishes, such as red snapper and several groupers are especially vulnerable to overfishing because they are typically long-lived, slow-growing, recruitment limited, and spatially restricted. Loss of spawning potential is especially important too.

Fishing operations, such as trawling, gill netting, purse seining, or hook and line are largely species indiscriminate (Darnell, 1990; NMFS, 1998; NOAA, 1998) so fishing results in large quantities of by-catch. By-catch is discarded for economic, legal or personal considerations and includes fish and invertebrates, as well as protected species, such as marine mammals, sea turtles and sea birds. Trawl by-catch from the shrimp industry is the predominant finfish by-catch issue in the Gulf of Mexico, where as many as 115 fish species are discards (NOAA, 1998). Recent estimates of annual by-catch ranges from almost 1 kg to 10 kg or more of finfish for each 0.5 kg of shrimp (McKinney, 1991). This estimate amounts to more than 46 billion fish per year, including 36 billion Atlantic croaker (*Micropogonias undulatus*), 5.5 billion seatrout (*Cynoscion* spp.), 1.8 billion longspine porgy (*Stentomus caprinus*), 1.5 billion spot (*Leiostomus xanthurus*), and 41 million red snapper (*Lutjanus campechianus*) (NOAA, 1998). Bottom trawls and dredges are extensively used (Pattillo et al., 1997; NOAA, 1998) which severely scour the bottom (NOAA, 1998). Nutrient concentrations in near-bottom water can be increased, and bottom sediment can be re-suspended, resulting in pollutant desorption and decreased dissolved oxygen levels. The U.S. Congress has responded to concerns by increasing requirements of the Marine Mammal Protection Act, the Endangered Species Act, and most recently, the Sustainable Fisheries Act to reduce or eliminate by-catch (NMFS, 1998).

COASTAL EROSION AND LANDFILL

Probably the greatest damage to fisheries is caused by the expansion into the shallow, productive coast by real estate, industrial, oil and agricultural interests (Hoese and Moore, 1998). Coastal development, dredging operations (Fig. 5), construction of inshore installations and other harbour and waterway improvements can cause changes in currents, salinity gradients and turbidity that can affect the habitats of the Gulf of Mexico (Galtsoff, 1964).

As the coast becomes increasingly urbanized, natural drainage is altered. Roads, roof tops, and other relatively impermeable surfaces increase runoff. Ditches along roads not only collect surface runoff, they can intercept subsurface flow and bring it onto the surface, bringing about accelerated storm runoff and higher peak flows in small drainage basins (Chamberlin et al., 1991). These activities are besieging wetland habitats along much of the Gulf coast, the most common activities including construction of canals and channels for water control and navigation, dredging, spoil disposal, impounding, draining and filling (Duke and Kruczynski, 1992). These impacts vary from state to state: in Texas, subsidence due to extraction of oil, gas and freshwater are primary factors; alterations of hydrodynamic flow due to construction of navigation channels, as well as subsidence, led in Louisiana; while dredging and fill operations predominate in Mississippi, Alabama and Florida. In the north-central Gulf, loss of wetland habitats that serve as nursery area is believed to be primarily the result of channelization, river control and coastal development. Loss of wetlands in the far western and eastern Gulf is believed to result primarily from urbanization and poor water management practices (Duke and Kruczynski, 1992; EPA, 1994b).

Tidal marshes may be vast (Fig. 6). In northwest Florida, tidal marshes are impounded to control salt marsh mosquitoes with minimal use of pesticides and ditching (Carlson, 1983; Haddad and Joyce, 1997). Since these mosquitoes oviposit on moist exposed soils of the high marsh, impoundment and flooding destroys the larval habitat. In some areas of Florida, entire marsh areas, some encompassing hundreds of acres, have been dammed and flooded

Fig. 5. Dredging activities used in northern Gulf of Mexico for navigational purposes.

Fig. 6. Section of the vast marsh lands bordering the United States coast of the northern Gulf of Mexico.

(Haddad and Joyce, 1997). Although this method decreases the use of pesticides, it also reduces biological productivity; marsh vegetation and mangroves die from the effects of constant submersion, and the estuarine-dependent larval and juvenile forms of marine species no longer have access to their prime nursery grounds. Impounding is not currently a major form of mosquito control on the Gulf coast, but as the population increases, so will pressure to expand this method of control.

Other activities such as construction of marinas, port facilities, real estate development, and conversion to other uses also contribute to the growing loss of wetlands habitat (Duke and Kruczynski, 1992). Also, natural phenomena such as subsidence, sea-level rise, and erosion are impacting large areas of coastal habitat along the Gulf. Rising sea level can increase water depth and lengthen the hydroperiod of wetlands causing changes in their dominant plant species or converting them to open water areas (Mulholland et al., 1997). Although the main effects of sea-level rise would be seen in coastal marshes, extensive areas of bottomland and swamp forests could be affected by increased flooding and saltwater intrusion (Keeland et al., 1995).

One of the purposes of stream channelization is to improve the movement of water downstream to control flooding in urban areas (Brinson, 1990). This is typically accomplished by deepening, widening, and straightening the channel. This results in sharper pulses in flow, which can initiate gullying and the transport of sediments downstream into Gulf estuaries. The resulting narrower, swifter, and deeper stream channel reduces habitat diversity, species richness of fish, and commercial catches of fish.

Urbanization and channelization lead to increased amounts of runoff laden with silt, whose high turbidity decreases the penetration of sunlight into the water (Butler, 1954; EPA, 1994a). Fine silt, deposited on the bottom, coats old shells, making them unsuitable as cultch for oyster spat. Unusually high freshwater inflows can also inhibit maturation and spawning (Stanley and Sellers, 1986) and may clog gills and interfere with filter feeding and respiration. When this sediment settles out it can smother oyster reefs.

Relative Sea-level Rise and Landform Alteration

One of the most critical problems facing coastal areas of the northern Gulf is the high rate of relative sea-level rise (RSLR) due to a combination of eustatic sea-level rise and subsidence (NOAA, 1985; EPA, 1994a; Day et al., 1995). Current evidence indicates that sea-level rise is leading to reduced coastal accretion, water-quality deterioration, coastal habitat destruction, decreased biological production, coastal erosion, and salt-water intrusion in a number of coastal areas. Coastal subsidence can also be caused by enhanced drainage, lowered freshwater input, groundwater withdrawal, navigation channels, and petroleum extraction. Although these operations do not always destroy coastal habitat directly, they do amplify tidal forces, which are generally low in the Gulf (Day et al., 1995; Johnston et al., 1995). Coastal wetlands of the GOM exist within a relatively narrow elevation range, and for these wetlands to retain their elevation relative to local water levels, the rate of regional subsidence, sea-level rise, and sediment transport out of these areas cannot be greater than the depositional rate of inorganic silts and organic detritus. If coastal wetlands do not accrete vertically at a rate equal to the rate of RSLR, they will become stressed due to water logging and ultimately will disappear.

The construction and maintenance of dams and canals, especially navigation canals, is a major cause of changes in sedimentation patterns and wetland loss, especially in high precipitation areas (>100 cm/yr) of the Gulf coast (Duke and Kruczynski, 1992; Orlando et al., 1993; Baird et al., 1996). All important rivers in the Gulf of Mexico have been dammed, causing a reduction in freshwater inflow, nutrients, the influx of suspended sediments, and increased influence of pulsing events such as storms and river floods which are responsible for coastal accretion, higher net biological production, and increased deltaic functioning (Day et al., 1995). The high degree of water management modifies seasonal precipitation and runoff variations that are the primary factors in determining the flows in many rivers (Mulholland et al., 1997). Lower baseflows cause changes in the dominant plant species in riparian and associated wetland areas as upland species that are semi-flood tolerant invade these habitats displacing truly hydrophytic species. Reservoirs also reduce the sediment load and affect delta habitat formation (Baird et al., 1996). Without adequate flushing, soil salinity can become elevated in riparian ecosystems.

Freshwater flushing rate, through its dilution of saline waters, is also a critical parameter governing biological processes in Gulf estuaries (Pattillo et al., 1997). Reductions of freshwater inflow have been linked to reduced harvest yields of penaeid shrimp, and affect filter feeding organisms such as oysters, clams, and other invertebrates by depriving them of an important food source (EPA, 1994a; Baird et al., 1996; NOAA, 1997a). In south Texas, oyster beds have almost disappeared. Damming also restricts the movement of migratory fishes, such as Alabama shad, American eel, and striped bass (Pattillo et al., 1997). Impoundments, consisting of dams and other systems of dikes and water-control structures (Lin and Beal, 1995) have led to a loss of wetlands that maintain water quality. For example, habitat changes have been attributed to water management and land-use practices in southern Florida (McIvor et al., 1994; Smith-Vaniz et al., 1995). These changes may be contributing to the decline in pink shrimp landings from the Dry Tortugas and may eventually affect lobster and other fishery landings due to the loss of habitat used by juveniles. Landform alterations that create conditions whereby nutrient-rich freshwater inflow is diverted to open water and bypasses these wetlands, can lead to conditions that intensify coastal blooms, which can create large zones of hypoxic bottom waters (Sklar and Browder, 1998).

Saltwater Intrusion

Saltwater intrusion related to canal construction is responsible for the loss of many acres of Louisiana coastal marshes. Freshets can have a negative impact on fisheries by providing low salinity habitat areas for estuarine species. When these species move out of the freshet to follow their preferred benthic food resources they are unable to withstand the change in salinity (Hoss and Thayer, 1993). Dredge and fill activities for navigation channels is probably the largest form of estuarine alteration and has been responsible for the majority of conversions of wetland habitats into open water habitats on the Gulf coast (Duke and Kruczynski, 1992). Approximately 45% of the nation's import and export trade passes through Gulf ports, which are usually located within the broad, shallow estuaries typical of the Gulf of Mexico (McKinney, 1991). Dredging is necessary to provide access channels to these ports. Consequently, the Gulf of Mexico is the most dredged region in the US; 75% of all dredged material disposal in the US occurs here, and 90% of that occurs between Mobile Bay and the Laguna Madre, which includes some of the Gulf's most productive waters.

Dredging

Commercial dredging removes substantial portions of natural oyster grounds in the estuaries along the Gulf. Sediment from channel dredging, storms and floods profoundly affects oyster communities causing their destruction through burial (Britton and Morton, 1989). Oysters must also contend with sediment of their own making in the form of faeces and pseudofaeces. If water circulation is limited, oyster reefs can be smothered as this material accumulates. Sustained heavy sediment loads in the water interfere with respiration and eventually smother oysters (Stanley and Sellars, 1986; Britton and Morton, 1989). In the past, oyster shell was dredged as a source of minerals for agricultural and farm animal nutrients and for building materials (EPA, 1994a) which depleted stocks severely. Between 1912 and 1964, the volume of oyster shell dredged from Texas bays, 190 million cubic meters, was greater than the volume dredged to create the Panama Canal, and 78% of it came from Galveston Bay. In Mobile Bay, over one million cubic meters of shell were removed from 1947 through 1968 (Schroeder et al., 1998). The depressions left by mining activities became persistent sites of relatively high salinity containing hypoxic to anoxic water. Today, because of ancillary environmental damage, oyster dredging has been almost eliminated along the Gulf Coast.

Navigation canals produced by dredging are frequently constructed wider than the widths specified on permit applications (EPA, 1994a). Erosion from a high volume of boat traffic can also cause canal widths to increase further, and increased open water from canal construction can result in flow changes and saltwater intrusion (Duke and Kruczynski, 1992) which has been responsible for the loss of many acres of Louisiana coastal marshes. Furthermore, canals are constructed in straight lines, whereas natural creeks and channels are serpentine resulting in major differences in water and sediment movement between canals and natural drainage systems. This can also affect seagrasses by altering or blocking water currents (Gore, 1992), and filling is equally devastating to seagrass meadows (Fonseca et al., 1992); the addition of only a few centimetres of sediment can completely bury seagrasses. Construction and maintenance of navigation channels pose a significant threat to fishery resources through habitat destruction and alteration (Lindall and Saloman, 1977). The amount of channelization by local interests is not known, but more than 7000 km of navigation channels are completed, under construction, or planned by the U.S. Army Corps of Engineers. Almost all require periodic dredging. Consequences of this that are particularly damaging include physical loss of habitat by creation of spoil islands, segmentation and isolation of bays, increased shoaling, increased saltwater intrusion and flushing time, alteration of tidal exchange and circulation patterns, increased turbidity, and destruction of submerged and emergent vegetation. Filled areas represent the most obvious form of estuarine alteration; most filling occurs in productive parts of the estuarine ecosystem.

EFFECTS FROM URBAN AND INDUSTRIAL ACTIVITIES

Other than natural disasters, the greatest threat is activities related to the extraction of oil and gas. These activities can result in detrimental impacts to onshore, nearshore, and offshore wetlands and other aquatic habitats (Duke and Kruczynski, 1992). Onshore oil and gas activities, which can affect habitats, include construction of pipelines and support facilities; nearshore activities include construction and use of navigation channels. While it is difficult to quantify a direct cause and effect relationship between petroleum extraction activities on the outer continental shelf and habitat loss, it should be noted that the rapid wetlands loss in the central area of the Gulf coast is confined to the most concentrated development of onshore and offshore oil and gas recovery efforts.

Much of the coastal marshes in Louisiana are subject to management by man in the form of levees and water-control structures to manipulate water levels (Herke, 1995). This marsh management was traditionally used to maintain water levels for boat traffic, but is now also used by public agencies and private entities to improve duck and alligator habitat. Private landowners sometimes use levees to prevent losing legal title to their land through its erosion or subsidence below mean low water. By using levees to convert the submerged areas to "fast lands" these areas can remain in private ownership; otherwise, such areas can revert to state ownership. In addition, marsh management is now recommended as a means to reduce marsh loss in Louisiana. However, studies have found that the extensive use of this practice significantly reduces fisheries production and offshore recruitment and may be having a detrimental effect on fisheries.

Mariculture

Increasing regulation of commercial fishing has led to a boom in penaeid shrimp or fish mariculture. The potential impacts of these mariculture ventures are substantial and include loss of wetland area, eutrophication of the receiving body, competition by accidentally released non-native species, and transmission of exotic diseases to native stocks (Hopkins et al., 1995). To address these concerns, a combination of regulatory measures and directed research are underway at the federal and state government levels. Along the Gulf coast, Texas is at the top of the list as a location for such ventures due to its 1425 mile coastline and the availability of coastal land that is not adequate for traditional agriculture crops (Martin, 1986). A variety of factors in other Gulf coast states is also responsible for Texas' popularity for mariculture (Martin, 1986). Sandy soils and permit regulation are major barriers to production in Florida. The minimal coastline available for mariculture limits projects in Mississippi and Alabama. In Louisiana, despite the availability of huge expanses of coastal property, the low elevation and marshy condition along the coast are not very suitable for typical mariculture operations. However, the State of Louisiana allows private entities with the proper permits to use natural marshes for mariculture provided these areas are under a marsh management plan, which usually involves the use of levees or other water level controls (Herke, 1995). By isolating marshes for mariculture purposes, natural fisheries and offshore recruitment are reduced due to the loss of these areas as nursery and production grounds.

Effects of Increasing Use

Millions of people depend on the Gulf of Mexico for their living and flock to its shores for entertainment and relaxation. As the population soars, so does the demand upon the natural resources of this area. Changes in land use and attitudes of people have altered natural fire regimes, which has played a major part in reshaping coastal ecosystems and coastal habitats (LaRoe et al., 1995). The demand for fresh potable water along the Gulf coast has increased with the population and presently, the Gulf of Mexico is deprived of 40 to 90% of spring runoff from 44 rivers due to upstream diversion and impoundment (Rozengurt, 1992). Water control structures block anadromous fish species from migrating to their spawning areas. Declines in populations of the Alabama shad *Alosa alabamae* are believed to be due in part to dams barring their spawning runs (Pattillo et al., 1997) and the loss or reduction of freshwater inflow can have detrimental effects on both recreational and commercial fishery populations (Zimmerman and Minello, 1993; Hammerschmidt et al., 1998). As saltwater intrudes into brackish and freshwater areas, the increased salinities that result can cause changes in species composition as well as increased predation and disease from organisms previously excluded by lower salinity waters. The American oyster, *Crassostrea virginica*, and Atlantic rangia, *Rangia cuneata*, move into formerly freshwater areas as brackish conditions develop (Pattillo et al., 1997). American oysters, present in areas with increasing salinities, become more susceptible to stenohaline organisms such as the predatory oyster drill, *Thais haemastoma*, and the infectious protozoan, *Perkinsus marinus*.

Loss of freshwater inflow and the discharge of pollutants contribute to the degradation of the water quality of the northern Gulf (Chesapeake Bay NPS Program from Nutrient Enrichment Action Agenda for GOM; NOAA, 1997b). Of particular concern is nutrient enrichment, whose sources are numerous and varied. Nutrient enrichment is a serious environmental concern because of the effects of oxygen depletion and noxious algal blooms (from Nutrient Enrichment Action Agenda for GOM; Handley, 1995). Urban and suburban land users contribute much higher nutrient loads, on a per acre basis, than other land users; therefore, rapid urbanization leads to intensified non-point source pollution (NOAA, 1997a; Chesapeake Bay NPS Prog from Nutrient Enrichment Action Agenda for GOM).

Diatoms, tiny planktonic algae, are thought to provide the primary energy source for traditional food webs that support top predators. The abundance of coastal diatoms is influenced by silicon supplies, whose Si:N atomic ratio is about 1:1 (Redfield Ratio). Diatoms out-compete other algae in a stable and illuminated water column of favourable silicate concentration. Anthropogenic enrichment of N and P, leading to long-term increases in these nutrient loadings, has also led to long-term declines in the Si:N and Si:P ratios (Turner and Rabalais, 1991a, from Nutrient Enrichment Action Agenda for GOM). This decline has particularly favoured non-diatom blooms and is a key factor associated with the global epidemic of novel toxic and harmful phytoplankton blooms and phylogenetic shifts in the phytoplankton biomass predominance in coastal seas (Smayda, 1989, 1990, 1991, from Nutrient Enrichment Action Agenda for GOM). In the northern Gulf of Mexico, riverine input of Si:N from the Mississippi River has decreased from 4:1 to approximately 1:1 over the last three decades (Turner and Rabalais, 1991b, from Nutrient Enrichment Action Agenda for GOM). These changes may have a major impact on Si availability, phytoplankton species availability, carbon flux and hypoxia. Phylogenetic shifts within phytoplankton communities may alter the food supply available to herbivorous organisms. Evidence suggests that smaller, "less desirable" flagellate and cyanobacterial-dominated communities are less acceptable as food for grazers. Zooplankton, the main consumers of whole diatoms and a staple of juvenile fish, are thus affected by these nutrient changes in a cascading series of interactions. Alternatively, changes in type and distribution of higher trophic level herbivores and predators may have a cascading effect down the food web to primary producers.

Effects of Fire

Fire is an important natural ecological force, and growing urbanization along the Gulf has caused changes in both fire frequency and occurrence (Wade et al., 1980; LaRoe et al., 1995). New roads making coastal areas more accessible to development, and the growing number of tourists, dramatically increase the risk of fire. As the coast becomes more developed, fire suppression is increased due to the construction of roads and canals, and by public pressure to protect urban areas, maintain aesthetic landscapes, and prevent reduction in air quality. This, coupled with the drainage of wetlands and other low-lying areas, allows fuel in the form of vast amounts of organic detritus to accumulate. If this material is not periodically burned off by small fires it builds up to a point that results in hotter, more intense fires that are more destructive and require more effort to stop. These changes in the natural frequency and intensity of fires are causing a loss of species diversity, loss or decline of fish and wildlife habitat, destruction of fertile organic soils, changes in surface waterflow patterns, and site degradation that opens areas to invasion by exotic species that are often harmful to an ecosystem.

A natural fire cycle is important in maintaining many coastal ecosystems by preventing the invasion of fire-intolerant species and eventual loss of the original ecosystem (Wade et al., 1980; LaRoe et al., 1995). Fires kill woody species encroaching on saltmarshes ensuring the continued dominance of grasses. Pine forests are fire-adapted communities that rely on fire to exclude hardwood species. Fires also maintain stands of cypress at the expense of any invading pine species. Alterations of natural fire frequencies caused by human influence can change fire from a beneficial influence to a detrimental one. Generally, mature stands of cypress and mangrove forests are little affected by fire due to water flow flushing ground litter from these areas and the wetness of any remaining materials. However, when flow is altered due to channelization, fill, or drainage projects, dry organic material can accumulate and cause intense fires that can destroy these communities. Such drained and burned sites in Florida are susceptible to invasion by undesirable exotic species such as melaleuca (*Melaleuca quinquenervia*) and Brazilian pepper (*Schinus terebinthifolius*) preventing recovery of native vegetation. The heat from these extreme fires can destroy the thick organic soils of wetlands, lowering land elevations and making an area more prone to flooding or open water formation.

The growing industrialization and population along the northern Gulf of Mexico is responsible for a growing amount of pollution entering the Gulf and affecting water quality. For decades, estuarine and marine environments have been used as major repositories of wastes (Lindall and Saloman, 1977; Kennish, 1992). A complex mixture of pollutants, including oxygen-demanding organic materials, pesticides petroleum products, silt, heat, radioactive substances, heavy metals, debris, and other deleterious substances contained in sewage effluent and storm water runoff, have chemically altered some coastal habitats to the extent that their productivity and usefulness are endangered or lost. Pollution of water bodies with poor water circulation (such as coastal lagoons with low turnover rates) is also a serious problem (EPA, 1994a). Further aggravating this problem is the reduction of freshwater inflow due to impoundment and/or diversion activities (Orlando et al., 1993). Without the flushing provided by freshwater pulses, these areas tend to concentrate incoming pollutants for prolonged periods. These systems can contain several important habitats that can be affected (EPA, 1994a). Although they are not always harmed by the indirect discharge of sewage effluents, the complex faunal communities that utilize them are devastated. Water bodies receiving polluted inflows become eutrophic, with frequent massive fish kills due to de-oxygenation, and associated health hazards destroy the value of these areas for recreation or fisheries. Furthermore, most contaminants settle to the bottom upon entering the marine environment where they can be incorporated into food webs by benthic animals. Coral reefs are especially susceptible to stresses caused by pollutants (Taylor and Bright, 1974).

Urban runoff also contains a variety of water pollutants (Lindall and Saloman, 1977; Baird et al., 1996). Concentrations of the pollutants range from the quality of drinking water to raw sewage. Street litter is a significant source of urban pollution with dust and dirt being the largest component by weight. Other components include refuse left over from poor waste disposal and collection activities, wastes from pet and other animals, yard litter, and construction debris. Sediment from construction sites, where protective vegetation has been removed, is often a concern. A significant source in urban areas originates from atmospheric deposition (Baird et al., 1996). This can occur as dry deposition of airborne particles, which may later be dissolved and carried by rainfall and runoff, or as rainfall deposition. Components delivered from atmospheric sources include metals and nutrients such as nitrogen and phosphorus.

The expansion of urban development, industry, and recreational use in this area have also given rise to large amounts of debris being improperly placed into the Gulf where it can affect water quality (EPA, 1993). Although, historically, most debris is plastic, it also includes wood, metal, rubber, paper, glass, tar, and cloth. Major land-based sources of marine debris include users of beaches, docks, and marinas, storm water runoff from urban areas, inadequate sewage systems, solid waste disposal, river-borne debris; and wreckage from aging docks and other water dependent structures. Debris is also generated at sea from different sources that include ships and offshore activities that typically involve maritime and petroleum activities. Debris in the Gulf can cause a variety of impacts to living marine resources (O'Hara, 1992). Several species of marine animals are known to become entangled in various forms of debris and some is also ingested by marine animals.

Effects of Petroleum Oil

Among the most serious pollution threats to the biota in estuaries is petroleum oil (NOAA, 1985; Kennish, 1992). The sensitivity of organisms to oil pollution depends on several factors such as age, size and maturity of the individual, time of the year, type of oil, and other factors. The early life stages of estuarine and marine animals are usually more sensitive to crude oil compounds than adults. Oil causes both acute, lethal and long-term, sublethal effects (Kennish, 1992), but in addition, oil dispersant materials applied during clean-up operations are toxic to a wide variety of marine organisms and their use can add to environmental damages caused by a spill.

Salt marshes, mangroves, mudflats, and other low-energy areas tend to trap oil effectively. Hydrocarbon contaminants accumulate in the sediments where they can have long-lasting adverse effects, remaining for years before being re-released back into the surrounding environment when the area is struck by, for example, hurricanes, storm surges or floods. Salt marshes are especially vulnerable to this because of their ability to hold oil. Bioturbating infauna (e.g. fiddler crabs) facilitate the movement of oil deeper into the sediments. Shallow-rooted plant species that are in greater contact with oil-contaminated sediments are initially more susceptible to damage than deeper-rooted plants, but this changes as the oil migrates deeper. Oil and chemical spills also impose significant impacts on seagrass habitat in the intertidal zone (Fonseca, 1992). Exposure to oil and chemical products will also incorporate carcinogenic and mutagenic substances into the food chain, which can induce mortalities that lower the diversity and functional value of the seagrass system.

Transportation of crude oil and petroleum products is a continual threat to the environmental health of the Gulf (Rainey, 1991). Most major oil spills occurring recently have been due to transportation of imported oil (Defenbaugh, 1990). The principal methods for moving oil are by pipeline, barge, and tanker. Import and export activities over 17 years up to 1991 had resulted in 91 oil spills greater than 1000 barrels. Approximately 75% of the accidental oil spills in the U.S. occur in coastal waters, primarily estuaries, enclosed bays, and wetlands. An examination of the locations and frequency of historic spills in the Gulf show that most occur near terminals in ports and harbours, and are caused by barging accidents. Nearly as many spills occur from the movement of tankers in Gulf coastal waters. Gulf spills also tend to be concentrated near the mouths of rivers and bays where ship traffic is usually more congested and navigation more hazardous (NOAA, 1985). The largest spills were located primarily in open waters and in harbours and ports. While large spills represented less than 3% of the spills in the northern Gulf between 1975 and 1980, their volume accounted for over 60% of the total of all oil spilled.

Approximately 45% of the total oil entering the ocean waters comes from urban runoff, polluted rivers, non-petroleum industries, and municipal wastes and effluents (NOAA, 1985; Kennish, 1992). Activities related to the transportation of oil account for another 33%, of which 25% is attributable to accidents and major spillages, and the remainder to normal operational losses. Another 20% to 25% is released from natural oil seeps on the seafloor. While major spills and accidents attract greater public attention, chronic oil pollution associated with routine operations of coastal oil refineries and oil installations, as well as discharges from industrial and municipal sources affect a greater area. Some oil is routinely released as part of normal operations (Ehler et al., 1983). Ships, for example, discharge oil when bilge water is pumped out, and during tank cleaning and ballasting. Tank cleaning and ballasting account for about 70% of operational discharges. Moreover, wastewaters released from refineries and offshore oil rigs contribute contaminants to offshore and coastal waters. These routine losses over time can greatly exceed losses from major accidents.

PROTECTIVE MEASURES

The loss of quantity and quality of the Gulf of Mexico and its resources has declined since the late 1970s due to passage and enforcement of federal and state laws and regulations (Duke and Kruczynski, 1992). Many federal agencies are mandated by legislative statutes to address habitat degradation issues and support protection and restoration efforts. These agencies include: U.S. Environmental Protection Agency, U.S. Department of Commerce, U.S. Department of the Interior, U.S. Department of Defense, U.S. Department of Agriculture, and U.S. Department of Transportation (EPA, 1994a). In addition to regulation of activities that can affect the Gulf's coastal waters, many agencies participate in wetland research and demonstration projects that are underway or about to begin. For example, the U.S. Fish and Wildlife Service is digitizing information on wetland acreage in Gulf coastal states, NOAA continues to collect data in the Status and Trends Program, and EPA has initiated a monitoring program in near coastal waters and estuaries (Ecological Monitoring and Assessment Program). The U.S. Army Corps of Engineers is designing and testing several freshwater diversion projects to improve the condition of wetland habitats along the lower Mississippi River and the use of dredged material for habitat development and restoration (Landin et al., 1989).

The legal and institutional framework in place to protect the Gulf of Mexico extends from the international level all the way down to individual user groups. One of the most important international laws that protects the Gulf is MARPOL. The U.S. joined 39 other nations to ratify Annex V of MARPOL, which came into effect in 1988. MARPOL V bans the dumping of plastics by vessels at sea and in navigable waters and also regulates the disposal of other types of solid waste in the marine environment (EPA, 1993; O'Hara, 1992). However, Annex V restrictions only apply to ships of countries which are signatory to the MARPOL Protocol, and public vessels are exempt from these rules.

MARPOL V is implemented in the U.S under the Marine Plastic Pollution Research and Control Act of 1987 (MPPRCA). The Act prohibits the disposal of plastic by vessels at sea, requires that the effects of plastic pollution on the marine environment be identified and reduced, and regulates the allowable distance from shore for the discard of all other solid waste materials. Enforcement of the Act is handled by the U.S. Coast Guard. This act also requires establishment of a public education program on plastic pollution and citizen pollution patrols.

The Fishery Conservation and Management Act was originally enacted in 1976 to establish a national program for the conservation and management of the fishery resources of the United States. The Act designates eight regional Fishery Management Councils (New England, Mid-Atlantic, South Atlantic, Caribbean, Gulf of Mexico, Pacific, North Pacific, and Western Pacific). Each regional council prepares Fishery Management Plans (FMPs) for those fisheries, both recreational and commercial, which they determine to require active management.

The Gulf of Mexico Fishery Management Council is based in Tampa, Florida, and meets every two months at various locations around the Gulf Coast. Fishery Management Plans in effect include Gulf and South Atlantic spiny lobster, Gulf of Mexico reef fish, Gulf of Mexico corals, Gulf of Mexico shrimp, Gulf of Mexico stone crab, Gulf of Mexico red drum, and Coastal Migratory Pelagic species.

The Fisheries Conservation Zone established by the original Magnuson Act in 1977 was superseded by a 200-mile Exclusive Economic Zone (EEZ) in 1983. The EEZ extends from the seaward boundary of each of the coastal states (generally 3 miles from shore), to 200 nautical miles from shore. The seaward boundaries of Texas, Puerto Rico, and the Gulf Coast of Florida are 9 nautical miles from shore.

On October 11, 1996, President Clinton signed the Sustainable Fisheries Act, which re-authorized and amended the Magnuson-Stevens Fishery Conservation and Management Act (NOAA, 1996). The stated purposes of the re-authorized Act are to prevent overfishing, rebuild overfished stocks, ensure conservation, facilitate long-term protection of essential fish habitats, and to realize the full potential of the Nation's fishery resources. Among its provisions is a new requirement that all federal fisheries management plans must be amended to include description, identification, conservation, and enhancement of Essential Fish Habitat (EFH). EFH is defined as "waters and substrate necessary to fish for spawning, breeding, feeding, or growth to maturity". Other provisions include scientifically sound stock assessments, and conservation measures in each Fishery Management Plan to prevent overfishing.

In 1972, Congress passed the Coastal Zone Management Act (CZMA) in order to establish a national policy to "preserve, protect, develop, and where possible, to restore or enhance, the resources of the Nation's coastal zone for this and succeeding generations" and to "encourage and assist the states to exercise effectively their responsibilities in the coastal zone through the development and implementation of management programs to achieve wise use of the land and water resources of the coastal zone". The National Estuarine Research Reserve System (NERRS) was created with the passage of the CZMA. The NERRS is a protected areas network of federal, state, and local partnerships, dedicated to fostering a system of estuary reserves that represents the wide range of coastal and estuarine habitats found in the United States and its territories (NOAA, 1996). Three estuaries in the Gulf of Mexico have been designated to receive support under this program, Weeks Bay, Alabama and Apalachicola Bay and Rookery Bay, both in Florida.

The National Marine, Protection, Research and Sanctuaries Act (MPRSA) of 1972, administered by the National Oceanic and Atmospheric Administration, regulates the ocean dumping of waste, provides for a research program on ocean dumping, and provides for the designation and regulation of marine sanctuaries (NOAA, 1999). The purpose of this Act is to prevent "unregulated dumping of material into the oceans, coastal, and other waters that endanger human health, welfare, and amenities, and the marine environment, ecological systems and economic potentialities." Material includes, but is not limited to: dredged material, solid waste, incinerator residue, garbage, sewage, sewage sludge, munitions, chemical and biological warfare agents, radioactive materials, chemicals, biological and laboratory waste, wrecked or discarded equipment, excavation debris, and industrial, municipal, agricultural, and other waste. Section 102 of this act authorizes the EPA to issue ocean-dumping permits for the transport to and disposal of materials in the oceans, excluding wastes regulated by the USCOE. The USCOE is authorized under Section 103 of MPRSA to issue permits for the ocean disposal of dredged material. Title III of MPRSA, known as the National Marine Sanctuaries Act (NMSA), is establishing a system of marine protected areas that includes areas of the Gulf of Mexico. NMSA charges the Secretary of the department of Commerce to identify, designate, and manage marine sites based on conservational, ecological, recreational, historical, aesthetic, scientific or educational value within significant national ocean and Great Lake waters. The Florida Keys National Marine Sanctuary was established in 1990 to provide protection for this area, which has one of the most diverse assemblages of underwater plants and animals in North America. Located along the border between the southeastern Gulf of Mexico and the Atlantic Ocean, this area contains mangrove fringe forests, seagrass meadows, hard-bottom regions, and coral reefs. In 1991, the Flower Garden Banks National Marine Sanctuary, an area of coral reefs and rich underwater faunal diversity off the coast of Texas, became the nations' tenth sanctuary and the first located exclusively in the Gulf.

In recognition of the fact that estuaries are unique and endangered ecosystems and that the issues relating to them are too complex to be handled by traditional pollution-control programs, the U.S. Congress established the National Estuary Program (NEP) in 1987 under Section 320 of the Clean Water Act (Crum, 1999). The NEP is intended to identify estuaries of national significance and to establish and oversee a process for improving and protecting their water quality and enhancing their living resources. The NEP currently includes 27 estuary projects nationwide, seven of which are in the Gulf of Mexico: Corpus Christi Bay and Galveston Bay in Texas, Barataria–Terrebonne Estuarine Complex in Louisiana, Mobile Bay in Alabama, and Tampa Bay, Sarasota Bay, and Charlotte Harbour in Florida.

On a national scale, spatial and temporal trends of chemical contamination and biological responses to that contamination have been monitored since 1984 by NOAA's National Status and Trends (NST) Program (NOAA, 1998b). The NST program is managed by the Center for Coastal Monitoring and Assessment in NOAA's Nation Ocean Service. The purpose of the program is to determine the current status of, and to detect changes in, the environmental quality of U.S. estuarine and coastal waters. Temporal trends are monitored through the Mussel Watch project that analyzes mussels and oysters collected annually throughout the coastal and estuarine United States. Spatial trends are described from chemical concentration measured in surface sediments collected by both the Mussel Watch and Benthic Surveillance Project from sites along the coastal U.S. Of the 37 estuaries monitored in the northern Gulf, biological resource impacts due to toxic algal species were reported in 25, while nuisance algae were reported in 22.

In addition to interagency activities, many Federal agencies have taken individual action to preserve coastal environments. The U. S. Army Corps of Engineers has been regulating activities in the nation's waters under the Rivers and Harbors Act originally passed by Congress in 1890. This Act, as amended in 1899, prohibits dumping of solid wastes in the navigable waters of the U.S. and their tributaries and regulates activities that could affect the course, location, condition or capacity of such waters. Prior to 1968, the primary thrust of the USCOE regulatory program was the protection of navigation. However, as the result of several laws and judicial decisions handed down since the 1960s, the program has evolved into one that considers the full public interest by balancing favourable impacts to the environment against the detrimental impacts.

Currently, the most comprehensive wetlands regulatory tool available to managers is the Section 404 Regulatory Program established by the Federal Water Pollution Control Act Amendments of 1972, which has the goal of restoring and maintaining the chemical, physical, and biological integrity of the Nation's waters (Johnston et al., 1992; Holmberg, 1988). The Act was further strengthened by amendments in 1977 and 1987. The United States Environmental Protection Agency (EPA) has promulgated regulations under this Act, commonly referred to as the 404 regulatory program, that make it unlawful to discharge dredged or fill material into waters of the United States without first receiving authorization unless the discharge is covered under an exemption. The EPA has the final authority for activities in coastal wetland areas. However, the U.S. Army, Corps of Engineers (ACE) actually administers the 404 regulatory program under the overview of the EPA, while the United States Fish and Wildlife Service is responsible for investigating potential impact to fish and wildlife resources during the 404 permitting process. Along the Gulf coast, administration of the 404 program is a cooperative effort between the ACE and an approved state coastal zone management program. These state programs were founded and are funded through the Coastal Zone Management Act. In many instances the guidelines of the state programs are more restrictive than the federal program.

Each of the five Gulf of Mexico States also has a regulatory framework for addressing habitat degradation. Dredging, filling, and construction in Florida state waters are regulated by the Florida Department of Environmental Regulation under the Warren S. Henderson Wetlands Protection Act of 1984. The Alabama Environmental Management Act passed by the state legislature in 1982 created the Alabama Environmental Management Commission (AEMC) and established the Alabama Department of Environmental Management (ADEM), which consolidated several smaller commissions, agencies, programs and staffs responsible for implementing environmental laws. Among other things, the AEMC is charged with developing the state's environmental policy and adopting environmental regulations. The ADEM administers all major federal environmental laws, including the Clean Water Act and federal solid and hazardous waste laws. Mississippi's Department of Environmental Quality, through the Office of Land and Water Resources (LWR), is charged with the conservation, management, protection, and encouragement of appropriate development of the water resources of the state. The LWR performs its duties through the study of state water resources, developing information regarding such resources, and maintaining databases of this information. The mission of the Louisiana Department of Natural Resources (DNR), created in 1976, is to preserve and enhance the nonrenewable natural resources of Louisiana through conservation, regulation, and management/exploitation. The DNR exercises regulatory and permitting functions through the offices of Conservation and Coastal Restoration and Management. In Texas, to maintain water quality standards, the Texas Natural Resources Conservation Commission provides Water Quality Certification under Section 401 of the Clean Water Act except for projects involving the exploration, production, and transportation of raw petroleum products. Certification for activities involving raw petroleum products is through the Railroad Commission of Texas.

Besides federal and state regulation, the ecological and societal functions of coastal areas and wetlands are becoming better known to resource managers and to the general public that has become increasingly more interested and active in environmental affairs. Better informed managers and increased public awareness have made a positive difference in management decisions concerning the disposition of habitat in these areas (EPA, 1994a).

REFERENCES

Armour, C.L., Duff, D.A. and Elmore, W. (1991) The effects of livestock grazing on riparian and stream ecosystems. *Fisheries* **16**(1), 7–11.

Baird, C., Jennings, M., Ockerman, D. and Dybala, T. (eds.) (1996) Characterization of nonpoint sources and loadings to the Corpus Christi Bay National Estuary Program Study. Rep. CCBNEP-05, Texas Natural Resource Conservation Commission, Austin, TX, 226 p.

Beccasio, A.D., Fotheringham, N., Redfield, A.E., Frew, R.L., Levitan, W.M., Smith, J.E. and Woodrow, Jr., J.O. (1982) Gulf coast ecological inventory: user's guide and information base. U.S. Fish Wildl. Serv. Biol. Rep. FWS/OBS-82/55, 191 p.

Bohnsack, J.A. (1992) Reef resource habitat protection: the forgotten factor, In *Stemming the Tide of Coastal Fish Habitat Loss*, ed. R.H. Stroud. National Coalition for Marine Conservation, Inc., Savannah, GA, pp. 117–129.

Brinson, M.M. (1990) Riverine forests. In *Ecosystems of the World. Vol. 14, Forested Ecosystems*, eds. A.E. Lugo, M. Brinson, and S. Brown. Elsevier, Amsterdam, The Netherlands, pp. 87–141.

Briggs, J.C. (1974) *Marine Zoogeography*. McGraw-Hill, New York, NY. 475 pp.

Britton, J.C. and Morton, B. (1989) *Shore Ecology of the Gulf of Mexico*. University of Texas Press, Austin, TX, 387 pp.

Butler, P.A. (1954) Summary of our knowledge of the oyster in the Gulf of Mexico. *U.S. Fish Wildl. Serv., Fish. Bulletin* **55**, 479–489.

Carlson, D.B. (1983) The use of salt-marsh mosquito control impoundments as wastewater retention areas. *Mosquito News* **43** (1) 6.

Chamberlin, T.W., Harr, R.D. and Everest, F.H. (1991) Timber harvesting, silviculture, and watershed processes, In *Influences of Forest and Rangeland Management on Salmonid Fishes and Their Habitats*, ed. W.R. Meehan. American Fisheries Society Special Publication 19, Bethesda, MD, pp. 1–15.

Cintron-Molero, G. (1992) Restoring mangrove systems. In *Restoring the Nation's Marine Environment*, ed. G.W. Thayer. Maryland Sea Grant, College Park, MD, pp. 223–277.

Continental Shelf Associates, Inc. and Texas A&M University, Geochemical and Environmental Research Group (1998) Northeastern Gulf of Mexico Coastal and Marine Ecosystem Program: Ecosystem Monitoring, Mississippi/Alabama Shelf; First Annual Interim Report, U.S. Dept. of the Interior, U.S. Geological Survey, Biological Resources Division, USGS/BRD/CR-1997-0008 and Minerals Management Service, Gulf of Mexico OCS Region, New Orleans, LA, OCS Study MMS 97-0037, 133 pp. + app.

Cowardin, L.M., Carter, V., Golet, F.C. and LaRoe, E.T. (1979) Classification of wetlands and deepwater habitats of the United States. U.S. Dept. of the Interior, U.S. Fish Wildl. Serv., Washington, D.C., USFWS/OBS-79/31. 131 pp.

Crum, W.B. (1999) National Estuary Program in the Gulf of Mexico. In *The Gulf of Mexico Large Marine Ecosystem*, eds. H. Kumpf, K. Steidinger and K. Sherman. Blackwell Science, Malden, MA, pp. 667–671.

Culliton, T.J., Warren, M.A., Goodspeed, T.R., Remer, D.G., Blackwell, C.M. and McDonough III, J.J. (1990) Fifty years of population change along the Nation's coast 1960–2010. U.S. Dept. Commerce, Rockville, MD, 41 p.

Dame, R.F. and Patten, B.C. (1981) Analysis on energy flows in an intertidal oyster reef. *Marine Ecology Progress Series* **5**, 115–124.

Darnell, R.M. (1985) Gulf of Mexico continental shelf—an ecological overview. *Oceans '85 Proceedings*, Vol. 2, pp. 1124–1127.

Darnell, R.M. (1990) Mapping of the biological resources of the Continental Shelf. *American Zoologist* 30, 15–21.

Darnell, R.M. (1991) Biology of the estuaries and inner Continental shelf of the northern Gulf of Mexico. In *The Environmental and Economic Status of the Gulf of Mexico*, ed. G. Flock, pp. 161–172. Gulf of Mexico Program Office, Stennis, MS.

Darnell, R.M. and Defenbaugh, R.E. (1990) Gulf of Mexico: Environmental overview and history of Environmental Research. *American Zoologist* 30, 3–6.

Day, J.W., Jr., Pont, D., Hensel, P.F. and Ibanez, C. (1995) Impacts of sea-level rise on deltas in the Gulf of Mexico and the Mediterranean: the importance of pulsing events to sustainability. *Estuaries* 18, 636–647.

Defenbaugh, R.E. (1990) The Gulf of Mexico—a management perspective. *American Zoologist* 30, 7–13.

Duke, T. and Kruczynski, W.L. (1992) Status and trends of emergent and submerged vegetated habitats, Gulf of Mexico, U.S.A. U.S. Environmental Protection Agency, Stennis Space Center, MS, 161 p.

Ehler, C.N., Basta, D.J. and LaPoint, T.F. (1983) Analyzing the potential effects of operational discharges of oil from ships in the Gulf of Mexico. In *Proceedings of the 1983 Oil Spill Conference (Prevention, Behavior, Control, Cleanup), February 28–March 3, 1983, San Antonio, TX*. American Petroleum Institute, Washington, DC. pp. 323–330.

EPA (Environmental Protection Agency) (1993) Marine debris action agenda for the Gulf of Mexico. Office of Water, Gulf of Mexico Program. Stennis Space Center, MS. EPA 800-K-93-002, 151 pp.

EPA (Environmental Protection Agency) (1994a). Habitat degradation action agenda for the Gulf of Mexico. Office of Water, Gulf of Mexico Program. Stennis Space Center, MS. EPA 800-K-93-002, 140 p.

EPA (Environmental Protection Agency) (1994b) Nutrient enrichment action agenda for the Gulf of Mexico. Office of Water, Gulf of Mexico Program. Stennis Space Center, MS. EPA 800-G-94-004, 161 p.

Fonseca, M.S. (1992) Restoring seagrass systems in the United States. In *Restoring the Nation's marine environment*, ed. G.W. Thayer. Maryland Sea Grant, College Park, MD, pp. 79–110.

Fonseca, M.S., Kenworthy, W.J. and Thayer, G.W. (1992) Seagrass beds: nursery for coastal species. In *Stemming the Tide of Coastal Fish Habitat Loss*, ed. R.H. Stroud, National Coalition for Marine Conservation, Inc., Savannah, GA, pp. 141–147.

Franceschini, G.A. and El-Sayed, S.Z. (1968) Effect of Hurricane Inez (1968) on the hydrography and productivity of the western Gulf of Mexico. *Sonderdruck aus der Deutschen Hydrographischen Zeitschrift* 21, 193–202.

Galtsoff, P.S. (1964) The American oyster. *U.S. Fish Wildl. Serv., Fish. Bulletin* 64, 1–480.

Gore, R.H. (1992) *The Gulf of Mexico*. Pineapple Press, Inc., Sarasota, FL, 384 pp.

Gunter, G. (1963) The fertile fisheries crescent. *Journal of the Mississippi Academy of Science* 9, 286–290.

Haddad, K.D. and Joyce, Jr., E.A. (1997) Management. In *Ecology and Management of Tidal Marshes, A Model From The Gulf of Mexico*, eds. C.L. Coultas and Y. Hsieh. St. Lucie Press, Delray Beach, FL, pp. 309–329.

Hammerschmidt, P., Wagner, T. and Lewis, G. (1998) Status and trends in the Texas blue crab (*Callinectes sapidus*) fishery. *Journal of Shellfish Research* 17, 405–412.

Handley, L.R. (1995) Seagrass distribution in the Northern Gulf of Mexico. In *Our Living Resources: A Report to the Nation on the Distribution, Abundance, and the Health of U.S. Plants, Animals, and Ecosystems*, eds. E.T. LaRoe, G.S. Farris, C.E. Puckett, P.D. Doran and M.J. Mac. U.S. Dept. Int., Washington, DC, pp. 273–277.

Hansen, D.J. and Wilson, Jr., A.J. (1970) Residues in fish, wildlife and estuaries. *Pesticides Monitoring Journal* 4(2), 51–56.

Herke, W.H. (1995) Natural fisheries, marsh management, and mariculture: complexity and conflict in Louisiana. *Estuaries* 18(1A), 10–17.

Hoese, H.D. and Moore, R.H. (1998) *Fishes of the Gulf of Mexico*. Texas A&M University Press, College Station, TX, 422 pp.

Holmberg, N. (1988) Protection as a form of management for estuarine wetlands: The section 404 regulatory program and its impacts on estuarine wetlands. In *The Ecology and Management of Wetlands, Vol. 2: Management, Use and Value of Wetlands*, ed. D.D. Hook. Timber Press, Portland, OR, pp. 45–49.

Hopkins, J.S., Sandifer, P.A., DeVoe, M.R., Holland, A.F., Browdy, C.L. and Stokes, A.D. (1995) Environmental impacts of shrimp farming with special reference to the situation in the continental United States. *Estuaries* 18, 25–42.

Hoss, D.E. and Thayer, G.W. (1993) The importance of habitat to the early life history of estuarine-dependent fish. *American Fisheries Society Symposia* 14, 147–158.

Johnston, J.B., Field, D.W. and Reyer, A.J. (1992) Disappearing coastal wetlands. In *Restoring the Nation's Marine Environment*, ed. G.W. Thayer, pp. 53–58. Maryland Sea Grant, College Park, MD.

Johnston, J.B., Watzin, M.C., Barras, J.A., Handley, L.R. (1995) Gulf of Mexico coastal wetlands: case studies of loss trends. In *Our Living Resources: A Report to the Nation on the Distribution, Abundance, and the Health of U.S. Plants, Animals, and Ecosystems*, eds. E.T. LaRoe, G.S. Farris, C.E. Puckett, P.D. Doran and M.J. Mac. U.S. Dept. Int., Washington, DC, pp. 269–272.

Keeland, B.D., Allen, J.A. and Burkett, V.V. (1995) Southern forested wetlands. In *Our Living Resources: A Report to the Nation on the Distribution, Abundance, and the Health of U.S. Plants, Animals, and Ecosystems*, eds. E.T. LaRoe, G.S. Farris, C.E. Puckett, P.D. Doran and M.J. Mac. U.S. Dept. Int., Washington, DC, pp. 216–218.

Kennish, M.J. (1992) *Ecology of Estuaries: Anthropogenic Effects*. CRC Press, Boca Raton, FL, 494 p.

Kenworthy, W.J. (1992) Conservation and restoration of the seagrasses of the Gulf of Mexico through a better understanding of their minimum light requirements and factors controlling water transparency. In *Indicator Development: Seagrass Monitoring and Research in the Gulf of Mexico*, ed. H.A. Neckles. U.S. Environmental Protection Agency, Office of Research and Development, Environmental Research Laboratory, Gulf Breeze, FL, pp. 17–31.

Kenworthy, W.J. and Haunert, D.E. (eds.) (1991) The light requirements of seagrasses: proceedings of a workshop to examine the capability of water quality criteria, standards and monitoring programs to protect seagrasses. NOAA Tech. Memo. NMFS-SEFC-287, Beaufort, NC, 181 pp.

LaRoe, E.T., Farris, G.S., Puckett, C.E., Doran, P.D. and Mac, M.J. (eds.) (1995) *Our Living Resources: A Report to the Nation on the Distribution, Abundance, and the Health of U.S. Plants, Animals, and Ecosystems*. U.S. Dept. Int., Washington, DC, 530 pp.

Landin, M.C., Webb, J.W. and Knutson, P.L. (1989) Long-term monitoring of eleven Corps of Engineers habitat development field sites built of dredged material, 1974–1987. U.S. Corps Engin. Tech. Rep. D-89-1, Vicksburg, MS.

Lin, J. and Beal, J.L. (1995) Effects of mangrove marsh management on fish and decapod communities. *Bulletin of Marine Science* 57, 193–201.

Lindall, W.N. and Saloman, C.H. (1977) Alteration and destruction of estuaries affecting fishery resources of the Gulf of Mexico. *Marine*

Fisheries Review 39(9), 1–7

Leipper, D.F. (1954) Marine meteorology of the Gulf of Mexico, a brief review. In Gulf of Mexico—Its Origin, Waters, and Marine Life, ed. P.S. Galtsoff, *U.S. Fisheries Bulletin* **55**, 89–98.

Lynch, S.A. (1954) Geology of the Gulf of Mexico. In Gulf of Mexico—Its Origin, Waters, and Marine Life, ed. P.S. Galtsoff. *U.S. Fisheries Bulletin* **55**, 67–86.

Martin, N. (1986) Redfish Farmers. *Texas Shores* **19**(3), 13–15.

McDonald, I.R. (1998) Natural oil spills. *Scientific American* **279** (5), 56–61.

McEachron, L.W., Matlock, G.C., Bryan, C.E., Unger, P., Cody, T.J. and Martin, J.H. (1994) Winter mass mortality of animals in Texas Bays. *Northeast Gulf Science* **13** (2), 121–138.

McIvor, C.C., Ley, J.A. and Bjork, R.D. (1994) Changes in freshwater inflow from the Everglades to Florida Bay including effects on biota and biotic processes: a review. In *Everglades: The Ecosystem and Its Restoration*, eds. S.M. Davis and J.C. Ogden. St. Lucie Press, Delray Beach, FL, pp. 117–146.

McKinney, L. (1991) Managing America's Sea—The Gulf of Mexico. In *The Environmental and Economic Status of the Gulf of Mexico*, ed. G. Flock. Gulf of Mexico Program Office, Stennis, MS, pp. 173–184.

McLellan, H.J. (1963) Some features of the deep water in the Gulf of Mexico. *Journal of Marine Research* **21**, 233–245.

Meador, M.R. and Layher, A.O. (1998) Instream sand and gravel mining. *Fisheries* **23**(11), 6–13.

Meehan, W.R. (1991) Introduction and overview. In *Influences of Forest and Rangeland Management on Salmonid Fishes and Their Habitats*, ed. W.R. Meehan. American Fisheries Society Special Publication 19, Bethesda, MD, pp. 1–15.

Mitsch, W.J. and Gosselink, J.G. (1993) *Wetlands*, 2nd edn. Van Nostrand Reinhold, New York, 722 pp.

Mulholland, P.J., Best, G.R., Coutant, C.C., Hornberger, G.M., Meyer, J.L., Robinson, P.J., Stenberg, J.R., Turner, R.E., Vera-Herrera, F. and Wetzel, R.G. (1997) Effects of climate change on freshwater ecosystems of the south-eastern United States and the Gulf Coast of Mexico. *Hydrological Processes* **11**, 949–970.

NMFS (National Marine Fisheries Service) (1998) Managing the Nation's Bycatch. National Oceanic and Atmospheric Administration, Washington, D.C., 174 pp.

NOAA (National Oceanic and Atmospheric Administration) (1985) Gulf of Mexico, Coastal and Ocean Zones, Strategic Assessment: Data Atlas. National Ocean Service, Rockville, MD, 182 pp.

NOAA (National Oceanic and Atmospheric Administration) (1996) National Estuarine Research Reserve System, a tour of the reserves. Sanctuaries and Reserves Division, Silver Spring, MD, 60 pp.

NOAA (National Oceanic and Atmospheric Administration) (1997a) The 1995 national shellfish register of classified growing waters. Office of Ocean Resources Conservation and Assessment, Strategic Environmental Assessments Division, Silver Spring, MD, 398 pp.

NOAA (National Oceanic and Atmospheric Administration) (1997b) NOAA's estuarine eutrophication survey, Vol. 4: Gulf of Mexico Region. Office of Ocean Resources Conservation and Assessment, Silver Spring, MD, 77 pp.

NOAA (National Oceanic and Atmospheric Administration) (1998) (on-line). Ecological effects of fishing, by S.K. Brown, P.J. Auster, L. Lauck, and M. Coyne. NOAA's State of the Coast Report. Silver Spring, MD.

NOAA (National Oceanic and Atmospheric Administration) (1998b) National Status and Trends Program for Marine Environmental Quality. National Status and Trends Program, Silver Spring, MD, 32 pp.

NOAA (National Oceanic and Atmospheric Administration) (1999) (on-line). National Sanctuaries Programs: History. Silver Spring, MD.

O'Hara, K.J. (1992) Marine debris: taking out the trash. In *Stemming the Tide of Coastal Fish Habitat Loss*, ed. R.H. Stroud. National Coalition for Marine Conservation, Savannah, GA, pp. 81–90.

Orlando, S.P., Jr., Rozas, L.P., Ward, G.H. and Klein, C.J. (1993) Salinity Characteristics of Gulf of Mexico Estuaries. NOAA/NOS Office of Ocean Resources Conservation and Assessment, Silver Spring, MD, 209 pp.

Pattillo, M.E., Czapla, T.E., Nelson, D.M. and Monaco, M.E. (1997) Distribution and abundance of fishes and invertebrates in Gulf of Mexico estuaries, Volume II: Species life history summaries. ELMR Rep. No. 11. NOAA/NOS Strategic Environmental Assessments Division, Silver Spring, MD, 377 p.

Pequegnat, W.E., Gallaway, B.J. and Pequegnat, L.H. (1990) Aspects of the ecology of the deep-water fauna of the Gulf of Mexico. *American Zoologist* **30**, 45–64.

Platts, W.S. (1991) Livestock grazing, In *Influences of Forest and Rangeland Management on Salmonid Fishes and their Habitats*, ed. W.R. Meehan. American Fisheries Society Special Publication 19, Bethesda, MD, pp. 389–423.

Rabalais, N.N. (1990) Biological studies of the southwest Florida shelf. *American Zoologist* **30**, 77–87.

Rainey, G. (1991) The risk of oil spills from the transportation of petroleum in the Gulf of Mexico. In *The Environmental and Economic Status of the Gulf of Mexico*, ed. G. Flock, pp. 131–142. Gulf of Mexico Program, Stennis Space Center, Stennis, MS.

Rezak, R., Gittings, S.R. and Bright, T.J. (1990) Biotic assemblages and ecological controls on reefs and banks of the northwest Gulf of Mexico. *American Zoologist* **30**, 23–35.

Rozengurt, M.A. (1992) Alteration of freshwater inflows. In *Stemming the Tide of Coastal Fish Habitat Loss*, ed. R.H. Stroud. National Coalition for Marine Conservation, Inc., Savannah, GA, pp. 73–80.

Ruffner, J.A. (1980) *Climates of the States*, 2nd edn. Vols. I and II. Gale Research Co., Detroit, MI, 1175 pp.

Schmitt, C.J. and Bunck, C.M. (1995) Persistent environmental contaminants in fish and wildlife. In *Our Living Resources: A Report to the Nation on the Distribution, Abundance, and the Health of U.S. Plants, Animals, and Ecosystems*, eds. E.T. LaRoe, G.S. Farris, C.E. Puckett, P.D. Doran and M.J. Mac. U.S. Dept. Int., Washington, DC, pp. 413–416.

Schroeder, W.W., Cowan, J.L.W., Pennock, J.R., Luker, S.A. and Wiseman Jr., W.J., (1998) Response of resource excavations in Mobile Bay, Alabama, to extreme forcing. *Estuaries* **21**, 652–657.

Shaw, S.P. and Fredine, C.G. (1956) Wetlands of the United States: their extent and value to waterfowl and other wildlife. U.S. Fish Wildlife Service, Circular No. 39. USGPO, Washington, DC, 67 pp.

Sklar, F.H. and Browder, J.A. (1998) Coastal environmental impacts brought about by alterations to freshwater flow in the Gulf of Mexico. *Environmental Management* **22** (4), 547–562.

Smith, G.B. (1976) Ecology and distribution of eastern Gulf of Mexico reef fishes. Fla. Mar. Research Publ. 19, Fla. Dept. Nat. Resources, Mar. Research Lab., St. Petersburg, FL, 78 pp.

Smith, R.L. (1980) *Ecology and Field Biology*. Harper and Row, New York. 835 pp.

Smith-Vaniz, W.F., Bohnsack, J.A. and Williams, J.D. (1995) Reef fishes of the Florida Keys. In *Our Living Resources*, pp. 279–284.

Stanley, J.G. and Sellers, M.A. (1986) Species profile: life histories and environmental requirements of coastal fishes and invertebrates (Gulf of Mexico)—American oyster. U.S. Fish Wildl. Serv. Biol. Rep. 82 (11.64), 25 p.

Taylor, D.D. and Bright, T.J. (1974) Preliminary reports of heavy metals in organisms of the West Flower Garden Bank. In *Biota of the West Flower Garden Bank*, eds. T.J. Bright and L.H. Pequegnat, pp. 58–63. Gulf Publishing Co., Houston, TX.

Tomczak, M. and Godfrey, J.S. (1994) *Regional Oceanography: An Introduction*. Pergamon, Oxford and New York, 422 pp.

Wade, D., Ewel, J. and Hofstetter, R. (1980) Fire in south Florida ecosystems. U.S. Dept. Agric. For. Serv., Gen. Tech. Rep. SE-17, Southeast. For. Exp. Stn., Asheville, NC, 125 p.

Ward, Jr., G.H. 1980. Hydrography and circulation processes of Gulf Estuaries. In *Estuarine and Wetland Processes*, eds. P. Hamilton and K.B. MacDonald. pp. 183–213. Plenum Publishing Corp, NY.

Wiseman, W.J. and Sturges, W. (1999) Physical oceanography of the Gulf of Mexico: processes that regulate its biology. In *The Gulf of Mexico Large Marine Ecosystem*, eds. H. Kumpf, K. Steidinger and K. Sherman. Blackwell Science, Malden, MA, pp. 77–92.

Zieman, J.C. and Zieman, R.T. (1989) The ecology of the seagrass meadows of the west coast of Florida; a community profile. U.S. Fish Wildl. Serv., Washington, D.C., Biol. Rep. 85 (7.25), 155 pp.

Zimmerman, R.J. and Minello, T.J. (1993) Watershed effects on the value of marshes to fisheries. In *Coastal Zone '93, Vol. I*, eds. O.T. Magoon, W.S. Wilson, H. Converse and L.T. Tobin, pp. 538–547. American Society of Civil Engineers Press, New York.

THE AUTHORS

Mark E. Pattillo
U.S. Army Corps of Engineers, Galveston District,
P.O. Box 1229,
Galveston, TX 77551-1229, U.S.A.

David M. Nelson
National Ocean Service,
1305 East-West Highway,
Silver Spring, MD 20910, U.S.A

Chapter 28

COASTAL MANAGEMENT IN LATIN AMERICA

Alejandro Yáñez-Arancibia

At the beginning of the new millennium Latin America has several new arrangements with international bodies, as well as a number of bi- or trilateral international agreements. These, with continuing economic and environmental globalization, are leading to different strategies for evaluating problems and for integrated coastal planning and management, both from national and regional perspectives.

Latin America is a mosaic of differing training and experiences, of differing cultural roots and resources, different social development, different ecosystems and ecological approaches. It covers several biogeographical regions with varied biodiversity and climatic zones, and has many pristine areas as well as highly degraded zones. This chapter describes the integrated coastal zone management of Latin America, using examples from several countries, and illustrates several of the coastal management techniques which are being used.

Variations in scope and progress made by different Latin American countries show the difficulty of creating a single integrated management approach. Nevertheless, at the opening of the new millennium, Latin America has achieved a stage of relative maturity in terms of integrated coastal zone management.

INTRODUCTION

More than ten years ago Sorensen and Brandani (1987), noted five criteria for comparing 19 coastal Latin American countries in terms of management of coastal resources and environments. This was a broad effort, but any comparison in Latin America at present only risks producing generalities which do not reflect any particular pattern and remains far from providing solutions for particular and country-specific problems. Nevertheless, these authors have rightly shown several benefits, including improvement of information transfer among nations, providing new entrants with the experiences of predecessors, assistance in defining the scope of projects, and helping aid agencies to set priorities as well as aiding the design of programs. However, new international forums and the economic and environmental globalization require different strategies for evaluating problems and for developing an integrated coastal management. At the opening of the new millennium, Latin America has several new arrangements under NAFTA, MERCOSUR, APEC, CARICOM as well as a number of bi- or trilateral international agreements.

It is timely to focus on the Latin American countries' responses to the requirement for integrated ocean and coastal management (Cicin-Sain, 1993a; Yáñez-Arancibia, 1999a). The purpose of this chapter is thus to provide some definitions and points of view on integrated coastal management and to provide an assessment of key issues and country experiences. Its purpose is not to resolve or reconcile divergent perspectives, but it is clear that there appears to be basic agreement on major concepts, even though there have been somewhat different approaches (Yáñez-Arancibia, 1999b).

DEFINITIONS

The Coastal Zone

For management purposes, the boundaries of a coastal area should be defined by the extent of biophysical, economic, and other social interactions. Islands or small nations such as the Lesser Antilles, or the Caribbean in general, can be viewed entirely as a 'coastal zone'. In most other countries, size puts a practical limit on the extent of a manageable area. In these cases, there are rarely clearly defined physical boundaries, in which case prominent physical landmarks or other physical criteria, political or administrative boundaries and selected environmental units are often used (Scura et al., 1992). In numerous Latin American countries, especially in academic institutions, the coastal zone is defined as a broad space of interactions between the sea, the land, the fresh-water drainage, and the atmosphere (Yáñez-Arancibia, 1986).

A more functional definition (Windevoxhel et al., 1999) has a flexibility which permits considerations of biological, biophysical, social, and economic aspects. In this, the coastal zone is understood to be the geographic space in which the principal interchanges of material and energy occur between the marine and terrestrial ecosystems.

Functional criteria must be established, so for the purpose of this chapter the coastal zone is taken to mean a broad eco-region with intense physical and biological interactions, where dynamic interchanges occur of energy and materials between land, fresh water, atmosphere, and adjacent sea. Typically, such areas include wetlands, coastal plains, low-river basins, mangroves, coastal dunes, coastal lagoons, estuaries and adjacent ocean (Yáñez-Arancibia, 1986; Lemay, 1997; Windevoxhel et al., 1999).

In practice, several countries have enacted legislation which sets aside a much narrower strip, usually between 20 and 200 m of land preceding the shoreline (or mean high tide) as public or under state jurisdiction (Lemay, 1997). For example, Mexico has a terrestrial–maritime zone of 20 m, Uruguay has one of the broadest legally defined coastal zones in Latin America at 250 m, Ecuador 8 m and the mangrove greenbelt, Brazil has 33 m, Colombia 50 m, Costa Rica has a 50 m public zone and a 0–200 m restricted zone, Venezuela 50 m and Chile 80 m. The delimitation of maritime boundaries is an important element of State sovereignty.

Integrated Coastal Zone Management

To some, the concept of coastal zone management represents either a panacea for every excess of the private sector or governmental agencies, or the solution to every unsolved coastal problem. In reality, a successful programme is based on a comprehensive and integrated planning process, which aims at harmonizing cultural, economic, and environmental values and balancing environmental protection and economic development with a minimum of regulation (CEP/UNEP, 1995). Management without an appropriate planning process tends to be neither integrated nor comprehensive, but rather a sectorial activity.

There is an emerging consensus, after UNCED (1992), that 'integrated coastal area management' (ICAM), is replacing the old 'integrated coastal zone management', but the two are used interchangeably. There is less consensus about adding the word "marine" to pair off with "coastal" or the word "planning" to pair off with "management" (CEP/UNEP, 1995; Chua, 1993). In practice, and here, the marine area is subsumed within the term coastal, and further assumes planning as a function in management. There are no shortages of definitions of any of these terms (Sorensen, 1993; Awosika et al., 1993).

Integrated coastal management (ICM) is, of course, a dynamic process by which decisions are taken for the use, development and protection of coastal areas and resources, to achieve goals established in cooperation with user groups and authorities. ICM recognizes the distinctive character of the coastal zone, is multiple-purpose-oriented, analyzes implications of development, conflicting uses, and interrelationships between physical processes and human activities, and promotes linkages and harmonization between sectoral, coastal and ocean activities (Knecht and Archer, 1993).

There are at least seven different kinds of integration, each of which has its own limits (Knecht and Archer, 1993; CEP/UNEP, 1995): (a) intergovernmental, (b) land–water interface, (c) intersectorial, (d) interdisciplinary, (e) institutional, (f) temporal, (g) managerial. It is to be noted that the inclusion of NGOs and the local public in both the planning and the management process is of vital importance, and Sorensen (1993) is correct when he says the essential ingredients are: (a) a coastal system perspective, and (b) a multisectoral approach.

The evolution of a 'modern' ICM has matured over recent decades (Vallega, 1992; Yáñez-Arancibia, 1999b). Rapid technological advances, and the growing consciousness of the importance of coastal management for national policies have had important implications. In the 1970s the objectives changed, so that environmental protection and preservation became objectives to be pursued, together with the exploitation and use of marine resources. In the 1980s ICM evolved toward multiple-use, requiring resolution of conflicts; then in the 1990s it evolved towards exploiting resources, protecting critical habitats as well as minimising conflicts between uses.

Sustainable Development

Whole books and extensive papers have been devoted to defining this term (Cicin-Sain, 1993a; Smith, 1997). The term is, in fact, so vague that it has been used not only by advocates of precaution to refer to environmental sustainability of economic activity but also by advocates of growth who refer to the sustainability of economic expansion.

In 1987 the World Commission on Environment and Development (WCED), attempted to define environmentally sustainable development, and phrases which have passed into common parlance are 'Sustainable development is development that meets the needs of the current generation without compromising the needs of future generations' (WCED, 1987).

Unfortunately, it is impossible to know what the preference of future generations will be. Logically, precaution would dictate the preservation of the natural environment in its unaltered state, and we thus arrive at the so-called "strong" definition which is: 'Sustainability is the economic development that does not compromise environmental integrity' (Smith, 1997).

Of course, the social component must be included, a concern which has been present for years. But it is only recently that sustainable development has assumed prominence as an important concept and philosophy to guide economic development and environmental management. Consequently, there is still some confusion, even among environmental professionals. For some, there is a tendency to view sustainable development in rather narrow terms, such as "new environmental technologies", or "population stabilization" (Muschetts, 1997). And yet, if countries are to adopt sustainable development as a central organizing principle, as they must to obtain life support systems, secure a healthy environment and promote widespread property, a multifaceted approach is necessary.

The real value of a comprehensive conception of sustainability is its ability to shed light on how to make the best use of all available opportunities. Making sustainability operational is really a matter of predicting and measuring it, far more than just defining it (Costanza and Patten, 1995). At the UNCED 92, Principle 3 characterised sustainable development as: 'The right to development must be fulfilled so as to equitably meet developmental and environmental needs of present and future generations' (UNCED, 1992). UNCED 92 Principle 4 further states 'in order to achieve sustainable development, environmental protection shall constitute an integral part of the development process and cannot be considered in isolation from it'.

These two principles, stated as part of Agenda 21, have profound implications for use and stewardship of natural resources, ecology and environment. Because of misunderstandings, there is a tendency in official gatherings and agency programs to focus upon areas of consensus and very specific missions (Muschetts, 1997). Controversies may be swept aside, hoping that continued economic growth and new technology will solve problems of poverty and environment. Table 1 includes some of the more fundamental, root causes, as well as economic, environmental and technology factors which are more frequently mentioned in sustainable approaches (see Muschett, 1997).

Thus many different definitions of "sustainable development" abound. Achieving it entails a continuous process of decision-making, so that there is never an end-state of sustainability, since the balance between development and environmental protection must constantly be readjusted. Further, sustainable development is a process of change in which the exploitation of resources, the direction of investments, orientation of technological development, and institutional change are made consistent with future as well as present needs" (Cicin-Sain, 1993b). This also underlines the concept of needs, especially those of the Third World,

Table 1

Elements of sustainable development (from Muschett, 1997)

- Population stabilization
- New technologies/technology transfer
- Efficient use of natural resources
- Waste reduction and pollution prevention
- Integrated environmental systems management
- Determining environmental limits
- Refining market economy
- Education
- Changes in perception and attitude
- Social and cultural changes

Table 2
Major coastal resource management issues in Central America. 3 = Priority, 2 = significant; 1 = minor, localised (from Foer and Olsen, 1992)

	Belize	Guatemala	Honduras	El Salvador	Nicaragua	Costa Rica	Panama
Degraded water quality in estuaries/lagoons	2	3	3	3	1	3	3
Losses in estuarine-dependent fisheries	1	3	3	3	2	3	3
Destruction of mangroves	1	3	3	3	1	3	3
Poor shorefront development practices	2	2	3	3	1	3	2
Degradation of scenic/cultural resources	1	2	3	2	2	3	2
Ocean storms and/or severe flooding	3	1	3	2	3	2	1
Destruction of coastal wetlands	1	2	2	3	1	2	2
Dams on major rivers	1	2	2	2	1	2	2

Table 3
Institutional issues in coastal resources management in Central America. 3 = Priority; 2 = Significant; 1 = Minor or localised (from Foer and Olsen, 1992)

	Belize	Guatemala	Honduras	El Salvador	Nicaragua	Costa Rica	Panama
Inadequate implementation of existing regulations	3	3	3	3	3	3	3
Lack of trained personnel	3	3	3	3	3	2	3
Overlapping jurisdictions/interagency conflict	2	3	3	3	2	3	3
Inadequate public support for existing management initiatives	2	3	2	2	3	2	3

and the idea of limitations on the environment's ability to meet them. Before the sustainable development concept, there was growing realization that the world is facing a series of environmental crises, some global in nature, which threaten the future viability of life on earth, and some more local, which threaten the attainment of both development and quality of life.

The Latin America and Caribbean Focus

Just like any large area, Latin America is a mosaic of different heritages, social development, ecosystems, biodiversity and climatic zones. It is a broad region, including both temperate and tropical latitudes, rainforests in the north to fjords in the South, including polar cold and coastal deserts. Moreover, it contains one of the highest reserves of fresh water, petroleum, minerals, forests, fisheries, fertile soils, fruits, cattle, food supplies, and natural or wild landscapes and seascapes. These all increase the challenge of ICM in this region.

For a number of reasons, not discussed here, Latin America has been generally poorly represented in ICM. For instance, Cicin-Sain (1993a) notes only one paper from Latin America, despite the fact that 10 Latin American countries have developing ICM projects (Sorensen, 1993), many addressing urgent problems (Table 2). In the light of this, it is appropriate to look for a general strategy towards financing regional ICM programmes (Lemay, 1996, 1997; Cicin-Sain and Knecht, 1998).

Meso America, the bridge between the two American hemispheres, has 6603 km of coastline which is 12% of the coastline of both Latin America and the Caribbean. This coast harbours 567,000 ha of mangroves (8% of the extent of the world's mangroves) which accounts for 7% of the region's forest cover. It also contains 1600 km of coral reef, especially in Belize and Mexico. The continental shelf, with a total area of 237,650 km^2, has rich fisheries, and the region has more than 1.1 million km^2 of exclusive economic zone (Windevoxhel, 1997). Meso America has many river basins, from Chetumal Bay in Mexico to Panama Bay, and at least 107 coastal wetlands of international importance. There is, fortunately, a growing level of awareness and concern about the deteriorating conditions in Central America's coastal resources (Foer and Olsen, 1992) (Table 3).

Perhaps the most important critical coastal habitats are the mangroves. The ratio of total mangrove area to total surface area and coastline length of each country (Table 4) shows that mangroves are the most important forest in several countries and thus should be given priority in management and conservation. Recent data confirm the relative distribution of mangroves of Pacific and Atlantic coasts of Latin America and the Caribbean (Table 5); only in the southern countries are mangroves not present.

The coastal and insular countries of the Caribbean are among the most heavily exploited because of their landscape attractions and resources. All constitute the Wider Caribbean Region CEP/UNEP 1995. The number of people in the region living near the sea is increasing, along with the size and densities of coastal cities. Throughout the entire region, coastal areas support major industrial complexes, trade centres and tourist resorts. Competition for space

Table 4

Recent estimates of mangrove cover and the respective percentage of total country's area and length of the coastline in Latin America and Caribbean (extracted from Lacerda, 1993)

	Area (ha)	% Country surface
Continental countries		
USA	190,000	0.02
México	524,600	0.27
Belize	73,000	3.10
Guatemala	16,040	0.15
Nicaragua	60,000	0.50
Honduras	121,340	1.08
Costa Rica	41,330	0.08
El Salvador	35,235	1.65
Panamá	171,000	2.22
Colombia	358,000	0.31
Ecuador	161,770	0.60
Peru	4,791	0.01
Venezuela	250,000	0.27
Guiana Francesa	5,500	0.06
Guyana	150,000	0.70
Suriname	115,000	0.70
Brazil	1,012,376	0.12
Insular countries		
Trinidad & Tobago	7,150	1.40
Jamaica	10,624	1.02
Cuba	529,700	4.80
Haiti	18,000	0.65
Republica Dominicana	9,000	0.20
Puerto Rico	6,500	0.71
Bahamas	141,957	10.18
Bermuda	20	<0.01
Guadeloupe	8,000	4.49
Martinique	1,900	1.73
Cayman Islands	7,268	27.60
Antilles*	24,571	–

*Includes only the islands from where reliable mangrove surveys have been reported (Anguilla, Antigua, Aruba, Barbados, Barbuda, Bonaire, Curacao, Dominica, Grenada & Grenadines, Montserrat, Nevis, St. Kitts, St. Lucia, St. Vincent, Turks & Caicos).

Table 5

Mangrove forest cover in the Atlantic and Pacific coasts of Latin America, including the Caribbean islands, compared to world mangrove forest areas (from Lacerda 1993)

	Mangrove area (ha)	% of the total
Atlantic Coast	2,143.356	52.8
Pacific Coast	1,154.289	28.5
Caribbean Islands	764.690	18.7
Total	4,062.335	(100)
Latin America, Caribbean	4,062.335	28.6
Africa	3,257.700	22.9
Southeast Asia	6,877.600	48.5
World total	14,197.635	(100)

Table 6

Coastal crises: common problems and solutions (from CEP/UNEP, 1995)

Issues and problems	Effective actions
Depletion of inshore commercial and recreational fisheries from overfishing	Sustainable fisheries management for long-term productivity; possible limited entry
Degradation of coastal habitats	Improved management of biodiversity
Damage of coastal areas from uncontrolled development	Control of coastal development through planning, zoning and permitting procedures
Beach damage from sand mining and vegetation removal for resort development	Management of tourism for minimal erosion and reduced environmental impact
Water pollution from oil, sewage, urban runoff and sediments	Improved control of watershed effluents and urban waste disposal practices
harbour and estuarine pollution, congestion, and siltation	"Special Area" management planning, including use of EIA process, for harbours, wetlands, estuaries, industrial sites, and urban areas
Nutrient pollution of coastal waters from agricultural fertilisers and pesticides	Development of non-point-source pollution control programme
Loss of coastal wetlands and estuarine habitats	Establishment of no-net-loss policy for wetlands
Loss of scenic landscapes and seascapes, and historic areas	Landscape management and easement strategies to protect scenic coastlines and historic sites

along continental shorelines in the western and more southerly reaches of the region, is almost as severe.

As a consequence, pollution along the more densely settled and heavily used segments of these coastlines has become pervasive, as discharges have arisen from both terrestrial and marine sources, including cruise ships and the oil industry. Table 6 summarises common problems in the Wider Caribbean. The high productivity (from higher nutrient levels from rivers, estuaries and local upwellings) of these coastal marine ecosystems is also threatened by the lack of an adequate management policy and enforcement framework in which to manage resources for ecologically sustainable development (UNEP, 1995). Habitat loss and

Table 7

Wider Caribbean coastline activity in selected states (from CEP/UNEP, 1995). x = not available, a = goods loaded, b = goods unloaded

	Length of marine coast (km)	Maritime area (x 1000 km²)		Urban population in large coastal cities (1000s)		Average annual volume of goods loaded and unloaded 1988-90 (1000 metric tons)			Annual petroleum production		Proven oil and gas reserves	
		Shelf to 200 m depth	Exclusive Economic Zone			Petroleum	Crude products	Dry cargo	Oil (x1000 metric tons) 1992	Gas (million cubic m) 1992	Oil (million metric tons) 1992	Gas (billion cubic m) 1992
				1980	2000							
Caribbean												
Antigua/Barbuda	153	x	x	x	x	0	61b	83	0	0	0	0
Bahamas	3542	85.7	759.2	x	x	10524	3702	3222	0	0	0	0
Barbados	97	0.3	167.3	100	146	107b	51b	573	0	0	0	0
Cayman Islands	160	x	x	x	x	1357	36b	117				
Cuba	3735	x	362.8	6628	8942	5850b	3821	14244	0	0	0	0
Dominica	148	x	20.0	x	x	x	5b	93	0	0	0	0
Dominican Rep.	1288	18.2	268.8	2787	5797	1630b	785b	4358	0	0	0	0
Grenada	121	x	27.0	x	x	x	22b	71	0	0	0	0
Guadeloupe	306	x	x	142	196	x	370b	1221				
Haiti	1771	10.6	160.5	1216	2845	x	11b	838	0	0	0	0
Jamaica	1022	40.1	297.6	1016	1689	1210b	1203	10122	0	0	0	0
Mártinique	290	2.4	x	217	279	231b	282	876				
Trinidad & Tobago	362	29.2	76.8	623	1110	6518	2670	5638	6922	5799	78	261
Central America												
Belize	386	x	x	x	x	0	106b	306	0	0	0	0
Costa Rica	1290	15.8	258.9	1050	225 8	464b	336b	2662	0	0	0	0
Guatemala	400	12.3	99.1	780	932	683	204	4232	0	0	0	0
Honduras	820	53.5	200.9	583	192 3	397b	204	1849	0	0	0	0
México	9330	442.1	2851.2	6529	950 1	71817a	7377	19833	85656	11370	5712	1926
Nicaragua	910	72.7	159.8	1166	283 7	495b	183b	1280	0	0	0	0
Panama	2490	57.3	306.5	989	174 9	1192b	441	1939	0	0	192.7	7
South America												
Colombia	2414	67.9	603.2	2926	392 6	9442	6901	15231	0	0	10	40
French Guiana	378	x	x	x			137b	273	0	0	0	0
Guyana	459	50.1	130.3	213	425	x	474b	1919	0	0	0	0
Suriname	386	x	101.2	140	216	x	615b	6185	0	0	0	0
Venezuela	2800	88.1	363.8	5158	932 4	58367a	23564	26768	427 28	537 5	966	765

environment degradation is the primary threat to the region's marine productivity. The real challenge lies in the improvement of coordinated support to planning, monitoring, managing, and restoring the coastal marine ecosystem in the Wider Caribbean Region (CEP/UNEP, 1995; Griffith and Ashe, 1993).

In general terms, coastal resources are often common property resources with open free access to all users. Free access often leads to excessive use, the "tragedy of the commons" (Bernal et al., 1999). The absence of exclusive-use rights is the source of both biological and economic waste and conflict. Management intervention is generally necessary to achieve and maintain desired levels of maximum sustained yield, as well as desired levels of quality of coastal resources. Coastal management programmes can prevent the loss of natural resources throughout development regulations, proper monitoring and enforcement. Most of the common problems in the coastal crises in Latin America, and suggested direction for solutions, are summarised in Table 7.

Activities that add further value to coastal resources in Latin America include recreation and tourism, which has now become a major source of domestic and foreign exchange. The intrinsic economic value of coastal resources represents a capital investment for humankind by nature. The goods and services derived from them may be viewed as the interest generated by the investment. Hence, destruction of the resource base means depletion of both capital and interest. The economic value of resources is a function of ecosystem health, and this principle is an important term of reference for management in Latin America (Yáñez-Arancibia, 1999b).

REGIONAL EXAMPLES

A recent collection of articles (Yáñez-Arancibia, 1999a) contains elements of the real framework for coastal management in Latin America. These can be summarised as follows.

One of the first contributions in Chile towards effective coastal zone management (Paskoff and Manriquez, 1999) examines the coastal fringe of central Chile from an ecosystem level. It integrates the legal framework for coastal occupation, and gives recommendations for harmonious development. The area is being affected by increasing demographic pressure, mainly related to a boom of seaside tourism within the framework of a growing market economy. In Chile, a large industrial fishery development has turned the country into the third fishing nation in the world. Bernal et al. (1999) review recent regulatory innovations which, for the first time, include new management tools that allocate resources in the form of Individual Fishing Quotas (IFQs) and Individual Transferable Quotas (ITQs). At the same time, Territorial Users Rights in Fisheries (TURFs) have been incorporated into law to enhance self-regulatory practices among artisanal fishermen.

Also in Chile, Alvial and Reculé (1999) give an analytical point of view showing that flexibility must be the key approach in management models because of the diverse and variable Chilean coastal zone and its fishery heterogeneity, oceanographic characteristics, sustained population and increase of both infrastructure and pollution. A coastal zone management plan for Chile, more than creating new regulations—many of which already exist—needs to set up an effective mechanism among different users, and generate mechanisms to resolve differences, provide adequate scientific and technical information, and establish simple mechanisms for education and citizen participation.

The wealth of experience obtained by Argentina and Uruguay is utilised through the reciprocal cooperation framework provided by the Treaty of the Rio de la Plata and its Maritime Front (González Lapeyre, 1999). This is a long-term successful management tool for binational fishery management in one of the world's major deltaic systems. Martínez and Fournier (1999) present essential elements which describe the genesis, development, results and status of the EcoPlata instrument, a Uruguayan binational, multi-institutional approach to integrated coastal zone management. Emphasis is made on planning, institutional scientific research, institutional arrangements and participants, leading to integrated management.

The main factors responsible for the degradation of the coastal areas in Brazil and their impact on coastal human population is described by Diegues (1999), who also analyses the main policies of the Brazilian government concerning the management of coastal areas. Shortcomings of the methodologies are explained, showing that for the solution of existing ecological and social problems, the establishment of marine protected areas is important.

Many problems identified along the Brazilian coast can be solved through adequate human education (Reis et al., 1999). A description of how these problems are identified, and the skills, knowledge and attitude of personnel needed to adequately perform their tasks are identified and presented; the authors feel that appropriate education through international training courses is an essential component of ICM.

The main economic activities of the "Special Management Zone" Bahia de Caraquez, include agriculture, aquaculture, fishing and tourism (Arriaga et al., 1999). These are managed under the framework of the Coastal Resources Management Project (CRMP) established in a cooperative program with the Coastal Resources Center of the University of Rhode Island, USA. In this case, binational cooperation had success in a number of initiatives in one of the most important mangrove areas in the world.

In Colombia, a project has been implemented by the government to rehabilitate the Ciénaga Grande de Santa Marta, a key mangrove ecosystem on the Caribbean coast (Botero and Salzwedel, 1999). This multi-institutional and bi-national initiative focused on the management of hydrological resources, of faunal and floral resources, the role of social development, and institutional strengthening.

Windevoxhel et al. (1999) analyse Central American coastal zones including those of Panamá, Costa Rica, Nicaragua, Honduras, Salvador, Belize and Guatemala, focusing on ecosystem health, landscapes and seascapes, and biodiversity. In this region, tourism is significant, but integrated coastal management has been limited by information gaps, restricted technical and financial capacity, and strong sectoralism. Population density, tourism, fishing, aquaculture and agriculture are analysed at the regional level, and the role of natural protected coastal areas as a management tool is found to be important here too. The participation of numerous international foundations is extensive in this region, many projects depending on international cooperation and inter-agency coordination for their success. Progress and experiences indicate a potential initiative towards the "Alliance for Sustainable Development" in Central America.

Yáñez-Arancibia et al. (1999a) describe the Atlantic coast of Guatemala. Here the focus is on an ecosystem framework for planning and management, which includes the most important ecological processes, economic development, environmental problems, and guidelines for its management. The approach permitted analysis of tourism, hunting and fishing, agriculture and livestock farming, urbanization, oil-related activities, ports and means of communication, and finally the integration of these activities to provide management recommendations. It was clear that an integrated management plan for the Atlantic coast of Guatemala must be grounded in a development strategy based on scientific knowledge, integrating the ecosystem with economics.

Further, the environment and its problems in the coastal zone of Campeche has been recently subjected to an analysis which took in commercial and artisanal fishing, maritime transport, agriculture and cattle grazing in lowland areas, urban expansion, building of highways, and tourism (Yáñez-Arancibia et al., 1999b). After developing seven study cases, a management approach was developed considering four main actions: promotion of institutional arrangements; strengthening of public awareness related to coastal resources management policies and capabilities; gathering, analysis and dissemination of information related to coastal resources development; and provision of technical solutions to coastal resource uses in conflict. Science played a significant role in the politics of the policy process, both in protecting key estuarine ecosystems and in the planning process (see Yáñez-Arancibia and Day, 1988; Yáñez-Arancibia et al., 1998).

The Mexican coast along the Gulf of Mexico and the Caribbean has several ecological and socioeconomic problems and suffers environmental impacts (Zárate Lomelí et al., 1999). Here the legal, institutional, and technical framework that is applied to the coastal management of the region is leading towards the definition and implementation of a Program of Integrated Management (PIM). In this, adjustments to the main instruments of national environmental policy are proposed, to be applied to the coastal zone in order to achieve sustainable development and the improvement in conditions of the communities.

CONCLUSIONS

The variation in scope and the amount of progress made by different Latin American countries show the difficulties experienced in some of creating integrated management approaches to ecosystems and resources. It also indicates the effort still required for sustainable development and resource utilization.

Management of all natural resources usually focuses on three goals: (a) overcoming conflicts associated with sectoral management; (b) preserving the productivity and biological diversity of coastal ecosystems; and (c) promoting an equitable and sustainable allocation of coastal resources. This is exactly the case, too, in Latin America (Lemay, 1997; Post and Lunding, 1996). While the objectives of any ICM program are specific to the coastal problems in a particular area, an equitable, transparent process of governance is also required. Coastal management relies on a variety of techniques to achieve its specific objectives, in the sense of Lemay (1997). Following these approaches, the 13 examples summarised above can be arranged within a scheme of 'Coastal Management Techniques' which are: (a) coastal management plans, (b) land-use zoning and setbacks, (c) marine protected areas, (d) management and restoration of coastal habitats, (e) coastal pollution control, (f) shoreline

Table 8

Some Latin America initiatives on coastal zone management

Chile	*Data Base System* supported by Fundación Chile and the Navy's Maritime Territory National Board.
Uruguay/ Argentina	*Binational Technical Commission of the Rio de la Plata and its Maritime Front.*
Uruguay	*EcoPlata Project,* supported by the Universidad de la República, Faculty of Science, Program of Marine Science and Atmosphere, and the International Development Research Centre (IDRC-Canada).
Brazil	*Núcleo de Apoio á Pesquisa sobre Populacóes Humanas e Areas Umidas Brasileiras UPAUB,* ex-Program of Research and Conservation of Wetlands in Brazil PPCAUB, University of Sao Paulo, supported by IUCN and Ford Foundation.
Brazil	*Train-Sea-Coast Programme,* supported by the United Nations Division for Ocean Affairs and the Law of the Sea DOALOS/UN, at the University of Rio Grande Foundation.
Brazil	*National Coastal Management Program.*
Ecuador	*Coastal Resources Management Program,* initially supported by an Agreement signed between the Government of Ecuador and the United States Agency for International Development USAID. Ministry of Mines and Energy and the Natural Resources Ecuador, and the University of Rhode Island CRC, USA.
Colombia	*Environmental Management Plan Programme,* developed for Technical Cooperation (GTZ).
Central America (Globally)	*IUCN's Mesoamerican Wetlands and Coastal Zone Conservation Programme,* through the Regional Direction Office IUCN/ORMA, Moravia Costa Rica; and some support from CATIE, TNT, WWF, DANIDA, NORAD, SRI, Royal Embassy of The Netherlands, The European Community, Organization of American States.
Costa Rica	*National Coastal and Marine Program.*
Barbados	*Coastal Conservation Program.*
Guatemala	*Integrated Tourism Management Plan for the Atlantic Coast,* supported by Guatemala Institute of Tourism (INGUAT), and UNDP Caribbean Environment Programme.
Guatemala/ Honduras	*Bilateral Plan for Border Development* (BIFINO), partially supported by the Organization of American States, and the United Nations Development Programme UNDP.
Mexico	*Programme of Ecology, Fisheries and Oceanography of the Gulf of Mexico EPOMEX,* started in July 1990 and finished in January 1997, supported by the Secretary of Public Education SEP, and Grants from the Organization of American States, United Nations Development Programme UNDP/CEP, the IUCN Central America, the WWF, the National Council for Science and Technology CONACYT, and mainly the government of the State of Campeche.

stabilization, (g) close access regimes, (h) capacity building, (i) inter-agency coordination, (j) community-based management, (k) conflict resolutions, (l) environmental assessment, and m) international cooperation (see Yánez-Arancibia 1999 b).

This classification is also useful in understanding the different stages of evolution and approach of any country. Most items in this table are the products of present or past programmes in each country, several of which have now been in existence for sufficient time to have made significant inroads into public policy (Table 8).

Other minor initiatives exist (UNEP, 1995). While it is difficult to isolate all the factors explaining the continuity of these programs, one key factor is obviously that there are clearly defined coastal problems. These programs have demonstrated an ability to evolve from an initial, rather restricted focus towards a more integrated and participatory approach. Analysing the perspective of some of these programs, Lemay (1997) pointed out that in every country the status of coastal ecosystems is only now starting to be documented, and reliable long-term data on resources or coastal water quality are almost non-existent. In general terms, coastal management in Latin America is mainly a collection of projects or isolated papers which may or may not support economic development priorities in the coastal zone.

Experiences connected with sustainable development in Latin America show that traditional sectoral approaches have not been effective in maintaining the productive value of coastal areas. While the role of public sector institutions in maintaining resources has gone through major shifts, responsibilities are not well articulated nor have incentives been introduced to ensure that private sector intervention addresses sustainability. The situation is now changing, new terms of reference are available, and international scenarios are encouraging sustainability. The old paradigms are staggering along, but at the opening of the new millennium, adoption of new ones means that Latin America will mature in terms of Integrated Area Management.

ACKNOWLEDGMENTS

To Dr. Charles Sheppard from University of Warwick, Coventry UK, for inviting me to produce this contribution, and for waiting for it with patience. Dr. John W. Day Jr from the Coastal Ecology Institute at Louisiana State University USA, and Dr. Michele H. Lemay from Interamerican Development Bank, Department of Sustainable Development, Division of Environment, Washington, both read the manuscript and made valuable suggestions.

REFERENCES

Alvial, A. and Reculé, D. (1999) Fundación Chile and the integrated management of the coastal zone. *Ocean and Coastal Management* **42** (2–4), 143–154.

Arriaga, L., Montaño, M. and Vásconez, J. (1999) Integrated management perspectives of the Bahía de Caráquez zone and Chone River estuary, Ecuador. *Ocean and Coastal Management* **42** (2–4), 229–242.

Awosika, L., Baromthanarat, S., Comforth, R., Mendry, M., Koudstall, R., Ridgley, M., Sorenson, S., De Vrees, L. and Westmacolt, P.S. (1993) Management arrangements for the development and implementation of coastal zone management programmes. World Coast Conference Organizing Committee. *International Conference on Coastal Zone Management. The Netherlands, 1–5 November.*

Bernal, P.A., Oliva, D., Aliaga, B. and Morales, C. (1999) New regulations in chilean fisheries and aquaculture: ITQs and territorial user rights. *Ocean and Coastal Management* **42** (2–4), 119–142.

Botero, L. and Salzwedel, H. (1999) Rehabilitation of the Cienaga Grande de Santa Marta, a mangrove-estuarine system in the Caribbean coast of Colombia. *Ocean and Coastal Management* **42** (2–4), 243–256.

CEP/UNEP (1995) Guidelines for integrated planning and management of coastal and marine areas in the wider caribbean region. UNEP (OCA)/CAR W.G. 17/3. Kingston Jamaica, 28–30 June 1995, 141 pp.

Cicin-Sain, B. (ed.) (1993a) Integrated coastal management. Special issue. *Ocean and Coastal Management* **21** (1–3), 377 pp.

Cicin-Sain, B. (1993b) Sustainable development and integrated coastal management. *Ocean and Coastal Management* **21** (1–3), 11–43.

Cicin-Sain, B. and Knecht, R. (1998) *Integrated Coastal and Ocean Management: Concepts and Practices.* IOC-UNESCO Publishing, Island Press Washington DC, 518 pp.

Costanza, R. and Patten, B.C. (1995) Defining and predicting sustainability. *Ecological Economics*, **15**, 193–196.

Chua, T.E. (1993) Essential elements of integrated coastal zone management. *Ocean and Coastal Management* **21** (1–3), 81–108.

Diegues, A.C. (1999) Human populations and coastal wetlands: conservation and management in Brazil. *Ocean and Coastal Management* **42** (2–4), 187–210.

Foer, G. and Olsen, S. (1992) *Central America's Coast Profile and an Agenda for Action.* University of Rhode Island CRC, Narragansett, USA, 294 pp.

Gónzalez Lepeyre, E. (1999) The Maritime Front of the Río de la Plata as an instrument for binational fisheries management. *Ocean and Coastal Management* **42** (2–4), 155–164.

Griffith, M.D. and Ashe, J. (1993) Sustainable development of coastal and marine areas in small island developing states: a basin for integrated coastal management. *Ocean and Coastal Management* **21** (1–3), 269–284.

Knecht, R. and Archer, J. (1993) "Integration" in the US coastal zone management programme. *Ocean and Coastal Management* **21** (1–3) 183–199.

Lacerda, D. (ed.) (1993) *Conservation and sustainable utilization of mangrove forest in Latin America and Africa regions. Part I. Latin America.* ISME-ITTO, Japan, 2, 272 pp.

Lemay, M.H. (1996) Financing integrated coastal management in Latin America and the Caribbean: directions for a strategy at Inter-American Development Bank. *Coastal and Marine Workshop IUCN World Conservation Congress*, October, 13 pp.

Lemay, M.H. (1997) Coastal and marine resources management in Latin America and the Caribbean: a strategy background paper. Inter-American Development Bank Document, Washington, D.C. 56 pp.

Martínez C.M. and Fournier, R. (1999) EcoPlata: an Uruguayan multi-institutional approach to integrated coastal zone management. *Ocean and Coastal Management* **42** (2–4), 165–186.

Muschett, F.D. (ed.) (1997) *Principles of Sustainable Development.* St. Lucie Press, Debray Beach, FL, 176 pp.

Naves, C. (ed.) (1989) *Coastlines of Brazil. Proceedings Coastal Zone'89 Symposium on Coastal and Ocean Management, Charleston South*

Carolina, 11–14 July 1989. American Society of Civil Engineers, New York, 296 pp.

Paskoff, R. and Manriquez, H. (1999) Ecosystem and legal framework for coastal management in Central Chile. *Ocean and Coastal Management* **42** (2–4), 105–118.

Post, J. and Lunding, C. (1996) *Guidelines for Integrated Coastal Zone Management*. Environmentally Sustainable Development Studies and Monograph Series No. 9. The World Bank, Washington, D.C.

Reis, E.G., Asmus, M.L., Castello, P.J. and Calliari, L.J. (1999) Building human capacity on coastal and ocean management—implementing the Train-Sea-Coast Programme in Brazil. *Ocean and Coastal Management* **42** (2–4), 211–228.

Scura, L.F., Chua, T.E., Pido, M.D. and Paw, J.N. (1992) Lessons from integrated coastal zone management: the Asian experience. In: Chua, T.E. and Scura, L.F. (ed). Integrative Framework and Methods for Coastal Area Management. *ICLARM Conference Proceedings* **37**, 1–68.

Smith, F. (1997) *Environmental Sustainability, Practical Global Implication*. St. Lucie Press, Boca Raton, FL, 287 pp.

Sorensen, J. and Brandani, A. (1987) An overview of coastal management in Latin America. *Coastal Management* **15**, 1–25.

Sorensen, J. (1993) The international proliferation of integrated coastal zone management efforts. *Ocean and Coastal Management* **21** (1–3), 45–80.

UNCED (1992) United Nations Conference on Environment and Development Agenda 21 Chapter 17, Rio de Janeiro, Brazil.

UNEP (1993) Common guidelines and criteria to the wider Caribbean region for the identification, selection, establishment, and management of protected areas of national interest. UNEP (OCA)/CAR W.G. 11/6 Second Meeting ISTAC-SPAW. French Guiana 3–5 May.

UNEP (1995) Workshop on Integrated Planning and Management of Coastal Areas in the Wider Caribbean. UNEP (OCA)/CAR W.G. 17/4. Kingston Jamaica, 28–30 June, 13 pp. Annex I–IV.

Vallega, A. (1992) *Sea Management: A Theoretical Approach*. Elsevier, London, 259 pp.

WCED (1987) Our Common Future. World Commission on Environment and Development. Oxford University Press, Oxford, UK, 400 pp.

Windevoxhel, N.J. (1997) Wetland and coastal zone conservation in Mexo-America. IUCN Wetlands Programme, Newsletter 15, pp. 20–2.

Windevoxhel, N.J., Rodríguez, J.J. and Lahmann, E.J. (1999) Situation of integrated coastal zone management in Central America: experiences of the IUCN wetlands and coastal zone conservation program. *Ocean and Coastal Management* (Special Issue), **42** (2–4), 257–282.

Yáñez-Arancibia, A. (1986) Ecología de la Zona Costera: Análisis de siete Tópicos. AGT Editorial, México D.F. 186 pp. 2nd ed. Reviewed 2000 300 pp (in press).

Yáñez-Arancibia, A. (ed.) (1999a) Integrated Coastal Management in Latin America, Special Issue. *Ocean and Coastal Management* 42 (2–4) 77–368.

Yáñez-Arancibia, A. (ed.) (1999b) Terms of reference towards coastal management and sustainable development in Latin America: introduction to Special Issue on progress and experiences. *Ocean and Coastal Management* 42 (2–4), 77–104

Yáñez-Arancibia, A. and Day, J.W. (eds). (1988) Ecology of Coastal Ecosystems in the Southern Gulf of Mexico: the Terminos Lagoon Region. Inst. Cienc. del Mar y Limnol. UNAM. Coastal Ecology Institute LSU, Organization of American States, Washington, DC, Editorial Universitaria México D.F., 518 pp.

Yáñez-Arancibia, A. and Lara-Domínguez, A.L. (eds.) (1999) Mangrove Ecosystems in Tropical America Instituto de Ecología A.C. Xalapa Unión Mundial de la Naturaleza UICN Costa Rica, NOAA/NMFS, USA, 380 pp.

Yáñez-Arancibia, A., Zarate Lomelí, D. and Terán, A. (1995) Evaluation of the coastal and marine resources of the Atlantic coast of Guatemala. UNEP Caribbean Environment Programme, Kingston, Jamaica, 34, 1–64.

Yáñez-Arancibia, A., Rojas Galaviz, J.L., Zárate Lomelí, D., Lara-Domínguez, A.L., Villalobos, G.J., Rivera, E. and Sánchez-Gil, P. (1998) El Ecosistema de Petenes en Campeche, Península de Yucatán, y Terminos de Referencia para su Manejo Integrado en la Zona Costera, México: Instituto de Ecología A.C. Xalapa, 280 pp. (submitted).

Yáñez-Arancibia, A., Zárate Lomelí, D., Gómez Cruz, M., Godinez Orantes, R. and Santiago Fandiño, V. (1999a) The ecosystem framework for planning and management the Atlantic coast of Guatemala. *Ocean and Coastal Management* **42** (2–4), 283–318.

Yáñez-Arancibia, A., Lara-Domínguez, A.L., Rojas Galaviz, J.L., Zárate Lomelí, D., Villalobos Zapata, G.J. and Sánchez-Gil, P. (1999b) Integrating science and management on coastal marine protected areas in the Southern Gulf of Mexico. *Ocean and Coastal Management* **42** (2–4), 319–344.

Yáñez-Arancibia, A., Lara-Domínguez, A.L., Rojas Galaviz, J.L., Zárate Lomelí, D., Villalobos Zapata, G.J. and Sánchez-Gil, P. (1999c) Integrated coastal zone management plan for Terminos Lagoon, Campeche, Mexico, pp. 565–592. In *The Gulf of Mexico Large Marine Ecosystem, Assessment, Sustainability, and Management*, eds. H. Kumpf, K. Steidinger and K. Sherman. Blackwell Science, Inc. Massachusetts, 710 pp.

Zárate Lomelí, D., Saavedra Vázquez, T., Rojas Galavíz, J.L., Yáñez-Arancibia, A. and Rivera Arriaga, E. (1999) Terms of reference towards an integrated management policy in the coastal zone of the Gulf of Mexico and the Caribbean. *Ocean and Coastal Management* **42** (2–4), 345–368.

THE AUTHOR

Alejandro Yáñez-Arancibia
Department of Coastal Resources,
Institute of Ecology A.C.,
Km 2.5 Antigua Carretera Coatepec, P.O. Box 63,
Xalapa 91000, Veracruz, México

Chapter 29

SOUTHERN GULF OF MEXICO

Felipe Vázquez, Ricardo Rangel, Arturo Mendoza Quintero-Marmol, Jorge Fernández, Eduardo Aguayo, A. Palacio and Virender K. Sharma

The southern, Mexican, coast of the Gulf of Mexico supports fishing, extensive cattle ranching, crop production, and industrial and oil production activities. Along the Mexican coast there are several different ecological regions. There are semi-arid, dry and humid coastal plains and hills (north west plain of the Yucatan Peninsula), large and mostly brackish lagoons (Madre, Tamiahua, Alvarado, Terminos) and several rivers (Grande, Panuco, Coatzacoalcos, Grijalva–Usumacinta). Along this long coastline there are extreme wet and dry climates. In September and October hurricanes occur, several in the past being very damaging. North winds blow from September to March when the temperature decreases to about 6°C, and this carries rain to most of the Gulf coast.

There are extensive mangrove areas and a few areas of offshore coral reefs in the Gulf (Lobos in Tamaulipas State; Isla Verde in Veracruz State; Cayo Arcas in Campeche Sound). Wetland and marsh areas are also extensive, especially around the coastal lagoons and estuaries, while extensive seagrass beds occur off the coast. For many groups of organisms, including mammals and birds, the region has a high biodiversity. However, development pressures have resulted in several problems. The activities of the extensive livestock industry which requires large areas of pasture has resulted in extensive deforestation along the Gulf coast with consequent problems of run-off. Rivers carry sediment, metals and pesticides into the Gulf waters, and over-fishing is evident for several species and in several areas of this coast. The addition of fecal coliforms to rivers, lagoons and seawater is also a great problem. New government initiatives are being developed which aim to tackle the problems of sustainable development in a region which has inadequate financial resources.

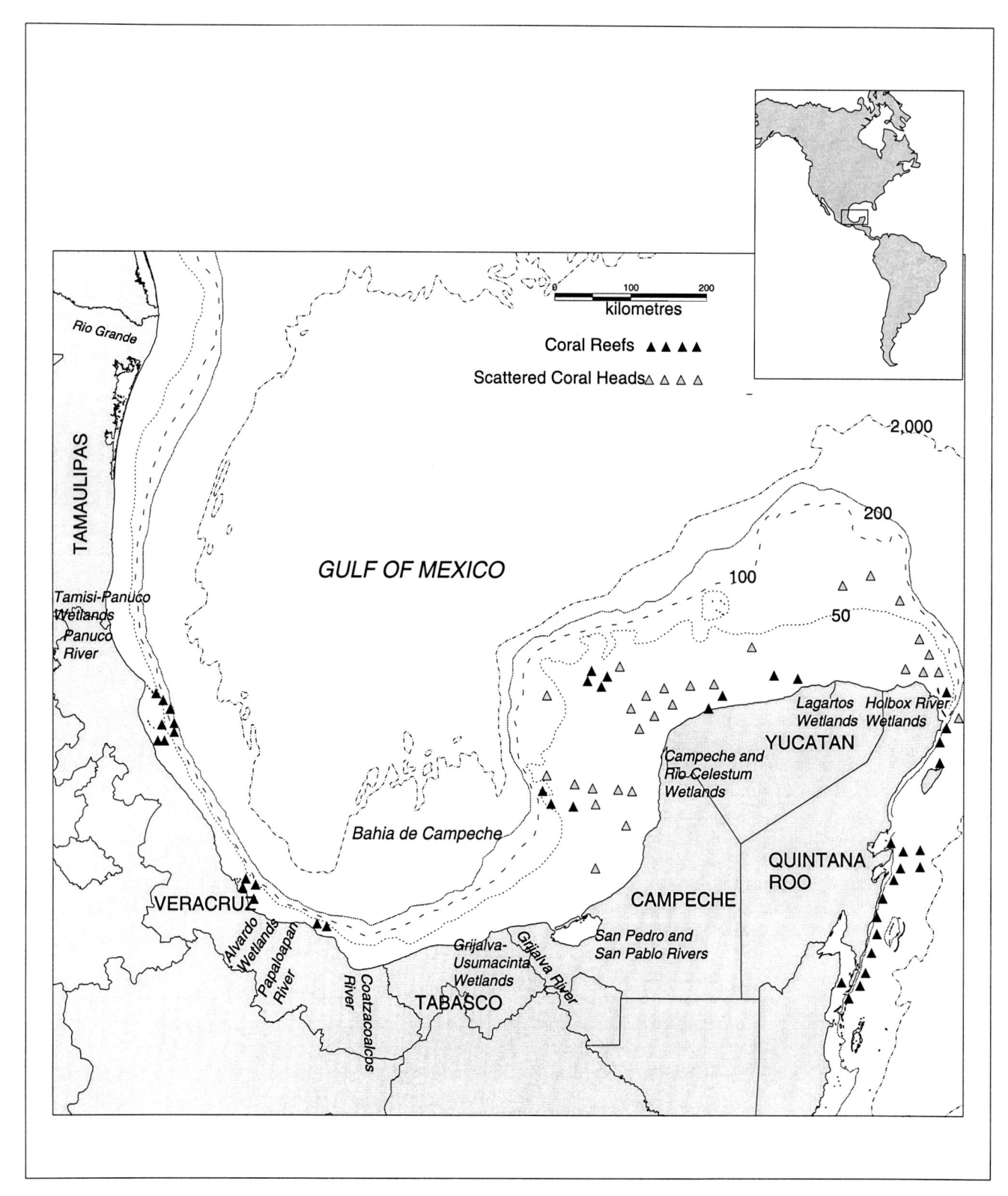

Fig. 1. Map of the Mexican coast of the Gulf of Mexico.

THE DEFINED REGION

The Gulf of Mexico is a basin partly separated from the adjacent Caribbean Sea (Fig. 1) with a horizontal area of 1600 million km^2. Together with the Caribbean Sea it forms the "American Mediterranean" and is frequently referred to as the Mexican Basin. It is deep, with a depth of 3688 m in the central area (Fig. 2). Bottom topography in the eastern Gulf of Mexico is dominated by a broad shallow shelf extending north of the Yucatan Peninsula, the Campeche Sound, while the western Gulf of Mexico is dominated by a narrow, shallow shelf from Tabasco to Tamaulipas states. Steep escarpments along the Mexican coast mark the continental slope.

Coral reefs, wetlands, mangroves and seagrass are important marine ecosystems in the Gulf. Diversity of some systems, reefs for example, may be less off Tamaulipas and Veracruz States when compared to the coral reefs of the Caribbean Sea, due both to natural conditions of marine currents, winds and suspended solids from rivers, and to human activities such as tourism, industrial, oil production and input of wastewater. Mangroves have a high diversity along the coast, and associated with these are several species of many invertebrate groups, notably crustaceans, molluscs, annelids and echinoderms as well as fishes.

ENVIRONMENTAL FRAMEWORK

There are two dominant semi-permanent circulation features in the Gulf of Mexico (Elliot, 1982): the intense Loop Current system in the eastern Gulf and an anticyclonic cell of circulation along the western boundary (Nowlin and McLellan, 1967). In the winter and early spring, wind stress maintains an area of cyclonic circulation in the Bay of Campeche. To the north of this cyclone is an anticyclone which is fed by those moving into the western Gulf of Mexico after they are pinched off from the Loop Current (Merrell and Morrison, 1981). Figure 3 (Fernández et al., 1993), illustrates the water masses of the Gulf. The interaction of these masses with the winds produces the climate of the southern area of the Gulf of Mexico. The principal seasons are a dry season (March–June), a rainy season (July–September) and a season of north winds (September–March).

The surface salinity in Campeche Sound maintains values of 36.4 to 36.6 ppt (Nowlin, 1972). In coastal areas during the rainy season it falls to 30 ppt, and at these times the plume of the Grijalva–Usumasinta and Coatzacoalcos rivers can extend seawards almost 90 km. In Campeche Sound, temperature and salinity have a great influence on the distribution of the marine ecosystems.

Dissolved oxygen in Campeche Sound throughout the year maintains values close to saturation level. pH values average 8.2 ± 0.1 in most surface waters (PEMEX-UNAM, 1998, 1999) and the upwelling in the northern part of the Peninsula of Yucatan (Secretaria de Marina, 1980) creates conditions which are rich with nutrients and beneficial for fishing. Off Punta Zapotitlan, Veracruz is located adjacent to another upwelling of lower intensity (PEMEX-UNAM, 1998; 1999).

Suspended solids from the Jamapa River, close to the coral reef off the Port of Veracruz, appears to be harmless to coral growth. At the present time, however, the area covered by this coral reef is decreasing. The increase of the suspended solids in the Jamapa River comes from erosion of soils provoked by destruction of the rainforest in the highlands of Veracruz State. Other rivers are similarly affected: during the rainy season in the Grijalva–Usumacinta system on the Tabasco coast, there is also a high discharge of suspended solids (West et al., 1985). The suspended solids, which are accompanied by high levels of nutrients, enter an area where there is extensive fishing, especially for shrimp (INP, 1994).

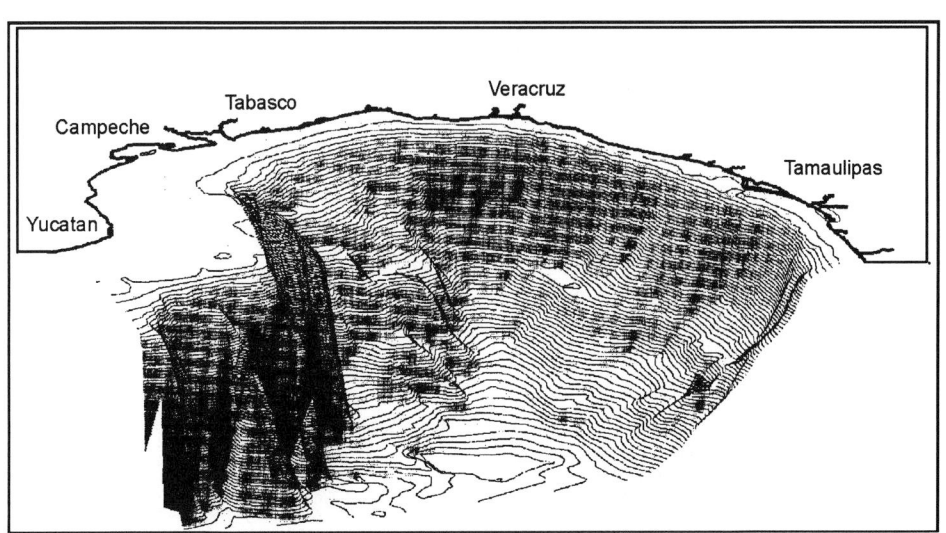

Fig. 2. Bathymetry of the Gulf of Mexico (Z value × 100, from SICORI, PEMEX).

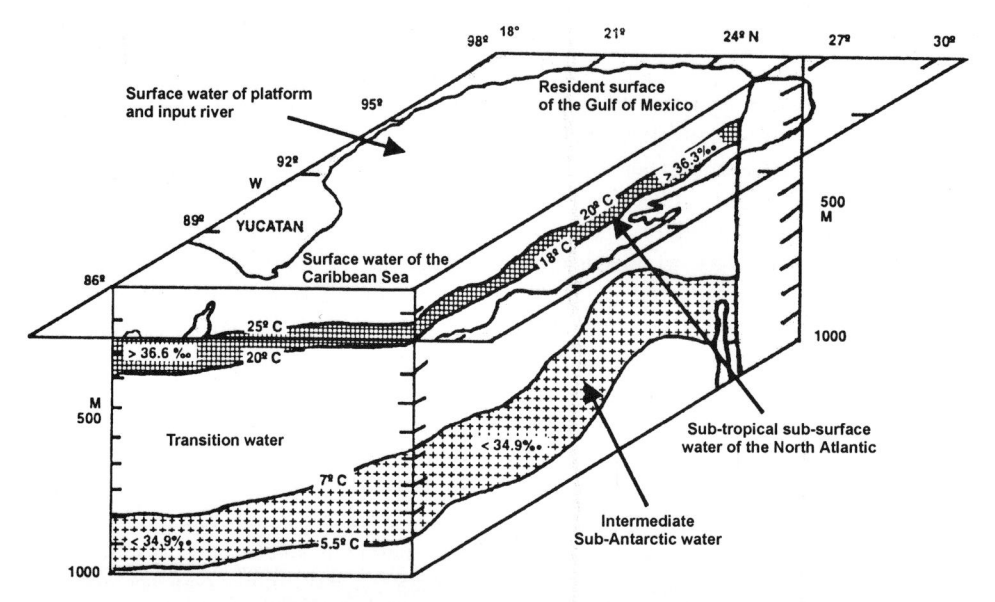

Fig. 3. Principal water masses of the Gulf of Mexico.

Fig. 4. Major recent hurricanes in the Gulf of Mexico: (A) Gilberto (09/88); (B) Roxanne (11/95); (C) Opal (11/95).

Table 1
Principal characteristics of the coral reefs in the Gulf of Mexico

Coral reef	Form	Major axis (m)	Width (m)	Name of emerged island	Distance from coast (km)
Blanquilla	Platform	730	500	SW	5
Medio	Platform	–	–	–	7.5
Lobos	Platform	630	300	Isla Lobos	11.25
Tuxpan	Platform	–	–	–	11.25
Anegada de Adentro	Platform	1870	500	–	7.5
La Blanquilla	Platform	875	625	El Peyote	4.22
Isla Verde	Platform	1120	125	Isla Verde	5.37
Isla Sacrificios	Platform	1000	500	Isla Sacrificios	1.42
Anegada de Afuera	Platform	4370	1250	–	16.25
Santiaguillo	Platform	375	250	Cayo, South	19.75
Anegadilla	Platform	625	125	–	20.5
Isla de En Medio	Platform	2250	1800	Isla de En medio	6.25
Blanca	Platform	875	500	–	2.62
Chopas	Platform	5000	1620	Isla Salmedina	3.25
Alacrán	Atoll	24670	13000	Islas Pérez, Chica, Pájaros, Desertora and Desterrada	130
Cayo Arenas	Platform	972	729	Cayo to the southeast	166.68
Triangulos Oeste	Platform	656	300	Cayo to the south	193.32
Triangulos Este	Platform	2370	547	Cayo to the southeast	187.05
Triangulos Sur	Platform	2000	1000	–	188.90
Cayo Arcas	Platform	2630	1540	Cayos Centro, Este and Oeste	127.68

In the South of the Gulf, hurricanes have an important effect, notably on coral reefs (Veracruz, Cayo Arcas; PEMEX-UNAM, 1998), on wetlands and on erosion (Veracruz, Tabasco and Campeche). Hurricanes Gilberto, Opal and Roxanne (Fig. 4) are amongst the most important meteorological disturbances in the last twelve years.

MAJOR SHALLOW WATER MARINE AND COASTAL HABITATS

The Gulf of Mexico is basically an area of terrigenous sedimentation, but despite this, reef formation is well developed in three areas (Carricart-Ganivet and Horta-Puga, 1993) (Fig. 5). The first is from Veracruz north to near Cabo Rojo, off Tamiahua lagoon, and near the outlet of the river Tuxpan. Further south is the Veracruz reef system which is divided in two groups by the river Jamapa–Atoyac; and thirdly there are reefs of the Bank of Campeche (Table 1). Abundant coral species include the substantial reef builders *Montastrea cavernosa*, *M. annularis*, *Diploria strigosa*, *D. clivosa*, *Acropora palmata*, *Siderastrea radians*, *Porites astreoides*, *P. porites*, *Colpophyllia natans* and *Agaricia agaricites*. The main impacts which these reefs of the southern Gulf of Mexico suffer are classified in Table 2.

There are several protected areas with important species whose status is threatened (Fig. 5). The national park "Arrecife Alacranes" is located in Yucatan. Twenty-four species of stony corals have been reported here, which grow in a windward barrier in the central lagoon and on the leeward or western coast. The barrier is constructed by *Acropora palmata* down to 10 m deep, while deeper there are associations of *Diploria*, *Montastrea*, *Porites* and *Agaricia*. Marine bird density is very high, especially of *Larus atricilla*, *Anus stolidus*, *Frigate magnificens*, *Sula dactylatra* and *S. leucogaster*. Reptiles include a small lizard (*Mabuya mabuya*) and three turtles: the green (*Chelonia mydas*), lute (*Coriaceous dermochelys*) and hawksbill (*Eretmochelys imbricata*) all of which are seen around the islands and nest on the beaches.

In Yucatan peninsula is Celestún creek, with species including monkey, leopard (*Leopardus wiedii*), ocelot (*L.*

Table 2
Impacts on some reef in the Gulf of Mexico (Chávez and Hidalgo, 1988)

Reef	Intensity						
	Point	Diffuse	Occasional	Chronic	Low	Medium	High
Florida Keys	*			*		*	
Flower Garden			*		*		
Blanquilla	*	*	*	*	*	*	
Lobos	*	*	*	*	*	*	
Isla Verde		*		*		*	*
Isla de En Medio	*	*		*		*	*
Arcas	*	*	*	*		*	*
Alacranes	*			*		*	

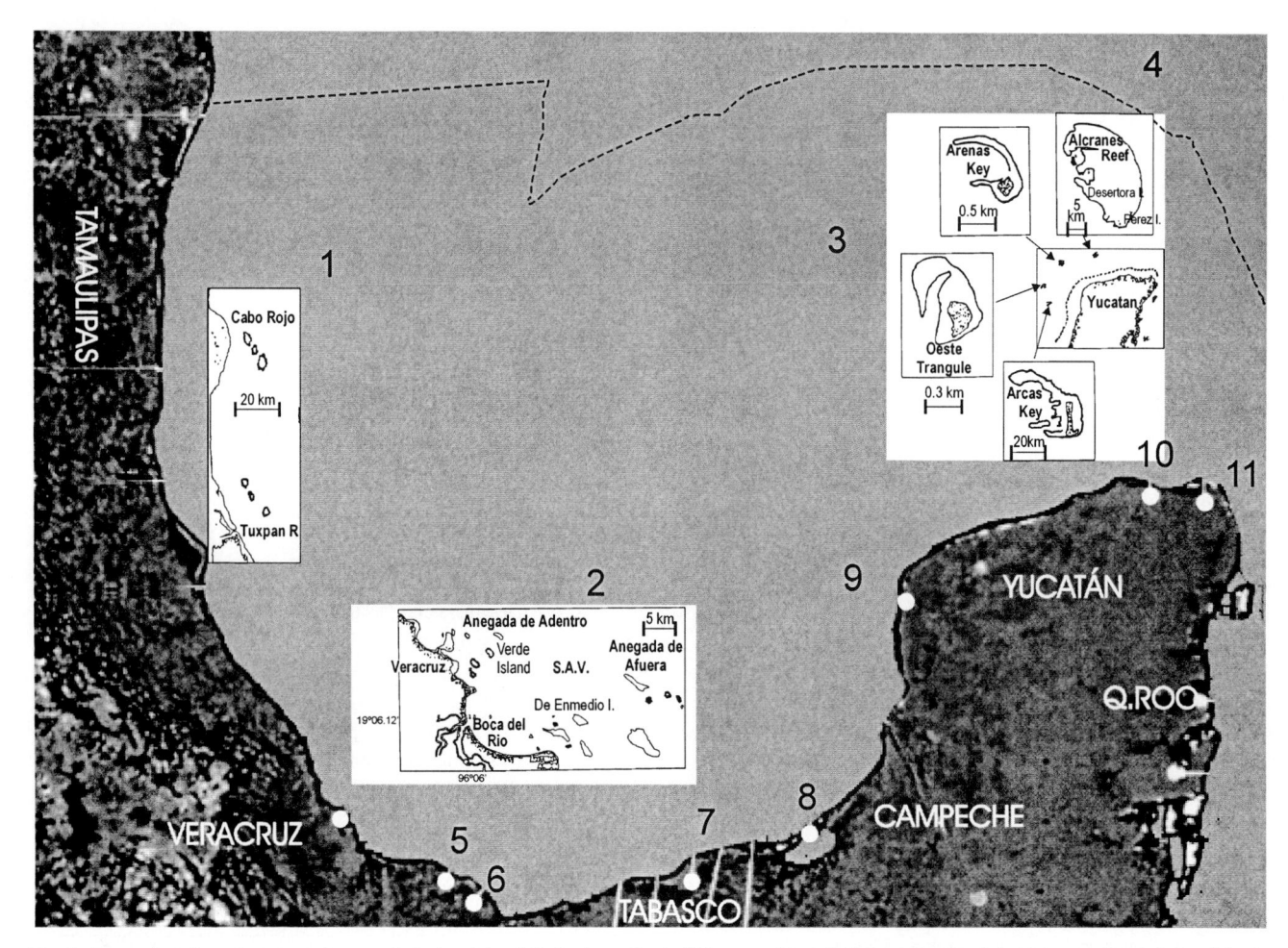

Fig. 5. Areas of reefs and protected zones. 1: Lobos Island. 2: Systems Reef of Veracruz Port. 3: Triangules Reef. 4: Alacranes Reef. 5: San Martin Volcano. 6: Santa Martha Mountain Range. 7: Centla Marsh. 8: Terminos Lagoon. 9: Celestum Ria. 10: Lagartos Ria. 11: Tum Blam.

Pardalis), and jaguar (*Panthera onca*). The area also contains flamingo (*Phoenicopterus ruber roseus*) and harbours several endemic species such as the rattle yucateca (*Campylorhynchus yucatanicus*) and hummingbird earwig (*Doricha eliza*); species of restricted distribution here include the quail, cotuí and American stork. It includes nesting sites for the leatherback turtle (*Coriaceous eretmochelys*), the swamp crocodile (*Crocodylus morelettiy*) and the boa (*Constrictor boa*).

The protected area of Terminos lagoon is also located in Campeche. Fauna which may be at risk include the stork jabiru, manatee, crocodile, badger, raccoon, ocelot, jaguar and marine turtles. In this area a large part of the fauna is Caribbean sub-tropical, representing the northern limit for several South American species. The area is also enriched by the presence of several Nearctic species that have emigrated from North America.

The state of Tabasco contains the swamps of Centla. There are few faunal inventories for this Reservation, and almost all species listings are inferred from similar regions of the Southeast or high watershed of the Grijalva–Usumacinta region. There are reported to be at least 60 species of fish, 85 of reptiles, 26 of amphibians, 103 of mammals and 264 species of birds in this important site.

In the state of Veracruz, the fauna and flora are also rich. Here, 102 species of mammals have been described, 49 species of amphibians, 109 reptiles, 561 birds, 437 species of fish in the coasts of the Sierra of Tuxtlas and 359 species of Lepidoptera.

Starting in 1960, the swamps of the coastal region have engendered great interest, not only because they support a high diversity of wild life (Ramamoorthy et al., 1993), but also because the swamps themselves contribute large quantities of nutrients to the marine system. Mexico has enormous swamp systems on its coasts, comparable with those of Brazil and Australia. This habitat is distributed from the coast of Belize in the Caribbean Sea to the Tamiagua lagoon in the Gulf of Mexico. The forests of the swamp of Terminos Lagoon and the plain of Tabasco are particularly striking, as are the systems of the Alvarado, Tampamachoco and Tamiahua lagoons in Veracruz (Menéndez, 1976; Rico-Gray and Lot, 1983). Trees in the swamps are bushy and do not exceed 5 m in height. The Madre lagoon, Tamaulipas, represents the limit of distribution in the Gulf of red mangrove and *L. racemosa*, while *A. germinans* continues on toward the south of Texas and Louisiana (Lot et al., 1975; Sherrod et al., 1986).

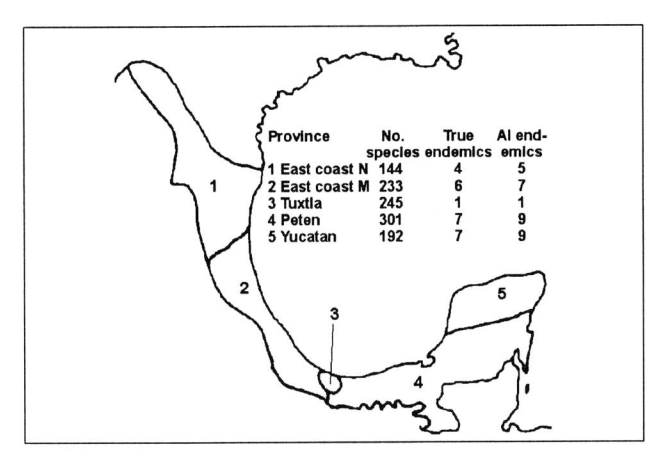

Fig. 6. Avian provinces, diversity and endemism in the coast of the Gulf of Mexico (Ramamoorthy et al., 1993).

Province	No. species	True endemics	Al endemics
1 East coast N	144	4	5
2 East coast M	233	6	7
3 Tuxtla	245	1	1
4 Peten	301	7	9
5 Yucatan	192	7	9

In the Gulf of Mexico there are remarkable forests of red mangrove in the lagoons of Terminos and El Vapor, Atasta, the San Pedro River and the outlet of the Grijalva river. Here these mangrove trees can reach 30 m in height. In Terminos lagoon, trees of *A. germinans* also reach 15 m (Jardel et al., 1987).

Molluscs

There are numerous studies on the molluscs of the Gulf of Mexico lagoons (García-Cubas, 1963, 1968, 1988; Reguero, 1990; García-Cubas et al., 1990) especially on those of economic importance, such as clams (*Rangia cuneata*) and oysters (*Crassostrea virginica*). Oil spills have led to numerous compensation claims, but in many cases, because there have been insufficient baseline data on water quality, compensation to fishermen may not reflect the true value of the resulting damage.

Fishes

As many as 41 orders (82% of the world's total) and 206 families (46.3% of the world's total) are represented in Mexican waters (Cohen, 1970; Nelson, 1984). Different species dominate in different parts, but the main families of fish in the estuarine and lagoon ecosystems of the Gulf of Mexico are the Carcharhinidae, Dasyatidae, Clupeidae, Engraulidae, Characiniidae, Ariidae, Atherinidae, Cichlidae, Mugilidae, Gobiidae, Bothidae, Soleidae, Lutjanidae and Poecillidae (Yañez, 1986; Resendez and Kobelkowsky, 1991). According to Fuentes-Mata (1991), lagoons of the Gulf with the greatest faunistic diversity are Terminos (with 118 species), Tuxpan-Tampamachoco (99), Laguna Madre (78) and Tamiahua (60).

Marine Turtles

Of the eight species of marine turtles that are known in the world, seven inhabit Mexican beaches. In 1986, 16 beaches throughout Mexico were declared Areas of Reservation and

Table 3
Principal birds in the Gulf of Mexico coast (SEMARNAP–CONABIO, 1997)

Common name	Scientific name	Distribution
Red-billed Pigeon	Columba flavirostis	Gulf of Mexico.
White-Tipped Dove	Leptotila verreuxi	Tamaulipas, Veracruz, Tabasco, Campeche, Yucatan
Aztec Parakeet	Aratinga nana	South of Tamaulipas, Veracruz, Tabasco, Campeche, Yucatan
White-fronted Parrot	Amazona albifrons	Tabasco, Campeche, Yucatan
Green Jay	Cyanocorax yncas	Tamaulipas, Veracruz, Tabasco, Campeche, Yucatan.
Western Bluebird	Sialia mexicana	Tamaulipas and north of Veracruz
Northern Mockingbird	Mimus polyglottos	Tamaulipas and north of Veracruz
Tropical Mockingbird	Mimus gilvus	Yucatan
Long-billed Thrasher	Toxostoma longirostre	Tamaulipas and north of Veracruz
Scrub Euphonia	Euphonia affinis	Central part of Veracruz, Tabasco, Campeche, Yucatan
Blue-gray Tanager	Thrampis episccopus	Veracruz, Tabasco, Campeche, Yucatan
Red-throated Ant-tanager	Habia fuscicauda	Veracruz, Tabasco, Campeche, Yucatan
Red (Common) Cardinal	Cardinalis cardinalis	Tamaulipas, Veracruz, Tabasco, Campeche, Yucatan
Blue-black Grassquit	Valatinia jacarina	Veracruz, Tabasco, Campeche, Yucatan
Red-winged Blackbird	Agelaiaus phoeniceus	Veracruz, Tabasco, Campeche, Yucatan
Great-Tailed Grackel	Quiscalus mexicanus	Veracruz, Tabasco, Campeche, Yucatan
Bronzed (Red-eyed) Cowbird	Molothrus aeneus	Veracruz, Tabasco, Campeche, Yucatan
Yellow-tailed Oriole	Icterus mesomelas	Veracruz, Tabasco, Campeche, Yucatan
Northern (Bullockis) Oriole	Icterus galbula	Winters in Yucatan and is migratory in Veracruz, Tabasco, Campeche.
House Sparrow	Passar domesticus	Veracruz, Tabasco, Campeche, Yucatan
American Flamingo	Phoenicopterus ruber roseus	Yucatan

Places of Refuge for turtle nesting, reproduction and growth; two of these are in the Gulf of Mexico: Nuevo Rancho, in Tamaulipas and Lagartos creek in Yucatan. In June 1990, an order went into effect prohibiting the capture and sale of marine turtles and of turtle products. The beaches of Nuevo Rancho, Tamaulipas are the only place in the world where the 'Lora' turtle breeds; and its range includes the southeast USA, specifically Texas and Louisiana and the coasts of Campeche.

Birds

Birds of the Gulf of Mexico coast have a high diversity (Fig. 6) (Table 3), and some parts of this area, such as the Laguna Madre, are considered to be migratory or biological corridors (Contreras-Balderas, 1993). The Laguna Madre in Tamaulipas is an especially good example (Contreras-Balderas, 1993). Of a total of 86 species of aquatic birds in 17 families, 38 have a distribution which includes the American continent; 23 are distributed in the Northern Hemisphere, 10 are Nearctic; seven neo-tropical, and seven are cosmopolitans and exotics. Nearly one half are residents, the rest being migratory or seasonal visitors.

Shrimp

The major shrimp species in the Gulf of Mexico are *Lithopenaeus setiferus*, *Farfantepenaeus aztecus* and *F. duorarum*. Shrimp is the main export species from Campeche Sound but production has diminished in the last few years. Fishermen are strongly influential, so environmental authorities point to the oil industry as being the only party responsible for the drop in production. Because the main fishing regions are located near to the coast (in 15 or 30 m depth) it will be necessary also to determine the impacts of the rivers with their loads of metals, organochlorine compounds and pathogens before determining the causes of the fishing decline (Boesch et al., 1987).

Deeper Benthos

In Campeche Sound, considerable work has been done on the benthos (Soto and Escobar, 1995; Sánchez, 1994; Cantú-Díaz and Escobar, 1992; Raz-Guzmán and Sánchez, 1992; Caso, 1979). Nevertheless little is known about the distribution and abundance of organisms near the oil platforms or of the effects of the industry on the fauna, though there appears to be a decrease in the abundance of organisms towards the area containing the platforms. It is not clear whether this effect was due to periodic contact with toxic substances in bleedwater, substrate disturbance due to currents around the platform legs, removal of substrate or other cause (Harper et al., 1980).

OFFSHORE SYSTEMS

Upwelling and the Yucatan Current

In the Gulf of Mexico, two major areas of primary production exist. The first is the Southern part of the Gulf, mainly Campeche Sound; the second is in the Northwest of the Gulf, along the Hondo River and Mississippi. Various studies have been carried out in Campeche Sound (Gómez-Aguirre, 1974; Licea, 1977; Licea and Santoyo, 1991; Moreno and Licea, 1994; PEMEX-UNAM, 1998, 1999). Diatoms form the dominant group near the coast where they comprise up to 100% of the total. Diatoms diminish offshore, contributing as little as 1% in some places. This pattern is only altered by patches of *Hemiaulus sinensis* and *Hemiaulus membranaceus* (Table 4). Naked dinoflagellates, phytoflagellates, together with cyanophytes and coccolithophores replace diatoms offshore and towards the continental platform. In the central area of this region is a domain of *Oscillatoria thiebautii*, restricted to the surface layer. Phytoplankton is at its most abundant during winter and spring, especially in coastal areas. The highest registered value is a million cells per litre.

Commercially important species of zooplankton in the southern Gulf of Mexico are the larva of the shrimp (Espinosa, 1997). The largest densities of larvae are obtained in summer and autumn, and the smallest in winter (Table 5). The maximum concentration of protozoa was observed near Lagoon of Machona and the Grijalva–Usumacinta system. Other dominant groups of zooplankton in Campeche Sound are Copepods, Chaetognathas, Ostracods, Appendicularians and Gastropods.

Table 4
Principal species of phytoplankton in the Gulf of Mexico

Diatoms		Dinoflagellates	
Bacteriastrum delicatulu	*Hemiaulus hauckii*	*Acutissimum*	*D. caudata*
Hyalinum	*H. membranaceus*	*Ceratium furca*	*D. tripos*
Chaetoceros affinis	*Leptocylindus danicus*	*C. fusus*	*Exauviella compressa*
Ch. coarctatus	*Nitzschia longissima*	*Massiliense*	*Goniaulax diegensis*
Ch. compressum	*N. bicapitata*	*Teres*	*Gymnodinium breve*
Ch. curvisetum	*N. pungens*	*Trichoceros*	*Prorocentrum micans*
Ch. decipiens	*Rhizosolenia alata*		*Peridinium depressum*

Table 5
Seasonal changes in the abundance of shrimp larva (organisms/100 m^3), Campeche Sound

Larval form	February (Winter)	August (Summer)	November (Autumn)
All larval forms	84.15	567.00	463.41
Protozoans	21.16	365.94	298.81
Mysis	1.87	27.72	39.13
Postlarval	61.12	173.34	125.47

In the Gulf of Mexico, the two main upwelling systems that enrich the plankton are in the north of Yucatan and near Punta Zapotitlan, Veracruz. The first is due to the waters flowing from the Yucatan channel (Secretaria de Marina, 1980). The second arises from the abrupt change of bathymetry from 800 m to 150 m off the coast of Punta Zapotitlan. That of Yucatan has been studied the most (Bogdanov, 1969; Bessonov et al., 1971; PEMEX-UNAM, 1998, 1999).

Fisheries

Tuna is represented by three species: *Thunnus albacares* (Yellowfin), *Thunnus atlanticus* (Blackfin) and *Thunnus thynnus* (Bluefin). These are distributed throughout the southern Gulf of Mexico, but the main capture area is in the west, in Campeche Sound and north of Yucatan. Size at capture is between 1.5 and 2.8 m, with a weight of between 50 and 176 kg. The most abundant species in the catch is *Bagre marinus*, which is distributed along the coastal area and which is caught mainly in southern Veracruz and Tabasco. Their capture size is 50–100 cm with a maximum weight of 2.3 kg.

Sharks captured are mainly *Rhizoprionodon terraenovae* (Atlantic sharp-nosed shark), *Sphyrna tiburo* (Bonnethead shark), *Mustelus nirrisi* (Florida smooth-hound) and *Carcharhinus falciformis* (silky shark). Their distribution is similar to that of the Gafftopsail. Its maximum size of capture is 112 cm. The northern net snapper (*Lutjanus campechanus*) has a distribution along the whole coast and is caught mainly in the north and northwest of the Yucatan Peninsula. Their capture size is between 60 and 100 cm with a weight average of 900 g. The distribution of flounder (*Paralichthys lethostigma, Syacium gunteri, Cyclopsetta chittendeni, Syacium papillosum*), *Paralichthys lethostigma* is similar to that of the sharks. Ninety percent of this catch is from Tamaulipas; their size average is 40 cm.

POPULATIONS AFFECTING THE AREA

The present total population of the coastal states is 13,212,559 (Fig. 7), which is 14.1% of the total population of Mexico (INEGI, 1998a,b,c; 1997a,b). Forty-five

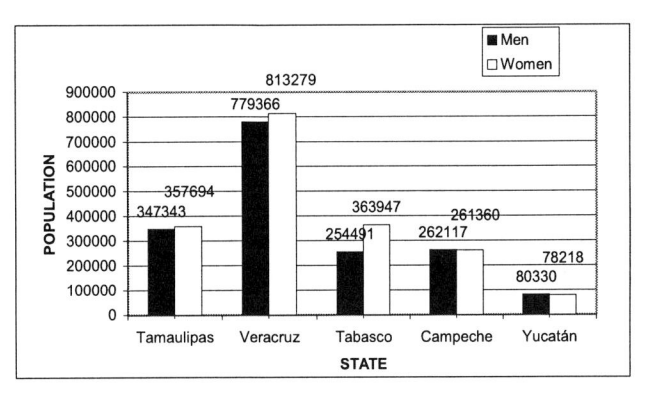

Fig. 7. Total population on the coast of the Gulf of Mexico, by State (1990).

municipalities have coasts in the Gulf with a population of 3,598,145 representing the 3.84% of the total population of Mexico. At present, the Gulf coasts do not have a dense population (Table 6), but in recent years there has been a strong tendency for this to rise.

Of the 2.8 million coastal inhabitants in 1980, 70% lived at an elevation of 1 to 5 m, mainly along the coast or at river borders very near their mouths (Ortíz-Pérez et al., 1996). Rain, tropical storms and hurricanes cause severe problems in these marginal places.

The major classes of coastal lands, are: forest, savanna, mangrove, tular and popal, agriculture and pasture. The major human land use is low technology but extensive cattle ranching. To support the cattle, many forest ecosystems have been changed to new pasture, so that this use of the land now accounts for 25% of the coastal zone. The dominant commercial monoculture crops are sugar-cane and bananas, located principally in Veracruz and Tabasco (Ortíz-Pérez et al., 1996).

Tourism here is poorly developed. Only the Ports of Veracruz, Coatzacoalcos and Tampico have sufficient infrastructure to develop these activities. Tourism in Merida, Yucatan (at 40 minutes from the beach) is the most developed and attracts international tourism. In Veracruz most tourism is national.

Historically the Gulf coasts (lagoons, estuaries) were used by the people for artisanal fisheries. The principal methods used are casting nets, small dragnets, lines with several fishhooks and 'fykes'. In the 1940s, fishing in coastal lagoons was in equilibrium with stocks. These artisanal

Table 6

Total population of Mexico and the coastal States of the Gulf of Mexico (modified from Ortíz-Pérez et al., 1996)

Area	1960	1970	1980	1990	1996*
Country	34,923,129	48,377,363	66,846,833	78,140,006	93,700,000
Coastal States	5,080,858	7,138,668	10,090,396	12,446,178	13,212,559
% Total Country	14.55	14.76	15.09	15.93	14.1
Coastal Zone	1,236,269	1,864,189	2,792,613	3,526,055	3,598,145
% Total Country	3.54	3.85	4.18	4.51	3.84
% Coastal States	24.33	26.11	27.68	28.33	27.23

fishing methods in general still do not have severe effects on the habitats which support them. Possibly the main adverse effects of the artisanal fishing are the locations of the nets in each lagoon, whether fishermen use fine mesh nets to catch small species (this practice is common when the resource is low) and fishing during the closed season. An increase in fishermen during the high season, and clandestine fishing are now causing over-exploitation of the resource and a near disappearance of some of the fish. More recently too, new technology, bigger boats with engines, industrial activities and pollution, together with an increase in the number of fishermen, have had more marked consequences.

At the beginning of the 20th century, the Gulf was characterised by extensive plantation agriculture of sugarcane, banana, and sisal among others, following development of the railway in southern Mexico. Later, commercial cattle raising became important, principally for internal markets (Ortíz-Pérez et al., 1996). In the earlier part of the century, Tamaulipas and Veracruz were important oil producers, but today Tabasco and Campeche produce 95% of Mexico's total oil production. Thus, for a long time the Gulf of Mexico was used for transportation, cattle, cloth and furniture from Europe until, at the beginning of this century, the oil industry became the most important.

As with many developing countries, Mexico does not carry out enough research to serve the needs of its people. At present there is isolated research in the lagoons, estuaries and seas. The Campeche Sound has had special attention for five years due to the oil industry. The lagoons of Carmen–Pajonal–Machona, Pom–Atasta and Terminos have research programs on water and sediment quality carried out by several universities.

RURAL FACTORS

On the Tamaulipas and Veracruz coasts especially, rivers transport large quantities of suspended solids due to inland deforestation (Beltrán, 1988; West et al., 1985). Inland forest clearance and erosion from the watershed is critical along the Gulf coast. In Tabasco the changes in the Grijalva–Usumacinta watershed is changing the delta of the Grijalva and San Pedro–San Pablo rivers (Psuty, 1965; Vázquez, 1994).

More so than fertilisers, pesticides are a critical problem here. The geographical regions of the coastal zone can be based on hydrology and pesticides use (Benítez and Bárcenas, 1996). The northern coast has the biggest arable crop and has a large runoff into the Laguna Madre and Rio Grande (29.7%). Pesticides are used in high volumes, are very toxic and persistent (Benítez and Bárcenas, 1996). In Tamaulipas, further south, a replacement of the cotton crop by sorghum and vegetables changed the volume and class of pesticides, while in Tabasco and Campeche in the 1980s, a rice program caused an increase of the pesticides (INEGI,

1998d; Benítez and Bárcenas, 1996). Other factors that increase the content of pesticides in particular zones are control of locust in the north of Campeche, and mosquito and malaria control.

COASTAL EROSION AND LANDFILL

In some parts of the Gulf coast, erosion and loss of beach is an important problem, and in Tabasco salinisation of some land causes problems as well. In Anton Lizardo (10 km from the Port of Veracruz) erosion is high and there has been a loss of over 100 m in the last eight years. East of the mouth of La Machona lagoon (Tabasco) there is a continuing strong erosion; in the last 20 years the coastline has receded more than 150 m (Vázquez, 1994). In the east of the Atasta Peninsula is another area with this problem, where north winds and hurricanes have accelerated the problem over the last six years, resulting in the loss of more than 100 m.

The Mexican coastal plain contains only a few sizable sections of geologically recent sedimentation. One of the most extensive of these occurs along the plains of the combined Mezcalapan and Usumacinta rivers (Tabasco). Numerous rivers course through this plain carrying sediments from the Chiapas–Guatemala highlands. Most of the

Table 7

Principal fishing species and production in the Gulf of Mexico (INP, 1994)

Species	Common name	Production (tons)
Rangia cuneata	Atlantic tiger lucine	1139
Thunnus albacares	Yellowfin tuna	3349
Loligo pealei	Atlantic long-finned	71
Lolliguncula Brevis	Western Atlantic brief squid	
Farfantepenaeus aztecus	Northern brown shrimp	24093
Farfantepenaeus duorarum	Pink shrimp	
Sicyonia Brevirostris	Rock shrimp	
Sicyonia dorsalis	Shrimp rock Lesser	
Lutjanus campechanus	Northern red snapper	3393
Callinectes sapidus	Blue crab	12921
Callinetctes bocourti	Blunt-tooth swimcrab	
Callinectes rathbunae	Sharp-tooth swimcrab	
Panulirus argus	Caribbean spiny lobster	614
Panulirus laevicauda	Smoothtail spiny lobster	
Mugil cephalus	Striped mullet	6015
Epinephelus morio	Red grouper	11368
Diapterus olisthostomus	Irish pompano, Irish mojarra	39823
Eucinostomus argenteus	Silver mojarra	
Crassostrea virginica	American cupped oyster	31729
Crassostrea rhizophorae	Mangrove cupped oyster	
Octopus vulgaris	Common octopus	13339
Centropomus undecimalis	Common anook	3996
Scomberomorus maculatus	Atlantic Spanish mackerel	7381
Carcharhinus falciformis	Silky shark	8444

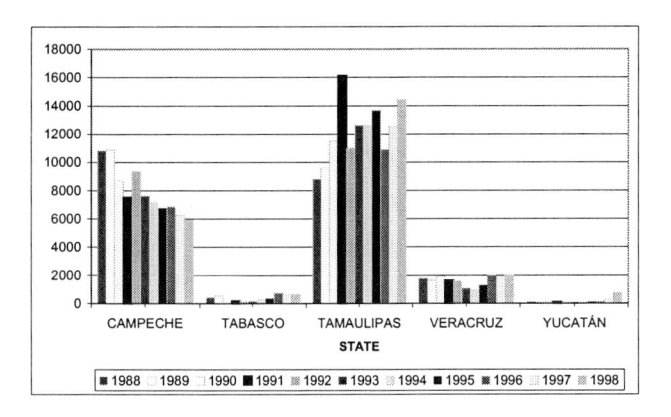

Fig. 8. Production of shrimp (tons live weight) in the states of the Gulf of Mexico (1988–1998).

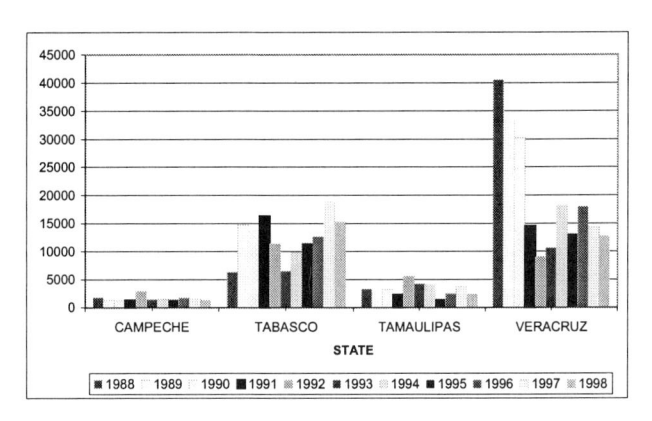

Fig. 9. Production of oyster (tons live weight) in the states of the Gulf of Mexico (1988–1998).

water is channelled coastward through the Grijalva river and others (Psuty, 1965), and the lowlands of Tabasco have been impacted over many years from landfill to build roads. Construction and channelling has resulted in drainage of some areas.

URBAN AND INDUSTRIAL ACTIVITIES

Artisanal and Non-industrial Uses

The most important artisanal and non industrial changes in the Gulf, are: forest, popal, tular, savanna and mangrove conversion to support cattle ranching; and popal, tular, savanna and mangrove felling. The felling of forest and mangrove on the Gulf coasts is a common activity and the wood is used largely for fuel. The coasts of Tabasco contain large palm plantations, which are important to the economy of the local people. Many years ago this activity was very attractive due to the profits, but in 1994 in the Campeche and Yucatan coasts, lethal yellow disease arrived from the Caribbean which destroyed all palm plantations in these states.

The principal effects of the forest and mangrove and felling are a loss of biodiversity along the coast of the Gulf of Mexico, erosion and loss of soils, and a loss of wetlands.

Industrial Uses

At present the important uses of Gulf coast are macro-scale fishing, raw materials transport, and the oil industry. Aquaculture is poorly developed in Gulf coasts; fortunately the Ecuador experience with shrimp ponds is remembered. Construction is also poorly developed.

The Gulf of Mexico and the Caribbean coastal zone, with its 2805 km of coastline, 771,700 km^2 of exclusive economic zone and a broad continental shelf, represents a significant fishing region, even though it is not as important in this respect as the Pacific coastal zone of Mexico (Ortíz-Pérez et al., 1996). The macro-scale fishing extends along the whole Gulf of Mexico coast. Several species are fished (Table 7). Shrimp have a great economic value and a considerable part of the production is exported to the USA. Shrimp in the Gulf of Mexico are distributed from coastal lagoons to depths of 100 m on the edge of the continental shelf, and their production during the last ten years is shown in Fig. 8. Tamaulipas has the biggest production of all the states of the Gulf of Mexico and this production in general has been maintained, while in Campeche production has decreased due to over-fishing, capture of the breeding adults in lagoons and estuaries, the oil industry, the lack of sustainable practices and the El Niño effect.

The oyster fisheries in the Gulf are also important. American oyster (*Crassotrea virginica*) is the primary species commercially exploited. The principal oyster fisheries are located in coastal lagoons (Ortíz-Pérez et al., 1996). Historical production (Fig. 9) of the oyster in Veracruz has decreased markedly over the years, due to over-fishing and pollution.

Shipping and Offshore Accidents

In the Gulf of Mexico are located the principal crude oil fields of the Western Hemisphere: East Texas, Gulf Coast and Mississippi Delta region, and Campeche Sound. The potential resources of these areas are around 2.24 and 21.99 billions of barrels of oil and 5.48 and 44.40 tcf of natural gas (Botello et al., 1996; Foote et al., 1983). Campeche Sound produces 80% of the total production of crude oil in Mexico. The port of Dos Bocas and Cayo Arcas (Fig. 10) are two major exporting points. The ports of Tampico and Veracruz transport acids, alkalis, glycols, sulphur, phosphorus, fertilisers, phenols and other raw materials.

Vessel traffic across the Gulf is high. As a result there are increased discharges of oil and sanitary wastes. In general, oil pollution is associated with offshore oil production, marine transportation, exploration, production and development, vessel cleaning, pipeline ruptures and trans-shipment accidents (Boesch and Rabalais, 1987). Four major oil spills have occurred in the Campeche Sound: Ixtoc-1 (06/03/1979), Abkatum 91 (1986), Yum II (10/10/1987) and Och-1B (08/20/1988) (PEMEX, 1980, 1991). Ixtoc-1 lost a total of more than 140 million gallons of oil (Golob and McShea, 1981),

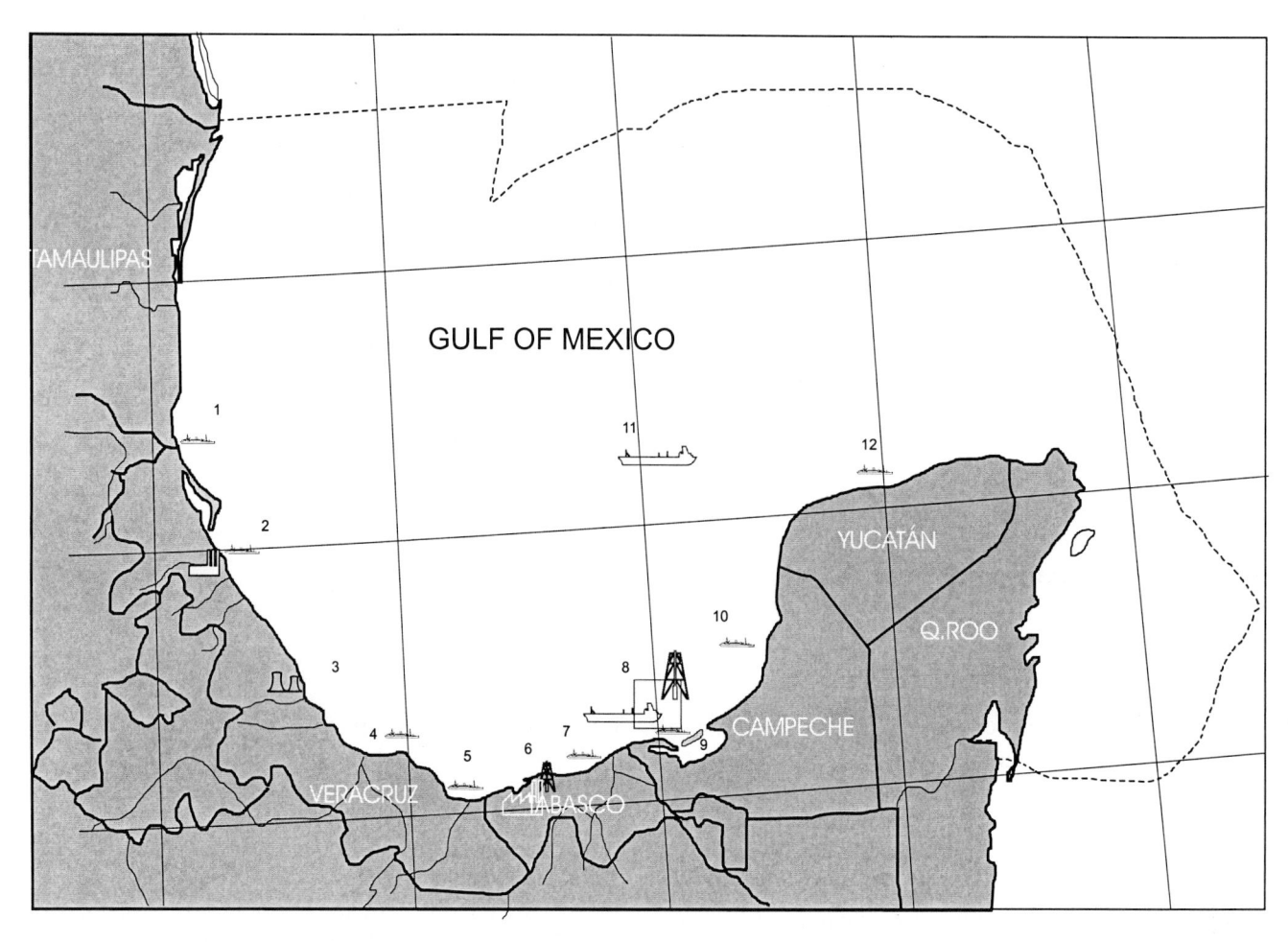

Fig. 10. Ports and industries along the Mexican coast. 1: Port of Tampico. 2: Port of Tuxpan and Electric Industry. 3: Laguna Verde Nuclear Power Station. 4: Veracruz Port. 5: Coatzacoalcos Port. 6: Industry and Oil Production. 7: Frontera Port. 8: Oil Exploration. 9: Cd. Del Carmen Port. 10: Lerma Port. 11: Crude Oil "Monobuoys". 12: Progreso Port.

and the impact of crude oil on the marine environment is well documented (Parker, 1974; Ray, 1981; Baker et al., 1981; NRC, 1985; Boesch and Rabalais, 1987; Witt, 1995). At present, a model of the oil spill has been developed with a focus in Campeche Sound (PEMEX-UNAM, 1999); this was calibrated with field measurements of current speed, current direction, pressure, temperature and salinity. It is believed that an oil spill of 1000 barrels from the marine platforms will take around five days to arrive at the coast of Tabasco or Campeche, depending on the meteorological conditions.

Only a few cities of the Gulf of Mexico coast have good sewerage systems (Matamoros, Ports of Tampico, Altamira, Veracruz and Coatzacoalcos); the others mainly have septic tanks. One of the main problems of the Gulf coast is the content of fecal coliforms in coastal waters and sediments.

EFFECTS OF HUMAN ACTIVITIES ON NATURAL PROCESS

The complexity of the components of these ecosystems means that activities in one area may have subtle repercussions in other seemingly unrelated components. This feature contributes to a lingering uncertainty about whether the effects of an activity are understood well enough to be predictive. Ecosystems may be highly connected to other ecosystems, particularly in the coastal ocean. Continental shelf ecosystems interact with coastal systems by environmental forces (e.g., runoff, storms, etc.) and movement of biota between them (Boesch and Rabalais, 1987).

One way to examine the relationship between the activities and the environmental factors is with environmental matrices. For this region, each activity can be divided into components and qualitatively correlated with the environmental factors in order to identify and evaluate the effects arising as a result of human activities or natural processes. Two factors are considered: the probability that the effect may occur and the duration of the effects. These cannot always be quantitatively expressed, but it is clear that they vary greatly among the issues (Boesch and Rabalais, 1987). Due to the diverse nature of the activities along the Gulf of Mexico, two examples are given: the offshore oil industry and the extensive cattle, agricultural and urban activities (Figs. 11 and 12). Thus, during petroleum exploration, seismic surveying and drilling have a probability of

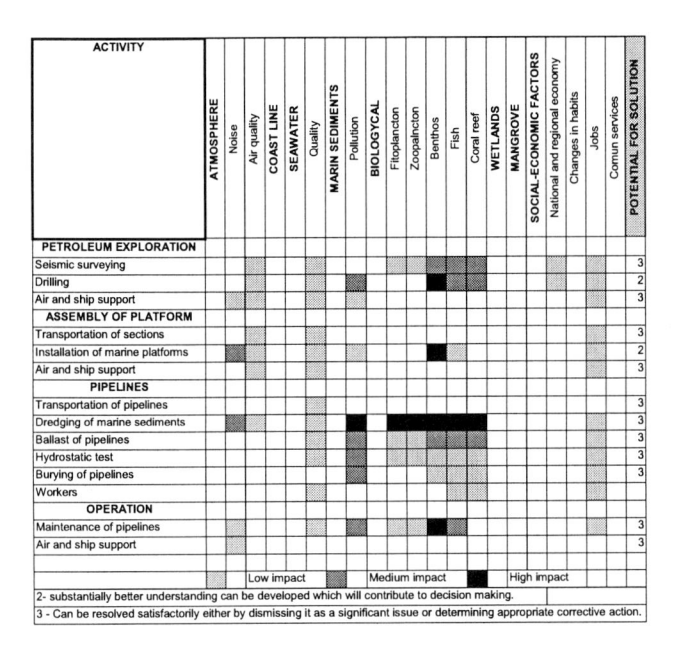

Fig. 11. Matrix of environmental impact to offshore oil industry in the Gulf of Mexico.

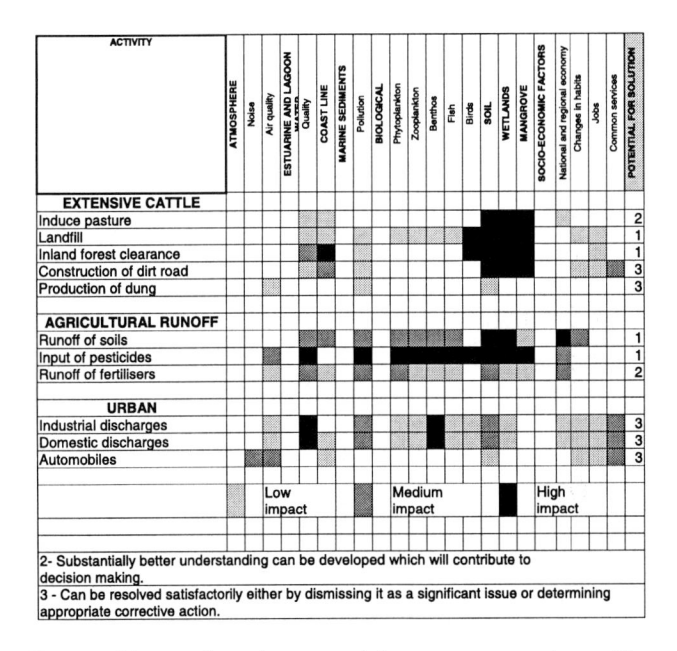

Fig. 12. Matrix of environmental impact to extensive cattle, agricultural runoff and urban activities in the Gulf of Mexico.

medium or high impact to the benthos, while during the assembly and attachment of a platform to the sea floor there can be a high impact to the benthos, perhaps for a short time. Dredging of marine sediments during pipeline laying has a high impact (see Fig. 11), while during operations, the maintenance of pipelines has high impact.

In the second example, pasture created to support the extensive cattle industry has low impact to the water quality of estuaries and lagoons, but has a high impact to soil, wetlands and mangrove (Fig. 12). At the same time, landfill and inland forest clearance have a high impact on distributions of birds, soils, wetlands and mangrove. The construction of dirt roads has a high impact to wetlands and mangrove, and the production of dung has a contribution to air quality due to methane liberation. In agricultural activities, pesticide input has a high impact to biological organisms, habitats and waters due to its persistence, lipid solubility and biomagnification (Fig. 12) (Wiemeyer, 1996). The input of pesticides has medium impact to air quality, national and regional economy, while runoff of fertilisers has medium impact to phytoplankton, soil, marine sediments, estuarine and lagoon water quality, and runoff of soils has medium impact to biological organisms but a high impact to soil and wetlands. Using this and geographic information systems, the present authors are processing spatial data into forms of information suitable for making decisions along the Gulf of Mexico coast (Demers, 1997).

PROTECTIVE MEASURES

In response to threats to the marine environments, Mexico started some policy initiatives in 1971, but the understanding of the environment by politicians at that time was poor. In 1981, the Public Works Law was issued, which required those planning the construction of buildings to consider environmental impacts (Ortíz-Pérez et al., 1996).

The Federal Protection of the Environmental Law was issued in 1982, to control air and water pollution. In 1988, the general law of Ecological Balance and Environmental Protection brought remarkable changes in the understanding of environmental problems by Mexican institutions and private industry. Under this law, sensible regulations have included conservation, restoration, and improvement of the environment, as well as protection of natural areas, flora and fauna. Also, the new regulations suggest a reasonable utilization of the natural elements and encourage estimation of the economic benefits of protecting ecosystems and preventing pollution (Ortíz-Pérez et al., 1996). Management of the coastal strip of the Gulf of Mexico and its natural resources calls for the involvement of several institutions and the enforcement of various legal instruments at federal, regional and local levels (Saavedra-Vázquez, 1996).

With the entry of Mexico to NAFTA (1995) OCDE (1994), Basilea (1992), Agenda 21 and IMO (1979) environmental regulation is now more rigorous. All applications for World Bank loans for development in areas with natural resources need environmental evaluation before the loan can be assigned.

The National Development Plan (1995–2000) is a strategy for environmental protection which will permit economic development of the country, but also promote sustainable growth. With this, developed a national strategy to achieve a global and regional balance between economic, social and environmental goals, and then to reduce the process of environmental damage and develop

Table 8
Some international agreements signed by Mexico

Agreement	Date
1. International Convention for the regulation of whale hunting	30 June 1949
2. International Convention for the prevention of pollution of hydrocarbons in water	20 July 1956
3. Constitutive Convention for Inter-governmental Maritime Consultative Organization	30 August 1970
4. Lw of the Sea. United Nations	18 March 1983
5. Basel Convention on the Trans-boundary Movement of Hazardous Wastes and their Disposal	May 1989
6. Adhesion for Economic Development and Cooperation Organization	15 April 1994
7. Agreement of Environment Program with Guatemala	31 October 1997
8. International Convention for Prevention of Pollution from Ships 1973; Protocol 1978 (MARPOL)	July 1992

environmental regulation (Saavedra-Vázquez, 1996; Plan Nacional de Desarroyo, 1995). The major programs in the National Development Plan related to the coastal zone are: (i) Program of Environment; (ii) Program of fishing and aquaculture; (iii) Hydraulic Program and (iv) Program of Forestry and Natural Resources. However, Mexico has insufficient funding and qualified personnel to permit its full implementation.

Programs of Sustainable Regional Development (Proders)

A major objective of the environmental policies of the present administration is to induce sustainability in national development. In this context they have established the 'Programs of Sustainable Regional Development' (Proders), directed to assist a group of high-priority regions of Mexico, the objective of which is to generate processes that allow a balance to be created between economic growth and a better quality of life, together with the conservation of natural resources. Also, 'Proders' seeks to coordinate environmental policies with other government institutions and with social organizations, academic, and non-governmental organizations.

From 1995, the SEMARNAP began to implement the Proders procedure in 21 high-priority regions. These regions have a total area of 292,177 km^2 and a total population of over 11 million inhabitants, mostly peasants. Of these regions, 19 are areas influenced by different indigenous groups, either with small properties or living in public land and communities. For the implementation of the regional programs, a commission was formed with the objective of coordinating efforts, and of increasing effectiveness at federal, state, regional and municipal levels.

Table 8 shows international agreements signed by Mexico relevant to the coastal zone. Because Mexico is not a country with a strong marine tradition, little interest exists in developing long-term studies that allow an understanding of the marine environment. As in other countries, socioeconomic factors have a priority in Mexico. An understanding of the diverse factors that control biodiversity will help achieve a sustainable development in the coastal area of the Gulf of Mexico, and the application of geographical systems of information, satellite images, elaboration of environmental models and the establishment of specific long-term monitoring will help to gain a better understanding and use of the coastal area.

The present deterioration is caused by both human and natural factors. Deforestation has been one of the main variables that has modified its coasts, and the presence of the petroleum and nuclear-electric industries will always pose possible impacts. The excessive development of the extensive cattle raising in the coastal region is another factor that needs to be controlled in the short term to avoid further deterioration.

ACKNOWLEDGMENTS

We thank Juan Fuentes and Roman Pérez for their help with figures. We are also grateful to PEP-RM-NE and PEP-RM-SO for the permission to use information. The preparation of this chapter was funded by UNAM (Project: 132).

REFERENCES

Baker, J.M., Moeso Suryowinoto, I., Broks, P. and Rowland, S. (1981) Tropical marine ecosystems and the oil industry; with a description of a post-oil spill survey in Indonesian Mangroves. In: *Petroleum and the Marine Environment*, ed. PETROMAR. EUROCEAN, pp. 679–703

Beltrán, J.E. (1988) *Petróleo y desarrollo, La Política Petrolera en Tabasco*. Gobierno del Estado de Tabasco, Villahermosa Tabasco. 247 pp.

Benítez, J.A. and Bárcenas, C. (1996) Patrones de uso de los plaguicidas en la Zona Costera del Golfo de México, pp. 155–167. In *Golfo de México, Contaminación e Impacto Ambiental: Diagnóstico y Tendencias*, eds. V.A. Botello, J.L. Rojas-Galavíz, J.A. Benítez and D. Zárate Lomelí. Epomex Serie Científica, 5. 666 pp.

Bessonov, N., González, O. and Elizarov, A. (1971) Resultados de las investigaciones cubano-soviéticas en el Banco de Campeche. Departamento de Hidroquímica de C.I.P., Cuba, pp. 317–323.

Boesch D.F., Butler, J.N., Cacchione, D.A., Geraci, J.R., Neff, J.M., Ray, J.P. and Teal, J.M. (1987) An Assessment of the long-term environmental effects U.S. offshore oil and gas development activities: future research needs, pp. 1–53. In *Long-Term Environmental Effects of Offshore Oil and Gas Development*, eds. F. Boesch and N.N. Rabalais. Elsevier, London and New York, 708 pp.

Boesch, F. and Rabalais, N.N. (ed.) (1987) *Long-term Environmental Effects of Offshore Oil and Gas Development*. Elsevier, London and New York, 708 pp.

Bogdanov, D.V. (1969) Some oceanographic features of the Gulf of Mexico and the Caribbean Sea. A. S. Bogdanov (Ed.) Soviet Cuban Fishery Research. U.S. Department of Commerce, Spring Geld. 13–35.

Botello A.V., Ponce-Vélez, G., Toledo, A., Díaz-González, G. and Villanueva, S. (1996) Ecología, Recursos Costeros y Contaminación en el Golfo de México, pp. 25–44. In *Golfo de México, Contaminación e Impacto Ambiental: Diagnóstico y Tendencias*, eds. V.A. Botello, J.L. Rojas-Galavíz, J.A. Benítez and D. Zárate Lomelí. Epomex Serie Científica, 5. 666 pp.

Cantú-Díaz Barriga, A. and Escobar, E. (1992) Isopods of the genus Excorellana Stebbing, 1904. (Crustacea, Isopoda, Corallanidae of the Southern Gulf of Mexico and the Mexican Caribbean, with a redescription of *Excorallana subtilis* (Hansen, 1890). *Gulf Research Reports* **8** (4), 363–374.

Carricart-Ganivet, J.P. and Horta-Puga, G. (1993) Arrecifes de Coral en México. In *Biodiversidad Marina y Costera de México*, eds. S.I. Dalazar-Vallejo and N. Emilia González. CONABIO-CIQRO, pp. 80–90.

Caso, M.E. (1979) Los equinodermos (Ateroidea, Ophiuroidea y Equinoidea) de la Laguna de Términos, Campeche. *An. Inst. Cienc. del Mar y Limnol. Pub. Esp.* 183 pp.

Cohen, D.M. (1970) How many recent fishes are there? *Proc. Calif. Academic Science*, 4th Ser. **38**, 314–346.

Contreras-Balderas, A.J. (1993) Avifauna de la Laguna Madre Tamaulipas. In *Biodiversidad Marina y Costera de México*, eds. S.I. Dalazar-Vallejo and N. Emilia González. CONABIO-CIQRO, 553–558.

Chávez, E.A. and Hidalgo, E. (1988) Los arrecifes coralinos del caribe noroccidental y Golfo de México en el contexto socioeconómico. *An. Inst. Cienc. del Mar y Limnol. Univ. Nal. Autón. México* **15** (1), 167–176.

Demers, M.N. (1997) *Fundamentals of Geographic Information Systems*. John Wiley & Sons, Inc. 486 pp.

Elliot, B.A. (1982) Anticyclonic Ring in the Gulf of Mexico. *Journal of Physical Oceanography American Meteorological Society* **12**, 1992–1309.

Espinosa (1997) Patrones de distribución espacio-temporal de los estadios larvarios de camarones Penaeus en la Sonda de Campeche, Tesis Maestría, Oceanografía Biológica y pesquera ICMyL, UNAM 69 pp.

Fernandez, E.A., Gallegos, A. and Zavala, J. (1993) Oceanografía Física de México. Zona Económica Exclusiva. *Ciencia y Desarrollo* **18** (108), 24–35.

Foote, R.Q., Martin, R.G. and Powers, R.B. (1983) Oil on gas potential of the maritime boundary region in the central Gulf of Mexico. *American Association of Petroleum Geologists Bulletin* **67** (7), 1047–1065.

Fuentes-Mata, P. (1981) Aspectos Biológico y ecológicos de la ictiofauna de la desembocadura del Rio Balsas, Mich-Gro Tesis, UNAM, Mexico.

García-Cubas Jr., A. (1963) Sistemática y distribución de los micromoluscos recientes de la Laguna de Términos, Campeche, México. *Bol. Inst. Geol., UNAM* **67**(4), 1–55.

García-Cubas Jr., A. (1968) Ecología y distribución de los micromoluscos recientes de Laguna Madre, Tamaulipas, México. *Bol. Inst. Geol., UNAM* **86**, 1–44.

García-Cubas, A. (1988) Características ecológicas de los moluscos de la Laguna de Términos, Cap. 16: 277–304. In: *Ecología de los ecosistemas costeros en el sur del Golfo de México: La región de la Laguna de Términos*, ed. A. Yañez-Arancibia and J.W. Day. Jr. Coastal Ecology. Instituto de Ciencias del Mar y Limnología, UNAM, LSU. Editorial Universitaria, México.

García-Cubas, A., Escobar de la Llata, F., González Ania, L.V. and Reguero, M. (1990) Moluscos de la Laguna Mecoacán, Tabasco, México: sistemática Ecológica. *An. Inst. Cienc. Del Mar y Limnol.* **17**(1), 1–30.

Golob, R.S. and McShea, D.W. (1981) Implications of the Ixtoc-I Blow-Out and Oil Spill. In *Petroleum and the Marine Environment*, ed. PETROMAR. EUROCEAN, 743–759 pp.

Gómez-Aguirre, S. (1974) Reconocimientos estacionales de la hidrología y plactón en la Laguna de Términos, Campeche, México (1965–1965). *An. Inst. Cienc. Del Mar y Limnol.* **1** (1), 61–81.

Harper, D.E., Jr., Potts, D.L., Salzer, R.R., Case, R.J., Jaschek, R.L. and Walker, C.M. (1980) Distribution and Abundance of Microbenthic and Meiobenthic Organisms, pp. 133–177. In *Environmental Effects of Offshore Oil Production, The Buccaneer Gas and Oil Field Study. Marine Science*, ed. B.S. Middleditch. Plenum Press, New York and London. 446 pp.

INEGI (1997a) *Anuario Estadístico del Estado de Campeche*. Instituto Nacional de Estadística, Geografía e Informática. 338 pp.

INEGI (1997b) *Anuario Estadístico del Estado de Veracruz*. Instituto Nacional de Estadística, Geografía e Informática. 958 pp.

INEGI (1998a) *Anuario Estadístico del Estado de Tabasco*. Instituto Nacional de Estadística, Geografía e Informática. 460 pp.

INEGI (1998b) *Anuario Estadístico del Estado de Tamaulipas*. Instituto Nacional de Estadística, Geografía e Informática. 462 pp.

INEGI (1998c) *Anuario Estadístico del Estado de Yucatán*. Instituto Nacional de Estadística, Geografía e Informática. 480 pp.

INEGI (1998d) *Estadística del Medio Ambiente, 1997*. Informe de la Situación General en Materia de Equilibrio Ecológico y Protección al Ambiente, 1995–1996. Instituto Nacional de Estadística, Geografía e Informática. 461 pp.

INP (1994) *Atlas pesquero de México*. Secretaría de Pesca. Instituto Nacional de Pesca. 243 pp.

Jardel, E.J., Saldaña, A.A. and Barreiro, M.T.G. (1987) Contribución al conocimiento de la ecología de los manglares de la Laguna de Términos, Campeche, México. *Ciencias Marinas* **13** (3), 1–22.

Licea, D.S. (1977) Variación Estacional del Fitoplacton de la Bahía de Campeche México (1971–1972). *FAO Fisheries Report* **200**, 253–257.

Licea, D.S. and Santoyo, H. (1991) Algunas características ecológicas de fitoplacton de la región Central de la Bahía de Campeche. *An Inst. Cienc. Del Mar y Limnol.* **18** (2), 157–167.

Lot, H.A., Vázquez-Yañez, C. and Méndez, F. (1975) Physiognomic and floristic changes near the northern limit of mangroves in the Gulf Coastal of Mexico. In *Proceeding of International Symposium on Biology and Management of Mangroves*, eds. G.E. Walsh, S.C. Snedaker and H.S. Teas. East–West Center Honolulu, Hawaii, pp. 56–61.

Méndez, F. (1976) Los Manglares de la Laguna de Sontecomapan, Los Tuxtlas, Veracruz: Estudio florístico ecológico. Tesis. Facultad de Ciencias, UNAM.

Merrell, W.J. Jr. and Morrison, J.M. (1981) On the Circulation of the Western Gulf of Mexico with Observations from April 1978. *Journal of Geophysical Research* **86** (C5), 4181–4185.

Moreno, L. and Licea, S. (1994) Morphology of three related Coscinodiscus Ehrenberg taxa from the southern Gulf of Mexico and coastal north Pacific of Mexico. In *Proceedings of the Eleventh International Diatom Symposium*. Memoirs of the California Academy of Sciences (17), pp. 113–127.

Nelson, J.S. (1984) *Fisher of the World*, 2nd edn. John Wiley & Sons, New York.

Nowlin, W. (1972) Winter circulation patterns and property distributions. In *Contribution on the Physical Oceanography of Gulf of Mexico*, eds. L.R.A. Capurro and L. Reid. Texas A & M. Univ. Oceanogr. Studies, Vol. 2. Gulf Publ. Co., Houston, TX, pp. 3–51.

Nowlin, Jr., W.D. and McLellan, H.J. (1967) A characterisation of the Gulf of Mexico waters in winter, Sears Foundation. *Journal of Marine Research* **25** (1), 29–59.

NRC (1985) *Oil in the Sea. Inputs, Fates and Effects*. National Research Council. National Academy Press, Washington D.C.

Ortíz-Pérez, M.A., Valverde, C. and Psuty, N.P. (1996) The impacts of sea-levels rise and economic development on the lowlands of the Gulf Coast. In *Golfo de México, Contaminación e Impacto Ambiental: Diagnóstico y Tendencias*, eds. V.A. Botello, J.L. Rojas-Galavíz, J.A. Benítez and D. Zárate Lomelí, pp. 459–470.

Parker, P.L. (1974) Effects of pollutants on marine organisms NFS/IDOE, Workshop on effects of populations on marine organisms. Sidney British Columbia, Canada, August 11–14. 46 pp.

PEMEX (1980) Informe de los trabajos realizados para el control del pozo Ixtoc-1, el combate del derrame de petróleo y determinación de sus efectos sobre el ambiente marino.

PEMEX (1991) Evaluación de la calidad del agua, sedimentos y algunos aspectos biológicos en el litoral del Golfo de México. Gerencia de Protección Ambiental PEMEX. 139 pp.

PEMEX-UNAM (1998) Diagnóstico actual de la calidad ambiental de la zona costera del Golfo de México (Sonda de Campeche, Zona costera de Atasta, Dos Bocas y sistema Lagunar de Tabasco) donde se localizan las plataformas petroleras y chapopoteras naturales. Informe Final campañas oceanográficas SGM-1 a 3.

PEMEX-UNAM (1999) Evaluación prospectiva para el programa de monitoreo continuo del efecto de la actividad petrolera en el Golfo de México. Monitoreo ambiental para el proyecto de modernización y optimización del Campo Cantarell, Informe Final de la Campaña Oceanográfica SGM-4.

Psuty, N. (1965) Beachridge development in Tabasco, Mexico. *Annals of the Association of American Geographers* **55**, 112–124.

Ramamoorthy, T.P., Bye, R., Lot, A. and Fa, J. (eds.) (1993) *Biological Diversity of Mexico. Origins and Distribution.* New York Oxford, Oxford University Press. 812 pp.

Ray, J.P. (1981) The effects of petroleum hydrocarbons on corals. In *Petroleum and the Marine Environment*, ed. PETROMAR. EUROCEAN, pp.705–726.

Raz-Guzmán and Sánchez, A.J. (1992) Registros adicionales de cangrejos braquiuros (Crustacea: Braquiura) de Laguna de Términos, Campeche. *An. Inst. Biol. Univ. Nal. Autón. México, Ser. Zool.* **63** (1), 29–45.

Reguero, M. (1990) Estructura de la comunidad de moluscos en lagunas costeras de Veracruz y Tabasco, México. Informe Interno. Trabajo de Investigación del Doctorado en Ciencias (Biología). Fac. Ciencias, UNAM.

Reséndez, A. and Kobelkowsky, A. (1991) Ictiofauna de los sistemas lagunares costeros del Golfo de México, México. *Universidad y ciencia*, **8** (15), 91–110.

Rico-Gray, U. and Lot, A.H. (1983) Producción de la hojarazca del manglar de la Laguna Mancvchs, Veracruz. *México Biótica* **3**, 295–301.

Saavedra-Vázquez, T.E. (1996) Normatividad en Zonas Costeras, pp. 605–640. In Golfo de México, Contaminación e Impacto Ambiental: Diagnóstico y Tendencias, eds. V.A. Botello, J.L. Rojas-Galavíz, J.A. Benítez, D. Zárate Lomelí. Epomex Serie Científica, 5. 666 pp.

Sánchez, A.J. (1994) Feeding habits of *Lutjanus apodus* (Osteichthyes:Lutjanidae) in Laguna Términos, to the Gulf of México. *Investigaciones Marinas* **15**(2), 125–134.

Secretaría de Marina (1980) Contribución al conocimiento de las características fisicoquímicas de las aguas del Caribe Mexicano. Química del Oceano. 99 pp.

SEMARNAP-CONABIO (1997) Aves canoras y de ornato. Instituto Nacional de Ecología, SEMARNAP-Comisión Nacional para el Conocimiento y Uso de la Biodiversidad, 177 pp.

Sherrod, C.L., Hoekaday, D.L. and McMillan, C. (1986) Survival of red mangrove, *Rhizophora mangle*, on the Gulf Mexico coastal of Texas. *Contribution in Marine Science* **29**, 43–59.

Soto, L.A. and Escobar, E. (1995) Coupling mechanisms related to benthic production in the South-western Gulf of Mexico. *EMBS Greece, Olsen & Olsen International Symposium Series*, pp. 233–242.

Vázquez, F. (1994) El Sistema Lagunar El Carmen-Pajonal-Machona del Estado de Tabasco: su hidrodinámica, la estabilidad de sus bocas y de su línea de costa. UNAM. 132 pp.

West, R.C., Psuty, N.P. and Thom, B.G. (1985) *Las Tierras Bajas de Tabasco, Gobierno del Estado de Tabasco*, Inst. De Cultura de Tabasco. 409 pp.

Wiemeyer, S.N. (1996) Other Organochlorine Pesticides in Birds. In *Environmental Contaminants in Wildlife*, eds. W.N. Beyer, G.H. Heinz and A.W. Redmond-Norwod. Lewis Publishers, pp. 99–115.

Witt, G. (1995) Polycyclic Aromatic Hydrocarbons in Water and Sediment of the Baltic Sea. *Marine Pollution Bulletin* **31** (3–12), 237–248.

Yañez-Arancibia, A. (1986) *Ecología de la Zona Costera, análisis de siete tópicos.* A.G.T. Editor, México, D.F.

THE AUTHORS

Felipe Vázquez
*Instituto de Ciencias del Mar y Limnología,
UNAM, Cd. Universitaria,
A.P. 70-305,
Mexico City, C.P. 04510 Mexico*

Ricardo Rangel
*Instituto de Ciencias del Mar y Limnología,
UNAM, Cd. Universitaria,
A.P. 70-305,
Mexico City, C.P. 04510 Mexico*

Arturo Mendoza Quintero-Marmol
*PEMEX-Exploracíon-Produccion-Región Marina Noreste, Calle 31, Esq. Periferica,
Cd. del Carmen, Campeche, C.P. 24170 Mexico*

Jorge Fernández
*PEMEX-Exploración-Producción-Región Marina Suroeste, Calle 33 S/N, Edif. Cantarell,
Cd. del Carmen, Campeche, C.P. 24170 Mexico*

Eduardo Aguayo
*Instituto de Ciencias del Mar y Limnología,
UNAM, Cd. Universitaria,
A.P. 70-305,
Mexico City, C.P. 04510 Mexico*

A. Palacio
*Instituto de Ingeniería,
UNAM, Cd. Universitaria,
Mexico City, C.P. 04510 Mexico*

Virender K. Sharma
*Chemistry Department,
Florida Tech.,
150 West University Blvd.,
Melbourne, FL 32901-6975, U.S.A.*

Chapter 30

THE PACIFIC COAST OF MEXICO

Alfonso V. Botello, Alejandro O. Toledo, Guadalupe de la Lanza-Espino and Susana Villanueva-Fragoso

The Mexican Pacific coast extends for about 8,000 km and covers eleven coastal states. It contains a large variety of coastal environments, derived from the interaction of diverse geological, biological, oceanographic and atmospheric processes. Because of its length, this description of the Mexican Pacific is based on seven important physiographic provinces, which together comprise a large number of coastal and marine environments dominated by particular oceanographic features (winds, currents, biological productivity, physico-chemical factors).

The variety of coastal type has given rise to a very high biodiversity. Also, the different provinces play an important role in migration patterns of coastal and marine species; they are important nesting and breeding areas, and parts are rich for fisheries, energy production and natural and mineral resources.

Dominant meteorological elements include annual tropical depressions due to the El Niño south oscillations (ENSO), which have marked effects on resources in that they provides the region with nutrient-rich water masses which greatly benefit the fisheries, mainly tuna.

Around 14% of the Mexican population (12 million people) inhabit the coastal states of the Mexican Pacific, most of them concentrated in small and medium sized coastal communities. The most populated area is the northwest (six million) followed by the central provinces (five million), whereas the south province has around one million inhabitants.

Very little industrial activity takes place on the Mexican Pacific littoral. Some of the more important industrial developments include giant salt exploitation sites in Guerrero Negro on the Gulf of California, agroindustrial and agricultural activities of the Imperial Valley of Mexicali and Northwest (Culiacan Valley), fishery processing industries of Guaymas and Mazatlán, and shrimp farms in Sinaloa. There are also some industrial harbours. The impact exerted by human activities has been caused by changes in the use of land, alteration of wetlands and coastal lagoons conversion of wetlands. Eutrophication of estuaries and coastal lagoons is important in several sites, as is loss of habitats due to large urban and tourist projects.

There are 21 natural protected areas totalling about 4,600,000 ha. Also the country is party to several international agreements for the protection and management of marine and coastal resources. However, Mexico has not undertaken much definite action for the effective conservation and protection of coastal and marine resources, due to the lack of legal instruments, and poor co-ordination between different governmental levels.

Seas at The Millennium: An Environmental Evaluation (Edited by C. Sheppard)
© 2000 Elsevier Science Ltd. All rights reserved

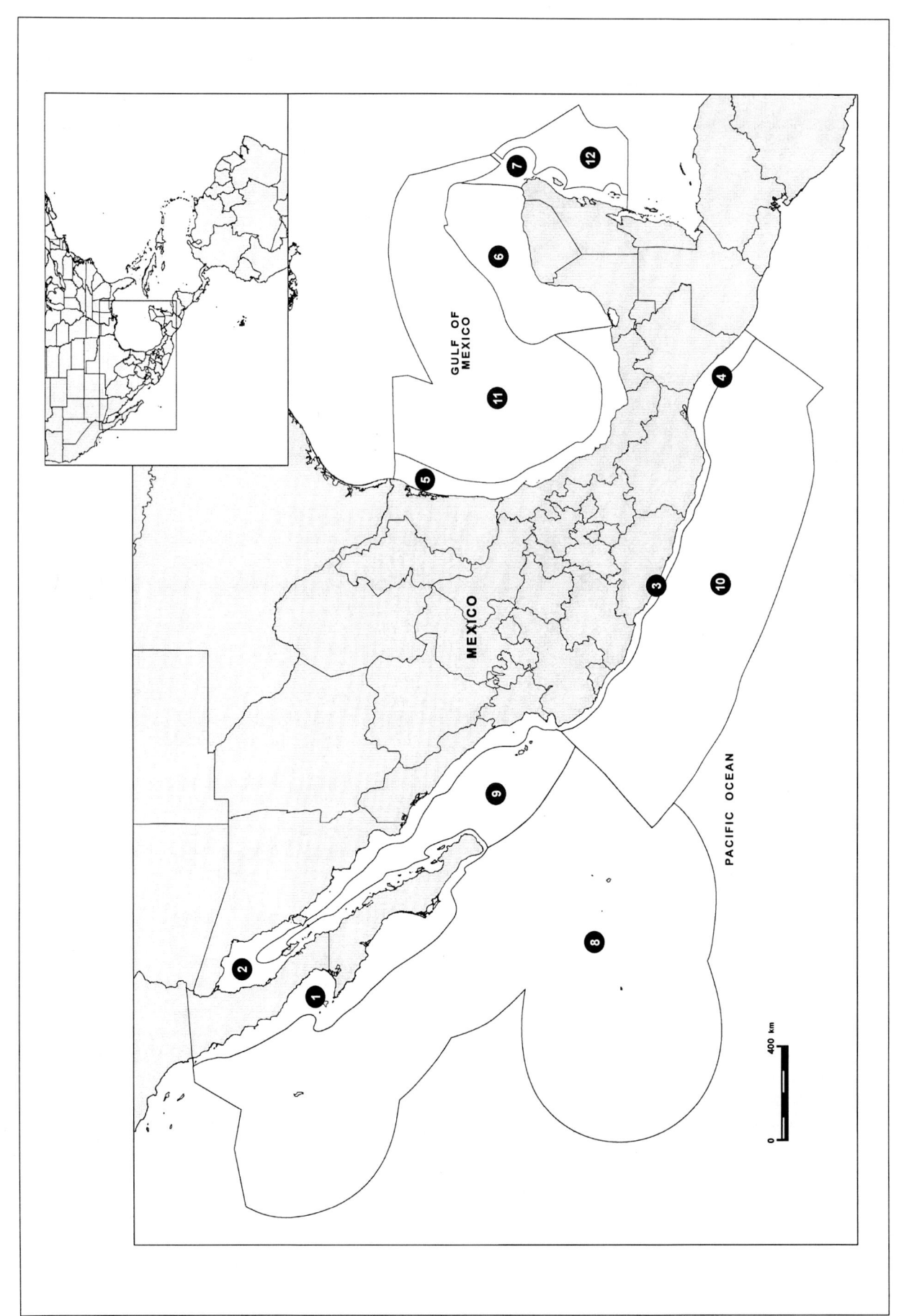

Fig. 1. Map of the Pacific coast of Mexico. Coastal Provinces: 1. Californiana. 2. Golfo de California. 4. Panámica. 5. Golfo Noroeste. 6. Plataforma de Yucatán. 7. Caribeña. Oceanic Provinces: 8. Pacífico Norte. 9. Mar de Cortés. 10. Pacífico Tropical Sur. 11. Golfo de México. 12. Mar Caribe.

INTRODUCTION

The Mexican coastline extends for 11,592 km along the Pacific Ocean, Gulf of California, Gulf of Mexico, and the Caribbean. Of the 31 states constituting the national territory, 17 are coastal, with 11 being located on the Pacific and Gulf of California, or western side, with the remaining six lying on the Eastern Gulf of Mexico and Caribbean side of the country (Fig. 1). The littoral states along the Pacific Ocean and Gulf of California are longer than those bordering the Gulf of Mexico and the Caribbean (Table 1).

The large extent of the Mexican littoral is a striking feature of the physiography of the country, and results in a great variety of coastal environment. The area has a high productivity and diversity, derived from the interaction of many different geological, biological, oceanographic, and atmospheric processes (Figs. 2 and 3).

Because of the large size of the Mexican Pacific littoral, the description of the Mexican Pacific is based on the regions and the physiographic provinces published by the National Institute of Statistics, Geography, and Informatics (INEGI, 1989) (Fig. 4).

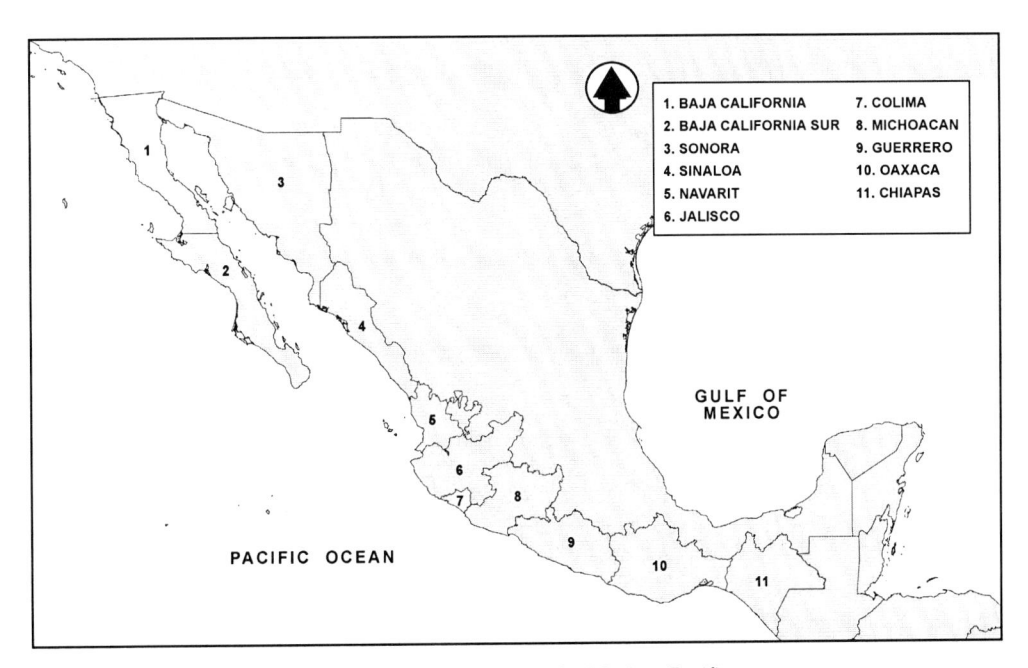

Fig. 2. Coastal states facing the Mexican Pacific.

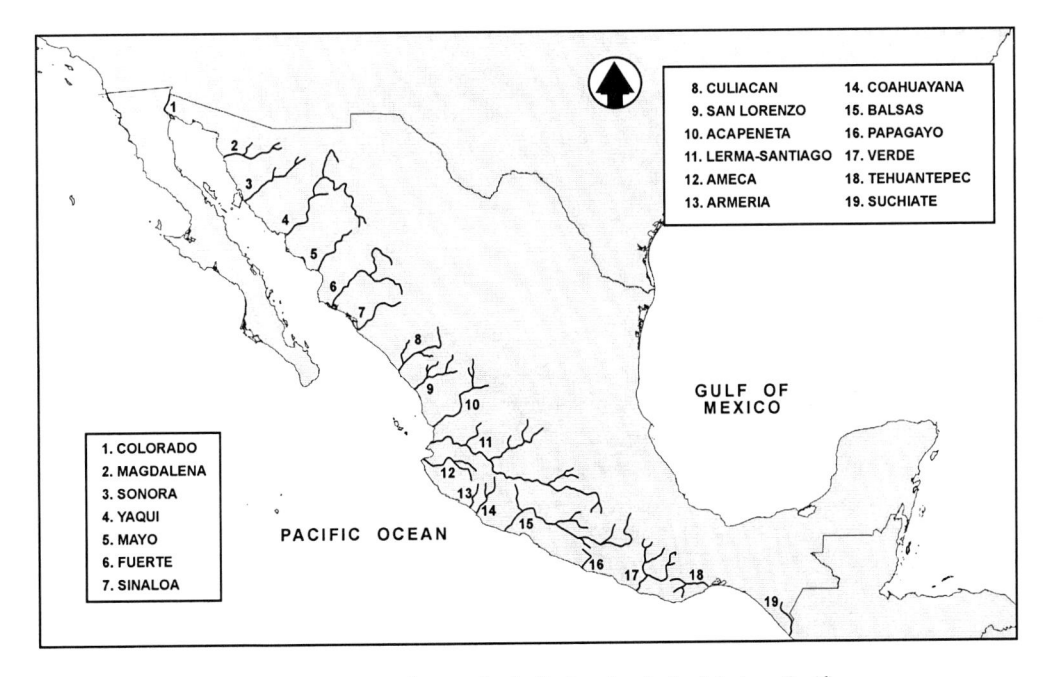

Fig. 3. Main rivers and watersheds discharging in the Mexican Pacific.

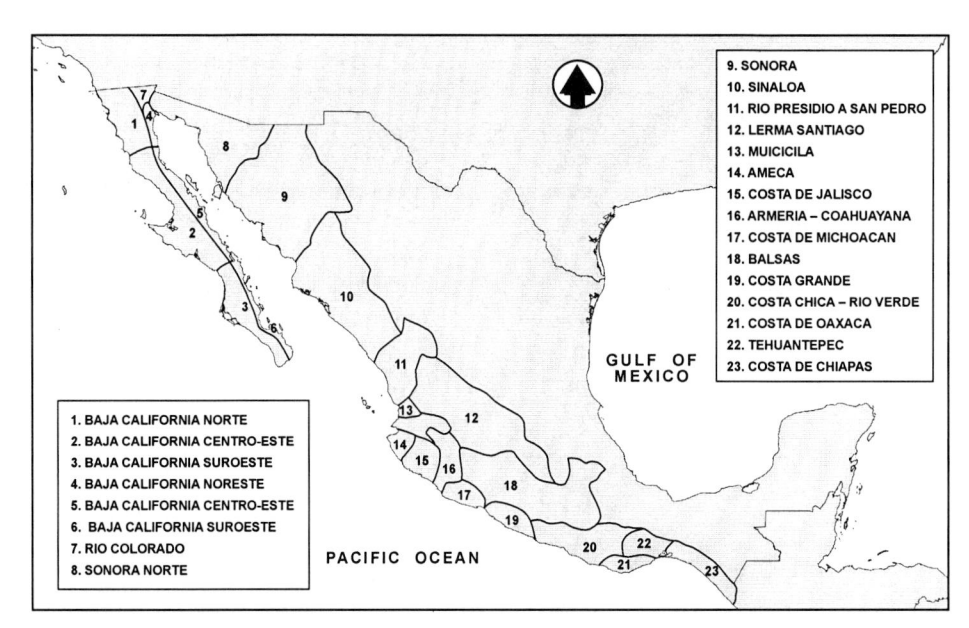

Fig. 4. Main coastal states and provinces in the Mexican Pacific.

Table 1

Length of the Mexican Pacific littoral (INEGI, 1989)

State	Length (km)
1. Baja California:	
Costa del Pacífico	880
Costa del Golfo de California	675
2. Baja California Sur:	
Costa del Pacífico	1400
Costa del Golfo de California	830
3. Sonora	1207
4. Sinaloa	640
5. Nayarit	300
6. Jalisco	341
7. Colima	139
8. Michoacán	246
9. Guerrero	484
10. Oaxaca	597
11. Chiapas	255

COASTAL AND OCEANIC PROVINCES

Californian Coastal Province

The Californian Coastal Province comprises the Pacific coast of the Baja California Peninsula stretching from the border with the USA to Cabo San Lucas, and includes the states of Baja California and Baja California Sur. It possesses 1600 km of coastline and is located between latitudes 32°31'58" to 22°52', and longitudes 117°07'31" to 109°53'. There are 16 coastal lagoons with high and intermediate relief, and many dry valleys with small basins within this area.

The climate is semi-hot, semi-arid, desertic, with an evaporation of 1400–2500 mm (annual average), and a precipitation of only 50–100 mm (annual average). Winds have marked seasonal patterns; winds blow predominantly towards the south during spring–summer, and towards the north during fall–winter.

The narrow continental shelf is, on average, less than 20 km; its widest portion is 50–70 km. In the central and southeast regions, surf energy is high and tidal energy generates high velocity currents. The main current is the California Current coming from the north, with an Equatorial Countercurrent towards 20°N. The California Current is classified as subarctic transition water whereas the Equatorial Current is part of the mass of the Equatorial Pacific.

Meteorological elements include annual tropical depressions associated with the "El Niño" phenomenon, or southern oscillation (ENSO). These features provide this region with nutrient-rich water, highly productive for fishery, especially for tuna, and changes to it have marked effects on resources.

There is volcanic-sedimentary-type substrate to the north with igneous and metamorphic rocks with alluvial deposits. In the central region are igneous-metamorphic rocks, and to the south lie alluvial deposits. Features include clay deposits, slimes, calcareous and siliceous skeletons, continental slope, trenches, marginal sea basins, isolated mountains, mountain crests, insular arches, abyssal plains, and the Patton rift and Popcorn Range. There are also major Basins (Colonett, San Quentel, San Isidro) and the Sebastián Viscaíno Bay.

Habitats include exposed areas, rocky beaches, cliffs, rocky platforms, coastal lagoons, vegetated dunes, and marshes. Lagoonal bodies are very productive, especially the macroalgal beds, cliffs, rocky platforms, protected beaches, and small islands. Sheltered marine areas are especially rich in seagrass beds, macroalgae, mangroves, and marshes, while the dunes contain rich sand-underbrush and xerophytic vegetation. The coast is an important transit zone for endemic and endangered species, providing feeding and nesting sites for migrating and resident birds, as well as for marine turtles. The grey whale as well as numerous fish species use this coast, due to its substantial areas of shelter.

Endemic birds, coral and abalone species are also found, although in comparison to the Gulf of California, endemic species are not abundant and have been affected by tourism and urban developments.

Environmental work in this area includes the use of species indicators for metal pollution. Substantial modification of marshes has been monitored using *Roscón plamoteado*, while some smaller pelagic species such as sardines and anchovies are important monitors because of their abundance, variability, and ecological role. Macroalgae, such as *Macrocystis pyrifera*, which are important for the maintenance of specific communities, are also monitored. The presence of *Pleuroncodes planipes* crayfish, which are characteristic of temperate–tropical transition zones, is important for the regional trophic chain, and is used to indicate natural and anthropogenic oceanographic changes.

North Pacific Oceanic Province

This region comprises the states of Baja California, Baja California Sur, Sinaloa, Nayarit, Jalisco, and Colima, limiting to the south with the oceanic provinces of the Sea of Cortés and South Tropical Pacific.

The climate is temperate in Baja California and semi-dry and hot to subhumid in Baja California Sur, with an annual rainfall of 100–200 mm in Baja California and of 300–500 mm in the Revillagigedo Islands (rain from July to December). Aeolian circulation in Baja California drives northeast winds, while in the Revillagigedo Islands winds to the south and southeast generate tropical storms. Meteorological phenomena in Baja California include cold fronts and El Niño effects.

The substrate in Baja California is formed by metamorphic rocks, while in the Revillagigedo Islands igneous rocks dominate. Slopes are steep. Tectonically Baja California is made up by the Pacific plate with telluric movements, and in the Socorro Islands, fumaroles provide main sources of energy.

The California Current dominates Baja California while the California and North Equatorial Currents dominate the Revillagigedo Islands. The main water masses in Baja California are superficial ones deriving from these currents. There are seasonal surges in the Revillagigedo Islands, though tides are semidiurnal in Baja California. Most areas have surf of high energy. Water depths in Baja California are 200–4100 m and steeply reach 3500 m around the Revillagigedo Islands.

The marine habitat of Baja California is mainly oceanic. Islands have beaches, cliffs and emerged rocks, whereas the Revillagigedo Islands in contrast are characterized by reefs. Baja California contains routes for migrating marine mammals, and is famously a reproduction site for the grey whale, as well as for seals (an endemic species of Baja California) and for the California sea lions, whereas the Revillagigedo Islands are a region for migrating tuna, finback whale, and sharks, and provide important nesting areas for birds and turtles.

Gulf of California Coastal Province

This province includes coasts of the states of Baja California, Baja California Sur, Sonora, Sinaloa and Nayarit, extending for 6500 km, between latitude 31°43′ to 20°24′, and longitude 114°43′ to 105°43′.

The climate to the north is semi-hot, and to the southwest it is semi-hot, desertic and mountainous.

The north and east coasts are semidesertic. The highest evaporation occurs in the north (2200–2400 mm), the central part has least (240–1600 mm) while evaporation is high at the mouth of the Gulf (1400–2000 mm). Precipitation in the north is 50–600 mm; in the centre and south it rises to 200–1200, while in the mouth it is only 160 mm per year. There is an aeolian circulation in the winter from the northwest and in the summer from the southeast (average velocity 5 m/s), and there is a hurricane season from August to October. The area experiences marked El Niño effects.

The substrate is of igneous, metamorphic rocks in the north, joined by alluvial sediments in the south. There are important phosphorite deposits. There are basins decreasing in depth towards the interior of the Gulf, while at the peninsular continental limit there are submarine canyons.

The East Pacific Plate and its transforming fault system dominate tectonics. The irregular shelf is not wider than 5 km, widening at La Paz to 20 km and at Sinaloa and Nayarit to 85 km, where it is then interrupted by the Frailes Canyon. Hydrothermal vents in the Guaymas basin provide the main sources of energy.

The main current is thermohaline, geostrophic and seasonal. In general, during the winter the flow is towards the south and in the summer to the north. Alternate cyclonic and anticyclonic gyres superficially join both coasts. The water masses come from the California Current and from the eastern subtropical Pacific. At greater depths, the water comes from a mix of subtropical, intermediate Antarctic and deep Pacific streams.

Tides are diurnal, semidiurnal and mixed, depending on the part of the Gulf. Wave energy is low, and depths vary from being deepest in the southern Pescadores Basin (3000 m), shallowing towards the High Gulf (15–50 m).

These habitats are highly diverse and very productive. Tidal areas and coastal lagoons are perhaps the richest, together with mangrove systems and coral reefs. Some of these diverse communities are currently being altered by agro-industrial wastes and tourist developments.

This area is important for the reproduction of the Monterrey sardine and contains both endemic species ("totoaba", *Totoaba macdonaldi*), endangered species such as "vaquita marina", (*Phocoena sinus*), and marine turtles. There are large regions of shrimp nurseries in Sinaloa and Nayarit and for anchovy in the northwest. The island region supports 80% of the sea lion populations of California, who give birth to 85% of their offspring here. It is also important for bivalve productivity.

Species benefiting from the high productivity of this area are, in general, the marine mammals, fishes, birds, marine turtles and molluscs (including mother-of-pearl). Numerous resident and migrating birds (seagulls and pelicans), marine turtles, plants and invertebrates all add to the richness of life in this region. The centre of this biological activity is probably located at Quino Bay, Guaymas. Yavaros, Grandes Islas region, and in coastal lagoons of Sinaloa.

Sea of Cortes Oceanic Province

This region includes coasts of the states of Baja California, Baja California Sur, Sonora, Sinaloa, Nayarit and parts of Jalisco, extending along the Gulf of California to its southern limit with the North Pacific Oceanic Province.

The climate is predominantly semi-hot and sub-humid with a rainfall of 125–400 mm. Dominant winds come from the northwest in fall and winter ("Nortes" season), and the area is subject to tropical storms, hurricanes, and El Niño effects.

The substrate is made up of igneous and sedimentary rocks, slimes and clays. Tectonic activity comes from the junction of the North-American and the Pacific plates with a Rift zone, such that hydrothermal chimneys are common.

The predominating currents are of the North Equatorial type, whereas water masses are tropical at the surface, and subtropical deeper. Deep water comes from the Pacific and Antarctic masses. Tides are semidiurnal, and the average depth of the province is 200–2300 m.

The habitat is oceanic. Most spectacular is the common humpback whale, found occupying the coastal and oceanic region of Banderas Bay for reproduction during the winter. These pelagic cetaceans point to abundant fishery resources.

Pacific Center Coastal Province

This includes coastal states of Jalisco, Colima, Michoacán, Guerrero, extending for 1512 km between latitude 20°24′ to 15°52′ and longitude 105°42′ to 95°00′.

The climate is hot, sub-humid to humid, with rains during the summer and an annual evaporation of 1400–2200 mm; precipitation averages 800–1600 mm. Trade winds ("vientos alisios") predominate from the southwest, changing to northwest at the equator. Annual average speed is 10 m/s.

There are tropical perturbations from May to November, when tropical storms, cyclones and hurricanes, as well as variable effects from El Niño, all affect the marine resources.

Substrates are igneous and metamorphic, with muddy sands, pelagic and terrigenous clays and detritus minerals. Submarine topography is of the east Pacific dorsal type, with zones of fractures. The shelf is very narrow, widening towards Jalisco and Michoacán. Oil is currently being prospected here.

Dominant currents are the Californian and North Equatorial. Surges occurring in the winter are produced by the cold winds running from the Pacific to the Gulf of Mexico. Tides are mixed and are semidiurnal. Energy is high with significant reflux velocities, and the least amplitude occurs near Lázaro Cárdenas, Michoacán. Water depth is 2000 m off Cape Corrientes, 4762 m off Colima, and 4000–4562 m on the coasts of Guerrero and Oaxaca.

The habitats of this coast include coastal lagoons, marshes, estuaries, mangroves, seagrass, coral reefs, coastal dunes, headlands and cliffs. There are large areas of seagrass and wetlands. Coastal dunes are colonised by *Atriplex* spp., *Suaeda* spp., red (*Rhizophora mangle*), white (*Laguncularia racemosa*), black (*Avicennia germinans*) and Chinese mangroves. It is well forested and contains palms. Its environmental characteristics are an important factor for a shrimp fishery which is carried out in the wetlands, and for tuna, swordfish, marlin, and sailfish offshore. Important marine turtle nesting areas occur here.

Mangroves species are used as indicator species here, whose height indicates environmental deterioration. Also seagrass condition is monitored. *Ceratrium* spp. is used to monitor red tides, while *Saliconia bigelovii* is used to indicate hypersalinity. Sea urchins (*Toxopneustes roseus*) and polychaetes indicate the quality of the general environment and *Thypha domingensis* is used to monitor eutrophication.

The centre of biological activity is probably Cape Corrientes (Piaxtla Point), a transitional zone that is temperate tropical.

Panamic Coastal Province

The Panamic Coastal Province comprises the states of Oaxaca and Chiapas, with 300 km of coast at latitude 15°52′ to 14°30′ and longitude 95°00′ to 92°00′.

The climate is hot, semi-arid and sub-humid, with an annual evaporation of 975–200 mm, and rainfall of 1200–5000 mm. Dominant winds are from the north and northwest; during the winter the Tehuantepec winds predominate (less than 10 m/s).

Meteorological elements include tropical perturbations from May to November, El Niño with variable effects on the resources, and Tehuantepec winds during the winter.

The substrate is comprised of igneous and metamorphic rocks; sediments are pelagic clays, middle and fine sands and coarse slimes. The continental shelf is wide, with an average width of 100 km at the east region of the

Tehuantepec Gulf, 4–6 km near Galeras Point (Oaxaca) and 50–80 km near Chiapas. Oil wells (exploration) and hydroelectric plants are sources of energy in this region.

The current is North Equatorial. Water masses are of tropical superficial water and intermediate water from the Antarctic. Surges occur in the winter, in the east region of the Tehuantepec Gulf; the winds in the region displace the superficial water mass, producing upwelling. Cyclonic and anticyclonic gyres occur with mixed and semidiurnal tides, and waves have high energy. Near Puerto Angel (Oaxaca), maximum depth is 4552 m.

Habitats and communities include bays, coastal lagoons, mangroves, sandy beaches, dunes, marshes, saltpetre beds, cliffs, and reefs. They also include seagrass systems, beaches, and wetlands containing red, white and black mangroves. Near the coast are low forests, grasslands, savannas and groves called "chaparrales".

Important environmental aspects include numerous turtle nesting beaches and large numbers of endemic species of fish, algae and corals, in a large variety of habitats. Important species include migrating and resident sea birds. The centre of biological activity is probably the Gulf of Tehuantepec.

Tropical South Pacific Oceanic Province

The tropical South Pacific Ocean Province comprises the states of Jalisco, Michoacán, Guerrero, Oaxaca, and Chiapas, whose extension is limited by the Central and Panamic Pacific coastal provinces.

The climate is hot and humid in the summer, and hot sub-humid to dry in the winter, with an average yearly rainfall of 1200–1400 mm. Important weather features include tropical storms, hurricanes, "nortes" in the winter, hurricanes in the summer, and El Niño effects.

The substrate is made up by igneous and sedimentary rocks, tuleithic basalt, slimes, clays and muds. The general topography is of the basin type with abyssal plain with fractures.

The currents are the California and North Equatorial, the Equatorial Countercurrent and the Costa Rica Current. There are mixed tides, high-energy surf, and depths range to 4000 m.

The habitat is oceanic. The whole province is affected by pollution with submarine waste, oil spills, and environmental damage caused by diverse shipping.

POPULATIONS AFFECTING THE AREA

Around 12 million people inhabit the coastal states of the Mexican Pacific, which represents 14% of the Mexican population. Most of them are concentrated in small or medium-sized coastal communities. The most populated region of the Mexican Pacific is the northwest, with a population of around six million Mexicans. About five million people inhabit the Central Provinces of the Mexican Pacific, whereas the South Panamic Province has only around one million inhabitants.

There are sparsely populated areas, such as the Baja California Peninsula where, due to the desert conditions and arid climate, agriculture is poor. The population is concentrated in some coastal cities of great importance due to their ties to the North American economy. These cities are Ensenada, Tijuana, La Paz, and the region of the San Lucas and San José Capes. The largest coastal populations are located in the Northwest Pacific and the Sea of Cortés. Most notable among these are Guaymas, Topolobampo, and Mazatlán. On the continental side of these provinces, in a large region of the Northwest coast, large hydro-agricultural facilities have been constructed that have exploited rivers, altering the water balance and polluting the coastal waters.

In the provinces of the Central Pacific, from Sinaloa to Oaxaca, there are about five million inhabitants, whose basic activities are fishery and agriculture. Except for the coastal states of Sinaloa and Nayarit, where irrigation is used for the cultivation of cereals, vegetables, sugar cane, and tobacco, the rest of the region depends on rainfall for its agricultural activities, and the irrigated areas are relatively small.

Further south, coastal populations engage in fishing in coastal lagoons and estuaries. The rural population lives on a subsistence economy, where cattle raising is of great importance, occupying increasingly more and more of the coastal plains. These are small communities of barely a couple of hundred inhabitants which are surrounded by larger communities of several thousand inhabitants. Puerto Angel, Salina Cruz, and Puerto Madero are the most important among these communities.

In the southernmost South Panamic Province, cattle raising, agriculture, and fishery predominate on the coastal plains and in the local waters. There is an irrigated agricultural region known as the Soconusco, adjacent to the border with Guatemala, where rivers have been exploited, and technologies with high energy consumption and agrochemical products are being used, risking the richness of the littoral systems.

Very little industrial activity is present on the Mexican Pacific coast. However, there are some relevant industrial developments: the giant salt exploitations in Guerrero Negro on the Gulf of California, the agroindustrial activities of the Imperial Valley of Mexicali, the irrigated agricultural fields of the Northwest, the fishery processing industries of Guaymas and Mazatlán, the industrial harbor of Lázaro Cardénas-Las Truchas (Michoacán), and the oil harbor in Salina Cruz (Oaxaca).

RURAL FACTORS AND FISHING

In the provinces of Coastal California, Gulf of California, Sea of Cortés, and North Pacific, there are important fisheries of benthic invertebrates such as shrimp, lobster, mussels, oysters, and snails (Cifuentes et al., 1990–1997). Among the smaller pelagic species, sardine and anchovies play an important role—both fisheries represent 30% of the entire national fishery production. Among the larger pelagic species, that of yellowfin tuna (*Thunnus albacares*), carp (*Katsuwonus pelamis*), bonito (*Sarda chilensis*), sailfish (*Isthiophorus platypterus*), swordfish (*Xiphias gladius*), blue marlin (*Makaria nigricans*), striped marlin (*Tretapturus andax*), and black marlin (*Makaria indica*) are the most important. Fishing is performed traditionally in the western coast of the peninsula, the mouth of the Gulf of California, Las Marias and Revillagigedo Islands, stretching into the coasts of Michoacán and Guerrero. Most of these fisheries are now exploited by foreign fishery fleets, especially by he Japanese fleets.

Undoubtedly, the shrimp fishery has reached its highest level among the exporting industrial activities of the region. At present, it represents the largest foreign currency exchange among the resources exploited from the Mexican seas. Industrial shrimp fishery in the Northwest started at the beginning of the century and acquired its greatest importance during the 1940s. Its deep-sea sites are located off the states of Sonora and Sinaloa, where a net towing fishery is practised. Different species are caught there, among them: the brown shrimp (*Penaeus californiensis*), blue shrimp (*Penaeus stylirostris*), white shrimp (*Penaeus vannamei*), and red shrimp (*Penaeus brevirostris*). The blue, white, and brown shrimps are usually caught in the coastal lagoons and littorals; around 60% of domestic catches are from these regions. Around 1500 vessels form the shrimp fleet, supported by 20 freezing plants and seven packing plants, whose infrastructure provides labour for a large part of the population. Because of the characteristics of the catching nets, a large amount of by-catch is also caught (7–12 kg per kg of shrimp), and this by-catch is only partially used profitably. This represents an evident waste and it has an important effect on both the trophic chains and on the biodiversity of the continental shelf.

Tuna, sardine, and anchovy stand out among the pelagic fisheries. The Mexican tuna fleet is one of the largest and most modern in the world and it operates mainly in the territorial seas of the East Pacific. The species mainly caught are the bluefin tuna (*Thunnus thynnus*), the yellowfin tuna (*Thunnus albacares*), and the albacore (*Thunnus alalunga*). Different fishing gears are used, such as circling net, the pole, the purse seine, the *almadraba* (Moorish for "trap"), the trawl line, and the fishing rod. A complete industrial infrastructure exists in this region of the Northwest Pacific for the catching, processing, canning, and marketing of the tuna fish. From its beginning (around 1930) this industry has focused on export, mainly to the North American markets. However, due to the embargo declared by the United States because of the incidental catch of dolphins, production since 1981 has been re-oriented towards domestic consumption, though to a lesser degree it still serves other countries.

The sardine fishery in the Northwest Pacific waters is another example of the economic and social relevance of the fisheries in this Mexican region. The species caught are: the Monterrey (*Sardinops sagax caerulea*), the "crinuda" (*Ophistonema libertate*), the Japanese (*Etreumess teres*), and the "big-mouth" (*Cetengragraulis mysticetus*). The former two represent 90% of the domestic catch, which at its peak reached 450 000 tons per year. Shrimp and fishing net vessels have been adapted for this fishery. The modern vessels are provided with electronic devices such as sonar, radio, automatic pilot, and radar. The industrial infrastructure is complemented on land with processing, preserving and canning plants, and marketing is efficient. It is one of the cheapest and most popular foods, accessible to large sectors of the population in its different forms. Several factors have caused a crisis to develop; most relevant being over-exploitation and the general economic crisis of the country, which has affected investments. From direct human nourishment it has recently been adopted as raw material for agricultural food, especially for fowl.

The anchovy fishery is relatively new in Mexican waters. It gained popularity with the decline of the sardine fishery. Despite its well-known nutritional value its use has mainly been to make fish flour, for which more than 95% of its production is used.

Mexico is among the regions with the highest diversity of algae and seagrasses in the world (Salazar-Vallejo and González, 1993). Its diversity is similar to that of the Red Sea, greater than that of the Mediterranean and that of South America, and can be compared to that of Japan. The hundreds of kilometres of the western coast of the Baja California Peninsula and the Gulf support rich populations of different species, among which stand out are *Macrocystis pyrifera*, *Gelidium robustum*, and *Gigartina canaliculata*. The extensive shallows bordering the hundreds of kilometres of coastal littoral sustain an important activity based on the exploitation of marine macroalgae, especially near the large islands of the north and centre of the Gulf of California, such as Angel de la Guarda Island, Tiburon Island, and Turner Island. Marine algae provides the fourth largest tonnage of the national fishery production, just behind the sardine, anchovy, and shrimp. Its production, estimated at several million tons (wet weight) per year, has been exported mainly to the USA and Japan for at least the last four decades. Its value as direct food for humans or animals, as raw material for food, and in the pharmaceutical, paper and clothing industries gives a strategic value to the Mexican seas.

There are some other important fisheries being carried out in the Northwestern Mexican Pacific, such as that of the lobster, the red urchin, the mussel and snail, octopus, abalone, oyster, and some other fish species, though these are relatively minor.

The artisanal shrimp fishery practised in the coastal lagoons, estuaries, and littorals poses other social, ecological, and technical problems that differ from those of deep-sea fisheries. For many years this type of fishery was reserved to community oriented organizations (Co-ops), which have used traditional catching techniques based on the use of diverse fishing gear. These include traps known as "tapos" or "chiqueros", which trap the shrimp in certain areas of the lagoons at their arrival in either postlarval or juvenile stages, and the people then wait for them to reach commercial sizes. Presently this activity is threatened by many problems, particularly environmental ones such as the over-exploitation of the resource, as well as by stresses produced by human activity in the lagoons, and by land-holding regulations. In the last years, this activity has been opened to private enterprises, which have focused their attention on highly productive areas to establish shrimp farms with international financial support. In many aspects, the situation endured by these artisanal fisheries, which supply a large proportion of the domestic demand, is a close reflection of the problems being experienced by Mexican society generally at the end of the century.

The Pacific Center Coastal Province comprises the south of Sinaloa, Nayarit, Jalisco, Colima, Michoacán, Guerrero and Oaxaca. This region supports a rich and varied marine and coastal fauna, due to the overlapping of the California and Panamic provinces with which it shares species. This diversity, enriched twice, makes this tropical Panamic region one of the most important fishing areas of Mexico, especially for some pelagic species such as the tuna, anchovy, sardine, carp, and shark, and for continental shelf species, such as the shrimp. About 250 different species have been identified on the continental shelf. Its littoral is characterized by the presence of important lagoonal systems, such as the Huizache-Caimanero, Teacapán-Agua Brava, and Mexcaltitán, and the bays of San Blas, Banderas, and Chamela. In these systems some 105 species have been identified, which sustain shore fisheries based on the shrimp and oyster, and some fin-fish species that satisfy the direct needs of some of the local communities as well as the demands of some regional markets. The trawling fisheries of the littoral are based on species such as the "mojarra", the red snapper, the porgy, and the grunt "ronco".

The Gulf of Tehuantepec possesses large fishery resources, including tuna, mackerel, sardine, saurel and carp. The shrimp fishery dominates coastal lagoons, to which fin-fishes of high commercial value, such as shark, red snapper, porgy, mojarra, stripped mullet, sea bass, as well as turtle, can be added. Most of these fisheries are barely developed, except for the shrimp and turtle fisheries. Tuna and sardine fisheries have only gained relevance in recent years.

Fisheries on the coasts of Guerrero and Oaxaca are mainly artisanal and are restricted to the coastal lagoons, where most of the rural communities catch shrimp and some fish species, basically for their own consumption or to supply regional markets. They use rustic crafts and wooden paddles, although lately they have been using fibreglass boats with outboard motors. Fishing gears consist of nets, casting nets, hammocks, conical shrimp nets, and traps known as "tapos". The latter is an ancient technique consisting of the construction of cages made with mangrove rods and synthetic fibre nets (CIESAS, 1985–1990). The beaches of Oaxaca, which are turtle nesting sites, are favoured sites for reproduction of several turtle species including the "golfina" (*Lepidochelys olivacea*) and the tringlada (*Dermoschelys coriacea coriacea*). Thousands of turtles arrive on the beaches every year and the turtle fishery has therefore been an important activity for the inhabitants of this region, who profit from its meat, oil, carapace, and eggs. Over-exploitation, fostered by commercial interests foreign to the region, has endangered these species, so their utilization has had to be regulated by establishing closed seasons and creating turtle conservation areas.

Cliffs, beaches, and bays with beautiful scenic views provide the region with a huge tourist potential. The latter has prompted governmental strategies towards developing large tourist projects, such as Ixtapa-Zihuatanejo, Acapulco, and recently, Huatulco and Puerto Escondido. These are located in areas with small and diverse villages which subsist on agriculture and low productivity shoreline fisheries. These megaprojects are isolated patches, in an environmental and social context, and are wealthy sites surrounded by poverty, causing marked contrasts.

There are, however, short-term plans for tourist developments aimed at integrating different centres with plans to link the scenic values of the beaches and bays with the cultural and archaeological riches of the region. Hence, one plan is to link Acapulco, Ixtapa-Zihuatanejo, Taxco and Mexico City, and another one to join Puerto Escondido and Huatulco with the city of Oaxaca.

Although the area has the least industrial development of the Mexican coast, other activities are starting to cause stresses to its biological richness, such as the oil harbour of Salina Cruz in the state of Oaxaca, and the agroindustrial emporium of the Soconusco in Chiapas. In Salina Cruz, activities related to oil and petrochemical products are beginning to affect the environmental health of the harbour and its underlying marine area. Polyaromatic hydrocarbons have been found in sediments, molluscs, and in fishes consumed by the inhabitants of the area (Botello et al., 1998). The Soconusco area, a rich agroindustrial region with diverse plantations (rice, banana, soy, coffee), requires the massive use of fertilizers and pesticides. These have been used without any control, and as such have altered the ecological balance. Further elements causing tension are the projects to establish shrimp farms in critical areas of the coastal plain.

These projects have been promoted by private entrepreneurs, the government, and international financing agencies. They leave the fate of the resources, as well as the nutrition and health of the rural communities, uncertain.

EFFECTS OF URBAN DEVELOPMENT AND INDUSTRIAL ACTIVITIES

The impact exerted by human activities has come about from several factors: changes in the use of the land, alteration of wetlands and coastal lagoons by construction of dams for irrigation purposes and to produce electricity, conversion of wetlands into shrimp farms, eutrophication of estuaries and coastal lagoons, loss of habitats due to urban and tourist developments, and pollution by municipal and agroindustrial wastes (Páez-Osuna et al., 1998a).

Populations are concentrated mainly in some medium-sized and small coastal communities: among the most relevant ones are Ensenada, La Paz, Guaymas, Mazatlán, Topolobampo, Manzanillo, Acapulco, Zihuatanejo, and Salina Cruz. These coastal communities have grown without coastal urban development ordinances, which means that besides producing agricultural water run-off charged with fertilizer and pesticide residues and other agro-industrial waste waters, the untreated municipal waters are also discharged into coastal systems, polluting not only fishery areas but also beaches and recreational areas.

The coastal zone of the Northwestern Mexican Pacific has been, in one sense, a huge experimental laboratory for the technological revolution of the 1960s; the "green revolution". Based on the control of rivers through dams and channelling, and the massive use of agrochemical products (fertilizers and pesticides), this activity has altered the regional hydraulic balance and has produced stresses due to the pollution of the waters that sustain biological productivity and fisheries.

Some activities linked to the development of high density tourist centres have also affected the environment, such as dredging for marinas and modification of habitats.

The accelerated population growth in Mexico has migrated steadily towards the coastal zone where anthropogenic activities were initially, and continue to be, directed at agriculture, particularly in areas adjacent to estuarine and lagoon environments. Among these, noteworthy are those of northwest Mexico within the Gulf of California (Sonora and part of Sinaloa) and the Mexican Pacific (south of Sinaloa and Nayarit).

The northwest area has a predominantly arid and semi-arid climate, with extreme periods of rainy and dry seasons and a marked fresh water deficit, as well as the lowest density of river basins dammed for agricultural and livestock management. However, there are lagoon areas that are fed by rivers whose flow is controlled by dams.

The coastal lagoons of the states of Sinaloa and Nayarit support fisheries that are important at the national level, but these have a variety of problems, including natural ones due to their geological evolution, and those caused by agricultural, livestock and aquacultural pollution.

Páez-Osuna et al. (1998a) compared the nitrogen and phosphorus discharges of agriculture (141,232 N ton/year and 35,272 P ton/year), the municipality (45,990 N ton/year and 16,097 P ton/year) and aquaculture (2866 N ton/year and 462 P ton/year). Although the northwest of Mexico represents little more than 20% of the territory it boasts a mechanized agriculture that produces more than 10 million tons of 68 different vegetables and seeds on more than 1.5 million hectares, helped with liberal use of fertilizers and pesticides (de la Lanza-Espino and Flores-Verdugo, 1998). The contribution of nitrogen and phosphorus from shrimp culture, commonly cited as important sources of such chemicals, represents only 1.5% and 0.9% respectively, of all sources of these fertilisers.

This aquaculture became important only recently, as although during the 1970s it was recognized as an important economic activity and a potential source of food, more than 20 years passed before it became developed. Up to 1997 there were 284 shrimp farms in the country with 18,685 ha of cages, canals and tanks. Sinaloa and Nayarit have 82% of the farms, with 78% of the surface area in working order (DGA, 1997).

The coastal area of the two states supports important urban centres (cities) far from the estuarine and lagoon environments, and settlements are only around 5000 to 50,000 inhabitants. Communal farmers and fishermen are also found along the margins, many of which discharge their drainage and water used for economic activities directly into the coastal bodies.

Case Studies

Sinaloa has eleven coastal systems, and Nayarit has three. However, some aquatic systems are noteworthy because of their size, environmental characteristics, and fisheries, as well as because of pollution problems and conflicting economic interests between adjacent lands used for different purposes. In Sinaloa these include Bahía de Navachiste, Ensenada del Pabellón-Altata and Huizache y Caimanero, and in Nayarit these are Teacapán-Agua Brava and Mezcaltitán. The environmental problems that have developed there will be analysed below.

Navachiste–San Ignacio–Macapule bays.

These bays have an area of 24,700 ha, located between 25°53'–25°11'N and 108°11'–108°55'W. They are separated from the sea by a barrier and communicate with it only via two canals. Two rivers drain into the system: the Sinaloa and the Arroyo Ocoroni, with two important dams (300 million m^3 and 2900 million m^3, respectively) that are used for irrigation (with 97,820 ha and 8650 ha, respectively). It has a dense and exuberant mangrove vegetation with the exception of the area adjacent to the agricultural lands. The mangrove species are *Rhizophora mangle*, *Avicennia germinans* and *Laguncularia racemosa* (with 14,200 ha), together with the marsh halophytes *Salicornia* sp. and *Sesuvium* sp., among others.

Fertile soils are located behind the water bodies as well as inland, with seed, cotton and vegetable crops enhanced mainly by irrigated (88,242 ha) and seasonal (60 ha) agriculture.

The region is annually subjected to cyclones and hurricanes, particularly in July and September, that cause flooding in agricultural and urban areas. In contrast, high evaporation rates generate saline soils. The water table is located at depths between 3 and 300 m and, as water is obtained from dams, the water table is under-exploited.

From the fisheries point of view, the area is mainly important for its shrimp and for certain commercially valuable fish. Migratory birds from Canada and the United States use the coastal area seasonally to rest and feed.

There are few studies on the impact generated by the various human activities in the area. Heavy metals (Cu, Hg, Zn) and pesticides (chlorinated and organophosphorous) have been recorded in shrimp showing that they are present in amounts smaller than the limit established by the national norms (NOM-029-SSAI-1993) (SEMARNAP-FAO, 1997).

There are 153 shrimp farms among the canals that form part of the coastal system, of which only 118 are in working order. Many of these are near mangroves that have deteriorated as a result of changes in the dynamics of the water circulation, and in some cases by logging, whereas in others there has been incipient growth along the margins of the culture ponds. International groups have recommended that this type of aquaculture take place away from this vegetation and with a better technology, in harmony with the environment.

Altata-Ensenada del Pabellón.

This coastal system is located between 107°28'–107°48'W and 24°19'–24°32'N and is formed by a bay to the west (Altata) and by an estuarine lagoon water body to the east (Ensenada del Pabellón), with an area of 36,000 ha (7500 ha and 28,500 ha respectively). Ensenada del Pabellón is formed by small marginal lagoons and has a major central body. The greater part of the water of the river Culiacán enters the system near the inlet to the sea, with canals to the north and northeast. This river provides water to both two nearby cities and to agriculture. The bigger city, Culiacán, is 40 km away and has approximately 300,000 inhabitants, and the smaller one, Navolato, is 20 km away with 50,000 inhabitants. Both cities are on the river margins. The terrestrial area surrounding Ensenada del Pabellón is agricultural (vegetables and sugar cane). Páez-Osuna et al. (1998a) calculated a surrounding agricultural area of 113,522 ha from which a variety of pesticides has reached and been recorded in the water. There are sediments and organisms containing concentrations which sometimes exceed the legal levels, with marked seasonal and spatial variations. The results show up in diverse ways such as environmental stress, changed ages and habitats of organisms, their tolerance, physiological states and taxa, among others. Among the chlorinated pesticides, endosulfan has reached levels of 140 ng/g in mussel tissue, DDT in clam tissue is 300 pg/g and chlorpyrifos in sediment is 7.6 ng/g. These are associated with vegetable crops and levels increase during the dry season (February). In spite of the organophosphorous compounds degrading more rapidly, some are persistent and have been detected in concentrations of <0.01 ng/g d.w. For example, parathion is preferred for sugar cane crops. This is a persistent organophosphorous compound and there is evidence that it is used commonly. However, chlorpyrifos is more toxic and sensitive in shrimp at levels of 10 ng/g. This pesticide is stable in marine systems and its use coincides with the arrival of postlarvae during the rainy season (Readman et al., 1992).

Apart from the chlorinated and organophosphorated pesticides, some fungicides contain heavy metals with a varying degree of toxicity, and these have been detected in fish, shrimp, clams, mussels and oysters. Notable are Cd with 20.8 $\mu g/g$ in shrimp, and Cu, Mn and Zn with a moderate degree of pollution in oysters (*Crassostrea corteziensis*) and mussels (*Mytella strigata*), with marked spatial, temporal and physiological variations. High contents of Cu (5.6 $\mu g/g$) are also notable in the fish *T. mozambica* and *Mugil curema* (6.3 $\mu g/g$) (Izaguirre-Fierro et al., 1992).

Increased sediment is also associated with agriculture. An increase of approximately 20 times the potential reserve of phosphorus and nitrogen that can be diffused into the water column as orthophosphates and ammonium has been recorded in the sediment of areas affected by drainage outlets, in comparison with unaltered areas. This high content of nutrients, in spite of having localised origins, is spread throughout the system by the tides at concentrations that cause eutrophication (de la Lanza-Espino and Flores-Verdugo, 1998). The sugar industry also has an effect in the form of organic residues that enrich the sediment; confirmed through changed ratios of TCH:TOC (Páez-Ozuna et al., 1998b).

Huizache y Caimanero lagoon system

Huizache y Caimanero is located 250 km from the Altata-Ensenada del Pabellón lagoon system between 22°50'–23°05'N and 105°55'–106°15'W. Both lagoon systems are in the state of Sinaloa and have a similar climate. Huizache y Caimanero is located between two rivers: the Presidio to the north and the Baluarte to the south. Both reach the sea directly, are markedly seasonal and communicate with the lagoons through long and winding canals. The canals were built in 1940 to improve the flow of fresh water to the northern sections of both lagoons.

The marsh of Huizache has an area of 14 km^2 and the lagoon of Caimanero an area of 51 km^2. During the dry season, the marsh dries up completely, whereas the lagoon loses 60% of its volume and reaches salinities of 60 PSU. Its accelerated geologic evolution has tended to silt it up converting it to land used for agriculture over the course of around 40 years. Small settlements of fishermen and communal farmers are unevenly distributed. This lagoon system was the first in Mexico to be environmentally managed by dredging marine inlets and keeping them open to allow the entrance of the shrimp postlarvae that sustain an important fishery. It must be mentioned that this has been classified as extensive aquaculture supported by local fishing aids called "tapos" (structures built with mangrove trunks in the canals that allow the postlarvae to enter but bar the adults so that they return to the sea to reproduce) (de la Lanza-Espino and García-Calderón, 1991).

The first efforts to establish shrimp culture in Mexico (semi-intensive) were carried out in this lagoon system, and a variety of aquacultural and technical problems resulted in a loss of area, changes in the circulation in freshwater bodies, a loss of postlarvae, an increase in silting and a decrease in the fishery. In view of these numerous problems, shrimp culture has not expanded in this lagoon system and only six farms are operational at present with low and irregular yields (DGA, 1997).

The proximity to agriculture has caused peripheral runoff containing fertilizers and agrochemicals to reach the sediment and, in consequence, the shrimp. Information indicates that the chlorinated pesticides apparently follow the annual cycle, associated with rainfall-evaporation, with maximum concentrations in July input by the rain and the atmosphere. The most important in the sediment are heptachlor (18.2 μg/kg), aldrin (6.9 μg/kg) and DDT (16.4 μg/kg) (Rosales et al., 1985). In 1983 low levels under the limit permitted, were quantified in fish, shrimp and molluscs. However, HCH (all isomers) with 2.5 ng/g, dieldrin with 10.5 ng/g, and DDT with 11.7 ng/g can now be detected (Rosales and Escalona, 1983).

As a result of the composition of agrochemicals, metals have also been recorded in benthic organisms such as shrimp with Cu (21 μg/g), Fe (133 μg/g) and Zn (84 μg/g) reaching maximum values similar to those of other areas, although with spatial, temporal and metabolic differences (Páez-Osuna and Ruiz-Fernández, 1995).

In spite of its economic importance, this lagoon system has had few studies with respect to pollution.

Teacapán–Agua Brava lagoon system

This complex system is located in the state of Nayarit between 22°04'–22°35'N and 105°20'–105°50'W, with a surface area of 40,000 ha and more than 1500 km^2 of tidal canals and mangroves. Four rivers drain into this system. From north to south they are the Cañas, Acaponeta, Rosamorada and Bejuco. The second is the most important and it is constant (although its volume decreases during the dry season), whereas the others are markedly seasonal and dry up completely during the dry season.

Its geological evolution has been very active. Silting up, changes in the river beds, marine transgressions and regressions, and the climate have created many water bodies of different sizes, different morphological types. Multiple beach barriers have remained in some places and have formed bars in others.

This estuarine system includes more than 150 lagoons, channels, streams, swamps, marshes, dew ponds and dams, many of which contain mangroves.

There are only 11 bodies bigger than 100 ha, but the largest is Agua Brava with a little more than 750 ha. This system communicates with the sea through two inlets, one to the north (Teacapán) 1660 m wide and 9 m deep, and an artificial inlet to the south (Cuautla) built in 1976, 200 m wide and 10–20 m deep. As a result of high flows, the inlet has expanded, and in 1998 it was 1500 m wide and 50 m deep; it is still unstable (SEMARNAP-FAO, 1995). As the area is hit by hurricanes and other tropical perturbations, the dynamics and speed with which it will expand further are unknown. This situation has significantly changed the distribution pattern of the salinity. Marine water has been allowed in which has increased the salinity of a previously mesohaline environment, and has affected the mangrove community with a quite obvious decrease of 24% (Flores-Verdugo, 1994).

This aquatic system has a semi-desert to warm sub-humid climate, with a rainfall gradient from north to south of 1000–1500 mm, an evaporation of 1900 mm, and a water deficit.

It has winds of the monsoon type from May to October and a season of hurricanes and tropical storms from June to November.

The El Niño South Oscillation phenomenon also affects this coastal area, not only by increasing the sea level but by modifying the pattern of seasonal winds and rains and thus the volume of the rivers and inundation areas, depending on the intensity of the phenomenon. When the phenomenon is strongest and lasts two years, the drought is greater during the first year and the rains are more abundant, but with an even greater imbalance during the second (de la Lanza-Espino, 1992).

The mangrove forest around this lagoon system is especially rich. It covers 108,113 ha (15% of the total in Mexico) together with that of the Mezcaltitán lagoon to which it is adjacent and with which it shares a common origin. It is mainly made up of *Laguncularia racemosa* (white mangrove) and *Rhizophora mangle* (red mangrove) (de la Lanza-Espino et al., 1996). This vast community has been fed by seasonal inundations from adjacent rivers or by those that drain into the lagoon system. However, between 1980 and 1990 it decreased by 24% as a result of diverse anthropogenic activities (Flores-Verdugo, 1994), including agriculture, aquaculture and the construction and filling in of the Presa Aguamilpa dam and the supphir "San Rafael", that altered not only the inundation area but also the sedimentation of the delta of the main river, increasing erosion (Ortíz-Pérez and Romo-Aguilar, 1994; de la Lanza-Espino et al., 1996).

The adjacent agricultural area in northwestern Nayarit reached 123,162 ha (42% of the state) in 1991 with a potential increase of 110,000 ha for irrigated agriculture (5%), seasonal agriculture (15%), livestock and forestry (19%), grassland (21%) and the rest for fisheries and aquaculture.

Nitrogenated and phosphorated fertilisers are used in agriculture for more than 19 crops, including seeds and vegetables. More than 4,500,000 tons/year of more than 10 formulas are used (SEMARNAP-FAO, 1995). Also, levels of faecal coliforms of 240,000 NMP/100 ml have been detected during the dry and rainy seasons in neighbouring localities in rivers such as the Acaponeta.

Other studies on trace metals in the coastal states of the Mexican Pacific show the mean concentrations of metals in sediments (Fig. 5). However, analyses are quite limited (Botello et al., 1994; Guerrero, 1993; Osuna-López and Páez-Osuna, 1986; Páez-Osuna and Osuna-López, 1987, 1990b; Osuna et al., 1990; Rosales et al., 1994). Average Pb concentrations of 55.00 $\mu g\ g^{-1}$ are reported for the areas of Chiapas, Gulf of California. The coastal lagoons of the states of Sinaloa, Nayarit, Oaxaca, show low concentrations (less than 50 $\mu g\ g^{-1}$), and hence could be considered as areas not polluted by Pb. Other metals, such as Cd and Cr, are reported with low concentrations, except for the northwest region of Baja California, where the highest values for these metals were found (Cd, 10.10 $\mu g\ g^{-1}$; Cr, 75.00 $\mu g\ g^{-1}$) and the Mitla Lagoon in Guerrero (91.98 $\mu g\ g^{-1}$).

Cu (176.18 $\mu g\ g^{-1}$), Co (43.29 $\mu g\ g^{-1}$), Ni (147.94 $\mu g\ g^{-1}$), and Zn (254.81 $\mu g\ g^{-1}$) show similar concentrations in all analysed systems, except for the mouth of the Gulf of California, which contains the highest values for these four metals (Páez-Osuna and Osuna-López, 1990a).

Average concentrations of toxic metals in the seven studied states of the Mexican Pacific show that Baja California (75.00 $\mu g\ g^{-1}$) and Sinaloa (59.00 $\mu g\ g^{-1}$) have the highest values for Cr, whereas large concentrations of Pb were found in the states of Guerrero (59.64 $\mu g\ g^{-1}$) and Chiapas (54.79 $\mu g\ g^{-1}$).

Mercury is the metal most extensively studied in Baja California, in the fish *Tilapia mossambica*, the clam *Corbicula fluminea*, and in mussels *Mytilus edulis* and *Mytilus californianus* (Gutiérrez-Galindo and Flores-Muñoz, 1986; Gutiérrez-Galindo et al., 1988, 1989). Values range from 0.05 to 0.49 $\mu g\ g^{-1}$ dry weight. These values are below those established by FDA (Nauen, 1983) as permissible for human consumption (2.5 $\mu g\ g^{-1}$).

Pb concentrations listed on Table 2 are below the maximal allowed levels according to FDA, except for that in the species *Mytilus strigata* (Páez-Osuna and Marmolejo, 1990) *Mugil curema*, *Tilapia mossambica* (Gutiérrez-Galindo et al. 1989), and *Crassostrea corteziensis* (Páez-Osuna et al., 1991) from the harbour of Mazatlán (Sinaloa). Values range from 10.00 to 15.00 $\mu g\ g^{-1}$, as well as for *C. corteziensis* (7.70 $\mu g\ g^{-1}$) from the Camichin estuary (Nayarit) (Páez-Osuna et al., 1988).

Cd concentrations in the species listed in Table 2 show a uniform distribution, except for the molluscs of the Navachiste lagoon of the state of Sinaloa (Páez-Osuna and Tron-Mayer, 1995), which show values higher than 8 $\mu g\ g^{-1}$. Cr, Cu, and Co present similar concentrations in diverse species of molluscs, crustacea, or fishes (Izaguirre-Fierro et al., 1992; Páez-Osuna et al., 1988, 1995; Páez-Osuna and Ruíz-Fernandez, 1995; Villanueva and Botello, 1996).

Studies in the state of Baja California have shown variations in the values among species and from one region to another; this could be ascribed to temporal fluctuations of metal levels in the aquatic biota, amount of metal introduced into the environment, weight changes occurring in the living organisms and the direct effects of temperature, salinity, and other characteristics altering seasonally, as well as the stability of the source of these metals (Pringle et al., 1968).

The high levels of metals in coastal areas are due mainly to the continuous and massive discharge of residual waters by coastal cities, by industrial, mining, tannery activities, galvanoplasty, and fertilizer runoffs. These levels vary according to the periodicity and magnitude of the discharges. Finally, atmospheric emissions from urban and industrial activities in the coastal regions are another important source of pollutants.

The information presented here shows an increased tendency of pollution with metals in coastal areas of the Mexican Pacific, mainly Pb, Cd, Cr, and Zn, metals that produce toxic effects on organisms living in Mexican coastal areas and hence deleterious effects on important fisheries of the region, such as those located in Sinaloa and Nayarit.

Finally, information on the presence of petroleum hydrocarbons is very limited and sparse for the Mexican Pacific in comparison with the data obtained in the Gulf of Mexico coasts.

Botello et al. (1998), in an extensive study carried out in the Port of Salina Cruz and adjacent areas in Oaxaca (Cen-

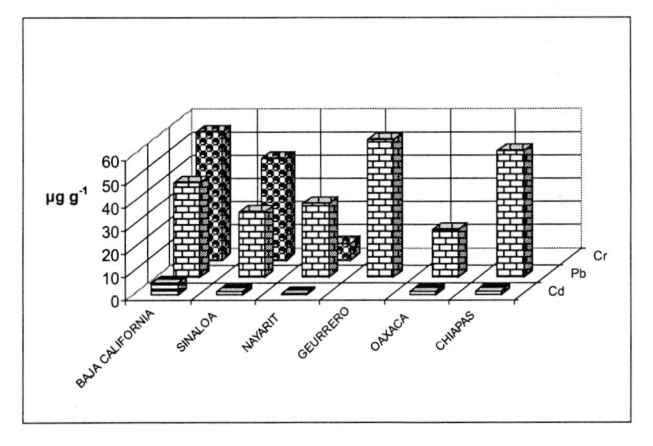

Fig. 5. Mean concentration of toxic metals in sediments ($\mu g\ g^{-1}$) from the Mexican Pacific.

Table 2

Mean values for metals in organisms ($\mu g\ g^{-1}$ dry wt) in coastal areas from the Mexican Pacific

State/organism	Lead	Cadmium	Chromium	Copper	Cobalt	Nickel	Zinc	References
Baja California								
Northwest Coast								
Panulirus inflatus		0.33	0.45	82.91	1.45	2.47	74.45	Páez-Osuna et al. (1995)
Penaeus californiensis		5.26		41.09		5.63	147.90	Páez-Osuna et al. (1995a)
Southern Gulf of California								
Saccrostrea iridescens		3.60		20.40		1.70	402.00	Páez-Osuna & Marmolejo (1990)
Mexicali Valley								
Tilapia sp.		N.D.	0.33	2.35			29.79	Gutierrez-Galindo et al. (1989)
C. carpio				3.36			69.82	Gutierrez-Galindo et al. (1989)
La Paz Harbour								
Penaeus californiensis	0.5–0.30	0.10–0.15		18–25			60–75	Mendez et al. (1997)
Sinaloa								
Navachiste Lagoon								
C. corteziensis	1.38	10.30	1.04	67.40	0.80	2.60	509.00	Páez-Osuna et al. (1991)
C. palmula	3.20	10.30	0.80	104.00	1.10	2.30	1190.00	Páez-Osuna et al. (1991)
Cerrito Habour								
Saccrostrea iridescens		3.60		20.00		1.70	402.00	Páez-Osuna et al. (1991)
Culiacán Valley								
Mugil curema	14.14	0.47	2.64	16.07	4.72	4.37	54.33	Izaquirre-Fierro et al. (1992)
Tilapia mossambica	15.21	1.12	2.21	35.09	2.82	4.78	378.67	Izaquirre-Fierro et al. (1992)
Mazatlán Harbour								
C. corteziensis	14.60	1.50	0.63	82.00	3.80	6.10	1620.00	Páez-Osuna and Marmolejo (1990a)
Mytella strigata	11.70	0.20		12.10	2.20	8.10	26.00	Páez-Osuna (1990b)
Penaeus vannamei		0.57	1.45	23.30	0.91	1.30	60.60	Izaguirre-Fierro et al. (1992)
Penaeus stylirostris		0.44	0.72	21.20	1.08	1.72	83.70	Páez-Osuna & Marmolejo (1990)
Altata-Ensenada from Pabellón Lagoon								
C. californiensis		2.50	0.90	9.95		8.11	377.00	Páez-Osuna et al. (1993)
C. subrugosa		2.88	2.13	50.28		9.83	434.17	Páez-Osuna et al. (1993)
Tellina sp.		6.00	2.70	119.00		3.80	806.00	Páez-Osuna et al. (1993)
C. palmula		8.20		150.00		2.90	943.00	Páez-Osuna et al. (1993)
C. corteziensis		3.90		147.00		1.90	727.00	Páez-Osuna et al. (1993)
Teacapan Lagoon								
Penaeus stylirostris	N.D.	0.44	0.72	21.20	1.08	1.72	83.70	Páez-Osuna and Ruíz 1995
Nayarit								
Camichin Estuary								
C. corteziensis	7.70	6.10	0.80	26.60	3.80	3.20	426.00	Páez-Osuna et al. (1995)
San Cristobal Estuary								
C. iridescens		1.49		25.89		2.10	628.33	Páez-Osuna et al. (1995)
Northwest Nayarit Continental Shelf								
Penaeus stylirostris	N.D.	0.61	0.29	36.60	0.42	0.72	74.60	Páez-Osuna and Ruíz (1995)
Colima								
Manzanillo Harbour								
Crassostrea iridescens				82.63		14.75	614.60	Guerrero (1993)
Oaxaca								
Salina Cruz Harbour								
C. iridenscens	0.36		0.14			0.70		Botello et al. (1994)
Penaeus stylirostris	0.22		N.D.			0.69		Botello et al. (1994)

tral Mexican Pacific), show important concentrations of PAH's in sediments and crustacean organisms (*Penaeus stylirostris*) from the study area. The major PAHs in sediments of the inner harbour were fluoranthene, benzo(a)anthracene, and benzo(b)fluoranthene in a range of concentration from 0.01 to 3.20 $\mu g\ g^{-1}$ dry weight.

Of paramount importance was the presence of PAHs in tissues of the crustacean *Penaeus stylirostris* and mollusc *Crassostrea iridiscens*. Chemical analysis showed the presence of pyrene, benzo(ghi)perylene, benzo(b)fluoranthene and benzo(a)pyrene in a range of concentration from 0.90 to 14.50 $\mu g\ g^{-1}$ dry weight.

These data confirm that oil pollution in the Salina Cruz Port and adjacent areas originates mainly from municipal effluents, direct discharges from the local refinery, spillages from fishing fleet vessels, port operations and presumably also atmospheric transport. The level of contamination may threaten important local fisheries.

Other studies on petroleum pollution were carried out in the coasts of Baja California, Sonora and Sinaloa states using bivalves as biological indicators of pollution (Páez-Osuna, 1998a).

The results obtained show very low levels of petroleum hydrocarbons, mainly PAHs, of which concentrations have increased in important ports such as Ensenada, Guaymas and Mazatlán.

Indeed, pollution by oil in the Mexican Pacific coast is not an important ecological issue compared with the levels found in the Gulf of Mexico coasts, where the biggest petrochemical plants in Latin America are located, as well as the main oil marine platforms producing 2×10^6 barrels of crude oil daily.

PROTECTIVE MEASURES

One of the most useful measures in the protection of natural resources in general and marine resources in particular, is the creation and maintenance of protected areas (McNeely, 1994; Heywood and Watson, 1995). It is estimated that at least 1% of the coastal zones around the world are protected under some form of measure compared with the 6% for the terrestrial surface. Marine Protected Areas (MPA) are coastal zones or oceanic zones designed, regulated and managed for the ecosystem conservation along with its functions and resources; they range from the small, highly protected reserves which sustain several species and maintain different natural processes, to bigger areas in which the conservation and development of different socio-economic activities are combined (De Fontaubert et al., 1996).

The MPA, if well planned, could achieve more than the protection of critical habitats and species alone, by providing a reference or base line for the application of adequate management, and by establishing and maintaining an interaction between science and politics. The majority of

Table 3

Protected Natural Areas in the Mexican Pacific

	Protected Natural Areas (Mexican Pacific)	Area (ha)
1	El Vizcaíno	2,546,790
2	San Ignacio	
3	Costa Oriental Vizcaino	
4	Isla Guadalupe	25,000
5	R.B. Archipiélago Revillagigedo	636,685
6	Cabo San Lucas, Cabo Pulmo	7,111
7	Bahía Loreto	
8	Islas del Golfo de California	150,000
9	Isla Rasa	61
10	Isla Tiburón	120,800
11	Alto Golfo de California	934,756
12	Cajón del Diablo	
13	Isla Isabel	194
14	Arcos de Vallarta	
15	Playas Ciuxmala, El Tecuán	
16	Mismaloya y Teopa	
17	Playa de Tierra Colorada	
18	Lagunas de Chacahua	14,187
19	Huatulco	11,890
20	Playa de Puerto Arista	
21	La Encrucijada	144,168
	Total	4,591,642

MPAs include zones where different uses of the resources coincide in a compatible manner; because of this it is essential to take into account traditional sustainable uses of the local communities (Table 3).

The only feasible alternative for the maintenance of the marine systems is for conservation strategies whose strong scientific bases are oriented towards the protection of the function of the areas, rather than of single or specific resources. In Mexico, there are 112 marine protected areas which cover approximately 12 million hectares; however, marine ecosystems are poorly represented (Table 3).

For the Mexican Pacific coasts there are 21 natural protected areas totalling 4,591,642 ha. These areas protect different species such as the grey whale, the seal, the turtle, the dolphin, marine birds and highly productive ecosystems like the mangroves and the wetlands.

Besides the Declaration of Natural Protected Areas, Mexico is included in different international agreements for the protection and management of marine resources, for example:

1. International Program for the reduction of the Incidental Captures in Trade Fishing Operations.
2. Intergovernmental Agreement on Dolphin Conservation.
3. International Whale Commission.
4. Co-operation Agreement on Wildlife Conservation.
5. Trilateral Agreement on the Conservation of wetlands and Migratory Birds.

6. Management Plan for Aquatic Birds of North America.
7. Cooperation Programme for the Biodiversity Conservation Mexico-USA.
8. Convention on International Trade of Endangered Species
9. Convention on Biodiversity

However, in spite of these legal instruments, Mexico has not developed definite action for the effective conservation and protection of its natural resources. This is due to the lack of focus, the lack of co-ordination between the departments of state, and the different governmental levels, and because of different views of the authorities in the application or interpretation of the laws.

Therefore, it has not been possible to establish an adequate basis for the protection of natural resources (marine and terrestrial) of the natural protected areas on the Mexican pacific coast, or for their balanced utilisation.

REFERENCES

Botello, A.V., Villanueva, F.S., Pica, G.Y., Díaz, G.G. (1994) Impacto sobre los sistemas acuáticos. Evaluación Geoquímica del Puerto de Salina Cruz. p. 183–207. In *Riqueza y Pobreza en la Costa de Chiapas y Oaxaca*. Centro de Ecología y Desarrollo A.C. México D.F.

Botello, A.V., Villanueva, F.S. Díaz, G.G. and Escobar, B.E. (1998) Polycyclic aromatic hydrocarbons in sediments from Salina Cruz Harbour and coastal areas, Oaxaca, México. *Marine Pollution Bulletin* 30, 554–558.

CIESAS (1985–1990) *Serie sobre los pescadores de México. Cuadernos de la casa Chata*. Varios volúmenes. Centro de Investigaciones y Estudios Superiores en Antropología Social. Museo Nacional de Culturas Populares.

Cifuentes, L., Torres-García, J.L. and Frías, M. (1990–1997) *El océano y sus recursos*. 12 volúmenes. Serie La Ciencia desde México. Secretaría de Educación Pública (SEP). Fondo de Cultura Económica (FCE). Consejo Nacional de Ciencia y Tecnología (CONACYT).

De Fountaubert, A.C., Downes, D.R., Agardy, T.S. (1996) *Biodiversity in the Seas: Implementing the Convention on Biology Diversity in Marine and Coastal Habitats*. IUCN Gland and Cambridge, Cambridge.

De la Lanza-Espino, G. and García Calderón, J.G. (1991) Sistema lagunar Huizache y Caimanero Sin. Un estudio socio ambiental, pesquero y acuícola. *Hidrobiológica* 1 (1), 1–35.

De la Lanza-Espino G. (1992) Variación climática de corto plazo y su trascendencia en una laguna costera. *Ciencia* 43, Num. Especial, 103–110.

De la Lanza-Espino, G., Sánchez, N., Sorani, V. and Bojórquez J.L. (1996) Características geológicas, hidrológicas y del manglar en la planicie costera de Nayarit, México. Investigaciones Geográfica. *Boletín del Instituto de Geografía* 32 ,33–54.

De la Lanza-Espino G., and Flores-Verdugo, F. (1998) Nutrient fluxes in sediment (NH_4^+ and $PO4^{-3}$) in NW coastal lagoon Mexico associated with an agroindustrial basin. *Water, Air and Soil Pollution*. 107 (1–4), 105–120.

DGA (1997) Anuario Estadístico de Pesca. Dirección General de Estadística y registros Pesqueros. SEMARNAP. 241 p.

Flores-Verdugo F. (1994) Humedales. *World Wildlife Fund*, 14–15.

Guerrero, C.Y.C. (1993) Evaluación de la concentración de metales pesados en el ostión de roca *Crassostrea iridescens*, agua y sedimento de la Bahia de Manzanillo, Col. Tesis de Licenciatura.

Facultad de Ciencias, Universidad Nacional Autónoma de México. México DF.

Gutiérrez-Galindo, E. and Flores-Muñoz, G. (1986) Disponibilidad biológica de mercurio en las aguas de la costa norte de Baja California. *Ciencias Marinas* 12 (2), 85–98.

Gutiérrez-Galindo, E., Flores-Muñoz, G. and Aguilar, F.S. (1988) Mercury in freshwater fish and clams from the Cerro Prieto geothermal field of Baja California, Mexico. *Bulletin Environmental Contamination Toxicology* 41, 201–207.

Gutiérrez-Galindo, E., Flores-Muñoz, G. and Rojas, R.V. (1989) Metales traza en peces del Valle de Mexicali, Baja California, Mexico. *Ciencias Marinas* 15 (4), 105–115.

Heywood, V.H. and Watson, R.T. (1995) *Global Biodiversity Assessment*. NEP. Cambridge University Press. Cambridge.

INEGI (Instituto Nacional de Estadística, Geografía e Informática), (1989) Datos básicos de geografía de México. Secretaría de Programación y Presupuesto, México.

Izaguirre-Fierro, G., Páez-Osuna, F. and Osuna-López, J.I. (1992) Metales pesados en los peces del Valle de Culiacán. *Ciencias Marinas* 18, 143–151.

Mc Neely, J. (1994). Introduction to a special issue on protected areas. *Biodiversity Conservation* 3, 387–389.

Mendez, L., Baudilio, A.E. and Magallón, F. (1997) Effects of stocking densities on trace metal concentration in three tissues of the brown shrimp *Penaeus californiensis*. *Aquaculture* 156, 21–34.

Nauen, C.E. (1983) Compilation of legal limits for hazardous substances in fish and fisheries products. Food and Agriculture Organization of the United Nations. Rome, October 1–100.

Ortiz-Pérez, M.A. and Romo Aguilar L. 1994. Modificaciones de la trayectoria meándrica en el curso bajo del Río Grande de Santiago, Nayarit, México. *Investigaciones Geográficas Boletin* 29, 9–23, Instituto de Geografía, Univ. Nal. Autón. de México.

Osuna, L.I. and Paéz-Osuna, F. (1986) Cd, Co, Cr, Cu, Fe, Ni, Pb y Zn en los sedimentos del puerto y antepuerto de Mazatlán. *Ciencias Marinas* 12 (2), 35–45.

Osuna, L.I., Zazuela, P., Rodríguez, H. and Páez, O.F. (1990) Trace metal concentrations in mangrove oyster (*Crassostrea corteziens*) from tropical lagoon environments, Mexico. *Marine Pollution Bulletin* 21 (10), 486–488.

Páez-Osuna, F. and Osuna-López, J.I. (1987) Acumulación de metales pesados en Mitla: Una laguna costera tropical. *Ciencias Marinas* 13, 97–112.

Páez-Osuna, F. and Osuna-López, J.I. (1990a) Aspectos genéticos de los sedimentos marinos de la boca del Golfo de California evidenciados por la geoquímica de sus metales pesados. *Geofísica Internacional* 29, 47–58.

Páez-Osuna, F. and Osuna-López, J.I. (1990b) Heavy metals distribution in geochemical fractions of surface sediments from the lower Gulf of California. *Anales Instituto Ciencias del Mar y Limnología* 17,287–298.

Páez-Osuna, F. and Marmolejo-Rivas, C. (1990a) Trace metals in tropical coastal lagoon bivalves, *Crassostrea corteziensis*. *Bulletin Environmental Contamination Toxicology* 45, 538–544.

Páez-Osuna, F. and Marmolejo-Rivas, C. (1990b) Trace metals in tropical coastal lagoon bivalves, *Mytella strigata*. *Bulletin Environmental Contamination Toxicology* 45, 545–551.

Páez-Osuna, F. and Ruíz-Fernandez, C. (1995) Comparative bioaccumulation of trace metals in *Penaeus stylirostris* in estuarine and coastal environments. *Estuarine and Coastal Shelf Science* 40, 35–44.

Páez-Osuna, F. and Tron-Mayer, L. (1995) Distribution of heavy metals in tissues of the shrimp *Penaeus californiensis* from the Northwest coast of Mexico. *Bulletin Environmental Contamination Toxicology* 55, 209–215.

Páez-Osuna, F., Zazueta, P.H. and Izaguirre, F.G. (1991) Trace metals in bivalves from Navachiste Lagoon, Mexico. *Marine Pollution Bulletin* 22 (6), 305–307.

Páez-Osuna, F. Guerrero-Galván, S.R. and Ruíz-Fernández, A.C. (1998a) The environmental impact of shrimp aquaculture and the coastal pollution in México. *Marine Pollution Bulletin* **36** (1), 65–75.

Páez-Osuna F., Bojórquez-Leyva H. and Green-Ruiz C., 1998b., Total carbohydrates: organic carbon in lagoon sediments as an indicator of organic effluents from agriculture and sugar-cane industry. *Environmental Pollution* **102**, 1–6.

Páez-Osuna, F., Osuna-Lopez, J.I., Izaguirre-Fierro, G., Zazueta-Padilla, H.M. (1993) Heavy metals in clams from a subtropical coastal lagoon associated with an agricultural drainage basin. *Bulletin Environmental Contamination Toxicology* **50**, 915–921.

Páez-Osuna, F., Izaguirre, G.F., Godoy, R.I., González, F. and Osuna-Lopez, J.I. (1988) Metales pesados en cuatro especies de organismos filtradores de la región costera de Mazatlán. Técnicas de extracción y niveles de concentración. *Contaminación Ambiental* **4**, 31–39.

Páez-Osuna, F., Pérez-González, R., Izaguirre-Fierro, G., Zazueta-Padilla, H.M. and Flores-Campaña, L.M. (1995) Trace metal concentrations and their distribution in the lobster *Panulirus inflatus* (Bouvier 1895) from the Mexican Pacfic coast. *Environmental Pollution* **90**, 163–170.

Pringle, B.H., Hissong, D.E., Katz, E.L. and Mulwaka, S.T. (1968) Trace metal accumulation by estuarine molluscs in marine animals. *Journal Sanitary Division Society* **94**, 455–475.

Readman, J.W. Kwong L.W., Mee L.D., Bartiocci J., Nilve G., Rodríguez-Solano J.A. and González-Farias, F. (1992) Persistent organophosphorus pesticides in tropical marine environments. *Marine Pollution Bulletin* **24** (8), 398–402.

Rosalez, M.T.L. and Escalona, R.L. (1983) Organochlorine residues in organisms of two different lagoons of Northwest Mexico. *Bulletin Environmental Contamination and Toxicology* **30**, 456–463.

Rosales, M.T.L., Escalona, R.L., Alarcón, R.M. and Zamora, V. (1985) Organochlorine hydrocarbon residues in sediments of two different lagoons of Northwest Mexico. *Bulletin Environmental Contamination and Toxicology* **35**, 322–330.

Rosales, H.L., Carranza, E.A. and Santiago-Perez, S. (1994) Heavy metals in rocks and stream sediments from the northwestern part of Baja California, Mexico. *Revista Internacional Contaminación Ambiental* **10**, 77–82.

Salazar-Vallejo S.I. and González N.E. (1993) Biodiversidad marina y costera de México. Comisión Nacional para el conocimiento y aprovechamiento de la Biodiversidad (CONABIO). Centro de Investigaciones de Quintana Roo. 865 p.

SEMARNAP-FAO (1995) Estudio Piloto para un Plan de Desarrollo Acuícola en el Sistema Lagunar Teacapán-Agua Brava, Nay. Proyecto UTF/MEX./035/Mex. "Modernización del Sector Pesquero" México: 161 PP.

SEMARNAP-FAO (1997) Componente Ambiental del Proyecto TCP/Mexico 4 555 Camaronicultura Rural para Nayarit y Sinaloa. México: 463 p.

Villanueva, F.S. and Botello, A.V. (1996) Presencia de metales en el puerto de Salina Cruz y áreas adyacente, Oaxaca. 6° Congreso Nacional de Geoquímica. Instituto Nacional de Geoquímica, A.C., México. *Acta INAGEQ* **2**, 301–306.

THE AUTHORS

Alfonso V. Botello
*Institute for Marine and Limnology Sciences,
National Autonomous University of Mexico,
Marine Pollution Laboratory, P.O. Box 70305,
México City 04510 D.F., México.*

Alejandro O. Toledo
*Institute for Marine and Limnology Sciences,
National Autonomous University of Mexico,
Marine Pollution Laboratory, P.O. Box 70305,
México City 04510 D.F., México.*

Guadalupe de la Lanza-Espino
*Institute for Biological Sciences,
National Autonomous University of Mexico,
Marine Ecology Laboratory, P.O. Box 70233,
México City 04515 D.F., México*

Susana Villanueva-Fragoso
*Institute for Marine and Limnology Sciences,
National Autonomous University of Mexico,
Marine Pollution Laboratory, P.O. Box 70305,
México City 04510 D.F., México*

Chapter 31

BELIZE

Alastair R. Harborne, Melanie D. McField and E. Kate Delaney

Belize, Central America, contains some of the most important marine resources in the Caribbean including a 220 km barrier reef, shelf lagoon, three offshore atolls and extensive mangroves. The distribution of benthic communities within these systems is influenced by a range of factors, such as depth, wave action and underlying geology, but there are limited effects from annual seasonal variation. More important are stochastic events including coral bleaching from increased seawater temperatures and solar irradiance, and hurricanes. Although the Belize Barrier Reef is known to be part of a highly linked "Meso-American" reef system which extends into Mexico and Honduras, national and international sources and sinks of larvae to benthic and pelagic communities are poorly understood.

Entrained larvae and pollutants are transferred via currents and must be managed on a regional scale. National problems stem from over-fishing, sedimentation, agricultural run-off and urban pollution, particularly sewage. The effects from these are still small compared to many areas of the Caribbean, largely because of a low population density and distance from the mainland to the reefs. However, the population and tourism industry are growing as is a desire to diversify the economy. These factors must be carefully managed to avoid losses in other economically productive areas, such as the major fisheries (lobster and conch) which are already presumed to be over-exploited. Similarly, agricultural practices have almost certainly caused damage to some near-shore lagoons by sedimentation and contamination. Mangroves have been lost during the expansion of coastal zone towns, and effects from aquaculture, oil exploration, coastal development and diver damage may increase significantly in the future.

Tourism and fisheries are two of Belize's biggest industries, and there are a range of initiatives to protect the coastal zone. These efforts, recently consolidated within a Coastal Zone Management Authority, have established adequate environmental legislation but enforcement and monitoring remain limited. A series of marine protected areas has been devised, and some have been given World Heritage status, but there is limited monitoring of the efficacy of these reserves. A holistic approach has been possible because Belize is a small country, impacts are limited and management measures are both relatively easy to implement and assisted by national and regional funding agencies.

Seas at The Millennium: An Environmental Evaluation (Edited by C. Sheppard)
© 2000 Elsevier Science Ltd. All rights reserved

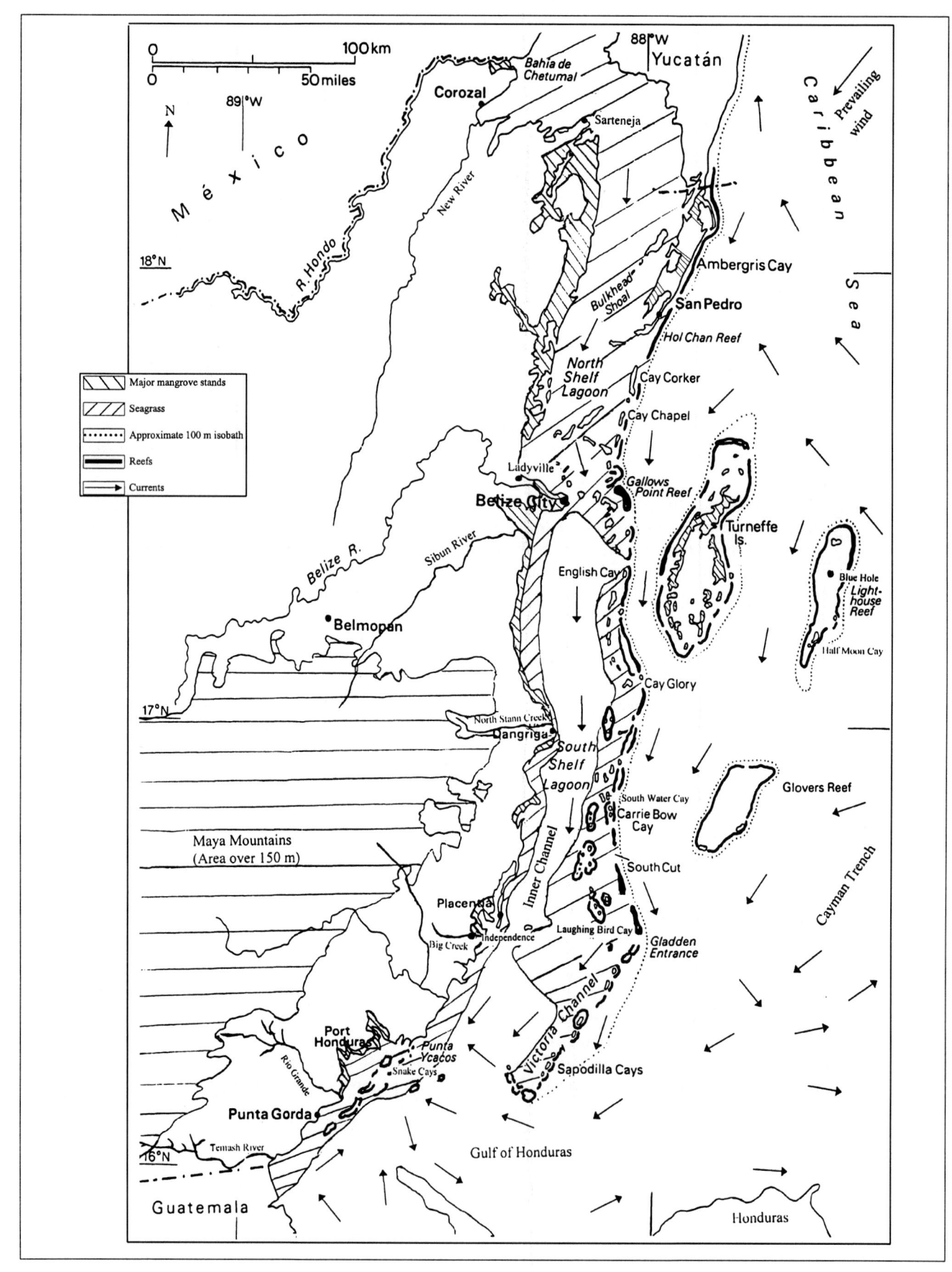

Fig. 1. Map of Belize. (Adapted from Craig, 1966; Heyman and Kjerfve, 1999; UNEP/IUCN, 1988; McField et al., 1996).

THE DEFINED REGION

Belize in Central America (Fig. 1) is small relative to many countries in the wider Caribbean but contains some of the most important marine resources in the region. The coastal zone of Belize is well known for the longest barrier reef in the western hemisphere but also includes three of the Caribbean's atolls, extensive mangrove forests, seagrass beds and estuarine systems. The significance of the coastal zone to Belize is highlighted by the fact that over 50% of the national territory is marine: there are 23,657 km^2 of territorial sea from a total of 46,620 km^2 (Hartshorn et al., 1984).

The continental shelf is 240 km long and varies in width between 13 and 48 km. The Belize Barrier Reef is nearly continuous for 220 km along this shelf from the Sapodilla Cays in the south to the Mexican border. The barrier reef system also encloses approximately 6000 km^2 of lagoon (Sedberry and Carter, 1993) and includes over 1000 cays (islands).

In addition to interactions with bordering countries, the important reefs of Belize are connected to most reefs in the wider Caribbean via currents that flow north from Brazil before entering the Gulf of Mexico. There is also a counter-clockwise gyre in the Gulf of Honduras. Such gross oceanographic patterns have led to a high similarity between biological communities throughout the region. These patterns also provide a conduit for facilitating the spread of waterborne pollutants and pathogens, an example of which was the devastating die-off of *Diadema antillarum* sea urchins in the early 1980s (reviewed by Lessios, 1988). Such region wide threats have led to management beyond geo-political boundaries, including the Meso-American Reef Initiative along nearly 1000 km of reefs from the Yucatan Peninsula to the Bay Islands (Cortés and Hatziolis, 1998).

Although Belize can, and has, been affected by other countries in the region, it also has significant effects on downstream areas. Cortés (1997) indicates that Belize is probably among the most important source areas of fish, coral and other invertebrate larvae in the area and this is supported by analysis of Caribbean reef connectivity (Roberts, 1997). However, the results of recruitment studies are equivocal, and one of the few studies carried out was on the lobster *Panulirus argus* (Glaholt and Seeb, 1992) which showed that Belizean lobsters are genetically different to those in Florida and that their larvae do not contribute significantly to Florida's population. Although further studies are required to investigate the origin of larvae in Belize, it seems likely that there are differences depending on species, and that the atolls may be more 'self-seeding' than the barrier reef system.

SEASONALITY, CURRENTS AND NATURAL ENVIRONMENTAL VARIABLES

Seasonality

Belize lies in the subtropical belt and experiences a dry and a wet season. There is also spatial variation across the country. Rainfall varies considerably between seasons (Heyman and Kjerfve, 1999) and noticeably affects salinity patterns. Hence, the southern section of the continental shelf often has a fresh water lens not usually found in the northern section. Water is known to be less saline around the Sapodilla Cays (Perkins, 1983) but generally there is a well mixed surface layer of isohaline and isothermal water to a depth of 50 m (James and Ginsburg, 1979). Seasonal changes in macro-algal and seagrass growth are well known in reef and lagoonal systems in the area, and indeed throughout the Caribbean. There is also an indication that the northern winds which often occur between November and March, as cold and wet air is pushed south by Arctic air masses, affect migration, location or recruitment of marine biota (Perkins, 1983; Sedberry and Carter, 1993).

Currents

Prevailing winds in Belize are from the north-east for 70% of the year (Koltes et al., 1998) and, since the area is micro-tidal, cause influential wind-driven currents. The predominant surface current inside the barrier reef is southerly, with a northerly current seaward of the three atolls and some westerly flow around their northern ends and within their lagoons (Perkins, 1983). The northerly current is a continuation of the current which flows west from Venezuela and north past Mexico. The hydrography of Belize is also influenced by the counter-clockwise surface gyre between Roatan (Honduras) and Glovers Atoll (Heyman and Kjerfve, 1999). Furthermore, deep, nutrient-rich oceanic waters occasionally enter the Gulf of Honduras from the Caribbean.

In Belize, 95% of waves are from the east and the remainder, from the west, have little mechanical effect on forereefs (Burke, 1982). Waves from the east are significantly modified by the atolls and this has a vital role in causing the variation in zonation seen between the three provinces of the barrier reef. For example, the zone of high-relief spur and groove formations in shallow water is only seen in areas of modified wave force.

Hurricanes

The effect of hurricane disturbance on reef geomorphology and benthic communities is well documented and is an important environmental factor in Belize. Hurricanes have been recorded since 1787 (Stoddart, 1963) but perhaps the best studied was Hattie in 1961. Within a swathe of heavy damage that crossed Turneffe Atoll and the central section of the barrier reef, Stoddart reported the removal of almost all trace of spur and groove formations on the barrier reef along with 80% of corals, virtually no living corals on the east side of Turneffe Atoll and a spectrum of coral resistance to mechanical damage from *Montastraea annularis* (most resistant) to *Acropora cervicornis* (least resistant). He also noted the higher levels of damage on the barrier reef than the atolls which may have been caused by the channel between them increasing wave height. Stoddart suggested a recovery period of 20–25 years but commented in 1969 that recovery was limited by mobile debris, algal competition and increased turbidity.

Subsequent hurricanes have provided further evidence of their importance in affecting the barrier reef and atolls. Hurricane Greta in 1978 caused significant lagoonward movement of *Acropora cervicornis* but this transport of living fragments is a key factor in its distribution (Rützler and Macintyre, 1982). Hurricane Mitch in 1998 caused 7 m seas for several days but there are currently little data on its effects, although Mumby (1999) reports that approximately 90% of living *Acropora palmata* was removed at some sites on Glovers Atoll, where a combination of bleaching and hurricane disturbance reduced coral recruit densities to 20% of levels recorded previously.

Bleaching

Also significant to the reefs are coral bleaching events caused when seawater temperatures rise. The loss of symbiotic zooxanthellae when temperature rises has often occurred on corals in Belize but there had not been a widespread, major bleaching event (McField, 1999) until 1995 when temperatures of over 29.4°C and high solar irradiance coincided with low wind speeds, affecting 52% of the corals (McField, 1999). Major reef builders, such as *Montastraea annularis* and *Agaricia tenuifolia*, were most affected and may cause long-term threats to reef integrity (Cortés and Hatziolis, 1998). More recently, in 1998 a second bleaching event occurred, and although few data are yet published, the effects on juvenile corals have been documented (Mumby, 1999).

THE MAJOR SHALLOW WATER MARINE AND COASTAL HABITATS

Geological Setting

Extensive work has been carried out on the geology of Belize, and reviews on various aspects are given by Stoddart (1962), Miller and Macintyre (1977), James and Ginsburg (1979), Precht (1993) and Macintyre and Aronson (1997). Belize's geology is controlled by passive plate location and it lies on the Yucatan continental block which split from the Nicaraguan-Honduras to form the Cayman Trench seaward of the three atolls and a series of submarine ridges. Two of these ridges are in deep water but the best developed ridge forms the southern edge of the barrier reef and continues as the base for Glovers Atoll and Lighthouse Reef. A fourth ridge underlies the central section of the barrier reef and Turneffe Atoll, while a fifth ridge is the base for the northern section of reef. The topography of the continental shelf reflects that of the mainland which has a flat lying north and a southern part uplifted by the Maya mountains. The shelf (lagoon) can be divided into three sections (southern, central and northern) based on relief and reef development. The distribution of recent sediments is significantly influenced by this topography.

Belize Barrier Reef

The Belize Barrier Reef and its associated ecosystems are a resource of immense social and economic importance to the country. Although commonly referred to as the Belize Barrier Reef, the reef complex can be followed as an ecological and geological unit along the Yucatan Peninsula for approximately 450 km (Craig, 1966). After the reef nearly merges with Ambergris Cay at Rocky Point, it parallels the Mexican coastline to Tulum and then consists of scattered coral heads to Cozumel. This direct association with other Central American countries, combined with the connections caused by currents carrying pollutants and entrained larvae, indicate the importance of treating Belize as part of a larger system inextricably linked with Mexico and the Gulf of Honduras (Gibson et al., 1998). For example, the currents which flow south along the continental shelf mean that the population of Chetumal (Mexico) may cause equal or greater effect to Belize's coastal zone than many of its own towns (McField et al., 1996).

The barrier reef is not continuous along its length but consists of linear segments separated by a series of channels. However, the reef has built up to within 20 cm of sea level along 57% of its length (Burke, 1982). Variation of reef characteristics has led Burke (1982) and later Macintyre and Aronson (1997) to describe three distinct provinces. The northern province, 46 km long from Rocky Point to Gallows Reef Point, contains 31% of the barrier platform. The central province, 91 km from Gallows Reef Point to Gladden Spit, contains 62% and finally the southern province is 10 km long from Gladden Spit to the Sapodilla Cays. A major influence on the forereef structures within these provinces is the wave energy reaching them after attenuation by the atolls.

The central province contains the best and most well studied reef development (Macintyre and Aronson, 1997). The central province has long wide sections of unbroken reef with three distinct structural features, two zones of spur and groove formations and a shelf edge coral ridge, in some areas (Burke, 1982). A transect across this reef profile at Carrie Bow Cay has been described in detail by Rützler and Macintyre (1982).

Atolls

Belize's three atolls, between 7 and 45 km from the barrier reef, are Glovers Atoll (132 km^2), Lighthouse Reef (126 km^2) and Turneffe Atoll (330 km^2) (Fig. 2). Their development has been more influenced by wave exposure than regional factors, such as sea-level rise, which has caused Glovers and

Fig. 2. View northwards along fringing reef on eastern side of Turneffe atoll.

Fig. 3. Fringing reefs around East Snake Cay.

Lighthouse to be more similar to each other than the more protected Turneffe (Gischler and Hudson, 1998). Glovers and Lighthouse have deep lagoons with numerous patch reefs and land area covering less than 3% of the atoll. In contrast, Turneffe has a land area of 22%, a shallow, poorly circulated lagoon and few patch reefs, except in the north of the atoll where there is little shelter from Lighthouse. The atolls have sides sloping into abyssal waters and their zonation is well known (Stoddart, 1962, and subsequent studies). Again, wave exposure has a key role and Lighthouse and Glovers have, for example, more *Acropora palmata* and pavements of *Lithothamnion*. Similarly, there are significant differences between the leeward and windward reef on each atoll.

Patch Reefs, Faroes and Cays

Throughout the Belizean shelf lagoon there are numerous patch reefs from small collections of coral heads to areas 80 m across (James and Ginsburg, 1979). These reefs are much more abundant in the southern lagoon than the north and support a wide variety of coral communities depending on shelf position, wave and current energy and depth (Precht, 1993). In addition to the patch reefs within the southern shelf lagoon, there are a series of rhomboid-shaped atoll-like features (faroes) which may be formed by submerged sand or rubble cays (James and Ginsburg, 1979). They have similar zonation patterns to the patch reefs and generally have steeply sloping sides with deep (often 15–30 m) channels and central lagoons (Miller and Macintyre, 1977).

Belize has over 1060 cays on the barrier reef and atolls (McField et al., 1996). Many are described in Stoddart et al. (1962, 1982).

Macro-algae

Most reefs within the Caribbean have experienced a dramatic decrease in coral cover and concomitant increase in macro-algae over the last two decades. This pattern has been attributed to a number of factors, particularly the mortality of *Diadema* urchins in the 1980s, removal of herbivorous fish and increased nutrients within the water column. Belize has been cited as an undisturbed system, but data show algal cover increasing from less than 10% to current levels of over 60% (e.g. McClanahan et al., 1999).

This ecological shift, however, has not been consistent in Belize and suggests a complex set of synergistic factors varying within the coastal zone. For example, on the barrier reef and Glovers Atoll patch reefs the changes paralleled the decimation of *Acropora* from white-band disease and show the role of coral mortality along with putative changes in herbivory and nutrients (McClanahan et al., 1999). The physico-chemical environment must also have an important role since the increased flushing on the forereef at Glovers Atoll seems to have limited disease and the increase in erect macro-algae is less apparent (McClanahan and Muthiga, 1998). In contrast, on the rhomboid reefs close to Carrie Bow Cay the result of *Acropora cervicornis* death is an alternative community state dominated by *Agaricia tenuifolia* (Aronson et al., 1998). Aronson et al. (1998) suggest that this shift seems to have been caused by intense herbivory by the urchin *Echinometra viridis*, reducing macro-algae and facilitating *Agaricia* recruitment.

Mainland Fringing Reefs

Reefs along the Belize mainland are limited by fluctuations in salinity and high turbidity. The only growth possible is in southern Belize where terrigenous sediments are removed most efficiently. Perkins (1983) reports some reef development along the mainland between Placencia and Punta Ycacos but fringing reefs are also known around the Snake Cays (Fig. 3). The mainland reefs are species-poor and support only resistant genera such as *Siderastrea* and *Porites*.

Lagoonal Shelf

The lagoonal shelf of Belize is dominated by seagrass beds, particularly *Thalassia*, which significantly modify the sediment regime and are important nursery and feeding grounds for many species. The lagoon is also a sink for estuaries from 16 major watersheds (McField et al., 1996) and has a coastal strip of terrigenous sediments bordered by sand beaches or mangroves. Trawls near Ambergris Cay indicate that the fish community is dominated by grunts (Haemulidae) and consists mainly of juveniles of reef species (Sedberry and Carter, 1993).

Mangroves

Belize has 783.16 km^2 of mangroves (Gray et al., 1990) and this is likely to represent 90–95% of historically known cover (McField et al., 1996). Mangroves fringe most of the coastline and brackish rivers and cover many cays, playing an important role in processes such as nutrient cycling and sediment trapping and acting as a nursery area. Gray et al. (1990) describe the main mangrove communities and their extent.

Biodiversity

The diversity of reefal, lagoonal and mangal systems within the coastal zone of Belize, combined with significant research efforts, has resulted in a good understanding of biodiversity compared to many countries in the Caribbean. The biodiversity within many taxa is regarded as high for the region with, for example, at least 94% of the zooxanthellate scleractinian species known from the Caribbean found here (Fenner, 1999) and over 50% of the Tubificidae (Erséus, 1990). Extensive fish species lists have been compiled for marine, brackish and freshwater systems with other studies documenting many additional taxa.

Belize also supports important populations of manatees (*Trichechus manatus*) and crocodiles (*Crocodylus acutus* and *C. moreletti*). Manatees, surveyed by O'Shea and Salisbury (1991) and Auil (1998), were found in larger numbers than any other Caribbean country because of the high quality habitat and low level of killing. In contrast, Platt and Thorbjarnarson (1997) indicates that densities of *C. acutus* are amongst the lowest reported for the region, probably caused by over-exploitation, habitat quality and competition with *C. moreletti*. Green (*Chelonia mydas*), hawksbill (*Eretmochelys imbricata*) and loggerhead (*Caretta caretta*) turtles are also known to nest in Belize and there is anecdotal evidence of leatherback (*Dermochelys coriacea*) and Kemp's ridley (*Lepidochelys kempi*) being present (McField et al., 1996).

OFFSHORE SYSTEMS

Deep Reef Habitats

The deep slopes of Belize are well studied and support reef organisms at great depths because of clear water. At Glovers Atoll, scleractinians and *Halimeda* were found at 100 m and crustose red algae at 250 m (reviewed by Stoddart, 1976). Beyond the photic zone, submersible studies, particularly by James and Ginsburg (1979), have shown at least two different profiles. In the southern barrier reef, the eastern side of Glovers and Turneffe Atolls and both sides of Lighthouse Reef the steep escarpment continues without interruption to over 1000 m because of the influence of the Cayman Trench. In contrast, the western side of Glovers and Turneffe Atolls and the central barrier reef changes at 300–400 m and then slopes much more gently. Within both profiles there are talus accumulations between 120–150 m followed by cliffs separated by gullied, sediment slopes from 150–200 m. Beyond 200 m there are fewer cliffs.

There is a true deep-reef fish fauna in Belize and Jamaica, although some of the juveniles of these species can be found at less than 50 m (Colin, 1974). Working on Glovers Atoll, Colin (1974) also found 60 species of reef fish between 50 and 305 m. This study indicated that the small, cryptic fauna was very poorly known.

Pelagic Waters

Offshore fisheries are limited throughout the Caribbean despite 80% of the water being deeper than 1800 m (UNEP/ECLAC, 1984). This is largely caused by the stable thermocline and onshore winds which limit upwellings and hence nutrient supply. However, the potential of deep sea fisheries beyond the barrier reef has not been fully explored, possibly because of a lack of capital funding. There have been proposals to further develop the fishery but extraction is currently by only a few fisherfolk with snapper reels (McField et al., 1996).

POPULATIONS AFFECTING THE AREA

Population and Demography

Belize is a diverse ethnic mix including Creoles, East Indians, Garifuna, Maya groups and Mennonites in addition to Latin Americans, Europeans and Chinese. The population in 1997 was 228,700 (EIU, 1998–99) giving a density of approximately 10 persons km^{-2} with an annual increase smaller than other Latin American countries. Belize City is the largest population centre and is an order of magnitude larger than any other town (Fig. 4).

Fig. 4. Belize City, divided by the Belize River.

Table 1 shows the main towns and cities in the coastal zone and highlights the fact that the proportion of the population in this area is reducing, partially caused by an increasing number of refugees (McField et al., 1996). Most of the cays support only shifting fishing populations or small tourist resorts but since many are either leased or in private ownership increasing development is likely. The balance of at least 50% of people living in rural areas and less than 50% of the population in the coastal zone is considered unusual compared to global averages and is a key factor in assisting coastal zone management (McField et al., 1996).

Table 1

Main population centres (greater than 1000 persons) within the coastal zone of Belize. Modified from McField et al. (1996). Original data from Central Statistics Office, Belmopan.

Town	Population		
	1980	1991	1994 (estimate)
Corozal Town	6899	6926	7644
Sarteneja	1005	1365	1433
Belize City	39771	44031	49122
San Pedro Town	1125	2001	2060
Ladyville	1810	2373	2664
Dangriga	6661	6565	7171
Independence	1474	1921	2115
Punta Gorda	2493	3956	4268
Others (total)	4413	5877	6422
Total (coastal zone)	65651	75015	82899
Total (Belize)	144857	194000	211000
Population in coastal zone (%)	45.3	38.7	39.3

Use of the Coastal Zone

Belize has a long history of artisanal use of the coastal zone, which can be traced to the Mayan Indians between 300 BC and 900 AD. The Mayans used cays in the lagoons as fishing stations, ceremonial centres and burial sites and utilised a range of fisheries, including conch, finfish, turtle eggs and manatees (Perkins, 1983). The first Europeans arrived in the late 1500s to harvest logwood, subsequently moving to the extraction of mahogany and then piracy. Garifuna also arrived from Roatan (Honduras) in the 17th century to fish and harvest timber. Belize was formerly known as British Honduras, a name which dates back to 1840, subsequently became a British colony in 1862 and gained independence in 1981.

The population of Belize has continued to place modest pressure on the coastal zone, with the principal uses being an export and artisanal fishery, aquaculture, tourism, small-scale shipping and oil exploration. However, independence from the UK in 1981 has increased the need to attain economic viability and reduce the dependence on imports and the country's natural resources will play a key role. Exploitation of these resources seems particularly likely through tourism but increasing the attractiveness of the country to tourists will place heavy demands on the coastal zone. Cortés (1997) lists the major threats to the reefs of Belize as fishing, sedimentation, tourism, agro-chemicals, sewage, solid wastes and dredging. Tourism and its associated demands can exacerbate all these other detrimental factors. It seems likely that sustainably managing the tourist industry and its associated infrastructure is vital to the integrity of coastal resources.

Assessing changes from anthropogenic impacts relies on baseline data for comparison. Apart from occasional collections of marine organisms the first studies were between 1920 and 1960. However, the first detailed studies and documentation of the barrier reef and atolls was by Stoddart (e.g. 1962) and Craig reviewed the fisheries of Belize in detail in 1966. These and other studies provide a baseline for a large amount of subsequent research and, although they do not contain quantitative data on ecological components such as the percent of live coral cover, are useful for assessing recent changes. However, Jackson (1997) indicates that despite the limited impacts in Belize, reefs in the 1960s were far from pristine and that data for a truly natural Caribbean reef are not available. Such problems with data interpretation may be aided by working on both ecological and palaeobiological time scales (e.g. Aronson et al., 1998). Future research will be assisted by significant habitat mapping work, including a national classification scheme (Mumby and Harborne, 1999), and quantitative ecological monitoring projects, by a range of Belizean and international agencies, NGOs and academic institutions carried out over the last decade.

RURAL FACTORS

Agriculture

Approximately 35% of the total land area of Belize is considered potentially suitable for agricultural use, although only between 10 and 15% is cultivated during any one year (Programme for Belize, 1995). Belize's commercial agriculture is concentrated on bananas, sugar and citrus fruit, although numerous crops such as rice, kidney beans, and maize are also grown. Sugar-cane cultivation is predominantly concentrated in the northern districts whilst banana plantation agriculture is located across the highly fertile alluvial plains of the south. Citrus production is less defined spatially, although the bulk of the industry is located throughout the southern Stann Creek and Toledo districts of Belize. Aside from commercial activities, the traditional system of "milpa" farming (shifting cultivation) utilises small plots of cleared tropical forest for rice, kidney bean and plantain cultivation across the foothills of the Maya Mountains.

Land clearance for commercial agriculture alters the drainage basin water and material flux to the coastal system, and affects water quality within the catchment (Holligan and de Boois, 1993). Relative to potential impacts on peripheral marine resources, the silt load in rivers, with associated nutrients and contaminants, may be considered the most important dimension of this process (Martin et al., 1980). Agricultural impacts on the riverine flux of material from the watershed to the marine system in Belize may be divided into two processes, (a) initial land clearance and (b) agricultural practices. The commercial cultivation of citrus, sugar cane and bananas in Belize involves the large-scale clearance of natural vegetation cover (King et al., 1993). In the case of citrus and bananas this vegetation is predominantly moist tropical forest on the southern coastal plains whilst further north savannah grasslands and scrub-land are cleared for sugar-cane production. Most deforestation is for agriculture rather than forestry which is largely for mahogany and pine (McField et al., 1996). It is reported that material mobilised as a result of climatological and hydrological interactions with such activities may potentially be conveyed via the river system to the peripheral coastal zone of Belize (Archer, 1994). Agricultural practices continue to add nutrients and sediments.

Potential Impacts on the Marine System

There is little quantitative information on direct impacts of agricultural activities on the Belizean barrier reef environment, and no long-term monitoring initiatives have been established to quantify the delivery process relative to rural anthropogenic processes. Such information would be essential for detailing temporal interactions between agrohydrological, geo-morphological and ecological processes within the Belizean coastal zone (IGBP, 1995).

However, although a direct link between rural anthropogenic activity and process change across the barrier reef system is, at best, difficult to establish, a number of significant off-site effects within the peripheral coastal zone have been identified (GESAMP, 1994). Perhaps the most significant potential impact of rural agricultural practice on the Belizean coastal zone is that of sedimentation and contamination within the near-shore environment. Adjacent to the banana and citrus growing regions of Belize, shallow marine habitats such as patch reefs and seagrass beds are located only 500–1000 m from the mainland. High suspended sediment concentrations or turbidity within the near-shore water column may reduce photosynthetic activity by restricting light penetration. In addition, effects may occur directly where sediment and associated contaminants settle on and smother coral polyps, restricting basic physiological functions. Accelerated algal growth related to increased nutrient and contaminant delivery may take place under certain conditions, however it should be noted that significant debate surrounds the cause and effect dimensions of this process. Further north in the citrus and sugar growing regions of Belize, agricultural interactions with the near-shore environment are likely to be significantly reduced, as the distances between farms and the marine ecosystem are greater.

Fluvial suspended material delivery has been reported to directly impact coral growth rates in Puerto Rico (Miller and Cruise, 1995), and in Belize, accelerated sediment, nutrient and contaminant delivery to the barrier reef system may likewise influence coral growth. However, investigations into the properties and application of agrochemicals used by banana and citrus industries have reported that even if pesticides, fertilisers and herbicides were to reach the barrier reef, their properties would have deteriorated significantly, reducing the strength of detection, quantification and source identification (Hall, 1994). Their effect on offshore atolls will be further reduced and may be negligible. Despite this research, more work is needed to examine possible effects on sensitive recruitment phases or via bioaccumulation.

Artisanal Fishing

Fishing has historically been a primary occupation for Belizeans and all fisheries are small-scale commercial operations (Perkins, 1983). Department of Fisheries statistics indicate that in 1998 there were approximately 350 boats and 1,900 fisherfolk but they are organised into five co-operatives and have significant political influence (McField et al., 1996). Marine products are export orientated and the wild-caught industry was worth approximately US$19.6 million in 1998, with 80% of the catch exported and 60% going to the USA.

The dominant fisheries are lobster (mainly *Panulirus argus*) and conch (mainly *Strombus gigas*) but significant amounts of finfish are caught, concentrating on higher quality species such as groupers (Serranidae) and snappers (Lutjanidae) (Gibson et al., 1998). There are also small fisheries for turtles, shrimp and stone crabs. Most fishing is conducted in the shallow waters on and inside the barrier reef and on the shallow reefs and lagoons of the atolls (Perkins, 1983).

There are direct threats to the populations of lobster, conch and grouper from over-fishing, with tourist demand a key factor. These fisheries were already considered close to their maximum sustainable yields in the early 1980s (Perkins, 1983) but modelling populations is difficult because catch and effort data are not collected systematically (McField et al., 1996) and because of visits by illegal alien fisherfolk. There is anecdotal evidence of decreasing catch per unit effort (King, 1997). Since most fisheries are exploited with traditional equipment, with the exception of shrimp trawling, indirect damage to benthic habitats is small-scale and limited to breakage from anchors, skin divers, nets and discarded gear (Gibson et al., 1998). The use of SCUBA, poisons and explosives is prohibited. Within the Caribbean there is evidence of over-fishing of herbivorous fish contributing to increased coverage of macro-algae, but evidence is equivocal in Belize and may be limited because of the concentration on higher value (piscivorous) species.

Lobster

Lobster have been harvested commercially in Belize since at least the 1920s when it was largely controlled by foreign interests. By 1995 fisherfolk were extracting 363,000 kg of lobster with an export market of US$8.8 million (McField et al., 1996). In addition, an estimated 23–45 kg of undersized lobster are caught and consumed locally on Caye Caulker alone (King, 1997). Most lobsters are caught by either skin divers using a hook and stick or traps (Hartshorn et al., 1984). These traps are generally wooden and based on a 1920s Canadian design but are increasingly made from oil drums (King, 1997).

Conch

Conch is the second most valuable fishery in Belize with catches around 180,000 kg (Appeldoorn and Rolke, 1996) worth exports of US$1.15 million (McField et al., 1996). Most conch are taken by skin divers in the back reef and seagrass beds where the aggregating behaviour of individuals makes them susceptible to exploitation (Perkins, 1983). Appeldoorn and Rolke (1996) highlighted the low density of adults in shallow habitats and there is evidence of increased populations in marine protected areas, both indicating over-exploitation. However, catches appear to be relatively consistent and the paradox could be caused by a deep, unfished stock so that catch (shallow water) may be independent of the spawning stock (Appeldoorn and Rolke, 1996).

Finfish

Finfish in Belize are generally caught for the domestic market and of the 114,000 kg caught in 1993–94 approximately 80% were consumed locally (McField et al., 1996). Hook-and-line fishing is dominant in Belize and this gear selects for piscivores so the catch is predominantly groupers and snappers (Koslow et al., 1994). There is also a seasonal fishery for estuarine species such as mullet (*Mugil* spp.) and some gill-nets for sharks (McField et al., 1996). The shark fishery is over-exploited but a surplus-production model for finfish provides evidence that there is capacity for further expansion, and current effort seems to be only 10% of levels that would maximise landings (Koslow et al., 1994). However, these results must be used with caution, particularly since it is difficult to model the effects of fishing on spawning aggregations which contribute a significant portion of the catch. At least six spawning aggregations are known in Belize, located at Rocky Point, Cay Glory, Gladden Entrance and the north-east corner of the three atolls (Carter and Sedberry, 1997). Fish are often caught before they spawn and some of the areas are thought to be over-exploited or no longer functional (McField et al., 1996).

Additional Fisheries

Belize's coastal zone supports a variety of localised fisheries including shrimp, turtles and crabs. Despite being a legally protected species, there are confirmed reports of manatees being killed for meat in the Port Honduras area and anecdotal reports from around Ambergris Cay, Sarteneja, Corozal and Dangriga (O'Shea and Salisbury, 1991; Auil, 1998). The penaid shrimp fishery supported 11 trawlers in 1988 which exploited stocks in Victoria Channel and the lagoon between Belize City and Placencia (McField et al., 1996). However, the fishery seems limited by a lack of knowledge, expertise and capital (Perkins, 1983) and has declined to a catch of 34,250 kg in 1995 (McField et al., 1996). Stone crabs (*Menippe mercenaria*) and blue crabs (*Callinectes sapidus*) are caught with baited traps and catches increase during the close season for lobster fishing (McField et al., 1996). Perkins (1983) reports 1,360 kg of turtle meat being sold in the early 1980s but this has declined because of increased regulations and declining populations.

COASTAL EROSION AND LANDFILL

The marine system and mainland activities are closely linked via the rivers and numerous watersheds on the mainland. Effects are larger in southern Belize because of the greater rainfall and have been studied and modelled in detail by Heyman and Kjerfve (1999) who concluded that any land-use decisions have ecological and economic effects on the marine resources. Currently development in the coastal zone is relatively low but is increasing, largely associated with the tourism industry, agriculture and aquaculture.

Effects on the Coastal Zone

Rivers throughout Belize introduce large amounts of terrigenous material close to the shore but there is little evidence of alteration to patterns of longshore drift by coastal development. Longshore currents are reworked by heavy surf and carried south by currents to form headlands and beach ridges (Miller and Macintyre, 1977). Belize Bight, north of Belize City, is also naturally occurring from longshore currents flowing northwards (Miller and Macintyre, 1977).

Deforestation and other detrimental land uses in Caribbean watersheds have also resulted in significant coastal erosion (Cortés and Hatziolis, 1998). Several areas in Belize have suffered from beach erosion, particularly the mouth of the Sibun River, Commerce Bight south of Dangriga and the mouth of Monkey River. Cay development has also caused some erosion and the establishment of piers on larger cays has affected beach profiles (McField et al., 1996). Changes to cay and mainland beaches caused by degradation of the reefs which dissipate much of the wave energy have not been documented.

Direct effects on coastal zone habitats from excavation or conversion have been limited. Perhaps the most dramatic change has been the expansion of Belize City since it was originally a small mangrove peninsula (Hartshorn et al., 1984). Construction around San Pedro (Ambergris Cay) has also involved land clearance, infilling and building unnatural beaches (McField et al., 1996) and uprooting seagrass beds to improve swimming (Perkins, 1983). Furthermore, there has been some mangrove clearance on the offshore cays and significant dredging and alteration of the landscape to facilitate a golf course on Caye Chapel, but data on changes to adjacent reefs are sparse.

Dredging in Belize should only be undertaken with the appropriate permits but enforcement is limited and there are known to be many cases of illegal activity (Gibson et al., 1998). However, an increasing number of permits for marine dredging have been granted and most aim to fill land for tourism or real estate development (McField et al., 1996). There is also some industrial dredging, such as that related to mining activities at the entrance of North Stann Creek River (McField et al., 1996) and the construction of the port at Big Creek (J. Gibson, pers. comm.).

EFFECTS FROM URBAN AND INDUSTRIAL ACTIVITIES

Initiatives by the Government of Belize to diversify the economy include efforts to increase revenue in all sectors including tourism, manufacturing, aquaculture and oil exploration. Currently most of the Caribbean has limited industrialisation and pollution of the coastal zone has not

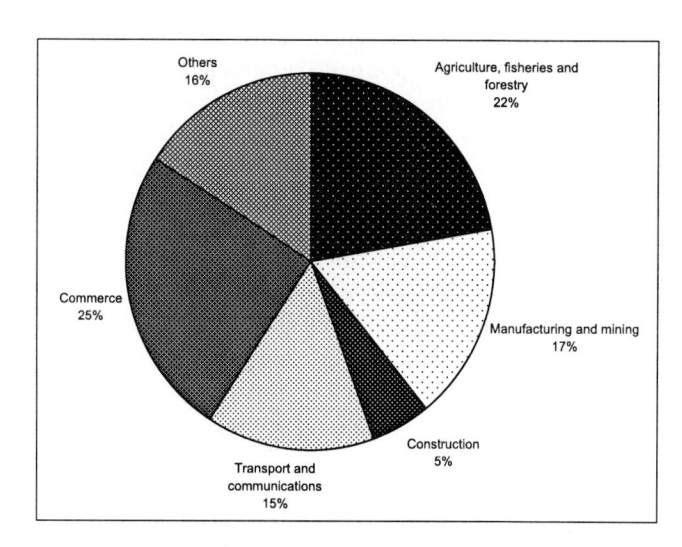

Fig. 5. Origins of gross domestic product in Belize in 1997. Data source: Economist Intelligence Unit. 1998. Country report. Jamaica, Belize, Organisation of Eastern Caribbean States (Windward and Leeward Islands).

reached levels seen in many more developed regions (UNEP/ECLAC, 1984). However, increasing development must be carefully monitored since prevailing currents rapidly move waterborne pollutants between countries (Davidson, 1990).

Value-added revenue from manufacturing has risen from US$40.7 million in 1980 to US$59.5 million in 1993 and, although this represents a decreasing proportion of GDP, since all major industries are located in the coastal zone this expansion threatens marine resources (McField et al., 1996). The origins of GDP in Belize in 1997 (Fig. 5), show the importance of marine resources. Management is assisted by the distance of the reef from the mainland (Perkins, 1983). Furthermore, the lack of a deepwater port limits economic expansion from large-scale coastal industry (O'Shea and Salisbury, 1991). Therefore, Belize has encouraged an open investment climate, such as the sale of cays, which is highly attractive to foreign investors (Katz, 1989).

Artisanal and Non-industrial Uses

There are numerous small-scale enterprises along the coastline but their effects on marine resources are limited. At the mouth of the Sibun River there has been some mining, by shovel, for quartz sand for cement (Perkins, 1983). Perkins also reports that some coral was mined for the streets of Belize City in the 1940s and 1950s but this has since ceased. No lime production from coral is known but bioprospecting seems likely to increase.

There is a small but growing trade in aquarium fish and approximately 27,000 fish and invertebrates, with a value of over US$40,000, were exported in 1994 (Carter and Sedberry, 1997). Further increases may reduce local populations of popular species. There is some sale of curios, particularly on Ambergris Cay, where tourists can purchase species such as tritons, helmet shells, cowries and black coral products from licensed and illegal collectors (Perkins, 1983).

Aquaculture and Fishing

Encouraging aquaculture is consistent with government aims to diversify the economy. Belize has a great potential for shrimp farm development on the abundant boggy lowland (Katz, 1989). Coastal areas and rivers are also suitable for raising non-native tilapia (*Oreochromis niloticus*). Currently there is little evidence that aquaculture is threatening the health of the coastal zone but there are concerns of nutrient enrichment, oxygen depletion, mangrove clearance, introducing alien species and the spread of disease from shrimp ponds (McField et al., 1996).

Small-scale aquaculture has included harvesting sponges, especially on Turneffe Atoll, and the alga *Eucheuma* which is used as a thickening agent (Perkins, 1983). Exploitation of *Eucheuma*, which may have a significant potential, has occurred on Turneffe Atoll and around Placencia and Hunting Cay. There have also been aborted attempts to introduce the American lobster (*Homerus americanus*) and to raise native lobsters in pens on Turneffe Atoll.

Pond-raised shrimp farming has increased dramatically over the last two decades and is the fastest growing fisheries sector. Approximately 90% of capital investment in aquaculture is for shrimp farms and there are at least six in Belize with exports of over 590,000 kg and income of US$5.25 million (McField et al., 1996) and increasing significantly. The ponds mainly raise *Penaeus vannamei* but *P. stylirostris* and *Macrobrachyum rosenbergii* are also used (Hartshorn et al., 1984).

Most fishing in Belize is small-scale and the small continental shelf may not be able to support an expanded, high-tech fishing industry (Perkins, 1983). However, there is a rapid expansion of longlining by Asian fleets in the Caribbean and this poses a threat to stocks of tuna, billfish and pelagic gamefish (Davidson, 1990). The overall catches for Belize since 1987 are presented in Fig. 6.

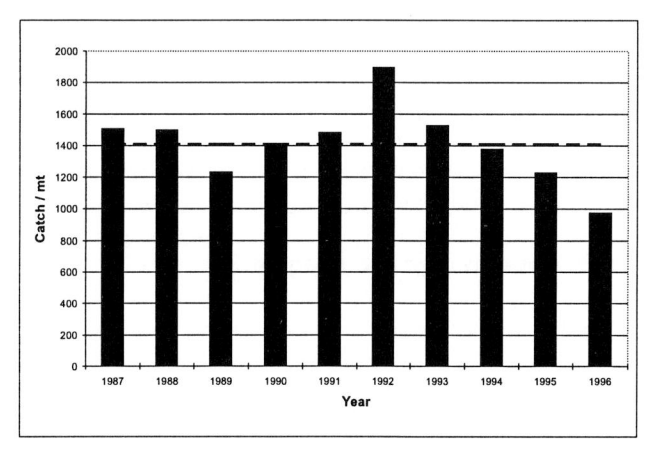

Fig. 6. Nominal catches from all fisheries in Belize. Dashed line represents mean catch. Data source: FAO, 1996. Fishery statistics capture production. FAO Yearbook Volume 82.

Tourism

Tourism is the largest generator of foreign exchange in Belize and 134,289 visitors arrived in the country during 1997 (EIU, 1998–99), attracted by its proximity to North America, being English speaking, and having a good climate. Most tourists are SCUBA divers and sports fishermen, and 77% snorkel or dive in the coastal zone (Gibson et al., 1998). The tourist centre is San Pedro but there are also many visitors to Caye Caulker, South Water Cay, Tobacco Cay, Placencia and atoll resorts.

Since 78% of hotel rooms are in the coastal zone (Gibson et al., 1998), increasing tourist infrastructure is a key factor contributing to urban pollution and habitat conversion. However, there are also direct effects on the reef and damage inflicted by divers, boats and anchors has been seen at all popular sites, although these are not thought to be severe (Gibson et al., 1998). Furthermore, as there may be over 14,000 sports fishermen arriving each year, some target species, such as bonefish and tarpon, may become over-exploited (McField et al., 1996). Tourists may also disturb manatees and bird colonies.

Industrial Effects

Although there are not many effects to the entire coastal zone from industrial activity, localised areas receive significant pollution. The main industrial plants in Belize are processing factories for sugar, citrus and seafood, rum distilleries, a brewery, soft drinks bottling, diesel electricity generators and garment factories (McField et al. 1996). Effluents and other wastes from these plants usually have large biochemical oxygen demands, are acidic, or contain chemicals such as sodium hydroxide. Around 950,000 tonnes of sugar cane are processed each year in Belize, along with the production of approximately 100,000 gallons of citrus concentrate.

Effluents and waste products are generally dumped straight into rivers and the lack of aeration ponds means that they do not meet environmental standards (McField et al., 1996). New River, close to many of the refineries and factories, is considered the most polluted in Belize with fish kills reported. Fish kills have also been recorded in the Belize River (Hartshorn et al., 1984). Data on the effects of a hydroelectric station on the Macal River are sparse.

Despite limited heavy industrial activity, a study by Gibbs and Guerra (1997) in the bottom muds of Belize City harbour found levels of cadmium, copper, lead and zinc sufficient to cause environmental problems to the biota. Concentrations of chromium were also tested but found to be at natural levels. Gibbs and Guerra (1997) hypothesise that cadmium, lead and zinc originate from a nail and battery factory sited on the Belize River, and copper is from antifouling boat paint and electrical and plumbing products.

Urban Pollution

Most Belize reefs are buffered from urban pollution by their distance from the mainland and channels of deep water. However, with population growth and increasing numbers of tourists, sufficient sewerage treatment is vital, and Belize City and San Pedro are the only coastal towns with a central sewerage system. Both of these plants have only secondary treatment via settlement ponds and release nutrients via mangrove stands and, although there is a possibility of localised eutrophication, they are thought to conform to WHO bacterial guidelines (McField et al., 1996). Nutrients are also thought to enter the marine environment via outfall pipes and septic tanks, particularly on the cays. Contamination by human waste from water craft is considerable (UNEP/ECLAC, 1984).

In addition, significant inputs of sewage from Mexican towns are carried into Belize by the south flowing currents. A study by Ortiz-Hernández and Sáenz-Morales (1999) examined the discharge from Chetumal which is estimated as 200 m^3 day^{-1} via a pluvial system, and showed that areas used for recreation have a concentration of faecal coliforms above levels given in Mexican legislation. However, this load, combined with the organic matter in the River Río Hondo, does not seem to increase the BOD of Chetumal Bay and provides evidence of self-depuration. Similarly, there is currently limited evidence of wide-scale nutrient enrichment in Belize but the effects on some benthic communities are known and are indicative of future changes if nutrient inputs are increased.

Lesser impacts are caused by litter from dumping by mainland populations, often in mangroves, and by boats (Hartshorn et al., 1984). Solid wastes from Belize City are dumped at a landfill site which receives around 920 m^3 week^{-1}, plus industrial waste, and may leach heavy metals into the coastal zone (McField et al., 1996). Drawing potable water has caused salt water intrusion and faecal contamination in urban areas, especially San Pedro (Hartshorn et al., 1984) and may lead to more common use of desalination plants.

Shipping and Offshore Impacts

Belize is considered to be a low risk shipping zone (UNEP/ECLAC, 1984) but most exports and imports are carried by sea and there is a constant threat of spills and groundings. There are three main ports at Belize City, Big Creek and Commerce Bight plus Esso's private dock which received 15 tankers of international fuel supplies in 1994 (McField et al., 1996). None of these ports have a deepwater dock and deep draft vessels anchor offshore and use barges. Although there are a minimum of 13 shipwrecks on the barrier reef there has been only minor damage from groundings and spills. Few hazardous substances are shipped though Belizean waters with perhaps only fertiliser, agro-chemicals and oil spills posing significant threats. Small oil spills have been seen and there are an increasing number of tar balls from ballast water and tanker washing on both international ships and local barges moving fuel to the cays (Gibson et al., 1998). Numbers of cruise ships are also increasing substantially and 13,661 passengers arrived in 1994 (McField et al., 1996).

Most oil is imported into Belize but there has been significant exploration. Perkins (1983) reports at least 40 exploratory wells, of which 12 are on the continental shelf, although all are capped and abandoned since none found commercially viable deposits despite the proximity to large Mexican and Guatemalan fields. In addition to shot-holes in seagrass beds from seismic testing, if deposits are found there are obvious threats to benthic habitats from drilling, construction and processing.

PROTECTIVE MEASURES

Management of Belize's coastal zone has evolved from the sectoral management of commercial fisheries and conservation of important bird species to the broader approach of ecosystem management and integrated coastal zone management. This integrated approach (see McField et al., 1996; Gibson et al., 1998) is critical as most of the current threats originate some distance from the valuable habitats. Conservation and sustainable management of marine resources are high priorities of the general population and of the many non-governmental organisations involved in resource management in Belize as well as of most government departments. This broad-based approach is largely responsible for the successes to date, although conflicting views and a general impatience with the slow pace of "sustainable development" are fostering a new international business climate seeking quick profits through massive development initiatives.

In 1990 a Coastal Zone Management Unit was established within the Fisheries Department, later assisted by the UNDP/GEF Coastal Zone Management Project in 1993. The Coastal Zone Management Act of 1998 consolidated efforts within the autonomous Coastal Zone Management Authority (CZMA), and assisted with implementation and research by an affiliated institute. Although no regulatory powers have yet been developed within the CZMA, it serves as the focal point of marine conservation planning, monitoring and research.

Policy Development and Integration

Several integrated committees exist which provide broad-based platforms to discuss policy development and the implementation of key programs. The board of the CZMA includes senior government representatives and can approve policies. Similarly, the Barrier Reef Committee was established as a national platform for review of the Conservation and Sustainable Use of the Meso-American Barrier

Reef System and the World Heritage Site. A Marine Protected Areas Committee fosters communication and exchanges among protected areas managers and advisory committees. Finally, a National Coral Reef Monitoring Working Group was formed to integrate and co-ordinate various reef monitoring efforts throughout the country.

Regulation of Development

Belize's relatively recent introduction to international tourism and commercial development means that many models are available as guides to assist with ensuring sustainable development. However, as developmental pressures increase, so does the potential for serious environmental degradation. Belize has adequate environmental legislation but lacks enforcement and monitoring capacity. For example, under the Environmental Impact Assessment (EIA) regulations of 1995, the Department of Environment enforces regulations and screens projects that may require EIAs. Similarly, the Land Utilisation Authority is responsible for Special Development Areas which are a form of strategic planning, providing for the zoning of land-use. A zoning plan for Belize's marine waters will ultimately be developed within an overall Coastal Zone Management Plan.

Regulation of Tourism

The Belize Tourist Board regulates the tourism industry, including the expanding cruise industry, which many view as a growing threat to ecologically sensitive areas. The Tourist Guide Regulations require that all tour guides meet standard levels of professional training and licences can be revoked for non-compliance with environmental or other regulations. However, dive guides are normally quite effective at "self-regulation" and have initiated a series of "conservation zones". Dive operators also play a major role in the installation and maintenance of mooring buoys. However, these initiatives may be jeopardised by pressure within to accommodate the mass-tourism market rather than current small-scale eco-tourism ventures.

Fisheries Regulation

The Department of Fisheries manages the fisheries industry, which includes aquaculture. No fishing is allowed on SCUBA and there are other gear restrictions, size limits and closed seasons. However, government resources are inadequate to patrol the waters of Belize or to fully enforce these regulations. Six marine reserves have been established to assist fisheries management by replenishing heavily exploited stocks, while also protecting essential habitats (coral reefs, seagrass beds and mangroves).

Control of Pollution

The Environmental Protection Act of 1992 provides the framework through which the Department of Environment enforces regulations preventing pollution. The EIA process further ensures proposed industrial activities take environmental protection measures into account during the planning stages. Although enforcement manpower is severely limited, the small-scale of Belize's industrial sector aids the identification and control of potential sources of pollution.

Marine Protected Areas

The establishment of marine and coastal protected areas, summarised in Table 2 and including some with World

Table 2

The marine and coastal protected areas of Belize. At least nine additional areas are proposed as protected areas. CR = Crown Reserve, FR = Forest Reserve, MR = Marine Reserve, NM = Natural Monument, NP = National Park, NR = Nature Reserve. Modified from McField et al. (1996) and Auil (1998).

Existing protected areas	Area (acres)	Date established	Date of completion of management plan
Bacalar Chico NP & MR*	15,117	1996	1994
Bird Cayes (various locations) CR	6,744	1977	–
Blue Hole NM*	1,023	1996	–
Burdon Canal NR	5,252	1992	–
Caye Caulker FR	160	1998	–
Caye Caulker MR*	10,618	1998	1996
Corozal Bay (manatee) WS*	177,762	1998	–
Deep River FR	77,499	1941	–
Gales Point (manatee) WS	9,095	1998	–
Glovers Reef MR*	81,175	1993	1988
Half Moon Caye NM*	9,771	1982	1995
Hol Chan MR*	4,035	1987	1986
Laughing Bird Cay NP*	10,119	1991	1994
Paynes Creek NP	31,676	1994	in prep.
Port Honduras MR*		2000	1998
Sapodilla Cays MR*	33,401	1996	1994
Sarstoon-Temash NP	41,898	1994	in prep.
Shipstern Nature Reserve (private)	18,852	1987	1990
South Water Caye MR*	78,374	1996	1993
Total	612,570		
Total with significant marine area (10 protected areas *)	421,395		

Heritage listing, has been an essential component of marine conservation efforts in Belize. Currently there are seven designated Marine Reserves, four National Parks, two Wildlife Sanctuaries, two Natural Monuments and one Nature Reserve. There are seven Crown Reserves, which are essentially bird sanctuaries on small cays, and one private reserve in the coastal zone. The role of NGOs and local community-based management is expanding and advisory committees are playing increasingly important roles. For example, establishment of some reserves has been assisted by provision of baseline data from international volunteer programmes (e.g. Coral Cay Conservation).

Marine protected areas (MPAs) are a useful tool for addressing a number of threats to coral reefs, particularly those related to tourism carrying capacities, over-exploitation of commercial species and the potential ecological benefits of increased herbivory. Zoning schemes enable multiple uses in these areas, including recreational diving, sports-fishing, and traditional small-scale fishing and full protection (no-take zones) in key areas. Over-success of tourism in parks can be a concern, as in Hol Chan, which receives over 30,000 visitors a year. The value of protected areas in promoting sustainable fisheries and in regulating tourism and other activities is now well documented and the Hol Chan Marine Reserve has been cited as an international model. Research at Hol Chan and Half Moon Caye has illustrated that MPAs can result in fish populations with significantly greater abundance and larger-sized individuals (Polunin and Roberts, 1993; Carter and Sedberry, 1997). Current research indicates that 30% of the coastal zone of Belize should be closed to fishing and the remainder managed by traditional methods.

Challenges

Although the conservation efforts of the last two decades have been successful, there are many challenges to sustaining them. MPAs often suffer from inadequate funding levels and rely on external financing. The Protected Area Conservation Trust raises money through a tourist departure tax and a percentage of park entrance fees. This fund should be supporting basic infrastructure and associated costs of managing protected areas, although a relatively small proportion of the funds are actually used for park management (versus more general environmental education, and community development projects). Recently there has also been organised opposition from some fishermen who are not convinced of replenishment reserves or complain that the existing park regulations are not adequately enforced. The existing piecemeal approach to MPA management might be better served by a unified Parks Service, as has been previously recommended (Programme for Belize, 1995). A committee has recently been formed to address this issue.

One of the most difficult challenges facing all marine conservation efforts in Belize is the lack of sustainable financing. While many international donor agencies have contributed greatly to these efforts, the reliance on such short-term project-based funding reduces the long-term national approach. The CZMA has addressed this issue (Coastal Zone Management Project, 1995) and will continue to seek sustainable revenue-generation, although many of the obvious sources have already been utilised. Like many countries, the government of Belize is facing increasing economic constraints and a variety of revenue generation strategies will be necessary.

ACKNOWLEDGEMENTS

The authors would like to thank Janet Gibson and Alex Page for reviewing this paper and Caroline Turnbull for collating reference lists. ARH wishes to thank the Department of Fisheries, Coastal Zone Management Project and University College of Belize for their support during fieldwork between 1986 and 1998. MDM is supported by the Elsie and William Knight Jr. Fellowship and Department of Marine Science (University of South Florida), with prior support and assistance from the Coastal Zone Management Project. EKD is supported by Fyffes, the Banana Growers Association of Belize and Department of Geography (University of Southampton).

REFERENCES

Appeldoorn, R.S. and Rolke, W. (1996) Stock abundance and potential yield of the queen conch resource in Belize. Report to CARICOM Fisheries Resource Assessment and Management Programme and Belize Fisheries Department.

Archer, B.A. (1994) United Nations Environment Project Regional Coordinating Unit Report on land-based sources of marine pollution inventories. UNEP.

Aronson, R.B., Precht, W.F. and Macintyre, I.G. (1998) Extrinsic control of species replacement on a Holocene reef in Belize: the role of coral disease. *Coral Reefs* 17, 223–230.

Auil, N. (1998) Belize manatee recovery plan. Project No. BZE/92/G31 Sustainable Development and Management of Biologically Diverse Coastal Resources—Belize.

Burke, R.B. (1982) Reconnaissance study of the geomorphology and benthic communities of the outer barrier reef platform, Belize. pp. 509–526. In The Atlantic Barrier Reef Ecosystem at Carrie Bow Cay, Belize I: Structure and communities, eds. K. Rützler and I.G. Macintyre. *Smithsonian Contributions to the Marine Sciences* 12.

Carter, J. and Sedberry, G.R. (1997) The design, function and use of marine fishery reserves as tools for the management and conservation of the Belize Barrier Reef. *Proceedings of the 8th International Coral Reef Symposium* 2, 1911–1916.

Coastal Zone Management Project (1995) Institutional development and sustainable financing mechanisms for coastal zone management in Belize. Project Report No.95/1. UNDP/GEF Sustainable Development and Management of Biologically Diverse Coastal Resources—Belize.

Colin, P.L. (1974) Observation and collection of deep-reef fishes off the coasts of Jamaica and British Honduras (Belize). *Marine Biology* **24**, 29–38.

Cortés, J. (1997) Status of the Caribbean coral reefs of Central America. *Proceedings of the 8th International Coral Reef Symposium* **1**, 335–340.

Cortés, J. and Hatziolos, M.E. (1998) Status of coral reefs of Central America: Pacific and Caribbean coasts. pp. 155–163. In *Status of Coral Reefs of the World: 1998*, ed. C.R. Wilkinson. AIMS, Australia.

Craig, A.K. (1966) Geography of fishing in British Honduras and Adjacent Coastal Areas. Technical Report No. 28, Coastal Studies Institute, Louisiana State University, USA.

Davidson, L. (1990) Environmental assessment of the Wider Caribbean Region. UNEP Regional Seas Reports and Studies No. 121. UNEP.

EIU (Economist Intelligence Unit) (1998–99) Country profile. Jamaica, Belize, Organisation of Eastern Caribbean States (Windward and Leeward Islands). Economist Intelligence Unit Limited.

Erséus, C. (1990) The marine Tubificidae (Oligochaeta) of the barrier reef ecosystems at Carrie Bow Cay, Belize, and other parts of the Caribbean Sea, with descriptions of twenty-seven new species and revision of *Heterodrilus, Thalassodrilides* and *Smithsonidrilus. Zoologica Scripta* **19** (3), 243–303.

Fenner, D. (1999) New observations on the stony coral (Scleractinia, Milleporidae, and Stylasteridae) species of Belize (Central America) and Cozumel (Mexico). *Bulletin of Marine Science* **64** (1), 143–154.

GESAMP (1994) Anthropogenic Influences on sediment delivery to the coastal zone and environmental consequences. GESAMP Reports and Studies No. 52.

Gibbs, R.J. and Guerra, C. (1997) Metals of the bottom muds in Belize City harbour, Belize. *Environmental Pollution* **98** (1), 135–138.

Gibson, J.P., McField, M.D. and Wells, S.M. (1998) Coral reef management in Belize: an approach through Integrated Coastal Zone Management. *Ocean and Coastal Management* **39**, 229–244.

Gischler, E. and Hudson, J.H. (1998) Holocene development of three isolated carbonate platforms, Belize, Central America. *Marine Geology* **144**, 333–347.

Glaholt, R.D. and Seeb, J. (1992) Preliminary investigations into the origin of the spiny lobster, *Panulirus argus* (Latreille, 1804), population of Belize, Central America (Decapoda, Palinuridea). *Crustaceana* **62** (2), 159–165.

Gray, D., Zisman, S. and Corves, C. (1990) *Mapping the Mangroves of Belize*. Department of Geography, University of Edinburgh.

Hall, M. (1994) Agricultural pollution on the coral reef in the Stann Creek district, Belize; an assessment of sources, effects and Government policy. Unpublished MSc thesis. Imperial College of Science, Technology and Medicine (University of London).

Hartshorn, G. and 16 others (1984) *Belize Country Environmental Profile*. Robert Nicolait & Associates Ltd.

Heyman, W.D. and Kjerfve, B. (1999) Hydrological and oceanographic considerations for Integrated Coastal Zone Management in Southern Belize. *Environmental Management* **24** (2), 229–245.

Holligan, P.M. and de Boois, H. (1993) Land–ocean interactions in the coastal zone (LOICZ) science plan. IGBP Report No. 25.

IGBP (1995) Land–ocean interactions in the coastal zone. Expert meeting on coral reef monitoring, research and management, Bermuda. October 1994. IOC-IUCN_LOICZ/WKSHP/94.4.

Jackson, J.B.C. (1997) Reefs since Columbus. *Proceedings of the 8th International Coral Reef Symposium* **1**, 97–106.

James, N.P. and Ginsburg, R.N. (1979) *The Seaward Margin of Belize Barrier and Atoll Reefs*. International Association of Sedimentologists. Special Publication No. 3. Blackwell Scientific Publications, Oxford, London, Edinburgh, Melbourne.

Katz, A. (1989) Coastal resource management in Belize: potentials and problems. *Ambio* **18** (2), 139–141.

King, R.B., Pratt, J.H., Warner, M.P. and Zisman, S.A. (1993) *Agricultural Development Prospects in Belize*. Natural Resources Institute Bulletin 48 Chatham, UK.

King, T.D. (1997) Folk management and local knowledge: Lobster fishing and tourism at Caye Caulker, Belize. *Coastal Management* **25**, 455–469.

Koltes, K.H., Tschirky, J.J. and Feller, I.C. (1998) In UNESCO (1998), pp. 79–94. CARICOMP—Caribbean Coral Reef, Seagrass and Mangrove Sites. Coastal Region and Small Island Papers 3, UNESCO, Paris.

Koslow, J.A., Aiken, K., Auil, S. and Clementson, A. (1994) Catch and effort analysis of the reef fisheries of Jamaica and Belize. *Fishery Bulletin* **92** (4), 737–747.

Lessios, H.A. (1988) Mass mortality of *Diadema antillarum* in the Caribbean: What have we learned? *Annual Review of Ecology and Systematics* **19**, 371–393.

Macintyre, I.G. and Aronson, R.B. (1997) Field guidebook to the reefs of Belize. *Proceedings of the 8th International Coral Reef Symposium* **1**, 203–222.

Martin, J.M., Burton, J.D. and Eisma, D. (eds.) (1980) River inputs to ocean systems. Proceedings of a SCOR/ACMRR/ECOR/IAHS/UNESCO/CMG/IABO/IAPSO Review and Workshop held at FAO headquarters, Rome (1979) with the collaboration of the Intergovernmental Oceanographic Commission and the support of the United Nations Environment Programme.

McClanahan, T.R., Aronson, R.B., Precht, W.F. and Muthiga, N.A. (1999). Fleshy algae dominate remote coral reefs of Belize. *Coral Reefs* **18**, 61–62.

McClanahan, T.R. and Muthiga, N.A. (1998) An ecological shift in a remote coral atoll of Belize over 25 years. *Environmental Conservation* **25** (2), 122–130.

McField, M.D. (1999) Coral response during and after mass bleaching in Belize. *Bulletin of Marine Science* **64** (1), 155–172

McField, M.D., Wells, S.M and Gibson, J.P. (eds.) (1996) State of the Coastal Zone Report. Belize, 1995. Coastal Zone Management Programme and Government of Belize.

Miller, J.A. and Macintyre I.G. (1977) Third International Symposium on coral reefs, Field guidebook to the reefs of Belize. The Atlantic Reef Committee, Florida.

Miller, R.L. and Cruise, J.F. (1995) Effects of suspended sediment on coral growth: evidence from remote sensing and hydrologic modelling. *Remote Sensing of the Environment* **53**, 177–187.

Mumby, P.J. (1999) Bleaching and hurricane disturbances to populations of coral recruits in Belize. *Marine Ecology Progress Series* **190**, 27–35.

Mumby, P.J. and Harborne, A.R. (1999) Development of a systematic classification scheme of marine habitats to facilitate regional management and mapping of Caribbean coral reefs. *Biological Conservation* **88**, 155–163.

Ortiz-Hernández, M.C. and Sáenz-Morales, R. (1999) Effects of organic material and distribution of faecal coliforms in Chetumal Bay, Quintana Roo, Mexico. *Environmental Monitoring and Assessment* **55**, 423–434.

O'Shea, T.J. and Salisbury, C.A. (1991) Belize—a last stronghold for manatees in the Caribbean. *Oryx* **25** (3), 156–164.

Perkins, J.S. (1983) The Belize Barrier Reef ecosystem: An assessment of its resources, conservation status and management. New York Zoological Society report.

Platt, S.G. and Thorbjarnarson, J.B. (1997) Status and life history of the American Crocodile in Belize. Final report to the UNDP-GEF Belize Coastal Zone Management Project.

Polunin, N.V.C. and Roberts, C.M. (1993) Greater biomass and value of target coral-reef fishes in two small Caribbean marine reserves. *Marine Ecology Progress Series* **100**, 167–176.

Precht, W.F. (1993) Holocene coral patch reef ecology and sedimentary architecture, northern Belize, Central America. *Palaios* **8** (5), 499–503.

Programme for Belize (1995) Towards a national protected areas system plan for Belize. Report to the Government of Belize and NARMAP.

Roberts, C.M. (1997) Connectivity and management of Caribbean coral reefs. *Science* **278**, 1454–1457.

Rützler, K. and Macintyre, I.G. (1982) The habitat distribution and community structure of the barrier reef complex at Carrie Bow Cay, Belize, pp. 9–66. In Rützler, K. and Macintyre, I.G. The Atlantic Barrier Reef Ecosystem at Carrie Bow Cay, Belize I: Structure and communities. *Smithsonian Contributions to the Marine Sciences* **12**.

Sedberry, G.R. and Carter, J. (1993) The fish community of a shallow tropical lagoon in Belize, Central America. *Estuaries* **16** (2), 198–215.

Stoddart, D.R. (1962) Three Caribbean atolls: Turneffe Islands, Lighthouse Reef, and Glover's Reef, British Honduras. *Atoll Research Bulletin* **87**, 1–151.

Stoddart, D.R. (1963) Effects of Hurricane Hattie on the British Honduras reefs and cays, October 30–31, 1961. *Atoll Research Bulletin* **95**, 1–142.

Stoddart, D.R. (1976) Structure and ecology of Caribbean coral reefs. *FAO Fisheries Report* **200**, 427–448.

Stoddart, D.R., Fosberg, F.R. and Spellman, D.L. (1982) Cays of the Belize Barrier Reef and lagoon. *Atoll Research Bulletin* **256**, 1–22.

UNEP/ECLAC (1984) The state of marine pollution in the Wider Caribbean region. UNEP Regional Seas Reports and Studies No. 36. UNEP.

UNEP/IUCN (1988) Coral Reefs of the world. Vol. 1. Atlantic and Eastern Pacific. IUCN Gland.

THE AUTHORS

Alastair R. Harborne
Coral Cay Conservation, 154 Clapham Park Road, London, SW4 7DE, U.K.

Melanie D. McField
Department of Marine Science, University of South Florida, 140 Seventh Ave South, St. Petersburg, FL 33701, U.S.A. (Correspondence address: P.O. Box 512, Belize City, Belize)

E. Kate Delaney
Department of Geography, University of Southampton, Southampton, SO17 1BJ, U.K.

Chapter 32

NICARAGUA: CARIBBEAN COAST

Stephen C. Jameson, Lamarr B. Trott, Michael J. Marshall and
Michael J. Childress

Nicaragua's Caribbean coastline is 450 km long and fringes the largest continental shelf in Central America. Coastal forests are still abundant despite large-scale clearing, and hardwoods abound in the coastal zone where exploitation has been difficult. Most rainfall drainage flows to the Caribbean. The coastal zone contains a great diversity of resources and ecosystems with great potential for generating wealth and a good quality of life.

Offshore, coral reefs vary from small patches and pinnacles to large, complicated platforms and well-defined belts and, apart from the narrow zone occupied by the coastal boundary current, are distributed across virtually the entire shelf. Their distribution is partly determined by river flow rates and proximity to land, and by the frequency of storms. There are also many coral cays. Seagrass beds are some of the most extensive in the Caribbean, if not the world, providing major feeding grounds for large populations of green turtles. Mangroves are also extensive, bordering an estimated 600 km^2 of lagoon shores.

Nicaragua's natural resources have not been adequately monitored but it is apparent that in many cases they have been degraded. Most of the population is involved in subsistence farming. The commercial and artisanal lobster fishery fleets are significant, and shrimp are the second most important fishery. Evidence exists of over-exploitation in some cases but, with proper management, the marine system has great potential both for Nicaragua and for the condition of the Caribbean Sea Large Marine Ecosystem. The government understands the value of these resources and has received excellent natural resource management guidance and technical assistance. Preventing the mismanagement and destruction of these resources and ecosystems, which has occurred in other Central American countries, now depends on the ability of government and the Nicaraguan people to work together.

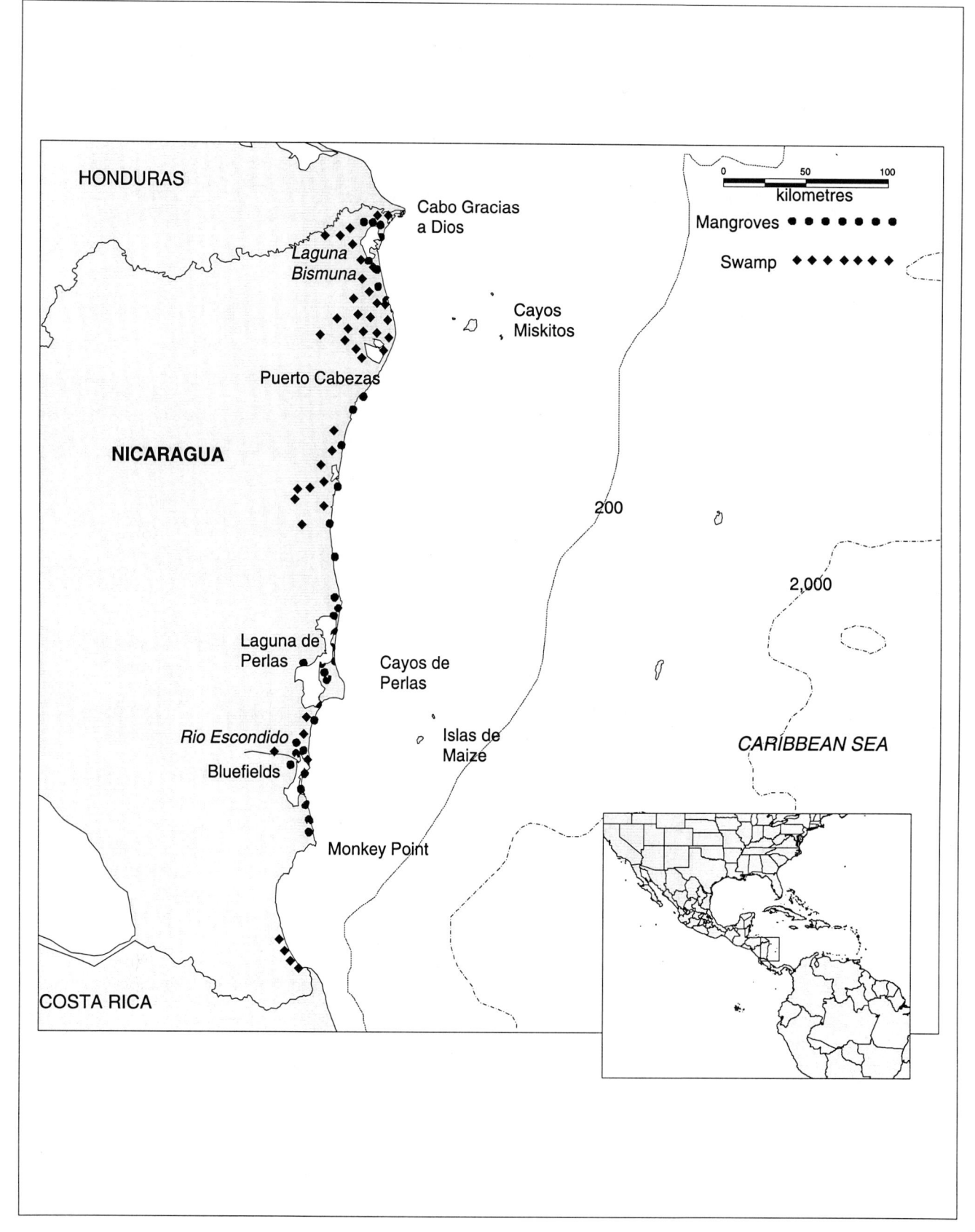

Fig. 1. Map of Caribbean coast of Nicaragua.

THE DEFINED REGION

The east coast of Nicaragua lies within the southern part of the Caribbean Sea. Nicaragua is the largest country in Central America (129,494 km^2), with a Caribbean coastline 450 km long (INPESCA, 1990) and the largest continental shelf in Central America (Fig. 1; and see Fig. 2 in Chapter 33). There are many coral reefs and many coral cays which are used as campsites from which turtles and lobsters are caught, as well as by pirate fishermen from other countries. The maritime claims of Nicaragua include a contiguous zone consisting of a 25-nm security zone, the continental shelf, and a territorial sea out to 200 nm.

The central mountains form the country's main watershed, and most drainage flows to the Caribbean. Rivers include the 425-mile long Coco River along the Nicaragua–Honduras border, the Rio Grande de Matagalpa, and the San Juan, Prinzapolka, Escondido, Indio and Maiz Rivers (Foer, 1992).

The coastal lowlands form the extensive and still sparsely populated area known as Costa de Mosquitos. These have an average width of 100 km, and are among the widest in Central America, and are divided into the North and the South Atlantic Autonomous Regions (RAAN and RAAS) covering 45% of Nicaragua's territory. The lowlands are hot and humid and their soils are generally leached and infertile. Pine and palm savannas predominate as far south as the Laguna de Perlas, after which tropical rainforests dominate. Fertile soils are found only along natural levees and floodplains of the numerous rivers and many lesser streams into the complex of shallow bays, lagoons, and salt marshes of the coast (Scott and Carbonell, 1986; LCRS, 1999).

Forests are abundant and, despite large-scale clearing, about one-third of the land in Nicaragua—or approximately 4 million hectares—was still forested in 1993. Hardwoods abound in the coastal zone but stands are mixed, making exploitation difficult. Some logging of mahogany, cedar, rosewood, and logwood for dyes takes place, and large stands of pine in the northeast support a small plywood industry (LCRS, 1999).

SEASONALITY, CURRENTS, NATURAL ENVIRONMENTAL VARIABLES

The climate in Nicaragua is warm (mean air temperature 25.3–26.6°C) and relatively humid. Annual mean air temperature at San Andrés Island was 27°C with a 10°C seasonal range (Geister and Diaz, 1997). Extreme temperatures occurred in March 1973 (17°C) and July 1971 (34.4°C).

Average annual rainfall is about 2500 mm (UNEP, 1988). Bluefields towards the south receives about 4480 mm (Koltes et al., 1997). Rainfall is seasonal and varies greatly between the northern and southern part (Ryan, 1992). May to October are the wettest months when the coast is subject to heavy flooding along all major rivers having good-sized catchment basins. Near the coast, floodwaters spill over onto the floodplains; riverbank agricultural plots are often seriously damaged, and savanna animals die in considerable numbers. Heavy rains accompanying mid-latitude cyclonic storms may sweep through, and nearshore waters are subject to considerable sedimentation at such times (Roberts and Murry, 1983).

Solar radiation data from Bluefields (López de la Fuente, 1994) are listed in Table 1. Tropical storms and hurricanes are quite common in summer. Hurricane Joan struck in 1988 causing more destruction to wildlife, natural resources and infrastructure than years of war. An estimated 10% of the rainforests (250,000 ha of broadleaf and mangrove forest) were blown down along the Caribbean in four hours (Norsworthy and Barry, 1989). When Joan hit Corn Island it destroyed 95% of the homes, crops and wells and crippled the fishing industry (Nietschmann, 1990). The wave surge lasted for nearly one week and wave heights greater than 15 m were generated. Surges re-suspended large sand banks and shifted them onto several extensive coral formations, badly damaging them (Ryan et al., 1997). In October 1998, Hurricane Mitch became the strongest storm in the western Caribbean for 10 years and Nicaragua suffered a major setback, affecting 870,000 people, or 18.2% of the population and killing 2863 people. The Caribbean shrimping fleet was largely unaffected, but Hurricane Mitch caused severe damage to infrastructure. Receding floodwaters left stagnant pools and the heightened risk of mosquito-borne diseases, cholera and leptospirosis. Agriculture and shrimp aquaculture were severely impacted, and total losses were estimated as over US$1.5 billion (USEM, 1998).

There is little seasonal variation in surface water temperatures (25.5–28°C). Trade winds blow steadily from the ENE at

Table 1

Solar radiation, PAR, and cloud cover data for Corn Island (López de la Fuente, 1994). Adapted from Ryan et al. (1997).

Month	Daily Radiation[1]	PAR[2]	Sunshine (h/d)	% Cloud Cover[3]
January	15.94	9,051	12.3	64.9
February	119.61	11,137	12.5	61.9
March	22.36	12,434	12.8	56.4
April	23.14	12,870	13.2	56.9
May	19.81	11,577	13.2	74.8
June	16.88	10,477	13.3	75.3
July	15.99	10,318	13.2	86.0
August	16.63	10,320	13.0	82.3
September	16.46	10,219	17.7	81.9
October	15.80	9,798	12.1	81.8
November	15.02	8,942	11.7	73.0
December	14.11	8,404	11.6	74.5

[1] Global radiation expressed in megaJoules/m^2 surface area.
[2] PAR = Phytosynthetically Active Radiation measurements expressed in μ-Einsteins/s/m^2, represents the amount of solar radiation available to perform phytosynthesis.
[3] Cloud cover is calculated from the average number of octaves of the sky that are clouded each month.

7–10 m/s (Roberts and Suhayda, 1983) and generate a westward flow with some upwelling (Milliman, 1976) and a seasonally variable Western Caribbean Gyre (Glynn, 1973). The winds and density gradients set up by riverine effluents produce a strong (>70 cm/s) north/south flowing coastal boundary current. This creates a 10-km wide zone with terrigenous sediments and high turbidity. Outside this zone, reef growth is not inhibited (Roberts and Murry, 1983). Tides are semi-diurnal and average less than 1 m in range.

Salinity variation within the Caribbean basin is greatest in the shallow estuaries and correlates with rainfall. Hypersaline conditions can persist for months or years where tidal flushing is reduced by shallow sand/mud banks but, more often, estuaries have reduced salinity. At San Andrés and Providencia Islands, salinity ranged between 34–36.3 ppt (Geister and Diaz, 1997), values which are typical of other offshore regions also.

There is little information on nutrient concentrations. Hallock et al. (1988) suggest that marine waters in the Nicaraguan Rise are enriched, and point to its extensive *Halimeda* banks as evidence. Waters overlying the Nicaraguan Rise are generally low in chlorophyll, but the acceleration of the Caribbean Current here brings the uppermost thermocline/nutricline up onto the bank edges and even to the sea surface (Hallock and Elrod, 1988). The waters of the Nicaraguan Rise are probably just too mesotrophic to support reef growth (Hallock, pers. com.).

MAJOR SHALLOW WATER MARINE AND COASTAL HABITATS

The coastal zone contains a great diversity of resources and ecosystems with great potential for generating wealth and a good quality of life for the Nicaraguan people (Ryan, 1996, 1999) including the largest remaining stands of coastal forest in the region (tropical pine and broadleaf forests), and numerous coral reefs (Ryan, 1992, 1994a), seagrass beds (Ryan, 1994b), extensive lagoons, mangrove wetlands and estuaries. The forests are important sources of timber, and provide important functions in flood and erosion control, and the self-purification of waters. The Nicaraguan Caribbean coastal zone, for pragmatic purposes, has been proposed (MAIZCo, 1997) as having a terrestrial limit of the inland extension of mangrove habitat and brackish waters, and a marine limit of the 12 nm of the territorial sea. Many of the areas are in excellent condition.

The continental shelf here is the broadest in all of Central America, reaching westward for 250 km in the north to 20 km wide near the border with Costa Rica in the south. The shelf deepens sharply to 20–40 m and maintains this depth to the abrupt shelf edge (Murry et al., 1982; Owens and Roberts, 1978). There are proven deposits of oil but these have been only marginally studied (Foer, 1992).

Coral reefs

Coral reefs vary from small patches and pinnacles to large, complicated platforms and well-defined belts and, apart from the narrow zone occupied by the coastal boundary current, are distributed across virtually the entire shelf. Only a few may be observed from the surface (Roberts and Murry, 1983). River flow rates, proximity to land and rivers, and the frequency of storms influence the distribution of reefs on the Nicaragua shelf (Ryan, 1993). There are several major coral reef areas and most can be divided into those on the nearshore shelf (to 25 km), central shelf (to the edge of the shelf) and those on the shelf edge. Reefs in the south are poorly formed, probably due to high inputs of freshwater and sediments from the Rio San Juan (Ryan and Zapata, in prep).

The sea urchin *Diadema antillarum* suffered mass mortality across the entire Caribbean in 1983–84. *Diadema* was still absent from Corn Island in 1995 (Woodley et al., 1997). However, some low densities in some shallow (<5 m) nearshore reefs occur and in 1996 Jameson (1998) found the shallow water population in the Miskitos Cays Marine Reserve to be slowly recovering.

Bleaching and disease has not been observed at the Corn Island CARICOMP monitoring site from 1992–1997 (Ryan et al., 1997).

The Miskito Coast Marine Reserve (or Miskito Coast Protected Area: MCPA) lies about 50 km offshore (Fig. 2). Jameson (1996a) found water turbidity relatively high (average visibility 10 m), caused by sedimentation from the Coco and Honduran rivers. Twenty-seven hard coral species and 12 gorgonian species were collected during a three-day survey of the MCPA (Jameson, 1998). Table 2 lists the dominant coral species in the MCPA. Considering the habitat availability and healthy conditions, all Caribbean coral species should be found in the MCPA.

Table 2

Dominant coral species found at reefs in the Miskito Coast Marine Reserve (Jameson, 1998)

Reef Name	Date	Location	Depth (m)	Dominant Coral Species
Nasa Reef	4/30/96	14°18.43'N, 82°57.20'W	1–2	*Acropora palmata, Montastrea annularis, Millipora complanata, Porites astreoides*
Uanvatkira Rock	5/1/96	14°25.30'N, 82°52.50'W	3–4	*Acropora palmata, Porites astreoides, Porites porites, Diploria clivosa*
North Miskito Reef	5/1/96	14°29.60'N, 82°41.80'W	1–4	*Acropora palmata, Millipora complanata*
Pinnacle southwest of Morrison & Dennison Reef	5/2/96	14°24.22'N, 82°57.16'W	10–15	*Montastrea cavernosa, Montastrea annularis, Porites porites, Porites astreoides, Agaricia agaricites*

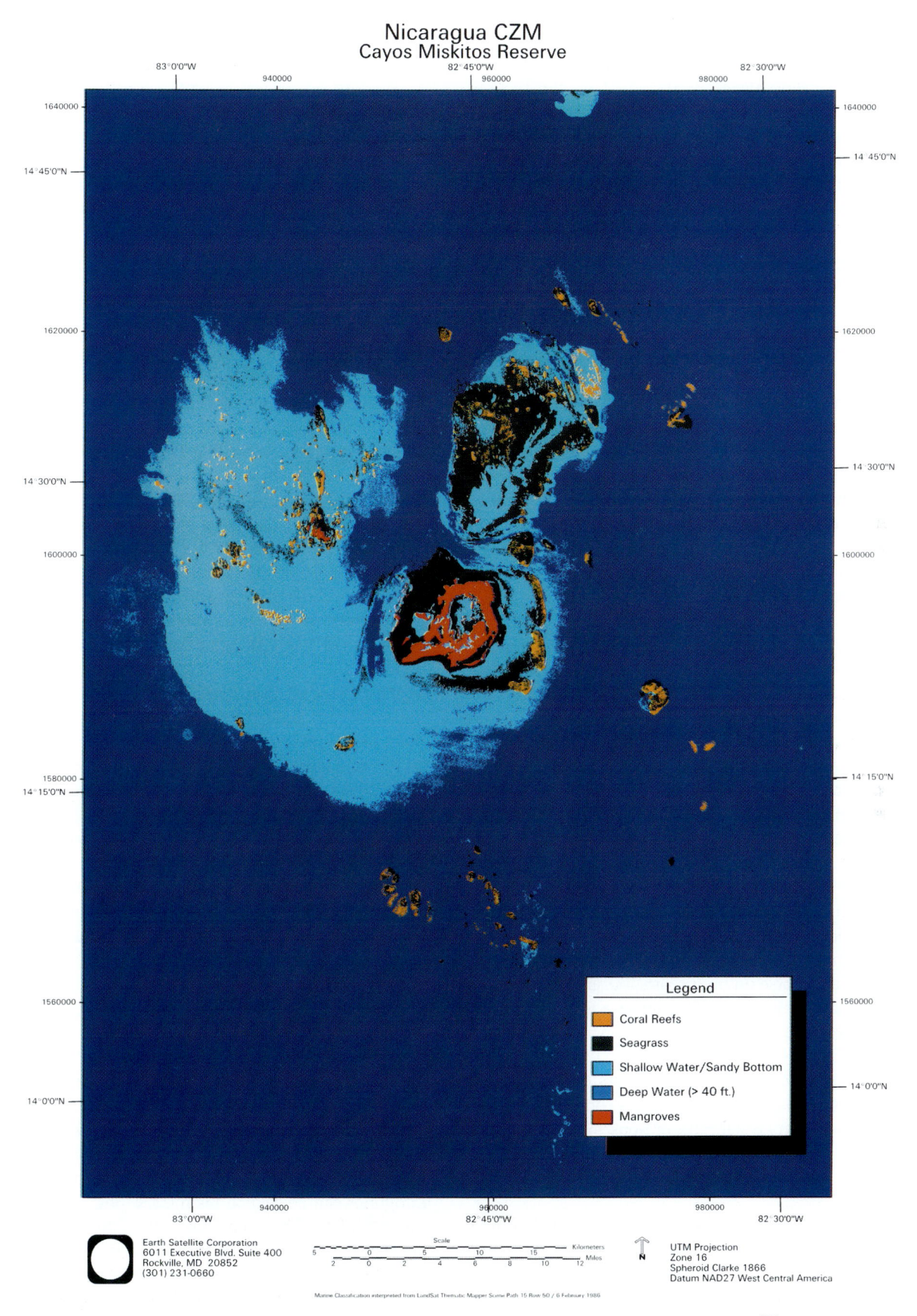

Fig. 2. Satellite photo of Cayos Miskitos (photo courtesy of Earth Satellite Corporation's global Landsat™ mosaic).

The Pearl Cays complex of shallow reefs lies close to shore on the edge of the turbid coastal boundary. The cays support thriving *Acropora palmata* on their windward eastern sides. Coralline algae are abundant on the shallow margins, and mangroves occupy the cay interiors. Part of the Pearl Cays platform, especially the eastern section, has complicated carbonate build-ups, many of which do not reach the surface (Roberts and Murry, 1983). About 27 coral species occur, as is typical for reefs of the western Caribbean.

Great Corn Island (maximum altitude 100 m) is volcanic with low coastal areas and numerous beaches between rocky headlands, with a vegetation primarily of coconut groves and mangroves. There is discontinuous but linear shallow reef around the north and southeast coast (Roberts and Suhayda, 1983), and a triple fringing reef around the north coast (Geister, 1977, 1983; Ryan et al., 1999). Seagrass beds occur also (Ryan, 1994b).

While most of the nearshore coral at Great Corn Island has undergone a major decline (less than 10% live coral cover) over the last decade, the percent live coral coverage (25.2%) at the deeper CARICOMP site is within the range of values reported for other Caribbean sites (Woodley et al., 1997).

Seagrass and Turtles

Seagrass beds (predominately *Thalassia testudinum*) are some of the most extensive in the Caribbean, if not the world. These seagrass beds are major feeding grounds for what is reported to be the largest green turtle (*Chelonia mydas*) population in the Atlantic (Carr et al., 1978; Mortimer, 1983). Algal flats (predominantly *Halimeda*) dominate the vegetative cover on the Nicaragua shelf (Phillips et al., 1982) and on the adjacent Nicaragua Rise (Hine et al., 1988).

Sea turtle research has been conducted on four species (Groombridge, 1982; Nietschmann, 1975, 1981; Bacon et al., 1984). Recently, Lagueux (1998) studied human use patterns and harvest trends. There is very little nesting by green and loggerheads *Caretta caretta* or the Leatherback *Dermochelys coriacea* despite the extensive areas of seagrass. Hawksbills are sparse on both coasts, but may be seen on many cays and reefs; about 100 animals a year and nearly all of the eggs laid are collected. In addition, nesting females are killed when they are found.

Conch seem to be scarce. Marshall (pers. comm.) reports none of the typical evidence of an active conch fishery (i.e., no shell piles around restaurants or houses).

Mangroves and Lagoons

Mangroves are extensive, and an estimated 600 km^2 line lagoon shores, the coast, and cover coral islands (Robinson, 1991). There are no broad mangrove wetlands, but there is a relatively thin mangrove band surrounding each lagoon (see Fig. 1). Each river has mangrove lining its banks. Both black (*Avicennia* spp.) and red mangroves (*Rhizophora* spp.) are present with few signs of destruction (Melnyk and Benge, 1996). Currently, mangroves are abundant and in good health near Puerto Cabezas as well.

Lagoons are critical habitats for fisheries in Nicaragua (Marshall, 1996), and many contain seagrasses, especially *Halodule*. The distribution of fish and shrimp in lagoons differs with the season. During the dry season *Gerres cinereus* was the most abundant fish caught over mud bottoms (Marshall et al., 1994), but fewer are caught in the wet season. Salinity changes also affect the distribution of fish and shrimp in the lagoons. During the dry season species were variably distributed across salinity zones (Marshall et al., 1994) and many species caught in the lagoons in April are not captured during the wet season in August (1996 data), when salinities fell to near zero. Salinities measured in passes during the wet season were slightly higher than the lagoonal salinities but were always less than 5 ppt (Marshall, 1996).

Depth preferences are also evident in the distributions of fish and shrimp. In trawls conducted at depths ranging from less than 1 to more than 10 m, species were distributed across the entire depth range or within more restricted depth zones. *Arius felis* and *Stellifer lanceolatus* were most abundant in depths from 4–8 m, *Bairdiella* spp. were abundant in depths from 5–6 m, while *Centropomus undecimalis* and *C. ensiferus* were most abundant in shallow water (1–2 m) (Marshall, 1996). This data will be useful in mapping and defining nursery grounds and protective habitats in the Miskito Coast lagoons and can also be used to assess the impact of commercial and artisanal fisheries on the recruitment of juveniles to the lagoons.

It was obvious from earlier dry season surveys (Marshall et al., 1994) that mangrove forest prop roots and seagrass beds are important fishery habitats here and that their protection should be a high priority. The results (Marshall, 1996) should be continued over several years and extended to all major lagoons and extended to monitor populations of fishery species and to determine environmental impacts of coastal trawling and other types of nearshore fisheries.

OFFSHORE SYSTEMS

Nutrient input from rivers, estuaries, and wind-induced upwelling all help make the waters off the northern coast of South America the Caribbean's most productive area. Most of the fishing is artisanal. Longliners fishing off Nicaragua are mainly converted shrimp trawlers that capture large pelagic fishes (tuna and swordfish) and bottom fish. Longlines can be several kilometres long and might have 100 to 500 baited hooks per kilometre. Most of the fishing is believed to be done by Korean vessels, though there are probably fewer than 10 vessels involved in the pelagic fisheries and many national vessels are being converted into lobster dive boats according to local reports. Longliners target primarily red snapper and yellowtail snapper.

Automatic jigging machines are a relatively new type of gear for Nicaragua, and vessels using this technique are also believed to be converted shrimp vessels. The product from jigging machines is said to be superior, since the fish are captured live and can be preserved rapidly (Jones, 1996a).

While much general offshore biological and fisheries information is available for the Caribbean coast of Nicaragua, little data has been collected in a systematic way that could provide a foundation for ongoing fishery management. Jones (1996b) provides recommendations and field data collection forms to help establish a solid base of statistical information for fisheries managers in Nicaragua.

POPULATIONS AFFECTING THE AREA

Nicaragua's natural resources have not been adequately or consistently monitored but it is apparent that in many cases they have undergone significant degradation. Generally, however, much of Nicaragua's Caribbean coastal resources are intact (Foer, 1992) and the country has the opportunity to prevent the widespread destruction which has occurred in other Central American countries. Until the 1960s most of these ecosystems were not highly impacted, but they are now experiencing increasing pressures. During the 1980s, the interruptions caused by the civil war allowed reestablishment of some communities, but the heavy exploitation and degradation has now resumed. The National Environmental Action Plan (NEAP, 1994) lists deforestation, soil erosion and water pollution as the most serious environmental problems.

About 7% of Nicaragua's 2.7 million inhabitants live in the Caribbean region (4% in coastal municipalities with a density of 6 inhabitants/km^2). On average only 21% of homes have access to sewage services (10% in the south and 33% in the north). About 48% of homes have no access to public sources of drinking water (62% in the south and 34% in the north). Unemployment is on average 27% (32% in the north) compared with 17% nation-wide (MAIZCo, 1997).

Most of the population is involved in subsistence farming. The people and products of Nicaragua's Caribbean coast have easier access and closer ties to international markets than they do to Managua and western Nicaragua. As this relatively unexploited area becomes increasingly accessible, it will become an important generator of income and employment to larger numbers of people, although the only fertile soils here are found along the natural levees and narrow floodplains of the numerous waterways. Ownership and control over the resources is a source of long-term conflict between the Miskito, Sumo, Rama, and Garifuna people who have lived here for centuries, and the Spanish, British and Nicaraguan Ladino colonists, and more recently, the large numbers of international resource pirates operating illegally in coastal waters (Foer, 1992; Vegas, 1996). The coasts are popular recreation destinations for Nicaraguans and the government recognizes the coral reefs as a potential tourist attraction, creating in 1991 a large marine park in the Miskito Cays and adjacent Caribbean coastal lagoons.

The eastern slopes of the central highlands are lightly populated with pioneer agriculturalists and small communities of Indians, but the numbers are growing rapidly and spreading eastward toward the coast and the region inhabited by the Miskitos, Sumo, Rama, and Creole peoples. This region, extending into Honduras, is called Yapti Tasba (Motherland) by the Indians, and has a population of over 260,000, made up of four territorially based indigenous nations and Nicaraguan (Ladino) immigrants. The autonomy of this region was the cause of the fighting between the indigenous groups and the Sandinistas that began in 1981; the people have fiercely defended their territory for centuries against Spanish, British and other would-be occupiers. Today, these lands are divided by the Nicaraguan government into two regions run by an elected government. The Rio Wangki (Rio Coco), which runs along the Nicaraguan–Honduran border, has historically been the major population concentration of Miskito communities. The largest town by far is Bluefields, with 41,000 people, followed by Corn Island with 4100. Eight communities in the RAAS have electric power.

Tourism

International tourism is currently a relatively minor activity in Nicaragua. The potential is quite significant although the majority of visitors head to the Pacific beaches. Nicaragua recognizes that it still has a fair amount of what is becoming an increasingly sought after and rapidly disappearing commodity, especially in Central America—wilderness and minimally impaired environments. The master plan for the development of tourism in Nicaragua contemplates ecotourism for the area (MITUR, 1996).

RURAL FACTORS

Indiscriminate deforestation takes place in the upper and middle areas of the Caribbean coastal basins as a result of slash-and-burn agriculture and the forestry industry. Hurricane Joan destroyed 50,000 ha of forest and opened the area up for agricultural and livestock husbandry (Foer, 1992). Because 90% of Nicaragua's watersheds drain towards the east coast through eleven major rivers (Ryan et al., 1997), increased deforestation has led to high erosion along many of the watersheds. The resultant sediment loads are believed to have killed several of the large nearshore reef complexes on the central Nicaraguan shelf (Ryan, 1992).

Fisheries

Fishing has long been a source of food for the domestic market, but the rich fishing grounds of the Caribbean side began to be exploited for export of shrimp and lobster only

> ## Cultures and Stakeholders
>
> Nicaragua's extensive Caribbean lowlands region has never been fully incorporated into the nation. This area, known as the Costa de Mosquitos, is isolated from western Nicaragua by rugged mountains and dense tropical rainforest with poor communications. Costeños (the indigenous people and Creoles native to the Caribbean lowlands) are also divided by history and culture from the whites and mestizos of the west, whom they call "the Spanish".
>
> The Caribbean lowlands were never part of the Spanish empire but were, in effect, a British protectorate beginning in the 17th century. In the mid-19th century, the US displaced Britain, and not until 1894 did the entire region come under Nicaraguan administration. Even then, continuing United States political weight, commercial activity, and missionary interest in the Caribbean lowlands eclipsed the weak influence of western Hispanic Nicaragua until World War II. As a result, costeños have not traditionally regarded themselves as Nicaraguans; they see Nicaraguan rule as an alien imposition and fondly recall the years they enjoyed under British and American tutelage. Costeños are more likely to speak English or an indigenous language at home than Spanish.
>
> The Caribbean lowlands are home to a multiethnic society of Miskito, Creoles, and mestizos, with small populations of Sumu, Rama, and Garifuna, an Afro-Carib group. The Miskito, the largest indigenous group, reflect the region's diverse ethnic history. Like the Sumu, they are linguistically related to the Chibcha of South America. Their culture and genetic heritage reflects contacts with Europeans. A Miskito monarchy, established over the region with British support in 1687, endured into the 19th century. The Miskito population is concentrated in northeastern Nicaragua, around interior mining areas and along the banks of several rivers. Honduras also has a large Miskito population. In modern times, they have survived by alternating subsistence activities with wage labour.
>
> The Creoles are the descendants of colonial-era slaves, Jamaican merchants and West Indian labourers. As British influence receded, Creoles displaced the Miskito at the top of the region's ethnic hierarchy and became the key colonial intermediary. Concentrated in the coastal cities, contemporary Creoles are mainly English-speaking, are urban, well educated, and amply represented in skilled and white-collar occupations. Creoles are disdainful of indigenous groups, over whom they maintain a distinct economic advantage. All Caribbean groups, however, share a resentment of the Hispanic elite.
>
> The expanding mestizo population in the Caribbean lowlands is concentrated in the region's western areas, inland from the Caribbean littoral. Since the 1950s, the expansion of export agriculture in the western half of the country has forced many dispossessed peasants to seek new land. On the Caribbean side of the central highlands, this has produced bitter clashes between mestizo pioneers and Miskito and Sumu agriculturalists, over what the indigenous people regard as communal lands.
>
> Within contemporary Caribbean lowlands society, a clear ethnic hierarchy exists. The indigenous groups occupy the bottom ranks and are the most impoverished, least educated, generally with the least desirable jobs. Above them, at successively higher ranks, are recently arrived poor mestizos, Creoles, and a small stratum of middle-class mestizos. Prior to 1979, Europeans or North Americans, sent to manage foreign-owned enterprises, were at the top of the hierarchy. In the mines, Miskito and Sumu work at the dangerous, low-wage, underground jobs; mestizos and Creoles hold supervisory positions; and foreigners dominate the top positions. Also prior to 1979, a special niche was occupied by a small group of Chinese immigrants, who dominated commerce.
>
> The last census data are from 1971. Since then, the region has experienced rapid natural increase and heavy migration of mestizos from the west. In the early 1980s, armed conflict in the region drove thousands of Miskito over the Honduran border, but as the violence ebbed in the late 1980s, refugees returned.

in the 1980s. A 1987 loan allowed Nicaragua to double the size of its fishing fleet to 90 boats but damage by Hurricane Joan in 1988 to processing plants, and the United States trade embargo in 1985, kept production levels far below the potential catch. Restoration of trade produced a surge in exports (LCRS, 1999).

The Nicaraguan lobster fishery is made up of both commercial and artisanal fleets, distinguished by size of the vessels. The artisanal fleet is limited to shallow waters surrounding the Miskito Cays and Corn Island where the catch accounts for about 8% of the total Nicaraguan lobster production. Much of the commercial effort occurs just outside the boundary of the reserve and it is believed the MCMR may account for as much as 50% of the Nicaraguan total. However, over 50% of the catch is composed of juveniles (Childress and Herrnkind, 1996).

Shrimp are the second most important fishery, and the artisanal and commercial fishers catch four species of penaeid shrimp. In Nicaragua, shrimp are harvested along the entire coastline (Klima, 1996).

Thirty-five species of reef fish surveyed at nine reef sites in shallow water (1–5 m) revealed that numbers of herbivores exceeded those of carnivores (Marshall, 1996). Scaridae (*Scarus* spp. and *Sparisoma* spp.) and a kyphosid (*Kyphosus* sp.) were the most abundant. Parrot fish, chubs and Pomacentrids were also abundant on most reefs. Snappers were present but not abundant, as were groupers and a wide range of other, typical reef species. Few fish of commercial importance and adequate size for reef fisheries were seen. Most existing and future reef fish fisheries would thus focus on the reefs and hard bottoms in deep water (>15 m) that surround the Miskito Cays.

The Nicaraguan Center for Hydrobiological Research estimates the fisheries biomass resource to be approximately 195,000 Mt, excluding pelagic resources outside the 12 mile territorial limit. It is estimated that 54,000 Mt of this is exploitable. In addition to the above, the estuarine systems have a theoretical biomass of a further 15,000 Mt, consisting of snook, white sea bass, mojarra, mullet and catfish (Ryan et al., 1993).

Much more information will be needed to preserve the sustainability of the reef fish resource as human demands grow. The Gulf Coast Fisheries Management Council (1989) and Bohnsack and Sutherland (1985) present information needs within the Gulf of Mexico generally, which are the same needs as exist for Nicaragua specifically. An abbreviated and slightly modified listing of the recommendations include: (1) identify optimum reef fish habitat; (2) identify environmental and habitat conditions that promote reef fish production; (3) determine the connections between estuarine habitats and reef fish recruitment; (4) identify areas of critical importance to reef fish, and (5) determine the impact of shrimp trawling in the Miskito Channel on reef fish populations.

Reef fish populations do not seem to be impacted by fishing. However, there is concern that the aquarium fishery developing at Corn Island may expand into the protected area (Marshall, 1996).

The fishery sector has not made great progress towards achieving diversification away from shrimp and lobster (MAIZECo, 1997). The use of intensive diving, chemical products to drive lobsters out from their hiding places, nets to block the entrances to coastal lagoons, drag nets and Jamaican-type traps, represent a threat to the integrity of the ecosystems (Jones, 1996a; MAIZCo, 1997).

Spiny lobster (*Panulirus argus*) is one of the most valuable species and economically is the most important fishery in Nicaragua (Marsh and Gallucci, 1996). Landings have been on the increase with a reported catch of over 3 million pounds in 1995, probably due to a fishing effort which has doubled since 1990. Populations seem to be maintaining themselves, albeit at a young age-class population structure (Childress and Herrnkind, 1996). The five principal risk factors identified in Nicaragua include: (1) mortality of berried females and juveniles, (2) incidental catch of lobsters in shrimp trawls, (3) illegal or pirate fishing, and (4) spatial concentration of commercial fishing effort. Management options and research priorities to address these risks are provided by Marsh and Gallucci (1996) and Childress and Herrnkind (1996).

Recruitment of juvenile shrimp into the fishery begins in April and continues at a high level to November (Klima, 1996). Artisanal, nearshore fishermen harvest almost exclusively white shrimp from between 3 and 35 m depth. Industrial fishermen harvest all four species, but mainly red shrimps. The fishery produced 4.9 million pounds of shrimp in 1995 but this declined to 3.6 million pounds in 1994 despite increased fishing effort (Klima, 1996).

Nicaragua has had a long practice of limiting the number of fishing vessels operating off the Caribbean coast. While Nicaragua has management issues in penaeid shrimp fisheries which are common worldwide, two primary risk factors which have been identified are the harvest of juveniles and industrial fishing in the coastal zone (Marsh and Gallucci, 1996; Klima, 1996).

Sea turtles are now considered threatened or endangered in many areas in the Caribbean as a result of over-exploitation. In the MCMR fishing pressure is very high. Green turtles are captured and held alive, to be sold as a meat substitute (Jameson, 1998).

EFFECTS FROM URBAN AND INDUSTRIAL ACTIVITIES

The nature and effects of industrial uses of the Nicaraguan Caribbean coast are minor compared to many countries in the region. Commercial lobster and shrimp plants are in Bluefields, and a lobster and scale-fish fishery and two seafood processing plants (which produce over 40% of Nicaragua's total seafood exports) provide the primary sources of income for residents of Great and Lesser Corn Islands (Ryan et al., 1997). Development has been hampered by lack of infrastructure and difficulties of access and communication. Most products are shipped by highway to and from ports in Costa Rica and Honduras. Nicaragua has three Caribbean seaports (El Bluff, El Rama and Puerto Cabezas), two of which are basically piers which handle limited cargo.

Water quality around the major coastal settlements has deteriorated due to industrial and shipping activities, fish processing plants, and human wastes, but there are few studies documenting conditions. The Bahia de Bluefields is the most degraded body of water, since residents dispose of sewage via septic tanks and latrines (UNEP, 1988). Industrial waste is not high at present (Table 3), though pesticides were high (Table 4) and a major problem is solid waste disposal in the lagoon (CIMAB, 1996). Wells have high coliform levels and the government has a limited program to distribute chlorine to well owners. The Escondido and Kukra rivers deposit large quantities of sediments into the Bay, diminishing its depth. The Bay averages only 1 m depth, and circulation is further restricted by Deer Island which forms a barrier to the ocean. Prevailing winds and wave action concentrate contaminants along the west side of the Bay where Bluefields is situated. Corn Island and Puerto Cabezas also experience water contamination (MAIZCo, 1997). Many of the Great Corn Island nearshore coral reefs and mangroves appear to be degraded due to subterranean discharges of sewage-contaminated groundwater originating from the island (Ruden, 1993; Ryan, 1994c; Broegaard, 1995; Ryan et al., 1997).

Small amounts of gold and silver in northeast Nicaragua provide much-needed export income. All mines were

Table 3
Average amounts of heavy metals in sediments in Bluefields, Nicaragua (CIMBA, 1996)

Year	Organic matter (μg/l)	Co (μg/l)	Cu (μg/l)	Fe (%)	Mn (μg/l)	Ni (μg/l)	Pb (μg/l)	Zn (μg/l)
1994	16.95	21	53	5.96	826	14	19	117
1995	14.30	21	54	6.33	670	19	18	111

Table 4
Organochlorine pesticides in waters and sediments in Bluefields, Nicaragua in 1995 (CIMBA, 1996)

Location	Lindane	Heptachlor	Aldrin	Dieldrin	DDT
Water (ng/l)	1.9	10.1	4.5	6.85	21.25
Sediments (ng/g)	1.8	3.9	0.8	6.4	8.8

nationalized in 1979 which, combined with hostilities, caused production of gold and silver to halve in the 1980s. Small amounts of copper, lead, and tungsten have been mined in the past, and the country has unexploited reserves of antimony, tungsten, molybdenum, and phosphate. The only important mining activities with potential for affecting coastal areas are the gold mines around Siuna and Bonanza, whose processing can release mercury and cyanide.

Exploratory drilling for petroleum will commence in the near future within the MCPA. The potential for direct damage to reefs and seagrass beds (and therefore fisheries) will depend on the drilling methods and practices utilized (Hudson et al., 1982).

PROTECTIVE MEASURES

There is presently a lack of coordinated support among the 38 Caribbean nations for marine ecosystem monitoring and management, and integrated management will be difficult to achieve because of differing cultures, education levels, and economic development. One proposed solution to the fisheries over-exploitation problem is the creation of marine fishery reserves, with each country contributing in proportion to its shelf area and potential biotic wealth (Richards and Bohnsack, 1990).

In Nicaragua, the 29,000 km² Miskito Coast Protected Area (MCPA) was established with the assistance of several U.S. environmental groups and the USAID/NOAA partnership (Trott, 1996). The Miskito Coast Marine Reserve (MCMR) component consists of an area defined by a 40 km radius around the big island of Miskito Cays (see Fig. 2). The terrestrial component consists of a 20 km coastal zone band between Cabo Gracias a Dios to the north and Wounta to the south. A draft management plan for the MCPA was released in 1995 (CCC, 1995), and Jain (1996) used this to develop a framework for general and specific objectives (Table 5). Jameson (1996b) outlines the benefits of adding UNESCO Biosphere Reserve designation to the MCPA and Harrington and Gallucci (1996) describe the benefits and steps required to implement co-management between the Nicaraguan government and the Miskito Indians.

Investigations regarding threats to mangroves should be carried out (Melnyk and Benge, 1996). Shrimp aquaculture has already destroyed many hectares of mangroves on the Pacific coast and the responsible agencies must ensure that this does not happen in the MCPA. The economic benefits and costs of shrimp farming should also be evaluated before any ponds are established (Janssen and Padilla, 1996).

In addition to the MCPA, there are 16 other protected areas. Several are located in the coastal strip of the MCPA itself and were declared protected areas in 1988. Another is not located in the coastal zone but serves as a buffer zone for coastal ecosystems.

There are four natural reserves in the southern region. The Indio-Maíz Reserve is 295,000 ha and the forestry reserve Wawasang is 327,000 ha. Others are considerably smaller. The Protected Areas Directorate has also identified the mouth of the Rio Grande (44,700 ha), Pearl Lagoon (18,600 ha), The Perlas Cays (2300 ha), and Lesser Corn Island (290 ha) as potential future reserves.

ICZM

Nicaragua's ICZM initiative resulted from its environmental action plan (PAA-NIC) approved in 1993. The plan states that "the patterns followed to exploit coastal natural resources are not sustainable and unless they are modified, will result in loss of biodiversity in the coastal zones and in a reduction of their economic potential" (MAIZCo, 1997).

In May, 1995 a Phase 1 project document was prepared (MAIZCo) that outlined the role of the government. MAIZCo was launched in January 1996. The program supports acquisition of knowledge and capacity building, and the first phase was implemented by the Ministry of the Environment and Natural Resources (MARENA).

In 1997 MARENA produced a proposal for the sustainable use of coastal resources (MAIZCo, 1997), made possible

Table 5
Summary of recommendations for the Miskito Coast Protected Area from Trott (1996).

Objectives	Summary of Recommendations
Management	1. Conserve the biodiversity, endangered species and ecosystems found within the reserve. 2. Conserve the condition and abundance of natural resources so they might be used in traditional ways by future generations 3. Improve the lives and economy of communities living in the reserve. 4. Establishment of the Miskito Coast Protected Area was an important first step in meeting these objectives
Lobster	1. Set minimum size of 135 mm tail length, 75 mm carapace length, 5 ounce tail weight. 2. Prohibit capture of gravid females. 3. Prohibit fishing in the nursery habitat surrounding Big Miskito Cay including the mangroves, seagrass, and fringing coral reefs to an approximate depth of 10 m. 4. Prohibit the use of hooks and spears by divers. 5. Traps must be made of wood and contain an escape gap of not less than 2.125 inches. 6. Traps may not be baited with turtle meat or sub-legal sized lobsters. 7. Traps should be placed only on sandy bottoms.
Shrimp	1. Prohibit shrimp trawling within the reserve. 2. Set minimum harvest size of 60 tails/pound within the reserve lagoons. 3. Preserve the lagoon environment.
Lagoon Fish	1. Set minimum mesh size of 5 inches for gill nets. 2. Prohibit block netting methods across lagoon passes. 3. Conduct studies to determine if closed seasons are necessary for some species. 4. Develop additional or alternate fisheries, such as crabs or prawns.
Reef Fish	1. Encourage hook and line fishing methods, and discourage destructive methods, especially chemical and explosive. 2. Prohibit the collection of aquarium trade reef fish and corals within the reserve. 3. Conduct studies to determine if closed seasons are necessary for some species. 4. Consider the establishment of a no-take protected area for reef fish.
General	1. While it is desirable to have input from all stakeholders, management decisions should be made by a small decision-making body. 2. Establish a method to limit fishing effort. 3. Prohibit industrial fishing vessels. 4. Establish sources of financial support. 5. License all vessels that operate within the reserve. 6. Apply for recognition as an International Biosphere Reserve. 7. Formalize the boundaries and policies to reduce fishing effort within the reserve buffer zone, in cooperation with the Ministry for the Development of Fisheries. 8. Implement programs in resource education, technical training, economic alternatives, scientific research, and control and surveillance. 9. Continue and expand monitoring programs, in cooperation with the Center for Research on Aquatic Resources, to collect data on catch and effort for lobsters, shrimp and lagoon fish. 10. Develop education programs to educate fishermen regarding the relationship between the proposed management measures and the biology of the fishery species. 11. Consider the development of eco-tourism activities. 12. Comply with the national recommendations regarding the harvest of marine turtles. 13. Create regulations under the "Ley General del Medio Ambiente y los Recursos Naturales" to accomplish the recommendations. 14. Enforcement of these regulations, with cooperation from the Navy, is critical to the welfare of the reserve.

through cooperation with the Netherlands and Denmark. The Pearl Lagoon Basin was selected for a pilot project to implement the methodology.

Many aspects of Nicaraguan law need modification to effectively manage the nation's coastal resources (see 'Nicaragua: Pacific Coast'). At this time, Nicaragua has no institutional or legal framework specifically relevant to ICZM and management of coastal resources is determined by specific sectoral laws which fail to provide an efficient or effective legal framework. Nicaragua is a party to many international agreements that provide other protective measures to ensure the wise use of its coastal resources.

The ICZM capacity is low. Institutions dealing with natural resources have a presence in the area, but the Autonomous Government lacks operational structures to fulfil its mandate. This makes inter-institutional co-ordination weak and controversial. In addition, the Regional Council's Commission on Natural Resources is not yet an active participant in the ICZM process and there is no other structure to take its place. NGOs have a coordination initiative but have not been effective.

Gradually, the different organizations have been improving their communication and resolving their conflicts. Joint co-ordination and agreements have been achieved, but in most cases, the duplication of tasks in the respective mandates have prevented their implementation thus far (MAIZCo, 1997). Mismanagement is mainly due to economic reasons and poor government management; government bodies have been severely criticized for lack of co-ordination, planning and political will (MAIZCo, 1997).

In conclusion, many of Nicaragua's Caribbean marine and coastal resources are minimally impaired and have great potential to improve the quality of life for the Nicaraguan people. Preventing mismanagement and destruction of these resources and ecosystems, which has occurred in other Central American countries, now depends on effective government and the ability of the Nicaraguan people to work together as a united nation.

ACKNOWLEDGEMENTS

We thank Serge Andrefouet, Vincent Gallucci, Pamela Hallock Muller, Cynthia Lagueux, Dulcie Linton, Enrique Lahmann, Yassy Naficy, Kennard Potts, Julie Robinson, Joseph Ryan, Arike Tomson, Leslie Upchurch and Mariska Weyerman for research assistance.

REFERENCES

Bacon, P., Berry, F., Bjorndal, K., Hirth, H., Ogren, L. and Weber, M. (1984) *The National Reports. Proceedings of the Western Atlantic Turtle Symposium, Vol. 3*. University of Miami Press, Florida.

Bohnsack, J.A. and Sutherland, D.L. (1985) Artificial reef research: a review with recommendations for future priorities. Bulletin of Marine Science 37(1), 11–39.

Broegaard, J.R. (1995) Corn Island: Las aguas oscuras bajo el paraiso. *WANI* 18, 45–49.

Carr, A., Carr, M. and Meylan, A. (1978) The ecology and migrations of sea turtles: 7, The West Caribbean green turtle colony. *American Museum of Natural History* 162, 1–42.

CCC (1995) Draft management plan for the Cayos Miskitos Marine Reserve. Caribbean Conservation Corporation, USA.

Childress, M.J. and Herrnkind, W.F. (1996) The Miskito Coast Marine Reserve spiny lobster population: Assessment and recommendations. In: Trott, L.B. (ed) Recommendations and reports for the management of fisheries in the Miskito Coast Marine Reserve of Nicaragua, Environmental Initiative of the Americas Fisheries Project October 1995 to September 1996. USAID Environment Center, Global Bureau, Washington, DC.

CIMAB (1996) Final reports-Nicaragua case study, UNDP/GEF Project on planning and management of heavily contaminated bays and coasts in the wider Caribbean. Centro de Ingerieria y Manejo Ambiental de Bahias, Havana, Cuba.

Foer, G. (1992) Profile of the coastal resources of Nicaragua. In: Foer, G. and Olsen, S. (eds) Central America's Coasts: Profiles and an Agenda for Action. The University of Rhode Island Coastal Resources Center, Naragansett, RI, 30 pp.

Geister, J. (1977) The influence of wave exposure on the ecological zonation of Caribbean coral reefs. *Proc. 3rd International Coral Reef Symposium, Miami* 1, 23–29.

Geister, J. (1983) Holocene West Indian Coral reefs: geomorphology, ecology and facies. *Facies* 9, 173–284.

Geister, J. and Diaz, J.M. (1997) A field guide to the oceanic barrier reefs and atolls of the southwestern Caribbean (Archipelago of San Andres and Providencia, Colombia. *Proc. 8th International Coral Reef Symposium* 1, 235–262.

Glynn, P.N. (1973) Aspects of the ecology of coral reefs in the Western Atlantic Region. In: *Biology and Geology of Coral Reefs, Vol II*, eds. O.A. Jones and R. Endean. Academic Press, New York, pp. 271–324.

Groombridge, B. (1982) *The IUCN Amphibia-Reptilia Red Data Book, Part 1: Testudines, Crocodylia, Rhynchocephalia*. IUCN, Gland Switzerland.

Gulf of Mexico Fishery Management Council (1989) Reef fish management plan. Gulf of Mexico Fishery Management Council, 881 Lincoln Center, 5401 West Kennedy Boulevard, Tampa, FL.

Hallock, P. and Elrod, J.A. (1988) Oceanic chlorophyll around carbonate platforms in the western Caribbean: Observations from CZCS data. *Proceedings of the Sixth International Coral Reef Symposium, Townsville, Australia, 8–12 August 1988*, Vol. 2, pp. 449–454.

Hallock, P.A., Hine, C., Vargo, G.A., Elrod, J.A. and Jaap, W.C. (1988) Platforms of the Nicaraguan Rise: Examples of the sensitivity of carbonate sedimentation to excess trophic resources? *Geology* 16, 1104–1107.

Harrington, G.A. and Gallucci, V. (1996) Analysis of the artisanal fisheries and the potential for co-management in the Miskito Cays Protected Area, Atlantic coast, Nicaragua. In: Trott, L.B. (ed) Recommendations and reports for the management of fisheries in the Miskito Coast Marine Reserve of Nicaragua, Environmental Initiative of the Americas Fisheries Project October 1995 to September 1996. USAID Environment Center, Global Bureau, Washington, DC.

Hine, A.C., Hallock, P., Harris, M., Mullins, H.T., Belknap, D.F. and Jaap, W.C. (1988) *Halimeda* bioherms along an open seaway: Miskito Channel, Nicaraguan Rise, S.W. Caribbean Sea. *Coral Reefs* 6, 173–178.

Hudson, J.H., Shinn, E.A. and Robin, D.M. (1982) Effects of offshore drilling on Philippine reef corals. *Bulletin of Marine Science* 32, 890–908.

INPESCA (1990) Diagnostico de la Actividad Pesquera de Nicaragua. Corporacion Nicaraguense de la Pesca, Ministerio de Economia y Comercio, Fondo Nicaraguense de Inversiones, Secretaria de Planificacion y Presupuesto, Managua.

Jain, M. (1996) Information and options for management of the Miskitos Cays Protected Area in Nicaragua. In: Trott, L.B. (ed) Recommendations and reports for the management of fisheries in the Miskito Coast Marine Reserve of Nicaragua, Environmental Initiative of the Americas Fisheries Project October 1995 to September 1996. USAID Environment Center, Global Bureau, Washington, DC.

Jameson, S.C. (1996a) Miskito Coast Reserve; Coral reef ecosystem survey and management recommendations. In: Trott, L.B. (ed) Recommendations and reports for the management of fisheries in the Miskito Coast Marine Reserve of Nicaragua, Environmental Initiative of the Americas Fisheries Project October 1995 to September 1996. USAID Environment Center, Global Bureau, Washington, DC.

Jameson, S.C. (1996b) Miskito Coast Reserve; Biosphere reserve concept analysis. In: Trott, L.B. (ed) Recommendations and reports for the management of fisheries in the Miskito Coast Marine Reserve of Nicaragua, Environmental Initiative of the Americas Fisheries Project October 1995 to September 1996. USAID Environment Center, Global Bureau, Washington, DC.

Jameson, S.C. (1998) Rapid ecological assessment of the Cayos Miskitos Marine Reserve with notes on the stony corals off Nicaragua. Atoll Res Bull, September 1998, No. 457, Smithsonian Institution, Washington, DC, 15 pp.

Janssen, R. and Padilla, J. (1996) Valuation and evaluation of management alternatives for the Paghiliao Mangrove Forest. CREED Working Paper 313, International Institute for Environment and Development, London.

Jones, C.R. (1996a) Fishing vessels and gear used in the Miskito Coast Protected Area, Nicaragua. In: Trott, L.B. (ed) Recommendations and reports for the management of fisheries in the Miskito Coast Marine Reserve of Nicaragua, Environmental Initiative of the Americas Fisheries Project October 1995 to September 1996. USAID Environment Center, Global Bureau, Washington, DC.

Jones, C.R. (1996b) Field program for fisheries management in the Cayos Miskitos, Nicaragua. In: Trott, L.B. (ed) Recommendations and reports for the management of fisheries in the Miskito Coast Marine Reserve of Nicaragua, Environmental Initiative of the Americas Fisheries Project October 1995 to September 1996. USAID Environment Center, Global Bureau, Washington, DC.

Klima, E.F. (1996) Shrimp management options in Nicaragua's protected area. In: Trott, L.B. (ed) Recommendations and reports for the management of fisheries in the Miskito Coast Marine Reserve of Nicaragua, Environmental Initiative of the Americas Fisheries Project October 1995 to September 1996. USAID Environment Center, Global Bureau, Washington, DC.

Koltes, K.H. and 21 others (1997) Meterological and oceanographic characterization of coral reef, seagrass and mangrove habitats in the wider Caribbean. *Proc. 8th International Coral Reef Symposium* **1**, 651–656.

Lagueux, C.J. (1998) Marine turtle fishery of Caribbean Nicaragua: human use patterns and harvest trends. PhD Dissertation, University of Florida, Gainesville, 215 pp.

LCRS (1999) Library of Congress Research Service Country Study http://lcweb2.loc.gov/cgi-bin/query/r?frd/cstdy:@field(DOCID+ni0077).

López de la Fuente, J. (1994) Atlas of Solar Radiation in Nicaragua. Report to the Swedish Development Assistance Program, Stockholm.

MAIZCo (1997) Action plan proposal for management of coastal zones in Nicaragua. Ministry of the Environment and Natural Resources, General Environment Directorate, Coastal Zones Centre, Managua, Nicaragua, 55 pp.

Marsh, T. and Gallucci, V. (1996) Stock assessment and management of lobster and shrimp fisheries in Nicaragua. In: Trott, L.B. (ed) Recommendations and reports for the management of fisheries in the Miskito Coast Marine Reserve of Nicaragua, Environmental Initiative of the Americas Fisheries Project October 1995 to September 1996. USAID Environment Center, Global Bureau, Washington, DC.

Marshall, M.J., Leopoldo, P.W. and Jerris, J.F. (1994) Lagoonal reconnaissance and near coastal fish surveys in the Miskito Coast Protected Area. Report to the Caribbean Conservation Corporation, Gainesville, Florida, USA.

Marshall, M.J. (1996) Lagoon fish surveys in the Miskito Coast Protected Area, Nicaragua. In: Trott, L.B. (ed) Recommendations and reports for the management of fisheries in the Miskito Coast Marine Reserve of Nicaragua, Environmental Initiative of the Americas Fisheries Project October 1995 to September 1996. USAID Environment Center, Global Bureau, Washington, DC.

Melnyk, M. and Benge, M. (1996) Mangroves. In: Trott, L.B. (ed) Recommendations and reports for the management of fisheries in the Miskito Coast Marine Reserve of Nicaragua, Environmental Initiative of the Americas Fisheries Project October 1995 to September 1996. USAID Environment Center, Global Bureau, Washington, DC.

Milliman, J.D. (1976) Caribbean Coral Reefs. In *Biology and Geology of Coral Reefs, Vol. 1*, eds. O.A. Jones and R. Endean. Academic Press, New York, pp. 1–50.

MITUR (1996) Master plan for the development of tourism in Nicaragua. Ministry of Tourism, Managua, Nicaragua.

Mortimer. J. (1983) The feeding ecology of the West Caribbean Green Turtle (*Chelonia mydas*) in Nicaragua. *Biotropica* **13**, 49–58.

Murry, S.P., Hsu, S.A., Roberts, H.H. and Owens, E.H. (1982) Physical processes and sedimentation on a broad, shallow bank. *Estuarine Coastal Shelf Science* **14**, 135–157.

NEAP (1994) National environmental action plan. Government of Nicaragua, Managua.

Nietschmann, B. (1975) Of turtles, arribadas and people. *Chelonia* **2** (6), 6–9.

Nietschmann, B. (1981) Following the underwater trail of a vanishing species—the Hawksbill Turtle. *National Geographic Society Research Reports* **13**, 459–480.

Nietschmann, B. (1990) Conservation by conflict in Nicaragua. *Natural History Magazine*, November 1990, New York.

Norsworthy, K. and Barry, T. (1989) Nicaragua: A country guide. The Interhemispheric Education Resource Center, Albuquerque, NM.

Owens, E.H. and Roberts, H.H. (1978) Variations of wave-energy levels and coastal sedimentation, eastern Nicaragua. *Proc 16th Coastal Engineering Conference, Hamburg, West Germany*, pp. 1195–1214.

Phillips, R.C., Vadas, R.L. and Ogden, N. (1982) The marine algae and seagrasses of the Miskito Bank, Nicaragua. *Aquatic Botany* **13**, 187–195.

Richards, W.J. and Bohnsack, J.A. (1990) The Caribbean Sea: A large marine ecosystem in crisis. In *Large Marine Ecosystems: Patterns, Processes and Yields*, eds. K. Sherman et al. American Association for the Advancement of Science, Washington, DC, pp. 44–53.

Roberts, H.H. and Murry, S.P. (1983) Controls on reef development and the terrigenous-carbonate interface on a shallow shelf, Nicaragua (Central America). *Coral Reefs* **2**, 71–80.

Roberts, H.H. and Suhayda, J.N. (1983) Wave-current interactions on a shallow reef (Nicaragua, Central America). *Coral Reefs* **1**, 209–214.

Robinson, S. (1991) Diagnostico preliminar de la situacion del medio ambiente en la region autonoma sur (RAAS). INDERA, May 1991, Managua.

Ruden, F. (1993) The hydrogeologic environment of an oceanic basaltic island. *Memoirs of the 24th Congress of the International Association of Hydrogeologists (Oslo, Norway)*, pp. 166–1182.

Ryan, J.D. (1992) Ecosistemas marinas y el manejo sostenible en la costa central del Caribe Nicaraguense. WANI #13, Universidad Centroamericana Press, 15 pp.

Ryan, J.D. (1993) Pastos Marinos del Caribe Nicaraguense. WANI 15, Universidad Centroamericana Press, 12 pp.

Ryan, J.D. (1994a) Ecosistemas de los arrecifes coralinos en la plataforma central de Nicaragua. WANI #14, Universidad Centroamericana Press, 13 pp.

Ryan, J.D. (1994b) Seagrass Meadows and Marine Plants on the Nicaraguan Caribbean Coast, (in Spanish) WANI #15, Universidad Centroamericana. 12 pp.

Ryan, J.D. (1994c) The Corn Island reef survey: Coral degradation patterns and recommended actions. Report to NORAD and IRENA (April), 42 pp.

Ryan, J.D. (1996) Ecosystems of Nicaragua's Coastal Zone (Spanish).

Final report Danida/MARENA, Managua.

Ryan, J.D. (1999). Coastal ecosystems of the Rio San Juan. Final Report prepared for Rìo San Juan Watershed GEF Project.

Ryan, J. and Zapata, Y. (in prep.) The Corals of Nicaragua. In *Coral Reefs of Latin America*, ed. J. Cortez.

Ryan, J., Gonzalez, L. and Parra, E. (1993) Plan de acciun del amedio ambiente Nicaraguense: recursos acu_ticos. DANIDA,/ASDI/ World Bank report.

Ryan, J., Miller, L., Zapata, Y., Downs, O. and Chan, R. (1997) Great Corn Island, Nicaragua. In *Caribbean Coastal Marine Productivity (CARICOMP): Coral Reef, Seagrass and Mangrove Site Characterizations*, ed. B. Kjerfve. UNESCO Technical Book Series, Paris.

Ryan, J., Miller, L., Zapata, Y. and Downs, O. (1999) A characterization of the coral reefs at the Great Corn Island, Nicaragua CARICOMP site. UNESCO Technical Book Series.

Scott, D.A. and Carbonell, M. (1986) *A directory of Neotropical Wetlands*. IUCN, Gland and Cambridge.

Trott, L.B. (ed.) (1996) Recommendations and reports for the management of fisheries in the Miskito Coast Marine Reserve of Nicaragua, Environmental Initiative of the Americas Fisheries Project October 1995 to September 1996. USAID Environment Center, Global Bureau, Washington, DC.

UNEP/IUCN (1988) *Coral Reefs of the World, Volume 1: Atlantic and Eastern Pacific*. UNEP Regional Seas Directories and Bibliographies. IUCN, Gland Switzerland and Cambridge, U.K./UNEP, Nairobi, Kenya xlvii + 373 pp, 38 maps.

USEM (1998) Economic effects of Hurricane Mitch on Nicaragua. U.S. Embassy, Managua, December 1998, Overview and summary, Managua, Nicaragua, U.S. Embassy Managua Econ/Commercial Section www.usia.gov/abtusia/posts/NU1/wwwhcom. html.

Vegas, A. (1996) Fisheries enforcement in the Nicaraguan Miskito Coast Marine Reserve. In: Trott, L.B. (ed.). Recommendations and reports for the management of fisheries in the Miskito Coast Marine Reserve of Nicaragua, Environmental Initiative of the Americas Fisheries Project October 1995 to September 1996. USAID Environment Center, Global Bureau, Washington, DC.

Woodley, J.D. and 29 others (1997) CARICOMP monitoring of coral reefs. *Proceedings of the 8th International Coral Reef Symposium* **1**, 651–656.

THE AUTHORS

Stephen C. Jameson
Coral Seas Inc. – Integrated Coastal Zone Management,
4254 Hungry Run Road,
The Plains, VA 20198-1715, U.S.A.

Lamarr B. Trott
National Oceanic and Atmospheric Administration,
National Marine Fisheries Service,
1315 East West Highway,
Silver Spring, MD 20910, U.S.A.

Michael J. Marshall
Coastal Seas Consortium,
5503 40th Avenue East,
Bradenton, FL 34208, U.S.A.

Michael J. Childress
Department of Biological Sciences,
Idaho State University,
Pocatello, ID 83209-8007, U.S.A.

Chapter 33

NICARAGUA: PACIFIC COAST

Stephen C. Jameson, Vincent F. Gallucci and José A. Robleto

Nicaragua is the largest country in central America, with a coastline of 305 km. Its relatively dry western slopes have attracted farmers since colonial times. The area is tropical, but relatively dry, and consequently supports a tropical fauna. By the mid-1960s cotton was cultivated on 80% of the arable land along the Pacific coast which led to widespread deforestation, erosion, and biocide contamination, and reduced the original vegetation cover by 90%. In 1994 the country had approximately 155,000 ha of mangroves, about 45% of which were located on the Pacific Coast, but this has been reduced by over half by deforestation.

The Gulf of Fonseca is a deep embayment and is one of the most important coastal ecosystems of the region, shared with Honduras and El Salvador. Several rivers drain into it, but these rivers now also deposit large quantities of sediments and pollutants arising from changes in drainage regimes and up-hill deforestation. It is heavily populated by almost 500,000 people; it is economically important because it is a place of trade and commercial activities between the three countries, and opportunities to exploit the natural resources of this rich environment have attracted many people who have in turn further compounded the degradation and loss of valuable natural resources. Shrimp aquaculture is very important here.

Several hundred species of demersal fish are commonly taken as by-catch in the shrimp fisheries, but do not sustain commercial fisheries. Significant fisheries on small coastal pelagic fish, such as the Central Pacific anchovetta and Pacific thread herring, have existed in the Gulf of Panama since the 1950s, but currently most of the production is used for fish meal and oil. The most valuable fish off Nicaragua are tuna (yellowfin, skipjack, and bigeye tuna) caught mainly by fleets from overseas. Serious population-related problems include pollution, lack of facilities for solid waste disposal, lack of basic infrastructure, and lack of integrated planning to prepare for natural hazards, problems related to mining, and cultural displacement of natives. Several areas of Nicaraguan law need modification to effectively manage the nation's coastal resources. Therefore, integrated management is needed to ensure that the potential value of the natural resources contributes to the sustainable development of the Pacific coast.

Some of Nicaragua's Pacific marine and coastal resources have been over-exploited, degraded or destroyed. However, they offer great potential to improve the quality of life for the Nicaraguan people and the condition of the Pacific Central American Coastal Large Marine Ecosystem. The government understands the value of these resources and has received excellent natural resource management guidance and technical assistance. Preventing mismanagement and destruction of these resources and ecosystems, which has occurred in other Central American countries, now depends on effective government and the ability of the Nicaraguan people to work together as a united nation.

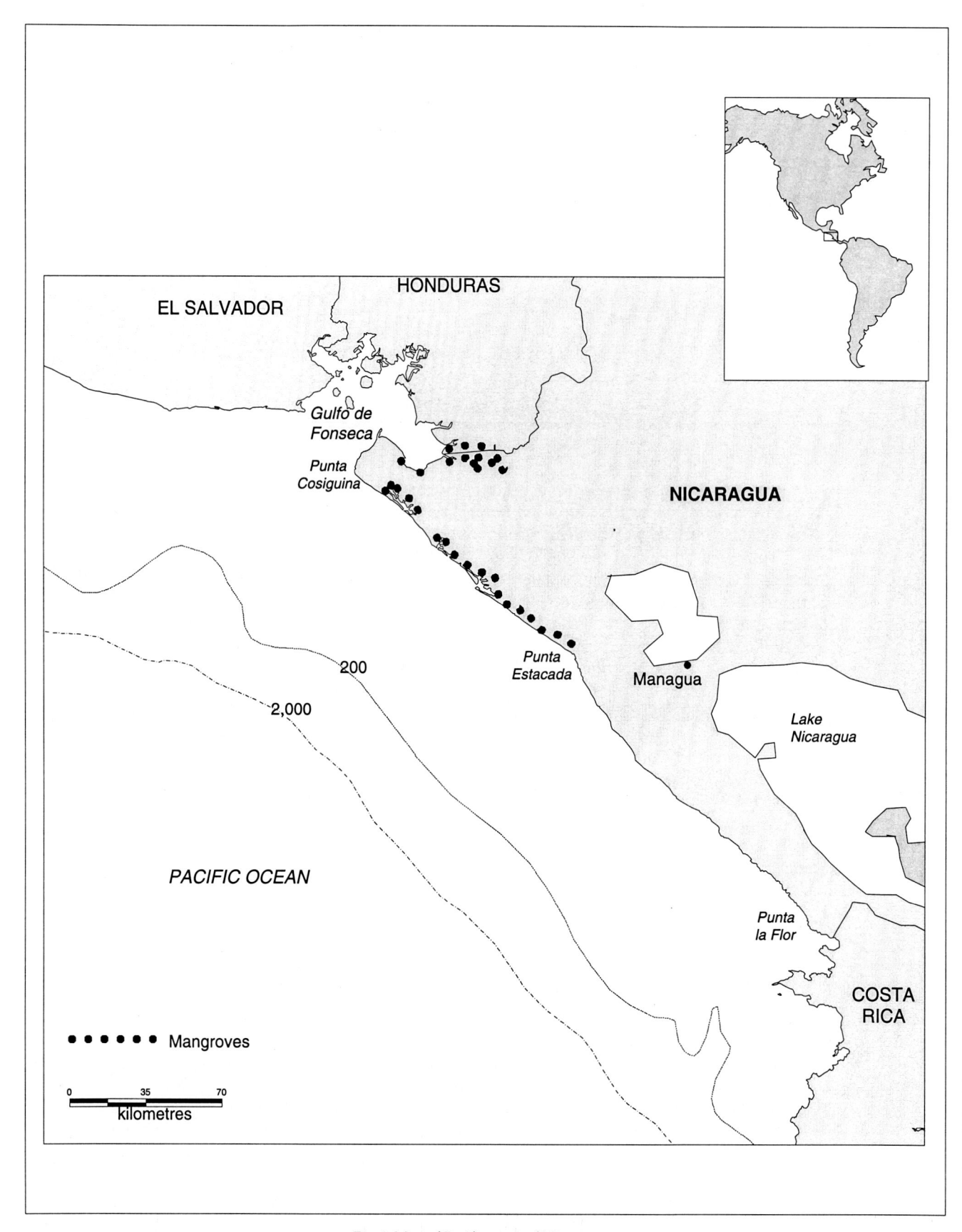

Fig. 1. Map of Pacific coast of Nicaragua.

THE DEFINED REGION

Nicaragua is the largest country in Central America (Fig. 1), covering 129,494 km^2. There are three major physiographic zones: Pacific lowlands; the wetter, cooler central highlands; and the Caribbean lowlands. The Pacific coastline is 305 km long (INPESCA, 1990) and the continental shelf is steep, reaching extreme ocean depths very near the coast. The Middle American Trench is located just 100 km offshore, and the adjacent, strong crustal subduction zone results in frequent and sometimes severe earthquakes. Nicaragua and adjoining states are some of the poorest of the Americas, and many impoverished people have migrated to the coast to make a meagre living from subsistence fishing and farming. In the past 20 years, the region has been torn by civil wars, but today the region's economy is diverse and developing. The main commercial products are agricultural and manufacturing, though there are also significant mining activities as well.

The coastal lowland is a narrow strip about 75 km wide. The Pacific volcanic chain is the most important geomorphological feature of the northwestern part and is formed by almost 40 volcanoes extending for about 300 km (Sánchez and Quiróz, 1998). Most of the area is flat, except for the volcanoes between the Golfo de Fonseca and Lago de Nicaragua, whose peaks lie just west of a large crustal fracture that forms a long, narrow depression and which is occupied in part by the largest freshwater lakes in Central America: Lago de Managua (56 km long and 24 km wide) and Lago de Nicaragua (about 160 km long and 75 km wide). These two lakes are joined by the Río Tipitapa, which flows south into Lago de Nicaragua, which in turn drains via the Río San Juan to the Caribbean Sea. The valley of the Río San Juan forms a natural passageway close to sea level across the Nicaraguan isthmus and was considered as a possible alternative to the Panama Canal in the past. The lowlands are fertile and highly enriched with volcanic ash, densely populated and well cultivated. West of the lake region is a narrow line of ash-covered hills and volcanoes that separate the lakes from the Pacific Ocean (Fig. 2).

The central highlands are rugged mountain terrain, composed of ridges 900 to 1800 m high. Its drainage is primarily toward the Caribbean, and those that flow west to the Pacific Ocean are steep, short, and intermittent

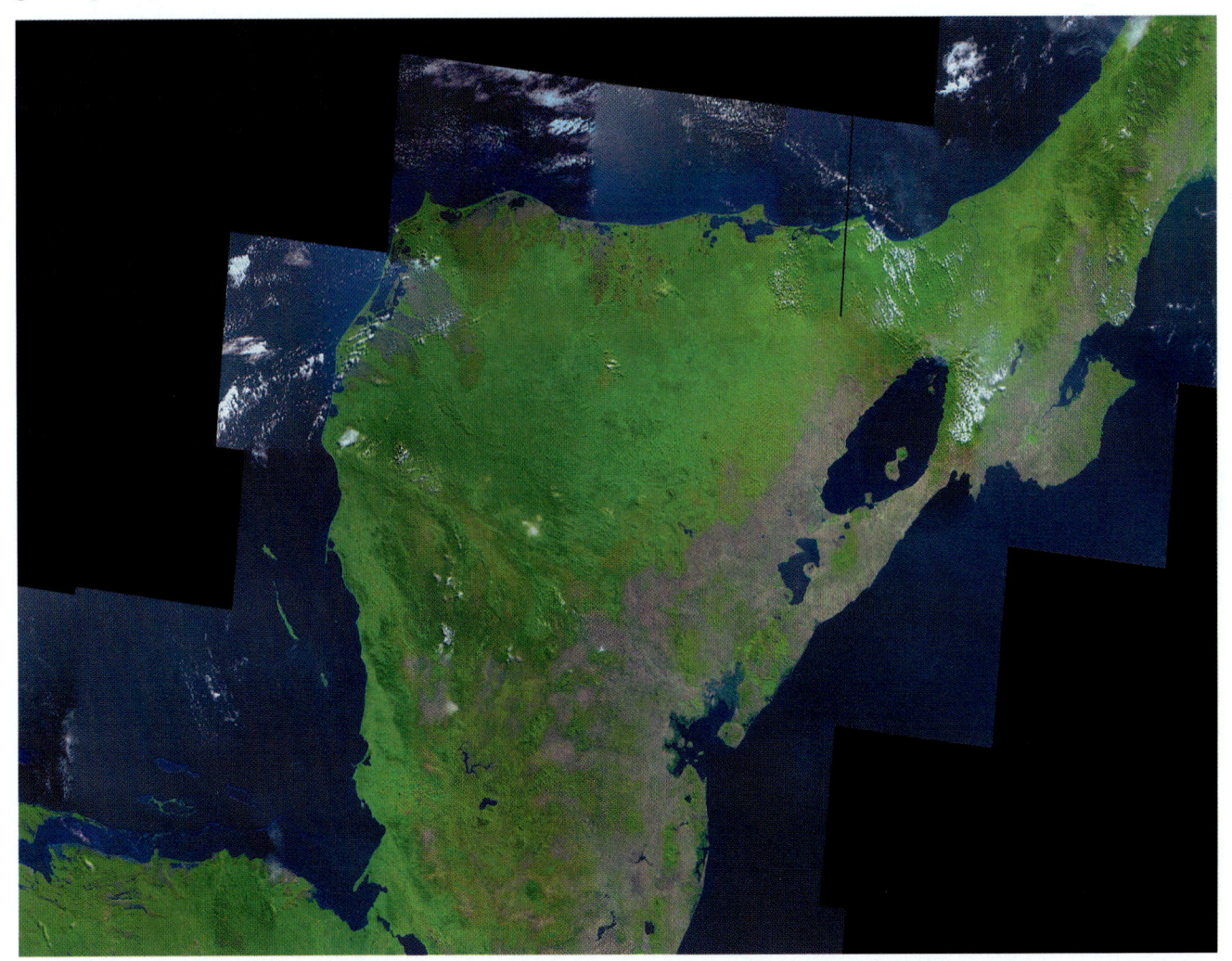

Fig. 2. The coasts of Nicaragua. Photo is from Earth Satellite Corporation's global Landsat^{TM} mosaic.

(PROTIERRA, 1995). The coastal zone might be defined as including the entire Nicaraguan Depression with its large freshwater lakes, but a large portion of the rain falling upon this basin area drains into the lakes, and ultimately eastward, and so will not directly and significantly impact coastal activities and features of the Pacific side (Foer, 1992). The most important rivers are the Rios Negro, Tamarindo and Estero Real, the latter carrying significant freshwater and sediments into the Gulf of Fonseca (Ryan, 1993; Quirós, 1998). Most of the rivers in the Pacific region show severe signs of pollution. Some of them are used as primary receptors of city sewage, industrial effluents and solid wastes (Ryan, 1993, Quirós, 1998), and because those rivers are located in agricultural areas they also receive a heavy load of sediments and pesticides (DANIDA, 1997). The Pacific coast of Nicaragua also has the largest number of groundwater aquifers in the nation (Sánchez and Quiróz, 1998). The relatively dry western slopes have drawn farmers from the Pacific region since colonial times and are now well settled (USEM, 1998).

SEASONALITY, CURRENTS, NATURAL ENVIRONMENTAL VARIABLES

The area falls within the Large Marine Ecosystem situated between the California Current and the Peru (Humboldt) Current. Monthly mean ocean temperatures remain above 26°C throughout the year, and consequently the area supports a tropical fauna. Water masses are influenced by the Peru (Humboldt) Current that moves northwards, joined by the North-Equatorial Counter Current and North-Equatorial Current. There are seasonal cold core eddies, interacting with warm core anticyclonic gyres off the Gulf of Fonseca as well (Brenes et al., 1998).

The region belongs to the dry tropic zone, with a savanna tropical climate. Temperature ranges between 25°C and 30°C (González, 1997; Ryan, 1993), and rainfall varies between 700 mm and 1500 mm per year. There are two very distinct seasons: the dry season (November–April) and the rainy season (May–October). Surface seawater has an average salinity of 34 ppt and a temperature above 25°C, varying from 21–23°C in the south (González, 1997), to 28°C in the north (Ryan, 1993; Gutiérrez et al., 1998). Depending on the season, there is a thermocline between 30–70 m depth. Temporary changes in the thermocline and currents allow deep, colder water masses to come to the surface during the annual upwelling. In general, upwelling waters have a low concentration of dissolved oxygen (Ryan, 1993).

MAJOR SHALLOW WATER MARINE AND COASTAL HABITATS

The resources in the coastal waters of the Pacific region are highly valued in terms of biomass (sea turtles, shrimp and an abundance of pelagic and demersal fish). The Pacific continental platform has an area of 5350 square miles (Ryan, 1993), and it includes a mixture of rocky, muddy and sandy bottoms (Ryan, 1993), some of which are suitable for trawling.

The coastal zone of Nicaragua has, for the pragmatic purposes required by the coastal zone action plan (MAIZCo, 1997), a terrestrial limit of the inland extension of mangrove habitat and brackish waters, and a marine limit of 12 nm. The Pacific is less diverse than the Caribbean coast, but has very important ecosystems none the less. Its integrity has been affected by over-use, contamination and degradation of the mangrove ecosystems and adjacent gallery forests and dry tropical forest. The southern part of the coast is less diverse than the north but is less impacted. The current value of the ecosystems of the north, however, remains significantly higher due to its intensive use and high population.

Even though the Pacific shelf of Nicaragua is considerably smaller than its Caribbean shelf, the estimated biomass is four times larger, partly due to enrichment from annual upwelling. The fishery biomass estimated for the Pacific coast is approximately 400,000 tonnes (MAIZCo, 1997). In addition, the mangrove ecosystems are highly productive.

Penaeid shrimp are the most important species for Pacific coastal communities. This product is exported, mainly to the USA, and provides substantial foreign exchange. Since the 1960s, total catch and catch per unit effort has been declining (Ryan, 1993). Shrimp aquaculture occupies 25,000 ha and in 1994, 23,000 tons of shrimp were produced (PRADEPESCA, 1995). At least 90% of the shrimp farms have been constructed on former mangrove or salt pond areas.

Shrimp fishing grounds are located from north San Juan del Sur to Casares beach, and from the Punta Ñata to Aserradores. The highest catches (Table 1) are produced between October and March. White shrimp is the most commonly caught species, comprising about 46% of the catch. The white shrimp is abundant from January to February and is found in depths less than 30 m. Red shrimp are second in abundance and is found deeper than 30 m (Ryan, 1993). The spiny lobster (*Panulirus gracilis*) is not an economically important resource. It was over-exploited during the 1970s and populations are recovering. Lobsters are usually caught by diving and up to a depth of 40 m.

Important mollusc species located in the estuaries include the black conch (*Anadara tuberculosa*), casco de burro (*Anadara grandia*), and the oyster (*Crassostrea* spp.). Black conch are found in muddy substrate within mangrove roots and have been heavily exploited, while the casco de burro is now considered endangered. Other molluscs such as mejillón (*Modiolus pseudolulies*), barba de hacha (*Mytella* spp.), and caracol (*Strombus galeatus*) are considered underexploited. The dardo squid (*Loliolopsis diomedae*) is frequently found between 50 and 150 m depth (Ryan, 1993).

The Pacific shelf is dotted with small isolated patches of solitary Pocilloporid corals and gorgonians, but there is a general absence of corals due to upwelling, noxious plankton blooms, river discharges, sedimentation, and El Niño events.

Table 1

Shrimp species of the Pacific coast of Nicaragua (adapted from Ryan, 1993)

Common name	Scientific name
White shrimp	*Litopenaeus occidentalis*
	L. stylirostris
	L. vannamei
Red shrimp	*L. brevirostris*
Brown shrimp	*L. californiensis*
Little shrimp	*Trachypenaeus byrdi*
	Xiphopenaeus riveti
	Protrachypeneus precipua

Table 2

Abundance in 1994 of mangrove species in the Estero Real, Nicaragua (adapted from Hurtado, 1994). Ht = height in meters, Ha = area of mangrove species in hectares, WAC = % of mangroves within specified Ha area, TAC = % of total Estero Real area covered by species.

Species present	Ht	WAC	Ha	TAC
Rizhophora mangle *R. racemosa*	15–20	70–100	1,470	6.27
R. mangle *R. racemosa*	15–20	40–70	1,858.5	7.92
R. mangle *R. racemosa*	6–15	70–100	3,488	14.88
R. mangle *R. racemosa*	6–15	40–70	3,856.5	16.45
Avicenia germinans *A. bicolor* *Conocarpus erectus* *Laguncularia racemosa*	0.5–6		12,774.5	54.48

In 1994 there were approximately 155,000 ha of mangroves (Agüero and González, 1997), about 45% of which are located on the Pacific Coast (Hurtado, 1994). Today, this has been reduced by deforestation (González, 1997). Estimates in 1983 of 60,000 ha and in 1988 of 70,000 ha on the Pacific coast have been reported (Foer, 1992), of which one third are in the Estero Real zone on the Gulf of Fonseca (Table 2) (Agüero and González, 1997; Hurtado, 1994). The most common species are the red mangrove (*Rhizopora mangle*, *R. racemosa*, and *R. harrisonii*), salt tree (*Avicenia bicolor*, *A. germinans*), botoncillo (*Conocarpus erectus*) and the pineapple mangrove (*Pelliciera rhizophorae*) (Agüero and González, 1997).

The largest area of mangroves extends from Puerto Sandino to the Gulf of Fonseca, with an estimated extension of 830 km^2 (PROTIERRA, 1995). Within this, Estero Real is one of the largest (600 km^2) areas (González, 1997). Most of the shrimp farms currently operating in Nicaragua are concentrated here (Hopper, 1988), on mariculture lands given in concession by the government to private investors and co-operatives. The Estero Padre Ramos is one of the best preserved parts and has a mangrove area of approximately 311 km^2 with potential for tourism (González, 1997). Other important sections are Aserradores-Corinto (60 km^2) with a very complex system of secondary estuaries, and Poneloya–Puerto Sandino (64 km^2) with tourism potential that provides nesting sites for aquatic birds.

At present, there are three protected mangrove areas totalling about 70,000 ha. The wildlife refuge of Río Escalante-Chacocente (4800 ha) is an area where Olive Ridley and Leatherback turtles are protected. It includes one of the last remaining tropical dry forests in the country, a vegetation which was once characteristic of the Nicaraguan Pacific.

Estuaries in the north remain connected to the Pacific ocean all year round, but those in the south Pacific (Managua, Carazo and Rivas) are usually only connected to the ocean during the rainy season.

The Gulf of Fonseca is shared with Honduras and El Salvador, and is fringed with mangroves and wetlands covering approximately 3200 km^2. It is a deep embayment and is one of the most important coastal ecosystems of the region (Cedeño, 1996). The Choluteca, Nacaome, Coascara and Negro drain into it, but these rivers now also deposit thousands of tonnes of sediments and pollutants arising from changes in drainage regimes and up-hill deforestation, resulting also in increased salinity in some of the estuarine systems which is affecting biodiversity and species distribution (Quirós, 1998).

Five species of sea turtles are found here (Groombridge, 1982; Nietschmann, 1975; Cáceres, 1998; CCC, 1998). The Olive Ridley (*Lepidochelys olivacea*) nests here in considerable numbers, and Chacocente beach is also a nesting site for small groups of Leatherback turtle (*Dermochelys coreacea*).

OFFSHORE SYSTEMS

Nutrient enrichment is not caused by the classic Ekman transport upwelling (Bakun et al., 1999), but by a combination of equatorial upwelling, open ocean upwelling driven by wind stress curl, by currents largely induced by the curvature of flow patterns, which is part of the mechanism underlying the 'Costa Rica Dome structure', and by episodic "downwind" coastal upwelling forced by winds blowing across the Isthmus from the Caribbean. In the latter, the mountainous barrier results in lower pressure over the Pacific side of the Isthmus, with the result that intense jets of wind blow directly down the pressure gradient through mountain passes and exert strong off-shore directed stress on the sea surface. As a result, distinctive upwelling plumes often extend offshore from three locations corresponding to major gaps in the mountains (Bakun et al., 1999).

An extensive oxygen-minimum layer exists. It is 1200 m thick off Mexico and Central America, thinning towards the equator. Off Nicaragua, its upper boundary is shallower than 50 m (Wyrtki, 1965).

Productivity and Fisheries

Several hundred species of demersal fish are commonly taken as by-catch in the shrimp fisheries, but do not sustain commercial fisheries. Significant takes of small coastal pelagic fish such as the Central Pacific anchovetta and Pacific thread herring have existed in the Gulf of Panama since the 1950s, but currently most of the production is used for fish meal and oil. The most valuable fish off Nicaragua are tuna (yellowfin, skipjack, and bigeye tuna) caught mainly by fleets from overseas (Ryan, 1993).

Primary productivity is relatively high (90–180 g C/m^2/year) while the secondary production in the first 100 m of the water column averages 200–500 mg/m^2. This produces two major areas with high fishery yields. One is an important feeding ground for tunas, located 400–500 nautical miles offshore and another is located within the Gulf of Papagayo near San Juan del Sur beach (Ryan, 1993).

Information about species distribution and biomass has been gathered by the Norwegian, Fritdjov Nansen, between 1986 and 1987 and by commercial fishing research from the Icelandic vessel FENGUR (Table 3). The most commercially important fauna are fish and crustaceans (Tables 4 and 5) found between 60 and 70 m. These consist of pelagic fish, small demersals (anchovies, clupeids, carangids, jurels, sawfish, and barracudas), squids, octopus, and red crab. Demersal resources include snappers, lobsters, and shallow water shrimp (Ryan, 1993).

Demersal fish off Nicaragua are classified into four habitat groups with respect to depth (Ryan, 1993): near-shore (<50 m); intermediate zone (50–100 m); open sea (100–200 m); and those from the continental shelf break (to 300 m). The most abundant and economically valuable demersal species are found close to 50 m depth (Table 5) and include snappers, roncadores, and mojarras.

In off-shore waters the most important commercial species are langostinos (*Pleuroncodes planiceps*) and big head shrimp (*Heterocarpus vicarius*), mantis shrimp (*Squilla* spp.) and the giant squid (*Dosidiscus gigas*) (Ryan, 1993). Small pelagic fish are found in the deeper waters (>150/200 m) of the continental slope, the most important being swords (*Trichiurus nitens*), argentina (*Argentina aliceae*) and serranos (*Epinephelus* spp.). Of these species argentina is the most abundant.

Big pelagic fish with high commercial value include yellow-fin tuna (*Thunnus albacares*) but starting in 1996, dorado (*Coryphaena hippurus*) became the most abundant pelagic species landed off Nicaragua (Hernández and Maradiaga, 1998). Other pelagic species captured by exploratory fishing during August 1995–August 1997 include: yellow-fin tuna (*Thunnus albacares*); bonito or barrilete tunas (*Corypaena hippurus*); white marlin (*Makaira mazara*); pink marlin (*Tetrapterus audax*); sword fish (*Xiphias gladius*); vela fish (*Istiophorus platypterus*); arenque (*Elagatis bipinnulate*); blue shark (*Prionace glauca*); gray shark (*Carcharhinus falciformis*); hammerhead shark (*Sphyrna lewini*); porous shark (*Carcharhinus*

Table 3

Estimated biomass of the most important marine resources of the Nicaraguan Pacific (adapted from Ryan, 1993)

Fishery resource	Biomass (tons)
Near-Shore	
1. Small Pelagics	72,000
2. Demersals	160,000
Commercial	30,000
Other	130,000
3. Crustaceans	4,000
Litopenaeus	3,000
Panulirus	1,000
Off-Shore	
1. Fish	180,000
Tuna	>15,000
2. Crustacean	127,000
Cephalopods	19,000
Total	341,000
Grand Total	577,000

Table 4

Summary of economically important marine resources with respect to depth and season (adapted from Ryan, 1993)

Marine resource	Depth (m)	Season
Tunas		
Barrilete (*Katsuwonus pelamis*)	Surface	April–June
Bonito	no data	March–April
Squids		
Coastal Zone	<50	Feb.–May
Oceanics	>100	Feb.–May
Shrimp		
Big head (*Heterocarpus vicarius*)	>100	no data
Demersals		
Snappers (*Lutjanidae* spp.)	<50	August
Pajaritas	50–100	May
Doncellas (*Hermianthus signifer*)	100–200	no data
Langostino (*Pleuronectoides planiceps*)	100–200	no data
Serranos (*Epinephelus* spp.)	100–200	no data
Palometa (*Peprilus triacanthus*)	no data	no data
Sharks (*Isurus* spp.)	no data	no data
Semipelagics		
Argentina (*Argentina aliceae*)	100–200	no data
Macarela	100–200	Aug.–Nov.
Sable (*Trichiurus nitens*)	100–200	Feb.
Sardineta	100–200	Aug.–Nov.
Pelagics		
Barracuda	0–50	May–Sept.
Scombridae spp.	0–50	no data
Carangidae spp.	0–50	no data
Clupeidae spp.	0–50	no data
Engraulidae spp.	0–50	August

Table 5

Scientific and common names of snapper species in the Pacific Ocean off Nicaragua with respect to the depths they inhabit and commercial minimum size (CMS) (adapted from Ryan, 1993).

Common name	Scientific name	Depth (m)	CMS (cm)
Pargo rojo	*Lutjanus jordani*	20–55	55
Pargo colorado	*L. colorado*	24	90
Pargo amarillo	*L. argentriventris*	52	63
Pargo negro	*L. novemfasciatus*	15–26	110
Huachinango	*L. peru*	23–75	36
Pargo rayado	*L. viridis*	no data	30
Pargo raicero	*L. aratus*	no data	55
Coconaco	*Hoplopargus guntheri*	no data	50
Pargo lunajero	*L. guttatus*	24	54

porosus); thresher shark (*Alopias vulpinus*); bull shark (*Carcharhinus leucas*), and wahoo (*Acanthocybium solandri*).

Artisanal fishing communities number about 38 (based on a Nicaraguan government survey; Anon., 1996), with about 950 artisanal fishing vessels and about 3770 artisanal fishermen. Over half of the fishermen associate with some type of organization. In many cases, intermediaries with connections to the market ("acopiadores") act as agents between the fishermen or their organization and the commercial operators.

Three types of vessels are in common use in the country: *Botes*, made of wood, often without a motor; *pangas*, wood or fibreglass with outboard motor; and *lanchas*, wood or fiber glass or metal with inboard motor and maybe navigation equipment. Pangas and lanchas are used primarily on the Pacific coast (Ryan, 1993). Incorrect gill net mesh size used by artisanal fishermen is the main cause of catching immature fish. Lost nets are also a growing problem, but there is no information on the number of nets lost or of mortality by this cause (Ryan, 1993).

Lobsters (*Panulirus gracilis*) are exploited almost entirely by artisanal fishermen. About 300,000 pounds of tails could be extracted annually without ill effects (Foer, 1992).

The Olive Ridley (Paslama) turtle (*Lepidochelis olivacea*) is mainly threatened by egg harvesting. Its eggs are sold in local markets, restaurants, bars, and exported illegally to Costa Rica and Honduras (Ryan, 1993).

POPULATIONS AFFECTING THE AREA

About 45% of Nicaragua's population live in the general Pacific region. It is ethnically homogenous but is marked by extremes in the distribution of power and wealth. By the year 2000 one-third of Nicaragua's total population and over two-thirds of the urban population is expected to reside in the capital (Foer, 1992). Managua will then be 13 times the projected size of León, the only other urbanized area expected to have over 100,000 people in the year 2000.

In 1998, the province of Managua had a population of about 1.2 million inhabitants (24% of the country's total) (USEN, 1999).

Serious population-related problems include pollution, lack of facilities for solid waste disposal, lack of basic infrastructure, and lack of integrated planning to prepare for natural hazards, problems related to mining, and cultural displacement of natives (USEN, 1999).

The Pacific coast of Nicaragua has been divided for management into two regions. The north Pacific region (León and Chinandega) which includes nine municipalities, and the south Pacific region (Managua, Carazo and Rivas) including seven municipalities (Gonzalez, 1997). Population density at national level is 43 inhabitants/km^2 but in the Pacific coastal municipalities it is 106 inhabitants/km^2 (nearly 20 times greater than on the Atlantic coast). Two thirds of homes have access to sanitary services, making it the region with the least problems in this respect, and only 14% of homes do not have access to drinking water. There is 22% unemployment in the Pacific coastal municipalities (MAIZCo, 1997). The main economic activities in the region (fishing, shrimp farming) generate a significant income, but the population does not benefit much in terms of social development since the benefits go directly to businessmen and the central government (MAIZCo, 1997).

Estero Real

Many of the natural resources of the Estero Real have been ecologically degraded (Quirós, 1998) and the area has extreme poverty, high unemployment, few economic alternatives, and lack of management and environmental education (DANIDA, 1997). Hurtado (1994) pointed to the lack of planing and management, a high demand for natural resources and very few options for alternative sources of income. MARENA (through a DANIDA-MANGLARES Project) has been working to establish a management plan for the Estero Real that will benefit local residents while sustaining the resource (Hurtado, 1994).

Gulf of Fonseca

The coastal zone of the Gulf of Fonseca is heavily populated, especially over the border in Honduras. It has been estimated that almost 500,000 people live around the Gulf (DANIDA, 1997). It is economically important because it is a place of trade and commercial activities between three countries, and opportunities to exploit the natural resources of this rich environment have attracted many people who have in turn further compounded the degradation and loss of valuable natural resources (DANIDA, 1997).

Over-exploitation of the coastal and marine resources is occurring for two different reasons. One segment of the population is over-exploiting the natural resources because they live in or near extreme poverty with few economic alternatives; they survive at the expense of exploiting their immediate natural environment. The other segment are the investors in large industrial shrimp farming complexes or agro-industries focused on non-traditional export crops. These do not create real alternative work opportunities, nor do they pass along many benefits to local residents. As a consequence of policies of the Nicaraguan government and of unsustainable management of the watersheds draining into the Gulf of Fonseca, degradation of the marine and coastal environment of the Gulf system has rapidly increased. There has been a significant reduction in mangrove cover, increases in pollution of surface, ground and sea water, and reduction in both the quality and quantity of marine and fishery resources. The increasingly restricted and scarce marine resources also have further marginalized traditional human users of mangroves, wetlands and marine resources (DANIDA, 1997).

Uneven social development is complicated by unbalanced development between the three countries that share the waters of the Gulf. Besides having the poorest population, Nicaragua is the main basic supplier of raw material for the other two countries and does not gain the full economic benefit from the exploitation of its resources (DANIDA, 1997). The traditional fishing industry of Salvador depends largely on the marine resources of Nicaragua and, likewise, Honduran fishermen harvest, mostly illegally, shrimp larvae from Nicaraguan estuaries. Endangered species are being taken out of the country to be sold in the international market. Because of lack of infrastructure, Nicaraguan shrimp products are processed in Honduras by Honduran seafood plants.

South Pacific Region

In the south Pacific region the coastal population depends more on the natural resources than in the north Pacific region. There are fewer opportunities, less infrastructure such as sewage facilities and there is 24% unemployment. The distribution of benefits derived from the development of natural resources is adequate but, like the north, not very equitable. The population has a low level of education. Unfortunately, the economic benefits generated by most development activities do not filter down to the local communities. The south seems, however, to be the most stable. Although ecosystem diversity is low, tourism perhaps provides the best opportunity for economic development here (MAIZCo, 1997).

RURAL FACTORS

By the mid-1960s cotton was cultivated on 80% of the total arable land of the Pacific coast (40% of all cultivated land in Nicaragua). This led to widespread deforestation, erosion, and biocide contamination of land and water, and reduced

---

Government in Nicaragua

Nicaragua takes its name from Nicarao, a chief of the indigenous tribe living around Lake Nicaragua. In 1524, the Spanish settled the region, establishing two towns: Granada on Lake Nicaragua and León east of Lake Managua. Nicaragua gained independence from Spain in 1821, briefly becoming a part of the Mexican Empire and then a member of a federation of independent Central American provinces. In 1838, Nicaragua became an independent republic.

Much of Nicaragua's recent politics has been characterized by rivalry between the Liberal elite of León and the Conservative elite of Granada, which often spilled into civil war. Due to differences over an isthmian canal, concessions to Americans in Nicaragua and a concern for what was perceived as a destabilizing influence, in 1909 the United States provided political support to Conservative-led forces. The United States maintained troops in Nicaragua until 1933 who were engaged in a running battle with rebel forces led by the Liberal general Sandino, who rejected an arrangement to end the fighting between Liberals and Conservatives.

After the departure of U.S. troops, Somoza and two sons who succeeded him, maintained close ties with the U.S. Their dynasty ended in 1979 with an uprising led by the Sandinistas. The latter established an authoritarian dictatorship, nationalized many private industries, confiscated private property, supported Central American guerrilla movements, and maintained links to international terrorists. The US suspended aid to Nicaragua in 1981, provided assistance to the Nicaraguan Resistance and later imposed an embargo on U.S.–Nicaraguan trade.

In response to pressure, the Sandinista regime agreed to nationwide elections in 1990, resulting in the presidency of Violeta Barrios de Chamorro. For seven years, her government achieved major progress. This was followed by the Aleman government, and Nicaragua is now a constitutional democracy with executive, legislative, judicial, and electoral branches of government.

---

the original vegetation cover by 90% (Foer, 1992). During the 1978–79 harvest, over 280 pesticide-related hospitalizations or emergency consultations were reported from San Vicente Hospital in León, and very high levels of DDT were found in human milk-fat samples (Swezey et al., 1986). There are no studies on pesticide levels in commonly eaten fish or shellfish in coastal waters (Foer, 1992).

Erosion has been increased by the intensive use of the land and inappropriate technologies. The levels of erosion exceed 50 Mt/ha/year in cotton areas (four times greater than permissible levels), and no terraces or drains are used on slopes of approximately 5%.

The greatest environmental problems were caused by hundreds of state farms created by the agrarian reform. Mangrove stands in the Estero Real region were felled to establish state banana plantations, causing losses in shrimp farming and fisheries. Eighteen miles of nesting beaches of the Pacific Ridley turtle were destroyed by landslides caused when the Ministry of Agriculture ordered the removal of coffee shade-trees in the Carazo region in a futile effort to control a spreading tree fungus (Nietschmann, 1990). Degradation of forests first started with the introduction of cotton and sugar cane, and subsequently cattle farming increased the problem. Today, forests suffer from progressive extraction of firewood, either as a commercial activity or for self-consumption, and now there are but a few pockets of dry tropical forests left in isolated areas of the Maribios volcanic range of mountains (González, 1997).

Fauna is under threat from poachers, and from habitat destruction. The high pressure caused by the diversified use of the mangrove ecosystems and the excessive exploitation of some commercially important species has led to the extinction of species in some areas (Ryan, 1993). The watershed drainage system supports only short-lived flow because rivers are poorly developed between the mountain range and the coast. Flows are very low and the use of water for consumption and as sewer discharge routes requires special attention. Some river basins are flat lands dedicated to intensive irrigation farming which has further lowered flows (MAIZCo, 1997). Ground water has been contaminated by natural and by anthropogenic causes (sewage, pesticides, etc.) in some areas (Quirós, 1998), and intrusion of marine water, particularly in the Chichigalpa–Chinandega–Corinto area has followed (Quirós, 1998).

Contamination of freshwater systems with nutrients and pesticides is related to discharges from agricultural activities carried out near the river basins. High volumes of sedimentation affect rivers (Quirós, 1998), and chemical contamination is highly concentrated in several areas. Industrial effluents and sewage have diminished the ecological potential of important water systems such as the Xolotlán Lake, Corinto Bay and other lagoons; it is considered that sedimentation as a result of deforestation and bad agricultural practices is the most serious threat to the rich aquatic biodiversity of Nicaragua (Ryan, 1993; González, 1997; DANIDA, 1997; Sepúlveda and Vreugdenhil, 1998).

EFFECTS FROM URBAN AND INDUSTRIAL ACTIVITIES

Industrial Fishing

This activity concentrates on a small number of species and is characterized by intensive exploitation in limited areas because of a lack of support and technical assistance, lack of financial support and incentives, and government organization problems (Ryan, 1993). Some species are being exploited well under their potential and, if harvested, could reduce the pressure on over-exploited species (MAIZCo, 1997). The shark fishery is an example of poor resource optimization. Shark can potentially offer the most by-products. However artisanal fishermen in Nicaragua capture shark only for their fins (Ryan, 1993).

It is difficult to partition the industrial fleet between the two coasts, so the following numbers are national. Some vessels may travel from one coast to the other via the Panama Canal. Annual catch is about 12,000 metric tons. Internal markets receive about 58% of the production and 42% is for external markets, the majority of which (95%) is lobster and shrimp. Of about 970 employees, 422 are women, 60% of whom work on processing the catch and 40% are administrative workers. About 29% of the men are employed directly in the fishing, 59% have "indirect" employment on the vessels and 12% are employed on administration (Ryan, 1993).

Fishing gear is not optimally designed for cost-effectiveness. Shrimp trawling nets waste resources because 90% of the total catch is by-catch (Ryan, 1993) which in most cases is discarded. After pressure from the US, shrimp trawlers started using nets with TEDS (turtle exclusion devices). To date, no formal evaluation of the results of TEDS has been made for Nicaragua (Ryan, 1993).

The role of crustaceans in the export market is critical. The population dynamics of shrimp, in particular, has significant biological problems that make the prediction of future production or catch difficult. There are two apparent factors affecting shrimp landings. In one case, white shrimp landings *Penaeus schmitti* are correlated with rainfall over a 22-year period, with no relationship with temperature. Over the same period, red shrimp *P. duoranrum* landings are correlated with water surface temperature, but not with rainfall (Anon., 1995). These observations so far simply deepen the mystery connected with shrimp populations. Shrimp production in 1994 was 14,030,000 pounds from the Pacific coast and 11,600,000 pounds from the Atlantic coast, excluding shrimp mariculture. The average annual catch per trip is 173 pounds. The maximum sustainable yield of Pacific stocks is approximately 1.8 million pounds of tails, or about 30–40 boats' worth of effort. In 1992 there were about 27 boats operating, capturing 0.7 million pounds (INPESCA, 1990). There is no closed season on the Pacific coast (Foer, 1992).

As well as shrimp species listed in Table 1, other shrimps of the family Pandalidae are fairly numerous and are potential future fisheries. The same may be true for the shrimp family Solenoceridae. Two other shrimp species that periodically appear in the catch are: *Xiphopenaeus riveti* and *Trachypenaeus byrdii* (Perez, 1993).

Shrimp Aquaculture

Shrimp aquaculture is very important here. In 1993 less than US$600,000 of shrimp were exported, while in 1994 more than US$7 million were exported from aquaculture alone. Seventeen communities registered 59 establishments in 1995, 32 of which are near Puerto Moraza'n. About 95% of the total cultured shrimp is exported. About 3450 ha are in production and 2000 ha are already designated for future expansion. About 1100 men and 125 women are employed:

3% work in laboratories, 87% working on production and 10% in administration. Government sources project employment for 6000 people. The average production cycle is about 3 months. Larvae are captured naturally from lagoons (about 30%) and the remainder are purchased. The primary shrimp species cultivated are *Penaeus vannamei* and *P. stylirostri*. There are several potential problems with shrimp aquaculture mainly connected with the destruction of mangroves and degradation of water quality. There is also the destruction of many small creatures of various species which are indiscriminately caught along with the shrimp post-larvae in the fine nets.

In some countries, development of shrimp aquaculture has decreased the availability of food for domestic consumption. Areas formerly productive for locally consumed species no longer can support these species and the shrimp are mostly all exported. Conversely, shrimp aquaculture can earn large sums of foreign currency, create new employment and provide incentive for the development of technical capacity and even a sense of stewardship for the environment. Careful consideration of these factors needs to be taken by the government during this aquaculture development process (Foer, 1992).

Other Exploited Species

A potential giant squid *Dosidicus gigas* is being investigated, as is that for a potential artisanal fishery for oysters in the Gulf of Fonseca. Sharks (*Isurus* spp.) are exploited because of the value of their fins (US$25–28/pound) (Ryan, 1993). Hernández and Maradiaga (1998) itemise other species currently being landed: *Prionace glauca, Carcharhinus falciformis, Sphyrna lewini, Carcharhinus porsus, Alopias vulpinus,* and *Carcharhinus leucas*. Snappers (*Lujtanus guttatus*) and other fish species still represent more than 80% of catches by artisanal fishermen, or secondarily by the industrial fleet as a shrimp by-catch (Ryan, 1993). Finally, the lobster, *Cervimunida johni*, and the Chilean lobster, *Pleuroncodes monodon*, which are the basis for major fisheries in Chile are both captured in Nicaragua occasionally.

Seaports and Shipping

Nicaragua has six seaports, all of which are operated by the Government Port Authority (ENAP). The most suitable for commercial shipping is the Port of Corinto located on the Pacific, 110 miles northwest of Managua. Corinto has a capacity of 1,516,900 tons annually and is presently upgrading its facilities to increase competitiveness. Puerto Sandino, also on the Pacific Coast, is primarily used for the import of crude petroleum. The remaining Pacific port of San Juan del Sur has limited capacity and uses barges to load and unload cargo (USEN, 1999). Ships currently discharge their liquid and solid wastes directly into the estuaries (DANIDA, 1997) because of a lack of services and the absence of controls.

Major Economic and Political Trends

Nicaragua has made progress since 1990 in consolidating democratic institutions and in fostering economic growth. The economy grew by 4% in 1998, down from 5% in 1997. The primary sector (agriculture, livestock and fishing) was the most affected, with the growth rate falling from 8.3 in 1997 to 4.2 in 1998. The secondary sector (manufacturing, construction and mining) also saw its growth rate slow from 5.7 to 4% from 1997 to 1998. Private investment, from both domestic and foreign sources, is rising and the private banking sector continues its expansion. Nevertheless, GDP per capita is only an estimated $441, the second-lowest in the hemisphere. The unemployment rate fell from 14.3% in 1997 to 12.3% in 1998. The combined rate of unemployment and underemployment was 25.5%.

For 1999, estimates for real GDP growth range from 5 to 6.3% (the higher figure is the government's estimate). This robust growth is being led by commerce, services, agriculture, and construction—residential, commercial and public. In March 1998, the IMF approved an Enhanced Structural Adjustment Facility for Nicaragua, and international support in the aftermath of Hurricane Mitch has been significant.

In May 1999, donor countries pledged $9 billion for Central American reconstruction, $2.5 billion of which is earmarked for Nicaragua. The United States and other donor countries in the Paris Club also deferred Nicaragua's debt payments until early 2001 (the foreign debt is more than $6 billion, one of the highest per capita debts in the world).

Nicaragua is essentially an agricultural country with a small manufacturing base, dependent on imports for most items. In recent years the government has liberalized foreign trade and eliminated most non-tariff barriers and foreign exchange controls. The U.S. remains Nicaragua's largest trading partner by far; two-way trade in 1998 totalled $790 million, figures which have risen steadily in recent years. Prospects look good for sustained rapid economic growth. However, success depends on the government's economic ability, its progress in resolving property disputes, its strengthening of the rule of law, and its steps to remove obstacles to private investment.

Agricultural production accounts for 26.7% of Nicaragua's GDP and two-thirds of exports. Tourism is increasing; 381,600 tourists visited Nicaragua in 1998, up 13.2% over 1997. The country expects to host over 450,000 visitors a year by the end of the decade, and tourism was Nicaragua's third most important source of foreign exchange in 1998, representing a $90 million market—an increase of 12.5% over 1997. The best prospects for investment in this sector exist outside of Managua, for example on the Pacific coast and in ecotourism areas (USEN, 1999).

REFERENCE

USEN (1999) Country Commercial Guide. US Embassy Nicaragua http://www.usia.gov/abtusia/posts/NU1/wwwhe08.html

Mining

Several companies have won concessions for new exploration. Some 124,100 troy ounces of gold were produced in 1998, a 54% increase over 1997, and 67,300 troy ounces of silver were produced in 1998, which represents a 98% increase over 1997. A small gold mine, the El Limón mine, operates north of León on the Pacific coast (USEN, 1999).

A 'Special Law for Exploration and Exploitation of Hydrocarbons' (Law No. 286), was signed in 1998. These new hydrocarbon (and electricity) laws open the door for increased private sector involvement in Nicaragua's energy sector and could provide significant commercial opportunities for international exploration companies (USEN, 1999).

PROTECTIVE MEASURES

Disorderly resource management has led to degradation and destruction of economically and ecologically important ecosystems. There is a lack of consistency and coherency among policies for the use of coastal resources and current legal and institutional frameworks contain vacuums in laws and regulations (Sepúlveda and Vreugdenhil, 1998). There is no existing comprehensive and integrated coastal zone management legislation. At the National Assembly, important coastal zone management laws are still waiting to be approved. Among them are the Fishery and Aquaculture Law, the Forestry Law, and the Hydrocarbons Law (DANIDA, 1997).

MAIZCo (1997) lists several areas of Nicaraguan law that need modification to effectively manage the nation's coastal resources.

1. Constitutional measures, laws and decrees relating to maritime territory are unclear.
2. Contradictions exist between institutions derived from central and regional government. The Autonomy Statute, Law No. 28, is not clear on the supremacy of the regional government over the institutions, resulting in a lack of co-ordination.
3. There is a lack of procedural norms to formalize the autonomous regime, preventing the latter from exercising their mandates effectively.
4. Organization of communal lands needs to be speeded up. These are lands located within specific administrative areas, and for coastal zones management plans to function, must be implemented at municipality level.
5. Forestry by-laws are inconsistent, there is little capacity for inspection of forestry development, and the by-laws do not establish administrative responsibilities.
6. Regarding fishing, there are no specific or clear legal organisational procedures, but rather a series of loose administrative measures.

The General Law on the Environment and Natural Resources fills in a few legal gaps on issues of territorial and coastal waters. Even with this, however, laws regulating forestry, fishing and aquaculture are still required. One of the main problems remains the lack of co-ordination among institutions, resulting in duplication and an inadequate allocation of roles. Co-ordinating efforts have been made, but existing mechanisms are inadequate. Also, the level of environmental education is weak. Existing information on natural resources, which is handled sectorally, is not accessible and, further, environmental education programmes do not take into account the cultural, social and economic concerns of the coastal population.

Environmental management capacity was examined by MAIZCo (1997). The north Pacific region had the highest capacity, with a great number of NGOs conducting projects along the coastal zone and in Estero Real in particular. Government offices concerned with natural resources are also present. These have a relatively high capacity to coordinate and organize among themselves (with the exception of MEDEPESCA). The south Pacific is less well off, with little coordination between agencies and NGOs. In several cases there is co-ordination, but in others management has difficulties. Although the municipality of Rivas is not in the coastal zone, it is where other central government offices are located, thus hindering coordination between coastal municipalities which have no institutional representation (MAIZCo, 1997).

Therefore, integrated management is needed to ensure that the potential value of the natural resources contributes to the sustainable development of the Pacific coast. In the north, the main objective will be to maintain the integrity of the ecosystems. The current value of those systems is high, but their ecological integrity is under serious threat, both in terms of the region's ecological potential and its contribution to the national economy. In the south Pacific, the possibilities for better development of coastal natural resources lie in fishing and tourism. In these sectors the integrity of the ecosystems must also be carefully maintained (MAIZCo, 1997).

Information on economically important aquatic resources is limited and needs to be upgraded and maintained. Research and monitoring also needs to be conducted on potential sources of conflict between coastal rural development and other land/sea uses. Information on how shrimp fisheries interact with other fisheries is also critical. Strengthening local administrations, which are poorly equipped to do proper fisheries monitoring and management, will be the first step in this process. Tourism should be developed and environmental impact requirements established and enforced so this can be done wisely. Finally, there needs to be a strengthened awareness among local people and governments of the importance of preserving ecosystem integrity and this should be translated into effective action (MAIZCo, 1997).

Recognizing this situation, the Government of Nicaragua (with the help of various donor countries) is undertaking efforts to organize and coordinate its institutions with other political and social organizations, so that together they can reach a consensus on the need for integrated management of the coastal zones, in accordance with the environmental policies outlined in Nicaragua's Environmental Action Plan published in PAANIC (1993) (MAIZCo, 1997).

The Nicaraguan Government should play an important role in the promotion of social development, by investing in infrastructure and encouraging private investment. MAIZCo (1997) clearly identifies the need for ICZM so that Nicaragua can optimize use of its coastal resources for the benefit of residents. Of critical importance in this process is the efficient and effective coordination of all organizations involved, especially government institutions, in the planning and implementation of actions involving the sustainable use of coastal resources. Consequently, in the process of developing this coordination, it is crucial that proposed ICZM strategies emphasize improving public administration. MAIZCo (1997) identifies five critical strategies for ICZM:

1. Strengthening technical and scientific capacity.
2. Creating adequate legal and institutional frameworks for ICZM.
3. Increasing and structuring the knowledge base.
4. Promoting the participation of the local population and civil society.
5. Creating a regional planning process.

In conclusion, many of Nicaragua's Pacific marine and coastal resources have great potential to improve the quality of life for the Nicaraguan people. The government understands the value of these resources and has received excellent natural resource management guidance and technical assistance. Preventing mismanagement and destruction of these resources and ecosystems, which has occurred in other Central American countries, now depends on effective government and the ability of the Nicaraguan people to work together as a united nation.

ACKNOWLEDGEMENTS

We thank Serge Andrefouet, Cynthia Lagueux, Enrique Lahmann, Yassy Naficy, Kennard Potts, Julie Robinson, Joseph Ryan, Arike Tomson, Leslie Upchurch and Mariska Weyerman for research assistance.

REFERENCES

Anon. (1995) Informe Bienal 1993–1994. Centro De Investigacion De Recursos Hidrobiologicos, Managua, Nicaragua.

Anon. (1996) Informe preliminar i censo nacional de la actividad pesquera y acaicola. Ministerio de Economia y Desarrollo Direccion Pesquero Mede Pesca. Manuaga, Nicaragua.

Agüero, M. and González, E. (1997) Diagnóstico y evaluación de alternativas de desarrollo de la camaronicultura en Estero Real, Nicaragua. Proyecto uso adecuado de los recursos naturales del manglar (DANIDA MANGLARES, Estero Real), 43 pp.

Bakun, A., Csirke, J., Lluch-Belda, D. and Steer-Ruiz, R. (1999) The Pacific Central American Coastal LME. In *Large Marine Ecosystems of the Pacific Rim: Assessment, Sustainability, and Management*, eds. K. Sherman and Q. Tang. Blackwell Science, Cambridge, MA.

Brenes, C., Hernández, A. and Gutiérrez, A. (1998) Sea surface thermohaline variations along the Nicaraguan Pacific coastal waters. *Tópicos Meteorológicos y Oceanográficos* 5 (1), 17–25.

Cáceres, G. (1998) Administración y manejo de tortugas marinas. In *Proceedings of the Sea Turtle National Workshop*. Universidad Centroamericana (UCA), January 20–22. Managua, Nicaragua, pp. 37–39.

CCC (1998) Tortugas marinas: Una guía educativa. Caribbean Conservation Corporation, USA, 29 pp.

Cedeño, V. (1996) Situación del medio ambiente marino costero en el litoral Pacífico. proyecto conservación para el desarrollo sostenible de Centroamérica. Managua.

DANIDA (1997) Socio-environmental study of the Gulf of Fonseca. Danida-Manglares Project, Nicaragua.

Foer, G. (1992) Profile of the coastal resources of Nicaragua. In Foer, G. and Olsen, S. (eds.), Central America's Coasts: Profiles and an Agenda for Action. The University of Rhode Island Coastal Resources Center, Naragansett, RI, 30 pp.

González, L. (1997) Diagnóstico ecológico de las zonas costeras de Nicaragua. Programa de manejo integral de las zonas costeras (MAIZCo). Ministerio del Ambiente y Recursos Naturales, Dirección General del Ambiente, Managua, 78 pp.

Groombridge, B. (1982) *The IUCN Amphibia-Reptilia Red Data Book, Part 1: Testudines, Crocodylia, Rhynchocephalia*. IUCN, Gland, Switzerland.

Gutiérrez, M.M., Cáceres, F.M., Díaz, M.J., Arancina, J.D., León, C.C., Gallo, M., Barrera, J.E., Aguilar, B., Sediles, E., Paniagua, C. and Cajina, O. (1998) Diagnóstico de la zoa costera del Pacífico norte de Nicaragua. Proyecto conservación para el desarrollo sostenible en América Central, Nicaragua, CATIE/OLAFO/Manglares, Nicaragua.

Hernández, A. and Maradiaga, J. (1998) La pesqueria de peces pelágicos en el Océano Pacífico de Nicaragua. Proyecto Pesca de Mediana Altura (PMA). Ministerio de Economía y Desarrollo, Managua, 125 pp.

Hopper, L.A. (1988) Proyecto de asistencia para la elaboración de un plan general de desarrollo de la camaronicultura marina en Nicaragua. Informe Técnico FAO TCP/NIC, Managua.

Hurtado, N. (1994) Estudio de caso: Manejo y uso adecuado de los recursos del manglar en Estero Real. In: *El ecosistema de manglar en América Latina y la cuenca del caribe: su manejo y conservación*, ed. D.O. Suman. Rosenstiel School of Marine and Atmospheric Science, Universidad de Miami, Miami, FL and The Tinker Foundation, New York, pp. 168–175.

INPESCA (1990) Diagnostico de la Actividad Pesquera de Nicaragua. Corporacion Nicaraguense de la Pesca, Ministerio de Economia y Comercio, Fondo Nicaraguense de Inversiones, Secretaria de Planificacion y Presupuesto, Managua.

MAIZCo (1997) Action plan proposal for management of coastal zones in Nicaragua. Ministry of the Environment and Natural Resources, General Environment Directorate, Coastal Zones Centre, Managua, Nicaragua, 55 pp.

Nietschmann, B. (1975) Of turtles, arribadas and people. *Chelonia* 2 (6). 6–9.

Nietschmann, B. (1990) Conservation by conflict in Nicaragua. *Natural History Magazine*, November 1990, New York.

PAANIC (1993) Plan de Acción Ambiental de Nicaragua. MARENA. Managua, Nicaragua.

Perez, M. (1993) Algunas especies de crustaceos con potencial pesquero en Nicaragua. Centro de Investigacion de Recursos Hidrobiologicos.

PRADEPESCA (1995) Situación actual y perspectivas del cultivo d Camarón en el istmo Centroamericano. Resumen del III Simposio

Centroamericano sobre Camarón Cultivado (ANDAH-FPX), Honduras, PRADEPESCA, Panamá.

PROTIERRA (1995) Recursos naturales renovables. León, Chinandega. CONAGRO/BANCO MUNDIAL, Managua, 37 pp.

Quirós, G. (1998) Diagnóstico del rstado de los recursos biofísicos, socioeconómicos e institucionales. Sistema Marino Costero Comisión Centroamericana de Ambiente y Desarrollo (CCAD), Unión Mundial para la Naturaleza (UICN), Managua, 68 pp.

Ryan, J. (1993) Diagnóstico y propuesta del plan de acción de recursos acuáticos. Plan de Acción Ambiental de Nicaragua (PAANIC), Managua, 240 pp.

Sánchez, J. and Quiróz, G. (1998) Diagnóstico del estado de los recursos socioeconómicos e institucionales del Golfo de Fonseca (PROGOLFO-NICARAGUA). In Proceedings of the workshop manejo y conservación de humedales con aplicación de sistemas de información geográfica, Universidad Centro Americana (UCA) 23 al 26 de Febrero de 1998, pp. 82–110.

Sepúlveda, N. and Vreugdenhil, D. (1998) Documento de trabajo: Una estrategia institucional para manejar las costas y los mares de Nicaragua. Programa de Apoyo Institucional para la DGCA, MARENA, Managua, Nicaragua, 112 pp.

Swezey, S.D., Murry, D. and Daxl, R. (1986) Nicaragua's revolution in pesticide policy. *Environment* **28**, 1.

USEM (1998) Overview and summary, Managua, Nicaragua, U.S. Embassy Managua Econ/Commercial Section, U.S. Embassy, Managua, December 1998, www.usia.gov/abtusia/posts/NU1/wwwhcom.html.

USEN (1999) Country Commercial Guide. US Embassy Nicaragua http://www.usia.gov/abtusia/posts/NU1/wwwhe08.html.

Wyrtki, K. (1965) Summary of the physical oceanography of the eastern Pacific Ocean. Institute of Marine Resources, Univ of Calif, San Diego, Ref. 65-10, 78 pp.

THE AUTHORS

Stephen C. Jameson,
Coral Seas Inc. – Integrated Coastal Zone Management,
4254 Hungry Run Road,
The Plains, VA 20198-1715, U.S.A.

Vincent F. Gallucci,
University of Washington,
School of Fisheries,
Seattle, WA 98195, U.S.A.

José A. Robleto,
University of Mobile,
Latin American Campus, San Marcos,
Carazo, Nicaragua

Chapter 34

EL SALVADOR

Linos Cotsapas, Scott A. Zengel and Enrique J. Barraza

The coastal zone of El Salvador is diverse and supports several different coastal and marine habitats, with the outer coastline dominated by sand (or mixed sand and gravel) beaches and rocky shorelines (rocky platforms and cliffs) and the estuarine areas dominated by mangroves and tidal flats. Rocky reefs and sand bottom habitats occur in nearshore marine waters. El Salvador's coastal and marine habitats, ecosystems, and their associated biota, have not been studied extensively. Practically the entire coastal zone area of El Salvador is inhabited, although few urban areas are present—the largest city is Acajutla (est. population 67,000). Fishing is a principal economic activity at both artisanal and commercial levels, and is mostly unregulated. Certain land-use and agricultural practices are possibly causing environmental impacts in the coastal zone. Examples include excessive deforestation, over-grazing, soil erosion, poor water resources management, and run-off of agricultural chemicals. Largely untreated wastewater discharges from industrial and domestic sources may also be a source of pollution in coastal and marine areas. In terms of sediment contamination, the immediate vicinity of highly industrialized areas, such as the major maritime port, are the most polluted, with other areas showing low or background levels of contaminants. Small oil spills near port areas have also been documented. The recently passed Environmental Law of El Salvador may provide a mechanism to protect the nation's coastal zone and marine resources from future degradation. A system of protected areas, including the nation's first marine protected site, will also serve in this regard.

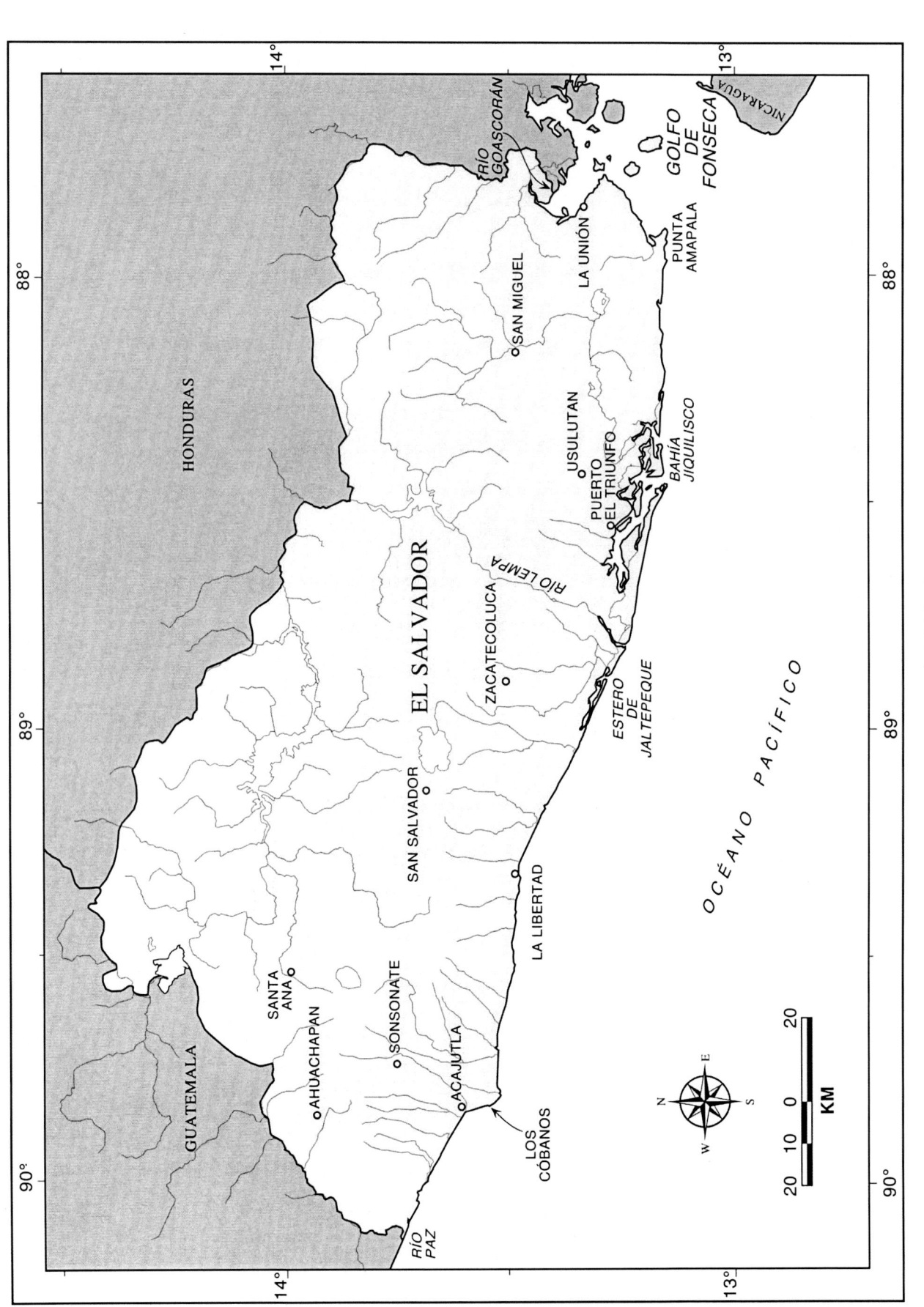

Fig. 1. Map of El Salvador.

THE DEFINED REGION

El Salvador is located on the Pacific coast of Central America, centred around 89°W, 13°30'N. Its coastline and associated ecosystems can be divided into three principal morphological or functional parts—the continental coastline located directly on the Pacific Ocean, the inshore bays and estuaries, and the offshore/pelagic zone. The coastline is predominantly oriented west-to-east, stretching for approximately 275 km from the border with Guatemala at Río Paz in the west, to Punta Amapala and the entrance to Golfo de Fonseca in the east (Fig. 1). From Punta Amapala the coastline turns north into Golfo de Fonseca, ending at the border with Honduras at Río Goascorán (a distance of roughly 45 km). Within the Golfo de Fonseca there are several small islands which are also part of El Salvador. Landward of the outer coast shoreline, there are three large estuarine systems; listed by decreasing size, they are the Golfo de Fonseca, Bahía de Jiquilisco, and Estero de Jaltepeque. There are also 17 smaller estuaries, some of which are open to the ocean only seasonally. The largest river in the country, Río Lempa, roughly bisects the coastline, discharging between the Estero de Jaltepeque and Bahía de Jiquilisco.

The coastal zone of El Salvador is diverse and supports several different coastal and marine habitats including mangrove forests, salt marshes, sand beaches, gravel beaches, tidal flats, rocky platforms and reefs, rock cliffs, and islands. Each habitat supports different types of biological communities. El Salvador's coastal and marine areas are relatively rich in biodiversity, a fact that is surprising given that the country is the most densely populated non-island nation in the Western Hemisphere, and that it has relatively few natural areas left. El Salvador's forest cover has been reduced to less than 10% of its original extent (3% of the country is currently forested), and though poorly documented, the deforestation which has taken place has also resulted in a decrease in biological diversity nationwide (Lovejoy, 1986). Komar (1997), reports that the country has only 3590 km^2 (359,000 ha) of forest or scrub habitats left. Based on a report by the World Resources Institute (1995), nationwide El Salvador has 458 species of mammals and birds per 10,000 km^2 (Costa Rica ranks first in the world with 615 species/10,000 km^2), and 3277 plant species per 10,000 km^2 (Colombia ranks first in the world with 10,735 species/10,000 km^2). Though similar specific comparisons are not available, marine and coastal biodiversity in El Salvador is likely similar to other Pacific coast areas in Central America.

Nearshore coastal and marine ecosystems are widely recognized for their ecological importance, being productive and diverse environments. However, these ecosystems are also vital components of both local and national economies. The economic value of ecosystems cannot be easily ascertained, though Costanza et al. (1997) have suggested that nearshore coastal and marine ecosystems (such as estuaries, tidal marshes, mangroves, seagrass beds, and coral reefs) which constitute about 6.3% of the earth's surface, make up 43% of the value of its ecological services. Despite this, El Salvador's coastal and marine habitats and ecosystems, like those in many tropical locations, have not been studied extensively.

SEASONALITY, CURRENTS, NATURAL ENVIRONMENTAL VARIABLES

El Salvador has a tropical climate with two seasons driven by rainfall patterns rather than temperature changes. Summer (the dry season), occurs from November to April and characteristically has winds out of the northeast. Winter (the rainy season), occurs from May to October and characteristically has easterly winds. The transition period between seasons typically lasts two to four weeks. The mean annual air temperature within the coastal zone is 28°C, though maximum temperatures in excess of 40°C may occur. The annual precipitation ranges considerably from region to region (915–2909 mm/yr) with a mean precipitation in the vicinity of 1700 mm/yr in the coastal areas (Almanaque Metereológico, 1992). During El Niño years, both temperature and rainfall averages may vary considerably. Generally, the El Salvador coastal zone is not subject to direct hits from tropical hurricanes and cyclones, though during Hurricane Mitch in 1998, very heavy rainfall (more than 0.70 m) occurred in a very short time-period (Williams, 1999). Heavy run-off associated with Hurricane Mitch resulted in significant changes (depositional and erosional) in certain coastal areas, such as the inlet associated with the mouth of the Río Lempa.

Some of the main environmental factors influencing nearshore marine habitats and biological resources include tides, temperature, salinity, nutrient availability, sediments, type of substrate, and major oceanic currents. El Salvador's coastline is meso-tidal, with tides in the 2.0–3.0 m range. This large tidal range partially accounts for the existence of relatively wide intertidal zones in some areas (extensive tidal flats, mangrove areas, rocky intertidal reefs, etc.). This results in the potential exposure of intertidal organisms to high temperatures and drying, a fact that becomes an important limiting factor on intertidal diversity. Lessmann (1986), reports that coastal water temperatures in El Salvador vary from 27–29°C, a fairly stable condition. However, intertidal pools and shallow slow-moving waters in some estuarine areas may reach water temperatures considerably higher. During a juvenile marine shrimp study, Sandifer et al. (1992) reported water temperatures of 43°C in Bahía de Jiquilisco.

Salinity in open marine waters typically ranges from 28–35 ppt, with the lower end of the range only observed in localized areas with estuarine influence. Within the estuaries, the typical range is 0-35 ppt depending on the position within the estuary, time of year (dry season vs. wet season), and river discharge. Much higher salinities can also occur in some locations during the dry season. Salinity readings up to 94 ppt were recorded by Sandifer et al. (1992) at Barrancones, within the Golfo de Fonseca, during the dry season, even though average salinity at the site was 31 ppt. The wide range in salinity within the estuaries of El Salvador is driven by their relatively small size and seasonal differences in river discharge: during the dry season stream flow is minimal and is similar to that in marine waters; during the wet season salinity drops to near fresh-water as riverine influence dominates. Estuarine areas with extremely shallow waters can experience greater than average seawater salinities in the dry season, due to high rates of evaporation. This is especially true for estuaries that lose connection to marine waters during the dry season, when ocean inlets close.

Little quantitative information is available on marine and estuarine sediments and sedimentation. During the rainy season, rivers carry large amounts of sediments, which not only affect the water quality of estuaries but also of nearshore marine waters. Sediment influx into the estuaries and nearshore areas results in increased turbidity and low light penetration, affecting the types of biological communities that can exist. It is suspected that increased turbidity due to sedimentation may limit a wider variety of hard corals and other attached biota from flourishing in certain areas, despite the fact that water temperatures are suitable for growth (Thurman and Webber, 1984).

Information on nearshore and offshore current patterns is very limited. Though localized nearshore currents vary, the dominant nearshore current along the coastline of El Salvador runs from east to west. It is also known that El Salvador lacks strong upwelling currents that have been observed in other locations in the region, such as in the neighbouring areas of the Tehuantepec and Papagayo Gulfs (Lluch-Cota et al., 1997).

THE MAJOR SHALLOW WATER MARINE AND COASTAL RESOURCES

The outer coastline of El Salvador is dominated by sand (or mixed sand and gravel) beaches and rocky shorelines (rocky platforms and cliffs) (RPI, 1998a,b). Rocky reefs and sand-bottom habitats occur in nearshore marine waters. Estuarine shorelines are dominated mainly by mangroves and tidal flats (RPI, 1998a,b). In some places, smaller stands of salt marsh occur inland of the mangroves, in sediments that are less often flooded and usually hypersaline. Subtidal estuarine habitats include mud and sand-bottom areas. Submerged seagrass beds and mollusc reefs, though perhaps present in some areas, are not known to be a major component of the subtidal estuarine habitats. The coastal beaches and estuarine habitats are associated with the coastal plain, and the rocky shorelines are part of several coastal mountain ranges. Sediments comprising the coastal plain are predominantly of volcanic origin. These features were formed during the low-stand of the sea during the Pleistocene period as large rivers eroded the volcanic terrain of El Salvador and carried vast amounts of sediment to the low-stand shoreline. Over time, these sediments were reworked into proto-barrier beaches that transgressed landward to their present position during the Holocene.

Mangroves and Reefs

Mangroves are a vital component of El Salvador's coastal and marine environment. Mangroves are used extensively by numerous fish and invertebrate species as nursery grounds, by coastal wildlife for nesting, feeding, roosting areas, and as general habitats for a variety of other species (Barraza and Vásquez, 1993). Six species of mangroves (*Rhizophora mangle*, *R. mucronata*, *Laguncularia racemosa*, *Avicennia nitida*, *A. bicolor*, and *Conocarpus erectus*) are reported in El Salvador. These species grow in muddy substrates that are sheltered from wave action and strong tidal currents. Figure 2 illustrates a generalized cross-section of the vegetation present in Bahía de Jiquilisco and depicts the relative location of each mangrove species.

Several estimates of mangrove areal coverage in El Salvador have been reported. Based on aerial photos, topographic maps, and aerial surveys, the extent of the mangroves in El Salvador in 1997 was calculated to be approximately 35,362 ha: 17,525 ha in the Bahía de Jiquilisco complex; 8530 ha in the Estero de Jaltepeque complex; 4666 ha in the Golfo de Fonseca area; and 4641 ha in other areas (RPI, 1998a,b). Jiménez (1994) reported a similar mangrove extent of 35,235 ha: 19,847 ha in the Bahía de Jiquilisco complex; 5385 ha in the Estero de Jaltepeque complex; 4657 ha in the Golfo de Fonseca area; and 5346 ha in other areas. Marroquín (1992) reported total mangrove cover of 34,424 ha in 1974 and 26,772 ha in 1989. Due to differences in methods, the three different studies cannot be directly compared to address changes in mangrove cover over time. However, the study by Marroquín (1992) did identify a 22% decline in mangrove cover from 1974 to 1989.

The most extensive nearshore rocky reef complex in El Salvador is found adjacent to Los Cóbanos, located south of the major port of Acajutla. The reef at Los Cóbanos is

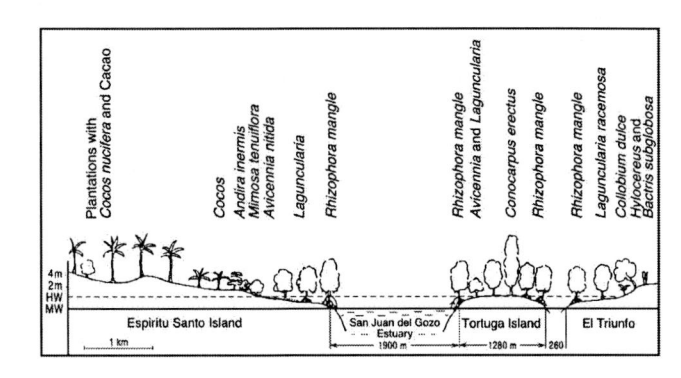

Fig. 2. Generalized cross-section of the vegetation present in Bahía de Jiquilisco (adapted from Gierloff-Emden, 1976).

estimated to be approximately 1500 ha. This rocky reef is a major aggregation area for fish and invertebrates. A variety of soft and hard corals are also found there (Lemus et al., 1994). Nearshore rocky reef areas, smaller in extent, can also be found in the eastern part of the country, just west of Punta Amapala and Playa Maculis, and in other locations. Rocky reefs in deeper waters are also present in other locations, based mainly on the types of fish and invertebrates harvested in these areas, but details about the types of reef present, or their extent, are not known at this time.

Extensive studies of the coastal and nearshore fauna and flora of El Salvador are limited. RPI (1998a,b) compiled published and expert information on the principal species of marine and coastal mammals, birds, fish, reptiles, and invertebrates found in El Salvador, and depicted known concentration areas and sensitive sites spatially in a set of coastal resource atlases, summarized below.

Mammals and Birds

Very few studies have been conducted on marine and coastal mammals. Generally speaking, the identity of what species are present in the country is known, but information on population dynamics, seasonal activities, etc. is not. The bottlenose dolphin (*Tursiops truncatus*) is reported to be the most common marine mammal in El Salvador, while two other species, *Stenella attenata* (spotted dolphin) and *S. longirostris* (spinner dolphin), are reported to be widespread in nearshore areas. Up to five species of whales may periodically be present offshore, but the regularity of whale occurrence and the importance of Salvadoran waters for these species is not known. The California sea lion (*Zalophus californianus*) has been reported several times in El Salvador, but it is suspected that these were accidental visitors (MINED, 1995).

Most mammals found in the coastal zone of El Salvador are small species not specifically associated with marine and coastal habitats. Although the common raccoon (*Procyon lotor*) occurs in a variety of coastal and inland habitats, it is particularly abundant in mangrove and estuarine areas in El Salvador. Also, fish eating bats such as the lesser bulldog bat (*Noctilio albiventris*) are often associated with mangrove forests. These bats feed on small fish found in estuarine waters.

Birds are probably the best studied of all the fauna of El Salvador. Detailed information exists on the life-history, seasonality, and population status, including nesting survey counts, of many coastal and marine species. Komar (1997), reports that 509 species of birds are known to occur throughout El Salvador, of which 310 are breeding residents, the balance being classified as migratory visitors, transients or vagrants. Of these, approximately 200 species occur in nearshore marine and coastal areas, including: pelagic seabirds, coastal diving birds, gulls and terns, waterfowl, wading birds, shorebirds, raptors, and passerine and other "non-water" birds (Komar et al., 1993). Komar et al. (1993) report that 70 of these species are threatened or endangered.

The three major estuaries, as well as the smaller estuaries, provide very important concentration and feeding areas for large numbers of birds and a large variety of species. The estuaries, as well as the ocean inlets associated with them, are probably the most important habitats for coastal birds. The mangrove forests provide protection, roosting and nesting sites to many water birds, while the inlets and intertidal flats provide important feeding grounds. Mangroves also provide critical living and nesting habitat for non-marine birds, such as the highly endangered white-fronted and yellow-naped parrots.

Several important areas for coastal and marine birds deserve specific mention. The estuary associated with Barra de Santiago (east of Bocana Zapote) contains an area locally referred to as "Colegio de las Aves". This complex of mangroves, tidal creeks, tidal flats, and small coastal streams is probably the single most important bird nesting area in El Salvador's coastal zone, with peak nesting activity reported from June to August (Komar et al., 1993).

A number of smaller colonial nesting sites also occur along the coast in various areas, often associated with small mangrove islets or the few small rocky islands that occur. Estero de Jaltepeque especially supports large numbers of wading birds, including reddish egret (*Egretta rufescens*), little blue heron (*Egretta caerulea*), and great blue heron (*Ardea herodias*). The islands in the Golfo de Fonseca include large concentrations of coastal diving birds (e.g., brown pelican, *Pelecanus occidentalis*) and pelagic seabirds (e.g., magnificent frigatebird, *Fregata magnificens*). The Río Lempa river mouth is an especially important area where numerous species of shorebirds, such as sanderling (*Calidris alba*), western sandpiper (*Calidris mauri*), and Wilson's plover (*Charadrius wilsonia*) concentrate. Nearly every river mouth and ocean inlet in the country would also be considered important shorebird areas.

Compared to coastal and estuarine species, the distributions of offshore marine birds (notably *Oceanodroma* species) are less well known. Particular concentration areas can vary greatly, both spatially and temporally, although certain regions may be more important than others. In El Salvador, and in most other offshore marine areas worldwide, more research into the distribution, life-history, and movements of pelagic marine birds (while they are at sea) is needed, especially with reference to potential effects of marine pollution and global climate change.

Reptiles

Reptiles found in nearshore and coastal areas of El Salvador include sea turtles, freshwater turtles, crocodilians, iguanas, and snakes (including sea snakes). Though in relatively low numbers, crocodiles, caimans, and alligators occur throughout appropriate habitats in El Salvador, particularly in mangroves, other coastal wetlands, estuarine waters, coastal rivers, ponds, and impoundments. Sea snakes, *Pelamis platurus*, are reported for the Golfo de Fonseca, and may also occur in other marine and estuarine areas. Relatively little is known about their distribution and life history.

The outer coast sand beaches of El Salvador are used as nesting areas by four species of marine turtle: Pacific green turtle (*Chelonia mydas agassizi*), leatherback (*Dermochelys coriacea*), hawksbill (*Eretmochelys imbricata*), and olive ridley (*Lepidochelys olivacea*). Nesting for the green turtle and the olive ridley is also reported to occur inside Bahía de Jiquilisco at Punta San Juan and Punta Rancho Viejo. At least one of the four species may be found nesting in any month of the year, except March and April when nesting is not common (Hasbún and Vásquez, 1991; RPI, 1998a,b). The most common of the sea turtles found in El Salvador is the olive ridley. Quantitative information on the numbers of sea turtles that nest along the coastline of El Salvador is not available, as regular coast-wide surveys are not conducted.

In addition to nesting, in-water concentrations for green turtles and olive ridleys are known for Canal Barillas, Estero el Coyajón, and other parts of Bahía de Jiquilisco. Though not well documented, in-water concentrations in other estuarine areas and in offshore waters may also occur. All sea turtles in El Salvador are protected species but, despite legal protections, eggs and perhaps adult turtles are still illegally harvested for individual consumption, and are even sold for local commercial utilization (e.g., sea turtle eggs are offered on the menus of some local, small seafood establishments).

Fishes

Numerous fish species inhabit El Salvador's marine and estuarine waters. Though general information on species presence and distribution is known, especially for harvested species, specific information on fish aggregation sites, spawning areas, nursery areas, population estimates, etc. is not well known. Important fisheries species include groupers, snappers, drums (corvinas, seatrout), snooks, mackerels, and sharks. Sharks are heavily targeted, mainly for the sale of fins to Asian markets. Based on information gathered from local fisheries experts, the distribution of most marine fish along the El Salvadoran coastline is generally uniform, with the exception of the rocky reef areas such as Los Cóbanos, where many additional reef-associates are found (RPI, 1998a,b). Another exception is the distribution of fish in El Salvador's estuaries, where certain species are apparently found predominantly or even exclusively in one estuarine system or another. It is possible, however, that major differences in estuarine fish distribution are the result of incomplete and sporadic surveys, rather than true biogeographic differences. In regards to the estuaries, many species occurring in El Salvador display the classic estuarine-dependency observed in other parts of the world, where young life-stages of marine species are found in estuarine nursery habitats such as mangrove swamps, tidal creeks, and coastal rivers. The fact that many estuarine-dependent marine species comprise important fisheries should provide an economic and cultural incentive to protect mangrove swamps and other critical estuarine habitats.

In addition to estuarine-dependent marine species, a few diadromous species also occur in El Salvador. One good example is the mountain mullet (*Agonostomus monticola*). Similar to other tropical montane regions, diadromous species in El Salvador spend much or all of their adult lives in freshwater streams, where they also spawn. Eggs or early-larval stages move downstream to coastal areas, where the larvae and juveniles use estuarine and marine waters for early-development, migrating back upstream to freshwater habitats as juveniles or sub-adults. This diadromous life-history strategy is a form of obligate amphidromy (see Holmquist et al., 1997; Radtke and Kinzie, 1996). Unlike anadromous species in temperate parts of the world, such as salmonids and sturgeon, tropical amphidromous species do not comprise major commercial fisheries (although harvests for local consumption may occur, especially during concentrated migratory "runs"). However, quite similar to anadromous fish, tropical amphidromous species could serve well as focus-species of conservation interest. Healthy populations of such species can be good indicators of overall watershed health and useful monitors of successful coupling between inland watersheds and ocean systems, since they require intact stream habitats and estuarine or marine waters to complete their life cycle. As with anadromous species, excessive water withdrawals, stream diversion, migratory barriers (dams), water pollution, improper land use, and loss of natural habitat in both inland and coastal areas can threaten the survival of these species.

Marine Invertebrate Resources

Most of the information available on marine invertebrates in El Salvador pertains to the major crustacean and bivalve species that are harvested. Penaeid shrimp make up the most important commercial fisheries resource in the country, although other species are harvested as well, mainly for local consumption. Penaeid shrimps in El Salvador are harvested in a marine trawl fishery, similar to other parts of the world. The penaeid shrimps *Penaeus vannamei* and *P. stylirostris* are the most important harvested species. Both these shrimps use estuarine mangrove habitats as nursery grounds. Juveniles occur year-round in the estuaries, but *P. vannamei* juveniles are more abundant during the rainy season from May to October (Sandifer et al., 1992). Out of three coastal areas studied (Barra de Santiago, Puerto Parada, and Barrancones), Barra de Santiago appeared to be the richest shrimp nursery grounds (Sandifer et al., 1992).

Similar to the mountain mullet, a complementary assemblage of amphidromous shrimps also occur as larvae and juveniles in the estuarine waters of El Salvador.

Diadromous shrimp such as *Macrobrachium* and *Atya* spend their adult lives in freshwater riverine habitats, often in mountainous areas, using estuarine and marine habitats for larval and juvenile development. The life histories of these shrimps, and their potential importance for watershed monitoring and conservation, match closely to those of the amphidromous fish described previously. Some of the larger species of *Macrobrachium* (some of which can be very large, on the order of 10–20 inches in total length) are harvested for local consumption. *Atya* may also be harvested for local consumption or bait.

Other important marine and coastal crustaceans include crabs such as *Callinectes* spp. and *Menippe frontalis*, both of which are harvested. The land crabs *Cardisoma crassum* and *Gecarcinus lateralis*, are important ecologically because of their burrowing activities in mangroves areas and along coastal rivers. Land crabs may also be harvested for local consumption, as they are in other Latin American countries. The spiny lobster, *Panulirus* sp., occurs in reef areas and along rocky shorelines, and is harvested in earnest wherever it is abundant.

Major bivalves species include a variety of oysters (*Ostrea* spp.), mussels (*Mytella* spp.), and blood arks or cockles (*Anadara* spp.). As a rule, harvested oysters are attached forms, growing on rocky substrate in marine waters, rather than reef-forming species associated with estuarine waters. These oysters are harvested by free-diving using mask and fins, for supply to local seafood establishments or for other local consumption. The blood arks occur in muddy estuarine areas and mangrove habitats, and are harvested by hand.

Additional studies on invertebrates are listed below, including species not covered here, such as polychaetes (Hartmann-Schröder, 1957, 1959; Molina-Lara and Vargas-Zamora, 1993; Barraza, 1994; Rivera-Muñoz and Ibarra-Portillo, 1995) and echinoderms (Barraza, 1993, 1994; Barraza and Carballeira, 1998).

OFFSHORE SYSTEMS

There are very few studies (especially recent studies) that have been conducted on the offshore ecosystems and pelagic communities specific to El Salvador. During the EASTOPAC cruises (1967–1968), some basic physical, chemical, and biological oceanographic data were collected including measurements of salinity, temperature, winds, nitrates, phosphates, chlorophyll, zooplankton, fish larvae, tuna distributions, and seabird counts (NOAA, 1970–1972). These cruises were not specific to El Salvador, but some of the data collected included areas of El Salvador's offshore territorial waters. It is known that the offshore waters of El Salvador do not experience major upwelling currents. Pelagic fish, including sharks, tuna, marlins, mackerels, etc., do occur and some of these species are actively fished. The Middle America Trench runs approximately 100–200 km offshore, mostly parallel to El Salvador's shoreline.

POPULATIONS AFFECTING THE AREA

Practically the entire coastal zone area of El Salvador is inhabited, and even within the mangrove forests there are some areas with small settlements. This is not surprising since El Salvador is the most densely populated country in Central America, and has the highest population density of any non-island nation in the Western Hemisphere (Lovejoy, 1986). There are four principal population centres directly on the coastline of El Salvador: Acajutla (est. pop. 67,000), La Libertad (est. pop. 55,000), La Unión (est. pop. 50,000), and Puerto El Triunfo (est. pop. 21,000). Acajutla and La Libertad are located on the continental coastline, La Unión in the Golfo de Fonseca, and Puerto El Triunfo in Bahía de Jiquilisco. These towns, as well as the numerous villages in the coastal areas, rely principally on fisheries (both artisanal and commercial) and agriculture as the main economic resources. In recent years tourism, notably beach homes and some hotel developments, has also occurred on the coast.

Agriculture is the predominant land use in El Salvador. Though agricultural development extends directly to the coast, the degree of urbanization is not great, except in major port areas, where much of the development is industrial in nature, related to port activities and the populations which develop around such areas. However, away from developed ports, even nearby in some cases, much of the immediate shoreline appears relatively natural (except for mostly single-family houses that are present throughout much of the shoreline), even if it is backed by agricultural lands rather than natural habitats. Other obvious land-use changes in many estuarine areas include salt production ponds, built in areas which naturally support mangrove, salt marsh, and salt flat habitats.

The main maritime port of El Salvador is the Port of Acajutla, which was built in the 1960s. The overwhelming percentage of goods imported and exported to or from El Salvador via maritime transport pass through this port. Smaller ports exist at La Unión, La Libertad, and at El Triunfo, though the last two are principally fishing ports. The major Port of Acajutla is discussed in more detail in later sections.

RURAL FACTORS

Agriculture, the predominant land use in El Salvador, may negatively impact coastal and marine ecosystems through the run-off of sediments, fertilizers, pesticides, and animal wastes to rivers and the coast (Foer and Olsen, 1992). Certain land-use and agricultural practices, such as excessive deforestation, over-grazing, soil erosion, poor water resources management, and over-application or misapplication of agricultural chemicals, may contribute to downstream impacts in coastal and marine areas. Run-off effects are most noticeable during the rainy season.

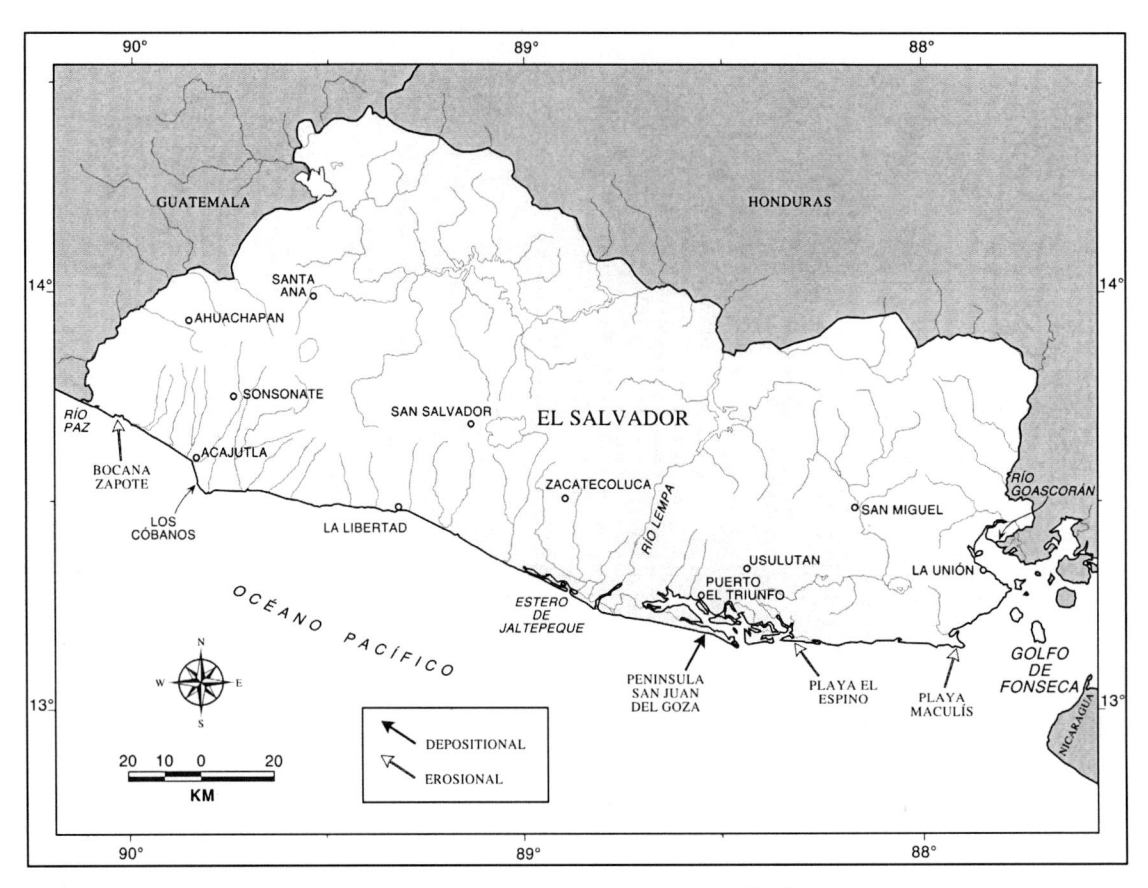

Fig. 3. Areas of coastal deposition and erosion in El Salvador.

Increased sedimentation can be a serious problem, especially in tropical areas that support reefs. For instance, during the rainy season, the reefs at Los Cóbanos are affected by increased sedimentation, resulting in reduced growth of corals and other attached biota. Although sedimentation on the reefs at Los Cóbanos has not currently been linked to agricultural practices, comparison of sedimentation inputs from rivers associated with differing levels of agricultural development should be examined in El Salvador. In regards to fertilizer and animal waste run-off, examination of nutrient levels, algal productivity, oxygen demand, and microbial contamination in agricultural watersheds and adjacent coastal areas have not been examined.

Though no systematic surveys have been conducted, pesticide concentrations in sediment samples from the mouths of two rivers that drain agricultural lands were examined in one recent study (RPI, 1995; Michel and Zengel, 1998). Concentrations of nine specific pesticides were all below detection levels, although detectable amounts of DDT and its degradation products were found (0.0–0.038 ppm dry weight, total DDT-related compounds) in both areas. Two of five sediment samples were equal to or below the effects-range low (ERL) value of 0.005 ppm for DDT, while three of five were between the ERL and the effects-range median (ERM) value of 0.046 ppm reported by Long et al. (1995). Concentrations below the ERL represent minimal-effects range, where ecological effects are expected to be rarely observed. Concentrations between the ERL and ERM represent a possible-effects range, where ecological effects would occasionally occur. Concentrations above the ERM, represent cases where effects are expected to regularly occur. Though this study provides some insight into pesticide contamination in estuarine sediments, it should be recognized that firm conclusions cannot be drawn from such limited sampling. Also important to consider, samples from the study cited above were collected in the dry season only. Samples collected during the wet season would also be needed to fully assess pesticide contamination from agricultural run-off.

COASTAL EROSION AND LANDFILL

There is evidence of active coastal erosion in several parts of El Salvador (RPI, 1998a,b). Recent field observations, including detailed beach profiles from 21 field stations, have identified erosional zones in several parts of the country including Bocana El Zapote, Playa El Espino, and Playa Maculís. Areas of coastal deposition have also been documented, such as at the Peninsula of San Juan del Gozo (Fig. 3). Beach profile stations should be revisited regularly, particularly after major storms such as Hurricane Mitch, and after major coastal development projects are completed, to assess changes in the shoreline over time.

EFFECTS FROM URBAN AND INDUSTRIAL ACTIVITIES

Artisanal and Non-industrial Uses of the Coast

Although extensive cover of mangroves still occurs in El Salvador, mangrove cutting and clearing are small-scale activities that can result in large cumulative impacts over time. Though a protected resource, mangroves continue to be harvested or removed to create opportunities for other land uses. Mangroves are cut mainly for use as fuel, although they are harvested for building materials as well. Mangroves are cleared to make way for agricultural expansion, salt pond creation, and to a lesser extent aquaculture. Mangroves are also cleared to make way for small settlements. It is expected that demands for fuel, building materials and land will exert increasing pressure on mangrove habitats in the future. The only study to directly compare changes in mangrove cover over time, reported a 22% loss of mangrove habitats between 1974 and 1989 (Marroquín, 1992).

Though clearing of mangrove areas for aquaculture in other parts of the world, including other Latin American countries, has resulted in drastic declines in mangrove cover, this has not been the case in El Salvador (at least not to date). Aquaculture, particularly shrimp culture, has not extensively developed in El Salvador. In the 1980s, marine shrimp culture was initially started near Los Cóbanos and in the Golfo de Fonseca. Total area under cultivation was relatively small (less than 100 ha). Promoted by FUSADES (Fundación Salvadoreña para el Desarrollo Económico y Social), an effort was made in the late 1980s and early 1990s to expand shrimp farming in the country. For several reasons, including the civil war that was ongoing during that time, few investors showed any interest in this and, as a result, there was no significant aquaculture development.

Clearing of mangrove areas for construction of salt production ponds (secondarily used for aquaculture) continues to occur, though this activity seems to have been less widespread in the last several years. The total area of all salt production ponds in 1997 was estimated to be 1,876 ha (RPI, 1998a,b). Some portion of this area represents the amount of mangroves lost directly to salt ponds, some of which were constructed directly on salt pans or flats, or on other habitats in addition to mangroves. Salt pond construction has occurred predominantly in the Golfo de Fonseca and Bahía de Jiquilisco areas. Several recently-developed salt ponds were detected in the northern part of Golfo de Fonseca in 1997 (RPI, 1998b).

Fisheries

Fishing is a principal economic activity at both artisanal and commercial levels. Artisanal and commercial fishing are discussed together in this section, since total harvests from both categories are similar. Penaeid shrimp are the most

Table 1

Summary of fisheries catch by major type for commercial and artisanal fisheries (adapted from CENDEPESCA, 1996).

Fishery type	Commercial catch (Mt)	Artisanal catch (Mt)
Snappers	–	333
Sea trout/Corvinas	–	232
Mackerels	–	156
Catfish	–	94
Sharks	–	759
Other fish	365	2188
Shrimp	4460	403
Other crustaceans	32*	532
Molluscs	*	601
Total	4857	5298

*Other crustaceans and mollusc catch is combined.

important commercial fisheries resource in El Salvador, accounting for almost all the commercial harvest and 48% of the total catch (artisanal and commercial) reported in 1995 (CENDEPESCA, 1996). The artisanal catch includes a much wider variety of species, and leads commercial fishing catch for all groups other than shrimp (Table 1).

Fisheries statistics indicate that a total of 10,155 Mt (4857 Mt by commercial fishing and 5298 Mt by artisanal fishing) of marine and estuarine fisheries products were landed in 1995 (CENDEPESCA, 1996). According to Article 25 of the General Law governing Fishing Activities of El Salvador, artisanal fishing is defined as fishing conducted from boats less that 10 m in length in water depths less than 40 fathoms, and that at least 70% of the product is consumed fresh (not processed in any way). Approximately 91% of the fisheries landed using artisanal methods are caught by individuals, with the balance landed by fishing cooperatives (CENDEPESCA, 1996). In 1995 the domestic commercial fishing fleet included 80 vessels, mainly engaged in the shrimp fishery (CENDEPESCA, 1996).

The current status of marine fisheries populations in El Salvador is not well known. It is not known if over-fishing is impacting marine fisheries resources, or if current harvests are relatively sustainable. It is likely that landings data to address this question exist, but unfortunately these data were not available at the time of writing. One item of both past and recent concern is the sustainability, or lack thereof, of the shark fishery (Villatoro Vaquiz, 1977). In contrast to the nationwide fisheries catch reported in Table 1, 1990–1994 artisanal catch statistics for the Port of Acajutla show that the shark catch on average was an order of magnitude higher than the catch of either snappers or corvina (CENDEPESCA data, reported in RPI, 1995). However, shark catch fluctuated by one to two orders of magnitude from year to year among independent fishermen, with an overall declining trend for the four years (the catch by fishing cooperatives stayed about the same). It is difficult to assess why shark harvests fluctuated so drastically—there may have been changes in market demand, numbers of fishermen, changes in fish abundance, etc.; however, the magnitude of annual change does raise questions. Observations at fishing ports near Acajutla in 1995 revealed that sharks dominated much of the catch. Dried shark fins were the primary product related to the shark fishery. Because of the emphasis on shark fins, it is possible that shark catches are drastically under-estimated (if mainly fins are landed and weighed, and not whole fish). If this is indeed the case, shark harvests might actually be very much higher than reported. Based on the world-wide interest and need for improved shark fisheries management and conservation, further examination of the sustainability of shark harvests in El Salvador is greatly needed.

In addition to population level fisheries effects, the loss of or damage to important fisheries habitat caused by destructive fishing practices is also of concern. Though prohibited, several destructive artisanal fishing practices are still in use, including the use of explosives in mangrove lagoons, and the poisoning of fish and invertebrates in coastal rivers. The use of nets and trawls on rocky reefs may also destroy corals and other attached organisms. The commercial fishing industry may also employ some practices that may adversely impact coastal habitats and species. Shrimp trawls, for example, sweep the seabed, and though their effect on benthic communities has not been properly assessed, this practice is likely to be destructive, at least in certain habitats. By-catch associated with shrimping may also be a serious concern. New rules established by the U.S. government have required that sea turtle excluder devices (TEDs) be used for any harvests of wild-caught shrimp that will be exported to the U.S. Though difficult to monitor or enforce, it is known that at least some Salvadoran shrimping vessels now utilize TEDs, hopefully reducing impacts to marine turtles while also reducing other by-catch.

Other Extracted Resources

Unregulated small-scale sediment removal (both sand and gravel) is common in the coastal zone of El Salvador, primarily for local construction purposes. Though this practice is not permitted, it continues to occur relatively unchecked. Targeted areas include river mouths, beaches, and sand dunes. Sediment removal is also actively practised on inland riverine point bars on the coastal plain, especially in the dry season when these areas are most accessible. Though it occurs inland, this practice robs coastal areas of sediment supply, though the magnitude of such effects is not known. Dams on inland rivers also deplete sediment supply to the coast, evidenced by the infilling of these dams by trapped sediments. Excessive depletion of sediment supply to the coast could be contributing to coastal erosion.

In the 1950s, shallow water corals at the Los Cóbanos reefs were locally exploited to produce cement, a practice which destroyed or degraded extensive patches of reef. An algal invasion followed, and most hard corals did not recover. There is not much data about coral species composition prior to coral mining activities, although Gierloff-Emden (1976) reported on observations of corals made in 1956. Currently, Los Cóbanos reef is a rocky reef dominated by soft corals (gorgonians) and other attached organisms with limited hard coral development. No coral or other reef mining activities are now known to occur in the country.

Industrial Uses of the Coast

Industrial development in coastal El Salvador is mainly limited to Acajutla, though there is some industrial development in other areas. In 1995, 27 industrial facilities were identified at the Port of Acajutla. These facilities were involved in the following activities: petroleum refining and storage; fertilizer storage; industrial chemical production and storage; power generation; foodstuff (grain, edible oils, molasses) processing and storage; aquaculture production and seafood processing; and light manufacturing (RPI, 1995; Michel and Zengel, 1998). Reported average daily storage of liquid products in Acajutla in 1995 totalled about 904,000 bbls (37,968,000 gallons), of which 78% (by volume) was petroleum-related, associated mainly with a refinery and two distribution centres. Petroleum-related storages included crude oil (43% by volume), diesel (21%), bunker C (16%), gasoline (10%), and other oil types (10%). Annual liquid marine transfers reported at the port and the adjacent marine transfer area for the refinery were also dominated by petroleum (91% by volume), of which 52% was crude oil, 35% diesel, 6% liquefied petroleum gas (LPG), 4% gasoline, 2% bunker C, and 1% other petroleum types.

Industrial wastewater discharges to estuarine and marine waters from Acajutla have also been inventoried (RPI, 1995, Michel and Zengel 1998). Reported industrial wastewater discharges totalled nearly 2,500,000 gallons per day in 1995. Major wastewater streams included industrial cooling water (52% by volume), shrimp pond circulation water (32%), industrial process water (8%), industrial sewage (8%), seafood processing (<1%), and industrial wash water (<1%). Stormwater run-off from industrial areas was not included in discharge estimates or totals, although it is certainly substantial. Discharges from petroleum-related facilities (including stormwater run-off in some cases) are treated by oil–water separation prior to release (although oil streamers were observed in one discharge area). Treatment of other wastewater prior to release is minimal, although at least one major facility was taking steps to reduce wastewater discharge through recycling and the use of mist eliminators. Another facility trucked wastewater offsite for treatment and disposal by land-farming techniques. Most industrial wastewater in Acajutla is discharged to a canal which drains directly to nearshore marine waters. Refinery wastewaters are discharged to a small stream which flows directly to the sea. Also worth mentioning are industrial and domestic wastewater discharges from inland areas which often flow to the coast via rivers. For instance, discharges from industries in the inland town of Sonsonate flow to the coast near Acajutla in the Sensunapan River. Wastewater inputs to marine waters from such sources have not been documented.

Marine Contamination

An initial assessment of marine contamination in the vicinity of Acajutla was conducted in 1995. Oyster (*Ostrea irridescens*) tissues and fine-grained sediments from several stations were analyzed for petroleum hydrocarbons (PAHs), chlorinated organic compounds, pesticides, and trace metals (RPI, 1995; Michel and Zengel, 1998). Results of the oyster analyses were compared to over 20 years of similar monitoring data from southern California (Mearns et al., 1991). Results of the sediment analyses were compared to effects-range low (ERL) values and effects-range median (ERM) values reported by Long et al. (1995).

Most oyster samples had only background levels of PAHs, although oysters collected close to the main industrial canal discharge were up to 100 times background levels. Only black muds from between the main piers of the industrial port area contained sediment PAHs at levels of environmental concern (slightly above the ERL of 0.55 ppm dry weight for low-molecular weight PAHs). Total PAHs at all other sites were well below ERL values. The distribution of PAH compounds in oysters and sediments indicated that the source of petroleum contamination was chronic petrogenic hydrocarbon releases (crude and refined oils) rather than pyrogenic hydrocarbons (from combustion of fossil fuels).

Concentrations of specific pesticides in oysters and sediments were below detection limits in most samples, the one exception being Heptachlor, detected at 0.146 ppm dry weight in one composite oyster sample from near the industrial canal discharge. DDT and its degradation products (DDE, DDD) were detected in several samples. Total DDT-related compounds in oysters typically ranged from one half to two times background levels of 0.05 ppm dry weight reported for southern California. One sample, however, had concentrations greater than ten times southern California background levels. Total DDT-related compounds in sediments from the main industrial port, and from a smaller fishing vessel harbour, were relatively high, all above the ERM value of 0.46 ppm dry weight. Values from nearby river mouths were lower, as noted previously.

Total PCBs were below detection levels in most oyster and sediment samples. PCBs in oyster tissues from one site were higher than background values, but well below median values from harbours and bays in southern California. However, PCBs in sediments from the main port area were quite high, well above ERM values of 0.18 ppm dry weight, indicating PCB levels of high environmental concern in the local vicinity of the port.

Trace metals concentrations in oysters and sediments varied widely. Copper and zinc concentrations were similar to levels normally associated with moderately polluted sites in the United States, somewhat surprising considering the current level of industrial development near Acajutla. Copper levels in sediments were above the ERL but well below the ERM at all sites, while zinc was either approaching or above the ERL at most sites, indicating that concentrations of both these metals may be approaching levels of ecological concern. It is possible that elevated levels of copper and zinc are related to natural sources, although this has not been investigated. In contrast to copper and zinc, levels of arsenic, cadmium, chromium, lead, and nickel were relatively low at most sites, and do not appear to be contaminants of major concern at this time. As with nearly all the contaminants examined, trace metals with low concentrations at most sites were generally elevated in the main port area. For example, lead concentrations in sediments were below detection limits at nearly all sites, except in the port area where concentrations approached the ERL. In addition to studies near Acajutla, trace metal concentrations in sediments at La Unión (in the Golfo de Fonseca) have also been examined (Barraza and Carballeira, 1998). Trace metals in this much smaller and less industrialized port area were reported to be below levels of ecological concern.

In conclusion, although several industrial contaminants were found to be at or approaching levels of ecological concern near Acajutla, sites with elevated concentrations of pollutants were mainly limited to the major industrial wastewater discharge and the main pier structures for the major marine port in the country. Chronic contamination outside of these limited areas is probably minor, although more extensive surveys and monitoring coast-wide and over time are needed to better assess the situation, especially as development proceeds along the coast in the future.

Cities

The coastal cities, towns, and villages of El Salvador do not currently have municipal sewage and wastewater treatment facilities. Human wastes are disposed of directly to septic tanks or latrines, and in some cases directly on the ground. Domestic wastewater is discharged to the ground or to small ditches and creeks that flow to coastal water bodies. In some areas, such as Acajutla and La Unión, community collection systems collect and pipe a proportion of domestic wastewaters directly to the coastal rivers, estuaries, or the ocean. Using estimates for the typical amounts of domestic wastewater generated per capita per day in Latin America, the daily volume of domestic wastewater discharged from a city the size of Acajutla would be roughly 1,000,000 gallons (RPI, 1995). Note that this figure does not include run-off from open latrines and septic tanks, or groundwater contamination from septic systems. Based on more precisely reported industrial wastewater discharges for Acajutla, domestic wastewater would account for 29% of total wastewater discharges to the marine environment

(RPI, 1995). Potential pollutants associated with domestic wastewaters include organic matter, nutrients, surfactants (soap), and microbial contaminants. The ecological and human health problems (from swimming, shellfish consumption, etc.) associated with untreated domestic wastewater discharges to coastal waters have not been investigated in El Salvador. Basic surveys of faecal coliform levels in coastal waters near major settlements would be valuable in assessing current risks.

Shipping and Offshore Accidents and Impacts

Catastrophic offshore spills have not occurred in El Salvadoran waters. However, two marine spills worth mentioning occurred in 1994. 400 bbls (16,800 gallons) of Venezuelan Recon oil were spilled in June 1994 during transfer operations at an offshore vessel off-loading area and pipeline associated with the refinery at Acajutla (Michel and Henry, 1997). A smaller release of No. 6 fuel oil occurred in December 1994 from the power generation plant at the Port of Acajutla (RPI, 1995; Michel and Henry, 1997).

During the June 1994 incident, about 30 bbls of chemical dispersants were applied to the spill in the surf zone and in shallow waters (4–6 m), effectively dispersing much of the oil, although nearby sand beaches and rocky shores were also oiled. To assess the exposure of water column and benthic marine resources to the oil, oysters (Ostrea irridescens) from subtidal rocky areas (3–9 m depth) were analyzed for PAH concentrations at 1 week and 4 weeks post-spill (Michel and Henry, 1997). Sediments were also collected and analyzed for PAHs. Two oyster samples from the area of dispersed oil contained total PAHs of 147 and 164 ppm dry weight during the first sampling period, compared with background levels of 1.0 ppm from nearby reference sites. Four weeks later, PAH levels had dropped by 94–98%, indicating continued depuration in oysters which had been exposed to the spilled oil. Half-lives for individual PAHs were calculated and were generally consistent with results from laboratory studies. Sediments did not appear to be significantly contaminated by the spilled oil.

In the area affected by the spill, there were no reports of oiled or dead birds, fish, invertebrates, or other organisms. Also, commercial shrimp fishing activities which take place in offshore waters beyond the spill area did not appear to be affected by the spill. No impacts to gear from other fisheries were reported either. Documented impacts or potential impacts associated with the June 1994 spill centred around oyster and finfish fisheries. Oysters from the spill area are harvested by local fishermen using free-diving methods. It was assumed that this activity was disrupted by the spill for at least several days. It is not known if oyster harvests were avoided for longer time periods based on health concerns, detection of tainted oysters by smell or taste, or other factors.

Fishing vessels from a nearby fishermen's cooperative stayed in port because of the presence of oil slicks. They did not fish for approximately 20 days thereafter because of a reported lack of fish. They reported impacts mainly in rocky areas where they usually catch snappers, groupers, corbina, parrotfish, and nuhuilla (RPI, 1995). After 20 days they began fishing elsewhere, travelling farther from port than usual to locate fish. They reported that snappers started coming back after three months post-spill, but that grouper had not returned as of March 1995. No biological studies were conducted to confirm impacts to finfish, and examination of annual catch records for the fishing cooperative were inconclusive. To estimate order of magnitude impacts to fisheries, Version 1.1 of the U.S. Department of Interior's Type A Natural Damage Assessment Model for Coastal and Marine Environments was used. The model was run using habitat types and species similar to those fishermen reported as impacted during and after the spill. After analysing several different model runs to best match conditions observed during the spill, results indicated total lost fisheries catch to be about 2000–4000 kg, including losses of potential recruits for up to 21 years post-spill. Though certainly subject to debate, the model results indicate that fisheries losses may have been somewhat less than those expected based on reports by local fishermen. It is of interest that fisheries losses estimated by the model were much less than the mean annual landings of 165,000 kg (1990–1993) reported for the fishing cooperative and independent fishermen in Acajutla. Also, the standard deviation for the annual mean was similar to losses estimated by the model (RPI, 1995).

PROTECTIVE MEASURES

The most comprehensive measures recently taken in the protection and preservation of El Salvador's coastal zone and marine resources is the federal Ley del Medio Ambiente (Environmental Law) which was passed on 4 May 1998 (Decreto No. 233, Diario Oficial Tomo No. 339, Numero 79). Several sections of this law specifically address coastal and marine resources (Table 2). Actual regulatory implementation of the Environmental Law and related laws could significantly improve the protection and conservation of El Salvador's coastal and marine areas.

Protected areas (parks, refuges, etc.) have existed in El Salvador for many years. Over the past several years, increased environmental awareness and a renewed interest in designating and managing protected areas has led to efforts to formalize protected area status for many sites, as well as the addition of other new protected sites. To improve protection and management of such areas, definitive legal boundary descriptions, land surveys, and titles for many of these areas have been proposed as a first step in determining the extent of these areas. As of 1998, there are at least 45 small protected areas which fall within 25 km of El Salvador's coastline (some of which may be informal or proposed protected areas) (Fig. 4). These are mostly

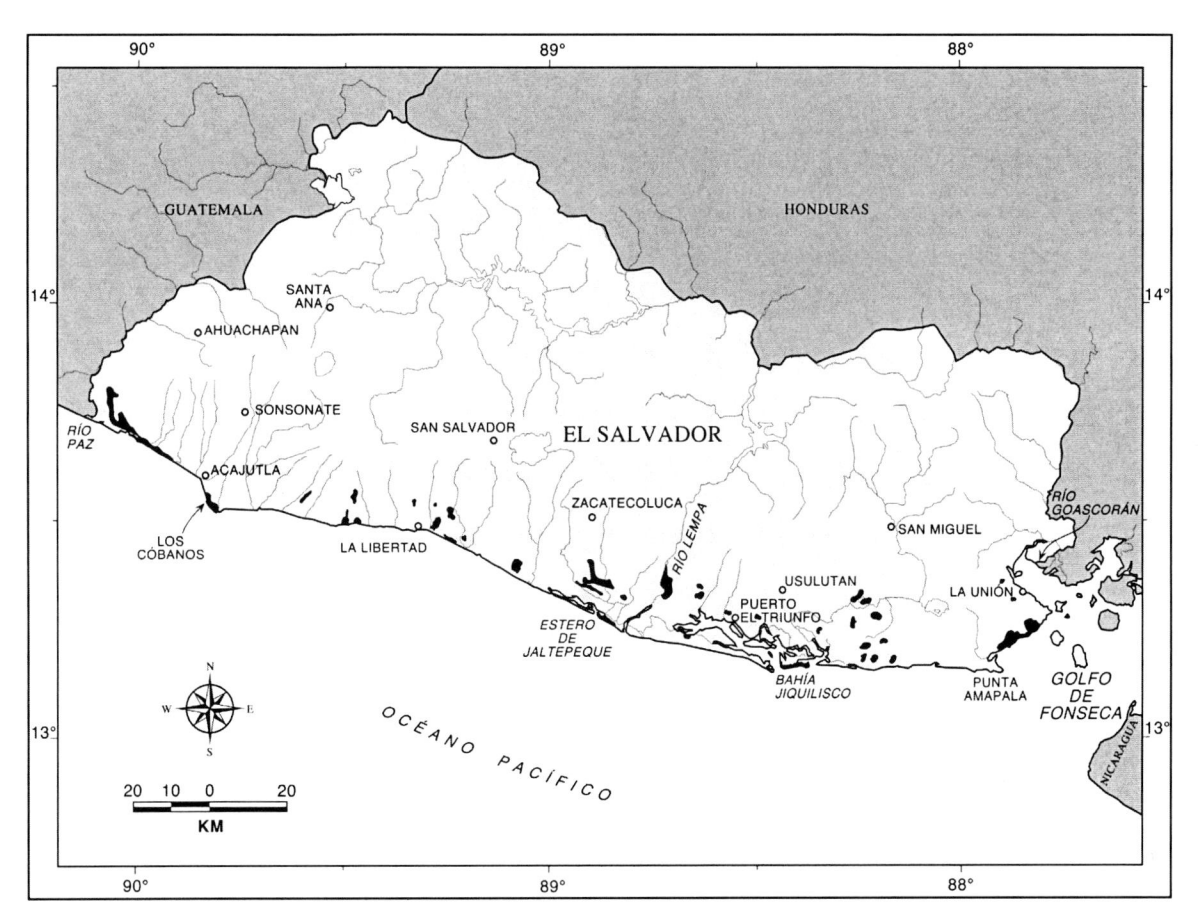

Fig. 4. Protected areas in El Salvador. Black shaded areas indicate the protected areas.

Table 2

Coastal and marine related sections of El Salvador's Environmental Law and related legislation

Protective action	Supporting legislation or program
Reduced coastal pollution from domestic and industrial sources	Environmental Law
Planned development in the coastal zone, including environmental reviews and permitting requirements	Environmental Law
Improved management of marine resource exploitation	Environmental Law and Marine Resources Plan
Reduced sedimentation in coastal areas	Economic incentives for inland reforestation
Revision of fisheries regulations, including new size and catch limitations	Proposal of new Fisheries Law
Prohibited collecting of marine reef fauna	Environmental Law

land-based, but do include significant areas of mangrove forests and other important coastal habitats. The Los Cóbanos reefs have recently been added to this list, although their status has not yet been finalized (official surveyed boundaries are still pending). Once established, this area will be the first marine-based protected area in the country. Considering the increasing interest in marine reserves world-wide, perhaps additional marine areas will be protected in the future.

REFERENCES

Almanaque Metereológico (1992) El Salvador.

Barraza, E.J. (1993) Comentarios de los equinodermos de la zona rocosa Solymar, La Libertad, El Salvador. Publicaciones Ocasionales No. 4. Museo de Historia Natural de El Salvador. 7 pp.

Barraza, E.J. and Vásquez, M. (1993) Estudio de la diversidad de camarones y peces del estero de San Diego, El Salvador. 1993. Publicaciones Ocasionales. Museo de Historia Natural de El Salvador. 15 pp.

Barraza, E.J. (1994) Guía ilustrada de algunos poliquetos (Annelida: Polychaeta) de Solymar, La Libertad, El Salvador. Boletín Técnico No. 2, Asociación Amigos del Arbol. 17 pp.

Barraza, E.J. and Carballeira, A. (1998) Una nota corta sobre los metales pesados de la bahía de La Unión, Golfo de Fonseca, El Salvador. Publicación Ocasional. No. 1. Ministerio de Medio Ambiente y Recursos Naturales, El Salvador. 8 pp.

CENDEPESCA - Ministerio de Agricultura y Ganaderia, Dirección General de Desarrollo Pesquero (1996) Anuario de Estadisticas Pesqueras, Año 1995, Vol. 21. 79 pp.

Costanza, R., d'Arge, R. and de Groot, R. (1997) The value of the world's ecosystem services and natural capital. *Nature* **387**, 253–260.

Foer, G. and Olsen, S. (1992) Las Costas de Centroamérica. Diagnósticos y Agenda para la Acción. Agency for International Development, Regional Office for Central America Programs. 290 pp.

Gierloff-Emden, 1976. *La Costa de El Salvador*. Minsiterio de Educación, Dir. de Publicaciones. San Salvador, El Salvador.

Hartmann-Schröder, G. (1957) Neue Armandia Arte (Opheliidae) aus Brasilien un El Salvador. *Beitr. Neotr. Fauna* **1**.

Hartmann-Schröder, G. (1959) Zur ökologie der polychaeten des Mangrove-Estero-Gebietes von El Salvador. *Beitr. Neotr. Fauna* **1**, 69–183.

Hasbún, C.R. and Vásquez, M. (1991) *Sea Turtle Nesting in El Salvador*. World Wildlife Fund. 59 pp.

Holmquist, J.G., Schmidt-Gengenbach, J.M. and Buchanan Yoshioka, B. (1997) High dams and marine-freshwater linkages: effects on native and introduced fauna in the Caribbean. *Conservation Biology* **12** (3), 621–630.

Jiménez, J.A. (1994) *Los Manglares del Pacífico Centroamericano*. Fundación UNA, Heredia, Costa Rica. 352 pp.

Komar, O. (1997) Avian diversity in El Salvador. The Wilson Bulletin. 33 pp.

Komar, O., Dueñas, C. and Rodriguez, W. (1993) Inventario de Aves Marinas de la Costa Salvadoreña. Secretaria Ejecutiva del Medio Ambiente (SEMA/CONAMA) Pub # 33: 71 pp plus maps.

Lemus, L.S., Pocasangre, J.A. and Zelaya, T.D. (1994) Evaluación del Estado Actual de la Distribución y Cobertura de los Arrecifes Coralinos en la Zona de Los Cóbanos, Departamento de Sonsonate. Universidad de El Salvador, Facultad de Ciencias Naturales y Matemática, Escuela de Biología. Tesis de Licenciatura. 40 pp.

Lessmann, K.W. (1986) Las Aguas Territoriales y la Morfología Litoral. In *Geografía de El Salvador*, pp. 123–247. Dirección de Publicaciones e Impresos de El Salvador.

Lluch-Cota, S.E., Alvarez-Borrego, S., Santamaría-del Ángel, E.M., Müller-Karger, F.E. and Hernández, S. (1997) The Gulf of Tehuantepec and adjacent areas: spatial and temporal variation of satellite-derived photosynthetic pigments. *Ciencias Marinas* **23** (3), 329–340.

Long, E.R., MacDonald, D.D., Smith, S.L. and Calder, F.D. (1995) Incidence of adverse biological effects within ranges of chemical concentrations in marine and estuarine sediments. *Environmental Management* **19**, 81–97.

Lovejoy, T.E. (1986) Species leave the ark one by one. In *The Preservation of Species*, ed. B.G. Norton, pp. 13–27. Princeton University Press, Princeton, NJ.

Marroquín, E. (1992) Diagnóstico de la situación actual y dinámica del deterioro del ecosistema estero-manglar. Secretaría Ejecutiva del Medio Ambiente. 51 pp.

Mearns, A.J., Matta, M., Shigenaka, G., MacDonald, D., Buchman, M., Harris, H., Golas, J. and Lauenstein, G. (1991) Contaminant trends in the southern California Bight, inventory and assessment. National Oceanic and Atmospheric Administration Technical Memorandum NOAA ORCA 62. Seattle, Washington.

Michel, J. and Henry, C.B. (1997) Oil Uptake and Depuration in Oysters After Use of Dispersants in Shallow Water in El Salvador. *Spill Science & Technology Bulletin* **4** (2), 57–70.

Michel, J. and Zengel, S.A. (1998) Monitoring of Oysters and Sediments in Acajutla, El Salvador. *Marine Pollution Bulletin* **36** (4), 256–266.

MINED (Ministerio de Educación de El Salvador) (1995) *Historia Natural y Ecológica de El Salvador Tomo I*, ed. F. Serrano. Editorial Offset, S.A. de C.V., Mexico, 398 pp.

Molina-Lara, O. and Vargas-Zamora, J.A. (1993) Estructura del macrobentos del estero de Jaltepeque, El Salvador. *Rev. Biol. Trop.* **42**, 165–174.

NOAA, National Marine Fisheries Service (1970–1972) EASTOPAC 1967-1968, A Cooperative Effort towards Understanding the Oceanography of the Eastern Tropical Pacific Ocean. Circular 330, Vol. 1–10, Washington D.C, USA.

Radtke, R.L. and Kinzie, R.A. (1996) Evidence of a marine larval stage in endemic Hawaiian stream gobies from isolated high-elevation locations. *Transactions of the American Fisheries Society* **125**, 613–621.

RPI (Research Planning, Inc.) (1995) Diagnóstico Ambiental en el Medio Costero Marino de la Zona de Acajutla. San Salvador, El Salvador, 75 pp.

RPI (Research Planning, Inc.) (1998a) Levantamiento y Mapeo de Indices de Sensibilidad Ambiental de la Linea Costero-marina entre las desembocaduras de los ríos Paz y Lempa, El Salvador (volumen 1). 65 pp.

RPI (Research Planning, Inc.) (1998b) Levantamiento y Mapeo de Indices de Sensibilidad Ambiental de la Linea Costero-marina entre las desembocaduras de los ríos Lempa y Goascoran, El Salvador (volumen 2). 75 pp.

Rivera-Muñoz, J.R. and Ibarra-Portillo, R.E. (1995) Estudio de los Poliquetos (Annelida: Polychaeta) del Estero de la Barra de Santiago, Ahuachapán, durante la Estación Lluviosa. Universidad de El Salvador, Facultad de Ciencias Naturales y Matemática, Escuela de Biología. Tesis de Licenciatura. 49 pp.

Sandifer, P.A., Dugue, R., Cotsapas, L., Malecha, S.R. and Ramos, J. (1992) Abundance of Juvenile Marine Shrimp in the Coastal Waters of El Salvador. 133 pp plus 54 Appendices. Report prepared for FUSADES (Fundación Salvadoreña para el Desarrollo Económico y Social), San Salvador, El Salvador.

Thurman, H.V. and Webber, H.H. (1984) *Marine Biology*. Bell & Howell Company, Columbus, OH, 446 pp.

Villatoro Vaquiz, O.A. (1977) Evaluacion de la poblaciones de tiburones en el El Salvador y comentarios para la sostenibilidad. Ministerio de Agricultura y Ganaderia, Centro de Desarrollo Pesquero, Division de Investigacion Pesquera, Nueva San Salvador, El Salvador. 70 pp.

Williams, A.R. (1999) After the Deluge. Central America's Storm of the Century. *National Geographic* **196** (5), 108–129.

World Resources Institute (1995) *World Resources 1994–1995*. Prepared by WRI in collaboration with UNEP and UNDP. Pub. No. WR94-95, Oxford University Press, New York.

THE AUTHORS

Linos Cotsapas,
Research Planning, Inc.,
1121 Park St., Columbia, SC 29201, U.S.A.

Scott A. Zengel,
Research Planning, Inc.,
1121 Park St., Columbia, SC 29201, U.S.A.

Enrique J. Barraza
Ministry of the Environment,
San Salvador, El Salvador

Chapter 35

JAMAICA

Marjo Vierros

Jamaica has a rich natural heritage, with a range of habitats and plant and animal species. The 795 km long coastline is highly irregular, with diverse ecosystems, including bays, beaches, rocky shores, estuaries, wetlands, cays, seagrass beds and coral reefs. Jamaica's shallow-water marine ecosystems suffer from a number of impacts, including severe overfishing, industrial and domestic pollution and sedimentation, much of which has its origins in the pressures caused by a fast-growing population.

These impacts are perhaps most clearly illustrated in the deterioration of Jamaica's coral reefs, caused by a combination of human and natural disturbances, including over-fishing of herbivorous fish, nutrient enrichment of the water, and a die-off of the grazing urchin *Diadema*. Long-term studies document a phase shift from a system dominated by corals to one dominated by fleshy macroalgae.

Hillsides in the coastal watershed are subjected to forest clearance, estimated to be occurring at a rate of 10,000 hectares per year, leading to run-off and the removal of 80 million tons of topsoil yearly from farms and forests around the country. In addition, coastal mangrove areas, wetlands and seagrass beds are being destroyed by development, as developers continue to show preference for coastal areas, especially along the island's North Coast. Seagrass beds are affected by dredging and filling for the construction of ports, buildings, large water-front industries, water channels, and roads. This has resulted in the destruction of large tracts of seagrasses in the vicinity of major ports and harbours.

There is appreciable drift of the population to urban areas, and some urbanised areas, especially around the capital, are affected by a wide range of domestic and industrial pollutants. Natural environmental variables causing marked effects include hurricanes which have had marked impacts on the coastal and shallow marine habitats, but there is also probably an impact from groundwater which is naturally nutrient enriched.

Jamaica's waters are severely overfished with a record number of 240,000 people reportedly earning a living from the island's 795 kilometres of coastline. Overfishing on the narrow coastal shelf became apparent as early as the 1960s, and today the local fishery supplies only a third of local demand.

Environmental conservation has been strengthened by legislation, such as the Natural Resources Conservation Act of 1994, and by increasingly active NGOs.

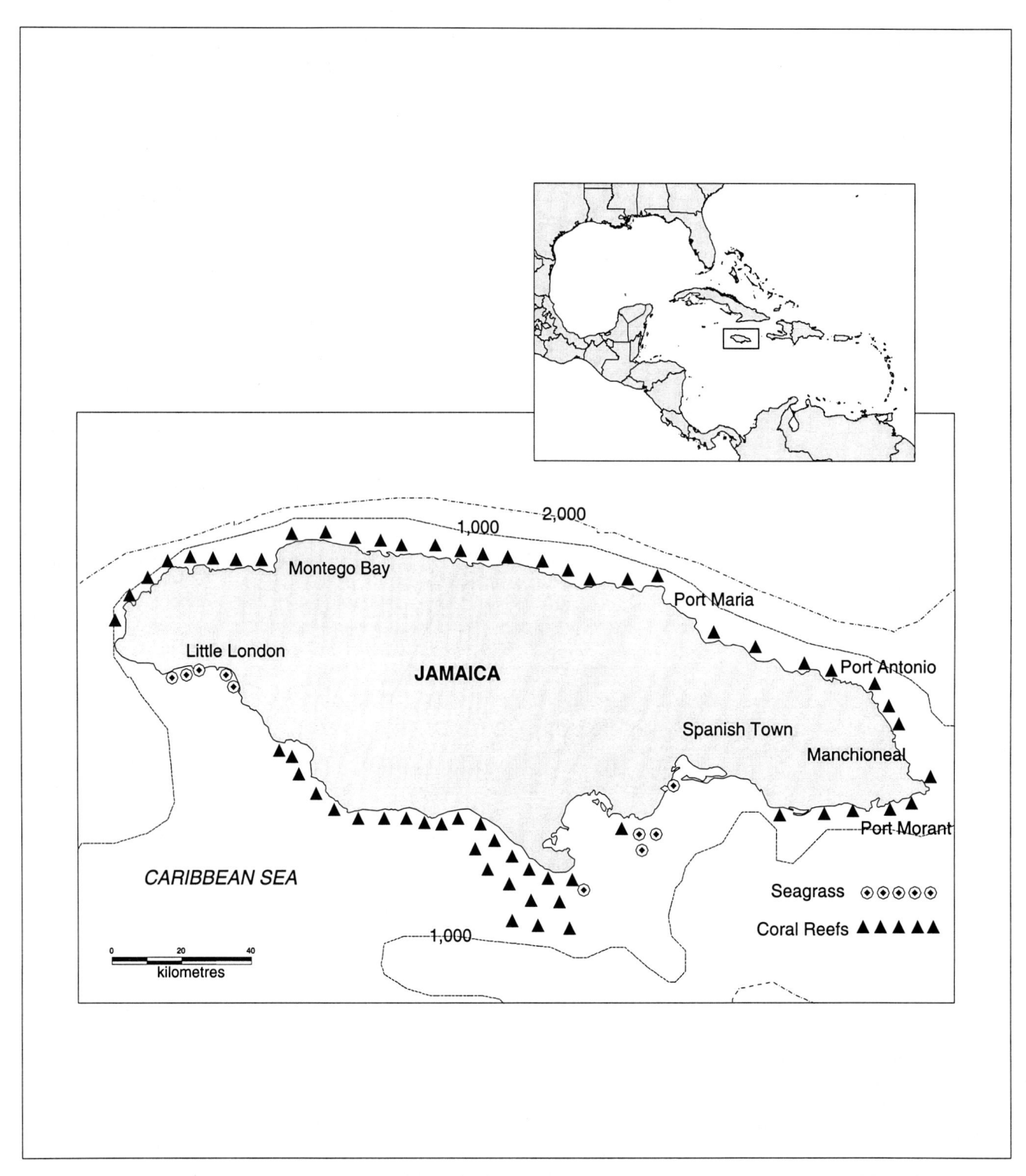

Fig. 1. Map of Jamaica, also showing distribution of shallow marine habitats. Note the narrow shelf on the north coast and the wider one on the south coast.

THE DEFINED REGION

Jamaica is located in the centre of the Caribbean Sea, 150 km south of Cuba and 161 km west of Haiti. It is the third largest island in the Caribbean, 235 km long and between 35 and 82 km wide, with a total land area of 10,939.7 km² (Fig 1). It has a highland interior with peaks and plateaux running the length of the island, surrounded by flat coastal plains. The high Blue Mountains dominate the landscape to the east, resulting in mountain rainforests and torrential streams, which transport coarse sediments to the coast. The highest point is Blue Mountain Peak at 2256 m. The valleys in this area are steep-sided and often heavily eroded. The western areas of the island consist of hills, plateaux and low mountains, and wide, sluggish rivers. Developed karst topography can be found in Cockpit Country, and to a lesser extent in the Dry Harbour and John Crow Mountains. Elsewhere the karst is less developed and the terrain takes the form of rolling hills, shallow sinkholes and ridges. The coastal plain is mostly less than 3.2 km wide, although coastal lowlands exist along the western and eastern ends of the island, as well as along areas of the south coast. Major swamp areas exist at the Upper Morass and the Great Morass in the southwest, and on Westmoreland Plain at the western end of the island.

Jamaica's 795 km of coastline is varied and irregular, indented with bays and extended by sand pits and bars. On the south coast, the low-lying shoreline is edged by long, straight cliffs, wetlands and black sand beaches. The north coast is rugged, with many white sand beaches, the most famous of which is probably the four mile stretch of beach along the west coast at Negril. Irregular cliffs formed by Blue Mountain foothills dominate the Portland coast, with well developed pocket beaches. Wave energy is very unevenly distributed around the island. The eastern, northeastern and southeastern coasts are exposed to high wave energy, while the western coast is very sheltered. The south coast has numerous sand cays, the most well known of which are located off Kingston. Much further offshore lie the larger Morant Cays and Pedro Cays.

The island, and especially the coastal areas, are under heavy development pressure. The population of the island has doubled in the last 30 years to about 2.5 million, of which 65% lives within 5 km of the coast. Industrial development is found in the southeast, around Kingston, and substantial tourism development exists on the north coast (Woodley et al., 1998) which is an internationally known tourist destination, and boasts white sand beaches and high-rise hotels in Montego Bay and Ocho Rios, as well as the famous Negril beach. More inland-oriented and intimate establishments in the Portland Parish are becoming increasingly popular. The Southwest coast from Milk River to Bluefield is another recreational and resort area, and is still relatively unexploited with open plains. The Black River area is an easily accessible network of lagoons, rivers and small lakes with an abundance of rare flora and fauna.

SEASONALITY, CURRENTS, NATURAL ENVIRONMENTAL VARIABLES

Jamaica's tropical maritime climate is modified by north and northeasterly tradewinds and by land–sea breezes. In low coastal areas average annual temperature is 27°C, while on Blue Mountain Peak it is 12°C. The coldest months are January and February, the warmest July and August. Rainfall is marked by monthly, annual and spatial variability; average rainfall for the entire island is 195.8 cm. The Blue Mountains and the northeast coast lie in the path of the tradewinds, receiving the highest annual rainfall of over 330 cm. Kingston, located in the lee of the mountain range, receives less than 127 cm annually. Water shortages are characteristic of the southern coastal lowlands, making irrigation necessary for agriculture. May and October are the wettest months, while March and June are the driest (NRCD, 1987).

The surface temperature of coastal water generally varies between 24°C in January and 29.8°C in July, although higher temperatures are also encountered. Tides vary in height between 20 and 36 cm, and are irregular, tending to double up so that they only occur once a day. The current along both coasts flows westward, increasing during the trade wind season between April and December (Wells, 1988).

Hurricanes and tropical storms periodically affect the island. From 1870 to Hurricane Gilbert in 1988, there were a total of 39 recorded hurricanes. Jamaican reefs had suffered little hurricane damage for more than 30 years until the onslaught of Hurricane Allen in 1980 (Woodley et al., 1998). Prior to 1980, Jamaica had a period of almost four decades without a major storm event. Hurricanes Allen and Gilbert, both category 5 storms, caused great damage along north Jamaica, especially at shallow coral reef sites, where the branching elkhorn and staghorn corals (*Acropora palmata* and *A. cervicornis*) especially suffered serious mortality (Hughes, 1994; Woodley et al., 1981).

THE MAJOR SHALLOW WATER MARINE AND COASTAL HABITATS

Jamaica has a rich natural heritage, with a range of habitats and plant and animal species, and ranks fifth amongst the islands of the world in the number of endemic species. Jamaica's coastline is highly irregular with diverse ecosystems, including bays, beaches, rocky shores, estuaries, wetlands, cays, seagrass beds and coral reefs (NRCA, 1997). Jamaica's shallow water marine ecosystems suffer from a combination of impacts, including overfishing, industrial and domestic pollution and sedimentation.

Charismatic marine species include manatees and sea turtles, which nest on the beaches and offshore cays. The population of manatees has not been surveyed since 1993, but it is generally believed that less than 100 individuals remain (NRCA, 1997). The American Crocodile (*Crocodylus acutus*) is found along the south coast. In addition, the mainland and surrounding cays support several breeding species of seabird (Wells, 1988).

Coral Reefs

Jamaica's coral reefs are among the best studied in the world. They may also be the longest directly observed submarine ecosystem, with data available since the 1950s, when T.F. Goreau and his associates initiated their studies (Goreau, 1992). Subsequent observations from Discovery Bay Marine Laboratory of the University of West Indies, as well as others, have added to the wealth of information that is now available about Jamaica's marine ecosystems.

According to Wells and Lang (1973) Jamaica is located at the centre of coral diversity in the Atlantic Ocean. Over 60 species of reef-building corals grow here, with fringing reefs occurring on a narrow, 1–2 km shelf along most of the north coast of Jamaica. Reefs also grow sporadically on the south coast on a much broader shelf of over 20 km wide (Hughes, 1994). In addition, reefs and corals can be found on the neighbouring banks of the Pedro Cays, 70 km to the south, and the Morant Cays, 50 km to the southwest (Woodley et al., 1998).

The reefs of Jamaica are recent structures, probably less than 5000 years old and superimposed on a succession of older, submerged shorelines and drowned reefs that mark the various eustatic low stands of the Pleistocene and Holocene sea levels (Goreau and Wells, 1967). The distribution of reef zones is strongly dependent on the reef profile. As a result, there are large variations in the distance of the reef crest from the shore along the narrow north coast shelf, which contributes to the variable width of the adjacent reef zones.

Jamaica's coral reefs have undergone large changes during the past thirty years. These changes present a case study in the devastating effects of combined human and natural disturbances on a reef system. Natural impacts, most importantly hurricane damage and *Diadema antillarum* die-off, started a major deterioration in Jamaica's coral reefs. Their recovery after these natural disturbances was prevented by chronic human disturbance, notably over-fishing, increased sediment and pollution runoff (Woodley et al., 1998). The phase shift from corals to algae occurred, showing how human and natural disturbances may reinforce one another (Woodley, 1995).

Goreau (1959) described a typical Jamaican reef zonation from early studies in Ocho Rios. The pattern shown, though typical in its time, has changed drastically since Hurricane Allen in 1980 and subsequent disturbances. The back reef region is composed of a shore zone and a lagoon or channel. The shore zone is composed of a rich scleractinian population on rocky shores to a depth of 3 m, while the lagoon zone is between 10 and 300 m wide with a sandy bottom, sparse coral populations and turtle grass *Thalassia testudinum*. Gorgonians, molluscs and echinoderms are abundant in this zone and the bulk of the bottom sediments are from foraminifera and *Halimeda* fragments.

The reef crest is composed of the rear zone, the reef flat and the *Acropora palmata* zone. The rear zone forms the inshore limit of the reef crest and rises abruptly from the sandy bottom of the lagoon at a depth of 2–3 m, and is the site of a rich and varied coral population. The reef flat is formed almost entirely of dead and unconsolidated colonies of branching corals (*Acropora palmata*). This zone averages 40 m wide and is less than 0.5 m deep. Exceptions are channels and pools where depth may increase to over 2 m. The reef flat gives way to a narrow zone populated almost entirely by large colonies of *Acropora palmata*, which take the full force of the surf. The reef slopes gently to depths between 5 and 7 m over a distance of 30–60 m, the lower *A. palmata* zone creating a kind of a moat since the reef becomes shallower again in the buttress zone. Coral density is much reduced here. Large areas of the bottom are covered in unconsolidated rubble. The deeper part of the zone has a more diverse coral population.

The buttress zone consists of spurs or buttresses of living coral, which project outward to deeper water, separated from each other by narrow canyons or grooves whose floors are 8–10 m deep. This zone extends from Port Antonio in the east to Montego Bay in the west, though there are some interruptions, and the width of the zone is variable.

The shallower forereef consists of large areas of staghorn coral, *Acropora cervicornis*. The zone is from 30 to 100 m wide and 8–15 m deep, broken by sandy tracts. There is high vertical relief. The lower forereef is characterized by *Montastrea annularis* and other massive corals, and dense coral may grow down to 70 m (Goreau and Wells, 1967).

It is possible that Goreau's descriptions of luxurious growth of *Acropora palmata* and *A. cervicornis* represented an atypical condition. According to Woodley (1992) the condition of the reef at that time followed a long hurricane-free period. The single longest period known with no hurricane activity is an interval of 36 years between 1944 and 1980, which is the time period of flourishing *Acropora* thickets, and also the time when most descriptions of these coral reefs were written. It is likely that these descriptions were in fact of an atypical situation.

The south coast reefs are very different. The area has a wide shelf with unconsolidated sediments. Corals form patch reefs without a buttress zone, and are surrounded by a sandy bottom. Many of the reefs are in poor condition, especially near Kingston Harbour and river mouths such as the Black River. Their corals are heavily stressed by polluted water and suspended river sediments. According to Goreau (1959) the south coast reefs may partly be erosional remnants of larger reefs, and the differences between reefs on Jamaica's north and south coast were due to differences in hurricane frequency, with the south coast being subject to more frequent hurricanes.

The classic north coast reef zonation patterns described by Goreau and colleagues in the 1950s–1970s no longer

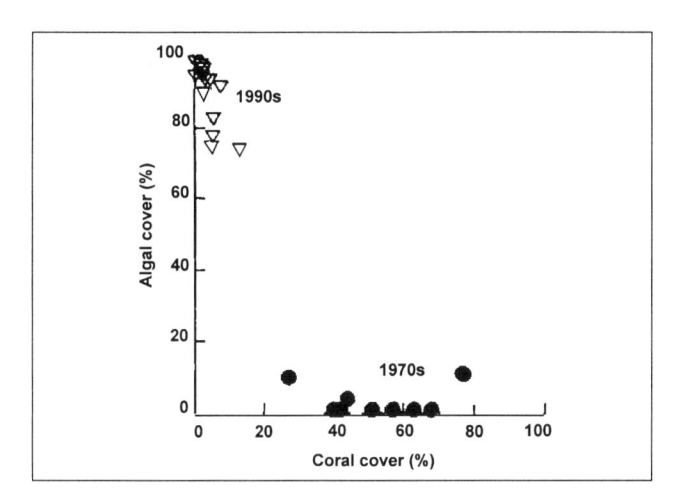

Fig. 2. Large-scale community phase shift from a coral-dominated to an algal-dominated system. The figure shows the change in percent cover of algae and coral between the 1970s and the 1990s (Hughes, 1994).

exist, and a striking phase shift has taken place from a coral dominated system to one dominated by algae (Hughes, 1994). The *Acropora palmata* and *A. cervicornis* zones (Goreau, 1959) are no longer recognizable (Hughes, 1993). These changes are due to multiple human and natural impacts. Woodley et al. (1998) measured coral cover on nine reef sites on the north coast in the late 1970s. The results averaged 52% coral cover at 10 m depth. In the 1990s the coral cover on these reefs had declined to 3%. During the same time period, the coverage of fleshy macroalgae had increased from 4% to 92% (Woodley et al., 1998). Similar results were also reported by other researchers (Fig. 2).

Reef deterioration was also evident from Goreau's (1992) report of the results of 40 years of ecological study at 11 sites around Jamaica. All of his sites showed clear reef degradation in terms of changing community composition, coral cover or growth rates. This degradation was not uniform. Some areas were highly affected by a combination of stress factors, while other areas were less impacted. Hellshire, for example, was reported to be the most intensely and longest stressed reef site in Jamaica, and the reason for its poor condition is the outflow from the extremely polluted Kingston Harbour. Similarly, other reefs close to major population centres, such as Ocho Rios, Montego Bay and Negril suffer from multiple stressors, most importantly sewage pollution, resulting in algal overgrowth. Sites in areas with low population densities, such as West Hanover showed minimum disturbance. The stress factors caused by human activities, such as sedimentation, sewage and industrial pollution, overfishing, bleaching, dredging and boats, anchor and diver damage, showed great variability between sites. The rate and character of observed reef degradation depended on the type, intensity and duration of stresses, and Goreau concluded that the net impact of all stresses is the accelerating decline of all reefs around the island.

Although the human stresses on coral reefs were already present in the 1970s and earlier, there appeared to be little visible coral deterioration prior to Hurricane Allen in 1980, which severely damaged the reefs of north Jamaica. These are normally sheltered and had not been impacted by a severe hurricane since 1917. Data collected by Woodley et al. (1981) during the weeks following the hurricane indicated that the damage to the reefs was patchy on several scales. Damage varied locally between reef zones and within zones. Damage was greatest in shallow sites and among branching species, while deeper than 10 to 15 m, encrusting and massive shaped colonies fared better (Hughes, 1994). Damage, especially in exposed areas, was caused by the force of the waves, recorded at Discovery Bay at over 12 m, and by the dislodged material carried by the waves. Dense stands of *Acropora palmata* and *Acropora cervicornis* were levelled and physical disturbance was recorded down to 50 m deep.

The hurricane exposed large amounts of substratum by abrasion, erosion and fracture. This increased the area available for recruitment and growth of sessile organisms. A bloom of the green algae *Trichosolen duchassaingii* occurred in shallow water within a week, but subsided after a month. Near total mortality of *Acropora cervicornis* occurred, and *A. palmata* did not fare much better. Hurricane Allen created overnight patterns of distribution and abundance strikingly different from pre-existing states (Woodley et al., 1981). Destruction of the tall branching *Acropora* species caused a severe reduction in reef structural complexity. The reef at Discovery Bay had been highly complex prior to the hurricane, with average spatial indices from 3.8 m/m at 3 m to 2.8 m/m at 10 m. After Hurricane Allen, the reef architectural complexity was reduced to 1.6 and 1.4 m/m respectively (Steneck, 1993). Coral cover was also reduced. In 1977, Hughes (1993) recorded coral covers between 47% and 70% near Discovery Bay, but after Hurricane Allen, this had fallen to 22–38%, and the relative abundance of species had changed significantly.

Within a few months, substantial coral recruitment began, and while recruitment by *Acropora* was minimal, other corals settled onto the free spaces and the reef began to recover. During the following 3–4 years coral cover increased slowly, rising from 22% in 1981 to 29% in 1984 at a study site in Rio Bueno (Hughes, 1993). However, the species composition changed; two weedy species, *Agaricia agaricites* and *Briarium asbestinum* more than doubled their cover from 1981 to 1983 (Hughes, 1993).

All this changed, however, when a mass mortality of *Diadema* took place in the Caribbean between 1982 and 1984, caused by a species-specific pathogen. The effects were seen in Jamaica in 1983 with far-reaching consequences to the coral reefs of the island. Jamaica had an unusually high abundance of *Diadema* prior to the die-off. Even though overfishing had been taking place since the 1960s, the abundant *Diadema* grazed down the algae, allowing corals to dominate. The high abundance of *Diadema* was itself likely to be due to overfishing of its predators (Hughes, 1994).

Diadema populations fell abruptly by mid-August 1983 by almost two orders of magnitude (Hughes, 1993). When the *Diadema* died, macroalgae grew over the reefs, smothering living hard corals and preventing new coral larvae from settling. Coral larval settlement failed and most of the adult colonies that survived Hurricane Allen have been killed by algal overgrowth (Hughes, 1994). Now several algae form extensive mats up to 10–15 cm deep, now covering up to 92% of the substrate compared with 4% previously.

The current depleted fish stocks are not sufficient to reduce algal abundance in the absence of *Diadema*. The most abundant coral on the forereef today is *Montastrea annularis*, but even this species has declined to 0–2% cover at depth of 10 m in 1993. Future hurricanes will likely reinforce rather than reverse the phase shift, as illustrated by the impact of Hurricane Gilbert in 1988 (Hughes, 1994).

However, reduced herbivory alone could not have caused the massive macroalgal blooms witnessed in many locations. Lapointe (1998) presents evidence that nutrient enrichment in Discovery Bay was a major contributing factor as well. The nutrient concentrations measured in Discovery Bay are some of the highest reported for coral reefs worldwide and help explain why such impressive standing stocks of macroalage have developed on the reef system.

The potential for eutrophication at Discovery Bay was noted by D'Elia et al. (1981), and is associated with increased population growth and sewage pollution. Nutrient concentrations further increased in the back reef area during the 1980s and spread offshore, elevating nutrient concentrations above thresholds found to sustain maximum productivity of macroalgae. Thus the phase shift from corals to macroalgae was directly linked to nutrient concentrations. It is likely that the naturally high concentrations of NO_3^- in groundwater in Jamaica have been increased by deforestation, agricultural and industrial development and sewage contamination. The locations of macroalgal-dominated habitats (Hughes, 1994) indicate large-scale coastal nutrient enrichment, not only from groundwater discharges, but also from river discharges. It appears that control of both fishing practices and nutrient reduction are needed to restore Jamaica's coral reefs (Lapointe, 1998).

A second major hurricane, Gilbert, impacted the island in 1988. This caused additional damage to corals, although less than Hurricane Allen. Hurricane Gilbert also reduced algal cover and biomass substantially. However, this effect was short-lived and the algal cover returned to previous proportions within a few weeks (Hughes, 1994). Many deeper sites lost almost half their coral cover during 1988. In 1993, in Discovery Bay, coral cover was less than 3% at many sites, while algal cover everywhere was greater than 90% (Hughes, 1993).

Jamaica's coral reefs continue to suffer from combined human-induced and natural stresses. With the growth of human populations, nutrient pollution has increased. This is particularly evident near Kingston, where the pollution plume from Kingston Harbour has contributed to increased coral mortality west of the harbour. Soil erosion, and sedimentation from rivers, has been a serious problem in Jamaica for 50 years. Healthy coral populations can be found around Port Royal Cays, where coral cover up to 20% can be found. Mass bleaching took place in Jamaica during 1987, 1989 and 1990, with considerable mortality, and wide-spread bleaching was also recorded in 1998 (Woodley et al., 1998).

Some coral recovery has taken place in a few locations. Mendes et al. (1999) report an increase in live coral cover from approximately 12% in 1989 to over 29% in 1999 at depths between 1 and 8 m on the fringing reef at Lime Cay, just south of the capital city of Kingston. This is presumed to be due to an increase in the abundance of *Diadema* in the shallow reef areas. Lime Cay became a part of a protected area in 1998, and is up-current of the polluted outflow from Kingston Harbour. Reef restoration efforts are also under way in Montego Bay, which is both severely impacted and one of the leading tourist destinations in Jamaica. Long-term restoration efforts include establishing strategic partnerships, integrated coastal zone management decision support modelling, watershed management, sewage treatment interventions, an alternative income programme, zoning and fisheries management, monitoring, education, volunteer and enforcement programmes (Jameson et al., 1999).

Watersheds and Wetlands

Jamaica has four large coastal wetland areas called morasses. The Negril Morass is located on the west coast, the Black River and Portland Bight Morasses are located on the south coast, and the Cape Morant morass is on the eastern end of the island. In addition, there are many smaller wetlands.

The Negril Morass is isolated from the sea by a solid sandy beach barrier with a few drainage outlets, which were converted into canals in the late 1950s. Hendry (1982) studied its evolution, and concluded that the peat has accumulated at the rate of the rising Holocene sea level. Coke et al. (1982) identified twelve vegetation types. The back of the coastal barrier is fringed with Sabal forest and the interior area is dominated by a *Cladium–Sagittaria* association and hummocky swamp. Typical mangrove vegetation of *Rhizophora, Avicennia, Conocarpus* and *Laguncularia* is found only in the northwestern part, particularly around the North River Canal. It is likely that construction of the North Canal resulted in the destruction of mangrove communities to the north of Bloody Bay (NRCA, 1996). The uncontrolled development on the banks of the South Negril River has quite likely placed further stress on existing mangroves in this area, due to the disposal of domestic waste. High faecal coliform levels measured at the mouth of the canal give an indication of the magnitude of the problem (NRCA, 1997).

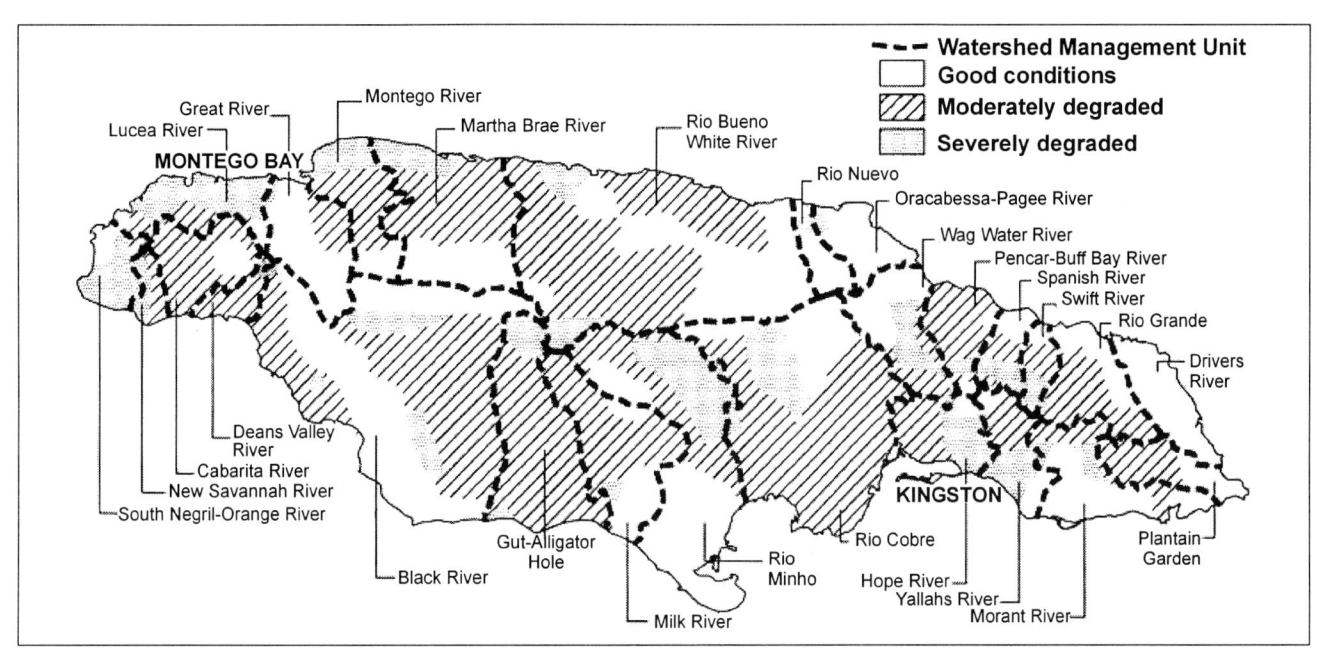

Fig. 3. The condition of Jamaica's watersheds (NRCA, 1997).

The Black River Morass is composed of three basins, connected to a fourth basin north of Santa Cruz by the Black River. The main basin area consists of a network of meandering rivers, which have a common outlet at the town of Black River. South of Black River, a beach barrier cuts the morass off from the sea. A vegetation map (Coke et al., 1982) distinguishes twelve vegetation types. The mangrove vegetation, which characterizes the lowest basin is dominated by a *Cladium–Conocarpus* assemblage, a *Scirpus oleneyi* zone close to Broad River and a zone of mangrove forest. The *Cladium–Conocarpus* zone represents a transition between freshwater swamp and mangrove, and occupies most of the area north of the lower part of Broad River and also large areas between the mangrove forest south of the river. Extensive mangrove forests are found between Broad River and the coast, and north of the river. These forests are dominated by *Rhizophora* and *Conocarpus*. The coastal mangrove fringe along the back of the coastal barrier is formed by *Rhizophora*, *Avicennia*, *Laguncularia* and *Conocarpus*. There are also important mangrove areas west of Black River. The most obvious human impact is found along the upper part of Black River, where the swamp forest has been cleared and the levee areas along the rivers are used for cultivation and cattle grazing. The banks of the lower river are affected by waves from tourist boats on crocodile-watching excursions.

Coastal wetlands fringe the coast of Portland Bight from Portland Ridge in the southwest to the southern end of Hellshire Hills, but are absent on the stretch of coast between Port Esquivel and Old Harbour Bay. This is also the area of strongest human impact, caused by shipping and effluents from harbours, agriculture and municipalities. There are no major rivers in the area. In addition to the wetlands, there are also patchy coral reefs and cays of great natural interest. In the bay between Rocky Point and Portland Ridge, the West Harbour basin is connected to the open Bight through a myriad of channels between mangrove stands. This is probably the ecologically most dynamic coastal zone in Jamaica. The area around Little Goat and Great Goat Island, and the nearby mangrove headland of Carbarita Point, have numerous lagoons and minor channels.

Each of Jamaica's 26 watershed management units have portions that are considered degraded, as can be seen in Fig. 3. Natural and plantation forests are usually located within the upper reaches of watersheds. Removal of these trees can have serious impacts on low-lying areas, including increased flooding, sedimentation, altered river courses, and reduction in aquifer recharge and available water supplies. It is commonly believed that Jamaica's rivers are drying up. While there may not be agreement on how many rivers no longer flow, there has been a trend towards both reductions in the flows of many rivers as well as increase in the intensity of flooding. Both are attributed to loss of forest cover in watersheds (NRCA, 1997).

OFFSHORE SYSTEMS

Offshore, the Pedro Bank to the southwest and Morant Cays to the southeast of Jamaica, are important fishing and bird breeding areas. The Morant Cays, located approximately 60 km southeast of Morant Point on mainland Jamaica, consist of four isolated low-lying islands with sparse vegetation and surrounding reefs and shoals. They are located on a ridge rising from the seabed at 1000 to 1500 m depth. The total area less than 100 m deep is 100 km^2. The topography to windward of the main reefs is steep, while

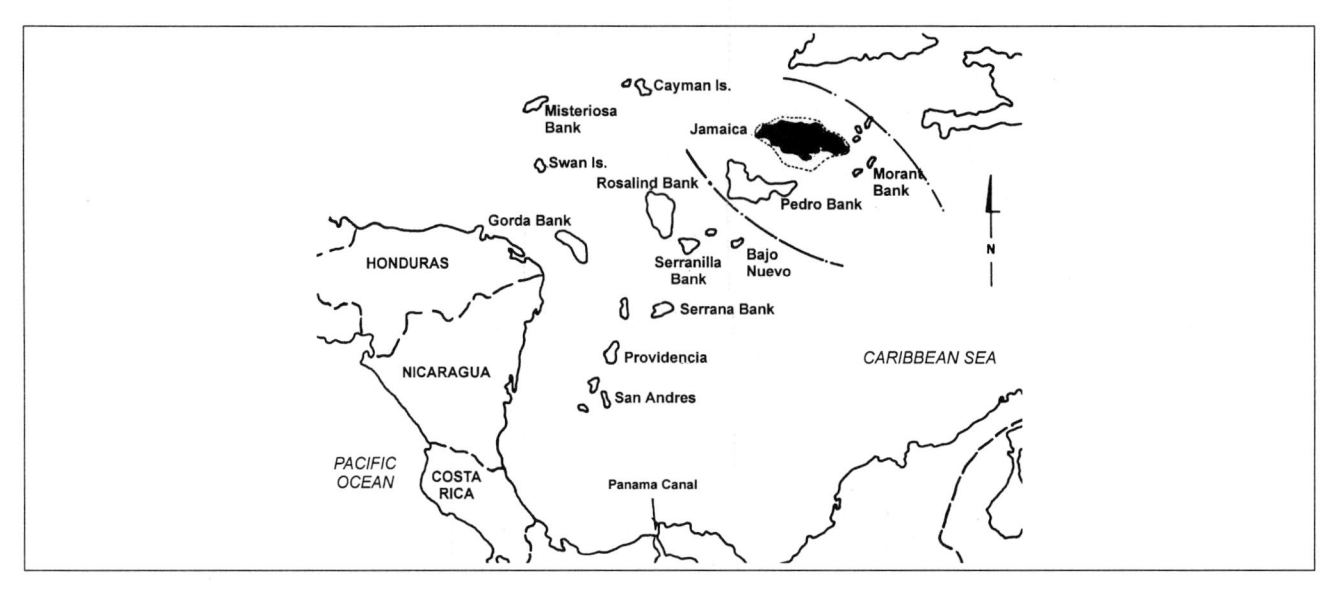

Fig. 4. Location of Jamaica's inshore and offshore fishing areas (NRCD, 1987).

large shallow shoal areas are found leeward. The reefs are exposed to heavy wave action and strong currents. Little is known about the reefs and the oceanographic conditions. There are nesting seabird populations on the island (Wells, 1988).

The Pedro Cays are a large shallow water bank about 100 km to the southeast of Jamaica. The total shelf area less than 50 m deep is about 8000 km^2, and rises from the seabed at 800 m or more. Most of the bank has water depths between 5 and 40 m, and the submarine topography is relatively flat. There are patches of corals and algae, coral rubble, sand and silt. Four low-lying cays are located on the southeast end of the Bank. Strong seas and currents are prevalent. Zans (1958) surveyed a section of the reef around North East Cay and found simple reefs, arising from a sandy platform at 6 m depth on both forereef and backreef (Wells, 1988).

Both of the banks are important offshore fishing areas (Fig. 4), with groups of fishermen resident on the cays. In addition to the finfish fishery, consisting primarily of trap fishing, conch harvesting also takes place on the Pedro Banks. The fishermen are serviced by carrier vessels from Kingston, with whom they trade for cash, fishing gear and supplies (Aiken, 1992). In 1984, there were approximately 350 fishermen in residence on the Pedro Banks and up to 100 in the Morant Cays (Wells, 1988). This number has increased considerably since, placing the area under a high degree of environmental stress. The fishing pressure in the area is increasing, and there is a need for management of both the terrestrial and marine areas. The banks' considerable distance from the mainland makes policing and enforcement difficult.

The deep habitats are generally not well known. Lang (1974) described those off Discovery Bay down to a depth of 300 m. According to these studies, the major biological zones between 55 and 300 m at Discovery Bay can be broken down into two main sections, the deep fore-reef (55–120 m) and the island slope (120–300 m) (Fig. 5). The primary reef framework between 55 and 70 m is constructed by hermatypic corals, which cover the upper surfaces of many ledges. Sclerosponges are abundant underneath ledges, while demosponges, crustose red algae and *Halimeda* spp. are common on vertical surfaces. Between 70 and 100 m the primary framework constructors are sclerosponges, which grow under ledges and caves. On vertical surfaces demosponges and crustose red algae predominate. Both hermatypic corals and halimeda are rare.

Between 100 and 120 m the base of the escarpment is constructed primarily of lithified reef sediments and debris. Demosponges and crustose coralline algae grow over vertical reef surfaces, while most horizontal and inclined surfaces are covered with unconsolidated reef sediments. On the island slope, the area between 120 and 200 m is made up of both hard substrates and unconsolidated sediment surfaces. Encrusting and lithistid demosponges, crustose red algae and ahermatypic corals cover the vertical surfaces of the hard substrates. On horizontal surfaces, stalked and stalkless crinoids, ahermatypic corals, gorgonians and antipatharians are common. The unconsolidated sediment surfaces contain a few epibenthic demosponges, stalked crinoids, starfish, urchins and hermit crabs. In occasional pockets of thick sediment, burrows and animal trails can be seen.

Between 190 and 300 m the vertical surfaces of limestone blocks are covered with encrusting and lithistid demosponges, ahermatypic corals, stylasterine corals and *Holopus rangi*. There are many apparently barren areas. Fauna on horizontal hard substrate surfaces includes ahermatypic corals, stalked crinoids and alcyonaceans. The predominantly fine-grained sediments contain large populations of infaunal organisms, including worms and cerianthid anemones. Hermit crabs and other motile scavengers and grazers are abundant (Lang, 1974).

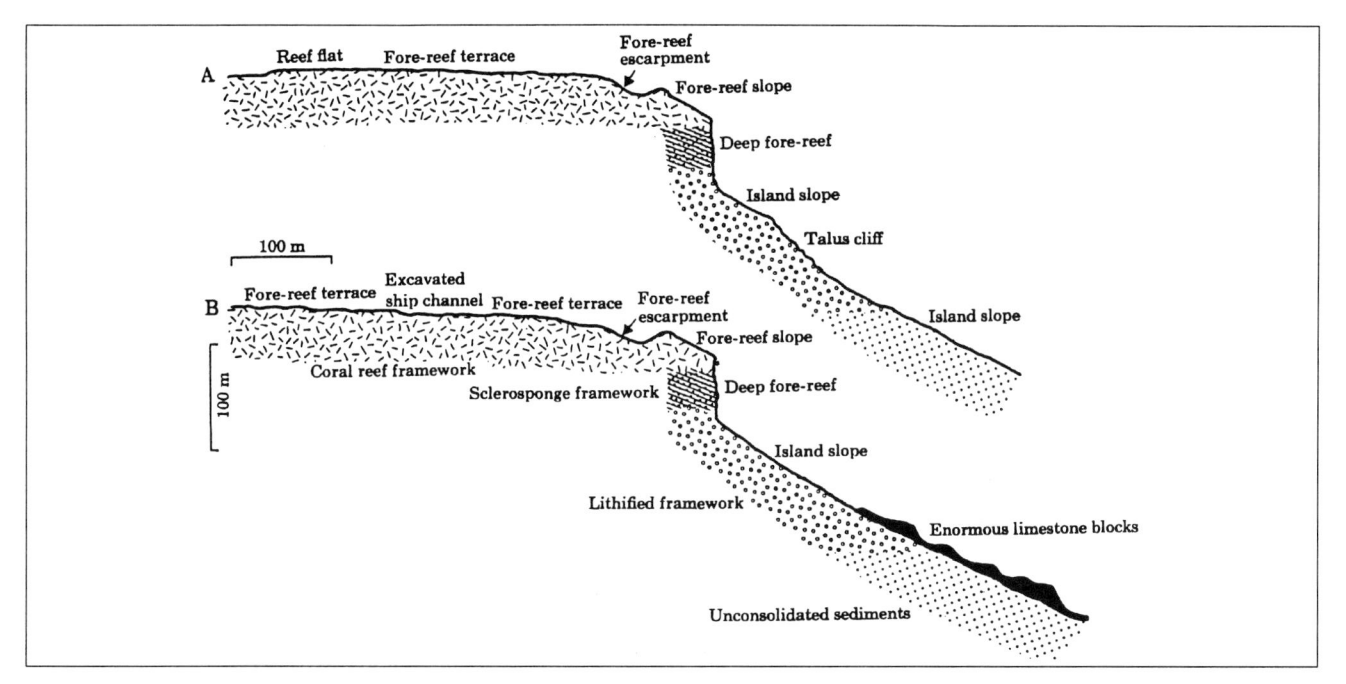

Fig. 5. Profile diagrams of two transects (A and B) off Discovery Bay harbour down to a depth of 300 m (Lang, 1974).

POPULATIONS AFFECTING THE AREA

Jamaica's population in 1999 numbered 2.6 million (World Population Reference Bureau, 1999) with an average population density of 230 persons km^{-2}. The density is unevenly distributed. At the end of 1996, approximately 50% of the population of Jamaica lived in urban areas, with Kingston and the parishes of St. Andrew and St. Catherine being home to over one million people, with a population density of 1528 persons km^{-2}. Approximately 65% of the total population lived within 5 km of the coast (NRCA, 1997).

There is a high incidence of rural to urban drift. Migration to the city often leads to poverty, as is evident in overcrowded tenements and an increasing number of squatter settlements. Population pressure in urban areas places a severe stress on the water supply, sewerage systems and garbage disposal (NRCA, 1997). The National Plan of Action on Population and Development (PIOJ, 1995) reports "In Jamaica, the urban system is characterized by the overwhelming preponderance of a few major urban centres including the Kingston Metropolitan Region (KMR). The KMR includes the Kingston and St. Andrew Metropolitan Area (KMA) and Portmore. The continued concentration of population in the KMR and other major urban centres poses specific economic, social, infrastructural and environmental challenges for the government." Evidence of this is seen in water shortages and the pollution of coastal waters by sewage and other sources near urban areas, including Kingston and Montego Bay.

Jamaica's population growth is similar to that of many developing countries. Before 1870, the population was less than half a million. By 1925, the population had doubled, and had done so again by 1975 (Hughes, 1994). It is expected that at the current rate of population the "doubling time" is 40 years, and that by the year 2010 the population will have reached 2.9 million, and by the year 2025 it will be 3.2 million (World Population Reference Bureau, 1999). However, migration away from the island at an estimated rate of about 18,000 per year helps keep the population growth rate at 1.1%. It has been estimated that this is sustainable and can be supported by the nation at satisfactory standards of living (PIOJ, 1995). Still, in 1996, approximately 26.1% of the population lived below the poverty line (NRCA, 1997).

There is a clear relationship between population and the environment. With the faltering economy and scarce jobs, more people turn to natural resources for their livelihood. A growing population with limited access to land and other natural resources often leads to squatting and farming on unsuitable areas, such as steep slopes, gullies and wetlands (NRCA, 1997). The population also places stresses on sewage disposal. Of the 112 sewage treatment facilities monitored in 1996, only 30% were operating satisfactorily (Environmental Control Division, 1997). The result is seen as deteriorating water quality in the vicinity of urban areas.

In addition to the resident population, a substantial number of tourists visit Jamaica every year. The tourism sector continues to be a major source of income for the country, bringing foreign exchange earnings and employment opportunities. In 1996 there were 180 hotels, 275 guesthouses, 978 resort villas, and 436 apartments (NRCA, 1997). The tourism industry places many demands on the environment, including the clearance of seagrasses from swimming beaches, the clearance of wetlands for development, and pollution caused by inadequate waste disposal.

RURAL FACTORS

Jamaica's reef fishery is severely overfished with a record number of 240,000 people reportedly earning a living from the islands 795 km of coastline (Neufville, 1999).

Overfishing on the narrow coastal shelf became apparent in the 1960s (Woodley et al., 1998). Fish catches are being reduced by increasing numbers of fishermen, poor fishing techniques, such as the use of fine mesh nets to trap immature fish, as well as illegal dynamiting and poisoning. The 1996 finfish catch was estimated to be 14,500 metric tons (NRCA, 1997).

The demand for fish and fish products is great and growing. The local marine fishery supplies only approximately one-third of the total demand. Additionally, unemployment is high and fishing is one of the few job opportunity options for a large segment of the coastal population. The fishery of Jamaica is essentially artisanal, and is mainly conducted by fishermen in small vessels. Fish pots are used extensively, as are nets and hook and line. Spearfishing, illegal dynamiting and poisons are used to a smaller degree. The fish traps in Jamaica are usually called Z-traps or Z-pots, and are generally covered with hexagonal wire mesh with a minimum aperture of 3.18 cm. The average Jamaican Z-type fish pot measures 230 × 120 × 60 cm (Aiken, 1992).

There is evidence that fish populations in Jamaica were already decimated more than 100 years ago (Jackson, 1997). Munro (1983) showed that by the late 1960s, fish biomass had been reduced by up to 80% on the narrow fringing reefs of the north coast. This was mainly the result of intensive artisanal fishtrapping. By 1973, the number of canoes deploying fish traps was well above sustainable levels. The taxonomic composition of fish had also changed alarmingly. Large predators, such as sharks, *lutjanids* (snappers), *carangids* (jacks), *ballistids* (triggerfish) and *serranids* (groupers) had virtually disappeared on the north coast. The herbivorous fish, such as *scarids* (parrotfish) and *acanthurids* (surgeonfish) were small. As a result, fully half of the species caught in traps are below minimum reproductive size. Further to this, because adult stocks on the north coast of Jamaica have been reduced for several decades, populations today may rely heavily on larval recruitment from elsewhere in the Caribbean. A similar decline in stocks has been seen recently on the south coast. A modernization of the fishing fleet with motorized canoes has made the wide shelf increasingly accessible, reflected in a halving of the catch per unit effort (Hughes, 1994). Herbivorous fish are now commonly served in restaurants (Steneck, 1993).

Fishing impacts coral reefs both directly and indirectly. Direct impacts include the removal of organisms and habitat damage from destructive fishing practices. Indirect impact includes removal of important components of the ecosystem, such as predators and herbivores, which could disrupt ecological relationships (Bohnsack, 1993). Significant differences have been shown between reefs in Jamaica under different levels of exploitation (Munro, 1983; Koslow et al., 1988). Knowlton et al. (1990) reported that predation from a snail (*Coralliophila*), a polychaete (*Hermodice*) and a damselfish (*Pomacentrus*) prevented recovery of staghorn corals (*Acropora*) after a hurricane. Fishing could also have potentially aggravated coral mortality by reducing the natural predators of these species. Although the precise role of fishing in these events is unknown, Jamaica is certainly intensively exploited (Munro, 1983; Koslow et al., 1988 in Bohnsack, 1993).

Jamaica is primarily an agricultural country, and agriculture employs 36% of the population. Yet agriculture presently contributes greatly to environmental degradation. The use of steep hillsides for agriculture is common, and the clearing of unstable slopes for cultivation results in serious soil erosion, especially when no proper soil conservation techniques are used. Eighty million tons of topsoil is lost yearly from farms and forests around the country. The use of slash-and-burn methods often causes forest fires. In addition, misuses of agricultural chemicals, including pesticides, herbicides and fertilizers are contributing to water pollution. Traces of agricultural chemicals are found in groundwater, rivers and the sea. The use of chemicals, including pesticides, herbicides and fertilizers in agriculture continues to increase (Fig. 6). The misuses of these chemicals are contributing to water pollution (NRCA, 1997).

Jamaica's approximately 267,000 ha of forests are under severe threat due to land clearing for cultivation, fuel wood, round log stakes and charcoal production. The extent of deforestation is not fully known (Evelyn, 1997). In 1995, deforestation was estimated to be occurring at a rate of 10,000 ha per year. Less than 6% (77,000 ha) of Jamaica's forests are relatively undisturbed. The remainder is badly disturbed secondary forest (169,000 ha) or plantations (21,000 ha). Efforts to improve the management of forests and watersheds have been made through the Forest Act of 1996 and the Water Resources Act of 1995 (NRCA, 1997).

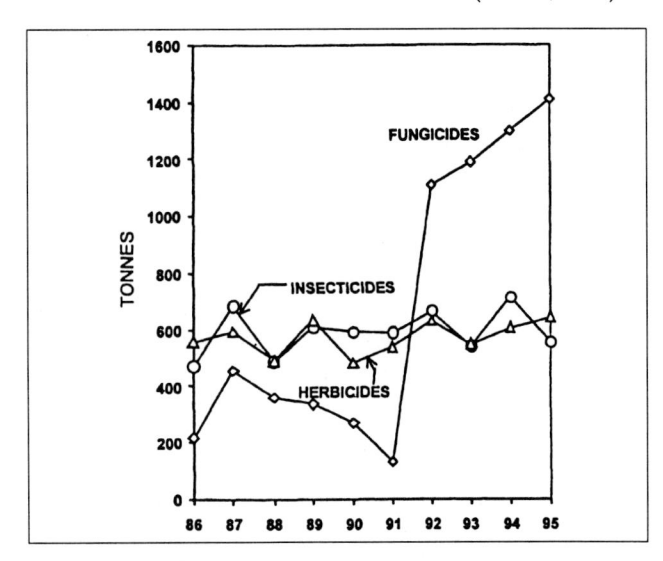

Fig. 6. Increase in the import of agricultural fungicides, insecticides and herbicides between 1986 and 1995 (NRCA, 1997).

The development of mariculture in Jamaica had its beginnings in the 1977 Oyster Culture Project which was a joint effort of the Ministry of Agriculture and the University of West Indies Department of Zoology. The project studied the culturing of mangrove oysters (*Crassostrea rhizophorae*) in Bowden Bay, St. Thomas and subsequently expanded to include Davis Cove at Green Island in Hanover, East Harbour at Port Antonio in Portland, and Bogue in St. James. Apart from the activities stemming from the Oyster Culture Project, there is only one other active mariculture enterprise in Jamaica, a privately owned tilapia farm utilizing seawater, producing approximately 3.4 million kg of fish. Other options for development in mariculture include sea moss culture, establishment of a marine shrimp hatchery, and the cage culture of finfish (NRCA, 1998).

COASTAL EROSION AND LANDFILL

The coastal environment is being altered to provide facilities for tourism and related uses. As a consequence, beach and coastline erosion is accelerating, aggravated by mining of sea sand. In addition, coastal mangrove areas, wetlands and seagrass beds are being destroyed by development. Developers continue to show preference for coastal areas, especially along the island's North Coast (NRCA, 1997).

Loss of sand from beaches may occur as a result of several factors. These include sand mining and quarrying of aggregate from river channels and beaches; physical development causing erosion as a result of an alteration in either wave pattern or in the pattern of inshore currents; wave wash from passing power boats; and storm damage. Illegal sand mining is a long-standing problem, which is continuing due to the lack of enforcement, and because offenders operate at night and are seldom seen. Shoreline alterations are undertaken by many developers in an attempt to improve their property. Frequently, because of a lack of knowledge of coastal engineering and coastal oceanography, these developments cause shoreline erosion and remedial measures have to be put in place by constructing groynes or other structures. In addition, erosion arising from the wash of power-boats demands that proper regulations for the use of power boats in near-shore waters are promulgated and enforced (NRCA, 1999).

One of the most common sources of damage to seagrasses has been dredging and filling for the construction of ports, buildings, large water-front industries, water channels, and roads. This has resulted in the destruction of large tracts of seagrasses in the vicinity of major ports and harbours, such as Kingston Harbour and environs, and Montego Bay. Ports and marinas must continuously dredge and dispose of spoil to keep important inlets and channels open. In addition to the direct dredging of seagrasses or the dumping of fill on top of the beds, this activity causes increases in water turbidity, causing declines in adjacent coral reefs and seagrasses (NRCA, 1998b).

Coastal areas typically have complex land use and ownership patterns, and are affected by an overlapping of responsibilities and management. Both physical and visual access to the shoreline is decreasing as a result of development. There is a perception that the best beaches have been taken over by hotels for their exclusive use. Coordination of both development and protection efforts can involve numerous stakeholders and needs a considerable amount of time and money. During 1995–1996, the NRCA prepared guidelines for dredging, marinas and small craft harbours, benthic structures, coastal protection and enhancement structures, and underwater pipelines and cables. These, in combination with the new NRCA Permit and Licence system should help reduce the negative impacts of coastal development (NRCA, 1997).

EFFECTS OF URBAN AND INDUSTRIAL ACTIVITIES

The disposal of solid, liquid and hazardous waste continues to pose environmental problems. An estimated 10,000 tonnes of hazardous waste, consisting mainly of waste oils, is generated in Jamaica yearly. Sewage generation is estimated at 455 million litres per day. Of this, only about 25% is collected and treated in conventional treatment systems and the remainder is disposed of using pit latrines, soakaways and septic tanks. About 51% of the Jamaican population still uses pit latrines. Sewage effluent and industrial waste are contaminating aquifers and surface waters at an increasing rate (NRCA, 1997).

Because of the cost of installing treatment plants, and the difficulties of maintaining them in good operational order, many coastal properties rely on soakaway pits, tile fields or septic tanks for sewage disposal. However, most of these properties are situated on porous limestone bedrock, with the result that sewage disposal systems frequently leach nutrients and sometimes bacteria into the adjacent coastal water (NRCA, 1999).

Water quality on Jamaica's recreational beaches has been monitored by several agencies and has been consistently good except for a few urban beaches where "spikes" associated with periods of heavy rains are experienced. With increasing urbanisation, and the associated growth of informal settlements, concerns have grown as to whether the existing coastal water monitoring arrangements are adequate (NRCA, 1999).

The sugar and rum industry are a major source of water pollution. These industries have made a commitment to rid the country of dunder by the early 21st century. Other industries, which have made commitments to anti-pollution investments include the cement company, coffee industry, and the power sector. Bauxite has long been the major mineral resource in Jamaica, and has contributed significantly to the country's economy. Bauxite mining produces atmospheric dust and noise pollution. An additional source of pollution has in the past been the disposal of red mud residues, which resulted in the seepage of caustic solutions in the groundwater. This practice has now been replaced by more environmentally friendly dry mud techniques (NRCA, 1977).

Oil pollution sometimes results from poor environmental practices at garages and other work places. At times, oil is deliberately dumped in storm gullies, rivers, streams, etc., and ultimately washed to sea. Oil washed onto the beach frequently results in the formation of "tar balls", which cling to the feet, damage clothing, damage fishing nets and cause fouling of fishermen's boats.

Due to the prevailing north-easterly trade winds, any ship disaster north or east of Jamaica might result in a major pollution incident on northern tourist beaches. A national oil spill emergency plan and the Office of Disaster Preparedness and Emergency Management, the Natural Resources Conservation Authority and the Jamaica Defence Force Coast Guard are responsible for implementing the Plan (NRCA, 1999).

Air pollution in areas of high population density, such as Kingston and Montego Bay continues to create major health problems with the local population. Air pollution is caused by vehicle emissions, emissions from industrial sources, and the burning of garbage (NRCA, 1977).

An extreme case of pollution caused by urban areas is illustrated by the case of Kingston Harbour. Kingston Harbour is widely regarded as one of the finest natural harbours of the world, and Port Bustamante, located in the northwest corner of the harbour, is one of the busiest trans-shipment ports in the Caribbean region. The deterioration of Kingston Harbour had begun to attract public comment and scientific investigation approximately 30 years ago. In 1970 and 1976, Goodbody and Wade, two scientists at the University of West Indies, carried out studies and published disturbing reports concerning changing ecological conditions in the harbour. These studies were followed by, for example, Webber and Rolf (1996), which tracked the influence of Kingston Harbour outflow to stations offshore and on the Hellshire coast.

Monitoring Kingston Harbour has revealed that major sources of pollution are sewage and industrial effluent discharged directly into the Harbour, or into the gullies and rivers that enter it. In response, the Kingston Harbour Rehabilitation Project was launched in 1996. The Project started by targeting specific pollution loads and developing institutional capacity to sustain a recovery effort (NRCA, 1997).

PROTECTIVE MEASURES

The Beach Control Act (1960) was for many years the main law controlling coastal development. This licensed construction or drainage works near the shore, but was easy to ignore (Woodley et al., 1998). The Natural Resources Conservation Department (NRCD) was established in 1975 as an attempt to address the need for an umbrella environmental management agency. The need for increased legislative authority, as well as for more comprehensive environmental management, led to the NRCD being transformed into the Natural Resources Conservation Authority (NRCA) in 1991. The expanded authority includes the power to request environmental impact assessments for projects, license discharges of trade effluent to the environment, require performance evaluation of pollution control facilities, and declare and manage national parks and other protected areas (NRCA, 1997). The Natural Resources Conservation Act (1994) greatly strengthened environmental management. The law also increased the staffing of the Natural Resources Conservation Authority (NRCA) in the Ministry of Environment and Housing (Woodley et al., 1998). Permits are now required from the NRCA for watershed development, land clearing of 10 ha of more for agricultural development, and clear cutting of forested areas of 3 ha or more on slopes greater than 25 degrees (NRCA, 1977).

Although the government has adopted a "polluter pays" policy, the costs of installing pollution control equipment and changing commercial and industrial processes, as well as the time required to implement such changes are a major challenge. The NRCA Environmental Permit and Licensing system, which came into effect in January 1997, will improve the monitoring of air and water emissions and ensure that both existing and new facilities come into compliance with standards. These standards include Trade Effluent Standards, Sewage Effluent Standards and Ambient Air Quality Standards. The NRCA, in keeping with the public "right to know" policy, provides full disclosure of Environmental Impact Assessments and is developing a Public Pollution Register (NRCA, 1997).

The Forest Act of 1996 and the Water Resources Act of 1995 are expected to result in improved management of forests and watersheds (NRCA, 1997). The Fishing Industry Act seeks to conserve and manage fisheries resources by addressing issues such as licensing fishermen using traps or pots, nets, spear guns and lines from boats. The new Draft Fisheries Bill presents a more comprehensive approach to fisheries management (NRCA, 1997). In 1998, the Council on Ocean and Coastal Zone Management was formed, creating a mechanism for improved integrated coastal zone management. The Council reports directly to the cabinet, and on it are represented all major sectors (Woodley et al., 1998).

Jamaica is a participant in several international and regional treaties and conventions (Table 1). Work continues on the development of the National Protected Area system. The "Policy for the National System of Protected Areas" was approved by Parliament in November 1997. The National Park Trust Fund was established in 1991 to finance the national protected area system. Planning studies were launched to develop Protected Area Systems in Black River and in Port Royal/Palisadoes areas. NGOs worked together

Table 1

International Conventions signed by Jamaica

- Convention on International Trade in Endangered Species of Wild Flora and Fauna (CITES)
- Montreal Protocol on Substances that Deplete the Ozone Layer
- Convention for the Protection and Development of the Marine Environment of the Wider Caribbean Region
- Convention on Biological Diversity
- Convention on Wetlands of International Importance especially as Waterfowl Habitats (Ramsar Convention)
- United Nations Framework Convention on Climate Change
- United Nations Convention on the Law of the Seas
- International Convention on the Prevention of Pollution from Ships
- Convention on the Prevention of Marine Pollution by Dumping of Wastes and Other Matter
- Convention concerning the Protection of the World Cultural and Natural Heritage
- Convention on Fishing and Conservation of the Living Resources of the High Seas.

with the NRCA towards establishment of protected areas in Negril, Port Antonio, Portland Bight, Ridge and Hellshire. The Negril and Green Island watersheds were declared Environmental Protection Areas in November 1997 (NRCA, 1997). However, recent reductions in government funding to the protected areas will pose difficulties for their effective management. The location of existing and proposed protected areas can be seen in Fig. 7.

The first International Coral Reef Initiative (ICRI) Regional Workshop for the Tropical Americas was hosted by Jamaica in 1995. At the meeting, the ICRI Framework for Action was adapted to the specific needs of the Caribbean, and individual countries were asked to hold national meetings to adopt national action plans. Jamaica held its national meeting in June 1997, and adopted the Jamaica coral reef action plan (JCRAP). The main strategy of JCRAP is to establish, under a co-management regime, a chain of marine protected areas around the island and on the inshore and offshore cays and banks. The Jamaican government has stated its intention to declare 14 national parks, marine parks and marine protected areas by the end of the decade. The Montego Bay Marine Park and the Negril Marine Park have already been officially declared. The government intends to delegate the management of these parks and protected areas to suitable NGOs (Espeut, 1999).

The Ocho Rios and Montego Bay Marine Parks were classic "paper parks" with no staff and no funding until 1989, when the Montego Bay Marine Park was revitalised with USAID funding. In 1998, new land/sea coastal management areas were created at Negril and the Portland Bight (Woodley et al., 1998). Jamaica has only two designated fish sanctuaries, one at Bogue Lagoon in Montego Bay and one at Bowden in Morant Bay (NRCA, 1997).

An island-wide coral reef monitoring programme, using photo and video transects is being undertaken by the NRCA in partnership with hotels, water sports operators and selected NGOs. This programme has begun with monitoring stations in the Montego Bay Marine Park. In 1996–1997 monitoring stations were established at a number of sites including Negril, Portland Bight, Hellshire, Ocho Rios, San San and Bluefields (NRCA, 1997). The CARICOMP monitoring site at Discovery Bay is soon to be joined by others at Portland Bight and Montego Bay. In addition, long-term monitoring for the effects of climate change is to be carried out for the Caribbean Planning for

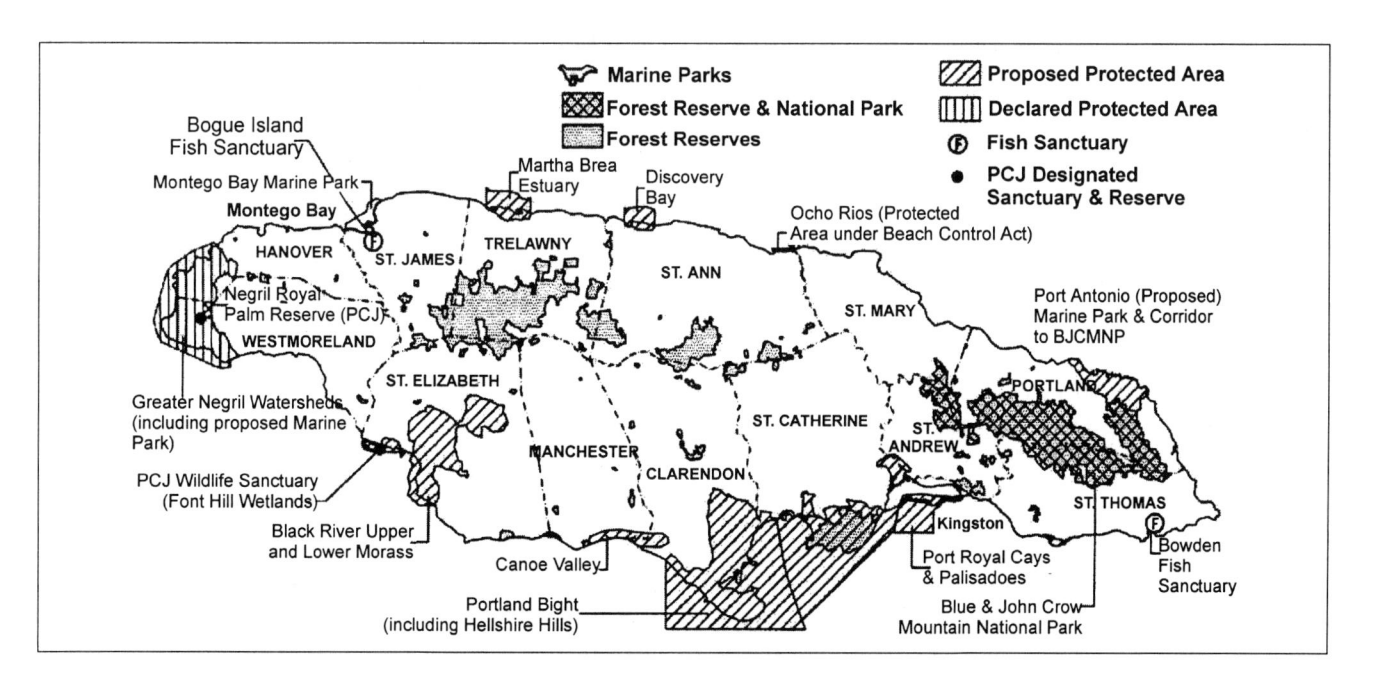

Fig. 7. The National System of Protected Areas, including both marine and terrestrial reserves (NRCA, 1997).

Adaptation to Climate Change project (CPACC) (Woodley et al., 1998). The University of West Indies carries out research in pollution assessment and mitigation, seagrass studies, reef monitoring, and reef fishery management.

Government efforts have increasingly been supplemented by NGO activities. Groups, such as the Negril Coral Reef Preservation Society, have formed all around the country, and are doing valuable work. Other groups include the Portland Environmental Protection Association, the St. Ann Environmental Protection Association, and Caribbean Coastal Area Management Foundation. The Fisheries Improvement Programme at the Discovery Bay Marine Laboratory is helping artisanal fishermen manage their own fishery resources (Woodley et al., 1998).

There has been a progression away from "top-down" control by a central authority to incorporate "bottom-up" participation by local resource users in a co-management framework. One example of such a management strategy can be seen at the Portland Bight Sustainable Development Area (PBSDA). This area, to be declared a protected area, is located on Jamaica's south coast just west of Kingston Harbour, and consists of 520 km^2 of land and 1350 km^2 of marine area. The land includes coastal forest, wetlands and the largest continuous stand of mangroves left in Jamaica. The marine area includes eight proposed fish sanctuaries. Co-management of the area was initiated by the establishment of the Portland Bight Fisheries Management Council, which included artisanal fishers, government representatives and Caribbean Coastal Area Management Foundation representatives. The council drafted fisheries regulations, which banned dynamiting, dragnetting, and the use of SCUBA for fishing; defined minimum mesh sizes for fish traps and nets; recommended the establishment of no fishing areas; and proposed a system of limited entry for fishers into the PBSDA. This action will result in a slow reduction in fishing effort and number of fishers over time. Annual fees were proposed for fishing within the PBSDA, and, once established, will contribute to the cost of fisheries management. It was also recommended that 50 fishers be appointed honorary game wardens and fisheries inspectors by the government. The proposed regulations were sent to each fishers' co-op and association for ratification. This process was fully participatory and resulted in a situation where both the Portland Bight fishers and the government own the regulations, which they helped to draft. The effort is still in its infancy, but thus far the Portland Bight Fisheries Management Council has accomplished real community participation in planning and decision making (Espeut, 1999).

Management plans are being developed for the most threatened species, and new legislation to facilitate efforts for species and biodiversity conservation is being proposed (NRCA, 1997). As an example, conch harvesting has been regulated since 1993. The 1996 catch was slightly below the 1.8 million quota set and there is also a closed season. Quotas will be reduced annually until the estimated sustainable yield of 1.5 million kg is reached. Most harvesting takes place on the Pedro Banks. Lobster harvesting is also regulated by an annual closed season during breeding, and there are size limits on the lobster that can be harvested (NRCA, 1997). In addition, there are restrictions on the trade of both conch and lobster under CITES, to which Jamaica is a signatory.

That these measures do not always work when confronted by the realities of economy, is illustrated by a July 1999 decision by the Jamaica Court of Appeals against the attempts by the Minister of Agriculture and the Natural Resources Conservation Authority (NRCA) to enforce restrictions on conch and lobster fishing. Two of the island's largest seafood exporters were refused conch fishing quotas for the 1998/1999 season by the Ministry of Agriculture and the NRCA. The exporters consequently took their case to the court. The Court of Appeal Judge wrote in his decision, "It is unthinkable to visualise that in the alleged delicate economic climate in Jamaica, a Minister of Government would seek to hinder the exportation of 731,483 lbs of conch, one third of the total national quota harvestable, with the consequent loss of product and foreign currency. This is, indeed, curious and cause for some query." There is no law in Jamaica, other than CITES, an international treaty, that requires conch exporters to get permits to export conch. The Fisheries Division gives licences for the harvesting of conch based on scientific assessments made annually by the Marine Science Unit at the University of West Indies. Permits are approved by the NRCA under the CITES treaty amounts recommended by the Fisheries Division. The court decision enraged environmentalists and was seen as a dangerous precedent on an island that has some of the most over-fished waters in the world (Neufville, 1999).

The main impediment to recovery of coral assemblages on Jamaican reefs is the presence of dense algal mats, which often occupy 90% or more of the space, and either prevent larval settlement or overgrow coral recruits. Storms may remove some of the algal biomass, but it is likely to recover too quickly to allow corals to compete. Herbivory could likely decrease algal cover. However, overfishing continues to be a reality in Jamaica. In addition, *Diadema* recruitment has been negligible at most sites and densities remain at very low levels (Hughes, 1993). There is a need for the regulation of fishing effort, creation of fish sanctuaries and community/government co-management of coastal resources for fishing and tourism. In addition, there is a need for environmental education, especially for adults (Woodley, 1995).

Efforts to control the deforestation problem are also in progress. In 1996, four tree nurseries were operated by the Forestry Department, each capable of producing 1,500,000 seedlings annually. However, the demand for seedlings and planting capacity is much lower. The 1997 target was to plant 400 ha. Other tree nurseries are being established for both fuelwood and reforestation purposes. Reforestation efforts focus on establishment of a few commercial tree species (typically Caribbean pine, mahoe and cedar), but they are poor substitutes for the biologically rich natural forests, which have many different species within a single stand (NRCA, 1997).

Jamaica's coastal and marine areas continue to face a variety of environmental stressors ranging from overfishing and nutrient pollution caused by increased human populations, to soil erosion caused by development and deforestation. Efforts to manage these problems are under way, and these efforts increasingly involve intersectoral collaboration and community participation. However, there are still many obstacles on the path to sustainable development. These include inadequate legislation, uncoordinated planning, inadequate levels of public awareness, illiteracy and poverty (NRCA, 1997). Reversing the trend of environmental degradation will take considerable time and effort. The government's willingness to take action is demonstrated by new policies, standards and guidelines for sewage and trade effluents, vehicle and stacks emissions and beach access and use policy, as well as the emerging National Protected Area System. It is also demonstrated by the preparation of the first Jamaica National Environmental Action Plan (JANEAP) in 1995, and additional legislation relating to water resources, pesticide regulation, forestry, fisheries, energy, watershed management, industrial development, and land use. There is also a growing environmental NGO movement and the increasing participation of private citizens in activities designed to reverse the trend of environmental degradation (NRCA, 1997).

REFERENCES

Aiken, K.A. (1992) Fisheries and Marine Conservation. Presentation for the Natural History Society of Jamaica.

Bohnsack, J.A. (1982) The effects of piscivorous predator removal on coral reef fish community structure. 1981 Gutshop: Third Pacific Technical Workshop Fish Food Habits Studies. Washington Sea Grant Publication, pp. 258–267.

Coke, L.B., Bertrand, R. and Batchelor, S. (1982) Macrophyte vegetation of the Negril and Black River Morasses, Jamaica. Botany Department, University of the West Indies, Kingston and Petroleum Corp. of Jamaica, 29 pp.

D'Elia, C.F., Webb, K.L. and Porter, J.W. (1981) Nitrate-rich groundwater inputs from Discovery Bay, Jamaica: A significant source of N to local coral reefs? *Bulletin of Marine Science* **31**, 903–910.

Environmental Control Division (1997) Annual Report. Ministry of Health, Kingston, Jamaica.

Espeut, P. (1999) The Jamaica Coral Reef Action Plan and the Portland Bight Sustainable Development Area. *InterCoast Network*, Spring Issue, pp. 14–15.

Evelyn, O.B. (1997) Deforestation in Jamaica. Report to the NRCA.

Goreau, T.F. (1959) The ecology of Jamaica reefs. I. Species composition and zonation. *Ecology* **70**, 275–279.

Goreau, T.F. and Wells, J.W. (1967) The shallow-water Scleractinia of Jamaica. Revised list of species and their vertical distribution range. *Bulletin of Marine Science* **17**, 442–53.

Goreau, T.J. (1992) Bleaching and reef community change in Jamaica: 1851–1991. *American Zoologist* **32**, 683–695.

Hendry, M.D. (1982) The structure, evolution and sedimentology of the reef, beach and morass complex at Negril, Western Jamaica. Report to Petroleum Corporation of Jamaica, P.O. Box 579, Kingston 10, Jamaica.

Hughes, T.P. (1993) Coral reef degradation: a long-term study of human and natural impacts. *Proceedings of the Colloquium on Global Aspects of Coral Reefs, Health, Hazards, and History.* University of Miami. pp. 208–213.

Hughes, T.P. (1994) Catastrophes, phase-shifts, and large-scale degradation of a Caribbean coral reef. *Science* **265**, 1547–1551.

Jackson, J.B.C. (1997) Reefs since Columbus. *Coral Reefs* **16** (5), 23–32.

Jameson, S.C., Huber, R.M. and Miller, M. (1999) Restoration of a valuable coral reef ecosystem: Reeffix Montego Bay, Jamaica. In: Proceedings from the International Conference on Scientific Aspects of Coral Reef Assessment, Monitoring, and Restoration, Ft. Lauderdale, Florida.

Knowlton, N., Lang, J.C. and Keller, B.D. (1990) Case study of natural population collapse: Post-hurricane predation on Jamaican staghorn corals. *Smithsonian Contributions to Marine Science* **31**, 1–25.

Koslow, J.A., Hanley, F. and Wicklund, R. (1988) Effects of fishing on reef communities at Pedro Bank and Port Royal cays, Jamaica. *Marine Ecology Progress Series* **43**, 2021–212.

Lapointe, B.E. (1998) Nutrient thresholds for bottom-up control of macroalgal blooms on coral reefs in Jamaica and southeast Florida. *Limnology and Oceanography* **42**(5) II: 1119–1131.

Lang, J.C. (1974) Biological zonation at the base of a reef. *American Scientist* **62**, 272–281.

Mendes, J., Woodley, J.D. and Henry, C. (1999) Changes in reef community structure on Lime Cay, Jamaica, 1989–1999: The story before protection. In: *Proceedings from the International Conference on Scientific Aspects of Coral Reef Assessment, Monitoring, and Restoration*, Ft. Lauderdale, Florida.

Munro. J.L. (ed.) (1983) Caribbean coral reef fishery resources. *ICLARM Studies and Reviews* 7. International Centre for Living Aquatic Resources Management, Manila Philippines. 276 pp.

Neufville, Z. (1999) Jamaican Court Chooses Commerce over Protection. Environment News Service, July 12, 1999.

NRCA (1996) National Policy for the Conservation of Seagrasses.

NRCA (1997) Jamaica—State of the Environment. Natural Resources Conservation Authority, Jamaica, 53 pp.

NRCA (1998) Jamaica National Environmental Action Plan. JANEAP 1998 Status Report. 197 pp.

NRCA (1998b) Mariculture, Draft Policy and Regulation. Natural Resources Conservation Authority, Coastal Zone Management Division, Jamaica, April 1998.

NRCA (1999) Beach Policy. Natural Resources Conservation Authority of Jamaica Publication.

NRCD (1987) Jamaica: Country Environmental Profile. 362 pp.

PIOJ (1995). Economic and Social Survey, Jamaica 1994, prepared by the Planning Institute of Jamaica, ISSN 0256-5013.

Steneck, R.S. (1993) Is herbivore loss more damaging to reefs than hurricanes? Case studies from two Caribbean reef systems. *Proceedings of the Colloquium on Global Aspects of Coral Reefs, Health, Hazards, and History.* University of Miami. pp. 220–224.

Webber, D.F. and J.C. Roff (1996) Influence of Kingston Harbour on the phytoplankton community of the nearshore Hellshire Coast, southeast Jamaica. *Bulletin of Marine Science* **59** (2), 245–258.

Wells, J.W. and Lang, J.C. (1973) Systematic list of Jamaican shallow-water scleractinia. *Bulletin of Marine Science* **23**, 55–58.

Wells, S.M. (1988) *Coral Reefs of the World. Vol. 1: Atlantic and Eastern Pacific.* UNEP/IUCN. 373 pp.

Woodley, J.D. (1992) The incidence of hurricanes on the north coast of Jamaica since 1870: are the classic reef descriptions atypical? *Hydrobiologica Reviews* **247**, 133–138.

Woodley, J.D. (1995) Tropical Americas Regional Report on the Issues and Activities Associated with Coral Reefs and Related Ecosystems. Prepared for the 1995 International Coral Reef Initiative Workshop, Dumaguete City, Philippines.

Woodley, J.D., Chornesky, E.A., Clifford, P.A., Jackson, J.B.C.,

Kaufman, L.S., Knowlton, N., Lang, J.C., Pearson, M.P., Porter, J.W., Rooney, M.C., Rylaarsdam, K.W., Tunnicliffe, V.J., Wahle, C.M., Wulff, J.L., Curtis, A.S.G., Dallmeyer, M.D., Jupp, B.P., Koehl, M.A.R., Niegel, J. and Sides, E.M. (1981) Hurricane Allen's impact on Jamaican coral reefs. *Science* **214**, 749–755.

Woodley, J.D., De Meyer, K., Bush, P., Ebanks-Petrie, G., Garzon-Ferreira, J., Klein, E., Pors, L. and Wilson, C. (1998) Status of coral reefs in the south-central Caribbean. In *Status of Coral Reefs of the World: 1998*, ed. C. Wilkinson. Australian Institute of Marine Science.

Zans, V.A. (1958) The Pedro Cays and Pedro Bank. Report on the survey of the Cays. 1955–1957. Geological Survey Dept. *Jamaica Bulletin* **3**, 47.

THE AUTHOR

Marjo Vierros
UNEP-CAR/RCU, 14–20 Port Royal St.
Kingston, Jamaica
Correspondence address: Rosentiel School of Marine and Atmospheric Science, University of Miami, Dept. of Marine Geology & Geophysics, 4600 Rickenbacker Causeway, Miami, FL 33149-1098, U.S.A.

Chapter 36

PUERTO RICO

Jack Morelock, Jorge Capella, Jorge Garcia and Maritza Barreto

Puerto Rico is part of a volcanic island platform that includes Puerto Rico and the Virgin Islands. Puerto Rico's prevailing weather is tropical. The trade winds blow consistently from east-northeast or east in the winter and from east-southeast in the summer. Stronger wind speeds are recorded during summer and winter than in spring and autumn.

The narrow island shelf can be described in terms of north, east, south and west provinces. The north is the narrowest and is marked by higher wave energy and more terrigenous sediments. The east, south and west are carbonate platforms with coral reefs and dominantly carbonate sediments. Each of these provinces has different physical energies.

Marine habitats are being diminished by excessive influxes of sediments and nutrients and by overfishing. During the past 50 years, more than 50 percent of the living coral has been lost and the rate of loss of reef areas has accelerated during the past 20 years. The high population density (>1000 people per square mile) and a shift of population to coastal areas has had a strong effect.

Although the problems of loss of habitat are generally recognized, very little has been done to protect the environment. Local resources for protection are meagre, and Federal (United States) efforts are directed to Florida, Hawaii and the Pacific in terms of coral reef preservation.

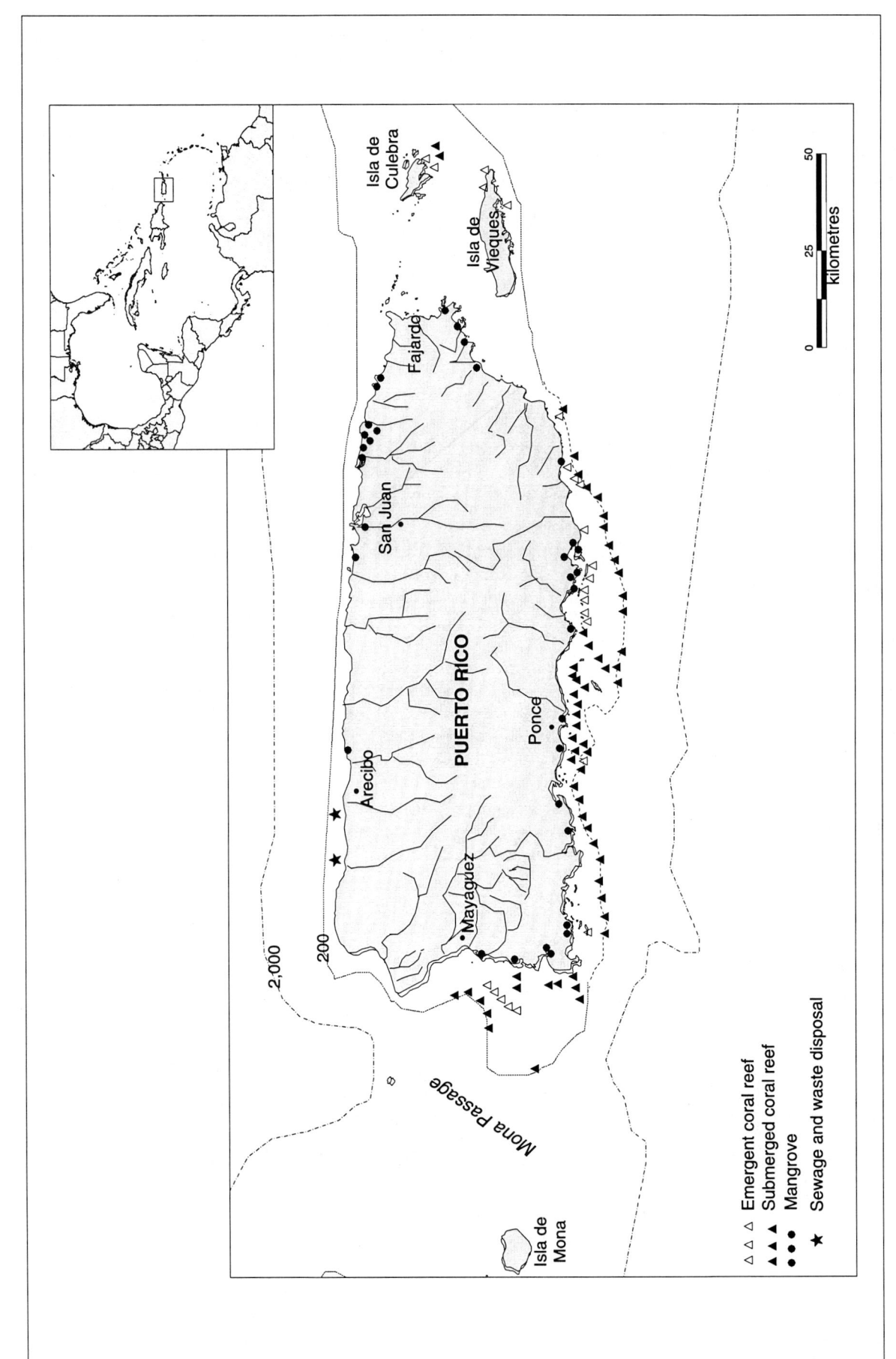

Fig. 1. The area between the coastline and insular shelf break varies from 0.3 to 10 km, with the only wide shelf on the southwest corner, south east of Ponce and the eastern platform. Coral reefs are shown for the west, south and east shelf, but they are too small to map on the north coast. The distribution of sewage discharge, and urban development is shown in relation to reef and mangrove areas. The three major cities are Greater San Juan, Mayagüez and Ponce. Arecibo, on the north coast, does not impact reef areas, but Fajardo on the east coast is a major factor in reef loss. Mangrove distribution is mapped from the U.S. Geological Survey topographic maps, drainage patterns are mapped from USGS data, and the coral reef distribution is from unpublished data of Morelock.

INTRODUCTION

The island was born at the leading edge of the Caribbean plate as the Puerto Rico Trench subduction zone developed. Limestones filled in between volcanic flows to form an island 35 by 110 miles. This occurred over the last 65 million years, but the actual shape and size of the island was essentially completed 40 million years ago. Active volcanism lasted from Cretaceous through Eocene time. The central mountain chain, a core of volcanic material and batholiths, lies south of the centreline, so that the coastal plain is wider on the north than the south, and most of the river drainage is to the north coast. Figure 1 defines the part of the platform occupied by Puerto Rico and the smaller islands of Vieques and Culebra. The insular shelf is limited in size and drops abruptly to deep water. Variations in the character of the habitat are controlled by wave and current energies, sediment types and sediment influx, and by bottom features. The north and northwest insular shelf supports a different marine ecosystem to that on the west, south and east shelves.

The basic structure of a central mountain volcanic core flanked by limestone deposits (with karst erosion patterns) is modified by clastic coastal plains and alluvial fans to the north and south forming three main physiographic units (Fig. 2). During part of the island's development, it was subjected to periods of intense deformation resulting in extensive folding and faulting (Morelock, 1978).

The north coast is Tertiary limestones and Recent deposits of alluvial plains, sand dunes, beachrock, and eolianites, except for volcanic material in the northeast. The limestones are cemented and only slowly erode by solution to form karst terrain in contrast to the Recent deposits which are unconsolidated or weakly cemented and therefore easily eroded.

The south coast has similar rock types except for intrusive igneous rocks at the eastern end, and a large area of alluvial fan deposits in the eastern part of this coast. Cretaceous and Tertiary limestones are more common in the western half of the south coast.

The east and west coasts are normal to the structural trends of the Island. The west coast has a pattern of rocky shorelines of limestone and volcaniclastic rocks, alternating with alluvial valley deposits, beaches, swamp and mangrove shorelines. On the east coast, the rocky shorelines are volcanic and intrusive igneous rocks alternating with the same pattern of modern alluvial deposits as is found on the west coast.

PHYSICAL PARAMETERS

Local, everyday weather in the Puerto Rico region is determined by the interplay between the land topography of this archipelago and the mid-latitude high pressure cell in the North Atlantic, westward travelling tropical waves, and cold fronts arriving from the north and northeast. The northeastern Caribbean lies along the northern edge of the Trade Wind belt which circles the globe from east to west; locally the Trade Winds are associated with the mid-latitude high pressure cell, whose centre is periodically displaced along the Bermudas–Azores latitude band. This clockwise rotating, high surface pressure system generates easterly winds (Trade Winds) over the region. The actual wind direction varies from northeast to southeast, depending on the geographical distribution of the isobars over the North and Tropical Atlantic.

The Wave Climate

The most frequent wave climate in our region consists of a background field of eastward-moving seas and swell generated by the Trade Winds, that is locally modified by bottom topography and by the behaviour of the wind as it hits the multiple islands. Most islands create shadow wind

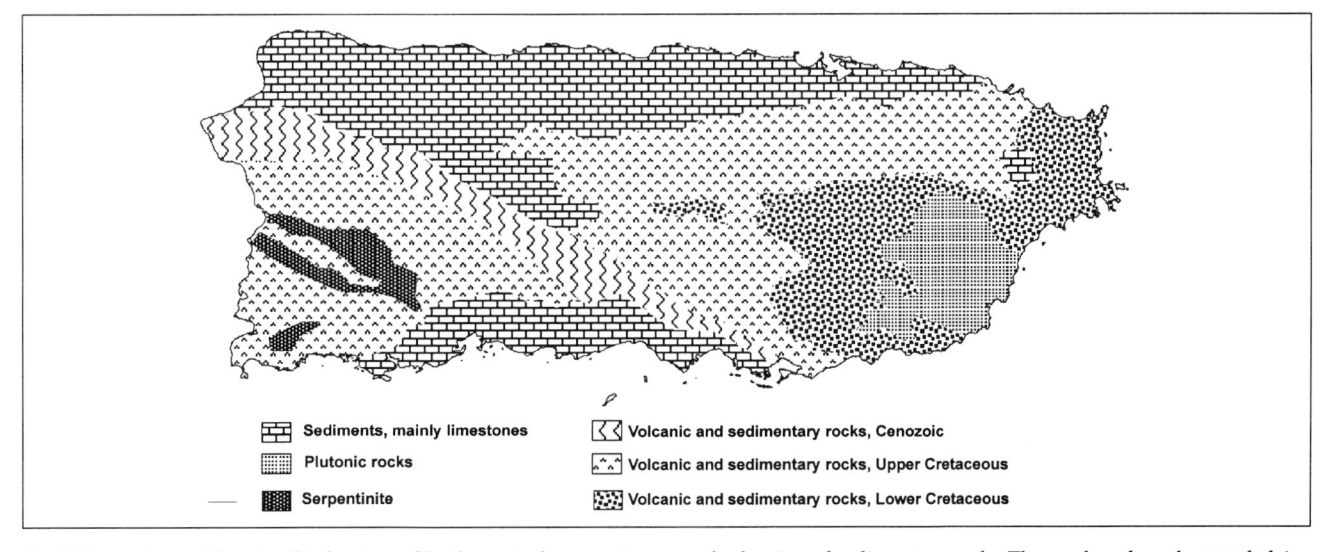

Fig. 2. The geology of the island is dominated by the central mountain range of volcanic and sedimentary rocks. The north and south coastal plains are limestones and alluvial deposits. Data for the geological map is from Garrison et al. (1972).

and wave regions over their leeward (western) waters. As the easterly waves approach the islands and are affected by the bottom, they refract towards shore creating a westward longshore current. Under steady easterly winds a windward-facing coastline receives wind and swell head on throughout the day, but the west coast is generally much calmer. A typical forecast under these conditions is for 1–2 m waves and swell throughout the region. Local winds close to the islands are also influenced by the diurnal land–sea heating cycle which often induces a strong sea breeze from the west during the afternoons and very calm conditions at night along the western coastline.

Westward-moving tropical waves are best known for the "low" part of the surface pressure cycle as the "highs" are indicative of dry windy weather. Tropical lows generally exhibit variable weaker winds, and increased cloudiness and precipitation. Wave conditions under a tropical low are usually calmer, except near centres of strong precipitation, which are accompanied by strong winds (and large seas) of short duration. During the hurricane season in the Atlantic, from June to November, high sea surface temperatures and favourable atmospheric conditions allow the development of these tropical lows into tropical depressions, tropical storms, and hurricanes. Hurricanes can develop 15–20 m waves near their centre and swell is radiated asymmetrically along their trajectory.

Whereas tropical waves move westward, mid-latitude weather systems propagate towards the east. Frontal systems over the North American continent result from the surface displacement of cold air masses from high latitudes (Alaska and Canada). These systems move towards the south and east and sustain strong winds along the cold–warm air boundary. As these systems exit the east coast of the U.S. towards the Atlantic, the strong winds generate high seas which decay into large swell waves that propagate southwards into the northern Caribbean. Large swell (12–16 s periods, breakers of 3–3.5 m but up to 4 m) is common along exposed northern Puerto Rico coastlines during the winter season (November–April). Arrival at the islands is easily predicted by following the quasi-periodic behaviour of the continental cold fronts. The arrival of a cold front generally means large swell along the north coast and flat seas along the south coast.

Surface waves in the ocean are generally divided into two major groups, seas and swell. Seas are formed by the local wind stress and are therefore strongly dependent on the speed and direction of the overlaying wind. Swell is formed far away and its speed and direction are in a straight line from its source, most noticeably far-ranging storm events. Whereas seas have periods of a few seconds and wavelengths in the order of several to tens of meters, swell peaks are spread hundreds of meters apart and, as any surfer knows, they travel in groups and there are only 10–15 s before the next big one in the group.

Long-term studies of the wave and currents patterns for Puerto Rico are rare, with most of the available data coming from the Summary of Synoptic Meteorological Observations (SSMO) and local consulting reports. Three wave regimes occur in Puerto Rico waters: (1) easterlies seas, (2) North Atlantic wave regime, and (3) Caribbean wave regime. The easterlies produce low energy seas that approach from the northeast to southeast affecting all of the island's waters, except for the leeward west coast. The North Atlantic wave regime consists of waves generated in middle latitudes, that can travel long distances from the generation location to Puerto Rican waters.

The west and north coasts have deep water wave heights from 1.2 to 1.8 m with occurrence of 25 to 10 percent respectively during a one year period (SSMO). Highest short period waves occur from May to September and long period waves are more frequent from January to July. Swells approaching from east and northeast are the most frequent with wave heights ranging from 0.3 to 3.6 m. Wave heights of more than 4 meters have a less than 5 percent occurrence. During storm conditions, the normal swells shift to higher amplitude waves.

Wave energy decreases toward the south and east coasts. Wave heights for the Caribbean wave regime are smaller than the North Atlantic, but higher than east coast seas (Lugo-Fernandez et al., 1994). Reduction in wave height is caused by an increase in island sheltering and because Caribbean and easterlies waves are lower energy. In eastern waters, more waves are of short period (less than 6 s) and low amplitude (from 0.3 to 0.92 m); generated by the combination of local sea breezes and easterlies. The strength of seas varies according to seasonal trends with the highest waves during summer and winter when the easterly breeze is stronger. Waves with periods of 6 s occur from October and December.

The combination of local shadow zones from the prevalent Trade Winds, wave refraction due to bottom topography, and the interference patterns created by multiple wave sources creates complex wave patterns between the closely spaced Virgin Islands. Even under the "normal" conditions treacherous waves may develop in narrow channels between the islands due to the focusing of wave energy by the bottom topography.

Winds and Hurricanes

Puerto Rico and the Virgin Islands are subject to easterly seas of 1 to 1.5 m under steady Trade Wind conditions, with shadow zones commonly found toward the western end of the individual islands. The east–west orientation of the PR-VI platform causes north–south gradients in the surface wave field so that under northeasterly winds the north coasts are exposed to wave action and the south coasts are protected. This reverses under southeasterly winds and is modified by the land–sea breeze diurnal cycle. During the winter months large, surfer-friendly, swell arrives every couple of weeks along the north coast and some of it squeezes through the exposed passages between the

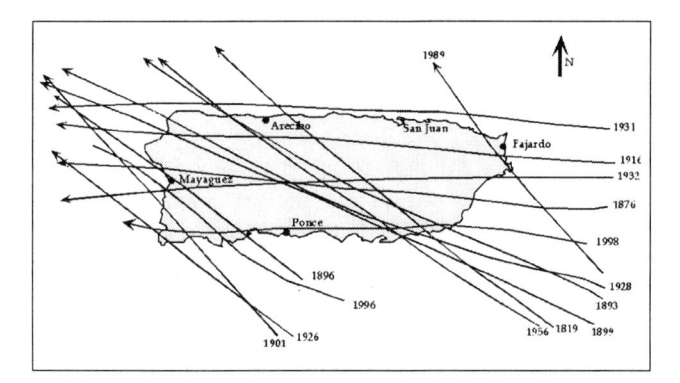

Fig. 3. In this figure, only the hurricane tracks passing over Puerto Rico are shown. Data are from Barrreto (1997) with additional tracks from personal experience.

islands. Extreme wave events accompany the arrival of tropical storms and hurricanes.

Hurricane recurrence studies to 1998 reported more than 130 hurricanes in the Tropical Atlantic zone with 21 coming within 60 miles of Puerto Rico (Neumann et al., 1978; pers. data to 1999). About 30 hurricanes have passed less than 400 km south or north of San Juan since 1940 (Fig. 3). Extratropical storm systems are significant energy contributors to the north, northeast and northwest coast of Puerto Rico. These storms are very well organized systems of low pressure generated in high latitudes of the Atlantic Ocean. The swells from these events have the capability to generate significant wave power upon reaching the coastal areas.

Water Bodies and Their Circulations

The Caribbean Sea is a semi-enclosed basin bounded by the Lesser Antilles to the east, the Greater Antilles (Cuba, Hispaniola, and Puerto Rico) to the north, and by Central America to the west. Passages between the various islands allow the inflow of North Atlantic, Tropical Atlantic, and Equatorial waters into the basin. The main outflow is through the Yucatan Channel between Cuba and the Yucatan Peninsula.

Caribbean waters are well stratified with depth, which means that at different depths the fluid is moving in different directions, according to the sources and sinks for each water mass. In the ocean around Puerto Rico (and this varies within the Caribbean), lies the Caribbean Surface Water, the local mixed-layer, whose lower boundary is known as the seasonal thermocline (technically it is the pycnocline but these two boundaries approximately coincide in depth); Subtropic Underwater to about 180 m; Sargasso Sea Water to about 325 m; Tropical Atlantic Central Water to just over 700 m; Antarctic Intermediate Water to 900 m; and North Atlantic Deep Water reaching the bottom. The island passages do not allow Atlantic bottom water to enter the Caribbean.

The structure and composition of the Caribbean Surface Water, that in which most human activity occurs, exhibit a well defined seasonal pattern. In the northeastern Caribbean Sea the depth of the thermocline reaches a maximum of close to 100 m in the spring (January–March) and a minimum in the order of 25 m in the fall (September–October). Density, temperature, and salinity follow the same seasonal pattern with temperatures ranging from 26 to 30°C and salinities from 36.3 to 34 PSU, respectively. The large range in offshore surface salinities is due to the northwards advection-mixing of South American riverine outflow in the eastern Caribbean Sea, specially from the Orinoco River; the seasonal surface salinity range is therefore narrower northwards into the North Atlantic. While the Orinoco effect creates a seasonal north–south surface salinity gradient in the eastern Caribbean, the Amazon River outflow becomes entrained in pools or eddies that, after a circuitous trajectory through the Tropical Atlantic, arrive at the Windward Islands as pools of green (high chlorophyll content, low salinity) water and enter the Caribbean from the east.

The mean circulation pattern of the wind-driven surface waters around the Puerto Rico–Virgin Islands shelf is in a west-southwest direction; these waters join the general western flow of the Caribbean towards Yucatan Strait. This archipelago is bounded to the east by Anegada Passage and to the west by Mona Passage, both of great strategic and economic importance to the region (see Fig. 1).

In the Caribbean Sea the meridional distribution of the zonal wind stress generates a circulation cell where deep waters upwell along the north coast of South America and surface waters (enriched by upwelling and by the Orinoco loading) are advected northwards into our region, specially during the fall season. Satellite images in the visible spectrum (CZCS and SEAWIFS) clearly show the meridional spreading of green water in the eastern Caribbean. The northward edge of the Orinoco plume does not extend far to the north of Puerto Rico. A persistent feature of the geostrophic flow south of Puerto Rico is the generally eastward transport in the upper 100 m, in a direction that is the opposite of the expected westward advection of Caribbean Sea waters. Eastward geostrophic flow is limited to near-surface waters, while deeper flow is generally westward. These observations are consistent with a net eastward geostrophic transport in the northeastern Caribbean, south of the Puerto Rico–Virgin Islands platform.

In the North Atlantic, the curl of the wind stress induces a large-scale Sverdrup transport towards the south that is then compensated by the intense northward-flowing Gulf Stream along the east coast of the U.S. The northeastern Caribbean receives part of this large-scale, climatological, southwestward transport. The convergence of these two distinct, Caribbean and North Atlantic, dynamical regimes defines our region as a boundary zone, with the edge of the green Orinoco plume often referred to by local researchers as the Caribbean Front.

There are no named current systems in the vicinity, which is not characterized by persistent extreme surface currents. The main axis of the Caribbean Current flows south of Puerto Rico, from the southeastern Antillean passages, through roughly the north–south centre of the Caribbean Basin, west of Jamaica, and out through the Yucatan Channel. Seasonal changes associated with the north-south excursion of the Inter Tropical Convergence Zone (ITCZ) result in maximum mean surface currents in the central Caribbean during the summer.

Superimposed on the mean circulation, tidal currents are the dominant component of the offshore currents; this is to be expected given the oceanic character of this region. The oscillatory, usually elliptical, tidal flows are mostly cancelled (vector averaged) in the calculation of the mean flow. In the open waters of Mona Passage typical peak tidal currents are in the order of 50–75 cm/s, corresponding to mean speeds over a tidal cycle of 25–30 cm/s, whereas the mean resultant velocity (the vector average) is only about 15 cm/s. The mean transport through Mona Channel is of 1–2 Sv into the Caribbean (1 Sv = 1 million cubic meters per second). The tidal current ellipses in surface and near-surface waters in Mona Passage are mixed semidiurnal and rotate in a clockwise sense. Tidal currents vary in magnitude in phase with the astronomical tidal forcing cycles of perigee–apogee and lunar declination. Due to the highly stratified nature of Caribbean waters, current speeds drop quickly with depth below the mixed layer.

Low frequency variability at time scales of weeks and months is additionally observed in current time series in the region. As an example of counterintuitive flows, the trajectory of a surface drifter deployed north of Puerto Rico in October 1980 started initially in a southwestward direction and then turned eastward along the north coast of Puerto Rico before describing a large anticyclonic gyre. This drifter passed very close to its initial position four months after deployment. Note that the long-term mean is in the "right" direction but the instantaneous velocities at short timescales are not.

During the passage of Hurricane George on 22 September 1998, currents of nearly 150 cm/s were measured at a depth of 34 m in the western Mona Passage.

Satellite altimetry and numerical modelling studies, have shown high levels of mesoscale activity superimposed on, or averaging into, the mean circulation of the Caribbean Sea and of the western Tropical Atlantic to the east. Just upstream from the eastern Caribbean, these mesoscale features consist of anticyclonic eddies that arrive at the eastern Caribbean from interaction of the equatorial long-wave field with the Brazil Current Retroflection. Model simulations show the dissipation of these eddies upon impact with the southeastern Antilles, in the vicinity of Barbados and the transfer of mass, energy, and vorticity into the Caribbean resulting in the spawning of new eddies west of the islands. Both cyclonic and anticyclonic eddies have been observed inside the Caribbean Basin where they are advected westward, in the direction of the mean flow. Modelling studies show anticyclonic eddies reaching all the way to the Yucatan Channel where they have a profound effect on the Loop Current in the Gulf of Mexico. Several mechanisms have been proposed for the formation of eddies inside the Caribbean Sea: island-flow dynamics, bottom-flow dynamics (due to the Aves Ridge and other bathymetric features), and eddy-boundary dynamics.

Coastal currents around Puerto Rico and the US Virgin Islands are mainly tidally and wind driven. Whereas the tide along north and western Puerto Rico is mainly semi-diurnal (two cycles per day), along the south coast the diurnal (one cycle per day) component predominates. The narrow and shallow shelf is in most places directly exposed to the open ocean, especially along the north coast. With the exception of bays and lagoons, coastal flows are steered by the coastline–shelf topography and are therefore east–west along the north and south coasts, north-south in Mona Passage, and variable on the shallow Virgin Island platform. Typical peak tidal speeds of 10–20 cm/s have been observed at numerous sites in the region; the mean vector velocity is usually less than 5 cm/s. The typical pattern is that of oscillatory currents parallel to the coastline.

The local wind stress, dominated by the easterly Trade Winds, pushes surface waters towards the west, the same direction as the large-scale offshore mean flow. Coastal geomorphic net drift indicators also show the westward dominance on the north coast (Morelock et al., 1985). However, during times of weak easterly winds near bottom waters are commonly observed to flow towards the west. This behaviour is known to occur along the north and south coasts of PR and has been attributed to (1) a reverse pressure gradient resulting from the action of the mean flow on the abrupt island topography, and/or (2) a mean eastward external geostrophic transport.

Specific examples for Mayagüez (west coast of Puerto Rico) and Guayanilla (a protected bay along the south coast) reveal the different diurnal vs. semidiurnal tidal regimes. The mean speeds (scalar means) of the currents in Mayagüez-Añasco Bay are always larger than the net speeds (resultant vector) as is typical of oscillatory currents such as tidally dominated flows. These currents flow back and forth along an axis and result in a small net transport towards the south. The net flow represents what is left over after subtracting the oscillatory tidal flow, while the mean flow is roughly half the tidal current amplitude. The situation in the MAB is that of a semidiurnal tidal current with a maximum amplitude of about 20 cm/s, typically 10–14 cm/s, that is amplified in one preferred direction along its principal axis of oscillation due to the presence of a large-scale mean flow (the net flow). A net south flow can be ascertained from the geomorphic data presented in Grove (1998).

The subsurface currents in Guayanilla Bay are tidally driven, following the prevailing diurnal tide with a weak semidiurnal component. Tidal current vectors rotate along a topographically flattened tidal ellipse in a clockwise sense. As expected, the tidal currents lag the tidal elevations by a 6 hour time lag. These tidal currents are bottom intensified and could well account for the high levels of turbidity observed in Guayanilla Bay. Residual near-bottom currents enter the bay along the eastern side of the channel and exit along the western side, suggesting a counterclockwise residual flow within the bay. Tidal volume exchange across the entrance channel exceeds previous estimates and is calculated in the average to be at least in the order of 15–40% of the volume of the bay. The mean observed currents are from Tallaboa Bay into Guayanilla Bay (east to west).

Tides

Tides throughout the northeastern Caribbean Sea exhibit a complex behaviour. Along the south coasts of PR and Vieques the tide is principally diurnal while the tide along the north and west coasts of PR is semidiurnal. The diurnal band actually extends south across most of the Caribbean and is surrounded by areas where the semidiurnal tide is stronger. This is further complicated as we approach Vieques due to the presence of the semidiurnal anticlockwise rotating amphidromic system, centred south of St. Croix. Accurate numerical prediction of the oceanic tide close to the islands becomes rather difficult due to the steep bathymetry of the Antillean Island Arc (and the lack of high resolution bathymetry) and the proximity of the M2, N2, and S2 amphidromes.

Vertical oscillations in the water column driven by the barotropic tide, known as internal tides, are observed to extend from the seasonal thermocline to the maximum observed depth. The amplitudes of these tidal oscillations are inversely proportional to the stability of the water column, resulting in a general increase in amplitude with increasing depth.

SHELF MORPHOLOGY AND SEDIMENTS

Composition of the coastlines can also be categorized as: (1) hard (resistant) composed of limestone or igneous rock; (2) semi-durable composed of eolianite or beachrock which is less cemented and more erodible than the first group; (3) erodible unconsolidated deposits such as beach, alluvial fan, alluvial plain, or dune. Another category, mangrove shorelines, lies between group 2 and 3, and often shows accretion. These mangrove shorelines are restricted to low energy conditions (Morelock and Trumbull, 1985; Barreto et al., 1994).

Insular shelf and slope morphology varies greatly in Puerto Rico with differences in shelf width, shelf and slope inclination, shelf break depth, and the extent of natural barriers. The shelf has inclinations ranging from 0.1 to 3.0 degrees and widths ranging from 0.3 to 21 km. The insular shelf is extremely small in comparison to continental shelves.

On the west coast, moderate shelf inclination varies from 0.1 to 0.5 degrees with shelf widths ranging from 0.4 to 6.6 km. On the north coast, the shelf inclination varies from 0.22 to 11 degrees. The shelf width ranges from 0.3 to 3.2 km.

The flattest shelves are found on the east and south coasts, where the platform is inclined from 0.1 to 0.7 degrees. On the east, shelf inclination varies from 0.1 to 0.3 degrees. On the south, shelf inclination is from 0.1 to 0.7 degrees. The south coast has a wider insular shelf ranging from 5 to 21 km.

The shelf break is in depths of 10–40 m around the island. On the west and north, the shelf break is at a depth of 10–40 m, on the east, the shelf break is 15–30 m deep, and the south coast shelf break ranges from 15 to 40 m.

The thin, recent, unconsolidated sediment cover of the Puerto Rico Insular Shelf is very diverse with little lateral continuity, because of large variations in physical and biological parameters controlling sedimentation. The north shelf is subjected to the highest wave energies, largest influx of river sediment, and has few coral reefs. Except for inner shelf zones, much of the sediment there contains enough silt and clay to be described as mud. The lower wave energy shelf around the rest of the island is dominated by coral reefs and mangrove forests that play important roles in the distribution of sediment. The dominant sediment type is calcareous skeletal sand. There are extensive white carbonate sand facies on the larger carbonate platforms of the west, south and east insular shelf. Sediment patches with more than 50% mud are present across the shelf at mouths of the major rivers. The upper insular slope is blanketed by a sandy mud containing <50% sand. The most important physical factors controlling the type of sediment are: wave energy, dilution by rivers and presence of reefs (Schneidermann et al., 1976). Other environmental parameters including bottom topography, depth, coastal configuration, coastal rock type, reefs and mangroves, affect carbonate sedimentation by controlling the terrigenous sediment supply.

Sediments on the insular shelf are mapped in three groups (Fig. 4):
– Carbonate sands, hardground and reef
– Terrigenous sands and muds with carbonate sands and hardground
– Terrigenous muds

Terrigenous sand facies are associated with river discharge or coastal erosion. At times, the river discharge disperses laterally (mainly to the west on the north coast) on the inner shelf, and at other times, the fluvial load is transported offshore during floods (Pilkey et al., 1978).

Mixed sediment facies are the result of river sediments mixed with local *in situ* carbonate production. Variability in the components of the mixture along the inner shelf depends on the proximity of local sediment sources and the physical forces that produced mixing.

Most of the hardgrounds such as eolianites, beach rock, dead coral reef and rock promontories are found along the inner shelf of Puerto Rico. Extensive and continuous

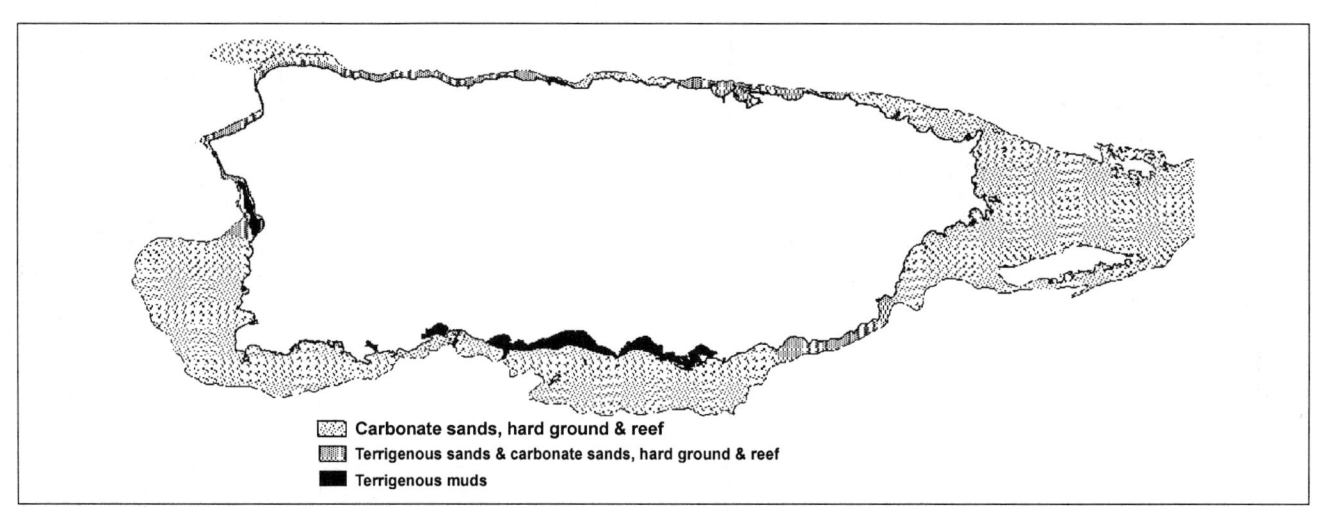

Fig. 4. The shelf sediments have been generalized into two sands and mud deposits. The pattern of sediments on the shelf has been generalized from a number of publications.

hardgrounds are present along the northwest and northeast inner shelf. These are eolianites, dead coral and beach rock with very little sand. Hardground constitutes the outer rocky shelf from La Parguera to Ponce, but the nearshore is sand and mud.

SHALLOW WATER AND COASTAL HABITATS

Three tropical habitats are present: coral reefs, *Thalassia* seagrass meadows, and mangrove forests. The rest of the insular shelf is sediment-covered or hardground. The general distribution of coral reefs has been mapped (Morelock, pers. data; Garcia et al., in press) and an image analysis program of mapping is being conducted by NOS-NOAA. The distribution of mangrove forests is mapped from USGS topographic sheets, but these are not current. Very little mapping has been done of the *Thalassia* meadows.

Mangrove

Mangrove forests can be found around most of the island, but the largest forests are at the east end of the north coast around to the north part of the east coast, and on the south coast wrapping around to the west coast (see Fig. 1). The mangroves form the shoreline in most areas, except for the north coast where they lie behind beach deposits or in interior river basins. Although no accurate measurement is available, Wadsworth (1968) reported more than 150 square miles of mangrove in seven of the northeast mangrove forests. Mangrove lagoons are fairly common around the island, forming distinctive habitats.

Reefs

Three types of reefs are recognized on the Puerto Rican shelf. Coral reefs are mostly found as fringing, patch, shelf and submerged shelf-edge reefs. Fringing reefs occur adjacent to land with little or no separation from shore. A low input of terrigenous sediment is important, and the best-developed fringing reefs occur off shorelines where rainfall is low, there is little relief, or the hillsides are stabilized by heavy vegetation. Clearing of natural vegetation has resulted in the loss of most of these reefs in Puerto Rico. Fringing coral reefs are found throughout most of the northeast, east and southern coastlines associated with erosional "rocky" features of the shelf. Patch reefs are smaller features, roughly equant in plan view. While many of these have reached sea level, many Puerto Rico patch reefs are submerged with the tops more than 4 m below the water surface. Most of the patch reefs in Puerto Rico are on the open shelf as pinnacles.

Submerged shelf-reefs are Caribbean platform margins that presently sit in water depths greater than 15 m after being flooded by rising sea level 6,000–10,000 years ago. Since then, they have not been able to offset the effects of deepening water and limiting base area, so many of them have been left behind (Fig. 5). While coral and other calcifying organisms occur along most of these margins, they do not produce carbonate at a rate sufficient for the reef to 'catch up' with sea level. The present influx of sediments and nutrients into reef areas is actually causing loss of the existing coral. Barrier reefs are separated from the shoreline by a moderately deep body of water—the lagoon, which has a characteristic sediment facies. These may form at the shelf edge, or may be located more inshore, usually on an antecedent break in slope. Barrier reefs are not present, but instead a common reef type in Puerto Rico has been called a shelf reef, which falls between the criteria for barrier or patch reefs (Morelock et al., 1977). They are similar to patch reefs in shape, but are usually larger, more linear, and are aligned in roughly shore-parallel sets (Fig. 6). The sediments behind these reefs are similar to those in front; no lagoonal type sediments occur.

Rock reefs are submerged hard substrate features of moderate to high topographic relief with low to very low

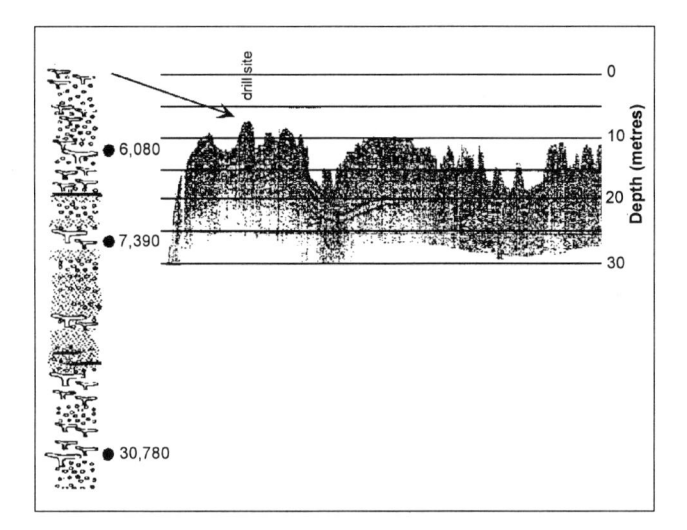

Fig. 5. The end of seismic line 12 at La Parguera crosses the shelf edge normal to the shelf break. The double ridge at the outer edge is Holocene *Acropora palmata* reef over an eroded Pleistocene karst surface. The inner ridge was cored for 19 m, showing a Holocene reef about 15 m thick over the older Pleistocene surface. The present surface of the reef is a thin layer of massive coral living at a water depth of 20 m. Ages are in years before present.

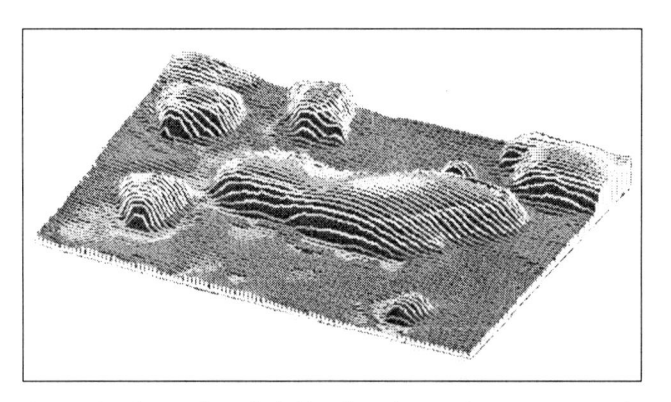

Fig. 6. This shows aligned shelf reefs with a gap between two reefs. Five patch reefs are shown around the left-hand shelf reef. The three western patch reefs and the two shelf reefs have crests at the water surface. The eastern patch reefs have a surface at about 8 m below surface. The sediment surface around the reefs averages 20 m deep.

coral cover, mostly colonized by turf algae and other encrusting biota. Coral colonies are abundant in some cases (e.g. *Diploria* spp., *Porites astreoides*, *Acropora palmata*), but grow mostly as encrusting forms, providing minimal topographic relief. These are coastlines subjected to high wave energy, abrasion and sedimentation stress. The underlying substrate may be any rock type, more commonly igneous rocks, eolianites, and beachrock. These reefs are important habitats for fish and macroinvertebrates since they usually are the only available structure providing underwater topographic relief in these areas. Some have developed on submerged rocky headlands and are characterized by the development of coralline communities adapted to grow under severe wave action and strong currents. They are mostly flat, eolianite platforms ranging in depth from 5 to 30 m, largely covered by turf algae, encrusting sponges and scattered patches of stony corals. Coral colonies are typically encrusting forms, perhaps an adaptation to the extremely high wave energy which prevails seasonally on the north coast.

Mud reefs are formed in shallow wave-protected areas where secondary frame builders such as *Thalassia* and *Porites* can grow upward while trapping sand and mud.

Corals grow throughout most of the insular shelf of Puerto Rico (see Fig. 1), yet the physical, climatological and oceanographical conditions which influence coral reef development vary markedly among the insular shelf segments. The shelf of the north and northwest coasts is narrow (<3 km) and shallow communities are subjected to mechanical abrasion and sedimentation produced by high waves, particularly during winter, as cold fronts from the North Atlantic reach the Caribbean Antilles. These are local rock reefs with coral cover of less than ten percent. They are too small areally to appear on the map. The west coast receives substantial sediment and nutrient loading from river discharge. The northeast coast has a wider shelf, partially protected from wave action by a chain of small emergent rock reefs aligned east–west between the main Island and the Island of Culebra. The northeast coast is upstream from the discharge of major rivers resulting in more appropriate conditions for coral reef development. The east coast is characterized by extensive sand deposits (unconsolidated) which constrain coral reef development, but scattered rock formations within this shelf section have been colonized by corals. Isla de Culebra and Isla de Vieques lie at the eastern boundary of the Puerto Rican shelf in clear waters that promote growth of coral.

The south coast has lower wave energy and the insular shelf is generally wider, with a broad carbonate shelf from Mayagüez on the west to Guanica on the south coast and from Ponce to Arroyo on the south coast. Rivers with smaller drainage basins discharge on the southeast coast and only small intermittent creeks discharge on the southwest coast, which has been classified as a semi-arid forest. The south coast also features a series of embayments and submarine canyons (Acevedo and Morelock, 1989). Small mangrove islets fringe the south coast and many of these provide hard substrate for coral development.

The shelf-edge drops off at about 20 m with an abrupt, steep (sometimes vertical) slope. At the top of the shelf-edge lies a submerged coral reef formation which gives protection to other reefs, seagrass and mangrove systems of the inner shelf. The southwest coast is relatively wide and dry, with many emergent and submerged coral reefs that provide adequate conditions for development of seagrass beds and fringing mangroves.

Modern shelf-edge reefs formed in Puerto Rico some 8000 years bp (see Fig. 5; Hubbard et al., 1996). Inner reefs, formed on top of submerged banks and sandy bottoms of the flooded erosional shelf are believed to be about 6000 years old. The rise in sea level associated with the last Pleistocene glaciation flooded the lower limestone ridges of

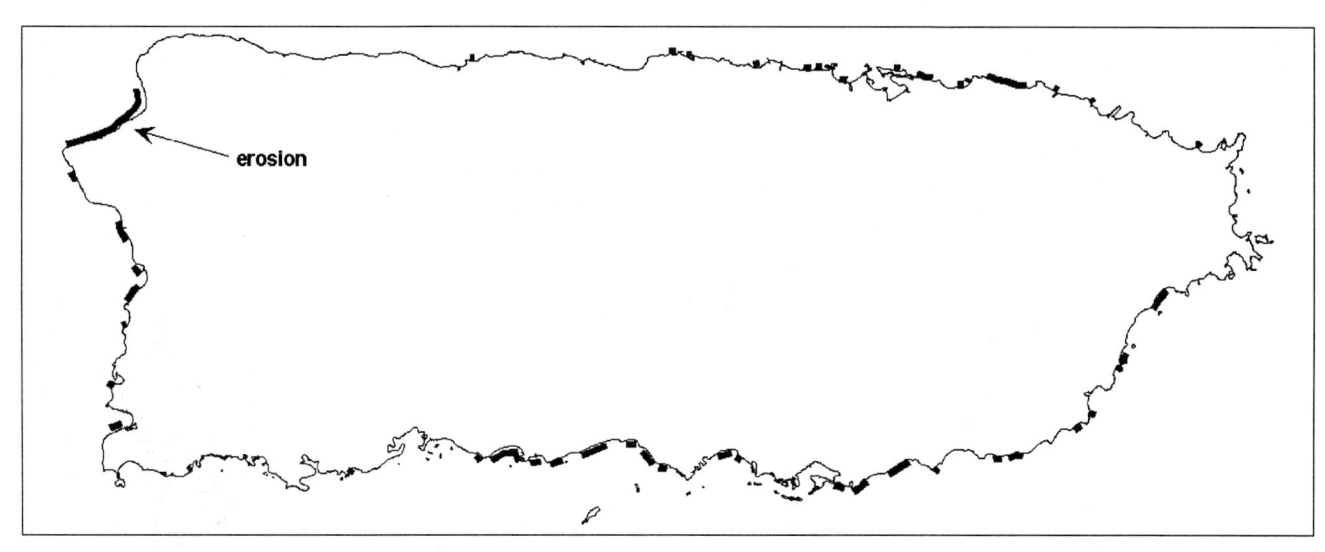

Fig. 7. Areas of erosion around Puerto Rico. Rates of erosion range from 0.1 to 1.1 m/year. Erosion was determined from five sets of aerial photographs taken in 1936, 1951, 1960, 1971, and 1987. Data are from Barreto (1997) and Morelock (1978, 1984).

the shelf, providing appropriate sites for coral growth and subsequent reef development (Glynn, 1973). Cross-shelf seismic profiles provided by Morelock et al. (1994) support the theory of Kaye (1959), which states that reefs on the southwest coast developed on drowned calcarenite cuestas formed as eolianite structures parallel to the coastline during the Wisconsin glacial period. Proper substrate, depth, and water transparency conditions in the southwest coast allowed for extensive development of coral reefs during the mid-Holocene period.

The outer shelf is irregular as a consequence of the karst bedrock surface and subsequent reef growth. Holocene and possibly Pleistocene deposits form a thin veneer over the late Pleistocene erosional surface. The shape and development of both the reef and the unconsolidated deposits have been strongly influenced by the morphology of the surface, the duration of inundation, the availability of sediment, and the energy of the environment. The shelf edge is formed by an almost continuous submerged reef. This reef is a double ridge along much of its length. The inner ridge is a platform surface 14–18 m below sea level; the crest of the narrow outer reef ridge is at 16–22 m. The valleys between the ridges range from 20 to 35 m deep.

POPULATION DEVELOPMENT AND LAND USE: EFFECTS FROM URBAN AND INDUSTRIAL ACTIVITIES

Only two land use studies have been done in Puerto Rico, one for DNER in 1970 and another for EPA in 1997. A general pattern of conversion of naturally vegetated areas into sugar cane farmland began with the Spanish colonization. A later pattern was conversion of mangrove coastal vegetation to coconut groves. This change of cover led to excessive sediment runoff and the loss of many areas of coral. Industrialization and urbanization that gained momentum after World War II has greatly increased the influx of sediments onto the shelf and brought a new problem, nutrient influx. The past decade has been marked by rapid urbanization of coastal areas with both a heavy increase of sediment influx from home construction and a heavy load on the existing sewer facilities. Visual examination of the shelf reefs and rock formations suggest that at least half of the existing living coral has been lost in the past five decades (Morelock, 1997).

Sewage and solid waste disposal have been a major problem in terms of environmental quality. Both ocean outfalls and interior discharge are shown in Fig. 1. This is an island of almost four million people occupying 3500 square miles—a population density of 1090 people per square mile.

Coastal erosion is not severe, but much of the coast does show erosion (Fig. 7) (Morelock, 1984). Movement of this sediment into coral reef areas has occurred to only a small extent. However, the high rate of erosion east of Jobos Bay has resulted in sediments covering the lower 20 feet of the fringing reef at Cayo Caribe.

However, very little has been done to preserve the natural features of the marine environment in Puerto Rico. One area has been designated a US National Estuarine Sanctuary (under NOAA supervision) and several areas have been designated as reserves by the Puerto Rico government.

REFERENCES

Acevedo, R. and Morelock, J. (1989) Effects of terrigenous sediment influx on coral reef zonation in southwestern Puerto Rico. *Proceedings of the Sixth International Coral Reef Symposium*, pp. 189–193.

Barreto, M. (1997) Shoreline Changes in Puerto Rico (1936–1993). Ph.D, University of Puerto Rico RUM.

Barreto, M., Morelock, J. and Vasquez, R. (1994) An Integrated mapping and databank system for coastal changes. Part A. West coast Puerto Rico. *Proc. of the Second Thematic Conference. Remote Sensing for Marine and Coastal Environments*.

Garcia, J., Morelock, J., Hernandez, E. and Goenaga, C. (In press). Coral Reefs of Puerto Rico. *Coral Reefs of Latin America*, ed. J. Cortez. Springer-Verlag, New York.

Garrison, L.E., Martin R.G. Jr., Berryhill H.L. Jr., Buell M.W. Jr., Ensminger H.R. and Perry R.K. (1972) Preliminary tectonic map of the eastern Greater Antilles region. Map I-732. Misc. Geologic Investigations: Map I-732, U.S. Geological Survey, Washington, D.C.

Glynn, P.W. (1973) Aspects of the ecology of coral reefs in the western Atlantic region. In *Biology and Geology of Coral Reefs. Biology 1*, eds. O.A. Jones and R. Endean, pp. 271–324. Academic Press, New York.

Grove, K. (1998) Nearshore sediment transport and shoreline change at Anasco Bay, Puerto Rico. Ph.D, University of Puerto Rico.

Hubbard, D.K., Gill, I.P., Burke, R.B. and Morelock, J. (1996) Holocene reef backstepping—southwestern Puerto Rico shelf. *Proceedings of the Eighth International Coral Reef Symposium*, preprint.

Kaye, C.A. (1959) Shoreline features and Quaternary shoreline changes, Puerto Rico U.S. Geological Survey Professional Paper 317-B. U.S. Geol. Surv., Washington, D.C.

Lugo-Fernandez, A., Hernandez-Avila M.L. and Roberts H.H. (1994) Wave-energy distribution and hurricane effects on Margarita Reef, southwestern Puerto Rico. *Coral Reefs* **13**, 21–32.

Morelock, J. (1978) *Shoreline of Puerto Rico*. Coastal Zone Program, Department of Natural Resources, San Juan, PR, 45 pp.

Morelock, J. (1984) Coastal Erosion in Puerto Rico. *Shore and Beach* 18–27.

Morelock, J. (1997) Status of coral reefs in Puerto Rico. *Coral Reef Workshop*, abstract.

Morelock, J., Hernandez-Avila, M., Schwartz, M.L. and Hatfield, D.M. (1985) Net shore-drift on the north coast of Puerto Rico. *Shore and Beach*, 16–21.

Morelock, J., Schneidermann, N. and Bryant W.R. (1977) Shelf reefs, southwestern Puerto Rico. In *Reefs and Related Carbonates—Ecology and Sedimentology*, eds. S.H. Frost, M.P. Weiss and J.B. Saunders, pp. 17–25. Studies in Geology 4. American Association Petroleum Geologists, Tulsa, Oklahoma.

Morelock, J. and Trumbull, J. (1985) Puerto Rico Coastline. In *The World's Coastlines*, ed. E. Bird. Van Nostrand Reinhold, New York.

Morelock, J., Winget, E. and Goenaga, C. (1994) Marine geology of the Parguera-Guanica guadrangles, Puerto Rico. USGS Misc. Map Series, U.S. Geological Survey, Washington, D.C.

Neumann, C.J., Cry, G.W., Caso, E.L. and Jarvinen, B.R. (1978) Tropical cyclones of the North Atlantic Ocean, 1871–1977. *Climatology Series 6-2*, U.S. Department of Commerce, Asheville, N.C.

Pilkey, O.H., Trumbull, J.V.A. and Bush, D.M. (1978) Equilibrium shelf sedimentation, Rio de La Plata shelf, Puerto Rico. *Journal of Sedimentary Petrology* **48** (2), 389–400.

Schneidermann, N., Pilkey, O. H. and Saunders, C. (1976) Sedimentation on the Puerto Rico insular shelf. *Journal of Sedimentary Petrology* **46** (1), 167–173.

Wadsworth, F.H. (1968) Conservation of the natural features of our estuarine zones. Unpublished report, U.S. Forest Service, San Juan, PR.

THE AUTHORS

Jack Morelock
University of Puerto Rico R.U.M.,
Department of Marine Sciences, P.O. Box 3200,
Lajas, Puerto Rico 00667

Jorge Capella
University of Puerto Rico R.U.M.,
Department of Marine Sciences,
Mayagüez, Puerto Rico

Jorge Garcia
University of Puerto Rico R.U.M.,
Department of Marine Sciences,
Mayagüez, Puerto Rico

Maritza Barreto
University of Puerto Rico Rio Piedras,
Geography Department,
Rio Piedras, Puerto Rico

Chapter 37

THE TURKS AND CAICOS ISLANDS

Gudrun Gaudian and Paul Medley

The Turks and Caicos Islands are geographically part of the Bahamian Banks, lying off the southeast coast of Florida and to the north of Hispaniola. They are surrounded by tropical deep waters. The coral reefs are mostly pristine, particularly away from centres of human habitation. The islands are sparsely populated overall, although numbers of people are seasonally inflated by tourists. Tourism is a major environmental concern, as housing, food, water, and extra energy, need to be provided. Measures are being developed by the government of the Turks and Caicos Islands to manage tourism sustainably. Other industries are very localised and thus show relatively small effects on the environment overall.

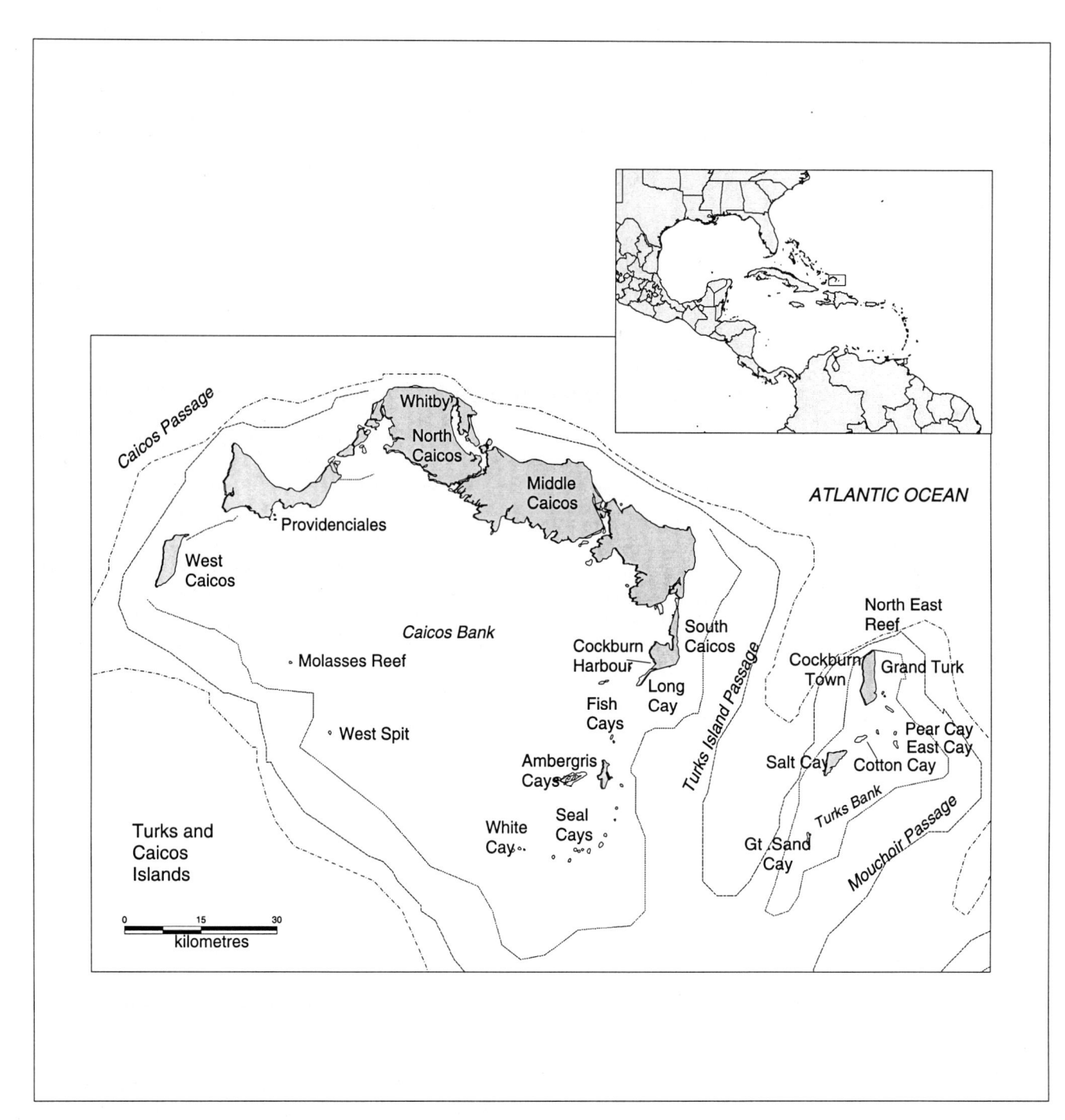

Fig. 1. Map of the Turks and Caicos Islands.

THE DEFINED REGION

The Bahamian Banks form an extensive archipelago of islands, cays and sandbanks separated by deep ocean channels, extending over 800 km from southern Florida to Hispaniola (Fig. 1). It is the only example of a large open ocean island system in the tropical Atlantic Ocean. The politically separate Turks and Caicos Islands (TCI) lie at the southern end, and consist of the Caicos Bank and Turks Bank. Between the TCI and the Dominican Republic lie two additional submerged banks, the Mouchoir Bank and Silver Bank. The shallow parts of these are extensive coral reefs. Some coral heads may emerge at extremely low tides, otherwise these banks are completely submerged.

The origins of the Bahamian Banks are still much debated, despite scientific drillings on cays and deep submarine surveys (Sealey, 1985). One theory suggests that the archipelago was once an extension of the Florida Plateau, but is now separated from the continental coast by an 80 km wide chasm. It is suggested that the Bahamian Bank is growing upwards from the deep seafloor by deposition of pure calcium carbonate (aragonite) which precipitates out of the seawater (Sealey, 1985). Whether this is the case or not, the vast deposits of aragonite on the Bahamian shelf are considered to be exceptionally interesting as they provide examples of the warm, shallow, lime-depositing seas which were characteristic of the past but which are atypical of modern marine conditions (Newell et al. 1951 in UNEP/IUCN, 1988).

The whole archipelago is surrounded by deep water, reaching a depth of over 3500 m in places, and can thus be pictured as a steep submerged mountain range with only the tips breaking the water surface. The individual banks are characterised by low-lying arid islands and cays, separated by shallow water ranging between 1 m and 20 m depth. Although each bank is different, a common pattern exists: the central part is flat with large islands lying along the north and east edges, and occasional low islands and rocks along the west and southern edge. The outer shelf surrounding the bank is usually less than 1.5 km wide and slopes seaward to 15–50 m before plunging steeply. On the seaward side of most islands is a lagoon and fringing reef, while the sides of islands facing inwards towards the centre of the bank are characterised by soft sediments and associated communities.

SEASONALITY, CURRENTS, NATURAL ENVIRONMENTAL VARIABLES

Shallow water sea temperatures range from 25°C in the winter (December to March) to 29°C in the summer (July to October), although higher temperatures of 31°C have been recorded in shallow lagoons with little water circulation. Unusually, temperatures as low as 22.5°C were recorded over the coral reefs off Grand Turk in early 1999 (B. Riggs, pers. comm.).

The strongest winds occur between December and March, with winds increasingly blowing from the northwest rather than the prevailing easterly trade winds, which are otherwise consistent most of the year.

The hurricane season starts on the 1st June and lasts until the 1st November, although devastation from hurricanes or tropical storms is not common. Between the years 1800 and 1995, hurricanes which have resulted in a significant loss of human life have hit the Bahamas chain ten times, including most recently in 1960 (Donna) and 1992 (Andrew). Anecdotal information indicates that sand dunes and bars that make up much of the islands shift permanently following hurricane activity, but there is no data to measure this. It has been reported (FAO/CFRAMP, 1998) that hurricanes have a strong influence on species which have nursery areas in shallow water, such as spiny lobster, which may adversely affect recruitment to that fishery and may exacerbate overfishing.

The Turks and Caicos Islands receive around 900 mm rain per year, most falling between May and October, with North Caicos receiving more rain than Grand Turk, as is reflected in the lushness of the vegetation.

The only major current in the archipelago is the Florida arm of the Gulf Stream, which flows between Little Bahamas Bank and the American mainland. Surface water currents, down to 10 m depth, are generated by strong winds during the winter season, and tidal currents are particularly noticeable at narrow inlets. The tidal variation is about 1 m, exposing the shallow reef flats and many sand banks at very low tides.

THE MAJOR SHALLOW WATER MARINE AND COASTAL HABITATS

Being an island archipelago, the whole Bahamian Bank including the Turks and Caicos Islands can be classified as coastal. The islands are made of limestone and fossilised sand dunes, and sand banks. Rain water rapidly drains through the limestone, collecting in subterranean lenses. There are no rivers or any other natural fresh water runoffs. Vegetation is limited to low scrub dominated by halophytes. In natural hollows, where leaf litter has been accumulating over time to form soil and retain water, there can be dense stands of natural forest.

The biological diversity of the shallow marine waters is comparatively high within the wider Caribbean region, as few reefs and shallow water habitats are affected by human activities relative to the large area of the banks and fringing reefs. About 30 species of scleractinian corals have been found on these reefs (Gaudian, 1995) and around 290 inshore marine fishes have been noted for the Bahamas as a whole (WCMC, 1999).

Mangroves

Mangrove stands, dominated by *Rhizophora mangle*, *Laguncularia racemosa* and *Conocarpus erecta*, grow in sheltered bays and creeks and can be extensive and dense. No quantitative estimates are available for the area of mangroves in the Turks and Caicos Islands, although dense stands of mangroves grow in South Creek on Grand Turk, and along the bank sides of all the Caicos Islands. The extent of these mangrove stands is limited by the availability of fresh water.

Coral Reefs

Reef areas are extensive. The Caicos Bank (6140 km^2) and the Turks Bank (324 km^2) are fringed by coral reefs. These reefs are close to land (up to 400 m) along the northern and eastern coasts, as well as along the western side of West Caicos. Elsewhere, an extensive shallow sandy bank, interspersed with coral outcrops and seagrass beds, divides the shore from the fringing reef.

Bahamian coral reefs have been studied extensively, including Hydro-Lab submersible studies down to 70 m at various locations (Bunt et al., 1981). The deepest distribution of scleractinian corals known in the Atlantic has been found off nearby San Salvador: *Agaricia grahamae* was observed to 119 m and *Montastrea cavernosa* to 113 m (Reed, 1985). There are many areas with fine, extensive and virtually untouched coral reefs in the Bahamas, descriptions of some being compiled in the *Coral Reef Directory* (UNEP/IUCN, 1988). About 30 species of corals have been recorded, of which the most important reef builders are *Montastrea annularis, M. cavernosa, Siderastrea siderea, Diploria labyrinthiformis, Porites porites, Acropora palmata* and *Agaricia agaricites* (Newell et al., 1959). There is relatively little quantitative information specifically on the reefs and adjacent marine environments of the TCI. A general description of the reefs has been given by Gascoine (1991) in a guide specifically designed for divers visiting the TCI but little else has been done since the compilation on the TCI in UNEP/IUCN (1988). The latter includes such grey literature as reports and briefing notes produced by various expeditions to the islands, for example Operation Raleigh (1986). An expedition by deep-sea submersible was conducted by Harbour Branch, Florida, off Grand Turk in 1994, in order to collect sponges for biomedical purposes. Sullivan et al. (1994) examined in detail the abundance patterns of scleractinian corals on the windward margin of the Caicos Bank, on the eastern side of South Caicos. Gaudian and Medley (1995) collected quantitative baseline data to characterise reef communities on Grand Turk, and assess diver impact on the benthos, which was found to be small, but measurable.

The most common reef-building corals on the shallow fringing reefs of the TCI are very similar to those for the rest of the Bahamas Bank noted above (Gaudian, 1995). The encrusting sponge *Cliona langae* was found to be common on the sites studied.

In the TCI, fringing reefs start at a depth between 10 and 20 m, depending on location, and plunge steeply over several small terraces into great depths. The water above the reefs is clear, as there are no rivers transporting sediments into the sea. Turbidity is higher during storms, when bottom sediments are resuspended by wave action, and during exceptionally calm weather, when plankton blooms in the surface waters. Because coral growth is restricted by the availability of light, hermatypic corals are generally found above 50 m on the vertical slopes of the fringing reef, being replaced by gorgonids and encrusting sponges in deeper water. Along exposed shores, branching *Acropora palmata* grows on the shallow reef and adjoining reef flat. Although large stands of *A. palmata* have been decimated by disease and storms elsewhere in the Caribbean, large healthy stands can still be found at Salt Cay, and the exposed sides of Grand Turk, South Caicos and some parts of the Caicos Bank. Elsewhere on the shallow upper fringing reef, the massive coral *Montastrea annularis* and *M. cavernosa* are dominant, being replaced by plates of agariciid corals on the steep slopes.

Large expanses of the shallow banks have sandy substrate, with small coral patch reefs where there is exposed hard substrate. These patch reefs are usually dominated by gorgonids, with several species of scleractinian corals and sponges. They are important habitats for the spiny lobster (*Panulirus argus*) fisheries.

Seagrasses, Sand and Soft Bottom Substrate

Typically islands slope gradually towards the centre of the bank, where is a build up of soft sediment. Towards the edge of the bank, there is substantial oolithic sand, with often only a light scattering of algae such as *Avrainvillea* sp., *Penicillus* sp., *Halimeda* sp., and *Acetabularia* sp. Seagrass beds are extensive, consisting of predominantly *Thalassia testudinum*, often coated with microscopic and macroscopic epiphytes, and interspersed with other seagrasses such as *Syringodium filiforme* and *Halodule wrightii*. Dominant algae within the seagrass beds are several species of *Halimeda, Caulerpa,* and *Sargassum*. Seagrass beds provide grazing for turtles, visiting manatees, fish and invertebrates. There is no quantitative or qualitative information on the infauna of the soft bottom substrates of the Banks. Near reefs, seagrass beds stabilise bottom sediments that otherwise could damage coral, and closer to the islands they help slow coastal erosion.

Some detailed habitat mapping of the eastern edge of the Caicos Bank was conducted in 1994 as part of *Panulirus argus* stock assessment and management (Department of Environment and Coastal Resources, unpublished data). Closer to the southern end of the Caicos Islands, the substrate becomes a muddy/sandy mix, fringed by mangrove roots and halophytes. These habitats are marked by abundant populations of queen conch (*Strombus gigas*), which is also considered vulnerable to over-fishing in the region, although populations in TCI appear abundant and stable (CFMC/CFRAMP, 1999).

Other Species

Marine turtles (*Caretta caretta, Cholera mydas, Dermochelys coriacea* and *Eretmochelys imbricata*) are vulnerable to coastal development as they breed on sandy beaches. In the TCI, green and hawksbill turtles are caught for local consumption (mainly as by-catch while diving for lobster and conch), although interest in turtle meat is waning. Since the whole archipelago provides many breeding beaches in remote locations, it is not clear to what degree turtles are threatened.

Visiting West Indian manatees have been seen on the Caicos Bank and North Creek of Grand Turk, as well as in some lagoons of the Bahamas. Manatees need flowing freshwater to survive and therefore may not be permanently resident in the Bahamian archipelago (Reynolds and Odell, 1991). From December to March, humpback whales (*Megaptera novaengilae*) migrate through the Turks Passage, between the Caicos Bank and Grand Turk, on their way to breeding and birthing grounds on the Silver and Mouchoir Banks.

OFFSHORE SYSTEMS

There are no major offshore systems such as upwellings. The Florida arm of the Gulf Stream flows eastward between Florida and the Bahamas. The tradewinds drive surface currents in an east–west direction. The pelagic waters support populations of swordfish, marlin and tuna and other pelagic species found throughout the tropical Atlantic.

POPULATIONS AFFECTING THE AREA

Considering the large area of the archipelago, it has a sparse resident population. This is greatly inflated annually by tourists, mainly concentrating on resort islands and cays. It is estimated that currently around 19,000 people live in the TCI, including non-nationals. Illegal immigrants (suggested at around 3000) from Haiti and the Dominican Republic add to this resident population. Expatriates are the biggest growth sector. The estimated area of the Turks and Caicos Islands landmass is 499.7 km^2 (Olsen, 1986), and the majority of the population is concentrated on Providenciales, one of the six islands and cays that is inhabited.

Providenciales has the biggest and fastest growth in development, both for residential and hotel building projects. In order to avoid pollution of the surrounding seas through sewage effluent, which in turn would affect tourism numbers, Providenciales has a number of sewage treatment plants and desalination plants. Most of these installations are part of the hotel complexes, but some are run by Government through aid-funded projects. The greatest pollution problem on Providenciales, and indeed on all the other inhabited islands and cays, is the lack of solid refuse treatment facilities. All the domestic and commercial rubbish is burnt in open sites, causing some air pollution and an environmental health risk. The potential effects of chemicals seeping from those dumps either into the ground water or into the ocean, washed there by rain, have not been studied.

The success of Providenciales as a tourism destination started in 1986 with the building of an international runway for large aeroplanes. Since then, the building of hotels and houses has increased, attracting many Turks Islanders from the smaller islands and cays to Providenciales, in search of work and a better lifestyle. Tourist numbers have increased dramatically over the last few years; up to 50,000 tourists now visit the islands, most of whom stay on Providenciales.

The waters around Providenciales are used for recreational purposes mostly by tourists (diving, snorkelling, sailing, etc.) and fishing activities by local fishermen to supply hotels (some finfish, lobster and conch), or for export (lobster and conch), as well as for domestic use.

Grand Turk has about 2500 residents, most of whom work for the Government; many have left for Providenciales and the private sector. Several hundred tourists a year visit Grand Turk, attracted by the good diving and small-scale operations of that island. There are only a few small hotels on the islands.

Traditionally South Caicos is the island from which most of the fishing for conch and lobster is conducted, although conch fishing has expanded from Providenciales. South Caicos depends almost solely on the fishing industry.

Fishing for local consumption has been a continuous activity since the arrival of the first settlers in the islands—the Taino indians in pre-Columbian times. This is documented by piles of conch shells and conch shards used for cooking left by these inhabitants. Export of dried conch was part of the regular trade between Haiti and the TCI. Fishing for conch and lobster for export to the USA began in the late 1960s following the introduction of freezer plants. Since 1966, catch and effort data on conch, lobster and finfish for export has been collected. There are about 200 full-time fishermen in the fishery, but details of their subsistence fishing for finfish and conch has not been recorded.

Quantitative investigations into the condition of several dive sites on coral reefs in Grand Turk, South Caicos and Providenciales was started in 1993, and regular monitoring is being carried out. Descriptive information of some of the dived reefs has been given by several expeditions in the 1980s (e.g. Operation Raleigh, 1986).

RURAL FACTORS

There is very little agriculture in the TCI and Inagua, as rainfall is low and the soil is generally poor. Anecdotal accounts suggest agriculture declined on some islands following major hurricanes which resulted in significant soil loss. Archaeological work in the TCI suggest that there was much more tree and shrub cover in the past, before the salt trade, cotton and sisal industry in the 18th and 19th centuries, implying that trees were felled for clearing.

Today, some small crops are grown for local consumption and for trade within families. Almost all food is imported from the United States, but in the past there was much more trade with Hispaniola, from where fruit and vegetables were brought by sailing boats.

Subsistence fishing for conch has been undertaken since the islands were first settled by Taino Indians and conch remains a traditional food for local residents. Conch are currently collected by snorkelling from small family-owned fishing boats. From before 1900 up to the early 1960s, conch were dried and exported to Haiti, and a small conch-drying cottage industry still exists, particularly on Grand Turk. Most conch is now exported frozen.

COASTAL EROSION AND LANDFILL

Since tourism has expanded into a major industry, much construction work is being conducted throughout the archipelago close to the beaches, for both hotel complexes and private residences. All the basic building materials are found within the islands; stone is mined from quarries and sand is dug up from beaches or mined from the bank. In the TCI, legislation stipulates beach setback and replanting to avoid erosion of soils. In general, damage to the marine environment as a result of construction work along the coast has been fairly small, partly due to the watchfulness of residents and partly because of the nature of these constructions. However, the sea fronts of some of the residences in Providenciales are gradually being washed away by longshore currents. This is a result of natural current activity and winter storm surge, unregulated building, and altered current patterns from sand mining further along the reef flat.

There is limited but unregulated harvesting of mangroves for charcoal production, carried out by immigrants from neighbouring countries, though charcoal-making has not been a traditional industry. As yet, there have been no large-scale environmentally damaging projects, although there is heavy pressure to develop the islands and unutilised space. Vigilance and local resistance have been able to stall some of these projects until appropriate Environmental Impact Assessments have been carried out. The problem is that much of the pristine landscape on Providenciales and Middle Caicos is allocated for development, leaving no areas set aside for wildlife.

The Department of Environment and Coastal Resources in the TCI is collecting baseline data on beach movement. This information is not published or compiled in reports.

EFFECTS FROM URBAN AND INDUSTRIAL ACTIVITIES

Artisanal and Non-industrial Uses of The Coast

In the adjacent Bahamas, queen conch (*Strombus gigas*) is fished by free diving, diving with compressors or collection from shallow water. Conch landings in 1997 were 1.43 million lbs, about the same as that from the TCI but from a much larger area. Conch are not heavily exploited in the Bahamas across all banks, but this does not exclude local depletion particularly around major population centres such as New Providence. In contrast, the TCI conch stock is probably fully exploited, but not overfished (CFMC/CFRAMP, 1999).

The spiny lobster (locally called crawfish) fishery of the Bahamas as a whole is the fourth largest spiny/rock lobster fishery in the world. The latest analyses of Department of Fisheries (DoF) catch and effort data, although not conclusive, appear to show that the spiny lobster resource is at least fully exploited (FAO/CFRAMP, 1998), and catches have not increased in the past four years despite increased fishing effort. The TCI spiny lobster fishery is more artisanal in nature with vessels only going on one-day trips. The lobster are caught by hooking, using mask and fins, from small boats. Experimental artificial lobster habitats have been deployed in some areas off South Caicos to provide additional shelter for spiny lobsters against natural predators and to increase the stock size. The catch is sold to the processing plants on South Caicos and Providenciales which then package the fish for export to the United States. Trolling off small boats is used to catch small tunas and other off-the-reef pelagic fish. Scalefish is mostly sold directly to the hotels or consumed locally.

The spiny lobster fisheries may be particularly vulnerable at present because these fisheries throughout the Caribbean are reaching the point of maximum exploitation (Cuba, Mexico, Florida/USA) or overfishing (Brazil, Central America) (FAO/CFRAMP, 1998) and it is possible that some areas of the Bahamas Bank including the TCI are supplied with larvae from these areas. However, Medley and Ninnes (1997) found a correlation between spawning stock size and recruitment after a delay of four years. The relationship awaits verification, but suggests that significant recruitment may be supplied from local stocks, and therefore local populations may be vulnerable to recruitment overfishing.

PROTECTIVE MEASURES

Both TCI are signatories to the World Heritage Convention, but they have not inscribed sites under the Convention. The TCI have signed the RAMSAR Convention and have established a RAMSAR site along the southern edge of the eastern Caicos islands, but the TCI—as a British Overseas Territory—still have to decide on the legal issue and responsibilities of becoming a signatory in relation to the British Government. The TCI are a signatory to the Cartagena Convention's Specially Protected Areas and Wildlife Protocol.

Several Marine Protected Areas (MPAs) have been set up to allow the wise and sustainable use of the coastal marine environment. Many of the designated areas still only exist on paper. The list of marine protected areas in

Table 1
Protected areas in the Turks and Caicos islands

	Size (ha)	Year
Area of Historical Interest		
HMS Endymion Wreck	2.6	1992
Salt Cay	176	1987
National Park		
Admiral Cockburn Land and Sea	431	1987
Chalk Sound	1460	1987
Columbus Landfall Marine	518	1992
East Bay Islands	3541	1987
Fort George Land & Sea	494	1987
Grand Turk Cays, Land and Sea	156	1987
North West Point Marine	1026	1987
Princess Alexandra Land and Sea	2644	1987
South Creek	74	1987
West Caicos Marine	397	1992
Nature Reserve		
Admiral Cockburn	431	1987
Bell Sound	1142	1975
North, Middle & East Caicos (RAMSAR Site)	54390	1992
Pigeon Pond and Frenchman's Cay	2392	1987
Princess Alexandra	182	1987
Vine Point and Ocean Hole	757	1992
Sanctuary		
Big Sand Cay	151	1987
French, Bush and Seal Cays	20	1987
Long Cay Sanctuary	80	1987
Three Mary Cays	13	1987

Table 1 has been derived from the WCMC Protected Areas Database, and the sites selected are based on the IUCN definition of marine protected areas (resolution GA 17.38, 17th General Assembly, IUCN), which describes them as "any area of intertidal or subtidal terrain, together with its overlying water and associated flora, fauna, historical and cultural features, which has been reserved by law or other effective means to protect part or all of the enclosed environment". This definition includes sites with only a very small subtidal or intertidal territory, which might otherwise be regarded as wholly terrestrial. Also, within this list are sites described as recommended and proposed, these are thought to have no current legal protection status.

A number of regulations exist to protect local fisheries. In the TCI there is a minimum carapace length for lobster of 83 mm, a minimum tail weight of 5 oz, and the same closed season. Use of SCUBA to catch lobster, conch and finfish remains illegal.

Because conch is listed in Appendix II of the Convention on International Trade in Endangered Species of Flora and Fauna (CITES) as a species that is at risk of over-exploitation, the TCI has set a total export quota of 600,000 lbs processed meat (682 t landings), which has successfully maintained catch rates and protected the fishery (CFMC/CFRAMP, 1999). There is a minimum shell length, but it is not enforced as conch are shelled ("knocked") at sea.

Illegal fishing remains a problem for all islands, particularly away from population centres where enforcement is non-existent. The amount and significance of illegal fishing is unknown. Marine protected areas in the TCI are listed in Table 1 (DECR, 1992). In order to raise the necessary funding to run and maintain the national park system, the TCI Government introduced an environment tax in 1997. The tax is raised on hotel beds used per day and is paid into a special account for national parks management.

REFERENCES

Bunt, J.S., Williams, W.T. and Chalker, B.E. (1981) Coral associations at depths of 45 to 125 feet in the Bahamian region. *Proceedings of the 4th International Coral Reef Symposium, Manila* 1, 707–714.

CFMC/CFRAMP (1999) Report on the queen conch stock assessment and management workshop. Belize City, Belize, 15–22 March 1999. In prep.

Deleveaux, V. and Higgs, C. (1995) A Preliminary Analysis of Trends in the Fisheries of the Bahamas Based on the Fisheries Census. Department of Fisheries, Bahamas.

DECR (1992) Maps of National Parks, Nature Reserves, Sanctuaries and Areas of Historical Interest as Listed in the National Parks Order 1992. Department of Environment and Coastal Resources, Grand Turk, TCI.

FAO/CFRAMP (1998) Report on the Caribbean Spiny Lobster Stock Assessment and Management Workshops: Section 2. Belize, 21 April–2 May 1997 and Mérida, 1 June–12 June 1998. FAO, In prep.

Gascoine, B. (1991) *Diving, Snorkelling and Visitor's Guide to the Turks & Caicos Islands*. Graphic Reproductions, Miami USA. 94 pp.

Gaudian, G. and Medley, M. (1995) Diver carrying capacity and reef management in the Turks and Caicos Islands. *Bahamas Journal of Science* 3 (1).

Gaudian, G. (1995) Sustainable environmental carrying capacity of the dive sites on Grand Turk, TCI. Report to the Department of Environment and Coastal Resources, Government of the Turks and Caicos Islands.

Holowesko, L.P. (1997) Environmental Preservation In: The Commonwealth of The Bahamas. *Islander Magazine* Issue 4 July 1997 (published by Skye International Teleservice Centre).

Medley, P.A.H. and Ninnes, C.H. (1997) A recruitment index and population model for spiny lobster (*Panulirus argus*) using catch and effort data. *Canadian Journal of Fisheries and Aquatic Sciences* 54, 1414–1421.

Newell, N.D., Rigby, J.K., Whiteman, A.J. and Bradley, J.S. (1951) Shoal water geology and environs, eastern Andros island, Bahamas. *Bulletin of the American Museum of Natural History* 97, 1–30.

Newell, N.D., Imbrie, P., Purdy, E.G. and Thurber, D.C. (1959) Organism communities and bottom facies, Great Bahamas Bank. *Bulletin of the American Museum of Natural History* 117 (4), 177–228.

Olsen, D.A. (1986) Fisheries assessment for the Turks & Caicos Islands. Consultant's report, FAO, Rome, 73 pp.

Operation Raleigh (1986) Report on the distribution of habitats and species of the north coast of Providenciales and Leeward Cays. Report on the TCI expedition, University of York, Part 1, 58 pp.

Reed, J.K. (1985) Deepest distribution of Atlantic hermatypic corals discovered in the Bahamas. *Proc. 5th Int. Coral Reef Symp., Tahiti* 6, 249–254.

Reynolds, J.E. and Odell, D.K. (1991). *Manatees and Dugongs*. Facts on File Ltd, Oxford.

Sealey, N.E. (1985) *Bahamian Landscapes. An Introduction to the Geography of the Bahamas*. Collins Caribbean, 96 pp.

Sullivan, K.M., Chiappone, M. and Lott, C. (1994) Abundance patterns of stony corals on platform margin reefs of the Caicos Bank. *Bahamas Journal of Science* **1** (3), 2–11.

UNEP/IUCN (1988) *Coral Reefs of the World. Vol. 1: Atlantic and Eastern Pacific*. UNEP Regional Seas Directories and Bibliographies. IUCN, Gland, Switzerland and Cambridge, UK/UNEP Nairobi, Kenya, xlvii + 373 pp, 38 maps.

WCMC (1999) Website information at www.wcmc.org.uk.

THE AUTHORS

Gudrun Gaudian
Sunny View, Main Street,
Alne, N. Yorks, YO61 1RT, U.K.

Paul Medley
Sunny View, Main Street,
Alne, N. Yorks, YO61 1RT, U.K.

Chapter 38

THE DUTCH ANTILLES

Adolphe O. Debrot and Jeffrey Sybesma

The six islands and offshore Saba Bank which make up the Dutch Antilles lie in the Caribbean Sea. The principal marine littoral habitats are coral reefs, seagrass beds, algal beds, mangroves and salt ponds, all of which have been only partially inventoried and mapped, so that few baseline data exist for most islands. Only scattered documentation exists on marine pollution loads and resource use pressure. Nevertheless, indications are that the current magnitude of coastal development and industrial pollution in Aruba, Curaçao and St. Maarten have already reached such proportions that large tracts of reef and related coastal habitats are being rapidly degraded. Major problem areas identified for the various islands are: pollution caused by oil refineries; sewage discharge and eutrophication of coastal waters; beach tar and litter contamination; erosion due to overgrazing and poor real-estate development practices; overfishing; municipal landfills which form a long-term threat; and unregulated coastal urbanization. Several islands have extensive areas which are vulnerable to hurricane wave damage and long-term sea-level rise.

Governmental efforts towards a more coherent environmental development policy only began in 1992 with the Rio Conference. In 1996 the Netherlands Antilles Ministry of Public Health and Environment published a framework for environmental policy, and in 1998 a legal framework was established for nature management and conservation. Recent decades have seen significant advances, particularly in nature conservation, which have largely been driven by "governmental NGO" national park management foundations. Nevertheless, on the whole, current environmental legislation remains deficient and fragmentary and fails to adequately address most major issues. Also, government investment in environmental matters remains minimal, notwithstanding high GNPs and standards of living by regional comparison. None of the islands have as yet developed any vision of the greater issue of population size which drives most economic, infrastructural and environmental problems.

A recent assessment by ECLAC (Economic Commission for Latin America and the Caribbean) identifies the establishment of good governance as the main priority for the implementation of sound environmental policy in the Netherlands Antilles. This is followed by institutional capacity building, and policy and legislation development, whereby public awareness and participation are recommended as key elements of strategy.

The chance that the various problems will effectively be dealt with in the near future differs between islands. We predict that the less populous islands (Saba, St. Eustatius, Bonaire) where population pressure is less, the economies are simpler and more ecotouristically oriented, and where government is less complex, show the greatest promise in being able to deal effectively with environmental issues. We further predict that issues such as rampant coastal urbanization, fishing pressure and pollution by the petroleum sector (and other large enterprises) are not likely to be dealt with effectively in the foreseeable future on most islands, notwithstanding the many management plans and the various initiatives towards policy development in these areas.

Seas at The Millennium: An Environmental Evaluation (Edited by C. Sheppard)
© 2000 Elsevier Science Ltd. All rights reserved

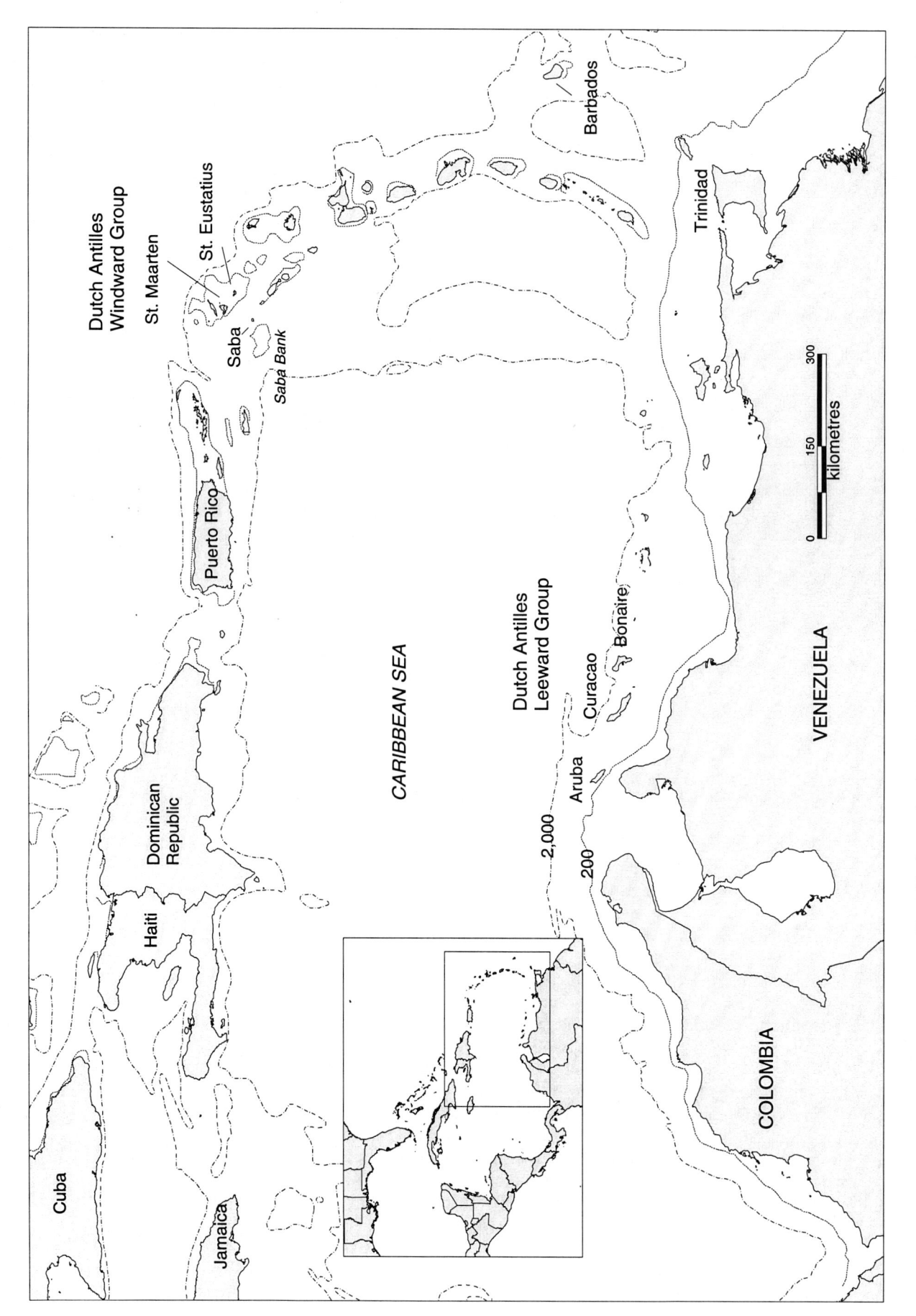

Fig. 1. The leeward and windward groups of the Dutch Antilles in the Caribbean Sea.

DEFINITION AND GENERAL DESCRIPTION OF THE REGION

The Dutch Antilles comprise the Netherlands Antilles and Aruba and consist of a leeward and windward island group, separated by about 900 km of open sea (Fig. 1). Neither of the two subregions here distinguished can be considered a main ecosystem in and of themselves but simply form part of a greater "Caribbean" or "tropical western Atlantic" ecosystem. Aruba, which forms part of the leeward island group, was formerly part of the Netherlands Antilles but has had a sovereign status within the Kingdom of the Netherlands since 1986. Throughout the 1980s, until a consultative referendum held in 1993 indicated otherwise, the ultimate disintegration of the remaining Netherlands Antilles into smaller island states had been considered inevitable. At present this matter is by no means settled yet. Some basic descriptive data for the six island territories and offshore Saba Bank can be found in Table 1.

Leeward Group

The leeward group is composed of the islands Curaçao, Bonaire, and Aruba (Fig. 2) which lie between 12–13°N and 68–70°W, 30–70 km off the coast of Venezuela in the southern Caribbean. With a total surface area of about 444 km², Curaçao is the largest of the three and is surrounded by well developed fringing coral reefs which constitute the principal nearshore marine habitat. The total number of inhabitants has been stable in recent decades at about 150,000, and urbanization is concentrated in the east-central third of the island around the industrial Schottegat harbour. The economy is principally supported by offshore banking, oil industry, ship repair and maintenance, tourism and trade (container trans-shipment). The Netherlands Antilles, of which Curaçao is by far the largest and most populous island, has a high per capita GNP by regional standards (US$ 11,032 in 1995) (CBS, 1996). Large areas of Curaçao have been deforested due to extensive livestock grazing and past agriculture, but today runaway urbanization and tourist subdivision development projects are the main threat to the remaining coastal wildland areas and reefs.

Bonaire is 288 km² in surface area and had 14,218 inhabitants in 1995, up more than 25% from 11,000 in 1991. The island is surrounded by fringing coral reefs which constitute the principal nearshore marine habitat. Urbanization is concentrated in the central part of the island around the anchorage of Kralendijk. The economy is principally supported by dive tourism. Limited oil trans-shipment, real-estate development and evaporative sea-salt production are other significant economic activities. Large areas of the island have been deforested due to extensive livestock grazing, wood harvest and seasonal agriculture, which remain significant to the present.

Table 1
Preliminary quantitative* overview of coastal habitat categories for the seven territories of the Dutch Antilles

	Aruba	Curaçao	Bonaire	St. Maarten	St. Eustatius	Saba	Saba Bank
Capital	Oranjestad	Willemstad	Kralendijk	Phillipsburg	Oranjestad	The Bottom	–
Surface area (km²)	190	444	288	34	21	13	2200[1]
Total inhabitants	83600	150000	14200	38000[2]	2000	1200	0
Population density (per km²)	440	338	49	1118	95	92	0
Littoral Habitats (approx. surface areas* in ha)							
Coral reefs	990	4560	4372	692	180	14	18958
Seagrass beds	3520	494	104	2799	82	56	?
Reefal algal beds	1247	2223	3335	?	?	?	?
Mangroves	292	55[3]	79	0	0	0	0
Saliñas	0	378	2178	60	0	0	0
Undeveloped coastline[4] (km)	35	90	77	5	20	16	0

*Habitat surface area estimates are strictly preliminary and indicative, as coastal and marine surveys have been only partial for most territories. Only secondary, largely qualitative sources were used. With one exception (see below), area estimates correspond exactly (were not rounded) to the habitat areas as sketched in Figs. 1 and 2.
(1) Surface of submerged areas up to 200 m depth. Other surface areas given in this row are simply island surface areas.
(2) Not included are approx. 25,000 unregistered inhabitants.
(3) This estimate (Debrot and de Freitas, 1991) is based on mangrove coverage as from aerial photographs, and is more precise than, and not comparable to, the other preliminary estimates presented.
(4) Includes only coastline which is either prohibitive for industrial and urban development (e.g. exposed coasts of the leeward islands), or for which government policy intent (at this time) is to designate it as protected coastline.

Aruba has a surface area of about 190 km² and is situated on the continental shelf of South America. Much of the leeward coast is dominated by a string of barrier islands and a sandy coastal lagoon. The sandy beaches have been the basis for the development of mass tourism which is the principal source of revenue. The second most important industry is the oil industry. The total population size was 83,600 by 1995 and population growth due to immigration is high (5% annually during 1991–1995) (DOW Aruba, 1995). As with the Netherlands Antilles, per capita GNP is high by regional standards. The aridity of the island and sparse vegetation are striking.

The leeward Dutch islands lie in an arid region of the Caribbean. Average annual rainfall varies between 567 mm for Curaçao and 425 mm for Aruba. The wind is quite persistent and blows from northeast to southeast (average direction: east) for more than 95% of the time. Predominant wind speeds are at levels 4 and 5 of the Beaufort scale, with average annual windspeeds of between 6.7 and 7.5 m/s. Monthly mean temperatures average about 27.5°C with monthly maxima and minima of about 31°C and 25.5°C, respectively (Meteorologische Dienst, 1982).

The average annual sea surface water temperature is 26.8°C and varies from 25.4°C in February to 28.1°C in September. The islands lie downstream one of the most productive fishing grounds of the Caribbean—the area of upwelling near Margarita, Venezuela (Sturm, 1991). Even though the waters are generally highly oligotrophic, the islands lie within an area in which seasonal upwelling occurs (Sturm, 1991; Gast, 1998). The principal current regime is that of a branch of the South Equatorial Current which flows up into the Caribbean Sea along the coast of Guyana and Venezuela (Schuhmacher, 1976). Eddies along the southwest coast are common, in which the longshore current direction at any given time or place may be contrary to the principal current direction. Mean tidal range is about 30 cm.

Windward Group

The windward group is composed of the islands of St. Maarten, St. Eustatius and Saba, and the Saba Bank, all of which lie at about 17–18°N and 63°W (Fig. 3), and which form part of the Leeward Islands of the Lesser Antilles. St. Maarten is the largest island and has a surface area of approximately 86 km², 34 km² of which falls under Dutch jurisdiction. The remainder of the island falls under French jurisdiction. The island is surrounded by shallow bank waters and displays extensive sandy beach development. In recent decades the population of Dutch St. Maarten has grown explosively from about 2700 in 1960 to more than 38,000 registered inhabitants in 1995 (Rojer, 1997). Unregistered inhabitants likely number in the tens of thousands, leading to extensive shanty-town development on the hillsides. Large areas of the island remain deforested due to extensive livestock grazing in the past, but unbridled urbanization, tourist resort development and sprawling shanty-towns are now the major new usurpers of natural habitat.

St. Eustatius has an area of about 21 km². The island is surrounded by relatively shallow bank waters and the coasts of the island are dominated by steep cliffs, while sandy beaches are rare. The total number of inhabitants is about 2000, and urbanization is largely concentrated in Oranjestad. Large areas of the island remain deforested due to extensive livestock grazing and past agriculture. Island income levels have risen significantly since the establishment of an oil terminal on the island in 1982.

Saba is about 13 km². The island is a steep dormant volcano rising from depths of 600 m extending to 870 m above sea level. The shores are steep and inaccessible and the island has only one significant beach which is of black volcanic sand (and which may be seasonal) and a couple of bays with cobble and rubble. The island is inhabited by approximately 1200 people, distributed among four villages. Large areas of the island are deforested due to over-grazing by livestock, though the summit contains good, original forest.

The Saba Bank is a large sunken atoll southwest of Saba, which lies totally within the potential Exclusive Economic Zones for which the Netherlands Antilles could claim national jurisdiction in accordance with the Law of the Seas Convention. The surface area of the bank (depths 8–200 m) is about 2200 km². There is no land and no habitation. While the majority of the bank lies at depths of 20–50 m, large parts lie at depths of 10–20 m and have extensive reef development. The bank appears to be the largest actively growing atoll in the Caribbean (Meesters et al., 1996).

The windward island group differs climatologically from the leeward group, most importantly in terms of rainfall and susceptibility to hurricanes. In the windward group the average annual rainfall is about 1100 mm for all three islands. Wind blows from between 060° and 120° direction for more than 80% of the time. The predominant wind speeds are at levels 3 and 4 of the Beaufort scale, with average annual windspeeds of about 5.4 m/s. Monthly mean temperatures average about 26.7°C with monthly maxima and minima of about 30°C and 24°C, respectively (Meteorologische Dienst, 1982). The average annual sea surface water temperature is 26.4°C and varies from 24.7°C in February to 27.9°C in September. The islands lie inside the hurricane belt and experience hurricane conditions on average once every 4–5 years. The principal current regime is that of the North Equatorial Current which flows westward.

MAJOR SHALLOW MARINE AND COASTAL HABITATS

The principal marine littoral habitats of the Dutch Antilles are coral reefs, seagrass beds, algal beds, mangroves and salt ponds, all of which have been only partially

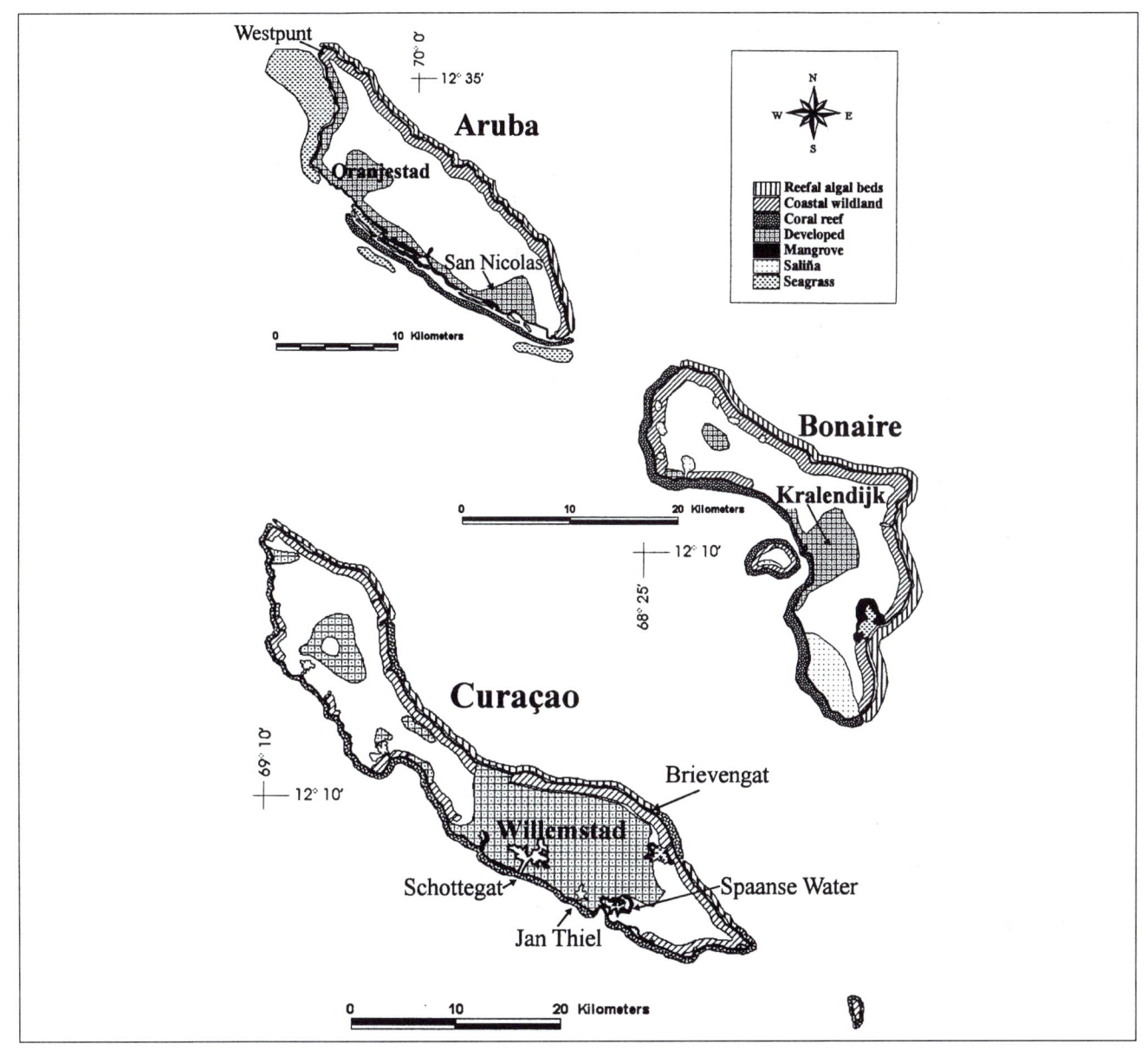

Fig. 2. The distribution of principal coastal and marine habitat areas in the leeward Dutch Antilles group.

inventoried and mapped, and few baseline data exist for most islands. Figures 2 and 3 provide maps of the leeward and windward groups, respectively, indicating the approximate distribution of principal marine habitats and (developed) population centres. Table 1 provides a rough quantitative assessment of the various habitats. However, surface area estimates are strictly preliminary as coastal and marine habitat surveys have been only partial for most territories. Also, natural shorelines and terrestrial coastal habitat are recognized as being integral components of the coastal ecosystem and are included in this review.

Coral Reefs/Reefal Algal Beds

Of the three leeward islands, it is especially Curaçao and Bonaire and their satellite islands which possess luxuriant, steep fringing reefs close to shore along the leeward coasts (to depths of about 60 m and generally within 100 m of the shore) (Fig. 2, Table 1). The good reef development may be ascribed to factors such as low rainfall and a steep reef profile, which define the oceanic nature of the reefs, as well as the fact that these islands lie outside the hurricane belt. Aruba has much more limited reef development (Table 1) as it is situated on the continental shelf and has much more sand in its shallow sublittoral areas. The nearness of the reefs to land on all islands makes them highly accessible and highly vulnerable to land-based impacts.

Along the northeast coasts of the islands wave energy impact is apparently too great to allow extensive coral growth in shallow areas, and instead the calcareous substratum is covered with large fields of the reefal alga *Sargassum* (Bak, 1975) down to depths of 10 m. These algal

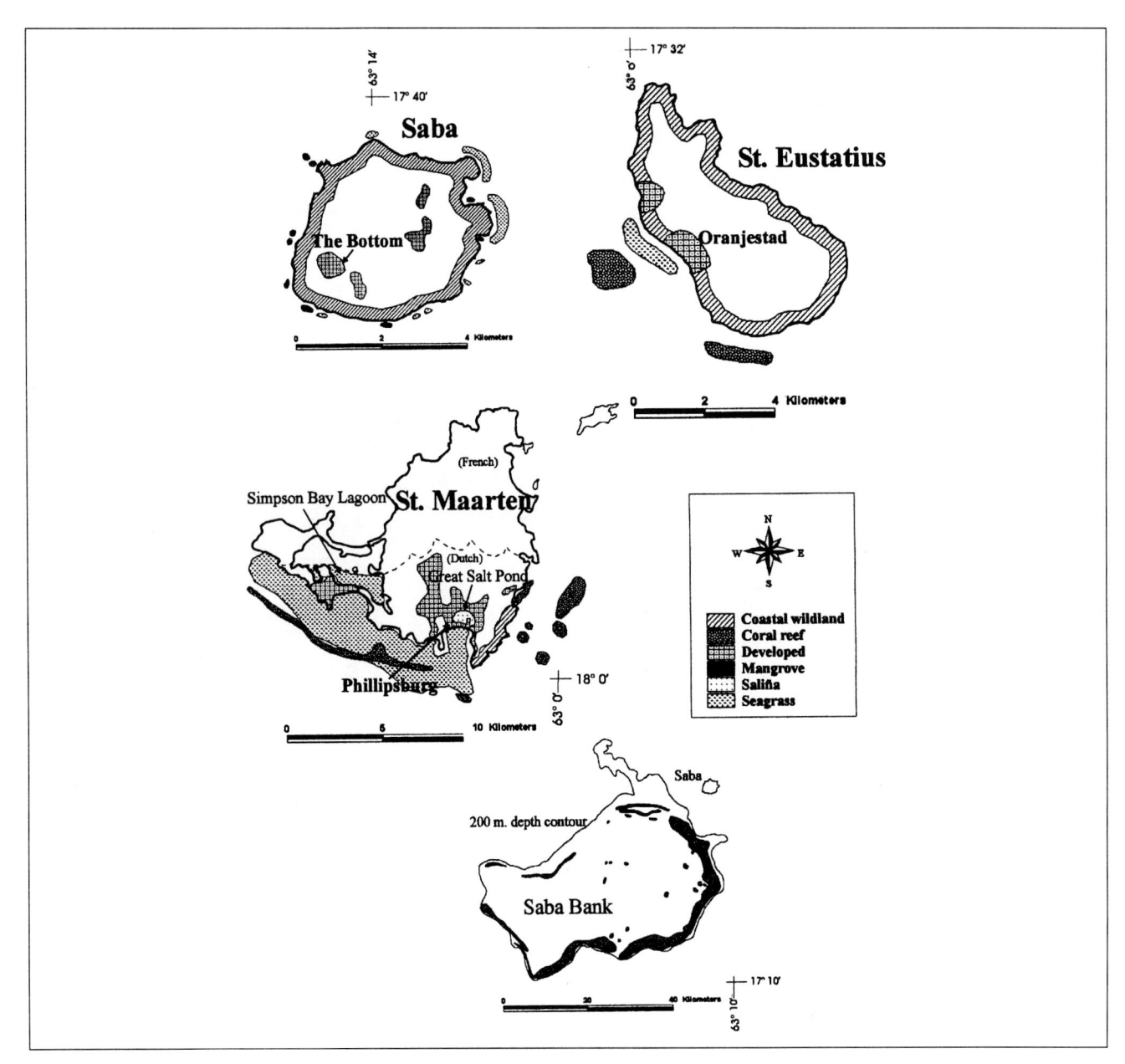

Fig. 3. The distribution of principal coastal and marine habitat areas in the windward Dutch Antilles group.

beds are a major, but little studied habitat of the leeward islands. A narrow band of coral reefs is typically found at depths of 10–30 m.

Algal beds in the windward group may also be of great ecological importance, and in St. Eustatius have been found to support significant populations of conch, *Strombus gigas* (Sybesma et al., 1993).

In the windward group, reef development around the islands is much more limited and patchy (Bak, 1975, 1977; van't Hof, 1985, 1989; Sybesma et al., 1993). Survey results by Sybesma et al. (1993), however, indicate that the existence of significant deep reef systems surrounding St. Eustatius cannot be excluded until more extensive surveys are carried out. An important exception to the generally reef-poor windward group is the Saba Bank (Table 1), where total reefal areas are estimated to be about 150 km^2 and luxuriant reef fronts are estimated to amount to 20–40 km^2 (Meesters et al., 1996). However, practically nothing is known about the reefs of the Saba Bank.

Mangroves

Mangrove habitat occurs almost exclusively in the shore zone of inland bays at the interface of the terrestrial and marine environments. As is typical of arid regions, where mangrove growth is limited due to low fresh water influx, zonation may be partial or incomplete, and mangrove vegetation typically forms a narrow fringe lining inland waters. By far the most common mangrove species is the red mangrove (*Rhizophora mangle*).

Mangrove habitat, which serves an important nursery function for many reef fish, is predominantly found in the leeward group (Table 1). In the windward group this habitat type is only found on the island of St. Maarten in tiny remnant patches. This habitat type was never abundant on St. Maarten but is now almost completely destroyed.

Seagrass Beds

Seagrass beds of the sheltered lagoons and shallow offshore areas are another habitat of major ecological importance, not only in terms of their fish nursery function, but also as habitat for the endangered conch (*Strombus gigas*) and green turtle (*Chelonia mydas*). The principal seagrass in Curaçao and Bonaire is *Thallassia testudinum*, commonly known as turtle grass. At depths of up to 4 m, it often forms dense cover on sandy and muddy bottoms in the inland bays.

In the windward group, seagrass beds are found in protected coves and lagoons, or in deeper water where wave action is less. Manatee grass *Syringodium filiforme* and turtle grass, *Thallassia testudinum*, are the principal two seagrass species in St. Maarten. In St. Eustatius, manatee grass is the only seagrass species of importance (Sybesma et al., 1993). The existence of extensive seagrass communities on the Saba Bank cannot be excluded.

Salinas

Salinas (salt lakes and salterns) are landlocked saline water bodies which basically represent remnants of former bays and generally do not have a direct opening to the sea. In the leeward islands these are found on Curaçao and especially Bonaire, but not Aruba (Table 1). Environmental conditions such as salinity and temperature fluctuate heavily on an annual basis, and these habitats are characterized by large fluctuations in aquatic life during the course of the year. Salinas (especially in Bonaire) are of great importance as foraging and nesting areas for the endangered Caribbean flamingo, *Phoenicopterus ruber ruber*. The salinas of Curaçao amount to feeding and resting habitat for about 300 flamingos year round (Debrot and de Freitas, 1999). In the windward group salina habitat is only found on St. Maarten but has almost completely been destroyed by filling.

Coastal Wilderness

In recent years, shorelines and terrestrial coastal habitats have been recognized as being integral components of the coastal ecosystem. In the Dutch Antilles access to the coast is critical to gecarcinid land crabs (*Gecarcinus lateralis, G. ruricola*), and the West Indian hermit crab (*Coenobita clypeatus*), which function as ecologically important terrestrial omnivores in the hinterlands but which depend on the sea for reproduction. Many shore birds and sea turtles all require undisturbed natural coastline for survival.

Coastal islets, which may or may not be inside lagoons, are found off all Dutch Antilles except St. Eustatius, and are of particular breeding importance for various species of terns. In Saba, coastal cliffs form important breeding habitat for red-billed tropicbirds, *Phaethon aethereus* (Walsh-McGehee and Lee, 1998). Coastal wilderness is also important to coastal resource users. Recent surveys indicate tourists in Curaçao attach great value to natural surroundings and landscapes in addition to other beach parameters. Boat user surveys indicate that coastal scenery is either important or very important to over 90% of respondents, while almost 75% preferred a natural landscape (van't Hof et al., 1995).

RURAL FACTORS, COASTAL EROSION, LAND ACCLAMATION AND EXCAVATION

Livestock Grazing and Erosion

Table 2 provides a qualitative overview of the magnitude of anthropogenic impacts to the marine environment in the Dutch Antilles.

Extensive, excessive and uncontrolled livestock grazing has impacted all islands for centuries. It has been a major factor in causing deforestation on all islands, and continues to be a problem except on Curaçao and St. Maarten, where livestock ranching declined dramatically in recent years as property prices have risen in light of expanding urbanization. Also, dramatic increases in crime and theft on these islands, particularly on Curaçao, have become a major blight to animal husbandry. On all islands, including Aruba, legislation has long existed banning livestock grazing in public areas (e.g. Bakhuis, 1990; Koopmans et al., 1994) but this has never been enforced anywhere except until recently inside the Christoffel Park, Curaçao. However, in September 1998 the government of St. Eustatius began with a program for removal of loose-roaming livestock (VOMIL, 1998).

Erosion due to the combined causes of deforestation, overgrazing and other deleterious land-use practices (e.g., large-scale clearing of land by bulldozers for building projects) remains serious on all islands. Modern stormwater management in which rain water is channelled off to sea, instead of allowed to penetrate the ground, also likely contributes to a greater sediment load to the nearshore environment. However, in the leeward islands rainfall is low and seasonal, which tends to temporally restrict sediment stress. The coastal areas of the leeward islands are also, to a large degree, rimmed by coralline rock and most runoff to sea is channelled through lagoons and salinas where sediments settle and accumulate. In this way, most of the potentially devastating erosion is trapped and does not reach the reef environment. Nevertheless, for Saba, van't Hof (1985) indicates heavy seasonal sediment input due to erosion (by goats and rainfall) as a limiting factor to reef development (Fig. 4).

Fig. 4. Rum Bay on Saba, 1997, an island also known as "The Unspoiled Queen". Deforestation, exacerbated by goat grazing on erosion-prone slopes, results in sedimentation which is detrimental to the fringing coral reefs of the island. Deforestation also contributes to global processes which indirectly harm the island's marine biota, such as global warming, sea-level rise and global changes in weather patterns. Photo: A. Rojer.

In Curaçao, Bonaire and Aruba, large parts of the coastline are composed of fossil coralline rock or volcanic rock. Coastal erosion is therefore hardly a problem. In the windward group Saba has no significant coastal erosion, partly because the shores are mainly cliffs, while on St. Maarten, hurricanes occasionally cause serious beach erosion. On St. Eustatius coastal erosion is a serious problem in the central third of the island which is composed of a thick packet of loose volcanic dust and where erosion is exacerbated by deforestation.

Land Reclamation and Excavation

In Curaçao coastal lagoons of Rif, Waaigat and various other areas of the Schottegat, all within the industrial and commercial centre of Willemstad have been filled, since the beginning of the century, thereby destroying significant amounts of lagoonal habitat. Filling and destruction of shallow lagoonal habitat continues in these areas as well as inside the Spaanse Water, which is rapidly being developed as an urban area.

Several muddy inland waters of the southwest coast of Curaçao, such as Piscaderabaai (1972) and Sta. Marthabaai which formerly had narrow, shallow entrances, have been dredged for improved access for vessels. This has translated into persistent sediment loads to adjacent reef areas and large-scale degradation of those reefs (e.g. Bak, 1978). Plans are being made to open up several saliñas (such as Jan Thiel, Rif St. Marie and Cas Abou), which will cause even greater persistent sediment loads to the reefs and destroy the saliña habitat which the endangered Caribbean flamingo visits for feeding.

In Curaçao dredging of bottom sediment regularly takes place inside the industrially developed Schottegat bay. Djohani and Klok (1988), Rijkswaterstaat (1992) and Ebbing (1997) document significant contamination for certain areas within the bay. The spoils are being dumped three miles off Willemstad harbour at depths of about 3–400 m.

In Aruba the only significant filling taking place is at the municipal dump which lies in the coastal lagoon system used for recreation and other purposes (Koopmans et al., 1994).

In Bonaire a lagoonal system excavated for hotel construction continues to discharge sediment onto adjacent reefs and has led to ruination of groundwater quality in adjacent coastal areas, ever since its excavation many years ago. In St. Maarten, significant filling of saliña and lagoonal habitat has occurred and continues to take place. The Great Salt Pond of St. Maarten, which was the largest saliña of the island, lies landwards from the capital of Phillipsburg and is subject to filling for urbanization and also contains the municipal landfill.

EFFECTS FROM URBAN AND INDUSTRIAL ACTIVITY

Artisanal and Non-industrial Uses

Fishing Activity

In 1988, the total number of fishing boats of 5 m or more in length for Curaçao was estimated at 182, and the number of professional fishermen at 120 full-time and 250 part-time fishermen (LVV, 1988). Total fish catch was estimated at 900–1200 metric tons per year (LVV, 1988) of which 10–15% constituted demersal, largely reef fish (LVV, 1988; FAO, 1991). The total value of fisheries production amounts to approximately ANG 6.3 million per year (LVV, 1992).

The most common fishing methods used are trolling, line fishing, seine fishing, and trap fishing. There is wide awareness among those who use the reef of a rapid decline in abundance and catches of valued reef organisms, and studies indicate that densities of large piscivores (groupers, snappers, jacks, barracuda) are low (van't Hof et al., 1995). The Agriculture and Fisheries Service of Curaçao (LVV)

Table 2

Overview and qualitative assessment of the magnitude of various deleterious factors with respect to the marine environment for the seven territories of the Dutch Antilles

	Aruba	Curaçao	Bonaire	St. Maarten	St. Eustatius	Saba	Saba Bank
Rural Factors							
Livestock grazing	P	M	P	M	P	P	NA
Soil erosion	P	P	P	P	P	P	NA
Coastal erosion	L	L	L	M	Pl	L	NA
Land acclamation and excavation	M	P	L	P	L	L	NA
Misc. Artisanal and Non-industrial Use							
Fishing pressure (demersal stocks)	L	P	L	P	M	L	P?
Recreational diving	L	L	M	L	L	L	NA
Recreational boats	L	Pl	L	P?	L	L	NA
Seabird disturbance	L	P	L	M?	L	L	NA
Beach sand mining	L	L	L	L	L	L	NA
Artificial beaches & replenishment	L	P	L	NA	NA	NA	NA
Industry							
Oil contamination	P	P	M?	M?	M?	L?	NA
Beach tar	P?	P	P	M?	M?	M?	NA
Industrial effluents	P	P	Pl	NA	NA	NA	NA
Sand mining offshore	L	L	NA	NA	NA	NA	NA
Groundings/anchor damage	L	M	L	P	M	NA	M
Cities							
Coastal development/urbanization	P	P	M	P	L	L	NA
Sewage	P	P	Pl	P	L	L	NA
Municipal dump	P	L	L	P	L	L	NA
Litter	P	P	M?	P?	P	L?	NA
Cumulative Degradation	P	P	M	P	M	L	L

P = problem level; Pl = problem level but localized; M = moderate; L = low; NA = not applicable, ? = likely but not certain.

considers the reef to be overfished and is alarmed by the continued overcapitalization of the fishery sector (Dol, 1995). Recent years have seen the development of the deleterious fishing practice of spanning gillnets clear across the openings of inland waters to target larger reef fish which enter the lagoons at night.

In 1998 total numbers of resident fulltime commercial fishermen for Bonaire were less than 20 (J.A. De Meyer, Bonaire Marine Park, pers. comm.). Roberts and Hawkins (1994a) indicate high concentrations of predatory groupers and snappers which can be interpreted as indicative of relatively low cumulative levels of fishing pressure.

As Aruba has a seafloor surface of about 2630 km^2 in the 50–200 m range, it has much greater access to demersal resources than the other Dutch Antilles (Guidicelli, 1980). Total fish production (1998) is estimated at about 250 metric tons per year. The island only has about 20 full-time fishermen, 660 part-time fishermen which fish "regularly" and more than 3000 occasional fishermen. The fleet consists of 265 locally built small vessels. The stocks are likely not yet overfished (B. Boekhoudt, LVV Aruba, pers. comm.).

In St. Maarten, reef fish populations lack large piscivores such as groupers and snappers (van't Hof, 1989). Fishermen ascribe apparent declines in the catches to factors other than excessive fishing (van't Hof, 1989). On St. Eustatius there are about 15 active fishermen, of which only four are full-time. The principal catch species are lobster, reef fish, pelagics and conch. About 300 fishing pots are deployed (Sybesma et al., 1993). The paucity of larger piscivores such as grouper and snappers is indicative of overfishing. In Saba, fishing pressure appears to be limited (van't Hof, 1985). In 1998 about five people fished regularly around the island and about 5–10 West Indian fish traps were in use (D. Kooistra, Saba Marine Park, pers. comm.).

In recent years the Saba Bank is increasingly being fished by non-Saban fishermen. Meesters et al. (1996) provide some data for the Saban fleet while Guidicelli and Villegas (1981) estimated total fish catch by all fleets to be about 1000 tons annually. Meesters et al. (1996) report that spearfishing on the bank is not uncommon and that recently spearfishermen and divers using hookah gear (divers with surface-compressed air) have harvested groupers, lobsters and conch from large areas of the bank, while average fish sizes in the catch appear to have decreased considerably. A large section of the Saba Bank is spared from finfish fishing due to ciguatera poisoning which renders the fish dangerous for consumption (Meesters et al., 1996).

Recreational Diving

Marine park and tourism offices have data which show consistent growth of dive tourism for Saba, Bonaire and Curaçao in recent years. The island with most divers is Bonaire. By 1998 total divers had increased to 29,000 annually, corresponding to annual revenues of about US$ 39 million (J.A. De Meyer, Bonaire Marine Park, pers. comm.). Roberts and Hawkins (1994a) indicate that at 1994 levels, diving activity at Bonaire dive sites, some of which were dived close to 6000 times per year, likely caused little lasting damage.

Recreational Boating

The Curaçao Ports Authority estimates that there may be a total of 2500–3000 boats in Curaçao, of which 1300–1800 would be pleasure boats (van't Hof et al., 1995). The largest concentration of recreational boaters in Curaçao can be found in Spaanse Water, where 993 vessels were counted in May 1992 (Debrot et al., 1998). In Bonaire a total of 575 vessels are registered, 295 as fishing boats and 168 as pleasure boats (J.A. Frans, Bonaire Marine Park, pers. comm.). In Aruba 265 locally built small vessels and 124 pleasure yachts were registered in 1998 (B. Boekhoudt, LVV Aruba, pers. comm.). In St. Maarten at times the number of vessels (largely foreign yachts) may number upwards of 1000 in Simpson Bay Lagoon alone (AIDEnvironment/EcoVision, 1996), which certainly must adversely affect water quality.

Disturbance of Seabird Habitat Areas

The Netherlands Antilles are of regional significance as a principal nesting area for a number of seabirds (e.g. van Halewijn and Norton, 1984). In Curaçao all principal seabird and shorebird sites (Debrot and de Freitas, 1991) are seriously impacted by disturbance or remain vulnerable to projected future building projects.

On Aruba, the internationally significant tern breeding islands lack formal status as protected habitat (Koopmans et al., 1994) but are closed annually to public access during the breeding season by government decree (R. de Kort, Parke Nacional Arikok, pers. comm.). Monitoring and enforcement is conducted on a nonstructural basis by volunteers. In Bonaire, breeding and foraging areas (for terns and upwards of 2000 flamingos year-round) are kept relatively free of human disturbance largely thanks to the salt company keeping the public out of the principal habitat area (saltworks area). On Saba and St. Eustatius, human disturbance of the regionally important tropicbird breeding colonies is not a problem due to the general inaccessibility of the coastal cliffs. For St. Maarten the significance of offshore islands for seabird breeding is likely but has not yet been confirmed.

Beach-sand and Coral-rubble Mining

The beaches of the leeward islands are used for nesting by a limited numbers of at least four species of endangered sea turtles. Illegal artisanal sand and gravel mining on public beaches in Curaçao has long been a recognized problem but enforcement has always been lacking. Even small-scale gravel mining may cause long-term damage to beach quality (for both sea turtle nesting and recreation) as the beaches of Bonaire and Curaçao, in particular, tend to be poorly endowed with sand. Today, illegal sand mining likely continues on all three islands of the leeward group but at very limited levels.

Touristic Artificial Beaches and Beach Replenishment

Due to the paucity of natural sandy beaches in certain areas of the coast of Curaçao, a trend has developed since 1986 to create artificial beaches for tourism using containment dams in areas where beaches were not formerly present. Sand replenishment on natural beaches also takes place regularly. These activities cause serious damage to sessile sea life due to initial smothering and prolonged sediment stress. All of at least seven artificial beaches in Curaçao require periodic replenishment as sand retention is poor. On Aruba artificial beach replenishment is rare as the beaches accumulate sand naturally. On Bonaire, beach replenishment is also rarely done as the island prides itself on its luxuriant reefs. On St. Maarten beaches are quite vulnerable to hurricanes. Nevertheless, natural recovery, which appears to be intimately tied to recovery of nearshore seagrass beds, is often relatively rapid (AIDEnvironment/Ecovision, 1996).

Artificial Dive Objects

In the past numerous objects have been intentionally sunk on the reefs of Curaçao (especially) and Bonaire as a dive attraction for tourists. In Curaçao the majority of such intentional sinkings have been failures due to faulty preparation and omitting to anchor the objects once sunk. As the reef is steep, many sunk vessels have simply disappeared into the depths leaving nothing but a large scar in the reef (Carmabi archives).

Industrial Use

Oil Contamination

Large oil refineries have been in operation since the early 1920s in Curaçao (1992 capacity: 237,000 bls d^{-1}) and Aruba (1985 capacity: 300,000 bls d^{-1}; 1995 capacity: 225,000 bls d^{-1}) and have brought significant oil pollution to the islands.

Fig. 5. Long-term accumulations of tar "rock" deposits (50–80 cm high) on a former rubble beach at Caracasbaai, Curaçao, 1993. Photo: I. Nagelkerken.

Total oil pollution load for the refinery in Curaçao is estimated at 1050 t/yr, accounting for well upwards of 90% of the total insular marine oil pollution load (Buth and Ras, 1992). In Curacao, as well as in Aruba, subterraneous oil contamination is severe (e.g., Koopmans et al., 1994), causing significant amounts of oil to leach out into the sea. In Curaçao, a large 62 ha impoundment inside the Schottegat harbour was used to dump approximately 2 million m^3 of unrefinable tar residues, where it remains to this date (H. Hoyer, Curaçao Environmental Service, pers. comm.). Each year during the hurricane season when the wind direction changes, oil slicks make their way out of the Schottegat, soiling ship hulls and hotel beaches.

Tar contamination of coastal beaches has been studied for Bonaire (Newton, 1987) and Curaçao (Debrot et al., 1995) and is generally very high. Massive tar contamination (Fig. 5) was found to have long-lasting deleterious effects on intertidal molluscs (Nagelkerken and Debrot, 1995), while chronic oil pollution has been found to seriously affect submerged coral communities in Aruba (Bak, 1987). Bonaire has had oil trans-shipment facilities since 1975 and St. Eustatius since 1982. Small oil spills appear to be rather frequent at the Statia Oil Terminal (once every two months or less) and dispersants appear to be used (Meesters et al., 1996).

Industrial Effluents

In Curaçao, the oil refinery is estimated to account for about 91% of the insular marine pollution load in terms of BOD$_5$ (estimated at 2800 t/yr) and 87% of the insular pollution load in terms of N (estimated at 330 t/yr) (Buth and Ras, 1992). A large fraction of the mangrove and seagrass habitat of Curaçao (those of the Schottegat and Piscaderabaai) are already polluted to the point at which their ecological functioning can be considered to be grossly impaired (Debrot and de Freitas, 1991). Baseline data on contaminants for various inland bays by Djohani and Klok (1988), Rijkswaterstaat (1992) and Ebbing (1997) show that bottom sediments of especially the Schottegat are contaminated with Cu, Pb, Zn, Ni, Cd, petroleum and other organics. Contaminated fish from inside the Schottegat are only caught for consumption on a limited scale. However, no studies have yet been done to assess contamination levels in these fish or foodfish species from the intensively fished nearby reef waters.

Some of the most serious and persistent marine contamination likely results from the spoils of the Curacao Dry Dock with annual production figures of about 150 vessels per year (Buth and Ras, 1992). At Brievengat industrial effluents are largely related to the soap industry, but the Willemstad paint factory (total annual paint production of about 2000 m^3) also discharges waste water at Brievengat. The large amount of cooling water discharged by the oil refinery is responsible for major thermal pollution of the Schottegat bay in Curaçao. The hot saline discharge water of the Curaçao water desalination plant has a major though localized effect on the coral community near the discharge point at Parasassa (Hoppe, 1982). Industrial cooling water production in Aruba (1995) is approximately as follows: the refinery 60,000 m^3 d^{-1}; the water and electricity plant 30,000 m^3 d^{-1} (DOW Aruba, 1995).

In Bonaire the principal industrial effluents are thermal wastewater from a desalination plant which causes localized damage to the reef (Bak and van Moorsel, 1979).

Offshore Sand Mining

On a few occasions in the past, offshore sand mining on an industrial scale has occurred along the windward coast of Curaçao at depths of about 30 m. More structural, industrial-level sand mining only occurs on Curaçao in the mouth of the St. Jorisbaai on the windward coast where sand tends to accumulate at shallow depths. The removal of sand at the entrance to this bay may enhance circulation within the bay but may also degrade beaches inside it, where sea turtle nesting has been observed in the past, and it may increase turbidity (C. Schmitz, pers. comm.). Some regular sea sand mining also still occurs on beaches at Bartolbaai at Wacawa in Curaçao. However, most sand and gravel in Curaçao today is produced by the Curaçao Mijnmaatschappij which mines and crushes fossil coral rock from the Tafelberg mountain at Sta. Barbara. On Aruba marine sand is periodically dredged off Westpunt from 15–20 m of water for use in road construction by the Aruba Dept. of Public Works (B. Boekhoudt, LVV Aruba, pers. comm.).

Anchor Damage by Ships and Ship Groundings

Postma and Nijkamp (1996) and Meesters et al. (1996) point to limited but uncontrolled and potentially damaging anchoring by large ships on the Saba Bank. In St. Eustatius uncontrolled anchoring by ships is believed to be destroying reefs and conch fishing grounds (J. Begeman, St.

Eustatius Marine Park, pers. comm.), whereas in St. Maarten ship groundings on the historic Proselyte reef are frequent. The Curaçao reefs have also experienced the effect of several boat groundings: in June 1995 a small freighter ran aground leading to the destruction of some 8500 m² of reefs (Debrot, 1995), whereas in 1998 about 2700 m² of reefs suffered damage due to temporary anchoring of a drill platform (Vermeij and Kardinaal, 1998). Damages have not been claimed by government and remain uncompensated.

Cities

Coastal Urbanization

Due to steep rocky shores and heavy wave and wind conditions on the windward coasts of the islands of the leeward island group, coastal urbanization in the leeward island group is practically limited to leeward coasts, where the principal coral reefs and lagoonal habitats are concentrated. At present, about 25% of the leeward coast of Curaçao is taken up by urban, suburban and industrial development. In the next decade this is to be expanded to about 70%, leaving only about 30% of the total leeward coastline as natural undeveloped coastline (Executive Council Curaçao, 1995). The final outcome is less certain though, as most of the "useless green patches" are highly disputed.

In Aruba 95% of the leeward coastline has already been developed and only about 5% can be considered natural or semi-natural (B. Boekhoudt, LVV Aruba, pers. comm.). In Aruba too, these areas remain under high pressure for development. Likewise, most coastline of St. Maarten is already developed or rapidly being developed. The only natural coastal habitat tentatively planned for the near future is 5 km of coastline on the east side of the island (Fig. 2). In Bonaire, coastal development is picking up but the most recent draft of the nature management plan allows for significant coastal wilderness. In Saba and St. Eustatius, in contrast, there should be no great pressure to urbanize the coastal zone within the near future due to the steep and unstable slopes of the coastal cliffs.

Hurricanes 'Luis' and 'Marilyn' in St. Maarten recently demonstrated how hurricanes can lead to the wide dispersal of large amounts of anthropogenic rubbish in the marine habitat (AIDEnvironment/EcoVision, 1996). In the leeward Dutch islands, hurricanes are much less frequent, but ultimately inevitable (Meteorologische Dienst, 1982). In the three leeward islands, as in St. Maarten, large areas of land are quite vulnerable to both hurricane wave force and long-term sea-level rise. Urbanization in the wrong areas, without building setbacks or with inadequate building codes thus constitutes a long-term liability to environmental integrity of nearshore habitats. In contrast, the natural coastline does not carry the high environmental liabilities associated with the urbanized coastline.

To address the issues of both hurricane risk and long-term sea-level rise requires long-term urban planning and building codes. Only Curaçao has made a major effort to structure and formalize urban and land-use planning, but the issues of long-term sea-level rise and hurricane vulnerability remain as yet totally unaddressed.

Sewage Discharge and Treatment

In Curaçao, significant parts of central Willemstad as well as a number of government housing subdivisions (38% of total sewage production) are presently on a sewage collection system. Of this about 75% is subject to primary and secondary treatment, whereas 25% is discharged into the sea (Buth and Ras, 1992). Large parts of the coastline are significantly eutrophied (Gast, 1998). While the treated waste water is used for crop irrigation and landscaping, the remainder of sewage production (62% of all households) is largely discharged into cesspools, which ultimately leach into and contaminate the groundwater. In Curaçao groundwater demonstrates elevated nitrate concentrations especially under urbanised areas (Louws et al., 1998).

In Aruba approximately 26% of all sewage produced is collected for secondary treatment, possibly one quarter of which (Buth and Ras, 1992), overflows into the sea during rain showers. In addition, about 6% of all sewage is discharged directly into the sea. The remainder (approximately 60% of all sewage) is discharged into individual cesspools and septic tanks. The paucity of useable groundwater in Aruba means that the use of septic tanks is much more prevalent than in Curaçao (DOW Aruba, 1995). Disposal of contaminated sewage sludge from waste water treatment plants is a problem in both Curaçao and Aruba.

Bonaire has no sewerage system. Most households discharge waste into cesspools, whereas most hotels have septic systems and reuse waste water for landscaping purposes. Roberts and Hawkins (1994b) find indications of eutrophication around the town of Kralendijk where the hotels are concentrated. In St. Maarten most residential areas use cesspools and only 400 homes are connected to sewers. Large amounts of sewage is discharged directly or indirectly over land into the sea and sewage discharge by visiting yachts in the Simson Bay lagoon and Great Bay harbour is a problem (Buth and Ras, 1992). Saba and St. Eustatius do not have sewerage systems and rely totally on cesspools and septic systems. No direct discharge takes place on these islands (Buth and Ras, 1992).

Municipal Dumping

In Curaçao, municipal dumping of solid waste in the sea ceased in 1978. Today only animal cadavers and pumped sewage sludge are dumped along the windward coast. All solid waste is disposed of at an inland landfill. Illegal littering and dumping are still a serious problem on the island but largely take place inland along rural dirt roads.

Fig. 6. Submerged recreational litter at Groot Knip public beach, Curaçao, 1994. Photo: I. Nagelkerken.

The first, but now closed, municipal landfill of Curaçao lies adjacent to a saliña of major importance as a feeding area for flamingos and a nesting area for several tern species (Voous, 1983). Contamination of this valuable wetland by leachate from the landfill has not yet been proven but would seem to be ultimately unavoidable (Buth and Ras, 1992).

In Aruba and in St. Maarten, present landfills are at sea level, in a backreef lagoon (Koopmans et al., 1994) and an abandoned salt pond, respectively. This makes these dumps at high risks of disturbance by hurricanes and allows ready percolation of contaminants into the surrounding sea water. These will likely develop into costly environmental problems for future generations. In Saba, where a small landfill was once used, solid wastes are now simply dumped into a coastal gorge.

Beach Litter and Submerged Litter

Recent surveys by Debrot et al. (1999) and Nagelkerken et al. (in prep.) indicate that the beaches of Curaçao are heavily littered (Fig. 6). On the leeward public beaches of Aruba recreational litter is also a major problem, with plastic cups, beverage bottles, cans and synthetic diapers being the principal items collected during recent annual beach clean-up events (B. Boekhoudt, LVV Aruba, pers. comm.). Buth and Ras (1992) and Sybesma et al. (1993) indicate beach and submerged litter problems in St. Eustatius, while submerged litter and all kinds of rubbish are particularly abundant in front of inhabited coastal areas and many hotels in Curaçao (Debrot, pers., observ.) and Bonaire (Roberts and Hawkins, 1994b).

Cumulative Environmental Impacts and Ecological Trends

The combined effect of deleterious factors such as discussed above may translate into significant declines in key species, large-scale species shifts, and losses in terms of biodiversity and ecological resilience. In this respect, Bak and Nieuwland (1995) and Debrot et al. (1998) document alarming long-term declines in coral communities of Curaçao and Bonaire. The losses could not be accounted for by any known natural causes and instead, coastal urbanization, eutrophication and artificial beach construction were implicated as most probable causes.

In recent years the reefs have also suffered various "natural" disasters such as mass mortalities of the sea urchin *Diadema antillarum* (Bak et al., 1984), mass mortalities of two dominant shallow-water corals, *Acropora palmata* and *A. cervicornis* (Bak and Criens, 1981), wide-scale mortalities of seafans (Nagelkerken et al., 1997), several major coral bleaching events (Meesters and Bak, 1993; CARICOMP, 1997), the most recent of which was in 1998 (Debrot, pers. observ.), booming populations of the ascidian *Trididemnum solidum*, which overgrows and kills corals (Bak et al., 1996) and mass recruitment of balloonfish *(Diodon holocanthus)*, in which the highest densities occurred on anthropogenically stressed reefs (Debrot and Nagelkerken, 1997).

For St. Maarten, van't Hof (1989) concluded that most nearshore reefs were moderately to heavily degraded while the seagrass beds appeared relatively healthy. In Saba the cumulative anthropogenic impacts on the marine environment appear limited at present (van't Hof, 1985).

PROTECTIVE MEASURES

Policy Development and Legislation

A position paper on the environment prepared for the 1992 Rio Conference on Environment and Development (UNCED) forms the starting point for a more coherent effort in environmental policy development and planning by the government of the Netherlands Antilles (Government of the Netherlands Antilles, 1992). Until that time issues regarding the protection and management of the (marine) environment were handled only on an *ad hoc* basis. In 1962 the National Parks Foundation of the Netherlands Antilles had been founded to handle park management and nature protection (Sybesma, 1992). While, among others, the organization succeeded in achieving semi-official park or protected status for various important marine areas in the nation, the organization lacked governmental status or even a formalized advisory role in decisionmaking.

A long-standing legal discussion in the Netherlands Antilles, which greatly inhibited policy development until the early 1990s, was that of the judicial competence of the central and island legislators to install marine environmental legislation. The problem was finally solved in 1998 through an amendment of the Constitutional Island Regulations Law which established both the central and the island legislator's competence to legislate in environmental matters. In general, the environment is the competence of the island government. Only when treaty requirements are laid upon the Netherlands Antilles does the central government have the competence to act. The central government, however, does remain responsible for the implementation of legislation regulating trade of endangered species. Since Aruba's autonomy in 1986, it has only one layer of government, which helps make the implementation of environmental policy and legislation easier.

In 1993 the central government of the Netherlands Antilles enacted a national Fishery Protection Law that regulates commercial fisheries in the territorial seas and the 200 nautical mile Economic Fisheries Zone (EFZ). Commercial fishing in Netherlands Antilles waters now requires a license. Some basic restrictions, rules and regulations are in place with regard to certain fishing methods, the use of certain materials and equipment, and certain species such as marine mammals, sea turtles, spiny lobster (*Panulirus argus*) and queen conch (*Strombus gigas*). Aruba enacted comparable fisheries legislation in 1992.

In 1996 the Department of Public Health and Environental Hygiene published a framework for an environmental policy in the Netherlands Antilles, in which focus was directed towards five major themes. These were (1) garbage and sewage, (2) oil, (3) tourism, (4) nature and (5) public awareness (VOMIL, 1996).

The national legal framework for nature management and conservation, enacted in 1998 in the Netherlands Antilles, obligates both the central government as well as the island governments to implement the nature treaties of which the Netherlands Antilles is party, as part of the Kingdom of the Netherlands. These treaties are the Ramsar or Wetland Convention, the Bonn or Migratory Species Treaty, the Convention on International Trade in Endangered Species of wild flora and fauna (CITES), the Spaw Protocol of the Cartagena Convention on the protection of the marine environment in the Wider Caribbean Sea and the Convention on Biological Diversity.

For Aruba, the most significant (recent) developments in terms of legislation for nature protection and environment in general have been the 1991 enactment of legislation regulating import and export of endangered species, and the 1992 amendment of the list of protected species to include, among others, several native endangered reptiles and raptors (Koopmans et al., 1994).

Maritime legislation has recently been enacted with regards to shipping in the Netherlands Antilles. The majority of articles follow from treaty obligations made under the authority of the IMO (van Rijn, 1992). The norms and regulations required under MARPOL have been enacted as well as the different liability conventions such as CLC and the Fund Convention. However, implementation is quite deficient.

With regards to fisheries matters, small-scale, but non-commercial fisheries within the territorial waters (12 nm) fall under each island's jurisdiction. However, only the island of Saba has implemented an island fisheries ordinance.

All islands have so-called environmental hindrance ordinances in place. This legislation requires all activities and installations that may cause hindrance by pollution of air, soil or water to be licensed by government. These licenses restrict and limit certain types of pollution and prohibit some activities that may harm the marine environment. In practice, though, the significance of this legislation to the environment is minimal for both Aruba and the Netherlands Antilles (Koopmans et al., 1994). Some islands have installed garbage legislation to guide the process of garbage collection and discharge. Dumping in the sea is prohibited on all islands, but the problems of waste disposal remains serious.

A notable recent piece of legislation is that regarding land-use planning in Curaçao (Executive Council Curaçao, 1995). Few matters affect the marine habitat more than land usage in the adjacent coastal zone. Other islands (incl. Aruba; Koopmans et al., 1994) have not started to formally address the issue of land-use planning but do have draft nature management plans which will, if enacted, partially address these matters.

With regard to sewage, only St. Maarten so far has legislation to regulate the disposal of sewage, legally prohibiting the discharge of untreated effluent directly into the sea. Work is under way by the Department of Public Works to eliminate all direct untreated ocean discharges in Curaçao by the year 2000 (Buth and Ras, 1992). In Bonaire, the government is trying to find funding for centralized water treatment because of evidence that nutrient leaching is causing damage to the reefs, while Aruba is hopeful in being able to implement a large sewage treatment plan within the near future (DOW Aruba, 1995).

A number of other laws are in draft form at present. One of these is a framework law regarding environmental pollution for the Netherlands Antilles. If enacted it will obligate all island governments to install local legislation that deals with the management of hindrance, garbage, sewage and other pollution of the environment. Important is the political notion that in general the marine environment should not be used as a dumping place for liquid and/or material waste. Another notable initiative is the comprehensive nature management law for Aruba (Koopmans et al., 1994), which at the time of writing is not yet enacted. Policy development in terms of draft legislation continues and (shelved) management plans abound, in both Aruba and the Netherlands Antilles.

Implementation and an Evaluation

Table 3 provides an overview of the current state of marine environmental legislation in the Netherlands Antilles, as well as the most relevant international treaties dealing with the protection of the marine environment. The table also provides a qualitative assessment of the sophistication of the legislation as based on such criteria as the range of issues addressed, the detail to which regulations are stipulated and the scope and level of possible sanctions.

As already made clear, international treaties often form the basis for policy development and implementation at both national and island levels. However, as of 1998, implementation of treaty requirements by means of national and island level legislation, remains largely deficient (Table 3). The existing legislation is, furthermore, generally deficient in that it fails to address many or most of the critical issues.

Notwithstanding the extraordinary efforts by various struggling management agencies, the generally deficient legislation is also generally coupled with deficient implementation due to a lack of funding and government support in enforcement. In certain cases the net effectiveness nevertheless appears to be adequate so long as nothing serious happens. Examples are the marine parks of Saba and Bonaire. An essentially unspectacular level of legislation and modest implementation due to the perennial shortage of funds, nevertheless, miraculously combine for satisfactory effectiveness simply due to the fact that environmental pressures are low.

A rare example of strong legislation which addresses most critical issues in the area of concern in a quite rigorous fashion, is the previously mentioned land-use planning legislation in Curaçao (Executive Council Curaçao, 1995). However, political and media support for the legislation are low. Government fumbling has compromised the effectiveness and even the continued existence of this legislation within one year after implementation.

Institutional capacity is critical to the implementation of environmental policy. The 1992 installation of a small division for environment and nature policy within the Netherlands Antilles Department of Public Health and Environmental Hygiene, the 1992 installation of VROM Aruba (Department for Urban Planning and Environment) and the 1989 installation of VROM St. Maarten constitute the main structural advances in (environmental) institutional capacity in recent years.

Law enforcement is critical to effective implementation of maritime and marine environmental legislation. However, enforcement of most environmental law is considered an incidental matter and remains minimal. At present marine park managers and their personnel are given special police authority and a notable exception to the rule (as long as funds last) is the rigorous enforcement of anti-spear fishing legislation along the coast of Curaçao, by the Curaçao Marine Park. Another notable exception is the assiduous enforcement of oil spills inside the Curaçao harbour areas by the Harbourmaster. The Department of Maritime Affairs of the Netherlands Antilles has the task of enforcing legislation regarding oil contamination by ships in the territorial waters of both the Netherlands Antilles and Aruba but enforcement is largely ineffective (Koopmans et al., 1994).

Since the installation of a coastguard for the Netherlands Antilles and Aruba, the regular police has fully withdrawn from enforcement in the marine environment. The mandate of the coastguard fortunately includes enforcement of environmental and fisheries regulations, even though these are not priority areas. The coastguard is adequately equipped with vessels and manpower for its task with respect to environmental and fisheries enforcement.

In recent years, sanctions in the Netherlands Antilles for environmental violations have been expanded from detention and fines to include administrative sanctions such as withdrawal of permits, forfeiture of deposits, penalties for non-compliance and seizure or confiscation of material.

Practical Management Measures

In addition to management implemented through and based on legal foundations as discussed above, practical implementation of several projects has taken place over the years which are of significant value to coastal resource conservation. These include the Flamingo Sanctuary, Bonaire, 1956; Washington Park, Bonaire 1969; Slagbaai Park, Bonaire, 1977; the Christoffel Park, Curaçao, 1978; the Curaçao Underwater Park, 1983; the 7-Boca Park, Curaçao, 1993; the Arikok National Park, Aruba, 1997—all of which provide actual, though non-legally based, protection to valuable coastal habitat.

Public attitude, in particular a propensity to litter and a disregard for nature, are among the greatest environmental challenges on the larger more industrialised islands. Therefore, public awareness and involvement are of strategic importance to long-term sustainable environmental management (ECLAC, 1998). The most significant recent programs in terms of public information and education in the Netherlands Antilles are a long-term anti-littering campaign led by the (government) foundation Korsou Limpi i Bunita and a structurally funded nature education project by the (government) Carmabi Foundation in Curaçao, involving upwards of 20,000 school children each year.

Prognoses and Prospects

In light of the current magnitude of problems, the protective measures taken to date are clearly insufficient.

Table 3

Qualitative overview of the current status of implementation of marine environmental policy and legislation in the Netherlands Antilles (i.e. excl. Aruba)

General theme	Specific theme	Treaty	Party (1)	National translation (2)	National implementation	Island translation (2)	Island implementation	Comb. effectiveness	Remarks
Marine environment legislation	general	1982 United Nations Convention on Law of the Sea (UNCLOS)	?	by means of several marine and maritime laws (b)	basic	marine environment ordinances [Curacao: 1976(w),Bonaire: 1984(s),Saba: 1987(s),St. Eustatius: 1996(s)	basic	?	
Marine pollution legislation	liability	1969 Civil Liability Convention plus 1976 Protocol	?	1996 national law on marine liability (s)	basic	N		?	difficult judicial procedures can now be avoided
	intervention	1969 Intervention Convention plus 1973 Intervention Protocol	1975; 1983	N		N		?	
	liability	1971 Fund Convention	?	1996 national law on marine liability (s)	basic	N		?	difficult judicial procedures can now be avoided
	dumping	1972 London Convention	1978	N		marine environment ordinances [Bonaire: 1984 (b); Saba: 1987 (b); St. Eustatius: 1996 (b)	weak	poor	polluted sludge is still dumped in Curacao
	intervention	1972 Collision Registration Convention (ColReg)	?			?		?	
	marine pollution by ships	1978 MARPOL Convention plus Annexes	1983	national law on pollution of the sea by ships (s)	weak	marine environment ordinances [Bonaire: 1984(b); Saba: 1987(b); St. Eustatius: 1996(b)	basic	poor	
	oil pollution	1983 Cartagena Convention; 1983 Oil Spill Protocol	1986; 1986	N		marine environment ordinances [Bonaire: 1984(w); Saba: 1987(w); St. Eustatius: 1996(w)	basic	poor	
	intervention	1990 Oil Preparedness,Response and Contingency Convention (OPRC)	?	N		island contingency plans (w)	?	poor	
	pollution from land	Cartagena Conv.; Land based sources of marine pollution Protocol (3)	1985	N	weak	marine environment ordinances [Bonaire: 1984(b); Saba: 1987(w); St. Eustatius: 1996(w)"	weak	poor	
Marine natural environment legislation	wetlands birds	1971 Ramsar or Wetlands Convention	1980	1998 national law on nature management and protection (b)	basic	marine environment ordinances [Curacao: 1976(w); Bonaire: 1984(s); Saba: 1987(s); St. Eustatius: 1996(s)	basic	moderate	
	trade	1973 CITES Convention	N	1998 national law on nature management and protection (b)	basic	marine environment ordinances [Curacao: 1976(w); Bonaire: 1984(s); Saba: 1987(s); St. Eustatius: 1996(s)	basic	moderate	

General theme	Specific theme	Treaty	Party (1)	National translation (2)	National implementation	Island translation (2)	Island implementation	Comb. effectiveness	Remarks
	migratory species	1979 Bonn Convention	1983	1998 national law on nature management and protection (b)	basic	marine environment ordinances [Curacao: 1976(w); Bonaire: 1984(s); Saba: 1987(s); St. Eustatius: 1996(s)"	basic	moderate	
	marine parks and species	1983 Cartagena Convention 1990 SPAW protocol (4)	1985; 1990	1998 national law on nature management and protection (b)	basic	marine environment ordinances [Curacao: 1976(w); Bonaire: 1984(s); Saba: 1987(s); St. Eustatius: 1996(s)	basic	good	
	biological diversity	1992 Biodiversity Convention	N	1998 national law on nature management and protection (b)	weak	marine environment ordinances [Curacao: 1976(w); Bonaire: 1984(s); Saba: 1987(s); St. Eustatius: 1996(s)	basic	poor	
Fisheries legislation	pelagic	Conventions on fisheries on the high seas and straddling stocks	?	1991 national fisheries law (b) (5)	basic	fisheries ordinance (Saba: 1996(s))	basic	?	only Saba has such legislation
	coastal					marine environment ordinances [Curacao: 1976(w); Bonaire: 1984(b); Saba: 1987(b); St. Eustatius: 1996(b)	weak	poor	Curacao: weak legislation but strong enforcement
Coastal zone legislation	coastal zone planning					Curacao: spatial development plan: 1995(s); tourism master plan (b)	weak	poor	only Curacao has these
	marine protected areas					local policy plans and different (concept) nature policy plans(w)	weak	good	good: only Bonaire,Saba,St. Eustatius
	sewage treatment					local policy plans(w)	basic	moderate	only Curacao and St. Maarten have these plans
	misc. env. impacts					local hindrance ordinances (s)	weak	poor	
Institutional capacity				Dept. Maritime Aff. (w),Dept. P. Health & Env. (w),Coast Guard (b),Research institute (b)	weak	island services e.g. Urban Plann. (Curacao: b; other isl.: w),LVV (w),Environmental Serv. (w); also subsidized NGO's (b)	weak	poor	

(1) N: no; ?: unknown/no data available; year of enactment given for those treaties for which we know the Netherlands Antilles to be a party.
(2) s: strong; b: basic; w: weak
(3) this protocol is still being negotiated
(4) this protocol is not yet into force
(5) including the 200 nm EFZ.

With few exceptions, all principal coastal habitats of the Dutch Antilles are under major environmental pressure and, on the whole, current environmental legislation remains weak in failing to address most key issues. The framework nature law enacted by the central government of the Netherlands Antilles in 1998 provides some hope that the islands within the constellation may be pressured into implementing significant legislation within the near future. On the positive side, it can be pointed out that various island governments have been working on proposals for implementation and/or legal reinforcement of various marine parks in an effort to boost dive tourism. Also, significant advances have been made in recent years in terms of nature conservation and management on the various islands, albeit ad hoc and fragmentary. A notable success for the governmental parks foundations has been the development of the concept of users paying for their use of the marine environment so as to be able to finance a minimum level of marine park management (Bonaire and Saba). Recent years have seen a dramatic growth in terms of membership and influence for the local environmental movement, which originated in the Netherlands Antilles in 1988 with the founding of Defensa Ambiental in Curaçao. In Aruba environmental activism has also grown significantly since the 1991 founding of the association Stimaruba (Koopmans et al., 1994).

Nevertheless, under the current political and socio-economic circumstances, government (at both national and insular levels) often has great difficulty in making effective decisions and implementing them sustainably. Not surprisingly, better governance has been identified as critical to the implementation of a sound environmental policy in the Netherlands Antilles (ECLAC, 1998). Coastal resource users are environmentally aware and express support for improved natural resource management (e.g. Debrot and Nagelkerken, 1999). Yet, environmental issues are rapidly politicised and polarised in terms of environment versus progress and environmental concerns generally receive unfavourable press. Government funding for environmental protection and management remains sparse and often non-structural, notwithstanding the regionally high per capita GNPs and standards of living, and the recognized importance of the environment to sustainable tourism. As a consequence, institutional capacity for policy and legislation development as well as implementation remains very limited and is identified as a high priority issue (ECLAC, 1998). Public awareness and participation are, furthermore, recommended as key elements of strategy for implementation (ECLAC, 1998).

None of the islands have as yet developed any vision on the greater issue of population pressure which is the principal driver of most economic, infrastructural and environmental problems. As long as this issue is not dealt with, and as long as environmental management is not integrated into socio-economic development plans, environmental problems will essentially remain unbounded and solutions will forever lag behind.

On the more populous islands (Aruba, Curaçao, St. Maarten), high population densities with the concomitant housing problems, economic problems, and infrastructural problems mean that, for the near future, environmental issues will continue to be under-regarded in terms of government priority. A strategy of economic dependence on the petroleum industry and other large enterprises will further mean that governments will continue to hesitate in demands for sound environmental practice. We predict that issues such as rampant coastal development, excessive fishing pressure and pollution by the petroleum sector and other large enterprises are not likely to be dealt with effectively in the foreseeable future, notwithstanding the many management plans and the various initiatives towards policy development in these areas. The less populous islands (Saba, St. Eustatius, Bonaire) where population pressure is less, economies are simpler and more eco-touristically oriented, and government is less complex and more effective, would seem to show the greatest promise in being able to deal with environmental problems before they grow out of hand.

Global issues such as climate change, and sea-level rise, which will have major consequences for several islands, lie largely beyond the scope of the direct influence of the government and populace. However, only through firm local action can the Dutch Antilles hope to take a credible stand on these issues. To date, none of the Dutch Antilles have undertaken any measures with respect to greenhouse gasses, or emission reduction (Koopmans et al., 1994; ECLAC, 1998).

ACKNOWLEDGMENTS

We thank L. Pors and K. Dekker, both of Carmabi, for drafting our maps and for calculating habitat surface areas. B. Boekhoudt of the Department of Agriculture and Fisheries is thanked for providing extensive preliminary data for Aruba. The following individuals are further thanked for providing additional information: G. Boekhoudt of VROM Aruba, J. Lue of DOW Aruba, R. de Kort of Parke Nacional Arikok, Aruba; D. Kooistra of the Saba Marine Park, J. Begeman of the St. Eustatius Marine Park, J. A. de Meyer of the Bonaire Marine Park and H. Hoyer of the Curaçao Environmental Service. Work on this review by the first author was made possible by the Island Government of Curaçao and the Central Government of the Netherlands Antilles through their annual subsidy to the CARMABI Foundation. Work by the second author was made possible by the Central Government of the Netherlands Antilles through their annual subsidy to the University of the Netherlands Antilles.

REFERENCES

AIDEnvironment/EcoVision (1996) Saint Maarten's natural resource after 'Luis' and 'Marilyn': an assessment of the environmental damage and related problems caused by two hurricanes. Dept. VROM, St. Maarten.

Bak, R.P.M. (1975) Ecological aspects of the distribution of reef corals in the Netherlands Antilles. *Bijdragen tot de Dierkunde* **45** (2), 181–190.

Bak, R.P.M. (1977) Coral reefs and their zonation in the Netherlands Antilles. *Stud. Geol.* **4**, 3–16.

Bak, R.P.M. (1978) Lethal and sublethal effects of dredging on reef corals. *Marine Pollution Bulletin* **9**, 14–16.

Bak, R.P.M. (1987) Effects of chronic oil pollution on a Caribbean coral reef. *Marine Pollution Bulletin* **18**, 534–539.

Bak, R.P.M., Carpay, M.J.E. and de Ruyter van Steveninck, E.D. (1984) Densities of the sea urchin *Diadema antillarum* before and after mass mortalities on the coral reef of Curaçao. *Marine Ecology Progress Series* **17**, 105–108.

Bak, R.P.M. and Criens, S.R. (1981) Survival after fragmentation of colonies of *Madracis mirabilis*, *Acropora palmata* and *A. cervicornis* (Scleractinia) and the subsequent impact of a coral disease. *Proc. IVth Intern. Coral. Reef Symp.* **2**, 221–227.

Bak, R.P.M., Lambrechts, D.Y.M., Joenje, M., Nieuwland, G. and van Veghel, M.L.J. (1996) Long-term changes on coral reefs in booming populations of a competitive colonial ascidian. *Marine Ecology Progress Series* **133**, 303–306.

Bak, R.P.M. and Nieuwland, G. (1995) Long-term changes in coral reef communities along depth gradients over leeward reefs in the Netherlands Antilles. *Bulletin of Marine Science* **56** (2), 609–619.

Bak, R.P.M. and van Moorsel, G.W.N.M. (1979) De invloed van het water en energie bedrijf (Bonaire) op het koraalrif. CARMABI Report, CARMABI, Curaçao.

Bakhuis, W.L. (1990) Wetgeving en natuurbeheer, pp. 5–14. In *Milieurecht Congres, Curaçao, September 1989*, eds. Q.B. Richardson and J. Sybesma. Antilliaanse Juristen Vereniging, Curaçao.

Buth, L. and Ras, J. (1992) *Inventory of the land-based sources of marine pollution, Netherlands Antilles, 1992.* RZZ (Council for Sea Research and Sea Activities, Curaçao.

CARICOMP (1997) Studies on Caribbean coral bleaching 1995. *Proc. VIIIth Intern. Coral Reef Symp.* **1**, 673–678.

CBS (Centraal Bureau voor de Statistiek, Nederlandse Antillen) (1996) Economisch profiel, Nederlandse Antillen. CBS, Curaçao.

Debrot, A.O. (1995) Dumped rice destroys reefs in Curaçao. *Marine Pollution Bulletin* **30**, 631.

Debrot, A.O. and de Freitas, J.A. (1991) *Wilderness areas of exceptional conservation value in Curaçao, Netherlands Antilles.* NCIN Mededelingen No. 26, University of Amsterdam, The Netherlands.

Debrot, A.O. and de Freitas, J.A. (1999) Avifaunal and botanical survey of the Jan Thiel Lagood Conservation Area, Curaçao. CARMABI Report, CARMABI, Curaçao.

Debrot, A.O. and Nagelkerken, I. (1997) A rare mass recruitment of the balloonfish (*Diodon holocanthus* L.) in the Leeward Dutch Antilles, 1994. *Caribbean Journal of Science* **33**, 284–286.

Debrot, A.O. and Nagelkerken, I. (1999) User perceptions on resource state and management options for the Curaçao coastal zone. Proc. 29th Meeting Assoc. Mar. Lab. Caribb. 58 (Abstract).

Debrot, A.O., Bradshaw, J.E. and Tiel, A.B. (1995) Tar contamination on beaches in Curaçao. *Marine Pollution Bulletin* **30**, 689–693.

Debrot, A.O., Kuenen, M.M.C.E. and Dekker, K. (1998) Recent declines in the coral fauna of the Spaanse Water, Curaçao, Netherlands Antilles. *Bulletin of Marine Science* **63**, 571–580.

Debrot, A.O., Tiel, A.B. and Bradshaw, J.E. (1999) A study of beach debris contamination in Curaçao, Netherlands Antilles. *Marine Pollution Bulletin* **38**, 795–801.

Djohani, R.H. and Klok, C. (1988) *Een onderzoek naar de waterkwaliteit van enkele baaien van Curaçao op basis van biologische en abiotische parameters.* CARMABI Report, CARMABI, Curaçao.

Dol, T. (1995) Groei visserij baart zorgen; Gerard van Buurt: "verzadigingspunt in verschiet". *Napa, Amigoe*, April 8, 1995, p. 1.

DOW Aruba (1995) Afvalwaterstructuurplan Aruba 1997–2010. DOW (Dept. Public Works), Aruba.

Executive Council, Curaçao (1995) Eilandelijke Ontwikkelingsplan Curaçao 1995. (2 Vols) DROV (Urban Planning and Housing Service), Curaçao.

ECLAC (Economic Commission for Latin America and the Caribbean) (1998) National implementation of the SIDS/POA: A Caribbean perspective. ECLAC, Trinidad.

Ebbing, J.H.J. (1997) Geochemische inventarisatie van enkele baaien op Curaçao. TNO-rapport NITG 97-238-B , Nederlandse Organisatie voor Toegepast-Natuurwetenschappelijk Onderzoek, The Netherlands.

FAO (1991) Project Proposal Development of the Pelagic Fishery in the Curaçao/Bonaire Area. Draft, 13 November, 1991.

Gast, G.J. (1998) Microbial densities and dynamics in fringing coral reef waters. Ph.D. dissertation, University of Amsterdam.

Government of the Netherlands Antilles (1992) National Environmental Report. VOMIL (Netherlands Antilles Department of Public Health and Environmental Hygiene), Curaçao.

Guidicelli, M. (1980) Programme for fisheries development and diversification in the southern Netherlands Antilles: Aruba, Curaçao and Bonaire. WECAF Rept. 32, UNDP, FAO, Rome.

Guidicelli, M. and Villegas, L. (1981) Program for fisheries development in Saba and St. Eustatius. WECAF Rept. 39, UNDP, FAO, Rome.

Hoppe, W. (1982) Invloed van vervuiling op het koraalrif van Curaçao met speciale referentie naar de KAE en COT. CARMABI Report, CARMABI Curaçao.

Koopmans, J.W.T., Sjak-Shie, E.L. and Thodé, G.A.E. (1994) Milieu en ruimtelijke ordening in Aruba. VAD, Oranjestad, Aruba.

Louws, R.J., Vriend, S.P. and Frapporti, G. (1998) De grondwaterkwaliteit van Curaçao. H_2O **26**, 788–791.

LVV (Dienst Landbouw Veeteelt en Visserij, Curaçao) (1988) Visserij Ontwikkelings Plan III. SEP, Curaçao.

LVV (Dienst Landbouw Veeteelt en Visserij, Curaçao) (1992) Kort overzicht van de agrarische en visserij sector, 1992. Dienst LVV, Curaçao.

Meesters, E.H. and Bak, R.P.M. (1993) Effects of coral bleaching on tissue regeneration potential and colony survival. *Marine Ecology Progress Series* **96**, 189–198.

Meesters, E.H., Bos, A. and Gast, G.J. (1992) Effects of sedimentation, lesion morphology, and lesion position on coral tissue regeneration. *Proc. VIIth Int. Coral Reef Symp.* **1**, 681–688.

Meesters, E.H., Nijkamp, H. and Bijvoet, L. (1996) Towards sustainable management of the Saba Bank. KNAP Project 96-1. Netherlands Antilles Dept. Public Health and Environmental Hygiene/ AIDEnvironment, Amsterdam.

Meteorologische Dienst (Meteorological Service of the Netherlands Antilles and Aruba) (1982) Beknopt overzicht van het klimaat van de Nederlandse Antillen. Meteorological Service, Curaçao.

Nagelkerken, I., Buchan, K., Smith, G.W., Bonair, K., Bush, P., Garzón-Ferreira, J., Botero, L., Gayle, P., Heberer, C., Petrovic, C., Pors, L. and Yoshioka, P. (1997) Widespread disease in Caribbean sea fans: I. Spreading and general chracteristics. *Proc. VIIIth Intern. Coral Reef Symp.* **1**, 679–682

Nagelkerken, I. and Debrot, A.O. (1995) Mollusc communities of tropical rubble shores of Curaçao: long-term (7+ years) impacts of oil pollution. *Marine Pollution Bulletin* **30**, 592–598.

Nagelkerken, I., Wiltjer, M., Debrot, A. O. and Pors, L.P.J.J. Submerged debris on the coral reef at recreational beaches in Curaçao, Netherlands Antilles (in prep.)

Newton, E. (1987) Tar on beaches, Bonaire, Netherlands Antilles. *Caribbean Journal of Science* **23**, 139–143.

Postma, T.A.C. and Nijkamp, H. (1996) Seabirds, marine mammals and human activities on the Saba Bank; field observations made during the Tydeman expedition, April–May 1996. AIDEnvironment, Amsterdam.

Rijkswaterstaat. (1992) Problematiek vervuild havenslib haven Willemstad Curaçao. Ministerie van Verkeer en Waterstaat, Rijswijk, The Netherlands.

Roberts, C.M. and Hawkins, J.P. (1994a) Report on the status of Bonaire's coral reefs 1994, part 2: effects of diving on coral reefs of Bonaire and status of the island's fish communities. Eastern Caribbean Center, Univ. Virgin Islands, St. Thomas, U.S.V.I.

Roberts, C.M. and Hawkins, J.P. (1994b) Report on the status of Bonaire's coral reefs 1994, part 1: critical management issues for coral reefs of Bonaire. Eastern Caribbean Center, Univ. Virgin Islands, St. Thomas, U.S.V.I.

Rojer, A. (1997a) Biologische Inventarisatie van St. Maarten. KNAP-project 96-10. CARMABI Report. CARMABI, Curaçao.

Schuhmacher (1976) *Korallenriffe; ihre Verbreitung, Tierwelt un Ökologie.* BLV Verlagsgesellschaft mbH, Munchen, Germany.

Sturm, M.G., de L. (1991) The living resources of the Caribbean Sea and adjacent regions. *Caribbean Marine Studies* **2** (1&2), 18–44.

Sybesma, J. (1992) The role of NGOs in managing national parks: The Netherlands Antilles National Parks Foundation as an example. *Proc. IVth World Congress on National Parks and Protected Areas, Caracas,* Workshop III-4.

Sybesma, J., van't Hof, T. and Pors, L.P.J.J. (1993) Marine area survey: an inventory of the natural and cultural marine resources of St. Eustatius, Netherlands Antilles. CHL Consulting Group/CARMABI, Curaçao.

van Halewijn, R. and Norton, R.L. (1984) The status and conservation of seabirds in the Caribbean. *ICBP Techn. Publ.* **2**, 169–222.

van't Hof, T., Debrot, A.O. and Nagelkerken, I.A. (1995) Curaçao Marine Management Zone: a plan for sustainable use of Curaçao's reef resources—a draft plan for consultation. CTDB/STINAPA Report, CARMABI, Curaçao.

van't Hof, T. (1989) Towards conservation of the marine environment St. Maarten/St. Martin; report of a preliminary reef survey. Stinapa, St. Maarten.

van't Hof, T. (1985) Saba Marine Park: a proposal for integrated marine resource management in Saba. STINAPA (Netherlands Antilles National Parks Foundation) Report, CARMABI, Curaçao.

van Rijn, A.B. (1992) Milieurecht in de Nederlandse Antillen. Leerstoel voor Milieu en Ontwikkeling, University of the Netherlands Antilles.

Vermeij, M.J.A. and Kardinaal, W.E.A. (1998) Coral reef damage at Lyhoek as a result of construction activities in Caracas Bay, Curaçao (N.A.). CARMABI Report, CARMABI, Curaçao.

VOMIL (Netherlands Antilles Department of Public Health and Environmental Hygiene) (1996) Contouren van het Milieu-en Natuurbeleid Nederlandse Antillen. VOMIL, Curaçao.

VOMIL (1998) *Nieuwsbrief.* Sectie Milieu en Natuur **3** (7), November 1998. VOMIL, Curaçao.

Voous, K.H. (1983) *Birds of the Netherlands Antilles.* De Walburg Pers, Zutphen, The Netherlands.

Walsh-McGehee, M. and Lee, D.S. (1998) The conservation status of tropicbirds in the West Indies (Abstract). Society of Caribbean Ornithology Meeting, Guadeloupe. *El Pitirre* **11** (2), 60–61.

THE AUTHORS

Adolpe O. Debrot
*Carmabi Foundation,
Piscaderabaai, P. O. Box 2090,
Curaçao, Netherlands Antilles*

Jeffrey Sybesma
*University of the Netherlands Antilles,
Jan Noorduynweg 111, P.O. Box 3059,
Curaçao, Netherlands Antilles*

Chapter 39

UK OVERSEAS TERRITORIES IN THE NORTHEAST CARIBBEAN: ANGUILLA, BRITISH VIRGIN ISLANDS, MONTSERRAT

Fiona Gell and Maggie Watson

The United Kingdom Overseas Territories (UKOT) of Anguilla, British Virgin Islands and Montserrat are in the Lesser Antilles, in the northeast Caribbean. All are small. Tropical storms and hurricanes are the most common causes of natural disturbance in this region, and for the Atlantic basin as a whole there has been an increase in the number of strong hurricanes since 1995. In Montserrat, volcanic activity is the outstanding environmental problem, and this has caused extensive destruction in the last few years.

Shallow, sheltered habitats support large areas of seagrasses, mostly of shallow beds of *Thalassium testudinum* (turtle grass). Coral communities are also extensive. Mangroves are reduced in extent, mostly as a result of piecemeal destruction for waterside developments although mangrove felling is illegal. The beaches of the islands are one of their most important tourist attractions.

The northern islands have seen a phenomenally rapid increase in tourism, which has replaced traditional industries such as fishing and salt extraction, and contributed to a large growth in population. By contrast, Montserrat's population has more than halved in the last five years, mainly because of volcanic activity. Economic development from the two main industries (tourism and financial services) has generally reduced the importance of fishing as a livelihood, but even so, some fish resources appear to be declining, in part because the growth of tourism has increased demand for fish products. As overseas territories of the UK, these islands fall within the UK jurisdiction for many laws and regulations, but retain independence in others. Generally, the small sizes of the islands have meant that environmental aspects have been under-resourced, though various small government departments and several NGOs are active in the region.

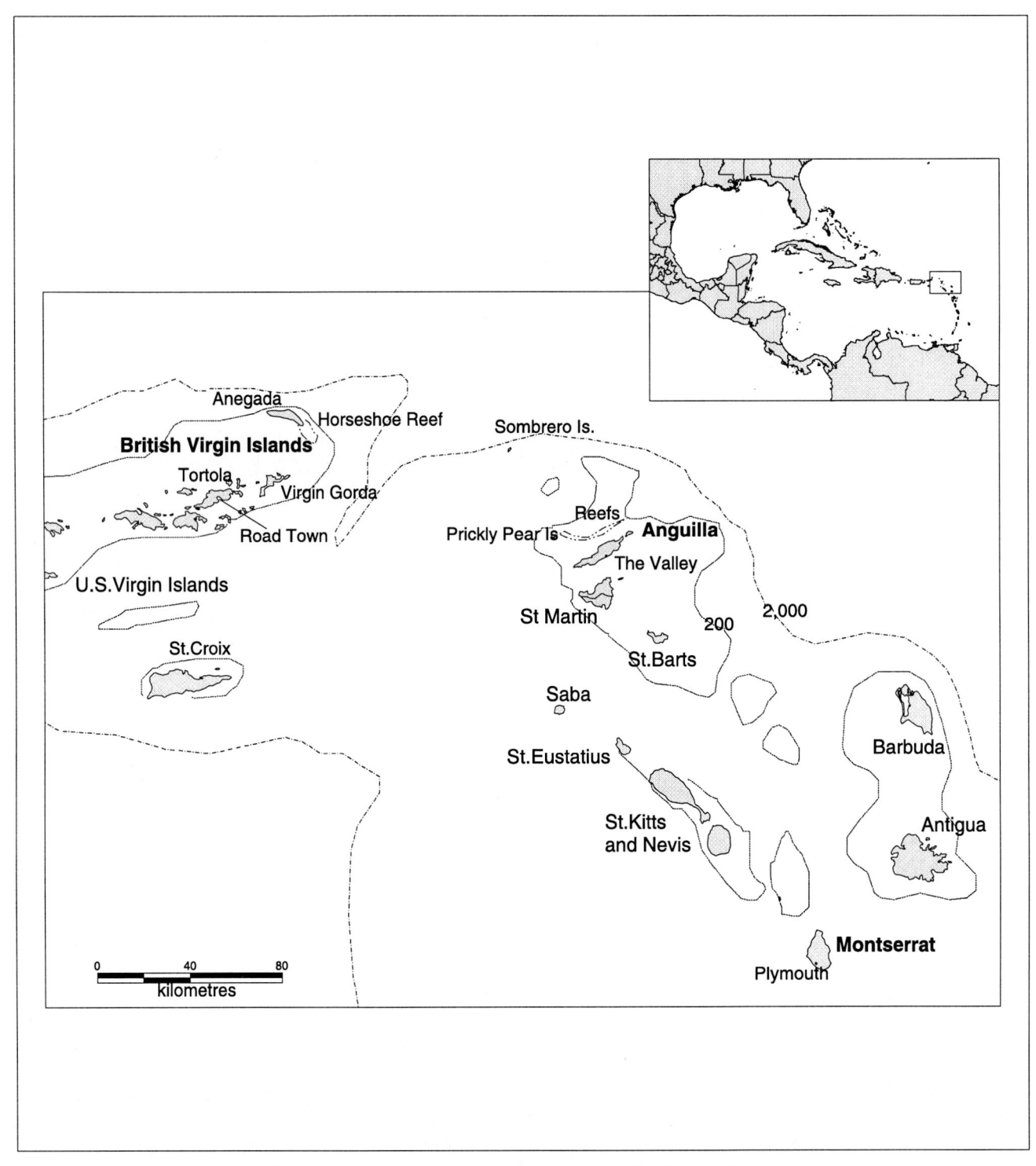

Fig. 1. Map of northeast Caribbean, showing the British Virgin Islands, Anguilla and Montserrat.

THE DEFINED REGION

The United Kingdom Overseas Territories (UKOT) of Anguilla, British Virgin Islands and Montserrat are in the Lesser Antilles, in the north of the Eastern Caribbean island arc (Fig. 1). The islands are small, and because of both this and their political diversity, most Caribbean Small Island Developing States (SIDS) are interdependent for natural resources such as fisheries.

The Territories considered here form part of the nation-state of the UK but are not represented in the UK parliament and have independent elected governments. The UK is responsible for defence, international relations and has some say in legislation (Pienkowski, 1999).

Anguilla was administered as a single federation with St. Kitts and Nevis but sought separation in the 1960s, came under direct UK administration in the 1970s and eventually became a separate British Dependent Territory in 1980. Anguilla is a Caribbean Community (CARICOM) observer and an associate of the Organisation of Eastern Caribbean States (OECS). It is a low island of 90.65 km^2 (highest point 65 m) rising from the Anguilla Bank. On the north coast cliffs are almost 30 m high and depths are 23–45 m within 1 km of shore, whilst the south coast is low, with 30–40 sandy bays. Anguilla has several small offshore cays and islands, including Sombrero, 61 km to the northwest.

The British Virgin Islands (BVI) comprise more than 50 islands and islets, covering 155 km^2, though only 16 islands are inhabited. Tortola is the largest and most heavily populated island. Virgin Gorda, Anegada and Jost Van Dyke are amongst the more important. BVI has been a UKOT since 1971 and is an associate member of OECS and CARICOM. The islands are made of Cretaceous sedimentary and metamorphic/volcanic rocks, rising to 543 m above sea level (Fig. 2), with the exception of Anegada to the north which is made from carbonates and reefs.

Montserrat lies 43 km south west of Antigua, with which it has strong historical links. Only 18 km by 11 km, it has an area of 102 km^2. A narrow coastal shelf drops quickly to nearly 200 m only 650 m offshore along the southern half of the island. In the north, the shelf slopes more gently and the 200 m contour is 4.6 km offshore (Bovey et al., 1986). The pear-shaped island rises to 915 m and is formed from exposed peaks of part of the Lesser Antillean Archipelago, the undersea mountain ridge that curves between Puerto Rico and Trinidad. The ridge was formed by subduction of the Atlantic plate beneath the Caribbean Plate during the Miocene period. After 350 years of dormancy, Montserrat became volcanically active in 1995, when the capital Plymouth was destroyed and approximately half the island became uninhabitable. Much may remain uninhabitable for the next decade.

SEASONALITY, CURRENTS, NATURAL ENVIRONMENTAL VARIABLES

In Montserrat, volcanic activity is the outstanding environmental variable. During the active phase from 1995 to early 1998, vegetation was completely lost in many areas, leading to severe erosion. Huge plumes of sediment entered the sea at several locations, some containing 150 g dry weight of sediment per litre of water (Brosnan, 1999). Most sediment settled out of the water column less than 300 m from shore, and the effects on reefs on the east and southwest of the island have been severe. The volcano has been quiet from March 1998, but in November 1999 there were reports of renewed activity.

Tropical storms and hurricanes are more common causes of natural disturbance in this region; an average of one direct hit every 20 years is often quoted. Twenty hurricanes passed between 10°N and 19°N at 60°W during 1950 to 1997, and in the last 90 years, roughly two thirds of intense Atlantic hurricanes crossed the 60°W meridian between 18°N and 25°N, closely approaching or passing over the northern part of the Eastern Caribbean island chain (Jones, 1999). For the Atlantic basin as a whole there has been an increase in the number of strong hurricanes since 1995 which may be part of a quasi-cyclic pattern alternating between active and quiet hurricane phases every 25–40 years. Such phases are probably related to surface temperatures in the Atlantic (Landsea, 1999). Hurricane-generated waves and high rainfall can devastate nearshore reefs and cause coastal erosion.

With the exception of these major disturbances, the climate of the islands is stable, with summer and winter temperatures around 29°C and 24°C respectively. Anguilla receives an average of 102 cm of rain a year—the lowest in the Leeward Islands. By contrast, rainfall on Montserrat varies between 107 cm near sea level to 205 cm at 365 m (Corker, 1986). Most rain falls in the summer months, with a dry period from January until around May. Northeast trade winds prevail.

The islands are influenced by the North Equatorial current which flows westwards from the open Atlantic. Anguilla experiences winter currents of 0.4–0.5 knots

Fig. 2. 'High' islands of the British Virgin Islands (photo C. Sheppard).

flowing 280–290°, whilst summer currents are slightly stronger (0.6–0.9 kt) and take a more northwesterly course between 300 and 310°. However, the Anguilla Bank is an irregular, relatively shallow platform and water flowing over the platform interacts with coastal currents to generate short-lived reversals in current direction (Towle, 1979). Tidal range is approximately 23 cm on average, with a maximum range of 30–60 cm at Spring tides. The tidal cycle is 14 hours, i.e. approximately semi-diurnal. For Montserrat, currents from the east normally flow around the northern and southern ends, creating longshore currents that converge at Bransby Point on the island's southwest central coast. Montserrat's west coast is a leeward shore, but is occasionally affected by Atlantic storms. Swells reach heights of 1.2 m approximately 10 times every winter and are expected to reach 3.7 m approximately once a year (Cambers, 1981). Currents around the BVI are poorly documented but are generally less than 0.5 kt.

MAJOR SHALLOW WATER COASTAL MARINE HABITATS AND BIODIVERSITY

Seagrass

Anguilla's seagrass beds cover 3400 ha (Olsen and Ogden, 1981) including an extensive bed at Crocus Bay (Wells, 1988) (Fig. 3). Large areas of seagrasses are found in sheltered bays around the BVI as well. As in Anguilla, most of the shallow beds are *Thalassium testudinum* (turtle grass), with some deeper beds of *Syringodium filiforme* (manatee grass) and some mixed beds. Tidal range is small, so intertidal seagrasses are limited. Montserrat has only three main seagrass beds. The largest (750 ha) is around the northern tip of island (IRF, 1993). The other two are on the east and west coasts (Jeffers, pers. comm., in IRF, 1993). There are three species of seagrass around the island (Brosnan et al., 1999) but available shallow water restricts the extent of this habitat.

Coral Reefs

Coral communities cover approximately 22% of the Anguillan shelf area (Olsen and Ogden, 1981). Living reefs are mostly on the north side of the island, with an extensive system running approximately parallel to the north coast several kilometres offshore. South coast reefs are more susceptible to hurricanes and in many places consist of a framework of dead *A. palmata*, with low diversity of living coral (Salm, 1980) Anguilla's reefs were badly damaged by Hurricane Donna in 1960, but in 1982 the 17 km long reef area along the southeast coast was considered to be the most important, largely unbroken, reef area in the Eastern Caribbean (Putney, 1982, in Wells, 1988). The reefs at Dog Island are reportedly in good condition and visitors are discouraged (Smith et al., 1997).

Extensive coral communities exist throughout the BVI with many well developed reefs. Horseshoe Reef (Anegada) covers approximately 77 km^2. Estimates of total coral reef area on several of the larger islands are given in Table 1. Figures reflect only shallow non-emergent reefs and therefore underestimate the total reef area which would include the numerous small patch reefs on the shelf area. The estimated area of *Acropora* communities destroyed by hurricanes, siltation, white band disease and other impacts is given in brackets (Table 1) (from Lettsome, 1998).

The cover estimates originated from a coastal inventory conducted in 1993 and may now have some inaccuracies (Lettsome pers. comm). More accurate and updated information on the distribution of habitats in the BVI will be available soon. An ongoing GIS project using satellite imagery and extensive ground truthing will begin in 2000, co-organized by the Conservation and Fisheries Department (CFD) and OECS. Some reefs, especially shallow water *Acropora*, were damaged by Hurricane Hugo in 1989, and Luis and Marilyn in 1995 also impacted the BVI (Smith et al., 1997). Most recently, high seas created by Hurricane Lenny (November 1999) damaged reefs on Virgin Gorda, the island closest to the storm centre.

Before the volcanic activity in Montserrat, coral communities were found in small patches interspersed with sand and sediment on the north, south and west coasts (Jeffers, pers. comm. in IRF, 1993). Lack of hard substrate, combined with high run-off, limits development of corals, and high nutrient input encouraged algal over-growth on many reefs. The Sustainable Ecosystems Institute (SEI) carried out reef surveys in 1995/96, and afterwards monitored the effects of the volcano. Brosnan et al. (1999) reported high coral diversity but small individual coral heads. The baseline data collected by SEI was from habitats already impacted by moderate fishing pressure and by natural environmental disturbances. They found 37 hard coral species, 17 gorgonians and other octocorals, 87 species of marine invertebrate, 37 algal species and 67 fish species. Fish abundance decreased significantly between 1995 and 1996, possibly as a result of Hurricane Luis. Volcanic sediments have had a severe impact on reef growth,

Table 1

Area of coral reefs on several islands and area of *Acropora* communities recently destroyed (Lettsome, 1998)

Location	Area of reef (ha)	Area recently destroyed (ha)
Tortola and Beef Island	987	159
Anegada	4589	382
Virgin Gorda and North Sound	743	341
Jost Van Dyke, Great and Little Tobago	540	314
Norman Island	709	25

Fig. 3. Large bays in Anguilla contain abundant seagrass. Coasts are typically well vegetated, even in this driest island. Anguilla is a low carbonate island; in the background is the high island of St Martin (divided into Dutch and French sectors) (photo C. Sheppard).

Fig. 5. Salt pans in Anguilla. Most low-lying islands have embayments where traditionally salt has been obtained by evaporation (photo C. Sheppard).

particularly on reefs in the east and southwest of island. Direct deposits of ash and waterborne sediment led to coral bleaching, an increase in coral diseases and the disintegration of large sponges, once very common. Some areas of coral were completely covered and in places an algal film developed over the sediment, preventing it from dispersing. Where ash plumes fell, 64% of *Agaricia* coral colonies bleached. In February 1997 monitoring revealed increased degradation of reefs in the south of the island (Brosnan, 1999).

Mangroves

In the 1950s, mangroves lined much of the southern coastline of Tortola in the BVI (Lettsome, 1998) (Fig. 4). Mangrove stands at Road Town, Sea Cows Bay, Nanny Cay and at other sites have since largely been displaced by marinas, tourism complexes and housing (Lettsome, 1998). Extensive stands remain around Beef Island, and Paraquita Bay with smaller stands at the west side of Fat Hog's Bay, including some on land recently donated to the National Parks Trust. There are small areas of mangrove on Anegada, Virgin Gorda and Jost van Dyke (Lettsome, 1998). Although mangrove felling is illegal, many areas have been lost through piecemeal destruction for waterside developments. The Conservation and Fisheries Department demonstrate the importance of mangroves through a number of awareness-raising activities such as local radio programmes, schools visits and replanting activities. For example, a replanting scheme on Tortola combines habitat restoration with education. However, recent storms, especially Hurricane Lenny in November 1999, have damaged many newly established propagules (Evans, pers. comm.). Around Montserrat, mangroves are quite limited. IRF (1993) notes only two sites, with the most important being approximately 6 ha at Fox's Bay on the west coast. On Anguilla a few areas of mangrove are found in enclosed bays (Wells, 1988).

Fig. 4. Red mangrove *Rhizophora* in the British Virgin Islands (at very low tide) is an important but diminishing habitat (photo C. Sheppard).

Fig. 6. The main income of these islands is tourism, attracted by water sports and beaches (Anguilla) (photo C. Sheppard).

Salt Ponds

Salt ponds are a significant habitat for resident and migratory birds in the islands (Fig. 5). In BVI, ponds are found on Tortola, Beef Island, Jost van Dyke, Anegada and Norman Island. However, they have suffered from development. Of 30 identified in a survey in 1984, only 10 remained in 1993 (Lettsome, 1998), with resultant losses of coastal protection (see below). In Anguilla, some salt ponds, for example Maunday's Bay Pond, have been developed (Wells, 1988) but 20 remained in 1990 (Pritchard, 1990).

Beaches

Beaches are one of the Caribbean's most important tourist attractions (Fig. 6). Anguilla has more than 30 spectacular beaches (Proctor, 1997, Pritchard, 1990) whilst Lettsome (1998) estimated almost 80 km of white sand in the BVI. However, development has caused conflicts of interest, and access to beaches for traditional uses has been identified as a serious problem on both islands (Cambers, 1997). Montserrat has extensive beaches, but a large proportion are in the unsafe south section of the island.

Biodiversity

The islands provide important habitat for both resident and migratory birds. According to the Anguilla National Trust, nearly 30% of the 120 bird species recorded on Anguilla are considered globally or regionally threatened or endangered. Outlying cays (including Sombrero) are especially important for seabirds, the most internationally significant facet of Anguilla's wildlife (Pritchard, 1990).

In BVI, roseate tern, frigate bird, brown boobies, least tern and noddy tern are amongst those protected from interference during their nesting season (CFD, 1997a). Caribbean flamingoes were once common in the British Virgin Islands, with thousands of pairs reported in the last century. The young were a popular source of food and so, by the middle of the 20th century, the species had become locally extinct (Lazell, 1999). In 1983 several agencies, including the BVI National Parks Trust and the Conservation Agency, began a programme to reintroduce the flamingo to Guana Island and Anegada. In 1995 five young were successfully fledged at Anegada. Transient Flamingos occasionally arrive, and the flock (now more than 20) is expected to expand further (Petrovic, 1998).

Turtles are found throughout the Lesser Antilles, and green, hawksbill and leatherback are all present. IRF (1993) records seven turtle nesting areas around Montserrat. Historically turtle fisheries were important in the BVI but the territory was never a commercial exporter so turtle have only been fished at a subsistence level (Eckert et al., 1992). Recently the fishery has declined but, unusually for such a developed economy, turtle fishing is still permitted outside the nesting season from December 1 to March 31 (Jarecki, 1996). Hawksbill and green turtles are caught using nets. Local environmental groups are opposed to the fishery but turtle meat is still sold in local supermarkets. Turtle fishing was unregulated in Anguilla in 1981, with catches of up to 15 per day (Olsen and Ogden, 1981). Now, years can go by without a successful nest (Anguilla National Trust, 1999).

The islands have a number of endemic species, such as the Anegada ground lizard, the endangered Anegada Rock Iguana and the Virgin Gorda Gecko in BVI. A benthic invertebrate study at Guana Island (BVI) will span three years from 1999, and may already have found new species. On Sombrero (Anguilla), the endemic black lizard *Ameiva corvina* is threatened by development. At least one other endemic species has been identified in ongoing EIAs (Petrovic pers. comm.). Montserrat has five regionally endemic birds, and one single island endemic—the Montserrat Oriole (*Icterus oberi*).

POPULATIONS AFFECTING THE AREA

For Anguilla and BVI, the most significant factor affecting human populations over the last few decades has been a phenomenally rapid increase in tourism, replacing traditional industries such as fishing and salt extraction. Anguilla's tourist industry flourished in the 1980s and 90s. In 1911 the population was just 4075, and by 1990, that had approximately doubled to 8000 (Proctor, 1997), but by 1999 the figure had increased again to an estimated 11510 (CIA, 1999). Visitor numbers increased dramatically from 17,561 in 1982 to 125,780 in 1995 (Proctor, 1997; IRF, 1996). This expansion led to a construction boom in the 1980s, and GDP was $165.1 million in 1995, of which 91% came from tourism.

The population in the BVI underwent a similar expansion to about 18,000 in 1999. Tourism, contributes 45% of GDP, which was US$183 m in 1997 (CIA, 1999). Visitor arrivals increased 103% between 1984 and 1994. In 1995 there were 65 hotel rooms per 1000 people, but a large part of tourism comes from live-aboard yachting holidays and from the cruise ship industry. In season, cruise liners can deposit more than 5000 visitors per day, and each year the islands are visited by an estimated 300,000 tourists (Miller and Louisy, 1995).

By contrast, Montserrat's population has more than halved in the last five years. In 1995 Montserrat had a population of 11,000, but after the eruption of the Soufriere Hills Volcano in 1995 and 1997, as many as 8000 people were evacuated. Although some people have begun to return, the present population is estimated at just 4500. The population is now concentrated in the north of the island.

In 1995 Montserrat's GDP was $58 m, with tourism accounting for 31% (CTO, 1995, in IRF, 1996). Tourism is only just starting again after the volcano, and the main activity now is construction. Montserrat received £59 million between 1995 and 1998 to aid the recovery process, and the UK government has pledged a further £75 million for 1999 to 2001 (FCO, 1999). The UK and the Government of Montserrat have formulated a comprehensive Sustainable Development Plan. The impact of so many people concentrated in the northern third of the island is impacting the environment and concentrating pressure on natural resources.

FISHING

Economic development has generally reduced the importance of fishing as a livelihood. At the same time the growth of tourism has increased demand for fish products. Throughout the northeast Caribbean the most important fishing gear are fish pots. These often target demersal or reef fish, and are placed in water from 5 to 30 m.

Anguilla

Although small-scale and artisanal, fishing was once a major industry for Anguilla. Most fishers use pots, but seines are used in the calmer summer months. Olsen and Ogden (1981) estimated the maximum sustainable yield (MSY) for demersal stocks to be 2740 tonnes for finfish and 230 tonnes of lobster and 230 tonnes of conch and other shellfish, with 80% of the production potential coming from the 22% of the shelf area containing coral reef communities. Although inshore stocks were heavily fished, shelf stocks were considered to be under-exploited. Government statistics record a conservative estimate of 280 fishers in 1980 (Government of Anguilla et al., 1992). In 1993, Mokoro Ltd estimated around 400 people were involved in the fishing industry, producing between 300 and 500 tonnes of fish, lobster and conch worth 3.3% of the 1991 GDP.

Nearshore resources have declined under increasing pressure. Export duties allow an estimate of lobster exports in 1979 of at least 28,364 kg, though the true value is probably much higher. In 1980, fishers were discarding scarids and snappers up to 2–3 kg as trash fish and fish pots used *Epinephelus striatus*, now commercially extinct in many parts of the Caribbean, as lobster bait (Salm, 1980). However by 1987, signs of overfishing were apparent for lobster and potfish (Stephenson, 1992). Unfortunately no landing data were kept.

Anguilla has an Exclusive Fisheries Zone (EFZ) that covers approximately 85,500 km^2. This zone extends a full 200 miles to the north, but is limited to the west by the British Virgin Islands, to the east by Antigua and to the south by St Maarten and St Bartholomew. Anguilla is on the migratory route of a number of species of large pelagic fishes, so there is also a seasonal fishery for species such as kingfish and dolphinfish, by domestic fishermen using small boats. Longlining for deep-swimming tuna has been tried by a few domestic fishermen, but most have gone back to trapping and handlines.

British Virgin Islands

Fishing was traditionally a major source of food and income in BVI. The islands are surrounded by a number of shallow areas and fishing banks, attracting demersal fish as well as invertebrates such as lobster and conch. One source of fishing pressure comes directly from tourists and casual fishers, many of whom take conch, lobster (often undersized) and fish for their own consumption. This is illegal but unenforceable. Management actions concentrate on education for this sector.

Commercial trap fishing in the BVI has increased over recent years, leading to a decline in fish populations and a change in catch composition (Smith et al., 1997). Pomeroy (1999) made an extensive study of the domestic fishery. He estimated 174 fishers mainly used traps or traps and handlines, with a few using seine nets, fishing rods, SCUBA (now illegal) or other fishing methods. Fish traps were set throughout the year, for an average of 40 weeks. Longline use was seasonal, from October to May. Seine nets are used all year round but particularly from November to March for jacks and from March to August for bonito (*Sarda sarda*), yellowtail snapper (*Ocyurus chyrsurus*) and other species. The total BVI commercial catch for 1998 was estimated at 819,329 kg, with a value of US$6,652,221 (Pomeroy, 1999). The total area of shelf available to BVI fishers is 3130 km^2, giving an estimated 259 kg/km^2 or 2.94 trips/km^2 for 1998. A major problem with the commercial nearshore fishery is ciguatera poisoning from predatory demersal fish. This limits the market for some nearshore fish and many businesses prefer to import fish or serve pelagic species.

The BVI has an Exclusive Fishing Zone of 84,000 km^2, stretching almost due north into the central Atlantic, but bounded on other sides by the United States Virgin Islands, Puerto Rico and Anguilla (CFD, 1997b). The islands lie in the migratory path of a number of large pelagic species (MRAG, 1993) and host an active sport fishery which targets blue marlin, mackerels, barracuda, wahoo and bonitos. However, the domestic sport fishing and longline industry is not well developed. In 1993 there were only two local longline vessels, which fished 15–20 miles offshore out of Anegada. In the same year, 82% of USVI sport fishing occurred in BVI waters under licences costing only $200 a year per boat (MRAG, 1993). A lot of fish were sold back to hotels in the BVI. Licence fees and penalties for illegal fishing have since been increased, but developing a stronger BVI industry could potentially produce a higher revenue than licensing foreign commercial vessels, with less impact on fish stocks.

Montserrat

Montserrat has a very small EFZ of only 6000 km^2 (MRAG, 1993) and a small coastal shelf of 140 km^2. The coastal shelf is too small to support large demersal stocks so pelagic stocks are important. Many of these pelagic species are migratory, providing a seasonal fishery from November to March. The main target species in the early 1990s were snapper, grouper, kingfish, barracuda and skipjack tuna, many of which can be ciguateric. Also important were conch, spiny lobster, some reef and demersal fishes, turtles and shark, which were all often taken undersize (Jeffers, pers. comm. in IRF, 1993).

In 1991 there were 250 fishermen (33% full time) fishing mainly with handlines and fish traps from 53 dory-style fishing vessels of 4.5–6 m (Jeffers, pers. comm in IRF, 1993). Fisheries development is also limited by a lack of safe harbours due to the geography of the island. Large vessels are forced to seek shelter in Antigua. Commercial longlining has been proposed but in 1993 there was only one local vessel capable of commercial long-lining, and this had stopped fishing (MRAG, 1993). Fish catches for 1997 were estimated at 46 tonnes (FAO, 1999).

Before the eruptions, the main threat to coral reefs of Montserrat was thought to be fishing, predominantly trap and spear-fishing. The concentration of people in the safe area at the north of the island has had a serious impact on the marine resources of this area. Fishing effort has been particularly intense here, and decreases in catches were noticed in 1997 (Brosnan, 1999).

COASTAL EROSION AND LANDFILL

Waves and excessive rain from hurricanes and tropical storms lead to erosion and sedimentation damage in nearshore waters. Even moderate rainfall over land cleared for development or agriculture can result in heavily sedimented water extending offshore more than 1 km (Cambers, 1999). Such effects are mitigated by coastal protection from salt ponds, mangroves and sand dunes. However, extended periods without major hurricane damage may encourage inappropriate development (e.g. mangrove felling or sand mining). Anguilla, BVI and Montserrat have all experienced construction booms of various intensities in recent years, with subsequent effects on the marine environment. During the 1980s, Anguilla underwent a phase of extremely rapid tourism development. The need for building sand encouraged indiscriminate beach mining; for example at Sile Bay extensive dunes up to 6 m high were heavily exploited, resulting in encroachment of the sea 46 m inland after Hurricane Luis in 1995. Sand mining was prohibited after 1994 except on licensed beaches. Sand is currently mined from areas at Windward point but this will be depleted within a couple of years (Hendry and Bateson, 1997). The government is committed to restricting beach mining, but offshore sand mining is being considered as an economical alternative to imports (Hendry and Bateson, 1997).

After Hurricane Donna in 1960, Anguilla had almost four decades without a severe hurricane. Hurricanes Luis and Marilyn (which struck only a week or so apart in 1995) caused some damage to both BVI and Anguilla (Smith et al., 1997) but were relatively small systems. However in November 1999 the centre of powerful Hurricane Lenny passed very close to Anguilla, devastating sea defences, infrastructure and many hotels. Several coastal areas were seriously eroded, including Bankie Banx Preserve, Covecastles and Sonesta Beach, and heavy rains caused floods 2–3 m deep in places. By contrast, although only approximately 80 km north of the hurricane centre, BVI suffered minimal damage.

Hurricane Hugo passed close to BVI in September 1989, and battered Puerto Rico. It generated heavy rainfall and waves approximately 4 m high. Beaches where sand had been mined were eroded 5–10 m inland and seawater encroached 30 m (Lettsome and Potter, 1997). Coastal reclamation projects suffered severe erosion and shallow *Acropora* reefs were impacted. The Beach Protection Ordinance prohibits beach mining without the permission of the Minister of Natural Resources, but this has proved ineffective in protecting these coastal resources (Lettsome, 1998). Between 1982 and 1996, 10416 m^3 of sand were legally mined, with 94% coming from Josiah's Bay, Fat Hog's Bay, Brewers Bay and Cane Garden Bay (Tortola) and Little Bay (Virgin Gorda). Eight dredging operators and one unlimited beach sand mining permit are not included in this total (Lettsome, 1998).

Hugo was also the most damaging hurricane to hit Montserrat in recent years. Sustained winds of 140 mph and gusts of 180 mph battered the island for 12 hours. Ninety eight percent of homes were damaged (20% completely destroyed) and total losses exceeded US$300 million (UNDP, 1989; Butler, 1991). The hurricane compounded erosion problems caused by hurricane David in 1979. Building for a growing tourist industry and a cultural switch from wooden to stone (with cement) houses in the 1960s had already resulted in serious beach excavation which was exacerbated by reconstruction attempts after both hurricanes (Cambers, 1999). Beaches suffered a mean erosion rate of 1.05 m/year between 1966 and 1990 (Cambers, 1990).

EFFECTS FROM URBAN AND INDUSTRIAL ACTIVITIES

The main industries on Anguilla are tourism and financial services. The impact of urban development on nearshore coastal resources (especially coral reefs) may have been lessened by the dry climate. Oxenford and Hunte (1990) concluded the reefs and seagrasses were in relatively good condition despite a decade of tourism expansion, but noted high macro algae cover and turbid water which are often associated with land-based pollution. They considered the minimal rainwater run-off an unlikely cause and attributed the conditions to slow recovery of *Diadema* (after the 1984 Caribbean wide die-off) and naturally fine sediments. Nonetheless, Proctor (1997) notes scrub clearance and dune reconfiguration for building space and development has increased the sedimentation load on the reefs through beach erosion and subsequent beach re-nourishment without adequate environmental controls.

The most significant industrial activity currently proposed for Anguilla is construction of a rocket launching site. The Anguillan Government has agreed to grant Beal Aerospace a 98-year lease for the remote islet of Sombrero. Only 388 m across, Atlantic swells often crash right over the island. Beal plans to launch communication satellites via the yet untested BA-2 rockets, carrying 800 tonnes of hydrogen peroxide and kerosene (Pearce, 1999). The company plans to launch 12 rockets a year, and industrial accidents are a significant threat. An environmental impact assessment has been heavily criticised as incomplete by the Anguilla National Trust, RSPB, American Bird Conservancy, BirdLife International and regional NGOs such as Island Resources Foundation. So far, many of the arguments have centred on terrestrial life including the endemic *Ameiva corvina* lizard. The Anguilla National Trust describes Sombrero as the most important seabird nesting site in the east Caribbean. Although too exposed for well developed shallow reefs, marine life is rich (Ogden et al., 1985) but poorly documented. The planned facility has not yet been licensed by the British National Space Agency, and discussions continue.

As on Anguilla, the two main industries in BVI are tourism and financial services. The most severe impacts on the marine environment stem from both the shortage of flat coastal land and population pressure intensified by high visitor numbers. Several coastal areas have been reclaimed in Road Town Tortola, and other smaller reclamations occur elsewhere in the Territory. Tortola produces an estimated 40 tonnes of garbage a day (Lettsome, 1998). At Pockwood Pond, one of Tortola's largest stands of red mangroves stood next to a power station, incinerator, cement plant, and rock quarry. The mangroves impeded plans for commercial expansion until an oil spill devastated the area. Rather than restore the habitat, permission was granted to reclaim the area for industrial use. The reclamation extended beyond the mangrove, and also affected nearby reefs (Petrovic, 1998).

Dredging to deepen harbours and boat channels, build marinas and reclaim land requires a permit, but a recognized problem is mining beyond the area of the permit in order to sell mined sand (Jarecki, 1996). Even within coastal bays, dredging operations are rarely screened.

Roads cut on steep hillsides to increase the value of undeveloped real estate generate substantial sediment run-off. Where development does proceed near the coast, topsoil cleared from the building site has sometimes been pushed down the slope to end up in the sea. Such impacts are localized in the BVI at present, but development is occurring at an alarming rate.

Sewage presents another disposal problem to small islands, and is compounded by the high number of visitors. Although a few hotels and commercial enterprises have installed small treatment works, such facilities in BVI are minimal (Petrovic, 1998). Most houses have septic tanks although a few coastal dwellings discharge directly into the sea. In BVI urban areas, drainage channels designed for rainwater are sometimes contaminated with sewage (Lettsome, 1998). At Cane Garden Bay, Tortola's flagship beach and an important anchorage, algal blooms, bacterial contamination and public opinion galvanized government into constructing the first public sewage treatment works. However, few yachts are equipped with holding tanks and even fewer marinas have pump-out facilities. In crowded marinas such as Wickhams Cay and Virgin Gorda Yacht harbour, sewage is a problem, compounded with pollution from boat yards including paint residues, solvents and leaking engine oil (Petrovic, 1998).

Before the volcanic activity, Montserrat was becoming one of the Caribbean's premier retirement destinations. Beginning in the 1960s, tourism had developed into the leading economic sector and the island was a cruise ship destination. Industrial impacts were moderate although the ship-to-shore pipeline transferring oil to the Texaco and Delta tank facilities has been indicated as a potential threat (IRF, 1993). Waste oil from private vehicles and generators is often disposed in storm drains or poured directly on the ground, where it washes into the sea (IRF, 1993).

LEGISLATION AND PROTECTIVE MEASURES

As UKOTs, Anguilla, BVI and Montserrat cannot be parties to international conventions and treaties in their own right. Where necessary or desirable, conventions to which the UK has acceded may be extended to the OTs. In most cases local legislation is also required. For example the requirement to designate a wetland on joining the Ramsar Convention on Wetlands of International Importance is technically already discharged by the UK (Pritchard, 1990). No initial provision was made for Anguilla or BVI in the Ramsar Convention, although Montserrat proposed a site (Pritchard, 1990). In 1994 BVI suggested the Western Salt Ponds of Anegada as a RAMSAR site (Lettsome, 1998) and this site was finally accepted in 1999. Twenty-six international conventions and treaties with an environmental theme have been extended to the BVI, but domestic regulations do not cover them all (Lettsome, 1998).

The UK Government has produced a White Paper dealing with biodiversity issues in overseas territories (FCO, 1999). Although UKOTs have 10 times more endemic species than the UK (FCO, 1999), most of the UK's financial contribution to international conservation is via Global Environment Fund (GEF), earmarked for developing countries and not available to UKOTs (Pienkowski, 1999). UKOTs are eligible for Darwin Initiative funds for biodiversity conservation from the UK government, but little work has been done. Finding money for resource management can be a challenge, especially where resource use does not contribute a large part to GDP. For example, fisheries administration might be expected to cost approximately 5% of annual production value, but the 1993 budget for the Anguillan Fisheries Department, although equalling 20% of the 1991 production value, was not enough to cover costs of elementary catch and effort recording (Mokoro Ltd, 1993). In the BVI, the combined budget of the Conservation and Fisheries Department and the National Parks Trust averaged only 1% of the overall budget between 1990 and 1996 (Lettsome, 1998). However, an important development in the region is a program to harmonize fishing regulations between OECS states, including harmonized fishery management legislation, common fisheries surveillance zones and common fishing zones (OECS, 1999).

Coastal Protection

Anguilla has several laws which can be used to regulate coastal resource use. There is a Beach Control Ordinance 1988 which prohibits damage to plants, shrubs and trees, and allows beaches to be declared as protected from sand mining. Seventeen beaches were so designated in 1988. The Cruising Permit Ordinance of 1980 applies to the coastal region, out to 3 nautical miles. Permit fees are charged, and anchoring is prohibited around Sandy Island, Prickly Pear Cays, Seal Island, Dog Island and Rendezvous Bay (Pritchard, 1990).

BVI drew up Coast Conservation Regulations in 1990 but they are not yet enacted. Similarly, the Land-use Planning Bill, legislation on ground water monitoring and control and the Parks and Protected Areas System Plan of 1986 are all still in the pipeline. The Conservation and Fisheries Department's jurisdiction over development projects is largely limited to the Land Development (Control) Ordinance which is administered by the Development Control Authority and Town and Country Planning (Lettsome, 1998). There is no legal requirement for EIAs before development in BVI, and where penalties exist, rapid economic development makes them less effective as a deterrent (Lettsome, 1998). No specific legislation exists to control the import, use, storage or disposal of toxic chemicals. However, many of these issues should be addressed under the new National Integrated Development Plan which provides a framework to promote, plan and rationalize the sustainable development of natural resources.

An area where BVI has pioneered conservation is in the use of mooring buoys to prevent anchor damage to coral reefs. The mooring system, now maintained by the National Parks Trust, was originally conceived, funded and operated by BVI Dive Operators in the 1980s. In addition, private enterprises operate moorings in several popular anchorages, demonstrating the potential for synergy between private enterprise and conservation.

Fisheries Legislation

Fishing within Anguilla's territorial waters and EFZ is regulated by the Fisheries Protection Ordinance of 1988, which repealed the Turtle Ordinance of 1984. Under the new legislation, regulations were set out in 1990, including sports fishing regulations (Pritchard, 1990). Anguilla has a policy of reserving most of the fish resources in its exclusive fisheries zone for domestic fishermen. In the past, foreign fishing vessels caused serious conflicts of interest, for example illegal Taiwanese longline vessels were caught hauling traps belonging to local fishermen in the EFZ. By 1993, Anguilla was reported to have effective surveillance and enforcement of shelf areas and of unlicensed foreign fishing (Mokoro, 1993). Foreign vessels are not totally excluded, but are only licensed north of 19°. Within the OECS framework, the Anguillan Government is an informal observer, and accepts OECS conditions for licensing foreign fishing vessels, but has yet to make a decision on common fishing zones.

The perception that large foreign fishing vessels deplete near and inshore resources to the detriment of local users is shared by BVI. Previously, United States longliners were licensed to target swordfish in the BVI EFZ, but since 1990 there has been a moratorium on large-scale foreign commercial long-line fishing, including several rejected applications from Taiwan. Under the BVI's recent Fisheries Act (1997), locals and visitors alike require a licence to take any marine life. Fishing using scuba, or spearguns attracts a US$15,000 fine. The Act provides for the declaration of fishing priority areas, marine protected areas, species protection, gear restrictions, and spatial or temporal closures. Local fisheries management areas can be set up where local users help devise by-laws. Under the law, anti-pollution measures can be enforced, and the Chief Fisheries Officer is mandated to ensure stocks are not overexploited, through implementation of a fisheries management plan. The Act sets out wide ranging stop, search and seize powers for enforcement officers, and sets fines up to US$500,000 in addition to confiscation of foreign vessels infringing the Act. However, enforcement procedures are not yet well worked out. In March 1999 all foreign vessel licences were suspended, but the resulting outcry from the neighbouring United States Virgin Islands led to temporary licences being issued from June 1999. Although the Act has a sound basis in sustainable management, its impact will be diminished unless sufficient resources are made available for enforcement. For example, collection of undersized lobster and conch, particularly by tourists and non-professional fishermen remains a problem.

Montserrat's fishing regulations were enacted in 1982, but were generally not enforced. More recent legislation has been drafted to correspond with OECS-harmonized fisheries legislation, with specific habitat-protection measures including provision for the adoption of MPAs (IRF, 1993).

Marine Protected Areas (MPAs)

Anguilla's Marine Parks Ordinance of 1982 empowered the Governor to designate MPAs and acquire private land. The legislation was amended in 1992 so as to restrict damaging activities and impose fines or imprisonment as penalties. After a history of discussions and plans for multi use/zoned marine protected areas in Anguilla which dated back to the early 1970s, five marine parks were established in 1993 (Smith et al., 1997). Four of these (Sandy Island, Prickly Pear—including the Seal Island reef system, Island Harbour and Dog Island) are designed to protect reefs and are managed by the Department of Fisheries.

The BVI has marine protected areas at the Wreck of the Rhone National Park, The Baths (Virgin Gorda), and Anegada's Horseshoe Reef. However, these are not 'no take' areas. Commercial fishing around the Rhone and the Baths is somewhat restricted because of day-to-day conflicts of interest with the tourist industry. Horseshoe Reef is off-limits to both commercial and recreational use. All anchoring is forbidden, and despite its isolation, few charter boats venture there because of the treacherous waters. Fishing was initially banned, but permits were later issued to Anegadian fishers. Pomeroy (1999) estimated the annual catch from Horseshoe reef as 47.7 tonnes, of which over 16.4 tonnes was lobster.

Montserrat has no designated MPAs although proposals have been put forward for the northernmost tip of the island (Bovey et al., 1986), Fox Bay in the west (IRF 1993) and an extensive zoning system encompassing fishery zones, recreational zones and no-take preserves (Brosnan et al., 1999). Since 1979, the mangroves and pond at Fox's Bay have been leased to the Montserrat National Trust as a bird sanctuary.

PROSPECTS AND PROGNOSES

The immediate outlook for the marine environment of Montserrat does not seem good. The impacts of the volcano on nearshore systems have been serious and it may take a long time for the ecosystems to recover. The concentration of the population in the north will inevitably increase the pressure on the marine resources less heavily impacted by the volcano, but it must be hoped that the sustainable development plan will be effective. Restricted access to the south coast may provide a *de facto* marine protected area which could help maintain the island's fisheries in the long term.

In the British Virgin Islands and Anguilla the future of the marine environment looks relatively good. Both territories are wealthy, reducing the pressure on marine resources through subsistence use. There is an awareness of environmental issues at all levels of society. The present financial reliance on tourism provides every incentive to maintain healthy marine environments as a major source of revenue, although development is rapid and in some places badly planned and implemented. However, a change in rules governing offshore financial service industries could decrease the revenue of these islands, and the tourist market is fickle. The loss of either one of these industries would lead to an increased dependence on marine resources, particularly nearshore fish and invertebrates. Revenue currently gained from schemes such as mooring buoy programmes would be lost. Government departments and NGOs must continue to work with local communities and tourism operators to implement long-term sustainable management strategies for coastal development and conservation. With these in place, the marine habitats of these beautiful islands should be sustained well into the new millennium.

ACKNOWLEDGEMENTS

This chapter would not have been possible without extensive resources made available by the Island Resources Foundation library, British Virgin Islands. Particular thanks to W. Dressler at IRF. Hugh Philpott at the Foreign and Commonwealth Office kindly commented on an earlier draft. ICLARM contribution number 1561.

REFERENCES

Anguilla National Trust (1999) Home Page http://web.ai/ant. Accessed 30 December 1999.

Bovey et al. (1986) Montserrat National Park: Ecological and cultural feasibility assessment. Joint effort of the Montserrat National Trust, World Wildlife Fund/UK, and the University of Alberta.

Brosnan, D.M. (1999) Ecological impacts of the Montserrat Volcano: Pictorial account of its effects on land and sea. SEI. www.sei.org/impacts.html. Accessed Dec. 1999.

Brosnan, D.M., Grubba, T.L.J., Backman, D.K., Boylon, K. and Moore, L.T. (1999) The coral reefs of Montserrat, West Indies: Diversity, conservation and ecotourism. SEI www.sei.org/coral.html. Accessed Dec. 1999.

Butler, P. (1991) Making a move on Montserrat. Rare Centre for Tropical Bird Conservation. Philadelphia

Central Intelligence Agency 1999 Central Intelligence Agency World Fact Book. www.odci.gov/cia/publications/factbook/. Accessed Dec. 1999.

Cambers, G. (1981) Sand resources in Montserrat. Prepared for Government of Montserrat by Caribbean Planning and Development Company, Barbados.

Cambers, G. (1990) Beach changes in Montserrat between 1966 and 1990. OECS/NMRU/German Technical Co-operation Programme. Castries, St. Lucia.

Cambers, G. (ed.) (1997) *Managing Beach Resources in the Smaller Caribbean Islands*. Papers presented at a UNESCO–University of Puerto Rico Workshop, 21–25 October 1996, Mayagüez, Puerto Rico. Coastal Region and Small Islands Papers, No. 1. UPR/SGCP-UNESCO, Mayagüez, 269 pp.

Cambers, G. (1999) Sand Mining: ICZM Perspectives in the Caribbean. *Reef Encounter* **26**, 14–16.

CFD (Conservation and Fisheries Department) (1997a) Conservation Locally in the BVI. October 1997. Conservation and Fisheries Department, Road Town, BVI.

CFD (Conservation and Fisheries Department) (1997b) Fisheries Development in the British Virgin Islands: Emerging Issues. A technical report for the sub-committee on productive sectors: National Integrated Development Strategy, Conservation and Fisheries Department, Ministry of Natural Resources and Labour, British Virgin Islands.

Corker, I. (1986) Montserrat: a resource assessment. Overseas Development Administration report no. P-164. Land Resources Development Centre, Surrey, UK.

Eckert, K.L., Overing J.A. and Lettsome, B.B. (1992) WIDECAST Seaturtle Recovery Action Plan for the British Virgin Islands. CEP Technical Report No. 15, UNEP Caribbean Environmental Programme, Kingston, Jamaica. 116 p.

FAO (1999) FAO Fisheries Department–Production of fish, crustaceans and molluscs (1988–1998). www.fao.org/WAICENT/FAOINFO/FISHERY/statist/summtab/z2l ist.asp. Accessed Dec. 1999.

FCO (Foreign and Commonwealth Office) (1999) Britain and the Overseas Territories—a modern partnership. White Paper.

Government of Anguilla, WTO/UNDTCD/UNDP (1992) Tourism and economic development in Anguilla: a tourism strategy for the nineties.

Hendry, M.D. and Bateson, R.I. (1997) Seeking sand source alternatives: an island case study. In *Managing Beach Resources in the Smaller Caribbean Islands. Coastal Region and Small Island Papers No. 1*, ed. G. Cambers, pp. 133. UPR/SGCP-UNESCO, UNESCO–University of Puerto Rico Workshop, 21–25 October 1996, Mayaguez, Puerto Rico, Mayaguez.

IRF (Island Resources Foundation) (1993) Montserrat environmental profile. Island Resources Foundation, St. Thomas, USVI. pp. 124.

IRF (Island Resources Foundation) (1996) Guidance for best management practices for Caribbean coastal tourism. Government of Montserrat, Island Resources Foundation and the Montserrat National Trust. IRF, United States Virgin Islands.

Jarecki, L.L. (1996) Linking tourism and nature conservation in the British Virgin Islands. A Case Study for International Postgraduate Course in Environmental Management for Developing countries, Dresden University of Technology.

Jones, D. (1999) An analysis of Eastern Caribbean hurricane tracks for 1950–1997 provides some interesting landfall statistics! http://www/caribwx.com/

Landsea, C.W. (1999) FAQ: hurricanes, typhoons and tropical cyclones. Part G. Tropical Cyclone Climatology. http://aoml.noaa.gov/hrd/tcfaqG.html#G4, Version 2.8 Modified 12 Aug. 1999.

Lazell, J. (1999) Guana Island—A Natural History Guide. The Conservation Agency. From an excerpt on the Guana Island web-page: www.guana.com.

Lettsome, B. and Potter, L. (1997) Sand mining in the British Virgin Islands—a second look. In *Managing Beach Resources in the Smaller Caribbean Islands. Papers presented at a UNESCO–University of Puerto Rico Workshop, 21–25 October 1996, Mayagüez, Puerto Rico. Coastal Region and Small Island papers, No. 1*, ed. G. Cambers (ed.), UPR/SGCP-UNESCO, Mayagüez, 269 pp.

Lettsome, B. (1998) The environment of the British Virgin Islands: emerging issues. National Integrated Development Plan. Sectoral Presentation. Conservation and Fisheries Department, Ministry of Natural Resources and Labour, Government of the British Virgin Islands.

Miller, E. and Louisy, P. (1995) H. Lavity Stoutt Community College Strategic Development Plan. Millrowe Consultants Ltd. Kingston, Jamaica.

Mokoro Ltd. (1993) Anguilla Strategic Review: interim report to the Anguillan Government.

MRAG (1993) Large pelagic fisheries in the Caribbean: Their role in the economies of UK dependent territories. Report to the Overseas Development Administration. Caribbean UK Dependent Territories: EFZ Management Study. MRAG.

OECS (1999) Fisheries in the OECS: the Way Forward. Organisation of Eastern Caribbean States. www.oecsnrmu.org/ Accessed Dec. 1999.

Ogden, N.B., Gladfelter, W.G., Ogden, J.C. and Gladfelter, E.H. (1985) Marine and terrestrial flora and fauna notes on Sombrero Island in the Caribbean. *Atoll Research Bulletin* **292**, 61–74.

Olsen, D.A. and Ogden, J.C. (1981) Management Planning for Anguilla's Fishing Industry. Draft Report prepared for Eastern Caribbean Natural Area Management Program. US Virgin Islands.

Oxenford, H.A. and Hunte, W. (1990) A survey of marine habitats around Anguilla, with baseline community descriptors for coral reefs and seagrass beds. Department of Agriculture and Fisheries, Government of Anguilla, 177 pp.

Pearce, F. (1999) Trouble in Paradise. *New Scientist* **20**, 18–19.

Petrovic, C. (1998) Environmental issues in the British Virgin Islands. *Islander Magazine* **5**, 25–30.

Pienkowski, M. (1999) Paradise Misfiled? Nature conservation in UK's Overseas Territories and the role of UK Government and NGOs. Ecos. UK Overseas Territories Conservation Forum. www.ukotcf.org/ Accessed Dec. 1999.

Pomeroy (1999) Economic analysis of the British Virgin Islands commercial fishing industry. Research Report. ICLARM Caribbean/Eastern Pacific Office/Conservation and Fisheries Department, Road Town, Tortola, BVI.

Pritchard, D. (1990) The Ramsar Convention in the Caribbean with special emphasis on Anguilla. Royal Society for the Protection of Birds Sabbatical Report, RSPB Sandy, Bedfordshire. UK.

Proctor, O. (1997) Destroying the goose that lays the golden egg. In *Managing Beach Resources in the Smaller Caribbean Islands. Papers presented at a UNESCO–University of Puerto Rico Workshop, 21–25 October 1996, Mayagüez, Puerto Rico. Coastal region and small islands papers, No. 1*, ed. G. Cambers, pp. 215–220. UPR/SGCP-UNESCO, Mayagüez, 269 pp.

Putney, A.D. (1982) Survey of Conservation Priorities in the Lesser Antilles. Final Report. Caribbean Conservation Association, Caribbean Environment Technical Report 1.

Salm, R.V. (1980) Anguilla. Coral Reefs and Marine Parks Potential. Consultancy report for selection and design of marine parks and reserves. ECNAMP unpubl. rep. Available from Island Resources Foundation.

Smith, A.H., Rogers C.S. and Bouchon, C. (1997) Status of Western Atlantic coral reefs in the Lesser Antilles. *Proc. 8th International Coral Reef Symposium* **1**, 351–356.

Stephenson, A. (1992) Anguilla: country profile 1991. Anguilla Government.

Towle, E.L. (1979) Survey of conservation priorities in the lesser Antilles. Trip Report, Anguilla Field Work July 19-21, 1979 and August 6-9 1979. Island Resources Foundation unpubl. rep., Island Resources Foundation, United States Virgin Islands.

UNDP/DEP (1989) Hurricane Hugo in Montserrat Reconnaissance Report on the Structural Damage. Sept. 1989. UNDP Barbados.

Wells, S. (1988) *Coral Reefs of the World. Vol. 1. Atlantic and Eastern Pacific*. IUCN/UNEP, pp. 1–6.

THE AUTHORS

Fiona Gell
ICLARM Caribbean and Eastern Pacific Office,
PMB 158, Inland Messenger Service,
Road Town, Tortola,
British Virgin Islands

Maggie Watson
ICLARM Caribbean and Eastern Pacific Office,
PMB 158, Inland Messenger Service,
Road Town, Tortola,
British Virgin Islands

Chapter 40

THE LESSER ANTILLES, TRINIDAD AND TOBAGO

John B.R. Agard and Judith F. Gobin

All the islands of the Lesser Antilles have coasts that border on both the tropical western Atlantic and the Caribbean Sea. Major sills in the passages between these islands control water flow into the Caribbean Sea from the Atlantic Ocean. Horizontal motion below the average sill depth of the Antillean Arc (1200 m) is almost stagnant. These islands are probably the most important physiographic features of the Caribbean Sea as they act as the gatekeepers to the integrity of the Caribbean marine environment. The coastal marine environments around the islands are generally oases of high production associated with shallow waters, coral reefs, mangrove swamps, estuaries and coastal lagoons surrounded by deep oligotrophic seas.

The oceanography of the southern Lesser Antilles is strongly influenced by the outflow of two of the world's largest river systems, the Amazon and the Orinoco. Superimposed on this regime are the periodic passage of large eddies of Amazon water from the Guyana Current. The marine production of offshore waters is generally low due to the relatively stable thermocline, which in the absence of significant upwelling prevents the mixing of nutrient-rich deep waters with surface waters. The main seasonal variation of the islands is due to rainfall. The passage of hurricanes are other periodic events that occasionally have significant impacts on the marine biota of these islands.

Penaeid shrimp dependent on estuarine conditions and muddy bottoms are the most valuable fishery resource harvested on the continental shelf between Trinidad and Venezuela. Pelagic fishes offshore (e.g. flyingfish, kingfish, dolphinfish, tuna, swordfish, sharks) and inshore (e.g. kingfish, jacks, herrings and anchovies) are the main commercial fisheries resource exploited among the islands in the area of mixed water stretching from the north coast of Trinidad to St. Vincent. In the clear blue oligotrophic waters from St. Lucia to the Virgin Islands the only significant fisheries are for lobsters and conchs inshore, and for tuna offshore. Only Trinidad and Tobago and the French islands of Martinique and Guadeloupe show any noticeable increase in fish catches from 1990 to 1995. Large commercial fishing vessels from several nations not indigenous to the sub-region frequently exploit the limited fish stocks within the Exclusive Economic Zone of these islands. In many cases, these vessels operate without the knowledge and consent of island governments. All of the islands have fisheries legislation but a shortage of trained personnel and the high cost of effective fisheries patrols in offshore as well as inshore waters and marine parks hinder their effective enforcement.

Seas at The Millennium: An Environmental Evaluation (Edited by C. Sheppard)
© 2000 Elsevier Science Ltd. All rights reserved

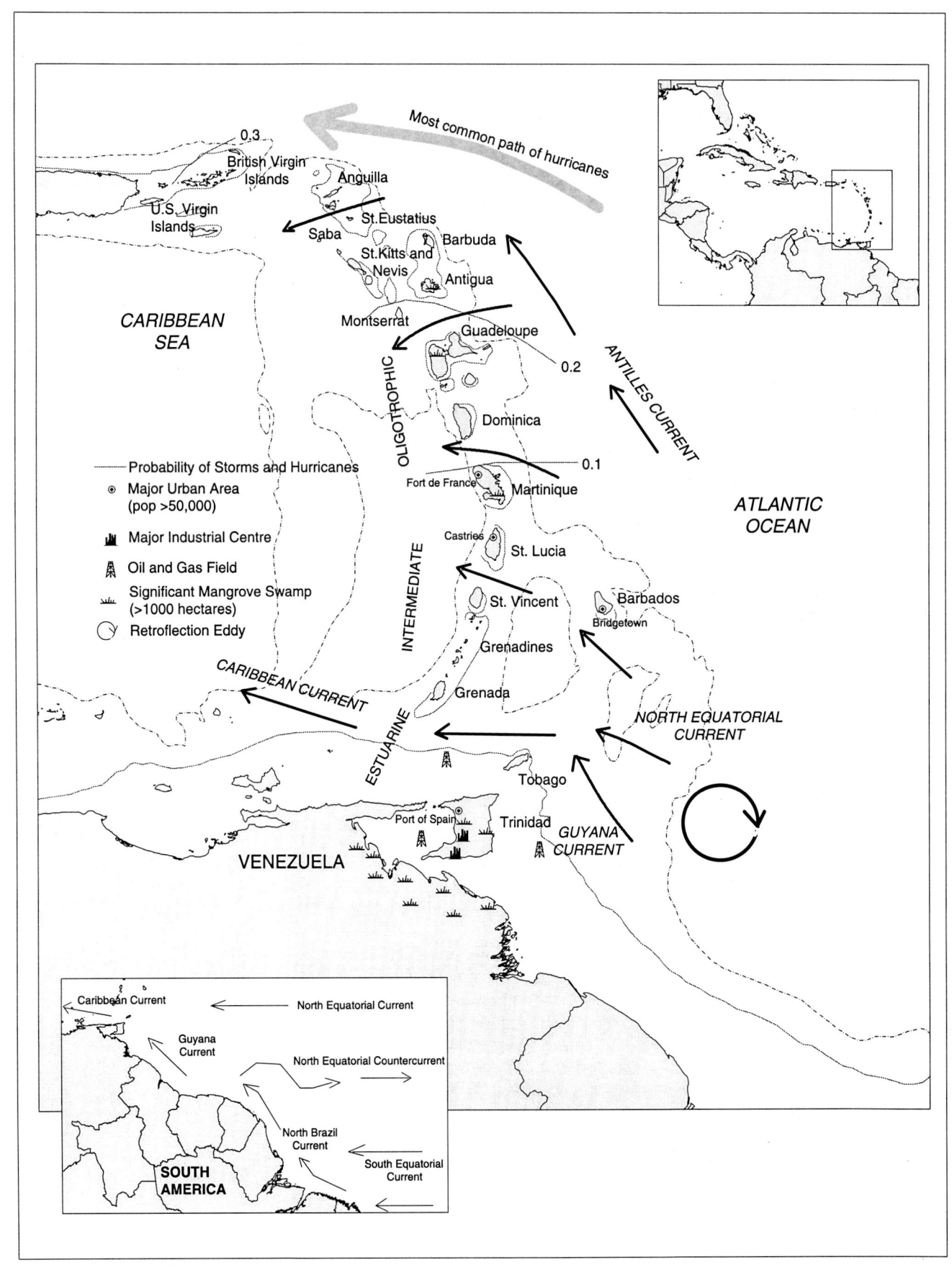

Fig. 1. Major features of coastal and marine environment of the Lesser Antilles, Trinidad and Tobago.

In these islands human impacts on the marine environment are significant because population density is high ranging from 83 km^{-2} in Anguilla to 614 km^{-2} in Barbados. Ongoing deforestation is a serious problem affecting the coastal zone in Trinidad and Tobago, Guadeloupe, Martinique, St. Lucia and the British Virgin Islands. Artisanal fishing methods such as trawling for shrimp, cutting mangrove roots to harvest oysters, and over-harvesting of edible sea urchins, lobsters and conch, also damage marine habitats. Beach sand mining is the major human-induced cause of coastal erosion in the Eastern Caribbean. Marine pollution from inadequately treated sewage effluents is a problem on every island because of the lack of adequately maintained centralised sewage treatment facilities.

The annual number of tourist visitors in individual islands is substantially greater than their resident population in 12 out of 14 instances, excluding only Dominica and Trinidad. Airport and marina construction to provide facilities for tourists have resulted in the filling in of coastal mangroves and increasing sedimentation in coral reef and seagrass areas. The development of heavy industry in the coastal zones of the various territories is very limited except for the island of Trinidad. Dense petrochemical-related shipping traffic passing through narrow straits around Trinidad and Tobago make this area a high-risk zone for marine pollution from shipping accidents.

The Lesser Antillean countries are signatories to several important international conventions and programs, which are geared to protect the marine and coastal environment. However, the record of implementing the provisions of these conventions is very poor. Further, there are few significant ongoing marine investigations in the sub-region except for those undertaken through the Caribbean Coastal Marine Productivity (CARICOMP) network of Marine Laboratories, Parks and Reserves. The islands will have to significantly increase their environmental protection efforts if they are to stem the tide of pollution and natural resource depletion.

THE DEFINED REGION

This chapter covers the Lesser Antillean arc of islands in the Eastern Caribbean, as well as the islands of Trinidad and Tobago which are geologically parts of the South American continent. The sub-region includes the following countries: Anguilla, Antigua and Barbuda, Barbados, British Virgin Islands, Dominica, Grenada, Guadeloupe, Martinique, Montserrat, Netherlands Antilles (Saba and St. Eustatius), St. Kitts (St. Christopher)-Nevis, St. Vincent and the Grenadines, St. Lucia, Trinidad and Tobago, and the U.S. Virgin Islands (Fig. 1). All the islands of the archipelago have coasts that border on both the tropical southwest Atlantic and the Caribbean Sea. Major sills in the passages between these islands are the controllers of water flow into the Caribbean Sea from the Atlantic Ocean (Fig. 1). Horizontal motion below the average sill depth of the Antillean Arc (1200 m) is almost stagnant. These islands are probably the most important physiographic features of the Caribbean Sea as they act as the gatekeepers to the integrity of the Caribbean marine environment from the Atlantic Ocean. The water forcing its way in the upper layer of the sea, through the narrow channels between the islands creates jet currents as well as large turbulent wakes and eddies. These currents may concentrate marine organisms as well as nutrients and pollutants in the wake of islands. The coastal marine environments around the islands are generally oases of high production associated with shallow waters, coral reefs, mangrove swamps, estuaries and coastal lagoons surrounded by deep oligotrophic seas. The wellbeing of these small islands is therefore intimately associated with the protection of their coastal ecosystems. The area seems amenable to the large marine ecosystem (LME) approach to managing marine resources. However, since this LME includes the territorial waters of 15 countries, the effectiveness of this management tool is dependent on the co-operation of all the bordering countries.

SEASONALITY, CURRENTS, NATURAL ENVIRONMENTAL VARIABLES

Ocean surface temperatures are about 27°C with seasonal fluctuations of no more than 3°C and a decrease of 10-15°C within the upper 200 m, beyond which there is little change. The main seasonal variation of the islands is due to rainfall. The seasons are caused by the annual displacement of the Inter-tropical Conversion Zone (ICTZ) northward during April to September and its return southward during January to March. Climatically the year is divided into a dry season lasting from about January to May and a wet season from June to December. There are differences in precipitation between the islands with rainfall being highest among the central islands of the Antillean arc from Anguilla to Dominica. The effect of local rainfall on the marine biota is generally insignificant in comparison with the dominant hydrographic regime.

The oceanography of the southern Lesser Antilles is strongly influenced by the outflow of two of the world's largest river systems, the Amazon and the Orinoco. Together they account for about 20% of fresh water discharges into the world's oceans. During February to May, the Guyana Current advects water of high primary productivity and sediment concentration from the Amazon River along the edge of the continental shelf of northern South America. It then curves left to join the North Equatorial Current as it enters the Caribbean Sea in a broad 150–200 km stream mainly between Tobago and Barbados (Fig. 1). During this period, the discharge of the Orinoco is low and among the islands, only Trinidad is under its influence.

During June to January, the North Brazil Current retroflects or veers offshore into the North Equatorial Counter Current taking about 60% of the annual discharge of Amazon water eastward toward Africa. Offshore in the Atlantic, this water mixes with the North Equatorial Counter Current before the lens of low salinity water enters the southeastern Caribbean about six months to a year later (Muller-Karger et al., 1988). These events seem to cause diminished flow between the islands of the eastern Caribbean by diverting a major source of water for the Guyana Current. The weakening of the Guyana Current permits the northwestward dispersal of Orinoco water from the Gulf of Paria towards the Antilles due to eastern Caribbean Ekman forcing (Muller-Karger and Varela, 1989). During the peak of the wet season from July to November, the discharge of the Orinoco may completely engulf Trinidad, Tobago, Grenada and St. Vincent (Fig. 2). The gradient of surface salinity may then range from 20 ppt near the Mouth of the Gulf of Paria between Trinidad and Venezuela, to 36 ppt in the oligotrophic waters of the Antilles Current around the Virgin Islands.

The retroflection of the North Brazil Current in the Atlantic also causes pieces of that current to break off and form eddies of up to 400 km diameter with swirl speeds of 17–84 cm/s (Richardson et al., 1994). Some eddies loop from the surface down to depths of 900 to 1200 m as they translate northwestward at a speed of 4–16 cm/s towards the Lesser Antilles. The relative shallowness (350–1000 m below sea level) and narrowness (220 km) of the gap between Barbados and Tobago may cause the Eddies to disintegrate as they try to pass between the islands into the Caribbean Sea (Fig. 1). They are occasionally large enough to engulf the island of Barbados with consequent sudden changes in salinity and phytoplankton species composition (Stansfield et al., 1995).

Thus, the gradient of oceanographic conditions around the islands constitutes the major overall environmental determinant of habitat boundaries. These include:

1. Estuarine conditions and dark green or brown turbid waters (seen as red and yellow in Fig. 2) around Trinidad and Tobago due to the influence of the Orinoco and Amazon Rivers. During the dry season in the early part of the year the system is affected by the periodic passage of large eddies in the Guyana Current, while during the later part of the year the

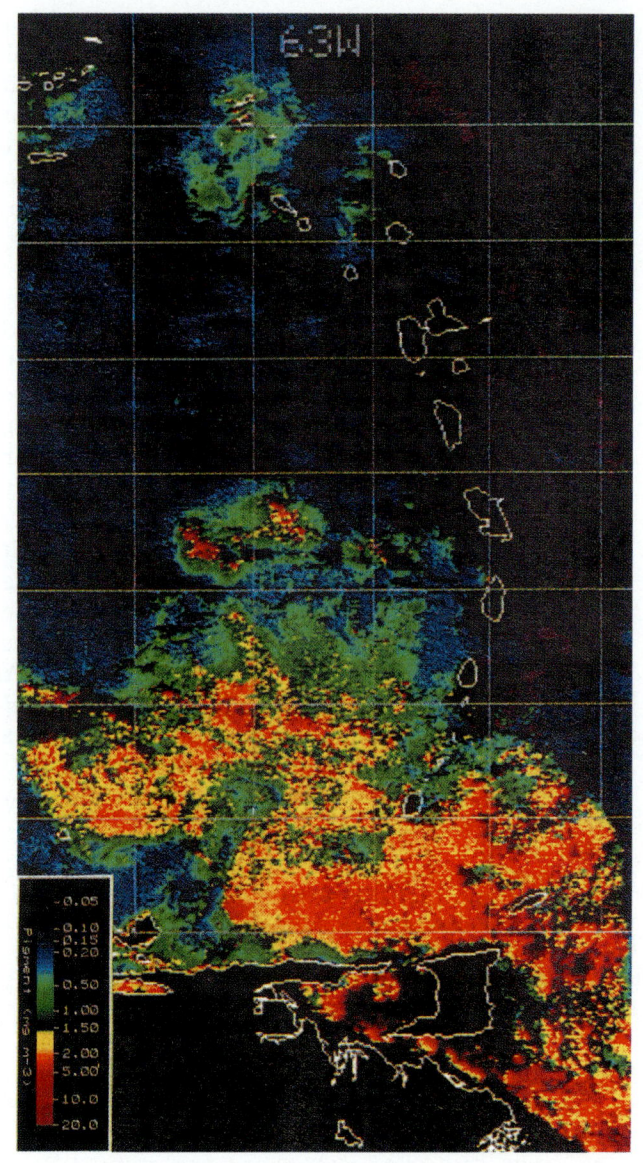

Fig. 2. CZCS composite satellite images of the Lesser Antilles and the Orinoco River plume during September 2–7, 1979. (Photo courtesy Frank Muller-Karger, University of South Florida).

major hydrographic influence is from the flood waters of the Orinoco River

2. Intermediate aquamarine waters (seen as blue-green in Fig. 2) from Grenada to St. Vincent, which during the first half of the year are formed by the mixing of water from the oligotrophic Atlantic North Equatorial Current with sediment and plankton laden Amazon River water from the previous season. Later in the year the area of intermediate water is formed by the mixing of the North Equatorial Current with the edge of the Orinoco River plume.
3. Oligotrophic clear bluish waters (seen as dark blue in Fig. 2) from the Atlantic North Equatorial Current and Antilles Current flowing into the Caribbean Sea above sill depth between the islands stretching from St. Lucia to the Virgin Islands.

Mangrove communities occur throughout the sub-region but attain their most extensive development in the estuarine conditions at the Trinidad end of the Antillean arc. The more southerly coral reefs at Trinidad have a reduced coral diversity dominated by species living at the limit of their tolerance to low salinity and suspended sediments. Conversely, coral reefs are more diverse and seagrass beds more extensive in the oligotrophic waters found from the Virgin Islands to St. Lucia.

Hurricanes and tropical storms may also have an impact on the marine biota of these islands. They form in the area between 5 and 10° off the equator over the tropical Atlantic Ocean where the surface temperature is in excess of 26.7°C. They derive their kinetic energy from latent heat of condensation and serve to transport this accumulated energy and precipitable water mass poleward. The Coriolis force is the cause of the rotation and they may typically extend from 100 km to as much as 1500 km in diameter at maturity. Their normal track takes them westward across the Atlantic between June and November where they typically pass on the Atlantic side of the Leeward Islands with the highest frequency occurring in September (Fig. 1). The probability of storms making landfall decreases sharply towards the Trinidad and Tobago end of the Lesser Antilles.

Coral reefs seem able to survive the other mainly temporary physical effects of average storms such as tidal surge; increased inundation with consequent reduced salinities and increased suspended solids and nutrients from land runoff. However, since 1960 hurricane intensity has increased giving rise to some of the most intense storms ever experienced in the region, e.g. David, Allen, Gloria, Gilbert, Hugo, Andrew and Georges. The most powerful hurricanes can affect coral reefs through the physical removal of live corals. Algae and pioneer coral genera such as *Agaracia*, *Porites*, *Favia* and *Millepora* may be the first to colonize the exposed substrate or broken coral fragments. Hurricanes may also favour the spread of branching genera of corals such as *Acropora*. The breaking off, scattering and re-growth of the numerous asexual recruits favours the spread of this genus and others with a similar growth form (Fong and Lirman, 1995). Destruction of these overhanging branching colonies also benefits slower-growing massive corals by allowing them more light for growth (Rogers et al., 1982). Other known ecological effects are the reduction of shelter for fish and other organisms, as well as the provision of new surfaces for colonisation by algae and invertebrates. In Martinique very strong hurricanes have also flattened but not uprooted the mangrove trees of entire swamps. Observation suggests that the increased surface area for settlement provided by the mangrove branches dipping into the water, favoured encrusting species such as oysters. Of greater concern in the islands is storm surge, which can erode up to 10 m of beach within an hour. Hurricanes may also bring another beneficial effect through the upwelling of nutrient-rich water along or near the track temporarily improving fisheries production.

THE MAJOR SHALLOW WATER MARINE AND COASTAL HABITATS

Except for Trinidad and Tobago, the marine production of the Lesser Antilles is constrained by the lack of a shallow continental shelf and the relatively stable thermocline. In the absence of significant upwelling, these factors prevent the mixing of nutrient-rich deep waters with surface waters. Coral reefs, mangrove swamps and seagrass beds have all solved the problem of obtaining nutrients under these conditions and are the three most important coastal habitats in the sub-region. There are many direct links between the extent and health of these habitats and the productivity of the inshore fisheries, which support human populations. The majority of bottom-dwelling fish species in the shallow nearshore waters of the Eastern Caribbean (more than 300 species, of which an estimated 180 species are landed for human consumption) are associated with coral reefs as adults (Towle and Towle, 1991). Many of these reef fishes as well as conch and lobsters, utilise mangrove swamps and/or seagrass beds as nursery habitats in their juvenile stages. Mangroves may act as an exporter of nutrients or traps for terrigenous materials and as such their removal adversely affects coastal water quality, frequently with deleterious consequences for adjacent coral reefs. Coral reefs and seagrass beds also protect coastal areas from erosion. Trinidad being on the continental shelf in the estuary of the Orinoco River is an exception to overall prevailing low nutrient conditions. This has given rise to exceptionally high primary productivity in the shallow Gulf of Paria. The muddy bottom of the area also has a diverse and highly productive benthic fauna, which supports the enhanced fisheries production of this area compared to the other islands. The Gulf of Paria coast of Trinidad also has some hydrocarbon adapted invertebrate fauna associated with natural oil seepage (Agard et al., 1993).

The small islands of the Lesser Antilles have relatively extensive coastlines in comparison to their land area and almost every coastline has mangroves (Table 1). The total area of mangroves in the Lesser Antilles is about 20,636 ha comprising seven species. The Red mangrove *Rhizophora mangle* is most common while *R. harrisonii* and *R. racemosa* are apparently restricted to Trinidad. The Black mangrove *Avicennia germinans* is widespread, whereas *A. schaueriana* is present on several islands but nowhere common. White mangrove *Laguncularia racemosa* is present on most islands but rarely forms large stands and Button mangrove *Conocarpus erectus* is a common component of wetland margins and littoral woodlands (Bacon, 1993).

The largest mangrove swamps are found where there are extensive river systems (e.g. Trinidad) or islands with low-relief coastal plains with substantial freshwater inflow (e.g. Guadeloupe). In this sub-region the endangered West Indian manatee (*Tricheus manatus*) which once ranged throughout the Lesser Antilles is now solely represented by

Table 1

Major features of the coastal zone of the Lesser Antilles

Country	Maritime Region			Marine Habitats		Coral reefs
	Length of coastline (km)	Shelf area (km^2)	EEZ area (km^2)	Mangrove sites		
				No.	Area (ha)	Present (P) or Absent (A)
Anguilla	–	–	–	10	270	P
Antigua and Barbuda	153	3,570	110,103	45	1,175	P
Barbados	97	300	167,384	14	20	P
Dominica	148	716	15,092	10	10	P
Grenada and the Grenadian Grenadines	121	1,600	27,440	28	248	P
Guadeloupe	306	1,650	26,200	15	8,000	P
Martinique	290	2,400	13,000	30	1,900	P
Montserrat	35	106	21,100	4	4	P
St. Kitts-Nevis	–	850	11,319	16	79	P
St. Lucia	105	520	16,121	18	157	P
St. Vincent and the Grenadines	–	1,800	32,585	17	50	P
Trinidad and Tobago	362	29,000	76,800	49	7,150	P
US Virgin Islands	–	–	–	21	978	P
British Virgin Islands	–	–	–	55	627	P
Total	1,617	42,512	517,144	287	20,636	–

– = Not available. Sources: (1) UN Statistical Yearbook 42nd Issue; (2) Bacon (1993).

about a dozen adults in the Nariva Swamp on the East Coast of Trinidad. In the smaller islands of the Eastern Caribbean spatial coverage by mangroves may be restricted (Table 1) and the trees often show poor development in the form of low coastal scrub (e.g. Barbados, Dominica, Montserrat, St. Vincent and the Grenadines. The main factors responsible for this are limited freshwater runoff, hypersaline conditions, wave exposure and seasonal hurricanes. The biomass of Caribbean mangroves including sites at Barbados and Trinidad, ranges from 1 to 19 kg/m^2 which appears to fall within the lower part of the global spectrum of biomass data (CARICOMP Program, 1997).

Almost every island in the sub-region has coral reefs (Table 1). Reef development is greatest on islands with low rainfall and little sedimentary runoff such as Antigua and Barbuda, St. Vincent and the Grenadines. Lesser Antillean reefs seem quite uniform with about 37 species of hermatypic scleractinian corals. Trinidad is an exception as it has a reduced coral biodiversity of about 17 species due to the high sediment load from the Orinoco River. Noticeably absent from Trinidad are common Caribbean coral genera such as *Madracis*, *Isophyllia* and *Mycetophyllia*. One of the major species responsible for maintaining the ecology of Caribbean coral reefs is the sea urchin *Diadema antillarium*. This species is herbivorous and a bioeroder. In 1983 the importance of this keystone species became apparent when *D. antillarum* were reduced to about 1% of their normal abundance. The postulated cause was the spread of a pathogen from Panama to the Lesser Antilles and the rest of the Caribbean. In the aftermath of the epidemic, algal overgrowth of corals caused major changes in the community ecology of coral reefs (Lessios, 1988).

Seagrass beds of *Thalassia testudinum* or turtle grass occur throughout the islands. Other species such as *Halodule wrightii* and *Syringodium filaforme* are frequently interspersed but nowhere common. The highest measured seagrass biomass in the sub-region is at Barbados (2900–3800 g/m^2) while the lowest is in Tobago (200–500 g/m^2) (CARICOMP, 1997a). Seagrass beds stabilise bottom sediments, retard coastal erosion and provide grazing for the green turtle (*Chelonia mydas*), manatees (*Tricheus manatus*) and parrotfish (Scaridae). Snappers (Lutjanidae), grunts (Scaridae), queen conch (*Strombus gigas*) and the edible sea urchin (*Tripneustes esculentus*) all forage in seagrass meadows. Other common benthically rooted algae are *Caulerpa*, *Halimeda*, *Penicillus*, *Rhipocephalus* and *Udotea*.

OFFSHORE SYSTEMS

The offshore hydrography of the region is dominated by a subsurface high saline water mass between depths of 100 and 200 m with a salinity maximum of 36.8 ppt and temperatures from 22 to 23°C (Kumar et al., 1991). This is referred to as subtropical underwater. Below this can be found Antarctic intermediate water at a depth of 700–800 m, a salinity maximum of 35.0 ppt and temperature of about 6–7°C. This water mass enters the Caribbean Sea through the Grenada passage after being carried by the Guyana current. A major offshore feature of the region is the intrusion of a lens of low salinity water (<33.5 ppt) from the Amazon discharge, with a vertical extent of up to 50 m depth between Tobago and Barbados. During the wet season the surface waters of the southeastern Caribbean are influenced by the freshwater discharge of the Orinoco River. There is no significant upwelling in the region although strong localised upwelling occurs seasonally along the north coast of Trinidad.

Water column primary production along the island arc is low ranging from 102 to 2026 (av. 391) mg C m^{-3} d^{-1} with highest values associated with the intrusion of Amazon water (Bhattathiri et al., 1991). Typically, diatoms of the genera *Navicula* and *Coscinodiscus* dominate phytoplankton in the estuarine conditions between Trinidad and Tobago. In the intermediate conditions extending from Tobago to Barbados and St. Vincent a mixed bloom of diatoms comprising *Chaetoceros* sp., *Thalassiothrix* sp., *Rhizosolenia faroensis* and *Skelotenema* sp. can usually be found. In the oligotrophic waters from St. Lucia to the Virgin Islands blue-green algae of the genus *Trichodesmium* dominate (Agard et al., 1996).

Due to the generally nutrient impoverished nature of offshore areas in the Caribbean Sea, shelf area and river influences are the major influences on fisheries production in the Eastern Caribbean. Penaeid shrimp which are dependent on estuarine conditions and muddy bottoms are the most valuable fishery resource harvested from the dark green or turbid brown water around Trinidad (Table 2). Pelagic fishes offshore (e.g. flyingfish, kingfish, dolphinfish, tuna, swordfish, sharks) and inshore (e.g. kingfish, jacks, herrings and anchovies) are the main commercial fisheries resource exploited among the islands in the area of mixed water stretching from Trinidad and Tobago to St. Vincent. In the clear blue oligotrophic waters from St. Lucia to the Virgin Islands the only significant fisheries are for lobsters, conchs and tuna offshore. Occasionally, some coral reef fishes are harvested for the marine aquarium trade. Sailfish and Blue marlins caught by charter boats are increasing in importance. Total fish catches landed for all the Lesser Antilles are relatively small at about 39,000 metric tons and show no indication of significant increase from 1990 to 1995 (Table 2). Expansion of fisheries exploitation in the region has traditionally been hindered by over-fishing of near-coastal waters by subsistence and other small scale fishers, reliance on small open wooden fishing boats and inadequate cold storage facilities. Another problem, which is currently being addressed, is increasing co-operation between neighbouring countries to facilitate stock assessment of shared stocks. Meanwhile large commercial fishing vessels from several nations not indigenous to the sub-region are exploiting the limited fish stocks in the EEZs of these islands. In many cases, these vessels operate without the approval of island governments. All of the islands have fisheries legislation in place but a shortage of trained personnel and the high cost of effective fisheries patrols in offshore as well as inshore waters and marine parks hinder their effective enforcement.

Table 2

Fisheries of the Lesser Antilles

Country	Fish catches (thousand metric tons)		% Change 1990–1995	Common marine species harvested
	1990	1995		
Anguilla	0.4	0.4	0	–
Antigua and Barbuda	0.9	0.5	–44.4	Lobsters
Barbados	3.0	3.3	10	Flyingfish, Dolphinfish, Kingfish, Halfbeaks, Sharks, Tuna, Blue Marlin
British Virgin Islands	1.4	1.0	–28.6	–
Dominica	0.4	0.8	100	–
Grenada	1.8	1.5	–16.7	Tuna, Sailfish, Jacks, Clupeoids, Groupers, Dolphinfish, Wahoo, Flyingfish
Guadeloupe	8.6	9.5	10.4	Dolphinfish, Tuna, Conchs, Turtles
Martinique	3.6	5.4	50.0	*Sarda sarda*, Kingfish, Tuna, Dolphinfish, Sharks
Montserrat	0.2	0.2	0	–
St. Kitts-Nevis	0.6	0.2	–66.7	–
St. Lucia	0.9	1.0	11.1	Tuna
St. Vincent and the Grenadines	9.0	1.5	–83.3	Dolphinfish, Tuna, Kingfish, Snappers, Groupers
Trinidad and Tobago	8.4	13.0	54.8	Kingfish, Shrimp, Jacks, Sharks, Clupeoids, Tuna, Sailfish
US Virgin Islands	0.7	0.9	28.6	Lobsters, Conchs, Blue Marlin, Sailfish
Total	39.9	39.2	–1.8	

Source: UN Statistical Yearbook 42nd Issue (data available as of 30 June 1997).

POPULATIONS AFFECTING THE AREA

Human beings adversely affect the marine environment through waste disposal and natural resource depletion. The magnitudes of these impacts are influenced by demographic characteristics such as population size, growth rate, density, distribution, financial income and access to sanitary facilities. The islands of the Eastern Caribbean are very small with surface areas of the major islands ranging from 96 to 5130 km^{-2}. In such small islands human impacts on the marine environment are significant because population density is high ranging from 83 km^{-2} in Anguilla to 614 km^{-2} in Barbados (see Table 3).

Table 3

Population and human settlements of the Lesser Antilles

Country	Latest census		Mid-1995[a] popn. estimate (000's)	Land surface area 1995 (km^2)	Annual rate of popn. increase 1990–95 (%)	Popn. density 1995 (km^{-2})	Population of largest urban city in 1995	
	Date	Popn.					City	Est. popn. (000's)
Anguilla	10.04.84	6,987	8	96	2.7	83	The Valley	0
Antigua and Barbuda	28.05.91	62,922	66	442	0.6	149	St. John's	24
Barbados	02.05.90	257,082	264	430	0.5	614	Bridgetown	123
Br. Virgin Is.	12.05.91	17,809	19	151	3.4	126	Road Town	8
Dominica	12.05.91	71,794	71	751	–0.1	95	Roseau	–
Grenada	30.04.81	89,088	91	344	0.2	267	St. George's	33
Guadeloupe	15.03.90	387,034	428	1,705	2.1	251	Pointe-a-Pitre	27
Martinique	15.03.90	359,579	379	1,102	0.9	344	Fort-de-France	104
Montserrat[b]	12.05.80	11,932	11	102	0.0	108	Plymouth	2
St. Kitts-Nevis	12.05.91	40,618	41	261	–0.5	157	Basse-Terre	12
St. Lucia	12.05.91	135,685	145	622	1.8	234	Castries	53
St. Vincent and the Grenadines	12.05.91	106,499	111	388	0.7	285	Kingstown	27
Trinidad and Tobago	02.05.90	1,234,388	1,306	5,130	1.4	255	Port of Spain	52
US Virgin Is.	01.04.96	101,809	105	347	0.6	303	Charlotte Amalie	13
Total			3,045	13,866	1.0	220		

[a]These figures involve the use of an average fertility hypothesis.
[b]Estimates changing rapidly due to evacuation of island because of volcanic activity.
Source: UN Department of Economic and Social Affairs, Population Division.

There are large differences in the rates of population change in these islands ranging from –0.5 in St. Kitts/Nevis to 3.4 in the British Virgin Islands (the former figure being largely due to emigration and the latter figure to immigration). It is noticeable that most of the islands (9 out of 14) have growth rates of less than 1.0% per year (Table 3). The average annual rate of population growth for the Lesser Antilles during the first half of the 1990s was about 1.0% as compared to 1.7% during the 1980s and higher values in the previous four decades. This suggests that the islands are approaching the mature stage of the demographic cycle with the population growth rate declining.

The total population of the Lesser Antilles in 1995 was estimated to be about three million persons (Table 3). Despite the general decrease in population growth rate, if present trends continue the total population size will grow by about 76,000 persons between 1995 and the year 2000. The populations of the islands are concentrated in the coastal zone and on average about 15% live in the capital cities. An exception is Barbados where about 46% of the population reside in the capital Bridgetown.

There is widespread concern that the disproportionate reliance of the island economies on the resources of the coastal environment makes them sensitive to the impacts of expanding human population. Even so, the history of marine investigations into the conditions of the area is short. Marine studies in the sub-region other than for fisheries purposes could be described as spasmodic and largely dependent on the transit of research vessels from metropolitan countries. These studies typically lasted no more than a few days. Some investigations were however more deliberate and for example Cruise P-6907 by the R/V *John Elliott Pillsbury* from RSMAS at the University of Miami took bottom trawls or dredges at more than 100 stations in the Lesser Antilles during 1969. From the late 1960s a few marine science and natural resource management oriented organisations developed in the Lesser Antilles. The largest of these is the Institute of Marine Affairs (IMA) located in Trinidad. These organisations are largely devoted to collecting baseline information relative to their individual islands and only a few regional studies have been done to-date. A notable exception is the ongoing fisheries assessment program conducted through the Organisation of Eastern Caribbean States and Canadian International Development Authority (OECS/CIDA). Similarly, the Caribbean Oceanographic Resources Exploration (CORE) which involved scientists from the Caribbean and India was able to collect physical and biological oceanographic baseline data for the sub-region. The most important ongoing initiative is the Caribbean Coastal Marine Productivity (CARICOMP) research and monitoring network of Marine Laboratories, Parks and Reserves. The network was established in 1990 and currently involves 25 sites in 16 countries in the Wider Caribbean. Three of these sites are in the Lesser Antilles at Saba, Barbados and Tobago. The network conducts a standardised, synoptic set of measurements of the structure, productivity, and associated physical parameters of relatively undisturbed coral reefs, seagrasses, and mangroves. The principal goals of the program are to determine the dominant influences on coastal productivity and to discriminate human disturbance from long-term natural variation in coastal systems over the full regional range of their distribution (CARICOMP, 1997b).

RURAL FACTORS

Agriculture in the eastern Caribbean is organised on three basic production systems: large-scale plantation agriculture; small-scale sedentary farm agriculture; and migratory or shifting agriculture (Gumbs, 1981). The dominant large-scale plantation agriculture generally produces export crops such as sugar, bananas and cocoa. The production of these crops is based on a mechanised monoculture system on flat land, which requires high inputs of fertilisers and pesticides. Total fertiliser consumption in the sub-region at 46.6 thousand metric tons is high and increasing in most countries (Table 4). More than half of this amount consists of nitrogenous and phosphate fertilisers, which may cause algal blooms when, washed into the marine environment. Runoff of soils and chemicals occurs during the rainy season when the land is laid bare. Degradation of the soil occurs over time and chemicals may find their way via waterways or directly into sensitive marine habitats. In one recorded incident, during 1978 mercury and DDT used as a seed dressing at an agricultural facility in Chaguaramas Trinidad washed down into the sea via a river causing a massive fish kill.

Small privately owned farms, the second most important of the agricultural systems in the sub-region, produce export crops in addition to food crops for the domestic market. Farms are frequently located on marginal lands and farming practices are often poor due to financial constraints and lack of access to modern technology. Poor knowledge of the ecological effects frequently results in the indiscriminate application of chemicals with consequent runoff to aquatic systems.

The most significant agricultural effect on the environment comes from slash-and-burn farming. In this method forest is clear-felled and the vegetation burned in the dry season, in order to plant crops in time for the approaching rainy season. The farmer makes minimal inputs to maintain this subsistence level of farming and there is little or no soil conservation. This form of shifting agriculture is usually practised on highly erodable soils on steep mountain slopes. It is the major cause of deforestation in the region. The problem is most acute in Trinidad and Tobago with deforestation taking place at an average rate of 3000 ha per annum (Table 5). Deforestation is also a serious problem in Guadeloupe, Martinique, the British Virgin Islands, St. Kitts and Nevis. In the other islands forest cover appears to be stable but in some cases such as Barbados and the U.S. Virgin Islands all the old growth forests have already been removed. Since Trinidad is in an estuarine area with naturally high, suspended sediments, turbidity in the coastal zone due to soil erosion from deforestation has less effect on coastal marine species, which are already tolerant to some degree. On the other hand, all of the other islands have numerous coral reefs, which may be easily smothered by eroded sediments from deforestation.

Table 4

Fertilizer consumption in the Lesser Antilles FC = fertilizer consumption.

Country	FC 1994/95 (thousands metric tons)				Change FC 1990–1995 (%)
	Nitrogen (N)	Phosphorus (P_2O_5)	Potash (K_2O)	Total FC	
Anguilla	–	–	–	–	–
Antigua and Barbuda	–	–	–	–	–
Barbados	1.5	0.2	1.0	2.7	0
Br. Virgin Is.	–	–	–	–	–
Dominica	2.0	1.3	1.3	4.6	35.3
Grenada	–	–	–	–	–
Guadeloupe	1.0	1.0	5.0	7.0	27.2
Martinique	3.0	2.0	9.0	14.0	–44.2
Montserrat	–	–	–	–	–
St. Kitts-Nevis	0.3	0.2	0.3	1.0	–11.1
St. Lucia	3.0	2.0	2.0	7.0	20.7
St. Vincent and the Grenadines	1.0	1.0	1.0	3.0	30.4
Trinidad and Tobago	3.0	1.0	2.0	6.0	30.0
U.S. Virgin Is.	1.0	0.3	0	1.3	0
Total	15.8	9.0	21.9	46.6	17.6

Source: UN Statistical Yearbook 42nd Issue (data available as of 30 June 1997).

Artisanal fishing methods can also damage marine habitats. For example, trawling for shrimp by scraping up the seabed in Trinidad has destroyed nursery grounds and the habitat of demersal species with a near collapse of inshore fisheries. In addition, the by-catch is discarded along with the juveniles of many non-target species. In Trinidad, oysters are harvested by cutting off the entire prop-roots of red mangrove on which they have settled (rather than scraping off the oysters). This practice systematically reduces the amount of suitable substrate available for oysters to settle with a consequent decline in the fishery. Over harvesting of lobsters and conch in several islands has led to a switch from relying on traps to employing dive teams whose efforts to remove animals from their hiding places often result in damage to reefs. In St. Lucia and elsewhere, Sea Moss (*Gracilaria* spp.) is harvested by pulling the entire plant from the substrate, preventing regeneration. This practice has led to a decline in wild stocks. Other threats to coral reef systems include collection of live coral specimens for sale as souvenirs.

Table 5

Deforestation in the Lesser Antilles

Country	Total forest (10³ ha)		Total forest (% of land)		Total change (10³ ha) 1990–1995	Annual change (10³ ha)	Arable land (10³ ha)	
	1990	1995	1990	1995			1990	1995
Anguilla	–	–	–	–	–	–	–	–
Antigua and Barbuda	9	9	20.5	20.5	0	0	8	8
Barbados	0	0	0	0	0	0	16	16
British Virgin Islands	5	4	33.3	26.7	–1	0	–	–
Dominica	46	46	61.3	61.3	0	0	7	5
Grenada	4	4	11.8	11.8	0	0	5	4
Guadeloupe	87	80	51.5	47.3	–7	–1	–	–
Martinique	40	38	37.7	35.8	–2	–	–	–
Montserrat	3	3	30	30	0	0	–	–
St. Kitts-Nevis	11	11	30.6	30.6	0	0	–	–
St. Lucia	6	5	9.8	8.2	–1	0	–	–
St. Vincent and the Grenadines	11	11	28.2	28.2	0	0	4	4
Trinidad and Tobago	174	161	33.9	31.4	–13	–3	74	74
US Virgin Islands	0	0	0	0	0	0	–	–
Total	396	372	348.6	331.8	–24			

Source: UN Statistical Yearbook 42nd Issue (data available as of 30 June 1997).

COASTAL EROSION AND LANDFILL

Beach sand mining is the major human-induced cause of coastal erosion in the eastern Caribbean. Throughout the sub-region, beach sand has been traditionally regarded as a free natural resource available for the taking by anyone. This attitude has had disastrous consequences for the sub-region, which governments have only been recently trying to address.

Beach sand mining (1990 est. 96,000 tons) in St. Vincent and the Grenadines has been the cause of severe beach erosion, flooding of coastal areas, loss of dunes and other habitats. In order to address the situation, the government decided as of 1st January 1995 to import sand from Guyana for all government projects and to implement controls on beach sand mining. This action was widely expected to cause an increase in the cost of construction and so there was a massive stockpiling of sand. In a two-month period the volume of sand mined from the beaches was 2.5 times the annual volume (Porter, 1997). This caused serious erosion, which was accentuated by tropical storm Iris in August 1995. Similarly in Tobago, beach sand mining has caused severe erosion and although the practice has been brought under a licensing system, few beaches have shown signs of recovery. In Grenada the story is similar, with beach sand mining continuing at the rate of up to 65,000 cubic yards per year. Another common practice is offshore dredging to re-establish hotel beaches removed by hurricanes. This has occurred in Anguilla, Barbados, St. Kitts/Nevis and St. Lucia. Dredge and filling of mangrove areas to construct Marinas and Resort developments are also now a common feature of the region.

EFFECTS FROM URBAN AND INDUSTRIAL ACTIVITIES

Artisanal and Non-industrial Uses of the Coast

In the Lesser Antilles, humpback whales (*Megaptera novaeangliae*) migrate annually to calving grounds in the Grenadines and also between the islands of Antigua and Anguilla. This has been the basis for the operation of two shore-based artisanal fisheries for whales. The island of Bequia in the Grenadines has a whaling industry dating back to 1875 and is allowed an aboriginal whaling quota of three humpback whales per year by the International Whaling Commission (Ward, 1987). Between 1950 and 1984, it has been estimated that between 52–70 Humpback whales were killed by the Bequia fishery, about 70% of them females. The harvest is strictly for local consumption. Occasionally Dolphins, Sperm Whales and Killer Whales are also taken. Another fishery for pilot whales (*Globicephala macrorhyncus*) is based in St. Vincent, with a few boats operating out of St. Lucia. Both whale fisheries are in rapid decline and the whaling quota has been reduced to two per year.

Hawksbill turtles (*Chelonia mydas*) are heavily exploited in the Grenadines and small turtles are captured with spearguns to be stuffed and sold as tourist curios. Stony corals and black corals are also sold to tourist as curios.

The white-spined sea urchin (*Tripneustes ventricosus*) is harvested in the Caribbean for its edible roe. The demand for the delicacy has led to severe over-harvesting especially in St. Lucia, Barbados, the Tobago Cays and the French islands. In 1990 the Government of St. Lucia introduced a

Table 6
Tourist arrivals and international tourism receipts

Country	Number of Tourist Arrivals (000's)				Tourism Receipts (000,000's US$)			
	1991	1995	Total change 1991–95	% Change 1991–95	1991	1995	Total change 1991–95	% Change 1991–95
Anguilla	31	39	8	25.8	35	48	13	5.4
Antigua and Barbuda	197	212	15	7.6	314	329	15	4.8
Barbados	394	442	48	12.2	460	680	220	47.8
British Virgin Islands	147	253	106	72.1	109	191	82	75.2
Dominica	46	60	14	30.4	28	33	5	17.9
Grenada	85	108	113	27.1	42	58	16	38.1
Guadeloupe	370	640	270	73.0	251	458	207	82.4
Martinique	315	457	142	45.1	255	384	129	50.6
Montserrat	17	18	1	2.4	12	14	2	16.7
St. Kitts-Nevis	83	79	–4	13.3	74	65	–9	–1.4
St. Lucia	158	232	74	46.8	173	268	95	29.5
St. Vincent and the Grenadines	52	60	8	15.4	53	57	4	3.8
Trinidad and Tobago	220	260	40	9.1	101	73	–28	–27.7
US Virgin Islands	470	454	–84	–3.4	778	821	43	5.5
TOTAL	2585	3314	729	28.2	2685	3479	794	29.6

Source: UN ECLAC.

co-management arrangement of the fishery with community groups. These groups are licensed for the harvesting season, in return for their observing minimum size limits and restrictions on harvest location (Smith and Berkes, 1991).

In St. Lucia the Mankote mangrove has been traditionally used as a source of wood for charcoal production. Over-harvesting and the declaration of the island mangroves as Marine Reserve Areas have led to the development of a fuel-wood reforestation project using *Leucaena* sp. as an alternative.

In Tobago, local boatmen use glass-bottomed boats to ferry tourists out to view the popular Buccoo Reef. Visitors are encouraged to disembark into the water on the reef and walk, frequently trampling corals under their sandalled feet. Current practice by Park authorities is to discourage the activity or confine it to already degraded areas.

Industrial Uses of the Coast

In the Lesser Antilles there are only two large scale industrial uses of the coast where significant ownership and profits are held extra-regionally, viz. tourism in all islands and oil and gas exploitation together with petrochemical production in Trinidad. The tourism resource mainly comprises pristine natural assets such as beaches and coral reefs or island culture as well as physical plant such as hotels and marinas. Most of the large hotels and resorts are owned by international hotel chains or groups of foreign investors. Although tourism is the major source of income for the sub-region, it often creates problems where it exceeds the carrying capacity of the coastal zone to provide facilities for tourists. Total tourism receipts in the sub-region of about 3.5 billion US$ during 1995 represent a 29.6% increase compared to 1991 (Table 6). Tourist arrivals of about 3.314 million persons to the region in 1995 were greater than the entire population (1995 est. 3.045 million, Table 6) of the Eastern Caribbean. The annual number of tourist arrivals in individual islands is substantially greater than their resident population in 12 out of 14 instances excluding only Dominica and Trinidad (cf. Tables 3 and 6). Further, tourist arrivals to the sub-region are increasing at the rate of about 5.6% or 145,800 persons per annum. One undesirable trend is that in Trinidad and Tobago although tourist arrivals have been increasing, tourism receipts have dropped. This it is suggested is due mainly to an increase in the influx of cruise ship passengers whose all-inclusive accommodations provide little return to the country.

In order to supply the needs of tourists, one of the unfortunate side effects of the extraordinary financial success of this industry is pollution and degradation of the coastal environment. Airport and marina construction have resulted in the filling in of coastal mangroves and increased sedimentation in coral reef and seagrass areas. Rapid development within the hotel construction industry has been responsible for increased beach sand mining to supply construction material and for resorts being built too close to the water. Coral reef damage is also a growing problem because of boats running aground or anchoring on them and visitors trampling on them or breaking off pieces for souvenirs. Spearfishing associated with the lucrative dive tourism industry is also another cause of depletion of reef communities. Hotel demand for high valued commodities such as lobsters and conch are the cause of them being over-harvested. Marine pollution from inadequately treated sewage effluents is a problem on every island because of the lack of adequate centralised sewage treatment facilities. This has led to the proliferation of poorly designed, ill-maintained package sewage treatment plants associated with individual resorts. There are no consistent regulations across the sub-region requiring yachts to have holding tanks for their sewage wastes and to use shore based waste collection facilities. This has allowed yachts to empty their sewage wastes directly into the coastal marine environment solely at their discretion. Fuel and oil waste spillage from ship refuelling activities is also a widespread problem. Associated with the increase in yacht traffic are the construction of numerous marinas and boat maintenance facilities in the islands. A problem of particular concern at these facilities is that throughout the region boats are being painted with antifouling paints containing extremely toxic organo-tin compounds, which have been banned for this use in North America and Europe.

The development of heavy industry in the coastal zones of the various territories is very limited except for the island of Trinidad. The basis of the enormous industrial development of the island is the exploitation of oil (1996 est. offshore production 36,300,380 barrels) and natural gas (1996 est. offshore production 804.1 MMcfgd) mainly by multinational energy companies. Using natural gas from 17 offshore fields as a feedstock, several energy intensive and world scale petrochemical plants have been established along the Gulf of Paria coast. These include several iron and steel, iron carbide, ammonia, methanol, urea, and liquefied natural gas plants among others. The scale of production is such that at full activation of the installed capacity, Trinidad and Tobago will be the world's largest exporter of ammonia (3.455 million t/yr) and methanol (2046 million t/yr) by the year 2000. Over a period of 14 years, several fish kills in the area have been traced back to industrial effluents especially ammonia (Heileman and Siung-Chang, 1990) and the risk of such incidents is likely to increase. Another impact of large-scale industrial development is that the scarcity of unoccupied flat coastal land for expansion puts pressure on mangrove areas as in Trinidad. An alternative approach to expansion has also resulted in the smothering of seagrass beds due to construction of an artificial island offshore.

Shipping and Offshore Accidents and Impacts

The exploitation of oil is also the source of widespread pollution in the south of the island of Trinidad. From 1993 to 1995 the mean number of reported oil spills to the marine environment was 215 per annum resulting in the net annual loss of about 5000 barrels of oil on average (unpublished data, Ministry of Energy, Trinidad and Tobago). To this must be added 100,000 barrels of oil per annum due to effluents, which find there way into the marine environment via rivers. A further 15,000 barrels of oil enter the marine environment in produced water from marine installations offshore. Together these inputs suggest pollution inputs to the marine environment of about 120,000 barrels of oil per annum. These inputs dwarf the rate of natural oil seepage and oil pollution of marine sediments extends along the entire western coast of Trinidad but is greatest near oil refineries (Agard et al., 1988). This pollution has resulted in the massive depletion of soft bottom benthic communities up to several kilometres away from the sources (Agard et al., 1993).

In the sub-region oil-bearing supertankers or cargo ships passing through certain high-risk zones for shipping accidents increase the risk of marine pollution in the area. Most at risk are the more than 1000 large vessels per annum passing between Trinidad and Tobago or through the narrow entrances to the Gulf of Paria. These vessels transport oil, gas and chemicals to industrial estates on the West Coast of Trinidad. In fact, one of the largest oil spills ever recorded from a tanker collision occurred on July 19, 1979 just 30 km northeast of Tobago. The collision of two fully laden supertankers the Atlantic Empress and the Aegean Captain resulted in oil spillage estimated at 90,000 t. The resulting oil slick driven by strong winds fortunately moved away from land into the Caribbean Sea. Another high-risk zone for shipping accidents is the narrow Anegada Passage between the Virgin Islands because it lies on a major shipping lane to the eastern seaboard of the United States. This shipping lane passes along the Antillean island arc and is the source of tar balls occasionally found along windward exposed beaches of the island chain. The likely source is oily ballast washings from petroleum tankers (Atwood et al., 1987/88).

Cities

The pollution impacts of cities in the sub-region are mainly due to chronic low-level discharges either directly or via rivers to the marine environment. The main point sources of aquatic pollution are from domestic sewage and industrial effluents. Of these sources, domestic sewage is the more important. Most urban areas have centralised sewage treatment facilities whereas rural areas tend to be served by septic tanks or pit latrines. In the islands the percentage of the population which resides in urban areas is generally high being less than 34% in only two islands, so that potentially most domestic sewage finds its way into treatment facilities. However, poor maintenance of these treatment facilities has led to localised inputs of sewage into the marine environment in some parts of the Eastern Caribbean. The percentage of domestic wastes currently treated is unknown, but the high proportion of poor or non-functional sewage treatment plants (17–100%) is a cause for concern (Table 7).

Industrial effluents are a more limited problem as, except for Trinidad, industrial activity in the islands is minor and largely based on sugar, alcohol, soft drinks and food processing. In most cases the small and medium-sized manufacturing industries are in urban areas and discharge their effluents into surface drains, rivers or the sea with little treatment. As a result, waste loads (especially BOD, total suspended solids and total nitrogen) from industrial sources are a major problem particularly in Barbados and on the island of Trinidad.

Table 7

Status of centralised sewage treatment facilities in urban areas in the Lesser Antilles

Country	Urban population (1995) (% of total pop.)	Operating conditions of sewerage treatment facilities		
		Good (%)	Poor (%)	Undetermined (%)
Anguilla	11.0	–	–	100
Antigua and Barbuda	35.8	47	48	5
Barbados	47.3	83	17	0
Br. Virgin Is.	56.0	–	–	100
Dominica	69.3	–	–	100
Grenada	35.8	80	20	0
Guadeloupe	99.4	–	–	100
Martinique	93.3	–	–	100
Montserrat	16.3	–	–	100
St. Lucia	37.2	46	54	0
St. Kitts-Nevis	34.0	75	25	0
St. Vincent and the Grenadines	48.1	0	100	0
Trinidad and Tobago	71.7	54	46	0
US Virgin Is.	–	–	–	100
Mean		64	44	

Sources: Urban population data UN ECLAC; Sewage Treatment Facilities data UNEP.

PROTECTIVE MEASURES

The management of the marine and coastal areas of the Lesser Antilles is characterised by piecemeal efforts outside the framework of integrated coastal area management plans. In every instance the legal framework exists but the approach usually taken is either not to designate marine protected areas (MPAs) at all or to designate them with little attempt at implementation of management. This phenomenon is referred to as 'paper parks'. Of the 16 sites so designated in the region together with 15 other coastal areas earmarked for protection (Table 8), only three appear to be fully managed at present. This suggests that about 90% of these areas may not be adequately managed. This assessment appears to be much worse than that reported a decade previously by the Organisation of American States. At that time the Marine Islands Nature Reserve in St. Lucia and the Barbados Marine Reserve were rated as fully managed, but more recent information suggests that they no longer warrant this status (van't Hof, 1994). Interestingly, the few areas which appear to be fully managed are operated either by an agency of a colonial government, as in the U.S. Virgin Islands, or by a non-governmental organisation as in Saba and the British Virgin Islands. Lack of implementation of protected area management has been attributed to a composite of lack of funding, lack of trained personnel and lack of public support (van't Hof, 1994). The region has the technical resources to solve these problems but most international assistance programs draw on human resources from outside the region. This approach slows the long-term solution, which should be based on developing

Table 8

Status of Marine Protected Areas in the Lesser Antilles

Country	Marine Protected Areas (1995)			Existing Marine Protected Areas
	No.	Area (ha)	Other coastal areas needing protection	
Antigua and Barbuda	2	2500	1	Palaster Reef National Park, Salt Fish Tail National Park
Barbados	1	250	1	Barbados Marine Reserve
British Virgin Is.	1	323	2	Wreck of the Rhone Marine Park
Dominica	2	–	0	Cabrits National Park, Soufriere Scott's Head Marine Reserve
Grenada	0	–	–	–
Guadeloupe	1	–	0	Grand Cul-de-sac Marine Natural Reserve
Martinique	1	–	0	Caravelle Littoral Conservation Area
Montserrat	1	–	0	Fox,s Bay Bird Sanctuary Private Reserve
Saba	1	–	0	Saba Marine Park
St. Kitts-Nevis	0	–	0	–
St. Lucia	2	5	1	Maria Islands Nature Reserve, Soufriere Marine Management Area
St. Vincent and the Grenadines	1	–	0	Tobago Cays Marine Park
Trinidad and Tobago	1	650	7	Buccoo Reef Marine Park
U.S. Virgin Islands	2	6429	3	Virgin Islands National Park, Buck Island Reef National Monument
TOTAL	16		15	

Source: OAS, IUCN.

indigenous capacity. It is difficult for individuals from outside the region to appreciate the requirements of marine resource management on the insular scale where less than a handful of professional staff are available on each island. If the limited expertise is pooled, then the area may be amenable to the large marine ecosystem (LME) approach to managing marine resources. However, since this LME includes the territorial waters of more than 14 countries, the effectiveness of this management tool is dependent on the co-operation of all the bordering countries.

The Lesser Antillean countries are also participants in several important international conventions and programs which are geared to protect the marine and coastal environment (Table 9). Due to its intensive maritime activities and sensitive marine environment, the wider Caribbean Region has been designated a "Special Area" under Annex V of the International Convention for the Prevention of Pollution from Ships – MARPOL 73/78. The Convention imposes severe waste-disposal restrictions on ships using the Caribbean Sea and requires ports to provide reception facilities on land. However, before special area status can be officially activated, the IMO's Marine Environment Protection Committee has to be notified that "sufficient and adequate" reception facilities exist in the region. These shore-based facilities must be able to accept the waste that will no longer be discharged into the sea. Governments are also required to enact domestic legislation to give effect to the provisions of the treaty. Although most of the countries in the Eastern and Wider Caribbean are signatories to the Convention, to date they have collectively done little to discharge their obligations under the treaty. Twelve months after they do, the regime can actually start being applied. Only then will ships transiting the Caribbean be legally prohibited from dumping virtually anything at all overboard.

Since 1990, most governments of the region have also been signatories to the Cartagena Convention. Under this Convention a Protocol on Specially Protected Areas and Wildlife (SPAW) for the Caribbean has been adopted as part of the Caribbean Environment Program (CEP). Similar provisions for the designation of protected areas and species are also found in the Convention on Biological Diversity (arising out of the 1992 Earth Summit in Rio de Janeiro) and the Convention on Wetlands of International Importance Especially as Waterfowl Habitat (Ramsar). A common theme in the sub-region is that the domestic legislation which all these Conventions require has not been enacted. The islands will have to significantly increase their environmental protection efforts if they are to stem the tide of pollution and natural resource depletion.

ACKNOWLEDGEMENTS

We thank Professor Kay Hale, Rosensteil School of Marine and Atmospheric Sciences, University of Miami for expert assistance in locating obscure literature. The Coastal Zone Colour Scanner (CZCS) satellite image of the Lesser Antilles in Fig. 2 was made available courtesy of Dr. Frank Muller-Karger, University of South Florida. Alicia Laurent provided assistance with data preparation. This work was supported by a University of the West Indies study and travel grant.

Table 9

Participation of Lesser Antillean countries in international conventions applying to marine and coastal environments

Country	RAMSAR	SPAW	Biodiversity	MARPOL
Anguilla (UK)	X	X	X	X
Antigua and Barbuda	X	X	X	
Barbados		X	X	X
British Virgin Is. (UK)	X	X	X	X
Dominica			X	
Grenada				
Guadeloupe (Fr.)	X	X	X	X
Martinique (Fr.)	X	X	X	X
Montserrat (U.K.)	X	X	X	X
Saba (Neth.)	X	X	X	X
St. Kitts-Nevis			X	
St. Lucia			X	X
St. Vincent and the Grenadines		X		X
Trinidad and Tobago	X	X	X	X
US Virgin Islands	X	X	X	X

RAMSAR: onvention on Wetlands of International Importance especially Waterfowl Habitat.
SPAW: Specially Protected Areas and Wildlife Protocol to the Convention for the Protection and Development of the Marine Environment of the Wider Caribbean Region (Cartagena Convention).
Biodiversity: United Nations Convention on Biological Diversity.
MARPOL: International Convention for the Prevention of Pollution from Ships.

REFERENCES

Agard, J.B.R., Boodosingh, M. and Gobin, J. (1988) Petroleum residues in surficial sediments from the Gulf of Paria, Trinidad. *Marine Pollution Bulletin* **19** (5), 231–233.

Agard, J.B.R., Gobin, J. and Warwick, R.M. (1993) Analysis of marine macrobenthic community structure in relation to natural and man induced perturbations in a tropical environment (Trinidad, West Indies). *Marine Ecology Progress Series* **92**, 233–243.

Agard, J.B.R., Hubbard, R.H. and Griffith, J.K. (1996) The relation between productivity, disturbance and the biodiversity of Caribbean phytoplankton: applicability of Huston's dynamic equilibrium model. *Journal of Experimental Marine Biology and Ecology* **202**, 1–17.

Atwood, D.K., Burton, F.J., Corredor, J.E., Harvey, G.R., Mata-Jimenez, A.J., Vasquez-Botello, A. and Wade, B.A. (1987/88) Petroleum pollution in the Caribbean. *Oceanus* **30** (4), 25–32.

Bacon, P.R. (1993) Mangroves in the Lesser Antilles, Jamaica and Trinidad and Tobago. In: *Conservation and Sustainable Utilization of Mangrove Forests in Latin America and African Regions, Part 1: Latin America*, ed. L.D. Lacerda. Mangrove Ecosystems Technical Reports, Vol. 2. International Society for Mangrove Ecosystems and International Tropical Timber Organization, pp. 155–193.

Bhattathiri, P.M.A., Wagh, A.B., Chow, B.A., Hubbard, R. and Mohammed, A. (1991) Primary production and phytoplankton biomass from the Caribbean Sea. *Caribbean Marine Studies* **2** (1&2), 45–53.

CARICOMP Program (1997) Structure and productivity of mangrove forests in the Greater Caribbean region. *Proc. 8th Int. Coral Reef Symp.* **1**, 669–672.

CARICOMP (1997a) Variation in ecological parameters of *Thallasia testiudinum* across the CARICOMP network. *Proc. 8th Int. Coral Reef Symp.* **1**, 663–668.

CARICOMP (1997b) Caribbean marine productivity (CARICOMP): A research and monitoring network of marine laboratories, parks, and reserves. *Proc. 8th Int. Coral Reef Symp.* **1**, 641–646.

Fong, P. and Lirman, D. (1995) Hurricanes cause population expansion of the branching coral *Acropora palmata* (Scleractinia): wound healing and growth patterns of asexual recruits. *Marine Ecology* **16** (4), 317–335.

Gumbs, F. (1981) Agriculture in the wider Caribbean. *Ambio* **10** (6), 335–339.

Heileman, L.I. and Siung-Chang, A. (1990) An analysis of fish kills in coastal and inland waters of Trinidad and Tobago, West Indies, 1976–1990. *Caribbean Marine Studies* **1** (2), 126–136.

Kumar, S.P., Murty, V.S.N., Khan, A.A., Jones, M.A., Wagh, A.B. and Desai, B.N. (1991) Hydrographic characteristics and circulation in the Caribbean Sea during April and May 1990. *Caribbean Marine Studies* **2** (1 & 2), 69–80.

Lessios, H.A. (1988) Mass mortality of *Diadema antillarum* in the Caribbean: What have we learned? *Annual Review of Ecology and Systematics* **19**, 371–393.

Muller-Karger, F.E., McClain, C.R. and Richardson, P.L. (1988) The dispersal of the Amazon's water. *Nature* **333** (6168), 56–59.

Muller-Karger, F.L. and Varela, R.J. (1989) Influjo del Rio Orinoco en el mar: observaciones con el CZCS desde el espacio. *Mem. Soc. Nat. La Salle* **49** (131–132), 361–390.

Porter, M. (1997) Sandmining in St. Vincent and the Grenadines after the landmark decision of 1994. In *Coastal Regions and Small Island Papers 1. Managing Beach Resources in the Smaller Caribbean Islands. Workshop Papers*, ed. G. Cambers. University of Puerto Rico, Sea Grant College Program.

Richardson, P.L., Hufford, G., Limeburner, R. (1994) North Brazil Current Retroflection Eddies. *Journal of Geophysical Research* **99**, 5081–5093.

Rogers, C.S., Suchanek, T.H. and Pecora, F.A. (1982) Effects of hurricanes David and Frederick (1979) on shallow *Acropora palmata* reef communities: St. Croix, U.S. Virgin Islands. *Bulletin of Marine Science* **32** (2), 532–548.

Smith, A.H. and Berkes, F. (1991) Solutions to the "Tragedy of the Commons": Sea urchin management in St. Lucia, West Indies. *Environmental Conservation* **18** (2), 28–31.

Stansfield, K.L., Bowman, M.J., Fauria, S.J. and Wilson, T.C. (1995) Water mass and coastal current variability near Barbados, West Indies. *Journal of Geophysical Research* **100**, 24819–24830.

Towle, J.A. and Towle, E.L. (1991) Environmental Agenda for the 1990's: A Synthesis of the Eastern Caribbean country profile series. Caribbean Conservation Association/Island Resources Foundation. 71 pp.

van't Hof, T. (1994) Resolving common issues and problems of marine protected areas in the Caribbean. Caribbean Conservation Association. 58 pp.

Ward, N.F.R. (1987/88) The whalers of Bequia. *Oceanus* **30** (4), 89–93.

THE AUTHORS

John B.R. Agard
Department of Life Sciences,
The University of the West Indies,
St. Augustine, Trinidad and Tobago

Judith F. Gobin
Institute of Marine Affairs,
Hilltop Lane, Chaguaramas,
Port of Spain, Trinidad and Tobago

Chapter 41

VENEZUELA

Pablo E. Penchaszadeh, César A. Leon, Haymara Alvarez,
David Bone, P. Castellano, María M. Castillo, Yusbelly Diaz,
María P. Garcia, Mairin Lemus, Freddy Losada, Alberto Martin,
Patricia Miloslavich, Claudio Paredes, Daisy Perez,
Miradys Sebastiani, Dennise Stecconi, Victoriano Roa and
Alicia Villamizar

The coast of Venezuela is very diverse in geological and topographical terms. The annual surface seawater temperature ranges between 20 and 29°C. Twelve upwelling zones have been identified, mainly related to the intensity of the Trade Winds. During upwelling events, the temperature decreases locally between 5 and 7°C. Regional temperature variations follow a seasonal pattern, colder during the dry season and warmer in the rainy season. The Orinoco River discharges an average of 36,000 m^3/s to the Atlantic Ocean and influences the salinity patterns, currents, suspended materials and nutrients in the Venezuelan Atlantic coast and the Caribbean Sea. The major shallow water marine and coastal habitats are sandy beaches, rocky shores, seagrass beds (mainly *Thalassia testudinum*), coral reefs, coastal lagoons and mangroves.

The coastal area of Venezuela has, in general terms, a high primary production due to the upwelling systems and from the nutrient supply by rivers and watersheds. This contributes to a diversified fishery. The main official statistics of marine and estuarine fisheries show a catch of around 400,000 mt/yr and the main products are sardine (*Sardinella anchovia*), mollusk bivalves (*Arca zebra*), and mugilids (*Mugil* spp.).

The main export of Venezuela is petroleum, and there are coastal areas with chronic oil contamination. There are also reports of contamination with heavy metals. Power plants, oil refineries, petrochemical plants, paper-mill facilities, and sewage discharge directly to the sea, and have impacted many areas of the coastal zone of Venezuela.

Venezuela has a legislation framework for coastal management, which includes National Parks, and other areas managed under Special Regulations. Nevertheless, protection of the coastal area is not efficient, since economic forces ignore these regulations.

Seas at The Millennium: An Environmental Evaluation (Edited by C. Sheppard)
© 2000 Elsevier Science Ltd. All rights reserved

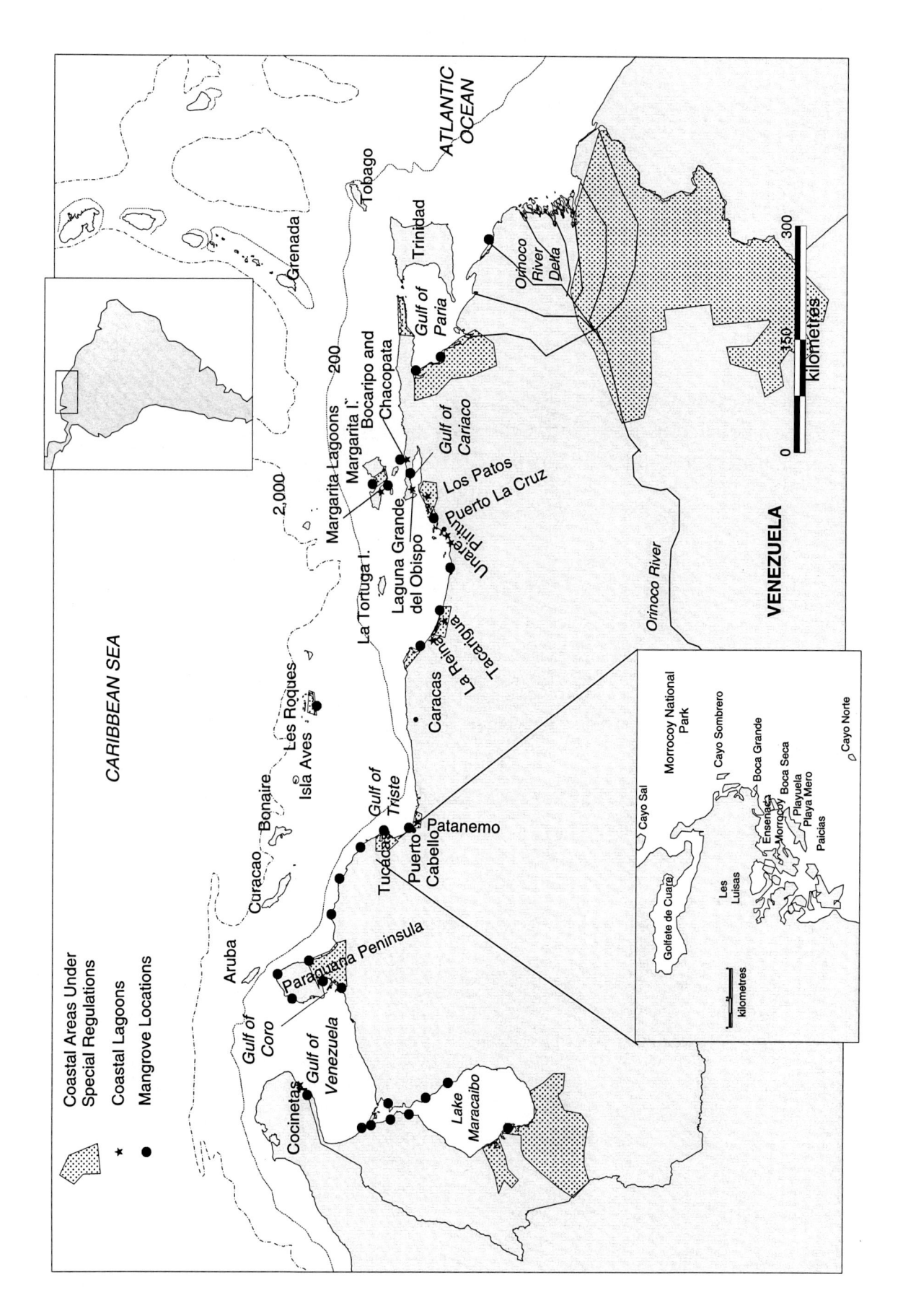

Fig. 1. The coastline of Venezuela.

INTRODUCTION

The coastal zone of the Venezuelan Caribbean is a populated area, subject to intense development pressure. Petroleum refineries, dredging, tourism, waste disposal, fisheries, import and export activities of raw and manufactured materials, are among the most common activities that take place along this fragile coast. The management of the coastal zone is therefore recognized to be of fundamental importance in the developmental strategy of the country.

The rational use and management of coastal habitats and marine resources are major challenges to the populations of the region. Conflicts over the use, over-exploitation of resources, and destruction of habitats are urgent problems that managers and decision makers face on a daily basis. Tempting short-term exploitation options in the past have led to the depletion of renewable natural resources. Solutions to these problems are not easy because of the lack of knowledge about how different coastal marine ecosystems work, what interactions exist between different systems, and how resistant individual as well as connected ecosystems are to the perturbations affecting them.

In the past few decades, Venezuela has entered a stage of technological and industrial development which poses a number of growing problems related to pollution and degradation of the environment, particularly marine ecosystems.

Industrial expansion and diversification, growing urban areas and the lack of adequate wastewater-treatment plants have resulted in the discharge of polluting agents into the rivers, lakes, coastal lagoons and the sea. It is necessary to add to this list the changes in the runoff regime of streams due to improper management of hydrographic basins. These basins are seriously affected by indiscriminate deforestation and erosion. As a consequence, their rivers are characterized by high sedimentation rates, presence of heavy metals, pesticides, and eutrophication. Examples of these systems are some beaches of the Central Littoral coast and Barlovento Zone, close to the openings of the Tuy, Guapo, and Curiepe rivers (Fig. 1) (Carrillo et al., 1987; Martin, 1987; Carrillo, 1989). Golfo Triste (Fig. 1) is a region characterized by an extensive seashore (685 km) starting at the Paraguana Peninsula and ending in Punta Peñon in Patanemo Bay. Along this coastline are two oil refineries, a petrochemical and a thermo-electrical plant, one pulp and paper-mill industry, the port of Puerto Cabello and docks, along with important human settlements. The coast exhibits contamination problems from oil pollutants as well as domestic and rural wastes. The rivers of the region are also seriously polluted and degraded. Special attention was given to Caño Alpargaton whose waters possess traces of mercury coming from a chlor-alkali plant (at present not in operation) of a petrochemical industry in Moron. Exhaustive work has been conducted to analyze the presence and bio-accumulation of mercury in the seawater, sediments, and biota (Iglesias and Penchaszadeh, 1983; Perez, 1988).

THE VENEZUELAN CONTINENTAL COASTLINE

Venezuela has approximately 2875 km of continental coastline (Fig. 1); 1927 km (67%) are in the Caribbean Sea and 948 km (33%) in the Atlantic Ocean. Ellenberg (1978) divided the coastline into the following eight regions, from the eastern side.

Atlantic zone: Formed by the coasts of Monagas State and Delta Amacuro, comprised mainly of the Orinoco delta with an extensive mangrove area of around 46,802 ha (Conde and Alarcon, 1993).

Sucre coast to Puerto La Cruz: Semi-arid hinterland of heavily folded Mesozoic limestones, with deep river valleys. It has a very narrow shelf but is well protected against wave attack and has sparse mangroves in bays. Corals are abundant but reefs are not important. The shore is subsiding tectonically and so there is a slow coastal retreat.

Anzoategui and Miranda States: This is a sub-humid coastal plain, which accumulates sediments from the sea and rivers. The shelf is very shallow and is up to 50 km wide. Coastal drift of sediments is very marked, so mangroves are absent on the direct ocean beach but are well developed around coastal lagoon systems of the coastal plain. Corals are not abundant. There is slow tectonic uplift and there is progradation of the shore from sediments.

Central Littoral to Puerto Cabello: Sub-humid to semi-arid hills, intensely folded and faulted, consisting of metamorphic rocks and a variety of sediments. The shelf is wider than 15 km. Wave exposure is not very great, but cliffs occur and mangroves and corals are sparse. Uplift of this coastal area is important in this region, though some erosion is evident. Bay formation is due to the accumulation of river sediments, and bays reach far into hinterland. Coast retreat is marked generally, but at some places there is progradation due to uplift.

Golfo Triste from Puerto Cabello to Punta Tucacas: Puerto Cabello-Tucacas: This has a wide coastal plain without mangroves or corals. Marine erosion dominates, and coastal drift is strong. Some small islands are abundant, most of them being characterized by a fringing reef on the wave-exposed side. There is slight tectonic uplift but the coast is retreating in most parts. *Tucacas-Punta Tucacas*: This sector has very shallow water, protected against direct wave exposure. There is strong accumulation of sediments by coastal drift. Areas of dense mangrove forests occur, and the sedimentary progradation is assisted by tectonic uplift.

Paraguana Peninsula: On its east side this is a semi-arid area with flat-lying limestone and a shallow shelf. There is strong wave exposure and coastal drift is towards the south. Mangroves are rare and there are fringing coral reefs in the southern part. At the southern side both accumulation and progradation of the coast is evident, while the northern side is characterized by marine erosion and retreat. On its west side, this is similar to the east side and emergence increases southwards. It has very shallow waters (Gulf of Venezuela). Mangroves and corals are sparse. The north is prograding while the south is eroding. On its south side, the coastal belt consists of lagoon-type and sandy accumulations. The water is not deeper than 8 m in the Gulf of Coro. Corals are abundant on rocks, but there are few reefs. Mangroves are locally important. The area shows emergence and progradation.

Gulf of Venezuela: Along the stretch between Quebrada El Volcan-Jutaipano the coastal plain consists of terrestrial and marine accumulations and the Gulf of Venezuela is shallow with little erosion from waves. Locally, mangroves are important and there are almost no corals. There is probably a slight emergence of the coast and some local progradation. Between Jutaipano and Castilletes is a semi-arid coastal plain with significant emergence. There is strong marine erosion, creating small cliffs. Rapid sediment transport by coastal drift ensures that the coast is refreshing in this area.

Maracaibo lake: An estuary of muddy sediments.

SEASONALITY, CURRENTS, NATURAL ENVIRONMENTAL VARIABLES

The surface temperature along the Venezuelan coastline is affected by cold-water plumes whose borders differ in form, size, and intensity depending on the season and wind intensity.

Annual surface seawater temperature ranges between 20 and 29°C with a daily thermal variation ranging up from 0 to 2.5°C. The first months of the year show the largest variation (Fig. 2). Between July and December water temperature rises and daily differences reach up to 2.5°C. These daily oscillations become less intense during upwelling. On the other hand, the yearly variations span 5°C if the upwelling waters are taken as a reference. The oceanic waters can have variations of up to 3°C. This 2°C difference between oceanic and coastal waters is due to upwelling.

The north–south oscillation of the Intertropical Convergence Zone (ITCZ) constitutes the main cause of seasonal changes in salinity and temperature of the Caribbean Sea surface waters (Müller-Karger and Varela, 1989).

The regional temperature variations follow a seasonal pattern. The waters are colder during the dry season, and warmer in the rainy season (Fukuoka, 1965; Okuda, 1974). When the Trade Winds reach maximum speeds, the cold water plumes become more intense and more detectable, having the effect of decreasing surface water temperature by mixing it with deep water.

Seasonal changes govern the variations in surface water temperature in the region. However, upwelling modifies the spatial–temporal temperature distribution of the Venezuelan coast.

Twelve upwelling zones have been identified on the southern coastline of the Caribbean Sea. During the upwelling events, the temperature decreases between 5 and 7°C. Thermal fronts occasionally show up in the surface water mixture limits.

West to the Goajira Peninsula, an important seasonal upwelling moves towards the northwest. The Gulf of Venezuela shows a marked difference in temperatures from east to west, due to the presence of the less dense and warmer waters of the Maracaibo Lake. To the west of the Gulf of Venezuela an upwelling focus is located near to Punto Fijo, Paraguana Peninsula. Near to Cabo San Roman, very cold waters (as low as 18°C) arise, influential towards the northwest of the Caribbean Sea. Less frequent is the upwelling at Puerto Cumarebo, which forms an almost parallel front to the Istmo and the Paraguana Peninsula. At Cabo Codera, the upwelling is most anomalous, since on occasions it has been reported while other upwelling fronts are not present. The eastern coast of Venezuela shows several important focal areas between Puerto La Cruz and Cumana, north to Araya and Margarita Island, and close to Carupano (Sucre State) (Figs. 1 and 2). All these sources act as one pulse, clearly dependent on the major wind intensity during the first six months of the year. Its intensity and propagation are considerable, covering up to 200,000 km², and contributing to the most productive area known in the Caribbean Sea. At the Paria Peninsula and north of Trinidad, cold waters are present with limited duration and variable intensity. Those waters that emerge at the Paria Peninsula extreme, sometimes form mesoscale eddies towards the northwest, probably due to the exit of less dense and warmer waters from the Gulf of Paria.

The Orinoco River

The Orinoco River runs over 2000 km from the Sierra Parima, Amazonas State, south of Venezuela to the Atlantic Ocean (Fig. 1), discharging an average of 36,000 m³/s of water, forming the third largest river of the world in terms of discharge following the Amazonas and Zaire rivers (Weibezahn, 1990). The Orinoco influences the salinity patterns, currents, amount of suspended material and nutrients in the Venezuelan Atlantic coast and the Caribbean Sea.

The extension of the Orinoco basin is 900,000 km² (Vasquez, 1989) and it is shared by Venezuela (71%) and Colombia (29%). Three main physiographic units occur in the watershed: the Guayana Shield, the Andes, and the Llanos (Lewis et al., 1995), which result in rivers with particular geochemical characteristics. Their classification is based on their optical features:

1. *Black water*: tea-coloured water with low pH and conductivity, and low concentration of suspended solids (mainly from the Guayana Shield).
2. *Clear water*: acid-neutral pH waters with low concentration of suspended solids (mainly from the Andes and the Guayana shield).
3. *White water*: basic-neutral pH with high concentration of suspended and dissolved solids (mainly from the Llanos).

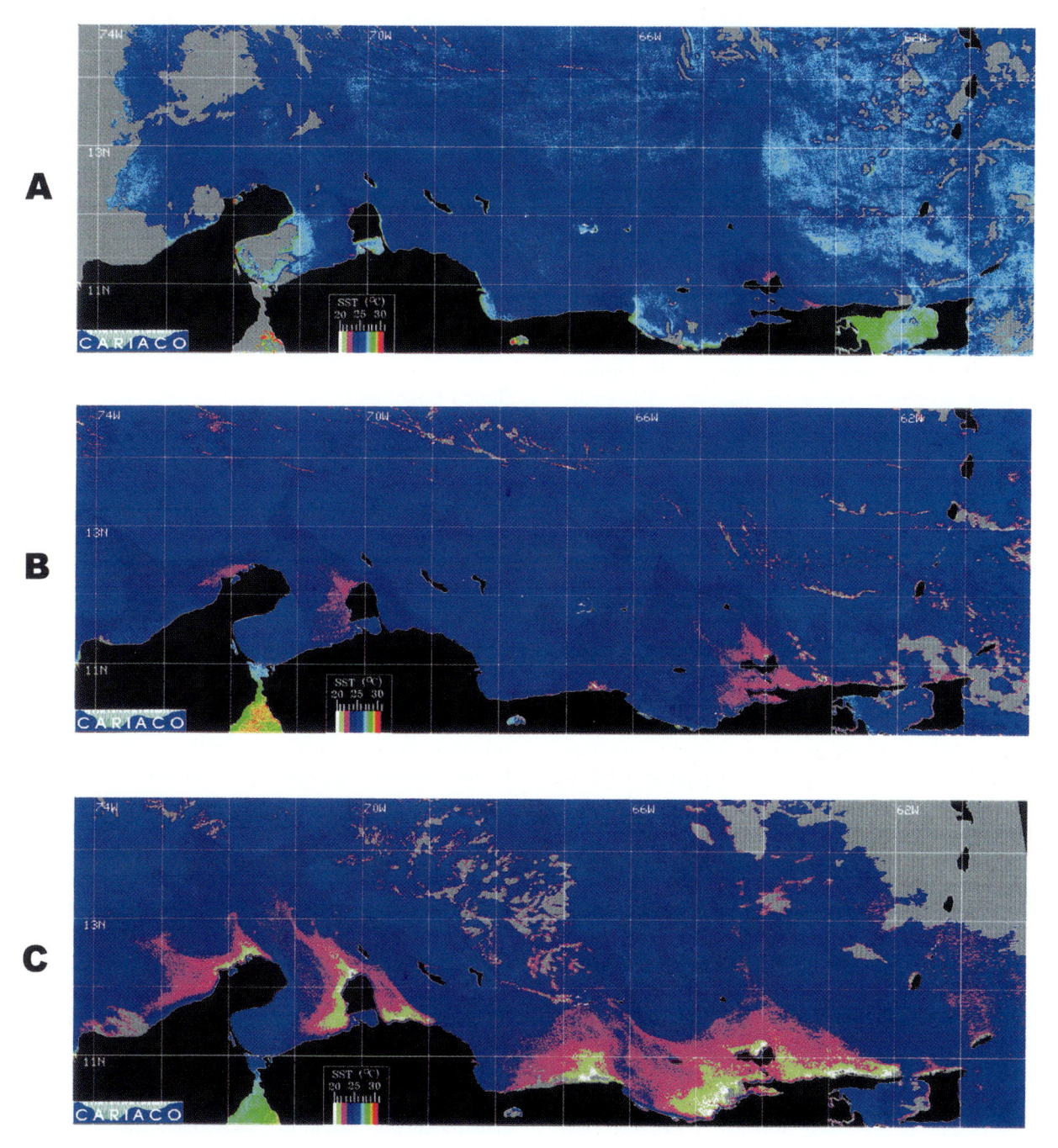

Fig. 2. AVHRR (Advanced Very High Resolution Radiometer) images of SST (sea surface temperature) for (A) January, (B) February and (C) April 1998.

Black and clear water rivers are found in the rivers Ventuari, Sipapo, Caura, and Caroni, which drain from the Guayana Shield while white water rivers (Guaviare, Meta, and Apure) run through the Llanos (sedimentary plains). Fifty percent of the water transported by the Orinoco is provided by the rivers draining the Guayana Shield. This only represents one third of the total watershed area (Lewis and Saunders, 1989). In contrast, the waters coming from the Andes and the Llanos have an important influence on the dissolved and suspended load. Nearly 90% of the suspended material is contributed by the Meta, Guaviare, and Apure rivers and only the Meta provides almost 50% of the total sediment load transported by the Orinoco River (Meade et al., 1990).

The total nitrogen concentrations in the Orinoco River at Ciudad Bolivar are high (460 μg/l), particularly when taking into account the fact that the Orinoco watershed is relatively undisturbed. This high concentration could be explained by the high rate of atmospheric nitrogen fixation conducted by the watershed vegetation. On the other hand, total phosphorous levels are relatively low (64.66 μg/l) as a result of a dilution effect produced by the flow of the rivers draining the Guayana Shield, having one of the lowest total P levels in the world (Lewis and Saunders, 1989; Lewis et al., 1995)

The seasonal fluctuation of the Orinoco River discharge has a strong influence on the transport of suspended and dissolved material into the southeastern Caribbean Sea and the Atlantic Ocean. This fluctuation is directly related to seasonal variations (rainy and dry season) in the watershed, since the Orinoco main stem has not been totally regulated. Only one of its main tributaries is dammed (Caroni River). Due to these factors, the Orinoco hydrography shows an unimodal curve with a large discharge change between a rainy and a dry season (Lewis and Saunders, 1989).

During the dry season (January–March) the Orinoco waters upon reaching the ocean circulate very close to the coast, entering the Gulf of Paria. During the rainy season the Orinoco waters flow towards the east covering the Continental Shelf and the east of Trinidad and Tobago, having an influence on the Caribbean Sea. These variations of flow affect the salinity of the coastal zone of the Orinoco Delta and the Gulf of Paria. Salinity levels below 35% occur near the coast in the Gulf of Paria during the dry season. This condition extends to the Atlantic Shelf of the Orinoco, Trinidad and Tobago, and the sub-Eastern Caribbean during the rainy season (Monente et al., 1994).

As expected, the seasonal variations in discharge also affect the transport of sediments into the coastal zone. According to Monente (1990), a higher sediment concentration is observed during the periods of high and falling water in the Gulf of Paria and the coastal marine side of the Delta. In addition, the dynamic of the suspended sediments in the sub-Eastern Caribbean is also influenced by the Amazon and the Orinoco River (Monente et al., 1994).

The Orinoco River Delta comprises a zone that opens to the Caribbean Sea, from Barrancas (Monagas State), with a perimeter of 300 km and a surface area estimated at 30,000 km^2. It functions like a wetland with a great variety of aquatic systems whose characteristics depend on a series of factors. The effects of tide and season greatly affect the physico-chemical characteristics of the waters.

The Delta has three main zones (MARNR, 1979):
1. *Upper Delta*. Higher lands and less floodable, with predominantly herbs and low arboreal communities along the riverbed.
2. *Mid Delta*. The land is mid-flooded and the herbal vegetation is substituted by woody communities of the Moriche Palm, and to a lesser degree by woods along the shores of water bodies.
3. *Low Delta*. This is highly flooded and mangroves dominate.

Until recently, the Orinoco River discharged its waters through the great Caños Manamo, Macareo and Boca Grande. In 1965, the construction of a dike to close Caño Manamo, which in normal conditions discharged 3500 m^3/s and up to 8000 m^3/s in flooding conditions, commenced for the purposes of navigation. The effect of tides on this zone is now noticeable, considering that the coast before and after the dike varies by 2 m in the dry season and up to 8 m in the rainy season in the Orinoco. The system has been transformed into a delta which is totally controlled by the tide variation. Additionally, the sediments periodically disposed in the bottom by the Orinoco River before the closing are now being leaked by the marine currents, causing a deepening in these and a major penetration of marine water.

THE MAJOR SHALLOW WATER MARINE AND COASTAL HABITATS

Sandy Beaches

According to wave energy, sediment origin and particle size the sandy beaches of Venezuela can be classified in three types: (1) dissipative beaches with fine and very fine sediments, rich in organic matter, (2) high energy beaches with coarse sediments and intermediate content of organic matter, and (3) low energy beaches of carbonate biogenic sediments with variable content of organic matter.

Dissipative beaches are not very diverse; they are characterized by few but highly abundant and dominant (both in number and biomass) bivalve species (*Tivela mactroides, Donax denticulatus,* and *Donax striatus*). *Tivela mactroides* is present in dense populations near the mouth of rivers, particularly where large amounts of terrigenous particles are transported in suspension, or in areas where disturbance of sediment produces large amounts of suspended material (McLachlan et al., 1996). Populations of *T. mactroides* reach high densities; Prieto (1983) reported mean densities at Playa Guiria (Sucre State) of 788–1024 ind. m^{-2} with a maximum biomass of 1229 g m^{-2} shell free wet weight. *T. mactroides* has a short life span and rapid growth. At Playa Guiria they reach 30 mm in 12 months, with a life span of 18 months (Prieto, 1983). Natural mortality is very high; it is very common to see dead and decaying individuals in the upper surf zone. There is a constant renewal of the population as a consequence of this high natural mortality. *Donax denticulatus* population is mostly distributed in the wash zone, with few individuals in the surf zone. There is also a pronounced sorting of individuals by size in the wash zone with large adults almost being entirely confined to the oxygen saturated zone, and small individuals generally being found in the unsaturated zone. This distribution is maintained throughout the tidal cycle and during wind-generated changes in water height. Etchevers (1975) evaluated the biomass of *D. denticulatus* in Playa la Restinga (Margarita Island) and estimated that there was 14.2 tons (wet weight including shells) in 18 km of beach. For *D. striatus*, the growth rate at these beaches was very rapid during the first months after the recruitment. The clams reached 50% of their asymptotic size in 2.5 months, 66% in 4 months and 90% in 8 months.

These beaches are used by fish as nursery grounds and food sources. Both *D. denticulatus* and *D. striatus* are important in the beach food chain as primary consumers of phytoplankton and detritus. In turn, they are eaten by a wide range of predators. *D. denticulatus* is eaten by a variety of fish including *Menticirrhus littoralis*, *Conodon nobilis*, *Trachinotus carolinus*, *T. goodei* and *Umbrina coroides* (Penchaszadeh, 1983; Riera, 1995), and by ghost crabs *Ocypode verreauxi*. The fishes *M. littoralis* and *T. carolinus* were found to eat both entire *Donax* as well as the siphons (Riera, 1995).

High energy beaches have a diverse taxonomic composition with no dominant species and the ichthyofauna associated with them is scarce. Figure 3 shows a typical zonation of a high energy beach in Quizandal, Carabobo State). Their supra-littoral is well developed and the most conspicuous species are *Excirolana brasiliensis* and *Ocypode quadrata*. In mid-littoral and infra-littoral the most conspicuous taxa are molluscs (*Mazatlania aciculata*, *Olivella verreauxi*, *Prunum prunum*, *Polinices hepaticus*, *Persicula interruptolineata*), Polychaetes and Echinoderms (*Mellita quinquiesperforata*) (Penchaszadeh et al., 1983).

Diversity is very low on the beaches of carbonate origin. The characteristic species are *Heterodonax bimaculatus* and *Excirolana mayana* (de Mahieu, 1984). Their fish fauna is often associated with nearby coral reefs and *Thalassia testudinum* beds.

Rocky Shores

Venezuelan rocky shores are areas of great diversity, since they show a variety of microhabitats. Several authors have contributed to the knowledge on the zonation and studies of the rocky shore ecosystems: Rodriguez (1959) in Margarita Island, Almeida (1974), Perez (1980) and de Mahieu (1984) in the west-central littoral (Golfo Triste) (Fig. 1). In these areas the tides are reduced to around 60 cm. The middle and supralittoral fringe limits depend more on the wind or even the wave mode than on the tide variations. Generally this fringe shows a low vertical development, but a very adequate horizontal development.

At the rocky fragments or segments of the littoral zone from the Patanemo Bay, Carabobo State and Tucacas, Falcon State (Fig. 1), three great areas can be distinguished: supralittoral, midlittoral and infralittoral. They are characterized by groups or associations of particular organisms:

The *Littorina* zone, located on the supralittoral spray and splash zone of major or minor depth, varies according to the locality and mode. The highest fringe, away and less exposed to the sea, is commonly distinguished by several species of mollusc. This community is represented by gastropods as *Littorina ziczac*, *L. nebulosa*, *L. mespillum*, *Nodilittorina tuberculata*, *Nerita versicolor*, *N. tessellata*, *N. peloronta*, *Tectarius muricatus*, and *Planaxis nucleus*. In this zone microscopic algae are attached to rock gauges giving a greenish or blackish aspect to the rock.

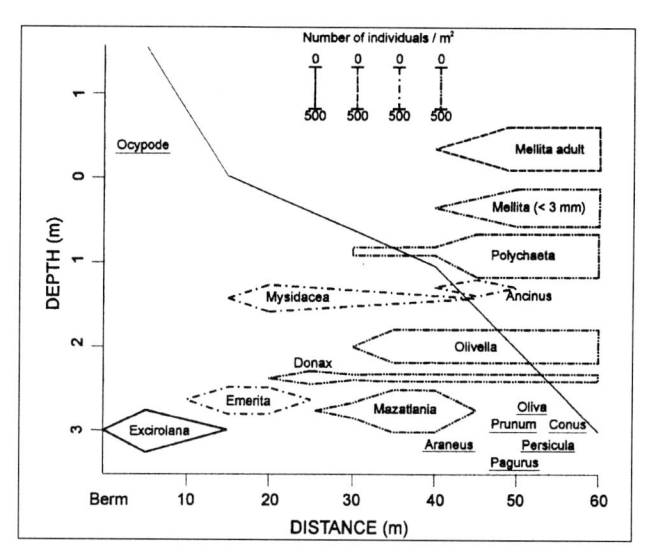

Fig. 3. Sandy beach community zonation at Quizandal, Venezuela (from Penchaszadeh et al., 1983).

The barnacle zone, located in the lower supralittoral, limits a relatively clear zone between the algal fringe (lower) and *Littorina* zone (higher). The barnacle zone is occupied by *Balanus* sp. and *Chthamalus* sp., but can occasionally be absent or inconspicuous. Abundant tidepools are present, housing algae such as *Padina gymnospora* and *Chaetomorpha aerea*, bivalves like *Isognomum radiatus* and *Perna perna*, gastropods like *Planaxis nucleus*, *Cittarium pica* and the anfineuraus *Chiton squamosus*, and *Acanthopleura granulata*. On the eastern coast of Venezuela, where an important phytoplankton primary production occurs, the midlittoral and infralittoral is colonized by the mussel *Perna perna* and in recent times has been invaded by *Perna viridis*, originally from the Indo-Pacific (Penchaszadeh and Velez, 1995).

The algal zone in the midlittoral is the more diverse of the littoral profile. The composition of algae varies depending on wave exposure. *Sargassum cymosum* and *Pterocladia* sp. on rocks, tend to displace other algae like *Ulva lactuca* and *U. fasciata* (Rodriguez, 1959; Gonzalez, 1977). Other species present are *Laurencia papillosa*, *L. obtusa* and *L. scoparia*, *Dictyota dichotoma*, *Ectocarpus breviarticulatus*, *Caulerpa racemosa*, *C. sertularioides*, *Galaxaura marginata*, *G. obstusa* and *Padina gymnospora*. A great variety of animals live on this algae fringe, particularly amphipods of the families Ampithoidae, Hyalidae, and Stenothoidae, and the molluscs *Fissurella barbadensis*, *F. nimbosa*, *F. nodosa*, *Planaxis nucleus*, *Diodora listeri*, *Acmaea antillarum*, *Thais haemastoma floridana*, *T. rustica*, *Purpura patula*, and *Cittarium pica*, among others.

Frequently it is possible to find a great number of sea urchins inside holes in the rocks, *Echinometra lucunter* being the most common. Numerous ophiurids like *Ophiothrix angulata* and *Ophioderma* sp. are also present. It is also possible to observe large patches of colonial anemones like *Zoanthus sociatus* and *Palythoa caribaeorum* (Bastidas and Bone, 1996).

The sublittoral zone is characterized by small aggregations of corals and hydrocorals like *Diploria clivosa, Acropora palmata, Millepora alcicornis, M. complanata,* and algae such as *Halimeda opuntia, Caulerpa racemosa, Laurencia papillosa,* and *Dictyota* sp.

Seagrasses

Seagrass beds are common along the Venezuelan coast. Many commercially important fish species spend part of their lives in the seagrass ecosystem as it is an important nursery ground. Unfortunately, seagrasses are being destroyed at an increasing rate due to the development of coastal resorts in the absence of regulatory legislation and adequate land reinforcement.

Eight seagrass species have been reported in Venezuela (Ganesan, 1989). *Thalassia testudinum* is the most widely distributed species on the western and central Venezuelan coast (Acosta, 1974) as well as in Margarita Island (Hambrook et al., 1979; Zieman et al., 1997; Perez, 1997). *T. testudinum* covers large areas in the sub-littoral zone along the eastern Venezuelan coast. It covers approximately 70% of the Cariaco Gulf and many square kilometers along the coast of Sucre State (Fig. 1). Its distribution is interrupted by reefs and small rocky beds only where benthic algae are the dominant macrophytes (Vera, 1978).

Venezuelan seagrasses grow in association with corals between the low tide and 10 m or more depth. Along stretches of the eastern Venezuelan coast, *T. testudinum* forms dense mono-specific beds in deeper water where soft substrate is frequent, while *Halodule wrightii* and *Syringodium filiforme* grow in shallow water. In these areas, small rocks, coral fragments, and shells provide the only hard substrate for algal patches that grow on the hard dead coraline substrate (*Halimeda* sp.), resulting in highly diverse community.

Productivity values reported by Zieman et al. (1997) are in the order of 1.8 $g/m^2/d$ in Morrocoy National Park (Golfo Triste) (Fig. 1) and 1.5 $g/m^2/d$ for Margarita Island.

The *Thalassia* beds house a highly diverse fauna of resident species as well as visitors, including a long list of invertebrates (polychaetes, molluscs, gastropods, crustacean) (Isea, 1994) and fish (*Acanthurus bahianus, Chaetodon capistratus, Eucinostomus gula, E. havana, Gerres cinereus, Haemulon flavolineatum, H. sciurus, Ocyurus chrysurus, Monacanthus ciliatus, M. tuckeri, Lactophrys quadricornis, Pomacentrus leucostictus, Sparisoma chrysopterum, S. radians, Sphyraena barracuda, Sphoeroides spengleri,* and *S. testudineus*) (Diaz, 1997).

Coral Reefs

The best developed coral reefs in the Venezuelan Caribbean are to be found around the islands offshore from the continental coast. The Archipielago de Aves (East and West), Archipielago de Los Roques, La Orchila, Isla de Aves, La Tortuga, and La Blanquilla all have diverse coral reefs, most of them in pristine condition (Fig. 1). Except for the Archipielago de Los Roques (which in spite of the fact that it is a National Park protected under special laws suffers intense fishing as well as other activities derived from tourism), all of these islands have either low population densities or occasional residents centred on fishery or seasonal tourist activities (Woodley et al., 1997). Otherwise they belong to the Venezuelan Navy (La Orchila and La Blanquilla) and are considered restricted areas. The reef fauna in these islands is similar to the rest of the Caribbean, with more than 40 scleractinian coral species. The most important in coverage and abundance are *Montastrea annularis, M. cavernosa, Colpophyllia natans, Diploria* spp., *Acropora palmata, A. cervicornis, Porites porites, P. astreoides, Madracis decactis, Siderastrea siderea,* and *Agaricia* spp. Thirty-five species of octocorals are the most representative of the area, mainly *Pseudopterogorgia americana, P. acerosa, Plexaura homomalla, P. flexuosa, Eunicea* spp., *Muricea atlantica, Gorgonia ventalina,* and *Telesto riisei*. They are distributed in some localities to a depth of 45 m. The sponge fauna is also varied and little studied. Alvarez and Diaz (1985) reported for the Archipielago de Los Roques 30 species of desmosponges to a depth of 35 m. They estimate that there should be at least 30 species more. This group, as in other Caribbean localities, is abundant at major depths.

The Margarita, Coche, and Cubagua islands, which are east of Venezuela (Fig. 1), also have some localities with well developed reefs. They show signs of increasing damage due to the intense pressure caused by continuous fishery, tourist, and urban activities.

The Continental Venezuelan Caribbean coastline has few localities with significant reef development. The presence of several rivers that carry great amounts of sediments to the sea is the explanation for this.

The reefs with major development in the continental coast are found at the Morrocoy National Park (Falcon State), San Esteban National Park, and Turiamo Bay (Carabobo State). San Esteban reefs border keys near the coast, extending down to a depth of 12 m. These continental reefs are however less diverse than those located at the oceanic islands. Morrocoy's reefs occupy a more extensive area in the internal lagoons, oceanic mouths entrance and oceanic keys. Thirty-five scleractinian and thirty-four octocorals species have been reported to a depth of 18 m. The reefs at Morrocoy National Park have been subject to a great deal of tourism and urban activity, and to the influence of the discharge of neighbouring rivers. In recent years these reefs have suffered several bleaching events (1987, 1990, 1994, and 1995) as a consequence of water warming and recently an event of invertebrate mass mortality (January, 1996). The latter was probably due to climatic anomalies (rainfall) and upwelling accompanied by an excessive phytoplankton proliferation, which is uncommon in this area.

The Turiamo Bay, located in the east-central coast, constitutes one of the continental locations in Venezuela where a diverse reef fauna can still be observed. A bordering reef has developed in the east and west coast of the bay, showing a diversity gradient rapidly diminishing in the north–south direction. The west coast of the bay is being affected by an accelerated process of sedimentation, probably generated by dredging activities at the mouth of the bay and by the leaching of the eroded coast. The eastern coast is in good condition and has 25 scleractinian species distributed near the mouth to a depth of 12 m.

The Mochima Bay (Sucre State), on the Venezuelan northeastern coast, is subject to an upwelling process, which is characteristic of this geographic area. Numerous localities with coral reef communities are present in Mochima Bay. Twenty-six species of scleractinian are present, most of them growing on sedimentary rocks, to a depth of 14 m near the mouth of the Bay (Pauls, 1982). As in the Turiamo Bay, Mochima has a diversity gradient that diminishes in the north–south direction.

During the last decades, the central coast from Puerto Frances to Carenero, including the Buche and Los Totumos Bays, has suffered an accelerated damage of its reef communities due to a growing urban process, and the construction of a fuel supplier terminal (Clamens, 1987). More than 80% of the invertebrate reef species reported in the 1980s have disappeared from the zone.

Mangroves

The Venezuelan mangroves cover an area of approximately 673,000 ha along the coastline (Conde and Alarcon, 1993) (Fig. 1). The diversity of the coastal environment is expressed in differences of topography, climate, wave strength (energy), amplitude of tides, sedimentation and salinity. Consequently, the mangrove distribution pattern also varies. The greatest area of mangroves—about 495,200 ha or 73.4% of the total in the country—is located in the Orinoco Delta and the Gulf of Paria (Fig. 1). Some 138,300 ha (20.5%) of mangroves are found in coastal lagoons of the west-central and east-central zones of the country. The other 15,000 ha (2.3%) of mangroves are found in the Gulf of Venezuela north to Maracaibo Lake (Fig. 1).

On the north coast of Venezuela the greatest beneficiaries of preserved mangrove areas are the traditional artisanal fishing communities. This is the case in Tacarigua de la Laguna (Fig. 1) in Miranda State, where those most prejudiced against mangrove protection are developers for urban land, seeking to expand areas for hotels, yacht clubs and secondary homes (FUDENA, 1988).

The mangroves in Venezuela have clear zones. The most seaward is formed by a fringe of *Rhizophora mangle*, considered pioneering plants. According to the water movement, the dense curved roots which are characteristic of this external mangrove zone, promote the sedimentation of suspended material producing a fine sand bottom. The roots are the substrate for a very diverse community, especially for the mangrove oyster *Crassostrea rhizophorae*.

Following this *Rhizophora* zone is an *Avicennia germinans* fringe, which is easily recognizable by its abundance of pneumatophores. Associated with this fringe is *Laguncularia racemosa*, followed by an internal fringe of *Conocarpus erectus*, species that prosper on firm sandy soil away from the tides (Pannier and Pannier, 1989; Conde and Alarcon, 1993).

Coastal Lagoons

An important group of coastal lagoons with a total surface of 26,900 ha exists in Venezuela, located in the region from the Goajira Peninsula to Sucre State and Nueva Esparta (Margarita Island and Coche). Distributed from west to east are the following lagoons: Cocinetas in the Goajira Peninsula; Boca de Caño in Falcon State; Patanemo in Carabobo State; La Salina, Laguna Grande, La Reina and Tacarigua in Miranda State; Unare and Piritu in Anzoategui State; Los Patos, Laguna Grande del Obispo, Bocaripo and Chacopata in Sucre State; Boca Chica, La Restinga, Los Portillos, Boca de Palo, Laguna de Raya, Punta de Piedras, Punta de Mangle, Las Maritas, El Morro de Porlamar, Caño El Cardon and Zaragoza in Margarita Island, and El Saco in the Island of Coche of Nueva Esparta State (Fig. 1).

All these lagoons are separated from the sea by sandy bars, with the exception of the Patanemo Lagoon (which has a small coral reef), the Laguna Grande de Obispo and the lagoon of Boca Chica (Ramirez, 1996).

Some of the Venezuelan coastal lagoons are legally protected by the Areas Bajo Regimen de Administracion Especial (ABRAE) (Areas Under Special Administration). This responds to the physical and natural conditions as well as the social and economic processes that have been developed for the use of these renewable natural resources.

Many of them have been used for the development of different activities that include an economy of subsistence, refuge, fisheries and aquaculture, urban activities and tourism.

The coastal lagoons are affected by a series of natural factors and anthropogenic activities. Among the natural factors are the tides, which affect water circulation and biological, physical, and chemical processes. The input of fresh water coming from the rivers also influences temperature and salinity, which causes a wider variation range in the hydrological and biochemical factors. Coastal lagoons are also affected by sediments coming from the drainage of the rain waters, that may in time cause the complete filling and disappearance of these ecosystems.

Among the anthropogenic effects are the constructions along their river-sides of industries, urban projects, aquaculture farms, tourism facilities and the abusive pruning of mangrove for lumber-use (this has considerably reduced the surfaces previously occupied). Further, there is a spread of electric lines, contamination by solid waste, pesticides, and sewage waters with proliferation of human discharges that are poured directly into the swamps.

Laguna de Tacarigua is the best studied coastal lagoon (Conde, 1996). It is a shallow (average depth: 1.2 m) estuarine lagoon with an approximate area of 72 km^2. The permanent, continuous, rectangular sandbar that isolates the lagoon from the Caribbean Sea is mainly formed of detrital material; its width ranges from 300 to 1000 m. In its western extreme, there is a narrow pass that connects this estuary to the open sea. Five embayments or zones, freely communicated by channels, can be delimited. The central part of the lagoon is dotted with red mangrove (*Rhizophora mangle*) islets. There is one important fluvial tributary, the Guapo River. This discharges continuously through the Madre Casañas Canal, and several intermittent seasonal creeks along the southern shore. The sedimentary load entering the lagoon from the river was estimated at 150 m^3/year. It has been estimated that every year 3×10^6 m^3 of sediments are transported by littoral currents towards the lagoon's inlet.

The sediments draining into the lagoon are dominated by alluvial and marine calcareous materials. The surficial sediments have a detrital origin, although there are neoformations of pyrite and evaporites, and most likely there are *in situ* formations of chlorite, calcite and aragonite. As a whole, muddy-clay fractions predominate in most of the sediments.

As a consequence of the shallow depth of the lagoon, water temperature fluctuates periodically and daily. Maximum temperatures were observed in August and September (33.8°C), while minimum temperatures were registered in February and April (26.4°C).

Tacarigua can be considered eurihaline-mixohaline. Salinity can vary locally and temporarily from hypersaline levels (54.0‰) at the eastern and central zones, where the freshwater influx is sporadic, to limnetic records close to the Guapo River discharge area. Salinity regimes depend on several factors: freshwater input, rainfall, intrusion of seawater, and evaporation.

Planktonic primary productivity at Laguna de Tacarigua is higher than the productivity estimated at other offshore and inshore localities on the Venezuelan coast, which average less than 10 mgC/m^3/h. Productivity is markedly seasonal, averaging 36.1 mgC/m^3/h; the maximum values coincide with rain peaks, during the middle and the end of the year, when the primary production reaches 187.76 mgC/m^3/h (Conde, 1996).

The presence, and abundance of, the American flamingo (*Phoenicopterus ruber*), an endangered species with few nesting sites in the Caribbean, has been monitored quite accurately since the early 1980s, although interspersed observations go back to the 1940s. The common bird species reported for Laguna de Tacarigua are shown in Table 1. The common fish species are shown in Table 2.

Until 1940, the American crocodile (*Crocodylus acutus*) was quite populous in Laguna de Tacarigua. In 1949 sightings of crocodiles were common. However, from 1940 to 1960 this species was relentlessly exploited by French and English companies which exported hides to Europe. The current population of this reptile is small.

Another common reptile that lives among the mangrove trees is the arboreal snake *Corallus hortulanus*. Among several species of terrestrial mammals that have been observed within the mangrove forests, and among the vegetation

Table 1

Common bird species reported for Laguna de Tacarigua (taken from Conde, 1996)

Species	Common name
Ajaia ajaia	Roseate spoonbill
Anas discors	Blue-winged teal
Anhinga anhinga	Anhinga
Anous stolidus	Brown noddy
Aramides axillaris	Rufous-necked wood-rail
Ardea cocoi	White-necked heron
Cochlearis cochlearis	Boat-billed heron
Dendrocygna bicolor	Fulvous whistling-duck
Egretta caerulea; E. tricolor	Egrets
Eudocimus albus	White ibis
Eudocimus ruber	Scarlet ibis
Himantopus hamatopus	Common stilt
Jabiru mycteria	Jabiru
Larus atricilla	Laughing gull
Mycteria americana	American wood ibis
Nyctanassa violacea	Yellow-crowned night heron
Pelecanus occidentalis	Brown pelican
Phaetusa simplex	Large-billed tern
Phalacrocorax olivaceus	Cormorant
Rallus longirostris	Clapperrail
Rynchops nigra	Black skimmer
Sterna fuscata	Sooty tern

Table 2

Common fish species in coastal lagoons in Venezuela, reported for Laguna de Tacarigua (taken from Conde, 1996)

Species	Common name
Arius spixii	Catfish
Selenapsis herzbergii	Catfish
Centropomus ensiferus	Snook
C. pectinatus	Snook
C. parallelus	Snook
C. undecimalis	Snook
Mugil brasiliensis	Mullet
M. curema	Mullet
M. liza	Mullet
Diapterus rhombeus	Mojarra
Eugerres plumieri	Mojarra
Gerres cinereus	Mojarra
Tarpon atlanticus	Atlantic tarpon
Caranx hippos	Crevalle jack
Elops saurus	Ladyfish
Trinectes maculatus brownii	Hogchoker

that surrounds the lagoon, are: the crab-eating racoon (*Procyon cancrivorous*), the prehensile-tailed porcupine (*Coendu prehensilis*), the ocelot (*Felis pardalis*), the capuchin or ring-tail monkey (*Cebus olivaceus*), the paca (*Agouti paca*), the agouti (*Dasyprocta aguti*), the brocket deer (*Mazama americana*), the capybara (*Hydrochaeris hydrochaeris*), and bats of the genus *Noctilio* and *Saccopteryx*.

VENEZUELAN FISHERIES

The trawl fisheries began to explore in Venezuela at the end of the 1940s but commercial development began in the 1950s with operations carried out in the Gulf of Venezuela.

The trawl system used in the initial stage utilised "Italian" or "Mediterranean" type crafts. Fleets were first settled down in Punto Fijo and Maracaibo after carrying out exploratory fisheries along the Venezuelan coast. They identified these localities as the most productive areas for species such as shrimps, cephalopods and flake fish. In 1963 the "Florida" or "American" type boats started fishing the region.

Worker conflicts between artisanal and trawl fisheries, have occurred. However, in practice both fisheries have worked alongside each other for a long period.

Without denying the occasional existence of worker conflicts among the artisanal and industrial fisheries, the problems that the artisanal fishermen confront cannot be attributed to the trawl fishery. The best evidence is that there is not a correlation (trawl:artisanal) with other areas, that is, the communities of fishermen who live in areas where trawl fishery is not practised do so under identical conditions to those who reside in areas of occasional or systematic interference (Gimenez et al., 1993). The national artisanal and industrial marine productions in 1995 were 331,610 mt and 108,454 mt respectively, with 373 trawl units and 15241 artisanal units (MAC, 1996).

The fish captured by the trawl fleet include a great number of commercial species such as *Lutjanus synagris, L. analis, L. griseus, Vomer setapinnis, Cynoscion similis, Macrodon ancyclodon, Micropogon furnieri, Carcharhinus acronotus, Orthopristis ruber, Priacanthus arenatus*, and *Sphyraena barracuda*. All these species are not necessarily fished in all the areas, so some of them are indicative species of specific areas and habitats. In general the areas reported with more unloading in 1989 were the Gulf of Venezuela and the Sucre-Margarita area with 12761 mt and 6690 mt respectively; in order of importance, the area from the Delta, Unare's platform and Golfo Triste. The tendency of the total yield has been to continuously increase since the early seventies, due to a high increase in the loading in the Gulf of Venezuela. On the other hand, the rest of the areas produce around 1000 mt and 5000 mt (Gimenez et al., 1993).

The catch of the trawl fishery was estimated at 63.5% of the total catch by volume, representing 4500 mt for 1977–79 in Golfo Triste (Penchaszadeh et al., 1984). The shrimp *Penaeus notialis* is the most important species of the marketable catch—its proportion in volume is 6.5 times greater during the night than during the day. The ratio between shrimp and the rest of the catch is 1:20. Many of the fish species that are exploited by the trawl fishery use the shallow sandy bottoms as nursery grounds (Penchaszadeh and Salaya, 1985), and fishing is carried out in shallow waters for species such as *Chloroscombrus chrysurus, Diapterus rhombeus* and *Gerres cinereus*. However, many other species are caught by the fisheries from deeper waters, as is the case with *Lutjanus synagris, Synodus foetens* and *Upeneus parvus* (Penchaszadeh et al., 1986).

The development of the canning industry in Venezuela primarily uses the sardine *Sardinella anchovia*. The capture of sardines in 1995 was 153,037 mt (MAC, 1995). These industrial parks began in the late 30s when the private industry installed plants in Cumana and Mariguitar.

From the technological point of view, the capture of the sardine is not an example of advanced technology since the fishing of that species is carried out with beach nets (beach chinchorros) operated by boats that constitute the "Sardine Fleet". However, it should be noted that this fishing system is highly productive and efficient. The productivity of the "Sardine Fleet" ranges from 100 to 500 mt. From the beginning of the artisanal fishing until the arrival of the trawl fishery, there existed a period called "The Sardine Period", so named due to the dominance of this resource in the total catch. The total Venezuelan fleet haul of tuna from the Pacific Ocean in 1995 was 72,835 mt (MAC, 1996), the trawl capture was 30,280 mt and snapper-grouper 4768 mt (MAC, 1996). The total shrimp production in 1995 was 10,786 mt (MAC, 1995).

Aquaculture

The first resource exploited by the Spaniards in Venezuela was the pearl-oyster *Pinctada imbricata* in Cubagua Island. The Spanish depleted the banks and exterminated the native people through slavery. Finally, an earthquake destroyed the Town of Nueva Cadiz in the late 17th century.

Since the 1960s, various efforts were made to culture the mussel *Perna perna* and the mangrove oyster *Crassostrea rhizophorae*, especially in Sucre State. Salaya et al. (1973) reported the results of the culture on suspended cords from a raft, with an annual crop of 15–17 mt. Rapid growth and extraordinary phytoplankton concentrations (due to upwelling) were suspected to be sufficient for a rapid growth of this activity. Nevertheless, the poor capacity to monitor water quality and prevent red tides, the lack of a proper system of refrigeration and distribution, and the lack of good sound political measures to promote this activity, caused the mollusc aquaculture in Venezuela to fail, even when technical and scientific knowledge was available (Velez, 1991).

Marine shrimp farming in Venezuela began in the mid 80s, when the first farms were installed in the country.

Today, 11 registered farms exist in the government offices that control aquaculture (Servicio Autonomo de los Recursos Pesqueros y Acuicolas: SARPA). These enterprises are located in the northwestern part of the country (Zulia and Falcon States) and in the northeast of the country (Anzoategui, Sucre and Nueva Esparta). In 1998, the coastal surface used for this activity extended to 1500 ha.

A series of steps and controls have been set up to try to minimize the effects of farming activity on the environment. Some of these involve studies of environmental impact aimed at verifying which are the effects of shrimp farming and to try to adjust the causes so that the impact can be minimized. In the same way, control and supervision plans have been established so that the development of the activity can be monitored.

The farming system in most of these establishments is semi-intensive but some assays have been done with intensive farming systems, and these have been very satisfactory. In semi-intensive farming the pools are rectangular and made of soil, with a surface of 2–10 ha each. Artificial air systems are not used, 40% of the total pool volume is replaced, supplementary feeding is done by concentrated pellets, organic and inorganic fertilizers are used in order to increment the primary water production and the culture density is 16–24 larvae/m^2. The cultivated shrimps are introduced species (native to the Pacific Ocean), *Penaeus vannamei* and *P. stylirostris*. The farming system is mostly a single culture of *P. vannamei*, or the combined culture of this species with 10–20% of *P. stylirostris*. The culture cycle has a duration of 115–145 days and at harvest the production yield goes from 1500 to 2500 kg/ha/cycle (Rosenberry, 1996) with a survival rate of about 60–90%, depending on the farm. During the year, 2.4 culture cycles are done.

Since 1988 the production has been increasing, rising from 1 mt in 1988, to 4600 mt in 1997. Industry's production rose between 1992 and 1997 by 267%, high for the region. The productivity is estimated to be 3860 kg/ha/year, a very high value compared with countries like Ecuador, for example, which in 1996 produced 923 kg/ha/year (Rosenberry, 1996). The labour impact of this industry is the provision of 1500 jobs during the whole year. At present, the complete production is exported mainly to the United States and some European countries. For the North American market the product is exported without the head, and with sizes of around 41–50 and 36–40 shrimp/lb. For the European market the shrimp goes with the head and the sizes are similar to those of the North American market.

POPULATIONS AFFECTING THE AREA

Venezuela has around 22 million inhabitants with a population density of 21 inhab/km^2; Distrito Federal (Fig. 1) has the highest with 1090 inhab/km^2. The total population in the coastal Districts has reached 5.6 million inhabitants (28.80% of the national total). In terms of distribution, nearly 40% of the coastal inhabitants are concentrated on the western coast, specifically by Maracaibo Lake. The second highest concentration is in the Central Region (16.3%), and the third highest the Eastern Region (15.6%).

The most important establishments along the coastline are: Maracaibo city (Zulia State) (1.2 million inhabitants), Cabimas (Zulia State); Punto Fijo and Coro (Falcon State); Puerto Cabello (Carabobo State); Ocumare de la Costa (Aragua State); La Guaira and Catia La Mar (Dtto. Federal); Barcelona-Puerto La Cruz (Anzoategui State); Cumana (Sucre State); Pedernales (Delta Amacuro State); and Porlamar (Nueva Esparta State). In the Federal Dependencies (offshore islands) live 2245 inhabitants, mainly distributed in Los Roques and Los Testigos (Fig. 1).

Ever since the Colonial Period, the coastline of Venezuela has been influential due to its accessibility, which favours the exploitation of pearls in the east (Cubagua and Margarita islands) and precious metals, cacao, and coffee on the Central coast. Since 1930, the population has been influenced by petroleum activities, which allowed the establishment of settlements around the western coast.

Along Venezuela's coastline, a variety of economical activities like fisheries and tourism take place. Fisheries occur mainly on the eastern coast and these contribute about 50% of the national production. Tourism is distributed along almost all the coastline, from Falcon State to Sucre State. Petroleum activities (exploration, extraction, and refinery) occur mainly on the western coast. Industrial activities are located around the Central Region especially in Carabobo and Aragua States.

The petroleum activities, as well as other industrial activities and their effluents, have contributed to the pollution of marine areas along the coast, especially in Zulia State (Table 3). The contamination by urban effluents is evident in the Distrito Federal's coast, which has increased its population without any urban planning to control the deposition of sewage and solid wastes. The Tuy River basin represents another problem for the coast. Several activities take place within this basin: agriculture, industries, sand-mines, and

Table 3

Main economic activities by states in the Venezuelan coastline

State	Activity
Zulia	Petroleum extraction, Petrochemical
Falcon	Petroleum refinery, Fishery, Tourism-recreation
Carabobo	Industrial, Tourism
Aragua	Tourism
Dtto Federal	Tourism-recreation, Services
Miranda	Fishery, Tourism
Anzoategui	Petroleum refinery, Tourism
Sucre	Fishery
Delta	Fishery, Petroleum exploration
Nueva Esparta	Tourism, Services, Commercial

sewage disposal. The Guaire River, a tributary to the Tuy River, receives untreated sewage waters from Caracas' 6 million inhabitants, and finally discharges into the Caribbean Sea at Miranda State (Fig. 1).

Anzoategui State has a coastal management problem due to industrial and commercial activities, tourism, fishery, oil extraction and refinery, mines, and urbanism.

COASTAL EROSION AND LANDFILL

Morrocoy National Park: Case Study

During the last two decades, the coral communities of the Parque Nacional Morrocoy, located in the northwest of Venezuela (Fig. 1), have been affected by different kinds of man-made disturbances. These began towards the end of the 1960s with the construction of houses and docks, and the disposal of garbage, sewage, and other materials into Chichiriviche Bay.

Weiss and Goddard (1977) considered that sewage pollution caused the greatest damage. The decade of the 70s began with the spread of buildings into all existing quays of the area and to many shoals, sand banks, and reef flats (Bone et al., 1993). After tremendous community pressure, in May 1974 the area was decreed a National Park by the Venezuelan Government and the buildings were completely removed within one year. Any further development of the coastal areas included within the Park was prohibited. However, even with protection, damage suffered by the reefs has increased. Reefs reported healthy in 1973 by Weiss and Goddard (1977), showed high dead coral cover values when first studied in 1979 by Bone (1980). Other reefs in the northern part of the Park had obvious symptoms of deterioration such as high dead coral coverage and the smothering of many living colonies by sediments (Bone et al., 1993). The ten most abundant species which showed relative dead coverage were *Montastrea annularis, Acropora palmata, A. cervicornis, Agaricia tenuifolia, Colpophillia natans, Porites porites, P. astreoides, Diploria clivosa, D. strigosa,* and *Millepora alcicornis*. Evidence suggests that the extent and degree of damage today is mainly due to a high and increasing sedimentation process that began in 1972. This recent kind of man-made disturbance has its origin on the mainland where poor management of agricultural land has created serious erosion problems, with sediments being carried into the Park waters by run-off, by several rivers and the action of marine coastal currents.

The most abundant species were the most efficient in removing sediments. Bone et al. (1993) found a similar pattern. They compared the community structure of Cayo Sal, a degraded reef with high turbidity, with Cayo Pescadores, a healthy reef with less turbidity. They found that the only parameters which could clearly express the differences between the two reefs were the damage index (the percentage of dead cover over the natural mortality of control reefs) and the number of species. Thus, species diversity and living coral cover may not necessarily be different in two reefs with very different levels of turbidity and sedimentation.

Conversion of Mangroves to Small-scale Shrimp Ponds

The coastal wetland of Hueque (1500 ha) is located north of the western coast of Falcon State (Fig. 1). It is characterized by the mangrove species *Rhizophora mangle, Laguncularia racemosa,* and *Avicennia germinans*. This coastal environment receives the waters of the Caribbean Sea through the openings of the Hueque and Guay Rivers, north of the wetland. This drainage is of vital importance for the ecological functioning of the coastal sector of this wetland.

The anthropogenic activity in the area has persisted without change for the last thirty years and is mainly extensive agriculture and cattle raising. In 1986 the construction of a shrimp industry began. Although it never functioned, the drainage and deforestation of the wetland was carried out through 200 ha, affecting the mangroves of the Guay River, obstructing the natural drainage by the presence of pools and roads, and by dredging the Guay River for navigation.

As a result, the area was modified in several ways: changes in the mangrove vegetation of the Guay and Hueque River, modification of the normal flux of tides, new openings in the mangrove forests by deforestation, accumulation of organic waste from dredging, decrease of leaching from the continental lands, and modification of the courses of the rivers (Villamizar, 1994; Fonseca, 1998).

The mangrove areas from the Hueque wetland have been considered by MARNR (1992) to be a wild fauna reserve, and its natural vegetation is protected. At this moment, it is being considered for classification as an Area Under Special Administration which should insure tougher measures to maintain the condition of this area.

EFFECTS FROM URBAN AND INDUSTRIAL ACTIVITIES

The Eastern Coastline

Since the main export of Venezuela is petroleum, special attention has been paid to oil pollution. Research has been carried out on the pollution levels by petroleum residues and aliphatic hydrocarbons in the northeastern region of Venezuela since 1980. Reports indicate areas with chronic contamination and areas that, even though free from extraction activities, show petroleum residues due to marine currents of the gulf that transport some of these hydrocarbons. Also, Bonilla (1982) demonstrated that hydrocarbon contamination was minimized in the Gulf of Cariaco.

Margarita Island

Even though Margarita Island is far away from the oil activities, residues of this industry have been detected with average concentrations reported in the order of 7.53±3.01 and 23.5±17.82 mg/l from samples collected in the months of January–April and June–September of 1986, respectively. Apparently these are transported by superficial currents that move toward the west (Cedeño, 1991).

Bergantin Bay

The impact created by the National Oil Company (PDVSA) on the Bergantin Bay is intense due to the presence of refinery effluents and from the industries loading docks. The continuous accumulation of hydrocarbons in the surface water of this bay has a marked seasonality and the concentrations are in the order of 0.4–6.25 mg/l, the highest values reported in January and May.

Pozuelos Bay

The hydrocarbon analysis carried out in the Pozuelos Bay and its adjacent areas show that the Neveri's outlet had a hydrocarbon concentration of 90 μg/l, and of 80 μg/l in the influence area of the effluents of the Corpoven-PDVSA refinery in Puerto la Cruz. The concentrations inside the Bay oscillated around 15 and 80 μg/l. These values were also observed away from the coast, which would be explained by the maritime traffic that includes the passage and anchorage of tankers near Borracha Island.

Area of Jose

Monitoring done in the area of Jose showed that the hydrocarbon concentration was highest during the months of April, May and July (1.4–1.87 mg/l); the rest of the months the concentrations were around 0.06 and 0.37 mg/l. This variation is probably primarily due to maritime transportation, since this area has no effluents. The Criogenic's dock is the most contaminated zone with a hydrocarbon concentration that varies throughout the year from 0.1 to 5.0 mg/l, depending on eventualities at the Criogenic complex. Secondly, variations could also be due to changes of superficial currents.

Pollution With Heavy Metals

Table 4 shows the results of the analysis performed for heavy metals along the coasts of the Anzoategui and Sucre States, which if compared to those reported by Muhamad (1992) for a non-contaminated zone, indicate an increase in some metal concentrations.

The Western Coastline

The main pollution problems reported for this area are: oil activities and urban activities (Maracaibo Lake), refinery (east coast of Falcon State), tourism and chemical (Morrocoy National Park), and agriculture, industry and ports (Golfo Triste) (Fig. 1).

Maracaibo Lake

Intense oil extraction, agriculture, cattle raising, and urban activity on the Maracaibo lake basin (Fig. 1) has submitted this ecosystem to high pressure. The Instituto de Conservacion del Lago de Maracaibo (ICLAM, created in 1981) which depends on the Ministerio del Ambiente y Recursos Naturales Renovables (MARNR), has compiled all information on contamination of the Maracaibo Lake. Control and management of water quality involves the study of the effluents coming from Maracaibo City, the Petrochemical complex el Tablazo, the tributary rivers (Chama, Motatan, Catatumbo, Escalante, Palmar, Machango, among others), and the accidental oil spills coming from the continuous extraction in the Maracaibo Lake (Parra, 1986).

Perhaps the most systematic information comes from the eutrophication survey done by PDVSA and reported by Parra (1986). In this report the appearance of algae proved to be a symptom of enrichment by sewage waters at various zones in the Maracaibo Lake. This information was confirmed by Satellite Images (Centro de Procesamiento Digital de Imágenes, CPDI).

Table 4

Heavy metal content (μg/g) in marine sediments for different zones in Anzoategui State and Golfo de Cariaco

	Superficial sediment (μg/g)					
	Cd	Cr	Cu	Pb	Ni	Zn
Not contaminated	1.0	20.0	10.0	5.0	10.0	110.0
Jose		0.88	57.49	14.34	11.5	108.22
Jose	1.54	14.42	15.16	16.72	39.42	102.41
Jose	2.81	37.96	20.00	9.67	30.42	82.26
Jose	0.86	57.49	14.20	13.26		107.76
Bergantin Bay		1.88	20.90	14.94	28.83	127.05
Bergantin Bay	2.39	25.17	11.96	21.86	27.44	92.83
Bergantin Bay			13.33		45.60	292.72
Pertigalete Bay			8.37		20.42	118.12
Pozuelos Bay			9.25		11.68	90.05
Guanta Bay			16.46		25.07	120.29
Barcelona Bay			13.15		13.07	108.21
Barcelona Bay			135.0	42.00	44.00	212.00
Pto. la Cruz-Pertigalete			84.0	49.00	78.00	155.00
Anzoategui State coasts			118.0	41.00	57.50	138.00
Lag. Chica-Golfo Cariaco					50.20	43.57
Punta Arenas					24.51	32.66

Table 5
Results of the analysis of marine sediments in Morrocoy National Park, Falcon State, Venezuela (Perez, 1988)

Station	Distance (m)	Al %	Fe %	Mn 10^{-2} %	Zn 10^{-3} %	Pb 10^{-3} %	Cd 10^{-4} %	Cu 10^{-4}%	Ni 10^{-3} %	Cr 10^{-3}%
Cayo Animas		<0.2	<0.1	<0.1	0.5	<0.1	0.6	3.1	0.1	<0.1
Caño Grande	2	*0.9	*1.1	*0.4	*4.1	0.3	1.4	8.1	1.2	*7.0
	50	0.5	0.4	<0.1	3.7	*0.9	*2.8	*10.5	1.0	5.2
	100	<0.2	0.2	0.2	2.2	0.5	1.8	5.2	0.5	1.6
Ensenada Morrocoy		<0.2	<0.01	<0.1	<0.2	<0.1	0.8	2.4	0.1	<0.1
Punta Tucacas	2	<0.2	<0.01	<0.1	<0.2	<0.1	0.5	2.1	0.1	<0.1
	100	<0.2	<0.01	<0.1	<0.2	<0.1	0.6	1.5	<0.05	<0.1
Las Luisas	2	<0.2	<0.01	<0.1	<0.2	<0.1	0.5	2.6	<0.05	<0.1
	50	<0.2	<0.01	<0.1	<0.2	<0.1	<0.8	2.1	0.1	<0.1
	100	<0.2	<0.01	0.2	0.5	0.3	0.6	2.6	0.1	<0.1

*Maximum concentration values found for each element.

Table 6
Relevant Venezuelan regulations related to planning on coastal areas. Source: Sebastiani and Villamizar (1991), updated by the same authors

Legislation	Year	Key Provisions
Ley Forestal de Suelos y Aguas (Law of Forestry, Soils and Waters	1966	Creation of national parks, natural monuments, forest regional reserves. It establishes a protection area of 50 m from each side of a navigable river and 25 m for a non-navigable river.
Ley Organica del Ambiente (Environmental Act)	1976	Land use planning. Creation, protection and conservation of natural parks, forest reserves, natural monuments, protected zones, natural wildlife refuges and sanctuaries. Activities that could cause environmental damage only can be authorized if there are guarantee procedures and norms to correct or diminish the damage.
Ley Organica de la Administracion Central (Central Administration Act)	1976	Creation of the Ministry of Environment and Natural Resources (Ministerio del Ambiente y de los Recursos Naturales Renobables). It is in charge of developing and applying programs of conservation, defense and regulation of land use in rural areas and National Parks.
Reforma Parcial del Reglamento de la Ley Forestal de Suelos y Aguas (Partial Reform of the Regulation of Forestry Law of Soils and Waters)	1977	Protection area of at least 50 m from margins of lakes or natural lagoons.
Ley Organica de Ordenacion del Territorio (Law of Territorial Planning)	1983	There must be plans to guide land use. There are areas that must have special regulation named Areas Bajo Regimen de Administracion Especial (ABRAE): National Parks, Protected Areas, Forest Reserves, Areas for Security and Defense, Reserves and Sanctuaries for Wild life, Natural Monuments, Zones for tourism, Areas under International Treaty.
Decreto 623 (Decree 623)	1989	An 80 m wide "Protected Coastal Zone" is established for which a land use plan must be developed by the Ministry of Environment and Natural Renewable Resources.
Decreto 1843 (Decree 1843)	1991	Protection of mangroves (all species) and vital areas associated to mangroves (corals, marine seagrasses, wetlands and other related ecotones). The Ministry of Environment and Natural Renewable Resources will establish the need for an environmental impact assessment, also this Ministry could allow, in specific situations, the intervention of mangrove areas and vital areas associated.
Ley Penal del Ambiente (Criminal Law of the Environment)	1992	Criminal acts in coastal areas are considered: discharge of pollutants, construction without required studies, beach degradation, hydrocarbons discharge, illegal fisheries, omission of accidents.

Morrocoy National Park

Several studies have been carried out in this Park which lies very close to the industrial activities at Golfo Triste (Fig. 1). It receives discharges by nearby rivers and is subject to high tourist pressure. Perez (1988) reports raised metal levels in numerous organisms, in water and in sediment at five localities of the Park (Table 5). These results show that the heavy metals analyzed are distributed in all the sampled zones and are below the levels reported for contaminated natural areas. Recently, MARNR (1994) tested the water quality at the Morrocoy National Park and concluded:

Table 7

Coastal areas under special regulations (ABRAE-Areas Bajo Regimen de Administracion Especial). Source: Sebastiani and Villamizar (1991), updated by the same authors and MARNR (1992).

Type of area under special regulation and activities allowed	Name of the area	State
Forest reserve. Sustainable forest exploitation, hydroelectricity, deforestation and plantations, use of water resources, exploitation of plants, animals and wildlife for commercial purposes, passive recreation, industrial activities with restrictions and restrictive sport hunting.	Guarapiche Imataca	Sucre and Monagas Bolivar and Delta Amacuro
Woods lot. Agriculture, deforestation and plantation, use of water resources, exploitation of plants and animals and wild life for commercial purposes, passive recreation, industrial use without restrictions, forestry, restricted human occupation.	Rio Guanipa	Sucre and Monagas
National Park. Educational, cultural and recreational activities. Use of water resources with severe restrictions.	Medanos de Coro Morrocoy San Esteban Henri Pittier Laguna de Tacarigua Mochima Paria Peninsula Archipielago de Los Roques Laguna de la Restinga Tuerepano Mariusa	Falcon Falcon Carabobo Aragua and Carabobo Miranda Sucre & Federal Dependencies Sucre Federal Dependencies Nueva Esparta Delta Amacuro Delta Amacuro
Natural monuments. Use of water resources with severe limitations and recreational activities.	Laguna de las Marites Las Tetas de Maria Guevara	Nueva Esparta Nueva Esparta
Protected zone. Sustainable Forest exploitation, hydroelectricity, reforestation and plantations, use of water resources, exploitation of plants and animals and wild life for commercial purpose, passive recreation, industrial activities with restrictions, deforestation for agriculture and urban activity with severe restrictions, human occupation and restrictive sport hunting.	Higuerote Cabos, Puntas and Lagoons of Margarita Island Territorial space next to the national coast. Central Littoral.	Miranda Nueva Esparta All the Territory of Miranda and Federal District.
Hydraulic reserves. Sustainable Forest exploitation, hydroelectricity, deforestation and plantations, mining with severe restrictions, exploitation of plants and animals and wild life for commercial purpose, passive recreation, restricted industrial activities, restricted wood cutting for agriculture and urban activity with, limited human occupation and limited sport hunting.	Sur del Lago de Maracaibo Zone	Zulia
Wildlife reserve. Extensive passive and wildlife utilization.	Cienaga de Juan Manuel, Aguas Negras and Aguas Blancas.	Zulia
Wildlife refuge. Research and extensive passive recreation	Cienaga de Juan Manuel Aguas Negras and Aguas Blancas. Cienaga Los Olivitos Cuare Isla de Aves	Zulia Zulia Falcon Federal Dependency
Zone for agriculture. Agricultural activities	Barlovento	Miranda

(1) the maximum heavy metal concentration registered for the upper sediments was higher than that observed in other regions of the country. It was also comparable with other marine coasts in the world that have been affected by domestic and/or industrial discharges (Cd, Pb, Cu, Fe, Ni, Cr).

(2) No pesticides were reported in sediments analyzed.

(3) Sixty-seven percent of the samples contained faecal coliforms in the different stations, indicating that molluscs collected or farmed there were not suitable for consumption.

(4) The highest dissolved and dispersed hydrocarbons, lubricants, and grease concentrations were reported after a holiday season, which indicates high embarkation activity in the marine area of the Park.

Golfo Triste

The strong industrial activities (power plant, refinery, petrochemical plant, and pulp-paper-mill industry) has made this zone a highly impacted area by hydrocarbons, heavy metals, organic compounds, and increase of water temperature. The effect on the marine environment at Punta Morón of the power plant has been thoroughly studied (Penchaszadeh and Losada, 1987). This study showed that the warm thermal plume varies in extension, form and direction, depending on the functioning of the power plant as well as a complex interaction of climatic and hydrological factors. It was established that the heated water is located in the surface layers only and does not go beyond three meters deep, so the bottom is not directly exposed to the increase in temperature.

Work in Golfo Triste examined environmental conditions of the marine coast and oil industry operations (Perez, 1980). These show that there are no significant amounts of dispersed-dissolved oil in the water column in the areas and its surroundings. There are two places where the hydrocarbon concentration on the sediments is high: around Puerto Cabello and El Palito.

Regarding the heavy metals in Golfo Triste, the Perez (1988) report established that the levels are high (see Table 5 for results on sediment samples at Morrocoy National Park).

PROTECTIVE MEASURES

In Venezuela, many strategies are being used to control intervention on coastal areas. These strategies are required, directly or indirectly by the law. Table 6 shows some of the existing regulations that enforce these strategies, and Table 7 and Fig. 1 show the coastal areas considered under special regulation regimes.

The legislation framework is based in coastal management of areas specifically assigned for protection. Nevertheless, the main problem is the absence of surveillance due to insufficient economic resources.

REFERENCES

Acosta, J.M. (1974) Estudio de las comunidades vegetales de la Bahia de Los Totumos. *Bol. Soc. Venez. Cienc. Nat., Caracas*. **31** (128/129), 79–112.

Almeida P. (1974) Distribucion de los moluscos en la costa centro-occidental de Venezuela (Patanemo-Punta Tucacas). Comparacion de los habitats litorales. *Mem. Soc. Cien. Nat. La Salle*. Caracas **34** (97), 24–52.

Alvarez, B. and Diaz, M.C. (1985) Las esponjas de un arrecife coralino en el Parque Nacional Archipielago de Los Roques: taxonomia y ecologia. Undergraduate Thesis. Escuela de Biologia. Universidad Central de Venezuela. Caracas. 216 p.

Bastidas, C. and Bone, D. (1996) Competitive strategies between *Palythoa caribaeorum* and *Zoanthus sociatus* (Cnidaria: Anthozoa) in a flat reef environment in Venezuela. *Bulletin of Marine Science* **59** (3), 543–555.

Bone, D. (1980) Impacto de las actividades del hombre sobre los arrecifes coralinos del Parque Nacional Morrocoy, Estado Falcon. Undergraduate Thesis. Universidad Central de Venezuela. Caracas. x + 122 p.

Bone, D., Losada, F. and Weil, E. (1993) Origin of sedimentation and its effect on the coral communities of a Venezuelan National Park. *Ecotropicos* **6**(1), 10–21.

Bonilla, J.R. (1982) Algunas caracteristicas geoquimicas de los sedimentos superficiales del Golfo de Cariaco. *Bol. Inst. Oceanogr. Univ. Oriente* **21**, 133–135.

Carrillo, R.J. (1989) Venezuela. Chapter 8, pp. 113–134. In *International Handbook of Pollution Control*, ed. E. Kormondy. Grenwood Press, New York. xvi + 466 pp.

Carrillo, R., Martin, A. and Poleo, R. (1987) Estudio de la contaminacion ambiental en Venezuela. Universidad Simon Bolivar. Caracas. 156 pp.

Cedeño, G. (1991) Contaminacion por residuos de petroleo e hidrocarburos alifaticos en algunas areas de la region nororiental de Venezuela. Trabajo de ascenso a titular. Universidad de Oriente. Cumana. 161 pp.

Clamens, S. (1987) Efectos de la sedimentacion, generada en la fase de construccion del proyecto S.A.M.M., Lagoven S.A. en el arrecife coralino y la pradera de *Thalassia testudinum* en Carenero, Dpto. Brion, Edo. Miranda. Undergraduate Thesis. Universidad Simon Bolivar. Caracas. 88 pp.

Conde, J.E. (1996) A profile of Laguna de Tacarigua, Venezuela: A tropical estuarine coastal lagoon. *Interciencia* **21**, 282–292.

Conde, J.E. and Alarcon, C. (1993) Mangroves of Venezuela. In *Conservation and Sustainable Utilization of Mangrove Forests in the Latin American and Africa Regions. Part I. Latin America*, ed. L.D. Lacerda, pp. 211–243. Mangrove Ecosystems Technical Reports Series Vol 2. The International Society for Mangrove Ecosystems (ISME) & The International Tropical Timber Organization. Okinawa, Japan.

de Mahieu, G.P. (1984) Milieu et peuplements macrobenthiques littoraux de Golfo Triste (Venezuela). Etudes experimentales sur sa pollution. These Grade de Docteur D'Etat-sciences. l'Université d'Aix-Marseille II. Faculte des Sciences de Luminy. France. 333 pp.

Diaz, Y. (1997) Relaciones troficas en la ictiofauna asociada a praderas de Thalassia en el P.N. Morrocoy. Undergraduate Thesis. Universidad Simon Bolivar. Caracas. xii + 120 pp.

Ellenberg, L. (1978) Coastal types of Venezuela—an application of coastal classifications. *Z. Geomorph. N.F.*, **22** (4), 439–456.

Etchevers, S.L. (1975) Notas ecologicas y evaluacion de la poblacion de chipi-chipi, *Donax denticulatus* en Playa La Restinga, Isla de Margarita, Venezuela. *Mem. II Simp. Latinoamericano de Oceanografia Biologica, Cumana*, pp. 233–249.

Fonseca, H. (1998) Comportamiento hidrologico del rio Hueque, Edo. Falcon y su influencia en la vegetacion de manglar. Master's Thesis. Universidad Simon Bolivar, Caracas. xx + 235 pp.

FUDENA (1988) *Manglares. La importancia economica de los manglares en la politica, planeamiento y manejo de los recursos naturales costeros*. Caracas. 37 pp.

Fukuoka, J. (1965) Condiciones meteorologicas e hidrograficas de los mares adyacentes a Venezuela. *Mem. Soc. Cien. Nat. La Salle*. Caracas. **25** (70–72), 11–38.

Ganesan, E.K. (1989) *A catalog of benthic marine algae and seagrasses of Venezuela*. Fondo Editorial CONICIT, Caracas. 247 pp.

Gimenez, C., Molinet, R. and Salaya, J.J. (1993) *La pesca industrial de arrastre*. Editorial Grupo Carirubana. Caracas. xxxi + 321 pp.

Gonzalez, A.C. (1977) Estudio fico-ecologico de una region del litoral central (Punta de Tarma), Venezuela. *Acta Botanica Venezuelica* **12**, 207–240.

Hambrook, J., Layrisse, M. and Colmenares, R. (1979) Contribucion al conocimiento de la comunidad de *Thalassia testudinum* en Punta Moron, pp. 269–283. In *Ecologia del ambiente marino costero de Punta Moron (Termoelectrica Planta Centro, Estado Carabobo, Venezuela). Informe Final de la Primera Fase*, ed. P.E. Penchaszadeh. CADAFE-Universidad Simon Bolivar, INTECMAR. Caracas. 343 pp.

Iglesias, N. and Penchaszadeh, P.E. (1983) Mercury in sea-stars from Golfo Triste, Venezuela. *Marine Pollution Bulletin* **14** (10), 396–398.

Isea, J.A. (1994) Variacion espacial y temporal de la epifauna movil asociada a las praderas de Thalassia testudinum. Undergraduate Thesis. Universidad Simon Bolivar. Caracas. iv + 101 pp.

Lewis Jr., W.M. and Saunders III, J.F. (1989) Concentration and transport of dissolved and suspended substances in the Orinoco River. *Biogeochemistry* **7**, 203–240.

Lewis Jr., W.M., Hamilton, S.K and Saunders III, J.F. (1995). Rivers of Northern South America, pp. 219–256. In: *Rivers of Northern South America*, eds. C.E. Cushing, K.W. Cummins and G.W. Minshall. Elsevier, Amsterdam.

MAC (1995) *Anuario estadistico agropecuario*. Ministerio de Agricultura y Cria. Caracas. xiii + 319 pp.

MAC (1996) *Estadisticas del subsector pesquero y acuicola de Venezuela. 1990–1995.* Año 1, numero 1. Servicio Autonomo de los Recursos Pesqueros y Acuicolas (SARPA). Ministerio de Agricultura y Cria. Caracas. 289 pp.

MARNR (1979) *Inventario Nacional de Tierras Delta del Orinoco y Golfo de Paria.* Serie de Informes Cientificos. Zona 2/IC/21. 97 pp.

MARNR (1992) *Areas Naturales Protegidas.* DGSPOA/ACM/01. 56 pp.

MARNR (1994) *Evaluacion de la calidad de las aguas del Parque Nacional Morrocoy.* Serie de informes tecnicos. Caracas. v + 75 pp.

Martin, A. (1987) Estudio integral de la contaminacion acuatica en la cuenca del rio Tuy (Edos. Aragua-Miranda). Master's Thesis. Instituto Universitario Politecnico de las Fuerzas Armadas Nacionales (I.U.P.F.A.N.), Caracas. xiv + 277 pp.

McLachlan, A., Dugan, J.E., Defeo, O., Ansell, A.D., Hubbard, D.M., Jaramillo, E. and Penchaszadeh, P. (1996) Beach Clam Fisheries. *Oceanography and Marine Biology: an Annual Review* **34**, 163–223.

Meade, R.H, Weibezahn, F.H., Lewis Jr., W.M., and Perez-Hernandez, D. (1990) Suspended-sediment budget for the Orinoco River, pp. 55–79. In *The Orinoco River as an Ecosystem*, eds. F.H. Weibezahn, H. Alvarez, and W.M. Lewis, Jr. Editorial Galac, Caracas. 430 pp.

Monente, J.A. (1990) Influencia del Orinoco en el Caribe, materia en suspension. *Mem. Soc. Cienc. Nat. La Salle.* Caracas. **49–50** (131–134), 257–271.

Monente, J.A., Pujos, M. and Jovanneau, J.M. (1994) Origen y composicion de las masas de agua que ingresan al Caribe suroriental. *Interciencia* **19**, 79–85.

Muhamad, S. (1992) *Toxic Metal Chemistry in Marine Environments.* Marcel Dekker, Inc. New York. 350 pp.

Müller-Karger, F. and Varela, R. (1989) Influjo del rio Orinoco en el Mar Caribe: Observaciones con el CZCS desde el espacio. *Mem. Soc. Cien. Nat. La Salle.* Caracas **186**, 361–390.

Okuda, T. (1974) Caracteristicas oceanograficas generales de la costa suroriental del Mar Caribe. *Cuadernos Azules*, Cumana **15**, 58–69.

Pannier, F. and Pannier, R.F. (1989) *Manglares de Venezuela.* Cuadernos Lagoven, Caracas. 67 pp.

Parra P.G. (1986) *La conservacion del Lago de Maracaibo. Diagnostico Ecologico y Plan Maestro.* LAGOVEN-PDVSA, Caracas. 86 pp.

Pauls, S.M. (1982) Estructura de las comunidades coralinas de la Bahia de Mochima, Venezuela. Master's Thesis. Instituto Oceanografico, Universidad de Oriente, Cumana, 124 pp.

Penchaszadeh, P.E. (1983) Subtidal sandy beach trophic structure in the area of Punta Moron, Venezuela, pp. 523–528. In *Sandy Beaches as Ecosysytems*, eds. A. McLachlan and T. Erasmus. W. Junk Publishers. The Hague. 757 pp.

Penchaszadeh, P.E., de Mahieu, G., Farache, V. and Lera, M.E. (1983) Ecology of the sandy beach gastropod Mazatlania aciculata in Quizandal (Carabobo, Venezuela), pp. 655–660. In *Sandy Beaches as Ecosystems*, eds. A. McLachlan and T. Erasmus. W. Junk Publishers. The Hague. 757 pp.

Penchaszadeh, P.E., Salaya, J.J., Guzman, R. and Molinet, R. (1984) *Estructura de la pesqueria de arrastre de Golfo Triste, region Centro-Occidental de Venezuela, con especial referencia al material de descarte o broza.* Universidad Simon Bolivar, Caracas, 164 pp.

Penchaszadeh, P.E. and Salaya, J.J. (1985) Estructura y ecologia trofica de las comunidades demersales en Golfo Triste, Venezuela, pp. 571–598. In *Recursos Pesqueros Potenciales de Mexico: la pesca acompañante del camaron*, ed. A. Yañez-Aranciba. UNAM, Mexico, Chapter 12.

Penchaszadeh, P.E., Salaya, J.J., Molinet, R. and De Feo, O. (1986) Aspectos del reclutamiento en las comunidades demersales en Golfo Triste, Venezuela. Taller TRODERP-IREP, COI-FAO; Ciudad del Carmen, Suppl. 44, 203–214.

Penchaszadeh, P.E. and Losada, F.J. (eds.) (1987) Ecologia del ambiente marino costero de Punta Moron y comunidades inscrustantes de Planta Centro, (Estado Carabobo, Venezuela). Informe final de la tercera fase. CADAFE-Universidad Simon Bolivar, INTECMAR, Caracas. 835 pp.

Penchaszadeh, P.E. and Velez, A. (1995) Presencia del mejillon verde *Perna viridis* (Linnaeus, 1758), originario de la region Indo-Pacifica, en el Oriente Venezolano. *Com. Soc. Malac. Urug.* **7**, 401–402.

Perez N.H. (ed.) (1980) Estudio ambiental marino costero de Golfo Triste y marco de referencia para evaluar efectos de operaciones petroleras. Informe final del contrato MARAVEN-Universidad Simon Bolivar, INTECMAR, Caracas. xvi + 142 pp.

Perez, D. (ed.) (1988) Linea base de referencia biologica en el ambiente marino costero del area de Golfo Triste. Informe final. PEQUIVEN-Universidad Simon Bolivar, INTECMAR, Caracas. 279 pp.

Perez, D. (1997) Variabilidad espacio-temporal del crecimiento, la productividad y las caracteristicas estructurales y demograficas de praderas de *Thalassia testudinum* en Venezuela. Trabajo de ascenso a titular. Universidad Simon Bolivar, Caracas. iv + 162 pp.

Prieto, A.S. (1983) Ecologia de *Tivela mactroides* (Born, 1778) (*Mollusca, Bivalvia*) en Playa Guiria (Sucre, Venezuela). *Bol. Inst. Oceanogr. Univ. Oriente.* **22**, 7–19.

Ramirez, P. (1996) *Lagunas Costeras Venezolanas.* Universidad de Oriente. Porlamar, Venezuela. ixx + 275 pp.

Riera, A. (1995) Relaciones troficas interespecificas en una comunidad ictica de una playa arenosa del Estado Falcon, Venezuela. Undergraduate Thesis, Universidad Simon Bolivar, Caracas. 63 pp.

Rodriguez, G. (1959) The marine communities of Margarita Island. Venezuela. *Bull. Mar. Sci. Gulf and Carib.* **9** (3), 237–280.

Rosenberry, B. (ed.) (1996) *World Shrimp Farming.* Shrimp News International.

Salaya, J., Y. Bauperthuy & J. Martinez. 1973. Estudio sobre la biologia, pesqueria y cultivo del mejillon, *Perna perna* (L), en Venezuela. *Informe Tecnico # 62*. Ministerio de Agricultura y Cria. Caracas. 52 p.

Sebastiani, M. and Villamizar, A. (1991) Coastal Wetlands Management in Venezuela, pp. 503–512. In *Coastal Wetlands*, eds. S.H. Bolton and O.T. Magoon. American Society of Civil Engineers, New York. xii + 515 pp.

Vasquez, E. (1989) The Orinoco River: a review of hydrobiologycal research. *Regulated Rivers: Research & Management* **3**, 381–392.

Vera, B. (1978) Introduccion al estudio taxoecologico de la comunidad de *Thalassia* en las aguas costeras de la region noroccidental del Estado Sucre, Venezuela. Undergraduate Thesis. Universidad de Oriente, Cumana. 103 pp.

Velez, A. (1991) Biology and culture of the Caribbean or mangrove oyster, *Crassostrea rhizophorae*, in the Caribbean and South America, pp. 117–125. In *Estuarine and Marine Bivalve Mollusks Culture*, ed. W. Menzel. CRC Press, Boca Raton, FL.

Villamizar, A. (1994) Analisis Geografico Historico y desarrollo estructural del manglar de Rio Hueque, Edo. Falcon. Master's Thesis. Universidad Simon Bolivar. Caracas. 250 pp.

Weibezahn, F.H. (1990) Hidroquimica y solidos suspendidos en el alto y medio Orinoco, pp. 150–210. In *The Orinoco River as an Ecosystem*, eds. F.H. Weibezahn, H. Alvarez and W.M. Lewis, Jr. Editorial Galac, Caracas. 430 pp.

Weiss, M.P. and Goddard, D.A. (1977) Man's impact on coastal reefs: an example from Venezuela, pp: 111–124. In *Reefs and Related Carbonates: Ecology and Sedimentation*, eds. S.H. Forst, M.P. Weiss and J.B. Saunders. American Association of Petroleum Geologists, Tulsa, OK.

Woodley, J.D, De Meyer, K., Bush, P., Ebanks-Petrie, G., Garzon-Ferreira, J., Klein, E., Pors, L.P.J.J. and Wilson, C.M. (1997) Status of coral reefs in the south central caribbean. *Proc. 8th Int. Coral Reef Symp.* **1**, 357–362.

Zieman, J., Penchaszadeh, P., Ramirez, J.R., Perez, D., Bone, D., Herrera-Silveira, J., Sanchez-Arguelles, R.D., Zuniza, D., Martinez, B., Bonair, K., Alcolado, P., Laydoo, R., Garcia, J.R., Garzon-Ferreira, J., Diaz, G., Gayle, P., Gerace, D.T., Smith, G., Oxenford, H., Parker, C., Pors, L.P.J.J., Nagelkerken, J. A., Van Tussenbroek, B., Smith, S.R., Varela, R., Koltes, K. and Tschirky, J. (1997) Variation in ecological parameters of *Thalassia testudinum* across the CARICOMP network. *Proc. 8th Int. Coral Reef Symp.* **1**, 663–668.

THE AUTHORS

Pablo E. Penchaszadeh
Universidad Simon Bolivar, Apartado 89000, Caracas 1080-A, Venezuela

César A. Leon
Universidad Simon Bolivar, Apartado 89000, Caracas 1080-A, Venezuela

Haymara Alvarez
Universidad Simon Bolivar, Apartado 89000, Caracas 1080-A, Venezuela

David Bone
Universidad Simon Bolivar, Apartado 89000, Caracas 1080-A, Venezuela

P. Castellano
Centro de Procesamiento de Imagenes Digitales (CPDI), Sartenejas, Caracas, Venezuela.

María M. Castillo
Universidad Simon Bolivar, Apartado 89000, Caracas 1080-A, Venezuela

Yusbelly Diaz
Universidad Simon Bolivar, Apartado 89000, Caracas 1080-A, Venezuela

María P. Garcia
Universidad Simon Bolivar, Apartado 89000, Caracas 1080-A, Venezuela

Mairin Lemus
Instituto Oceanografico de Venezuela, Universidad de Oriente, Cumana, Venezuela.

Freddy Losada
Universidad Simon Bolivar, Apartado 89000, Caracas 1080-A, Venezuela

Alberto Martin
Universidad Simon Bolivar, Apartado 89000, Caracas 1080-A, Venezuela

Patricia Miloslavich
Universidad Simon Bolivar, Apartado 89000, Caracas 1080-A, Venezuela

Claudio Paredes
Universidad Simon Bolivar, Apartado 89000, Caracas 1080-A, Venezuela

Daisy Perez
Universidad Simon Bolivar, Apartado 89000, Caracas 1080-A, Venezuela

Miradys Sebastiani
Universidad Simon Bolivar, Apartado 89000, Caracas 1080-A, Venezuela

Dennise Stecconi
Centro de Procesamiento de Imagenes Digitales (CPDI). Sartenejas, Caracas. Venezuela

Victoriano Roa
Universidad Simon Bolivar, Apartado 89000 Caracas 1080-A, Venezuela

Alicia Villamizar
Universidad Simon Bolivar, Apartado 89000 Caracas 1080-A, Venezuela

Chapter 42

THE CARIBBEAN COAST OF COLOMBIA

Leonor Botero and Ricardo Alvarez-León

The Colombian Caribbean contains several tropical marine ecosystems (coral reefs, sea grass and calcareous algal beds). In addition it supports others in the northern area which are subjected to upwelling, where seasonal anomalies in the water temperature create special structural and functional characteristics of the biotic communities.

In oceanic areas, coralline and volcanic islands form the archipelago of San Andrés and Providencia which contains very large and diverse coral reef formations. Major environmental problems in the Colombian Caribbean relate to urban development (sewage and solid wastes disposal), unplanned tourism, port development and road construction without proper studies and mitigating measures. Colombia has many environmental regulations which are hard to implement due to conflicting social, cultural and economic conditions. The national government and regional agencies are currently developing policies and plans for coastal management in which natural as well as socio-economic and socio-cultural factors are being integrated.

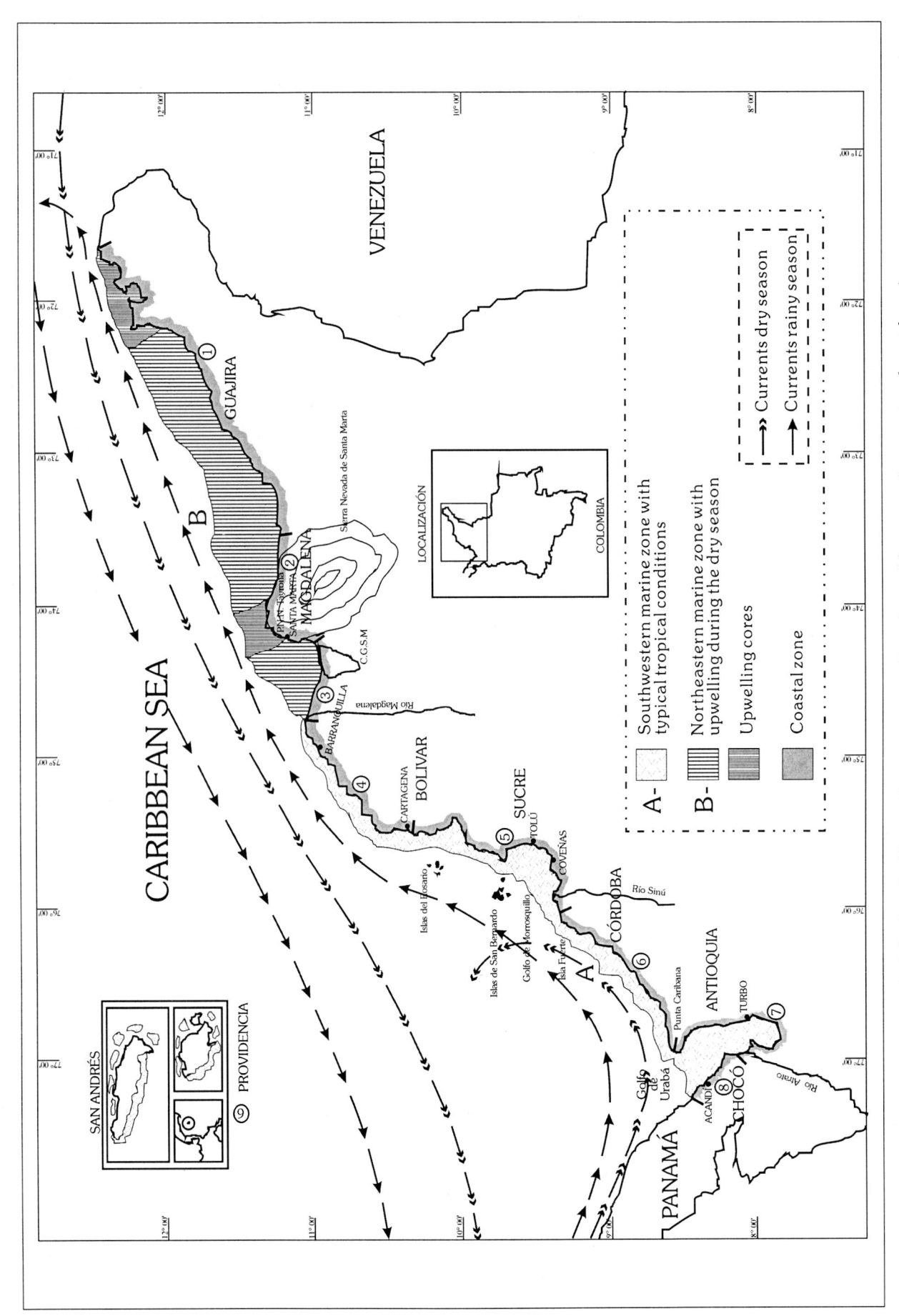

Fig. 1. The Colombian Caribbean with main oceanographic and coastal features, major currents and coastal provinces.

THE DEFINED REGION

The Colombian Caribbean marine area covers approximately 540,876 km² of the southern Caribbean Sea, along a coastline of 1600 km, from Cabo Tiburón, bordering Panama, to the border with Venezuela in the Guajira Península (Fig. 1). Included in this area is the archipelago of San Andrés and Providencia, a set of oceanic islands, atolls and coralline banks aligned north–northeast along the Nicaraguan Rise between 80°–82°W and 12°–14.5°N (Díaz et al., 1996). Although closer to Central America than to South America, these have been part of Colombia since 1822. A wide variety of tropical marine and coastal ecosystems are found along the Colombian coast, such as coralline and rocky reefs, seagrass and algal beds, mangroves, estuaries and coastal lagoons.

The Colombian Caribbean marine area can be divided into two main oceanographic categories (Alvarez-León, 1993) (Table 1). There is a Southwest zone from Golfo de Urabá, bordering Panama, to the mouth of the Magdalena River in the Department of Atlántico. This includes the marine areas of the archipelago of San Andrés and Providencia, and has conditions which are typical of tropical seas. Secondly there is a Northeast zone from a few km east of the mouth of the Magdalena River, in the Department of Magdalena, to the tip of the Guajira Peninsula, with seasonal oceanographic conditions similar to those of subtropical waters. This difference is due to ocean–atmosphere interactions when east tradewinds predominate; these provide the necessary energy to generate such anomalies as upwelling in some parts of the northeastern Colombian Caribbean.

SEASONALITY, CURRENTS AND NATURAL ENVIRONMENTAL VARIABLES

The climate of the Caribbean coast of Colombia is characterized by a bimodal wet–dry seasonality. A strong, dry season from December to April, is followed, in some areas, by a short rainy season in May and June. In July and August dry conditions predominate and from September to November there is maximum precipitation throughout the region. Between Santa Marta and the northernmost limits of the Guajira department, the climate is very dry with a precipitation of 500–1000 mm/yr. From the mouth of the Magdalena River to the Golfo de Urabá, precipitation ranges between 1000 and 1500 mm/yr while on the southernmost part of this Gulf precipitation is around 2000–2500 mm/yr. Relative humidity oscillates between 80 and 98% with an average value of 88% (Alvarez-León, 1989; HIMAT 1992). The duration and intensity of the dry season in the Colombian Caribbean is largely determined by the predominance and strength of the northeast tradewinds, while during the rainy season, winds from the southwest blow along most of the coast (Bula-Meyer, 1990a).

Average temperatures for Colombian Caribbean waters range between 28 and 30°C. Tidal ranges are only 0.3–0.5 m (Alvarez-León, 1989; Giraldo, 1994). Currents are governed mainly by strong east tradewinds which generate the Caribbean Current which has a western direction, or by southwest-west winds which generate the weak Colombian Countercurrent that flows east from the Golfo de Urabá (Fig. 1) (Bula-Meyer, 1990a; Donoso, 1990; Giraldo, 1994). During dry seasons, when tradewinds from the northeast are blowing at their maximum speed, the Caribbean Current may reach as far west as Golfo de Morrosquillo, restricting the action of the Colombian Countercurrent to the southwestern section of the Colombian Caribbean. When northeast tradewinds decrease their speed during the rainy seasons, the Colombian Countercurrent may reach as far east as Riohacha in the Guajira Department.

Continental fresh water flows into the Colombian Caribbean mainly from three major rivers: Magdalena, Atrato and Sinú. The Magdalena River, which is the largest river in Colombia with a length of 1,500 km, drains a basin of approximately 257,438 km² along a considerable part of the Colombian Andes. Daily measurements between 1940 and 1993 indicate an annual water discharge of 7106 m³s⁻¹ and a sediment transport of 138.92×10^6 t yr⁻¹ (Restrepo and

Table 1
Characteristics of the Colombian Caribbean, Southwest and Northeast of the Magdalena River mouth.

Characteristic	Southwest zone	Northeast zone
Continental platform	Wide	Narrow
Coastal topography	Flat with low elevations	Abrupt with elevations up to 5775 m a.s.l.
Average precipitation	Around 800 mm during the dry season and 1200 mm during the wet season	Less than 200 mm during the dry season and around 500 mm during the wet season
River discharge	Magdalena, Atrato and Sinú (year-round)	Small, seasonal rivers
Upwelling and/or outwelling	Outwelling mainly during rainy season	Upwelling during dry season
Climate	Semi-arid and semi-humid	Dry-arid
Surface water temperature	Warm, tropical (>28°C), little seasonal variation	Cooler (°C), drastic seasonal changes
Average sea-water salinity	≤35 ppt	≥36 ppt
Coral reefs and seagrass beds	Well developed, down to 50 m depth	Restricted and little developed, down to 20–30 m depth

Kjerfve, 2000). The Atrato River, the second largest river in Colombia, drains a basin of 35,700 km², discharges 2740 m³ s⁻¹ of water into the Caribbean Sea and transports approximately 11.26×10^6 t yr⁻¹ of sediment. The Sinú River empties into the Golfo de Morrosquillo and drains 10,180 km². Measurements from 1963 to 1993 indicate an annual water discharge of 373 m³ s⁻¹ and a sediment transport of 6×10^6 t yr⁻¹. Restrepo and Kjerfve (2000) estimate annual water and sediment discharges from Colombian rivers into the Caribbean to be 333.77 km³ yr⁻¹ and 163.17×10^6 t yr⁻¹, respectively.

In the archipelago of San Andrés and Providencia, northeast and east–northeast tradewinds are predominant most of the year, with average speeds ranging between 4 and 7 m s⁻¹ (Díaz et al., 1996). Strong wave energy generated by these tradewinds is an important factor that largely controls the geomorphological characteristics of the archipelago as well as its sedimentological regime and biological communities. During the second half of the year, storms with winds of up to 20 m s⁻¹ from the West and Northwest can occur sporadically (Díaz et al., 1996). The archipelago is located in the hurricane belt of the Caribbean and thus is periodically affected by them, with considerable damage both to the coral reefs and to the infrastructure of the islands.

Upwelling System

A large part of the Northeastern Colombian Caribbean (north of the Magdalena River mouth) (Table 1) is characterized by a coastal upwelling of deep (150–200 m) 'sub-tropical underwater' mainly during the long and short dry seasons (Perlroth, 1968; Bula-Meyer, 1977; Fajardo, 1979), when northeast tradewinds blow at average speeds of 10 m s⁻¹. This water mass has a relatively higher nutrient concentration (9–12 times richer in NO_3), lower temperature (22–25°C) and higher salinity (36.5 ppt) than normal Caribbean surface waters (Bula-Meyer, 1985). The relatively low level of inorganic nutrients, as compared to other upwellings in Perú or California is due to specific conditions of the Subtropical Underwater which upwells from areas already impoverished in these nutrients (Corredor, 1977, 1979). Phytoplankton productivity along this coastal upwelling has been estimated and although values are much lower than those of Perú or California, they are significantly higher than those found for normal Caribbean waters (Corredor, 1979).

Upwelling is the cause of special environmental conditions that characterise the coastal waters along Magdalena and Guajira departments, where two main cores have been identified, one across from the northern Guajira Peninsula and the second across Cabo de la Aguja and Tayrona National Park (Wust, 1964; Gordon, 1967; Bula-Meyer, 1985). The last of these has been described mainly by the physical–chemical conditions of surface waters and their effect on the macroalgal communities of the area (Bula-Meyer, 1985). The latter are vigorous and rich in species diversity. Upwelling, together with strong water agitation and abundant rocky substrata down to 10–25 m deep, causes several anomalies to the typical Caribbean flora during the dry seasons. These include blooms of benthic macroalgae, an increase in species number, growth 2–4 times greater than that for the same species in other Caribbean areas, atypical zonation of several species, an absence of certain characteristic genera and species of the Caribbean (*Penicillus, Turbinaria, Cymopolia*) and the presence of certain subtropical and warm temperate genera and species (*Plocamium, Porphyra, Acrosorium, Ectocarpus confervoides, Dictyopteris hoytii*) (Bula-Meyer, 1977, 1989–1990, 1985).

Several other correlations between upwelling and biota have been identified (Antonius, 1972; Werding and Erhardt, 1976; Guillot and Márquez, 1975; Meyer and Macurda, 1976; Caycedo, 1977; Alvarez-León et al., 1995). The existence of only small, coastal fringing reef in the internal parts of bays and inlets, and the lack of true coral reefs near Santa Marta and Guajira have been attributed in part to the cold temperature generated by the upwelling.

During rainy seasons, coastal upwelling recedes, due to a weakening of the northeast tradewinds. The Colombian Countercurrent can then reach Tayrona National Park carrying substantial amounts of suspended sediments and nutrients from the Magdalena River. This generates phytoplankton blooms, water turbidity and decreases in salinity by 2–4 ppt relative to upwelling values (Bula-Meyer, 1985). Paradoxically, the biomass of most macroalgal populations decreases dramatically, probably due to temperature increase as hypothesized by Bula-Meyer (1989–1990).

SHALLOW WATER MARINE ECOSYSTEMS AND COASTAL HABITATS

The Colombian Caribbean Sea contains most types of marine ecosystems and habitats of the Western Tropical Atlantic. These include, coral reefs, seagrass beds, subtidal algal beds, mangrove wetlands, estuaries, coastal lagoons, sandy and muddy subtidal and intertidal flats and rocky shores. The region (both marine and terrestrial) includes eight National Parks and five other protected areas, sanctuaries or forest reserves covering approximately 2000 km² and including all of the above-mentioned marine ecosystems as well as different terrestrial ones (Alvarez-León, 1989; Sánchez-Páez and Alvarez-León, 1997).

Coastal Morphology

Based on CORPES-CA (1992) and Molina et al. (1998), the coastal and marine areas of the Colombian Caribbean can be categorized as having the following sectors or provinces (Fig. 1):

(1) *Guajira Peninsula*: from the mouth of the Palomino river to the northern end of the Guajira Department, the coastal zone is mainly desert. The northern section of coast is very rocky, while the middle and southern sectors are

made of continental and marine clastic deposits. The absence of topographical barriers allows for strong wind and wave action on the coastal edge. Hurricane tails and periodic strong waves also contribute to erosion of this coast. It is a very wide continental platform with extensive macroalgal beds (Schnetter, 1981) with some endemic marine fauna (Díaz, 1990).

(2) *Sierra Nevada de Santa Marta*: from the mouth of the Palomino River to the northeast of Ciénaga city the coast is made of igneous and metamorphic rocks with a hilly and mountainous topography, some of which extends into the sea forming small bays and inlets with sandy beaches, seagrass beds and narrow fringing coral reefs. The continental platform is very narrow, due to which accretional sedimentary processes are absent. There is low vulnerability to wave action and erosional processes due to strong lithological consistency.

(3) *Magdalena River Delta*: from Ciénaga city to Bocas de Ceniza–mouth of the Magdalena River there is a low coastal zone, subject to floods and made of terrigenous silt, clay and sands. The large sand bar "Isla de Salamanca" contains a variety of environments such as mangroves, salt and flood plains, coastal lagoons, sand dunes and beaches. It is very vulnerable to wave and wind action, and is subject to erosional processes. The continental platform and slope is traversed by submarine canyons probably built by the old mouth of the Magdalena River.

(4) *Central Caribbean*: from Bocas de Ceniza–mouth of the Magdalena River to Cartagena, tertiary sedimentary rock hills alternate with fluvial quaternary deposits. There is considerable sediment movement from the mouth of the Magdalena River. The continental platform, made of terrigenous sands and sandy silts, ranges in width from 5 to 30 km. The littoral zone is dominated by sandy beaches subjected to strong wave action and interrupted by rocky cliffs and sand bars across the entrances of coastal lagoons such as Ciénaga de la Virgen. Mud diapirism is common.

(5) *Cartagena–Punta de Piedra and Isla Fuerte*: this is a coast of tertiary sedimentary rocks mixed with different quaternary deposits. Two important deltas are located in this sector: Canal del Dique is a river-dominated delta protected from wave and tidal action by the bay of Barbacoas. The Tinajones Delta (Sinú river) is formed both by wave and fluvial energy. In this sector, large accretion processes are due to fast development of deltas. The continental platform is covered by terrigenous sediments and coral reefs, the latter being found around the archipelagos of Islas del Rosario, Islas de San Bernardo and Isla Fuerte.

(6) *Punta de Piedra (Isla Fuerte)–Punta Caribana*: this area is lithologically constituted by lodolites and turbidites. Erosion is critical in this area and affects several human settlements. The continental platform is wide here and is made of terrigenous sediments. Mud diapirism is common both on the platform and on the continent.

(7) *Urabá Gulf*: the landscape of this region is defined mainly by alluvial plains and mangrove forests. The gulf

Fig. 2. Bolivar Cay in the oceanic archipelago.

shows the largest sedimentary accretion found on this Caribbean coast and is related to the rapid progradation of the Atrato River through its deltaic system.

(8) *Darien*: this is a mountainous zone made mainly by tertiary volcanic rocks organized into discontinuous cliffs, separated by alluvial valleys. There is very low coastal erosion. The continental platform is made of terrigenous silts and bioclastic sands with coral formations along the rocky littoral zone.

(9) *Island areas*: islands include the oceanic islands San Andrés and Providencia, with cays and banks, the whole forming one extended archipelago (Figs. 1 and 2). The origins of the islands and banks are associated with volcanic activity and exuberant coral reef development. Reefs are the dominant factor in the present marine environment of this area, forming wide barriers and rings on reduced platforms around the islands. The complex processes of growth, erosion, sedimentation and alteration of current and wave patterns of coralline formations, have shaped the submarine relief and promoted the presence of other associated environments such as reef lagoons, seagrass beds and sand flats with calcareous algal beds.

Coral Reefs

The largest and most important coral reef formations are located in the oceanic area of the Colombian Caribbean around the islands and cays of archipelago of San Andrés and Providencia, more than 700 km offshore, north of the Colombian continental coast and to the east of the Nicaraguan platform (Fig. 1). Coral formations include two barrier reefs offshore from the two major islands, San Andrés and Providencia, as well as patch and fringing reefs, atolls and banks (Díaz et al., 1996). The coralline system around these islands is one of the largest in the Atlantic (approximately 5000 km^2 including coral reefs, seagrass beds and sedimentary bottoms) and the reefs support approximately 45 species of scleractinian corals, 163 macroalgal species, 118 species of sponges and 40 gorgonaceans (Diaz et al., 1996). Recent studies (Díaz et al., 1995; Díaz et al., 1996; Garzón-Ferreira and Kielman, 1993; Garzón-Ferreira et al., 1996; Garzón-Ferreira, 1997) indicate a considerable and general degradation of coral reefs in this area, similar to that occurring in many other parts of the Caribbean Sea. Degradation is observed not only in reefs subjected to intense anthropogenic activity (San Andrés is one of the most densely populated islands in the world with 1548 inhabitants km^{-2}) but is also occurring in remote and isolated reefs of the archipelago complex (Garzón-Ferreira, 1997).

Due to substantial river discharge and the dominance of sedimentary bottoms, coral reefs are scarce along the continental platform (Garzón-Ferreira, 1997). Less than 10% of this continental fringe has coralline development. The largest and most developed coral formations are found in the southwest (Table 1) around the archipelago of Islas del Rosario (southwest of Cartagena), Islas de San Bernardo (across Golfo de Morrosquillo) and Isla Fuerte (Fig. 3). In Golfo de Urabá, fringing reefs are limited to the northwestern coast close to the Panamá border, due to fresh water and sediment outflow from the Atrato river. In general, the southwestern zone of the continental Caribbean is influenced by periodic outflows of freshwater, nutrients and sediments through the Atrato, Sinú and Magdalena river mouths, as well as by increasing tourism and other anthropogenic actions which affect the health of coral ecosystems. In the northeastern zone (Table 1) of the continental Caribbean coast, coral formations are found mainly along the Santa Marta and Tayrona National Park coastline. Here, cold upwelling waters, lower water transparency and hard substrata which is generally limited to approximately 25 m depth, restrict coral formations to small, shallow fringing reefs (Garzón-Ferreira and Cano, 1991). These fringing reefs also suffer from anthropogenic activities in the coastal zone such as sewage discharge, tourism and blast fishing, as well as freshwater, sediment and nutrient riverine outflow during the rainy seasons. Coral formations along the Guajira platform are limited to small patch and fringing reefs in restricted locations (Solano, 1994).

Almost all symptoms and factors relating to coral reef degradation which have been reported in the literature, have been observed during the last 20 years in the Colombian Caribbean (Garzón-Ferreira, 1997). Live coral coverage has decreased to an average level of 20–30% of available hard substrata. Massive coral mortality has occurred, as in other Caribbean areas, during the late 1970s and 1980s. During the 1990s, some degradation continued to occur, although massive episodic mortalities have not been reported except for the die-off of *Gorgonia ventalina* about a decade ago (Garzón-Ferreira and Zea, 1992; Garzón-Ferreira and Kielman, 1993; Nagelkerken et al., 1997a,b).

Coral reef health monitoring in Colombia began in 1992 when the marine research centre INVEMAR joined the CARICOMP international program and implemented a permanent monitoring site in the continental Caribbean coast. Monitoring sites have been extended now to two other localities, one in the area of Islas del Rosario and the other one in the oceanic reefs of the San Andrés and Providencia archipelago (J. Garzón, unpublished report).

Seagrass Beds

Although not very well studied, seagrass beds are found throughout large portions of the Colombian Caribbean, from the Guajira Península in the north, to the Golfo de Urabá in the southwestern limits, as well as around islands and cays in the oceanic archipelago of San Andrés and Providencia. The most conspicuous beds are located in Bahía Portete and Cabo de la Vela (Guajira department), bays and inlets of Tayrona National Park and Santa Marta area (Magdalena department), around Islas del Rosario in the Cartagena area, Islas de San Bernardo in the Golfo de Morrosquillo, Golfo de Urabá and areas around the islands and cays of the archipelago de San Andrés y Providencia. They are absent in areas influenced by strong freshwater discharges from large rivers (Magdalena, Sinú and Atrato). *Thalassia testudinum* is the most abundant and frequent species, found generally to a depth of 10 m. Other common species are *Syringodium filiforme, Halodule wrightii, Halophila decipiens, and Halophila baillones* (Laverde, 1994).

Important resources associated with seagrass beds (turtles, conchs, sea stars) have been depleted by man. Several beds have been damaged due to activities associated with urban, industrial or tourism development such as dredging or landfill in the Santa Marta area while others, far from urban centres, are well preserved (as in Tayrona National Park, Rosario islands, Guajira Peninsula and San Andrés and Providencia Archipiélago).

Macroalgae

This coast is rich in macroalgae, with approximately 450 species out of about 600 reported in the tropical and subtropical American Atlantic (Bula-Meyer, 1990b). This high diversity can be attributed in part, to old environmental

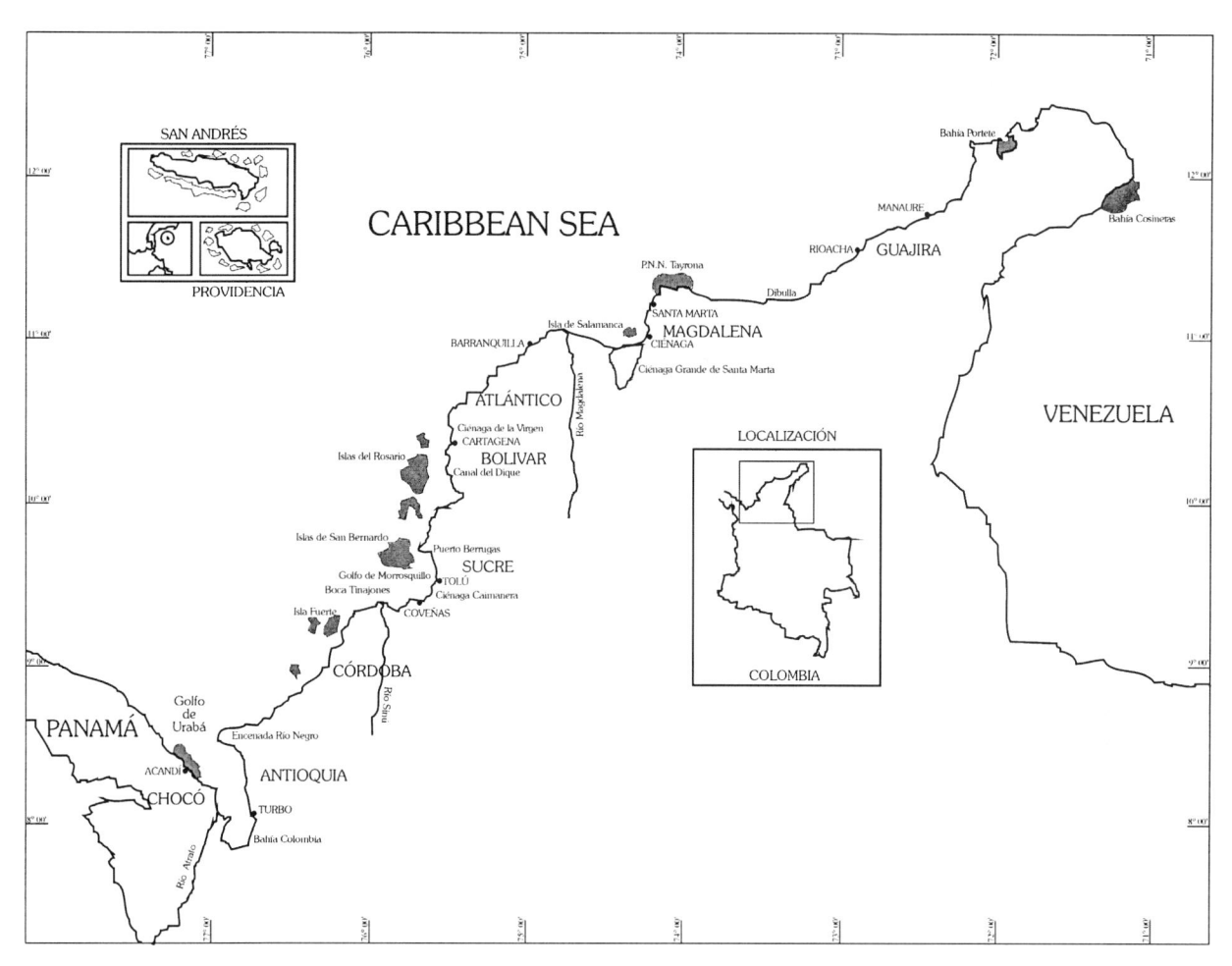

Fig. 3. Location of main coral and seagrass ecosystems.

diversity and stability (Bula-Meyer, 1990). The greatest species richness is found in the Santa Marta and Tayrona National Park area. The latter presents several outstanding factors relating to macroalgae, such as the presence of an endemic genus of brown alga, *Cladophyllum* (Bula-Meyer, 1980), and very large sizes in certain genera such as *Sargassum* sp. which can reach heights of 3–4 m in beds of several thousand square metres. Islas del Rosario, Islas de San Bernardo, Guajira Peninsula and the oceanic islands San Andrés and Providencia also show rich macroalgal flora.

Mangroves

Approximately 1530 km^2 of the Colombian Caribbean coast are covered by mangrove forests (Sánchez-Páez et al., 1997) composed mainly of *Rhizophora mangle*, *Avicennia germinans* and *Laguncularia racemosa*. In some areas, *Conocarpus erecta* and *Pelliciera rhizophorae* are also present. Generally, tree-height does not exceed 25 m due to predominantly arid conditions. The largest mangrove wetlands are found in the deltas of the Magdalena (including Canal del Dique), Sinú and Atrato rivers (Fig. 4). Smaller mangrove wetlands are found in the islands of San Bernardo and Rosario as well as in Ciénaga de la Virgen (close to Cartagena), Tayrona National Park, Bahía Portete (in the Guajira Peninsula) and Old Providence Island. Rehabilitation of mangroves is an important activity here.

OFFSHORE SYSTEMS

Importance of Fisheries

In the southwestern sector, river discharge supplies considerable nutrient enrichment into the pelagic environment, although turbidity associated with riverine waters prevents high photosynthesis of phytoplankton (CORPES-CA, 1992). The discharge of the Magdalena river is the most important continental source for fertilization of surface waters of the Colombian Caribbean, affecting an area larger than the Exclusive Economic Zone (CORPES-CA, 1992). The mixture of outwelled nutrients with marine waters causes the formation of "fronts" with higher pelagic productivity in which planctivorous organisms concentrate. There is not much information on the pelagic resources of the southern Colombian Caribbean although some records exist of the presence and artisanal catches of pelagic species, specifically *Tarpon atlanticus*. However, recently this population has been severely depleted due to destruction or

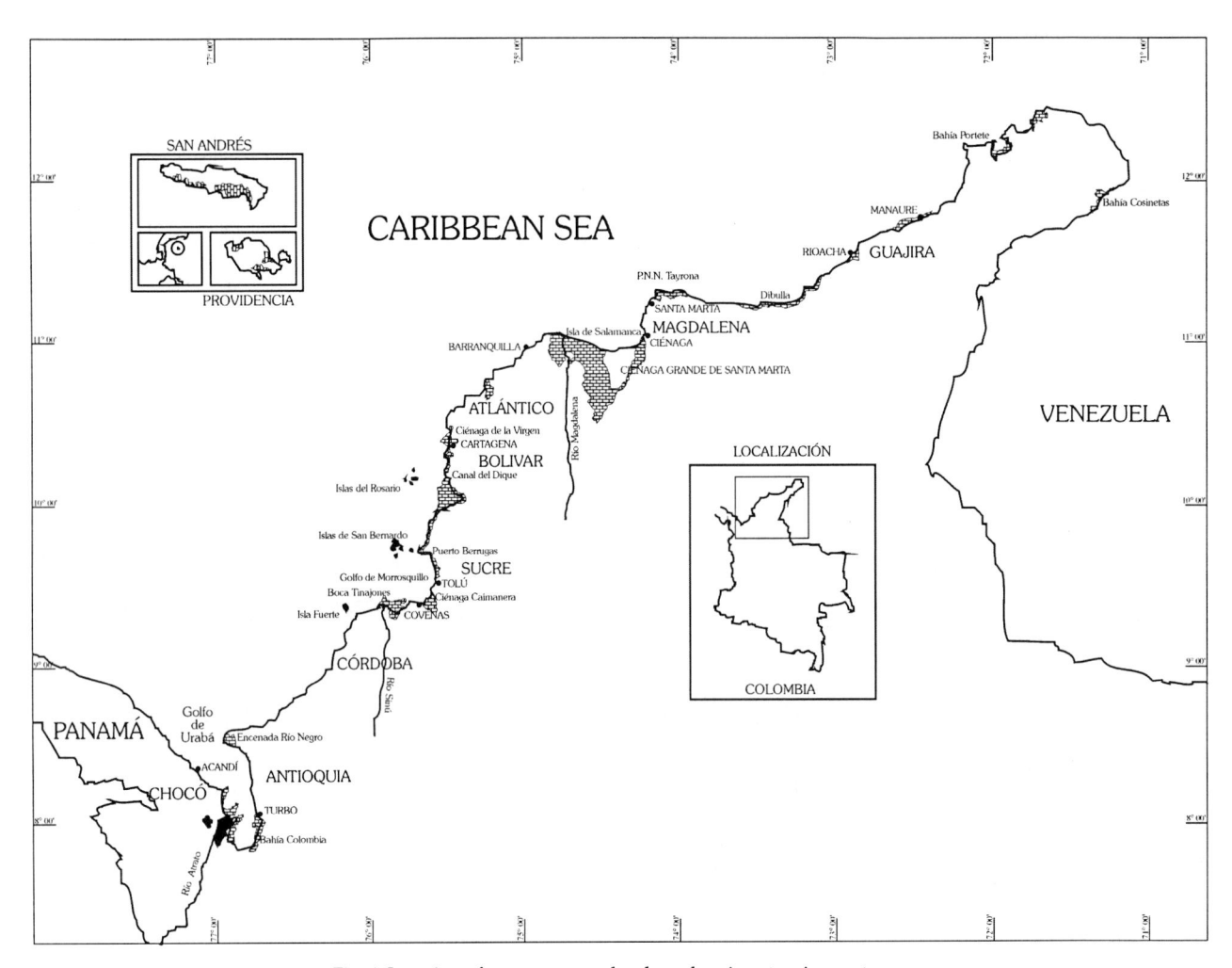

Fig. 4. Location of mangrove wetlands and main estuarine systems.

deterioration of coastal lagoons which were nursery areas for this species. Also common along this southern pelagic zone are sierras, carites and macabí (*Scomberomorus maculatus, S. ragalis, Caranx hippos, C. latus, C.crysos* and *Elops saurus*) (CORPES-CA, 1992). These species are not subjected to large-scale fisheries but are mostly captured by local, artisanal fishermen. Among small pelagics, *Opisthonema oglinum*, is abundant and widely distributed. Demersal resources are represented mainly by shrimps of the genus *Penaeus*, on which an important fishing industry is based.

In the northeastern sector, fish resources are mostly related to upwelling as described above. Small pelagics of the family Clupeidae are thought to be of potential importance for fisheries, mainly *Opisthonema oglinum, Sardinella anchovia* and *Harengula clupeola*. Associated with these planctivorous species are some relatively abundant predators, such as *Thunnus albacares, T. atlanticus, Auxis thazard, Scomber colias, S. japonicus, Decapterus macarellus, D. punctatus, Euthynnus alleteratus, Coryphaena hippurus* and *Elegatis bipinnulata*. Squids, such as *Loligo plei, L. pealei* and *Illex* sp. are another potential resource for the fisheries in this area although almost no research has been done on their populations (CORPES-CA, 1992; Alvarez-León et al.,

1995). Demersal resources mainly include shrimps of the genus *Penaeus* (*P. duorarum, P. brasiliensis, P. notialis* and *P. subtilis*), snappers and pagoras, mainly *Lutjanus synagris* and *Micropogonias furnieri*, with maximum catches from Cabo de la Vela in the Guajira Peninsula (Alvarez-León and Lesser-Mehr, 1986).

Exploratory and commercial fishing cruises show larger catches per unit (both in weight and size of the individuals) in this northern area compared to the southwestern zone.

In general, industrial fisheries use towing vessels of the Florida type, with escape mechanisms for marine turtles. The most exploited resources since the 1970s are shrimps (*Penaeus notialis, P. brasiliensis, P. schmitti* and *Xiphopenaeus kroyeri*) from the shallow continental platform, although since the 1980s catches have decreased considerably. Shrimps are packed for export under strict quality control in processing plants in Cartagena. Demersal fishes are caught mainly as a by-product (ranfaña) of the shrimp fishing activity although a few vessels specialize in fish catches, mainly snappers, groupers, corvinas and sharks. Lately, tuna (*Thunnus atlanticus, T. thynnus, Euthynnus alleteratus*), fished with long lines and purse seines, has become a very important product for export to the United States,

Mangroves in the Magdalena River Delta

The Magdalena river delta comprises a very large, estuarine–lagoonal system (1280 km^2) known as Ciénaga Grande de Santa Marta. It is the largest of its kind in the Caribbean area and historically is the main source of fish and shellfish for the north coast of Colombia. Extensive mangrove forests covered 520 km^2 of this deltaic area up to the late 1960s. Three stilt villages are located in the two main lagoons of the system (Fig. 1). Anthropogenic activities during the last 40 years, mainly alterations to the hydrological regime through freshwater diversion, construction of roads, dikes and berms, resulted in salinisation of mangrove soils to salinities higher than 100 ppt for more than six months a year (Botero, 1990).

As a consequence of soil hypersalinisation, massive forest degradation started in the late 1960s and, up to the present day, almost 70% of the original forest is dead (Gónima et al., 1996). Besides mangrove mortality, a reduction in the diversity and abundance of fish, bird and invertebrate fauna has been reported, as well as contamination of water, sediments and organisms with pesticides, heavy metals and pathogenic bacteria (Botero and Mancera-Pineda, 1996). A progressive and significant increment in water salinity, seston concentration and eutrophication has also been documented to 1994 (Botero and Mancera-Pineda, 1996; Mancera-Pineda and Vidal, 1994)). During the last six years (1994–1999) the Colombian government has been implementing a rehabilitation project for the area, mainly through diversion of fresh water back into the system, management of forest and fisheries resources and social development programmes. Fresh water from the Magdalena river is flowing back into the system (approximately 163 m^3/s) and mangrove regeneration is clearly visible (Botero and Salzwedel, 1999). Several institutions involved in research and environmental management are currently monitoring the mangrove rehabilitation in terms of forest structure, nutrient exchanges, fish and shellfish populations and biodiversity. The Ministry of the Environment has been supporting and enhancing several other projects in which diverse planting methods are being devised for restoration of small-scale degraded areas. Hypocotyls, seedlings, shoots and stems, from natural forests and from experimental green houses, are obtained and planted with the active participation of the local native communities.

Fig. 1. Stilt village in the mangrove–lagoonal system Ciénaga Grande de Santa Marta.

European Union and Japan. Other abundant species caught with tuna are *Coryphaena hippurus, Tetrapterus albidus, Makaria nigricans, Xiphias gladius* (sword fish), *Scomberomorus cavalla* (sawfish) and several shark species such as *Galeocerdo cuvieri, Prionace galuca, Carcharodon carcharias* and *Isurus oxyrhychus*.

POPULATIONS AFFECTING THE AREA

According to the latest census, the Colombian Caribbean population reached 7,088,990 in 1993, which is 19.75% of the total Colombian population. Coastal inhabitants number approximately 2,800,000. The largest and most important population centres are the cities of Barranquilla, Cartagena and Santa Marta, where most tourism, industrial and port activities take place. The population on San Andrés and Providencia Archipiélago is approximately 50,000, San Andrés Island having one of the highest population densities (1850 inhab/km²) in the Caribbean basin.

Fishing, tourism, ports and navigation (recreational and commercial) are the main activities in the marine areas, but since 1985 shrimp aquaculture has been developing steadily. Through several large rivers, the Caribbean coastal and marine areas receive many of the run-off substances and environmental impacts occurring inland, including from the Andean region which has a large human population and intensive urban and industrial activities. The Magdalena River alone discharges annually approximately 138 million tons of sediment into the Caribbean basin together with considerable amounts of nutrients, fertilizers and pesticides.

RURAL FACTORS

Artisanal Fishing

Artisanal fishing is an important activity sustaining inhabitants of small towns and villages located around the most productive ecosystems of the coast. Recent statistical information of fish catches indicates a decrease in fish resources mainly due to overfishing and habitat degradation (Sánchez-Páez et. al., 1997). Most fish products are destined for national or local (family) and tourism consumption and are sold in the cities of Cartagena, Barranquilla and Santa Marta. Mangroves and coastal lagoons are fished mainly for oysters (*Crassostrea rhizophorae*), mullets (*Mugil incilis, M. liza*), mojarras (*Eugerres plumieri, Gerres cinereus, Diapterus rhombeus*), catfish (*Arius proops., Cathorops spixii*), juvenile

> ### Sierra Nevada de Santa Marta
>
> Although not a marine ecosystem, Sierra Nevada de Santa Marta (SNSM), the highest coastal mountain in the world with a height of 5775 m a.s.l., exerts an important influence on the coastal zone of the Colombian Caribbean, especially around the area of Santa Marta, Tayrona National Park, Ciénaga Grande de Santa Marta and part of the Guajira Península. SNSM is very rich and diverse in terrestrial and freshwater ecosystems and is considered a Biosphere Reserve. At least 15 rivers originate in the ice caps of this mountain, draining different types of highlands, forests and agricultural lands and finally flowing into the Caribbean Sea or the estuarine lagoon Ciénaga Grande de Santa Marta. Due to colonization and related activities in the Sierra, these rivers carry increasing amounts of sediments, pesticides and organic matter which eventually reach the marine or estuarine zones. The very abrupt coastal topography of Tayrona Park and Santa Marta area is a direct consequence of the presence of this coastal mountain (Hernández-Camacho et al., 1998).

tarpons (*Tarpon atlanticus*) and snooks (*Centropomus undecimalis, C. ensiferus*), crabs (*Callinectes boucurti, C. sapidus*) and shrimps (*Penaeus* spp. and *Xiphopeneus kroyeri*). Red snappers (family Lutjanidae), groupers (family Serranidae), lobsters (*Panulirus argus*) and queen conch (*Strombus gigas*) are important resources associated with coral reefs, rocky bottoms and seagrass beds around the reefs of Islas del Rosario and Islas de San Bernardo, on the platform of the Guajira Peninsula (which is well known for its lobster catches) and from the oceanic islands (Alvarez-León and Lesser-Mehr, 1986; Gutièrrez and Valderrama, 1994).

A variety of artisanal fishing methods are used ranging from line hooks, gill nets, dragnets, casting nets and traps, to more harmful ones such as blast fishing, bolicheo and zangarreo. In the latter, areas are surrounded with gill nets and the water is hit with poles or agitated with an outboard engine to frighten fish or generate anoxic conditions due to removal of anoxic bottom sediments. This is done in coastal lagoons and mangrove areas. These methods obviously have consequences both for the environment and the associated living resources. Although blast fishing, bolicheo and zangarreo are illegal fishing practices, local authorities are unable to control them.

Aquaculture

Shrimp aquaculture started here in 1985, and today shrimp ponds with *Penaeus vannamei* and *P. stylirostris* cover a total area of approximately 1900 ha (19 km^2) in the southwest. No mangrove areas have been affected by this as farms have been located on sand flats and on high grounds or terrestrial lands. A total annual production of 6000 tons is exported to the United States, Europe and Japan. The productivity of Colombian shrimp mariculture is relatively high, approximately 2000 kg/ha/yr (Aguilera, 1999). In the last four years, shrimp mariculture has made important scientific and technological advances related to disease prevention and control. Through family selection procedures, the industry has now specific virus-resistant groups from which all larvae used by farmers to seed their ponds is now being produced. This means that Colombia is now self-sufficient regarding shrimp larvae production and can thus avoid import from other countries or wild catches, both of which are known sources of viral diseases to cultured shrimp.

Fish mariculture is still at the research and laboratory level with a few projects presently aimed at inducing reproduction in red snapper (*Lutjanus analis*) or on feeding and growth rate experiments with wild-caught snapper and grouper juveniles. Bivalve (oysters and scallops) mariculture has been developed at a research scale, with good prospects for artisanal or industrial applications, especially with *Crassostrea rhizophorae*, based on seed collection in the wild (H. Rodríguez, unpublished information).

Land Uses

According to IGAC (Instituto Geográfico Agustín Codazzi) 18% of the land along the Colombian Caribbean has agricultural potential, but various limitations mean that only 4% is economically suitable. Around 38% of the land is capable of forestry, but most of it should be dedicated just to conservation. Although only 32% of the land is fit for rearing cattle, the reality is that almost 72% of the land in the Caribbean region of Colombia is used primarily for this activity (CORPES-CA, 1992) with obvious environmental consequences to soil, forest and fresh water. Large palm oil, banana and other fruit plantations in the foothills of the Sierra Nevada de Santa Marta contribute to fertilizer and pesticide runoff into the coastal zone, especially into the Ciénaga Grande de Santa Marta (Plata et al., 1993). Increasing human settlement on this mountain, combined with agricultural activities and tree felling, promotes erosion around river beds, thus increasing sediment loads arriving at the coastal zone. Although legislation and measures are being implemented, to control anthropogenic activities in the middle and highlands, there is still much to be done in order to rehabilitate forests and watersheds on this large and important coastal mountain.

COASTAL EROSION

Table 2 gives quantitative information on the degree of erosion and accretion in the littoral zones. Of the 1600 km of Colombian Caribbean coastline, 12% shows accretion, 16% strong erosion and 72% is stable or variable. Highest accretion is linked to areas where deltas, sand spurs and large beaches are being formed. The greatest erosional processes (in Guajira, Isla de Salamanca, Punta Piedra-Punta Caribana

Table 2
Degree of erosion or accretion in the littoral zone of the Colombian Caribbean

Province	Erosion		Accretion	
	km	%	km	%
I. Guajira Península	156	56	22	11
II. Sierra Nevada de Santa Marta	14	5	7	3
III. Magdalena River Delta	28	10	0	0
IV. Central Caribbean	11	4	24	12
V. Cartagena–Punta de Piedra (Isla Fuerte)	18	6	55	26
VI. Punta de Piedra (Isla Fuerte)–Punta Caribana	17	6	17	8
VII. Urabá Gulf	29	11	82	40
VIII. Darien	5	2	0	0

and Urabá Gulf) with possible consequences to industrial, urban or tourist areas, are mainly related to natural factors such as wave and wind energy, subsidence and/or sea-level rise, or to structures such as jetties, wave breaks or sites of sand and gravel extraction along beaches and river beds.

EFFECTS FROM URBAN AND INDUSTRIAL ACTIVITIES

Tourism

The Caribbean coast of Colombia is presently the main tourist area in the country. Although some ecotourism has been developing in recent years, the truth is that a large part of the coastal and marine habitat degradation can be attributed to disorganised and uncontrolled tourism activities. Harmful effects can be observed mainly in beach and coral reef areas close to main population centres where garbage accumulation, coral extraction and destruction, sediment and sewage discharges have had a significant effect. Governmental agencies in charge of environmental regulation are increasingly trying to protect the existing National Parks and Reserves as well as to increase their area, especially highly productive and diverse systems such as mangroves, coral reefs and coastal lagoons.

Port Activities

Significant traffic goes through ports on this coast. In the last ten to fifteen years, coal exports from mines in the Guajira Península and inland have significantly increased in number and size. Unfortunately, port construction is being done without adequate planning and without taking into account other uses of the coastal and marine zone. Coal export activities do not coexist well with tourism, fishing and urbanization, and have a negative impact on the tourism industry as a result of coal dust emissions in the air, water and beaches, as well as increasing cargo shipping close to tourist areas.

Crude oil is exported through a large marine facility in Golfo de Morrosquillo. Until 1998, crude oil was pumped to a large static ship tanker located in the middle of the Gulf from where oil would be distributed to ships. The static tanker was recently removed from the Gulf and oil is now pumped directly to ships entering and leaving the area periodically. So far, only small oil spills (100–200 gallons) have occurred as a consequence of these operations.

Urban Development

Several areas of this coast have poor sanitary conditions due to discharges of raw domestic sewage into the sea. No sewage treatment plants are available in the main urban centres. An estimated 26,300 tons/yr of sewage enters the Colombian Caribbean from urban and municipal areas. In addition, the sea receives, via several large rivers, many products from inland towns and industries, mainly in the Andean region with its large human population and intensive industries. The Magdalena River alone discharges annually approximately 138 million tons of sediment into the Caribbean basin, together with considerable amounts of nutrients, fertilizers, pesticides and heavy metals.

PROTECTIVE MEASURES

There are many regulations regarding natural resources and environments but only recently has the government started to develop specific policies and plans for integrated coastal zone management in which natural as well as social, economic and cultural conditions are taken into account. Probably the most noteworthy case is that of the rehabilitation programme of Ciénaga Grande de Santa Marta (PROCIENAGA, 1995) in which the national and local government have invested considerable amounts of financial resources and effort. It already shows positive signs of mangrove recovery. The programme has also led to preliminary fisheries management and control while convincing the native fishermen that the only way to guarantee a sustainable resource is to implement management measures and strategies. The National Parks Authority, together with the regional environmental agencies, are increasingly controlling impacts, especially in coral reef and mangrove areas through stricter enforcement of regulations. Several of the most polluting industries located around the Bay of Cartagena have been required either to decrease their pollutant load into the Bay or to close down.

REFERENCES

Aguilera, M.M. (1999) Los cultivos de camarón en el Caribe colombiano. *Aguaita Uno*, 24–38.

Alvarez-León, R. and Lesser-Mehr, E.S. (1986) Aspectos sobre el reclutamiento delosrecursos demersales en las costas colombianas. In A. Yañez-Arancibia and D. Pauly (eds.), IOC/FAO Workshop on recruitment in tropical coastal demersal communities, Rep. 44, pp. 107–122. Mexico.

Alvarez-León, R. (1989) Los ecosistemas marinos del Caribe colombiano. Bull. Inst. Geol. Bassin d'Aquitaine, Bordeaux **45**, 131–143.

Alvarez-León, R. (1993) Mangrove Ecosystems of Colombia. In: Lacerda, L.D. (ed.) Conservation and sustainable utilization of mangrove forests in the Latin America and Africa regions. pp. 75–113. ITTO/ISME Project PD 114/90 (F). 272 pp.

Alvarez-León R., Aguilera-Quiñones, J., Andrade-Amaya, C.A. and Nowak, P. (1995) Caracterización general de la zona de surgencia en la Guajira colombiana. Rev. Acad. Colombiana de Ciencias Exactas Fisicas y Naturales **75**(19), 679–694.

Antonius, A. (1972) Occurrence and distribution of stony corals (Anthozoa and Hydrozoa) in the vicinity of Santa Marta, Colombia. Mitt. Inst. Colombo-Alemán Invest. Cient. **6**, 89–103.

Botero, L. (1990) Massive mangrove mortality in the Caribbean coast of Colombia. Vida Silvestre Neotropical **2** (2), 77–78.

Botero, L. and Mancera-Pineda, E. (1996) Síntesis de los cambios de origen antrópico ocurridos en los últimos 40 años en la Ciénaga Grande de Santa Marta (Colombia). Revista de la Academia Colombiana de Ciencias Exactas, Físicas y Naturales **20** (78), 465–474.

Botero, L. and. Salzwedel, H. (1999) Rehabilitation of the Cienaga Grande de Santa Marta, a mangrove–estuarine system in the Caribbean coast of Colombia. Ocean and Coastal Management **42**, 243–256.

Bula-Meyer, G. (1977) Algas marinas bénticas indicadoras de un área afectada por aguas de surgencia frente a la costa Caribe de Colombia. An. Inst. Inv. Mar. Punta Betín **9**, 45–71.

Bula-Meyer, G. (1980) *Cladophyllum schnetteri* a new genus and species of Sargassaceae (Fucales, Phaeophyta) from the Caribbean coast of Colombia. Bot. Marina **23**, 555–562.

Bula-Meyer, G. (1985) Un nuevo núcleo de surgencia en el Caribe colombiano detectado en correlación con las macroalgas. Boletin Ecotrópica **12**, 3–25.

Bula-Meyer, G. (1989–1990) Altas temperaturas estacionales del agua como condición idsturbadora de las macroalgas del Parque Nacional Tayrona, Caribe colombiano. An. Inst. Inv. Mar. Punta Betín **19/29**, 9–22.

Bula-Meyer, G. (1990a) Oceanografía. In *Caribe Colombia*, ed. M.C. Jimeno, pp. 100–114. Fondo José Celestino Mutis–FEN, Colombia. 270 pp.

Bula-Meyer, G. (1990b) Macroflora marina. In *Caribe Colombia*, ed. M.C. Jimeno, pp. 135–154. Fondo José Celestino Mutis–FEN, Colombia. 270 pp.

Caycedo, I.E. (1977) Fitoplancton de la Bahía de Nenguange (Parque Nacional Tayrona), Mar Caribe, Colombia. An. Inst. Inv. Mar. Punta Betín **9**, 17–44.

CORPES-CA (1992) El Caribe colombiano: realidad ambiental y desarrollo. Rapidoffset Ltda. Santafé de Bogotá. 275 p. + anexos.

Corredor, J.E. (1977) Aspects of phytoplankton dynamics in the Caribbean Sea and adjacent regions. FAO Fish. Rep., 200, pp. 101–104.

Corredor, J.E. (1979) Phytoplankton response to low level nutrient enrichment through upwelling in the Colombian Caribbean Basin. Deep-Sea Research **26**, 731–741.

Diaz, J.M. (1990) Malacofauna subfósil y reciente de la Bahía de Portete, Caribe colombiano, con notas sobre algunos fósiles del Terciario. Boletin Ecotrópica **23**, 1–22.

Díaz, J.M., Garzón-Ferreira, J. and Zea, S. (1995) Los arrecifes coralinos de la Isla de San Andrés (Colombia): estado actual y perspectivas para su conservación. Acad. Colomb. Cien. Exac. Fis. Nat., Colec. Jorge Alvarez Lleras **7**, 150 p.

Díaz, J.M., Díaz-Pulido, G., Garzón-Ferreira, J., Geister, J., Sánchez, J.A. and Zea, S. (1996) Atlas de los arrecifes coralinos del Caribe colombiano. I. Complejos Arrecifales Oceánicos. INVEMAR, Santa Marta, Serie de Publicaciones Especiales no. 2.

Donoso, M.C. (1990) Circulación de aguas en el Mar Caribe. Memorias VII Seminario Nacional de Ciencias y Tecnologías del Mar, Cali. pp. 345–356.

Fajardo, G.E. (1979) Surgencia costera en las proximidades de la Península de la Guajira. Bol. Cient. CIOH **2**, 7–19.

Garzón-Ferreira, J., Zea, S. and Díaz, J.M. (1996) Coral health assessment in four southwestern Caribbean atolls. Abstr. 8th. Intern. Coral Reef Symp., Panamá: 68.

Garzón-Ferreira, J. and Zea, S. (1992) A mass mortality of *Gorgonia ventalina* (Cnidaria: Gorgoniidae) in the Santa Marta area, Caribbean coast of Colombia. Bull. Marine Science **50** (3), 522–526.

Garzón-Ferreira, J. and Cano, M. (1991) Tipos, distribución, extensión y estado de conservación de los ecosistemas marinos costeros del Parque Nacional Natural Tayrona. Manuscr. VII Concur. Nal. Ecol., FEN/INVEMAR, Bogotá/Santa Marta (Colombia), 82 p.

Garzón-Ferreira, J. and Kielman, M. (1993) Extensive mortality of corals in the Colombian Caribbean during the last two decades. In *Proc. Colloq. Global Aspects of Coral Reefs: Health, Hazards and History*, ed. R. Ginsburg, pp. 247–253. RSMAS/Univ. Miami.

Garzón-Ferreira, J. (1997) Arrecifes coralinos: un tesoro camino a la extinción? Colombia Ciencia y Tecnología **15** (1), 11–19.

Giraldo, L. (1994) Estado actual del conocimiento de la oceanografía física del Caribe y Pacífico colombiano. In *Memorias del Taller de Expertos sobre el estado del Conocimiento y Lineaminetos para la Estrategia Nacional de Biodiversidad*, pp. 269–278. CCO/DNP/ENB Minca, Magdalena, Colombia, 311 p.

Gónima. L., Mancera-Pineda, J.E. and Botero, L. (1998). Aplicación de imágenes digitales de satélite al diagnóstico ambiental de un complejo lagunar estuarino tropical: Ciénaga Grande de Santa Marta, Caribe colombiano. INVEMAR, Santa Marta, Serie de Publicaciones Especiales no. 4.

Gordon, A.L. (1967) Circulation of the Caribbean Sea. Journal of Geophysical Research **72**, 6207–6223.

Guillot, G.H. and Márquez, G.E. (1975) Estudios sobre los tipos de vegetación marina bentónica en el litoral del Parque Nacional Tayrona, costa Caribe colombiana. Tesis Profesional. Depto de Biología, Univ. Nacional de Colombia, 116 pp.

Gutiérrez, B.F. and Valderrama, M. (1994). La pesca artesanal en Colombia. In: Memoria Oceanográfica, Conmemoración de los 25 años de creación de la Comisión Colombiana de Oceanografía, pp. 86–89. Boletín Especial CCO, Santafé de Bogotá, Colombia.

Hernández-Camacho, J.I., Sánchez-Páez, H., Rodríguez-Mahecha, J.U., Castaño-Uribe, C., Cano-Correa, M. and Mejía, I.Y. (1998) *El Sistema de Parques Naturales de Colombia–30 años: Espacios estratégicos y sagrados*. UAAESPNN-Ministerio del Medio Ambiente. Santafé de Bogotá, Colombia, 497 pp.

HIMAT (1969–1992) *Registros de diferentes estaciones del Caribe colombiano. Parámetros: precipitación, temperatura, caudales, transporte de sedimentos y mareas*. Instituto Colombiano de Meteorología y Adecuación de Tierras. Santafé de Bogotá, Colombia.

Laverde, C.J. (1994) Estado del conocimiento de las praderas de fanerógamas marinas en Colombia. In *Memorias del Taller de Expertos sobre el Estado del Conocimiento y Lineaminetos para la Estrategia Nacional de Biodiversidad*, pp. 132–141. CCO/DNP/ENB Minca, Magdalena, Colombia, 311 pp.

Mancera-Pineda, E. and Vidal, L.A. (1994) Florecimineto de microalgas relacionado con maortandad masiva de peces en el complejo lagunar Ciénaga Grande de Santa Marta (Colombia). An. Inst. Inv. Mar. Punta Betín **23**, 103–117.

Meyer, D.L. and Macurda, D.B. (1976) Distribution of shallow-water crinoids near Santa Marta, Colombia. Mitt. Inst. Colombo-Alemán Invest. Cient. **8**, 141–123.

Molina, L.E., Pérez, F., Martínez, J.O., Franco, J.V., Marín, L., González, J.L. and Carvajal, J.H. (1998) Geomorfología y aspectos erosivos del litoral Caribe colombiano. *Publicaciones geológicas especiales del INGEOMINAS* **21**, 1–74.

Nagelkerken, I., Buchan, K., Smith, G.W., Bonair, K., Bush, P., Garzón-Ferreira, J., Botero, L., Gayle, P., Heberer, C., Petrovic, C., Pors, L. and Yoshioka, P. (1997a) Widespread disease in Caribbean Sea Fans: I. Spreading and general characterisitics. *Proc. 8th Int. Coral Reef Symp.* **1**, 679–682.

Nagelkerken, I., Buchan, K., Smith, G.W., Bonair, K., Bush, P., Garzón-Ferreira, J., Botero, L., Gayle, P., Harvell, C.D., Heberer, C., Kim, K., Petrovic, C., Pors, L. and Yoshioka, P. (1997b) Widespread disease in Caribbean Sea Fans: II. Patterns of infection and tissue loss. *Mar. Ecol. Progr. Ser.* **160**, 255–263.

Perlroth, I. (1968) Distribution of mass in the near surface waters of the Caribbean. Nat. Oceanogr. Data Center Progress Rep., p. 72. Nov. 1–15.

Plata, J., Campos, N.H. and Ramírez, G. (1993) Flujo de compuestos organoclorados en las cadenas tróficas de la Ciénaga Grande de Santa Marta. *Caldasia* **2**, 199–204.

PROCIENAGA (1995) Plan de manejo ambiental de la subregión Ciénaga Grande de Santa Marta 1995–1998. CORPAMAG, INVEMAR, CORPES C.A., GTZ.

Restrepo, J.D. and Kjerfve, B. (2000) Water and sediment discharges from the western slopes of the Colombian Andes with focus on Río San Juan. *Journal of Geology* **108**, 17–33.

Sánchez-Páez, H. and Alvarez-León, R. (1997) Zonificación y categorías de manejo para las áreas silvestres costeras de Colombia: La representatividad de los ecosistemas de manglar. Taller sobre Areas Costeras y Marinas Protegidas CEPAL/UICN/FAO/GTZ/CORPAMAG/ PROCIENAGA del I Congreso Latinoamericano de Parques Nacionales y otras areas Protegidas. Santa Marta, Colombia.

Sánchez-Páez, H., Alvarez-León, R., Pinto-Nolla, F., Sánchez-Alférez, A.S., Pino-Rengifo, J.C., García Hansen, I. and Acosta-Peñaloza, M.T. (1997) Diagnóstico y zonificación preliminar de los manglares del Caribe en Colombia. In: Sánchez-Páez, H. y R. Alvarez-León (eds). Proy. PD 171/91 Rev. 2 (F) Fase 1 Conservación y manejo para el uso múltiple y el desarrollo de los manglares en Colombia, MMA/OIMT. Santa Fé de Bogotá D.C. (Colombia), 511 pp.

Schnetter, R. (1981) Aspectos de la distribución regional de algas marinas en la Costa atlántica de Colombia. *Rev. Acad. Col. Cien. Exact. Fis. Nat.* **15** (57), 63–74.

Solano, O.D. (1994) Corales, formaciones arrecifales y blanqueamiento de 1987 en Bahía Portete (Guajira, Colombia). *An. Inst. Invest. Mar. Punta Betín* **23**, 149–163.

Werding, B. and Erhardt, H. (1976) Los corales (Anthozoa e Hidrozoa) de la Bahía de Chengue en el Parque Nacional Tayrona (Colombia). *Mitt. Inst. Colombo-Alemán Invest. Cient.* **8**, 45–57.

Wust, G. (1964) *Stratification and Circulation in the Antillean Caribbean Basins. Part I.* Columbia University Press, 201 pp.

THE AUTHORS

Leonor Botero
COLCIENCIAS, Trans. 9A # 133-28, Santafé de Bogotá, Colombia

Ricardo Alvarez-León
Universidad de la Sabana, Depto. Ciencias de la Vida, Campus Universitario Puente del Común, Edif. E-2, of. 232, Chía, Cundi., Colombia

Chapter 43

THE PACIFIC COAST OF COLOMBIA

Alonso J. Marrugo-González, Roberto Fernández-Maestre and Anders A. Alm

The Colombian Pacific coast is characterised by abundant rain, a high tidal range of up to 5 m, low salinity and tsunamis. Being tropical it has warm temperatures but an absence of hurricanes. The area has two seasons: a rainy and a dry season. There is a general south–north coastal drift along a coast which is made up of a wide mixture of sandy areas, both eroding and accreting, and cliffs, both sedimentary and igneous. Several different water masses converge in this area, some of which carry high nutrients and hence generate a high productivity. Marine habitats are varied and include reefs and beaches, which grade into extensive mangrove forests and into lowland coastal forests. Some of the latter are still extensive, though large areas have been removed, built upon or otherwise impacted in recent years. Most of the area has low pollution contamination from industry or towns, and maintains a very rich biodiversity and high productivity.

However, the area is very poor in human terms. There is enormous under-development, and poverty and health problems are severe compared with other areas of the country. Urban centres likewise are comparatively poorly developed. However, this coast is a peaceful area in comparison with other parts of the country that have suffered from serious and complex confrontations between governmental forces, drug dealers, different insurgent groups and private justice groups. This is because of the difficult climatic and geographic conditions of the area, and because of the general poverty of the region.

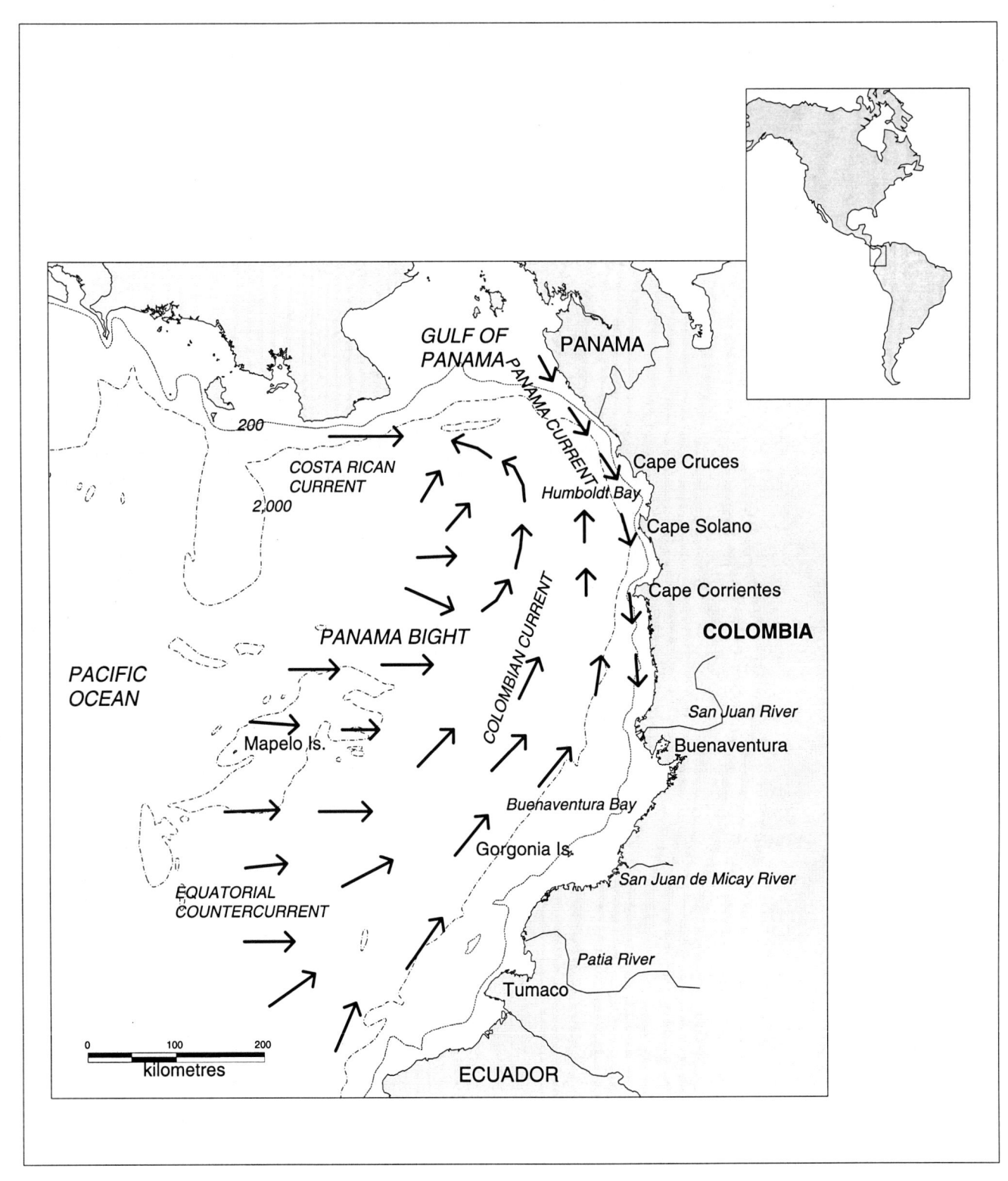

Fig. 1. Map of the Pacific coast of Colombia.

THE DEFINED REGION

The Colombian Pacific coast (Fig. 1) is about 1300 km long, and runs between the Panamá border in the north and the Ecuador border in the south. Politically it is divided into four departments: Nariño, Cauca, Valle del Cauca and Chocó. The region is characterised by high precipitation during the whole year, and by the abundance of mangrove forests and swamps. The Pacific continental shelf is narrow in this region (around 25 nautical miles), giving a shelf area of 5602 square nautical miles between shore and the 200 m depth contour.

SEASONALITY, CURRENTS, NATURAL ENVIRONMENTAL VARIABLES

The Colombian Pacific coast lies in an area of low atmospheric pressure in which the trade winds from the northern and southern hemispheres converge, forming the Intertropical Confluence Area. This area has warm rising winds with high humidity, forming a wide cloudy front which produces between 2000 and 7000 mm of rain each year. This front oscillates north and south during the year, giving rise to the two seasons and to seasonal variations of the region's major oceanographic conditions (Giraldo, 1994; Arias, 1994).

The waters of the Colombian Pacific are relatively warm and have low salinity near shore. The salinity can vary from 20 ppt near the coast, rising to oceanic values of 33.5 ppt further offshore. Except for the upwelling area in the Gulf of Panama (between 7.5 and 9.0°N), temperatures mostly range from 25 to 26°C. The characteristic water masses in the region have been identified as Central Equatorial Water, Subsurface Equatorial Water, Intermediary Antarctic Water and Pacific Deep Water (Andrade, 1992). The distribution of these water masses is correlated with the dissolved oxygen and nutrient content in the water column. The vertical profiles of these parameters are well understood: their conditions are altered periodically during the presence of the El Niño phenomenon, which greatly influences the region's climatic and oceanographic behaviour. The El Niño consists of an increase of water temperature, a substantial increase of the sea level due to the presence of Kelvin waves, a decrease of rainfall, a proliferation of vortices for the whole area and an increase in the flow of the Colombian Current. In general terms, it has been established that, under normal conditions, the highest concentrations of nutrients are in the area close to Gorgona island and in the coastal area of the Gulf of Panamá, where upwelling phenomena are induced by the behaviour of the wind. Off Cape Marzo, another zone of high nutrient concentration, associated with a thermal dome, has been identified around 79°W (Giraldo, 1994).

There is a seasonal water upwelling cycle in the Panamá Bight. In February the intensity of the northerly winds is maximum, the sea level lowers, a deep layer upwelling begins, the photic layer becomes rich in nutrients and the planktonic biomass increases to its maximum. May is relatively calm, the sea level rises, there is minimum upwelling of the deep layers, the photic layer becomes poor in nutrients and fishing activity increases to benefit from high levels of productivity. In October the intensity of the southern winds is maximum, elevating the sea level further, the water upwelling practically disappears and the nutrient concentration in the photic layer falls, reducing fishing activity to a minimum. Upwelling has a greater intensity in the first three months of the year when the northeast winds are stronger (Giraldo, 1994).

The system of equatorial currents in the Pacific has a clearly defined structure in the open ocean, but this structure loses its definition closer to the American continent. It affects the waters of the Panamá Inlet. The effects of the east-bound Equatorial Countercurrent disappear between 85 and 90°W. This is intensified from May to December, when it partially joins the circulation system of the Panama Inlet and the Costa Rican current. It disappears temporarily between the months of February and April, allowing the northerly trade wind to push tropical waters of the inlet, which initiates the Panama current. The Colombian current is notable for bringing water with low salinity northwards, and it interacts with the Chocó current which moves southward along the coast (Giraldo, 1994; Cantera, 1994b).

The mix of currents in the Colombian Pacific creates several series of vortices, some of which are large enough and intense enough to induce upwelling. The most important of these seems to be located in front of Cape Marzo, 79°W.

Tides in the Colombian Pacific coast are of semidiurnal type and reach an approximate maximum amplitude of 5 m with a tidal range that increases from south to north. They are classified as high mesotidal to macrotidal. Tides are also an important factor in the coastal dynamics and have a direct influence on the coastal ecosystems. There are no hurricanes on the Pacific coast, but there are tsunamis. Occasionally, the surf is increased by winds associated with open sea storms. Waves are generally 0.5 to 2 m high with intervals from 10 to 15 seconds. The coastal drift is south–north (Cantera, 1994b).

El Niño phenomena have dramatic effects on the tropical and subtropical marine ecosystems of the oriental Pacific. Their direct effects are reflected, in the first instance, by alterations in the composition of the phytoplanktonic communities. Also the increase of sea surface temperature associated with El Niño causes severe bleaching and death of corals, and has an important impact on the populations of other associated organisms in the tropical oriental Pacific. The coral reefs of the Colombian Pacific have not been exempt from such effects (Zapata, 1992; Cantera, 1994b).

THE MAJOR SHALLOW WATER MARINE AND COASTAL HABITATS

The Colombian Pacific zone has one of the highest vegetation covers of the country, but it is calculated that 86% of the mangrove along the Colombian Pacific has been impacted to date. The native plant association, dominated by the mangrove *Mora megistosperma*, has been heavily exploited. Mangrove forests represent 7.5% of the total Colombian Pacific exploitable forest (Fig. 2). In the north coast around Baudó Sierra, small mangrove forests exist in Nuquí and Coquí, and riverine mangrove systems develop along the Juradó River, at the mouths of the Putumia, Curiche, Baudó and Docanpadó Rivers and towards the internal part of the Tumaco Inlet. There are fringing mangroves in the Utría Inlet, La Chunga and La Aguara, and in the Pichidó isthmus. Mangroves are also developed in Juradó, towards the delta of the San Juan River, and in the mouths of Torogomá, Charambirá and around Choncho island (Buenaventura Bay). They are also developed in Virudó, Catripe, and in the intertidal areas of the Bongo River and towards the inner parts of Málaga Bay, between Cape Soldado (Buenaventura Bay) and in the shallow Cajambre River Mouth, as well as between Yurumangui and Guapi Rivers. Mangrove forests extend about 30 km inland between Iscuandé River and Tumaco Inlet. This whole area is dominated by freshwater and sediments from the Tapaje, Satinga, Sanquianga and Patía Rivers. Of approximately 350,000 ha of mangroves that were estimated to exist in the 1950s, some 281,300 hectares were left by the year 1973, 86% of which were considered to have been impacted. A preliminary study by INDERENA (National Institute of Renewable Natural Resources) and of the CVC (Valle del Cauca Regional Corporation), calculated that this extent had further decreased to some 110,000 hectares in 1992. The present exploitation of mangrove is so disorganised that there is no good and reliable estimate of the area covered today (Cantera, 1994b).

Shore birds are diverse. The 105 marine and shore bird species described so far (visitors and residents) include more than 90% of the taxa that exist in marine habitats of the intertropical area of America. Compared with the total number of resident species in Colombian coasts (18), the number of migratory species (31) is high. The other species are of unknown status (Naranjo, 1994).

The reefs of the eastern Pacific are characterised by their small size, discontinuous occurrence, and low development in areas of water upwelling which are subject to high river discharges. This is reflected in their low diversity of species. In the Colombian Pacific there are basically three areas with substantial development of coral reefs. These are Gorgona Island, Utría Inlet and Malpelo Island. The biggest reefs in Gorgona are true fringing reefs. Malpelo does not possess true coral reefs but supports coral communities on the rocks, and smaller coral communities have also been reported from the north coast of Chocó, from Cape Corrientes towards the Panama frontier (Zapata, 1994) They have not been described or studied.

Among the more interesting local animal species there are: panther, otter, armadillo, weasel, peccary, anteater, whale (occasionally), dolphins, eagles, sparrows, hawks, turtles, crocodiles, iguanas, snakes, boas, octopi and numerous crustaceans. The Gorgona National Park is the main breeding area of humpback whales in the South American Pacific. Blue whales may be seen off Colombia as are the turtles known locally as "golfina" (*Lepidochelys olivacea*), "canal" (*Dermochelys coriacea*), "black" (*Chelonia agassizzi*) and "tortoise shell" (*Eretmochelys olivacea*) (Capella et al., 1994; Gonzalez et al., 1994).

Along the north coast are cliffs formed from basalts, especially at Cape Marzo, Humboldt Bay, Cape Cruces, Cape Solano, Utría Inlet and Cape Corrientes. Along the rest of the north coast, Cupica Bay, Solano Bay and Tibugá Inlet, the coast is formed by basic green vulcanites and ultramaphics. In the region towards the south of Cape Corrientes, cliffs are discontinuous formations (Cantera, 1994a).

OFFSHORE SYSTEMS

According to the Interamerican Tuna Commission, the greatest fishery is in the Panamá Bight and lasts from April to May, while from September to October it is at its minimum. Three important banks have been identified in the Colombian Pacific, two of those located off Tumaco Inlet and the other in the vicinity of the Malpelo Island. The catch is made up of tuna, anchovy and shrimp (Giraldo, 1994).

POPULATIONS AFFECTING THE AREA

The Pacific coast has a population estimated in 1992 to be 817,000 inhabitants, distributed through 32 municipalities. As many as 73.5% are located in urban centres such as Buenaventura (30%) and Tumaco (14%). The rate of growth has been estimated as being 2% above the national rate of increase of 1.8%. The rural population is located in dispersed communities of shrinking size (Fig. 3) due to the high migration, especially towards Buenaventura, Tumaco, Quibdó and some town outside the region (Calero et al., 1994).

The public health situation in the Colombian Pacific is precarious (Table 1). The average rate of infant mortality reaches 110 children for each thousand born alive. These rates are four or five times higher than the national average, and are among the highest in the world, similar to Congo (115), but higher than India (90), Haiti (94), Bolivia (106) and Bangladesh (108) (DNP, 1992). Gastrointestinal infections due to the lack of sewage systems and a lack of drinkable water in most of the municipalities maintain a high morbi-mortality index. Water is extracted from artisanal wells that do not comply with sanitary requirements (Calero et al., 1994). This region is the poorest in Colombia and is severely under-developed compared to other regions regarding health, education and infrastructure.

Fig. 2. Large estuarine areas with mangroves.

Fig. 3. Coastal fishing village.

Table 1

Social indicators of the Colombian Pacific Coast black communities (Corporacion Juridica Libertad, 1997; Comunidades Negras y Derechos Humanos en Colombia, 1997

Indicator	
Percentage of the population of the Colombian Pacific Coast	90%
Life expectancy (years)	55
Population without basic needs	80%
Annual income per capita ($)	500–600
Families with income below the minimum wage ($120)	79%
Families with monthly income below $40	70%
Families with monthly income below $4	39%
Population living in extreme poverty	60%
Population without access to health services	70%
Doctors for each 10,000 inhabitants	1.6
Housing with public services	29%
Running water in municipal districts	48%
Sewer system in municipal districts	10%
Running water in rural areas	13%
Sewer system in rural areas	2%
Garbage collection facilities	10%
Availability of primary school education	77%
Availability of secondary school education	38%
Infant mortality	11%

The Colombian Pacific Coast was a peaceful area until President Samper (1994–1998) announced the construction of an interoceanic channel in the region. Since then, attacks, murder, arbitrary detention and harassment of local people from paramilitary groups, police and military agents have increased (Corporacion Juridica Libertad, 1997). For some years, a national association of artisanal fishermen (ANPAC, Asociacion Nacional de Pescadores Artesanales de Colombia), has denounced the activities of industrial fishing in restricted areas, such as beaches and river outlets, as well as the contamination of waters and bays by industrial wastes. These activities contravene the rights of artisanal fishermen and adversely affect the survival of marine species that use these areas. In protected areas such as the National Park of Sanquianga, equipment of artisanal fishermen is confiscated, although larger ships continue with their activities (Comunidades Negras y Derechos Humanos en Colombia, 1997).

The Colombian Pacific has been characterised by development activities such as timber exploitation which supplies nearly 60% of the national market, and by its commercial aquaculture. The year 1993 stands out for its industrial and artisanal marine production, when the area produced 89.4% of the national total, including 25.2% of white shrimp and all of the *Tilapia* production (Calero et al., 1994).

Man's use of mangrove swamps is very old. Since its discovery during colonial times—and still in the present—the Colombian Pacific coast has suffered from uncontrolled exploitation. It has been subject to periodic extractive periods which have depleted resources without improving the quality of life of the coastal inhabitants. Gold, platinum, banana, tannins, dwarf fan palm, coconut, wood and shrimp culture, are only some examples of those intense extractive activities which have resulted in the severe under-development still seen today. At the moment, many extractive activities are being carried out in mangrove forests with neither regulation nor control (Cantera, 1994b).

RURAL FACTORS

The climatic and soil conditions at the coast do not make significant agricultural development possible as the soils are practically devoid of nutrients. However, there are some suitable areas for cultivation of the African palm which is planted in monocultures. Some mining of gold and platinum takes place (Calero et al., 1994).

COASTAL EROSION AND LANDFILL

About 40% of the 1300 km of the Pacific coast is steep-worked igneous rocks, mainly volcanic rocks of the Baudó

Table 2
Erosive tendency of the Colombian Pacific coast

Coast with high erosion	8.7%
Coast with medium erosion	36.2%
Coast with very low to stable erosion	51.5%
Coast with accretion	3.6%

Table 3
Contribution of the Pacific Coast to national production

Activity	Contribution
Forest volume	198 million m^3
Forest reserve	17%
Sawn wood	42%
Wood pulp	70%
Platinum ($850,000)	82%
Gold ($49 million)	18%
Silver ($200,000)	13%
Marine fishing potential	140,000 tons annually
Marine fishing	89%
White shrimp (marine aquaculture)	25%
Thilapia (marine aquaculture)	100%

Sierra and sedimentary rocks in the areas near Buenaventura and Tumaco Inlet. The remaining 60% forms a low coast characterised by great geomorphic variety (accretion and erosion). Features which stand out in various places are cliffs, barrier islands, old and recent alluviums, tidal lands, mangrove swamps, intertidal plains, tombolos, reefs, flood plains, alluviums of abandoned channels, old beach deposits and tertiary hills.

The differential tendency of this coast to erode has been classified (Table 2). The Colombian Pacific is an active coast with a varied physiography of rocky and wide coastal plains. The morphology and evolution of the different types of coast are a function of several factors. There is a humid tropical climate (maximum annual precipitations of 8–9 m) which supplies huge quantities of sediments to the coast from the Andes. The coast is also tectonically active; interactions between the Andean structures and several systems of transversal fractures create a coast line formed by more or less subsidential blocks. There is also the erosion from a tidal system with medium to low wave energies. Differences in the contributions and in the relative importance of waves and tides generate the different degrees of erosion seen. Also, comparisons between maps from 1847 and radar images from 1992 (IGAC-INTERA) show important morphological variations in the coastal strip over this time period. Seismic subsidence of the land is thus another important factor causing coastal erosion. Accretion in some areas also occurs, mainly due to tidal deposits caused by wave diffraction and swells associated with storms, and local changes in the contribution of fluvial matter and storms, some originating from El Niño phenomena (Correa et al., 1996; Martinez et al., 1990).

EFFECTS FROM URBAN AND INDUSTRIAL ACTIVITIES

Artisanal and Non-industrial Uses of the Coast

The Pacific coast is important to the country in economic terms (Table 3). Very few of the coastal or marine organisms living in the sections of coasts which are dominated by cliffs are used by the population. Some molluscs species on the cliffs of the northern coast are consumed by the residents, but in the southern part it has been observed that some fishermen destroy the cliff to extract shrimps which are then used as bait. The fishing in estuaries is generally more important to the coastal population. Open-sea species are generally fished for consumption in the interior part of the country, or for export (Cantera, 1994a).

The still extensive mangrove forests of the Colombian Pacific coast are used in several ways, including fishing, hunting, recreation, wood extraction for construction or charcoal, subsistence agriculture and livestock grazing. The main agriculture products are rice, banana, coconut, "chontaduro", yucca, "papachina", corn, sugar cane, cocoa, rubber, African palm and some fruit-bearing trees. The most common species of livestock are pigs, hens and, on a smaller scale, cattle (Cantera, 1994b).

The fisheries in areas near mangrove forests are mainly artisanal, and the fish caught are used by the fisherman for direct consumption or for selling in small local markets. It is considered that approximately 10,000 to 15,000 artisanal fishermen live in the Colombian Pacific coast, in some 69 fishing communities (see Fig. 3). Three types of fishing activity are carried out in the mangrove forests: white fishing (families Mugilidae, Centropomidae, Scianidae, Ariidae and Gerridae), the fishing and gathering of crustaceans (shrimps and crabs) and the extraction and gathering of molluscs, mainly bivalves but including gastropods on a smaller scale. Mangrove species are cut for firewood and for conversion into charcoal by sawmills. The main forms of subsistence activity at present are the logging of forests, mainly "Natales" and "Machares", and the Naidí palm exploitation. Timber-yielding species are used in local construction (pilings and beams of houses, furniture, crafts, tools, etc.) or sold as logs to the sawmills. The red mangrove is also used locally as medicine for curing throat infections and for treating hair (Cantera, 1994b).

The mangrove ecosystems offer an ideal place for ecotourism, and provide some alternative revenue for the regional economy. However, several difficulties inhibit this, the main ones being difficult and expensive accommodation, difficulty of adapting construction to the local environment, ingress by water in intertidal areas and the presence of illnesses like cholera, malaria and leishmaniasis (Cantera, 1994b).

Hunting also takes place, the species that are most pursued being deer (*Odocoileus* sp.) peccary or ñeque, guagua, guatín, babillas, otter, iguanas, tigrillo (*Procyon carnivorus, Felis pardalis*), panther (*Panthera onca*) and armadillo. These are hunted during night with lamps. "Ratas espinosas" and "chuchas" (*Chironectes minimus* and *Didelphis marsupialis*) are also captured using traps (Cantera, 1994b). These, together with poaching of turtle nests, trawling for shrimps, habitat deterioration by the construction of tourist infrastructure on nesting beaches, and marine contamination, are factors that have contributed to the dramatic decrease of the natural populations of these animals in Colombia, placing many of them in the category of endangered species (Amorocho, 1990).

The mangrove forests also provide the coastal inhabitants with vegetable fibre, shells and animal skeletons for local crafts, which is an important economic activity. Aquaculture in brackish waters of mangrove swamps is a recent activity in the Colombian Pacific coast and it has so far been tried only with marine shrimps and fresh water fishes (*Tilapia*), with more or less promising results. These activities, however, are encouraging the destruction of mangrove forests in the Colombian Pacific, whose deterioration is also due to inadequate watershed management in the rivers and growing contamination with solids and domestic waste waters (Cantera, 1994b).

Minerals of strategic importance here include bauxite, manganese, cobalt, tin, chromium, nickel and petroleum. However, mining is contaminating rivers with mercury. Studies near the San Juan River in the central part of the Pacific coast found mercury concentrations above 0.5 mg/kg in fish and 0.15 mg/l in urine samples of the inhabitants. Mercury concentrations over 0.02 mg/l were found in blood of people older than 15 years. The concentration in humans should be below 0.0005 mg/l (Comunidades Negras y Derechos Humanos en Colombia, 1997).

Cities

Buenaventura and Tumaco stand out for their high contributions of organic material due to fishing industries, sawmills, shrimp farming and processing, and the fish flour and palm oil factories. In the last few years regional corporations have initiated controls to reduce the loads of industrial levels of organic matter. Now, some industries like the Tumaco palm oil factories have treatment plants for their waste (Calero et al., 1994).

Studies carried out in the Pacific coast indicate that the contamination generated by the gas stations, landing stages and docks is transient and disperses away from the discharge point until reaching normal levels (Marrugo-Gonzalez, 1990; 1994). Pollution studies in the Colombian Pacific showed that the dissolved oxygen in most stations was above 4.0 mg/l, which is the permissible minimum value needed for flora and fauna survival in marine and estuarine waters according to ordinances of the Ministry of Health. In the water column there was no contamination by cadmium or copper. Possible zinc contamination was detected. Cadmium in sediments, bound to carbonates, in spite of being high, is not biologically available due to the water pH. There is no contamination by copper in the sediments (Niño and Panizzo, 1990; Hernández et al., 1994).

The state of conservation of the Colombian oceanic systems is relatively good. Given the great width of the oceanic areas, they present a high capacity for recovery in the event of any potential polluting incident (Arias, 1994).

In general the quality of the Colombian Pacific water is still not markedly affected by pesticides, hydrocarbons or organic matter. Contamination tends to increase in time and space, and the bivalves of Buenaventura Bay and in the sector between Puente El Pindo and El Morro (Tumaco Inlet) do have high hydrocarbons levels in their tissues and are potentially dangerous for human consumption. Concentrations of hydrocarbons and pesticides are relatively high in Tumaco Inlet, which, due to its fishes and aquaculture resources, should be an area of special environmental management (Calero et al, 1994).

Industrial Uses of the Coast

Colombian oceanic waters are far from providing a significant part of the economy of the country. This does not mean that they are not being taken advantage of, and particularly in the Pacific, foreign ships specialised in fishing large pelagics and highly migratory species, use the Colombian Exclusive Economic Zone illegally. The main reason for this abnormal situation has to do not only with the lack of an effective means of control (coast guards), but mainly with the absence of a national fishing fleet for the benefit of the country. However, a recent increase in catch should be highlighted (Arias, 1994).

Most of the industrial fishing is carried out in coastal areas. The main species used in this type of exploitation are: prawns, shrimps and fish like anchovy (*Cetengraulis mysticetus*) and "plumuda" (*Opisthonema libertate*), mainly for fish flour (Cantera, 1994b).

According to the Boletín Estadístico Pesquero (Fishing Statistical Bulletin) of INPA (1993), fish catch has increased 8.3 times from 1985 to 1993 due to tuna and sardine capture—now very important components of the fishing economy. As for prawn and shrimp capture, the statistics show a 90% decrease, while for the cultivation there is a notable increase in production of approximately 77 times between the years 1985 and 1992. There was a decrease in aquaculture production in 1992 and, bearing in mind the increasing area of land dedicated to shrimp farming, one can assume that there was a problem of mortality. This may be due to the increased levels of hydrocarbons and pesticides (Boletín Científico No. 5 of CCCP) or to diseases in the ponds. The Boletín Estadístico Pesquero of INPA, shows a positive balance of trade in the last seven years, which indicates the importance of these resources in the gross national product of the country. It is convenient that the mangrove swamp ecosystems and the areas dedicated to shrimp farming receive special management to guarantee the water quality.

Besides the fisheries of species which migrate to the estuaries, molluscs such as oysters, mussels and clams are also economically important. The latter are mainly "piangua" and "sangara" (Cantera, 1994b). The number of important species in the country as a whole is about 281, of which 126 are from the Pacific. Fishes represent 74% of this total, crustaceans 17% and molluscs 9%. Most of the fish species are benthic, followed by estuarine, coastal pelagic and oceanic pelagic species. Crustaceans are mainly caught in the benthic shelf, coast and estuarine habitats. For molluscs, the main species are from the coast and estuaries (including the mangrove forests) (Valderrama, 1994).

Colombian marine commercial landings in 1993 were 93,259 tons, of which 89% came from the Pacific Ocean. The tunas, yellow fin tuna, *Thunnus albacares*, bonito, *Katsuwonus pelamis*, and anchovy, *Cetengraulis mysticetus* (Engraulididae), are the most important species representing 85% of the marine landings. The "plumuda", *Opisthonema* sp., in the Pacific is under-exploited, but it has good potential. Other important species are the medium-sized coastal pelagic fishes (*Caranx* spp., *Trachinotus* spp., *Chloroscombrus chrysurus*, *Seriola* spp., *Selene* spp.). Although their numbers are not very high, these resources could be subject to increased fishing efforts. Within the group of demersal fishes, species under fishing pressure are: *Lutjanus* spp., *Epinephelus* spp., *Haemulon* spp., *Pomadasys* spp., *Anisotremus* spp., *Mycteroperca* spp). It has been considered that the fishing potential of this group is still not entirely maximised.

The fishing of "cherna café", *Mycteroperca xenarca*, is slightly above its maximum sustainable yield. It is considered that sharks are still a under-exploited resource (Carcharhinidae, Lamnidae and other families). However sharks are a fragile resource needing careful management. As for crustaceans, the shallow waters shrimps (Penaeidae) and white shrimp, *Penaeus occidentalis* are the most outstanding species and are highly overexploited resources. Also, the fishing of "tiger" shrimp, *Trachypenaeus byrdi*, and "tití" shrimp, *Xiphopenaeus riveti*, are at the limits of their maximum sustainable yields. There are potentials for deep waters resources (continental shelf and slope) such as *Solenocera agassizii*, and *Heterocarpus* spp. The crab, *Callinectes arcuatus*, is under-exploited. With respect to molluscs, the squids (Loliginidae and Ommastrephidae), *Loligo gahi* and *Lolliguncula panamensis*, are those of highest commercial interest. The giant squid of the slope and deep oceanic waters, *Dosidicus jigs*, may offer future potential (Valderrama, 1994).

PROTECTIVE MEASURES

The legislative framework for the management of fisheries resources in the country is the 1990 13th Law or General Statute of Fishing. Under this law, the INPA has established some management actions and it has reaffirmed others implemented by INDERENA (Valderrama, 1994). Some of its measures include:

(a) *Fishing quotas and determination of maximum sustainable yields, such as in the case of shallow waters shrimps*. Annual fishing quotas have been established by resource or type of resource based on available abundance or biomass information.

(b) *Regulation of fishing methods*. In particular, for the industrial fishing of shrimp, tuna, anchovy, lobster and molluscs, regulations have been established. There are general regulations that prohibit specific methods, or general methods which impact the aquatic environments or fishing resources, such as fishing with dynamite or physical modification of the aquatic environment.

(c) *Reserve areas*. In the Pacific coast, the Tumaco Inlet is an reserve area set aside for exclusive fishing by the residents in the inlet.

(d) *Bans*. Bans on fishing for some species have been established.

(e) *Limitation of the fishing effort*. Given the overexploitation of the shallow water shrimp resource in the Pacific, free access to the activity has been limited by not authorising new fishing licences. Also, the licenses of the fleets that stop fishing are not renewed.

Other Measures

The INPA has begun the creation of a Gene Bank of Fishing and Aquarian Resources. An agreement was signed by the International Fisheries Genetic Bank, IFGB. This technical cooperation agreement began activities in 1995.

In the Colombian Pacific coast several flora and fauna sanctuaries have been created:

(a) the Sanquianga National Park, which is an estuarine system located on the coast of the department of Nariño (2°40'N and 78°28'E). It has an approximate area of 80,000 ha (Rubio, 1990);

(b) the Gorgona Island National Park 02°57'N, 78°12'W) which includes several small islands located around a larger one (Gorgonilla, El Horno, El Viudo);

(c) the Utría Inlet National Park, located in Utría Inlet. Protection of fauna and flora in general was initiated some decades ago because several international juridical instruments were subscribed to with that purpose. However, an analysis of these juridical instruments and their influence in the legislation of our country reveals the low level of real participation of Latin America in general, and certainly of Colombia in particular, in the processes of implementation of these instruments. This situation is paradoxical given that these countries are owners of the greatest biodiversity on the planet (Rengifo, 1994).

The "Plan Nacional de Contingencia Contra Derrames de Hidrocarburos en Aguas Marítimas" (National Contingency Plan Against Hydrocarbons Spills in Marine Waters) is an operative technical document that establishes, delimits and identifies the contamination of vulnerable areas; it provides basic information about the affected area including available infrastructure, personal, equipment, etc. and the resources at risk; it defines responsibilities and functions; it rationalises the use of the available personnel, equipment and materials, and suggests lines of action to cope with incidents (Calderon and Gutiérrez, 1994).

The climatic characterisation of this region is quite well documented in several papers, from older ones such as West (1957) to several new studies. Due to recent high interest in the country related to the conditions of the Pacific (El Niño phenomena), in recent years a network of meteorological stations have contributed valuable climatic information related to possible effects of global warming (Cantera, 1994b).

To reinforce the policy of environmental protection the Colombian government created the Ministry of Environment in the mid-1990s. This ministry has now taken charge of the functions of environmental protection that until now have been under the responsibilities of INDERENA.

REFERENCES

Amorocho, D. (1990) Biología Reproductiva en la Tortuga Golfina (lepidochelys olivacea) en la Playa Larga, El Valle (Chocó). In: *Memorias del VII Seminario Nacional de Ciencias y Tecnologías del Mar, Cali, Colombia.*

Andrade, G., Ruiz, J.P. and Gómez, R. (1992) Biodiversidad, Conservación y Usos de Recursos Naturales. Colombia en el Contexto Internacional. Presencia, Bogotá, 126 pp.

Arias, F.A. (1994) Contribución para Definir el Estado del Conocimiento de los Sistemas Oceánicos Colombianos con Énfasis en la Parte Biológica. In: *Estado del Conocimiento y Lineamientos para una Estrategia Nacional de Biodiversidad en los Sistemas Marinos y Costeros Colombianos*, Minca (Magdalena, Colombia).

Calderón, D.M. and Gutiérrez, J.A. (1994) Plan Nacional de Contingencia Contra Derrames de Hidrocarburos. In: *IX Seminario Nacional de Ciencias y Tecnologías del Mar y Congreso Latinoamericano en Ciencias del Mar, Medellín, Colombia.*

Calero, L.A., Marrugo-González, A.J. and Casanova, R.F. (1994) Diagnóstico de la Contaminación Marina Enfocado a la Parte Social y Económica del Pacífico Colombiano. In: *IX Seminario Nacional de Ciencias y Tecnologías del Mar y Congreso Latinoamericano en Ciencias del Mar, Medellín, Colombia.*

Cantera, J.R. (1994a) Biodiversidad de Acantilados Rocosos en el Pacífico Colombiano: Estado de su Conocimiento. In: *Estado del Conocimiento y Lineamientos para una Estrategia Nacional de Biodiversidad en los Sistemas Marinos y Costeros Colombianos*, Minca (Magdalena, Colombia).

Cantera, J.R. (1994b) El Ecosistema de Manglar en el Pacífico Colombiano: Estado de su Conocimiento. In: *Estado del Conocimiento y Lineamientos para una Estrategia Nacional de Biodiversidad en los Sistemas Marinos y Costeros Colombianos*, Minca (Magdalena, Colombia).

Capella, J., González, L.F. and Bravo, G.A. (1994) Residencia y Retorno Reproductivo de Ballenas Jorobadas, megaptera novaeangliae, en el Pacífico Colombiano. In: *IX Seminario Nacional de Ciencias y Tecnologías del Mar y Congreso Latinoamericano en Ciencias del Mar, Medellín, Colombia.*

Comunidades Negras y Derechos Humanos en Colombia. Internet. www.nodo50.ix.apc.org/sodepaz21art9.htm (31 March 1997).

Corporación Jurídica Libertad. Diagnóstico y Situación de Derechos Humanos en el Chocó-Colombia. Internet. www.derechos.org/libertad/choco. html (17 April 1997)

Correa, I.D., Gayet, J. and Vernette, G. (1996) El Litoral Pacífico Colombiano: Aproximación a los Contextos Estructurales e Hidrodinámico. In: *Memorias del X Seminario Nacional de Ciencias y Tecnologías del Mar.* Santafé de Bogotá D.C.

DNP (Departamento Nacional de Planeacion) (1992) Plan Pacífico, Santafé de Bogotá: DNP, 89 P.

Giraldo, L. (1994) Estado Actual de Conocimiento de la Oceanografia Fisica del Caribe y Pacífico Colombianos. In: *Estado del Conocimiento y Lineamientos para una Estrategia Nacional de Biodiversidad en los Sistemas Marinos y Costeros Colombianos*, Minca (Magdalena, Colombia)

González, L.F., Haase, B., Capella, J., Bravo, G.A, Féliz, F., Lyrholm, T., and Gerrrodette, T. (1994) Movimiento de Ballenas Jorobadas megaptera novaeangliae en la Costa Oeste de Suramérica. In: *IX Seminario Nacional de Ciencias y Tecnologías del Mar y Congreso Latinoamericano en Ciencias del Mar, Medellín, Colombia.*

Hernández, L.B., de Villaveces, M.C. and Báez, M.C. (1994) Estudio y Evaluación de la Contaminación por Metales Traza en Zonas del Pacífico Colombiano. In: *IX Seminario Nacional de Ciencias y Tecnologías del Mar y Congreso Latinoamericano en Ciencias del Mar, Medellín, Colombia.*

INPA (Instituto Nacional de Pesca y Acuicultura) (1993) Boletín Estadístico Pesquero, Santa Fé de Bogotá.

Marrugo-González, A.J. (1990) Estudio de la Contaminación por Hidrocarburos en el Litoral Pacífico Colombiano. In: *Memorias del VII Seminario Nacional de Ciencias y Tecnologías del Mar, Cali, Colombia.*

Marrugo-González, A.J. (1994) Estudio de la Contaminación Marina por Hidrocarburos en Áreas Críticas de la Costa Pacífica Colombiana Etapa II. In: *IX Seminario Nacional de Ciencias y Tecnologías del Mar y Congreso Latinoamericano en Ciencias del Mar, Medellín, Colombia.*

Martínez, J.O., González, J.L., Carvajal, J.H. and Escobar, L.M. (1990) Problemas Geológicos Asociados a la Línea de Costa del Pacífico Colombiano: Geomorfología y riesgos geológicos. In: *Memorias del VII Seminario Nacional de Ciencias y Tecnologías del Mar, Cali, Colombia.*

Naranjo, L.G. (1994) Evaluación del Estado del Conocimiento de las Aves Marinas y Playeras en Colombia. In: *Estado del Conocimiento y Lineamientos para una Estrategia Nacional de Biodiversidad en los Sistemas Marinos y Costeros Colombianos*, Minca (Magdalena, Colombia).

Niño, M.C. and Panizzo, L. (1990) Estudio Evaluativo de Cadmio y Zinc en Sedimentos Superficiales de la Bahía de Buenaventura. In: *Memorias del VII Seminario Nacional de Ciencias y Tecnologías del Mar, Cali, Colombia.*

Rengifo, A.J. (1994) Aspectos Jurídicos para la Preservación, Conservación y Utilizacion de la Biodiversidad Marina y Costera. In: *Estado del Conocimiento y Lineamientos para una Estrategia Nacional de Biodiversidad en los Sistemas Marinos y Costeros Colombianos*, Minca (Magdalena, Colombia).

Rubio, E.A. and Estupiñan, F. (1990) Ictiofauna de Parque Nacional Natural Sanquianga, un Análisis de su Estructura y Perspectivas para su Manejo. In: *Memorias del VII Seminario Nacional de Ciencias y Tecnologías del Mar, Cali, Colombia.*

Valderrama, M. (1994) El Recurso Pesquero Maritimo: Diversidad, Estado de Aprovechamiento, Potencialidad y Ordenacion.

Lineamientos Generales. In: *Estado del Conocimiento y Lineamientos para una Estrategia Nacional de Biodiversidad en los Sistemas Marinos y Costeros Colombianos,* Minca (Magdalena, Colombia).

West, R. (1957) *The Pacific Lowlands of Colombia: A Negroid Area of the American Tropics.* Louisiana University, Louisiana, 278 pp.

Zapata, L.A. (1992) Contribución al Conocimiento de la Biología, Hábitos Alimenticios y Crecimiento en la Carduma (*C. mystycetus*) en el Pacífico Colombiano. Thesis. Universidad del Valle. Cali, Colombia.

Zapata, L.A. (1994) Las Comunidades y Arrecifes Coralinos del Pacífico Colombiano. In: *Estado del Conocimiento y Lineamientos para una Estrategia Nacional de Biodiversidad en los Sistemas Marinos y Costeros Colombianos,* Minca (Magdalena, Colombia).

THE AUTHORS

Alonso J. Marrugo-González
Universidad de Cartagena,
Facultad de Ciencias Químicas y Farmacéuticas,
Zaragocilla, AA 1661,
Cartagena, Colombia

Roberto Fernández-Maestre
Universidad de Cartagena,
Facultad de Ciencias Químicas y Farmacéuticas,
Zaragocilla, AA 1661,
Cartagena, Colombia

Anders A. Alm
Universidad de Cartagena,
Facultad de Ciencias Químicas y Farmacéuticas,
Zaragocilla, AA 1661,
Cartagena, Colombia

Chapter 44

PERU

Guadalupe Sanchez

The coast of Peru extends in a fairly straight line for over 3080 km. Most of the coastal zone is arid semi-desert, though the northern end has landscapes with streams and great expanses of vegetation, having notable stands of mangroves with a rich diversity. The continental shelf is up to 65 miles in the central section, but is much narrower in the southern and northern ends of the coast.

Climate is influenced by the Peruvian or Humboldt Current whose cool temperature condenses moisture in the equatorial air masses, producing fog and clouds in the central coastal region. The upwelling off the Peruvian coast is exceptionally important, with marked effects on sea level, climate and on productivity. During El Niño events, the upwelling is suppressed, and the pelagic ecosystem suffers radical changes: warm waters which are poor in nutrients flow in and inhibit high primary production, and modify the composition of the plankton. The herbivorous pelagic fish which feed on the plankton are reduced as a "tropicalization" occurs due to the migration of different species from equatorial waters. Copepods which are the dominant group during normal years and which are herbivorous, decline and are replaced by warm-water species which generally are carnivorous predators of eggs and fish larvae. Peak production of anchovy and sardine have been over 9 million tons, especially related to the first fishery resource.

There has been a marked population growth since the 1950s. Habitats in the coastal zone in Peru have suffered alterations from both natural processes and various activities resulting from the increase in population and increasing demand for land. Heavy use of pesticides has produced toxic residues, although the main pollution is based on domestic and industrial residual wastes that contaminate beds of rivers, altering the quality of fresh waters and the Peruvian marine coast, but new laws and several protected areas are now encouraging environmental considerations in new development.

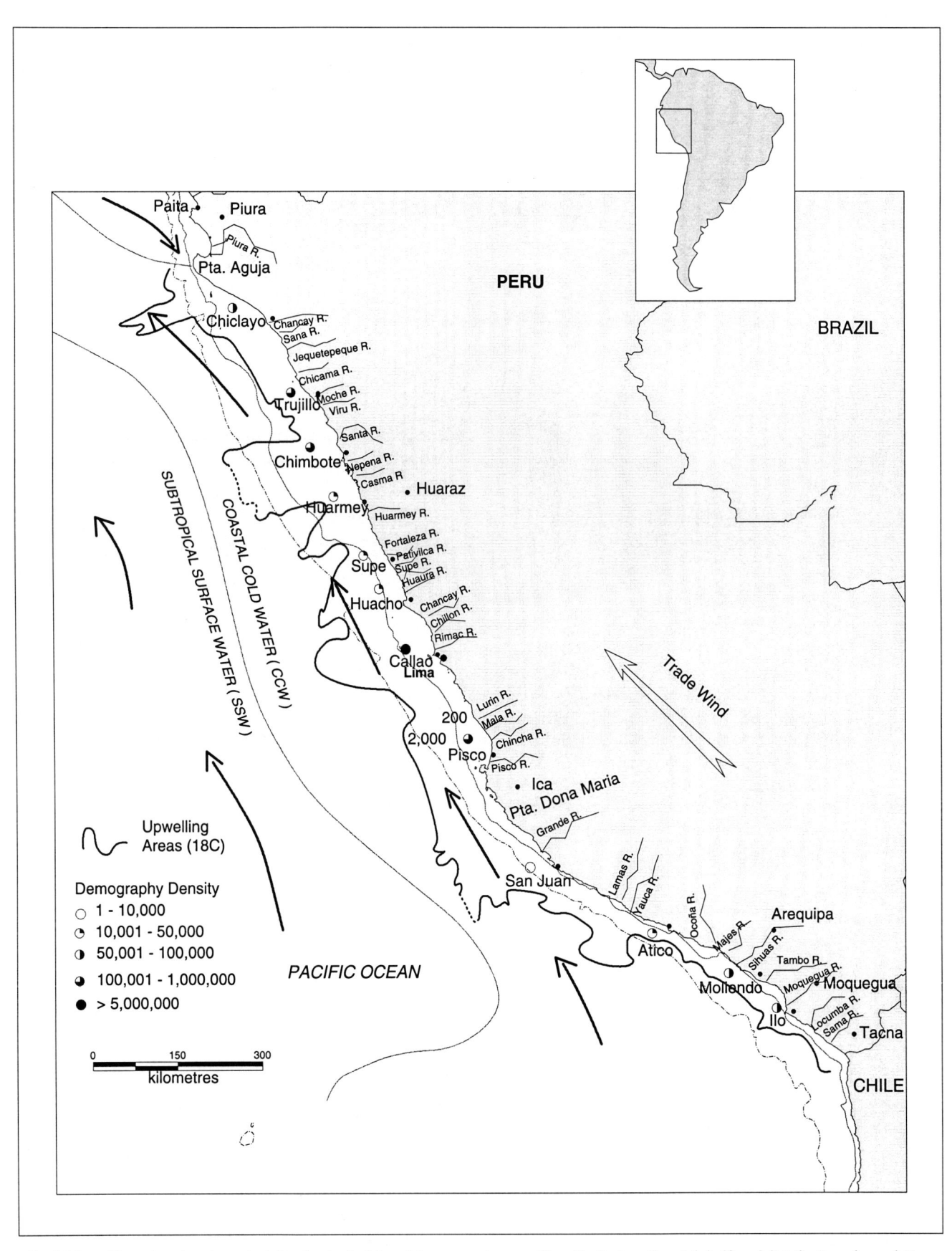

Fig. 1. Map of Peru showing rivers draining the Andes Mountains, currents, upwelling, limit of continental shelf, and distribution of population.

CHARACTERISTICS OF THE PERUVIAN COAST

Peru is located in the central-western part of South America and covers about 1,285,216 km² (Figs. 1 and 2). Of this, 94 km² are islands. The coast extends from the Department of Tumbes (03°23'42.5"S) to Tacna (18°21'34.8"S) and is a long and mostly straight line of 3080 km, whose geological characteristics are typical of western South America. The coastal zone is bordered to the east by the Andes mountains, which run parallel along the littoral about 100 km inland. This littoral zone represents 10.6% of the national territory, but contains most of the population.

The coastal zone is arid semi-desert, shaped by subaereal and fluvial erosion. In this region, there are both marine and fluvial terraces, and valleys are generally narrow. Rivers are low in winter and increase in summer when rains are intense in the Andean zone (Fig. 1).

The continental shelf is up to 65 miles in the central section, but much narrower in the southern and northern ends of the coast. These characteristics are important factors in the distribution of species and of the substantial fishery resources.

Characteristics of the Coastal Zone

Peru forms part of the pronounced westerly projection of the South American continent, between the Gulf of Guayaquil and the Island Lobos de Tierra. The littoral of the northern department of Tumbes, bordering Ecuador, differs from the rest of the Peruvian coast by having mangroves. Here, the Amotape mountains and low hills back the coastal zone. Further south are found the best beaches, which extend for nine miles to the cove of La Cruz. The coast of Piura is desert and semidesert, and is more than 100 km wide. On this coast, the mountains of Paita and Illescas extend to 390 m and 515 m, respectively, constituting (together with the Amotape mountains) the remains of the mountain chain of the ancient coast or the zone of the western massives (Ministerio de Marina, 1982).

South of the bay of Paita is the Sechura inlet, in whose southern end is the port of Bayóvar and the oil terminal from a pipeline which originates from the northeast Peruvian tropical forest. Six miles offshore is the largest Peruvian island, Lobos de Tierra. Less than 11 km long, the island is considered to be part of the ancient mountain chain of the coast.

From this point southwards, the coast is typically arid. The coast of Lambayeque region is mainly one of vast plains, cut by the rivers of La Leche, Chancay and Zaña. Another island, Lobos de Afuera, actually two small, adjacent islands each approximately 5 km long, is used as a base for fishing activities.

Further south, the morphology of the coastline has a pattern which is continuously repeated along the Peruvian seashore. This pattern consists of a series of rocky promontories, each of which protects a port that is located northward of a cliff, and south of these promontories are straight sandy or shingle beaches several kilometres long.

The desert between Trujillo and Santa river is becoming a fertile area due to irrigation projects which use Santa river waters. Further south still, the coast that previously was one of long and open coves with sandy beaches, has increasing numbers of embayments, many of the latter surrounded by coastal cliffs which are extensions of hills that reach the coast. In this area, called Ancash, marine erosion has created several deep bays, many islands and sheltered areas, including those used by the ports of Chimbote and Samanco, the former being the main fishing port of Peru.

In the department of Lima, the Andes mountains lie close to the sea, forming a succession of high cliffs except where several rivers drain, forming important valleys which support intense agricultural activity. The desert features of the Lima region include numerous pampas with sand dunes, and a coast characterized by numerous coves, promontories and headlands. Lima and Callao (Fig. 3) lie on alluvial plains, with several terraces that reach the shoreline, and the large San Lorenzo Island is offshore here.

High cliff with a narrow platform of abrasion, and sandy beaches, characterise much of the central Peruvian coast. Occasionally there are marshes fed by rivers and by surplus water from irrigation. The coastline of Ica department then differs in several morphological aspects; it is desert with mineral and non-metallic deposits. The uniformity of the coast is broken by the Peninsula of Paracas which is important geologically, resulting from tectonic collapse of the fragmentary mountain chain on the coast. Paracas bay has beautiful beaches, fronting a flat, sandy interior and some hills with heights of up to 200 m.

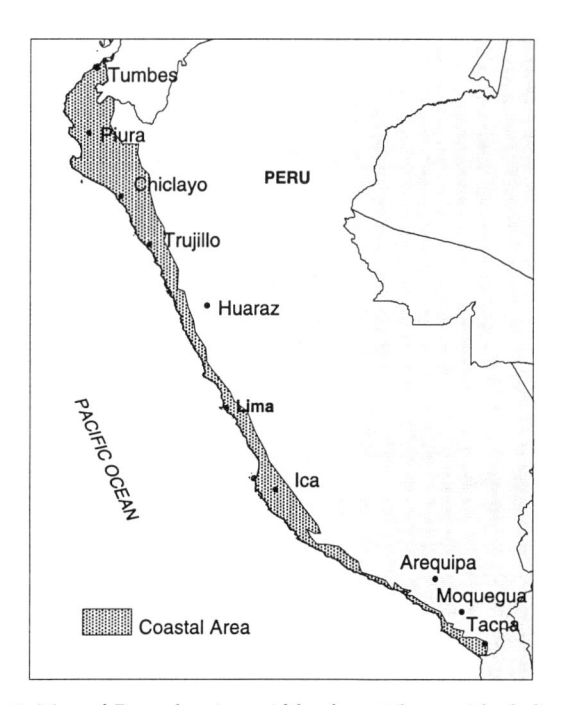

Fig. 2. Map of Peru showing width of coastal zone (shaded) and provinces.

Fig. 3. A stony beach in La Punta peninsula, Callao.

Along a section of 230 miles the coast then continues in a southeast direction, and includes San Nicolas and San Juan ports. This coast is initially rocky and high, as far as the Ocoña river where it becomes flatter with sandy, shingle beaches. From here, the coast is very sinuous, with cliffs, up to Matarani bay. Then, from Mollendo to the border with Chile, the coast is flat and mainly sandy, but with areas of cliffs.

Marine and Coastal Characteristics

The Peruvian coast has 53 rivers, though most flow only intermittently. The flanks of most are almost vertical walls and canyons, cutting the coastal belt until reaching the sea. Most rivers increase in volume in summer, from December to April, due to rains in the highlands. In winter, most river water does not reach the sea. Eight of them have an important volume of water all year, Tumbes being the only navigable river of the coast.

Coastal Climate

Climate is influenced by the Peruvian or Humboldt Current (Fig. 1). Its cool temperature condenses moisture in the equatorial air masses that blow from the sea toward the coast, producing fog and clouds in the coastal region, between sea level and 500 m high. This characteristic is almost exclusive to the Peruvian Current influence zone and is not seen very often north of 5°S where the climate is more temperate and warm.

The main factors that condition the climate include:
- an atmosphere dominated by the Pacific anticyclone and by the southeast tradewinds, a wet air mass that blows onto dry soil with very little or no vegetation;
- an almost uniform, mild temperature, but with cold air during winter due to the dampness. In the central and southern coast the temperature does not fall below 11°C;
- a sea with cold temperatures due to the Peruvian Current and to emerging deep waters, following the Peruvian Current direction;
- important rivers, especially in the north.

The Peruvian coast has notable contrasts of climate. The northern end has landscapes with streams and great expanses of vegetation, originally being a typically tropical landscape. The central and southern coasts show a 'wet desert' climate, with the exception of oases and mountains veiled in mists known as "lomas" and where, thanks to winter rainfall, vegetation grows near the desert. This zone normally extends from between 100 and 500 m above sea level and up to approximately 15 km inland. An example of this coastal feature is Lomas de Lachay, in Lima, which has a very rich fauna and flora.

An important aspect in the coastal zone of Ica (south-central province), is intense eolic sand and dust transport. Excessive warming of the land generates dust storms that frequently last many hours, moving at more than 60 km per hour, taking huge yellowish dust clouds inland. These sand storms, called Paracas, due to the direction from where they come, deposit their dust in the foothills of the Andes to more than 2000 m above sea level.

Currents

The Peruvian Current, or Humboldt, is a cold-water mass that moves northward from the Antarctic. It flows from the central coast of Chile to the northern coast of Peru (Punta Aguja, 05°S), where it turns toward northwest and becomes the South Equatorial Current. The Peruvian Current has two components: the Coastal Current flows near the coast at a speed of 0.2 to 0.3 knots before it changes and becomes the South Equatorial Current; and the Peruvian Oceanic Current which flows at 0.5 to 0.7 knots. It has a high biological productivity and, in surface layers, a high nutrient turnover.

The Subtropical Surface Current that flows far offshore becomes closer inshore mainly when there is a weakening of the tradewind, and has a great influence on the southern area. It has a warm temperature, and salinities higher than 35.1 psu. The Surface Equatorial Current flows north of 6°S, mainly in the spring and summer months, and its fluctuations are related to the displacement of the "equatorial front".

The anticyclonic circulation in the Pacific and the Eastern Pacific upwelling strongly affects Peru. Offshore there is a considerable zonal gradient, related to the coastal upwelling, that increases the movement of cold waters toward the surface, so that south of 6°S the temperature increases westward. Northward 4°S there is a southern gradient related to the equatorial front that develops between the months of April and December.

In summer and autumn, the most intensive gradients occur near the coast, with the 24±C isotherm located parallel to the coast, and with a noteworthy "warm tongue" projecting southward. In winter and early spring the isotherm is not parallel with the coastline, and it is during this period that the Peruvian Current is intensified and flows northwest (Zuta and Guillen, 1970).

Salinity and Dissolved Oxygen

Salinity of the coastal waters is influenced by the subtropical salty waters along almost all the coast, and by lower salinity waters from the equatorial region above 5°S.

Waters from the coastal upwelling generally have salinities less than 34.9 psu in the south and less than 35 psu near the north. This difference is important to the anchovy fishery, a species very sensitive to variations in salinity. Concerning the vertical distribution of salinity, the maxima are found at the surface, and as a rule salinity decreases with depth, except where Subantarctic Waters meet the Subsurface Equatorial Waters.

Off the Peruvian coast, average oxygen concentration in the surface layer has a greater range in summer and autumn than in winter and spring. It also increases further offshore. Minimal values, <2 ml/l, are found in the areas of upwelling, and maximum values up to 7 ml/l occur in areas where photosynthesis is rapid.

There is an oxygen minimum layer where the oxygen falls abruptly to values of less than 0.5 ml/L, generally between 50 and 800 m deep. Below this, and close to the salinity minimum related to the extension of the Antarctic Intermediate Waters, oxygen content increases quickly with depth.

The Coastal Upwelling

The upwelling off the Peruvian coast has a singular importance because of its effects on sea level, on climate and on productivity (Fig. 4). El Niño is an interaction between the ocean and the atmosphere that occurs in the Pacific. It produces an anomalous warming of surface waters of the Pacific Equatorial off the coasts of Peru and Ecuador. These global climatic changes modify the marine environment, causing different degrees of effect depending on the intensity of the event. The El Niño event has marked ecological consequences in this region; often these consequences are strongly negative, mainly for economy of the countries located in the Pacific basin (Arntz and Fahrbach, 1996).

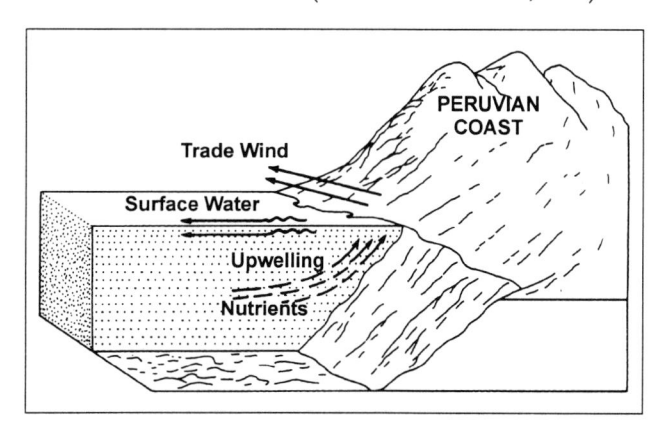

Fig. 4. Diagram of the Peru upwelling mechanism (after Wyrtki, 1963; Sánchez and Soldi, 1988).

The upwelling allows waters from subsurface layers to reach the surface and to move far from their origin through divergent horizontal flows (Wyrtki, 1963), and it creates considerable changes in the vertical distribution of the salinity and nutrients in the surface layer. The location of the continental shelf and extent of the upwelling zone is shown in Fig. 1. The area affected varies in area and distribution according to the season. In summer it extends for 800 miles long by 5 miles wide, covering a total of 12,000 square miles, while in winter the area of upwelling may become three times greater.

BIOLOGICAL DIVERSITY

There are approximately 759 species of fishes from the Peru marine system, which belong to 402 genera and 169 families. Of these, 73 species are economically important. There are two provinces. The first is north of 6°S, the so-called Provincia Panameña of warm waters that have an abundant flora and tropical fauna with high diversity (Sánchez and Orozco, 1997). The second is the Peruvian–Chilean province, whose characteristics are temperate, and where species may be extremely abundant, e.g. the Peruvian anchovy *Engraulis ringens*, the sardine *Sardinops sagax sagax*, the jack mackerel *Trachurus pictaratus murphy* and the mackerel *Scomber japonicus* (Fig. 5).

There are 400 species of crustaceans, 5% being commercially valuable prawns of the genus *Penaeus*, the crab or "jaiva" of the families Xanthidae, Portunidae or Canceridae. Important also are microcrustaceans which are food sources for the important fishes, and of these there are known to be 61 species of copepods, 11 euphausids, 3 cladocera and 40 amphipods. A variety of other invertebrates include the molluscs with 971 species, echinoderms with 61 species, approximately 300 polychaetes, 5 salps, 11 chaetognats, 52 foraminifers, and others. Also, there are approximately 201 species of macroalgae, several used for production of agar and alginates. This high productivity supports good populations of marine mammals (Fig. 6).

Wetlands and Protected Areas

Wetlands provide a series of products for the subsistence of the rural inhabitants. Some of the most notable wetlands in the Peruvian coast include mangroves in bays and lagoons of the north. Tumbes contains mangroves from 03°30′ to 03°59′S, covering 2972 ha, with very notable species biodiversity. This area has a uniform aspect of sandy beaches with vegetation, resembling a jungle, and access is difficult. The National Sanctuary of the Mangroves of Tumbes has the objective to protect the mangrove ecosystem. Furthermore, the National Park Amotape, the National Forest of Tumbes and the Biosphere Reserve of the northeast all include valuable and productive ecosystems (INRENA, 1996).

Engraulis ringens (J): Peruvian anchovy

Sardinops sagax sagax (J): Pacific sardine, Pilchard

Trachurus picturatus murphy (N): Jack mackerel, Inca scad

Scomber japonicus H: Snake mackerel

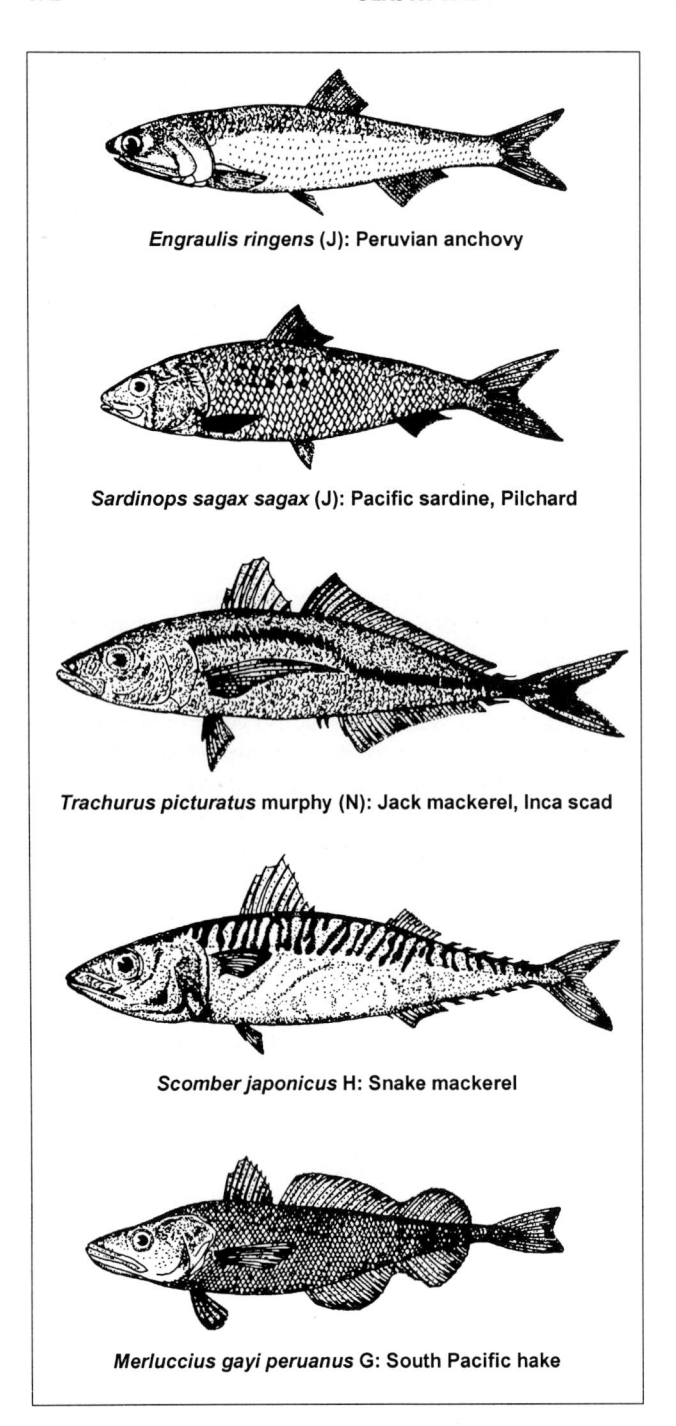

Merluccius gayi peruanus G: South Pacific hake

Fig. 5. Commercially important fish in Peru.

The mangrove species are *Rhizophora mangrove*, and the red mangrove *Avicennia tomentosa*. Mollusc and crab populations are rich, as are several marine birds such as *Phalacrocorax olivaceus* and several species of herons (Peña, 1971). Prawns are of great importance in this zone. These live in the estuaries and include species of commercial importance: *Penaeus occidentalis* "white prawn", *Penaeus stylirostris* "white prawn". In this zone, aquaculture of *P. vannameis* is being developed.

The swamps of Villa cover 396 ha, located in the district of Chorrillos, Lima, between 12°13'S and 77°01'W. This was

Fig. 6. Fur-seals inhabiting islands and peninsulas of the central and southern coast of Peru.

declared as the 'Reserved Zone of Villa Swamps', to protect the flora and fauna. The National Reserve of Paracas, near Pisco-Paracas covers 335,000 ha, 117,406 of which are terrestrial and 217,594 ha are marine. It is located in the district of Paracas between 14°08'S and 76°15'W and continues for 72 km. The Reserve has a rich biodiversity in a variety of habitats and extensive beaches, all influenced by the Peruvian Cold Current which brings high primary productivity. In Pisco-Paracas, more than 330 species of macrobenthic invertebrates have been identified (Acero and Tarazona, 1995). Independence Bay, a special geographical feature, is especially rich (Paredes et al., 1988). Four species of marine turtles (Zeballos, 1991) also occur here, with several marine mammals, including two species of fur seals, otters, dolphins and whales.

The National Sanctuary of Mejia Lagoons in the province of Islay (17°19'S and 71°51'W) covers 691 ha. The area includes small coastal "lomas" along the coastline of the Sanctuary, and receives water from the Tambo river. Along both banks are lagoons which provide shelter for several bird species in danger of extinction and provide refuges for migrant species. It is also the habitat of several endemic species. A common vegetation is "totora" *Typa* sp. and rushes *Scirpus* sp.

Biological Effects of El Niño

During El Niño, the pelagic ecosystem suffers radical changes. The coastal upwelling weakens, allowing warm waters which are poor in nutrients to rise and inhibit high primary production. The composition of the plankton is modified, diatoms commonly disappear and in their place the dinoflagellates become an important part of the trophic chain. Herbivorous pelagic fish feed on the plankton, and in turn are part of the diet of guano birds, fur seals and whales. A "tropicalization" due to the migration of species from equatorial waters to the south, and of oceanic waters toward the coast, reduces both biomass and production.

Zooplankton species are also affected. There is a decrease in endemic species of the Cold Coastal Waters, which are replaced with species of oceanic or sub tropical waters. During El Niño, the biomass of copepods, which are the dominant group during normal years and which are herbivorous, falls considerably and they are replaced by species from warm waters which generally are carnivorous predators of eggs and fish larvae. During the El Niño of 1997–1998, the volumes of plankton decreased considerably, reaching an average of 0.62 ml/m^3 between June and July of 1997 (Chang et al., 1997).

As already noted, the Peruvian anchovy is one of the most important species for the industrial fishery. It inhabits the Peruvian Current and reacts sensitively to warm water. Another important species is the sardine which inhabits oceanic waters jointly with the jack mackerel, mackerel and others. In an El Niño event these species enter coastal waters, replacing the anchovy. In the major El Niño of 1997–98, the anchovy migrated southward to waters in the north of Chile where weak upwellings allowed them to obtain their required food, even though this affected their reproductive activity. In the earlier El Niño of 1982–83, the anchovy showed a similar behaviour but without great success; on that occasion they failed to reproduce.

Demersal species, such as hake, move from their normal locations to others on the upper continental slope which retain good oxygen concentrations. In coastal, shallow and intertidal water, the influence of El Niño is notable, affecting the benthic communities too. Crustaceans and molluscs of the intertidal zone may suffer high mortality from the negative impacts of El Niño, and banks of the mussel *Aulacomya ater* may be eliminated.

Another aspect of El Niño is the immigration of subtropical and tropical species. Crustaceans of economic importance include *Xiphopenaeus riveti* and *Penaeus occidentalis* which moved southwards as far as Ilo in 1982–83 and to Callao in 1997–98. A mollusc favoured by El Niño is the scallop *Argopecten purpuratus*.

FISHERIES

Pelagic Fisheries

According to FAO (1994), the fishery in the world reached a total of 103 million tons in 1994, based mainly on continental fishery activity, aquaculture and the Peruvian anchovy. Pelagic species like the anchovy, sardine, jack mackerel and mackerel (Fig. 5) sustain the industrial pelagic fishery, the first two species being used for the manufacture of fish meal. Fluctuations in catches are based on fluctuations of the marine environment.

In the 1980s, landings of anchovy and sardine increased until 1994 when 9.17 million tons were taken, followed in 1995 and 1996 by a decline in catches as a consequence of cold surface temperatures (Fig. 7). However, in 1996 a landing of 8.08 million tons between both species was reached. With the El Niño of 1997–98, catch of both species totalled 6.47 million tons (Bouchon and Ñiquen, 1997).

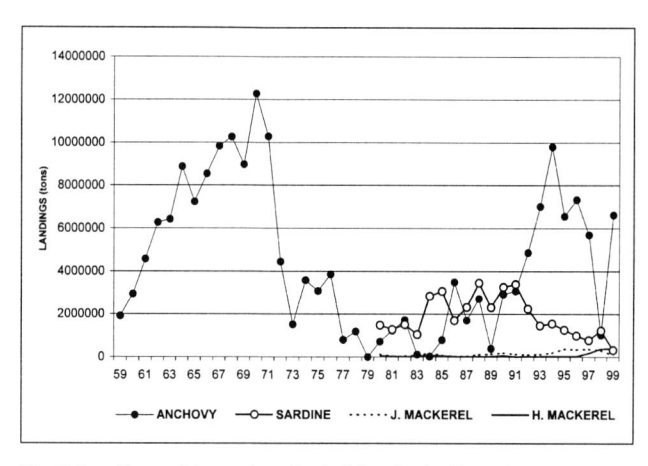

Fig. 7. Landings of the main pelagic fishes in the Peruvian ecosystem. Source: Bouchon and Ñiquen, 1997)

Demersal Fishery

Several important species include the hake *Merluccius gayi peruanus* that inhabits deep areas between 100 and 200 m (Guevara et al., 1996) including the edge of the shelf and upper part of the continental slope in areas associated with subsurface currents (see Fig. 5). This species is distributed mainly north of 10°S (Samame et al., 1983).

Recently the fishery of this species has increased, from 56,000 tonnes in 1993 to 192,000 tonnes in 1996. In 1997, the landings were reduced by the presence of El Niño but in 1998 there was a recovery of the fishery in the northern coast of Peru. Other demersal species include *Cynoscion analis* "common Peruvian weakfish", *Trachinotus paitensis* "paloma pompano", *Paralabrax humeralis* "Peruvian rock seabass", "smoothhounds" and *Prionotus stephanophrys* "blackfin gurnard". Altogether, landings of these species have started to decline in recent years.

A different group of species are targeted by the artisanal fisheries (Fig. 8) and some new species and invertebrate species are also taken (Fig. 9).

COASTAL POPULATIONS AND THE MAIN SOURCES OF POLLUTION

The coast of Peru has a population of 13,277,079 inhabitants according to the last census of 1993 (see Fig. 1). This is a marked population growth from the 1950s, and growth has accompanied a favourable development in the economy. The latter has allowed the development of programs of conservation and restoration of cases of environmental deterioration which resulted from the often disordered growth of its cities (Sanchez and Muñoz, 1995).

Activities on the coast are very diverse (Fig. 10), causing pollution of different types and with different degrees of intensity. In some cases the sources of contamination have been identified and evaluated; domestic pollution is the most notorious.

Paralabrax humeralis (V): Peruvan rock seabass

Paralichthys adspersus (S): Fine flounder

Odontesthes regia regia (H): Peruvian silverside

Trachinotus paitensis (C): Paloma pompano

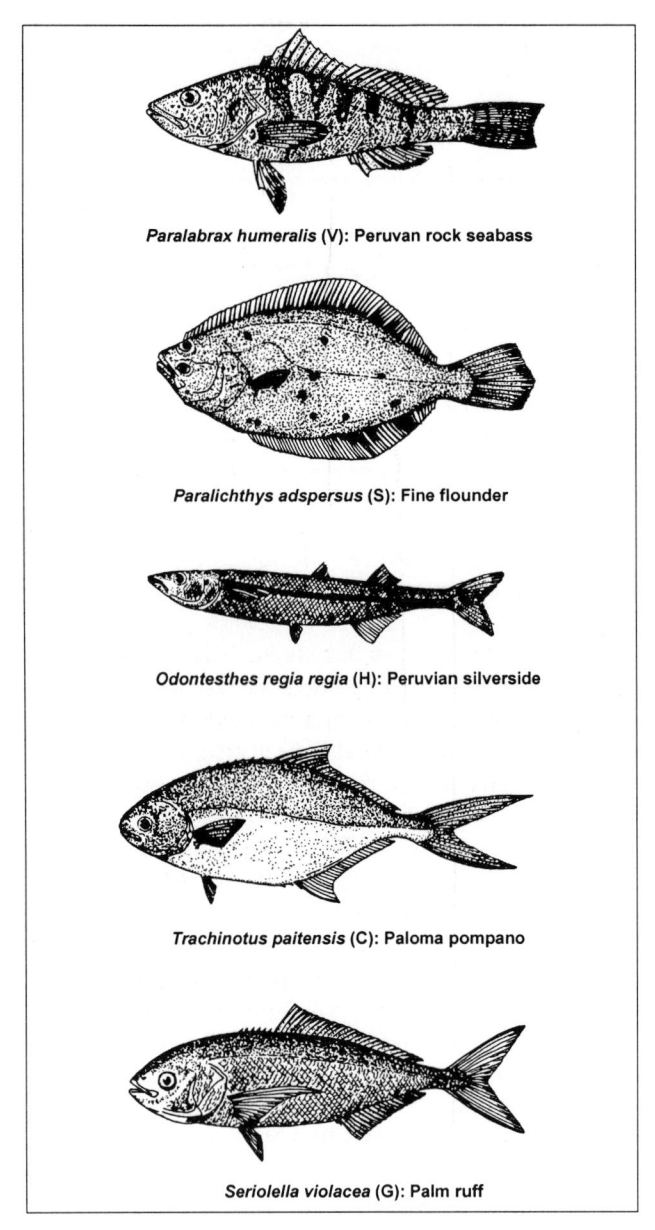

Seriolella violacea (G): Palm ruff

Fig. 8. Fishes important to the artisanal fisheries.

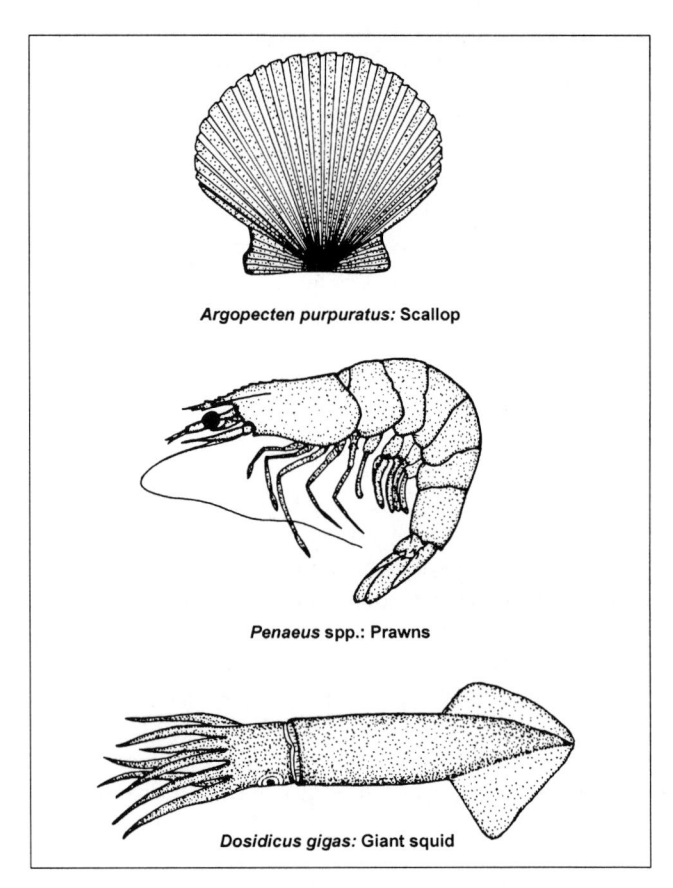

Argopecten purpuratus: Scallop

Penaeus spp.: Prawns

Dosidicus gigas: Giant squid

Fig. 9. Invertebrate fishery species in Peru.

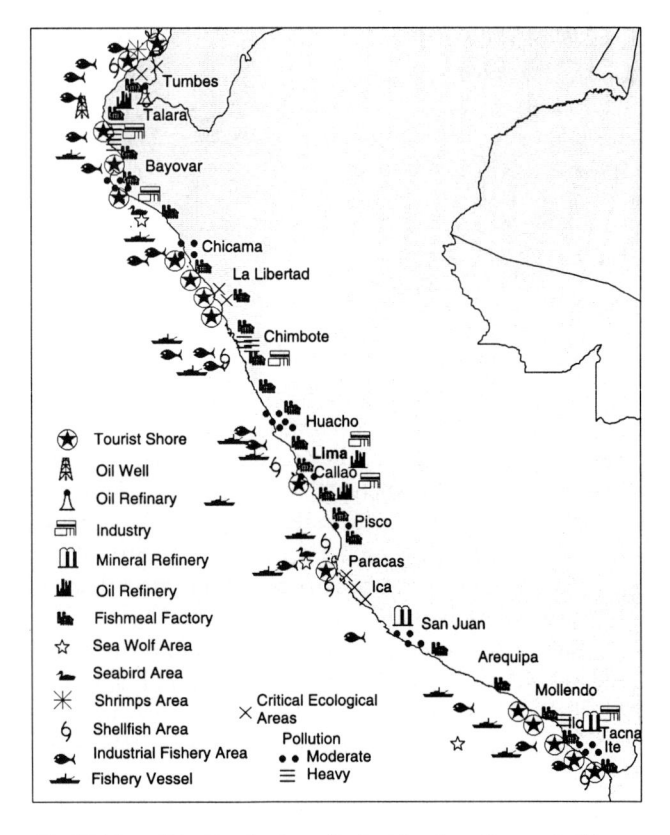

Fig. 10. Map of the distribution of industries along the coast of Peru.

Habitat Degradation

Habitats in the coastal marine zone in Peru have suffered alterations from both natural processes and various activities resulting from the increase in population and the demand for land. As a result, pollution of rivers, wetlands and the sea has arisen from industrial, mining and agricultural wastes. Residues from tanneries, paper mills, mining and metallurgical factories are highly toxic, and the fishing waste in Chimbote, Paracas and Paita strongly degrade the quality of the marine environment in those locations. Occasional increases in the intensity of rains causes overflows of rivers, which lead to flooding of vast alluvial plains that stay dry the rest of the year.

Heavy use of pesticides produces toxic residues that contaminate beds of rivers, altering the reproduction of

The Main Populated and Industrial Areas

The city of Tumbes in the province of the same name is located in the north of Peru, with a population of 74,601. Activities are artisanal fisheries, agriculture and trade. There is intensive culture of prawns in the delta of the Tumbes river where 9000 ha is worked by 60 companies. Agriculture is mainly rice, followed by fruit. The Province and city of Talara is north of Lima. Exploration and exploitation of petroleum on the coast and continental shelf is intense, as is artisanal fishing. Talara is a fishing port supplying the interior also. Greatest landings come from jack mackerel, mackerel, drums, hake and mullet, as well as giant squid and prawns.

The city of Paita has 42,491 inhabitants, the main activity of whom is fishing, both industrial and artisanal. Paita has three industrial areas containing fish processing and fishmeal plants, and a wheat milling plant. The Province of Trujillo contains the important port of Salaverry. The region also contains one of the most extensive irrigation projects on the coast. The Province has a population of 639,554, many of whom are involved in leather manufacture, canning of crops or production of sugar, paper, liquors, pottery and handmade products. Sugar cane is extensive. Fishing activity is important mainly in the district of Chicama where Peruvian anchovy, sardine, jack mackerel and mackerel are landed.

The city of Chimbote is the main fishing port of the Peruvian coast and has 278,271 inhabitants. In the 1960s, Chimbote was known as the First Fishing Port of the World, receiving millions of tonnes of anchovy and sardine, and contains 30 fishmeal, oil and cannery factories. Some steel, electro-ceramic, and trading exists, along with ship repair. Substantial amounts of cotton and rice are grown.

Lima, the capital of Peru, is located on the central coast. It has 350,000 inhabitants and a floating population of more than 1,500,000. Activities are very diverse, covering 60% of the industrial production and 90% of the commercial, financial and administrative services. Callao is the main marine port. The population of the region is approximately 6.5 million inhabitants (30% of the total population in the country) with an annual growth of 2.4%, due mainly to high migration from rural areas.

The Province of Pisco is 249 km south of Lima. The Bay of Paracas, the National Reserve of Paracas and the Bay of Pisco are important areas. The latter has substantial fishing and agriculture, and its beaches are used for vacation. Pisco has an industrial area, whose wastes are discharged to the sea. The Bay of Paracas has intense artisanal and industrial fishing.

The Province of Ilo has the most important fishing port in the south and two important maritime ports: Ilo and Ilo-Mining. The Province has a population of 52,182, some of whom work in two copper mines. On the coast, industrial activities are smelting and refining. Annual production of copper is of 300,000 metric tons. Mine tailings were transported 100 km, and flow by gravity to the coast and are now disposed of in a tailing trench. Industrial fishing produced 121,688 tons of fishmeal and 15,311 tons of oil in 1993. Agriculture is mainly of olives.

aquatic birds and freshwater species like the river shrimp. Expansion of agriculture has resulted in the loss of large areas of wetlands, such as areas surrounding the lagoons in the national Sanctuary of Mejia which lost approximately 3000 ha of wetlands in the 1990s (INRENA, 1996).

Domestic and Industrial Pollution

There are 53 main hydrographic basins in Peru. One of the most important rivers is the Rimac river because it is the main source of water for Lima, the capital. Its quality has deteriorated because of the rapid and chaotic growth of the population, especially in the littoral, producing very high levels of biological and chemical pollution. This occurs fundamentally because of illegal discharges in many points of its course, discharges of tailings from the mineral works, and from point discharge of domestic wastes from both banks.

Furthermore, industrial and domestic waste waters are discharged through various pipes, with an approximate volume of 18 m^3/s. It has been estimated that in such areas about 72.2% of the total organic load in the water is released in this way, representing 89,510 tonnes BOD per year (Sánchez and Muñoz, 1995).

The principal industries which discharge residual waters to the sea are fishing, mining and oil. Thus, the pollutant load from fishing industries has been quite severe in areas such as Pisco-Paracas, Chimbote and Paita. However, legal measures have recently been made to minimize their impact, especially where the coastal form and low speeds of the currents do not allow for good dispersion and dilution.

The mining industry and related activities have highly toxic residues that have caused serious damage in some coastal water bodies. For example, in the Rimac river there is no longer any kind of aquatic fauna due to tailings of mineral concentrates from mining in the western mountains. This effect is also migrating to the marine environment where there is deterioration in the outlet area.

Mining

In 1995, the Ministry of Energy and Mining identified a total of 24 enterprises with 30 plants involved in the deposition of residues, tailings and dross. Among these, 3 plants had discharges to the marine environment. Their final products are principally the minerals of baritine ground, bentonite ground, Portland cement, copper, lead, zinc, cadmium, sulphuric acid, white cement and non-metallic products, coal, iron, gold, concentrates of copper and molybdenum, copper–silver blist and calcium carbonate. Ite bay in the south is one of the areas most affected by mine tailings. Over more than 30 years the copper mines have discharged their tailings into the coastal area, and at the end of 1996 the discharge was 90,000 metric tons of tailings per day into Playa Inglesa, a beach in the Department of Tacna. These tailings are deposited in pools, which is a better alternative than direct discharge to the sea.

Oil Hydrocarbons

The oil activity in Peru, which includes exploration, exploitation (in the sea and on the land), processing and distribution, is carried out in the coastal littoral mainly in the north, including in the Peruvian tropical forest. Most extraction is done on the continental shelf, the crude oil and gas being taken to large storage tanks to be transported by ships to the refineries near Lima. Oil from the forest is led to the northern port of Bayovar through the Transandean pipeline, and from this port it is moved by tankers to the refinery at La Pampilla.

Loading and unloading of crude and refined oil is accomplished by submarine lines or docks. In these areas there is a high risk of oil spills, and because of this, the National Contingency Plan has been implemented to combat oil spills by means of widespread coordination between the Navy and participating institutions.

This Contingency Plan has been executed several times between 1990 to 1995. In 1990, 14,000 barrels of kerosene were spilled in the Bay of Supe when a vessel grounded. There was serious discomfort to the population near the coast, because of the smell which lasted for several days. Artisanal fishing was affected because it altered the distribution of species, and strongly disturbed benthic communities on rocky substrates and affected beaches. The National Contingency Plan was activated, and the spill resulted in temporary prohibition of the artisanal fishing and some research to evaluate the effects of the incident.

Other spills have involved 438 barrels of crude oil on the beaches of Conchan caused by breakdown of the submarine line, and a spill of diesel on the beaches of Ventanilla. In both these cases too, the National Contingency Plan was activated at the local level, and technical research was conducted to evaluate effects.

Agriculture

Important agricultural products grown on the coast include rice, which in 1993–94 was planted on 117,300 ha, mainly in the north where major rivers exist for irrigation. Sugar cane occupies 120,200 ha of the best irrigated lands in the country. Both use pesticides which, in Peru, are widely marketed. At present, approximately 548 chemical products of biological and synthetic origin are sold for controlling crop diseases, most of this variety and quantity being used on the Peruvian coast and forest edge (SENASA, 1994).

The use of pesticides, especially organochlorines, increased by 100% between 1982 and 1987. However, the higher use did not create higher crop yields, but it did cause economic expense and adverse ecological impacts, such as the sterilization of soils.

Coastal Erosion

The main erosion processes are deflation, dune fields, river and rain erosion, slides, mud flow or "huaycos", vions and intense laminar erosion (Alvarez, 1989).

These cause destruction of vulnerable ecosystems, such as mangroves in Tumbes, where between 1.5 and 2.0 million m^3/year of sediments is carried by the rivers when the warm El Niño event occurs. It is noteworthy that on the coast of Tumbes, intensive deposition of sediments causes constant morphological change and growth of new land (ONERN, 1989).

Added to this are contributions of development and mining. In the active deltas of Pisco and Locumba rivers, deltaic sedimentation comes from 30 million m^3/year of tailings (until December of 1996), which has increased the land by 1 to 2 km seaward along 20 km of shoreline, with submarine sedimentation of 1 to 20 m thickness extending 15 km offshore.

Erosion may also take place, however, such as occurred in the north coast of Trujillo following the installation of rocky breakwaters that altered the coastline.

PERUVIAN LEGISLATION ON ENVIRONMENTAL PROTECTION

Peru has in its legislation, laws for the protection of the atmosphere and the sustainable development of its geographical areas. In the 1993 Political Constitution of Peru, there is a decree stating that all persons have the right to enjoy an appropriate environment for the development of life.

The Environmental and the Natural Resources Code, Decree Law No. 613, contains articles which require Environmental Impact Studies (EIA) in various circumstances. The Organic Law of Municipalities has the objectives of formulating, approving, executing and supervising the integration of development plans.

The Framework Law for growth of Private Investment deals with the conservation of the atmosphere and the sustained use of natural resources. For that purpose, clear rules for the protection of the environment and for reduction of environmental pollution will be given.

On the other hand, each sector has established rules for sustainable development, which guarantee the protection of natural resources and the economic and social growth of the people of Peru. Sectors like Agriculture, Energy and Mines, Fishery and Industries have also established rules for conducting Environmental Impact Studies.

REFERENCES

Acero, R. and Tarazona, J. (1995) Informe Nacional. Areas Costeras y marinas Protegidas del Perú. INRENA. Informe para la Reunión del Grupo Ad-hoc sobre Areas costeras y Marinas Protegidas. CPPS. Lima.

Arntz, W. and Fahrbach, E. (1996) EL Niño, Experimento Climático de la Naturaleza. Fondo de Cultura Económica, Mexico, 312 pp.

Alvarez, J. (1989) La erosión en la zona costera continental peruana. En el seminario-Taller sobre la erosion de la zona costera del Pacífico Sudeste. CPPS-PNUMA.

Bouchon, M. and Ñiquen, M. (1997) Estadísticas de la pesquería pelágica en la costa peruana. Informe Interno DGIRH. Inst. Mar Perú.

Chang, F., Delgado, E. and Villanueva, P. (1997) Caracteristicas del fitoplancton con enfasis en los indicadores biologicos durante el Crucero Oceanografico 9706-07. Informe Interno. DOB. Inst. Mar Peru.

FAO (1994) Anuario Estadístico de Pesca, capturas y desembarques No. 78.

Guevara, R., Castillo, R. and Gonzales, A. (1996) Aspectos metodológicos de la evaluación directa de la merluza (*Merluccius gayi peruanus*) con el método del Area Barrida (Cr.BIC SNP-1, 9505-06). *Inf. Inst. Mar Perú* **117**, 8–15.

Instituto del Mar del Peru (1997) Aporte al conocimiento de la diversidad biológica marina de la Reserva Nacional de Paracas, Ica, Perú. Informe a la CPPS-PNUMA.

INRENA (Instituto Nacional de Recursos Naturales) (1996) Estrategia Nacional para la conservación de Humedales en el perú. Programa de Consevación y Desarrollo Sostenido de Humedales, Lima. Perú.

Ministerio de Marina (1982) Derrotero de la Costa del Perú. Vol. I and II. Direc. De Hidrografía y Navegación de la Marina. HIDRONAV-34.

ONERN (Oficina Nacional de Evaluacion de Recursos Naturales) (1989) Delta Tumbes. En el Seminario-Taller sobre erosión de la zona costera del pacífico Sudeste. CPPS-PNUMA.

Paredes, C., Tarazona, J., Canahuire, E., Romero, L. and Cornejo, O. (1988) invertebrados Macro-Bentonicos del Area de Pisco, Peru. In H. Salzwedel y A. Landa (ed.) Recursos y Dinamica del ecosistema de Afloramiento Peruano. *Boletino Instituto del Mar del Peru-Callao*, Vol. Extraordinario, 121–132.

Peña, M. (1971) Zonas de distribución de los bivalvos marinos del Peru. An. Cient. Univer. Nacional Agraria. Nº 9(1-2):46-55. Lima.

Samame, M., Espino, M., Castillo, J., Mendieta, A. and Damm, V. (1983) Evaluación de la población de merluza y otras especies demersales en el área de Puerto Pizarro-Chimbote. (Cr.BIC Humboldt 8103-04, marzo–abril 1981)). *Boletino Instituto del Mar del Perú* **5** (5),111–191.

Sanchez, G. and Muñoz, A. (1995) Contaminación Marina en el Peru proveniente de Fuentes de Origen terrestre. Informe de Consultoría. Informe Nacional. CPPS-PNUMA.

Sanchez, G. and Orozco, R. (1997) Diagnostico Regional sobre las Actividades Realizadas en Tierra que afectan los ambientes Marino, Costero y Dulceacuicolas asociados al Pacifico Sudeste. Informe de Consultoría. Informe Nacional. CPPS-PNUMA.

Sanchez, de B.G. and Soldi, H. (1988) Survey and Monitoring of Marine Pollution in Peru. In: Regional Co-operation for Environmental Protection of Marine and coastal areas of the Pacific Basin, UNEP 1988. *Regional Seas Report and Studies* **97**, 99–114.

SENASA (Servicio Nacional de Sanidad Agraria) (1994) Plaguicidas Agrícolas y Sustancias Afines registradas en el Ministerio de Agricultura. Informe del Servicio Nacional de Sanidad Agraria, Perú.

Wyrtki, K. (1963) The horizontal and vertical field of motion in the Peru Current. *Bulletin of the Scripps Institution of Oceanography* **8**, 313–346.

Zeballos, J. (1991) Situacion de las Tortugas Marinas en el Perú. Informe Interno DGIRH. Instituto del Mar del Perú.

Zuta, S. and Guillén, O. (1970) Oceanografía de las Aguas Costeras del Perú. *Boletín Instituto del Mar del Perú* **2** (5), 157–324.

THE AUTHOR

Guadalupe Sanchez
*Instituto del Mar del Peru,
Callao, Peru*

Chapter 45

THE CHILEAN COAST

Ramón B. Ahumada, Luis A. Pinto and Patricio A. Camus

The Chilean coast extends from 18°30" to 57°30"S, along the western coast of South America. This region of the Pacific is under the influence of a sub-Antarctic current system, known as the Humboldt Current. Dominant coastal south-southwest winds cause seasonal upwelling near the coast. Upwelled water masses rich in nutrients are responsible for an elevated primary production in coastal waters which can sustain an annual fisheries of seven million tons.

The Chilean coast is a straight, west-facing shoreline with few embayments. A desert dominates the north part of the country, a transitional region is in the centre and temperate forest develops in the south. In the north, industrial development has occurred including mining, exploitation of marine resources and the installation of fuel terminals. Dumping and disposal of mining tailings on the coast of Chañaral is considered one of the major environmental impacts of mining activities in the country. Agriculture, manufacturing and heavy industry appears in the central part of Chile. Between 34° and 42°S, timber, pulp and cellulose industries together with dairy farms are important. Recent development in coastal areas has expanded the oil and fishmeal industries, the chemical industry, oil refineries, the loading of mineral ores, and general loading of goods—all exerting pressure on the few semi-enclosed embayments. The impact of these activities is local but very intensive, causing the disappearance of local fauna.

The morphology of the southern coast beyond 42°S is complex and formed by fjords, channels and islands. Recently, this region has increased its economic development mainly by the creation of salmon farms, exploitation of timber, and tourism. This area has a low-density population, but this could change drastically in the new millennium. Mining of zinc and lead along the Aysen Sound are potential sources of contamination in what can still be perceived as a pristine area.

The effects of "El Niño" events—the regional manifestation of a large-scale ocean–atmosphere fluctuation (Southern Oscillation)—on the Peruvian and Chilean coasts are varied and dependent on their intensity, and are often perceived as 'positive' or 'negative' according to their impact on human populations or economic activities. These effects can also be classified as biological or physical ones, which are usually interdependent. To date, the real ecological or evolutionary impact of El Niño on coastal species or communities remains largely unknown.

Environmental conditions on the coast of Chile are strongly influenced by the nature of the water masses. The subantarctic water mass (SAAW) regulates the coastal climate while the subsurface equatorial water mass (ESSW) brings nutrients and low oxygen to the surface waters during upwelling events. Introduction of high-nutrient, low-oxygen waters to shallow embayments causes high primary production to increase, high sedimentation rates generating sub-oxic bottom waters. These ecosystems are highly susceptible to contamination by high loads of organic matter, creating eutrophic or even hypertrophic conditions. Saltmarshes are an example of subsystems where hypertrophication has occurred as a result of pollution (Rudolph and Ahumada, 1987; Ahumada et al., 1989).

Seas at The Millennium: An Environmental Evaluation (Edited by C. Sheppard)
© 2000 Elsevier Science Ltd. All rights reserved

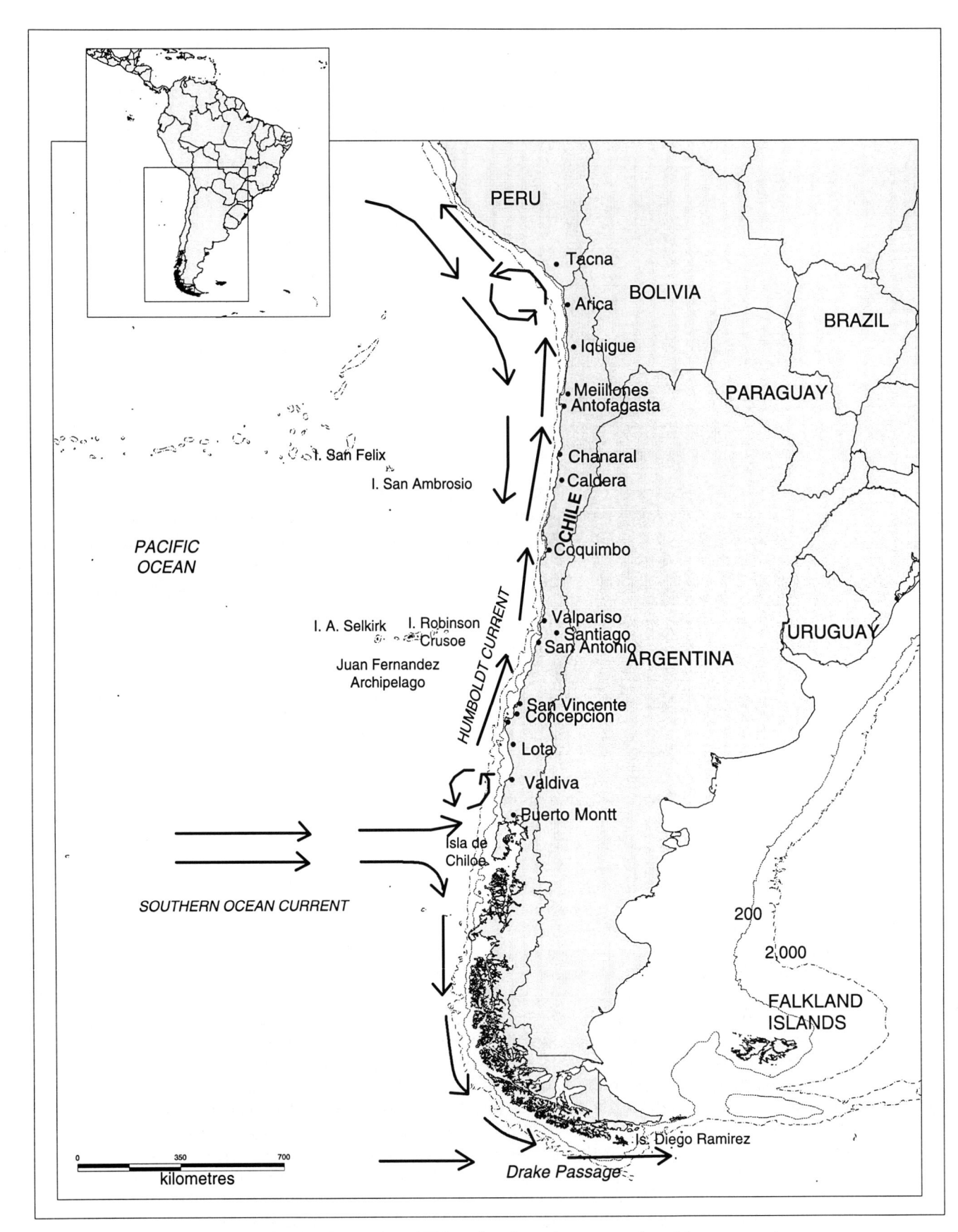

Fig. 1. Map showing the coastline of Chile and a schematic representation of the sea currents.

The large-scale patterns of distribution and abundance of marine species in Chile has four major determinants. The first is a massive cooling effect along the continental coast associated with the combined influence of subantarctic waters and upwelling processes, which confer a predominantly cold character on the biota. The second is the latitudinal variation of climatic and topographic features of the coast, which is important in the differentiation of the present main biogeographic regions. The third is a possible spatial differentiation of the biological consequences of El Niño (and La Niña) on coastal populations and communities, in which the northern Chilean area would have experienced greater disturbance effects at historical time scales. The fourth is a factor of a different nature such as the geographical variation in the occurrence and strength of anthropogenic impacts due to the marked spatial differences in the distribution and concentration of both coastal human populations and industrial activities along the coast.

The creation of the National Environmental Commission and recent environmental regulations have recently begun to help public awareness about environmental issues. Consultants have gained experience in understanding the need to harmonize economic development with nature conservation. Slowly, these new ideas are helping in the preservation of the coastal environments in Chile.

THE PHYSICAL SETTING

The Chilean coast extends in latitude from 18°30" to 57°30"S, along the western coast of South America. This region of the Pacific is under the influence of a sub-Antarctic system current, known as the Humboldt Current. The West Wind Drift current (WWDC), located on the sub-Antarctic zone at 42°S, reaches the continent, creating: (a) an equator-ward surface coastal current (the Peru–Chile current) and an ocean current, both flowing separated by an Equatorial Countercurrent; and (b) a poleward flow known as the Cape Horn current. The water mass carried by the WWDC is defined as a Sub-Antarctic water mass (SAAW) (Fig. 1).

The Chilean coastline located between 18°S and 41°30"S, is an extended and straight, west-facing shoreline, with few embayments along the coast. The continental shelf is narrow in the northern zone and has a depth of 100–150 m. The shelf off Valparaiso is an exception, where a hang-shelf appears eight nautical miles wide and 800 m in depth. The widest part of the shelf in this coast is found off Talcahuano, being 25 nautical miles wide and 150 m deep, and flanked by two submarine canyons—the Itata canyon and the Bio-Bio canyon—both originating from active rivers during the Quaternary period. In the southern area as far as Puerto Montt (42°S), the continental shelf is incipient and narrower than off Talcahuano.

The morphology of the southern coast beyond 42°S is complex and formed by fjords, channels and islands, produced by glacial erosion occurring over the past 12,000 years, after glacier lobes covered the continent (Clapperton, 1994). This region extends to Diego Ramirez island (56°30"S) on the Cape Horn.

The ocean floor extends from the continental margin and is characterised by a deep trench, approximately 5000 km long from the equator to 46°S, off the Penas Gulf. The Chilean trench located at the boundary of the Nazca Plate, converges towards the South American plate. At 46°S, the Chilean submarine mountain range reaches the continent shaping the boundary of the plates of Nazca, South America and Antarctic, known as the Triple Junction. Four systems of oceanic islands are associated with the Pacific region of Chile: (a) the islets of San Félix and San Ambrosio; (b) the Sala and Gómez islet; (c) the Juan Fernández Archipelago (or Robinson Crusoe); and (d) Pascua island (or Easter Island).

WATER MASSES

Several water masses (Fig. 2) are described in relation to the dynamics of the South Eastern Pacific:

The Subtropical Water Mass (STW), characterised by superficial, oceanic, warm and saline waters which conform to the gyre of the central ocean and seasonally move to the northern coast of Chile and southern Peru. During "El Niño" events, the STW reaches the coast as a superficial water mass between 0 and 50 m deep. Changes in the physical and chemical features of the seawater induce deep changes on intertidal and subtidal coastal species. Temperature increases +3 or +4°C at the coast in a layer 30–50 m deep. The coastal upwelling process is depressed or ceases, and often exotic species are advected to or reach the coast, while nutrients are depleted and fish migrate from the coast or experience mass mortality.

The Sub-Antarctic Water Mass (SAAW), characterised by medium temperatures and low salinity water (Table 1). This is more dense than the Subtropical Water Mass, and when these two water masses converge on the coast, the latter (STW) takes place on the surface. The flow of the SAAW is

Table 1
Physical and chemical features for the Pacific East Boundary water masses

Water mass	Depth (m)	Temp. (°C)	Salinity (UPS)	Density	D. oxygen (ml/l)	Nitrate (μM)	Nitrite (μM)	Ammonia (μM)	Phosphate (μM)	Silicate (μM)
STW	0–100	>20	>35.00	<25.3	Saturation	>5.0	0.5	0.3	>0.5	>10.0
SAAW	0–300	12–17	33.8–34.6	<26.4	Saturation	1.5	0.3	0.1	0.1	3.0
ESSW	300–600	8–12	34.5–34.9	<26.8	>1.0	>35.0	1.0	10.0	3.5	50.0
IAAW	700–1400	6–7	34.3–34.4	<27.3	2.0–4.0	>40.0	0.3	0.1	>2.5	<80.0
PDW	1500–5000	2–5	34.5–34.7	<30.0	4.0–6.0	>40.0	0.5	0.1	>3.0	>100.0

Data source: Silva and Konow, 1975; Reid, 1973; Ahumada, 1989.

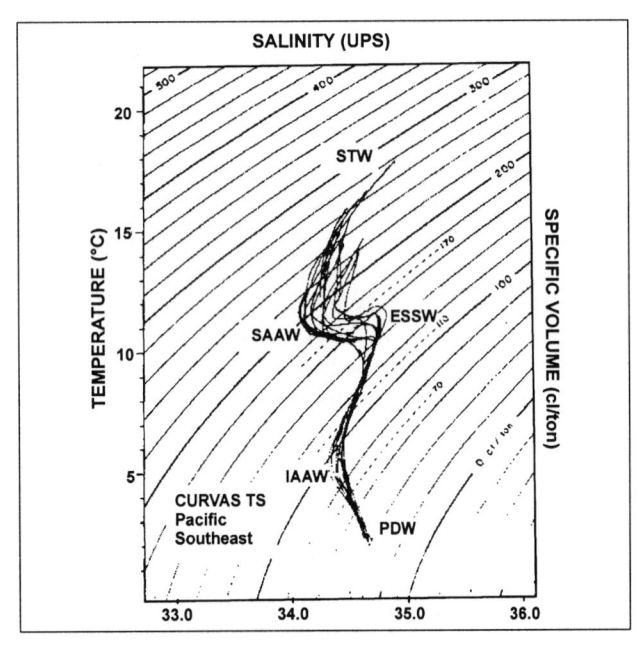

Fig. 2. Temperature–salinity diagram showing the dominant water masses off the coast of Chile (taken from Atlas Oceanográfico para la Educación, Servicio Hidrográfico de la Armada de Chile).

equator-ward and has two branches: the oceanic current and the coastal current, jointly referred to as the Humboldt Current System. Usually, this water mass moves between 0 and 300 m deep along the coast, with some temperature and salinity changes due to the exchange of heat and water vapour with the atmosphere. Figure 2 shows the T–S curve for the SAAW indicated by a star in a curve near the subantarctic region.

The Equatorial Subsurface Water Mass (ESSW) has a poleward flow and is characterised by low temperature and medium salinity. The relevant feature of this water is its low content of dissolved oxygen (less than 1 ml/l). The oxygen is depleted by remineralization of organic matter produced as detritus that "rains" from the euphotic zone. Respiration processes consume the dissolved oxygen and produce new nutrients or oxidative nutrients, distributing them along the coast during an upwelling event. This water mass is responsible for the elevated primary production in coastal waters.

The Intermediate Antarctic Water Mass (IAAW) with an equator-ward flow is characterised by low temperature and low and constant salinity. This water mass appears between 500 and 1500 m in depth, and moves equator-ward.

The Pacific Deep Water Mass (PDW) is characterised by low temperature and saline waters with a poleward flow. These waters have an ascending movement to the south and reach the surface at the Antarctic divergence.

In the coastal zone, the surface water SAAW dominates throughout the year. From 22°S and during the summer, the STW spreads causing the thinning of the SAAW as a coastal current. During "El Niño" events, the STW may reach the coastline, producing several ecological changes as discussed in the next section.

LARGE AND MESOSCALE NATURAL VARIABILITY

Long-term Variations: El Niño Southern Oscillation (ENSO)

El Niño events are of particular importance. An early sign that usually allows recognition of the occurrence of an El Niño phenomenon is the mass mortality and stranded fish observed on the north and central coasts of Peru during the month of December. The onset of the event is associated with a mesoscale ocean–atmosphere phenomenon beginning in May/June, developing from July until November, and reaching the coast in December and January/February of the following year.

El Niño, the regional manifestation of a large-scale ocean–atmosphere fluctuation (Southern Oscillation), is brought about by intensification and relaxation of easterlies or southeast trade winds. The Southern Oscillation has been associated with the variation of the atmospheric pressure system in the Pacific ocean: the high pressure at the Pacific anticyclone and the low pressure at the Indonesia cyclone. A weakening of the Pacific anticyclone produces a warm event known as the negative face of the Southern Oscillation or ENSO. The magnitude of the southeast trade relaxation, and its timing in relation to regular regional seasonal relaxation, determines the strength of the resulting El Niño event. The mechanism

Table 2
Years of occurrence of El Niño along the South West coast of South America (after Quinn and Neal, 1983) using the Chile Rainfall index.

	Year													
	1934	1936	1941	1944	1953	1958	1965	1968	1972	1978	1983	1987	1991	1997
Intensity	W	W	S	W	S	M	W	S	VS	W	VS	W	W	VS
S.M.T.A.										+2	+7	+3	+2	+4
Lat. South					22°				30°	18°	33°	25°	22	?

S.M.T.A = Surface Mean Temperature Anomaly off Callao (Perú). Lat. South = effect along shore.
Intensity scale: W = weak; M = moderate; S = strong; VS = very strong.

suggested by Quinn and Neal (1983) establishes that before El Niño, strong southeast trades and equatorial easterlies intensify the Equatorial Current, coinciding with the build-up in the sea level and an accumulation of warm water in the Western Pacific. As soon as the wind relaxes, the accumulated water flows eastward producing a change in the sea level in South America. It occurs in the form of an internal equatorial Kelvin wave that spreads along the coast of South America to both hemispheres. In the southern hemisphere an abnormal warming of surface waters occurs, associated with changes in the circulation patterns. To date, different indices have been developed to measure and forecast the magnitude of ENSO events. The most important indices are the anomaly of atmospheric pressure in Tahiti and Darwin (Southern Oscillation index, SOI), the South tropical rainfall in Chile (Quinn and Neal, 1983) and the mean of sea surface temperature anomaly.

According to Quinn and Neal (1983), El Niño events have a mean frequency of 3.2 years, being of varying intensity and somewhat erratic in time. It was previously thought that El Niño was an aperiodic and heterogeneous process, likely due to the different time intervals (seven or eleven years) between two or three strong events and to their "normal" or "abnormal" features according to their suggested intensity (Table 2). To date, however, El Niño is better conceived as a complex, multi-levelled phenomenon integrating events with different recurrence intervals associated with different intensities, whose predictability still remains elusive.

A new index called the Multivariate ENSO Index (MEI) is used to predict the event when it shows the first signs, i.e., when the process begins to show some anomalous climatic features. Comparison of different indices does not always provide the same results regarding El Niño years (Fig. 3). According to the MEI, there was a strong signal of the development of an El Niño for 1998. During July and October 1997, the index revealed an intensity similar to that of El Niño 82–83, followed by a clear decrease from November–December 1997 to January 1998 (Fig. 4), and the model predicted a return to normal conditions by May 1998, as apparently occurred. The effects of El Niño events on the Peruvian and Chilean coasts are varied and dependent on their intensity, and are often perceived as "positive" or

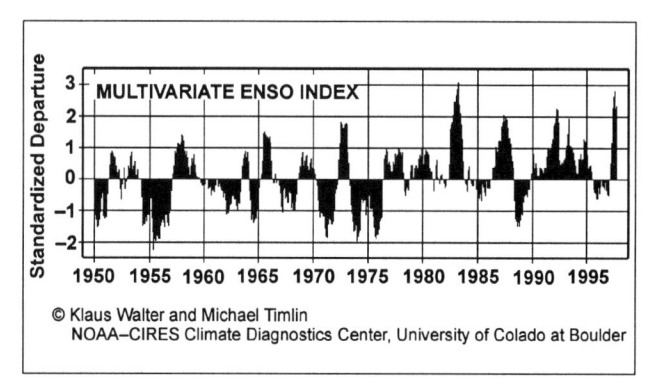

Fig. 3. Multivariate ENSO Index (MEI) intensities for a time series of 47 years (after Walter and Timblin, Climate Diagnostics Center Web page).

"negative" according to their impact on human populations or economic activities. Likewise, these effects can also be classified as biological or physical ones, which are usually interdependent. Among the several physical alterations due to El Niño events are: (a) warming of the surface ocean waters and coastal waters—changes in temperature up to +6°C along the coast of Perú have been detected; the positive temperature anomaly decreases poleward; (b) warming of the air masses up to 2 or 3°C over the normal temperature; (c) deepening of the thermocline, as much as 270 m over the average (Guillén et al., 1985); (d) weakening of upwelling events; (e) intrusion of the STW to the coast, producing a series of ecological changes; (f) rise of sea level up to 30 cm; and (g) different climate changes occurring in the southern hemisphere before, during and after an El Niño event, such as heavy and episodic rainfall in Chile and Peru, alternating with increased atmospheric temperature (e.g., +3°C).

Among the biological effects of El Niño are, for instance: (a) migration of adults and/or death of juvenile and larval fish species; (b) mortality of some macroalgal species on the coast due to nutrient depletion, secondarily generating habitat loss for several associated species; (c) bathymetric migration of certain coastal fish to reach for appropriate temperature conditions; (d) decrease of the sea lion population due to the migration or death of their food resources; and (e) increased growth or development of several opportunistic species, some of them of economic importance (e.g.,

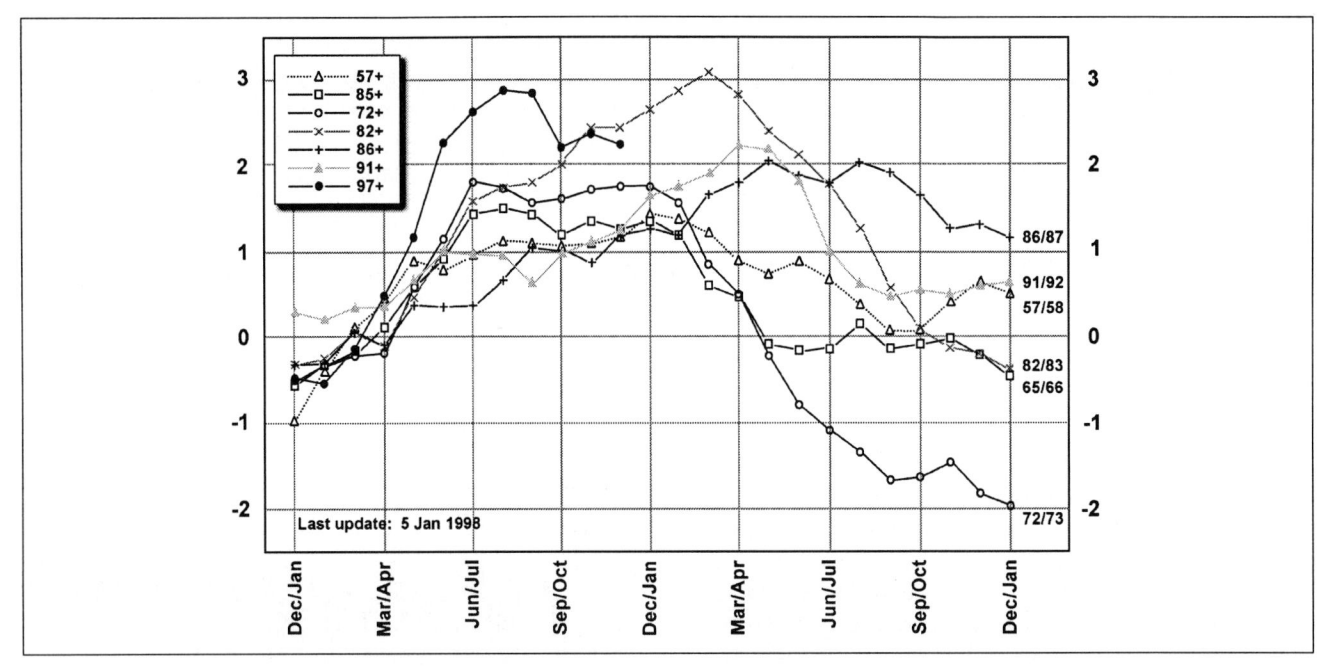

Fig. 4. MEI showing monthly variation for the six strongest El Niño events.

Octopus vulgaris and *Argopecten purpuratus*). More generally, these biological effects can be described as breeding or recruitment failures and abundance fluctuations of varying extent. One of the most dramatic impacts occurs when mass mortality finally leads to local extinction in some particular species, altering their geographical population structure (and consequently changing the structure of gene flow and dispersal) or even their latitudinal distribution range. Some species do not always recover to their pre-El Niño conditions, and the occurrence of a new event can have unpredictable consequences if coupled with prior ones. To date, the real ecological or evolutionary impact of El Niño on coastal species or communities remains largely unknown.

Seasonal Variation

Strong differences in seasonal climatic regimes are present along the extensive coastline of Chile. From latitude 18°30"S to 30°S, no large differences between winter and summer exist. Winter is usually a dry season with a mean rainfall of 1 mm in Arica (≈20°S) to near 100 mm in Coquimbo (≈30°S). For Arica the value corresponds to the annual mean for the past five years.

A prominent feature of this region—as often observed in the western part of continents near oceanic gyres—is the presence of deserts. In Chile, the Atacama desert, one of the driest deserts in the world, is part of the so-called arid diagonal of South America. The Atacama desert is a major factor influencing the distribution and abundance of terrestrial biota on a large spatial scale, although its importance for marine biota remains enigmatic. Rivers originating in the mountains of the Andes range, carve long and deep

Table 3

Characteristic rivers with various flow regimes associated to different climatic regions from North to South of Chile. Flow was measured at the mouth of the river and the geographical positions are those taken on the coast.

River	Surface (km^2)	Length (km)	Flow (m^3 s^{-1})	Latitude South
Lluta	3400	147	2.29	18°25"
Loa	33570	440	2.70	21°20"
Copiapó	18407	162	1.90	27°20"
Elqui	9657	75	7.13	29°53"
Limarí	11760	64	7.34	30°42"
Aconcagua	7163	142	39.00	32°48"
Maipo	15380	250	92.30	33°36"
Rapel	14177	60	162.00	33°55"
Maule	20295	240	467.00	35°18"
Bío-Bío	24029	380	899.00	36°49"
Imperial	12054	55	240.00	38°40"
Toltén	7886	123	572.00	39°15"
Valdivia	9902	150	687.00	39°51"
Bueno	17210	130	570.00	40°20"
Baker	26726	170	875.00	47°15"

transversal valleys with few of them reaching the coast (Table 3). From latitude 30°S to 41°S, the mean annual rainfall increases from 100 mm up to 2200 mm, such as in the city of Valdivia. In this region, rivers are fed by windward slope rain and Andean melt water. Rivers carry an adequate quantity of water to reach the coastal zone; however, freshwater reaching the shore causes only local dilution of the seawater, without major consequences on

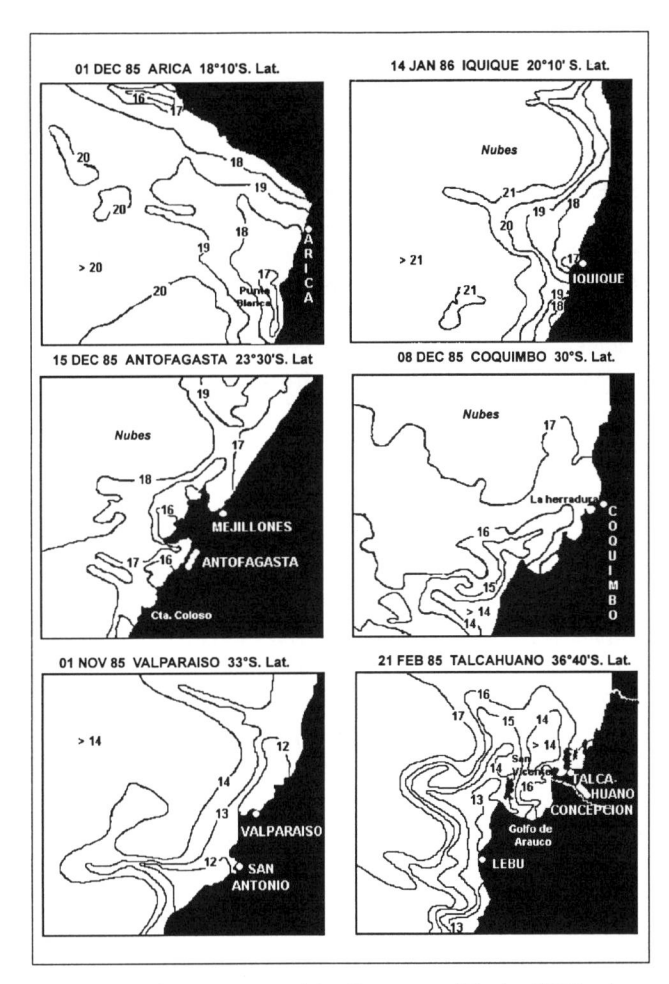

Fig. 5. Coastal zones reported for Fonseca and Farías (1987) where important and frequent upwelling events occur during some season of the year.

water masses. In contrast, south of latitude 42°S, several rivers flow to the fjords and channels creating an estuarine system where the dilution of seawater becomes important. Salinity reaches 32–33 UPS in this system. Brandshorst (1971) called this the "Channel Water mass", and recognised it as a local water mass extending up to Gulf of Arauco (latitude 37°S).

Before and during the warm period of El Niño events, the rainfall can produce substantial changes in river flow. At this time, heavy rainfall in the Andean range, known as the "Bolivian winter", causes important flooding of dry riverbeds with the destruction of roads and fields. In some cases, water reaches the ocean but most of the water is drained to the desert causing several plant species to flourish and flowers to blossom in the Atacama desert. The flowering desert extends from latitude 26° to 30°S.

During winter, north and northwest winds dominate along all of the coast. This pattern is more conspicuous when it reaches latitude 33°S. The north wind brings rain and cold weather. The south wind dominates from September to March. The onset of the dominance is alternated with north and west winds. During this period, upwelling processes begin to take place throughout the region. The ESSW intensifies and spreads to the coast, and the intense and persistent winds produce upwelling events lasting between five and seven days. Throughout the year, when south winds dominate, the probability of upwelling occurrence is high. Along the coast, during spring and summer, major upwelling zones occur (Fonseca and Farías, 1987); all of these areas are associated with high primary production (Fig. 5).

Knowledge about the ESSW is probably the most important element in understanding the upwelling events on the Chilean coast and other coastal processes (see Table 1). The ESSW is known as the Chile–Peru Current and originates in the Subsurface Countercurrent called the Cromwell Current, which flows eastward. This current accumulates as a subsurface water in the northern coast of Peru, reaching 800 m in thickness and a poleward flow as a subsurface current. The low oxygen content of the ESSW ($O_2 < 1$ ml l^{-1}) results from the consumption of oxygen by the detritus rain falling from the surface waters reaching a neutral buoyancy at the density of this water. The continuous input of organic matter and its remineralisation, produce high nutrient concentrations (some of the highest values reported in the ocean), high levels of dissolved organic matter, high levels of trace metals (the bioaccumulation is 10^3 over the water value) and anoxic processes (such as dissimilative denitrification). Some of these processes have been described by Dugdale et al. (1977) off the Peruvian coast.

COASTAL MARINE ECOSYSTEMS

An Oceanographic Perspective

It is not easy to define and establish clearly recognisable ecosystem boundaries in Chile. Nevertheless, when considering general systems it is possible to follow Smith (1970) and Ellenberg (1973) who defined an ecosystem as "a functional unit with recognisable boundaries and an internal homogeneity". In such a unit, organisms interact with their abiotic environment forming a system that could be defined by the extent and relationship between the flows of energy and matter, the degree of homogeneity of environmental conditions (e.g., limit values for relevant physical parameters), and/or the concurrent spatial distribution of its constituting species. Once the physical limits of the system are approximately defined, it may be treated as an open system, assuming a certain degree of "self-maintenance" or even "self-regulation". On this basis, it is possible to distinguish four major Chilean ecosystems (Bernal and Ahumada, 1985).

South Pacific Eastern Margin Ecosystem

South Pacific Eastern Margin Ecosystem or Humboldt Coastal Current Ecosystem, extends from latitude 41°S to

18°30"S. This system is influenced by the Sub-Antarctic Water (SAWW) along the coast. An important change in the conservative parameters are determined by the extensive movement and the surface condition of this water. However, in general the water temperature and salinity of the SAWW is less than that of the surrounding waters (i.e., 18°C and 34.8 UPS). During spring and near the city of Antofagasta, subtropical waters move to the coast causing the thinning of subantarctic waters.

Coastal Ecosystems

Upwelling ecosystem: Along the coast there are zones where the subsurface waters upwell seasonally. These waters provide nutrients to the surface waters. During this process, the temperature decreases and a temperature discontinuity defines the area of upwelling. Changes in physical, chemical, biological and geological variables define this particular ecosystem. South winds causing upwelling, dominate along the coast in spring and therefore upwelling events are more frequent between September to March along the northern and central coast of Chile. Typical upwelling areas along the Chilean coast have been reported by Fonseca and Farías (1987) using satellite imaging (Fig. 5). The more conspicuous changes on the coastal zone during upwelling are the dissolved oxygen content of waters below 15 m deep, high nutrient concentration and changes in the pH and Eh of the waters and sediments. The coastal upwelled waters rich in preformed nutrients sustain annual landed fisheries of seven million tons, approximately 20% of the total landed fisheries of the world.

Embayment ecosystem: From latitude 18°30"S to 41°50"S, there are several embayments of tectonic origin, most of them open to the North. This kind of coastal system could be defined as an ecosystem on its own with clear boundaries and characteristic dynamics (i.e., as a residence time). In the Chilean coast several of these embayments are open to the North. During the upwelling season the ESSW or oxygen minimum water spreads to the interior of these embayments. This feature produces an increase of nutrients, aeration of the surface water, high primary production and high sedimentation rates (in which nearly 50% of the particulate organic carbon production falls to the sediments (Ahumada, 1991) where it undergoes remineralization. In all of these embayments, the sediments are anoxic, rich in organic matter, and active sulphate reduction occurs. On the other hand, during the winter time, the dominant water mass is the SAAW, saturated in oxygen, and low in nutrients and salinity.

Central South Pacific Gyre Ecosystem

Here two different ecosystems are recognized.

Pelagic oceanic Ecosystem: The oceanic system on the Central South Pacific is not well known. Its main oceanographic features have been described following some national and international oceanographic expeditions. The most complete studies have been carried out on the first 100 km near the coast and surrounding the oceanic islands, or in relation with oceanic resources such as tuna, cephalopods, swordfish and cuttle fish.

Islands or Archipelago Ecosystems: This kind of ecosystem refers to the oceanic islands. Nevertheless, the different island groups are widely-spaced units with relatively high degrees of geographical isolation and endemism, and their respective biotas lack a strong compositional affinity and therefore they cannot be considered as a single system.

Sub-Antarctic Ecosystem

Oceanic Ecosystem: This corresponds to the area off the Chilean fjords and channels. Here, the West Wind Drift current is locally called Cape Horn Current. This region extends from the Subtropical Convergence to the Antarctic Convergence and is dominated by Subantarctic Waters. The surface water temperature fluctuates between 8–12°C.

Estuarine Ecosystem: An estuary is a semi-enclosed coastal zone where freshwater drainage produces a permanent or semi-permanent dilution of the seawater. Detection of the change in salinity allows to define the boundaries of the estuary. The following types of estuaries constitute this group: (1) area adjacent to the mouth of a river (coastal plain); (2) fjord and channel system (deep basin); and (3) littoral systems such as salt marshes or coastal lagoons (bar-build estuary) (Pritchard, 1968).

Estuaries type (1) and (3) can be found north of latitude 42°S, whereas type (2) is located south of latitude 42°S. This extensive region is an area with a low population density, of only a few inhabitants per square kilometre. Pickard (1971) classified the region into three areas based on its hydrographic, geological and climatic characteristics, which align simply into northern, central and souther fjords. The salmon culture industry has been developed in the northern fjords whereas the others are is pristine, with little human impact.

Antarctic Ecosystem

Geophysical studies have established that glaciation in Antarctic began 20 million years earlier than in the Arctic, creating a longer period of cold acclimation in Antarctica and different strategies to resist the cold temperature (Dunbar, 1977). The ecosystem within the physical boundaries of the Antarctic continental mass to the south and the Antarctic Convergence to the north is an old system, as reflected in the diversity and abundance of the bottom fauna and mid-water fish stocks (Hedgpeth, 1977). The quantitative and qualitative features of the basic processes of the antarctic systems is demonstrated by the distribution of the dominant crustacean herbivore: *Euphausia superba*. This species is considered because of its abundance as an important resource of the open ocean. Estimates of the potential yield range from 50 millions tons upwards (Gulland, 1977).

The Antarctic ecosystem is clearly described by Steele (1974) when he says "the phytoplankton of the open sea is eaten nearly as fast as it is produced generating a highly productive zooplankton pelagic system that supports populations of whales, penguins, and seals, and possibly abundant intermediate populations of fishes and cephalopods. The animals living at the sea bottom depend on the herbivores faeces for their food supply". From this point of view, it is believed this system is extremely fragile and should not be disturbed by human activities (Hedgpeth, 1977).

A Biological Perspective

Overall, the flora and fauna of the Chilean coast may be considered as a temperate biota. In general, diversity tends to be lower than in other areas at the same latitude due to the surface cooling produced by subantarctic waters. Upwelling of cool waters also decreases the temperature of areas adjacent to the shore. Thus, the above factors seem to facilitate the latitudinal expansion of several species along the coast, and in turn their ecological dominance in particular environments. This is especially remarkable in the case of littoral rocky habitats, where large-sized kelps such as *Lessonia trabeculata* in the shallow subtidal and *Lessonia nigrescens* in the intertidal appear as the most representative species of the Chilean littoral seascape (e.g., see Ojeda and Santelices, 1984; Camus and Ojeda, 1992). Under these circumstances, and at a large spatial scale, it is common to observe relatively low species numbers, but with a large number of individuals per species. From a community perspective, nonetheless, the species richness and composition of the littoral biota may exhibit an important spatial variation (beta diversity) along the coast, and it is possible to find highly diverse communities alternating with poor ones within a same region (Camus, 1998). Additionally, no clear relationship has been found between diversity and latitude for a variety of taxa, a lack of trend that could also be related to the relatively small temperature variations observed on a geographical scale.

Regarding the extent and limits of the Chilean marine biota, different biogeographical studies tend to agree with respect to the major distributional patterns. A comprehensive analysis was performed by Brattström and Johanssen (1983) on the distribution and abundance of 240 benthic invertebrate species pertaining to 13 taxa, and their results were compared with ten early zoogeographical propositions. These authors recognised two major biogeographical units along the Chilean coast, both broadly provinces of temperate nature. Despite the fact that names and exact boundaries of these units can vary depending on the researcher, we provide some traditional groupings that roughly correspond to Brattström and Johanssen's (1983) units. These are: (a) a Northern warm-temperate region ranging from Perú to latitude 30°S, mostly termed the Peruvian Province, and (b) a Southern cold-temperate region extending from latitude 42°S down to 56°S, usually termed the Magellanic Province. Between the two regions, both these authors and prior researchers recognise a transitional area (from ca. 30°S to 42°S), although it is not clear whether the area constitutes a zoogeographical province on its own, and there is some disagreement about its biological nature and limits, particularly regarding the southern limit of the "Peruvian" province.

Despite their differences, most studies concur on the existence of three areas, and their limits and extensions agree with the geographical distribution of river outflow and climate features. A remarkable feature of the northern area is the small number of tropical species inhabiting the coast, whether macroalgae (e.g., Santelices, 1980) or invertebrates (e.g., Brattström and Johanssen, 1983), apparently due to the reduced surface temperature as compared to other equivalent latitudes in the world (a difference of −10°C as suggested by Viviani, 1979). Noteworthy, during some El Niño events (particularly stronger ones), the northern Chilean coast can be invaded by a variety of exotic species of tropical origin, a transient phenomenon with supposedly slight consequences on the structure and organization of littoral communities (see Glynn, 1988, for a review). In the past fifteen years, however, Chilean biologists have been increasingly concerned with the potential impact of El Niño as a historical, large-scale disturbance on present-day biogeographical patterns (Tomicic, 1985; Camus, 1990), and also on many ecological processes in populations and communities of both marine and terrestrial systems (for a review, see Jaksic, 1998). For instance, rocky intertidal communities from northern Chile are usually severely disturbed during El Niño events, and they exhibit a much higher degree of beta diversity than those in central Chile, suggesting that El Niño might promote a regional differentiation in diversity patterns (Camus, 1998). In fact, large-magnitude events such as El Niño 1982–83 caused the local extinction of several populations of some important intertidal species in the northern area, such as the kelp *Lessonia nigrescens*, among others, without a noticeable impact on their central Chilean populations (Castilla and Camus, 1992). A major result of this phenomenon is the modification of the geographical population structure of these species, which not only triggers processes of extinction–recolonization dynamics on a regional scale (Camus, 1994a), but also generates complex interactions between large- and small-scale factors during the recovery process (Camus, 1994b), whose consequences are not clearly understood.

With regard to the biogeography of the pelagic system, patterns are slightly similar to those observed in the coast. However, even though it would be possible to recognise three main areas, namely, the Northern, Central and Southern regions, separation appears to be less marked than it is along the coast as distributions near the boundaries are not so simple and several taxa exhibit a gradient between regions. Indeed, a zoogeographical classification proposed by Antezana (1981), on the basis of the distribution of euphausiid fauna, recognizes only two main regions: (a) a large Peru–Chilean province, ranging from 3 to 42°S and composed by four districts; and (b) a Magellanean province, ranging from 42° to ca. 60°S and composed of two districts.

The Oceanic Island system not only appears as a heterogenous habitat but also each group of islands has its own fauna. In fact, specialists on marine fish (Pequeño, pers. comm.) recognise that each one has a different fauna with little overlap among them. Nevertheless, Easter Island, the most distant Chilean island offshore exhibits the most relevant differences and it is possible to recognise its fauna (including the benthic one) as Polinesic (Arana, pers. comm.).

South of latitude 41°30"S, within the zone of fjords and channels, Pickard (1971) recognized three zones separated according to certain oceanographic features. Currently, new systematic oceanographic studies are being carried out in the area in order to understand the dynamics and environmental features of the region.

Table 4 shows these marine habitats which can be separated and characterised by their constituent species. Many of these species are present or at least occur in other systems, but not always with the characteristic abundance found in their main region. On the other hand, some species are distributed across many systems, likely due to the reduced temperature gradient along the coast. On this basis, some representative species of the different habitats were selected to give an idea of the biological distribution and diversity of these areas.

As a concluding remark on the present, large-scale patterns of distribution and abundance of marine species in Chile, we may highlight four major determinants. First, a massive cooling effect along the continental coast associated with the combined influence of subantarctic waters and upwelling processes, which confer a predominantly cold character to the biota. Second, the latitudinal variation of climatic and topographic features of the coast, which would have been important in the differentiation of the present main biogeographic regions. Third, a possible spatial differentiation of the biological consequences of El Niño (and La Niña) on coastal populations and communities, where the northern Chilean area would have experienced a greater disturbance effect on a historical time scale. And

Table 4

List of dominant species in the different ecosystem of west Pacific border along the Chilean Coast

Habitat	Northern system	Central system	South system	Oceanic system	Oceanic island system*
Nearshore algae	Lessonia nigrescens	Lessonia nigrescens	Lessonia spp.	—	Zonaria stipitata
	Hipnea cenomyce	Gelidium chilense	Chordaria linearis	—	Sargassum skottsbergii
	Chondrus canaliculatus	Mazzaella spp.	Gigartina spp.	—	Cladophora socialis
Nearshore invertebrate	Jehlius cirratus	Jehlius cirratus	Comasterias lurida	—	Palinurus pascuense
	Heliaster helianthus	Perumytilus purpuratus	Choromytilus chorus	—	Jasus frontalis
Coastal fishes	Semicossyphus maculatus	Sciaena spp.	Eleginops maclovinus	Helicolenus lengerichi	Scorpis chilensis
	Scartichthys gigas	C. geniguttatus	Galaxias spp.	Parona signata	Girellops nebulosus
	Kyphosus analogus	M. viridis	Normanichthys crockeri		Blennidae, Labridae
	Anisotremus scapularis	Cilus gilberti			Serranidae
Pelagic fishes	Engraulis ringens	Strangomera bentinki	Clupea fuegensis	Scomberesox saurus	Seriola lalandii
	Sardinox sagax	Sardinox sagax	Thyrsites atun	Sardinox sagax	Sardina de JF.
	Truchurus murphyi	Truchurus murphyi	Truchurus murphyi	Truchurus murphyi	Pseudocaranx chilensis
	Sarda chilensis	M. magelanicus	M. magellanicus	Thunnus albacares	Thunnus albacares
	Seriolella violacea	Seriolella porosa	Seriolela cerulea	Seriola sp.	Thunus albacares
	Scomber japonicus	Scomber japonicus	Lepidotus chilensis	Xifias gladius	
Demersals fishes	Merlucius gayi	Merlucius gayi	Merlucius australis	Micromesistius australis	Nuraenichthys chilensis
Benthic fishes	Paralichthys adspersus	Paralichthys microps			
	Genypterus maculatus	Genipterus maculatus	Genipterus chilensis		Gymnothorax spp
	Genipterus chilensis	Genipterus chilensis	Genipterus blacodes		Paralichthys spp.
		Hippoglossina macrops			

This table was elaborated in collaboration with Dr. G. Pequeño and Prof. J Chong.

fourth, a factor of different nature such as the geographical variation in the occurrence and strength of anthropogenic impacts due to the marked spatial differences in the distribution and concentration of both coastal human populations and industrial activities along the coast (see next section). While the operation of this last determinant in the system is new in historical terms, the first three are much more ancient, widespread and complex in their interaction. Present topographic features of the coast and cold oceanic conditions would be a result of a chain of processes beggining in the Oligocene with the opening of the Drake passage and the apparent origin of the Humboldt current system (Brundin, 1989). Later, during the Quaternary period where Andean orogenesis and coastal uplift events occurred with important effects on coast line topography, there were important variations in sea level and temperature (see *Revista Chilena de Historia Natural* 1994, for a multidisciplinary analysis of the Quaternary in southern South America). Apparently, oceanic features stabilized only during the Holocene period, and therefore the present cold conditions and the large-scale biological patterns could also be of recent origin (Camus, 1990, and references therein). Whether the present-day patterns and conditions will be modified by the ongoing global climate change is a matter of speculation. However, many authors suggest that the outcome would be more related to a progressive intensification of El Niño and La Niña events than to a gradual warming (Jaksic, 1998), and to date the magnitude and direction of potential biological changes remain unpredictable.

HUMAN COASTAL ACTIVITY

Chile has a population of approximately 13,200,000 inhabitants. Although nearly 22% of the population live on coastal cities, only two cities have more than 500,000 inhabitants, four between 250,000 and 125,000, and the rest have fewer than 100,000 inhabitants (1992 Census). In 1974, the Chilean government divided the country into 13 administrative regions, all of them having different natural resources and industrial activities (Table 5).

To understand and classify the nature of coastal pollution entering the aquatic environment by anthropogenic activities, it is necessary to know more about the geography of the country. The coast is narrow and flanked by a chain of coastal mountains with an elevation of about 800–1200 m above sea level (masl). Further inland a central plateau with an elevation of 700–900 masl appears and extends 100–150 km before reaching the Andean range. This mountain range ascends up to 4000–6000 masl.

The relief is the main factor controlling the water regime flowing west of the Andes. Rivers are initially turbulent, slowing their flow in the central plateau and finally modelling the coastal range as they move toward the ocean.

Rainfall is poor or absent in the north between latitude 18°S and 30°S, where the Atacama desert crosses transversal valleys formed by rivers. In the northern part of Chile, the major activity is mining. The principal cities are also the main ports and developed because of the increase of mineral production. Some transversal valleys, where rivers flow from the Andes, are occupied mainly by natives on a subsistence agriculture. The rivers occasionally terminate on the coast and if so, they carry little water.

Between latitude 30°S and 34°S, rainfall increases from 100 mm yr^{-1} to 900 mm yr^{-1}. Between latitude 34°S and 56°S increases to 1000 mm yr^{-1} with a maximum of 3000 mm yr^{-1}.

In the northern part of Chile, rivers have a low flow, with water originating on the Andean range. On the other hand, the rivers in the Central region toward the South, have a regime based on melting snow and lacustrine outflow, with an important rainwater contribution at the

Table 5

Census of population by regions, number of inhabitants on the coastal areas and population density by region from north to south

Region	Population inhabitants	Area (km^2)	Density (hab/km^2)	Rate of growth 10 years (%)	Coastal inhabitants	Population (%)
Tarapacá	341,000	56,698	5.8	2.40	321,000	94.1
Antofagasta	407,000	126,444	3.2	1.92	270,000	66.4
Atacama	231,000	75,523	3.1	2.58	33,000	14.5
Coquimbo	502,000	40,656	12.4	1.96	248,000	49.3
Valparaíso	1,374,000	16,396	83.8	1.35	719,000	52.4
Metropolitan	5,170,000	15,349	333.8	1.97	—	—
El Libertador Gral. B. O'Higgins	688,000	16,365	42.1	1.73	5,400	0.8
Maule	834,000	30,301	27.5	1.42	49,800	6.0
Bio Bío	1,730,000	36,929	46.8	1.39	510,000	29.5
Araucanía	775,000	31,858	24.3	1.10	14,400	1.9
Los Lagos	953,000	66,997	14.2	1.23	257,000	27.0
Aysén	82,000	109,025	0.8	2.37	10,000	12.3
Magallanes and Antarctic territory	143,000	1,382,033	0.1	0.84	138,000	96.5

central plateau and superficial runoff. In this context, the human activity has been developed on different regions near the rivers and along the embayments.

In the Northern region of the country, the main economic activity is related to the mining of metal and non-metal resources. Gold, silver, zinc, lead, iron, manganese, molybdenum and copper have been the main products obtained from metallic mining at the Andes mountain range and the central plateau since the past century. Elements such as nitrate, iodate, phosphate and carbon are the main non-metallic minerals which are mined.

The industrial activity of the coastal zone is related to: (1) transport of minerals from the mining of copper, tails that reach the sea and shipment of the concentrate; (2) industrial fisheries, landing fish at the port, their industrial processes and by the activity of intensive marine aquaculture, (3) transport of petroleum, fuel terminals and oil refineries; (4) steel factories, industrial metal–mechanical processes and metal smelting; (5) a cellulose industry, and (6) the discharge of urban wastewaters on the coast.

Development of the industrial activity in Chile is of recent history involving the export of raw materials, after the depression of 1929. The idea of "substituted industrialisation" and "poles of development" were promoted by the government (del Valle, 1985). "Substituted industrialisation" refers to the creation of enterprises administered by the state, for buying and processing the natural resources to obtain a product that incorporates aggregated value. The idea was to develop the mining, agricultural and service goods. "Poles of development" corresponded to the localisation of new enterprises on cities far away from the capital, to create a centre of development of the country. Prior to these measures, early industrial development was related to the extraction of raw materials.

Mining

Exploitation of nitrate salts ($NaNO_3$ and KNO_3), gold, silver and copper were probably the first mining activities in Chile, although their development was slow and small in volume. The exception was the intense mining of sodium nitrate that provoked environmental problems due to clear cutting the "tamarugo" forest to be used as fuel. *Prosopis tamarugo*, a unique tree species from the northern desert known as the "Pampa of the Tamarugal", was driven almost to extinction as an indirect result of the mining of sodium nitrate. Similar indirect effects of mining have threatened other woody plants in northern Chile.

The main anthropogenic activity related to mining is located in the north region of the country. In the so-called "Great Mines", mining of copper is operated by 35 companies, 28 of which are located north of 30°S. The other seven are distributed between latitude 30°S and 34°S. The mining of iron and tin is located around 30°S. Manufacture of steel is accomplished on San Vicente Bay (≈36°S). Finally, mining of lead and zinc takes place in the south region (≈46°S), in the Aysen fjord. As a rule, mining of metals is located on the Andean range and its products are transported to the coast where it is exported overseas. The risk of pollution comes from infiltration of liquid tailings, transport of concentrated copper powder and accidents during seaport loading activities. Table 6 shows the main metallic mining activities carried out in Chile during the last 50 years, with the present production as a way to assess pollution risks of the coastal areas.

Mine tailings obtained from copper extraction are deposited in salt mines which are not permeable basins. For the extraction of 100 ton of copper mineral, it is necessary to remove 100 ton of sterile rock materials. The rock mineral contains 1.2% of copper, which is ground, dissolved with 300 ton of water and reacted with up to 2 tonnes of sulphuric acid. A chemical flotation process is used to obtain copper and molybdenum and 400 tonnes of tailings. The latter is a semi-liquid residue or waste originating from the extraction process. To mitigate the impact of harmful mine tailings, the waste is deposited in basins, thus creating sedimentation lagoons. The risk of percolation and contamination of groundwater is high and driven by the pH of the tails (≈11) and the impermeability of the basin. In the desert, it is possible to find dry salty lagoons with impermeable basins that can been used to deposit copper mining tails and recover the water.

Export of copper mining products for 1995 were as follows: refined copper (48%), blister copper in bars (2%), concentrated fine copper powder (44%), other forms of copper (5.5%). Therefore, the pollution risk by copper is mainly associated with concentrated copper during its transportation and shipping at seaport (Table 7).

Fishing Industry

The fishing industry is another group of activities developed on the coastal zone. The major product is fishmeal, which includes about 85% of all fish captured. This activity begun in 1960 in the northern part of Chile, in Iquique, expanding later to the near localities of Arica and Pisagua. The first economic crisis borne by the fishing

Table 6

Statistics of mining activities of specific metal resources in Chile

	Number of companies	Initial year exploitation*	Total production at 1996 (TMF)
Copper	35	1929	3,141,000
Iron	2	1953	9,081,481
Gold	6	1929	51.8
Silver	3	1841?	1,129.9
Molybdenum	30	1980	17,415
Lead	1	1903	1,089
Zinc	1	1924	35,625

*Date when an important industrial development occurs.

Table 7
Industrial activity having an environmental impact in the coastal zone

Industrial activity	Geographic zone	Initial year	Type of industry	Locality	Current production (ton-year in 1997)
Fishery	I Region	1960	Fishmeal	Arica	522,481
				Iquique	849,200
	II Region	1976	Fishmeal	Tocopilla	159,939
	II Region	1976	Fishmeal	Mejillones	167,708
	III Region	1980	Fishmeal	Caldera	121,632
	IV Region	1975	Fishmeal	Coquimbo	
	V Region	1970	Fishmeal	San Antonio	298,271
	VIII Region	1970	Fishmeal	Talcahuano	1,016,908
				San Vicente	741,966
				Coronel	1,114,812
Marine Aquaculture	III Region	1985	Algae, molluscs		
	IV Region	1985	Scallops culture	Guanaqueros	
	VIII Region	1975	Salmon culture	River	951
	IX Region	1980	Salmon culture	River, Lakes	
	X Region	1980	Salmon culture	Puerto Montt	78,056
				Chiloe	47,425
	XI Region	1980	Salmon culture	Seno Aysen	13,432
	XII Region	1990	Salmon culture	Punta Arenas	1,635
Mining	I Region	1994	Concentrate Cu	Puerto Iquique	
	I Region	1994	Concentrate Pb	Puerto Arica*	
	II Region	1990	Concentrate Cu	Cta. Coloso, Antofagasta	
	II Region	1991	Concentrate Pb	Antofagasta*	
	III Region	1992	Concentrate Cu	Chañaral, B. Calderilla	
	III Region	1992	Pellet Fe	Chañaral	
	IV Region	1970	Pellet Fe		
	VIII Region	1949	limestone	San Vicente	
	XII Region	1949	limestone	Isla Guarello	
Mining Tails	III Region	1980	Cu	Chañaral	
	III Region	1986	Cu	Chañaral, Cta Hedionda	
	III Region		Fe	Chañaral, Huasco	
	IV Region		Fe	Coquimbo, Guayacan	
	XI Region		Zn	Seno Aysen	
Cellulose	VII Region		Cellulose	Constitución	
	VIII Region			Cellulose	Rio Bío Bío
	VIII Region			Cellulose	Rio Bio Bio
	VIII Region			Cellulose	Rio Bio Bio
	VIII Region			Cellulose	Rio Bio Bio
	VIII Region	1985		Cellulose	Golfo Arauco.
Oil Refinery	V Region	1950	Hydrocarbons	Bahía Quintero	
	VIII Region	1965	Hydrocarbons	Bahía San Vicente	
Oil Terminals	All the coast		Hydrocarbons	Seaports	
Oil Platform	XII Region		Magellan Strait		
Municipal waters	In all the port		Organic matter	Coastal cities	
Municipal waters	1992 still without treatment			Arica, Iquique, La Serena	
Submarine pipes				Viña del Mar, Bahia Concepción	

*Lead originates in Bolivia.

Table 8

Major oil spills (i.e., more than 10 ton) during the last 20 years along the coast of Chile

Ship name or cause	Year	Site of accident	Type of hydrocarbons	Impact and mitigation
B/T Napier	1973	Chiloé, Guamblin	30,000 ton crude oil	Without report of impact.
B/T Metula	1974	Magellan Strait	52,000 ton crude oil	40 km of coast exposed to the asphaltic fraction. Beach cleaning.
B/M Astrapatagonia	1975	Southern Channels	1,000 ton fuel oil	Without report of impact.
B/M Northern Breeze	1975	Quintero Bay	440 ton diesel/fuel oil	Beach cleaning
B/T Cabo Tamar	1978	San Vicente Bay	17,000 ton crude oil	4 km of coast with asphaltic fraction. Beach and salt marsh cleaning.
OBO Valparaiso	1987	Concepción Bay	Not evaluated	Light and polar fractions of crude oil were partitioned into seawater*
Breakage of pipeline and fire	1993	San Vicente Bay	Diesel fuel	Loss of small ships and general damage.

activity was in 1965, due to El Niño events of 1959 and 1964. These events affected the reproduction and recruitment of different fish species in the following years. The fish stock is increasingly facing overfishing and catastrophic collapse.

From 45 initial fishery industries producing fish meal and fish oil in Iquique, only six were operating in 1965 and ten in the entire northern region at the end of this century. In 1972, the fishery industry expanded to the central part of Chile (i.e., San Antonio and Talcahuano) and began a new stage of development. The fisheries industries landed seven percent of the world total. Marine aquaculture began as an artisanal activity in fjords and channels in the southern part of Chile in 1970, with the culturing of oysters and mussels. This artisanal culture was replaced in 1985 by industrial salmon culture. This was intense in some regions, which brought some problems resulting from over-feeding, and this type of activity is the main source of large quantities of organic matter, grease, oil, vitamins and hormones which affect, in diverse ways, the marine environment. In the most affected regions, more than 300 aquaculture centres exist. During the last five years, the fish industry has reported the capture of 94% of the total fish landed in the country with a value of 650 million US dollars. Artisanal fisheries with 5% of the total capture is worth 300 million dollars, and the aquaculture industry with only 2% of the total capture has a value of 350 million dollars (SERNAP, 1998). It is surprising to discover a similarity in cash value in these three vastly different areas.

Petroleum

Along the coast there are ca. 40 oil terminals, 2 petroleum refineries and 2 oil platforms. Crude oil, fuel oil, diesel oil, and several refined products are the main petroleum products transported along coastal waters. Oil spills greater than 10 m^3 are shown in Table 8. Accidental spills smaller than 5 m^3 are frequent in oil terminals and seaports. This low but continuous input of oil causes a major impact on these ports. High-risk or "critical" areas as defined by CPPS-PNUMA (1985) are Antofagasta, Valparaiso, Talcahuano and Punta Arenas. A recent study of different sedimentary hydrocarbons and organic matter done at San Vicente bay identifies a coke gas plant, discharges of fish industries and petroleum maritime terminals as the source for hydrocarbon contaminants (Mudge and Seguel, 1998).

These results are consistent with the general hydrographic dynamic of San Vicente bay (Ahumada, 1992, 1995; Arcos et al., 1993). Studies of hydrocarbon and organosulphur contaminants in sediments of coastal zones are relevant to evaluate changes in the ecosystem caused by anthropogenic activity. Similar studies to those carried out along the Northeast Pacific (Pinto and Leif, 1991; Prahl et al., 1996) are starting to be carried out locally (Mudge and Seguel, 1999).

Industrial Wastewater

The best protected embayments have been used as seaports, and around them a series of industrial activities have developed. There are several examples along the coast: Mejillones, Calderilla, La Herradura, Concepcion, etc. There is only one iron and steel factory in Chile, located in San Vicente Bay, on the VIII Region. Around this factory several related industries exist, producing different types of wire, steel balls for mills, cement, dockyards.

Cellulose Industry

The processing of cellulose uses two different systems: the Kraft chemical process using a sulphur solution and the physical process. Major industries use the Kraft system requiring large amounts of water. Two species of tree are utilized: pine and eucalyptus, both species being introduced to Chile and requiring a mean rainfall greater than 700 mm y^{-1}. This type of industry is located on the south part of Chile, next to big rivers and where large areas of pine and eucalyptus trees are planted. There are nine cellulose industries, four Kraft processing plants located along the Bio Bio river, one in Constitucion and another in Arauco along the coast, disposing their liquid waste into the river and the coastal zone, respectively. Among the major pollutants of this activity are organic mater (i.e., BOD_5), resin acids, totals solids, dioxins, sulphides, mercaptans, chloride, phenols, etc.

Municipal Wastewaters

Municipal waste is normally discharged into rivers and lakes when the cities are located in the interior of the country or, in coastal cities, waste is discharged into the sea. There are 113 major towns that discharge their wastewaters into rivers and about 41 that discharge them into the sea. In any of these coastal cities more than 10 discharge points can exist, usually involving no treatment. In recent years, the wastewater discharge to the sea is changing from a superficial dumping of wastewater in the intertidal zone to a submarine pipe. The volume of domestic wastewater discharged in Chile in 1992 was estimated in 1.8×10^6 m^3 per day. Half of this volume is discharged in Santiago to the Mapocho river, and transported to the coast by the Maipo river. Two indexes are used in Chile to evaluate the possible impact of the effluent which is rich in organic waste to the reception water body: the 5-day Biochemical Oxygen Demand (BOD_5) and the Chemical Oxygen Demand (COD). Both of them are based on the oxidation of organic matter content per unit of volume of waste and offer a projection of the impact on the oxygen consumption by the discharged waters. Some effluents typically have high oxygen demand values: for example BOD_5 for municipal waste is 300 to 800 ml l^{-1}; for fishmeal waste it is between 3000 to 5000 ml l^{-1}; for a cellulose plant it is 1500 ml l^{-1}; and for a dairy product plant it may be 4000 ml l^{-1} (Rudolph, pers. comm.).

Case Studies

Three cases studies are analysed to assess environmental problems on the coastal zone, each one corresponding to the worst case scenario in relation to a specific type of contamination.

Mining: Lessons for the Future

Copper mining is the main mining activity in Chile. Molybdenum, gold, silver and rhenium are other metals obtained from copper mines. An annual extraction of 1,250,000 tons of copper, produces 15,000 tons of molybdenum and significant quantities of gold, silver and rhenium. For this purpose about 100,000,000 tons of minerals must be removed together with an equal tonnage of sterile materials. The process requires 300,000,000 tons of water, 2,000,000 tons of sulphuric acid, and chemical additives for flotation producing 100,000,000 tons of tailings (Corvalán, 1985).

Two copper mines located on the Andes mountains in the III Region, discharge their liquid tailings into the Rio Salado basin: one of them begun its production on 1929 and the other begun on 1959. The sediments are transported 150 km reaching the coast in Chañaral, a small city of 41,450 inhabitants (census of 1992). For forty-six years, the tailings reached the coast accumulating in the sediments of Chañaral Bay, changing its depth. Between 1962 and 1969 the coastline moved 130 m in the north and 100 m in the southern area of the bay. About 250 million tons of tailings were deposited as sediment into the bay, affecting its subtidal region. Fine dry sediment was carried by the wind and became a nuisance over the city. During 1975, the tail waste was discharged on Caleta Palitos, a fishing village about 10 km north of Chañaral Bay. For 12 years another 100 million tons of tail waste built new deposits of tailing sediments on Caleta Palitos.

Castilla (1983) assessed the tail-waste impact over the biological communities on the beaches around Chañaral. He reported only two species in the contaminated area: *Bateus truncatus* (green shrimp) in tidepools and *Enteromorpha compressa*, an algae, both very abundant. *E. compressa* exposed to high levels of copper was studied by Correa et al. (1996) to define whether the mechanism of resistance for this algae has an adaptative nature or is controlled by heredity. According to García (1985 the impact of tail sediments extended 50 km along the coast of Chañaral. Several beaches silted up and the people of Chañaral in 1987 complained to the mining company about the pollution of about 100 km of coast. The civil suit was accepted by the Supreme Court, constituting the first legal settlement of the citizens for an environmental case in Chile.

In 1992 the Division El Salvador (CODELCO) constructed a big lagoon of tailings. A long channel of 70 km connects the extraction processing site to the lagoon. The residual liquid is adjusted to flow at a pH ≈ 11, to maintain the metal elements as complexes in the tail. The accumulation lagoon of tailings is 130 ha in area and is 10 m deep (Fig. 6). In this lagoon, the sediments settle and the supernatant water (clear water) is pumped back to Rio Salado. Some of this water is used to irrigate experimental cultures of native plants of the area. The content of metal in the clear water is low and the flora and fauna that use the water is monitored for metals according to environmental laws.

Fishmeal Industry: Effect on the Coastal Zone

There are two important environmental impacts caused by the fishmeal industry. The first is related to the resource. The total capture of fish in Chile during 1996 was 6,725,734 tons. The percentage was used as follows: 85.6% was converted to the industrial fishmeal and oil; 4.3% was frozen; 4.2% was canned, 4.1% was dried or smoked and 1.6% was left as cold/fresh fish.

The fishmeal industry produces a residual liquid with a high oxygen requirement. The intensive salmon culture

Table 9

Characteristic parameters of a fish industry effluent

Type of industry	Parameter	Concentration	Unit	Flow (l s^{-1})	Charge (kg in 8 h)
Fishmeal					
Discharged water	BOD$_5$	2000	ml O$_2$ l^{-1}	66	5431
	COD	4000	ml O$_2$ l^{-1}		10861
	Ammonia	5.0	μM		0.13
	Grease and oil	0.5	g l^{-1}		950
Processing residual waters	BOD$_5$	2600	ml O$_2$ l^{-1}	150	16046
	COD	3600	ml O$_2$ l^{-1}		22217
	Ammonia	20	μM		1.21
	Grease and oil	0.2	g l^{-1}		864
Canning					
	BOD$_5$	3500	ml O$_2$ l^{-1}	10	1440
	COD	6000	ml O$_2$ l^{-1}		2468
	Ammonia	20	μM		0.81
	Grease and oil	2.0	g l^{-1}		57.6

*Normal oxygen saturation content of sea water necessary to oxidate the organic matter of emissary discharge of 8 hours (calculated).
Data obtained after 1991–1992 and 1993, during the program of impact mitigation for the Industries of Talcahuano, San Vicente and Coronel.

also has an environmental impact due to the excessive quantity of fish faecal pellets and unconsumed food pellets that settle at the bottom sediments.

Fishmeal production in 1997 was 1,227,561 tonnes, about 12.4% lower than 1996, and the production of oil in 1997 was 291,981 tonnes, or 29.5% lower than in 1996. Pelagic fish captured are species such as anchovies (*Engraulis ringens*) with 1,757,499 tonnes annually and Jack mackerel (*Trachurus murphii*) with 2,917,064 tonnes. Other species used for fishmeal are: common sardine (*Clupea bentinki*) and Merluza de cola (*Macroronus magellanicus*).

The second environmental impact is related to the industrial-liquid waste. Liquid discharges have a high content of organic matter which includes digested liquids, blood and ground fish, all of which are dumped into the water during the transport of fish from the area of fishing to the seaport. The main content of the organic waste depends on the time of transport, temperature, size of the ship hold and size of fish. Secondly there is the residual liquid from the industrial processes, which contain solids, ammonia, organic matter and grease material (Table 9). For years the fishmeal industry dumped their residual waste directly in coastal waters producing a local hypertrophic area near the fishing port. Under these circumstances, dissolved oxygen is consumed during degradation of organic matter creating an anoxic environment. Under anaerobic conditions, other terminal electron acceptors such as nitrate and sulphate are used in the metabolisation of organic matter. These microbially-mediated reactions produce ammonia and sulphide respectively, which are very harmful to the ecosystem. In all sites where fishmeal residue is discharged, an oxygen-depleted black sediment appears. In these areas, a film of oil and grease in the surface water further obstructs normal oxygen exchange between the air and sea water creating an extremely harsh environment for living organisms.

The actual organic discharge for the fishing industry is shown in Table 9, corresponding to the sum of discharged fish waters plus the residual process waters. Estimated total oxygen requirement (BOD$_5$) is obtained assuming a saturation value of 7.25 mg O$_2$ l^{-1}, for local waters with a salinity of 34,400 UPS and a temperature of 13.4°C. The last column of the table shows the effect of fish discharge from the boats in terms of the BOD$_5$ requirement in terms of water volume.

The residual waters of effluents dumped to the bay require the total amount oxygen contained in ≈2315 m^3 of seawater. In each embayment there are about eleven fish industries. Recently the fishing discharge has been done with recirculating water coming from the industrial process.

Environmental impacts of the fishmeal production has been localised on beaches near harbours and industrial areas. Among the major sites environmentally impacted we find the coastal zones of Arica, Iquique, Mejillones, Taltal, Calderilla, San Antonio, Talcahuano, San Vicente, Coronel y Lota. One of the major chronic contaminated areas is the Rocuant saltmarsh of Talcahuano (Rudolph and Ahumada, 1987; Ahumada et al., 1989). Around this saltmarsh, nine fishing industries are located spilling their wastewater into this water body. All of the aquatic fauna was replaced by bacteria and microciliates, and the water has become suboxic with anoxic sulphide producing sediments.

Municipal Wastewaters: The Submarine Disposal Solution

Municipal wastewaters are an important contaminant along the whole country. Until 1985, most wastewater was dumped directly into rivers, lakes and along the seashore on the intertidal or near the infralittoral zone in all of the coastal cities. Starting in 1990, environmental laws restrained disposal of wastewaters. Major pipelines are being constructed to carry wastewaters to the coast and dumped through submarine conduits offshore. Design for these pipelines is based on the LT-90 index, corresponding to the time when 90% of bacteria have been destroyed. Apparently, the country has favoured the construction of submarine disposal pipelines to wastewater treatment. Due to cost, these submarine pipes were in most cases inside an embayment, and all of them have been placed at shallow depths of less than 30 m. Experience has demonstrated that wastewater treatment and the construction of submarine pipes are complementary tools when cost and maintenance are taken into account. Data is still lacking to demonstrate the best option or the best investment on this type of infrastructure. Potential damage to the benthic fauna due to the toxicity of residues in the area has not been fully addressed. Recent studies indicate that flat fish present in Concepcion Bay show lesions on the skin attributed to contaminants (Leonardi and Tarifeño, 1996, 1997).

ADVANCES IN CONTROL AND POLLUTION ABATEMENT

Two periods are distinguished in relation to environmental coastal protection in Chile. Initially, some legal ordinance existed dispersed throughout the legislation. Responsibility for avoiding environmental problems has been difficult to achieve since many different organisations had a degree of interest in the matter but no technical expertise or funding to carry out their mission. This permissive period was therefore due to a lack of regulations.

In 1987, based on Navigation laws, Regulation 12600-550 was established to oblige all industries discharging liquid effluents into the sea to carry out environmental impact assessments. Some of these studies have been partially published (Arcos et al., 1993; Ahumada, 1995a,b). This Regulation provided a grace period of five years to finish the study and present a proposal to mitigate the impacts and to generate a monitoring program for three consecutive years with seasonal sampling in the receptor water body. During this period a lot of experience was gained and, after five years in 1992, more than the 90% of the industries were observing this rule.

In 1990, the National Environmental Commission was created as a legal support department to the government on environmental issues. In 1993, the Basis for the Environmental Law was written and several regulations approved thereafter. Finally, the Environmental Impact Assessment document was incorporated into all industrial or environmental projects to be carried out in Chile. One of the important aspects of these documents recommended that an environmental monitoring program ought to be carried out seasonally by the industry to detect any environmental change caused by their activity. When undesired changes occur, the company is obliged by law to present an abatement program to minimize those changes.

REFERENCES

Ahumada, R. (1991) Producción y destino de la biomasa fitoplanctónica en un sistema de bahías, en Chile Central: una hipótesis. *Biologia Pesquera* **18**, 53–66.

Ahumada, R. (1992) Patrones de distribución de metales traza (Cr, Ni, Zn, Cu, Cd y Pb) en sedimentos superficiales de Bahía San Vicente, Chile. *Revista de Biologia Marina Valparaíso* **27**(2), 265–282.

Ahumada, R. (1994) Nivel de concentración y bioacumulación de metales pesados (Cd, Cr, Cu, Fe, Hg, Pb y Zn) en tejidos de organismos bénticos de Bahía San Vicente. *Revista de Biologia Marina Valparaíso* **29**(1), 2–18.

Ahumada, R. (1995a) Bahías: Areas de uso múltiple un enfoque holístico del problema de la contaminación. *Ciec. Tecnol. Mar, CONA*, Número Especial: 59–68.

Ahumada, R. (1995b) Programa de Vigilancia del contenido de metales traza (As, Cd, Cu, Hg, Mo, Pb, Se y Zn) en sedimentos marinos de Caleta Coloso. *Cienc. y Tec. del Mar, CONA*, Número Especial: 89–100.

Ahumada, R. (1998) Metales traza (Ba, Cd, Co, Cr, Cu, Ni, Pb, V y Zn) en los sedimentos del Seno Aysen: línea base y alteraciones ambientales. *Ciec. Tecnol. Mar, CONA* **21**, 75–88.

Ahumada R., Rudolph, A. and Martínez, V. (1983) Circulation and fertility of waters in Concepcion Bay. *Estuarine and Coastal Shelf Science* **16**, 95–105.

Ahumada, R., Troncoso, A., Rudolph, A., Morrillas J. and Contreras, T. (1989) Coloración roja producida por bacterias: Marismas Rocuant, Talcahuano. *Bol. Soc. Biol. Concepción* **60**, 7–16.

Ahumada, R., Matrai, P. and Silva, N. (1991) Phytoplankton biomass distribution and relationship to nutrient enrichment during an upwelling event off Concepcion, Chile. *Bol. Soc. Biol. Concepción (Chile)*, **62**, 7–19.

Antezana, T. (1981) Zoogeography of euphausiids in the Southeastern Pacific Ocean. Memorias del Seminario sobre Indicadores Biológicos del Plancton. UNESCO, Montevideo, pp. 5–24.

Arcos D., Furet, L., Carrasco, F., Nuñez, S. and Vargas, F. (1993) Eutroficación del ambiente marino de Chile central: efectos inducidos por la evacuación de residuos industriales líquidos. *Invest. Mar., Valparaíso* **21**, 43–50.

Bernal, P. and Ahumada, R. (1985) Ambiente Oceánico. pp. 55–106. In: *Medio Ambiente en Chile*, ed. Fernando Soler. Centro de Investigación y Planificación del Medio Ambiente (CIPMA). Ediciones Universidad Católica. 413 pp.

Bernal P., Ahumada, R., González, H., Pantoja, S. and Troncoso, A. (1989) Flujo de energía en un modelo trófico para la Bahía de Concepción. *Biologia Pesquera* **18**, 4–13.

Boje, R. and Tomczak, M. (1978) Ecosystem analysis and definition of bounderies in upwelling regions, pp. 1–11. In *Upwelling Ecosystems*, eds. R. Boje and M. Tomczak. Springer-Verlag. New York.

Brandhost, W. (1971). Condiciones oceanográficas estivales frente a la costa de Chile. *Rev. Biol. Mar., Valparaíso* **14** (3): 45–84.

Brattström, H. and Johanssen, A. (1983) Ecological and regional zoogeography of the marine benthic fauna of Chile. *Sarsia* **68**, 289–339.

Brundin, L.Z. (1989) Phylogenetic biogeography. In *Analytical Biogeography. An Integrated Approach to the Study of Animal and Plant Distributions*, eds. A.A. Myers and P.S. Guiller. Chapman and Hall, London. pp. 343–369.

Camus, P.A. (1990) Procesos regionales y fitogeografía en el Pacífico Sudeste: el efecto de "El Niño-Oscilación del Sur". *Rev. Chil. de Hist. Nat.* **63**, 11–17.

Camus, P.A. and Ojeda, F.P. (1992) Scale-dependent variability of density estimates and morphometric relationships in subtidal stands of the kelp *Lessonia trabeculata* in northern and central Chile. *Marine Ecology Progress Series* **90**, 193–200.

Camus, P.A. (1994a) Dinámica geográfica en poblaciones de *Lessonia nigrescens* Bory (Phaeophyta) en el norte de Chile: importancia de la extinción local durante eventos El Niño de gran intensidad. *Rev. Inv. Cien. y Tec., Ser Cien. Mar* **3**, 58–70.

Camus, P.A. (1994b) Recruitment of the intertidal kelp *Lessonia nigrescens* Bory in northern Chile: successional constraints and opportunities. *Journal of Experimental Marine Biology and Ecology* **184**, 171–181.

Camus, P.A. (1998) Estructura espacial de la diversidad en ensambles sésiles del intermareal rocoso en Chile centro-norte: la diversidad local como un resultado de determinantes de multiescala. Ph.D. Thesis, P. Universidad Católica de Chile, Santiago. 262 pp.

Castilla, J.C. and Nealler, E. (1978) Marine environmental impact due to mining activities of El Salvador copper mine, Chile. *Marine Pollution Bulletin* **9** (3), 67–70.

Castilla, J.C. (1983) Environmental impact in sandy beaches of copper mine tailing at Chañaral, Chile. *Marine Pollution Bulletin* **14** (12), 459–464.

Castilla, J.C. and Camus, P.A. (1992) The Humboldt–El Niño scenario: coastal benthic resources and anthropogenic influences, with particular reference to the 1982/83 ENSO. In Payne A.I.L., K.H. Brink and K.H. Mann (eds.). Benguela Trophic Functioning. *South African Journal of Marine Science* **12**, 111–119.

Carrera, M.E., Valenta, P., Ahumada, R. and Rodríguez, V. (1993) Determinación Voltamétrica de Metales trazas en la columna de agua y sedimentos en Bahía de Concepción. *Rev. Biol. Mar. Valparaíso* **28** (1), 151–163.

Clapperton, C.M. (1994) The quaternary glaciation of Chile: a review. *Revista Chilena de Historia Natural* **67**, 369–387.

Correa J., González, P., Sanz, P., Muñoz, J. and Orellana, M.C. (1996) Copper–algae interactions: interritance or adaptation?. *Environmental Monitoring Assessment* **40**, 41–54.

Corvalán, J. (1985) Recursos no Renovables, pp. 165–181. In *Medio Ambiente en Chile*, ed. F. Soler Rioseco. Centro de Investigación y Planificación del Medio Ambiente (CIPMA). Ediciones Universidad Católica de Chile. 413 pp.

Del Valle, A. (1985) Energía. pp. 183–223. In *Medio Ambiente en Chile*, ed. F. Soler Rioseco. Centro de Investigación y Planificación del Medio Ambiente (CIPMA). Ediciones Universidad Católica de Chile. 413 pp.

Djurfeldt, L. (1989) Circulation and mixing in a coastal upwelling embayment; Gulf of Arauco, Chile. *Continental Shelf Research* **9** (11), 1003–1016.

Dugdale, R.C., Goering, J.J., Barber, R.T., Smith, R.L. and Packard, T.T. (1977) Denitrification and hidrogen sulfide in the Perú upwelling region during 1976. *Deep-Sea Research* **24**, 601–608.

Dunbar, M.J. (1977) The evolution of polar ecosystem, pp. 1063–1076. In: *Adaptations within Antarctic Ecosystems*, ed. G.A. Llano. Smithsonian Institution. Washington DC. 1252 pp.

Ellenberg, H. (ed.) (1973) Ziele und stand der ökosystemforschung. In *Ökosystemforschung*. Springer, pp. 1–31.

Fonseca, T.R. and Farías M. (1987) Estudio del proceso de surgencia de la Costa Chilena utilizando percepción remota. *Investigacion Pesquera (Chile)* **34**, 33–46.

García, R. (1985) Estudio sobre contaminación por aguas de relave de minería de El Salvador en el litoral costero de Chañaral. Estudio. Dirección General del Territorio Marítimo y Marina Mercante, Chile. 75 pp.

Glyn, P.W. (1988) El Niño–Southern Oscillation 1982–1983: nearshore population, community, and ecosystem responses. *Annual Review of Ecology and Systematics* **19**, 309–345.

Guillen O., Lostaunau, N. and Jacinto, M. (1985) Características del fenómeno "El Niño" 1982–83. *Boletín Inst. Mar Peru-Callao*, vol. Extraordinario, 9–21

Gulland, J.A. (1977) Antarctic marine living resource, pp. 1135–1144. In: *Adaptations within Antarctic Ecosystems*, ed. G.A. Llano. Smithsonian Institution. Washington DC. 1252 pp.

Hedgpeth, J.H. (1977) The Antarctic marine ecosystem, pp. 1–10. In: *Adaptations within Antarctic Ecosystems*, ed. G.A. Llano. Smithsonian Institution. Washington DC. 1252 pp.

Jaksic, F.M. (1998) The multiple facets of El Niño/Southern Oscillation in Chile. *Rev. Chil. de Hist. Nat.* **71**, 121–131.

Leonardi, M. and Tarifeño, E. (1996) Efecto de la descarga de aguas servidas por un emisario submarino en los lenguados, *Paralichthys microps* (Gunther, 1881) y *Paralichthys adpersus* (Steindachner, 1867) en la bahía de Concepción, Chile: evidencias experimentales. *Rev. Biol. Mar., Valparaíso* **31**(1), 23–44.

Leonardi, M. and Tarifeño, E. (1997) Evaluación de las enfermedades de peces asociadas a las descargas de aguas servidas en los lenguados *Paralichthys microps* (Günther, 1881) y *Paralichthys adpersus* (Steindachner, 1867), como bioindicadores de la polución ambiental costera en la bahía Concepción. Parte II. Histopatologías. *Resumenes XVII Congreso Ciencias del Mar*. Universidad de Chile. 115–116.

Margalef, R. (1978) What is an upwelling ecosystem? pp. 12–14. In *Upwelling Ecosystems*, eds. R. Boje and M. Tomczak. Springer-Verlag, New York.

Mudge, S. and Seguel, C.G. (1997) Trace organic contaminants and lipid biomarkers in Concepcion and San Vicente Bays. *Bol. Soc. Chil. Quim.* **42**, 005–015.

Mudge, S. and Seguel, C.G. (1999) Organic contamination of San Vicente Bay, Chile. *Marine Pollution Bulletin* **38** (11), 1011–1021.

Ojeda, F.P. and Santelices, B. (1984) Ecological dominance of *Lessonia nigrescens* (Phaeophyta) in central Chile. *Marine Ecology Progress Series* **19**, 83–91.

Peterson, W.T., Arcos, D., McManus, G.B., Dam, H., Bellantoni, D., Johnson, T. and Tiselius, P. (1988) The Nearshore Zone during Coastal Upwelling: Daily Variability and Coupling between Primary and Secondary Producction off Central Chile. *Progress in Oceanography* **20**, 1–40.

Pickard, G. (1971) Some physical oceanographic features of inlets of Chile. *Journal of the Fisheries Research Board of Canada* **28**, 1077–1106.

Pinto, L.A. and Leif, R.N. (1991) Occurrence and distribution of elemental sulphur and organosulphur compounds in Washington (USA) continental slope sediments. In *Organic Geochemistry. Advances and Applications in the Natural Environment*, eds. D.A.C. Manning et al. Manchester University Press, Manchester, pp. 252–255.

Prahl, F.G., Pinto, L.A. and Sparrow, M.A. (1996) Phytane from chemolytic analysis of modern marine sediments: A product of desulfurization or not? *Geochimica et Cosmochimica Acta* **60**, 1065–1073.

Pritchard, D.W. (1968) Estuarine circulation pattern. In: *Estuaries*, ed. J.P. Lauff. 480 pp.

Quinn, W.H. and Neal, V.T. (1983) Long-term variations in the Southern Oscillation, El Niño, and Chilean subtropical rainfall. *Fisheries Bulletin, U.S.* **81**, 363–374.

Reid, J.L. (1973) Transpacific hydrographic sections at Lats. 43°S and 28°S, the SCORPIO Expedition III. Upperwater and a note on southward flow at middepth. *Deep-Sea Research* **20** (1), 51–68.

Rudolph, A. (1995) Coastal environmental alterations caused by fishing industry from organic matter disposal: a case of study. *Cienc. Tec. Mar, CONA*, No. Especial: 69–78.

Rudolph, A. and Ahumada, R. (1987) Intercambio de nutrientes entre una marisma con una fuente carga de contaminantes orgánicos y las aguas adjacentes. *Bol. Soc. Biol. Concepción, Chile* **58**, 151–169.

Santelices, B. (1980) Phytogeographic characterization of the temperate coast of Pacific South America. *Phycologia* **19**, 1–12.

SERNAP (Servicio Nacional de Pesca) (1998) *Anuario Estadístico de Pesca 1997*. Ministerio de Economía, Fomento y Recostrucción. Chile. 307 pp.

Silva, N. and Konow, D. (1975) Contribución al conocimiento de las masas de agua en el pacífico suroriental. Expedición Krill. Crucero 3–4, julio–agosto 1974. *Rev. Com. Perm. Pacífico Sur* **3**, 63–75.

Silva, N., Maturana, J., Sepúlveda, I. and Ahumada, R. (1998) Remineralización de la materia orgánica en la Región norte de los fiordos y canales del sur de Chile. *Ciec. Tecnol. Mar CONA* **21**, 49–74.

Silva, N. and Neshyba, S. (1979) On the southmost extension of the Peru–Chile Undercurrent. *Deep-Sea Research* **26**, 1387–1393.

Silva, N., Sievers, H.A. and Prado, R. (1995) Características oceanográficas y una proposición de circulación para algunos canales australes de Chile, entre 41°20′S y 46°40′S. *Rev. Biol. Mar., Valparaíso* **30**(2), 207–254.

Smith, F.E. (1970) Analysis of ecosystems. In *Analysis of Temperate Forest Ecosystems*, ed. D.E. Reischle. Springer Verlag, pp. 7–18.

Steele, J.H. (1974) *The Structure of Marine Ecosystems*. Harvard University Press. Cambridge. Massachusetts.

Tomicic, J.J. (1985) Efectos del fenómeno El Niño 1982–83 en las comunidades litorales de la Península de Mejillones. *Investigacion Pesquera* (Chile) **32**, 209–213.

Troup, A.J. (1965). The Southern Oscillation. *Quarterly Journal of the Royal Meteorological Society* **91**, 490–506.

Viviani, C.A. (1979) Ecogeografía del litoral chileno. *Studies on Neotropical Fauna and Environment* **14**, 65–123.

THE AUTHORS

Ramón B. Ahumada
Facultad de Ciencias,
Universidad Católica de la Santísima Concepción,
Campus San Andrés, Paicaví 3000, Casilla 297,
Concepción, Chile

Luis A. Pinto
Facultad de Ciencias,
Universidad Católica de la Santísima Concepción,
Campus San Andrés, Paicaví 3000, Casilla 297,
Concepción, Chile

Patricio A. Camus
Facultad de Ciencias,
Universidad Católica de la Santísima Concepción,
Campus San Andrés, Paicaví 3000, Casilla 297,
Concepción, Chile

Chapter 46

TROPICAL COAST OF BRAZIL

Zelinda M.A.N. Leão and José M.L. Dominguez

This chapter provides a general overview of the tropical coast of Brazil, with emphasis on the marine realm. The described region extends for approximately 3000 km, and has three different sectors (northern, northeastern and eastern), each one with distinctive characteristics.

The north and northeast sectors have a predominantly semi-arid climate, whereas the eastern sector is tropical and humid. Northeasterly trade winds occur mostly in the north sector, but in the northeast and east sectors southeasterly and easterly trade winds are more important. Three factors controlled sedimentation along the coast of tropical Brazil during Late Quaternary time: the sea-level history that played an important role in the evolution of the coastal zone and its related ecosystems over the last 5000 years; the sediment supply that is primarily regulated by the local relief and climate; and the climate itself, which is the major control of the large and active dune fields found in the northern sector.

Coral reefs are one of the most prominent marine ecosystems of tropical Brazil, particularly because of the unique character of its low-diversity coral fauna which is rich in endemic species; it is a relic fauna from the Tertiary, which forms unusual mushroom-shaped coral pinnacles. They include the southernmost coral reef communities of the Atlantic, and a small atoll.

A transition from siliciclastic dominant sediments on the coastline, to pure carbonates toward the middle and outer shelves, characterises the continental margin of tropical Brazil. Development was generally low until recently due to lack of roads and infrastructure, but new development has led to expansion in the area. Uncontrolled urban development, associated with heavy industrialisation and consequent pollution—uses which are not appropriate for these coastal marine ecosystems—and the accelerated deforestation of the Atlantic Rainforest are, today, the major threats to the tropical Coast of Brazil.

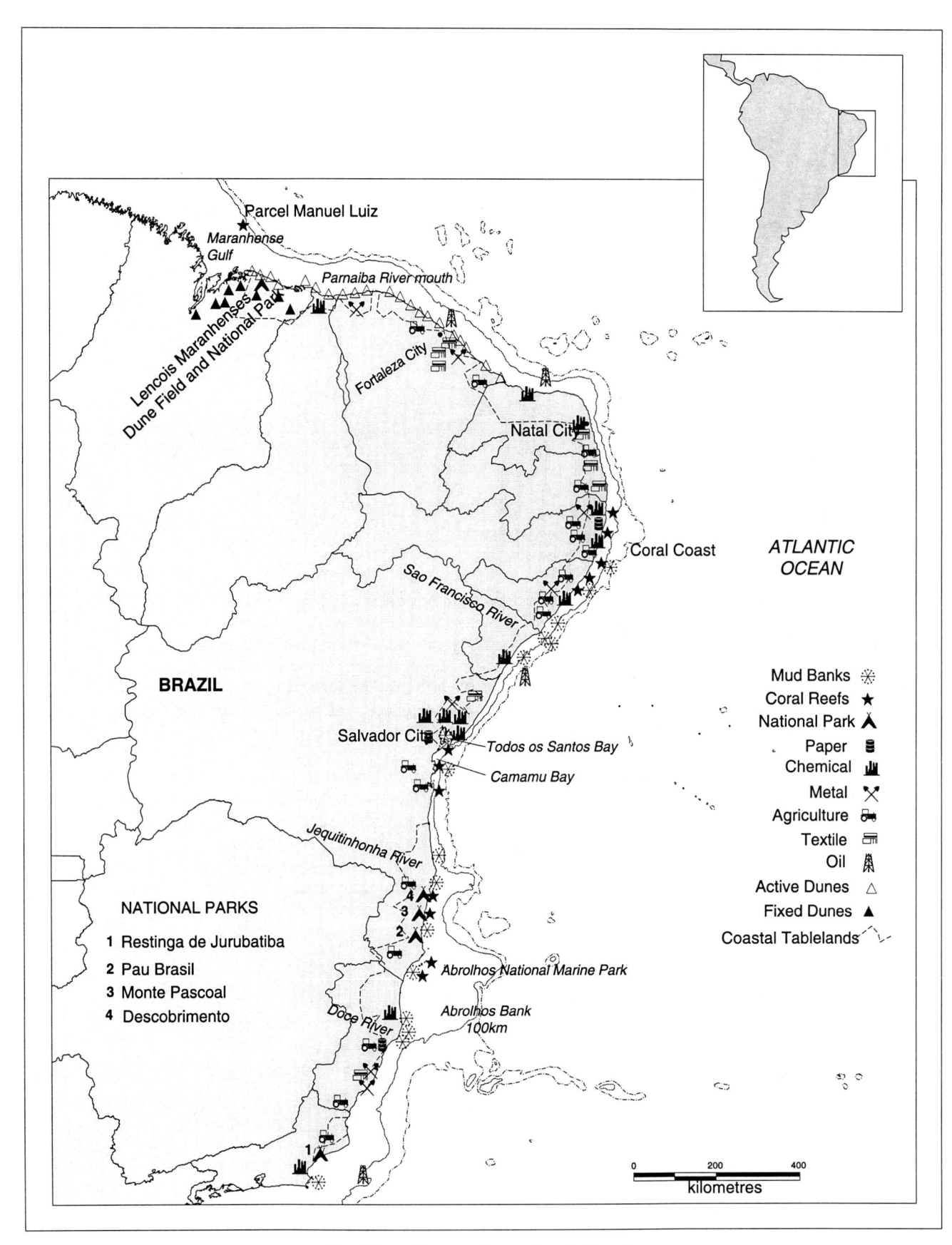

Fig. 1. Map of the coastal area of tropical Brazil.

THE REGION

The tropical coast of Brazil extends from the Maranhense Gulf (2°00'S) to the Paraíba do Sul coastal plain (21°50'S) (Fig. 1a–c). It comprises three sectors: the northern (2–5°S), the northeastern (5–12°S) and the eastern sectors (12–21°S). There are three major geomorphologic provinces within these regions:
1. A Precambrian basement hinterland dominated by rounded hills, with altitudes varying from 50 to 500 m.
2. The Tablelands—this is the most important and extensive morphological unit, almost continuously bordering the shoreline. They have an extremely flat surface with deeply incised flat-bottom valleys. This area is comprised of unconsolidated Late Tertiary alluvial sediments, mostly debris flow deposits, but also including fluvial channel and lacustrine deposits, named the Barreiras Formation. This represents an important reservoir of sediment to the coastal zone, particularly in those regions where active sea cliffs are present.
3. The Quaternary plain—Quaternary deposits of various origins (beach-ridge plains, wetlands and coastal dune fields) are distributed discontinuously along the shoreline and separated from the tablelands by a line of fossil sea cliffs.

The continental shelf, in Tropical Brazil, varies considerably in shape and width. It is mostly very narrow, with an average width of 50 km. In the southern section the shelf widens, particularly in the Abrolhos Bank region (see Fig. 1c), as a result of damming of sediment by volcanic seamounts. The shelf break is located at an average depth of 80 m. Sedimentation in the inner shelf, up to a depth of 20 m, is dominantly siliciclastic. In the middle and outer shelves carbonates dominate, except near major rivers where siliciclastic sedimentation can reach the outer shelf. Terrigenous mud accumulation is also restricted to these areas (Milliman and Barreto, 1975). Carbonate mud is locally deposited on the offshore inter-reefal zones.

OCEANOGRAPHIC PARAMETERS

Coastal Climate

The general atmospheric circulation pattern is controlled by two elements (Bigarella, 1972): (i) air masses generated in the South Atlantic high pressure cell, and (ii) advances of polar air masses. The South Atlantic is devoid of hurricanes, so only the above two elements, associated with the Intertropical Convergence Zone, define climate here. The northeastern and eastern coasts are therefore dominated by the southeasterly and easterly trade winds, whereas in the northern coast, northeasterly trades prevail. Along the eastern coast, a divergence zone of the trade winds occurs and northeastern winds blow to the south of this zone. A seasonal variation in the position of the South Atlantic high-pressure cell produces a north–south oscillation of the divergence zone between 10 and 20°S. This zone moves northward during summer and southward during winter. As a result, easterly and southeasterly winds dominate the coast north of 13°S, year-round, with speeds ranging from 5.5 to 8.5 m/s (US Navy, 1978). South of 13°S the easterly and southeasterly winds blow during fall and winter (April to September), and the northeasterly winds prevail during spring and summer (September to February); in this area the wind speed rarely surpasses 5.5 m/s (US Navy, 1978). The Antarctic polar front moves northward across the South American continent, east of the Andes Mountains as great anti-cyclones, and splits into two branches. The eastern branch moves along the coast towards the Equator and can reach as far as 10°S during winter but rarely reaches latitudes lower than 15°S in summer (Dominguez et al., 1992). The advance of this polar front also generates additional south-southeasterly winds, which reinforce the southeasterly winds generated by the anti-cyclone high-pressure cell. Gale force winds (25 m/s) have been measured along the advance of these polar fronts (Bandeira et al., 1975).

Climate in the north-northeastern coast of Brazil is classified as semi-arid, whereas in the eastern coast it is of the tropical humid type. The coastal zone from 4 to 6°S has at least four to five dry months during the year (Nimer, 1989). This extended dry season has favoured extensive dune development in this part of the coast.

Sea-surface temperature (SST) is the most conservative parameter along the Brazilian tropical coast, varying in the north-northeastern coast from 30°C during summer and fall (February to May) to 28°C from the end of winter to the beginning of summer (August to December). In the eastern coast, SST varies from 30°C (February to May) to 27°C (July and August). Minimum temperature, however, shows a marked decrease from north to south. On the north-northeastern coast, it decreases from 25°C during summer and fall, to 23°C during winter and spring; on the eastern coast, during winter, the minimum temperature can reach as low as 21°C (US Navy, 1978).

The wave pattern is conditioned by variation in the trade winds, and is related to movements of the offshore high-pressure centres. The Brazilian coast is mainly dominated by sea waves (locally generated waves with periods less than 7 s), and those with heights above 1 m account for more than 50% of observations (US Navy, 1978). These are the kinds of waves that Larcombe et al. (1995) found to be more effective in increasing turbidity of water in their four months of measurements, near the city of Townsville, Australia. In the Brazilian northern and northeastern coasts, the NE and E-SE waves dominate year round, with the eastern waves more important from January to May (summer–fall) and from September to November (spring). The southernmost part of the northeastern coast and the eastern coast, on the other hand, are dominated by eastern waves during the whole year. Northeast waves are only important from

November to February (summer), whereas southeast waves occur from March to August (winter).

Tides on the continental shelf are semi-diurnal. Due to the large latitudinal extent of the shelf, two different areas are defined (Hayes, 1979): (i) upper mesotidal in the northern and northeastern coasts, and (ii) lower mesotidal to microtidal in the eastern coast. The most conspicuous effect of the tidal component is observed in the northern coast, where it enhances the northwestward flow of the Brazilian Current (the North Brazilian Current), and periodically produces an intensification of this drift.

The Brazilian Current (BC) and the North Brazilian Current (NBC) are the main surface currents on the Brazilian continental margin (Stramma, 1991; Silveira et al., 1994). They originate from the South Equatorial Current at about 5 to 6°S and flow toward south (BC) with an average velocity of 50 to 70 cm/s, and to the north and northwest (NBC) attaining velocities of 30 cm/s. Data from the Atlas de Cartas Piloto (DHN, 1993) show that between 10 and 13°S, during July and August (austral winter), a reverse flow to the north can occur. North of 5°S, the North Brazilian Current becomes stronger as a result of combining with the South Equatorial Current.

Sedimentation

Major controls on sedimentation along the coast and in the adjacent continental shelf include sea-level history, sediment supply and climate.

Two important transgressive episodes affected the coastal zone during the Late Quaternary (Dominguez et al., 1992): (i) the Penultimate Transgression, which reached a maximum of 8±2 m above the present level around 123,000 years BP and the Last Transgression, which reached a maximum about 5100 years BP, when sea level was positioned 5±2 m above the present sea level. This sea-level history has played a pivotal role in the evolution of the coastal zone and its ecosystems during the last 5000 years. During the maximum of the Last Transgression most of the coastal zone was inundated. Estuaries, bays, lagoons and barrier islands were the most important coastal environments, and rivers drained into these and not directly into the open ocean. This was the time of maximum expansion of the mangrove forests and of optimum development for coral reefs on the adjacent shelf. During the following drop in sea level, vast areas of estuaries and lagoons disappeared and reef tops were exposed. Progradation of the shoreline, as a result of increased sediment supply from rivers and the drop in sea level itself, brought the shoreline and the coral reefs close together, or simply buried them with siliciclastic sediments. Increased turbidity resulted in additional stress to coral reefs and other inner shelf ecosystems.

Sediment supply varies considerably along Brazil's tropical coast and is controlled by area, relief and climate (Dominguez and Bittencourt, 1996). The rivers emptying into the coast between São Francisco to the Paraíba do Sul

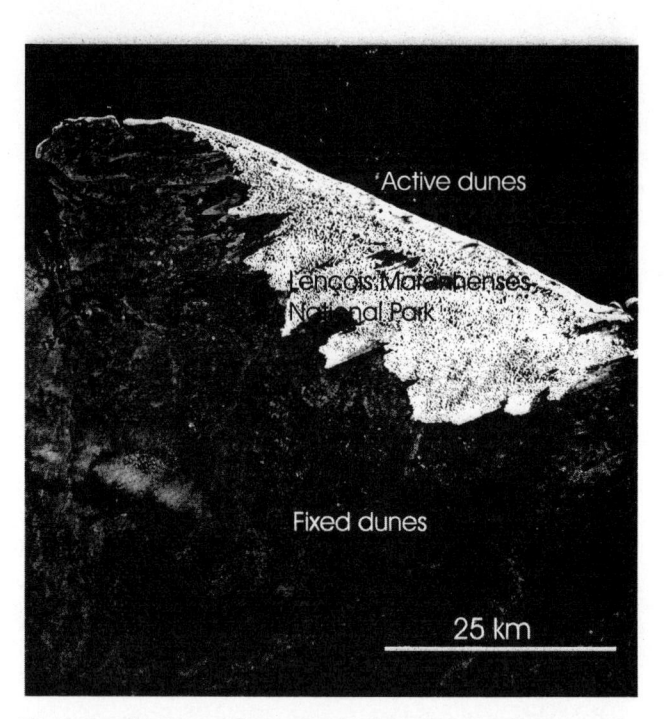

Fig. 2. Satellite image illustrating the Lençois Maranhenses on the Northern Sector of the Tropical Coast of Brazil.

river mouths are characterised by drainage basins that are larger, have higher mean altitude and wetter climate than along the coast between the São Francisco river mouth to the Maranhense Gulf. These differences in sediment supply are reflected directly in the distribution of Quaternary deposits along the whole coastal zone (see Fig. 1). Quaternary deposits resulting from progradation of the shoreline are abundant, south of the São Francisco River mouth, whereas north of this river the tablelands reach the present day shoreline, forming active sea cliffs.

Besides its effect on the sediment supply to the drainage basins, in the coastal zone itself climate controls the distribution of eolian deposits. In Tropical Brazil, active coastal dunes are restricted to those areas where more than three consecutive dry months occur during a year. These conditions are met along the northern coast where large dune fields are present. The Lençois Maranhenses dune field (Fig. 2), is the largest coastal dune complex in the entire South American continent.

MAJOR MARINE HABITATS

Continental Shelf

Carbonate sediments dominate the entire tropical Brazilian middle and outer shelves. Bioclastic carbonate gravel and sands (free-living non-articulated coralline red algae—maërl—*Halimeda*, benthic Foraminifera and mollusc debris), are also an important constituent in the inner shelf in many areas of the shelf (Coutinho, 1980; Dominguez and Leão, 1994; Testa, 1997; Testa and Bosence, 1998, 1999).

More commonly, the inner shelf is a typical mixing zone of siliciclastic and carbonate sediments, the former from river discharges, coastal erosion, and lower sea-level stands, and the carbonates from *in situ* growth and transport of calcareous organisms, such as red and green algae. Near the São Francisco River mouth (10°30'S), the largest river on east-northeastern Brazil, the carbonate sediment production is interrupted, probably due to water turbidity (Tiltenot et al., 1994). Also, the inner shelf of the eastern sector region, between the Jequitinhonha (15°00'S) and Doce (19°40'S) rivers, is influenced by river discharges, and here plumes of fine sediments are seen to advance some 50 km offshore. In these areas bioclasts occur only on the middle and the outer shelves, and the main carbonate sediments are mollusc shells, benthic Foraminifera tests, debris of calcareous algae, bryozoans, echinoids and, more rarely, coral gravel. Coral reefs occur along most of the carbonate province.

Coral Reefs

The Brazilian coral reefs form structures significantly different from most of the well known coral reef models in the world, as (i) they have a characteristic initial growth form of mushroom-shaped coral pinnacles, (ii) they are built by a very low-diversity coral fauna, rich in endemic species, and in which major reef builders are archaic forms, remnant of an ancient coral fauna dating back to the Tertiary, (iii) encrusting coralline algae has an important role in the construction of the reef structure, and (iv) the nearshore bank reefs are surrounded and even filled with muddy siliciclastic sediments. Thus they are a reef ecosystem ecologically unique and are, also, economically valuable for fisheries and ecotourism.

The reefs are formed by the coalescence of isolated columns called chapeirões that grow from the bottom of the ocean in a mushroom-like shape; their base is narrow and the top expands laterally. There are chapeirões of different heights and widths, in widely diversified stages of growth. When they are closely spaced, which usually occurs in the reefs nearest to shore, adjacent chapeirões fuse together at their tops, forming large compound reef structures called bank reefs, which have horizontal tops, are somewhat irregular and which can become completely uncovered during low tides. In these intertidal reef flats, large coral heads are truncated by erosion, and they alternate with numerous small pools, some shallow and sandy, but others rather deeper and with rocky bottoms. Irregular meandering channels of varied depths connect these pools with surrounding waters.

Corals, millepores and coralline algae build the rigid frame of the reefs in Brazil. The coral fauna comprises 18 species, almost half of them endemic to Brazilian waters (Belém et al., 1986). Among these endemic species some have affinities to the Miocene European corals and some are related to the Eocene Caribbean species (Laborel, 1969a,

1969b). These archaic species were preserved during Pleistocene low stands of sea level, in a refugium provided by the sea mountains off the Abrolhos Bank (Leão, 1982). Three milleporids and one stylasterid (Laborel, 1969a; Hetzel and Castro, 1994) represent the hydrocorals. The Brazilian hermatypic corals and hydrocorals were first described in the last century by Verrill (1868), later by Laborel (1969a, 1969b), and more recently have been examined in some detail (Amaral, 1994; Amaral et al., 1997; Castro, 1994; Echeverria et al., 1997; Maÿal and Amaral, 1990; Pires et al., 1992; Pitombo et al., 1988; Villaça and Pitombo, 1997).

Brazilian reefs can be grouped into various reef types: (i) nearshore bank reefs that comprise small discontinuous reef structures, adjacent to the beach, and of variable but often elongate forms, (ii) isolated bank reefs off the coast, of widely variable sizes (<10 m to >20 km) and shapes (elongate, circular, semi-arched), which due to the lateral discontinuity of their structures are different from the classical examples of barrier reefs, (iii) fringing reefs, more or less continuous, formed by encrustation of calcareous organisms on the rocky outcrops that usually occur bordering the coast of islands; (iv) isolated open-sea coral pinnacles, which are giant chapeirões that grow from the bottom and can reach the sea surface, (v) superficial reefs that are coral–algal constructions of no great thickness developed mostly above lines of beach rock, (vi) one small atoll, and (vii) drowned reefs.

These reefs are distributed into three major sectors along the tropical coast of Brazil. The northern coast apparently contains only one coral bank, 18 km long by 6 km wide, called Parcel de Manuel Luis, located about 80 km off the coast of São Luiz (State of Maranhão), and monospecific coral aggregates that occur along the coast of the state of Ceará (Coura, 1994; Laborel, 1969a, 1969b; Maida and Ferreira, 1997). The northeastern coast has several types of reef structures: (i) Rocas atoll located 267 km off the coast of Natal (Rio Grande do Norte State), (ii) coastal isolated bank reefs of varied shapes and dimensions forming one, two or three lines of reefs along the coast of the states of Rio Grande do Norte, Paraiba, Pernambuco and Alagoas, and which may be associated with lines of beach rock, and (iii) coral constructions bordering the shores of the oceanic islands. Examples are seen in the Fernando de Noronha Archipelago, about 345 km east off the coast of Rio Grande do Norte, in the São Pedro and São Paulo Archipelago about 500 km northeast from Fernando de Noronha (Kikuchi and Leão, 1998; Laborel 1969a, 1969b; Maida and Ferreira, 1997; Pires et al., 1992; Secchin, 1986; Testa, 1996, 1997), and off the eastern coast in three different sectors of the state of Bahia (see Fig. 1).

North of Salvador City, discontinuous coral bank reefs occur for about 20 km in very shallow waters; superficial coral–algal reefs, with thickness less than 1 m, developed above a line of beach rock, and drowned reefs occur in depths ranging from 20 to 70 m (Kikuchi and Leão, 1998;

The Biological Reserve of "Atol das Rocas"

Ruy K. P. Kikuchi

Departamento de Ciências Exatas, Universidade Estadual de Feira de Santana, BR-116, Campus Universitário, Feira de Santana, 44031-160, Bahia, Brazil

Rocas is the only atoll in the Southwestern Atlantic, and the smallest in the world (Fig. 1). It was the first Marine Biological Reserve created in Brazil, and the only activity allowed is scientific research. It was discovered in 1503 when a sailing boat struck it and sank, and from this point onwards, the atoll was regarded as hazardous.

The atoll lies on the western side of a flat-topped seamount, 260 km east of the city of Natal. Monthly average rainfall is 860 mm and ESE winds dominate, blowing usually at between 6 and 10 m/s. Tides are semi-diurnal with a maximum range of 2.7 m. The atoll lies in the Southern Equatorial Current which originates on the African coast. Water temperature averages 27°C, though inner pools may reach 39°C. Water visibility during good weather exceeds 20 m.

A reef front, reef flat and a lagoon are clearly distinct. On the windward side the reef front is a nearly vertical wall to depths of about 15 m, from which is a flat surface colonised by fleshy and coralline algae, corals and sponges, extending outwards another 1 km. On the lee side a spur-and-groove system develops to depths of 18 m. The reef flat is a ring 100 to 800 m wide interrupted by pools and two channels, with sandy deposits that are exposed at low tide. Two islets occur on the western side: the Cemitério Islet, which has a cross-bedded beachrock cliff, 1.5 m high, on its northeast side, and the 3 m high Farol Islet which has ruins of a 19th century lighthouse. Old reef spits rising 2 to 3 m above the reef flat surface, called "rocas" in Portuguese, give the atoll its name.

Soft algae and an association of coralline algae and vermetid gastropods mainly cover the reef surface. The corals *Siderastrea stellata, Montastrea cavernosa, Madracis decactis, Agaricia agaricites, Porites astreoides, Porites branneri, Favia gravida* and *Mussismilia hispida* (Echeverría et al., 1997) occur, mainly in the lagoon, in pools and in grooves of the reef front. *Siderastrea stellata* strongly dominates. The low diversity and cover of corals may have allowed coralline algae to develop more substantially here, and an 11 m thick reef framework core, primarily constructed by coralline algae, has accreted at a rapid average rate of 2.8 m/ky (Kikuchi and Leão, 1997).

The Biological Reserve of "Atol das Rocas" extends to the seamount top at the 1000 m isobath, covering 360 km^2. Created in 1978, conservation activities began in 1990. The first temporary station was established under the auspices of the Marine Turtle Foundation (Fundação Pró-TAMAR) and of the Manatee Project (Projeto Peixe-Boi Marinho–IBAMA). At the end of 1993, a more permanent station was built. It takes about 26 hours to reach the atoll from Natal. Research teams, composed of two park rangers (from IBAMA) and six scientists, students or volunteers, rotate every 25 days.

REFERENCES

Echeverría, C.A., Pires, D.O., Medeiros, M.S. and Castro, C.B. (1997) Cnidarians of the Atol das Rocas. *Proc. 8th Int. Coral Reef Symp.* **1**, 443–446.

Kikuchi, R.K.P. and Leão, Z.M.A.N. (1997) Rocas (Southwestern Equatorial Atlantic, Brazil): an atoll built primarily by coralline algae. *Proc. 8th Int. Coral Reef Symp.* **1**, 731–736.

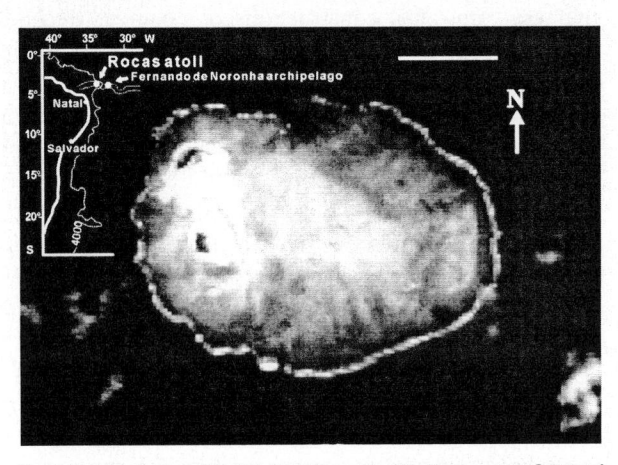

Fig. 1. Satellite image of the "Atol das Rocas" on the Northeastern Sector of the Tropical Coast of Brazil. Scale bar = 1 km.

Leão, 1996; Leão et al., 1988, 1997; Nolasco and Leão, 1986). In the Todos os Santos Bay, shallow fringing reefs, more or less continuous, border islands (Araujo et al., 1984; Leão, 1996; Leão et al., 1988).

Southern Bahia contains the largest and the richest area of coral reefs along the whole Brazilian coast. Its northern part is the least known area, but fringing reefs border the shores of islands and of Camamu bay. These reefs are more or less continuous structures, whose tops become completely exposed during low spring tides (Leão, 1996). North to the Abrolhos Bank, small bank reefs, with varied shapes and dimensions, may have grown on submerged strings of beach rock lying at about 10 m deep, but southward, where the continental shelf widens, forming the Abrolhos Bank, is the largest and richest coral reef area of Bahia (Fig. 3). These reefs form two arcs: a coastal arc located a few kilometres from the coastline is composed of bank reefs of varied shapes and dimensions, and an outer arc bordering the east side of the Abrolhos Archipelago, formed by isolated chapeirões in water deeper than 20 m. Incipient fringing reefs border the shores of the five islands that form the Abrolhos Archipelago (Castro, 1994, Leão, 1982, 1994, 1996, Leão and Ginsburg, 1997, Leão et al., 1988, Pitombo et al., 1988).

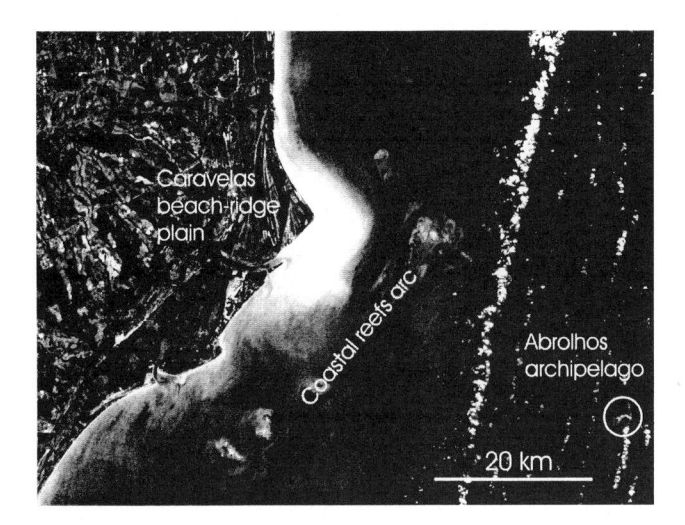

Fig. 3. Satellite image of the Eastern Sector of the tropical coast of Brazil illustrating the Caravelas beach-ridge plain and the Abrolhos coastal reefs.

Bays

Todos os Santos Bay is the biggest and the most important bay in the entire tropical coast of Brazil (Fig. 4). Its structure is inherited from an aborted-rifted basin formed during the South America–Africa separation, in Cretaceous Time. It contains about 35 islands, and although the bay receives about 200 m³/s, of freshwater drainage, this volume is two orders of magnitude less than the estimated volume of tidal water that enters through the main bay opening; its characteristics are clearly marine (Lessa et al., submitted). Muddy bottom sediment predominates in the northern portion, whereas its southern portion has an accumulation of medium to very coarse sands. Bioclastic sediments cover large areas of the bottom. Since its discovery, in the 16th century, the bay of Todos os Santos has been heavily used, from mining of calcareous sands, to petroleum exploration, overfishing, industrial sewage and intensive tourism.

MAJOR COASTAL HABITATS

The Atlantic Rainforest (The Maritime Forest)

When Portuguese settlers arrived in Brazil, the Atlantic Rainforest covered an area of about 1 million km², or 12% of the whole Brazilian territory. Today, this forest is reduced to 5% of its original distribution. In tropical Brazil the most important remnants of this maritime forest are located in the southern part of the State of Bahia, between the Jequitinhonha and the Caravelas strandplains (Fig. 5). The

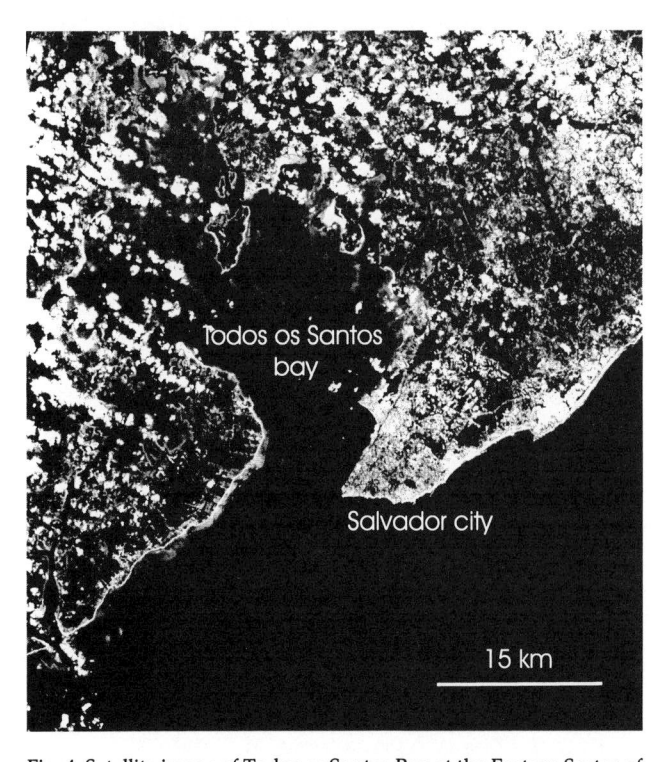

Fig. 4. Satellite image of Todos os Santos Bay at the Eastern Sector of the tropical coast of Brazil.

Fig. 5. Satellite image of the southern part of the Eastern Sector of the tropical coast of Brazil illustrating the coastal tablelands and remnants of the Atlantic Rainforest. Landsat-TM5 image.

The Abrolhos National Marine Park

Zelinda M.A.N. Leão

The National Marine Park of Abrolhos protects part of the largest and the richest coral reefs of Brazil, and of the South Atlantic. It comprises two isolated areas, which include nearshore isolated bank reefs, offshore volcanic islands with fringing reefs, and giant coral pinnacles. The Abrolhos reefs are ecologically unique, and the most characteristic reef growth form, the *chapeirões,* attain their maximum development here, and support communities that thrive in a periodically inhospitable muddy environment.

There are five islands, surrounded by fringing reefs (Fig. 1). The most visited and largest is Santa Barbara (Leão et al., 1994), approximately 1 km long. At its north and south sides are gravelly, sandy beaches, and fringing reefs develop along two thirds of its shore. Other islands are Redonda Island which has a reef fringe on its southeast, Siriba Island, Sueste Island (500 m long) and the small Guarita Island (100 m across), which is made of volcanic rocks with no sandy beaches. Reef communities grow on its slope. Several are separated by very shallow (<4 m) water. They are arranged in a ring, though this is not evidence of the remains of an old caldera.

Seventeen stony corals occur, six of which (*Mussismilia braziliensis, M. hispida, M. hartti, Siderastrea stellata, Favia gravida* and *Favia leptophylla*) are endemic species in Brazil. Among these, *M. braziliensis* is common but shows the greatest geographical confinement, occurring only along the coast of Bahia. *Siderastrea stellata* and *Favia gravida* are similar to Caribbean species and are the most common corals in the shallow intertidal pools. More cosmopolitan corals are minor components in Abrolhos. Two of the three millepores in Abrolhos are Brazilian endemics also: *Millepora braziliensis* and *M. nitida*. The distribution of these and soft corals are quite well known (Castro 1989, 1990, 1994; Belém et al., 1982).

The algal flora is abundant on the reefs. The crustose coralline algae *Lithothaminion, Lithophyllum, Sporolithon* and *Porolithon* are the most abundant genera (Figueiredo, 1997) and are major reef-framework builders. Fleshy algae dominate some reefs around the islands (Amado Filho et al., 1997), but on further offshore reefs they diminish, possibly due to higher herbivore activity (Coutinho et al., 1993). Most fishes identified in Abrolhos are related to the Caribbean fauna (Nunam, 1979). Thus the Abrolhos region is the southernmost area of the Atlantic inhabited by a coral reef fish fauna. About 39% of the species are herbivorous (Scaridae, Acanthuridae, Kyphosidae), 54% are omnivorous (Haemulidae, Balistidae, Pomacanthidae, Lutjanidae, Pomacentridae) and 7% are carnivorous (Serranidae, Carangidae, Sphyraenidae) (Telles, 1998).

Marine turtles visit Abrolhos for feeding and reproduction. In summer time *Caretta caretta* (the loggerhead) and *Chelonia midas* (green turtles) lay their eggs on the sandy beaches, while *Eretmochelys imbricata* (hawksbill turtles) feeds on invertebrates from the reefs. The humpback whale *Megaptera novaeangliae* migrates from subantarctic waters to the warmer shallows here. Sea birds like *Sula dactylatra* (blue-faced booby), *Sula leucogaster* (brown booby), *Fregata magnificens* (frigate bird), *Sterna fuscata* (sooty tern), *Anous stolidus* (brown nooddy), and *Phaeton aethereus* (red-billed tropicbird) all nest on the islands.

The Abrolhos National Marine Park was created in 1983, by Presidential Decree, under the jurisdiction of the Brazilian Institute of Environment and Natural Renewable Resources (IBAMA). It covers about 900 km^2 comprising two areas (see Fig. 1 of main chapter). It includes less than one fourth of the total area of reefs in Abrolhos. The park was not implemented until 1988, when a Triennial Plan (1989–1991) was prepared based on the experience of the first 15 months of park activity (Gonchorosky et al., 1989). The plan was divided into three main lines of action: environmental protection, environmental education and support for scientific research. The management plans with conservation programs are already in action (IBAMA/FUNATURA 1991). In the Abrolhos Archipelago, landing is only allowed in Redonda and the Siriba islands under the supervision of a Park technician. In Sueste and Guarita islands, landing, anchorage and diving are forbidden. Santa Barbara belongs to the Brazilian Navy and landing there is only permitted with an official authorization.

Fig. 1. Aerial photograph of the Abrolhos Archipelago on the Eastern Sector of the Tropical Coast of Brazil, facing to the east. Top left is the Santa Barbara Island, bottom left the Redonda Island, middle of the photo is the Siriba Island and top right the Sueste Island.

REFERENCES

Amado-Filho, G.M., Andrade, L.R., Reis, R.P., Bastos, W. and Pfeiffer, W.C. (1997) Heavy metal concentrations in seaweed species from the Abrolhos reef region, Brazil. *Proc. 8th Int. Coral Reef Symp.* **2**, 1843–1846.

Belém, M.J.C., Castro, C.B. and Rohlfs, C. (1982) Notas sobre *Solanderia gracilis* Duchassaing & Michelin, 1846 do Parcel de Abrolhos, Bahia. Primeira ocorrência de Solanderiidae (Cnidaria, Hidrozoa) no litoral brasileiro. *An. Acad. Bras. Ciênc.* **54** (3), 585–588.

Castro, C.B. (1989) A new species of *Plexaurella* Valenciennes, 1855 (Coelenterata, Octocorallia), from the Abrolhos reefs, Bahia, Brazil. *Rev. Bras. Biol.*, Rio de Janeiro, **49** (2), 597–603.

Castro, C.B. (1990) A new species of *Heterogorgia* Verrill, 1868 (Coelenterata, Octocorallia) from Brazil, with comments on the type species of the genus. *Bulletin of Marine Science* **4** (2), 411–420.

Castro, C.B. (1994) Corals of Southern Bahia. In: *Corals of Southern Bahia*, eds. B. Hetzel and C.B. Castro. Editora Nova Fronteira, Rio de Janeiro, pp. 161–176.

Coutinho, R., Villaça, R.C., Magalhães, C.A., Guimarães, M.A., Apolinario, M. and Muricy, G. (1993) Influência antrópica nos ecossistemas coralinos da região de Abrolhos, Bahia, Brasil. *Acta Biologica Leopoldensia* **15** (1), 133–144.

Gonchorosky, J., Sales, G., Belém, M.J.C. and Castro, C.B. (1989) Importance, establishment and management plan of the Parque Nacional Marinho dos Abrolhos, Brazil. In: *Coastlines of Brazil*, ed. C. Neves. Coastlines of the World. American Society of Civil Engineers, New York, pp. 185–194.

IBAMA/FUNATURA (1991) *Plano de Manejo do Parque Nacional Marinho dos Abrolhos*. Instituto Brasileiro do Meio Ambiente e dos Recursos Naturais Renováveis/Fundação Pró-Natureza, Brasília. Aracruz Celulose S.A., 96 pp.

Leão, Z.M.A.N., Telles, M.D., Sforza, R., Bulhões, H.A. and Kikuchi, R.K.P. (1994) Impact of tourism development on the coral reefs of the Abrolhos area, Brazil. In *Global Aspects of Coral Reefs: Health, Hazards and History*, compiled by R.N. Ginsburg. Rosenstiel School of Marine and Atmospheric Science, University of Miami, Florida, pp. 254–260.

Nunam, G.W. (1979) The zoogeographic significance of the Abrolhos area as evidenced by fishes. Ms. Thesis, Rosenstiel School of Marine and Atmospheric Science, University of Miami, Florida, 146 p.

Telles, M.D. (1998) Modelo trofodinâmico dos recifes em franja do Parque Nacional Marinho dos Abrolhos, Ba. Tese de Mestrado, Fundação Universidade do Rio Grande, 150 pp.

---

great majority of the Brazilian fauna and flora which are listed as endangered species, are endemic to this Atlantic Rainforest.

Restinga

This term is used in Brazil to describe the biological communities that thrive in sandy coastal deposits (dunes and beach-ridge plains) accumulated as a result of shoreline progradation. The vegetation, comprising grasses, shrubs and small trees, is adapted to soils that are virtually devoid of nutrients, and they exhibit xeromorphic characteristics allowing them to survive in this relatively dry sandy substrate. The salt spray is their most important source of nutrients. The most important Restinga in tropical Brazil are associated with the most prominent beach-ridge plains, such as the São Francisco, Jequitinhonha, Caravelas, Doce and Paraíba do Sul (Fig. 6) strandplains.

Wetlands

In tropical Brazil wetlands are very limited. Freshwater wetlands are restricted to flat-bottom valleys incised in the tablelands, or occur in low-lying areas in the Quaternary plains which acted as coastal lagoons during the maximum of the last Transgression. In these freshwater wetlands, emergent vascular macrophytes (grasses and *Juncus*) dominate. Mangrove forests occur in the lower segments of rivers. The most important mangrove forests in Brazil occur in the Parnaíba river strandplain, in the Todos os Santos and Camamu bays, and in the Caravelas strandplain. The dominant species in all these areas is *Rhizophora mangle*.

MAJOR ENVIRONMENTAL CONCERNS AND PRESERVATION

Until the 1980s most of the tropical coast of Brazil was virtually inaccessible. Occupation was concentrated in the major state capitals. The remaining coastal zone was dotted with small villages inhabited by local fishermen, and the area experienced economic stagnation. As a result of several plans to develop tourism activity in the region, massive federal and state government investments were made to build highways which provided easy access to the coastal region. Occupation of the coastal zone has dramatically increased since then, exerting diverse and numerous stresses on the coastal ecosystems. Figure 1 shows locations of major agri-industrial activities which directly affect the coastal and adjacent marine environments. Heavy industrialisation occurs in Fortaleza, Recife and Salvador regions. Major impacts to the marine realm are probably mostly associated with agri-industrial activities (i.e. sugarcane and

Fig. 6. Satellite image illustrating the Paraiba do Sul beach-ridge plain at the southern portion of the tropical coast of Brazil.

timber), paper mills, mineral and chemical industries and oil exploration. Most agricultural activity takes place in the tablelands which, by their morphological nature, offer very favourable conditions for intensive, mechanised farming practices. Carbonate materials (shells and coral fragments) have been mined for many years from Todos os Santos Bay, although this practice has recently been stopped. Studies are presently under way to evaluate the feasibility of exploiting calcareous algae, sand and gravel from the inner and middle shelves. Additional stresses to the marine environment come from increased sediment, due to the removal of the Atlantic Rainforest, and the disposal of industrial and urban effluents.

Fortunately the areas where the major reef complexes occur correspond to regions where development is still incipient, although nearby urban centres are experiencing accelerated growth, and tourist development is increasingly placing a toll on the environment. Today, oil production in Brazil comes, mostly, from offshore basins. The Campos basin, located in the south of this area is, nowadays, the most important producing area. In the rest of the region, the major human impacts are related to development of the coastal zone for recreational and tourist uses. These activities have resulted in widespread elimination of the Restinga ecosystem, which in most places has been replaced by developments or by coconut cultivation. This last activity dates back to the 18th century.

The institutions concerned with preservation of the coastal marine ecosystems are fairly new. In the north, a state marine park established in 1991 protects the offshore Manuel Luiz Bank, but it does not prevent the coral bank from being crossed by large ships en route to São Luiz harbour, which exposes the reefs to oil spills and shipwrecks. In the northeastern sector, an environmental marine protected area was recently designated (1998) that covers a great part of the coast of the states of Pernambuco and Alagoas; this is called the Coral Coast Environmental Protected Area, and it is the largest area of shallow coastal reefs so far created in Brazil. Offshore, the Biological Reserve of Atoll das Rocas and the Fernando de Noronha National Marine Park, have made lobster fishing illegal in the atoll, and have been controlling marine tourism in the Fernando de Noronha Archipelago. The Fernando de Noronha National Marine Park was created in 1988, and is located about 80 nautical miles eastward of the Rocas atoll.

In the east region, one national and two municipal marine protected areas have been created. The oldest is the Abrolhos National Marine Park that protects the richest area of coral reefs in Brazil. The two municipal areas are the Pinaunas Reef Environmental Protected Area at the entrance of the Todos os Santos Bay, designated in 1997, and the Recife de Fora Municipal Marine Park, created in 1998, about 5 nautical miles off the south coast of the state of Bahia. It is a small reef with an area of approximately 11 square miles that has been heavily used for marine tourism. Along the coast of the state of Bahia even reefs not included in park areas are still protected by law, given that the Bahia State Constitution declares that coral reefs are areas of permanent protection. However, enforcement is not yet effective.

Special Government Programs or Non Governmental Institutions protect distinctive faunal elements. For example the TAMAR Project–IBAMA, which protects marine turtles, has several stations along the whole tropical coast. The "Baleia Jubarte" Institute, responsible for the study of and concerns about the humpback whales, has its headquarters in South Bahia, and the Manatee Project, on the coast of the State of Alagoas, has been making efforts to protect this endangered animal.

In the coastal zone itself there are only five national parks. Four are located in the eastern sector. Three of them (Pau Brasil, Monte Pascoal and Descobrimento National Parks) were created with the purpose of protecting remnants of the Atlantic Rainforest. The Restinga de Jurubatiba National Park protects an extensive area of the Restinga ecosystem, and the Lençois Maranhenses National Park in the northern sector coincides with a large dune system (see Fig. 2).

REFERENCES

Amaral, F.M.D. (1994) Morphological variation in the reef coral *Montastrea cavernosa* in Brazil. *Coral Reefs* 13, 113–117.

Amaral, F.M.D., Silva, R.S., Mauricio-Da-Silva, L. and Solé-Cava, A.M. (1997) Molecular systematic of *Millepora alcicornis* Linnaeus, 1758 and *M. braziliensis* Verrill, 1868 (Hydrozoa: Milleporidae) from Brazil. *Proc. 8th Int. Coral Reef Sym.* 2, 1577–1580.

Araujo, T.M.F., Leão, Z.M.A.N. and Lima, O.A.L. (1984) Evolução do recife de coral da ilha de Itaparica determinada a partir de dados geológicos e geofísicos. Congr. Bras. Geol. 33, *Anais*, Rio de Janeiro, SBG, 1, pp. 159–169.

Asmus, H.E. (1970) Banco de Abrolhos. Tentativa de interpretação genética. PETROBRÁS/DEXPRO/DIVEX. Unpub. Report, Rio de Janeiro.

Bandeira A.N., Jr., Petri, S. and Suguio, K. (1975) Projeto Rio Doce. Petróleo Brasileiro S.A. Internal Report, 203 pp.

Belém, M.J.C., Rohlfs, C., Pires, D. and Castro, C.B. (1986) S.O.S. Corais. *Rev. Ciência Hoje* 5 (26), 34–42.

Bigarella, J.J. (1972) Eolian environments—their characteristics, recognition and importance. In *Recognition of Ancient Sedimentary Environments*, eds. J.K. Rigby and W.L. Hamblin. SEPM, Spec. Publ. 16, pp. 12–14.

Castro, C.B. (1994) Corals of Southern Bahia. In: *Corals of Southern Bahia*, eds. B. Hetzel and C.B. Castro. Editora Nova Fronteira, Rio de Janeiro, pp. 161–176.

Cordani, U.G. (1970) Idade do vulcanismo no Oceano Atlântico Sul. *Bol. Inst. Geoc. Astron.* Univ. São Paulo, 1, 9–76.

Coutinho, P.N. (1980) Sedimentação continental na plataforma continental Alagoas-Sergipe. *Arq. Ciênc. Mar* 21, 1–18.

Coura, M.F. (1994) *Contribuição ao plano de manejo do Parque Estadual Marinho do Parcel de Manuel Luiz, Ma-Brasil*. Monografia de Especialização. Universidade Federal do Maranhão, 87 pp.

DHN (1993) *Atlas de Cartas Piloto*. Ministerio da Marinha. Diretoria de Hidrografia e Navegação.

Dominguez, J.M.L. and Leão, Z.M.A N. (1994) Contribution of sedimentary geology to coastal environmental management of the Arembepe region, state of Bahia, Brazil. *14th Intern. Sedim. Congress, IAS Abstract*, J14–J15.

Dominguez, J.M.L. and Bittencourt, A.C.S.P. (1996) Regional assessment of long-term trends of coastal erosion in Northeastern Brazil. *An. Acad. Bras. Ci.* **68** (3), 355–371.

Dominguez, J.M.L., Bittencourt, AC.S.P. and Martin, L. (1992) Controls on Quaternary coastal evolution of the east-northeastern coast of Brazil: roles of sea-level history, trade winds and climate. *Sedimentary Geology* **80**, 213–232.

Echeverría, C.A., Pires, D.O., Medeiros, M.S. and Castro, C.B. (1997) Cnidarians of the Atol das Rocas. *Proc. 8th Int. Coral Reef Symp.* **1**, 443–446.

Figueiredo, M.O. 1997. Colonisation and growth of crustose coralline algae in Abrolhos, Brazil. *Proc. 8th Int. Coral Reef Symp.* **1**, 689–694.

Hayes M.O. (1979) Barrier island morphology as a function of tidal and wave regime. In: S.P. Leatherman (ed.), *Barrier Islands from the Gulf of St. Lawrence to the Gulf of Mexico.* Academic Press, New York, NY, pp. 1–27.

Hetzel, B. and Castro, C.B.C. (1994) *Corals of South Bahia.* Editora Nova Fronteira, R.J., 202 pp.

Kikuchi R.K.P. and Leão, Z.M.A.N. (1998) The effects of Holocene sea level fluctuation on reef development and coral community structure, Northern Bahia, Brazil. *An. Acad. Bras. Ci.* **70** (2), 159–171.

Laborel, J.L. (1969a) Madreporaires et hydrocoralliaires recifaux des côtes brésiliennes. Systematique, ecologie, repartition verticale et geographie. *Ann. Inst. Oceanogr. Paris* **4**, :171–229.

Laborel, J.L. (1969b) Les peuplements de madreporaires des côtes tropicales du Brésil. *Ann. Univ. d'Abidjan, Ser. E, II,* Fasc. 3, 260 pp.

Larcombe, P., Ridd, P.V., Prytz, A. and Wilson, B. (1995) Factors controlling suspended sediment on inner-shelf coral reefs, Townsville, Australia, *Coral Reefs* **14**, 163–171.

Leão, Z.M.A.N. (1982) Morphology, geology and developmental history of the southernmost coral reefs of Western Atlantic, Abrolhos Bank, Brazil. Ph.D. Dissertation, Rosenstiel School of Marine and Atmospheric Science, University of Miami, Florida, U.S.A., 218 pp.

Leão, Z.M.A.N. (1994) The coral reefs of southern Bahia. In *Corals of Southern Bahia,* eds. B. Hetzel and C.B. Castro. Editora Nova Fronteira, Rio de Janeiro, pp. 151–159.

Leão, Z.M.A.N. (1996) The coral reefs of Bahia: morphology, distribution and the major environmental impacts. *An. Acad. Bras. Ciênc.* **68** (3), 439–452.

Leão, Z.M.A.N. and Ginsburg, R.N. (1997) Living reefs surrounded by siliciclastic sediments: the Abrolhos coastal reefs, Bahia, Brazil. *Proc. 8th Int. Coral Reef Symp.* **2**, 1767–1772.

Leão, Z.M.A.N., Araujo, T.M.F. and Nolasco, M.C. (1988) The coral reefs off the coast of eastern Brazil. *Proc. 6th Int. Coral Reef Symp.* Australia, **3**, 339–347.

Leão, Z.M.A.N., Kikuchi, R.K.P., Maia, M.P. and Lago, R.A.L. (1997) A catastrophic coral cover decline since 3,000 years B.P., Northern Bahia, Brazil. *Proc. 8th Int. Coral Reef Sym.* **1**, 583–588.

Lessa, G., Dominguez, J.M.L., Bittencourt, A.C.S.P. and Brichta, A. (submitted). The tides and tidal circulation of Todos os Santos Bay, Northeastern Brazil: a general characterisation. *An. Acad. Bras. Ci.*

Maida. M. and Ferreira, B.P. (1997) Coral reefs of Brazil: an overview. *Proc. 8th Int. Coral Reef Symp.* **1**, 263–274.

Maÿal, E. and Amaral F.M.D. (1990) Ecomorfose em alguns escleractínios da costa pernambucana. *Trab. Oceanogr. Univ. Fed. Pe.* **21**, 239–251.

Milliman, J.D. and Barreto, H.T. (1975) Continental margin sedimentation off Brazil, Part 1. Background. *Contribution to Sedimentology* **4**, 1–10.

Nimer, E. (1989) *Climatologia do Brasil.* Instituto Brasileiro de Geografia e Estatística, Rio de Janeiro, 421 pp.

Nolasco, M.C. and Leão, Z.M.A.N. (1986) The carbonate buildups along the northern coast of the state of Bahia, Brazil. In *Quaternary of South America and Antarctic Peninsula,* ed. J. Rabassa. Balkema Pub., Vol. 4, pp. 159–190.

Pires, D.O., Castro, C.B., Migotto, A.E. and Marques, A.C. (1992) Cnidários bentônicos do Arquipélago de Fernando de Noronha, Brasil. *Bol. Mus. Nac. Zool.,* R.J. (354), 1–21.

Pitombo, F., Ratto, C.C. and Belém, M.J.C. (1988) Species diversity and zonation pattern of hermatypic corals at two fringing reefs of Abrolhos Archipelago, Brazil. *Proc. 6th Int. Coral Reef Symp.* Australia, **2**, 817–820.

Secchin, C. (1986) *Abrolhos—Parque Nacional Marinho.* Edit. Cor/Ação. Rio de Janeiro. 128 pp.

Silveira, I.C.A. da, Miranda, L.B. and Brown, W.S. (1994) On the origins of the North Brazil Current. *Journal of Geophysical Research* **99** (C11), 501–512.

Stramma, L. (1991) Geostrophic transport of the South Equatorial Current in the Atlantic. *Marine Research* **49**, 281–294.

Testa, V. (1996) Quaternary sediments of the shallow shelf, Rio Grande do Norte, NE Brazil. PhD Dissertation, Royal Holloway University of London, 411 pp.

Testa, V. (1997) Calcareous algae and corals in the inner shelf of Rio Grande do Norte, NE Brazil. *Proc. 8th Int. Coral Reef Symp.* **1**, 737–742.

Testa, V. and Bosence, D.W.J. (1998) Carbonate-siliciclastic sedimentation on a high-energy, ocean-facing, tropical ramp, NE Brazil. In *Carbonate Ramps: Oceanographic and Biological Controls, Modelling and Diagenesis,* eds. V.P. Wright and T. Burchette. Geological Society of London, Special Publication 149, pp. 55–71.

Testa, V. and Bosence, D.W.J. (1999) Biological and physical control on the bedform generation in the Rio Grande do Norte inner shelf, Brazil. *Sedimentology* **46**, 279–301.

Tiltenot, M., Jennerjahu, T., Irion, G. Morais, J.O. and Brichta, A. (1994) Sedimentological and geochemical indicators of river supply along the Brazilian continental margin. *14th Int. Sedim. Cong. IAS Abstract:* D77.

US Navy (1978) *US Navy Marine Climatic Atlas of the World, Vol. IV: South Atlantic Ocean.* US Navy, Washington, DC, 325 p.p

Verrill, A.E. (1868) Notes of the radiate in the Museum of Yale College, with descriptions of new genera and species. 4. Notes of the corals and echinoderms collected by Prof. C.F. Hartt at the Abrolhos reefs, Province of Bahia, Brazil. *Connecticut Acad. Arts Sci. Transact.* **1** (2): 351–371.

Villaça R. and Pitombo, F.P. (1997) Benthic communities of shallow-water reefs in Abrolhos, Brazil. *Rev. Bras. Oceanogr.* **45** (1/2), 35–43.

THE AUTHORS

Zelinda M.A.N. Leão
Laboratório de Estudos Costeiros,
Centro de Pesquisa em Geofísica e Geologia,
Universidade Federal da Bahia,
Rua Caetano Moura 123, Federação,
Salvador, 40210-340, Bahia, Brazil

José M.L. Dominguez
Laboratório de Estudos Costeiros,
Centro de Pesquisa em Geofísica e Geologia,
Universidade Federal da Bahia,
Rua Caetano Moura 123, Federação,
Salvador, 40210-340, Bahia, Brazil

Chapter 47

SOUTHERN BRAZIL

Eliete Zanardi Lamardo, Márcia Caruso Bícego, Belmiro Mendes de Castro Filho, Luiz Bruner de Miranda and Valéria Aparecida Prósperi

---

Brazil has a large and extended coastline. Due to the history of the country and the facilities the littoral zone offers, the coastal zone is overpopulated, with 50% of the total population of the country living there. The coast of Brazil has many ports, industries and tourism activities, which intensify in the southern part of the country, which is the focus of this chapter. The Southern Brazil coast, here defined from Abrolhos (Bahia) to Chuí (Rio Grande do Sul) is distinct from the rest of the country's coast due to its geographic characteristics and intense development.

Southern Brazil contains many different ecosystems such as sandy beaches, beautiful islands, mangroves, forests and some lagoon complexes. Associated with them is a variety of flora and fauna that make this region pleasant and unusual, attracting many tourists. In addition this region is considered the most developed part of the country, encompassing large cities, many industries, petroleum, harbour, fishing and agricultural activities, among others. As a consequence, some of the coastal and shallow marine systems have been polluted with heavy metals and organic compounds which come from domestic and industrial effluents. Some ecosystems which are already intensively degraded include Vitória, Guanabara and Santos Bays, as well as the coastal lagoons of Rio Grande do Sul.

In the last 20 years environmental concerns have been increasing and some attempts to reverse this situation have been undertaken. The government created some laws, Secretariats and Environment Agencies to deal with several problems, and a few programmes have been created to monitor and evaluate the ecosystems. Interactions with public and non-governmental organizations are also taking place to protect such ecosystems and, where possible, help them to recover.

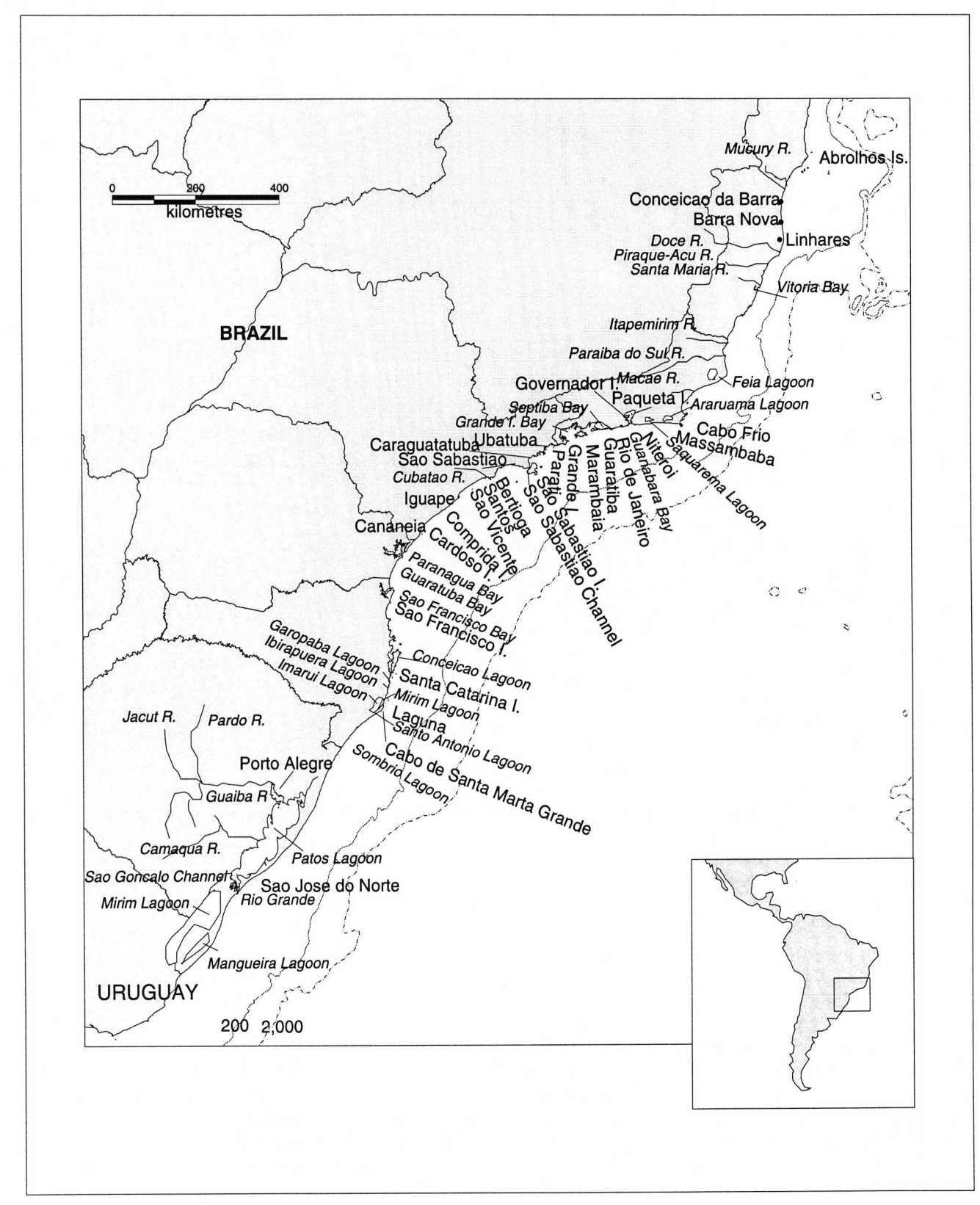

Fig. 1. Map of southern Brazil. The area covered is the southern and southeastern states.

INTRODUCTION

Brazil is the fifth largest country in the world in terms of area, and extends between 4°52'45"N and 33°45'10"S. The length of the coastline is around 7500 km, but this estimate reaches 9200 km when the indentations from all the estuarine and delta entrances are included. Brazilian coastal ecosystems include extensive estuaries, salt marshes, coral reefs, mangroves, coastal forest and lagoons, beaches and cliffs. A great variety of fauna and flora is found here, many of which are used by the local population according to their needs and technology. Brazil's coastal zone is being threatened by overpopulation, agricultural and industrial activities. The average demographic density in coastal regions is 87 inhabitants/km², five times higher than the national average of 17/km² (MMA, 1995). The reason for this is intense tourism and many industries particularly chemical, petroleum and cellulose. Reis et al. (1999) determined, from a questionnaire about Brazilian coastal problems, that the most common concerns are pollution, coastal erosion, inappropriate occupation of coastal areas, deforestation and reduction of environmental resources.

Castro and Miranda (1998) have divided the coast into six different sections based on physical environments and geographic characteristics. This chapter briefly describes the status of Southern Brazil, which is formed by the East, Southeast and South sections defined by these authors (Fig. 1).

The Southern region encompasses the South of Bahia State (BA), and the States of Espírito Santo (ES), Rio de Janeiro (RJ), São Paulo (SP), Paraná (PR), Santa Catarina (SC) and Rio Grande do Sul (RS). This area is very important in terms of finance and politics as well as in terms of fauna and flora diversity. It contains the two most important oil terminals, DTCS (Dutos e Terminais Centro Sul) in SP and GEBIG (Terminal Maritimo de Petróleo da Ilha Grande) in RJ, the two biggest ports of the country, Port of Santos (SP) and Rio de Janeiro (RJ), and another important Port of Tubarão (SC) (CDRJ, 1993). The population of these six states excluding the area south of Bahia is 73,841,861 accounting for 47% of Brazil's population (IBGE, 1999).

East Coast

The East Coast, from Abrolhos Bank (18'S, BA) to Cabo Frio City (23°S, RJ), features well developed sandy coastal plains with high cliffs (sedimentary and granitic rocks) intermixed with lowlands. Large estuaries, marine sedimentation and some sandstone reefs characterize the region (Reis et al., 1999). The climate is predominantly tropical.

Southeast Coast

The Southeast coast extends from Cabo Frio City (23°S) to Cabo de Santa Marta Grande (28.7°S). The mainland is formed from elevated granitic and gneiss rocks. This Precambrian structure is called "Serra do Mar" and it reaches the coast, creating an irregular coastline with cliffs, small bays and many islands. The northern limit, Cabo Frio, is characterized by coastal upwelling which makes it unique in Brazil and forms the biogeographic limit of many coastal taxa. The southern limit of this area, Cabo de Santa Marta Grande (SC), is the limit of mangroves systems along the Brazilian coastline. Most of this region has a subtropical climate, although the Rio de Janeiro coast is typically tropical (Reis et al., 1999; Diegues, 1999).

South Coast

Located between Cabo de Santa Marta Grande (28.7°S, SC) and Chuí (34.45°S, RG), the South Coast is a large coastal plain, reaching 120 km wide at some points. It has multiple beach ridges, dune fields and an extensive lagoon system. A barrier island 640 km long extends along most of this 740 km coastline and separates the ocean from the Mirim and Patos lagoon complex. The climate in this region is temperate (Reis et al., 1999).

Historical Setting

At the beginning of the 16th century, American settlers arrived in Brazil and colonization started from the coast and progressed inland. The first villages developed in the littoral zone and developed into commercial centres that remain today. Europeans also established themselves, and at that time, the economy was based on commerce and shipborne merchandise. The geography of the area was ideal for anchoring ships. The Portuguese settled in 18 centres in the 16th Century and of these, only São Paulo City was not located on the coast. The more productive areas became harbours, generating the first cities and later, important regional nuclei. However, this colonization process was not homogeneous in terms of population evenness.

During colonization, the important centres of Rio de Janeiro and São Paulo were established. Rio de Janeiro was the production and supply zone for the mining areas. The main agricultural product was sugar cane and its fermentation products. São Paulo's coastal area provided harbours at the Santos Estuary, São Sebastião and Ubatuba, but the easiest way to reach the inland plateau was through the Santos estuary. Thus the São Paulo littoral zone was centred on the cities of Santos and São Vicente (Santos estuary), and became a centre for exploration of Pau-Brasil and sugar cane cultivation. These activities were not as successful as they were in the Northeast of Brazil, so the settlers started a subsistence agriculture, cattle ranching and also captured local Indians for the slave trade (SMA, 1996).

During the 20th century the centres grew markedly; in 1900 the total population was 17 million inhabitants, which increased to 147 million in 1991 (MMA, 1995). The population is still not evenly distributed and the difference between North and South is dramatic. The Brazilian coastline population density ranges from 2/km² in some northern regions up to 586/km² in Rio de Janeiro and 79/km² in São Paulo (MMA, 1996).

In the late 1950s, the Brazilian coastal population changed with urbanization, tourism and industrialization. Industries depending on imported products were established close to ports, generating job opportunities. Today, five metropolitan regions with 15% of the total population live at the coast, and around 50% of the total population, or 70 million inhabitants, now live within 200 km of the sea.

PHYSICAL DESCRIPTION

East Coast: Abrolhos Bank to Cabo Frio

This region extends from 18°S to 23°S and has complex topographic features. From a width of 190 km in the Abrolhos Banks area, the continental shelf narrows southward, reaching 80 km in the Campos region (Castro and Miranda, 1998).

Slope and continental shelf water masses in this region are the result of mixing between three water bodies: tropical water, which is warm and salty (>20°C, >36 ppt) transported southward by the Brazil Current, the South Atlantic Central Water which is cold and less saline (<20°C, <36 ppt), and Coastal Water characterized by high temperatures and low salinity due to runoff and river discharges. On the outer shelf, the vertical mixing between the tropical water and South Atlantic Central Water is dominant. Due to frequent coastal upwelling events in the southern region of the Abrolhos-Campos region, the South Atlantic Central Water is advected onshore (Cacciari et al., 1993; Castro and Miranda, 1998.).

The flow of the Brazil Current in the banks region is not well documented, but there are indications that a significant part flows through the channels of the Abrolhos Banks. Between 16° and 19°S, the westernmost branch flows southward far from the coast, off the 3000 m isobath (Stramma et al., 1990), and wind stress is an important forcing mechanism for currents in this wide shelf. From Abrolhos to Cabo Frio, there is evidence that the Brazil Current flows southward along the shelf break, along a width of 70 km with speeds up to 0.7 m s^{-1} (Miranda and Castro, 1982). The associated meridional volume transport computed in relation to the 240 cl ton^{-1} isanosteric surface ($\cong$ 500 decibar surface) was 5.5 Sv (1 Sv = 10⁶ m³ s^{-1}). In the anticyclonic shear region a northward shallow counter flow with a surface speed of almost –0.2 m s^{-1} and a volume transport of –1.3 Sv is observed. Further south, near the Cabo Frio region (23°S), Evans and Signorini (1985) using the PEGASUS profiling system, measured speeds up to 0.5 m s^{-1} associated with the Brazil Current.

Southeast Coast: South Brazil Bight

The South Brazil Bight is up to 200 km wide, extending from Cabo Frio to Cabo de Santa Marta Grande (Fig. 1). Its

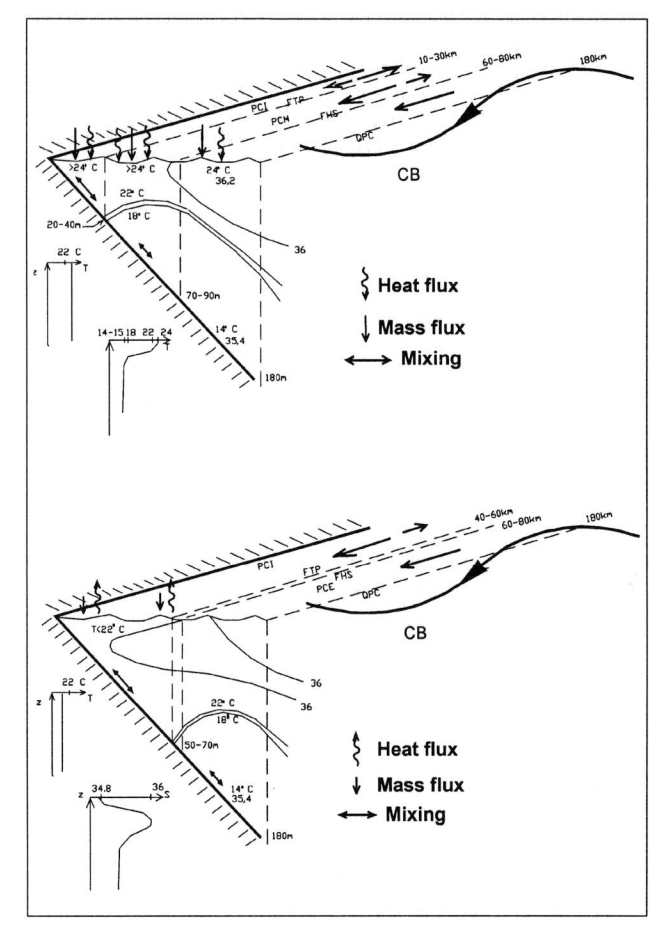

Fig. 2. Schematic representation of the physical characteristics in the South Brazil Bight for summer (top) and winter (bottom). PCI: Inner Shelf; PCM: Middle Shelf; PCE: Outer Shelf; FTP: Bottom Thermal Front; FHS: Surface Saline Front; QPC: Shelf Break; CB: Brazil Current; Fluxo de calor: Heat flux; Fluxo de massa: Mass flux; Mistura: Mixing (from Castro, 1996).

topography is smooth, with isobaths approximately parallel to the coastline.

Three water bodies have been identified here, especially during summer (Castro, 1996): Inner Shelf (IS), Middle Shelf (MS) and Outer Shelf (OS) (Fig. 2). The IS outer limit changes seasonally: it reaches the 20–40 m isobath during summer (10–30 km from the coast) and reaches the 50–70 m isobath during winter (40–80 km from the coast). The bulk stratification parameter (difference between the near bottom and the near surface conventional density) shows that during summer the smallest values are usually located in the narrow Inner Shelf. During winter, the zone of low stratification extends offshore to the 60–70 m isobath. Nevertheless, the IS stratification during summer is highly variable, and freshwater runoff is important.

The IS and the MS are separated by a bottom thermal front associated with intrusions towards the coast of cold slope waters identified as South Atlantic Central Water. Intrusion is greater during summer (Castro, 1996), when the MS extends offshore to the 70–90 isobaths (60–80 km from the coast). During winter the MS is very narrow, almost collapsing. The main characteristic of the MS is the high stratification due to the presence of a seasonal thermocline. During summer, near-surface and near-bottom temperatures in the MS are about 22–24°C and 14–15°C, respectively. There is high stratification during summer, while during winter the thermocline disappears.

The MS is separated from the Outer Shelf by a surface saline front located in the transition region between two upper layer water masses: the Mid Shelf Coastal Water and the Tropical Water. Stratification in the OS does not vary much during the year; values of 2.0 and 3.0 for the bulk stratification are frequently observed in the region.

Estimates show that were it not for the South Atlantic Central Water intrusions during summer, the MS would be almost homogeneous and without a dominant seasonal thermocline (Castro, 1996).

Meteorological forcing in the South Brazil Bight has high along-coast coherence, with perturbations having time scales of 4–12 days. As a result, coastal subtidal sea-level fluctuations with the same time scales show significant coherence for the whole region (Castro and Lee, 1995). Coastal sea-level response to wind forcing can be explained by a first mode barotropic continental shelf wave that propagates equatorward from the southern part of the South Brazil Bight.

Inner and mid-shelf currents are mainly wind-forced (Castro and Miranda, 1998). The kinetic energy contribution from the tidal currents (M2 is dominant) to the total current is about 20% for the along-shelf component and 40–50% for the cross-shelf component. Inner Shelf mean flow is probably northeastward during summer and southwestward during winter, while Middle Shelf mean flow is southwestward during the whole year. Subtidal oscillations in the current field have time scales of 3–12 days and are well correlated with the wind. Inner and mid-shelf waters respond mainly barotropically to the forcing mechanisms, especially during winter, and as a consequence, total currents present small vertical shear. During winter, there are along-shelf intrusions of cold water advected from the south which take place especially in the Middle Shelf and in some years reach the northern part of the Bight (Campos et al., 1996).

The Brazil Current continually forces poleward motions on the outer-shelf, and eventually meanders into the mid-shelf, leaving eddies which perturb locally the wind-forced circulation. Currents near the shelf break are predominantly to the southwest with mean speeds of about 0.5 m s^{-1}, reaching peaks higher than 1.0 m s^{-1}.

South Coast: Southern Brazilian Shelf

This shelf region extends from Cabo de Santa Marta to Chuí. The coastline lies NE–SW and the topography is smooth (see Fig. 1). The main characteristic of the Southern Brazilian Shelf is the relatively large amount of brackish water. Fresh water is discharged mainly by the Prata River. Patos Lagoon (Fig. 3), a coastal lagoon about 32°S, also

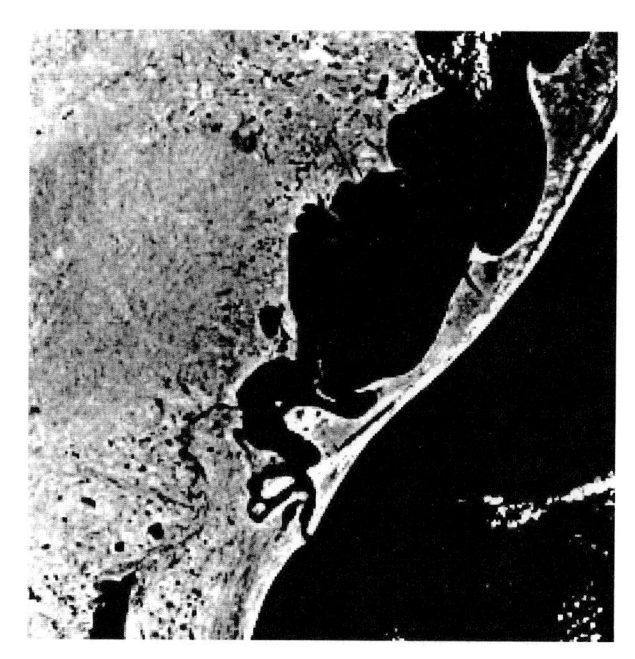

Fig. 3. Satellite image of southern part of the huge Patos lagoon, Rio Grande do Sul State.

contributes fresh water to the coastal region. The influence of the northward-flowing Malvinas Current transporting water masses of subantarctic origin is felt during the winter in the region (Miranda, 1972).

In addition to the water masses observed in the South Brazil Bight, the Southern Brazilian shelf is influenced by the Subantarctic Water (Miranda, 1972; Miranda and Castro, 1979), with low salinities and temperatures. This has greatest influence during winter, in the mid-shelf.

THE MAJOR SHALLOW WATER MARINE AND COASTAL HABITATS

Coral reefs of Brazil are limited to tropical areas, mainly located at the Northeast of the country. Sandy beaches are common from the Northeast to the South region where large dunes and barrier islands form. Mangroves are distributed from Oiapoque (Amapá) to Laguna (28.5°S, SC), occupying an area of 25,000 km². The most productive areas are the estuaries and mangroves, and this productivity favours economic activities, but some birds, fish, crustacean, molluscs, and mammals here are already endangered species. The Southeast coast has some large islands as well as 'restingas': areas of vegetation that grow on recent sandy coastal sediments which contain complex communities. Salt marshes are dominant in the South (Diegues, 1999).

East Coast (Abrolhos Bank to Cabo Frio)

Between the Abrolhos region and Mucuri Estuary (State limit between BA and ES) some coral reefs can be found with traces of Atlantic forest and mangroves. Abrolhos region is a unique island group very rich in flora and fauna.

From August to October, whales come to breed, attracting many tourists. Unfortunately, some animals are becoming endangered at Mucuri Estuary: examples include *Falco peregrinus* (Peregrine falcon) and *Ara ararauana* (Canindé macaw) (MMA, 1996; Diegues, 1999).

Espírito Santo State (ES) coast has rich mangrove areas. This state has small and sparse natural vegetation, in particular grass, cactus and bromeliads, and its islands are concentrated in the south. The most important mangroves are Conceição da Barra (Conceição da Barra City), Barra Nova (São Mateus City) and Piraquê-Açú River. In addition, lagoons, dune vegetation, beaches, and restingas, including the important Restingas of Guarapari, Jacarenema (Vila Velha City) and Itapemirim, are also abundant. Vitória Bay is located in the central part of the state and receives waters mainly from the Santa Maria River; this bay contains important industrial activities. Another important ecosystem in this region is the mouth of the Doce River, located at Linhares City. Some examples of endangered species on Linhares grassy marshes are *Crypturellus noctivagus* (Yellow-legged tinamou), *Myrmecophagus tridactyla* (Great anteater), *Lutra enudris* (Otter), and *Dermocheluys coriacea* (Trunk turtle) (MMA, 1996; Diegues, 1999).

The north coast of Rio de Janeiro State (RJ) is located between ES and Cabo Frio (RJ). It has the biggest lagoon in the RJ, the coastal Feia Lagoon, the Macaé River, by the important delta of Paraíba do Sul River. Erosion is very evident in some parts of this littoral zone. Some mangroves occur in the river mouth (MMA, 1996).

Southeast Coast (Cabo Frio to Cabo de Santa Marta Grande)

The Southeast coast encompasses several different habitats, including lagoons, mangroves and restingas around Cabo Frio. In the centre, the Atlantic Forest becomes denser with some littoral plains and, in the south, there is an imposing estuary–lagoon complex. Along the coast the Serra do Mar mountain chain almost reaches the coastline, creating cliffs and bays (Diegues, 1999).

In the Cabo Frio region, stabilized and mobile dunes and restingas isolate the ocean from the lagoons. In some places the lagoons are more than 2 km wide. Here one can find a restinga almost untouched by humans, Restinga de Massambaba, and two large coastal lagoons, Saquarema and Araruama Lagoons (MMA, 1996). Cabo Frio is characterized by coastal upwelling which limits the geographic distribution of many tropical species. Cabo Frio has a high diversity of macroalgae flora (Yoneshigue-Valentin and Valentin, 1992).

There are three large bays on the Rio de Janeiro coast: Guanabara, Grande Island and Sepetiba Bays. Guanabara Bay is the most important and, as seen from Sugar Loaf, is a symbol of Rio de Janeiro City. It has several islands, the largest being Governador, Fundão and Paquetá. Around the bay lies the metropolitan area of Rio de Janeiro City, including Niterói City. In Sepetiba and Grande Island Bay, hundreds of small islands can be found, with Grande Island the largest one. Sepetiba Bay is located 60 km to the west of Rio de Janeiro City and is isolated from the sea by the Restinga of Marambaia, a very well developed example of restinga in this state. Mangroves occur in some parts of Guanabara Bay, Guaratiba, Angra dos Reis and Parati (MMA, 1996).

São Paulo State has 2 large bays, Caraguatatuba and Santos Bay. Around Caraguatatuba Bay there are three "recreation" cities: Ubatuba, Caraguatatuba and São Sebastião, which receive thousands of visitors each summer. The very active São Sebastião Channel is located south of Caraguatatuba Bay, encompassing the most important oil terminal in Brazil: Dutos e Terminais Centro Sul. The second largest Brazilian Island, São Sebastião Island, separates São Sebastião Channel from the Atlantic Ocean. São Sebastião City extends from Caraguatatuba Bay southward, and also has intense tourism. The beaches are irregularly distributed, intermittent with rocky shores and cliffs. Mangroves here are seriously degraded by urban expansion, and several oil spills have occurred at São Sebastião Channel. The main species found here are *Rhizophora mangle* (Red mangrove), *Avicennia schaueriana* (Black mangrove), and *Languncularia racemosa* (White mangrove), but there are also *Hibiscus* sp. (Hibiscus), *Spartina ciliata* (Cordgrass) and *Acrostichum danaefolium* (Leather fern). Some of these are now locally almost extinct. Orchids, bromeliads and other small endemic plants make the local vegetation unique. Many birds, bats, marsupials, monkeys, rodents and reptiles also live in the Atlantic Forest along the São Paulo coast (SMA, 1996).

The central part of São Paulo is formed by Santos Bay where some mangroves, several islands, extensive channels and river mouths can be found, especially in Santos, São Vicente and Bertioga Cities. Cubatão River is an important river in this bay. In its basin is a large industrial complex which discharges effluents causing water and soil degradation. Mangroves between Santos and Bertioga are under constant threat of oil spillage from pipelines as well as harbour activities. Some significant pollution incidents have already occurred in the area, but mangroves have survived, even though degraded. As the Atlantic Forest is located some distance from the coast here, the whole littoral zone is populated, degrading restingas, mangroves and the Atlantic forest. Long sandy beaches with rocky shores, large estuaries, bays, islands and coastal plains form the littoral zone from the central part of São Paulo toward the south as far as Cabo de Santa Marta Grande (MMA, 1996).

Another important system in the Southeast coast is the estuarine-lagoon complex of Iguape-Cananéia-Paranaguá, located between the south of São Paulo and Paraná States, with an area of about 115 km². It is considered the most important littoral ecological system of Southern Brazil. This complex system and Guaratuba Bay are examples of extensive mangroves and salt marshes. The mangrove genera *Rhizophora, Avicennia* and *Languncularia* are well adapted to this region, which is dominated by *Spartina* spp. (Cordgrass) and *Crinum* sp. (Amaryllis) further inland. In addition, *Typha domingensis* (Cattail) associated with *Hedychium coronarium* (White hedychium) are abundant in these regions which are regularly flooded by fluvial waters (MMA, 1996). The Cardoso and Comprida Islands, located at the estuary of Cananéia, form an impressive restinga. Several species of Rhodophyta and Chlorophyta are recorded in the mangroves of Cardoso Island. Unfortunately, between Iguape-Paranaguá several vertebrate residents are already endangered: *Bubo virginianus* (Great horned owl), *Cebus apella* (Capuchins or Ring tail monkey) and *Tapirus terrestris* (Brazilian tapir or Lowland tapir). Low stands of *Avicennia schaueriana* (Black mangrove) and *Rhizophora mangle* (Red mangrove) terminate at Florianópolis (27.5°S, SC), but *Laguncularia racemosa* (White mangrove) extends southward to the poleward limit of mangroves at the mouth of the Araranguá River (29°S) (Diegues, 1999).

Santa Catarina State has many restingas and mangroves, interspersed with cliffs and sandy beaches. At the south of the state is a large lagoon complex with mobile dunes. Examples are the lagoon complexes of Imaruí, Mirim and Santo Antonio (Imaruí and Laguna Cities), Ibirapuera and Garopaba lagoons (Imbituba and Garopaba Cities). Santa Catarina also has a very large island, the Santa Catarina Island, which encompasses the large Conceição Lagoon with a major mangrove ecosystem. Santa Catarina Island is located close to the southern limit of mangrove distribution in Brazil. Another mangrove system can be found at Babitonga Bay, which is formed mainly by Cubatão River. The most southerly true mangroves on the Atlantic coast occur at Laguna (28.5°S, SC), where there is a mean annual temperature of 19.4°C, and where *Laguncularia racemosa* grows to no more than 2 m (MMA, 1996).

South Coast (Cabo de Santa Marta Grande to Chuí)

The South Coast is a large coastal plain, formed mainly by a long barrier island (640 km). This island is isolated from the mainland by a large lagoon system and contains well-developed and straight beaches, dunes, wetlands, and some restingas. Most beaches along the coast are exposed, but an exception occurs in the northernmost part of the barrier where rocky headlands made up of sandstone and basalt reach the coast and provide sheltered beaches (MMA, 1996; Reis et al., 1999).

Tidal marshes are dominant on the Southern coast of Santa Catarina and Rio Grande do Sul, and extensive salt marsh formation occurs at the margins of the coastal lagoon estuaries. The most important are associated with Patos, Mirim and Mangueira lagoons (Diegues, 1999). A large restinga occurs in São José do Norte.

Patos Lagoon (Fig. 3) has an area around 10,000 km² and with its interconnected lagoons, forms the largest and unique coastal lagoon complex in Southern Brazil. It is located at 30°S bordering Uruguay, and is 290 km long and 60 km width. In the North, the lake receives water from the Jacuí, Guaíba, Gravataí and other rivers that together with Camaquã River form a very important hydrographic basin. Patos Lagoon is connected to another big lake—Mirim Lagoon—through the São Gonçalo channel near Pelotas City. It is also connected to the Atlantic Ocean through a channel 740 m to 1.5 km wide. The expanding industrial district of Rio Grande and the big Port of Rio Grande, adjacent to this channel, discharges untreated effluents. The estuary of the lagoon is a very important breeding and nursery zone for much of the coastal fauna, which migrates through the channel and represents a significant percentage of the national fishery resources. Thus, discharge of pollutants into this system has biological and economic implications.

ACTIVITIES

Brazil has five metropolitan areas (out of a total of nine) on the coast, with two located in the southern part of the country: Rio de Janeiro with 9.7 million and Santos with 1.2 million inhabitants. In addition, most of the littoral states have their capitals on the coastline, with high demographic growth. Industries such as oil, chemical, mining and fertilizers have developed their activities in coastal cities due to good communications and transportation facilities. The overpopulation on the coast has brought serious environmental consequences including expanding slum areas, heavily concentrated pollution and extensive coastal habitat degradation.

Tourism is also an important economic factor, but is an unplanned and unregulated activity, affecting the coastal area and natural resources. Few programmes to guide tourists have been developed to prevent continuous environment alteration. Fishing is another significant economic activity along the coast and marine environment. The total catch is 565,000 t y^{-1}, being 40% artisanal fishermen and 60% industrial production. This is mainly concentrated between Rio de Janeiro and Rio Grande do Sul States (Diegues, 1999).

East Coast (Abrolhos Bank to Cabo Frio)

Along the southern coast of Bahia and the whole Espírito Santo, activities are basically artisanal fishery, sugar cane cultivation, sand mining in restinga areas and tourism. The main species caught are *Penaeus brasiliensis* and *P. schmitti* (White shrimps), *Lutjanus purpureus* (Southern red snapper), *Epinephelus* sp. (Grouper), *Thunus albacares* (Yellowfin tuna) and *Hemiramphus* sp. (Halfbeak). Calcareous algae exploitation has been developed in this region, and is under investigation because the real consequences of this activity to the marine environment are not yet understood. Some industries are also located here, such

Fig. 4. Vitória Bay in Espirito Santo State.

as Petrobrás (petroleum exploitation), Disa (sugar and alcohol distillation), Samarco Minerações (iron), Braspérola S.A. (textile industry) and Aracruz Celulose S.A. that uses an extensive area of eucalyptus plantation for paper pulp production (MMA, 1996).

Vitória Bay encompasses a huge harbour (Port of Tubarão) supporting the mining and iron industries from the Companies Vale do Rio Doce, Siderúrgica Nacional de Tubarão and Ferro e Aço (Fig. 4). There are another five ports located around the Vitória region including the Ports of Capuaba and Praia Mole. The Port of Tubarão is considered one of the world's most modern and efficient bulk shipment ports. Its ships annually handle 50 million tons of bulk, mainly ore, but it also handles mixed cargo, and giant carriers up to 300,000 TDW can anchor there (SSA, 1999). The port presents an oil spill risk for the fragile surrounding regions, such as Santa Maria River estuary, Vitória Bay and the Comboio Reserves, where marine turtles lay their eggs. Heavy metal mining also occurs in the Vitória Bay area (MMA, 1996).

Oil drilling has become another important economic activity in Brazil since 1973, and the main drilling along the southern coast is located around Campos City (north of RJ) which is exploited by PETROBRÁS (Brazilian Petroleum Cia) (Diegues, 1999). On the east coast tourism and urban regions are expanding (MMA, 1996).

Southeast Coast (Cabo Frio to Cabo de Santa Marta Grande)

Tourism is present throughout the Southeast coast of Brazil. With the exception of the big urban areas, commercial fishing is spread throughout the coast.

In Arraial do Cabo City there are salt extraction, sand mining and agricultural activities as well as a large chemical industry, the Nacional de Álcalis Cia. The cities in the Cabo Frio region are growing fast, with neither urban nor environmental conservation plans. The population density is 105.5 inh/km^2 (MMA, 1996).

Rio de Janeiro and Santos metropolitan areas are two giant urban centres, with 1343 and 641 inh/km^2, respectively (MMA, 1996). There are many chemical, petroleum and metallurgy industries in the surrounding areas, causing environmental degradation. Slums are increasing and the sewage treatment system has been deficient, causing pollution of coastal rivers, estuaries, lagoons and bays.

Approximately 2000 industries occupy the Rio de Janeiro metropolitan area. The principal ones are oil and its derivatives, metallurgy, plastic and synthetic rubber. The drained area around Guanabara Bay encompasses the second biggest industrial park, with eight petroleum terminals, 12 shipyards, and two petroleum refineries. These refineries are responsible for 17% of the national output, and an additional 6000 industries are located in the same drainage basin. The port of Rio de Janeiro is the second most important port in Brazil and around 2000 commercial ships dock there annually (CDRJ, 1993; Niencheski and Baungarten, 1998). Examples of big industries in Rio de Janeiro are Dow Chemical, Petrobrás and Petroflex. However, Rio de Janeiro City still has the biggest preserved urban forest: Parque Nacional da Tijuca (Tijuca National Park).

The littoral zone in south of Rio de Janeiro and north of São Paulo supports harbour operations, mining and oil terminal activities besides tourism and fishing. The Grande Island Bay together with the Sepetiba Bay form a complex coastal ecosystem which is very important ecologically. Sepetiba Bay is one of the most important fishing areas of Rio de Janeiro State but is threatened by the growing industrial activities (mainly manufacturing) that have established during the last 20 years. These will increase even more when a planned large seaport is constructed and more industrial investment introduced. In Grande Island and Sepetiba Bays there is the shipyard Verolme, the Nuclear Power Plant of Almirante Alvaro Alberto (also known as the Nuclear Power Plant of Angra dos Reis), and the maritime petroleum terminal of Grande Island Bay (GEBIG, Petrobrás). At Mangaratiba City, there is a mining terminal located at Guaiba Island. There are other intense activities including marine and terrestrial oil pipelines, oil tankers anchorage, and the Port of Sepetiba (MMA, 1996). The Port of Sepetiba is an alternative route to Santos and the Rio Grande area and handles approximately 30 million tons of bulk shipping annually (SSA, 1999). Unfortunately, the impact of all these activities and the pollution of ocean beaches and lakes are reducing the flux of tourists, mainly to the Rio de Janeiro metropolitan area.

In this region, there is some subsistence agriculture. The attractive small bays and beaches, interspersed with rock shores and Atlantic forest, make it a beautiful region. Tourism is more intense during the summer, with many people using the area as a second home (mainly those from São Paulo and Rio de Janeiro). High real estate prices push the local population away from the beaches to live in poorly constructed houses close to the mountains of Serra do Mar.

In São Sebastião City is the Port of São Sebastião and the oil terminal Almirante Barroso, now known as DTCS. It was constructed due to the development and growth of the industrial complex at Santos Bay. It is the most important oil terminal in Brazil, and up to 55% of Brazil's oil passes through this complex. Frequent accidents have occurred, disturbing fisheries and tourism activities as well as mangroves and other sensitive ecosystems. This area is conducive to this kind of activity due to the channel formed between the Island and the City of São Sebastião. São Sebastião Channel is narrow (2 to 6 km wide) but very deep in the middle (40 m) and well sheltered, making it the ideal place for shipping operations. In spite of these intense activities, the levels of oil in the water and sediments are low, mainly due to the strong currents in the area (Zanardi et al., 1999a,b).

Another large industrial complex in the Southeast coast is at Santos-Cubatão, the industrial complex of Cubatão. In 1988, around 1,100 industries were located in this area. The Cubatão complex is one of the largest Petrochemical Parks of the country and, in 1985, its industries produced 3% of the Brazilian gross domestic product (CETESB, 1999). Some oil, fertilizers and chemical industries as well as an iron production company are established there and are potential polluters. Among them, the most important industries are Presidente Bernardes Petrochemical Refinery, Companhia Petroquímica Brasileira (Brazilian Petrochemical Cia), COSIPA (São Paulo Siderurgic Cia) and Ultrafértil (fertilizer). The Port of Santos is located at Santos Estuary, which involves oil, mineral, transportation, fishing and sand activities. It is the largest port in Brazil in terms of general cargo and container movement. In 1995 it handled over 35 million tons and around 50% of the total container shipping volume in the country (SSA, 1999).

The estuarine-lagoon system of Iguape-Cananéia, Comprida Island and Ribeira Valley forms the southernmost littoral zone of São Paulo. In this region there are only small towns and the environment is still well preserved with only 38,081 inhabitants in 3287 km^2 of total area (MMA, 1996). The dominant activities are subsistence agriculture and fisheries. The Ribeira Valley is formed by the Ribeira do Iguape River and around it are grown banana, tea and passion fruit. In this area calcareous sand and lead mines exist, and some environmental accidents have already occurred with lead.

In Paraná and Santa Catarina States the main agricultural activities are vegetables, banana and rice, but subsistence agriculture is also developed. A large variety of industries, from textiles and food to computer facilities are located here. Artisanal and industrial fisheries and tourism are strong economic activities in the region, and shrimp is the main product at the lagoon complex of Laguna (28.5°S) (MMA, 1996). In this region the Ports of Paranaguá and São Francisco do Sul have annual shipping volumes of 15 and 9 million tons respectively, of which 90% are bulk. The main exporting product is soya bean, and the Port of Paranaguá also receives large shipments of automobile parts and grain (SSA, 1999).

The coast here is an important fishing region, responsible for 75% of the national catch. The main Brazilian fishery resource is *Sardinella brasiliensis* (Brazilian Sardinella) representing 45% of the total production. The major part of this resource is captured between Cabo Frio (23°S) and Santa Catarina Island (27°S), even though fishing extends from Cabo de São Tomé (22°S) to Cabo de Santa Marta Grande (28.7°S). The artisanal sector is only 25% and the main fish captured are *Mugil* sp. (Mullets), *Centropomus* sp. (Snook) and *Cynoscion* sp. (Weakfish). The industrial sector accounts for the remaining 75% of production using trawling, encircling nets and long-line fishing techniques (Diegues, 1999). In the Southeast Coast, juveniles and subadults of the following Sciaenids are dominant in bottom trawls: *Paralonchurus brasiliensis* (Banded ground drum), *Micropogonias furnieri* (Whitemouth croaker), *Isopisthus parvipinnis* (Shortfin corvina), *Stellifer brasiliensis* (Canganguá), and *S. rastrifer* (Rake stardrum). *M. furnieri* (Whitemouth croaker) is commercially exploited in the Cabo Frio region. Toward the south, the density is low but constant and consistent throughout the year. Adults and juveniles of *Mugil curema* (White mullet) are abundant throughout estuaries in São Paulo State, but this species gradually decreases in abundance toward the south. *Anchoviella lepidentostole* (Broadband anchovy) is caught in beach seine in the estuarine systems of Santos and São Vicente, and Clupeids (Sardines) are important in the estuarine system of São Paulo State (Vieira and Musik, 1994).

Penaeus brasiliensis (White shrimp) is present along the whole coast of Brazil but most abundant at Cabo Frio, Santos and Cananéia. *Penaeus paulensis* (White shrimp), also exists (Valentini et al., 1991). The shrimp *Xiphopenaeus kroyeri* (Atlantic sea bob) is found at Barra de Itabapoana, Barra de São João and between Parati (24°S) and Laguna (28.5°S). The cephalopods (Squids) *Loligo sanpaulensis* and *L. plei* are the by-catch on artisanal shrimp fisheries at several places, being especially abundant at Arraial do Cabo (Cabo Frio region) and along the Santa Catarina littoral zone (Haimovici and Perez, 1991). The principal fishing harbours are located at Santos, Rio de Janeiro and Itajaí.

South Coast (Cabo de Santa Marta Grande to Chuí)

Because of the many lagoon systems in the region, artisanal fisheries and tourism are intense. Urban centres are small and well distributed. There are several agricultural products such as rice, bananas, pineapples, strawberries, peaches and corn, among others, just a little farther from the coast. The runoff from these areas brings agrochemical products to the coastal waters contaminating the sensitive lagoon ecosystems. There is coal mining on the coast and a large industrial centre in the Patos Lagoon, including fertilizer and petroleum industries together with harbour activities (MMA, 1996). Rio Grande City is known as the

"Gateway to Mercosul" and borders Argentina and Uruguay. The Port of Rio Grande, around this city, has an annual shipping volume of approximately 10 million tons and almost 90% are bulk cargo (SSA, 1999). It encompasses four terminals: a petroleum terminal, grain and container terminals, and a vegetable oil terminal.

The Patos Lagoon region contains seven dominant species of demersal fish: *Micropogonias furnieri* (Whitemouth croaker), *Paralonchurus brasiliensis* (Banded ground drum), *Menticirrhus americanus* (Southern kingcroaker), *Lycengraulis* sp. (Anchovy), *Urophycis brasiliensis* (Codling) and the Ariids (Sea catfish) *Netuma barba* and *Netuma planifrons*. The *Netuma barba* (Salmon sea catfish) is the most important for the artisanal fisheries. The quantity and diversity of these species vary according to the seasons. The Pteropod *Limacina retroversa* is the dominant species during winter, showing the influence of the Malvina Current. During the summer the presence of *Creseis virgula* and *Limacina trochiformis* indicate Platform Tropical and oceanic waters, respectively. In the shallow area of the estuary and adjacent coastal region the atherinids (Silverside) *Xenomelaniris brasiliensis* and *Odontesthes bonariensis*, *Mugil liza* (Lebranche mullet) and the South American endemic *Jenynsia lineata* are dominant in beach seines (Vieira and Musik, 1994). During the summer, *Sympterygia acuta* (Skate), *Rhinobatus horkelii* (Brazilian guitarfish) and *Mustelus fasciatus* (Striped dogfish) are the most common species captured in beach seines for commercial use (Chao et al., 1982). Clupeids make only modest contributions to beach seine catches in the Patos lagoon estuary.

Pomatomus saltatrix (bluefish) is the principal pelagic species caught by commercial fishing at Rio Grande do Sul. *Micropogonias furnieri* (Whitemouth croaker) is around eight times more abundant here than the Southeast Coast, and the concentration of individuals varies during the year, affected by the position of the Subtropical Convergence. The Engraulids (Anchovies) *Anchoa marinii*, *Lycengraulis* sp. and *Engraulis anchoita* constitute up to 32% of mid-water trawl catches in the Patos lagoon, but are less important in bottom trawl and beach seines (Chao et al., 1982).

DEGRADATION

The primary problems on the Brazilian coastline are the destructive use of coastal resources and water pollution. Mangrove deforestation, landfill and domestic effluents from urban areas lower the water quality of rivers, streams and even the water table. In addition, natural and human-induced coastal erosion, mining activities for sand, heavy minerals and calcium carbonate, and overfishing contribute to the degradation of the coast. Tourism, an important economic activity, affects these ecosystems mainly during summer vacations (Reis et al., 1999). Many highways have been constructed along the coast, damaging beaches, mangroves and the Atlantic Forest. An important example is the road that links many coastal capitals, BR 101, built in 1970.

East Coast (Abrolhos Bank to Cabo Frio)

This area has the same problems noted above. In addition, unplanned urbanization of beaches and industrial pollution are increasing and deforestation has caused significant sediment deposition on rivers in the East Coast region. Some systems are already degraded, such as estuaries, mangroves and restingas which receive urban waste effluents and agrochemical runoff from the agricultural areas.

The north of East Coast is damaged by the presence of Aracruz Celulose S.A., by paper pulp production and by an extensive eucalyptus plantation. Since this industry started, fishing activities have decreased, fishermen are impoverished and there is no environmentally sensitive tree cutting plan (Diegues, 1999). The waters and marine sediments of the river close to this company (Piraque-Açú River) show evidence of pollution and some fish species are already eliminated (Niencheski and Baungarten, 1998). Since this river flows into Vitória Bay, it is a potential source of contamination to the estuary. Maiambá Lagoon is also a threatened system because it receives industrial waste from Samarco Mineração company (MMA, 1996). Mangrove swamps, islands and estuaries in the estuarine region of Vitória are critically degraded because the region receives sewage and domestic effluents from urban complexes and industries that produce detergents, phosphates, caustic soda and iron products. In addition, irregular urban occupation and sand mining damage these sensitive ecosystems as well (MMA, 1996; Diegues, 1999). The dragnet fishery also contributes to the degradation of sensitive regions like mangroves and beaches (MMA, 1996). The National State Park of Itaúnas, at Conceição da Barra, with 3150 ha, encompasses mangroves, restingas, dunes and beaches and is an effort to preserve these systems.

In the southern part of the East Coast there is the petroleum basin of Campos, a potential source of oil contamination in the area. Coastal erosion has been occurring at the mouth of Paraíba do Sul and Macaé Rivers, and sediments from the continental platform of Paraíba do Sul River showed Hg contamination (MMA, 1996; Niencheski and Baungarten, 1998). Paraíba do Sul River is used as a supply for drinking water by most cities around the Rio de Janeiro metropolitan area, and is constantly under threat of accidental industrial spillages. The local population occupies restingas, and urban sewage and byproducts from sugar cane distillation are dumped directly to the water bodies, further degrading these ecosystems (MMA, 1996).

Southeast Coast (Cabo Frio to Cabo de Santa Marta Grande)

This region has similar problems, but others are also common. Some examples are coastal flooding, pluvial erosion due to unregulated urbanization, pollution of coastal lagoons and conflict between fishing and tourist activities (Reis et al., 1999).

Deforestation and sand mining from dunes and beaches are destroying these ecosystems in the Cabo Frio area. In addition, lagoons and beaches are polluted due to untreated sewage. Sediments and organisms were analysed at Arraial do Cabo region through the National Program MOMAM (Marine Environment Monitoring) coordinated by the Brazilian Navy and the Institute of Sea Studies Almirante Paulo Moreira (IEAPM). The objective of this programme was to track the pollutants in the region. Results showed low levels of heavy metals considered characteristic of the natural environment. The presence of petroleum hydrocarbons and chlorinated organics indicated these compounds are reaching the area, even though the concentrations are not yet high (MOMAM, 1997).

Guanabara Bay is the most important Brazilian coastal bay with an area of 384 km^2. It is situated in the overpopulated Rio de Janeiro metropolitan area and receives 2 millions m^3 of sewage and industrial effluents. Rio de Janeiro, Duque de Caxias, São Gonçalo, Niterói, and several other small cities are located along its margins. The estimated population in this area is 10 million inhabitants and around 70% of its domestic sewage is dumped untreated directly into the bay. The local population is increasing 1.1% annually, making the establishment of the necessary infrastructure even more difficult. The port of Rio de Janeiro, the second most important port in Brazil, contributes to the increase in pollution in the bay (CDRJ, 1993; Niencheski and Baungarten, 1998). Around Guanabara Bay, the Atlantic forest has been deforested to give way to unplanned cities which are spreading over 'preserved' areas such as lagoons and river banks, mangroves, and restingas, contaminating and increasing sediment deposition in these ecosystems (MMA, 1996).

Although pollution control plans were instigated by FEEMA (Rio de Janeiro State Environmental Agency) in 1979, the water in Guanabara Bay has become seriously polluted. In 1991, only 15% of the huge domestic and industrial effluents dumped into the bay had some treatment, even though many sewage treatment plants had been constructed in the past 20 years. Eighteen t d^{-1} of petroleum hydrocarbons reaches the bay and it is estimated that 85% comes from urban runoff. Large amounts of nutrients cause water eutrophication close to the coast. Around 2×10^5 t y^{-1} of suspended materials enter Guanabara Bay. Suspended solids, organic matter, heavy metals, and hydrocarbons have been accumulating in the bottom sediments of Guanabara Bay for many years. As a consequence of these discharges, the fisheries yield decreased to 10% compared to that of 30 years ago (FEEMA, 1990).

The western margin of the Bay is in the worst condition and is effectively dead. Untreated sewage runoff enters the bay and the low capacity of water renewal makes the nutrient levels very high. The result is bad water quality with average dissolved oxygen of 3.1 mg l^{-1} near the bottom, anoxic bottom mud, mean faecal coliform of 1140 count ml^{-1} and excessive levels of ammonia and phosphate. Between 1980 and 1990 an increase of faecal pollution was observed based on coliform data. The levels were much higher than the standard allowed by sanitary quality regulation. The concentration of chlorophyll in this part of the bay exceeds 130 g l^{-1} as compared to 57 g l^{-1} for the average of the entire bay (FEEMA, 1990).

Some studies showed high concentrations of cadmium, lead, zinc, copper, chromium and mercury in the waters and sediments of Guanabara Bay (Lacerda et al., 1988; Perin et al., 1997). The fish, mussels and crustaceans are already contaminated by Hg (Diegues, 1999). The Cr concentration in suspended material and barnacles *Balanus* sp. in this bay is four and three times, respectively, higher than those found in a nearby area. Moreira and Pivetta (1997) found 0.7 to 20 mg kg^{-1} of Hg in sediments rich in organic matter, generally under anoxic conditions. Sediments also contain polyaromatic hydrocarbons (PAHs) with a geochronologic correlation with the beginning of the industrial period. The concentration of total PAHs ranged from 1.57 to 18.44 μg g^{-1}, exceeding the values recommended by GESAMP. Among the highest concentrations of compounds, BaP (Benzo-(a)pyrene), a carcinogenic compound, was present at 0.05 to 2.24 μg g^{-1}. The beaches, mangroves and estuaries around Guanabara Bay are considered critically degraded (Diegues, 1999). The air pollution in this Rio de Janeiro metropolitan area causes detrimental health effects and even premature mortality.

Sepetiba Bay is a semi closed area of 305 km^2, receiving water from the important Itaguaí and Guandu rivers that account for 75% of the total input. Its drainage basin is 2,400 km^2 and the population is estimated at 1.7×10^6 inhabitants. It receives domestic sewage, industrial effluents and in some areas agriculture discharges. The metallurgic industrial park along the northern coast contributes high levels of heavy metals that can already be identified in some sections of the bay. Even though Sepetiba Bay is generally cleaner than Guanabara Bay, the data show that Sepetiba water quality is rapidly decreasing. Dissolved oxygen is significantly declining and some algae growth is already evident (IBGE, 1999). In addition, this bay has eutrophication, toxic and faecal coliform problems that come from industrial pollution and a rapidly growing population. Among organisms from different taxonomic groups, oysters (*Cassostrea brasiliana*) and macroalgae had the highest Cd and Zn concentrations. Some organisms had concentrations 4 to 25 times higher than those from nearby clean areas. Metal contamination is already comparable with Guanabara Bay and during the past 15 years some studies have classified Cd and Zn as the main contaminants in sediments and biota of this bay (Lima et al., 1986; Lacerda et al., 1988; Amado Filho et al., 1999).

The Nuclear Power Plant Almirante Alvaro Alberto, oil terminals and pipelines, the oil tankers' anchorage, and the Port of Sepetiba release their oil and heavy metal residues to the Grande Island Bay and Sepetiba Bays (MMA, 1996). Between 1981 and 1990, 2432 oil tankers shipped in these bays and 53 oil spills were registered in this period. Hydrocarbons in beach and mangroves sediments were 0.7 to 90.5 $\mu g\ g^{-1}$, and the concentrations found in waters were in the range of 0.4–2.2 $\mu g\ l^{-1}$ 'Arab oil equivalents' (Melges-Figueiredo et al., 1993). Recently, sediments were analyzed and found to be contaminated by aliphatic and aromatic hydrocarbons, mainly at Sepetiba Bay (Melges-Figueiredo, 1999). Some studies have shown high concentrations (equal to or higher than the acceptable level) of Zn, Mn and Pb metals in Sepetiba Bay, especially in suspended material. Its waters contained mainly Cr and Cd and the bottom sediment was also contaminated by heavy metals (Lacerda et al., 1988; Niencheski and Baungarten, 1998).

The Jacarepaguá lagoon system (Jacarepaguá, Tijuca and Camorim lagoons) is located in the southeast coast of Rio de Janeiro. People living in the surrounding area use the entire system as a food source, but it is a potential zone of eutrophication with some red tides already observed. Jacarepaguá Bay receives the waste from cellulose, plastics, beverages and drug industries causing oxygen depletion and high sediment deposits. Animal mortality was already verified as a consequence of this pollution (Niencheski and Baungarten, 1998). High levels of Zn were already recorded for this area, but no Hg was found in Jacarepaguá lagoon. Particulate metals have been accumulating in the trophic chain and the steel chemical industries located around the lagoon may be responsible (Fernandes et al., 1993).

The coastal area between Rio de Janeiro City and Santos was deforested during the construction of the coastal road BR 101, known here as Rio-Santos. Unplanned urbanization started on its edges and allowed expansion of the fishing villages, unregulated exploration of mangrove resources and degradation of the ecosystems. Now, erosion is a major problem, causing huge sediment deposition and affecting more and more mangroves and other marine ecosystems. In addition, this area became a landslide and flooding zone, putting the population at risk. This coastal region has significant and indiscriminate tourism, leading to degradation of beaches, coastal rocks and mangroves (MMA, 1996).

At São Sebastião City, in the São Sebastião Channel, where the DTCS oil terminal is located, fisheries, tourism and indeed the entire environment, with its mangroves, beaches and coastal rocky areas, have to put up with frequent oil spillages. The DTCS has been in operation since 1967 and 145 accidents have occurred between 1985 and 1994 (Poffo et al., 1996). The worst case was that of the "Brazilian Marina" tanker in 1978, where 6000 m³ of oil spilled into the Channel (Weber and Bícego, 1991; Poffo et al., 1996). The oil spread northwards on the strong currents in the area (up to 4 knots), reaching many beaches and damaging shellfish cultures and also shrimp and fisheries. As a result of these constant oil inputs, petroleum hydrocarbons have been consistently measured in water and sediments in the São Sebastião Channel and adjacent areas. Weber and Bícego (1991) analyzed total petroleum hydrocarbons in water and found 0.19 to 8.52 $\mu g\ l^{-1}$ Carmópolis oil equivalents, during November 1985 to August 1986. A monitoring programme between October 1993 to April 1995, reported lower concentrations of petroleum hydrocarbons in water, ranging from 0.15 to 4.9 $\mu g\ l^{-1}$ Carmópolis oil equivalents (Zanardi et al., 1999b). Typical aromatic and aliphatic petroleum hydrocarbons were also found in water and sediments from this region (Ehrhardt et al., 1995; Zanardi et al., 1999a). Even though the concentrations are generally low, the amount of oil these systems receive during each event is significant. In May, 1994, when a pipeline ruptured, 2700 m³ of crude oil spilled into the São Sebastião Channel and adjacent areas. On that occasion, the concentration of oil in the water reached 49.6 $\mu g\ l^{-1}$ Carmópolis oil equivalents and the bottom sediments were impregnated visibly with oil. The water rapidly recovered to normal concentrations, but the sediments still contained typical petroleum hydrocarbons 7 months later (Zanardi, 1996; Zanardi et al., 1999b). Despite the high toxicity of the oil spilled, the rocky shore populations were not affected, probably due to the high currents and wind speeds in the region (Lopes et al., 1997).

Around Santos City is an estuary encompassing mangroves, islands, channels and estuaries. Urban sewage and industrial effluents intensively degrade most of these ecosystems. The Santos Estuary and Bay have serious environmental problems mainly due to the industrial complex of Cubatão and the Port of Santos, responsible for 19% of the total oil accidents that occur at São Paulo State (Awazu et al., 1985).

The Cubatão complex on the Cubatão River Basin extends toward Santos Estuary. Some industries used to use the basin as a final receptor for their effluents with inadequate treatment. Around 100,000 kg month^{-1} of several pollutants, such as Zn, phenol and Hg, used to be discharged into the estuary waters. In addition, the effluents from the harbour have been dumped directly into the estuary for several years, leading to contamination of the water, sediment and biota. Studies during the 1980s showed that even though some industrial effluent sources were controlled, the area was contaminated. High concentrations of metals and chlorinated organics were found in sediments, water and aquatic organisms such as fish and crabs (Boldrini et al., 1989). Mangroves are also altered due to heavy metals contamination, solid residues, changed water courses, frequent oil spills, and industrial and domestic effluents (Lamparelli et al., 1993). Because of the intense harbour activities, more than 5000 tons of sediment must be dredged every year, and this material has usually been disposed of in the ocean. Results of ecotoxicological and chemical work showed that the sediment and dredged material were mainly contaminated with PAHs (particularly benzo(a)pyrene) and metals such as Hg, Ni, Cu, Pb, and Zn. Some samples were also mutagenic for *Salmonella typhirurium* (bacteria) and impaired embryo development of sea urchin *Lytechinus variegatus* (Prósperi et al., 1998).

Petroleum hydrocarbons in Santos Estuary showed high concentrations close to the industrial complexes of Cubatão (Bícego, 1988). Total PAHs in sediments of this area had very high concentrations varying from 0.08 to 42.39 $\mu g\ g^{-1}$ (Nisighima, 1999), at levels that can affect benthic organisms. Bivalves sampled during the International Mussel Watch Programme (IMW 1991–1992) were used to assess the occurrence of selected chlorinated hydrocarbons and PAHs along the Brazilian coastline. The highest concentrations of these compounds were measured in Santos and Guanabara Bay (Taniguchi et al., 1999).

There are 1,208,776 inhabitants around Santos Bay, including Santos, São Vicente, Guarujá, Cubatão, Praia Grande and Mongaguá Cities, and during the summer vacations, the populations doubles (MMA, 1996). Some investment in basic sanitation, sewage collection and construction of a submarine emissary has been instigated. These improvements have contributed to increased seawater quality and these waters are still being monitored by CETESB (São Paulo State Environmental Agency).

In addition, this dense industrial complex has polluted the air and soil. Until 1984, Cubatão discharged, daily, thousands of tons of pollutants into the air and of this, 250 tons was dust (CETESB, 1999). In the same year, workers' and residents' health was threatened due to the intense air, soil and water pollution. In order to improve the quality of life, an environmental control programme, called the Cubatão Pollution Control Project, was created to reduce pollution to acceptable levels within five years. This project included the co-operation of state and local governments, technicians, environmental agencies, public opinion and financial support. Industries also participated in the project and 23 of them represented the 320 air–water–soil pollution source industries. Immense effort was taken to reduce, monitor and control the pollution. In 1984, 62 control programmes were established to fight air pollution, 11 for water and 35 for suspended particles. These programmes consisted of equipment, construction work and production procedures necessary to reduce the emissions to the standards proposed by CETESB. After their programme had been implemented, CETESB estimated that industries spent close to U$ 550 million in pollution control equipment during 1983–96. As a result, a permanent monitoring system was implemented and between 1984 and 1989, the air pollution was reduced 92.4% for fluorides, 97.4% for ammonium hydroxide, 84.5% for sulphur dioxide, 22.2% for nitrogen oxides and 92% for particles. The water quality improved significantly and the Cubatão River was restored to good condition (even the fish returned to its waters). The government kept working to restore the ecosystems around Santos Bay. In 1989 and 1990, 3 billion native tree and bush seeds were thrown into the inaccessible Serra do Mar using helicopters and planes, in order to restore 60 km^2 of mountain vegetation damaged by the pollution (CETESB, 1999). Despite these efforts many mangrove stands have been cleared due to the urban development (MMA, 1996; Diegues, 1999).

Several islands and river deltas form the estuarine–lagoon complex of Iguape/ Cananéia/ Paranaguá, which is considered a very important ecological system in the southern littoral zone of Brazil. These ecosystems are being destroyed by pollution from agriculture and uncontrolled tourism, mainly because the sewage treatment is not enough to deal with the increasing demand. The worst contribution to the degradation comes from the Port of Paranaguá (considered the third most important in Brazil), due to ignorant handling of chemical products. In addition, overfishing (mainly shrimps) and unplanned urbanization (only 15.8% of the population around Paranaguá has sewage collection) are altering the systems (MMA, 1996).

The coal-mining complex in Santa Catarina, located between Florianópolis and Cabo de Santa Marta Grande, is causing environmental problems for mangroves, beaches and coastal lagoons. The effluents have high acidity, iron oxide and some heavy metals such as Cr, Ni and Zn, affecting artisanal fisheries. Petroleum residues from harbour activities, unregulated tourism and over-fishing are also contributing to degrading the bays, estuaries and islands on the coast. Restingas, dunes and mangroves have been filled to facilitate urban expansion, generating real estate speculation and increasing water pollution due to domestic sewage. Babitonga Bay has been receiving heavy metals from metal-mechanic industries and the agrochemicals used to cultivate rice are reaching and damaging the lagoon-complex of Laguna (SC).

South Coast (Cabo de Santa Marta Grande to Chuí)

The primary problems related to the South Coast of Brazil are urbanization, coastal erosion, sand mining, overfishing and water contamination from industry and agriculture residues (Reis et al., 1999). The Porto Alegre metropolitan area is located north of the Patos Lagoon and includes a petrochemical complex, leather, metallurgical and chemical industries. These urban–industrial residues are degrading the coastal lagoons and salt marshes (Diegues, 1999). Rio Grande City is located south of Patos Lagoon, and all of its domestic and industrial sewage, including effluents from a fertilizer industry, harbour activities and petroleum refinery are discharged into the lagoon (MMA, 1996). Some studies in this area showed high concentrations of ammonium, phosphates, Cu, Zn and Cd in water and sediments and a trend of increasing Pb, Cu and Mn levels around the estuary mouth (Niencheski and Baungarten, 1998). Many agricultural activities surround the lagoon complexes and the resulting agrochemical products contaminate these systems. Blooms of *Microcystis aeroginosa* and coliforms in this lagoon are related to the high levels of nutrients, mainly ammonium (Niencheski and Baungarten, 1998). Deforestation and exhaustive drainage from lagoons to irrigate plantations are other common problems of the Brazilian South Coast (MMA, 1996).

ENVIRONMENTAL LAWS

The Brazilian Government became involved in coastal preservation and management during the 1970s when degradation of ecosystems increased due to industrialization and urban growth. Diegues (1999) gives a detailed review of Brazilian government and social concerns with environment protection. Table 1 shows the environmental organizations and programmes in chronological order.

The first organization responsible for the environment was the Secretariat for the Environment (SEMA), created in 1973. In 1974, the government implemented the Interministerial Commission for Marine Resources (CIRM) to coordinate research and management on marine resources. A Secretariat for this commission was created in 1979 (SECIRM) and chaired by the Navy Ministry. In 1981, through a national law (Law 6938), a National Council on the Environment (CONAMA) was established and formed by society members and several government sectors. CONAMA is responsible for the main policies related to the environment. As many countries have been exploiting their marine resources, the UNCLOS Laws of the Sea (1982), was created to set limits within each country. In January, 1993, through Law 8617, the Brazilian Congress defined its territorial sea as 12 miles from each coastal state and the economic exclusive zone (EEZ) as 200 miles (Carvalho and Rizzo, 1994; Diegues, 1999).

Coastal management is supported by the Federal Constitution, which defined the Coastal Zone, Amazon Forest, Atlantic Forest, Serra do Mar and Pantanal Matogrossense, as National Property (1988). The Federal government makes general laws, states adapt and detail them according to their needs and the Municipal governments search for local interest (Carvalho and Rizzo, 1994). In the same year (1988), the Brazilian government implemented the PNGC (National Plan of Coastal Management) through Law 7661, with the objective of promulgating a National Programme of natural coastal resources use. This programme involves: a national system of coastal management information; implementation of a programme to define zones (deliberated by the Environmental Agency of the States); incentive for government and society to work together in planning and monitoring programmes (SMA, 1996; Freire et al., 1996). The Ministry of Environment, Water Resources and Legal Amazon (MMA) supervises the activities related to PNGC. The Environmental and Natural Resources Institute (IBAMA), created in 1989, collaborates with this Ministry to formulate and coordinate the Environmental National Politic (Diegues, 1999).

According to Freire et al. (1996), economic, social and environmental differences cause problems in the occupation and use of coastal natural resources. Most of the littoral regions are already damaged, fragmented and different from their original character, mainly in the southeast coast, where intense human activities began a long time ago. The government implemented the Studies on Continental

Table 1

Main national programs and environmental organizations related to marine ecosystems

Date	Organization	Description
1973	SEMA	Secretariat for the Environment
1974	CIRM	Interministerial Commission for Marine Resources
1979	SECIRM	Secretariat of Interministerial Commission for Marine Resources
1981	CONAMA Law 6938	National Environment Council dealing with National Environmental policy
1986	CONAMA	Approved the first legislation requiring environment impact analysis for large projects
1987	PGC	Coastal Management Program—published by CIRM
1988	Constitution of 1988	Atlantic Forest/Coastal Zone declared as important areas for management and development
1988	Project LEPLAC	Geophysical data to establish the limits of Brazilian economic exclusive zone
1988	PNGC - Law 7661	National Plan of Coastal Management
1989	IBAMA	Brazilian Institute of the Environment and Renewable Natural Resources
1992	UNCED-92	International Congress occurred at Rio de Janeiro—Government and social organization actively participated
1992	MMA	Ministry of the Environment, Water Resources and Legal Amazon (incorporated SEMA)
1993	Law 8617	Territorial sea (12 miles each State) and Economic Exclusive Zone (200 miles) defined
1994	Project REVIZEE	Evaluation of potential resources of sea
1995	GERCO	National Program of Coastal Management

Platform (LEPLAC) in 1988, and the Project of Evaluation of Resources Alive at EEZ (REVIZEE) that started in 1994. In addition to Coastal Management, these studies are responsible for consolidating a system to plan and provide leadership for marine and coastal zone resources.

In 1995, the National Programme of Coastal Management (GERCO) proposed de-centralization giving more initiative to States and Municipalities, according to their different interests and situations, even though most decisions are taken at Federal level. The State environmental organizations are responsible for laws, implementation of quality control plans and resources. Among other responsibilities, they give licences and supervise potentially polluting activities, and control and establish the state pattern of environmental quality.

In São Paulo, 2 main organizations protect coastal environments. São Paulo State Environment Secretariat (SMA) is responsible for specific projects and maps out coastal ecosystems. São Paulo State Environmental Agency (CETESB) is responsible, among other activities, for monitoring water quality of the state rivers; air quality monitoring and the diagnosis of coastal areas, including water, sediment and organism contamination. CETESB also acts in emergency situations such as oil spills, where they manage containment and cleaning procedures, and develop some environmental education programmes such as Projeto Praia Limpa (Clean Beach Project).

Rio de Janeiro State Environmental Agency (FEEMA) is responsible for the coastal management system and for mapping the littoral zone in Rio de Janeiro. They also have an Emergency plan at Guanabara Bay to remove pollution from the Bay. The Environment Institute of Paraná (IAP) linked to SEMA (Secretariat for the Environment) is responsible for water quality control of rivers, reservoirs and coastal waters of the state, and air quality of the metropolitan region of Curitiba. They also promote the programme Baía Limpa (Clean Bay) to sustain and develop the socio-economic processes of the Paraná coastal region. Other agencies responsible for environmental quality in Southern Brazil are the State Secretariat of Environmental Affairs in Espírito Santo, the Environment Foundation (FATMA) in Santa Catarina and the State Foundation for Environment Protection (FEPAM) in Rio Grande do Sul.

However, the Coastal Management Law has not been entirely successful. The production of maps was considered a priority instead of preparation and implementation of plans. Continuous changes and misunderstandings between government institutions at federal, state and municipal levels have led to delays in implementation. In addition, ill-considered regulation laws ignored the needs of the coastal population. As a result, the support of the local communities for these management plans is weak, making it more difficult to protect the ecosystems (Diegues, 1999).

These preservation areas are called environmental conservation units. These areas are divided into two groups: units of direct and indirect use (or full protection). Direct use involves environmental protection areas, national forests and extractivist reserves. Units of indirect uses are the National Parks, biological and ecological reserves, and ecological stations.

There are 29 conservation areas on the Brazilian coastline. Rio de Janeiro has the highest number of environmental conservation units, but only 7.5% of the coastal zone is protected. On the other hand, Paraná State, which has one of the smallest coasts of Brazil, has 56% of its area protected by four Federal conservation units. São Paulo, also with four Federal conservation units, covers 12.8% of its coastal zone while Rio Grande do Sul has the lowest percentage of area protected by Federal law, just 1.8% (Carvalho and Rizzo, 1994).

Oceanographic and other non-governmental institutions or organizations have supported conservation units, environmental education and research activities on the Brazilian southern coast. Some universities have developed scientific research in coastal regions to increase knowledge of marine ecosystems and the most important are listed. In São Paulo, the Oceanographic Institute of the University of São Paulo and the Marine Biology Centre (CEBIMAR) is linked to the University of São Paulo and the Federal University of Campinas. Some universities in Rio de Janeiro are also working on environmental concerns, and the two most important are Fluminense Federal University (UFF) and the State University of Rio de Janeiro (UERJ). Also there is the Sea Studies Centre of the Federal University of Paraná, the Federal University of Santa Catarina and the University of Vale do Itajaí. Other institutions include the Centre of Coastal, Limnologic and Marine Studies (CECLIMAR) linked to the Federal University of Rio Grande do Sul and the University Foundation of Rio Grande (FURG).

The most active environmental groups working in marine ecosystem protection in Southern Brazil are: Pró-Tamar (marine turtle protection), SOS Mata Atlantica (protecting Atlantic Forest against deforestation), MAQUA (aquatic mammals protection) and NEMA (environmental monitoring, working at Peixe Lagoon and Ecologic Station of Taim). These organizations, and several others, have environment education programmes and work with the local population teaching them subsistence alternatives other than marine resources exploitation. These groups are also promoting projects to preserve, protect and recover environments. Some of these projects are: restoration of degraded areas, reduction of clandestine garbage deposits, programmes to reintroduce wild animals back into the system, selective garbage programmes and tourism vs. ecology programmes.

Finally, coastal zone handling has been discussed between the public, local communities and several government sectors. However, many problems still persist in reconciling the use of the coast and its natural resources with maintaining and improving its environmental health.

ACKNOWLEDGMENTS

We would like to thank Aluísio Rebelo Rocha (Neca) for producing the map and Dr. Catherine Clark and Cynthia Moore for English review.

REFERENCES

Amado Filho, G.M., Andrade, L.R., Karez, C.S., Farina, M. and Pfeiffer, W.C. (1999) Brown algae species as biomonitors of Zn and Cd at Sepetiba Bay, Rio de Janeiro, Brazil. *Marine Environment Research* **48**, 213–224.

Awazu, L.A.M., Serpa, R.R. and Aventurato, H. (1985) Análise histórica da ocorrência de acidentes ambientais no Estado de São Paulo. In: 13° Congresso Brasileiro de Engenharia Sanitária e Ambiental. Maceió, AL. 22 pp.

Bícego, M.C. (1988) Contribuição ao Estudo de hidrocarbonetos biogênicos e do petróleo no ambiente marinho. *Dissertação de Mestrado*. Instituto Oceanográfico da Universidade de São Paulo. 156 pp.

Boldrini, C.V., Eysink, G.G.J. and Martins, M.C. (1989) Avaliação preliminar da contaminação por metais pesados na água, sedimento e organismos aquáticos do Rio Cubatão. Relatório Técnico—CETESB. 28p + anexos.

Cacciari, P.L., Harari, J. and Pereira, J.E.R. (1993) Identificação e distribuição das massas de água e da corrente de superfície sobre a plataforma continental e o talude continental da Bacia de Campos no verão e inverno de 1992. In: Programa de Monitoramento Ambiental Oceânico da Bacia de Campos, RJ, L.R. Tommasi, ed. FUNDESPA, São Paulo, Brazil, pp. II.11–II.21. (Technical report in Portuguese)

Campos, E.J.D., Ikeda, Y., Castro, B.M., Gaeta, S.A., Lorenzzetti, J.A. and Stevenson, M.R. (1996) Experiment Studies Circulation in the Western South Atlantic. *EOS Transactions, American Geophysical Union*, 77 (27), 253–259.

Carvalho, V.C. and Rizzo, H.G. (1994) Ministério do Meio Ambiente, dos Recursos Hídricos e Amazônia Legal, Brasília. A zona costeira brasileira: subsídios para uma avaliação ambiental. Brasília, MMA, 211 pp.

Castro, B.M. (1996) Correntes e massas de água da plataforma continental norte de São Paulo. Tese de Livre-Docência, Instituto Oceanográfico da Universidade de São Paulo. 248 p.

Castro, B.M. and Lee, T.N. (1995) Wind forced sea level variability on the Southeast Brazilian Shelf. *Journal of Geophysical Research* 100 (C8), 16045–16056.

Castro, B.M. and Miranda, L.B. (1998) Physical Oceanography of the Western Atlantic Continental Shelf located between 4 N and 34 S. In *The Sea*, Vol. 11, ed. A.R. Robinson and K.H. Brink, pp. 209–250.

CDRJ (Companhia das Docas do Rio de Janeiro) (1993) Relatório Estatístico Anual—1992. Rio de Janeiro. RJ, Brazil.

CETESB (1999) In *Cubatão: A change of air* [Online]. Available: www.cetesb.br, Sept. 1999.

Chao, L.N., Pereira, L.E., Vieira, J.P., Bemvenuti, M.A. and Cunha, L.P.R. (1982) Relação preliminar dos peixes estuarinos da Lagoa dos Patos e região costeira adjacente, Rio Grande do Sul, Brasil. *Atlantica, Rio Grande* 5 (1), 67–75.

Diegues, A.C. (1999) Human populations and coastal wetlands: conservation and management in Brazil. *Ocean and Coastal Management* 42, 187–210.

Ehrhardt, M.G., Weber, R.R. and Bícego M.C. (1995) Caracterização da fração orgânica dissolvida nas águas do Porto de São Sebastião e Praia do Segredo, Canal de São Sebastião, São Paulo, Brasil. *Bolm. Inst. Oceanogr.* 11, 81–86.

FEEMA (Fundação Estadual de Engenharia do Meio Ambiente) (1990) Projeto de recuperação gradual do ecossistema da Baía da Guanabara, Rio de Janeiro. 364 pp.

Fernandes, H.M., Cardoso, K., Godoy, J.M.O. and Patchineelam, S.R. (1993) Cultural impact on the geochemistry of sediments in Jacarepaguá Lagoon, Rio de Janeiro, Brazil. *Environmental Technology Letters* 14, 93–100.

Freire, O.D.S., Pereira, L.G.G. and Lima, R.J.C. (1996) Ministério do Meio Ambiente, dos Recursos Hídricos e Amazônia Legal, Brasília. Macrodiagnóstico da zona costeira do Brasil na escala da união. Brasília, MMA. 277 pp.

Evans, D.L. and Signorini, S.R. (1985) Vertical Structure of the Brazil Current. *Nature* 315, 48–50.

Haimovici, M. and Perez, J.A.A. (1991) Abundância e distribuição de cefalópodos em cruzeiros de prospecção pesqueira demersal na Plataforma Externa e Talude Continental do Brasil. *Atlantica, Rio Grande* 13 (1), 189–200.

IBGE (1999) In *Censo populacional 1996* [Online]. Available: www.ibge.gov.br. Sept 1999

Lacerda, L.D., Souza, C.M.M. and Pestana, M.H.D. (1988) Geochemical distribuition of Cd, Cu, Cr and Pb in sediments along the southeastern Brazilian coast. In: *Metals in Coastal Environments of Latin America*, eds. U. Seeliger, L.D. Lacerda and S.R. Patchineelam, pp. 86–89. Spriger, Berlin.

Lamparelli, C.C., Moura, D.O., Rodrigues, F.O. and Vicent, R.C. (1993) Biomonitoramento de ecossistemas aquáticos e de transição—Manguezais. São Paulo, Relatório Técnico CETESB, 18 p. + anexos.

Lima, N.R.W., Lacerda, L.D., Pfeiffer, W.C. and Fiszman, M. (1986) Temporal and spatial variability in Zn, Cr, Cd and Fe concentrations in oyster tissues (*Crassostrea braziliana* L.) from Sepetiba Bay, Brazil. *Environmental Technology Letters* 7, 453–460.

Lopes, C.F., Milanelli, J.C.C., Prósperi, V.A., Zanardi, E. and Truzzi, A.C. (1997) Coastal monitoring program of São Sebastião Channel: assessing the effects of "Tebar V" oil spill on rocky shore populations. *Marine Pollution Bulletin* 34 (11), 923–927.

Melges-Figueiredo, L.H., Herms, F., Pereira, S., Hamacher, C., Meniconi, M., Carreira, R., Costa, E., Lima, A., Araújo-Guerra, L., Cunha, A. and Almeida, D. (1993) Hydrocarbon distribution in the coastal environment of Ilha Grande Bay under influence of a Maritime Petroleum Terminal. *Third Latin American Congress on Organic Geochemistry, Manaus.* Extended Abstracts, pp. 90–93.

Melges-Figueiredo, L.H. (1999) Investigação das contribuições antrópicas e naturais em sedimentos costeiros utilizando-se hidrocarbonetos marcadores. PhD Dissertation, Departamento de Química, Pontifícia Universidade Católica, Rio de Janeiro, RJ, Brazil. 151 pp.

Miranda, L.B. (1972) Propriedades e variáveis físicas das águas da plataforma continental do Rio Grande do Sul. Tese Doutorado, Instituto de Física da Universidade de São Paulo, 127 pp.

Miranda, L.B. and Castro, B.M. (1979) Aplicação do diagrama T-S estatístico volumétrico à análise das massas de água da plataforma continental do Rio Grande do Sul. *Boletim do Instituto Oceanografico*, S Paulo 28 (1), 185–200.

Miranda, L.B. and Castro, B.M. (1982) Geostrophic flow conditions of the Brazil Current at 19°S. *Ciência Interamericana* 22 (1), 44–48.

MMA (Ministério do Meio Ambiente, Recursos Hídricos e da Amazônia Legal) (1995) Ecossistemas brasileiros e principais macrovetores do desenvolvimento: subsídios ao planejamento e gestão ambiental. Brasília (BR). MMA. 108 pp.

MMA (Ministério do Meio Ambiente, Recursos Hidricos e da Amazônia Legal) (1996) Perfil dos Estados Litorâneos do Brasil: Subsídios e implantação do Programa Nacional de Gerenciamento Costeiro. 301 pp.

MOMAM (Monitoramento do Ambiente Marinho) (1997) Relatório sobre as análises dos poluentes realizadas em 1996/1997. Ministério da marinha/ IEAPM. 82 pp.

Moreira, J.C. and Pivetta, F. (1997) Human environmental contamination by mercury from industrial uses in Brazil. *Water Air and Soil Pollution* 97, 241–246.

Niencheski, L.F. and Baungarten, M.G. (1998) Avaliação do Potencial Sustentável dos recursos Vivos na Zona Econômica Exclusiva. Oceanografia Química: Levantamento Bibliográfico e Estado atual do Conhecimento. MMA, CRM, FEMAR. 169 pp.

Nisighima, N.F. 1999. Aplicação da Cromatografia a Líquido de Alto Desenpenho (HPLC) na purificação e separação de hidrocarbonetos de sedimentos das regiões de Santos e Cananéia, São Paulo, Brasil. *Dissertação de mestrado*. IOUSP. 89 pp.

Perin, R., Fabris, G., Manante, S., Rebelo-Wagener, A., Hammacher, C. and Scotto, S. (1997) A five-year study on the heavy-metal pollution of Guanabara Bay sediments (Rio de Janeiro, Brazil) and evaluation of the metal bioavailability by means of geochemical speciation. *Water Research* 31, 3017–3128.

Poffo, I.R.F., Nakasaki, A., Eysink, G.G.J., Heitzman, S.R., Cantão, R.F., Midaglia, C.L.V., Caetano, N.A., Serpa, R.R., Aventurato, H. and Pompéia, S.L. (1996) Dinâmica dos vazamentos de óleo no Canal de São Sebastião. São Paulo, Relatório Técnico CETESB, 2 volumes.

Prósperi, V.A., Eysink, G.G.J. and Saito, L.M. (1998) Avaliação do grau de contaminação do sedimento ao longo do canal de navegação do Porto de Santos. Relatório Técnico CETESB. 33 pp.

Reis, E.G., Asmus, M.L., Castello, P.J. and Calliari, L.J. (1999) Building human capacity on coastal and ocean management—implementing the Train-Sea-Coast Programme in Brazil. *Ocean and Coastal Management* **42**, 211–228.

SMA (Secretaria do Meio Ambiente) (1996) Macrozoneamento do Litoral Norte: plano de gerenciamento costeiro. São Paulo: SMA. Série Documentos. 202 pp.

SSA (Seaports of South America) (1999) In: *Brazil* [Online]. Available: www.seaportsinfo.com.saonly.html (Sept. 1999).

Stramma, L., Ikeda, Y. and Peterson, R.G. (1990) Geostrophic transport in the Brazil Current region north of 20°S. *Deep-Sea Research* **37**, 1875–1886.

Taniguchi, S., Montone, R.C., Weber, R.R., Lara, W.H. and Sericano, J.L. (1999) The International Mussel Watch Programm in Brazil. Assessment of Coastal Marine Pollution. *Marine Pollution Bulletin* in press.

Valentini, H., Incao, F.D., Rodrigues, L.F., Rebelo Neto, J.E. and Rahn, E. (1991) Análise da pesca do Camarão-Rosa (*Penaeus brasiliensis* and *P. paulensis*) nas regiões sudeste e sul do Brasil. *Atlantica, Rio Grande* **13** (1), 143–157.

Vieira, J.P. and Musik, J.A. (1994) Fish faunal composition in warm-temperate and tropical estuaries of western Atlantic. *Atlantica, Rio Grande* **16**, 31–53.

Weber, R.R. and Bícego, M.C. (1991) Survey of petroleum aromatic hydrocarbons in the São Sebastião Channel, SP, Brazil, November 1985 to August 1986. *Boletim do Instituto Oceanografico* São Paulo, **39**: 117-121.

Yoneshigue-Valentin, Y. and Valentin, J.L. (1992) Macroalgae of the Cabo Frio Upwelling Region, Brazil: Ordination of communities. In *Coastal Plants Communities of Latin America*, ed. U. Seeliger, pp. 31–50. Academic Press.

Zanardi, E. (1996) Hidrocarbonetos no Canal de São Sebastião e na Plataforma Interna Adjacente—Influência do Derrame de Maio de 1994. Dissertação de Mestrado. Instituto Oceanográfico da USP, São Paulo, 112 pp.

Zanardi, E., Bícego, M.C., Miranda, L.B. and Weber, R.R. (1999a) Distribution and origin of hydrocarbons in water and sediment in São Sebastião, SP, Brazil. *Marine Pollution Bulletin* **38** (4), 261–267.

Zanardi, E., Bícego, M.C. and Weber, R.R. (1999b) Dissolved/Dispersed Petroleum Aromatic Hydrocarbons in the São Sebastião Channel, SP, Brazil. *Marine Pollution Bulletin* **38** (5), 410–413.

THE AUTHORS

Eliete Zanardi Lamardo
University of Miami – RSMAS,
Dept. Marine and Atmospheric Chemistry,
4600 Rickenbacker Causeway,
Miami, FL 33149, U.S.A.

Márcia Caruso Bícego
Universidade de São Paulo,
Dept. Oceanografia Física do Instituto Oceanográfico,
Pca do Oceanográfico, 191,
Cidade Universitária, SP, 05508-900, Brazil

Belmiro Mendes de Castro Filho
Universidade de São Paulo,
Dept. Oceanografia Física do Instituto Oceanográfico,
Pca do Oceanográfico, 191,
Cidade Universitária, SP, 05508-900, Brazil

Luiz Bruner de Miranda
Universidade de São Paulo,
Dept. Oceanografia Física do Instituto Oceanográfico,
Pca do Oceanográfico, 191,
Cidade Universitária, SP, 05508-900, Brazil

Valéria Aparecida Prósperi
CETESB – Companhia de Tecnologia de Saneamento Ambiental,
Setor de Ictiologia e Bioensaios com organismos aquáticos,
Av. Prof. Frederico Hermann Jr., 345,
Alto de Pinheiros, SP 05489-900, Brazil

Chapter 48

THE ARGENTINE SEA: THE SOUTHEAST SOUTH AMERICAN SHELF MARINE ECOSYSTEM

José L. Esteves, Nestor F. Ciocco, Juan C. Colombo, Hugo Freije,
Guillermo Harris, Oscar Iribarne, Ignacio Isla, Paulina Nabel,
Marcela S. Pascual, Pablo E. Penchaszadeh, Andrés L. Rivas and
Norma Santinelli

The Southeast South American shelf marine ecosystem (SSASME) extends over the entire continental shelf off the eastern shores of Argentina, Uruguay and southeastern Brazil (23° to 55° latitude south). It is one of the widest in the world with smooth relief. The La Plata River basin system drains into this marine ecosystem providing fresh waters and nutrients that support a rich mix of coastal and marine fauna and flora. The La Plata River is the second largest river in South America, after the Amazon.

The Malvinas/Falklands Current also contributes nutrients to the highly productive marine ecosystem that extends along the edge of the shelf-break, sustaining large populations of invertebrates, fish, marine birds and mammals. There is a strong interdependence between the shelf/slope zone, located between 170 and 850 km from the coast, and coastal activities. Fish catches near the shelf-break support land-based processing plants, and marine birds and mammals that feed in the area for almost six months each year, return to the coast to breed in large colonies which, as valuable tourist attractions, contribute significantly to the regional economy.

Areas that are most valuable in terms of global biodiversity are the shores of Patagonia. Besides providing resting and breeding sites for marine birds and mammals, this coast has significant coastal wetlands that are used as feeding and resting areas by migratory shorebirds.

At present, more than one third of the total Argentinean population (~12,000,000) is settled 80 km around the ports of Buenos Aires and La Plata and the population densities are inversely proportional to latitude. In Patagonia, population density is estimated at one inhabitant per square kilometre, except for the valleys of the rivers Colorado, Negro and Chubut.

In the northern section of this coastal system, the La Plata River carries significant levels of pollution which have little-studied effects on the marine ecosystem. In Patagonia, although industrial activity is growing along the coast, it is incipient and coastal water is of high quality and free from industrial, agricultural and, in most cases, urban pollutants.

Clear jurisdictions and effective administration based on coastal zone management planning and a harmonious interrelationship between the provincial and national governments are the major challenges for the long-term protection of the biological diversity and health of the Southeast South American shelf marine ecosystem.

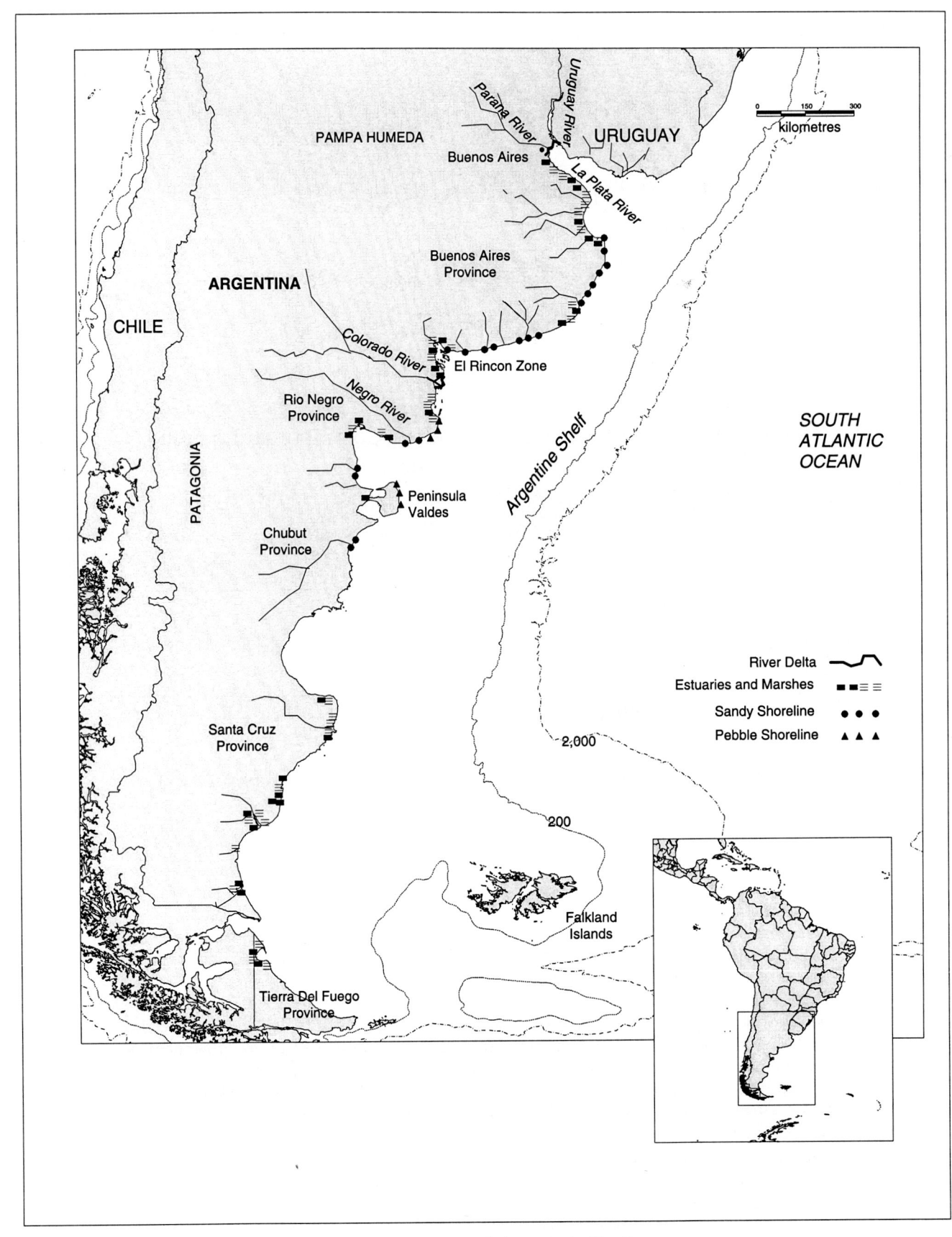

Fig. 1. Map of Argentina, including principal landforms.

THE DEFINED REGION

The Southeast South American shelf marine ecosystem (SSASME) extends over the entire continental shelf (23° to 55°S) along the eastern shore of Argentina, Uruguay and southeastern Brazil. This shelf is one of the widest in the world with smooth relief. The continental margin of Argentina is over 3500 km long, averaging 400 km in width to the shelf-break. The width of the platform increases toward the south from 170 km in front of the Buenos Aires coast to 850 km at the latitude of the Malvinas/Falkland Islands. Seventy percent of its area has a depth of more than 70 m and there are two gulfs (Nuevo and San Matías) where the maximum depths exceed the external shelf border depths (180 and 200 m, respectively). The region of the shelf-break is characterized by a marked continental slope with transversal canyons. The origin of this continental margin goes back more than 140 million years to the Late Jurassic–Early Cretaceous, when the opening of the South Atlantic took place. Most of it is a tectonically passive margin characterized by eustatic changes associated with thermal relaxation and crustal sag (Urien and Zambrano, 1996). Different morphostructural features are recognized north and south of 49°S. A typical continental passive margin of Atlantic type developed to the north, along which most of the Argentine continental shelf extends. The southern section is composed of two physiographic units: the Malvinas/Falkland plateau and the Scotia arc formed by a northern and southern ridge and a volcanic island arc, a prolongation of the Cordillera Austral, which is part of the Scotia plate. In this region a number of isolated and poorly extended island platforms are present (Parker et al., 1996). Their seaward limit is defined by an abrupt increase in slope to 200 m (Parker et al., 1997).

In the open sea, the Brazil Current and the Falklands/Malvinas Current are the major ocean flows and impose an eastern boundary to the SSASME. While the former provides sub-tropical and oligotrophic waters, the latter contributes cold nutrient-rich water of sub-antarctic origin. The encounter of these two currents defines a Confluence Zone (CZ) of mixed waters between 35° and 40°S. Significant physical and biological processes depend upon the seasonal pattern of current intensification and latitudinal movement of the CZ. Other oceanographic features, such as frontal zones, upwelling and low-salinity coastal waters, contribute to the hydrological structure. Finfish and invertebrate species of both commercial value and key ecological significance are distributed over a wide latitudinal range, which reveals the ecological extent of the SSASME and the intimate connection between the sub-tropical and sub-antarctic realms. Degradation of coastal water quality and anthropogenic perturbations to living marine resources are also significant throughout this system (Bisbal, 1995).

The main concerns for coastal management in Argentina are erosion, industrial pollution, and impacts from urban development, principally in the Buenos Aires metropolitan region and, to a lesser extent, along the Province of Buenos Aires. The remaining 72% of the coast to the south, from Patagonia to Tierra del Fuego, has localized problems associated with towns. This urban and rural split suggests that Argentina should have different organizational arrangements and strategies for managing its coastal resources and environments.

Argentina's coastal zone (Fig. 1) could be subdivided into four well defined sectors:

1. *Río de La Plata basin system*. This is, after the Amazon, the second largest river in South America. With a surface of 30,000 km², a huge drainage basin (more than 3 million km²), and freshwater runoff (16,000–28,000 m³/s), it receives about 90 million t/yr of particulate matter from the temperate and tropical regions of South America (Fig. 2a).

2. *Buenos Aires Province* is characterized by a high rate of precipitation (800 mm/year) and rivers from the plain with low flow rates. They flow over 400,000 km² of lands—known as "pampa húmeda" (wet pampas)—given over to extensive production of cereals and cattle breeding (Fig. 2b). This constitutes an important source of anthropogenic substances and could affect the coastal area. The "El Rincón" region (The Corner) is an unusual feature of the Argentinean coast (Fig. 2c), comprising some 2500 km² of a very complex system of tidal channels separating mudflats, salt marshes, and islands. The three main channels are Bahía Verde, Bahía Falsa and Bahía Blanca—the two first sites being almost pristine environments.

2. *The Argentinean Patagonia* is located in the south of the American continent. It is bordered by over 3000 km of coast and covers an area of 787,000 km²; its northern limit is the Colorado River and its western is the Andes mountains. The climate is arid and semi-arid in the east

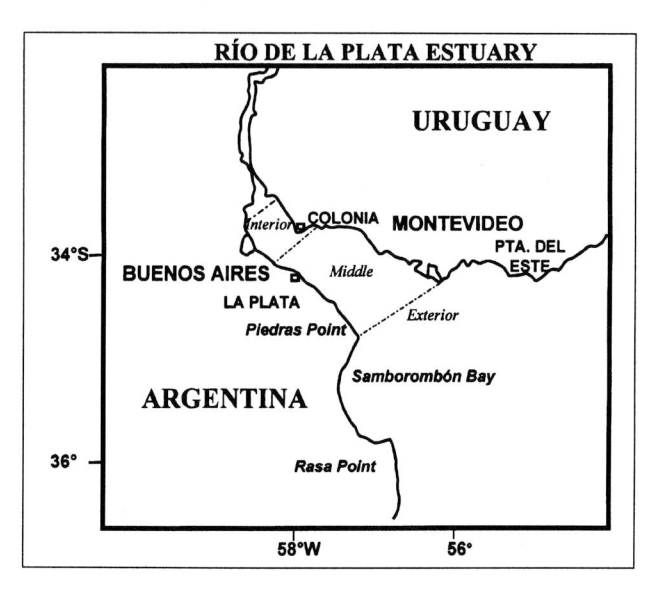

Fig. 2a. Rio de la Plata estuary.

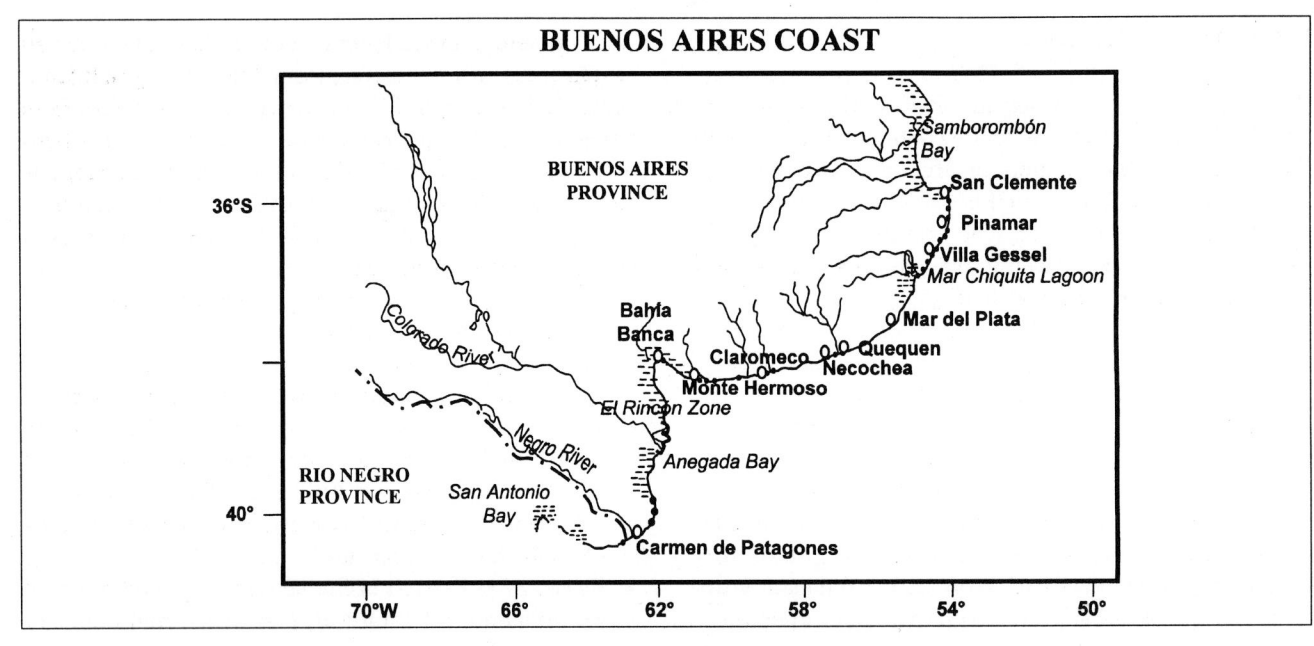

Fig. 2b. Province of Buenos Aires coast.

and centre of the Patagonian plateau, with precipitation values of approximately 150 mm/yr on the coast (Fig. 2d). The most important rivers that flow to the Atlantic Ocean are the Colorado River (130 m^3/s), the Negro River (1000 m^3/s), the Chubut River (50 m^3/s), and the Santa Cruz river (750 m^3/s) (Ferrari Bono, 1990). The Malvinas/Falkland current inputs nutrients to the highly productive marine ecosystem that extends along the edge of the shelf-break, sustaining large populations of invertebrates, fish, marine birds and mammals.

4. *Tierra del Fuego Island.* The large Island of Tierra del Fuego (52°S to 56°S and 63°W to 75°W) is part of an archipelago, shared with the Republic of Chile, at the

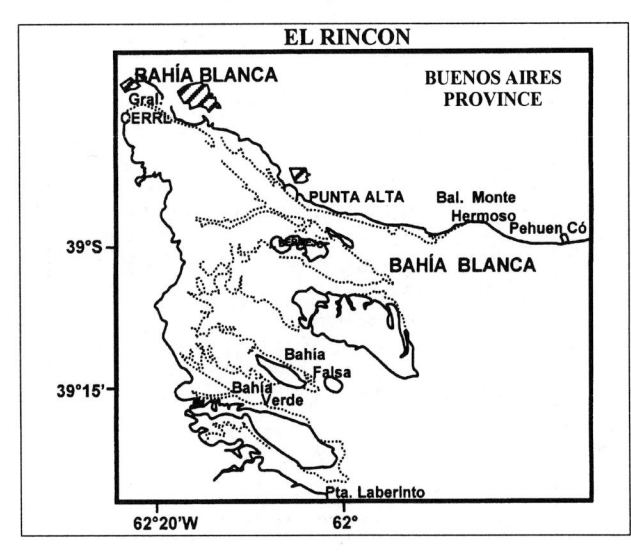

Fig. 2c. El Rincón coast.

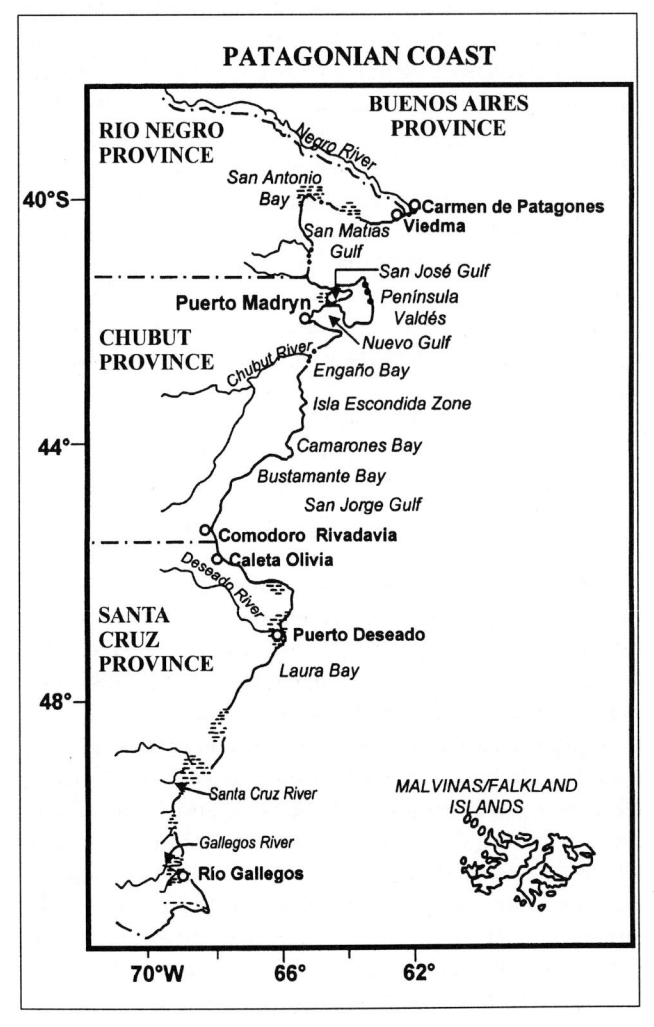

Fig. 2d. Patagonia coast.

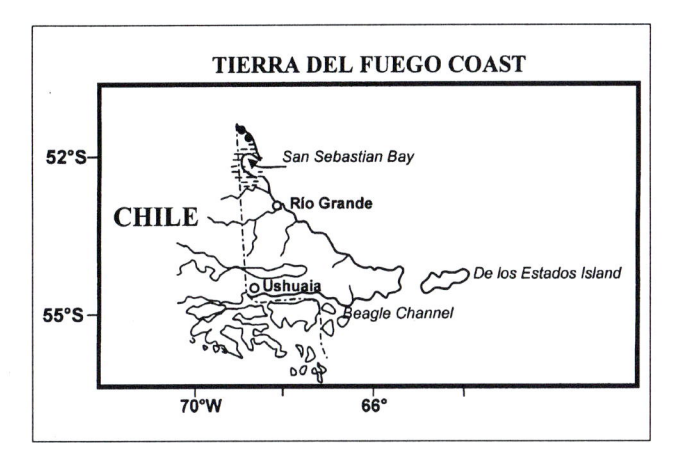

Fig. 2e. Tierra del Fuego coast.

southern tip of the continent (Fig. 2e). It has more than 600 km of coast line with an area of 21,200 km². Two different environments can be identified: (i) the oriental coast on the Atlantic Ocean, with temperate–cold and semi-arid climate, values of precipitation in the order of 350 mm/yr and a water temperature of between 0.2°C in winter and 6°C in summer; the tidal range is up to 10 m; (ii) the south coast on the Beagle Channel, with humid and temperate–cold climate, values of precipitation of 600 mm/yr and a water temperature of between 4.5°C in winter and 9°C in summer and a tidal range of less than 1 m. Tierra del Fuego represents a special environmental area for wildlife, and includes species from continental regions and from the Atlantic, Pacific and Southern Oceans (Goodall et al., 1993).

SEASONALITY, CURRENTS, NATURAL ENVIRONMENTAL VARIABLES

Water washing the continental shelf of Argentina is of subantarctic origin, coming from the north area of the Drake Passage and the Malvinas/Falkland current. The exchanges of heat and mass with the atmosphere alter the surface characteristics and continental discharges locally modify the salinity, especially in the areas of the Magellan Strait and Río de la Plata estuary and—with less significance—the mouths of the rivers Negro and Santa Cruz.

The temperature, forced by the superficial flow of heat, shows a noticeable annual cycle (Podestá et al., 1991; Rivas, 1994). Seasonal fluctuation controls the density variability which causes summer stratification. Mean superficial temperatures during mid-summer varies from 22°C in front of La Plata river, to 6°C to the south of the Malvinas/Falkland islands, and from 12°C to 2°C during mid-winter, respectively (Hoffmann et al., 1997).

Seasonal variations of salinity are small. There is a path of low salinity ($S < 33.2$ practical salinity unit or psu) stretching along the coast from the Magellan Strait, which when reaching the San Jorge gulf turns NNE, increasing its salinity (up to 33.7 psu) and continues to the Río de La Plata estuary (Guerrero and Piola, 1997) (Fig. 3). At the shelf-break, the salinity reaches values typical of sub-antarctic water of the Malvinas/Falkland current in the southern sector ($S = 34.1$ psu) and of the subtropical Brazilian current in the northern sector ($S = 35.0$ psu). Apart from that, the discharge of fresh water from La Plata river generates an area of diluted water connected to the north with Los Patos lagoon (Brazil).

There are three systems on the platform (Carreto et al., 1995; Martos and Piccolo, 1988): (1) the coastal system, where tidal and wind mixing maintain the homogeneity of the water column throughout the year; (2) the system of subantarctic shelf waters, characterized by the typical cycle of development and breakdown of the seasonal thermocline; and (3) the Malvinas/Falkland system presenting colder and more saline waters. These systems are frequently separated by quasi-permanent fronts. On the inner-shelf, several coastal fronts are identified, these being haline—e.g., the marine front off La Plata river (Ottman and Urien, 1965, 1966; Guerrero et al., 1997), the front off "El Rincón" (Guerrero and Piola, 1997) and the front along the coast of the Santa Cruz province (Krepper and Rivas, 1979)—or thermal—e.g. the tidal front located to the northeast of Península Valdés (Carreto et al., 1986).

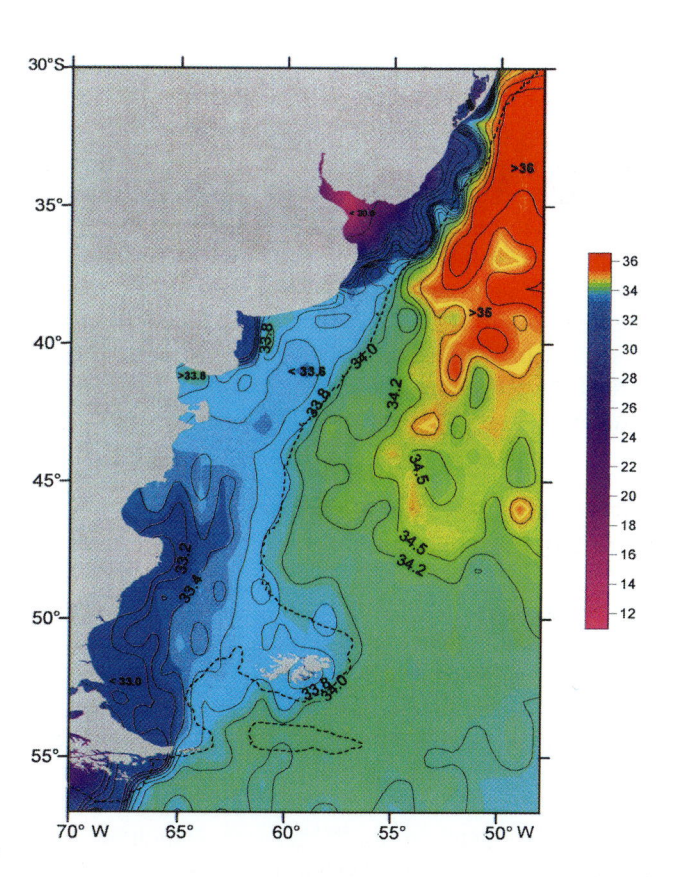

Fig. 3. Surface salinity distribution on the Argentine Basin (adapted from Guerrero and Piola, 1977, with permission from Boschi, ed.).

In the shelf-break area, the boundary between the shelf water and the Malvinas/Falkland sub-antarctic water is referred to as the shelf/slope front. Available information (Proyecto de Desarrollo Pesquero, 1968/71) shows evidence of upwelling processes in this area. This front is topographically trapped, persistent and robust. It is associated with high biological productivity and may be important in inhibiting the cross-shelf exchange of particulate matter. However, little is known of its dynamic behaviour or of the mechanisms regulating exchange of heat, salt, momentum and nutrients between platform and shelf-break waters.

Circulation

Knowledge of the circulation of the area has been studied mainly based on numerical simulations; there exists only one long-term *in situ* current measurement (Rivas, 1997). All the numerical models implemented in the area (Piola and Rivas, 1997) determine a NNE flow but they disagree on the estimation of intensity. Models that have analyzed the annual cycle agree on circulation intensification during winter, mainly associated with changes of wind intensity and direction. Some differences are observed in relation to the circulation direction in the Buenos Aires Province shelf, north of 41°S. A controversial hypothesis (see Piola and Rivas, 1997) suggests the existence of a summer countercurrent with a south-southwest direction named "Deriva Cálida Costera" (Coastal Warm Drift; Balech, 1971).

There is an intensification of tidal currents in the Patagonian coastal areas (Glorioso and Flather, 1995, Rivas, 1997). These currents keep over 80% of their total kinetic energy. This is because the Patagonian platform presents favourable conditions for the resonance of the semidiurnal component, resulting in a clear difference between tidal amplitudes recorded in the north and south of 41°S. While an equinoctial height of less than 0.8 m was recorded on the coast of La Plata river, in some locations of South Patagonia tides reach up to 13 m. Dissipation of tidal energy usually takes place on continental shelves and it is estimated that 8.5% of global dissipation takes place in the eastern coast of South America (Miller, 1966).

THE MAJOR SHALLOW WATER MARINE AND COASTAL HABITATS

Sandy Beaches

The sandy beach community in the northern Buenos Aires province is mainly composed of the yellow clam *Mesodesma mactroides*, the coquina clam *Donax hanleyanus*, several endemic southern atlantic gastropods such as *Buccinanops duartei*, *Olivancillaria vesica* and *O. uretai*, the isopod *Cirolana argentina*, the anomuran *Emerita brasiliensis*, amphipods and polychaetes (Olivier and Penchaszadeh, 1971). *Mesodesma mactroides* develops dense populations, with 14.2 kg/m^2 (wet weight) recorded in 1968. The second most abundant species is *Donax hanleyanus* (Penchaszadeh and Olivier, 1975). Both bivalves show large fluctuations in their densities and, since the early nineties, the yellow clam has been declining severely.

Rocky Shores

Except for the intrusion into the sea of the Tandilia formation, there is no true rocky shore between the La Plata river and northern Patagonia. On northern Argentinean hard substrates, the general rule is low diversity of macro and megafauna. In the Mar del Plata area, the supralittoral fringe is exclusively occupied by the pulmonate gastropod *Siphonaria lessoni* (Olivier and Penchaszadeh, 1968), while the midlittoral is covered mainly by the mussel *Brachidontes rodriguezi*, in impressively high densities, up to 175,000 ind./m^2 (20 kg/m^2 (ww) (Penchaszadeh, 1973a). Common algae here are, as epizootic on Brachidontes, *Ulva lactuca*, *Porphyra umbilicalis*, *Chaetomorpha* cf. *antennina* and *Enteromorpha intestinalis*. In tidal pools, the coralinacean algae *Bostrychia* develops dense stands, almost monopolizing the substratum.

There are very scarce *Mytilus edulis platensis* among the Brachidontes, while other animals are the amphipods *Hyale grandicornis* and *Allorchestes* sp., polychaetes and nemerteans.

The upper midlittoral community is characterized by a belt of the red algae *Hildenbrandtia lecannelieri*, which hosts a diverse blue-green algae such as *Calothrix crustacea*, *Brachytrichia quoyi*, *Phormidium corium*, *Schizothrix cacicola*, *Microcoleus tenerrinus*, *Kyrtuthrix maculatus*. A typical belt of cirripedians in the upper midlittoral—which characterizes this community almost everywhere else in the world—was completely lacking in natural conditions up to the late sixties, when *Balanus* sp. and *Chtamalus* sp. then became established (Penchaszadeh, 1973).

The sea-star *Patiria stellifer* is the main benthic predator of *Brachidontes* in its lower vertical distribution, and *Micropogonias furnieri* is the most important fish having *Brachidontes* in its diet (Olivier et al., 1968).

In northern Patagonia, the midlittoral fringe is dominated by another Mytilid, *Perumytilus purpuratus*, and in the south the community diversity becomes higher (Olivier et al., 1966).

Infralittoral Characteristic Communities

There are, in northern Patagonia, some very characteristic sublittoral communities. For instance, the *Aulacomya ater* banks are composed of a mytilid which develops high density populations at 6–10 m depth in the San Jose and Nuevo Gulfs. A mean density of 2400 ind/m^2 was reported for the Punta Loma Bank, in the Nuevo Gulf (Penchaszadeh et al., 1974). Two infaunal bivalves are also present in these banks, *Venus antiqua* and *Eurhomalea exalbida* which are preyed upon by the volutid gastropod *Odontocymbiola magellanica*. Common algae in the *Aulacomya* bank are *Dictyota* sp. and *Codium* spp.

There are over 150 species of macroalgae distributed along the coast of the three provinces of Rio Negro, Chubut and Santa Cruz, of which at least four are of commercial interest (Piriz and Casas, 1996). Between them, one well represented sublittoral community in Patagonia is the kelp forests of *Macrocystis pyrifera* (Boraso and Kreibohm, 1980; Hall, 1980).

Circalittoral Banks of *Mytilus edulis platensis*

A peculiar community of mussels develops in dense banks in depths of 36–58 m off the Province of Buenos Aires, in sand-shell mixed bottoms (Penchaszadeh, 1971; Penchaszadeh et al., 1974). It presents the highest diversity of invertebrates recorded for these latitudes. *Mytilus* recruits are predated by the echinoid *Arbacia dufresnei* and *Pseudechinus magellanicus*, the gastropods *Calliostoma* spp. and *Tegula patagonica*, and by the chitons *Chaetopleura tehuelcha*. The sea-star *Astropecten brasiliensis* predates on mussels up to 6 months old (22 mm in shell length), *Mytilus* representing 83% of the digested bivalve mollusc diet in number of items in sea-star stomach contents (Penchaszadeh, 1973b). When adult, this mussel is prey for a variety of gastropods, among them *Trophon geversianus*, *T. laciniatus* and *T. varians*, and various species of Volutidae. The mussel bank is visited by a variety of fish such as *Discopyge tschudii*, *Psammobatis scobina*, *Squatina argentina*, *Raja cyclophora*, *Paralichthys pagonicus*, *Etropus lingimanus*, *Prionotus nudigula*, *Porichthys porossissimus*, *Dules auriga*, *Conger* sp. and *Pinguipes fasciatus*. A complex food web develops (Penchaszadeh, 1979), and the success of the annual recruitment season (August–November) depends on the availability of filamentous or cord-shaped substrata, in the form of hydrozoans, colonial tunicates or polychaete tubes. Massive settlement in the adult banks is needed to avoid complete destruction by predators (Penchaszadeh, 1983).

Mud-flats, Salt-marshes and Coastal Lagoons

The principal estuaries and salt marshes on the Argentine coast are the Río de La Plata estuary, Mar Chiquita coastal lagoon, El Rincón and Anegada bay (Province of Buenos Aires); San Antonio bay (Province of Rio Negro); Península Valdés and Bustamante bay (Province of Chubut); Ría Deseado and Ría Santa Cruz (Province of Santa Cruz) and San Sebastián bay in Tierra del Fuego (Schnack, 1985). These areas are inhabited by large concentrations of invertebrates which include bivalves and polychaete worms. These form a valuable food resource for migratory shorebirds.

The Mar Chiquita coastal lagoon in Buenos Aires is a 46 km² shallow water estuarine environment running parallel to the coast, mainly impounded by a sedimentary barrier and connected to the open sea with one tidal inlet (Fasano et al., 1983). The watershed encompasses approximately 10,000 km² of extensive agricultural areas drained by three artificial channels and two creeks. The small access channel behaves like an estuary with variation in salinity mainly due to wind and tidal input.

The only seagrass inhabiting the Mar Chiquita lagoon is *Ruppia maritima* (Ruppiaceae). Waterfowl that forage on these patches in high densities include mainly swans (*Coscoroba coscoroba*, *Cygnus melancoryphus*), coots (*Fulica armillata*, *F. leucoptera*) and ducks (*Anas georgica*, *A. platalea*, *A. flavirostris*, *A. sibilatrix*, *A. versicolor*) (Bortolus et al., 1998). The lagoon is surrounded by large salt marshes dominated by the cordgrass *Spartina densiflora*, which is probably the main source of productivity and the basis of this estuarine food chain (Olivier et al., 1972).

The Rincón region near Bahía Blanca in Buenos Aires Province supports significant biological activity, based on high primary productivity due to a diatom winter bloom (Freije et al., 1980). The region is considered valuable as a breeding and nursery area for crustacea (Mallo, 1984) and fish (Lopez Cazorla, 1987) which inhabit the surrounding waters. Many species of shorebirds migrate through or live in these wetlands.

The South Western Atlantic burrowing crab (*Chasmagnathus granulata*) is one of the most abundant macroinvertebrates on marshes and estuarine environments from southern Brazil to northern Patagonia. It is distributed in almost all the zones of marshy intertidal areas in the soft bare sedimentary mudflats and in areas with *S. densiflora* (Spivak et al., 1994). Extremely well adapted to exposure to atmospheric air, this crab occupies the uppermost part of the intertidal area, frequently several hundred meters from the water edge in marshy areas. Their burrowing activity oxygenates the marsh, enhances soil drainage, and modifies sediment and meiofaunal abundance. These crabs are mainly deposit feeders in mud flats but are herbivorous in the *Spartina*-dominated areas (Iribarne et al., 1997). Evidence suggests that the extensive burrow beds built by the crabs between the marshes and the open estuary act as a large macrodetritus retention area, reducing the amount of organic matter exported from marshes (Botto and Iribarne, 1998).

Shorebirds are one of the main wildlife features of the coastal wetlands of Argentina. These include red knot (*Calidris canutus*), sanderling (*C. alba*), white-rumped sandpiper (*C. fuscicollis*), Hudsonian godwit (*Limosa haemastica*), greater yellowlegs (*Tringa melanoleuca*), lesser yellowlegs (*T. flavipes*), American golden plover (*Pluvialis dominica*), black-bellied plover (*P. squatarola*), semipalmated plover (*Charadrius semipalmatus*) and ruddy turnstone (*Arenaria interpres*). All these species breed in the northern part of North America and use South America as a wintering area (Martinez, 1993; Blanco and Canevari, 1995; Botto et al., 1998). The main areas of non reproductive concentration are located in Tierra del Fuego (Blanco and Canevari, 1995). A few species of shorebirds breed on the coast or in the hinterland of Patagonia. These include the Magellanic plover (*Pluvianellus socialis*), the two-banded plover (*Charadrius falklandicus*) and the rufous-chested dotterel (*Xonibyx modestus*). All three species migrate up the coast as far as Buenos Aires in the winter months of April to September. Migratory shorebirds concentrate in large numbers in a few locations, that in general are shared by several species. These areas are enormously valuable but, given their small distance from oil-producing areas (e.g., Bustamante bay and San Sebastian bay), they are extremely vulnerable.

Unique Patagonian Environments

Areas on the Argentine coast that are most valuable in terms of global biodiversity are the shores of Patagonia. They are used as resting and breeding sites by marine birds and mammals and the coastal wetlands are used as feeding and resting areas for migratory shorebirds.

The major significance of these wildlife areas is their biomass. Although there are only 16 species of marine birds (Table 1), 3 pinnipeds and 15 cetaceans that breed in Patagonia (Table 2), they number in tens or even hundreds of thousands. Colonies of marine birds and mammals are not evenly distributed along the coast of continental Patagonia. There are long stretches of coast with very scarce wildlife. Furthermore, often several species breed next to one another in relatively small areas, usually on isolated points and islands. This "patchy" distribution is probably related to the existence of nearby food resources or to predator avoidance. Most colonies are concentrated in three areas (Península Valdés, Cabo Dos Bahias to Bustamante bay and Puerto Deseado to Laura bay).

As well as birds and mammals that breed on these shores, many others migrate through, or feed in, the south-western Atlantic. These include about forty species of marine birds including penguins, albatrosses, petrels and gulls (Canevari et al., 1991) and over forty species of seals, dolphins, beaked whales and baleen whales (Lichter, 1992).

Marine Mammals

There are 75 colonies of sea lions and fur seals in the provinces of Rio Negro, Chubut and Santa Cruz. The southern sea lion (*Otaria flavescens*), has an estimated population of between 75 and 80,000 animals (Reyes et al., 1996; Dans et al., 1996). The southern fur seal (*Arctocephalus australis*) numbers 20,000 animals. The southern elephant seal (*Mirounga leonina*) gathers to breed and molt mainly on the shores of Peninsula Valdés, Chubut (Lewis, 1996). This is the only area where this species is known to be increasing in number. Its population is estimated at 40,000 adults with an annual production of 12,000 pups (Campagna et al., 1996).

Table 1
Marine birds that breed in Patagonia

Common name	Scientific name
Magallanic penguin	*Spheniscus magellanicus*
Rockhopper penguin	*Eudyptes chrysocome*
Southern giant petrel	*Macronectes giganteus*
Imperial cormorant	*Phalacrocorax atriceps*
Rock cormorant	*Phalacrocorax magellanicus*
Guanay cormorant	*Phalacrocorax bougainvillii*
Red-legged cormorant	*Phalacrocorax gaimardi*
Neotropic cormorant	*Phalacrocorax olivaceus*
Kelp gull	*Larus dominicanus*
Dolphin gull	*Larus scoresbii*
Band-tailed gull	*Larus atlanticus*
South American tern	*Sterna hirundinacea*
Royal tern	*Sterna maxima*
Cayenne tern	*Sterna eurygnatha*
Great skua	*Catharacta antarctica*
Chilean skua	*Catharacta chilensis*

Table 2
Marine mammals that breed in Patagonia

Common name	Scientific name
Pinnipeds	
Southern sea lion	*Otaria flavescens*
South american fur seal	*Arctocephalus australis*
Southern elephant seal	*Mirounga leonina*
Cetaceans	
Southern right whale	*Eubalaena australis*
Killer whale	*Orcinus orca*
Dusky dolphin	*Lagenorhynchus obscurus*
Commerson's dolphin	*Cephalorhynchus commersonii*
Peale's dolphin	*Lagenorhynchus australis*
Hourglass dolphin	*Lagenorhynchus cruciger*
Bottlenose dolphin	*Tursiops truncatus*
Common dolphin	*Delphinus delphis*
Risso's dolphin	*Grampus griseus*
False killer whale	*Pseudorca crassidens*
Long-finned pilot whale	*Globicephala melas*
Southern right-whale dolphin	*Lissodelphis peronii*
Franciscana dolphin	*Pontoporia blainvillei*
Spectacled porpoise	*Australophocoena dioptrica*
Burmeister's porpoise	*Phocoena spinipinnis*
Marine otter	*Lutra provocax*

Several species of dolphins including the abundant dusky and common dolphins are found off large sections of the coast of Argentina, while others such as Commerson's dolphin are restricted to relatively small ranges (Pedraza et al.,1996). Twenty-eight species of cetaceans (21 small dolphins, porpoises and breaked whales, 7 large whales) are found in waters around Tierra del Fuego, where the most numerous cetacean seen from shore is the Commerson's dolphin (Goodall et al., 1993).

A significant portion of the world population of southern right whales (*Eubalaena australis*) breeds in the southwestern Atlantic in waters around Península Valdés and adjoining bays in Chubut. The total number of individuals in Península Valdés is close to 2000, of which 400 to 600 gather in these waters each season.

Marine Birds

There are 72 breeding colonies in a total of over 150 breeding and roosting sites on the coast of Rio Negro, Chubut and Santa Cruz; each one contains between one and 8 species. Many of these sites are shared with colonies of sea lions and elephant seals. The abundance of different species of marine birds is uneven. Magellanic penguins (*Spheniscus magellanicus*) with a population of 893,000 pairs surpass by at least one order of magnitude the next most abundant species, the kelp gull (*Larus dominicanus*), with 72,500 pairs and the imperial cormorant (*Phalacrocorax atriceps*) with 35,000 pairs. The following species have low populations: Olrog's gull (*Larus belcheri*), dolphin gull (*Larus scoresbii*), red legged cormorant (*Phalacrocorax gaimardi*) and rockhopper penguin (*Eudyptes chrysocome*) (Yorio and Harris, 1997).

There are 210 species of birds in Tierra del Fuego; about 95 of these are found in San Sebastian Bay (Goodall et al., 1993).

OFFSHORE SYSTEMS

In the shelf/slope front, the little available data indicate a sector of maximum concentration of chlorophyll and nitrate in the Argentine Sea, over a path stretching from 37° to 50°S to the west of the shelf break. Figure 4 shows chlorophyll concentration estimates from data of the Coastal Zone Color Scanner (CZCS) (Podestá, 1997). Although a higher abundance of plankton has been observed in shelf breaks of other parts of the world, the length of this stretch of high concentrations of pigments (about 1500 km) and its temporal and spatial continuity, makes this a unique phenomenon (Podestá, 1997). The offshore system is also associated with species of commercial importance, which support an important fishery. Among others, the following species are caught: Argentine hake (*Merluccius hubbsi*), southern hake (*Merluccius australis*), southern blue whiting (*Micromesistius australis*), Hoki (*Macruronus magellanicus*), Kingclip (*Genypterus blacodes*), Shortfin squid (*Illex argentinus*) and Longfin squid (*Loligo* sp.). This region is also the feeding area of species of marine mammals whose coastal breeding and moulting sites have become an important tourist resource, particularly the Patagonian elephant seal (*Mirounga leonina*). Campagna et al. (1995; 1998), using a geographic location time-depth-recorder (GLTDR), demonstrated that all female southern elephant seals

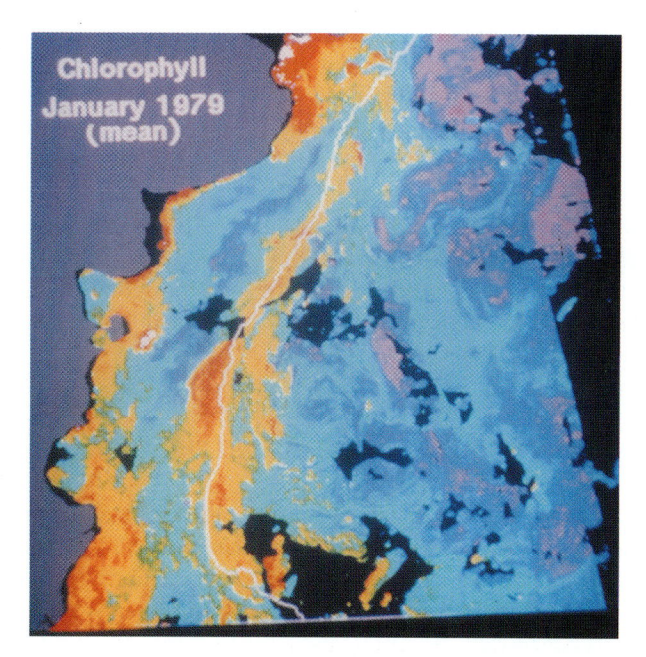

Fig. 4. Chlorophyll concentration along the Argentine shelf-break (Podesta, 1997, with collaboration agreement between Servicio Meteorológico Nacional, Argentina, and Miami University, USA).

studied cross the continental shelf in seven days or less, spending 89% of the recorded time at sea over deep water. The females concentrated their foraging efforts in pelagic temperate waters of the southwest Atlantic Ocean, between 36° and 46°S and up to 1200 km away from land.

Prince et al. (1992) showed that wandering albatrosses from South Georgia flew in areas only 240 km off the coasts of southern Brazil and Uruguay, and that they always flew to feed on the shelf-break zones; in some of the trips they travelled along this shelf-break zone for more than 2000 km. Distribution and abundance of anchovy are related to the areas of upwelling along the shelf break (Bakum and Parrish, 1991). Several mechanisms may cause the upwelling along the shelf break (Podestá, 1989).

There is a strong interdependence between this area and coastal activities. In the case of fisheries, the catches carried out near the shelf break support land-based processing plants; on the other hand, marine birds and mammals which feed in the area for almost six months, return to the coast each year to breed. These areas are valuable tourist attractions, as such contributing significantly to the regional economy.

Offshore fisheries seem to produce a strong impact on the fauna, through accidental mortality by fishing nets and high catch volume of target and discard species. The sea lion (*O. flavescens*), for instance, tangles with fishing gear. The most affected individuals are apparently males and the capture level is estimated between 1 and 2% of the total population per year in the south of Chubut (Crespo et al., 1997). Both the Dusky and the Commerson's dolphin have been found dead in trawl nets. Marine birds are also affected, mainly the Magellanic penguin, followed by the

Sooty shearwater (*P. griseus*), while the Imperial cormorant (*Phalacrocorax atriceps*) and the Black-browed Albatross (*D. melanophrys*) are only occasionally caught. Incidental capture by longline fishing vessels indicates that the most affected species in the Patagonian sector of the continental shelf is the Black-browed albatross. The annual mortality of these birds in the southwest Atlantic is estimated around 2400 and 8400 individuals. Populations of wandering albatross have decreased apparently by incidental catch caused by longline fishing vessels in the shelf-break area (Prince et al., 1992).

HUMAN POPULATIONS AFFECTING THE AREA

The population densities are in inverse ratio with latitude in Argentina (Fig. 5). At present, more than one third of the total Argentinean population is settled in the northern 80 km (12 million people) around the ports of Buenos Aires and La Plata (Fig. 2a). Approximately half of the Uruguayan population lives in its northern-shore cities (1.5 million), including Montevideo and Colonia. This estuary has been recognized as one of the most important pollutant input sources to the coastal system (C.A.R.P., 1989; Pucci, 1991). The largest urban and industrial centres of Argentina lie along its path, and large quantities of waste are discharged daily into Rio de la Plata via small tributaries and untreated effluents (see the Rio de la Plata Estuary box).

The largest coastal city of the Buenos Aires province is Mar del Plata, a city of 500,000 inhabitants. During the summer season (December to March), the affluence of people from Buenos Aires and other cities, (up to four million tourists), increases the environmental impact on more than twenty coastal cities and towns. Besides summer tourist activities, fishing is the most important activity given that a large part of the fish catches from the Argentinean shelf (40%) is landed in Mar del Plata port. Agriculture is also an important activity in all the coastal area. The land is characterized by its high content of organic matter that makes it suitable for the extensive culture of wheat, oats, linen, sunflower, soya and potatoes. Agriculture is one of the subsections with the largest growth of added value within the primary sector of the region. Production of wheat, corn and sunflower has shown a 134% increase during the period 1993–1995. The major pollution problem in this area is related to lack of complete treatment of domestic sewage. This problem is restricted to the Mar del Plata area, but given that the main activity is beach-related tourism, the solution of this problem should be immediate.

The city of Bahía Blanca is located 7 km from the seaside and together with several nearby towns has a growing population of 350,000 inhabitants. Since 1982 a "Petrochemical Sector" has been developed, including a factory of nitrogenated fertilizers which will produce 1,200,000 t/yr of urea. Medium-size industries and service enterprises are located near the port area. A pre-treatment plant for domestic sewage (1.5 m^3/s) is beginning operations to remove solids under 0.75 mm.

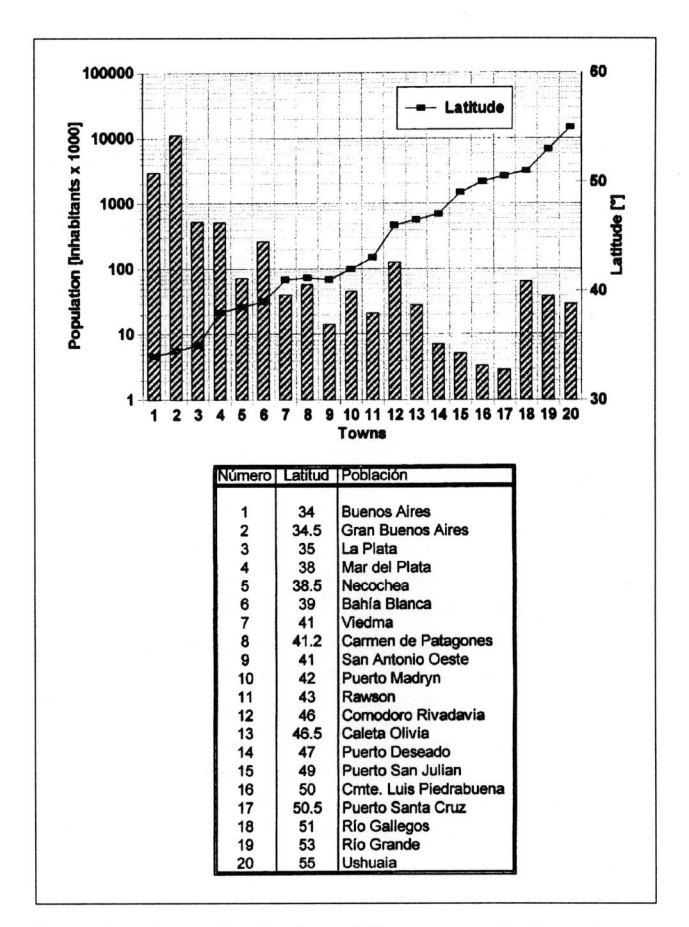

Fig. 5. Population density from different towns in Argentina vs. latitude (Population Census, 1991).

Número	Latitud	Población
1	34	Buenos Aires
2	34.5	Gran Buenos Aires
3	35	La Plata
4	38	Mar del Plata
5	38.5	Necochea
6	39	Bahía Blanca
7	41	Viedma
8	41.2	Carmen de Patagones
9	41	San Antonio Oeste
10	42	Puerto Madryn
11	43	Rawson
12	46	Comodoro Rivadavia
13	46.5	Caleta Olivia
14	47	Puerto Deseado
15	49	Puerto San Julian
16	50	Cmte. Luis Piedrabuena
17	50.5	Puerto Santa Cruz
18	51	Río Gallegos
19	53	Río Grande
20	55	Ushuaia

The evolution of Patagonian settlements followed European colonization. In 1779 Carmen de Patagones was founded, 30 km from the mouth of the River Negro on the Atlantic. About fifty years later, Punta Arenas (Chile) was founded on the West sector of the Magellan Strait; but it was only in 1865 that a Welsh colony settled in the lower valley of the Chubut river. Other towns on coastal Patagonia were founded after 1885. Population density is estimated at one inhabitant per square kilometre and, except for the valleys of the rivers Colorado, Negro and Chubut, the rural population is very low. Only Comodoro Rivadavia has over 100,000 inhabitants. Pollution problems are observed in coastal cities but have not been fully evaluated. Recently, several cities have been assessed (Esteves et al., 1996, 1997a, 1997b, 1998) determining that their impact is directly related to the geographical features of the area. Cities located inside gulfs, bays or coves have higher impact levels than those located on open sea. Industrial activity is restricted mainly to Puerto Madryn in the Nuevo gulf (light metals and fishing activities); Comodoro Rivadavia and Caleta Olivia in the San Jorge gulf (oil loading); Río Gallegos (oil and carbon) and Río Grande (oil). The impact of these activities is local. However,

The Río de la Plata Estuary

Juan C. Colombo

*Environmental Chemistry and Biogeochemistry, Facultad de Ciencias Naturales y Museo,
Universidad Nacional de La Plata, Paseo del Bosque s/n, 1900, La Plata, Argentina*

SYSTEM CHARACTERISTICS

The Río de la Plata is located between Uruguay and Argentina at 34–36°S latitude and 55–58°W longitude (Fig. 2a) in a temperate-humid zone (mean temperature 16°C, rainfall 900 mm). The estuary, discovered in 1516, was the main route of colonization of southern South America, and more than one third of the total Argentinian population (ca. 12 million people) is settled in the first 100 km of the south shore which includes the principal ports of Buenos Aires and La Plata. Approximately half of the Uruguayan population (ca. 1.5 million) lives in northern-shore cities, principally in Montevideo and Colonia.

The Río de la Plata is a funnel-shaped coastal-plain estuary 300 km long, 30–220 km wide and 0.5–25 m deep with a total surface area of 30,000 km^2 (Balay, 1961; Urien, 1972). Its drainage basin covers more than 3 million km^2 of tropical and temperate rainy areas of Brazil (46%), Argentina (30%), Paraguay (13%), Bolivia (6.4%) and Uruguay (4.6%). The Paraná (3780 km) and Uruguay rivers (1790 km), main collectors of the basin, carry to the Río de la Plata 500–880 km^3 fresh water and about 40, 90 and 8 million tons of dissolved solids, suspended solids and total organic carbon per year, respectively (Degens et al., 1991). The particulate charge is largely carried by the Paraná river, principally in suspended form (75–150 mg/l) but also as bedload transport (~6 kg fine sand/day at Paraná Guazu). This massive contribution of material feeds a vast delta of 18,000 km^2.

Structurally, the Río de la Plata resulted from the tilting of the pre-Cambric Brazilian crystalline shield which dominates the high and rocky Uruguayan north shore. The Argentinian south shore is formed by reworked quaternary sediments (Pampean loess), and is low and silty with large mud deposits (Ottman and Urien, 1966; Parker, 1990). The material supplied by the Paraná (silty clay with low sand) and the Uruguay rivers (chiefly sand), and relict sand from the last Holocene transgression, complete the sedimentary sources of the Río de la Plata. Overall, sediments are dominated by silts and clays, with a higher contribution of riverine sands in the upper sector, and of finer material in the middle and external zones which also present sand shoals transported from the Argentine shelf (Biscaye, 1972; CARP, 1989).

The shallow depth and strong freshwater discharge (16–28 × 10^3 m^3/s) of the estuary restrict the penetration of seawater. The upper reaches are thus essentially freshwater environments. The presence of sand shoals in the middle (Ortiz Bank) and outer area (Archimedes and English Banks) deflects freshwater discharge via the south channel to the Samborombón Bay, and through the deeper Northern Channel to the Brazilian coast. The penetration of marine waters is favoured by southeast winds which restrict the discharge of freshwater, raising the level of the estuary and causing disastrous floods on the Argentine shore. The outer area of the Río de la Plata is estuarine with steep salinity gradients (1–5 to 25–30‰) and a double-layer circulation in the deeper northern channel (Ottman and Urien, 1965; CARP, 1989). The tide penetrates all along the estuary (0.3–1 m) having a semi-diurnal uneven regime with higher amplitude on the south shore as a result of Coriolis and average current velocities of 5–60 cm/s (Lanfredi et al., 1979).

The Río de la Plata is usually divided in three sectors: (1) Upper zone, extending from the Paraná and Uruguay rivers to Buenos Aires (Punta Quilmes)-Colonia, freshwater area dominated by the Paraná delta, (2) Intermediate zone, extending to Punta Piedras-Punta Brava (Montevideo), chiefly freshwater with occasional saline intrusion, and (3) Outer sector, down to Punta Rasa-Punta del Este, mixing zone with steep salinity gradients. The Río de la Plata is a partially mixed estuary with stratification in the deepest channels. In the mixing zone (0.5–5‰ salinity), resuspension of bottom material, particle trapping and flocculation of fines and the strong ionic gradient generate a high turbidity zone with suspended solid concentrations of 150–500 mg/l (Bazán and Janiot, 1991) and up to 2 g/l on the Argentinian side (Colombo, 1998 unpublished results). The increase of ionic strength in this area also appears to favour the release of phosphate from bottom sediments (Pizarro et al., 1992).

The environmental gradients are naturally reflected by the composition of the biological community. The upper and middle sectors are dominated by typical freshwater species, including green algae, diatoms and cyanophyceae (Frenguelli, 1941; Guarrera, 1959; Roggiero, 1988; Gomez and Bauer, 1998), oligochaetes, bivalves, especially the invasive *Corbicula* and *Limnoperma* sp. (Ituarte, 1984; Pastorino et al., 1993), crustaceans (cladocerans, copepods, crabs) and fishes (principally detritivorous and catfishes). Eurihaline diatoms and crustaceans are common in the intermediate zone. In the outer third, truly marine and eurihaline plankton, macroalgae and fish predominate (Cousseau, 1985). The Río de la Plata is moderately productive; chlorophyll-a concentration range from 0.3 to 15 mg/m^3. Extraordinarily high values have been reported in freshwater euthrophic embayments (60–250 mg/m^3) dominated by diatoms and cyanophyceae, and in the outer zone during marine diatom blooms (CARP, 1989).

HUMAN IMPACT

In spite of the large dimensions and dilution capacity of the estuary, the vast urban–industrial zone developed in the first 100 km of the Argentinian coast has produced a significant impact in the 0.5–2 km coastal band (AGOSBA-OSN-SHIN, 1992). Massive amounts of polluted waters and toxic materials are discharged daily to the estuary via small tributaries which act as open sewers, or directly as untreated effluents. The major sewers of Buenos Aires city, for example, discharge about 2 million m^3/day of crude effluent 2.5 km from the coast. Sixty km south, La Plata city also discharges crude sewage directly into the estuary. This massive organic load produces anoxic or sub-oxic situations in tributaries and, depending on temperature and currents, in shallow coastal embayments. Low oxygen and high organic carbon and conductivity values (<1–2 mg O$_2$/l, >10–20 mg DOC/l, >0.5 mS/cm) characterize polluted tributaries, but the strong freshwater discharge and turbulence of the estuary rapidly attenuate the impact 3–5 km offshore in the upper sector (8–9 mg O_2/l, 5–10 mg DOC/l and <0.15 mS/cm, respectively) (Colombo et al., 1994).

Introduced nutrients and organic inputs are significant in mass balance terms. The urban contributions of dissolved phosphorus and nitrogen have been estimated as 29 and 23% of the total load, respectively, in the freshwater sector (Pizarro and Orlando, 1984). Mean dissolved phosphate and nitrate values in this area are ≈0.002–0.05 and 0.4–0.6 and up to 0.2 and >10 mg/l, respectively, near the coast (Pizarro and Orlando, 1984; AGOSBA-OSN-SHIN, 1992). The nutrient discharge of the Río de la Plata appears to have some fertilizing effect in adjacent coastal waters. Over the continental shelf, Río de la Plata waters meet warm subtropical waters from the southward-flowing Brazilian current and cold, nutrient-rich subantarctic waters from the northward-flowing Malvinas current, developing complex and productive frontal systems. The existence of a north–south elongated offshore eutrophication zone has been deduced from satellite images (Szekielda et al., 1983) and recurrent phytoplankton blooms occur in this area (Carreto et al., 1986; Gayoso, 1996) which constitute the spawning and feeding grounds of several economically important species such as the anchovy, the hake and the white croaker.

The anthropogenic impacts are also reflected by the hydrocarbon contents of waters and sediments. Reported concentrations for the water column are 0.01–24 μg/l, with higher values on the south shore (CARP, 1989). Hydrocarbon composition of bottom sediments (Colombo et al., 1989) indicates abundance of petrogenic hydrocarbons in tributaries and nearby coastal areas affected by petrochemical effluents, tanker ballast washings and spillages. Offshore sediments show the contribution of vascular-plant hydrocarbons and pyrogenic aromatic hydrocarbons transported from major fossil fuel combustion sources. Dissolved organochlorines also reflect the higher impact in the coastal fringe compared to offshore stations (Colombo et al., 1990; Janiot et al., 1991). Bivalves and fishes contain abundant chlorinated pesticides (BHCs, Chlordanes, DDTs) and PCBs. Asiatic Clams, used as sentinel organisms along 150 km of the Argentinian coast, showed an order of magnitude decrease of PCB levels with increasing distance to the Buenos Aires-La Plata urban area (Colombo et al., 1995). From 1986 to 1993, Asiatic Clams from the Río de la Plata showed sustained levels of PCBs and lindane, and a marked decrease of trans-Chlordane (60%) and DDT (90%), possibly reflecting a reduction of the pesticide load to the estuary. Isolated reports of Uruguayan Asiatic Clams indicate a lower degree of contamination (Farrington and Tripp, 1995).

PCBs, dioxins and furans have been recently analyzed in Río de la Plata bivalves and fishes (Colombo et al., 1997; 2000). Owing to their detritivorous-feeding habits, bioaccumulation of organic pollutants is very high in the abundant bottom-dwelling fishes of the Río de la Plata. Carps (*Cyprinus carpio*) and "Sabalos" (*Prochilodus lineatus*), a dominant iliophagous species which comprises 86% of the total fish catch in the estuary, have high concentrations of petrogenic hydrocarbons, PCBs, dioxins and furans, exceeding the tolerance limits set for human consumption (Colombo et al., 2000). Trace metal levels in Río de la Plata fish and bivalves are relatively low (Marcovecchio and Moreno, 1992; Bilos et al., 1998) reflecting their distinct bioaccumulation process and environmental dynamics. In contrast to suspended particles and sediments, which are enriched in anthropogenic metals (Cr, Cu, Zn) in the urban sector of the estuary, clams do not show clear geographical trends (Bilos et al., 1998).

STATE OF KNOWLEDGE

The Río de la Plata Estuary is ranked among the top 10 major systems on earth, yet several important aspects are poorly understood including the biogeochemistry of natural and anthropogenic compounds. The effects of dissolved and particulate xenobiotics on the rich biological resources of the estuary and adjacent continental shelf are completely unknown. A preliminary model performed with data collected in the upper and intermediate area yield interesting results. Taking an average freshwater discharge of 1.7 km^3/day with a conservative suspended solid concentration of 75 mg/l and a total particle load of 128,000 tons/day, our trace metal and organic determinations yield the following discharges: *dissolved phase*, PCBs and DDTs < 1 kg/day each, Cu, Cr, Mn and Fe, 8, 7, 14 and 760 tons/day respectively; *particulated phase*, PCBs and DDTs ~3 kg/day each, Cu, Cr, Mn and Fe, 7.5, 30, 105 and 5700 t/day, respectively. An important, and still unknown, proportion of this freshwater load would be retained through complexation and particle deposition in inner sedimentary basins (e.g. Samborombon Bay). The other probable major destination of particulated xenobiotics is the southern continental shelf of Brazil, e.g. the "mud well" (Pozos de fango), a 40 m-deep depositional area located in front of the Río Grande.

REFERENCES

Administración General Obras Sanitarias Buenos Aires-Obras Sanitarias de la Nación-Servicio de Hidrografía Naval (AGOSBA-OSN-SIHN) (1992) Río de la Plata. Calidad de aguas, Franja Costera Sur, 115 pp.

Balay, M.A. (1961) El Río de la Plata entre la atmósfera y el mar. Publ. No. 621, Servicio de Hidrografía Naval.

Bazán, J.M. and Janiot, L.J. (1991) Zona de máxima turbidez y su relación con otros parámetros del Río de la Plata. Informe Técnico No. 65/91. Servicio de Hidrografía Naval.

Bilos, C., Colombo, J.C. and R. Presa, M.J. (1998) Trace metals in suspended particles, sediments and Asiatic Clams (*Corbicula fluminea*) of the Río de la Plata estuary, Argentina. *Environmental Pollution*, in press.

Biscaye, P.E. 1972. Strontium isotope composition and sediment transport in the Río de la Plata estuary. In *Environmental Framework of Coastal Estuaries*, ed. B.W. Nelson. The Geological Society of America, Memoir 133, pp. 349–357.

Carreto, J.I., Negri, R. and Benavides, H.R. (1986) Algunas características del florecimiento del fitoplancton en el frente del Río de la Plata. I : Sistemas nutritivos. *Revista Investigación Desarrollo Pesquero* **5**, 7–29.

Comisión Administradora del Río de la Plata (CARP) (1989) Estudio para la evaluación de la contaminación en el Río de la Plata. SHIN-SOHMA, 422 pp.

Colombo, J.C., Pelletier, E., Brochu, C., Khalil, M. and Catoggio, J.A. (1989) Determination of hydrocarbons sources using n-alkane and polyaromatic hydrocarbon distribution indexes. Case study: Río de la Plata Estuary, Argentina. *Environmental Science and Technology* **23**, 888–894.

Colombo, J.C., Khalil, M.F., Arnac, M., Horth, A.C. and Catoggio, J.A. (1990) Distribution of chlorinated pesticides and individual polychlorinated biphenyls in biotic and abiotic compartments of the Río de la Plata, Argentina. *Environmental Science and Technology* **24**, 498–505.

Colombo, J.C., Bilos, C., R. Presa, M.J. and Schroeder, F. (1994) Contaminación química en el Río de la Plata. Evaluación del impacto de efluentes urbano-industriales mediante monitoreo electrónico, químico y biológico. *Gerencia Ambiental* **6**, 420–451.

Colombo, J.C., Bilos, C., Campanaro, M., R. Presa, M.J. and Catoggio, J.A. (1995) Bioaccumulation of polychlorinated biphenyls and chlorinated pesticides by the Asiatic Clam *Corbicula fluminea*: its use as sentinel organism in the Río de la Plata estuary, Argentina. *Environmental Science and Technology* **29**, 914–927.

Colombo, J.C., Brochu, C., Bilos, C., Landoni, P. and Moore, S. (1997) Long term accumulation of individual PCBs, dioxins, furans, and trace metals in Asiatic Clams from the Río de la Plata Estuary, Argentina. *Environmental Science and Technology* **31**, 3551–3557.

Colombo, J.C., Bilos, C., Remes Lenicov, M., Colautti, D., Landoni, P. and Brochu, C. (2000) Detritivorous fish contamination in the Río de la Plata estuary. A critical accumulation pathway in the cycle of anthropogenic compounds. *Canadian Journal of Fisheries and Aquatic Sciences*, in press.

Cousseau, M.B. (1985) Los peces del Río de la Plata y su frente marítimo. In *Ecología de comunidades de peces en estuarios y lagunas costeras: hacia una integración de ecosisitemas*, ed. A. Yañez. Univ. Nac. Autónoma de Mexico, 24, pp. 515–534.

Degens, E.T., Kempe, S. and Richey, J.E. (1991) Summary: Biogeochemistry of Major World Rivers. In *Biogeochemistry of Major World Rivers*, eds. E.T. Degens, S. Kempe and J.E. Richey. SCOPE 42, John Wiley and Sons Ltd., pp. 335–347.

Farrington, J.W. and Tripp, B.W. (1995) International Mussel Watch Project. Initial Implementation Phase, Final Report. NOAA Technical Mem. NOS ORCA 95.

Frenguelli, J. (1941) Diatomeas del Río de la Plata. *Revista del Museo de La Plata, Botánica* **15**, 213–235.

Gayoso, A.M. (1996) Phytoplankton species composition and abundance off Río de la Plata (Uruguay). *Archives Fisheries Marine Research* **44**, 257–265.

Gomez, N. and Bauer, D.E. (1998) A study about the phytoplankton from the Southern Coastal Fringe of the Río de la Plata (Buenos Aires, Argentina). *Hydrobiologia*, in press.

Guarrera, S.A. (1946) Contribución al conocimiento de las Chlorophyceae del Río de la Plata. *Revista de la Administración Nacional del Agua* **110**, 21–140.

Ituarte, C.F. (1984) Aspectos biológicos de las poblaciones de *Corbicula largillierti* Philippi (Mollusca Pelecypoda) en el Río de la Plata. *Revista Museo de La Plata* **XIII**, 231–247.

Janiot, L.J., Orlando, A.M. and Roses, O.E. (1991) Niveles de plaguicidas clorados en el Río de la Plata. *Acta Farmacéutica Bonaerense* **10**, 15–23.

Lanfredi, N.W., Schmidt, S.A. and Speroni, J. (1979) Cartas de corrientes de marea (Río de la Plata). Informe Técnico No. 03/79, Servicio de Hidrografía Naval.

Marcovecchio, J.E. and Moreno, V.J. (1992) Evaluación del contenido de metales pesados en peces de la Bahía de Samborombon. *Frente Marítimo* **12**, 139–146.

Ottmann, F. and Urien, C.M. (1965) Le melange des eaux douces et marines dans le Río de la Plata. *Cahiers Océanographiques* **XVII**, 703–713.

Ottmann, F. and Urien, C.M. (1966) Sur quelques problèmes sédimentologiques dans le Río de la Plata. *Revue Géographie Physique et Géologie Dynamique* **3**, 209–224.

Parker, G. (1990) Estratigrafía del Río de la Plata. *Asociación Geológica Argentina Rev.* **XLV** (3–4), 193–204.

Pastorino, G., Darrigran, G., Martín, S.M. and Lunaschi, L. (1993) *Limnoperma fortunei* (Dunker, 1857) (Mytilidae), nuevo bivalvo invasor en aguas del Río de la Plata. *Neotrópica* **39**, 34.

Pizarro, M.J. and Orlando, A.M. (1984) Distribución de fósforo, nitrógeno y silicio disueltos en el Río de la Plata. Publicación H.625, Servicio de Hidrografía Naval.

Pizarro, M.J., Hammerly, J., Maine, M.A. and Suñe, N. (1992) Phosphate adsorption on bottom sediments of the Río de la Plata. *Hydrobiologia* **228**, 43–54.

Roggiero, M.F., 1988. Fitoplancton del Río de la Plata I. *Lilloa* **XXXVII** (1), 137–152.

Szekielda, K.-H., Piatti, L. and Legeckis, R. (1983) Turbidity zones over the Río de la Plata region as monitored with satellites. *SCOPE/UNEP, Sonderband* **55**, 183–192.

Urien, C. (1972) Río de la Plata Estuary Environments. In *Environmental Framework of Coastal Estuaries*, ed. B.W. Nelson. The Geological Society of America, Memoir 133, pp. 213–233.

---

hydrocarbon dispersion occurs in the coastal area each time an oil spill of a certain intensity is produced during ship-loading operations. Hydrocarbon (HC) concentrations in intertidal superficial sediment from Patagonian coastal zone varied between undetected and 737.6 $\mu g/g$ of dry sediment for aromatic HC while aliphatic HC concentrations ranged from 0.06 to 1304.7 $\mu g/g$ of dry sediment (Commendatore et al., 1996). High concentrations were observed in the San Jorge Gulf where crude oil production is the most important economic activity. Furthermore, the highest aromatic and aliphatic hydrocarbon concentrations were also present in isolated spots at the north of this Gulf. This accumulation could be produced by prevalent west–southwest winds and marine currents that carry the hydrocarbons spilled in Comodoro Rivadavia and Caleta Olivia ports, and from tanker ballast washing. The levels were more important in harbour areas than in nearby zones. Lower or undetected levels of HC were found in sites with low or non-anthropogenic activity.

The highest metal levels in intertidal surface sediments from the Patagonian coastal area of Argentina, were generally confined to areas adjacent to anthropogenic activities (Gil et al., 1988; 1989). Mercury was not detected in any sample, and cadmium was only detected in Bahía San Antonio, where extremely high values of lead, zinc and copper were also measured (Commendatore et al., 1996). An old metal-rich mineral accumulation waste dump close to this bay was identified as the main pollution source.

In Tierra del Fuego Province the most important human establishments are located in Ushuaia (nearly 30,000 inhabitants) and Río Grande (38,000 inhabitants). Ushuaia was founded in 1893 and during the last ten years the population increased from 18,000 inhabitants to approximately 30,000. Two activities dominate the coastal economy: (i) factories for the industrial assembly of electronic devices and (ii) heavy shipping traffic in the local port. Petroleum extraction is the main industrial activity in Río Grande. Amin et al. (1996) demonstrate that, even if the minimum values of the total levels present in Ushuaia Bay agreed with those reported as a natural background, a slight increase in metal concentration exists associated with human activities.

Chronic oil pollution remains a problem, from oil production and probably from washing of bilge and shipping in the Strait of Magellan; these cause tarballs, leading to oil and oiled birds on beaches (Goodall et al., 1993).

RURAL FACTORS

Buenos Aires Province is an important agricultural area and so the corresponding soils are strongly treated with fertilizers and pesticides. Thus, knowledge of the impact on the coastal zone is important. There are very few details on the effect of fertilizers and pesticides in extensive agricultural areas on the Pampas coastal zone. All evidence suggests that there is increased input of sediments due to erosion during farming activities. The concentration of metals (Cu, Zn, Cd, Pb, Cr, Mn, Fe) in the Mar Chiquita lagoon sediments showed an increasing trend starting at the lagoon mouth. Nevertheless, the levels are close to the values recognized as background contents of natural systems (Marcovecchio, 1996). Similar results were obtained from analysis of Cu, Cd and Pb from suspended particulate matter sampled at several lagoon sites. The results were comparable to the concentrations measured in non contaminated rivers and seawater.

However, the lagoon is suffering one of the most striking effects of the invasion of a species in the Argentine coast. The reef-building serpulidae polychaete (*Ficopomatus enigmaticus*) is a species that forms large structures similar in shape to coral reefs (Obenat and Pezzani, 1994; Olivier et al., 1968). They live in eurihaline waters and are cosmopolitan, probably originating in Australia and introduced to Argentina before the 1950s. As a consequence, they have largely changed the water circulation, habitat structure and sediment dynamics in the lagoon. Approximately 25% of the area of the lagoon is now occupied by reefs. It is interesting to note that this lagoon is the only southwest Atlantic site where this species has developed such an important population. The construction of a bridge during the mid 1950s that reduced the width of the inlet channel from 254 m to 80 m diminished the saltwater circulation inside the lagoon, providing optimum habitat for the development of this species.

Ferrer et al. (1994) analyze the trace metal transference from the continent to the marine coastal sediments in the littoral area of Mar del Plata. Metal concentrations in soils of southeastern Buenos Aires Province were always lower than the standards internationally accepted as characteristic of non-impacted soils (Poblet et al., 1994).

Along the northern coast of the Bahía Blanca estuary, the Sauce Chico River drains an area of 1600 km^2 and the Napostá Grande Creek drains 920 km^2, both increasingly affected by intensive agriculture. The main urban settlements and activities are also located on this side of this estuary.

Existing information on pesticides in coastal Patagonia shows that concentrations of all types of pesticide analyzed during the International Mussel Watch Project (Farrington and Tripp, 1995) did not exceed 10 ng/g (0.01 μg/g) and most were present at low or non-detectable values. Pesticide concentrations in marine birds and mammals remain low or not detected (Gil et al., 1997).

COASTAL EROSION AND LANDFILL

Along all Argentine coastlines there is evidence of the Holocene transgressive–regressive cycle, leaving cheniers and estuaries. While most of the coast forms are an accretion type, there are also very common erosive forms, mainly to the south, with cliffs more than 50 m high, probably related to the regional epirogenic uplift of Patagonia (Feruglio, 1949/1950; Rutter et al., 1989). In the coastal plain, beach ridges have been deposited during the last 4000 years (Fasano et al., 1983). These forms are common in mesotidal as well as in macrotidal zones with large wave energy (Isla, 1996). The Pampa and Patagonia continental shelves are characterized by postglacial submergence and by effects of storm-dominated processes north of Mar del Plata city (Isla, 1996).

In the Buenos Aires Province, erosion is natural and caused mainly by the effects of storms coming from the south ("sudestadas"). The cliffs of Mar del Plata are receding at a rate of up to 2 m/yr. At the popular bathing resorts of Villa Gesell or Pinamar beach erosion rate is of the order of 1 m/yr. However, in the area of Mar Chiquita (at the boundary between the cliffs and barrier coast), the erosion rate exceeds 7 m/yr.

The construction of harbours (Mar del Plata, Quequen) produced an obstruction to the drift of the order of 500,000 to 1,000,000 m^3/yr to the north. Those beaches immediately to the north of these harbours, constructed at the beginning of the century, were severely affected by coastal erosion, and many disappeared. Constructed groin fields then produced a more effective obstruction to the littoral drift; more beaches disappeared to the north and were artificially recovered with different grain sizes. Beach nourishment practices have never been carried out.

Coastal landfills are not common on the Argentine coast. The shoreline of the La Plata river in the neighbourhood of Buenos Aires city has been reclaimed with landfills. Salt marshes close to San Clemente del Tuyu (Samborombón bay) and Bahia Blanca were also subjected to landfill.

On the Buenos Aires alluvial coast, there is constant dredging at harbour entrances (Buenos Aires, Mar del Plata, Bahía Blanca). The navigation channel of Bahía Blanca is nearly 100 km long and supports ship traffic to several ports: Arroyo Pareja (where two anchored buoys transfer crude oil and derivatives); Naval Base Pto. Belgrano (main naval facility of the Argentine Navy);

Ingeniero White (export of cereals, pellets and sunflower oil); Posta de Inflamables (liquefied gas, petrochemical and distillation products); Puerto Galván (cereals, fertilizers and general ships cargo).

Oil is extracted at sea close to the coast of Comodoro Rivadavia and in or near the Magellan Strait (mostly by the Chilean Republic). Further offshore, there is extraction of crude oil in the northern coast of Tierra del Fuego (San Sebastian bay).

There is no offshore extraction of other minerals. Sand for construction is usually extracted from beaches close to cities. Due to erosion problems, sand mining was forbidden at several localities on the Buenos Aires coast, though it continues in others (Isla, 1996) and along the Patagonian coast. Mining authorities estimate monthly extraction volumes of 9000 m^3 of beach sand and 23,000 m^3 of dune sand.

EFFECTS FROM URBAN AND INDUSTRIAL ACTIVITIES

Commercial Fisheries

Fish catches declared by the national fleet—ca. 800 vessels, approximately 56% coastal fleet, 19% ice vessels and 25% freezer vessels (Registro Nacional de Buques)—have grown significantly over the last thirty years, particularly during the 1990s. The 200,000 t/yr caught in the late 1960s increased to 1,339,614 t in 1997 (Anon., 1973; Redes, 1998). Over 90% of the catch is exported. Catches carried out by freezer vessels (57–60.7% of the total caught in 1994–96) exceeded those by ice vessels. Worldwide, Argentina went in six years from 33rd place (1989: 475,529 t) to 22nd (1995: 1,148,761 t) (FNI, 1991; Redes 1997a).

Industrial Fisheries

The basis of the Argentine fishery is the Argentine hake *Merluccius hubbsi*. This species lives between 80 and 800 m deep and between 55° and 35°S; exceptionally up to 23°S (Otero et al., 1982; Angelescu and Prenski, 1987; Cohen et al., 1990). Reports from the National Agriculture, Livestock and Fishery Secretariat (Anon., 1997) state that Argentine hake represented 53.9% of the total catch and 45% of exported volume in 1985–96. At least since the 1980s, calls for cautious use of this species have been made (Otero and Verazay, 1988). Recently, symptoms of acute damage to this resource were detected (FAO, 1992; Bezzi et al., 1994) because catches surpassed maximum permissible levels. During the second half of 1997, the biological closed season was significantly extended in area and time.

Until 1982, foreign fleets (mainly Polish, Soviet, Spanish, Taiwanese, Japanese and East German) caught Argentine hake, Southern hake (*Merluccius australis*), Southern blue whiting (*Micromesistius australis*) and Antarctic cod (*Notothenia rossii*) in the Malvinas/Falklands area (over 2 million tons among the four species during the 1977–81 period; Csirke, 1987). The Argentina–UK war forced the decrease of operations and the search for alternative resources (Malaret, 1986; Bisbal, 1993). As a result, came massive catches of Shortfin squid (*Illex argentinus*) and Longfin squid (*Loligo sp.*) by foreign fleets (Caddy, 1994; Rodhouse et al., 1995). On average, 507,500 t/yr of squid were caught in the Southwest Atlantic during 1982–92, with a maximum of 762,000 t in 1989 (FAO, 1992). In 1987 (764,000 t) the first evidence of over-exploitation appeared (Csirke, 1987) and in 1990, alternative steps to reduce the fishing effort were suggested (FNI, 1990). Squid captures within the Exclusive Economic Zone (EEZ) never exceeded 50,000 t until 1992, but beginning in that year they suddenly climbed as a result of renting of foreign jigging vessels. In 1993–97 1,270,263 tons of *I. argentinus* were captured in the EEZ, with a record of 389,903 t in 1997. During the same period, total catches of *I. argentinus* in Argentina–UK (EEZ + FICZ: Falkland Islands Interim Conservation and Management Zone, U.K.) reached 1,762,391 t, with 537,226 t recorded in 1997.

Other species are being exploited near their maximum established levels (Southern blue whiting; 84,000–110,000 t/yr) or are under-exploited (Hoki, *Macruronus magellanicus* and Argentine anchovy, *Engraulis anchoita*) (Anon., 1997). Kingclip catches, *Genypterus blacodes*, frequently exceed 20,000 t/yr. In 1994–96 trotline vessels caught over 45,000 tons of Patagonian toothfish, *Dissostichus eleginoides* (FICZ, FOCZ: Falklands Outer Conservation Zone (Argentina/U.K.), Antarctic waters and South of the EEZ) (Otero et al., 1982; Malaret, 1986; Redes, 1997b).

Among crustaceans the shrimp, *Pleoticus muelleri*, is the most important and is the species with the highest exportation value. A large-scale fishery started in 1982 (Boschi, 1986), concentrated in coastal waters between 43–48°S and annual captures have sharp fluctuations (range: 1083–24,397 t/yr; maximum in 1992, Anon., 1997).

Since 1995, the Patagonian scallop *Zygochlamys patagonica* supports a new fishery (depth: 60–120 m); (main fishing grounds: 39–44°C; catches 1996: 36,952 t) (Ciocco et al., 1998).

Semi-industrial and Artisanal Fisheries

In the North of the EEZ (Buenos Aires coast), the "yellow fleet" (coastal ships) catch Whitemouth croaker (*Micropogonias furnieri*), Striped weakfish (*Cynoscion striatus*) and Argentine anchovy (*Engraulis anchoita*) among others. This last resource (around 20,500 t/year) provides about 2000 jobs and a revenue of around US$ 6,500,000/yr. Among the bivalves, the mussel (*Mytilus edulis platensis*) and the yellow clam, *Mesodesma mactroides*, fisheries were important in Buenos Aires in 1950–1980.

During recent decades in the San Matias Gulf, around 12 trawling coastal ships operating from San Antonio Oeste have caught Argentine hake, Kingclip and other species.

Argentine hake supports a new artisanal fishery by boats with trotlines. The dredge shellfishery from this gulf caught over 8000 t of Tehuelche scallop, *Aequipecten tehuelchus*, caught by 9–18 stern trawlers in 1969–72. Then, this fishery collapsed due to over exploitation and rebound during the 1980s. The fishery has been closed since 1995 (Ciocco et al., 1998).

On the Chubut Province coast, around 10,000 t/year of Argentine hake are seasonally caught by 25–30 trawling coastal ships, representing 1000–3000 seasonal jobs for Rawson city and its neighbourhood. Bivalve molluscs, gastropods, octopods, some crustaceans and coastal fish support small scale fisheries of regional importance. The San José Gulf shellfishery catches 500–1000 t/year of Tehuelche scallop (*Aequipecten tehuelchus*), ribbed mussel (*Aulacomya ater*), clam (*Ameghinomya antiqua*) and mussel (*Mytilus edulis platensis*) involving 150–200 jobs and valued at 1–2.5 million US$/year. The Tehuelche scallop fishery collapsed in 1995 (Ciocco et al., 1998). The Argentine stiletto shrimp (*Artemesia longinaris*) (20–44°S) is an occasional target of the coastal trawling fleet. The Southern king crab (*Lithodes santolla*) supports artisanal fisheries in the San Jorge Gulf and the Beagle Channel (Anon., 1997).

Data from the coastal trawling ships operating between 41°S and 52°S during 1993–96 showed over 100 species caught, and individuals from 83 species of commercial importance are discarded at sea (Caille et al., 1997).

Marine Aquaculture in Argentina

Development of marine aquaculture in Argentina has been a slow and difficult process, unlike other countries of Latin America, such as Chile or Ecuador, where the development was explosive. Intensive fishing activities of molluscs in the North Patagonian San Matías and San José Gulfs caused a permanent ban on extractive methods. The collapse of these fisheries enhances the importance of aquaculture projects.

Other than certain isolated attempts at oyster culture carried out during the first half of the century, experimental aquaculture began in the middle of the 1970s and focused on mollusc aquaculture in the North Patagonian gulfs (Pascual and Zampatti, 1998).

Aquaculture has been focused on three native molluscs: the Tehuelche scallop (*Aequipecten tehuelchus*), the Puelche oyster (*Ostrea puelchana*) and the mussel (*Mytilus edulis platensis*) (Olivier et al., 1972; Penchaszadeh et al., 1974).

Scallop production through aquaculture was not encouraged for a long time because of two coastal shell fisheries, the bottom dredge fishery in San Matías Gulf and the diving fishery in San José Gulf. The experimental scallop and Puelche oyster culture are based on the use of spat collection, hatchery and growth at commercial scale using Spanish-style culture ropes and sub-tidal trays (Orensanz et al., 1991). Levels of collection have been insufficient. Hatchery, however, has been successful. Puelche oyster reaches commercial size in 28–30 months, a similar period to other oyster species. The oyster culture industry has recently begun in San Matías Gulf (Río Negro Province). The pilot production stage has been accomplished for the mussel species. Small culture developments of this species are being carried out in San José, Nuevo and San Matías Gulfs (Ciocco, 1995), and in Claromecó (Buenos Aires Province).

At present, the yellow clam (*Mesodesma mactroides*), the Japanese oyster, (*Crassostrea gigas*) introduced in Argentina in 1982, and the southern mussel (*Mytilus chilensis*) are also being investigated.

In 1997 the first official hatchery for mollusc spat production was inaugurated in the Province of Río Negro. The success of this project may extend aquaculture along the shoreline of the provinces of Buenos Aires, Río Negro and Chubut where the massive production of seed gives the producer a better chance of success. Puelche oyster seeds, mostly destined for export, are being produced, and technology for massive production of scallop and yellow clam seed is being tested.

Fish culture is in its early stages. Some species of commercial importance, such as *Paralichtys* species were identified. Experimental hatcheries have begun in the area of Mar del Plata, carried out with *Paralichthys patagonicus*. Experiences with salmonids were carried out with hatcheries of rainbow trout (*Onchorhynchus mykiss*), coho salmon (*O. kisutch*) and chinook salmon (*O. tshawystcha*) (Pascual and Orensanz, 1996).

Two crustaceans of potential culture are the Patagonian shrimp, *Pleoticus muelleri*, and Argentine stiletto shrimp, *Artemesia longinaris*. The former has a high value and supports an important fishery along the Patagonian coastline. This species has been studied for several years (Scelzo, 1987).

Environments: Characteristics and Adaptability

The Argentine coast is very exposed. The lack of protected sites, such as bays or creeks, necessitates the use of open-sea culture techniques. These characteristics aside, in most available sites, the use of structures such as rafts, sticks or superficial lines in shallow waters requires the use of long lines and heavy infrastructure. Techniques are complicated, expensive and require the use of ships and diving. The use of inter- or subtidal racks, inundated ponds or culture on bottoms and beaches are valid alternatives. Entanglement of right whales in experimental spat collecting gear in San José gulf occurred in 1995. There is, however, a high quality of coastal waters, free from industrial, agricultural and in most cases, urban pollutants.

Harmful Algae

Toxicity in molluscs of the Argentine coasts has affected coastal fisheries (Tehuelche scallop, mussel, ribbed mussel, clams), exceeding on many occasions the maximum level authorized for human consumption (80 μgSTXeq./100 g). Harmful effects of the PSP toxins on fish were determined as well. Evidence of massive mortality of mackerel (Montoya et al., 1996), and PSP accumulation in mackerel livers (Carreto et al., 1993) were found. This suggests that toxins accumulate through the food chain. In coastal Patagonia, PSP toxicity is associated with natural phenomena and oceanographic conditions. This occurs, for instance, when cold water upwelling at the beginning of the summer facilitates blooms of *Alexandrium tamarense* in the Nuevo gulf (Esteves et al., 1992).

Biological Species Introduction

Accidental biological introductions are poorly documented. The North Pacific barnacle *Balanus glandula* presence was detected in Mar del Plata intertidal zone in the late 1960s (Spivak and L'Hoste, 1976) and reported later in San José and Nuevo Gulfs (Gómez Simes, 1993). The Asian kelp *Undaria pinnatifida* was detected in Nuevo Gulf in 1992 and has shown good adaptation (Casas and Piriz, 1996). The Asian mussel *Limnoperma fortunei* is invading the Río de la Plata estuary (Pastorino et al., 1993).

Several examples of intentional biological introductions have been reported (Pascual and Orensanz, 1996). In 1809–1904 in Blanca Bay trials were carried out with the American Oyster, *Crassostrea virginica*, and the European oyster, *Ostrea edulis*. In 1981, Japanese Oyster *Crassostrea gigas* culture assays were carried out in San Blás Bay. In 1990 the Chilean oyster *Tiostrea chilensis* was introduced in the Chubut Province. Probably also intentional was the introduction of the clam *Corbicula largillierti*, an invasive species common in the Río de la Plata estuary (Ituarte, 1984). Salmonid introductions for culture trials (mainly rainbow trout, *Oncorhynchus mykiss* and coho salmon, *Oncorhynchus kisutch*) have been reported since 1990 in several places of the Patagonian coastline (Pascual and Orensanz, 1996).

Effects of Coastal Effluent Inputs

Urban and industrial pollution is a local problem in coastal Patagonia. No treatment of liquid effluents exists or else is deficient. This generates eutrophication. San Antonio bay is located in an area of important primary and secondary productivity. Its beaches are used by migratory birds, such as knots and sandpipers, to feed and rest. Evidence of eutrophication due to urban pollution was found with high concentrations of nitrogen, phosphorus and silica. Among the phytoplankton, species characteristic of eutrophic waters include *Nitzschia longissima*, *Prorocentrum micans* and species of the toxic *Pseudonitzschia* (Esteves et al., 1996). In the city of Puerto Madryn, urban and industrial effluents flow into the sea at 0.20 m³/s, characterized by high concentrations of nitrogen, phosphorus, bacteria, and low concentrations of dissolved oxygen, generating eutrophication in the north coast (Esteves et al., 1997a). The seaweed *Ulva lactuca*, is particularly efficient and has displaced other algae. Among the dinoflagelates, the toxic species *A. tamarense*, is so abundant that it appears sometimes as the only algae species. When cities are located on the marine front or in areas with important currents, efficient dilution of waters occurs, minimizing the impact of pollutants. This is the case of the Chubut river mouth in Engaño bay (Esteves et al., 1997b). The city of Puerto Deseado shows growing human impact. In some beaches located near this city, bacteria concentrations are over the limits permitted by international legislation (Esteves et al., 1998). Metals such as lead, copper and zinc were significantly elevated in the port area. Hydrocarbons in sediments show highest values around the pier, generated by shipping activity.

Solid waste is not efficiently treated. It is dumped and buried in open-air sites, sometimes only protected with wire fences. It is an inappropriate means of disposal because of the meteorological conditions of the area, characterized by frequent regular or strong west wind. This wind blows waste out to sea and affects wildlife. Sea lions are frequently observed with plastic bands on their necks, and marine birds or mammals with plastic bags in their stomachs. Garbage accumulations also affect populations of marine birds, especially the Kelp gull, and decrease biodiversity (Yorio et al., 1996). Feeding by scavengers also spreads different types of pollutants.

Ships operate mainly from the harbours of Buenos Aires (commercial, crops), La Plata (commercial), Mar del Plata (fishing), Necochea (crops, fishing), Bahia Blanca (crops, military), San Antonio (fruits and juice), Puerto Madryn (commercial, industrial, fishing), Comodoro Rivadavia (crude oil, fisheries), Caleta Olivia (crude oil), Puerto Deseado (fisheries), Río Gallegos (mining and crude oil) and Ushuaia (commercial).

Oil Spill

Offshore petroleum activity is not yet very extensive. The most important oil spill accidents on the coast were produced by the "Metula" oil tanker (August 1974) in the Magellan Strait (Chile) where more than 53,500 tons of crude oil were spilled (Hann, 1975; Schwarz, 1978); an oil "pavement" still remains (18 years later) in nearby Chilean tidal inlets (Goodall et al., 1993). The "San Jorge" oil tanker at the La Plata river (February 1997) spilled nearly 30,000 tons of crude oil, affecting numerous beaches and marine birds. In May 1972, 3000 tons of crude oil were spilled into the La Plata river, due to a collision between the Thien Chee and Royston Grange ships (Moreno, pers. communication). In 1962, a ship ("Ameghino") exploded at the La Plata harbour and in 1997 there was another ship explosion at the Bahia Blanca harbour. In September 1982, 15 km of kelp coast were covered by crude oil at Bahia Bustamante. In September 1991, 17,000 Magellanic penguins died along 750 km of the Chubut Province coast due to an oil spill; the ship responsible remains unidentified.

Chronic oil pollution is a regional problem. It is a significant mortality factor for adult Magellanic penguins along the coast of Argentina and may be reducing population numbers. The use of offshore buoys on very rough coasts (Bahia San Sebastian, Caleta Olivia and Comodoro Rivadavia) is a dangerous practice. The transport of crude oil along the Parana River (from Buenos Aires to Asuncion, Paraguay) is also dangerous.

The Argentine continental shelf has not been exhaustively explored for oil (Turic et al., 1996). Offshore exploration was initiated in 1969 in the Salado basin with negative results. The north sector of the San Jorge gulf has been extensively studied. At the moment, the southern basins and mainly the area that surrounds the Malvinas/

Table 3

Laws and agreements undertaken by Argentina

Law or Decree	Objective	Observations
Law 21,353 (1976)	International Convention to prevent pollution of waters by hydrocarbons.	OILPOL 54.
Law 21,947 (1979)	Convention on the prevention of pollution of the seas through the spillage of waste matter and other substances.	London, 1969.
Law 22,190 (1980)	Law of Navigation and on the prevention and surveillance of the pollution of waters or other components of the marine environment by agents originating from shipping and naval devices.	The enforcement authority is the Argentine Coast Guard.
Law 22.344 (1980)	Convention on the International Trade of Endangered Species of Wild Fauna and Flora.	CITES. Washington, 1981. The Enforcement Agency is the Secretariat of Natural Resources and Sustainable Development (SRNDS).
Law 23,456 (1986)	International Agreement relating to involvement on the high seas in the event of accidents that cause pollution by hydrocarbons.	
Law 23.918 (1991)	Convention on the Conservation of the Migratory Species of Wild animals.	
Law 23.919 (1991)	Convention relating to wetland areas of international importance, especially as the habitat of aquatic birds.	Ramsar, 1991. Aims at the conservation of the bird life that nests in humid areas, estuaries and wetlands of significance.
Law 24,089 (1992)	Convention for the Prevention of Pollution from Shipping.	MARPOL 73/78. Annexes I, II, IV and V are in force in the country.
Law 24,292 (1993)	Convention on the preparation, response and cooperation in the matter of pollution by hydrocarbons (O.P.R.C.).	
Law 24,375 (1994)	Agreement on the Protection of Biological Diversity.	Establish systems of protected areas in order to guarantee conservation *in situ*.
Law 24,543 (1995)	Convention of the United Nations on the Law of the Sea.	Sovereign rights over the Exclusive Economic Zone (EEZ) up to two hundred nautical miles.
Decree 89.180/4 1	Hemispheric Convention for Conservation.	Protection of the flora and fauna and scenic beauties of the countries of the Americas.
Decree 1979/78	Man and Biosphere Program (MAB) of UNESCO.	National Committee.

National Law or Decree	Objective	Action
The National Constitution. Reform of 1994	grants the basic right to a healthy environment, suitable for human development.	establishes the need to prescribe the minimum prerequisites of protection.
Law 24,051 (1991)	The Hazardous Waste Law.	A complex case of conflicts by overlapping of jurisdictions.
Law 23,879 (1990)	Environmental Impact Evaluation.	
Marit. Decree No. 11/97	New navigation routes for crude oil vessels.	Oil spill prevention. Marine biodiversity protection.
Marit. Decree No. 10/97	Special protection zones in the argentine littoral.	Toxic spill prevention. Marine biodiversity protection.
Law 24922 (1998)	Fish Resources Management. Fishing license by individual transferable quota system.	It could favour ice vessel activity and job creation.
Secretariat of Agriculture, Fisheries and Foodstuffs. Resolution No. 245/91 (1991)	Protection of the Argentine hake reproductive area from Isla Escondida (43°30'S to 44°30'S and 64°W to the coast).	This Resolution was recently amplified in area and time, forbidding all kinds of fishing.

Falkland Islands represent the most promising sites for oil discovery.

PROTECTIVE MEASURES

International Agreements

The Argentine Republic has ratified various international agreements concerning the control of marine pollution, biodiversity, etc. (Table 3).

Clear and effective administration based on coastal zone management planning and a harmonious interrelationship between Provinces and National Governments represent the major challenge for the long-term protection of the biological diversity and health of the SSASME. Overlapping spheres of authority are not only seen between the Federal and Provincial Governments, but also between administrative bodies on the same level within each State organisation. The reason why this situation exists so often in the environmental field lies in the impossibility of allocating responsibility among often many intervening government departments.

Specific cases for each Province exist, in which a particular ecosystem is protected, in some cases by laws from the nation; in other cases through provincial laws or municipal ordinances (Table 4). Environmental impact assessments, for instance, although mandatory by law, are, for the most part, mere formalities and, in many cases, ineffective in avoiding anthropogenic environmental damage governed by economic need.

In spite of broad legislation which protects the environment, there is a deficiency in the fulfilment of these laws. Control inefficiency, lack of technological facilities adapted to new environmental problems, lack of budget, and scarce or poorly trained human resources, produce this deficiency.

Table 4
Provincial and municipal ordinances

Law or Decree	Objective	Action
Law 5965	Province of Buenos Aires	Regulate the pollution of waters and atmosphere.
1996	Province of Buenos Aires	Biosphere Reserve of Mar Chiquita Lagoon, designated by the Man and the Biosphere. (MAB) Bureau of UNESCO.
Law 2669 (1993)	Province of Río Negro	Provincial system of protected areas.
Law 1503	Province of Chubut	Hazardous waste and prohibition of the entry of waste from other jurisdictions.
Law 2161 (1985)	Province of Chubut	Provincial system of protected areas. Include Península Valdés and San José Gulf Marine Park.
Law 3742 (1992)	Province of Chubut	Provincial System of conservation of the Tourist Heritage, incorporating the Reserves and protected areas.
Law 786/72	Province of Santa Cruz	Legal framework for the protection of natural areas.
Law 1451	Province of Santa Cruz	Prohibition of the pollution of public, surface or underground water and full power to impose restrictions and easements on private property. However, it is insufficient for the regulation of pollution in coastal zones.
Law 2472	Province of Santa Cruz	Restrictive regime for the entry of hazardous waste of any kind into the territory of the Province.
Law 2951	Province of Santa Cruz	Law of coastal defense, with prohibitions of spillage of waste without authorization.
Law 55 (1993)	Province of Tierra del Fuego	Protection of the environment.
1992	Province of Tierra del Fuego	Protected Area between Cabo Espíritu Santo and the mouth of Ewan river. Marine biodiversity protection. With the advice of the Western Hemisphere Shorebird Reserve Network and other groups.

REFERENCES

Amin, O., Ferrer, L. and Marcovecchio, J. (1996) Heavy metal concentrations in littoral sediments from the Beagle Channel, Tierra del Fuego, Argentina. *Environmental Monitoring Assessment* **41** (3), 219–231.

Angelescu V. and Prenski, L. (1987) Ecología trófica de la merluza común del Mar Argentino. (Merluccidae, *Merluccius hubbsi*). Parte 2: Dinámica de la alimentación analizada sobre la base de las condiciones ambientales, la estructura y la evaluación de los efectivos en su área de distribución. *Contr. Cient. Inst. Nac. Inv. y Des. Pesq. (INIDEP), Argentina* **561**, 205 p.

Anon. (1973) Producción Pesquera de la República Argentina, 1943 a 1972. Compendio de Publicaciones del Ministerio de Agricultura y Ganadería de la Nación. (Buenos Aires, Argentina).

Anon. (1997) Flota pesquera argentina, capturas marítimas totales y exportaciones del sector pesquero argentino, 1991–1996. Secretaría de Agricultura, Ganadería, Pesca y Alimentación. Buenos Aires, Argentina. 300 pp.

Bakum A. and Parrish, R.H. (1991) Comparative studies of coastal pelagic fish reproductive habitats: the anchovy (*Engraulis anchoita*) of the southwestern Atlantic. *ICES Journal of Marine Research* **48**, 343–361.

Balech E. (1971) Notas históricas y críticas de la oceanografía biológica argentina. Servicio de Hidrografía Naval, H-1027, 57 pp.

Bezzi S., G. Cañete, M. Pérez, M. Renzi, and H. Lassen, 1994: Report of the INIDEP Working Group on Assessment of Hake (*Merluccius hubbsi*). North of 48°S (southwest Atlantic Ocean). *Documento Científico Inst. Nac. Inv. y Des. Pesq. (INIDEP), Argentina* **3**, 5–28.

Bisbal, G.A. (1993) Fisheries management on the Patagonian shelf. A decade after the 1982 Falklands/Malvinas conflict. *Marine Policy* **17** (3), 213–229.

Bisbal, G.A. (1995) The southeast South American shelf large marine ecosystem. *Marine Policy* **19** (1), 21–38.

Blanco, D. and Canevari, M. (1995) 3. Situación actual de los chorlos y playeros migratorios de la zona costera Patagónica (prov. de Río Negro, Chubut y Santa Cruz). Plan de manejo Integrado de la zona costera patagónica. *Fundación Patagonia Natural* (Puerto Madryn, Argentina) **3**, 1–26.

Boraso, A.L. and Kreibohm, I. (1980) Observaciones preliminares sobre la reproducción de *Macrocystis pyrifera* en las costas argentinas. *Contrib Cient. CENPAT* **30**, 1–9.

Boschi, E.E. (1986) La pesquería del langostino del litoral patagónico. Cuadernos de Redes. *Revista de la Industria Pesquera Argentina* **5**, 20–26.

Bortolus, A., Iribarne, O. and Martinez, M. (1998) Relationship between waterfowl and the seagrass *Ruppia maritima* in a southwestern Atlantic coastal lagoon. *Estuaries* **21** (3) 710–717.

Botto, F. and Iribarne, O. (1998) The effect of the burrowing crab *Chasmagnathus granulata* on the benthic community of a SW Atlantic coastal lagoon. *Journal of Experimental Marine Biology and Ecology* **241**, 263–284.

Botto, F., Iribarne, O., Martinez, M., Delhey, K. and Carrete, M. (1998) The effect of migratory shorebirds on the benthic fauna of three SW Atlantic estuaries. *Estuaries* **21** (4) 700–709.

Caddy, J.F. (1994) Cephalopod and demersal finfish stocks: some statistical trends and biological interactions. The 3rd International Cephalopod Trade Conference, Venecia, Italia. FAO, Globefish, 28 pp.

Caille G., González, R., Gosztonyi, A. and Ciocco, N. (1997) Especies capturadas por las flotas de pesca costera en Patagonia—Programa de biólogos observadores a bordo—1993–1996. Informes Técnicos del Plan de Manejo Integrado de la Zona Costera Patagónica. *Fundación Patagonia Natural* (Puerto Madryn, Argentina) **27**, 1–21.

Campagna, C., Le Boeuf, B.J., Blackwell, S.B., Crocker, D.E. and Quintana, F. (1995) Diving behaviour and foraging location of female southern elephant seals from Patagonia. *Journal of Zoology, London* **236**, 55–71.

Campagna, C., Lewis, M. and Quintana, F. (1996) Tendencia poblacional y distribución del elefante marino del sur en la Península Valdés. Informes Técnicos del Plan de Manejo Integrado de la Zona Costera Patagónica. *Fundación Patagonia Natural* (Puerto Madryn, Argentina) **14**, 1–23.

Campagna, C., Quintana, F., Le Boeuf, B.J., Blackwell, S.B. and Crocker, D.E. (1998) Diving behaviour and foraging ecology of female southern elephant seals from Patagonia. *Aquatic Mammals* **24** (1), 1–11.

Canevari, M., Canevari, R., Carrizo, G., Harris, G., Mata, R.J. and Straneck, R.J. (1991) *Nueva guía de las aves argentinas*. Fundación Acindar. 908 pp.

CARP (Comisión Administradora del Río de la Plata) (1989) *Estudio para la evaluación de la contaminación en el Río de la Plata*. SHIN-SOHMA, 422 pp.

Carreto, J.I., Benavides, H.R., Negri, R.M. and Glorioso, P.D. (1986) Toxic red tide in the Argentine Sea. Phytoplankton distribution and survival of the toxic dinoflagellate *Gonyaulax excava* in a frontal area. *Journal of Plankton Research*, **8** (1), 15–28.

Carreto, J.I., Akselman, R., M.O. Carignan, A.D. Cucchi Colleoni and M. Pájaro, 1993: Presencia de veneno paralizante de moluscos en hígado de caballa de la región costera bonaerense. *INIDEP, Doc. Cient.*, **2**, 53–59.

Carreto, J.I., Lutz, V.A., Carignan, M.O., Cucchi Colleoni, A.D. and De Marco, S.G. (1995) Hydrography and chlorophyll *a* in a transect from the coast to the shelf-break in the Argentinian Sea. *Continental Shelf Research* **15** (2/3), 315–336.

Casas, G.N. and Piriz, M.L. (1996) Surveys of *Undaria pinnatifida* (Laminariales, Phaeophyta) in Golfo Nuevo, Argentina. *Hydrobiologia* **326/327**, 213–215.

Ciocco, N. (1995) 1. Marisquería mediante buceo en el Golfo San José; 2. Primeras experiencias privadas de cultivo de bivalvos en los golfos San José y Nuevo. Informes Técnicos del Plan de Manejo Integrado de la Zona Costera Patagónica. *Fundación Patagonia Natural* (Puerto Madryn, Argentina) **2**, 46 pp.

Ciocco, N.F., Lasta, M.L. and Bremec, C. (1998) Pesquerías de bivalvos: mejillón, vieiras (tehuelche y patagónica) y otras especies. Cap. VI. En *El Mar Argentino y sus recursos Pesqueros*, ed. Boschi, 2, pp. 142–166. INIDEP, Mar del Plata.

Cohen, D.M., Inada T., Iwamoto, T. and Scialabba, N. (1990) FAO Species Catalogue. Gadiform fishes of the world (Order Gadiformes). An annotated and illustrated catalogue of cods, hakes, grenadiers and other gadiform fishes known to date. *FAO Fisheries Synopsis* **10** (125) 442 pp.

Commendatore, M., Gil, M., Harvey, M.A., Colombo, J.C. and Esteves, J.L. (1996) Evaluación de la contaminación por hidrocarburos y metales en la zona costera patagónica. Informes Técnicos del Plan de Manejo Integrado de la Zona Costera Patagónica. *Fundación Patagonia Natural* (Puerto Madryn, Argentina) **21**, 1–47.

Crespo, E., S. Pedraza, S. Dans, N. García, M. Koen Alonso, L. Reyes and M. Coscarella, 1997: Interacciones operacionales entre mamíferos marinos y pesquerías de arrastre en el norte y centro de Patagonia. Informes Técnicos del Plan de Manejo Integrado de la Zona Costera Patagónica. *Fundación Patagonia Natural* (Puerto Madryn, Argentina) **30**, 1–28.

Csirke, J. (1987) The Patagonian Fishery Resources and the Offshore Fisheries in the South-West Atlantic. *FAO Fisheries Technical Papers* **286**, 78 p.

Dans, S., Crespo, E., Pedraza, S., González, R. and García, N. (1996) Estructura y tendencia de los apostaderos de lobos marinos de un pelo en el norte de Patagonia. Informes Técnicos del Plan de Manejo Integrado de la Zona Costera Patagónica. *Fundación Patagonia Natural* (Puerto Madryn, Argentina) **13**, 1–21.

Esteves, J.L., Santinelli, N., Sastre, V., Diaz, R. and Rivas, O. (1992) A dinoflagellate bloom and P.S.P. production associated with upwelling in Golfo Nuevo, Patagonia, Argentina. *Hydrobiologia* **242**, 115–122.

Esteves, J.L., Solís, M., Sastre, V., Santinelli, N., Gil, M., Commendatore, M. and Raies, C.G. (1996) Evaluación de la contaminación urbana de la bahía de San Antonio (Provincia del Río Negro). Informes Técnicos del Plan de Manejo Integrado de la Zona Costera Patagónica. *Fundación Patagonia Natural* (Puerto Madryn, Argentina) **20**, 1–26.

Esteves, J.L., Solís, M., Santinelli, N., Sastre, V., González Raies, C., Hoffmeyer, M. and Commendatore, M.G. (1997a) Evaluación de la contaminación urbana de la Bahía Nueva (Provincia del Chubut). Informes Técnicos del Plan de Manejo Integrado de la Zona Costera Patagónica. *Fundación Patagonia Natural* (Puerto Madryn, Argentina) **31**, 1–32.

Esteves, J.L., Solís, M., Gil, M., Santinelli, N., Sastre, V., González Raies, C., Hoffmeyer, M. and Commendatore, M.G. (1997b) Evaluación de la contaminación urbana de la Bahía Engaño (Provincia del Chubut). Informes Técnicos del Plan de Manejo Integrado de la Zona Costera Patagónica. *Fundación Patagonia Natural* (Puerto Madryn, Argentina) **35**, 1–29.

Esteves, J.L., Solís, M., Gil, M., Santinelli, N., Sastre, V., Commendatore, M., Ocariz, H. and Gonzalez Raies, C. (1998) Evaluación de la contaminación urbana de la Ría de Deseado (Provincia de Santa Cruz).Informes Técnicos del Plan de Manejo Integrado de la Zona Costera Patagónica. *Fundación Patagonia Natural* (Puerto Madryn, Argentina) **36**, 1–50.

Fasano, J.L., Isla, F.I. and Schnack, E.J. (1983) Un analisis comparativo sobre la evolución de los ambientes litorales durante el Pleistoceno tardio-Holoceno. Laguna de Mar Chiquita (Buenos Aires) - Caleta Valdes (Chubut). Simposio Oscilaciones del Nivel del Mar Durante el Ultimo Hemiciclo Deglacial en la Argentina (IGCP) *Universidad Nacional de Mar del Plata. Actas*, 27–47. Mar del Plata, Argentina.

FAO (1992) Examen de la situación de los recursos pesqueros mundiales. Parte I: recursos marinos. FAO Circular de Pesca FIRM/C710 (Rev. 8, Parte 1) (Es), 120 pp.

Farrington, J.W. and Tripp, B.W. (1995) International Mussel Watch Project. Initial Implementation Phase. Final Report. NOAA Technical Memorandum NOS ORCA 95. 63 pp and 5 Append.

Ferrari Bono, B. (1990) La potencialidad del agua. Recursos hídricos continentales de la Patagonia Argentina. *Ciencia Hoy* 2 (7), 54–67.

Ferrer, L., Poblet, A., Scagliola, M. and Marcovecchio, J. (1994) Accumulation of trace metals in coastal marine sediments: the continental transference. In *6th International Conference, Delphi, October 1994*, ed. Varnavas, pp. 1–3.

Feruglio, E. (1949–1950) *Descripción Geológica de la Patagonia*. Dirección de Yacimientos Petrolíferos Fiscales, Tomo I: 1–334, Tomo II: 1–349, Tomo III: 1–431. Buenos Aires, Argentina.

FNI (Fishing News International) (1990) Southwest Atlantic *Illex* squid: fishing effort can not be sustained. *Fishing News International* 29 (6), 14–15.

FNI (Fishing News International) (1991) World Catch. *Fishing News International* 31 (3), 30–33.

Freije, R., Zavatti, J., Gayoso, A. and Asteasuain, R. (1980) Producción primaria, pigmentos y fitoplancton del Estuario de Bahía Blanca. (1) Zona Interior: Puerto Cuatreros. *Contr. Cient. IADO* 46.

Gil, M.N, Esteves, J.L. and Harvey, M.A. (1988) Metal content in bivalve molluscs from the San José and Nuevo Gulfs, Patagonia, Argentine. *Marine Pollution Bulletin* 19 (4), 181–182.

Gil, M.N., Sastre, V., Santinelli, N. and Esteves, J.L. (1989) Metal content in seston from the San José Gulf, Patagonia, Argentina. *Bulletin of Environmental Contamination and Toxicology* 43, 337–341.

Gil, M., Harvey, M.A., Beldoménico, H., García, S., Commendatore, M., Gandini, P., Frere, E., Yorio, P., Crespo, E. and Esteves, J.L. (1997) Contaminación por metales y plaguicidas organoclorados en organismos marinos de la zona costera patagónica. Informes Técnicos del Plan de Manejo Integrado de la Zona Costera Patagónica. *Fundación Patagonia Natural* (Puerto Madryn, Argentina) 32, 1–28.

Glorioso, P.D. and Flather, R.A. (1995) A barotropic model of the currents off SE South America. *Journal of Geophysical Research* 100, 13427–13440.

Gómez Simes, E. (1993) *Balanus glandula* Darwin, 1854 (Cirripedia, Operculata) en los golfos Nuevo y San José, Chubut, Argentina. III Jornadas Nacionales de Ciencias del Mar. Puerto Madryn, Argentina. p. 94.

Goodall, N.R., Schiavini, A.C. and Benegas, L.G. (1993) The presence of mammals and birds along the northeastern coast of Tierra del Fuego. 1as. Jornadas Argentinas de preservación del recurso agua en la industria petrolera. Mendoza, 16/19 Noviembre de 1993. pp. 1–28.

Guerrero, R.A. and Piola, A.R. (1997) *Masas de agua en la Plataforma Continental, en: El Mar Argentino y sus Recursos Pesqueros, Tomo I: Antecedentes históricos de las exploraciones en el mar y las características ambientales*, ed. E.E. Boschi. INIDEP. I, pp. 107–118.

Guerrero, R.A., Acha, M.E., Framiñan, M.E. and Lasta, C. (1997) Physical oceanography of the Río de la Plata estuary. *Continental Shelf Research* 17 (7), 727–742.

Hall, M. (1980) Evaluación de los recursos de *Macrocystis pyrifera*. Costa de la Provincia del Chubut. *Contrib Cient. CENPAT* 31, 6 pp.

Hann, R.W. (1975) Follow-up field study of the Oil Pollution from the tanker "Metula". Environmental Engineering Division. Texas A&M University. 57 pp.

Hoffmann, J.A., Núñez, M.N. and Piccolo, M.C. (1997) Características climáticas del Océano Atlántico Sudoccidental. In: *El Mar Argentino y sus recursos pesqueros, Tomo I: Antecedentes históricos de las exploraciones en el mar y las características ambientales*, ed. E.E. Boschi. INIDEP I, pp. 163–193.

Iribarne, O., Bortolus, A. and Botto, F. (1997) Between-habitats differences in burrow characteristics and trophic modes in the southwestern Atlantic burrowing crab *Chasmagnathus granulata*. *Marine Ecology Progress Series* 155, 132–145.

Isla, F.I. (1996) Morphology and sedimentology of the southwest Atlantic coastal zone and continental shelf from cabo Frio (Brazil) to Peninsula Valdes (Argentina). *Explanatory text of the Atlas*, pp. 62–70, eds. L.R. Martins and I.C.S. Correa. CECO-UFRGS, Brazil.

Ituarte, C.F. (1984) Aspectos biológicos de las poblaciones de *Corbicula largillierti* Philippi (Mollusca Pelecypoda) en el Río de la Plata. *Revista Museo de La Plata* XIII, 231–247.

Krepper, C.M. and Rivas, A.L. (1979) Análisis de las características oceanográficas de la zona austral de la plataforma continental argentina y aguas adyacentes. *Acta Oceanographica Argentina* 2 (2), 55–82.

Lewis, M. (1996) El elefante marino del Sur. Biología de la especie, descripción general de la agrupación de la Península Valdés y protocolos de trabajo. Informes Técnicos del Plan de Manejo Integrado de la Zona Costera Patagónica. *Fundación Patagonia Natural* (Puerto Madryn, Argentina) 16, 1–29.

Lichter, A. (1992) Huellas en la arena, sombras en el mar. Los mamíferos marinos de la Argentina y la Antártida. Ediciones Terra Nova, Buenos Aires, 284 pp.

López Cazorla, A. (1987) Contribución al conocimiento de la ictiofauna marina en el área de Bahía Blanca. Tesis Doctoral. Universidad Nacional de La Plata. 247 pp.

Malaret, A.E. (ed.) (1986) Impacto ecológico y económico de las capturas alrededor de las Malvinas después de 1982. *Contr. Inst. Nac. Inv. y Des. Pesq.* (INIDEP), Argentina, No. 513, 115 p.

Mallo, J.C. (1984) Desarrollo larval y cultivo en laboratorio del camarón marino *Peisos petrunkevitchi* (Crustacea, Decapoda, Sergestidae). Tesis Doctoral. Universidad Nacional de La Plata. 175 pp.

Marcovecchio, J. (ed.) (1996) Pollution processes in coastal environments. *International Conference on Pollution Processes in Coastal Environments. Mar del Plata, Argentina*. 430 pp.

Martinez, M.M. (1993) Las aves y la limnología. In: *Conferencias de Limnología*, eds. A. Boltovskoy and H.L. López. Instituto de Limnologia. "Dr. R.A. Ringuelet". La Plata, pp. 127–142.

Martos, P. and Piccolo, M.C. (1988) Hydrography of the Argentine continental shelf between 38° and 42°S. *Continental Shelf Research* 8 (9), 1043–1056.

Miller, G.R. (1966) The flux of tidal energy out of the deep ocean. *Journal of Geophysical Research* 71, 2485–2489.

Montoya, N.G., Akselman, R., Franco, J. and Carreto, J.I. (1996) Paralytic shellfish toxins and mackerel (*Scomber japonicus*) mortality in the Argentine Sea. In: *Harmful and Toxic Algal Blooms*, eds. T. Yasumoto, Y. Oshima and Y. and Fukuyo. IOC (UNESCO), pp. 417–420.

Obenat, S. and Pezzani, S.E. (1994) Life cycle and population structure of the polychaete *Ficopomatus enigmaticus* (Serpulidae) in Mar Chiquita Coastal Lagoon, Argentina. *Estuaries* 17, 263–270.

Olivier, S.R., de Paternoster, I.K. and Bastida, R. (1966) Estudios biocenóticos en las costas del Chubut (Argentina). I. Zonación biocenológica de Punta Pardelas (Golf Nuevo). *Bol. Inst. Biol. Mar* 10, 75 pp. Mar del Plata, Argentina.

Olivier, S.R., Bastida, R. and Torti, M.R. (1968) Sobre el ecosistema de las aguas litorales de Mar del Plata. Niveles tróficos y cadenas alimentarias pelágico-demersales y bentónico-demersales. *Ser. Hidr. Naval. Buenos Aires*, H 1025, 1–46.

Olivier, S.R. and Penchaszadeh, P.E. (1968) Evaluación de los efectivos de almeja amarilla *Mesodesma mactroides*, Desh., en las costas de la Provincia de Buenos Aires. *Proy. Des. Pesq. FAO Ser. Inf. Tecn.* 8, 1–19.

Olivier, S.R. and Penchaszadeh, P.E. (1971) Estructura de la comunidad, dinámica de la población y biología de la almeja amarilla (*Mesodesma mactroides* Desh, 1854), en Mar Azul (Pdo. Gral. Madariaga, Buenos Aires, Argentina). Capítulo I, Ecología. *Proy. Des. Pesq. FAO Ser. Inf. Tecn.* 27, 1–35.

Olivier, S.R., Escofet, A., Penchaszadeh, P.E. and Orensanz, J.M. (1972) Estudios ecológicos en la región estuarial de Mar Chiquita (Buenos Aires, Argentina). I. Las comunidades bentónicas. Anales de la Comisión de Investigaciones Científicas, Tomo CXCIII, 237–262.

Orensanz, J.M., Pascual, M.S. and Fernandez, M.E. (1991) Scallop resources from the southwestern Atlantic (Argentina). In *Scallops: Fisheries and Aquaculture*, ed. S. Shumway, pp. 981–999. Elsevier, Amsterdam.

Otero, H.O., Bezzi, S.I., Renzi, M.A. and Verazay, G. (1982) Atlas de los recursos pesqueros demersales del Mar Argentino. *Contr. Inst.Nac.Inv.y Des.Pesq. (INIDEP), Argentina*, No. 423, 248 pp.

Otero, H.O. and Verazay, G. (1988) El estado actual del recurso merluza común (*Merluccius hubbsi*) y pautas para su manejo pesquero. *Publ. Com. Técn. Mix. Fr. Mar.* **4**, 7–24.

Ottman, F. and Urien, C.M. (1965) Les melange des eaux douces et marines dans le Río de la Plata. *Cahiers Oceanographiques* **17** (10): 703–713.

Ottman, F. and Urien, C.M. (1966) Sur quelques problémes sédimentologiques dans le Río de la Plata. *Revue de Géographie Physique et de Géologie Dynamique* **8** (3), 209–224.

Parker, G., Violante, R.A. and Paterlini, M.C. (1996) Fisiografía de la plataforma continental. In: *Geología y recursos naturales de la Plataforma Continental Argentina*, Chapters 1 (1–16). Statement of the XIII Argentine Geological Congress and the III Petroleum Exploration Congress, Buenos Aires.

Parker, G., Paterlini, M.C. and Violante, R.A. (1997) El fondo marino, en: El Mar Argentino y sus Recursos Pesqueros, Tomo I: Antecedentes históricos de las exploraciones en el mar y las características ambientales, ed. E.E. Boschi. INIDEP. I: 65-87.

Pascual, M.S. and Orensanz, J.M. (1996) Introducciones y transplantes de especies marinas en el litoral patagónico. Informes técnicos del Plan de Manejo Integrado de la Zona Costera Patagónica (Puerto Madryn, Argentina) No. 9: 1-16.

Pascual, M.S. and Zampatti, E.A. (1998) El Cultivo de Moluscos Bivalvos. En *El Mar Argentino y sus recursos Pesqueros*, ed. E.E. Boschi. INIDEP, Mar del Plata, II, pp. 167–194.

Pastorino, G., Darrigran, G., Martín, S.M. and Lunaschi, L. (1993) Limnoperma fortunei (Dunker, 1857) (Mytilidae), nuevo bivalvo invasor en aguas del Río de la Plata. *Neotrópica* **39**, 34.

Pedraza, S., Schiavini, A., Crespo, E., González, R. and Dans, S. (1996) Estimación preliminar de la abundancia de algunas especies de pequeños cetáceos del Atlántico Sudoccidental. Informes Técnicos del Plan de Manejo Integrado de la Zona Costera Patagónica. *Fundación Patagonia Natural* (Puerto Madryn, Argentina) **17**, 1–11.

Penchaszadeh, P.E. (1971) Estudios sobre el mejillón (*Mytilus platensis* d'Orb), en explotación comercial del sector bonaerense, Mar Argentino. I. Reproducción, crecimiento y estructura de la población. FAO. CARPAS 5/D Tec., 12: 1–15.

Penchaszadeh, P.E. (1973a) Ecología de la comunidad del mejillín *Brachidontes rodriguezi* en el mediolitoral rocoso de Mar del Plata (Argentina): el proceso de recolonización. *Physis, Sr. A* **84**, 51–64.

Penchaszadeh, P.E. (1973b) Comportamiento trófico de la estrella de mar *Astropecten brasiliensis*. *Ecologia. Buenos Aires* **1**, 45–54.

Penchaszadeh, P.E. (1979) Estructura de la comunidad y procesos que la determinan en bancos circalitorales de mejillón *Mytilus platensis*. UNESCO, Sem. Bentos Atlántico Suroccidental, Montevideo, Uruguay, pp. 131–148.

Penchaszadeh, P.E. (1983) Ecología larvaria y reclutamiento del mejillón del Atlántico suroccidental, *Mytilus platensis* d'Orbigny. *Cahiers Biol. Mar. París* **XXI** (2), 169–180.

Penchaszadeh, P.E., Burgos, M.A. and Scilingo, M.A. (1974) Situación actual de los conocimientos biológicos y ecológicos de la cholga (*Aulacomya ater*, Mollusca, Mytilidae). Perspectivas de su explotación y cultivo. Actas II Reunión del Grupo Intern. de Coord. para el Océano Austral, Buenos Aires, julio/1974.

Penchaszadeh, P.E. and Olivier, S.R. (1975) Ecología de una población de "berberecho", *Donax hanleyanus* Phil. en Villa Gesel, Argentina. *Malacologia, USA* **15** (1), 133–146.

Piola, A.R. and Rivas, A.L. (1997) Corrientes en la Plataforma Continental. En *El Mar Argentino y sus Recursos Pesqueros, Tomo I: Antecedentes históricos de las exploraciones en el mar y las características ambientales*, ed. E.E. Boschi. INIDEP I, pp. 119–132.

Piriz, M.L. and Casas, G. (1996) Macroalgas de interés comercial en las costas del sur de Chubut y norte de Santa Cruz. Informes Técnicos del Plan de Manejo Integrado de la Zona Costera Patagónica. *Fundación Patagonia Natural* (Puerto Madryn, Argentina) **26**, 1–36.

Poblet, A.W., Osterrieth, M. and Marcovecchio, J.E. (1994) Lead and Copper distribution in soils from southeastern Buenos Aires Province (Argentina). 6th International Conference, Delphi, October 1994, ed. S.P. Varnavas. pp. 1–3.

Podestá, G.P. (1989) Migratory pattern of Argentine Hake *Merluccius hubbsi* and oceanis processes in the southwestern Atlantic Ocean. *Fishery Bulletin, U.S.* **88** (1), 167–177.

Podestá, G.P. (1997) Utilización de datos satelitarios en investigaciones oceanográficas y pesqueras en el Océano Atlántico Sudoccidental. En *El Mar Argentino y sus Recursos Pesqueros. Tomo 1: Antecedentes históricos de las exploraciones en el mar y las características ambientales*, ed. E.E. Boschi, 222 pp.

Podestá, G.P., Brown, G.O. and Evans, R. (1991) The annual cycle of satellite-derived sea surface temperature in the southwestern Atlantic Ocean. *Journal of Climate* **4**(4), 457–467.

Prince, P.A., Wood, A.G., Barton, T. and Croxall, J.P. (1992) Satellite tracking of wandering albatrosses (*Diomedea exulans*) in the South Atlantic. *Antarctic Science* **4** (1), 31–36.

Proyecto de Desarrollo Pesquero (1968–1971) Datos y resultados preeliminares de las campañas Pesquería. Serie de Informes Técnicos, *Proy. Des. Pesq.*, publicaciones 10/I a 10/XI.

Pucci, A.E. (1991) Estado del medio ambiente marino costero de la República Argentina. *J. Invest. Contam. De las Aguas, IOC-UNESCO* 39–44.

Redes (1997a) *Revista de la Industria Pesquera Argentina* **96**, 18–19.

Redes (1997b) *Revista de la Industria Pesquera Argentina* **98**, 108–112.

Redes (1998) *Revista de la Industria Pesquera Argentina* **100**, 24–25.

Reyes, L., Crespo, E. and Szapkievich, V. (1996) Distribución y abundancia de lobos marinos de un pelo en el centro y Sur de Chubut, Argentina (1996). Informes Técnicos del Plan de Manejo Integrado de la Zona Costera Patagónica. *Fundación Patagonia Natural* (Puerto Madryn, Argentina) **10**, 1–24.

Rivas, A.L. (1994) Spatial variation of the annual cycle of temperature in the Patagonian shelf between 40 and 50° of south latitude. *Continental Shelf Research* **14** (13/14), 1539–1554.

Rivas, A.L. (1997) Current-meter observations in the Argentine Continental Shelf. *Continental Shelf Research* **17** (4), 391–406.

Rodhouse, P.G., Barton, J., Hatfield, E.M. and Symon, C. (1995) *Illex argentinus*: life cycle, population structure and fishery. *ICES Marine Science Symposia* **199**, 425–432.

Rutter, N., Schnack, E.J. and del Rio, J. (1989) Correlation and dating Quaternary littoral zones along the Patagonian coast, Argentina. *Quarterly Science Reviews* **8**, 213–274.

Scelzo, M. (1987) Posibilidades de cultivo de camarones y langostinos marinos en Argentina. Panorama de la Acuicultura en la Argentina. SECYT. Delegación Regional Patagonia (CRUB-Univ. Nac. del Comahue). *Cuadernos Universitarios* **17**, 23–34.

Schnack, E.J. (1985) *Argentine: The World's Coastline*, eds. E.C.F. Bird and M.L. Schwartz. Van Nostrand Reinhold Co., New York.

Schwarz, J.F. (1978) *El Caso Metula*. Ed. Inst. de Public. Navales. Buenos Aires. 183 pp.

Spivak, E. and L'Hoste, S.G. (1976) Presencia de cuatro especies de *Balanus* en la costa de la provincia de Buenos Aires. Distribución y aspectos ecológicos. Mimeo. 11 pp.

Spivak, E., Anger, K., Luppi, T., Bas, C. and Ismael, D. (1994) Distribution and habitat preferences of two grapsid crab species in Mar Chiquita Lagoon (Province of Buenos Aires, Argentina). *Helgolander Meeresunters* **48**, 59–78.

Turic, M.A., Nevistic, A.V. and Rebay, G. (1996) Geologia y recursos naturales de la plataforma continental. XIII Congreso Geologico Argentino y III Congreso de Exploracion de hidrocarburos. Buenos Aires, 1996. *Relatorio*, 405–423.

Urien, C.M. and Zambrano, J.J. (1996) Estructura del margen continental. In: *Geología y recursos naturales de la Plataforma Continental Argentina*, Chapter 3 (29–65). Statement of the XIII Argentine Geological Congress and the III Petroleum Exploration Congress, Buenos Aires.

Yorio, P., E. Frere, P. Gandini and M. Giaccardi (1996) Uso de basurales urbanos por gaviotas: magnitud del problema y metodologías para su evaluación. Informes Técnicos del Plan de Manejo Integrado de la Zona Costera Patagónica. *Fundación Patagonia Natural* (Puerto Madryn, Argentina) **22**, 1–22.

Yorio, P. and Harris, G. (1997) Distribución reproductiva de aves marinas y costeras coloniales en Patagonia: relevamiento aéreo Bahía Blanca, Cabo Vírgenes (Noviembre, 1990). Informes Técnicos del Plan de Manejo Integrado de la Zona Costera Patagónica. *Fundación Patagonia Natural* (Puerto Madryn, Argentina) **29**, 1–20.

THE AUTHORS

José L. Esteves
CENPAT-CONICET, Bv. Brown 3000, (9120) Puerto Madryn, Chubut, Argentina

Nestor F. Ciocco
CENPAT-CONICET, Bv. Brown 3000, (9120) Puerto Madryn, Chubut, Argentina

Juan C. Colombo
Química Ambiental y Bioquímica, Facultad de Ciencias Naturales y Museo, Universidad Nacional de La Plata, Paseo del Bosque s/n, (1900) La Plata, Argentina

Hugo Freije
Universidad Nacional del Sur, Química Ambiental, Av. Alem 1253, (8000) Bahía Blanca, Argentina

Guillermo Harris
Fundación Patagonia Natural, Marcos A. Zar 760, (9120) Puerto Madryn, Chubut, Argentina

Oscar Iribarne
Universidad Nacional de Mar del Plata, Biologia, CC 573, Correo Central, (7600) Mar del Plata, Argentina

Ignacio Isla
Universidad Nacional de Mar del Plata, Centro de Geología de Costas

Paulina Nabel
Museo Argentino de Ciencias Naturales-CONICET, Av. A. Gallardo 470, (1405) Buenos Aires, Argentina

Marcela S. Pascual
Universidad Simon Bolivar, Apartado 89000, Caracas 1080-A, Venezuela

Pablo E. Penchaszadeh
Museo Argentino de Ciencias Naturales-CONICET, Av. A. Gallardo 470, (1405) Buenos Aires, Argentina

Andrés L. Rivas
CENPAT-CONICET, Bv. Brown 3000, (9120) Puerto Madryn, Chubut, Argentina

Norma Santinelli
Universidad Nacional de la Patagonia, Belgrano 504, (9100) Trelew, Chubut, Argentina

Chapter 49

THE GULF OF GUINEA LARGE MARINE ECOSYSTEM

Nicholas J. Hardman-Mountford, Kwame A. Koranteng and
Andrew R.G. Price

The Gulf of Guinea Large Marine Ecosystem (LME) lies between the Bijagos Islands (Guinea-Bissau) and Cape Lopez (Gabon). It is generally defined as the area influenced by the flow of the Guinea Current. The coastal area is characteristically low lying and interspersed with marshes, lagoons and mangrove swamps. The region has a monsoon climate with high precipitation and almost constant monthly temperatures. Many rivers flow into the Gulf of Guinea, giving warm, low salinity coastal waters, except during the upwelling seasons in the central part of the Gulf. Mangroves are found around the major river mouths in the Gulf of Guinea, especially in the Niger Delta. Some corals are present in coastal and offshore areas, but true reefs are absent. Turtles, marine mammals and seabirds are also present. A number of fish communities are present in coastal and offshore waters.

The Gulf of Guinea is the most densely settled coastal area in Africa and is highly impacted by human activities. Mangroves, which constitute an important resource for coastal populations, are damaged by over-exploitation and pollution of water bodies from urban run-off. Forest clearance in rural areas is another major problem, causing topsoil erosion. Artisanal and industrial fisheries and aquaculture are an important source of employment and food in the region and shallow coastal waters appear fully or over exploited. Other anthropogenic activities include onshore and offshore oil production, damming of major rivers, port development and landfill. Such activities have serious effects on marine and coastal environments and can contribute to coastal erosion. A number of protected areas now exist and some environmental legislation is in place. However, enforcement is difficult, mainly due to constraints on financial, physical and human resources.

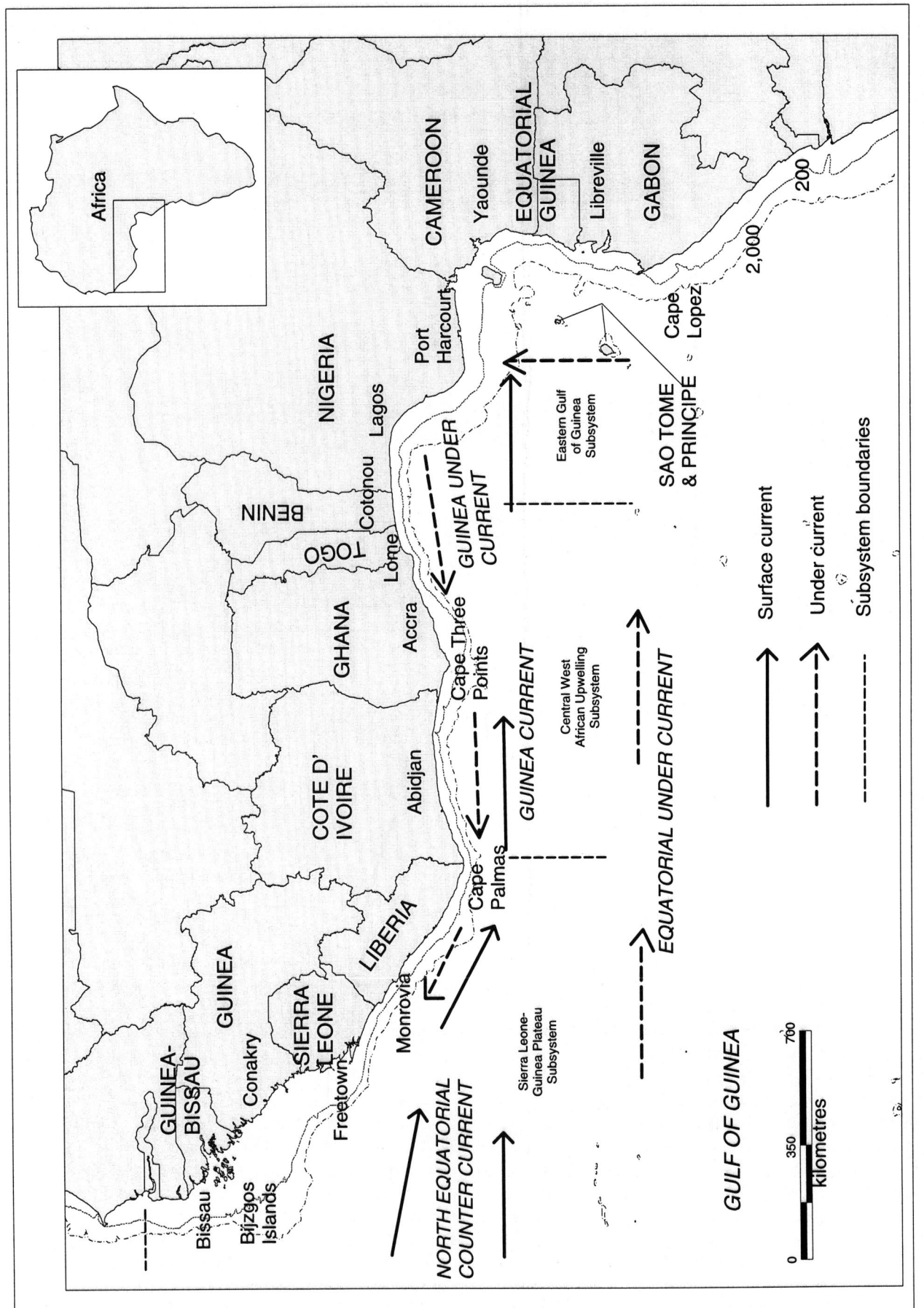

Fig. 1. Map of the Gulf of Guinea Large Marine Ecosystem showing political boundaries, place names and LME subsystems. Currents are adapted from Binet and Marchal (1993).

THE DEFINED REGION

The coastal marine environment of West Africa has been classified into three Large Marine Ecosystems (LMEs): the Canary Current LME (NW Africa), the Gulf of Guinea LME (Central West Africa) and the Benguela Current LME (SW Africa) (Binet and Marchal, 1993).

The Gulf of Guinea LME lies between the Bijagos Islands (Guinea-Bissau, ~11°N) and Cape Lopez (Gabon, ~1°S). This area includes the maritime waters of 12 coastal states (Guinea-Bissau, Guinea, Sierra-Leone, Liberia, Côte d'Ivoire, Ghana, Togo, Benin, Nigeria, Cameroon, Equatorial Guinea and Gabon) as well as the island states of Equatorial Guinea (Bioko, formerly Fernando Po) and Sao Tome and Principe. The system is generally defined by the flow of the Guinea Current so is sometimes referred to as the Guinea Current LME. It is bounded to the north by the Canary Current and to the south by the South Equatorial Current (Binet and Marchal, 1993).

Tilot and King (1993) divide the Gulf of Guinea LME into three subsystems, each defined by its particular characteristics, which nevertheless contribute to the functioning of the ecosystem as a whole and interact with each other. These subsystems are:

1. Sierra Leone and Guinea Plateau: from the Bijagos Islands (Guinea-Bissau) to Cape Palmas (Liberia/Côte d'Ivoire). This area is characterised by the largest continental shelf in West Africa and has large riverine inputs, giving thermal stability. It is also in the seasonal passage of the Inter-Tropical Convergence Zone (ITCZ).
2. Central West African Upwelling: from Cape Palmas to Cotonou (Benin). This thermally unstable subsystem is characterised by seasonal upwelling of cold, nutrient-rich, subthermocline water, which dominates its annual cycle and drives the biology of the subsystem. Variability in upwelling strength leads to variability in productivity.
3. Eastern Gulf of Guinea: from Cotonou to Cape Lopez (Gabon), including the offshore islands of Bioko and Sao Tome and Principe. This area is characterised by thermal stability and a strong picnocline. Its productivity depends on nutrient input from land drainage, river flood and turbulent diffusion through a stable pycnocline. Variability of river input depends on climatic fluctuations of the monsoon.

The coastline of West Africa came into existence when the South American landmass broke away from Africa during the early Cretaceous period. They gradually drifted apart to form the Atlantic Ocean (Allersma and Tilmans, 1993).

The coastline of the Gulf of Guinea is generally low lying and interspersed with marshes, lagoons and mangrove swamps. Lagoons range from small on the rocky coasts to the large Ebrie Lagoon, adjacent to the port of Abidjan, Côte d'Ivoire and the large Keta Lagoon on the Volta Delta in Ghana. Inlets connect the rivers, lagoons and marshes to the ocean. Behind the coast, estuaries branch into numerous winding tidal creeks through mangrove swamps. Two large delta systems punctuate the coastline. The complex Niger–Benue river system of Nigeria forms the expansive Niger Delta, the brackish-water sector of which consists of about 10,000 km^2 of estuaries, intertidal swamps, rivers and winding saline creeks (Ssentongo et al., 1986). In Ghana, the much smaller Volta River also forms a delta. The creeks of the Volta connect the estuary to the large lagoons eastward of it. The most seaward extent of the coastal zone is composed of sandy, low beach ridges and spits. A few rocky areas exist along the coastline. Steep coasts with small bays and narrow beaches occur between Cape Palmas and Fresco and from Cape Three Points almost to the mouth of the Volta (Allersma and Tilmans, 1993). Other rocky areas include the Freetown peninsula in Sierra Leone and Mount Cameroon.

The continental shelf of the Sierra Leone–Guinea Plateau is the largest in West Africa, especially off Guinea-Bissau, covering approximately 53,000 km^2 (Tilot, 1993). In the central and eastern Gulf of Guinea, the continental shelf is narrow. Shelf widths are 20–25 km along the coast from Côte d'Ivoire to Cameroon, except between Cape Three Points and the Volta Delta, where it reaches 80 km wide in parts, and around the Niger Delta, where it has a width of 50–65 km (Allersma and Tilmans, 1993).

Major geomorphic features of the continental shelf in the Gulf of Guinea include bathymetric undulations of sand ridges, canyons, gullies, dead Holocene coral banks, pockets of hard ground and rocky bottoms (Awosika and Ibe, 1998). Submarine canyons exist off the Vridi canal (Trou Sans Fond), in Côte d'Ivoire and off west Nigeria (Avons Deep). Indications of similar structures are reported off the Volta Delta, off the west coast of the Niger Delta (Mahin Canyon) and off the Calabar estuary, Nigeria (Allersma and Tilmans, 1993).

SEASONALITY, CURRENTS, NATURAL ENVIRONMENTAL VARIABLES

Seasons and Climate

Climate

The Gulf of Guinea is in an equatorial humid zone—a humid tropical climate with almost constant monthly temperatures and a relatively large amount of precipitation. Precipitation is generally much more important than evaporation (Tilot, 1993; Allersma and Tilmans, 1993).

The annual climatic cycle is dominated by the Saharan atmospheric depression that begins around April and is caused by increased solar insolation warming the continental landmass (Tilot, 1993). The low-pressure zone increases rapidly and migrates past 20°N in July. This depression plays a fundamental role in both climatology and oceanography, generating the monsoon, which is responsible for most of the region's rainfall (Tilot, 1993).

Seasons

Four seasons are seen in the Gulf of Guinea region: a short cold season from December to January, a long warm season from February to June, from July to September is a long cold season and finally in October and November there is a short warm spell (Longhurst, 1962; Koranteng, 1995).

Air Temperatures

The temperature variability in the Gulf of Guinea is small. In the Sierra Leone–Guinea Plateau area the mean annual surface air temperature is 24.5°C with annual variation of 3°C (Tilot, 1993). In the central and eastern Gulf, the average daily maximum varies between 27 and 29°C in August–September and 31–33°C in February–March. The average daily minimum varies from 21–22°C in August to 23–24°C in March (Allersma and Tilmans, 1993), however, temperatures can vary from 17 to 34°C near sea level (Ssentongo et al., 1986; Hearn, 1998).

Winds

In the central and eastern Gulf, the wind is a persistent southwesterly monsoon, modified by land and sea breezes in the coastal area. Speeds vary between 0.5 m/s (night) and 1.5–2 m/s (day) along the coast of Côte d'Ivoire and increase to between 0.5–2.5 m/s (night) and 2–6 m/s (day) in Nigeria. Storms are very rare. Weaker line squalls with heavy rain and strong winds of short duration occur occasionally (Allersma and Tilmans, 1993). The Sierra Leone–Guinea Plateau area also has monsoon winds blowing from April to October. During the rest of the year winds in this region are northeasterly maritime tradewinds.

A major feature of the climate system in the tropical Atlantic Ocean is the Inter-Tropical Convergence Zone (ITCZ), a region of atmospheric instability at the convergence of the northern and southern tradewinds. This frontal zone separates the northern dry, heavy, continental air masses from the southern humid, lighter, marine air masses. Latitudinal migration of the ITCZ generates seasons over the tropical Atlantic Ocean and African landmass (Binet and Marchal, 1993).

During the boreal winter, there are some occurrences of the hot, dry, northeasterly harmattan wind when the ITCZ deviates from its normal southerly position at 5–7°N (Allersma and Tilmans, 1993).

Rain

Although rainfall is generally high throughout the Gulf, there are considerable differences in the amount and seasonal distribution of precipitation. Rainfall greater than 1500 mm per year feeds tropical rainforests in Côte d'Ivoire and western Ghana and east from Cotonou. There are two maxima (May–June and October–November). However, in the eastern Gulf the monsoon climate leads to one long rainy period (Allersma and Tilmans, 1993). Cameroon has the highest rainfall with Mount Cameroon receiving 11,000 mm per year. This is mainly due to orographic effects of the imposing volcano and the perpendicular orientation of the coast to the main oceanic flows. In general, rainfall in Cameroon's mangrove swamps is 4000–6000 mm per year. In the south, around Kribi, rainfall is about 2400 mm per year (Appolinaire, 1993).

The central coast from Takoradi to Cotonou is known as the Accra dry belt, with only 2–3 months receiving more than 100 mm of rain (Allersma and Tilmans, 1993). It is thought that this low rainfall is due to stabilisation of the atmosphere by cold, upwelled surface waters.

Rivers

Many rivers flow into the Gulf of Guinea (Fig. 2). The Niger is the largest and has many distributaries in a typical delta carrying its water across the coast. Almost twenty estuarine

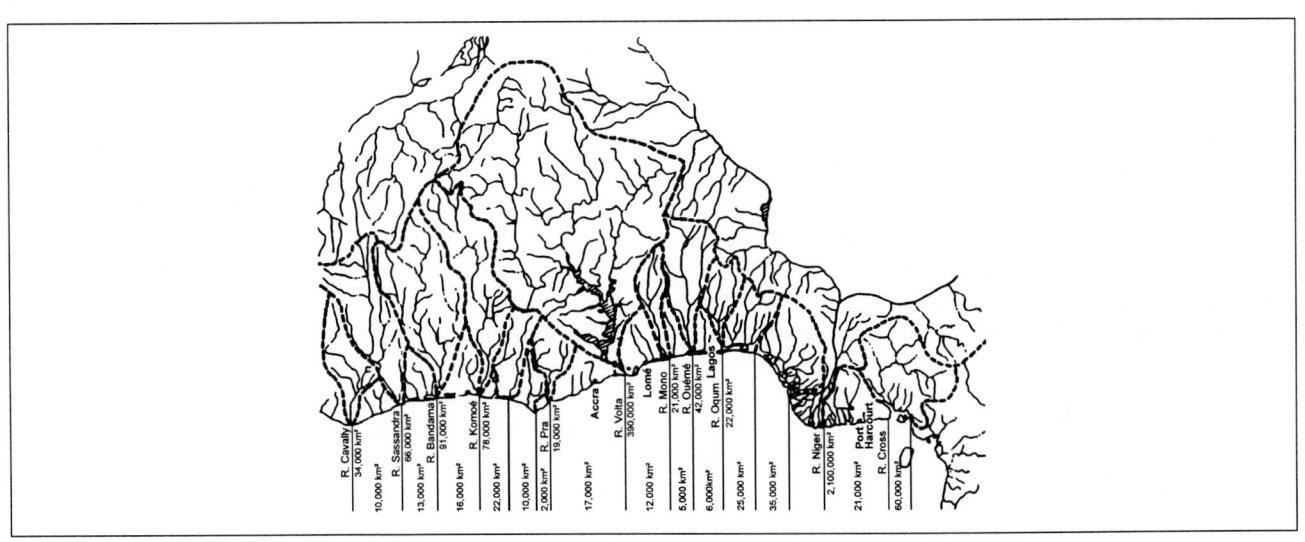

Fig. 2. Map showing the rivers and coastal catchment areas of the Gulf of Guinea from Côte d'Ivoire to Cameroon (Source: Allersma and Tilmans, 1993).

mouths interrupt the barrier beaches that separate mangrove swamps from the sea. The Volta is much smaller but still has a distinct delta, which protrudes from the coast, discharging on its western flank. Most others have no delta. In the eastern Gulf, the Cross River opens into a large estuary at the Nigerian port of Calabar. This estuary stretches across the border between Nigeria and Cameroon (Allersma and Tilmans, 1993). In Cameroon, a number of rivers pour into the Rio del Rey and Cameroon Estuary (Appolinaire, 1993). 145 billion m³ of water annually pour into the Gulf of Guinea from Cameroon rivers alone (Price et al., 1997). Many small rivers flow into the ocean between the mouths of the larger rivers (Allersma and Tilmans, 1993).

River flow is seasonal; small coastal rivers carry little water during the long dry season. Some only open to the sea during the wet season. The larger ones show a peak around August–October and low discharges during the rest of the year. Evaporation in coastal lagoons and marshes reduces outflow to the sea, which may become negative (Allersma and Tilmans, 1993).

Sedimentary Inputs

The main supply of sediments to the Gulf of Guinea comes from rivers and coastal erosion (Allersma and Tilmans, 1993). Relatively little is known about sediment discharges from rivers within the Gulf. Twelve major rivers, including the Volta, Niger and Benue, contribute over 92 million tonnes of sediment annually into the Gulf (Mahé, 1997; Folorunsho et al., 1998). These rivers have a total catchment area of 3.497×10^6 km². Transparency is low and turbidity high owing to the quantity of organic and continental matter suspended in the water. The percent sand in the total load is estimated at 10–15%, depending on conditions (Allersma and Tilmans, 1993). Table 1 shows the sediment contribution from rivers and coastal areas from Côte d'Ivoire to Nigeria.

The direction of sediment transport through inlets may be reversed during dry periods, leading to deposition of some fine sediments in lagoons, estuaries and marshes. Quantities involved are considerable and increase with the tidal prism of the system. This process helps maintain the morphological equilibrium of the coastal plain and shore (Allersma and Tilmans, 1993).

During the 1970s and 80s, river inputs decreased in the region (Tilot, 1993), coinciding with the period of the sub-Saharan drought (Lamb, 1982). This has resulted in reduced flows of almost all rivers opening into the Gulf of Guinea (Mahé, 1997).

Oceanography/Marine Hydrology

Sea Water Quality/Structure

The high precipitation and numerous rivers in the eastern Gulf of Guinea result in large masses of warm (>24°C), low salinity (<35‰) water, called Guinean waters, mixing with the Tropical Surface Waters (TSW) and circulating throughout the Gulf of Guinea. These are invariably shallow and rest on colder water, their extent varying greatly throughout the year. Guinean waters are permanent in the Bight of Biafra (UNEP/IUCN, 1988).

Table 1

Sediment supply from rivers and coastal areas in the Gulf of Guinea from Côte d'Ivoire to Nigeria. The coast has been divided into five physiographical areas, for which total amounts have been derived. The corresponding rivers and coastal catchment areas are shown in Fig. 2. The low sediment yield of the Niger Delta is a result of deposition in the inner delta as well as inclusion of large areas of desert and parts of the Cameroon Mountains in its catchment. (Source: Allersma and Tilmans, 1993)

Catchments	Catchment area (1000 km²)	Length of coast (km)	Load (1000 t/ year)	Yield (t/km²/ year)	Sand (million m³/year)
Cavally	34	–	2400	70	0.23
Coastal	10	170	500	50	0.05
Sassandra	66	–	4600	70	0.45
Coastal	13	120	700	50	0.08
Bandama	91	–	6400	70	0.6
Coastal	16	135	800	50	0.08
Komoé	78	–	4700	60	0.45
Coastal	22	120	1300	60	0.13
Coastal	10	75	700	70	0.06
Côte d'Ivoire	**340**	**620**	**22100**	**65**	**2.13**
Coastal	2	60	100	50	0.04
Pra	19	–	1300	70	0.13
Coastal	7	255	300	40	0.1
West and Central Ghana	**28**	**315**	**1700**	**61**	**0.27**
Volta	390	–	15000	40	1.0
Coastal	12	150	500	40	0.06
Mono	21	–	1300	60	0.15
Coastal	5	80	300	60	0.03
Ouémé	42	–	2100	50	0.2
Coastal	6	90	300	50	0.03
Ogun	22	–	1100	50	0.1
Coastal	25	95	0	0	0
Volta to West Nigeria	**523**	**415**	**20600**	**39**	**1.57**
Coastal	35	180	0	0	0
Niger	2100	150	40000	19	2.5
Coastal	21	220	0	0	0
Niger Delta	**2156**	**550**	**40000**	**19**	**2.5**
Cross	60	70	7500	125	0.7
Total	**3117**	**1970**	**91900**	**30**	**7.14**

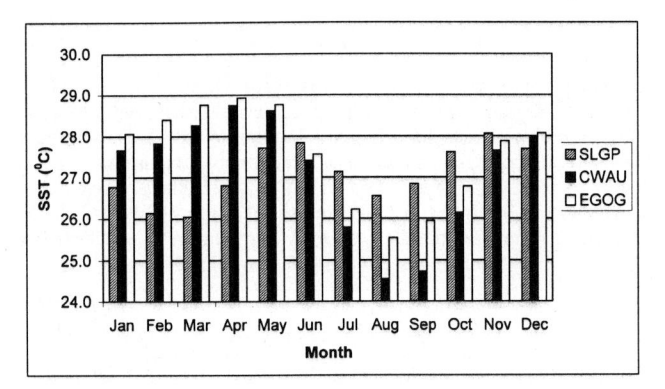

Fig. 3. Mean monthly SST for each of the Gulf of Guinea subsystems for the years 1950 to 1990. SLGP = Sierra Leone and Guinea Plateau, CWAU = Central West African Upwelling, EGOG = Eastern Gulf of Guinea (Source: COADS dataset).

The Tropical Surface Waters overlie colder South Atlantic Central Water (Longhurst, 1962). Stratification becomes enhanced during the warm seasons, especially the long warm season. The depth of the thermocline can vary seasonally from approximately 10–60 m (Longhurst, 1962; Koranteng, 1998).

In the central and eastern areas of the Gulf, sea surface temperature (SST) varies between 27 and 29°C outside of the upwelling seasons (Allersma and Tilmans, 1993) but can drop to below 22°C at the coast during the major upwelling. Figure 3 shows the mean monthly SST for each of the Gulf of Guinea subsystems taken from the COADS dataset (Woodruff et al., 1987).

Tides/Waves

Oceanic conditions, tides and waves are fairly constant along the whole coast, the only variation being the orientation of the shore with respect to wave direction. A semi-diurnal tide with an average range of about 1 m occurs almost simultaneously along most of the coast. This range is somewhat higher in the eastern Gulf. Coastal currents caused by these tides are weak (Allersma and Tilmans, 1993).

Waves build and maintain the coastal barrier, separating the lagoon from the sea. The force of river discharge, tides and incident waves, leads inlets to open into the sea (Allersma and Tilmans, 1993). The tidal wave is modified when it enters inlets, lagoons and estuaries and stronger tidal streams occur in these waters. In the tidal estuaries of Cameroon, waters can reach up to 40 km upstream (Price et al., 1997). The meeting of fresh and saline water gives rise to density currents (Allersma and Tilmans, 1993).

Littoral Transport and Marine Sedimentology

Littoral transport is the main mode of sand displacement along the coast. Fine sediments are carried in suspension while coarse material forms the beaches and adjacent seabed. Mud settles in less turbulent waters offshore, lagoons and swamps. The persistence and power of the perpetual wave action, particularly Atlantic Swell, leads to high transport rates. Along rocky parts of Côte d'Ivoire and Ghana the transport capacity of approximately 0.85 million m^3 per year far exceeds the supply of sand. The long, concave, sandy coasts between Cape Palmas and the Niger Delta carry an eastward littoral transport, the rate of which decreases in an easterly direction. Côte d'Ivoire receives sand from the rocky coast between Cape Palmas and Fresco and some rivers, while the coastline between the Volta Delta and west Nigeria receives sand from the west and central Ghana coast and Volta River. The gradual decrease in transport rate towards the eastern Gulf indicates a general accretion of the coasts. At the Niger Delta, sand moves away from the main mouth of the Niger River and along its two flanks. Interruption of the coast by a large number of tidal inlets probably reduces transport to less than 0.5 million m^3 per year, east or west. Unstable inlets shift in the direction of littoral transport. At the eastern end of Côte d'Ivoire and on both sides of the Niger Delta, areas of little transport and accumulation occur (Allersma and Tilmans, 1993).

Main losses from the littoral flow occur through offshore transport and the effects of relative sea level rise (Allersma and Tilmans, 1993). The underwater canyon of Trou Sans Fond comes within less than 1 km of the coast at Abidjan. This, combined with the sharply bending coastline, leads to seaward diversion of sediments and change in littoral transport of about 400,000 m^3 per year. This is not seen in other similar areas of the region, e.g. Avons Deep off west Nigeria (Allersma and Tilmans, 1993).

Ocean Currents

The Guinea Current dominates the oceanography of the LME. This current is an eastward, superficial flow, fed by the North Equatorial Counter Current (NECC) off the Liberian coast. The Guinea Current is quite shallow, having an average depth of 15 m near the coast and 25 m offshore (Binet and Marchal, 1993). Generally it flows barely near the coast except near promontories (Allersma and Tilmans, 1993). Although the location of the NECC changes seasonally according to the position of the ITCZ, the position of the Guinea Current remains fairly constant (Binet and Marchal, 1993). However, it is reinforced by the monsoon and can be modified by the harmattan. Underneath the Guinea Current flows the westward Guinea Under-Current. This originates in the Bight of Biafra as a return branch of the Equatorial Under-Current. The Guinea Under-Current can be observed at the surface in the Bight of Biafra (Longhurst, 1964) before it sinks under the Guinea Current as it flows westward (Binet and Marchal, 1993).

Upwelling

One of the major features of interest in the Gulf of Guinea LME is the seasonal upwelling seen in the central subsystem. There are two periods of upwelling per year. The major upwelling coincides with the long cold season (July to September) and the minor during the short cold season (December to January) (Longhurst, 1964; Koranteng, 1998). During this period SST falls, surface salinity and nutrient levels increase and dissolved oxygen levels generally fall (Houghton and Mensah, 1978; Mensah and Koranteng, 1988). Upwelling can also occur during the harmattan in the boreal winter (Allersma and Tilmans, 1993). There is further discussion of upwelling in the section on Offshore Systems.

Productivity and the Seasonal cycle

Sources of nutrient input to the Gulf of Guinea vary between subsystems. In the Sierra Leone–Guinea Plateau and eastern Gulf, productivity relies upon terrestrial runoff from land drainage and river flow during rainy periods. In the central upwelling subsystem, cold, nutrient-rich water, upwelled from below the thermocline, is the most important input (Houghton and Mensah, 1978; Binet and Marchal, 1993; Tilot and King, 1993). Primary production is initially stimulated by nutrient input from the first rains (June), then by the major upwelling and finally by the flood of larger rivers (September–October). Phytoplankton biomass is, therefore, at a maximum between June and September (Anang, 1979; Binet and Marchal, 1993) before declining due to nutrient depletion of surface layers and reduced external supply. During this period, production is solely dependent on regenerated nutrients, except during the minor upwelling season when low-intensity, short-lived upwellings take place (Binet and Marchal, 1993). The zooplankton biomass follows the same seasonal pattern but with a two week lag time (Binet, 1976). However, during the period of high productivity (June to September), zooplankton biomass appears to be correlated much more with rainfall than upwelling (Binet and Marchal, 1993). Most fish spawn at this time (Mensah and Koranteng, 1988).

MAJOR SHALLOW WATER MARINE AND COASTAL HABITATS

Mangroves

Mangroves, mainly *Rhizophora* sp., *Conocarpus* sp., *Avicennia* sp., *Mitragyna inermis* and *Laguncularia* sp., are a feature of many coastal areas in the Gulf of Guinea (Fig. 4). The largest areas are found in the Niger Delta, which is the third largest mangrove forest in the world and the largest in Africa (Ssentongo et al., 1986). Mangroves are also present around other river mouths and coastal lagoons. The total spatial mangrove cover for the region is 21,300–45,700 km² (WCMC, 1997; GEF, 1996/1997). Table 2 shows the breakdown by country.

A description of mangrove zonation in the Gulf of Guinea is provided by Saenger and Bellan (1995). However,

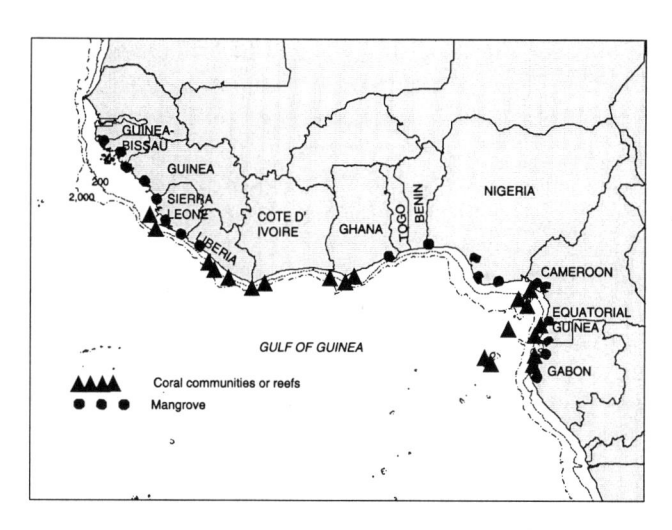

Fig. 4. Distribution of coral communities and mangroves in the Gulf of Guinea LME (Sources: mangroves WCMC 1998, corals UNEP/IUCN, 1988).

Table 2
Estimated spatial cover of mangroves in countries of the Gulf of Guinea (Source: WCMC 1997, GEF, 1996/1997).

Country	Mangrove Area (km²)
Benin	30
Cameroon	3060
Côte d'Ivoire	20
Equatorial Guinea	200
Gabon	2500
Ghana	20–100
Guinea	2230
Guinea-Bissau	2366
Liberia	200
Nigeria	9700–33280
Sao Tome and Principe	–
Sierra Leone	1000–1710
Togo	–

this is broadly generalised, as clear zonation is either absent at many sites or not well studied.

In Ghana, mangrove distribution is closely related to the presence of lagoons and estuaries along the coastline. Estimated cover is approximately 10,000 ha and occurrence is generally sparse, but good stands are present near Assiama, at Iture and in the Volta Delta, where the best stands occur (GEF, 1996/1997). Approximately 30% of Cameroon's coastline is occupied by mangroves covering some 300,000 ha (Kelleher et al., 1993), located in two main areas at the mouths of the major estuaries (Price et al., 1997).

Mangroves are associated with moderate or high biological diversity and abundance of associated species. Large mollusc populations form the basis for fish and avian food chains and are a major human food source. Crustaceans are also prevalent and mangroves form important areas for shrimp reproduction. Shallow bays, inlets and channels, which form an integral part of the mangrove system, are primary fish habitats. Euryhaline species spawn offshore but use estuarine mangrove areas as larvae, juveniles or adults. Mullets (Mugilidae), carangids, gerrids, lagocephalids, clupeids (e.g. Sardinella spp.), snappers (Lutjanidae), drums and croakers (Sciaenidae), sea catfishes (Aridae), groupers (Serranidae, *Epinephelus* sp.) and tarpons (Elopidae, *Megalops* sp.) all come into this category (Pauly, 1976; Koranteng, 1995). Use of mangroves can be for food, but is mainly linked to reproduction. Mangroves also provide good shelter for immature fish hiding from predators (Tilot, 1993). Estuarine fish, such as gobids, seabreams (Sparidae), syngnathids, eleotrids, grunts (Pomadasydae) and some cichlids (*Tylochromis jentinki*, *Tilapia* sp.), spend most of their life in brackish water but migrate to the open sea or continental waters at certain life stages. Some mangroves may still harbour crocodiles (Tilot, 1993).

Corals

Very little has been written about corals in the Eastern Atlantic. The following is taken from (UNEP/IUCN, 1988).

No true reefs exist along the West African coast but there are a number of rich coral communities. These generally form in very shallow protected coves, outside which the number and size of colonies decrease abruptly. In open waters, hermatypic species are limited to depths less than 20 m although some exceptions occur in the offshore islands of the eastern Gulf. Species found in the region include two endemic oculinid corals: *Schizoculina africana* and *S. fissipara*. These are adapted to the very low salinities of Guinean waters, and are absent outside these waters, e.g. at Annobon. Colonial shallow water dendrophyllids are among the most abundant corals in West Africa, covering vertical rocky surfaces with brilliantly coloured populations. Their taxonomy is still confused and it is unclear how many species are restricted to the area. The genus *Astrangia* is well represented. More tolerant species of *Millepora* and hermatypic corals occur in low salinity, mainland, littoral waters. *Madracis pharensis* appears to be abundant everywhere throughout the region. *Siderastrea radians* is found along the mainland coast and both *S. radians* and *S. siderea* are found in the islands. There are three species of *Porites* in the region: *P. astreoides*, *P. porites* and *P. bernhardi*. *P. bernhardi* is endemic to West Africa from Liberia to Gabon and in the islands, but is absent from Sierra Leone northwards. A faviid population has also been found in Gabon and the islands. *Monastrea cavernosa* is found only in the islands and *Acropora* is absent from the region. A noticeable amount of reef construction occurs in a few areas along the mainland coast and around the islands. Locations of coral communities in the Gulf of Guinea are shown in Fig. 4.

Marine Reptiles, Mammals and Birds

Of the seven species of marine turtles remaining in the world, five are represented throughout the Gulf of Guinea. These are: *Chelonia mydas* (green turtle), *Caretta caretta* (loggerhead), *Eretmochelys imbricata* (hawksbill), *Lepidochelys olivacea* (olive ridley) and *Dermochelys coriacea* (leatherback). The status of each of these is presented in Table 3.

Poikao Island in the Bijagos archipelago has the largest turtle breeding ground in West Africa (Tilot, 1993). The offshore islands of the eastern Gulf of Guinea are also important nesting areas for turtles. Greens, Hawksbills, Olive Ridleys and Leatherbacks lay their eggs on Bioko; Greens, Hawksbills and Leatherbacks breed on Sao Tome and Principe and Leatherback turtles, at least, nest on Annobon. Bioko is probably the most important island in terms of number of species and nesting individuals, however, nesting places are presently restricted to barely 20 km along the southern coastline (Castroviejo et al., 1994). Eggs are laid here during the dry season (November–February) (Hearn, 1998). On Principe, peak reproductive activity is in December (Hearn, 1998).

The Humpbacked dolphin *Trichechus senegalensis* and the West African manatee *Sousa teuszii* are found throughout the Gulf of Guinea. Both species appear on the IUCN Red List of Threatened Species, the West African manatee being classified as vulnerable. Data is insufficient for the Humpbacked dolphin but it is classified as highly endangered under CITES (Donoghue and Wheeler, 1994; WCMC, 1996).

The Gulf of Guinea forms part of the West African flyway, one of the major annual bird migration routes between breeding and wintering areas. Many of the migrant bird species come from Europe, as there are many important wintering sites for marine and coastal birds situated along the shores of the tropical eastern Atlantic. The Bijagos Islands are the most important breeding ground for Palearctic shorebirds in the Gulf, with a breeding population of over 1 million (Schwarz, 1992). Other parts of the Sierra Leone–Guinea Plateau region are important bird wintering areas. It is estimated that the area between Sierra Leone and Ghana holds about 700,000 waders in winter (Smit and Piersma, 1989). According to Altenburg (1987), knowledge of the number of wintering waders between Ghana and Angola is relatively poor. However, a conservative estimate is about 300,000 birds.

Table 3

Status of marine turtles in the Gulf of Guinea according to IUCN Red List Classification. (Source: WCMC, 1996).

Species	Common Name	IUCN Red List Classification
Chelonia mydas	Green turtle	Critically Endangered
Caretta caretta	Loggerhead turtle	Endangered
Eretmochelys imbricata	Hawksbill turtle	Critically Endangered
Lepidochelys olivacea	Olive Ridley turtle	Endangered
Dermochelys coriacea	Leatherback turtle	Endangered

Islands

Bijagos Archipelago

The Bijagos Islands in Guinea-Bissau constitute an archipelago of about 80 islands formed by the ancient delta of the Rio Geba and the Rio Grande de Buba. The coastal environment follows a gradual cline from brackish water and mudflat areas to rock and beach fringed offshore islands. It has very diverse tidal habitats: mudflats and mangrove forests, silicate and shell beaches and intertidal rocky areas (Schwarz, 1992). It is extremely rich in organic material and plankton and has great biological diversity. The highest species diversity is found in the Orango Islands (Schwarz, 1992). The archipelago is also highly productive because it benefits from high freshwater inputs, mainly from the Geba and Cacheu rivers, precipitation and associated coastal run-off. Offshore, strong currents and a seasonal, southerly extension of the Senegalese Upwelling support this.

Important fish species around Bijagos include mullets *Mugil cephalus*, shads *Ilisha africana*, sea catfishes *Arius heudeloti*, baracudas *Sphyraena sphyraena*, groupers esp. *Serranus aeneus*, snappers *Lutjanus ageneus*, corvinas *Olithus brachyganthus* and numerous cartilaginous fishes (*Raja, Dasyatis, Carcharinus, Sphyrna* sp.) (Tilot, 1993). Coastal sciaenids form the major part of catches in the archipelago (Domain, 1988).

The islands also form important breeding and nursery grounds for fish and crustaceans, including spawning grounds for the following species: demersals *Macroamphosus* sp., *Sparus caerusostictus*, coastal sparids *Pagellus bellotti, Dentex congoensis, D. angolensis,* and pelagics *Pomatomus saltatrix, Sardinella maderensis* and *Decapterus rhonchus* (Tilot, 1993).

As well as being the second most important West African breeding ground for Palearctic shorebirds (after the Banc d'Arguin in Mauritania), the Bijagos archipelago is home to the largest manatee population in West Africa. Additionally, the low salinity waters around the islands are one of the few places where the hippopotamus *Hippopotamus amphibius* is adapted to the sea. The freshwater turtle *Pelosius subniger* is also present in these waters (Tilot, 1993).

OFFSHORE SYSTEMS

Upwelling

The coastal upwelling observed in the region occurs twice a year. There is a long upwelling period between June and October and a short one in January. According to Bakun (1978), the Guinea Current region is similar to other eastern ocean boundary upwelling areas in the appearance of cool sea temperatures near the coast, productive coastal fisheries and a zone of low rainfall on the adjacent coast. Also the eastward flowing Guinea Current and westward flowing undercurrent give the system a structure of surface and subsurface circulation similar to other eastern ocean boundary upwelling areas (Roy, 1995). There are, however, several important differences between the Guinea Current region and other eastern boundary systems. These include the zonal rather than meridional trend of the coast, the influence of a rather narrow, intense, coastwise current and a rather unusual lack of correspondence, on a seasonal time scale, between sea-temperature features attributable to upwelling and features in the overlying wind stress field (Bakun, 1978). These differences have made understanding the system extremely complex and a mechanism for driving the upwelling has not yet been agreed upon, although a number of hypotheses have been postulated. These include wind driven Ekman upwelling (Verstraete, 1970), intensification of the Guinea Current (Ingham, 1970), dynamic interaction of the Guinea Current with Cape Palmas and Cape Three Points (Marchal and Picault, 1977) and remote forcing from the western Atlantic (Moore et al., 1978; Servain et al., 1982). It now appears that the mechanism is probably a combination of these factors (Roy, 1995).

The winter upwelling has received less attention than the summer upwelling. Intensification of the Guinea Current in January and February may contribute to the upward movement of the thermocline associated with this upwelling (Morlière, 1970). The wind may also be an important contributor to the winter upwelling (Roy, 1995).

Interannual Variability in Upwelling

An intensification of the winter upwelling has been observed with a continuous decrease of SST starting in the mid-1970s (Pezennec and Bard, 1992; Koranteng and Pezennec, 1998). An increase in SST difference between coastal and offshore areas has also been observed over the same period (Pezennec and Bard, 1992). Decomposition of the SST signal into various components showed that the long-term trend was one of SST increase, perhaps linked to global warming (Mendelssohn and Roy, 1994; Koranteng, 1998).

Along the Ghana–Côte d'Ivoire coastal system a constant intensification of the wind stress appears to have taken place over the last 30 years (Roy, 1995). An intensification of the coastal westward undercurrent is also thought to have occurred during the 1980s (Binet and Marchal, 1993), however, this is based on current observations collected during 1983 and 1984, a period of unusual conditions in the Atlantic due to one of the strongest El Niños this century occurring in 1983 (see Hisard et al., 1986). No *in situ* current data is available since.

Atlantic warm events (Atlantic Niños), thought to be related to Pacific El Niños, have been noted to occur periodically in the western Atlantic, particularly in 1984 (Philander, 1986; Hisard, 1986). These cause an incursion of warm water from the equatorial region into the Gulf of Guinea, which travels down the coast as far as the Benguela

Table 4

Synopsis of benthic communities in the Gulf of Guinea (Source: Longhurst, 1958)

Community	Habitat	Characteristic species
1. Shallow Soft Deposit Communities		
Venus community	Shelly sands	*Aloidis sulcata, Lathyrus filosus, Clavatula* spp., *Paguristes* spp., *Branchiostoma* spp., *Glycimerus* spp.
Amphioplus community	Clean, shelly and sandy muds	*Amphioplus congoensis, Ochetostoma mercator, Sipunculus titubans, Callianassa guineensis, Aloidis dautzenbergi*
— Estuarine sub-community		*Clavatula coerulea, Callianassa balssi, Alpheus pontederiae, Acrocnida semisqamata*
— Offshore sub-community		*Callianassa guineense, Sipunculus titubans, Automate evermanni, Sternaspis scutata*
Tellina Iso-community	Clean sand	*Terebra* spp., *Donax rugosa, Albunea intermedia, Cirolana* spp.
Pachymelania Community	Upper reaches of creeks, and upper estuarine regions	*Pachymelania aurita, Aloidis trigona, Iphigenia truncata, Clibinarius cooki*
Littoral Crab community		
— Ocypoda sub-community	Sandy beaches of the open coast	*Ocypoda cursor, Ocypoda africana*
— Grapsid sub-community	Marshes and mangrove flats of the estuarine region	*Sesarma* spp., *Sarmatium curvatum, Uca tangeri*
— Cardiosoma sub-community	Sandy areas of mangrove flats	*Cardiosoma armatum*
2. Shallow Hard Substrate Communities		
Inshore reef community	Reefs	*Alcyonium* sp., *Leptogorgia* sp., *Balanus amphitrite, Pteria atlantica, Echinometra lucunter*
Intertidal hard substrate community		
— Open coast rock sub-community	Rocky shores	*Nodilittorina meleagris, Littorina punctata, Chthamalus dentatus, Siphonaria pectinata, Balanus amphitrite, Palythoa monodi, Fissurela nubecula*
— Estuarine rock sub-community		*Littorina cingulifera, Ostrea tulipa, Chama gryphina*
— Mangrove sub-community		*Littorina angulifera, Balanus rhizophorae*
3. Deep Soft Deposit Communities		
Deep Shelf Community	Shelly sand and shelly mud below 80 m	*Leptometra celtica, Hyalonoecia tubicola, Ophiothrix tomentosa, Cidaris cidaris, Pennatula phosphorea, Cuspidaria cuspidata, Stichopus regalis*
Continental Slope Community	Occurs below the continental edge	*Geodia* sp., *Umbellula huxleyi, Funiculina quadrangularis, Stereomastis sculpta, Polycheles typhlops, Ophiacantha abyssicola, Aereosoma hystrix*
4. Deep Hard Substrate Communities		
Yellow Coral Community	Rocky ground	*Dendrophyllia* sp., *Ophiothrix tomentosa, Avicula hirundo*
Continental Edge Marl Community	Occurs on very soft rock at the continental edge	*Pholadidae loscombiana, Ibla* sp., *Lima excavata, Phascolosoma antillarum*

Current in the south (Hisard, 1986). The warm conditions suppress the upwelling of cold, nutrient-rich water and thereby affect seasonal productivity cycles and fish catches. It has also been noted that Atlantic Niños tend to occur the following year to ENSO events, suggesting that the two phenomena are linked (Bakun, 1996).

Global climate change could affect the productivity of upwelling areas in the Gulf of Guinea by causing an intensification of alongshore wind stress on the ocean surface leading to accelerated upwelling (Bakun, 1990). Whether this would improve fish production by providing more nutrients or reduce it by causing greater dispersion of phytoplankton is uncertain.

Continental Shelf Benthic Communities

The benthic fauna of the West African continental shelf has not been sufficiently studied. However, it has been shown that there is a faunistic barrier around 80 m deep and the benthic fauna are classified as follows (Longhurst, 1958; Buchanan, 1957):

1. shallow soft deposit communities
2. shallow hard substrate communities
3. deep soft deposit communities
4. deep hard substrate communities

These benthic communities and their characteristic species are summarised in Table 4.

Plankton Productivity

Plankton communities in the Gulf of Guinea have not been extensively studied. Investigations, conducted mainly off Ghana and Côte d'Ivoire, have shown that phytoplankton cell counts, chlorophyll *a* concentrations and primary productivity rates are high during the major upwelling period (>1000 mg C/m^2/day) and low during the non-upwelling period (<700 mg C/m^2/day) (Anang, 1979). Dinoflagellates form the main components of the phytoplankton population during the non-upwelling period and diatoms dominate at other times. The dinoflagellates consist of such genera as *Peridinium*, *Ceratium*, *Prorocentrum* and *Dinophysis* and the diatom flora include *Skeletonema*, *Nitzschia* and *Thalassiosira* (Anang, 1979). Investigations from Sierra Leone have shown that diatoms are more abundant in inshore waters, while dinoflagellates are more frequent offshore (Aleem, 1979).

The following zooplankton groups have been identified in samples collected from the region: Copepoda, Ostracoda, Cladocera, Decapoda, Larvacea, Thaliacea, Chaetognatha and larvae of bottom invertebrates. Copepods outnumber any other taxonomic group of zooplankton. During the upwelling season, up to 88% of the zooplankton organisms in the coastal area are copepods. In the open sea, the percentage is generally much smaller (20–40%). *Calanoides carinatus* is the most abundant copepod (Mensah, 1995; Greze et al., 1969). Swarms of Salps and Pyrosomids (Thaliacea) are often observed in the deep-sea parts of the Gulf of Guinea (Le Borgne, 1983).

Patterns of variability in zooplankton biomass appear to be the same interannually, though absolute values vary monthly and annually. Underlying the fluctuations is a general declining trend in abundance of zooplankton (Mensah, 1995). It is unclear whether the decline is associated with variation in species composition of the plankton. This observation has serious implications for pelagic fisheries, especially sardinellas which feed directly on zooplankton.

Fish Populations

The distribution and abundance of pelagic fish in the Gulf of Guinea depends mainly on the environment (Bard and Koranteng, 1995) whilst depth and type of substrate are important factors that influence fish community structure (Koranteng, 1998). Demersal fish and crustacean populations in the region are known to be uniform and stable.

From the results of the Guinean Trawling Survey (Williams, 1968), the fish communities identified in the subregion are summarised as follows (from Longhurst, 1969):
A. Sciaenid community
B. Eurybathic or thermocline species
C. Lutjanid community
D. Sparid community (shallow element and deep element)
E. Deep shelf community
F. Continental slope community

Recent investigations (Koranteng, 1998) have confirmed these species groupings and also shown that a number of species remain faithful to their assemblages over time. It appears that there are clear faunal discontinuities around 30 m, 100 m and 200 m deep with the first ecotone closely related to depth and thermocline, the second to drastic shelf drop, and the third to division between shelf and slope fish assemblages. It also seems that the dynamics of the assemblages are influenced by physico-chemical parameters of the water masses, mainly temperature, salinity and dissolved oxygen (Koranteng, 1998).

Fishery resources in the Gulf of Guinea are usually classified into small pelagic species, large pelagic species, coastal demersal species and deep-water demersal species (Koranteng et al., 1996). Some of the most important fish species exploited in the Gulf of Guinea are listed in Table 5.

Three commercially important penaeid shrimps occur in the Bight of Biafra. These are *Penaeus notialis* (pink shrimp), *Parapenaeopsis atlantica* (Guinea shrimp) and *Palaemon hastatus* (estuarine white shrimp). The shrimps use the sea as well as the bays, estuaries and mangrove swamps during their life cycle (Ssentongo et al., 1986, Ajayi and Anyanwu, 1997). Cephalopods are also important commercially. Species include cuttlefish *Sepia officinalis*, squid *Loligo* sp., and to some extent octopus *Octopus vulgaris* (Tilot, 1993).

Pelagic Variability

Significant changes have occurred in Gulf of Guinea fish stocks, most notably the collapse of the *Sardinella aurita* fishery in the central upwelling region in 1973, coinciding with the proliferation of triggerfish *Balistes carolinensis*. There was then a rapid decline in abundance of triggerfish in the late 1980s, coinciding with a rise in abundance of cuttlefish *Sepia officinalis* and globefish *Lagocephalus laevigatus* (Koranteng, 1998). It is not known to what extent these species shifts were due to fishing pressure or environmental forcing.

Variability of river inputs, depending upon climatic fluctuations of the monsoon, generally determines the distribution, life cycle and sizes of fish species in the thermally stable subsystems of the Gulf of Guinea. Since 1970, river inputs have decreased, as has the seasonal equatorial upwelling (Mahé, 1991). It has been observed that generally abundant and comparatively large migrating pelagic fishes are displaying sedentarism and dwarfism (Marchal, 1991) and adapting to scarcer nutrient inputs by living in deeper, colder waters and consuming less oxygen (Longhurst and Pauly, 1987). This is presumed to be an adaptive response to environmental conditions and has been seen in other species from other areas (Tilot, 1993).

Whale Migrations

Fin whales, Humpback whales and some toothed whales migrate to the waters of the Gulf of Guinea from Antarctica

Table 5

Commercial fish species in the Gulf of Guinea (Source: Koranteng et al., 1996).

Resource	Families	Species
Small (coastal) Pelagic	Clupeidae (small pelagics: sardines, bonga shad etc.)	Sardinella aurita Sardinella maderensis Ethmalosa fimbriata
	Scombridae (mackerel)	Scomber japonicus
	Engraulidae (anchovies)	Engraulis encrasicolus
	Carangidae (scads)	Decapterus rhoncus
Large Pelagic	Thunnidae (tuna)	Thunnus albacares Thunnus obesus Katsuwonus pelamis Euthynus alletteratus Istiophorus albicans Xiphias glaudius Makaira nigricans Tetrapturus albidus
Coastal demersal	Sparidae (porgies, seabreams)	Pagellus bellottii Sparus caeruleostictus Dentex canariensis
	Haemulidae (grunts)	Pomadasys incisus Pomadasys jubelini Brachydeuterus auritus
	Sciaenidae (croakers, drums)	Pseudotolitus spp. Umbrina spp.
	Lutjanidae (snappers)	Lutjanus fulgens Lutjanus agennes
	Mullidae (mullet) Serranidae (groupers)	Pseudupeneus prayensis Epinephelus spp.
	Polynemidae (threadfins)	Galeoides spp.
	Penaeidae (shrimp)	Parapenaeopsis atlantica Penaeus notialis
Deep water demersal	Sciaenidae (croakers, drums)	Penteroscion mbizi
	Ariommatidae Geryonidae (deep sea crabs)	Ariomma bondi Geryon maritae
	Penaeidae (shrimp)	Parapenaeus longirostris

(Jefferson et al., 1993; Elder and Pernetta, 1991). This is probably to breed, as breeding whales have been caught in the Bight of Biafra in the vicinity of Bioko (Irvine, 1947). Occasionally, a large whale is sighted, stranded in shallow waters in the Gulf of Guinea or washed ashore dead. Unfortunately, since such an occurrence is rather infrequent, no accurate records are kept.

POPULATIONS AFFECTING THE AREA

Nearly all major cities, harbours, airports, industries, extensive agricultural plantations and other socio-economic infrastructures in the region are located in coastal areas (GEF, 1996). In fact, the coastal zone of the Gulf of Guinea is the most densely settled in Africa (Saenger et al., 1998). Some population indicators are given in Table 6 and Table 7 shows the degree of development in each country.

Uses of the Marine Environment

The large population inhabiting the coastal zone of the Gulf of Guinea is dependent on the lagoons, estuaries, creeks and inshore waters surrounding them (Saenger et al., 1998). Rivers and lagoons serve as important waterways for the transportation of goods and people. They are also important sources of animal protein in the form of fish and shellfish. Above all, the water systems represent a source of domestic water supply for both rural and urban communities (Wandan and Zabik, 1996). Unfortunately, pollution from residential and industrial sources has affected the waters of the Gulf of Guinea, resulting in habitat degradation, loss of biological diversity and productivity, and degenerating human health (Saenger et al., 1998).

Mangroves of the Gulf of Guinea are a particularly important resource for the coastal communities. They are used for firewood, fish smoking, building materials, salt production, oysters and fisheries and medicinal purposes. However, overuse and pollution have severely damaged them (Saenger et al., 1998, GEF, 1996/1997). Urban expansion and industrial growth has led to these mangroves becoming much reduced in extent with several species that were once present being no longer found. In many instances mangrove areas have been reduced to saline grasslands of Paspalum vaginatum. Sewage treatment facilities are very limited throughout the region and raw sewage is discharged both into coastal lagoons and the rivers flowing into them. It is estimated that 186 m^3 of untreated sewage flow into Ebrie Lagoon from Abidjan each day (Saenger et al., 1998). This, combined with the limited tidal water exchange of lagoons, has led to widespread eutrophication (Saenger et al., 1998).

RURAL FACTORS

Agriculture is important to all countries in the region, both at subsistence and commercial level. A number of rural and agricultural practices impact the marine and coastal environment. The use of chemical fertilisers and pesticides has markedly increased with the development of commercial agriculture and the need to improve food production and protect human health against insect-borne diseases (Ibe et al., 1998). Although organochlorine-based pesticides are still used, awareness of their danger has spread so the majority are now organo-phosphorous and carbamate based (Portman et al., 1989). Run-off of these chemicals may reach surface or groundwater where they may persist for long periods (Wandan and Zabik, 1996). Investigations of PCBs have shown they exist at a background level but are

Table 6

Demographic information for Gulf of Guinea countries (Sources: CIA, 1997 and Scowcroft, 1995)

Country	Population size (millions)	Population growth rate	GDP (per capita)	% Urban population in coastal cities
Benin	5.90 (1997 est.)	3.31% (1997 est.)	$1440 (1996 est.)	100
Cameroon	14.67 (1997 est.)	2.86% (1997 est.)	$1230 (1996 est.)	54
Côte d'Ivoire	14.99 (1997 est.)	2.35% (1997 est.)	$1620 (1996 est.)	84
Equatorial Guinea	0.44 (1997 est.)	2.57% (1997 est.)	$800 (1995 est.)	0
Gabon	1.19 (1997 est.)	1.47% (1997 est.)	$5400 (1996 est.)	0
Ghana	18.10 (1997 est.)	2.21% (1997 est.)	$1530 (1996 est.)	72
Guinea	7.41 (1997 est.)	1.1% (1997 est.)	$950 (1996 est.)	100
Guinea-Bissau	1.18 (1997 est.)	2.33% (1997 est.)	$950 (1996 est.)	100
Liberia	2.60 (1997 est.)	6.92% (1997 est.)	$1100 (1995 est.)	100
Nigeria	107.13 (1997 est.)	3.05% (1997 est.)	$1380 (1996 est.)	20
Sao Tome & Principe	0.15 (1997 est.)	2.54% (1997 est.)	$1000 (1995 est.)	NA
Sierra Leone	4.89 (1997 est.)	3.54% (1997 est.)	$980 (1996 est.)	100
Togo	4.74 (1997 est.)	3.54% (1997 est.)	$970 (1996 est.)	100

Table 7

Major industries and sources of employment in Gulf of Guinea countries (Source: CIA, 1997)

Country	Industries	Labour force
Benin	Textiles; cigarettes; beverages & food; construction materials; petroleum.	NA
Cameroon	Petroleum and gas production & refining; food processing; light consumer goods; textiles; lumber.	NA
Côte d'Ivoire	Foodstuffs & beverages; wood products; oil extraction & refining; automobile assembly; textiles; fertilisers; construction materials; electricity.	Agriculture approx. 85%
Equatorial Guinea	Petroleum & gas production; fishing; sawmilling.	Majority subsistence agriculture, forestry and fishing.
Gabon	Food & beverage; textile; lumbering & plywood; cement; petroleum and gas extraction & refining; manganese, uranium & gold mining; chemicals; ship repair.	Agriculture 65%; industry & commerce; services.
Ghana	Mining; lumbering; light manufacturing; aluminium; food processing.	Agriculture & fishing 54.7%; industry 18.7%; sales & clerical 15.2%; professional 3.7%; services, transportation & communications 7.7%
Guinea	Bauxite, gold, diamonds; aluminium refining; light manufacturing and agricultural processing industries.	Agriculture 80.0%; industry & commerce 11.0%; services 5.4%; civil service 3.6%
Guinea-Bissau	Agricultural products processing, beer, soft drinks.	Mainly farming and fishing.
Liberia	Rubber processing; food processing; construction materials; furniture; palm oil processing; iron ore, diamonds.	Agriculture 70.5%; services 10.8% industry & commerce 4.5%; other 14.2%
Nigeria	Crude oil and gas extraction and refining; coal, tin, columbite; palm oil, peanuts, cotton, rubber, wood, hides & skins; textiles; cement & other construction materials; food products; footwear; chemicals, fertilisers; printing; ceramics; steel.	Agriculture 54%; industry, commerce & services 19%; government 15%
Sao Tome & Principe	Light construction; textiles; soap; beer; fish processing; timber.	Majority subsistence agriculture and fishing
Sierra Leone	Mining (diamonds, bauxite, rutile); small-scale manufacturing (beverages, textiles, cigarettes, footwear); petroleum refining.	Agriculture 65%; industry 19%; services 16%
Togo	Phosphate mining; agricultural processing; cement; handicrafts; textiles; beverages.	Agriculture 64%; industry 9%; services 21%; unemployed 6%

not a problem yet (IOC, 1985). Pesticide concentrations in marine fish are low but much higher levels are found in freshwater fish. Organo-phosphorous pesticides generally break down quickly in the aquatic environment around where they are used, so are not expected to reach the marine environment (Portman et al., 1989). However, recent work in Ghana has revealed the rising importance of pesticides in coastal waters (Joiris et al., 1997).

Inorganic, especially nitrate and phosphate based, fertilisers are being used on an increasing scale. Substantial quantities of nutrients originating from domestic and agricultural effluents, which are used in primary production, are carried to the sea through river outflows. As yet these fertilisers do not appear to be causing eutrophication in marine waters but, coupled with sewage pollution, they could be a serious threat to lagoons (Portman et al., 1989).

Slash-and-burn is a traditional farming practice in several Gulf of Guinea countries. Large areas of forest are cleared and the bush is burnt, rendering the soil amenable to erosion. After torrential rainfall, topsoils are washed away. The main cause of soil erosion, however, appears to be deforestation in the interests of exploiting timber resources (Portman et al., 1989). In Ghana, there is general over-exploitation of mangroves for fuel wood. The high-density rural population along the coast relies on this fuel wood as a source of energy. Exploitation at the Volta Delta is very alarming. Another serious threat is the clearing of mangrove areas for salt mining. It is not possible for mangroves to regenerate afterwards, since the practice makes soil dry and compact. In some areas, such as the Songor Lagoon and Densu Delta (west of Accra), this practice has reduced mangrove vegetation to a few scattered clumps. Pressure on the land for residential development, with the increasing population in the coastal zone, has led to reclamation of some mangrove areas (GEF, 1996/1997).

From a biodiversity perspective, the mangrove palm *Nypa fruticans* is a significant problem. It has recently become distributed throughout the coastal area of the Gulf of Guinea, invading and replacing native mangrove species. Although *N. fruticans* is known from the fossil record from throughout the Niger delta, the current populations were introduced to Nigeria early this century from Singapore. Since then, it has spread throughout the Niger, Imo, Bonny and Cross Rivers. While the spread has been slow, it appears to be accelerating, facilitated by local villagers who value its thatching properties. Most recently, it has been reported to be growing in and around the mouth of the Volta River in Ghana (Saenger et al., 1998).

Coconuts (*Cocos* spp.) are important and the most conspicuous crops along the coast of several Gulf of Guinea countries. In Ghana, they constitute an industry worth US$8.8 million (Overfield et al., 1997). In the last two decades, the lethal yellowing disease called the Cape St. Paul Wilt has seriously threatened this important industry. Infected coconut trees die leaving the shoreline bare and prone to coastal erosion as the binding effect of the crops' roots in the sandy beach is removed.

Aquaculture

In the Gulf of Guinea countries, fish is a relatively cheap source of animal protein. In sub-Saharan Africa, aquaculture dates back only to the late forties and early fifties when colonial administrations sought alternatives for the diversification of rural economies and improvement of the animal protein component in the diet of the rural inhabitants (Delince and Obiekezie, 1996). In recent years, decline in marine fish landings has led to the intensification of aquaculture in the region. The principal species cultured are tilapias (*Tilapia* sp., *Sarotherodon* sp.) and clarias (*Clarias, Heterobranchus* and *Chrysichthys* spp.). Various forms of fish farming practices are found in the Gulf of Guinea area; these range from small subsistence ponds in rural areas, to large commercial farms. Cage culture in lagoons is practised in Côte d'Ivoire and in Benin traditional brackish water brush park fish farming or "acadjas" is still practised. This method involves dumping twigs at selected areas in the waterbody, which serve as fish aggregating devices. Fish that aggregate in the area are harvested periodically. The "acadja" method has generated serious environmental concerns including deforestation and siltation of waterbodies. In addition, eutrophication in the Weija reservoir in Ghana, that supplies water to parts of Accra, has been blamed on this aquacultural practice (M. Entsua-Mensah, pers. comm.). Mariculture is not a common practice in the region, although oyster farms exist in Sierra Leone and Nigeria.

COASTAL EROSION AND LANDFILL

Rocky coasts in Côte d'Ivoire and Ghana both receive small supplies of sediments, less than the littoral transport capacity and are, therefore, eroding. Along sandy coasts sediment supply predominates and much more sediment has accumulated, however, accretion lifts the shore to increasingly deeper water. In combination with sea level rise and sediment losses, this will ultimately lead to equilibrium or even recession of the shore (Allersma and Tilmans, 1993).

Although some coastal erosion is natural, it is not clear how much has been contributed to it through anthropogenic activities like the damming of rivers (for hydroelectric or irrigation purposes) and riverbed mining for sand, reducing the amount of sediment reaching the sea. Some instances suggest these activities have a very severe impact, such as the partial disappearance of the town of Keta following the construction of the Akosombo dam on the Volta River, Ghana and coastal erosion in nearby Togo and Benin (Portman et al., 1989). Abban (1986), however, points out that the Volta River could never have replaced the amounts presently being eroded. In some areas it appears that reduced freshwater flow has caused the extension of sand spits, as seen across the mouth of the Volta, and construction of the Cotonou dam in Benin has resulted in less erosion of the harbour area and reduced sedimentation on the coast (Portman et al., 1989). Similarly, removal of sand and gravel for construction and land reclamation have led to erosion near Freetown, Sierra Leone, but similar activities near Lomé, Togo, have had no impact because the sand was dredged from a deposition zone (Portman et al., 1989).

The consequences of harbour development, stabilisation and regulation of inlets and other coastal modifications, which have interrupted coastal sediment transport, are much clearer: sand has accumulated on the western side of structures and there is erosion on their eastern sides. The severe coastal erosion, which reduced the width of Victoria beach, Lagos, by 2 km, was a result of the construction of a breakwater (Elder and Pernetta, 1991). Sand is regularly supplied to Victoria Beach (3 million m³ every four years) to help counteract the erosion. Similar erosion of 0.5 km occurred at Escaros, Nigeria after breakwaters were completed in 1964 and at Abidjan after the Vridi canal was opened in 1950. This erosion eventually cut through a road (Portman et al., 1989). These processes continue to influence the coast in both directions (Allersma and Tilmans, 1993). Erosion of barrier beaches will eventually lead to the loss of the highly productive lagoons they shield (Elder and Pernetta, 1991).

Sea level rise also leads to considerable erosion. This may explain the currently increasing erosion of beaches. Anthropogenic sea level rise may upset the entire balance of sediments for the whole coast (Allersma and Tilmans, 1993).

Landfill is also becoming a problem throughout the Gulf of Guinea, especially reclamation of coastal marshland areas. In Nigeria alone thousands of square kilometres have been lost to this practice. For example, in 1984 extensive dredging at Lagos and deposition of the spoil in adjoining mangrove swamps led to high suspended solids in most embayments and severe damage to the oyster fishery (Portman et al., 1989). Infilling for industrial use, combined with other industrial practices, has destroyed breeding and nursery grounds of commercial fishes (Ezenwa and Ayinla, 1994).

EFFECTS FROM URBAN AND INDUSTRIAL ACTIVITIES

Artisanal and Non-industrial Uses of the Coast

Artisanal fisheries are the most important West African fishing sector, accounting for over 70% of the total marine catch (Tilot and King, 1993). The most developed artisanal fisheries are close to the upwelling areas on the coasts of Côte d'Ivoire, Ghana and Togo. Important marine fisheries also occur in non-upwelling areas, such as the continental shelf areas off Sierra Leone and the Bijagos archipelago (Chaveau, 1991). Coastal brackish-water fisheries in creeks, estuaries and intertidal mangrove swamps form another important resource.

In artisanal fisheries, several gears are used, ranging from cast nets to purse seine-like encircling gears. Wicker cages and traps are also used in lagoon areas. Fishing is mainly from canoes. Some of the groups fishing here have adopted adaptive fishing strategies to cope with the unstable nature of the resources (Tilot and King, 1993).

Throughout West Africa, artisanal fisheries play a major role in the provision of fish for domestic consumption. For most countries in the region, artisanal fish catch constitutes over 60% of the total domestic catch. Small pelagic species are most important in terms of quantities landed and are mainly for local consumption. Large pelagic species are mainly exported out of the region and demersal fish species are the most valuable.

Artisanal fisheries generally employ more people than industrialised fisheries because the latter use labour-saving technology. Artisanal fishing tends to be labour intensive, using traditional technologies that are often well suited to the task and can be very productive (Tilot and King, 1993). Occasional (or part-time) fishermen make up a large proportion of labour in artisanal fisheries (Tilot and King, 1993). Thus, fishing is also an important secondary source of income for many people.

Illegal fishing using dynamite and pesticides such as DDT is still widely practised throughout the region. Undoubtedly this has adverse effects on aquatic fauna and mangroves (Saenger and Bellan, 1995). In Cameroon, hunting and trapping in shark nets threatens sirenians and cetaceans, despite some local protection (Price et al., 1997). Other artisanal activities include the hunting of turtles on beaches, e.g. the southern beaches of Bioko (Hearn, 1998). Turtle shells are sold at or near areas of coastal tourism in Cameroon (Price et al., 1997).

Industrial Fishing

Industrial fishing is relatively recent in the Gulf of Guinea and accounts for 30% of catches by West African fleets (Tilot and King, 1993). They mainly catch high value demersal species. Shrimp fishing vessels tend to operate in the coastal sector. Industrial vessels are predominantly nationally owned or part of joint ventures so there are few private West African fishing fleets. Intra-regional agreements exist between some countries, permitting fishing in neighbouring countries' waters (Tilot and King, 1993).

Foreign fleets have also been present on a large scale since the 1960s owing to high productivity in the region. These are primarily from Europe, Russia and ex-eastern bloc countries, the Korean Republic and Japan. EC countries represent the largest fleet in the region and have fishing agreements with many of the nations (Tilot and King, 1993).

Tuna resources are exploited by baitboats and purse seiners. These vessels are mainly of Ghanaian, Ivorian, Korean, Japanese, Spanish and French nationalities and fish throughout almost the entire LME, from close inshore to as far as the Equator to the south (Koranteng et al., 1996).

This level of industrial fishing makes the activity an important source of employment throughout West Africa. It is also an inexpensive and vital protein source in a region for which food security is a problem. Some of the more densely populated Gulf of Guinea countries import fish for domestic consumption from ex-Soviet and East European fleets, as well as from less densely populated West African countries, such as Morocco, Mauritania and Senegal (Tilot and King, 1993).

Industrial fishing has severe environmental impacts including substrate modification. This may have the greatest impact where inshore fisheries, e.g. *Penaeus* sp., operate in important nursery grounds for other species, such as near mangroves, lagoons and estuaries. Trawling can cause resuspension of bottom sediments resulting in physical smothering of benthic organisms. Also, pollutants, such as anti-fouling paints and hydrocarbons from fuel oil, are introduced into the marine environment by fishing vessels. Impacts can be seen on non-target species from by-catch and the incidental mortality of many species, including marine mammals, turtles and seabirds, from lost or discarded fishing gear (Tilot and King, 1993). In general, offshore demersal stocks seem to be under-exploited in the Gulf of Guinea whereas those in shallow coastal waters appear either fully exploited or over-exploited (Koranteng et al., 1996). As traditional fishery resources decline, industrial vessels tend to exploit the same grounds as artisanal fishers.

Onshore Oil Production

In the Niger Delta, 23 out of 62 oil fields are within mangrove areas. Oil terminals are spread throughout the delta while numerous seismic lines and oil pipelines criss-cross the mangroves (Shell, 1998; Saenger et al., 1998). Oil spills are common; between 1970 and 1982 alone, there were 1581 oil spills involving a total of 2 million barrels (Saenger et al., 1998). Since 1989, Shell Petroleum Development Company (SPDC) has recorded an average of 221 spills per year in their operational area alone. These have involved a total of 7350 barrels of oil a year, although the majority of spills involve less than eight barrels (1 tonne) (Shell, 1998). While most spills are small, they tend to occur within mangrove waterways. As a result, surface waters become contaminated and undrinkable, localized fisheries production declines and, in many instances, local inhabitants are forced to emigrate to other areas (Saenger et al., 1998). One of the major issues of a conflict between SPDC and the Ogoni people of the Niger Delta has been environmental degradation (PIRC, 1996). Despite this, the mangrove forest is the least disturbed of the Niger Delta's ecological zones: only 5–10% of it has been lost to urban growth and industrial development, including oil company activities (Shell, 1998).

Main threats to the survival of both endemic and migrant birds in the Gulf of Guinea include oil spills. In the islands and coastal zone of the Bight of Biafra, pollution from activities connected with the oil industry could lead to additional risk facing avian populations.

Toxic Waste Trade

International trade in toxic wastes is vast—an estimated 3.7 million tonnes were exported to developing countries between 1986 and 1988. In 1988, investigations into a spate of poisoning symptoms among residents of Koko in Nigeria revealed that the problem was caused by toxic chemicals, dumped by an Italian waste-disposal company, leaking from a nearby site. The Italian government was forced to remove the toxic material (Elder and Pernetta, 1991).

Cities

Urban Waste

With a few exceptions, all pollution in the coastal area of the Gulf of Guinea comes from land-based sources. Industrial pollution has not been well studied in the Gulf of Guinea area.

Abidjan is built on the side of Ebrie Lagoon. Pollution problems there are caused by improper pouring out of residuary waters into the natural environment of the lagoon, with no preliminary processing. The degree of flushing in the lagoon has helped minimise the impact (Portman et al., 1989). However, the constant supply of biodegradable compounds has led to considerable eutrophication, especially in areas with very low rates of water renewal, such as bays (GEF, 1996/1997). At least four embayments regularly become anoxic due to the pollution load (Dufour and Slépoukha, 1975).

Urban wastes are not just a problem in Côte d'Ivoire, but along the entire coast. For example, at Nouake Lagoon in Cotonou, they pile high on the shore spilling into the lagoon itself (GEF, 1996). Abattoir wastes, industrial wastes, solid rubbish, medical wastes (including syringes), organic wastes, industrial chemical wastes (including dyes and alkalis from the textile industry), and residual oils are all thrown into this lagoon or directly into the Bight of Benin (IOC, 1994).

In Ghana, most research has involved only site-specific one-time water quality sampling. Available studies show serious pollution of surface waters especially off large townships where industrial effluents, discharged into nearby drains, end up in waterbodies (Biney, 1990). Studies have shown that domestic and industrial pollutants create biological oxygen demand and concentrate toxic chemicals in coastal waters. Lagoons tend to be most affected because of their limited water exchange (IOC, 1994). An associated problem is siltation from factory effluent. This has considerably reduced the depth of some affected lagoons to the extent that fishing activities have been halted. It has also reduced the effect of the tidal action of the sea, leading to the lagoons becoming bodies of foul water. In Ghana, out of 16 lagoons studied, 12 were polluted, two grossly so (Korle and Chemu) (Portman et al., 1989). Korle lagoon is so polluted it is no longer biologically productive from a fishery viewpoint. Although other lagoons are currently regarded as only slightly polluted, the expected large increases in population size, urbanisation and industrial development could significantly increase the risk to them (IOC, 1994). According to IOC (1994), if there is no increase in pollution control by 2010, or in many cases much earlier, most coastal areas will have become biologically inactive with sterile, unusable areas around them.

In most towns and villages in the coastal region, sewage is discharged untreated to rivers, streams, lagoons and coastal waters via short outfalls. In Lagos, Nigeria, untreated sewage is collected from depots around the city and dumped into Iddo lagoon. Construction of sewage treatment facilities and longer discharge pipes is now underway in Abidjan, Accra, Lagos, Douala and Libreville (Portman et al., 1989).

A major health risk, due to sewage pollution of coastal waterways, is the presence of pathogenic bacteria. In the harbour at Lagos, pathogens included faecal bacteria, *Salmonella typhi* and *E. coli*. These were found in water, sediments and edible shellfish. Faecal coliforms have also been found in stretches of the Volta Estuary (Portman et al., 1989). Typhoid, cholera, hepatitis, polio and dysentery are all endemic in the region and organisms which cause these diseases must occur in waters receiving sewage inputs (Portman et al., 1989). An additional problem is that coliforms and *Pseudomonas* sp. from bodies of water around Port Harcourt, Nigeria, have been seen to show antibiotic resistance (Sokari et al., 1988).

Cameroon, like other countries in the Gulf of Guinea, is becoming increasingly contaminated by industrial and domestic waste. Pollution off the Cameroon coast is particularly of concern because it is the convergence point for major ocean currents. Thus, not only is the dispersal of pollutants off the coast slow, but contaminants from other areas, transported by ocean currents, will accumulate along the coastline (Price et al., 1997).

Port Development

Access from the ocean has always been difficult along the West African coast because of the high waves and lack of large river mouths (Allersma and Tilmans, 1993). Some protection could be found east of large promontories along rocky parts of the coast so settlements sprung up and eventually the first modern harbour (Takoradi) was built. Some deep inlets and estuaries were good for navigation, but increasing traffic and the size of vessels required artificial improvement (Allersma and Tilmans, 1993).

Presently, large artificial harbours with breakwaters have been constructed in the region, e.g. Takoradi, Sekondi and Tema (Ghana), Cotonou (Benin), Lomé (Togo), Lagos (Nigeria). Main engineering works connected with port development in the region include:

- Regulation of inlets with jetties and dredging (e.g. Vridi canal, Côte d'Ivoire and Lagos Harbour, Nigeria).
- Construction of large harbours with breakwaters extending into the sea (e.g. Takoradi, Sekondi, Tema, Ghana; Lomé, Togo; Cotonou, Benin).
- Construction of sea defences, such as revetments, sea walls and groynes, at various locations.
- Extraction of sand for construction at various locations and replenishment (e.g. Victoria Beach, Lagos, Nigeria, see erosion section).
- Dredging in some estuaries (e.g. in the Niger Delta).

Some of these developments have led to sand accumulating on their western side and erosion on their eastern side (Allersma and Tilmans, 1993). In Cameroon, recent extension work on the port at Douala has damaged mangroves in the Wouri Estuary (Saenger et al., 1998). Unfortunately future extensions are planned.

Dams

Nearly all main rivers in the Gulf of Guinea have been dammed in at least one location. Reservoirs and dams are used to regulate flow, generate electric power and divert water for irrigation. These reduce the total amount of water and seasonal flow variation (Allersma and Tilmans, 1993; Koranteng, 1998). In some areas, regular flooding in the wet season has been eliminated so several lagoons, which used to be refilled during the floods, have been lost (Portman et al., 1989). Water is lost via extra evaporation in reservoirs and through evapo-transpiration in irrigated areas (Allersma and Tilmans, 1993). In some cases, where large lakes have been formed, e.g. the Volta lake in Ghana, local climate has changed (Portman et al., 1989).

There are two main downstream influences of dams on the sediment load:
1. Reduction in supply of material from upstream because of entrapment in the reservoir;
2. Reduction of the transport capacity of modified flow downstream.

Rough estimates indicate a reduction to 60% of the original transport of sand being supplied to the sea by the Volta after construction of the Akosombo dam (Allersma and Tilmans, 1993). Influence of the Kanji dam on supply by the Niger River is much smaller because it does not influence the Benue River, which contributes more than half the water and about two-thirds of sediments (Allersma and Tilmans, 1993). Across the Niger Delta, however, it is estimated that construction of dams has resulted in a 70% loss of sediment catchment area (Portman et al., 1989). Erosion of the riverbed downstream of the dam replaces some entrapped sediments (Allersma and Tilmans, 1993). Reduction in the load of fine sediments is more difficult to estimate because it originally moved mainly as wash-load in the lower part of rivers. Its reduction may be 50% or more in smaller rivers, but again, much less in the Niger (Allersma and Tilmans, 1993). On the other hand, deforestation and other destruction of vegetation may have increased sediment yield in some catchments. Quantitative information is not available (Allersma and Tilmans, 1993).

Reduced seasonality of flow patterns may also affect the stability of inlets through coastal barriers (Allersma and Tilmans, 1993) and the extent of intrusion of the estuarine salt wedge inland (Portman et al., 1989). This can have important ecological effects including destruction of mangroves and rainforest areas by the intrusion of salt water (Portman et al., 1989). Reduction in nutrients reaching the coastal zone has led to a highly significant loss of fish

Cameroon: Conservation Concerns

A.R.G. Price and R. Klaus

Department of Biological Sciences, University of Warwick, Coventry, CV4 7AL, UK

The Cameroon coast is some 400 km in extent and characterised by sand and rock beaches, estuaries with mangroves and swamps, backed by luxuriant freshwater vegetation. Subtidal areas are principally sedimentary, with occasional seagrass, and there is limited coral on basalt rock (Klaus et al., 1997; Price et al., 1997a,b,c,d).

COASTAL ENVIRONMENTAL ASSESSMENT

A recent coastal survey was undertaken at 36 sites, supported by WWF-Cameroon (Fig. 1), which showed moderate impacts from construction, tar and oil pollution, beach/driftwood, general refuse, solid waste and pollution. Refuse is widespread and moderately abundant along the coastal zone. Beach oil and tar are generally less of a problem. Relatively low mean concentrations of tar balls were also recorded from a quantitative study on three beaches (0.31 g m^{-2}, 4.88 g m^{-2}, and 0.11 g m^{-2}) by Gabache et al. (1998).

The abundance of major coastal ecosystems and species groups was also determined, and is of value for identifying candidate sites for a protected area system.

Offshore environmental assessment has been undertaken. From obervations at 77 sites, nine major habitats were recognised, mud (>80% mud in sediment) being predominant. In addition, Advanced Very High Resolution Radiometer (AVHRR) and Coastal Zone Colour Scanner (CZCS) satellite imagery were obtained to show sea surface temperature and primary productivity from chlorophyll concentrations along the Cameroon coast. Upwelling is of great significance to primary productivity and the fisheries, but its influence in Cameroon needs to be clarified.

MAIN ENVIRONMENTAL CONCERNS AND ISSUES

More general coastal environmental issues have been identified and include: (1) high coastal turbidity, although the extent to which this is natural or has increased as a result of soil erosion brought about by deforestation needs further study; (2) fishery problems which include non-sustainable harvesting (e.g. use of dynamite and poisons) and also open access to most fishing areas; (3) insufficient information to judge the pros and cons of increased 'ecotourism'; (4) likely oil pollution and other impacts from the Cameroon/Chad pipeline; (5) inadequate knowledge on the extent of trade in threatened and conservationally important species (i.e. hawksbill turtles, corals); (6) difficulties in upholding national environmental legislation and international agreements; and (7) an inadequate system of marine protected areas.

Follow-up activities undertaken by WWF-Cameroon have included analysis of coastal issues relating to the oil industry, including the Chad/Cameroon pipeline (Price et al., 2000); and development of a coastal zone management/governance framework, linking together key WWF information and other national/regional data (e.g. GEF/LME and EU fisheries projects), and to help ensure more sustainable use of coastal resources. Other requirements include: questionnaires or interviews to determine local perceptions of coastal environmental and governance related issues; greater resolution of habitats in selected areas; the influence of spatio-temporal variations in upwelling and implications for primary productivity and fisheries and further analysis of coastal protected area needs and initiation of a system plan.

REFERENCES

Gabache, C.E., Folack, J. and Yongbi, G.C. (1998) Tar ball levels on some beaches in Cameroon. *Marine Pollution Bulletin* 36 (7), 535–539.

Klaus, R., Price, A.R.G., and Abbiss, M. (1997) Enhanced satellite images of Cameroon coastal zone. Preliminary survey of the marine and coastal ecosystem of Cameroon. Output 2. WWF, Cameroon, 11 pp.

Price, A.R.G., Klaus, R. and Abbiss, M. (1997a) Compilation and assessment of existing information. Preliminary survey of the marine and coastal ecosystem of Cameroon. Output 1. WWF, Cameroon, 38 pp.

Price, A.R.G., Abbiss, M., Klaus, R., Kofani, M. and Webster, G. (1997b) Field data based on ground-truthing. Preliminary survey of the marine and coastal ecosystem of Cameroon. Output 3. WWF, Cameroon, 18 pp.

Price, A.R.G., Klaus, R., Abbiss, M., Webster and G. Kofani, M. (1997c) Cameroon's marine and coastal ecosystems, threats and management requirements. Preliminary survey of the marine and coastal ecosystem of Cameroon. Output 4. WWF, Cameroon, 12 pp.

Price, A.R.G., Klaus, R., Abbiss, M., Webster and G. Kofani, M. (1997d) WWF Project final report. Preliminary survey of the marine and coastal ecosystem of Cameroon. Output 4. WWF, Cameroon, 8 pp.

Price, A.R.G., Klaus, R., Sheppard, C.R.C., Abbiss, M.A., Kofani, M. and Webster, G. (2000) Environmental and bioeconomic characterisation of coastal and marine systems of Cameroon, including risk implications of the Chad–Cameroon pipeline project. *Aquatic Ecosystem Health and Management* 3 (1), 137–161.

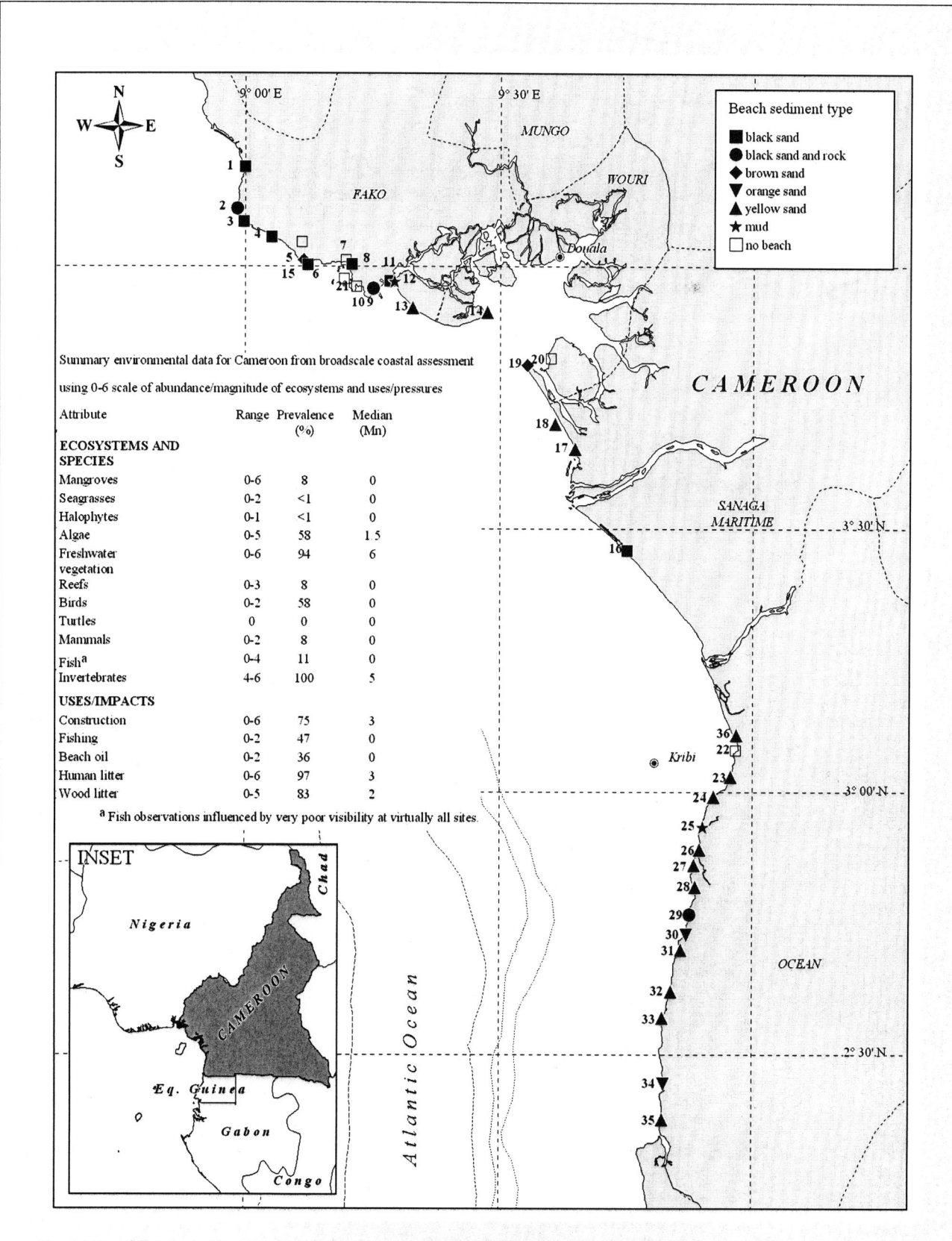

Fig. 1. Map of Cameroon showing the location of coastal study sites, their major beach sediment type and summary environmental data (from Price et al., 2000).

Summary environmental data for Cameroon from broadscale coastal assessment using 0-6 scale of abundance/magnitude of ecosystems and uses/pressures

Attribute	Range	Prevalence (%)	Median (Mn)
ECOSYSTEMS AND SPECIES			
Mangroves	0-6	8	0
Seagrasses	0-2	<1	0
Halophytes	0-1	<1	0
Algae	0-5	58	1.5
Freshwater vegetation	0-6	94	6
Reefs	0-3	8	0
Birds	0-2	58	0
Turtles	0	0	0
Mammals	0-2	8	0
Fish[a]	0-4	11	0
Invertebrates	4-6	100	5
USES/IMPACTS			
Construction	0-6	75	3
Fishing	0-2	47	0
Beach oil	0-2	36	0
Human litter	0-6	97	3
Wood litter	0-5	83	2

[a] Fish observations influenced by very poor visibility at virtually all sites.

Beach sediment type:
- black sand
- black sand and rock
- brown sand
- orange sand
- yellow sand
- mud
- no beach

catches in some parts of the Niger Delta (Portman et al., 1989). Seasonal nutrient pulses to the ocean are required for the productivity cycle.

Shipping and Offshore

Oil pollution in the marine environment of the Gulf of Guinea may be a potential problem because of both oil production and maritime transport. Drilling activities of oil exploration surveys lead to contamination of drill cuttings and blanketing of benthos by drilling muds, however, these effects are generally localised around drilling platforms (Portman et al., 1989). Pipeline development can have serious environmental effects. A recent survey indicated the proximity of a proposed Cameroon–Chad pipeline route to hard corals growing on rocky subtidal substrates (Price et al., 1997).

Both Nigeria and Gabon export large amounts of their oil. Annually, 706 million tonnes are transported to Europe and America (Portman et al., 1989). The wide sea lanes mean the number of accidents involving tankers has been low compared to elsewhere in the world, with only three major accidents up to 1981 (Portman et al., 1989). There have, however, been a total of 10 shipping accidents involving tankers, including the suspicious loss of the "fully laden" *Salem* in 1980. Smaller accidents are estimated to occur at a rate of 1.5 spills per year of 1000 tonnes within 50 miles of land and 0.26 spills per year of 334 tonnes greater than 50 miles from land (Portman et al., 1989).

There have been three well blowouts in the region. These include the Funiwa 5 blowout in 1980, which spilt more than 400,000 barrels of oil into coastal waters of Nigeria and caused mortality of some benthic communities and mangrove areas (Portman et al., 1989).

Tar balls have been noted on a number of beaches with elevated levels found in the eastern part of the Gulf of Guinea. Several kilograms per square metre were found at some sites in Nigeria and Cameroon (Asuquo, 1991; Enyenihi and Antia, 1986; Okonya and Ibe, 1986; IOC, 1985). Levels in Côte d'Ivoire are an order of magnitude lower (IOC, 1985). Tar balls have even been found in areas remote from oil production (Portman et al., 1989). This pollution is attributed to oil exploration and transportation in the region (IOC, 1985), especially port operations (Portman et al., 1989). Small-scale oil losses are a common feature of most port operations due to spills and tank washings (Portman et al., 1989).

Heavy metals are a potential problem related to oil production and pollution, however, levels in marine species appear to be background. Heavy metals in fauna do not

Table 8

International Agreements and Conventions signed up to by Gulf of Guinea countries. Countries marked with * have signed the Agreement/Convention but not ratified it. All other named countries are party to the Agreement/Convention (Source: CIA, 1997).

Name of International Agreement/Convention	Gulf of Guinea countries signed up to the Agreement/Convention
Biodiversity	Benin, Cameroon, Côte d'Ivoire, Equatorial Guinea, Gabon*, Ghana, Guinea, Guinea-Bissau, Liberia*, Nigeria, Sao Tome and Principe*, Sierra Leone, Togo
Climate change	Benin, Cameroon, Côte d'Ivoire, Gabon*, Ghana, Guinea, Guinea-Bissau, Liberia*, Nigeria, Sao Tome and Principe*, Sierra Leone, Togo
Desertification	Benin, Cameroon*, Côte d'Ivoire*, Equatorial Guinea*, Gabon*, Ghana*, Guinea*, Guinea-Bissau, Nigeria, Sao Tome and Principe*, Sierra Leone*, Togo
Endangered species	Benin, Cameroon, Côte d'Ivoire, Equatorial Guinea, Gabon, Ghana, Guinea, Guinea-Bissau, Liberia, Nigeria, Sierra Leone, Togo
Environmental modification	Benin, Ghana, Liberia*, Sao Tome and Principe, Sierra Leone*
Hazardous wastes	Côte d'Ivoire, Guinea, Nigeria
Law of the sea	Benin*, Cameroon, Côte d'Ivoire, Equatorial Guinea*, Gabon*, Ghana, Guinea, Guinea-Bissau, Liberia*, Nigeria, Sao Tome and Principe, Sierra Leone, Togo
Marine dumping	Côte d'Ivoire, Gabon, Liberia*, Nigeria
Marine life conservation	Ghana*, Liberia*, Nigeria, Sierra Leone
Nuclear test ban	Benin, Cameroon*, Côte d'Ivoire, Equatorial Guinea, Gabon, Ghana, Guinea-Bissau, Liberia, Nigeria, Sierra Leone, Togo
Ozone layer protection	Benin, Cameroon, Côte d'Ivoire, Gabon, Ghana, Guinea, Liberia, Nigeria, Togo
Ship pollution	Côte d'Ivoire, Equatorial Guinea, Gabon, Ghana, Liberia, Togo
Tropical timber 83	Cameroon, Côte d'Ivoire, Gabon, Ghana, Liberia, Togo
Tropical timber 94	Cameroon, Côte d'Ivoire, Gabon, Ghana, Liberia, Togo
Wetlands	Côte d'Ivoire, Gabon, Ghana, Guinea, Guinea-Bissau, Togo
Whaling	Ghana, Nigeria, Sierra Leone

appear to be a problem in the region so far. The highest levels of mercury were found in Yellowfin Tuna (*Thunnus albacores*), but these were not high enough to be of concern and were less than those found in the Mediterranean (IOC, 1985). Little data exists regarding heavy metal concentrations in sediments.

PROTECTIVE MEASURES

Marine and Environmental Legislation

Many Gulf of Guinea countries are signatories to a number of international environmental agreements. These are given in Table 8.

Even though some of these agreements are not wholly or even partly marine (e.g. Framework Convention on Climate Change, Tropical Timber Agreement), they have implications for the marine environment. For example, conservation of forests will help reduce soil erosion and coastal sedimentation, which, if excessive, can be harmful to photosynthetic marine ecosystems. Unfortunately, neither Nigeria nor Cameroon, the two countries with the largest mangrove areas, are party to the Convention on Wetlands of International Importance especially as Waterfowl Habitat (Ramsar Convention of 1971). In countries that are signatories to the convention (e.g. Ghana), Ramsar sites are identified, delineated, protected and studied (Price et al., 1997). Some countries have also developed national environmental legislation.

Although a considerable body of laws and agreements has been adopted, implementation and enforcement is often problematic due to constraints relating mainly to physical and human resources (Price et al., 1997).

Table 9

Number of protected areas and estimated total area having protected status, including proposed protected status, for each of the Gulf of Guinea countries (Source: WCMC, 1997, Koranteng, 1995)

Country	No. of Protected Areas	Total Area Protected
Benin	1	100
Cameroon	2	4600
Côte d'Ivoire	3	330
Equatorial Guinea	4	1500
Gabon	4	6600
Ghana	4	–
Guinea	3	–
Guinea-Bissau	3	–
Liberia	2	2004
Nigeria	–	–
Sao Tome & Principe	–	–
Sierra Leone	2	150
Togo	1	9

Protected Areas

In Ghana, Muni and Sakumo lagoons, Densu delta and the Keta-Songor wetlands have been declared as Ramsar sites (Koranteng, 1995). Other protected areas in the sub-region are indicated in Table 9. Lagoons and coastal wetlands at Fresco, Aby (Côte d'Ivoire), Songor lagoon complex, Anlo-Keta (Ghana) and coastal lagoons and wetlands of the Mono River (Togo/Benin) have been proposed as marine protected areas (Schwarz, 1992).

Gulf of Guinea LME Projects

The UNIDO/GEF Gulf of Guinea LME Project was set up involving countries from the region in an attempt to manage the marine environment of the Gulf of Guinea, across political boundaries. Member countries are Benin, Cameroon, Côte d'Ivoire, Ghana, Nigeria and Togo. Possibly Equatorial Guinea, Sao Tome and Principe and Gabon would join soon. Scientists from these countries are engaged in collaborative research with a view to meeting the specific objectives of the project. These objectives are given as:
1. to strengthen regional institutional capacities to prevent and remedy pollution of the Gulf of Guinea LME and associated degradation of critical habitats,
2. to develop an integrated information management and decision making system for ecosystem management,
3. to establish a comprehensive programme for monitoring and assessment of the living marine resources, health and productivity of the LME,
4. to prevent and control land-based sources of industrial and urban pollution, and develop national and regional strategies and policies, including forging regional Conventions and Protocols for the long-term management and protection of the Gulf of Guinea LME.

To be successful, the LME management concept must be coupled with management of the adjacent Marine Catchment Basins (GEF, 1996).

Another 'LME project' running in the region is the EU/INCO DC Project: *Impacts of environmental forcing on marine biodiversity and sustainable management of fisheries in the Gulf of Guinea.* It is a collaborative research project between EU and Gulf of Guinea scientists with the following general objectives:
1. to assess the impacts of upwelling and other forms of environmental forcing on marine biodiversity and the dynamics of artisanal and industrial fisheries,
2. to develop and implement an information and analysis system for the sustainable management and governance of fisheries resources in the Gulf of Guinea.

A realisation of the large spatio-temporal scales at which marine processes occur and the high dependence of human populations on ecosystem health has led to these collaborative programmes, showing the acceptance among scientists in the Gulf of Guinea region of the need to share data and work together to manage their marine environment.

CONCLUSION

The Gulf of Guinea LME has areas of high biodiversity, productive fisheries and extensive mineral reserves. It is also the most densely settled coastline in Africa. Thus, there is a high degree of coupling between the human population and the natural resources and environmental dynamics of the coastal and marine environment. For example, environmental fluctuations lead to variability in food production while increasing human populations and industrial development lead to over-exploitation of natural resources and pollution. Despite the environmental problems outlined in this chapter, advances are being made in the Gulf of Guinea, especially through the collaborative LME programmes. These are providing a better understanding of natural environmental variability, together with a move towards sustainable use of the marine environment.

REFERENCES

Abban, J.F. (1986) Coastal erosion in the West African region. In: C.A. Biney (ed.) Proceedings of the National Workshop on the Joint FAO/IOC/WHO/IAEA/UNEP Project on Monitoring of Pollution in the Marine Environment of the West and Central African Region, WACAF/2, Accra, Ghana, 31 July 1985. Institute of Aquatic Biology Technical Report No. 110, pp. 43–119.

Ajayi T.O. and A.O. Anyanwu (1997) Marine fisheries of Nigeria: Recent investigations, resource evaluation, state of exploitation and management strategies. Prepared for the CECAF (Committee for Eastern Central Atlantic Fisheries) Working Party on Resources Evaluation, Accra, Ghana, 24–26 Sept. 1997.

Aleem, A.A. (1979) Marine Microplankton from Sierra Leone (West Africa). *Indian Journal of Marine Science* 8 (4), 291–295.

Allersma, E. and Tilmans, W.M.K. (1993) Coastal Conditions in West Africa—a Review. *Ocean and Coastal Management* 19, 199–240.

Altenburg W. (1987). Waterfowl in West African coastal wetlands. A summary of current knowledge of the occurrence of waterfowl in wetlands from Guinea Bissau to Cameroon and a bibliography of information sources. WIWO report 15, Zeist. 72 pp.

Anang E.R. (1979). The seasonal cycle of the phytoplankton in the coastal waters of Ghana. *Hydrobiologia* 2 (1), 33–45.

Appolinaire, Z. (1993) Mangroves of Cameroon. In *Conservation and Sustainable Utilization of Mangrove Forests in Latin America and Africa Regions. Part II, Africa,* ed. E. Diop. International Society for Mangrove Ecosystems and International Tropical Timber Organisation. pp. 193–209.

Asuquo, F.E. (1991) Tar balls on Ibeno-Okposo beach of south-east Nigeria. *Marine Pollution Bulletin* 22 (3), 150–151.

Awosika L.F. and A.C. Ibe (1998). Geomorphic features of the Gulf of Guinea shelf and the littoral drift dynamics. In *Nearshore Dynamics and Sedimentology of the Gulf of Guinea.* Proc. First IOCEA Cruise, IOC/LME Gulf of Guinea, eds. C.A. Ibe, L.F. Awosika and K. Akapp. IOC/UNIDO/GEF, pp. 25–33.

Bakun, A. (1978) Guinea Current Upwelling. *Nature* 271: 147–150.

Bakun, A. (1990) Global Climate Change and Intensification of Coastal Ocean Upwelling. *Science* 247: 198–201.

Bakun, A. (1996) *Patterns in the Ocean: ocean processes and marine population dynamics.* California Sea Grant, La Jolla, CA, USA.

Bard F.X. and K.A. Koranteng (eds.) (1995). *Dynamics and Use of Sardinella Resources from Upwelling off Ghana and Ivory Coast.* Proceedings of Scientific Meeting, Accra 5–8 Oct., 1993. ORSTOM Editions, Paris.

Binet, D. (1976) Biovolumes et poids secs zooplanctoniques en relation avec le milieu pélagique au-dessus du plateau ivoirien. *Cahiers ORSTOM Série Océanographie* 9, 247–266.

Binet, D. and Marchal, E. (1993) The Large Marine Ecosystem of the Gulf of Guinea: Long-Term Variability induced by Climatic Changes. In *Large Marine Ecosystems—Stress Mitigation and Sustainability,* eds. K. Sherman, L.M. Alexander and B. Gold. American Association for the Advancement of Science. pp. 104–118.

Biney, C. (1990) Review of characteristics of freshwater and coastal ecosystems in Ghana. *Hydrobiologia* 208, 45–53.

Buchanan J.B. (1957). The bottom fauna communities across the continental shelf off Accra, Ghana (Gold Coast). *Proceedings of the Zoological Society London* 130, 1–56.

Castroviejo, J., Juste, B.J., Perez del Val, J., Castelo, R., Gil, R. (1994) Diversity and status of seaturtle species in the Gulf of Guinea islands. *Biodiversity and Conservation* 3(9), 828–836.

Chaveau, J.P. (1991) Les variations spatiales et temporelles de l'environnement socio-économique et l'evolution de la pêche maritime artisanale sur les côtes ouest-africaines. Essai d'analyse en longue période: XV–XX siècle. In *Pêcheries Ouest-Africaines, Variabilité, Instabilité et Changement,* eds. P. Cury and C. Roy. ORSTOM Editions. pp. 14–25.

CIA (1997) World Factbook '97. http://www.odci.gov/cia/publications/factbook/index.html.

Delince, G. and Obiekezie, A. (1996) Sustainable aquaculture in West and Central Africa. ACP-EU Fisheries Research Initiative. Proc. Second Dialogue Meeting, Western and Central Africa, the Comoros and the European Union. Dakar, Senegal. pp. 119–135.

Domain, F. (1988) Rapport des campagnes de chalutages du N.O. André Nizery au large des côtes de Guinée-Bissau. Institute de Recherche Agronomique de Guinée, Ministère français de la coopération.

Donoghue, M. and Wheeler, A. (1994) *Dolphins: Their Life and Survival.* David Bateman Ltd, Auckland, New Zealand.

Dufour, P. and Slépoukha, M. (1975) L'oxygène dissous en lagune Ebrié: Influence de l'hydroclimat et des pollutions. *Documents Scientifique C.R.O. Abidjan,* ORSTOM 6(2), 75–118.

Elder, D. and Pernetta, J. (1991). *Oceans: A Mitchell-Beazley World Conservation Atlas.* In association with IUCN: the World Conservation Union. Mitchell Beazley Publishers, London.

Enyenihi, U.K. and Antia, E.E. (1986) Tar balls as indicator of crude petroleum pollution of the beaches of Cross River State, Nigeria. Workshop Report. IOC (41), Annex V.9, 8 pp.

Ezenwa, B.I. and Ayinla, O.A. (1994) Conservation strategies for endangered fish breeding and nursery grounds within the coastal wetlands of Nigeria. *Aquatic Conservation: Marine and Freshwater Ecosystems* 4(2), 125–133.

Folorunsho R., Awosika, L.F. and Dublin-Green, C.O. (1998) An assessment of river inputs into the Gulf of Guinea. In *Nearshore Dynamics and Sedimentology of the Gulf of Guinea,* Proc. First IOCEA Cruise, IOC/LME Gulf of Guinea, eds. C.A. Ibe, L.F. Awosika and K. Aka. IOC/UNIDO/GEF, pp. 177–186.

GEF (1996) Gulf of Guinea Large Marine Ecosystem Project. UNIDO, UNDP, NOAA, UNEP.

GEF (1996/1997) GOG LME Newsletter. No. 6, Oct 1996–Mar 1997.

Greze V.N., Gordejeva, K.T. and Shmeleva, A.A. (1969) Distribution of zooplankton and biological structure in the Tropical Atlantic. In *Proceedings of the Symposium on the Oceanography and Fisheries Resources of the Tropical Atlantic.* Results of ICITA and GTS, Abidjan, Ivory Coast, 20–28 October 1966. UNESCO Publications, Paris. pp. 85–90.

Hearn, G. (1998) Bioko Island Website: An emerging site for African biodiversity research, conservation, and ecotourism. http://www.beaver.edu/Bioko/.

Hisard, P. (1986) El Niño response of the tropical Atlantic Ocean during the 1984 year. *International Symposium on Long Term Changes in Marine Fish Populations*, pp. 273–290.

Hisard, P., Hénin, C., Houghton, R., Piton, B. and Rual, P. (1986) Oceanic conditions in the tropical Atlantic during 1983 and 1984. *Nature* **322**, 243–245.

Houghton R.W. and Mensah, M.A. (1978) Physical Aspects and Biological Consequences of Ghanaian Coastal Upwelling. In *Upwelling Ecosystems*, eds. R. Boje and M. Taniczak. Springer-Verlag, Berlin. pp. 167–180.

Ibe C.A., Awosika, L.F. and Aka, K. (eds.) (1998) *Nearshore Dynamics and Sedimentology of the Gulf of Guinea*. Proc. First IOCEA Cruise. IOC/LME Gulf of Guinea. 225 pp.

Ingham, M.C. (1970) Coastal upwelling in the Northwestern Gulf of Guinea. *Bulletin of Marine Science* **20**, 1–34.

IOC (1985) First Workshop of Participants in the Joint FAO–IOC–WHO–IAEA–UNEP Project on Monitoring of Pollution in the Marine Environment of the West and Central African Region (WACAF/2-Pilot Phase). Dakar, Senegal, 28 Oct.–1 Nov. 1985. Workshop Report. IOC 41. UNESCO.

IOC (1994) IOC Regional Workshop on Marine Debris and Waste Management in the Gulf of Guinea. Lagos, Nigeria, 14–16 Oct. 1994. Workshop Report. IOC 113. UNESCO.

Irvine, F.R. (1947) *The Fish and Fisheries of the Gold Coast*. HM Service, Gold Coast.

Jefferson, T.A., Leatherwood, S. and Webber, M.A. (1983). FAO species identification guide. Marine Mammals of the World. Rome, FAO. 320 pp.

Joiris C.R., Otchere, F.A. Azokwu, M.I. and Ali, I.B. (1997) Total and Organic Mercury, and Organochlorides in a Bivalve from Ghana and Nigeria. In *The Coastal Zone of West Africa: Problems and Management*, eds. S.M. Evans, C.J. Vanderpuye and A.K. Armah. Penshaw Press, U.K. pp. 89–92.

Kelleher, G., Bleakely, C., Wells, S. (eds.) (1993) *A Global Representative System of Marine Protected Areas. Vol II: Wider Caribbean, West Africa and South Atlantic*. GBRMPA,World Bank, IUCN. 93 pp.

Koranteng K.A. (1995) Fish and Fisheries of Three Coastal Lagoons in Ghana. Global Environment facility/Ghana Coastal Wetlands Management Programme. GW/A.285/SF2/31.

Koranteng K.A. (1998) The impacts of environmental forcing on the dynamics of demersal fishery resources of Ghana. PhD Thesis, University of Warwick. 377 pp.

Koranteng, K.A. and Pezennec, O. (1998) Variability and Trends in Environmental Time Series along the Ivorian and Ghanaian Coasts. In *Global versus Local Changes in Upwelling Systems*, eds. M.H. Durand, P. Cury, R. Mendelssohn, C. Roy, A. Bakun and D. Pauly. ORSTOM Editions, Paris. pp. 167–177.

Koranteng K.A., McGlade, J.M. and Samb, B. (1996) Review of the Canary and Guinea Current Large Marine Ecosystems. In: *ACP-EU Fisheries Research Initiative. Proceedings of the Second Dialogue Meeting, Western and Central Africa, the Comoros and the European Union*. Dakar, Senegal, 22–26 April 1996. ACP-EU Fisheries Research Report (2), Brussels. pp. 61–84.

Lamb, P.J. (1982) Persistence of sub-Saharan drought. *Nature* **299**, 46–48.

Le Borgne, R. (1983) Note on swarms of Thaliacea in the Gulf of Guinea. *Océanographie Tropicale* **18**(1), 49–54.

Longhurst A.R. (1958) An ecological survey of the West African marine benthos. Colonial Office Fishery Publication No. 11. 102 pp.

Longhurst, A.R. (1962) A review of the oceanography of the Gulf of Guinea. *Bulletin de l'Institut Francaise d'Afrique Noire* **24**(A)3, 633–663.

Longhurst, A.R. (1964) The coastal oceanography of Western Nigeria. *Bulletin de l'Institut Francaise d'Afrique Noire* **26**(A)2, 337–402.

Longhurst A.R. (1969). Species assemblages in tropical demersal fisheries. In *Proceedings of the Symposium on the Oceanography and Fisheries Resources of the Tropical Atlantic*. Results of ICITA and GTS, Abidjan, Ivory Coast, 20–28 October 1966. UNESCO Publications, Paris. pp. 147–168.

Longhurst, A.R. and Pauly, D. (1987) *Ecology of Tropical Oceans*. Academic Press Inc, London. 407 pp.

Mahé G. (1997). Freshwater yields to the Atlantic Ocean: local and regional variations from Senegal to Angola. Presented at the 1994 CEOS workshop on Global Versus Local Changes in Upwelling Systems, Monterey, California, USA.

Mahé, G. (1991) La variabilité des apports fluviaux au Golfe de Guinée utilisée comme indice climatiques. In *Pêcheries Ouest-Africaines, Variabilité, instabilité et changement*, eds. P. Cury and C. Roy. ORSTOM Editions. Paris. pp. 147–161.

Marchal, E. (1991) Nanisme et sedentarité chez certaines espèces de poissons pélagique: deux aspects d'un même réponse à des conditions défavorables. In *Pêcheries Ouest-Africaines, Variabilité, instabilité et changement*, eds. P. Cury et C. Roy. ORSTOM Editions, Paris. pp. 192–200.

Marchal, E. and Picault, J. (1977) Répartition et abondance évaluées par échointégration des poissons du plateau continental ivoiro-ghanéen en relation avec les upwelling locaux. *Journal de Recherche Océanografique* **2**, 39–57.

Mendelssohn, R. and Roy, C. (1994) Changes in mean level, seasonal cycle and spectrum in long term coastal time series from various Eastern Ocean Systems. 26th Liège International Colloquium on Ocean Hydrodynamics, May 2–6; 1994.

Mensah, M.A. (1995) The occurrence of zooplankton off Tema during the period 1969–1992. In *Dynamics and Use of Sardinella Resources from Upwelling off Ghana and Ivory Coast*, eds. F.X. Bard and K.A. Koranteng. ORSTOM Editions, Paris. pp. 290–304.

Mensah M.A. and Koranteng, K.A. (1988) A review of the oceanography and fisheries resources in the coastal waters of Ghana. Marine Fisheries Research Report No. 8, Fisheries Research and Utilization Branch, Tema, Ghana. 35 pp.

Moore, D.W., Hisard, P., McCreary, J.P., Merle, J., O'Brien, J.J., Picault, J., Verstraete, J.M. and Wunsch, C. (1978) Equatorial adjustment in the eastern Atlantic Ocean. *Geophysical Research Letters* **5**, 637–640.

Morlière, A. (1970) Les saisons marines devant Abidjan. *Documents Scientifiques C.R.O. Abidjan* **1**, 1–15.

Okonya, E.C. and Ibe, A.C. (1986) Stranded pelagic tarball loadings on Badagry Beach, Nigeria. Workshop Report. IOC 41, Annex V.10, 6 pp.

Overfield D.A., Adam, M., Ghartey, N., Arthur, R., Duhamel, R. and Willoughby, N. (1997) The economic and social consequences of Cape Saint Paul wilt disease in Ghana and its implications for coastal management issues. In *The Coastal Zone of West Africa: Problems and Management*, eds. S.M. Evans, C.J. Vanderpuye and A.K. Armah. Penshaw Press, U.K.

Pauly, D. (1976) The biology, fishery and potential for aquaculture of *Tilapia melanotheron* in a small West African lagoon. *Aquaculture* **7**, 33–49.

Pezennec, O. and Bard, F.X. (1992) Importance écologique de la petite saison d'upwelling ivoiro-ghanéenne et changements dans la pêcherie de *Sardinella aurita*. *Aquatic Living Resources* **5**, 249–259.

Philander, S.G.H. (1986) Unusual conditions in the tropical Atlantic Ocean in 1984. *Nature* **322**, 236–238.

Picault, J. (1983) Propogation of the seasonal upwelling in the eastern tropical Atlantic. *Journal of Physical Oceanography* **13**(1), 18–37.

PIRC (1996) Controversies Affecting Shell in Nigeria: Report to Clients. http://www.pirc.co.uk/shellmar.htm.

Portman, J.E., Biney, C., Ibe, A.C. and Zabi, S. (1989) State of the marine environment in the west and central African region. *UNEP Regional Seas Reports and Studies*, No. 108. UNEP.

Price, A.R.G., Klaus, R. and Abbiss, M. (1997) Compilation and assessment of existing information. Preliminary survey of the marine and coastal ecosystem of Cameroon. Output 1. WWF, Cameroon, 38 pp.

Roy, C. (1995) The Cote d'Ivoire and Ghana Coastal Upwellings: Dynamics and Changes. In *Dynamics and Use of Sardinella Resources from Upwelling off Ghana and Ivory Coast*, eds. F.X. Bard, and K.A. Koranteng. ORSTOM Editions, Paris. pp. 346–361.

Saenger, P. and Bellan, M.F. (1995) The mangrove vegetation of the Atlantic coast of Africa. A review. Laboratoire d'Ecologie Terrestre, Université de Toulouse III. 58 pp.

Saenger, P., Santiane, Y., Baglo, M., Isebor, C., Armah, A.K. and Nganje, M. (1998) The Gulf of Guinea Project: Managing Mangroves to Protect Biodiversity in West Africa. http://brooktrout.gso.uri.edu/ICmangsaenger.html.

Schwarz, B. (1992) Identification, establishment and management of specially protected areas in WACAF region: national and regional conservation priorities in terms of coastal and marine biodiversity. Report prepared by IUCN for UNEP, OCA/PAC, April, 1992.

Scowcroft, C.P. (1995) Africa: Background Paper. ICLARM Research Planning Workshop, Cairo, 23–27 Sept. 1995.

Servain, J., Picault, J. and Merle, J. (1982) Evidence of remote forcing in the equatorial Atlantic Ocean. *Journal of Physical Oceanography* **12**, 457–465.

Shell (1998). Shell in Nigeria. http://www.shellnigeria.com/

Smit C.J. and T. Piersma (1989) Numbers, midwinter distribution, and migration of wader populations using the East Atlantic Flyway. In *Flyways and Reserve Network for Water Birds*, eds. H. Boyd and J.-Y. Pirot. IWRB Special Publication No. 9. pp. 24–60.

Sokari, T.G., Idiebele, D.D., Ottih, R.M. (1988) Antibiotic resistance among coliforms and *Pseudomonas* spp. from bodies of water around Port Harcourt, Nigeria. *Journal of Applied Bacteriology* **64**, 355–359.

Ssentongo G.W., E.T. Ukpe and T.O. Ajayi (1986). Marine Fishery Resources of Nigeria: A review of exploited fish stocks. CECAF/ECAF Series 86/40 (En), Rome, FAO.

Tilot, V. (1993) Description of the Different Large Marine Ecosystems of West Africa. IUCN Marine Program.

Tilot, V. and King, A. (1993) A Review of the Subsystems of the Canary Current and Gulf of Guinea Large Marine Ecosystems. IUCN Marine Programme.

UNEP/IUCN (1988). *Coral Reefs of the World Volume 1: Atlantic and Eastern Pacific*. IUCN, Gland, Switzerland and Cambridge, UK / UNEP, Nairobi, Kenya.

Verstraete, J.M. (1970) Etude quantitative de l'upwelling sur le plateau continental ivoirien. *Documents Scientifiques C.R.O. Abidjan* **1**(3), 1–17.

Wandan, E.N. and Zabik, M.J. (1996) Assessment of surface water quality in Côte d'Ivoire. *Bulletin of Environmental Contamination and Toxicology* **56**(1), 73–79.

WCMC (1996) Animal Redlist: Result of Red List country enquiry for Gabon. http://www.wcmc.org/.

WCMC (1997) Marine Programme. http://www.wcmc.org.uk/marine/

WCMC (1998) Forest Programme. http://www.wcmc.org.uk/forest/

Williams F. (1968) Report on the Guinean Trawling Survey, Organisation of African Unity Scientific and Technical Research Commission (99).

Woodruff, S.D., Slutz, R.J., Jenne, R.L. and Steurer, P.M. (1987) A Comprehensive Ocean–Atmosphere Data Set. *Bulletin of the American Meteorological Society* **68**, 1239–1249.

THE AUTHORS

Nicholas J. Hardman-Mountford
*Centre for Coastal and Marine Sciences,
Plymouth Marine Laboratory,
Plymouth, U.K.*

Kwame A. Koranteng
*Marine Fisheries Research Division,
Ministry of Food and Agriculture,
Ghana*

Andrew R. G. Price
*Dept. of Biological Sciences,
University of Warwick,
Coventry, U.K.*

Chapter 50

GUINEA

Ibrahima Cisse, Idrissa Lamine Bamy, Amadou Bah, Sékou Balta Camara and Mamba Kourouma

The Guinea Exclusive Economic Zone covers about 71,000 km² and has a high diversity of marine fauna and flora. The coast has numerous estuaries, mangrove stands and beaches, and the offshore habitat is dominated by sediments, high proportions of which are terrigenous silts which arrive from heavy run-off during periods of high seasonal rainfall.

The upwelling which brings nutrient enrichment to more western countries terminates around the western border of Guinea, though some enrichment of these waters does occur. Plankton are enriched, and support an industrial and artisanal fishery which, while not large in global terms, brings valuable protein, employment and foreign exchange to the people of the coastal region. Both pelagic and demersal fish species are important, together with cephalopods and shrimp. Except for tuna boats, most of the fishery is for demersal species, and catches have been increasing considerably through the 1990s.

People of the coastal zone depend mainly on small-scale agriculture as well as artisanal fishing. Rice plantation and some shrimp farms occupy increasing amounts of estuaries and land once occupied by mangroves, and significant deforestation has taken place. While coastal erosion is not severe overall, it has resulted in important losses, especially in Conakry. Urban and village development has been unregulated, and there is very little industrial or domestic pollution infrastructure or control, with the result that high levels of several contaminants have been recorded in marine species or sediments. There is some more recent progress towards environmental management and pollution control.

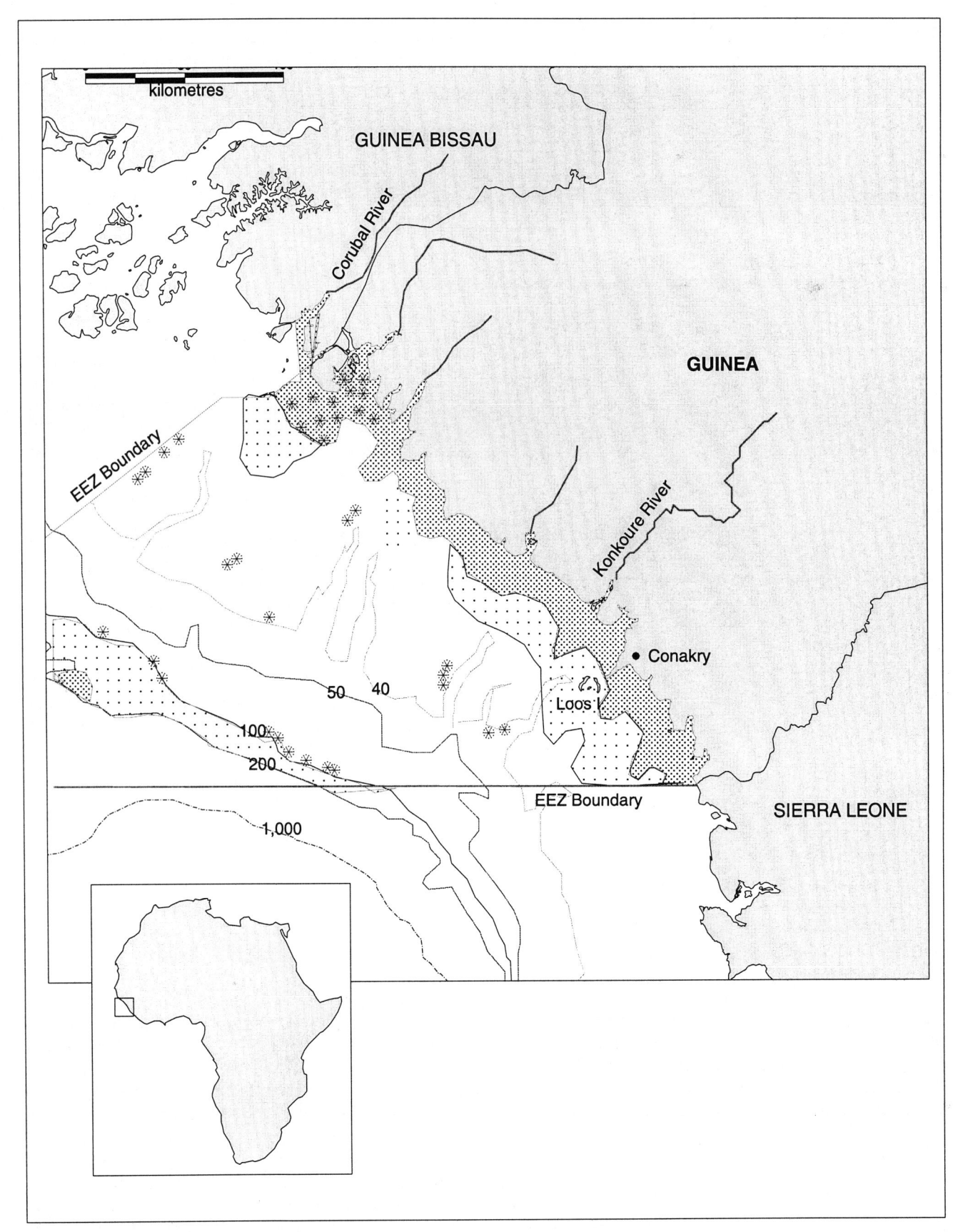

Fig. 1. Map of Guinea. Stars indicate positions of rocky outcrops, areas of sparse dots are those with 0–50% terrigenous sediment, areas of denser dots are areas with >50% terrigenous sediment.

THE DEFINED REGION

The Republic of Guinea in West Africa has an area of 245,857 km² with a 1997 population of 7,164,000 (Fig. 1). Its Atlantic shoreline is 300 km long, with a tropical climate characterised by two seasons, a rainy and a dry season. The annual rainfall on the coast averages 4000 mm, which is considerably higher than that of the interior. The annual rainfall variations impose constraints on agriculture, since much may fall in huge and occasional showers.

The continental shelf is 43,000 km², the largest of the west African coast (Rougeron, 1996) and it reaches its widest extent in the north. The Guinean EEZ has an area of 71,000 km² (Postel, 1955; Arseny and Guerman, 1985; Domain and Bah, 1993). Below 20 m depth its slope is mostly gentle, with most of the shelf area falling in the 20–100 m depth band (Table 1). Two capes are prominent: the peninsula of Conakry which is further extended by Loos Island, and Cape Verga. Towards the border with Guinea-Bissau, many islands are strung along the coast; these are surrounded by rocks which makes the zone difficult for trawling.

The continental shelf is cut almost in half by submarine canyons, 'fossil valleys', marking the river entrances. The sea bottom is composed of silts and sands, containing terrigenous particles originating from rivers and the watershed. Below 20 m, sandy bottoms dominate. Around the old valleys are outcrops of hard substrates, some in the north being characterised by the presence of 'ridings', outcrops 5 m in height.

SEASONALITY, CURRENTS, NATURAL ENVIRONMENTAL VARIABLES

Along the coast it rains from May to November. In the early 1990s, precipitation ranged from 3015 to 4210 mm in Conakry, with values showing important yearly variations. This heavy rainfall is the main cause of considerable salinity variations in inshore waters, whereas trade winds influence the northwest African upwelling and cold water intrusion in the northwest of the EEZ.

Many rivers, most of which are north of Conakry (the Konkoure, the Fatala and Rio Nunez) bring large quantities of silt-laden fresh water. Flow is directly linked to the rainfall on the slope, and increases rapidly from June to August, then decreases in December. Throughout the water column, salinity drops, and the turbidity may be clear up to 100 miles offshore. In the rainy season, winds are generally southwest, and warm waters flow to the north of the continental shelf. Surface temperatures are 28–29°C. Deep temperatures decrease towards the open sea, and in deep waters temperatures decrease slightly with depth.

In the dry season, a low surface salinity region near shore persists, though this is less remarkable than in the rainy season. Values are typically 32.0 ppt at the coast, rapidly reaching 35.5 ppt offshore. Bottom salinity is higher in the dry season and even at the coast it remains more than 32.8 ppt. Isohaline curves usually lie parallel to the coastline. Guinean continental shelf waters are located between northern upwelling waters and southern warm waters (Wauthy, 1993). Blown by the wind from the north, the front between these two bodies moves on an annual basis. The maximum southern extension of the upwelling waters occurs during the first half of March. As an example, in February 1993, surface temperature in the south was 27°C, which decreased to a minimum value of 23°C in the northwest. This marked cooling shows the influence of the upwelling in this season, and is an important influence on the pelagic biota.

Tides are generally semidiurnal. Tidal ranges may be more than 4 m, inducing strong currents of 70 cm s^{-1} on the bottom and 90 cm s^{-1} at the surface (Allen and Salomon, 1980; Keita 1998). These tidal currents play an important role in freshwater mixing and influence where sediment deposition occurs.

In January–February, the Canaries current affects the region, decreasing the temperature. In November, there is generally an anti-cyclonic circulation. The upwelling is active during the period of January–May, and weakens from June to November.

THE MAJOR SHALLOW WATER MARINE AND COASTAL HABITATS

There are three main zones: an inshore or coastal band (0–20 m depth), a middle shelf (20–60 m) and the outer shelf (60–200 m). These differ in their relief, their dynamics and their sediment characteristics. The inshore zone is most influenced by tides. Most of this area is covered with muddy deposits, but in the region of Verga-Cape, Kaloum peninsula and Loos Island there are sands, rocks and gravel. Organically rich sediments, such as are created in coastal mangroves, are an important part of the sediments in this zone. The middle shelf is a large plateau, fairly dynamic and affected by sediments from the estuaries and by mobile muds. The outer shelf is largest in the north. In the shallowest part of this zone there are sometimes belts of sand lying parallel to the coast whose width may reach 100–250 m; they reach almost 2 m high, extending for 6 km in length.

Because of the high rains, the mixing of sea with fresh waters causes the development of a biologically rich front.

Table 1
Areas of different depths of the Guinean continental shelf

Depth	0–10	10–20	20–40	40–100	100–200	Total
Area (km²)	5339	6498	18134	10679	2267	42919
%	12	15	42	25	5	

The ecosystem which develops provides rich artisanal fishing which is important to a large part of the population. Phytoplankton is rich in the inshore region (Kouzmenko and Haba, 1988). Sponges are common throughout the area to 200 m depth (Camara, 1998), though the main invertebrates are polychaetes, gastropods and bivalves, cephalopods, crustaceans, echinoderms and annelids (Keita et al., 1995). The main pelagic fish in the Guinean EEZ are *Sardinella aurita, S. maderensis, Decapterus punctatus, D. rhoncus, Scomber japonicus,* and *Ethmalosa fimbriata.* Also present but less exploited are *Engraulis encrasicolus,* the Carangidae and small Thonnidae.

About 207 demersal species from 87 families have been identified in a trawling survey (Domain, 1989). The principal demersal fish communities are those dominated by the Sciaenidae, which are mainly found in the coastal zone and near estuaries with muddy bottom and sandy-marl. A community with 11 species of sparidae is concentrated offshore, mainly from 20 to 50 m on sandy bottoms and rocky outcrops (Morize et al., 1995). A continental upper slope community dominated by nine species is found between 100 and 200 m.

Reptiles occur along the coast, around the islands, especially in estuaries, swamps and mangroves. There are several tortoises and turtles, including the leatherback. Dolphins and Baleen whales are also seen (Bah et al., 1996).

OFFSHORE SYSTEMS

Upwelling

The flow of waters rich in nutrients from the continental slope is caused by wind-driven currents and by a density current (Keita et al., 1998). This brings to the surface deep waters which are not only rich in nutrients but also rich in oxygen. The maximum phytoplankton abundance in the frontal zone has a biomass of 5–8 g m^{-3}. Zooplankton concentrations follow this. The horizontal boundary to the upwelling is also influenced and strengthened by raised concentrations of terrigenous substances whose nutrients add to those from the upwelling. The period of intense activity is February, March, and April. In April the speed of the current varies from 33 cm s^{-1} in the north to 20 cm s^{-1} in the southwest and 12 cm s^{-1} in the southeast (Keita, 1998).

The location where fishing takes place confirms existing information about the geographical distribution and movement of the fish. Fishing takes place in the first quarter when the pelagic fish population (small sardines, chinchards) migrates south, following the movement of cold waters from the Senegal–Mauritanian upwelling. This stops in the northwestern part of the EEZ. The pelagic fishing effort in Guinean waters is restricted by this boundary, and the pelagic resources in Guinean waters are relatively modest, being 8 to 10 times less than in Senegal. The only significant concentrations are small sardines found all year near the coast, and sardine and chinchards which temporarily come from the north with the cold water of the Senegalese upwelling in the dry season, in 80 to 200 m depth.

Fishing

The pelagic fish catch in 1996 was 50,000 to 200,000 tons. Most are caught in the northwestern part of the region on the continental shelf, at the end of the dry season when the productivity is maximum due to the upwelling. This limited quantity is due to the fact that the enrichment mechanism and the development of planktonic food is much less in Guinea compared to that off Senegal, Mauritania and Morocco.

Three types of fishing exist in the coastal zone: traditional fishing, industrial fishing and shrimping. Traditional fishing is an old activity, practised all along the coast by riverside populations, mostly the Bagas and the Sousous. This sector has strongly developed over the last decade. There has been an increase in the number of canoe landing stages and an increase in the number of canoes (estimated at 2343), a greater proportion of which are motorised, resulting in a production estimated today at 50,000 tons, shared between pelagic and demersal species (Morize, 1995). The local production supplies the home market, but recently methods have been developed using freezer canoes which look for specific species for foreign markets.

This fishing is diverse in its practices, under the influence of migratory nationals and foreigners such as Ghanaians, Leonese and Senegalese (Boujou, 1992). To aid this sector, there are several agencies such as the canoe motorization centre which imports and sells fishing equipment and which trains and equips local mechanics, the National Fishing Supervision Center which looks after, controls and supervises fishing in order to maintain compliance with regulations, the Fish Smoking Center which increases the quality and production of smoked fish and which also especially helps women. Finally there is the National Halieutic Sciences Center of Boussoura (CNSHB) which is in charge of evaluating resources and developing fishing research programs.

Fishing is a varied socio-cultural activity. Six kinds of canoes are used: kourous, gbankegnis, yolis, botis, salans and the flimbotes. The three types of fishing methods found along the coast are mesh nets, circle mesh nets, line and hooks (Domalain et al., 1989). The coast has over one hundred landing stages whose quality and infrastructure are varied and sometimes inadequate.

Industrial Fishing

This is subject to regulations. Since 1985 Guinea has had a fishing Code which was revised in 1990 before being published in 1995. Five types of licences are defined: pelagic fishing, demersal fishing, cephalopod fishing, shrimping and tuna fishing. Two fishing areas are defined also: demersal fishing nearer shore and pelagic fishing further offshore. Net sizes are regulated, being 70 mm for drawn nets for fish and cephalopod trawls, 50 mm for bottom shrimp trawls, and 30 mm for pelagic fish.

Industrial fishing boats can freeze catches and stay at sea for more than two months, seldom landing in Conakry. Except for tuna boats, the industrial fishing is now dominated by demersal fishing and trawlers. The demersal trawlers possess demersal fishing licences, and pelagic fishing boats now are only a small part of the total flotilla.

In accordance with the annual fishing plan, 119 fishing licences were issued in 1996. Of these, 82 exploited the continental shelf resources and 37 were tunny boats. The size of the flotilla allowed to exploit Guinean waters decreased between 1995 and 1996. Effort decreased from 4229 to 3617 fishing days, a decrease of 14%, and cephalopod fishing decreased from 6623 to 5320 fishing days, a decrease of about 20%. At the same time, the pelagic fishing effort increased from 138 to 653 fishing days, and the days worked by shrimping boats rose from 1008 to 1853 fishing days. The changes appear to be the consequence of a development of deep shrimp stock exploitation along the edge of the continental shelf. Overall, industrial fishing performance increased 30% to 30,000 tons (against 23,000 tons previously) due to greater effort by pelagic trawlers.

Importance of Maritime Fishing

Guinea benefits from its EEZ, which contributes to food security, improves the social and economic living conditions of the people and contributes to the economy. According to Chavance (1997) production from local fishing was estimated to be 52,000 tons in 1995. The total number of local fishers is about 900 persons and from this number we can estimate that the fishing sector generates nearly 36,000 jobs or 1.3% of the Guinean active population (FAO, 1996). The value of this production was estimated at $US 18,700,000 in 1991, of which about $US 4,000,000 was exported, providing a trade balance of sea products of about $US 1,000,000. The fishing sector also contributes to earnings through fishing licences and fees.

Shrimp Culture

This is carried out in the Koba plain, which is prone to flooding, by a company called SAKOBA with French assistance. Shrimps are also caught between 20 and 40 m on the continental shelf.

POPULATIONS AFFECTING THE AREA

The population of the coastal zone is estimated at 1,963,384 with an average density of 61.94/km².

Guinea is a developing country, with a 52% schooling rate and average life expectancy of 47 years. Its health system has been restructured, but still suffers from a lack of essential materials. There is limited access to drinking water in certain zones: some areas have one water pump for 100 inhabitants. It is said that the coastal population has lived in this region since Neolithic times, and certainly, European navigators witnessed these populations in the 14th century. Besides agriculture, pig and chicken farming is practised in some parts.

There has been fishing supervision since the creation of C.N.S.P. in 1990. There is a swamp mangrove project at Dubreka, and a mangrove research center at Boussoura (Conakry).

There are various pressures on the natural resources along the coast apart from hunting and fishing. Shrimp farming is practised, rice is grown in swamp and mangrove areas as well as in estuaries, salt production is carried out in swamp and mangrove zones, and significant deforestation occurs through wood cutting. This wood is used for home consumption, for salt production, fish drying and house building. According to Yansané (1998) 18,000 tons of wood a year is transported to Conakry. There are some food and agribusiness industries, building material industries, and tanneries and soap factories. For most, there are no waste treatment facilities, so most of the factories discharge their wastes directly into the sea or into drains.

Coastal erosion is not a critical problem in most areas. According to Camara Morlaye et al. (1998) coastal erosion and subsequent flooding has destroyed some farming areas, and has also caused the disappearance of Conakry's most important beaches. Erosion is influenced by the fairly high tides which reach 4.5 m in the port of Conakry, 4.1 m in the Tamara Island and 5 m in Dubreka (Camara et al., 1998). Erosion is increased by heavy rains which deposit 4000 mm a year and, in some seasons, strong winds of 2.5 to 6 m s^{-1}.

In addition, deforestation of mangroves in particular has exaggerated the effects of coastal sediment transportation, increasing the exposure of the shores to winds and waves. In some areas too, urbanisation near the shore has increased run-off.

Guinea is now facing great difficulties in domestic and industrial waste management. Most of the coastal zone is considered as a rubbish tip.

Unregulated development by inhabitants along the coast of Conakry has resulted in a proliferation of different sources of pollution from diverse origins. Most of the pollutants reach the sea via run-off, much of which passes through agricultural regions which are regularly treated by pesticides, as well as carrying household and industrial pollutants as well. It is very difficult to evaluate precisely the volume of the pollutants carried by the run-off waters. Rivers collect dissolved substances, though some of the harmful elements are natural in origin (Camara et al., 1998a). The latter include not only nutrients but also some heavy metals from natural rock erosion. The most important activity is undeniably artisanal fishing. Fishers' camps or villages are situated along the entire coast.

Table 2

Classification of wastes found during a trawling campaign to measure dumped materials (Keita et al., 1998).

Designation	Composition (%)	Average weight per trawling (g)
Plastics	30	223
Metal containers	22	170
Other metal objects	8	50
Resins and derivatives	10	80
Synthetic and wrapping materials	8	45
Wood objects	30	150
Total	100	668

Table 3

Microelement concentrations in Guinean coastal waters (in $\mu g/l$)

	Fe	Mn	Cu	Zn	Pb	Ni
0 meters	1.62	0.19	0.48	0.59	0.16	0.65
	0.48	0.07	0.38	0.37	0.10	0.29
	2.10	0.26	0.86	0.96	0.24	0.94
30 meters	1.69	0.18	0.51	0.60	0.13	0.53
	0.35	0.06	0.28	0.24	0.05	0.19
	2.03	0.24	0.79	0.84	0.18	0.72

Industrial development has accelerated recently in Guinea but many industries still use old procedures which generate pollution. The Guinean Bauxite Company extracts 11,000 to 13,000 tons of bauxite a year, and this covers large areas with dust, which is then carried directly into coastal waters. An aluminium factory at Fria creates wastes which in the past were poured into the Konkoure river which then carries them to the coast of Conakry, but more recently, these wastes have been ponded. About 80 km from Conakry is a food processing factory whose wastes, with high organic content, are discharged without treatment.

In Conakry large quantities of pollutants are discharged or dumped into the sea (Table 2). The condition of maritime waters is worsening due to spilled and discharged hydrocarbon products, which form a true black tide along the coast. This was particularly severe after incidents in 1991 when fish were found to have ingested solid tar (Keïta et al., 1998) and in some places water pH fell below 5.0. Offshore detergent levels were about 200–300 g/l, and sediments contained high levels of several metals (Table 3).

There is also pollution from maritime transport and navigation. The port of Conakry receives nearly 500 to 600 ships per year, which dump 1000 to 1500 tons of wastes per year. Also from 1992 to 1993, leaking installations strongly affected the fauna and flora, pollution coming from ships as well. Dredging of Conakry's port is also a source of pollution.

PROTECTIVE MEASURES

In 1990 the National Center of Supervision and Fishing Protection (C.N.S.P) was created to apply protection and supervision of maritime fisheries. It has elaborated and promulgated a maritime fishing code and a code for the protection and evaluation of the environment. Articles in the code of the environment create an administration in charge of resources and management, and these prohibit dumping, leaking, waste release and other forms of pollution in Guinean continental waters.

Guinea is party to the London convention of 1954 on hydrocarbons, regulating and placing responsibilities for accidents and dumping, and has an emergency plan for dangerous situations that can lead to pollution in the sea or adjacent zones.

REFERENCES

Allen and Salomon (1980) Role sédimentologique de la mareé dans les estuaires à fort marnage. 26ème Congrès Géologique International, Paris, 1980, Resumé Vol. II.

Arseny, B. and Guerman Zoyev (1985) Les ressources en poisson de la Z.E.E. Guinéenne et les perspectives de leur exploitation.

Bah, M. et al. (1996). *Monograph on the Biodiversity*. D.N.E., Conakry, Guinea.

Berrit, G.R. (1962) Contribution to the knowledge of seasonal variations in the gulf of Guinea. Surfaces' observations along the navigation line. Part 2: Regional studies. *Cahiers Oceanographiques*, C.C.O.E.C. **14** (9), 633–643.

Bouju, S. (1992) Migrate fishers along the coast of Guinea from the 18th century up to now. Doc. Scientific, National Halieutic Sciences Center of Boussoura (C.N.S.H.B.), No. 16.

Camara, S., Bangoura, K., Magassouba, M. et al. (1998a) The potential causes of the Guinean coastal waters, pollution and strategy for cleaning the coastal line. *Bulletin of the Rogbane Center*, No. 12 (October).

Camara, S. et al. (1998b) Analysis of the biological diversity of the navies and coastal ecosystem of Guinea. Identification of priorities for their conservation. Projet GUI/97/A/1G/99.

Camara M. Morlaye et al. (1998) Plan de gestion intégrée de la zône côtière guinéenne (Baie de Sangaréah et de Tabounsou). ICAM (Projet Régional PNUE/FAO).

Chavance, P. et al. (1997) *Atlas of Guinea Maritime Fishing*. ORSTOM/C.N.S.H.B.

Domain, F. (1989) Rapport des Campagnes de Chalutages du N/O André Nizéry dans les eaux Guinéennes de 1985 à 1988. Doc. Scientifique. C.R.H.B. No. 5.

Domain, F. and Bah, M.O. (1993) *The Sedimentologic Map of the Guinean Continental Shelf*. Editions ORSTOM, Paris, 15 pp.

Domalain, G. et al. (1989) The Guinea canoes registration, I: Peninsula of Conakry and the Loos Island. Doc Scientific No. 6, C.R.H.B.

FAO (1996) Report: FAO/TCP/RAF/2381, 1996 sub-Regional Commissions for fisheries. Round-table conference. Scientific and technical report of the conjuncture 1997. C.N.S.H.B., Conakry, Guinea.

Fontana, A. and Lootvoet, B. (1994) The diagnostic studies of the Guinean maritime fishing sector. Doc. ORSTOM, Guinea.

Keita, Ansoumane et al. (1998) Identification, human grading, the analysis of the durability of the resources systems on the navy. Biological diversity and the principal cause of pressures.

Keita, A. et al. (1995) Plancton et bioproductivité de la zône côtière guinéenne. Atelier régional sur la gestion intégrée du littoral. ICAM. UNESCO No. 121.

Keita, M.L. et al. (1998) Fronts and upwelling of the coastal zone, in the prospect for the rationalisation of the Halieutic resources. *Bulletin of the Rogbane Center*, No. 12. C.E.R.E.S.C.O.R., Guinea.

Keita, M.L. and Bangoura, K. (1998) *The Health of the Oceans*. C.E.R.E.S.C.O.R. Guinea.

Kouzmenko, V.L. and Haba, B. (1988) Phytoplancton (composition et distribution); production primaire, chlorophylle atlantique tropicale. Région Guinéenne. Naukova Dunka, Kiev.

Morize, E. et al. (1995) End of studies reports. Project of protection and supervision of the Guinean ZEE "Scientific aspect". Doc. Multigr., ORSTOM/C.N.S.H.B., 137 pp.

Postel, E. (1955) Les faciés bionomiques des Côtes de la Guinee française. *Rapp. Cons. Int. Expl. Mer* 137, 10–13.

Rougeron (1996) The study of spatial sharing of the demersal resources in the Guinean continental shelf. Relation with benthos and physical environment. D.E.A Thesis. The National Agronomic High School of Rennes, 27 pp.

Wauthy, B. (1983) Introduction á la climatologie due golfe de Guinée. *Océanographie Trop.* **18** (2).

Yansané, A. (1998) Le schéma Directeur d'amenagement de la mangrove de Guinée (SDAM) et sa mise en oeuvre. Bull. du Centre de Rogbané, No. 12, Conakry, Oct. 1998.

THE AUTHORS

Ibrahima Cisse
The National Center of Halieutic Sciences of Boussoura,
B.P. 3060,
Conakry, Republic of Guinea

Idrissa Lamine Bamy,
The National Center of Halieutic Sciences of Boussoura,
B.P. 3060
Conakry, Republic of Guinea

Amadou Bah,
The National Center of Halieutic Sciences of Boussoura,
B.P. 3060,
Conakry, Republic of Guinea

Sékou Balta Camara
The National Center of Halieutic Sciences of Boussoura,
B.P. 3060,
Conakry, Republic of Guinea

Mamba Kourouma,
The National Center of Halieutic Sciences of Boussoura,
B.P. 3060,
Conakry, Republic of Guinea

Chapter 51

CÔTE D'IVOIRE

Ama Antoinette Adingra, Robert Arfi and Aka Marcel Kouassi

The Ivoirian oceanic zone is bordered to the north by the Gulf of Guinea shoreline, stretching from Cape Palmas (7°30W) to Cape Three Points (2°W). The continental shelf is 25–30 km wide, with a surface of about 16,000 km². The shoreline, with a length of 530 km, is formed by sandy beaches, shaping a wide arch open to the Atlantic Ocean. A major morphological feature, the "Trou Sans Fond" Canyon cuts the continental shelf in front of Abidjan; there, depths over 1000 m are rapidly reached a few kilometres offshore. Large rivers (Cavally, Sassandra, Bandama and Comoé) drain the country from the north to the south and flow into the ocean either directly or via a lagoon.

The current pattern is dominated by the Guinean Current flowing eastward in the upper layer (average speed and maximum velocity of about 0.5 kt and 2 kt, respectively) and by the Ivoirian Undercurrent running westward in the subsurface layer, (average speed 0.4 kt). Waves from the open sea are very energetic and the swell originating from the South Atlantic Ocean produces a permanent surf parallel to the coastline. Tides are semi-diurnal with diurnal inequality and an amplitude ranging from 0.8 to 1 m. Coastal upwellings occur seasonally along the shoreline, from July to September (major event) and in January (minor event). The prevailing coastal winds are the Monsoon Trade winds blowing from southwest to south-southwest with a speed of about 3–4 m s^{-1}. The climate is governed by the latitudinal displacement of the Inter-Tropical Convergence Zone (ITCZ) separating a humid air mass of oceanic origin (Monsoon period) and a dry air mass of continental origin (Harmattan season). The major rainy season (54% of the annual rainfall, ranging from 1500 to 2200 mm yr^{-1}) occurs generally from May to July, the minor rainy season (16%) occurs from October to November. The major dry season starts in December and ends in March and the minor one between August and September.

The coastline encompasses a variety of coastal habitats including lagoons, estuaries, mangroves, swamps and humid zones. These critical habitats providing spawning grounds for numerous fish, molluscs, birds, manatees and other life forms are now undergoing rapid destruction as a result of intense human activities. Deriving mainly from the history of the country's first contact with European seafarers, nearly all major infrastructures in the country are located in the coastal area. Pollution from these various sources affects the water and their natural living resources. Environmental degradation including critical habitat destruction and loss of biodiversity are among the major impacts. Concerns about the deterioration of the coastal and marine environment of the Côte d'Ivoire coupled with the experiences gained from the country's participation in several regional and international conventions have led to the preparation of a National Environmental Action Plan and also to the vote by the Parliament of a new Outline Law on Environment.

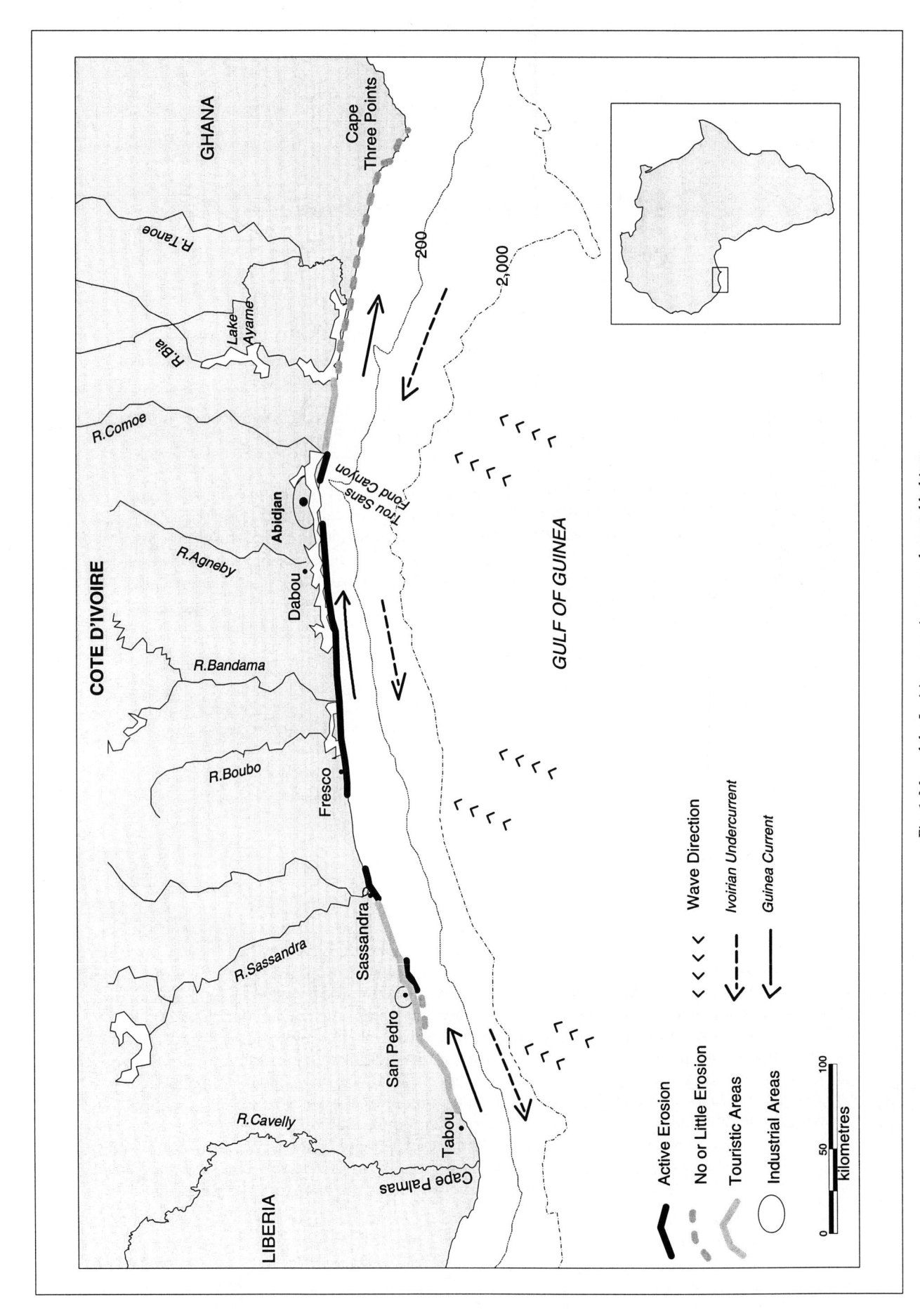

Fig. 1. Map of the Ivoirian oceanic area and coastal habitats.

THE DEFINED REGION

Two syntheses of the studies conducted along the coastal areas of Côte d'Ivoire exist: one mainly of shallow areas of Ebrié Lagoon (Durand et al., 1994), the other concerns other coastal areas (Le Lœuff et al., 1993). A third synthesis (a special issue of *Océanographie Tropicale*, 1983, number 18) describes some pelagic studies conducted in the Gulf of Guinea, near the Equator.

Geological Context and Geographical Limits

The Ivoirian oceanic area is bordered to the north by the Gulf of Guinea coastline, to the west by the Cape Palmas (7°30W, Liberian border) and to the east by the Cape Three Points (2°W, located in Ghana). To the south, the continental slope delimits a narrow continental shelf (25–30 km wide, surface about 16,000 km²). The slope is smooth until –120 to –150 m, then increases sharply (Martin, 1973). A major morphological feature, the "Trou Sans Fond" Canyon cuts the continental shelf in front of Abidjan: there, depths over 1000 m are rapidly reached a few kilometres offshore. Côte d'Ivoire has a land area of 318,000 km² and a shoreline of 530 km. The littoral is made of a series of sandy beaches and forms a wide arch open to the Atlantic Ocean. It can be subdivided into two parts (Fig. 1):
- To the East of Fresco, it is a flat coast, with sandy and monotonous structures of sedimentary origin (Quaternary). Several lagoons (submersed fluviatile basins) are separated from the sea by a littoral bar, built and maintained by waves and currents.
- To the West of Fresco, it is a more complex structure, where the metamorphic basement reaches the sea. Rocky capes with low cliffs alternate with sandy bays.

Water Inputs, From the Continent and From the Ocean

Five major rivers flow into the Gulf of Guinea (Table 1):
- the Cavally River drains a forest area from Mount Nimba (Liberia) to the coastline;
- the Sassandra and the Bandama rivers drain the country from north to south. Both rivers rise in the northern region of the Côte d'Ivoire, which is dominated by savannah vegetation. Dam construction for power production (Buyo on the Sassandra River, Kossou and Taabo on the Bandama River) markedly decreases the freshwater flow from these two rivers;
- the Comoé River also rises in the savannah zone of Burkina Faso. Its natural mouth (Grand Bassam inlet) was closed after the opening of the Vridi Canal and now, the river flows seaward through the Ebrié Lagoon;
- two dams (Ayamé I and II) regulate the Bia River. Its mouth opens in the north shore of the Aby Lagoon, itself opened to the ocean at the Assinie Pass.

Table 1

Main river characteristics. Flow data correspond to the 1980–1996 period

	Draining surface (km²)	Length (km)	Average flow (m³ s⁻¹)	Minimum flow (10⁹ m³ yr⁻¹) and record year	Maximum flow (10⁹ m³ yr⁻¹) and record year
Cavally	30,000	700	390	3.8 (1992)	30.1 (1966)
Sassandra	75,000	650	355	6.2 (1984)	33.8 (1962)
Bandama	97,000	1050	173	2.0 (1978)	22.4 (1957)
Comoé	78,000	1160	121	0.8 (1983)	14.9 (1968)
Bia	10,000	290	na	0.8 (1986)	5.2 (1964)
Tanoé	16,000	385	na	1.8 (1958)	7.6 (1968)

na: Not available.

These large rivers are characterised by high flow during the flood season and by low flow during the dry season. All of them have permanent openings to the Gulf of Guinea, and their floods strongly influence the marine system. Several coastal rivers reach the ocean via a lagoon (Tanoé River into the Aby Lagoon, Mé River and Agnéby River into the Ebrié Lagoon, Boubo River into the Grand Lahou Lagoon). Other minor forest rivers have their mouths west of Fresco. In most cases, they have a low flow during the dry season; then, the ocean transport rapidly builds a sandy ridge, closing the mouth and forming little lagoons. They flow again seaward during the flood season.

The coastal ocean along the Côte d'Ivoire shoreline has a morphological, geographical and dynamical unity, under a seasonal continental influence and under a permanent oceanic influence. It is widely open to the Atlantic Ocean, but its functioning is driven by local forces. Biological processes governed by the structure characterise both the pelagic and the benthic systems. Major changes occur in July and in January, when cold and rich deepwater reaches the surface.

SEASONALITY, CURRENTS AND NATURAL ENVIRONMENTAL VARIABLES

Ecological processes in the area are largely governed by seasonal or periodic phenomena, occurring over a wide frequency range. These processes include lateral hydrodynamics (waves and currents), vertical oscillations (thermocline migrating up and down), continental water inputs (during the flood) and climatic variations (solar radiation, winds). Some aperiodic phenomena (e.g. strong winds) may sometimes affect the upper layer of the water column.

Currents and Hydrodynamics

Currents

Dominated by two systems, the current pattern is simple. The Guinean Current (GC) flows eastward in the upper

layer (0–30 m depth), with an average speed of about 0.5 kt and a maximum velocity close to 2 kt. The Ivoirian Undercurrent (IU) runs westward in the subsurface layer, with an average speed of 0.4 kt. Both current systems are permanent, and their movements at the surface induce a powerful coastal drift. However, superficial circulation is spatially and temporally affected by high variability (Colin, 1988):
- horizontally, the GC can extend southward, but then its speed and intensity decrease rapidly;
- vertical inversions occur from January to March and in October, as a result of a southward shift of the GC allowing the IU to reach the surface;
- the GC reaches its maximum intensity from May to August and from December to February. The IU velocity is high from July to November and from February to April.

Waves

These are from the open sea and often very energetic. The swell originates during the austral winter from the South Atlantic (50–60°S), and produces permanent surf parallel to the coastline. Breaking waves induce littoral transport, favoured by surface currents (Fig. 1). In the western part of the littoral, erosion is weak (the load of sand carried is 200 km^3 yr^{-1}), since the shore is rocky and capes alternate with bays. High erosion characterises the central zone, and the beach ridge is actively eroded (800 km^3 yr^{-1}). Since the opening of Vridi Canal and the construction of a protection pier, transport between the central and the eastern zones is interrupted. Most of the sand coming from the west either goes into the Trou Sans Fond Canyon or contributes to the growth of beaches adjacent to the western side of the pier. Erosion is very active in the eastern part of the littoral between the Vridi Canal and Grand Bassam (400 km^3 yr^{-1}). Consequently, these processes modify the shoreline, move the river mouths and destroy coastal buildings. They are also a threat to San Pedro and Abidjan's harbours and for the sites located east of Vridi, industrial (oil refinery, airport) as well as tourist facilities such as hotels along the beach.

Tides

These are semi-diurnal with diurnal inequality. Tidal amplitude is low, ranging from 0.8 to 1 m. The average seawater level is a good indicator of the occurrence of the upwelling events. It is low in January and from July to September, high between May and June and from October to November. Seawater level fluctuation is around 15-cm ± the tide signal.

Wind

Monsoon Trade winds blow 10 months a year from south-west to south-southwest. They are weak (3–4 m s^{-1}) and regular, characterised by diel rhythms. Their speed increases during the boreal summer (4–6 m s^{-1}). In January and February, Northeast Trade winds (also called Harmattan) blow from north-northeast to northeast. Despite their reduced speed, the enormous sandy and dusty load they carry in the atmosphere induces high nebulosity.

Coastal Upwellings

These occur seasonally along the shoreline, from July to September (major event) and in January (minor event). The thermocline moves upward under the combined action of winds and currents. Intensification of zonal wind along the shoreline causes a Kelvin wave trapped at the Equator. The eastern border of the basin reflects this wave as secondary Rosby and Kelvin waves, which in turn induce the upward movement of the thermocline. During the boreal summer, local wind speed increases. Its direction rotates slightly eastward, and becomes more parallel to the coast. Therefore, prevailing winds contribute more to the vertical movement, while the GC speed increases during the same period. Coastal morphology and dynamic processes enhance locally this phenomenon (cape effect). More intense and more lasting major events are observed between Tabou and Sassandra. In this part of the coast, minor upwellings always occur. From Fresco to Abidjan, the main upwelling decreases in intensity and in duration and the minor event is generally weak, sometimes absent. East of Abidjan, cooling is reduced during the main event, and the minor upwelling sometimes occurs. Interannual variability of these events is high, and nutrient enrichment varies considerably (Arfi et al., 1993b). For several years, cooling has increased in the western part of the littoral, particularly during the minor cold event.

Climate

The climate of the coastal zone is governed by the latitudinal displacement of the Inter-Tropical Convergence Zone (ITCZ). Alternation of rainy and dry seasons is regulated by zonal and seasonal variations of the ITCZ along the coastline. The ITCZ separates two air masses: the humid air mass of oceanic origin (Monsoon period) and the dry air mass of continental origin, (Harmattan season).

In the coastal area, heavy rains occur generally from May to July (Fig. 2a), but the rainy season begins sometimes in April. During this wet period, humid air goes northward, and rainfall represents 54% of the total for the year. August and September are dry and cool: this is the short dry season. Rains come again in October and November, when humid air goes southward. This short rainy season provides on average 16% of the total rainfall. December to March is the main dry and hot season, with a short cool event (January or February) during the Harmattan period. Rainfall is higher in the western part of the shore (about 1800–2200

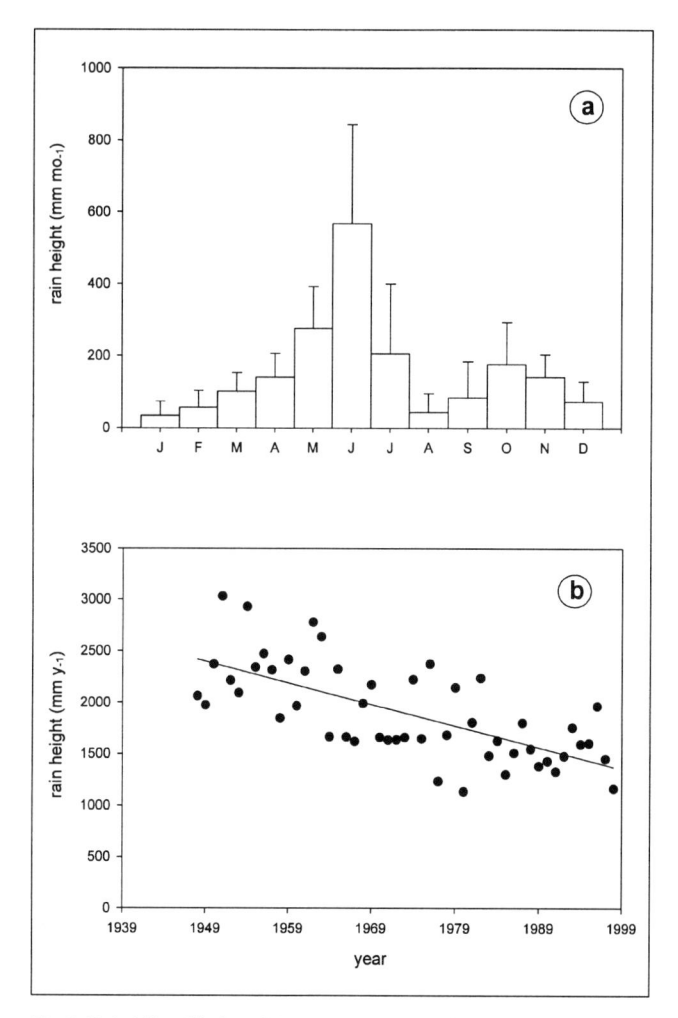

Fig. 2. Rainfall at Abidjan (1948–1998). (a) Monthly rains (average and standard deviation) and (b) annual rain with first-order fit (rain = −20.9 * year +3423, $n = 51$, $r = 0.68$).

mm yr^{-1}) than in the central zone, where low values are usually observed (1200–1500 mm yr^{-1}). In the eastern zone, rainfalls range from 1500 to 1800 mm yr^{-1}. Since the 1970s, the intensity of the main rainy season has decreased, while the duration of the short rainy season has been increasing. In recent years, the two seasons showed comparable rainfall. Both for the main rainy season (the lowest values ever recorded in Abidjan) and the short rainy season (heavy rains until December), 1998 was an atypical year. Low rainfall was already observed on occasion in the past decades, but such years, rare in the past, are now more common (Fig. 2b). If the 1950s and 1960s were rather rainy, the decrease observed from the 1970s has now resulted in a large deficit in the coastal area. Now, only the western extremity of the littoral shows annual rainfall higher than 1800 mm. From Sassandra to Grand Lahou, rainfall is lower than 1400 mm yr^{-1}.

Air temperature shows little variation around 26°C, with an average change of 4 and 8°C, respectively, daily and monthly. The sky is often cloudy, particularly during the rainy seasons. Storms are frequent, since the littoral is under the transit line of storms originating from the eastern extremity of the Gulf of Guinea. Tornadoes often occur in March and April.

Effect of Hydroclimatic Variables on the Biological Compartments

Two seasonal factors strongly affect marine and coastal communities living along the shore: floods, enhancing continental influence, and upwellings, enhancing oceanic influence. Freshwater inputs (run-off and river flows) are high from April to July and from October to November. Their consequences on the marine environment are linked to their nature (decrease in salinity), their particulate load (high turbidity) and their dissolved load (increase in nutrient concentrations). Nationally, these effects are limited to the first metres of the water column, and the river plumes in the ocean extend rarely more than a few kilometres southwards. Ocean hydrodynamic factors disrupt the front between continental and marine waters, ensuring rapid mixing. For rivers having their mouth in a lagoon (Bandama, Bia, Tanoé and Comoé), the mixing with brackish water modifies their characteristics. Therefore, the freshwater influence is weakened. Deep seawater inputs extend to the superficial and euphotic cold and nutrient-rich waters, and the whole continental shelf is influenced when upwellings occur. Enrichment is higher in the western zone of the littoral than in the eastern zone and is correlated with upwelling intensity and duration.

In the 0–80 m layer, large seasonal variations govern ecological factors (temperature, salinity, water transparency and nutrient concentrations). Annual cycles are defined from a coastal station in front of Abidjan:

A. After the flood, salinity increase is linked to the reduction of continental outputs (low water for the major rivers, local dry season) and to deep water inputs during the minor cold event.

B. From mid-May to June, salinity decreases when Guinean waters (GW) are present. GW are a combination of oceanic water and freshwater originating from the coastal area of West Africa, carried eastward by the Guinean Current. From October to December, floods induce a marked salinity decrease: the hydrological situation is comparable to the GW sequence, although decrease in salinity is less pronounced.

C. The local rainy season ends when the thermocline reaches the surface. From July to September, upwellings induce sharp water cooling and an increase of surface salinity.

Coastal upwellings establish themselves when surface temperatures fall below 26°C. During this period, surface salinity is high, contrasting with the desalting events linked to the floods. Water transparency decreases in June, when the GW invade the coastal area. Light attenuation is high from June to October (average Secchi depth: 9 m), low from November to May (18 m). The euphotic layer is 20–25 m during the high turbidity period, 35–40 m during the low

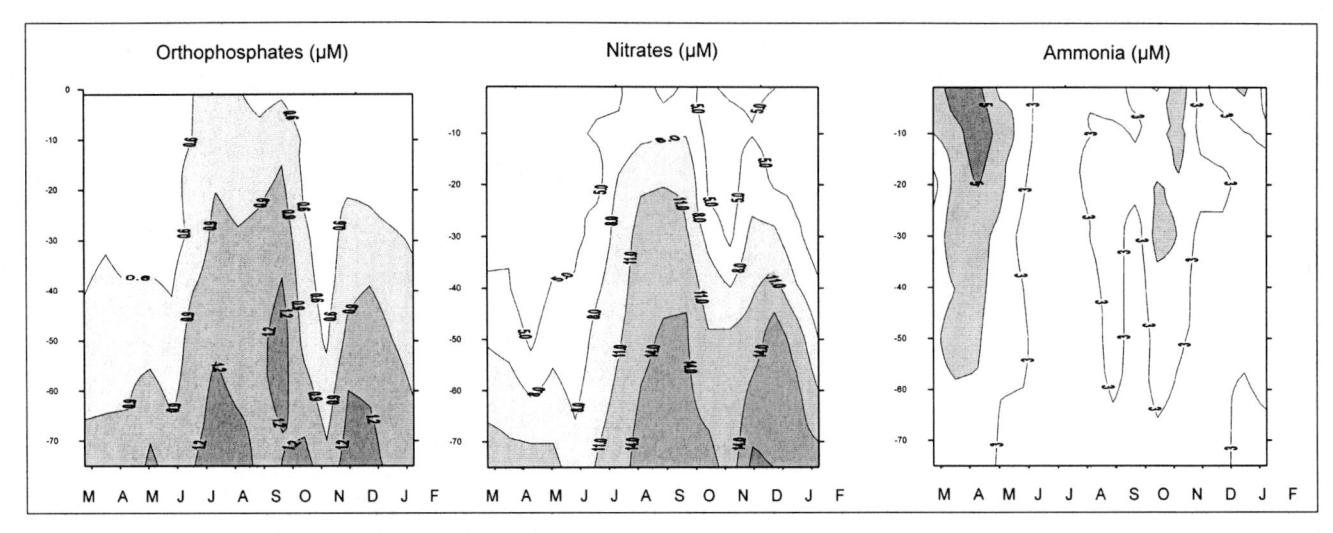

Fig. 3. Contour plot of nutrient annual cycle with depth at the Abidjan coastal station.

turbidity period. Turbidity has decreased between the periods 1966–1971 (average Secchi depth: 10 m, euphotic layer: 27 m) and the 1992–1997 (respective values: 14 m and 36 m). Turbidity is related to local rains, and the decreasing rainfall during the past decades now reduces continental influence. Dam effects complete this phenomenon, since a large part of freshwater particles is trapped in the reservoirs and does not reach the sea. Therefore, the effects of the GW intrusion seem to be less effective today.

Phosphate enrichment of the euphotic layer is clear from mid-June to mid-October (Fig. 3), when the 0.6 μM contour line reaches the surface. Transient incursion of phosphate-rich water above the –30 m level is observed from mid-December to mid-January. Nitrates show the same annual pattern: the average concentration near the bottom (14.4 μM) is four times higher than near the surface (3.7 μM). At these two levels, there is no obvious seasonal cycle, while such a cycle is clear between –20 and –50 m, with a marked concentration increase from mid-June to mid-October. The euphotic layer shows higher concentrations during the upwelling situation than during the sequence under continental influence. Ammonia shows an opposite pattern, with high concentrations from March to May and from October to December. These high values reflect active mineralisation process in the water column that occurs simultaneously with the input of organic matter of continental origin.

THE MAJOR SHALLOW WATER MARINE AND COASTAL HABITATS

The coastline encompasses a variety of coastal habitats including coastal lagoons, estuaries, mangroves, swamps and humid zones. The more characteristic coastal habitats are the lagoon systems. They combine brackish and shallow ecosystems, mangrove and estuaries, in a geographical continuum starting with freshwater conditions and ending at the seashore. Swamps and humid zones cover large areas, mainly along lagoon shores (Fig. 4).

Mangrove

Coastal wetlands are located along the banks of lagoons and estuaries (Nicole et al., 1994). The dominant species are *Rhizophora racemosa*, *Avicennia germinans* and *Conocarpus erectus* (respectively called red, white and grey mangrove). These three species do not coexist in all areas, and *R. racemosa* is usually the main species. It grows well in low salinity zones and it is observed both at the water's edge and further inland. *A. germinans* survives in higher salinity, while *C. erectus*, rather rare, grows at the interface between mangrove and forest. They are accompanied by other

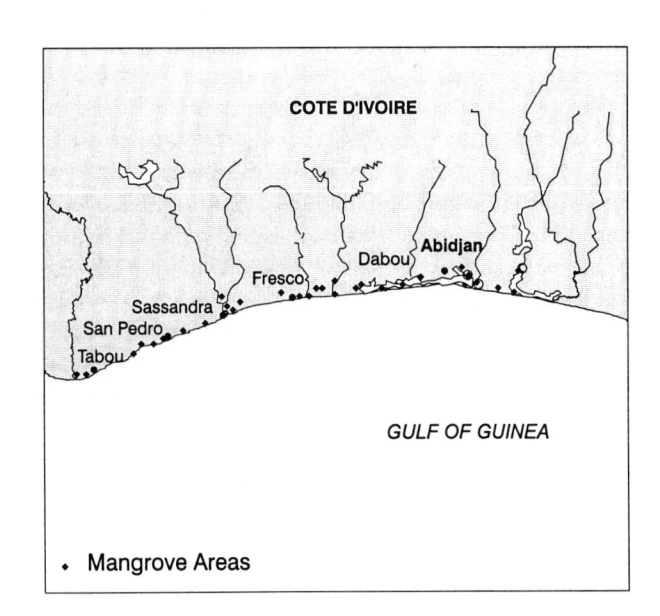

Fig. 4. Mangrove habitats in Côte d'Ivoire.

species, like *Drepanocarpus lunatus, Hibiscus tiliaceus, Dalbergia ecastaphyllum, Acrostichum aureum, Phoenix reclinata, Pandanus candelabrum, Panicum repens* and *Paspalum vaginatum*.

Lagoons

Several shallow systems are observed from Fresco to Assinie (Fig. 5). Most of the scientific efforts have focused on the Ebrié Lagoon. Some studies were conducted on the Aby Lagoon, and the Grand Lahou Lagoon is poorly known.

These three systems communicate by artificial canals: Asagny Canal links Grand Lahou and Ebrié lagoons, while Assinie Canal links Ebrié and Aby lagoons. Narrow and shallow, these waterways were once used for transportation but are no longer maintained.

Physical Framework

The Ebrié system (523 km², 120 km long, 1–7 km width, average depth 4.8 m, maximum depth 28 m) stretches parallel to the shoreline. Several bays (half-closed or

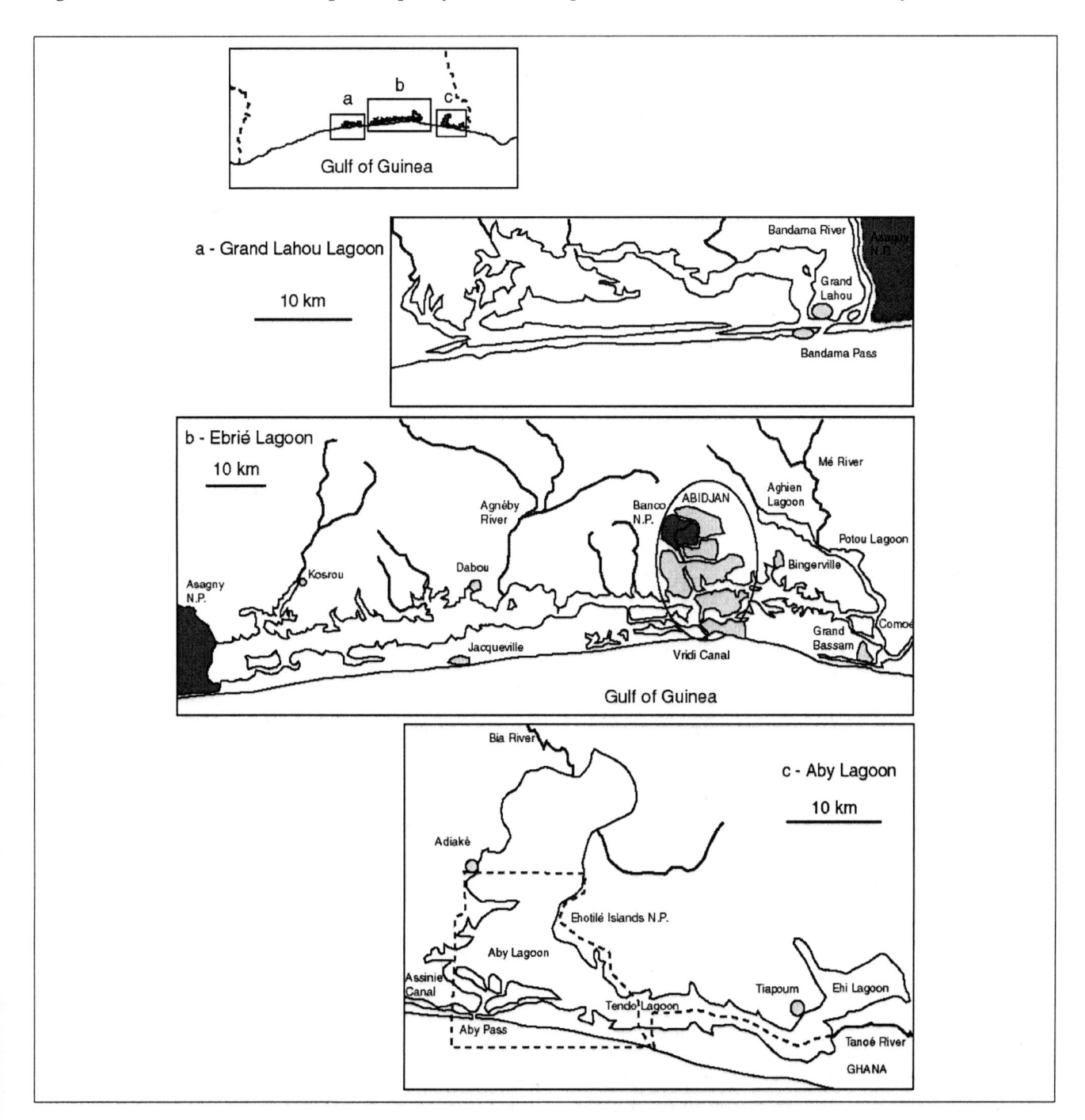

Fig. 5. Sketches of Côte d'Ivoire lagoons.

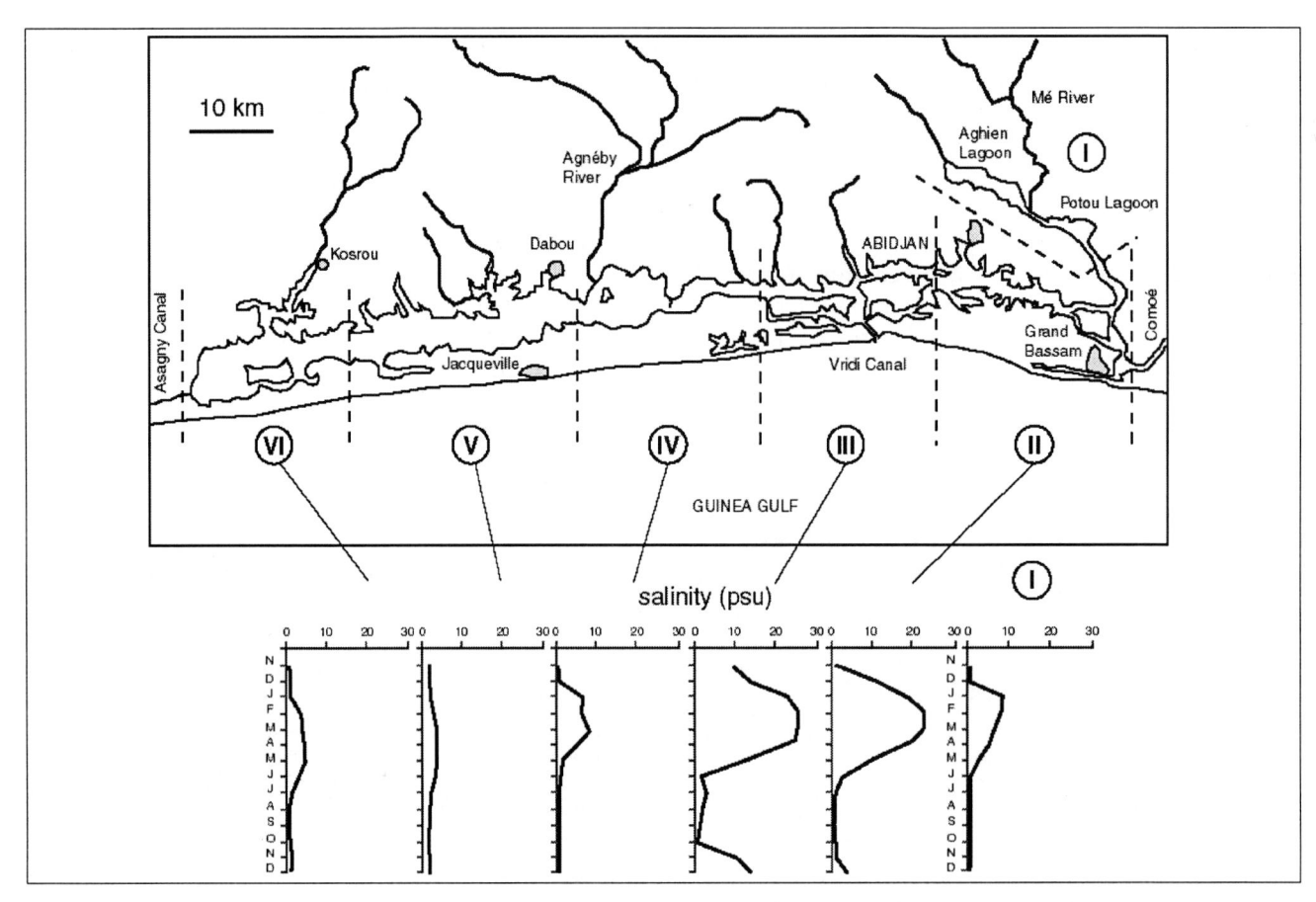

Fig. 6. Zonation in Ebrié lagoon as defined from salinity seasonal variations.

opened perpendicular to the main axis) and secondary basins (Aghien and Potou Lagoons, 43 km²) complete a system with a complex morphology. The artificial Vridi Canal, a unique link between the lagoon and the ocean, allows permanent communication with the Gulf of Guinea and makes Abidjan a secure harbour. Freshwater inputs into the lagoon (rainfall and river flows) range between 2.3 and 22.3×10^9 m³ yr^{-1}. The minimum value was recorded in 1983, in intense drought situation. The average input is estimated to 6.3 10^9 m³ yr^{-1}. The solid inputs average 0.4 10^6 t yr^{-1}, but the seaward export of the particulate load is estimated to 10% of this figure (Tastet and Guiral, 1994).

Hydrological Zonation

Continental and marine influences are permanently in opposition in the lagoon. From May to December when there is local rainfall and then river floods, the whole lagoon is desalted; the more intense are the annual freshwater inputs, the more lasting will be the phenomenon. From January to April, continental inputs are reduced, but evaporation is at its maximum in a context of dominant oceanic influence. Salinity is highest in the estuarine part of the lagoon. From this seasonal alternation and the morphological structure, a zonation (Fig. 6) has been proposed (Pagès et al., 1979). Zones V and VI are oligohaline, with maximum salinity close to 5 psu and limited amplitude. Zones I and IV are mesohaline (maximum around 9 psu, and minimum close to 0, linked to the flood intensity). Zones II and III are polyhaline. Their highest salinity values are comparable (23 and 27 psu, respectively), but during the flood season, salinity is very low in sector II, and more variable in sector III which is more an estuary under tide influence. Sectors I, V and VI are rather confined, with low turn-over rates, while sectors II, III and IV have high turnover rates. Lagoon waters are characterised by high nutrient concentrations (Dufour, 1994a). In most parts of the north shore, wind-induced resuspension is intense where depths are low, contributing to water turbidity and sediment instability (Arfi et al., 1993a; Arfi and Bouvy, 1995).

Plankton

Phytoplankton biomass and productions are relatively high in areas characterised by high turnover rates, while the more confined areas feature reduced algal activity (Dufour, 1994b). Cyanobacteria, diatoms and chlorophytes prevail in the algal communities (Iltis, 1984). Abidjan's polluted bays show often algal blooms, sometimes characterised by very high numbers of cyanobacteria and euglena (Arfi et al., 1981). Marine or brackish forms characterise the lagoon zooplankton, but the dominant community, with high

Table 2

Main hydrobiological features over the Côte d'Ivoire continental shelf (after Binet, 1979)

	J	J	A	S	O	N	D	J	F	M	A	M
Rainfall	Main rainy season		Short dry season		Short rainy season		Main dry season				Tornadoes	
River regime	Coastal river floods			Main river floods			Low waters					
Upwelling			Main cold events					Short cold events				
Surface temperature		Cold				Warm		Cold			Warm	
Surface salinity	Low		High			Low			High			
Guinea Current thickness (m)	20		10			15		10			30	
GC velocity (kt)	1.4		0.4			0.5		0.5			1.0	
Ivoirian Undercurrent location	Coastal		South								South	
IU speed (kt)	0.8		0.3	0.8				0.6			0.4	
Primary productivity	Moderate		High			Low		Moderate			Low	

Acartia clausi numbers, is not very diverse (Arfi et al., 1987; Pagano and Saint-Jean, 1994). Freshwater and oceanic communities show low numbers. The main assemblage (dominated by cladocerans and cyclopoïds) is observed in both extremities of the lagoon, while the second one (*Lucifer*, calanoids, chaetognaths, salps and medusae) is more diversified and is observed in the estuarine part during high salinity periods. Transition communities show high rotifer numbers.

Benthos

Four main assemblages are defined, all dominated by molluscs (Zabi and Le Lœuff, 1994). In the estuarine area, *Crassostrea gasar* and *Brachyodontes tenuistriatus* are dominant in the upper levels, accompanied by euryhaline crustaceans and polychaetes. *Anadara senilis* and *Tagelus angulatus* prevail in the second assemblage, associated with crustaceans. They share the same environment as the first community, but in less energetic and deeper areas. *Pachymelania aurita* and *Congeria ornata*, seconded by polychaetes dominate the benthos in the eastern part of the lagoon, particularly in low salinity environments. This assemblage covers large areas in the 0.5–2 m strata where microphytobenthos is very abundant (Plante-Cuny, 1977). The fourth assemblage, dominated by *Corbula trigona* and *Iphigenia truncata*, is observed in the whole western part of the lagoon, with very high numbers of *C. trigona*.

Fishes and Fisheries

With 153 species described (Daget and Iltis, 1965; Albaret, 1994), the Ebrié lagoon has relatively high diversity. Morphological complexity, biotope variety and simultaneous presence of marine, brackish and desalted conditions explain this diversity. Several species of continental origin can support moderate salinity variations (*Chrysichtys nigrodigitatus*, *C. maurus*, *C. auratus*, *Clarias ebriensis* and *Hemichromis fasciatus*) but some species are dependant on more specific environments (e.g. *Tilochromis jentinki jentinki*, characteristic of estuarine conditions). Most of them are marine species adapted to brackish situations (*Liza grandisquamis*, *Pomadasys jubelini*, *Ethmalosa fimbriata*, *Trachinotus teraia* and *Pseudolithus elongatus*), which reproduce in the lagoon or which have one development phase in the ocean. Other marine fishes can make transient incursions into the brackish environment. They are accompanied by a wide group of secondary species which have been observed in particular occasions or locations, which contribute to the community diversity. Around 80% of the species are carnivorous, feeding largely on crustaceans, penaeids and mysids, but the more abundant fishes like *E. fimbriata* are adaptable and opportunistic. Fisheries collapsed in the eighties, and the collective fishery has now considerably decreased. Artisanal fishery is now focused on shrimps and *E. fimbriata*. Increasing pollution in the estuarine area (induced by the growing importance of Abidjan sewage and non-point pollution) limits the fisheries in the lagoon.

OFFSHORE SYSTEMS

The annual cycle in offshore areas reflects the same hydrobiological features as those observed in the continental shelf (Table 2); these conditions drive marine community development.

Benthic Assemblages

Benthic assemblages of the continental shelf are characteristic of sandy sediment, and the bottom composition (grain size, organic richness) is the main determinant. From the shoreline to the slope break, four ecological zones have been defined (Le Lœuff and Intès, 1993).
- The infralittoral area lies from the surf zone to around –30 m. Very high hydrological variability is observed (Table 3), with large seasonal variations (at a depth of 10 m, 10°C in temperature, around 3 psu in salinity, but differences are higher at the surface where salinity is close to 7 psu.
- Between –30 and –65 m, the coastal circalittoral zone is permanently under the influence of the thermocline. Seasonal temperature differences are still high (around 10°C), but salinity variations are limited (less than 1.5 psu). During the warm season, temperature exhibits high frequency variability, induced by the vertical movement of the cold layer.
- From –65 m to the slope, the circalittoral zone is hydrologically stable. Diurnal and seasonal changes are limited, owing to the fact that the South Atlantic Central Water is permanently present.
- Below the shelf break there is a biological break-point, but the deeper community remains in hydrological continuity with the circalittoral zone.

From studies conducted from 1966 to 1973, Le Lœuff and Intès (1993) have defined seven benthic assemblages in this continental shelf. Sand and silty sand communities characterise the infralittoral biota. Fifty-three species were found in the first assemblage, forty-five in the second. Sand assemblage (52 species), silty sand (118 species) and sandy silt communities (92 species) are observed in the coastal circalittoral. A coarse and silty sediment assemblage (53 species) is observed in the circalittoral zone. The deep external margin community (21 species) characterises the slope. These assemblages are largely dominated by polychaetes, crustaceans and molluscs, and echinoderms can be sometimes abundant. Carnivorous and detritivorous species are dominant, followed by limivorous and filter-feeders. These communities show analogies with those described in Ghana or in Sierra Leone. In the upper levels (infralittoral and circalittoral), communities show the highest abundance and diversity during the warm season (February to April) and during the main upwelling (August to October). The two rainy seasons are characterised by a decrease in abundance and in diversity.

Phytoplankton

Reyssac in the sixties and Dandonneau in the seventies described algal communities and their productivity, summarised in Sevrin-Reyssac (1993). Nutrient inputs during upwelling events allow relatively high algal growth (4–10 10^6 cell l^{-1}, chlorophyll concentrations higher than 1 mg m^{-3}, and average primary productivity of 1 g C m^{-2} d^{-1}). These values usually decrease southward. During the warm season, phytoplankton abundance is low (chlorophyll concentrations lower than 0.2 mg m^{-3}, average primary productivity lower than 0.3 g C m^{-2} d^{-1}), and the same decreasing trend toward the open sea is observed. Dinoflagellates are abundant and diverse in warm conditions (158 species, 65 species of which are *Ceratium*). *Gymnodinium splendens* can induce red tides. Cyanobacteria (*Oscillatoria*) are often present in warm water (>27°C), but disappear rapidly with the intrusion of cold waters. Some diatoms show affinity to warm waters (*Biddulphia sinensis*, *Hemiaulus membranaceus*), but most of them (several species of the genera *Chaetoceros*, *Bacteriastrum*, *Rhizosolenia* and *Coscinodiscus*) bloom in upwelling situation. In these conditions, diversity is low.

Zooplankton

Zooplankton studies, conducted in the seventies (Seguin, 1970; Binet, 1979; Le Borgne and Binet, 1979), are summarised by Binet (1993). Zooplankton communities are perturbed by hydrological instability, such as cold water intrusion during the warm season or periodic freshwater inputs. On the other hand, the main cold event induces a lasting and stable sequence. Zooplankton development is favoured, but the coastal areas are not among the richest: high numbers are often encountered above 60–100 m depths. Open sea assemblages are dominated by copepods, followed by ostracods (though sometimes they can be as abundant as the copepods), appendicularians and chaetognaths. These organisms have their maximum abundance during the main cold event, but they are also well represented during the minor upwelling: their abundance is closely related to the phytoplankton biomass. Salps, pteropods and cladocerans (*Evadne tergestina*) are very

Table 3

Descriptive statistics for temperature (°C), salinity (psu), and Secchi depth (m) at the coastal station near Abidjan (1992–1997 data). The euphotic depth (m) is calculated after a linear regression between Secchi disk values and light attenuation coefficients ($n = 43, r = 0.84, p < 0.001$, Arfi, unpublished data).

	Min.	25%	Median	75%	Max.
Temperature (–10 m)	18.9	23.2	26.4	27.9	29.9
Salinity (–10 m)	32.59	34.71	35.15	35.54	36.01
Temperature (–50 m)	15.9	18.3	19.8	22.0	26.4
Salinity (–50 m)	34.73	35.69	35.79	35.90	36.15
Secchi depth	4	9	13	18	30
Euphotic depth	14	25	32	39	55

abundant during the short cold event. Large crustaceans (*Lucifer faxonii*, mysids, euphausids) and larvae of benthic decapods reach their peak in February, June and from October to December. Most of the species encountered above the continental shelf are eurytherm. During the warm season, carnivorous species dominate the plankton; then, diversity is high but numbers are low. Brackish species (*Acartia clausi*, *Paracalanus parvus*) are observed in the Gulf of Guinea during the flood season. In upwelling conditions, herbivorous plankton (dominated by *Calanoides carinatus*) show high numbers during the algal bloom, then are replaced by omnivorous species (*Temora turbinata*, *Centropages chierchiae*).

Pelagic Fisheries

The pelagic ichthyofauna is characterised by marked variability (Marchal, 1993), linked to climatic variation (Cury and Roy, 1987) but also to the fisheries' performance (Ecoutin et al., 1993). Fisheries are based on clupeids (*Sardinella aurita*, *S. maderensis*), exploited by semi-industrial and by artisanal techniques (Pezennec et al., 1993). The first one, operated mainly from Abidjan, was very active until 1973, when there was a severe decrease in abundance of *S. aurita*. The second one, still important, is operated from several hundreds of canoes, scattered all along the coast. Estimates of annual catches range from 20 to 30,000 metric tons, and Ghana and Côte d'Ivoire share probably the same sardinella stock. Abidjan is also the home of an important tuna fishery with more than 150,000 metric tons processed yearly. But the waters above the continental shelf represent a small part of these total catches (Amon-Kothias and Bard, 1993; Stretta et al., 1993). Three species (*Thunnus albacores*, *Katsuwonus pelamis* and *Thunnus obesus*) represent the main catches (10,000 to 20,000 metric tons) in the coastal area. Catches are very seasonal, with high values from July to December, probably related to migration patterns.

Demersal Fisheries

A relatively homogeneous community of sciaenids (*Galeoides decadactylus*, *Pomadasys jubelini*, *Brachydeuterus auritus*, *Pseudolithus senegalensis*, and *Cynoglossus canariensis*) is observed above the 10–50 m depth. A sparid community (*Dentex angolensis* and *Pagellus bellotii*), associated with sharks, triglids and groupers, is observed above the 50–120 m depth (Caverivière, 1993). Trawlers and canoes undertake demersal fishing. Total catches range between 5000 and 10,000 metric tons. The bulk of these catches is carried out in the western part of the continental shelf (San Pedro and Sassandra areas). A shrimp fishery is based on *Penaeus notialis*, with an annual catch estimated to 600 metric tons yr^{-1} (Lhomme and Vendeville, 1993). The bulk of that resource is located west of the large river inlets, above the 25–50 m depth. Spawning occurs in November in the ocean, followed by larval development in brackish waters. Migration and recruitment (lasting between 3.5 and 4 months) occur in February in the ocean (Garcia, 1977). Crayfishes are also exploited but the resource is not known.

POPULATION

Of the 12.6 million inhabitants of Côte d'Ivoire, more than 4 million people live in the coastal cities of Abidjan, Grand-Bassam, Jacqueville, Grand-Lahou, Sassandra, San Pedro and Tabou. Coastal population is projected to reach 9 million in 2015, with a growth rate estimated to 4% (national growth rate: 3.7%; average population density: 37.7 inhabitants km^{-2}). Abidjan is presently the economic capital, with about 3.0 million inhabitants. It represents 21% of the total country population and 51% of the total urban population (Table 4).

Before colonisation, indigenous populations (Krou, Nzima, Alladjan, Ebrié, Ahizi, Avikam) populated the coastal zone. Their major activities were centred on small-scale farming and subsistence fisheries using rudimentary tools (Le Lœuff et al., 1993). Development of coastal areas started with colonisation, which introduced profit-earning agriculture (palm tree and coconut), trading and shipping along the coastline. These colonial activities attracted more and more people, especially fishermen from the neighbouring countries (Ghana, Benin, Togo and Liberia, Surgy, 1965) and natives from forest and savanna areas. Trading between European seafarers and Africans took place in several cities (Assinie, Grand-Lahou, Sassandra and Tabou).

The opening of the Vridi Canal in 1950, followed by the construction of Abidjan harbour, gave a major boost to Côte d'Ivoire's economy. Traditional fishing gave place progressively to trawling and tuna fisheries, with the introduction of purse seiners. The number of trawlers rapidly increased from 12 in 1954 to 40 in 1959. Substantial further investments were made, like ice factories, canneries, cold-storage

Table 4

Population trend of the city of Abidjan

Year	Population	Rate
1912	1400	12%
1920	5370	
1934	17,000	
1945	46,000	10%
1950	65,000	
1955	125,000	
1960	180,000	9.3%
1963	254,000	
1970	550,000	11.6%
1975	951,000	
1979	1,415,000	3.8%
1988	1,929,000	
1998	2,500,000	(estimate)

and fish meal industries. Today, Abidjan is the largest tuna and container port of West Africa.

The construction of Abidjan harbour also caused industrial development: more than 60% of the industries of the country are located in the coastal zone or near Abidjan (tourism, oil refinery and offshore oil and gas exploration and exploitation). Abidjan harbour contributes respectively to 96% and 66% of the country's import and export. Its activities represents 90% of the sea traffic of the country and 75% and 40% of that of the neighbouring landlocked countries, Burkina Faso and Mali, respectively. San Pedro harbour, constructed in 1971, represents 10% of Côte d'Ivoire's port activities.

The presence of the country's major industries along the coast has been a factor in the rapid population increase. These industries offer employment opportunities, attracting workers from the landlocked states. Table 4 presents the evolution of Abidjan's population from 1921 to 1994.

RURAL FACTORS

Côte d'Ivoire's economy is based on agriculture, of which coffee, cocoa, palm oil, rubber, bananas, coconut and pineapple productions were considered by the government as the "spear head" of the national economy. Annual production was reported at 868,000 tons for coffee, cocoa and palm oil, 195,000 tons for pineapple, 195,000 tons for bananas, 67,000 tons for rubber, and 55,000 tons for coconut. The rapid population increase and development of agro-industrial activities have exerted a considerable threat on the aquatic environment. Most of the large-scale agricultural plantations located in the coastal zone use large quantities of chemical fertilisers and phytosanitary products, such as insecticides, fungicides, nematocides, raticides and herbicides. Run-off of fertilisers and pesticides from farm lands into rivers and coastal marine ecosystems induces eutrophication and chemical contamination of water and fishes. Organochlorine pesticides have been found in the tissues of several marine fish species (*Pagellus bellotii*, *Epinephelus aeneus*, *Cynoglossus canariensis*, *Pseudotolithus senegalensis*, *Sphyraena sphyraena* and *Penaeus notialis*).

Many fisheries of the country are artisanal and based mainly in the coastal zone. Population pressures have increased consumption and demands and have led to destructive fishing methods. The use of poisonous substances such as DDT is very common (Marchand and Martin, 1985), inducing mass destruction of fish, coastal environment pollution and health hazard to people.

Although there is a mesh size regulation, lack of enforcement has caused some alteration to the ecological balance of coastal lagoons. Trawling has become dominant in an area formerly dominated by traditional fishermen, in response to increasing demand for fish and fish products. These operations are largely unregulated (or do not conform to existing regulations), with illegal mesh size resulting in destructive fishing, including undersized fish catches.

The open ocean of this region seems yet to be largely unaffected either by man-induced damage or by over-exploitation of natural resources. But living resources are endangered by foreign fleets that "poach" fish from the oceanic area. This fishing pressure has an adverse impact on small-scale operations conducted by local fishing fleets, and artisanal fishermen have noticed a marked decrease in catches of large pelagic and migratory species.

COASTAL EROSION AND LANDFILL

To accommodate the population increase in coastal cities, new housing projects are undertaken, inducing collection of construction materials along the coastline for building purposes. Mining of sand, gravel and other construction materials in the coastal zone is a common practice. Sand and gravel mining from lagoons and from estuaries tends to destroy natural habitats and decrease the amount of fluvial sediment input to the coastline, thereby accelerating shoreline retreat. Sand extraction directly from beaches seriously depletes the sediment pool available and beach retreat is either induced or accelerated.

Increasing awareness of revenue potentially generated by tourism has also led to increased building of tourism facilities on beaches along the coast (Assinie Mafia, Grand-Bassam, Sassandra, San Pedro, Grand-Béréby, etc.). Construction activities in the coastal zone loosen the sediment binding by removing the surface revetments and increasing rainwater runoff. Thus, soil erosion is enhanced. On the other hand, structures constructed on the coast by strengthening soil, may lead to decreased sediment supply to the shoreline. The opposite problem of increased siltation and sediment starvation along the coast depends on the local physiographic conditions.

Besides the increased threat of erosion, mining of construction materials from the coastal zone tends to disrupt fragile ecosystems (e.g., mangroves) and affect their productivity. Several rivers (mainly the Sassandra and Bandama Rivers) are dammed in at least one location. Construction of these dams has resulted in loss of catchment area for sediment, due to the effective entrapment of particles in the reservoirs. Sometimes this loss of sediment input is blamed for coastal erosion that has occurred since the construction of these dams. The severe coastal erosion of Grand-Lahou area is attributed to this process (Abé, 1995). Dam construction has caused the decrease of freshwater input in the lower estuarine reaches of the Bandama River and the alteration of intrusion extent of the estuarine salt wedge inland, with ecological effects on the flora and fauna.

Harbour construction for national and international trade was found to have a direct negative impact on the environment. Erosion at Port-Bouet beach, estimated to be 3 m yr^{-1}, is attributed to the construction of the port of Abidjan (Koffi et al., 1993).

EFFECTS OF URBAN AND INDUSTRIAL ACTIVITIES

Waste Disposal

Although modern industries have had tremendous positive effects on the economy of the coast, they have also largely contributed to the pollution of water bodies along the coast. Near Abidjan, the Ebrié lagoon is heavily polluted by industrial discharges (Broche and Peschet, 1983; Métongo et al., 1993; Dufour et al., 1994). Estimations of pollution loads generated by industries in the Abidjan area have been undertaken (Kouassi et al., 1995). Food manufacturing and textile production are the dominant sources of industrial pollution, producing approximately 85% of the waste volume and 95% of the pollution load. There is no significant production of toxic substances in the area, although pesticide, glue and wood preservative industries have generated some toxic organic effluents.

Domestic effluents include direct sewage and sludge discharges from septic tanks. Only two cities located in the coastal area (Abidjan and San Pedro) have a sewerage system which is even partly effective. The other coastal cities do not yet have either proper drainage facilities or sewage treatment plants. Now, most discharges are made to the waters immediately inshore or to estuaries, or, even worse, to lagoons, most of which have only very poor exchange with the open coastal waters (Métongo et al., 1993).

Dufour and Slepoukha (1975) have provided data on sewage volume discharged to the Ebrié lagoon compared with the volume of receiving waters. In the area immediately around Abidjan, the volume of domestic sewage discharged each year was equivalent to about 18% of the total lagoon volume (estimated to 2.6×10^9 m^3). Domestic sewage accounts for just over half the total effluent volume. For the whole area, the quantity of domestic and industrial wastewater discharged annually is twice the lagoon volume. The Ebrié Lagoon is seasonally subject to marked salinity variations (0–28 psu) and this flushing effect probably helps to minimise impacts. Nevertheless, at least four embayments become regularly anoxic due to the pollution load.

There are few proper facilities to dispose of sewage in many coastal communities, and arrangements for collection of household rubbish are limited or non-existent. As a result, populations use the estuarine and coastal waters or the beaches as dumping places. Some of this rubbish accumulates on the beaches and the scale of the problem is positively correlated with population density. The most commonly encountered forms of litter are plastics, metal cans, and less readily degraded forms of household and industrial refuse.

Toxic wastes are mainly generated by chemical industries. Total production of industrial solid wastes in the coastal area is not known. However, the city of Abidjan concentrates over 60% of the 100,000 metric tons of waste produced by industry, of which one third is eliminated with domestic refuse at the landfill sites. The rest is either recycled or incinerated in the open air (Kouassi et al., 1995).

Oil and Gas Exploration and Exploitation

A major shipping route for oil tankers and other bulk carriers bound for the Middle East to Europe passes offshore from Côte d'Ivoire. The sea-lanes are very wide and the number of accidents involving tankers has been low compared to elsewhere in the world. Nevertheless, beaches are not free of oil pollution. Much of the oil residue found along the shore is due to spills or tank washings from tankers visiting ports in the region, although other sources are important.

Now, exploration, exploitation and refining of oil and gas proceed along the littoral of the country. Although the impacts of these activities are not known, the possibility exists that these activities, while contributing to economic development, routinely introduce a variety of pollutants to the coastal zone and ocean. Pollution includes hydrocarbons from occasional spills, from chronic low-level releases associated with leaking valves, corroded pipelines and ballast water discharges. Construction of pipe networks for hydrocarbon and gas exploitation and transportation, on or near the coast, constitutes visible structural modification of the coastal zone that has adverse effects on coastline migration.

Depletion and Degradation of Coastal Habitats

A major consequence of the population shift toward coastal areas is the increasing degradation rate of coastal habitats. These habitats are lost or damaged by a wide range of activities, including discharge of raw or poorly treated sewage, dredging and filling, discharge of industrial effluents, erosion and overfishing. Specific threats to particular habitats are outlined below.

Lagoons and Estuaries

Lagoons and estuaries are among the most productive of all coastal waters. They serve the special needs of migrating nearshore and oceanic species that require shallow protected habitats for breeding or as sanctuary for their larval stages. Lagoons and estuaries—and the productive source of protein they provide—are threatened by urban encroachment, pollutants of various kinds, siltation and overfishing. Urban development such as that occurring around the Ebrié Lagoon has resulted in habitat (nurseries, sanctuaries, etc.) degradation. Domestic sewage, garbage, and waste fuel are major causes of the decline in productivity of lagoons near urban areas (Albaret and Charles-Dominique, 1982; Kouassi et al., 1990). Industrial effluents, agricultural runoff, and increased sedimentation from poor upstream land and water management schemes are also contributing factors.

For a decade, free floating macrophytes (*Eicchornia crassipes, Pistia stratiotes, Salvinia molesta*) have invaded coastal sites, drifting with freshwater. Most of the large reservoirs are colonised (Ayamé I and II, Taabo and Buyo), as are the rivers. Large rafts of *E. crassipes* and associated species are carried seaward, and then run aground on the beaches. On the large scale at which this happens now in Côte d'Ivoire, these weeds strongly affect water quality, and impede water uses such as transportation, fisheries and other harbour activities.

Mangroves

In most coastal lagoons (especially the Fresco Lagoon), mangroves are severely affected as they are continuously harvested for fuel-wood. Mangroves are also affected by erosion, by changes in salinity and through the construction of canals (Carmouze and Caumette, 1982; Zabi, 1982). These canals were intended for use as transport routes. Their construction had the immediate side effect of increasing suspended solids in the water, which can lead to degradation of benthic fauna. This is followed by more permanent damage as the hydrological regime and salt intrusion alters and spoils banks, and impedes land run-off. Now, water-weeds, like Water Hyacinth (*Eicchornia crassipes*) invade these canals and contribute to their slow filling. Most of the large vertebrates inhabiting the mangrove have almost disappeared, like manatee (*Trichenus senegalensis*), pigmy hippopotamus (*Choeropsis liberiensis*) and crocodiles (*Crocodylus niloticus* and *C. cataphractus*). Forest buffaloes (*Syncerus caffer nanus*) and forest elephants (*Loxodonta africanus cyclotis*) are endangered like several other mammals (otters, chimpanzees) by uncontrolled poaching and deforestation. Mangroves of Côte d'Ivoire have poor bird populations compared to adjacent areas. Developing rice fields could provide additional wetland habitats to waders.

Aquaculture and Fisheries

There are several aquaculture ventures in lagoon regions. There is potentially a risk from sewage or industrial pollution either directly through damage to their stocks or indirectly through adverse effects on the quality of their products (Adingra et al., 1997). Mass fish mortalities occur near urban areas as a result of pollution in the Ebrié Lagoon (Zabi, 1982; Carmouze and Caumette, 1985) while harmful fishing practices in the Aby, Ebrié and Grand-Lahou lagoons are frequent.

Cultural and Historic Sites

Côte d'Ivoire's coastal areas include many cultural and historic sites. Several water bodies serve as objects of worship or are considered to be abodes of gods. Though an intangible resource, such cultural attributes are part of the heritage of riverine populations. This is very important since many inhabitants are followers of traditional religions. The cultural aspects of some of the water bodies are used as a vehicle for encouraging people participation in the management of their resources. Nevertheless, Grand Bassam (classified as an historic site by UNESCO), Assinie and Grand-Lahou are examples of historic cities destined to disappear if nothing is done to stop severe coastal erosion occurring along the eastern part of the coastline.

PROTECTIVE MEASURES

Present measures to control and minimise degradation of marine and coastal environments of Côte d'Ivoire are included in the report on the National Environmental Action Plan (NEAP). The NEAP was elaborated to improve environmental management through strengthening institutional capacity, formulation of standards for environmental quality, development of economic incentives to promote environmental management and establishment of national environmental data management systems. Also envisaged are developments of Integrated Coastal Area Management, biodiversity preservation and integrated management of water resources.

Table 5

International conventions adhered to in the coastal and marine environment

Conventions Place and Adoption Dates	Ratification Date
Convention on International Trade of sites on Endangered Wild Fauna and Flora Species (1975), Washington, 1973	1994
Convention on Biological Diversity, Rio de Janeiro, 1992	1994
International Convention for Prevention of Pollution by Ships (MARPOL) London, 1973	Date to be specified
Convention related to Ozone Layer Depletion: Vienna, 1988; Montreal Protocol, 1987; London Amendment, 1990	1993
Convention on Climate Changes, Rio de Janeiro, 1992	1994
Convention related to Cooperation on Protection and Exploitation of Sea Geographical coastal areas of West and Central African Region (WACAF) Abidjan, 1981	1983
Protocol related to Cooperation on Fight against Pollution in case of Critical Situation	1983
RAMSAR Convention related to Humid Areas of International Importance aiming at Guaranteeing Strengthened Protection of Stay and Nestling Places of some Migratory Species	1993
Bâle Convention on the Control of Transboundary movements of Dangerous Wastes and their Destruction	
Bamako Convention on the Prohibition of Importing to Africa Dangerous Wastes	

Some legislation existed in Côte d'Ivoire but problems arose with implementation and enforcement because many provisions were not specific and penalties were obsolete. Since 1996, a new Outline-Law on the Code of Environment (Le Code de l'Environnement Law No. 96-766 of October 3, 1996) has been voted by the parliament. This new legislation with its pursuance law under preparation, takes into account all environmental aspects including public health, pollution, natural resource management, environmental impact assessment, etc.

Côte d'Ivoire is engaged in many regional initiatives ranging from generally conceived political organisations to highly specialised ones. In the area of marine management and protection, the country has ratified several international conventions (Table 5).

REFERENCES

Abé, J. (1995) Etude comparative de la dynamique sédimentaire aux embouchures des fleuves du littoral ivoirien. *Proc. Int. Conf. "Coastal Change 95"*, Bordomer-IOC, Bordeaux, pp. 347–363.

Adingra, A.A., Guiral, D. and Arfi, R. (1997) Impacts of lagoonal bacterial pollution on an aquacultural site (Ebrié lagoon, Côte d'Ivoire). In *African Inland Fisheries Aquaculture and the Environment*, ed. K. Remane. Publ. FAO, pp. 207–220.

Albaret, J.J. (1994) Les poissons, biologie et peuplements. In *Environnement et ressources aquatiques de Côte d'Ivoire. Tome 2. Les milieux lagunaires*, eds. J.R. Durand, P. Dufour, Guiral and S.G. Zabi. Editions de l'ORSTOM, Paris, pp. 239–279.

Albaret, J-J. and Charles-Dominique, E. (1982) Observation d'un phénomène de maturation sexuelle précoce chez l'Ethmalose *Ethmalosa fimbriata*, Bowdich dans une baie polluée de la lagune Ebrié, Côte d'Ivoire. *Doc. Sci. Centr. Rech. Océanogr. Abidjan* **13**, 23–31.

Amon-Kothias, J.B. and Bard, F.X. (1993) Les ressources thonières de Côte d'Ivoire. In *Environnement et ressources aquatiques de Côte d'Ivoire. Tome 1. Le milieu marin*, eds. P. Le Lœuff, E. Marchal and J.B. Amon Kothias. Editions de l'ORSTOM, Paris, pp. 323–352.

Arfi, R. and Bouvy, M. (1995) Size, composition and distribution of particles related to wind induced resuspension in a shallow tropical lagoon. *Journal of Plankton Research* **17**, 557–574.

Arfi, R., Dufour, P. and Maurer, D. (1981) Phytoplancton et pollution. Premières études en baie de Biétri (Côte d'Ivoire). Traitement mathématique des données. *Oceanologica Acta* **4**, 319–329.

Arfi, R., Pagano, M. and Saint-Jean, L. (1987) Communautés zooplanctoniques dans une lagune tropicale (lagune Ebrié, Côte d'Ivoire). Variations spatio-temporelles. *Rev. Hydrobiol. Trop.* **20**, 21–36.

Arfi, R., Bouvy, M. and Guiral, D. (1993a) Wind induced resuspension in a shallow tropical lagoon. *Estuarine, Coastal and Shelf Sciences* **36**, 587–604.

Arfi R., Pezennec O., Cissoko S. and Mensah M. (1993b) Evolution spatio-temporelle d'un indice caractérisant l'intensité de la résurgence ivoiro-ghanéenne. In *Environnement et ressources aquatiques de Côte d'Ivoire. Tome 1. Le milieu marin*, eds. P. Le Lœuff, E. Marchal and J.B. Amon Kothias. Editions de l'ORSTOM, Paris, pp. 111–122.

Binet, D. (1979) Le zooplancton du plateau continental ivoirien. Essai de synthèse écologique. *Oceanologica Acta* **2**, 397–410.

Binet, D. (1993) Zooplancton néritique de Côte d'Ivoire. In *Environnement et ressources aquatiques de Côte d'Ivoire. Tome 1. Le milieu marin*, eds. P. Le Lœuff, E. Marchal and J.B. Amon Kothias. Editions de l'ORSTOM, Paris, pp. 167–193.

Broche, J. and Peschet, J.L. (1983) Enquête sur les pollutions actuelles et potentielles en Côte d'Ivoire. In *Réseau National d'Observation de la qualité des eaux marines et lagunaires en Côte d'Ivoire*, eds. P. Dufour P. and J.M. Chantraine. Paris, ORSTOM et Ministère de l'Environnement, 451 pp.

Carmouze, J.P. and Caumette, P. (1985) Les effets de la pollution organique sur les biomasses et activités du phytoplancton et des bactéries hétérotrophes dans la lagune Ebrié (Côte d'Ivoire). *Rev. Hydrobiol. Trop.* **18**, 183–212.

Caverivière, A. (1993) Les ressources en poissons démersaux et leur exploitation. In *Environnement et ressources aquatiques de Côte d'Ivoire. Tome 1. Le milieu marin*, eds. P. Le Lœuff, E. Marchal and J.B. Amon Kothias. Editions de l'ORSTOM, Paris, pp. 427–488.

Colin, C. (1988) Coastal upwelling events in front of the Ivory Coast during the FOCAL program. *Oceanologica Acta* **11**, 125–138.

Cury, P. and Roy, C. (1987) Upwelling et pêche des espèces pélagiques de Côte d'Ivoire: une approche globale. *Oceanologica Acta* **10**, 347–357.

Daget, J. and Iltis, A. (1965) Poissons de Côte d'Ivoire. *Mém. IFAN* **74**, 385 pp.

Dufour, P. (1994a) Du biotope à la biocénose. In *Environnement et ressources aquatiques de Côte d'Ivoire. Tome 2. Les milieux lagunaires*, eds. J.R. Durand, P. Dufour, D. Guiral and S.G. Zabi. Editions de l'ORSTOM, Paris, pp. 93–108.

Dufour, P. (1994b) Les microphytes. In *Environnement et ressources aquatiques de Côte d'Ivoire. Tome 2. Les milieux lagunaires*, eds. J.R. Durand, P. Dufour, D. Guiral and S.G. Zabi. Editions de l'ORSTOM, Paris, pp. 109–136.

Dufour, P. and Slepoukha, M. (1975) L'oxygène dissous en lagune Ebrié: Influence de l'Hydroclimat et des Pollutions. *Doc. Sci. Cent. Rech.Océanogr. Abidjan* **6**, 75–118.

Dufour, P., Kouassi, A.M. and Lanusse, A. (1994) Les Pollutions. In *Environnement et ressources aquatiques de Côte d'Ivoire. Tome 2. Les milieux lagunaires*, eds. J.R. Durand, P. Dufour, D. Guiral and S.G. Zabi. Editions de l'ORSTOM, Paris, pp. 309–334.

Durand, J.R., Dufour, P., Guiral, D. and Zabi, S.G. (eds.) (1994) *Environnement et ressources aquatiques de Côte d'Ivoire. Tome 2. Les milieux lagunaires*. Editions de l'ORSTOM, Paris, 546 pp.

Ecoutin, J.M., Delaunay, K. and Konan, J. (1993) Les pêches artisanales maritimes. In *Environnement et ressources aquatiques de Côte d'Ivoire. Tome 1. Le milieu marin*, eds. P. Le Lœuff, E. Marchal and J.B. Amon Kothias. Editions de l'ORSTOM, Paris, pp. 537–549.

Garcia, S. (1977) Biologie et dynamique des population de crevettes roses (*Penaeus duorarum notialis* Perez-Farfante) en Côte d'Ivoire. *Trav. Doc. ORSTOM* **79**, 271 pp.

Iltis, A. (1984) Biomasses phytoplanctoniques de la lagune Ebrié (Côte d'Ivoire). *Hydrobiologia* **118**, 153–175.

Koffi, K.P., Affian, K. and Abé, J. (1993) Contribution à l'étude des caractéristiques morphologiques de l'unité littorale de Côte d'Ivoire, Golfe de Guinée. Cas du périmètre littoral de Port-Bouët. *J. Ivoir. Océanol. Limnol.* **2**, 43–52.

Kouassi, A.M., Guiral, D. and Dosso, M. (1990) Variations saisonnières de la contamination microbienne de la zone urbaine d'une lagune tropicale estuarienne - Cas de la ville d'Abidjan (Côte d'Ivoire). *Rev. Hydrobiol. Trop.* **23**, 179–192.

Kouassi, A.M., Kaba, N. and Métongo, S. (1995) Land based sources of pollution and environmental quality of the Ebrié lagoon waters. *Marine Pollution Bulletin* **30**, 295–300.

Le Borgne, R. and Binet, D. (1979) Dix ans de mesures de biomasse de zooplancton à la station côtière d'Abidjan: 1969–1979. *Doc. Sci. Centre Rech. Océanogr. Abidjan* **10**, 165–176.

Le Lœuff, P. and Intès, A. (1993) La faune benthique du plateau continental de Côte d'Ivoire. In *Environnement et ressources aquatiques de Côte d'Ivoire. Tome 1. Le milieu marin*, eds. P. Le Lœuff, E. Marchal and J.B. Amon Kothias. Editions de l'ORSTOM, Paris, pp. 195–236.

Le Lœuff, P., Marchal, E. and Amon Kothias, J.B. (1993) *Environnement et ressources aquatiques de Côte d'Ivoire. Tome 1. Le milieu marin.* Editions de l'ORSTOM, Paris, 589 pp.

Lhomme, F. and Vendeville, P. (1993) La crevette rose *Penaeus notialis* (Pérez Farfante, 1967) en Côte d'Ivoire. In *Environnement et ressources aquatiques de Côte d'Ivoire. Tome 1. Le milieu marin*, eds. P. Le Lœuff, E. Marchal and J.B. Amon Kothias. Editions de l'ORSTOM, Paris, pp. 489–520.

Marchal, E. (1993). Biologie et écologie des poissons pélagiques côtiers du littoral ivoirien. In *Environnement et ressources aquatiques de Côte d'Ivoire. Tome 1. Le milieu marin*, eds. P. Le Lœuff, E. Marchal and J.B. Amon Kothias. Editions de l'ORSTOM, Paris, pp. 237–269.

Marchand, M. and Martin, J.L. (1985) Détermination de la pollution chimique (hydrocarbures, organochlorés, métaux lourds) dans la lagune d'Abidjan (Côte d'Ivoire) par l'étude des sédiments. *Océanogr. Trop.* **20**, 25–39.

Martin, L. (1973) Carte sédimentologique du plateau continental de Côte d'Ivoire. ORTSOM–CRO Abidjan, notice explicative no. 48, 19 pp and 3 maps.

Métongo, S.B., Kouassi, A.M. and Kaba, N. (1993) Evaluation qualitative et quantitative de la pollution marine en Côte d'Ivoire. Contrat de Recherches CRO/OMS. Editions CRO Abidjan, 100 pp.

Nicole, M., Egnankou Wadja, M. and Schmidt, M. (1994) A preliminary inventory of coastal wetlands of Côte d'Ivoire. IUCN, Gland, Switzerland, viii + 80 pp.

Pagano, M. and Saint-Jean, L. (1994) Le zooplancton. In *Environnement et ressources aquatiques de Côte d'Ivoire. Tome 2. Les milieux lagunaires*, eds. J.R. Durand, P. Dufour, D. Guiral and S.G. Zabi. Editions de l'ORSTOM, Paris, pp. 155–188.

Pagès, J., Lemasson, L. and Dufour, P. (1979) Eléments nutritifs et production primaire dans les lagunes de Côte d'Ivoire. *Doc. Sci. Centre Rech. Océanogr. Abidjan* **3**, 1–30.

Pezennec, O., Marchal, E. and Bard, F.X. (1993) Les espèces pélagiques côtières de Côte d'Ivoire. Ressources et exploitation. In *Environnement et ressources aquatiques de Côte d'Ivoire. Tome 1. Le milieu marin*, eds. P. Le Lœuff, E. Marchal and J.B. Amon Kothias. Editions de l'ORSTOM, Paris, pp. 387–426.

Plante-Cuny, M.R. (1977) Pigments photosynthétiques et production primaire du microphytobenthos d'une lagune tropicale, la lagune Ebrié (Abidjan, C.I.). *Cahiers ORSTOM, Océanogr.* **15**, 3–25.

Seguin, G. (1970) Zooplancton d'Abidjan (Côte d'Ivoire). Cycle annuel (1963–1964). Etude qualitative et quantitative. *Bull. IFAN, sér. A*, **32**, 607–663.

Sevrin-Reyssac, J. (1993) Phytoplancton et production primaire dans les eaux marines ivoiriennes. In *Environnement et ressources aquatiques de Côte d'Ivoire. Tome 1. Le milieu marin*, eds. P. Le Lœuff, E. Marchal and J.B. Amon Kothias. Editions de l'ORSTOM, Paris, pp. 152–166.

Stretta, J.M., Petit, M. and Slépoukha, M. (1993) Les prises de thonidés et leur environnement au large de la Côte d'Ivoire. In *Environnement et ressources aquatiques de Côte d'Ivoire. Tome 1. Le milieu marin*, eds. P. Le Lœuff, E. Marchal and J.B. Amon Kothias. Editions de l'ORSTOM, Paris, pp. 353–385.

Surgy, A. De (1965) *Les pêcheurs de Côte d'Ivoire. Tome I: Les pêcheurs maritimes* (3 fascicules), CNRS-CNDCI-FAN, 224 pp.

Tastet J.P. and Guiral, D. (1994) Géologie et sédimentologie. In *Environnement et ressources aquatiques de Côte d'Ivoire. Tome 2. Les milieux lagunaires*, eds. J.R. Durand, P. Dufour, D. Guiral and S.G. Zabi. Editions de l'ORSTOM, Paris, pp. 35–58.

Zabi, S.G. (1982) Les peuplements benthiques liés à la pollution en zone urbaine d'Abidjan (Côte d'Ivoire). *Oceanologica Acta* suppl. 4: 441–455.

Zabi, S.G. and Le Lœuff, P. (1994) La macrofaune benthique. In *Environnement et ressources aquatiques de Côte d'Ivoire. Tome 2. Les milieux lagunaires*, eds. J.R. Durand, P. Dufour, D. Guiral and S.G. Zabi. Editions de l'ORSTOM, Paris, pp. 189–227.

THE AUTHORS

Ama Antoinette Adingra
Centre de Recherches Océanologiques, BP V18, Abidjan, Côte d'Ivoire

Robert Arfi
Centre de Recherches Océanologiques, BP V18, Abidjan, Côte d'Ivoire

Aka Marcel Kouassi
Centre de Recherches Océanologiques, BP V18, Abidjan, Côte d'Ivoire

Chapter 52

SOUTHWESTERN AFRICA: NORTHERN BENGUELA CURRENT REGION

David Boyer, James Cole and Christopher Bartholomae

Southwestern Africa is strongly affected by the Benguela Current, which is one of the world's four major eastern boundary currents, influencing the coastal environments of western South Africa, Namibia and southern Angola. A major effect is the upwelling of cool, nutrient-rich water along the coastal edge of the continental shelf, creating the necessary conditions for a highly productive food chain, culminating in large fish stocks.

The coastal margin of the Northern Benguela is bounded by the extremely arid Namib Desert, whose one source of permanent fresh water is the Kunene River. The low population density has resulted in very low human impacts.

The offshore region supports large fish stocks, the most important being small pelagics, horse mackerel and hakes. Namibia has a large and economically important fishing industry. Heavy fishing pressure, particularly prior to independence in 1990, resulted in depletion of many of the living marine resources.

Since independence, the re-building of depleted stocks has been encouraged through strictly enforced management controls, mainly by limited Total Allowable Catches. Several stocks are now showing signs of a sustained recovery, and annual catches have been allowed to increase. Other stocks, notably pilchard, remain depleted.

The remoteness of this region and low population density ensures that much of the region will remain in a pristine state into the new millennium and the foreseeable future. Development along the coastline, if managed responsibly, need not result in any long-term degradation of vulnerable habitats. The greatest political and scientific challenge will be the re-establishment of fully recovered fish stocks. Achieving this goal will involve continued improvement of stock assessment techniques and a better understanding of the natural processes which control the population dynamics of species in the region.

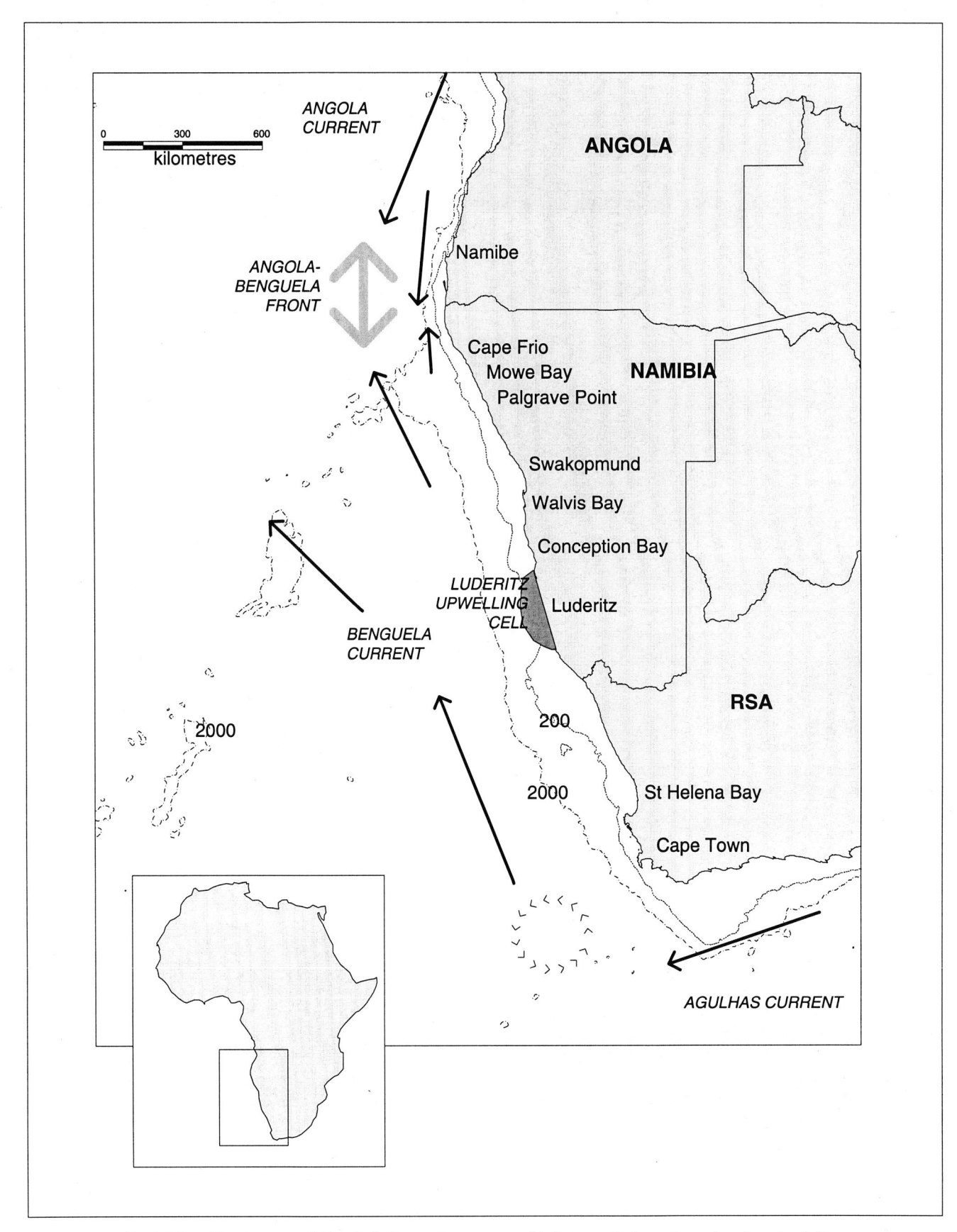

Fig. 1. Map of Southwestern Africa, including main oceanographic features, bathymetry and surface circulation.

THE REGION

The Benguela Current is one of the world's four major eastern boundary currents. It is characterised by predominately equatorward flow, high levels of Ekman-driven coastal upwelling, and a highly productive coastal ecosystem. The region is bounded at both ends by warm water currents (Fig. 1). To the north are the tropical Angola Basin and the southerly flowing Angola Current, while to the south the Agulhas Current flows round the southern tip of Africa from the tropical Indian Ocean. Both boundaries exhibit strong thermal fronts.

Strong perennial upwelling off Lüderitz (27–28°S) effectively separates the Northern Benguela from the Southern Benguela. A northwesterly moving tongue of upwelled, turbulent water with little vertical stratification acts as a semi-permanent environmental barrier to the longshore transport of pelagic fish eggs and larvae (O'Toole, 1977; Agenbag, 1980; Boyd and Cruickshank, 1983; Agenbag and Shannon, 1988). Fish stocks that occur in deeper waters, e.g. hakes (*Merluccius capensis* and *M. paradoxus*) and mesopelagic fish, are less affected, however, and form continuous populations with those in the Southern Benguela (Mas-Riera et al., 1990).

This chapter focuses on the Benguela Current Large Marine Ecosystem (Sherman, 1991) from approximately 28°S northwards. The region covers around 150,000 km² taking the edge of the continental shelf as the offshore boundary. The coastal upwelling generally extends between 150 and 200 km from the coast, though upwelled water may mix with surface oceanic water a further 600 km into the South Atlantic (Lutjeharms and Stockton, 1987). The northern boundary is dynamic, and is marked by a strong thermal front with warm Angolan water, usually between 14° and 17°S (e.g. Shannon, 1985; Lutjeharms and Meeuwis, 1987).

The Namib Desert forms the coastal margin (Fig. 2). Mean annual rainfall is only around 15 mm per year; during this century widespread annual rainfalls of greater than 100 mm have only been recorded in 1934, 1976, and 1978 (Lancaster et al., 1984).

Most previous reviews of the Benguela System have been on the southern part where most research has been focused (e.g. Chapman and Shannon, 1985; Shannon and Pillar, 1986). This review attempts to focus on the much less well known northern Namibia and southern Angola region. Its 'nearshore' region includes the littoral and surf zone, where the low number of people living on the coast provide little impact. The offshore region, from the surf zone to the offshore boundary of the system, has high productivity and important commercial fish stock, and is focused upon here.

PHYSICAL ENVIRONMENT

The Benguela Current forms the eastern limb of the South Atlantic subtropical gyre (Stramma and Peterson, 1991). From the Cape of Good Hope it skirts equatorward until offshore of Lüderitz where it deflects to the northwest (see Fig. 1). The prevailing winds are responsible for frequent Ekman transport and the upwelling of cool nutrient-rich water along the edge of the continental shelf.

The intensity of upwelling in the Northern Benguela is far from uniform, with tremendous spatio-temporal variability according to local bathymetry, fluctuations in wind, coastally trapped waves, and periodic intrusions of tropical water (Boyd, 1987) (Fig. 3). Strong upwelling is generally associated with headlands and where the continental shelf is narrow. Conversely, upwelling tends to be sluggish along concave stretches of coastline and where the continental shelf is wide.

For detailed information on the region's oceanography and meteorology see Nelson and Hutchings (1983), Shannon (1985), and Boyd (1987); for chemical processes see Chapman and Shannon (1985) and for sedimentology of the area see Rogers and Bremner (1991).

Meteorology and Surface Winds

The prevailing southerly and southeasterly winds are primarily governed by the south Atlantic anticyclone. In the Northern Benguela the anticyclonic flow of air is entrained northwards along the coast by a thermal barrier of hot air rising off the Namib Desert.

The main seasonal and latitudinal trends in upwelling favourable winds (i.e. longshore winds) may be seen in Fig 4. The strongest winds are found where the coastline changes orientation from broadly northwesterly to more northerly, at Lüderitz and Cape Frio, and the weakest between Möwe Bay and Walvis Bay, where the coastline is concave. Shorter-term diurnal (caused by sea-breezes) and 'event scale' variability are superimposed over these seasonal trends, the latter being 5–12-day cycles of southerly components (Boyd, 1987).

Fig. 2. The hyper-arid coastal edge of the Northern Benguela region (photo D. Boyer).

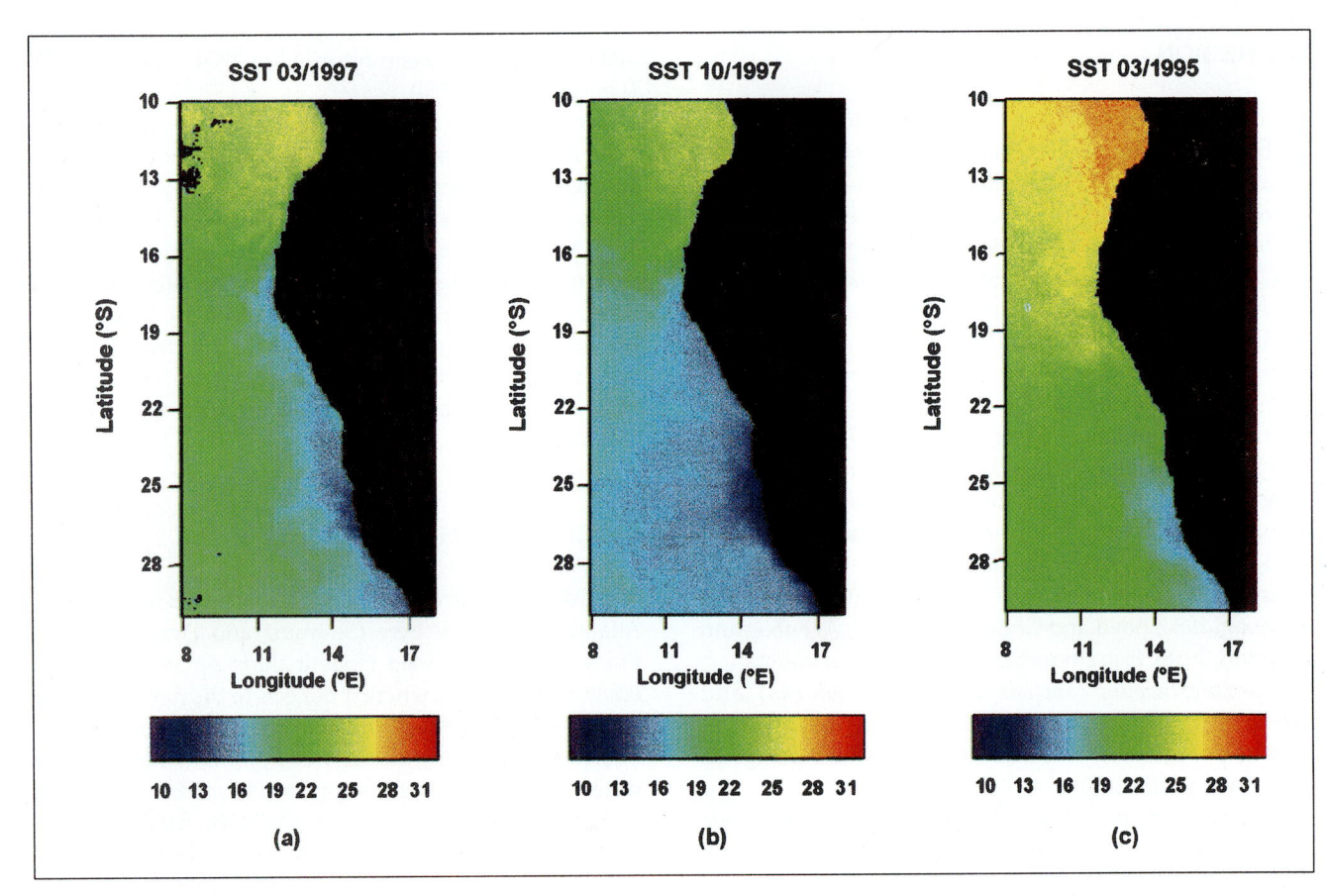

Fig. 3. SST off Namibia and southern Angola during (a) summer, (b) spring, and (c) 1995 Benguela Niño.

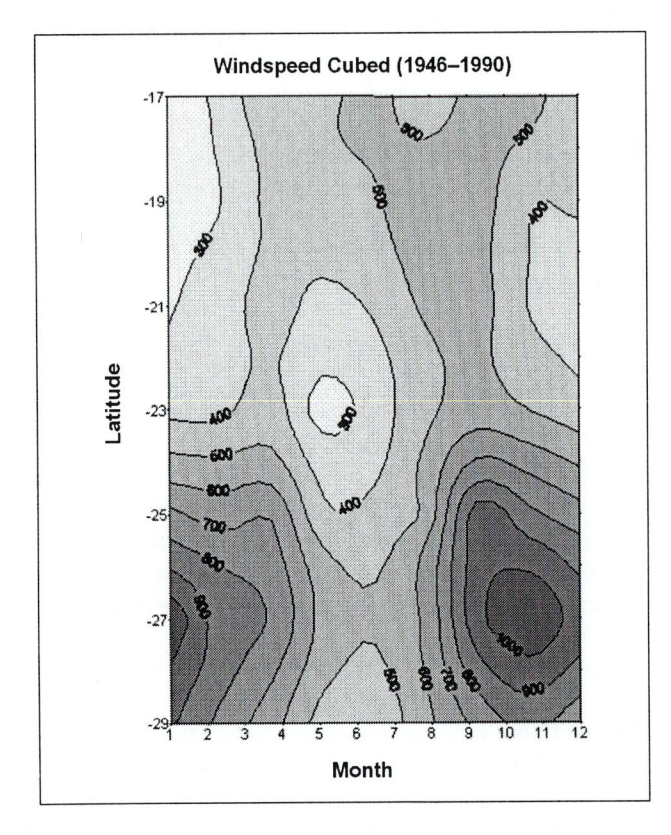

Fig. 4. Monthly wind-stress along the Namibian coast.

Macroscale Currents

The Benguela Current originates as northward flow near the Cape of Good Hope. Off Lüderitz the main geostrophic flow starts to head northwest, separating from the coast and beginning to widen. The current accelerates over areas with steep bathymetry and meanders over bathymetric plains (Shannon, 1985). It is mainly fed by the South Atlantic Current, but also receives some input from the Agulhas Current and from sub-Antarctic surface water (Shannon and Taunton-Clark, 1989).

South of 23°S the predominant offshore currents coincide with the anticyclonic wind fields. North of 23°S the situation is more complicated. Moroshkin et al. (1970) conclude that a divergence zone exists offshore of 20°S (see Fig. 1). Near shore, the wind-driven upwelling currents are northwesterly flowing until deflected offshore at about 18°S, near the frontal region between cool upwelled water and warm, saline, tropical Angolan water. Close inshore and to the north of the front there is a pronounced southerly, warm Angolan Current.

Sub-surface, there is a polarward undercurrent flowing along the edge of the continental shelf as far as the Cape of Good Hope, often with low oxygen concentrations (Nelson and Hutchings, 1983; Shannon, 1985). Further inshore, there are often subsurface intrusions of saline Angolan water over the central Namibian shelf during summer/autumn (Boyd et al., 1987).

Upwelling Activity and Mesoscale Features

Nutrient enrichment from upwelling plays an over-riding role in shaping the region's ecosystem and productivity. Upwelling intensity is closely linked to the magnitude of the southerly wind and local bathymetry. Diurnal and 'event' scale pulsing drive high levels of short-term upwelling variability, which are superimposed onto seasonal and inter-annual trends. The main upwelling cell is found off Lüderitz, with others at Cape Frio, Palgrave Point and Conception Bay.

Remote forcing from the equatorial Atlantic has also been implicated in the propagation of coastally trapped waves which suppress cold water upwelling, and during Benguela Niño years, is responsible for strong poleward intrusions of Angolan water which 'primes' upwelling by advecting cool, deep, offshore water onto the coastal shelf (Boyd and Agenbag, 1985; Shannon, 1985; Boyd, 1987). Seasonal variation in upwelling is greater off northern and central Namibia due to latitudinal differences in the forcing of the system. Between 24° and 28°S, upwelling is reasonably consistent throughout the year, whereas north of 24°S there is a marked contrast between vigorous upwelling activity in winter and spring, and the more sluggish conditions during summer and autumn.

The width of the coastal upwelling regime on average ranges between 150 and 200 km, and is separated from surface South Atlantic waters by a highly convoluted thermal front characterised by eddies and long upwelling filaments (e.g. Lutjeharms and Stockton, 1987; Kazmin et al., 1990). These upwelling filaments penetrate far out into the South Atlantic into a 'filamentous mixing area', which may extend a further 600 km or so offshore of the main upwelling front.

The vertical structure of the water column is a function of wind-driven turbulence, solar heating of the surface layers, and intrusions of warmer water masses. Consequently, the frequency with which thermoclines are found in the region varies with area and season. Shallow thermoclines are most frequently found during summer in sheltered areas, whereas deep thermoclines may occur in more exposed regions due to the inward advection of warmer water masses (e.g. Du Plessis, 1967; Boyd, 1987).

Northern Namibian Region (15–19°S)

This region is dominated by the meeting of upwelled and tropical waters at the Angola–Benguela front, where temperature gradients can be as high as 4°C per 1° latitude (Shannon, 1985). The dynamic balance between the southerly flowing Angola Current and the northwesterly flow of upwelled water determines the position of the front. It is furthest north during winter (Lutjeharms and Meeuwis, 1987; Lutjeharms and Stockton, 1987), and furthest south

Fig. 5. A frontal zone showing high concentrations of phytoplankton (photo D. Boyer).

during summer and autumn (November to April). The midpoint of the front can usually be found anywhere between 10°S and 20°S, although on average it is found between 14° and 17°S (Shannon, 1985; Meeuwis and Lutjeharms, 1990).

Surface eddies are caused by the frontal shear between the Angolan Current and the upwelled water currents as they turn offshore, and upwelling filaments often develop off Cape Frio and the Kunene estuary in winter and spring (May to October). During summer and early autumn (December to March), 'tongue-like' poleward intrusions of warm Angolan water along the coast, and high levels of vertical stratification are common (Meeuwis and Lutjeharms, 1990). Onshore surface flow associated with frontal eddies may also occur at this time of the year (Shannon et al., 1987; Boyd et al., 1999) (Fig. 5).

Central Namibian Region (19–24°S)

Moderate upwelling favourable winds and the wide shallow Swakop shelf (which prevents the upwelling of water from below 150–200 m) characterise this region. It has

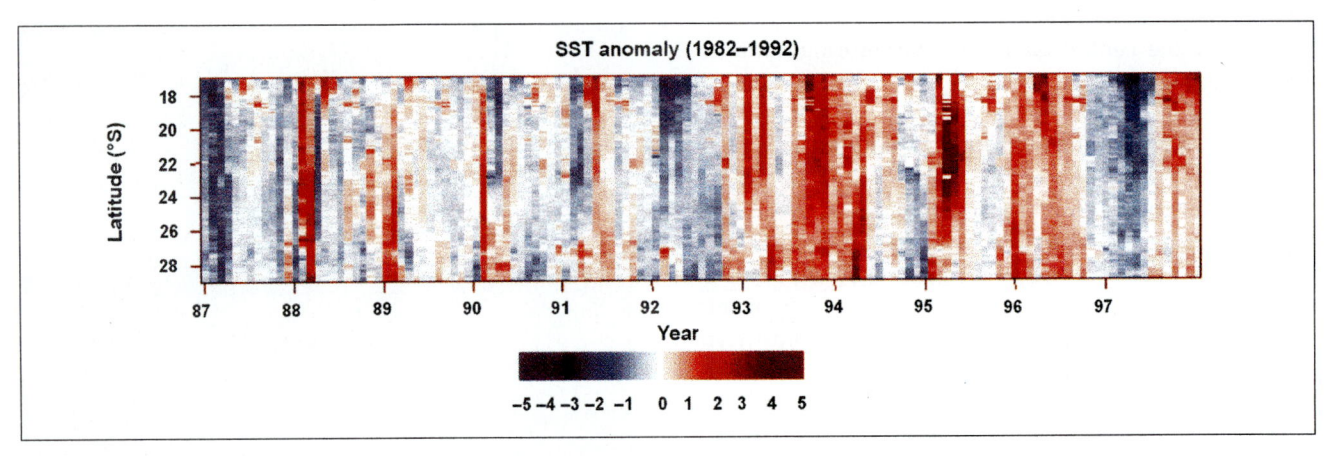

Fig. 6. SST anomalies along the Namibian coast (20–50 km offshore).

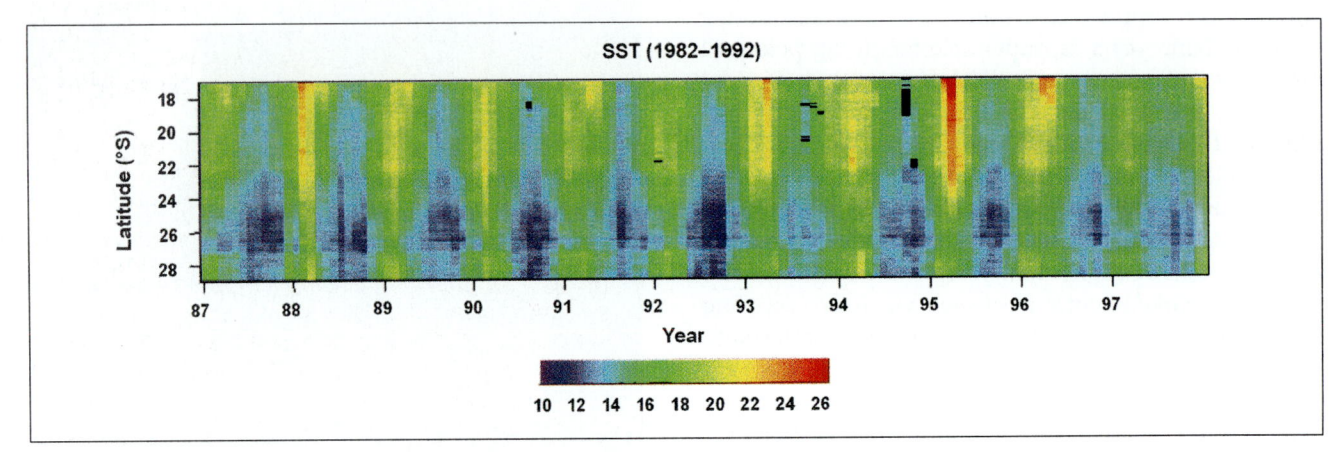

Fig. 7. SST along the Namibian coast (20–50 km offshore).

relatively low intensity upwelling (Boyd, 1987), and shallow thermoclines are common in summer and early autumn (November to March) (Du Plessis, 1967; Boyd, 1987). At Palgrave Point and Conception Bay, steeper, narrower continental shelves promote greater upwelling.

Onshore and longshore flow of warmer waters is common during summer and early autumn, when upwelling activity reaches a seasonal low. Intrusions of tropical Angolan water come from the north (Boyd et al., 1987), whereas intrusions of oceanic water come from the west and northwest (O'Toole, 1980). Intrusions of warmer water are normally associated with a relaxation in upwelling. There are exceptions, for example, during Benguela Niño years when warm water intrudes from the north, even if there has been no relaxation in longshore windstress (Shannon et al., 1986), and O'Toole (1980) observed that when upwelling is strong off both Cape Frio and in the Lüderitz region oceanic water can be drawn into the central Namibian region.

The area between 21 and 23°S is of interest as a 'semi-permanent convergence zone' between the Lüderitz and the northern/central Namibian coastal regimes. Little is known about surface flow dynamics in the region, but some kind of semi-enclosed circulation system is likely (Barange and Boyd, 1992), perhaps as a result of strong cyclonic wind shear in the region (Bakun, 1996).

Lüderitz Upwelling Region (24–28°S)

This region is one of the world's most intense and consistent upwelling regions (Bakun, 1996), by virtue of its strong southerly winds and narrow, deep continental shelf (Boyd, 1987). Upwelling is common around 26°S, and water typically rises from depths of 250–350 m. A northwesterly moving tongue of freshly upwelled, turbulent water with little vertical stratification acts as a 'semi-permanent' environmental barrier between the Northern and Southern Benguela (Boyd and Cruickshank, 1983; Agenbag and Shannon, 1988).

Eddies and filaments are common features of the upwelling front off Lüderitz. Occasionally 'superfilaments' may extend over 1000 km into the South Atlantic. They result from the combined action of intense upwelling and 'offshore entrainment' of the filament by warm water rings, which have been advected north after 'budding' off from the Agulhas Current (Lutjeharms et al., 1991). Tongues of warm water are also occasionally advected up the southwest African coast. With few exceptions, these intrusions have little effect on upwelling activity north of about 30°S.

Interannual Variability

There are high levels of interannual variability in upwelling. From SST measurements the existence of a weak 8–10-year cycle between alternating cool and warm periods has been found (Taunton-Clark and Shannon, 1988), cool periods being associated with enhanced cold water upwelling and vice versa for warm periods (Figs. 6 and 7). SSTs may vary 5°C or more from the long-term mean.

Benguela Niños

Anomalously warm years are known as Benguela Niños (Shannon et al., 1986), and have occurred several times this century (Taunton-Clark and Shannon, 1988), with especially strong events recorded in 1963, 1984 and 1995 (Stander and De Decker, 1969; Gammelsrød et al., 1998).

Like Pacific El Niños, Benguela Niños are thought to originate from anomalous atmospheric conditions in the western tropical Atlantic. They usually last six months or more, beginning around January and February, and may be thought of as an extension of the seasonal summer warming in the northern and central regions. The Angola–Benguela front is displaced south, and warm, highly saline water advects southwards as far as 25°S (see Fig. 3c). These intrusions deepen the thermocline with the result that when upwelling does occur, it will often be from the nutrient-impoverished layers above the thermocline, and the severe nutrient impoverishment leads to marked declines in productivity (Boyd et al., 1987). These often seem to be preceded by cool years (e.g. 1963 and 1984 Niños). The observation that Benguela Niños occur approximately a year after Pacific El Niños has led to speculation that atmospheric conditions over South America provide a 'teleconnection' between the Pacific and Equatorial Atlantic (Bakun, 1996).

Oxygen Concentrations

Low-oxygenated (<0.5 ml l^{-1}) bottom water off central and northern Namibia is a characteristic seasonal feature of coastal upwelling processes. This occurs especially during summer and autumn, when upwelling is reduced. The decay of organic matter leads to a decrease of oxygen levels near the seabed. This area of anoxic water is usually concentrated in a coastal band downstream of the main upwelling cell off Lüderitz. It is particularly pronounced between Conception Bay and Cape Cross within the 100 m isobath. With the onset of winter and spring, upwelling intensifies and oxygen levels tend to rise again. The effects can persist well beyond the normal seasonal period, possibly due to an influx of oxygen-poor water from the Angola Dome region

---

Red Tides

The Northern Benguela region experiences harmful algae blooms. Dead and dying fish, as well as other forms of marine life, are occasionally observed in the sea and on the beaches between Conception Harbour and the Ugab River. Copenhagen (1953) attributed these mortalities to "anoxia, anoxia plus hydrogen sulphide poisoning and poisonous plankton". Brongersma-Sanders (1957) was the first to suggest that mass mortalities in the Walvis Bay region may be a consequence of a particular dinoflagellate species (*Gymnodinium* sp.). Since then several species of poisonous algae have been identified in the northern Benguela e.g. *Gymnodinium galatheanum*, *Gymnodinium galatheanum*, *Peridinium triquetrum*, *Goniaulax tamarensis* (now *Alexandrium tamarense*) and *Alexandrium catenella*.

Dinoflagellate blooms in the northern Benguela were especially noticeable in 1993 and 1994, when regional changes in climate produced shifts in wind direction, resulting in calmer conditions with reduced upwelling and an increase in dinoflagellate blooms along much of the coast. Harmful algae blooms are usually considered as localised inshore events, although it is increasingly being recognised that they may have more far-reaching effects, particularly through the impact on recruitment of a number of fish species.

REFERENCES

Brongersma-Sanders, M. (1957) Mass mortality in the sea. Memoirs of the Geological Society of America 67, 941–1010.

Copenhagen, W.J. (1953) The periodic mortality of fish in the Walvis region: a phenomenon in the Benguela Current. Investl Rep. Div. Fish. S. Afr. 14. 35 pp.

(Bubnov, 1972). For example, during much of 1993 and 1994, and to a lesser extent during 1997–1998, the oxygen levels in the water close to the seabed were found to be persistently low over a wide area (Fig. 8). In the summer, oxygen levels were as low as 0.25 ml l^{-1} over much of the coastal waters between Cape Frio and Walvis Bay. During 1994 the vertical extent of the low-oxygen layer ranged between 50 and 250 m in thickness.

Cape hakes have a high degree of tolerance to hypoxic waters (Woodhead et al., 1996) and show behavioural adaptations to permit survival in this apparently unfavourable environment (Huse et al., 1998). Nevertheless, young fish may be displaced further offshore, where they are subjected

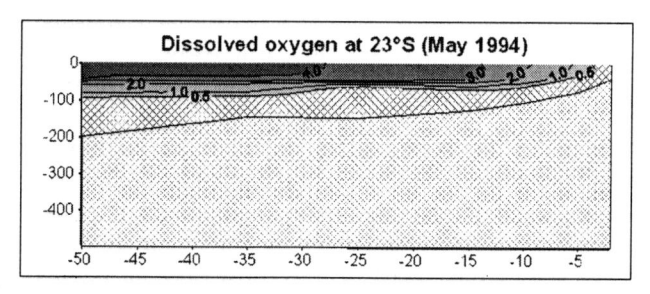

Fig. 8. Oxygen section off Walvis Bay (23°S) showing a thick layer oxygen poor water (<0.5 ml l^{-1}) above the seabed.

to increased predation by older hake (Hamukuaya et al., 1998) and undergo mass mortalities (Woodhead et al., 1996).

Most Namibians living along the central coast are familiar with so-called "sulphur eruptions" because of the characteristic rotten eggs smell accompanied by a turquoise discoloration of the ocean. These eruptions are most common during the summer months and are caused by anaerobic decomposition of high levels of organic matter in the sediments. This decomposition strips the surrounding water of dissolved oxygen, and can cause heavy mortalities of near-shore marine life.

Relevance to Living Resources

These oceanographic processes play a major role in influencing productivity. This particularly affects pelagic fisheries, since conditions influence survival of eggs, larvae, and juvenile fish (Cole and McGlade, 1998a). Links between oceanographic conditions and the pilchard and anchovy stocks have been found with respect to spawning (Parrish et al., 1983; Le Clus, 1990, 1991; Cole, 1997), larval abundance and distribution (O'Toole, 1977; Badenhorst and Boyd, 1980; Hewitson, 1987; Olivar, 1990), adult distribution (Crawford and Shannon, 1988; Gammelsrød et al., 1998) and recruitment success (Shannon et al., 1988; Cole, 1999).

Bakun's (1996) Triad theory describes three classes of oceanographic process important to survival of these juvenile stages. Enrichment of the food chain will primarily result from upwelling activity; concentration of food will occur in frontal regions or across thermoclines; and retention of eggs and larvae in suitable nursery habitats will result from either inward intrusion of warmer water masses (e.g. Barange and Boyd, 1992; Boyd et al., 1999; Cole and McGlade, 1998b) or from mesoscale features leading to semi-enclosed circulation.

THE MAJOR SHALLOW WATER MARINE AND COASTAL HABITATS

The 1500 km coastline between Lüderitz and Baìa dos Tigres, is relatively uniform. Sandy beaches account for about 60% of the shore, with 10% rocky beaches and 28% mixed sandy and rocky areas. The remaining 2% are lagoons (estimated from Campbell, 1993). The whole coast receives high-energy wave action, mainly from the south-southwest, which causes considerable longshore drift. Tides are semi-diurnal, with ranges of up to 2 m.

Species diversity and endemism is low, but abundance can be very high. Most of the Northern Benguela coastline falls in the temperate zoogeographical Namib province and has similar fauna to the Namaqua province to the south (Bally, 1983; Emmanuel et al., 1992). The area north of 17°S forms a transition zone with the tropical Angolan system (Field and Griffiths, 1991).

With few exceptions, the coast has been unchanged by human activities. A large part of it is both heavily protected or is inaccessible. The handful of settlements and towns were only established in the last 100 years.

Bays and Mudflats

Northerly longshore drift has created several sand-spits, sheltered bays and mudflats. The most important are Sandwich Harbour, Walvis Bay, and Baìa dos Tigres. A natural rocky inlet at Lüderitz provides shelter at the southern limit. Lüderitz and Walvis Bay are both natural harbours with well developed fish and cargo handling facilities. Sandwich Harbour and Baìa dos Tigres are in remote areas where access is difficult; currently they are uninhabited.

In Walvis Bay 35 km^2 of mudflats have been developed into salt pans, which produce around 500,000 tonnes of salt per annum.

Both salt pans and mudflats provide rich feeding grounds for birds (Williams, 1987; Noli-Peard and Williams, 1991). About 90,000 birds regularly use the Walvis Bay wetlands, of which, on average, 45% are Palaearctic migrants and 50% intra-African migrants, including 50,000 flamingos (Fig. 9). Sandwich Harbour generally supports about half these numbers, although up to 170,000 birds have, on occasion, been recorded there (Simmonds, 1996).

Eight bird species found at Walvis Bay and Sandwich Harbour are listed by Robertson et al. (1998) as having a conservation status of vulnerable or more endangered (Table 1). Also, more than 1% of the world breeding

Table 1

Species of birds classified as vulnerable*, endangered** or critically endangered***, that occur in the coastal or marine habitats (adapted from Robertson et al., 1998)

Common name	Scientific name	Main habitat
Great crested grebe***	Podiceps cristatus	Bays and mudflats
White pelican**	Pelecanus onocrotalus	Bays and mudflats
Greater flamingo**	Phoenicopterus ruber	Bays and mudflats
Lesser flamingo**	Phoenicopterus minor	Bays and mudflats
African black oystercatcher*	Haematopus moquini	Bays and mudflats
Chestnutbanded plover*	Charadrius pallidus	Bays and mudflats
Hartlaub's gull*	Larus hartlaubii	Bays and mudflats
Caspian tern*	Hydroprogne caspia	Bays and mudflats
Jackass penguin***	Spheniscus demersus	Islands and neritic
Cape gannet**	Morus capensis	Islands and neritic
Crowned cormorant**	Phalacrocorax coronatus	Islands and coastal
Damara tern[a]**	Sterna balaenarum	Coastal
Swift tern*	Sterna bergii	Coastal

[a]The Damara tern is endemic to the Northern Benguela (Robertson et al., 1998).

Fig. 9. Flamingos at Sandwich Harbour (photo D. Boyer).

population of 18 species of birds are found in these bays and mudflats (Williams, 1987; Simmonds, 1996). As a result both Sandwich Harbour and Walvis Bay lagoons were designated Wetlands of International Importance under the Ramsar Convention in 1995 (Curtis et al., 1998), and together are considered to be the most important wetland area in southern Africa. The proximity of the Walvis Bay mudflats to urban and industrial developments makes them potentially vulnerable to pollution.

River estuaries

The only perennial river is the Kunene, though there are ten more which are ephemeral and which may flow for a few days every few years, although some, such as the Swakop River, may be dry for a decade or more. The Kunene estuary supports small populations of resident and migratory birds and various estuarine and marine fish species. It is the southern limit of the distribution of the Nile crocodile (*Crocodylus niloticus*) and Nile soft-shelled terrapin (*Trinyx triunguis*) on the west coast of Africa. Green turtles (*Chelonia mydas*) also occur in the waters of the river mouth (Griffin and Channing, 1991). The area surrounding the estuary is currently uninhabited.

Islands

Twelve islands occur towards the southern end of the Northern Benguela coast. These range from the 90 ha Possession Island to several of less than one ha. None possess fresh water and all support little, if any, terrestrial vegetation. Due to their location in the productive waters of the Northern Benguela, and the protection they provide from terrestrial predators, a number host avian breeding colonies, including several endangered species: African penguins, Cape gannets (Fig. 10), crowned cormorants and the southern African endemic bank cormorant (*Phalacrocorax neglectus*).

These islands are all situated off remote parts of the coast, and are rarely visited. In the 19th century, however, a thriving guano mining industry stripped many metres of this substrate from several of the islands; a total of around 250,000 t. This deprived the penguins, in particular, of sheltered nesting burrows and probably contributed to the decline of the African penguin population by about 70% in the past century (Crawford et al., 1995). Extensive egg collection around the turn of the 20th century also reduced the penguin populations. More recently, however, collapses in the pilchard and anchovy stocks (important prey for the penguins) are probably the main cause of the continued population decline.

Beaches

Sandy and rocky beaches are both characterised by lower species diversity and a moderate to high biomass compared to other beaches of southwest Africa (McLauchlan, 1985; Donn and Cockroft, 1989; Sakko, 1998). No endemic organisms have been recorded (Bally, 1983). Shore invertebrates are dominated by filter feeders which exploit high concentrations of organic particulate matter, especially diatoms.

Fig. 10. A Cape gannet rookery on Mercury Island (photo H. Boyer).

Fig. 11. Surf angling is a popular sport for tourists (photo D. Boyer).

Table 2

Fish species of the surf zone that are caught for recreational or commercial purposes (calculated from Kirchner, 1998 and Kirchner et al., in press).

Common name	Scientific name	Approximate annual catch (t)
Kob	*Argyrosomus inordorus*	500
Steenbras	*Lithognathus aureti*	100
Galjoen	*Dichistius capensis*	50
Dassie	*Diplodus sargus*	20

Table 3

Estimated biomass of fish removed by commercial fishing since the 1940s

Species	Total catch (t)
Pilchard (*Sardinops ocellatus*)	15 million
Horse mackerel (*Trachurus capensis*)	12 million
Hakes (*Merluccius capensis* and *M. paradoxus*)	12 million
Anchovy (*Engraulis capensis*)	4 million
Other species	<1 million

Mussels are the most common filter feeders, especially the white mussel *Donax serra* on sandy shores and the brown mussel *Perna perna*, bisexual mussel *Semimytilus algosus* and Mediterranean mussel *Mytilus galloprovincialis* on rocky areas.

The alien Mediterranean mussel has become established here and, by the end of the 1990s, extended to northern Namibia (B. Currie, pers. comm.) after arriving, probably, in ballast water in Cape Town in the 1970s (Hockey and van Erkom Schurink (1992).

The traditional tourist coastal pursuits of surf and sun are less popular in this region than in many areas because of the cold water and generally cool, foggy climate. Even the most accessible beaches are generally only visited by anglers, and most beaches are in a near-pristine state. The compacting effects of vehicles driving on the beach are insignificant compared to the pounding of the surf. However, diamond mining in the south can have profound effects on local coastal habitats.

Surf Zone

Recreational angling from the beach, or small boats, is a popular coastal activity (Fig. 11). Four species are caught (Table 2), although several shark species are also targeted, including spotted gully shark (*Triakis megalopterus*) and bronze whaler (*Carcharhinus brachyurus*).

A few commercial linefish vessels also target kob, creating a multi-user fishery that is difficult to regulate. A lack of adequate controls, and an increase in recreational fishing has resulted in signs of over-fishing over the last two decades (Kirchner, 1998).

Only one endemic organism has been recognised in the northern Benguela: the disc lamp shell *Discinisca tenuis* (phylum Brachiopoda), which occurs in the surf zone. This benthic filter feeder occurs sub-tidally and is often present in continuous layers for hundreds of metres on the seabed.

OFFSHORE SYSTEMS

The upwelling of nutrient-rich waters into the sunlit surface layers results in a high mean primary productivity of around $2 \, g \, C \, m^{-2} \, d^{-1}$ (Shannon and Pillar, 1986). This is comparable to or higher than other coastal upwelling systems, and is similarly associated with low species diversity and few endemic species. For example, of the 185 species of diatom recorded in Namibian waters (Kruger, 1980) only four are endemic to the Benguela, and none to the northern part alone (Sakko, 1998). Likewise, 243 species of copepods (the most abundant zooplankton in the region) have been recorded, but none are endemic (Carola, 1994).

Fishing, in terms of both its commercial value and environmental impact, is the most important human activity in the region. The expansion in industrial fishing since the end of World War II has had a considerable impact on the population levels of targeted species, and has affected the community structure and trophic functioning of the ecosystem.

Commercial trawling started around 1900, but the first large-scale fishing operation began with the introduction of purse seiners soon after World War II. For a number of years, the Northern Benguela supported some of the world's largest fisheries. During the late 1960s, catches of pilchard peaked at over one million tonnes per annum (Fig. 12), whilst hake and horse mackerel catches have each been close to 500,000 t for many years (Figs. 13 and 14). The fishing industry currently accounts for over 10% of Namibian GDP, and an annual catch of around 700,000 t places Namibia in the top twenty fishing nations.

The fisheries of this region, like other coastal upwelling systems, fall into three main groups; epipelagic shoaling fish (mainly pilchard and anchovy), horse mackerel and hakes (Parrish et al., 1989). Together they account for more than 95% of the total Namibian catch in terms of tonnage and revenue (Table 3). Other living resources include a range of demersal finfish and benthic crustacea. Complete lists of all species recorded to date in Namibian and Angolan waters are given in Bianchi et al. (1993) and Fischer et al. (1981), respectively.

Epipelagic Zone

The sunlit waters of approximately the upper 50 m form the epipelagic zone in the region. Phytoplankton abundance is high and variable. High concentrations of diatoms are

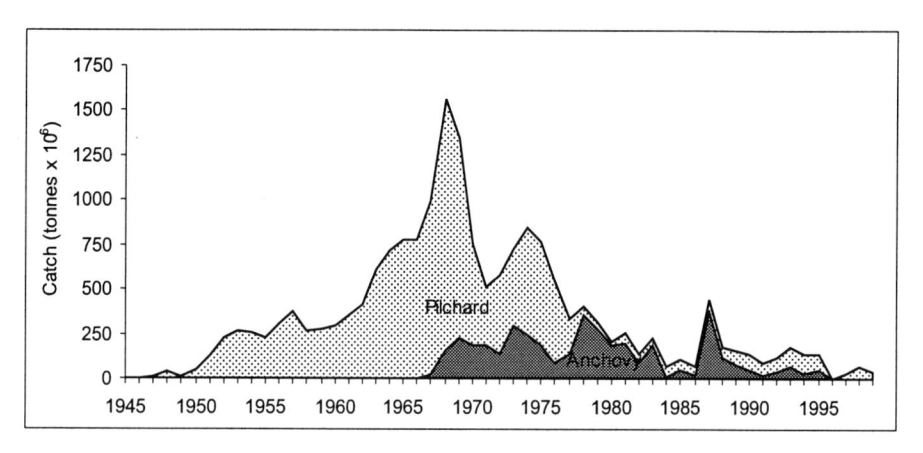

Fig. 12. Purse seiner catches of pilchard and anchovy in the Northern Benguela.

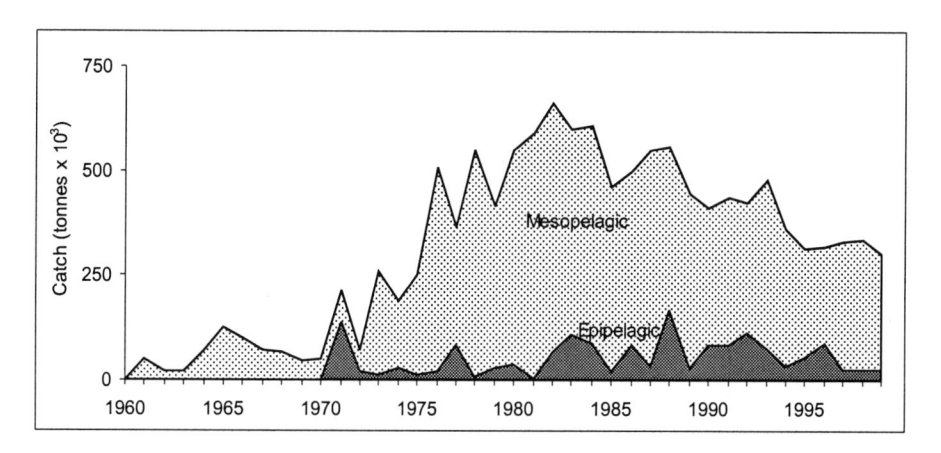

Fig. 13. Purse seiner catches of epipelagic horse mackerel and midwater trawler catches of mesopelagic horse mackerel in Namibian waters.

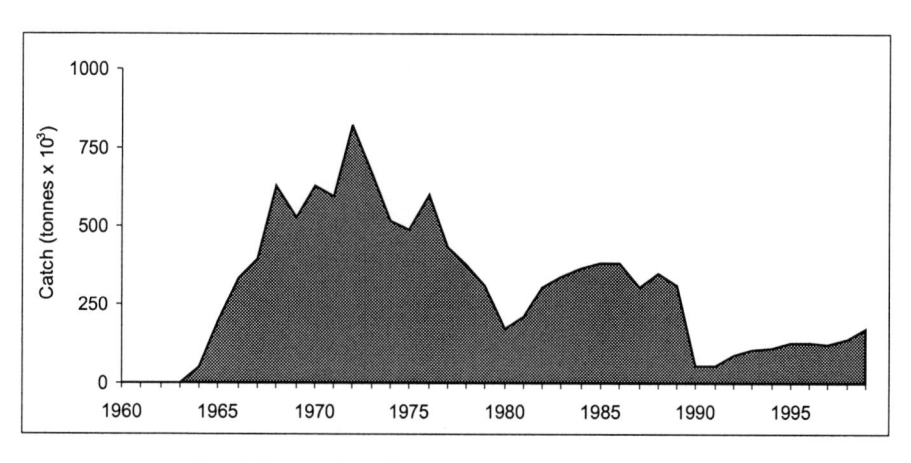

Fig. 14. Demersal trawl catches of hake in Namibian waters.

associated with cold water upwelling activity, whilst microflagellates and dinoflagellates tend to dominate when upwelling is suppressed and the water column is vertically stratified (Pieterse and van der Post 1967; Shannon and Pillar, 1986; Mitchell-Innes and Pitcher, 1992). Copepods and euphausiids dominate the zooplankton community.

Pilchard, anchovy and juvenile horse mackerel are, in terms of biomass, the main epipelagic fish. Traditionally pilchard has been the backbone of the pelagic fishing industry (see Table 3), although in recent decades anchovy and horse mackerel have assumed greater importance. Sardinella (*Sardinella aurita* and *S. madarensis*) are harvested in southern Angola, but these are essentially fish of the tropics that occur at the Angola–Benguela front and are not discussed further.

The Southern Benguela and Northern Benguela both have stocks of pilchard and anchovy separated by the Lüderitz upwelling cell (Boyd and Cruickshank, 1983), but are not genetically distinct (Grant, 1985). The exchange of adults, juveniles and eggs/larvae across this 'barrier', during the rare periods when there is little or no upwelling off Lüderitz, is thought to prevent the stocks from becoming totally reproductively isolated from each other (Boyd and Badenhorst, 1980; Cruickshank, 1984; Hewitson, 1987).

Fishing Activity

In the 1950s, soon after the start of commercial fishing, the region's pilchard stocks were around 5 million tonnes (Butterworth, 1983). Successive years with good recruitment during the late 1950s and early 1960s caused the population to increase to more than 10 million tonnes by the mid-1960s (Crawford et al., 1987), after which a series of stock crashes resulted in a biomass of only a few thousand tonnes by 1996. Similarly, catches have fallen from a peak of 1.4 million tonnes in 1968 to a low of 2400 t in 1996.

The stock crashes all showed a similar pattern; a run of years with poor recruitment combined with sustained fishing pressure, resulting in a crash of an order of magnitude or more. Following each crash the stock partially recovered, but full recoveries have been thwarted by continued or increased fishing pressure.

Juvenile horse mackerel catches doubled when the pilchard stock was collapsing in the 1970s. This may have been due to either a redirection of fishing effort towards horse mackerel, a real increase in the size of the horse mackerel stocks, or most likely, a combination of both.

The failure of the pilchard stock to recover even to a fraction of its former levels of abundance is most likely because the adult spawning biomass has remained at low levels, primarily by fishing, thus affecting recruitment. In addition, fewer age-classes in the population and a reduction in distribution has further decreased the likelihood of large recruitment.

Environmental Variability

Pilchard and anchovy have similar life-history traits in terms of high fecundity, batch spawning, rapid growth, and relatively short life spans (see Blaxter and Hunter, 1982, for review). These traits allow populations to respond rapidly to favourable conditions. The population sizes of these stocks are highly variable, on both year-to-year and decadal time-scales. This natural population variability occurs in the absence of fishing pressure, as confirmed by long-term records from clupeoid scale deposits in bottom sediments and guano deposits (Shackelton, 1987, 1988).

Carnivorous zooplankton may also have a large impact on the population dynamics of these small pelagics. Over the past two or three decades there has been a dramatic increase in the abundance of two jellyfish species: *Chrysaora hysocella* and *Aequorea aequorea*, although any impact they have remains unmeasured. Jellyfish are significant predators of zooplankton and ichthyoplankton in other parts of the world, so it is possible that they may influence the size of pelagic fish stocks in the Northern Benguela (Gibbons et al., 1992; Fearon et al., 1992).

Changes in the population dynamics of fish stocks have shown a remarkable degree of synchrony in many upwelling systems, both between species in the same system and between different systems. For example, regime shifts, in which the reduction in biomass of one species occurs simultaneously with the increase in abundance of another, are common (Lluch-Cota et al., 1997). Whether they are driven exogenously by, for example, environmental changes favouring one species over another, or through direct competition between species, is not always clear. The Northern Benguela seems to be unusual in this respect because, despite pilchard being depressed for more than two decades, horse mackerel seems to have only partly replaced pilchard, while anchovy has remained at comparatively low levels throughout. It is also possible that the fishing pressure has simply been too great to allow any species to fill this vacant niche. Similarly, changes in abundance of the clupeoid stocks of the Northern Benguela have shown little correspondence with the global synchrony in abundance of many other clupeoid stocks around the world (Kawasaki, 1983; Bakun, 1996).

Mesopelagic Zone

This zone extends from the epipelagic zone to a metre or two above the seabed. Adult Cape horse mackerel, bearded gobies (*Sufflogobius bibarbatus*) and several species of lantern fish (Myctophidae) are classified as 'mesopelagic'. Typically, they undergo extensive diurnal vertical migrations; at night they ascend into the epipelagic zone, although possibly not for feeding which occurs mainly during the day near the bottom.

Horse mackerel is the only mesopelagic subjected to heavy fishing pressure. Although large stocks of gobies over the central Namibian shelf have attracted the attention of the fishing industry, attempts to catch and process this species have, to date, not proved viable.

Horse mackerel have a wide distribution, being found from the Lüderitz upwelling cell to southern Angola, and from the surf zone to beyond the shelf break. The highest concentrations, however, are found between 17 and 21°S. A second stock occurs in the Southern Benguela, but transfer between the two stocks is believed to be slight, and for management purposes the two are treated separately.

A second species, the Cunene horse mackerel (*Trachurus trechae*), occurs in northern Namibia and southern Angola.

The Commercial Fisheries

The large-scale pelagic fishery first started in the late 1940s. Purse seiners initially targeted pilchard, and catches peaked in the 1960s. This was followed by the first of several collapses in the stock. Anchovy and horse mackerel then became important, suggesting species replacement, but anchovy catches have since declined too. A management strategy in the 1970s was to heavily fish anchovy in the hope that this would stimulate pilchard recovery, but all three species have remained consistently low.

Anchovy catches averaged around 200,000 tonnes per annum until 1983, after which they declined considerably. A brief increase in 1987 due to an influx of eggs and larvae from South Africa (Hewitson, 1988) was temporary, and now this stock is severely depleted.

Juvenile horse mackerel catches partly compensated for the lack of pilchard and anchovy, but the industry is very over-capitalised. When fully operational this industry employed more than 5000 people, so the declining stocks had severe social consequences. Midwater trawlers attained peak catches of 600,000 t in the early 1980s, using subsidised vessels from the Soviet bloc who purchased the products, but now both vessels and catches have declined.

Cape hake and deepwater hake are the major demersal fishery. Annual hake catches reached over 500,000 t, but since Independence, strict limits reduced this to 55,000 t, increasing through the 1990s to about 200,000 t per annum. Namibian monkfish was initially a by-catch, but is now important at around 15,000 t per annum. Commercially viable catches of orange roughy were made in the mid 1990s and catches peaked at 15,000 t. This species is highly vulnerable to overfishing; controls have failed, and catches fell to less than 3000 t in 1999.

Two high value crustaceans are caught. Rock lobster (*Jasus lalandii*) catches reached 10,000 t per annum but declined dramatically in the late 1980s. Current catches are limited to a few hundred tonnes which may be promoting a slow recovery. Deep-sea red crab (*Chaceon maritae*) catches peaked in 1983 at 10,000 tonnes, but have also declined to 2500 tonnes per annum.

SECONDARY EFFECTS OF FISHING

Reduced food for predators, by-catch of non-commercial species and destruction of habitat are all serious concerns. Since commercial fishing commenced Namibian penguins have declined by more than 95%. This may have stabilised (Crawford et al., 1995), but the species is critically endangered (Robertson et al., 1998). The gannet population has declined by almost 80% (Crawford et al., 1991), also due to prey depletion (Robertson et al., 1998). Indeed, 20% of coastal bird species have suffered substantial declines, largely due to loss of prey.

Drift-net fishing and beam trawling are not permitted in Namibia, and long-lining is not common. By-catch remains important. Since Independence, it has been illegal to discard commercially valuable fish, which has provided a strong incentive to avoid unwanted fish. Also, to discourage capture of valuable by-catch, punitive levies are imposed.

REFERENCES

Bailey, G.W., Beyers, C. de B. and Lipschitch, S. (1985) Seasonal variation of oxygen deficiency in waters off southern SWA in 1975 and 1976 and it's relation to catchability and distribution of Cape rock lobster *Jasus lalandii*. *South African Journal Marine Science* **3**, 197–214.

Crawford, R.J.M., Ryan, P.G. and Williams, A.J. (1991) Seabird consumption and production in the Benguela and western Agulhas ecosystems. *South African Journal Marine Science* **11**, 357–375.

Crawford, R.J.M., Williams, A.J., Hofmyer, J.H., Klages, N.T.W., Randall, R.M., Cooper, J., Dyer, B.M. and Chesselet, Y. (1995) Trends of African penguin (*Spheniscus demersus*) population in the 20th century. *South African Journal Marine Science* **16**, 101–118.

Hewitson, J.D. (1988) Spatial and temporal distribution of the larvae of the anchovy *Engraulis capensis* Gilcrest in the northern Benguela region. Colln. scient. Pap. int. Commn. SE. Atl. Fish, 15(2), pp. 7–17.

Robertson, A., Jarvis, A.M., Brown, C.J. and Simmons, R.E. (1998) Avian diversity and endemism in Namibia: patterns from the Southern African Bird Atlas Project. *Biodiversity and Conservation* **7** (4), 495–512.

There is some overlap in the distributions of these two species, particularly between 16 and 17°S. Morphologically both species are very similar and species are not distinguished in catches. Aggregations frequently contain both species, but the proportion of Cape horse mackerel increases towards the south and Cunene horse mackerel to the north.

Twelve million tonnes of Cape horse mackerel have been removed by fishing from the Northern Benguela system since the 1960s, more than any other species apart from pilchard and hake. Despite this, the stock is still in a robust state, supporting catches of around 400,000 t per annum throughout the 1990s. The reasons for this unusual resilience remain unclear.

Demersal zone

Hakes are commercially the most important demersal fish in the region, with a total of 12 million tonnes harvested to date. Two species occur off Namibia. Cape hake is found primarily in waters of less than 380 m depth, while deepwater hake occurs closer to the shelf break between 150 m and 800 m depths and beyond. Both species show a positive relationship between size/age and depth (Payne, 1989). Cape hake is mainly distributed off central and northern Namibia, while the deepwater hake population is thought to be an extension of the Southern Benguela stock. A doubling in biomass of deepwater hake off Namibia during the 1990s suggests an expansion of the Southern Benguela

Marine Mammals

Seals (*Arctocephalus pusillus*) were first harvested on a large scale in the 17th century by sealers from the Northern Hemisphere. By the start of the 20th century the seal population of the entire Benguela region had been reduced to around 100,000 animals (David, 1989). The population has now recovered to about 2 million, partly due to controls on harvesting and, more recently, a decline in the value of the seal products. Sustainable harvesting of seals is permitted in Namibia, with some 20,000 to 40,000 seal pups and 3000 to 5000 bulls harvested annually. The main products are: pelts, animal feed and bull genitalia which are exported the Far East.

The Namibian seal population is estimated to be one million individuals of which around half breed at Wolf and Atlas Bay, one quarter at Cape Cross and the remainder on islands and minor mainland colonies. Non-breeding, ephemeral rookeries occur near Conception Bay, Sandwich Harbour, Walvis Bay and Cape Frio.

Seals are estimated to consume more than 1 million tonnes of fish per annum in the region (Wickens et al., 1992). Despite much debate over the role of seals in the Benguela system, it is not clear to what extent seals affect the size of commercial fish stocks, mainly because a large portion of their diet is composed of non-commercial fish species (David, 1980). Indeed, the fact that the seal population has increased in the last 50 years, in spite of the depletion of the nearshore pelagic fisheries, has been attributed to their ability to switch to other prey.

Large whales were exploited during the 18th and 19th centuries; the name "Walvis" is Afrikaans for whale. By the mid-19th century the whale stocks had been depleted, and today whales are only sighted infrequently. The most common are probably Southern right (*Eubalaena australis*), humpback (*Megaptera novaeanglaie*) and Minke (*Balaenoptera acutorostrata*) whales. Aerial surveys off South Africa suggest that at least the abundance of the Southern right whale is increasing (Best and Ross, 1989).

Many species of dolphin and small whales occur in the Northern Benguela (Ross and Best, 1989). Of these, the dusky dolphin (*Lagenorhynchus obscurus*), bottlenose dolphin (*Tursiops truncatus*) and long-finned pilot whale (*Globicephalus melas*) are the most common. The Benguela or Heaviside's dolphin (*Cephalorynchus heavisidii*), an endemic to the Benguela system, occurs close to the shore along the Namibian and western South African coastlines. Due to its limited distribution this species may be considered vulnerable.

Fig. 1. Cape Cross: one of several large seal colonies (photo D. Boyer).

REFERENCES

Best, P.B. and Ross, G.J.B. (1989) Whales and whaling. In *Oceans of Life off Southern Africa*, eds. A.I.L. Payne and R.J.M. Crawford. Vlaeberg Publishers, Cape Town, pp. 315–338.

David. J.H.M. (1989) Seals. In *Oceans of Life off Southern Africa*, eds. A.I.L. Payne and R.J.M. Crawford. Vlaeberg Publishers, Cape Town, pp. 288–302.

Ross, G.J.B. and Best, P.B. (1989) Smaller whales and dolphins. In *Oceans of Life off Southern Africa*, eds. A.I.L. Payne and R.J.M. Crawford. Vlaeberg Publishers, Cape Town. pp. 303–314.

Wickens, P.A., Japp, D.W., Shelton, P.A., Kriel, F., Goosen, P.C., Rose, B., Augustyn, C.J., Bross, C.A.R., Penney, A.J. and Krohn, R.G. (1992) Seals and fisheries in southern Africa—competition and conflict. *South African Journal of Marine Science* 12, 773–790.

---

stock, given that there is no evidence of recruitment in the northern part of the stock (Hampton et al., 1999).

A third species of hake (*Merluccius polli*) occurs in southern Angola, as far south as the border with Namibia. Given its generally low population density, however, there is no targeted fishing on this species.

Namibian hake catches remained between 500,000 t and 600,000 t in the 1970s and 1980s, despite clear signs that the stock was being overfished. When Namibian authorities took over the management of the fish resources at Independence in 1990, catches were initially reduced to less than 100,000 t per annum then, as the stock started showing signs of recovery towards the end of the 1990s, increased to 200,000 t.

Hakes are opportunistic feeders, becoming increasingly piscivorous and cannibalistic with age (Punt et al., 1992). They feed largely on myctophids and gobies. In addition, Cape hake are thought to prey heavily on deepwater hake, especially in view of the fact that there is extensive overlap in the distribution of large Cape hake and smaller deepwater hake (Botha, 1985). Hake are thought to consume several million tonnes of fish annually, giving them an important role in the overall trophic functioning of the system.

A number of other species are harvested by industrial trawling, or have been in the past. Unfortunately the population dynamics of many of these species are poorly known, and while the effect of fishing on their population status can often be guessed, the impact on the wider community structure is unknown. For example, west coast sole (*Austroglossus microlepis*) and kingklip (*Genypterus capensis*) once supported small but important fisheries, but are rarely caught today. Annual catches of monkfish (*Lophius vomerinus*), were as high as 15,000 t in the 1980s and declined to a tenth of that in the early 1990s. In recent years catches have risen again to around 15,000 t per annum.

A recent addition to the Namibian fishery is orange roughy (*Hoplostetus atlanticus*). Orange roughy form dense aggregations at the shelf break between 500 m and 1200 m depths. Within just five years of the discovery of these aggregations, the stock was showing serious signs of depletion.

EFFECTS OF URBAN AND INDUSTRIAL ACTIVITIES

Due to the aridity of the coast and the immediate hinterland, the 1500 km coastline supports a permanent population of little more than 100,000 people. There are only four towns of note, ranging in size from 4000 to 45,000 people. Walvis Bay and Lüderitz are both industrial harbours, whilst Swakopmund and Henties Bay are primarily tourist centres.

The major commercial activities in this region, aside from fishing, are tourism, oil and gas prospecting, and diamond mining. The main shipping lanes are well offshore, and the high-energy ocean environment of the Northern Benguela ensures that any local adverse affects from these activities are quickly dissipated and absorbed into the system.

Fishing Industry and Ports

Walvis Bay and Lüderitz both host large modern fishing fleets and fish processing factories, and around 800 cargo vessels visit Walvis Bay annually. Recent developments to improve Walvis Bay's transport links to the interior are likely to increase the volume of cargo handled by the port.

Organic pollution emanates from the fishing factories, as do hydrocarbons and heavy metals from the various shipping and ship-repair activities. Although this pollution is relatively small-scale and localised, it occurs close to the important wetlands. Any increase in the level of background pollution, or a major spill, could potentially have serious consequences for this habitat.

Mining

Diamonds were found on ancient beach terraces in southern Namibia in 1908, which resulted in much of the southern part of the Namibian coastline and adjacent areas being closed to public access for most of the 20th century. As a result, apart from those regions directly affected by mining, this area has been largely protected from human disturbance and qualifies as one of the best-preserved habitats in Namibia. In 1986 much of the Namib Desert between Walvis Bay and Lüderitz changed status and is now part of a national park, the Namib Naukluft Park, and hence is still protected from most forms of human activity.

Currently, shore-based and offshore mining concessions extend from the Orange River to 26°30'S and therefore extend outside of the Northern Benguela system. The environmental impacts of diamond mining are generally highly localised. Obviously, the environment in the immediate vicinity of mining is totally altered, and sedimentation from tailings disposal may affect some habitats, but these effects are likely to occur within a relatively close distance of the mining operations.

The Kudu gas field 170 km due west of Oranjemund was discovered in 1974 and has proven to be a significant gas reserve with potential reserves of at least 5 trillion cubic feet of gas. Liquid petroleum reserves have not been discovered, despite extensive exploration.

Tourism

Coastal tourism is largely concentrated around Swakopmund and Henties Bay, although beach angling occurs at several camps in the north as well as along the 250 km stretch of coastline that is open to public access in the central region. In addition there are a number of small nature conservation and tourist camps along the coast.

Currently around 100,000 people visit the Namibian coastline annually. While tourism is expanding rapidly, the remoteness of Namibia, the limited range of activities available and the harsh climate limits the level of tourist activity.

Aquaculture

Given that local markets are small, and that the coastline offers few protected sites, opportunities for aquaculture are limited. Some ventures exist in Walvis Bay and Lüderitz lagoons, involving raft-cultured oysters (*Ostrea edulis* and *Crassostrea gigas*), black and bisexual mussels and seaweeds (*Gracilaria gracilus*). Environmental impacts are highly localised, and the escapement of exotic species, such as oysters, into the wild is unlikely, due to the hostile conditions outside of the lagoons.

THE NEXT MILLENNIUM

Due to the remoteness of the Northern Benguela system, and the paucity of fresh water, much of the system remains in a near-pristine state. The region emerges from the 20th century, however, with the legacy of excessive and uncontrolled fishing, resulting in most fish stocks being fully exploited or depleted.

Opportunities for further expansion of urban centres or industrial activities in the region are limited largely by the scarcity of water. Despite this, the coastal population

Table 4

Status of selected International Conventions, Agreements and Codes of Conduct related to the marine environment (pers. comm. Wolfgang Scharm, Ministry of Fisheries and Marine Resources, Arnt Aidijervi, Directorate of Maritime Affairs)

	Status	Date
United Nations Convention on the Law of the Sea (UNCLOS)	Ratified	10/12/82
FAO Code of Conduct	*	
Agreement for the Implementation of the Provisions of the United Nations Convention on the Law of the Sea of 10 December 1982 Relating to the Conservation and Management of Straddling Fish Stocks and Highly Migratory Fish Stocks	Ratified	1998
South East Atlantic Fisheries Organisation (SEAFO)	Under negotiation	
International Convention on the Conservation of Atlantic Tunas – ICCAT	Signed	1999
Convention for the Conservation of Antarctic Marine Living Resources: CCMALR	Acceded	1999
International Whaling Commission		
Convention on Biological Diversity: Rio de Janeiro	Signed Ratified	12/6/92 12/3/97
Convention on Wetlands of International Importance especially as Waterfowl Habitat: Ramsar		1995
Convention on International Trade in Endangered Species of Wild Fauna and Flora: CITES		1991
FAO/ICES Code of Practice for Consideration of Transfer and Introduction of Marine and Freshwater Organisms	*	
Climate change: Rio de Janeiro	Ratified	
Agreement to Promote Compliance with International Conservation and Management Measures by Fishing Vessels on the High Seas	Accepted	1998
International Convention for the Prevention of Pollution from Ships: MARPOL	Ratification in preparation	
Convention for the Prevention of Marine Pollution by Dumping Waste and Other Matters: LDE	Ratification in preparation	
International Convention on Oil Pollution Preparedness, Response and Co-operation: OPRC	Ratification in preparation	
International Convention Relating to Intervention on the High Seas in Cases of Oil Pollution Casualties	Ratification in preparation	
Convention on the International Maritime Organisation	Ratification in preparation	

*This has no legal basis, merely states an intent.

doubled in the four years between 1990 and 1994, and further increases are expected. In addition, various coastal developments have been proposed, for example a naval base near Walvis Bay and a hydro-electric dam several hundreds of kilometres upstream from the mouth of the Kunene River. The former may add further pressure to the already threatened lagoon of Walvis Bay whilst the latter would alter the flood pattern of the lower Kunene, and in turn would affect the estuarine system (Simmonds et al., 1993).

Fisheries Management, Past, Present and Future

The early years of the Northern Benguela fishery were targeted almost exclusively at the pelagic stocks, primarily pilchard. Virgin (and hence robust) stocks and limited fishing effort ensured that fishing was initially sustainable. As sardine stocks collapsed elsewhere in the world, fishing vessels and factories were transferred to the Northern Benguela region. Due as much to lack of knowledge as anything else, total allowable catches (TACs) were allowed to increase until, in 1968, the first of a series of collapses of the pilchard stock occurred. By 1990, the pelagic stocks of the Northern Benguela were a fraction of their former abundance.

During the late 1960s, increases in the number of deep-sea stern trawlers, mostly of European origin, which were fishing outside of Namibia's territorial boundaries, led to uncontrolled fishing pressure on the resources, mainly the horse mackerel and demersal stocks. This prompted the formation of the International Commission for the South East Atlantic Fishery (ICSEAF) in 1969 with the specific task of managing the fishery. At the time, 17 countries were full members of ICSEAF, most being engaged in active fishing and research. Regardless of the public pronouncements of responsible management, ICSEAF allowed nations to increase their fishing capacity almost at will and, not surprisingly, the overfishing continued (Moorsom, 1984).

The new Namibian authorities considered ICSEAF as an irrelevant management forum, whose main role was to legalise the plunder of Namibian resources. Therefore at Independence, the 200 n.m. exclusive economic zone was declared in terms of UNCLOS (Table 4) and all foreign fishing vessels were required to leave.

The Sea Fisheries Act provided the legal framework for the "conservation of marine ecology and the orderly exploitation, conservation, and promotion of certain marine resources", the emphasis being on the utilisation of Namibia's resources on a sustainable basis and to the benefit of Namibians. A fisheries White Paper outlines the government's policy towards fisheries. The underlying theme is somewhat contradictory; to encourage fishing while at the same time promoting the rebuilding of stocks.

Control is exercised through TACs in conjunction with limited access. All offloading of catches must take place in Walvis Bay or Lüderitz under stringent surveillance. In addition a strong and stable government has meted out severe punishment to offenders as a deterrent to illegal fishing.

Angola has been plagued by civil war for more than 25 years and this has limited the development of the country's own fishing fleet. However, while many of Angola's fish resources show clear signs of over-exploitation, others may in fact be under-utilised (Hampton et al., 1999). One of the consequences of the war has been an almost complete breakdown in fisheries management, surveillance and control. Fishing by foreign fleets, nominally under Angolan registration, continues (including Namibian vessels fishing for sardinella), while a small research and management cadre attempt the impossible task of monitoring and controlling the fishery with very limited resources at their disposal.

Namibia, in contrast, has many factors favouring successful and sustainable utilisation of its living marine resources. The country has a strong legal framework backed by political will, only two harbours to monitor, friendly neighbours, and a fairly discrete physical system with limited sharing of stocks with other countries. Many international conventions, agreements and codes of conduct have either been signed or ratified, giving Namibia the legal (and moral) basis with which to manage its own resources.

However, the biggest problem is lack of knowledge of the biological, environmental and ecosystem processes which inhibit the rational and optimal exploitation of the marine resources. Uncertainties in stock definition, abundance and basic biological parameters, a lack of understanding of recruitment processes and linkages between stock dynamics and the environment all contribute to imprecise recommendations that lack conviction (Hampton et al., 1999).

During the final years of this century several research programmes between the three nations of the Benguela system (South Africa, Namibia and Angola) and various collaborating donor nations (primarily Norway and Germany) have been initiated. Programme such as the Benguela Environment Fisheries Interaction and Training (BENEFIT) programme, ENVIFISH, Benguela Current Large Marine Ecosystem (BCLME) programme and others aim to address these problems.

Other Developments and Commercial Activities

As has been noted, most of the Northern Benguela coastline is protected and, in general, the nearshore region is in a pristine state. Specific areas have been totally altered, mainly by mining, but these are small in relation to the overall extent of the region.

The Namibian Cabinet approved an environmental assessment policy in 1994, but so far this lacks formal legislative support. However, the Namibian Constitution requires that development projects undergo an environmental assessment before they can be approved. It is uncertain if and when the Cabinet will pass the current version of the Environmental Management Act, but constitutionally some legislation in this field will need to be approved in the near future.

The mining sector has its own legislation, which demands certain requirements to be met before developments can be undertaken. The Petroleum and Mineral Acts of 1991 and 1992 set certain criteria for environmental impact assessment before mining activities are carried out. For example, all exploratory licensees must complete a full EIA prior to exploratory drilling or development of any commercial discovery. Oil spill contingency plans have been developed by the Government Action Control Group—a cross-sectoral committee to co-ordinate any oil spill or related disasters.

The legislation for controlled development that will take cognisance of, and strive to alleviate negative environmental impacts is either in place or very nearly so. Whether future generations will benefit from Namibia's policy of sustainable development remains firmly in the hands of the decision-makers.

REFERENCES

Agenbag, J.J. (1980) General distribution of pelagic fish off South West Africa as deduced from aerial fish spotting (1971–1974 and 1977) and as influenced by hydrology. *Fisheries Bulletin of South Africa* 13, 55–67.

Agenbag, J.J. and Shannon, L.V. (1988) A suggested physical explanation for the existence of a biological boundary at 24°30'S in the Benguela system. *South African Journal of Marine Science* 6, 119–132.

Badenhorst, A. and Boyd, A.J. (1980) Distributional ecology of the larvae and juveniles of the anchovy Engraulis capensis Gilcrest in relation to the hydrological environment off South West Africa, 1978/79. *Fisheries Bulletin of South Africa* 13, 83–106.

Bakun, A. (1996) Patterns in the ocean. California Sea Grant College System / Centro de Investigaciones Biológicas del Noroeste, México. 323 pp.

Bally, R. (1983) Intertidal zonation on sandy beaches of the west coast of South Africa. *Cahiers de Biologie Marine (Paris)* 24, 85–103.

Barange, M. and Boyd, A.J. (1992) Life history, circulation and maintenance of Nyctiphanes capensis (Euphausiacea) in the northern Benguela upwelling system. In Benguela Trophic Functioning, eds. A.I.L. Payne, K.H. Brink, K.H. Mann and R. Hilborn. *South African Journal of Marine Science* 12, 95–106.

Bianchi, G., Carpenter, K.E., Roux, J-P., Molloy, F.J., Boyer, D. and Boyer, H.J. (1993) FAO Species Identification Sheets for Fishery Purposes. The Living Marine Resources of Namibia. FAO, Rome, 250 pp.

Blaxter, J.H.S. and Hunter, J.R. (1982) The biology of clupeiod fishes. *Advances in Marine Biology* 20, 1–223.

Botha, L. (1985) Occurrence and distribution of Cape hakes Merluccius capensis Cast. and M. paradoxus Franca in the Cape of Good Hope area. *South African Journal of Marine Science* 3, 179–190.

Boyd, A.J. (1987) The oceanography of the Namibian shelf. Ph.D. thesis, University of Cape Town: 295 pp.

Boyd, A.J. and Agenbag, J.J. (1985) Seasonal trends in the longshore distribution of surface temperatures off Southwestern Africa 18–34°S, and their relation to subsurface conditions and currents in the area 21–24°S. *International Symposium on the Most Important*

Upwelling Areas off Western Africa (Cape Blanco and Benguela). Consejo Superior de Investigaciones Cientificas. Instituto de Investigaciones Pesqueras Barcelona 1985 **1**, 119–148.

Boyd, A.J. and Badenhorst, A. (1980) A review of some aspects of anchovy recruitment and associated environmental influences in ICSEAF Divisions 1.4 and 1.5. *Collection of Scientific Papers of the International Commission of the South East Atlantic Fisheries* **7** (2), 31–37.

Boyd, A.J. and Cruickshank, R.A. (1983) An environmental basin model for the west coast pelagic fish distribution. *South African Journal of Science* **79**, 150–151.

Boyd, A.J., Salat, J. and Masó, M. (1987) The seasonal intrusion of relatively saline water on the shelf off northern and central Namibia. In The Benguela and Comparable Ecosystems, eds. A.I.L. Payne, J.A. Gulland and K.H. Brink. *South African Journal of Marine Science* **5**, 107–120.

Boyd, A.J., Largier, J.L., Nelson, G., Sundby, S., Iita, A., Filipe, V. and Mouton, D. (1999) Near surface current regimes in the Benguela and southern Angola system in winter. Poster at the SAMMS Symposium, Cape Town, RSA.

Bubnov, V.A. (1972) Structure and characteristics of the oxygen minimum layer in the Southeastern Atlantic. *Oceanography* **12** (2), 193–201.

Butterworth, D.S. (1983) Assessment and management of pelagic stocks in the southern Benguela region. In Proceedings of the Expert Consultation to Examine Changes in Abundance and Species Composition of Neritic Fish Resources, San Jose, Costa Rica, April 1983, edited by G.D. Sharp and J. Csirke, FAO Fisheries Report No.291, 2, pp. 329–405.

Campbell, E.E. (1993) Coastal geomorphology, dunes, beach fauna and flora and seaweeds of Namibia. In: Environmental Data Workshops for Oil Contingency Planning: Centre for Marine Studies, UCT.

Carola, M. (1994) Checklist of the marine planktonic copepoda of southern Africa and their worldwide geographic distribution. *South African Journal of Marine Science* **14**, 225–253.

Chapman, P. and Shannon, L.V. (1985) The Benguela ecosystem Part II. Chemistry and related processes. *Oceanographic and Marine Biology Reviews* **24**, 65–170.

Cole, J.F.T. (1997) The surface dynamics of the Northern Benguela upwelling System and its Relationship to Patterns of Clupeoid Production. Ph.D. thesis, University of Warwick. 208 pp.

Cole, J.F.T. (1999) Environmental conditions, satellite imagery, and clupeoid recruitment in the northern Benguela upwelling system. *Fisheries Oceanography* **8** (1), 25–38

Cole, J.F.T. and McGlade, J. (1998a) Clupeiod population variability, the environment and satellite imagery in coastal upwelling systems. *Reviews in Fisheries and Fish Biology* **8**, 445–471.

Cole, J.F.T. and McGlade, J. (1998b) Temporal and spatial patterning of sea surface temperature in the northern Benguela upwelling system: possible environmental indicators of clupeoid production. In Benguela Dynamics: Impacts of Variability on Shelf-Sea Environments and their Living Resources, eds. S.C. Pillar, C.L. Moloney, A.I.L. Payne and F.A. Shillington. *South African Journal of Marine Science* **19**, 143–157.

Crawford, R.J.M., Shannon, L.V. and Pollock, D.E. (1987) The Benguela ecosystem Part VI. The major fish and invertebrate resources. *Oceanographic and Marine Biology Reviews* **25**, 353–505.

Crawford, R.J.M. and Shannon, L.V. (1988) Long-term changes in the distribution of fish catches in the Benguela. In *Proceedings of the International Symposium on Long Term Changes in Fish Populations. Vigo, 1986*, eds. T. Wyatt and M.G. Larrañeta, pp 449–480.

Crawford, R.J.M., Williams, A.J., Hofmyer, J.H., Klages, N.T.W., Randall, R.M., Cooper, J., Dyer, B.M. and Chesselet, Y. (1995) Trends of African penguin (*Spheniscus demersus*) population in the 20th century. *South African Journal of Marine Science* **16**, 101–118.

Cruickshank, R.A. (1984) Anchovy recruitment off South West Africa south of Walvis Bay. *Collection of Scientific Papers of the International Commission of the South East Atlantic Fisheries* **11** (1), 67–75.

Curtis, B., Roberts, K.S., Griffin, M. Bethune, S., Hay, C.J. and Kloberg, H. (1998) Species richness and conservation of Namibian freshwater macro-invertebrates, fish and amphibians. *Biodiversity and Conservation* **7** (4), 447–466.

Donn, T.E. and Cockroft, A.C. (1989) Macrofaunal community structure and zonation of two sandy beaches on the central Namib coast, South West Africa/Namibia. *Madoqua* **16**, 129–135.

Du Plessis, E. (1967) Seasonal occurrence of thermoclines off Walvis Bay, South West Africa, 1959–1965. *Investigational Reports of the Marine Research Laboratory of South West Africa* **13**, 35.

Emmanuel, B.P., Bustemante, R.H., Branch, G.M., Eekhout, S and Odendaal, F.J. (1992) A zoogeographical and functional approach to the selection of marine reserves on the west coast of South Africa. *South African Journal of Marine Science* **12**, 341–354.

Fearon, J.J., Boyd, A.J. and Schülein, F.H. (1992) Views on the biomass and distribution of *Chrysaora hysoscella* (Linné, 1766) and *Aequoea aequorea* (Forskål, 1775) off Namibia, 1982–1989. *Scientia Marina* **56** (1), 75–85.

Field, J.G. and Griffiths, C.L. (1991) Littoral and sublittoral ecosystems of Southern Africa. In *Ecosystems of the World 24: Intertidal and Littoral Ecosystems*, eds. P.H. Nienhuis and A.C. Mathieson. Elsevier, Amsterdam, pp. 323–346.

Fischer, W., Bianchi, G. and Scott, W.B. (1981) FAO Species Identification Sheets for Fishery Purposes. Eastern Central Atlantic; Fishing areas 34, 47 (in part). FAO, Rome, 7 vols.

Gammelsrød, T., Bartholomae, C.H, Boyer, D.C., O'Toole, M.J., Filipe, V.L.L. (1998) Intrusions of warm surface layers along the Angolan–Namibia coast in February–March 1995: The 1995 Benguela Niño. In Benguela Dynamics: Impacts of Variability on Shelf-Sea Environments and their Living Resources, eds. S.C. Pillar, C.L. Moloney, A.I.L. Payne and F.A. Shillington. *South African Journal of Marine Science* **19**, 41–56.

Gibbons, M.J., Stuart, V., and Verheye, H.M. (1992) Trophic ecology of carnivorous zooplankton in the Benguela. In Benguela Trophic Functioning, eds. A.I.L. Payne, K.H. Brink, K.H. Mann and R. Hilborn. *South African Journal of Marine Science* **12**, 421–437.

Grant, W.S. (1985) Population genetics of the Southern African pilchard, *Sardinops ocellata*, in the Benguela upwelling system. *Int. Symp. Upw. W. Afr. Inst. Inv. Pesq., Barcelona*, (1): 5510–562.

Griffin, M. and Channing, A. (1991) Wetland-associated reptiles and amphibians of Namibia—a national review. *Madoqua* **17**, 221–225.

Hampton, I., Boyer, D.C., Penney, A.J., Pereira, A.F. and Sardinha, M. (1999) Integrated overview of fisheries of the Benguela Current region. BCLME Thematic Report.

Hamukuaya, H., O'Toole, M.J. and Woodhead, P.M.J. (1998) In Benguela Dynamics: Impacts of Variability on Shelf-Sea Environments and their Living Resources, eds. S.C. Pillar, C.L. Moloney, A.I.L. Payne and F.A. Shillington. *South African Journal of Marine Science* **19**, 57–59.

Hewitson, J.D. (1987) Spatial and temporal distribution of larvae of the anchovy, *Engraulis capensis* Gilcrist in the northern Benguela region. M.Sc. thesis, University of Port Elizabeth.

Hockey, P.A.R. and Van Erkom Schurink, C. (1992) The invasive biology of the mussel *Mytilus galloprovincialis* in southern Africa. *Transactions of the Royal Society of South Africa* **48**, 123–139.

Huse, I., Hamukuaya, H., Boyer, D., Davies, S., Malan, P. and Stromme, S. (1998) The diurnal vertical dynamics of Benguela hakes and their potential prey. In Benguela Dynamics: Impacts of Variability on Shelf-Sea Environments and their Living Resources, eds. S.C. Pillar, C.L. Moloney, A.I.L. Payne and F.A. Shillington. *South African Journal of Marine Science* **19**, 365–376.

Kawasaki, 1983. Why do some fishes have wide fluctuations in their numbers? Biological basis of fluctuation from the viewpoint of evolutionary ecology. In Proceedings of the Expert Consultation to Examine Changes in Abundance and Species Composition of Neritic Fish Resources, San Jose, Costa Rica, 18–29 April 1983, eds. G.D. Sharp and J. Csirke. *FAO Fish Reports*, 2 (293), 731–777.

Kazmin, A.S., Legeckis, R. and Fedorov, K.N. (1990) Evolution of the temperature field in the Benguela upwelling using ship and satellite measurements. *Soviet Journal of Remote Sensing* 7 (3), 427–444.

Kirchner, C.H. (1998) Population dynamics and stock assessment of the exploited silver kob (*Argyrosomus inodorus*) in Namibian waters. Ph.D. Thesis. University of Port Elizabeth, RSA, pp. 448.

Kirchner, C.H., Sakko, A.L. and Barnes, J.I. (in press). The economic value of the Namibian recreational and surf fishery. *South African Journal of Marine Science*, in press.

Kruger, I. (1980) A checklist of South West African marine phytoplankton, with some phytogeographical relations. *Fisheries Bulletin of South Africa* 13, 31–53.

Lancaster, J., Lancaster, N. and Seely, M.K. (1984) Climate of the central Namib Desert. *Madoqua* 14, 5–16.

Le Clus, F. (1990) Impact and implications of large-scale environmental anomalies on the spatial distribution of spawning of the Namibian pilchard and anchovy populations. *South African Journal of Marine Science* 9, 141–159.

Le Clus, F. (1991) Hydrographic features related to pilchard and anchovy spawning in the northern Benguela system, comparing three environmental regimes. *South African Journal of Marine Science* 10, 103–124.

Lluch-Cota, D.B., Hernández-Vázquez, S. and Lluch-Cota, S.E. (1997) Empirical investigation on the relationship between climate and small pelagic global regimes and El Niño-Southern Oscillation (ENSO). FAO Fish. Circ. No. 934. Rome, Italy.

Lutjeharms, J.R.E. and Meeuwis, J.M. (1987) The extent and variability of South-east Atlantic upwelling. In The Benguela and Comparable Ecosystems, eds. A.I.L. Payne, J.A. Gulland and K.H. Brink. *South African Journal of Marine Science* 5, 35–49.

Lutjeharms, J.R.E., Shillington, F.A. and Duncombe Rae, C.M. (1991) Observations of extreme upwelling filaments in the southeast Atlantic ocean. *Science* 253, 774–776.

Lutjeharms, J.R.E. and Stockton, P. (1987) Kinematics of the upwelling front off southern Africa. In The Benguela and Comparable Ecosystems, eds. A.I.L. Payne, J.A. Gulland and K.H. Brink. *South African Journal of Marine Science* 5, 35–49.

Mas-Riera, J., Lombarte, A., Gordoa, A. and MacPherson, A. (1990) Influence of the Benguela upwelling on the structure of demersal fish populations off Namibia. *Marine Biology* 104, 175–182.

McLauchlan, A. (1985) The ecology of two sandy beaches near Walvis Bay. *Madoqua* 16, 129–135.

Meeuwis, J.M. and Lutjeharms, J.R.E. (1990) Surface thermal characteristics of the Angola–Benguela front. *South African Journal of Marine Science* 9, 261–279.

Mitchell-Innes, B.A. and Pitcher, G.C. (1992) Hydrographic parameters as indicators of the suitability of phytoplankton populations as food for herbivorous copepods. In Benguela Trophic Functioning, eds. A.I.L. Payne, K.H. Brink, K.H. Mann and R. Hilborn. *South African Journal of Marine Science* 12, 355–365.

Moorsom, R. (1984) *Exploiting the Sea* (A Future for Namibia: 5). Catholic Institute for International Relations. London. 123 pp.

Moroshkin, K.V., Bubnov, V.A. and Bulatov, R.P. (1970) Water circulation in the Eastern South Atlantic Ocean. *Oceanology* 10, 27–34.

Nelson, D.M and Hutchins, L. (1983) The Benguela upwelling area. *Progress in Oceanography* 12, 333–356.

Noli-Peard, K.R. and Williams, A.J. (1991) Wetlands of the Namib coast. *Madoqua* 17, 147–153.

Olivar, M.P. (1990) Spatial patterns of ichthyoplankton distribution in relation to hydrographic features in the Northern Benguela region. *Marine Biology* 106, 39–48.

O'Toole, M.J. (1980) Seasonal distribution of temperature and salinity in the surface waters off South West Africa, 1972–74. *Investl Rep. Sea Fish. Inst. South Africa*, 121–125.

O'Toole, M.J. (1977) Investigations into the early life history of the South West African anchovy *Engraulis capensis* Gilcrest. In Investigations into some important fish larvae in the South East Atlantic in relation to hydrological environment. PhD thesis, University of Cape Town, Paper 3, 38 pp.

Parrish, R.H., Bakun, A., Husby, D.M. and Nelson, C.S. (1983) Comparative climatology of selected environmental processes in relation to eastern boundary current pelagic fish reproduction. In: Proceedings of the expert consultation to examine changes in abundance and species composition of neritic fish resources, San Jose, Costa Rica, 18–29 April 1983, eds. G.D. Sharp and J. Csirke. *FAO Fish Rep.* 2 (293), 731–777.

Parrish, R.H., Serra, R. and Grant, W.S. (1989) The monotypic sardines, Sardina and Sardinops; their taxonomy, distribution, stock structure, and zoogeography. *Canadian Journal of Fisheries and Aquatic Science* 46 (11), 2019–2036.

Payne, A.I.L. (1989) Cape hakes. In: *Oceans of Life off Southern Africa*, eds. A.I.L. Payne and R.J.M. Crawford. Vlaeberg Publishers, Cape Town, pp. 136–147.

Pieterse, F. and Van der Post, D.D. (1967) Oceanographical conditions associated with red-tides and fish mortalities in the Walvis Bay region. *Investigational Reports of the Marine Research Laboratory of South West Africa* 14, 125 pp.

Punt, A.E., Leslie, R.W. and Du Plessis, S.E. (1992) Estimation of the annual consumption of food by Cape hake *Merluccius capensis* and *M.paradoxus* off the South African west coast. In Benguela Trophic Functioning, eds. A.I.L. Payne, K.H. Brink, K.H. Mann and R. Hilborn. *South African Journal of Marine Science* 12, 611–634.

Rogers, J. and Bremner, J.M. (1981) The Benguela Ecosystem. Part VII. Marine–geological aspects. *Oceanography and Marine Biology Annual Reviews* 23, 105–182.

Robertson, A., Jarvis, A.M., Brown, C.J. and Simmons, R.E. (1998) Avian diversity and endemism in Namibia: patterns from the Southern African Bird Atlas Project. *Biodiversity and Conservation* 7 (4), 495–512.

Sakko, A.L. (1998) The influence of the Benguela upwelling system on Namibia's marine biodiversity. *Biodiversity and Conservation* 7 (4) 419–433.

Shackelton, L.Y. (1987) A comparative study of fossil fish scales from three upwelling regions. *South African Journal of Marine Science* 5, 79–84.

Shackelton, L.Y. (1988) Fossil pilchard and anchovy scales—indicators of past fish populations off Namibia. In *Proceedings of the International Symposium on Long Term Changes in Fish Populations, Vigo, 1986*, eds. T. Wyatt and M.G. Larrañeta, pp 58–68.

Shannon, L.V. (1985) The Benguela Ecosystem. Part I. Evolution of the Benguela, physical features and processes. *Oceanographic and Marine Biology Annual Reviews* 23, 105–182.

Shannon, L.V. and Pillar, S.C. (1986) The Benguela ecosystem Part III. Plankton. *Oceanographic and Marine Biology Reviews* 24, 65–170.

Shannon, L.V. and Taunton-Clerk, J. (1989) Long-term environmental indices for the ICSEAF area. *Selected Papers of the International Commission on SE Atlantic Fisheries* 1, 5–15.

Shannon, L.V., Boyd, A.J., Brundrit, G.B. and Taunton-Clark, J. (1986) On the existence of an El Niño-type phenomenon in the Benguela system. *Journal of Marine Research* 44 (3), 495–520.

Shannon, L.V., Agenbag, J.J., and Buys, M.E.L. (1987) Large and mesoscale features of the Angola-Benguela front. In The Benguela and Comparable Ecosystems, eds. A.I.L. Payne, J.A. Gulland and K.H. Brink. *South African Journal of Marine Science* 5, 11–34.

Shannon, L.V., Crawford, R.J.M., Brundrit, G.B. and Underhill, L.G. (1988) Responses of fish populations in the Benguela ecosystem to environmental change. *Journal du Conseil, Conseil International pour l'Exploration de la Mer* 45, 5–12.

Sherman, K. (1991) The large marine ecosystem concept: research and management strategy for living marine resources. *Ecological Applications* 1 (4), 349–360.

Simmonds, R.E. 1996. Namibian Wetland Birds. In *African Waterfowl Census 1996*, eds. T. Dodman and V. Taylor. The International Waterfowl and Wetlands Research Bureau. Slimbridge, UK. pp 107–114.

Simmonds, R.E., Braby, R. and Braby, S.J. (1993) Ecological studies of the Cunene River mouth; avifauna, herpetofauna, water quality, flowrates, geomorphology, and the implications of the Epupa Dam. *Madoqua* **18**, 163–180.

Stander, D.H. and de Decker, A.H.B. (1969) Some physical and biological aspects of an oceanographic anomaly off South West Africa in 1963. *Investigational Reports of the Marine Research Laboratory of South West Africa* **81**, 46 pp.

Stramma, R.G. and Peterson, L. (1991) Upper-level circulation in the South Atlantic ocean. *Progress in Oceanography* **26**, 1–73.

Taunton-Clark, J. and Shannon, L.V. (1988) Annual and interannual variability in the SE Atlantic during the 20th Century. *South African Journal of Marine Science* **6**, 97–106.

Williams, A.J. (1987) Conservation management of the Walvis Bay wetland with particular reference to coastal bird numbers and their conservation significance. Report to Walvis Bay Round Table No. 37, Namibia.

Woodhead, P., Hamukuaya, H., O'Toole, M.J., Strømme, T., Saetersdal, G. and Reiss, M.R. (1996) Catastrophic loss of two billion juvenile Cape hake recruits during widespread anoxia in the Benguela Current. ICES International Symposium: Recruitment of Exploited Marine Populations: Physical–Biological Interactions. Baltimore, Maryland, USA. Sept. 1997. 18 pp.

THE AUTHORS

David Boyer
National Marine Information and Research Centre,
P.O. Box 912,
Swakopmund, Namibia

Christopher Bartholomae
National Marine Information and Research Centre,
P.O. Box 912,
Swakopmund, Namibia

James Cole
2 Dolphin Cottage, 31 Penny St.
Portsmouth PO1 2NH, U.K.

INDEX

Page numbers in bold refer to tables, in italics to illustrations. Roman numerals indicate volume number.

abalone, recreational harvesting, South Africa II-139
abalone fishery II-27, II-137
— Australia II-586
— dive fishery, Victoria Province II-668
— Tasmania II-655–6
Aboriginal people, Australia II-583–4, II-619–20
— dugongs a festive food for II-619
— southern and northern Nullabor Plain II-682
aboriginal subsistence whaling III-74
Abrolhos Archipelago and the Abrolhos Bank, Brazil *I-732*
— coral reefs *I-720*, *I-724*, *I-726*
Abrolhos Bank–Cabo Frio, Southern Brazil I-734
— Brazil Current I-734
— environmental degradation I-740
— eucalyptus plantations for pulp production I-738
— — industrial pollution from pulp production I-740
— major activities I-737–8
— shallow water marine and coastal habitats I-735–6
— Vitória Bay
— — environmental degradation I-740
— — oil spill risk for surrounding fragile areas I-739
Abrolhos National Marine Park, Brazil *I-720*, I-726–7, *I-728*
abundance, change in may be due to new competitor I-257
abundances, damped oscillations in III-275
accidental introductions *see* alien/accidental/exotic/introduced organisms/species
accretion, Gulf of Guinea coast I-778
acid mine drainage, Tasmania II-654
acid sulphate soil conditions, associated with mangrove clearance II-369, II-395
acid sulphate soil run-off
— a critical issue in eastern Australia II-638
— Great Barrier Reef region II-620–1
adaptive management III-400
Aden, Gulf of *II-36*, II-38, II-47–61, *II-48*
— coastal erosion and landfill II-58
— defined II-49
— effects from urban and industrial activities II-58–9
— major shallow water marine and coastal habitats II-51–4
— offshore systems II-54–5
— populations affecting the area II-55–6
— ports **II-55**
— production higher II-42
— protective measures II-59–60
— — conventions signed relating to the Marine Environment **II-59**
— rural factors II-56–8
— seasonality, currents, natural environmental variables II-49–51
Aden port II-55
— sand and mud flats II-53
Adriatic Sea *I-268*
— agricultural pollution load I-275
— circulation and water masses I-270
— described I-269–70
— differences between east and west coasts I-269, I-272
— effects of urban and industrial activities I-277–8
— — lagoon ports and canal harbours I-277
— nuisance diatom bloom III-298

— salinity and tidal amplitude I-269
— seasonality, currents and environment variables I-270–1
adult biomass, minimum viable, required for stock replacement III-382–3, *III-383*
adverse health effects, of toxic chemicals II-456–8
— coplanar PCBs most suspect contaminants II-457
Aegean Sea I-233–52
— Aegean basins I-235
— — partitioned by islands I-237
— behaviour of water masses I-236–7
— coastal erosion and landfill I-245
— defined I-235–6
— — Miocene—present evolution I-235
— effects from urban and industrial activities I-245–9
— human populations I-241–2
— northern, Turkish, sediment decreased through dam construction III-352
— offshore systems I-240–1
— protective measures I-245–9
— rural factors I-242–5
— seasonality, currents and natural environmental variables I-236–7
— shallow water ecosystems and biotic communities I-237–40
aerial surveys, for bird distributions III-109
Africa, Eastern, coral bleaching and mortality events III-48
Africa, southwestern I-821–40
— Benguela Current Large Marine Ecosystem I-823
— commercial fisheries I-833
— effects of urban and industrial activities I-835
— major shallow water marine and coastal habitats I-828–30
— the next millennium I-835–7
— — fisheries management, past, present and future I-836–7
— — oil spill contingency plans I-837
— — other development and commercial activities I-837
— offshore systems I-830–5
— physical environment I-823–8
— status of selected Convention, Agreement and Codes of Conduct **I-836**
African dustfall
— an ecological stressor in the Florida Keys I-410
— potential cause of other Caribbean coral diseases I-410
Agalega (Mascarenes) II-255
— coconut plantations II-257
— population II-260
— upwelling increases nutrients II-259–60
Agenda 21 III-351
— and promotion of ICZM III-352
— — evolution of III-352
— on Protection of the Oceans I-58
— on rights and duties of fishing III-158
— using two UNCED principles I-459
aggregate extraction
— offshore sites, Irish Sea I-95–6
— sand for beach nourishment I-53–4, I-96
— sand and gravel, off UK coasts I-53, I-74
aggregate mining, wadi beds II-27
agrarian reform, Nicaragua, created environmental problems I-538–9
agricultural pollution II-376
— Malacca Strait II-338
— Mauritius II-261
— problem of, Adriatic and Tyrrhenian basins I-275–6

— southern Spain I-174
— Sri Lanka II-184
agriculture
— American Samoa, Tutuila Island II-770
— Argentina I-758
  — Buenos Aires Province, fertilisers and pesticides I-762
— Australia II-584
  — poor agricultural practices II-584
— in the Bahamas I-423–4
— Bangladesh II-292
— Belize I-508
  — commercial, effects of I-508
  — potential impacts on the marine system I-508
— Borneo II-374–5
— Cambodia, traditional II-576
— Chesapeake Bay
  — best management practices I-347
  — contributes N and P to pollution I-345
— China's Yellow Sea coast, increased output II-492
— coastal
  — Mozambique II-107
  — subsistence, Gulf of Aden II-57–8
— coconut plantation, Marshall Islands II-781
— Colombian Pacific Coast I-682
— the Comoros II-247
— Coral, Solomon and Bismark Seas region II-435
— Côte d'Ivoire, basis of the economy I-816
— creating problems, Andaman, Nicobar and Lakshadweep Islands II-193
— eastern Australian region II-636–8
  — central and southern New South Wales II-638
  — northern New South Wales II-637–8
  — south-east Queensland II-637
— effects of in Tanzania II-88–9
— El Salvador I-551
  — poor land-use and agricultural practices I-551–2
— extensive cattle industry I-479
— Fiji Islands, and its impact II-758
— Great Barrier Reef region II-620–1
  — land clearing II-620–1
  — sedimentation and nutrients II-621
— Gulf of Guinea, use of agrochemicals I-784–6
— Gulf of Papua II-600–1, II-604
— Hawaiian Islands II-801
— Hong Kong, declining II-542
— influenced by water availability, Arabian Peninsula southern coast II-25–6
— intensive
  — Greece and Turkey I-242
  — Guadalquivir valley I-174
— Jamaica I-568
— Lesser Antilles I-635
  — export crops I-635
— Madagascar II-123–4
  — hill rice planting leads to soil loss II-123
— the Maldives **II-208**
  — environmental effect of II-207
— Marshall Islands II-778
— Mascarene Region
  — Mauritius II-260–1
  — Reunion II-261
  — Rodrigues II-261
— Mexican Pacific coast I-489
  — and the green revolution I-492
— Mozambique coast, inappropriate techniques II-109
— N and P to Baltic Sea I-104, *I-104*
— New Caledonia II-729, **II-729**
— Nicaragua
  — Caribbean coast I-523
  — Pacific coast I-538–9, I-540
— Peru, use of organochlorine pesticides I-696

— the Philippines II-412–13
— plant nutrients from I-93
— poor farming practice leads to estuarine sedimentation and turbidity II-136
— Samoa II-716
— Sea of Okhotsk, poorly developed II-469
— the Seychelles
  — deforestation for coconut plantations II-238
  — poor agricultural practice and unwise development activity II-238–9
— Somalia
  — few crops 179, II-77
  — nomad/semi-nomad II-77
— southern Brazil
  — subsistence I-738, I-739
  — sugar cane cultivation I-737–8
— Southern Gulf of Mexico, plantations I-476
— in the Sundarbans II-153
— Taiwan west coast II-503
— Tonga II-716
— Turks and Caicos Islands, small amount only I-591–2
— Vanuatu
  — and its impacts II-743
  — Pilot Plantation Project II-743
— Vietnam, rice growing II-563
— western Indonesia II-396
agrochemical pollution
— Malacca Strait II-318–19
— south coast, southern Brazil I-739
agrochemicals
— use of
  — China II-492
  — Vanuatu II-743
— western Indonesia II-396
Agulhas Bank II-135, II-137
Agulhas current *I-822*, I-824, *II-101*, *II-134*, II-135
Agulhas Gyre *II-101*
Agulhas Province *II-134*
air masses
— main winter transport routes over the Arctic *I-13*
— transport of from industrialised areas to the Arctic I-5
air pollution
— by ships, prevention of III-337
— Chesapeake Bay I-341
— locally high, Jamaica I-570
— Rudnaya River Valley, eastern Russia II-485
— Sea of Japan II-476–7
  — industrial contributions II-476–7
  — sources of II-476
— southern Brazil
  — Rio de Janeiro metropolitan area I-741
  — Santos Bay I-743
— through forest burning, Borneo II-375
air temperatures
— Côte d'Ivoire I-809
— Gulf of Guinea I-776
Air-Ocean Chemistry Experiment (AEROCE) I-224, I-229
airborne remote sensing, use of multiple digital video cameras III-284
airmasses, contribution to pollution over the Sea of Japan *II-476*
airport construction, Bermuda, effects of I-228
airports, environmental problems of II-214
Airy's theory, to predict energy in a wave III-312
Al Batinah *II-18*, II-19, II-26
— coastal erosion acute II-28
— dams across wadis keep sediment from coast II-27
— a wadi plain II-27
Alaska, protection for high relief pinnacles III-386
Alaska Coastal Current *I-374*, I-375
— and Ekman drift I-378
Alaska Current *I-374*, I-375, *III-180*
Alaska, Gulf of I-373–84, III-128

— anomalous along-shore flow III-184
— baseline studies, pollutants I-381-I-383
— climatic changes III-179–86
   — changes in physical properties III-181-3
   — impact of El Niño events III-183-4
   — implications of observed changes III-184-5
   — physical properties III-181
   — possible biological effects: salmon as an example III-185-6
— effects from urban and industrial activities I-383
— the *Exxon Valdez* oil spill I-382
— Forest Practices Act I-383
— geographic setting I-375
— human populations I-381
— major shallow water marine and coastal habitats I-375
— overfishing of virgin groundfish stocks, change to fishing other species III-120-1
— oxygen depletion impacts fisheries III-219
— physical oceanography I-375, I-378-9
   — shelf hydrography and circulation vary seasonally I-378
— primary productivity and nutrient cycles I-379-80
   — biomass doubling round perimeter, hypotheses I-380
— rural factors I-383
— 'strip-mining' of Pacific Ocean Perch III-120, III-121
— tidal propagation III-188-90
— trophic shift in marine communities I-376-8
Alaska Gyre *III-180*
Alaska Peninsula *I-374*
Alaskan Stream *I-374*, I-375, *III-180*
albacore I-141
albatrosses, wandering I-757
Albermarle Sound *I-352*, I-360
— environmental problems for bottom communities I-360
Albermarle—Pamlico Estuarine System *I-352*, I-353, I-362
— low benthic diversity and high seasonality in abundances I-360
— supports extensive shellfisheries I-360
Albufeira Lagoon *I-152*, I-154
— phytoplankton community I-156
— purification of mussels I-158
— RAMSAR site I-162, *I-163*
Aleutian Islands *I-374*
Aleutian Low Pressure region/system I-376, II-465
— linked to Alaskan Basin and shelf I-375, I-378
Alexander Archipelago *I-374*, I-375
algae I-590, I-649, I-754
— Andaman and Nicobar Islands II-192
— benthic, degraded, Black Sea I-291
— blue-green I-106
— brown I-47, I-70, I-109, I-125, I-159, I-290, I-669, II-23, II-42, II-664
— calcareous, maërl I-69–70, I-140
— calcareous, exploitation of, Abrolhos Bank–Cabo Frio, Southern Brazil I-737
— coralline II-370, II-711, II-741
   — crustose II-72, II-796
— drift, long-range dispersal, Western Australia II-695
— encrusting I-21
— epigrowth I-21
— epiphytic II-665, III-2
   — reduce diffusion of gases and nutrients to seagrass leaves II-701
— fucoid I-70
   — sensitive to oil III-275
— Galician I-137
— Great Barrier Reef region II-615
   — cross-shelf difference in communities II-615
— green I-70, I-71, I-194
   — *Halimeda* II-615
— Gulf of Maine, comparable to northern Europe I-311
— Gulf of Mannar II-164
— marine
   — in the Azores I-205-6
   — Fiji Islands II-754
   — Marshall Islands II-779

— Mexican Pacific coast, diversity of I-490
— red I-47, I-70, I-109, I-125, I-159, I-290, I-754
   — endemism, Australian Bight II-678
— reefal I-599–600
— the Seychelles II-236
— The Maldives II-204
— toxic
   — in English Channel I-69
   — Irish Sea I-90
— Vietnam II-564
— Wadden Sea, changes in I-49
algal assemblages, Western Australia II-695
algal beds
— green, New Caledonia II-726
— Hawaiian Islands II-796
algal blooms I-743, I-821, III-198
— Baltic Proper I-126, I-128
— Baltic Sea I-125
— Bay of Bengal II-275
— Black Sea, intensifying I-290
— detection of III-298-300
   — coccolothophores III-298
   — cyanobacteria III-298-300
   — red and white tides (HABs) III-298
— English Channel I-69
— entrainment blooms II-20
— harmful (HAB) I-827
   — and fish kills II-320
   — Hong Kong coastal waters II-541
   — increase in I-20
   — may be associated with ENSO events II-408
   — Yellow Sea II-495
   — *see also* harmful algal blooms (HABS)
— lakes, central New South Wales II-642
— novel III-259
— nuisance, creating anocic/hyppoxic disturbances III-219
— potentially harmful II-10–11, II-13
— red, Zanzibar II-92
— toxic
   — Australia's inland waters II-585
   — Carolinas coast I-356
   — increase may be due to ballast water I-26
   — Portuguese coastal waters I-160
— toxic and nuisance
   — Irish Sea I-90
   — North Sea I-53
algal mats I-131
— cyanophyte mat formers I-357
— may cause anaerobic condition I-53
algal reefs, Red Sea II-41
algal ridges, Fangataufa atoll, French Polynesia *II-818*
algal turf I-205, I-583
— Red Sea II-42
algal-vermetid reefs (boilers) I-227
alien/accidental/exotic/introduced organisms/species I-26, II-587, II-588–9, II-623, II-641
— accidental introduction of III-86
— Adriatic and Tyrrhenian seas I-282
— Argentine coastal waters I-765
— the Azores I-209
— Baltic Proper I-110
— Baltic Sea I-129
— Bay of Bengal II-279–80
— Black Sea I-290, I-291
— carried in ballast water II-319, III-227
— a coastal transboundary issue III-355
— a concern in Hong Kong II-542
— effects on seabirds III-114
— English Channel I-70
— Gulf of Maine I-312
   — and Georges Bank I-310

— Hawaiian Islands II-802
— Irish Sea I-89
— a major impact on the Tasmanian marine environment II-657–8
— Mediterranean mussel, southwestern Africa I-830
— the Mediterranean via the Suez Canal I-257–8
— North Sea I-47–8
— Victoria Province, Australia II-661, II-669
— Western Australia II-701
— *see also* new, novel occurrences and invasive disturbances
Alisios Winds *see* Trade Winds
alkyl lead I-95
alluvial fans II-37
alluvial plains, Borneo II-366
Altata-Ensenda del Pabellón, Mexican Pacific coast I-493
— pesticides causing environmental stress I-493
aluminium contamination, from bauxite mining II-395–6
Amazon River water, entering the Caribbean I-579
American flamingo, at Laguna de Tacarigua, Venezuela I-652
American Samoa II-765–72
— coastal habitats II-767–9
— coral reefs II-712, II-767–8
    — degradation of II-712
— effects from urban and industrial activities II-770–1
— environment II-767
— offshore systems II-769
— population II-714, II-769
    — must import to support population II-769
— protective measures II-771
    — marine protected areas II-771, **III-771**
— rural factors II-770
Americas, Pacific Coast, coral bleaching and mortality events III-53–4
americium I-94
Amirantes Bank *II-234*, II-235
ammonia, exports from Trinidad and Tobago I-638
amnesic shellfish poisoning (ASP) III-221
Amnesic Shellfish Toxin contamination, Scottish scallop fishery I-90
Amoco Cadiz oil spill
— change to *Fucus* III-272
— migration of oil layers downward within beach sediments III-270–1
amphidromic points *I-44*, I-45
amphipods, highly sensitive to oil III-274–5
Amursky Bay, eastern Russia
— chemical pollution **II-484**
— decline in annual biomass II-484
— ecological problems II-483–5
— the ecosystem is decaying II-484–5
— heavy metals in bottom sediments II-484
— metal concentrations, high in suspended matter II-484
— organic substances in wastewater II-483
anchialine pools, Hawaii II-796
anchor damage
— fishing and diving II-806
— ship groundings, Florida Keys I-409
— to corals I-605–6, I-637
anchovy
— beach seining, southern Brazil I-739
— Peru, replaced by sardines during an El Niño event I-693
anchovy fishery I-141
— Black Sea
    — collapse of I-290
    — Turkey I-292
— larval fishery, Taiwan Strait II-504–5
— Mexican Pacific waters I-490
— Peru I-691, I-693
— southwestern Africa I-832, I-833
Andaman Islands *II-270*
Andaman and Nicobar Center for Ocean Development (ANCOD) II-280
Andaman, Nicobar and Lakshadweep Islands II-189–97
— biodiversity II-192

— climate and coastal hydrography II-191
— coastal ecosystems II-191–2
— conservation measures II-195–6
— fish and fisheries of the Andamans II-192
— impacts of human activities on the ecosystem II-193, II-195
— islands of volcanic origin II-191
— Lakshadweep Islands II-194–5
— national parks and wildlife sanctuaries II-195
— population II-193
Andaman Sea *II-270*, *II-298*, *II-310*
— coral reefs II-302
— poor reef status II-303
— surfaces water influenced by freshwater continental runoff II-191
Andros Island *I-416*, I-431
Angola, development limited by civil war I-837
Angola Current *I-822*, I-824
Angola–Benguela front *I-822*, I-825
— southward displacement I-827
anguilla I-226, *I-616*, I-617
— beaches I-620
— climate I-617
— coastal protection I-624
— coral reefs I-618
— Exclusive Fisheries Zone I-621
— hurricane damage I-622
— impact of urban development I-622–3
— increased tourism I-620
— mangroves I-619
— marine parks I-625
— rocket launching site, environmental impact assessment criticised I-623
— seagrass beds I-618, *I-619*
— sustainable yields for fisheries I-621
— tourism I-620, I-622
Anguilla Bank I-618
Anjouan (Ndzouani/Johanna)
— coconut and ylang-ylang II-246
    — fringing reefs II-246
— reef front changed by quarrying II-248
— young island II-246
Annapolis Royal, Nova Scotia, tidal energy project III-316
Anole, Lac and Lac Badana *II-67*, II-68
— ecological profiles *II-76*
anoxia I-337, III-366
— due to decay of large plant biomass III-259
— from HABs III-298
— summer, Thessaloniki I-241
anoxic basins, confined, for dumping of contaminated sediments I-27
anoxic water
— Oslofjord *I-27*
— shallow-silled fjords I-17
anoxic/hypoxic disturbances III-219–21, *III-220*
— fish kills due to hypoxia III-220
Antarctic Circumpolar Current *II-581*, II-649, II-650, *II-674*
Antarctic Convergence II-583
Antarctic ecosystem, diversity and abundance in I-706–7
Antarctic Intermediate Water I-579, I-632
— Mascarene Region II-255
Antarctic polar front I-721
Antarctic Treaty system **III-335**, III-342
— Madrid Protocol III-342
Antarctica
— birds of III-108
— the grave of whaling III-75
anthropogenic disturbance, imposed on seagrasses III-10–11
anthropogenic influences, Norwegian coast I-24
anti-cancer/anti-infective agents, natural or modelled on natural products III-38–9
antibiotics, used in marine fish cages III-369
anticyclones
— Azores I-188, I-203

— Azores—Bermuda High I-439
— Bermuda High I-223
— Great Australian Bight II-675
— Mascarene II-116
— North Pacific High I-378
— Siberian High I-287, I-376, II-465
— south Atlantic I-823
antifouling paints I-788
— *see also* marine antifoulants; organotins; tributyltin (TBT)
Antilles Current I-419, *I-628*
Apo Reef *II-406*
Aqaba, Gulf of *II-36*, II-39
— minerals as pollutants *II-36*, II-39
aquaculture I-52, I-162, I-366, I-835, II-558–9, III-368–9
— Adriatic coasts I-275
    — lagoons and small lakes I-277
— Asian, significance of III-166
— Australia II-587
— the Bahamas, unsuccessful I-427
— Cambodia II-576–7
— can be a perfect environment for epidemics III-221
— Chilean fjords I-712
— the Comoros II-249
— Côte d'Ivoire I-818
— diseases, from and among facilities III-226
— east coast, Peninsular Malaysia II-352–3
    — changing water quality II-353
— effective planning and management to control pollution and disease III-170
— French Polynesia II-820
    — raising barramundi II-820
— Galicia I-144
— global, overview III-166
— Godavari-Krishna delta II-171
— limited in the Mascarenes II-262
— Malacca Straits, vulnerable to oil spill damage II-319–20
— marine, Tasmania II-656
— marine, Argentina I-764–5
    — biological species introduction I-764
    — environments: characteristics and adaptability I-764
    — harmful algae I-764–5
— marine, overview III-166–71
    — culture facilities III-170
    — fish aquaculture, development of III-168–9
— Marshall Islands II-781–2
— New Caledonia II-729
— oyster farming, eastern Australia II-641
— the Philippines, issues associated with II-413
— ponds in marine and brackish water, high demands on water III-368–9
— scenarios for opportunities to secure and increase production III-171
— seaweed, Fiji Islands II-754
— small-scale II-438
— sustainable II-375
— sustainable, Chinese and Thai experience III-171–7
    — national perspectives on issues and challenges III-171–2
    — national support for III-172
— The Maldives II-207
— Tyrrhenian coast I-279
— Venezuela I-653–4
— Victoria Province, Australia II-668–9
— west coast of Malaysia II-339
— Western Australia II-700
— western Indonesia II-394–5
    — high fluxes of nutrients and sediments into nearshore waters II-394–5
— Yellow Sea coastal waters II-494
— *see also* mariculture; shrimp aquaculture; shrimp farming
aquarium fish
— from Mauritius II-263

— trade in I-510
aquarium fish collection
— Cambodia II-576
— Great Barrier Reef II-617
— Hawaiian Islands II-803
— Marshall Islands II-784
— western Indonesia II-400
aquarium fish trade, New Caledonia II-734
aquasports, non-consumptive resource use II-141
aquatic ecosystems, atmospheric pathway a significant pathway of pollutant and nutrient fluxes III-201, III-207
aquatic organisms, farming of *see* aquaculture
aquatic plant farming III-167–8
aquatic vegetation
— decline of, Tangier Sound I-345, I-347
— improving, Chesapeake Bay I-347
Arabian Gulf II-1–16, *II-18*
— autumn seabird breeding season III-110
— coral bleaching and mortality events III-47
— corals little affected by Gulf War oil spills III-274
— defined II-3–4
— development issues II-9–11
— evaporation in excess of precipitation II-3
— future marine studies II-13
— major coastal habitats and biodiversity II-6–9
— marine studies II-3–4
— natural environmental variables II-4–6
— oil II-3
    — oil spills and the Gulf War II-11–13
— protective measures II-13
Arabian Gulf Co-operative Council (AGCC), Marine Emergency Mutual Aid Centre (MEMAC) II-29
Arabian Oryx Sanctuary II-31
Arabian Plate II-37
Arabian Sea
— abundance of meso-pelagic fish II-24–5
— coral bleaching and mortality events III-47–8
— crossed by major trading routes II-25
— a cyclone-generating region II-20
— erosion–deposition cycle II-27
— extreme marine climate II-20–1
— floor of II-19
— low pressure, ambient-temperature water flushing removes oil from mangroves III-277
— mangrove stands II-23
— northern, no freshwater inflow from Arabian Peninsula II-23
— suboxic conditions II-21
Arcachon Bay, French coast, serious environmental impacts of TBT III-250
arctic haze I-13, I-14
Arctic Mediterranean I-33
Arctic Monitoring and Assessment Programme (AMAP) I-9, I-26
— POPs ubiquitous III-361
areal surveillance, in fisheries management III-160
Argentina
— co-operation with Uruguay I-463
— laws and decrees undertaken by **I-766**, I-767
— main coastal management concerns I-751
— provincial and municipal ordinances **I-767**
Argentine coast
— estuaries and salt marshes I-755–6
    — Mar Chiquita coastal lagoon I-755
Argentine Sea *see* Southeast South American Shelf Marine Ecosystem
Argentinian continental shelf
— circulation I-754
— coastal system I-753
— Malvinas/Falklands system I-753
— subantarctic shelf waters system I-753
arid soil extraction, for beach regeneration, southern Spain I-177
Arkona Basin I-101, I-106
— deterioration of oxygen conditions I-105

armouring
— of beaches
    — and oil penetration III-269–70
    — and oil persistence III-276
Arrow oil spill, effect on *Fucus* spp. III-272
arsenic II-12
arsenic contamination, Bangladeshi groundwater II-293
artificial islands, for airport construction I-54
artificial reef structures I-347–8
artisanal fishing, impact of, New Caledonia II-730–1
Aruba *I-596*, I-597
— coastal and marine habitats *I-599*
— coastal urbanization I-606
— fishery I-603
— landfill I-602
— limited reef development I-599
— mass tourism I-598
— sewage discharge I-606
ASEAN Council on Petroleum (ASCOPE) II-325
Ashmore Reef National Nature Reserve, Western Australia II-696
Asia Minor Current I-236
Asian developing regions: persistent organic pollutants in the seas II-447–62
— coastal waters II-449–54
— open seas II-454–60
asphalts, formation of III-270
astronomical tidal forcing III-190, III-193
Aswan Dam, effects of I-256
Atacama Desert I-704
Atchafalaya River *I-436*, I-443
Athens *I-234*, I-241
— primary sewage treatment I-245–6
Atlantic basin
— and ENSO events I-223–4
— and tropical cyclones I-224
Atlantic Conveyor, postulated reversal of III-86–7
Atlantic Niños, related to Pacific El Niños, effects of I-781–2
Atlantic Rainforest
— Abrolhos Bank–Cabo Frio, Southern Brazil I-735
— Brazilian tropical coast I-725–7
Atlantic Water I-255
— entering the Tyrrhenian circulation I-272, I-274
— flowing into the North Sea I-45, *I-45*
atmosphere—surface fluxes, determined using surface analysis methods III-204
atmospheric nuclear experiments, fallout from I-11
'atmospheric particles'
— described III-199
— origins of III-199
— primary and secondary particles III-199
— size distribution in space and time III-199–200
    — can change as a result of physical and chemical processes III-200
Australia
— an island continent II-581
— biogeography II-581
— coastal erosion and landfill II-585
— continental shelf II-582
— coral bleaching and mortality events III-52–3
    — Western Australia (1998), variable III-3
— degradation of the Great Barrier Reef III-37
— EEZ II-581
— effects of urban, industrial and other activities II-586–9
— human populations affecting the area II-583–4
— laws protect flatback turtle III-64
— major shallow water marine and coastal habitats II-582
— mangroves III-20
— marine biogeographical provinces *II-675*
— offshore systems II-583
— protective measures II-589–91
    — general marine environmental management strategies II-590

— International arrangements and responsibilities II-590
— marine protected areas II-590
— oceans policy II-590–1
— a regional overview II-579–92
— status of scientific knowledge of the marine environment II-584
— rural factors II-584–5
— seasonality, currents, natural environmental variables II-581–2
— status of the marine environment and major issues II-591
Australia, eastern: a dynamic tropical/temperate biome II-629–45
— biogeography II-631
— coastal erosion and landfill II-639
— effects of urban and industrial activities II-639–42
— human populations affecting the area II-635–6
— major shallow water and coastal habitats II-632–5
— protective measures II-642–3
    — evaluation of protected areas II-643
    — legislation and responsibilities II-642
    — marine protected areas II-643
    — protected species II-643
— rural factors II-636–8
— seasonality, currents, natural environmental variables II-631–2
— status of the marine environment II-643
Australia, northeastern, Great Barrier Reef region II-611–28
— biogeography II-613
— coastal erosion and landfill II-622
— effects of urban and industrial activities II-622–4
    — protective measures II-624–7
— major shallow water marine and coastal habitats and biota II-614–19
— offshore systems II-619
— populations affecting the area II-619–20
— rural factors II-620–1
— seasonality, currents, natural environmental variable II-613–14
Australian Sea Lion II-687
— on IUCN Red List II-681
— threats to recovery of II-685–6
AVHRR imagery
— detection of cyanobacteria blooms in the Baltic III-299–300
    — algorithm for bloom detection III-299
— sea surface temperature (SST), Venezuela *I-647*
— for trends in water clarity III-296
The Azores I-201–19
— climate I-203–4
— coastal erosion and landfill I-211
— effects from urban and industrial activities I-212–13
— major shallow water marine and coastal habitats I-205–9
    — importance of harbours I-209
— offshore systems I-209–10
— populations I-210–11
— protective measures I-213–17
    — in word but not deed I-214–15
— the region defined I-203
— rural factors I-211
— seasonality, currents, natural environmental variables I-203–5
Azores anticyclone I-188, I-203
Azores Current I-204–5, *I-204*
Azores Microplate I-203
Azores—Bermuda High I-439
Azov, Sea of *I-286*, I-287
— suffering from hypoxia I-291

Bab el Mandeb *II-36*, II-38
— flow through not clear II-39–40
— traversed by oil tankers II-59
back water effect, Bangladesh II-273
bacteria
— on the Southern Californian shoreline I-399
— sulphur-oxidizing III-263, *III-263*
— zooplankton and phytoplankton as reservoirs for III-221
bacterial decomposition, Arabian Sea II-24

Baffin Bay *I-6*, I-7, I-8
— increased mercury in upper sediments I-9
Bahama Bank Platforms, calcium carbonate I-419
Bahama Banks
— biota of Caribbean origin I-418
— fossil reefs, used to date sea-level change I-417
— islands formed during sea-level lowstands I-417
— origins of debated I-589
The Bahamas I-415-33
— agriculture I-423-4
— aquaculture I-437
— artisanal and commercial fisheries I-424-5
— coral bleaching and mortality events III-54
— effects from urban and industrial activities I-427-30
— Landsat TM–seagrass biomass relationship III-286
— major shallow water marine and coastal habitats I-419-22
— Marine and Coastal National Parks of the Bahamas **I-431**
— offshore systems I-422-3
— origin of I-417-18
— other fisheries resources I-426-7
— population I-423
— protective measures I-430-2
— seasonality, currents, natural environmental variables I-418-19
— tourism and its effect on the population I-423
Bahamas National Trust I-432
Bahamas National Trust Park *I-416*
Bahamas Reef Environment Education Foundation I-432
Bahamian Banks *see* Bahama Banks
Bahamian Exclusive Economic Zone (EEZ) I-422
Bahrain *II-2*
— coral bleaching and mortality events III-47
Baie de Seine *I-66*, I-70
— dumping of metals I-75
— PCBs and organochlorines in meio- and macrofauna I-75
Baie de Somme *I-66*
— marsh with ponds I-69
baiji (Yangtze river) III-90
— will be affected by Three Gorges Dam III-97
Baird's beaked whale III-76, III-92, III-93
Baja California I-486
— marine habitat I-487
Baja California Sur I-486, I-487
Bajuni Islands (Archipelago) *II-67*, II-68, *II-75*
— coral carpet development II-72
— development of shelf since isotope stage 5e *II-73*
— fringing reefs II-72, II-74
— shows features of barrier island complex II-69, *II-73*, *II-75*
— — build-up of sand bodies II-69, *II-71*
Bajuni Sound
— coral knobs, patch reefs and seagrass beds II-74
— intertidal abraded flats facing channels II-77
— mixture of habitats *II-71*, II-75, II-76, *II-78*
"balance of nature" III-86
baleen (rorqual) whales III-74, III-76
— estimates of numbers III-82-3
— "safe catch" limit calculations III-79
ballast water
— dumping of II-295
— environmental risks from II-319, III-226-7
Baltic ecosystem, pollution sensitivity of I-129-30
Baltic Proper *I-122*
— biota
— — alien species I-110
— — birds and mammals I-109
— — main coastal and marine biotopes I-108-9
— — pelagic and benthic organisms I-109
— central parts permanently stratified I-123
— chemical munitions dumped in I-117
— eutrophication I-103-8
— — biological effects of I-109-10
— — inputs from land and atmosphere I-103-4
— — oxygen I-104-6
— — temporal and spatial variability in nutrients I-106-8
— major oceanic inflows I-101-2, *I-102*
— — environmental conditions dependent on I-124
— natural immigrants I-108
— persistence of anoxic zones I-105-6
— reasons for problems I-101
— shallow banks providing spawning and nursery areas I-124
— standing stock and carbon flow I-126, *I-127*
Baltic Sea *I-100*, *I-122*, III-205
— anoxic areas III-258
— basins *I-121*/I-123
— cyanobacteria blooms III-299
— — investigation using AVHRR imagery III-299
— described I-123
— freshwater immigrants I-125
— including Bothnian Sea and Bothnian Bay I-121-33
— — coastal and marine habitats I-124-8
— — effects of pollution I-130-1
— — environmental factors affecting the biota I-123-4
— — fish and fisheries I-128-9
— — pollution sensitivity of the Baltic ecosystem I-129-30
— — protective measures I-132
— southern, eutrophication of coastal inlets III-260-3
— southern and eastern regions I-99-120
— — biota in the Baltic Proper I-108-10
— — condition of the Baltic Proper I-103-8
— — environmental pollutants I-112-17
— — fish stocks and fisheries I-110-12
— — regional setting I-101-3
— a young sea I-101, I-108
Baltic Water *I-18*, I-23
band ratio, for retrieval of chlorophyll content from spectral radiance III-297
Bangladesh II-285-96
— coastal environment and habitats II-289-91
— cyclones and storm surges II-288
— fisheries resources II-294
— legal regime II-295
— Naaf estuary Ransar site II-289
— need for integrated coastal management II-295-6
— offshore system and fisheries resources II-292-5
— physical setting of the Bay of Bengal II-287
— population and agriculture factors in the coastal areas II-291-2
— seasonality, currents, and natural environmental variables II-287-8
bank reefs, Brazil I-723, I-723-4
banks and shoals
— Madagascar II-118
— Mascarene Plateau, habitats poorly known II-256
Bar al Hickman *II-18*
Barbados Marine Reserve I-639
Barcelona Convention I-249, I-262, III-363
Barents Sea, effects of collapse of capelin stock III-129
barnacles I-71, I-194
— barnacle belt, Spanish north coast I-140
— bioindicators of Cu, Zn, and Cd concentrations I-213
— Deltaic Sundarbans II-156
— Mauritius II-258
Barnegat Inlet *I-322*
barramundi fishery, Gulf of Papua II-602-3, II-607
barrier island lagoons I-437
barrier island-sound systems, diverse communities I-361
barrier islands I-353
— provide shelter for seagrass communities I-358
barrier reefs II-370
— Andros Island I-421
— Belize *I-502*, I-503, I-504
— — affected by hurricanes I-504
— — Barrier Reef Committee I-512-13
— — coral mortality and macroalgae increase I-505

— system may be damaged by sediment, nutrients and contaminants I-508
— Fiji Islands II-756, **II-756**
— French Polynesia II-817
— Madagascar II-118
— Mauritius II-258
— New Caledonia II-725
— off northeast Kalimantan II-388
— San Andrés and Providencia Archipelago I-668
— *see also* Great Barrier Reef
basin water exchange processes I-21
basking sharks, Irish Sea I-91
Basque Country
— heavy industry decline, leaving environmental damage I-142
— recovery of coastal water quality I-142
Bass Strait *II-648*, *II-662*, II-663
— local sea floor pollution from offshore rigs II-669
— oscillatory tidal currents II-650
— species richness II-651
— tidal currents II-664
— topography II-663
Bassian Province II-649
bathing water quality I-51, I-76
— Aegean Sea I-248
— Israeli coast I-261
— Turkish Black Sea coast I-302–3
Bay of Biscay Central Water, off Ortegal Cape I-139
Bay of Fundy, Canada, tidal barrage III-316
bays, mudflats and sand spits
— southwestern Africa I-828–9
— saltpans I-828
— Walvis Bay and Sandwich Bay, Ramsar wetlands I-829
Bazaruto Islands, Mozambique II-104–5
— fishing a key activity II-109
— shallow and shelf waters II-104
beach angling I-835
beach armour, or beach nourishment III-66
beach erosion
— Australia II-585
— the Bahamas I-428
— Belize I-510
— Fiji Islands II-760
— human-induced I-211
— and sand drift, NSW, case study II-640
— South Florida I-428
— south-east Queensland and New South Wales II-639
— Tanzania, possible reasons for II-93
— buffer zone principle II-93
— through sand mining II-782
beach forest vegetation II-777
beach formation, and longshore drift, east coast of Madagascar II-117
beach loss, in the Comoros II-248
beach mining *see* sand mining
beach nourishment/replenishment I-604, II-356, III-352
— Gulf of Guinea I-787
— sand for I-53–4, I-96, I-176, I-366, I-428
beach sands, quarrying of II-248
beach seining, Great Barrier Reef region II-623
beach vegetation, east coast, Peninsular Malaysia II-349
beach-rocks habitat, possible origin of I-273
beaches
— Adriatic, more severe storm patterns I-276
— Anguilla I-620
— artificial, for tourists I-604
— Curaçao, littered I-607
— impermeable layers affect oil penetration and persistence III-270
— loss of sand from, Jamaica I-569
— Malacca Straits II-312
— Maldivian II-203
— not meeting EU standard, Irish Sea coast I-95
— oil contamination of III-364

— and soft substrates, Victoria Province, Australia II-665–6
— Southern California, recreational shoreline monitoring I-399
— Sri Lanka, squatters on II-183
— Turkish Black Sea coast I-302, **I-302**, **I-303**
— *see also* sandy beaches
beaked whales III-92
Beaufort Sea, summer ice thickness and residual circulation III-190, *III-190*
Beaufort's Dyke *I-84*, I-85
bêche-de-mer fishery, Coral, Solomon and Bismark Seas region II-437
bêche-de-mer production, New Caledonia II-730
Belfast Lough *I-84*
Belgium, offshore wind farm planned III-309
Belize I-501–16
— after the 1998 bleaching and hurricane Mitch III-55
— coastal erosion and landfill I-509–10
— Coastal Zone Management Authority I-512, I-514
— Coastal Zone Management Unit I-512
— effects from urban and industrial activities I-510–12
— geology of I-504
— major shallow water marine and coastal habitats I-504–6
— Marine Protected Areas Committee I-513
— National Coral Reef Monitoring Working Group I-513
— offshore systems I-506
— populations affecting the area I-506–7
— population and demography I-506–7
— use of the coastal zone I-507
— protective measures I-512–14
— challenges I-514
— policy development and integration I-512–13
— regulation of development I-513
— rural factors I-508–9
— seasonality, currents and natural environmental variables I-503–4
— source area for fish, coral and other larvae I-503
Belize Barrier Reef *I-502*, I-503, I-504
Belize City, habitat destruction during growth of I-510
Belle Tout Lighthouse, Beachy Head, relocation of III-353
beluga whales III-76, III-90
— contaminant-induced immunosuppression III-95
— hunted in Arctic and sub-Arctic III-92
Bengal, Bay of (northwest coast) and the deltaic Sundarbans II-145–60
— effects from urban and industrial activities II-154–6
— major rural activities and their impact II-153–4
— major shallow water marine and coastal habitats II-149–51
— protective measures II-156–9
— captive breeding programmes II-158–9
— conservation of biological resources II-156–7
— conservation policies II-157–8
— seasonality, currents and natural environmental variables II-148–9
— social history and population profile II-151–3
Bengal, Bay of II-269–84
— chemical features of the water II-274
— coastal habitats and biodiversity II-274–7
— defined II-271
— effects from urban and industrial activities II-279–80
— marine fisheries II-277–8
— mining, erosion and landfill II-279
— natural environmental variables II-271–4
— new millennium: need for east coast zone management authority II-280–1
— physical setting II-287
— populations affecting the area II-278
— protective measures II-280
— Coastal Ocean Monitoring and Prediction System, Indian Coast II-280
— public awareness II-280
— rural factors II-278–9
Bengal Deep Sea Fan II-287
Bengal tiger II-150, II-290
— Project Tiger II-156, II-157

Bengkali Strait II-311
Benguela Current *I-822*, I-825
— eastern boundary current I-823
— fish kills, from upwelling, blooms and oxygen depletion III-129
— red tides III-218
Benguela ecosystem *II-134*
— upwelling II-135
Benguela Environment Fisheries Interaction and Training Programme I-837
Benguela Niño years I-825, I-826, I-827
benthic assemblages, offshore, Côte d'Ivoire I-814
benthic biomass, South China Sea coast **II-554**
benthic communities
— affected by turbidity and oxygen depletion III-259
— Gulf of Guinea I-782, **I-782**
— Irish Sea
  — linked to sediment type *I-86*, I-89
  — threats to I-89, *I-94*
— macrofaunal, English Channel bed I-71
— offshore, Carolinas coast I-362
— Portuguese coastal waters, temporal changes due to natural causes I-159–60
benthic fauna
— Aegean Sea I-240
— Baltic Proper, adverse changes below the halocline I-110
benthic microbial communities, effects of eutrophication on, the Bodden III-261–2
benthic monitoring III-243
benthic organisms
— as biomonitors of radionuclides I-116
— West Guangdong coast II-554
benthic species, high mortality rate from trawling III-123
Benthic Surveillance Project (USA) I-452
benthic vegetation, Portuguese coastal waters I-159–60
benthos
— Campeche Sound I-474
— Côte d'Ivoire I-813
— infaunal, filtering ability, Chesapeake Bay I-342
— Marshall Islands II-780
— Southern California Bight, effects of anthropogenic inputs I-395–6
Bergen *I-18*, I-27
Bering Sea *I-374*
— by-catch as an issue III-146
— decline in some marine mammal populations through the pollock fishery III-129
— Eastern, change in groundfish species composition III-128
— residual currents III-190
Bering Sea ecosystem, example of cumulative and cascading impacts III-98–9
Bermuda
— acid rain I-229
— Bermuda Atlantic Time-series (BATS) program I-224–6
— conservation laws I-230
— Hydrostation S program I-224–5
— threats to reef environment I-228
Bermuda High I-223
Bermuda Platform I-227
Biddulphia sinensis, an Asian introduction I-26
Bien Dong Sea II-563
— great natural resource potential II-567
— *see also* South China Sea
Bijagos Archipelago, Guinea-Bissau I-781
— breeding and nursery ground, fish and crustaceans I-781
— diverse tidal habitats I-781
Bikini Atoll, nuclear tests at, costs of clean up II-783
Bimini *I-416*
— deep water sport fishing I-422
— mangroves stunted I-422
— sand mining/dredging I-428
bioaccumulation
— of heavy metals II-156

— in seabirds and fish I-212
— of pesticides, northern Gulf of Mexico I-444
— in the Southern California Bight I-394
  — of DDTs and PCBs in seabirds I-398
  — effects of in fish I-397
  — in marine mammals I-399
— of TBT III-250
bioavailability, of contaminants in sediments I-25, I-26
biocides
— antifouling, the ideal III-249
— in marine antifoulant paints III-248
biodiversity II-336
— Andaman, Nicobar and Lakshadweep Islands II-192, **II-193**
— Anguilla I-620
— Arabian Gulf II-6–7
— Australia II-583
— of the Baltic Sea I-124–5
  — marine species decrease to the north I-124–5
— Bay of Bengal **II-274**
— benefits to tourism and recreation II-326
— British Virgin Islands I-620
— coral reefs III-34
  — endangered III-36–7
— Coral, Solomon and Bismark Seas region II-431
— Deltaic Sundarbans II-150, **II-150**
— English Channel
  — fauna I-70–1
  — flora I-69–70
— Fiji Islands II-757
  — marine **II-754**, II-762
— Godavari-Krishna delta II-170
— Great Barrier Reef II-613
— Gulf of Aden II-49
— Gulf of Guinea LME I-794
— Gulf of Mannar II-164, **II-164**
— high
  — macroalgae and invertebrates, the Quirimbas II-102, II-104
  — New Caledonia II-726, **II-726**
— hunted in Arctic and sub-Arctic III-92
— increase from fjord head to coast I-22
— Lakshadweep Islands II-194
— Malacca Straits II-312
— marine
  — Australian Bight II-678–9
  — Hawaiian Islands II-794
  — southern Australia II-582
  — Vanuatu **II-741**
  — Vietnam II-564
  — western Indonesia, threats to II-389–90
— Palk Bay **II-166**
— Palk Bay–Madras coast II-169
— Patagonian shores I-756
— Peru I-691–3
  — biological effects of El Niño I-692–3
  — wetlands and protected areas I-691–2
— the Philippines II-408–10
— potential loss of in the Comoros II-251
— range of effects of fishing III-378–9
— rocky shores, eastern Australian region II-633
— South African shores II-135
— in the Sundarbans II-291
— super-K species III-39
— and system integrity, mangroves III-28
— in terms of higher taxa, mine tailings III-241–2
— threatened by by-catch III-136
— western Indonesia II-389–90
— Wider Caribbean I-589
— within the Belize coastal zone I-506
— Yellow Sea, loss of II-497
biodiversity management, Great Barrier Reef Marine Park II-625–6
biodiversity recovery, mine tailings III-241

bioerosion II-69
— contributes to destruction of the reef matrix II-397
— the Maldives, after coral reef bleaching event II-214
biofilms
— as indicators for eutrophication III-263
— and microorganism habitats III-261
— oxygen supply to III-263
— photoheterotrophic III-263
— toxic III-219
biogenic species, effects of oil spills III-273–4
biogeochemistry, around Bermuda I-225
biogeography
— Australia II-581
— eastern Australian region II-631
— equilibrium theory of III-380
— Fiji Islands II-753
— Vanuatu II-739
— Western Australian region II-695
biological communities
— and coastal stability III-352
— effects of oil spills III-272–5
    — alterations in pattern of succession and dominance III-275
    — on consumers (predators and herbivores) III-274
    — on prey species III-274
    — on sensitive species with localized recruitment III-274–5
    — on structuring communities III-2734
biological factors, the key to the Sundarban coast II-148
biological invasions see alien/accidental/exotic/introduced organisms/species
biological oxygen demand (BOD)
— direct increases in through nutrient loading I-367
— high, leads to hypoxia/anoxia I-364
biological production
— increased, Baltic Sea I-130
— Portuguese coastal waters I-153–4
biology, evolutionary, and "punctuated equilibrium" III-86
biomagnification I-5
— of Cd and Hg, Greenland I-9, *I-10*
— of mercury in the biota, Jakarta Bay II-396
"biophile" elements I-116
biophysical features, gradients in
— Hawaii II-795
— the Maldives II-202, **II-202**
biopollution see alien/accidental/exotic/introduced organisms/species
bioregions, Tasmania II-651
biosphere, value of ecosystem services III-395
Biosphere Reserves
— Gulf of Mannar II-163
— Mananara Nord, Madagascar II-127
— Nancowrie Biosphere Reserve II-196
— Odiel saltmarshes I-170
— Rocas Atoll, Brazil I-724
— Sikhote-Alin Biosphere Reserve, eastern Russia II-481–2
— Sundarban Biosphere Reserve II-147, II-156–7
biota
— influence on oil persistence III-271
— Madagascan, effects of climate on II-116–17
    — migratory patterns of some species II-116
    — shallow-water assemblages differ from North to South II-116
biotoxin and exposure disturbances III-218–19, *III-220*, **III-220**
biotoxins
— causing mortalities III-218
— cyanobacterial, implicated in chronic diseases III-218
— effects of direct exposure to III-219
birds
— Asian, organochlorine pollution in II-452, *II-454*
— commercially reared III-225
— Doñana National Park as breeding site and migratory stop I-171–2
— El Salvador I-549
— Gulf of Mexico Coast *I-473*, **I-473**, I-474
— land and sea, Marshall Islands II-778

— Mai Po marshes, Hong Kong, high species diversity II-539
— migratory
    — disease among III-225
    — exposed to HCHs and PCBs in India II-452
    — Sarawak II-368
— *see also* marine birds; seabirds; shorebirds
Biscay, Bay of, weak circulation I-138
Biscayne Bay *I-406*, I-407, III-2, III-9
Biscayne National Park I-412
Bismark Sea *II-426*
Bitter Lakes, kept Suez Canal salinity high II-38
bivalve molluscs, particle-feeding III-240
bivalve mortalities III-226
bivalves
— Aegean I-239
— Bothnian Bay I-125
— filter feeding activity reduced by brown tides III-219–20
— Portuguese coastal waters, some bacterial problems I-158
— southern Baltic I-109
BKD (kidney disease) I-35
Black River Morass, Jamaica I-565
— mangroves I-565
Black Sea I-285–305
— anoxic interface I-287
— climate I-287
— coastal development expected to continue I-303
— effects of recurring hypoxic conditions III-219
— environmental policy difficult to implement I-304
— land-based pollution I-296–303
— major shallow water marine and coastal habitats and offshore systems I-290–4
— residence time of waters I-303
— seasonality, currents, natural environmental variables I-287–9
— southern Black Sea, Turkey I-294–6
— surface circulation I-287–8, *I-288*
    — upper layer general circulation *I-289*
— world's largest anoxic water mass I-287
Black Sea Environmental Programme I-290, I-294–5
— Black Sea Action Plan I-295, I-303
— financing of I-295
Black Sea Water I-236, I-243
blacklisting, of fishing vessels III-160–1
blast fishing see destructive fishing
blast fishing, Indonesian seas II-382
blue crab pot fishery I-362
— Pamlico and Neuse estuaries I-360
Blue Mountains, Jamaica, effects of I-561
blue mussels I-52, I-131
— Baltic Proper I-126
— Greenland, high lead concentrations I-12
blue whales III-82, III-82–3
— secretly killed III-77, III-86
BOD see biological oxygen demand (BOD)
Bodden, southern Baltic Sea
— changes of ecosystem structure and function following eutrophication III-264, **III-265**
— characteristics of III-260
    — high filter and buffer capacity III-260
— eutrophication of III-260–1
    — buffer capacity exhausted III-261
    — causes of III-260
    — effects on benthic microbial communities III-261–2
    — investigations of the impact on the nitrogen cycle III-262–3
— Nordrügensche Bodden, effects of increasing eutrophication III-261–2
— remediation possibilities III-263–4
bolide impact, effects of Chesapeake Bay I-337
"Bolivian winter" I-705
Bonaire *I-596*, I-597
— coastal development I-606
— coastal and marine habitats *I-599*

— dive tourism I-597
— excavation for construction, ruined groundwater quality I-602
— fishery I-603
— Flamingo Sanctuary and Washington Park I-609
— sewage discharge I-606
— turtle grass I-601
Bonifacio Strait I-273
Bonn Convention III-340
Borneo II-361–79
— erosion and landfill II-375
— habitat types II-364–5
— human populations II-373–4
— major coastal habitats II-366–72
— management objectives for marine protected areas II-377
— marine conservation areas II-376–8
— natural environmental variables II-365–6
— offshore systems II-372–3
— regional extent II-363–5
— rural factors II-374–5
— urban and industrial pollution II-375–6
Bornholm Basin I-101
— deterioration of oxygen conditions I-105
Bornholm Deep I-102, I-106
Bosphorus I-198, *I-286*
— connects Black Sea to the Mediterranean I-287
Boston *I-308*, I-312, I-313
Boston Harbor I-314, I-317
— a history of lead I-315
Bothnian Bay *I-100*, *I-122*, I-123, I-126
— heavy metals I-131
— standing stock and carbon flow I-126, *I-127*
Bothnian Sea *I-100*, *I-122*
— fresh water species I-125
— heavy metals I-131
— standing stock and carbon flow I-126, *I-127*
bottlenose whales III-76, III-83, III-92
bottomland forest wetlands, Gulf Coast, USA I-440
boulder/cobble shores, the Azores I-207
Boundary Current, North and East of Madagascar *II-114*
boundary-layer flow III-206
— *see also* internal boundary layers
bowhead whales III-76
— catch limits III-83
— distinct stocks III-81
— occasional Canadian aboriginal kills III-81
brackish-water conditions, adaptation to in Baltic Sea I-125
Brahmaputra River *II-146*
braided channels, Somalia
— biota in II-77
— channel levees II-77, *II-78*
— encrusted hard bottoms *II-76*, II-77
Brazil
— degradation of coastal areas I-463
— education to solve some problems I-463
— mangroves III-20
— Special Management Zone, Bahia de Caraquez I-463
Brazil Current *I-722*, *I-751*, *I-753*
— Abrolhos Bank–Cabo Frio region I-734
Brazil Current–Falklands/Malvinas Current confluence zone I-751
Brazil, southern I-731–47, I-744–5
— activities I-737–40
— degradation I-740–3
— east coast I-733
   — Abrolhos Bank–Cabo Frio I-734
   — habitats I-735–6
— environmental laws I-744–5
   — territorial sea defined I-744
— historical setting I-733–4
— major shallow water marine and coastal habitats I-735–7
— physical description I-734–5
— south coast I-733

— habitats I-737
— large coastal plain I-737
— Southern Brazil Shelf I-735
— southeast coast I-733
   — habitats I-736–7
   — South Brazil Bight I-734–5
Brazil, tropical coast of I-719–29
— major coastal habitats I-725–7
— major environmental concerns and preservation I-727–8
— major marine habitats I-722–5
— oceanographic parameters I-721–2
— the region I-721
— Rocas Atoll I-723
   — Biosphere Reserve I-724
— Southern Bahia, coral reefs I-724
breaking wave forces, increased with rising sea level III-190
bridges, may obstruct water flow in estuaries II-135
brine pools
— contain hydrogen sulphide I-442
— hot, Red Sea II-39
brine rejection, due to freezing I-33
British Indian Ocean Territories (BIOT)
— included in UK's ratification of conservation and pollution Conventions II-231
— *see also* Chagos Archipelago, Central Indian Ocean
British Virgin Islands *I-616*, I-617
— Coast Conservation Regulations I-624
— coral communities I-618
— effects from urban and industrial activities I-623
— Exclusive Fishing Zone I-621
— expansion of tourism I-620
— hurricane damage I-622
— mangroves I-619
— marine protected areas I-625
— mooring system I-624
— National Integrated Development Plan I-624
— tourism I-620
Brittany coast *I-66*, I-69, I-74
Broad River Estuary *I-352*, I-354
brominated chemical, widespread contamination by III-362
brown tides I-439, III-1298
— anoxic impact III-219–20
— cause persistent economic fisheries losses III-220
— high bivalve mortality III-220
Browns Bank *I-311*
Brunei *II-362*, II-365
— population II-374
Bryde's Whale II-676, III-78
bubble burst activity 202
— change in deposition velocity due to III-202–3, *III-203*
Buckingham Canal, southeast India, a health hazard II-170
Buenaventura Bay *I-678*
— high hydrocarbon levels in bivalves I-683
Buenos Aires Province I-751
— coast *I-752*
— coastal erosion I-762
— effects of harbour construction I-762
Burmeister's porpoise III-90
burning
— of grassland, Madagascar, leads to topsoil loss II-123–4
— of secondary forest, produced air polluting haze, Borneo II-375
— *see also* fire
burrowing animals, effects on oil spills III-271
Busc Busc Game Reserve II-80
Busc Busc, Lac *II-67*, II-68
by-catch I-445, II-57, II-59
— black-browed albatross as I-758
— by-catch reduction programs III-140–1
   — reductions in BPUE III-141
   — reductions in effort III-140–1
— as a component of fishing mortality III-140

— creating conservation problems III-136
— defined III-137
— and discard mortality III-123–5
    — reasons for discard III-123
— Great Barrier Reef region II-623
— history of the issue: some early examples III-141–7
    — coastal gillnets and seabirds III-146
    — discards in shrimp and prawn trawls III-143
    — gillnets and cetaceans III-144–5
    — high seas drift nets III-145
    — longlines and sea turtles III-146
    — longlines and seabirds III-145–6
    — Northeast Pacific groundfisheries III-146–7
    — shrimp–turtle problem III-143
    — trawls and cetaceans III-145
    — tuna–dolphin problem III-141–3
— and incidental take III-379
— includes many dolphins III-91
— indirect results of III-379
— a main fisheries issue III-161
— of non-target organisms III-368
— not incorporated in most fisheries management models III-367
— originally ignored, now important III-136–7
— prawn trawling Torres Strait and Gulf of Papua II-604
— problems and solutions III-135–51
    — by-catch classification: why is it useful? III-139–40
    — definitions III-137
    — into the 21st century III-147–8
    — reasons for discarding III-137–8
    — regulations and guidelines III-138–9
— of seabirds III-113
— shrimp fishing/trawlers I-490, I-536, I-539, I-554, II-121
— solutions to III-147
— squids, southern Brazil I-739
— and technology III-148
— wasteful I-3
by-catch-per-unit effort (BPUE), reductions in III-141
— deployment and retrieval changes III-141
— management action III-141
— training III-141
by-catch-reduction devices (BRDs) III-144, III-145, III-146
bycatch quotas III-156
Bylot Sound, Thule, nuclear weapon accident, plutonium contamination I-11

Cabo de Santa Marta Grande-Chui, southern Brazil
— environmental problems I-743
    — primary problems I-743
— major activities I-739–40
    — tourism I-739
— marine and coastal habitats I-737
— Port of Rio Grande I-740
Cabo Delgado, dividing point for South Equatorial Current II-101
Cabo Frio–Cabo de Santa Marta Grande, southern Brazil I-736–7
— bays
    — Rio de Janeiro coast I-736
    — Sao Paulo State I-736
— environmental degradation I-738
    — Cubatao Pollution Control Project I-743
    — from unregulated urbanisation I-740
    — in Guanabara Bay I-741
    — Santos estuary I-742–3
    — in Sepetiba Bay I-741
— estuarine–lagoon complex, Iguape–Cananéia–Paranaguá
    — important littoral ecological system I-736–7
    — subsistence agriculture and fisheries I-739
— major activities I-738–9
— mangroves degraded, south of Caraguatatuba Bay I-736
cachelot see sperm whales/whaling
Cadiz Bay *I-168*
— commercial port I-176
— saltmarshes in decline I-177
— urban–industrial development environmentally hazardous I-174
Cadiz, Gulf of *I-168*
— connects Atlantic and Mediterranean trading ports I-181
— contamination related to heavy industrial activity I-179, **I-180**, *I-181*
— meeting of water masses I-169
— monitoring quality of coastal waters I-179
— rich fishing grounds I-173
— use of traditional fishing techniques I-175–6
Cadiz–Tarifa arc
— rocky I-172
— rugged bottoms, Tarifa I-172
cadmium (Cd) I-5, I-75, I-260–1, I-301
— Baja California I-495, **I-496**
— concentrations in the Azores I-213
— in Greenland seabirds and mammals I-9, *I-10*, I-14
— levels unsafe in Norwegian mussels and fish liver I-25
— off Cumbrian coast I-94
— in whales I-230
caesium (Cs), conservative behaviour I-13
caesium-137(^{137}Cs)
— from Chernobyl I-11
— from Sellafield
    — by long-distance marine transport I-11
    — in Irish Sea water I-94, **I-95**
— outflow from the Black Sea I-249
— Sea of Japan II-479
Caicos Bank *I-588*
— reef areas I-590
Caicos Passage *I-416*, I-417
calamari fishery, Victoria Province, Australia II-668
calcification, corals
— enhanced by algal turf communities III-38
— and nutrient uptake III-35–6, *III-36*, III-38
— physiology of III-35, *III-36*
California Current *I-374*, *I-386*, I-486, I-487, I-488, I-489, I-543, III-113, *III-180*
— affected by El Niño and La Niña events I-387
California sea lion I-399
Californian Coastal Province, Mexico I-486–7
— environmental importance of I-487
— features of I-486
— habitats I-486–7
— monitoring work I-487
Calvados Coast *I-66*, I-69
Calvert Cliffs *I-336*, I-346
Cambodia II-299
Cambodian Sea II-569–78
— coastal deterioration due to erosion and landfill II-576
— coastal habitats II-573–4
— coastal population II-575
— defined local marine environment II-571
— effects of the rural sector II-575–6
— effects of urban and industrial development II-576–7
— offshore habitat II-574–5
— physical and chemical conditions in surface waters II-572–3, **II-573**
— protective measures II-577–8
    — limited perception of environmental impact II-577
    — status of marine environment and habitats protection measures **II-577**
— seasonal variability of the natural environment II-571–3
Cameroon
— conservation concerns I-790, *I-791*
    — coastal environment assessment I-790
— increasing waste contamination I-789
— rivers I-777
Campeche Sound *I-468*
— oil industry I-476
— primary production I-474
Canadian Archipelago I-5, *I-6*

Canary Current I-204, *I-204*, I-223
Canary Islands I-185–99
— climate I-188
— development pressures and protective measures I-196–8
    — ecosystem characteristics and impacts on coastal communities **I-197**
— fishing resources I-195–6
— marine ecosystems I-193–5
— physical and geological background I-187–9
    — geological evolution of I-187
— seasonality, currents and natural environmental variables I-189–92
Canary Islands Counter Current I-189
Canary Islands Stream I-189
canneries, Fiji and American Samoa II-718, II-770
canning industry, Venezuela I-653
Cantabrian Sea *I-136*
— summer subsurface chlorophyll maximum I-141
Cantabrian Shelf, sand and silt *I-136*, I-137
Cap-Breton Canyon *I-136*
Cape Cod *I-308*, *I-311*, I-314
Cape Fear Estuary I-353, I-362
— source of chronic BOD load I-364
— well flushed I-363
Cape Fear River *I-352*, I-354, I-361, I-364
— turbidity, faecal coliforms and BOD, correlation I-364, **I-365**
Cape Hatteras *I-352*
— associated with geographic division of plankton I-356
Cape Horn Current I-701
Cape Johnson Trench II-426
Cape Sable *I-308*
capelin I-376
— industrial fishery based on I-18
capelin fishery, collapse of affected small cetaceans III-97
Capelinhos Mountain, Faial I-203, I-207
captive breeding programmes, Sundarbans
— estuarine crocodile II-158
— horseshoe crabs II-158–9
— Olive Ridley turtle II-158
carbon dioxide (CO_2)
— anthropogenic, elevated III-38
— atmospheric
    — increase in III-188
    — reduces oceanic $CaCO_3$ supersaturation III-36
— disposal at sea to mitigate climate change III-366
— from conversion of bicarbonate III-35, III-35–6
carbon dioxide pollution, transferred to deep sea I-7
carbonate platforms, the Bahamas I-417
carbonate sediments
— dominate tropical Brazilian middle and outer shelves I-722–3
— open shelf, Australian Bight II-681
carbonates, biogenic and chemically precipitated, Borneo II-366
Cardigan Bay *I-84*
— shore communities I-88
— Special Area of Conservation I-90–1
Cariaco Gulf *I-644*
— *Thalassia* beds I-650
Caribbean, collapse of coral reefs III-37, *III-37*
Caribbean and Atlantic Ocean, coral bleaching and mortality events III-54–6
Caribbean Basin, eddies within I-580
Caribbean Coastal Marine Productivity (CARICOMP), research and monitoring network I-634
Caribbean Current I-580, I-617–18, *I-628*, I-665
Caribbean Lowlands, Nicaragua I-524
— ethnic hierarchy I-524
— home to Miskito, Creoles and mestizos I-524
Caribbean Oceanographic Resources Exploration (CORE) I-634
Caribbean Sea I-579
Caribbean Small Island Developing States I-617
Caribbean Surface Water, density, temperature and salinity I-579
Caribbean–North Atlantic convergence I-497, **I-497**

Carlsberg Ridge II-49
Carolina Coasts, north and south I-351–71
— better management on non-point source runoff essential I-367
— coastal rivers, problems of I-361
— environmental concerns in estuarine and coastal systems I-363–6
— flora and fauna of the coastal waters I-354–63
— limits needed on nutrient inputs to rivers and estuaries I-367
— physical setting I-353–4
    — sources of pollutants I-354
— prognoses for the future I-366–7
carrying capacity III-86
Cartagena Convention
— protocol on Specially Protected Areas and Wildlife (SPAW) I-608, I-640
— SPAW protocol III-68
— and whale protection III-85
Carysfort Reef
— continuing decline in deep and shallow waters I-410, I-412
— coral vitality: long-term study I-410–12
— reaching stage of ecological collapse I-412
— sediments of I-409
"cascade hypothesis" III-99
catchment impacts, Tasmanian coastline II-654
categorical correlation matrix, from HEED survey *III-217*
catfish, mysterious mortalities III-223
Cay Sal Bank I-417, I-421
Cayman Islands, coral bleaching and mortality events III-54
cays
— Belize I-505
    — shifting populations I-507
CC:TRAIN III-327–8
— operates under TRAIN-X principles III-327
— role of III-327
— Vulnerability and Adaptation Assessment (V&A) COURSE III-327–8
CDOM (coloured dissolved organic matter) III-297
cellulose industry, Chile, disposal of liquid waste I-712–13
Central Adriatic I-269
— surface circulation I-271
— water column divisions I-270–1
Central America
— analysis of coastal zones I-463
— described I-460
— institutional issues in coastal resource management I-460
— major coastal resource management issues **I-460**
Central Bass Strait Waters II-664
central Bight water mass, salinity of II-676
Central Equatorial Water I-679
central south Pacific gyre ecosystem I-706
Central South Pacific Ocean *see* American Samoa
cephalopods, important in Gulf of Thailand II-306
'Certain Persistent Organic Pollutants', negotiations on III-340
cetaceans
— caught by gillnets III-144–5
— caught in midwater trawls III-145
— in the English Channel I-70
— Great Australian Bight II-680
— Gulf of Aden II-54
— increased strandings in the North Sea I-50
— organochlorine residue levels, western Pacific II-455, *II-459*
— Patagonian coast I-756–7
— PCB concentrations II-455, *II-459*
— Southern California I-399
— vulnerable to undersea noise II-685
— *see also* dolphins; whales
Chagos Archipelago, Central Indian Ocean II-221–31
— available for defence purposes, with conservation provisions II-230
— biogeographic position in the Indian Ocean II-224–5
— a British Indian Ocean Territory II-223
— geographical and historical setting II-223–4

— importance of Chagos II-231
— major shallow water marine and coastal habitats II-226–9
— offshore systems II-229–30
— population, urban and industrial activities II-230
— a pristine environment II-230
— protective measures II-230–1
— reef studies II-223
— seasonality, currents, natural environmental variables II-225–6
Chagos Bank *see* Great Chagos Bank
Chain Ridge *II-64*, II-67
chalk cliffs, erosion of I-68
Challenger Bank I-227
Chang Yun Ridge, Taiwan Strait, effects on pollutants II-501–2
chank fishery, Sri Lanka II-183
Channel Islands I-71
Char Bahar *II-18*, II-24
Charleston Harbor *I-352*, I-353
chemical cargoes, loss of from shipping III-364
chemical contamination/pollution
— Amursky Bay, eastern Russia **II-484**
— Bay of Bengal II-295
— English Channel, and its impacts I-74–6
— Faroes I-31, I-36
— northern New South Wales catchments II-637
— Norway I-24–5
— southern Brazil, Paranaguá Port I-743
chemical energy, from natural oil seeps I-442
chemical industry, southern Brazil I-738
chemical spills, toxic I-72–3
chemical warfare agents, dumped in the Baltic Proper I-117
chemical wastes, Bangladesh, disposal of II-295
Chernobyl
— contamination from I-11, I-55
— in Baltic seawater I-116–17
Chesapeake Bay I-335–49, III-207
— C and D canal *I-336*, I-339
— coastal erosion and landfill I-346
— the defined region I-337–9
— eutrophication and remediation III-205, III-206
— land use *I-338*
— largest estuarine system in the USA I-337
— major shallow water marine and coastal habitats I-341–3
— offshore systems I-343–4
— pesticides affected eelgrass III-9
— populations affecting the area I-344–5
— principal rivers entering I-337, I-339, **I-339**
— protective measures I-346–8
— nutrient management plans for manures I-347
— pollution clean up I-346
— removal of eelgrass by rays III-8
— rural factors I-345
— seasonality, currents, natural environmental variables I-339–41
Chesapeake Bay ebb tidal plume I-344
— USEPA Chesapeake Bay Program I-346
Chesapeake Bay ecosystem, loss of keystone species leads to ecosystem shift III-227
Chesil Beach *I-66*, I-68
Chichester Harbour *I-66*, I-74
Chile
— industrial fishery development I-463
— possibilities for coastal management plan I-463
— small cetaceans as fish bait III-92–3
Chile–Peru Current I-705
Chilean Coast I-699–717
— advances in control and pollution abatement I-715
— coastal marine ecosystems I-705–9
— major determinants of distribution/abundance of marine species I-708–9
— human coastal activity I-709–15
— large and mesoscale natural variability I-702–5
— long-term variations: El Niño Southern Oscillation I-702–4

— seasonal variations I-704–5
— physical setting I-701
— oceanic islands I-701
— water masses I-701–2
Chilean Coast oceanic islets I-701
Chilean Trench I-701
Chilka Lake, India, and environmental problem II-279
China
— aquaculture
— allocation and utilisation of natural resources III-174
— changes in living standards and consumer preferences III-172–3
— changing scenarios in development III-172–5
— freshwater aquaculture expansion III-173
— production diversification III-174–5
— provincial expansion III-173–4
— short comings in the sector III-175
— coastal population, Yellow Sea II-491
— increased agricultural output, Yellow Sea coast II-492
— legislation and regulations concerning water quality III-175, **III-175**
— mortalities in mariculture attributed to red tides III-219
— oil spills II-495
— species used in aquaculture III-171, III-174–5
China Sea, effects of sea level rise, cases considered III-190, *III-192*, III-193
chlorinated compounds, Great Barrier Reef region II-623
chlorinated polycyclic aromatic hydrocarbons (Cl-PAHs), in the Baltic I-114
chlorobiphenyl congeners, decreased concentrations, Baltic Sea I-114
chlorofluorocarbons (CFCs) III-188
chlorophyll
— Argentine Sea I-757, *I-757*
— concentrations, Côte d'Ivoire I-814
— deep chlorophyll maximum, Levantine Basin I-256
— southern Aegean Sea I-240
chlorophyll *a* III-297
chlorophyll content
— Case 2 waters, determination of III-297
— distinguished from CDOM III-297
chlorophytes I-821
Chocó Current I-679
Chokoria Sundarbans II-289, II-290
Christiaensen Basin, New York Bight, highly contaminated I-325
Christmas Island II-583, II-696
Chrysochromulina leadbeteri bloom I-26
Chrysochromulina polylepis bloom (spring 1988) I-20
Chumbe Island, Tanzania
— Marine Park II-94
— environmental education programme II-94
Chwaka Bay, Tanzania, effects of herbicide use II-89
ciguatera poisoning, Mascarene Region II-261–2, II-264
Ciguatoxic Fish Poisoning (CFP) III-221
circulation
— Aegean I-236
— Argentinian continental shelf I-754
— Bay of Bengal, monsoonal II-273, *II-273*
— English Channel I-67
— fresh water-induced, off estuaries I-46
— Gulf of Mexico, Loop Current system I-469
— largest Baltic Sea basins I-123
— northern Gulf of Mexico I-439
— Yellow Sea and East China Sea II-489
circulation patterns
— Coral, Solomon and Bismark Seas Region II-428, *II-428*
— the Maldives II-201
— Polynesian South Pacific II-816, *II-816*
CITES III-340
— Convention on Migratory Species III-68
cities
— Guangdong II-559

— Indonesia **II-384**
— the Philippines II-416–17
clams, Maputo Bay, *Vibrio* contamination II-109
clay–oil flocculation, reduces oil retention in fine sediments III-271
clear water, assists remote-sensing III-284
cliffs, high, central Peruvian coast I-689
climate
— Arabian Gulf II-4
　— winter and summer monsoons II-4
— Arabian Sea coastal areas II-20
— Bahamas archipelago I-418–19
— Chesapeake Bay I-339–41
　— tornadoes and hurricanes I-340–1
— and coastal hydrography, Andaman and Nicobar Islands II-191
— coastal, tropical Brazil I-721
— Coral, Solomon and Bismark Seas Region II-428
— Côte d'Ivoire I-808–9
— east coast, Peninsular Malaysia II-347
— eastern Australian region II-631
— effect on Madagascar's biota II-116–17
— El Salvador I-547
— Fiji II-753
— Florida Keys, hurricanes I-408
— Great Australian Bight II-675
— Great Barrier Reef region II-613
— Gulf of Guinea I-775–6
— Hawaiian Islands II-795
— leeward and windward Dutch Antilles *I-596*, I-598
— Mexican Pacific Coast I-486, I-487, I-488, I-489
— Nicaragua I-519, I-534
— northern Gulf of Mexico I-439
— the Philippines II-407, *II-407*
— Red Sea, driven by migration of the Inter-Tropical Convergence Zone II-39
— Sea of Okhotsk, similar to arctic seas II-465
— and seasonal rainfall, Chagos Islands II-225
— seasonal variation, Chilean coast I-704–5
— the Seychelles
　— controlling factors II-235
　— humid tropical II-235
— Somalian Indian Sea coast
　— bi-modal rainfall II-65–6
　— temperature II-66
— Vanuatu II-739
— (weather), Borneo II-365
— (weather), French Polynesia II-815
— western Sumatra II-386
— Xiamen region, China II-515
climate change III-369–70, III-394
— and changes in marine mammals and seabirds I-9
— and changes in thermohaline circulation I-33
— combined with stressors, possible effects of III-230
— and coral reef degradation III-44
　— *see also* coral reef bleaching
— Global Seagrass Declines and Effects of Climate Change III-10–11
— and greenhouse gas emissions III-47
— Gulf of Alaska III-179–86
— monitoring for Caribbean Planning for Adaptation to Climate Change project, Jamaica I-571–2
— and the North Sea I-47
— past, west coast, Sea of Japan II-483
— and sea level change, effects on coastal ecosystems III-187–96
— and warming, Chagos Islands II-225
Climate Change Convention III-327
— implementing the challenges III-327
Climate Change Project (GEF), the Maldives II-215, II-216
climate disturbances, significant III-224
climate influences, act alongside global-scale environmental change III-225
climate shift, Gulf of Alaska I-377
cloud cover, limiting factor for remote sensing III-286

co-management
— in fisheries III-159
— Fisheries Master Plan, Mozambique II-111
— from management to co-management: the *Pomatomus saltatrix* fishery II-138
co-occurring biological anomalies III-214, III-217
coagulation, of particles III-200
coal mining
— Great Barrier Reef hinterland II-622
— southern Brazil I-739, I-743
coastal accretion
— Borneo II-366, II-375
— western Indonesia II-384
coastal area management, integrated, need for III-170–1
coastal area management systems, traditional, Coral, Solomon and Bismark Seas region II-442–3
coastal areas
— changes, tides and long waves through sea level rise III-188
— propagation of tides in III-188
— sea breezes III-207
— showing signs of eutrophication III-199
— Venezuela
　— areas under Special Regulation **I-658**
　— relevant planning regulations **I-657**
Coastal Biodiversity Action Plans I-77
coastal cold water *I-688*
coastal construction
— Hawaiian Islands II-805
— may provide additional solid substratum II-699
— Xiamen region II-519
coastal currents, Puerto Rico and US Virgin Islands I-580
coastal data coordination III-355
coastal defences
— English Channel coast I-78–9
— soft and hard, loss of, Mozambique II-108
coastal development
— in the Comoros II-247–8
— impacts of, the Bahamas I-427–9
— Western Australia, leading to habitat loss and alienation II-699
coastal dunes, heavy metal mining II-639
coastal ecosystems
— Andaman and Nicobar Islands II-191–2
— changes in, Xiamen region II-524–5
　— Maluan Bay II-524
　— Tong'an Bay II-524–5
— complex, Grande Island Bay with Septiba Bay, southern Brazil I-738
— Coral, Solomon and Bismark Seas region II-430
— effects of climate change and sea level on III-187–96
　— coastal effects III-188
　— effects on coastal storm surges and estuarine flood risk III-190, III-193–4
　— effects on tides and tidal currents III-188–90
　— other effects III-194–5
— Marshall Islands, threatened II-784
— Palk Bay II-166–7
— Palk Bay–Madras coast II-168–9
　— lagoon ecosystem II-169
— southeast India, degradation of II-172
— Taiwan, problems of ignorance and lack of public awareness II-511
coastal environment, and habitats, Bangladesh II-289–91
— beaches II-289
— mangroves II-290–1
— Matamuhuri delta and coastal islands II-289
— St Martin's Island II-289–90, **II-290**
— seagrasses II-290
Coastal Environment Program, the Philippines II-419
coastal erosion
— American Samoa II-770
— Ancash, Peru I-689

— Australia
    — and landfill II-585
    — and sea-level change II-585
— the Azores I-211
— the Bahamas I-428–9
— Bay of Bengal II-272
— Brunei II-375
— Cambodia II-576
— Chesapeake Bay I-337
— China's Yellow Sea coast, causes of II-493
— Colombian Caribbean Coast I-672–3, **I-673**
— Colombian Pacific Coast I-682
— Côte d'Ivoire I-816
— and deposition, El Salvador I-552, *I-552*
— east coast, Peninsular Malaysia II-353, II-356–7
— English Channel
    — English coast I-68–9
    — French coast I-69
— Great Australian Bight II-684
    — problem of uncontrolled vehicle access II-684
— Gulf of Guinea
    — and sediment supply I-777
    — through anthropogenic activities I-786
— Gulf of Mexico I-476
— Gulf of Thailand, west coast II-307
— Hawaiian Islands
    — now a socio-economic problem II-805
    — through subsidence II-804–5
— and landfill
    — Adriatic Sea I-276
    — Argentine coastlines I-762–3
    — around the North Sea I-53–4
    — Belize I-509–10
    — Bermuda I-228–9
    — Chesapeake Bay I-346
    — eastern Australian region II-639
    — Great Barrier Reef region II-622
    — Gulf of Aden II-58
    — Hong Kong II-542–3
    — Irish Sea Coast I-93
    — Jamaica I-569
    — Lesser Antilles I-636
    — Madagascar II-124–5
    — Marshall Islands II-782–3
    — Mozambique II-108
    — northern Gulf of Mexico I-445–8
    — northern Spanish coast I-144–5
    — Oman II-27–8
    — The Seychelles II-239
    — South Western Pacific Islands II-716–18
    — southern Spanish coast I-176–7
    — Sri Lanka II-184
    — Torres Strait II-603
    — Turks and Caicos Islands I-592
    — Tyrrhenian Sea I-276–7
— landfill, and effects from urban and industrial activities, West Australia II-700–2
— and landfill, Fijian Islands II-760
— Malacca Straits II-317
— The Maldives II-208–10
    — from pleasure boats II-209–10
— Mediterranean coast of Israel I-258–9
— New Caledonia II-731
— northern end of Sumatra II-339
— Peru I-696
— Portuguese coast I-161
— potentially a serious problem in the Mascarenes II-262
— Puerto Rico I-584
— Tanzania II-93
— Tasmania II-655
— Turks and Caicos Islands I-592
— Vanuatu II-744
— Victoria Province, Australia II-667
— western Indonesia, effects of shrimp ponds on II-395
— western Taiwan II-506
— Xiamen region II-517, **II-517**
coastal flooding, and sea level rise III-190
coastal forests
— Madagascar east coast, alleviate mangrove problem II-125
— Malacca Straits II-312
— Tanzania II-88, II-89–90
coastal habitats
— affected by livestock II-26
— American Samoa II-767–9
    — changes to II-770
— Australian Bight II-678
— Bay of Bengal II-274–7
    — loss of on east coast of India II-279
— and biodiversity, Arabian Gulf II-6–9
— Borneo II-366–72
    — coral reefs II-370–2
    — mangroves II-366–9
    — rocky shores II-369
    — sandy shores II-369–70
    — seagrass and algae II-370
— Brazilian tropical coast I-725–7
    — Abrolhos National Marine Park I-726–7
    — Atlantic Rainforest (maritime forest) I-725–7
    — restinga I-727
    — wetlands I-727
— Cambodia II-573–4
    — Botum Sakor National Park II-573–4
    — Kampot Bay habitat II-574
    — Koh Kong Bay II-573
    — Kompong Som semi-enclosed bay habitat II-574
— Côte d'Ivoire, depletion and degradation of I-817–18
— Dutch Antilles **I-597**
    — under environmental pressure I-612
— Gulf of Thailand II-301–4
— Oman and Yemen, effects of cold nutrient rich upwellings II-50
coastal lagoons I-645
— Adriatic Sea I-272, I-276
    — intrusion of allochthonous species I-282
— Baltic I-124
— and basin estuaries, Sri Lanka II-178–9
— and coastal lakes, Baltic Proper I-109
— Côte d'Ivoire I-811, *I-811*
    — Ebrié lagoon I-811–12
    — and estuaries, depletion and degradation of habitats I-817–18
— Curaçao, habitat destruction I-602
— and estuaries, Tasmania, depauperate II-653
— forming New River Estuary I-353
— Gulf of Guinea coast, suffering from eutrophication I-788–9
— Mar Chiquita coastal lagoon, Argentina I-755
    — changes made by alien reef-builder I-762
    — experimental aquaculture hatcheries I-754
    — heavy metals I-762
— Mexican Pacific Coast I-482, I-486, I-490
    — Huizache y Caimanero lagoon system I-493–4
    — Teacapán—Agua Brava lagoon system I-494–5, I-497
— Muthupet Lagoon, Palk Bay II-167
— Nicaragua, Caribbean coast I-522
    — depth preferences, fish and shrimp I-522
— Pulicat Lake, Palk Bay–Madras coast II-169
— saltwater, Coral, Solomon and Bismark Seas region II-430
— Venezuela I-651–3
    — affected by natural and anthropogenic factors I-651
    — Laguna de Tacarigua I-652
    — legal protection for some I-651
— west Taiwan II-502
    — industrial parks II-509
coastal lowlands, Nicaraguan Pacific coast I-533

coastal management
— Argentina, main concerns I-751
— Latin America, regional examples I-463–4
— rational, lack of, Adriatic and Tyrrhenian seas I-281
— southern Brazil I-744
    — active environmental groups I-745
    — Coastal Management Law not entirely successful I-745
    — National Programme of Coastal Management, proposed de-centralization I-744
    — preservation and conservation areas I-745
    — Sao Paulo I-744–5
— southern Brazil Rio de Janeiro State Environmental Agency I-745
— techniques, classification of I-464–5
— working with nature III-352–3
— Xiamen region, an integrated approach to II-525–32
coastal management, in the future III-349–58
— coastal management and policy evolution III-350–2
— data and inclusivity III-355
— financing III-356
— integration in III-350
— legal issues III-353–5
— physical systems and management III-352–3
— relocation of historic buildings III-353
— solitary waves III-355–6
— success needs political will III-351
— transboundary issues III-355
coastal management and planning
— effectiveness depends on access to relevant information III-356
— legal issues III-353–5
coastal management professionals, training programmes for III-329–30
coastal management programmes, integration with national climate change action plans III-330
coastal management schemes, at the planning stage III-351–2
coastal marine areas, self-management of II-442
coastal and marine ecosystems
— Malacca Straits II-312–14
— nearshore, El Salvador, ecological important I-547
coastal marine ecosystems, Chile I-705–9
— biogeography of the pelagic system I-707–8
— a biological perspective I-707–9
— south Pacific eastern margin I-705–7
coastal and marine problems, frequently transboundary III-355
coastal marine waters
— eastern Korea, wastewater pollution II-478
— western coast, Sea of Japan II-477–8
    — Amur lagoons II-477
    — Northern Sakhalin II-477
    — pollution
        — in Amursky, Nakhodka and Ussuriysky Bays II-478
        — from ore mining and chemical production, Zolotoy Cape–Povorotny Cape II-477–8
    — southern region, sporadic water pollution II-478
coastal models, for areas associated with sea-level rise III-195
coastal morphology, Colombian Caribbean coast I-666–7
coastal plains, fertile, Al Batinah and Salalah II-26
coastal platform, raised I-170
coastal pollution, Chile, related to geography I-709
Coastal Protected Areas (Proposed), Mozambique II-110, **II-110**
coastal protection, Sri Lanka II-184
coastal reclamation I-93
coastal reef terrace, Somalia, diversified ahallow marine environments II-69, *II-70*
coastal reefs, Western Australia, macroalgal and invertebrate communities II-693
coastal region, Southern Gulf of Mexico, recognition of importance of swamps I-472
Coastal Regulation Zones (CRZ), West Bengal II-157–8, **II-158**
Coastal Resource Management Program, the Philippines II-419
coastal resources
— Australian Bight II-682

— east coast, Peninsular Malaysia, conservation legislation II-355–6
— southern Brazil
    — destructive use of I-740–3
    — economic, social and environmental differences causing problems I-744
— Tasmania, commercial usage of II-655–6
— use in Coral, Solomon and Bismark Seas region II-434
coastal seas
— Coral, Solomon and Bismark Seas region
    — fisheries II-440
    — impacts of large urban areas II-439–40
    — industrial-scale impacts II-439
    — shipping and offshore accidents and impacts II-440–1
    — village-level impacts II-438
coastal and shallow water habitats, Puerto Rico I-582–4
coastal squeeze, causing habitat reduction III-353
coastal states, responsibility for pollution by seabed activities III-338
coastal terrestrial vegetation, Madagascar, varies II-119
coastal uses, Hawaiian Islands II-802–4
Coastal Warm Drift I-754
coastal waters
— Carolinas, flora and fauna of
    — benthic microalgae I-357–8
    — coastal benthic invertebrate communities I-360–2
    — estuarine and coastal finfish communities I-362–3
    — estuarine phytoplankton I-354–6
    — estuarine zooplankton I-359
    — macroalgae I-356–7
    — marine phytoplankton I-356
    — marine zooplankton I-360
    — offshore benthic communities I-362
    — seagrasses and other rooted submersed aquatic vegetation I-358–9
— Coral, Solomon and Bismark Seas region
    — impacts of land use II-435–6
    — subsistence and artisanal fisheries II-436–7
    — threats to sustainability II-437–8
— Norwegian, constituents of I-20
— West Guangdong coast, water quality II-555
— Xiamen region
    — bacterial pollution II-522–3
    — mainly in good condition II-521–2
    — Maluan Bay sediments a secondary pollution source II-524
    — oil pollution II-523
    — organic pollution, eutrophication and red tides II-522
    — sea dumping and disposal of solid waste II-523–4
    — threatened by rapid economic development II-522
coastal wave energy development
— current and future prospects III-315
    — Osprey 2000 III-315
— enclosed water column devices III-314, *III-315*
— location advantages and disadvantages III-314
— tapered channel concept III-314–15
coastal wetlands
— Argentina I-755–6
— Black Sea, changes in ecology of I-293
    — anthropogenic influences resulting in loss/degradation of I-293
— Côte d'Ivoire I-810–11
— Hueque, Venezuela, drainage and deforestation I-655
— loss of
    — northern New South Wales II-638
    — Yellow Sea II-493–4
— loss of, Southern California I-388
    — affecting seabirds I-398
— lost to landfill, Hawaiian Islands II-805
— mangrove destruction, the Bahamas I-428
— northern Gulf of Mexico, potential loss of I-446
— west Taiwan II-502–3
— *see also* khawrs; swamps; wetlands
coastal wilderness, Dutch Antilles I-601
Coastal Zone Management Act (USA) I-451–2

— National Estuarine Research Reserve System I-452
Coastal Zone Management Plan (CZMP), Sri Lanka II-185–6
— Special Area Management Plans II-186
coastal zone, Somali coast
— alterations of II-80
— geomorphic features II-68–72
    — Bajuni barrier islands II-69, *II-73*, *II-75*
    — braided channelized coast II-69–72
    — coastal reef terrace II-69, *II-70*
    — Merka Red Dune Complex II-68–9, *II-70*
    — rivers and alluvial plains II-68
— suffering degradation through lack of protective measures II-80
coastal zones I-4
— Argentina, main divisions I-751–2
— artisanal and non-industrial uses
    — El Salvador I-553
    — Madagascar II-125
    — Mozambique II-108–9
    — Taiwan II-509
— atmospheric deposition to III-205–7
    — atmospheric flow in, and atmosphere—surface exchange III-206–7
    — impacts in III-205–6
— Bay of Bengal
    — Calcutta and Howrah a significant detriment to II-148
    — mangroves II-149
— Belize I-503
    — aquaculture and fishing I-511
    — artisanal use of I-507
— and non-industrial use I-510–11
    — assessment of anthropogenic impacts I-507
    — effects of coastal erosion and landfill I-510
    — management of I-512
    — protected areas **I-513**
    — tourism I-511
— Brazilian tropical coast, for recreation and tourism I-728
— Cambodia traditional agriculture II-576
— Chile
    — flora and fauna I-707
    — industrial activity I-710–15
    — passage of rivers through I-704–5
    — southern marine habitats I-708, **I-708**
    — upwelling ecosystems and embayment ecosystems I-706
    — upwellings *I-705*
— Colombian Caribbean Coast, influence of Sierra Nevada de Santa Marta I-672
— community-based management, Hawaiian Islands II-809
— the Comoros, urban and industrial impacts II-251
— Côte d'Ivoire
    — cultural and historic sites I-818
    — mining of construction materials from I-816
— definitions of I-458
— east coast, Peninsular Malaysia
    — assessment of liquid and airborne pollution II-351
    — changes in catchment land use II-351
    — development of tourism II-351
    — relevant legislation and guidelines **II-356**
— eastern Australian region agriculture in II-636–8
— eastern Taiwan Strait II-501
— El Salvador I-547
    — coastline meso-tidal I-547
    — industrial uses I-554–5
— Guinea, pressure on natural resources I-801
— industrial effects I-511
— Integrated Management Plan for the Coastal zones of Brunei II-379
— legal definition in need of revision III-354
— Lesser Antilles, major features **I-632**
— Madagascar
    — cities II-125
    — industrial uses II-125
    — shipping and offshore accidents II-125
— Mozambique
    — industrial uses II-109
    — Mecúfi Coastal zones Management Projects II-111
    — a priority area II-110
    — Xai-Xai Sustainable Development Centre for the Coastal zones II-111
— need to understand physical systems III-352
— New Caledonia, potential effects of mining discharges II-731
— Nicaragua, Caribbean coast I-520
    — recreational and tourist destinations I-523
— Nicaragua, Pacific coast I-543
— Peru
    — arid semi-desert I-689
    — characteristics of I-689–90
    — climate influenced by Peruvian current I-690
    — 'wet desert' climate I-690
— pressure on and degradation of III-295
— shallow, human impact on III-7
— southeast India, domestic sewage a problem II-172
— southern Brazil
    — urbanization, tourism and industrialization I-734
    — very little is protected I-745
— Southern Gulf of Mexico, land classification I-475
— Sri Lanka
    — Coastal zones Management Plan (CZMP) II-185–6
    — pressures on II-181
— Taiwan
    — abused by public and private sector II-511
    — coastal land subsidence II-507
    — longshore currents II-506
— Tanzania II-85
— Tasmania
    — recreational use of II-655
    — urban development II-655
— tropical, sensors relevant for mapping III-284
— tropical Brazil, transgressive episodes I-722
— Venezuela, development pressure I-645
— Victoria, Australia
    — indigenous peoples II-666
    — white settlers II-666
coastal/marine structures
— effects on beach morphology I-258–9, I-276
— *see also* dikes and breakwaters
coastguards, for the Dutch Antilles I-609
coastline
— Gulf of Guinea, low-lying and swampy I-775
— Jamaica I-561
— Puerto Rico, shelf morphology I-581
— Venezuela
    — economic activities I-654–5
    — influential because accessible I-654
    — regions of I-645–6
— West Africa I-775
coastline modification
— Coral, Solomon and Bismarck Seas region II-438
— French Polynesia II-821
coasts
— braided and channelized, southern Somalia II-69–72
— erosion, protection reduces sediment availability III-352
— North Sea, urban and artisanal use of I-54
Cobscook Bay, marine ecosystem bibliography I-316
coccolithophores, and algal bloom detection III-298
cockle harvesting/fisheries I-51–2, I-73
— disrupts the environment I-92
coconut tree replanting, Marshall Islands, benefits vs. impacts II-785
Cocos (Keeling) Atoll II-583
cod I-35, I-49, I-313
— Baltic Sea I-110
    — growth conditions less favourable I-128
    — concentrations of PCBs and DDT in **I-37**

"cold pool", connection to Georges Bank I-324
cold water dome, persistent, East China Sea II-508
cold-water plumes, Venezuelan coast I-646
collisional tectonics, Aegean area I-235
Colombia
— coral bleaching and mortality events III-54
— rehabilitation of key mangrove system I-463
Colombia, Caribbean Coast I-663–75
— coastal erosion I-672–3
— effects from urban and industrial activities I-673
— offshore systems I-669–71
— populations affecting the area I-671
— protective measures I-673
— rural factors I-671–2
— seasonality, currents, natural environmental variables I-665–7
    — bimodal wet–dry seasonality I-665
— shallow water marine ecosystems and coastal habitats I-666–9
Colombia, Pacific Coast I-677–86
— coast is tectonically active I-682
— coastal erosion and landfill I-681–2
— effects from urban and industrial activities I-682–4
— major shallow water marine and coastal habitats I-679–80
— offshore systems I-680
— populations affecting the area I-680–1
— protective measures I-684–5
    — Gene Bank of Fishing and Aquarian Resources I-684
    — National Contingency Plan against Hydrocarbon Spills in Marine Waters I-685
— rural factors I-680–1
— seasonality, currents, natural environmental variables I-679
Colombian Current *I-678*
colonial powers, and the Malacca Strait II-337
colonisation, of North Sea still going on I-45
Commission de l'Océan Indien (COI) *see* Indian Ocean Commission
common dolphins, northern stock III-91
Common Fisheries Policy I-57
Common Market for Eastern and Southern Africa II-80
community-based coastal resources management, the Philippines II-418
Comores *see* Comoros Archipelago
Comoros Archipelago *II-114*, II-243–52, *II-244*
— effects from urban and industrial activities II-251
— major shallow water marine and coastal habitats II-245–6
    — Anjouan II-246
    — Grande Comore II-245
    — Mayotte II-246
    — Moheli II-246
— offshore systems II-246–7
— populations affecting the area II-247
— protective measures II-251–2
    — regulations for environmental protection and management II-251
    — watershed improvements II-252
— seasonality, currents, natural environmental variables II-245
— threats to the environment II-247–51
conch fishery I-591, I-635, I-637
— Belize I-509
— Jamaica I-572
conservation
— efforts in Belize I-512, I-514
— of whale stocks III-75–7
— Yellow Sea area, inhibiting factors II-495–6
conservation measures/policies
— Andaman, Nicobar and Lakshadweep Islands II-195–6
— for the Sundarbans II-157–8
Conservation Sensitive Management System (CSMS), analysis of coastal geomorphological sensitivity III-353
construction, and infilling of coastal mangroves I-637
consumption, and population III-396
contaminants
— chemical, safety limits in Norwegian fish and shellfish I-25

— chemical, secondary sources I-24
— entering the Southern California Bight
    — in biota I-394
    — DDT contamination I-390
    — processes undergone I-390–1
    — in sediments I-392–4
    — in the water column I-391–2
— Gulf of Maine I-314–16
— input, transport and biological responses of, North Sea I-56
— levels in Greenland marine ecosystem I-5
    — future trends I-14–15
    — indirect evidence for I-9
— in the marine environment I-450
— Norway, within reach of tidal activity I-28
contamination I-158
— bacterial, Tagus estuary I-158
— biocide I-538
— chemical, from sewage, Arabian Gulf II-10
— Colombian Pacific coast, mainly transitory I-683
— marine, El Salvador I-555
— toxic, North Carolina estuaries I-360–1
— *see also* mercury contamination; microbiological contamination
continental collision II-19
continental seas, western Indonesia, uniqueness of II-383
continental shelf
— Norwegian I-23
    — bottom fauna communities I-23
    — bottom substrate I-23
— *see also* Argentinian continental shelf; South Brazil Bight
Continental Shelf Alternative (CSA) sites I-327
Convention on Biological Diversity I-608, III-341
Convention on the Conservation of Antarctic Marine Living Resources III-342
Convention for the Protection of the Black Sea against Pollution I-294
Convention on the Protection of the Marine Environment of the Baltic Sea Area (Helsinki Convention) I-101
Cook Inlet *I-374*, I-375
copepods I-814, I-831, II-54
— diapausing populations II-24
— Gulf of Alaska I-379–80
— North Carolina I-359
— Southern Bight I-48
copper (Cu) I-131
— Azorean amphipods I-213
— surficial sediments, Penobscot Bay *I-316*
copper levels, Island Copper Mine III-239
copper mining, Chile I-710
— lessons for the future, a case study I-713
— lessons for the future, case study, complaints about pollution and construction of new tailings lagoon I-713
— *see also* Island Copper Mine, Canada
copra, from Chagos Islands II-223, II-229
coprostanol, west Taiwan coast II-506, II-509
coquinas II-77
coral assemblages, Arabian Sea II-22
coral atolls
— American Samoa II-769
— Belize I-504–5
— Borneo II-370
— Chagos Archipelago II-223
— Lakshadweep Islands II-194
— Marshall Islands II-775
— the Seychelles II-236
    — peripheral reefs II-236
— Western Australia II-693
    — shelf edge
coral barrier, reef beaches, Cambodia II-574
coral bleaching I-419
— August 1998 I-421
— Belize reefs I-504
— related to water temperature I-409–10

coral carpets, Somali coast II-72
coral cays
— Borneo, Pulau Sangalaki II-371
— western Indonesia II-388–9
coral collection, Hawaiian Islands II-803
coral colonies, use of high-resolution airborne methods for status of III-288
coral communities
— Aruba, affected by oil pollution *I-596*, I-597
— British Virgin Islands I-618
  — effects of hurricanes I-618
— coastal reef terrace, Somalia II-69, *II-70*
— eastern Australian region II-633
— Gulf of Aden II-52
  — natural stressors II-52
— Gulf of Thailand II-304
— offshore islands, Taiwan Strait II-504
— southern Somali coast II-72
— West African coast I-780
— on Yemeni black basalt effusions II-51
coral degradation, tourist areas, Gulf of Thailand II-304
coral diseases
— global epidemic of III-226
— Hawaiian Islands II-807
coral harvesting, prohibited, the Bahamas I-430
coral mining II-315
— affecting reefs of Pulau Seribu II-397
— the Comoros II-248
  — uses of corals II-249
— contributes to reef and forest degradation, Tanzania II-91
— El Salvador I-554
— illegal, western Indonesia II-391
— Madagascar II-125
— the Maldives II-208
  — reefs show little sign of recovery II-208, *II-209*
  — sea-level rise effects exacerbated by II-215
— Mozambique II-108
— *see also* coral quarrying
coral mortality
— Colombian Caribbean Coast I-668
— due to Crown-of-Thorns starfish outbreaks II-616
coral pinnacles
— Bajuni Islands II-74
— Brazil I-723
coral quarrying, Gulf of Mannar II-165
coral reef areas
— no-take reserves III-386
— US Marine Protected Areas III-386
coral reef bleaching III-37
— 1998 event
  — Madagascar II-116
  — Sri Lanka II-180
— American Samoa II-712
— Chagos Archipelago
  — evidence of cover decline pre-1998 II-228, *II-228*
  — massive coral mortality after 1998 event II-228, *II-228*
— the Comoros
  — 1983 in Mayotte II-249
  — 1998 event, mass mortality II-249
— and coral death III-38
— Coral, Solomon and Bismark Seas region II-429
— Fiji II-711
— followed by coral disease III-224
— French Polynesia
  — 1991, 1994 and 1998 bleaching events II-822
  — problem of remoteness from coral recruitment II-822
— Great Barrier Reef II-617
— Gulf of Thailand II-304
— Hawaiian Islands II-807
— the Maldives II-202–3
  — 1998 event very extensive II-214

— consequences of on the socio-economic welfare of communities II-214
— Marshall Islands II-777
— and mortality, 1998 event III-43–57
  — Arabian Region III-47–8
  — Caribbean and Atlantic Ocean III-54–6
  — Central and Eastern Pacific Ocean III-53
  — Chagos Archipelago III-49
  — East Asia III-52
  — Indian Ocean III-48–9
  — interpretations and conclusions III-46–7
  — mechanisms of III-44
  — Pacific coast of the Americas III-53–4
  — Pacific Ocean, Northwest and Southwest III-52–3
  — Singapore, Thailand, Vietnam III-52
  — Southeast Asia III-49–52
— and mortality, Socotra II-52
— the Philippines II-408
— recent El Niño years, time too long for reef tolerances III-224
— several periods, Andaman Sea II-303
— the Seychelles
  — 1998 event II-236
  — massive mortalities II-237
— stress thought to be main reason with other factors II-137
— Vanuatu II-742
— western Indonesia II-389
coral reef communities
— east coast, Peninsular Malaysia II-348
— Moorea Island, French Polynesia II-819
coral reef ecosystems
— Andaman and Nicobar Islands II-191–2
— remote sensing of III-287–8
  — degrees of sophistications III-287
  — future challenges III-288
  — reef habitat maps III-287–8
  — representation of individual habitats III-287–8
  — use of colour aerial photography III-287
coral reef fish II-117, II-180
— The Maldives II-202
coral reef habitats, Sri Lanka II-180
coral reef microcosms, abundance in III-40
Coral Reef Monitoring Project, USEPA I-412
Coral Reef Rehabilitation and Management Project (COREMAP), western Indonesia II-393, II-399–400
coral reef species, Tanzania II-86–7
Coral Reef Symposia III-44
— late awareness of degradation III-37
coral reef zones
— Bermuda Islands I-227
  — rim and terrace reefs I-227–8
coral reefs II-693, III-33–42
— affected by hurricanes I-631
— American Samoa II-712, II-767–8
  — recovering from series of natural disturbances II-768
  — reef fish assemblage II-768
— Anguillan shelf I-618
— Australia II-582
  — vulnerable to eutrophication and sedimentation II-585
— the Bahamas I-421
  — near-shore health of I-429
— Bay of Bengal II-277
— biodiversity of III-34
  — based in calcium framework building III-35
— Borneo II-370–2
  — Berau Barrier reef system II-370
  — erosion of, Kota Kinabalu Bay and Tunku Abdul Rahman Park II-376
  — reef flats II-370–1
— Brazilian I-722, I-723–4
  — differ from well known coral reef models I-723
  — formed by coalescence of 'chapeiroes' I-723

— reef types I-723
— Cambodia II-574
— cf. rainforests **III-34**
— Chagos Islands II-223
   — biological patterns on the reef slopes II-227–8
   — changes over twenty years II-228–9
   — ecology of II-226–7
   — island ecology II-229
   — reef flats, algal ridges, spur and groove systems II-227
   — submerged banks and drowned atolls II-223, **II-224**
— Colombian Caribbean Coast I-668
      — best development in southwest I-668
      — general degradation I-668
— Colombian Pacific Coast I-680
— conservation, an international priority III-40–1
— damaged, Morrocoy National Park, Venezuela I-655
— decline in coral cover III-37
— degradation
   — early III-37
   — primary bases for III-40
— 'design' makes for vulnerability to changing environmental conditions I-413
— destructive fishing practices III-123
— Dutch Antilles
   — natural disasters I-607
   — and reefal algal beds I-599–600
— east coast, Peninsular Malaysia II-355
   — effects of creating Marine Parks II-355
— effects of sewage not well documented II-804
— endangered III-36–7, III-41
   — loss through destructive fishing methods III-37
   — lost through siltation and eutrophication III-37
— Fiji Islands II-755–7
   — reef provinces **II-756**
   — reef types II-756, **II-756**
   — studies on Suva reef II-756
— Florida Keys I-407–8
   — Carysfort Reef I-410–12
   — coral diseases I-410, *I-411*
   — degradation from agricultural and urban factors I-409
   — Florida Bay Hypothesis I-407
   — recruitment low I-410
   — stressed by environmental change I-408–9
— fossil, Red Sea coastline II-38
— French Polynesia II-817–19
   — Fangataufa atoll gastropod assemblage studies II-817
   — reef monitoring networks II-824
   — reef restoration schemes II-824
   — status of II-823
— Great Barrier Reef region II-616–17
   — Crown-of-Thorns Starfish outbreaks II-616–17
   — little evidence of long-term decline II-616
   — pressures and status II-616
— Gulf of Aden II-52
— Gulf of Mannar II-163–4
   — and the Gulf islands II-164
— Gulf of Mexico
   — impacts on some reefs **I-471**
   — principal characteristics **I-471**
— and hard bottoms, northern Gulf of Mexico I-442
   — decline of I-445
— Hawaiian Islands II-795, II-796–7
   — benthic reef life II-799–800
   — better management needed in northwestern islands II-810
   — Kane'ohe Bay, reef restoration II-796, II-809
   — monitoring of II-796
— indicators of oceanic health and global climate change I-413–14
— Jakarta Bay, once beautiful now almost destroyed II-397
— Jamaica I-562–4, I-566, *I-567*
   — changes in I-562
   — *Diadema* mass mortality I-563–4
   — differences in I-562
   — hurricane damage I-562, I-563
   — impacts of fishing I-568
   — impediments to recovery I-572
   — reef deterioration I-563
   — reef zonation I-562, I-562–3
   — some recovery I-563, I-564
— Kenya, result of predator overfishing III-128
— Lesser Antilles I-631
— Madagascar II-118
   — ancient II-118
— Malacca Strait II-336
— the Maldives II-202–3
   — need to keep reefs healthy II-218
   — protected from mining by tourism II-217
— and marine environments, pharmaceuticals from III-39
— Marshall Islands II-778–9
— Mauritius II-258
— natural perturbations III-37
— natural products, identification and extraction from III-40
— New Caledonia II-725–6
— Nicaragua, Caribbean coast I-520–2
   — *Diadema* mass mortality I-520
   — Miskito Coast Marine Reserve I-520, *I-521*
   — reef fish I-524, I-525
— Norwegian coast I-21–2
— Palk Bay II-166–7
— Papua New Guinea coastline II-429
— the Philippines II-408–9
   — effects of overfishing, sedimentation and destructive fishing II-415
   — primary productivity II-409
   — reef health II-408–9
— primary productivity III-35
— Puerto Rico I-582–4
   — fringing reefs I-582
   — patch reefs I-582
   — shelf reefs I-582, I-583–4
— Queensland Shelf II-619
— Quirimba Archipelago II-102
— Red Sea II-37, II-40–1
   — alignment of II-39
   — coral distributions II-41
   — effects of oil pollution II-43
   — fringing reefs II-40
— Reefs at Risk analysis III-44
— risk criteria classification evaluates potential risk from ports and harbours II-417–18
— role of nutrients in degradation III-38
   — sensitivity of to N, P and CO_2 III-38
— the Seychelles II-236
   — threatened II-236
— Singapore, smothered by siltation II-339
— social and economic value III-38, III-40–1
— Solomon Islands II-429–30
— Somali coast II-65
   — fringing reefs II-65
   — shelf and fringing II-72–4, II-80
— South Western Pacific Islands II-711–12
— Southern Gulf of Mexico I-469
   — and protected zones I-471–2, *I-472*
   — Vera Cruz reef system I-471
— Sri Lanka II-180
— Straits of Malacca II-314
— and submerged banks, Lakshadweep Islands II-194
   — degradation from siltation and sponge infestation II-194
— Taiwan II-503, II-504
— Taiwan Strait, increasingly threatened II-504
— Tanzania II-85–7
   — degraded sites II-86
   — restoration project, Dar es Salaam II-89

— Thailand
    — Andaman Sea II-302
    — Gulf of Thailand II-302-3
— threats pre-1998 III-44
— Torres Strait II-597
— true, Bar Al Hackman II-22
— Turks and Caicos Islands I-590
— Vanuatu II-741-2
    — condition/special features of coral communities **II-742**
    — status of II-742
— Venezuela I-650-1
    — coastline reefs, less diverse and under pressure I-650
    — Mochima Bay I-651
    — on offshore islands, pristine condition I-650
    — Turiamo Bay, diverse reef fauna I-650-1
— Vietnam II-565
— vulnerable to oil spills and their effects III-273-4
— Western Australia II-696
— western Indonesia II-388-9
— western Sumatra II-387
— *see also* coral atolls; fringing reefs; patch reefs; reefs
Coral Sea II-426, II-427, *II-612*
— Chesterfield Islands and Bellona reef II-711
— nutrient and sediment loads II-614
— western, circulation in II-614
Coral Sea Basin II-427
Coral Sea Coastal Current II-428, *II-594*
Coral Sea Island Territories, inclusion in Great Barrier Reef Marine Park? II-627
Coral Sea water II-581
Coral, Solomon and Bismark Seas Region II-425-46
— coastline change II-438
— human impacts on coastal seas II-438-41
— land and sea use factors impacting on coastal waters II-435-8
    — inability to manage stocks for sustainability II-437
— offshore systems II-431-3
— people, development and change II-433-5
    — inadequate information for establishing any form of baseline II-435
— provisions for the management and protection of coastal seas II-441-3
    — community-based management II-442-3
    — national administrative and legal arrangements II-441
    — protected species, habitats and areas II-441-2
    — regional cooperation II-442
— seasons, currents, seismicity, volcanicity and cyclonic storms II-428-9
— shallow water marine and coastal habitats II-429-31
coral-rubble mining I-604
coralligenous formations, Tyrrhenian Sea I-273-4
coralline coast, Mozambique *II-100*, II-101, II-104
corals
— Andaman and Nicobar Islands II-191-2
— Arabian Gulf II-7-8
— Brazilian, some endemics I-723
— Chagos Islands
    — most diverse site in the Indian Ocean II-225
    — soft corals II-227
    — a stepping stone for corals II-224-5
— collection for aquarium and shell trade II-438
— the Comoros, used in building II-249
— deep-sea, little studied III-379
— diversity of, Spratly Islands II-364
— French Polynesia II-819
— Great Barrier Reef, diversity and species assemblages II-616
— Gulf of Oman II-22
— hermatypic I-566, I-590, I-723, II-633
    — the Bahamas I-421
— Maldivian, zooxanthellate and azooxanthellate II-202
— Marshall Islands, great biodiversity II-778
— Papua New Guinea reefs II-431

— physiology of calcification III-35
    — calcification and nutrient uptake III-35-6
    — scleractinian corals III-35
— scleractinian
    — Andaman Sea II-302
    — Fiji II-711
    — Gulf of Mannar II-164
    — Lakshadweep Islands II-194
    — Madagascan, affinities of II-117
    — Palk Bay II-167
    — Vanuatu II-741
    — Venezuela I-650
    — western Indonesia II-389
— the Seychelles, growth affected by southeast trade winds II-236
— species diversity, Grand Récif of Toliara II-118
— Sri Lanka II-180
— stony II-796
    — east coast of Taiwan II-508
    — Fiji Islands II-757
— Taiwan II-504
— Turks and Caicos Islands I-589, I-590
— Vanuatu, similarities with Great Barrier Reef II-741
— varying growth conditions, Puerto Rico I-583
— Western Australia II-696
Coriolis effect III-311
Coriolis Force I-631, II-274, II-288, III-316
Coro, Gulf of *I-644*, I-646
corrales (stockyards), Spanish fishing technique I-176
'Corriente de Navidad' I-139
Corsica
— Lavezzi nature reserve I-280-1
— transboundary park in the Bocche di Bonifacio I-281
Costa Rica Dome Structure I-535
Costa Rican Current I-489, *I-678*, I-679
Côte d'Ivoire I-805-20
— coastal erosion and landfill I-816
— construction of Abidjan harbour, importance of I-815-16
— effects of urban and industrial activities I-817-18
— geological context and geographical limits I-807
— major shallow water marine and coastal habitats I-810-13
— offshore systems I-813-15
— population I-815-16
— protective measures I-818-19
    — International conventions, coastal and marine environment **I-818**
    — National Environmental Action Plan I-818
— rural factors I-816
— seasonality, currents, natural environmental variables I-807-10
— water inputs, from continent and ocean I-807
Cotentin Peninsula *I-66*
Cox's Bazaar sand beach II-289
crab fishery, and yields, Island Copper Mine III-244
crabs, pelagic II-54-5
Crepidula fornicata, altered benthic habitats I-47-8
Cretan Sea *I-234*, I-235
crocodiles I-505
Cromwell Current I-705
cross-correlation analyses III-217
Crown-of-Thorns starfish II-303
— a potential environmental problem II-214
Crown-of-Thorns starfish outbreaks II-711, II-712
— causes indeterminate II-617
— east coast, Peninsular Malaysia II-355
— French Polynesia II-823
— Great Barrier Reef II-616-17
— Hawaiian Islands II-807
— Marshall Islands II-777
cruise ship discharge, the Bahamas I-430
crustacean farming III-169, III-170, *III-170*
crustacean fishery
— Alaska, collapse of I-380

— KwaZulu-Natal coast II-137
crustaceans I-34, I-172, II-586
— Côte d'Ivoire I-815
— El Salvador I-550–1
— Faroes I-34, I-35
— fishing for I-73
— glacial relict I-125
— Gulf of Maine I-311–12
— PAHs in I-497
— Peru I-691
— Spanish north coast I-139, I-140
— Tagus estuary I-157
— Vietnam II-564
cryptomonads I-355
crystalline rock habitats, Sri Lanka II-180
cultivated land, decline in, Guangdong II-558
cultural convergence, Sundarbans II-152
cultural evolution, and cultural adaptation III-396
Curaçao *I-596*, I-597
— beach replenishment I-604
— coastal and marine habitats *I-599*
— coastal urbanization I-606
— fringing reefs I-599
— land-use planning I-608, I-609
— livestock decline I-601
— national parks I-609
— reef considered to be overfished I-602–3
— reefs damaged by ship groundings I-606
— sewage discharge I-606
— turtle grass I-601
Curaçao Dry Dock, contamination from I-605
curio trade, marine life collection for 251, II-91, II-108, II-125, II-212
— marine ornamentals trade II-393
— Marshall Islands II-782
Curonian Lagoon I-109
— fishery I-111
currents
— affecting the Bahamas I-418, *I-418*
— Arabian Sea, mirror seasonal wind direction II-20
— Belize, affected by prevailing winds I-503
— Cambodian Sea II-571, *II-572*
— Côte d'Ivoire coast I-807–8
— eastern Australian region II-631
— Great Australian Bight II-676
— Gulf of Aden II-50
    — development of gyres II-50
— Gulf of Mexico I-437
— influencing Marsha, Islands II-775–6
— in the Malacca Straits II-311
— Mexican Pacific coast I-487
— offshore, Taiwan Strait II-501–2
— Sea of Okhotsk II-465–6
— South China Sea II-552
    — surface water patterns II-366, *II-366*
— surface, Java Sea II-383–4
— through the Fijian Groups II-754
— Torres Strait II-596
— Xiamen coastal waters II-516
    — residual currents in the Outer Harbour II-516
Currituck Sound *I-352*, I-360
customary fishing rights
— Fiji II-758, II-761
    — more interest in marine protected areas now II-762
customary marine practices/law
— Coral, Solomon and Bismark Seas region II-434
— Marshall Islands II-779, II-786
Cuulong Project, Mekong Delta II-299
Cuvier's beaked whale III-92
Cuvumbi Island, Bajuni Islands, reef front and reef flat II-74, *II-78*
cyanide, and the live reef fish food trade II-373, II-392–3
cyanide fishing

— Indonesian Seas II-392–3
— the Philippines II-415
cyanobacteria I-743, I-821, III-3, III-221
— blooms III-298–9
— and marine mass mortalities III-218
— The Maldives II-204
cyanophytes II-10–11
— picoplanktonic, Neuse Estuary I-355
Cyclades *I-234*, I-235
cyclone shelters, Bangladesh II-288
cyclones I-439, II-596
— affecting Madagascar II-116, II-117
— Bay of Bengal II-148
    — and storm surges II-288
— and coral cover II-616
— formation of, Coral, Solomon and Bismark Seas region II-429
— French Polynesia, related to abnormal El Niños II-815–16
— Gulf of Thailand II-301
— and high rainfall, Fiji II-753
— influence on Great Barrier Reef II-613
— the Maldives II-201
— Mascarene Plateau II-256
— Mexican Pacific coast I-488, I-493
— occasional
    — Mozambique II-102
    — Sri Lanka II-177
— the Philippines II-407
— South West Pacific Islands
    — damage by associated waves II-710
    — effects of II-709, II-710
— Vanuatu II-739–40
    — damage by II-740
— west Seychelles, infrequent II-235–6
Cyprus *I-254*
cytochrome P450, from *Exxon Valdez* oil spill I-382
cytochrome P450-1A values, elevated, Prince William Sound III-274
cytochrome P450-aromatase systems, inhibited by TBT III-249

dabs I-49, I-70
Dahlak Islands *II-36*
Dall's porpoise III-90
— caught in Japanese salmon drift-net fishery III-379
— hunted in Japanese waters III-94
— possible endocrine disruption II-456–7, *II-460*
Dalmation coast, tourism I-277
dam building
— on major rivers, contributing to coastal erosion II-108
— protests against II-306
Damperien Province, Western Australia II-694
dams
— adverse effect on mangroves III-19
— and canals, effects of construction and maintenance of I-446–7
— effects of, Côte d'Ivoire I-810, I-816
— effects of, Gulf of Guinea main rivers I-789, I-792
— in estuaries, South Korea II-493
— hydroelectric power, reduce salmon populations I-128
— responsible for hydrological change, east coast, Peninsular Malaysia II-353
— Taiwan rivers, increasing coastal erosion II-506
— Tasmania, regulation of freshwater flow II-654
— a threat to freshwater fish II-697
Danish water, total nitrogen input to III-201
Danube River, increased loads of organic and inorganic pollutants I-303
Dardanelles *I-234*, I-243
data mining III-212
— and data models III-214–15, **III-215**, **III-216**
— for disturbance indicator types and pathogen toxin and disease combinations III-214
— and other research, to retrospectively derive new time series III-216

Davis Strait *I-6*, I-7
— cod abundance and temperature I-8
— increased Hg in upper sediments I-9
— trawler fishery for Greenland halibut I-8
Daymaniyat Islands *II-18*, II-22
— Daymaniyat Islands National Nature Reserve II-30
DDT II-545
— air and surface seawater, worldwide II-454, *II-457*
— in animals from the Greenland seas I-9, I-11
— in Asian developing region waters II-449
— contamination in Southern California I-390, I-392
    — total DDT of most concern I-393, **I-393**
    — well preserved in anoxic deep basin sediments I-393
— Faroes I-37–8, **I-37**
— Norwegian west coast, still an environmental problem I-25
— recent use, Vietnam II-565
— in river waters, El Salvador I-552
— as seed dressing I-635
— still reaching Baltic Sea via precipitation I-131
de la Mare, Dr William, on the New Management Procedure (IWC) III-78
debris, non-biodegradable, a danger to sea turtles III-67
decision making process, for coastal management, greater inclusiveness more sensible III-355
Declaration on the Protection of the Black Sea Environment I-294
deep water formation
— Greenland Sea I-7
— Tyrrhenian Sea I-272
deep water renewal, Norwegian fjords I-21
deep-sea smelt, Okhotsk Sea II-466, II-468
deepwater dumpsite 106 *I-322*, I-327
deforestation I-241, I-427–8
— Australian Rainforest II-585
— Belize I-508, I-510
— Borneo II-374–5
— catchment areas of Malacca Strait II-338
— causing soil erosion I-174–5
— in the Comoros II-247
— effects of, Andaman and Nicobar Islands II-193
— from mining, New Caledonia II-731
— Great Barrier Reef region catchments II-620–1
— Guangdong II-558
— Guinea I-801
— Gulf of Guinea I-786
— Jamaica I-568
    — reforestation I-572–3
— Kamchatka peninsula II-469
— and land conversion, cause of high sedimentation rates II-384
— Lesser Antilles I-635, **I-636**
— logging in Gulf of Papua watersheds II-604
— Mozambique II-109
— Nicaragua I-523, I-538
— the Philippines II-412
    — reduction in forest cover II-413
— round Sea of Japan II-475
— and soil erosion, Hawaiian Islands II-801–2
— in southeast India, affects coastal zone II-165
— southern Brazil
    — of Atlantic forest I-741
    — of mangroves I-740, I-741
    — for Rio-Santos road I-742
— Southern Gulf of Mexico
    — and erosion I-476
    — impact of I-479
— Sumatra and Kalimantan II-395
— Tanzania II-89–90
— through livestock grazing, Dutch Antilles I-601, *I-602*
— west coast Taiwan II-503
degradation
— southern Brazil I-740–3
    — main problems I-740

Delaware Bay *I-322*, I-329, *I-336*
Deltaic Sundarbans *II-146*, II-147
— brackish waters support phytoplankton, macrobenthic algae and zooplankton II-150
— effects of seasonal changes II-149
— Hugli estuary, carries industrial discharges from Haldia region II-155
— important morphotypes II-147
— major rural activities and environmental impact II-153–4
— monsoon period II-149
— physico-chemical characteristics II-149, **II-149**
— population extremely poor II-152
— post-monsoon periods II-149
— pre-monsoon period II-148–9
— rich in natural resources II-152
— a unique ecosystem II-156
denitrification I-106, III-262
— water-column, Arabian Sea II-24
Denmark
— Action Plan for Offshore Wind Farms in Danish Waters III-307
— acts on behalf of Faroes I-36
— monitoring and modelling of offshore wind energy technology III-308
    — Wind Atlas Analysis and Application Program (WA^{S}P)
— offshore wind energy production III-307–8
    — Tunø Knob installation III-308
    — Vindeby installation III-307–8
— pilot offshore wind energy projects III-306
Denmark Strait *I-6*, I-7, I-9
D'Entrecasteau Basin *II-426*
D'Entrecasteaux Channel *II-648*
— localised cooling II-651
deposition processes, from the atmosphere III-200–1
Derjugin's Basin *II-464*, II-465
desalination
— Arabian Gulf
    — by-products II-10
    — multistage flash evaporation plants II-10
    — and power plants II-10
— Oman II-28–9
— Red Sea, saline discharges from desalination plants II-43
desalination plants, Bay of Bengal, effects of saline discharge II-279
desert
— Namibia I-823
— Peru I-689–90
developed countries, marine reserves in III-386
developing countries
— a chance for coastal zone development with fewer problems III-357
— marine reserves in III-386
— prawn/shrimp fisheries, high utilisation of catch, poor utilisation of species III-143
development pressure
— English Channel coasts I-71
— Venezuelan coastal zone I-645
Dhofar *II-18*
— excessive cutting of firewood II-26
— limestone cliffs II-19, II-21
Diadema mass mortality I-410, I-421, I-503, I-520, I-607, I-632
— co-incidental with El Niño conditions III-223
— Jamaica I-563–4
diamonds, coastal deposits, South Africa II-142
diarrhetic shellfish poisoning (DSP) I-69, I-160
diatom blooms I-155
— Baltic Sea I-125
— Irish Sea I-89
— spring and autumn, English Channel I-67
— toxic, causing debilitating illness III-218
diatom sediment records, Portuguese shelf I-160
diatoms I-355, I-692, I-743, I-783, I-814, I-821, I-830–1, II-24, III-221
— Australia II-583

— Bay of Bengal II-275
— Campeche Sound I-474
— discriminated against III-259
— in northern Gulf of Mexico I-449
— West Guangdong coast II-553
dibutyltin (DBT) II-449
Diego Garcia, Chagos Archipelago *II-221*
— military development imported alien flora II-229
— occasional storms from cyclone fringes II-225
— only inhabited island II-223
— recreational fishing II-230
diffusional transfer 202
diffusiophoresis III-201
diffusive attenuation coefficient
— and water quality III-296–7
    — K-maps III-296
dikes and breakwaters, construction interrupts sand drift I-176
dinoflagellate blooms I-827
— toxic, French Polynesia II-821
dinoflagellates I-692, I-783, I-814, I-831, II-320
— Australia II-583
— ichthyotoxic I-363–4
— increased number in the Arabian Gulf *II-12*, II-13
— toxic I-355, II-304, II-587, III-218, III-221
    — benthic, produce tumour-promoting agents III-219
    — *Pfiesteria piscicida* III-223
— winter-blooming I-355
dioxin
— discharge from old Norwegian magnesium plant I-25
— remobilization of I-25
discards I-146–7, **I-148**
— at sea, percentage of total catch III-377
— and by-catch III-123–5
— fate of III-127
— reasons for discarding III-137–8
disease
— an ecological opportunist III-225
— can devastate aquaculture ponds III-369
— chronic conditions within endangered populations, monitoring of III-227
— increase in extent and impact of III-225
— novel, possible introduction through aquaculture III-369
— reflect perturbations with ecosystems III-225
— transfer via ballast water III-355
— trophodynamically acquired, a global issue III-221
— water-related, Vanuatu II-745
disease disturbances III-225–6, **III-226**
diseases, water-borne, Marshall Islands II-781
Disko Bay, Greenland, shrimp fishing I-8
dispersants *see* oil dispersants
dissolved inorganic carbon (DIC), non-Redfield ratio depletion I-225–6
dissolved organic N (DON), varied origin III-199
dissolved oxygen I-240
— Campeche Sound I-469
— depletion of III-199
— and fish kills I-364
— Gulf of Aden II-51
— levels in the Cambodian Sea II-572–3
— variable, Canary Islands I-191, *I-192*
disturbance
— keystone-endangered and chronic cyclical III-227–8
— as a regular feature of an ecosystem III-227
disturbance categories/types III-217–28
— derivation III-217
— grouping of III-217
disturbance regimes III-227
— better understanding of the natural history of III-230
dive tourism
— artificial dive objects I-604
— Bonaire I-597

— Dutch Antilles I-604
— effects of on Borneo islands II-377–8
— Hawaiian Islands II-806
— SCUBA diving
    — East Kalimantan II-377
    — Sabah II-377
— Sri Lanka II-180
— Vanuatu II-742
diving and snorkelling, impacts from II-213
Djibouti *II-36*, II-49, II-55, *II-64*
— use of mangroves II-58
Dnestr River, pollutants carried I-303
Dogger Bank I-46, I-47
— plankton I-48
dogwhelks
— imposex I-25, I-36–7, I-55, I-75
— severe impact of TBT III-250–1
— southeast England, the Dumpton Syndrome III-252
Doldrums *see* Inter-Tropical Convergence Zone (ITCZ)
dolphin mortality
— by-catch is to some degree controllable III-139
— reduction in through changed purse-seining procedure III-141
— through tuna fishing III-142–3
dolphins I-90, I-834, II-24, II-205, II-368, III-90–1
— caught for shark bait II-57
— declining, Black Sea I-291
— Great Barrier Reef II-618–19
— Gulf of Guninea I-780
— illegal hunting in Black Sea III-94
— Java Sea II-363
— Marshall Islands II-780
— and mercury contamination III-95
— Patagonian coast I-756
— Sri Lanka II-180
— *see also* named varieties
dominance, pattern of altered by oil spills III-275
Doñana National Park, southern Spain *I-168*
— controversy over infrastructure improvements around I-173
— sand and marsh ecosystems I-171–2
Donax denticulatus, wash zone, dissipative beaches I-648
Dover *I-66*
Dover Straits *I-66*
— shipping through I-72
downwelling
— Ekman downwelling I-223
— nearshore, Alaskan Gulf I-376
dragnet fishery, southern Brazil, east coast, causing degradation I-740
drainage, a development issue, Arabian Gulf II-9
Drake Passage I-753
dredge and fill
— Gulf coastal plain, USA I-445
    — navigation channels I-447
— Jamaica, damaging seagrasses I-569
— of mangrove swamps III-19
dredge fishery, Victoria Province, Australia II-668
dredge spoil, dumped I-77
dredging I-623, II-124, II-546, II-642, II-655
— Belize I-510
— in Chesapeake Bay I-348
— Curaçao I-602
— damaging to seagrass beds III-7, III-8–9
— destructive III-379
— detrimental effects of I-145
— and dumping, in the Comoros II-247–8
— and earth-moving activities, Marshall Islands II-782–3, *II-782*
— effects of, Tagus estuary I-161
— environmental impacts of, southern Spanish coast I-176–7
— and filling
    — for construction purposes I-601–2
    — Jamaica, damaging seagrasses I-569

— followed by reclamation, French Polynesia II-821
— and habitat destruction III-379–80
— harbour entrances and navigation channels, Buenos Aires I-762–3
— instream, degradation and erosion caused by I-444
— Kuwait II-9–10
— and land fill, Hawaiian Islands II-805
— and landfill, Red Sea II-43
— Malacca Strait II-339
— Maputo and Beira II-108
— northern Gulf of Mexico I-447–8
— now a necessity, Malacca Straits II-317
— to improve Gulf of Aden ports 58
drift netting, most popular, east coast, Peninsular Malaysia II-352
Driftnet Ban, United Nations III-140, III-145
drill cuttings, contaminated, discharge of III-363
drilling fluids, oil-based, toxicity of III-363
drilling waste management I-23–4
"drowned river system" *see* Chesapeake Bay
drowned river valleys, Tasmania II-653
Drupella snails, coral-eating II-696
dry deposition I-13
— measurement techniques III-203–5
    — field measurement of, current status III-205
    — use of surrogate surfaces III-204–5
    — wind tunnel experiments III-204
— modelling frameworks and algorithms III-202–3
    — current modelling uncertainties III-202–3
    — mathematical treatment of physical processes III-202
— of particles to water surfaces, processes and consequences III-197–209
    — atmospheric deposition to the coastal zone III-205–7
    — atmospheric particles III-199–200
    — deposition processes III-200–1
    — gas deposition and role of particles III-205
    — importance of in nutrient fluxes III-201
    — nutrient fluxes and aquatic cosystem responses III-198–9
dry deposition velocities III-200–1, III-202, III-204–5
— based on eddy correlation III-205
Dry Tortugas *I-406*
— corals little diseased I-410
— decline in pink shrimp landings I-447
— extreme low temperature, and death of corals I-410
— showing only small decline I-412
Dry Tortugas National Monument I-412
— redesignated as Dry Tortugas National Park I-413
Dry Tortugas National Park, coral bleaching (1998) III-56
duck plague virus III-225, III-227
dugongs II-54, II-90, II-123, II-336, II-431, II-574, II-587, II-601, II-615
— Bazaruto Islands II-104
— eastern Australian region II-637
— endangered in Gulf of Mannar II-165, II-166
— Great Barrier Reef
    — entanglement problems II-619
    — pressures and status II-619
— in the Mayotte lagoon II-250
— Red Sea II-42–3
— Sri Lanka II-180
— Vanuatu II-741
— western Indonesian seas II-401
dumping
— at sea III-338–9, III-342, III-365–6
    — of dredge spoils III-365
    — some prohibitions achieved III-365
— Baltic Proper, of chemical warfare agents I-117
— Irish Sea
    — blast furnace spoil I-94
    — munitions I-96
— New York Bight I-323, I-324
— North Sea
    — of industrial waste and sewage sludge finished I-53, I-55
    — of munitions I-52

— *see also* waste dumping
Dumping Convention **III-334**
Dumpton Syndrome III-252
Dungeness *I-66*, I-68
Durvillia habitat, exposed coasts, Tasmania II-652
Dutch Antilles I-595–614
— Aruba *I-596*, I-597
— coral bleaching and mortality events III-54
— definition and description I-597–8
— effects from urban and industrial activity I-502–7
    — artisanal and non-industrial uses I-602–4
    — cities I-606
    — cumulative impacts and ecological trends I-607
    — industrial use I-604–6
— environmental hindrances ordinances I-608
— environmental institutional capacity I-609
— Fishery Protection Law I-608
— government environmental protection and management funding, sparse I-612
— law enforcement is critical I-609
— Leeward Group *I-596*, I-597–8
— major shallow marine and coastal habitats I-598–601
— maritime legislation I-608
— National Parks Foundation of the Netherlands Antilles I-607–8
— overview of implementation of marine environmental policy and legislation I-609, **I-610–11**
— protective measures I-607–12
    — implementation and an evaluation I-609
    — policy development and legislation I-607–9
    — practical management measures I-609
    — prognoses and prospects I-609, I-612
— rural factors, coastal erosion, landfill and excavation I-60–2
— Windward Group *I-596*, I-598
dynamite fishing
— cessation of II-95
    — case study Mtwara, Tanzania II-87, II-95
— Gulf of Thailand II-304

earthquake activity
— the Philippines II-408
— Vanuatu II-744
— Xiamen region II-516–17
East African Coastal Current *II-64*
— important for larval dispersal and downwelling II-85
— influenced by the monsoons II-85
East Arabian Current II-50
— behaviour of II-20
East Australian Current II-581, *II-581*, II-614, II-619, II-631, *II-632*, II-649–50, II-664
— giving distinctive ecosystems around the Kent Group II-651
— varies with El Niño/Southern Oscillation cycle II-650
East China Sea *II-474*, *II-500*, II-508
East Greenland Current I-5, I-7
East India Coastal Current (EICC) II-274
East Madagascar Current *II-101*, II-245
East Pacific Plate I-487
East Sakhalin Current II-466
Easter Island I-708
eastern Atlantic flyway, migratory feeding, North Sea coasts I-49
Eastern North Atlantic Central waters I-153
Ebrié lagoon, Côte d'Ivoire I-811–12
— artificial Vridi canal I-812
— fish I-813
— freshwater inputs I-812
— hydrological zonation I-812
— physical framework I-811–12
— pollution in I-817
— sewage discharged to I-817
echinoderms I-35, I-239, III-242
Ecklonia radiata habitat II-652
ecological catastrophe, Aznalcóllar mining spillage disaster I-178

ecological problems, and their causes, west coast, Sea of Japan, Amursky Bay II-483–5
ecological resources
— exploitation of III-366–9
   — aquaculture III-368–9
   — marine capture fisheries III-366–8
ecological tariffs III-401
ecological tax reform III-401
economic development, and coastal management III-351
economic incentives, to achieve economic goals III-401
economic income, skewed III-396
ecosystem, defined I-705
ecosystem health information III-212
ecosystem instability and collapse, processes typically leading to III-227
ecosystem management, and the reserve concept III-377–8
ecosystem modelling I-50
ecosystems
— changes in structure and function following eutrophication III-263, **III-264**
— connectedness of I-478
— dynamics of III-259
— effects of fisheries on III-117–33
— measuring impact of by-catch on III-144
— refers to processes and functions III-259
— response to increased nutrient loads unpredictable III-366
— retrogression III-259–60
— vulnerable species targetted by opportunistic microorganisms III-227
ecosystems services, value of III-395
ecotourism I-4, II-89, II-95, III-376
— and Brazilian coral reefs I-723
— Fiji II-762
— Great Barrier Reef, Australia II-586
— Marshall Islands II-780
— Midway Atoll II-801
— a non-consumptive resource use II-141
— possible in the mangrove ecosystems of Pacific Colombia I-682–3
— potential for, Mayotte II-251
eddies I-46, I-630
— in Caribbean Basin I-580
— mesoscale, in cold water at Paria Peninsula I-646
— surface, Angola-Benguela front I-825
— SWODDIES I-138
— *see also* Sitka Eddy
eddy accumulation techniques, dry deposition measurement III-203–4
eddy correlation/covariance techniques, dry deposition measurement III-203
eddy currents, Canary Islands I-189
eddy diffusion, between deep and upper waters I-391
eddy systems I-85, I-223
eelgrass, Carolinas I-358
effluent disposal pipes
— KwaZulu-Natal II-142
— using Algulhas Current for dispersal, South Africa II-142
effluents
— from shallow-well injection, migration of I-409
— polluting Norwegian coastal waters I-24
Egypt *II-36*
— dive tourism II-44
— Red Sea coast, coastal pollution by oil II-43
Ekaterina Strait *II-464*, II-465
Ekman downwelling I-223
Ekman transport II-274
— drives coastal upwelling off Oman–Yemen coast II-19
El Niño
— affects Mexican Pacific coast I-486, I-488, I-489
— affects Nicaraguan Pacific coast I-534
— biological effects of, Peru I-692–3

— change to subtropical and tropical species I-693
— causes of I-702
— a complex, multi-level phenomenon I-703
— described I-679
— dramatic effects of I-679
— in El Salvador I-547
— *see also* Atlantic Niño; Benguela Niño years
El Niño events II-776.
— affecting the California Current I-387
— affecting Gulf of Alaska mixed layer depth III-182
— biological effects of I-703–4
— bring drought, Coral, Solomon and Bismark Seas Region II-428–9
— and coral bleaching III-38
— extreme
   — and the 1997–98 coral bleaching event III-44–6
   — areas of bleaching III-46
   — impact on Indian and Southeast Asian winds III-44
   — interpretation and conclusions III-46
— impacts on Northeast Pacific Ocean III-183–4
   — effect of 1997–8 event III-183–4
— importance of I-702
— mechanism governing strength of I-702–3
— northern Chilean coast invaded by exotics I-707
— physical alterations due to I-703
— position of Subtropical Water Mass I-701
El Niño Southern Oscillation (ENSO)
— effects of I-223–4
— effects on climate and oceanography of Fiji Islands II-753–4
— effects on the Great Barrier Reef region II-613–14
— long-term variations I-702–4
— periodic influence in Australia II-582, II-632
El Salvador I-545–58
— coastal erosion and landfill I-552
— effects from urban and industrial activities I-553–6
— geography I-547
— major shallow water marine and coastal resources I-548–51
— offshore systems I-551
— populations affecting the area I-551
— protective measures I-556–7
   — Ley del Medio Ambiente (Environmental Law) I-556
   — protected areas I-556–7
— rural factors I-551–2
— seasonality, currents, natural environmental variables I-547–8
elasmobranchs, sensitive to ecological change, North Sea I-49
Electronic Chart Display and Information Service (ECDIS), Malacca Straits II-325
Elefsis Bay *I-234*, I-237
— decrease in invertebrate abundances I-239
— industrial pollution I-245
— metal pollution I-246
elephant seals I-756, I-757
Eleuthera *I-416*
EMECS (Environmental Management of Enclosed Coastal Seas) I-56
Emperor Seamounts
— fossil coral reefs II-800
— relationship to Hawaiian Islands II-794
enclosed shores, Hong Kong II-538
endangered species
— Abrolhos Bank–Cabo Frio, Southern Brazil I-736
— consequences of climatic irregularities III-224
— leatherback, green turtles and loggerheads III-62
— loss or imminent loss, Black Sea I-291, **I-292**
— Olive Ridley and Hemp's Ridley critically endangered III-63
— Philippines II-410
— sea turtles III-143
— vaquita III-90
endangered species protection, western Indonesia II-400–1
endemic species
— Coral, Solomon and Bismark Seas region II-431
— Hawaiian Islands II-797, II-800
— Marshall Islands II-779

endemism
— Australian Bight II-678–9
— islands of the southwest Indian Ocean II-251
— low, reefal communities, French Polynesia II-819
— southern African marine biota II-135
— southern Australia II-581
— Tanzanian coastal forests II-88
endocrine disrupters
— potential harm from III-86
— TBT III-249
energy
— from the oceans III-303–21
    — possibility of combined technologies III-311
— from seagrasses, detrital and direct grazing pathways III-3
engineering structures, causing beach erosion II-639
English Channel I-65–82
— anthropogenic impacts on I-74–7
— coastal and marine habitats I-68–71
— defined I-67
— formation of I-67
— lies at boundary of Boreal and Lusitanian biogeographical regions I-68
— physical and biological environment I-67–8
— protective and remediation measures I-77–9
    — International Conventions and Agreements, and EC Directives **I-78**
— urban and rural populations I-71–4
Eniwetak Atoll, nuclear tests II-783
ENSO events
— 1997/1998, impacted mammals and seabirds in several regions III-225
— 1998 event II-236
— affected by climate change III-370
— effects of, Vanuatu II-740
— effects on seabirds III-110
— influence on New Caledonia II-725
— influencing sea surface temperatures II-384
— Marshall Islands II-776–7
    — coral bleaching II-777
    — Crown-of-Thorns starfish II-777
    — sea-level rise II-777
— the Philippines, manifestations of II-408
— South Western Pacific Islands, affect oceanography and climate II-710
— Teacapán—Agua Brava lagoon system I-494
entanglement
— of mammals and seabirds in floating synthetic debris III-125–6
— usually fatal for small cetaceans III-145
entanglement nets (jarife), Tanzania II-90
environmental assessment, for newer Paupuan mines II-439
environmental coastal protection, Chile I-715
environmental concerns
— Brazilian tropical coast I-727–8
— coral reefs, Hawaiian Islands II-807
— in the Maldives II-206–7, **II-206**
— mariculture, Hawaiian Islands II-803
— Marshall Islands II-780, **II-781**
    — related to cross-sectoral issues II-784–5
environmental conservation, restoration and improvement, Gulf of Mexico I-479
environmental criteria, concepts in II-202
environmental degradation
— Andaman and Nicobar Islands II-193
— Cabo Frio–Cabo de Santa Marta Grande, southern Brazil I-738
— China II-492
— coastal forested watershed, Cambodia II-575
— early, the Seychelles II-239
— east coast, Peninsular Malaysia
    — of coral reefs II-355
    — from tourism II-353–4
— Godavari-Krishna delta II-171–2

— Hong Kong II-543–6
— Lakshadweep Islands II-194
— Palk Bay–Madras coast II-169
— parts of North Spanish coast I-142
— the Philippines II-418
— potential, Tasmanian marine fish farms II-641
— Sabah coral reefs II-371, *II-372*
— southern Brazil I-740–3
— Taiwan's marine and coastal environment II-510
— Tasmanian estuaries II-654–5
— through land reclamation, West Guangdong II-557
— Western Australia, metropolitan areas II-697
environmental effects, of Australian fishing II-587
environmental impact assessment, Hawaiin Islands II-808
environmental impacts
— of alien species in Irish Sea I-89
— Chilean fishmeal industry I-713–14
— of coastal settlements, eastern Australian region II-640
— of coastal zone industries, Chile **I-711**
— and ecological trends, Dutch Antilles I-607
— from repetitive trawling II-624
— industrial fishing, Gulf of Guinea I-787–8
— of intensive fishing
    — on the Irish Sea environment I-92
    — on Irish Sea fisheries I-92
— of mining activities, New Caledonia II-731
— nickel mining, New Caledonia II-713
— of tourism II-212–13, **II-213**
environmental issues
— Australia II-586
    — marine environment **II-589**
— Mascarene Region II-265
— south-east Queensland II-641–2
environmental management
— poor, Vietnam II-566
— sound, expected of companies and organisations III-351
— Torres Strait and the Gulf of Papua II-607–8
environmental management strategy, Marshall Islands II-787
environmental matrices
— Gulf of Mexico
    — extensive cattle pasture I-479
    — offshore oil industry I-478–9
environmental opportunities and prospects, Marshall Islands II-787–8
environmental pollution, in the Bay of Bengal II-154–5
environmental problems
— Cambodia, no clear perception of II-576, II-577
— critical, Xiamen region II-521–4
    — bacterial pollution II-522–3
    — oil pollution II-523
    — organic pollution, eutrophication and red tides II-522
    — pollution from pesticides II-524
    — sea dumping and disposal of solid waste II-523–4
    — sediment deterioration and secondary pollution II-524
— Malacca Straits II-317–21
— Marshall Islands, related to population pressure II-780
— some east coast Indian cities II-278
environmental projects, French Polynesia II-824
environmental quality criteria, use in Norwegian fjords and coastal waters I-27
Environmental Risk Assessment, Malacca Strait II-320–1, *II-321*
— analysis of likelihood of adverse effects II-320–1
— retrospective analysis of decline in key habitats II-320, **II-321**
Environmental Sensitivity Index (ESI) III-269, **III-270**
environmental stress, and altered benthos, Southern California I-396
environmental variability, in fish stocks, southwestern Africa I-832
environments, deep, Red Sea II-39
EOS satellite, with colour sensor MODIS III-295
epi-continental seas I-45
epiphytes, seagrass, productivity of III-2
Equatorial Counter Current I-486, I-489, II-201

Equatorial Countercurrent *I-678*, I-679, I-701, *II-581*, II-614, II-619, *II-724*, II-767, II-775, II-780
Equatorial Current, and El Niño I-703
Equatorial Subsurface Water Mass, low dissolved oxygen content I-702
Equatorial Under Current *I-774*
equity and fairness, in fisheries management 157–8, III-155
— inter-generational equity III-157
— issue of relative deprivation and historical participation III-157
equity theory in fisheries management III-157-8
Eritrea *II-36*
EROD induction I-56
erosion
— of beaches through sand mining I-636
— due to hurricanes I-622
— Dutch Antilles I-601–2
— Nicaraguan Pacific coast I-538
erosion–accretion cycles, Hawaiian Islands II-804–5
erosion–deposition cycles
— Arabian Sea II-27
— Gulf of Aden II-58
— Inhaca Island II-105
ERTS-1 satellite, early recognition of cyanobacteria blooms III-299
ESA satellite ENVISAT, with ocean colour sensor MERIS III-295
Escheria coli
— Hong Kong bathing waters II-541
— Penang and Selangor II-319
Espichel, Cape *I-152*, I-153
Essential Fish Habitat (EFH) I-451
estuaries
— east coast, Peninsular Malaysia II-349
— eastern Australia II-634
— Gulf of Maine I-310, *I-311*
— New York Bight, altered condition I-327
— North Sea, importance of I-49
— northeastern Australia II-614
— northern Gulf of Mexico
    — freshwater flushing rate a critical parameter I-447
    — tidal and subtidal oyster reefs I-441–2
— South Africa II-135–6
    — greatest threat will be lack of water II-136
    — resource use II-136
— southwestern Africa I-829
— Taiwan Strait II-502, II-503
— Tasmania II-653
    — degree of sedimentation II-654
    — southern, tannin-stained II-652
estuarine crocodiles
— Great Barrier Reef II-618
— Solomon Islands, depleted II-431
estuarine ecosystem, Palk Bay–Madras coast II-169
estuarine habitats
— Channel coast of England I-68
— Xiamen region, Jiulongjiang estuary II-518
estuary wetland ecosystems, North China, affected by oil and agriculture II-494
EU
— Habitats Directive, omits to take natural change into account III-354–5
— MAST programme I-56
— nature conservation, a shared competency I-57
EU/EC
— Convention on International Trade in Endangered Species (CITES) I-608
— Migratory Species Treaty (Bonn Convention) I-608
EU/EC Directives **I-78**
— on Bathing Water Quality I-76, I-95
— concerning waste-water treatment I-145
— Conservation of Natural Habitats and of Wild Fauna and Flora I-52, I-57, I-77, I-96–7, I-162
— Conservation of Wild Birds I-52, I-57, I-77, I-162, I-214
— on the control of nitrates I-53
— for protection of European Seas **I-57**
— Shellfish Hygiene I-76
— on Shellfish Waters I-76
— to increase urban wastewater treatment, too many derogations I-55
Eucla Bioregion, Australian Bight II-678, II-679
euphotic layer, Côte d'Ivoire I-810
Euphrates, River II-3
Europe
— current and planned offshore wind farms III-309–11, *III-310*
— wind resources predicted III-308, *III-309*
European Environment Agency, arrangements for protecting North East Atlantic in the 1990s **I-57**
European Union, funding for off shore wind energy technology III-307
eustatic movements, tilting European landmass I-47
eutrophication I-137–8, I-174, I-278, II-542, II-770, II-771, III-366
— and algal blooms I-20
— areas of French channel coast contributing to I-74
— in Australian coastal waters II-585
— Baltic Proper I-103–8
    — biological effects of I-109–10
    — rivers bring pollutants I-103
    — and supersaturation of oxygen I-104–5
— Baltic Sea I-130–1
— Bermuda I-228
— Black Sea I-290
    — changing phytoplankton community composition I-290
— of coastal areas, western Indonesia II-396
— concepts for remediation III-263
— cultural, Western Australia II-700
— in fjords with restricted circulation I-24
— from aquaculture I-144
— from fish farming I-26
— Guanabara Bay, southern Brazil I-741
— Gulf of Guinea, coastal waters I-784, I-788–9
— Gulf of Thailand II-307
— Hawaiian Islands II-804
— Irish Sea, possible near English coast I-87
— in its broadest sense III-198–9
— lagoon, Vanuatu II-745
— leads to increase in N and P III-259
— local, Red Sea II-43
— of marine waters: effects on benthic microbial communities III-257–65
    — coastal inlets of the southern Baltic Sea III-260–3
    — definition and sources of pollution III-258
    — effects of on marine communities III-259–60
— in most Carolinas estuaries I-363
    — estuarine fish kills I-363–4
    — measures for reduction I-366
    — more stringent N and P reductions needed I-366
— North Korean artificial lakes II-493
— North Sea I-53
— northern Gulf of Mexico, growing problem I-443, I-450
— in Norwegian coastal waters, dependent on transboundary load I-24
— and nutrients, English Channel I-74
— Patagonia I-765
— the Philippines II-417
— potential for, Jamaica I-564
— in a range of aquatic environments III-205–6
— secondary effects I-108
— some North Carolina estuaries I-356, I-363
Euvoikos Gulf *I-234*, I-237, I-244
— affected by wastes I-246
evaporation
— in excess of precipitation
    — Arabian Gulf II-3
    — Red Sea II-39

— Gulf of California Coastal Province, Mexico I-487–8
evolutionary mechanisms, allowing toxic species to spread III-219
Exclusive Economic Zones (EEZs) I-2
— Andaman and Nicobar Islands II-191
— Azores Archipelago I-203
— the Bahamas I-422
— Cambodia II-571
— and fisheries management III-158
— Lakshadweep Islands II-194
— Mozambique II-105
— North Sea I-51
— the Philippines II-408
— South Africa II-137
— USA I-451
    — utilization of economic resources I-365–6
exotic species/introductions *see* alien/accidental/exotic/introduced organisms/species
exports, from Faroese fishery I-35–6
extinction—recolonization dynamics, Chilean coast I-707
extreme events, The Maldives II-203
Exuma Cays *I-416*, I-419
— patch reefs I-429–30
— shallow marine habitat I-421
Exuma Cays Land and Sea Park
— evidence for positive effect of protective measures I-431–2
— regulations I-431
Exuma Sound I-417, I-428
Exxon Valdez oil spill I-382, III-275
— biological effects at low levels of aromatic compounds III-364
— characteristics of III-272–3
— decrease in *F. gardneri* III-272
— oiled mussel beds III-271, III-274
— part of severity of effects on *Fucus* due to high-pressure-hot-washing III-275
Eyre Bioregion, Australian Bight II-678

faecal coliforms I-743
— areas of Gulf of Guinea coast I-789
— contaminates freshwater sources, Marshall Islands II-781
— Dar es Salaam and Zanzibar II-92, II-93
— Hong Kong bathing beaches II-541
— Johore Strait II-335
— Maputo Bay II-109
— Morrocoy National Park I-658
— Peninsular Malaysia west coast II-319, II-340
— Sepetiba Bay, southern Brazil I-741
— Suva region, Fiji II-760–1
    — Suva harbour II-719
— Sydney beaches II-642
— Xiamen coastal waters II-522–3
faecal contamination, Durban beaches II-142
faecal pollution, Guanabara Bay, southern Brazil I-741
Fal estuary *I-66*, I-74, I-75
— closure of shellfishery I-69
Falklands/Malvinas Current I-751
FAO
— catch statistics reveal "fishing down" of food chains III-368
— Code of Conduct for Responsible Fisheries III-138, III-341–2
— Code of Conduct for Responsible Fishing III-159
— data base to maintain records of high seas fishing vessels III-160
— international plan to reduce incidental catch of seabirds in longline fisheries III-146
Farasan Islands *II-36*, II-38, II-41
farming, 'modernization' of III-7
Faroe Islands I-31–41
— coasts and shallow waters I-33–5
— mariculture I-35
— oceanic climate I-33
— offshore resources I-35–8
— pilot whales hunted III-92
— protective measures and the future I-38–40

— typical upper and deeper layer water flows I-33, *I-34*
faroes I-505
'faros', Maldivian mini-atolls II-201
Federal and Islamic Republic of Comoros (FIRC) II-247
Federal Water Pollution Control Act Amendments (USA), Section 404 Program I-452–3
feral animals, impact of Australian Bight coast II-684
fertiliser pollution, from Mozambique's upstream neighbours II-109
fertiliser run-off, Madagascar, impacting coral reefs II-123, II-124
fertilizer consumption, Lesser Antilles I-635, **I-635**
Ficopomatus enigmatus, positive effects of I-48
Fiji Islands *II-706*, II-707, *II-708*, II-751–64
— agriculture and fisheries II-715–16
— biogeography II-753
— coastal erosion and landfill II-760
— coastal modifications II-717
— coral reefs II-711
    — degraded by pollution II-711
— effects of urban and industrial activities II-760–1
— island groups II-753
— major shallow water marine and coastal habitats II-754–8
— marine environment critical II-762
— offshore systems II-758
— overfishing of marine invertebrate species II-759
— population II-714, II-758
— protective measures II-761–2
    — private sanctuaries II-761
— rural factors II-758–60
— seasonality, currents, natural environmental variables II-753–4
— sensitive and endangered marine species II-757
Fiji plateau II-753
fin whales, depleted III-78
financial and technology flows, global, importance for environmental decision-making III-346
Findlater Jet *II-18*, II-19
finfish communities, estuarine and coastal, Carolinas coast I-362–3
finfish farming III-170
finfish fishery, Belize I-509
Finisterre Cape *I-136*, I-139
Finland, Gulf of *I-100*, *I-122*
fire, effects of, northern Gulf of Mexico I-449–50
fire cycle
— natural, important in maintaining coastal ecosystems I-449
    — changes in detrimental I-449
fish
— the Azores I-207
— Baltic Sea
    — butyltins in **I-115**
    — larval stages in *Fucus* belt I-125
    — mixture of freshwater and marine species I-128
— Belize lagoonal shelf I-506
— Cambodia II-575
— coastal, eastern Australian region II-634
— Côte d'Ivoire, Ebrié Lagoon I-813
— deep water, "boom and bust" cycles III-120
— El Salvador I-550
    — estuarine nursery habitats I-550
— in English Channel I-70
— fauna, Australian Bight II-679
— Great Barrier Reef II-617
— Gulf of Guinea
    — commercial species **I-784**
    — Guinean Trawling Survey I-783
    — populations I-783
— Gulf of Mannar II-164–5
— Lesser Antilles I-633
— life history traits (*r*- and *k*-selection) III-119
— Mauritius II-259
— nearshore, Hawaiian Islands, depleted II-803
— North Aegean Sea I-238
— North Sea I-49

— North and South Carolina coasts I-362–3
— offshore, Fiji Islands II-758
— organochlorine pollution in, Asian waters II-449–51
— possible extinction of groupers and Humphead wrasse II-373
— problem of size-selective fishing III-121–2
    — changes in size structure, community level III-121–2
— South Aegean Sea I-238
— Southern California, pollution effects on I-397
— southwestern Africa
    — mesopelagic I-832–3
    — surf zone I-830
— species diversity, Gulf of Mexico I-473
— species and groups, Vietnam II-564
— Tagus estuary, nursery grounds I-157
— target species, ecologist-market collision III-158
— vulnerable species, fishing of III-120
Fish Aggregating Devices (FADs) II-372
— New Caledonia coastal fishery II-730
— used in the Comoros II-249
fish barrages, intertidal, Madagascar II-126
fish communities
— natural factors for change I-398
— Southern California, impacts of anthropogenic pollution I-397–8
fish diseases, from pollution I-397
fish diversity
— Baltic, marine, decreases northwards I-109
— demersal, soft grounds, Spanish north coast I-141
fish farming
— Faroes I-35
— Norway I-26
    — and expanding industry I-28
    — mass mortalities in fish cages I-20
— potential for, east coast of India II-278
— round North Sea I-52
— southern Spain I-177
fish farms, marine, Yellow Sea II-494
fish mortality, Tanzania II-89
fish plant, Marshall Islands, benefits vs. impacts II-785
fish processing at sea, dumping of organic material III-127
fish recruitment I-92
fish stock assessments III-156
— and Fishers' insights III-157
— relationship between present and future stocks III-156
— validity of single-species models in multi-species ecosystems III-156–7
fish stocks
— Adriatic Sea I-277
— affected by commercial fisheries III-97
— Arabian Gulf, dwindling II-11
— assessments in the Maldives II-210
— Baltic Sea I-110–12
    — freshwater species I-111
— Bangladesh **II-292**
— Chesapeake Bay I-342–3
    — coastal ocean spawners I-343
— collapsing I-3, III-119–21, III-367
— commercial fisheries, Western Australia **II-698**
    — Fish Habitat Protection Areas II-699
— Coral, Solomon and Bismark Seas region
    — coastal, information on II-434–5
    — fisheries management measures II-442
— declining
    — English Channel I-76
    — Gulf of Maine and Georges Bank I-310
— demersal fish, Gulf of Aden II-54
— dolphin stocks III-42
— evaluation of global status of III-377
— Gulf of Guinea, significant changes in I-783
— Gulf of Maine and Georges Bank I-312
— Gulf of Thailand III-128
— increased off Alaskan coast I-376–7

— Jamaica I-564
    — reefs overfished I-568–9
— Lesser Antilles, shared assessment of I-633
— Malacca Strait II-337
— Malacca Straits II-314
    — declining II-320
— New Caledonia II-730
— New York Bight, overfished I-330
— North Sea, management by EU I-57
— northern Spanish coast *I-147*
    — over-exploited and depleted I-146
    — sardine stock critical I-146
— off coasts of Oman and Yemen II-27
— over-exploited III-367
    — Irish Sea I-91
— Queen Conch I-425
— Sea of Okhotsk, some decrease in II-470
— some declines, Victoria Province II-668
— southern Spain, over-exploited I-173
— Tanzania II-88
Fish Stocks Agreement III-334, III-340, III-341, III-343
— *see also* Straddling Stocks Agreement
fisheries III-366–8
— Adriatic Sea I-277, **I-278**
— Aegean Sea, overfished I-244
— American Samoa, changing local fishing patterns II-770
— Andaman Islands, poor development of II-192, **II-193**
— Andaman and Nicobar Islands, deep-sea, development of recommended II-196
— Anguilla
    — artisanal I-621
    — Fisheries Protection Ordinance and Turtle Ordinance I-624
— Arabian Gulf II-8
— Arabian Sea II-25
    — artisanal II-26–7
    — industrial II-29
— Argentina
    — commercial I-763
    — industrial I-763–4
    — offshore, impact on fauna I-757–8
    — semi-industrial and artisanal I-763–4
— Australia II-586–7
    — Australian Fishing Zone II-586
    — commercial II-584
    — decline through overfishing II-587
— the Azores I-209–10
    — demersal and pelagic *I-209*, I-210
    — little regulation I-214
— the Bahamas
    — commercial and artisanal I-422, I-424–5
    — other resources I-426–7
    — protective regulations I-430–1
    — use of poisonous substances prohibited I-430
— Baltic I-111–12
— Bangladesh II-292, **II-293**
    — resources II-294
— Bay of Bengal
    — artisanal and industrial II-293
    — brackish water II-278
    — marine II-277–8
— Belize
    — artisanal I-508–9, I-511
    — regulation of I-513
— British Virgin Islands
    — commercial trap fishing I-621
    — moratorium on large-scale foreign fishing I-624
— Cambodia II-574
    — foreign poachers II-576
    — freshwater II-575
    — inland, Mekong River II-575
    — inshore, depleted II-576

— multi-fishing practices II-576
— Canary Islands I-195–6
   — cetacean fishery I-195
   — tuna I-196
— Carolinas I-362–3
   — fisheries monitoring programs I-363
   — viable, stressed and overfished I-363
— Chesapeake Bay, decline in I-343, I-346
— Chile I-710–12
   — cash values I-712
   — new management tools I-463
— co-management in III-159
— Colombian Caribbean Coast I-669–71
   — artisanal I-671–2
   — industrial I-670–1
   — related to upwellings in northeast I-670
   — *Tarpon atlanticus*, depleted population I-669–70
— Colombian Pacific Coast
   — artisanal rights protected I-681
   — commercial landings I-684
   — industrial I-683
   — industrial fishing denounced I-681
   — in the mangrove forests I-682
   — protective measures I-684
— the Comoros, artisanal II-249
— Coral, Solomon and Bismark Seas region II-440
   — problems best solved by community-based initiatives II-442
   — subsistence and artisanal II-434, II-436–7
   — tuna fishery II-432, *II-433*
— Côte d'Ivoire I-816
   — changes in I-815–16
   — demersal I-815
   — pelagic I-815
— declining yields III-377
— destructive II-374, II-387, II-438, II-566
   — the Comoros II-249
   — Indonesian Seas II-392–3
— destructive fishing techniques II-377, II-503, III-37, III-122–3
   — the Philippines II-415
   — *see also* cyanide; dynamite fishing
— discards by ocean region **III-367**
— Dutch Antilles I-602–4
— east coast, Peninsular Malaysia II-357
   — artisanal II-352–3
   — main pelagic species II-352
— eastern Australian region
   — commercial II-640–1
   — commercial and recreational II-637
— Ebrié lagoon, Côte d'Ivoire I-813
— effects on ecosystems III-117–33
— El Salvador I-550, I-552–3
   — affected by oil spills I-556
   — artisanal I-553
— English Channel I-73–4
   — impacts arising from fishing I-76–7
— Faroes I-35–6
   — collapse then increase I-35
— Fiji
   — commercial catch data II-760
   — commercial/artisanal II-715–16
   — overfishing II-760
   — overfishing of marine invertebrates II-759
   — subsistence, impacts of II-758
— and the Fish Stocks Agreement III-341
— French Polynesia
   — lagoonal species II-819
   — subsistence and commercial II-819–20
— Godavari-Krishna delta II-170–1
— Great Australian Bight
   — commercial II-682–3, II-684
   — inshore commercial fishing II-682–3

— Great Barrier Reef region
   — commercial II-623
   — live food-fish export II-623
   — recreational II-621
— Greenland I-8
— Guangdong
   — closed fishing seasons II-560
   — outstripping sustainable capacity II-557–8
— Guinea
   — artisanal I-800, I-801
   — Fish Smoking Center I-800
   — industrial I-800–1
   — National Fishing Supervisions Center I-800
   — pelagic I-800
   — traditional I-800
— Gulf of Aden II-56–7
   — artisanal II-56
   — industrial, illegal and uncontrolled II-59
— Gulf of Guinea
   — artisanal I-787
   — illegal fishing I-787
   — industrial I-787–8
— Gulf of Maine and Georges Bank I-312–13
— Gulf of Mannar II-164–5
   — indiscriminate use of small nets II-165
— Gulf of Thailand II-305–6
   — conflict between subsistence and light-luring fishermen II-306
   — decline in the resource II-305
   — dominant demersal fish groups II-305
   — dominant pelagic fish groups II-306
   — mortalities of shellfish and finfish III-218–19
   — multi-gear, multi-species II-305
— Hawaiian Islands
   — commercial II-802–3
   — derelict fishing gear II-804
   — resistance to licensing/monitoring II-809–10
   — subsistence, artisanal and recreational II-802
— Hong Kong II-538
   — seriously affected by urban development II-543
— impact on seabirds III-111
— impact on small cetaceans III-97–8
— Indonesian II-391–4
   — artisanal II-391, II-394
   — commercial exploitation of offshore fisheries II-391
   — high fishing pressure on major stocks II-391
   — reef fisheries II-393
— industrial
   — effects of III-112
   — sequential overfishing III-120–1
— intensive fisheries management III-367
— Irish Sea I-91–2
   — artisanal I-91, I-93
   — comparison with North Sea I-92, **I-92**
— issues along the Black Sea Coast I-292, **I-293**
   — need for cooperative action I-292
— and its problems I-3
— Jamaica
   — management plans I-572
   — reef I-566, I-568
— Japan, drive fisheries, small cetaceans III-93
— Java Sea, pelagic catch II-391
— Lakshadweep Islands II-194
— large incidental capture of sea turtles III-66
— late development of, Somalia II-80
— Lesser Antilles **I-633**, I-636–7
   — artisanal methods can damage marine habitat I-635
   — ongoing assessment I-634
   — overfishing of nearshore waters I-633
— long-lived species replaced by shorter-lived species III-367
— Madagascar
   — artisanal II-121, II-122–3, II-125

— commercial II-121–2
— offshore, estimated potential 120, **II-120**
— Malacca Strait, mainly artisanal II-338
— Malacca Straits, capture fisheries and coastal mariculture II-314
— the Maldives II-210–12
 — coral reef fisheries II-211–12
 — lagoonal bait fisheries II-210–11
— management for multispecies, marine harvest refugia I-244–5
— Marshall Islands II-783–4
 — artisanal II-784
 — maximum sustainable yield not yet reached II-783–4
 — pelagic fisheries II-784
— Mascarene Plateau
 — artisanal II-261–2
 — banks fishery II-256
 — effects of artisanal fisheries II-262
 — potential for longline and deep-water fishery II-260
 — some foreign tuna fishery II-260
— may negatively affect survival of marine mammals and birds III-128–9
— Mediterranean, Israeli I-257
— Mexican Pacific coast I-489–91
 — artisanal I-490–1, I-491
 — benthic invertebrates I-489–90
 — foreign fleets I-490
— Montserrat I-621–2
— Mozambique II-105, II-107
 — artisanal II-107
— New Caledonia II-728
 — artisanal II-729–31
 — Beche-de-Mer and Trochus exploitation II-730
 — professional II-729–30
 — subsistence reef fisheries II-715, II-729–30
— New York Bight
 — "foreign fishing" I-330
 — Magnuson Act I-330
 — recreational I-323, I-329
— Nicaraguan Caribbean coast I-522, I-523–5
 — automatic jigging machines I-523
— Nicaraguan Pacific coast I-534
 — artisanal fishing I-537
 — big pelagic fish I-536
 — demersal fish I-536
 — economically important marine resources **I-536**
 — estimated biomass **I-536**
 — industrial I-539
— North Sea I-51–2
 — cockle fishery I-51–2
 — detritus from catch-processor ships III-127
 — shellfish I-51
— North Spanish coast
 — artisanal I-144
 — demersal I-138
 — Galicia I-142
 — negative impacts I-146
 — pelagic I-141
— northern Gulf of Mexico I-445
 — threatened by effects of navigation channels I-447–8
— Norwegian coast I-25–6
— Oman, artisanal II-21, II-26–7
— Palk Bay II-167
— Peru
 — artisanal I-693, *I-694*
 — demersal I-693
 — pelagic I-693
— the Philippines
 — coastal, affected by loss of mangroves II-414
 — intrusion of commercial fishers into municipal waters II-416
 — offshore II-411
— processing waste III-137
— purse-seine
 — by-catch from different methods III-124, *III-124*
 — dolphin mortality III-124
 — reduction in dolphin mortality III-141
 — tuna mortality, Eastern Tropical Pacific III-94
 — tuna, Pacific, incidental take III-379
— recreational I-323, I-329, I-362, I-397, I-430–1
— Red Sea II-42
 — artisanal II-43
 — reef fisheries II-43
— rejects and marketable catch III-137
— Sabah, reef, overfished II-372
— Samoa II-714
 — inshore II-716
— Sarawak II-372
— Sea of Okhotsk
 — commercial, ecosystem effects II-469–70
 — intensive II-470
— selective fishing III-136
 — may result in ecosystem imbalance III-139
 — Norway III-138
— the Seychelles
 — artisanal II-238
 — foreign vessels charged for fishing II-238
 — reef fisheries II-236
— small-scale, more important than large-scale commercial II-140
— South Africa
 — artisanal and subsistence II-139–41
 — commercial II-136–8
 — optimal utilisation of catch II-138
 — recreational angling II-138–9
 — target switching common II-137
— South Western Pacific Islands, Industrial II-718–19
— southern Brazil I-737
 — artisanal I-737, I-739
 — industrial I-739
— Southern California Bight
 — commercial I-396–7
 — recreational I-397
— Southern Gulf of Mexico I-475
 — artisanal I-475–6
 — species and production **I-476**
— southern Portugal
 — artisanal, Tagus and Sado estuaries I-161
 — coastal waters I-159
— southern Spain I-172
 — major economic activity, Gulf of Cadiz I-172
 — overexploited I-175
— southwestern Africa I-830, I-833
 — commercial trawling I-830, **I-830**
 — hake I-833–4
 — purse seiner catches *I-831*
 — secondary effects of fishing I-833
— Spratly Islands, pressure on recently increased II-364
— Sri Lanka
 — chank and sea cucumber II-183
 — coastal II-182
 — marine II-181–2
 — marine ornamental II-182–3
— Tagus and Sado estuaries I-159
— Taiwan Strait II-503
 — artisanal fishing II-506
 — "bull-ard" fishery II-504–5
 — coral reef II-504
 — decline in fishermen II-505
 — seasonal grey mullet fishery II-503–4
— Tanzania
 — artisanal II-90–1
 — commercial II-90
— target and non-target catches III-137
— Tasmania II-654
 — commercial II-655–6

— individual transferable quotas (ITQs) II-656
— offshore II-653
— Tonga, subsistence reef fisheries II-716
— Torres Strait and Gulf of Papua
    — commercial II-604
    — subsistence and artisanal II-601–3
— tropical seas, many low-value species in catches II-372
— Tyrrhenian Sea I-279
— Vanuatu
    — artisanal II-715
    — commercial II-744
    — community-based management schemes II-745–6, **II-746**
    — impacts of II-743–4
    — major area for development II-744
    — undeveloped II-744
— Venezuela I-653–4
    — worker conflicts, artisanal and trawl fisheries I-653
— Victoria Province
    — commercial II-667–8
    — specialised bay and inlet fishery II-667–8
— Western Australia, commercial II-697–9
— Xiamen region, capture fisheries II-519
— Yemen, artisanal II-56
— Yuzhnoprimorsky region II-480
— *see also* by-catch; discards; recreational fishing
fisheries agreements, regional III-340
fisheries law, global III-158–9
— international law, based on UNCLOS III-158–9
    — national approaches to III-158
— regional agreements III-158
fisheries management
— for artisanal and subsistence fisheries, South Africa II-140–1
— based on ecosystem principles III-377
— binational I-463
— and closed areas III-378
— designed for fish conservation III-157
— emergence of "epistemic community" in III-163
— emerging global fisheries laws III-158–9
— exploitation rate, implementation error in III-389
— Great Barrier Reef region, evaluation of II-624
— Hawaiian Islands II-803–4
    — need for restrictions II-810
— incentive structures III-155–6
— issues related to enforcement III-160–1
    — blacklisting III-160–1
    — flag state responsibility III-161
— issues related to legitimacy of III-155, III-157
    — substantive and process legitimacy III-155
— issues related to surveillance III-159–60
    — observer programs III-160
— limits to exploitation and target exploitation rates III-382
— Malacca Strait, little information to base sustainable use on II-341
— the Maldives, for single species fisheries II-211–12
— ocean dynamics and marine habitat changes becoming important III-185–6
— participation in III-159
— past, present and future, southwestern Africa I-836–7
— precautionary approach III-79
— social aspects of management issues III-161–2
    — restrictions on effort III-161
    — restrictions on technique III-161
— as a social science problem III-153–64
    — equity and fairness III-157–8
    — "governance" and "institutional embeddedness" III-155
    — scientific realism III-156–7
— Sri Lanka **II-186**
— Torres Strait and the Gulf of Papua II-605–7
Fisheries and Oceans Canada, view of passage of El Niño signal, Northeast Pacific III-184–5
fisheries regulations, Madagascar
— conservation of valuable species II-126

— management and policies II-126
— new local control law (GELOSE) II-126
fisheries resources, New Caledonia, not safe from certain threats II-731
fisheries science
— and Fishers' insights III-157
— a mandated science III-156
fisheries scientists I-2
Fisheries Sector Program (FSP), the Philippines II-418–19
fisheries—ecosystem interaction model III-118–19, *III-118*
Fishery Conservation and Management Act (USA) I-451
fishery systems, sources of uncertainty in analysis of III-389
fishing
— as a business III-155
— ecosystems effects of III-378–80
    — by-catch and incidental take III-379
    — debris and ghost fishing III-380
    — habitat destruction III-379–80
— illegal, "patterned deviant behaviour" III-155
— regulatory discards seen as perverse III-156
fishing discards *see* discards
fishing fleet, global, overlarge III-155
fishing fleets, influencing government over international laws III-158
fishing gear
— cause of seabird and marine mammal mortalities II-656
— derelict, dangerous nuisance II-804
— destructive II-27
— subsistence fishing II-436
fishing mortality
— directed, of target organisms III-118–22
    — density-dependent responses III-121
    — loss of genetic diversity III-122
    — overfishing III-119
    — population collapses III-119–21
    — population size III-118–19
    — size-selective fishing III-121–2
— indiscriminate III-122–6
    — by-catch and discard mortality III-123–5
    — caused by lost gear, ghost fishing III-125–6
    — from physical impacts and destructive practices III-122–3
— natural, indiscriminate changes in III-126–9
    — changes mediated by biological interactions: competition and predation III-127–9
    — changes mediated by dumped 'food subsidies' III-127
    — changes mediated by habitat degradation III-126
— underestimation of, and population depletion III-126
fishing practices
— concerns about direct and indirect effects of on marine ecosystems III-352
— destructive I-365, I-366, I-554
fishing pressure, above optimum, Irish Sea I-92
Fishing Reserves, Mascarene region II-265, **II-265**
fishing techniques
— controls on economically inefficient III-161
— potential for environmental impacts II-29
fishmeal, Chile
— effect on coastal zone I-713–14
— environmental impacts of I-713–14
— product of fishing industry I-710, I-712
fjord circulation
— creates environmentally significant gradients I-19
— two-layer flows I-19
fjord complexes, Greenland I-8
fjords
— as contaminant traps I-19
— described I-19
— Norwegian coast I-17, I-19
    — changes in shallow water communities I-22
    — deep, contain arctic bottom fauna I-20
    — deep water circulation I-19
    — main fluvial input at head I-19

— southern, stagnant bottom water I-19
— west coast, large and deep I-19
flash floods II-23
flatback turtles III-63–4
— Great Barrier Reef II-618
— restricted range III-64
— a vulnerable species III-64
Fleet lagoon *I-66*, I-68
Flinders Current *II-581*, *II-674*, II-676
Flinders Region, Victoria Province II-663
Flindersian Province
— southern Australia, marine fish fauna II-679
— Western Australia II-694
flood and coastal hazard maps, Hawaiian Islands II-805
flood risk, and rising sea level III-190–4
flood warning system, Firth of Clyde III-352
Flores Sea *II-382*
— atolls, reefs damaged by destructive fishing practices II-389
Florida
— marine reserves, key design principles III-390
— use of zoning III-389–90
Florida arm, Gulf Stream I-589
Florida Bay *I-406*, I-407, *I-436*
— effects of cyanobacteria III-218
— influx of freshwater, nutrients and sediments I-409
— seagrass die-off
— shown by SPOT XS imagery III-286
— a trend for coastal waters in the new Millennium? III-14–16
— western, importance of to Florida Keys reef tract III-16
Florida Current I-407
Florida Escarpment *I-436*
Florida Keys I-405–14, *I-436*
— agriculture and urban factors I-409
— area of sediment influx I-409
— coral bleaching and mortality events III-56
— coral vitality: long-term study of Carysfort reef I-410–12
— Florida Keys National Marine Sanctuary and Protection Act I-412
— global stresses I-409–10
— grim picture for the future of the reefs I-413
— industrial stresses I-409
— islands composed of calcium carbonate rock I-408
— localized ecological reef stress I-410
— major shallow water, marine and coastal habitats I-407–8
— populations affecting the area I-408
— protective measures I-412–13
— multi-use zoning concept I-413
— stress to the environment and reefs I-408–9
— USEPA coral reef monitoring project I-412
Florida Keys Coral Reef Disease Study I-412
Florida Keys National Marine Sanctuary I-408, I-409, I-452
Florida Keys Reef Ecosystem: Timeline **I-413**
Florida Loop Current I-437, I-439
— carries effects of Mississippi floods I-409, *I-410*
Florida Straits *I-416*, *I-436*, I-437, I-439
Flower Garden Banks *I-436*
— marine sanctuary I-442, I-452
— no real coral bleaching III-56
flushing time, North Sea I-46
fly ash disposal, China II-495
fog and cloud, Peruvian coast I-690
Fonseca, Gulf of *I-532*, I-537–8, *I-546*
— coastal diving birds I-549
— industrial shrimp farming I-538
— over-exploitation of resources I-537–8
Food and Agricultural Organisation *see* FAO
food chains
— energy transfer by zooplankton I-359
— marine, Greenland, heavy metals in I-9
food pollution, Asian developing countries and OCs in fish II-450
food webs, detrital III-3
forest clearance *see* deforestation

forestry
— Australia, environmental considerations II-584
— northern Spanish coastal area, causing soil loss I-142–3
forests
— Chesapeake Bay catchments I-345
— El Salvador I-547
— Gulf coastal plain, USA, conversion to farmland I-444
— Nicaragua I-519, I-520
— degradation of I-539
Forth Estuary Forum
— economic appraisal of the Forth Estuary III-351
— economic development a 'Flagship Project' III-351
Forth Kuril Strait *II-464*, II-465
Frailes Canyon I-487
Fram Strait *I-6*
Framework Convention on Climate Change III-340
France
— influence in French Polynesia II-823–4
— La Rance tidal energy project III-304, III-316
— nitrogen inputs to the English Channel and North Sea I-74
— offshore wind farm planned III-309
— prohibition of organotin-based paints on small boats III-251
franciscana (La Plata river dolphin) III-90
Fraser Island II-634
Freez Strait *II-464*, II-465
freezing, northern Gulf of Mexico, causes mass mortalities I-439
French Polynesia II-813–26
— effects of human activities II-820–3
— exploitation of resources II-819–20
— islands described II-815
— major shallow water marine and coastal habitats II-817–19
— offshore systems II-817
— population and politics II-815
— protective measures II-823–4
— Marine Area Management Plans II-824
— reef restoration schemes II-824
— seasonality, currents, natural environmental variables II-815–17
fresh water, lacking, Namibia I-835
freshet (spring flow), Chesapeake Bay I-340
freshwater
— a critical resource
— the Comoros II-251
— the Mascarene Region II-264–5
— dilution of, Hong Kong coastal waters II-537
— a future problem
— Marshall Islands II-780, II-781
— New Caledonia II-729
— reduced in estuaries, Victoria Province, Australia II-666–7
— resources scarce, Xiamen region II-518
freshwater coral kills, Hawaiian Islands II-807
freshwater flushing
— important for water bodies with poor circulation I-449–50
— rate a critical parameter in northern Gulf of Mexico estuaries I-447
freshwater influx
— affects seasonal patterns, Yellow Sea II-489
— Florida Bay I-409
— Gulf of Alaska I-378
— northern Gulf of Mexico I-437, I-443
— deprived of most of spring runoff I-448
— to Cambodian Sea II-571
freshwater lens, Belize continental shelf I-503
freshwater supplies, the Bahamas I-429
freshwater systems, contamination by nutrients and pesticides I-539
freshwater turtles, east coast, Peninsular Malaysia II-350
fringing reefs II-277
— Andaman and Nicobar Islands II-191, *II-192*
— Borneo II-370
— Brazil I-723
— Colombian Pacific Coast, Gorgona Island I-680
— Comoros Islands II-246
— many threats to II-248

— east coast, Peninsular Malaysia II-350
— Fiji Islands II-756
— French Polynesia II-817, *II-818*
— inner and outer, Tanzania II-85–6
— islands in Torres Strait II-597
— Madagascar II-118
— mainland Belize I-505–6
— Mauritius II-258
— the Philippines II-408
— Puerto Rico I-582
— San Andrés and Providencia Archipelago I-668
— the Seychelles II-236
— western Indonesia II-389
frontogenesis, subtropical I-223
Fucus I-70
— Baltic Sea, decreased through eutrophication I-130–1
— lead and zinc in, near major Greenland mines *I-12*
— response to oil spills III-273
Fucus serratus, scarcity of in Faroes I-33
Fucus vesiculosus
— community changes
   — disappearances associated with fish/fisheries decline I-131
   — southern Baltic I-110
— hard bottoms, Baltic Sea I-125
Fujeirah *II-18*, II-19
— construction of oily waste reception facilities II-29
Fuma Island, Bajuni Islands, sandy tail *II-75*, II-76
Fundy, Bay of *I-308*, I-311, *I-311*
— cetacean–gillnet interaction III-144–5
fur seals, Patagonian coast I-756

Galapagos Islands, coral bleaching and mortality events III-53
Galicia *I-136*, I-137
— dense coastal population I-142
Ganges River *II-146*
— mouths of *II-286*
gap-fraction analyses, used with remote sensing III-286
garbage disposal, poor, Sri Lanka II-183
garfish, Baltic Sea I-111
gases, solubility of in water, and removal by wet deposition III-200
Gdansk Basin I-101, I-106
Gdansk Deep I-101
— cyclic behaviour, phosphate and nitrate concentration I-108
Gdansk, Gulf of, anthropogenic heavy metals in sediments I-115
Gelidium sesquipedale beds
— Cantabrian coast I-140
— Portuguese coast I-159
Gene Bank of Fishing and Aquarian Resources, Colombia I-685
genetic diversity, loss of in exploited fish populations III-121, III-122
Geneva Convention, addresses military threats to the marine environment III-342
geological features, South Western Pacific Islands region II-712
geology, Gulf of Thailand II-299
Georges Bank
— closed areas III-386
   — as larval source sites III-381, *III-382*
— and "cold pool" I-324
— nontidal surface circulation *I-324*
— primary production over I-310
Georges Bank (R. Backus and D. Bourne) I-309
Georges Basin I-309, *I-311*
Geosphere Biosphere Programme (IGBP), scientific definition of coastal zone III-295
German Bight
— cadmium budget I-54
— effects of oxygen depletion I-53
— organic pollutants in I-54
Ghana, mangroves I-779
ghost fishing III-380
— and fishing mortality III-125
giant clam culture, Marshall Islands II-781, II-782

giant clam fishery, Fiji, overexploited II-759
giant clam hatchery, South Sulawesi II-401
giant clams, endangered, the Philippines II-410
Gibraltar Strait *I-168*, I-169, I-172
gillnets
— and cetaceans III-144–5
— coastal
   — give high small cetacean by-catch III-94–5
   — and seabirds III-146
— drifting/driftnets, high incidental catches III-124–5
— and ghost fishing III-125
gillnetting
— by-catch from II-57
— a danger to dugongs and dolphins II-180
— and decline in Australian dugongs II-587
— prohibited in Gulf of Alaska I-383
— prohibited in Southern California I-397
— Western Australia II-697
Gippsland Lakes
— Victoria
   — adjusting to changed environment II-667
   — affected by high mercury levels II-669
glaciation, Antarctic I-706
glass-eels I-157
Global Climate Change (GCC) models, predict increase in extreme events III-47
global codification, learning-based approach to III-346–7
global commons, and sustainability III-360
Global Coral Reef Monitoring Network II-435
Global Ocean Observation System I-56
Global Plan of Action (GPA) III-344, III-345
global stresses, Florida Keys I-409–10
global warming
— an abdication of intergenerational responsibilities III-370
— potential effects on oceans III-98
— and range extension I-88
— response to climate change III-181–2
— and whale distribution III-86–7
Global Waste Survey III-339, III-345
— waste connectivity III-338–9
Glovers Atoll (Reef) *I-502*, I-504–5
— deep reef habitats I-506
GNP versus sustainable welfare III-395
GOA *see* Alaska, Gulf of
Godavari and Krishna deltaic coast II-170–3
— biodiversity II-170
— climate and coastal hydrography II-170
— environmental degradation II-171–2
— fish and fisheries II-170–1
— geological features II-170
— impacts of human activities on the ecosystem II-172
— mangrove ecosystem II-170
— suggestions and recommendations II-172–3
Golfe Normano-Breton, gradient in zooplankton type and biomass I-70
Gorgona Island National Park, Colombia I-681, I-684
Gotland Basin I-101, I-106
Gotland Deep *I-101*, I-102
gradient techniques, dry deposition measurement III-203
Grand Comore (Ngazidja/Ngazidia)
— establishment of a coelacanth park II-251
— small reefs and restricted seagrass beds II-245
— a volcanic island II-245
Grand Turk *I-588*, I-590
— tourists I-591
gravel habitats, sensitive to disturbance III-123
Great Australian Bight II-673–90
— coast of geological significance II-678
— coastal erosion and landfill II-684
— effects from urban and industrial activities II-684–6
— major shallow water marine and coastal habitats II-678–81

— need for greater research and conservation management II-686
— offshore systems II-681
— populations affecting the area II-682
— protective measures II-686–8
    — Great Australian Bight Marine Park II-686–8
    — research II-688
— rural factors II-682–4
— seasonality, currents, natural environmental variables II-675–7
— a single demersal biotone II-675
— Yalata Aboriginal Land Lease II-682
Great Australian Bight Marine Park II-686–8
— Benthic Protection Area II-687
— Commonwealth Marine Park II-687
— Marine Mammal Protection Area II-687
— potential pressures on marine conservation **II-688**
— Whale Sanctuary, Head of Bight II-686
Great Australian Bight Trawl Fishery II-682, II-683
Great Bahama Bank *I-416*, I-421
Great Barrier Reef II-595, *II-612*, III-41
— degradation of III-37
Great Barrier Reef Marine Park II-595
— aboriginal interests in II-620
— does not fisheries management II-624
— a model in large marine ecosystem conservation II-626
— protective measures II-624–7
    — biodiversity conservation II-625–6
    — evaluation of the management II-626–7
    — fisheries management II-626
    — human and financial resources II-624–5
    — impact of tourism II-626
    — management framework II-624
    — management mechanisms II-625
    — shipping and oil spills II-626
    — water quality II-626
— tourism II-622
Great Barrier Reef Marine Park Act II-620
Great Chagos Bank *II-221*
— peat on Eagle Island II-229
— seismic tremors II-225–6
— world's largest atoll II-223
Great Corn Island, volcanic with reefs I-522
Great Lakes Fisheries Commission III-154
Great South Channel I-309, *I-311*
Great Whirl *II-18*, *II-48*, II-49, II-50, II-66
— high speed of currents II-54
Greater Antilles, coral bleaching and mortality events III-54, III-56
Greece
— protected areas I-250
— turtle protection not carried through III-355
green turtle nesting sites, Hawaiian Islands II-799
green turtles I-780, II-54, II-180, II-250, II-289, II-336, II-350, II-598, II-779, II-784
— an endangered species III-62
— Chagos Islands II-229
— death through exposure to okadaic acid III-219
— "edible turtle" III-62
— fibropapilloma incidence on, and turtle mortality III-227
— Fiji, endangered II-757
— Great Barrier Reef II-618
— growth slow III-62
— Hawaiian Islands, with fibropapillomatosis II-799
— The Maldives II-205
green urchins I-312
greenhouse gases
— atmospheric III-188
— commitments to reductions in III-304
greenhouse scenarios, for Australia II-585
greenhouse warming III-369
Greenland icecap, increase in Pb, Cd, Zn and Cu in snow and ice cores I-9
Greenland Sea *I-6*, I-7

Greenland seas I-5–16
— future threats for the marine environment I-14–15
— oceanography I-7–8
    — bedrock shorelines I-8
    — ocean surface circulation *I-7*
— pollutant sources I-11–14
— population and marine resources I-8–9
— present contaminant levels I-9–11
Greenland—Scotland Ridge, acts as partial barrier I-33
Grenada *I-644*
Grenadines, whaling I-636
grey seals I-49–50, III-95
— Cornwall and Scilly Isles I-70
— Irish Sea I-91
grey whales III-74, III-76
— census of III-79
— numbers of III-81
— for subsistence III-83
Grotius, freedom of the oceans concept (1609) I-2
groundings, shipping, South Western Pacific Islands II-718
groundwater
— the Azores I-207
— the Bahamas, vulnerable to contamination I-429
— pollution of, Borneo II-376
groundwater contamination, Nicaraguan Pacific coast I-539
groundwater extraction, Taiwan, causing subsidence and saline intrusion II-507, II-509
groundwater pollution
— Gulf of Cadiz I-174
— round Sea of Japan II-477
grouper fishery, the Maldives II-211–12
growth rates, seagrasses III-2
Guadalcanal Island II-427
— effects of earthquakes II-429
— small-scale aquaculture trial II-438
Guadalquivir River *I-168*
— estuary fishing I-175
— regulation of fishing I-183
Guadimar River *I-168*
— spillage from the Aznalcóllar mining disaster I-178
guano mining I-829
Guatemala, focus on an ecosystem framework for planning and management I-463–4
Guban coastal plain *II-64*
Guinea I-797–803
— continental shelf I-799
— fishing benefits from its EEZ I-801
— major shallow water marine and coastal habitats I-799–800
    — main zones I-799
— offshore systems I-800–1
— populations affecting the area I-801–2
— protective measures I-802
    — National Center of Supervision and Fishing Protection I-802
    — party to London Convention I-802
— seasonality, currents, natural environmental variables I-799
Guinea Current *I-774*, I-778, I-781, I-807–8
Guinea, Gulf of I-773–96
— coastal erosion and landfill I-786–7
— defined region I-775
    — geomorphic features of continental shelf I-775
    — LME divided into subsystems I-775
— effects from urban and industrial activities I-787–93
— islands I-781
— major shallow water marine and coastal habitats I-779–81
— offshore systems I-781–4
— populations affecting the area I-784
— protective measures I-793–4
    — Gulf of Guinea LME projects I-793–4
    — marine and environmental legislation I-793
    — protected areas I-793
— rural factors I-784–6

— seasonality, currents, natural environmental variables I-775–9
Guinea Under Current *I-774*, I-778, I-781
Guinea waters I-777
Gulf of Aqaba
— continuation of the Red Sea rift II-37
— sea-floor spreading II-37
Gulf Area Oil Companies Mutual Aid Organisation II-29
Gulf of California Coastal Province, Mexico I-487–8
— habitats I-487–8
Gulf Coast Fisheries Management Council I-525
Gulf Coast, USA, habitats of I-440
Gulf of Guinea LME projects I-793–4
Gulf of Maine Point Source Inventory I-316
Gulf of Mexico Aquatic Mortality Network III-228
— multi-jurisdictional monitoring III-229, III-231
Gulf of Mexico Fishery Management Council I-451
Gulf of Mexico states (USA), regulations addressing habitat degradation I-452
Gulf Stream I-189, I-204, *I-222*, I-223, I-227, I-356, I-362, I-419
— meanders, off the Carolinas coast I-353
— transient effects I-324
Gulf War
— oil spills II-11–13
  — oil penetration aided by burrowing animals III-271
Gush Dan outfall, sewage sludge I-259, I-261
Guyana Current I-630, I-632
Gyrodinium aureolum blooms I-69, I-90

habitat changes, small, dramatic effects on whales III-87
habitat degradation III-130
— by fishing methods, effects of III-126
— and change, greatest effects on inshore small cetaceans III-96–7
— Coral, Solomon and Bismark Seas region, localised concerns II-437–8
— leads to decline in fish populations III-126
— Peruvian coast I-694–5
habitat destruction III-379–80
habitat diversity, northern Gulf of Mexico I-438
habitat infilling, North Sea coasts I-54
habitat loss
— American Samoa II-770
— coastal, through new building I-173
— east coast, Peninsular Malaysia II-353
habitat modification, can damage sea turtle populations III-66
habitats I-3
— restoration, Chesapeake Bay I-347
HABS *see* harmful algal blooms (HABs)
haddock I-35, I-49, I-313
Hadramout *II-18*
Haifa *I-254*
Haifa Bay I-255
— coastal zone outside I-261
— heavy metal monitoring I-262
— land-based pollution sources I-259–61
    — nutrients I-259–60
    — toxic metal and organic pollutants I-260–1
Hainan Current II-537, II-538
Hainan Island *II-550*, II-551
Hajar Mountains, Oman *II-18*, II-21
hake I-313
— Argentine Sea I-757
— in Peruvian fisheries I-693
— silver, contaminants in I-315
— southwestern Africa I-833, *I-833*–4
hake fishery, Argentina I-763–4
Halaniyat Islands *II-18*
— some coral cover II-22
halibut fisheries, individual quota system, US Pacific Northwest III-155
haline stratification
— Baie de Seine I-67

— off Lancashire–Cumbrian coasts I-86
haling, catch per unit whaling effort (CPUE) indices III-79
halocline, Black Sea, generation of I-288–9
halogenated hydrocarbons (HHCs) III-95
harbo(u)r porpoises I-50, III-90
— by-catch in the North Atlantic III-94
— English Channel I-70
— Irish Sea I-90
harbour seals I-49–50
harbours
— and marinas, accumulation of TBT in sediments III-250
— TBT contamination not reduced III-254
hard bottom habitats
— Norwegian coast I-21
    — gradients of change I-22
— Southern California Bight I-388
— Tyrrhenian Sea I-273
hard bottoms I-442
hard substrate communities, Bahamas Archipelago **I-420**
hard-bottom outcrops, Carolinas coast I-353, I-362
hard-bottom/coral reef communities, Bahamas Archipelago **I-420**
hardgrounds, Puerto Rican shelf I-581–2
hardshores, the Azores I-205, *I-205*
— intertidal zonation pattern I-205, *I-206*
harmful algal blooms (HABs) III-98, III-213, III-298
— cause of mass lethal mortality disturbances III-223
— as indirect cause of morbidity and mortality III-218
— profit from wetland and mangrove destruction III-221–2
harp seals
— effect of commercial fisheries on III-97
— hunted in Greenland I-8
Hawai'i
— marine reserves III-386
— no-fishing 'kapu' zones III-386
Hawai'i Coastal Zone Management Plan II-808
Hawaiian Islands National Wildlife Refuge II-798
Hawaiian Islands (USA) II-791–812
— challenges for the new millennium II-809–10
    — better management in the northwestern islands II-810
    — community-based coastal area management II-809
    — controls over fishing II-810
    — preservation of natural tourism amenities II-810
    — public educations II-809–10
— coastal uses II-802–4
— geological origin of II-793–4
— islands listed **II-794**
— major shallow-water marine and coastal habitats II-795–800
— offshore systems II-800–1
— populations affecting the area II-801
— protective measures II-807–9
    — development plans II-808
    — environmental impact assessment II-808
    — environmental legislation II-807–8
    — marine environmental restoration II-809
    — marine protected areas II-808–9
— region defined II-793–4
— rural factors II-801–2
— seasonality, currents, natural environmental variables II-794
— urbanization factors II-804–7
Hawaiian Ocean Resources Management Plan II-808
hawksbill turtle I-780, II-205, II-229, II-250, II-350, II-779, II-784, II-799, III-62–3
— critically endangered III-62–3
— endangered, Fiji II-757
— Great Barrier Reef II-618
— harvested for tortoise shell III-62
hazardous substances, targets planned by North Sea states III-362–3
hazardous waste
— Marshall Islands, nuclear testing II-783
— shipments of II-320
HCHs II-545

— air and surface seawater, worldwide II-454, *II-456*
— in the English Channel I-75
heat flux, across Line-P, Northeast Pacific III-184
heavy metal contamination
— Dutch Antilles I-605
— riverine and coastal environment, western Indonesia II-396
— West Guangdong and Pearl River delta II-556, **II-556**
heavy metal mining I-738
— Australia II-639, II-641
heavy metal pollution I-279
— Amursky Bay, eastern Russia II-484
— Anzoategui State and Cariaco Gulf, Venezuela I-656
— from mining, Greenland I-11–12
— Johore Strait II-335
— Liverpool Bay I-95
— Malacca Straits II-318
— Morrocoy National Park I-657–8, **I-657**
— northern Spanish coast I-145, **I-146**
— Odiel saltmarshes I-170–1
— Pago Pago harbour, American Samoa II-712
— southern Brazil
  — Guanabara Bay I-741
  — and mangroves I-742
  — Sepetiba Bay I-741, I-741–2
— Tanzania II-92
— west Taiwan, and "green oysters" II-509–10
heavy metals
— affecting small cetaceans III-95
— Arabian Gulf
  — in sediments II-12
  — in sewage II-10
— Australia II-588
— Bay of Bengal
  — in cultured prawn tissue II-155
  — and estuaries II-155
— Bothnian Bay and Bothnian Sea I-131
— coastal and river waters, Gulf of Cadiz **I-180**
— dune mining of, KwaZulu-Natal II-142
— found in coral skeletons II-390
— from pesticides and fungicides, in shellfish I-493
— Great Barrier Reef region II-623
— in harbour muds, Belize City I-511
— in the Hoogly (Hugli) river II-278
— incorporated into estuarine trophic web I-179
— long-range airborne transport of I-115
— Norway I-24–5
— off west coast, Peninsular Malaysia II-341
— offshore, Guinea I-802, **I-802**
— one-hop contaminants I-13
— Patagonian coast I-765
  — localised I-761
— a potential problem, Gulf of Guinea I-792–3
— in sediments, Teacapán—Agua Brava lagoon system I-495, *I-495*
— in shrimps I-493, I-494
— in surficial sediments, Gulf of Thailand II-305, **II-305**
— Suva Harbour, Fiji II-719
Hector's dolphin III-91
— incidental mortality III-94
HEED (Health Ecological and Economic Dimensions) approach III-214
HELCOM Convention, on Baltic Marine Environmental Protection strategy I-132
Helsinki Conventions on the Protection of the Marine Environment of the Baltic Sea Area III-339
herpesvirus, ducks, geese and swans III-225
herring I-49, I-90, I-313
— Baltic Sea I-110, I-110–11, I-128
— Vistula Lagoon I-112
herring fishery, decline in Irish Sea I-91
hexachlorocyclohexane isomers *see* HCHs
Himalayas, and the Bay of Bengal II-287

Hiri Current II-619
Honduras, coral bleaching and mortality events III-54
Honduras, Gulf of *I-502*, I-503
Hong Kong II-535–47
— coastal erosion and landfill II-542–3
— coastline and coastal waters II-537
— effects from urban and industrial activities II-543–6
— Hong Kong Special Administrative Region II-537
— hydrography II-537
— Mai Po marshes II-539–41
— major shallow water marine and coastal habitats II-537–8
  — mixed subtropical fauna and flora II-537–8
— offshore systems II-538
— populations affecting the area II-538–42
— problems in Victoria Harbour II-542
— protective measures II-546
  — oil pollution control ordinances II-546
  — Water Control Zones II-546
— rural factors II-542
— Sewerage Master Plans II-546
Hoogly (Hugli) river, heavily polluted II-155, II-278–9
Hormuz, Straits of II-3, *II-18*, II-19
— corals in Musandam II-22
— important shipping lanes II-28
— inflowing water enriches the Gulf II-6
— water flow through II-4, *II-5*
horse mackerel I-141
— southwestern Africa I-832–3, *I-833*
horseshoe crabs
— captive breeding programmes, Sundarbans II-158–9
— potential source of bioactive substance II-159
hotspot sediments
— environmental implications of I-25
— Southern California Bight I-392
Houtman Abrolhos reefs, Western Australia II-696
Hudson Canyon *I-322*
Hudson Shelf Valley *I-322*, I-326, *I-326*
— up- and down-welling aids sediment transport I-323–4
Huelva *I-168*
— commercial port I-176
— urban-industrial development environmentally hazardous I-174
— waste dumped into the estuary I-179, **I-179**
Huizache y Caimanero lagoon system I-493–4
— fertilizers and agrochemicals from runoff found in shrimps I-494
— lagoon environmentally managed I-494
— marsh I-493–4
— shrimp culture I-494
human activities
— accelerating eutrophication, and its results III-198–9
— direct and indirect effects of seabirds III-111, III-114
— effects of, French Polynesia II-820–3
— increased spread rate of organisms and changed meaning of distance III-336–7
human breast milk
— contaminated with organochlorines
  — Hong Kong II-545
  — India and China II-452
Human Development Index, low, Papua New Guinea and the Solomon Islands II-434
human impacts, on coastal sediments III-352–3
human-assisted redistributions *see* alien/accidental/exotic/introduced organisms/species
Humber estuary I-54
— seasonal change between nutrient source and sink I-53
Humboldt Current *I-700*, I-701
— *see also* Peruvian Current
Humboldt Current ecosystem *see* south Pacific eastern margin ecosystem
humpback whales III-76
— Bermuda Islands I-226–7
— Hawaiian Islands II-800

— recovery of III-82
hurricane damage, and coral bleaching, Belize III-55
hurricanes I-340–1, I-408
— affecting Belize I-503–4
— Anguilla, British Virgin Islands and Montserrat, damage from I-622
— the Bahamas I-418–19
— cause ecological effects to Gulf of Mexico habitats I-439
— destructiveness of III-224
— dispersing anthropogenic rubbish in the marine habitat I-606
— Dutch Antilles, and oil slick movement I-605
— effects on Carolinas coast I-364
— El Salvador I-547
— Gulf of Mexico I-471, *II-470*
— intensity of increased I-631
— Jamaica I-561
    — damaging the coral reefs I-562, I-563, I-564
— most common path, Lesser Antilles *I-628*
— Nicaragua, Joan and Mitch **I-519**
— northeast Caribbean I-617
— Pacific Mexican coast I-487, I-488, I-489, I-493
— Puerto Rico I-578, I-579, I-580
— Turks and Caicos Islands I-589
— windward Dutch Antilles I-598
hydrocarbon pollution
— Australia II-588
— from tanker off Mozambique II-109
hydrocarbons
— enhanced values near refineries I-75
— surface marine sediments, Taiwan Strait II-510
hydroelectric power
— for heavy industry I-24
— regulation of river flow for I-19
hydrogen sulphide (H_2S) I-105–6, *I-105*, I-287
— in prawn ponds II-154
hydrography, Hong Kong II-537
hydrography and circulation, Gulf of Thailand
— forcing mechanisms in coastal seas II-300
— understanding of limited II-299–300
hydrologic cycle III-394
hydrology, Xiamen region, China II-515–16
hydrothermal vents, Guayamas basin I-487
hygroscopicity, of particles III-200
hyper-nutrification II-320
hypersaline conditions, south Texas and South Florida I-438
hypersalinisation, Magdalena Delta soils I-671
hyposalinity, Jiulongjiang estuary surface water II-518
hypoxia
— Black Sea I-201, I-303
— Florida Keys, near rivers and deltas I-408
— New York Bight I-325

ice cover
— Baltic Proper I-102
— in Chesapeake Bay I-340
— Greenland seas I-5
— winter, Sea of Okhotsk II-466
ice scouring, winter I-311
icefjords, Greenland I-8
— halibut fishing I-8
Iceland, ITQ system III-162
ICES *see* International Council for the Exploration of the Sea (ICES)
ICES
— Seabird Ecology working group III-112
— studies on ghost fishing III-125
— study group on bottom trawling III-126
ICZM *see* Integrated Coastal Zone Management (ICZM)
image data sources III-294–5
immigrants, behaviour of, Levantine basin I-257–8
IMO *see* International Maritime Organization (IMO)

imposex
— as an indicator of environmental contamination III-253
— causal link to TBT III-249
— dogwhelks I-25, I-36–7, I-55, I-56, I-158, III-250–1, *III-251*
— from tributyltin
    — Malacca Strait II-339
    — Singapore and Port Dickson II-319
— genetic aspects: the Dumpton Syndrome III-252
— occurrence in southwest England *III-251*
— whelks I-210–11
impoundments, lead to loss of wetland I-445–6, I-447
incentive structures, in fisheries management III-155–6
increased prevalence theory III-226
Index of Sustainable Economic Welfare (ISEW) III-395
India
— conservation policies for the Sundarbans II-157
— production of DDT and HCHs II-449
— Southeast II-161–73
    — Godavari and Krishna deltaic coast II-170–3
    — Gulf of Mannar II-163–6
    — Palk Bay II-166–8
    — Palk Bay-Madras Coast II-168–70
Indian Monsoon Current II-274
Indian NE Monsoon current *II-64*
Indian Ocean
— anticyclonic gyre in Mascarene region II-255
— biogeographic position of the Chagos Islands II-224–5
— Central, coral bleaching and mortality events III-48–9
— sever coral bleaching III-224
— Southern, coral bleaching and mortality events III-48
— western, oceanic platform II-235
Indian Ocean Commission (COI)
— joined by the Comoros II-247
— Madagascar a member II-127
— pilot ICZM operation, Mauritius and Reunion II-266
Indian Ocean Whale Sanctuary II-29, II-180
Indian—Australian plate margin, upthrust II-712
individual quota systems III-155, III-161
— best management tool for many fisheries III-162
individual transferable quotas (ITQs)
— ideal but complex and can cause problems III-162
— will lead to concentration of access rights III-162
Indo-Pacific beaked whale II-54
Indo-West Pacific Marine Province, Pan-Tethyan origin II-707
Indonesia
— coral bleaching and mortality events III-52
— Lembata Island, only subsistence traditional whaling operation III-84
— mangroves III-20
— marine conservation target II-401
— national management systems and legislations and the Malacca Straits II-321–3
— National Oil Spill Contingency Plan II-323
— national parks programme II-342
— northeastern, great species diversity II-389
Indonesia, Western, Continental Seas II-381–404
— land-based processes affecting Western Indonesian seas II-395–9
— major marine and coastal habitats II-385–90
— marine resource extraction II-390–5
— a microtidal region II-384
— oceanography II-383–5
— prospects for the future II-403
— protective measures II-399–403
    — endangered species protection II-400–1
    — Integrated Coastal Zone Management II-399–400
    — marine protected areas II-401, **II-402**
    — PROKASIH Program II-400
Indonesian Throughflow, Coral, Solomon and Bismark Seas Region II-428
Indus river dolphin III-90
industrial activities, Malacca Strait II-339

industrial centres, Russian sector, Sea of Japan basin **II-475**
industrial crowding, Calcutta II-155
industrial development
— coastal zone, Cambodia, no serious environmental pollution II-576–7
— Great Barrier Reef hinterland, loss of coastal habitat and water quality II-622
— limited, Coral, Solomon and Bismark Seas region II-439
— poor in Sundarbans II-153
— Tasmania II-654
— Victoria Province coast II-669
— Xiamen region II-518–19
industrial discharges/effluent
— Cabo Frio–Cabo de Santa Marta Grande, southern Brazil I-741
— Hawaiian Islands II-805–6
— Lesser Antilles I-638–9
— Madras, discharged into the estuaries II-169–70
— and marine pollution, Dutch Antilles I-605
— Tanzania II-92
industrial diversification II-263
industrial pollution
— east coast, Peninsular Malaysia II-351
— Fiji II-745
— Gulf of Guinea coast I-788–9
— Hong Kong II-542
— Malacca Straits II-317, **II-318**
— New Caledonia II-733
— Pago Pago harbour, American Samoa II-771
— Peru I-695
— Sarawak II-376
— southern Brazil, east coast I-740
— Sri Lanka II-184
industrial waste III-339
— dumping of, Tasmania II-654
— poorly handled, the Philippines II-417
— Tasmania
— disposal in coastal waters II-656
— dumping of II-654, II-656
industrialisation, at an early stage, Vietnam II-565
industry
— Argentine coast, Bahía Blanca, petrochemicals I-758
— around the North Sea I-52
— Brazilian tropical coast, effects of I-727–8
— causing pollution, Peruvian coast I-693, I-694–5, *I-694*
— development of round the Aegean I-241
— eastern Australian region, central New South Wales II-636
— El Salvador I-554–5
— English Channel coasts I-72
— expansion and diversification, Venezuelan coastal zone I-645
— growth of, Côte d'Ivoire I-815–16
— Guinea, uses old procedures I-802
— Gulf of Maine I-313
— development and change I-314
— heavy, Noumea and Suva II-718
— marine, Guangdong II-559, **II-559**
— Mexican Pacific Coast I-489
— causing environmental stress I-491
— Nicaraguan Caribbean coast I-525
— northern Spanish shoreline, effects of I-145–6
— Oman
— Muscat and Salalah II-25
— new developments, Sur and Sohar II-25, II-27
— and sources of employment, Gulf of Guinea countries **I-785**
— southern Brazil
— impinging on shallow water coastal and marine habitats I-736, I-737
— industrial complexes I-737–9
— Trinidad I-638
— Vanuatu II-745
infaunal biodiversity assessment, Island Copper Mine III-237
— large infaunal species III-242–3

— juveniles, seen infrequently III-242–3
— sampling design and procedures III-237–8, III-239
— similarity analyses of species and abundance data III-239
— tailings rate tolerable to infauna III-239
infections, in humans, from casual exposure to water III-219
informal settlements, South African coast, a pollution problem II-142–3
infrastructure
— causing coastal modification, Samoa II-717
— interrupting coastal sediments, Gulf of Guinea I-786–7
— leading coastal occupation, Brazilian tropical coast I-727
— the Maldives
— interference with natural erosion-deposition cycles II-208
— interfering with sand movement II-203
— Mascarene Islands, interferes with sand movement II-262
Inhaca and Portuguese Islands Reserves, Mozambique II-105
— slow soil recovery from slash and burn agriculture II-107
— turtle nesting beaches II-105
inorganic nutrients
— Neuse Estuary I-366
— Pamlico, Neuse and New River I-363–4, **I-363**
inshore habitats, Australian Bight II-678
— biogeographical regions II-678
Institute of Marine Affairs (IMA), Trinidad I-634
institutional mechanisms
— established under IMO and UNEP III-345
— importance of III-344–5
integrated coastal area management *see* Integrated Coastal Zone Management (ICZM)
integrated coastal management
— need for in Bangladesh II-295–6
— Xiamen region
— land—sea integration II-531–2
— movement towards II-528
Integrated Coastal Zone Management (ICZM) I-79
— and Agenda 21 III-350
— definitions I-458–9
— finance for III-356
— is it achievable? III-356–7
— must include economic development III-351
— needs to develop in Madagascar II-129
— Nicaragua I-541–2
— critical strategies for I-542
— MAIZCo I-526, I-528, I-534
— Nicaraguan Caribbean coast I-526–8
— the Philippines II-418
— a process not a solution III-350
— western Indonesia II-399–400
— CEPI projects II-399
integrated coastal zone management projects, Tanzania II-94
— constraints on development and implementation of II-96
— programmes attempting to put ICZM into practice II-94–5, **II-94**
— Tanga Coastal Zone Conservation and Development Programme II-94
Integrated Ecological Economic Modelling and Assessment (SCOPE project)
— basic framework III-398
— steps in III-399
integrated marine and coastal management (ICAM) III-341, III-344
integration, different kinds I-459
Inter-American Tropical Tuna Commission (IATTC) I-680
— Tuna-Dolphin Program, training skippers III-142
Inter-Tropical Convergence Zone (ITCZ) I-629, I-679, II-39, II-49, II-66, II-347, II-725, II-753, II-780
— Côte d'Ivoire I-808
— Gulf of Guinea I-776
— over Chagos Islands II-225
— and seasonal changes in the Caribbean sea surface waters I-646
Intermediate Antarctic Water Mass I-702
Intermediate Arctic Water I-679
internal boundary layers III-206–7
— concept III-206

International Commission for the South East Atlantic Fishery, and overfishing I-836
International Conferences on the Protection of the North Sea I-26
international conservation organisations, in the South Western Pacific Islands II-720
International Convention for the Conservation of Atlantic Tuna (ICCAT) III-158
— relies on trade restrictions III-159
International Convention on the Long-Range Transport of Air Pollutants I-132
International Convention for the Prevention of Pollution of the Sea by Oil (OILPOL) III-337
International Convention for the Regulation of Whaling III-340, III-341
International Coral Reef Initiative (ICRI), Regional Workshop for the Tropical Americas, Jamaica I-571
International Council for the Exploration of the Seas (ICES) I-8, I-50–1
— Geneva Convention III-76
— maximal/maximum sustainable catches/yields (MSY) III-75, III-81, III-118
International Geosphere-Biosphere Program(IGBP), GLOBEC program III-369–70, III-394
International Maritime Dangerous Goods Code III-343
International Maritime Organization (IMO)
— complete ban on TBT recommended I-75
— Great Barrier Reef a 'Particularly Sensitive Area' II-616
— Gulf of Aden a 'special area' II-59
— lack of oil reception facilities leads to pollution III-345
— Marine Environment Protection Committee I-640
— oil as a serious pollutant from shipping accidents III-336
— prevention of maritime pollution by non-traditional pollutants III-343
— see also MARPOL Convention
International Mussel Watch Programme/Project
— bivalves sampled, southern Brazil I-743
— Patagonian coast I-762
International North Sea Conferences I-57
International Ocean Institute Training Programmes III-328–9
International Pacific Halibut Commission III-158
international rivers, create problems for North Sea I-58–9
International Safety Management (ISM) Code III-343
International Seabed Authority, prevention of pollution from seabed mining III-338
International Whaling Commission (IWC) III-341
— aboriginal subsistence whaling management procedure (1982) III-83–4
    — a new procedure requested (1995) III-84
    — a persistent cause of bad feeling III-84
— beluga stocks III-90
— considered best body to govern small cetacean catches III-99
— Decades of Cetacean Research III-82
— effectiveness of III-99
— Indian Ocean Whale Sanctuary II-29
— limits finally to be set for all catches III-83
— mandate for the whaling industry III-77
— New Management Procedure III-78
— proposals for sustainable catch limits III-78
— Revised Management Procedure III-79, III-80
    — DNA fingerprinting to reveal sub-populations III-80
    — process error III-80
— and smaller cetaceans III-76
— Sodwana Declaration II-29
— UN requested ten year moratorium III-78
intertidal areas
— English south coast, important to migrating wildfowl I-71
— intertidal rocky communities, Spanish north coast I-140
    — coasts exposed to moderate wave action I-140
    — estuarine coasts I-140
— Long Island and New York, poor intertidal fauna I-326
— Southern Californian Bight, unique I-388

— Tyrrhenian Sea, 'trottoir' I-273
Intertropical Confluence Area see Inter-Tropical Convergence Zone (ITCZ)
intoxication events, Madagascan artisanal fishery II-123
invertebrate communities, coastal benthic, North and South Carolina I-360–2
invertebrate fisheries/gleaning
— Mozambique II-108–9
— South Africa II-139
    — catch per unit effort (CPUE) II-139
— Tanzania II-90
invertebrates
— coral reef, Marshall Islands II-778
— in the Deltaic Sundarbans II-150
— Great Barrier Reef II-617
— Lake Tyres, Victoria Province II-665
— land, Marshall Islands II-778
— marine, El Salvador I-550–1
— Spanish north coast I-139–40, I-140
    — Cantabrian shelf megabenthos I-141
Ionian Sea *I-268*
Iran *II-2*, II-21
— desalination plant at Bushehr II-10
— southern, Makran coast II-19
Iraq *II-2*
— devastated by Gulf War II-11
— projects possibly detrimental to the Gulf
    — drainage of marshes in South II-9
    — Third River (Main Outfall Drainage) *II-6*, II-9
Ireland, offshore wind farm being considered III-309
Irish Sea I-83–98
— coastal erosion and landfill I-93
— effects from urban and industrial activities I-93–6
— major benthic marine and coastal habitats I-87–9
    — range limit for some northern and southern species I-88
— natural environmental variables I-85–7
— offshore systems I-89–92
— populations affecting the area I-92–3
— protective measures I-96–7
— region defined I-85
— rural factors I-93
— topography and sediments I-85, *I-86*
Irish Sea Coast
— artisanal and non-industrial use I-93–4
— environmental impact of cities I-94–5
— industrial uses I-94
— ports and shipping I-95
Irminger Current I-7
Irminger Sea, redfish exploited I-8
Irrawaddy dolphin III-91–2
irrigation
— afalaj system, Arabia II-26
— and drainage, development poor, Somalia II-79
— Gulf coastal plain, USA, growing problem I-443–4
— and water shortage, southern Spain I-175
island communities, Great Barrier Reef region II-615
— high floral biodiversity II-615
Island Copper Mine, Canada, effect of mine tailings on the biodiversity of the seabed III-235–46
— discharge of tailings III-236
— no contamination problem from bioactivation of trace metals III-237
— recovering and unaffected stations III-240–1
'Island Domains', South Western Pacific II-707
islands
— and beaches
    — Marshall Islands II-777
    — The Maldives II-203
— offshore
    — and beaches, Hawaiian Islands II-797
    — and coral reefs, east coast, Peninsular Malaysia II-350

ISM Code *see* International Safety Management (ISM) Code
isostatic rebound I-375
Israel, coast of, and the Southeast Mediterranean I-253–65
— coastal erosion I-258–9
— effects of land-based pollution I-259–61
— natural characteristics I-255–6
— Nature Reserves and National Parks Authority I-262
— population I-258
— protective measures I-262–3
    — National Masterplan for the Mediterranean coast I-261
— shallow marine habitats, the Red Sea invaders I-256–8
ITCZ *see* Inter-Tropical Convergence Zone (ITCZ)
IUCN, and Project Tiger II-157
Ivittuut, Greenland *I-6*, I-12
Ivoirian Undercurrent I-808
Ivory Coast *see* Côte d'Ivoire
IWC *see* International Whaling Commission (IWC)
Izmir I-241
Izmir Bay *I-234*, I-236, I-237, I-244
— dinoflagellate red tides I-238
— eutrophication spreading I-244
— industrial development I-246–7
— wastes dumped with no treatment I-247

Jakarta Bay
— heavy metal loading II-396, **II-398**
— 'Where Have All the Reefs Gone'? the demise of Jakarta Bay and the final call for Pulau Seribu II-397–8
Jakarta Mandate on Marine and Coastal Biodiversity III-341
— integrated marine and coastal management (ICAM) III-341
Jamaica I-559–74
— coastal erosion and landfill I-569
— coral bleaching and mortality events III-55
— effects of urban and industrial activities I-569–70
— Environmental Protection Areas, Negril and Green Island watersheds I-571
— geography of I-561
— macroalgal mats smother old reefs III-223
— major shallow water marine and coastal habitats I-561–5
— Montego Bay and Negril Marine Parks I-571
— offshore systems I-565–6
— populations affecting the area I-567
— Portland Bight Fisheries Management Council I-572
— Portland Bight Sustainable Development Area I-572
— protective measures I-570–3
    — Council on Ocean and Coastal Zone Management I-570
    — Environmental Permit and Licensing system I-570
    — International Conventions participation I-570, **I-571**
    — National System of Protected Areas *I-571*
    — Natural Resources Conservation Authority I-570
— rural factors I-568–9
— seasonality, currents, natural environmental variables I-561
— stresses causing collapse of many reef species III-223
Jamaica coral reef action plan I-571
Jangxia Creek, China, tidal energy project III-316
Japan
— acquisition of whaling technology III-74
— against the Southern Ocean Whale Sanctuary III-85–6
— banned organotin-based paints completely III-251, III-253
— coral bleaching and mortality events III-52
— directed hunts of small cetaceans III-93–4
— mistaken killing of dolphins III-97
— resumed Antarctic whaling III-77
— still killing minke whales III-78
— whale meat
    — continued search for III-78
    — insatiable demand for III-78
Japan, Sea of II-473–86
— climate change II-483
— ecological problems and their causes II-483–5
    — Amursky Bay II-483–5

— Rudnaya River Valley II-485
— landscapes and tourism II-482–3
— natural resources, species and protected areas II-480–1
— protected areas II-481–2
— state of marine, coastal and freshwater environment II-476–80
Java Sea *II-362*, II-363, II-363–4, *II-382*, II-552
— fish landings related to the monsoon II-372
— fishery production II-363
— fragile II-363
— monsoonal climate II-383–4
— pelagic resource base considered heavily exploited II-393
jellyfish I-832
jet drops and film droplets, cause enhancement to particle deposition III-202
Jiangsu Coastal Current II-489
Jinjira *see* St Martin's Island, Bangladesh
John Pennecamp Coral Reef State Park I-412
Johore Strait II-311, II-334–5
— East Strait II-334
— effect of causeway II-334
— increase in sewage and industrial waste discharges II-335
— low wave energy II-334
— poor water quality II-334, II-335
— seagrass beds II-334
— Sungei Buloh Nature Park II-335
— West Strait, problems II-334
Jordan Basin I-309, *I-311*
Juba-Lamu embayment *II-67*
Jutland Current *I-18*, I-23

Kakinada sand spit, Godavari and Krishna deltaic coast II-172, II-272
Kalimantan
— East, SCUBA diving II-377
— Northeastern *II-362*
    — coastal habitats II-365
— transmigration programme II-374
Kamchatka *II-464*
— "eastern channel", west coast II-466
Karimata Strait *II-382*
karst I-561
— formation I-417
Kattegatt *I-100*, *I-122*
kelp II-52
— Chile I-707
— Faroes, forests of *Laminaria hyperborea* I-33–5
— Galician coast I-140
— giant, extensive beds, Tasmanian waters II-652
— Gulf of Maine I-312
— Irish Sea I-89
— Norwegian coast I-21
    — harvesting causes public concern I-26
    — kelp forests denuded I-21
— response to oil spills III-273
— Southern California Bight I-388, I-395
kelp forests III-35
Kelvin wave dynamics III-184
Kelvin waves, and El Niño I-703
Kemp's ridley turtle III-63
— critically endangered III-63
Kent Group ecosystems, Tasmania II-651
Kerch Strait I-287, I-298
Kerguelen Islands II-583
Key Largo *I-406*, I-408
— reefs showing physical or biological stress I-410
Key Largo Formation I-408
— migration of effluent through I-409
Key Largo National Marine Sanctuary I-412
Key West *I-406*, I-408
khawrs
— brackish coastal wetlands II-23, *II-52*
— larger, may host mangroves II-53

— periodic opening to sea II-23, II-52–3
— roads may interfere with the process II-27
Kilinailau Trench *II-426*
Kishon River, Israel
— carrying nitrogen and phosphorus I-259
— heavy metal contamination in the estuary I-260–1
Kislaya, Bay of, tidal energy project III-316
Kizilirmak delta, Turkey, conservation area I-293
knowledge-based tools in coastal management III-355
Korea, disappearance of tidal marsh zone II-490
Korea Strait II-475
Korean Peninsula, Yellow Sea coast population II-491–2
Kosi Bay system, KwaZulu-Natal, estuarine resource use II-136
krill, Antarctic, food for seabirds III-108
Krishna-Godavari deltaic coast *II-162*, II-163
Kruzenshtern Strait *II-464*, II-465
Kuril Basin *II-464*, II-465
Kuril Islands *II-464*
Kuroshio Current II-489, II-516, II-537
— eastern Taiwan II-501, *II-502*, II-508
Kuwait *II-2*
— coral islands II-7, *II-8*
— coral reef fish fauna II-8
— devastated by Gulf War II-11
— dual-purpose power/desalination plants II-10
— hydrographic changes due to Third River project II-9
— landfill II-9–10
— mariculture II-8–9
— nutrients offshore II-6
— offshore circulation anomaly II-4
Kuwait Action Plan II-29
Kuwait Institute for Scientific Research, research related to pollution by petrochemical industries II-3–4
KwaZulu-Natal
— degraded state of estuaries II-135–6
— from management to co-management: the *Pomatomus saltatrix* fishery II-138
— importance of subsistence and artisanal fishermen II-140
— intertidal harvesting, Maputaland Marine Reserve II-140
— invertebrate fishery/gleaning II-139
— well managed II-139
— *see also* South Africa
Kyoto Agreement III-304
Kyoto Declaration, and Plan of Action III-139

La Hague, radionuclides transported from *I-14*, I-54–5, I-76
La Niña I-708, I-709
— strong
— associated coral bleaching III-46
— bleaching south of typhoon path III-47
La Perouse seamount II-257
Labrador Current *I-352*, I-356
Lac Badana National Park II-80
Laccadives *see* Lakshadweep Islands
lagoonal habitats, important in Mascarene Region II-256
lagoonal shelf, Belize I-506
lagoons
— east coast, Peninsular Malaysia II-349
— Marshall Islands, water residence time II-776
Laguna de Tacarigua, Venezuela I-652–3
— American crocodile I-652
— common birds and fish **I-652**
— eurihaline-mixohaline I-652
— planktonic productivity I-652
Laguna Madre *I-436*
lake waters, airborne spectrometer studies, and chlorophyll content III-297–8
Lakshadweep Islands II-20, *II-190*, II-194–5
— biodiversity II-194
— climate and coastal hydrography II-194
— conservation recommendations II-195
— environmental degradation II-194
— low numbers of seabirds recorded II-204
— population and tourism II-194
Lakshadweep (Laccadive)—Maldives—Chagos Ridge II-201, II-223
— formation of II-223
— spread of coral biodiversity along II-202
Laminaria, harvested for alginate I-70
Laminaria hyperborea forests I-33–4
land, an increasingly scarce resource, the Comoros II-251
land clearance, for agriculture, impact of silt, nutrients and contaminants I-508
land degradation
— Australia II-585
— Fiji II-758
land mosses, as biomonitors I-116
land reclamation
— for agriculture and spoil disposal I-144
— Coral, Solomon and Bismark Seas region II-438
— effects of, Western Port Bay, Victoria II-667
— Guangdong coast II-557
— and habitat loss, Singapore II-334
— large-scale, Hong Kong II-542–3, II-546, *II-546*
— Malé's artificial breakwater II-208–9
— pros and cons, the Maldives II-208
— small-scale, Tasmania II-655
— Taiwan II-505
— an earlier policy II-510
— effects of II-509
— for urban development, western Indonesia II-398, *II-399*
— west coast of Peninsular Malaysia II-339
— west coast Taiwan II-503
land subsidence, caused by artesian wells II-307
land tenure rights
— Marshall Islands II-786
— vs. governance system, Marshall Islands II-785
land use, Colombian Caribbean Coast I-672
land-slips, and associated sediment plumes, show a dynamic landscape II-429
land-use practices, environmental threat, the Comoros II-247
landfill
— American Samoa II-770
— Belize I-510
— biodiversity loss and monetary loss I-145
— Curaçao I-602
— Djibouti, impacting mangroves II-58
— due to urbanization, Arabian Gulf II-9–10
— and habitat loss, major concern, Cambodia II-576
— impacting some lowlands, Gulf of Mexico I-477, I-479
— Kota Kinabalu, Sabah II-375
— the Mascarenes II-263
— New Caledonia II-731
— North and South Korea II-493
— not common in Argentina I-762
— obvious form of estuarine alteration I-448
— a problem throughout the Gulf of Guinea I-787
— Sarawak II-375
Langstone Harbour *I-66*, I-174
Lanzarote, Los Jameos del Agua, endemic invertebrates I-195
Laperuz Strait *II-464*, II-465
Large Marine Ecosystems (LMEs) I-2
— Arabian Gulf II-1–16
— Arabian Sea *II-64*
— Australia *II-580*
— basis for assessing the health of III-230–1
— Bay of Bengal II-287
— Benguela Current LME I-823–37
— East African Marine Ecosystem II-85
— Gulf of Guinea I-773–96
— health impacted by morbidity, mortality and disease events III-218

— Lesser Antilles, Trinidad and Tobago I-627–41
— many becoming stressed III-218
— Mexico, northern Gulf of I-437
— Red Sea II-35–45
— Somali Coastal Current II-63–82
— South Western Pacific Islands Region II-707, II-709
— West Iberian large marine ecosystem I-137
Latin America, coastal management in I-457–66
— coastal zone defined I-458
— integrated coastal zone management defined I-458–9
— the Latin America and Caribbean focus I-460–3
— regional examples I-463–4
— some initiatives on coastal zone management **I-464**
— sustainable development defined I-459–60
Law of the Sea *see* UNCLOS (UN Convention on the Law of the Sea)
lead (Pb)
— Baja California I-495, **I-496**
— in Boston Harbor sediments I-315
— contamination by in Tagus estuary I-158
— decrease in Greenland snows I-14
— from petroleum combustion I-329
— in Greenland marine mammals I-9
— in the North Atlantic III-361
lead pollution
— Greenland I-12
— ships' paint I-37
lead and zinc mining, Chile I-710
leatherback turtles I-780, II-205, II-350, **II-355**, III-61–2
— considered endangered by IUCN III-62
Leeuwin Current II-581, *II-581*, *II-674*, II-676, II-677, II-694, *II-695*
— introduces an Indo-Pacific element II-676
— linked to population dynamics of West and South Australia's commercially important pelagic species II-676, II-679
Leeuwin Under Current *II-674*
Lesser Antilles
— coral bleaching and mortality events III-56
— defined I-629
— economies over reliant on coastal environment I-634
— southern, influence of Amazon and Orinoco rivers I-629–30
Lesser Antilles, Trinidad and Tobago I-627–41
— coastal erosion and landfill I-636
— effects from urban and industrial activities I-636–9
— major shallow water marine and coastal habitats I-631–2
— offshore systems I-632–4
— populations affecting the area I-633–5
— protective measures I-639–40
— rural factors I-635–6
— seasonality, currents, natural environmental variables I-629–31
Levantine Basin *I-254*, I-255
— deep chlorophyll maximum I-256
Levantine Intermediate Water I-255, I-256
— entering the Tyrrhenian Sea I-272
— silicates in I-271
Levantine Sea *I-234*, I-241
Levantine Surface Water I-255
life, beginning on Earth III-394
light availability, and seagrass growth I-358
light-stress-induced mortality, seagrasses III-15
Lighthouse Reef *I-502*, I-504–5
Ligurian Sea *I-268*
limestone, used for cement, Sri Lanka II-181
limestone mountains, Arabian coast II-19, II-21
limestone platforms, Red Sea, foundation for "Little Barrier Reef" II-40
limpets I-71
— harvesting of, the Azores I-206, I-210
lindane, Seine estuary I-75
liquefied natural gas (LNG), Indonesia II-390
Lisbon *I-152*
Lisbon embayment I-153
literacy rate, Gulf of Mannar II-165

litter
— conspicuous source of pollution, Fiji Islands II-761
— dumping of I-512
— and floating wastes, Borneo II-375–6
— and marine debris, western Indonesia II-398–9
— ocean and beach, Australia II-588, II-640, II-684, **II-685**
Little Andamans, reported deforestation II-193
Little Bahama Bank *I-416*, I-421
littoral communities, Faroes, response to wave exposure I-33
littoral transport
— Côte d'Ivoire I-808
— sand, Gulf of Guinea I-778
Littorina littorea, lacking in Faroes I-33
live reef fish food trade II-373, II-392–3
— Cambodia II-576
— Coral, Solomon and Bismark Seas region II-440
— Great Barrier Reef region II-623
Liverpool Bay *I-84*, I-85
— discharges to I-95, **I-95**
livestock grazing, and erosion I-601–2
livestock rearing
— changing pattern of, Oman and Yemen II-26
— Gulf of Aden coasts II-57–8
lobster fishery II-80
— Belize I-509
— Lesser Antilles I-635, I-637
— Madagascar, under local control II-126
— Nicaragua I-524
— Pacific coast I-537
— Yemen II-59
lobster habitats, artificial I-425
local institutions, learning from III-400
loggerhead turtles II-205, II-574, III-63, III-355
— developmental migrations III-63
— Great Barrier Reef II-618
logging
— Alaska, leads to habitat degradation I-381, I-383
— Cambodia II-575
— effects of
— Andaman and Nicobar Islands II-193, II-195
— Coral, Solomon and Bismark Seas region II-435–6
— Gulf of Papua II-607
— probably a threat to coastal and marine environments II-604
Lombok Strait *II-382*, II-383
— tanker traffic II-390
London Dumping Convention II-524, III-338, III-365
— 1996 Protocol III-339
— evolution of III-338–9
— greater prohibition of dumping III-339
— need for more comprehensive perspective on waste management III-339
— permitted amounts of dredged materials **III-365**
Long Bay *I-352*
Long Beach *I-322*
long-distance transport of pollutants
— atmospheric
— Baltic Sea I-132
— lead in the southern Baltic I-116, I-214
— one-hop or multi-hop pathways I-13
— to the Sargasso Sea I-229
Long-Range Transboundary Air Pollution Convention III-340
longline fisheries
— incidental capture of sea turtles a problem for III-146
— Madagascar II-122
— and seabirds III-145–6
— some important by-catch III-125
longshore drift, Belize I-510
longshore sand transport, Israeli coast I-256
Looe Key National Marine Sanctuary I-412
Lophelia banks III-379
Lord Howe Island II-635, II-636, II-643

Louisade Archipelago *II-426*
Louisade Plateau *II-426*
Louisiana–Texas Slope and Plateau *I-436*
low energy environments
— biological effects aligned with oil persistence III-275–6
— oil dynamics III-271
lowland evergreen forest, Cambodia II-573–4
Lüderitz upwelling cell *I-822*, I-825
— eddies, filaments and superfilaments I-826
— separates Northern Benguela from Southern Benguela I-823
Lyme Bay *I-66*

Maamorilik, Greenland *I-6*
— sources of lead and zinc pollution identified I-12
Macao *II-550*
mackerel I-49, I-313
— Baltic Sea I-111
Macquarie Harbour, Tasmania, affected by mining pollution II-654
Macquarie Island II-583
macroalgae III-3
— the Bahamas I-429
— Baltic Proper I-109
— benthic I-70
    — Baltic, changes in I-109–10
— Carolinas coast
    — colonization limited I-357
    — offshore I-362
    — rich communities occur on rocky outcrops I-357
    — a transition zone I-356–7
— Colombian Caribbean Coast I-668–9
— communities, Gulf of Aden II-52
— eutrophication causing community changes I-130–1
— Great Barrier Reef region II-615
— increase in Belize reefs I-505
— marine I-490
— northern Patagonia I-755
— Norwegian coast I-21
— red, disappearance from Wadden Sea creeks I-48–9
— reef habitats, Tasmania II-652
— southern Arabia, luxuriant during the Southwest Monsoon *II-22*, II-23
— spring bloom, Gulf of Aden II-50, II-54
— Western Australia II-695, II-701
    — habitat alienation and fragmentation II-699
    — habitat loss II-699
— western Sumatra II-387
macroalgal blooms I-666
macroalgal mats I-53
macrobenthic assemblages/fauna
— intertidal sedimentary areas I-47
— North Sea I-48
macrobenthos II-277
— southern Yuzhnoprimorsky region, reduction in density of II-481
Macrocystis pyrifera, adjacent to *Lessonia* and*Phyllospora* habitats, Tasmania II-652
macrofauna
— intertidal zone, English Channel I-71
— North Sea, northern and southern species I-48
— Portuguese coastal waters I-159
    — some limits on life-span and size I-159
macromolluscs, Chagos Islands II-227–8
macrophytes
— free floating, Côte d'Ivoire reservoirs and rivers I-818
— growth restricted, the Bodden III-261
— replaced by nuisance algal species I-53
Madagascar II-101, II-113–31, *II-244*
— coastal erosion and landfill II-124–5
— effects from urban and industrial activities II-125
— islets and islands II-119
— major shallow water marine and coastal habitats II-117–19
— offshore systems II-120

— populations affecting the area II-120–3
— protective measures II-125–9
    — Environmental and other legislation II-126–7
— recommendations and prognosis II-129
— rural factors II-123–4
— seasonality, currents, natural environmental variables II-115–17
Madagascar Current *II-114*
Madeira Current I-204, *I-204*
maërl I-69–70
— commercial exploitation I-70
maërl vegetation I-69–70, I-140
Mafia Island, Tanzania *II-84*, II-85
— Multi User Marine Park II-86, II-94
    — unsustainable octopus harvesting II-90–1
Magdalena River
— discharges fertilize the Colombian Caribbean I-669
— mangroves in the delta I-671
    — rehabilitation project I-671
— sediment movement from I-667
Magdalena River Basin, Colombia I-665–6
Magellan Strait I-753, I-762
magellanic penguin, affected by oil spills I-765–6
Mai Po marshes, Hong Kong II-539–41
— Deep Bay area
    — conversion of mangroves to fishponds II-539
    — habitat loss a threat II-540
    — over-wintering site for cormorants II-539
    — shrimp ponds (Gei Wais) II-540
— mudflats II-539
— protective measures II-540
— a RAMSAR Site II-539
    — threats to II-540
— Site of Special Scientific Interest II-539
Maine, Gulf of and Georges Bank I-307–20, I-330
— boreal waters I-309
— circulation of the Gulf I-309, I-310, *I-311*
    — nontidal surface circulation I-324, *I-324*
— continual addition of new species I-312
— early settlement and development I-313, I-314
— effects from urban and industrial activities I-314–17
— forestland, development of I-313
— growth of environmental regulation I-313–14
— Gulf of Mine Habitat Workshop I-318
— hydrological regions I-310
— major shallow water marine and coastal habitats I-310–12
— natural environmental variables: currents, tides, waves and nutrients I-310
— offshore systems I-312
— population affecting the area I-312–14
— principal basins I-309
— urban areas I-313
— Working Group on Human Induced Biological Change I-318
— workshops and Proceedings I-309
    — primary research goals and tasks identified **I-318**
Maine, Gulf of, cetacean–gillnet interaction III-144–5
MAIZCo (ICZM), Nicaragua I-526, I-528, I-534
Makaronesy archipelagoes, physical data **I-187**
Makassar Strait *II-382*, II-383
— tanker traffic II-390
Malacca Strait *II-310*, *II-332*, *II-382*
— diluted by river discharges II-333
— including Singapore and Johore Straits II-331–44
    — coastal population II-337
    — general environmental setting II-333–4
    — impact of human activities II-337–40
    — protective measures and sustainable use II-341–2
    — water quality II-340–1
— main problems and issues II-341–2
— marine and coastal habitats II-335–7
— organotins II-449
— pre-European trading route II-337

— tanker traffic II-390
Malacca Straits II-309–29
— conclusions and recommendations II-325–7
— critical environmental problems II-317–21
— natural environmental conditions II-311–14
    — climatology and oceanography II-311–12
    — coastal and marine ecosystems II-312–14
    — geography II-311
    — populations affecting the area II-311
    — topography II-311
— ports, trade and navigation II-316–17
— protective measures II-321–5
    — coordination of the management of the Straits II-325
    — international legal regime governing the Straits II-324–5
    — national management systems and legislations II-321–4
— ratification by littoral states of international conventions **II-324**
— resource exploitation, utilization and conflicts II-314–17
— total net economic value, marine and coastal resources **II-326**, II-327
Malacca Straits Demonstration Project (MSDP) II-325
Malacca Straits Strategic Environment Management Plan needed II-327
malaria control I-476
Malaysia II-299
— approaches to prevent and control marine pollution II-323
— coral bleaching and mortality events III-51
— Kuala Selangor Nature Park II-342
— legislation on environmental protection II-323
— mangroves II-312
— Matang managed mangrove forest III-25
— Matang Mangrove Forest Reserve II-342
— move to fisheries management II-341
— National Oil Spill Contingency Plan II-323
— Peninsular, coastal plains and basins on west coast II-311
— regulation
    — of land-based pollution II-323
    — of toxic and hazardous wastes II-323
    — of vessel-related marine pollution II-323
— swine farming source of agricultural waste II-318–19
— water quality monitoring programme II-340–1, **II-341**
Malaysia, Peninsular, East Coast of II-345–59
— coastal erosion and land reclamation II-353
— effects from urban and industrial activities II-353–4
— impact on habitats and communities II-354–5
— major shallow-water marine and coastal habitats II-348–50
— populations affecting the area II-350–1
— protective measures II-355–7
    — 1985 Fisheries Act II-355–6
    — coastal erosion II-356–7
    — prognosis II-357
    — turtle hatcheries II-356
— rural factors II-351–3
— seasonality, currents, natural variables II-347–8
— the shallow seas II-350
Malaysia and the Philippines, coral bleaching and mortality events III-52–3
The Maldives II-199–219
— coastal erosion and landfill II-208–10
— effects from urban and industrial activities II-210–15
— major shallow water marine and coastal habitats II-202–5
— offshore systems II-205
— part of Lakshadweep (Laccadive)—Maldives—Chagos Ridge II-201
— populations affecting the area II-206–7
— protective measures II-215–18
    — carrying capacity, sustainability and future prospects II-217
    — environmental legislation and related measures **II-215**, II-216
    — environmental restoration II-217
    — initiation of Protected Area system II-216–17
    — multidisciplinary Environmental Impact Assessment (EIA) II-216
    — national and regional development plans II-216
— rural factors II-207

— seasonality, currents, natural environmental variables II-201–2
Malé Declaration II-216
Maluan Bay, Xiamen region
— chemical oxygen demand (COD) II-522
— ecosystem changes II-524–5
Malvinas/Falkland Islands *I-750*, I-751, I-752, I-753, I-766–7
— fishing industry I-763
— squid fishery I-763
mammals
— North Sea I-49–50
— as pests, Marshall Islands II-778
man
— and changes in marine mammals and seabirds I-9
— interference with river flows I-365
Man and Biosphere Programme (UNESCO), research project, Moorea and Takapoto reefs II-817
managed coastline retreat I-54, I-79, III-353
management I-3–4
management practices, customary, Madagascan marine and coastal resources II-126
management problems, Xiamen region II-526–8
— conflicting uses in marine waters II-526–7
— lack of knowledge and information II-527
— new management measures II-528–9
— transboundary problems II-527–8
management structures and policy, South Africa
— advisory bodies II-143
— legislation II-143
— principles II-143
Managua, Lake *I-532*, I-533
manatee grass I-601
manatees I-505, I-509, I-591, I-631
mandated science, fisheries management as III-156
mangrove communities, Borneo, mixed, extending up river valleys II-367
mangrove crab culture, sustainable community aquaculture II-375
mangrove deforestation II-438
— Mozambique II-109
— promotes coastal erosion II-108
— Torres Strait and Gulf of Papua II-605
mangrove destruction, West Guangdong II-559
mangrove ecosystem management, sustainable basis for needed, Malacca Strait II-341–2
mangrove ecosystems
— Godavari-Krishna delta II-170
    — impact of deforestation and prawn seed collection II-172
    — indiscriminate exploitation II-171
— interaction with other ecosystems III-18, III-24
— need to improve sustainable use III-29
— Palk Bay–Madras coast II-168–9, II-170
— rehabilitation of III-26–7
    — concern for the human factor III-28
    — criteria and practical considerations III-27
    — goals III-27
    — need for III-26–7
    — replanting programmes III-27
— species poor, but support biodiversity III-24
mangrove forests
— Borneo
    — distribution factors II-366–7
    — successional communities on accreting shores II-367
— Brazilian tropical coast I-722, I-727
— Colombian Pacific Coast I-679–80
    — exploitation of I-680, I-681
    — fisheries in I-682
    — uses of I-682–3
— Gulf of Papua II-597–8
    — lack low salinity species II-598
    — mangrove tree species, Fly River delta II-597
    — pristine II-597
— Indian Sundarbans II-147

— managed, Sundarbans the first II-148
— Palk Bay, areas cleared for salt pans II-167–8
— Palk Bay–Madras coast, reduced by human activity II-169
— the Philippines
  — loss of and loss of coastal productivity II-414
  — provide nursery grounds II-408
— Sundarbans
  — home of the Bengal tiger II-150
  — under serious threat II-151
  — zoned on the tidal flats II-149–50
— Thailand, use for shrimp farming III-175–6
— western Indonesia, reduced by clearing II-385, II-388
mangrove logging, Cambodia II-575
mangrove palm, Gulf of Guinea, a significant problem I-786
mangrove shrimp ponds III-21–2, III-25
— effects of acid release III-25–6
mangroves III-17–32
— Abrolhos Bank–Cabo Frio, Southern Brazil I-735, I-736
— aerial roots III-18, *III-19*
— American Samoa
  — limited occurrence II-768
  — loss of II-768
— Andaman and Nicobar Islands II-192
  — conservation activities initiated II-195–6
  — degraded sites, restoration of II-196
— Australia II-582
— the Bahamas I-421–2
  — typical zonation I-421–2
— Bangladesh II-290–1
  — land area changes and biodiversity II-290–1
  — *see also* Sundarbans
— Bay of Bengal II-277
  — depletion in Orissa II-279
— Belize I-506
  — some clearance I-510
— biodiversity and human communities III-27–8
— Borneo II-366–70
  — clearing for shrimp farms II-368–9
  — fauna II-368
  — non-conversion uses II-369
  — for wood chip industry II-369
— and braided channels *II-71*, II-76–7
  — low-energy intertidal environment II-76–7
— British Virgin Islands I-619
— Cabo Frio–Cabo de Santa Marta Grande, southern Brazil
  — degraded I-742–3
  — estuarine–lagoon complex I-736–7
  — lost to urban development I-743
  — seriously degraded I-736
— Central America, critical coastal habitat I-460
— clearance for aquaculture III-368
— coastal lagoons, Côte d'Ivoire I-818
— Colombian Caribbean Coast I-669
  — rehabilitation important I-669
— Coral, Solomon and Bismark Seas region II-430, II-438
  — species diversity II-431
— Côte d'Ivoire I-810–11
— degradation of, the Comoros II-248
— development limited, northwest Arabian Sea and Gulf of Oman II-23
— distribution *III-18*
  — patterns of III-20
— Dutch Antilles I-600–1
— east coast, Peninsular Malaysia II-348
  — destroyed by development, Pulau Redang II-353
  — estuaries, lagoons and mainland II-349
  — rate of destruction alarming II-354
— eastern Australian region II-632–3
  — loss of II-639
— ecological values III-22–4
  — importance to fish populations III-23

— mangrove litter III-23
— productivity III-22–3
— stabilisation of exposed land III-23–4
— El Salvador I-548, I-553
— Fiji Islands II-755
  — cleared for development II-760
— the future III-38–9
— Guinea, deforestation I-801
— Gulf of Aden II-53
— Gulf Coast, USA I-440
— Gulf of Guinea I-779–80
  — Ghana I-779
  — importance of I-779–80
  — over-exploited I-786
— Gulf of Mannar II-164
— Gulf of Thailand II-301
— Hawaiian Islands II-797, II-802
— Iran, Saudi Arabia and Bahrain II-7
— Jamaica
  — Black River Morass I-565
  — Negril Morass I-564
— killed by hypersalinity II-135
— Koh Kong Bay, Cambodia
  — may now be seriously degraded II-576
  — now cleared for shrimp farming II-577
  — pristine II-573
— Lesser Antilles I-631, I-632, I-637
— long term oil retention in sediments III-271
— Madagascar II-118–19
  — change in cover II-118–19, II-123
  — exploitation of II-125
  — harvesting regulations II-126
  — majority on the west coast II-118
— Magdalena River Delta I-671
— Malacca Strait II-335–6
  — conversion for aquaculture no longer valid II-342
— Malacca Straits II-312
  — a natural resource being lost II-314–15
— Malaysia, loss through land reclamation II-312
— The Maldives II-204
  — high species richness II-204
— Marshall Islands II-777, II-778, II-779
— Mauritius II-258
— Mayotte Island II-246
— Mexican Pacific coast I-488, I-492
— need for easier availability of existing knowledge III-28
— negative effects of oil spills III-273
— New Caledonia II-726
— Nicaraguan Caribbean coast I-522
— Nicaraguan Pacific coast I-535
  — contamination and degradation of the ecosystem I-534
  — protected areas I-535
  — reduction in, Gulf of Fonseca I-538
— northeastern Australia II-614–15
  — reclamation and draining threats II-615
— and other habitats, Tanzania II-87–8
  — mangrove harvesting II-89–90
  — restoration of, Dar es Salaam II-89
  — use of mangrove timber **II-88**
— patterns of use III-24–6
  — benefits from mangroves **I-25**
  — misuse through international agencies III-26
  — pressures for change III-24–5
  — recreation and ecotourism III-26
  — shrimp aquaculture III-21–2, III-25
— Peru I-691–2
— the Philippines II-409
  — source of fishery and forest products II-409
— present extent and loss III-19–22
  — areal statistics III-20, **III-20**
  — causes of loss **III-21**

— eastern and western groups III-19–20
— loss to shrimp farming III-21–2
— problems of human pressure III-18–19
— Puerto Rico I-582
  — clearance of, effects I-584
— the Quirimbas II-102–3
— red, Gulf of Mexico I-472–3
— Red Sea II-40
  — hard-bottom (reef) and soft-bottom mangals II-41
— rehabilitation of key system, Colombia I-463
— remote sensing of III-285–6
  — detection of change in resources III-285
  — future challenges III-286
  — mangrove leaf area index (LAI) III-285–6
— role in sediment stabilisation II-414
— Sinai Peninsula II-41, *II-41*
— Somalia II-71, *II-71*, II-77
— South Western Pacific Islands II-710
— spawning and nursery grounds II-335
— Sri Lanka II-179
  — damaged by shrimp aquaculture II-179
— Sumatra, zonation depending on tidal regime II-312
— Turks and Caicos Islands I-589–90
  — harvesting of mangroves I-592
— Vanuatu, species diversity II-740
— Venezuela
  — conversion to shrimp ponds I-655
  — mainly in Orinoco Delta and Paria Gulf I-651
— Victoria Province, Australia II-665
— Vietnam II-564, II-565
  — effects of destruction of II-566
— west coast Taiwan II-503
— Western Australia II-695
— western Indonesia II-385–8, **II-388**
— western Sumatra II-387
  — very productive II-388
— Xiamen region II-518
Mangueira Lagoon, southern Brazil *I-732*, I-737
Mannar, Gulf of II-163–6
— climate and coastal hydrography II-163
— fish and fisheries II-164–5
— human population and environmental degradation II-165
  — islands affected by new harbour at Tuticorin II-165
— impact of human activities on the ecosystem II-165–6
  — effects of industries and the power station II-165
— main marine ecosystems II-163–4
— National Marine Park and a Marine Biosphere Reserve II-163
Manning Shelf Bioregion, Australia II-634
Manus Basin *II-426*
Manus Trench *II-426*
Maracaibo Lake *I-644*, I-646
— ecosystem under extreme pressure I-656
— population round I-654
Maria Island, Tasmania
— changes in oceanographic climate recorded II-651
— Marine and Estuarine Protected Area II-658–9
mariculture I-52, I-69, I-319, I-362, II-314, II-339
— American Samoa II-770
— culturing of mangrove oysters I-569
— Gulf of Thailand II-306
— Hawaiian Islands, research and development II-803
— Kuwait II-8–9
— northern Gulf of Mexico I-448
— potential for, eastern Russia II-481, II-482
— Taiwan II-505
— Tanzania II-91–2
— Xiamen region II-519, **II-520**
marinas, sources of contaminants I-72
marinculture, Guangdong, main methods II-558
marine antifoulants III-247–56
— biocide-free 'non-stick' coatings, for the future? III-254

— effectiveness of regulations: measuring and monitoring TBT in the environment III-253–4
— leaching during normal operations III-364
— new self-polishing copolmer paints III-248
— organotin-based paints banned on smaller boats III-251
— TBT
  — ban on, finding safe alternatives III-254–5
  — environmental impacts of III-250–1
  — persistence of in the environment III-249–50
— TBT-based, legislative control of III-251, III-253
  — regulations have reduced contamination III-253
— *see also* tributyltin (TBT)
marine biota
— Azores I-203
— Chilean I-707
  — Peruvian Province and Magellanic Province I-707
— Southern California Bight, effects of anthropogenic activities I-394–9
marine birds
— Patagonian coast, Argentina I-756, I-757
  — breeding **I-756**
marine circulation
— Arabian Gulf II-4
— development of two-gyre system
  — Arabian Gulf II-4
  — Gulf of Aden II-50
  — Somali Indian Sea Coast II-66
marine climate data, accessibility of III-216
marine and coastal communities, Côte d'Ivoire, affected by seasonal factors I-809–10
marine coastal and estuarine ecosystems, trophic status categories III-258
marine and coastal habitats
— Baltic Sea I-124–8
  — biodiversity of I-124–5
  — hard-bottom communities I-125–6
  — pelagic communities I-126, I-128
  — soft-bottom communities I-126
— Chesapeake Bay I-341–3
— continental seas, western Indonesia II-385–90
— English channel coast I-68–9
  — loss of coastal habitats I-74
  — protective and remediation measures I-77–9
— French channel coast I-69
— Irish Sea
  — intertidal habitats I-87–9
  — sub-tidal habitats I-89
— Lesser Antilles, Trinidad and Tobago I-631–2
— Malacca Strait II-335–7
— offshore, English Channel I-69
— and offshore systems, Black Sea I-290–4
— Sargasso Sea and Bermuda Islands I-227–8
  — coral reef zones I-227–8
  — two inshore nutrient zones I-228
— shallow
  — Adriatic Sea I-272–3
  — Tyrrhenian Sea I-273–4
— shallow, Dutch Antilles I-598–601
  — coastal wilderness I-601
  — coral reefs/reefal algal beds I-599–600
  — mangroves I-600–1
  — saliñas I-601
  — seagrass beds I-601
— shallow, southern Spain I-170–2
  — Eastern sector I-172
  — Western sector I-170–2
  — Western sector described I-170–2
— shallow water
  — Australia II-582
  — the Azores I-205–9
  — the Bahamas I-419–22

— Bay of Bengal II-149–51
— Belize I-504–6
— Chagos Archipelago II-226–9
— Colombian Pacific Coast I-679–80
— the Comoros II-245–6
— Coral, Solomon and Bismark Seas region II-429–31
— Côte d'Ivoire I-810–13
— east coast, Peninsular Malaysia II-348–50
— eastern Australian region II-632–5
— El Salvador I-548–51
— Fiji Islands II-754–8
— French Polynesia II-817–19
— Great Australian Bight II-678–81
— Great Barrier Reef region II-614–19
— Guinea I-799–800
— Gulf of Aden II-51–4
— Gulf of Alaska I-375
— Gulf of Guinea I-779–81
— Gulf of Maine and Georges Bank I-310–12
— Hawaiian Islands II-795–800
— Hong Kong II-537–8
— Jamaica I-561–5
— Marshall Islands II-777–9
— Mascarene Region II-256–9
— Mozambique II-102–5
— New Caledonia II-725–6
— New York Bight I-325–7
— Nicaraguan Caribbean coast I-520–2
— Nicaraguan Pacific coast I-534–5
— North Sea I-47–50
— northern Gulf of Mexico I-440–2
— Oman 21–4
— the Philippines II-408–10
— Red Sea II-40–2
— Sea of Okhotsk II-467–8
— the Seychelles II-236–7
— Somali Indian Ocean coast II-72–7, *II-78*
— South Western Pacific Islands II-710–12
— southeast South American shelf marine ecosystem I-754–7
— southern Brazil I-735–7
— Southern Gulf of Mexico I-471–4
— southwestern Africa I-828–30
— Sri Lanka II-178–80
— Taiwan Strait II-502–3
— Tanzania II-85–8
— Tasmanian region II-651–3
— The Maldives II-202–5
— Torres Strait and Gulf of Papua II-597–8
— Turks and Caicos Islands I-589–91
— Vanuatu II-740–2
— Venezuela I-648–53
— Victoria Province, Australia II-664–6
— Western Australia II-695–6
— Xiamen region II-517–18
— Spanish north coast I-139–40
   — intertidal rocky communities I-140
   — soft-bottom communities I-139–40
   — subtidal rocky communities I-140
   — wetlands and marshes I-139
— Vietnam II-564–5
   — littoral habitat II-564–5
marine and coastal protected areas, Madagascar II-127
marine communities
— changes in structure anddiversity III-127
— effects of eutrophication on III-259–60
— Gulf of Alaska, trophic shift in I-376–8
   — ocean/climate variability I-375
— major increase in cod and ground fish I-375–6
— population changes of shrimp and forage fish I-375
marine conservation, Vanuatu, traditional and modern practices II-745–6, **II-746**, **II-747**

marine conservation areas, Borneo II-376–8
marine conservation and resource management, scope of III-376
marine ecosystem health
— concept III-213
— marine epidemiological model III-213
— tracking of HABs III-213
marine ecosystem health as an expression of morbidity, mortality and disease events III-211–34
— basis for assessing the health of large marine ecosystems III-230–1
— categories of disturbance III-217–28
   — anoxic/hypoxic disturbances III-219–21
   — biotoxins and exposure disturbances III-218–19, *III-220*, **III-220**
   — disease disturbances III-225–6
   — keystone-endangered and chronic cyclicaldisturbances III-227–8
   — mass lethal mortality disturbances III-223–4
   — new, novel occurrences and invasive disturbances III-226–7
   — physically forced (climate/oceanographic) disturbances III-224–5
   — trophic-magnification disturbances III-221–2
— data assimilation methods III-215–17
— disturbance type derivation III-217
— Gulf of Mexico Aquatic Mortality Network III-228
— HEED approach III-214
— network for developing standards and achieving consensus III-229–30
— survey methods III-214–15
marine ecosystems
— Canary Islands I-193–5
   — *Cymodocea–Caulerpa* communities I-195
   — deepwater species rise at night I-194
   — mesolittoral area I-194
   — pelagic system I-193–4
   — rocky and sandy seabeds I-195
   — sublittoral area I-194
   — supralittoral area I-194
— and coastal habitats, shallow water, Colombian Caribbean Coast I-666–9
— Gulf of Mannar II-163–4
— long-term concerns about direct and indirect effects of fishing practices III-352
— rocky strata I-194–5
— sustainability of human activities on III-359–73
   — exploitation of ecological resources III-366–9
   — global commons III-360
   — global trends III-369–71
   — introduction of hazardous substances and radioisotopes to the marine environment III-360–6
   — towards sustainability III-371
— western Sumatra III-386–7
   — disturbance and stress II-387
marine environment
— annual input of petroleum hydrocarbons **III-268**
— global legal instruments (at year 2000) III-331–48
   — Antarctic regime III-342
   — from present to future III-342–7
   — Law of the Sea III-332, III-336
   — marine organisms III-340–1
   — marine pollution III-336–40
   — protection from military activities III-342
   — "soft law" instruments III-339, III-344
   — taking stock III-332, **III-333–5**
— introduction of elevated nutrient loads III-366
— introduction of hazardous substances and radioisotopes to III-360–6
   — contminants from shipping III-364–5
   — dumping of wastes at sea III-365–6
   — land-based sources III-361–3
   — operational discharges from the offshore oil and gas sector III-363–4
— protection of from military activities III-342

— bilateral agreements III-342
— regional protection programmes III-363
marine environment strategies, Australia II-590
Marine Environmental Act, Faroes I-38
marine environmental agreements, global
— instruments and their institutional arrangement III-344-5
— move away from binding commitments leads to increased vulnerability III-345
marine environmental protection, legislative power of national governments I-57-9
Marine Environmental Protection Committee (MEPT: IMO), ban on TBT proposed III-365
marine environmental restoration, Hawaiian Islands II-809
marine environmental science, multinational training programmes in III-323-30
— CC:TRAIN III-327-8
— International Ocean Institute Training Programme III-328-9
— synthesis III-329-30
— TRAIN-COAST-SEA III-325-7
— UN training programmes, TRAIN-X strategy III-324-5
marine epidemiological information system, information flows *III-229*
Marine and Estuarine Protected Areas (MPAs), Tasmania II-658
marine fishery reserves I-526
marine habitats, smothered I-383
marine habitats, Brazilian tropical coast I-722-5
— bays I-725
— continental shelf I-722-3
— coral reefs 723-4
marine habitats, shallow
— the Bahamas I-421-2
— coral reefs I-421
— mangroves I-421-2
— seagrass I-421
— Israeli shelf, Red Sea invaders I-256-8
— Tagus and Sado coastal waters I-157-8
— marine biological resources I-157-8
— plankton I-157
— Tagus salt marshes I-157
marine harvesting, attempts to limit impacts of III-378
marine mammals
— abnormalities and perturbations due to toxic chemicals II-456
— accumulate toxins in blubber I-91
— Aegean Sea I-239
— Africa, southwestern I-834
— Alaska, food-limited I-377
— Australian Bight II-680
— Baltic Proper I-109
— Bazaruto Islands II-104-5
— Chesapeake Bay I-343
— and coastal mammals, El Salvador I-549
— contamination by and bioaccumulation of persistent organic organochlorines II-455-6
— Great Barrier Reef II-618-19
— hunted in Greenland I-8-9
— importance of polynyas to I-7
— Jamaica I-561-2
— little affected by radionuclides III-96
— loss of, Java Sea II-364
— Maldives II-205
— Marshall Islands II-779, II-780
— Patagonian coast, Argentina I-756-7
— breeding **I-756**
— the Philippines II-410
— Portuguese coastal waters I-159
— and reptiles, Hawaiian Islands II-800, **II-800**
— small, west coast, Sarawak II-368
— Southern California, effects of anthropogenic activities I-399
— Sri Lanka II-180
— susceptible to chemicals accumulation III-361
— threatened in Hong Kong waters II-542

marine monitoring, Southern California Bight, unable to assist environmental management I-400
Marine National Parks
— Andabar and Nicobar Islands **II-196**
— Madagascar II-127
— Mozambique
— Bazaruto, dugongs and turtles II-110
— Inhaca and Portuguese Islands II-110
— the Quirimbas may be next II-110
— the Seychelles II-239-40, **II-240**
— Curieuse and Sainte Anne II-240, II-241
marine nature reserves
— French Polynesia II-824
— Skomer Island *I-84*, I-96
— Strangford Lough *I-84*, I-96
marine organisms
— common, eastern coast of Taiwan II-508
— harmful effects of increased exposure to UV-B III-370
— legal instrument **III-334-5**
— migration circuit concept III-378
— protection of III-340-1
— TBT, accumulation of III-250
— TBT, toxic to III-250
— Yuzhnoprimorsky region
Marine Park of the North Sporades *I-234*, I-250
Marine Parks
— east coast, Peninsular Malaysia II-356, II-357
— Italy
— Adriatic coast I-280
— Tyrrhenian coast I-280
— Jamaica I-571
— Tanzania II-86, II-94
— further park proposed for Mnazi Bay, Mtwara II-94, II-95
— western Indonesia, zoning schemes II-401
marine plants, Aegean Sea I-239
marine pollution III-336-40
— atmospheric pollution III-340
— Australia II-588-9
— Bay of Bengal II-280
— by persistent organic organochlorines II-455-6
— dismantling of ships III-339
— dumping III-338-9
— from present to future III-3427
— learning-based approach to global codification III-346-7
— from wastes and sewage, Vanuatu II-747-8
— Hong Kong, land-based origin II-542
— Israeli legislation against I-261-2
— Java Sea II-398
— land-based III-339-40
— legal instruments **III-333-4**
— Malacca Strait III-339
— Malacca Straits, sea-based sources II-319-20
— oil and chemical spills II-319-20
— oily discharges II-319
— TBT II-319
— Mozambique II-109
— North Sea, changes in approaching control of I-59
— regulation of land-based sources, Sumatra II-322
— risk of, Lesser Antilles I-638
— sea-based, Indonesian legislation II-322
— seabed activities: peaceful exploration and exploitation III-336, III-338
— shipping III-336
— South China Sea II-354
— Sri Lanka, land-based sources II-184-5
— western Indonesia II-396-8
— Xiamen region
— management of II-532
— monitoring of II-531
marine pollution parameters
— West Guangdong coast II-554-5

— conventional water quality parameters II-555
— pollution sources and waste products II-554–5
— trace toxic organic contaminants in the Pearl River II-555–6
marine populations, dispersal patterns III-380–1
— early life dispersal distance III-381
— larval stage duration III-381
— potential dispersal ranges III-381
marine protected areas
— the Azores I-214, **I-215**
  — proposed **I-216**
— the Comoros II-251
— Djibouti II-55
— east coast of Sumatra **II-322**
— insufficient, Strait of Malacca II-342
— Irish Sea, slow process I-96
— a limited success rate, Tanzania II-93–4
— Sweden I-132
Marine Protected Areas (MPAs)
— American Samoa II-771, **II-771**
— Anguilla and British Virgin Islands I-625
— Australia II-590, II-686
  — Great Australian Bight Marine Park II-686–8
— Belize **I-513**, I-514
— Borneo, management objectives for II-377
— classification of III-377
— eastern Australian region II-643
— Fiji Islands, proposed **II-761**
— Hawaiian Islands **II-803**, II-808–9
  — need to extend coverage II-810
— healthy communities of endangered species at risk from severe storms III-224
— Lesser Antilles I-639–40, **I-639**
— Mexico I-497
— New Caledonia II-734, **II-734**
— Papua New Guinea II-441
— the Philippines II-419
— restrict fishing effort III-161
— South West Pacific Islands II-719–20
— Terminos Lagoon, Southern Gulf of Mexico I-472
— Turks and Caicos Islands I-592–3
— Vanuatu II-746–7
— Victoria Province, Australia II-670
— western Indonesia II-401, **II-402**, II-403
  — many only "paper parks" II-401
marine reptiles, Aegean Sea I-240
marine reserves
— assessing the effectiveness of III-384–9
  — regional summaries of reserve implementation III-386
  — within-reserve effects III-384–5
— demarcate boundaries
  — simply and reliably III-383–4
  — useful III-390
— design of III-380–4
  — dispersal III-380–1
  — edge effects III-383–4
  — minimum viable biomass III-382–3
  — number of III-384
  — optimum size III-383
  — reserve size and number III-381–2
— effectiveness depends on public acceptance, understanding and compliance III-390
— failures in resource management, reasons for III-378
— fine filter and coarse filter III-380
— framework for design of 'no-take' reserves and networks evolving III-376
— global distribution of *III-376*
— guidelines for the development of III-380
— Madagascar II-127
— Mayotte, Longogori Reserve (S passage) II-250
— need a robust approach to management III-377
— no-take zones III-376, III-377, III-378, III-386

— oceanographic setting crucial for egg and larval stages III-380–1
— placement in current/counter current and gyre systems III-381, *III-381*
— potential benefits of III-384
— potential to meet diverse management objectives III-376
— predicting the effects of III-386–9
  — dynamic pool models III-387, III-389
  — logistic models III-387
  — spatial harvesting models (yield per recruit) III-387, III-388
— and resource management III-375–92
  — closed areas and fisheries management III-378
  — dealing with uncertainty III-389
  — ecosystem effects of fishing III-378–80
  — ecosystem management and the reserve concept III-377–8
— social and economic considerations III-389–91
  — enforcement and compliance III-390
  — user participation in management process III-390–1
  — zoning III-389–90
— South Africa, Maputaland Marine Reserve II-140
marine resources
— Australia II-584
— exploitation of, Coral, Solomon and Bismark Seas region II-434
— Greenland
  — fishery I-8
  — hunting I-8–9
— Guangdong II-556
— utilised in the Bahamas **I-424**
marine scientists I-2
marine species
— and habitats, protection of, duplication or complementary **I-58**
— timing of reproduction and reproductive success II-51
— Xiamen coastal waters **II-517**
marine transgression, Kenya and Somalia II-67
Marine Turtle Specialist Group (IUCN) III-64–5
marine turtles, Fiji Islands II-757
marine zonation scheme, Xiamen region II-529–30, *II-530*, **II-530**
marine zoning III-389–90
— *see also* Marine Protected Areas (MPAs); marine reserves
Marmara Sea *I-234*, I-243
MARPOL II-324
— Gulf of Oman designated a 'Special Area' II-28, II-29
— Malacca Straits, possible designation as a 'Special Area' II-325
— special status for Wider Caribbean area I-640
MARPOL Convention I-38, I-55, I-147, I-248, I-262, I-451, II-658, III-343
— importance of III-337
— operational discharges regulated under III-364
marsh and estuarine systems, central Gulf of Mexico I-438
marsh vegetation, persistent oil effects III-276
Marshall Islands II-773–89
— changes in the country's sociopolitical status II-775
— coastal erosion and landfill II-782–3
— cultural and historical resources II-779
— degrading of Ebeye's natural resources II-783
— effects from urban and industrial activities II-783–4
— ENSOs II-776–7
— geographical gradients in physical features **II-776**
— key environmental and marine regulations **II-786**
— major shallow water marine and coastal habitats II-777–9
— need for greater level of protection II-785
— offshore systems II-779–80
— other environmental concerns II-784–5
  — community and individual conflicts linked to land tenure and governance II-785
  — conflicts from different user interests II-785
  — greater incorporation of environmental concerns in development planning II-785
  — limited understanding on cross-sectoral issues II-784
— overall assessment of environmental governance II-788
— populations affecting the area II-780
— protective measures II-786–8

— institutions II-786
— international legislation and regional programmes II-787
— land tenure and customary marine practices II-786
— national legislations II-786
— proposed environmental policies and strategies II-788
— responses to environmental issues: assessment and status II-787–8
— US Army, Kwajalein Atoll (USAKA) procedures and standards II-786–7
— rural factors II-781–2
— seasonality, currents, natural environmental variables II-775–6
— species, habitats and sites of conservational interest II-779
marshes
— coastal, Louisiana, managed by man I-448
— continental, Guadalquivir River I-170
— eastern Russia II-482
— *see also* coastal wetlands; saltmarsh; tidal marshes; wetlands
Martinique, effects of hurricanes I-631
Mascarene Anticyclone II-116
Mascarene Basin II-255
Mascarene Plateau II-255
— an obstacle to deep water flow II-255
— close to a tidal amphidrome II-256
— internal wave generation II-256
Mascarene Region II-116, II-253–68
— coastal erosion and landfill II-262
— effects from urban and industrial activities II-262–5
    — artisanal and non-industrial uses of the coast II-262–3
    — cities and sewage discharges II-264
    — freshwater II-264–5
    — light industry II-263
    — sand mining and lime production II-263
    — shipping, offshore accidents and impacts II-265
    — tourism II-264
— major shallow water marine and coastal habitats II-256–9
— offshore systems II-259–60
— populations affecting the area II-260
— protective measures II-265–6
— rural factors II-260–2
— seasonality, currents, natural environmental variables II-255–6
Mascarene Ridge *II-254*
Masirah Island *II-18*
— some coral cover II-22
mass lethal mortality disturbances III-223–4
— many reports coincide with climate extremes III-223–4
mass mortalities, cause ecosystem collapse and reorganization round a new stable state III-223
"maszoperie" I-110
Matang Mangrove Forest Reserve, Malaysia II-342
Mauritius *II-254*, II-255
— agriculture II-260–1
    — deforestation causes top soil loss II-260–1
— artisanal fishing II-261
— cities and sewage discharges II-264
— coral reefs II-258
    — soft corals II-258
    — spur and groove zone II-258
— eutrophication in the lagoons II-261
— Fishing Reserves II-265, **II-265**
— fishponds (barachois) II-262
— habitat diversity II-257–9
— mangroves II-258
— Nature Reserves II-265, **II-265**
— population II-260
— Round Island, free from introduces mammals and plans II-259
— sand mining and lime production, effects of II-263
— Southeast Trades drive most winds II-256
— stone-crushing plants, environmental effects of II-263
— sugar mills and textile plants II-263
— Terre Rouge Bird Sanctuary II-258
— tourism II-264

maximal/maximum sustainable catches/yields (MSY) III-75, III-81, III-118
Mayotte (Maore/Mahore) *II-244*
— coconut, bananas and ylang-ylang grown for export II-246
    — Iris Bank II-246
— forest more luxuriant now II-247
— fringing reef II-246
— introduction of sewage treatment II-251
— lagoon a series of hydrologic basins II-248–9
— land-based pollutants and the lagoon II-251
— Marine Reserve of Longogori II-250
— reef threatened by silting II-248
— remained with France II-247
— seagrass beds II-248
— sedimentation in the lagoon affect fishing II-249
Mediterranean
— development of marine aquaculture III-169
— offshore wind energy development slower III-311
— receives water from the Black Sea I-22
Mediterranean Action Plan (MAP) I-249
Mediterranean Climate, Great Australian Bight II-675
Mediterranean fauna
— poor in animal species I-240
— zooplankton I-240–1
Mediterranean Sea III-205
Mediterranean Water (MW) I-139, I-153, I-169, I-205
MEDPOL National Monitoring Programme (Greece) I-249
meiobenthos II-277
Mellish Plateau *II-426*
Menai Strait *I-84*, I-96
mercury contamination
— Colombian Pacific coast I-683
— Haifa Bay I-260, I-262
— Kalimantan rivers II-396
— southern Brazil, east coast I-740
mercury (Hg) I-5, I-95, I-116, I-793, II-390
— accumulation in the Everglades I-444
— in Azorean seabirds, fish and cephalods I-212
— Baja California I-495, **I-496**
— in cinnabar mine tailings, the Philippines II-416
— evidence of increasing atmospheric concentrations I-14
— in Faroese pilot whale meat I-31, I-36, I-39
— in Greenland seabirds and mammals I-9, *I-10*
— low levels in Arabian Gulf II-12
— a multi-hop contaminant I-13
— in seabirds of German North Sea coast I-49
— as seed dressing I-635
— in surficial sediments, Gulf of Thailand II-305, **II-305**
Merka Formation II-68
Merka Red Dune Complex II-68–9, *II-70*
— potential for a national glass industry II-79
MESA New York Bight Atlas Monograph Series I-323, I-325
Meso America *see* Central America
Messina Strait I-273, I-281–2
— animal communities I-281
    — Atlantic affinity species I-281
    — endemic species I-282
    — relict species I-281
— intense tidal currents I-281
Mestersvig, Greenland *I-6*
— pollution from lead—zinc mine I-12
metal biomagnification, absent in southern Baltic food chain I-116
metal pollution
— Baltic Sea I-114–16
    — atmospheric and riverine fluxes I-115–16
    — biota I-116
— Russian Far East II-476
    — of surface waters II-477
— *see also* heavy metal pollution
metal sequestration, Tagus salt marshes I-158

metals
— English Channel
— generally low I-74–5
— higher near estuaries and inshore I-75
— entering North Sea from mining and industry I-54
— in land-based pollution III-361
methane
— emissions from the Bodden III-262
— increased in the atmosphere III-188
methane gas hydrates, Blake Plateau I-367
methylmercury I-114, III-95
— in Faroese pilot whale meat I-36
metropolitan areas, located near estuaries, problems associated with climate change and sea level rise III-194
Metula oil spill, Chile III-276
— asphalt formation III-270
Mexican Basin, and Sigsbee Deep I-442
Mexican Pacific coast
— case studies I-492–7
— Altata-Ensenda del Pabellón I-493
— Huizache y Caimanero lagoon system I-493–4
— Navachiste—San Ignacio—Macapule bays I-492–3
— Teacapán—Agua Brava lagoon system I-494–5, I-497
Mexican Pacific coastline I-485, **I-486**
Mexico
— coastal zone of Campeche, analysis of environment and its problems I-464
— coasts have ecological and socioeconomic problems I-464
— and Gulf of Mexico, coral bleaching and mortality events III-56
— Pacific coast, coral bleaching and mortality events III-53–4
Mexico, Gulf of
— growth of hypoxic zone III-219
— problem of incidental mortality of juvenile red snappers III-143–4
Mexico, northern Gulf I-435–56
— coastal erosion and landfill I-445–8
— defined I-437–8
— eastern sector I-437
— effects from urban and industrial activities I-448–51
— interactions
— with North Atlantic Ocean I-437
— with waters and biota of the Caribbean I-437
— a large marine ecosystem I-437
— major shallow water marine and coastal habitats I-440–2
— offshore systems I-442–3
— populations affecting the area I-443
— rural factors I-443–5
— seasonality, currents, natural environmental variables I-438–40
— western and central sectors I-437
Mexico, Pacific coast I-483–9
— Californian Coastal Province I-486–7
— effects of urban development and industrial activities I-492–7
— case studies I-492–7
— North Pacific Ocean Province I-487–8
— Pacific Center Coastal Province I-488–9
— population I-489
— protective measures I-497–8
— included in international agreements I-497–8
— rural factors and fishing I-489–91
— Sea of Cortes Oceanic Province I-488
— Tropical South Pacific Ocean Province I-489
Mexico, Southern Gulf I-467–82
— coastal erosion and landfill I-476–7
— effects of human activities on natural processes I-478–9
— environmental framework I-469–71
— major shallow water marine and coastal habitats I-471–4
— offshore systems I-474–5
— populations affecting the area I-475–6
— protective measures I-479–80
— international agreements signed by Mexico **I-480**
— National Development Plan (1995–2000) I-479–80
— programs of sustainable regional development (Proders) I-480

— rural factors I-476
— urban and industrial activities I-477–8
micro-organics, sediment contamination by slight in the English Channel I-75
microalgae
— benthic I-357–8
— primary production I-357–8
— toxic I-356
microalgal blooms, increased due to due to seagrass die-off III-15
microbes
— oil degrading, in cyanobacterial mats II-12
— oxydation of organic material III-261
microbial diseases, and prawn mortality II-154
microbiological contamination
— English Channel
— impact on bathing water quality I-76
— impact on shellfish I-76
Micronesia, Marshall Islands a part of II-775
Micronesia, Federated States of, coral bleaching and mortality events III-52
microphytobenthos, North Sea I-48
Middle American Trench I-533
Middle Atlantic Bight I-323
— nontidal surface circulation *I-324*
Midway Atoll National Wildlife Refuge II-798
migration circuit concept, marine organisms III-378
migrations, seasonal, Okhotsk Sea II-466
military activities, protection from **III-335**
military usage
— French Polynesia II-822
— Hawaiian Islands II-804
— Marshall Islands II-783
Mindanao Current II-407
mine tailings
— copper mining, polluting Chilean coast I-713
— construction of large lagoon for tailings I-713
— effect of on the biodiversity of the seabed, Island Copper Mine, Canada III-235–46
— after mine closure sustainable ecological succession soon established III-245
— biodiversity in terms of higher taxa III-241–2
— crab fishery and yields III-244
— data set III-237–8
— habitat change since mine closure III-241
— large infaunal species III-242–3
— species evennness III-240–1
— species richness III-238–9
— tailings deposition levels affecting fauna III-239–40
— time- and cost-effectiveness of the benthos surveys III-243–4
— placement of for minimal and reversible environmental losses III-236
— risk of groundwater contamination I-710
mine tailings deposition, what is the tolerable rate? III-239–40
mineral extraction, areas of interest on Tyrrhenian sea floor I-279
mineral resources, Somalia II-79
mineral springs, west coast, Sea of Japan II-483
minerals
— coastal sands, Sri Lanka II-181
— Colombian Pacific coast I-683
— deep-sea deposits, Mascarene Basin II-260
— Godavari basin II-170
— Malacca Straits II-316
— Palk Bay–Madras coast II-168
— South Africa II-142
Minimata disease II-396, II-398
mining
— Alaska I-381
— Australia II-585
— Chile I-710
— lessons for the future, a case study I-713
— coastal, the Philippines II-415–16

— problems from cinnabar mines, Palawan II-415–16
— wastes and tailings serious threats to the marine environment II-415
— diamonds, Namibia I-835
— Great Barrier Reef hinterland II-622
— Greenland, and heavy metal pollution I-11–12
— Gulf of Papua II-607
   — fate of sediments from II-603
— New Caledonia, and its effects II-731–2, *II-732*
— Nicaragua I-525–6, I-541
— Papua New Guinea II-439
   — problem of mine discharges II-439
   — tailings discharged direct to the sea II-439
— Peru I-695
— runoff from increasing sedimentation and turbidity I-444–5
— small-scale, the Philippines II-414–15
   — discharge of untreated mine tailings II-414–15
— southern Spain
   — Aznalcóllar disaster I-178
   — problems of drainage from I-178
— Sumatra, surface and submarine II-395
— Tasmania, impact on coastal waters II-654
mining pollution I-710
minke whales III-78
— catch limits III-83
— counting of III-79–80
— number estimates III-82
   — probably more than one biologically distinct population
Miskito Coast Marine Reserve, Nicaragua I-526
— corals I-520, *I-521*
— recommendations for **I-527**
Mississippi River *I-436*, I-443
— deltaic marshes I-440
— results of 1993 extreme flooding III-219
— silt and clay from I-437
Mnemiopsis, effects of introduction to the Black Sea I-290, I-304
Mogadishu Basin *II-67*
Moheli (Moili/Mwali), good soils II-246
molecular markers, used to identify contaminant sources I-392
mollusc aquaculture I-653
— Patagonia I-764
mollusc diversity, Sunda Shelf II-389
mollusc farming III-167, *III-168*
molluscan fauna, larger Norwegian fjords I-20–1
molluscs I-34, I-172, II-586
— carrying bacteria, Sri Lanka II-184
— culturing of II-278
— economically important, Colombian Pacific Coast I-684
— fishing for I-73
— lagoons, Gulf of Mexico I-473
— Nicaraguan Pacific coast I-534
— Spanish north coast I-139
— Tagus and Sado estuaries I-158
— Vietnam II-564
Monin—Obukhov theory 202
monitoring and enforcement, becoming easier on the high seas III-390
monitoring programmes, lacking, Yellow Sea II-496
monk seals
— Hawaiian Islands II-799, II-800
— Mediterranean Sea, decline in III-129
monsoons II-116, II-537
— affecting southeast India II-163, II-166
— affecting Tanzania II-85
— Bay of Bengal II-148
   — causes high concentrations of heavy metals in Bay region II-155
   — currents and gyres II-273–4, *II-274*
— Borneo II-365
— Cambodia II-571
— Coral, Solomon and Bismark Seas Region II-428
— east coast, Peninsular Malaysia II-347

— intermonsoon changeover period II-347
— winds, waves and currents II-347–8
— Gulf of Thailand, influence of II-300, II-301
— Indian system influences climate of northwestern Arabian Ocean II-19–20
   — effects of Southwest Monsoon II-19
— influences in Malacca Straits II-311
— influencing northern Maldivian islands strongly II-201
— and the Intertropical Convergence Zone II-49
— and the Malacca Strait II-333
— the Philippines II-407
— South China Sea coast II-551
— and the Sri Lankan climate II-177
— Vietnam II-564
— winds affect northern and northwestern Madagascar II-117
— winter and summer, Arabian Gulf II-4
Mont Saint Michel Bay
— important bird location I-71
— marshes I-69
Montauk Point *I-322*, I-323
Montego Bay, Jamaica *I-560*
— reef restoration efforts I-546
Monterrey sardine I-488
Montserrat *I-616*, I-617, I-618, I-623
— climate I-617
— coral communities
   — before volcanic activity I-618
   — effects of volcanic activity I-618–19
— endemic birds I-620
— fishing I-621–2
   — regulations not enforced I-624
— hurricane damage I-622
— loss of population due to volcanic activity I-620–1
— mangroves limited I-619
— seagrass beds I-618
— Sustainable Development Plan I-621
— volcanically active *I-616*, I-617
mooring system, British Virgin Islands I-624
moorings, the Bahamas I-431
morbilli virus *see* phocine distemper viruses
Morecambe Bay *I-84*
— shore communities I-88
Moreton Bay, Queensland
— environmental problems II-642
— high biodiversity *II-630*, II-634
— residential marinas II-639
Morondava, Madagascar, problems of land loss II-124–5
morphological abnormalities, indicator of system health III-225
Morrocoy National Park, Venezuela *I-644*
— much damage due to sedimentation I-655
— pollution estimates I-657–8
— subject to man-made disturbances I-655
Morrosquillo, Golfo de *I-664*, I-665, I-666
mother-of-pearl shell II-44
motor vehicles, emissions from I-345
Mozambique II-99–112
— coastal divisions II-101
— coastal erosion and landfill II-108
— effects from urban and industrial activities II-108–9
— high tidal range II-II-101
— major shallow water marine and coastal habitats II-102–5
   — Bazaruto Islands II-104–5
   — Inhaca and Portuguese Islands Reserves II-105
   — Quirimba Archipelago II-102–4, II-245
— offshore systems II-105, II-107
— participation in International Conventions **II-105**
— population II-107
— protected by Madagascar II-101
— protective measures II-109–11
   — Framework Environmental Law (1997) II-110
— rural factors II-107–8

— seasonality, currents, natural environmental variables II-101–2
— tropical humid to subhumid climate II-101
Mozambique Channel II-101, *II-114*, II-251
— carried high volume of crude oil traffic II-109
Mozambique Current II-101, *II-101*, *II-114*, II-245
Mozambique Gyre *II-101*, II-105
mucilaginous aggregates, effects of I-278
mud flats *see* soft shores
mud reefs I-583
muddy bottoms, Gulf of Papua II-597
mudflats
— and accreting mangroves, Perak and Selangor II-336
— extensive, Malacca Straits II-312
— Hong Kong II-539
Multiple Marine Ecological Disturbances (MMEDs) III-212
— episodic events and co-occurring anomalies III-214
— HEED approach III-214
— HEED database and GIS III-229–30
— indicators of decline in ecosystem health III-230
— observational reports, additional information for III-215–16
— pooling of co-occurring biological disturbance data, resulting evaluations III-216–17
— use of marine ecosystem health framework III-230–1, *III-230*
Multiple Marine Ecological Disturbances (MMEDs) program
— Health Ecological and Economic Dimensions (HEED) of III-215
— HEED system III-215, III-229
— scale in aggregation of anomaly indicators III-212
multiple-use management model, Great Barrier Reef Marine Park, criticism of II-616
Multivariate ENSO Index (MEI) I-703, *I-704*
munitions, dumping of in Beaufort's Dyke I-96
Murat Bioregion, Australian Bight II-678, II-679
Musandam *II-18*
Muscat *II-18*
— Qurm National Nature Reserve
    — at risk from urban development II-30
    — mangroves at risk II-28
mussel beds, protect sequestered oil from weathering III-271
mussel cultivation
— Albufeira lagoon I-156
— Galicia I-144
Mussel Watch Project I-452
— and contamination by organotins II-451, **II-453**
mussels I-34–5, I-51, I-290, III-167
— adapted to hypersaline conditions I-442
— Albufeira lagoon, purification needed I-158
— Island Copper Mine dock, bioaccumulation in III-244
— marine, Scope-for-Growth I-56
— metal contamination, Tagus estuary I-158
— response to oil spills III-273
— Venezuela I-649
Muthupet Lagoon, Palk Bay II-167
Mytilus edulis platensis I-754
— circalittoral banks, Buenos Aires Province I-755

N/P ratios
— Aegean Sea I-240
— Baltic Sea I-130
— trophic zone, Baltic Proper I-106
Namib Desert I-823
Namib Naukluft National Park I-835
Namibia
— claimed her EEZ at independence I-836
— diamonds I-835
— favourable factors for a sustainable fishery I-837
— the fishing industry I-830
— hake catches I-834
— Kudu gas field I-835
— Namib Naukluft National Park I-835
— other development and commercial activities I-837
— ports, Walvis Bay and Lüderitz I-835
— "sulphur eruptions" I-828
— sustainable harvesting of seals I-834
Nancowrie Biosphere Reserve, Andabar and Nicobar Islands II-196
Nares Strait I-5, *I-6*, I-13
narwhals III-76, III-90
Nassau Grouper fishery I-425
National Coastal Erosion Study, east coast, Peninsular Malaysia II-356
National Estuary Program (USA) I-452
National Marine Fisheries Service (NMFS; USA), publishes bycatch statistics III-159
National Marine, Protection, Research and Sanctuaries Act (USA) I-452
National Parks
— Colombian Pacific I-684
— Fiji Islands II-761–2
— Gulf of Cadiz I-182, **I-182**
— Lac Badana National Park, Somalia II-80
— Masoala National Park, Madagascar II-127
— Southern Gulf of Mexico, "Arrecife Alacranes" I-471
Natura 2000 network I-52, I-77
natural capital depletion tax III-401
natural events, hazards and uncertainty, Marshall Islands II-776
natural hazards, Xiamen region II-516–17
natural resource depletion I-634
natural resource extraction, Alaska I-381
natural resource regions, Russian East Coast II-480–1
— Severoprimorsky region II-480
— Yuzhnoprimorsky region II-480–1
Nature Reserves, Mascarene region II-265, **II-265**
Nauru Basin *II-426*
Navachiste—San Ignacio—Macapule bays, Mexican Pacific coast I-492–3
— dams I-492
— mangroves I-492
navigation channels, problems created by I-447–8
Nazareth Bank *II-254*, II-255
Nazca Plate I-701
nearshore habitats, Hong Kong II-538
needs, concept of, in Third World I-459–60
Negril Morass, Jamaica I-564
— hummocky swamp I-564
nehrungen *see* sand barriers, southern Baltic coast
nematodes I-48
Netherlands, offshore wind energy production in III-308
neurotoxic shellfish poisoning (NSP) III-221
Neuse River Estuary *I-352*, I-366, III-205
— picoplanktonic cyanophytes I-355
— productivity pulses and algal blooms I-354
— zooplankton abundance and planktonic trophic transfer I-359
New Amsterdam anticyclone *see* Mascarene Anticyclone
New Britain Trench II-427
New Caledonia *II-706*, *II-707*, *II-708*, II-723–36, II-729
— coastal erosion and landfill II-731
— coastal modifications II-717
— coral reefs II-711
— effects from urban and industrial activities II-731–4
— the lagoon II-715
— major shallow water marine and coastal habitats II-725–6
    — offshore systems II-726–7
— Marine Protected Areas (MPAs) II-719
— nickel mining II-729
    — case study II-713
— populations affecting the area II-713–14, II-727–9
    — activities affecting the sea II-728–9
— protective measures II-734–5
    — Marine Protected Areas (MPAs) II-734, **II-734**
    — reef-monitoring project II-734
    — species-specific regulations II-734
— rural factors II-729–31

— seasonality, currents, natural environmental variables II-725
— subsistence reef fisheries II-715
— ZoNéCo Programme II-727
New Guinea Basin *II-426*
New Guinea Coastal Undercurrent II-428
New Ireland Basin *II-426*
new, novel occurrences and invasive disturbances III-226–7, III-230, III-259
New Providence Island *I-416*
— coastal erosion I-428–9
— perturbations over the last fifty years I-430
New River Estuary *I-352*, I-366
New South Wales
— central, major habitats II-635
— fisheries management system, gives security within adaptive management III-400
— share system conceptual framework relevant to other fisheries III-400–1
New York Bight I-321–33
— Bight restoration plans I-330–1
— dump sites in I-323
— major physical, hydrographic and chemical factors I-323–5
— major shallow water marine and coastal habitats I-325–7
— "new ways forward" I-331
— New York harbor
    — anthropogenic impacts I-327
    — clean up efforts I-327
— offshore systems I-327
— opposing uses I-331
— populations and conditions affecting the Bight I-328–9
— resources at risk I-329–31
— some habitat improvement I-330
New York Metropolitan area, growth of landfill and dumping of wastes I-328
New Zealand, no-take areas III-386
New Zealand Fur Seal II-681
Newfoundland, collapse of Atlantic cod fishery III-120
NGO activities, Jamaica I-572
NGO participation, increased due to "soft law" instruments III-344
Nicaragua
— central highlands, drainage of I-533–4
— coasts *I-533*
— cotton cultivation, after-effects of I-538
— government in I-538
    — laws awaiting approval I-541
— major economic and political trends I-540
— Pacific volcanic chain I-533
Nicaragua, Caribbean coast I-517–42
— coastal resource management need modification of law I-527–8
— effects from urban and industrial activities I-525–6
— geography of I-519
— ICZM I-526–8
— major shallow water marine and coastal habitats I-520–2
— natural reserves I-526
— offshore systems I-522–3
— population I-523
    — eastern slopes of the central highlands I-523
— protective measures I-526
— rural factors I-523–5
— seasonality, currents, natural variables I-519–20
Nicaragua, Lake *I-532*, I-533
Nicaragua, Pacific coast I-531–42
— effects from urban and industrial activities I-539–41
— geography/geology of I-533–4
— major shallow water marine and coastal habitats I-534–5
— marine resources used by Honduras and Salvador I-538
— offshore systems I-535–7
— populations affecting the area I-537–8
— protective measures I-541–2
— rural factors I-538–9
— seasonality, currents geography/geology of I-534

Nicaraguan Center for Hydrobiological Research, estimate of fisheries biomass I-525
nickel mining
— New Caledonia II-713, II-729
    — destructiveness of II-713, II-731
    — environmental effects of processing still largely unknown II-733
Nicobar Islands *II-190*, *II-270*
Niger, river and delta I-776–7
— oil a source of conflict I-788
— onshore oil production in the Delta I-788
— sand movement in the delta I-778
Nigeria, mangroves III-20
Nile Delta, retreat of I-258
nitrate concentrations, Maria Island II-651
nitrate contamination/pollution I-174
nitrate enrichment, euphotic layer, Côte d'Ivoire I-810
nitrate salts, mining of Chile I-710
nitrates I-255–6
— increased load to Baltic Sea I-130
— and seagrass and other aquatic vegetation I-358
nitrification III-262
nitrogen
— anthropogenic
    — in Baltic Proper I-104, *I-104*
    — emissions to the atmosphere III-199
    — load to Chesapeake Bay I-345
— increased load to Baltic Sea I-130
— load to Chesapeake Bay I-344–5
— organic, and phosphorus at depth in Bay of Bengal II-274
nitrogen concentration, Orinoco River I-647
nitrogen cycle, investigations of the impact of eutrophication on III-262–3
nitrogen fixation, important in Baltic waters I-128
nitrogen flux, riverine, increase in III-366
nitrous oxide, increased in the atmosphere III-188
no-take reserves III-376, III-377, III-378, III-386
— for fisheries management, coral reefs III-386
NOAA
— environmental survey baseline investigation, DWD 106 I-327
— National Status and Trends Program I-452
noise pollution, effect on marine mammals III-97
nomadic jellyfish, an immigrant I-157
Nord-Pas de Calais coast I-67
Nordostrundingen *I-6*
Norfolk Island II-635, II-636, II-643
— catastrophic soil erosion II-638
Normandy *I-66*
"nortes" I-489
North Aegean Trough I-235
North Atlantic, Sverdrup transport in I-579
North Atlantic Central Water, South of Finisterre Cape I-139
North Atlantic Current I-5, *I-18*, I-20, I-23, I-204, *I-204*, I-223
North Atlantic Deep Water I-579
North Atlantic Oscillation I-223
— and exceptional North Sea conditions I-45
North Atlantic water, southern Spain I-169
North Brazilian Current I-630, I-722
— retroflection of causes eddies I-630
North Carolina
— demersal marine zooplankton I-360
— eutrophication in some estuaries I-356
— toxic contamination I-360–1
North Channel *I-84*, I-85
North East Monitoring Program I-315
North Equatorial Counter Current I-630, II-407, II-428
— Gulf of Guinea *I-774*, I-778
North Equatorial Current I-223, I-487.I-488, I-489, I-617–18, *I-628*, *II-406*, II-407
North Equatorial Drift *I-222*
North Equatorial Pacific Current II-775, II-779–80, II-795, II-797

North Inlet *I-352*, I-356, I-359
— salt marsh estuary I-353
North Korea, famine due to deforestation, soil erosion and natural disasters II-492
North Loyalty Basin *II-426*
North New Hebrides Trench *II-426*
North Pacific Current *I-374, III-180*
North Pacific Fisheries Management Council, observer program on all larger and some smaller fishing vessels III-147
North Pacific Groundfish Observer Program (NPGOP) III-160
North Pacific High I-378
North Pacific Oceanic Province, Mexico I-487–8
North Pacific Pressure Index (NPPI) I-376
North Pacific Subtropical Anticyclonic Gyre II-407, II-795
North Sea I-43–63
— atmospheric nitrogen deposition assessment III-201
— cessation of some polluting inputs I-56
— climate I-45
— coastal erosion and landfill I-53–4
— cyclonic circulation I-45–6
— disposal of dredged material in I-55
— duplication of responsibility for **I-58**
— effects of urban and industrial activities I-54–6
— eutrophication I-53
— general decline in populations of marine mammals and seabirds III-129
— groundfish assemblage, changes in III-128
— major shallow water marine and coastal habitats I-47–50
— partitioning of I-46
— populations affecting the North Sea I-50–3
— protective measures I-56–9
    — international arrangements affecting protection of **I-56**
    — protection at subregional level **I-58**
— reduction of nutrient input by bordering countries agreed III-220–1
— region defined I-45
— seabirds and fisheries in III-112
— seasonality and natural environmental variables I-45–7
North Sea management
— need to increase management plans I-59
— new system is required I-59
North Sea Task Force I-57
— Monitoring Master Plan I-75, I-77
Northeast Asia Regional Global Observing System (NEARGOOS) II-497
Northeast Atlantic, high by-catch mortality rates III-125
Northeast Pacific shelf, detritus from catch-processor ships III-127
Northeast Providence Channel *I-416*
Northern Adriatic I-269
— communities of I-172
— described I-269
— formation and circulation of water masses I-271
— primary production high in offshore systems I-174
— salinity I-271
northern Benguelan current region *see* Africa, southwestern
northern fur seals, enzyme induction by PCBs II-457, *II-460*
northern Gulf of Mexico shelf, pulses of shrimp, crabs and fish I-438
Northern New South Wales, major habitats II-634
Northwest Arabian Sea and Gulf of Oman II-17–33
— coastal erosion and landfill II-27–8
— geography and geology II-19
— major shallow water marine and coastal habitats II-21–4
— offshore systems II-24–5
— populations II-25
— protective measures II-29–31
— rural factors II-35–7
— seasonality, currents and natural environmental stresses II-19–21
— urban and industrial activities II-28–9
Northwest Atlantic Fisheries Organization (NAFO) I-8
Northwest Pacific Action Plan II-496
Northwest Providence Channel *I-416*

Northwestern African Upwelling I-190
Norway
— climate modifies fjord morphology I-19
— coastal wave energy development III-315
— data on contamination in organisms and bottom sediments I-27
— "no discards" policy forces selective fishing III-138
— use of antibiotics in fish cages III-369
— Whaling Act (1929) III-77
— whaling technology III-74–5
— wild fish may contain antibiotic residues III-369
Norwegian Atlantic Current *I-19*
Norwegian coast I-17–30
— anthropogenic influences I-24–6
— environmental setting I-19
— major shallow water marine and coastal habitats I-21–3
— monitoring programmes, environmental quality criteria and protective measures I-26–8
— natural environmental variables I-19–21
— north—south community gradient less than expected I-22–3
— offshore systems I-23–4
Norwegian Coastal Current *I-18*, I-20, I-23
Norwegian Sea, depletion of rorquals III-74
Norwegian Trench I-17
nuclear fuel reprocessing plant, La Hague I-72, I-76
nuclear power, India II-279
nuclear power plants/stations I-54, I-72
— southern Brazil I-738
Nuclear Test Ban Treaty III-342
nuclear waste dump sites, Sea of Japan II-478, *II-480*
nucleation III-200
Nullarbor National Park, Australian Bight II-682, II-684
Nullarbor Plain, Great Australian Bight region II-682
nursery areas, restriction of fishing in III-378
nutricline, Levantine basin I-256
nutrient burden, reduction of III-258–9
nutrient changes, drastic, signs of in the Bodden III-260–1
nutrient concentrations, reef and lagoonal water, French Polynesia II-816
nutrient discharges, to North Sea, changes in I-53
nutrient effects, of seagrasses III-2
nutrient enrichment
— of coastal waters of Florida reef tract I-409
— Cockburn Sound, Western Australia, effects of II-700–1
— from equatorial and open ocean upwelling I-535
— from sewage effluents, The Maldives II-204
— from upwelling, southwestern Africa I-825
— New York Bight I-324–5
— of northern Gulf of Mexico I-449
— problems caused by II-212
— a threat to coral reefs III-36
nutrient fluxes
— and aquatic ecosystem responses III-198–9
— importance of dry deposition processes III-201
nutrient loading, anthropogenic III-366
nutrient loads
— Chesapeake Bay I-344–5
    — reduction in I-346
— high
    — brought by the Mississippi I-437, I-438, I-443
    — introduction of to the marine environment III-366
— increased, North Carolina I-358
— Yellow Sea coastal waters II-495
nutrient ration, effects of alteration III-259
nutrient reduction, Norwegian obligation I-24, I-26
nutrient supply
— Northeast Pacific
    — effected by shallow mixed layers III-184
    — vulnerable to climatic change III-171
— reduced in Californian coastal wasters III-185
nutrients
— causing eutrophication, Hong Kong II-542

— from fertilisers, Australia II-584
— anthropogenic inputs, to the English Channel I-74
— Arabian Gulf II-5–6, **II-5**
— Argentine Sea I-757, *I-757*
— Baltic Proper, temporal and spatial variability in I-106–8
— calcification and uptake of III-35–6
— concentrations around Tasmania II-651
— concentrations of, Canary Islands I-191–2
   — and nitrite I-191–2, *I-193*
— from fish farming III-369
— from the land, oceans a sink for III-394
— Gulf of Alaska shelf, from deep water I-379, I-380
— high, nearshore, Hawaiian Islands II-795
— horizontal gradients, Tagus and Sado embayments I-155, *I-156*
— input into Haifa Bay I-259–60
— inputs during upwelling events and high algal growth, Côte d'Ivoire I-814
— Irish Sea
   — increasing I-87
   — sources of inputs I-93, **I-93**
— large increase of to Baltic Sea I-130
— low
   — in Australian waters II-582
   — off Central-eastern Australia II-632
— Red Sea II-39
— reduction of input to North Sea by bordering countries agreed III-220–1
— role of in coral reef degradation III-38
— and salinity, off Madagascar II-115–16
— sources of input into Gulf of Guinea I-779
— to Bay of Biscay from Cantabria I-143–4
— transferred from sea to land by birds, Chagos Islands II-229

ocean drift netting, banned II-719
ocean dumping II-322
ocean physics, and foraging seabirds, North Pacific III-113
ocean resources, intergenerational and interspatial effects of use of III-397
ocean species, common, affected by coastal pollution and ocean dumping I-330
Ocean Station Papa, Gulf of Alaska *III-180*
— deviation from normal salinity III-182, *III-182*
— mid-winter mixed layer depth trend III-182, *III-182*, III-185
ocean temperature zones, Australia II-581
ocean temperatures, effects of changes in III-182
Ocean Thermal Energy Conversion (OTEC) III-304
ocean thermal energy plants II-279
— India, impacts of II-279
oceanic islands, associated with Chile I-701
— habitats and faunas I-708
oceanic mixed layer, Gulf of Alaska III-182, III-185, III-814
oceanic productivity, may decrease with climate change III-369
oceanic swell, Chagos Islands, resistance of spur and groove system II-227
oceanographic conditions, as determinants of habitat boundaries, Lesser Antilles I-630
oceanography
— Australia II-583
— of the Bahamas I-419
— Case 1 and Case 2 waters III-294
— eastern Australian region II-631
— Fiji Islands II-754
— in Greece I-242
— Marshall Islands II-779–80
— Vanuatu II-740
oceanography/marine hydrology, Gulf of Guinea I-777–9
— littoral transport and marine sedimentology I-778
— ocean currents littoral transport and marine sedimentology
— productivity and the seasonal cycle I-779
— sea water quality/structure I-777–8
— tides/waves I-778

— upwelling I-778–9
oceans
— biological divisions I-2
— biome scheme I-2
— common property and open access characteristics III-397
— ecological, economic and social importance of III-393–403
   — ecological importance III-394
   — economic importance of III-395
   — social importance III-396–7
   — sustainable governance III-397–402
— human impacts on III-360
— impediments to ecological or scientific divisions I-2–3
— low in nitrates and phosphate, Australia II-583
— missing areas, reasons for I-3
— political divisions I-2
— unique problems III-397
Oceans Policy, Australia II-590–1
OCs *see* organochlorines
octachlorostyrenes (OCSs) I-54
Oculina reefs, Florida
— experimental reserve III-390
— mostly destroyed by fishers III-379
Odiel River *I-168*
— estuary receives high metal load I-171, I-174
Odiel saltmarshes I-170–1
ODP oceanographic surveys, of the Somali Basin II-67
off shore wind energy III-304–11
— access issues III-306
— foundations, design concepts III-306
— problems of grid connection III-306
— prospects for the future III-308–11
   — new technology and research needs III-311
— review of current technology III-304–7
— status at the millennium III-3078
— suitable off shore areas constrained III-307
— wind farms offshore III-305–6
   — size of installation III-306
offshore habitat, Cambodia II-574–5
offshore petroleum installations, Norway, monitoring of bottom conditions I-27
Offshore Pollution Liability Agreement (northwest Europe) III-338
offshore systems
— Adriatic Sea I-274
— Aegean Sea I-240–1
   — deep-water fauna I-240–1
— American Samoa II-769
— Australia II-583
   — geomorphology II-583
   — oceanography II-583
   — offshore territories II-583
— the Azores I-209–10
— the Bahamas I-422–3
   — deep water channels and V-shaped canyons I-422
— Belize deep reef habitats I-506
   — pelagic waters I-506
— Borneo II-372–3
— Chagos Archipelago II-229–30
— Chesapeake Bay I-343–4
— Colombian Caribbean Coast, importance of fisheries I-669–71
— Colombian Pacific Coast I-680
— Comoros Archipelago II-246–7
   — Geyser and Zélée Bank II-246–7
— Coral, Solomon and Bismark Seas region II-431–3
   — the environment II-431–2
   — tuna fisheries II-432, *II-433*
— Côte d'Ivoire I-813–15
   — benthic assemblages I-814
   — phytoplankton I-814
   — zooplankton I-814–15
— eastern Australian region
   — Elizabeth and Middleton Reefs II-635

— Lord Howe Island II-635
— Norfolk Island II-635
— El Salvador I-551
— Fiji Islands II-758
— and fisheries resources, Bangladesh II-292–5
— French Polynesia II-817
— Great Australian Bight II-681
  — variable abundance of pelagic fish II-681
— Guinea I-800–1
  — fishing I-800
  — industrial fishing I-800–1
  — upwelling I-800
— Gulf of Aden II-54–5
— Gulf of Alaska I-380
— Gulf of Guinea I-781–4
  — interannual variability in upwelling I-781–2
  — upwellings I-781
— Gulf of Maine and Georges Bank I-312
— Gulf of Papua II-598–9
— Gulf of Thailand II-304–5
— Hawaiian Islands II-800–1
  — deep-sea cobalt manganese crusts II-800
— Hong Kong II-538
— Jamaica I-565–6
  — deep habitats not well known I-566
  — Discovery Bay I-566, *I-567*
  — Morant Cays I-565
  — Pedro Cays I-566
— Lesser Antilles I-632–3
  — low salinity lens of Amazon discharge, Tobago to Barbados I-633
— Madagascar II-120
— The Maldives II-205
— Marshall Islands II-779–80
  — oceanography II-779–80
— Mascarene Region II-259–60
— Mozambique II-105, II-107
— New Caledonia II-726–7
  — pelagic zone II-736–7
  — ZoNéCo programme II-727
— New York Bight I-327
— Nicaraguan Caribbean Coast I-522–3
— Nicaraguan Pacific coast I-535
  — oxygen-minimum layer I-535
— northeastern Australia II-619
— northern Gulf of Mexico I-442–3
  — neritic province I-442
  — oceanic province I-442
— Northwest Arabian Sea and Gulf of Oman II-24–5
— the Philippines II-410–11
  — offshore fisheries II-411
  — oil and gas II-411
  — productivity low II-410
  — upwelling and internal waves II-411
— Portuguese coastal waters I-159
— Red Sea II-42–3
— Sea of Okhotsk II-468
  — major and permanent zones of vertical intermixing II-468
  — mesopelagic layer II-468
— South Western Pacific Islands II-712–13
  — pelagic communities II-712–13
— southeast South American shelf marine ecosystem I-757–8
— Southern Gulf of Mexico I-474–5
  — fisheries I-475
  — upwelling and the Yucatan Current I-474–5
— southern Spain I-172
— southwestern Africa I-830–5
  — demersal zone I-833–5
  — environmental variability I-832
  — epipelagic zone I-830–2
  — fishing activity I-832

  — mesopelagic zone I-832–3
— Spanish North coast I-140–1
— Tanzania II-88
— Tasmania II-653
— Tyrrhenian Sea I-274–5
— Vanuatu II-743
— Victoria Province
  — pelagic system II-666
  — slope communities II-666
— West African continental shelf benthic communities I-782, **I-782**
  — fish populations I-783
  — pelagic variability fish populations
  — plankton productivity I-783
  — whale migrations I-783–4
— Western Australian region II-696
— Yellow Sea II-490–1
  — resident and migratory species II-491
  — spawning, nesting and nursery area II-490
offshore wave energy conversion systems III-313–14
— research now limited III-314
oil
— biodegradation increased with clay–oil flocculation III-271
— effects on near shore populations and communities III-271–5
  — effects on biological communities III-272–5
  — general effects III-271–2
— mutagenic effects of long-term exposure III-278
— penetration affected by viscosity and type III-269
— persistence of on shores III-268–71
  — beach wetting: adhesive properties of oil III-270
  — dynamics of in low-energy environments III-271
  — oil-contaminated sandy beaches III-270–1
  — permeability III-269–70
— stickiness of, possible effects on III-270
— total entering northern Gulf of Mexico I-450–1
— trapped in low energy areas I-450
oil contamination
— low levels Gulf of Aden beaches II-59
— southern Brazil I-740
oil deposits, Nicaragua I-520
oil dispersants I-556
— toxic I-76, I-450
— toxicity of III-276
oil exploration I-40
— in Greenland Seas I-14
— Nicaragua, Caribbean coast I-526
oil exploration and production
— Argentina I-766–7
— Gulf of Guinea I-792
— offshore, Australia II-588
— Peru I-696
— southern Brazil I-738
— Western Australia II-702
oil exports
— Colombian Caribbean Coast I-673
— Nigeria and Gabon I-792
oil and gas
— exploration and exploitation, Côte d'Ivoire I-817
— the Philippines II-411
— Trinidad I-638
oil and gas exploration
— drilling waste piles on Norwegian seabed I-17, I-23–4
— Irish Sea I-96
— Mozambique coast II-109
— North Carolina coast I-366–7
— North Sea I-51
  — exploitation increasing I-52–3
— and production, Gulf of Thailand II-304–5
— South Africa II-142
oil and gas fields
— Bass Strait II-669
— east coast, Peninsular Malaysia II-354

oil and gas industry
— Australia II-584
— Sakhalin Shelf II-470–1
oil and gas installations
— offshore
— far-field effects possible III-364
— marine pollution from III-360
— operational discharges from III-363–4
— other chemicals in use, possible effects of III-364
oil and gas potential, Somalia II-79
oil and gas production
— Adriatic Sea I-278
— Malacca Straits II-315–16, *II-315*
— North Sea, environmental impacts at all stages I-55–6
— northern Gulf of Mexico I-448
— offshore, pollution by III-338
— transport of crude oil and environmental threat I-450
— Yemen II-59
oil and gas resources, South China shelf II-557
oil industry, offshore, Gulf of Mexico I-478–9
oil installations, offshore, Brunei II-372
oil persistence III-278
— effects may change as oil weathers III-271
— from the Refinería Panama storage tank rupture III-273
oil pipelines II-603–4, II-605
— environmental effect, Gulf of Guinea I-792
— southern Brazil, effects of rupture I-742
oil pollution II-376
— Black Sea I-294
— chronic
— Argentinian coast I-766
— western Indonesia, from oil refineries and production facilities II-390
— Côte d'Ivoire beaches I-817
— Dutch Antilles I-604–5
— False Bay, South Africa II-142
— from shipping III-364
— Gulf of Guinea I-792
— Gulf of Oman
— from routine tanker operations II-28
— worsening II-29
— heavy near Chittagong and Chalna, Bangladesh II-294–5
— increasing, South China Sea II-354
— Jamaica I-570
— and loss of seabirds III-113
— marine ecosystem at Toamasina (Madagascar) threatened II-125
— of marine sediments I-638
— Mexican Pacific coast I-497
— northern Gulf of Mexico I-450–1
— not avoided by small cetaceans III-96
— Sakhalin Shelf II-471
— Saronikos Gulf I-245
— sensitivity of organisms to I-450
— southern Brazil, Sepetiba Bay I-741–2
— Sri Lanka II-185
— Straits of Hormuz II-28
— Sydney Harbour and Botany Bay II-641
— Tierra del Fuego I-762
— Tyrrhenian Sea coasts I-279
— Venezuela
— eastern coastline I-655–6
— western coastline I-656–9
— Xiamen coastal waters II-523
oil production
— Argentina I-761, I-763
— Chile I-712
— Guangdong II-557
— Gulf of Papua, pipelines for delivery of II-603–4
— Mexico I-476
— Niger Delta, onshore I-788
— northern Gulf of Mexico *I-443*

— Northern Sakhalin, causing serious concern II-477
— northwestern Arabian Sea II-28
— onshore, Wytch Farm, Dorset I-72
— western Indonesia II-390
— effects of increase in II-390
oil refineries
— Aden II-59
— Aruba and Curaçao I-604–5
— Fawley I-72
— Malacca Straits *II-315*, II-316
— Mogadishu II-80
— South Korea II-495
— southern Brazil I-738
oil revenues, invested in infrastructure II-28
oil seeps
— Gulf of Alaska I-381
— resulting from salt tectonism, northern Gulf of Mexico I-442, I-451
oil slicks
— Bay of Bengal II-280
— from ballast water I-72
oil spill contingency planning, Coral, Solomon and Bismark Seas region II-441
oil spill response equipment, for the Malacca Straits II-325
oil spills I-512, I-638, II-354, II-417
— accidental, risk of I-14
— acute
— lethal and sublethal effects III-272
— sublethal and chronic effects from III-278
— Aegean Sea I-248
— affect sea turtles III-67
— Alaska, *Exxon Valdez* I-382
— amount spilled *III-268*
— Arabian Gulf II-3
— and the Gulf War II-11–13
— Argentina I-765–7
— "Metula" spill I-765
— "San Jorge" spill I-765
— Australia II-587–8
— Bay of Cadiz I-181
— Black Sea, Nassia disaster I-294
— Chilean coast **I-712**
— differences in lead to different responses III-272–3
— direct causes of mortality III-272
— El Salvador I-556
— endangering western Indonesian coastlines II-388
— English Channel I-72, I-75–6
— Amoco Cadiz, long term effects I-76
— Torrey Canyon, major damage from oil dispersants I-76
— Galician coast I-146
— Great Australian Bight II-686
— Gulf of Aden II-59
— Gulf of Mexico I-473
— high potential for, western Indonesian seas II-390, **II-390**
— Hong Kong, mainly minor II-542
— and illegal discharges, North Sea I-55–6
— increasing, Marshall Islands II-783
— Malacca Strait II-339
— Malacca Straits II-319–20
— serious impact on fragile ecosystems II-319
— Standard Operation Procedures (SOP) II-322–3
— Maracaibo Lake I-656
— Mozambique Channel II-251
— Niger Delta I-788
— northern Gulf of Mexico I-450
— Norway, few I-26
— off Fujeirah II-28
— persistence on beaches III-168–71
— persistent oil effects and their causes III-275–6
— Peru I-696
— a risk for Mozambique II-109
— *Showa Maru* II-334

— small, Sri Lanka II-185
— South China Sea (western) II-559
— southern Brazil
    — Sao Sebastiao City I-742
    — Sepetiba Bay I-741–2
    — southeast coast, and degraded mangroves I-736
— Tasmania II-657
— timing of important III-269, III-270, III-272
— Tobago I-638
— Torres Strait, *Oceanic Grandeur* II-605
— treatment effects III-276–7
    — bioremediation III-277
    — dispersants III-276
    — injuries due to III-275, III-276
    — manual removal of oil III-276–7
    — problems with use of heavy machinery III-277
    — sand-blasting and high/low-pressure-water techniques III-277
— Xiamen coastal waters II-523
— Yellow Sea II-495
oil tanker traffic
— Malacca Strait II-339
— Malacca Straits II-316–17
— Sepetiba Bay, southern Brazil I-741–2
oil terminals, southern Brazil, Sao Sebastiao City I-739, I-742
oil-well fires II-11
okadaic acid III-40, III-219
Okhotsk, Sea of II-463–72
— currents II-465–6
— effects from urban and industrial activities II-469–71
— major shallow-water marine and coastal habitats II-467–8
— offshore systems II-468
— populations affecting the area II-468–9
— protective measures II-471
    — fishery regulations II-471
    — poaching and overfishing II-471
— rural factors II-469
— seasonality, currents, natural environmental variables II-466–7
    — seasonal changes in biota II-466
— winds drive winter water movement II-465
Okhotsk shelf, productive fish area II-467
Old Bahama Channel *I-416*, I-417
Olive Ridley turtle I-780, II-180, II-205, II-289, III-63
— captive breeding programmes, Sundarbans II-158
— critically endangered III-63
— distinctive nesting behaviour III-63
— threatened I-537
Oman *II-2, II-18*
— agriculture II-26
— construction of regional fishing harbours II-27
— coral communities
    — damaged by nets *II-30*
    — limiting factors II-22
— increase in beach tar II-28
— industrial diversification, new industrial development, Sohar and Sur II-25, II-27, II-28
— limestone mountains *II-18*
    — unique vegetation II-21
— network of conservation areas proposed II-30
    — national nature reserves II-30
— population II-25
— Ra's Al Hadd National Scenic Reserve II-30
— Ra's Al Junayz National Nature Reserve II-30
— seabirds III-110
— upwellings along coast affect local weather II-50
Oman, Gulf of *II-18*
— current flow *II-18*, II-20
— defined II-19
— fish biodiversity II-24
— low-energy environments II-21
— sea water temperature fluctuations II-21, II-22, *II-22*
— shallow water marine and coastal habitats II-21–4

— corals, reefs and macroalgae II-22–3
— seagrasses II-23–4
— turtles II-24
— *wadis, khawrs* and mangroves II-23
Ontong Java Plateau *II-426*
open ocean habitats I-224–7
— anguilla I-226
— biogeochemistry of the area round Bermuda I-224–6
— humpback whales I-226–7
— *Sargassum* community I-226
open sea banks I-47
Operation Raleigh I-590
optical remote sensing, governing processes involved III-295, *III-296*
orcas ("killer whale") III-76
— high levels of PCBs III-96
— widely distributed III-91
organic carbon, New York Bight, sources of I-324–5
organic matter/material
— Baltic Proper, sinks to soft bottoms I-128
— Baltic Sea I-112–14
— delivered to Baltic Proper **I-103**, I-104
organic pollutants
— Aegean Sea **I-248**
— Baltic Sea I-131
— Faroes I-37–8
— Nervión estuary, northern Spain I-145–6
— Sargasso Sea I-229
organic pollution
— Adriatic Sea I-270
— in the English Channel I-75
— Hong Kong coastal waters II-543
    — trace contaminants II-544
— Sea of Japan *II-479*
— West Guangdong II-559
    — Pearl River mouth II-555
— west Taiwan II-509
— Xiamen region II-522
organisms, sensitivity to oil III-272
organochlorine burdens
— Arctic beluga III-362
— slowing recovery of Baltic Sea seal populations III-361
organochlorine pollution
— fish from Asian waters II-449–51
— oceanwide II-454
organochlorine residues
— in Hoogly (Hugli) river sediments II-279
— in rivers and marine biota, Malaysia II-319, II-338, II-341
organochlorines
— in Australia's marine environment II-588
— Jakarta Bay II-398
— in Liverpool Bay I-95
— southern and western Baltic I-112, *I-113*
    — DDT in herring and perch I-112, *I-114*
— in USA dolphins III-96
— use of in China II-492
organochlorines, persistent
— Asian developing countries *II-448*, II-449, *II-453*
— contamination and bioaccumulation in marine mammals II-455–6
    — toxic effects II-456–8
— contamination in North and South Pacific II-454
— global fate II-455
— Hong Kong II-542
— major pollution sources now II-455
— river and estuarine sediments, Asian developing region II-449, *II-451, II-550*
— temporal trend of contamination II-458–60
    — in Antarctic minke whales II-458, *II-460*
— West Guangdong coast II-555–6
organophosphate compounds I-493
organotin pollution, in fish II-451, *II-454*

organotins
— in anti-fouling paints I-146
— Asian developing regions II-449
— in USA dolphins III-96
— *see also* tributyltin
Orinocco effects I-579
Orinoco Basin
— extent of and physiographic units I-646
— geochemical characteristics of rivers I-646–7
Orinoco River
— changed tidal effects after closure of Caño Manamo I-648
— delta *I-644*
   — functions like a wetland I-648
   — main zone I-648
— effects of fluctuation in discharge I-648
— influences of, Venezuelan Atlantic coast and Caribbean Sea I-646–8
Orinoco River Plume I-630
Ortegal Cape *I-136*, I-139
Oslo *I-18*
— contamination of harbour sediments I-27
Oslo and Paris Commissions (OSPARCOM) I-43, III-221
— monitoring contaminants in sediments, biota and waters I-77–8
— monitoring programmes (JMP and JAMP) I-27, I-77–8
OSPAR Convention I-26, I-57, **I-57**, I-147, I-162
— hazardous substance targets III-362–3
— stronger controls on off shore oil and gas III-362–3
Otway Region, Victoria Province II-663
over-exploitation
— of coastal fish resource, Sri Lanka II-182
— littoral fish, Canary Islands I-196
— Mexican artisanal shrimp fisheries I-490, I-491
— Northwest Pacific sardine fishery I-490
— of timber, Vanuatu II-743
— turtle fishery I-491
— Venezuelan coast zone I-645
over-harvesting, of renewable resources, Marshall Islands II-784
overcapacity, world's fishing fleets III-367, III-377
overfishing I-3, I-312, III-119, III-130, III-367
— Adriatic Sea I-278
— Carolinas I-366
— and catch decline, Mozambique II-107
— Colombia
   — Caribbean Coast I-671
   — Pacific Coast, shrimps I-684
— in the Comoros II-248
— Coral, Solomon and Bismark Seas region, boom-and-bust cycles II-437
— Malacca Straits II-320
— 'Malthusian overfishing' III-123
— of marine invertebrates, Fiji Islands II-759
— North Sea I-52
— northern Gulf of Mexico I-445
— Oman and Yemen II-27
   — with industrial methods II-29
— of predators III-128–9
— of salmon and seatrout, in the English Channel I-73–4
— South Western Pacific Islands II-716, II-719
— technique restrictions not effective against recruitment overfishing III-161
— threatens artisanal fishing, Belize I-509
— Vanuatu II-747
overgrazing
— effects of I-444
— Hawaiian Islands II-802
— Oman and Yemen II-26
overgrazing/overcropping eastern Australian region II-637
overharvesting, of fish resources, Palk Bay–Madras coast II-170
Owen Fracture Zone *II-64*, II-67
oxygen deficiency, Baltic Proper, affecting cod spawning I-110
oxygen depletion III-259

— between Mississippi and Sabine rivers I-443
— effects of, North Sea I-53
Oxygen Minimum Layer, Arabian Sea II-24
oyster banks, Chesapeake Bay I-341
oyster beds
— importance of, Carolinas coast I-362
— Texas, disappearing I-447
oyster culture, Venezuela I-653
oyster farms
— Korea II-494
— West Guangdong II-558–9
oyster fishery, small commercial, KwaZulu-Natal II-139
oyster habitat, Chesapeake Bay, destruction of III-126
oyster harvesting
— Chesapeake Bay I-341–2
— northern Gulf of Mexico I-442
oyster reefs, northern Gulf of Mexico I-441–2
oyster shell, for agricultural uses I-447
oysters III-167
— communities destroyed through burial I-447
— decline in production, Tagus estuary I-162
— dragnets and dredging damage seagrass beds III-8–9
— effects of TBT III-250
— El Salvador I-551
   — contamination in I-555
— farming of, Normandy and Brittany coasts I-73
— as indicators of environmental contamination III-253
— rock oysters II-52
— Texas, declines in linked to salinity perturbations I-439–40
— *see also* pearl oyster culture; pearl oyster fishery
ozone, tropospheric, Azores I-204
ozone depletion III-370–1
— a possible impact on small cetaceans III-98

Pacific basin, detection of mid-1970s regime shift III-224–5
Pacific Center Coastal Province, Mexico I-488
— habitats I-488
— important fishing area I-491
— varied marine and coastal fauna I-491
Pacific continental platform I-534
Pacific Deep Water I-679, I-702, II-583
Pacific Ocean
— Central and Eastern, coral bleaching and mortality events III-53
— equatorial current system I-679
— north, surface layer is fresh III-181
— Northwest, coral bleaching and mortality events III-52
— Southwest, coral bleaching and mortality events III-52–3
Pacific Ridley turtle, destroyed by landslides I-539
Pagassitikos Gulf *I-234*, I-236, I-237, I-244
— nutrient rich I-244
— some chlorinated biphenyls I-247–8
Pago Pago harbour, American Samoa
— dredging and filling causing reef loss II-717
— heavy metal pollution II-712, II-771
— industrial activities II-770–1
— loss of habitats II-770
PAH monitoring, Prince William Sound and Cook Inlet I-381
PAHs *see* polycyclic aromatic hydrocarbons (PAHs)
Palk Bay *II-162*, II-163, II-166–8
— climate and coastal hydrography II-166
— coastal ecosystems II-166–7
— fish and fisheries II-167
— geological features II-166
— human population and environmental degradation II-167
— Muthupet Lagoon II-167
— Vedaranyam wildlife sanctuary II-167
Palk Bay–Madras coast II-168–70
— biodiversity II-169
— climate and coastal hydrography II-168
— coastal ecosystems II-168–9
— environmental degradation II-169

— affected by decrease in freshwater II-169
— fish and fisheries biodiversity II-169
— geological features II-168
— impacts of human activities on the ecosystem II-169–70
Pamlico River Estuary *I-352*
— dinoflagellates I-355
— freshwater eelgrass I-359
— productivity pulses and algal blooms I-354
Pamlico Sound *I-352*, I-353
— environmental problems for bottom communities I-360
Panama Bight *I-678*
— seasonal upwelling cycle I-679
— tuna/anchovy/shrimp fishery I-680
Panama Current *I-678*
Panama, Gulf of *I-678*
Panamic Coastal Province, Mexico I-488–9
— habitats and communities I-489
"pantry" reserves, Marshall Islands II-786
Papua, Gulf of *II-426*, *II-594*
— agriculture II-600–1
— fisheries management II-606–7
— low level of development II-599
— water and sediment discharge to II-596, **II-596**
Papua New Guinea (PNG) II-427, **II-427**, II-595
— coral bleaching and mortality events III-53
— coral reefs in good conditions II-431
— effects of logging II-435
— impact of land use on coastal waters II-435–6
— level of social and economic development lo II-434
— mangroves II-430
— population and demography II-433–4, **II-434**
— research providing some information on marine biota II-435
— seagrass species diversity II-431
— sedimentary coast, backed by mountains II-427–8
Papuan Barrier Reef II-430, II-597
paralytic poisoning, and saxitoxin III-40
Paralytic Shellfish Poisoning (PSP) I-69, I-160, I-765, II-320, II-339, II-418, II-495, II-543, III-221
Paria, Gulf of *I-644*
— dry season circulation of Orinoco waters I-648
Paria Peninsula, cold waters present I-646
Paris Convention, Protection of Marine Pollution from Land-based Sources III-339
particles
— hygroscopic growth of, and deposition velocities III-203
— removal by wet deposition III-200
Pas de Calais, zooplankton in waters off I-70
pastoral nomadism, Somalia II-77, II-80
Patagonian coast, Argentina I-751–2, *I-752*
— eutrophication I-765
— evolution of settlements I-758, I-761
— hydrocarbon concentrations I-761
— intensification of tidal currents I-754
— northern, characteristic sublittoral communities I-754–5
— unique environments I-756–7
  — marine birds I-757
  — marine mammals I-756–7
patch reefs I-505, II-7–8, II-65, II-191
— Glovers Atoll I-505
— lagoonal
  — Bermuda I-227, I-228
  — Mauritius II-258
— Madagascar II-118
— northern Kalimantan and eastern Sabah II-371
— Tanzania II-85–6
Patella, affected by oil dispersants I-76
Patos Lagoon, southern Brazil *I-732*, I-737
— fishes of I-740
PCBs I-54, I-279, I-392, I-555
— in animals from the Greenland seas I-9, I-11, I-14
— in Asian developing region waters II-449
— in Bergen fish and shellfish I-27
— in blubber II-455
— concentrations in Arctic species III-361, *III-362*
— coplanar PCBs considered more toxic II-457
— estimated loads in the global environment II-455, **II-455**
— in the Gulf of Maine I-317
— high levels in Irish Sea mammals I-95
— Hong Kong coastal waters II-544, **II-545**
— in marine sediments, Vietnam II-565–6
— marine transport of from European waters I-14, *I-14*
— pollution from, Thailand II-449
— reasons for persistence of III-362
— reduction in positively affecting Baltic seal populations I-131
— uniform distribution of in air and surface seawater II-454, *II-458*
— in whale blubber I-36
Peale's dolphins III-91, III-93
Pearl Cays complex, corals I-522
pearl oyster and Chank beds, Gulf of Mannar II-164
pearl oyster culture II-207
— French Polynesia, problems of II-820
— Marshall Islands II-781, II-782
— Solomon Islands II-438
pearl oyster fishery
— Arabian Gulf, in decline II-11
— Fiji, overexploited II-759
— French Polynesia II-820
pearl oysters II-598
Pearl River Delta *II-550*
— phytoplankton II-553
— tributaries II-551
Pearl River mouth
— irregular semi-diurnal tides dominate II-552
— and the Pearl River II-551
pearl shell fishery II-602
peat, Sri Lanka II-181
peat swamp forests, Malacca Strait II-336, II-342
peat swamps
— east Sumatra II-312
— Malacca Strait II-335
pelagic organisms
— Irish Sea I-90–1
— West Guangdong coast II-554
Pemba *II-84*, II-85
Penobscot Bay *I-308*, I-314
Pentland Firth, Scotland, considerable tidal currents III-318
peroxyacetylnitrate (PAN) I-205
Persian Gulf *see* Arabian Gulf
persistent organic compounds, contribute to seal decline I-50
persistent organic pollutants (POPs) I-5, I-39–40, I-381, III-360, III-361
— Asian developing regions II-447–62
— atmospheric transport to colder regions III-361
— decline very slow III-362
— Greenland I-9, I-11, *I-11*
  — from Europe and Russia I-13
  — may affect human health I-11
— toxicity risk to small cetaceans III-95–6
— transport and deposition in Europe I-56
Peru I-687–97
— biodiversity I-691–3
  — Peruvian—Chilean province I-691
  — Provincia Panameña I-691
— characteristics of the coast I-689–91
— coastal populations and the main sources of pollution I-693–6
— collapse of anchovy fishery III-119, III-129
— decline in guano birds III-111, III-129
— direct and indirect small cetacean catches III-93
— legislation on environmental protection I-696
— main populated and industrial areas I-695
— National Contingency Plan (oil spills) I-696
Peru (Humboldt) Current I-486, I-487, I-488, I-489, I-543

Peruvian Current
— high productivity I-692
— influences coastal climate I-690
— northward-flowing I-690
pesticide pollution
— and coastal ecosystems, the Seychelles II-239
— from Mozambique's upstream neighbours II-109
— Peru I-694–5
— Southern Gulf of Mexico I-476, I-479
— Xiamen coastal waters II-524
pesticides
— causing environmental stress, Altata-Ensenada del Pabellón I-493
— discharged to Black Sea I-301–2
— El Salvador
    — in oysters I-555
    — in river waters I-552
— golf courses, Hawaiian Islands II-806
— Gulf of Guinea I-784–5
— in land-based pollution III-361
— loss of from shipping III-364
— Malaysian waterways II-341
— organochlorine, Vietnam coastal waters II-565
— and other chemicals used in aquaculture III-369
— in shrimps I-493
— use and abuse, the Philippines II-412–13
— use round Sea of Japan II-475
— use of in Tanzania II-88–9
— used in Greece, concentration in the marine environment I-243–4
Peter the Great Gulf
— changes in bottom communities II-481
— radioactivity in II-470–80
Peter the Great Marine Reserve, endangered II-477
petrochemicals, southern Brazil I-739
petroleum, Venezuela's main export I-655
petroleum hydrocarbons, in Arabian Gulf sediments II-12
petroleum refineries *see* oil refineries
Pfisteria, ichthyotoxic I-363–4, I-366
Phaeocystis I-69, I-90, I-356
pharmaceuticals
— from marine environments and coral reefs III-39
— natural products and chemicals III-38–9
— natural products from coral reefs III-40
Philippine Tuna Research Project II-410
Philippines, The II-405–23, II-416–17
— the area and its natural environmental variables II-407–8
— coastal erosion and landfill II-413–14
— coral bleaching and mortality events III-51–2
— effects from urban and industrial activities II-414–18
— major shallow-water marine and coastal habitats II-408–10
— mangroves
    — clearance for aquaculture III-368
    — lost to fishponds III-21–2
— offshore systems II-410–11
— populations affecting the area II-411–12
— protective measures II-418–20
    — Coastal Environment Program (CEP) II-419
    — Coastal Resource Management Program II-419
    — community-based coastal resources management (CB-CRM) II-418
    — Fisheries Sector Program II-418–19
    — legalities, utilization, conservation and management of the coastal areas **II-420**
    — marine protected areas II-419
— rural factors II-412–13
— seimically active II-408
phocine distemper virus (PDV) (1988)
— killed seals on English south coast I-71
— reduced seal numbers, North Sea I-50
phocine distemper viruses III-225
phosphate I-255
— increased load to Baltic Sea I-130

— a limiting factor I-240
— release of, the Bodden III-262–3
— seawater, Vietnam II-565
phosphate enrichment, euphotic layer, Côte d'Ivoire I-810
phosphate mining, Makatea, Tuamotu Archipelago II-822
phosphoric acid manufacture I-72
phosphorite ore, Onslow Bay I-366
phosphorus
— from phosphate rock processing I-94
— increased load to Baltic Sea I-130
— as a limiting nutrient I-106
— load to Chesapeake Bay I-344–5
phosphorus level, low, Orinoco River I-647–8
photic layer, Panama Bight I-679
photoinhibition II-42
photosynthesis III-394
— anoxigenic III-263
phthalates, in coastal sediments, west Taiwan II-509
Phyllophera meadows, Black Sea, decrease in I-291, I-304
physical environmental anomalies, make entire populations vulnerable III-225
physically forced (climate/oceanographic) disturbances III-224–5
phytohydrographic associations, Arabian Gulf II-6–7, **II-7**
phytoplankton
— Carolinas
    — estuarine I-354–6
    — marine I-356
— Guinea I-800
— Gulf of Guinea I-779
— Gulf of Thailand II-304
— harmful effects of increased exposure to UV-B III-370
— key roles of III-198
— Malacca Strait II-333
— Malacca Straits II-312
— offshore, Côte d'Ivoire I-814
— species diversity, Vietnam II-564
— West Guangdong coast II-553
— wide range in UV-B sensitivity, effects of III-370–1
phytoplankton assemblages, Australia II-583
phytoplankton biomass
— and biodiversity, Pearl River mouth II-553–4
— coastal, Sunda Shelf II-384
— nutrient control of, Tuamoto archipelago II-817
phytoplankton blooms I-244, II-50, II-700
— Baltic Proper
    — changes in composition and dominance I-109
    — spring I-106
— Canary Islands I-192
— Irish Sea I-89
— Norway, spring and summer I-20
— Spanish north coast I-140, I-141
phytoplankton communities, Mascarene Region, nutrient-limited II-259
phytoplankton ecology, Bay of Bengal II-275, *II-276*
phytoplankton growth
— North Spanish coast I-138
— Tagus and Sado coastal waters I-155–6
— western English Channel I-67
phytoplankton processes, Spanish north coast, modified by oceanographic processes I-141
picoplankton, becoming dominant, the Bodden III-261
pilchard fishery
— Australian Bight II-683
    — may affect seabirds II-685
— southwestern Africa I-831
    — failure of I-832
— Victoria Province, Australia II-668
pilchards, killed by herpes virus II-587
pilot whales III-76, III-91
— hunted, Faroe Islands III-92
— long-finned I-31, I-36

Pinatubo, Mount, effects of eruptions II-408
pinnipeds, effects of decline in III-99
Pinus pinea forests, southern Spain I-170, I-181
Pitt Bank *II-222*
plaice I-49
— Baltic, stock decreased I-111
plankton
— abundance of, Argentine Sea I-757
— in the Black Sea I-290
— Côte d'Ivoire I-812–13
— eastern Australian region II-634
— North Sea I-48
— Peru, changed during El Niño I-692
— Tagus estuary I-157
— West Guangdong coast **II-554**
plankton assemblages, southeastern Taiwan II-508
plankton productivity, Gulf of Guinea I-783
planktonic communities, French Polynesian lagoons II-816–17
planktonic food web, model for, Takapoto atoll II-817
planktonic systems, Irish Sea I-89–90
Plantagenet Bank I-227
plate boundary, diffuse, through Chagos area II-226
platform reefs II-388
— the Seychelles II-236
plutonium (Pu) I-94
— fall out from nuclear tests I-117
PO_4 levels, affected by Tagus and Sado river discharges I-155, *I-157*
Po basin
— agricultural and industrial pollutants from I-275
— evolution of Po delta I-276
poaching, and marine reserves III-390
Poland
— discharged partly treated sewage to River Vistula I-115
— effect of increased standard of living I-112
— environmental contamination from mining I-114–15
— pollutants to Baltic Proper I-103–4
"poles of development", Chile I-710
politics, in fisheries management III-154
pollack I-313
pollock fishery
— Alaska III-147
— Bering Sea III-129
pollutant layers, elevated III-200
pollutants
— atmospheric transport of I-13
— baseline studies, Gulf of Alaska I-381, I-383
— brought by river to Black Sea I-287, I-298–9
— from Turkish Black Sea coast *I-200–301*, **I-300**
— Carolinas coast, sources of I-354
— carried to Baltic Proper by rivers I-103
— direct discharge into estuaries, Portugal I-162
— dispersion and deposition processes in the coastal zone III-206
— entering the Bay of Bengal II-278
— environmental, Baltic Sea I-112–17
— global redistribution of *III-361*
— higher sensitivity of Baltic populations I-129
— increase susceptibility to infection III-225
— and the internal boundary layer III-207
— Malacca Strait II-340
— marine current transport of I-13–14, *I-14*
— Palk Bay–Madras coast II-169
— and pathogens, waterborne, spread of in the Caribbean I-503
— reduction of phosphorus and nitrogen to North Sea I-24
— sea ice transport of I-14
— Southern California Bight
   — largest reduction from publicly owned treatment works I-389–90
   — multiple source discharges I-388–9, *I-389*
   — reductions in I-389
— *see also* long-distance transport of pollutants
polluter pays principle II-323, II-327, III-339, III-401

— Jamaica I-570
pollution
— Aegean Sea I-245–9
— affecting eelgrass beds III-7–8
— affecting small cetaceans III-95–6
— airborne, Azores I-204
— atmospheric, from ships III-340
— Bahamas, effects on water quality and near-shore habitat I-429–30
— of beaches, Sri Lanka II-183
— Belize
   — control through the Environmental Protection Act I-513
   — from industry I-511
   — urban I-512
— Black Sea, sources of I-287, I-297, **I-297**, I-298–9, **I-298**, *I-299*
— Cabo Frio–Cabo de Santa Marta Grande, southern Brazil I-738–9
— definition and sources of III-258
— degrading Gulf of Guinea coastal waters I-784
— effects of in the Baltic Sea I-130–1
   — eutrophication I-130–1
   — heavy metals I-131
   — organic pollutants I-131
   — pulp mill industry I-131
— effects of climate change and sea level rise III-195
— effects of land-based sources, Israel I-259–61
   — Haifa Bay I-259–61
— entering the northern Gulf of Mexico I-449–50
— entering rivers and coastal seas, western Indonesia II-396
— estuarine and marine, South Africa II-142–3
— from fishponds, the Philippines II-414
— from ships, Coral, Solomon and Bismark Seas region II-440–1
— from shrimp farming and agrochemicals, Bangladesh II-294
— in the Great Barrier Reef region II-623–4
— Guinea I-801, I-802
— Gulf of Alaska, mainly from long-distance transport I-381
— Gulf of Mannar II-165
— impacting on seabirds I-49
— industrial
   — and domestic, Peru I-694, I-695
   — Tanzania II-92
— Kingston Harbour, Jamaica I-570
— land-based
   — contributing to marine pollution III-339–40, III-340, III-360, III-361–3
   — regulation of III-362–3
   — slower treatment of III-339–40
   — threat to coral reef biodiversity II-389
— land-based, Turkish Black Sea coast I-296–303
   — from city sewerage systems I-300–1, *I-301*
   — from rivers I-299–300
   — monitoring I-291–302
   — pesticides and PCBs I-301–2
— localised, Coral, Solomon and Bismark Seas region II-439
— Malacca Strait
   — faecal coliform count II-340
   — Indonesian side not systematically monitored II-341
   — land-based II-340
   — sea-based II-340
— Malacca Straits II-317–20
   — agricultural waste II-318–19
   — coliform contamination II-319
   — land-based II-317–18, **II-318**
   — main problem areas II-325–6
   — sea-based sources II-319–20
— moderate in Canary Islands I-198
— Nicaraguan Pacific coast I-534
— organic I-835
— Papeete, Tahiti II-817
— point-source, Fiji II-760
— release to marine environment from point and non-point sources III-258
— Saronikos Gulf I-238

— and seabird deaths III-114
— secondary, from Maluan Bay sediments, Xiamen region II-524
— the Seychelles, from habitation and farms II-239
— Taiwan Strait, effects of tidal currents II-502
— through shipping operations and accidents I-146
— Venezuela I-654
— vessel-sourced, reduction of III-343
— Western Australia II-697
pollution abatement, some progress, North Sea countries I-52
pollution hotspots
— identified in Baltic I-132
— the Philippines II-418
pollution prevention programmes
— Yellow Sea
    — economic problems a major impediment to II-496
    — land-based, obstacles to II-495–6
pollution-contamination distinction difficult III-258
polybrominated diphenylethers (PBDEs) I-36
— presence in marine mammals III-362
polychaetes I-157, I-239
— Faroes I-34
— first to colonize mine tailings III-240
— Spanish North coast I-139, I-140
polychlorinated biphenyls *see* PCBs
polycyclic aromatic hydrocarbons (PAHs) I-56, I-392, I-555, II-417, III-361
— Arabian Gulf II-12
— attributed to oil seeps I-381
— in fjords I-24
— Great Barrier Reef region II-623
— Hong Kong coastal waters II-544, **II-545**
    — posing a risk to ecosystems and seafood consumers II-545
— multi-hop contaminants I-13
— in Norwegian shellfish I-25
— Prince William Sound, delayed effects I-382
— in sediments, Baja California I-497
— southern Brazil
    — Guanabara Bay I-741
    — Santos Estuary I-743
— Taiwan Strait II-510
— *see also* PAH monitoring
polynyas
— Baffin Bay I-8
— Greenland Sea I-7
— Sea of Okhotsk II-466
Pomeranian Bay I-109, I-115
Poole Bay *I-66*
POPs *see* persistent organic pollutants
population
— Adriatic and Balkan coastlines I-275
— Aegean, ancient cultures, modern cities I-241–2
— Alaska I-381
— American Samoa II-769
    — population growth rates II-769, *II-769*
— Andaman, Nicobar and Lakshadweep Islands II-193
    — and increasing environmental degradation II-193
    — Lakshadweeps II-194
— Anguilla, British Virgin Islands and Montserrat I-620–1
— Arabian Gulf II-9
— Argentinian coast I-758, I-761–2
— around the Irish Sea I-92–3
— around the North Sea, effects of I-50–3
— around the Sea of Japan II-475
— Australia II-586
    — Aboriginal peoples II-583, II-682
    — general community II-584
    — indigenous communities II-583–4
— the Azores I-210
— the Bahamas I-423
— Baltic catchment I-50
— Bangladesh II-291–2, **II-292**

— Belize I-506–7
— Borneo II-373–4
— Brazil I-733
— Carolinas coast, growing I-354
— Chesapeake Bay area
    — Europeans I-344
    — growth and development I-344
    — Native Americans I-344
    — sprawl development I-345
— coastal
    — Cambodia II-575
    — surrounding the Yellow Sea II-491–2
    — Vietnam II-563
— Colombian Pacific Coast I-680–1
— Comoros Archipelago II-247
— and consumption III-396
— Côte d'Ivoire I-815–16
    — coastal cities I-815
    — indigenous population I-815
— and demography
    — Papua New Guinea II-433–4, **II-434**
    — Solomon Islands II-433–4, **II-434**
— Dutch Antilles I-612
    — leeward group I-597, I-598
— east coast, Peninsular Malaysia II-350–1, *II-351*
    — urban growth rate II-351, **II-351**
— eastern Australian region II-635–6
    — New South Wales II-636
    — south-east Queensland II-636
— El Salvador I-551
— English Channel coasts I-71
— Faroes I-35
— Fiji II-758
— Florida Keys I-408
— Great Australian Bight region II-682
    — European colonisation II-682
— Great Barrier Reef region II-619–20
    — indigenous people II-619–20
    — trends in II-620
— Greenland I-8
— growing beyond sustainable limits II-173
— Guinea coastal zone I-801–2
— Gulf of Aden states, rural and poor II-56, **II-56**
— Gulf of Guinea coast I-784
    — country demographic information **I-785**
— Gulf of Maine I-314
— Hawaiian Islands II-801
    — decline of native population II-801
    — demographic patterns II-801
    — tourism trends II-801
— Hong Kong II-538, II-541
— Huelva and Cadiz, southern Spain I-172
— human
    — growth of III-396
    — total impact of III-396
— and human settlements, Lesser Antilles I-634, **I-634**
— Iranian Gulf of Oman coast II-25
— Israel I-258
— Jamaica I-567
    — development pressure I-561, I-567
— Madagascar II-120–1
    — increase in major coastal towns II-121
— Marshall Islands
    — coastal uses and environmental issues II-780
    — demographic patterns II-780
— Mascarene Region
    — cities and sewage discharges II-264
    — coastal II-260
— Mexican Pacific coast I-489
    — in medium and small communities I-492
— Mozambique, trend to urbanisation II-107

— New Caledonia
  — distribution of II-728
  — structure II-727–8
— of the New York Bight area I-328–9
— Nicaraguan Caribbean coast I-523
— Nicaraguan Pacific coast I-537–8
  — Estero Real, degraded natural resources I-537
  — Gulf of Fonseca I-537–8
  — population-related problems I-537
— northern coast of Spain I-142, *I-143*
— northern Gulf of Mexico I-443, I-448
— Norway I-17, I-24, I-28
— Oman II-25
— Palk Bay coast II-167
— Pearl River delta and West Guangdong II-556–8
— Peruvian coast I-693
  — cities I-695
— the Philippines II-411–12
  — coastal population depends on coastal fisheries II-414
  — growth rate II-412
  — rural–urban migration II-412
— Poland and Lithuania I-103–4
— and politics, French Polynesia II-815
— rates of change, Lesser Antilles I-634
— Red Sea coastline, mainly major ports and cities II-43
— regions round Sea of Okhotsk II-468–**II-469**
  — Sakhalinsky region, unique II-468
— round the Malacca Straits II-311
— rural, Cambodia II-575
— the Seychelles II-237–8
— Singapore II-337
— Somalia 77
  — migration to cities II-77, *II-79*
— South Western Pacific Islands II-713–14
  — population trends II-713
— southeast India, Gulf of Mannar II-165
— southern Black Sea coast, Turkey **I-296**, I-297
— southern Brazil
  — cities growing fast I-738
  — metropolitan areas I-737, I-738
  — Rio de Janeiro and Santos I-738
  — Santos Bay I-743
— Southern California I-388
— Southern Gulf of Mexico I-475
— Southern Portugal I-160–1
— Sri Lanka II-177, II-181
— of the Sundarbans II-153
  — scheduled castes/tribes II-153
— Tanzania II-85, II-88
— Tasmania II-653–4
— The Maldives II-206–7
  — demographic patterns II-206
— Torres Strait and Gulf of Papua II-599–600
— Turks and Caicos Islands I-591
— Tyrrhenian coasts I-275, I-279
— UAE II-25
— Vanuatu II-743
  — dual economic structure II-743
— Venezuela I-654–5
— Victoria, Australia II-666
  — indigenous people II-666
— Vietnam II-566
— west Taiwan coast II-505
— Western Australia II-696–7
— Yellow Sea, high densities inhibit conservation II-495
— Yemen II-25
population biomass, minimum viable in a marine reserve III-382–3
population density, Australia II-586
population growth rates, Oman and Yemen II-25
porpoises III-90
— *see also* Dall's porpoise ; harbo(u)r porpoises

port activities, Colombian Caribbean Coast I-673
port development
— Gulf of Guinea I-789
— Xiamen region II-519, II-521
  — deep harbours II-519
Portland, Maine *I-308*, I-313
Portsmouth Harbour *I-66*
— loss of salt marsh I-74
Portugal, Tagus and Sado estuaries I-151–65
— benthic vegetation I-159–60
— coastal erosion and landfill I-161
— the defined region I-153–4
— dredging I-161–2
— natural environmental variables and seasonality I-153–6
— offshore systems I-159
— populations affecting the area I-160–1
— protective measures I-162, **I-163**
— rural fishing I-161
— shallow marine habitats I-157–8
— upwelling effect on fisheries I-160
— urban and industrial effects I-162
Portuguese coastal waters I-153
Posidonia oceanica meadows I-274
power station effluent temperature, Hawaiian Islands II-805–6
power stations, conventional, North Sea coasts I-54
prawn culture
— Grand Bahama I-437
— the Sundarbans II-152–3, II-153–4
  — causing deterioration of coastal water bodies II-154
  — ecological crop loss II-154
prawn fishery
— Borneo, linked to mangroves II-368
— highest discard/catch ratios III-143
— large discard III-125
— northern New South Wales II-638
prawn trawling fishery
— Great Barrier Reef region II-623
  — effects on benthic communities II-623–4
— Torres Strait and Gulf of Papua II-604, II-606
Preah Sihanouk National Park, Cambodia II-571
precautionary action, principle of III-371
precautionary principle
— and equity III-157
— in Straddling Stocks Agreement III-158
precipitation
— Colombian Caribbean Coast I-665
— Gulf of Alaska, large freshwater flux I-378
— increasing III-188
— Norwegian coast I-19
primary production
— Adriatic Sea I-270
— Andaman and Nicobar Islands II-191
— Baltic Proper I-108
— Bay of Bengal coastal waters II-275, *II-276*
— by phytoplankton, usually N and P limited III-198
— Cambodian Sea II-573
— Campeche Sound I-474
— Canary Islands I-192, I-193
— Chagos Islands II-230
— coastal areas, Coral, Solomon and Bismark Seas region II-431–2
— coral reefs III-35
  — the Philippines II-409
— Gulf of Aden II-55
— Gulf of Cadiz I-172
— Gulf of Guinea I-779
— high, northern Gulf of Thailand II-302
— high, Sunda Shelf II-384–5
— important, Tasmania II-654
— inner New York Bight I-324, *I-325*
— Irish Sea, related to stratification of water masses and distribution of fronts I-89–90

— Izmir Bay I-244
— Malacca Strait II-334
— Malacca Straits II-312
— the Maldives II-202
— Norwegian coast, strong seasonality I-20
— and nutrient cycles, Gulf of Alaska I-379–80
  — nutrient source probably deep ocean I-379
— pelagic, Baltic Proper I-126
— percentage from the sea III-394
— Red Sea, low II-42
— stimulated by macronutrients III-259
— Tagus estuary I-157
— water column, Lesser Antilles I-633
— western English Channel I-67
primitive earth III-394
Prince William Sound *I-374*, I-375, I-380
— pollution of I-381
— southern, zooplankton community influenced by advection from the Alaskan Shelf I-380
— *see also Exxon Valdez* oil spill
principle components analysis, for grouping of disturbance types III-217
produced water
— containing mercury, Gulf of Thailand II-305
— effects of oil and chemicals in III-363
— from oil and gas production I-24, I-55
productivity
— Arabian Sea
  — and the Northeast Monsoon II-20
  — and the Southwest Monsoon II-19
— high, Gulf of Paria, Trinidad I-631
— and Nicaraguan Pacific coast fisheries I-536
— primary and secondary, of seagrasses III-2–3
— Somali Current LME II-66
PROKASIH (Clean Rivers Program), western Indonesia II-400
property rights, importance in regulation of the coast III-354
property rights institutions, sophisticated II-400
property rights regimes, and the oceans III-396–7
protected areas
— Borneo, mangroves in II-367–8
— Gulf of Guinea I-793
— Marshall Islands, resistance to II-788
— Turks and Caicos Islands **I-593**
Protected Natural Areas, Mexican Pacific I-497, **I-497**
protected species
— eastern Australian region II-643
— Madagascar II-17
— status and exploitation of, the Comoros II-249–50
  — coelacanths II-250
proton secretion III-35
— and nutrient uptake III-36
Providenciales, TCI
— pollution problem I-591
— surrounding waters, fishing and tourism I-591
— tourist destination I-591
Prymnesium parvum I-26
public health, Colombian Pacific coast I-680, **I-681**
Puerto Rico I-575–85
— coral bleaching and mortality events III-56
— geology I-577, *I-577*
— physical parameters I-577–81
— population development and land use:: effects from urban and industrial activities I-584
— shallow water and coastal habitats I-582–4
— shelf morphology and sediments I-581–2
— US National Estuarine Sanctuary I-584
Pulicat Lake, Palk Bay–Madras coast II-169, II-278
purse-seine fishery
— Madagascar II-122
— South Africa II-136–7
Puttalam Lagoon, Sri Lanka II-179

Puttalam Lagoon–Dutch Bay–Portugal Bay system, Sri Lanka, seagrass beds II-179
pyrite and evaporites, Laguna de Tacarigua I-652

Qatar *II-2*
— coral bleaching and mortality events III-47
queen conch I-590, I-592
Queen Conch fishery I-425
Queensland Plateau *II-426*, II-619
Queensland Trough *II-426*
Quirimba Archipelago, Mozambique
— fishing techniques II-107–8
— marine/coastal habitats II-102–4
— seagrass fishery, Montepuez Bay II-106
  — marema (basket traps), use of II-106
  — seagrass preferences II-102–4
— seagrass the most abundant habitat II-104
— source of productivity for South East Africa II-103–4
  — net primary productivity calculations II-103
  — unspoilt mixture of mangrove, seagrass and coral reefs II-103
quota systems, based on total allowable catch (TAC) III-161–2

radioactive discharges, Irish Sea I-94
radioactive oceanographic tracers, round the Azores I-204
radioactive pollution, Sea of Japan II-478–80
radioactive waste, stored in north Russia, cause for concern I-14–15
radionuclides
— ^{137}Cs, before and after Chernobyl I-116–17
— Black Sea, from Chernobyl disaster I-294
— from La Hague *I-14*, I-54–5, I-76, I-117
  — and Sellafield I-54–5, I-117, III-362
— in Greenland seas I-11
  — long-distance marine transport of I-13
— natural III-362
— not affecting small cetaceans III-96
— via Black Sea Water I-248–9
rainbow trout I-35
rainfall
— Coral, Solomon and Bismark Seas Region II-428
— Côte d'Ivoire I-808–9
— Dutch Antilles I-598
— Guinea I-799
— Gulf of Guinea I-776
  — Accra dry belt I-776
— Hawaiian Islands II-795
— Marshall Islands II-775
— Pearl River watershed II-551
— South West Pacific Islands II-709
— Torres Strait II-595
— Vanuatu II-739
— within the Fijian Group II-753
— *see also* precipitation
rainy seasons, Tanzania II-85
raised beaches II-349
Raleigh Bay *I-352*
RAMSAR Convention I-147, I-431, I-608, I-793, II-658, **III-334**, III-340
— TCI signed up to I-592
— Walvis Bay and Sandwich Bay wetlands, Namibia I-829
— Western Salt Ponds of Anegada (British Virgin Islands) accepted I-623
RAMSAR sites I-52, I-96, I-148
— Albufeira Lagoon I-162, *I-163*
— Inagua National Park, the Bahamas I-431
Ras al Hadd *II-18*, II-19
Ras Caseyr *II-64*
Ras Muhammed marine park *II-36*, II-44
Raso, Cape *I-152*, I-153
Ratak Submarine Ridge II-779
Recife de Fora Municipal Marine Park, Brazil I-728

recreational angling I-830, **I-830**, II-138–9
— conflict with commercial interests II-138–9
recreational boating, Dutch Antilles I-604
recreational fisheries, management of III-161
recreational fishing
— Australia II-587
— Great Barrier Reef region II-621
    — management of II-626
— Hawaiian Islands II-803
— New Caledonia II-729
— Tasmania II-655
— Victoria Province II-668
— Western Australia II-697
recreational industry, New York I-328
Red Sea II-35–45
— biogeographic position II-38
— coral bleaching and mortality events III-47
— endemic species II-38
— extent II-37
— geographical and historical setting II-37–8
— major shallow-water marine and coastal habitats II-40–2
— offshore systems II-42–3
— oil contamination throughout II-43
— population, urban and industrial activities II-43–4
— protective measures II-44
— receives continual supply of larvae for the Indian Ocean *II-36*, II-38
— seasonality, currents, natural environmental variables II-39–40
— turnover time II-40
Red Sea invaders, southeastern Mediterranean I-256–8
Red Sea rift II-37
Red Sea—Gulf of Aden water exchange II-39
red tides I-90, I-244, I-439, I-742, II-376, III-198, III-199, III-298
— Bay of Bengal II-275
— Benguela Current III-218
— and fish kills III-206
— Hong Kong II-542
    — impact on fish production II-543
— Izmir Bay I-238
— North Benguela region I-827
— Pearl River mouth II-553
— the Philippines II-418
— Rías Bajas I-141
— Straits of Malacca II-320
— Xiamen coastal waters II-522
— Yellow Sea II-495
Redfield ratios, P:N:Si, Arabian Gulf II-6
reed field, Liaohe Delta II-494
reef ecosystems, decline in abundance from poor fishing practices III-126
reef fish
— Marshall Islands II-778–9
— problem of size-selective fishing III-121
reef flats
— central Red Sea II-40, *II-40*
— Chagos Islands II-227
reef gleaning, Fiji Islands II-759
reef habitats, shallow, Tasmania II-652
reef and lagoonal water, French Polynesia II-816
reef systems, marine reserves, percentage adult population protected III-382
reefs
— artificial I-183
— Belize, buffered from urban pollution I-512
— deep reef habitats I-506
Refinería Panama storage tank rupture III-278
— long-term effects on physical structure of mangrove forest III-273, III-274
— mortality of subtidal corals III-273–4
Reflagging Agreement III-158, III-161
— requirements of III-160

regime shift, Gulf of Alaska shelf I-375
Regional Organisation for the Conservation of the Environment of the Red Sea and the Gulf of Aden (PERGS) II-59
Regional Organisation for the Protection of the Marine Environment (ROPME) II-3, II-29, II-44
Regional Seas Program III-340
— provision for protection against land-based pollution III-339
Relative Penis Size Index (RPSI) III-250
relative sea-level rise, and landform alteration, northern Gulf of Mexico I-446–7
relaxed eddy accumulation (REA) III-204
— application to particle measurement III-205
remineralisation processes, enhanced, the Bodden III-261
remittances, important in South Western Pacific Islands II-714
remote sensing
— applications of III-294
— Brazilian tropical coast *I-725*
— Coastal Zone Color Scanner information
    — Chagos Islands II-230
    — The Maldives II-202
— could aid seagrass research III-12
— in fisheries management III-160
— modern definition III-294
— provides large-scale synoptic data III-284
— satellite imaging of Chilean coast upwellings *I-705*, I-706
— and sea bottom types III-297
— of tropical coastal resource III-283–91
    — coral reef systems III-87–8
    — economic considerations III-288–9
    — mangroves III-285–6
    — sensors relevant to mapping tropical coastal zones III-284
    — tropical seagrass ecosystems III-286–7
    — types of data achievable III-284
— used to classify and quantify coastal marine habitats, Mauritius II-259
— *see also* AVHRR imagery
renewable energy, from the oceans III-304
reproductive disturbance
— in Baltic biota I-114
— in breeding colonies of seabirds, Southern California I-398
— from *Exxon Valdez* oil spill I-382
reproductive failure
— from organic chemicals III-95
— seabirds, and fish population collapse III-129
reptiles
— El Salvador I-549–50
— Great Barrier Reef II-618
reservoirs
— southern Spain, Guadiana and Guadalquivir basins I-175
— *see also* dams
residence time
— Baltic Sea water I-103
— fjord basin water I-21
residual currents, complex, Irish Sea I-85, *I-86*
resource depletion, the Philippines II-418
resource exploitation, French Polynesia II-819–20
resource management, USA potentially conflicting paradigms III-350
resource utilisation conflicts, Xiamen region **II-518**
resources
— non-renewable, South Africa II-142
— right of access to (South Africa) II-141
restingas
— Abrolhos Bank–Cabo Frio, Southern Brazil I-736
— Brazil I-727
— Santa Catarina State, southern Brazil I-737
retroflection eddies *see* eddies
Reunion *II-254*, II-255
— agriculture II-261
— artisanal fishing II-261–2
— ciguatera poisoning outbreaks II-261–2
— Fishing Reserves **II-265**, II-266

— steep volcanic surfaces support corals II-259
— tourism II-264
Revillagigedo Islands, Mexico I-487
— seasonal surges I-487
— some coral bleaching (1998) III-53–4
Rhine, River I-58–9
— Rhine Action Programme for rehabilitation of I-59
Rhine water, effects on southern North Sea I-46
Rhodos gyre I-236
Rías Bajas, Galicia I-137, I-138, I-141
Riau Archipelago, Sumatra II-311, II-337
— coral reefs II-314
— seagrass beds II-314, II-336
Riga, Gulf of *I-122*
right whales III-74, III-76, III-81
Rim Current, Black Sea I-287
Rincón region, Buenos Aires Province, significant biological activity I-755
ringed seals I-8, III-95
Rio de Janeiro metropolitan area
— industries and port I-738
— Jacarepaguá lagoon systems, receives industrial waste I-742
— pollution in I-741
— Guanabara Bay, sewage and industrial effluents I-741
Rio de la Plata basin system I-751
Rio de la Plata estuary *I-751*, I-759–61
— anthropogenic impacts I-760
— environmental gradients I-759
— formation of I-759
— human impact I-759–60
— impact of urban-industrial zone I-759
— nutrient discharge I-760
— pollutants I-760
— sectors of I-759
— state of knowledge I-760
— system characteristics I-759
Rio Declaration, on transgenerational responsibility III-360
river deltas, New Guinea coastline, support pristine mangrove forests II-597
river dolphins, India, DDT, PCBs and HCHs in II-452
river inflow
— Colombian Caribbean I-665
— Côte d'Ivoire I-807
— and run-off I-809
— Guinea I-799
— Gulf of Guinea I-776–7
river pollution, Russian Far East II-477
river run-off
— Bay of Bengal II-271, **II-272**, II-275
— lessened through irrigation withdrawals II-287
— and river impacts II-273
— Gulf of Thailand, freshwater discharge in addition to II-301
— in Malacca Strait II-333
river and wave material transport, east coast, Peninsular Malaysia II-582
rivers
— Chilean coast, various flow regimes I-704–5, **I-704**
— Gulf of Guinea coast, downstream effects of damming I-789, I-792
— influence on southern Spain I-169–70
— polluted, Venezuelan coastal zone I-645
— Southeast Asia, the most turbid II-306
Rocas atoll, Brazil I-723
— biological reserve I-724
rock lobster fishery II-56
— artisanal, Torres Strait II-602
— Australia II-586
— Gulf of Papua II-604
— South Africa II-137
— Tasmania II-655, II-656
— Victoria Province, Australia II-668
rock and surf angling II-138

rocky coastlines
— Irish Sea I-87–8
— communities of I-88
rocky reefs
— eastern Australian region II-634
— subtidal II-633
— El Salvador I-548–9
— Los Cóbanos the most extensive *I-546*, I-549, I-552
— habitats of, Australian Bight II-679
— Louisiana coast I-421
— Puerto Rico I-582–3
— Victoria Province, subtidal II-664
rocky and sandy seabeds, Canary Islands I-195
rocky shores
— Borneo II-666
— eastern Australian region II-633
— exposed, persistence of oil on III-276
— and headlands, east coast, Peninsular Malaysia II-349
— Malacca Strait II-336
— northern Argentina, macro and megafauna I-754
— Venezuela I-649–50
— algal zone I-649
— barnacle zone I-649
— *Littorina* zone I-649
— microhabitats I-649
— Victoria Province II-664
— western Sumatra II-387
Rodrigues *II-254*, II-255
— agriculture II-261
— artisanal fishing important II-262
— lagoons heavily silted II-259
— mangroves II-259
— population II-260
Rodrigues Bank *II-254*, II-255
Rompido Sand Cliffs *I-168*
Rudnaya River Valley, eastern Russia
— degradation/decay of ecosystems taking place II-485
— health of population requires improvement II-485
— polluted surface and groundwaters II-485
— a pre-crisis situation II-485
runoff
— agricultural, western Indonesia II-396
— annual, mainland Norway I-19–20
— and biota, Papua New Guinea coast II-597
— extensive, Tahiti and Moorea II-821
— increased by impervious surfaces I-445
— polluting Taiwan's coastal environment II-507, **II-510**
— silt-laden, increases turbidity I-446
— soil, Guangdong II-558
— urban, contains pollutants I-450
— *see also* soil runoff; stormwater runoff; surface runoff
Rupat Strait II-311
rural factors
— affecting the Aegean I-242–5
— affecting southern Spanish coastal zones I-174–6
rural land use
— impacts on western Indonesian seas II-395–6
— agriculture II-396
— deforestation II-395
— mining II-395–6
— *see also* land use
rural-urban migration I-567
— the Philippines II-412
— Tanzania II-92–3
Russians, whaling III-75

S:N ratio, Irish Sea I-87
Saba Bank *I-596*, I-598
— fishing by non-Saban fishermen I-603–4
— reefal areas I-600
Saba Island *I-596*, I-598, I-603

— coastal and marine habitats *I-600*
— sediments, due to erosion limiting factor for reef development I-601, *I-602*
Sabah *II-362*
— coastal habitats II-365
— illegal immigrants II-373
— population density II-373–4
— problem of Sipadan Island II-377
— reef destruction II-371, *II-372*
— SCUBA diving II-377
— Semporna Islands Park (proposed) II-377
  — current and potential threats to **II-378**
  — management plan for II-377
— Tunku Abdul Rahman Park, impacts on II-376
Sabellid worm colonies, Yemen II-51
sabkha
— described II-8
— Oman
  — at risk from rising sea levels II-28
  — Bar Al Hickman II-21
Saccostrea build-ups II-76, II-77
Sado estuary *I-152*, I-154
Saharan low pressure zone I-775
St Bees Head *I-84*
St Brandon Bank *II-254*, II-255
— lagoons with sandy floors II-257
— spur and groove regions with fish II-256–7
St Brandon Islands *II-254*, II-255
— artisanal fishing II-262
— seabirds and turtle nesting sites II-257
— temporary settlements II-260
St Brandon Sea *II-254*, II-257
St. Eustatius *I-596*, I-598, I-603
— coastal erosion I-602
— coastal and marine habitats *I-600*
— deep reef systems I-600
— manatee grass I-601
St. Helena Sound *I-352*, I-354
St. Lucia
— coral bleaching and mortality events III-56
— fuel-wood reforestation I-637
— Marine Islands Nature Reserve I-639
St Lucia system, KwaZulu-Natal, periodic hypersalinity in II-135
St. Maarten *I-586*, I-598, I-603
— coastal and marine habitats *I-600*
— decline in livestock I-601
— filling of saliñas and lagoons I-602
— manatee and turtle grasses I-601
— urbanization and tourist developments I-598
Saint Malo, Golfe de *I-66*
St Martin's Island, Bangladesh, focus for ecotourism II-289–90
saithe I-35, I-49
Sakhalinsky Bay *II-464*
Salalah *II-18*, II-19
saliñas, Dutch Antilles I-601
saline intrusion
— from over abstraction of groundwater II-26
— into aquifers, the Philippines II-414
— Marshall Islands II-781
— some Gulf of Aden coasts II-58
— Taiwan II-507
— to groundwater around Zanzibar II-93
salinisation
— Gulf of Mexico I-476
— of land II-584
salinity
— Aegean Sea I-236
— affecting Baltic biota I-123
— Arabian Gulf II-5
— Baltic Sea
  — and cod reproduction I-128
  — increase in I-111
— Bass Strait II-664
— in Bay of Bengal II-287
— Bien Dong Sea II-563, *II-563*
— Black Sea I-289
— Cambodian Sea II-571
— Campeche Sound I-469
— Canary Islands I-190–1, *I-191*, *I-192*
— changes at the deep salinity minimum, North Pacific III-182
— Chesapeake Bay I-339
— coastal waters, Xiamen region, China II-516
— Colombian Pacific I-679
— dry season, coastal zone of Gulf of Paria and Orinoco delta I-648
— El Salvador estuaries and open water I-547–8
— French Polynesia II-816
— Great Australian Bight II-677
— hypersalinity, Red Sea II-37–8
— impact on fisheries II-117
— Irish Sea I-86–7
— lagoon and open water, Vanuatu II-740
— lower near the Mississippi I-437
— Malacca Straits II-311–12
— Marshall Islands II-776
— near-bottom, Alaskan Gulf I-378–9
— negative anomaly, Alaskan coast III-183
— New York Bight I-323
— North Sea I-45, I-47
— ocean and lagoon, South Western Pacific Islands II-710
— off Guinea coast I-799
— Palk Bay II-166
— of Peruvian coastal waters I-691
— South China Sea II-348, II-552
— surface, Argentine Basin *I-753*, *I-753*
— surface water, Papua New Guinea coast II-596
— variation in estuaries, correlates with rainfall I-520
— variations, Côte d'Ivoire I-809
— Western Coral Sea II-614
— Yellow Sea II-490
salinity gradients
— Gulf of Suez II-39
— horizontal, Baltic Proper I-102
salinity stress I-364
salinity trend, North Pacific III-182
salinity–NO$_3$ relationship, northern Alaskan Gulf I-379
salmon
— Atlantic I-35, I-313
  — Baltic Sea I-110, I-111, I-128
  — farmed III-168
— Pacific
  — change in salmon survival rate in the open Pacific III-185
  — Okhotsk Sea II-467
  — possible biological effects of climatic change III-185–6
— sockeye, Alaska, accumulating PCBs and DDT I-381, I-383
salt diapirs I-442
salt extraction
— Gulf of Guinea I-786
— Southern Brazil I-738
salt flats, Lac Badana channel, Somalia II-77
salt pans
— Mauritius II-262
— Palk Bay, mangroves cleared for II-167–8
salt ponds
— Anguilla *I-619*, I-620
— British Virgin Islands I-620
— Cambodia II-574
— El Salvador I-553
salt tectonics, effects of I-442
salt wedge, Chesapeake's main tributaries I-337
salt-marsh plants, hybridisation of I-47
saltmarsh 24, I-49, I-358
— decreasing, Chesapeake Bay I-337

— eastern Australian region II-632
    — loss of II-639, II-640
— English Channel coasts I-68, I-69, *I-69*
— Gulf Coast, USA I-440
— Gulf of Maine I-311
— impact of dredging I-176
— lost in Cantabria and the Basque country I-144
— lost to infilling I-54
— Mar Chiquita coastal lagoon, Argentina I-755
— North Spanish coast I-139
— southern Spain I-170
    — Gulf of Cadiz I-177
— Sri Lanka II-180
— Tagus estuary I-158
— Tasmania II-653
— Victoria Province, Australia II-665
saltwater intrusion
— Belize I-512
— into coastal aquifers I-443, I-447
    — changes caused by I-448
Salvage Islands *I-186*, I-187
Salwa, Gulf of *II-2*
Samoa
— coastal modifications, serious impacts II-717
— coral reefs
    — cyclone physical destruction II-712
    — degraded II-712, II-714
— cyclone damage II-714
— population II-714
Samoa Group *II-706*, II-707, *II-709*
— agriculture and fisheries II-716
    — overfishing II-716
— coral reefs II-712
    — American Samoa II-712
    — Samoa II-712
— Marine Protected Areas (MPAs) II-720
— *see also* American Samoa
San Andrés and Providencia Archipelago
— coral reefs I-668
— strong wave energy I-666
— volcanic origin with reef developments I-667
San Jorge Gulf I-753
— crude oil production I-761
San Matias Gulf, fishing I-763–4
San Salvador Islands, mangroves stressed I-422
sand
— oolithic I-590
— white carbonate, Puerto Rico I-581
sand accumulation, Gulf of Mannar, Palk Strait and Palk Bay II-181
sand barriers, southern Baltic coast III-260
sand dunes I-170
— active, Brazilian tropical coast I-722
— Cabo Frio region, southern Brazil I-736
— embayments behind, nursery grounds I-326
— front edge recession I-429
— Great Australian Bight II-675
    — Yalata dunes II-678
— Mexican Pacific coast I-487
— removal of to enlarge beaches I-276
— scrub landscape, mammals in, Doñana National Park, southern Spain I-171–2
— with xerophytic grasses II-51
sand and gravel extraction
— east coast of England I-53
— El Salvador I-554
— English Channel I-74
    — impacts from I-77
— New York Harbor I-330
sand loss, from beaches I-176, I-177
sand mining I-604, I-786
— Anguilla I-622
— by suction, east coast, Peninsular Malaysia II-355
— causing beach erosion I-636
— the Comoros, forbidden II-248
— for construction, Israel I-258
— and coral mining, Marshall Islands II-782
— Côte d'Ivoire I-816
— and dredging, in the Bahamas I-428
— from beaches II-27
— Jakarta Bay II-391
— and lime production, Mauritius II-263
— Malacca Straits II-316
— the Maldives II-208
— the Mascarenes II-262
— offshore, Curaçao I-605
— Orissa, for heavy minerals II-279
— the Philippines II-415
— river, east coast, Peninsular Malaysia II-353
— Southern Brazil I-737, I-738
    — causing environmental degradation I-740, I-741
— Suva Reefs II-717
— Todos os Santos Bay, Brazil I-725
— Turks and Caicos Islands I-592
— Vanuatu II-744–5
sand spits/sand banks I-170, I-828–9
— east coast, Peninsular Malaysia II-349
— *see also* Kakinada sand spit, Godavari and Krishna deltaic coast
sand vegetation, Doñana National Park, southern Spain I-171
sand-eels I-49
sandstone/beach rock habitats, Sri Lanka II-180
sandy beaches
— Arabian Sea coast II-21
— the Azores I-207
— Borneo II-369–70
— Buenos Aires Province I-754
— Cabo Frio–Cabo de Santa Marta Grande, southern Brazil I-736
— east coast, Peninsular Malaysia II-349
— eastern Australian region II-633
— French channel coast I-69
— Gulf of Aden, deposits of ilmenite and rutile II-58
— Hawaiian Islands, colourful I-797
— high-energy, Yemeni and Somali coasts II-51
— loss of and beach nourishment I-176
— Malacca Strait II-337
— Marshall Islands II-777
— Mauritius II-258
— New York Bight I-325
— and oil spills III-269
— oil-contaminated III-270–1
— Oman II-21
— regeneration of causes serious impacts I-181
— replenished by wadis II-52
— southwestern Africa I-828
    — and rocky beaches I-829–30
— Taiwan II-502
— Venezuela I-648–9
    — dissipative beaches I-648–9
    — high energy beaches, zonation of I-649, *I-649*
— west coast of Malaysia II-311
— western Sumatra II-387
— Xiamen region, polluted II-519
— Yellow Sea II-490
— *see also* beach nourishment/replenishment
sandy bottom habitats I-548
sandy bottoms, southern Somali coast *II-70*, II-72, II-74
Sanganeb Atoll II-40
Santa Maria
— depauperate palagonitic tuff I-206–7
— limpet harvesting I-206
Santa Monica Bay *I-386*
Sarawak
— population II-374

— Pulau Bruit National Park II-368
— well managed coastal National Parks and Protected Areas II-376–7
sardine, horizontal migration of II-466–7
sardine fishery I-141, II-56
— beach seining II-26
— Northwest Pacific I-490
— Pacific I-396–7
— Peru I-693
— Portuguese I-160
— South Africa, collapse of II-137
— Venezuela I-653
sardinella fishery, southern Brazil I-739
Sardinia, 'Smeralda' Coast I-278, I-279
Sargasso Sea, defined I-223
Sargasso Sea and Bermuda I-221–31
— coastal erosion and landfill I-228–9
— effects from urban and industrial activities I-229
— major open ocean habitats I-224–7
— major shallow water marine and coastal habitats I-227–8
— populations I-228
— protective measures I-230
— seasonality, currents, natural environmental variables I-223–4
Sargassum community I-226
— displaced benthos I-226
Sargassum decurrens, distribution of, Australia II-695
Saronikos Gulf *I-234*, I-235–6, I-247
— industrial pollution I-245–6
— pollution in, affecting the plankton I-238
— seagrass beds and algae I-239
— water masses in I-237
satellite imagery, uses of III-287
satellite remote sensing III-394
— of the coastal ocean: water quality and algal blooms III-293–302
    — areas of application III-295
    — critical bottom depth III-297
    — semianalytical atmospheric radiation transfer models III-295
    — sensor measures upwelling radiance III-297
    — water quality parameters III-295–8
— coasts of III-294
— *see also* AVHRR imagery; remote sensing
satellite-based vessel monitoring systems III-160
Saudi Arabia *II-2*, *II-36*
— coral islands and patch reefs II-7–8
Saya de Malha Bank *II-254*, II-255
— sand, coral and green algae II-256
scallop fishery
— Argentina, collapse of I-764
— New Caledonia II-734
— Scotland I-90
scallop and Puelche oyster culture, Argentina I-764
scavengers, food subsidies from by-catch/discard/ and processing dumping III-127
scheduled castes/tribes
— Lakshadweep Islands II-194
— Sundarbans II-153
science and the seas III-394
scientific realism, in fisheries management III-155, III-156–7
Scilly Isles I-71
Scotian Shelf *I-311*
Scotland
— early attempts to determine effects of fishing on fish populations III-376
— legislation relating to the coastal and marine environment III-353
Scottish scallop fishery, Amnesic Shellfish Toxin contamination I-90
sea breezes, Borneo II-365
sea cliff retreat I-144, I-211
sea colour remote sensing III-294
— need for development III-298
— not as advanced as that for ocean colour III-300
Sea of Cortes Oceanic Province, Mexico I-488
— oceanic habitat I-488

sea cucumber fishery
— the Maldives II-212
— Sri Lanka II-183
sea cucumbers II-90, II-123, II-784
— exploitation of, Mozambique II-108
— Madagascar II-123, II-126
sea horses
— endangered, Palk Bay II-168
— Gulf of Mannar II-165
sea ice transport, of pollutants I-14
sea level
— anomalous rise, Alaskan coast, a propagating Kelvin wave III-84
— high seasonal oscillation, Bay of Bengal II-287
— history of, Brazilian tropical coast I-722
— rising III-188
— *see also* sea-level rise
Sea Moss, harvesting of I-635
sea otters, effected by *Exxon Valdez* oil spill III-274
sea salt extraction, southern Spain I-177
sea surface temperature (SST)
— Aegean Sea I-236
— American Samoa II-767, *II-767*
— Arabian Gulf and Straits of Hormuz II-4–5
— the Bahamas I-419
— Bay of Biscay I-14
— Black Sea I-289
— Brazilian tropical coast I-721
— Cambodian Sea II-572
— Canary Islands, spatial and temporal differences I-190*I-191*, *I-192*
— the Comoros II-245
— continental seas, western Indonesia II-384
— Dutch Antilles I-598
— east coast, Peninsular Malaysia II-348
— eastern Australian region II-631–2
— effect of Southwest Monsoon off Arabian coasts II-19
— French Polynesia II-816
— Great Australian Bight II-677
— Gulf of Guinea I-778
— Gulf of Thailand II-300
— higher, affecting tropical cyclones III-369
— increasing, central Gulf of Alaska III-182
— Indian Ocean and Southeast Asia (1998) *III-45*
    — 'hot spot migration III-44–6
— Irish Sea I-85–6
    — slight rise in I-86, *I-87*
— lagoon and open water, Vanuatu II-740
— long-term increase, Gulf of Guinea, linked to global warming? I-781
— Madagascar II-115
— Malacca Straits II-311
— Maldive Islands II-201
— New Caledonia II-725
— of the Nicaraguan Caribbean coast I-519–20
— off southern Angola *I-824*
— Somali coast II-66
— South China Sea II-407–8
— South China Sea coast II-552
— South Western Pacific Islands II-710
— Spanish north coast I-137
— Sri Lanka II-178
— sub-surface anomaly, Line-P, Northeast Pacific III-183–4
— Tasmania II-650–1
— varies with season, Yellow Sea II-490
— Venezuela
    — AVHRR images *I-647*
    — and upwellings I-646
— Victoria coastal waters II-664
— warmer, effects of III-195
— Western Coral Sea II-614
sea swell, North Spanish coast I-138
sea turtle conservation, Mozambique II-111

sea turtle fibropapilloma disease III-67
sea turtles 59–71, I-522, I-525, I-535, I-591, II-21, II-44, II-250, II-600, II-601, II-797
— Abrolhos National Park I-726
— Bahamas I-426
— breeding grounds, Chagos Islands II-229
— conservation priorities III-67–8
  — as flagship species for conservation III-68
  — international and regional levels III-68
  — national level III-67–8
— Coral, Solomon and Bismark Seas region II-431
— evolutionary history III-60
— Great Barrier Reef II-618
— Gulf of Guinea, status of I-780, **I-780**
— Gulf of Mexico I-473–4
— Hawaiian Islands II-799
— Java Sea I-364
— Lesser Antilles I-620, I-636
— life cycle III-60–1
  — delayed maturity III-61
  — pelagic migration drifting III-61
  — reproduction III-60–1
— living, biological status of 61–4
— and longline fisheries III-146
— Maldives II-205
— migrations, growth and population structure III-61
— modern research needs and tools III-64–5
  — information networks III-64–5
  — nesting beach studies III-64
  — use of PIT tags III-64
— Mozambique II-105
— nesting ground harvests difficult to control III-65
— the Philippines II-410
— protected by Indonesian Law II-400–1
— protected in El Salvador I-550
— Red Sea II-42
— Sri Lanka II-180
— Sunda Shelf II-400–1
— Tanzania II-87
— threats to species survival, historical and modern III-65–7
  — damage to nesting and foraging habitat III-66–7
  — direct harvest III-65
  — fisheries mortality III-65–6
— vulnerable to fisheries III-379
— Yucatan I-471
— *see also* named varieties
sea urchin harvest I-636–7
sea urchins
— effects on reefs, French Polynesia II-819, II-822
— Mauritius II-257–8
sea water temperatures, Turks and Caicos Islands I-589
sea waves, Brazilian tropical coast I-721
sea-bed activities, peaceful exploration and exploitation III-336, III-338
sea-bed current stress II-596
sea-bed exploration, the Philippines II-417
sea-bed and sediments, English Channel I-67, *I-68*, I-69
sea-floor spreading II-66
— Gulf of Aqaba II-37
sea-level rise
— adverse effect on coastal ecosystems III-98
— French Polynesia, effects of II-823
— Gulf of Guinea I-787
— impact on coastal habitats, northern Gulf of Mexico I-446, I-446–7
— Marshall Islands vulnerable to II-777
  — significant points II-777
— problem of Bay of, Bengal II-281, II-288, II-295
— puts reclaimed land at risk I-145
— South Western Pacific Islands, effects on II-716–17
— threat taken seriously in the Maldives II-214–15
— *see also* relative sea-level rise

sea-salt production, China II-494
sea-surface microlayer, enrichment of I-391–2
seabirds III-105–15
— Aegean Sea I-239–40
— American Samoa II-769
— Antarctic III-108
— "Arrecife Alacranes" national park, Yucatan I-471
— Australian Bight II-679–80
— the Azores
  — mercury in I-212
  — roosting/nesting sites I-207
  — some protection for I-214
— Chagos Islands, high diversity II-229
— El Salvador
  — nesting and migratory I-549
  — plentiful, "Colegio de las Aves" I-549
— English Channel, offshore, migrating and breeding species I-71
— evolution and history III-106–7
  — derivation III-106
  — Tertiary development III-106–7
— exploitation and conservation III-111–14
  — loss to growing populations III-111, III-114
  — ocean physics-foraging seabirds relationship, North Pacific III-113
  — seabirds and fisheries in the North Sea III-112
  — value to older cultures 111
— Faroese, concentrations of PCBs and DDT in **I-37**
— Fiji Islands II-758
  — threats to II-757
— food and fisheries III-111
— Great Australian Bight, threats to II-684–5
— Great Barrier Reef II-618, II-679–80
— habitat disturbance, Dutch Antilles I-604
— Hawaiian Islands II-798–9, **II-798**
  — threatened or endangered **II-799**
— and longline fisheries III-145–6
  — development from the Australian zone III-146
— of Madagascar II-120
  — breeding sites II-120
— the Maldives II-204–5
  — socio-economic importance of II-204
— Marshall Islands II-779
— modern III-107
  — main groups **III-106**
  — Miocene climax III-106
  — shorebirds/waders III-107
— movements and measurements III-107–11
  — breeding counts, "apparently occupied nest" III-109
  — coastal birds, fluctuating distribution III-109
  — complexities of breeding seasons III-109–11
  — distributions III-107
  — effects of ENSO events III-110
  — erratic productivity and mortality III-109
  — foraging ranges and feeding areas III-108
  — migrations III-107, III-110–11
  — seasonal fluctuations at sea III-109
— nesting and migratory, El Salvador I-549
— North Sea I-49
  — supported by fishing fleet discards I-51
— plentiful, El Salvador, "Colegio de las Aves" I-549
— reduced numbers of, Greenland I-8–9
— southern Baltic I-109
— Southern California
  — dramatic declines in I-398
  — effects of major impacts I-398
— and waders, Gulf of Aden coasts II-54
— *see also* fishing mortality; marine birds
seafood
— consumption, Torres Strait II-600, II-601
— health risks from, Southern California I-399–400
seagrass I-71, I-194, I-228, I-239, I-311, II-290

— Arabian Gulf II-7
— Australia II-582
  — major human-induced declines II-700, **II-700**
  — vulnerable to eutrophication and sedimentation II-585
— the Bahamas I-421
— Borneo II-370
— Carolinas I-361–2
  — and other rooted submersed aquatic vegetation I-358–9
  — remapping of I-366
— Coral, Solomon and Bismark Seas region II-430
— damaged by dredging and filling I-569
— distinguished from macroalgae, problem in remote sensing III-286
— east coast, Peninsular Malaysia II-348, II-349–50
  — threats to identified II-354–5, II-357
— eastern Australian region II-633
  — dieback, south-east Queensland II-637
— eastern sector, northern Gulf of Mexico I-438
— English Channel I-70
— epiphytes/epiphyte communities I-357–8
— Fiji Islands II-754–5
— Global Declines and Effects of Climate Change III-10–11
  — some losses documented III-10
— global status of III-1–16
  — declines and effects of climate change III-10–11
  — ecosystems services III-2–3
  — planning, management, policy, goals III-9, III-11
  — research priorities III-12
  — worldwide decline III-7–9
— Great Barrier Reef region II-615
— Guidelines for the Conservation and Restoration of in the USA and adjacent waters, a synopsis III-8
— Gulf of Aden II-53
— Gulf of Mannar, affected by trawling II-165
— Gulf of Thailand II-301
— Hawaiian Islands II-796, II-797
— high species diversity, Papua New Guinea and Torres Strait II-431
— impact on of oil and chemical spills I-450
— Jask and Char Bahar, Iran coast II-24
— Koh Kong Bay, Cambodia II-573
— lagoonal, Lakshadweep Islands II-194
— Malacca Strait II-336
— management possibilities, planning, policy and goals III-9, III-11
— Mar Chiquita coastal lagoon, Argentina I-755
— Marshall Islands II-779
— Mexican Pacific coast I-490
— North Sea, 1930s 'wasting disease' I-49
— Oman II-23
— Papua New Guinea coast II-598
— Red Sea II-42
— sediment accumulation and stabilization III-2
— Tasmania II-652
— The Maldives II-204
— Vanuatu II-740–1
— Victoria Province, Australia II-664–5
  — die-back, possibly due to catchment erosion II-667, II-668
  — epiphytic algae II-665
— wasting disease III-9, III-14
— Western Australia II-693, II-696
  — Kimberley coast II-694
— western Sumatra II-387
seagrass beds I-195, I-590
— Anguilla I-618, *I-619*
— Cambodia II-573, II-574
— Chesapeake Bay I-341
— Colombian Caribbean Coast I-668
— damaged by boat propellers III-9
— Dutch Antilles I-601
— El Salvador I-548
— fluctuations in salinity, water temperature and turbidity bad for I-441
— Johore Strait II-334
— in lagoons, Mauritius II-258
— Lesser Antilles, Trinidad and Tobago I-631
— little damaged by hurricanes III-2
— losses from low oxygen III-220
— Madagascar II-119
— Malacca Straits II-312, II-314
— New Caledonia II-726
— Nicaragua, Caribbean coast I-522
— northern Gulf of Mexico, high diversity of I-440–1, *I-441*
— nursery areas I-421, I-440–1
— nursery function III-3
— the Philippines II-410
  — conversion and utilization of II-413–14, **II-413**
— Quirimbas, increasing fishing pressure II-102
— role of, Coral, Solomon and Bismark Seas region II-430
— and sandy bottoms, southern Somali coast *II-70*, II-72, II-74, II-76
— the Seychelles II-236
— shelter function III-3
— South Western Pacific Islands II-711
— Sri Lanka II-179–80
— Tanzania II-8708
— Torres Strait II-598
  — seagrass abundance on reefs II-598
— Venezuela I-650
— West Florida Shelf I-438
— western Indonesia II-388
  — intertidal II-388
seagrass communities
— Australian Bight II-679
— Hong Kong, threatened II-541–2
seagrass dynamics, SPOT XS, Landsat TM and airborne methods all needed III-286–7
seagrass ecosystems
— affected by natural environmental impacts III-8
— divisions of animal community III-3
— nitrogen a rate-limiting factor III-3
— tropical, remote sensing of III-286–7
  — future challenges III-287
seagrass meadows *see* seagrass beds
seagulls, colonies dependent on groundfish trawling discards III-127
sealions I-756
seals I-834, II-797
— common I-109
  — French channel coast I-70
  — north east Irish coast I-91
— in the English Channel I-70–1
— Great Australian Bight II-680–1
— Northwest and Southwest Atlantic, effects of population recovery III-129
seamount fisheries, Emperor Seamounts II-802
seamounts
— the Comoros II-245
— fished and unfished, benthic biomass III-379
— Mascarene Region II-255, II-257
— offshore Tasmania II-653, II-659
— Vening Meinnesz Seamounts II-583
seasonality
— Belize I-503
— El Salvador I-547
— Jamaica I-561
— Nicaraguan Pacific coast I-534
— North Sea, temperature and salinity variability I-47
— Norwegian coastal ecosystem I-20, I-23
seawalls/shore structures, to protect the land, Hawaiian Islands II-805
seawater density, Gulf of Alaska III-181
seawater flooding, western Indonesia, carries contaminants and bacteria II-398
seaweed aquaculture, Fiji Islands II-754
seaweed farming III-167–8
— Tanzania II-91–2
seaweed harvesting, Norway I-26

seaweeds
— alien I-47
— Andaman and Nicobar Islands II-192
— brown and red, diversity decline, North Sea I-47
— Cadiz–Tarifa arc I-172
— cultivated, Yellow Sea II-494
— east coast of Taiwan II-508
— the Philippines, species diversity II-409
Secchi disk depth
— retrieved from satellite data III-296
— and water quality III-296
sediment accumulation, and stabilization, due to seagrasses III-2
sediment contamination, reduced in Gulf of Maine harbors I-317
sediment depletion, an increasing problem in coastal systems III-352
sediment drift I-47
sediment grain size, critical to permeability of oil III-269
sediment load
— high, western Indonesian rivers II-384, II-395
— south-east Queensland II-637
sediment mobility, from loss of aquatic vegetation III-259
sediment sinks, North Sea I-46
sediment supply
— Brazilian tropical coast I-722, I-728
— to Gulf of Guinea I-777
sediment transport
— alteration by coastal development, Western Australia II-699
— Bay of Bengal II-271–2, *II-272*
— into coastal zone from Orinoco River I-648
— patterns altered by harbour and reclamation engineering II-493
sedimentary basins, fault-controlled, cutting Somali coastline *I-67*, II-66
sedimentation
— affecting Andaman Sea coral reefs II-303
— between Rio and Santos I-742
— Borneo
    — a controlling factor in mangrove development II-367
    — high rates of, reducing coral cover and fish abundance II-366, *II-371*
    — a major concern II-375
— Brazilian tropical coast I-722
— can be serious in El Salvador I-552
— Chilka Lake II-279
— coastal, from deforestation, Australia II-666–7
— and contamination, near-shore environment, Belize I-508
— discharge into coastal waters, damage caused, Hawaiian Islands II-801–2
— excessive, Kuwait, and coral bleaching II-10
— Florida Keys, smothering coral I-408
— from logging, threat to Coral, Solomon and Bismark Seas region coral reefs II-436
— Great Barrier Reef region II-620–1
— Gulf of Mexico I-471
— high rates, western Indonesia II-384
— impact on Madagascan coral reefs II-123–4
— lagoonal, from mining, New Caledonia II-731
— Malacca Straits, a growing problem II-317
— Nicaraguan Pacific coast I-539
— Norwegian fjords I-21
— patterns in the Jiulongjiang estuary II-518
— Peruvian deltas I-696
— recent, Mexican coastal plain I-476
— reducing coral cover and fish abundance Borneo II-371, *II-371*
— Vanuatu, threatens some reefs II-742
— Western Australia, increased, carrying pollutants II-701
sedimentology, marine, Gulf of Guinea I-778
sediments
— accumulation of TBT in III-250
— Bay of Bengal II-273
— carbonates, West Florida shelf I-438
— containing agrochemicals, lagoons, French Polynesia II-821
— from rivers systems during the monsoon, Sri Lanka II-178
— Gulf of Aden, settle in deeper water II-53
— Gulf of Cadiz I-170, *I-171*
— high nutrient content I-493
— importance of availability of III-352
— influx to estuaries, El Salvador I-548
— Irish Sea bed I-89
— lagoonal, New Caledonia II-726
— marine
    — colonized by microorganisms III-261
    — trace metals in, Greenland I-9
— pulsed into Chesapeake Bay I-340–1
— shelf, Puerto Rico I-581–2
— size important in benthic community stratification II-665–6
— smothering coral polyps II-193
— Southern California Bight
    — contamination in I-392–4
    — quality assessed by toxicity studies I-396
— surface, Gulf of Maine, metal concentrations in I-315
— transport and deposition of III-352
— trapped by mangroves I-421
sediments loads, Fijian rivers II-758
sei whales III-82, III-82–3
— depletion of III-78
seine netting
— Montepuez Bay, Quirimba Island II-106
— Tanzania, destructive to reefs II-90
seismicity
— Coral, Solomon and Bismark Seas region II-429
— the Philippines II-408
— Vanuatu II-744
— Xiamen region II-516–17
— *see also* earthquake activity
seismology, sea level and island ages, Chagos Islands II-225–6
selenium (Se), detoxifying mercury in Greenland I-9
Sellafield, UK
— radioactive discharges to Irish Sea I-94
— radionuclides transported from I-11, I-13, *I-14*
sensors, satellite and airborne **III-284**
set-back lines, in coastal planning III-353
Setúbal embayment I-153
sewage
— affects coral reefs III-38
— in Bangladeshi rivers II-293
— Borneo
    — island treatment systems II-377–8
    — mainly reaches rivers untreated II-376
— causing problems in the Maldives II-212
— a disposal problem for small islands I-623
— domestic
    — and organic pollution, western Indonesia II-396
    — source of marine contamination, Coral, Solomon and Bismark Seas region II-440
— effluent discharged to deep injection wells I-429
— entering Aden Harbour II-58
— from Mexican towns, carried into Belize by currents I-512
— historically discharged contaminants from I-392
— impact of, Hawaiian Islands II-804, II-809
— inadequate facilities, Fiji Islands II-760–1
— Lesser Antilles I-638
— limited treatment, Gulf of Guinea I-784
— Madras, discharged into the estuaries II-169
— Malacca Strait, from the Sumatran coast II-340
— in Metro Manila II-416
— polluting Turkish Black Sea coast I-300–1, **I-301**
— Rio de la Plata estuary, effects of I-759
— seeps can lead to nutrient enrichment II-135–6
— and sludge discharges, Côte d'Ivoire I-817
— a threat to potable water supplies, Tanzania II-93
— Torres Strait and Gulf of Papua II-605
— treatment demanded in Canary Islands I-197
— untreated, problems of discharge to sea, Marshall Islands II-783

— and water use, the Bahamas I-429–30
sewage discharge
— direct to sea, east coast, Peninsular Malaysia II-351
— and high BOD, Malacca Straits II-317
— to groundwater lens, Bermuda I-228
— and treatment, Dutch Antilles I-606, I-608
— untreated
    — Gulf of Guinea I-789
    — to Irish Sea I-95
sewage disposal
— American Samoa II-770
— and deteriorating water quality, Jamaica I-567
— French Polynesia II-821–2
— into coastal waters, southeast India II-165
— Jamaica I-569
    — and deteriorating water quality I-567
— new scheme implemented, Hong Kong II-543
— a priority, New Caledonia II-733
— proper system needed for Lakshadweep Islands II-195
— Puerto Rico I-584
— Vanuatu, a concern II-745
sewage outfalls
— Southern California
    — contamination of sediments near I-393–4
    — improvement of condition of benthos near I-395–6
sewage pollution
— Cabo Frio–Cabo de Santa Marta Grande, southern Brazil I-741, I-742
— the Comoros II-251
— effects of, Gulf of Aden II-58
— impacting on coral reef building algae, Tanzania II-93
— Maputo Bay II-109
— Morrocoy National Park, Venezuela I-655
— of urban shorelines, Madagascar II-125
— Zanzibar, in nearshore waters II-93
sewage sludge
— disposed of in New York Bight I-327
    — ocean disposal ended I-330
— dumped at sea I-77
— dumping in Irish Sea I-96
sewage treatment
— Tasmania II-657
— Victoria Province, Australia II-669
    — facilities, Port Phillip Bay II-669
sewage treatment facilities
— lack of, Red Sea countries II-43
— Port Phillip Bay, Victoria Province, Australia II-669
sewage treatment plants
— poor design and maintenance I-638, **I-639**
— Tasmania, some improvement in II-657
sewage and waste management
— Australia II-586
— Great Barrier Reef Marine Park II-626
sewage wastes, reaching the Basque coast I-145
The Seychelles II-233–41
— coastal erosion and landfill II-239
— effects from urban and industrial activities II-239
— inner granitic islands II-235
    — Precambrian II-235
— major shallow water marine and coastal habitats II-236–7
— offshore systems II-237
— outer coralline islands and atolls II-235
    — high limestone islands II-235
— populations affecting the area II-237–8
— protective measures II-239–41
    — network of marine protected areas 239–40, **II-240**
    — ratification of international Conventions II-239, **II-240**
— rural factors II-238–9
— seasonality, currents, natural environmental variables II-235–6
— Seychelles National Land Use Plan II-238–9
Seychelles Bank II-235

shallow water ecosystems and biotic communities, Aegean Sea I-237–40
— benthic fauna I-238–9
— fish fauna I-238
— marine mammals I-239
— marine plants I-239
— marine reptiles I-240
— plankton I-237–8
— sea birds I-239–40
— zooplankton communities I-237
shallow water habitats
— Southern China coastal waters II-552–4
    — intertidal zone II-552–3
— Yellow Sea II-490
shark capture, Southern Gulf of Mexico I-475
shark fins, from by-catch, Marshall Island II-784
shark fishery
— Belize I-509
— El Salvador I-550, I-553–4
— Hawaiian Islands II-803
— Mozambique, linked to possible dugong population collapse II-104
— Nicaraguan Pacific coast I-539, I-540
— Yemeni II-57
sharks II-363
— as by-catch III-147
— Chagos Islands, drop in numbers II-228–9
— in the English Channel I-70
Sharm el Sheik *II-36*, II-44
Shatt al Arab *II-2*
— low oxygen saturation II-5
— nitrates, silicates and phosphates II-6
Shaumagin Islands *I-374*
Shebeli alluvial plain *II-67*, *II-70*
Shebeli River, Somalia *II-64*, II-68
shelf benthos
— eastern Australian region II-633–4
— Great Barrier Reef region II-615–16
shelf reefs, Puerto Rico I-582, I-583–4
shelf seas, vulnerable III-295
shelf-edge atolls, Western Australia II-696
Shelikof Strait *I-374*
— lower, interannual variation in copepod biomass I-380
Shelikov Bay *II-464*
shell deposits, Sri Lanka II-181
shellfish, spring mass mortalities, Taiwan II-506
shellfish culture plots, TBT in II-319
shellfisheries
— Albermarle—Pamlico Estuarine System I-360
— Canary Islands I-195
— Carolinas coast
    — barrier islands/coastal rivers region I-361–2
    — Northern Carolina, closed due to high bacterial counts I-364–5
— closure of I-76
— decline in Chesapeake Bay I-346
— Gulf of Cadiz I-175–6
— Irish Sea I-91
— New York Bight I-329
    — overharvested I-327
— northern Spanish coastal communities I-144
— San José Gulf I-764
— shell fish farming, Adriatic Sea I-277
shells, as a form of currency II-436–7
shifting/slash-and-burn cultivation II-88, II-435, II-601
— Gulf of Guinea countries I-786
— Lesser Antilles I-635
shingle areas, Channel coast of England I-68
shipping
— Australia II-587
— eastern Australian region II-641
— in the English Channel I-72–3

— Great Barrier Reef lagoon, inner and outer routes II-622–3
— Gulf of Aden II-55
— impacts of Bay of Bengal II-279–80
— impacts on the Gulf II-13
— intensive activity, South China Sea II-559
— Malacca Strait II-339
— Malacca Straits II-316–17, *II-316*
— and marine pollution III-336
    — inputs of contaminants from III-364–5
    — LOS legal framework III-333–4, III-336
— New Caledonia II-728
— New York Bight I-323
— offshore accidents and impacts
    — Lesser Antilles I-638
    — New Caledonia II-733–4
    — Tasmania II-657–8
— and offshore impacts
    — Belize I-512
    — El Salvador I-556
— potential source of marine pollution, Vietnam II-566
— and seaports, Nicaraguan Pacific coast I-540
— South Western Pacific Islands II-718
— Torres Strait II-605
— Victoria coast II-669
shipping accidents, potential for, the Maldives II-213
ships, dismantling of, potential pollutants III-339
shipwrecks, Madagascan coast II-125
shoalgrass, Carolinas I-358
shore birds
— coastal wetlands of Argentina I-755–6
— Colombian Pacific Coast I-680
— migratory
    — east coast, Peninsular Malaysia II-350
    — Hawaiian Islands II-798
shoreline erosion II-193
Shoreline Management Plans (SMPs) I-59
— England and Wales III-351–2
shorelines, effects of sea level rise III-188
shrimp aquaculture
— Colombian Caribbean Coast I-672
— developing at the expense of mangroves, around Beira II-109
— east coast of India
    — collapse of II-278
    — disease problem II-280
— El Salvador I-553
— environmental consequences cause great concern II-394
— extensive, Bangladesh II-292
— Guinea I-801
— Gulf of Mannar II-166
— Gulf of Thailand II-306
— Malacca Straits II-314
— mangrove clearance for, Rufiji Delta, Tanzania II-89, II-91
— Mexico I-492, I-493
    — extensive aquaculture I-494
— New Caledonia II-732, II-733
— a pollution source, Yellow Sea II-494
— Sri Lanka
    — damaging mangroves II-179
    — major pollution source II-184
— western Indonesia II-394
shrimp farming III-169, *III-170*, III-368
— Asia and Latin America III-368
— Belize I-511
— Cambodia II-576–7
— Colombian Pacific Coast I-683–4
— mangrove clearance for, Borneo II-368–9, *II-368*
— marine, Venezuela I-653–4
— Nicaraguan Caribbean coast I-526
— Nicaraguan Pacific coast I-535, I-538, I-539–40
shrimp fisheries II-27
— Arabian Gulf II-7, II-8, II-11

— Argentina I-763
— Brunei II-372
— Colombian Caribbean Coast I-670
— Côte d'Ivoire I-815
— deep Water, Fiji II-758
— Greenland I-8
— Guinea I-801
— Gulf of Guinea I-787
— Gulf of Thailand II-306
— highest discard/catch ratios III-143
— incidental capture of sea turtles III-66
— large discard III-125
— Madagascar
    — increasingly regulated II-126
    — more productive to the west II-117
    — substantial by-catch II-121
    — variable yields II-121
— Mexican Pacific coast I-490, I-493
    — artisanal I-490–1
— Morecambe Bay I-91
— Nicaraguan Caribbean coast I-524
— Nicaraguan Pacific coast
    — fishing gear not optimally designed I-539
    — penaeid and white shrimp I-534
— northwest Pacific Ocean, high discard rate III-137
— Pamlico Sound I-362–3
— Sofala Bank, Mozambique, changes in II-109
— South Carolina I-363
— southeast coast, Brazil I-739
— Southern Gulf of Mexico I-474
— and water temperature, Alaska I-377
shrimp nurseries I-488
shrimp ponds, western Indonesia II-388
shrimp trawling
— Gulf of Suez II-43
— Norway I-25
shrimp viral diseases III-226
shrimps, amphidromous and diadromous, El Salvador I-550–1
Si:N and Si:P ratios, decline in, northern Gulf of Mexico I-449
Siberian High I-376
— dominant over Black Sea in winter I-287
Siberian High Pressure Core II-465
Sicily, fishing ports I-279
Sierra Leone—Guinea Plateau, continental shelf I-775
Sierra Nevada de Santa Marta, Colombia I-672
Sikhote-Alin, eastern Russia, endemic species II-481
silicates
— Baltic Proper I-106
— indicators of eutrophication in Baltic Proper I-108
sills, associated with fjords I-19
Sinai Peninsula *II-36*
— mangroves II-41, *II-41*
Singapore *II-310*
— coastal modification II-317, II-339
— coral reefs, stressed II-314
— decline in fish catch II-320
— habitat loss severe II-334
— heavy metal pollution II-341
    — Keppel harbour II-318
— improvements in water quality II-341
— industrial centre II-339
— no regulation framework on the environment II-323–4
— Port, provides all major port services II-317
— Prevention of Pollution at Sea Act (1991) II-324
— regulation of land-based pollution II-324
— and sustainable development II-342
Singapore Strait *II-310*, II-311, II-333, II-334
— and South China Sea II-334
Singapore, Thailand, Vietnam, coral bleaching and mortality events III-52
SIORJI growth triangle II-337

Sites of Special Scientific Interest (SSSIs) I-96
Sitka Eddy I-375
Skagerrak *I-100*
Skomer Island *I-84*, I-96
Slovenia and Croatia, special nature reserves I-280
sludge dumping II-543
— New York Bight I-324
— Western Australia II-700
slumping, fjords I-21
small cetaceans III-89–103
— adressing the issues III-99–100
— by-catch in coastal gillnets III-94–5
— classification III-90–2
    — *odontoceti* (dolphins and porpoises) III-90
— and environmental change III-96–8
— human impacts III-92–8
    — direct and indirect catches III-92–5, III-99
    — pollution III-95–6
— populations in a fairly healthy state III-98
— vulnerable to bioaccumulation of toxins III-98
Small Island Developing States
— development problems II-265
— Federal and Islamic Republic of the Comoros II-247
— Sustainable Development of, Conference II-216
Snake Cays, fringing reefs I-505, *I-505*
social choice theory, conventional III-398
Socotra II-49, *II-49*, *II-64*
— artisanal fishery II-56
— coral communities II-52
— effective community management of traditional fisheries II-60
— no regulation on waste disposal II-58
— seabird and raptor nesting site II-54
— seagrass beds II-53
— terrestrial diversity II-49
— use of beach coral debris II-58
Socotra Archipelago II-49
— mangrove forests II-53
— to become a Biosphere Reserve? II-60
Socotra Eddy/Gyre *II-18*, *II-48*, II-50, II-66
sodium cyanide, fish asphyxiant III-123
soft bottom habitats, Malacca Strait II-337
soft corals II-742
soft shores
— eastern Australian region II-633
— Hong Kong II-538
soft-bottom benthos, Western Australia II-695
soft-bottom communities
— Baltic Sea I-126
— Spanish north coast I-139–40
— Tahiti II-819
— Tyrrhenian Sea I-274
soft-bottom habitats
— Southern California Bight I-388
— Taiwan Strait II-502–3
soft-bottom systems
— Carolinas coast I-353
    — diverse communities I-362
soft-sediment communities, Bahamas Archipelago **I-420**
soft-sediment habitats
— Tasmania II-652
— Victoria Province, Australia II-665–6
soil degradation, Lesser Antilles I-635
soil erosion I-142–3, I-174–5
— Australia II-585, II-637
— due to prawn culture II-153
— Fiji II-715, II-758
— Hawaiian Islands II-801–2
— high islands, French Polynesia II-821
— Jamaica I-568
— Mascarenes II-261, II-262
— northern end of Sumatra II-339

— a problem in China II-491
— round Sea of Japan II-475
— sediments raising river beds II-558
— though agriculture, the Philippines II-412
soil runoff, many causes, American Samoa II-770
solar radiation
— and heating of the Baltic Sea I-123–4
— Nicaragua **I-519**
— reduced by Gulf War smoke II-11
— southern Spain I-169
sole I-92, I-162
Solent *I-66*, I-72
solid waste disposal
— American Samoa, improvement in II-770
— inappropriate, Argentina I-765
— Marshall Islands *II-782*, II-783
— not understood, Cambodia II-576
solid wastes
— Côte d'Ivoire I-817
— dumped and burnt II-58, II-262
— dumping of
    — Bangladesh II-293–4
    — municipal, Dutch Antilles I-606–7
— poor disposal of
    — Fiji II-761
    — Vanuatu II-745
— a problem
    — Gulf of Guinea coast I-788–9
    — in the Maldives II-212
    — Metro Manila II-416
solitary waves, coastal erosion and habitat destruction issues III-355–6
Solomon Islands *II-426*, II-427, **II-427**
— concerns about pesticide pollution II-438
— coral reefs in good conditions II-431
— environmental assessment of the only mine II-439
— impact of land use on coastal waters II-435–6
— level of social and economic development low II-434
— mangroves II-430
— plantation agriculture II-435
— population and demography II-433–4, **II-434**
— positive effects of earthquake damage to reefs II-429
— rainforest logging II-435
— small ecosytems knowledge II-435
— tuna fishery II-423
    — "Tuna 2000" policy for sustainability II-432
Solomon Sea *II-426*
Solomon Strait *II-426*
Solway Firth *I-84*
Somali Current, seasonal reversal of II-50, *II-64*, II-66
Somali Natural Resources Management Programme II-60
Somali Plain *II-64*
Somalia II-49
— focus on peace and socio-economic considerations II-60
— Indian Ocean Coast of II-63–82
    — effects from human activities and protective measures II-80
    — Late Pleistocene to present-day event sequence II-72
    — major shallow water marine and coastal habitats 72–7, *II-78*
    — natural environmental parameters II-65–6
    — population and natural resources II-77–80
    — present day features of the coastal zone II-68–72
    — structural framework II-66–7
— location and extent II-65
— modern setting of the coast II-72
— poverty of II-56
— very low level of urbanisation and industrialisation II-58
Songo-Songo Archipelago, Tanzania, reef species richness differences II-86
Soudan Bank *II-254*, II-255
South Aegean volcano arc I-235
South Africa II-133–44

— the coastline II-135
— estuaries II-135–6
— management structures and policy II-143
— pollution and environmental quality II-142–3
— resources use II-136–42
South American Plate I-701
south Atlantic anticyclone I-823
South Atlantic Central Water
— Gulf of Guinea I-778
— South Brazil Bight I-735
South Atlantic Current I-824
South Atlantic high pressure cell I-721
South Atlantic subtropical gyre I-823
South Australia Current *II-674*
South Brazil Bight I-734–5
— bottom thermal front 734
— Inner Shelf, Middle Shelf and Outer shelf water bodies I-734–5
South Caicos, fishing centre I-591
South China Sea *II-310*, II-334, *II-346*, II-357, *II-362*, II-363, *II-382*, II-571
— coastal areas west of Hong Kong II-551
— fish landings related to the monsoon II-372
— marine pollution II-354
— monsoon generation II-365
— presence of thermocline and halocline II-348
— salinity II-348, II-552
— sea surface temperature (SST) II-407–8, II-552
— surface currents II-407
— surface water patterns II-366, *II-366*
— typhoon tracks I-365, *II-366*
— water quality **II-350**
South China Sea Current II-552
South China Sea warm current II-516
South East Anatolia Project(Turkey), may deprive Gulf of river flow and nutrients II-9, II-13
South East Asia, mangrove loss III-21
South East Asian—Great Barrier Reef Region, Indo-Pacific marine species diversity II-753
South East Queensland, major habitats II-634
South Equatorial Current I-598, *II-84*, II-101, *II-101*, II-254, II-255, II-428, *II-581*, II-614, II-619, *II-724*, II-767, II-775
— dominant round Madagascar *II-114*, II-115
— seed stock derives from further East II-117
— the Seychelles lie within II-236
South Florida Slope *I-436*
South Korea
— increase in agrochemical use II-493
— industrialization and urbanization II-492
South Pacific Central Water II-631
south Pacific eastern margin ecosystem I-705–7
— Antarctic ecosystem I-706
— central south Pacific gyre ecosystem
— islands/archipelago ecosystems I-706
— pelagic oceanic ecosystem I-706
— coastal ecosystem I-706
— sub-Antarctic ecosystem I-706
South Pacific Forum Fisheries Agency II-442, III-160
South Pacific Marine Region II-739
South Pacific Regional Environmental Program II-442, II-824
South Pacific subtropical gyre II-614
south Tasmanian bioregion, greater number of endemics II-652
South West Pacific Islands Region II-705–22
— aid projects short-lived II-720
— biogeography II-707–9
— coastal erosion and landfill II-716–18
— effects of urban and industrial activities II-718–19
— island groups *II-706*
— major shallow water marine and coastal habitats II-710–12
— new environmental initiatives at community level II-720
— offshore systems II-712–13
— protective measures II-719–20

— evaluation of marine environmental protection measures II-720
— marine protected areas II-719–20
— modern conservation practices II-719
— rural factors II-714–16
— seasonality, currents, natural environmental variables II-709–10
South Western Atlantic burrowing crab I-755
Southampton Water
— refinery discharges I-75
— vegetation loss I-74
Southeast Asia, coral bleaching and mortality events III-49–52
Southeast South American Shelf Marine Ecosystem I-749–71
— coastal erosion and landfill I-762–3
— effects from urban and industrial activities I-763–7
— human populations affecting the area I-758–62
— major shallow water marine and coastal habitats I-754–7
— offshore systems I-757–8
— protective measures I-767
— region defined I-751–3
— rural factors I-762
— seasonality, currents, natural environmental variables I-753–4
Southeast Trade Winds II-116, II-428, II-710, II-753
Southern Adriatic I-269
— beach-rocks habitat I-273
— central Oceanic community I-274
— surface circulation I-271
— water column divisions I-270–1
Southern Bluefin Tuna fishery, Australian Bight II-682, II-683
Southern Brazil Shelf I-735
Southern California I-385–404
— anthropogenic inputs
— distribution and fate of I-390–4
— and human contributions I-388–90
— biogeographic provinces and habitats I-387–8
— DDT contamination in I-390
— effects of anthropogenic activities on marine biota I-394–9
— geography and oceanography I-387
— human health concerns I-399–400
— monitoring and management of I-400
Southern California Bight *I-386*
— impaired water bodies I-392
— loss of wetlands affects fish nursery areas and migratory birds I-399
— variations in oceanic environment I-387
Southern China, Vietnam to Hong Kong II-549–60
— effects of urban and industrial activities II-558–9
— major shallow water biota II-552–4
— marine pollution parameters II-554–6
— populations affecting the area II-556–8
— protective measures II-559–60
— closed fishing seasons II-560
— local legislation II-559–60
— seasonality, currents, natural variables II-551–2
southern New South Wales, major habitats II-635
Southern Ocean Current *I-700*
Southern Oscillation I-702
Southern Pelagic Province of Australia II-675
southern right whale
— Australian Bight II-676, II-680, II-683
— disturbance and threats to II-685
Southern Shark fishery, Australian Bight II-682, II-683
southeast Asian archipelago, a zone of megabiodiversity II-389
Southwest Monsoon drift-current II-552
Soya Current II-466
Spain, North Coast I-135–50
— coastal erosion and landfill I-144–5
— effects from urban and industrial activities I-145–7
— lies in Northeast Atlantic Shelf and Eastern Canaries Coastal provinces I-137
— major shallow water marine and coastal habitats I-139–40
— populations affecting the area I-142
— protected coastal areas *I-143*, I-148

— protective measures I-147–9
  — application of Coastal Law I-147–8
  — Law for the Conservation of Natural areas and the Wild Fauna and Flora I-148
  — preservation of marine habitats I-149
  — protected areas being degraded I-148
— region defined I-137–8
— rural factors I-142–4
— seasonality, currents and natural environmental variables I-138–9
Spain, North West, Basques originated whaling III-74
Spain, southern, Atlantic coast I-167–84
— climate I-169
— coastal erosion and landfill I-176–7
— effects from urban and industrial activities I-177–83
— major marine and coastal shallow water habitats I-170–2
— offshore systems I-172
— populations affecting the area I-172–4
— protection measures I-181–3
  — Guiding Plan for Use and Management of La Breña and Barbate saltmarshes I-182–3
  — law on Andalusian Territory Regulation I-182
  — programmes and actions affecting whole coast I-183
— rural factors I-174–6
— seasonality, currents and natural environmental variables I-169–70
Spartina anglica
— a hybrid I-47
— spread of I-68
spawning, herring and mackerel, North Sea I-49
spawning stock biomass (SSB)
— Irish Sea, decline in I-92
— percentage preservation III-382
— Southern Bluefin Tuna III-367
— to give maximum sustainable yield (MSS) III-157
Special Areas of Conservation (SACs) I-52, I-77
— marine I-96–7, *I-97*
Special Management Areas, Hawaiian Islands II-808–9
Special Protected Areas (SPAs) I-52, I-77, I-96, I-148, I-162
Specially Protected Natural Territories, eastern Russia II-481–2
— black fir—hardwood forest ecosystem II-482
— "Borisovskoye plato" natural preserve II-482
— "Kedrovaya pad" nature reserve II-482
— Lazovsky Reserve II-482
— Sikhote-Alin Biosphere Reserve II-481–2
species, diversity of functional groups buffers impacts of stressors III-227–8
species abundance, effects of oil spills on III-271
species abundance and distribution, effects of variation in climate, temperature and other factor I-67–8
species diversity
— and distribution, Faroes I-33
— high, northern Australia II-581
— macroalgae, high Carolinas coast I-357
species evenness, mine tailings, Island Copper Mine III-240–1
species extinction, coral reefs III-37
species richness
— of corals in the Chagos Islands II-224–5, II-227
— mine tailings
  — patterns emerging III-238–9
  — under impact and during recovery III-238
— Oman beaches II-21
— Red Sea II-38
species vulnerability, to oil spills III-271
sperm whales/whaling III-74, III-76
— Azores III-74
— numbers indeterminate III-81
spilled oil, persistence of on shores and its effects on biota III-267–81
— effects of oil on nearshore populations and communities III-271–5
— effects of treatments III-276–7
— influences of biota on oil persistence III-271
— oil persistence on shores III-268–71
— recovery of biota and the importance of persistent oil 1275–6

— synthesis III-277–8
  — common themes III-278
Spiny Lobster fishery I-424–5
— Hawaiian Islands II-803
— Nicaragua I-525
— rocky coasts, Arabian Sea II-27
— South Caicos I-591
  — vulnerability to recruitment overfishing I-592
— Turks and Caicos Islands I-592
sponge harvesting I-426, I-511
sponges II-633
— encrusting I-590
— Guinea I-800
— populations, Palk Bay II-167
sport fishing I-621, II-138, II-213, II-260, II-806
— a problem I-197
sprat, Baltic Sea I-110, I-128
Spratly Islands *II-362*, II-364
— disputed claims to II-364
— Layang Layang Island, SCUBA diving II-377
spur and groove structures/systems
— Chagos Island reefs II-227
— Mascarene Region II-256–7, II-258
— outer reef slopes, Tanzania II-85
squid I-90, II-137
squid fishery, Argentina I-763
^{90}Sr, levels of fallout decreased in Greenland seas I-11
Sri Lanka II-175–87, *II-270*
— coastal and marine shallow water habitats II-178–80
— coastal resources management II-185–6
— direct and indirect small cetacean catches III-93
— geographical setting II-177
— marine pollution II-184–5
— marine resource use and populations affecting the area II-181–3
— non-living resources II-180–1
— protected marine fish II-183
— regulatory route in coastal planning and management III-353
— rural and urban factors affecting the coastal environment II-183–4
— seasonality, currents, natural environmental variables II-177–8
  — climate and rainfall II-177
  — currents and tides II-178
— shipping activities and fishery harbours II-185
Sri Lanka Turtle Conservation Project II-180
SST *see* sea surface temperature
stagnation periods
— Baltic Proper I-106, I-108
  — and formation of hydrogen sulphide I-105–6, *I-105*
Standard Operation Procedures (SOP), for oil spill response, Malacca Straits II-322–3
State Offshore Island Seabird Sanctuaries (Hawaii) II-797
Steller sea lion, decline in III-98–9, *III-99*
Stono River Estuary *I-352*, I-354
storm surges I-631
— associated with typhoons III-190, III-193–4, *III-193*, *III-194*
— and estuarine flood risk, effects of climate change and sea level rise 193–4, III-190
— from cyclones, Fiji II-710
— from tsunami waves, Vanuatu II-744
— Mascarene Plateau II-256
— North Sea I-46
— return periods III-194, *III-195*
— Xiamen region II-517
storm waves, Victoria coast II-663–4
storm-water management, problems, Dutch Antilles I-601
stormwater runoff
— contributes nutrients and heavy metals to coastal waters, New South Wales II-642
— increase in I-445
— nutrients in II-586
— polluted
  — Tasmania II-657

— Victoria Province II-669
Straddling Stocks Agreement III-158
— important principles III-158–9
Straits of Malacca *see* Malacca Straits
strandflat I-19
Strangford Lough *I-84*, I-96
stratification
  — of Caribbean waters I-579
  — near-surface waters, Southern California Bight I-391
  — north Pacific Ocean III-181
  — North Sea I-46
  — seasonal
    — New York Bight I-323
    — North Spanish coast I-139
  — Tagus estuary I-153–4
stratospheric Quasi-Biennial Oscillation (QBO) I-224
stream channelization, impact of I-446
stresses
  — causing coral reef deterioration I-563
  — favours domination by smaller organisms III-227
  — Florida Keys
    — global I-409–10
    — industrial I-409
    — localized, mass mortality of *Diadema* I-410
    — to the environment and reefs I-409
striped bass (rockfish) I-348
— juvenile habitat I-342–3
— spawning stock restored I-342
striped dolphin, hunted in Japanese waters III-91, III-94
sturgeon I-342
sub-Antarctic ecosystem I-706
Sub-Antarctic Water I-701–2, I-706, I-735, II-664
sub-cellular damage, in fish, from chlorinated hydrocarbons I-397
sub-Saharan drought, reduced flows of Gulf of Guinea rivers I-777
'sub-tropical underwater' I-666
Subarctic Boundary *III-180*, III-181
— represents abrupt change in stratification III-181
Subarctic Current *III-180*
submarine canyons *I-136*, I-137, I-170, I-667, I-701, I-799
— Australia II-583
— Bay of Bengal II-271
— and shelf valleys I-325–6, *I-326*
— Taiwan Strait II-501
— "Trou sans Fond" canyon, Côte d'Ivoire I-805, *I-806*, I-807, I-808
submarine mining, placer tin, causing high turbidity II-395
subsidence
— coastal
  — northern Gulf of Mexico I-446
  — Somali coast II-66, II-67
— Colombian Pacific coast, seismic I-682
— marginal Mesozoic, Somali coast II-66
— Sucre coast, Venezuela I-645
— through groundwater extraction, Taiwan's littoral zone II-507
subsidence phenomena, Po delta I-276
subsidies
— political attempt to create legitimacy III-155
— some necessary and justifiable III-155–6
"substituted industrialisation", Chile I-710
Subsurface Equatorial Water I-679
subtidal habitats
— Irish Sea
  — hard substrates I-89
  — soft substrates I-89
— rocky communities, Spanish north coast I-140
Subtropic Underwater I-579
Subtropical Convergence II-583, II-653
Subtropical Maximum Salinity Water II-255
subtropical mode water, formation of, Sargasso Sea I-223
Subtropical Surface Current *I-688*, I-690
subtropical underwater I-632
Subtropical Water Mass I-701

succession, pattern of altered by oil spills III-275
Sudan *II-36*
Suez Canal *I-254*, *II-36*
— and "Lessepian migration" II-38
— opened the Mediterranean to Red Sea migrants I-257–8
— *see also* Red Sea invaders
Suez, Gulf of *II-36*, II-37, II-39
— coastal pollution by oil II-43
Sulawesi Sea *II-362*, *II-382*
— coastal habitats II-365
sulphate, to the Sargasso I-229
Sulu Sea *II-362*
— circulation II-407
— coastal habitats II-365
— internal waves II-411
Sumatra
— accelerated natural erosion II-338
— central, alluvial coastal plain II-311
— east, peat swamps II-312
— maintenance of a mangrove buffer belt II-342
— mangroves II-312
  — converted to aquaculture ponds II-314–15
— massive deforestation II-338
— North, traditional fisheries management II-341
— sources of industrial pollution II-317–18
— western, marine ecosystems of II-386–7
sunbelt development, northern New South Wales II-636
Sunda Shelf II-312, II-333
Sunda Shelf region II-383
— Asian floral and faunal realm II-389
— exposed during Pleistocene lowstands II-383
— river discharge on Indonesian portion of II-384, **II-385**
Sunda Strait *II-382*
Sundarban Biosphere Reserve II-147, II-156–7
— Project Tiger II-156, II-157
Sundarban ecosystem II-148, II-290
Sundarbans
— Bangladeshi *II-146*, II-147
— British occupation II-151
— cultural convergence II-152
— early settlers II-151
— a hostile environment for settlement II-151–2
— Indian *II-146*, II-147
— name derivation II-290
— natural protection by II-290
— a necessity for Bangladesh II-291
— salinity intrusion close to II-287
— spread of colonization II-152
— *see also* Deltaic Sundarbans
sunlight penetration, reduced by nuisance micro- and macro-algae III-219
superficial reefs, Brazil I-723
surf, high energy I-487.I-486
surf zone, southwestern Africa I-830
Surface Equatorial Current I-690
surface runoff
— and river flow, Texas I-437
— source of pollutants to the Southern Californian Bight I-390, **I-391**
  — stormwater discharges, little regulation and control of I-390
surface temperature, mean global, rising III-188
surface waves, direction and velocity, Bay of Bengal II-148
surfaces, rainfall-impervious, effects on flood frequency/intensity I-341
surrogate surfaces, particle collection on III-204–5
surveillance, in fisheries management III-159–60
suspended particulate matter, Haifa Bay I-261
suspended sediments/solids
— estuarine waters, Perak and Johore II-338
— flowing into the Aegean I-242–3
— off southern Spanish coast I-175
sustainability
— adaptive management III-400

— in coastal management III-350, III-356
— Coral, Solomon and Bismark Seas region, threats to II-437–8
— of current and future practices III-360
— gauged against certain criteria III-360
— of human activities on marine ecosystems III-359–73
— requires precautionary approach to ecosystems management III-400
sustainable development I-465
— application to society III-350–1
— in coastal zones III-351, III-356
— definitions I-459–60
— elements of **I-459**
— Xiamen coastal waters, measures for II-531
sustainable energy, from the oceans III-304
Sustainable Fisheries Act (USA) I-451
sustainable governance, of the oceans III-397–402
— the deliberative process in governance III-397–8
— moving from public opinion to public judgement III-397
— use of "visions" III-397–8
— the deliberative process in governance two-tier social decision structure III-398, *III-398*
— integrated ecological–economic modelling III-398–9
— property rights regimes III-397
— conflict-resolution mechanisms III-399
— design principles III-399
— new III-399–401
— taxes and other economic incentives III-401–2
— ecological tax reform III-401
— shifting tax burden to ecological damage and consumption of non-renewables III-410–12
sustainable regional development, programs for, Mexico I-480
Suva Harbour, Fiji, pollution in II-719
Svalbard I-13, I-27
— fjords I-19
swamp forests
— Cambodia II-573
— Coral, Solomon and Bismark Seas region II-430–1
— Gulf of Mexico I-472
swamp lands, filled and canalised, no longer trap sediments II-136
swamps
— freshwater
— Coral, Solomon and Bismark Seas region II-430–1
— Gulf of Papua II-598
— Gulf of Mexico, Centla I-472
— Peru, 'Reserved Zone of Villa Swamps' I-692
Swatch of No Ground II-289
Sweden, offshore wind farm development III-309–10
swell waves I-190, I-618
— the Bahamas I-419
— Gulf of Aden II-50
— reaching Puerto Rico I-578
swine industry, Carolinas, a significant environmental threat I-366–7
synthetic drilling muds, impacting on marine benthic communities III-363
synthetic organic chemicals, in land-based pollution III-361–2
Szczecin Lagoon (Oder Haff) I-109
— fishery I-111
— metal pollution I-115

Tadjora, Golfe de II-49
— mangroves II-53
Tagus embayment, frontal boundary I-155
Tagus estuary *I-152*, I-153–4
— disposal of wastes in I-162, **I-162**
— effects of fertilizers and pesticides I-162
— flow patterns at the mouth I-154
— two distinct regions I-153
Tagus and Sado coastal waters I-155–6
Tagus and Sado coastal zone I-153
— protected areas I-162, *I-163*, **I-163**

Taiwan
— coral bleaching and mortality events III-50
— eastern II-508
Taiwan Current II-537, II-538
Taiwan Shoal *II-500*, II-501
Taiwan Strait II-499–512
— coastal erosion and landfill II-506–7
— effects from urban and industrial activities II-507–10
— major shallow water marine and coastal habitat II-502–3
— offshore systems II-503–5
— populations affecting the areas II-505
— protective measures II-510–11
— coastal zone management II-510–11
— instability of coastal policy II-510
— protection of coastal resources II-511
— rural factors II-506
— seasonality, currents, natural environmental variables II-501–2
Tangier Sound *I-336*
— decline of aquatic vegetation I-345, I-347
Tanker Safety and Pollution Prevention Conference III-337
Tanzania II-83–98
— coastal erosion II-93
— effects from urban and industrial activities II-92–3
— offshore systems II-88
— population II-88
— protected areas and integrated coastal management II-93–6
— rural factors II-88–92
— seasonality, currents, natural environmental variables II-85
— shallow water marine and coastal habitats II-85–8
— signatory to international conventions supporting ICZM **II-94**, II-95
— Southern, local community training and education II-95
Tanzanian Coastal Management Partnership II-95
tar
— on Aegean coasts I-248
— Bay of Bengal II-280
— contamination, Dutch Antilles beaches I-605
— on Israeli beaches I-262–3
— in the Sargasso Sea I-229
tar balls
— Belize I-512
— Gulf of Guinea beaches I-792
— Jamaica I-570
— Malacca Straits II-319
— Sri Lankan beaches II-185
Taranto, Gulf of *I-268*
Tasman Bay, destruction of coralline grounds III-126
Tasmania, catchment management policy II-655
Tasmanian Province II-649
Tasmanian region II-647–60
— coastal erosion and landfill II-655
— effects from urban and industrial activities II-655–8
— major shallow water marine and coastal habitats II-651–2
— offshore systems II-653
— populations affecting the area II-653–4
— protective measures II-658–9
— international conventions II-658
— Marine and Estuarine Protected Areas II-658–9
— state legislation II-658
— rural factors II-654–5
— seasonality, currents, natural variables II-649–51
Tasmanian Wilderness World Heritage Area II-653
Tatarsky Strait *II-464*, II-465, II-475
Taura Syndrome, spread by shrimp-eating birds? III-226
TBT *see* tributyltin (TBT)
TCI *see* Turks and Caicos Islands
Teacapán—Agua Brava lagoon system I-494–5, I-497
— adjacent agricultural area uses N and P fertilisers I-495
— estuarine system I-494
— mangroves I-494–5
^{99}Tc, increase in due in Greenland Seas from Sellafield I-11

technetium (Tc), conservative behaviour I-13
tectonic activity
— the Azores I-203, I-211
— extensional, Aegean Sea I-235
— northwestern Arabian Sea II-19, II-27
— off the Mexican Pacific coast I-487, I-488
Tehuantepec, Gulf of, fishery resources I-491
Tehuantepec winds I-488
'teleconnection', Pacific and Equatorial Atlantic, maybe? I-827
temperature, and the water column, Gulf of Alaska III-181
Terceira (Azores), destruction and creation of marshes I-207–9
Terminos Lagoon *I-468*
terraces, Somali coast II-72
terrestrial vegetation, Marshall Islands II-777–8
Texas Louisiana Shelf *I-436*
Thailand II-299
— butyltin compounds in sediments **II-452**
— mangrove loss, uncontrolled conversion to shrimp ponds III-21
— "shifting aquaculture" III-368
— shrimp farming III-175–6
— — move to more intensive methods III-176
— — regulations III-177, **III-177**
— — source of land for III-175–6
— species used in aquaculture III-171
Thailand, Gulf of II-297–308, *II-310*
— coastal erosion, land subsidence and sea-level rise II-307
— coastal habitats II-301–4
— defined II-299
— depletion through human intervention III-120
— effects from urban and industrial activities II-307
— geological description II-299
— offshore systems II-304–5
— physical oceanography II-299–300
— populations affecting the area II-306
— — rural factors II-306–7
— protective measures II-307
— ray and shark species reduced III-120
— seasonality, currents, natural environmental variables II-301
Thermaikos Gulf *I-234*, I-235, I-237, I-244
— increased phytoplankton abundance I-238
— organophosphorus pesticides present I-244
— river input I-244
— sewage pollution I-246
thermal pollution
— Curaçao I-605
— damaging to seagrass beds III-9
— discharges from desalination plants II-10, II-28
— Indonesian power plants II-390
thermal stratification
— around Bermuda I-225
— English Channel I-67
— summer
— — Irish Sea I-85
— — Spanish north coast I-140
thermal vents, deep sea, South Western Pacific Islands region II-712
thermal winds, Gulf of Aden II-50
thermohaline circulation, and climate change III-369–70, *III-370*
Thessaloniki *I-234*, I-244
— summer anoxia I-241
Third River (Main Outfall Drainage), Iraq *II-6*
— may seriously impact Gulf ecosystem II-9, II-13
— purpose of II-9
threshold hypothesis, for economic growth and welfare III-395
tidal amplitude
— Faroes I-33
— La Rance I-72
tidal barrages III-316–18
— construction will change the environment III-318
— double basin systems *318*, III-317
— electricity generation
— — ebb generation III-316–17, III-318
— — flood generation III-317
— — two-way generation III-317
— high costs of, but 21st century development likely III-319
— single basin schemes *317*, III-316
tidal bars *II-71*, II-76
tidal current generation III-318–19
— current and future prospects III-319
— SeaFlow Project III-319
— vertical axis and horizontal axis turbines III-318
— — problem of fixing III-319
tidal currents
— coastal waters, Xiamen region, China II-516
— Guinea coast I-799
— Puerto Rico I-580
— — Guayanilla Bay I-580
— — Mayagüez-Añasco Bay I-580
— strong
— — Irish Sea I-85
— — Yellow Sea II-490–1
— Taiwan Strait II-502
tidal energy III-304, III-315–19
— affects height of oil deposition III-269
— current and future prospects III-319
— harnessing the energy in tides III-316–19
— public perception of III-318
tidal flats
— clay–oil flocculation, reduces oil retention in fine sediments III-271
— vulnerable to oil spills III-271
— Xiamen region II-517–18
— Yellow Sea II-490
tidal marshes
— Cabo de Santa Marta Grande-Chui, southern Brazil I-737
— northwest Florida, impounded for mosquito control I-445–6
tidal power, electricity generation I-54
tidal range
— Baltic Sea I-124
— Galicia I-138–9
tidal regimes, effects of, Madagascar II-117
tidal residual, in homogeneous and stratified water III-188
tidal residual circulation, magnitude of III-188–90
tidal surges I-487, I-488, I-489
tidepools, with algae I-649
tides
— Chagos Islands II-225
— — water accumulates oxygen II-225
— Colombian Pacific Coast I-679
— and currents, affected by monsoons, Sri Lanka II-178
— diurnal and semi-diurnal, Puerto Rico I-580, I-581
— east coast, Peninsular Malaysia II-347
— extracting energy from III-315–16
— — Spring and Neap tides III-316
— harnessing the energy in III-316–19
— — tidal barrage methods III-316–18
— — tidal current generation III-318–19
— Indian Ocean II-66
— influences on III-316
— local tidal currents, Tanzania II-85
— Red Sea, annual, importance of II-39
— and tidal currents, effects of climate change and sea level rise III-188–90
Tierra del Fuego Island I-752–3, *I-753*
— whales I-756–7
Tierra del Fuego Province, coastal economy I-761–2
tiger reserves, India II-156, II-157
Tigris, River II-3
Tilapia, in aquaculture, Bahamas I-427
timber exploitation/harvesting
— Colombian Pacific Coast I-681
— problems from I-444
timber industry, Sakhalin, profitable but overharvesting II-469
TINRO Basin *II-464*, II-465

Tiran, Strait of *II-36*, II-37
titanium dioxide processing I-72
titanium—magnesium deposits, Yuzhnoprimorsky region II-480
Tivela mactroides, dissipative beaches I-648
Tobago *I-628*, *I-644*
Todos os Santos Bay, Brazil *I-720*
— mining of calcareous sand I-725
— Pinaunas Reef Environmental Protected Area I-728
Toliara region, Madagascar
— barrier reef II-116, II-118
    — signs of over-exploitation II-123
— changes to the Grand Récif, 1964–1996 II-123, II-124
— long swell II-116
— types of fishing and target species II-123
Tonga *II-706*, *II-707*, *II-709*
— agriculture and fisheries II-716
— at risk from sea-level rise II-716
— coastal modifications II-717
— coral reefs II-711–12
— Marine Protected Areas (MPAs) II-720
— population II-714
Tongatapu
— coral reefs II-711
    — degraded II-712
— Fanga'uta lagoon II-712
Tongue of the Ocean *I-416*, I-417, I-428
— V-shaped canyons I-422
Tordesillas, Treaty of (1494) I-2
Torres Strait *II-426*
— commercial use of marine resources II-600
— coral reefs II-597
— defined II-595
— environmental management II-608
— fisheries
    — artisanal II-602
    — management of II-606
    — problems of over-exploitation II-602
    — reef II-601
    — sedentary resources II-602
    — subsistence II-601
— high indigenous population II-600
— high seagrass species diversity II-431
— indigenous fishing rights protected II-584
— low level of development II-600
— population mainly on "Inner islands" II-599
— tidal circulation II-596
Torres Strait Baseline Study II-600
Torres Strait and the Gulf of Papua II-593–610
— coastal erosion and landfill II-603
— effects from urban and industrial activities II-603–5
— major shallow water marine and coastal habitats II-597–8
— offshore systems II-598–9
— populations affecting the area II-599–600
— protective measures II-605–8
    — environmental management II-607–8
    — fisheries management II-605–7
    — international agreements II-608
— rural factors II-600–3
— seasonality, currents and natural environmental variables II-595–7
Torres Strait Islands II-583
Torres Strait Treaty (Australia–Papua New Guinea) II-595, II-605–6
— Torres Strait Protected Zone (TSPZ) II-606
Torrey Canyon oil spill, increase in *F. vesiculosus* III-272, III-274, III-275
Total Allowable Catch (TAC)
— for exploited Baltic fish stocks I-112
— southwestern Africa I-836
— under CAP I-92
total suspended matter (TSM) III-297
tourism I-215, II-43–4
— Adriatic coasts I-277
— Alaska I-381, I-383

— Anguilla I-620, I-622
— Argentine coast I-758
— around the North Sea I-50, I-51
— Australia II-584
    — marine and coastal II-586
— awareness of, Côte d'Ivoire I-816
— the Bahamas, effects on the population I-423
— Belize I-511, I-514
    — demands on the coastal zone I-507
    — regulation of I-513
— Bermuda I-228
— British Virgin Islands I-620
— Canary Islands I-188–9, I-197
— coastal
    — development for unlikely to meet high standards, Mozambique II-108
    — Malacca Straits II-316, II-326
    — Xiamen region II-519, **II-520**, **II-521**
— and coastal pollution, Sri Lanka II-185
— Colombian Caribbean Coast I-673
— the Comoros II-251
— and coral reefs II-315
— coral reefs important for II-398
— east coast, Peninsular Malaysia II-353–4
— and economic value of coral reefs III-38, III-41
— English Channel coasts I-72
— Fiji Islands II-714, II-758
    — impacting on turtles II-757
— Florida Keys I-408
— French Polynesia, importance of II-823
— Godavari–Krishna delta II-172
— Great Australian Bight II-683–4
— Great Barrier Reef II-622
— Gulf of Maine and Georges Bank I-314
— Gulf of Mannar, affects coastal water quality II-166
— Hawaiian Islands II-797, II-806–7
    — boating, surfing and submarines II-807
    — development of Waikiki and shoreline erosion II-805
    — diving and snorkelling II-806
    — golf courses II-806
    — infrastructure II-806
    — sport fishing II-806
— impact of, Great Barrier Reef Marine Park II-626
— Irish Sea coast I-93
— Jamaica, demands on the environment I-567
— Lakshadweep Islands II-194
— and landscapes, west coast, Sea of Japan II-482–3
— Lesser Antilles I-637–8, **I-637**
— the Maldives
    — boating, fishing and other effects II-213
    — consideration of carrying capacity II-217
    — impacts from diving and snorkelling II-213
    — impacts from infrastructure II-212–13
    — many positive effects II-217
— Marshall Islands II-780
    — infrastructures and activities II-783
— Mauritius and Reunion, already well-developed II-264
— Mexican Pacific coast
    — development plans I-491
    — potential for I-491
— Namibia I-830
— New Caledonia, a developing sector II-732
— Nicaragua, Caribbean coast I-523
— North and South Carolina I-354
— Palk Bay area II-168
— potential, Nicaraguan Pacific coast I-535
— and recreation, Latin America I-462–3
— returning to Montserrat I-620
— round the Tyrrhenian Sea I-278–9
— the Seychelles II-238, II-240
    — Sainte Anne Marine National Park II-241

— small industry, Coral, Solomon and Bismark Seas region II-439
— southern Brazil I-737, I-738, I-739
    — destructive side of I-740
    — indiscriminate I-742
— Southern Gulf of Mexico, poorly developed I-475
— Southern Portugal I-161
    — pressure of, Gulf of Cadiz I-179, I-181
— southern Spain I-173, I-177
— southwestern Africa I-835
— Taiwan Strait coral reefs II-504
— Tanzania II-92
— Tonga II-714
— Turks and Caicos Islands I-591
— Tyrrhenian coast I-275
— upward trends in The Maldives II-206
— Venezuela I-654
tourism trends, Hawaiian Islands II-801
Townsville Trench *II-426*
toxic materials, accumulation in parts of Chesapeake Bay I-348
toxic residues, Peruvian coastal waters I-695
toxic waste
— deliberate dumping of II-59
— Indonesia, cradle-to-grave approach II-322
toxic waste trade, Nigeria I-788
toxicity
— in Argentine coast molluscs I-764–5
— of harmful algal blooms III-98
toxin bioaccumulation, from algal blooms III-199
toxins
— defensive III-39
— dinoflagellate III-40
trace elements
— atmospheric input into North Sea I-56
— Bay of Bengal II-274
trace metal pollution, the Philippines II-416
trace metals
— Faroes I-36–7
— Gulf of Maine I-314
— Hong Kong coastal waters II-544
— oysters and sediments, El Salvador I-555
— in sediments I-26
trade, Somalia II-80
trade agreement, affect fisheries management III-146
Trade Winds I-188, I-591, III-369
— Marshall Islands II-775
— Mexican Pacific coast I-488
— and the Northwestern African Upwelling I-190
— and the Puerto Rican wave climate I-577–8
— tropical coast of Brazil I-721
tragedy of the commons I-462
— and fishing management III-154
TRAIN-SEA-COAST III-325–7
— integrated management of coastal and marine areas III-325
— network III-326
    — capabilities of III-326
    — central unit responsibilities III-326
    — development units for specific training priorities III-326–7
    — range of training courses under development III-327
— UN/DOALOS (UN Division for Ocean Affairs and the Law of the Sea) III-325
    — programme of action III-325
TRAIN-X strategy (UN)
— main elements of III-324
— methodology III-325, III-329
— training networks III-324–5
— *see also* CC:TRAIN; TRAIN-SEA-COAST
trans-oceanic floating debris, even in Chagos Islands II-230
transboundary pollution
— affecting Mozambique II-109
— air pollution, Sea of Japan II-476
— oil spills in Malacca Straits II-325
— Xiamen region II-527–8
— Yellow Sea II-496
transboundary straddling stocks/species, the Philippines II-410
transgressive–regressive cycles, Holocene, Argentine coastlines I-762
Transkei coast
— artisanal and subsistence collectors II-140
— overexploitation of mussels II-141
transport, the Maldives, environmental concerns II-213–14
trap fisheries I-397
trawl fisheries
— eastern Bass Strait II-668
— restricted, Western Australia II-697
— southern Brazil I-739
— Venezuela I-653
trawl nets, by-catch excluders II-587
trawling
— affecting Black Sea biota I-290
— banned, by Indonesia II-320, II-338, II-341
— beam trawls, North Sea III-368
— bottom trawling
    — unselective III-124
    — very destructive III-379
— causing damage to the Norwegian continental shelf I-25
— destructive III-122–3
    — Gulf of Aden II-59
— detrimental effect on the English Channel I-76, I-77
— ecological impact of I-178
— and habitat destruction III-379–80
— North Sea I-51, *I-51*, I-52
— pelagic trawls more selective III-124
— restricted in southern Spain I-183
— Southern California Bight I-397
— trawls and dredges scour bottoms I-445
trawling exclusion zones, Gulf of Alaska I-383
treaties, traditional, substituted by global plans and programs of action III-344
Treaty of the Rio de la Plata and its Maritime Front I-463
tributyltin (TBT) I-72, I-95, II-12, II-339, II-449
— in Australia II-588, II-669, II-701
— in the Baltic I-114, I-131
— currently only from larger vessels III-364–5
— degradation of III-249–50
— and endocrine disrupter III-249
— environmental impacts of III-250–1
— high concentrations, Malacca Straits 319
— Hong Kong II-542
— and imposex I-25, I-36–7, I-55, I-75, I-210–11, III-364
— major antifoulant III-248
— a moiety III-248
— as part of free association paints III-248
— persistence of in the environment III-249–50
— in Prince William Sound I-381
— Suva Harbour, Fiji II-719, II-745
— Tagus and Sado estuaries I-158
Trinidad *I-628*
Trinidad and Tobago, estuarine conditions and turbid waters I-630
Triste, Gulf of *I-644*
— coastal retreat I-645
— contamination by oil, domestic and rural wastes I-645
— great industrial impact I-659
— high heavy metal levels I-659
Trobriand Trough *II-426*
trochus fishery
— Fiji, overexploited II-759
— New Caledonia II-730
— Torres Strait II-602
Trochus shell II-784, II-820
Tromelin seamount II-255, II-257, II-260
Trondheim *I-18*
— contaminated harbour I-27–8
trophic cascade model III-127

— cases of top-down controls in community structure III-128–9
trophic-magnification disturbances III-221–3
trophodynamic disturbances
— affect habitat supporting organisms III-221–2
— and habitat lost, significant portion of lost GDP III-222–3
Tropical Atlantic Central Water I-579
Tropical Cyclones I-223, I-224
— impact on the Sargasso Sea and Bermuda I-224
tropical habitats, Puerto Rico I-582–4
tropical lows, Puerto Rico I-578
Tropical South Pacific Oceanic Province, Mexico I-489
— oceanic habitat I-489
tropical storms I-488, I-489, I-617, I-631
— "Agnes" (1972), effects of I-341, I-346
— impacts on Chesapeake Bay I-340–1
— Xiamen region II-517
Tropical Surface Waters I-777
tropicalization, of Peruvian coastal waters I-692
troposphere, thermal stratification of, Canary Islands I-188
'trottoir', Tyrrhenian Sea I-273
"Trou sans Fond" canyon, Côte d'Ivoire I-805, *I-806*, I-807, I-808
trout, farming of III-168
Tsesis oil spill III-272
tsunamis II-744
— Colombian Pacific Coast I-679
— Coral, Solomon and Bismark Seas region II-429
— Fiji Islands II-753
Tubataha Reef *II-406*
tuna fisheries
— the Azores I-209
— Borneo II-373
— Canary Islands I-196
— Coral, Solomon and Bismark Seas region II-432, *II-433*
— French Polynesia II-819
— Gulf of Guinea I-787
— long line and purse-seine, Colombian Caribbean Coast I-670–1
— longline, Bay of Bengal II-277
— Madagascar II-121–2
— the Maldives II-210
— Marshall Islands II-784
— Mexico I-490
— Mozambique II-105, II-109
— Nicaraguan Pacific coast I-536
— the Philippines II-411
— recreational, Chagos Islands II-230
— South Western Pacific Islands II-718–19
— Southern Gulf of Mexico I-475
tuna–dolphin problem III-141–3
turbidity I-548
— affects kelp beds I-395
— changes in alter biological processes II-193
— coastal, Cameroon I-790
— Côte d'Ivoire I-809–10
— due to seagrass die-off III-14–15
— in fjords I-21
— harmful effects
  — in estuaries I-365
  — on habitats I-445
— high, inshore, northern Gulf of Mexico I-437
— Kuwait, caused by landfill II-9–10
— offshore, Guinea I-799
— a problem in Bermuda I-228–9
— Tagus estuary I-154
— of water, related to maximum depth of living reef II-397
— of water above seagrasses, Western Australia II-700
turbidity currents, responsible for Bahamian V-shaped canyons I-422
Turbo shell II-820
turbot, Baltic predator I-111
turbulence, thermally-induced or mechanically generated III-207
turbulent transport 202

Turkey
— Black Sea fishing industry I-292
— the southern Black Sea I-294–6
  — characteristics and population of the coast I-295–6
  — development of the coastal areas I-296
  — land-based pollution I-296–303
  — rural factors affecting the coast I-296
Turks Bank *I-588*
— reef areas I-590
Turks and Caicos Islands I-587–94
— artisanal and non-industrial uses of the coasts I-592
— coastal erosion and landfill I-592
— coastal habitat map III-289
— defined I-589
— major shallow water marine and coastal habitats I-589–91
— offshore systems I-591
— populations affecting the area I-591
— protective measures I-592–3
  — regulations to protect local fisheries I-593
— rural areas I-591–2
— seasonality, currents, natural variables I-589
Turneffe Atoll (Island) *I-502*, I-504–5
turtle eggs
— collection now banned, east coast, Peninsular Malaysia II-355
— eaten, Socotra II-57
Turtle Excluder Devices (TEDs) II-127, III-66, III-141, III-143
turtle fisheries I-620
— Mexican Pacific coast I-491
turtle grass I-562, I-601, I-632
— die-off in Florida Bay, a trend for coastal waters in the new Millennium? III-14–16
turtle hatcheries, east coast, Peninsular Malaysia II-356
Turtle Island *II-406*
turtle nesting beaches II-24, II-54, II-350, II-377
— the Comoros II-250
— Mozambique II-105
— St Brandon Islands II-257
— St Martin's Island, Bangladesh II-289
turtle nesting sites I-550
— West Africa I-780
turtlegrass II-598
Tweed-Moreton Bioregion, Australia II-634
Twofold Region, Victoria Province II-663
typhoon shelters, Hong Kong, polluted *II-543*, II-544–5
typhoons II-483
— Cambodia II-571
— Marshall Islands II-776
— present and future effects III-190, III-193, *III-194*
— South China Sea II-552, *II-552*
— Xiamen region II-516, II-517, **II-517**
— Yellow Sea II-489, II-490
Tyrrhenian Sea *I-268*, I-271–2
— algal species of tropical origin I-282
— Atlantic Water in I-272
— Central-Southern, abyssal fauna I-274–5
— coastal erosion I-276–7
— coastline I-270
— deep waters I-272
— effects of urban and industrial activities I-278–9
— effects of winds I-272
— limits I-270
— nutrient availability I-271
— shipping activity intense I-275, I-279
— typical Mediterranean biocenoses I-273

UAE *see* United Arab Emirates
UK
— commitment to offshore wind energy III-310
— 'insensitive structures', potential for legal action III-354
— many laws relating to the coast III-353

— NERC models., southern North Sea I-50
— nutrient input into the English Channel I-74
UK government, White Paper, biodiversity issues in overseas territories I-623–4
UK Overseas Territories in the Northeast Caribbean: Anguilla, British Virgin Islands, Montserrat I-615–26
— Darwin Initiative funds I-623
— legislation and protective measures I-623–5
— prospects and prognoses I-625
UN Conference on Environment and Development (UNCED) III-343
— Agenda 21 III-158
— and the London Dumping Convention reviews III-338–9
— Preparatory Committee, process-oriented criteria III-346
UN Convention on Biodiversity I-147, I-162
UN Convention on the Law of the Sea (LOS: UNCLOS) *see* UNCLOS (UN Convention on the Law of the Sea)
UN Framework Convention on Climate Change *see* Climate Change Convention
UN/DOALOS (UN Division for Ocean Affairs and the Law of the Sea), and the TRAIN-SEA-COAST programme III-325–6
UNCHE III-343
UNCLOS (UN Convention on the Law of the Sea) I-56, I-147, III-158, III-332, **III-333**, III-336
— adopted by the Philippines II-408–10
— authorises littoral states to undertake enforcement measures II-324–5
— basis for international fisheries law, little detail on application III-158
— requirements of ships transiting Malacca Straits II-324
— *see also* Reflagging Agreement; Straddling Stocks Agreement
UNEP
— global conference on sewage III-345
— need to accurately assess program effectiveness III-346
— progress with global 'POPs' Convention III-363
— Regional Seas Convention for East Africa (Nairobi Convention) II-127
— Regional Seas Programme I-2, III-339, III-340
    — Kuwait Action Plan I-2
    — Mediterranean Action Plan (MAP) I-249
    — Red Sea Action Plan II-44
    — South Pacific Region I-2
Unguja Island II-85
unique environments, Patagonian coast, Argentina I-756–7
UNITAR (UN Institute for Training and Research)
— approach of I-327
    — country team approach I-327
— Climate Change Programme training packages III-328
— development of CC:TRAIN III-327
— regional partners III-327
United Arab Emirates *II-2, II-19*
— coral bleaching and mortality events III-47
— special protection proposed for Khawr Kalba II-31
— *see also* Fujeirah
United Joint Group of Experts on the Scientific Aspects of Marine Pollution (GESAMP), definition of pollution III-258
uplift, Venezuelan coast I-645
upper water column characteristics, the Maldives II-201–2
upwelling index, Alaskan coast I-378, I-380
upwellings II-614
— act as barrier to gene flow and marine organism distribution II-51
— Arabian Sea II-19, II-66
    — cool and nutrient rich II-19
    — open-ocean and coastal II-19
    — stimulate phytoplankton II-24
— Arabian Sea system II-50
— attractive to seabirds III-113
— Bay of Bengal II-274
— Benguela ecosystem II-135
— Cambodia II-571, II-572, II-574
— Chagos Islands II-230
— Colombian Caribbean Coast I-666

— causes special environmental conditions I-666
— recedes in rainy season I-666
— restrict coral formations I-668
— Colombian Pacific Coast I-679
— Côte d'Ivoire I-808, I-809
— equatorial and open ocean I-535
— Galician and Cantabrian coasts I-139, I-141
— Great Australian Bight, provide nutrients to surface waters II-676, II-677
— Guinea coast I-800
— Gulf of Aden II-50–1, II-54
— Gulf of Alaska I-376
    — from Ekman pumping I-380, I-381
— Gulf of Guinea I-781
    — central subsystem I-778–9
    — intensification of winter upwelling I-781
    — interannual variability in I-781–2
— intensity may reduce with global warming III-369
— Long Island and New Jersey coastlines I-324
— Madagascan coast II-115
— the Maldives II-202
— North Sea I-46
— Northern Benguela I-823
— and nutrient enrichment, Messina Strait I-281
— off Luzon, South China Sea II-411
— off Peru I-690
    — importance of in El Niño events I-691
— off West Greenland coast I-7
— Oman coast II-19, II-50, II-54
— Sea of Okhotsk II-468
— shelf-break area, Argentinian continental shelf I-754
— shelf-edge, Carolinas coast I-353
— Somali coast II-50, II-51
— southeast coast, southern Brazil I-733
— southern Baltic I-111
— Southern Gulf of Mexico I-475
— southern Portuguese coast I-153, I-155
    — effects on fisheries I-160
— southern Spanish coast I-172
— southwestern Africa
    — central Namibian region I-825–6
    — effects of remote forcing from the equatorial Atlantic I-825
    — interannual variability I-827
    — low-oxygenated bottom water, northern Namibia I-827
    — Lüderitz upwelling cell *I-822*, I-823, I-825, I-826–7
    — northern Namibian Region I-825
    — wind-driven I-824
— upwelling ecosystem, Chilean coast I-706
— Venezuelan coast I-646
— Vietnam II-564
— Yemeni coast II-54
uranium, in the Hoogly (Hugli) river II-278–9
urban development
— coastal, Dutch Antilles I-606
— Colombian Caribbean Coast I-673
— Côte d'Ivoire, encroaching on lagoons and estuaries I-817
— Hong Kong
    — pressure on local marine environment II-543
    — rapid habitat loss II-541–2
— poorly planned, impact of, Coral, Solomon and Bismark Seas region II-439–40
— Tasmania II-654
    — pressure on the marine environment II-657
— Torres Strait and Gulf of Papua II-605
— unplanned, southern Brazil I-740, I-742
— Victoria Province, Australia II-669, II-670
urban, environmental and health problems, the Maldives II-212
— sewage-related problems II-212
urban impacts
— Australia II-586
— Noumea, New Caledonia II-733

urban and industrial activities
— Gulf of Mexico I-477–8
    — artisanal and non-industrial uses I-477
    — industrial uses I-477
    — shipping and offshore accidents I-477–8
— Yellow Sea II-494–5
    — aquaculture and coastal industries II-494
    — oil and oil spills II-495
    — wastewater and solid waste discharges II-494–5
urban and industrial activities, effects of
— American Samoa
    — habitat loss II-770–1
    — industrial activities and impacts II-770–1
    — other urban activities and impacts II-770
— Argentine coast I-763–7
    — commercial fisheries I-763
— Bay of Bengal II-279–80
— British Virgin Islands I-623
— Cambodia II-576–7
— Colombian Caribbean Coast I-673
— Colombian Pacific Coast I-682–4
    — artisanal and non-industrial uses of the coast I-682–3
    — cities I-683
    — industrial uses of the coast I-683–4
— the Comoros II-251
— Côte d'Ivoire I-817–18
    — depletion and degradation of coastal habitats I-817–18
    — oil and gas exploration I-817
    — waste disposal I-817–18
— Dutch Antilles I-502–7
— east coast, Peninsular Malaysia II-353–4
— eastern Australian region II-639–42
    — cities II-640
    — coastal settlements II-640
    — commercial fisheries II-640–1
    — industrial uses II-640
    — mining and dredging II-641
    — ports and shipping II-641
    — regional issues II-641–2
— Fiji Islands II-760–1
— Great Australian Bight II-684–6
— Great Barrier Reef region II-622–4
— Gulf of Aden II-58–9
— Gulf of Guinea I-787–93
    — artisanal and non-artisanal coastal use I-787
    — cities I-788–9
    — dams I-789, I-792
    — industrial fishing I-787–8
    — onshore oil production I-788
    — shipping and offshore I-792–3
    — toxic waste trade I-788
— Gulf of Thailand II-307
— Hong Kong II-543–6
    — anthropogenic contamination widespread II-543
— Lesser Antilles, Trinidad and Tobago I-636–9
— Madagascar II-125
— the Maldives II-210–15
— Marshall Islands II-783–4
— the Mascarene Region II-262–5
— Mozambique II-108–9
— New Caledonia II-731–4
    — artisanal and industrial uses of the coast II-731–2
    — cities II-732–3
    — shipping and offshore accidents and impacts II-733–4
— the Philippines II-414–18
    — artisanal and non-industrial uses of the coast II-414–15
    — cities II-416–17
    — industrial uses of the coast II-415–16
— Sea of Okhotsk II-469–71
— the Seychelles II-239
— shipping and offshore accidents and impacts II-417–18
    — pollution hot spots II-418
— South Western Pacific Islands II-718–19
— southwestern Africa I-835
— Taiwan Strait II-507–10
— Tanzania II-92–3
— Tasmania
    — artisanal and non-industrial coastal uses II-655
    — commercial usage of coastal resources II-655–6
— Torres Strait and Gulf of Papua II-603–5
    — artisanal and non-industrial coastal uses II-603
    — cities II-605
    — fisheries II-604–5
    — industrial uses II-603–4
    — shipping and offshore accidents, impacts II-605
— Vanuatu II-745
— Venezuela I-655–9
    — eastern coastline I-655–6
    — western coastline I-656–9
— Victorian Province, Australia II-667–9
    — commercial fisheries II-667–8
    — industrial uses of the coast II-669
    — recreational fisheries II-668
    — shipping and offshore accidents and impacts II-669
    — urban use (cities) II-669
— Western Guangdong coast II-558–9
    — artisanal and non-industrial uses II-558–9
    — cities II-559
    — industrial uses II-559
    — shipping and offshore accidents and impacts II-559
— western Indonesian II-396–9
urban, industrial and other activities, effects of, Australia II-586–9
urban pollution
— Borneo II-375
— Patagonia, and eutrophication I-765
urban sewage, a pollutant I-24
urban waste, a problem along entire Gulf of Guinea coast I-788–9
urbanisation
— adjoining an estuary, creates problems II-135–6
— of Carolinas coast I-365
— coastal and waste disposal, Tanzania II-92–3
— effects of, Hawaiian Islands II-804–7
— expansion allows debris to enter northern Gulf of Mexico I-450
— Florida Keys I-408
— Great Barrier Reef catchment II-622
— Pearl River delta II-554–5, II-558, II-559
— the Philippines
    — major industrial regions II-417
    — rapid, and informal settlements II-416
— rapid
    — east coast of India II-278
    — effects of I-449
— South Western Pacific Islands II-718
— Southern California I-388
— Vietnam, impact on the marine environment II-566
Uruguay, EcoPlata I-463
USA
— and adjacent waters, Guidelines for the Conservation and Restoration of seagrasses, a synopsis III-8
— Coastal Zone Management Act
    — CZM evaluation model III-347
    — independent evaluation of program effectiveness III-346–7
— eastern, atmospheric nitrogen pathway to watersheds III-201
— Endangered Species Act II-798
— marine reserves III-386
— property rights important in regulation of the coast III-354
— stringent environmental standards for uses of Kwajalein atoll II-786–7
user conflict
— conflict resolution over use of Aliwal Shoal, South Africa II-141, II-142
— from increased resource use II-142

Ushant *I-66*
USSR
— catch reports false, collusion with Japan III-77, III-86
— natives of eastern Siberia, need for whale meat III-83
UV radiation, increase in III-370

Vanuatu *II-706*, II-707, *II-708*, II-717, II-737–49
— agriculture II-714–15
— biogeography II-739
— coastal erosion and landfill II-744–5
— coral reefs II-711
   — cyclone damage II-711
— customary law and reef ownership II-741
— cyclone-prone II-710
— effects of urban and industrial activities II-745
— island groups II-739
— major shallow water marine and coastal habitats II-740–2
— Marine Protected Areas (MPAs) II-719
— offshore systems II-743
— population II-713
— population affecting the area II-743
— protective measures II-745–7
   — marine conservation II-745–6, **II-746**, **II-747**
   — Marine Protected Areas (MPAs) II-746–7
— rural factors II-743–4
— seagrasses II-711
— seasonality, currents, natural environmental variables II-739–40
— unsatisfied demand for fresh fish II-744
vaquita, Gulf of California
— endangered III-90
— incidental catch III-94
Vas Deferens Sequence Index (VDSI) III-250
Vedaranyam Wildlife Sanctuary, Palk Bay II-167
Venezuela I-643–61
— coastal erosion and landfill I-655
— the continental coastline I-645–6
— effects from urban and industrial activities I-655–9
— fisheries I-653–4
— major shallow water marine and coastal habitats I-648–53
— populations affecting the area I-654–5
Venezuela, Gulf of *I-644*, I-645, I-646
Venice, Gulf of *I-268*
— salinity I-271
Vening Meinesz Seamounts II-583
vermetid reefs *I-66*, I-255, I-256–7
vermin and insect infestations, Tanzanian coast II-88–9
vertebrates
— accumulation of TBT in III-250
— endangered, southern Brazil I-737
— Gulf of Aden II-54
Victoria Province, Australia II-661–71
— coastal erosion and landfill II-667
— effects from urban and industrial activities II-667–9
— effects of removal of native vegetation from catchments II-666–7
— Fisheries Management Plans II-670
— Gippsland Lakes, adjusting to changed environment II-667
— human populations affecting the area II-666
— major shallow water marine and coastal habitats II-664–6
— marine species, distribution patterns II-663
— offshore systems II-666
— problems in marine environment II-670
— protective measures II-670
   — endangered and vulnerable species II-670
— rural factors II-666–7
— seasonality, currents, natural variables II-663–4
Victorian Biodiversity Strategy II-670
Vietnam II-299
— coral bleaching and mortality events III-52
— mangroves lost to fishponds III-22, III-25
Vietnam and adjacent Bien Dong (South China Sea) II-561–8
— biodiversity II-564

— impacts from development II-566
— legislation II-566–7
   — Environmental Protection Law II-566–7
   — species in need of protection II-567, **II-567**
— marine and coastal habitats II-564–5
— physical parameters II-563–4
— regional setting II-546
— river systems and estuaries II-566
— water quality II-565–6
Virginian Sea *see* Middle Atlantic Bight
Vistula Lagoon I-109
— fishery I-111
volcanic islands, Mascarene region *see* Mauritius; Reunion; Rodrigues
volcanic pinnacle, off Tasmania II-583
volcanicity
— Coral, Solomon and Bismark Seas region II-429
— latent, American Samoa II-767
— the Philippines II-408
Volterra—Hjort formulation III-79
Vulnerability Index *see* Environmental Sensitivity Index (ESI)

Wadden Sea I-45, I-53
— black and white spots III-263
— decline in eelgrass beds III-7–8
— disappearance of red macroalgae I-48–9
— long-term changes from bottom trawling noted III-126
— suspended sediment I-46
— and *Zostera marina* III-6–7
   — decrease in area suitable for re-establishment III-7
   — large-scale decline III-6
wadi systems, show past erosional processes II-37
wadis
— Al Batinah II-27
— and development of khawrs II-23
— Gulf of Aden coast, flow from percolates into groundwater II-52
Wake Island II-775
— extinction of Wake Rail II-778
— war-time activities II-782
war-time effects, Marshall Islands II-782–3
warming events, El Niño, Northeast Pacific Ocean III-183–4, *III-183*
waste disposal I-634
— an environmental problem, Jamaica I-569
— Côte d'Ivoire I-817–18
— Faroes I-38–9
waste dumping
— English Channel I-77
   — regional, global and European regulation I-77, **I-78**
waste management
— difficulties of, Guinea I-801
— Vanuatu, by industry, minimal II-745
waste plastic pellets, dangers of, Arabian Gulf II-12
wastes
— disposal of, Bangladesh II-293–4
— domestic, dumpsite for, New Caledonia II-733
— dumping of
   — at sea III-365–6
   — Suva Harbour, Fiji II-719
   — in Xiamen coastal waters II-523–4
— Guangdong Province **II-555**
— Halong City, Vietnam II-566
— Turkish industry, input to Black Sea **I-298**, I-302
— urban and industrial, disposal of, Tasmania II-655
— *see also* chemical waste; hazardous waste; industrial waste; radioactive waste; solid waste; toxic waste; urban waste
wastewater
— direct discharge into coastal waters
   — Chinese Yellow sea II-494–5
   — West Guangdong coast II-555
— domestic and industrial, New York Bight I-323
— El Salvador

— domestic, untreated, discharged to marine environment I-555–6
— industrial discharges I-554–5
— entering Amursky Bay II-483
— Hong Kong, arrives in coastal waters II-538
— industrial, Chile I-712–13
— municipal, Chile I-713
— final discharge to the sea I-713
— submarine disposal solution I-715
— problems of discharge to sea, Marshall Islands II-783
— released to the Yellow Sea, Korea II-495
— reuse of, Oman and UAE II-29
— Tasmania, polluting II-657
— treatment seen as a priority, Mauritius II-263, II-264
— urban, causing contamination, southern Portugal I-179
— used for irrigation, Curaçao I-606
wastewater discharge
— major issue in Noumea, New Caledonia II-733
— Peru I-695
— to sea, Gulf of Aden II-58
Wastewater Management for Coastal Cities (C. Gunnerson and J. French) I-331
wastewater pollution, Arabian Gulf II-10
wastewater treatment
— Chesapeake Bay
— biological nutrient removal I-347
— improving I-347
— Norway I-27
water
— clean, needed for aquaculture III-170
— Colombian Pacific, from artisanal wells I-680
— potable, demand for, Gulf coast, USA I-448
— resources in Somalia II-79
water bodies and their circulations, around Puerto Rico I-579–81
water clarity/transparency
— Andaman Sea II-302
— decreasing, Baltic Sea I-130
water column habitats, Hawaiian Islands II-796
water column stability, Gulf of Alaska III-181
water masses
— Bien Dong Sea II-563–4, *II-564*
— transformation and spreading of II-564
— Chilean coast I-701–2
— Gulf of Mexico *II-470*
water pollution
— local, Turkish Black Sea coast I-302
— Pearl River delta II-554–5
— the Philippines II-416–17
— Sea of Japan II-477
— sugar and rum industry, Jamaica I-569–70
— threatens Mai Po marshes RAMSAR site II-540
water quality
— American Samoa, a concern II-770
— and aquaculture III-172
— declining, Australia's inland waterways and lakes II-585
— degraded, areas of Victoria coast, Australia II-669
— Great Barrier Reef Marine Park II-626
— impaired
— Adriatic coasts I-278
— northern Gulf of Mexico I-449, I-449–50, I-450
— and industry, Nicaraguan Caribbean coast I-525
— Jakarta Bay and Pulau Seribu, influence of Java on II-397
— Jamaican beaches I-569
— Jiulongjiang estuary II-518
— Malacca Strait II-340–1
— marine, often exceeding standards, east coast, Peninsular Malaysia II-351–2, **II-351, II-352**
— New York Bight I-328–9
— off Carolinas coast I-353
— Papeete Harbour, Tahiti II-817
— parameters, optical sensing III-295–8

— bottom depth and reflectance III-297
— poor
— central New South Wales metropolitan areas II-642
— creeks and groundwater, some areas of French Polynesia II-822
— northern New South Wales rivers II-637
— some improvement, Cubatao, southern Brazil I-743
— South China Sea **II-350**
— Southern Californian shoreline I-399
— Vietnam II-565
— west coast of Taiwan II-509–10
— West Guangdong coast II-555
— Western Australia II-699
— declining II-700–1
water quality parameters, optical sensing
— chlorophyll pigment, coloured dissolved organic matter and total suspended matter III-297–8
— retrieval of chlorophyll content from spectral radiance III-297
— secchi disk depth and diffusive attenuation coefficient III-296–7
water residence time, lagoonal waters, French Polynesia II-816
water scarcity, and dam building, side effects II-306
water surfaces, dry deposition to III-201–2
water temperature, Baltic Proper, seasonal variations I-102–3
water transit time, English Channel I-67
water weeds, invading mangrove canals, Côte d'Ivoire I-818
water-column nitrate inhibition I-358
water-leaving radiance, satellite data of, measurement of diffuse attenuation coefficient III-296
water-purification technology, Tyrrhenian coasts, Arno valley I-280
watershed management planning II-496
watershed management units, Jamaica I-565
wave climate
— deep wave heights I-578
— deep-water, off Fiji II-758
— Puerto Rico
— generated by Trade Winds I-577–8
— swell waves I-578
wave energy III-311–15
— current and future prospects III-315
— extraction of from waves III-311–12
— and persistence of oil III-269
— Puerto Rico I-578
— the resource III-313–15
wave energy programme, Chennai, India II-279
wave exposure, a key role in Belize atolls I-505
wave height, increase in I-46, I-68
wave intensity, and community structure I-22
wave power III-304
wave (swell) surges
— Marshall Islands II-776
— post-hurricane **I-519**
wave-power devices
— absorber mode III-313, *III-314*
— attenuator mode III-314
— design criteria III-313
— enclosed water column devices III-313
— flexible membrane devices III-313
— relative motion devices III-313
— tethered buoyant structures III-313
waves
— energy in is kinetic III-311, III-312
— from Antarctic winter storms, Hawaii II-795
— generation of at sea III-311–12, *III-312*
— Great Australian Bight, west coast swell environment II-675–6
— growth of through differential pressure distribution III-311, *III-312*
— Gulf of Guinea I-778
— interception of by an energy converter 312
— Levantine basin I-256
— modified by atolls, Belize I-503
— permanent surf, Côte d'Ivoire I-808
— the Philippines, generated by North Pacific storms II-407
— Significant Height III-311

— wind-induced, east coast, Peninsular Malaysia II-347
weather effects, the Comoros II-245
West African flyway, Gulf of Guinea a part of I-780
West African manatee I-780
West Bengal
— felling of mangrove forests II-151
— structure of Coastal Regulation Zone (CRZ) II-157–8, **II-158**
West Florida Shelf *I-436*
— seagrass beds I-438
West Greenland
— changes in marine climate I-7
— decline in marine mammal and seabird stocks I-8–9
— lower level of POPs than East Greenland I-13
West Greenland Current I-7
West Guangdong Province coast II-551
— intertidal habitats II-552–3
— phytoplankton and red tide organisms II-553–4
West Iberian large marine ecosystem I-137
West Kamchatka Current II-466
West Kamchatka Shelf, productive fish area II-467
West Wind cold water mass II-676
West Wind Drift Current (WWDC) I-701
Western Atlantic Ocean Experiment (WATOX) I-224, I-229
Western Australian region II-691–704
— coastal erosion, landfill and effects from urban and industrial activities II-700–2
— geomorphology of the coast II-693–4
— Kimberley coast, ria system II-694
— major shallow water marine and coastal habitats II-695–6
— offshore systems II-696
— populations affecting the area II-696–7
— protective measures II-702
　　— Western Australian Marine Parks and Reserves Authority II-702
— rural factors II-697–700
— seasonality, currents, natural environmental variables II-694–5
Western Central South Pacific Current II-619
Western Indian Ocean Marine Scientists Association (WIOMSA) II-95–6
Western Somali Basin II-67
wet deposition I-13, I-56, I-223, III-200
wetland forest, Cambodia II-574
wetlands
— Bangladeshi coast II-289
— drained, Tasmania II-655
— forested, Gulf Coast, USA I-440
— Gulf of Guinea, some protected areas I-793
— Jamaica
　　— Black River Morass I-565
　　— Negril Morass I-564
— loss of
　　— American Samoa II-770
　　— northern Gulf of Mexico I-445–6, I-448
　　— to agriculture, Peru I-695
　　— to urban development *II-718*
— Mai Po marshes, Hong Kong II-539–41
— and marshes, North Spanish coast I-139
— miniature, unique, the Azores I-207, *I-208*, I-215
　　— destruction and creation of I-207–9, I-211
— and protected areas, Peru I-691–2
　　— National Sanctuary of the Mangroves of Tumbes I-691
　　— National Sanctuary of Mejia Lagoons I-692
— reclamation of, Sri Lanka II-183–4
— tropical Brazil I-727
— Victoria Province, Australia II-666
Whale Research Programme under Special Permit (JARPA), Japan III-82
Whale Sanctuaries III-84–5
— entire Southern Ocean a sanctuary III-84–5
— further suggestions III-85
— in part of Antarctic Pacific sector III-77, III-84

— proposed for Indian Ocean III-78
whale watching II-251, III-85
— Great Australian Bight II-683
— south east Queensland II-636
whales I-343, I-591, I-834, II-24, II-205, II-582
— in the Azores I-213–14
— baleen (rorqual) whales III-76
— bottle-nosed III-76, III-383
— Colombia I-680
— Great Australian Bight II-680
— Great Barrier Reef II-618–19
— individual, tags and radio tags III-80
— Marshall Islands II-780
— Mexican Pacific coast I-487, I-488
— migrations of II-116
　　— Gulf of Guinea I-783–4
— Mozambique waters II-105
— in the Southern California Bight I-399
— Sri Lanka II-180
— *see also* individual types
whales and whaling III-73–88
— assumption about whales as food competitors not sound III-86
— in a changing world III-86–7
— competition III-85–6
— conservation III-75–7
　　— Mørch's memorandum III-75
　　— quotas controversial III-77
— historical perspective III-74–5
　　— expansion of whaling fleets III-75
　　— factory ships III-75
　　— modern whaling III-74
　　— sale of whaling technology to Japan III-74
— how many? what is happening? III-80–3
— just lookin' III-85
— nations withdrawing from whaling III-78
— 'the orderly development of the whaling industry' III-77–8
　　— installation of on-ship freezers III-77
— sanctuary III-84–5
— 'scientific' whaling III-78
— sharing resources III-84
— subsistence, and indigenous rights III-83–4
— visual counting and acoustic listening III-79–80
— whales in a changing world III-86–7
whaling I-3, I-36, I-636
— regulation of III-341
whiting I-49
Wider Caribbean I-460–2
— coastal crises I-461–2, **I-461**
— coastline activity **I-462**
— lack of port reception facilities for garbage III-345
— pollution along heavily urbanized/industrialized coasts I-461
widgeon-grass, Carolinas I-358
wildfowl and waders (shorebirds) III-107, III-110
Wildlife Management Areas
— Gulf of Papua II-608
— Papua New Guinea II-441
Wildlife Reserve, Kure Atoll, Hawaiian Islands II-798
Wilkinson Basin I-309, *I-311*
wind power III-304
wind shear III-305
wind speed, varies with height over different roughnesses III-305, *III-305*
wind stress, and residual flows in the English Channel I-67
wind turbines
— lifetimes of III-306
— modified to operate in sea conditions III-305
— new technology for installation in deeper water III-311
— power output of III-304–5, *III-305*
wind waves I-190
winds
— Aegean, summer Etesians I-236

— affecting the Bahama I-418, *I-418*
  — hurricanes I-418–19
— Arabian Gulf, Shamal and Kaus II-4
— Azores I-203–4
— Coral, Solomon and Bismark Seas Region II-428, *II-428*
— Côte d'Ivoire
  — monsoon I-808
  — Northeast Trade/Harmattan I-808
— Gulf of Aden, controlled by monsoons II-49–50
— Gulf of Alaska I-378
— Gulf of Guinea I-776
— Levantine basin I-256
— Mexican Pacific coast I-486, I-487
— monsoonal
  — Gulf of Thailand II-300
  — may cause damage, Palk Bay II-167
— Northern Benguela I-823
— northern Gulf of Mexico I-439
— northwest Arabian Sea II-19
— prevailing, Red Sea, and extreme sea breezes II-39
— Puerto Rico I-578–9
— *see also* hurricanes; Trade Winds; tropical storms
Winyah Bay *I-352*, *I-353*
women, trained to restore degraded mangrove forest, Dar es Salaam II-89
World Commission on Environment and Development I-459
World Conference on Fisheries Management and Development III-161
World Heritage Convention, TCI signatories to I-592
World Trade Organization (WTO), hindrance to environmental protection III-159

xerophytic shrub vegetation I-170
Xiamen region, China II-513–33
— clean up of Yuan Dang Lagoon II-531
— coastal waters of II-515
— critical environmental problems II-521–4
— ecosystem changes II-524–5
— geography of II-515
— local agencies with marine waters mandates II-526
— major shallow water marine and coastal habitats II-517–18
— protective measures II-525–32
  — a functional marine zonation scheme II-529–30
  — institutional arrangements for marine environment management II-526
  — land—sea integration II-531–2
  — laws and regulations II-525–6, II-529
  — major management problems II-526–8
  — Marine Management Co-ordination Committee (MMCC) II-528–9
  — marine pollution management II-532
  — monitoring, surveillance and emergency preparedness II-531
  — preparedness and response systems II-532
  — present limitations and recommendations II-531
— resource exploitation, utilisation and conflicts II-518–21
— seasonality, currents, natural environmental variables II-515–17
— a Special Economic Zone II-519
— towards integrated coastal management II-528
  — new management issues II-528–9
  — strategic management plan (SMP) and its implementation II-528

Yellow River, modern source of sediment to the Yellow Sea II-493
Yellow Sea *II-474*, II-487–98
— coastal erosion and landfill II-493–4
— main rivers entering II-489
— offshore systems II-490–1
— physical parameters and environmental variables II-489–90
— population II-491–2
— protection and conservation measures II-495–7
  — current conservation and marine protection measures II-496–7
  — factors inhibiting conservation II-495–6
— rural factors II-492–3
— shallow water habitats II-490
— shift away from demersal fish III-128
— urban and industrial activities II-494–5
Yellow Sea Large Marine Ecosystem (YSLME) programme II-496–7
Yellow Sea Warm Current II-489
Yellowfin fishery II-56
Yemen *II-18*, *II-36*, II-43, II-49
— beaches and mudflats important for seabirds and waders II-54
— dense and flourishing coastal algal community II-51
— fishing subsidies II-43
— limestone cliffs II-19
— new licensing round for oil and gas exploration II-59
— population II-25
— seasonal rainfall on southern mountains II-50

Zagros Mountains *II-18*, II-19
zährte, in southern Baltic rivers I-111
Zanzibar *II-84*, II-88
— problems of tourism expansion II-92
Zeehan Current II-649, *II-649*, II-650, *II-674*
Zhejiang—Fujian Coastal Current II-489, II-516
zinc pollution, Greenland I-12
zinc tolerance I-75
zinc (Zn) I-162
— Azorean amphipods I-213
zoogeographic provinces, South African shore II-135
zooplankton
— Antarctic I-707
— Bay of Bengal II-275–7
— Black Sea I-290–1
— Carolinas
  — estuarine I-359
  — marine I-360
— English Channel I-70
— Guinea upwellings, Senegal—Mauritanian, cold waters from I-800
— Gulf of Alaska I-379–80
— Gulf of Guinea I-779, I-783
— Irish Sea I-89–90
— Malacca Straits II-312
— the Maldives II-202
— North Spanish coast I-138, I-141
— offshore, Côte d'Ivoire I-814–15
— Peru, affected by El Niño I-692–3
— Southern Gulf of Mexico I-474
— southwestern Africa I-830, I-832
— Tagus estuary I-157
— Tyrrhenian Sea I-274
— Vietnam II-564
— West Guangdong coast II-554
zooplankton community, Deltaic Sundarbans II-150
Zostera I-70
— Amursky Bay, degradation of II-481
— *Z. marina*
  — propagation of from seed III-4–5
  — recovery of I-70
  — and the Wadden Sea III-6–7